Contents

D1087276

Metals Handbook® Ninth Edition

Volume 4
Heat Treating

Prepared under the direction of
the ASM Handbook Committee

Planned, prepared, organized
and reviewed by
the ASM Heat Treating Division
Council

Planning Committee

Lester E. Alban
James R. Easterday
Daniel S. Zamborsky

William H. Cubberly, Director of
Reference Publications
Vicki Masseria, Managing Editor
Craig W. Kirkpatrick, Associate
Editor
Bonnie Sanders, Editorial/Production
Assistant

 AMERICAN SOCIETY FOR METALS
METALS PARK, OHIO 44073

Copyright © 1981
by the
AMERICAN SOCIETY FOR METALS
All rights reserved

No part of this book may be reproduced, stored in a retrieval system, or transmitted, in any form or by any means, electronic, mechanical, photocopying, recording, or otherwise, without the written permission of the copyright owner.

First printing, November 1981
Second printing, June 1983
Third printing, September 1984
Fourth printing, August 1985
Fifth printing, February 1988

Metals Handbook is a collective effort involving thousands of technical specialists. It brings together in one book a wealth of information from world-wide sources to help scientists, engineers and technicians solve current and long-range problems.

Great care is taken in the compilation and production of this volume, but it should be made clear that no warranties, express or implied, are given in connection with the accuracy or completeness of this publication, and no responsibility can be taken for any claims that may arise.

Nothing contained in the Metals Handbook shall be construed as a grant of any right of manufacture, sale, use or reproduction, in connection with any method, process, apparatus, product, composition or system, whether or not covered by letters patent, copyright or trademark, and nothing contained in the Metals Handbook shall be construed as a defense against any alleged infringement of letters patent, copyright or trademark, or as a defense against any liability for such infringement.

Comments, criticisms and suggestions are invited, and should be forwarded to the American Society for Metals.

Library of Congress Cataloging in Publication Data

American Society for Metals
Heat treating.
(Metals handbook; 9th ed., v. 4)
Includes bibliographical references and index.

1. Metals—Heat treatment. I. American Society
for Metals. Handbook Committee. II. American Society
for Metals. Heat Treating Division. III. Series.
TA459.A5 9th ed., vol. 4 [TN672] 669s 81-12692
ISBN 0-87170-010-7   [671.3′6] AACR2

Printed in the United States of America

Foreword

For the first twenty years of its life, the American Society for Metals focused solely on heat treating. Launched in 1913 in Detroit as the Steel Treaters Club, the group amalgamated with the American Steel Treaters Society to form the American Society for Steel Treaters in 1920. A final name change to American Society for Metals in 1933 completed the transition from a society concerned primarily with heat treating to a society with far-ranging interests in a variety of metals topics.

Although almost a half-century has passed since this momentous change, heat treating has remained an area of deep interest and concern to ASM members. Indeed, heat treating is one of the foundations of the metals sciences and, therefore, of the Society.

Continued leadership of ASM in addressing this fundamental topic is demonstrated by the publication of Volume 4 of the 9th Edition of the Metals Handbook. An earlier volume in the 8th Edition dealt, in part, with heat treating. New Volume 4 reflects a significant expansion in subject coverage, consistent with a growing technology.

Mere volume of information, however, is by no means the sole criterion. With leadership comes a responsibility to match, and hopefully exceed, the usefulness and high technical standards established by the earlier volume. In this respect, new Volume 4 is innovative in a way that contributes to the scope and comprehensiveness of information presented. Every single article has been planned, organized, prepared, and reviewed by members of the ASM Heat Treating Technical Division, expert contributors with "hands on" experience in the many facets of heat treating technology.

The traditional quality of the Metals Handbook depends heavily on the efforts and talents of many people. In this volume, members of the Heat Treating Technical Division played a major role, but a full and rich share of the credit should go to the members of the ASM Handbook Committee for their continuing efforts, to individual contributors, and to the editorial staff. For their valuable contributions, we express our thanks and our gratitude.

John B. Giacobbe
President

Allan Ray Putnam
Managing Director

The Ninth Edition of Metals Handbook
is dedicated to the memory of
TAYLOR LYMAN, A. B. (Eng.), S. M., Ph. D.
(1917-1973)
Editor, Metals Handbook 1945-1973

Preface

Since the days of the ASM parent organization, American Society for Steel Treating, and the publication of its first Handbook in 1924, heat treating has played a very significant and ever-broadening role within the technical community. Traditionally, ASM has served as the leader in the gathering and dissemination of heat treating information in support of engineering and manufacturing needs. The publication of this Handbook on heat treating, Volume 4, Ninth Edition, is the latest offering in fulfillment of ASM's continuing responsibility.

Over the years, interest and activity in heat treating have continued to increase, both in volume and scope. This becomes evident by comparing the size of the 1948 Metals Handbook, containing approximately 100 pages of heat treating related articles to the 1964 Volume 2, which included 300 pages of heat treating information. Volume 4 is the largest yet, consisting of 800 pages of practical and easy to use information on the subject of heat treating.

Part of this increase results from the inclusion of new subject matter. Articles on fluidized bed processes, vacuum carburizing, microprocessors, laser and electron-beam heat treating are some of the new subject matter that reflects recent advancements in technology. In addition, sections on gas carburizing, furnace atmospheres, flame hardening, and heat treating of aluminum have been expanded appreciably. This new and updated information provides the reader with a comprehensive, useful, and technically reliable reference to current heat treating processes and related equipment.

One major innovation in the preparation of Volume 4 is the manner by which the contents were planned, organized, prepared, and reviewed. For the first time, these activities were put in motion and accomplished through the efforts of the ASM Heat Treating Technical Division. Members of the Heat Treating Division Council, each reflecting expertise in his particular area of interest, responded by supervising and participating in the preparation of appropriate manuscripts. Refinements and corroboration of their input was effected through review by a peer group—also members of the Heat Treating Division. The end result of this concerted effort, Volume 4, reflects the practical "hands on" experience and technical competence of these expert contributors.

As in past volumes of the Handbook, ASM expresses appreciation for the collective experience, skill, knowledge and dedication of the contributors, reviewers, and consulting editors to produce a book of such stature. They have earned the gratitude of the entire metallurgical community for their unselfish efforts.

James R. Easterday
Coordinator
Metals Handbook Activity
Heat Treating Division

Policy on Units of Measure

By a resolution of its Board of Trustees, the American Society for Metals has adopted the practice of publishing data in both metric and customary U.S. units of measure. In preparing this Handbook, the editors have attempted to present data primarily in metric units based on Système International d'Unités (SI), with secondary mention of the corresponding values in customary U.S. units. The decision to use SI as the primary system of units was based on the aforementioned resolution of the Board of Trustees, the widespread use of metric units throughout the world, and the expectation that the use of metric units in the United States will increase substantially during the anticipated lifetime of this Handbook.

For the most part, numerical engineering data in the text and in tables are presented in SI-based units with the customary U.S. equivalents in parentheses (text) or adjoining columns (tables). For example, pressure, stress and strength are shown in both SI units, which are pascals (Pa) with a suitable prefix (see the description of SI at the back of the volume), and in customary U.S. units, which are pounds per square inch (psi). To save space, large values of psi have been changed to kips per square inch (ksi), where one kip equals 1000 pounds. Some strictly scientific data are presented in SI units only.

To clarify some illustrations that depict machine parts described in the text, only one set of dimensions is presented on artwork. References in the accompanying text to dimensions in the illustrations are presented in both SI-based and customary U.S. units.

On graphs and charts, grids correspond to SI-based units, which appear along the left and bottom edges; where appropriate, corresponding customary U.S. units appear along the top and right edges. Some previously published charts, particularly histograms depicting statistical distribution of values of mechanical properties, could not be redrawn because of the absence of the original data points; these have been reproduced in their original forms, with SI equivalents on the top and right edges.

Data pertaining to a specification published by a specification-writing group may be given in only the units used in that specification or in dual units, depending on the nature of the data. For example, the typical yield strength of aluminum sheet made to a specification written in customary U.S. units would be presented in dual units, but the thickness ranges listed in that specification might be presented only in inches.

Data obtained according to specified test methods for which the specification implies a particular system of units are presented in the units of that system. Wherever feasible, equivalent units are also presented.

Conversions and rounding have been done in accordance with ASTM Standard E-380, with careful attention to the number of significant digits in the original data. For example, an annealing temperature of 1575 °F contains three significant digits (and possibly only two), because few commercial heat treatment systems can control the temperature of an entire load of parts within a spread of 10 °F. In this instance, the equivalent temperature would be given as 860 °C, or perhaps 850 °C depending on the degree of accuracy meant to be conveyed in the conversion; the exact conversion to 857.22 °C would not be appropriate. In many instances (especially in tables and data compilations), temperature values in °C and °F are alternatives rather than conversions.

The policy on units of measure in this Handbook contains several exceptions to strict conformance to ASTM E380; in each instance, the exception has been made to improve the clarity of the Handbook. Two examples of such exceptions are reporting temperature in °C rather than K and reporting stress intensity in $MPa\sqrt{m}$ rather than $MNm^{-3/2}$.

SI practice requires that only one virgule (diagonal) appear in units formed by combination of several basic units. Therefore, all of the units preceding the virgule are in the numerator and all units following the virgule are in the denominator of the expression (and no parentheses are required to prevent ambiguity).

Handbook Committee, Officers and Trustees

Members of the ASM Handbook Committee (1980-1981)

Martin G. H. Wells
(Chairman 1980-; Member 1976-)
Colt Industries

Gunvant N. Maniar
(Chairman 1978-1980;
Member 1974-1980)
Carpenter Technology Corp.

Robert Clark Anderson
(1978-1981)
Anderson & Associates

Donald R. Betner
(1974-1980) General Motors Corp.

Alexander V. Bublick
(1979-) United Technologies Corp.

Thomas D. Cooper
(1981-) Air Force Wright
Aeronautical Laboratories

Jerry L. Dassel
(1977-1980)
Kaiser Aluminum & Chemical Co.

Donald C. Engdahl
(1979-) The Boeing Co.

Philip H. B. Hamilton
(1977-1981)
Dominion Engineering Works, Ltd.

Dick W. Hemphill
(1977-) Dana Corp.

Jack A. Hildebrandt
(1974-) Clark Equipment Co.

George C. Hsu
(1980-)
Reynolds Metals Company

Lawrence J. Korb
(1978-) Rockwell International

Roger Mack
(1977-1980)
Abar Corporation

James L. McCall
(1977-)
Battelle Columbus Laboratories

Paul J. Mikelonis
(1978-1981)
General Casting Corp.

James H. Mikoda
(1979-1980)
Burgess Norton Manufacturing Co.

Peter Patriarca
(1978-1981)
Union Carbide Corp.

John E. Scheer
(1977-1980)
Standard Pressed Steel
Technologies, Inc.

Sonny G. Sundaresan
(1979-) Copeland Corp.

Charles N. Tanton
(1979-) Ford Motor Co. (retired)

George E. Wood
(1980-) IBM Corporation

viii

Officers and Trustees of the American Society for Metals

John B. Giacobbe
President and Trustee
Superior Tube Company

David Krashes
Vice President and Trustee
Massachusetts Materials Research, Inc.

Raymond L. Smith
Past President and Trustee
Michigan Technological University (retired)

Edward E. Slowter
Treasurer
Battelle Memorial Institute (retired)

Trustees

George E. Dieter
University of Maryland

F. Keith Lampson
Marquardt Company

John W. Pridgeon
Special Metals Corporation

George H. Bodeen
Lindberg Corporation

Edward J. Dulis
Colt Industries, Crucible Research Center

Harold D. Kessler
Cabot Corporation

Harold L. Gegel
Air Force Wright Aeronautical Laboratories

James F. Schumar
Argonne National Laboratory

William F. Zepfel
Sidbec-Dosco Ltd.

Allan Ray Putnam
Managing Director

Previous Chairmen of ASM Handbook Committee

R. S. Archer
(1940-1942) (Member, 1937-1942)

L. B. Case
(1931-1933) (Member, 1927-1933)

E. O. Dixon
(1952-1954) (Member, 1947-1955)

R. L. Dowdell
(1938-1939) (Member, 1935-1939)

J. P. Gill
(1937) (Member, 1934-1937)

J. D. Graham
(1965-1968) (Member, 1961-1970)

J. F. Harper
(1923-1926) (Member, 1923-1926)

C. H. Herty, Jr.
(1934-1936) (Member, 1930-1936)

J. B. Johnson
(1948-1951) (Member, 1944-1951)

R. W. E. Leiter
(1962-1963)
(Member, 1955-1958, 1960-1964)

G. V. Luerssen
(1943-1947) (Member, 1942-1947)

Gunvant N. Maniar
(1978-1980) (Member, 1974-1980)

W. J. Merten
(1927-1930) (Member, 1923-1933)

N. E. Promisel
(1955-1961) (Member, 1954-1963)

G. J. Shubat
(1972-1975) (Member, 1966-1975)

W. A. Stadtler
(1968-1972) (Member, 1962-1972)

Raymond Ward
(1976-1978) (Member, 1972-1978)

D. J. Wright
(1964-1965) (Member, 1959-1967)

ix

ASM Heat Treating Division Council

Gregory L. Serangeli
Chairman
Manager of Welding Development
Caterpillar Tractor Co.

Norman N. Breyer
Professor & Chairman
Department of Metallurgical and
    Materials Engineering
Illinois Institute of Technology

Charles G. Cloern
Vice President, Marketing
Ipsen Industries

James G. Conybear
Product Manager
Thermal Systems Technical Center
Midland-Ross Corp.

Jon L. Dossett
Vice President—Metallurgy
United States Gear Corp.

Robert W. Foreman
Director
Research & Development
Park Chemical Co.

Jack A. Hildebrandt
Corporate Metallurgist
Clark Equipment Co.

Dante Iacovoni
Vice President—Marketing
Lindberg
A Unit of General Signal

Norman O. Kates
Vice President—Technology
Lindberg Corp.

John M. Kelso
Chief Metallurgist
Benedict-Miller, Inc.

W. James Laird, Jr.
Vice President—Marketing
Research & Development
Upton Industries, Inc.

Robert F. Miller, Jr.
President
Vacuum Heat Treating Co., Inc.

Gordon L. Peterman
Ft. Worth Division
General Dynamics

David Scarrott
President
Scarrott Metallurgical Co.

Ross B. Shingledecker
Metallurgical Director
Manufacturing Services
Ladish Co.

Yancey E. Smith
Research Manager
Climax Molybdenum Co.

Donald R. Wensing
Chief Metallurgist
SKF Industries, Inc.

Author
Committees

Heat Treating
of Steel

Robert W. Foreman
Co-Chairman
Director
Research & Development
Park Chemical Co.

Yancey E. Smith
Co-Chairman
Research Manager
Climax Molybdenum Co.

Charles A. Apple
Engineer
Homer Research Laboratories
Bethlehem Steel Corp.

John W. Balai
Metallurgical Engineer
Holcroft
Division of Thermo Electron Corp.

Domenic A. Canonico
Director
Metallurgical and Materials
   Laboratory
C-E Power Systems
Combustion Engineering Inc.

Casimir F. Dombkowski
Senior Metallurgist
New Departure Division
General Motors Corp.

Stephen Floreen
Research Fellow
Research & Development Center
International Nickel Company, Inc.

David E. Gaylord
President
Progressive Heat Treating Co.

Bhupendra K. Gupta
Senior Project Development
   Engineer
Science & Technology
International Harvester Co.

Jack E. Haflinger
Chief Metallurgist
Lamson & Sessions Co.

Daniel J. Hayes
Metallurgical Engineer
United States Steel Corp.

Walter E. Heyer
Administrative General Manager
Heat Treating Equipment Division
Selas Corporation of America

Karl Heinz Kopietz
Marketing Manager
Heat Treating Products
Henry E. Sanson & Sons, Inc.

Gilbert M. Lahr
Chief Metallurgist
Detroit Diesel Allison Division
General Motors Corp.

W. James Laird, Jr.
Vice President—Marketing
Research & Development
Upton Industries, Inc.

Inna Lazarev
Supervisor Technical Service
Heat Treating Department
E. F. Houghton & Co.

William A. Leeper
Chief Metallurgical Engineer
Thornhill-Craver Co.

Richard J. Light
Chief Process Metallurgist
Surface Division
Midland-Ross Corp.

John S. Maxson, Jr.
Vice President
Procedyne Corp.

Norman C. McClure
Chief Metallurgist
Plant 2
Commercial Steel Treating Corp.

Quentin D. Mehrkam
Senior Vice President
Ajax Electric Co.

Anthony Meszaros
Metallurgical Supervisor
Park Chemical Co.

Jerrold Meyer
Division Metallurgist
Roller Bearings Division
SKF Industries, Inc.

Nicholas P. Milano
Director of Metallurgy
Illinois Gear
Wallace Murray Corp.

Edward R. Mueller
Director of Research
Tenaxol, Inc.

Daniel J. Olah
Materials Engineering Manager
Borg & Beck Division
Borg-Warner Corp.

T. V. Philip
Specialist
Tool and Alloy Metallurgy
Carpenter Technology Corp.

George L. Schiel
Chief Metallurgist
Metlab Co.

Duane J. Schmatz
Principal Staff Engineer
Scientific Laboratories
Ford Motor Co.

Edwin J. Schneider
Supervisor
Flat Rolled Products
Jones & Laughlin Steel Corp.

Paul R. Slimmon
Supervisor
Homer Research Laboratories
Bethlehem Steel Corp.

Romeo L. Suffredini
Chief Metallurgist
Winchester Arms Division
Olin Corp.

Peter Vernia
Senior Research Engineer
Metallurgy Department
General Motors Research
    Laboratories

Dwight A. Wilkinson
Chief Metallurgist
Saginaw Steering Gear Division
General Motors Corp.

Samuel L. Williams
General Engineer
Rock Island Arsenal

Case Hardening of Steel

Charles A. Stickels
Co-Chairman
Principal Staff Engineer
Research Staff
Ford Motor Co.

Donald R. Wensing
Co-Chairman
Chief Metallurgist
SKF Industries, Inc.

Donald R. Barber
President (retired)
Heatbath Corp.

T. Bell
Hanson Professor of Industrial
    Metallurgy
Department of Metallurgy &
    Materials
University of Birmingham

Larry E. Byrnes
Manager—Manufacturing Research
Federal Mogul Corp.

James G. Conybear
Product Manager
Thermal Systems Technical Center
Midland-Ross Corp.

Jon L. Dossett
Vice President—Metallurgy
United States Gear Corp.

Robert W. Foreman
Director—Research & Development
Park Chemical Co.

Kenneth D. Gladden
Senior Development Engineer
Caterpillar Tractor Co.

George Greene
Metallurgist
Industrial Truck Division
Eaton Corp.

Donald N. Guy
Project Manager
Technology Center
Lindberg Heat Treating Co.

Robert L. Hughes
Chief Metallurgist
Fairfield Manufacturing Co., Inc.

Karl Heinz Kopietz
Marketing Manager
Heat Treating Products
Henry E. Sanson & Sons, Inc.

W. James Laird, Jr.
Vice President—Marketing
Research & Development
Upton Industries, Inc.

Quentin D. Mehrkam
Senior Vice President
Ajax Electric Co.

Burton R. Payne, Jr.
President
Payne Chemical Co.

Dale A. Poteet, Jr.
Division Engineer
John Deere Component Works

Joseph A. Riopelle
District Manager/Chicago Office
Surface Division
Midland-Ross Corp.

Ronald D. Rogers
Chief Engineer
Materials Engineering
Automotive Operations
Rockwell International

John A. Swift
Consultant, Metallurgy
Henry E. Sanson & Sons, Inc.

Theodore B. Wilk
Vice President
Sales & Technology
The A. F. Holden Co.

William G. Wood
Vice President
Technology/Research &
    Development
Kolene Corp.

Heat Processing Equipment

Dante Iacovoni
Co-Chairman
Vice President—Marketing
Lindberg
A Unit of General Signal

Ross B. Shingledecker
Co-Chairman
Metallurgical Director
Manufacturing Services
Ladish Co.

Roger C. Anderson
Manager—Research &
    Development Laboratories
Ipsen Industries

Ed R. Byrnes, Jr.
Manager
Vacuum Equipment Marketing
Ipsen Industries

Douglas J. Cleary
Vice President and General
    Manager
Stanwood Corp.

Francis M. Fahrenwald
Fahrenwald Consulting

William L. James
President
Fennell Corp.

Norman O. Kates
Vice President—Technology
Lindberg Corp.

James Kelly
Director of Technology
Rolled Alloys

Fred W. Klag
Vice President
The Alloy Engineering Co.

W. James Laird, Jr.
Vice President—Marketing
Research & Development
Upton Industries, Inc.

Arthur L. LaMasters
Marketing Director
Cleveland Alloy Casting Co.

John W. Smith
Manager of Engineering
Holcroft Division
Thermo Electron Corp.

John E. Stein
Vice President—Manufacturing
Resisto-Loy Company, Inc.

Donald J. Tillack
Technical Sales Engineer
Huntington Alloys, Inc.

Dennis M. Wagen
Executive Vice President
Stanwood Corp.

Furnace Control Instrumentation

Ross B. Shingledecker
Chairman
Metallurgical Director
Manufacturing Services
Ladish Co.

Fred J. Bartkowski
Vice President
Marshall W. Nelson & Associates,
    Inc.

Roger G. Blocks
President
Chem-Al, Inc.

Jeffrey W. Boswell
Project Engineer
Sunbeam Equipment Corp.

Don G. Ensweiler
President
Heat Process Associates, Inc.

Paul L. Huber
Manager of Sales
Sunbeam Equipment Corp.

John E. O'Neil
Industry Manager
Fabricated Materials Market
Leeds & Northrup Co.
Unit of General Signal Corp.

Raymond Ostrowski
Sales Manager
Protection Controls, Inc.

Theodore K. Thomas
Business Unit Advertising Manager
Process Control Division
Honeywell Inc.

G. B. Zuber
Sales Manager
Instrument Division
Mine Safety Appliance Co.

Furnace Atmospheres
and Carbon Control

Ross B. Shingledecker
Chairman
Metallurgical Director
Manufacturing Services
Ladish Co.

Robert N. Blumenthal
Professor
Marquette University

Douglas H. Clingner
Heat Treat Metallurgist
Fairfield Manufacturing Co., Inc.

James R. Dale
Manager—Controlled Atmospheres
AIRCO Industrial Gases

Raymond L. Davis II
Carbon Control Instruments

Robert F. Gunow
President
Apt Consulting & Engineering

J. H. Kline (retired)
Heat Treat Department Metallurgy
Central Alloy District
Republic Steel Corp.

Max H. Priddy
Regional Manager
Selas Corporation of America

Ralph Puerta
Applied Research & Development
Air Products and Chemicals, Inc.

Lewis H. Shaefer
Marketing Manager
Anarad, Inc.

Wilfred G. Shedd
Co Owner & Secretary-Treasurer
Metallurgical Processing, Inc.

Localized
Heat Treating

Carl Fiorletta
Supervisor
Electron Beam Heat Treating
Systems
Sciaky Bros., Inc.

Ole Sandven
Chief Metallurgist
Avco Everett Metalworking Lasers

Nelson Stevens
Sales Manager
Lindberg/Cycle-Dyne
A Unit of General Signal Corp.

Heat Treating
of Cast Irons

Daniel S. Zamborsky
Chairman
Corporate Metallurgist
Warner & Swasey Co.

Lyle R. Jenkins
Vice President—Technology
Wagner Castings Co.

Edward F. Ryntz, Jr.
Senior Staff Research Engineer
General Motors Research
Laboratories

Kenneth E. Spray
Metallurgy Section Head
Material Corporate Laboratory
Clark Equipment Co.

Richard L. Ward
Metallurgist and Manager of
Quality Control
Bendix Industrial Group
Division of Warner & Swasey Co.

Heat Treating of Tool Steels

W. James Laird, Jr.
Chairman
Vice President—Marketing
Research & Development
Upton Industries, Inc.

Carl J. Oxford, Jr.
Vice President—Technology
National Twist Drill & Tool
Division
Lear Siegler, Inc.

Percy Rawcliffe
Manager—Metallurgical and
Physical Laboratories
Morse Cutting Tools
Division of Gulf & Western
Manufacturing

Carl Reichel
Plant Metallurgist
Drill & End Mill Division
Augusta Plant
TRW, Inc.

Ronald F. Spitzer
Chief Metallurgist
Bearings Division
TRW, Inc.

Daniel S. Zamborsky
Corporate Metallurgist
Warner & Swasey Co.

Heat Treating of Stainless Steels and Heat-Resisting Alloys

Daniel Rapoport
Chairman
Technical Center
Howmet Turbine Components Corp.

Ed R. Byrnes, Jr.
Manager—Vacuum Equipment
Marketing
Ipsen Industries

Kenneth L. Crooks
Senior Staff Engineer
Research & Technology
Armco, Inc.

Ranes P. Dalal
Senior Development Engineer
Technical Center
Howmet Turbine Components
Corp.

Matthew J. Donachie, Jr.
Senior Assistant Materials Project
Engineer
Pratt Whitney Aircraft Group
United Technologies Corp.

Dennis Macha
Supervisor—Metallurgical Services
Technical Center
Howmet Turbine Components Corp.

Robert N. Peterson
Manager of Metallurgical Services
Enduro Division
Republic Steel Corp.

Arthur D. Schwartz
Manager of Materials Performance
Engineering
Aircraft Engine Business Group
General Electric Corp.

Heat Treating of Nonferrous Metals

Jack A. Hildebrandt
Chairman
Corporate Metallurgist
Clark Equipment Co.

Sven E. Axter
Specialist Engineer
Boeing Commercial Airplane Co.

Larry J. Barker
Senior Research Associate
Kaiser Aluminum and Chemical
Corp.

Emmett N. Bossing
Chief Metallurgist
Teledyne Cast Products

Alan H. Braun
Manager—Research &
Development
Wellman Dynamics Corp.

Roger V. Carter
Manager—Metals Technology
Boeing Commercial Airplane Co.

Sidney L. Couling
Senior Research Metallurgist
Battelle Columbus Laboratories

Robert E. Droegkamp
Manager—Research &
Development
Fansteel Metals, Inc.

E. B. Fernsler (retired)
Technical Service Manager
Huntington Alloys, Inc.

Carl J. Gaffoglio
Manager—Applications
Engineering
Copper Development Association,
Inc.

Arthur L. Geary
Senior Metallurgist
Nuclear Metals, Inc.

William B. Hampshire
Technical Service/Metallurgy
Lead Industries Association, Inc.

William H. Heil
Supervisor, Quality Assurance
Timet
Division of Titanium Metals
Corporation of America

Walter Herman
Technical Director
Viking Metallurgical Corp.

David S. Hibbard
Development Engineer
Research & Technical Center
Anaconda Industries

Louis E. Huber, Jr.
Plant Metallurgist
Kawecki-Berylco Industries, Inc.
Division of Cabot Corp.

John H. Hull
Chief Metallurgist
Grafton Plant
Wyman-Gordon Co.

Nicholas C. Jessen, Jr.
Group Leader—Nuclear Division
Y-12 Plant
Union Carbide Corp.

Ralph J. Kotfila
Lead Engineer—Technology
McDonnell Aircraft Co.

Henry A. Kuchek
Senior Research Engineer
Dow Chemical Co.

Robert B. Leholm
Materials Engineer
McDonnell-Douglas Corp.

Joseph B. Long
Consultant

Theodore J. Louzon
Metallurgical Engineer
Bell Laboratories

J. Howard Mendenhall
Technical Associate
Olin Brass
Olin Corp.

Richard G. O'Rourke
Senior Quality Engineer
Brush Wellman, Inc.

Henry J. Profitt
Technical Director
The Light Metal Testing Division
Haley Industries Ltd.

Donald G. Schmidt
Consultant
R. Lavin & Sons, Inc.

Robert F. Schmidt
Technical Director
Colonial Metals Co.

Charles J. Scholl
Chief Product Metallurgist
Wyman-Gordon Co.

Stanley Shapiro
President
Revere Research, Inc.

James T. Staley
Technical Manager
Alloy Technology Division
Alcoa Technical Center
Alcoa Laboratories

Donald J. Tillack
Technical Sales Engineer
Huntington Alloys, Inc.

Allen B. Townsend
Development Consultant
Union Carbide Corp.

Patrick A. Tully
Manager—Metallurgy
Ampco Pittsburgh Corp.

R. Terrence Webster
Principal Metallurgical Engineer
Teledyne Wah-Chang Albany

Paul C. Wilson
Chief Engineer
Arwood Corp.

Edward D. Zysk
Technical Director
Carteret Operation
Engelhard Industries

Heat Processing of Powder Metallurgy Parts

W. James Laird, Jr.
Chairman
Vice President—Marketing
Research & Development
Upton Industries, Inc.

J. Howard Beck
President
BTU Engineering Corp.

Howard E. Boyer
Consultant

A. J. Craig, Jr.
Chief Materials Engineer
Homelite/Textron

Donald L. Dyke
Engineering Manager
Sintered Specialties

Donald Grendon
Sales Engineer
Drever Co.

Erhard Klar
Manager—Research &
Development
Chemical/Metallurgical Division
Glidden Metals
SCM Corp.

Kenneth H. Moyer
Product Development Engineer
Hoeganaes Corp.

George Otto
Supervisor—Process Engineering
The Maytag Co.

A. Thomas Sibley
Program Manager—P/M
Applications
Air Products & Chemicals, Inc.

Sang-Kee Suh
Senior Research Engineer
Ford Motor Co.

Contents

Heat Treating of Steel

Stress-Relief Heat Treating of Steel

By Domenic A. Canonico
Director of Metallurgical and
Materials Laboratory
C-E Power Systems
Combustion Engineering Inc.

STRESS-RELIEF HEAT TREAT-ING is used to relieve stresses that remain locked in a structure as a consequence of a manufacturing sequence. This definition separates stress relief heat treating from postweld heat treating in that the goal of postweld heat treating is to provide, in addition to the relief of residual stresses, some preferred metallurgical structure or properties (Ref 1 and 2). For example, most ferritic weldments are given postweld heat treated to improve fracture toughness of heat-affected zones. Moreover, austenitic and nonferrous alloys are frequently postweld heat treated to improve resistance to environmental damage.

Stress-relief heat treating is the uniform heating of a structure or portion thereof to a suitable temperature below the transformation range (Ac$_1$ for ferritic steels), holding at this temperature for a predetermined period of time, followed by uniform cooling (Ref 2 and 3). Care must be taken to ensure uniform cooling, particularly when a component is composed of variable section sizes. If the rate of cooling is not constant, new residual stresses can result that are equal to or greater than those

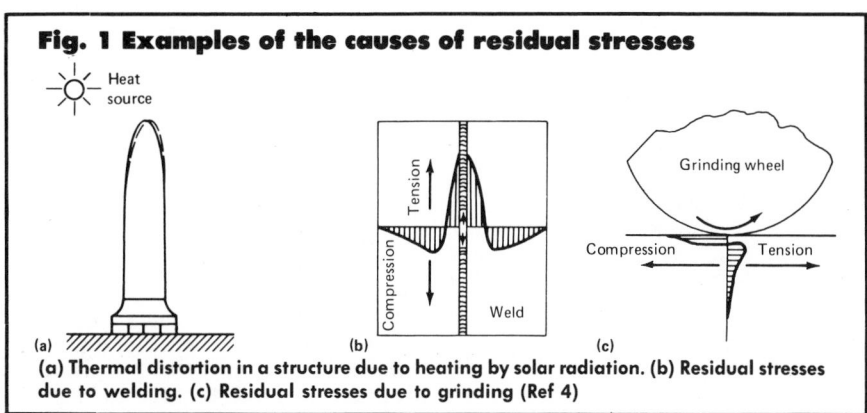

Fig. 1 Examples of the causes of residual stresses

(a) Thermal distortion in a structure due to heating by solar radiation. (b) Residual stresses due to welding. (c) Residual stresses due to grinding (Ref 4)

for which the heat treating was intended to relieve.

Stress-relief heat treating can reduce distortion and high stresses from welding that affect service performance. The presence of residual stresses can lead to stress corrosion cracking near welds and in regions of a component that has been cold strained during processing.

Residual stresses in a ferritic steel cause significant reduction in resistance to brittle fracture. In a material that is not prone to brittle fracture,

such as an austenitic stainless steel, residual stresses can be sufficient to provide the stress necessary to promote stress corrosion cracking even in environments that appear to be benign (Ref 4).

There are many sources of residual stress; they can occur during processing of the material from ingot to final product form (Ref 4 and 9). Residual stresses can be generated during rolling, casting or forging; during forming operations such as shearing, bending and machining; and during fabrication, in particular, welding. Residual stresses are present whenever a component is stressed beyond its elastic limit and plastic flow occurs. Bending a bar during fabrication at a temperature where recovery cannot occur (cold forming, for example), will result in one surface location containing residual tensile stresses, whereas a location 180° away will contain residual compressive stresses (Ref 5). Quenching of thick sections results in high residual compressive stresses on the surface of the material. These high compressive stresses are balanced by residual tensile stresses in the internal areas of the section (Ref 6).

Grinding is another source of residual stresses; these can be compressive or tensile in nature, depending on the grinding operation. Although these stresses tend to be shallow in depth, they can cause warping of thin parts (Ref 7).

The cause of residual stresses that has received the most attention in the open literature is welding. The residual stresses associated with the steep thermal gradient of welding can occur on a macroscale over relatively long distances (reaction stresses) or can be highly localized (microscale) (Fig. 1). Welding usually results in localized residual stresses that approach levels equal to the yield strength of the material at room temperature.

A number of factors influence the relief of residual stresses, including level of stress, permissible (or practicable) time for their relief, temperature, and metallurgical stability.

The relief of residual stresses is a time-temperature related phenomenon (Fig. 2), parametrically correlated by the Larson-Miller equation:

$$\text{Thermal effect} = T(\log t + 20)(10^{-3})$$

where T is temperature (Rankin) and t is hours. It is evident in Fig. 2 that similar relief of residual stresses can be achieved by holding a component for longer periods of time at a lower temperature. For example, holding a piece at 595 °C (1100 °F) for 6 h provides the same relief of residual stress as heating at 650 °C (1200 °F) for 1 h.

Relief of residual stresses represents typical stress-relaxation behavior, in which the material undergoes microscopic (sometimes even macroscopic) creep at the stress-relief temperature. Creep-resistant materials, such as the chromium-bearing low-alloy steels and

the chromium-rich high-alloy steels, normally require higher stress-relief heat treating temperatures than conventional low-alloy steels. Typical stress-relief temperatures for low-alloy ferritic steels are between 595 and 675 °C (1100 and 1250 °F). For high-alloy steels, these temperatures may range from 900 to 1065 °C (1650 to 1950 °F).

For high-alloy steels, such as the austenitic stainless steels, stress relieving is sometimes done at temperatures as low as 400 °C (750 °F). However, at these temperatures, only modest decreases in residual stress are achieved. Residual stresses can be significantly reduced by stress-relief heat treating those austenitic materials in the temperature range from 480 to 925 °C (900 to 1700 °F). At the higher end of this range, nearly 85% of the residual stresses may be relieved. Stress relief heat treating in this range, however, may result in sensitizing susceptible material. This metallurgical effect can

lead to stress-corrosion cracking in service (Ref 8). Frequently, solution-annealing temperatures at around 1065 °C (1950 °F) are used to achieve a reduction of residual stresses to acceptably low values.

Some copper alloys may fail by stress-corrosion cracking due to the presence of residual stresses. These stresses are usually relieved by mechanical or thermal stress-relief treatments. Stress-relief heat treating tends to be favored because it is more controllable, less costly, and also provides a degree of dimensional stability. Stress-relief heat treating of copper alloys is usually carried out at relatively low temperatures, in the range from 200 to 400 °C (390 to 750 °F) (Ref 9).

Resistance of a material to the reduction of its residual stresses by thermal treatment can be estimated with a knowledge of the influence of temperature on its yield strength. Figure 3 provides a summary of the yield strength

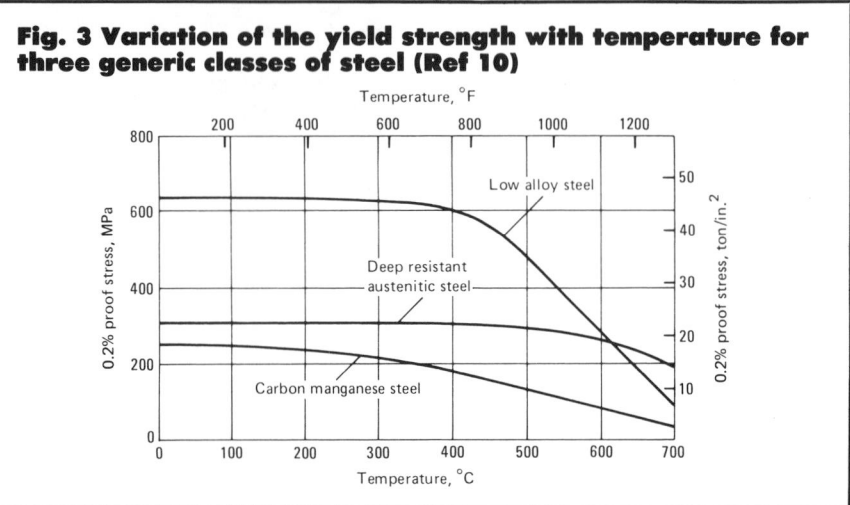

Fig. 2 Illustration of the relationship between time and temperature in the relief of residual stresses in steel (Ref 3)

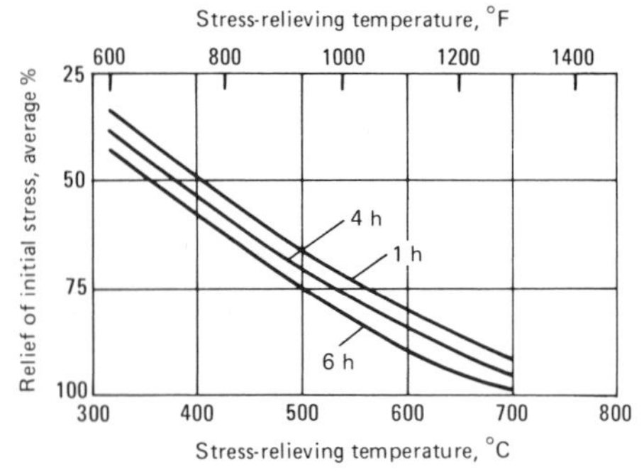

Fig. 3 Variation of the yield strength with temperature for three generic classes of steel (Ref 10)

to temperature relationship for three generic classes of steels. The room temperature yield strength of these materials provides an excellent estimate of the level of localized residual stress that can be present in a structure. To relieve the residual stress requires that the component be heated to a temperature where its yield strength approaches a value that corresponds to an acceptable level of residual stress. Holding at this temperature can, through the reduction of strain due to creep, further reduce the residual stress. Uniform cooling after residual-stress heat treating is mandatory if these levels of residual stress are to be maintained.

REFERENCES

1. The Metallurgical Effects of Residual Stresses, by N. Bailey: in *Residual Stresses,* The Welding Institute, 1981, p 28-33
2. *Metallurgy and Weldability of Steels,* by C. E. Jackson *et al.:* Welding Research Council, 1978
3. Fundamentals of Welding: *Welding Handbook,* 7th Ed., Vol 1, American Welding Society, 1976
4. *Defects and Failures in Pressure Vessels and Piping,* by Helmut Thielsch: Reinhold Publishing Corporation, NY, 1965, p 311
5. *Mechanical Metallurgy,* 2nd Ed., by G. E. Dieter: McGraw-Hill, Inc., 1976
6. *Residual Stresses and Fatigue in Metals,* by J. O. Almen and P. H. Black: McGraw-Hill Book Company, 1963
7. Machining: *Metals Handbook,* 8th Ed., Vol 3, American Society for Metals, 1967, p 260
8. Properties and Selection: Stainless Steels, Tool Materials and Special Purpose Metals: *Metals Handbook,* 9th Ed., Vol 3, ASM, 1980, p 47-48
9. Properties and Selection: Nonferrous Alloys and Pure Metals: *Metals Handbook,* 9th Ed., Vol 2, ASM, 1979, p 255-256
10. Thermal Stress Relief and Associated Metallurgical Phenomena, by C. G. Saunders: in *The Welding Institute Research Bulletin,* Vol 9, No. 7, Part 3, 1968

SELECTED REFERENCES

1. Welding and Brazing: *Metals Handbook,* 8th Ed., Vol 6, American Society for Metals, Metals Park, OH, 1971, p 213
2. Classification and Nomenclature of Internal Stresses, by E. Orowan: in *Symposium on Internal Stresses in Metals and Alloys,* The Institute of Metals, London, 1948, p 47-59
3. Stress Relieving of Weldments, by E. R. Parker: *Welding Journal,* Vol 36, No. 10, Oct 1957, p 433-S
4. The Effect of Residual Stresses on Fracture, by J. D. Harrison and R. H. Leggatt: in *Residual Stresses,* The Welding Institute, 1981, p 17-20
5. Residual Stresses in Welded Plates, by N. R. Nagaraja and L. Tall: *Welding Journal,* Vol 40, No 10, 1961, p 468-S

Normalizing of Steel

By Samuel L. Williams
General Engineer
Rock Island Arsenal

NORMALIZING OF STEEL is a heat treating process that is often considered from both thermal and microstructural standpoints. In the thermal sense, normalizing is an austenitizing heating cycle followed by cooling in still or slightly agitated air. Typically, the work is heated to a temperature about 55 °C (100 °F) above the upper critical line of the iron – iron carbide phase diagram, as shown in Fig. 1; that is, above Ac_3 for hypoeutectoid steels and above A_{cm} for hypereutectoid steels. To be properly classed as a normalizing treatment, the heating portion of the process must produce an austenitic phase (face-centered cubic crystal structure) prior to cooling. Typical normalizing temperatures for many standard steels are given in Table 1.

Normalizing is also frequently thought of in terms of microstructure. The areas of the microstructure that contain about 0.8% carbon are pearlitic (lamellae of ferrite and iron carbide). The areas that are low in carbon are ferritic (body-centered cubic atomic structure). In hypereutectoid steels, proeutectoid iron carbide (iron carbide in excess of that within the pearlite structure) can be present in the microstructure. Air-hardening steels are excluded from the class of normalized steels because they do not exhibit the "normal" pearlitic microstructure that characterizes normalized steels.

Uses. A broad range of ferrous products can be normalized. All of the stan-dard low-carbon, medium-carbon, and high-carbon wrought steels can be normalized, as well as many castings. Austenitic steels, stainless steels and maraging steels either cannot be normalized, or usually are not normalized.

The purpose of normalizing varies considerably. Normalization may increase or decrease the strength and hardness of a given steel in a given product form depending on the thermal and mechanical history of the product. Actually, the functions of normalizing may overlap with or be confused with those of annealing, hardening and stress relieving. Improved machinability, grain-structure refinement, homogenization, and modification of residual stresses are among the reasons for which normalizing is done. Homogenization of castings by normalizing may be done in order to break up or refine the dendritic structure and facilitate a more even response to subsequent hardening. Similarly, for wrought products, normalization can obliterate banded grain structure due to hot rolling, as well as large grain size or mixed large and small grain size due to forging practice. The details of normalizing treatments applied to three

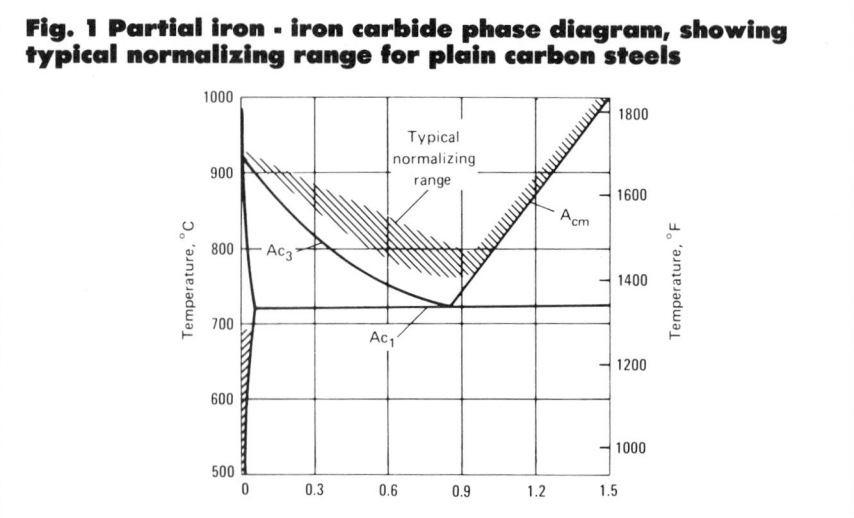

Fig. 1 Partial iron - iron carbide phase diagram, showing typical normalizing range for plain carbon steels

Table 1 Typical normalizing temperatures for standard carbon and alloy steels

Based on production experience, normalizing temperature may vary from as much as 27 °C (50 °F) below to as much as 55 °C (100 °F) above indicated temperature. The steel should be cooled in still air from indicated temperature.

Grade	Temperature °C	°F	Grade	Temperature °C	°F
Plain carbon steels			**Standard alloy steels (continued)**		
1015	915	1675	4817	925	1700
1020	915	1675	4820	925	1700
1022	915	1675	5046	870	1600
1025	900	1650	5120	925	1700
1030	900	1650	5130	900	1650
1035	885	1625	5132	900	1650
1040	860	1575	5135	870	1600
1045	860	1575	5140	870	1600
1050	860	1575	5145	870	1600
1060	830	1525	5147	870	1600
1080	830	1525	5150	870	1600
1090	830	1525	5155	870	1600
1095	845	1550	5160	870	1600
1117	900	1650	6118	925	1700
1137	885	1625	6120	925	1700
1141	860	1575	6150	900	1650
1144	860	1575	8617	925	1700
			8620	925	1700
			8622	925	1700
Standard alloy steels			8625	900	1650
1330	900	1650	8627	900	1650
1335	870	1600	8630	900	1650
1340	870	1600	8637	870	1600
3135	870	1600	8640	870	1600
3140	870	1600	8642	870	1600
3310	925	1700	8645	870	1600
4027	900	1650	8650	870	1600
4028	900	1650	8655	870	1600
4032	900	1650	8660	870	1700
4037	870	1600	8720	925	1700
4042	870	1600	8740	925	1700
4047	870	1600	8742	870	1600
4063	870	1600	8822	925	1700
4118	925	1700	9255	900	1650
4130	900	1650	9260	900	1650
4135	870	1600	9262	900	1650
4137	870	1600	9310	925	1700
4140	870	1600	9840	870	1600
4142	870	1600	9850	870	1600
4145	870	1600	50B40	870	1600
4147	870	1600	50B44	870	1600
4150	870	1600	50B46	870	1600
4320	925	1700	50B50	870	1600
4337	870	1600	60B60	870	1600
4340	870	1600	81B45	870	1600
4520	925	1700	86B45	870	1600
4620	925	1700	94B15	925	1700
4621	925	1700	94B17	925	1700
4718	925	1700	94B30	900	1650
4720	925	1700	94B40	900	1650
4815	925	1700			

typical production parts are given in Table 2, which also lists the reasons for normalizing and gives some of the mechanical properties obtained in the normalized and tempered condition. Comparisons of typical hot rolled or annealed mechanical properties versus typical normalized properties are presented in Table 3. Depending on the mechanical properties required, normalizing may be substituted for conventional hardening when the size or shape of the part is such that liquid quenching might result in cracking, distortion or excessive dimensional changes. Thus, parts that are of complex shape or that incorporate sharp changes in section may be normalized and tempered, provided that the properties obtained are acceptable.

The rate of heating generally is not critical for normalizing; on an atomic scale, it is immaterial. In parts having great variations in section size, however, thermal stress can cause distortion.

Time at temperature is critical only in that it must be sufficient to cause homogenization. Sufficient time must be allowed for solution of carbides, if present, and/or for movement of alloy atoms to obtain a desired final structure. Generally, time sufficient for complete austenitization is all that is required. One hour at temperature, after the furnace recovers, per inch of part thickness, is considered to be very liberal. Parts often can be austenitized adequately in much less time (with a saving in energy).

The rate of cooling significantly influences both the amount of pearlite and the size and spacing of the pearlite lamellae. At higher cooling rates, more pearlite forms, and the lamellae are finer and more closely spaced. Both the increased amount of pearlite and the greater fineness of the pearlite result in higher strength and higher hardness. Conversely, lower cooling rates result in softer parts. The effect of mass on hardness (via its effect on cooling rate) is illustrated by the data in Table 4. In any part having both thick and thin sections, the potential exists for variations in cooling rate, and thus for variations in strength and hardness as well as an increase in the probability of distortion or even cracking. Cooling rate sometimes is enhanced purposely with fans, to increase strength and hardness of parts or to decrease the time required, following the furnace operation, for sufficient cooling of parts to permit convenient handling.

After parts have cooled evenly to "black heat" below Ar_1 (the parts are no longer red, as when they were removed from the furnace), they may be water or

Table 2 Typical applications of normalizing and tempering

Part	Steel	Heat treatment	Properties after treatment	Reason for normalizing
Cast 50-mm (2-in.) valve body, 19 to 25 mm (¾ to 1 in.) in section thickness	Ni-Cr-Mo	Full annealed at 955 °C (1750 °F), normalized at 870 °C (1600 °F), tempered at 665 °C (1225 °F)	Tensile strength, 620 MPa (90 ksi); 0.2% yield strength, 415 MPa (60 ksi); elongation in 50 mm or 2 in., 20%; reduction in area, 40%	To meet mechanical-property requirements
Forged flange	4137	Normalized at 870 °C (1600 °F), tempered at 570 °C (1060 °F)	Hardness, 200 to 225 HB	To refine grain size and obtain required hardness
Valve-bonnet forging	4140	Normalized at 870 °C (1600 °F) and tempered	Hardness, 220 to 240 HB	To obtain uniform structure, improved machinability and required hardness

oil quenched to decrease the total cooling time. In heavy sections, cooling of the center material to "black heat" may require considerable time. Thermal shock, residual thermally induced stress, and resultant distortions are factors to be considered. The microstructure remains essentially unaffected by the increased cooling rate, provided that the entire mass is below the lower critical temperature, Ar_1, although changes involving precipitates may occur.

Carbon Steels. Table 1 lists typical normalizing temperatures for some standard grades of carbon steel. These temperatures can be interpolated to obtain values for carbon contents not listed.

Steels containing 0.20% C or less usually receive no treatment subsequent to normalizing. However, medium-carbon or high-carbon steels are often tempered after normalizing to obtain specific properties, such as a lower hardness for straightening, cold working or machining. Whether or not tempering is desirable depends primarily on carbon content and section size. Table 3 presents typical mechanical properties of selected carbon and alloy steels in the hot rolled, normalized and annealed conditions. A low-carbon or medium-carbon steel of thin section may be harder after normalizing than a high-carbon steel of large section size subjected to the same treatment.

Alloy Steels. For alloy steel forgings, rolled products and castings, normalizing is commonly used as a conditioning treatment before final heat treatment. Normalizing also refines the structures of forgings, rolled products and castings that have cooled nonuniformly from high temperatures.

Table 1 lists typical normalizing temperatures for some standard alloy steels.

Alloy carburizing steels, such as 3310 and 4320, usually are normalized at temperatures higher than the carburizing temperature, to minimize distortion in carburizing and to improve machining characteristics. Carburizing steels of the 3300 series sometimes are double normalized with the expectation of minimizing distortion; these steels are tempered at about 650 °C (1200 °F) for intervals of up to 15 h to reduce hardness to below 223 HB for machinability. Carburizing steels of the 4300 and 4600 series usually can be normalized to a hardness not exceeding 207 HB, and therefore, need not be tempered for machinability.

Hypereutectoid alloy steels, such as 52100, are normalized for partial or complete elimination of carbide networks, thus producing a structure that is more susceptible to 100% spheroidization in the subsequent spheroidize annealing treatment. The spheroidized structure provides improved machinability and a more uniform response to hardening.

Forgings

When forgings are normalized before carburizing or before hardening and tempering, the upper range of normalizing temperatures is used. However, when normalizing is the final heat treatment, use is made of the lower range of temperatures.

Furnaces. Either batch-type or continuous furnaces may be used for normalizing steel forgings. In a continuous furnace, forgings to be normalized are usually placed in shallow pans, and a pusher mechanism at the loading end

of the furnace transports the pans through the furnace. Furnace burners, located on both sides of the furnace, fire below the hearth, and combustion products rise along the walls of the work-zone muffle and exhaust into the roof of the furnace. No atmosphere control is used. Combustion products enter the work zone through ports lining both sides of the entire hearth. A typical furnace is 9 m (30 ft) long and has 18 gas burners (or 9 oil burners) on each side. For purposes of temperature control, such a furnace is divided into three 3-m (10-ft) zones, each having a vertical thermocouple extending into it through the roof of the furnace.

Processing. Small forgings are usually normalized as received from the forge shop. They are placed or piled loosely on the pans to a maximum depth of 75 mm (3 in.). A typical furnace has five pans in each of the three furnace zones. Heating is adjusted so that the work reaches normalizing temperature in the last zone. After passing through the last zone, the pans are discharged onto a cooling conveyor. The work, while still in the pans, is cooled in still air to below 480 °C (900 °F); it is then discharged into tote boxes, in which it cools to room temperature. Total furnace time is approximately 3½ h, but during this period the work is held at the normalizing temperature for only 1 h.

Normalizing of large open-die forgings usually is performed in batch-type furnaces pyrometrically controlled to narrow temperature ranges. Forgings are held at the normalizing temperature long enough to allow complete austenitizing and carbide solution to occur (usually one hour per inch of section thickness), and then are cooled in still air.

Table 3 Properties of selected carbon and alloy steels in the hot rolled, normalized and annealed conditions

AISI grade(a)	Condition or treatment	Tensile strength MPa	ksi	Yield strength MPa	ksi	Elongation(b), %	Reduction in area, %	Hardness, HB	Izod impact strength J	ft·lb
1015	As rolled	420	61	315	46	39.0	61	126	111	82
	Normalized at 925 °C (1700 °F)	425	62	325	47	37.0	70	121	116	85
	Annealed at 870 °C (1600 °F)	385	56	285	41	37.0	70	111	115	85
1020	As rolled	450	65	330	48	36.0	59	143	87	64
	Normalized at 870 °C (1600 °F)	440	64	345	50	35.8	68	131	118	87
	Annealed at 870 °C (1600 °F)	395	57	295	43	36.5	66	111	123	91
1022	As rolled	505	73	360	52	35.0	67	149	81	60
	Normalized at 925 °C (1700 °F)	485	70	360	52	34.0	68	143	117	87
	Annealed at 870 °C (1600 °F)	450	65	315	46	35.0	64	137	121	89
1030	As rolled	550	80	345	50	32.0	57	179	75	55
	Normalized at 925 °C (1700 °F)	520	76	345	50	32.0	61	149	94	69
	Annealed at 845 °C (1550 °F)	465	67	340	50	31.2	58	126	69	51
1040	As rolled	620	90	415	60	25.0	50	201	49	36
	Normalized at 900 °C (1650 °F)	590	86	375	54	28.0	55	170	65	48
	Annealed at 790 °C (1450 °F)	520	75	355	51	30.2	57	149	44	33
1050	As rolled	725	105	415	60	20.0	40	229	31	23
	Normalized at 900 °C (1650 °F)	750	109	425	62	20.0	39	217	27	20
	Annealed at 790 °C (1450 °F)	635	92	365	53	23.7	40	187	17	13
1060	As rolled	815	118	485	70	17.0	34	241	18	13
	Normalized at 900 °C (1650 °F)	775	113	420	61	18.0	37	229	13	10
	Annealed at 790 °C (1450 °F)	625	91	370	54	22.5	38	179	11	8
1080	As rolled	965	140	585	85	12.0	17	293	7	5
	Normalized at 900 °C (1650 °F)	1010	147	525	76	11.0	21	293	7	5
	Annealed at 790 °C (1450 °F)	615	89	375	55	24.7	45	174	6	5
1095	As rolled	965	140	570	83	9.0	18	293	4	3
	Normalized at 900 °C (1650 °F)	1015	147	500	73	9.5	14	293	5	4
	Annealed at 790 °C (1450 °F)	655	95	380	55	13.0	21	192	3	2
1117	As rolled	485	71	305	44	33.0	63	143	81	60
	Normalized at 900 °C (1650 °F)	465	68	305	44	33.5	54	137	85	63
	Annealed at 860 °C (1575 °F)	430	62	280	41	32.8	58	121	94	69
1118	As rolled	520	76	315	46	32.0	70	149	109	80
	Normalized at 925 °C (1700 °F)	480	69	320	46	33.5	66	143	104	76
	Annealed at 790 °C (1450 °F)	450	65	285	41	34.5	67	131	106	79
1137	As rolled	625	91	380	55	28.0	61	192	83	61
	Normalized at 900 °C (1650 °F)	670	97	395	58	22.5	49	197	64	47
	Annealed at 790 °C (1450 °F)	585	85	345	50	26.8	54	174	50	37
1141	As rolled	675	98	360	52	22.0	38	192	11	8
	Normalized at 900 °C (1650 °F)	705	103	405	59	22.7	56	201	53	39
	Annealed at 815 °C (1500 °F)	600	87	355	51	25.5	49	163	34	25
1144	As rolled	705	102	420	61	21.0	41	212	53	39
	Normalized at 900 °C (1650 °F)	665	97	400	58	21.0	40	197	43	32
	Annealed at 790 °C (1450 °F)	585	85	345	50	24.8	41	167	65	48
1340	Normalized at 870 °C (1600 °F)	835	121	560	81	22.0	63	248	93	68
	Annealed at 800 °C (1475 °F)	705	102	435	63	25.5	57	207	71	52
3140	Normalized at 870 °C (1600 °F)	890	129	600	87	19.7	57	262	54	40
	Annealed at 815 °C (1500 °F)	690	100	425	61	24.5	51	197	46	34
4130	Normalized at 870 °C (1600 °F)	670	97	435	63	25.5	60	197	86	64
	Annealed at 865 °C (1585 °F)	560	81	360	52	28.2	56	156	62	46
4140	Normalized at 870 °C (1600 °F)	1020	148	655	95	17.7	47	302	23	17
	Annealed at 815 °C (1500 °F)	655	95	415	61	25.7	57	197	55	40
4150	Normalized at 870 °C (1600 °F)	1155	168	735	107	11.7	31	321	12	9
	Annealed at 815 °C (1500 °F)	730	106	380	55	20.2	40	197	25	18
4320	Normalized at 895 °C (1640 °F)	795	115	465	67	20.8	51	235	73	54
	Annealed at 850 °C (1560 °F)	580	84	425	62	29.0	58	163	110	81
4340	Normalized at 870 °C (1600 °F)	1280	186	860	125	12.2	36	363	16	12
	Annealed at 810 °C (1490 °F)	745	108	470	69	22.0	50	217	51	38

(a) All grades are fine grained except for those in the 1100 series, which are coarse grained. (b) In 50 mm or 2 in.

Table 3 (continued)

AISI grade(a)	Condition or treatment	Tensile strength MPa	ksi	Yield strength MPa	ksi	Elonga-tion(b), %	Reduction in area, %	Hardness, HB	Izod impact strength J	ft·lb
4620	Normalized at 900 °C (1650 °F)	575	83	365	53	29.0	67	174	133	98
	Annealed at 860 °C (1575 °F)	510	74	370	54	31.3	60	149	94	69
4820	Normalized at 860 °C (1580 °F)	755	110	485	70	24.0	59	229	110	81
	Annealed at 815 °C (1500 °F)	680	99	465	67	22.3	59	197	93	69
5140	Normalized at 870 °C (1600 °F)	795	115	470	69	22.7	59	229	38	28
	Annealed at 830 °C (1525 °F)	570	83	295	43	28.6	57	167	41	30
5150	Normalized at 870 °C (1600 °F)	870	126	530	77	20.7	59	255	32	23
	Annealed at 825 °C (1520 °F)	675	98	355	52	22.0	44	197	25	19
5160	Normalized at 860 °C (1575 °F)	955	139	530	77	17.5	45	269	11	8
	Annealed at 810 °C (1495 °F)	725	105	275	40	17.2	31	197	10	7
6150	Normalized at 870 °C (1600 °F)	940	136	615	89	21.8	61	269	36	26
	Annealed at 815 °C (1500 °F)	665	97	410	60	23.0	48	197	27	20
8620	Normalized at 910 °C (1675 °F)	635	92	355	52	26.3	60	183	100	74
	Annealed at 870 °C (1600 °F)	535	78	385	56	31.3	62	149	112	83
8630	Normalized at 870 °C (1600 °F)	650	94	430	62	23.5	54	187	95	70
	Annealed at 845 °C (1550 °F)	565	82	370	54	29.0	59	156	95	70
8650	Normalized at 870 °C (1600 °F)	1025	149	690	100	14.0	45	302	14	10
	Annealed at 795 °C (1465 °F)	715	104	385	56	22.5	46	212	29	22
8740	Normalized at 870 °C (1600 °F)	930	135	605	88	16.0	48	269	18	13
	Annealed at 815 °C (1500 °F)	695	101	415	60	22.2	46	201	40	30
9255	Normalized at 900 °C (1650 °F)	935	135	580	84	19.7	43	269	14	10
	Annealed at 845 °C (1550 °F)	775	112	485	71	21.7	41	229	9	7
9310	Normalized at 890 °C (1630 °F)	905	132	570	83	18.8	58	269	119	88
	Annealed at 845 °C (1550 °F)	820	119	440	64	17.3	42	241	79	58

(a) All grades are fine grained except for those in the 1100 series, which are coarse grained. (b) In 50 mm or 2 in.

Axle-Shaft Forging. In forging an axle shaft made of fine-grain 1049 steel, only one end of the forging bar was heated to upset the wheel-flange section. When the part was examined in cross section from the flanged end to the cold end, the following metallurgical conditions were revealed:

The hot worked flanged area of the axle exhibited a fine-grain structure as a result of the hot working at the forging temperature (approximately 1100 °C or 2000 °F). However, a section adjacent to the flange, which also had been heated to the forging temperature but which had not been hot worked, exhibited a coarse-grain structure. Nearer the cool end of the shaft, a zone that reached a temperature of about 700 °C (1300 °F) exhibited a spheroidized structure. The cold end of the shaft retained its initial fine grain size throughout the forging operation.

In subsequent operations, this shaft was to be mechanically straightened, machined, and induction hardened. Because of the mixed grain structure,

these operations posed several problems. The coarse-grain area adjacent to the flange was extremely weak in the transverse direction, and there was a possibility that fracture would occur if this section were subjected to a severe straightening operation. The spheroidized area would not respond adequately to induction hardening, because the solution rate of this type of carbide formation was too sluggish for the relatively rapid rate of induction heating. Furthermore, the mixed metallurgical structure would present difficulties in machining. Consequently, normalizing was required in order to produce a uniformly fine-grain structure throughout the axle shaft prior to straightening, machining and induction hardening.

Low-Carbon Steel Forgings. In contrast to the medium-carbon axle shaft discussed in the preceding paragraphs, forgings made of carbon steels containing 0.25% C or less are seldom normalized. Only severe quenching from above the austenitizing temperature will have any significant effect on their structure or hardness.

Structural Stability. Normalizing and tempering is also a preferred treatment for promoting the structural stability of low-alloy heat-resistant alloys, such as AMS 6304 (0.45 C, 1 Cr, 0.5 Mo, 0.3 V), at temperatures up to 540 °C (1000 °F). Wheels and spacer rings used in the "cold" ends of aircraft gas turbine engine compressors are typical of parts subjected to such treatment to promote structural stability.

Effects on Mechanical Properties. Differences in mechanical properties obtained by normalizing and tempering and by quenching and tempering result from differences in the rate of cooling from the austenitizing temperature, and hence are functions of the hardenability of the steel and the section size of the part. For air cooling, just as for liquid quenching, a larger section size requires a higher alloy content if a given hardness is to be maintained. Attainment of the same hardness in a normalized part as in a liquid-quenched part also requires a higher alloy content, to compensate for the smaller hardening response elicited

Table 4 Effect of mass on hardness of normalized carbon and alloy steels

All data are based on single heats. Sources: data for 3310, 3140 and 4063 are from *Modern Steels and Their Properties,* 6th Ed., Bethlehem Steel Corp., 1966; all other data are from *Modern Steels and Their Properties* (Handbook 3310), Bethlehem Steel Corp., Sept 1978.

Grade	Normalizing temperature °C	°F	13(1/2)	25(1)	50(2)	100(4)
Carbon steels, carburizing grades						
1015 925		1700	126	121	116	116
1020 925		1700	131	131	126	121
1022 925		1700	143	143	137	131
1117 900		1650	143	137	137	126
1118 925		1700	156	143	137	131
Carbon steels, direct-hardening grades						
1030 925		1700	156	149	137	137
1040 900		1650	183	170	167	167
1050 900		1650	223	217	212	201
1060 900		1650	229	229	223	223
1080 900		1650	293	293	285	269
1095 900		1650	302	293	269	255
1137 900		1650	201	197	197	192
1141 900		1650	207	201	201	201
1144 900		1650	201	197	192	192
Alloy steels, carburizing grades						
3310 890		1630	269	262	262	248
4118 910		1670	170	156	143	137
4320 895		1640	248	235	212	201
4419 955		1750	149	143	143	143
4620 900		1650	192	174	167	163
4820 860		1580	235	229	223	212
8620 915		1675	197	183	179	163
9310 890		1630	285	269	262	255
Alloy steels, direct-hardening grades						
1340 870		1600	269	248	235	235
3140 870		1600	302	262	248	241
4027 905		1660	179	179	163	156
4063 870		1600	285	285	285	277
4130 870		1600	217	197	167	163
4140 870		1600	302	302	285	241
4150 870		1600	375	321	311	293
4340 870		1600	388	363	341	321
5140 870		1600	235	229	223	217
5150 870		1600	262	255	248	241
5160 860		1575	285	269	262	255
6150 870		1600	285	269	262	255
8630 870		1600	201	187	187	187
8650 870		1600	363	302	293	285
8740 870		1600	269	269	262	255
9255 900		1650	277	269	269	269

by air cooling compared with that brought about by liquid quenching.

Hardness is not the only property that is affected by differences in cooling rate. Other mechanical properties may be expected to differ from the normalized and tempered condition to the quenched and tempered condition, even when surface hardness is about the same for both treatments. For example, although tensile strength will be about the same, yield strength, elongation, and reduction in area will be lower in the normalized and tempered condition than in the quenched and tempered condition.

Multiple normalizing treatments are employed (*a*) to obtain complete solution of all lower-temperature constituents in austenite by the use of high initial normalizing temperatures (for example, 925 °C or 1700 °F), and (*b*) to refine final pearlite grain size by the use of a second normalizing treatment at a temperature closer to the Ac$_3$ temperature (for example, 815 °C or 1500 °F) without destroying the beneficial effects of the initial normalizing treatment.

Locomotive-axle forgings made of carbon steel to AAR Specification M-126, Class F (ASTM A236, Class F), containing 0.45 to 0.59% C and 0.60 to 0.90% Mn, are double normalized to obtain a uniformly fine grain structure along with other exacting mechanical-property requirements. Forgings made of a low-carbon steel (0.18% C) with 1% Mn intended for low-temperature service are double normalized to meet subzero impact requirements.

Bar and Tubular Products

Frequently, the finishing stages of hot-mill operations employed in making steel bar and tube produce properties that closely approximate those obtained by normalizing. When this occurs, normalizing is unnecessary and may even be inadvisable. Nevertheless,

Table 5 Typical mechanical properties of normalized alloy steel sheet

Grade	Thickness mm	in.	Tensile strength MPa	ksi	Yield strength(a) MPa	ksi	Elongation(b), %	Hardness, HRC
4130	4.9	0.193 835		121	585	85	14	25
4335(c)	4.6	0.180 1725		250	1240	180	8	48
4340(c)	2.0	0.080 1860		270	1345	195	7	50

(a) At 0.2% offset. (b) In 50 mm or 2 in. (c) Modified: 0.40% Mo, 0.20% V.

the reasons for normalizing bar and tube products are generally the same as those applicable to other forms of steel.

There is an additional reason, however, for normalizing bar and tube products. When these products are cold finished by a sequence of cold reductions with high subcritical anneals between passes, some spheroidization occurs. In such instances, the product is sometimes normalized before the last cold reduction pass. Normalizing eliminates the spheroidization that earlier passes and anneals may have generated and restores the pearlitic structure beneficial to machinability in low-carbon and medium-carbon grades of carbon or alloy steel.

Tubes are easier to normalize than bars of equivalent diameter, because the lighter section thickness of tubes permits more rapid heating and cooling. These advantages help minimize decarburization and promote more nearly uniform microstructures in tube products.

Furnaces. Continuous furnaces of the roller-hearth type are widely used for normalizing tube and bar products, especially in long lengths. Batch-type furnaces or other types of continuous furnaces are satisfactory if they provide some means for rapid discharge and separation of the load to permit free circulation of air around each tube as it cools. Continuous furnaces should have at least two zones, one for heating and one for soaking. Cooling facilities should be ample, so that uniform cooling can proceed until complete transformation has occurred. If tubes are packed or bundled during cooling from a high temperature, the purpose of normalizing is defeated and a semiannealed or a tempered product results.

Generally, protective atmospheres are not used in roller-hearth continuous furnaces for normalizing bar or tube products. The scale that forms during normalizing is removed by acid pickling or abrasive blast cleaning.

Alloy Steels. Although the principles involved in normalizing alloy steel bar, tube and pipe products are the same as those for normalizing carbon steels, application of these principles is sometimes more complex for alloy steels. Some alloy grades require more care in heating, to prevent cracking from thermal shock. They also require longer soaking times because of lower austenitizing and solution rates. For many alloy steels, rates of cooling in air

to room temperature must be carefully controlled. Certain alloy steels are forced-air cooled from the normalizing temperature in order to develop specific mechanical properties. This is a normalizing treatment only in the microstructural sense discussed near the beginning of this article.

Castings

In industrial practice, steel castings may be normalized in car-bottom, box, pit or continuous furnaces; the same heat treating principles apply to each type of furnace.

Loading. Furnaces are loaded with castings in such a manner that each casting will receive an adequate and uniform heat supply. This may be accomplished by stacking castings in regular order or by interspersing large and small castings so that load concentration in any one area is not excessive. At normalizing temperatures, the tensile strength of steel is greatly reduced, and heavy unequal sections may become distorted unless bracing and support are provided. Accordingly, small and large castings may be arranged so that they support each other.

Loading Temperature. When castings are charged, the temperature of the furnace should be such that the thermal shock will not cause metal failure. For the higher-alloy grades of steel castings, such as C5, C12 and WC9, a safe furnace temperature for charging is 315 to 425 °C (600 to 800 °F). For lower-alloy grades, furnace temperatures may be as high as 650 °C (1200 °F). For cast carbon steels and low-alloy steels with low carbon contents (low hardenability), castings may be charged into a furnace operating at the normalizing temperature.

Heating. After the furnace has been charged, the temperature is increased at a rate of approximately 220 °C/h (400 °F/h) until the normalizing temperature is reached. Depending on steel composition and casting configuration, a reduction in the rate of heating to approximately 28 to 55 °C/h (50 to 100 °F/h) may be necessary to avoid cracking. Extremely large castings should be heated more slowly, to prevent development of extreme temperature gradients.

Soaking. After the normalizing temperature has been reached, castings are "soaked" at this temperature for a period of time that will ensure complete austenitization and carbide solution. The duration of the soaking

period may be predetermined by microscopic examination of specimens held for various times at the normalizing temperature.

Cooling. After the soaking period, the castings are unloaded and allowed to cool in still air. Use of fans, air blasts or other means of accelerating the cooling process should be avoided.

Sheet and Strip

Hot rolled steel sheet and strip (about 0.10% C) are normalized primarily to refine grain size, minimize directional properties and develop desirable mechanical properties. Uniformly fine equiaxed ferrite grains normally are obtained in hot rolled sheet and strip by finishing the final hot rolling operation above the upper transformation temperature. However, if part of the hot rolling operation is performed on steel that has transformed partially to ferrite, the deformed ferrite grains usually will recrystallize and form abnormally coarse-grain patches during the self-anneal induced by coiling or piling at temperatures of 650 to 730 °C (1200 to 1350 °F). Also, relatively thin hot rolled material, if it is inadvertently finished well below the upper transformation temperature and coiled or piled while it is too cold to self-anneal, may possess directional properties. These conditions are unsuitable for some types of severe press drawing applications and may be corrected by normalizing.

Normalizing also may be used to develop high strength in alloy steel sheet and strip if the products are sufficiently high in carbon and alloy contents to enable them to transform to fine pearlite or martensite when cooled in air from the normalizing temperature. In general, the hardened material is tempered to attain an optimum combination of strength and ductility. Typical mechanical properties of normalized 4130, modified 4335 and modified 4340 steel sheet are given in Table 5.

Processing. The normalizing operation consists of passing the sheet or strip through an open, continuous furnace in which the material is heated to a temperature approximately 55 to 85 °C (100 to 150 °F) above its upper transformation temperature, 845 or 900 °C (1550 to 1650 °F), thus obtaining complete solution of the original structure with the formation of austenite, and then air cooling the material to room temperature.

Furnace Equipment. Normalizing furnaces are designed to heat and cool sheets singly or two in a pile. They are built in the form of long, low chambers, and usually comprise three sections: a preheating zone (12 to 20% of the total length); a heating, or soaking, zone (about 40% of the total length); and a cooling zone, which occupies the remaining 40 to 50% of the length.

Heating Arrangements. Normalizing furnaces usually are heated with gas or oil and do not employ protective atmospheres. Therefore, sheets are scaled during heat treatment. Burners are arranged along each side of the heating zone; they usually are above the conveyor, but occasionally are both above and below it. The furnace roof, which is higher in the preheating and soaking zones than in the cooling zone, is usually built in sections. In most furnaces, both the preheating zone and the cooling zone are heated by the hot gases from the heating zone. However, both of these zones may be equipped with burners for more accurate temperature control. Air is excluded by regulating the draft to maintain a slight pressure within all zones.

Conveyor-Type Furnaces. In modern furnaces of the conveyor type (the only type suitable for treating short lengths), sheets are carried through each of the three zones on rotating disks made of heat-resistant alloys. These disks have polished surfaces, which prevent them from scratching the sheets, and are staggered to ensure uniform heating. The disks are mounted on water-cooled shafts, which are driven by variable-speed motors through chains and sprockets or shafts and gears. These furnaces may be up to 2.5 m (100 in.) wide and from 27 to 61 m (90 to 200 ft) long. Fuel consumption is 2.3 to 5.2 million kJ per tonne (2.0 to 4.5 million Btu per ton) of steel treated, and production rates vary from 2.7 to 10.9 tonnes (3 to 12 tons) per hour.

Normalizing in a three-zone conveyor-type furnace equipped with pyrometric controls is a relatively simple operation. If scratching of sheets is to be avoided, the sheets are brought to the charging table and hand laid, one or more at a time, on a rider or conveyor sheet. Heavy sheets are normalized singly, but lighter sheets may be stacked two in a pile. To control heating and retard scaling, single sheets may be laid on a rider sheet and covered with a cover sheet. Sheets are carried by disk-rollers into the preheating zone, where they absorb heat rapidly because of the large temperature differential between the sheets and the interior of the furnace and because of the large surface-to-volume ratio. As the sheets become heated and the temperature differential is reduced, the rate of heat absorption slackens. After traveling 4½ to 6 m (15 to 20 ft), the sheets enter the soaking zone at a temperature several degrees below the normalizing temperature. Heating is completed in the soaking zone, which is maintained at a constant temperature, and sheets are held at the required temperature for a time sufficient to convert the microstructure to austenite before they are passed into the cooling zone. The sheets emerge from the cooling zone at a temperature that can be varied between 150 and 540 °C (300 and 1000 °F), and are conveyed for a short distance on the runout table, where, after being cooled rapidly in air, they are carefully removed from the rider sheet. The trip through such a furnace is carried out at a uniform speed of 0.03 to 0.10 m/s (5 to 20 ft/min) and requires 5 to 20 min to complete.

Catenary Furnaces. The catenary, or free-loop, type of furnace is designed for continuous normalizing of cold reduced steel unwound from coils; it does not have rolls or any other type of conveyor for supporting the material passing through the heating zone. The heating zones of catenary furnaces range in length from 6 to 15 m (20 to 50 ft). The preheating and cooling zones usually are shorter than those in conveyor-type furnaces, and for some kinds of work may be omitted entirely. At their exit ends, catenary furnaces may incorporate pickling or other descaling equipment for removing surface oxides formed on the steel during normalizing.

Annealing of Steel

By the ASM Committee on
Annealing of Steel*

ANNEALING is a generic term denoting a treatment that consists of heating to and holding at a suitable temperature followed by cooling at an appropriate rate, primarily for softening of metallic materials. It also is applied to produce desired changes in other properties or in microstructure. Steels may be annealed to facilitate cold working or machining, to improve mechanical or electrical properties, or to promote dimensional stability. The choice of an annealing treatment that will provide an adequate combination of such properties at minimum expense often involves a compromise. Terms used to denote specific types of annealing applied to steels are descriptive of the method used, the equipment used or the condition of the material after treatment. Many of these terms are discussed in the following sections or are defined in the glossary of annealing terminology at the end of this article.

Basic Concepts†

Critical Temperatures. The critical temperatures that must be considered in discussing annealing of steel are those that define the onset and completion of the transformation to or from austenite. For a given steel, the critical temperatures depend on whether the steel is being heated or cooled. Critical temperatures for the start and completion of the transformation to austenite during heating are denoted, respectively, by Ac_1 and Ac_3 for hypoeutectoid steels and by Ac_1 and Ac_{cm} for hypereutectoid steels. These temperatures are higher than the corresponding critical temperatures for the start and completion of the transformation from austenite during cooling, which are denoted, respectively, by Ar_3 and Ar_1 for hypoeutectoid steels and by Ar_{cm} and Ar_1 for hypereutectoid steels. These critical temperatures converge to the equilibrium values Ae_1, Ae_3 and Ae_{cm} as the rates of heating or cooling become infinitely slow. Figure 1 illustrates the positions of the Ae_1, Ae_3 and Ae_{cm} lines on the equilibrium phase diagram for plain carbon steels. The presence of other alloying elements will also have marked effects on these critical temperatures.

Table 1 provides critical temperatures for selected steels, measured at heating and cooling rates of 28 °C/h (50 °F/h). The equilibrium critical temperatures generally lie about midway between those for heating and cooling

at equal rates. Because annealing may involve various ranges of heating and cooling rates in combination with isothermal treatments, the less specific terms A_1, A_3 and A_{cm} are used here in discussing the basic concepts.

Annealing Cycles. In practice, specific thermal cycles of an almost infinite variety are used to achieve the various goals of annealing. These cycles fall into several broad categories that can be classified according to the temperature to which the steel is heated and the method of cooling used. The maximum temperature may be below the lower critical temperature, A_1 (subcritical annealing); above A_1 but below the upper critical temperature, A_3 in hypoeutectoid steels or A_{cm} in hypereutectoid steels (intercritical annealing); or above A_3 (full annealing).

Because some austenite is present at temperatures above A_1, cooling practice through transformation is a crucial factor in achieving desired microstructures and properties. Accordingly, steels heated above A_1 are subjected either to slow continuous cooling or to isothermal treatment at some temperature below A_1 at which transformation to the desired microstructure can occur in a reasonable amount of time. These

*Charles A. Apple, *Chairman,* Engineer, Homer Research Laboratories, Bethlehem Steel Corp.; Casimir F. Dombkowski, Senior Metallurgist, New Departure Div., General Motors Corp.; Jack E. Haflinger, Chief Metallurgist, Lamson & Sessions Co.; Daniel J. Hayes, Metallurgical Engineer, United States Steel Corp.; Richard J. Light, Chief Process Metallurgist, Surface Div., Midland Ross Corp.; George L. Schiel, Chief Metallurgist, Metlab Co.; Edwin J. Schneider, Supervisor, Flat Rolled Products, Graham Laboratories, Jones & Laughlin Steel Corp.; Paul R. Slimmon, Supervisor, Alloying and Heat Treatment, Homer Research Laboratories, Bethlehem Steel Corp.; Dwight A. Wilkinson, Chief Metallurgist, Saginaw Steering Gear Div., General Motors Corp.
† Parts of this and succeeding sections have been adapted from Ref. 1

Fig. 1 Fe-Fe₃C phase diagram, showing the temperature range of interest for annealing plain carbon steels

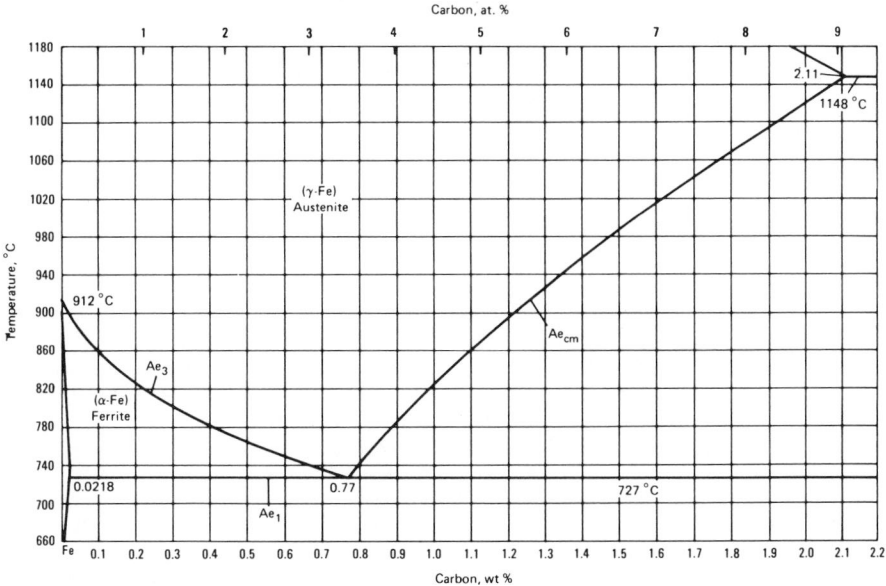

The equilibrium transformation temperatures Ae₁, Ae₃ and Ae_cm are labeled on the diagram.

Fig. 2 Schematic representation of some basic annealing schedules for a hypoeutectoid steel

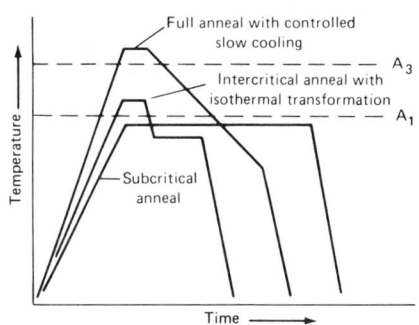

Table 1 Approximate critical temperatures for selected carbon and low-alloy steels

Steel	Critical temperatures on heating at 28 °C/h (50 °F/h)				Critical temperatures on cooling at 28 °C/h (50 °F/h)			
	Ac₁		Ac₃		Ar₃		Ar₁	
	°C	°F	°C	°F	°C	°F	°C	°F
1010	724	1335	877	1610	849	1560	682	1260
1020	724	1335	846	1555	816	1500	682	1260
1030	727	1340	813	1495	788	1450	677	1250
1040	727	1340	793	1460	757	1395	671	1240
1050	727	1340	768	1415	741	1365	682	1260
1060	727	1340	746	1375	727	1340	685	1265
1070	727	1340	732	1350	710	1310	691	1275
1080	729	1345	735	1355	699	1290	693	1280
1340	716	1320	777	1430	721	1330	621	1150
3140	735	1355	766	1410	721	1330	660	1220
4027	727	1340	807	1485	760	1400	671	1240
4042	727	1340	793	1460	732	1350	654	1210
4130	757	1395	810	1490	754	1390	693	1280
4140	732	1350	804	1480	743	1370	679	1255
4150	743	1370	766	1410	729	1345	671	1240
4340	724	1335	774	1425	710	1310	654	1210
4615	727	1340	810	1490	760	1400	649	1200
5046	716	1320	771	1420	732	1350	682	1260
5120	766	1410	838	1540	799	1470	699	1290
5140	738	1360	788	1450	727	1340	693	1280
5160	710	1310	766	1410	716	1320	677	1250
52100	727	1340	768	1415	716	1320	688	1270
6150	749	1380	788	1450	743	1370	693	1280
8115	721	1330	838	1540	788	1450	671	1240
8620	732	1350	829	1525	768	1415	660	1220
8640	732	1350	779	1435	727	1340	666	1230
9260	743	1370	816	1500	749	1380	713	1315

generalized treatments are illustrated schematically in Fig. 2. Under certain conditions, two or more such cycles may be combined or used in succession to achieve the desired results. The success of any annealing operation depends on proper choice and control of the thermal cycle, based on the metallurgical principles discussed in the following sections.

Heating Below A₁. Subcritical annealing does not involve formation of austenite. The prior condition of the steel is modified by such thermally activated processes as recovery, recrystallization, grain growth, and agglomeration of carbides. The prior history of the steel is, therefore, an important factor.

In as-rolled or forged hypoeutectoid steels containing ferrite and pearlite, subcritical annealing can adjust the hardnesses of both constituents, but excessively long times at temperature may be required for substantial softening. The subcritical treatment is most effective when applied to hardened or cold worked steels, which recrystallize readily to form new ferrite grains. The rate of softening increases rapidly as the annealing temperature approaches A₁. Cooling practice from the subcritical annealing temperature has very little effect on the established microstructure and resultant properties.

A more detailed discussion of the metallurgical processes involved in subcritical annealing is provided in Ref 2.

Heating Above A₁. Austenite begins to form when the temperature of the steel exceeds A₁. In hypoeutectoid steels, the equilibrium structure in the intercritical range between A₁ and A₃ consists of ferrite and austenite, and above A_3 the structure becomes completely austenitic. However, the equilibrium mixture of ferrite and austenite is not achieved instantaneously. Undissolved carbides may persist, especially if the austenitizing time is short or the temperature is near A_1, causing the austenite to be inhomogeneous. In hypereutectoid steels, carbide and austenite coexist in the intercritical range between A_1 and A_{cm} and the homogeneity of the austenite depends on time and temperature. The degree of homogeneity in the structure at the austenitizing temperature is an important consideration in the development of annealed structures and properties. The more homogeneous structures developed at higher austenitizing temperatures tend to promote lamellar carbide structures on cooling, whereas lower austenitizing temperatures in the intercritical range result in less homogeneous austenite, which promotes formation of spheroidal carbides.

Decomposition of Austenite. Austenite formed when steel is heated above the A_1 temperature transforms back to ferrite and carbide when the steel is cooled below A_1. The rate of austenite decomposition and the tendency of the carbide structure to be either lamellar or spheroidal depend largely on the temperature of transformation. If the austenite transforms just below A_1, it will decompose slowly. The product then may contain relatively coarse spheroidal carbides or coarse lamellar pearlite, depending on the composition of the steel and the austenitizing temperature. This product tends to be very soft. However, the low rate of transformation at temperatures just below A_1 necessitates long holding times in isothermal treatments, or very low cooling rates in continuous cooling, if maximum softness is desired. Isothermal treatments are more efficient than slow continuous cooling in terms of achieving desired structures and softness in the minimum amount of time. Sometimes, however, the available equipment or the mass of the steel part being annealed may make slow continuous cooling the only feasible alternative.

As the transformation temperature decreases, austenite generally decomposes more rapidly, and the transformation product is harder, more lamellar and less coarse than the product formed just below A_1. At still lower transformation temperatures, the product becomes a much harder mixture of ferrite and carbide, and the time necessary for complete isothermal transformation may again increase. Temperature-time plots showing the progress of austenite transformation under isothermal (IT) or continuous cooling transformation (CCT) conditions for many steels have been widely published (Ref 3 and 4) and illustrate the principles just discussed. These IT or CCT diagrams may be helpful in design of annealing treatments for specific grades of steel, but their usefulness is limited because most published diagrams represent transformation from a fully austenitized, relatively homogeneous condition, which is not always desirable or obtainable in annealing. However, transformation diagrams (either IT or CCT) that represent cooling from a specific austenitizing treatment may be developed by use of the techniques discussed in Ref 1. Such diagrams provide the information necessary for design of effective annealing schedules.

Cooling After Transformation. After the austenite has been completely transformed, little else of metallurgical consequence can occur during cooling to room temperature. Extremely slow cooling may cause some agglomeration of carbides and, consequently, some slight further softening of the steel, but in this regard such slow cooling is less effective than high-temperature transformation. Therefore, there is no metallurgical reason for slow cooling after transformation has been completed, and the steel may be cooled from the transformation temperature as rapidly as feasible in order to minimize the total time required for the operation.

If transformation by slow continuous cooling has been used, the temperature at which controlled cooling may be stopped depends on the transformation characteristics of the steel. However, the mass of the steel or the need to avoid oxidation are practical considerations that may require retarded cooling to be continued below the temperature at which austenite transformation ceases.

Effect of Prior Structure. The finer and more evenly distributed the carbides in the prior structure, the faster the rate at which austenite formed above A_1 will approach complete homogeneity. The prior structure, therefore, can affect the response to annealing. When spheroidal carbides are desired in the annealed structure, preheating at temperatures just below A_1 sometimes is used to agglomerate the prior carbides in order to increase their resistance to solution in the austenite on subsequent heating. The presence of undissolved carbides or concentration gradients in the austenite promotes formation of a spheroidal, rather than lamellar, structure when the austenite is transformed. Preheating to enhance spheroidization is applicable mainly to hypoeutectoid steels but also is useful for some hypereutectoid low-alloy steels.

Austenitizing Time and Dead-Soft Steel. Hypereutectoid steels can be made extremely soft by holding for long periods of time at the austenitizing temperature. Although the time at the austenitizing temperature may have only a small effect on actual hardness (such as a change from 241 to 229 HB), its effect on machinability or cold forming properties may be appreciable.

Long-term austenitizing is effective in hypereutectoid steels because it produces agglomeration of residual carbides in the austenite. Coarser carbides promote a softer final product. In lower-carbon steels, carbides are unstable at temperatures above A_1 and tend to dissolve in the austenite, although the dissolution may be slow.

Steels that have approximately eutectoid carbon contents generally form a lamellar transformation product if austenitized for very long periods of time. Long-term holding at a temperature just above the A_1 temperature may be as effective in dissolving carbides and dissipating carbon-concentration gradients as is short-term holding at a higher temperature.

Guidelines for Annealing

The metallurgical principles discussed above have been incorporated by Payson (Ref 1) into the following seven rules, which may be used as guidelines for development of successful and efficient annealing schedules:

Rule 1: The more homogeneous the structure of the as-austenitized steel, the more completely lamellar will be the structure of the annealed steel. Conversely, the more heterogeneous the structure of the as-austenitized steel, the more nearly spheroidal will be the annealed carbide structure.

Rule 2: The softest condition in the steel is usually developed by austenitizing at a temperature less than 56

°C (100 °F) above A_1 and transforming at a temperature (usually) less than 56 °C (100 °F) below A_1.

Rule 3: Because very long times may be required for complete transformation at temperatures less than 56 °C (100 °F) below A_1, allow most of the transformation to take place at the higher temperature, where a soft product is formed, and finish the transformation at a lower temperature, where the time required for completion of transformation is short.

Rule 4: After the steel has been austenitized, cool to the transformation temperature as rapidly as feasible in order to minimize the total duration of the annealing operation.

Rule 5: After the steel has been completely transformed, at a temperature that produces the desired microstructure and hardness, cool to room temperature as rapidly as feasible, to decrease further the total time of annealing.

Rule 6: To ensure a minimum of lamellar pearlite in the structures of annealed 0.70 to 0.90% C tool steels and other low-alloy medium-carbon steels, preheat for several hours at a temperature about 28 °C (50 °F) below the lower critical temperature (A_1) before austenitizing and transforming as usual.

Rule 7: To obtain minimum hardness in annealed hypereutectoid alloy tool steels, heat at the austenitizing temperature for a long time (about 10 to 15 h), then transform as usual.

These rules are applied most effectively when the critical temperatures and transformation characteristics of the steel have been established and when transformation by isothermal treatment is feasible.

Annealing Temperatures

For many annealing applications, it is sufficient simply to specify that the steel be cooled in the furnace from a designated annealing (austenitizing) temperature. Temperatures and associated Brinell hardnesses for simple annealing of carbon steels are given in Table 2, and similar data for alloy steels are presented in Table 3.

Heating cycles that employ austenitizing temperatures in the upper ends of the ranges given in Table 3 should result in pearlitic structures. Predominantly spheroidized structures should

Table 2 Recommended temperatures and cooling cycles for full annealing of small carbon steel forgings

Data are for forgings up to 75 mm (3 in.) in section thickness. Time at temperature usually is a minimum of 1 h for sections up to 25 mm (1 in.) thick; ½ h is added for each additional 25 mm (1 in.) of thickness

Steel	Annealing temperature °C	°F	Cooling cycle(a) °C From	To	°F From	To	Hardness range, HB
1018	855–900	1575–1650	855	705	1575	1300	111–149
1020	855–900	1575–1650	855	700	1575	1290	111–149
1022	855–900	1575–1650	855	700	1575	1290	111–149
1025	855–900	1575–1650	855	700	1575	1290	111–187
1030	845–885	1550–1625	845	650	1550	1200	126–197
1035	845–885	1550–1625	845	650	1550	1200	137–207
1040	790–870	1450–1600	790	650	1450	1200	137–207
1045	790–870	1450–1600	790	650	1450	1200	156–217
1050	790–870	1450–1600	790	650	1450	1200	156–217
1060	790–845	1450–1550	790	650	1450	1200	156–217
1070	790–845	1450–1550	790	650	1450	1200	167–229
1080	790–845	1450–1550	790	650	1450	1200	167–229
1090	790–830	1450–1525	790	650	1450	1200	167–229
1095	790–830	1450–1525	790	655	1450	1215	167–229

(a) Furnace cooling at 28 °C/h (50 °F/h)

Fig. 3 Heating and cooling cycle for process annealing of a load of coiled low-carbon steel sheet

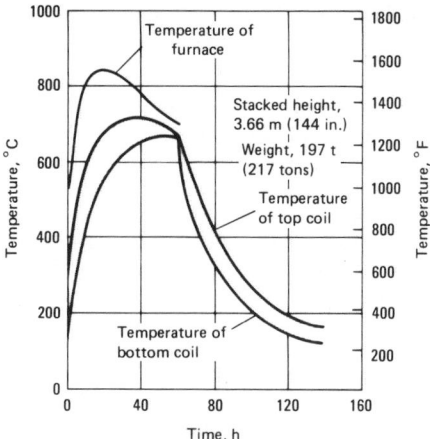

be obtained when lower temperatures are used.

When an alloy steel is annealed to obtain a specific microstructure, greater precision is required in specifying temperatures and cooling conditions for annealing. Table 4 presents, for a variety of standard alloy steels, typical schedules for such annealing operations.

In isothermal annealing to produce a pearlitic structure, particularly in forgings, an austenitizing temperature as much as 70 °C (125 °F) higher than that indicated in Table 4 may be selected in

order to decrease the austenitizing time.

For most steels, as indicated in Table 4, annealing may be accomplished by heating to the austenitizing temperature and then either (a) cooling in the furnace at a controlled rate or (b) cooling rapidly to, and holding at, a lower temperature for isothermal transformation. Both procedures result in virtually the same hardness; however, considerably less time is required for isothermal transformation.

Uniformity of Temperature. One potential contribution to the failure of an annealing operation is a lack of knowledge of the temperature distribution within the load of steel in the furnace. Furnaces large enough to anneal 18 t (20 tons) of steel at a time are not uncommon. The larger the furnace, the more difficult it is to establish and maintain uniform temperature conditions throughout the load, and the more difficult it is to change the temperature of the steel during either heating or cooling.

Furnace thermocouples indicate the temperature of the space above, below or beside the load, but this temperature may differ by 28 °C (50 °F) or more from the temperature of the steel itself (see Fig. 3), especially when the steel is in a pipe or box or when bar or strip is packed in a dense charge in a quiescent atmosphere. When these conditions exist, the distribution of temperature throughout the load during heating and cooling should be established by

Table 3 Recommended annealing temperatures for alloy steels (furnace cooling)

Steel	Annealing temperature °C	°F	Hardness (max), HB
1330	845–900	1550–1650	179
1335	845–900	1550–1650	187
1340	845–900	1550–1650	192
1345	845–900	1550–1650	...
3140	815–870	1500–1600	187
4037	815–855	1500–1575	183
4042	815–855	1500–1575	192
4047	790–845	1450–1550	201
4063	790–845	1450–1550	223
4130	790–845	1450–1550	174
4135	790–845	1450–1550	...
4137	790–845	1450–1550	192
4140	790–845	1450–1550	197
4145	790–845	1450–1550	207
4147	790–845	1450–1550	...
4150	790–845	1450–1550	212
4161	790–845	1450–1550	...
4337	790–845	1450–1550	...
4340	790–845	1450–1550	223
50B40	815–870	1500–1600	187
50B44	815–870	1500–1600	197
5046	815–870	1500–1600	192
50B46	815–870	1500–1600	192
50B50	815–870	1500–1600	201
50B60	815–870	1500–1600	217
5130	790–845	1450–1550	170
5132	790–845	1450–1550	170
5135	815–870	1500–1600	174
5140	815–870	1500–1600	187
5145	815–870	1500–1600	197
5147	815–870	1500–1600	197
5150	815–870	1500–1600	201
5155	815–870	1500–1600	217
5160	815–870	1500–1600	223
51B60	815–870	1500–1600	223
50100	730–790	1350–1450	197
51100	730–790	1350–1450	197
52100	730–790	1350–1450	207
6150	845–900	1550–1650	201
81B45	845–900	1550–1650	192
8627	815–870	1500–1600	174
8630	790–845	1450–1550	179
8637	815–870	1500–1600	192
8640	815–870	1500–1600	197
8642	815–870	1500–1600	201
8645	815–870	1500–1600	207
86B45	815–870	1500–1600	207
8650	815–870	1500–1600	212
8655	815–870	1500–1600	223
8660	815–870	1500–1600	229
8740	815–870	1500–1600	202
8742	815–870	1500–1600	...
9260	815–870	1500–1600	229
94B30	790–845	1450–1550	174
94B40	790–845	1450–1550	192
9840	790–845	1450–1550	207

placing thermocouples among the bars, forgings or coils. Regulation of the furnace during the annealing operation should be based on the temperatures indicated by these thermocouples, which are in actual contact with the steel, rather than on the temperatures indicated by the furnace thermocouples.

Spheroidizing

Steels may be spheroidized—that is, heated and cooled to produce a structure of globular carbides in a ferritic matrix—by the following methods:

1 Prolonged holding at a temperature just below Ae_1*
2 Heating and cooling alternately between temperatures that are just above Ac_1 and just below Ar_1
3 Heating to a temperature above Ac_1, and then either cooling very slowly in the furnace or holding at a temperature just below Ar_1
4 Cooling at a suitable rate from the minimum temperature at which all carbide is dissolved, to prevent reformation of a carbide network, and then reheating in accordance with method 1 or 2 above (applicable to hypereutectoid steel containing a carbide network).

The rates of spheroidizing provided by these methods depend somewhat on prior microstructure, being greatest for quenched structures in which the carbide phase is fine and dispersed. Prior cold work also increases the rate of the spheroidizing reaction in a subcritical spheroidizing treatment.

For full spheroidizing, austenitizing temperatures either slightly above the Ac_1 temperature or about midway between Ac_1 and Ac_3 are used. If a temperature slightly above Ac_1 is to be used, good loading characteristics and accurate temperature controls are required for proper results; otherwise, it is conceivable that Ac_1 may not be reached and thus that austenitization may not occur. Because time and temperature affect austenitization and thereby influence the number of undissolved carbides from which nucleation

*It is difficult to remain consistent with designations for critical temperatures. When heating with prolonged holding is being discussed, the critical temperatures of interest should be the equilibrium temperatures Ae_1 and Ae_3. Terminology becomes more arbitrary in discussions of heating and cooling at unspecified rates and for unspecified holding times.

and coalescence of the spheroids occur, close control of temperature is necessary. For example, if it is determined that spheroidization of a given steel will require an austenitizing temperature of 750 °C (1385 °F), a deviation of 11 °C (20 °F) may easily result in an incompletely spheroidized structure.

The spheroidized structure is desirable when minimum hardness, maximum ductility or (for high-carbon steels) maximum machinability is important. Low-carbon steels are seldom spheroidized for machining, because in the spheroidized condition they are excessively soft and "gummy", cutting with long, tough chips. When low-carbon steels are spheroidized, it is generally to permit severe deformation. For example, when 1020 steel tubing is being produced by cold drawing in two or three passes, a spheroidized structure will be obtained if the material is annealed for ½ to 1 h at 690 °C (1275 °F) after each pass. The final product will have a hardness of about 163 HB. Tubing in this condition will be able to withstand severe deformation during subsequent cold forming.

As with many other types of heat treatment, hardness after spheroidizing depends on carbon and alloy contents. Increasing the carbon or alloy content, or both, results in an increase in the as-spheroidized hardness, which generally ranges from 163 to 212 HB (Table 4).

Process Annealing

As the hardness of steel increases during cold working, ductility decreases and additional cold reduction becomes so difficult that the material must be annealed to restore its ductility. Such annealing between processing steps is referred to as "in-process" or simply "process" annealing. It may consist of any appropriate treatment. In most instances, however, a subcritical treatment is adequate and least costly, and the term "process annealing" without further qualification usually refers to an in-process subcritical anneal. It is often necessary to specify process annealing for parts that are cold formed by stamping, heading or extrusion. Hot worked high-carbon and alloy steels also are process annealed to prevent them from cracking and to soften them for shearing, turning or straightening.

Process annealing usually consists of heating to a temperature below Ae_1,

Table 4 Recommended temperatures and time cycles for annealing of alloy steels

Steel	Austenitizing temperature °C	°F	Conventional cooling(a) Temperature °C From	To	°F From	To	Cooling rate °C/h	°F/h	Time, h	Isothermal method(b) Cool to °C	°F	Hold, h	Hardness (approx), HB
To obtain a predominantly pearlitic structure(c)													
1340 830	830	1525	735	610	1350	1130	11	20	11	620	1150	4.5	183
2340 800	800	1475	655	555	1210	1030	8.3	15	12	595	1100	6	201
2345 800	800	1475	655	550	1210	1020	9.1	15	12.7	595	1100	6	201
3120(d) 885	885	1625	. . .	. . .	. . .	. . .	. . .	. . .	. . .	650	1200	4	179
3140 830	830	1525	735	650	1350	1200	11	20	7.5	660	1225	6	187
3150 830	830	1525	705	645	1300	1190	11	20	5.5	660	1225	6	201
3310(e) 870	870	1600	. . .	. . .	. . .	. . .	. . .	. . .	. . .	595	1100	14	187
4042 830	830	1525	745	640	1370	1180	11	20	9.5	660	1225	4.5	197
4047 830	830	1525	735	630	1350	1170	11	20	9	660	1225	5	207
4062 830	830	1525	695	630	1280	1170	8.3	15	7.3	660	1225	6	223
4130 855	855	1575	765	665	1410	1230	20	35	5	675	1250	4	174
4140 845	845	1550	755	665	1390	1230	14	25	6.4	675	1250	5	197
4150 830	830	1525	745	670	1370	1240	8.4	15	8.6	675	1250	6	212
4320(d) 885	885	1625	. . .	. . .	. . .	. . .	. . .	. . .	. . .	660	1225	6	197
4340 830	830	1525	705	565	1300	1050	8.3	15	16.5	650	1200	8	223
4620(d) 885	885	1625	. . .	. . .	. . .	. . .	. . .	. . .	. . .	650	1200	6	187
4640 830	830	1525	715	600	1320	1110	7.6	15	15	620	1150	8	197
4820(d)	. . .	. . .	. . .	. . .	. . .	. . .	. . .	. . .	. . .	605	1125	4	192
5045 830	830	1525	755	665	1390	1230	11	20	8	660	1225	4.5	192
5120(d) 885	885	1625	. . .	. . .	. . .	. . .	. . .	. . .	. . .	690	1275	4	179
5132 845	845	1550	755	670	1390	1240	11	20	7.5	675	1250	6	183
5140 830	830	1525	740	670	1360	1240	11	20	6	675	1250	6	187
5150 830	830	1525	705	650	1300	1200	11	20	5	675	1250	6	201
52100(f)	. . .	. . .	. . .	. . .	. . .	. . .	. . .	. . .	. . .	. . .	. . .	. . .	. . .
6150 830	830	1525	760	675	1400	1250	8.4	15	10	675	1250	6	201
8620(d) 885	885	1625	. . .	. . .	. . .	. . .	. . .	. . .	. . .	660	1225	4	187
8630 845	845	1550	735	640	1350	1180	11	20	8.5	660	1225	6	192
8640 830	830	1525	725	640	1340	1180	11	20	8	660	1225	6	197
8650 830	830	1525	710	650	1310	1200	8.4	15	7.2	650	1200	8	212
8660 830	830	1525	700	655	1290	1210	8.4	15	8	650	1200	8	229
8720(d) 885	885	1625	. . .	. . .	. . .	. . .	. . .	. . .	. . .	660	1225	4	187
8740 830	830	1525	725	645	1340	1190	11	20	7.5	660	1225	7	201
8750 830	830	1525	720	630	1330	1170	8.4	15	10.7	660	1225	7	217
9260 860	860	1575	760	705	1400	1300	8.4	15	6.7	660	1225	6	229
9310(e) 870	870	1600	. . .	. . .	. . .	. . .	. . .	. . .	. . .	595	1100	14	187
9840 830	830	1525	695	640	1280	1180	8.4	15	6.6	650	1200	6	207
9850 830	830	1525	700	645	1290	1190	8.4	15	6.7	650	1200	8	223
To obtain a predominantly ferritic and spheroidized carbide structure													
1320(d) 805	805	1480	. . .	. . .	. . .	. . .	. . .	. . .	. . .	650	1200	8	170
1340 750	750	1380	735	610	1350	1130	5.5	10	22	640	1180	8	174
2340 715	715	1320	655	555	1210	1030	5.5	10	18	605	1125	10	192
2345 715	715	1320	655	550	1210	1020	5.5	10	19	605	1125	10	192
3120(d) 790	790	1450	. . .	. . .	. . .	. . .	. . .	. . .	. . .	650	1200	8	163
3140 745	745	1370	735	650	1350	1200	5.5	10	15	660	1225	10	174
3150 750	750	1380	705	645	1300	1190	5.5	10	11	660	1225	10	187
9840 745	745	1370	695	640	1280	1180	5.5	10	11	650	1200	10	192
9850 745	745	1370	700	645	1290	1190	5.5	10	11	650	1200	12	207

(a) The steel is cooled in the furnace at the indicated rate through the temperature range shown. (b) The steel is cooled rapidly to the temperature indicated and is held at that temperature for the time specified. (c) In isothermal annealing to obtain pearlitic structure, steels may be austenitized at temperatures up to 70 °C (125 °F) higher than temperatures listed. (d) Seldom annealed. Structures of better machinability are developed by normalizing or by transforming isothermally after rolling or forging. (e) Annealing is impractical by the conventional process of continuous slow cooling. The lower transformation temperature is markedly depressed, and excessively long cooling cycles are required to obtain transformation to pearlite. (f) Predominantly pearlitic structures are seldom desired in this steel.

soaking for an appropriate time and then cooling, usually in air. In most instances, heating to a temperature between 11 and 22 °C (20 and 40 °F) below Ae₁ produces the best combination of microstructure, hardness and mechanical properties. Temperature controls are necessary only to prevent heating the material above Ae₁ and thus defeating the purpose of annealing.

When process annealing is performed merely to soften a material for such operations as cold sawing and cold shearing, temperatures well below Ae₁ normally are used and close controls are unnecessary.

In the wire industry, process annealing is used as an intermediate treatment between the drawing of wire to a size slightly larger than the desired finished size and the drawing of a light reduction to the finished size. Wire thus made is known as "annealed in process" wire. Process annealing is used also in the production of wire sufficiently soft for severe upsetting, and to permit drawing the smaller sizes of low-carbon and medium-carbon steel wire that cannot be drawn to the desired small size directly from the hot rolled rod. Process annealing is more satisfactory than spheroidize annealing for a material that, because of its composition or size (or both), cannot be drawn to finished size because it either lacks ductility or does not meet physical requirements. Also, material that is cold sheared during processing is process annealed to raise the ductility of the sheared surface to a level suitable for further processing.

Examples. A change to process annealed material was required, to overcome fabrication problems in two production operations involving 1040 and 1045 steels. Spheroidized 1040 steel was too soft for cold shearing; and when spheroidized 1045 steel was subsequently cold headed into stub shafts, excessive machining problems were encountered because of burrs and poor finish.

However, a spheroidized material may be necessary because of the sequence or severity of forming operations, or because of the mechanical properties desired in the end product. Thus, to eliminate cracks developed in process annealed 1035 steel during cold working for producing large-head carriage bolts, the maximum ductility provided by spheroidizing was required. In another operation, involving 1038

steel, spheroidizing was necessary so that the hardness of the end product was in the specified range of 78 to 88 HRB.

Annealed Structures for Machining

Different combinations of microstructure and hardness, considered together, are significant in terms of machinability. For instance, Fig. 4 shows that a partially spheroidized 5160 steel shaft was machined (by turning) with much less tool wear and better surface finish than the same steel in the annealed condition with a pearlitic microstructure and a higher hardness. Based on many observations, optimum microstructures for machining steels of various carbon contents are usually as follows:

Carbon, %	Optimum microstructure
0.06 to 0.20	As rolled (most economical)
0.20 to 0.30	Under 75-mm (3-in.) diam, normalized; 75-mm diam and over, as rolled
0.30 to 0.40	Annealed, to produce coarse pearlite, minimum ferrite
0.40 to 0.60	Coarse lamellar pearlite to coarse spheroidite
0.60 to 1.00	100% spheroidite, coarse to fine

The type of machining operation is also a factor. For example, certain gears were made from 5160 steel tubing by the dual operation of machining in automatic screw machines and broaching of cross slots. The screw-machine operations were easiest with thoroughly spheroidized material, but a pearlitic structure was more suitable for broaching. A semispheroidized structure proved to be a satisfactory compromise.

Semispheroidized structures can be achieved by austenitizing at lower temperatures, and sometimes at higher cooling rates, than those used for achieving pearlitic structures. The semispheroidized structure of the 5160 steel tubing mentioned above was obtained by heating to 790 °C (1450 °F) and cooling at 28 °C/h (50 °F/h) to 650 °C (1200 °F). For this steel, austenitizing at a temperature of about 775 °C (1425 °F) results in more spheroidization and less pearlite.

Medium-carbon steels are much more difficult to spheroidize fully than high-carbon steels such as 1095 and 52100. In the absence of excess carbides to nucleate and promote the spheroidizing reaction, it is more difficult to achieve complete freedom from pearlite in practical heat treating cycles.

At lower carbon levels, structures consisting of coarse, blocky pearlite in a ferrite matrix often are found to be the most machinable. In some alloy steels, this type of structure can best be achieved by heating to temperatures well above Ac₃ to establish a coarse austenite grain size, then holding below Ar₁ to allow coarse, lamellar pearlite to form. This process sometimes is referred to as "cycle annealing" or "lamellar annealing". For example, forged 4620 steel gears were heated rapidly in a five-zone furnace to 980 °C (1800 °F), cooled to 625 to 640 °C (1160 to 1180 °F) in a water-cooled zone and held at that temperature for 120 to 150 min. The resulting structure—coarse, lamellar pearlite in a ferrite matrix—had a hardness of 140 to 146 HB. (Ref 5)

Types of Furnaces

Furnaces for annealing are of two basic types: batch furnaces and continuous furnaces. Within either of these two types, furnaces can be further classified according to configuration, type of fuel used, method of heat application, and means by which the load is moved through, or supported in, the furnace. Other factors that must be considered in furnace selection are cost, type of annealing cycle, required atmosphere, and physical nature of parts to be annealed. In many cases, however, the annealing cycle used is dictated by the available equipment.

Batch-type furnaces are necessary for large parts such as heavy forgings and often are preferred for small lots of a given part or grade of steel and for the more complex alloy grades requiring long cycles. Specific types of batch furnaces include car-bottom, box, bell and pit furnaces. These furnaces may be controlled manually or may be equipped with programmed controllers that permit automatic operation.

Continuous furnaces such as roller-hearth, rotary-hearth and pusher types are ideal for isothermal annealing of large quantities of parts of the same grade of steel. These furnaces can be designed with various individual zones, allowing the work to be consecutively

Fig. 4 Correlation of annealing practice with surface finish and tool life in subsequent machining of 5160 steel

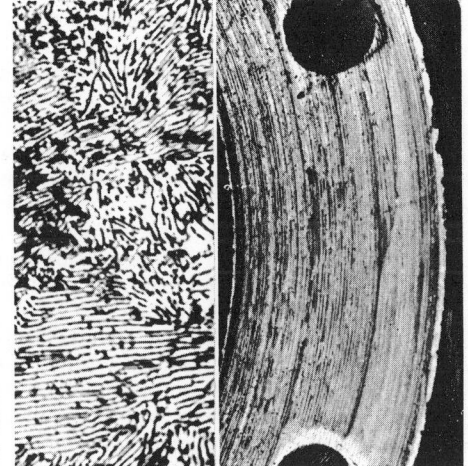
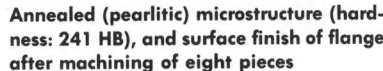

Annealed (pearlitic) microstructure (hardness: 241 HB), and surface finish of flange after machining of eight pieces

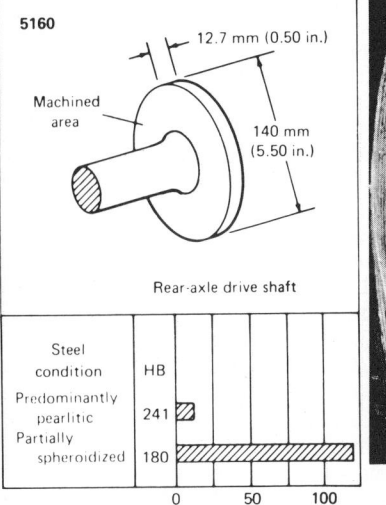

Rear-axle drive shaft

Steel condition	HB					
Predominantly pearlitic	241					
Partially spheroidized	180					

Tool life between grinds, min

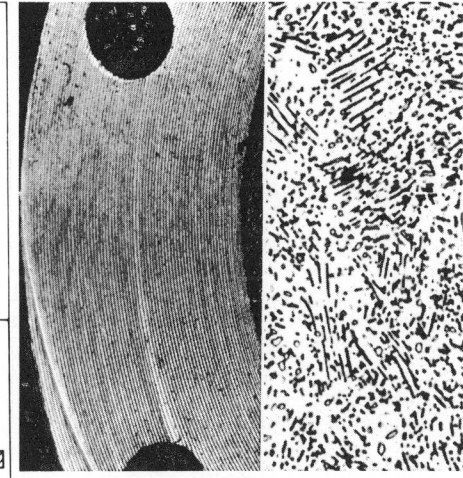

Partially spheroidized microstructure (hardness: 180 HB), and surface finish of flange after machining of 123 pieces

brought to temperature, held at temperature, and cooled at the desired rate. For more detailed discussion of the types of furnaces available for annealing, see Ref 6.

Furnace Atmospheres

Electric furnaces used with air atmospheres, and gas furnaces used with atmospheres consisting of the products of combustion, cannot be regulated for complete elimination of oxidation of the steel being treated. Only atmospheres independent of the fuel are generally considered satisfactory for clean or bright annealing. Excessive oxidation during annealing usually is prevented by the use of controlled atmospheres in conjunction with a suitable furnace that is designed to exclude air and combustion gases from the heating chamber. The gases and gas mixtures used for controlled atmospheres depend on the metal being treated, the treatment temperature and the surface requirements of the parts being annealed. The need to eliminate decarburization as well as oxidation is often a significant factor in the selection of annealing atmospheres.

The gas most widely used as a protective atmosphere for annealing is exothermic gas. This gas is inexpensive, the raw materials for making it are readily available and the results obtained with it are generally excellent. Hydrocarbon gases such as natural gas, propane, butane and coke-oven gas are commonly burned in an exothermic-gas producer, creating a self-supporting, heat-producing combustion reaction. A commonly used exothermic gas mixture contains 15% H_2, 10% CO, 5% CO_2, 1% CH_4 and 69% N_2. This gas is used for bright annealing of cold rolled low-carbon steel strip. It will decarburize medium-carbon and high-carbon steels, however, because of the carbon dioxide and water vapor it contains.

Exothermic gas sometimes is refrigerated to reduce its moisture content, particularly in geographic areas where the temperature of the water used for cooling is high. When decarburization of workpiece surfaces must be prevented, water vapor and carbon dioxide must be completely removed from the gas. Purified exothermic gas, with its carbon dioxide and water vapor removed, has many applications in heat treatment of steel without decarburization.

Purified rich exothermic gas, formed by partial combustion, is used for short-cycle annealing and process annealing of medium- and high-carbon steels of the straight-carbon and alloy types. For long-cycle batch annealing, however, this gas is unsuitable, because its high carbon monoxide content results in soot deposits on the work and because of the possibility of surface etching as a result of the relatively long time for which the work is in the critical low-temperature range where gas reactions can occur. In short-cycle annealing, these effects are minimal and the high-CO gas is then desirable because of its high carbon potential. The fairly lean purified gas formed by more complete combustion is used for long-cycle annealing of medium- and high-carbon steels of the straight-carbon and alloy types, and for batch and continuous annealing of low-carbon steel strip for tin plating.

Other atmospheres commonly used in annealing include endothermic-base, dissociated ammonia, vacuum, and nitrogen-base atmospheres, which consist of nitrogen plus additions such as hydrogen, methane, methanol and carbon monoxide. For more complete information, see the article on furnace atmospheres in this volume.

Annealing of Sheet and Strip

In terms of total tonnage of material processed, annealing of sheet and strip during production of steel-mill products represents the major use of annealing. Because such annealing is done to prepare the material for further processing (such as additional cold rolling or fabrication into parts) and because the temperatures employed are usually below the A_1 temperature, the more specific terms "subcritical annealing" and "process annealing" are appropriate, although common practice is to use the term "annealing" without qualification.

In annealing of sheet and strip, two techniques predominate: the batch process and the continuous process. In the batch process (also called "box annealing"), coils or cut lengths of sheet are placed on an annealing base and covered with containers that are sealed to hold the appropriate atmosphere. A furnace is then placed over the covered steel. A protective atmosphere is introduced within the inner covers to protect the steel from oxidation, and is circulated through the coils by use of fans and convector plates. Heating is provided by the outer furnace and may be done either through use of radiant tubes or by direct firing. The charge is heated to the required temperature and held for a period of time that will result in the desired properties. The outer furnace is then removed, and the coils are allowed to cool under the inner covers. When the temperature has been reduced to the point where oxidation of the steel will not occur, the inner covers are removed and the steel is forwarded for further processing.

In the continuous process, steel coils are uncoiled and drawn through a furnace where they are subjected to the annealing cycle under a protective atmosphere. After the sheet or strip has been cooled and removed from the furnace, further in-line processing (such as hot dip galvanizing) may be done, or the steel may be cut into sheets. In general, however, the steel is recoiled and then forwarded as in the batch process.

In addition to the obvious differences in equipment, the batch process and the continuous process differ considerably in several other ways. Batch annealing may require up to a week because of the large mass of material being treated, whereas continuous annealing is accomplished in about five minutes. Differences are also evident in the temperatures employed, with the batch process generally being conducted at lower temperatures. Because in batch annealing it is difficult to ensure that the temperature is uniform throughout the charge (which may consist of several hundred tons of steel), the continuous process offers the potential of more uniform properties. The short annealing times of the continuous process, however, frequently result in hardness levels slightly higher than those of similar material annealed by the batch process.

Cold Rolled Plain Carbon Sheet and Strip. The usual method of manufacturing cold rolled sheet and strip is to produce a hot rolled coil, pickle it to remove scale (oxide) and cold roll it to the desired final gage. Cold rolling may reduce the thickness of the hot rolled material in excess of 90%, which increases the hardness and strength of the steel but severely decreases its ductility. If any large amount of subsequent cold working is to be done, the ductility of the steel must be restored.

Annealing of cold rolled steel normally is designed to produce a recrystallized ferrite microstructure from the highly elongated, stressed grains resulting from cold work. During heating of the steel, and in the first segment of the holding portion of the cycle, the first metallurgical process to occur is *recovery*. During this process, internal strains are relieved (although little change in the microstructure is evident), ductility is moderately increased and strength is slightly decreased.

As annealing continues, the process of *recrystallization* occurs, and new, more equiaxed ferrite grains are formed from the elongated grains. During recrystallization, strength decreases rapidly, with a corresponding increase in ductility. Further time at temperature causes some of the newly formed grains to grow at the expense of other grains; this is termed *grain growth* and results in modest decreases in strength and small (but often significant) increases in ductility. Most plain carbon steels are given an annealing treatment that promotes full recrystallization, but care must be taken to avoid excessive grain growth, which can lead to surface defects (such as "orange peel") in formed parts.

The rates at which the metallurgical processes noted above proceed are functions of both the chemical composition and the prior history of the steel being annealed. For example, small amounts of elements such as aluminum, titanium, niobium, vanadium and molybdenum can decrease the rate at which the steel will recrystallize, making the annealing response sluggish and therefore necessitating either higher temperatures or longer annealing times to produce the same properties. Although the presence of these alloying elements is generally the result of deliberate additions intended to modify the properties of the sheet (as in the case of aluminum, titanium, niobium and vanadium), some elements may be present as residual elements (molybdenum, for example) in quantities great enough to modify the response to annealing. Conversely, larger amounts of cold work (greater cold reductions) will accelerate the annealing response. Therefore, it is not possible to specify a single annealing cycle that will produce a particular set of mechanical properties in all steels; the chemical composition and the amount of cold work also must be taken into account.

Cold rolled plain carbon steels are produced to a number of different quality descriptions. Commercial quality (CQ) steel is the most widely produced and is suitable for moderate forming. Drawing quality (DQ) steel is produced to tighter mechanical-property restrictions for use in more severely formed parts. Drawing quality special killed (DQSK) steel is produced to be suitable for the most severe forming applications. Typical properties of these grades may be found on page 156 in Volume 1 of this Handbook. Structural quality (SQ) steel is produced to specified mechanical properties other than those for the above three grades.

Typical annealing cycles for all possible combinations of composition, cold reduction and grade cannot be listed here. However, typical batch-annealing temperatures range from 620 to 690 °C (1150 to 1270 °F) for the coldest point in the charge. Cycle times vary with the grade desired and the size of the charge, but total times (from the beginning of heating to removal of the steel from the furnace) can be as long as one week.

Continuous-annealing cycles are of shorter duration and are conducted at higher temperatures than batch-annealing cycles. In some applications, the annealing temperature may exceed A_1. Typical cycles are 40 s at 700 °C (1290 °F) for cold rolled commercial quality sheet and 60 s at 800 °C (1470 °F) for drawing quality special killed sheet. Most continuous annealing of cold rolled sheet includes an overaging treatment designed to precipitate carbon and nitrogen from solution in the ferrite and to reduce the likelihood of strain aging. Overaging for 3 to 5 min at 300 to 450 °C (570 to 840 °F) accomplishes the desired precipitation of carbon and nitrogen.

Batch annealing and continuous annealing differ slightly in the properties they produce. Typical average properties of batch-annealed and continuous-annealed commercial quality plain carbon steel are as follows:

Annealing process	Yield strength MPa	ksi	Elongation, %
Batch.........	210	30.4	43.0
Continuous.....	228	33.0	41.7

High-strength cold rolled sheet and strip are growing in importance due to their high load-bearing capacities. Strength of sheet and strip can be increased through modifications of chemical composition and/or selection of different annealing cycles, but these methods result in decreased ductility. Plain carbon steels, produced by conventional techniques, may be batch annealed or continuous annealed under conditions that result only in recovery or partial recrystallization. Typical batch-annealing cycles of this type employ soak temperatures of 425 to 480 °C (800 to 900 °F) and various soak times. High-strength low-alloy (HSLA) steels containing alloying elements such as niobium, vanadium and titanium also may be produced as cold rolled grades. The additional alloying produces a stronger hot rolled steel, which is strengthened even more by cold rolling. Cold rolled HSLA steels may be recovery annealed to produce higher-strength grades or recrystallization annealed to produce lower-strength grades. Successful production of cold rolled HSLA steel requires selection of the appropriate combination of steel composition and hot rolled strength, amount of cold reduction and type of annealing cycle. Table 5 presents typical properties after recovery or recrystallization annealing, as appropriate, for a family of cold rolled sheet products employing titanium as the principal strengthening element.

Hot dip galvanized products are produced on lines that process either preannealed (batch annealed) or full hard coils. Lines for processing full hard coils incorporate an in-line annealing capability so that annealing and hot dip galvanizing can be accomplished in a single pass through the line. This in-line annealing, like continuous annealing of uncoated steel, generally results in slightly higher strength and slightly lower ductility

Table 5 Typical properties of titanium-strengthened cold rolled steel

Minimum yield strength MPa	ksi	Yield strength MPa	ksi	Tensile strength MPa	ksi	Elongation, %
275	40	325	47	415	60	30
345	50	380	55	490	71	26
415	60	455	66	545	79	23
485	70	525	76	605	88	19
550	80	615	89	670	97	17
690	100	745	108	800	116	12
825	120	885	128	905	131	10
965	140	1040	151	1050	152	5

than batch annealing. Maximum strip temperatures are below the A_1 temperature for commercial quality steel, but temperatures in excess of 845 °C (1550 °F) are required for DQSK grades. Galvanizing of preannealed steel results in properties similar to those of ungalvanized material.

The atmosphere in a continuous galvanizing line, in addition to protecting the sheet from oxidation, must remove any oxides present on the strip to promote metallurgical bonding between the steel and the zinc.

Tin mill products are distinguished from their cold rolled sheet mill counterparts chiefly by the fact that they are produced in lighter gages (0.13 to 0.38 mm, or 0.005 to 0.015 in.) and by the fact that some of them are coated with tin or chromium and chromium oxide for corrosion resistance. The sequence used for processing single-reduced tin mill products is similar to that for cold rolled sheet—that is, pickling, cold reducing, annealing and temper rolling of hot rolled coils. Double-reduced products are cold rolled an additional 30 to 40% following annealing (this step replaces temper rolling). Whereas much of the tonnage produced in tin mills is batch annealed, a considerable amount is continuous annealed (facilities for continuous annealing currently are more prevalent in tin mills than in sheet mills).

Because tin mill products traditionally have been produced at facilities separate from sheet mills and because applications for these products are different from those for cold rolled sheet, tin mill products have been assigned separate designations for indicating the mechanical properties developed during annealing. A list of these temper designations is given in Table 6.

Open-coil annealing, which is done in batch furnaces, involves loose rewinding of a cold reduced coil to pro-

Table 6 Temper designations for steel tin mill products

Designation	Hardness aim, HR30T
Batch (box) annealed products	
T-1	52 max
T-2	50 to 56
T-3	54 to 60
T-4	58 to 64
T-5	62 to 68
T-6	67 to 73
Continuously annealed products	
T-4 CA	58 to 64
T-5 CA (TU)	62 to 68
T-6 CA	67 to 73
Double-reduced products	
DR-8	73
DR-9	76
DR-9M	77
DR-10	80

vide open spaces between successive laps. This allows the controlled atmosphere gases to be drawn between the laps, providing faster and more uniform heating and cooling than are obtained with tightly wound coils. In addition, by control of the hydrogen content and dew point of the atmosphere, decarburizing conditions can be established. The carbon content of the steel can thereby be reduced to low levels for such materials as enameling steel and electrical steel.

Loose rewinding of coils for open-coil annealing is done on a turntable having a vertical mandrel. As the coil is wound, a twisted wire spacer is inserted between the laps. This spacer remains in the coil during annealing and is removed after the coil has been removed from the furnace. The coil is then tightly rewound and is ready for temper rolling.

Annealing of Steel Forgings

Annealing of forgings is most often performed to facilitate some subsequent operation—usually machining or cold forming. The type of annealing required is determined by the kind and amount of machining or cold forming to be done as well as the type of material involved. For some processes it is essential that the microstructure be spheroidal, while for others spheroidal structures may not be necessary or even desirable.

Annealing of Forgings for Machinability. In many cases, a structure suitable for machining can be developed in low-carbon steel forgings by transferring the forgings directly from the forging operation to a furnace heated to a proper transformation temperature, holding them at this temperature for a time sufficiently long to permit all the austenite to transform, then cooling in air. In this process, the effective austenitizing temperature is the finishing temperature of forging, not the initial forging temperature. This process is capable of producing reasonably uniform structures in forgings of uniform sections. However, in forgings shaped such that some portions are cooler than others, this difference in finishing temperature will cause the structures to be dissimilar. This process generally will not produce a spheroidal structure except in high-alloy steels containing large amounts of carbide-forming elements. If a lamellar structure is suitable for subsequent operations, however, this process can minimize energy usage and lower costs by reducing processing and handling time.

In many instances where the product or subsequent process requires a more consistent hardness, forgings can be subcritical annealed by heating to a temperature between 11 and 22 °C (20 and 40 °F) below Ae_1, holding sufficiently long (determined by degree of softening required), and then cooling in air (or equivalent). Care should be taken to maintain the temperature below Ae_1 to prevent formation of austenite, which would require a much lower cooling rate.

In forgings produced from higher-carbon steels with or without significant amounts of alloying elements, a spheroidal structure generally is preferable for high-speed machining operations. Direct transfer of high-carbon

steel forgings to a furnace for transformation sometimes can be used as the preliminary step of an annealing cycle and as a means of preventing the possibility of cracking in deep-hardening steel parts, but seldom will produce satisfactory properties alone. Most annealing of high-carbon steel forgings is done either in a batch furnace or in a continuous tray pusher furnace. Typical schedules for spheroidizing 52100 steel in a batch furnace are as follows:

- Austenitize by holding at least 2 h at 790 °C (1450 °F), furnace cool at 17 °C/h (30 °F/h) to 595 °C (1100 °F), then air cool.
- Austenitize by holding at least 2 h at 790 °C (1450 °F), cool as rapidly as practical to 750 °C (1380 °F), cool at 6 °C/h (10 °F/h) to 675 °C (1250 °F), then air cool.
- Austenitize by holding at least 2 h at 790 °C (1450 °F), cool as rapidly as practical to 690 °C (1275 °F), transform isothermally by holding at this temperature for 16 h, then air cool.

In all instances, the load should be distributed to promote uniform heating and cooling. Use of circulating fans in the furnace chamber will greatly aid in producing a product that is uniform in both hardness and microstructure.

A typical continuous furnace for annealing steel forgings might consist of five or six zones. An example of a specific spheroidize annealing treatment in such a furnace is given in the next section.

Annealing of Forgings for Cold Forming and Re-forming. If a steel forging or blank requires further cold forming, it may be necessary to soften it in order to enhance its plastic-flow characteristics. In general, this type of annealing is done only to the extent that the forming operation requires—that is, to satisfy dimensional, mechanical and tool-life requirements as well as to prevent cracking and splitting. Much intermediate annealing is done successfully, but cold forming processes are best performed on parts with totally spheroidized microstructures—especially for parts made of high-carbon steels.

In one plant, both 5160 and 52100 steels have been successfully spheroidized with a common cycle in a six-zone tray pusher furnace. In this cycle, the temperatures in the six zones are 750, 750, 705, 695, 695 and 680 °C (1380, 1380, 1300, 1280, 1280 and 1260 °F).

Time in each zone is 150 min. This process yields 5160 steel forgings with hardnesses of 170 to 190 HB and 52100 steel parts with hardnesses of 175 to 195 HB, both suitable for cold or warm restrike operations.

Low-carbon steels generally can be cold formed successfully after being heated to temperatures near A_1 and then being cooled through 675 °C (1250 °F) at a controlled rate. In one plant, 5120 steel annealed 1 to 2 h at 745 °C (1375 °F) and slow cooled has been cold formed successfully. Large quantities of 1008, 1513, 1524, 8620 and 8720 steels are being cold formed after annealing cycles consisting of 1 to 6 h at 720 °C (1325 °F) followed by slow cooling. The severity of the forming operation, as well as the grade of steel and history of the part, determines the extent of annealing required. Batch furnaces, continuous tray pusher furnaces and continuous belt furnaces are being used successfully to perform these types of annealing operations on low-carbon steels.

Any part that contains significant stresses resulting from cold forming or restrike operations should be reviewed for some type of stress-relief process. Stress relieving usually is done by means of time-temperature cycles that result in slight reductions in hardness. These cycles often consist of 1 h at 425 to 675 °C (800 to 1250 °F).

Annealing to Obtain Pearlitic Microstructures. Forgings—especially plain and alloy high-carbon steel forgings—sometimes are isothermally annealed to produce a pearlitic microstructure that is preferred for a subsequent process. In steels that are to be induction hardened, for example, the carbide distribution of a fine pearlitic structure offers excellent preparation for optimum control in selective hardening while producing a reasonably machinable core structure. Isothermal annealing to obtain fine pearlite can be performed in batch or continuous furnaces; however, temperature control and uniformity are more critical than in conventional slow cooling cycles, because a particular microstructure and a particular hardness level usually are desired. In one plant, a continuous belt-type furnace is used for isothermal annealing of 1070 steel forgings. The forgings are uniformly heated for 30 min at 845 °C (1550 °F), cooled to 675 °C (1250 °F) and held for 20 min, then rapidly cooled. The microstructure produced is essentially fine lamellar pearlite with a hardness of 219 to 228 HB. The hardness and the structure can be modified by adjusting the transformation temperature.

Hardness After Annealing. Figures 5 and 6 present data on distribution of hardness after annealing. These data are based on production annealing of forgings, bars, tubes and rings made of 1045, 4140, 4340, 8640H and 52100 steels. The details of each annealing cycle used are given also.

Annealing of Bar, Rod and Wire

Significant tonnages of bar, rod and wire are subjected to thermal treatments that decrease hardness and prepare the material for subsequent cold working and/or machining. For low-carbon steels (up to 0.20% C), short-time subcritical annealing often is sufficient for preparing the material for further cold working. Steels with higher carbon and alloy contents require spheroidizing, to impart maximum ductility.

Most carbon and alloy steel coiled products can be successfully spheroidized in accordance with rules 2 and 3 (see "Guidelines for Annealing"). In batch annealing, it is helpful to use higher-than-normal temperatures (such as 650 °C, or 1200 °F) during initial heating for purging, because the higher initial temperature promotes a lower temperature gradient in the charge during subsequent heating into the temperature range between A_1 and A_3. Use of a higher purge temperature also promotes agglomeration of the carbides in the steel, which makes them more resistant to solution in the austenite when the charge temperature is finally elevated. These undissolved carbides will be conducive to the formation of a spheroidal rather than a lamellar structure when transformation is complete.

A knowledge of the temperature distribution in the furnace and in the load can be a major factor in achieving a good, consistent response to spheroidization. Temperature distribution and control are much more critical in batch and vacuum furnaces, which may handle loads of up to 27 t (30 tons), than in continuous furnaces, in which loads of only 900 to 1800 kg (2000 to 4000 lb) may be transferred from zone to zone. Test thermocouples should be placed strategically at the top, middle and bottom (inside and outside) of the charge

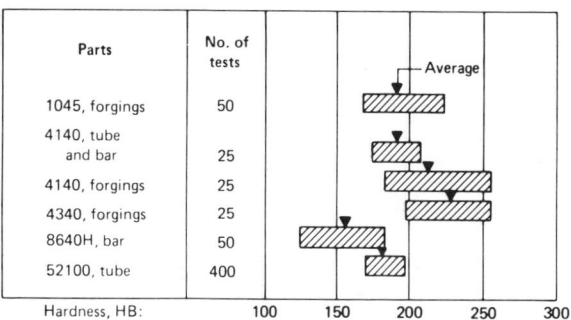

Fig. 5 Variation in Brinell hardness of annealed plain carbon and low-alloy steels

Parts	No. of tests
1045, forgings	50
4140, tube and bar	25
4140, forgings	25
4340, forgings	25
8640H, bar	50
52100, tube	400

Hardness, HB: 100 150 200 250 300

1045 steel forgings were heated at 790 °C (1450 °F) in a batch-type muffle furnace, furnace cooled at 11 °C/h (20 °F/h) to 650 °C (1200 °F), then air cooled. Specified maximum hardness was 207 HB. Hardness was measured on polished flash line; data cover a four-year period.

4140 seamless tubes and bars were annealed in a continuous car-bottom furnace to produce a predominantly lamellar structure. Bar diameter ranged from 47.6 to 203 mm (1.875 to 8.000 in.); tube wall thickness ranged from 16.0 to 35.0 mm (0.629 to 1.379 in.). The steel was austenitized at 885 °C (1625 °F), furnace cooled to 760 °C (1400 °F), furnace cooled at 11 °C/h (20 °F/h) to 635 °C (1175 °F), and air cooled.

4140 steel forgings for automotive transmissions were annealed in a batch furnace. Forgings were held 5 h at 675 °C (1250 °F). Hardness was measured on polished flash line. Specified hardness range was 170 to 241 HB.

4340 steel forgings for aircraft piston engines were annealed in a batch furnace. Forgings were held 8 h at 650 °C (1200 °F). Hardness was measured on polished flash line. Specified hardness range was 170 to 241 HB.

8640H hot rolled steel bars, 17.5 mm (11/16 in.) in diameter, cold heading quality, in coils, were spheroidize annealed to produce minimum hardness. Data represent as-received material.

52100 steel seamless tubes were austenitized at 790 °C (1450 °F), rapidly furnace cooled to 750 °C (1380 °F), cooled at 6 °C/h (10 °F/h) to 695 °C (1280 °F), and air cooled.

during development of cycles. In spheroidizing, to minimize formation of pearlite on cooling, it is important to ensure that no part of the charge be allowed to approach A_3. Conversely, if temperatures only slightly above A_1 are used and temperature controls are inaccurate because of poor placement of thermocouples, it is probable that the A_1 temperature will not be attained and that no austenitization will occur.

Table 7 gives typical mechanical properties that can be obtained in hypoeutectoid plain carbon steels by spheroidizing in accordance with rules 2 and 3. Recommended temperatures and times for lamellar and spheroidize annealing of hypoeutectoid alloy steels are presented in Table 4.

Prior cold working increases the degree of spheroidization and provides even greater ductility. For example, 4037 steel in the as-rolled condition normally can be spheroidized to a tensile strength of about 515 MPa (75 ksi). If, however, the material is drawn 20% and then spheroidized (referred to as "spheroidize annealed in-process"), the

resulting tensile strength will be around 470 MPa (68 ksi).

Although prior cold work can enhance response to annealing, caution must be observed in spheroidizing cold worked plain carbon steels with 0.20% C or less. Unless a reduction of at least 20% is applied, severe grain coarsening may be observed after spheroidizing. Such grain coarsening is the result of a critical combination of strain and annealing temperature peculiar to the steel and may severely impair subsequent performance.

In the wire industry, a wide variety of "in-process" annealing operations have been evolved for rendering coiled material suitable for further processing that may require formability, drawability, machinability or a combination of these characteristics. One large wire mill reports current use of 42 separate and distinct annealing cycles, the majority of which represent compromises between practical considerations and optimum properties. For example, annealing temperatures below those that might yield optimum softness some-

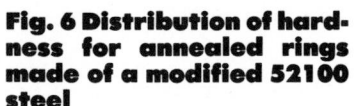

Fig. 6 Distribution of hardness for annealed rings made of a modified 52100 steel

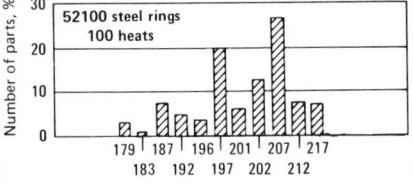

Rings were heated for 3 h at 790 °C (1450 °F), rapidly cooled to 725 °C (1340 °F), cooled at 8 °C/h (15 °F/h) to 695 °C (1280 °F), and air cooled. Hardness measurements were taken on rings located at extreme positions in furnace load. Treatment resulted in a spheroidized structure. Composition: 0.90 to 1.05 C, 0.95 to 1.25 Mn, 0.50 to 0.70 Si, 0.90 to 1.15 Cr

times must be used in order to preclude scaling of wire coils, which often can occur even in controlled-atmosphere furnaces. Even slight scaling may cause the coil wraps to stick together, which can impede coil payoff in subsequent operations.

Some of the terms used to describe various in-process annealing treatments are in common usage throughout the wire industry, whereas others have been developed within specific plants or mills. No attempt will be made here to list or define all the names that refer to specific treatments.

"Patenting" is a special form of annealing that is peculiar to the rod and wire industry. In this process, which usually is applied to medium- and higher-carbon grades of steel, rod or wire products are uncoiled, and the strands are delivered to an austenitizing station. The strands are then cooled rapidly from above A_3 in a molten medium—usually lead at about 540 °C (1000 °F)—for a period of time sufficient to allow complete transformation to a fine pearlitic structure. Both salt baths and fluidized beds have also been used for this purpose. This treatment increases substantially the amount of subsequent wiredrawing reduction that the product can withstand and permits production of high-strength wire. Successive drawing and patenting steps may be employed, if necessary, in order to obtain the desired size and strength level.

Table 7 Typical mechanical properties of spheroidized plain carbon steels

| | Tensile strength | | | |
| | Hot rolled | | Spheroidized | |
Steel	MPa	ksi	MPa	ksi
1010	365	53	295	43
1018	450	65	365	53
1022	470	68	385	56
1030	585	85	415	60
1038	600	87	485	70
1045	675	98	515	75
1060	860	125	550	80
1065	910	132	600	87
1524	510	74	450	65
1541	710	103	540	78

Austenitizing for patenting can be accomplished in oil, gas, or electric furnaces; in high-temperature lead or salt baths; or by induction or direct resistance heating. As an alternative to quenching in molten lead, continuous air cooling often is employed. Such "air patenting" is less expensive than "lead patenting" but results in coarser pearlite and often more proeutectoid ferrite, a microstructure that is less desirable from the standpoint of drawing high-strength wire.

Annealing of Plate

Plate products are occasionally annealed to facilitate forming or machining operations. Annealing of plate usually is done at subcritical temperatures, and long annealing times generally are avoided. Maintaining adequate flatness can be a significant problem in annealing of large plates.

Annealing of Tubular Products

Tubular products known as "mechanical tubing" are used in a variety of applications that can involve machining or forming. For these products, which are made from various grades of steel, annealing is a common treatment. In most annealing cycles, subcritical temperatures and short annealing times are used to reduce hardness to the desired level. High-carbon grades, such as type 52100 for bearings, generally are spheroidized to facilitate machining. Tubular products manufactured in pipe mills are rarely annealed. These products normally are used in the as-rolled, the normalized or the quenched and tempered condition.

Annealing Terminology

Box annealing: Annealing of a metal or alloy in a sealed container under conditions that minimize oxidation. In box annealing of ferrous alloys, the charge is heated slowly to a temperature usually below the transformation range but sometimes within or above it, and is then cooled slowly. This process sometimes is referred to as "close annealing" or "pot annealing".

Bright annealing: Annealing in a protective medium to prevent surface discoloration.

Finish annealing: A subcritical annealing treatment applied to cold worked low- or medium-carbon steel. Finish annealing, which is a compromise treatment, lowers residual stresses, thereby minimizing the risk of distortion in machining while retaining most of the benefits to machinability contributed by cold working.

Flame annealing: Annealing in which the heat is applied directly by a flame.

Full annealing: Heating to and holding at a temperature above the upper critical temperature to obtain full austenitization, followed by either slow cooling or isothermal transformation below the lower critical temperature. Specific structures and properties obtained depend on the composition and starting structure of the steel and on the particular time-temperature cycle employed.

Intercritical annealing: Any annealing treatment that involves heating to, and holding at, a temperature between the upper and lower critical temperatures to obtain partial austenitization, followed by either slow cooling or holding at a temperature below the lower critical temperature.

Intermediate annealing: Annealing of wrought metals or alloys at one or more stages during manufacture and before final treatment.

Isothermal annealing: Complete or partial austenitization of a ferrous alloy, followed by cooling to and holding at a temperature at which austenite transforms to a relatively soft ferrite-carbide aggregate.

Process (or "in-process") annealing: Any annealing operation applied for the purpose of restoring ductility for further or subsequent cold work. When used without further qualification, the term usually refers to a subcritical treatment.

Recovery annealing: A subcritical annealing operation that provides relief of residual stresses and some softening or recovery of ductility in cold worked steel. Temperatures employed are below those that would promote formation of new grains through recrystallization, and the degree of softening that occurs is less than that obtained by recrystallization annealing.

Recrystallization annealing: Annealing of a cold worked metal or alloy to produce a new grain structure without a phase change.

Spheroidize annealing (or "spheroidizing"): Any annealing treatment designed specifically to produce a spheroidal or globular form of carbide in the steel.

Stress relieving: Any annealing treatment will tend to reduce residual stresses, but only those treatments applied specifically for the purpose of reducing stresses in heat treated or cold worked steels are termed "stress relieving". The temperatures employed in most stress-relief treatments are below those necessary for complete recrystallization.

Subcritical annealing: Annealing at a temperature below the lower critical temperature. Subcritical annealing often is used to restore ductility between cold forming or drawing operations, in which case it often is referred to as "process annealing".

REFERENCES

1. The Annealing of Steel, by P. Payson: series of articles in *Iron Age,* June and July 1943, subsequently published as a 62-page booklet by Crucible Steel Co. of America

2. Annealing Heat Treatments, by B. R. Banerjee: *Metal Progress,* Nov 1980, p 59

3. *Atlas of Isothermal Transformation and Cooling Transformation Diagrams:* American Society for Metals, Metals Park, OH, 1977

4. *Atlas of Continuous Cooling Transformation Diagrams for Engineering Steels,* by M. Atkins: American Society for Metals, Metals Park, OH, in cooperation with British Steel Corp., 1980

5. Annealing and Carburizing Close Tolerance Driving Gears, by W. Snyder: *Metal Progress,* Oct 1965, p 121

6. *The Making, Shaping and Treating of Steel,* 9th Ed., edited by H. E. McGannon: United States Steel Corp., 1971

Austenitizing Temperatures for Hardening Carbon and Low-Alloy Steels

By Bhupendra K. Gupta
Senior Project Development
Engineer
Science and Technology
International Harvestor Co.

TEMPERATURES recommended for austenitizing carbon and low-alloy steels prior to hardening are given in Table 1 (for direct-hardening grades) and Table 2 (for carburized steels). Table 2 is applicable to carburized steels that have been cooled slowly from the carburizing temperature and are to be furnace hardened in a subsequent operation.

For most applications, the rate of heating to the austenitizing temperature is less important than other factors in the hardening process, such as maximum temperature attained throughout the section, temperature uniformity, time at temperature and rate of cooling. The thermal conductivity of the steel, the nature of the furnace atmosphere (scaling or nonscaling), thickness of section, method of loading (spaced or stacked), and the degree of circulation of the furnace atmosphere all influence the rate of heating of the steel part to the required temperature selected from Tables 1 and 2.

The difference in temperature rise within thick and thin sections of articles of varying cross section is a major problem in practical heat treating operations. When temperature uniformity is the ultimate objective of the heating cycle, this is more safely attained by slowly heating than by rapidly heating. Furthermore, the maximum temperature in the austenite range should not exceed that required to achieve the necessary extent of solution of carbide. The temperatures listed in Tables 1 and 2 conform with this requirement. When heating with significant cross-section variations, provisions should be made for slower heating to minimize thermal stresses and distortions.

Scaling, decarburization and other undesirable surface reactions related to furnace heating are dependent on furnace atmosphere. If these reactions are to be minimized or avoided, furnace heating must be performed in a protective atmosphere. A description of furnace atmospheres suitable for hardening is given in the section of this volume entitled "Furnace Atmospheres and Carbon Control". That section also discusses atmosphere generation and control. Information on austenitizing in molten salt baths, especially when austenitizing is to be followed by austempering in salt, is given in the article entitled "Austempering of Steel". See also the article on liquid carburizing.

Other subjects related to the hardening of carbon and low-alloy steels, and dealt with in separate articles, are as follows: quenching, induction hardening, flame hardening, austempering of steel and martempering of steel. A discussion of quenching from the carburizing temperature is given in the article on gas carburizing.

Table 1 Austenitizing temperatures for direct-hardening carbon and alloy steels (SAE)

Steel	Temperature °C	Temperature °F	Steel	Temperature °C	Temperature °F	Steel	Temperature °C	Temperature °F
Carbon steels			1146	800-845	1475-1550	50B50	800-845	1475-1550
			1151	800-845	1475-1550	50B60	800-845	1475-1550
1025	855-900	1575-1650	1536	815-845	1500-1550	5130	830-855	1525-1575
1030	845-870	1550-1600	1541	815-845	1500-1550	5132	830-855	1525-1575
1035	830-855	1525-1575	1548	815-845	1500-1550	5135	815-845	1500-1550
1037	830-855	1525-1575	1552	815-845	1500-1550	5140	815-845	1500-1550
1038(a)	830-855	1525-1575	1566	855-885	1575-1625	5145	815-845	1500-1550
1039(a)	830-855	1525-1575				5147	800-845	1475-1550
1040(a)	830-855	1525-1575				5150	800-845	1475-1550
1042	800-845	1475-1550	**Alloy steels**			5155	800-845	1475-1550
1043(a)	800-845	1475-1550	1330	830-855	1525-1575	5160	800-845	1475-1550
1045(a)	800-845	1475-1550	1335	815-845	1500-1550	51B60	800-845	1475-1550
1046(a)	800-845	1475-1550	1340	815-845	1500-1550	50100	775-800(c)	1425-1475(c)
1050(a)	800-845	1475-1550	1345	815-845	1500-1550	51100	775-800(c)	1425-1475(c)
1055	800-845	1475-1550	3140	815-845	1500-1550	52100	775-800(c)	1425-1475(c)
1060	800-845	1475-1550	4037	830-855	1525-1575	6150	845-885	1550-1625
1065	800-845	1475-1550	4042	830-855	1525-1575	81B45	815-855	1500-1575
1070	800-845	1475-1550	4047	815-855	1500-1575	8630	830-870	1525-1600
1074	800-845	1475-1550	4063	800-845	1475-1550	8637	830-855	1525-1575
1078	790-815	1450-1500	4130	815-870	1500-1600	8640	830-855	1525-1575
1080	790-815	1450-1500	4135	845-870	1550-1600	8642	815-855	1500-1575
1084	790-815	1450-1500	4137	845-870	1550-1600	8645	815-855	1500-1575
1085	790-815	1450-1500	4140	845-870	1550-1600	86B45	815-855	1500-1575
1086	790-815	1450-1500	4142	845-870	1550-1600	8650	815-855	1500-1575
1090	790-815	1450-1500	4145	815-845	1500-1550	8655	800-845	1475-1550
1095	790-815(a)	1450-1500(b)	4147	815-845	1500-1550	8660	800-845	1475-1550
			4150	815-845	1500-1550	8740	830-855	1525-1575
Free-cutting carbon steels			4161	815-845	1500-1550	8742	830-855	1525-1575
1137	830-855	1525-1575	4337	815-845	1500-1550	9254	815-900	1500-1650
1138	815-845	1500-1550	4340	815-845	1500-1550	9255	815-900	1500-1650
1140	815-845	1500-1550	50B40	815-845	1500-1550	9260	815-900	1500-1650
1141	800-845	1475-1550	50B44	815-845	1500-1550	94B30	845-885	1550-1625
1144	800-845	1475-1550	5046	815-845	1500-1550	94B40	845-885	1550-1625
1145	800-845	1475-1550	50B46	815-845	1500-1550	9840	830-855	1525-1575

(a) Commonly used on parts where induction hardening is employed. All steels from SAE 1030 up may have induction hardening applications. (b) This temperature range may be employed for 1095 steel that is to be quenched in water, brine or oil. For oil quenching, 1095 steel may alternatively be austenitized in the range 815 to 870 °C (1500 to 1600 °F). (c) This range is recommended for steel that is to be water quenched. For oil quenching, steel should be austenitized in the range 815 to 870 °C (1500 to 1600 °F).

Table 2 Reheating (austenitizing) temperatures for hardening of carburized(a) carbon and alloy steels (SAE)

Steel	Temperature °C	Temperature °F	Steel	Temperature °C	Temperature °F	Steel	Temperature °C	Temperature °F
Carbon steels			1527 760-790		1600-1650	4626 815-845		1500-1550
1010 760-790		1400-1450				4718 815-845		1500-1550
1012 760-790		1400-1450	**Free-cutting carbon steels**			4720 815-845		1500-1550
1015 760-790		1400-1450	1109 760-790		1400-1450	4815 800-830		1475-1525
1016 760-790		1400-1450	1115 760-790		1400-1450	4817 800-830		1475-1525
1017 760-790		1400-1450	1117 760-790		1400-1450	4820 800-830		1475-1525
1018 760-790		1400-1450	1118 760-790		1400-1450	8115 845-870		1550-1600
1019 760-790		1400-1450				8615 845-870		1550-1600
1020 760-790		1400-1450				8617 845-870		1550-1600
1022 760-790		1400-1450	**Alloy steels**			8620 845-870		1550-1600
1513 760-790		1600-1650	3310 790-830		1450-1525	8622 845-870		1550-1600
1518 760-790		1600-1650	4320 830-845		1525-1550	8625 845-870		1550-1600
1522 760-790		1600-1650	4615 815-845		1500-1550	8627 845-870		1550-1600
1524 760-790		1600-1650	4617 815-845		1500-1550	8720 845-870		1550-1600
1525 760-790		1600-1650	4620 815-845		1500-1550	8822 845-870		1550-1600
1526 760-790		1600-1650	4621 815-845		1500-1550	9310 790-830		1450-1525

(a) Carburizing is commonly carried out at 900 to 925 °C (1650 to 1700 °F); slow cooled and reheated to given austenizing temperature.

Quenching of Steel

By the ASM Committee on
Quenching of Steel*

QUENCHING OF STEEL is the rapid cooling of steel from a suitable elevated temperature. This generally is accomplished by immersion in water, oil, polymer solution or salt, although forced air is sometimes used. As a result of quenching, production parts must develop an acceptable as-quenched microstructure and, in critical areas, mechanical properties that will meet minimum specifications after the parts are tempered.

The effectiveness of quenching depends on the cooling characteristics of the quenching medium as related to the ability of the steel to harden. Thus, results may be varied by changing the steel composition or the agitation, temperature and type of quenching medium. The design of the quenching system and the thoroughness with which the system is maintained contribute to the success of the process. The design of the part also contributes to the mechanical properties and the distortion that will result from a particular quench.

The rate of heat extraction attributable to a quenching medium is greatly modified by the manner or condition in which the quenching medium is used. These modifications have resulted in the assignment of specific names to various quenching methods, such as direct quenching, time quenching, selective quenching, spray, fog and interrupted quenching.

Direct Quenching

Direct quenching is the most widely used method of treating steel. When carburized work is quenched from the carburizing temperature, or from a slightly lower temperature, the term "direct quenching" is used to distinguish this method from the more indirect practice of carburizing, slow cooling, reheating and quenching. Direct quenching practice is relatively simple and economical, and distortion of carburized parts is usually less frequent with direct quenching than with reheating and quenching, particularly on smaller parts.

Time Quenching

This form of quenching is used when the cooling rate of the part being quenched has to be changed abruptly at some time during the cooling cycle. The change in cooling rate may comprise either an increase or a decrease, depending on which is needed to attain desired results. The usual practice is to lower the temperature of the part by quenching in a medium (for instance, water) for a short time until the part has cooled below the nose of the time-temperature transformation curve, and then to remove the part and quench it in a second medium (for instance, oil), so that it cools more slowly through the martensite transformation range. In many applications, the second medium is still air.

Time quenching is most often used for minimizing distortion, cracking and dimensional change. It should be used with caution because successful application depends greatly on the skill of the operator.

Selective Quenching

This process is used when areas of a part are selected to remain relatively unaffected by the quenching medium. This can be accomplished by insulating the area to be protected or by allowing the quenchant to contact only those areas of the part that are to be quenched.

Spray Quenching

Streams of quenching liquid are directed at high pressure, up to 825 kPa (120 psi), to local areas of the workpiece in spray quenching practice. The cooling rate is fast and uniform over the entire temperature range of the quenching cycle because of the large volume of coolant used and because all of the coolant makes direct contact with the part being quenched. The velocity of the stream removes any vapor bubbles and creates spray droplets, which are available for heat transfer. However, low pressure spraying, in effect a flood-type flow, is preferred with certain polymer quenchants.

*Yancey E. Smith, *Chairman*, Research Manager, Climax Molybdenum Co.; Bhupendra K. Gupta, Senior Project Development Engineer, International Harvester; Karl Heinz Kopietz, Marketing Manager, Heat Treating Products, Henry E. Sanson & Sons, Inc.; Inna Lazarev, Supervisor Technical Service, Heat Treating Department, E. F. Houghton & Co.; William A. Leeper, Chief Metallurgical Engineer, Thornhill-Craver Co.; Nicholas P. Milano, Director of Metallurgy, Illinois Gear, Wallace Murray Corp.; Anthony Meszaros, Metallurgical Supervisor, Park Chemical Co.; Edward R. Mueller, Director of Research, Tenaxol, Inc.

Fig. 1 Temperature gradients and other major factors affecting the quenching of a gear

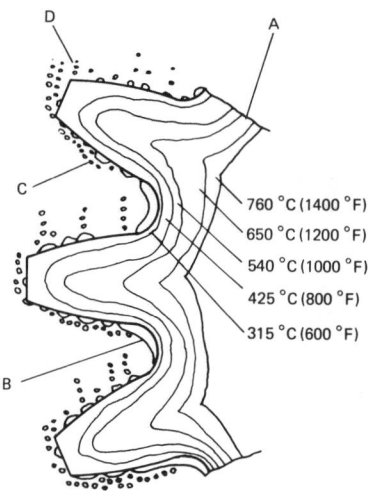

760 °C (1400 °F)
650 °C (1200 °F)
540 °C (1000 °F)
425 °C (800 °F)
315 °C (600 °F)

A, flow of heat from hot core of gear—temperature and flow rate vary with time; B, vapor blanket stage still exists due to large source of heat and poor agitation; C, trapped vapor bubbles condensing slowly; and D, vapor bubbles escaping and condensing. The gear was quenched edgewise in a quiescent volatile liquid. See text.

Fig. 2 Typical surface and center cooling curves, indicating the stages of heat transfer from a hot solid to a cold liquid

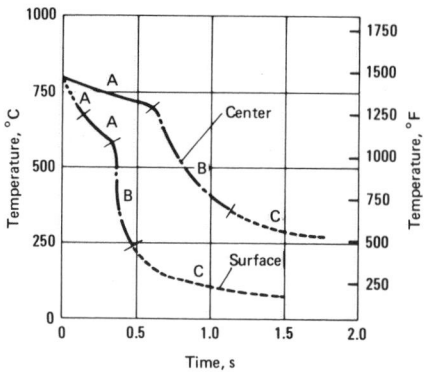

See text for discussion.

Fog Quenching

This procedure utilizes a fine fog or mist of liquid droplets and the gas carrier as cooling agents. Although similar to spray quenching, fog quenching is less effective, because the quenching mist or fog is not readily adapted to rapid removal or replacement by the cooler fog or mist once it has been heated by contact with the part being quenched.

Mechanism of Quenching

Several factors are involved in the mechanism of quenching: (a) internal conditions of the workpiece that affect the supply of heat to the surface, (b) surface and other external conditions that affect the removal of heat, (c) the heat-extracting potential of the quenching fluid in the quiescent state at normal fluid temperatures and pressures ("standard" conditions), and (d) changes in the heat-extracting potential of the fluid brought about by "nonstandard" conditions of agitation, temperature, or pressure. These factors are illustrated in Fig. 1 for a furnace heated gear that is quenched edgewise in a quiescent volatile liquid.

The effect of irregular configuration of the gear on the flow of heat from within the gear to the quenching area is demonstrated in part A of Fig. 1. High temperature persists near the surface at the roots of the teeth where large vapor bubbles are trapped. If the gear were induction or flame heated (and thus had a uniformly thin heated layer conforming to the irregular contour of the gear), heat supply to the quenching area would be more consistent, and quenching would progress more rapidly because heat also would flow simultaneously to the cold metal underlying the heated exterior.

A quiescent liquid experiences unavoidable movement as a result of the action of immersion, the turbulence of boiling, and convection currents. This minimum agitation eventually dissipates the accumulated heat to the surrounding large body of liquid, but local volumes of liquid become heated excessively, or may even vaporize, and this may affect the quenching action.

Volatile quenching liquids produce some vapor at all operating temperatures. Above the boiling point, the supply of vapor becomes so plentiful that an envelope of gas is formed around the surface of the workpiece. This enve-lope, or "vapor blanket", is maintained by radiated heat for as long as the heat is available (part B in Fig. 1). The temperature above which a total vapor blanket is maintained is called the "characteristic temperature" of the liquid.

At lower temperatures, the vapor consists of bubbles, which vary in size depending on the relation of boundary tension of the liquid, the gas and the solid at the temperatures concerned. In one liquid, numerous small, easily detached bubbles may be formed (part D in Fig. 1), whereas large, adherent bubbles, fewer in number, may be formed in another (part C in Fig. 1); this is referred to as the "bubble size characteristic" of the liquid. For any volatile liquid, the mechanical trapping of vapor bubbles (part C) will greatly retard the transfer of heat at the affected location.

Cooling Curves

The most useful way of accurately describing the complex mechanism of quenching is to develop a "cooling curve" for the quenching fluid under controlled conditions. A cooling curve test is sensitive to the factors that may affect the cooling ability of the quenchant, because the test simulates conditions of actual practice.

Cooling curves may be developed by quenching, from an elevated temperature, a test piece of the same steel of which the parts are made, in a sample of the quenching fluid. Sometimes an austenitic stainless steel specimen is

Fig. 3 Transformation diagrams and cooling curves for 8630 steel, indicating the transformation of austenite to other constituents as a function of cooling rate

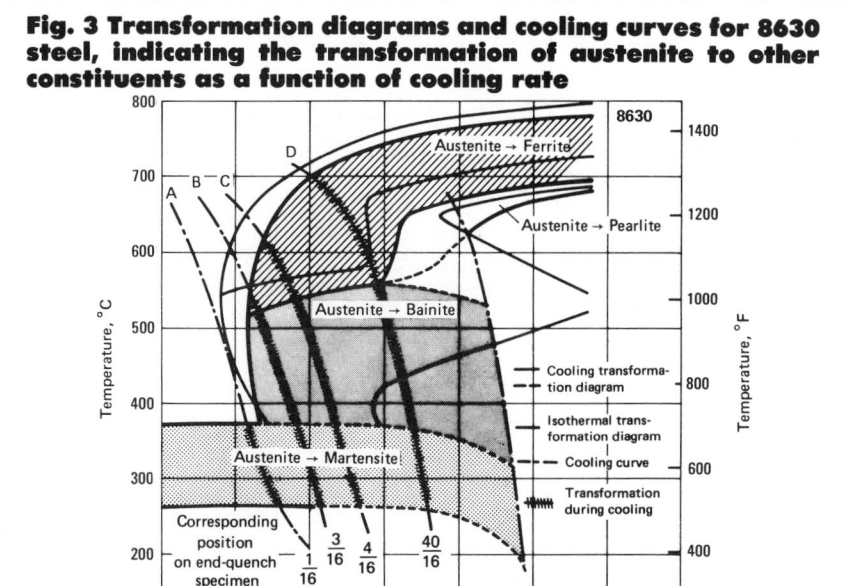

used, to avoid soaking or the necessity for a protective atmosphere. A high speed recorder is used for plotting temperature changes as measured by one or more thermocouples embedded in the test piece. The resulting time-temperature curve indicates the heat transfer characteristics of the quenching fluid.

The typical surface and center cooling curves shown in Fig. 2 graphically describe the four stages of heat transfer from a hot solid to a cold liquid.

Stage A′ in Fig. 2 illustrates the first effects of immersion. Sometimes called the "initial liquid contact stage", this stage is characterized by the formation of vapor bubbles that precedes the establishment of an enveloping vapor blanket. Stage A′ lasts for only about 0.1 s and is relatively unimportant in the evaluation of heat transfer characteristics. It is detectable only when extremely sensitive equipment is used, and it cannot be detected when the liquid is viscous or contains entrained gases, or when the bath is operated at a temperature near the boiling point of the liquid.

Stage A, called the "vapor blanket cooling stage", is characterized by the Leidenfrost phenomenon—namely, the formation of an unbroken vapor blanket that surrounds the test piece. It occurs when the supply of heat from the surface of the test piece exceeds the

amount of heat needed to form the maximum vapor per unit area of the piece. This stage is one of slow cooling, because the vapor envelope acts as an insulator and cooling occurs principally by radiation through the vapor film. This stage is not detectable in aqueous solutions of nonvolatile solutes (at about 5% concentration) such as potassium chloride, lithium chloride, sodium hydroxide or sulfuric acid. Cooling curves for these solutions start immediately with stage B.

When (a) saturated solutions of barium hydroxide, calcium hydroxide or other slightly soluble materials, (b) solutions containing finely dispersed solids, or (c) colloidal solutions in water are used, films are deposited on the test piece during the A stage, which results in the prolongation of both the A and C stages. This condition usually causes a more violent action in stage B. Solutions of some colloids or gels, such as polyvinyl alcohol, gelatin, soap and starch (but not water glass), form an envelope of gel *outside* the vapor blanket formed in stage A. The presence of this gel envelope prolongs the A stage and succeeding stages.

Stage B, the "vapor transport cooling stage", which produces the highest rate of heat transfer, begins when the temperature of the surface metal has been reduced somewhat and the continuous vapor film collapses; violent boil-

ing of the quenching liquid then occurs and heat is removed from the metal at a very rapid rate, largely as heat of vaporization. The boiling point of the quenchant determines the conclusion of this stage. Size and shape of the vapor bubbles are important in controlling duration of stage B, as well as the cooling rate developed within it.

Stage C is called the "liquid cooling stage"; the cooling rate in this stage is slower than that developed in stage B. Stage C begins when the temperature of the metal surface is reduced to the boiling point (or boiling range) of the quenching liquid. Below this temperature, boiling stops and slow cooling takes place thereafter by conduction and convection. The difference in temperature between the boiling point of the liquid and the bath temperature is a major factor influencing the rate of heat transfer. Viscosity also affects cooling rate in stage C.

Significance of Cooling Curves. The same mechanism is involved in cooling a test piece to evaluate a quenchant as is involved in quenching an actual part in a heat treating operation. For all or any selected portions of the curve, cooling curve information can be translated into cooling rate (in degrees per second), if desired. Although a cooling curve relates only to the size and material of the test piece, the thermocouple location, and conditions of the quenching liquid under which a test was performed, cooling curve data developed under one set of conditions can be translated to other conditions, as well as to H-values, by the application of heat transfer formulas.

Agitation is externally produced movement of the quenching liquid relative to the part, either by stirring the liquid or moving the part, or both in combination. This activity has an extremely important influence on the heat transfer characteristics of the quenching liquid. It causes an earlier mechanical disruption of the vapor blanket in stage A and produces smaller, more frequently detached vapor bubbles during the vapor transport cooling stage (stage B). It mechanically disrupts or dislodges gels and solids, whether they are on the surface of the test piece or suspended at the edge of the vapor blanket, thus producing faster heat transfer in liquid cooling (stage C). In addition to the above effects, agitation also brings cool liquid to replace heat-laden liquid.

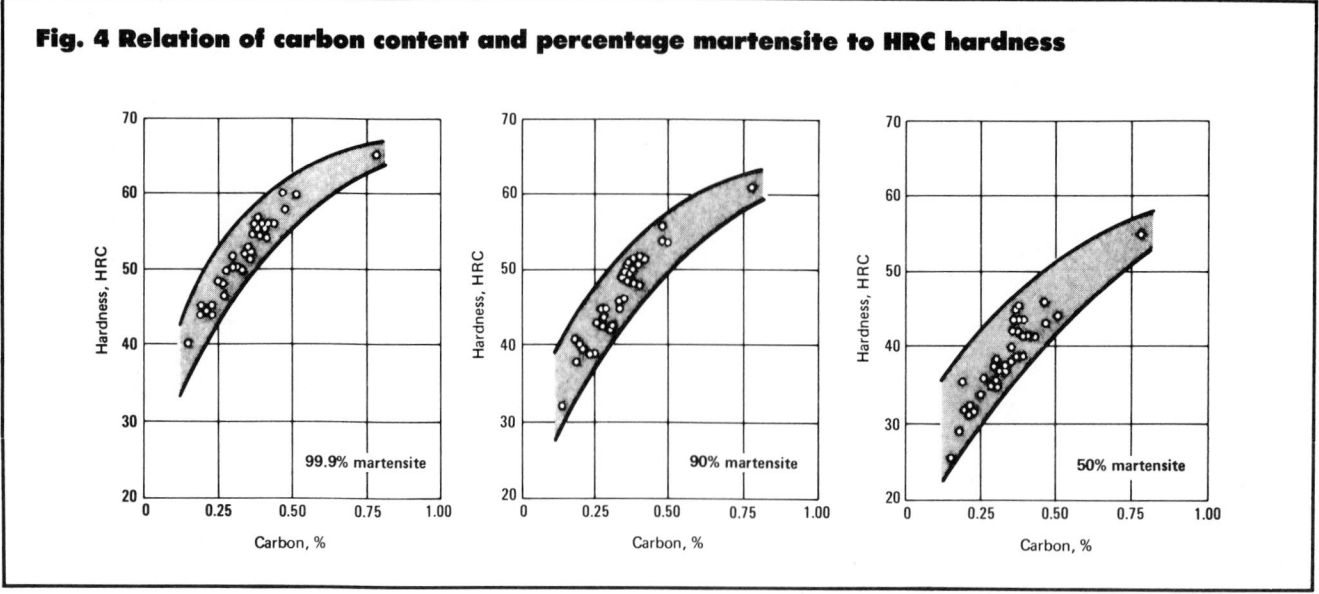

Fig. 4 Relation of carbon content and percentage martensite to HRC hardness

Temperature of Quenchant. The temperature of the liquid may markedly affect its ability to extract heat. As an example, water temperature is very important because it loses its cooling power as it approaches its boiling point. In oil, this effect is not as pronounced because oil becomes less viscous as the temperature is increased. This "thinning" of the oil offsets the temperature rise by a substantial amount.

Workpiece Temperature. Increasing the temperature of the test piece has relatively little effect on its ability to transfer heat to the quenchant. The rate of heat transfer may be increased simply because a greater temperature difference exists. The most noticeable change in ability to transfer heat probably comes from the more rapid oxidation of the surface of the test piece at higher temperatures. This can either increase or decrease the heat transfer ability, depending on the thickness of the oxide developed.

Metallurgical Aspects

Steel is quenched to control the transformation of austenite to desired microconstituents. The microstructures that may be secured are indicated in Fig. 3. Martensite is the as-quenched microstructure usually desired. As indicated by curve A in Fig. 3, to obtain the maximum amount of martensite, the cooling rate must be fast enough to avoid the nose of the time-temperature transformation (TTT) curve of the steel

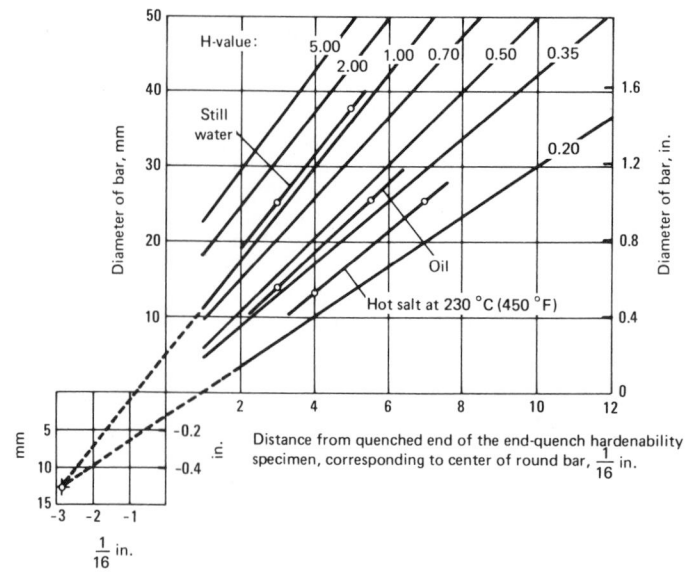

Fig. 5 Grossmann chart relating bar diameter, hardenability of steel and severity of quench

H-value (severity of quench): 5.00, strong brine quench, violent agitation; 2.00, poor brine quench, no agitation; 1.00, poor water quench, no agitation; 0.70, strong oil quench, good agitation; 0.50, good oil quench, good agitation; 0.35, good oil quench, moderate agitation; and 0.20, poor oil quench, no agitation

being quenched. If the cooling rate is not fast enough to miss the nose of the TTT curve (curves B, C and D in Fig. 3), some transformation to bainite, pearlite or ferrite will take place, with a corresponding decrease in the amount of martensite formed and the hardness developed.

Carbon Content and Hardenability. The maximum hardness obtainable in a steel quenched at a sufficient rate to avoid the nose of the TTT curve depends on the carbon content. The cooling rate (quenching efficiency) necessary to obtain a fully martensitic structure depends on the hardenability

Table 1 Effect of agitation upon the effectiveness of quenching

| Circulation or agitation | H-value or quenching power | | |
	Oil	Water	Caustic soda or brine
None	0.25 to 0.30	0.9 to 1.0	2
Mild	0.30 to 0.35	1.0 to 1.1	2 to 2.2
Moderate..............	0.35 to 0.40	1.2 to 1.3	...
Good..................	0.4 to 0.5	1.4 to 1.5	...
Strong	0.5 to 0.8	1.6 to 2.0	...
Violent	0.8 to 1.1	4	5

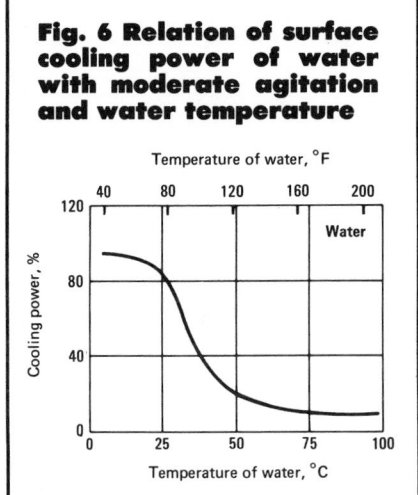

Fig. 6 Relation of surface cooling power of water with moderate agitation and water temperature

of the steel and the section thickness of the part. The relation of carbon content and percentage martensite to hardness is shown in Fig. 4.

Depending on the carbon content and hardenability of the steel, the cooling rate should be fast enough so that a high percentage of martensite will be produced in critically stressed areas of the part. Lower percentages of martensite are often acceptable in areas subject to lower stresses in service. Higher percentages of martensite in the as-quenched structure will produce higher fatigue and impact properties after tempering.

Cooling Rates. When carbon steel is quenched from the austenitizing temperature, a cooling rate equal to or greater than about 55 °C/s (100 °F/s), measured at 705 °C (1300 °F), is necessary for avoiding the nose of the TTT curve. The entire cross section of the piece must cool at this rate to obtain the maximum amount of martensite. Under ideal conditions, water provides a cooling rate of about 2760 °C/s (5000 °F/s) at the surface of steel cylinders 13 mm (1/2 in.) in diameter by 100 mm (4 in.) long. This rate of cooling decreases rapidly below the surface. Thus, for carbon steel, only light sections with a high ratio of surface area to volume can be fully hardened throughout the cross section.

When water or brine is used for quenching, a high temperature gradient is developed between the surface and the center of the part being quenched. This temperature gradient produces greater distortion and increases the hazard of cracking in all but simple, symmetrical shapes. Under favorable conditions, still quenching in a fast oil provides a cooling rate at the surface of a stainless steel cylinder 13 mm (1/2 in.) in diameter by 100 mm (4 in.) long of 2040 °C/s (3700 °F/s) between 830 and 545 °C (1530 and 1010

°F), compared with 2760 °C/s (5000 °F/s) for the surface of a similar specimen in water. Quenching in oil results in a much lower temperature gradient from surface to center, decreased distortion, and less probability of cracking.

With heavy sections, the cooling rate is limited by the rate of heat conduction from the interior to the surface. Rapid cooling of the center of an extremely large section is impossible by any quenching method, because of the "mass effect". Therefore, when deep hardening of a heavy section is mandatory, it is necessary to use an alloy steel with higher hardenability.

Tests and Evaluation of Quenching Media

The testing of quenching media can be classified into two categories: tests of hardening power (metallurgical response) and tests of cooling power (thermal response). The two properties are not the same, because hardening power also is related to the size, microstructure and composition of the steel being quenched.

Hardening-Power Tests. The final criterion for selection of a quenching medium is its hardening power, that is, its ability to develop a specified hardness in a given material-section size combination, under certain conditions of quenchant, agitation and temperature. The major factors influencing the hardness obtained on quenching can be summarized as follows:

- Nature and concentration (if applicable) of quenchant
- Temperature of quenchant
- Agitation and mass flow of quenchant
- Material composition, structure and thermal history of the metal component
- Section size, geometry, and surface condition of the metal component.

Jominy End-Quench Test (ASTM A255). This test is widely used to determine hardenability of steels. It consists of heating a specimen 25 mm (1 in.) round by 100 mm (4 in.) long to the desired hardening temperature, and then withdrawing heat from one end of the bar by the use of a water jet quench. The bar is then ground along its length to 0.4 mm (0.015 in.) below the surface, and hardness readings of the specimen are taken 1.6 mm (1/16 in.) apart. Hardenability is expressed in terms of the distance from the quenched end to a given hardness.

Immersion Quench Test. Because of the importance of agitation in quenching, the evaluation of hardening power of quenchants may be achieved using an immersion quench of the test piece. The hardness readings taken directly indicate the quenching capability, and test results of end quenching and immersion quenching can be easily correlated.

Cooling Power Tests. Because hardening power tests are time consuming and relatively difficult to perform, industry has developed simple, reproducible cooling power tests for the evaluation of quenching media. The four tests most commonly used are:

- Cooling curve test (or basic thermocouple test)
- The magnetic test
- The hot wire test
- The interval method, 5-s test.

Cooling Curve Test. The most useful method to determine the cooling power of a quenchant is by the basic

Fig. 7 Cooling curves showing effect of temperature on cooling power

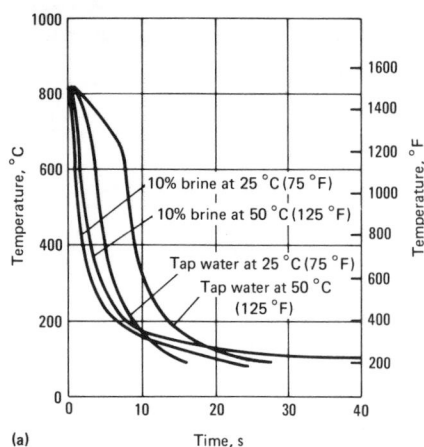

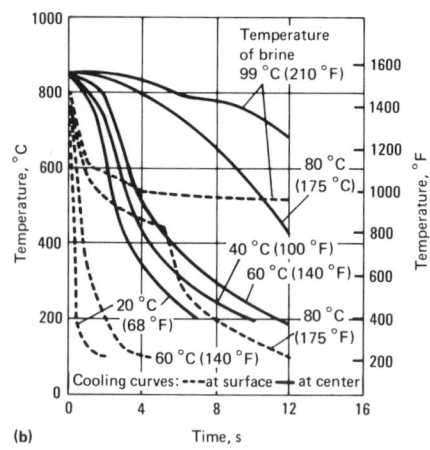

(a) Center cooling curves for water and 10% NaCl solution in quenching 18-8 stainless steel specimens 13-mm (1/2-in.) diam by 64 mm (2 1/2 in.). No agitation. (b) Center and surface cooling curves for 5% NaCl solution at 0.9 mm/s (3 ft/s) in quenching 0.95% C steel specimens 13-mm (1/2-in.) diam by 50 mm (2 in.)

cooling curve test, wherein cooling rates are established throughout the complete quenching process. The test is performed by quenching specimens, with thermocouples embedded at various points, and following the cooling process by using a temperature measuring device. Specimens utilized can be made of stainless steel, silver, nickel or carbon steel. Cylindrical or spherical shapes are most common. A specimen of relatively small mass may be desired for slow quenching fluids, whereas faster quenching fluids may require larger mass, or a recorder having a very high response rate. The use of a silver probe is believed to have an advantage over the other metals because it is easy to prepare, free from transformation effects, and resistant to corrosion. Its high conductivity averages out localized surface thermal conditions while high sensitivity is retained.

The latest test equipment is capable of the simultaneous recording of temperature versus time, and temperature versus cooling rate curves. Depending on the quenching media, the cooling curves developed depict two or three stages of quenching as well as the point of recalescence in hardenable steels. Molten salts and metals have only two stages of quenching, that is, they have no vapor phase.

When analyzing a cooling curve, the analyst should recognize that the heat transfer in metals is measured by the

Fig. 8 Relation of hardness and brine concentration when still-quenching end-quench specimens in 99 °C (210 °F)

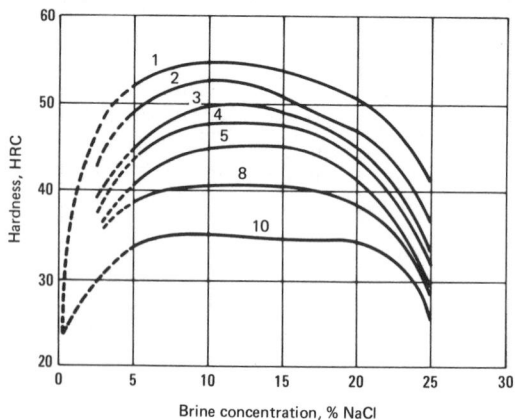

Numbers = distance from quenched end, 1/16 in.

expression kps, where k is the conductivity, p is the density, and s is the specific heat of the metal. The value of kps is 0.972 for silver, 0.49 for nickel and 0.308 for steel. Therefore, the temperature gradient in the test piece is much steeper in steel than in nickel, and it is steeper in nickel than in silver. Correspondingly, the location of the thermocouple must be more precise in a steel test piece than in a silver test specimen, and instruments must be capable

of a faster response for silver than is required for steel.

A surface-located thermocouple is capable of disclosing peculiarities of the cooling curve that are obscured by a center-located thermocouple. Moreover, the time factor is lengthened, and the temperature of occurrence of characteristic points on the curve appears to be higher when measurement is made at the center rather than the surface. If a surface thermocouple is used with a

Table 2 Relation of brine density to brine concentration

Salt, %	Specific gravity	Degrees Baumé	Salt g/l	Salt lb/gal
Sodium chloride solutions				
4	1.0268	3.8	41.1	0.343
6	1.0413	5.8	62.4	0.521
8	1.0559	7.7	84.5	0.705
9	1.0633	8.7	95.9	0.800
10	1.0707	9.6	107.1	0.894
12	1.0857	11.5	130.3	1.087
Sodium hydroxide solutions				
1	1.0095	1.4	10.1	0.0842
2	1.0207	2.9	20.4	0.1704
3	1.0318	4.5	31.0	0.2583
4	1.0428	6.0	41.7	0.3481
5	1.0538	7.4	52.7	0.4397

Fig. 9 Relation of hardness and distance from end of specimen quenched in water and brine

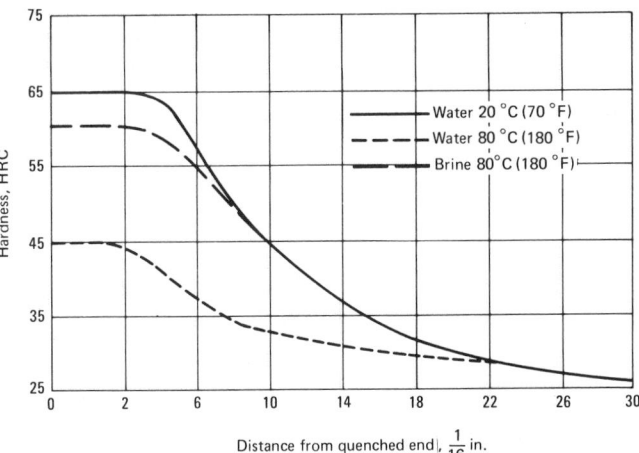

Water 20 °C (70 °F)
Water 80 °C (180 °F)
Brine 80 °C (180 °F)

Hardness, HRC

Distance from quenched end, $\frac{1}{16}$ in.

Cooling power of brine is greater than water at 80 °C (180 °F).

10-mm (0.4-in.)-diam silver test piece, the recorder must be capable of accurately registering rates of temperature change of at least 540 °C (1000 °F) in 0.1 s throughout the temperature range being explored.

The "characteristic temperature" (the lowest temperature at which a vapor envelope can be maintained) and the boiling point are unaffected by the temperature, the heat conductivity, or the mass of the test piece when the cooling curve is taken at the surface of the piece. This is not true when the curve is taken at any point below the surface. Surface thermocouple locations should always be used to obtain fundamental data, although other, more convenient, locations may be used for control purposes. The most common deviations from the above "standard" conditions of performing the cooling curve test are variations in temperature of the test piece or the quenching medium and in the degree of agitation. Over-all, the cooling curve test is an excellent research and quality control tool, and provides for a great degree of comparative accuracy. Standardization of the test would facilitate the comparison of results from different sources.

The Magnetic Test. This test makes use of the properties of metals that lose their magnetism when heated above a Curie point, and regain their magnetism when cooled below this temperature. The magnetic test was devised to provide a means of comparing the heat extraction rates of oil, molten salt, water or other quenching media. The test method involves the heating of a 22-mm (⅞-in.) diam nickel sphere, weighing approx 50 g (1.8 oz) to 835 °C (1625 °F) in either air or a controlled atmosphere. (A chromium-diffused nickel sphere can be used to eliminate the need for a protective atmosphere to avoid marked alteration of surface finish.) After uniformity of temperature is attained, the sphere is quenched in a 200-ml sample of quenchant located within a magnetic field. As the nickel sphere cools through the Curie temperature, it becomes magnetic and is attracted to the magnetic field. The time required for the nickel sphere to cool from 835 °C (1625 °F) to the Curie temperature, which is 355 °C (670 °F), is a measure of the cooling power of the quenching medium. The Curie temperature of nickel is below the nose of the isothermal transformation diagrams for most steels. Thus, with this method, quenching media can be compared for steels as well as for other metals. The faster the cooling power of the medium, the shorter the time required for the nickel to regain its magnetism.

Other alloys with different Curie points can be used to establish additional points on a cooling curve. A modification of the magnetic quench test has been used to study the influence of circulation and heat on the cooling power of quenching oils. A device known as the electronic quenchometer is installed directly in the quenching system so that results are obtained from the actual quenching process conditions of media, temperature and agitation. To compare the results of the magnetic test with the actual quenching power for certain steels, tests were made by quenching AISI 1046 steel bars in various oils. The test bars, 22-mm (⅞-in.) diam by 75-mm (3-in.) long, were quenched from an endothermic gas atmosphere after austenitizing at 815 °C (1500 °F). Transverse hardness tests correlated favorably with magnetic test results.

The Hot Wire Test. This method of quenchant evaluation consists of heating a nichrome or cupron wire (of standard gage and electrical resistance) by means of an electrical current, in a small quantity (100 to 200 ml) of quenchant. The quenchant tested usually is at quenching temperature, and the wire is supported by two copper or brass electrodes. Heating of the wire is

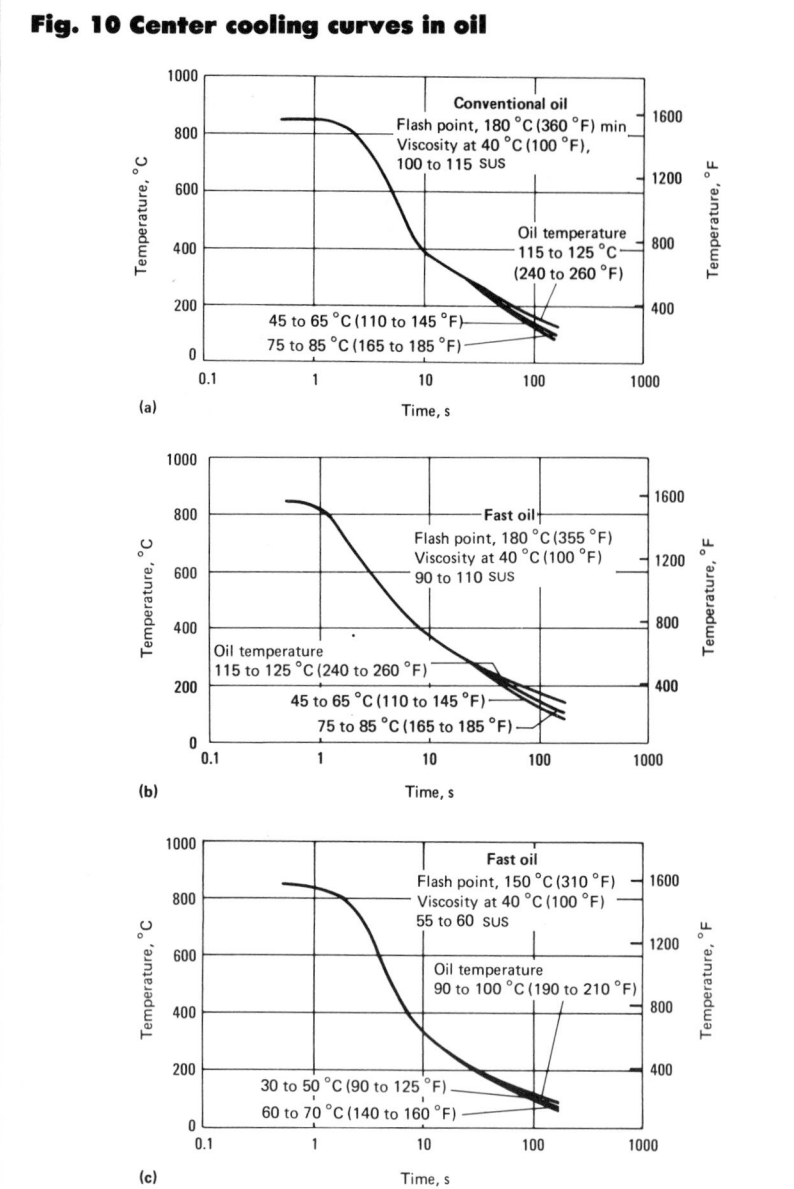

Fig. 10 Center cooling curves in oil

(a)

Conventional oil
Flash point, 180 °C (360 °F) min
Viscosity at 40 °C (100 °F),
100 to 115 SUS

Oil temperature
115 to 125 °C
(240 to 260 °F)

45 to 65 °C (110 to 145 °F)
75 to 85 °C (165 to 185 °F)

(b)

Fast oil
Flash point, 180 °C (355 °F)
Viscosity at 40 °C (100 °F)
90 to 110 SUS

Oil temperature
115 to 125 °C (240 to 260 °F)

45 to 65 °C (110 to 145 °F)
75 to 85 °C (165 to 185 °F)

(c)

Fast oil
Flash point, 150 °C (310 °F)
Viscosity at 40 °C (100 °F)
55 to 60 SUS

Oil temperature
90 to 100 °C (190 to 210 °F)

30 to 50 °C (90 to 125 °F)
60 to 70 °C (140 to 160 °F)

Center cooling curves for type 304 stainless steel specimens still quenched in a conventional oil and in two fast oils, showing effect of type and temperature of oil on cooling characteristics. Thermocouples were placed at geometric centers of specimens, which were 13 mm (0.5 in.) in diameter and 100 mm (4 in.) long. See text for discussion of significance of these data.

(8.8 oz) is heated to 815 °C (1500 °F) and quenched for 5 s. The oil sample is then stirred to ensure temperature equalization throughout the bath, and the temperature rise is noted to the nearest tenth of a degree Celsius. This process is repeated for a series of test bars. Finally, a bar of the same metal and size is fully quenched in a second 2-l sample of oil, and the rise in temperature is noted. Quenching power of the oil is computed according to the following equation:

$$\frac{A}{B} \times 100 = \% \text{ quench speed}$$

where A represents the average rise in oil temperature for the 5-s quench bars, and B represents the maximum temperature rise of the oil sample for the fully quenched specimen.

The 5-s test is used for determining gross changes in quenching oil because it is expedient and requires no special equipment. A duration of 5 s in quenching, however, constitutes only a comparison in the higher temperature region of the quench, and this may be misleading because it encompasses only a portion of the cooling curve—namely, the vapor blanket and vapor transport stages.

Evaluation of Severity for Quenching of Steels

The ability of a quenching medium to extract heat from a hot steel workpiece can be expressed in terms of the "H-value" (the severity of quench). If the H-value of still water is taken as 1.0, the H-values of oil, water and brine are as presented in Table 1.

The relations between bar diameter, inherent hardenability, and H-value may be plotted as shown in Fig. 5, which is referrred to as a "Grossman" chart. The H-values of a material are plotted as such: If a 25-mm (1-in.) -diam part is quenched in oil with "good agitation" (0.50 H-value), the chart shows that, reading across the 25-mm (1-in.) -diam horizontal line to the 0.50 H-value and then down, a value of 8 mm ($^5/_{16}$ in.) on the end-quench hardenability specimen is obtained. This means that the center of the bar will have the same hardness as that shown at 8 mm ($^5/_{16}$ in.) from the quenched end of a standard hardenability specimen made from the steel being quenched.

The application of the chart is limited because the quenchants involved are

accomplished by steadily increasing the current by means of a rheostat. The cooling power of the quenchant is indicated by the maximum current reading, as measured by an ammeter. Quenchants capable of extracting heat faster permit the passage of higher currents through the wire and thus register higher current values. This test resembles the 5-s test in that it compares quenchants only in the higher temperature range of quenching.

Interval Test. This method, also known as the 5-s test, is a rapid method of comparing the cooling power of quenchants. In this test, a 2-l (0.5-gal) sample of an oil is placed in an insulated container, and the oil temperature is noted. A bar of metal (usually, stainless steel) weighing about 250 g

Fig. 11 Center cooling curves in oil

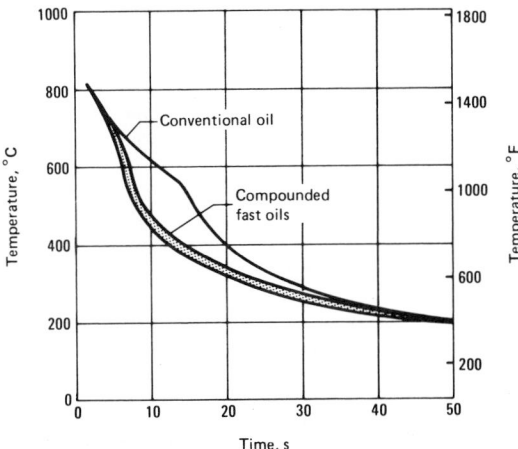

Center cooling curves for austenitic stainless steel quenched in conventional oil and compounded oils with fast quenching capacities. All oils at 52 °C (125 °F). Specimens were 13 mm (0.5 in.) in diam by 64 mm (2.5 in.) long.

Fig. 12 Comparison of cooling power of conventional and fast quenching oils

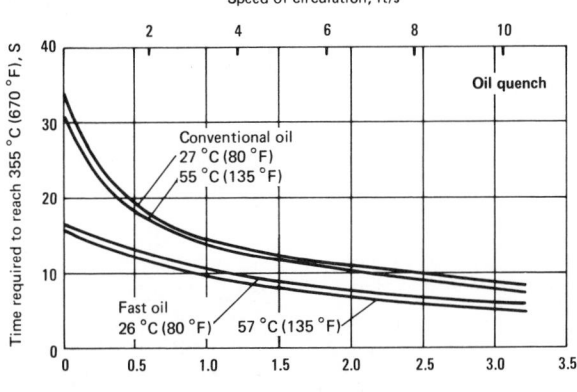

Effects of circulation and temperature of quenchant on the heat extraction rate of conventional and fast oils. A 25-mm (1-in.) nickel ball was heated in a reducing atmosphere and quenched to the Curie temperature 355 °C (670 °F).

described in general terms (for example, "very good oil quench—good agitation"). An added complication is the fact that what appears to be good agitation in a quenching system that has no parts in it may be poor agitation when a load of parts is immersed in the bath.

A simple test for evaluating the cooling power of a quench bath under any condition of loading is based on the use of Fig. 5. If the hardenability of the steel and the bar diameter are known, it is possible to plot one point on the quench-severity curve. For example, if

the 8-mm (5/16-in.) point on the end-quench specimen shows a hardness of 45 HRC, and a 25-mm (1-in.) -diam bar of the same steel quenched in the bath to be evaluated has a center hardness of 45 HRC, the result indicates a quench severity of 0.50. Regardless of the hardness obtained at the center of the 25-mm-diam test bar, it can be plotted on the chart of Fig. 5 at its correct location. For example, if the center hardness had been 48 HRC, and if the hardenability curve for the bar of steel tested had shown 48 HRC at 6 mm (1/4 in.), the 6-mm line and the 25-mm-

diam line would intersect at a quench severity of 0.70.

As shown in Fig. 5, for oil quenches the various severity lines of the Grossmann chart converge at a point outside the normal range of the chart. This point of intersection can be used as a second point in drawing any new quench-severity lines. An actual example is illustrated in Fig. 5 by the line labeled "Oil". The two experimental points on this line were obtained by quenching a bar of steel of known hardenability having two different diameters, 25 and 12.5 mm (1 and ½ in.). The use of a bar with two diameters offers an excellent means of checking results because it provides three points on the line. The results of three different operating commercial quenches, as determined with stepped test bars or bars having two different diameters, are plotted in Fig. 5.

The principal advantage of this test is that the specimen can be treated along with other work just as if it were a production part; thus, the quench is evaluated under the conditions of temperature and agitation that actually prevail in the production quenching of a load of parts. The test bar should be of such diameter that the as-quenched hardness at the center of the bar will fall on the sloping portion of the hardenability curve for the steel being quenched. The test can be performed most conveniently with a shallow-hardening steel because this type of steel permits the use of smaller diameter test bars, which are easier to prepare and to section after being quenched. The use of shallow-hardening steel in no way detracts from the validity of the test results, because severity of quench is an inherent characteristic of the bath and is not affected by part size or hardenability.

Quenching Media

Many different media have been used for quenching. Most are included in the list that follows, but some of these are used only to a very limited extent.

- Water
- Brine solutions (aqueous)
- Caustic solutions
- Oils
- Polymer solutions
- Molten salts
- Molten metals
- Gases, including still or moving

Table 3 Typical properties of commercially available quenching and martempering oils

Type of quenching oil	No.	API gravity	Flash point °C	°F	Pour point °C	°F	Viscosity at 40 °C (100 °F), SUS	Saponification	Ash, %	Water, %
Conventional, no	1	33	155	315	−12	10	107	0.0	0.01	0.0
additives	2	27	185	365	−9	15	111	0.0	0.03	0.0
Fast, with speed	3	33.5	190	370	−12	10	95	0.0	0.05	0.0
improvers	4	35	160	320	−4	25	60	0.0	0.20	0.0
Martempering, without	5	31.1	235	455	−9	15	329	0.0	0.02	0.0
speed improvers	6	28.4	245	475	−9	15	719	0.0	0.05	0.0
	7	26.6	300	575	−7	20	2550	0.0	0.10	0.0
Martempering, with	8	28.4	230	450	−9	15	337	2.0	1.1	0.0
speed improvers	9	27.8	245	475	−9	15	713	2.2	1.1	0.0
	10	25.5	300	570	−7	20	2450	2.5	1.4	0.0
ASTM test.....................		D287	D92	D92	D97	D97	D445, D2161	D94	D482	D95, D1533

Fig. 13 Comparison of water, oil and emulsions of water and soluble oil

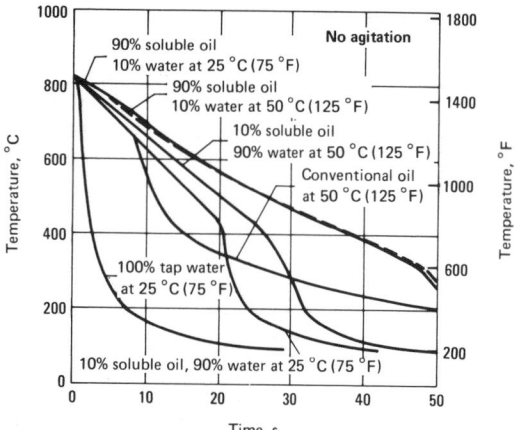

Center cooling curves for still-quenched 18-8 stainless steel specimens 13 mm (0.5 in.) in diameter by 64 mm (2.5 in.) long, indicating comparative cooling characteristics of plain water and soluble oil, at temperatures of 25 and 50 °C (75 and 125 °F)

Fig. 14 Advantage of low-viscosity quenching oil over high-viscosity oil

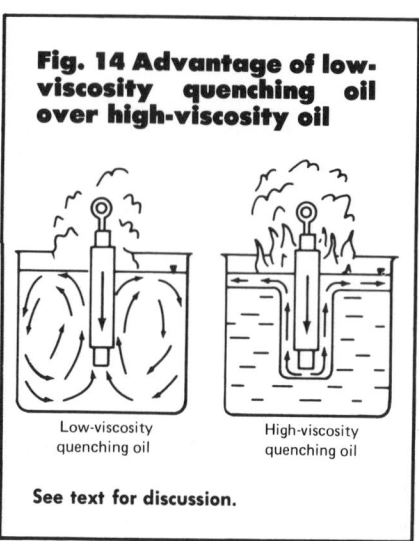

Low-viscosity quenching oil

High-viscosity quenching oil

See text for discussion.

- Fog quenching
- Dry dies, commonly water cooled.

Water

As a quenching medium, plain water approaches the maximum cooling rate attainable in a liquid. Its other advantages are that it is inexpensive and readily available, is easily disposed of without attendant problems of pollution or health hazard, and is an effective means of breaking scale from the surface of steel parts that are quenched from furnaces in which protective atmospheres have not been employed.

Water, therefore, is used wherever it is practical, that is, where the drastic quench afforded by water does not result in excessive distortion, or cracking of the workpiece. Water is widely used for quenching nonferrous metals, aus-tenitic stainless steels and other metals that have been given a solution treatment at elevated temperature. Plain water also is used to a considerable extent for rapid cooling of quench-hardenable steels where formation of vapor pockets with resulting soft spots is not objectionable.

One disadvantage of plain water as a quenchant is that its rapid cooling rate persists throughout the lower temperature range, in which distortion or cracking is likely to occur. Consequently, water usually is restricted to the quenching of simple, symmetrical parts made of the shallower-hardening grades of steel (plain carbon or low-alloy). Another disadvantage of using plain water is that its vapor blanket stage may be prolonged. This prolongation, which varies with the degree to which the complexity of the part being quenched encourages vapor entrapment and with the temperature of the quench water, results in uneven hardness and unfavorable distribution of stress. This may cause distortion, cracking or soft spots. Water quenched steel parts may rust unless immediately treated with a rust preventive.

To obtain reproducible results by water quenching, the temperature, agitation and contamination must be controlled. Water at a temperature of 15 to 25 °C (55 to 75 °F) will provide uniform quenching speed and reproducible results. As indicated in Fig. 6, the surface cooling power of water decreases rapidly as water temperature increases. (This behavior also is demonstrated in Fig. 7.) Hot water has a low cooling power because, as the boiling point is approached, the cooling action resembles that of steam.

Agitation is especially important in water quenching because it disperses

Fig. 15 Effects of water contamination

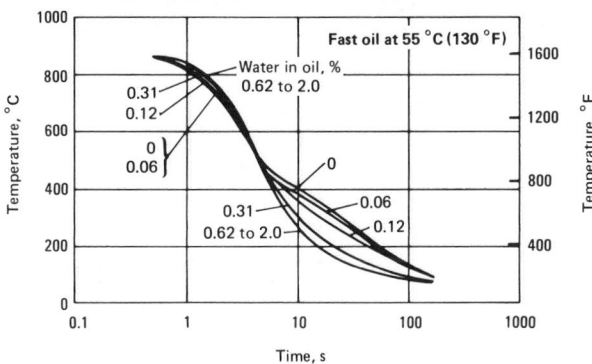

Effects of water contamination on the quenching power of fast oil at 55 °C (130 °F) in quenching type 304 stainless steel specimens 13 mm (0.5 in.) in diam by 100 mm (4 in.) long. Oil was not circulated, and specimens were not agitated. Thermocouples were located at the geometric center of specimens.

Fig. 16 Effect of boiling off water in oil

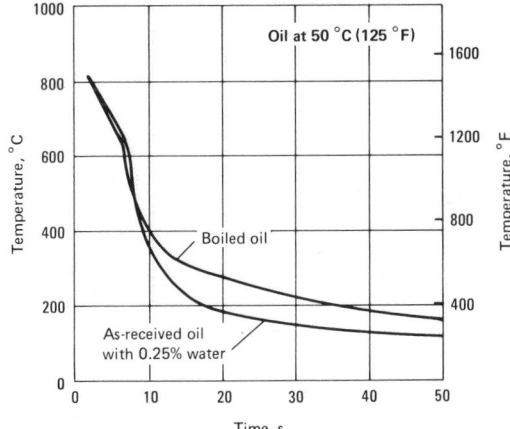

Center cooling curves for austenitic stainless steel specimens 13 mm (0.5 in.) in diam by 64 mm (2.5 in.) long, showing the effects of quenching in as-received oil containing 0.25% water and in the same oil after it had been freed of water contamination by being boiled at 120 °C (250 °F) for 5 h. Quenching temperature of oil in both conditions was 52 °C (125 °F).

vapor bubbles from parts and directs cooler water against the workpiece.

Contamination of water quench baths by dissolved heat treating salts is likely to increase the cooling rate, because salts effectively reduce the duration of the vapor blanket stage. However, contaminants such as soaps, algae, slimes or emulsion-formers reduce the cooling rate by trapping the steam or vapor film and thus prevent cooler water from contacting parts quickly enough to produce uniform results.

Insoluble salts that are carried over from activated salt baths into the quench bath may interfere with the operation of a closed pumping system or reduce the volume of available water in the quench tank. Periodic desludging or replacement of the quench water is therefore recommended.

Brine Solutions

The term "brine", as applied to quenching, refers to aqueous solutions containing various percentages of salt (such as sodium chloride or calcium chloride).

Advantages and Disadvantages. Brine offers the following advantages compared with plain water or oil, for quenching.

- Cooling rate is faster than that of water for the same degree of agita-

tion, or less agitation is required for a given cooling rate.
- Temperature is less critical than for water, thus requiring less control (note cooling curves in Fig. 7).
- Possibility of soft spots from steam pockets is less than in water quenching.
- Distortion is less severe than in water quenching.
- Heat exchangers are needed less often for the cooling of brine baths than they are for water or oil quench baths.
- Complex baffling patterns in the quench tank and propeller agitation, both of which are often necessary with oil or water as quenchants, are not always required with a brine quench, but may be required for steels with very low hardenability.

Ordinarily, the disadvantages of brine quenching will not prevent its use, because brine quenching is used only where oil or water quenchants do not provide desired results. Some of these disadvantages are:

- The corrosive nature of brine requires that, for reasonable service life, the quench tank, pumps, conveyors and other equipment in constant contact with the brine solution either be protected from corrosion by coating, plating or sheathing; or they must be made of corrosion resistant metals (such as copper-bearing or stainless steels).
- A hood may be needed to carry off the corrosive fumes that emanate from brine baths, to protect nearby machinery or delicate equipment from attack.
- Cost is higher than for water, largely because of the cost of the corrosion inhibitors that must be used after cleaning.
- Labor cost is increased because of the necessity for testing to control solutions and for providing a safe environment.

Cooling Rates. At 20 °C (70 °F) and moving at 0.9 m/s (3 ft/s), a solution containing 5 wt% sodium hydroxide can cool the surface of a cylinder 12.5 mm (1/2 in.) in diameter and 50 mm (2 in.) long from 870 to 200 °C (1600 to 390 °F) in 0.31 s. Under the same conditions, a 5% solution of sodium chloride requires 0.48 s, and plain water requires 1.2 s. However, a sodium chloride solution is safer, less costly and easier to handle than a sodium hydroxide solution, and may be preferred even

Fig. 17 Hardness values and changes in dimensions of 1085 steel test specimen after quenching in water, fast oil and conventional oil

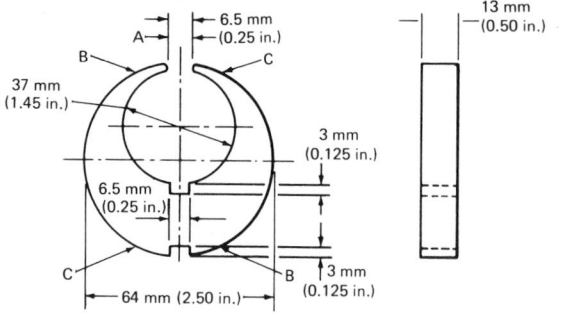

Quenching medium	Hardness, HRC		
	Maximum	Minimum	Variation
Water	67.0	63.0	4.0
Fast oil	66.0	63.0	3.0
Conventional oil	65.5	43.0	22.5

	Dimensional change					
	Gap A		Diameter B		Diameter C	
	mm	in.	mm	in.	mm	in.
Water	0.3404	0.0134	0.2591	0.0102	0.2946	0.0116
Fast oil	0.0533	0.0021	0.0813	0.0032	0.0610	0.0024
Conventional oil	0.0559	0.0022	0.0965	0.0038	0.0965	0.0038

Fig. 18 Effect of type of quenching oil on variations in center-to-pitch-line dimension

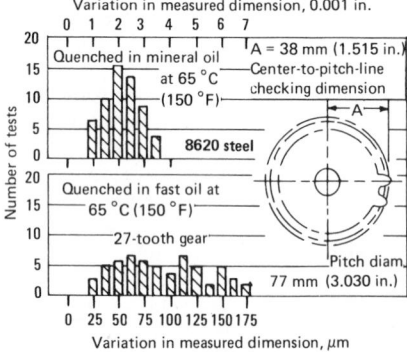

For transmission counter gears of 8620 steel, paraffin-base mineral oil had viscosity at 40 °C (100 °F) of 86 to 110 sus. Gears, carburized and direct quenched from 870 °C (1600 °F), were tempered at 155 °C (310 °F) after quenching, and had a hardness of 59 to 63 HRC.

Fig. 19 Cooling curves for polyvinyl alcohol solutions

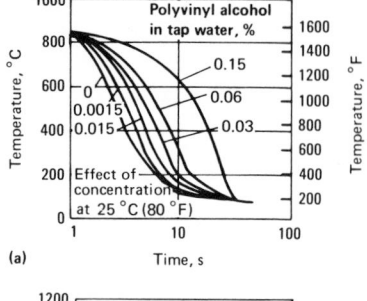

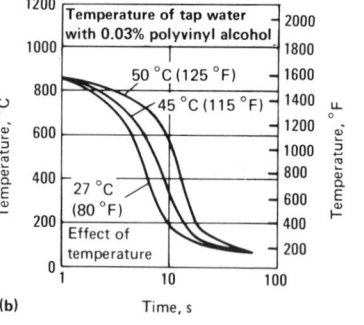

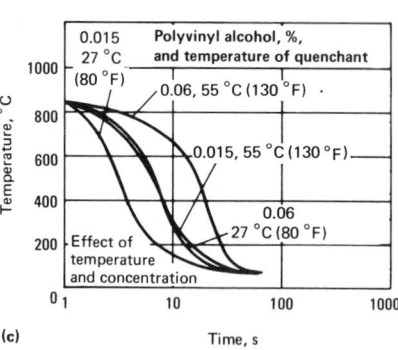

Center cooling curves for type 304 stainless steel specimens 13 mm (½ in.) diam by 100 mm (4 in.) long quenched in still tap water at 25 °C (80 °F) (or at other temperatures shown) containing various concentrations of polyvinyl alcohol. Thermocouple was placed in geometric center of each specimen. Water hardness, 130 ppm

though the latter provides faster cooling rates.

Brine is more widely used for small tool hardening applications, whereas caustic quenches are used for mass production processes.

Effect of Concentration. In conventional quenching, it has been found that a 10% NaCl solution is the most effective from the viewpoint of hardening. The concentration of the brine can be determined in several fashions, as in Table 2.

Although brine solutions (NaCl) of up to 24% concentration are optimum for reducing the vapor stage, such high concentrations generally are impractical. The value of the 10% NaCl solution in developing maximum surface hardness in conventional quenching is illustrated in Fig. 8. It should be noted that, although end quench values are used in this figure, the specimen was not quenched as in the standard end-quench test but was quenched in still water with only the end face contacting the water to simulate commercial quenching practice.

Effect of Temperature. The cooling power of brine solutions is not critically affected by small variations in operating temperature. Although such solutions can be used at or near the

Table 4 Comparison of the cooling power of commercially available quenching and martempering oils according to magnetic quenchometer test results

Types of quenching oil	Oil sample	Viscosity at 40 °C (100 °F), SUS	Flash point, °C	°F	Quenching duration from 885 °C (1625 °F) to 355 °C (670 °F)			
					At 27 °C (80 °F)		At 120 °C (250 °F)	
					Ni-ball	Chromized Ni-ball	Ni-ball	Chromized Ni-ball
Conventional	1	102	190	375	22.5	27.2	...	...
	2	105	195	380	17.8	27.9	...	...
	3	107	170	340	16.0	24.8	...	...
Fast.........................	4	50	145	290	7.0	(a)	...	...
	5	94	170	335	9.0	15.0	...	...
	6	107	190	375	10.8	17.0	...	...
	7	110	185	370	12.7	19.6	...	...
	8	120	190	375	13.3	17.8	...	...
Martempering, without speed improvers	9	329	235	455	19.2	27.6	18.4	22.1
	10	719	245	475	26.9	29.0	25.1	30.4
	11	2550	300	575	31.0	32.0	31.7	32.8
Martempering, with speed improvers	12	337	230	450	15.3	(a)	12.8	(a)
	13	713	245	475	16.4	17.9	14.0	15.6
	14	2450	300	570	19.7	17.0	15.1	15.4

(a) Not available

Fig. 20 Effect of concentration on cooling rate

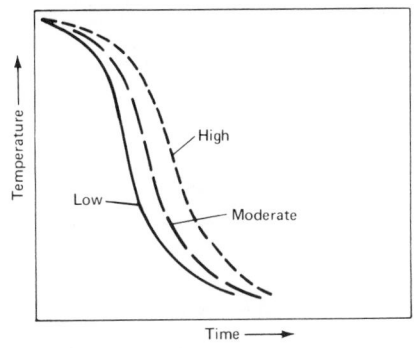

Fig. 21 Effect of temperature on cooling rate

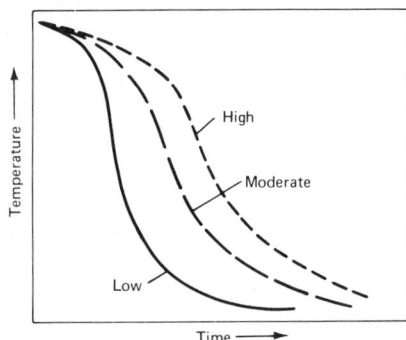

Fig. 22 Effect of agitation on cooling rate

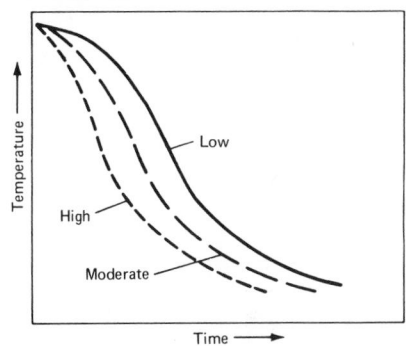

temperature of boiling water, they provide maximum cooling power at about 20 °C (70 °F). The relatively small effect of temperature on cooling power of brine compared with water is demonstrated in Fig. 9.

Effect of Contamination. Sludge and scale should be removed periodically, because both act to clog pumps and recirculating systems as well as to reduce the volume and cooling rate of the brine quench bath. Excess water reduces the solution strength and quenching power of the brine.

Caustic Solutions

Aqueous solutions of 5 to 10% sodium hydroxide (NaOH) often are used for quenching. The performance of such solutions is similar to brine solutions—

sharing about the same advantages as a quenchant. Its main shortcoming is its high alkalinity which is harmful to human skin.

Proprietary Salt Mixes

Proprietary salt mixes that have the same quenching effect as brine or caustic solutions have become commercially available recently. When used at 5 to 15% in aqueous solutions, they provide rust protection for the unwashed quenched parts, and they do not harm human skin.

Oil Quenching

Quenching oils can be divided into several distinctive groups. Based on

their composition, quenching effect and use temperature, quenching oils are categorized as conventional, fast, martempering, or hot quenching. An additional group of oil-based quenchants are oil-in-water emulsions. These find only limited use because of their unpredictable quenching characteristics, which are completely different from the characteristics of water-free quenching oils.

All modern quenching oils are based only on mineral oils, usually paraffin-base, and do not contain any fatty oils formerly used as additives to quenching oils or as quenching oils themselves. One reason for this is the much better aging stability of mineral oils; another is the fact that certain well-suited fatty oils are now unavailable.

Fig. 23 Cooling rate of a polyacrylate quenchant as a function of concentration and temperature

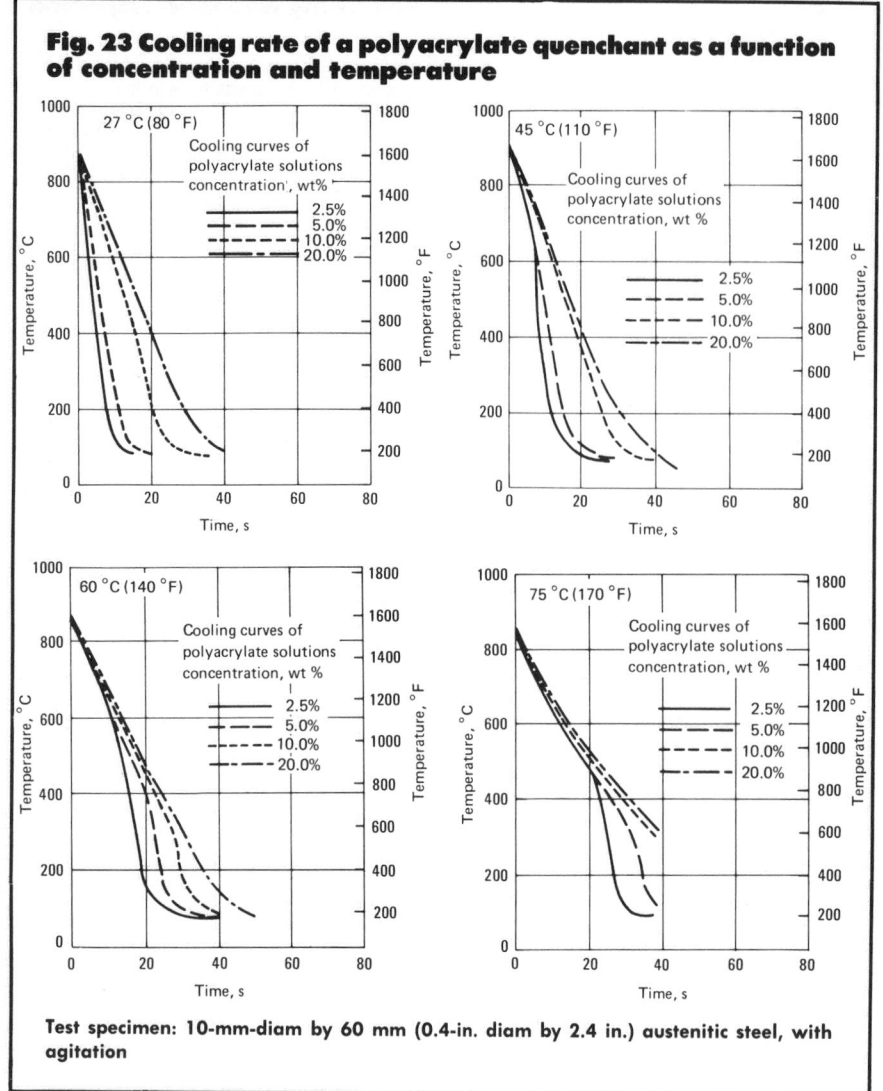

Test specimen: 10-mm-diam by 60 mm (0.4-in. diam by 2.4 in.) austenitic steel, with agitation

Fig. 24 Cooling rates of different polymer quenchants at 20% concentration at 25 and 60 °C (80 and 140 °F)

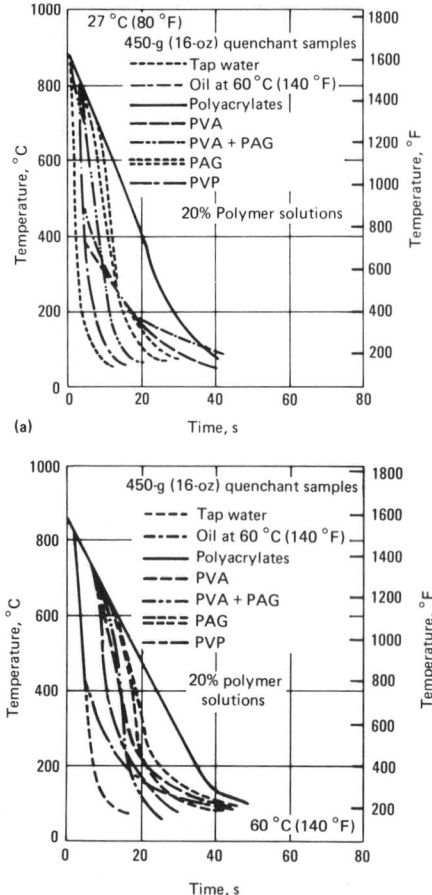

Test specimen: 10-mm diam by 60 mm (0.4-in. diam by 2.4 in.) scaleproof austenitic steel, medium agitation

Table 5 Comparison of the ranges of cooling power of commercially available quenching and martempering oils according to magnetic quenchometer test results using pure nickel balls

| Type of quenching oil | Quenching duration from 885 °C (1625 °F) to 355 °C (670 °F) | | | |
	At 65 °C (150 °F)	At 120 °C (250 °F)	At 175 °C (350 °F)	At 230 °C (450 °F)
Conventional	14 to 22	14 to 22	. . .	. . .
Fast	7 to 14	7 to 14	. . .	. . .
Martempering, without speed improvers	18 to 34	18 to 34	22 to 38	Approx 47
Martempering, with speed improvers	14 to 20	13 to 18	16 to 22	Approx 33

Sperm oil, which had been used successfully as quenchant for extremely bright quenching of very small workpieces, is now banned from use and from importation into the United States.

Conventional quenching oils are mineral oils, sometimes containing antioxidants; however, they are free from additives that alter their quenching effects. The typical viscosity of conventional quenching oils is 100 to 110

sus at 40 °C (100 °F), but it can reach about 200 sus at 40 °C (100 °F).

Fast quenching oils are mineral oil blends, usually with a viscosity between 50 and 110 sus at 40 °C (100 °F), but for the most part between 85 and 105 sus at 40 °C (100 °F). They contain especially developed proprietary additives that provide faster quenching effects; in addition they can be compounded further with antioxidants, wetting agents and other additives.

Martempering or hot quenching oils are solvent-refined paraffin-type mineral oils of very good thermal and oxidation stability. They are used at temperatures between about 95 °C (200 °F) and 230 °C (450 °F) for modified and

Fig. 25 Cooling curves obtained in quenching 4130 steel tubing in gas, oil and still air (normalizing)

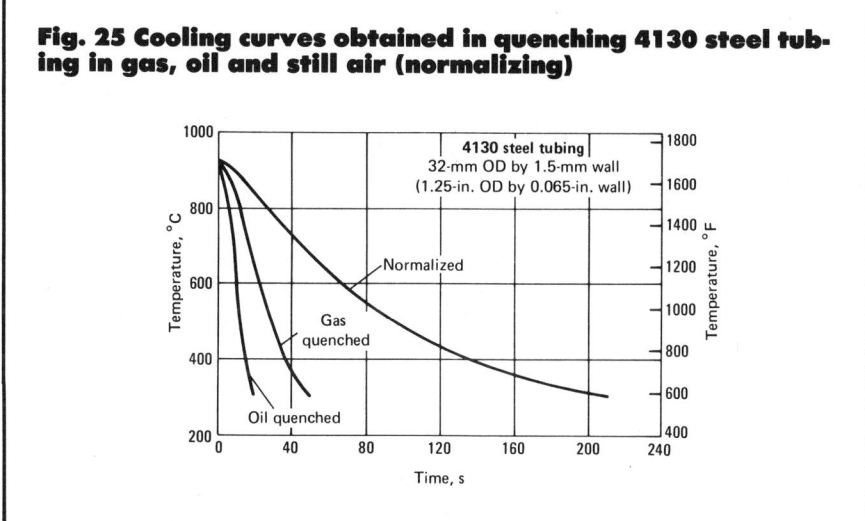

Fig. 26 Differences in Brinell hardness of forged 1095 steel disks, 100-mm (4-in.) thick, after oil quenching, gas quenching (forced air) and cooling in still air (normalizing)

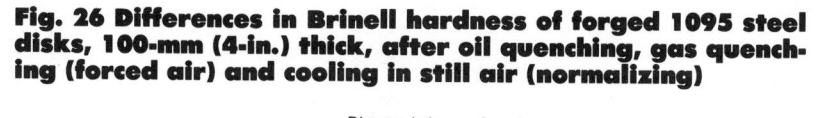

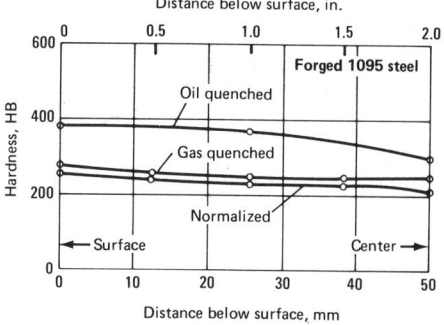

actual martempering of ferrous metals. Martempering oils may contain antioxidants to improve their aging stability. They are also available with comparatively high quenching effects, even at high use temperatures, because of the additions of very effective speed improvers.

Emulsions of soluble oils, usually employed as coolants for metal-working such as grinding, cutting and sometimes forming, are used for quenching with concentrations of 3 to about 15%. Such quenching emulsions are similar in characteristics to water-based quenchants, particularly when compared to synthetic polymer solutions.

Typical physical and chemical properties of several commercially available quenching oils are listed in Table 3, which also shows the appropriate

ASTM test methods for the determination of the properties listed. These tests are used by vendors for quality control on new quenching oils. They may be applied also to the routine evaluation of used quenching oils after a certain time-span of usage. Oils from different producers may differ in general properties and quenching effects. These differences result from variations in the compositions of base oils and the additives used.

Cooling Characteristics. The ideal quenchant would have a high initial quenching effect during vapor-phase and boiling-range periods but would cool slowly through the final convection range (liquid-cooling phase). Cold water, and especially the aqueous solutions of inorganic salts, show the highest initial quenching speeds, but they

also quench very fast at the end of the quenching process, that is, during the convection phase. Thus, they are restricted to quenching simple shapes and steels of comparatively low hardenability. For other workpieces, they would cause either intolerable degrees of distortion, or warpage and high quench crack rate. All quenching oils have considerably lower quenching effects than water or aqueous inorganic salt solutions. However, the heat extraction is more uniform and particularly slow at the end of the cooling cycle, that is, during the convection range. As a result, the dangers of distortion or cracking are diminished decisively.

Conventional quenching oils exhibit a comparatively long vapor-phase period, during which period the quenching speed is very low. The rate of cooling becomes higher during the boiling range, followed again by very slow cooling in the convection range. Thus, the quenching power of conventional quenching oils is far less than that of water and often inadequate for steel of lower hardenability.

Fast quenching oils show a high initial quenching speed, in some situations approaching the initial speed of water, followed by fast cooling during the boiling range. The cooling rate in the convection range is usually about the same for fast and conventional quenching oils. However, some fast quenching oils contain special additives that cause a faster quench speed in the convection range that results in a much better through-hardening effect than attainable with regular fast quenching oils; see fast oil (c) in Fig. 10.

Figure 11 shows the spread of cooling curves of eight different compounded (speed-improving additive-containing) fast quenching oils as a shaded band, and gives a curve for one typical conventional additive-free 100 SUS at 40 °C (100 °F) mineral quenching oil. These cooling curves, which show clearly the difference in quenching effects of conventional and fast quenching oils, were recorded without agitation and using a cylindrical austenitic stainless steel test probe of 13 mm (0.5 in.) diameter by 64 mm (2.5 in.) length, containing a thermocouple in its geometrical center. All oils were at a temperature of 50 °C (125 °F).

Table 4 shows the results of magnetic quenchometer tests in quenching durations, in seconds from 885 °C (1625 °F)

Fig. 27 "Wind tunnel" used in gas quenching large turbine-rotor forgings

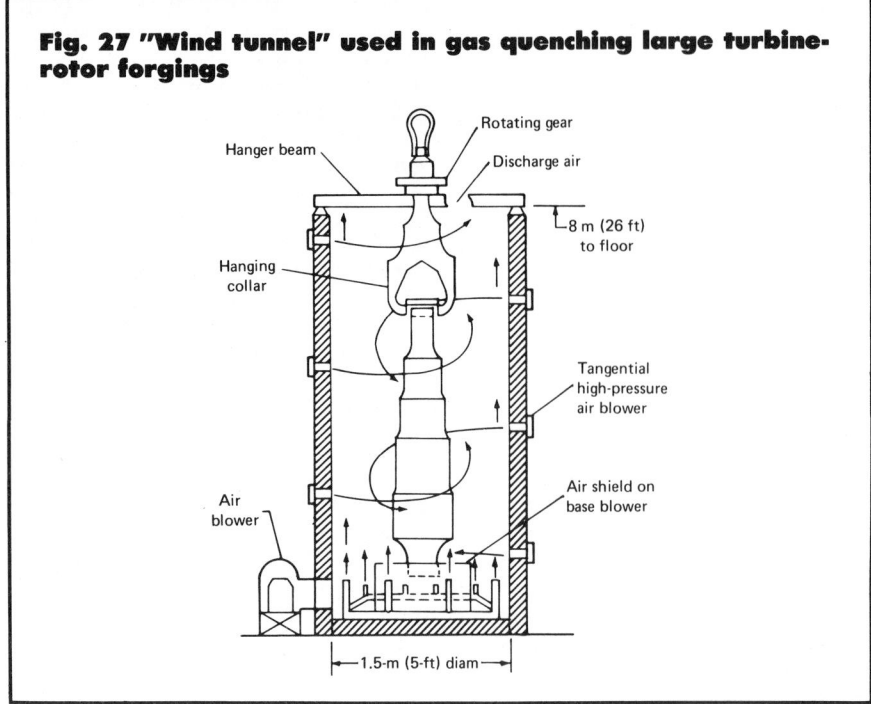

Rotating gear

Hanger beam

Discharge air

8 m (26 ft) to floor

Hanging collar

Tangential high-pressure air blower

Air blower

Air shield on base blower

1.5-m (5-ft) diam

Fig. 28 Surface cooling curves for blocks made of types T1 and A2 tool steels quenched from austenitizing temperatures by cooled nitrogen in a vacuum furnace

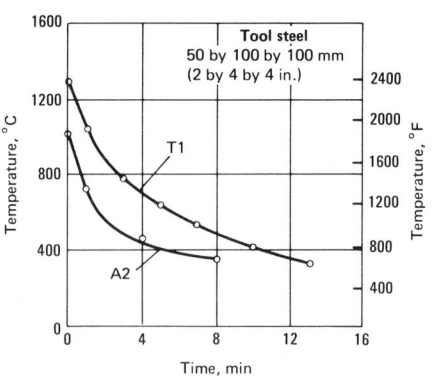

Tool steel
50 by 100 by 100 mm
(2 by 4 by 4 in.)

T1

A2

Time, min

Table 6 Power requirements for propeller agitation of quenching liquids

Volume of liquid in tank		Power required for velocity of approx 15 m/min (50 ft/min)			
		Conventional oil		Water or brine	
l	gal	W/l	hp/gal	W/l	hp/gal
190 to 3000	50 to 800	0.985	0.005	0.788	0.004
3000 to 7600	800 to 2000	1.182	0.006	0.788	0.004
7600 to 11400	2000 to 3000	1.182	0.006	0.985	0.005
Over 11400	Over 3000	1.380	0.007	0.985	0.005

to 355 °C (670 °F), of 22 mm (⅞ in.) diameter nickel balls (48 to 52 g). These results are listed for commercially available conventional and fast quenching oils as well as for martempering oils without and with additions of speed improvers. Conventional and fast quenching oils usually are tested at 25 °C (80 °F); the permissible range is 20 to 25 °C (70 to 80 °F). Martempering oils, which are used at higher temperatures, are commonly tested at 120 °C (250 °F). The test balls are made of either pure nickel or chromized pure nickel. The latter type was developed to approach the conditions and results obtained with pure nickel balls heated to 885 °C (1625 °F) under reducing atmospheres. This magnetic test, using the chromized nickel balls, is specified in ASTM D 3520, but it is so cumbersome that the much simpler plain nickel ball test usually is preferred.

Table 4 compares test results of the two magnetic quenchometer test methods. The already discussed difference in quenching effect between conventional and fast quenching oils is clearly indicated by either variant of the test. Table 4 also shows that martempering oils with speed improvers have, at 120 °C (250 °F), quenching effects as fast as the slower-quenching, low-temperature fast oils. As shown in Table 5, the martempering oils quench as fast as conventional 100 sus quenching oils when used at about 175 °C (350 °F). Thus, these modern fast quenching martempering oils permit modified or actual martempering of steel parts that are either thicker or lower in hardenability than previously permissible when using the slower quenching martempering oils without speed-improving additives.

The typical ranges for the magnetic

quenchometer speeds of the different quenching oil groups are listed in Table 5 as a function of their actual use temperature. Table 5 clearly shows the well-known fact that the quenching effect of quenching oils does not change much with changing oil temperature. Significant losses of quenching power occur only at temperatures over about 150 °C (300 °F), where martempering oils are preferred.

The insignificant effect of temperature changes and the influence of increased agitation on the cooling power of one conventional and one fast oil, as determined by a modification of the magnetic quenchometer test, are shown in Fig. 12.

The cooling curves in Fig. 10 prove also that the cooling characteristics of quenching oils are not changed over a wide range of operating temperatures. The only differences are found in the last stage of cooling, when the temperature of the quenched part approaches the oil temperature. During this stage, when heat is extracted only by convection, the cooling speed decreases with higher quenching oil temperatures. These differences can be important when quenching critical parts. Slower cooling speeds cause reduced temperature gradients between surface and core and result in lower internal stresses. This is especially beneficial while the steel part is passing through the temperature range of martensite formation, below 260 °C (500 °F), because it minimizes distortion and lowers the probability of quench cracking.

Fig. 29 Effect of flow rate of quenching oil on cooling time of scale-free bars of 9445 alloy steel

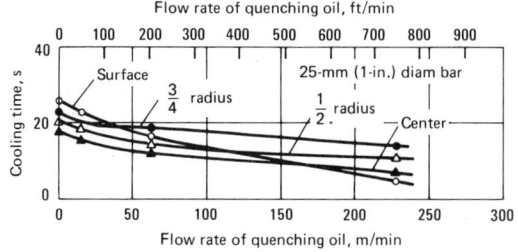

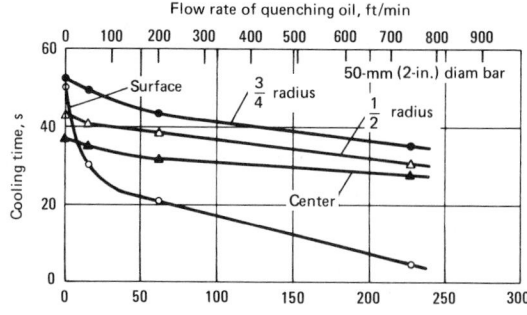

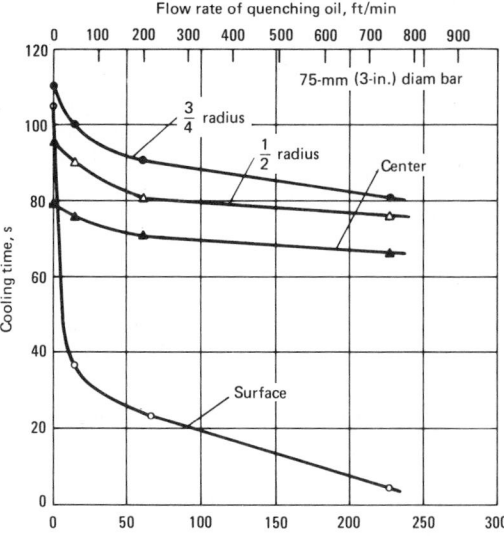

Bars were heated in an exothermic atmosphere, and were quenched in a mineral oil at 50 °C (120 °F). Saybolt universal viscosity of the oil at 40 °C (100 °F) was 79 s.

As indicated by the comparison of cooling curves presented in Fig. 13, emulsions of water and soluble oil combine the poorest features of water and oil quenchants. The center-cooling curves of an oil-in-water emulsion, consisting of 10% soluble oil and 90% water at 25 °C (75 °F) and 50 °C (125 °F), show that extremely stable vapor envelopes are formed at the beginning of the quenching cycle. Thus, at the higher bath temperature, the initial severity of the quench is reduced to below that of a conventional oil. Cooling in the latter stage is essentially the same as in water. Water-in-oil emulsions containing 90% soluble oil and 10% water have much lower cooling rates than conventional oil. For these reasons, oil and water emulsions gener-

ally are inferior to other quenchants. They are rarely used for quenching, although sometimes they may be used for spray quenching of induction or flame-heated parts. In these instances, the better-suited aqueous polymer quenchants would be more desirable.

The only legitimate use for oil-in-water emulsions is the final cooling of quenched and tempered fasteners to provide a black surface and rust protection. Such emulsions commonly contain about 3 to 10% of the soluble oil and are used at temperatures between about 30 °C (85 °F) and 80 °C (175 °F).

Oil Temperature. As shown in the previous section, the quenching effects of conventional and fast quenching oils are almost independent of their use temperatures. For practical reasons, these oils are generally maintained at temperatures between 40 and 95 °C (100 and 200 °F), most often at 50 to 70 °C (120 to 160 °F). Higher or lower oil temperatures are possible without any change in the as-quenched hardness of the hardened work, but are rarely utilized. Higher temperatures will cause faster aging of the oil and can lead to increased fuming. For safety purposes, the maximum use temperature should be at least 50 °C (120 °F) below the flash point of the oil. Lower oil temperatures may result in more distortion because of the faster quenching effect of the oil during its convection range, and may increase the fire hazard because of the higher viscosity of the cold oil.

Figure 14 shows the advantage of a more fluid quenching oil. Any given oil will have a lower viscosity at higher temperatures, and higher viscosity at lower temperatures. When a hot part is immersed into a low-viscosity oil, the fire hazard is less even if the oil's flash point is comparatively lower. This effect occurs because heat extracted from the part is distributed rapidly through the oil by the "thermosyphon" effect. Thus, oil contacting the hot metal at the air interface does not reach and exceed its flash point. Quenching oils with comparatively high viscosities—for example, cold oils—do not distribute heat as rapidly. Instead, a comparatively thin layer of oil is heated at the part surface and gets hotter as it flows upward. Thus, the oil can reach the air-oil interface at a temperature that is higher than its flash point. The oil temperature may even pass the fire point, resulting in immediate ignition.

In one test, two similar parts, each weighing 80 kg (175 lb) and at about 870 °C (1600 °F), were partially immersed in separate baths containing 180 kg (395 lb) of oil at 20 °C (70 °F) with no circulation. One oil had a flash point of 205 °C (400 °F) with a viscosity of 120 SUS at 50 °C (120 °F); it ignited immediately. The other tank contained oil with a much lower flash point of 120 °C (250 °F) but a lower viscosity of 46 SUS at 50 °C (120 °F); it took 20 s to ignite. Thus, two precepts are apparent: first, it is recommended that cold, viscous oils be preheated to decrease their viscosities, thereby diminishing the likelihood of ignition; second, higher-flash-point oils should be preferred. Typical conventional or fast quenching oils have minimum flash points of 125 °C (260 °F) and average flash points of 175 °C (350 °F).

Martempering oils are designated to be used at higher temperatures permitting modified or actual martempering of ferrous metals. The typical use temperature ranges for commercially available martempering oils are as follows:

Viscosity(a)	Flash point(b), °C	Temperature range, °C In open air	Under protective atmosphere
250-550	220	95-150	95-175
700-1500	250	120-175	120-205
2000-2800	290	150-205	150-230

(a) At 40 °C (100 °F) SUS. (b) Minimum

Fig. 30 Curves for identical cooling times in end-quench hardenability specimens and round bars quenched in oil

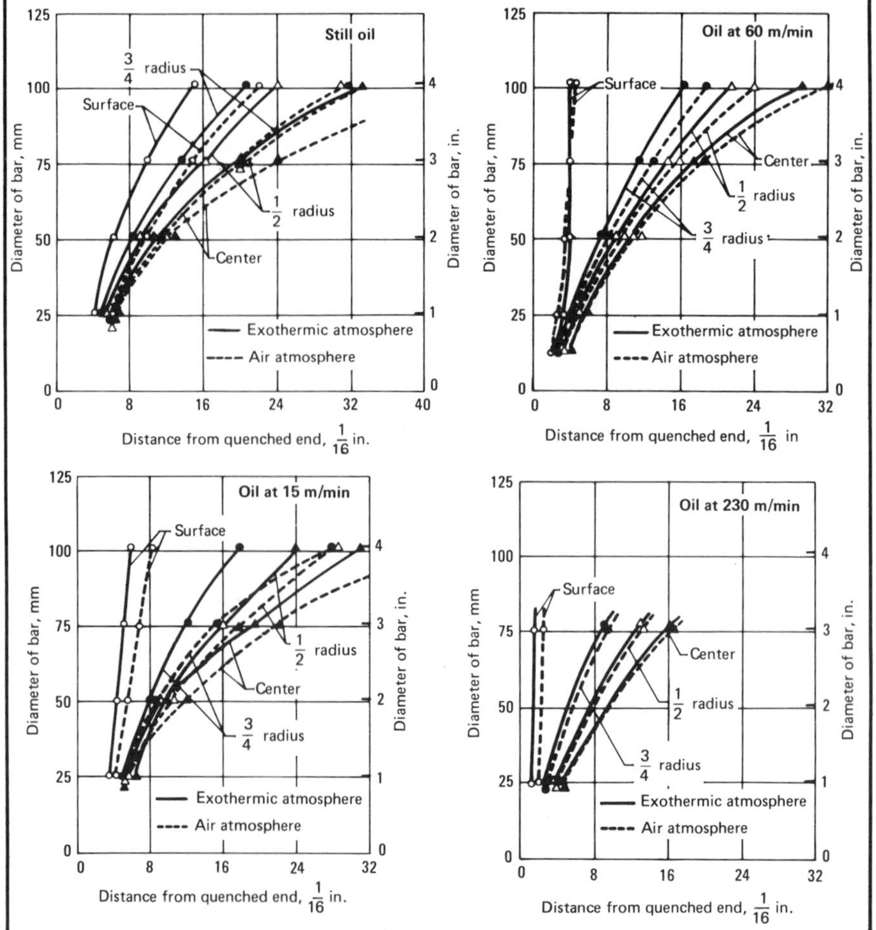

Mineral oil used for quenching had a Saybolt universal viscosity at 40 °C (100 °F) of 79 s.

Martempering oils, when covered by a protective (inert, neutral or reducing) atmosphere in an integral quenching chamber of a furnace with controlled atmosphere, are largely protected from oxidation by air. Thus, they may be used at higher temperatures and closer to their flash points than similar oils used in open air.

The dragout loss of a quenching oil depends mainly on its viscosity, which is dependent on its temperature. Because oils of higher viscosities usually are used at higher temperatures, the actual viscosities and the dragout losses of different quenching oils do not vary much. The following table shows the viscosities of one conventional and one fast quenching oil as well as two martempering oils as a function of the temperature. The recommended-use

temperature ranges for the four oils appear in bold-faced type below:

Oil temperature, °C	Conventional	Viscosity of oil, SUS Fast		Martempering
40	**100**	**95**	720	2630
65	**55**	**60**	190	555
95	**40**	**40**	85	180
120	35	35	**55**	90
150	30	35	**45**	**60**
175	...	...	**40**	**45**
205	...	...	35	**40**
230	...	...	...	35

When these oils are used at the middle of their use temperature ranges, their viscosities vary only from 40 to 60 SUS.

Quench loads will determine the temperature rise of the quenching oil, but uniform hardening usually can be maintained despite large temperature increases. However, the danger of fire must be considered when the oil temperatures approach the flash point. Parts with high ratio of surface area to volume increase the fire hazard because they transfer heat more rapidly to the oil, thus increasing the fire hazard. Such parts require lower quenching oil temperatures, larger oil volumes or oils of higher flash point.

The recommended capacity of an oil quench tank can be estimated by either of the following generalized rules:

• Volume of oil in liters equals weight of quench load in kg ÷ 10
• Volume of oil in gallons equals weight of quench load in pounds

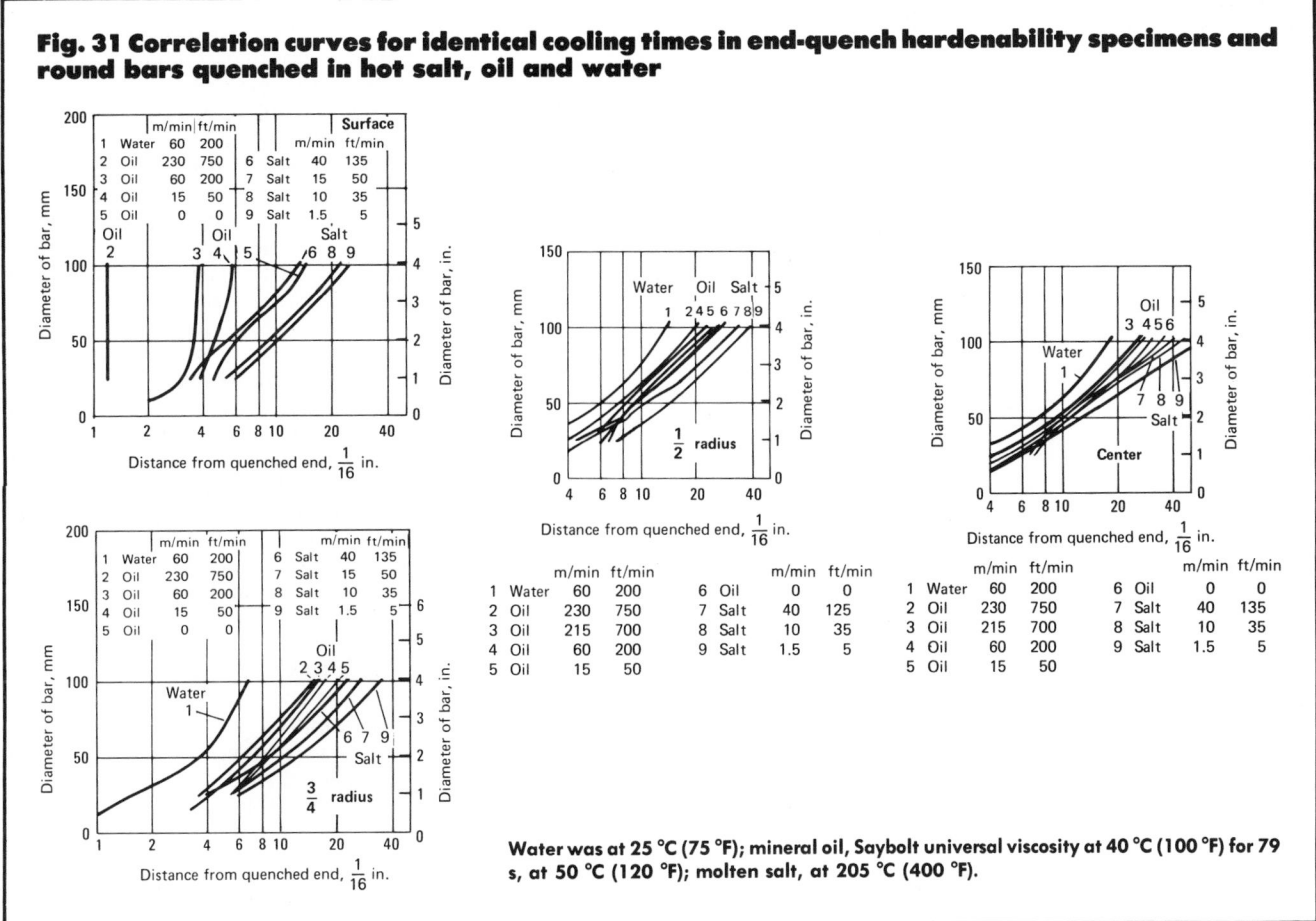

Fig. 31 Correlation curves for identical cooling times in end-quench hardenability specimens and round bars quenched in hot salt, oil and water

Water was at 25 °C (75 °F); mineral oil, Saybolt universal viscosity at 40 °C (100 °F) for 79 s, at 50 °C (120 °F); molten salt, at 205 °C (400 °F).

The two variations of the rule deliver somewhat different results, but the difference is only 17%. Thus, either of the rules can be used for a first approximation.

For very large, massive parts, the oil volume can be much lower than estimated using above rules. Such parts transfer the heat comparatively slowly to the oil. However, it is necessary to install larger coolers and provide fast circulation from tank to cooler in order to maintain the oil temperature at a safe, low level.

To minimize oxidation, all quenching oils should have as little contact with air as possible. It is recommended that the surface of open-air quench baths be kept as small as possible. For oil baths under controlled atmospheres, external exposure to the air should be as short as possible.

Quenching oils should be agitated (*a*) to distribute heat quickly and uniformly and (*b*) to maintain a uniform temperature, which provides a uniform quenching effect throughout the tank.

Only mechanical agitation should be used (pumps or impellers). Strong agitation should be used only when work is actually quenched. *Never bubble compressed air through oil for agitation!*

Heating elements used for preheating quenching oils should have less than 1.5 W/cm² (10.0 W/in.²) to prevent local overheating, which would cause not only unnecessary aging of the oil (polymerization) but also the formation of an insulating layer of oil-coke on the heater with the possible result of heating element burn-outs.

The quenching system, particularly heating elements and coolers, should not comprise copper or copper alloys. These metals act as catalysts, accelerating the oxidation and polymerization (aging) of mineral oils. Use of steel, ferrous castings, stainless steel, nickel-plated or tin-plated materials is recommended instead. Existing copper or brass equipment should be tin plated.

Water Contamination. Water in quenching oils is very dangerous, par-

ticularly in martempering oils used at temperatures above 100 °C (212 °F). Not only does it cause nonuniform or insufficient as-quenched hardness of the work, but also it creates heavy foaming and increased fire hazard because oil foam catches fire very easily. Oil foam that develops in an integral-quenching chamber can cause an explosion if the foam level reaches the hot furnace chamber.

The effect of water in quenching oils is shown in Fig. 15 in the form of center cooling curves. These results were recorded by quenching a cylindrical stainless steel probe in a fast quenching oil free of water and after water additions of 0.06 to 2.0%. At water concentrations below 0.12%, the initial quenching speed was increased, whereas higher water concentrations caused a more stable vapor blanket around the probe resulting in a lower initial quenching speed.

All water-containing oil samples quenched faster than the water-free fast oil during boiling and convection ranges. Because of the higher cooling speeds in the convection range, which overlaps with the martensite formation range of most common steels, increased distortion and an increased likelihood of cracking can be expected when using water-contaminated quenching oils.

Entrapment of water in localized sections of an oil bath can result in an explosion from the sudden formation of large volumes of steam. However, except where oil and water are distributed in stratified layers under stagnant conditions, there is little danger of explosion. With normal circulation of the bath, water content can be high without danger of explosion.

Water can be removed from an oil bath by several methods: (a) raising the temperature above the boiling point of the water (Fig. 16), (b) allowing the water to settle and draining it off, and (c) passing the bath through a centrifuge. However, used quenching oils may sometimes emulsify the water content. Such water-containing oils cannot be treated by using the method described in (b). The other two methods will work.

Surface Conditions of Quenched Work. In order to attain clean quench hardening, the quenching oil must retain its nonstaining features. This is best accomplished by maintaining low quenching oil temperatures to minimize oxidation or oil degradation. On the other hand, higher temperatures usually will minimize distortion.

Carbonaceous deposits on quenched parts (sludging) are symptomatic of oil breakdown. Normally, an oil does not reach this stage of degradation without

Fig. 32 Pitot tube calibration chart

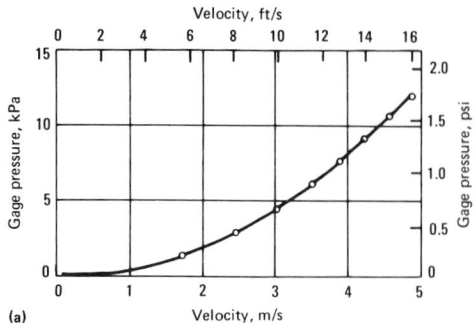

(a)

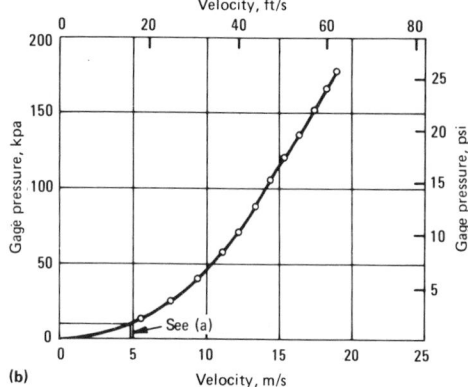

(b)

Chart for calibration of Pitot tube, assuming a Pitot tube factor, K, of 1.0. Velocity equals $K\sqrt{64.4\ h}$, where h is the pressure head, in feet of water. (a) Low-velocity and (b) full range.

Fig. 33 Center cooling curves showing the effect of scale on cooling curves, for 1095 carbon steel and 18-8 stainless steel quenched without agitation in fast oil

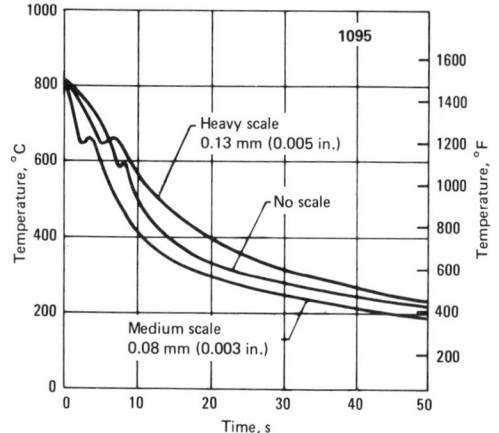

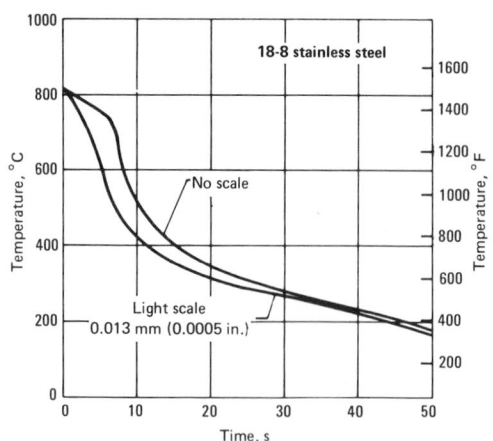

Oil was at 50 °C (125 °F) for the 1095 steel and 25 °C (75 °F) for the 18-8. Specimens were 13 mm (½ in.) diam by 64 mm (2½ in.) long.

Fig. 34 Effect of mass and section size on cooling curves in water quenching

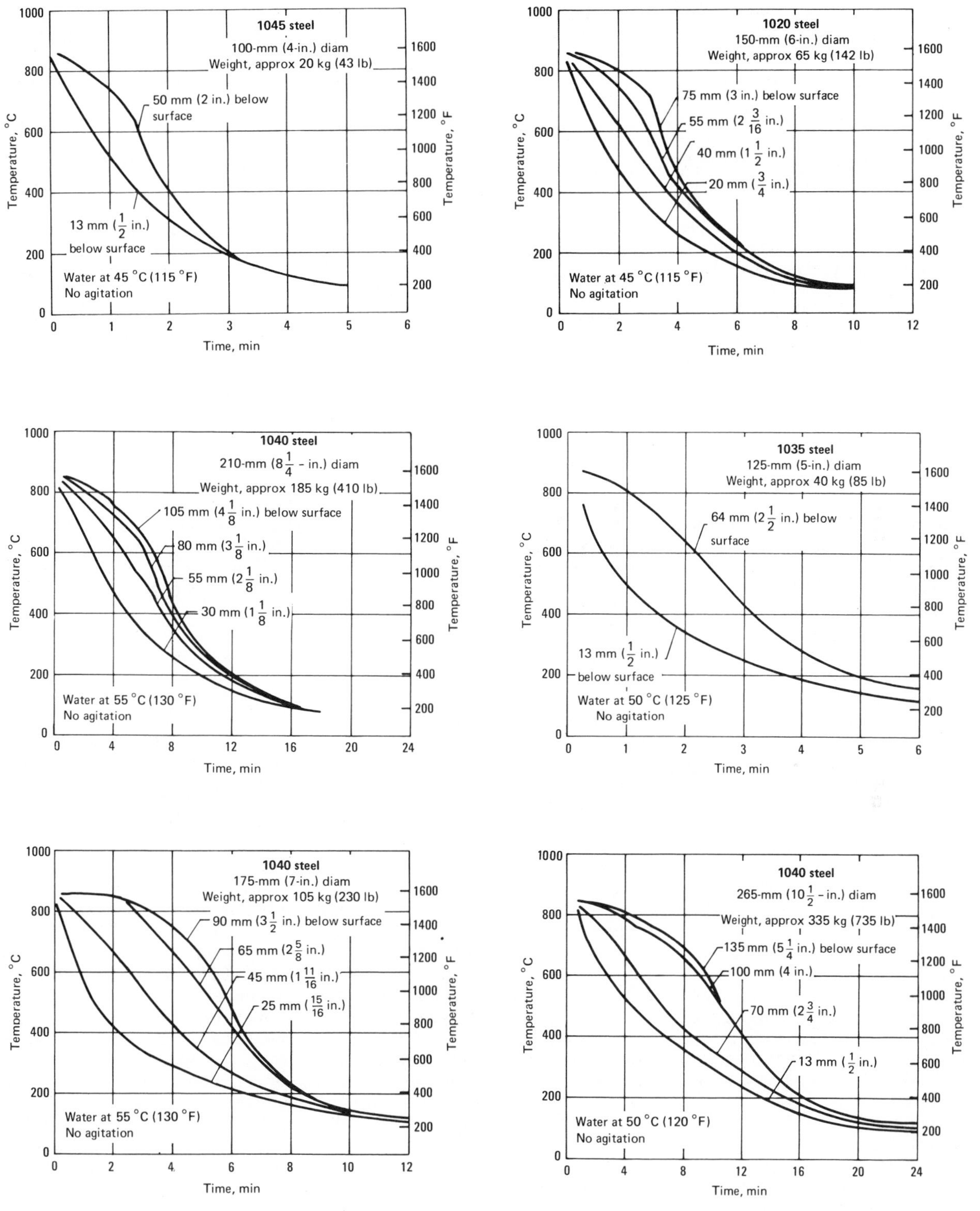
1045 steel
100-mm (4-in.) diam
Weight, approx 20 kg (43 lb)
50 mm (2 in.) below surface
13 mm (1/2 in.) below surface
Water at 45 °C (115 °F)
No agitation
1020 steel
150-mm (6-in.) diam
Weight, approx 65 kg (142 lb)
75 mm (3 in.) below surface
55 mm (2 3/16 in.)
40 mm (1 1/2 in.)
20 mm (3/4 in.)
Water at 45 °C (115 °F)
No agitation
1040 steel
210-mm (8 1/4 - in.) diam
Weight, approx 185 kg (410 lb)
105 mm (4 1/8 in.) below surface
80 mm (3 1/8 in.)
55 mm (2 1/8 in.)
30 mm (1 1/8 in.)
Water at 55 °C (130 °F)
No agitation
1035 steel
125-mm (5-in.) diam
Weight, approx 40 kg (85 lb)
64 mm (2 1/2 in.) below surface
13 mm (1/2 in.) below surface
Water at 50 °C (125 °F)
No agitation
1040 steel
175-mm (7-in.) diam
Weight, approx 105 kg (230 lb)
90 mm (3 1/2 in.) below surface
65 mm (2 5/8 in.)
45 mm (1 11/16 in.)
25 mm (15/16 in.)
Water at 55 °C (130 °F)
No agitation
1040 steel
265-mm (10 1/2 - in.) diam
Weight, approx 335 kg (735 lb)
135 mm (5 1/4 in.) below surface
100 mm (4 in.)
70 mm (2 3/4 in.)
13 mm (1/2 in.)
Water at 50 °C (120 °F)
No agitation
Temperature, °C
Temperature, °F
Time, min

detection, except in unprotected systems. Carbonaceous deposits are objectionable because they are difficult and costly to remove. However, they do not adversely affect hardening of the steel because they deposit after parts have reached a temperature below the range where rapid cooling is required.

When sludging occurs in installations that employ carbonaceous-protective atmospheres, the problem is complicated because carbon fallout may be the initial source of carbonaceous deposition. When quenching oil is contaminated by carbon fallout, it is sometimes difficult to determine other possible causes of carbon deposition.

Control of Quenching Oils. Routine analyses to determine the state of a used quenching oil should be performed about every three months. Used quenching oils are usually checked as follows:

- Crackle test (water, qualitative)
- Water content (quantitative, only if above test is positive) by ASTM test method D95
- Viscosity at 40 °C (100 °F) by ASTM test methods D445, D2161
- Sludge test by ASTM test method D2273
- Magnetic quenchometer speed, using plain nickel balls, at 25 °C (80 °F) for conventional and fast quenching oils; at 120 °C (250 °F) for martempering oils

Selection of Quenching Oil. The quenching of steel involves unsteady heat flow and kinetics of solid-state transformation; both of these are difficult to handle quantitatively. Therefore, the selection of oils for specific applications must be based largely on trial and error.

In many plants, selection of quenching oil is based on experience with similar parts and steels. When such experience is not available, it will be necessary to utilize one or more of the test methods previously described.

The most important criterion in selecting a quenching oil is that it provide a cooling rate that will harden the part without cracking it. Other considerations are degree of distortion, cost for the finished part and microstructure produced.

If a steel has sufficient hardenability to be fully hardened by being quenched in a conventional oil, there will probably be no advantage in using the more costly fast oils—as indicated below:

Example 1. In one plant, 4140 and 4135 steels were heat treated using both conventional and fast oils. Specimens 19 mm (³/₄ in.) in diam of 4140 were quenched in both types of oil from 855 °C (1575 °F), after which they were tempered at 595 °C (1100 °F). Average mechanical properties for several tests were:

Property	Conventional oil
Tensile strength	1035 MPa
	150 ksi
Yield strength	940 MPa
	136 ksi
Elongation in 50 mm or 2 in.	19.75%
Reduction in area	59.1%
Charpy impact	43 J
	32 ft·lb
Hardness	32 HRC

Property	Fast oil
Tensile strength	1035 MPa
	150 ksi
Yield strength	940 MPa
	136 ksi
Elongation in 50 mm or 2 in.	19.25%
Reduction in area	59.4%
Charpy impact	46 J
	34 ft·lb
Hardness	31.5 HRC

Tubular sections of 4135 steel, 120 mm (4.75 in.) OD, 70 mm (2.75 in.) ID, were quenched in both conventional and fast oils. A comparison of hardness traverses through the 25 mm (1 in.) wall sections showed uniform values of 46 to 47 HRC.

In other applications, the steel may be so low in hardenability that full hardening is difficult to obtain even in a water quench. Under these conditions, it is unlikely that a fast oil will offer sufficient advantage to solve the problem. However, between these extremes there are many applications in which it is advantageous to make discriminating selection, and fast oils often will prove advantageous.

In many instances, oil selection is based entirely on practical results and plant experience with various quenching oils. Specific instances of this nature are detailed in Examples 2 to 4.

Example 2. Transmission shifter shafts, 14.3 mm (0.562 in.) diam by 91 mm (3.6 in.) long, made of 1050 steel were carbonitrided at 815 °C (1500 °F) and quenched in conventional oil (86 to 110 SUS) at 40 °C (100 °F), after which they were tempered at 250 °C (480 °F). They were loaded vertically in baskets (1800 pieces per load), using a stabilizing screen. A minimum hardness of 50 HRC was required. Nonuniform hardness (within individual parts and from part to part) was encountered and caused excessive rejection.

Replacing the conventional oil with a fast oil solved the problem. Although the initial cost for the fast oil was 43% more than for conventional paraffin-base oils, cost of the heat treated parts was decreased because rejects were eliminated.

Example 3. Nonuniform hardness (50 to 60 HRC) was experienced in heat treating spring clips and fasteners made of 1064 steel and having section thicknesses from 0.25 to 1.55 mm (0.010 to 0.062 in.). The parts were heated in a shaker-hearth furnace at 870 °C (1600 °F) and continuously dropped at a rate of 90 kg/h (200 lb/h) into a 1890-l (500-gal) quenching tank. Temperature of the conventional quenching oil was maintained at 49 to 52 °C (120 to 125 °F) and circulation was 570 l/min (150 gal/min). Changing from the conventional oil to a fast oil resulted in consistent hardness values of 60 to 64 HRC.

Example 4. A minimum hardness of 62 HRC was needed on steel-mill rolls, 100 mm (4 in.) in diam by 815 mm (32 in.) long, made of 52100 steel. Water quenching developed sufficient hardness, but it also resulted in cracking of the 50-mm (2-in.) -diam, 100-mm (4-in.) -long bearing areas at the ends of the rolls. Quenching in a conventional oil resulted in hardness values of only 48 to 52 HRC. By using a fast oil, it was possible to develop hardness values of 64 to 65 HRC without cracking.

Greater distortion is sometimes associated with the use of fast oils, as indicated by these contrasting examples.

Example 5. One laboratory made use of modified Navy "C" test specimens of 1085 steel (Fig. 17) for a comparison of the distortion resulting in that material from quenching it from the austenitizing temperature in water, conventional oil and fast oil. As the average results for distortion and hardness (tabulated in Fig. 17) indicate, the fast oil produced slightly less distortion than the conventional oil while providing hardness values comparable to those obtained in water quenching.

Example 6. Figure 18 shows that significantly greater distortion oc-

Fig. 35 Effect of mass and section size on cooling curves in oil quenching

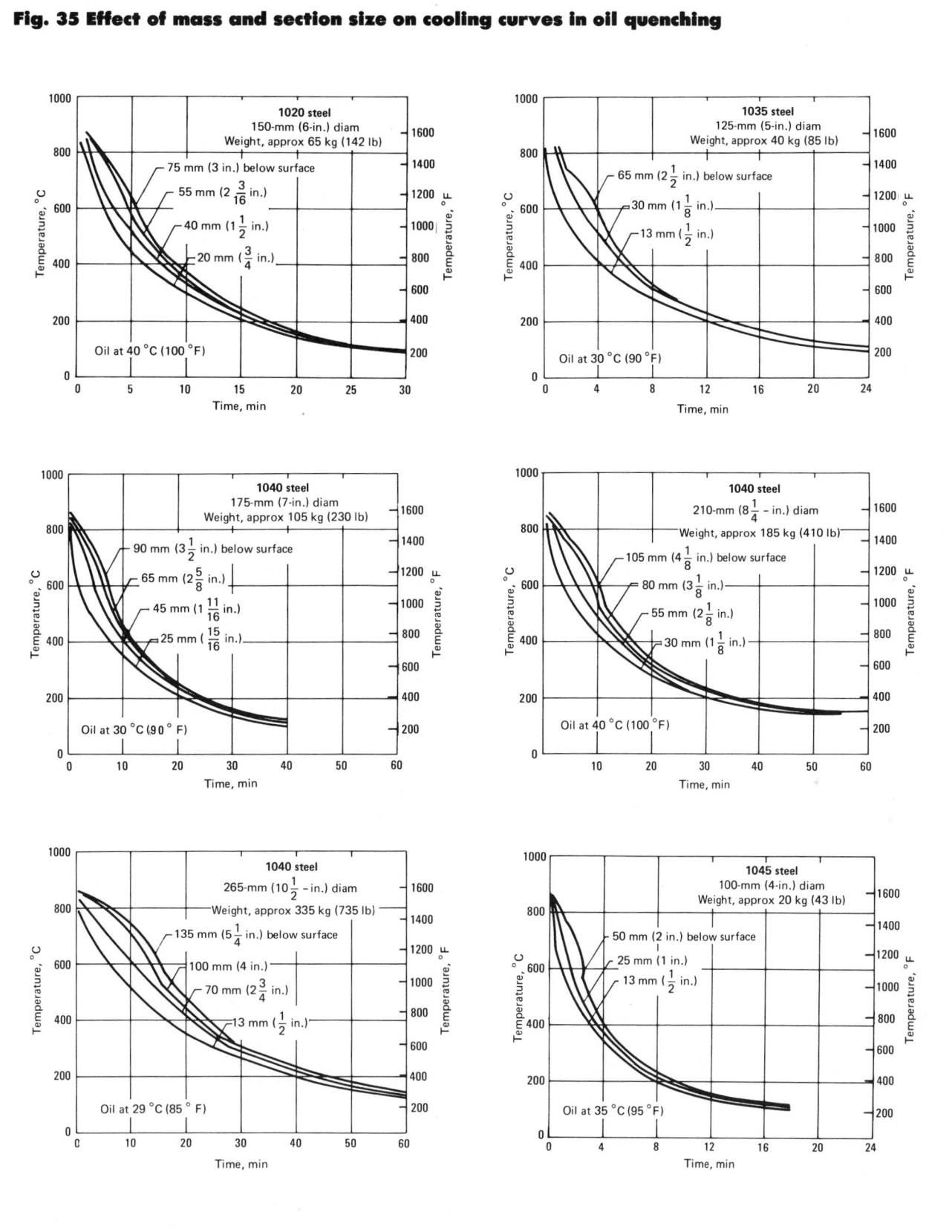

curred in carburized 8620 steel transmission gears when quenched in a fast oil than when quenched in a straight mineral oil; temperature of both oils was 65 °C (150 °F). Results plotted in Fig. 18 represent 60 tests for each type of oil.

In selecting a quenching oil, its contribution to the cost of the finished part must be considered. Aside from their initial cost (and the cost of replacing dragout), quenching oils can influence over-all cost by: (a) limiting the type of steel of which parts can be made; (b) determining the percentage of rejected parts that will require reworking; and (c) determining the extent of cleaning after quenching.

Although it is more common to fit the oil to the steel being treated than to fit the steel to the oil, there have been instances in which fast quenching oils permitted changing to steels of lower hardenability, with accompanying reductions in cost. One such situation is described in the following example.

Example 7. In one plant, truck ring gears were being made from 3310 steel and pinion gears from 4815 steel. It was found that, by using a fast quenching oil in place of conventional oil, 4718 steel could be used for both parts without sacrifice of properties. This resulted in a saving of about 15%, on a per-kilogram (per-pound) basis, in material costs.

In the following example, a fast oil proved to be cheaper, even though its initial cost was higher, than a conventional oil.

Example 8. In a plant employing a 3000-l (800-gal) quenching tank, 1500 l (400 gal) of conventional oil was consumed in quenching 29 500 kg (65 100 lb) of steel. This amounted to 19 kg of steel per liter (161 lb of steel per gallon) of oil. For the same system and type of work, only 1135 l (300 gal) of fast oil was consumed in treating 45 600 kg (100 500 lb) of steel—40 kg/l (335 lb/gal) of oil. Thus, despite the higher initial cost of the fast oil, the oil cost per kilogram or pound of steel treated was reduced by about 5.3%.

The lowered oil consumption was mainly the result of decreased dragout. In addition, the conventional oil formed more sludge, which resulted in higher maintenance cost and increased the cost of cleaning the heat treated parts.

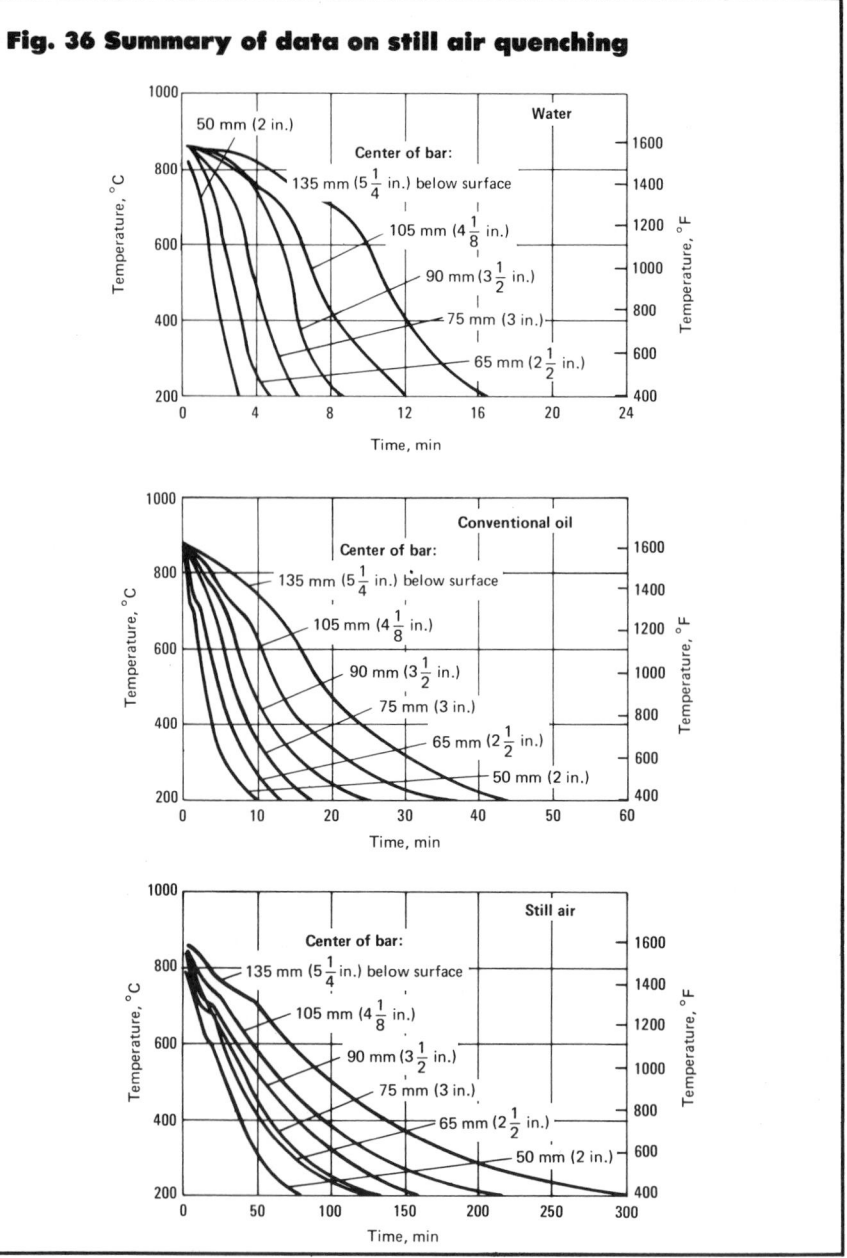

Fig. 36 Summary of data on still air quenching

Polymer Solutions

The technology of heat treating has seen dramatic growth during the current century. In the past, the principal quenchants were water, perhaps modified by some special additive such as inorganic salts, and naturally occurring oils. With the discovery of petroleum oil, hydrocarbon products became a major medium for quenching. It has been established that certain water-soluble organic polymers are useful in modifying the cooling characteristics of water. Whereas the patent literature has described a larger number of potential polymer candidates, in reality only four classes of product have achieved commercial prominence in the U.S. as polymer quenchants. In general, it has been an objective of polymer quenchants to provide cooling characteristics intermediate between water alone and petroleum oils.

Polyvinyl Alcohol (PVA). The use of aqueous solutions of polyvinyl alcohol (PVA) as quenching media was first described in U.S. Patent 2 600 290, which was issued in 1952. Polyvinyl alcohol was discovered in Germany and was introduced commercially into the U.S. in 1939. The chemical formula for PVA is

$$CH_3-CH-(CH_2-\overset{\overset{\displaystyle OH}{|}}{CH})_n$$

with OH above the first CH as well:

$$CH_3-\overset{\overset{\displaystyle OH}{|}}{CH}-(CH_2-\overset{\overset{\displaystyle OH}{|}}{CH})_n$$

Although PVA can be regarded as a polymer of vinyl alcohol, in reality all PVA resins are made by the hydrolysis of polyvinyl acetate. The extent of hydrolysis, which can govern commercial applications, may vary from partial (87 to 89%) through fully hydrolyzed (95%) to "super" hydrolyzed (99.7%). Furthermore, the water solubility and quenching characteristics of PVA resins (all of which are solids) will vary with the molecular weight of the polymer.

Cooling Characteristics. PVA was introduced in the mid-1950's as an additive to water to modify its cooling rate. As the curves in Fig. 19 (a) indicate, only slight variations in solution concentration are needed to produce changes in cooling characteristics of PVA solutions. At concentrations of less than 0.01%, the cooling characteristics at room temperature are only modestly different from those of water alone. With such small concentration variations, close control of PVA solutions is necessary. Control is complicated by the fact that quenched parts can become coated with an insoluble layer of resin, thus reducing the bath concentration. Maintaining an "effective" concentration requires specific control measures.

Polyalkylene glycols (PAG), or polyalkylene glycol ethers, were first introduced as a family of commercial products in the early 1940's. As shown in the diagram below, these materials are formulated by the random polymerization of ethylene and propylene oxides (although higher alkylene oxides and/or aryl oxides might be used also). Although block polymerizations of these same oxides are possible, these derivatives are less attractive as quenchants.

By varying the molecular weights and the ratio of oxides, polymers having broad applicability may be produced. Certain of the higher molecular-weight products were shown to have

utility as metal quenchants when used in aqueous solution (U.S. Patent 3 230 893). Proper selection of the polymer composition, and its molecular weight, provides a PAG product that is completely soluble in water at room temperature. However, the selected PAG molecules exhibit the unique behavior of inverse solubility in water—that is, water insolubility at elevated temperatures. This phenomenon provides the unique mechanism for cooling hot metal by surrounding the metal piece with a polymer-rich coating that serves to govern the rate of heat extraction into the surrounding aqueous solution. As the metal part approaches in temperature the temperature of the quenchant itself ("C" stage, Fig. 2), the PAG polymer coating dissolves to again provide a uniform concentration in the quenchant bath.

In the section describing the cooling characteristics of water, one of the disadvantages cited for plain water is that the vapor blanket stage (stage A in Fig. 2) may be prolonged. This prolongation encourages vapor entrapment that may result in uneven hardness and unfavorable distribution of stress, which in turn may cause cracking and/or distortion. By using polyglycol quenchants, uniform wetting of the metal surface results, thereby avoiding the unevenness and accompanying soft spotting. In fact, selection of the proper PAG quenchant can provide accelerated wetting so that the cooling rates achieved are faster than water and approach those achieved by brines. Thus, "brine quenching" is possible without the hazards and corrosiveness attendant with the use of salts or caustic solutions.

U.S. Patent 3 475 232 teaches that the addition of water-soluble alcohols, glycols or glycol ethers with 2 to 7 carbon atoms also can improve the wetting characteristics of PAG quenchants. Control of a multicomponent system then becomes more complex.

Whereas rusting can be a drawback when quenching with water alone, particularly where recirculation of treated water is not employed, solutions of polyglycol quenchants may be inhibited to provide corrosion protection of the quench-system components. Corrosion inhibition of quenched parts will be of short duration, so that specific protec-

tion should be provided following the tempering operation.

Cooling Characteristics. In the application of PAG quenchants for heat treating, three principal parameters are recognized to control the rate of cooling:

- Quenchant concentration
- Quenchant temperature
- Quenchant agitation

The influence of polymer concentration on cooling rates is illustrated by the cooling curves shown in Fig. 20. No specific concentrations are listed because the shape of the cooling curve, as well as the change in rate with concentration, will vary with the selection of the PAG quenchant. The slower rates of cooling achieved at the higher concentrations reflect the thickness of the polymer layer that surrounds the heated part during quenching. The polyglycol quenchants also are less sensitive to minor changes in polymer concentration, which is a recognized deficiency of polyvinyl alcohol and the other "film-forming" polymer quenchants.

Just as water exhibits a marked decrease in cooling capability as its temperature is elevated (Fig. 6), this same loss is translated to the aqueous solutions of PAG quenchants. The curves shown in Fig. 21 are illustrative of the general trends that would occur with changes in bath temperature; more detailed data would require specific identification of the particular PAG quenchant employed.

The use of no agitation with a polyglycol quenchant would be unusual. In general, low to moderate agitation is essential (a) to ensure that adequate replenishment of polymer occurs at the hot metal surface; and (b) to provide uniform heat transfer from the hot part to the surrounding reservoir of cooler quenchant. Vigorous agitation may be essential for achieving a rapid rate of cooling (for example, with a low hardenability steel) to avoid undesired transformation. Figure 22 clearly illustrates that, as agitation is increased, the cooling curves shift to more rapid rates.

Control Measures. The refractive index of oxyalkylene glycol polymer solutions (in the range employed for quenching) is essentially linear with concentration. Thus, the refractive index of a PAG quenchant solution serves as a measure of product concentration. Industrial model optical refractometers that employ an arbitrary scale may be

$$HO-(CH_2-CH_2-O)_n-(CH_2-\overset{\overset{\displaystyle CH_3}{|}}{CH}-O)_m-H$$

Fig. 37 Cooling curves for 64 mm (2½ in.) long austenitic stainless steel specimens of various diameters

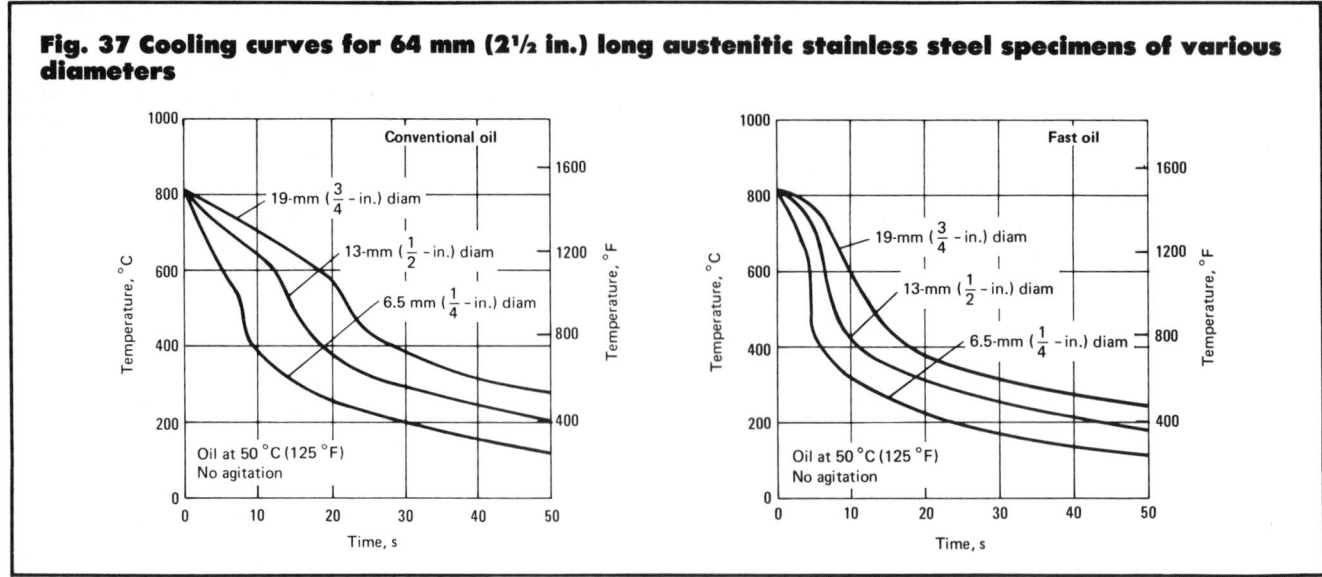

Fig. 38 Cooling curves for 100-mm (4-in.) long drill rod cylinders of various diameters

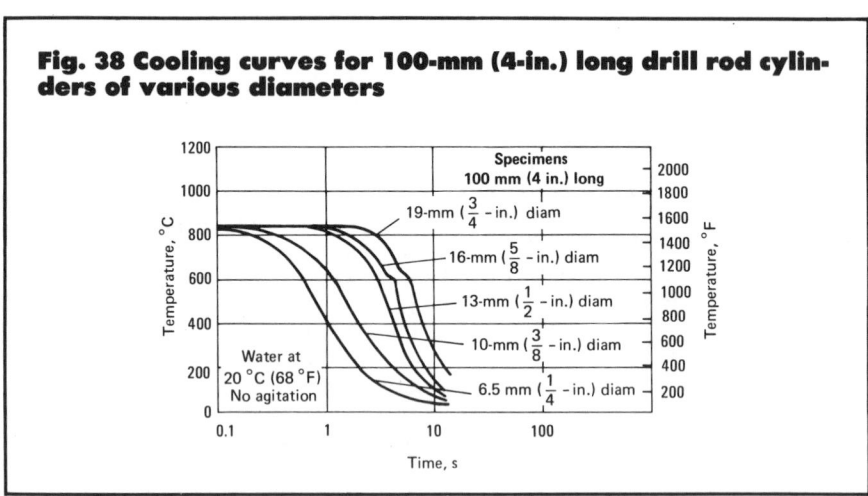

calibrated. Whereas such instruments prove invaluable for day-to-day monitoring of the quenchant concentration, the refractometer also will register other water-soluble components that are introduced to the used quenchant. When the indicated refractometer reading begins to provide erroneous numbers, some other analytical test is required to define the "effective" quenchant concentration. With polyglycol quenchants, kinematic viscosity measurements (which are correlated with concentration) have proven to be most useful.

As required, additional analytical tests for pH, inhibitor level and conductance may be useful adjuncts to a successful monitoring program. If the level of contaminants in the polyglycol quenchant becomes excessive—where these contaminants may be, in part, the same undesirable constituents that are detrimental to water alone, or oil—quenchant recovery can be effected thermally. By heating the quenchant solution (in whole or in part) above the separation temperature, a more-dense polymer-rich layer is obtained. Much of the water-soluble contamination can be withdrawn with the supernatant water layer. Solid contaminants such as scale or carbon would require settling, filtration and/or centrifugation.

Because PAG quenchants are highly bioresistant, the addition of a bactericide to the as-supplied quenchant is not required. Further, biochemical activity

in use is traceable not to the polyglycol polymer itself but to the introduction of nutrient contaminants. Microbiological treatment such as is employed with other aqueous metal working fluids generally will keep under control this foreign biological activity.

Polyvinylpyrrolidone (PVP) is derived from the polymerization of N-vinyl-2-pyrrolidone. PVP is a water-soluble polymer characterized by its unusual complexing and colloidal properties and by its physiological inertness (Ref 1 and 2). PVP has been available in the U.S. as a white, free-flowing powder, manufactured in four molecular-weight grades. Its structure is shown in the following diagram:

$$\begin{bmatrix} CH_2-CH_2 \\ CH_2 \quad C{=}O \\ N \\ -CH-CH_2- \end{bmatrix}_n$$

Solutions of PVP in water were first introduced as quenchants in 1975, coinciding with the issuance of U.S. Patent 3 902 929. The patent defines the molecular weight range for the pyrrolidone polymer, the quantity of polymer recommended for a solution concentrate (generally about 10% polymer solids) and the preferred use of a rust inhibitor and a bactericidal preservative.

Quenchant Variables. As with other polymer-type quenchants, concentration, bath temperature and agitation all play a role in establishing the cooling characteristics. By comparison, the quenching rates tend to be faster during the stable film and nucleate and boiling stages, but are slower during the convection stage. Because PVP does not have inverse solubility in water, only very small amounts of polymer film are retained on quenched parts at quenching temperatures from 30 °C (85 °F) to near boiling. Thus, a broader working range of temperatures for quenching can be employed.

Optical refractometer readings will provide initial control of concentration, but backup with viscosity measurements is strongly recommended. Means for removal of impurities by ultrafiltration was recently patented (U.S. Patent 4 251 292). This can be done without interrupting the quenching process.

Polyacrylates. The most recent addition to commercially available polymer quenchants in the U.S. is a product comprising an aqueous solution of sodium polyacrylate. Introduction was made at an ASM Heat Treating Conference/Workshop in May, 1977. The following reaction illustrates that the polymer may be achieved from the direct polymerization of sodium acrylate, or the alkaline hydrolysis of some polyacrylate ester.

$$\left[\begin{array}{c} -CH_2-CH- \\ | \\ C=O \\ | \\ O\ Na \end{array} \right]_n$$

By utilizing the salt of an alkali metal, in this instance sodium, the polymer is provided solubility in water.

Polyacrylate quenchants represent a class of quenchants completely different from the PVA, PAG or PVP types. The latter polymers fall into the type characterized as nonionic, that is, not ionizable or neutral. Polyacrylate quenchants are considered anionic, which is negatively charged. The charged character of the polymer imparts another dimension to the quenchant, strong polarity. The strong polarity provides water solubility but is also suspected of causing the polymer to operate by a different mechanism of heat extraction.

Unlike other polymer quenchants, polyacrylate solutions do not split on heating and do not form plastic films on the surface of the hot work. Their slower rate of cooling is based on the molecular weight of the polyacrylate and the subsequent viscosity of its solutions. By varying the molecular weights of the polymer, a whole family of quenchants can be designed covering a full range of applications from the fast quenching of water to the slow cooling of oils.

The quenching effect of the polyacrylate quenchants is a function of the three basic parameters: (a) polymer concentration, (b) bath temperature, and (c) bath agitation. The effect of polymer concentration and temperature for one of the commercially available polyacrylate quenchants is shown in Fig. 23.

The cooling curves of the polyacrylate solutions can be almost straight, which is the result of the extended vapor phase and reduced heat extraction during the boiling phase. This unique property of the polyacrylate quenchants allows their applications for hardening of crack-prone parts made of high hardenability steels. Applications of this kind usually are unobtainable with any other polymer quenchants or require much higher

concentrations of the polymer. Comparison of cooling curves of a polyacrylate quenchant with those of water, conventional quenching oil and a few typical polymer quenchants is shown in Fig. 24.

As can be seen in Fig. 24, the cooling rate of the polyacrylate solutions are distinctly slower than those of any other polymer quenchants. This illustrates their advantages as quenchants in particular for applications requiring a slow quenching effect.

With increasing polymer concentration and bath temperature, the cooling rate of the solutions can be slowed to the extent that many ferrous metals do not transform to martensite at all, but form bainite or pearlite. This "nonmartensitic" quenching can be utilized for new and unique heat treating procedures which are beneficial in many respects—for example, cost, energy, safety and environmental control. Some applications of the polyacrylate solutions for nonmartensitic quenching are as follows:

- Deep carburized parts, such as bearing races, balls and rollers, are usually double hardened in oil to obtain

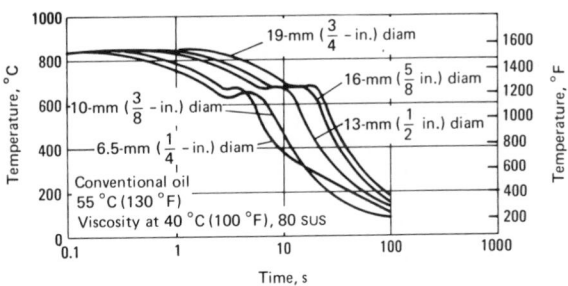

Fig. 39 Effect of mass and section size on center cooling curves in still quenching 100-mm (4-in.) long commercial drill rod cylinders of various diameters in water

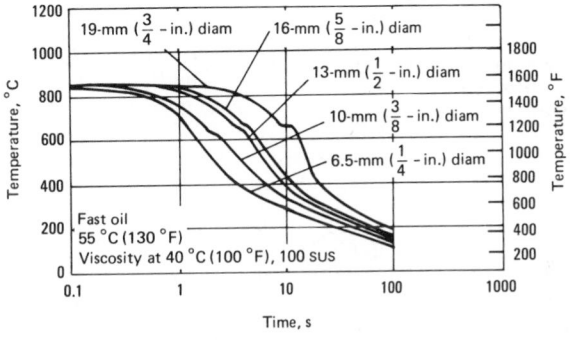

Fig. 40 Tank for continuous quenching

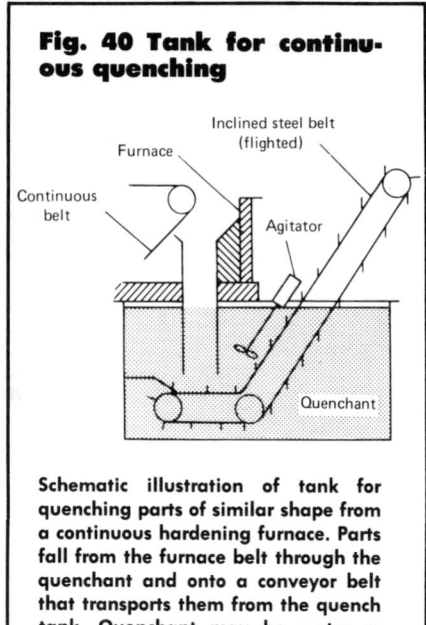

Schematic illustration of tank for quenching parts of similar shape from a continuous hardening furnace. Parts fall from the furnace belt through the quenchant and onto a conveyor belt that transports them from the quench tank. Quenchant may be water or oil.

Fig. 41 Temperature-controlled overflow tank for water quenching

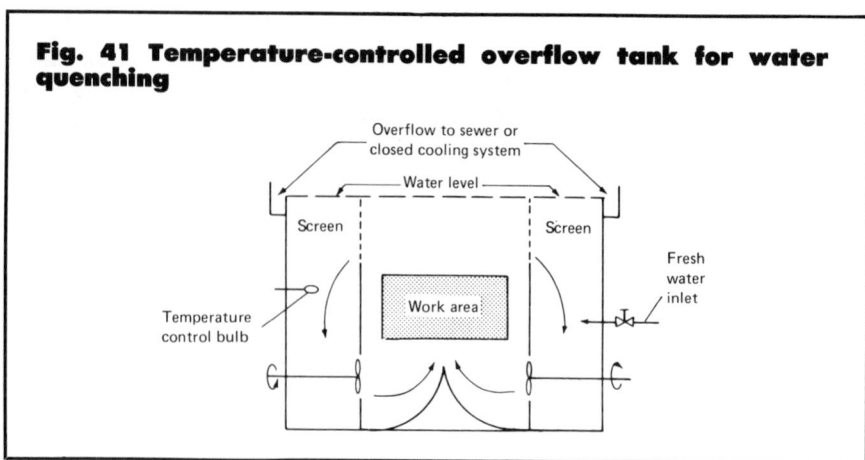

grain refinement. The first quenching can be done in a polyacrylate solution.

- Direct quenching of plain carbon steel with a high carbon content in polyacrylate solutions to obtain similar mechanical properties as obtained by quench and temper or austempering, such as automotive sway bars or railroad rails made of SAE 1070 to 1090
- Rod or wire patenting in polyacrylate solutions, instead of the commonly used lead or salt bath process at 510 to 565 °C (950 to 1050 °F)
- Direct quenching of hot formed parts in polyacrylate solutions to obtain good machinability without the usual necessary tempering process subsequent to hardening, in particular for steels of low hardenability, such as low-alloy carburizing steels
- Cooling of hot formed parts in polyacrylate solutions to prevent excessive scaling and decarburization during slow cooling in air, whereby practically the same microstructure is obtained as during air cooling
- Large concast steel slabs are water quenched to shorten cooling duration, thus permitting inspection shortly after casting. This is possible only for slabs of plain carbon steel with up to 0.2% carbon. Alloy steels and steels with higher carbon content have to be air cooled, which

requires several days. They would crack when water quenched. Now, the addition of polyacrylates to the slab quench water allows faster cooling of high carbon and alloy steel slabs.

The polyacrylate quenchants also can be used for quenching aluminum alloys after solution treatment to minimize distortion and warpage.

As any other polymer quenchants, the polyacrylate solutions require the use of agitation, the degree of which depends on the specifics of applications. As a general rule, a high degree of agitation is recommended for hardening operations, whereas minimal agitation is usually sufficient for nonmartensitic quenching.

The concentration control of the polyacrylate quenchants is based on their kinematic viscosities. To take into account the influence of contaminations, they also should be periodically checked with other laboratory methods including cooling curves analysis. This service usually is provided by a quenchant supplier.

Gas Quenching

Gas quenching is used to provide a cooling rate that is faster than that obtained in still air and slower than that obtained in oil.

In gas quenching, the austenitized workpiece is placed directly into the gas quenching zone or chamber, and heat is rapidly extracted from the metal by a fast-moving stream of gas. The cooling rate of the metal is related to the surface area and mass of the part and to the type and velocity of the cooling gas. The cooling rate can be adjusted and controlled by altering the last two of these variables, thus providing a degree of flexibility that can be obtained with very few other quenching media.

Recirculation. During the quench period, large volumes of relatively cold gases are directed through nozzles or vanes to achieve the impingement of high-velocity gas on the surface of the work load. After absorbing heat from the material being processed, the gases are cooled by being passed over water-cooled or refrigerated coils. Recirculating fans return the chilled gases to the high-velocity nozzles, through which they are again directed at the work to absorb more heat.

The gas quenching unit may be designed for either batch or continuous processing, and it is gastight, to prevent infiltration of air or loss of gas pressure. Various gases, ranging from air to complex mixtures, may be used for cooling, depending on process requirements. Protective atmospheres are commonly used, not only to produce bright work but also to increase the heat transfer rate between the gas and the work, thus bringing about an increase in the rate of cooling.

Applications. In some applications, quenching in still air is too slow, and oil quenching is undesirable because of distortion, cost factors, handling problems or the final hardness obtained. In these applications, gas quenching constitutes a useful compromise between two extremes.

Example 9. Gas quenching is used to advantage in hardening aircraft tubing made of low-alloy steel. Tubing with up to 3-mm (1/8-in.) wall thickness is satis-

Fig. 42 Layout showing the elements of a quenching system used with a four-row continuous gas carburizing furnace

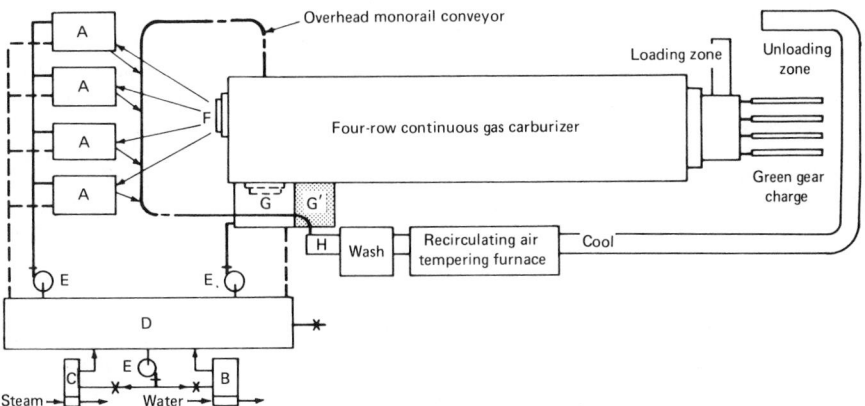

A, press quenches; B, cooler; C, heater; D, storage tank; E, pumps, strainers; F, slot door; G, atmosphere-hooded quench; G¹, open portion of G; and H, washer loading platform

factorily quenched in a continuous process. A comparison of the cooling curves obtained in quenching 4130 steel tubing 32-mm (1.25-in.) OD by 1.7-mm (0.065-in.) wall in gas, oil, and still air (normalizing) is presented in Fig. 25.

Example 10. When a steel that is not air hardenable because of its chemical composition or section size, or both, is quenched in oil, gas or still air, considerable differences in hardness are obtained. Brinell hardnesses, surface to center, obtained by quenching 100-mm (4-in.) -thick disks forged from 1095 steel in each of the three different media, are shown in Fig. 26.

Example 11. When cooled in still air, a 1095 steel collector ring with 100-mm (4-in.) maximum section thickness failed to develop satisfactory hardness. The necessary increase in hardness was obtained by forced-air cooling. The comparative hardness data obtained with these quenching techniques are as follows:

Distance below surface		Hardness, HB, when cooled in	
mm	in.	Still air	Forced air
3	1/8	247	272
6.5	1/2	235	248
22	7/8	229	248
32	1 1/4	229	248
41	1 5/8	229	248
50	2	217	248
60	2 3/8	217	255
70	2 3/4	229	255

Gas quenching is commonly used in the hardening of large forgings, to obtain more uniform cooling of heavy sections and to prevent cracking and thermal instability. Generally, as the size of the forging increases, it becomes more difficult to secure reasonable uniformity of properties throughout heavy sections. Assuming that the steel selected possesses adequate hardenability, it may still be necessary to gas quench to obtain desired properties. Gas quenching also will produce uniform mechanical properties in large parts of intricate shape and varying section thickness. For these parts, fixturing usually is required during the quenching cycle.

Example 12. Large turbine-rotor forgings are gas quenched in the "wind tunnel" apparatus shown schematically in Fig. 27. As each forging is rotated, six tangential high-pressure blowers create a convection current of air in the tunnel. A large blower, located at the base of the rotor and suitably deflected from the end of the shaft, forces a constant stream of air over the surfaces of the forging and out of the top of the tunnel. This method of quenching has been used to process hundreds of forgings, yielding acceptable mechanical properties from surface to center with virtually no distortion or cracking.

Hardening Tool Steel. Gas quenching has been used to obtain maximum hardness in various grades of air-hardening tool steel, two of which

are discussed in the following example.

Example 13. Tool steels of types A2 and T1, in the form of solid blocks 50 by 100 by 100 mm (2 by 4 by 4 in.), were successfully austenitized and gas quenched with cylinder nitrogen in a vacuum furnace. A sealed fan in the roof of the furnace chamber chilled the nitrogen by forcing it over the water-cooled walls of the chamber; the cooled gas was then admitted to the chamber at 69 kPa (10 psig) and directed at the steel. As indicated in Fig. 28, the A2 steel cooled from 1010 to 345 °C (1850 to 650 °F) in 8 min, and the T1 from 1290 to 345 °C (2350 to 650 °F) in 13 min. These rates were sufficient for maximum hardness.

Gas quenching also has been employed to develop the desired mechanical properties in other steel products, such as thin sheet and small-diameter wire.

Fog Quenching

In fog quenching, heat is rapidly extracted from the metal by a fast-moving stream of gas that contains water droplets. The cooling capacity of the "fog" is derived from both the absorption of heat by the gas and the heat of vaporization of the water. The addition of water droplets (or fog) to an air stream can increase its cooling capacity by a factor of 4 1/2. Fog quenching is most effective in the lower temperature ranges.

An outgrowth of gas quenching, fog quenching normally is substituted for liquid quenching in an effort to minimize distortion. Although less severe than a liquid quench, fog quenching is more severe than a plain gas quench of equivalent velocity. In general, the types of parts that are fog quenched are similar to those that are considered suitable for gas quenching.

Agitation

The most important influences on the cooling rate of steel in various quenching media are:

- Agitation (flow rate) of the quenching liquid
- Surface oxidation of the workpiece
- Mass and section size of the workpiece.

The various quenching media available do not automatically supply all the cooling rates desired, particularly in the range of quenching speeds between those obtainable in water and in still oil. The rate of cooling by oil quenching can be increased by altering the rate of oil flow past the work or by agitation of the work in the oil.

Oil Flow. Measurements of the rate of flow from pipes into tanks or from mechanical stirrers usually have shown too much variation to provide quantitative information on the effects of oil flow rate. However, flow rate data such as those plotted in Fig. 29, 30 and 31 obtained in specially designed equipment, are of more general applicability. The data in Fig. 29 were obtained on scale-free bars of 9445 alloy steel heated in an exothermic atmosphere and quenched in a mineral oil at 50 ° C (120 °F), viscosity at 40 °C (100 °F), 79 sus. Figure 30 shows the relationship between bar diameter and equivalent locations on the end-quench hardenability specimen, both for scale-free bars and for those scaled by heating in air. For the data in Fig. 29, 30 and 31, the following arbitrary criteria were used:

Position in quenched bar	Temperature range °C	°F
Surface	730 to 315	1350 to 600
3/4-radius	730 to 370	1350 to 700
1/2-radius	730 to 425	1350 to 800
Center	730 to 480	1350 to 900

Fig. 43 Schematic of a typical installation for high-volume batch quenching of carburized or hardened parts on trays

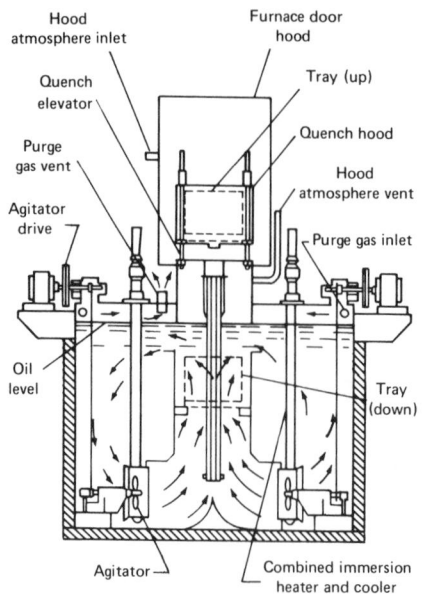

Directional vanes in the oil stream distribute the oil flow uniformly. Unit contains combined heating and cooling elements and provision for blanketing the surface of the oil with an inert gas atmosphere. Radiant tube is used for heating and cooling.

The smaller temperature ranges used for 3/4-radius, 1/2-radius and the center of the bar distort the relationships plotted in Fig. 29, 30 and 31; nevertheless, it is evident that increasing the rate of flow has the greatest influence on the cooling rate at the surface, and that the surface of a 75-mm (3-in.) bar can be cooled almost as fast as the surface of a 25-mm (1-in.) bar when the flow rate of the oil is above about 150 m/min (500 ft/min). It is evident also that the major advantage from agitation lies in the range between still oil and a flow rate of 60 m/min (200 ft/min), which is equivalent to an average hand agitation of the piece.

Molten Salt. Figure 31 shows the relation between molten salt flow and the cooling rates of various positions in scale-free bars in a marquenching salt at 205 °C (400 °F). Also shown in Fig. 31 are similar curves obtained from water and oil quenching, thus providing comparisons among three media giving a wide range of cooling rates. The data

are plotted to give correlation curves for identical cooling times in quenched round bars and standard end-quench hardenability specimens. The temperature ranges for the various positions are the same as those tabulated above. The bars were of medium-carbon 9400 steels, heated in an exothermic atmosphere.

Note that small-diameter bars (and the surface of all bar sizes shown) may be cooled faster in rapidly flowing salt at 205 °C (400 °F) than in still or mildly agitated oil at 50 °C (120 °F). However, as bar size increases, subsurface regions of salt quenched bars cool more slowly than those quenched in still oil. This limits the section size or steel composition that will provide a fully martensitic structure by hot salt quenching.

Water and Brine. When a 13-mm (1/2-in.)-diam cylinder, heated to 800 °C (1470 °F), is lowered into a flowing stream of water moving at 0.9 m/s (3 ft/s), the upstream surface of the cylinder cools to 205 °C (400 °F) in about 2 s, while the downstream surface cools to the same temperature in about 6 s. Cooling rates of the side surfaces will range between 2 and 6 s.

Quenching in a moving stream of brine or sodium hydroxide solution causes more nearly uniform, but faster, cooling of all surfaces.

Factors Controlling Agitation. In all agitated quenching baths, the degree and character of agitation vary from point to point within the bath; these variations are even more pronounced in spray quenching. Although accurate description or measurement of agitation is difficult, the principal factors that control agitation are well known. These include the general shape of the bath, location of the work, direction of flow currents, type of agitator, flow rate and power consumed. In spray or jet quenching, additional factors are encountered, such as the shape, arrangement and placement of the spray head in relation to the work, the pressure, velocity and size of the jets, and the total volume of quenching fluid used per unit of time.

Quenching velocities depend primarily on the mode of agitation. Very low velocities, not exceeding 0.9 m/s (3 ft/s), are encountered in immersion by gravity. Intermediate velocities, ranging from 1.1 to 1.8 m/s (3.5 to 6.0 ft/s), are achieved in hand quenching with an up-and-down, circular, or "figure 8" movement over 510 mm (20 in.) of trav-

Fig. 44 Equipment for uniform quenching of a thin-wall cylindrical part

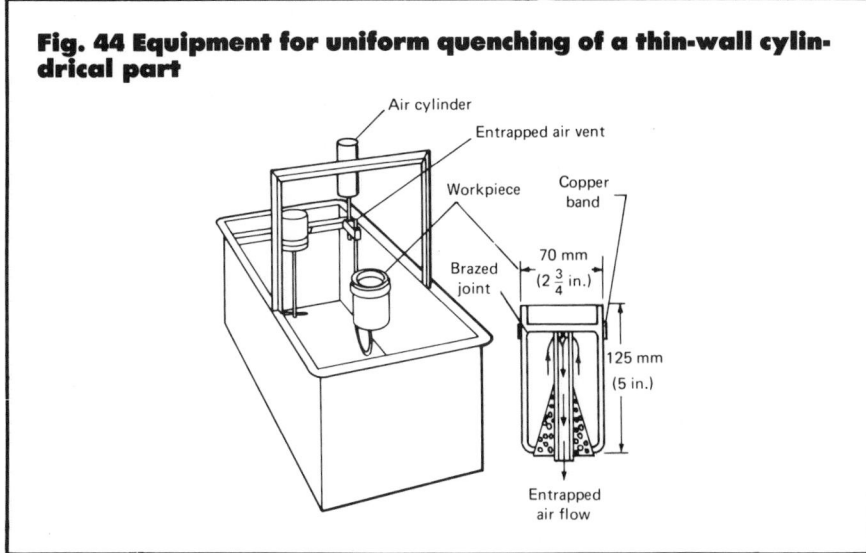

Fig. 45 Oil quenching tank with four top-entry propeller-type agitators used for quenching bar stock

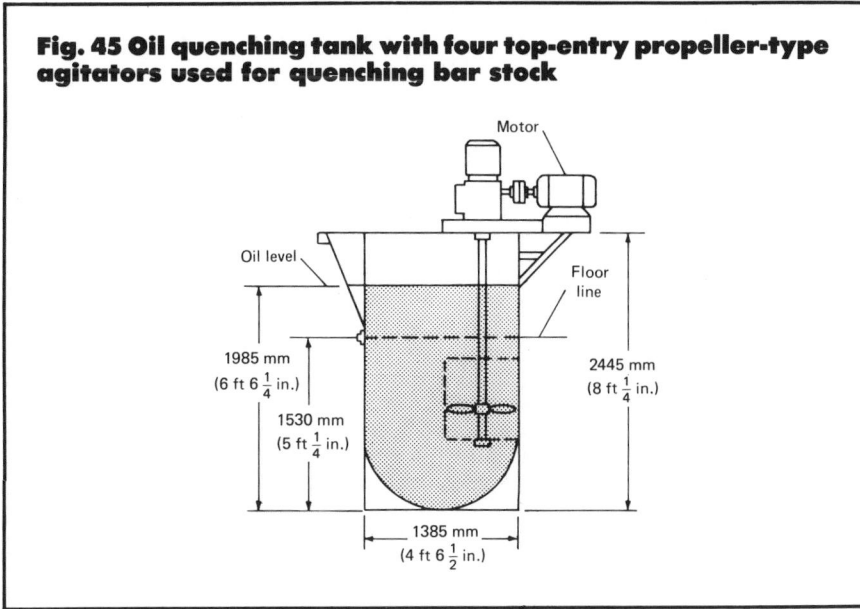

The velocity of the fluid as it strikes the tube is converted to a pressure head that is measured by the pressure gage. The pressure head (h), in feet of water, is converted into velocity (v), in feet per second, by the equation:

$$v = K\sqrt{2gh}$$

where K is the Pitot tube constant (most often 1.0, or close to unity) and g is the acceleration of gravity (980 cm/s^2 (32.2 ft/s^2). The relationships are shown in the Pitot tube calibration charts in Fig. 32.

Turbulent agitation in a quenching bath takes the form of a multitude of swirling eddy currents. This action is usually accompanied by a systematic gross movement within the quench tank caused by the position of the propeller or jet source of agitation and by the shape of the tank. Turbulent agitation is desirable for uniform cooling of those shapes that do not lend themselves to a complete washing of all surfaces with a lamellar, or streamline, flow. A multitude of turbulent eddy currents often will cause sufficiently uniform average movement over all surfaces of an irregular shape to produce adequate quenching.

Surface Oxidation

The effects of surface oxidation have been evaluated by magnetic testing, observation of high-speed motion pictures and cooling curve measurements.

Magnetic testing, which was described earlier in this chapter, has been used to determine the effect of surface oxidation on the rate of heat transfer from nickel specimens to the quenching medium. Specifically, the heat transfer rate of nickel balls heated to 885 °C (1625 °F) in an oxidizing atmosphere was compared to that of similar balls heated to the same temperature in a protective, reducing atmosphere. Rates of heat transfer from 915 to 355 °C (1675 to 670 °F) as shown in Tables 4 and 5 indicate that the balls heated in an oxidizing atmosphere cooled faster than the balls heated in a reducing atmosphere.

High-speed motion-picture techniques have been used to reveal the influence of oxide coatings on the quenching rate of steel in water containing polyvinyl alcohol, trisodium phosphate or carboxyl methylcellulose. When a steel specimen that had been heated in a protective atmosphere was

el. Spray quench rings are usually operated between 4.6 and 30 m/s (15 and 100 ft/s); special applications sometimes use velocities as high as 150 m/s (500 ft/s).

Measurement of Velocity. The Pitot tube is useful for measuring unidirectional velocity, such as that of a lamellar or jet stream, although it is not suitable for measuring turbulent (multidirectional) flow.

The tube can be constructed by drawing 6.5-mm (¼-in.) glass tubing to an inside diameter of approximately 0.4 mm (¹/₆₄ in.) and grinding the drawn end flat and square with the tube axis.

It may also be made of metal, provided the tube opening is knife-sharp and square with the axis of the tube. A suitable bourdon gage or a manometer is connected by rubber or transparent plastic hose to the tube. If precise measurements are required, a correction must be made for any quenchant column height in excess of the level of the upper opening of the Pitot tube.

The axis of the Pitot tube must be accurately aligned parallel to, and in the center of, the stream being measured. Alignment is obtained by exploring the stream with the tube, searching for the highest pressure reading.

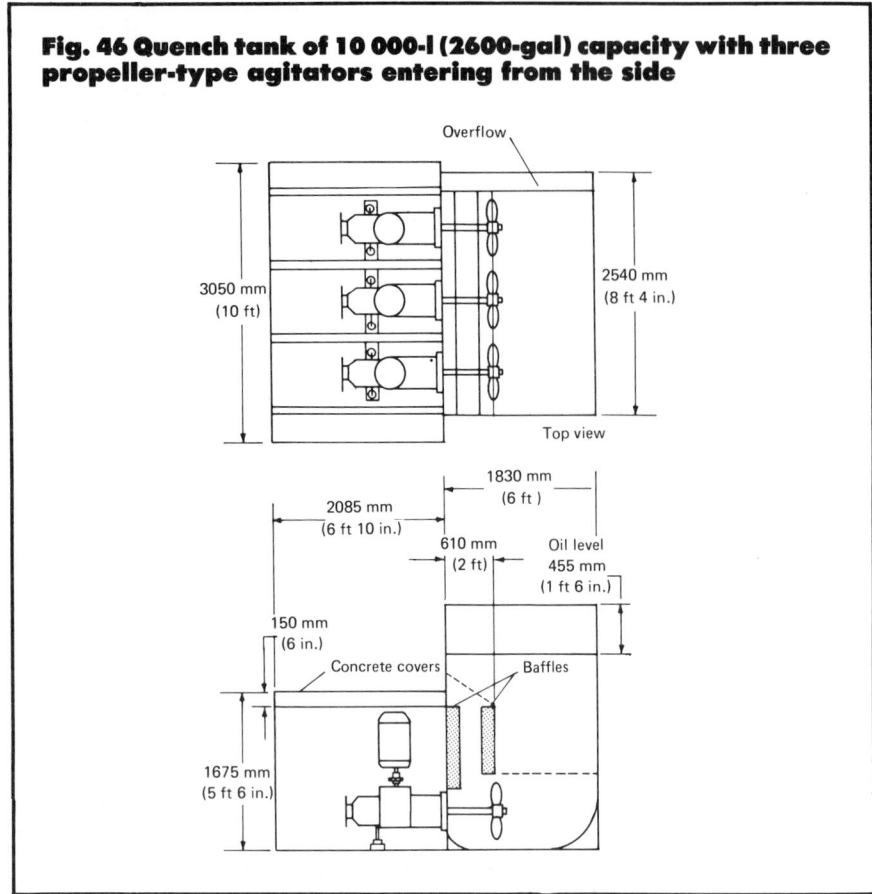

Fig. 46 Quench tank of 10 000-l (2600-gal) capacity with three propeller-type agitators entering from the side

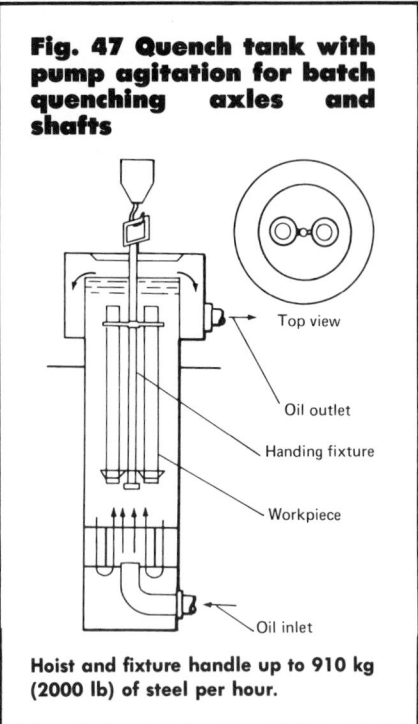

Fig. 47 Quench tank with pump agitation for batch quenching axles and shafts

Hoist and fixture handle up to 910 kg (2000 lb) of steel per hour.

submerged in the liquid, it was noted that a visible envelope, or cocoon, surrounded the specimen, preventing contact between the liquid and the steel surface. Finally, after a delay of many seconds, the envelope burst and quenching action commenced. The total quenching time in a 0.3% solution of polyvinyl alcohol in water was 37 s, a slower quenching rate than was obtained with conventional quenching oils. When an identical specimen that had been heated in an oxidizing atmosphere was quenched in the same polyvinyl alcohol solution, immediate contact was made between the solution and the specimen surface, and the total quenching time was only 2.3 s. This is comparable to the rate obtained in plain water.

Cooling curves, such as those in Fig. 33, also indicate the effect of an oxide scale on quenching characteristics. These curves were obtained by still quenching in fast oil. A scale not more than 0.08 mm (0.003 in.) deep increases the rate of cooling of 1095 steel as compared to the rate obtained on a speci-

men without scale. However, a heavy scale, 0.13 mm (0.005 in.) deep, retards the cooling rate. A very light scale, 0.013 mm (0.0005 in.) deep, also increased the cooling rate of the 18-8 stainless steel over that obtained on a specimen without scale.

Mass and Section Size

The cooling curves in Fig. 34 and 35 demonstrate the effect of mass and section size on the cooling curves of carbon steel in water and in oil. Figure 36 is a summary of Fig. 34 and 35, and includes data based on quenching in still air, for the center positions of the bars of various diameters. The combined effects of mass and quenching medium for cooling small sections are shown in Fig. 37, 38 and 39.

Quenching Systems

Equipment requirements for quenching may vary widely. A small plant making machine parts might each day require the hardening of only a few

simple carbon steel parts weighing about 1.4 kg (3 lb) apiece. For such an application, the quenching "system" would comprise a barrel of water, with piping to the water supply and a drain to the sewer. Handling equipment would consist simply of a pair of tongs. As quantity of work to be quenched and complexity of workpieces increase, however, various other items of equipment must, of course, be added to the quenching system.

For a complete quenching system, the following functional equipment usually is required and installed: (a) work tank or machines, (b) facilities for handling the parts quenched, (c) quenching medium, (d) equipment for agitation, (e) coolers, (f) heaters, (g) pumps and strainers or filters, (h) quenchant supply tank, (i) equipment for ventilation and for protection against hazards, and (j) equipment for automatic removal of scale from tanks.

Continuous Quenching. A tank such as that illustrated schematically in Fig. 40 is used commonly for continuous quenching of similar parts that can be dropped into the quenching liquid. In this installation, the parts fall from the furnace belt through a chute into the quenching medium and onto another belt that conveys them out of

Fig. 48 Typical oil-cooling system employing an underground storage tank

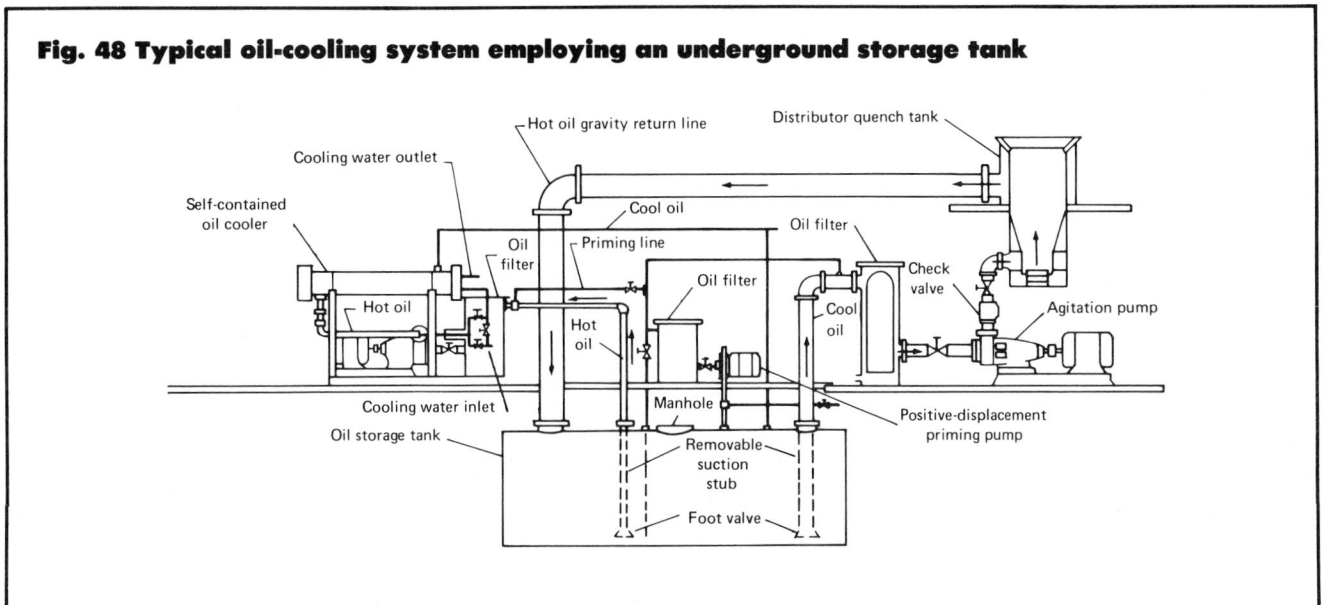

the quench. Agitation is provided by an impeller directed at the lower end of the quench chute. If the mass of work entering the tank is large, a means of controlling the temperature of the quenchant would be added. This design of quench system is usually limited to symmetric and simple shapes because free fall from the furnace belt can result in distortion and nonuniform quenching.

Water Quenching System. A simple arrangement for water quenching, incorporating a system for controlling water temperature, is shown in Fig. 41. In this system, a supply of fresh water is continuously introduced directly into the bath and is allowed to overflow to the sewer. Temperature control of the water is provided by a self-operated temperature regulator (which opens and closes at predetermined settings, these depending on the volume of parts that are being processed and on the water temperature desired).

The system is intended for use where water is plentiful and inexpensive. Where water is scarce, however, this system could be connected to a central or closed water-cooling system that would condition the water for reuse.

Oil Quenching Systems for Continuous Carburizers. Additional controls at the quenching operation are required for mass production of parts such as finished gears through a continuous carburizer. In a large automotive plant manufacturing rear and

front axle assemblies, the quenching of gears, stem pinions, and differential side gears and pinions requires special loading fixtures for batch-quenched parts and special quenching presses for ring gears. Because low cost per unit must be maintained throughout the entire heat treating process, quenching is integrated with heating, and handling facilities are arranged accordingly.

The clean, soft gears, fully machined except for finish grinding, are charged into a continuous pusher-type gas carburizer. Workpieces are loaded on alloy trays with fixtures that permit direct quenching.

Figure 42 shows the arrangement of the equipment that may comprise the quenching system for an application such as described above. The principal elements of this system are:

- Quenchant storage-supply tanks and pumps of sufficient capacity to maintain an efficient heat transfer from the gears
- Coolers and heaters to maintain the temperature of the quenchant for consistent quenching results
- Filters in the system to minimize free carbon and other foreign elements that may affect the rate of heat transfer or interfere with the efficient operation of quenching presses
- Agitation of the quenchant to obtain uniform quenching and minimize distortion

- Special alloy fixtures and handling facilities to transfer workpieces to the quench
- Quenching oil selected to provide the desired cooling rate
- Quenching presses for processing ring gears.

A typical batch-quenching tank that can be used for high-volume production of work on trays is illustrated in Fig. 43. Atmosphere protection, agitation, temperature control and other features are included in this arrangement. The pressure of the atmosphere gas must be greater than that of the outside air; otherwise, air will enter the chain-drive unit opposite the agitator motors, and the revolving chain will carry air into the oil and cause oxidation, especially when hot oil is being used.

Special techniques for quenching are often expensive, but they are sometimes required in order that specified results can be attained. The following example describes one such special technique.

Example 14. In Fig. 44, cylindrical parts, 70 mm (2¾ in.) in outside diameter by 125 mm (5 in.) long, were made from 1030 steel. Each part was a brazed assembly with one end open and the other end closed. The parts were preheated at 760 °C (1400 °F), austenitized in neutral salt at 870 °C (1600 °F), and quenched individually, open end down, in a 3 to 6% caustic solution at 20 to 25 °C (70 to 80 °F).

Nonuniform hardness and occasional excessive out-of-roundness were encountered. The nonuniform hardness was attributed to entrapped air in each part, which prevented uniform quenching. This was remedied by installing a vent pipe in the quenching tank. By quenching the parts individually over this pipe, the air was displaced and the quenchant reached all areas uniformly (Fig. 44), and uniform hardness was obtained.

Agitation Equipment

The agitation of a quenchant can be obtained in several ways. In conventional quench tanks, circulation of the quenching medium is usually provided by

- Pumps
- Passage of the workpiece through the quenching medium
- Manual or mechanical movement of the workpiece
- Mechanical propellers

The selection of any type of agitation method is dependent on the tank design, type of quenchant, volume of quenchant, the part design and the severity of quench required.

Pumps are commonly employed because they provide a controllable means for directing the quenchant. Also, the flow of the quenchant can be readily divided to provide circulation to more than one location in the tank. When the quenching medium is oil and a cooling system is employed, the pump used to recirculate oil to the cooling system also may be used to provide agitation. Recirculating pumps may be preferred for use in a "floating fountain" quench that is capable of removing heat quickly from an internal cavity of a workpiece.

Gravity fall of the hot workpiece through the quenchant usually is employed for quenching lightweight parts that have a large ratio of area to unit weight. If pumps or mechanical agitators were used, such workpieces probably would drift in the tank, and it would be difficult to handle the load with a conventional elevator-conveyor discharging quench tank.

Movement of the Workpiece. During the quenching of steel, it usually is desirable to obtain the most rapid cooling when the steel temperature is above about 540 °C (1000 °F). In this temperature range, the part is usually enveloped in the vapor blanket and the cooling rate is slowest. To accelerate cooling in this temperature range and to remove undesirable scale, rapid relative motion of the part in the quenchant is necessary. On small, low-production items, this can be accomplished by moving the part, or a small basket or tray of parts, through the quenchant by hand, in a "figure-8" motion. Workpieces may also be driven mechanically with respect to the quenching medium. For example, shafts are sometimes rotated in the quenching medium to produce the effect of agitation.

Where a variety of shapes and sizes is to be quenched, however, propeller agitation is most desirable, because it produces a turbulent motion.

Propellers. Aside from providing effective agitation, propellers, which are self-contained mechanical agitators, are compact, require no piping and can be easily removed for maintenance. Propellers must be properly located in the quench tank in order to function effectively.

Propellers are usually placed near the bottom of the quench tank to produce the most desirable agitation. A liquid accelerated by a propeller flows in a helical motion, in the same direction as the rotation of the propeller blades. This high-velocity stream moves across the bottom of the tank and spreads out as it moves away from the propeller. Upon striking the opposite wall, the stream is diverted upward and in the direction of rotation of the propeller. This produces a general rotation of the liquid, which is partially disrupted by the return cycle of the liquid to the propeller. The general motion of the quenchant in the tank is, therefore, a swirling motion and an up-and-down motion.

Propeller agitators may be either top-entry or side-entry units, as indicated in Fig. 45 and 46. Side-entry units are usually placed below the floor level, to conserve floor space. Top-entry units require more floor space but less excavation for installation.

After the size of the quench tank is determined, the horsepower required for agitation may be calculated from the data shown in Table 6. These power requirements are minimums to produce a velocity of about 15 m/min (50 ft/min). Higher horsepower is required for increased agitation.

Variables Affecting Agitation

Effect of Velocity. Tests have shown that quenching oil moving at a velocity of 23 m/min (75 ft/min) will increase the depth of hardening over that obtained at lesser velocities in sections up to 100 mm (4 in.) in diameter. In heavier sections, it is possible that no advantage can be gained by increasing the rate of agitation, because the rate of heat transfer through the workpiece becomes the limiting factor. However, the practical limit on agitation is that which produces foaming on the surface of the oil. Foaming is undesirable and indicates that air is being entrapped.

Number of Agitators. The number of units required depends on the total horsepower required and the horsepower of available units. Two or more smaller units provide more uniform agitation and more versatility than one large unit. The velocity of the quenchant can be varied by using two-speed motors or variable-pitch propellers, or a combination of both.

Relation of Tank Design to Agitation. A properly designed tank and two or more propellers usually will provide uniform agitation throughout the working area. Sometimes baffles are required to direct the quenchant flow uniformly throughout the tank. On top-entry units, guide vanes usually are installed in the draft tube to prevent the quenchant from vortexing.

Example 15. An oil quench tank used primarily for quenching bar stock is illustrated in Fig. 45. The four agitators develop a total of 100 hp. The horsepower per gallon for this particular tank is 0.04. The velocity is 60 m/min (200 ft/min), and the turnover is 113 500 l/min (30 000 gal/min). A similar tank holds 13 600 l (3600 gal) and has two 50-hp motors driving the propellers. The velocity is 110 m/min (360 ft/min), and the turnover of the quenchant is 196 000 l/min (51 800 gal/min); the horsepower per gallon is 0.028.

Example 16. A 10 000-l (2660-gal) oil quenching tank with three 25-hp side-entry agitators is shown in Fig. 46. This amounts to 0.0282 hp per gal and produces a velocity of 20 m/min (66 ft/min), with a turnover of 51 100 l/min (13 500 gal/min).

The tanks illustrated in Fig. 45 and 46 are designed with baffles and curved bottoms to obtain uniform and smooth flow of oil. Baffles are used in the return path to direct the flow of return oil to the input side of the propeller with a minimum of turbulence.

With these design features, a higher-velocity flow or more general circulation can be obtained with a given input horsepower. Without these design features, high agitation can spill the quenchant from the tank.

Example 17. A typical tank with pump agitation for batch quenching of axles and shafts in oil is illustrated in Fig. 47.

Design of Quench Tanks

To conform with present-day requirements, quenching systems are integrated into heat treating lines, and the automated lines are normally used to process many types of parts. It is seldom economically feasible to design a quench tank for only one type of part. One of the major goals of design is to obtain as much flexibility in the quenching system as possible without appreciably affecting the cost of the unit.

The design of a quench tank is primarily based on such factors as the kilograms (pounds) of steel quenched per hour and size of the part, as well as its shape, weight, section thickness, grade of steel and properties needed. A design based solely on kilograms (pounds) per hour may lead to miscalculations and unacceptable results.

Some practical suggestions for the design of quench tanks are as follows:

- Time necessary to "quench out" the part treated should be measured, and allowance should be made for this cycle when the hourly rating or capacity of the work tank is determined. When this factor is considered and the mechanism determined that provides for the desired quenching technique, the size of the enclosure of the work tank follows. This volume of medium then is used in other calculations concerning equipment and in the selection of size; for example, the volume of medium influences the selection of the size of cooler and whether a supply tank is necessary.
- Adequate space should be provided around the workpiece to get the full benefit of circulation and maximum heat absorption from the quenched parts.
- At best, it is difficult to control distortion caused by thermal strains when quenching. Therefore, the hot and plastic workpiece should not be mutilated and deformed by allowing it to fall on the chute or quench conveyor until sufficient quenching has taken place to form a hard crust that

will protect the part. On the other hand, if the part is large and roughly machined so that deformation in handling is not serious, premature destruction of the chute and conveyor mechanism is likely to result.

- The work tank should be accessible to facilities for maintenance and cleaning.
- If removal of scale from the tank is a serious problem, the use of a controlled atmosphere furnace should be considered.
- Adequate ventilation should be provided to protect workmen from fumes and to improve working conditions.
- If the quenching medium is oil, protection against fire hazard should be provided.
- Use of clean-out plates below the solution level of the work tank should be avoided, if possible. Gaskets are difficult to maintain and will leak unless many studs are used on closely spaced centers.
- Mechanisms for use in tanks for oil quenching may be of almost any material.
- Special selection of materials for mechanisms is usually necessary if brine or caustic is used. A high cost of maintenance will usually result.
- Alloy trays and fixtures of the 25Cr-12 Ni or 35Ni-15Cr type can be quenched in oil quite satisfactorily, but not in water, caustic or brine.
- If water, caustic or brine must be used, the use of special materials for pumps, coolers and other equipment should be considered.
- Wherever possible, the design of the work tank should provide for flexibility and control of such conditions as: (a) time cycle of quench, (b) volume of circulation, and (c) uniformity of bath conditions.
- It should not be expected that agitation will solve all problems. Excessive agitation that will blow light parts off the conveyor, and thereby lose them, should not be employed. A medium that has a higher speed of quenching should be used instead.
- A quenching bath should not be agitated so violently that foaming occurs on the bath surface. This means that air pockets are forming and that the maximum benefit of circulation is not being obtained.

Fixtures

Fixtures are used to provide support or restraint, as discussed in the following.

Support fixtures are widely used to minimize distortion during quenching. They may vary in design from a simple tray or rack to complex compartmented baskets and special holders. Distortion of parts such as shafts is minimized if they are hung or supported vertically during heating and quenching.

Circular parts, such as large rings, may be supported on flat surfaces during heating and quenching. However, this practice may cause nonuniform hardness, because it restricts the flow of quenchant. A preferred practice is to support such members on fixtures that have radial ribs, machined or ground flat, to permit free flow of the quenching medium to all areas.

Distortion of thin-wall circular members is sometimes corrected by inserting pins inside the ring to force the small diameter outward prior to tempering. The pins are a part of a turnbuckle assembly that permits adjustment. To ensure acceptable roundness after tempering and removal of the pin assembly, an over-correction of about 50% is usually required. However, corrective methods such as this are practical only for small-scale production.

Restraint fixturing is costly and is used primarily for highly specialized applications. Notable examples of parts that require restraint during heating and quenching are rocket and missile casings, or other large components with thin wall sections. For such components, two or more external restraining bands may be used. An articulated internal fixture also may be required. It is built up of numerous cast or stamped pieces pinned together to provide many points of support and a free flow of quenchant to the inner surface of the workpiece. Articulated fixtures can accommodate workpieces of various diameters and lengths. The whole assembly is hung in the furnace and then lowered into the oil quench tank.

Cold Die Quenching

Thin disks, long slender rods and other delicate parts that distort excessively when they are quenched in conventional liquid media, often can be quenched between cold dies with virtually no distortion.

Example 18. Large, thin thrust washers, which were blanked from cold rolled steel and contained machined oil grooves, developed considerable distortion as a result of blanking and machining stresses. A dimensional variation of 1.3 mm (0.050 in.) was common, and even larger variations sometimes occurred. Nevertheless, these washers had to be flat to within 0.13 mm (0.005 in.) after hardening.

To ensure the required flatness, the washers were squeezed between a pair of water-cooled die blocks immediately after they left the hardening furnace. The die blocks provided the necessary quenching action while maintaining flatness. When water-cooled beryllium copper die blocks were used in this application, the rate of cooling approached that obtained in water quenching.

Cold die quenching is limited to parts with a large surface area and a relatively small mass, such as washers, rods of small diameter, and thin blades.

Press Quenching

Commercial quenching presses are designed for controlled quenching of ring gears and other round, flat or cylindrical parts, to permit heat treating with minimum distortion.

Cooling Systems

As quenching proceeds in a given medium, heat is removed from the workpieces, and unless a cooling system is provided, the temperature of the quenching medium rises. Uniform results in the hardening operation depend in part on the control of quenchant temperature. Accordingly, coolers (and sometimes heaters) are required to achieve the desired temperature control. A typical oil-cooling system is shown schematically in Fig. 48.

Many types of heat exchangers are used for cooling quenching liquids, and each offers certain advantages. The rate of heat transfer through the cooling unit always depends on the velocity of the cooling medium passing through it. As the velocity of the cooling medium increases, the rate of heat transfer increases. Therefore, the efficiency of the cooling unit depends on an adequate velocity of flow, which in turn requires an adequate pumping facility.

The maintenance of cooling equipment is related to its design. Thus, if the cooling unit is capable of maintaining a uniform quenching oil temperature, the possibility of fouling is minimized. To facilitate maintenance, cooling equipment should always be designed for easy cleaning.

The shell-and-tube heat exchanger is a type of oil cooler that consists of a series of copper or steel tubes enclosed in a steel shell. The quenching liquid is circulated through either the tube section or the shell section, with the remaining section reserved for circulating the cooling liquid. City water or well water is often used as the coolant; with larger heat loads, an evaporative water-cooling unit may be employed to conserve water.

In general, disposable water is used when the total quenching load is less than 910 kg/h (2000 lb/h). However, the use of disposable water will depend on the water-consumption costs as compared to the initial and operating cost of the evaporative unit. Usually, the evaporative unit does not provide water at as low a temperature as can be expected from well water or the normal city source. Therefore, it normally requires a larger heat exchanger.

The shell-and-tube heat exchanger requires very little maintenance. If properly designed, the tubes may be removed for cleaning. The heat exchanger is flexible in its application and can easily be designed to comply with various specified conditions of velocity.

Maintenance costs are related to the number of mechanical units involved. Pumps are essential for water circulation, and in some types of towers, motor-driven fans, used for air circulation, are another maintenance item.

Selection of Cooling System. To select the optimum cooling system for the quenching circuit, the maximum demand or heat load must be determined from the following data:

- Weight of steel quenched during a specified period
- Specific heat of steel
- Temperature of steel entering the quench
- Temperature of steel leaving the quench
- Temperature of quenching liquid entering the work tank
- Temperature of quenching liquid leaving the work tank
- Characteristics of the quenching medium such as specific heat, specific gravity, viscosity, flash point, boiling point and point of vaporization of toxic solutions.

From the above data, and with full knowledge of costs and acceptable metallurgical results, proper cooling equipment can be selected for the quenching system.

Storage or Supply Tanks

For uniform hardness and minimum distortion in batch quenching, experience has indicated that the rise in temperature of the quenchant should be maintained within as narrow a range as possible—preferably within 5 °C (10 °F). This requires a large volume of quench liquid to limit the instantaneous temperature rise in the quench tank; this volume can be supplied by an auxiliary storage tank in the system. In addition to providing sufficient quenching liquid, the storage tank is an economical means of linking equipment such as pumps, coolers, heaters and filters, and it provides a means for the continual reuse of quenching liquids.

Design of the storage tank depends on individual requirements and conditions. Fundamentally, the tank should be of the closed type and sufficiently large to supply the necessary quench capacity for optimum temperature control when three-quarters full. The tank should be equipped with one or more manhole covers and numerous other openings on the top for drainage connections from the quench tanks. The ends of the tank should be provided with capped valve outlets for tapping the liquid supply. These outlets should conform in size with pipe outlets of normal size. A sight gage and a drain valve also should be attached to the storage tank.

There are advantages in having tanks positioned above the floor of a well-ventilated basement room that has the capacity to house pumps, coolers, filters or other equipment required in large quenching systems. This will permit gravity feed of the quench liquid to the pumps, eliminate any possibility of loss of pump prime, and allow insulation of the storage tank to minimize heat loss where temperature control of the liquid is desirable. However, there also are advantages in using underground storage tanks (notably, saving of space). The oil-cooling system shown in Fig. 48 employs an underground storage tank.

Heaters

Storage tanks containing water or brine may be heated economically by steam or hot water pipes or by radiant tubes inserted in the tanks. The optimum storage temperature range for water or brine is 13 to 25 °C (55 to 75 °F). When steam or hot water is used for heating quenching oils, there is always danger of contamination that will change the quenching characteristics of the oils, cause foaming and create danger of explosion.

When fuel-fired radiant tubes are used they should have the characteristic of burning deep in the tube, thus providing a cold leg to a distance well below the height of the quenching liquid. The products of combustion must be well vented. For oil storage tanks (or quenching tanks) the preferred temperature range is about 50 to 70 °C (120 to 160 °F), depending upon the type of quenching oil.

Immersion-type electrical heating elements also are used for heating the oil in either quenching or storage tanks. A choice between immersion electric heaters and radiant-tube burners is a matter of economics and convenience.

Pumps

Primarily, pumps are installed for one or both of two reasons:

- It is necessary to maintain a specified flow of the quenching liquid through the heat exchanger; usually, this is a continuous flow and need be only enough to compensate for heat transfer.
- Agitation is necessary in the quenching tank; quite often, a second pump is installed for this purpose.

Centrifugal pumps are used wherever possible, because wearing is minimized and the initial cost usually is lower. However, where a storage tank is employed, it usually is necessary to raise the oil from the storage tank to the quenching tank.

Because a centrifugal pump is not self-priming, it is customary to install a small positive-displacement pump for priming the centrifugal pump when necessary. Because a continuous-duty pump is usually small, this is often a positive-displacement pump and is used both for continuous circulation when necessary and for priming the centrifugal pump used in agitation.

By the installation of valves, it is usually feasible to connect the pump in such a way that it can be used for emptying or filling the quenching tank or tanks. Valves also should be installed to simplify the repair of pumps.

Centrifugal pumps are not designed for great suction-lift operation and, therefore, should be installed as near the oil supply as possible. Consideration should be given also to the resistance to flow set up by the piping system, so that the pump itself may be properly sized. If resistance of the piping is considerably less than anticipated, the motor of the pump is likely to become overloaded.

Mechanical Conditioning of Quenching Oils

Mechanical conditioning of quenching oils involves the removal of the following contaminants:

- Scale
- Carbonaceous materials, which may be (a) products of oil oxidation (sludge) or (b) carbon fallout, encountered in protective-atmosphere installations
- Other insoluble solids, such as sand
- Water
- Soluble compounds, such as carbon dioxide.

These contaminants can be removed by filtering, evaporation or draining. The solids can best be removed by appropriate bypass filters. The choice of filtering medium for removing solids is important. The most commonly used filtering media are the waste-pack, mineral-wool and cellulose types that must be replaced after their filtering ability has been exhausted. Clay filtering media are more expensive than the above types but can be reused after exhaustion, by suitable regeneration. However, the regeneration will not remove scale or sand. Clay media should be carefully selected when fast quenching oils are to be filtered, because it is possible to remove the additive along with the undesirable carbonaceous materials.

Magnetic filters or traps and strainers are useful in removing scale and other foreign materials. These types of filters can be easily cleaned and returned to service. They are especially helpful for preventing premature filter clogging and for protecting pumps.

Water can be removed by filtering or centrifuging, but these methods are expensive and are rarely used. Usually, bulk water is removed by draining and suspended water is removed by heating. Carbon dioxide is removed by heating.

Maintenance of Quenching Installations

Because quenching baths vary widely in design, shape, size and method of operation, it is not feasible to establish standard procedures for maintenance. However, typical schedules for maintaining large quenching installations are as follows:

Oil Quenching

Daily

- Check oil level in quench tanks.
- Check oil temperature.
- Check oil filter pressure.
- Check oil pumps and oil flow.

Weekly

- Check quenching rate of oil in production system.
- If oil filters are not included in oil system, check for solids in oil.
- Check oil temperature controller and control setting.

Monthly

- Drain quench tanks and remove sludge.

Semiannually

- Check heat exchanger coils, pipes and pumps.
- Replace oil filters when necessary.
- Check screens ahead of oil filter.
- Check oil storage tank for sludge, water leaks and general condition.
- Calibrate oil-temperature gage.
- Check oil for contamination.

Water Quenching

Daily

- Check water temperature.
- Check water pressure.
- Check water circulation.

Weekly

- Drain quench tanks and remove sludge.
- If water is recirculated, make necessary chemical addition to prevent calcium compounds from building up in tubes.

Brine Quenching

Daily

- Check brine temperature.
- Check brine concentration, and adjust as required.

Weekly

- Drain tanks and remove sludge.

- Check pumps and tank condition.
- Check quenching fixtures for signs of deterioration.

Quenching of Induction Hardened Parts

A great many induction hardening applications employ water as the quenching medium. Other media, such as conventional quenching oil, soluble oil, compressed air, water-base solutions of polyvinyl alcohol or brine, are occasionally used. Because water provides greatest ease in handling, minimum installation and maintenance of equipment, and greatest safety, it is used unless metallurgical considerations indicate the necessity for one of the other quenching media.

Quenching of Flame Hardened Parts

The proper application of a suitable quench in flame hardening is as important as proper heating. The quench must remove heat at a rate that will produce the desired structure and assist in controlling the depth of the hardened case.

Safety Precautions

Fire hazards always attend the use of quenching oils. The causes of quenching-oil fires and the methods of extinguishing them should be thoroughly understood.

One of the most common causes of fire in open quench tanks—and one of which operating personnel should be made aware—is ignition of the oil surface from a partially submerged load. Some aids to the prevention of such fires are:

- Careful design of hoists or conveyors used to lower the work load into the quench
- Periodic inspection of chains, sprockets and other components that may fail
- Having available an alternate power supply for use in the event that a power failure occurs during a quenching cycle.

The ignition of oil overflowing from a tank can produce a most damaging type of fire. When an oil quench tank is located adjacent to a furnace or any other source of ignition, special precautions are required, such as:

- Providing inside overflow drains, to prevent overfilling, and outside drains, to prevent the spread of burning oil
- Providing a system for detecting the presence of water in the oil (several detectors are commercially available). If the oil temperature approaches 120 °C (250 °F) with water present, foaming can result in overflow that exceeds the drain capacity.

A third hazard arises when oil is heated by the work load to a temperature above its flash point. A temperature less than 10 °C (50 °F) below the flash point is considered dangerous. To minimize the temperature rise, several methods may be employed, including the use of: (a) cooling coils in the tanks, (b) external heat exchangers, (c) more agitation, (d) a larger tank, or (e) oil with a higher flash point.

Extinguishing Oil Fires. A planned program for extinguishing quenching oil fires should include:

- A quick method of extinguishing the fire without contaminating the oil, such as by smothering it with a tank cover or by the use of a carbon dioxide system
- An auxiliary method giving longer-lasting protection, such as foam, dry chemical, or draining the tank
- Periodic training of personnel in fire prevention and fire extinguishing.

Facilities to drain oil from the tank to a safe location constitute one method of preventing serious fires.

Tank covers or lids can be used effectively to smother a fire in a small tank. These covers can be actuated by heat or, from a safe distance, manually.

There are two general types of carbon dioxide systems: (a) high-pressure systems, in which the gas is stored at room temperature; and (b) low-pressure systems, in which the gas is stored refrigerated to −20 °C (0 °F.)

The primary effectiveness of carbon dioxide lies in its ability to reduce the supply of oxygen at the surface of the oil to the point at which it cannot support combustion. In some systems, carbon dioxide has a cooling effect as it sublimes from a solid "snow" to a gas.

The advantages of carbon dioxide are that it does not contaminate the quenching oil or require cleanup. Disadvantages are the short duration of the protection afforded and the storage costs for a large installation.

Fire-fighting foam is a mass of fine, heat-resisting bubbles. It smothers an oil fire by floating over the oil surface and setting up into a stiff, long-lasting blanket. There are two general types of foam: chemical and mechanical (or air); both are equally effective.

An advantage of foam is its long-lasting protection. When a quench load has only partially submerged, or when the fire has heated surrounding metal, the protection must last until these sources of re-ignition are eliminated. Some disadvantages of foam are the cleanup problem after use and that it can cause oil over 120 °C (250 °F) to boil or foam unless the extinguisher is discharged several feet above the tank surface.

The dry chemical extinguisher uses mainly sodium bicarbonate discharged through heads with high-pressure nitrogen. The extinguishing action, advantages and disadvantages are similar to those of the foam type. Dry powders have been known to contaminate quenching oil.

REFERENCES

1. *Kirk-Othmer Encyclopedia of Chemical Technology*, John Wiley & Sons, Inc., Somerset, NJ 08873
2. *Handbook of Water-Soluble Gums and Resins*, edited by Robert L. Davidson, McGraw-Hill Book Co., New York, NY

SELECTED REFERENCES

- H. E. Boyer, Quenching Mediums and Equipment, ASM, 1979
- Application of Hardenability Concepts in Heat Treatment of Steels, by D. Doane, *Journal of Heat Treating*, Vol 1, No. 1, American Society for Metals, 1979
- Hardenable Carbon and Alloy Steels, in *Metals Handbook*, 9th Ed., Vol 1, American Society for Metals, 455, 1978
- *Quenching of Steels*, 40th National Metal Congress, Cleveland, OH, Oct. 27-31, 1958, ASM 1958
- *Principles of Heat Treatment*, 5th Ed., by M. A. Grossmann and E. C. Bain, ASM, 1964
- *The Quenching of Steels*, by H. J. French, American Society for Steel Treating, 1930
- Technology Reports, 1 through 17, by M. Togaya and I. Tamura, Osaka University, Osaka, Japan, 1952-1961; Japanese Metal Society, Vol 22, No. 12, 1958
- K. G. Speith and H. Lange, Mitt Kaiser Wilhelm Inst Eisenfor, Vol 17, p 175, 1935

- W. Peter, *Arch. Eisenhuettenw.*, Vol 21, p 395, 1950
- A. Rose, *Arch. Eisenhuettenw.*, Vol 13, p 345, 1940
- H. J. French, *Trans. ASST*, Vol 17, p 646, 1930
- *Hardenability of Alloy Steels*, by M. A. Grossmann, M. Asimow and S. F. Urban, American Society for Metals, 1939, p 124-190
- *Some Tests on Quenching Oil*, by T. F. Russell, Iron Steel Inst. (London), Spec. Rept. No. 24, 1939, p 283
- W. E. Jominy and A. L. Boegehold, *Trans. ASM*, Vol 26, 1938, p 574-600
- N. B. Pilling and T. D. Lynch, *Trans. AIME*, Vol 62, 1920, p 669

- D. J. Carney, *Trans. ASM*, Vol 46, 1954, p 882-924
- D. J. Carney and A. D. Janulionis, *Trans. ASM*, Vol 43, 1951, p 480-496
- A. R. Troiano and L. J. Klingler, *Trans. ASM*, Vol 44, 1952, p 775-785
- M. J. Sinnott and J. C. Shyne, *Trans. ASM*, Vol 44, 1952, p 758-774
- "Symposium on the Hardenability of Steel", by T. F. Russell, Iron Steel Inst. (London), Spec. Rept. No. 36, 1946, p 25-33
- E. W. Weinman, R. F. Thomson and A. L. Boegehold, *Trans. ASM*, Vol 44, 1952, p 803-844

- H. Scott, *Trans. ASM*, Vol 22, 1934, p 68, "Hardenability of Alloy Steels", American Society for Metals, 1939, p 251-258
- "An Evaluation of the Hardening Power of Quenching Media for Steel", by E. J. Eckel, R. W. Mayfield, G. W. Weush and F. A. Rough, Univ. Illinois Eng. Exp. Sta. Bull. No. 389, June 1951
- F. Wever and A. Rose, *Stahl Eisen*, Vol 74, 1954, p 749
- "Improved Quenching of Steel by Propeller Agitation", U.S. Steel Corp., 1957
- Crosby, Fiske and Forster, "NFPA Handbook of Fire Protection", 1954

Tempering of Steel

By the ASM Committee on
Tempering of Steel*

TEMPERING OF STEEL is a process in which previously hardened or normalized steel is heated to a temperature below the transformation range and cooled at a suitable rate, primarily to increase ductility and toughness. Steels are tempered by reheating after hardening to obtain specific values of mechanical properties and to relieve quenching stresses and ensure dimensional stability. Tempering usually follows quenching from above the critical temperature; however, tempering also is used to relieve the stresses and reduce the hardness developed during welding, and to relieve stresses induced by forming and machining.

Principal Variables

Variables associated with tempering that affect the microstructure and the mechanical properties of a tempered steel include temperature; time at temperature; cooling rate from the tempering temperature; and composition of the steel, including carbon content, alloy content and residual elements. In a steel quenched to a microstructure consisting essentially of martensite, the iron lattice is strained by the carbon atoms, producing the high hardness of quenched steels. On heating, the carbon atoms diffuse and react in a series of distinct steps that eventually form Fe_3C or an alloy carbide in a ferrite matrix of gradually decreasing stress

level. The properties of the tempered steel are determined primarily by the size, shape, composition and distribution of the carbides that form, with a relatively minor contribution from solid-solution hardening of the ferrite. These changes in microstructure usually decrease hardness, tensile strength and yield strength but increase ductility and toughness.

Under certain conditions, hardness may remain unaffected by tempering or may even be increased as a result of it. For example, tempering of a hardened steel at very low tempering temperatures may cause no change in hardness but may achieve a desired increase in yield strength. Also, those alloy steels that contain one or more of the carbide-forming elements (chromium, molybdenum, vanadium and tungsten) are capable of "secondary hardening"—that is, they may become somewhat harder as a result of tempering.

The tempered hardness values for several quenched steels are presented in Table 1. Temperature and time are interdependent variables in the tempering process. Within limits, lowering temperature and increasing time usually can produce the same result as raising temperature and decreasing time. With few exceptions, tempering is done at temperatures between 175 and 700 °C (350 and 1300 °F) and for times from 30 min to 4 h.

Tempering Temperature. Several empirical relationships have been made between the tensile strength and hardness of tempered steels such that measurement of hardness is commonly used to evaluate the response of a steel to tempering. Figure 1 shows the effect of tempering temperature on the hardness, tensile and yield strengths, elongation, and reduction in area of a plain carbon steel (AISI 1050) held at temperature for 1 h. Note that both room temperature hardness and strength decrease as the tempering temperature is increased. Ductility at ambient temperatures, as measured by either elongation or reduction in area, increases with tempering temperature.

Whereas elongation and reduction in area increase continuously with tempering temperature, toughness, as measured by a notched-bar impact test, varies with tempering temperature for most steels as shown in Fig. 2. Tempering at temperatures from 260 to 320 °C (500 to 610 °F) decreases impact energy to a value below that obtained at about 150 °C (300 °F). Above 320 °C, impact energy again increases with increasing tempering temperature. Both plain carbon and alloy steels respond to tempering in this manner. The phenomenon of impact-energy minima centered around 300 °C (570 °F) is called "500 °F temper embrittlement" or "blue brittleness". This phenomenon will be discussed more thoroughly in a subse-

*Peter Vernia, *Chairman,* Senior Research Engineer, Metallurgy Department, General Motors Research Laboratories; John W. Balai, Metallurgical Engineer, Holcroft, Division of Thermo Electron Corp.; Duane J. Schmatz, Principal Staff Engineer, Scientific Laboratories, Ford Motor Co.

Table 1 Typical hardnesses of various carbon and alloy steels after tempering

Data were obtained on 25-mm (1-in.) bars adequately quenched to develop full hardness.

Grade	Carbon content, %	°C: 205 °F: 400	260 500	315 600	370 700	425 800	480 900	540 1000	595 1100	650 1200	Heat treatment
Carbon steels, water hardening											
1030	0.30	 50	45	43	39	31	28	25	22	95(a)	Normalized at 900 °C (1650 °F); water quenched
1040	0.40	 51	48	46	42	37	30	27	22	94(a)	from 830-845 °C (1525-1550 °F); average dew
1050	0.50	 52	50	46	44	40	37	31	29	22	point, 16 °C (60 °F)
1060	0.60	 56	55	50	42	38	37	35	33	26	Normalized at 885 °C (1625 °F); water quenched
1080	0.80	 57	55	50	43	41	40	39	38	32	from 800-815 °C (1475-1500 °F); average dew
1095	0.95	 58	57	52	47	43	42	41	40	33	point, 7 °C (45 °F)
1137	0.40	 44	42	40	37	33	30	27	21	91(a)	Normalized at 900 °C (1650 °F); water quenched
1141	0.40	 49	46	43	41	38	34	28	23	94(a)	from 830-855 °C (1525-1575 °F); average dew
1144	0.40	 55	50	47	45	39	32	29	25	97(a)	point, 13 °C (55 °F)
Alloy steels, water hardening											
1330	0.30	 47	44	42	38	35	32	26	22	16	Normalized at 900 °C (1650 °F); water quenched
2330	0.30	 47	44	42	38	35	32	26	22	16	from 800-815 °C (1475-1500 °F); average dew
3130	0.30	 47	44	42	38	35	32	26	22	16	point, 16 °C (60 °F)
4130	0.30	 47	45	43	42	38	34	32	26	22	Normalized at 885 °C (1625 °F); water quenched
5130	0.30	 47	45	43	42	38	34	32	26	22	from 800-855 °C (1475-1575 °F); average dew
8630	0.30	 47	45	43	42	38	34	32	26	22	point, 16 °C (60 °F)
Alloy steels, oil hardening											
1340	0.40	 57	53	50	46	44	41	38	35	31	Normalized at 870 °C (1600 °F); oil quenched
3140	0.40	 55	52	49	47	41	37	33	30	26	from 830-845 °C (1525-1550 °F); average dew
4140	0.40	 57	53	50	47	45	41	36	33	29	point, 16 °C (60 °F)
4340	0.40	 55	52	50	48	45	42	39	34	31	Normalized at 870 °C (1600 °F); oil quenched
4640	0.40	 52	51	50	47	42	40	37	31	27	from 830-845 °C (1525-1575 °F); average dew
8740	0.40	 57	53	50	47	44	41	38	35	22	point, 13 °C (55 °F)
4150	0.50	 56	55	53	51	47	46	43	39	35	Normalized at 870 °C (1600 °F); oil quenched
5150	0.50	 57	55	52	49	45	39	34	31	28	from 830-870 °C (1525-1600 °F); average dew
6150	0.50	 58	57	53	50	46	42	40	36	31	point, 13 °C (55 °F)
8650	0.50	 55	54	52	49	45	41	37	32	28	Normalized at 870 °C (1600 °F); oil quenched
8750	0.50	 56	55	52	51	46	44	39	34	32	from 815-845 °C (1500-1550 °F); average dew
9850	0.50	 54	53	51	48	45	41	36	33	30	point, 13 °C (55 °F)

(a) Hardness, HRB

quent section on temper embrittlement.

Tempering Time. The diffusion of carbon and alloying elements necessary for the formation of carbides is temperature and time dependent. The effect of tempering time on the hardness of an 0.82% carbon steel tempered at various temperatures is shown in Fig. 3. The changes in hardness are approximately linear over a large portion of the time range when the time is presented on a logarithmic scale. Rapid changes in room-temperature hardness occur at the start of tempering in times less than 10 s. Less rapid, but still large, changes in hardness occur in times from 1 to 10 min, and smaller changes occur in times from 1 to 2 h. For consistency and less dependency on variations in time, components generally are tempered for 1 to 2 h. The levels of hardness produced by very short tempering cycles, such as in induction tempering, would be quite sensitive to both the temperature achieved and the time at temperature.

By use of an empirical tempering parameter developed by Holloman and Jaffe (Ref 2), the approximate hardnesses of quenched and tempered low- and medium-alloy steels can be predicted. The parameter is $T (c + \log t)$, where T is temperature in degrees Kelvin, t is time in seconds and c is a constant that depends on the carbon content of the steel. Reasonably good correlations are obtained except when significant amounts of carbide-forming elements or large amounts of retained austenite are present.

Cooling Rate. Another factor that can affect the properties of a steel is the cooling rate from the tempering temperature. Although tensile properties are not affected by cooling rate, toughness (as measured by notched-bar impact testing) can be decreased if the steel is cooled slowly through the temperature range from 375 to 575 °C (705 to 1065 °F), especially in steels that contain carbide-forming elements. Elongation and reduction in area may be affected also. This phenomenon is

Fig. 1 Effect of tempering temperature on room-temperature mechanical properties of 1050 steel

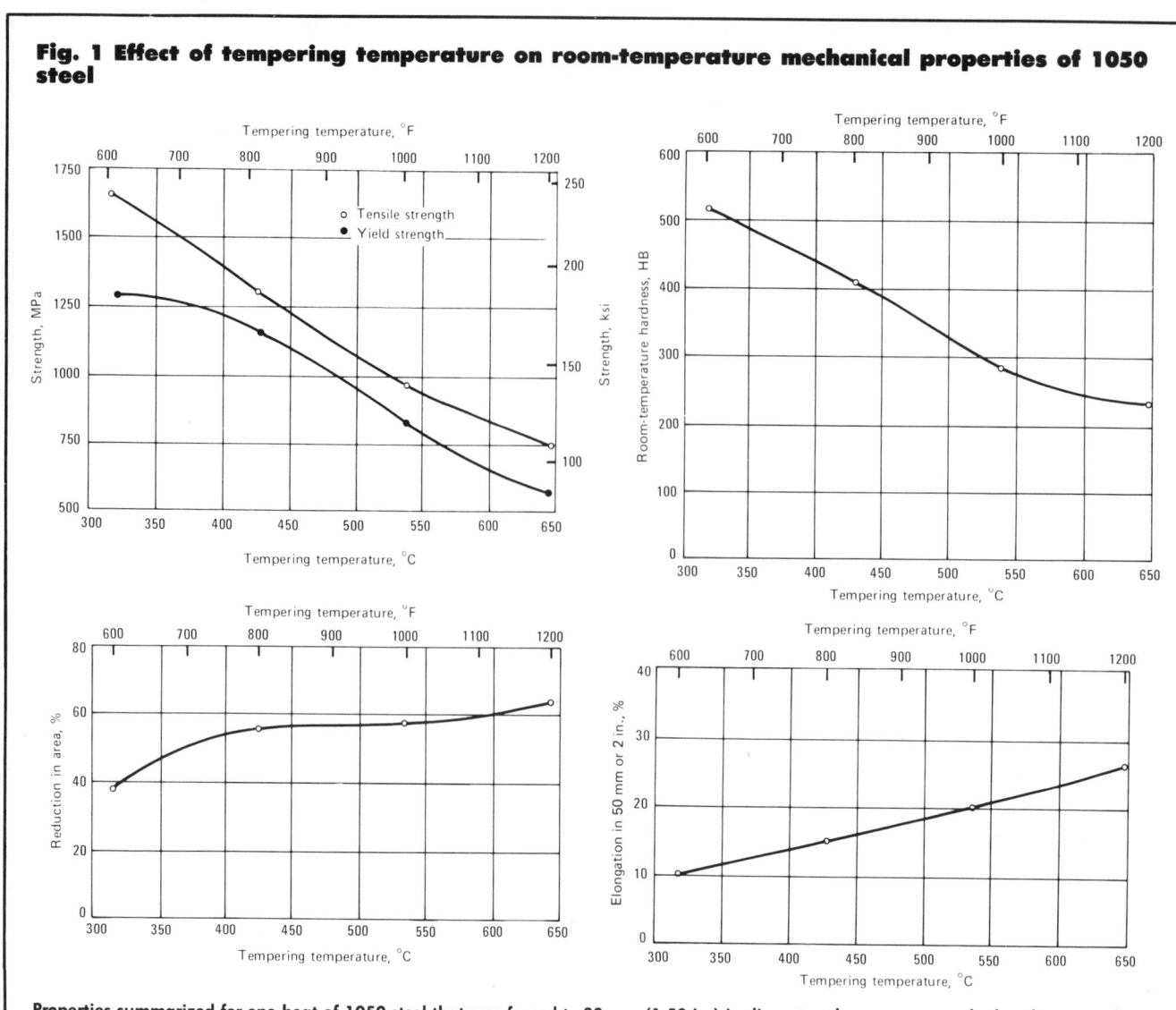

Properties summarized for one heat of 1050 steel that was forged to 38 mm (1.50 in.) in diameter, then water quenched and tempered at various temperatures. Composition of heat: 0.52 C, 0.93 Mn

called "temper embrittlement" and is discussed further in a subsequent section of this article.

Carbon Content

The effect of carbon content on the properties of tempered steel is illustrated in Fig. 4, which presents hardness data for 14 carbon steels that were tempered at temperatures ranging from 205 to 705 °C (400 to 1300 °F) and for times from 10 min to 24 h. The principal effect of carbon content is on as-quenched hardness. Figure 5 illustrates the relationship between carbon content and the maximum hardness

that can be obtained upon quenching. The relative difference in hardness as compared with as-quenched hardness is retained after tempering. Figure 6 shows the combined effect of time, temperature and carbon content on the hardness of four carbon-molybdenum steels of different carbon contents. Figure 7 shows the hardness of these steels after tempering for 1 h, as a function of tempering temperature; the effect of carbon content is clearly evident.

Alloy Content

The main purpose of adding alloying elements to steel is to increase the

steel's ease of hardenability—that is, its ability to form martensite on quenching from above its critical temperature. The general effect of alloying elements on tempering is a retardation of the rate of softening, especially at the higher tempering temperatures. Thus, to obtain a given hardness in a given period of time, alloy steels require higher tempering temperatures than do carbon steels. Alloying elements can be characterized as carbide-forming or non-carbide-forming. Elements such as nickel, silicon, aluminum and manganese, which have little or no tendency to occur in the carbide phase, remain essentially in solution in

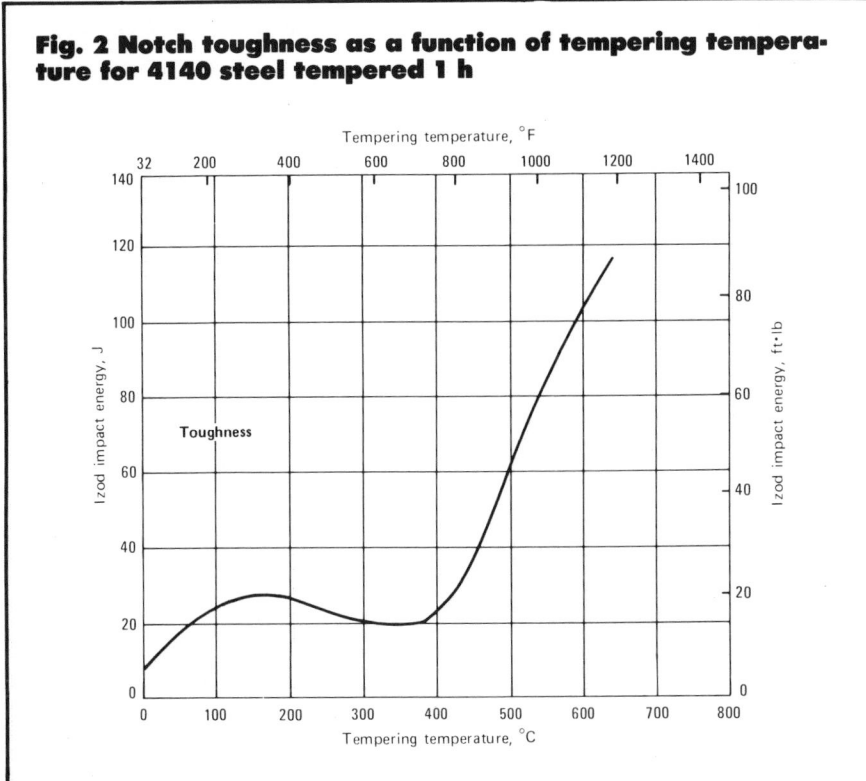

Fig. 2 Notch toughness as a function of tempering temperature for 4140 steel tempered 1 h

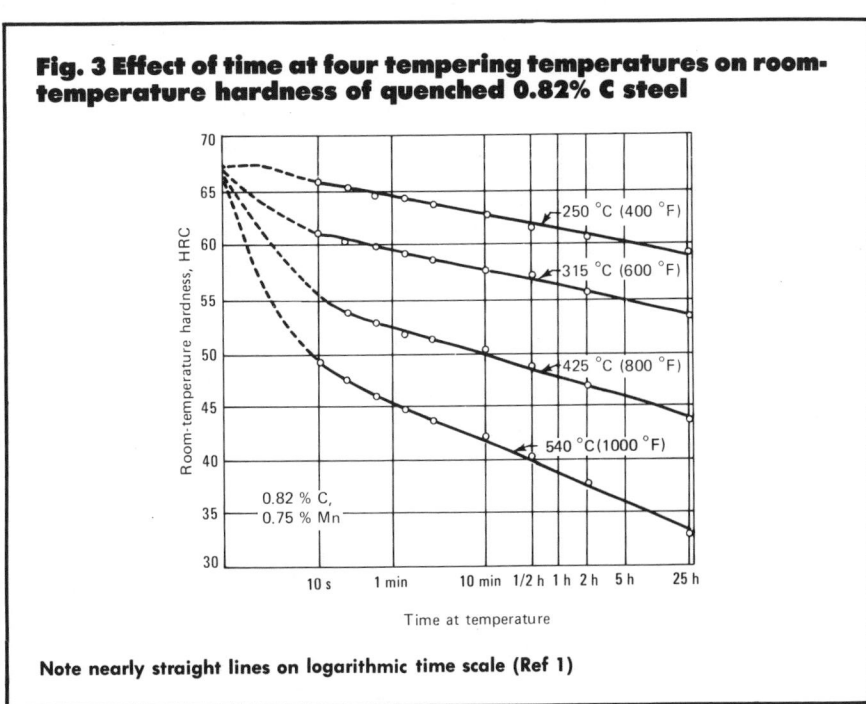

Fig. 3 Effect of time at four tempering temperatures on room-temperature hardness of quenched 0.82% C steel

Note nearly straight lines on logarithmic time scale (Ref 1)

ments is minimal at low tempering temperatures where Fe_3C forms; however, at higher temperatures, alloy carbides are formed and hardness decreases slowly with tempering temperature. Under certain conditions, such as with highly alloyed steels, hardness may actually increase. This latter effect is known as "secondary hardening".

The effect of molybdenum content on the tempering behavior of a 0.35% carbon steel is shown in Fig. 8. As the alloy content increases, the magnitude of the secondary-hardening effect increases. Synergistic effects of various combinations of alloying elements can occur: chromium tends to produce secondary hardening at a lower temperature than does molybdenum; and the combination of chromium and molybdenum produces a rather flat tempering curve, with the peak hardness occurring at a somewhat lower temperature than when only molybdenum is present. H11 steel is a widely used hot working die steel that contains nominally 0.35% C, 5% Cr, 1.5% Mo and 0.4% V. Figure 9 shows the room-temperature hardness of H11 as a function of tempering temperature. A very flat tempering curve results because of the specific combination of the three carbide-forming elements.

Tool Steels and Stainless Steels. Extensive data on tempering of tool steels (including H11) and martensitic stainless steels are given in the articles on heat treating of tool steels and heat treating of stainless steels and heat-resisting alloys.

Other Alloying Effects. Alloying elements produce a number of other effects besides providing ease of hardening and secondary hardening. The higher tempering temperatures used to temper alloy steels presumably permit greater relaxation of residual stresses and improve properties. Furthermore, the hardenability of alloy steels may permit a less drastic quench to be used so that as-quenched stresses are lower and cracking prior to tempering is minimized. The higher hardenability of alloy steels may also permit the use of a lower carbon content to achieve a given strength level but with improved ductility and toughness.

Residual Elements. Residual elements—those not intentionally added to a steel—can cause embrittlement. The elements that are known to cause embrittlement are tin, phosphorus, antimony and arsenic. A discussion of the

the ferrite and have only a minor effect on tempered hardness. Hardening due to the presence of these elements occurs mainly through solid-solution hardening of the ferrite. The carbide-forming elements, chromium, molybdenum, tungsten, vanadium, tantalum, niobium and titanium, retard the softening process by formation of alloy carbides. The effect of the carbide-forming ele-

Fig. 4 Influence of tempering temperature on room-temperature hardness of quenched carbon steels (Ref 3)

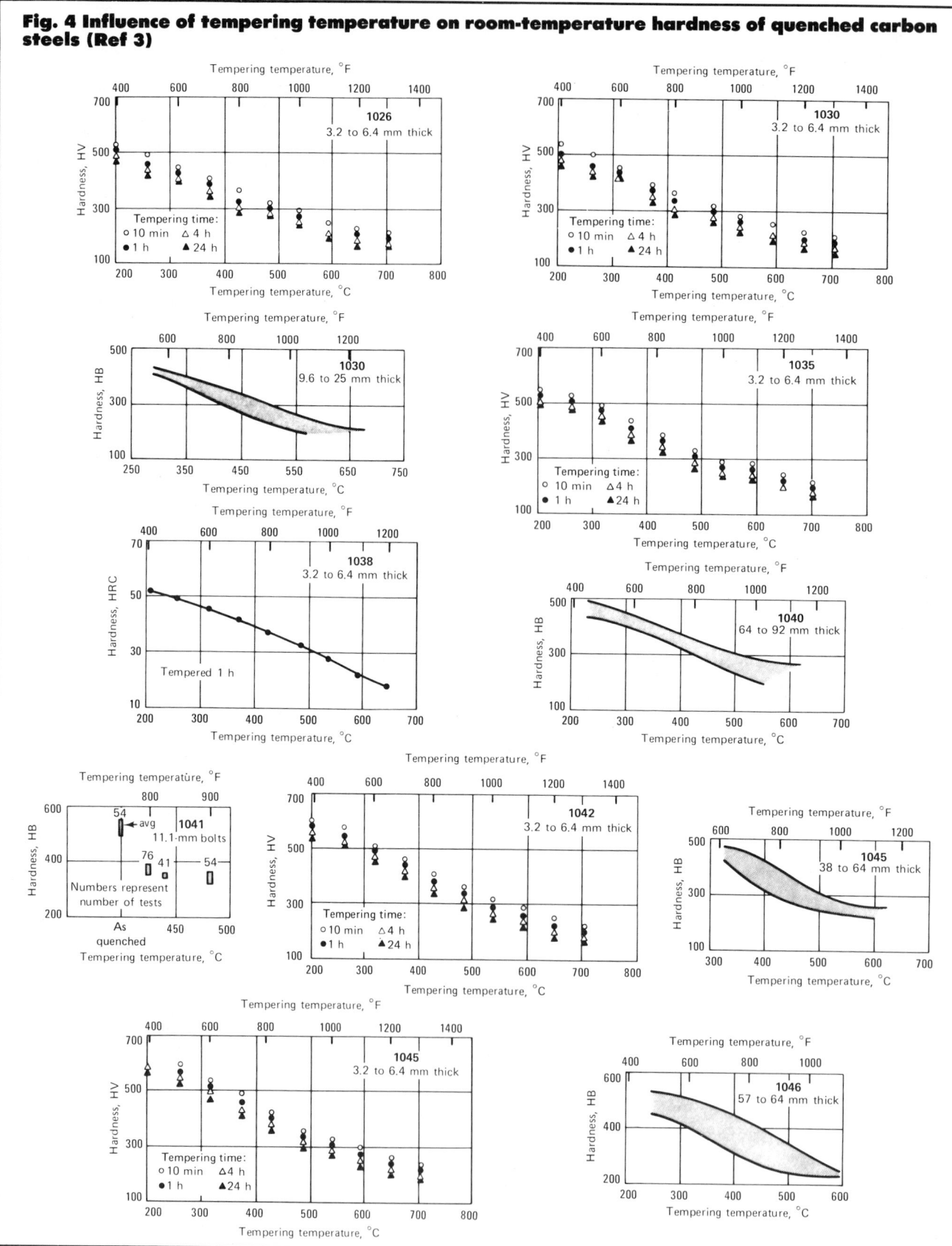

Fig. 4 (continued)

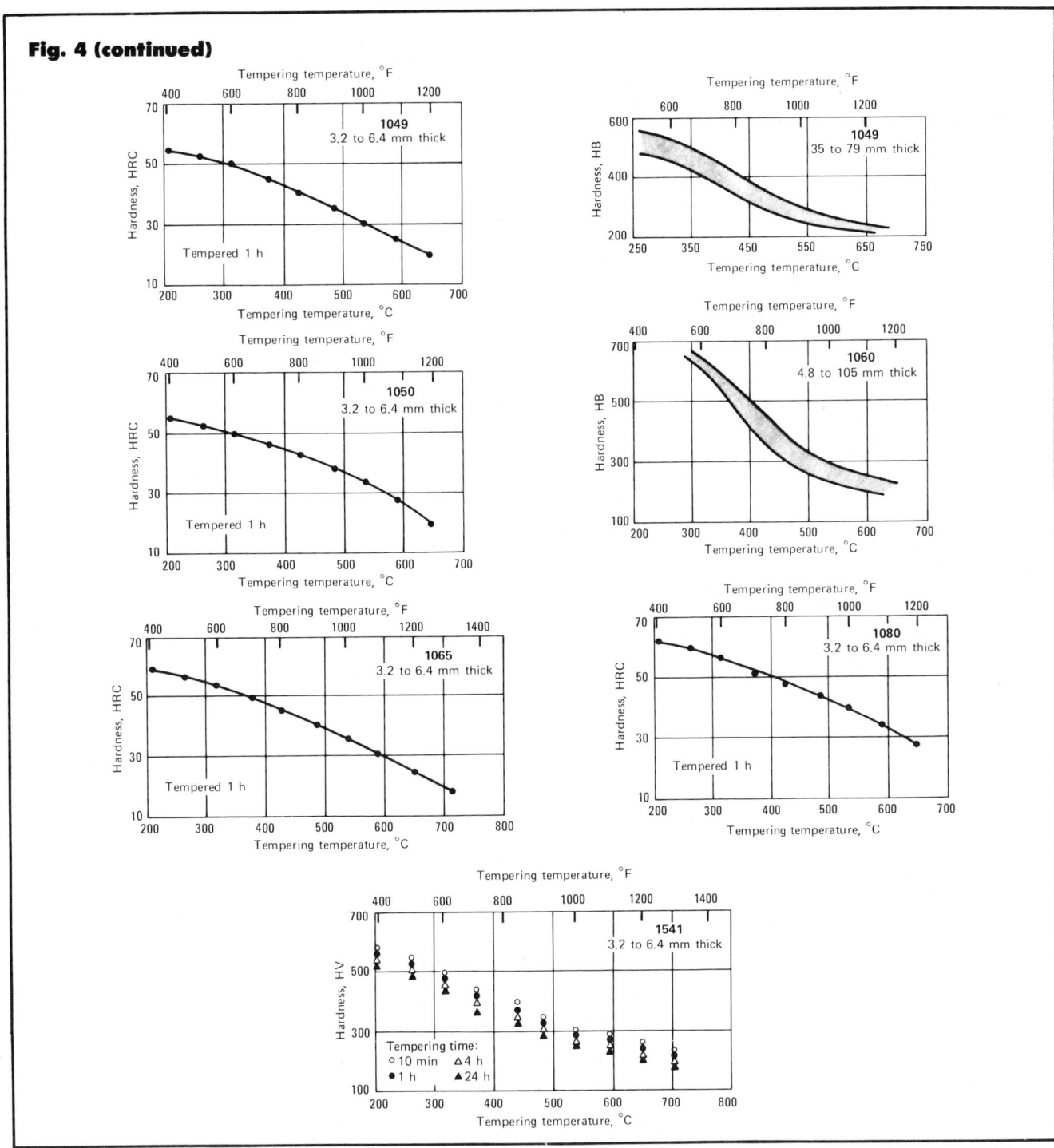

specific effects of these elements can be found in the section on temper embrittlement in this article.

Tempering Procedures

Tempering can be accomplished by soaking entire parts in the furnace for periods of time sufficient to bring the tempering mechanism to the desired point of completion or by selective heating of certain portions of the part to achieve toughness or plasticity in those areas.

Bulk processing may be done in convection furnaces or in molten salt, hot oil or molten metal baths. Selection of the type of furnace depends primarily on number and size of parts and on desired temperature. Table 2 gives temperature ranges, most likely reasons for use, and fundamental problems of these four types of equipment.

Selective tempering techniques are used to soften specific areas of fully hardened parts or to temper areas that were selectively hardened previously.

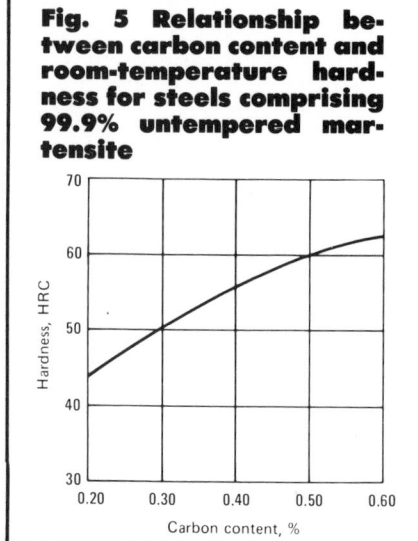

Fig. 5 Relationship between carbon content and room-temperature hardness for steels comprising 99.9% untempered martensite

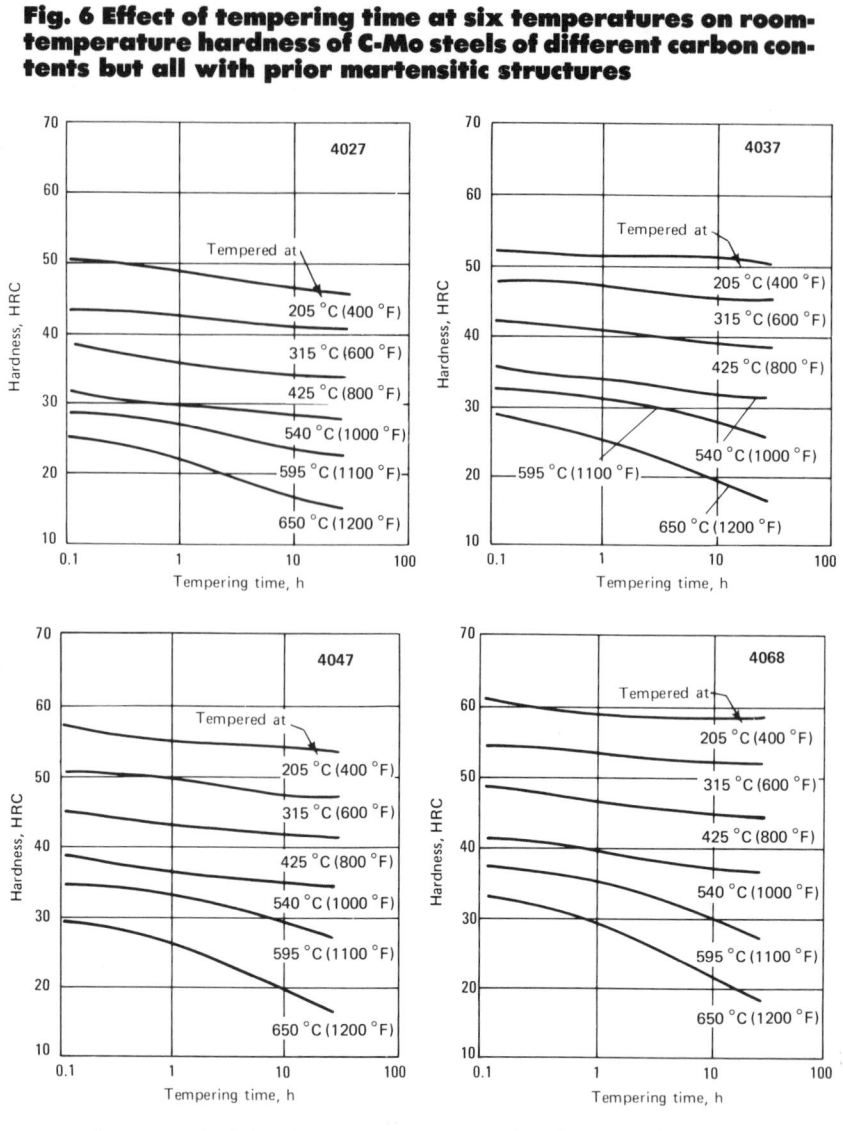

Fig. 6 Effect of tempering time at six temperatures on room-temperature hardness of C-Mo steels of different carbon contents but all with prior martensitic structures

4068 is former standard classification. See p 133 in Vol 1 of this Handbook.

The purpose of this treatment is to improve the machinability, the toughness or the resistance to quench cracking in the selected zone.

Induction and flame tempering are the most commonly utilized selective techniques because of their controllable local heating capabilities. Immersion of selected areas in molten salt or molten metal can be accomplished, but with somewhat less control.

Special processes occasionally are employed to achieve specific properties such as those derived from steam treating or use of protective atmospheres.

The tempering mechanism in certain steels is enhanced by cyclic heating and cooling. A particularly important procedure employs cycles between subzero temperatures and the tempering temperature to increase the transformation of retained austenite. The term used for this procedure, "multiple tempering", also is applied to procedures that employ intermediate thermal cycles to soften parts for straightening prior to the actual tempering operation designed to achieve the desired degree of toughness and plasticity.

Equipment for Tempering

Steel usually is tempered in either an air (convection) furnace or a salt bath (Fig. 10). Molten metal baths, oil baths, and flame or induction heating units are used also.

Convection Furnaces. The most commonly employed tempering method utilizes the recirculating or forced-air convection furnace, and the equipment most commonly used in conjunction with convection furnaces includes continuous belt conveyor, roller rail or dog beam pusher systems. Batch equipment such as box or pit types are used also.

Forced recirculating air is the most common and efficient method of tempering because it lends itself to a wide selection of furnace designs to accommodate a variety of products and capacities. Moreover, the metallurgical results are very good in terms of price per unit weight of yield.

Convection furnaces generally are designed for tempering temperatures of 150 to 750 °C (300 to 1380 °F). For temperatures up to 550 °C (1020 °F), recirculated hot air is supplied to the product from a chamber separate from the work-holding area, to avoid uneven heating by radiation. For temperatures of 550 to 750 °C, either forced convection or radiant heating is used depending on the metallurgical requirements of the product. To obtain closer control of metallurgical properties, the recirculated forced hot air is employed; but for greater efficiency, radiant heating is used, because transfer of radiant heat is greater as the temperature approaches 750 °C (1380 °F).

The most important phase of convection-furnace design is determination of the proper amount of forced-air supply. The objective of the blower is to furnish

Table 2 Temperature ranges and general conditions of use for four types of tempering equipment

| Type of equipment | Temperature range | | Service conditions |
	°C	°F	
Convection furnace	50 to 750	120 to 1380	For large volumes of nearly common parts; variable loads make control of temperature more difficult
Salt bath	160 to 750	320 to 1380	Rapid, uniform heating; low to medium volume; should not be used for parts whose configurations make them hard to clean
Oil bath	Up to 250	Up to 480	Good if long exposure is desired; special ventilation and fire control are required
Molten metal bath	Above 390	Above 735	Very rapid heating; special fixturing is required (high density)

Fig. 7 Effect of carbon content and tempering temperature on room-temperature hardness of four molybdenum steels

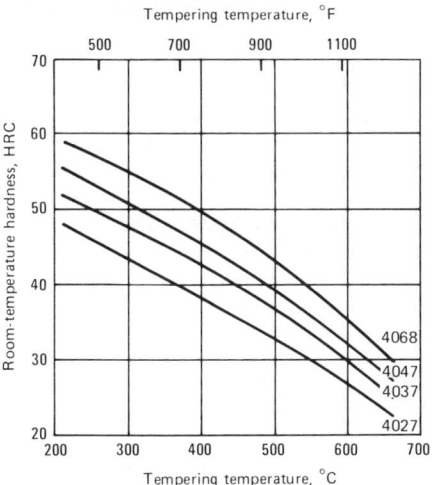

Tempering time: 1 h at temperature. 4068 is former standard classification. See p 133 in Vol 1 of this Handbook.

Fig. 8 Influence of molybdenum content on softening of quenched 0.35% C steels with increasing tempering temperature (Ref 1)

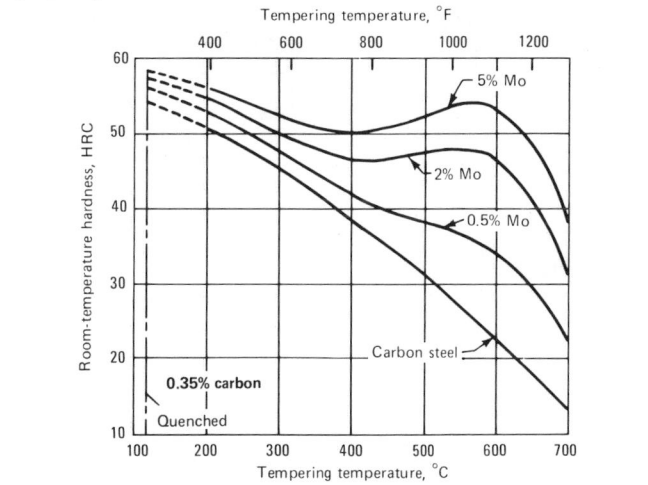

Fig. 9 Variation of room-temperature hardness with tempering temperature for H11 Mod steel

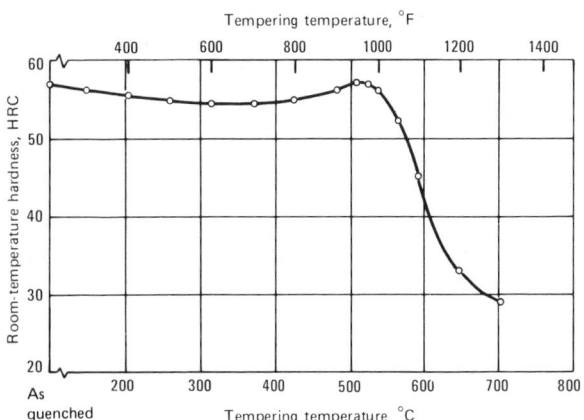

All specimens air cooled from 1010 °C (1850 °F) and double tempered 2 + 2 h at temperature

enough hot air to the complete work area so that it is efficiently used to heat the product in the shortest time thermophysically allowed. The type of product and material being processed will dictate the forced-air supply required, which is measured at the operating temperature. Consultation with fan manufacturers can help achieve maximum efficiency of heat transfer.

Heat for the furnace can be supplied by electricity, gas or oil. In most furnace designs, a dual heat-source capability can be built in, such as gas and electricity. This allows for more than one choice of utility when there is a shortage or a cost advantage, of one or the other.

Temperature control is accomplished

by positioning a thermocouple at the hot-air side of the recirculating system close to the product. When this technique is used, there is minimum danger of overheating, and loads of various sizes can be handled. This method also allows the duration of processing (holding time) to be varied by movement of the thermocouple location—but only within the limits of the furnace size

(and/or conveyor speed for continuous-type furnaces). Temperatures generally are held within ±5 °C (±9 °F).

Efficient use of a continuous furnace cannot be attained when production quantities are small or when parts vary in size, shape and mechanical requirements; the batch furnace is better suited for work of this type. When a continuous furnace is used for such

applications, production time is lost in raising or lowering furnace temperature. Sometimes, when the process variables must be changed, a dummy load must be placed in the furnace to accelerate a desired reduction of temperature, or production must be stopped until the temperature is stabilized.

Salt bath furnaces may be employed for tempering at 160 °C (320 °F) and above. Good heat transfer and natural convection in the bath promote uniformity of workpiece temperature.

All moisture must be removed from parts before they are immersed in the molten salt, because hot salt reacts violently with moisture. If dirty or oily parts are immersed in the bath, the salt will become contaminated and require more frequent rectification.

All parts tempered in salt must be cleaned soon after being removed from the bath, because the salt that clings to them is hygroscopic and may cause severe corrosion. Parts with small or blind holes from which salt is difficult to clean should not be tempered in salt.

Salt bath compositions and operating temperature ranges presented in Table 3 pertain to baths in common use for tempering and are classified according to Military Specification MIL-S-10699A (Ordnance).

Class 1 and class 2 salts are reasonably stable and seldom require rectification. If chlorides are added by carryover from a higher-temperature bath, they will cause an increase in the viscosity of the bath. Chlorides can be removed by filtering through fine screens or by cooling and settling out the insoluble chlorides as a sludge. Occasionally, carbonates become excessive. These can be removed by reaction with dilute nitric acid-base rectifiers. Upper temperature limits must not be exceeded, or salt will become highly oxidizing, even toward alloy steels.

Class 3 salts seldom require rectification. However, their high melting

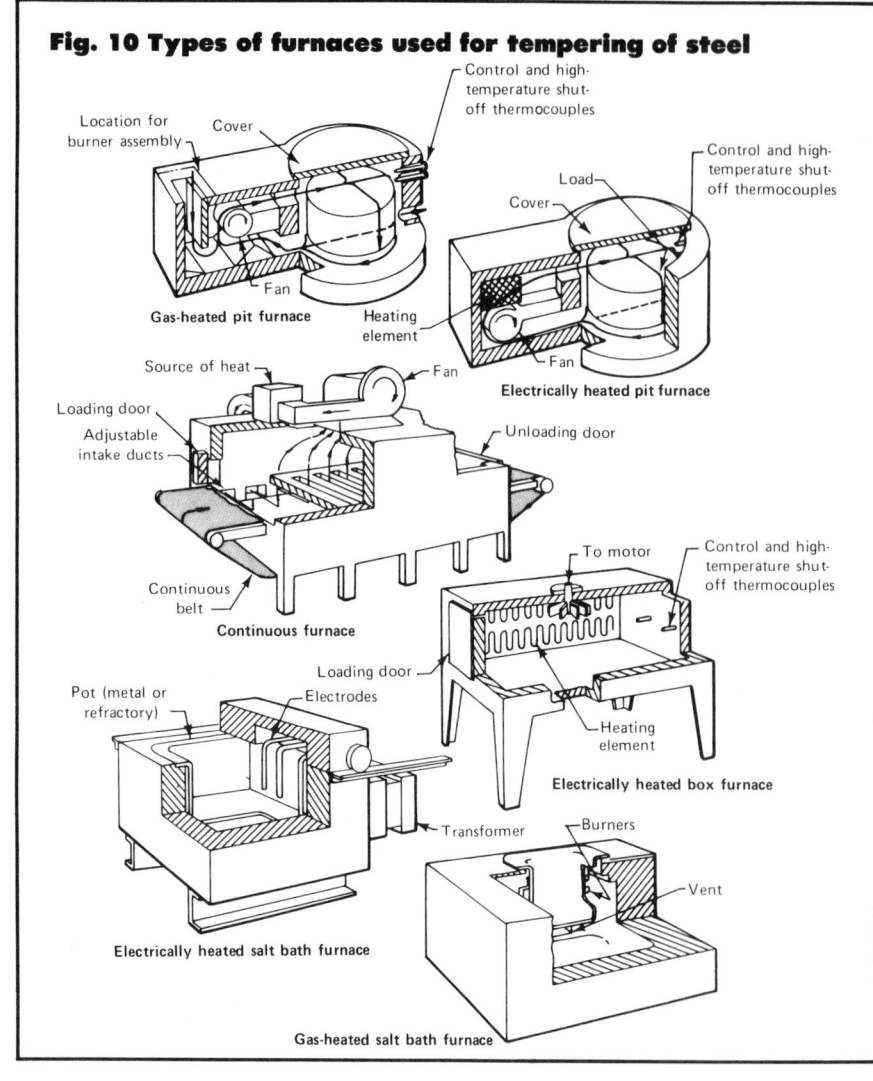

Fig. 10 Types of furnaces used for tempering of steel

Gas-heated pit furnace

Electrically heated pit furnace

Continuous furnace

Electrically heated box furnace

Electrically heated salt bath furnace

Gas-heated salt bath furnace

Table 3 Compositions and operating temperatures for salt baths used in tempering
Source: Military Specification MIL-S-10699A (Ordnance)

Class	Sodium nitrite	Sodium nitrate	Potassium nitrate	Sodium carbonate	Sodium chloride	Potassium chloride	Barium chloride	Calcium chloride	Operating temperature °C	°F	Fuming temperature °C	°F
1	37-50	0-10	50-60	...	...	...	...	...	165-595	325-1100	635	1175
2	...	45-57	45-57	...	...	...	...	...	290-595	550-1100	650	1200
3	...	...	...	45-55	...	45-55	...	...	620-925	1150-1700	938	1720
4	...	...	...	...	15-25	20-32	50-60	...	595-900	1100-1650	940	1725
4A	...	...	...	...	10-15	25-30	40-45	15-20	550-760	1025-1400	790	1450

Fig. 11 Variations in room-temperature hardness of three carbon steels after production tempering

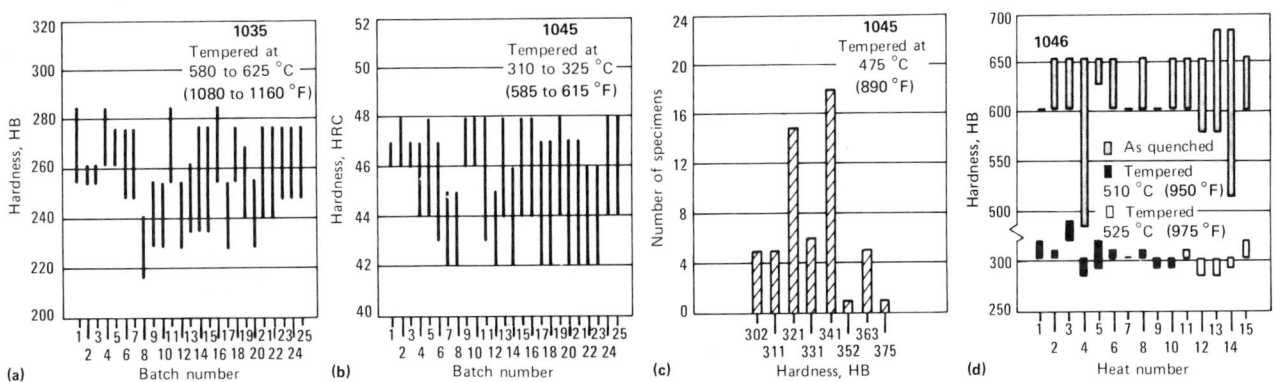

(a) Automotive steering-arm forgings made of fine-grain 1035 steel. Section thickness varied from 16 to 29 mm (⅝ to 1⅛ in.). Forgings were austenitized at 825 °C (1520 °F) in oil-fired pusher conveyor furnace, held 45 min, quenched in water at 21 °C (70 °F), and tempered 45 min at 580 to 625 °C (1080 to 1160 °F) in oil-fired link-belt furnace to required hardness range of 217 to 285 HB. Hardness was checked hourly with a 5% sample; readings were taken on polished flash line of 29-mm (1⅛-in.) section. Survey of furnace revealed temperature variation at 605 °C (1120 °F) of +8, −4 °C (+15, −7 °F). Data represent forgings from four mill heats of steel and cover a six-week period. (b) Woodworking cutting tools forged from 1045 steel. Section of cutting lip was hardened locally by gas burners that heated the steel to 815 °C (1500 °F). Tools were oil quenched and tempered at 305 to 325 °C (585 to 615 °F) for 10 min in electrically heated recirculating-air furnace to a desired hardness range of 42 to 48 HRC. Data were recorded during a six-month period and represent forgings from 12 mill heats. (c) Plate sections, 19 to 22 mm (¾ to ⅞ in.) thick, of 1045 steel were water quenched to a hardness range of 534 to 653 HB and tempered 1 h at 475 °C (890 °F) in continuous roller-hearth furnaces. Data represent a two-month production period. (d) Forged 1046 steel heated to 830 °C (1525 °F), quenched in caustic, and tempered for 1 h to a hardness range of 285 to 321 HB. Forgings were heated in a continuous belt-type furnace and individually dump quenched in agitated caustic. Forgings weighed 9 to 11 kg (20 to 24 lb) each; maximum section, 38 mm (1½ in.).

Fig. 12 Variations in room-temperature hardness of two alloy steels after production tempering

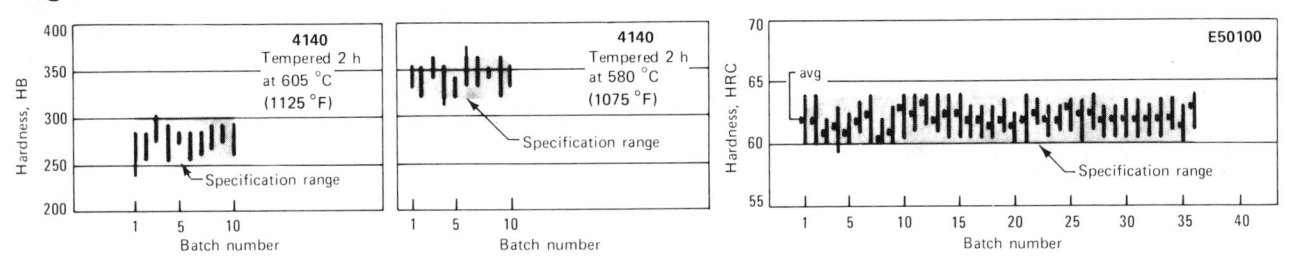

(a) Valve bonnets, 75 mm (3 in.) in diameter, made of 4140 steel from one mill heat. Parts were heated at 870 °C (1600 °F), oil quenched, and tempered at 605 °C (1125 °F) to a hardness specification of 255 to 302 HB. (b) Valve segments made of 4140 steel from one mill heat. Section thickness varied from 13 to 25 mm (½ to 1 in.). Parts were heated at 870 °C (1600 °F), oil quenched, and tempered at 580 °C (1075 °F) to a hardness of 321 to 363 HB. (c) Needle rollers made of E50100 steel wire varying in diameter from 2.2 to 2.6 mm (0.086 to 0.103 in.). Data represent 36 batches of parts (nine mill heats of steel) tempered to a hardness specification of 60 to 64 HRC. Each batch represents a minimum furnace load of 270 kg (600 lb).

points (about 560 °C, or 1040 °F) severely restrict their useful temperature range. Also, they are decarburizing to steels at temperatures above about 700 °C (1300 °F).

Class 4 salts, which are all-chloride neutral salts, are quite stable. They seldom require rectification but are restricted to temperatures above 590 °C (1100 °F).

Class 4A salts are close relatives of class 4 salts but contain calcium chloride, which lowers the minimum working temperature to 550 °C (1025 °F). The upper limit for these salts is more restricted than that for class 4 salts.

All of the salts for these baths are commercially available. The reader is referred to the military specification cited above for the chemical and other control procedures applicable to the various bath compositions. For further information, see the articles on martempering, on austempering and on heat treating of tool steels.

SAFETY PRECAUTION: *Introduction of cyanide salts or other reducing agents into nitrite tempering baths will cause violent explosions.*

Fig. 13 Variations in room-temperature hardness of two cast steels after production tempering

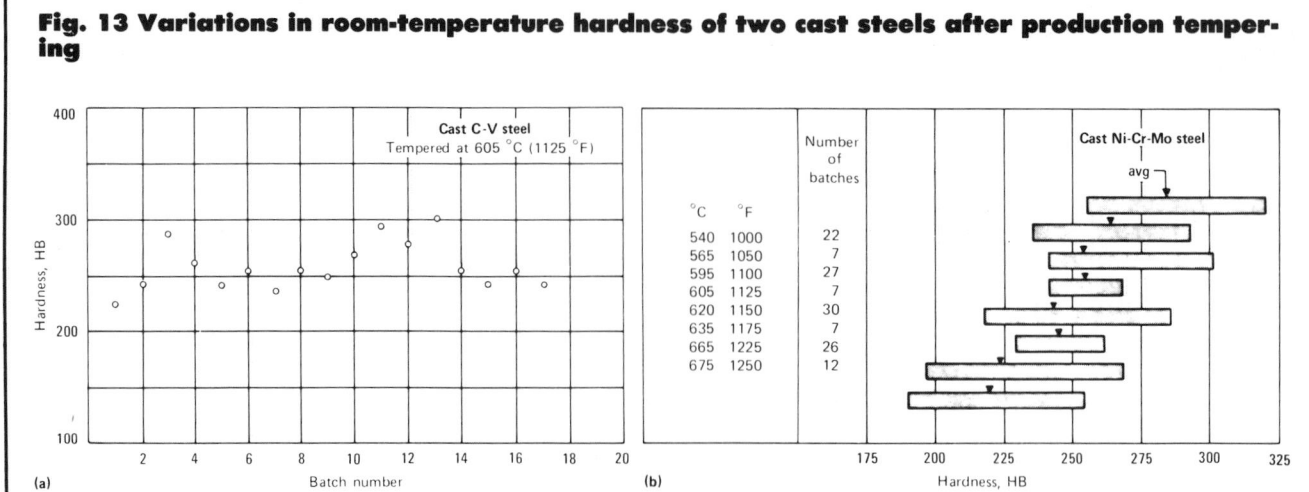

(a) Cast C-V steel containing 0.25 to 0.30 C, 0.70 to 0.80 Mn, 0.08 to 0.12 V and 0.40 to 0.60 Si, 25 mm (1 in.) thick, water quenched from 870 °C (1600 °F) and then tempered at 605 °C (1125 °F). (b) Cast Ni-Cr-Mo steel containing 0.28 to 0.33 C, 0.70 to 0.90 Mn, 0.40 to 0.70 Ni, 0.40 to 0.60 Cr, 0.15 to 0.25 Mo and 0.40 to 0.60 Si. Spread of hardness readings for various tempering temperatures

Oil bath equipment for tempering may be similar in design to that used for salt baths, or a steel tank over large hotplate-type burners will serve satisfactorily. Submerged electric heating elements may also be employed. Stirring is essential for temperature uniformity and satisfactory oil life. Simple, oven-type temperature controls may be employed, but localized overheating must be avoided to prevent fire and rapid decomposition of the oil. A standard thermometer of the proper range may be used to check the temperature of the oil.

Low-temperature tempering in a hot oil bath is a simple and inexpensive method that is especially useful for holding work at temperature for long periods of time. The practical temperature limit is about 120 °C (250 °F) without special ventilation and fire-protection equipment and about 250 °C (480 °F) with such precautions, which may include extremely efficient ventilators or inert-gas blanketing systems. When a tempering temperature above 205 °C (400 °F) is required, a salt bath is usually preferable to an oil bath.

Oils for tempering must resist oxidation and have a flash point well above the operating temperature. The most commonly used oils are high-flash-point paraffinic oils with antioxidant additives. Oils used for martempering (see the article on martempering of steels) are also satisfactory for tempering.

Molten metal baths for tempering have largely been replaced by salt baths. When employed, commercially pure lead, which melts at about 327 °C (620 °F), has proved to be the most generally suitable of all metals and alloys. For special applications, however, lead-base alloys having lower melting points are used.

Lead oxidizes readily. Although lead itself will not adhere to clean steel, adherence of lead oxide to steel surfaces is a problem, especially at higher tempering temperatures. Within the range of temperatures usually employed, a film of molten salt will protect the surface of the lead bath, and the work will be easily cleaned. Above 480 °C (900 °F), granulated carbonaceous material, such as charcoal, may be used as a protective cover.

Because of its high thermal conductivity, lead is useful for rapid local heating and selective tempering. A typical application is the tempering of a ball joint. The part is carburized and quenched to a minimum case hardness of 59 HRC and a core hardness of 30 to 40 HRC. The thread and taper then are tempered in lead to produce a maximum case hardness of 40 HRC.

Because of the high specific gravity of lead, parts tempered in molten lead will float unless held down by fixtures. All parts and fixtures must be dry when immersed in the bath, to prevent formation of steam in, and resultant

violent expulsion of, the molten lead. Precautions also must be taken to protect personnel from industrial lead poisoning; hoods and ventilating equipment are required.

Temperature Control. For either gas or electric heat, properly adjusted potentiometers of the on-off type will control the tempering temperature within ±6 °C (±10 °F) at the thermocouple location. With proportioning controls, these instruments can maintain temperatures within ±1 °C (±2 °F) at the thermocouple location.

Selection of Tempering Equipment

Selection of equipment for tempering is based principally on (a) temperature requirements and (b) quantity and similarity of the work to be treated. Temperature requirements are dictated by prior heat treatment and by the properties to be developed by tempering.

Process Control

Variation in hardness after tempering is most frequently the result of differences in prior microstructure, as discussed previously. When prior microstructure is the same, control of temperature is the most important factor in control of the tempering process.

In general, control of tempering temperature to within ±3 °C (±5 °F) is

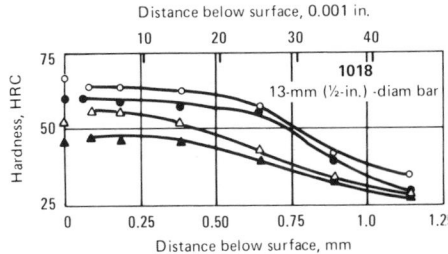

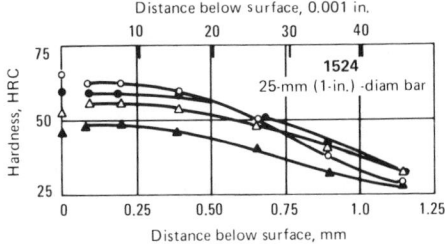

Fig. 14 Effect of tempering on room-temperature hardness of carburized cases

Carburized for 4.5 h at 925 °C (1700 °F), oil quenched, and tempered: ○ as quenched; ● tempered at 205 °C (400 °F); △ tempered at 315 °C (600 °F); ▲ tempered at 425 °C (800 °F)

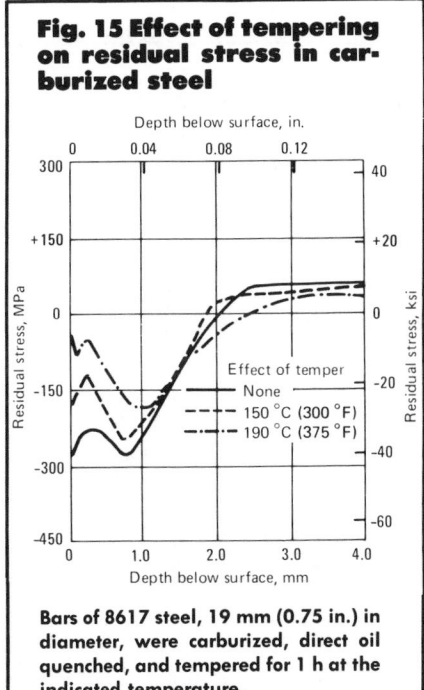

Fig. 15 Effect of tempering on residual stress in carburized steel

Bars of 8617 steel, 19 mm (0.75 in.) in diameter, were carburized, direct oil quenched, and tempered for 1 h at the indicated temperature.

adequate and is within the practical limits of most furnace and molten-bath equipment. Temperature variations are seldom permitted to exceed ±6 °C (±10 °F) unless mechanical-property requirements are correspondingly broad.

Examples of the range of variation in hardness obtained after tempering for a variety of wrought and cast steel parts are presented in Fig. 11, 12 and 13.

Tempering of Special Microstructures

Carburized Components. Tempering of carburized parts represents a special instance in that the combination of toughness, strength, hardness, residual stress and retained austenite all must be considered when selecting tempering time and temperature. Core properties cannot always be controlled by tempering when trying to achieve maximum case properties, and a favorable compressive residual stress pattern may be retained only at the expense of over-all toughness.

The effect of tempering on the hardness of carburized cases is shown in Fig. 14, and the influence of tempering on residual stress is illustrated in Fig. 15. The transformation of retained austenite and the resultant change in the relative volume of case and core are primarily responsible for the change in residual stress as a function of temperature. The effect of retained austenite on the performance of components is still a controversial issue. Reduction of retained austenite is apparently desirable for resistance to grinding abuse and to provide dimensional stability, but retained austenite appears to be beneficial for contact-fatigue durability.

Nonmartensitic Structures. Tempering of microstructures other than martensite and retained austenite also represents special applications of tempering. Reactions of structures containing substantial amounts of lower bainite are relatively similar to that of martensite in terms of the phenomena associated with carbide growth and coalescence.

Upper bainite and fine pearlite formed by controlled or relatively slow cooling simply respond by carbide growth and eventual ferrite recrystallization. The softening associated with tempering in such instances is shown in Fig. 16.

Impact properties of normalized and tempered and hardened and tempered structures at nearly equal hardness are shown in Fig. 17.

Induction Tempering

Extensive production experience has demonstrated the commercial success of induction heating for tempering in several types of applications. This subject is discussed in the article on induction hardening and tempering.

Special Tempering Procedures

Selective Tempering. Selective or localized tempering is applied to parts in which adjacent areas must have significantly different hardnesses. Chisels, punches, the upset ends of cold formed rivets, and the threaded portions of carburized parts are typical examples. Localized tempering is also employed in preheating and postheating of weld areas when a lowering of the hardness in the heat-affected zone is desired.

Selective tempering entails heating a restricted area to the required tempering temperature without heating the remainder of the part to this tem-

Fig. 16 Effect of prior microstructure on room-temperature hardness after tempering

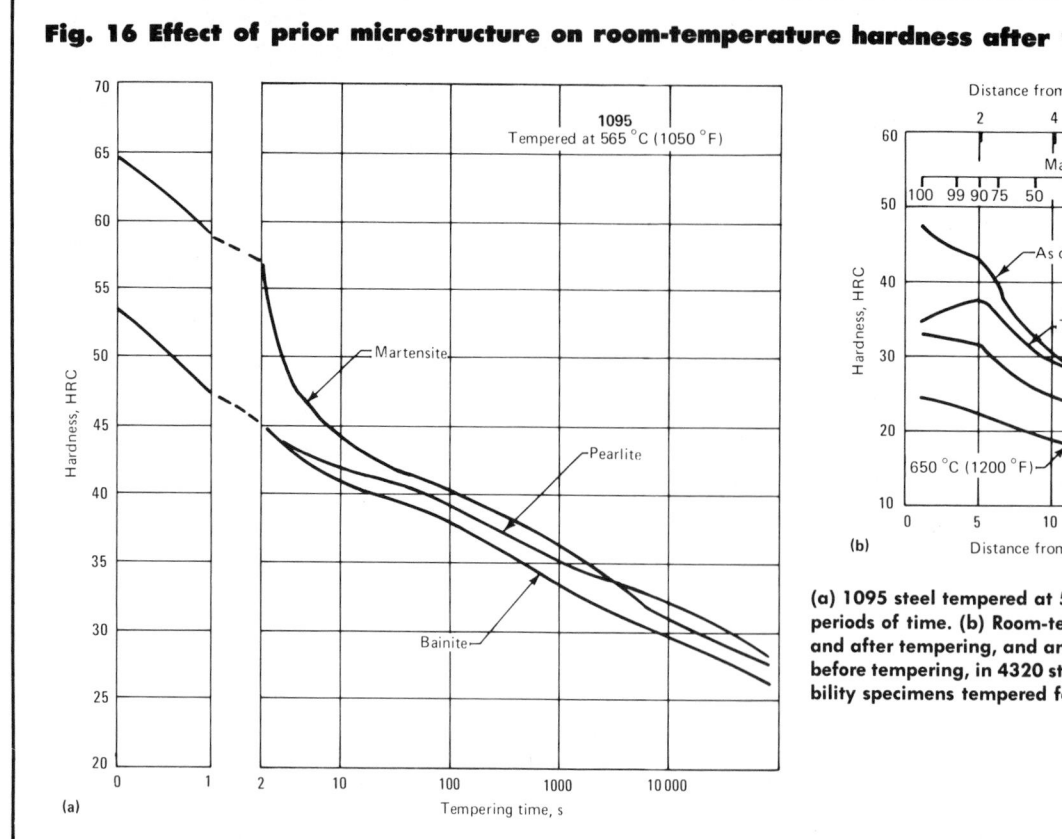

(a) 1095 steel tempered at 565 °C (1050 °F) for various periods of time. (b) Room-temperature hardness before and after tempering, and amount of martensite present before tempering, in 4320 steel end-quenched hardenability specimens tempered for 2 h

perature. Induction heating coils, special flame heads, and immersion in lead or salt baths are employed to achieve this selective heating. Selective tempering also is done by use of defocused lasers or electron beam equipment. Induction and flame techniques generally are used in high-volume production applications and are easiest to control. Deeper penetration is achieved with low-frequency (3 to 10 kHz) induction heating and salt immersion than with the other techniques.

To obtain rapid heating for selective tempering by immersion in salt or lead baths, it is usually necessary that the bath temperature be considerably higher than the desired tempering temperature. Consequently, the immersion time becomes the controlling factor in obtaining the desired tempering results. Lead, because of its higher rate of heat transfer, is more effective than salt.

Other selection factors, such as ease of fixturing, part configuration and cost, also influence the choice of equipment for selective tempering. In induction tempering, the same heating system can be used for both hardening and tempering.

Example 1. An upholsterer's tack hammer, forged from 1086 steel, is hardened on all surfaces to 53 to 60 HRC and then is tempered in salt at 190 °C (375 °F). This treatment provides the desired combination of hardness, toughness and magnetic properties for the horseshoe-magnet end of the hammer. However, the striking head must be selectively tempered in salt at 260 °C (500 °F) to produce the working hardness of 50 to 55 HRC.

Example 2. A chain pipe wrench handle forging, made of 4053 steel, is fully hardened and tempered for 1 h at 355 °C (675 °F) to produce an over-all hardness of 47 to 52 HRC. This is an ideal hardness for the wrench teeth but does not provide sufficient toughness for the I-beam section of the handle. This section is tempered further, selectively, by induction heating for 1 min at 480 °C (900 °F) to produce a hardness of 40 to 48 HRC.

The effect on hardness of the very short tempering times normally associated with selective heating is compared with the effect of furnace tempering in Fig. 18.

Multiple tempering is principally used (a) to relieve the quenching and straightening stresses in irregularly shaped carbon and alloy steel parts and thereby lessen distortion, (b) to eliminate retained austenite and improve dimensional stability in such parts as bearing components and gage blocks, and (c) to improve yield and impact strengths without decreasing hardness. The following examples illustrate these principal applications of the process.

Example 3. A six-throw, seven-bearing, counterweighted diesel-engine crankshaft weighing 80 kg (175 lb) was distorted in rough machining to such a degree that cold straightening was required. The straightening operation induced additional stresses, which caused severe distortion in final machining. The problem was solved by first tempering the 1046 steel shaft at 455 °C (850 °F) to a hardness of about 321 HB, which allowed hot straightening. The shaft was then retempered at 480 to 540 °C (900 to 1000 °F), depending on the composition of the particular heat, to produce a hardness of 269 to 302 HB and relieve residual stresses.

Example 4. In manufacture of a gage block of W1 tool steel (final hard-

Fig. 17 Effect of microstructure on notch toughness

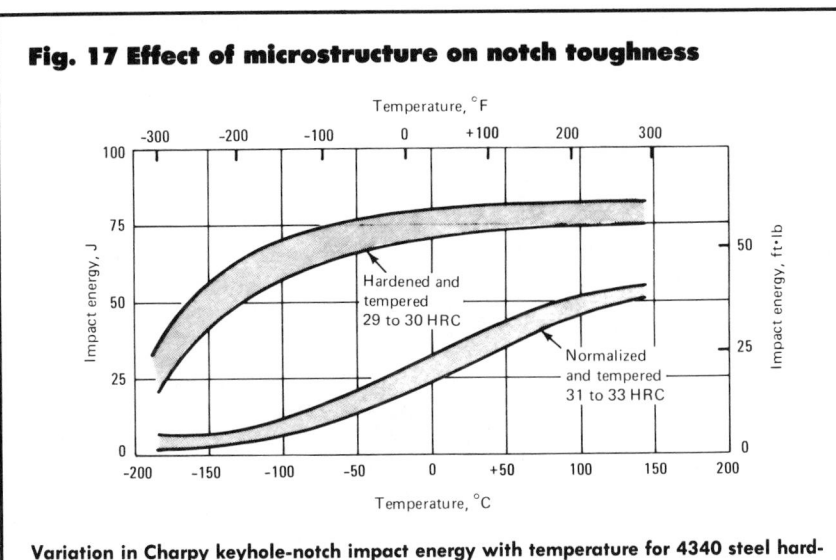

Variation in Charpy keyhole-notch impact energy with temperature for 4340 steel hardened and tempered to 29 to 30 HRC or normalized and tempered to 31 to 33 HRC

Fig. 18 Variations of room-temperature hardness with tempering temperature for furnace and induction heating

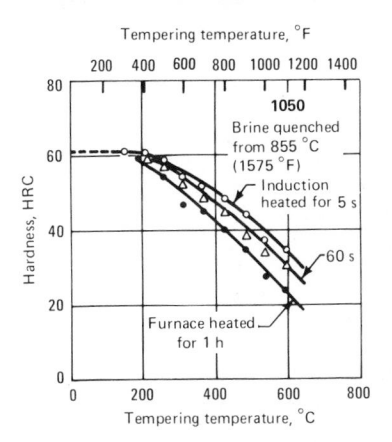

Fig. 19 Effect of temper embrittlement on notch toughness

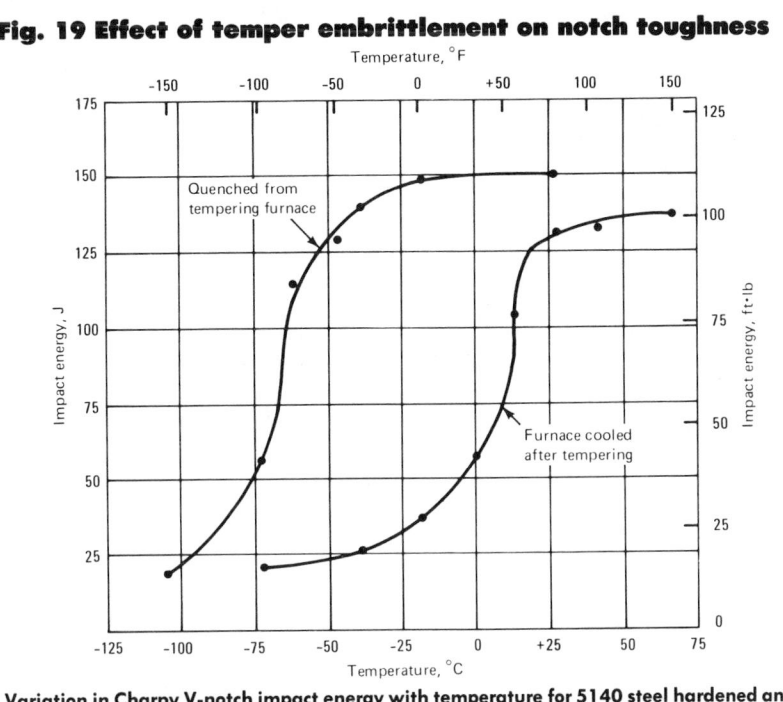

Variation in Charpy V-notch impact energy with temperature for 5140 steel hardened and tempered at 620 °C (1150 °F); one series of specimens was quenched from the tempering temperature, and the other was furnace cooled. Slow cooling of susceptible steels causes temper embrittlement (Ref 4)

ness, 65 to 66 HRC), the block is quench hardened after rough machining. It is then subjected to three consecutive cycles each consisting of cold treatment at −100 °C (−145 °F) for 1 h followed by tempering at 70 °C (160 °F) for 1 h. This minimizes the retained austenite and enhances dimensional stability before the block is finish ground.

Use of Fixtures

Many high-strength steel components having tensile strength greater than 1720 MPa (250 ksi) are finish machined before final heat treatment. To minimize distortion and satisfy stringent dimensional requirements, some of these components, such as cylinders, pressure vessels and thin parts, are held in fixtures during both hardening and tempering or during tempering only.

External rings, internal mandrels, jacks, screws, weights, wedges, dies and other mechanical devices are used to assist in dimensional correction.

Example 5. A welded pressure vessel made of 4135 steel, 380 mm (15 in.) in outside diameter by 1.8 m (6 ft) long by 3.2 mm (0.125 in.) thick, was found to be 1.3 to 3.8 mm (0.050 to 0.150 in.) out-of-round over its entire length when measured in the as-quenched condition. A tempering fixture consisting of a series of steel rings 125 mm (5 in.) wide reduced maximum out-of-roundness to 1.3 mm (0.050 in.) after tempering at 455 °C (850 °F) for 2½ h.

Steels that are susceptible to cracking at room temperature after hardening are given a low-temperature tempering treatment (a "snap draw") immediately after hardening and prior to final tempering in fixtures.

Cracking in Processing

Because of their carbon or alloy contents, some steels are likely to crack if they are permitted to cool all the way to room temperature during or immediately following the quenching operation. This likelihood is caused by generation of high tensile residual stresses during quenching, due to thermal gradients, abrupt changes in section thickness, decarburization or other hardenability gradients. Accordingly, for carbon steels containing more than 0.4% C and alloy steels containing more than 0.35% C, it is recommended that the parts be transferred to tempering furnaces before they cool below 100 to 150 °C (212 to 300 °F). Alternately, many heat treating operations use quenching oil for the tempering operation (martempering) or to avoid cooling below 125 °C (255 °F). Steels that are known to be sensitive to this type of cracking include 1060, 1090, 1340, 4063, 4150, 4340, 52100, 6150, 8650 and 9850.

Other carbon and alloy steels are generally less sensitive to this type of delayed quench cracking but may crack as a result of part configuration or surface defects. These steels include 1040, 1050, 1137, 1144, 4047, 4132, 4640, 8632, 8740 and 9840. Some steels, such as 1020, 1038, 1132, 4130, 5130 and 8630, are not sensitive.

Before being tempered, parts should be quenched to room temperature to ensure transformation of most of the austenite to martensite and to achieve maximum as-quenched hardness. Austenite retained in low-alloy steels will, upon heating for tempering, transform to an intermediate structure, reducing over-all hardness. However, in medium- to high-alloy steels containing austenite-stabilizing elements (nickel, for example), retained austenite may transform to martensite upon

cooling from tempering, and thus such steels may require additional tempering (double tempering) for relief of transformation stresses.

Temper Embrittlement

When carbon or low alloy steels are slow cooled from tempering above 575 °C (1065 °F), or are tempered for extended times between 375 and 575 °C (705 and 1065 °F), a loss in toughness occurs that manifests itself in reduced notched-bar impact strength compared to that resulting from normal tempering cycles and relatively fast cooling rates (see Fig. 19). The cause of temper embrittlement is believed to be precipitation of compounds containing trace elements such as tin, arsenic, antimony and phosphorus along with chromium and/or manganese. The intergranular nature of the fracture suggests that the embrittlement occurs at the prior austenite grain boundaries. Although manganese and chromium cannot be restricted, reduction of the other elements and quenching from above 575 °C are the most effective remedies for this type of embrittlement. It also has been found that small amounts of molybdenum retard embrittlement.

A second form of embrittlement, called "blue brittleness", occurs when steels are tempered at about 300 °C (570 °F) after being hardened. This is illustrated in Fig. 19. Generally, for steels containing potent carbide formers such as chromium, tempering between 200 and 370 °C (390 and 700 °F) should be avoided.

Hydrogen Embrittlement

Selection of tempering temperature, and the resultant hardness or plasticity, must include consideration of the potential problem of hydrogen embrittlement should the part be exposed to hydrogen through electroplating, phosphating or other means, or if environmental conditions cause cathodic absorption of hydrogen during service.

Generally, the restricted notch ductility of steels with hardness above 40 HRC presents ideal conditions for development of stress concentrations in parts containing notches or defects that would, in the presence of relatively low hydrogen concentrations, lead to failure of parts at stresses far below the nominal tensile strength of the material. Thus, tempering should be carried out to achieve hardness below 40 HRC if the part will be subjected to relatively high stresses and probable exposure to hydrogen.

REFERENCES

1. *Alloying Elements in Steel,* by E. C. Bain and H. W. Paxton: American Society for Metals, 1966, p 185, 197
2. Time-Temperature Relations in Tempering Steels, by J. H. Holloman and L. D. Jaffe: *Transactions of AIME,* Vol 162, 1945, p 223-249
3. R. A. Grange and R. W. Baughman: *Transactions of ASM,* Vol 48, 1956, p 165-167
4. The Effect of Quench-Aging on the Notch Sensitivity of Steel, by J. R. Low, Jr.: *Welding Research Council Research Reports,* Vol 17, 1952, p 253s-256s

Martempering of Steel

By the ASM Committee on
Martempering of Steel*

MARTEMPERING is a term used to describe an elevated-temperature quenching procedure aimed at reducing cracks, distortion or residual stresses. It is not a tempering procedure, as the name implies, and is more properly termed "marquenching".

Martempering of steel (and of cast iron) consists of: (a) quenching from the austenitizing temperature into a hot fluid medium (hot oil, molten salt, molten metal or a fluidized particle bed) at a temperature usually above the martensitic range (M_s point); (b) holding in the quenching medium until the temperature throughout the steel is substantially uniform; and then (c) cooling (usually in air) at a moderate rate, to prevent large differences in temperature between the outside and the center of the section. Formation of martensite occurs fairly uniformly throughout the workpiece during cooling to room temperature, thereby avoiding formation of excessive amounts of residual stress. The microstructure after martempering is essentially primary martensite, which is untempered and brittle for most applications. After being air cooled to room temperature, martempered parts are tempered in the same manner as if they had been conventionally quenched.

Figure 1 shows the significant difference between conventional quenching (a) and martempering (b). The advantage of martempering lies in the reduced thermal gradient between surface and center as the part is quenched to the isothermal temperature and then is air cooled to room temperature. Residual stresses developed during martempering are lower than those developed during conventional quenching, because the greatest thermal variations occur while the steel is in the relatively plastic austenitic condition and because final transformation and thermal changes occur throughout the part at approximately the same time. Martempering also reduces or eliminates susceptibility to cracking.

Although martempering is used primarily to minimize distortion, eliminate cracking and minimize residual stresses, it also greatly reduces the problems of pollution and fire hazard provided that nitrate-nitrite salts are employed rather than martempering oils. This is especially true where nitrate-nitrite salts are recovered from wash waters with systems that provide essentially no discharge of salts into drains. Any steel part or grade of steel responding to oil quenching can be martempered to provide similar physical properties. The quenching severity of molten salt is greatly enhanced by agitation and by water additions to the nitrate-salt bath. Both techniques are particularly beneficial in heat treating of carbon steels having limited hardenability. Table 1 compares the properties obtained in 1095 steel by martempering and tempering with those obtained by conventional quenching and tempering.

In many instances, martempering eliminates the need for quenching fixtures that would be required for minimizing distortion during conventional quenching, and thus reduces the cost of tooling and handling. However, changing from conventional quenching to

Table 1 Mechanical properties of 1095 steel heat treated by two methods

Specimen number	Heat treatment	Hardness, HRC	Impact energy J	Impact energy ft·lb	Elongation(a), %
1	Water quench and temper	53.0	16	12	0
2	Water quench and temper	52.5	19	14	0
3	Martemper and temper	53.0	38	28	0
4	Martemper and temper	52.8	33	24	0

(a) In 25 mm or 1 in.

*Quentin D. Mehrkam, Chairman, Senior Vice President, Ajax Electric Co.; Robert W. Foreman, Director, Research and Development, Park Chemical Co.; W. James Laird, Jr.,Vice President, Marketing; Research & Development, Upton Industries, Inc.; Jerrold Meyer, Division Metallurgist, Roller Bearings Division, SKF Industries, Inc.; Daniel J. Olah, Materials Engineering Manager, Borg & Beck Division, Borg-Warner Corp.; Romeo L. Suffredini, Chief Metallurgist, Winchester Arms Division, Olin Corp.

Fig. 1 Time-temperature transformation diagrams with superimposed cooling curves showing quenching and tempering

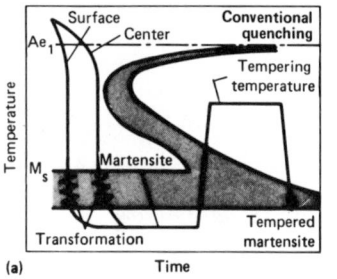 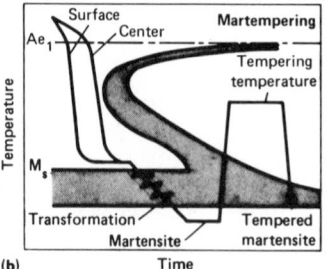 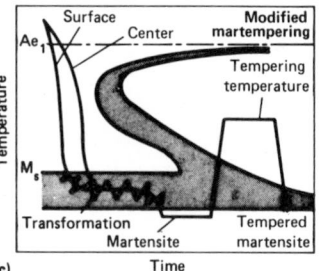

(a) Conventional process. (b) Martempering. (c) Modified martempering

martempering may require that dimensional variations in individual parts be studied before preheat-treatment dimensions are established.

Modified martempering differs from "standard" martempering only in that the temperature of the quenching bath is below the M_s point (Fig. 1c). The lower temperature increases the severity of quenching. This is important for steels of lower hardenability that require faster cooling in order to harden to sufficient depth, or when the M_s is high and some bainite is detrimental to the finished part. Thus, modified martempering is applicable to a greater range of steel compositions than is the standard process.

Although hot oil is invariably the quenchant employed for modified martempering at 175 °C (350 °F) and lower, molten nitrate-nitrite salts (with water addition and agitation) are effective at temperatures as low as 175 °C. Due to their higher heat-transfer coefficients, molten salts offer some metallurgical and operational advantages.

Martempering Media

Molten salt and hot oil are both widely used for martempering. Several factors must be considered when choosing between salt and oil. Operating temperature is the most common deciding factor. Oils are widely used for martempering at up to 205 °C (400 °F), and sometimes at temperatures as high as 230 °C (450 °F). Molten salt is used for martempering in the range from 160 to 400 °C (320 to 750 °F).

Composition and Cooling Power of Salt. A salt commonly used for martempering is composed of 50 to 60% potassium nitrate, 37 to 50% sodium nitrite, and 0 to 10% sodium nitrate. It

Fig. 2 Cooling curves for 1045 steel cylinders quenched in salt, water and oil

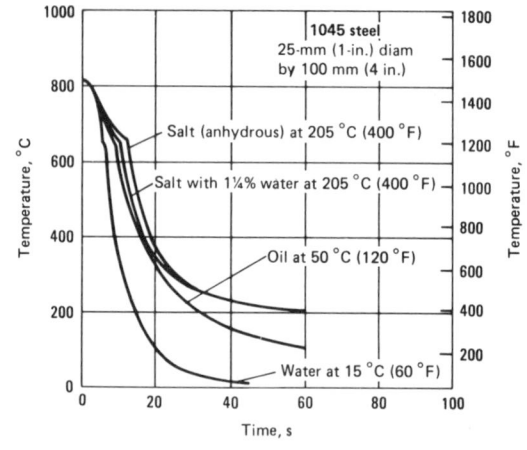

Thermocouples were located at centers of specimens.

melts at about 140 °C (280 °F) and may be used within a working range of 165 to 540 °C (325 to 1000 °F). Higher-melting-point (less costly) salts can be used for higher operating temperatures. These salts are comprised of 40 to 50% potassium nitrate, 0 to 30% sodium nitrite and 20 to 60% sodium nitrate.

The cooling power of agitated salt at 205 °C (400 °F) is about the same as that of agitated oil in conventional oil quenching. Addition of water to salt increases its cooling power, as indicated by the cooling curves for 1045 steel in Fig. 2 and by the hardness values for 1046 steel in Fig. 3. The cooling power of salt is compared with the cooling power of water, and of three types of oil, in Fig. 3.

Advantages of salt in comparison with oil for martempering are as follows:

- The viscosity of salt changes only slightly over a wide temperature range.
- Salt retains chemical stability, so that replenishment usually is required only to replace dragout loss. This is not always the case, however. In an installation for high-volume martempering of cylinder liners, for instance, breakdown of the salt did occur, forming a carbonate. This occurred on quenching from an endothermic atmosphere into an open-top salt quench bath at 245 °C (475 °F). The same operation conducted with a nitrogen atmosphere in the heating furnace does not produce such breakdown.
- Salt has a wide operating temperature range.
- Salt is easily washed from the work with plain water.

Fig. 3 Effects of quenchant and agitation on hardness of 1046 steel

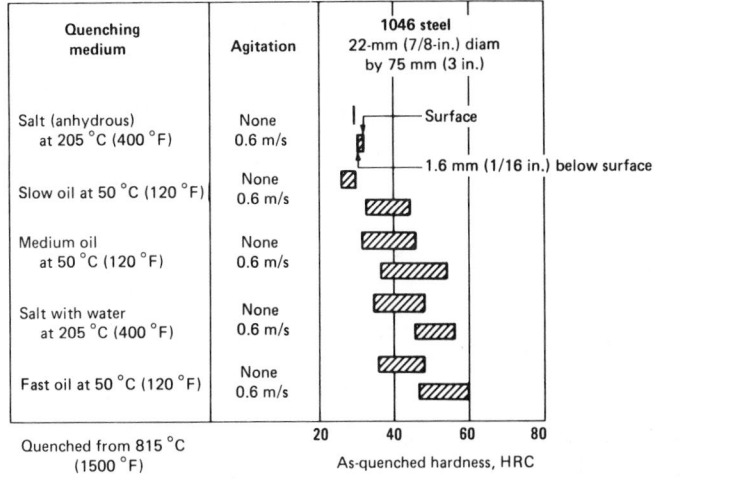

Quenching medium	Agitation	1046 steel 22-mm (7/8-in.) diam by 75 mm (3 in.)
Salt (anhydrous) at 205 °C (400 °F)	None 0.6 m/s	
Slow oil at 50 °C (120 °F)	None 0.6 m/s	
Medium oil at 50 °C (120 °F)	None 0.6 m/s	
Salt with water at 205 °C (400 °F)	None 0.6 m/s	
Fast oil at 50 °C (120 °F)	None 0.6 m/s	

Quenched from 815 °C (1500 °F)

As-quenched hardness, HRC

Table 2 Physical properties of two oils used for martempering of steel

Property	Value, for oil with operating temperature of:	
	95 to 150 °C (200 to 300 °F)(a)	150 to 230 °C (300 to 450 °F)
Flash point (min), °C (°F)	210 (410)	275 (525)
Fire point (min), °C (°F)	245 (470)	310 (595)
Viscosity, SUS, at:		
38 °C (100 °F)	235 to 575	...
100 °C (210 °F)	50.5 to 51	118 to 122
150 °C (300 °F)	36.5 to 37.5	51 to 52
175 °C (350 °F)	...	42 to 43
205 °C (400 °F)	...	38 to 39
230 °C (450 °F)	...	35 to 36
Viscosity index (min)	95	95
Acid number	0.00	0.00
Fatty-oil content	None	None
Carbon residue	0.05	0.45
Color	Optional	Optional

(a) Temperature range for modified martempering

- Less time is required for workpieces to attain temperature equalization in salt.
- No unusual disposal problems attend the rinsing operation.
- Salt is relatively easy to handle in powder form and easy to clean up if spilled; however, it is necessary to keep salts separated.

Disadvantages of salt in comparison with oil include the following:

- The minimum operating temperature of salt is 160 °C (320 °F).
- Quenching from cyanide-base carburizing salts is hazardous, because of

possible explosion.
- Explosion and splatter of high-temperature salt can occur if wet or oily parts are immersed.
- Potentially explosive reactions also occur with sooty atmospheres in atmosphere furnaces connected to martempering salt quenches.
- Quench salt can be contaminated by high-temperature neutral salt used for heating. Sludging operations are required to maintain quenching severity.

Oils for Martempering. Physical properties of two oils commonly used for martempering are listed in Table 2.

Such oils are compounded especially for martempering and, in comparison with conventional quenching oils, offer higher rates of cooling during the initial stage of quenching.

Quenching oil requires special handling when used in the temperature range from 95 to 230 °C (200 to 450 °F). To prolong its life, the oil should be maintained under a protective atmosphere (reducing or neutral). Deterioration is accelerated when oil is exposed to air at elevated temperatures. For example, for every 10 °C above 60 °C (every 18 °F above 140 °F), the rate of oxidation is approximately doubled. This causes formation of acid and sludge, which may affect both hardness and color of workpieces.

Example 1. A manufacturer of transmission components and axles martempers carburized parts at a rate of 115 kg/h (250 lb/h) in a 7550-l (2000-gal) oil tank that is completely covered by a vestibule hood and that is located directly beneath the discharge door of the continuous carburizing furnace. The vestibule contains a protective atmosphere consisting of carburizing gases that emanate from the furnace. The vestibule is equipped with an elevator that lowers the tray of parts into the martempering oil.

The ambient temperature in the vestibule immediately above the oil is 89 to 92 °C (193 to 197 °F) when the furnace discharge door is closed. The temperature of the oil is controlled at 150 °C (300 °F). However, during high-production quenching, the temperature of the martempering oil will rise to as high as 165 °C (325 °F). The atmosphere in the vestibule protects the carburized parts and the martempering oil from oxidation. To avoid possible oil fires, the furnace discharge door is closed before the workload is lowered into the martempering oil.

The martempering oil in this tank has not been replaced in several years of continuous operation (24 h per day, 7 days per week); make-up oil is added to the tank at a rate of about 755 l (200 gal) per month.

Bypass or continuous types of filter units with suitable filtering media (clay, cellulose cartridge or waste cloth), or with centrifugal filtering, help prolong oil life and maintain clean work. Filtering units are effective and relatively inexpensive. Oils should be circulated at a rate no lower than 0.9 m/s (180 ft/min) to break up excessive vapor that is formed during quenching.

Continuous operation of oils above 205 °C (400 °F) causes excessive breakdown. Once every one to six months, the oil should be subjected to complete physical and chemical testing to determine its condition. Such testing usually includes determinations of flash point, viscosity, degree of oxidation, contamination and cooling power.

Advantages of oil in comparison with salt are as follows:

* Oil can be used at lower temperatures.
* Oil is easier to handle at room temperature.
* Less dragout loss occurs with oil.
* Oil is compatible with all austenitizing salts.

Disadvantages of oil include the following:

* Oil is limited to a maximum operating temperature of 230 °C (450 °F).
* Oil deteriorates with use, thus requiring closer control.
* When martempered in oil, workpieces require more time to attain temperature equalization.
* Oil, whether hot or cold, is a fire hazard.
* Soap or an emulsifier is necessary to wash off the oil. Washers must be drained and refilled periodically. Oily wastes create disposal problems.

Safety Precautions

In operation of a martempering installation (salt or oil), the precautions appropriate for operating any other hot liquid bath are applicable. The following operator safety precautions are recommended:

* Operators of equipment should be thoroughly familiar with equipment manufacturer's instructions and with company safety recommendations before attempting to operate or service the equipment.
* Operators should wear gloves, face shields and protective clothing, as required. Safety equipment contaminated with oil or nitrate-nitrite salts should be discarded.

Nitrate-Nitrite Salts. Several precautions should be taken in operating nitrate-nitrite salt baths:

* All racks, fixtures and cleaning tools must be thoroughly washed clean of salts (hot water preferred), and thoroughly dried, before being reused;

otherwise, the presence of salt could contaminate the austenitizing furnace, and residual water could cause spattering of molten salt.

* All salt bath furnaces should be labeled so that the proper salt is used. Cyanide-containing salts should not be combined with nitrate-nitrite salts, because such mixtures result in explosive reactions. Good housekeeping is essential to prevent accidents. Martempering salts must be stored in closed, well-marked containers. These salts are hygroscopic and will absorb water if exposed. Yellow, orange or red coloring is commonly used for identification of nitrate-nitrite salts. Cyanide-containing materials may be identified by white, blue or gray. Color coding is not universally practiced but is strongly recommended. Everyone handling these materials should be familiarized with the explosion hazard of intermixing. For a discussion of attendant hazards and recommended practices, see the sections of this article entitled "Salt Contamination" and "Selection of Austenitizing Equipment".

* When water is added to a nitrate-nitrite salt to increase its quenching severity, the water should be trickled or atomized onto the surface of the bath and should never be introduced below the surface or under pressure. Otherwise, spattering or eruption can occur. Equipment manufacturers should be consulted as to recommended procedures and hardware.

* The salt bath should be equipped with a high-temperature limit controller to prevent the bath from exceeding 595 °C (1100 °F). At higher temperatures, nitrate salts decompose and may cause fire or explosion. Although nitrate-nitrite salts do not burn, they are oxidizing and will support combustion. Therefore, oxidizable materials should not be stored near the bath, and combustible materials should not be introduced into the bath unless designed for the purpose. In the event of a fire, a carbon dioxide extinguisher should be used. Sand is also effective in smothering floor fires. To avoid explosive spread of the molten medium, *water should never be used to extinguish a fire in or around a molten salt (or molten metal) bath.*

* Soot and carbonaceous materials should not be allowed to accumulate

on the surface of the molten bath.

Hot oil also requires safety precautions. Among the hazards that accompany its use are: (*a*) fire; (*b*) overflow, due to the increase in the volume of the oil during heating; (*c*) explosion, when an atmosphere cover is used over the quench tank; and (*d*) contamination by water, which can result in fire. Equipment normally used for protection against these hazards includes:

* Alarm system on temperature-control instrument. (Recommended maximum operating temperature is 55 °C, or 100 °F, below the flash point of the quenching oil.) Often, an additional system is added for maximum temperature alarm.
* An oil-level indicator
* If the oil is heated by gas-fired tubes, a safety control system to prevent firing when the air cooling system is off
* Expansion system to accommodate the change in volume that occurs when the oil is heated from room temperature to the operating temperature. A suitable system is a small tank equipped with an overflow return line from the main quench tank. The main tank is equipped with a small pump to return the oil from the expansion tank. The capacity of the pump is usually about 20 l/min (5 gal/min).
* A safety control to prevent oil from being heated unless it is agitated
* A carbon dioxide or foam system, installed over the quench tank for fire protection
* A water-detection system, with automatic alarm, that monitors the oil for water content.

Caution should be used in installing ventilating systems and any other system that could cause addition of water to the hot oil. When an atmosphere is used over the quenching medium, the general safety precautions outlined in the article "Gas Carburizing" should be followed.

Suitability of Steels for Martempering

Alloy steels generally are more adaptable than carbon steels to martempering. In general, any steel that is normally quenched in oil can be martempered. Some carbon steels that are normally water quenched can be martempered at 205 °C (400 °F) in sections

Fig. 4 Temperature ranges of martensite formation in 14 carbon and low-alloy steels

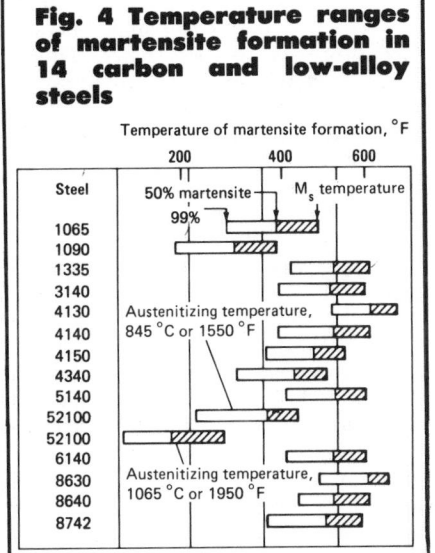

Fig. 5 Time-temperature transformation diagrams for 1034 and 1090 steels

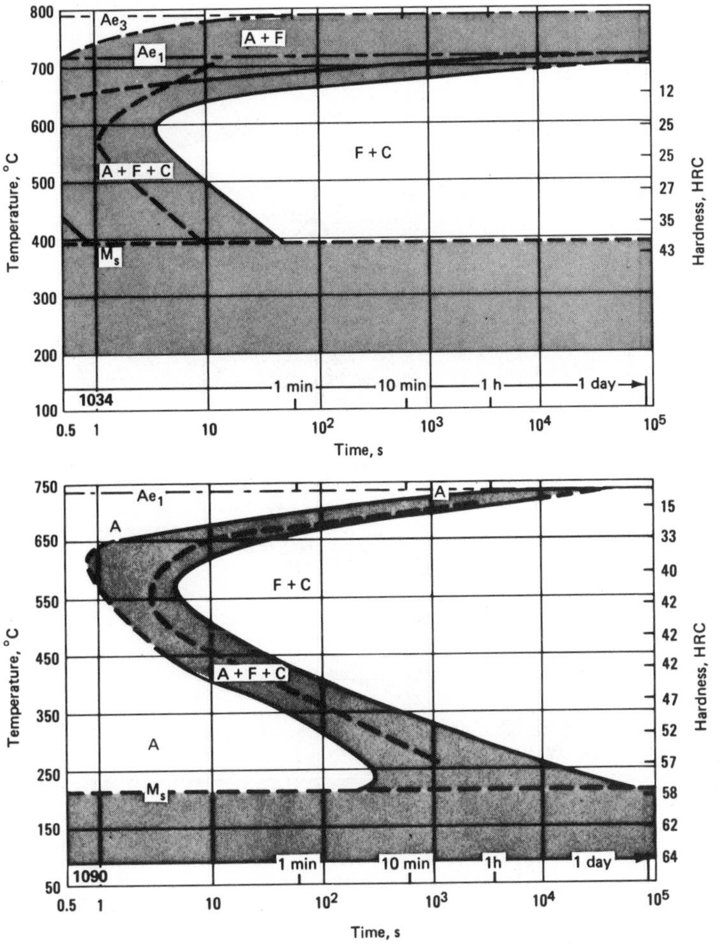

The 1090 steel was austenitized at 885 °C (1625 °F) and had a grain size of 4 to 5.

thinner than 5 mm (³/₁₆ in.), using vigorous agitation of the martempering medium. In addition, thousands of gray cast iron parts are martempered on a routine basis.

The grades of steel that are commonly martempered to full hardness include 1090, 4130, 4140, 4150, 4340, 4640, 5140, 6150, 8630, 8640, 8740, 8745, SAE 1141 and SAE 52100. Carburizing grades such as 3312, 4620, 5120, 8620 and 9310 also are commonly martempered after carburizing. Occasionally, higher-alloy steels such as type 410 stainless are martempered, but this is not a common practice.

Success in martempering is based on a knowledge of the transformation characteristics (TTT curves) of the steel being considered. The temperature range in which martensite forms is especially important. Figure 4 shows the martensite temperature ranges for 14 carbon and low-alloy steels. Two trends may be observed in these data: (a) as carbon content increases, the martensite range widens and the martensite transformation temperature is lowered; and (b) the martensite range of a triple-alloy (Ni-Cr-Mo) steel is usually lower than that of either a single-alloy or a double-alloy steel of similar carbon content.

Any steel that is to be martempered successfully must contain sufficient carbon or alloying additions to move the nose of the TTT curve to the right, thus permitting sufficient time for

quenching of workpieces past the nose of the TTT curve.

The TTT diagrams for a hypoeutectoid steel (1034), which is discussed later, and a hypereutectoid steel (1090) are shown in Fig. 5. The diagram for 1090 steel is the simplest form of transformation diagram, because no proeutectoid constituents (free ferrite or free carbide) are involved in the transformation at temperatures above that corresponding to the nose of the curve. The speed of transformation at the nose is related to the hardenability of the steel; when the nose of the TTT curve is far to the left on the diagram, the steel has lower hardenability; when the nose is to the right, the steel has higher hardenability. To achieve full hardening during quenching, the cooling curve of the steel must pass to the left of the

curve farthest to the left on the diagram. In production, some loss in as-quenched hardness is usually accepted in order to achieve minimum distortion.

A TTT diagram for a hypoeutectoid low-alloy steel (5140) suitable for martempering is shown in Fig. 6. The chromium in this steel causes the characteristic shape of the TTT curve near 540 °C (1000 °F).

The TTT diagram for an extremely high-hardenability steel (4340) is also shown in Fig. 6. The combined effect of nickel, chromium and molybdenum on hardenability is illustrated in this diagram. These elements cause a double nose on the TTT curve. The nose that occurs at about 480 °C (900 °F) is more significant in martempering than the one occurring near 650 °C (1200 °F).

Fig. 6 Time-temperature transformation diagrams for 4340 and 5140 steels

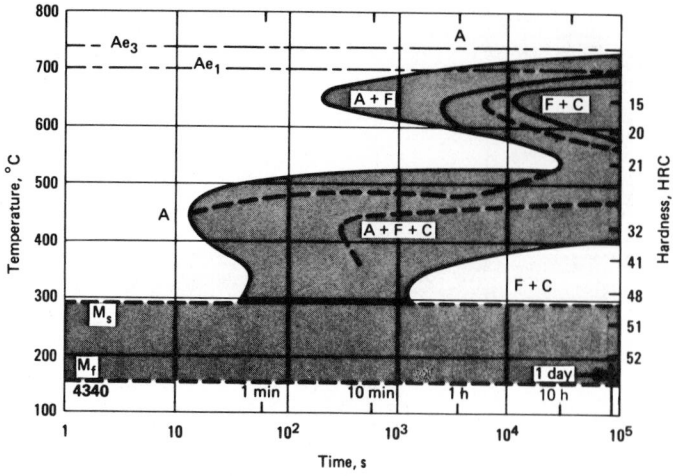

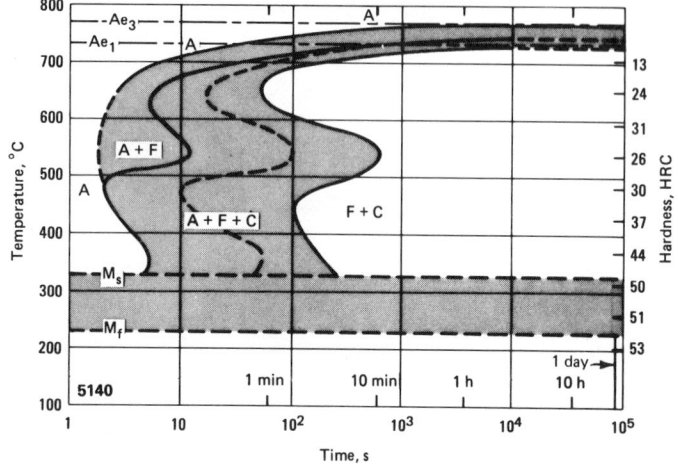

The austenitizing temperature for both steels was 845 °C (1550 °F); grain size was 7 to 8 for 4340 and 6 to 7 for 5140.

Steels having such high hardenability are easily martempered to fully martensitic structures.

Low-carbon and medium-carbon steels 1008 through 1040 are too low in hardenability to be successfully martempered, except when carburized. The TTT curve for the 1034 steel in Fig. 5 is characteristic of a steel that is unsuitable for martempering; except in sections only a few thousandths of an inch thick, it would be impossible to quench the steel in hot salt or oil without encountering upper transformation products.

Borderline Grades. Some carbon steels higher in manganese content, such as 1041 and 1141, can be successfully martempered in thin sections. Low-alloy steels that have limited ap-

plications for successful martempering are listed below (the lower-carbon grades are carburized before martempering):

1330 to 1345	4520
4012 to 4042	5015 and 5046
4118 to 4137	6118 and 6120
4422 and 4427	8115

Most of the above alloy steels are suitable for martempering in section thicknesses of up to 16 or 19 mm (⅝ or ¾ in.). Martempering at temperatures below 205 °C (400 °F) will improve hardening response, although greater distortion may result tha in martempering at higher temperatures.

Effect of Mass. The limitation of section thickness or mass must be considered in martempering. With a given

severity of quench, there is a limit to bar size beyond which the center of the bar will not cool fast enough to transform entirely to martensite. This is illustrated in Fig. 7, which compares the maximum diameter of bar that can be hardened by martempering, oil quenching and water quenching for 1045 steel and five alloy steels of various hardenabilities.

For some applications, a fully martensitic structure is unnecessary and a center hardness 10 HRC units lower than the maximum obtainable value for a given carbon content may be acceptable. By this criterion, maximum bar diameter is 25 to 300% greater than the maximum diameter that can be made fully martensitic (see lower graph in Fig. 7). Nonmartensitic transformation products (pearlite, ferrite and bainite) were observed at the positions on end-quenched bars corresponding to this reduced hardness value, as follows:

1045	15% pearlite
8630	10% ferrite and bainite
1340	20% ferrite and bainite
52100	50% pearlite and bainite
4150	20% bainite
4340	5% bainite

The influence of mixed structures such as these on the mechanical properties of the steel would have to be determined for each application.

Steels selected for martempering must be judged on hardenability and section size. To form the same amount of martensite, for a given section size, the carbon content or alloy content, or both, must be somewhat higher for martempering than for conventional (uninterrupted) quenching.

Control of Process Variables

The success of martempering depends on close control of variables throughout the entire process. It is important that the prior structure of the material being austenitized be uniform. Also, use of a protective atmosphere (or salt) in austenitizing is required, because oxide or scale will act as a barrier to uniform quenching in hot oil or salt.

The process variables that must be controlled in martempering include austenitizing temperature, temperature of martempering bath, time in martempering bath, salt contamina-

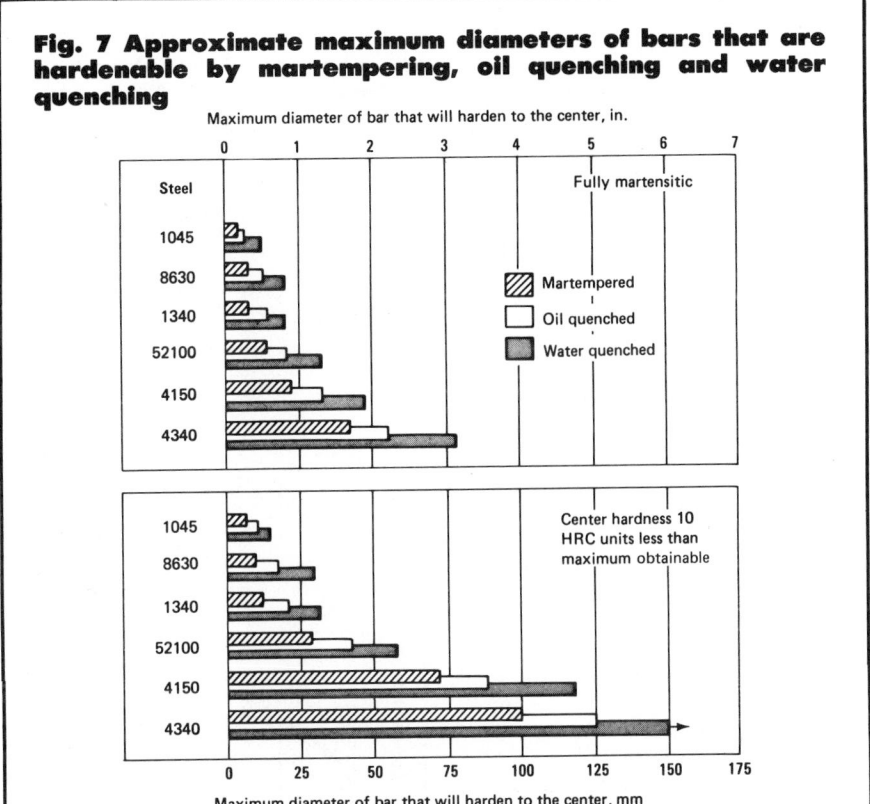

Fig. 7 Approximate maximum diameters of bars that are hardenable by martempering, oil quenching and water quenching

Maximum diameter of bar that will harden to the center, in.

Steel — Fully martensitic

- Martempered
- Oil quenched
- Water quenched

1045
8630
1340
52100
4150
4340

Center hardness 10 HRC units less than maximum obtainable

1045
8630
1340
52100
4150
4340

Maximum diameter of bar that will harden to the center, mm

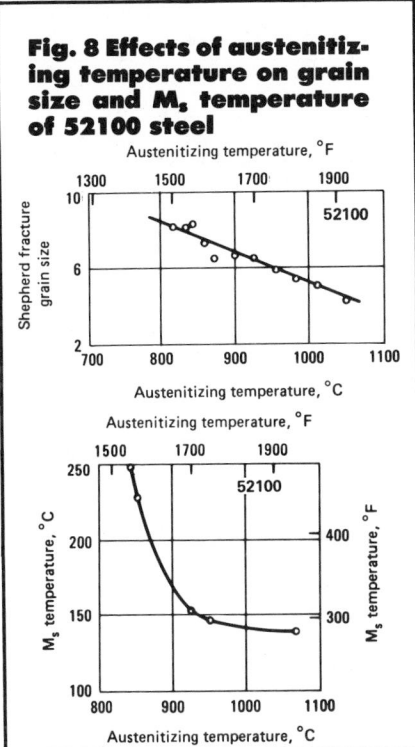

Fig. 8 Effects of austenitizing temperature on grain size and M_s temperature of 52100 steel

Austenitizing temperature, °F

Shepherd fracture grain size

52100

Austenitizing temperature, °C

Austenitizing temperature, °F

M_s temperature, °C

52100

M_s temperature, °F

Austenitizing temperature, °C

tion, water additions to salt, agitation, and rate of cooling from the martempering bath.

Austenitizing temperature is important because it controls austenitic grain size, degree of homogenization, and carbide solution, and because it affects the M_s temperature, which is important in establishing martempering procedures. As indicated in Fig. 8 for 52100 steel, an increase in austenitizing temperature lowers M_s temperature and increases grain size.

Temperature control during austenitizing is the same for martempering as for conventional quenching: a tolerance of 8 °C (15 °F) is common. The austenitizing temperatures most commonly used for several different steels are indicated in Table 3.

In most instances, austenitizing temperatures for martempering are the same as those for conventional oil quenching. Occasionally, however, medium-carbon steels are austenitized at higher temperatures prior to martempering, to increase as-quenched hardness.

For carburized parts, low austenitizing temperatures usually yield better size control during martempering. To obtain minimum dimensional changes,

the lowest austenitizing temperature that will yield satisfactory core properties should be used. The ratio of case depth to core, as well as the prior processing history of the steel (such as forging, rolling or drawing), can be controlling factors also, particularly for critical section shapes and sizes. Austenitizing temperatures that have been used for several carburizing steels are indicated in Table 3.

Salt Contamination. When parts are carburized or austenitized in a salt bath, they can be directly quenched in an oil bath operating at the martempering temperature. However, if the parts are carburized or austenitized in salt containing cyanide, they must *not* be directly martempered in salt, because the two types of salts are not compatible, and explosions can occur if they are mixed. Instead, one of two procedures should be used: either (a) air cool from the carburizing bath, wash, reheat to the austenitizing temperature for case and/or core in a chloride bath, and then martemper; or (b) quench from the cyanide-containing bath into a neutral chloride rinse bath maintained at the austenitizing temperature, and then martemper.

If the latter method is used, it is

essential to control the amount of cyanide buildup in the neutral rinse. When tests indicate more than 5% cyanide in the chloride rinse, part of the salt should be discarded and the remainder diluted with new salt.

All fixtures must be thoroughly cleaned after martempering, to prevent transfer of quenching salt to either cyanide baths or neutral chloride baths. Mixing of cyanide with nitrate-nitrite salts will cause explosion. A chloride bath that is contaminated with nitrate-nitrite salts will produce pitting and decarburization of parts immersed in it.

Temperature of the martempering bath varies considerably, depending on composition of workpieces, austenitizing temperature, and desired results. In establishing procedures for new applications, many plants begin at 95 °C (200 °F) for oil quenching, or at about 175 °C (350 °F) for salt quenching, and progressively increase the temperature until the best combination of hardness and distortion is obtained. Martempering temperatures (for oil and salt) that represent the experience of several plants are listed in Table 3.

Time in the martempering bath depends on section thickness and on the type, temperature, and degree of agitation of the quenching medium. The

Table 3 Typical austenitizing and martempering temperatures for various steels

Grade	Austenitizing temperature °C	°F	Martempering temperature Oil(a) °C	°F	Salt(b) °C	°F
Through-hardening steels						
1024 870		1600	135	275	. . .	. . .
1070 845		1550	175	350	. . .	. . .
1146 815		1500	175	350	. . .	. . .
1330 845		1550	175	350	. . .	. . .
4063 845		1550	175	350	. . .	. . .
4130 845		1550	. . .	. . .	205 to 260	400 to 500
4140 845		1550	150	300	. . .	. . .
4140 830		1525	. . .	. . .	230 to 275	450 to 525
4340, 4350 815		1500	. . .	. . .	230 to 275	450 to 525
52100 855		1575	190	375	. . .	. . .
52100 845		1550	. . .	. . .	175 to 245	350 to 475
8740 830		1525	. . .	. . .	230 to 275	450 to 525
Carburizing steels						
3312 815		1500	. . .	. . .	175 to 190	350 to 375
4320 830		1525	. . .	. . .	175 to 190	350 to 375
4615 955		1750	190	375	. . .	. . .
4720 845		1550	. . .	. . .	175 to 190	350 to 375
8617, 8620 925		1700	150	300	. . .	. . .
8620 855		1575	. . .	. . .	175 to 190	350 to 375
9310 815		1500	. . .	. . .	175 to 190	350 to 375

(a) Time in oil varies from 4 to 20 min, depending on section thickness. (b) Martempering temperature depends on shape and mass of parts being quenched; higher temperatures in range (and sometimes above range) are used for thinner sections and more intricate parts.

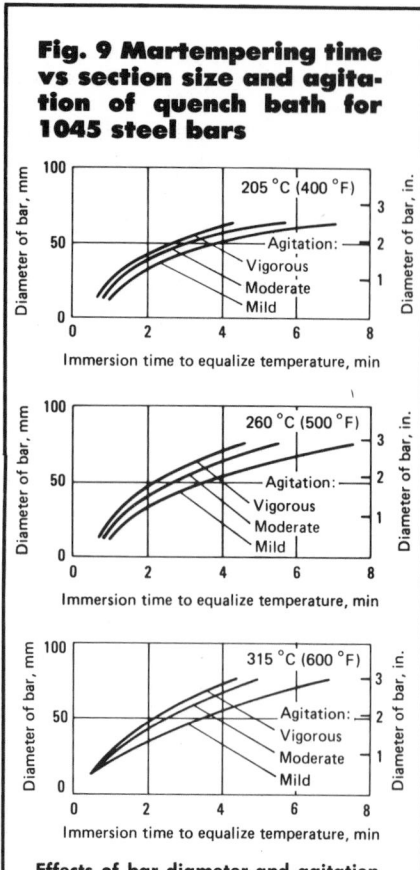

Fig. 9 Martempering time vs section size and agitation of quench bath for 1045 steel bars

Effects of bar diameter and agitation of quench bath on time required for centers of 1045 steel bars to reach martempering temperature when quenched from a neutral chloride bath at 845 °C (1550 °F) into anhydrous nitrate-nitrite martempering salt at 205, 260 and 315 °C (400, 500 and 600 °F). Length of each bar was three times the diameter.

effects of section thickness and of temperature and agitation of the quench bath on immersion time are indicated in Fig. 9.

Because the object of martempering is to develop a martensitic structure with low thermal and transformation stresses, there is no need to hold the steel in the martempering bath for extended periods of time. Excessive holding lowers final hardness, because it permits transformation to products other than martensite. In addition, stabilization of austenite may occur in medium-alloy steels that are held for extended periods of time at the martempering temperature.

The martempering time for temperature equalization in oil is about four to five times that required in anhydrous salt at the same temperature. For example, for a 25-mm- (1-in.-) diam bar martempered in salt at 205 °C (400 °F) with moderate agitation, the customary immersion time is about 1½ to 2 min, whereas about 8 to 10 min is required for the bar to attain temperature equalization in oil at 205 to 220 °C (400 to 425 °F). The immersion time required in salt can be reduced by as much as 50% by addition of 0.5 to 2% water.

Water Additions to Salt. Addition of water to increase the quenching severity of martempering salt usually is made by directing a fine spray of water onto the molten-salt surface. A protective shroud surrounds the water spray to prevent spattering (see previous section on safety precautions, in this article). The turbulence of the salt carries the water into the bath without spattering or hazard to the operator.

Water is continuously evaporating from the bath surface, and the rate of evaporation increases during quenching of hot work. Thus, it is necessary to add water periodically to maintain the water concentration and a uniform quenching severity. The amount of water to be added varies with the operating temperature of the salt, as indicated by the following recommended concentrations:

Temperature °C	°F	Water concentration, %
205	400	½ to 2
260	500	½ to 1
315	600	¼ to ½
370	700	¼

At present, there is no known means of automatically controlling the concentration of water in molten nitrate-nitrite salt. Water usually is controlled

at the discretion of the operator, who will add water as needed. On the basis of experience, however, it is possible to anticipate the need for water; here, the addition of water may be simplified by use of timers that can be adjusted to time the frequency and duration of water additions.

The presence of water is visually detectable by the operator, because steam is released when the hot work is immersed into the nitrate-nitrite salt. The steam causes a visible mounding of the salt above the quench area, and there is a characteristic sizzling caused by the vapor phase.

Besides visual appearance, periodic hardness checks of the work will indicate the activity of the bath. A more accurate determination can be made by

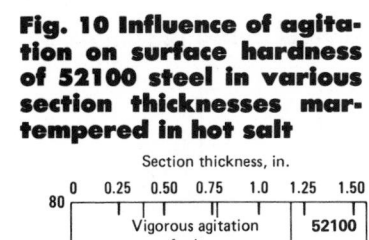

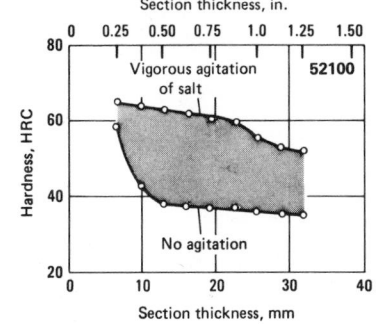

Fig. 10 Influence of agitation on surface hardness of 52100 steel in various section thicknesses martempered in hot salt

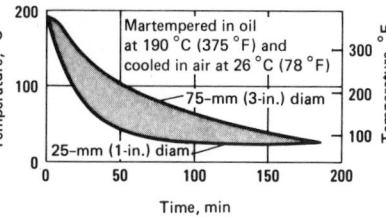

Fig. 11 Effect of section thickness on time required for air cooling of steel after martempering at 190 °C (375 °F)

Temperature measurements were made at the surface. Times can be reduced by approximately 30% by forced fan cooling.

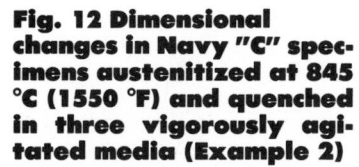

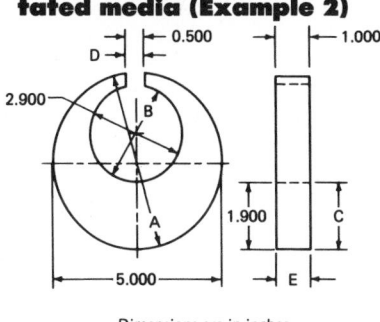

Fig. 12 Dimensional changes in Navy "C" specimens austenitized at 845 °C (1550 °F) and quenched in three vigorously agitated media (Example 2)

Dimensions are in inches

removing a small quantity of the salt and weighing it accurately before and after dehydrating it by heating to 370 to 425 °C (700 to 800 °F). Another method is to determine the freezing point of a small sample and then refer to a published curve that relates freezing point to water content for that specific salt.

Agitation of the martempering salt or oil considerably increases the hardness obtainable for a given section thickness, in comparison to that obtainable in still quenching. This is demonstrated in Fig. 10, which presents data for 52100 steel that relate section thickness, hardness and agitation.

In some instances, the rapid cooling produced by the most vigorous agitation increases distortion. Thus, mild agitation often is used in combination with water addition to obtain minimum distortion without sacrificing hardness.

Cooling from the martempering bath ordinarily is done in still air, to avoid large differences in temperature between the surface and the interior of the steel. Forced-air cooling by means of fans is occasionally used on sections over 19 mm (³/₄ in.) thick, but caution is required if the part varies in section thickness or has more surface exposed on one section, such as on threads or serrations, because objectionable amounts of distortion can occur on rapid cooling through the martensite range. Generally, cooling of workpieces in cool oil or water after removal from the martempering bath is considered undesirable, because cooling can re-establish thermal gradients and unequal stress patterns that can increase distortion.

Cooling time varies with the mass and density of the charge, the maximum section thickness of the workpiece, and the ambient air temperature. Usually, production loads of 365 to 815 kg (800 to 1800 lb) from either continuous or batch-type furnaces will require 2½ to 5 h to reach room temperature. Figure 11 shows the effect of section thickness. As indicated, the 25-mm- (1-in.-) diam sections were completely cooled in about half the time required for complete cooling of the 75-mm- (3-in.-) diam sections.

After being cooled to room temperature, martempered parts usually can be held at room temperature for several hours, and sometimes days, without risk of cracking, because residual stresses are low compared to those in conventionally quenched parts. Holding of parts at room temperature also permits more nearly complete transformation of steels in which transformation is sluggish.

Dimensional Control

In most instances, distortion is significantly less in martempered parts than in parts hardened by uninterrupted (conventional) quenching. However, prior processing often has a significant effect on distortion, regardless of the heat treating method used. Thus, for some applications, martempering may fail to solve distortion problems because excessive dimensional changes occurred during heating prior to martempering.

Occasionally, it is necessary to consider the effects of fabricating stresses that occur during forging, stamping, rolling and machining. When the work-

piece is heated, these stresses can cause a significant amount of distortion. A process anneal at 650 to 705 °C (1200 to 1300 °F) after rough machining or forming usually will relieve such stresses. Any resulting change in the size or shape of the part can then be corrected by finish machining prior to austenitizing and martempering. During heating, distortion also can occur as the result of temperature differences in a part having both light and heavy sections. This condition often is corrected by preheating at 650 to 705 °C (1200 to 1300 °F) prior to austenitizing.

Relatively large parts that require extreme flatness often must be press quenched. For example, a gear 178 mm (7 in.) in diameter, 13 mm (½ in.) thick at the rim and with a 6.4-mm (¼-in.) web could not be martempered with an acceptable degree of flatness. A large ring gear with internal teeth and a thin wall is another part for which martempering has been replaced by press quenching to obtain dimensions within acceptable limits.

The following examples describe specific situations in which distortion problems have been encountered. In some instances, the effects (on identical parts) of martempering and of oil quenching are presented for comparison.

Example 2. Fig. 12 shows data for nine Navy "C" specimens, made of a high-carbon alloy steel containing 0.95 C, 0.30 Si, 1.20 Mn, 0.50 W, 0.50 Cr and 0.20 V, were austenitized for 40 min at 845 °C (1550 °F); three of the specimens were subjected to conventional quenching in oil at 60 °C (140 °F), three were martempered for 2 min in salt at 205 °C (400 °F), and three were martempered for 2 min in salt at 245 °C (475 °F). Each quenching medium was agitated vigorously. All specimens were tempered to 63 to 64 HRC before being measured for dimensional changes.

The data accompanying Fig. 12 indicate that the martempered specimens—especially those quenched in salt at 245 °C (475 °F)—exhibited less distortion in every dimension. These test results indicate that lower stresses result from martempering than from conventional oil quenching.

Example 3. Bearing races, 215 mm (8³⁄₈ in.) in OD by 190 mm (7½ in.) in ID by 130 mm (5⅛ in.), made of 52100 steel, were austenitized in chloride salt at 850 °C (1560 °F) for 25 min, then martempered in salt at 230 °C (450 °F) for 2½ min and air cooled to room temperature. The resulting hardness was 63 to 64 HRC. This treatment produced an average growth of 0.08 mm (0.003 in.) and an average distortion (out-of-roundness) of 0.25 mm (0.010 in.). Prior to heat treatment, the machined races had an average out-of-roundness of 0.18 mm (0.007 in.). Thus, the average increase in out-of-roundness was only 0.08 mm (0.003 in.).

In this application, martempering reduced grinding time from 50 to 7 min per race by permitting a reduction in the grinding stock required for conventionally oil quenched parts.

Example 4. Rods 6.4 mm (¼ in.) in diameter by 255 mm (10 in.) long, with a 3.2-mm (⅛-in.) hole near one end, were made from oil-hardening drill rod. The requirements were a hardness of 60 to 62 HRC and maximum warpage of 0.25 mm, or 0.010 in. (indicator reading between centers).

The rods were heated by being suspended in a salt bath at 805 °C (1480 °F), and then were quenched in oil at 55 °C (135 °F). This treatment produced the required hardness, but warpage was excessive (up to 0.76 mm, or 0.030 in.). The problem was solved by replacing oil quenching with martempering in unagitated salt at 175 °C (350 °F).

Example 5. Figure 13 shows a thin-wall ring gear of 8625 steel, with a flange on one end, that had to be carburized to a depth of 1.3 mm (0.050 in.) and hardened. After carburizing followed by martempering in oil at 190 °C (375 °F), out-of-roundness ranged from 0.43 to 0.66 mm (0.017 to 0.026 in.), which was not acceptable. The problem was solved by martempering and then placing the ring gears over a plug

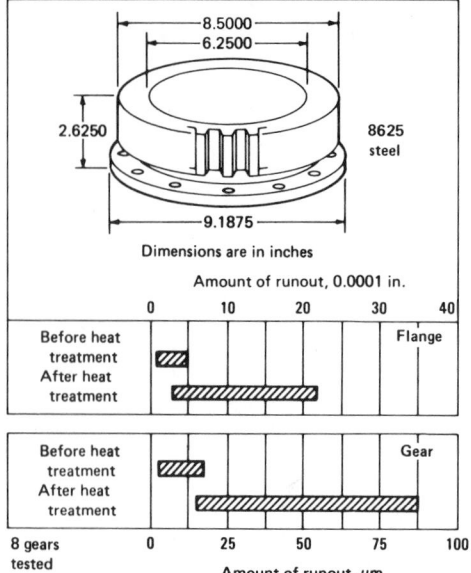

Fig. 13 Control range for 8625 steel ring gears (Example 5)

Dimensions are in inches

Eight ring gears made of 8625 steel were carburized to a depth of 1.3 mm (0.050 in.), martempered in oil at 190 °C (375 °F) and placed on a plug for cooling to room temperature. Measurements are total indicator readings.

machined to the final inside diameter of the gear for cooling to room temperature. This produced acceptable parts having a maximum runout of 0.09 mm (0.0035 in.). The runout for eight gears tested before heat treating and after plug cooling is indicated in Fig. 13.

Example 6. Figure 14 indicates the dimensional changes that occurred when different types and sizes of gears made of 8625H steel were carburized at 925 °C (1700 °F) to a depth of 1.0 to 1.5 mm (0.040 to 0.060 in.) and then martempered in oil at either 165 or 190 °C (325 or 375 °F).

Example 7. Figure 15 shows the dimensional changes that were encountered in carburizing and martempering seven different sizes of automatic-transmission gears made of 8620H steel. The gears were martempered in oil at 150 °C (300 °F). Shrinkage of the largest gear is associated with lower core hardness in the considerably heavier section of the gear.

Example 8. Figure 16 indicates the effects of various combinations of carburizing and quenching methods on the dimensions of 25-tooth reverse-idler gears for power-grader transmissions. All gears were carburized to a depth of 0.76 to 1.0 mm (0.030 to 0.040 in.). The smallest dimensional changes occurred

in gears that were liquid carburized and then martempered in salt at 205 °C (400 °F); the greatest changes occurred in those that were gas carburized and quenched in agitated oil at 45 °C (110 °F).

The effect of stress relieving on out-of-roundness also is indicated in Fig. 16. The data for bore, teeth and runout are for gears that were stress relieved before heat treating.

Examples 9 and 10. Figures 17 and 18 are histograms of distortion data, in terms of total indicator readings, for various shafts made of 8625 steel. The shafts depicted in Fig. 17 were machined from bar stock, whereas those in Fig. 18 were forged. All shafts were carburized at 925 °C (1700 °F) and martempered in oil at 165 °C (325 °F). Also, all shafts were in the vertical position during heat treating—some suspended, and some supported on one end—as indicated in Fig. 17 and 18.

Straightening After Martempering. Parts such as shafts, chain-saw guides, washers and springs require straightening or re-forming after heat treatment. This is sometimes not feasible after normal quenching and tempering. However, after such parts are removed from a martempering bath, straightening is readily accomplished, either by hand pressure or in a die

Fig. 14 Dimensional changes in carburized and martempered gears of 8625H steel (Example 6)

Gear No.	Before treating		After treating		Weight		Change in dimension, 0.0001 in. — Size over pins
	mm	in.	mm	in.	kg	lb	
1	99.95 to 100.00	3.9349 to 3.9370	100.02 to 100.06	3.9378 to 3.9395	2.15	4.75	
2	119.85 to 119.91	4.7184 to 4.7208	119.90 to 119.97	4.7205 to 4.7234	2.38	5.25	
3	163.59 to 163.65	6.4407 to 6.4429	163.69 to 163.76	6.4444 to 6.4473	2.15	4.75	
4	176.53 to 176.63	6.9501 to 6.9538	176.57 to 176.71	6.9517 to 6.9569	4.31	9.50	
5	234.19 to 234.28	9.2200 to 9.2235	234.17 to 234.30	9.2194 to 9.2245	4.54	10	
1	0.008 to 0.023	0.0003 to 0.0009	0.043 to 0.061	0.0017 to 0.0024	2.15	4.75	Rolling
2	0.005 to 0.018	0.0002 to 0.0007	0.018 to 0.084	0.0007 to 0.0033	2.38	5.25	
3	0.020 to 0.041	0.0008 to 0.0016	0.025 to 0.061	0.0010 to 0.0024	2.15	4.75	
4	0.008 to 0.043	0.0003 to 0.0017	0.010 to 0.046	0.0004 to 0.0018	4.31	9.50	
5	0.013 to 0.038	0.0005 to 0.0015	0.018 to 0.081	0.0007 to 0.0032	4.54	10	Change in dimension, μm

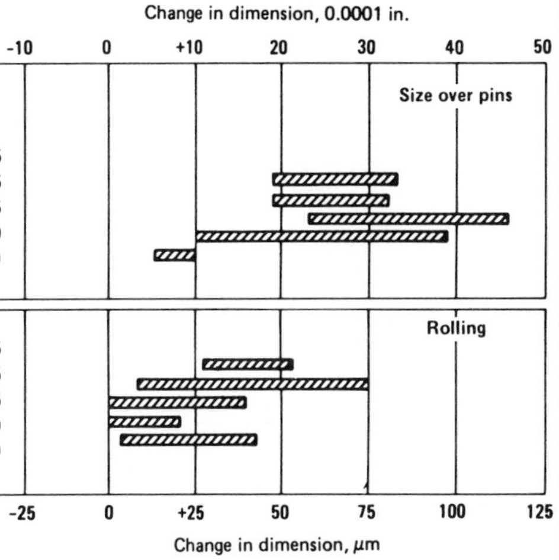

All gears were carburized at 925 °C (1700 °F) to a depth of 1.0 to 1.5 mm (0.040 to 0.060 in.) and martempered directly from the carburizing temperature in oil at 165 °C (325 °F). Gears 1, 2 and 5 were measured over 7.32-mm- (0.288-in.-) diam pins; gear 3 over 3.66-mm- (0.144-in.-) diam pins; and gear 4 over 8.78-mm- (0.3456-in.-) diam pins.

Fig. 15 Dimensional changes in carburized and martempered automatic-transmission gears made of 8620H steel (Example 7)

Gear No.	Before treating		After treating		Change in dimension, 0.0001 in. — Size over 7.32-mm- (0.288-in.-) diam pins
	mm	in.	mm	in.	
1	153.77 to 153.86	6.0540 to 6.0575	153.86 to 153.95	6.0575 to 6.0610	
2	153.70 to 153.77	6.0513 to 6.0538	153.77 to 153.87	6.0540 to 6.0580	
3	166.32 to 166.34	6.5480 to 6.5490	166.40 to 166.43	6.5513 to 6.5525	
4	191.75 to 191.81	7.5493 to 7.5515	191.85 to 191.92	7.5530 to 7.5560	
5	200.25 to 200.28	7.8840 to 7.8850	200.34 to 200.38	7.8873 to 7.8888	
6	213.10 to 213.13	8.3898 to 8.3910	213.18 to 213.23	8.3930 to 8.3950	
7 (a)	317.58 to 317.73	12.5030 to 12.5090	317.55 to 317.70	12.5020 to 12.5080	
1	0.000 to 0.000	0.0000 to 0.0000	0.000 to 0.033	0.0000 to 0.0013	Out-of-roundness
2	0.000 to 0.005	0.0000 to 0.0002	0.025 to 0.051	0.0010 to 0.0020	
3	0.000 to 0.008	0.0000 to 0.0003	0.000 to 0.030	0.0000 to 0.0012	
4	0.000 to 0.000	0.0000 to 0.0000	0.005 to 0.025	0.0002 to 0.0010	
5	0.000 to 0.008	0.0000 to 0.0003	0.005 to 0.020	0.0002 to 0.0008	
6	0.000 to 0.005	0.0000 to 0.0002	0.013 to 0.038	0.0005 to 0.0015	
7	0.000 to 0.013	0.0000 to 0.0005	0.000 to 0.018	0.0000 to 0.0007	Change in dimension, μm

(a) Size over 12.2-mm- (0.480-in.-) diam pins

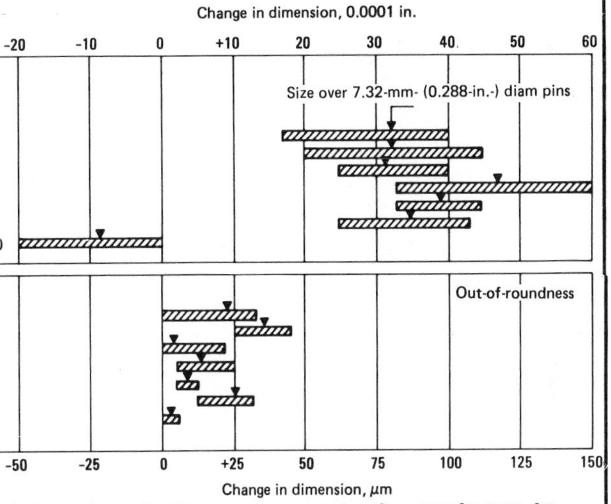

All gears were carburized at 925 °C (1700 °F) and then martempered directly from the carburizing temperature in oil at 150 °C (300 °F).

press, while the parts are still essentially austenitic.

When parts are clamped between plates held at approximately 150 °C (300 °F), the parts cool to the temperature of the plate. At this temperature, depending on the M_f temperature of the steel, the transformation to martensite may be only partial, but it is usually near enough to completion for the part to set permanently before it is removed from the clamped position. At this temperature, the film of salt (melting point, about 145 °C, or 290 °F) adhering to the part will not freeze, and so

removal of the part from the clamps or die can be accomplished without difficulty. The transformation of martensite is completed during subsequent air cooling to room temperature, and the part is within dimensional tolerance.

Forming After Martempering. Difficult-to-form materials, such as hot work die steels (H11) and some martensitic stainless steels used for missiles, rockets and high-speed aircraft, can be accurately formed and hardened by austenitizing, martempering and hot forming to shape immediately after they are extracted from the quench

bath, if the martempering temperature is above the martensite transformation temperature for the specific alloy being treated. During hot forming by rolling, forging, drawing or extruding, the metal consists of metastable austenite, which transforms to martensite on subsequent cooling to room temperature.

Parts of both simple and complex shape have been formed in this manner; after being air cooled to room temperature in or out of the forming die, the parts will accurately maintain their as-formed dimensions and yet be in the fully heat treated condition,

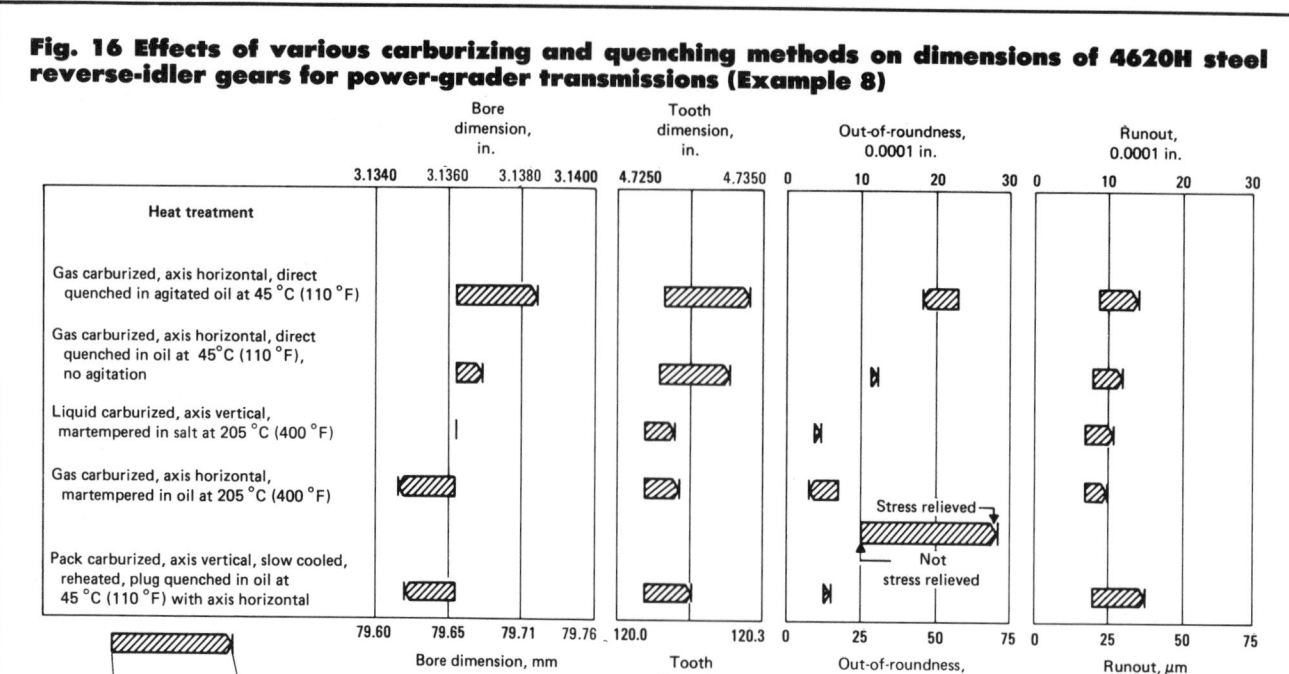

Fig. 16 Effects of various carburizing and quenching methods on dimensions of 4620H steel reverse-idler gears for power-grader transmissions (Example 8)

Gears were carburized to a depth of 0.8 to 1.0 mm (0.030 to 0.040 in.) and quenched to a hardness of 58 to 63 HRC.

Table 4 Typical applications of martempering steel parts in salt

Part	Grade	Maximum section thickness mm	in.	Weight kg	lb	Martempering conditions Temperature of salt °C	°F	Minimum time in salt, min	Required hardness, HRC
Compliant tube	4130	0.8	0.03	0.11	0.25	160(a)	320(a)	5	50(b)
Thrust washer	8740	5.1	0.20	0.05	0.1	230	450	1	52 min(b)
Chain link	1045	5.6	0.22	0.11	0.25	205(c)	400(c)	1	45 to 50(b)
Cotton-picker spindle(d)	Type 410	6.4	0.25	0.05	0.12	315	600(c)	1½	44 to 48(b)
Accessory drive shaft	9310 (e)	6.4	0.25	0.45	1.0	190	375	2½	90 (15N scale)
Clutch-adjustment nut	8740	7.6	0.30	0.14	0.3	230	450	2	52 min(b)
Seal ring	52100	7.6	0.30	0.18	0.4	190	375	10	65(b)
Spur pinion	3312 (e)	7.6	0.30	0.23	0.5	175	350	1½	90 (15N scale)
Internal gear	4350	8.9	0.35	0.36	0.8	245	475	2	54 min(b)
Dual gear(f)	4815 (e)	9.4	0.37	2.13	4.7	260	500	2	62 to 63(b)
Drive coupling	4340	10.2	0.40	0.27	0.6	230	450	2½	52 min(b)
Spline shaft	8720 (e)	10.2	0.40	0.50	1.1	190	375	2½	90 (15N scale)
Arbor sleeve	1117L (e)	10.2	0.40	0.59	1.3	205	400	3	...
Screw-machine spindle	8620 (e)	10.2	0.40	6.35	14.0	205	400	3	...
Driving barrel	4350	12.7	0.50	0.45	1.0	245	475	3	48 to 52(g)
Bearing race(h)	52100	12.7	0.50	13.2	29.2	220	425	2½	63 to 64(b)
Hog knive	9260	15.2	0.60	8.16	18.0	175(j)	350(j)	15	62(b)
Landing-gear spring	6150	19.1	0.75	14.7	32.5	260	500	2¾	56 to 57(b)
Internal gear	1117L (e)	25.4	1.00	1.36	3.0	205	400	3	...
Spur pinion gear(k)	4047	25.4	1.00	16.4	36.2	230(c)	450(c)	3	50 to 52(m)
Screw-machine sprocket	8620 (e)	38.1	1.50	9.07	20.0	205	400	3	...

(a) Salt contained 1½% water. (b) As quenched. (c) Salt contained water. (d) 6.4-mm (¼-in.) diam by 203 mm (8 in.) long. (e) Carburized. (f) 124-mm (4⅞-in.) OD by 32-mm (1¼-in.) ID by 102 mm (4 in.). (g) Final. (h) 224-mm (8¹³⁄₁₆-in.) ID by 251-mm (9⅞-in.) OD. (j) Salt contained 1% water. (k) 19-mm (¾-in.) OD by 92-mm (3⅝-in.) ID by 140 mm (5½ in.). (m) As-quenched hardness of teeth

Fig. 17 Distortion in martempering (Example 9)

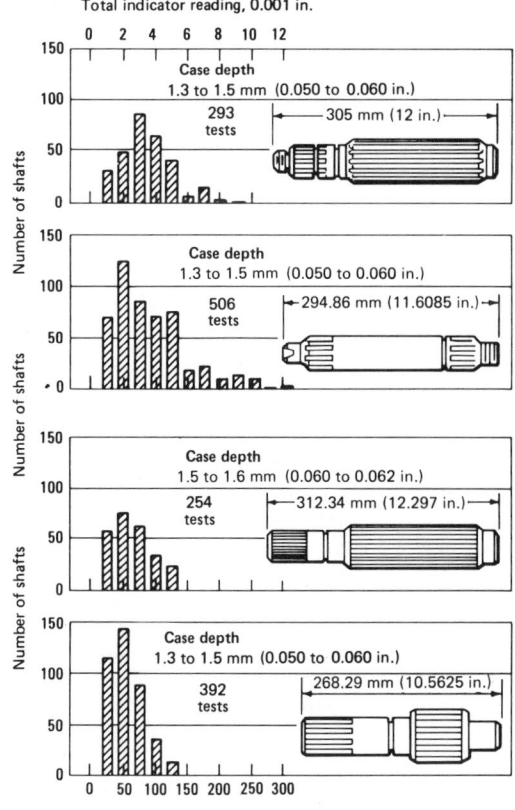

Total indicator reading, 0.001 in.

Histograms of distortion data on shafts of 8625 steel (machined from bar stock) after carburizing at 925 °C (1700 °F) and martempering in oil at 165 °C (325 °F). Shafts were heat treated in the vertical position: the top two shafts were suspended vertically from the threaded end; the lower two shafts were supported on one end.

requiring only a subsequent tempering operation.

This method of heat treating and forming has been successfully applied to sheet and plate that require nearly perfect flatness. It also has been used for forming type 420 stainless steel sheet into cups that require close dimensional tolerances and hardness values of 55 to 58 HRC. Working of metastable H11 that results in deformation of 58 to 94% at 480 °C (900 °F) produces a 19 to 32% increase in tensile strength.

Applications

Typical applications of martempering in salt and oil are indicated in Tables 4 and 5, which list and describe steel parts commonly treated and give details of martempering procedures and hardness requirements. From these tables, it is evident that martempering is used for parts of diverse shapes, weights, section sizes and steel compositions.

Although about a third of the applications in Table 4 and most of the applications in Table 5 are carburized parts, it is believed that in current industrial practice the tonnage of martempered through-hardening steels markedly exceeds that of martempered carburized steels. However, martempering is especially appropriate for carburized parts (particularly for splined shafts, cams and gears), because these parts generally are more difficult to grind and more costly to fabricate, and are made to closer dimensions, than are parts made of through-hardening steels.

Special adaptations of martempering are sometimes employed to achieve desired characteristics and to solve specific problems. Such modified techniques, however, usually require that all conditions be closely controlled, lest still greater problems result. One special technique that has been employed is described in the following example.

Example 11. A problem was encountered in obtaining a sufficient depth of hardness in forged balls made of 52100 steel, which ranged in diameter from 25 to 64 mm (1 to 2½ in.). Investigation ruled out both conventional oil quenching and martempering, because of low hardness penetration and the existence of quenching pearlite (an intermediate transformation product) in the microstructure.

This problem was solved by preceding martempering with a brief, timed quench in agitated brine at 23 °C (74 °F). Heat treatment of the balls then consisted of:

- Austenitizing in salt at 855 °C (1575 °F) for times ranging from 15 min for 25-mm- (1-in.-) diam balls to 50 min for 64-mm- (2½-in.-) diam balls
- Quenching in agitated brine at 23 °C (74 °F) for times ranging from 15 s for 25-mm-diam balls to 40 s for 64-mm-diam balls, and removing parts while hot (above 100 °C, or 212 °F) to accelerate evaporation of moisture
- Martempering in salt at 165 °C (325 °F) for 8 min (all diameters)
- Air cooling
- Tempering at 140 °C (285 °F) for 3 h (all diameters). By use of this treatment, the parts were successfully hardened to the desired depth. Surface hardness values were as follows:

Diameter of ball		Surface hardness, HRC
mm	in.	
27	1¹⁄₁₆	64.0-64.5
29	1⅛	64.5-65.5
32	1¼	63.5-64.5
33	1⁵⁄₁₆	64.0-64.5
35	1⅜	63.0-64.0
38	1½	63.0-63.5
41	1⅝	63.5-64.5
43	1¹¹⁄₁₆	63.0-64.0
44	1¾	63.5-64.5
48	1⅞	63.5-64.0
49	1¹⁵⁄₁₆	63.5-64.0
54	2⅛	61.5-63.5

Selection of Austenitizing Equipment

Austenitizing of steel prior to martempering may be done in virtually any

Table 5 Typical applications of martempering steel parts in oil

Part	Grade	Maximum section thickness mm	in.	Outside diameter mm	in.	Weight kg	lb	Carburizing temperature, °C	Depth of case μm	0.001 in.	Quenching temperature, °C	Temperature of martempering oil(a)°C	Surface hardness, HRC
Sleeve	52100	3.2	0.125	...	...	0.1	1/4	...	...	...	790	165	58-59
Spacer plate	1065	3.2	0.125	...	...	0.1	1/4	...	...	...	790	165	56-57
Bushing	1117	4.8	0.1875	51.0	2.009	0.2	1/2	910	1015-1220	40-48	910	190	58-62
Bushing	1117	6.4	0.25	76.3	3.0034	0.6	11/4	910	1015-1220	40-48	910	190	55-60
Shifter rail	1018	9.5	0.375	...	...	1.0	21/8	845(b)	255-455	10-18	845	165	55-60
Shifter rail	1018	9.5	0.375	...	...	1.4	31/8	845(b)	355-610	14-24	845	165	55-60
Spur gear	8620	12.7	0.5	320.6	12.620	12.7	28	925	1145-1525	45-60	845	150	55-60
Helical gear	4620H	19.1	0.75	331.5	13.050	16.9	37.2	925	760-1015	30-40	845	150	58-63
Herringbone gear	4820	19.1	0.75	283.2	11.150	16.3	36	925	1145-1525	45-60	845	150	55-61
Shifter rail	1141	25.4	1.0	25.4	1.0	0.9	17/8	885(b)	455-660	18-26	885	165	45-50
Spiral bevel gear	4620	25.4	1.0	210.6	8.29	5.1	11.25	925	1015-1270	40-50	845	150	55 min
Helical pinion	8617H	25.4	1.0	35.8	1.409	0.4	0.9	925	510-710	20-28	845	150	58-63
Spur gear	8625	31.8	1.250	83.8	3.300	4.3	93/8	925	1525-1725	60-68	925	190	58-62
Spur gear	4817H	34.0	1.340	186.7	7.350	8.6	19	925	1400-1780	55-70	845	150	58-63
Spur gear	8625	38.1	1.500	165.1	6.500	2.5	51/2	925	1525-1725	60-68	925	190	58-62
Splined shaft	8625	39.7	1.564	39.7	1.564	2.7	57/8	925	1400-1980	70-78	925	190	58-62
Spur gear	8625	44.4	1.750	108.0	4.250	3.5	73/4	925	1525-1725	60-68	925	165	58-62
Splined shaft	8625	44.4	1.750	44.4	1.750	2.0	41/2	925	1525-1725	60-68	925	190	58-62
Splined shaft	8620	50.8	2.000	50.8	2.000	5.1	111/4	925	1525-1725	60-68	925	165	58-62
Splined shaft	8625	65.0	2.559	65.0	2.559	6.8	15	925	1525-1725	60-68	925	165	58-62
Spur gear	8625	84.7	3.3343	245.5	9.667	11.9	261/4	925	1525-1725	60-68	925	190	58-62

(a) Minimum time in oil, 5 min. (b) Carbonitriding temperature

standard furnace. This phase of the operation has been successfully accomplished in furnaces ranging from small, simple box furnaces to large, fully automated, high-production installations. Both atmosphere-controlled furnaces and molten salt baths are widely used. Fluid beds are also being used to austenitize loads prior to martempering.

The choice of austenitizing equipment depends mainly on availability, shape and size of workpieces, production requirements, and permissible distortion.

Work that is austenitized in a gaseous atmosphere may be oxidized while being transferred to an oil or salt martempering furnace. However, no special considerations are necessary when salt media are used in both the austenitizing and martempering furnaces.

Chloride salts are preferred for austenitizing before martempering in salt, because chloride salts are compatible with, and are easily separated from, the martempering bath. The composition and characteristics of a typical chloride salt medium are as follows: composition, 45 to 55% sodium chloride and 45 to 55% potassium chloride; melting range, 650 to 675 °C (1200 to 1250 °F); working range, 705 to 900 °C (1300 to 1650 °F).

If it is necessary to austenitize in a bath that contains cyanide, such as a liquid carburizing bath, the work should be transferred to a neutral salt (chloride type) rinse at the austenitizing temperature prior to martempering in salt. Direct quenching from a noncyanide-type liquid carburizing bath is permissible (see the article entitled "Liquid Carburizing", in this volume).

Salt bath furnaces for austenitizing (and neutral rinsing, if used) are most commonly of the submerged- or immersed-electrode type, although externally heated pots also are satisfactory.

Selection of Martempering Equipment

The furnace used for martempering is essentially a heat exchanger. Its basic functions are (a) to absorb heat from the work being quenched, and then (b) to dissipate this heat to the surroundings, to maintain a constant temperature.

In its simplest form, the martempering furnace consists of a steel pot that contains the oil or nitrate-nitrite salt and that is heated internally or externally. Such a simple furnace can be successfully used for martempering in limited production quantities.

For continuous production, more complex equipment is needed to maintain optimum quenching conditions.

Fuel-fired (usually gas-fired) immersion tubes, electrodes or immersion heaters, located across the back wall and sides of the furnace, are also used for internal heating. Occasionally, furnaces are externally heated by fuel or electricity, but such furnaces are limited to relatively small installations because they are difficult to control.

With internal heating, furnace size is unlimited and is based on production requirements. Sizes may range from 0.06 m³ (2 ft³) to lengths of more than 18 m (60 ft) and depths of more than 14 m (45 ft).

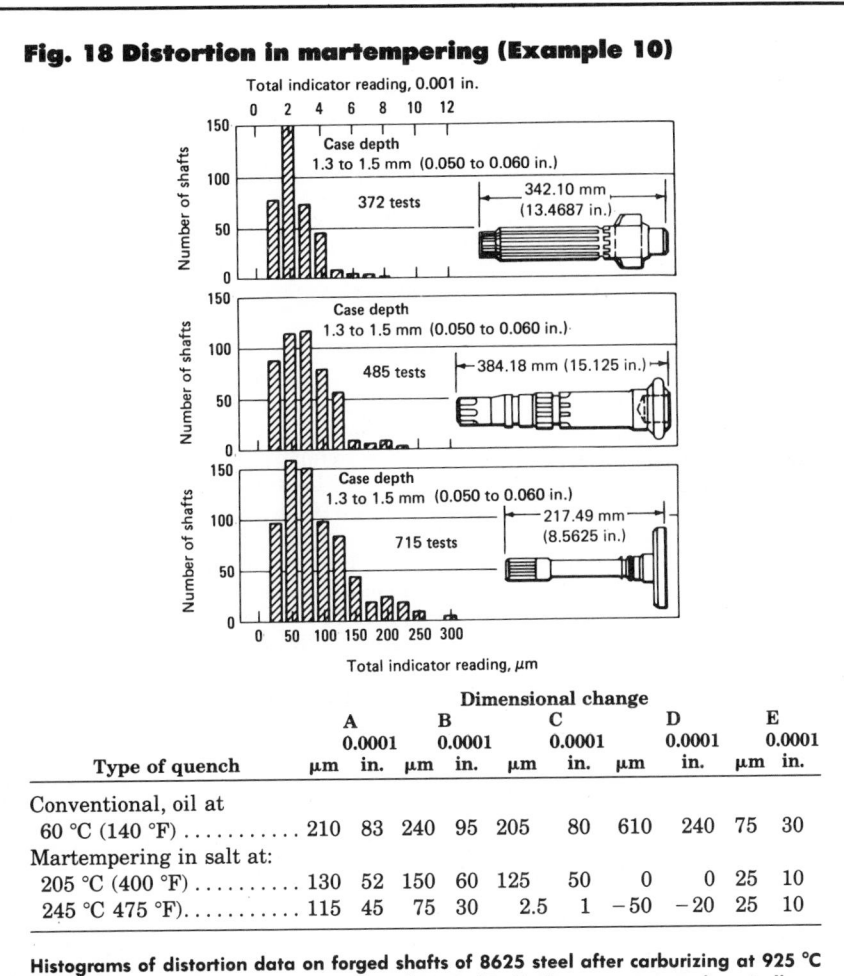

Fig. 18 Distortion in martempering (Example 10)

	Dimensional change									
	A		**B**		**C**		**D**		**E**	
	0.0001		0.0001		0.0001		0.0001		0.0001	
Type of quench	μm	in.	μm	in.	μm	in.	μm	in.	μm	in.
Conventional, oil at 60 °C (140 °F)	210	83	240	95	205	80	610	240	75	30
Martempering in salt at: 205 °C (400 °F)	130	52	150	60	125	50	0	0	25	10
245 °C 475 °F)	115	45	75	30	2.5	1	−50	−20	25	10

Histograms of distortion data on forged shafts of 8625 steel after carburizing at 925 °C (1700 °F) and martempering in oil at 165 °C (325 °F). Shafts were supported vertically on one end during heat treatment.

Table 6 Equipment requirements for oil martempering carburized parts made of 4024 and 4028 steels (Example 12)

Production requirements

Production rate . . . . 455 kg/h (1000 lb/h)
Weight of each piece . . . . 70 g to 1.0 kg (0.15 to 2.2 lb)
No. of pieces per hour . . . . Variable

Equipment requirements

Capacity of quench tank . . . . 18 925 l (5000 gal)
Type of oil . . . . Mineral oil (viscosity at 99 °C or 210 °F, 110 sus)
Temperature of oil . . . . 190 °C (375 °F)
Agitation . . . . High and low, as required

The operating temperature range for martempering usually is 165 to 595 °C (325 to 1100 °F). Temperature is measured by one or more thermocouples (depending on the size of the furnace), which are connected to controlling pyrometers that automatically control temperature within ±3 °C (±5 °F) by actuating the heating or cooling systems as required.

In continuous production, heat input from the work usually exceeds heat losses by radiation. Therefore, arrangements for cooling as well as for heating the martempering medium are required. To supplement heat losses by radiation from the surface of the bath, the exterior surface of the pot may be designed with cooling fins so that additional heat can be removed by forcing air through the cooling chamber between the pot and the casing walls. To increase heat dissipation, atomized water may be added to the stream of air. The mixture is then passed through a heat exchanger that is placed in the bath.

Agitation of the molten salt greatly improves the rate at which heat is extracted from the work. The furnace can be provided with a propeller-type pump that delivers the molten salt to the quench header into which the hot work is placed for quenching. By directing the flow of salt upward or downward through this quench header, effective control of the quenching severity of the salt is maintained, particularly if the speed of the pump can be varied. The salt also may be agitated by propeller mixers, centrifugal pumps or air bubblers located to produce effective agitation in the quench area. The use of air bubblers is not recommended, because they are not efficient and may cause carbonate buildup in the bath.

The furnace may have a second chamber in which the contaminating chlorides of barium, sodium and potassium that may be carried over from the austenitizing salt bath are separated by gravity from the nitrate-nitrite martempering salt. Contaminated salt is continuously circulated through the separating chamber, and, with a drop in salt temperature, more chlorides are precipitated from solution and settle to the bottom of the separating chamber. Because clarification is continuous, a uniform quenching condition is provided at all times. When work is austenitized in an atmosphere-controlled furnace, this chamber is not required.

The equipment required for martempering in hot oil is essentially the same as that for martempering in salt. Although the operating temperature range of oil baths is lower (95 to 230 °C, or 200 to 450 °F), the problem of maintaining a constant bath temperature is the same as for salt.

Equipment requirements for several specific operations are given in Tables 6 through 12 (Examples 12 through 18).

Example 12. The equipment required for martempering miscellaneous carburized parts (70 g to 1.0 kg, or 0.15 to 2.2 lb, per piece) made from 4024 and 4028 steels, at a rate of 455 kg/h (1000 lb/h), is listed in Table 6. Oil at 190 °C (375 °F) was the martempering medium used in this installation.

The parts were carburized at 915 to

Table 7 Equipment requirements for oil martempering carburized parts made of 8617 steel (Example 13)

Production requirements

Weight of load	455 kg (1000 lb) net
Weight of each piece	1.5 kg (3.3 lb)
No. of pieces treated per hour	75

Equipment requirements

Capacity of quench tank	7570 l (2000 gal)
Type of oil	Mineral oil with additives (viscosity at 38 °C (100 °F), 250 sus)
Temperature of oil	150 °C (300 °F)
Agitation	Direct flow(a)

(a) Agitation provided by two 3.7-kW (5-hp) motors driving 455-mm (18-in.) propellers at 370 rpm, causing the oil to flow at a rate of 915 mm/s (36 in./s)

Table 8 Equipment requirements for salt martempering automotive transmission shafts of 5040 steel (Example 14)

Production requirements

Weight of each piece	1.1 kg (2.5 lb)
Pieces per fixture	14
Production per hour:	
No. of loaded fixtures	10.7
No. of pieces	150
Weight of pieces	170 kg (375 lb)

Equipment requirements

Martempering furnace	Steel salt pot with resistance immersion heaters (70 kW)
Size of chamber(a)	1.22 m by 510 mm by 560 mm (48 by 20 by 22 in.)
Size of chloride-separating chamber	380 by 815 by 940 mm (15 by 32 by 37 in.)
Capacity of salt pot	3630 kg (8000 lb)
Type of salt	Nitrate-nitrite
Operating temperature	260 ± 3 °C (500 ± 5 °F)
Agitation	One 2.2-kW (3-hp) 150-mm (6-in.) propeller pump
Cooling system(b)	One 0.25-kW (⅓-hp) blower (25.5 m³/min, or 900 ft³/min)

(a) Total depth of salt was 940 mm (37 in.). (b) Cooling capacity of system was 215 kg/h or 475 lb/h (gross) from 845 to 230 °C (1550 to 450 °F) without exceeding 230 °C.

925 °C (1680 to 1700 °F) to a depth of 0.5 to 0.75 mm (0.020 to 0.030 in.) in a radiant-tube, gas-fired, three-row continuous pusher furnace with automatic quenching facilities. Parts were quenched from the final zone at 895 to 905 °C (1640 to 1660 °F). The carburizing atmosphere consisted of endothermic gas provided by a gas generator and 4.8 m³/h (170 ft³/h) of natural gas.

Example 13. In the operation for which equipment requirements are detailed in Table 7, oil at 150 °C (300 °F) was used as the martempering medium. In this operation, a 455-kg (1000-lb) load of 1.5-kg (3.3-lb) parts made of 8617 steel was quenched from an automatic batch-type furnace.

These parts were carburized to a depth of 1.0 mm (0.040 in.) in a radiant-tube, gas-fired batch furnace with automatic quenching facilities. The parts were carburized at 925 °C (1700 °F) and cooled in the furnace to a quenching temperature of 845 °C (1550 °F). The carburizing atmosphere consisted of endothermic gas and natural gas.

Example 14. Table 8 lists details of equipment required for salt martempering 1.1-kg (2.5-lb) transmission shafts, made of 5040 steel, at a rate of 170 kg/h (375 lb/h). The complete treatment was as follows:

- Austenitize for 35 min in neutral chloride salt at 845 °C (1550 °F).
- Martemper for 5 min at 260 °C (500 °F).

- Air cool (30 min) to 65 to 95 °C (150 to 200 °F).
- Temper for 45 min at 425 °C (800 °F).
- Air cool (5 min) to 95 to 120 °C (200 to 250 °F).
- Wash, rinse and dry.

Hardness after tempering and cooling to room temperature was 40 to 42 HRC (required hardness was 38 to 42 HRC).

Example 15. Table 9 lists equipment requirements for salt martempering 0.9-kg (2-lb) gears made of 6150 steel, at a rate of 128 pieces per hour. The gears were austenitized in a 60-kV·A submerged-electrode salt pot capable of heating 180 kg/h (400 lb/h) to 870 °C (1600 °F). The austenitizing pot measured 455 by 380 by 760 mm (18 by 15 by 30 in.) and contained 180 kg (400 lb) of alkali-chloride salt at an operating temperature of 845 °C (1550 °F). Table 9 gives martempering requirements.

Example 16. Details of a salt martempering furnace used in a commercial heat treating plant for quenching up to 180 kg/h (400 lb/h) are given in Table 10. This salt bath is capable of being operated at up to 400 °C (750 °F) and therefore can also be used for austempering.

Example 17. Table 11 shows details of a salt martempering bath capable of cooling 210 kg/h (465 lb/h) from 845 to 260 °C (1550 to 500 °F). This specific installation is used exclusively for heat treating piston rings made of 52100 steel.

Example 18. The equipment used in one installation for martempering aircraft landing-gear parts made of 4330 steel is detailed in Table 12. Following martempering, these parts are tempered at 425 °C (800 °F), which results in a hardness of 37 to 42 HRC.

Maintenance of Equipment

Lack of an established maintenance schedule may result in loss of process control or damage to the equipment, or both.

Salt System. Because martempering baths vary widely in design, shape, size and method of operation, it is not feasible to set forth a standard maintenance schedule. Manufacturers' recommendations for specific equipment should be followed; however, typical

Table 9 Equipment requirements for salt martempering gears made of 6150 steel (Example 15)

Production requirements

Weight of each piece . . . . 0.9 kg (2 lb)
Pieces per furnace load 32
Production per hour(a):
No. of pieces 128
Net work load . . . . 116 kg (256 lb)
Gross furnace load(b) . . . . 152 kg (336 lb)

Equipment requirements

Martempering furnace . . . . Immersion-heated salt pot(c)
Size of salt pot . . . . 610 by 380 by 840 mm
(24 by 15 by 33 in.)
Capacity of salt pot . . . . 270 kg (600 lb)
Type of salt . . . . Nitrate-nitrite (2% water added)
Quenching capacity of salt pot . . . . 180 kg/h (400 lb/h)
Operating temperature . . . . 205 °C (400 °F)
Agitation . . . . Air-operated stirrer

(a) Cycle time, 15 min. (b) Work plus fixtures: each fixture had an empty weight of 9.1 kg (20 lb) and contained eight gears. (c) Salt pot rated at 21 kV·A (3 phase, 60 cycle, 220 to 440 V) for heating to temperature range of 175 to 400 °C (350 to 750 °F); 0.37-kW (½-hp) blower (3 phase, 60 cycle, 220 V) used for cooling by driving room-temperature air between wall of pot and exterior shell of furnace

Table 10 Equipment requirements for salt martempering a variety of steel parts (Example 16)

Martempering furnace . . . . Steel salt pot
Method of heating . . . . 100-mm (4-in.) immersion tube
fired by natural gas(a)
Rated heat input . . . . 38.4 kW (131 000 Btu/h)
Operating temperature range(b) . . . . 205 to 400 °C (400 to 750 °F)
Capacity of salt pot . . . . 1725 kg (3800 lb)
Type of salt . . . . Nitrate-nitrite
Size of chloride-separating chamber(c) . . . . 205 mm by 1.07 m (8 by 42 in.)
Agitation method . . . . 0.19-kW (¼-hp) propeller mixer
Cooling method(d) . . . . Air through immersion tube

(a) Gas rated at 39.12 MJ/m³ (1050 Btu/ft³). (b) Temperature automatically controlled to ±3 °C (±5 °F). (c) Depth of salt, 760 mm (30 in.). (d) Cooling capacity, 180 kg/h or 400 lb/h (gross) from 845 to 260 °C (1550 to 500 °F) without exceeding 260 °C

schedules for maintaining a salt bath are as follows:

Each 8-h shift:

- Check instruments and thermocouples against a standard.
- Check neutrality of austenitizing bath (if salt containing cyanide is used for austenitizing, the cyanide content should be less than 2%).
- Remove sludge from martempering bath; mechanical separators (filter baskets or pans) eliminate the need for this operation.
- Check salt level.
- Check agitation of bath, and adjust as needed.

Weekly:

- Lubricate all moving parts.
- Remove sludge or contamination from the surface of the immersion heaters (radiant tubes or electrodes)

and from the walls, bottom, and top of the furnace.

Monthly:

- Check operation of all moving parts, such as blowers and pumps; adjust belt tension and alignment of shafts.
- Check electrical contacts of all contactors and relays, and repair as required; examine all electrical devices for proper operation.
- Remove fallen parts or debris from the quench header or quench area, to avoid fouling.
- Check heating and cooling facilities.

Semiannually or annually:

- Remove the salt from the furnace and check the condition of the pot, pumps and heating system.
- Clean and repair all electrical parts,

such as contactors, relays, motors and motor starters (it is especially important to remove condensed salt from all terminals and transformers).
- Clean and repair all moving parts, and lubricate as required.

Oil System. Following is a typical procedure for maintaining a high-temperature (175 to 205 °C, or 350 to 400 °F) oil-quench system:

Daily:

- Observe oil-temperature indicators every hour; verify the indicated temperature with a potentiometer once a day.
- Check oil level in sight gage, to ensure proper level and function of the automatic make-up unit.
- Check closed system frequently for proper pressure of the atmosphere blanket over the quenching oil; pressure of this blanket should be equal to furnace-atmosphere pressure.
- Check oil agitation by observation through sight doors and by noting if the pump shaft is operating at the proper speed or, preferably, by monitoring the load on the pump motor.
- Check condition of oil on parts emerging from quench. Undue discoloration or varnishing may indicate deterioration of oil.
- Visually check the performance of gas-fired immersion-heating tubes.

Weekly:

- Check condition of oil visually and by viscosity testing; record findings on graph to note trends.
- Check speed of pump shaft with tachometer, to ensure consistent oil flow.
- Check performance of temperature-control devices through "on-off" range, to ensure positive control.
- Check thermocouple.
- Check for proper operation and elevations of quenching mechanism or elevator.
- Check and clean pilot lights.
- Check and clean electrodes in ignition system of gas burners.
- Check safety control on gas lines for heating tubes.
- Check make-up oil supply.
- Check motor-operated venting systems for proper operation, and remove carbon buildup to prevent jamming.

Table 11 Equipment requirements for salt martempering piston rings made of 52100 steel (Example 17)

Piston rings were treated for a hardness of 52 to 55 HRC and minimum distortion. Treatment consisted of austenitizing for 45 min at 845 °C (1550 °F), martempering for 2½ min at 260 °C (500 °F), air cooling, and then tempering for 1 h at 370 °C (700 °F).

Production requirements

Production rate(a):

Gross . . . . 210 kg/h (465 lb/h)

Net . . . . 68 kg/h (150 lb/h)

Equipment requirements

Martempering furnace . . . . Immersion-heated steel salt pot(b)

Size of work chamber(c) . . . . 915 by 455 mm (36 by 18 in.)

Size of chloride-separating chamber(d) . . . . 380 by 785 mm (15 by 31 in.)

Capacity of salt pot . . . . 1950 kg (4300 lb)

Type of salt . . . . Nitrate-nitrite

Operating temperature of salt pot(e) . . . . 260 °C (500 °F)

Agitation . . . . One 2.2-kW (3-hp), 150-mm (6-in.) propeller pump

Cooling system . . . . One 0.25-kW (⅓-hp) blower (25.5 m³/min, or 900 ft³/min)

Cooling capacity . . . . 210 kg/h or 465 lb/h (gross) from 845 to 260 °C (1550 to 500 °F)

(a) Heavy mandrels were used as fixtures to retain shape of piston rings, thus accounting for wide difference between gross and net weights. (b) Resistance immersion heaters (60 kW). (c) Depth of salt, 760 mm (30 in.). (d) Depth of salt, 1.04 m (41 in.). (e) Automatically controlled to ±3 °C (±5 °F)

Table 12 Equipment requirements for salt martempering aircraft landing-gear parts of 4330 steel (Example 18)

Parts were treated as follows: austenitize for 45 to 60 min at 845 °C (1550 °F), martemper for 5 to 7 min at 205 °C (400 °F), air cool to room temperature, and then temper at 425 °C (800 °F) to produce a hardness of 37 to 42 HRC.

Production requirements

Production per hour . . . . One load of 270 kg (600 lb)

Equipment requirements

Martempering furnace . . . . Immersion-heated steel salt pot(a)

Size of work chamber(b) . . . . 1.5 by 1.9 m (60 by 75 in.)

Capacity of salt pot . . . . 21 850 kg (48 200 lb)

Type of salt . . . . Nitrate-nitrite

Operating temperature of salt pot(c) . . . . 205 °C (400 °F)

Agitation . . . . Two 2.2-kW (3-hp), 180-mm (7-in.) propeller agitators

Cooling system . . . . Natural draft

Cooling capacity . . . . 455 kg/h or 1000 lb/h (gross) from 845 to 205 °C (1550 to 400 °F)(d)

(a) Resistance immersion heaters (120 kW). (b) Depth of salt, 4.72 m (186 in.). (c) Temperature automatically controlled to ±3 °C (±5 °F). (d) Maximum temperature rise to 215 °C (415 °F)

Semiannually:

- Drain oil from, and clean, quench tank.
- Operate, and inspect functioning of, mechanical elements, such as elevator, oil pumps and adjustable oil deflectors.

- Check gas-fired heating tubes.
- Inspect V-belts for pump drive and agitators; replace if necessary.
- Check condition of temperature-measuring system.
- Determine condition of oil by physical and chemical testing of various properties.

Racking and Handling

The techniques for handling parts to be martempered may be similar to those for conventional oil quenching. However, racking and fixturing can be simplified for some martempering applications, because distortion is less.

Example 19. Extensive fixturing was required during conventional oil quenching of shaftlike parts made of 52100 steel. These parts were about 180 mm (7 in.) long and had a major diameter of 25 mm (1 in.). The fixtures weighed about the same as the work load.

A change from conventional quenching to martempering for 5 min in salt at 245 °C (475 °F) eliminated the need for the heavy, expensive fixtures and made it possible to hold distortion within the required limits. For martempering, the parts were placed vertically in simple fabricated baskets. This practice also resulted in a greater "payload".

The manner in which workpieces enter the bath generally is less critical in martempering than in conventional oil quenching. For example, large flat parts do not dish as much when martempered. However, the shape of each part must be considered individually, and some trial and error often is necessary before an optimum handling technique can be developed.

Nesting of small parts can be a problem, and development of a handling method that will result in uniform quenching often requires experimentation. A technique that proved successful for one application is described below.

Example 20. Flat blades that were not well suited to fixturing or wiring had a tendency to nest, which caused nonuniform quenching during martempering. This problem was overcome by use of a pump that directed a heavy flow of molten salt upward through a perforated metal basket, thus keeping the parts separated. This technique requires that the flow of salt be regulated to the floating characteristics of the workpieces.

For fixturing of parts, one of the following considerations may apply:

- Long, slender parts should be suspended.
- Symmetrical parts, such as bearing races and cylinders, can be stacked and supported on a rack or grid.
- Flat parts, such as circular saw blades, mower blades and clutch

plates, are best supported on horizontal slotted rods that provide the necessary separation.

- Coils of wire are supported either vertically on a spider-type grid or horizontally on support rods.
- Small parts can be loaded into a perforated ladle or basket and then dumped into the quench to obtain intimate quenching of all parts.
- Fixture design should be simple, free of welds (if possible) and easy to maintain. For example, fixtures supporting vertical stacks of bearing races should be removable for periodic grinding to maintain flatness.

Proper spacing of workpieces, to permit good flow of the quenchant around each part, is an important consideration in martempering. Also, the combined weight of workpieces and fixtures must be limited to the extent that the heat they contain is insufficient to cause a sharp increase in the temperature of the quenchant. In this regard, the size of a salt bath furnace is determined not only by the physical size of the work, but also by design requirements such as salt-separating systems, a means for agitation of the salt, and sufficient area for dissipation of heat through sidewalls.

Washing the Work

Regardless of the martempering medium used, the work load should not be washed until transformation is complete (all portions of workpieces should be near room temperature).

Martempering salts are completely water-soluble, and any hot-water soaking tank or spray washer will remove all salt from accessible areas. Cold water can be used, but its washing action is much slower.

The speed of washing will depend on how much hot water unsaturated with salt flows over the salt-coated surface. Therefore, agitation will increase the washing action, and a device such as an open-impeller sump pump can be used in a soaking tank to direct a stream of hot water into recessed areas. Steam jets may be used as supplementary equipment to remove salt from sections of intricately shaped parts with difficult-to-reach recesses for which conventional washing equipment is inadequate.

Quenching oils often present washing problems. Martempering oil adheres more tenaciously to workpieces than does conventional quenching oil, because of its higher viscosity—as high as 1200 Saybolt Universal Seconds (sus) at 38 °C (100 °F), compared with 100 sus at 38 °C for conventional oil.

Basically, the washing equipment is the same for both types of quenching oils, but a heavy-duty cleaner must be used to remove martempering oil. Several proprietary heavy-duty silicate-alkaline cleaners are available for cleaning parts martempered in hot oil. Vapor degreasing, and steam cleaning without detergent, may be used in special applications. (For additional information, see the article on selection of cleaning process, in Vol 5 of this Handbook.)

Austempering of Steel

By the ASM Committee on
Austempering of Steel*

AUSTEMPERING is the isothermal transformation of a ferrous alloy at a temperature below that of pearlite formation and above that of martensite formation. Steel is austempered by being:

- Heated to a temperature within the austenitizing range, usually 790 to 870 °C (1450 to 1600 °F)
- Quenched in a bath maintained at a constant temperature, usually in the range of 260 to 400 °C (500 to 750 °F)
- Allowed to transform isothermally to bainite in this bath
- Cooled to room temperature, usually in still air.

The process is described in detail by the inventors, E. S. Davenport and E. C. Bain, in U.S. Patent 1 924 099. The fundamental difference between austempering and conventional quenching and tempering is shown schematically in Fig. 1. Austempering of steel and hardenable grades of cast iron offers several potential advantages:

- Increased ductility or notch toughness at a given hardness (Table 1)
- Reduced distortion, which lessens subsequent machining time, stock removal and cost
- The shortest over-all time cycle to through harden within the hardness range of 35 to 55 HRC, with resulting savings in energy and capital investment.

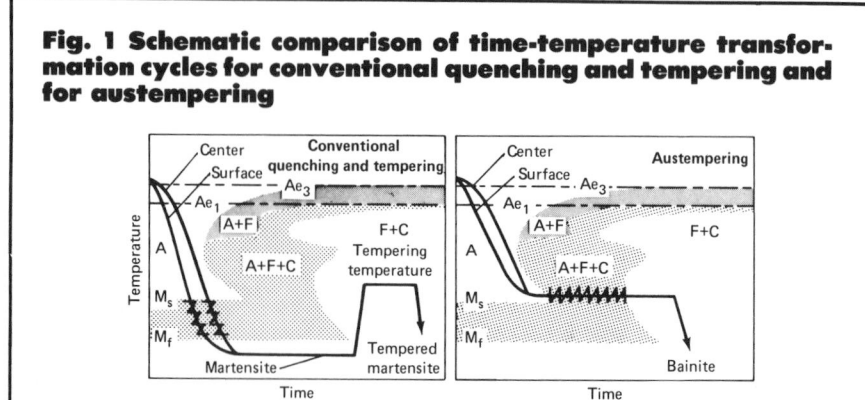

Fig. 1 Schematic comparison of time-temperature transformation cycles for conventional quenching and tempering and for austempering

For true austempering, the metal must be (a) cooled from the austenitizing temperature to the temperature of the austempering bath fast enough so that no transformation of austenite occurs during cooling, and (b) held at bath temperature long enough to ensure complete transformation of austenite to bainite. Modifications of these procedures, constituting departures from true austempering, are discussed in the section on modified austempering, later in this article.

Quenching Media for Austempering

Molten salt is the quenching medium most commonly used in austempering, because: (a) it transfers heat rapidly; (b) it virtually eliminates the prob-

lem of a vapor phase barrier during the initial stage of quenching; (c) its viscosity is uniform over a wide range of temperature; (d) its viscosity is low at austempering temperatures (near that of water at room temperature), thus minimizing dragout losses; (e) it remains stable at operating temperatures and is completely soluble in water, thus facilitating subsequent cleaning operations; and (f) the salt can be easily recovered from wash waters so that there is no discharge to drain.

Formulations and characteristics of two typical salt quenching baths are given in Table 2. The high-range salt is suitable for austempering only, whereas the wide-range salt may be used for austempering, martempering, and modifications thereof.

Water Additions to Salt. The

*Quentin D. Mehrkam, *Chairman,* Senior Vice President, Ajax Electric Co.; Robert W. Foreman, Director, Research & Development, Park Chemical Co.; David E. Gaylord, President, Progressive Heat Treating Co.; Walter E. Heyer, Administrative General Manager, Selas Corp. of America; Gilbert M. Lahr, Chief Metallurgist, Detroit Diesel Allison

quenching severity of a nitrate-nitrite salt can be increased significantly by careful addition of water. Agitation of the salt is necessary to disperse the water uniformly. Periodic water additions are needed to maintain required water content (see article on martempering). Water can be added with complete safety as follows:

- Water may be misted at a regulated rate into a vigorously agitated area of the molten bath.
- In installations where the salt is pump-circulated, returning salt is cascaded into the quench zone. A controlled fine stream of water may be injected into the cascade of returning salt.
- The austempering bath may be kept saturated with moisture by introducing steam directly into the bath. The steam line should be trapped and equipped with a discharge to avoid emptying condensate directly into the bath.

Water should never be added to a salt bath from a pail or dipper. Oil is seldom used for austempering, because of its chemical instability and resulting change in viscosity at austempering temperatures. Because of its persistent vapor phase, oil is a slower quenchant than salt at elevated temperatures and presents a fire hazard.

Steels for Austempering

The selection of steel for austempering must be based on transformation characteristics as indicated in time-temperature transformation (TTT) diagrams. Three important considerations are (a) the location of the nose of the TTT curve and the time available for bypassing it, (b) the time required for complete transformation of austenite to bainite at the austempering temperature, and (c) the location of the M_s point.

As indicated in Fig. 2, 1080 carbon steel possesses transformation characteristics that provide it with limited suitability for austempering. Cooling from the austenitizing temperature to the austempering bath must be accomplished in about 1 s to avoid the nose of the TTT curve, and thus prevent transformation to pearlite during cooling. Depending on the temperature, isothermal transformation in the bath is completed within a period ranging from a few minutes to about 1 h. Because of the rapid cooling rate re-

quired, austempering of 1080 can be successfully applied only to thin sections of about 5 mm (0.2 in.) maximum.

5140 low-alloy steel is well suited to austempering, as indicated by the TTT curve for this steel shown in Fig. 2. Approximately 2 s is allowed in which to bypass the nose of the curve, and transformation to bainite is completed within 1 to 10 min at 315 to 400 °C (600 to 750 °F). Parts made of 5140 steel or of other steels with similar transformation characteristics are adaptable to austempering in larger section sizes than are feasible for 1080 steel, because of the greater time allowed for bypassing the nose of the curve.

In addition to the steels previously indicated (1080 and 5140), steels adaptable to austempering include:

- Plain carbon steels containing 0.50 to 1.00% C and a minimum of 0.60% Mn
- High-carbon steels containing more than 0.90% C and, possibly, a little less than 0.60% Mn
- Certain carbon steels (such as 1041) with a carbon content of less than 0.50% but with manganese content in the range from 1.00 to 1.65%
- Certain low-alloy steels (such as the series 5100 steels) containing more than 0.30% C; the series 1300 to 4000 steels with carbon contents in excess of 0.40%; and other steels, such as 4140, 6145 and 9440.

Some steels, although they have sufficient carbon or alloy content to be hardenable, are borderline or impractical for austempering because (a) transformation at the nose of the TTT curve starts in less than 1 s, thus making it virtually impossible to quench other than thin sections in molten salt without forming some pearlite, or (b) they require excessively long periods of time for transformation. A typical example of a steel belonging to the first category is 1034, whose transformation characteristics are shown in Fig. 2. Transformation characteristics for 9261 (also shown in Fig. 2) indicate no difficulty in quenching past the nose of the curve, but the time required for isothermal transformation to bainite (about 24 h) is excessive. Other alloy steels having similarly excessive transformation times include those in series 4300, 4600 and 4800.

Austenitizing Temperature. As the austenitizing temperature of a high-carbon steel increases, the M_s temperature decreases because of more complete solution of carbon. The direct effects of alloying elements on the M_s point are much less pronounced than the effect of carbon. However, carbide-forming elements (such as molybdenum and vanadium) can tie up the carbon as alloy carbides and prevent complete solution of carbon. The approximate M_s temperature, in degrees Centigrade, of a completely austeni-

Table 1 Mechanical properties of 1095 steel heat treated by three methods

Specimen No.	Heat treatment	Hardness, HRC	Impact strength J	Impact strength ft·lb	Elongation in 25 mm or 1 in., %
1	Water quenched and tempered	53.0	16	12	· · ·
2	Water quenched and tempered	52.5	19	14	· · ·
3	Martempered and tempered	53.0	38	28	· · ·
4	Martempered and tempered	52.8	33	24	· · ·
5	Austempered	52.0	61	45	11
6	Austempered	52.5	54	40	8

Table 2 Compositions and characteristics of salts used for austempering

	High range	Wide range
Sodium nitrate	45 to 55%	15 to 25%
Potassium nitrate	45 to 55%	45 to 55%
Sodium nitrite	· · ·	25 to 35%
Melting point (approx)	220 °C (430 °F)	150 to 165 °C (300 to 330 °F)
Working temperature range	260 to 595 °C (500 to 1100 °F)	175 to 540 °C (345 to 1000 °F)

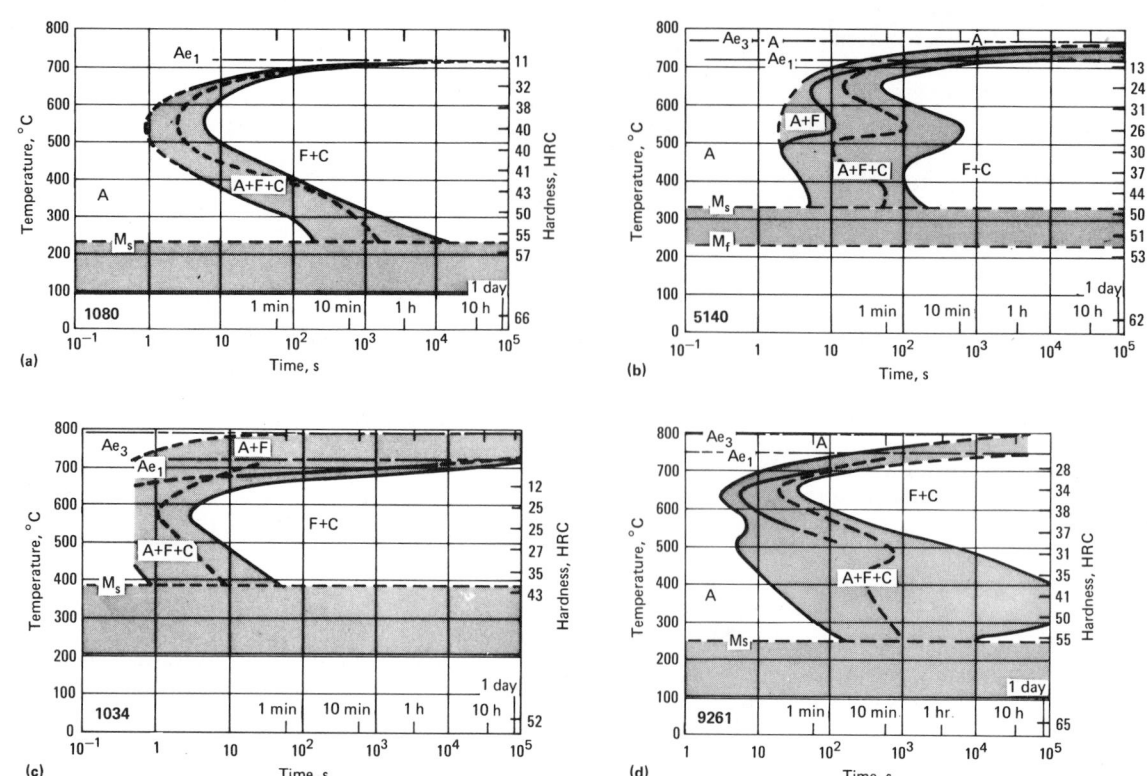

Fig. 2 Transformation characteristics of 1080, 5140, 1034 and 9261 steels, in relation to their suitability for austempering

1080, limited suitability for austempering because pearlite reaction starts too soon near 540 °C (1000 °F); 5140, well suited to austempering; 1034, impossible to austemper because of extremely fast pearlite reaction time at 540 to 595 °C (1000 to 1100 °F); 9261, not suited to austempering because of slow reaction to bainite at 260 to 400 °C (500 to 750 °F). See text for discussion of steel selection.

tized steel can be calculated by means of the following formula:

$$M_s = 538 - (361 \times \%C) \\ - (39 \times \%Mn) \\ - (19 \times \%Ni) \\ - (39 \times \%Cr)$$

Expressed in degrees Fahrenheit, the formula is:

$$M_s = 1000 - (650 \times \%C) \\ - (70 \times \%Mn) \\ - (35 \times \%Ni) \\ - (70 \times \%Cr)$$

Austenitizing temperature has a significant effect on the time at which transformation begins. As the austenitizing temperature is increased above normal (for a specific steel), the nose of the TTT curve shifts to the right, because of grain coarsening. In Fig. 2, for example, approximately ³/₄ s is allowed for quenching 1080 steel in order to avoid the nose of the curve. However,

this is based on an austenitizing temperature of 790 °C (1450 °F); higher austenitizing temperatures move the TTT curve to the right, allowing more time before transformation begins.

Practical use is sometimes made of this phenomenon in order to process compositions or section sizes that would otherwise be borderline for austempering. However, the coarser grain sizes resulting from higher austenitizing temperatures may be detrimental to some desired properties. Therefore, it is recommended that standard austenitizing temperatures be given preference for austempering. If experience with specific compositions and parts proves that advantages can be gained from the use of a higher temperature and that no harm will be incurred from grain coarsening, higher austenitizing temperatures may be employed.

Section Thickness Limitations

Maximum thickness of section is important in determining whether or not a part can be successfully austempered. For example, a fully bainitic structure can be obtained almost as easily in a 3-mm (1/8-in.) thick part that is 0.6 m (24 in.) long by 50 mm (2 in.) wide as in a 3-mm (1/8-in.) thick part that is only 75 mm (3 in.) long by 25 mm (1 in.) wide, assuming that all other conditions are equal and that proper facilities for handling the longer part are available.

For 1080 steel, a section thickness of about 5 mm (0.2 in.) is the maximum that can be austempered to a fully bainitic structure. Carbon steels of lower carbon content will be restricted to a proportionately lesser thickness. Lower carbon steels containing boron, howev-

Table 3 Hardness of various steels and section sizes of austempered parts

Steel	Section size mm	Section size in.	Salt temperature °C	Salt temperature °F	M_s temperature(a) °C	M_s temperature(a) °F	Hardness, HRC
1050	3(b)	0.125(b)	345	655	320	610	41 to 47
1065	5(c)	0.187(c)	(d)	(d)	275	525	53 to 56
1066	7(c)	0.281(c)	(d)	(d)	260	500	53 to 56
1084	6(c)	0.218(c)	(d)	(d)	200	395	55 to 58
1086	13(c)	0.516(c)	(d)	(d)	215	420	55 to 58
1090	5(c)	0.187(c)	(d)	(d)	· · ·	· · ·	57 to 60
1090 (e)	20(c)	0.820(c)	315(f)	600(f)	· · ·	· · ·	44.5 (avg)
1095	4(c)	0.148(c)	(d)	(d)	210(g)	410(g)	57 to 60
1350	16(c)	0.625(c)	(d)	(d)	235	450	53 to 56
4063	16(c)	0.625(c)	(d)	(d)	245	475	53 to 56
4150	13(c)	0.500(c)	(d)	(d)	285	545	52 max
4365	25(c)	1.000(c)	(d)	(d)	210	410	54 max
5140	3(b)	0.125(b)	345	655	330	630	43 to 48
5160 (e)	26(c)	1.035(c)	315(f)	600(f)	255	490	46.7 (avg)
8750	3(b)	0.125(b)	315	600	285	545	47 to 48
50100	8(c)	0.312(c)	(d)	(d)	· · ·	· · ·	57 to 60

(a) Calculated. (b) Sheet thickness. (c) Diameter of section. (d) Salt temperature adjusted to give maximum hardness and 100% bainite. (e) Modified austempering; microstructure contained pearlite as well as bainite. (f) Salt with water additions. (g) Experimental value

Fig. 3 Effect of section thickness on the hardness of austempered carbon and alloy steels

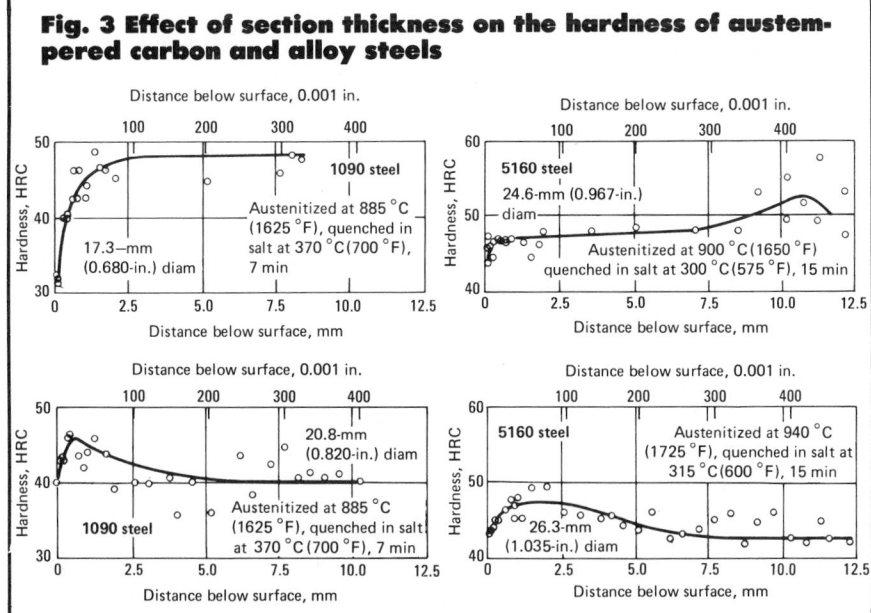

The 5160 steel was quenched in agitated salt containing some water. HRC values were converted from microhardness readings taken with a 100-g load. Low values of surface hardness result from decarburization. High hardness at center of 24.6-mm (0.967-in.) 5160 is due to segregation.

er, can be successfully austempered in heavier sections. In some alloy steels, section thicknesses up to about 25 mm (1 in.) can be austempered to fully bainitic structures.

Nevertheless, sections of carbon steel significantly thicker than 5 mm (0.2 in.) are regularly austempered in production when some pearlite is permissible in the microstructure. This is demonstrated in Table 3, which lists section sizes of austempered parts made of various steels.

The effects of section thickness on the hardness of austempered 1090 and 5160 steels are shown in Fig. 3. Hardness remains reasonably consistent to the center of a 17-mm (0.680-in.) diameter of 1090, but it becomes erratic when the diameter is increased to 21 mm (0.820 in.). A similar difference is evident for 5160 steel when the diameter is increased from 24.6 to 26 mm (0.967 to 1.035 in.).

The lower surface hardnesses indicated for 1090 and 5160 were the result of surface decarburization. The high core hardness of the 24.6-mm (0.967-in.) diameter of 5160 is attributed to chemical segregation in the center portion of the bar.

Applications

Austempering usually is substituted for conventional quenching and tempering for either or both of two reasons: (a) to obtain improved mechanical properties (particularly higher ductility or notch toughness at a given high hardness), and (b) to decrease the likelihood of cracking and distortion. In some applications, austempering is less expensive than conventional quenching and tempering. This is most likely when small parts are treated in an automated setup wherein conventional quenching and tempering comprise a three-step operation—that is, austenitizing, quenching and tempering. Austempering requires only two processing steps: austenitizing and isothermal transformation in an austempering bath.

The range of austempering applications generally encompasses parts fabricated from bars of small diameter or from sheet or strip of small cross section. Austempering is particularly applicable to thin-section carbon steel parts requiring exceptional toughness at a hardness near 50 HRC.

In austempered carbon steel parts, reduction in area is usually much higher than in conventionally quenched and tempered parts, as indicated in the following tabulation for 5-mm (0.180-in.)-diam bars of 0.85% C plain carbon steel:

Austempered mechanical properties

Tensile strength	1780 MPa (258 ksi)
Yield strength	1450 MPa (210 ksi)
Reduction in area	45%
Hardness.....................	50 HRC

Quenched and tempered mechanical properties

Tensile strength	1795 MPa (260 ksi)
Yield strength	1550 MPa (225 ksi)
Reduction in area	28%
Hardness.....................	50 HRC

The mechanical properties of sway bars made of 1090 steel and hardened by these two processes are listed in Table 4 (see also Table 1).

It is more important that austempered parts possess desired mechanical properties than that they have a 100% bainitic structure. In Table 3, it is evident from the hardness values indicated that several of the austempered steels have mixed structures. Higher-than-normal hardness indicates that some martensite has formed, and below-normal hardness indicates the presence of some pearlite. The formation of pearlite is more common and results from a quenching speed too slow for complete avoidance of the nose of the TTT curve.

In industrial austempering practice, a sizable percentage of applications are successful with less than 100% bainite. In fact, 85% bainite has been found to be satisfactory for some applications. Obviously, austempering is often "modified" to some degree in commercial application, and whether or not the metallurgical properties obtained conform to those obtained in true austempering is at least partially ignored if the treated parts meet service requirements. However, variations in hardenability from heat to heat may give rise to erratic results from variations in the amount of pearlite when borderline

Table 4 Comparison of typical mechanical properties of austempered and of oil quenched and tempered sway bars of 1090 steel

Property(a)	Austempered at 400 °C (750 °F)(b)	Quenched and tempered(c)
Tensile strength, MPa (ksi)	1415 (205)	1340 (200)
Yield strength, MPa (ksi)	1020 (148)	895 (130)
Elongation, %	11.5	6.0
Reduction of area, %	30	10.2
Hardness, HB	415	388
Fatigue cycles (d)	105 000(e)	58 600(f)

(a) Average values. (b) Six tests. (c) Two tests. (d) Fatigue specimens 21 mm (0.812 in.) in diameter. (e) Seven tests; range, 69 050 to 137 000. (f) Eight tests; range, 43 120 to 95 220

Table 5 Typical production applications of austempering
Parts listed in order of increasing section thickness

Part	Steel	Maximum section thickness mm	in.	Parts per unit weight kg	lb	Salt temperature °C	°F	Immersion time, min	Hardness, HRC
Plain carbon steel parts									
Clevis	1050	0.76	0.030	770/kg	350/lb	360	680	15	42
Follower arm	1050	0.76	0.030	412/kg	187/lb	355	675	15	42
Spring	1080	0.79	0.031	220/kg	100/lb	330	625	15	48
Plate	1060	0.81	0.032	88/kg	40/lb	330	630	6	45 to 50
Cam lever	1065	1.02	0.040	62/kg	28/lb	370	700	15	42
Plate	1050	1.02	0.040	0.1 kg	¼ lb	360	675	15	42
Type bar	1065	1.02	0.040	141/kg	64/lb	370	700	15	42
Tabulator stop	1065	1.22	0.048	440/kg	200/lb	360	680	15	45
Lever	1050	1.27	0.050	···	···	345	650	15	45 to 50
Chain link	1050	1.52	0.060	573/kg	260/lb	345	650	15	45
Shoe-last link	1065	1.52	0.060	86/kg	39/lb	290	550	30	52
Shoe-toe cap	1070	1.52	0.060	18/kg	8/lb	315	600	60	50
Lawn mower blade	1065	3.18	0.125	0.3 kg	⅔ lb	315	600	15	50
Lever	1075	3.18	0.125	24/kg	11/lb	385	725	5	30 to 35
Fastener	1060	6.35	0.250	110/kg	50/lb	310	590	25	50
Stabilizer bar	1090	19.05	0.750	4.5 kg	10 lb	370	700	6 to 9	40 to 45
Boron steel bolt	10B20	6.35	0.250	100/kg	45/lb	420	790	5	38 to 43
Alloy steel parts									
Socket wrench	6150	···	···	0.06 kg	⅛ lb	365	690	15	45
Chain link	Cr-Ni-V(a)	1.60	0.063	110/kg	50/lb	290	550	25	53
Pin	3140	1.60	0.063	5500/kg	2500/lb	325	620	45	48
Cylinder liner	4140	2.54	0.100	3 kg	7 lb	260	500	14	40
Anvil	8640	3.18	0.125	0.3 kg	¾ lb	370	700	30	37
Shovel blade	4068	3.18	0.125	···	···	370	700	15	45
Pin	3140	6.35	0.250	100/kg	45/lb	370	700	45	40
Shaft	4140(b)	9.53	0.375	0.1 kg	¼ lb	385	725	15	35 to 40
Gear	6150	12.70	0.500	0.9 kg	2 lb	305	580	30	45
Carburized steel parts									
Lever	1010	3.96	0.156	6.8 kg	15 lb	385	725	5	30 to 35(c)
Shaft	1117	6.35	0.250	66/kg	30/lb	385	725	5	30 to 35(c)
Block	8620	11.13	0.438	132/kg	60/lb	290 to 315	550 to 600	30	50(c)

(a) Contains 0.65 to 0.75% C. (b) Leaded grade. (c) Case hardness

conditions are involved in modified austempering.

Table 5 presents processing data for a number of specific parts made of various plain carbon, alloy and carburizing steels; these data are representative of austempering practice in at least a dozen different manufacturing plants.

Equipment and Processing Procedures

Austenitizing prior to austempering can be performed in molten salt baths, in one of several types of standard heat treating furnaces (including box, retort and shaker-hearth furnaces), or in special furnaces (such as direct-fired tunnel types). In the types of austenitizing furnaces that employ a gaseous atmosphere, atmosphere control is extremely important, because the formation of scale on parts will inhibit the rapid heat exchange that is essential in a successful austempering operation.

Molten salt baths for austenitizing are widely used for batch-type operation; they can also be mechanized for continuous operation. Heating in salt for batch-type operation is advantageous because the film of adhering salt prevents scaling during transfer from the austenitizing operation to the austempering bath. Salt bath furnaces heated by either immersed or submerged electrodes are most common, although externally heated pots also are used.

Because most austempering is performed in molten salt, the salt bath used for austenitizing must be compatible with the austempering salt. Chloride salts are recommended for austenitizing; the composition and characteristics of a typical chloride salt bath for austenitizing are as follows:

Sodium chloride . . . . 45 to 55%
Potassium chloride. . . . . 45 to 55%
Melting temperature . . 650 to 675 °C
    (1200 to 1250 °F)
Working temperature
    range. . . . . 705 to 900 °C
    (1300 to 1650 °F).

Austenitizing salt baths should not contain cyanide, because it reacts violently with quenching salts. Carbonate compositions and mixtures containing barium salts are not recommended, because they are difficult to separate from quenching salts.

Shaker-hearth furnaces with atmosphere control are widely used for austenitizing, especially where large quantities of small parts are being processed. In a typical automated installation, the workpieces enter the shaker from a feeder and, after being heated to the austenitizing temperature, are dropped through a chute into the austempering bath, thus avoiding exposure in air and the possibility of scaling. The parts are transported through the austempering bath on variable-speed mesh or plate belt-type conveyors at a speed that allows complete transformation before they are automatically transferred to a constant-speed conveyorized washing and drying setup.

Feeders for shaker-hearth furnaces vary widely in design, depending on the size, shape and fragility of workpieces. In one type of equipment, parts are fed by gravity from the hopper onto a reciprocating hearth, on which they are transported through the shaker-hearth furnace and then into the austempering furnace. This mechanism is used for small round or flat parts that will not be distorted by the vibrating action. Other methods of feeding to the moving hearth include "Walmil"-style ram volumetric feeders, weigh loaders, and special magnetic conveyor-and-bin combinations which disentangle parts that bundle.

Rotary retort furnaces are widely used for products in the saw-chain, concrete, explosive drive and air driven nail industries and in similar applications where both hardness and ductility are prime factors in the end product and reasonably large quantities of similar small parts are being processed without the threat of damage by the tumbling action of the retort. One of the advantages of the rotary retort, which advances the work by means of the internal spiral, is the absolute control of the retention time within the heated chamber. This control ensures uniformity of metallurgical properties of the product. Excellent atmosphere control is also attained, because the work progresses through a sealed retort, is sealed at the quench chute by the quench media, and the process has a restricted part entry area. Consistent production rates and uniformity of results are obtained by controlling the loading rate with a weigh loader that is fed from a storage hopper by discharging the workpieces to a loading elevator that in turn transports the load through a chute into the partially sealed entry of the retort.

Belt furnaces may use either a cast-link or mesh-type part-carrier belt. In either instance, the belt is totally contained within the heated chamber, with the work being fed into the chamber by means of either a shaker or vibrating hearth onto the moving surface of the belt.

The mesh-belt type of furnace is primarily used to process relatively thin sheet-metal stampings that would become distorted if loaded directly onto a heavy cast-link belt or would become tangled or distorted if processed through a rotary retort. Continuous automated parts feeding is accomplished by use of specialized magnetic conveyor feeders. These feeders strip the parts that tend to tangle from one edge of the pile in the feeding bin and carry them up to a vibratory trough. The trough in turn feeds the workpieces to the hearth that enters the furnace.

The cast-link belt types are used to handle heavier parts that would not be subject to distortion upon loading onto the hot cast-link belt. Continuous feeding is accomplished with Walmil volumetric feeders, automated orienters, or simple bin and vibratory trough types, depending upon part configuration. Each feeding system delivers parts to the hearth, which enters the furnace.

Belt-type furnaces ensure controlled retention time for repeatability of metallurgical properties of the product. The integrity of the atmosphere is controlled by sealing the furnace. A flexible seal about the shaker or vibrator hearth feeding into the chamber restricts atmosphere exhaust and the opening for parts entry. The discharge chute into the molten salt also is sealed.

Conveyorized transport through the austempering salt bath and wash tanks as described for shaker hearth furnaces is also employed with both the rotary retort and belt furnace systems.

Direct-fired tunnel-type furnaces are employed for heat treating large quantities of similar parts that must be hung from hooks or fixtures to minimize distortion. In a typical installation, the hooks or fixtures carrying the work travel in front of the operators and can be easily loaded. The work is conveyed through the tunnel-type heating zone, automatically removed after heating and dropped into the austempering bath, and then conveyed to washing and drying. During austenitizing, parts are in a protective atmosphere provided by the controlled gas-to-air ratio of the burners. The entire operation is automatic except for the hand loading of parts onto hooks.

Austempering Furnaces. Simple salt baths, either fuel-fired or electrically heated, are suitable for low-production austempering. Circulation can be provided by a propeller-type mixer. Most of these furnaces have no provision for cooling and depend on heat losses to equal the heat input acquired from workpieces.

Sufficient production quantities warrant the use of furnaces specifically designed for austempering. These furnaces usually provide for automatic handling. The salt bath units are heated with immersion heaters, immersed electrodes, or gas-fired immersion tubes. To maintain a constant temperature, heat is dissipated by forced air that cools the pot sidewall or by immersed plate coils. Both heating and cooling are automatically controlled. A salt-separating system is required to separate chloride salts dragged over from the austenitizing furnace. Chlorides, scale and other contaminants are removed by:

- *Gravity separation:* providing a separate chamber for the suspensions to settle; settled solids are then removed by manual desludging or sludge pans.
- *Continous filtration:* suspended contaminants are continuously passed through filter baskets; the baskets are periodically removed and dumped.

To provide uniform quenching, one or more propeller agitators are used to agitate the salt.

Racking and handling procedures for austempering are not necessarily different from those used for other heat treating operations. Racking or other means of keeping parts separated is often essential to the process. To minimize distortion, delicate parts must be handled with particular care.

For heating in tunnel-type or other special furnaces, a standard form of racking (Fig. 4) is usually employed. (Provision must also be made for lowering of parts into the austempering bath.) Parts similar to those illustrated in Fig. 4 may be batch treated by utilizing fixtures such as those shown in Fig. 5(a). Another common type of rack, useful for handling disks or rings, is shown in Fig. 5(b). The well-known Christmas tree-type of rack is used for parts that can be hung. Some parts are strung on wires that can be suspended into the austenitizing and quenching media. Heavier parts, or parts that cannot be hung, may be placed in baskets provided that rapid and uniform quenching can be obtained. Parts must not be stacked together, because this will result in nonuniform quenching.

Racks, baskets and other fixtures may be made of carbon steel, but if they are expected to last for many cycles it is usually more economical to make them of a heat-resisting alloy such as 35Ni-18Cr or another nickel-base alloy.

Washing and Drying. Salts used in the austempering bath are water soluble, and no special equipment is needed for their removal. Normal practice is to use clean, hot water, with agitation, for dissolving and removing salt from austempered parts. Cold water may be used, but its dissolving action is much slower. Hot water may not suffice to dissolve all salt from parts with deep blind holes, small crevices, laps, hidden corners, or ledges. If it is not completely removed, salt may bleed from such crevices or joints long after heat treating. Quenching into hot water directly from the austempering salt bath is often utilized to blow out trapped salt. Large parts may be cleaned with high-pressure steam. However, rapid cooling may cause distortion.

Equipment requirements for specific austempering operations are detailed in the three production-based examples that follow.

Example 1. Table 6 compares equipment required for two different methods of austenitizing and austempering small mechanical stops and levers. Parts of this type weighing 0.007 kg (0.015 lb) were processed at the rate of 4000/h by being austenitized at 845 °C (1550 °F) in an electrically heated salt-pot furnace and austempered in salt at 370 °C (700 °F). Parts weighing 0.002 kg (0.005 lb) were processed at the rate of 3700/h by being austenitized at 870 °C (1600 °F) in a controlled-atmosphere shaker-hearth furnace and austempered in salt at 360 °C (680 °F).

Example 2. Table 7 gives details of equipment required for austempering parts made of 1080 steel, at the rate of 1500/h. These parts, which varied in weight from 1.1 to 1.8 kg (2.4 to 3.9 lb), were austenitized at 925 °C (1700 °F) before being austempered.

Example 3. Table 8 lists details of the equipment comprising a manually operated monorail line for preheating, austenitizing, austempering, washing and rinsing, and corrosion protection,

Fig. 4 Method of continuously transporting parts on fixtures through an austenitizing furnace and into an austempering bath

Workpiece

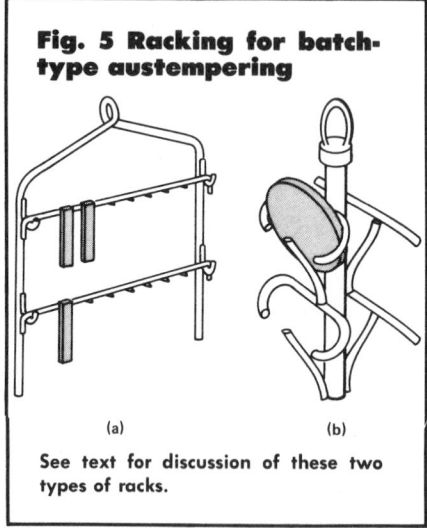

Fig. 5 Racking for batch-type austempering

(a)   (b)

See text for discussion of these two types of racks.

and indicates production capacities for processing tubes, plates and disks made of 1035 and 1060 steels.

Control of Process Variables

The achievement of desired results from austempering processes depends on control of the bath temperature and the immersion time of the work being processed, and on agitation of the bath.

Temperature control of austempering baths is important because:

- Bath temperature determines the

Table 6 Equipment requirement for two methods of austenitizing and austempering mechanical stops and levers

Requirements	Method 1	Method 2
Production requirements		
Weight of each piece, kg (lb).................................	0.007 (0.015)	0.002 (0.005)
Weight of each load, kg (lb)	12 (27)	(a)
Number of pieces per hour	4000	3700
Equipment requirements		
Type of austenitizing furnace..........................	Salt pot(b)	Shaker hearth(c)
Size of furnace, mm (in.)	380 by 760 by 560 (15 by 30 by 22)	305 by 2900 (12 by 114)(d)
Heat input, kJ(Btu) per hour	290 155 (275 000)(e)	90 210 (85 500)(e)
Amount of chloride salt, kg (lb)	770 (1700)	...
Operating temperature, °C (°F).........................	845 (1550)	870 (1600)
Type of austempering furnace	Salt bath(f)	Salt bath(g)
Size of furnace, mm (in.)	380 by 915 by 760 (15 by 36 by 30)(h)	610 by 1830 by 1220 (24 by 72 by 48)
Amount of nitrate-nitrite salt, kg (lb).....................	1770 (3900)	2720 (6000)
Heat input, kJ (Btu) per hour.........................	215 240 (204 000)	70 900 (67 200)
Operating temperature, °C (°F)...................	370 (700)(j)	360 (680)
Agitation...............................Propeller-type pump		...
Cooling system	Blower(k)	...

(a) Shaker hearth. (b) Electric furnace with internally heated ceramic pot. (c) Electrically heated furnace with endothermic atmosphere (dew point, −1 °C or 30 °F) generator. (d) Size of hearth. (e) At 845 °C (1550 °F). (f) Salt heated with electric resistance elements. (g) Electrically heated radiant tubes. (h) Quench area; salt-separating chamber, 305 by 710 by 1040 mm (12 by 28 by 41 in.). (j) Temperature of salt-separating chamber, 315 °C (600 °F). (k) Blower between pot and furnace wall, pyrometer controlled.

hardness and other properties obtained in the part being heat treated.

- Unintended freezing of a salt bath can result in unwarranted expense.
- At temperatures exceeding 455 °C (850 °F), nitrate-nitrite salt mixtures will cause pitting attack of steel pots and, more importantly, steel workpieces.
- *Local temperatures above 595 °C (1100 °F) anywhere in a nitrate-nitrite salt bath can cause violent exothermic reactions in the salt that may result in fire or explosion even after the source of furnace heat has been shut off.* Equipment manufacturers should be consulted as to their recommendations regarding safe *average* bath temperatures.

The potential hazard of nitrate-nitrite salt mixtures overheating as the result of failure of a thermocouple or control instrument must be recognized, and appropriate safeguards must be provided. A second instrument and thermocouple are recommended. These may be set 28 °C (50 °F) higher than the desired operating temperature and arranged to take over in the event of malfunction.

Usually, a variation in bath temperature of ±6 °C (±10 °F) is permissible. However, variations greater than ±6 °C have been known to cause unacceptable variations in the hardness of austempered parts.

Time in the bath should be sufficient to allow complete transformation. Allowing parts to remain in the bath for longer than the required time will increase the cost of the operation, but it is not harmful to mechanical properties (Table 9). In one instance, parts that had fallen from racks were unintentionally held at 265 °C (510 °F) for several months. Subsequent tests revealed that their properties were not harmed by the prolonged immersion.

Bath agitation can be a significant variable in austempering, because it affects quenching speed. Mechanical stirring, pumping and air agitation have all proved successful for specific operations. Pumping provides the greatest amount of salt movement, although it requires a more expensive installation. In many applications, the other methods have provided a sufficiently rapid quench. The suitability of stirring as compared to pumping is indicated in the following example.

Example 4. Mechanical stirring devices were originally used in the quenching system of a rotary retort furnace used for austempering (Fig. 6a). Although the pattern of agitation could be varied by changing the location and altitude of the impellers, this did not provide for displacement of the static salt within the chute; consequently, parts failed to quench adequately, because the static salt allowed parts to remain in clusters. Changing the angles of the mechanical stirrers failed to eliminate inadequate quenching.

Table 7 Equipment for austempering parts made of 1080 steel
Parts austenitized at 925 °C (1700 °F)

Production requirements	
Weight of each piece, kg (lb)	1.1 to 1.8 (2.4 to 3.9)
Production per hour	1500 pieces
Equipment requirements	
Austempering furnace	Submerged fuel-fired salt pot
Size of furnace, m³ (ft³)	6.6 (233)
Nitrate-nitrite salt, kg (lb)	11 340 (25 000)
Temperature, °C (°F)	345 to 360 (650 to 675)
Agitation	Pump, directed at delivery chute
Cooling	Forced air through burner tube

Table 8 Equipment requirements for austenitizing, austempering and corrosion protection of parts made of 1035 and 1060 steels

Equipment comprises a manually operated monorail heat treating line.

Production requirements

	Fabricated disk	Rectangular tube	Rectangular plate
Part	Fabricated disk	Rectangular tube	Rectangular plate
Steel	1035	1060	1060
Section thickness, mm (in.)	12.5 (0.492)	1.4 (0.055)	0.8 (0.032)
Weight of part, kg (lb)	1.1 (2.5)	0.2 (0.425)	0.01 (0.024)
Load weight (gross), kg (lb)	163 (360)	35.7 (78.75)	7.3 (16)
Number of pieces per h	432	720	3900
Preheating	705 °C (1300 °F); 10 min	705 °C (1300 °F); 5 min	705 °C (1300 °F); 5 min
Austenitizing	845 to 870 °C (1550 to 1600 °F); 10 min	830 °C (1525 °F); 10 min	830 °C (1525 °F); 10 min
Austempering	425 °C (800 °F); 10 min	345 °C (650 °F); 6 min	330 °C (630 °F); 6 min

Equipment requirements

Preheating furnace	Immersed-electrode salt bath, 0.45 by 0.6 by 1.6 m (18 by 24 by 62 in.)
Amount of chloride salt, kg (lb)	590 (1300)
Austenitizing furnace	Immersed-electrode salt bath, 0.45 by 0.6 by 1.6 m (18 by 24 by 62 in.)
Amount of chloride salt, kg (lb)	590 (1300)
Austempering furnace	Gas-fired salt bath, 0.6 by 1.2 by 1.8 m (24 by 48 by 72 in.)
Amount of nitrate-nitrite salt, kg (lb)	2270 (5000)
Heat input, kJ (Btu) per h	738 570 (700 000)
Agitation	Two 150-mm (6-in.) impellers
Two washing and rinsing tanks	Gas-fired, hot water; each 0.5 by 0.9 by 1.2 m (20 by 36 by 48 in.)
Capacity, each tank	570 l (150 gal)
Heat input, each tank, kJ (Btu) per h	316 530 (300 000)
Tank for corrosion protection, m (in.)	0.5 by 0.9 by 1.2 (20 by 36 by 48)

The lower section of the quench chute was redesigned for pump agitation (Fig. 6b). The output of a 946-l/min (250-gal/min) pump was directed upward into the chute at an angle of 60° above the horizontal from two oppositely inclined multiple orifice plates located at the bottom of the chute. Salt entering the chute was returned to the bath through a perforated section of the chute located just below the normal salt level. A skirt, extending outward from the chute above the perforated section and draped on all sides to a level below the salt level, maintained the atmosphere seal.

The output of a second 946-l/min (250-gal/min) pump was directed downward into the receiving basket at an angle of 27° below the horizontal. The salt entering the basket was returned to the bath through the perforated bottom and walls of the basket and over the top rim of the basket. A throttle valve was installed in the discharge line of each pump to control the discharge velocity at the orifice plates. A crossover line with shutoff valve made it possible to supply salt to the two pairs of orifice plates, independently, from either or both pumps. The degree of agitation could then be adjusted for maximum quenching to match the mass of the parts being processed.

Experience has proved that the out-put from one pump to all four orifice plates provides adequate agitation for all parts being treated. Full flow is used for heavier parts 3 mm (0.125 in.) thick, and an estimated half flow is used for lighter parts 1.5 mm (0.060 in.) thick. Rough adjustment of the pump throttle valve is made by observing the stream of salt discharged from a hole drilled in the manifold piping (this hole serves primarily to drain these lines when the pumps are shut off). Final adjustment is based on the contour of the surface of the bath as formed by the parts in the baskets. A markedly dished (concave) contour, with the parts sloped upward against all sides and corners of the basket, indicates that agitation is sufficient to avoid clustering of parts. A feathering of the parts to an acute angle against the sides of the basket indicates excessive agitation that could

Table 9 Effect of austempering time on hardness of three steels

Steel	Austempering temperature °C	°F	Rockwell hardness(a)						
			Austempering time, min						
			30	60	90	120	240	300	360
1095 (b)	230(c)	450(c)	91	90	90	90	90	90	90
	265(d)	510(d)	90	89	89	89	89	89	89
			Austempering time, min						
			1	2	5	10	20	40	80
8735 (e)	260(f)	500(f)	51	51	49	49	48	48	47
	315(f)	600(f)	49	45	46	46	46	46	46
	370(f)	700(f)	40	39	38	38	38	38	37
8750 (g)	260(f)	500(f)	58	56	53	51	52	52	51
	315(f)	600(f)	58	52	48	47	47	47	47
	370(f)	700(f)	54	42	39	39	39	38	39

(a) Rockwell 15-N hardness values for the 1095 steel; Rockwell C hardness values for the 8735 and 8750 steels. (b) Steel contained 0.90% C; specimen thickness, 0.25 mm (0.010 in.). Each hardness value is an average of 12 specimens; range of test values did not exceed one point on Rockwell 15-N scale. (c) Time for 100% transformation was 170 min. (d) Time for 100% transformation was 85 min. (e) Steel contained 0.37% C; specimen size, 16 by 32 by 2 mm (0.622 by 1.250 by 0.087 in.). (f) Time for complete transformation was 5 to 10 min. (g) Steel contained 0.49% C; specimen size, 25 by 25 by 3 mm (1 by 1 by 1/8 in.)

result in loss of parts over the basket rim.

Dimensional Control

Parts can usually be produced with less dimensional change by austempering than by conventional quenching and tempering. Austempering may be the best way to hold close tolerances without extensive straightening or machining after heat treatment.

Example 5. In one application, a U-shape part was formed from 1095 steel strip, 0.25-mm (0.010-in.) thick. The open end of the U varied within 0.25 mm (0.010 in.) after forming. When these parts were oil quenched from 800 °C (1475 °F) and tempered at 260 °C (500 °F), the open end of the U varied over a range of 1.3 mm (0.050 in.). However, when the parts were austempered for 90 min at 265 °C (510 °F), the dimensional spread was decreased to about 0.8 mm (0.030 in.).

Austempering has often proved helpful in minimizing the need for expensive straightening operations; a notable example is stabilizer bars.

Example 6. Table 10 compares the dimensional changes that occurred in stabilizer bars as a result of oil quenching and tempering with those that resulted from austempering. About 20% of these bars require straightening when oil quenched. This percentage is sharply reduced when austempering is used. With a new die setup, about 1 to 5% will require straightening for the first 3000 pieces while the dies are being trimmed for precise fit. Once the correct fit and setup are established, straightening will drop to less than 1/2%. This will carry through until the die wear becomes appreciable (approximately 40 000 pieces), at which point the amount of straightening required begins to increase.

The data in Fig. 7 and 8 further indicate the degree of improvement in dimensional stability attainable by austempering as compared to water or oil quenching and tempering.

Austempering Problems and Solutions

Problems encountered in austempering, together with their solutions, are discussed in the following examples.

Example 7. Threaded fasteners made of 1060 steel, designed to be embedded in concrete by explosive discharge, exhibited insufficient ductility

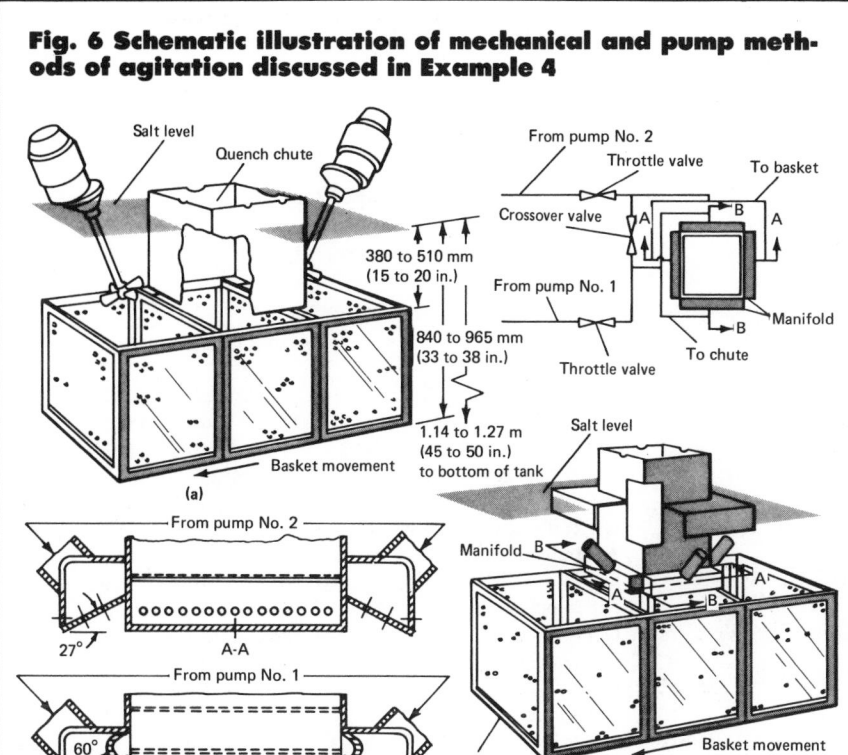

Fig. 6 Schematic illustration of mechanical and pump methods of agitation discussed in Example 4

(a) Mechanical stirring. (b) Pump agitation. This example illustrates a problem of salt bath quenching efficiency in which extensive redesign of the quenching system proved to be necessary. After redesign, the system could be used at full flow for heavier parts 3 mm (0.125 in.) thick and at about half flow for parts about half this thickness. Accompanying text indicates methods of adjusting the pump throttle valve for control.

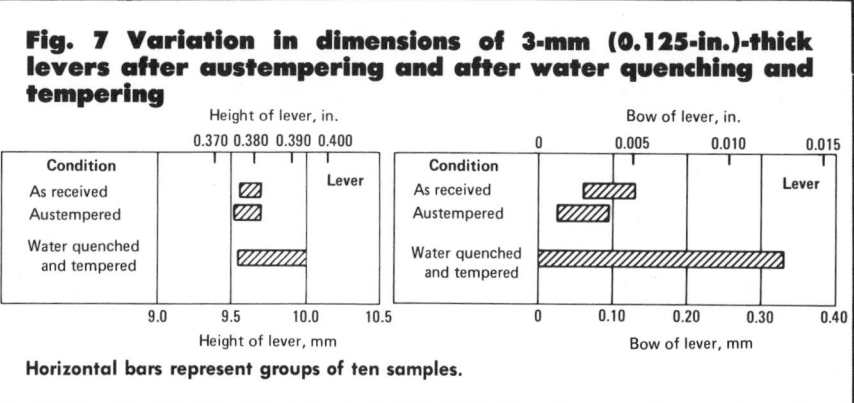

Fig. 7 Variation in dimensions of 3-mm (0.125-in.)-thick levers after austempering and after water quenching and tempering

Horizontal bars represent groups of ten samples.

after austempering. These parts were hardened by being heated to 845 °C (1550 °F) and quenched in a salt bath at 290 °C (550 °F) for 30 min. The lack of ductility was attributed to insufficient carbide solution. Parts with acceptable ductility after austempering were obtained by normalizing before final austenitizing and austempering.

Example 8. Lawn mower blades, approximately 3 mm (1/8 in.) thick, made of 0.50 to 0.60% C steel exhibited low hardness after austempering. Low hardenability due to lower-than-normal manganese content proved to be the cause. The manganese content desired for this application should be near the high side of the allowable

Table 10 Effects of oil quenching and tempering and of austempering on dimensions of stabilizer bars

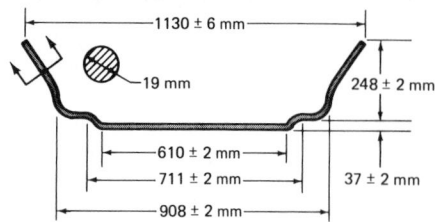

Specified dimension			Process	Measured dimension(a)					
				High		Low		Average	
mm		in.		mm	in.	mm	in.	mm	in.
1130 ±6	44$\frac{1}{2}$ ± $\frac{1}{4}$		OQ & T	1133	44$\frac{5}{8}$	1127	44$\frac{3}{8}$	1130	44$\frac{1}{2}$
			Austemper....	1130	44$\frac{1}{2}$	1126	44$\frac{5}{16}$	1127	44$\frac{3}{8}$
908 ±2	35$\frac{3}{4}$ ± $\frac{1}{16}$		OQ & T	911(b)	35$\frac{7}{8}$(b)	905(b)	35$\frac{5}{8}$(b)	910	35$\frac{13}{16}$
			Austemper....	910	35$\frac{13}{16}$	910	35$\frac{13}{16}$	910	35$\frac{13}{16}$
711 ±2	28 ± $\frac{1}{16}$		OQ & T	714(b)	28$\frac{1}{8}$(b)	711	28	713	28$\frac{1}{16}$
			Austemper....	713	28$\frac{1}{16}$	711	28	711	28
610 ±2	24 ± $\frac{1}{16}$		OQ & T	614(b)	24$\frac{3}{16}$(b)	611	24$\frac{1}{16}$	613(b)	24$\frac{1}{8}$(b)
			Austemper....	611	24$\frac{1}{16}$	610	24	611	24$\frac{1}{16}$
248 ±2	9$\frac{3}{4}$ ± $\frac{1}{16}$		OQ & T	249	9$\frac{13}{16}$	246	9$\frac{11}{16}$	248	9$\frac{3}{4}$
			Austemper....	248	9$\frac{3}{4}$	246	9$\frac{11}{16}$	246	9$\frac{11}{16}$
37 ±2	1$\frac{15}{32}$ ± $\frac{1}{16}$		OQ & T	38	1$\frac{1}{2}$	36.5	1$\frac{7}{16}$	38	1$\frac{1}{2}$
			Austemper....	38	1$\frac{1}{2}$	38	1$\frac{1}{2}$	38	1$\frac{1}{2}$
2(c)	$\frac{1}{16}$ (0.0625)(c)		OQ & T	1.3	0.050	0.13	0.005	0.8	0.032
			Austemper....	1.5	0.060	0.25	0.010	0.9	0.036

(a) Data represent measurements made on 12 samples of bars processed by each method. (b) Out of specification. (c) Arm-to-arm parallel

Fig. 8 Variation in pitch length of 2-mm (0.080-in.)-thick link plates after austempering and after oil quenching and tempering

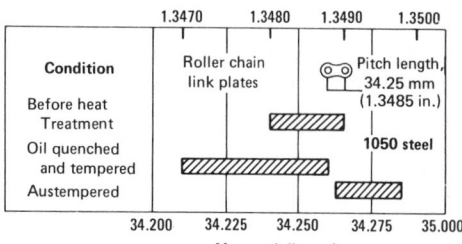

All link plates were austenitized at 855 °C (1575 °F) for 11 min: austempered link plates were held in salt at 340 °C (640 °F) for approximately 1 h, a time dictated by convenience in processing but not required to attain complete transformation.

range (0.60 to 0.90%), but the steel for these blades contained less than 0.50% Mn.

The problem was solved by first quenching the blades in a lower-temperature bath (just above M_s for $\frac{1}{2}$ min and then transferring them to the normal austempering bath at 315 °C (600 °F) and holding for $\frac{1}{2}$ h. This technique has been successfully applied to many carbon steels that were borderline for austempering because of low hardenability in relation to the section being quenched.

Example 9. A small percentage of parts made of medium-carbon and low-alloy steels had low hardness after austempering. The parts had been heated in a rotary retort furnace. Microscopic examination disclosed high-temperature transformation products, caused by under-quenching.

Investigation disclosed that this condition resulted when the parts were quenched in clusters rather than in a continuous sequence. Clustering resulted in a slow cooling rate at the center of the clusters. The problem was solved by redesigning the submerged portion of the quench chute to allow for high-velocity multiple jets of molten salt that could be directed upward into the chute and downward into the receiving basket (see Example 4 and Fig. 6).

Example 10. Nonuniform hardness was found in some parts hardened in an automated austempering cycle. The parts, made of 1050 steel, were 13 mm ($\frac{1}{2}$ in.) wide, 125 mm (5 in.) long and 1 mm (0.040 in.) thick. Heat treatment consisted of austenitizing at 845 °C (1550 °F) in a neutral salt, followed by quenching to 345 °C (650 °F) and holding for 15 min. The problem was traced to chloride contamination of the nitrate-nitrite quenching bath, a condition that caused the quenching bath to be sluggish. Inspection of the filtering screens indicated that they were not functioning adequately.

The problem was solved by establishing a procedure for reducing the bath temperature to 290 °C (550 °F) twice a week and thus precipitating the chlorides to a screen at the bottom of the bath for removal. This procedure eliminated the need for filtering screens and provided uniform quenching.

Example 11. Parts that required austempering to obtain ductility at high hardness were excessively brittle. These parts were made of 1055 steel and were hardened by being austenitized in a shaker-hearth furnace at 840 °C (1540 °F) for 8 min and quenched in a nitrate-nitrite salt at 370 °C (700 °F) for 10 min. Analysis of material and processing disclosed that the brittleness resulted from superficial nitriding, which was caused by the escape of nitrogenous gases from the austempering salts to the quench chute, where they combined with the furnace atmosphere. Increasing the furnace temperature to 860 °C (1580 °F) and reducing furnace time to 3 min solved the problem.

Example 12. Chain-saw components, made of low-alloy Cr-Ni-V steel containing 0.65 to 0.75% C, exhibited excessive brittleness in a torque test.

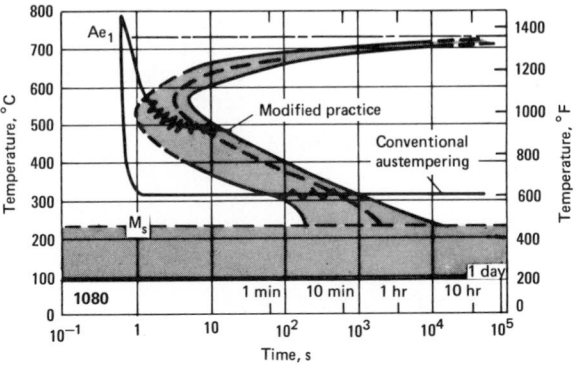

Fig. 9 Time-temperature transformation diagram for 1080 steel, showing difference between conventional and modified austempering

When applied to wire, the modification shown is known as "patenting".

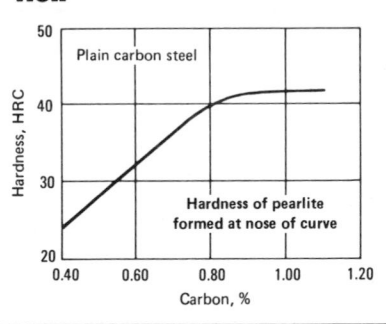

Fig. 10 Effect of carbon content in plain carbon steel on the hardness of fine pearlite formed when the quenching curve intersects the nose of the time-temperature diagram for isothermal transformation

The parts had been austenitized in a shaker-hearth furnace at 830 °C (1525 °F) and quenched in an agitated salt bath at 290 °C (550 °F). Examination disclosed excessive martensite and undissolved carbides in the microstructure. Over-spheroidization prior to hardening was determined to be the cause of brittleness. This resulted in the formation of abnormally large carbides that were not dissolved during the 11-min furnace cycle, thus lowering the carbon content of the matrix and raising the M_s temperature. This caused partial transformation to martensite instead of 100% lower bainite. The problem was solved by prehardening and tempering prior to the regular heat treating cycle.

Modified Austempering

As mentioned previously, modifications of austempering practice that give rise to mixed structures of pearlite and bainite are quite common in industrial practice. The amounts of pearlite and bainite may vary widely in different modifications of processing.

Patenting, a treatment used in the wire industry, is a significant and useful form of modified austempering, in which austenitized wire or rod is continuously quenched into a bath maintained at 510 to 540 °C (950 to 1000 °F) and held in the bath for periods ranging from 10 s (for small wire) to 90 s (for rod). Patenting provides a combination of moderately high strength and high ductility. As indicated in Fig. 9 by the

line designated "Modified practice", the process varies from true austempering in that the quenching rate, instead of being rapid enough to avoid the nose of the TTT curve, is sufficiently slow to intersect the nose, which results in the formation of fine pearlite.

Similar practice is usefully employed in applications involving plain carbon steels when a hardness between about 30 to 42 HRC is desirable or acceptable. The hardness of plain carbon steel quenched at a rate that intersects the nose of the TTT curve will vary with carbon content (Fig. 10).

Modified practices can be applied to parts having sections thicker than are normally considered practicable for austempering. These practices, however, also are subject to certain metallurgical limitations, such as the following which apply to the modified austempering of 1080 steel:

• The part to be quenched must be of sufficient mass or weight that cooling to the quench bath temperature cannot be accomplished in the time (about ¾ s) allowed for avoiding the nose of the TTT curve; otherwise, the part will undergo true austempering and may be harder than desired. The temperature of the quenching medium should not be raised above 370 °C (700 °F) in an effort to retard the cooling rate, or temper brittleness may result. For small parts, isothermal annealing at 565 °C (1050 °F) is preferred to austempering for obtain-

ing hardness in the range of 30 to 42 HRC.

• The part to be quenched is subject to a maximum, as well as a minimum, weight limit. If more than about 20 s is required for the center of the part to transform, or if there is a rise in bath temperature, some upper bainite may form, resulting in a variable hardness and mixed structure.

• The lower temperature limit of the quench bath depends on the weight of the part. For pieces weighing 1 to 2 kg (2 to 4 lb), the lower limit of bath temperature is about 330 °C (625 °F).

From the above it is evident that modified practices are limited by several critical factors and that some trial and error is necessary in developing an optimum cycle for parts of specific composition and section thickness.

Another form of modified austempering also entails the use of a special technique but produces results that are similar to those obtained in true austempering. This modification is employed for parts that, because of their size and the type of steel of which they are made, are difficult to quench rapidly enough to avoid the nose of the TTT curve before transformation begins. In such applications, the cooling rate can be increased by first quenching the part into a bath that is slightly above the M_s temperature. For 1080 steel, this is about 260 °C (500 °F), as shown in Fig. 9. The part is held at this temperature for only the brief time necessary to obtain temperature equalization throughout the section, and is then transferred to the austempering bath at a higher temperature and allowed to transform isothermally in the normal manner (see Example 8).

Washing and Drying

Plain water, preferably warm or hot, is most suitable for cleaning austempering salts from parts. A single overflowing tank can be used. In automated systems, it is common practice to use two conveyorized tanks, the first heated by incoming parts and the second separately heated. Fresh water is added to the second tank, with the overflow going to the first tank. In some cases, the concentrated salt solution is pumped back to the salt tank with available salt-recovery systems to recover both the salt and the water. This recovery reduces salt and water consumption and alleviates the disposal problem. The parts should be dried by warm forced air or treated with a rust preventive immediately after washing.

Safety Precautions

In operating an austempering installation, the same precautions should be taken that are appropriate for operation of any other hot liquid bath. These include the wearing of gloves and face masks. Several other special precautions are unique to operation of nitrate-nitrite baths:

- *Local temperatures above 595 °C (1100 °F) anywhere in a nitrate-nitrite salt bath can cause violent exothermic reactions in the salt that may result in fire or explosion even after the source of furnace heat has been shut off.* (Overheating might be caused by instrument failure or by overloading of the bath with hot workpieces.) Equipment manufacturers should be consulted as to their recommendations regarding safe *average* bath temperatures.
- *Nitrate-nitrite mixtures are oxidizing and should not be intermixed with materials that are readily oxidized. Cyanide and cyanide-bearing salts are especially incompatible with nitrate-nitrite salt mixtures.* Austempering should never be done directly from an austenitizing bath containing cyanide. When it is desirable to austemper cyanided or liquid carburized parts, they should first be immersed in a bath of neutral salt. This bath should be operated at the desired austenitizing temperature and controlled to prevent the accumulation of more than 2% sodium cyanide or potassium cyanide.
- When austempering is carried out in the same area where cyanide-containing salts also are used, the salt pots and salt containers should be color coded. Yellow, orange or red are commonly used for identification of nitrite and nitrite-containing salts. Cyanide-containing materials may be identified by white, blue or gray. *Anyone handling these materials should be made aware that explosions can occur if they are intermixed.*
- Finely divided carbonaceous materials should not be used. For example, it is common practice to use graphite, coal or powdered coke as an insulating cover on lead baths. *Under no circumstance should these be employed as coverings for nitrate-nitrite salt baths.* Accidental contamination of austempering baths with accumulations of carbon black, which may build up at the discharge end of gas carburizing or carbonitriding furnaces, must also be avoided. Where work is to be austempered directly from such furnaces, an opening for frequent cleanout should be provided.

Cold Treating of Steel

By Norman C. McClure
Chief Metallurgist—Plant 2
Commercial Steel Treating Corp.

COLD TREATING of steel is a supplemental treatment that may be used for such purposes as enhancing transformation of austenite to martensite and improving stress relief of castings and machined parts. Generally, 1 h of cold treatment is adequate for each inch of cross section.

Hardening and Retained Austenite

Whenever hardening is to be done during heat treating, complete transformation from austenite to martensite generally is desired prior to tempering. From a practical standpoint, however, conditions vary widely and 100% transformation rarely, if ever, occurs. Cold treating may be useful, in many instances, for improving the percentage of transformation and thus enhancing properties.

During hardening, martensite develops as a continuous process from start (M_s) to finish (M_f) through the martensite-formation range. Except in a few highly alloyed steels, martensite starts to form at well above room temperature. In many instances, transformation is essentially complete at room temperature. Retained austenite tends to be present in varying amounts, however, and, when considered excessive for a particular application, must be transformed to martensite and then tempered.

Cold Treating vs Tempering. Immediate cold treating without delays at room temperature or at other temperatures during quenching offers the best opportunity for maximum transformation to martensite. In some instances, however, there is a risk that this will cause cracking of parts. Therefore, it is important to ensure that the grade of steel and the product design will tolerate immediate cold treating rather than immediate tempering. Some steels must be transferred to a tempering furnace when still warm to the touch, to minimize the likelihood of cracking. Design features such as sharp corners and abrupt changes in section create stress concentrations and promote cracking.

In most instances, cold treating is not done before tempering. In several types of industrial applications, tempering is followed by deep freezing and retempering without delay. For example, such parts as gages, machineways, arbors, mandrils, cylinders, pistons, and ball and roller bearings are treated in this manner for dimensional stability. Several freeze-draw cycles are used for critical applications.

Cold treating is also used to improve wear resistance in such materials as tool steels, high-carbon martensitic stainless steels and carburized alloy steels for applications where the presence of retained austenite may result in excessive wear. Transformation in service may cause cracking and/or dimensional changes that can promote failure. In some instances, more than 50% retained austenite has been observed. In such cases, no delay in tempering after cold treatment is permitted, or cracking can develop readily.

Process Limitations. In some applications in which explicit amounts of retained austenite are considered beneficial, cold treating might be detrimental. Moreover, multiple tempering, rather than alternate freeze-draw cycling, is generally more practical for transforming retained austenite in high speed and high-carbon/high-chromium steels.

Hardness Testing. Lower than expected HRC readings may indicate excessive retained austenite. Significant increases in these readings as a result of cold treatment indicate conversion of austenite to martensite. Superficial hardness readings, such as HR15N, may show even more significant changes.

Precipitation-Hardening Steels. Specifications for precipitation-hardening steels may include a mandatory deep freeze after solution treatment and prior to aging.

Shrink Fits. Cooling the inner member of a complex part to below ambient temperature may be a useful way of providing an interference fit. Care must be taken, however, to avoid the brittle cracking that may develop when the inner member is made of heat treated steel with high amounts of retained austenite, which converts to martensite on subzero cooling.

Stress Relief

Residual stresses often contribute to part failure, and residual stresses frequently are the result of temperature changes that produce thermal expansion and phase changes—and, consequently, volume changes.

Under normal conditions, temperature gradients produce nonuniform dimensional and volume changes. In castings, for example, compressive stresses develop in lower-volume areas, which cool first, and tensile stresses develop in areas of greater volume, which are last to cool. Shear stresses develop between the two areas. Even in large castings and machined parts of relatively uniform thickness, the surface cools first and the core last. In such cases, stresses develop as a result of the phase (volume) change between those layers that transform first and the center portion, which transforms last.

When both volume and phase changes occur in pieces of uneven cross section, normal contractions due to cooling are opposed by transformation expansion. The resulting residual stresses will remain until a means of relief is applied. This type of stress develops most frequently in steels during quenching. The surface becomes martensitic before the interior does. Although the inner austenite can be strained to match this surface change, subsequent interior expansions place the surface martensite under tension when the inner austenite transforms. Cracks in high-carbon steels arise from such stresses.

The use of cold treating has proved beneficial in stress relief of castings and machined parts of even or nonuniform cross section. The following are features of the treatment:

- Transformation of all layers is accomplished when the material reaches −84 °C (−120 °F).
- The increase in volume of the outer martensite is somewhat counteracted by the initial contraction due to chilling.
- Re-warm time is more easily controlled than cooling time, thus allowing equipment flexibility.
- The expansion of the inner core due to transformation is somewhat balanced by the expansion of the outer shell.
- The chilled parts are more easily handled.
- The surface is unaffected by low temperature.
- Parts that contain various alloying elements and that are of different sizes and weights can be chilled simultaneously.

Advantages of Cold Treating

Unlike heat treating, which requires that temperature be precisely controlled to avoid reversal, successful transformation through cold treating depends only on the attainment of the minimum low temperature and is not affected by lower temperatures. As long as the material is chilled to −84 °C (−120 °F), transformation will occur; additional chilling will not cause reversal.

Time at Temperature. After thorough chilling, additional exposure has no adverse effect. When heat is used, holding time and temperature are critical. In cold treatment, materials of different compositions and of different configurations may be chilled at the same time even though each may have a different high-temperature transformation point. Moreover, the warm-up rate of a chilled material is not critical as long as uniformity is maintained and gross temperature-gradient variations are avoided.

The cooling rate of a heated piece, however, has a definite influence on the end product. Formation of martensite during solution heat treating assumes immediate quenching to ensure that austenitic decomposition will not result in the formation of bainite and cementite. In large pieces comprising both thick and thin sections, not all areas will cool at the same rate. As a result, surface areas and thin sections may be highly martensitic, and the slower-cooling core may contain as much as 30 to 50% retained austenite. In addition to incomplete transformation, subsequent natural aging induces stress and also results in additional growth after machining.

Aside from transformation, no other metallurgical change takes place as a result of chilling. The surface of the material needs no additional treatment. The use of heat frequently causes scale and other surface deformations that must be removed.

Equipment for Cold Treating

A simple home-type deep freezer can be used for transformation of austenite to martensite. Temperature will be approximately −18 °C (0 °F). In some instances, hardness tests can be used to determine if this type of cold treating will be helpful. Dry ice placed on top of the work in a closed, insulated container also is commonly used for cold treating. The dry ice surface temperature is −78 °C (−109 °F), but the chamber temperature normally is about −60 °C (−75 °F).

Mechanical refrigeration units with circulating air at approximately −87 °C (−125 °F) are commercially available. A typical unit will have the following dimensions and operational features: chamber volume, up to 2.7 m³ (95 ft³); temperature range, 5 to −95 °C (40 to −140 °F); load capacity, 11.3 to 163 kg/h (25 to 360 lb/h); and thermal capacity, up to 8870 kJ/h (8400 Btu/h).

Although liquid nitrogen at −195 °C (−320 °F) may be employed, it is used less frequently than any of the above methods because of its cost.

Heat Treating of Ultrahigh-Strength Steels

By T. V. Philip
Specialist, Tool and Alloy Metallurgy
Carpenter Technology Corporation

ULTRAHIGH-STRENGTH STEELS are heat treated by use of equipment and techniques similar to those employed for heat treating constructional alloy steels. The ultrahigh-strength steels ordinarily are quenched and tempered to specified hardnesses, but for critical applications it may be necessary to pull tensile specimens to ensure that a required combination of strength and ductility has been achieved. In still other instances, it may be necessary to conduct impact or fracture-toughness tests to ensure that a required level of resistance to brittle fracture has been attained.

Ultrahigh-strength steels are described in Volume 1 (Ninth Edition) of this Handbook. In that article, only those commercial structural steels capable of a minimum yield strength of 1380 MPa (200 ksi) are considered; and general characteristics, heat treatment and properties are presented for three types of alloys from among the broad class of ultrahigh-strength constructional steels. These three alloy types are (a) medium-carbon low-alloy steels, (b) medium-alloy air-hardening steels and (c) high-alloy hardenable steels.

The present article gives a brief description of the heat treatment and associated properties of the same types of steels, basically reproduced from the earlier article. Chemical compositions are given in Table 1. For a more detailed discussion of these ultrahigh-strength steels, the reader is referred to the original article.

Medium-Carbon Low-Alloy Steels

The medium-carbon low-alloy steels considered in this article are types 4130, 4140, 4340, 6150, 8640 and two modifications of 4340—namely, 300M and D-6a. These steels are readily hot forged. To avoid stress cracks resulting from air hardening, the forged part should be slowly cooled in a furnace or in an insulating medium. Prior to machining, usual practice is to normalize at 870 to 925 °C (1600 to 1700 °F) and temper at 650 to 675 °C (1200 to 1250 °F) or, if the steel is a deep air-hardening grade, to anneal by furnace cooling from 815 to 845 °C (1500 to 1550 °F) to about 540 °C (1000 °F). These treatments impart moderately hard microstructures suitable for machining. A very soft spheroidized structure can be obtained by full annealing. Such a structure is less well suited for machining than the normalized-and-tempered structure. However, for severe cold forming operations such as spinning, deep drawing and wiredrawing, the soft and ductile spheroidized structure is preferred. If blanks for parts are produced by flame cutting, they are annealed before being formed or machined. Welded parts, especially if complex, are stress relieved, or hardened and tempered, immediately after welding.

Table 1 Compositions of ultrahigh-strength steels described in this article

Designation or trade name	C	Mn	Si	Composition(a), % Cr	Ni	Mo	V	Co
Medium-carbon low-alloy steels								
4130	0.28-0.33	0.40-0.60	0.20-0.35	0.80-1.10	...	0.15-0.25	...	...
4140	0.38-0.43	0.75-1.00	0.20-0.35	0.80-1.10	...	0.15-0.25	...	...
4340	0.38-0.43	0.60-0.80	0.20-0.35	0.70-0.90	1.65-2.00	0.20-0.30	...	...
300M	0.40-0.46	0.65-0.90	1.45-1.80	0.70-0.95	1.65-2.00	0.30-0.45	0.05 min	...
D-6a	0.42-0.48	0.60-0.90	0.15-0.30	0.90-1.20	0.40-0.70	0.90-1.10	0.05-0.10	...
6150	0.48-0.53	0.70-0.90	0.20-0.35	0.80-1.10	...	...	0.15-0.25	...
8640	0.38-0.43	0.75-1.00	0.20-0.35	0.40-0.60	0.40-0.70	0.15-0.25	...	...
Medium-alloy air-hardening steels(b)								
H11 Mod	0.37-0.43	0.20-0.40	0.80-1.00	4.75-5.25	...	1.20-1.40	0.40-0.60	...
H13	0.32-0.45	0.20-0.50	0.80-1.20	4.75-5.50	...	1.10-1.75	0.80-1.20	...
9Ni-4Co steels								
HP 9-4-20	0.16-0.23	0.20-0.40	0.20 max	0.65-0.85	8.50-9.50	0.90-1.10	0.06-0.12	4.25-4.75
HP 9-4-30	0.29-0.34	0.10-0.35	0.20 max	0.90-1.10	7.0-8.0	0.90-1.10	0.06-0.12	4.25-4.75

(a) Phosphorus and sulfur contents may vary with steelmaking practice. Usually, these steels contain no more than 0.035 P and 0.040 S; 9Ni-4Co steels are specified to have 0.10 max P and 0.10 max S. (b) ASTM A681; composition ranges utilized by some producers are narrower.

4130

Type 4130 is a water-hardening alloy steel of low-to-intermediate hardenability.

Heat Treatments. The following standard heat treatments apply to type 4130 steel:

- *Normalize:* Heat to 870 to 925 °C (1600 to 1700 °F) and hold for a period that depends on section thickness; air cool. Tempering at 480 °C (900 °F) or above is often done after normalizing to increase yield strength.
- *Anneal:* Heat to 830 to 860 °C (1525 to 1575 °F) and hold for a period that depends on section thickness or furnace load; furnace cool.
- *Harden:* Heat to 845 to 870 °C (1550 to 1600 °F) and hold, then water quench; or heat to 860 to 885 °C (1575 to 1625 °F) and hold, then oil quench. Holding time depends on section thickness.
- *Temper:* At least ½ h at 200 to 700 °C (400 to 1300 °F); air cool or water quench. Tempering temperature and time at temperature depend mainly on desired hardness or strength.
- *Spheroidize:* Heat to 760 to 775 °C (1400 to 1425 °F) and hold 4 to 12 h; cool slowly.

Properties. Table 2 summarizes the typical properties obtained by tempering water-quenched and oil-quenched 4130 steel bars at various temperatures. Because 4130 steel has low hardenability, section thickness must be considered when heat treating to high

Table 2 Typical mechanical properties of heat treated 4130 steel

Tempering temperature °C	°F	Tensile strength MPa	ksi	Yield strength MPa	ksi	Elongation in 50 mm or 2 in., %	Reduction in area, %	Hardness, HB	Izod impact energy J	ft·lb
Water quenched and tempered(a)										
205	400	1765	256	1520	220	10.0	33.0	475	18	13
260	500	1670	242	1430	208	11.5	37.0	455	14	10
315	600	1570	228	1340	195	13.0	41.0	425	14	10
370	700	1475	214	1250	182	15.0	45.0	400	20	15
425	800	1380	200	1170	170	16.5	49.0	375	34	25
540	1000	1170	170	1000	145	20.0	56.0	325	81	60
650	1200	965	140	830	120	22.0	63.0	270	135	100
Oil quenched and tempered(b)										
205	400	1550	225	1340	195	11.0	38.0	450	...	...
260	500	1500	218	1275	185	11.5	40.0	440	...	...
315	600	1420	206	1210	175	12.5	43.0	418	...	...
370	700	1320	192	1120	162	14.5	48.0	385	...	...
425	800	1230	178	1030	150	16.5	54.0	360	...	...
540	1000	1030	150	840	122	20.0	60.0	305	...	...
650	1200	830	120	670	97	24.0	67.0	250	...	...

(a) 25-mm (1-in.) diam round bars quenched from 845 to 870 °C (1550 to 1600 °F). (b) 25-mm (1-in.) diam round bars quenched from 860 °C (1575 °F)

hardness or strength. Effects of mass on typical properties of heat treated 4130 steel are indicated in Table 3.

4140

Type 4140 steel is similar in composition to 4130 except for a higher carbon content, which imparts greater hardenability and strength. When any ultrahigh-strength steel is heat treated to high strength levels, it is subject to hydrogen embrittlement during operations such as acid pickling and cadmium or chromium electroplating.

Ductility can be restored in thin sections by baking for 2 to 4 h at 190 °C (375 °F), and in sections more than 38 mm (1½ in.) thick by baking for 23 h at 190 °C.

Heat Treatments. The following standard heat treatments apply to 4140 steel:

- *Normalize:* Heat to 845 to 900 °C (1550 to 1650 °F) and hold for a period that depends on section thickness; air cool.
- *Anneal:* Heat to 845 to 870 °C (1550

Table 3 Effects of mass on typical properties of heat treated 4130 steel

Round bars oil quenched from 845 °C (1550 °F) and tempered at 540 °C (1000 °F); 12.83-mm (0.505-in.) diam tensile specimens cut from center of 25-mm diam bar, and from mid-radius of 50- and 75-mm diam bars

Bar size		Tensile strength		Yield strength		Elongation in 50 mm or 2 in., %	Reduction in area, %	Surface hardness, HB
mm	in.	MPa	ksi	MPa	ksi			
25	1	1040	151	880	128	18.0	55.0	307
50	2	740	107	570	83	20.0	58.0	223
75	3	710	103	540	78	22.0	60.0	217

Table 4 Typical mechanical properties of heat treated 4140 steel

12.7-mm (¹/₂-in.) diam round bars, oil quenched from 845 °C (1550 °F)

Tempering temperature		Tensile strength		Yield strength		Elongation in 50 mm or 2 in., %	Reduction in area, %	Hardness, HB	Izod impact energy	
°C	°F	MPa	ksi	MPa	ksi				J	ft·lb
205	400	1965	285	1740	252	11.0	42	578	15	11
260	500	1860	270	1650	240	11.0	44	534	11	8
315	600	1720	250	1570	228	11.5	46	495	9	7
370	700	1590	231	1460	212	12.5	48	461	15	11
425	800	1450	210	1340	195	15.0	50	429	28	21
480	900	1300	188	1210	175	16.0	52	388	46	34
540	1000	1150	167	1050	152	17.5	55	341	65	48
595	1100	1020	148	910	132	19.0	58	311	93	69
650	1200	900	130	790	114	21.0	61	277	112	83
705	1300	810	117	690	100	23.0	65	235	136	100

Table 5 Effects of mass on typical properties of heat treated 4140 steel

Round bars oil quenched from 845 °C (1550 °F) and tempered at 540 °C (1000 °F); 12.83-mm (0.505-in.) diam tensile specimens cut from center of 25-mm diam bars, and from mid-radius of 50- and 75-mm diam bars

Diameter of bar		Tensile strength		Yield strength		Elongation in 50 mm or 2 in., %	Reduction in area, %	Surface hardness, HB
mm	in.	MPa	ksi	MPa	ksi			
25	1	1140	165	985	143	15	50	335
50	2	920	133	750	109	18	55	202
75	3	860	125	655	95	19	55	293

to 1600 °F) and hold for a period that depends on section thickness or furnace load; furnace cool.

- *Harden:* Heat to 830 to 870 °C (1525 to 1600 °F) and hold; oil quench. (For water quenching, which is rarely used, hardening temperatures are 815 to 845 °C, or 1500 to 1550 °F.) Holding time depends on section thickness.
- *Temper:* At least ¹/₂ h at 175 to 230 °C (350 to 450 °F) or 370 to 675 °C (700 to 1250 °F); air cool or water quench. Tempering temperature and time at temperature depend mainly on desired hardness. To avoid blue brittleness, 4140 usually is not tempered

between 230 and 370 °C (450 and 700 °F).

- *Spheroidize:* Heat to 760 to 775 °C (1400 to 1425 °F) and hold 4 to 12 h; cool slowly.

Properties. Table 4 summarizes the mechanical properties obtained by tempering oil-quenched 4140 steel at various temperatures. Because 4140 is not a deep-hardening steel, section size should be considered, especially when specifying heat treatment for high strength levels. The effects of mass on hardness and tensile properties of 4140 steel are shown in Table 5.

4340

Type 4340, the most popular steel in this class, is a deep-hardening steel. In thin sections, the steel is air hardening, although in practice it is usually oil quenched. When 4340 is heat treated to tensile strengths greater than about 1400 MPa (about 200 ksi), it is subject to hydrogen embrittlement. Parts exposed to hydrogen, such as in pickling or electroplating, should be baked at 185 to 195 °C (365 to 385 °F) for at least 8 h, and for 23 h if thicker than 38 mm (1¹/₂ in.), as soon as possible after the pickling or plating operation.

Heat Treatments. The following standard heat treatments apply to 4340 steel:

- *Normalize:* Heat to 845 to 900 °C (1550 to 1650 °F) and hold for a period that depends on section thickness; air cool.
- *Anneal:* Heat to 830 to 860 °C (1525 to 1575 °F) and hold for a period that depends on section thickness or furnace load; furnace cool.
- *Harden:* Heat to 800 to 845 °C (1475 to 1550 °F) and hold 15 min for each 25 mm (1 in.) of thickness (minimum, 15 min); oil quench to below 65 °C (150 °F); or quench into fused salt at 200 to 210 °C (390 to 410 °F), hold 10 min, then air cool to below 65 °C (150 °F).
- *Temper:* At least ¹/₂ h at 200 to 650 °C (400 to 1200 °F); air cool. Temperature and time at temperature depend mainly on desired strength or hardness.
- *Spheroidize:* Preheat to 690 °C (1275 °F) and hold 2 h, increase temperature to 750 °C (1375 °F) and hold 2 h, cool to 650 °C (1200 °F) and hold 6 h, furnace cool to about 600 °C (1100 °F), and finally air cool to room temperature. An alternative schedule is to heat to 730 to 750 °C (1350 to 1375 °F) and hold several hours, then furnace cool to room temperature.
- *Stress relieve:* After straightening, forming or machining, parts may be stress relieved at 650 to 675 °C (1200 to 1250 °F).

Properties. Through hardening of 4340 steel can be achieved by oil quenching round sections up to 75 mm (3 in.) in diameter, and by water quenching larger sections (up to the limit of hardenability). The influence of section size on tensile properties of oil-quenched and water-quenched 4340 is indicated by the data in Table 6.

Hardness of type 4340 as a function

Table 6 Effects of mass on mechanical properties of 4340 steel

Data from *Alloy Digest* and from A. M. Hall, Sr., "Introduction to Today's Ultrahigh-Strength Structural Steels", STP 498, American Society for Testing and Materials, Philadelphia, 1971

Section diameter		Tensile strength		Yield strength		Elongation in 50 mm or 2 in., %	Reduction in area, %	Hardness, HB
mm	in.	MPa	ksi	MPa	ksi			
Oil quenched and tempered(a)								
13	½	1460	212	1380	200	13	51	...
38	1½	1450	210	1365	198	11	45	...
75	3	1420	206	1325	192	10	38	...
Water quenched and tempered(b)								
75	3	1055	153	930	135	18	52	340
100	4	1035	150	895	130	17	50	330
150	6	1000	145	850	123	16	44	322

(a) Austenitized at 845 °C (1550 °F); tempered at 425 °C (800 °F). (b) 75-mm (3-in.) diam bar austenitized at 800 °C; 100- and 150-mm (4- and 6-in.) diam bars austenitized at 815 °C (1500 °F). All sizes tempered at 650 °C (1200 °F). Test specimens taken at mid-radius

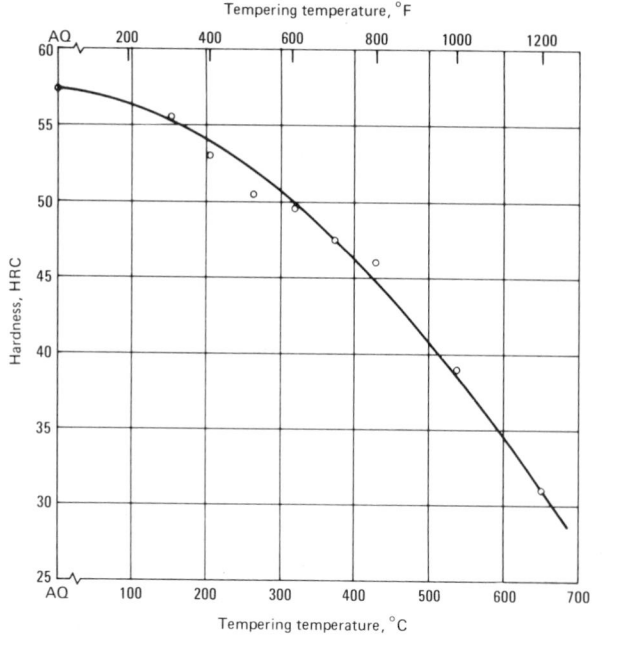

Fig. 1 Variation for hardness with tempering temperature of 4340 steel

All specimens oil quenched from 845 °C (1550 °F) and tempered 2 h at temperature

of tempering temperature is plotted in Fig. 1. Typical mechanical properties of oil-quenched 4340 are given in Table 7. Additional data on mechanical properties (notch toughness and fracture toughness) of this steel tempered to different hardnesses are given in Table 8.

300M

Alloy 300M is basically a silicon-modified (1.6% Si) 4340 steel, but is slightly higher in carbon and molybdenum and also contains vanadium. The steel exhibits deep hardenability. Because of its high silicon and molybdenum contents, 300M is more prone to decarburization than the steels so far described; and during heat treating, care should be exercised to avoid decarburization. When heat treated to strength levels higher than 1380 MPa (200 ksi), 300M is susceptible to hydrogen embrittlement, and thus parts should be properly baked after plating (same baking procedure as that used for type 4340).

Heat Treatments. The following standard heat treatments apply to 300M steel:

- *Normalize:* Heat to 915 to 940 °C (1675 to 1725 °F) and hold for a period that depends on section thickness; air cool. If normalizing is to enhance machinability, recharge into a tempering furnace at 650 to 675 °C (1200 to 1250 °F) before the steel reaches room temperature.

- *Harden:* Austenitize at 855 to 885 °C (1575 to 1625 °F). Oil quench to below 70 °C (160 °F); or quench in salt at 200 to 210 °C (390 to 410 °F) and hold 10 min, then air cool to 70 °C or below.

- *Temper:* Two to four hours at 260 to 315 °C (500 to 600 °F); double tempering recommended. This tempering procedure produces the best combination of high yield strength and high impact properties. Tempering outside the range given above results in severe deterioration of properties.

- *Spheroidize:* Heat to about 775 °C (1430 °F) and hold for a period that depends on section thickness or furnace load. Cool to 650 °C (1200 °F) at a rate no faster than 5.5 °C/h (10 °F/h), then cool to 480 °C (900 °F) no faster than 11 °C/h (20 °F/h), and finally air cool to room temperature. The same schedule is recommended for annealing.

Properties. Variations in hardness and mechanical properties of 300M with tempering temperature are presented in Table 9. This steel has deep hardenability, so that heat treated bars 75 mm (3 in.) in diameter have essentially the same tensile properties as bars 25 mm (1 in.) in diameter. Reductions in tensile strength, ductility and impact strength, however, are observed in heat treated bars 145 mm (5¾ in.) in diameter. Variations in properties of 300M with section size are presented in Table 10.

D-6a and D-6ac

Ladish D-6a was designed primarily for use at room-temperature tensile strengths of 1800 to 2000 MPa (260 to 290 ksi). It is deeper hardening than 4340. The alloy is called D-6a when produced by air melting in an electric furnace, and D-6ac when produced by air melting followed by vacuum arc remelting. Except for some improvements in mechanical properties of D-6ac due to melting practice, the characteristics of the two steels are identical.

Table 7 Typical mechanical properties of 4340 steel

Oil quenched from 845 °C (1550 °F) and tempered at various temperatures

Tempering temperature °C	°F	Tensile strength MPa	ksi	Yield strength MPa	ksi	Elongation in 50 mm or 2 in., %	Reduction in area, %	Hardness HB	HRC	Izod impact energy J	ft·lb
205	400	1980	287	1860	270	11	39	520	53	20	15
315	600	1760	255	1620	235	12	44	490	49.5	14	10
425	800	1500	217	1365	198	14	48	440	46	16	12
540	1000	1240	180	1160	168	17	53	360	39	47	35
650	1200	1020	148	860	125	20	60	290	31	100	74
705	1300	860	125	740	108	23	63	250	24	102	75

Table 8 Notch toughness and fracture toughness of 4340 steel tempered to different hardnesses

Hardness, HB	Equivalent tensile strength(a) MPa	ksi	Charpy V-notch impact energy J	ft·lb	Plane-strain fracture toughness MPa√m	ksi√in.
550	2040	296	19	14	53	48
430	1520	220	30	22	75	68
380	1290	187	42	31	110	100

(a) Estimated from hardness

Table 9 Typical mechanical properties of 300M steel

Round bars, 25 mm (1 in.) in diameter, oil quenched from 860 °C (1575 °F) and tempered at various temperatures

Tempering temperature °C	°F	Tensile strength MPa	ksi	Yield strength MPa	ksi	Elongation in 50 mm or 2 in., %	Reduction in area, %	Charpy V-notch impact energy J	ft·lb	Hardness, HRC
90	200	2340	340	1240	180	6.0	10.0	17.6	13.0	56.0
205	400	2140	310	1650	240	7.0	27.0	21.7	16.0	54.5
260	500	2050	297	1670	242	8.0	32.0	24.4	18.0	54.0
315	600	1990	289	1690	245	9.5	34.0	29.8	22.0	53.0
370	700	1930	280	1620	235	9.0	32.0	23.7	17.5	51.0
425	800	1790	260	1480	215	8.5	23.0	13.6	10.0	45.5

Table 10 Effects of mass on tensile and impact properties of 300M steel

Round bars, normalized at 900 °C (1650 °F), oil quenched from 860 °C (1575 °F), and tempered at 315 °C (600 °F)

Bar diameter mm	in.	Tensile strength MPa	ksi	Yield strength MPa	ksi	Elongation in 50 mm or 2 in., %	Reduction in area, %	Charpy V-notch impact energy when tested at +21 °C (+70 °F) J	ft·lb	−46 °C (−50 °F) J	ft·lb	−73 °C (−100 °F) J	ft·lb
25	1	1990	289	1690	245	9.5	34.1	30	22	26	19	24	18
75	3	1940	281	1630	236	9.5	35.0	26	19	19	14	12	9
150	5¾ ..	2120	308	1800	261	7.3	22.3	12	9	9	7	7	5

Heat Treatments. The following standard heat treatments apply to D-6a and D-6ac steels:

- *Normalize:* Heat to 870 to 955 °C (1600 to 1750 °F) and hold for a period that depends on section thickness; air cool.

- *Anneal:* Heat to 815 to 845 °C (1500 to 1550 °F) and hold for a period that depends on section thickness or furnace load; furnace cool to 540 °C (1000 °F) at a rate no faster than 28 °C/h (50 °F/h), then air cool to room temperature. Hot forgings may

be annealed as follows for maximum machinability: Immediately after forging, charge the parts into a 650 °C (1200 °F) furnace and hold 12 h, increase temperature to 900 °C (1650 °F) and hold for a period that depends on section size, cool to 650 °C (1200 °F) and hold 10 h, and finally air cool to room temperature.

- *Harden:* Austenitize at 845 to 900 °C (1550 to 1650 °F) for ½ to 2 h. Sections no larger than 25 mm (1 in.) in thickness or diameter may be air cooled. Larger sections may be oil quenched to 65 °C (150 °F); or salt quenched to 205 °C (400 °F), and then air cooled. For optimum dimensional stability, "aus-bay" quench into a furnace or salt bath at 525 °C (975 °F), equalize the temperature, then quench in an oil bath held at 60 °C (140 °F), or quench in 205 °C (400 °F) salt and air cool. The cooling rate during quenching significantly affects fracture toughness. For high fracture toughness, especially in heavy sections, austenitize at 925 °C (1700 °F), "aus-bay" quench to 525 °C (975 °F), equalize temperature, and oil quench to 60 °C (140 °F).

- *Temper:* Immediately after hardening, temper 2 to 4 h in the range 200 to 700 °C (400 to 1300 °F), depending on desired strength or hardness.

- *Spheroidize:* Heat to 730 °C (1350 °F) and hold 5 h, increase temperature to 760 °C (1400 °F) and hold 1 h; furnace cool to 690 °C (1275 °F) and hold 10 h; furnace cool to 650 °C (1200 °F) and hold 8 h; air cool to room temperature.

- *Stress relieve:* Heat to a temperature from 540 to 675 °C (1000 to 1250 °F) and hold for 1 to 2 h; air cool.

Properties. Typical room temperature hardness of D-6a steel bar as a function of tempering temperature is plotted in Fig. 2; other typical mechanical properties of D-6a bar are given in Table 11. Tensile properties of heat treated D-6ac billet material are given in Table 12.

6150

Type 6150 can be considered an ultrahigh-strength steel, although as a constructional steel it is not as popular as the other steels in this class. It is a shallow-hardening steel. Parts made of 6150 can be readily welded; after welding, parts should be normalized, then hardened and tempered to the desired hardness.

Fig. 2 Variation for hardness with tempering temperature for D-6a steel

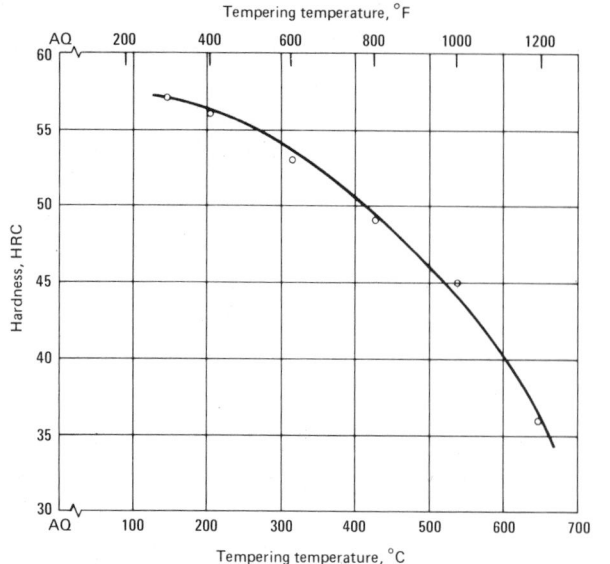

Tempering temperature, °F

Hardness, HRC

Tempering temperature, °C

All specimens oil quenched from 845 °C (1550 °F) and tempered 2 h at temperature

Table 11 Typical mechanical properties of D-6a steel bar

Normalized at 900 °C (1650 °F), oil quenched from 845 °C (1550 °F) and tempered at various temperatures

Tempering temperature °C	°F	Tensile strength MPa	ksi	Yield strength MPa	ksi	Elongation in 50 mm or 2 in., %	Reduction in area, %	Charpy V-notch impact energy J	ft·lb
150	300	2060	299	1450	211	8.5	19.0	14	10
205	400	2000	290	1620	235	8.9	25.7	15	11
315	600	1840	267	1700	247	8.1	30.0	16	12
425	800	1630	236	1570	228	9.6	36.8	16	12
540	1000	1450	210	1410	204	13.0	45.5	26	19
650	1200	1030	150	970	141	18.4	60.8	41	30

Table 12 Typical tensile properties of double tempered D-6ac billet

Austenitized 1 h at 900 °C (1650 °F), quenched in fused salt at 205 °C (400 °F) and held 5 min, then air cooled to room temperature. Tempered 1 h at 205 °C; second temper, 4 h at indicated temperature

Second tempering temperature °C	°F	Tensile strength MPa	ksi	Yield strength MPa	ksi	Elongation in 50 mm or 2 in., %	Reduction in area, %
480	900	1686.5	244.6	1540.3	223.4	11.1	40.0
510	950	1652.7	239.7	1519.7	220.4	13.2	44.1
540	1000	1613.4	234.0	1483.8	215.2	13.7	47.2

Heat Treatments. The following heat treatments apply to 6150 steel:

- *Normalize:* Heat to 870 to 955 °C (1600 to 1750 °F) and hold for a period that depends on section thickness; air cool.

- *Anneal:* Heat to 845 to 900 °C (1550 to 1650 °F) and hold for a period that depends on section thickness or furnace load; furnace cool.

- *Harden:* Austenitize at 845 to 900 °C (1550 to 1650 °F); oil quench.

- *Temper:* At least ½ h at 200 to 650 °C (400 to 1200 °F). Tempering temperature and time at temperature depend on desired final hardness or strength.

- *Austemper:* Austenitize in a salt bath at 845 to 900 °C (1550 to 1650 °F); quench in a salt bath at 230 to 315 °C (450 to 600 °F), hold 20 to 30 min, then oil quench or air cool to room temperature.

- *Martemper:* Austenitize in a salt bath at 845 to 870 °C (1550 to 1600 °F); quench in a salt bath at 230 to 260 °C (450 to 500 °F), equalize, then air cool or quench to room temperature. Temper to desired hardness.

- *Spheroidize:* Heat to 800 to 830 °C (1475 to 1525 °F), hold until heated through, furnace cool to 650 °C (1200 °F) and hold several hours, then cool slowly to room temperature.

Properties. Typical mechanical properties of small-diameter round sections of 6150 tempered at various temperatures are given in Table 13. Hardness and Izod impact energy as functions of tempering temperature are plotted in Fig. 3. The effects of section size on tensile properties and hardness are given in Table 14.

8640

Type 8640 steel was especially designed for maximum hardenability and best combination of properties with minimum alloying additions. It is an oil-hardening steel, but may be water hardened if precautions are taken to prevent cracking.

Heat Treatments. The following standard heat treatments apply to 8640 steel:

- *Normalize:* Heat to 870 to 925 °C (1600 to 1700 °F) and hold for a period that depends on section thickness; air cool.

- *Anneal:* Heat to 845 to 870 °C (1550 to 1600 °F) and hold for a period that depends on section thickness or furnace load; furnace cool.

- *Harden:* Austenitize at 815 to 845 °C (1500 to 1550 °F); quench in oil or water.

- *Temper:* At least ½ h at 200 to 650 °C (400 to 1200 °F). Temperature and time at temperature depend on desired hardness.

- *Spheroidize:* Heat to 705 to 720 °C (1300 to 1325 °F) and hold several hours; furnace cool.

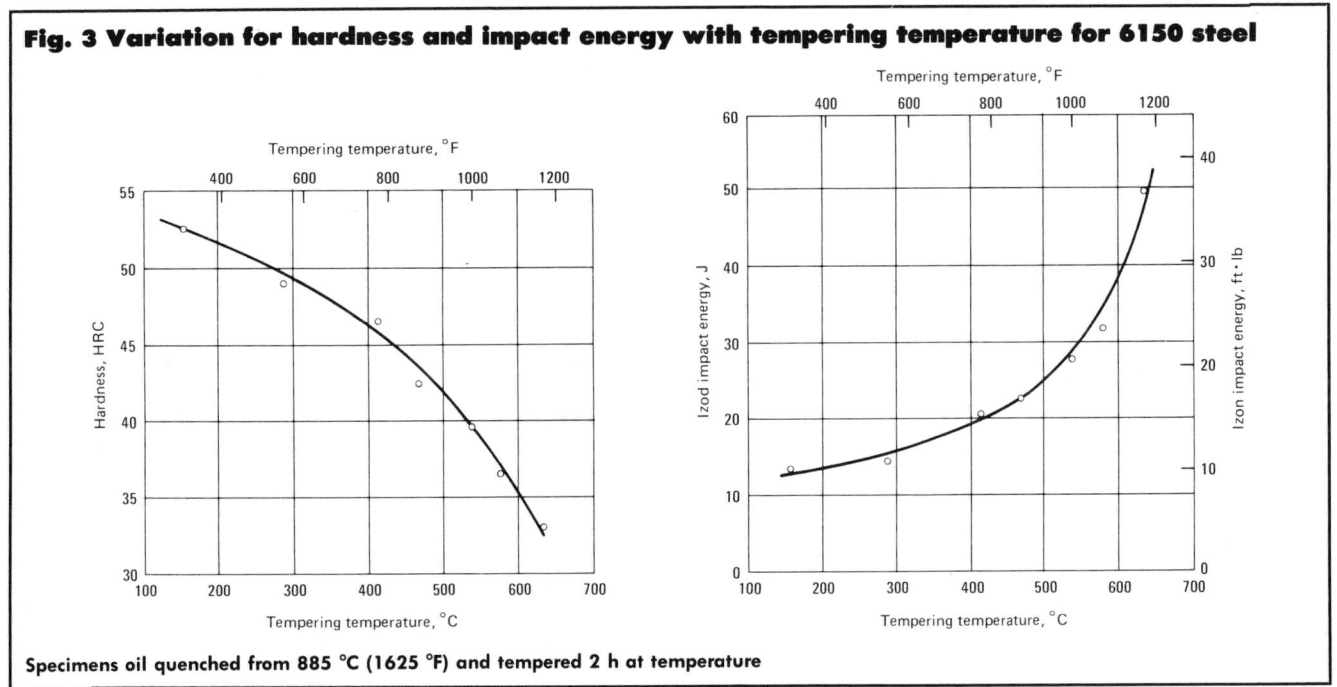

Fig. 3 Variation for hardness and impact energy with tempering temperature for 6150 steel

Specimens oil quenched from 885 °C (1625 °F) and tempered 2 h at temperature

Table 13 Typical room-temperature tensile properties of heat treated 6150 steel

Tempering temperature °C	°F	Tensile strength MPa	ksi	Yield strength MPa	ksi	Elongation in 50 mm or 2 in., %	Reduction in area, %	Hardness, HB	Izod impact energy J	ft·lb
Round bars, 14 mm (0.55 in.) in diameter(a)										
205	400	2050	298	1810	263	1	5	610	...	...
260	500	2070	300	1810	263	4	12	570	...	...
315	600	1950	283	1720	250	7	27	540	...	...
370	700	1770	257	1620	235	10	37	505	9	7
425	800	1585	230	1490	216	11	42	470	14	10
480	900	1410	204	1340	195	12	44	420	16	12
540	1000	1250	182	1210	175	13	46	380	20	15
595	1100	1150	167	1080	157	16	47	350	28	21
Round bars, 25 mm (1 in.) in diameter(b)										
425	800	1570	228	1450	210	10	37	461	...	...
480	900	1360	197	1210	175	11	41	401	...	...
540	1000	1180	171	1030	150	12	45	341	...	...
595	1100	1030	150	875	127	15	50	302	...	...
650	1200	920	133	760	110	19	55	262	...	...
705	1300	810	118	660	96	23	61	235	...	...

(a) Normalized at 870 °C (1600 °F), oil quenched from 860 °C (1575 °F) and tempered at various temperatures. (b) Oil quenched from 860 °C and tempered at various temperatures

Properties. Variations in typical properties of heat treated round sections of 8640 with tempering temperature are given in Table 15. Variations in properties with section size (mass effect) are given in Table 16.

Medium-Alloy Air-Hardening Steels

Heat treatments for the ultrahigh-strength steels H11 Modified (H11 Mod) and H13, which are also known as 5% Cr hot work die steels, are discussed in this section. These steels are similar in composition, heat treatment and many properties. They have deep hardenability and can be hardened through in large sections by air cooling. Air hardening results in minimal residual stresses after hardening. Both H11 Mod and H13 are secondary-hardening steels, and thus develop optimum properties when tempered at temperatures above the secondary-hardening peaks at about 510 °C (950 °F). These high tempering temperatures provide substantial stress relief and stabilization of properties so that the steels can be used advantageously at elevated temperatures. This also enables heat treated parts to be warm worked, or preheated for welding, at temperatures as high as 55 °C (100 °F) below the prior tempering temperature. Both H11 Mod and H13 steels are subject to hydrogen embrittlement, and parts should be baked after any exposure to environments where hydrogen may be absorbed. Baking for 24 h or longer at 190 °C (375 °F) or higher is recommended.

Because H11 Mod and H13 are air-hardening steels, forged parts must be cooled slowly after forging to prevent stress cracking. After forging, parts should be charged into a furnace at about 790 °C (1450 °F), soaked until

Table 14 Effects of mass on typical properties of heat treated 6150 steel

Round bars, oil quenched from 830 °C (1525 °F) and tempered at 540 °C (1000 °F); 12.83-mm (0.505-in.) diam tensile specimens taken from center of 25-mm bars and from mid-radius of 50- and 75-mm diam bars

Bar size		Tensile strength		Yield strength		Elongation in 50 mm or 2 in., %	Reduction in area, %	Hardness, HB
mm	in.	MPa	ksi	MPa	ksi			
25	1	1185	172	1040	151	14	45	341
50	2	1170	170	1030	149	13	48	341
75	3	1090	158	950	138	13	47	331

Table 15 Typical room-temperature mechanical properties of 8640 steel

Tempering temperature		Tensile strength		Yield strength		Elongation in 50 mm or 2 in., %	Reduction in area, %	Impact energy		Hardness	
°C	°F	MPa	ksi	MPa	ksi			J	ft·lb	HB	HRC
Round bars, 13.5 mm (0.53 in.) in diameter(a)											
205	400	1810	263	1670	242	8.0	25.8	11.5(b)	8.5(b)	555	55
315	600	1585	230	1430	208	9.0	37.3	15.6(b)	11.5(b)	461	48
425	800	1380	200	1230	179	10.5	46.3	27.8(b)	20.5(b)	415	44
540	1000	1170	170	1050	152	14.0	53.3	56.3(b)	41.5(b)	341	37
650	1200	870	126	760	110	20.5	61.0	96.9(b)	71.5(b)	269	28
Round bars, 25 mm (1 in.) in diameter(a)											
425	800	1382	200.5	1230	179	10	46	27(c)	20(c)	415	44
480	900	1250	181	1120	162	13	51	41(c)	30(c)	388	42
540	1000	1070	155	940	137	17	56	54(c)	40(c)	331	36
595	1100	1020	148	910	132	16	57	73(c)	54(c)	302	32
650	1200	865	125.5	760	110.5	20	61	83(c)	61(c)	269	28

(a) Oil quenched from 830 °C (1525 °F) and tempered at indicated temperature. (b) Izod. (c) Charpy V-notch

Table 16 Effects of mass on typical properties of heat treated 8640 steel

Oil quenched from 830 °C (1525 °F) and tempered at 540 °C (1000 °F)

Bar size		Tensile strength		Yield strength		Elongation in 50 mm or 2 in., %	Reduction in area, %	Surface hardness, HB
mm	in.	MPa	ksi	MPa	ksi			
25	1	1070	155	940	137	17	56	331
50	2	910	132	770	112	18	57	293
75	3	860	125	710	103	19	58	277

Table 17 Typical longitudinal mechanical properties of H11 Mod steel

Air cooled from 1010 °C (1850 °F); double tempered, 2 + 2 h at indicated temperature

Tempering temperature		Tensile strength		Yield strength		Elongation in 50 mm or 2 in., %	Reduction in area, %	Charpy V-notch impact energy		Hardness, HRC
°C	°F	MPa	ksi	MPa	ksi			J	ft·lb	
510	950	2120	308	1710	248	5.9	29.5	13.6	10.0	56.5
540	1000	2010	291	1675	243	9.6	30.6	21.0	15.5	56.0
565	1050	1850	269	1565	227	11.0	34.5	26.4	19.5	52.0
595	1100	1540	223	1320	192	13.1	39.3	31.2	23.0	45.0
650	1200	1060	154	850	124	14.1	41.2	40.0	29.5	33.0
705	1300	940	136	700	101	16.4	42.2	90.6	66.8	29.0

the temperature is uniform, and then slowly cooled, either in the furnace or in an insulating medium such as ashes, lime, mica or silocel. When the forgings have cooled, they should be spheroidize annealed. Weldments, especially heavy-section weldments, should be cooled slowly in a furnace heated to the preheating temperature, or in an insulating medium, immediately after welding. After being cooled, weldments should be given a full spheroidizing anneal.

H11 Mod

This steel is a modification of the martensitic hot work die steel H11, the significant difference being that H11 Mod has a slightly higher carbon content.

Heat Treatments. The following standard heat treatments apply to H11 Mod steel:

- *Normalize:* Generally not recommended. For effective homogenization, heat to about 1065 °C (1950 °F), soak 1 h for each 25 mm (1 in.) of thickness, air cool. Anneal immediately after the steel reaches room temperature. NOTE: There is a possibility that H11 Mod may crack during this treatment, especially if the surface is significantly decarburized.
- *Anneal:* Heat to 845 to 885 °C (1550 to 1625 °F) in a furnace, preferably one with controlled atmosphere, and hold to equalize temperature; cool very slowly in the furnace to about 480 °C (900 °F), then more rapidly to room temperature. This treatment should produce a fully spheroidized microstructure free of grain-boundary carbide networks.
- *Harden:* Preheat to 760 to 815 °C (1400 to 1500 °F), then raise the temperature to 995 to 1025 °C (1825 to 1875 °F) and hold 20 min plus 5 min for each 25 mm (1 in.) of thickness (minimum, 25 min); air cool in still air. This can be conveniently done in a neutral salt bath or a controlled-atmosphere furnace. For a few applications, oil quenching from the low end of the hardening temperature range may be done. Air cooling, which produces less distortion than oil quenching, is more commonly employed.
- *Temper:* At the secondary hardening peak temperature of about 510 °C (950 °F) for maximum hardness and strength, or preferably above the secondary-hardening peak to temper

Fig. 4 Variation for hardness with tempering temperature for H11 Mod steel

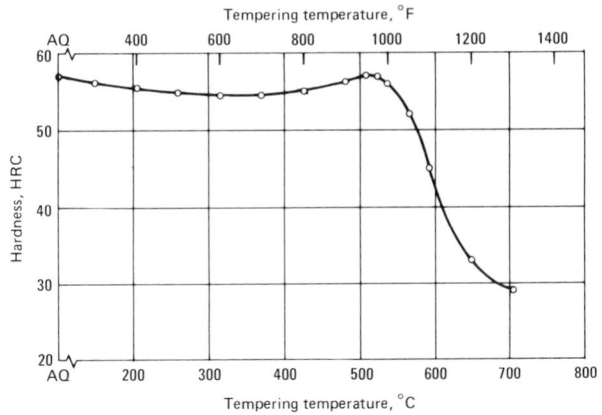

All specimens air cooled from 1010 °C (1850 °F) and double tempered, 2 + 2 h at temperature

Table 18 Effect of billet size and melting method on typical transverse strength and ductility of H11 Mod steel

Air cooled from 1010 °C (1850 °F); triple tempered, 2 + 2 + 2 h at 540 °C (1000 °F)

Billet size	Melting method	Tensile strength MPa	ksi	Reduction in area, %
150 by 150 mm (6 by 6 in.)......	Air	1965	285	16.1
	VAR	1985	288	25.7
300 by 300 mm (12 by 12 in.)....	Air	1972	286	7.2
	VAR	2013	292	19.7

Table 19 Typical room-temperature longitudinal mechanical properties of H13 steel

Round bars, oil quenched from 1010 °C (1850 °F) and double tempered, 2 + 2 h at indicated temperature

Tempering temperature °C	°F	Tensile strength MPa	ksi	Yield strength MPa	ksi	Elongation in 4 D, %	Reduction in area, %	Charpy V-notch impact energy J	ft·lb	Hardness, HRC
527	980	1960	284	1570	228	13.0	46.2	16	12	52
555	1030	1835	266	1530	222	13.1	50.1	24	18	50
575	1065	1730	251	1470	213	13.5	52.4	27	20	48
593	1100	1580	229	1365	198	14.4	53.7	28.5	21	46
605	1120	1495	217	1290	187	15.4	54.0	30	22	44

back to a lower hardness or strength with improved ductility and toughness. A minimum of 1 h at temperature should be allowed, but parts preferably should be double tempered (2 h at temperature, cool to room temperature, then 2 h more at temperature). Triple tempering is more desirable, especially for critical parts. For high-temperature applications, parts should be tempered at a temperature above the maximum service temperature to guard against unwanted changes in properties during service.

- *Stress relieve:* Heat to 650 to 675 °C (1200 to 1250 °F); cool slowly to room temperature. This treatment is often used to achieve greater dimensional accuracy in heat treated parts by stress relieving rough-machined parts, then finish machining, and finally heat treating to the desired hardness.

- *Nitride:* For increased wear resistance, finish machined and heat treated parts may be nitrided. The nitriding operation can be considered as the second temper of a double tempering operation. The parts should be gas or liquid nitrided at about 525 °C (980 °F). The nitrided case depth depends on time at temperature. For example, gas nitriding in 20 to 30% dissociated ammonia for 8 to 48 h normally produces a case depth of about 0.2 to 0.35 mm (0.008 to 0.014 in.).

Properties. Variation of hardness with tempering temperature for H11 Mod is plotted in Fig. 4. Variations in typical room-temperature longitudinal mechanical properties with tempering temperature are given in Table 17. As an indication of the deep air hardenability of this steel (to depths greater than 305 mm, or 12 in.), the transverse tensile strength and ductility obtained in large billets of air-melted and vacuum-arc-remelted H11 Mod are given in Table 18, which also shows the improvement in ductility that results from vacuum arc remelting. Each value is the average of four tests, two from the top and two from the bottom of the ingot.

H13

The main difference in composition between H11 Mod and H13 is the higher vanadium content of the latter (see Table 1); this leads to a greater dispersion of hard vanadium carbides, which results in higher wear resistance. H13 parts may be nitrided for additional wear resistance. Also, H13 has a slightly wider range of carbon content than H11 Mod. Depending on the producer, the carbon content of H13 may be near the high or low side of the accepted range, and for a given heat treatment the properties will vary correspondingly.

Heat Treatments. The following standard heat treatments apply to H13 steel:

- *Normalize:* Not recommended for H13. Some improvement in homogeneity can be obtained by preheating to about 790 °C (1450 °F), heating slowly and uniformly to 1040 to 1065 °C (1900 to 1950 °F) and holding 1 h for each 25 mm (1 in.) of

Fig. 5 Variation for hardness with tempering temperature for H13 steel

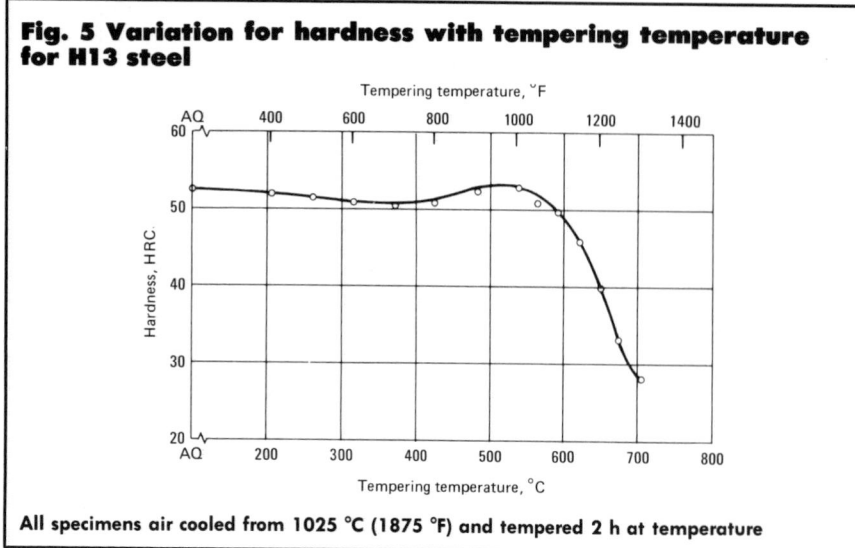

All specimens air cooled from 1025 °C (1875 °F) and tempered 2 h at temperature

Table 20 Room-temperature mechanical properties of HP 9-4-30 steel

| Property | Typical value for hardness of | | Minimum value(c) |
	49-53 HRC(a)	44-48 HRC(b)	
Tensile strength	1650-1790 MPa	1520-1650 MPa	1520 MPa
	(240-260 ksi)	(220-240 ksi)	(220 ksi)
Yield strength(d)	1380-1450 MPa	1310-1380 MPa	1310 MPa
	(200-210 ksi)	(190-200 ksi)	(190 ksi)
Elongation in 4 D	8-12%	12-16%	10%
Reduction in area	25-35%	35-50%	35%
Charpy V-notch impact energy	20-27 J	24-34 J	24 J
	(15-20 ft·lb)	(18-25 ft·lb)	(18 ft·lb)
Fracture toughness (K_{Ic})	66-99 MPa$\sqrt{m}$	99-115 MPa$\sqrt{m}$	...
	(60-90 ksi$\sqrt{in.}$)	(90-105 ksi$\sqrt{in.}$)	

(a) Oil quenched from 845 °C (1550 °F), refrigerated to −73 °C (−100 °F) and double tempered at 205 °C (400 °F). (b) Same heat treatment as (a) except double tempered at 550 °C (1025 °F). (c) For sections forged to 75 mm (3 in.) or less in thickness (or to less than 0.016 m², or 25 in.², in total cross-sectional area), quenched to martensite and double tempered at 540 °C (1000 °F)

thickness, and then air cooling. Just before or just as the steel reaches room temperature, it should be recharged into a furnace and given a full spheroidizing anneal. NOTE: There is a risk of cracking during this treatment, especially if done in a furnace where the atmosphere is not controlled to prevent surface decarburization.

- *Anneal:* Heat uniformly to 860 to 900 °C (1575 to 1650 °F) in a controlled-atmosphere furnace, or with the part in a neutral compound, to prevent decarburization, and hold to equalize temperature; cool very slowly in the furnace to about 480 °C (900 °F), then cool more rapidly to room temperature. This treatment should result in a fully spheroidized microstructure free from grain-boundary carbide networks.

- *Harden:* Preheat to 790 to 815 °C (1450 to 1500 °F), then raise the temperature uniformly to 995 to 1025 °C (1825 to 1875 °F) and soak 20 min plus 5 min for each 25 mm (1 in.) of thickness (minimum, 25 min); air cool in still air. For a few applications, oil quenching from the low side of the hardening temperature may be done, but at the risk of distortion or cracking. Air cooling is preferred, and usually is done from the high side of the hardening temperature range.

- *Temper:* At the secondary-hardening peak of about 510 °C (950 °F) for maximum hardness and strength, but preferably at a higher temperature to temper back to a lower level of hardness or strength with improved toughness and ductility. Double tempering—2 h at temperature, air cool, then 2 h more at temperature—is recommended; for critical parts, triple tempering may be desirable.

- *Stress relieve:* Heat to 650 to 675 °C (1200 to 1250 °F) and soak 1 h or more; cool slowly to room temperature. This treatment often is used to achieve greater dimensional accuracy in heat treated parts by stress relieving rough-machined parts, then finish machining, and finally heat treating to the desired hardness.

- *Nitride:* Finish-machined and heat treated parts may be nitrided. Because it is carried out at about the normal tempering temperature, nitriding can serve as the second temper in a double tempering treatment. The nitrided case depth depends on time at temperature. For example, gas nitriding at 510 °C (950 °F) for 10 to 12 h produces a case depth of 0.10 to 0.13 mm (0.004 to 0.005 in.), whereas gas nitriding for 40 to 50 h results in a case depth of about 0.3 to 0.4 mm (0.012 to 0.016 in.). For selective nitriding, copper plating is preferred for stopping off areas that are not to be nitrided; stop-offs containing lead should be avoided, because lead has been found to embrittle H13 steel.

Properties. The properties presented in this section are for H13 with a carbon content in the middle of the composition range (for composition range, see Table 1). Somewhat different properties should be expected when the carbon content is near the high end or the low end of the range.

Variation of hardness with tempering temperature for H13 is plotted in Fig. 5. Typical room-temperature longitudinal mechanical properties of bars tempered to different hardness levels are given in Table 19. H13 has deep hardenability, although its hardenability is slightly less than that of H11 Mod. For example, an H13 bar 330 mm (13 in.) in diameter and 2743 mm (108 in.) long, when fast air cooled from 1010 °C (1850 °F), had an as-quenched hardness of 45 HRC.

9Ni-4Co Steels

During the 1960's, Republic Steel Corporation introduced a family of four HP 9-4 (9Ni-4Co) steels having high fracture toughness when heat treated to very high strength levels. Among these, HP 9-4-20 and HP 9-4-30 are commercially available. They nominally contain 0.20 and 0.30% C, respectively (see Table 1 for chemical compositions). With the increase in carbon content, the attainable strength increases but with corresponding decreases in toughness and weldability. The high nickel content of 9% provides deep hardenability and toughness, and the 4% Co prevents retention of excessive austenite in heat treated parts. The carbide-forming elements, chromium and molybdenum, also impart hardenability, but the amounts of these carbide formers are adjusted to provide a fairly flat response to tempering without pronounced secondary hardening and its attendant reduction in toughness.

HP 9-4-20, although it has good weldability and fracture toughness, cannot develop a yield strength of 1380 MPa (200 ksi), which was selected as the criterion for ultrahigh-strength steels discussed in this article. Therefore, only HP 9-4-30 is discussed below.

HP 9-4-30

HP 9-4-30 steel has deep hardenability and can be fully hardened to martensite in sections up to 150 mm (6 in.) thick. Heat treated parts can be readily welded without any preheat or postheat treatment. After welding, parts may be stress relieved at about 540 °C (1000 °F) for 24 h.

Heat Treatments. The following heat treatments apply to HP 9-4-30 steel:

- *Normalize:* Heat to 870 to 925 °C (1600 to 1700 °F) and hold 1 h for each 25 mm (1 in.) of thickness (minimum, 1 h); air cool.
- *Anneal:* Heat to 620 °C (1150 °F) and hold 24 h; air cool.
- *Harden:* Austenitize at 830 to 860 °C (1525 to 1575 °F) and hold 1 h for each 25 mm (1 in.) of thickness (minimum, 1 h); water or oil quench. Complete the martensitic transformation by refrigerating at least 1 h at -87 to -60 °C (-125 to -75 °F); let warm to room temperature.
- *Temper:* At 200 to 600 °C (400 to 1100 °F), depending on desired strength; double tempering preferred. The most widely used tempering treatment is double tempering (2 h at temperature, air cool, then 2 h more at temperature) at a temperature from 540 to 575 °C (1000 to 1075 °F).
- *Stress relieve:* Usually required only after welding of restrained sections. Heat to 540 °C (1000 °F) and hold 24 h; air cool to room temperature.

Properties. Room-temperature mechanical properties of HP 9-4-30 double tempered at three different temperatures are presented in Table 20. The data for material double tempered at 540 °C (1000 °F) represent the minimum mechanical properties for this condition; properties listed for the other conditions may be considered typical.

REFERENCES

The data presented in this article were extracted from numerous sources as referenced in the original article by this author on Ultrahigh-Strength Steels, Metals Handbook, 9th Ed., Vol 1, p 421 to 443.

Heat Treating of Maraging Steels

By Stephen Floreen
Research Fellow
International Nickel Company, Inc.

MARAGING STEELS are ultrahigh-strength steels that differ from conventional steels in that they are not hardened by carbon content. Instead these steels are strengthened by precipitation of intermetallic compounds produced by age hardening a matrix of very-low-carbon martensite. Carbon, in fact, is an impurity in maraging steels and is kept at the lowest possible concentration.

Compositions of the common grades of maraging steel are given in Table 1. Grades have been developed that provide specific levels of yield strength ranging from 1030 to 3450 MPa (150 to 500 ksi).

The absence of carbon and the use of intermetallic precipitation to achieve hardening produce several unique characteristics of maraging steels that set them apart from conventional steels. Hardenability is of no concern. The low-carbon martensite formed after annealing is relatively soft, about 30 to 35 HRC. During age hardening there is only a very slight dimensional change. Thus, fairly intricate shapes can be machined in the soft condition and then hardened with a minimum of distortion.

Physical Metallurgy

Phase transformations in maraging steels are illustrated in Fig. 1. Plotted on the metastable diagram (Fig. 1a) are the austenite-to-martensite transformation on cooling and the martensite-to-austenite reversion on heating. The equilibrium diagram (Fig. 1b) shows that at higher nickel contents the equilibrium phases at low temperatures are austenite and ferrite.

In the metastable diagram, no phase transformations occur until the M_s temperature is reached and martensite is formed. Even very slow cooling of heavy sections produces only martensite, with no problems of low hardenability due to the large section size. The remaining alloying elements in the steel do, of course, alter the M_s temperatures significantly from those shown in Fig. 1, but the characteristic independence of cooling rate is not altered.

Table 1 Nominal compositions of commercial maraging steels

Grade	Ni	Composition(a), % Mo	Co	Ti	Al
18 Ni (200)	18	3.3	8.5	0.2	0.1
18Ni (250)	18	5.0	7.75	0.4	0.1
18Ni (300)	18	5.0	9.0	0.65	0.1
18Ni (350)	18	4.2(b)	12.5	1.6(b)	0.1
18Ni (Cast)	17	4.6	10.0	0.3	0.1

(a) For all grades, carbon content is no more than 0.03%. (b) Some producers use a combination of 4.8% Mo and 1.4% Ti, nominal.

For most grades of maraging steel, M_s temperatures are on the order of 200 to 300 °C (390 to 570 °F) and the steel is fully martensitic at room temperature.

Age hardening is produced by heat treating for several hours at temperatures typically on the order of 480 °C

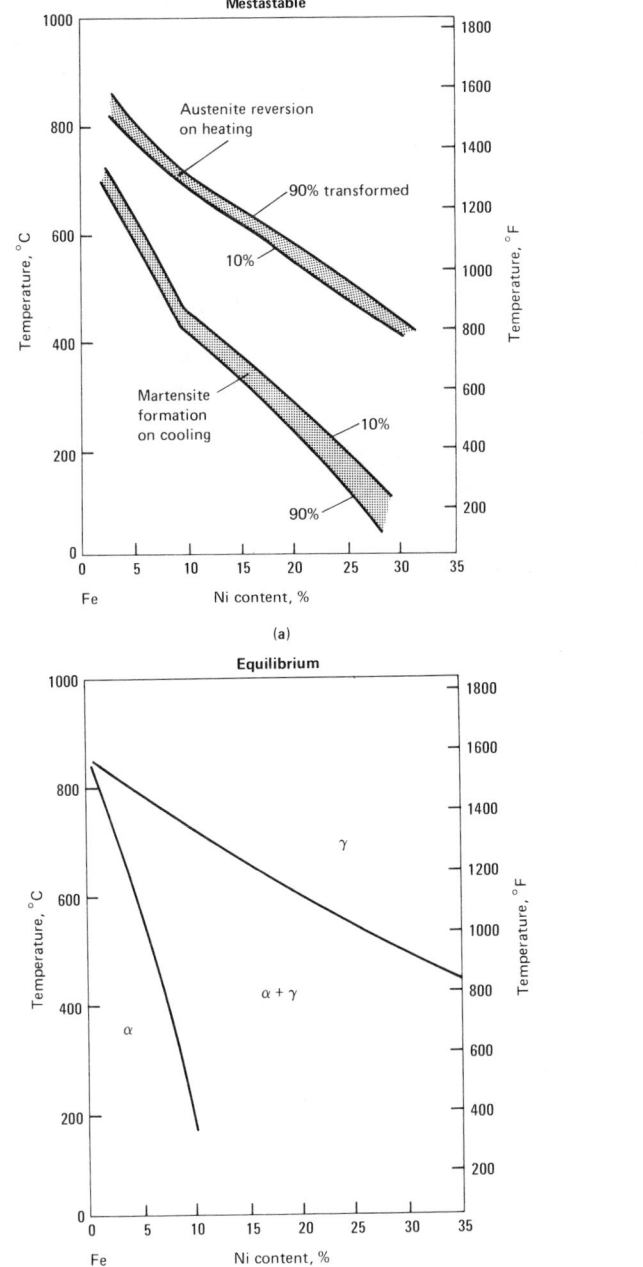

Fig. 1 Metastable and equilibrium phase relationships in the Fe-Ni system

aging steels usually is the result of (a) overaging by coarsening of the precipitate particles, and (b) reversion to austenite. Very substantial amounts of austenite, on the order of 50%, can eventually be formed.

Maraging steels normally contain little or no austenite after standard heat treatments, but sometimes austenite is deliberately formed. For example, if maraging steel is to be used in an application where overaging in service is expected, as in tooling for die casting of aluminum, the material will be slightly overaged beforehand. This minimizes overaging in service and resultant surface tensile stresses. Extreme overaging to form large amounts of reverted austenite also has been employed as an intermediate treatment to enhance response to cold working or to minimize the effects of thermal gradients during hot working and subsequent storage of extraordinarily heavy sections.

Overaged maraging steels are expected to show good resistance to brittle fracture and stress-corrosion cracking. Unfortunately, there appears to be considerable heat-to-heat variability in these characteristics when overaging heat treatments are used. Therefore, it is difficult to recommend specific overaging treatments that will produce consistent mechanical properties. In general, if a specific yield strength is required, it is better to use a maraging steel designed to produce that yield strength by conventional aging than to use an overaged higher-strength alloy.

Heat Treating

Conventional heat treatments for the standard grades of maraging steel are given in Table 2. Alloys with higher titanium contents are susceptible to formation of TiC films at austenite grain boundaries after holding at temperatures on the order of 900 to 1100 °C (1650 to 2000 °F). These films can severely embrittle the alloy when it is subsequently age hardened, leading to low-energy fractures along prior austenite grain boundaries. Prolonged annealing in this temperature range should be avoided for all compositions.

Solution Treatment. Maraging steels normally are solution annealed (austenitized) one hour for each 25 mm (1 in.) of section size. Atmosphere control may be necessary to minimize surface damage. Ordinarily, dry hydrogen or dissociated ammonia atmospheres

(900 °F). During this stage the equilibrium phase diagram becomes important. That is, with prolonged holding at 480 °C, the structure tends toward the equilibrium phases, primarily ferrite and austenite. Fortunately, the precipitation reactions that cause hardening proceed much more rapidly than the

reversion reactions that produce austenite and ferrite. Thus, very substantial increases in strength are achieved before reversion takes place. With longer aging times or higher temperatures, hardness will reach a maximum and then start to drop.

Overaging. Softening in aged mar-

Table 2 Heat treatments and typical mechanical properties of standard 18Ni maraging steels

Grade	Heat treatment (a)	Tensile strength MPa	ksi	Yield strength MPa	ksi	Elongation in 50 mm or 2 in., %	Reduction in area, %	Fracture toughness MPa√m	ksi√in.
18Ni (200)	A	1500	218	1400	203	10	60	155-200	140-220
18Ni (250)	A	1800	260	1700	247	8	55	120	110
18Ni (300)	A	2050	297	2000	290	7	40	80	73
18Ni (350)	B	2450	355	2400	348	6	25	35-50	32-45
18Ni (Cast)	C	1750	255	1650	240	8	35	105	95

(a) Treatment A: solution treat 1 h at 820 °C (1500 °F), then age 3 h at 480 °C (900 °F). Treatment B: solution treat 1 h at 820 °C (1500 °F), then age 12 h at 480 °C (900 °F). Treatment C: anneal 1 h at 1150 °C (2100 °F), age 1 h at 595 °C (1100 °F), solution treat 1 h at 820 °C (1500 °F) and age 3 h at 480 °C (900 °F)

are used. The cooling rate after annealing is of little consequence because it has no effect on either microstructure or properties. It is essential, however, that the steel be cooled to room temperature before it is age hardened. If this is not done, the steel may contain untransformed (retained) austenite and may be much softer than expected.

Age hardening normally is done at 455 to 510 °C (850 to 950 °F) for 3 to 12 h. In typical treatments at 480 °C (900 °F), grades 18Ni(200), 18Ni(250) and 18Ni(300) are held 3 to 6 h and grade 18Ni(350) is held 6 to 12 h. The 350 grade also is aged for 3 to 6 h at 495 to 510 °C (925 to 950 °F). For applications such as die casting tooling, aging at temperatures on the order of 530 °C (985 °F) is employed.

The standard age-hardening heat treatments listed in Table 2 produce 0.04% contraction in length in the 18Ni(200) grade, 0.06% contraction in the 18Ni(250) grade and 0.08% contraction in both the 18Ni(300) and 18Ni(350) grades. These very small dimensional changes during hardening allow many maraging steel components to be finish machined in the annealed condition. The finished parts then can be hardened without further machining. When greater dimensional accuracy is required, an allowance for contraction is readily made.

Cleaning After Heat Treatment

For removal of oxide films formed by heat treatment, grit blasting is the most efficient technique. Maraging steels can be chemically cleaned by pickling in sulfuric acid or by duplex pickling, first in hydrochloric acid and then in nitric-plus-hydrofluoric acid. As with conventional steels, care must be taken to avoid overpickling.

Case Hardening of Steel

Gas Carburizing

By the ASM Committee on
Gas Carburizing, Carbonitriding and
Nitriding*

GAS CARBURIZING, in current commercial practice, uses carbon from hydrocarbon gases and easily vaporized hydrocarbon liquids to produce a hard surface layer on steel parts. Gas carburizing is often referred to as "case carburizing". The main function of gas carburizing is to provide an adequate supply of carbon for absorption and diffusion into the steel.

Carbon Sources

Gaseous Sources. Gases most commonly used are natural gas, "manufactured" gas and certain propanes (Table 1). Butane is used infrequently. Where the demand for carbon is low, as in carbon restoration, endothermic generator gas is a suitable source of carbon.

Natural gas and propane are the preferred sources when available in high purity forms. The most desirable form of propane is derived from natural gas, rather than from petroleum. Propane obtained as a by-product of oil refining frequently contains excessive amounts of ethylene, propylene and other unsaturated hydrocarbons that break down rapidly to oily soot or coke. Because they often have a high carbon dioxide content, city gas, producer gas and other manufactured gases usually are undesirable sources of carbon for carburizing. These gases normally are blends of coke oven gas, water gas and natural

Table 1 Principal gases used as sources of carbon in gas carburizing

Constituent	Pittsburgh	Natural gas Kansas City	Indianapolis	Coke oven gas	Commercial (normal) butane	Commercial propane
Composition, vol %						
CO_2	0.8	0.4	2.2	. . .	. . .	
O_2	. . .	. . .	0.4	0.8	. . .	. . .
N_2	0.8	8.4	6.0	8.1	. . .	. . .
CO	. . .	. . .	. . .	6.3	. . .	. . .
H_2	. . .	. . .	. . .	46.5	. . .	. . .
CH_4 (methane)	83.4	84.1	87.2	32.1	. . .	. . .
C_2H_6 (ethane)	15.8	6.7	6.0	. . .	. . .	. . .
C_2H_4 (ethylene)	. . .	. . .	. . .	3.5	. . .	2.5
C_6H_6 (benzene)	. . .	. . .	. . .	0.5	. . .	. . .
C_4H_{10} (butane)	. . .	. . .	. . .	. . .	93.0	. . .
C_3H_8 (propane)	. . .	. . .	. . .	. . .	7.0	96.0
C_4H_{10} (isobutane)	. . .	. . .	. . .	. . .	. . .	1.5
Specific gravity	0.61	0.63	0.61	0.44	1.95	1.52

gas, which are maintained by law at a constant Btu heating value but have a variable chemical composition that may change frequently in a 24-h period. This produces nonuniform carburizing; consequently, this type of gas should not be used unless more desirable gas is unavailable.

For uniform carburizing, circulation of furnace gases is necessary. Because hydrocarbon gases provide large quantities of available carbon, relatively small flows of gas are required. The circulation resulting from gas flow alone is not always sufficient to produce uniform carburizing. It is usually essential that the furnace have high-volume forced circulation of the gases to all parts of the work load.

A common commercial practice is to use an endothermic gas or a purified exothermic gas as a carrier that is enriched with one of the hydrocarbon gases. The ratio of carrier gas to hydro-

*Charles A. Stickels, *Chairman*, Principal Staff Engineer, Research Staff, Ford Motor Co.; Larry E. Byrnes, Manager—Manufacturing Research, Federal Mogul Corp.; Jon L. Dossett, Vice President—Metallurgy, United States Gear Corp.; Robert L. Hughes, Chief Metallurgist, Fairfield Manufacturing Company, Inc.; Kenneth D. Gladden, Senior Development Engineer, Caterpillar Tractor Co.; Dale A. Poteet, Jr., Division Engineer, John Deere Component Works; Joseph A. Riopelle, District Manager/Chicago Office, Surface Division, Midland-Ross Corp.; Ronald D. Rogers, Chief Engineer, Materials Engineering, Automotive Operations, Rockwell International

carbon gas varies widely in industry but is usually in the ratio of 8-to-1 to 30-to-1. The ratio used depends on the types of carrier and hydrocarbon gases, furnace size and condition, amount of circulation, and the work surface.

Although the use of carrier gases results in improved circulation because of the larger volume used, the additional forced circulation provided by a fan is usually needed to obtain maximum uniformity of carburizing and furnace temperature, particularly with batch furnaces, in which the load is densely packed.

Liquid Hydrocarbon Sources. Liquids also are used as sources of carburizing gas. These liquids are usually proprietary compounds and range in composition from pure hydrocarbons such as terpenes, dipentene or benzene to oxygenated hydrocarbons such as alcohols, glycols or ketones. When a liquid is used, it normally is fed in droplet form to a target plate in the furnace, where it volatilizes almost instantaneously. The vapors dissociate thermally to provide a carburizing atmosphere containing carbon monoxide, carbon dioxide, methane and water vapor. The flow of liquid to the target plate is adjusted either manually or automatically to obtain the desired carbon potential. In a tight furnace, careful adjustment of the flow of liquid permits accurate control of carbon potential.

Forced fan circulation serves to distribute the atmosphere evenly throughout the furnace and to promote temperature uniformity. The fan recirculates the "spent" gas, which acts as a carrier gas and dilutes the rich gas to prevent excessive sooting.

Liquid sources are most suitable for batch-type furnaces in sizes up to about 0.9 m (3 ft) in diameter and about 1.8 m (6 ft) deep. They also may be used in locations where a supply of natural or manufactured gas is not readily available or is supplied in widely varying compositions.

Because they do not require external piping or gas generator equipment, the initial cost of liquid installations is comparatively low. However, as raw materials, the liquids normally are more costly than the carburizing gases. When the use of gas requires a carrier-gas generator, the over-all cost for gas may exceed that for liquids, because of the cost of operating and maintaining the generator. Automatic control is established more slowly with liquids, and in larger furnaces, liquids produce more soot, primarily because of inefficient cracking and the absence of a closely controlled carrier gas or diluent.

The sampling of atmospheres for analysis and the various control systems used are discussed in greater detail in the sections on gas-carburizing atmospheres, carbon concentration

gradients and surface carbon content later in this article.

Carrier Gases

Most gas carburizing furnaces employ one of several carrier gases to dilute and react with the hydrocarbon gas used as the principal source of carbon. The four common carrier gases, classes 201, 202, 302 and 402, are listed in Table 2. Class 302 gas is most commonly used today, because of its ease of use. Class 102 gas also is listed but is seldom used.

Class 102 gas, when used alone, is decarburizing, so its successful use depends entirely on the addition of hydrocarbon gas. Under these conditions, the ratio of hydrocarbon gas to carrier gas may be approximately 1 to 1, which results in excessive deposition of free carbon (soot or coke) on the workpiece and the furnace.

Class 202 gas offers a moderate range of control, some carbon availability for carburizing, and continuous operation with dew points of -40 °C (-40 °F) or lower.

Class 302 (endothermic) gas is generally the preferred carrier gas for use in gas carburizing furnaces and is the most widely used. It offers a broad range of carbon control, a moderate amount of carbon availability for car-

Table 2 Principal carrier gases used in gas carburizing

Class	Method of preparation	Air to gas ratio(a)	Composition(a), vol % N2	CO(b)	CO2	H2(b)	CH4	Dew point °C	°F	Fuel gas required(c) m³	ft³	Nature of atmosphere
102	Exothermic base (rich mixture)	6.0	71.5	10.5	5.0	12.5	0.5	(d)	(d)	4.4	155	Combustible; toxic; medium reducing
201	Prepared nitrogen base with lean mixture	9.0	97.1	1.7	...	1.2	...	-40	-40	3.8	135	Noncombustible; inert
202	Prepared nitrogen base with rich mixture	6.0	75.3	11.0	...	13.2	0.5	-40	-40	4.5	160	Combustible; toxic; medium reducing
302	Endothermic base (completely reacted and cooled to eliminate breakdown of CO into $C + CO_2$)	2.5	39.8	20.7	...	38.7	0.8	-4 to -21(e)	$+25$ to -5(e)	5.7(f)	200(f)	Combustible; toxic; very reducing
402	Charcoal base	...	64.1	34.7	...	1.2	...	-29	-20	5.7 kg Charcoal	12.5 lb	Combustible; toxic; very reducing

(a) Analyses based on 1055 kJ (1000 Btu) natural gas requiring 9.6 volumes of air for complete combustion. For high-H2 artificial gas, multiply the quoted ratio of air to gas by 0.5; for medium-H2, high-CO artificial gas, by 0.4; for propane, by 2.5; for butane, by 3.2. (b) If made with artificial gas, the CO will be slightly lower and H2 somewhat higher. With propane and butane, the reverse will be true. (c) Cubic metres (cubic feet) of 1055 kJ (1000 Btu) natural gas required to make 28 m³ (1000 ft³) of atmosphere. For high-H2 artificial gas, multiply value shown by 2.0; for medium-H2, high-CO artificial gas, by 2.5; for propane, by 0.4; for butane, by 0.3. (d) Room temperature (cooling by tap water). May be reduced to $+4$ °C ($+40$ °F) by refrigeration, or to -46 °C (-50 °F) by adsorbents. (e) Dew point is varied by changing the ratio of air to gas going to the generator. Most carburizing is done with dew point (at the generator) from -12 to -4 °C ($+10$ to $+25$ °F). (f) Plus 7 m³ (250 ft³) of fuel gas per 28 m³ (1000 ft³) of prepared atmosphere, for heating retort

burizing and, when operated with dew points of -1 °C ($+30$ °F) and above, continuous operation without weekend shutdowns for burnout. Lower temperatures may be used, down to -7 °C ($+20$ °F), but periodic burnout may be necessary.

Class 402 gas may be used as a carrier provided the equipment is designed to give the required control and a high cost for maintenance can be accepted.

Nitrogen gas is available from liquefaction plants, as a by-product from oxygen plants, or from modified exothermic generators as a class 201 atmosphere. Nitrogen alone is neutral, but in commercial form, it usually contains one or more minor impurities that may be oxidizing or decarburizing.

The inertness of pure nitrogen presents a problem of carbon control when used as a carrier for hydrocarbon gas. Because of the great capacity of hydrocarbons to supply carbon to steel, the ratio of nitrogen carrier gas to hydrocarbon gas must be large—from about 50-to-1 to 100-to-1. Even more important, the hydrocarbon gas must be controlled within limits of about 0.1% to maintain carburizing potential within the close ranges of $\pm 0.05\%$ carbon.

Additional information on the generation and use of the carrier gases discussed here is given in the article on furnace atmospheres in this volume.

Equipment

Gas carburizing furnaces vary widely as to physical construction, but they can be divided into two major categories: batch and continuous. The two types differ largely in their method of handling the work. In a batch-type furnace, the work is charged and discharged as a single unit or batch; in the continuous furnace, workpieces enter and leave the furnace as units in a continuing stream. The continuous furnace generally is favored for large-volume production of similar parts with total case depth requirements of 0.4 to 3 mm (0.015 to 0.120 in.).

Pit-Type Batch Furnaces. Pit-type carburizing furnaces consist essentially of two parts: the furnace, which is placed in a pit and extends to floor level or slightly above, and a cover or lid, which extends upward from floor level. These furnaces should be as gastight as possible; a positive pressure should be maintained within the furnace. It must be equipped with a fan to circulate gases.

Table 3 Typical equipment requirements for pit carburizing, reheating, quenching, and tempering large gears

Production requirements

Number of gears per 16-h day	6
Weight of each gear	Over 90 kg (200 lb)
Total depth of case	1.5 to 1.8 mm (0.060 to 0.070 in.)

Equipment requirements

Carburizing furnace:
Dimensions (internal)	0.9 m (3 ft) diam by 1.2 m (4 ft) deep
Floor space	2.4 by 4.6 m (8 by 15 ft)
Source of heat	Electricity (125 kV·A)
Operating temperature:	
Carburizing cycle	925 °C (1700 °F)
Cooling cycle prior to air cooling	790 °C (1450 °F)

Endothermic gas generator:
Floor space	2.1 by 3.0 m (7 by 10 ft)
Source of heat and atmosphere gas	Natural gas
Gas output	42 m³/h (1500 ft³/h)
Operating temperature	1010 °C (1850 °F)

Reheating furnace:
Diameter of rotary hearth	3.7 mm (12 ft)
Floor space	6.1 by 10.7 m (20 by 35 ft)
Source of heat	Natural gas
Heat input	1 583 000 kJ/h (1 500 000 Btu/h)
Reheating temperature of work	815 °C (1500 °F)

Quenching press:
Internal diameter	0.9 m (3 ft)
Floor space	1.7 by 2.8 m (66 by 109 in.)
Temperature of quenching oil	45 °C (115 °F)
Oil capacity	1136 l/min (300 gal/min)
Services:	
Air	1380 kPa (200 psi)
Electricity	220 V

Tempering furnace:
Floor space	2.4 by 4.6 m (8 by 15 ft)
Source of heat	Electricity
Heat input (max)	75 kW
Operating temperature (max)	760 °C (1400 °F)

Other items: cleaning equipment, quenching oil storage facility, heat exchanger for heating quenching oil, and handling equipment such as crane slings

Pit furnaces are available over a wide range of weight capacities and are well adapted to the use of devices for automatic carbon control. They are particularly suited to the processing of parts that must be cooled in the furnace. Depending on their shape, parts may be stack loaded, supported by holding fixtures, or retained in baskets. In the typical process situation, parts must not make contact (other than point contact) with each other if differences in case depth that are caused by shielding are to be avoided.

Carburized loads up to a certain size may be removed from pit-type furnaces at carburizing temperature and quenched. However, direct quenching is usually not feasible when large loads and large furnaces are involved.

An additional disadvantage of the pit-type furnace is that, if the work is to be direct quenched, the load must be moved from the atmosphere of the furnace into air before quenching. Although exposure in air is relatively brief, it results in the formation of an adherent black scale on the steel that, for many applications, must be removed by dilute mineral acids or by grit blasting. Thus, parts that must remain bright and scale-free after furnace treatment, such as parts with internal threads, are processed in horizontal batch furnaces and quenched under a cover of protective atmosphere.

Example 1. Table 3 lists equipment used in one plant for carburizing, cooling, reheating, press quenching and tempering of large gears. These gears weigh more than 90 kg (200 lb) each and are carburized to a depth of 1.5 to 2 mm (0.060 to 0.080 in.). Pit-type furnaces are well suited for carburizing parts that are cooled and reheated prior to quenching.

Horizontal batch furnaces are ideal for gas carburizing small parts requiring light case depths and direct quenching. These furnaces are well suited to handling batches of less than 900 kg (2000 lb) and, when equipped with an integral oil quenching system, can be made part of a continuous production line. These furnaces may also be equipped to cool loads slowly from the carburizing temperature. They can economically handle parts in quantities that are too small to justify the use of continuous equipment.

The ability to quench automatically under protective atmosphere is an outstanding advantage of the horizontal batch furnace over other types. Also, compared to the shaker-hearth furnace, it can produce deeper cases, primarily because of forced circulation, at a lower cost. It can handle parts that do, and parts that do not, require special stacking, placement or fixturing in the furnace. Because parts are not exposed to air, they can be produced almost entirely free of surface oxides. Gears, for example, can be carburized in horizontal batch furnaces without risk of oxide contamination or decarburization; they can easily be fixtured to avoid nicking and to minimize warpage. Flat pieces with sufficiently large holes can be suspended from rods to permit vertical heating and quenching, thus ensuring uniformity of case depth and minimizing the amount of straightening required after treatment.

Horizontal furnaces are versatile because they can accommodate many different types of loads and case depth requirements. With a high volume of forced, recirculated furnace atmosphere, they heat loads rapidly and provide a uniform case depth. When such an atmosphere is well agitated, satisfactorily uniform results can be obtained even with dense loads.

Rotary-retort batch furnaces are designed to handle loads of 45 to 680 kg (100 to 1500 lb). They are best suited for carburizing relatively small parts that may be indiscriminately loaded.

Seldom equipped with a circulating fan, this type of furnace consists of a gastight cylindrical retort that revolves within a firebrick-and-steel enclosed shell and is mounted on trunnions to facilitate loading and unloading of parts. The inner surface of the retort may contain a number of ribs designed to impart tumbling action to the parts being carburized, thus permitting greater exposure of the surface of the parts to the carburizing gases. The retort revolves on rolls and is turned continuously by a variable-speed drive. Loading and unloading are accomplished by removal of the front head. A chute may be used to convey the product from the furnace to the quenching medium. More information on use of rotary retorts as continuous furnaces is included in the section on continuous furnaces.

Parts loaded in a rotary-retort furnace must not be so heavy that the tumbling action will introduce nicks, if nicking is objectionable. Also, the parts must not be so fragile that the tumbling action and loading will cause undue distortion. Parts with outside threads, for example, might be damaged beyond use should the loading and tumbling action injure the threads.

Except for small parts of simple design, such as pins, bushings and rollers, rotary-retort batch furnaces operated noncontinuously are not well suited for parts that are to be directly quenched without subsequent reheating and hardening, because a load cannot be uniformly metered into the quenching medium when it is discharged from the retort. Differences in microstructural characteristics of the quenched parts result from this inability; where uniformity of heat treatment is desired, these differences may be unacceptable.

Rotary-hearth furnaces also may be used as batch-type units for uniform and controlled carburizing of parts such as spiral bevel gears, rings, sleeves, races and disks, that are individually loaded and die quenched. However, its use is recommended for low-volume production only.

Parts can be loaded at approximately the anticipated quenching rate. When the furnace is full (10 to 100 pieces, depending on part size), it is allowed to rotate until the carburizing cycle is completed, where the parts are subsequently quenched individually. For precision carburizing, this scheme ensures uniform heating of parts, uniform carburizing and close control of both

factors. Because the hearth rotates during the entire carburizing cycle (usually several hours), no part can remain near a cold spot (such as the furnace door) for an extended period of time. Gas composition and diffusion cycles are readily controlled, and furnace temperature can be lowered uniformly prior to quenching, to help reduce distortion. Except for normal checking, the furnace can be left unattended during the carburizing cycle.

Rotary-hearth furnaces are often used for secondary procedures, such as reheating previously carburized parts, because carburizing in one step can be inefficient.

Continuous Furnaces. Continuous carburizing furnaces are generally preferred for production loads exceeding 180 kg/h (400 lb/h) and requiring the same case depth, or for loads of sufficient size that require 24-h continuous operation with a minimum number of changes in required case depth. Continuous furnaces permit the individual quenching of larger parts and the batch quenching of smaller parts. Some types are equipped to provide cooling under cover of protective atmospheres.

Continuous furnaces can produce uniform carburizing results if parts are (a) uniformly loaded in terms of their weight and cross-sectional area, (b) provided with an adequate supply of carburizing gas with good circulation, and (c) uniformly quenched. In most continuous carburizing furnaces, circulating fans are used to improve atmosphere circulation and thereby promote more uniform carburizing results.

Continuous furnaces, in common with other types of carburizing furnaces, must be constructed so that air infiltration and contamination of the carburizing atmosphere by the quenching medium are held to a minimum. In some types, the furnace shell itself provides the necessary seal; with this construction, all seams and joints must be welded gastight. In other continuous furnaces, an alloy retort or muffle is used to provide an atmosphere seal. In furnaces that include direct quenching after carburizing, and in which the quenching medium also provides an atmosphere seal, it is often necessary to provide a means of preventing the quenching vapors from contaminating the furnace atmosphere. This is particularly true when the quenchant is a water solution.

Several types of furnaces are used for continuous gas carburizing, including the shaker-hearth, roller-hearth, rotary-retort and pusher furnaces.

Shaker-hearth furnaces use a reciprocating shaker motion to move the work along the hearth; this motion may be regulated to control the time cycle and case depth. Heating is efficient and confined mainly to the work load. Parts may be fed into the furnace by hand or by means of automatic metering and may be individually quenched. Use of this type of furnace is generally limited to lightweight parts that are to be carburized to case depths of 0.3 mm (0.010 in.) or less. The furnace hearth must be kept smooth, clean and at the proper level. Either the time cycle of work going through the furnace or uniformity of case depth should be checked at frequent intervals to detect any unwanted change in the force exerted by the shaker mechanism. When these precautions are adequately observed and the necessary atmosphere and temperature controls are provided, the shaker-hearth furnace can produce satisfactorily uniform case depth.

Unless a specially designed hearth, such as a corrugated hearth, is provided, heavy, flat parts are not suited to processing in shaker-hearth furnaces, because of the difficulty of obtaining adequate and uniform case on the part surface making contact with the hearth. Delicate parts and parts with fine threads may be mutilated by the action of the conventional shaker hearth. Without a special hearth, balls and cylinders will not move and progress uniformly in this type of furnace.

Two aspects of the design of shaker-hearth furnaces require special attention:

- Hearth plates should be of suitable weight to respond to the reciprocating motion of the shaker mechanism. This action propels the parts intermittently forward and ensures exposure and equal treatment to that portion of the part that rests on the hearth.
- Adequate exhaust facilities should be provided to handle quenching oil fumes. These fumes are highly carburizing and, unless properly disposed of, will interfere with control of the carbon potential of the atmosphere. They also will soil the surfaces of parts being processed.

Rotary-retort continuous furnaces are used to carburize the same types of small parts that can be handled in a rotary-retort batch furnace. The advantage of the continuous rotary furnace is that it can be automatically loaded and unloaded, thus eliminating the need for removing and replacing the head.

The inside of the retort is provided with a mechanized spiral rib throughout its length. The spiral rib can move the work load in either the forward or reverse direction. The frequency and duration of cycles of forward and reverse motion can be varied over any desired range. By this means, furnace length can be minimized and a reasonable agitation or tumbling action of the parts is obtained. The tumbling action provides for a better uniformity of case depth.

Continuous retorts cannot be tilted at will. Parts are fed at the front end and automatically issue from slots in the rear of the retort directly into the quenching medium. Because the front end of the furnace must be open to allow continuous charging, sufficient carburizing gas must be fed into the furnace to prevent the admission of outside air. These furnaces are suitable for carburizing to case depths of 0.4 to 2.5 mm (0.015 to 0.100 in.).

Pusher-type continuous furnaces are by far the most widely used continuous carburizing units. Construction usually consists of a gastight welded shell with radiant tubes for heating. The work is pushed through on trays with or without fixtures and, after completion of the carburizing cycle, may be quenched or cooled slowly. Circulating fans are almost always used for more uniform temperature and carburization. Most pusher-type furnaces are built with purging vestibules at the charge and discharge ends to reduce contamination of the atmosphere by air. In many instances, washing and tempering equipment is incorporated to provide a fully automated heat treating line.

Continuous pusher-type furnaces are designed for high-volume production. For example, they may have more than one row of work in process at the same time.

This type of equipment is quite versatile, and depending on the size and shape of parts and on permissible distortion, parts may be randomly loaded and quenched in an open tank. Alternately, parts may be removed individually from the furnace trays for plug or press quenching. Furnaces can be designed to provide a variety of special equipment and process requirements, such as number of trays, tray size, atmosphere control, atmosphere recirculation, temperature-control zones and quenching facilities.

Because of their size and more intricate charge and discharge mechanisms, continuous pusher-type furnaces are more expensive than the other types commonly used for carburizing. Consequently, production volume must be large enough so that the high equipment costs can be amortized during the life of the furnace.

In common with other atmosphere furnaces, continuous pusher-type furnaces must be capable of maintaining a positive pressure and must be able to exclude the products of combustion from the furnace chamber.

Example 2. Table 4 presents details of equipment components for continuous carburizing, tempering and washing of 9.9-kg (21.8-lb) gears in one plant. The cycle shown is typical of many high-production operations where the work is cooled to a predetermined temperature after carburizing, before direct quenching.

Safety Precautions

Because gaseous carburizing media are highly toxic and highly flammable and form explosive mixtures readily, a safety program should be established and rigidly followed. This program should be based on the training of operators and on preventive maintenance. A sound safety program protects the operator from injury, protects equipment and buildings, and prevents loss of work material; it can be simple and yet adequate to reach these objectives.

All the safety equipment available cannot substitute for the properly trained operator; most safety devices are no more than warning devices.

Important areas to be covered when training operating personnel include:

- Hazards involved
- Procedure for gassing the furnace. Usually, safety equipment will not permit introduction of atmosphere when furnace temperature is below 760 °C (1400 °F). Special safety procedures are essential if a combustible atmosphere is introduced into chambers at temperatures below 760 °C.
- Procedure for removing atmosphere
- Procedure for purging of chambers

and vestibules. Vestibules are usually equipped with explosion seals, and the proper use of these seals must be carefully explained because of the vulnerability of the vestibules to explosion.

- Procedures in emergencies such as power and equipment failures. Operators should be subject to trial runs, to ensure proper training for emergencies.
- Fire—proper consideration must be given to the type and location of fire-fighting equipment, and to various types of fires.
- Precautions in handling various hazardous gases and liquids.

Handling of Atmospheres. The main guideline pertinent to the safe handling of gas carburizing atmospheres includes knowledge of:

- The difference between inert and combustible gases
- Most prepared atmospheres contain more than 4% H_2 or 12.5% carbon monoxide, the low limit for explosion in air.
- Air-gas mixtures with small amounts of oxygen may ignite. According to the International Critical Tables, the flammability limits of hydrogen, carbon monoxide and methane in air are as follows: hydrogen, 9.0 to 68.6% H_2 (requires minimum of 6.59% O_2); carbon monoxide, 13.0 to 77.6% CO (requires minimum of 4.7% O_2); methane, 5.5 to 13.6% CH_4 (requires minimum of 18.1% O_2).
- The minimum ignition temperature of hydrogen or carbon monoxide is close to 595 °C (1100 °F).
- The introduction of a carburizing atmosphere into any furnace or part of a furnace at less than 760 °C (1400 °F) constitutes an explosion hazard.
- On the other hand, any gas mixture commonly used for carburizing can be safely put into a furnace at 760 °C (1400 °F) or above.
- To purge a furnace requires at least five times as much purging gas as the volume of the furnace itself.
- Before any repairs are made within a furnace, disconnect the gas lines, freshly purge the furnace, and blow air through the furnace continuously.
- Carbon monoxide is extremely toxic; unburned furnace or generated gases should not be exhausted into a room.
- If power or gas fails, shut down the

Table 4 Typical equipment requirements for continuous carburizing and tempering of 9.9-kg (21.8-lb) gears

Production requirements

Gears per day(a)	1155
Weight of each gear (8620 steel)	9.9 kg (21.8 lb)
Total depth of case	1.3 to 1.4 mm (0.050 to 0.055 in.)

Equipment requirements

Carburizing furnace:
Dimensions (internal)	1.8 by 8.2 m (6 by 27 ft)
Floor space	2.7 by 13.1 m (9 by 43 ft)
Source of heat	Natural gas(b)
Heat input	1 864 000 kJ/h (1 767 000 Btu/h)
Operating temperature:	
Carburizing cycle	940 °C (1725 °F)
Cooling cycle prior to quenching	870 °C (1600 °F)

Endothermic gas generator:
Floor space	1.8 by 2.7 m (6 by 9 ft)
Source of heat and atmosphere gas	Natural gas(b)
Heat input (includes gas generation)	801 000 kJ/h (759 000 Btu/h)
Operating temperature	1040 °C (1900 °F)

Tempering furnace:
Floor space	2.1 by 4.9 m (7 by 16 ft)
Source of heat	Natural gas(b)
Heat input	205 000 kJ/h (194 000 Btu/h)
Operating temperature (max)	260 °C (500 °F)

Other:
Trays per hour	3.85(c)

(a) Estimated for 20-h day. (b) 37 MJ/m³ (1000 Btu/ft³). (c) That is, one tray every 15½ min. Work trays, 457 by 610 mm (18 by 24 in.) holding 15 pieces per tray. Two rows of trays in pusher-type furnace

furnace promptly. If valuable work in process must be protected, switch to a reserve tank of protective gas and light the gas curtains at all doors.

- Be aware that the National Fire Prevention Association (NFPA) has developed safety standards specifically pertaining to safety of furnaces and related equipment. These worthwhile standards should be obtained and studied.

The use of a combustible atmosphere gas is attended by the greatest potential danger during start-up and shutdown of the equipment and during gas or power failures. When a gas or power failure occurs, the safest practice is to purge all chambers with an inert gas (bottled nitrogen is the most practical), metered at no more than the flow rate of the carrier gas. (In many modern systems, the purge is performed automatically, triggered by a pressure switch.) If inert gas is not available, the combustible gas must be immediately burned out of both vestibules of continuous carburizing furnaces.

The advantage of emergency standby power cannot be overstated. Because relatively little power (about 3 kW) is required to operate a 68-m³/h (2400-ft³/h) endothermic gas generator at full

capacity, even a small emergency power unit would permit quick restoration of carrier gas to two or three continuous carburizers or a dozen large pit furnaces, if the regular power source failed.

Continued gas flow considerably decreases the danger of explosion (regardless of furnace temperature), prevents damage to the work, and permits ready resumption of operation after the regular power supply is restored and the furnace is again brought to operating temperature. However, this procedure must be done under the continuous supervision of a well-qualified individual.

Starting and Stopping Furnace Operation

The best method of starting up a furnace is to purge air from the furnace to begin operation, or to purge combustible atmosphere from the furnace on shutting down, with at least five volumes of inert gas (nitrogen or carbon dioxide) from an outside storage tank. This is especially desirable for furnaces operating below 760 °C (1400 °F). A general procedure is outlined below for information *only*. In actual practice, *always* follow manufacturers' written instructions. Do *not* attempt to follow this brief guide in actual operations, because no general listing of this type can contain adequate safety provisions for your specific equipment.

Low-temperature furnaces operating at temperatures cooler than 760 °C or 1400 °F can best be operated by purging the chamber with five volumes of inert gas, before starting up or shutting down. A typical operating procedure may include the following steps.

To start up

- Open doors; bring furnace to temperature, above 760 °C (1400 °F).
- Place portable pilot light (torch) at gas entrances inside the chamber.
- Start flow of atmosphere gas.
- When gas ignites, remove pilots.
- Close doors almost completely.
- Light permanent flame curtains at entrance and exit.
- When gas coming from furnace ignites, furnace is ready for operation.

To shut down

- Keep furnace at temperature; maintain flame curtains.
- Slowly open entrance and exit doors wide.
- Shut off atmosphere gas.
- When clear of flame, furnace is purged.

Horizontal batch and continuous furnaces with no cooling chambers can best be operated by following a schedule similar to the one listed below.

To start up

- Open entrance and exit doors slightly. Heat furnace uniformly to 760 °C (1400 °F).
- Light pilots at gas entrances inside the chamber.
- Heat to operating temperature.
- Start flow of gas.
- When flame appears at both doors, heat treating operations can start.

To shut down

- Do not allow any part of furnace to cool below 760 °C (1400 °F).
- Open all doors.
- Turn off atmosphere gas.
- When clear of flame, furnace is purged.
- Keep furnace doors open, to prevent the accumulation of combustible gas from the furnace brickwork.

Propane. Many carburizing units use propane either for heating or as an enriching gas in carburizing. Propane gas is heavier than air; therefore, if leaks are present in the propane system, the gas will settle out in low places such as the pits under furnaces.

Insurance companies have records of many fires and explosions caused by propane leakage. When propane is used in heat treating, a continuous alarm system should be employed. A device of this type continuously samples the atmosphere in furnace pits and other designated areas. If propane leaks occur, an alarm is sounded before the mixture reaches an explosive composition, so that the gas can be diluted with air through the use of circulating fans or exhaust vents.

Preparation and Handling of Parts

While parts are being transferred from the machine shop to the carburizing furnace, they should be handled carefully to avoid nicks and surface damage, which are costly to correct after hardening is completed.

Cleaning. All parts should be thoroughly cleaned before they are charged into the furnace. Trays should be degreased by washing, although some users burn off organic matter in a furnace. Some of the contaminants that must be removed from parts to be carburized are listed below, together with their effects:

- Sulfur-bearing oils or sulfur compounds react in furnace atmosphere to decrease the carburizing action and also the life of heat-resistant alloy parts and protective copper plating.
- Iron oxides must be removed. Many gears are only partially machined, leaving hubs and webs with as-forged surfaces. If these are carburized without cleaning and are not subsequently grit blasted, the oxides become flaky and spongy and fall off in assemblies, damaging bearings and contaminating lubricants.
- Alkaline solutions are sometimes used for rust prevention on machined parts that are stored for a short time before carburizing and also to remove oil from the parts before carburizing. Residues from such solutions collect soot from the furnace and give a dark, dirty appearance to parts coming from the quench. Alkaline residues will also adversely affect heat-resistant furnace alloys. Parts that require protection by alkaline powder during storage should be rinsed thoroughly in uncontaminated hot water and allowed to dry before being charged into the carburizing furnace.
- Quenching salts may leave residues that are difficult to remove from trays, fixtures and work containers. If the same fixtures and containers are later reused in furnaces, the carburizing atmosphere will be affected deleteriously.
- Water in small amounts will upset the thermochemical balance of the furnace atmosphere. To avoid this, parts entering the furnace should be thoroughly dry.
- Trichloroethylene residues, from improper cleaning procedures, can cause severe etching of the work.

Loading Methods. To obtain maximum net load that can be carburized and quenched uniformly, the method of supporting the work in the furnace by trays, baskets, screens, spacers or other fixtures should be worked out carefully. Contact between the work and trays or baskets, contact between parts, and overly dense loads all result in uneven case depth and quench, and should therefore be minimized or avoided. The size and shape of parts will determine the method of loading for proper gas circulation and, for parts that are quenched directly on fixtures, will determine uniformity of hardening in the quench.

Process Variables

Successful operation of the gas carburizing process depends on control of three principal variables: temperature, time and atmosphere composition.

Effect of Temperature. The maximum rate at which carbon can be added to steel is limited by the rate of diffusion of carbon in austenite. This diffusion rate increases greatly with temperature; the rate of carbon addition at 925 °C (1700 °F) is about 40% greater than at 870 °C (1600 °F).

The temperature most commonly used for carburizing is 925 °C (1700 °F). This temperature permits a reasonably rapid carburizing rate without excessive deterioration of furnace equipment, particularly of heat-resistant alloys. The carburizing temperature is sometimes raised to 955 or 980 °C (1750 or 1800 °F), for certain deep case requirements. For shallow-case carburizing in which case depth must be kept within a specified narrow range, lower temperatures are frequently used, be-

Table 5 Values of case depth calculated by the Harris equation

Time (t), h	Case depth(a), after carburizing at:					
	870 °C (1600 °F)		900 °C (1650 °F)		925 °C (1700 °F)	
	mm	in.	mm	in.	mm	in.
2	0.64	0.025	0.76	0.030	0.89	0.035
4	0.89	0.035	1.07	0.042	1.27	0.050
8	1.27	0.050	1.52	0.060	1.80	0.071
12	1.55	0.061	1.85	0.073	2.21	0.087
16	1.80	0.071	2.13	0.084	2.54	0.100
20	2.01	0.079	2.39	0.094	2.84	0.112
24	2.18	0.086	2.62	0.103	3.10	0.122
30	2.46	0.097	2.95	0.116	3.48	0.137
36	2.74	0.108	3.20	0.126	3.81	0.150

(a) Case depth, mm = $0.635 \sqrt{t}$ (case depth, in. = $0.025 \sqrt{t}$) for 925 °C (1700 °F); $0.533 \sqrt{t}$ ($0.021 \sqrt{t}$) for 900 °C (1650 °F); $0.457 \sqrt{t}$ ($0.018 \sqrt{t}$) for 870 °C (1600 °F). For normal carburizing (saturated austenite at the steel surface while at temperature)

cause case depth can be more accurately controlled with the slower rates of carburizing obtained with lower temperatures.

For consistent results in carburizing, the temperature must be uniform. Uniformity on various locations throughout the work load depends on furnace design, load density, recirculation and heating rate. With high density of load, batch furnaces should have effective recirculating fans.

For a given load density, the difference in temperature between the outer and inner portions of the load may be high at carburizing temperatures, because the outside of the load is heated primarily by radiation and the rate of heating is rapid at carburizing temperatures. Also, the ability of the recirculated gas to decrease such differences in temperature is limited because of the low heat capacity of the gas at these temperatures.

At carburizing temperatures of 845 to 870 °C (1550 to 1600 °F), a given difference in the time required for different parts of the load to reach temperature has a lesser effect on case depth.

Other causes of large temperature differences are a high ratio of surface area to volume of individual parts, and high heat input. Nothing can be done about the ratio of surface to volume, but heat input can be regulated automatically with a heat-input controller, to minimize temperature differentials throughout the load.

For best control in batch furnaces, the thermocouple should be placed so that it reaches control temperature before any part of the charge. In continuous furnaces, the thermocouple should be as close as possible to the work without interfering with the flow of work through the furnace. For easy access in checking procedures, the thermocouple and protection tube usually are located through the sidewall of the continuous furnace. Because the first zone of a continuous furnace is a heating zone, the temperature-control thermocouple of this zone should be placed near the last part of the zone, to ensure against overheating the work. The thermocouple in the carburizing zones should be approximately in the middle of the zone. When the temperature of the last zone is lower than the carburizing temperature, for quenching purposes, optimum control is obtained by having the control thermocouple near the discharge end of the zone.

If Chromel-Alumel thermocouples (commonly used at carburizing temperature) are allowed to come in contact with reducing gases such as those found in the carburizing chamber, their accuracy is rapidly destroyed. To prevent this contact, thermocouple protection tubes are required.

Virtually all carburizing furnaces now in use have satisfactory automatic temperature control devices. However, errors in measurement of temperature do occur occasionally. Because of the value of the machined parts contained in large furnaces, in addition to the value of the equipment itself, frequent temperature checks are advisable.

Effect of Time. F. E. Harris developed a formula for the effect of time and temperature on case depth for normal carburizing (*Metal Progress,* Aug 1943), which can be shown in English units as:

$$\text{Case depth} = \frac{31.6 \sqrt{t}}{10^{(6700/T)}}$$

where case depth is in inches; t is time at temperature, in hours; and T is the absolute temperature, in degrees Rankine (Fahrenheit + 460).

The equivalent metric formula, using SI units, is:

$$\text{Case depth} = 660 \cdot e^{-8287/T} \cdot \sqrt{t}$$

where case depth is in millimetres, t is time in hours, and T is temperature in degrees Kelvin (Celsius + 273).

For a specific carburizing temperature, the relationship becomes simply:

$$\text{Case depth} = K\sqrt{t}$$

Case depth, mm = $0.635\sqrt{t}$ for 925 °C
(Case depth, in. = $0.025\sqrt{t}$ for 1700 °F)
Case depth, mm = $0.533\sqrt{t}$ for 900 °C
(Case depth, in. = $0.021\sqrt{t}$ for 1650 °F)
Case depth, mm = $0.457\sqrt{t}$ for 870 °C
(Case depth, in. = $0.018\sqrt{t}$ for 1600 °F)

Values of case depth calculated for times of 2 to 36 h at three common carburizing temperatures are given in Table 5.

When carburizing is purposely controlled to produce surface carbon concentrations somewhat less than saturated austenite, case depth will be slightly less than the equation shows.

In addition to the time at carburizing temperature, several hours may be required for bringing large workpieces to operating temperature. For work quenched directly from the carburizer, the cycle may be further lengthened to allow time for the work to cool from carburizing temperature to a quenching temperature of about 845 °C (1550 °F). Although some diffusion of carbon from case to core occurs during this time, diffusion is slower than it would be at the carburizing temperature. This period may be used deliberately as a moderate diffusion period, to lower the carbon concentration at the surface by maintaining an atmosphere of low carbon potential in contact with the work during this time.

Harris also developed a method for calculating the carburizing time and diffusion time to produce a carburized case of predetermined depth and carbon concentration at the surface, which can be shown as:

$$\text{Carburizing time} = \text{total time}\left(\frac{C - C_i}{C_o - C_i}\right)^2$$

and

$$\text{Diffusion time} = \text{total} - \text{carburizing time}$$

where total time, in hours, is calculated from the equation in Table 5; C is the final desired surface carbon concentration; C_o is the surface carbon concentration at the end of the carburizing cycle; and C_i is the concentration of carbon at the core.

This method is most adaptable to batch-type equipment. Successful application requires a well-conditioned furnace and a carrier gas of low carbon availability. All carbon additions are assumed to be made during the carburizing cycle while carrier gas and hydrocarbon gas are both supplied to the chamber. After the hydrocarbon addition to the carrier gas has been discontinued and the diffusion cycle started, it is assumed that no further carbon additions to the steel are made, either from the carrier gas or from hydrocarbon gas evolved from the furnace lining. It further is assumed that no carbon is removed from the steel during diffusion in an atmosphere consisting only of carrier gas. The assumption is most nearly valid when diffusion times are relatively short.

The Appendix contains a detailed discussion of the diffusion characteristics of carbon and nitrogen in iron.

Gas-Carburizing Atmospheres.
The furnace atmosphere used for gas carburizing consists of an endothermic carrier gas enriched with methane or propane. The main constituents of the atmosphere are CO, N_2, H_2, CO_2, H_2O and CH_4. Of these constituents, N_2 is inert, acting only as a diluent. The amounts of CO, CO_2, H_2O and H_2 present are very nearly the proportions expected at equilibrium from the reversible reaction:

$$CO + H_2O \rightleftarrows CO_2 + H_2 \qquad (Eq\ 1)$$

given the particular ratios of C, O and H in the atmosphere. Methane is invariably present in amounts well in excess of the amount expected if all the gaseous constituents were in equilibrium.

Although the sequence of reactions involved in carburizing is not known in detail, it is known that carbon can be added or removed rapidly from steel by the over-all reversible reactions:

$$2\ CO \rightleftarrows C\ (in\ Fe) + CO_2 \qquad (Eq\ 2)$$

and

$$CO + H_2 \rightleftarrows C\ (in\ Fe) + H_2O \qquad (Eq\ 3)$$

A carburization process based solely on the decomposition of Co would require large flow rates of atmosphere gas to produce appreciable carburizing. As an example, loss of just 0.47 g (0.017 oz) of carbon from a cubic metre of endothermic gas at 925 °C (1700 °F) is sufficient to reduce the ratio of CO to CO_2 from 249 to 132 and the carbon potential from 1.25 to 0.8%. Loss of 0.47 g (0.017 oz) of carbon represents about the same amount present in a steel part of 100 cm^2 (16 in.^{2}) surface area carburized to a depth of 1 mm (0.04 in.).

Methane or propane enrichment of endothermic gas provides the primary source of carbon for carburizing by slow reactions such as:

$$CH_4 + CO_2 \rightarrow 2\ CO + 2\ H_2 \qquad (Eq\ 4)$$

and

$$CH_4 + H_2O \rightarrow CO + 3\ H_2 \qquad (Eq\ 5)$$

which reduce the concentrations of CO_2 and H_2O, respectively. These reactions regenerate CO and H_2, thereby directing the reactions of Eq 2 and 3 to the right. Reactions in Eq 4 and 5 do not approach equilibrium. With methane contents typical of carburizing atmospheres, thermodynamic calculations show that, if the reactions did approach equilibrium, atmosphere CO_2 contents and dew points would be much lower than the values customarily observed. The sum of reactions in Eq 2 and 4 and in Eq 3 and 5 is reduced to:

$$CH_4 \rightarrow C\ (in\ Fe) + 2\ H_2 \qquad (Eq\ 6)$$

Thus, with constant CO_2 content and constant dew point, the net atmosphere composition change when carburizing is a reduction in methane content and an increase in hydrogen content. In most commercial operations, atmosphere flow rates are high enough and the rate of methane decomposition is low enough to prevent significant hydrogen buildup during a carburizing cycle. Carburization rates measured in CH_4-H_2 atmospheres have shown that, if the furnace-atmosphere methane content is high enough, above 1% at 925 °C (1700 °F), for example, some carburizing can occur by direct decomposition of methane on the steel surface, according to the reaction in Eq 6 (Ref 1).

Carbon-potential control during carburizing is achieved by varying the flow rate of the hydrocarbon enrichment gas and maintaining a steady flow of endothermic carrier gas. As a basis for regulating the enrichment

gas, the concentration of some constituent of the furnace atmosphere is monitored; for example, water vapor content by dew-point measurement, carbon dioxide content by infrared gas analysis or oxygen potential by zirconia oxygen sensor. The first two quantities provide measures of carbon potential according to the reactions of Eq 2 and 3. Oxygen potential is related to carbon potential by the reaction:

$$C\ (in\ Fe) + \tfrac{1}{2}\ O_2 \rightleftarrows CO \qquad (Eq\ 7)$$

When the carbon dioxide content of the atmosphere remains relatively constant, both carbon dioxide and oxygen potential provide good measures of the carbon potential. If the furnace atmosphere is sooting, the hydrogen content of the atmosphere rises, and the fraction of carbon monoxide is reduced correspondingly. Sooting also results in both lower carbon dioxide content and lower oxygen potential corresponding to a given carbon potential. The dew point corresponding to a given carbon potential remains nearly constant, however, even if the atmosphere is diluted with hydrogen as a result of sooting.

Calculation of Equilibrium Compositions.
Gas carburizing is a nonequilibrium process; that is, the gaseous constituents of the atmosphere are not fully in equilibrium with one another, and the atmosphere is not in equilibrium with the steel being carburized. Nevertheless, several important reactions—reactions in Eq 1 to 3, for example—approach equilibrium rapidly enough to permit predictions of the rate of carburizing from the atmosphere composition. Thus, the same case carbon gradient can be expected from different furnaces with different atmosphere gas flow rates when the following factors are held constant:

- Carbon potential, as inferred from CO_2, H_2O or oxygen potential measurements
- Carburizing time
- Carburizing temperature.

The equilibrium gas composition resulting from reactions of certain proportions of hydrocarbon gas and air is computed as follows:

- Six gaseous species are assumed to be present in the reacted gas: CO, CO_2, H_2, H_2O, CH_4 and N_2. The partial pressures of five of these are unknowns that can be determined; the sixth can be found by subtraction after the others become known.

- The C/H and O/N ratios are fixed by the nature of the hydrocarbon and the composition of air. In addition, by using the air/hydrocarbon ratio, three equations can be found relating the five unknown partial pressures.
- Two equilibrium relations, the reactions in Eq 1 and 4, for example, provide two additional equations, thus determining all the partial pressures.

Because explicit expressions for the unknowns cannot be written, trial-and-error computation methods must be used, necessitating the use of a computer (Ref 2). After the equilibrium composition of the gas is known, its carbon potential is found by writing the equilibrium relationship for Eq 2:

$$K_2 = \frac{a_c P_{CO_2}}{P^2_{CO}} \qquad \text{(Eq 8)}$$

where P_{CO} and P_{CO_2} are partial pressures of CO and CO_2, respectively; a_c is activity of carbon ($a_c = 1$ when the atmosphere is in equilibrium with pure carbon); and K_2 is the equilibrium constant for the reaction in Eq 2.

The expression K_2 can be computed from the Gibbs free energy of formation of CO and CO_2 at the temperature of interest (free energy values are tabulated in Ref 3):

$$\Delta F^0_{CO_2} - 2\Delta F^0_{CO} = -RT\ln K_2 \quad \text{(Eq 9)}$$

where ΔF^0_{CO} and $\Delta F^0_{CO_2}$ are free energies, T is absolute temperature, R is the gas constant, and K_2 is the equilibrium constant calculated in Eq 8.

The following expression relates the carbon activity and the carbon content of austenite:

$$\log a_c = \frac{2300}{T} - 0.920 + \frac{17\,950\,w}{T\,(100 - w)}$$
$$+ \log\left(\frac{4.65\,w}{100 - 5.65\,w}\right) \qquad \text{(Eq 10)}$$

where T is temperature in degrees Kelvin and w is % carbon in austenite. Combining Eq 8 and 10 gives a relation between carbon potential (that is, the equilibrium carbon content in austenitic iron), and carbon monoxide and carbon dioxide contents. Because the fractional change in carbon monoxide content is small even for large changes in carbon dioxide content, it usually is not necessary to monitor both gases to estimate the carbon potential. However, because endothermic gas derived from propane has a carbon monoxide

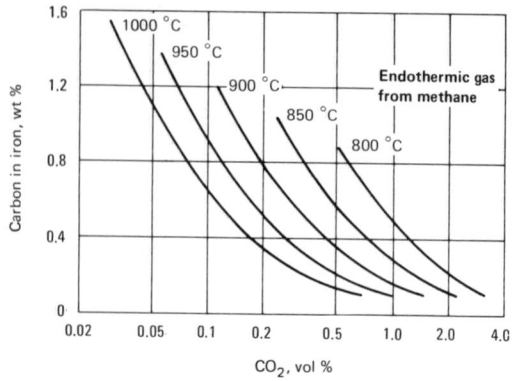

Fig. 1 Relationship between CO_2 content and carbon potential for endothermic gas from methane

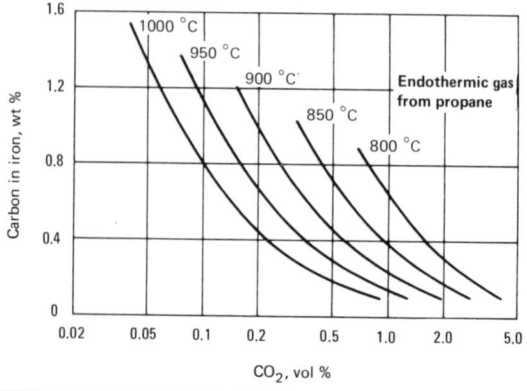

Fig. 2 Relationship between CO_2 content and carbon potential for endothermic gas from propane

content of about 23% while that derived from methane has a carbon monoxide content of about 20%, the carbon dioxide contents corresponding to a given carbon potential are higher for propane than for methane. Figures 1 and 2 show the relationship between carbon dioxide content and carbon potential for endothermic gas atmospheres derived from methane and propane, respectively.

The oxygen potential (or oxygen partial pressure), as measured by a zirconia oxygen sensor, is related to the carbon activity by the following equilibrium equation for the reaction given in Eq 8:

$$a_c = \frac{P_{CO_½}}{K_7 P_{O_2}} \qquad \text{(Eq 11)}$$

where P_{O_2} is the oxygen partial pressure and K_7 is the equilibrium constant. In carburizing atmospheres, the

oxygen partial pressure is approximately 10^{-19} to 10^{-25} atm.

The voltage output of a zirconia oxygen sensor, with air as a reference gas, is a function of the absolute temperature and the oxygen partial pressure according to the following expression:

$$\text{emf (volts)} = 4.9593 \times$$
$$10^{-5}T \log_{10}(P_{O_2}/0.209) \quad \text{(Eq 12)}$$

where T is absolute temperature and P_{O_2} is partial pressure in atmospheres. Combining Eq 10, 11 and 12 yields a relation between carbon potential and emf. Because this relation also depends on the carbon monoxide content of the atmosphere (Eq 11), emf measurements corresponding to a certain carbon potential for endothermic gas atmospheres derived from methane (Fig. 3) differ from those for atmospheres derived from propane (Fig. 4).

The water vapor content of the atmo-

Fig. 3 emf measurements of endothermic gas from methane

Fig. 4 emf measurements of endothermic gas from propane

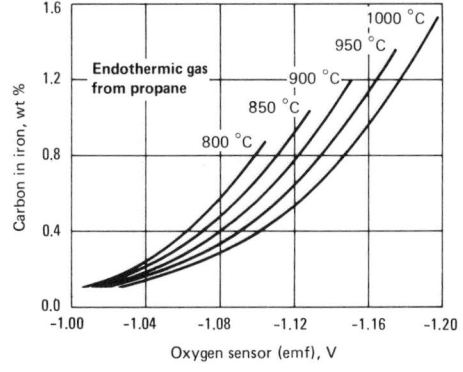

Fig. 5 Relationship between dew point and carbon potential for endothermic gas derived from methane

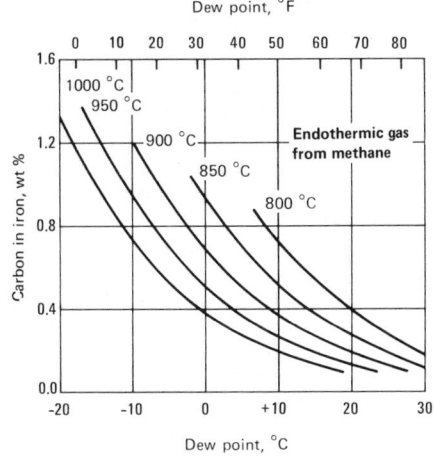

sphere is related to the carbon potential by the reaction in Eq 3. The equilibrium relation is

$$a_c = \frac{K_3 P_{CO} P_{H_2}}{P_{H_2O}} \qquad \text{(Eq 13)}$$

where K_3 is the equilibrium constant.

For atmospheres derived from any particular hydrocarbon, the product of the carbon monoxide and H_2 contents varies little for large changes in water vapor content. The water vapor content of the atmosphere gas usually is measured by determining its dew point. An equation relating dew point in degrees Celsius and P_{H_2O} in atmospheres is

$$\text{Dew point, } °C = \frac{5422.18}{(14.7316 - \ln P_{H_2O})} - 273.16$$

$$\text{(Eq 14)}$$

Equations 10, 13 and 14 can be combined to produce relations between dew point and carbon potential plotted in Fig. 5 and 6 for endothermic gas atmospheres derived from methane and propane, respectively.

If large volumes of parts with large surface areas are processed or if the furnace atmosphere is sooting, the rate of hydrogen production as described in Eq 6 may be sufficient to change significantly the relationships shown in Fig. 1 to 6. When an atmosphere becomes diluted with hydrogen, the carbon potential corresponding to a given carbon dioxide content or oxygen sensor emf is lowered. The relationship between dew point and carbon potential is not greatly affected, however. If there is no possibility that other diluents, such as nitrogen, have been added to the atmosphere, an abnormally low atmosphere carbon monoxide content (17% or less for endothermic gas derived from methane, for example) is an indication that the atmosphere is being diluted with hydrogen.

Other Carburizing Atmospheres. In recent years, in an effort to reduce the consumption of hydrocarbon gases used for furnace atmospheres, a number of alternatives to conventional gas carburizing atmospheres have been proposed. Some of these alternatives produce atmospheres similar in composition to conventional endothermic gas-base atmospheres, while others are radically different in composition. Among the former are (a) atmospheres formed from blends of propane, methane or other hydrocarbon gases and air introduced directly into the carburizing furnace, and (*b*) atmospheres formed within the furnace from methanol and nitrogen in 1 to 2 molar proportions. Among the latter are nitrogen-base atmospheres containing methane or propane with or without the addition of small amounts of air, carbon dioxide or water vapor.

For atmospheres with ratios of C, H, N and O similar to conventional atmospheres, the atmosphere carbon potential can be determined from CO_2, oxygen potential or dew point measurement (Fig. 1 to 6) if sufficient time is allowed for the gases to react within the furnace. That is, atmosphere gas flow rates must be low. Because reaction rates are faster at higher temperatures, higher carburizing temperatures are preferred.

For nitrogen-base atmospheres that are radically different in composition from conventional atmospheres, the equilibrium gas composition corresponding to a certain carbon potential can be computed (Ref 2). To relate carbon potential to a single measurement, the amount of nitrogen dilution must be held constant. When the nitrogen content is 80% or more, however, equilibrium carbon dioxide and dew point values for carburizing atmospheres are extremely low and difficult to measure. For many of these systems, it is better to avoid control based on gas composition and instead to adopt the same strategy used for vacuum carburizing; that is, carburize in a rich atmosphere for a short time to produce a carbon-saturated surface, then follow with a diffusion treatment in pure nitrogen to allow the carbon to penetrate and to decrease the surface carbon content. Lower-than-normal atmosphere flow rates are usually used in these systems as well.

Carbon Concentration Gradients and Surface Carbon Content

The carbon concentration gradient of carburized parts is influenced by carburizing temperature and time, type of cycle (various combinations of carburizing and diffusion times), carbon potential of the furnace atmosphere and the original composition of the steel. Compositions of the steels most often carburized or carbonitrided are given in Table 6.

In carburizing, carbon is transferred from the carburizing atmosphere sur-

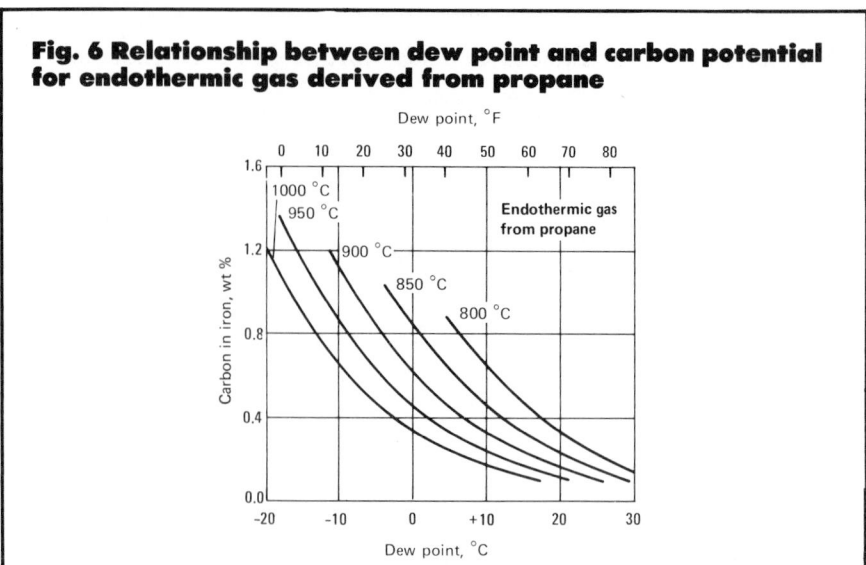

Fig. 6 Relationship between dew point and carbon potential for endothermic gas derived from propane

Table 6 Composition of steels for carburizing or carbonitriding

Steel	C	Mn	P(a)	S(a)	Composition, % Si	Ni	Cr	Mo
Carbon steels								
1010	0.08-0.13	0.30-0.60	0.035	0.045	· · ·	· · ·	· · ·	· · ·
1018	0.15-0.20	0.60-0.90	0.035	0.045	· · ·	· · ·	· · ·	· · ·
1019	0.15-0.20	0.70-1.00	0.035	0.045	· · ·	· · ·	· · ·	· · ·
1020	0.17-0.23	0.30-0.60	0.035	0.045	· · ·	· · ·	· · ·	· · ·
1022	0.18-0.23	0.70-1.00	0.035	0.045	· · ·	· · ·	· · ·	· · ·
1039	0.37-0.44	0.70-1.00	0.035	0.045	· · ·	· · ·	· · ·	· · ·
1524	0.19-0.25	1.35-1.65	0.035	0.045	· · ·	· · ·	· · ·	· · ·
1527	0.22-0.29	1.20-1.50	0.035	0.045	· · ·	· · ·	· · ·	· · ·
Resulfurized steels								
1113	0.13 max	0.70-1.00	0.07-0.12	0.24-0.33	· · ·	· · ·	· · ·	· · ·
1117	0.14-0.20	1.00-1.30	0.040	0.08-0.13	· · ·	· · ·	· · ·	· · ·
Alloy steels								
3310	0.08-0.13	0.45-0.60	0.025	0.025	0.20-0.35	3.25-3.75	1.40-1.75	· · ·
3310H	0.07-0.13	0.30-0.70	0.035	0.040	0.20-0.35	3.20-3.80	1.30-1.80	· · ·
4023	0.20-0.25	0.70-0.90	0.035	0.040	· · ·	· · ·	· · ·	0.20-0.30
4027	0.25-0.30	0.70-0.90	0.035	0.040	· · ·	· · ·	· · ·	0.20-0.30
4118	0.18-0.23	0.70-0.90	0.035	0.040	0.20-0.35	· · ·	0.40-0.60	0.08-0.15
4118H	0.17-0.23	0.60-1.00	0.035	0.040	0.20-0.35	· · ·	0.30-0.70	0.08-0.15
4140	0.38-0.43	0.75-1.00	0.035	0.040	0.20-0.35	· · ·	0.80-1.10	0.15-0.25
4320	0.17-0.22	0.45-0.65	0.035	0.040	0.20-0.35	1.65-2.00	0.40-0.60	0.20-0.30
4620	0.17-0.22	0.45-0.65	0.035	0.040	0.20-0.35	1.65-2.00	· · ·	0.20-0.30
4815	0.13-0.18	0.40-0.60	0.035	0.040	0.20-0.35	3.25-3.75	· · ·	0.20-0.30
4820	0.18-0.23	0.50-0.70	0.035	0.040	0.20-0.35	3.25-3.75	· · ·	0.20-0.30
5120	0.17-0.22	0.70-0.90	0.035	0.040	0.20-0.35	· · ·	0.70-0.90	· · ·
5140	0.38-0.43	0.70-0.90	0.035	0.040	0.20-0.35	· · ·	0.70-0.90	· · ·
8617	0.15-0.20	0.70-0.90	0.035	0.040	0.20-0.35	0.40-0.70	0.40-0.60	0.15-0.25
8617H	0.14-0.20	0.60-0.95	0.035	0.040	0.20-0.35	0.35-0.75	0.35-0.65	0.15-0.25
8620	0.18-0.23	0.70-0.90	0.035	0.040	0.20-0.35	0.40-0.70	0.40-0.60	0.15-0.25
8620H	0.17-0.23	0.60-0.95	0.035	0.040	0.20-0.35	0.35-0.75	0.35-0.65	0.15-0.25
8720	0.18-0.23	0.70-0.90	0.035	0.040	0.20-0.35	0.40-0.70	0.40-0.60	0.20-0.30
8822H	0.19-0.25	0.70-1.05	0.035	0.040	0.20-0.35	0.35-0.75	0.35-0.65	0.30-0.40
9310	0.08-0.13	0.45-0.65	0.025	0.025	0.20-0.35	3.00-3.50	1.00-1.40	0.08-0.15

(a) Maximum composition, %

rounding the steel part to the part surface. The carbon then diffuses slowly into the bulk of the part, establishing a carbon concentration gradient below the surface. The driving force for the carburizing reaction is called the carbon potential. Within the steel part, the high carbon surface has a higher carbon potential than the low-carbon interior; thus, carbon tends to move from the surface toward the center. Similarly, the carburizing atmosphere has a higher carbon potential than does the surface of the steel. If during processing the atmosphere carbon potential should fall below the carbon potential at the steel surface, carbon will be removed from the steel.

In thermodynamic terminology, the carbon potential would be defined as the partial molal free energy of carbon. However, it is customary in heat treating to identify carbon potential with carbon content of iron. The carbon potential of a gaseous mixture at a certain temperature is then defined as the carbon content of iron in equilibrium with that atmosphere at that temperature. Carburizing atmospheres frequently have carbon potentials higher than the solubility limit of carbon in iron. In such cases, it is better to refer to the carbon activity of the atmosphere. Carbon activity can be defined in terms of the oxygen potential, or carbon monoxide and carbon dioxide contents of the atmosphere when the gaseous constituents of the furnace atmosphere are in equilibrium.

At any temperature, there is a limit to the amount of carbon that can be dissolved in austenite. This limit, the carbon solubility limit in iron, is shown as the A_{cm} line in an iron-carbon phase diagram (Fig. 7). The addition of alloying elements to iron tends to reduce the carbon solubility limit; approximate limits are shown for several alloys in Fig. 7. In a carburizing atmosphere with a carbon potential well above the solubility limit, the carbon content of steel surfaces will exceed, in time, the solubility limit. The austenite at the surface is saturated in carbon at the solubility limit, and the rest of the carbon is present as carbide, $(Fe, M)_3C$, in low-alloy steels.

A particular carbon concentration gradient developed during carburizing depends on the following factors:

- The carbon potential of the furnace atmosphere. The higher the potential, the more rapidly carbon is supplied to the steel, and the steeper the

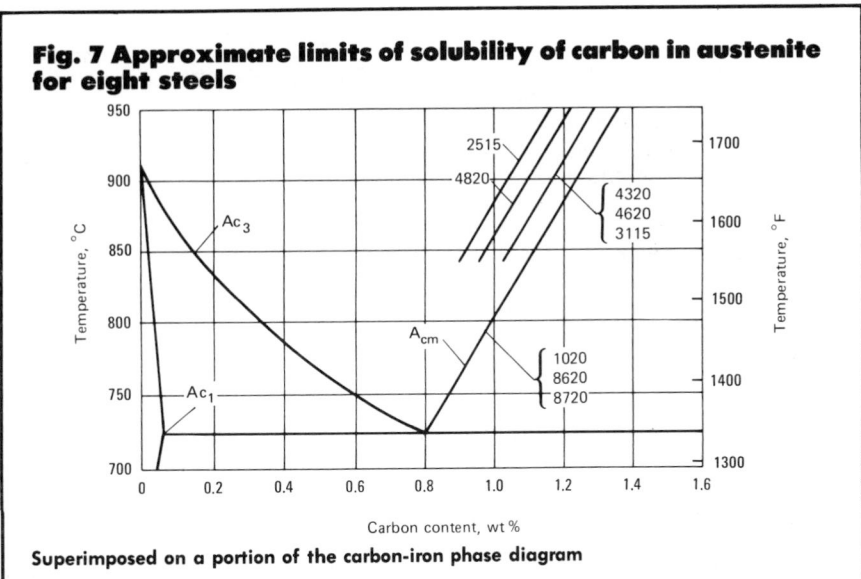

Fig. 7 Approximate limits of solubility of carbon in austenite for eight steels

Superimposed on a portion of the carbon-iron phase diagram

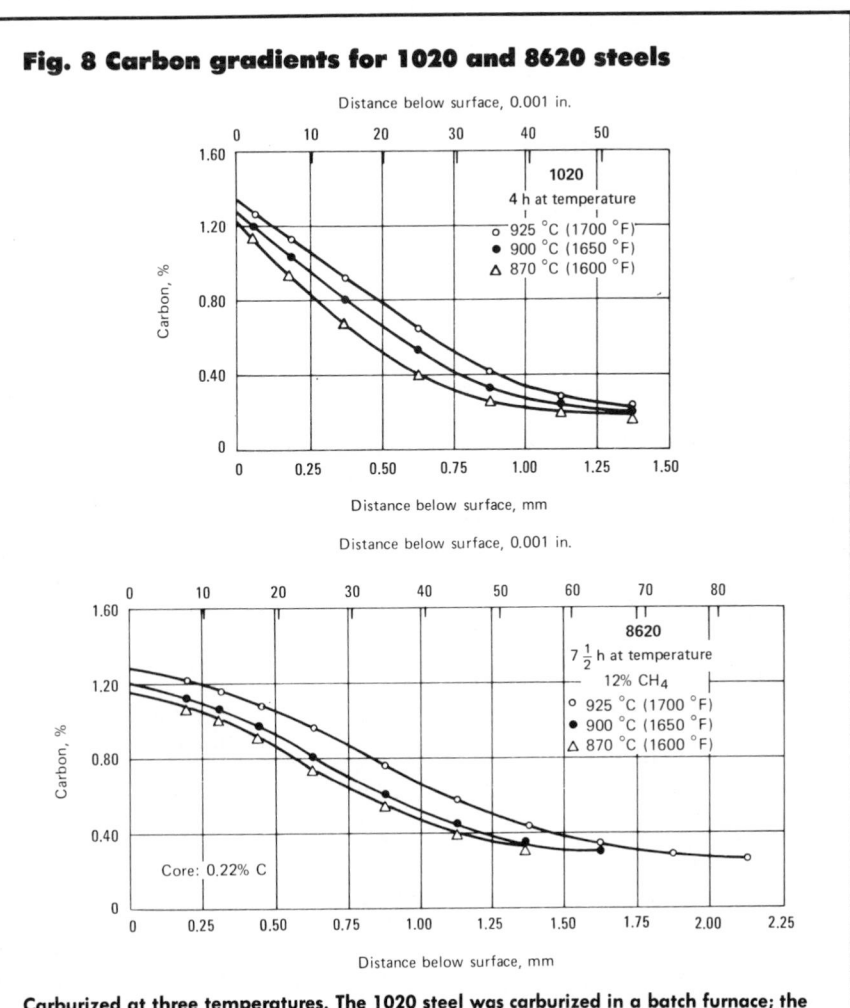

Fig. 8 Carbon gradients for 1020 and 8620 steels

Carburized at three temperatures. The 1020 steel was carburized in a batch furnace; the 8620 steel, in a recirculating pit furnace

carbon concentration gradient at the steel surface.

- The reaction rate at the steel surface. The nature and concentration of the molecular species present in the furnace atmosphere affect reaction rates. For example, the surface reaction is faster in an endothermic gas-base atmosphere than in a methane-hydrogen atmosphere.
- The carburizing temperature—increasing the temperature increases the reaction rate at the steel surface, and more importantly, it increases the rate of carbon diffusion within the steel. The carbon potential of an endothermic gas atmosphere with fixed proportions of carbon, hydrogen, nitrogen and oxygen increases as the temperature decreases.
- The carburizing time—the depth of carbon penetration increases as the time increases.
- Alloy content of the steel exerts its primary influence on the case carbon content and has a secondary influence on the rate of diffusion of carbon within the steel.

Normal Carbon Gradients. A normal carbon gradient is one produced by maintaining saturated austenite at the surface of the steel during the entire cycle. Figure 8 illustrates the influence of carburizing temperature on the carbon gradient for normal carburizing of a 1020 steel in a batch furnace. Comparable data also are given for 8620 steel carburized for 7½ h in an atmosphere containing 12% methane. Carburizing temperatures for both steels were 870, 900 and 925 °C (1600, 1650 and 1700 °F).

Carbon-gradient curves obtained in a typical batch furnace giving saturated austenite at the surface are shown in Fig. 9 for four steels after 4 h of carburizing at 870 and 925 °C (1600 and 1700 °F).

Typical influences of methane content, carbon potential and time on the normal carbon gradient in 1022 steel are shown in Fig. 10 and 11.

Modified Carbon Gradients. Various modifications of normal carbon gradients are obtained by control of temperature, time and atmosphere composition. The following are typical examples of zone control in continuous carburizing furnaces.

Example 3. Figure 12 illustrates two types of gas carburized cases obtained on 4027 steel in the same continuous furnace operating under identical

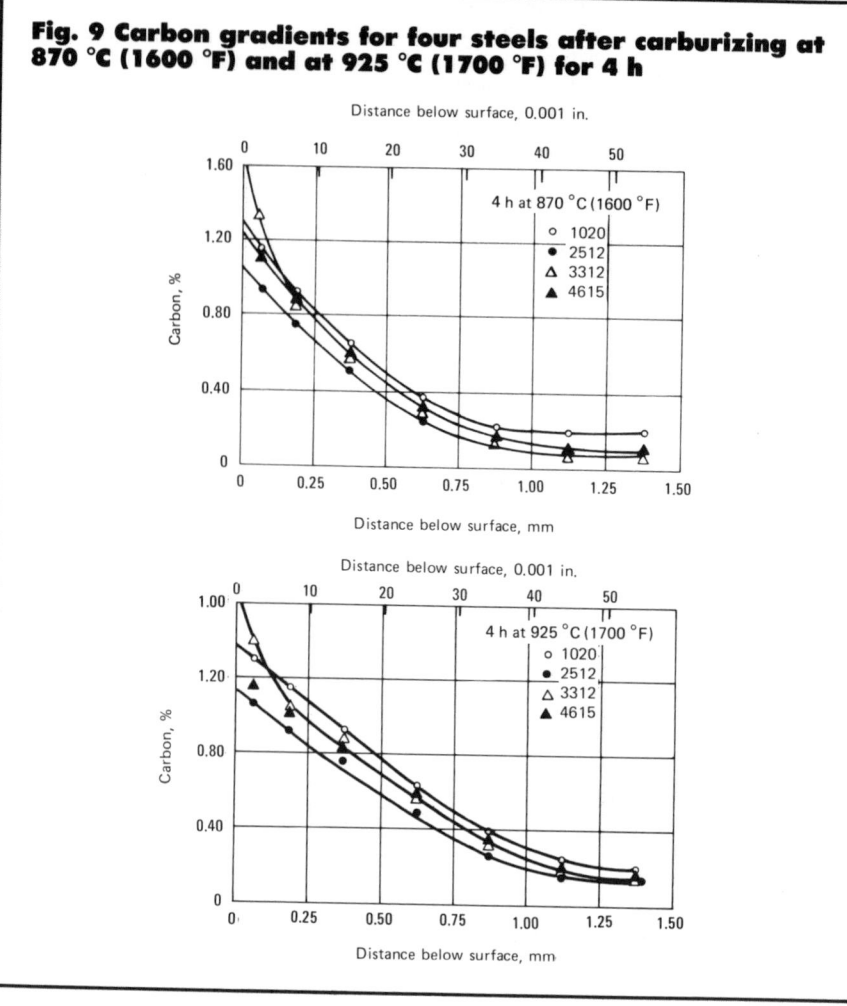

Fig. 9 Carbon gradients for four steels after carburizing at 870 °C (1600 °F) and at 925 °C (1700 °F) for 4 h

time-temperature cycles, but with different conditions of atmosphere control. The case with the higher carbon content at the surface was produced without consideration of high demand for carbon early in the carburizing cycle and low carbon demand during the final part of the cycle. A uniformly distributed mixture of carrier gas plus hydrocarbon gas was admitted at various points along the length of the furnace. The ratio of carrier gas to hydrocarbon gas was high enough to ensure minimum soot deposition during the carburizing and cooling portions of the cycle.

The other case shown in Fig. 12 was produced under identical conditions, except for the distribution of atmosphere within the furnace; the total amount of hydrocarbon gas required was admitted at several points in the carburizing zones, with the greatest amount added in the first of these zones. Carrier gas without enrichment was added in the diffusion and cooling zones.

Example 4. The 4815 steel for which data are shown in Fig. 13 was carburized in a three-zone continuous pusher-type furnace. Carbon potential was manually controlled. Details of the treatments used to obtain the three carbon gradients are given in the table accompanying Fig. 13. Additional examples are given in the section on commercial practice for carrier gas plus hydrocarbon gas.

Surface Carbon Content. As indicated in Fig. 12, control of carbon gradient also affects surface carbon concentration. Surface carbon content has a pronounced effect on the properties of the steel and must be controlled to obtain optimum results.

Carbon content in solution determines the amount of austenite that will be retained after quenching in a given steel composition. The amount of retained austenite is a function of dissolved carbon content, and because it affects the hardness of the quenched case, it is important to control it, in order to meet specifications for surface hardness. The carbon potential, carburizing cycle, quenching temperature and rate, diffusion cycle, and steel composition affect the amount of austenite retained after quenching.

Carbon gradient is related to case hardness, as shown by the data in Fig. 14 for specimens of 1024 steel. These specimens were carburized for 2¼ h at 900 °C (1650 °F) in an atmosphere containing 10% methane. They were oil quenched from 845 °C (1550 °F) and achieved a maximum surface hardness of about 62 HRC. However, a hardness of about 66 HRC was obtained at a depth of 0.25 mm (0.010 in.) below the surface, indicating the presence of some retained austenite at the surface.

It is generally agreed that retained austenite in amounts less than about 15% is not detrimental to pitting fatigue strength, when the austenite is finely dispersed. Part of the reason why small amounts of retained austenite are apparently not detrimental is that its presence allows mating surfaces to conform slightly faster and to spread the load more evenly, thus reducing local areas of high stress. In fact, substantially higher retained austenite content is desirable to resist pitting— 25 to 30%, and occasionally up to 40%. As long as hardnesses greater than 59 HRC can be maintained, high retained austenite seems to be beneficial for bearings.

The effect of surface carbon content on pitting fatigue strength has been subject to controversy for many years. Rolling-contact fatigue tests have indicated that pitting fatigue strength increases with increasing hardness for medium-carbon and high-carbon steels.

The importance of controlling surface carbon content in parts subject to bending fatigue is disputed, although most test work indicates that bending fatigue strength increases as retained austenite content decreases.

The effect of surface carbon content on resistance to wear and scuffing is too complicated to warrant a general conclusion. Excessive retained austenite permits surfaces to deform plastically under heavy loads, resulting in ripples, or "orange peel".

Statistical data relating to the control of surface carbon content are given

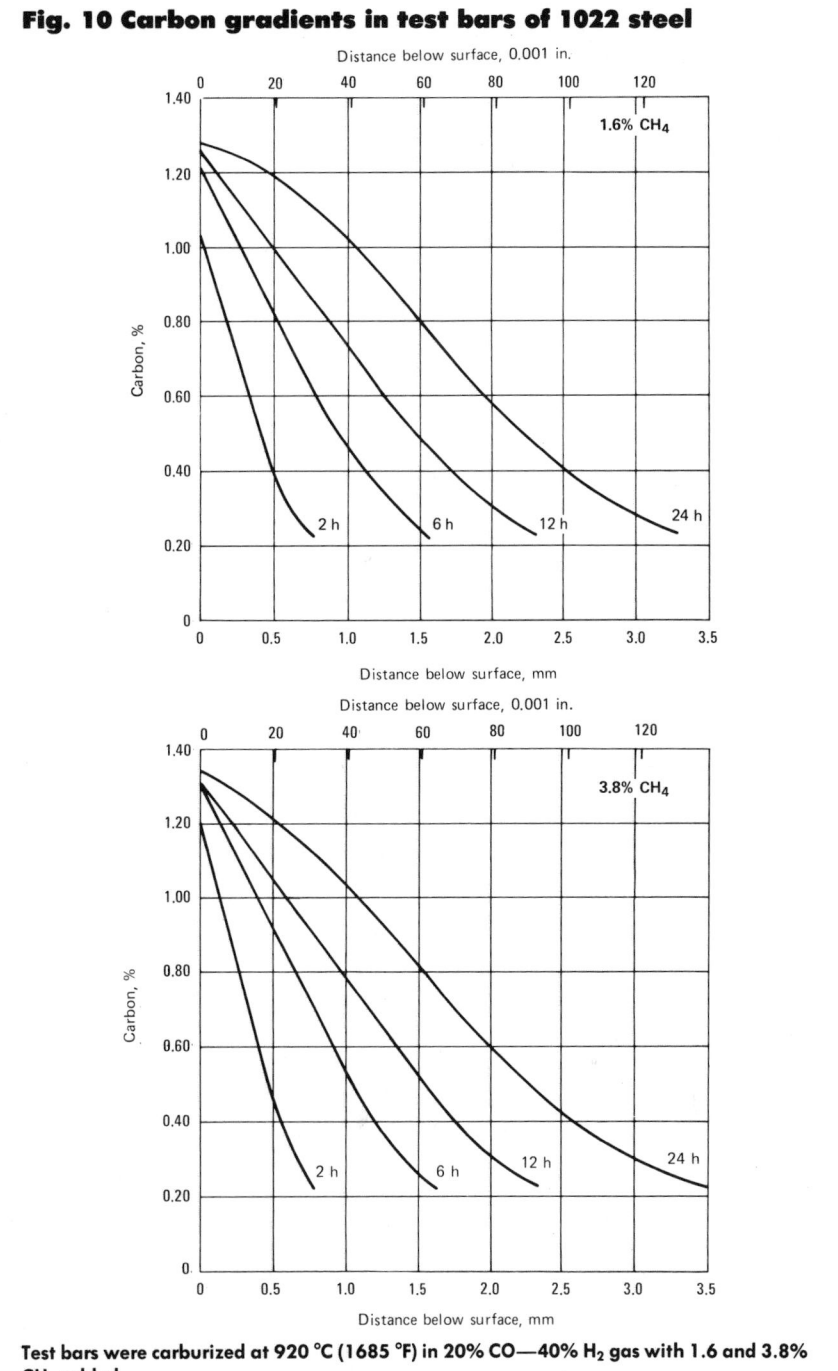

Fig. 10 Carbon gradients in test bars of 1022 steel

Test bars were carburized at 920 °C (1685 °F) in 20% CO—40% H₂ gas with 1.6 and 3.8% CH₄ added

in Fig. 15 to 18 (Examples 5 to 18) for several plain carbon and low-alloy steels. These data represent practice in eight different plants, thus involving several different types of furnaces, steel compositions and "aim" carbon contents. Control varied from good to poor, as the examples indicate.

Example 5. Data in Fig. 15 compare carbon concentrations 0.25 mm (0.010 in.) below the surface, obtained in gas carburizing 25 batches in each of three similar batch furnaces. The 75 tests for work from the three furnaces show a large majority of the concentrations to be within ±0.05% carbon of the 0.90% carbon aim. Parts were carburized at 925 °C (1700 °F). Automatic dew point controllers were used. A carrier gas with a dew point of +2 °C (35 °F) was

Fig. 11 Carbon gradients in 1022 steel

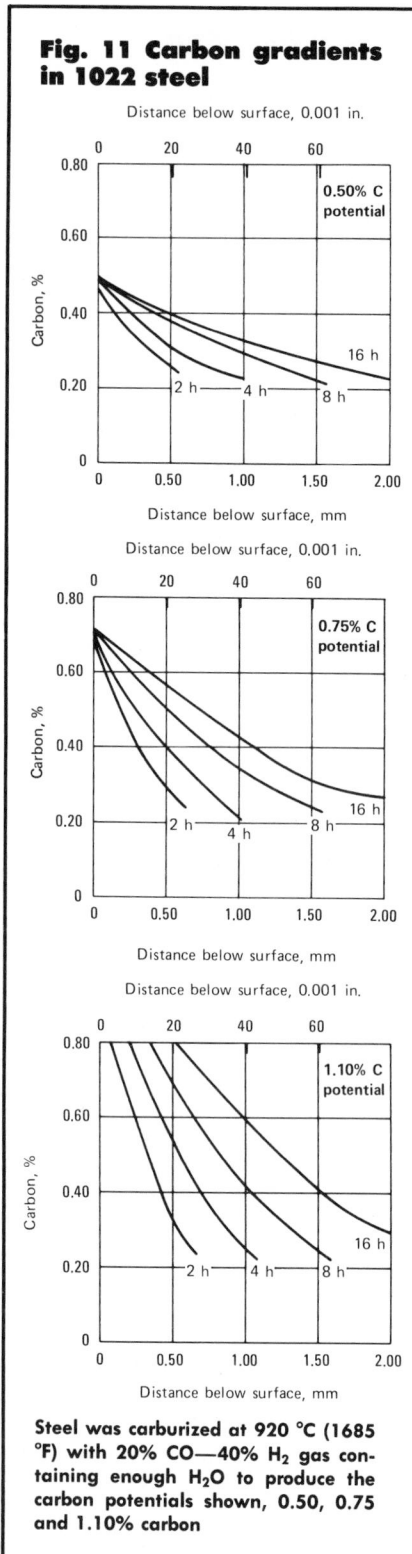

Steel was carburized at 920 °C (1685 °F) with 20% CO—40% H₂ gas containing enough H₂O to produce the carbon potentials shown, 0.50, 0.75 and 1.10% carbon

Fig. 12 Effect of zone control on carbon gradient in 4027 steel carburized in a continuous furnace

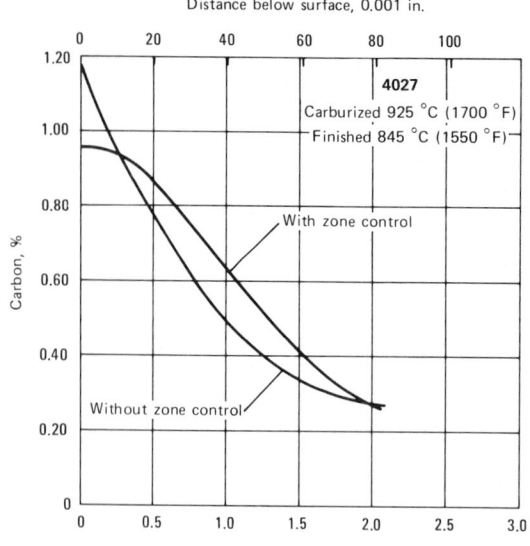

Total time, 10.3 h; zone temperatures, 900, 925, 925 and 845 °C (1650, 1700, 1700 and 1550 °F)

enriched with natural gas to maintain a dew point of −6 °C (22 °F) in the furnace.

Example 6. Data in Fig. 16 compare surface carbon concentrations in 50 specimens of rock bit cutters of 4815 steel carburized in two pit-type furnaces to an aim of 0.75% carbon. Furnaces were operated simultaneously, using the same gas generator. Each furnace was 0.9 m (36 in.) in diam by 1.8 m (72 in.) deep and contained a load of approximately 1360 kg (3000 lb). A carburize-diffuse cycle was used. An automatic dew point controller was employed on the generator but not on the furnace.

Example 7. Figure 16 shows surface carbon concentration, first 0.08-mm (0.003-in.) depth of cut, in 19 specimens of 4820 steel carburized with production parts in a two-row continuous furnace. The desired carbon content was 0.75 to 0.80%, with a specification of 0.70 to 0.90%. All results are within specification, and most of the values are within desired limits. Specimens were carburized at 925 °C (1700 °F) using a diffusion cycle, quenched from 815 °C (1500 °F) in oil at 60 °C (140 °F), tempered in lead at 620 °C (1150 °F) for 5 min, wire brushed and liquid-abrasive cleaned. Atmosphere at the charge end of the furnace was automatically controlled at a dew point of −15 °C (+5 °F) and at the discharge end at +3 °C (37 °F). Endothermic gas enriched with straight natural gas was

used as the carburizing medium; air was added at the discharge end.

Example 8. Data in Fig. 16 were obtained for 26 specimens of 4320 steel under the same carburizing and test conditions as for Example 7, except that the dew point was −1 °C (30 °F) at the discharge end and specimens were tempered at 650 °C (1200 °F). The desired carbon concentration was 0.85 ± 0.05%.

Example 9. Figure 16 shows results of 50 tests that represent variations in carbon concentration for continuous carburizing of 4027 steel. These results were obtained over a 3- to 4-year period of operation using an automatic dew point controller.

Example 10. Data in Fig. 16 were obtained for 32 specimens of 8620 steel under the same carburizing and test conditions as shown in Example 8 for carburizing 4320. However, the desired carbon concentration was 0.90 ± 0.05%.

Example 11. The aim carbon concentration was 0.90 ± 0.05% for parts carburized at 925 °C (1700 °F) in a horizontal batch-type furnace. The data shown in Fig. 16 were results of 12 tests based on the surface carbon content, first 0.08-mm (0.003-in.) depth of cut, of 25-mm (1-in.) rounds of 8620 steel carburized with the production parts,

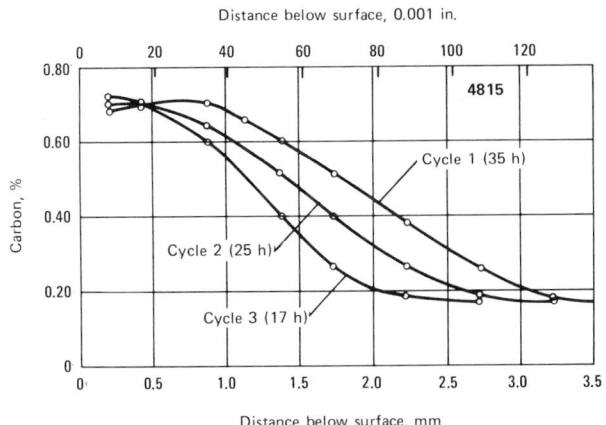

Fig. 13 Carbon gradients for three-zone continuous pusher-type carburizing furnace

Condition	Zone 1	Zone 2	Zone 3
Cycles 1, 2 and 3			
Temperature, °C (°F)	920 (1690)	920 (1690)	910 (1670)
Atmosphere	Methane	Methane	Carrier gas; air
Cycle 1 target: 2.8 mm (0.110-in.) case at 0.25% C			
Hours in zone 8		14	13
Carbon potential, % C 1.25		1.25	0.80(a), 0.55(b)
Cycle 2 target: 2.3 mm (0.090-in.) case at 0.25% C			
Hours in zone 6		10	9
Carbon potential, % C 1.25		1.25	0.80(a), 0.60(b)
Cycle 3 target: 1.8 mm (0.070-in.) case at 0.25% C			
Hours in zone 4		7	6
Carbon potential, % C 1.25		1.25	0.80(a), 0.75(b)
(a) At center of zone. (b) At discharge			

Conditions in each zone were as tabulated. Carbon potential was controlled manually.

quenched in oil from 845 °C (1550 °F), tempered in lead at 650 °C (1200 °F), wire brushed and liquid-abrasive cleaned.

The carburizing medium was endothermic gas plus natural gas for enrichment. No air was added. A dew point of −15 °C (+5 °F) was used through the carburizing cycle, −12 °C (+10 °F) for the diffusion cycle (925 °C or 1700 °F), and −3 °C (+27 °F) for the 845 °C (1550 °F) cycle that preceded quenching.

Example 12. Data in Fig. 17(a) represent carbon content at 0.2 mm (0.007 in.) below the surface. Actual surface carbon concentration ranged from 0.85 to 1.03%. The steel was carburized in a batch-type furnace using a hot-wire analyzer for automatic control of atmosphere. Depth of case to 50 HRC was 1.1 to 1.3 mm (0.043 to 0.052 in.); actual depth required was 1.14 to 1.27 mm (0.045 to 0.050 in.); total depth ranged from 1.9 to 2.1 mm (0.076 to 0.081 in.).

Example 13. Data in Fig. 17(b) also represent carbon content at 0.2 mm (0.007 in.) below the surface. Actual surface carbon concentration ranged from 0.84 to 1.19%. The steel was carburized in a batch-type furnace with manual control of atmosphere. Depth of case to 50 HRC was 0.4 to 0.6 mm (0.015 to 0.023 in.); actual depth re-

quired was 0.4 to 0.65 mm (0.015 to 0.025 in.); total depth ranged from 0.65 to 0.8 mm (0.025 to 0.031 in.).

Example 14. Carburizing and test conditions in Fig. 17(c) were the same as those used in Example 13. In this instance, surface carbon concentration ranged from 0.89 to 1.48%, and depth of case to 50 HRC was 0.65 to 0.95 mm (0.025 to 0.037 in.); actual depth required was 0.75 to 0.9 mm (0.030 to 0.035 in.). Total depth ranged from 0.86 to 1.35 mm (0.034 to 0.053 in.). (The same batch-type furnace was used to obtain the data for Examples 12, 13 and 14.)

Example 15. Data in Fig. 17(d) represent carbon concentration results of 27 tests taken at 8-h intervals from a tray pusher-type continuous furnace using a dew point controller. Desired carbon was 1.0%.

Example 16. Carbon concentration data in Fig. 18(a) were obtained by using round-bar test specimens that were placed at several locations throughout each charge of production parts of the same composition as the specimens. Specimens were carburized in a vertical pit-type furnace using one fan and endothermic atmosphere with methane as the enriching gas. The 8620 steel was carburized at 955 °C (1750 °F), without automatic atmosphere control and using a diffusion cycle consisting of flowing endothermic gas only during the last third of the carburizing cycle. The 3310 steel was carburized at 925 °C (1700 °F), with automatic atmosphere control but no diffusion cycle.

Example 17. Round bars were used to obtain the carbon concentration data shown in Fig. 18(b) for 4620 and 8620 steel. Both steels were carburized at 955 °C (1750 °F) using endothermic gas plus methane as an enriching gas. A diffusion cycle was used for the 8620 steel by shutting off the methane for the last third of the carburizing cycle. The 4620 steel was carburized without a diffusion cycle, and the carbon concentration ranged predominantly below the aim of 1.0%, while results for the 8620 steel were largely above the desired concentration.

Example 18. Data in Fig. 18(c) were obtained on round test specimens distributed throughout loads in continuous pusher-type furnaces. Endothermic gas plus methane as enriching gas was used. The 8620 specimens were carburized at 925 °C (1700 °F) without automatic atmosphere control or a diffusion

Fig. 14 Carbon and hardness gradients for 1024 steel

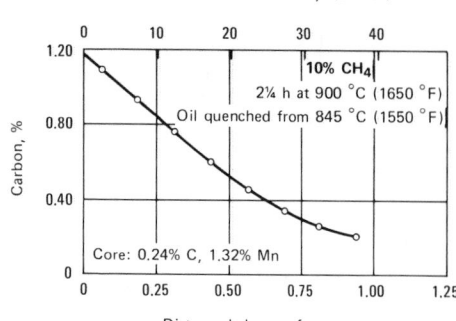 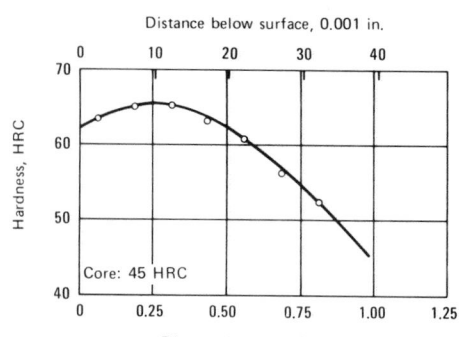

Carburized in a recirculating batch furnace under normal conditions to produce saturated austenite at the surface of the steel. Note the effect of retained austenite in decreased surface hardness. Specimens, 25 mm (1 in.) in diam by 150 mm (6 in.) long

Fig. 15 Variation of carbon content 0.25 mm (0.010 in.) below the surface for 1020 steel carburized in 3 similar batch-type furnaces

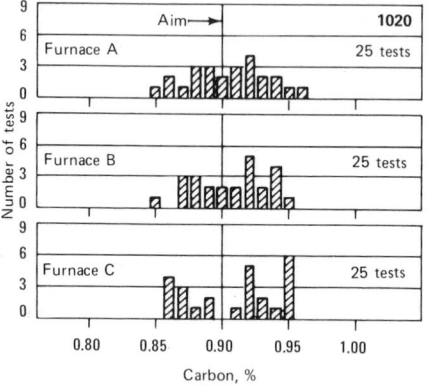

Fig. 16 Variation of surface (or near-surface) carbon content in alloy steels carburized under different conditions

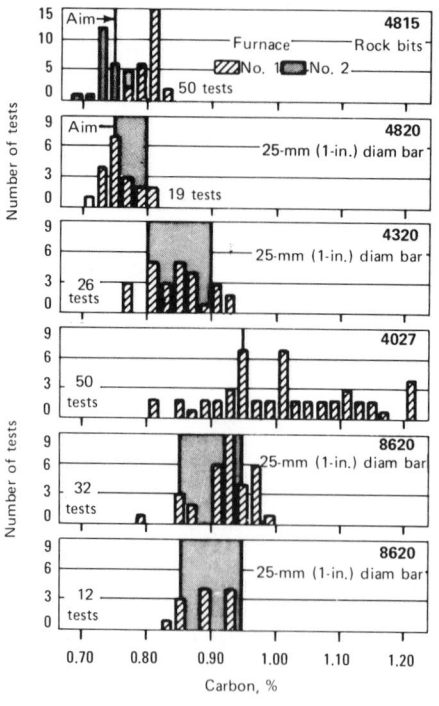

Fig. 17 Variation of surface (or near-surface) carbon content in 9310 and 3310 alloy steels carburized under different conditions

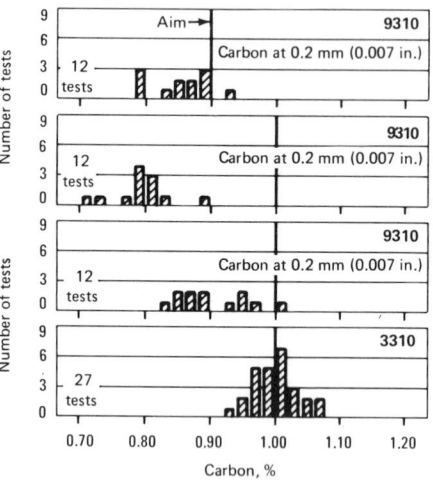

cycle, in a furnace with three fans. The 3310 specimens were carburized at 925 °C (1700 °F) in a furnace equipped with two fans, using automatic atmosphere control but no diffusion cycle. Surface carbon content may vary significantly as a function of location in the furnace, as indicated in the following example.

Example 19. Figure 19 indicates a variation in surface carbon from top to bottom of a pit-type furnace used for carburizing 4620 steel bearing races. Aim carbon content was 1.00%. In this example, it is apparent that surface carbon content was consistently higher in the bottom portion of the furnace, possibly indicating a slightly different operating temperature in this area.

Low Surface Carbon. Most steels can be carburized by the carbon satura-

tion-carbon diffusion type of cycle to produce surface carbon content well below saturation (for example, from 0.90 to 1.00% carbon).

There is a strong trend in present carburizing practice toward surface carbon concentrations of eutectoid composition or slightly higher. With the leaner alloy steels now being used, it

becomes increasingly important to utilize the full hardenability of the steel. Maximum hardenability of the alloy carburizing steels most commonly used is obtained at carbon concentrations near 0.8%. The excess carbides commonly found with high carbon concentrations promote the formation of transformation products other than martensite, and may also remove part of the carbide-forming elements from the austenite, thus affecting the true hardenability.

Low carbon concentrations at the surface also permit the economy of direct quenching of work from the carburizing furnace, instead of reheating. Direct quenching of carburized parts

Fig. 18 Variations of near-surface 0.13 to 0.25 mm (0.005 to 0.010 in.) carbon content of alloy steels carburized in different types of furnaces

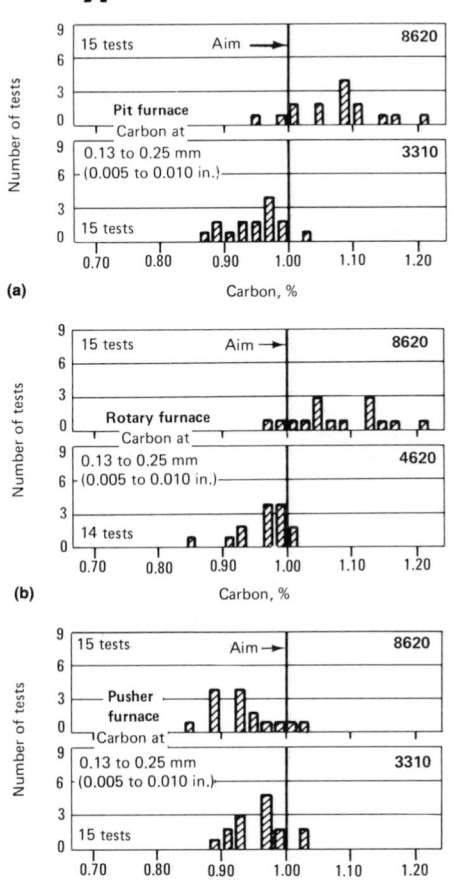

(a)

(b)

(c)

Fig. 19 Range of surface carbon with respect to position in furnace for 4620 steel bearing races

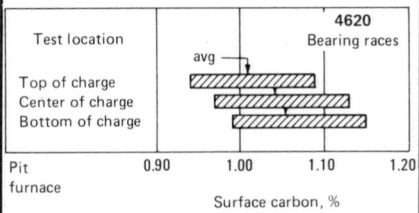

Carburized in a pit furnace 762 mm (30 in.) in diam by 914 mm (36 in.) deep. Open load of races was carburized for 7 h in natural gas atmosphere; 3½-h diffusion cycle followed. Surface carbon aim was 1.00%. Each bar on the chart represents 14 heats of steel.

Fig. 20 Reproducibility of case depths in 8620 steel

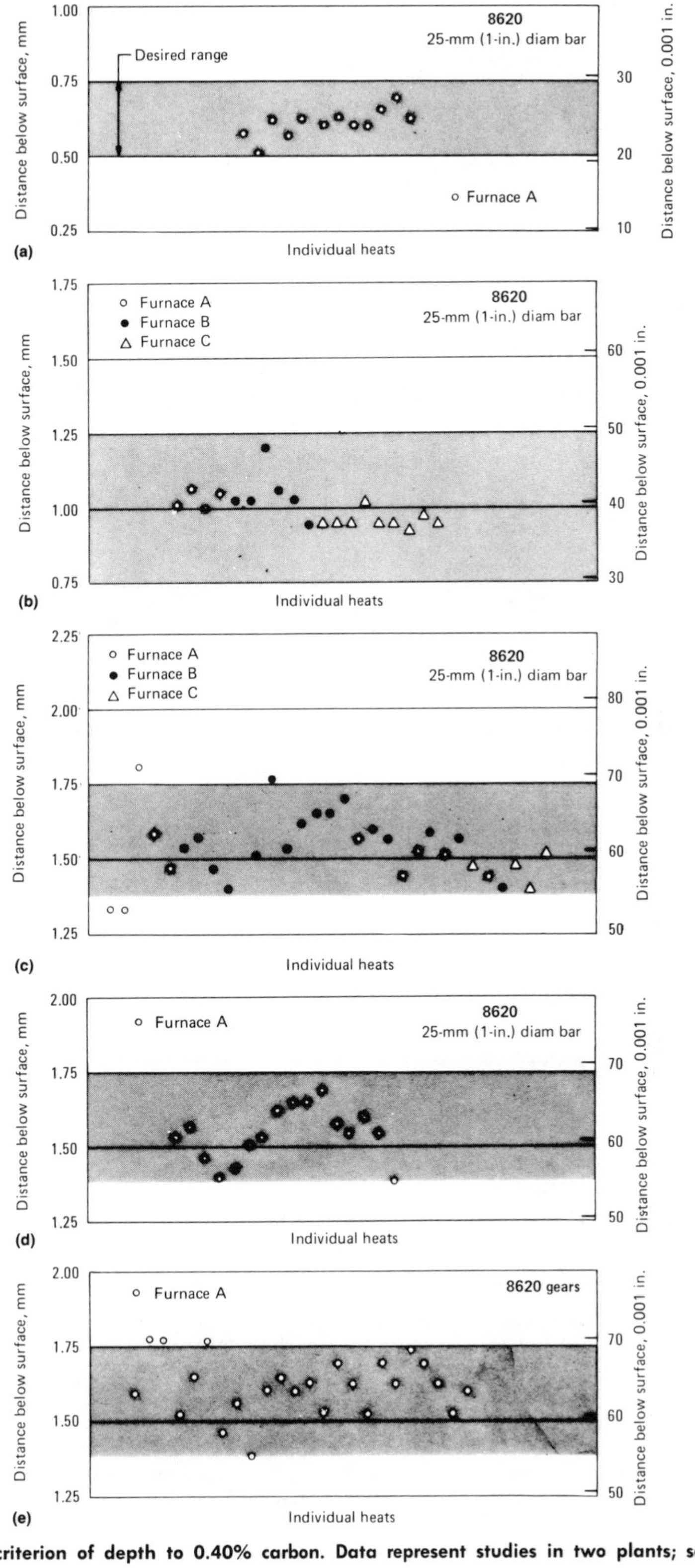

(a)

(b)

(c)

(d)

(e)

Using criterion of depth to 0.40% carbon. Data represent studies in two plants; see text.

Fig. 20 (continued)

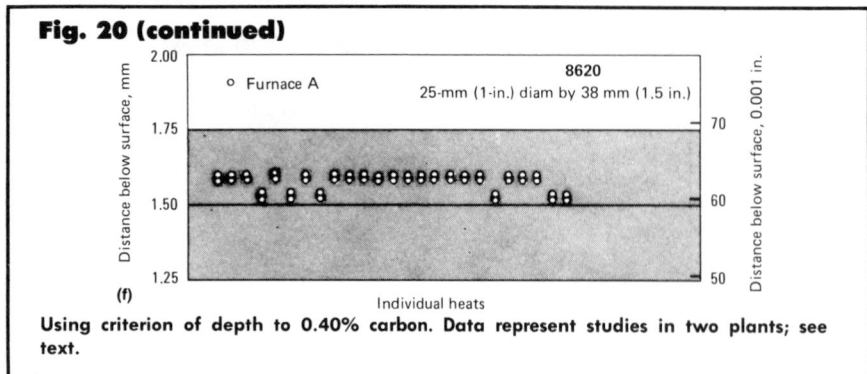

(f) Using criterion of depth to 0.40% carbon. Data represent studies in two plants; see text.

Fig. 21 Reproducibility of case depth in 4027 steel

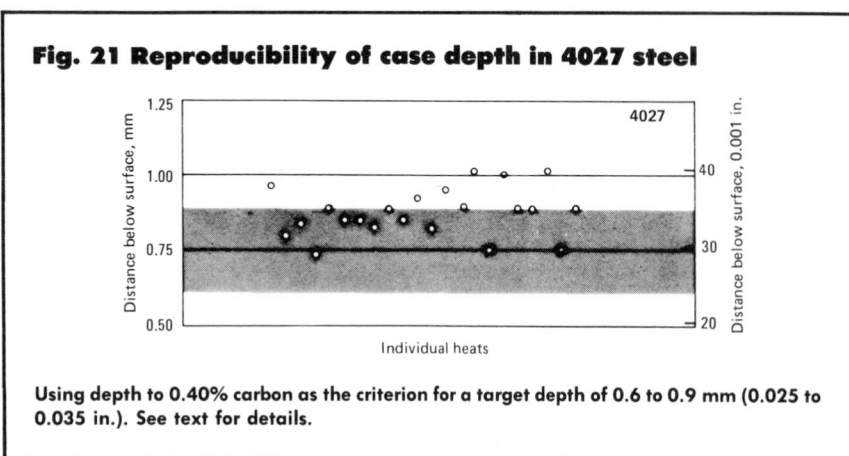

Using depth to 0.40% carbon as the criterion for a target depth of 0.6 to 0.9 mm (0.025 to 0.035 in.). See text for details.

Fig. 22 Reproducibility of case depth in 4620 steel

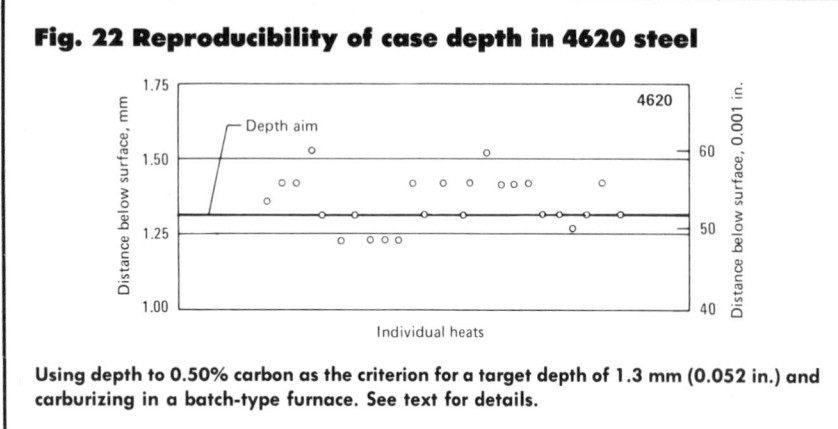

Using depth to 0.50% carbon as the criterion for a target depth of 1.3 mm (0.052 in.) and carburizing in a batch-type furnace. See text for details.

Fig. 23 Variation in carbon concentration of 4815 rock bit cutters

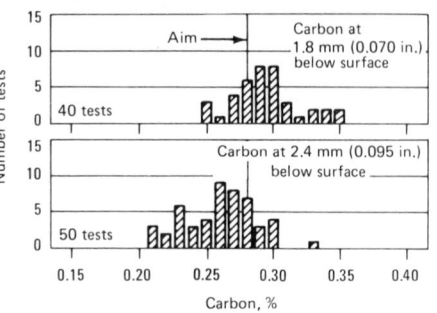

Variation in carbon concentration at 1.8 and 2.4 mm (0.070 and 0.095 in.) below surface of 4815 steel rock bit cutters carburized at 925 °C (1700 °F). Pit-type furnace, 914 mm (36 in.) in diam by 1830 mm (72 in.) deep, employed a carburizing-diffusion cycle. Dew point controlled at the generator only. Each test represents one load of approximately 1360 kg (3000 lb).

to exist, excessively low surface carbon concentrations and shallow case depths on the work in these areas will result.

Control of Case Depth

As preceding sections have shown, the depth of a carburized case in a given steel is a function of temperature, time at temperature and carbon potential. In commercial practice, case depth is controlled within certain tolerances, depending on the intended service of the carburized parts, the type and condition of the carburizing furnace, the cycles employed, and the limitations of control equipment. The end use of carburized parts is the principal determinant of how wide or narrow a range of case depths is acceptable. Parts to be used in less critical applications do not require close control of case depth, and the increased cost of ensuring close control would be unjustified.

Case Depth Variation. The data shown in Fig. 20 to 22 (Examples 20 to 27) provide a summary of the reproducibility of case depths within prescribed limits for test specimens and parts made of three different alloy steels. Operating details are given in each example.

Example 20. In Fig. 20(a), case depth was determined from a carbon-gradient curve secured by carburizing 25-mm (1-in.) rounds of 8620 with production parts at 925 °C (1700 °F) for 1¾

having high surface carbon concentrations favors the retention of austenite.

When surface carbon concentrations near eutectoid composition are desired, a multiple-manifold arrangement may be used for continuous furnaces, by which part of the carrier gas and all of the hydrocarbon gas are introduced in the front portion of the carburizing zone, where the carbon demand is high. Only carrier gas is supplied to the other zones, and the carbon potential is adjusted to give the desired final surface carbon concentration.

This "starved carbon" method requires a longer cycle for several reasons; for a given carbon content of the core, the diffusion rate of carbon in austenite decreases with a lowering of surface carbon concentration. Also, good recirculation of gases is essential. The atmosphere adjacent to the work quickly becomes depleted of available carbon, and if stagnant areas are allowed

Fig. 24 Comparison between depth of case to 0.40% carbon and to 50 HRC of 8620 steel

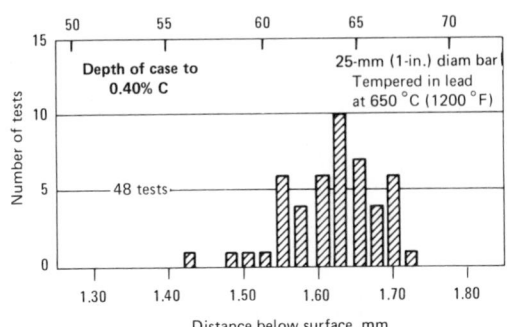

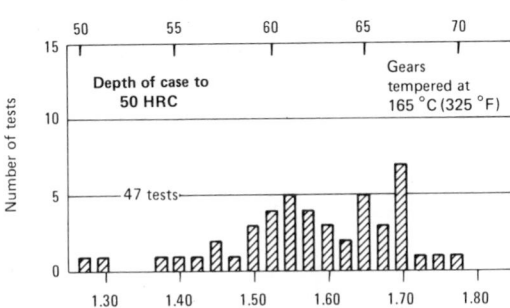

Carburized at 925 °C (1700 °F), oil quenched from 845 °C (1550 °F). Test specimens 25 mm (1 in.) in diam were processed with production gears in a two-row continuous gas carburizing furnace employing a diffusion cycle and an atmosphere of endothermic gas enriched with natural gas; air was added at the discharge end. Effective case depth to 50 HRC was measured on a microhardness traverse taken midway between the ends of a cross section of the gear tooth at the junction of the involute profile and root fillet.

Fig. 25 Comparison of case depth measurements made on the same gears in two different plants

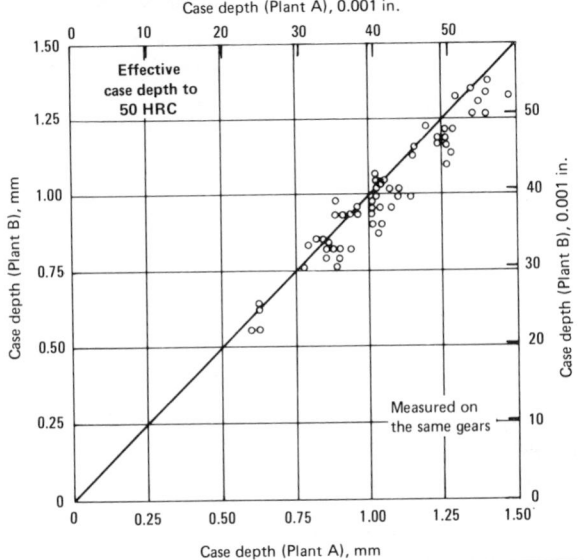

Fig. 26 Comparison of case depth measurements on the same specimens of 8620 steel in five different laboratories

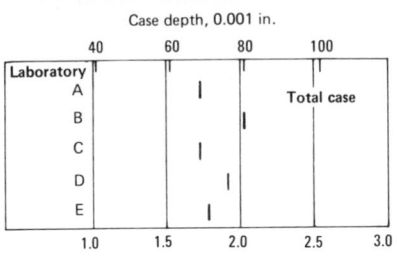

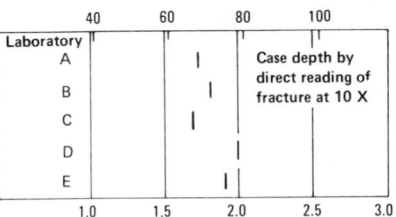

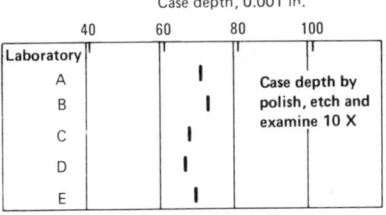

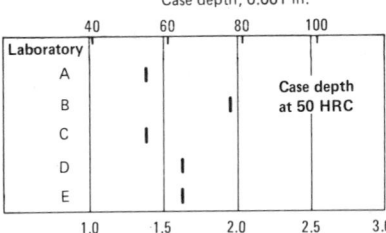

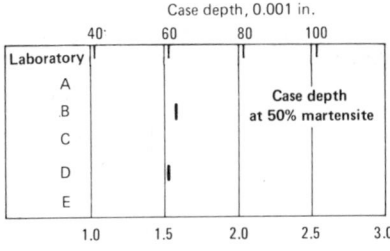

Sample 1: 8620 steel with core hardness, 30 HRC; surface, 0.90% carbon

Fig. 27 Comparison of case depth measurements on the same specimens of 8620 steel in five different laboratories

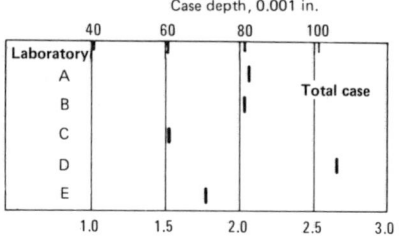

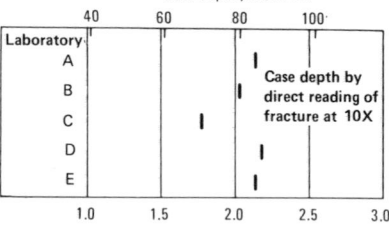

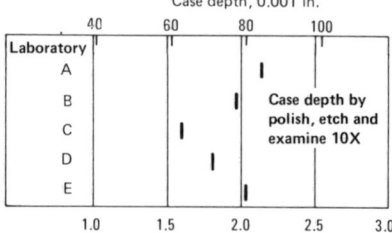

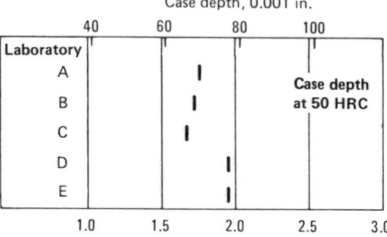

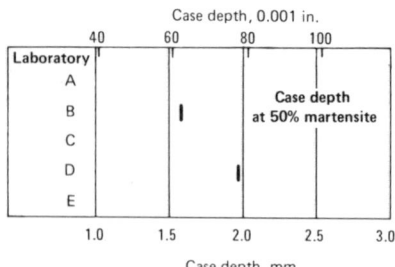

Sample 2: 8620 steel with core hardness, 40 HRC; surface, 0.90% carbon

Table 7 Efficiency of carbon utilization in a batch-type carburizing furnace

Total time, h	Total liquid l	pt	Initial carbon in liquid kg	lb	Total carbon in steel kg	lb	Total carbon absorbed, %
0.5	0.12	0.25	0.059	0.131	0.016	0.035	27.0
1.0	0.24	0.50	0.119	0.262	0.025	0.055	21.0
2.0	0.47	1.00	0.238	0.524	0.034	0.076	14.5
3.0	0.71	1.50	0.357	0.786	0.044	0.097	12.3
4.0	0.95	2.00	0.476	1.05	0.049	0.109	10.4
5.0	1.18	2.50	0.594	1.31	0.058	0.128	9.8
7.0	1.66	3.50	0.830	1.83	0.068	0.151	8.2

Load, 45 kg (100 lb); surface area, 929 000 mm^2 (1440 in.^{2})

Table 8 Efficiency of carbon utilization in a batch-type carburizing furnace

Total time, h	Total liquid l	pt	Initial carbon in liquid kg	lb	Total carbon in steel kg	lb	Total carbon absorbed, %
6.0	3.9	8.25	1.87	4.12	0.26	0.57	14
8.0	5.8	12.25	2.78	6.12	0.31	0.69	11
11.0	7.1	15.00	3.40	7.50	0.38	0.84	11

Load, 955 kg (2100 lb) gross, 675 kg (1490 lb) net; surface area, 9 000 000 mm^2 (14 000 in.^{2}). Work basket was 635 mm (25 in.) in diam by 914 mm (36 in.) deep. Load consisted of rock bit cutters randomly packed. Carburized at 925 °C (1700 °F). Surface carbon desired, 0.70 to 0.80%. Diffusion period from eighth to eleventh hour

h in a horizontal batch-type furnace, then equalizing the temperature at 845 °C (1550 °F) before quenching in oil. A diffusion type of carburizing cycle was used, with a dew point of −15 °C (+5 °F) for the first portion of the cycle at 925 °C (1700 °F), a dew point of −12 °C (+10 °F) for the diffusion cycle at 925 °C (1700 °F), and finally a dew point of −3 °C (+27 °F) for the 845 °C (1550 °F) portion of the cycle. The carburizing medium was endothermic plus straight natural gas as the enriching gas, without addition of air.

Test pieces were tempered in lead at 650 °C (1200 °F) after quenching, then wire brushed and liquid-abrasive cleaned. The target depth to 0.40% carbon was 0.5 to 0.75 mm (0.020 to 0.030 in.). With 0.65 mm (0.025 in.) as the average, the distribution was:

0.65 ± 0.025 mm
(0.025 ± 0.001 in.) = 75%
0.65 ± 0.050 mm
(0.025 ± 0.002 in.) = 92%
0.65 ± 0.075 mm
0.025 ± 0.003 in.) = 100%

Example 21. In Fig. 20(b), test pieces similar to those described in Example 20 were used to obtain case depth data (to 0.40% carbon) from three different batch-type furnaces using endothermic with straight natural gas as

the enriching gas, without addition of air. The loads were carburized at 925 °C (1700 °F), using a diffusion-type cycle and maintaining a dew point of −15 °C (+5 °F) for 3 h at 925 °C (1700 °F), a dew point of −12 °C (+10 °F) for the diffusion portion of the cycle, and finally −3 °C (+27 °F) for the 1-h portion of the cycle at 845 °C (1550 °F) prior to quenching.

The three different furnaces had the following characteristics as to size, atmosphere control and rated load:

Furnaces A and B
- Size: 0.75 by 1.2 m (30 by 48 in.); 0.46 m (18 in.) high
- Atmosphere control: Dew cell; manual reset
- Rated gross load: 680 kg (1500 lb)

Furnace C
- Size: 0.9 by 1.8 m (36 by 72 in.); 0.6 m (24 in.) high
- Atmosphere control: Infrared analyzer controlling carbon dioxide; automatic reset
- Rated gross load: 1590 kg (3500 lb)

The maximum variation recorded in 5 years was 0.9 to 1.2 mm (0.036 to 0.047 in.); target, 1 mm (0.040 in.). Distribution of the case depth was:

1.0 ± 0.025 mm
(0.040 ± 0.001 in.) = 32%

1.0 ± 0.050 mm
(0.040 ± 0.002 in.) = 47%
1.0 ± 0.075 mm
(0.040 ± 0.003 in.) = 90%
1.0 ± 0.100 mm
(0.040 ± 0.004 in.) = 95%
1.0 + 0.175 or − 0.100 mm
(0.040 + 0.007 or − 0.004 in.) = 100%

Example 22. For a required case depth (to 0.40% carbon) of 1.4 to 1.8 mm (0.055 to 0.070 in.), the same three furnaces as in Example 21 produced the results plotted in Fig. 20(c). Operation and method of testing were the same as for the shallower depths except that the total cycle at 925 °C (1700 °F) was 10.5 h; 5 h with a − 15 °C (+5 °F) dew point,

and 5.5 h with a dew point of − 12 °C (+10 °F). Equalizing time at 845 °C (1550 °F) was 1 h with a dew point of − 3 °C (+27 °F). Maximum case depth variation for 4 years was 1.4 to 1.7 mm (0.056 to 0.066 in.), with the following distribution:

1.5 ± 0.025 mm
(0.060 ± 0.001 in.) = 22%
1.5 ± 0.050 mm
(0.060 ± 0.002 in.) = 48%
1.5 ± 0.075 mm
(0.060 ± 0.003 in.) = 75%
1.5 ± 0.100 mm
(0.060 ± 0.004 in.) = 89%
1.5 ± 0.125 mm
(0.060 ± 0.005 in.) = 97%
1.5 ± 0.150 mm
(0.060 ± 0.006 in.) = 100%

Example 23. A horizontal batch-type furnace, operated the same as described in Examples 21 and 22 except that diffusion time with a − 12 °C (+10 °F) dew point was 6 h, produced the distribution of case depth shown graphically in Fig. 20(d) and tabulated as:

1.55 ± 0.025 mm
(0.061 ± 0.001 in.) = 31%
1.55 ± 0.050 mm
(0.061 ± 0.002 in.) = 50%
1.55 ± 0.075 mm
(0.061 ± 0.003 in.) = 63%
1.55 ± 0.100 mm
(0.061 ± 0.004 in.) = 81%
1.55 ± 0.125 mm
(0.061 ± 0.005 in.) = 100%

In this instance, the target depth (to 0.40% carbon) was 1.4 to 1.8 mm (0.055 to 0.070 in.). Test pieces of the type described in Example 20 were utilized for obtaining the data.

Example 24. A case depth of 1.4 to 1.8 mm (0.055 to 0.070 in.) (to 0.40% carbon) was specified for gears that were carburized at 925 °C (1700 °F) in a continuous furnace. The variation in case depth possibly is greater in a continuous furnace than in a batch furnace, as suggested by results of 24 tests (Fig. 20e). These data were obtained from tests made on actual gears, one per week for 24 weeks.

Example 25. Batch-type carburizing in the same plant as in Example 24 pro-

Table 9 Operating conditions for the use of undiluted hydrocarbon liquids in carburizing type 4815 steel rock bit cutters in pit retort furnace

Depth of case to 0.20% C		Carburizing			Diffusion			Total time, h
mm	in.	Time at temperature, h	Liquid l/h	pt/h	Time at temperature, h	Liquid l/h	pt/h	
1.5	0.060	4¼	0.9	2	1	0.2 to 0.5	½ to 1	5¼
2.0	0.080	6½	0.9	2	3	0.2 to 0.5	½ to 1	9½
2.5	0.100	7½	0.9	2	7½	0.2 to 0.5	½ to 1	15
2.8	0.110	9	0.9	2	9	0.2 to 0.5	½ to 1	18
3.0	0.120	12	0.9	2	9	0.2 to 0.5	½ to 1	21

Work container was 635 mm (25 in.) in diam by 914 mm (36 in.) deep. Gross weight of load, 955 kg (2100 lb); net weight, 675 kg (1490 lb). Cutters randomly packed. Time shown is time at 925 °C (1700 °F), as measured by calibrated check-thermocouple. Desired surface carbon (aim) was 0.70 to 0.80%.

Table 10 Operating conditions for carburizing type 9310 steel power-train gears in pit retort furnace using either propane or undiluted hydrocarbon liquids
Carburizing temperature, 925 °C (1700 °F); retort was 635 mm (25 in.) in diam by 914 mm (36 in.) deep.

Effective case depth to 0.40% C, mm	Time at temperature, h Carburizing	Diffusing	Heating Propane m³/h	ft³/h	Liquid l/h	pt/h	Carburizing Propane m³/h	ft³/h	Liquid l/h	pt/h	Diffusing Propane m³/h	ft³/h	Liquid l/h	pt/h	Cooling(a) Propane m³/h	ft³/h	Liquid l/h	pt/h
Manual control, for case depths under 0.9 mm (0.035 in.)(b)																		
0.3	1	¼	0.08-0.17	3-6	0.35	¾	0.11	4	0.35	¾	0.06	2	0.2	½	0.06	2	0.06	⅛
0.5	1	½	0.08-0.17	3-6	0.35	¾	0.11	4	0.35	¾	0.06	2	0.2	½	0.06	2	0.12	¼
0.6	1¼	¾	0.08-0.17	3-6	0.35	¾	0.11	4	0.35	¾	0.06	2	0.2	½	0.06	2	0.12	¼
0.7	1½	1	0.08-0.17	3-6	0.35	¾	0.11	4	0.35	¾	0.06	2	0.2	½	0.06	2	0.12	¼

Automatic control, for case depths over 0.9 mm (0.035 in.)

Effective case depth to 0.40% C, mm	Time at temperature, h Carburizing	Diffusing	Heating Propane m³/h	ft³/h	Liquid l/h	pt/h	Carbon-potential setting Carburizing	Diffusing	Cooling Propane m³/h	ft³/h	Liquid l/h	pt/h
0.08-1.0	7½	2	0.08-0.17	3-6	0.35-0.47	¾ to 1	1.15	0.90	0.08	3	0.12	¼
1.0-1.3	10	3	0.08-0.17	3-6	0.35-0.47	¾ to 1	1.15	0.90	0.08	3	0.12	¼
1.3-1.5	12	4	0.08-0.17	3-6	0.35-0.47	¾ to 1	1.15	0.90	0.08	3	0.12	¼
1.5-1.8	20	5	0.08-0.17	3-6	0.35-0.47	¾ to 1	1.15	0.90	0.08	3	0.12	¼
1.8-2.0	24	6	0.08-0.17	3-6	0.35-0.47	¾ to 1	1.15	0.90	0.08	3	0.12	¼

(a) Furnace cool to 760 °C (1400 °F), then cut off carburizing medium. For case depths under 0.9 mm (0.035 in.), continue furnace cooling to 595 °C (1100 °F), and hold for 2 to 6 h (depending on subsequent machining practice). If parts are to be machined prior to hardening, hold at 595 °C for 6 h, then furnace cool to 425 °C (800 °F). For case depths over 0.9 mm (0.035 in.), continue furnace cooling to 595 °C, hold for 6 h, continue furnace cooling to 425 °C and remove load. (b) Carburizing cycles based on 140 to 230 m² (1500 to 2500 ft²) of carburizing surface area; add scrap metal when loads are less than 140 m² (1500 ft²); for over 230 m² (2500 ft²), increase carburizing medium during carburizing and cooling.

duced case depths within an extremely close range, as indicated in Fig. 20(f). Case depths were determined by carburizing 25-mm (1-in.) -diam by 38-mm (1½ -in.) -long test pieces with 175-to-200-kg (390-to-450-lb) charges of gears, using dew point control. These test pieces were cut in half, polished, nital etched, and then examined microscopically.

Example 26. Figure 21 shows case depth variation (to 0.40% carbon) obtained over a period of 1 to 2 years from carburizing 4027 steel at 925 °C (1700 °F) in a continuous furnace with automatic dew point control. As indicated, the total variation was approximately 0.7 to 1.0 mm (0.029 to 0.040 in.); target depth was 0.6 to 0.9 mm (0.025 to 0.035 in.).

Example 27. In Fig. 22, depth to 0.50% carbon was the criterion, with a target depth of 1.3 mm (0.052 in.) for 4620 steel. Carburizing was done in a pit-type batch furnace 0.75 m (30 in.) in diam by 0.9 m (36 in.) deep. This unit used natural gas as a carburizing medium and was operated at 930 °C (1720 °F).

Case depth may also be expressed in terms of a given carbon content at a specified distance below the surface, as indicated in the following example.

Example 28. Figure 23 shows a statistical plot of results from tests taken at depths of 1.8 and 2.4 mm (0.070 and 0.095 in.) below the surface of carbu-

Table 11 Effect of carburizing temperature on grain size

Steel	Grain size before carburizing	Grain size after carburizing at:			
		925 °C (1700 °F)	980 °C (1800 °F)	1010 °C (1850 °F)	1040 °C (1900 °F)
1-h cycle					
1117	8	6 to 7	6 to 7	...	4 to 6
4615	8	8	8	...	8
8620	8	7	7	...	4 to 7
9315	8	8	8	...	8
3-h cycle					
1117	8	5 to 7	4 to 7	...	4 to 7
4615	8	8	8	...	8
8620	8	7	7	...	3 to 7
9315	8	8	8	...	8
		7.5-h cycle(a)		**4-h cycle(b)**	
1018	6 to 7	5 to 7	...	3	...
1022	6	7	...	4 to 6	...
1024	6	4 to 7	...	3	...
4617	8	5 to 7	...	5 to 7	...
8620	8	5 to 7	...	5 to 7	...

(a) Total depth of case (to 0.25% C) varied from 1.75 to 1.9 mm (0.069 to 0.075 in.). (b) Total depth of case (to 0.25% C) varied from 1.8 to 1.9 mm (0.072 to 0.077 in.).

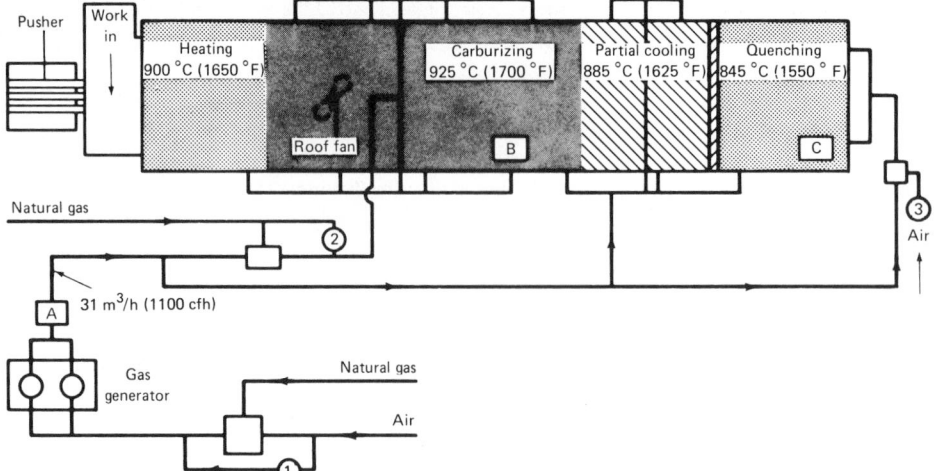

Fig. 28 Atmosphere manifolding for automatic dew point control of carbon potential of the atmosphere in a four-zone carburizing furnace

	Gas generator	Carburizing zone	Quenching zone
Dew point	−4 °C (+25 °F)	−9 C (+16 °F)	1 °C (34 °F)
Sampling location	A	B	C
Control valve	1	2	3
Fixed flow, natural gas	1.42 m³/h (50 ft³/h)		
Variable flow, natural gas	0 to 3.54 m³/h (0 to 125 ft³/h)		
Variable flow, air	0.14 to 0.85 m³/h (5 to 30 ft³/h)		

Typical carbon concentrations produced in the steel are 0.89% carbon at a depth of 0.13 mm (0.005 in.), 0.85% carbon at 0.25 mm (0.010 in.) and 0.80% carbon at 0.38 mm (0.015 in.).

Fig. 29 Carburizing conditions and carbon gradient produced using 8620H steel

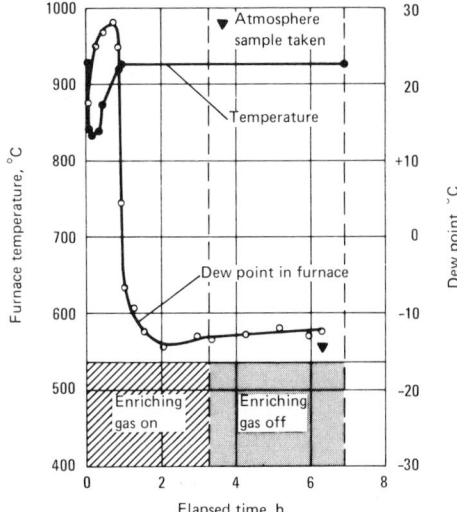

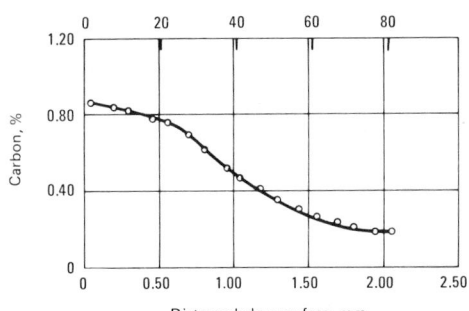

Load: ring gears, 231 kg (510 lb) net, 390 kg (860 lb) gross. **Furnace:** metallic-retort pit furnace, electrically heated. **Carrier gas:** 2.8 m³/h (100 ft³/h) of endothermic gas throughout the cycle, including 5 h and 59 min at temperature. **Enriching gas:** 0.3 m³/h (12 ft³/h) of natural gas; addition started when load was placed in furnace; flow of natural gas stopped after 2½ h at temperature, or 42% of at-temperature cycle. **Generator dew point:** −6 to −4 °C (+22 to +25 °F) throughout cycle. **Heating chamber pressure:** 5 to 8 mm (0.20 to 0.31 in.) water column. **Carburizing temperature:** 925 °C (1700 °F). **Cooling method:** Slow cooled from 925 °C (1700 °F) in cooling pit. **Analysis of atmosphere near end of cycle:** 20.8% CO, 0.4% CO_2, 34.0% H_2, 0% O_2, 0.8% CH_4, rem N_2

(a) Furnace temperature and atmosphere conditions for carburizing. (b) Carbon gradient produced by the cycle shown in (a)

rized 4815 steel rock bit cutters. Aim carbon content at both depths was 0.28%.

Measurement of Case Depth. Several methods for measuring case depth are in common use. Because the different methods do not measure the same location in a carburized case, confusion and misunderstanding may result if the method is not specifically defined. When the design of a part requires a certain case depth, it is important that the method of measurement be specified clearly; otherwise, the heat treater is at a loss to know what case depth to produce in the part.

It is not uncommon for engineers to specify carburizing by a phrase such as "case depth to be 1.25 mm (0.050 in.)", which leaves the heat treater in doubt as to what is wanted. One interpretation of such a statement could produce a case needlessly deep and expensive; another interpretation could yield a dangerously shallow case that would fail in service. When case depths are properly specified, with a notation of the method of measurement, there can be no misinterpretation of what is wanted. "Total case depth", "effective case depth to 50 HRC", and "case depth to 0.40% carbon" are specific descriptions that cannot be misinterpreted and are consistent with sound engineering practice.

The various methods of measuring case depth have been developed to meet particular needs for different kinds of work. There are chemical, mechanical and visual (macroscopic and microscopic) methods for measuring case depth. Test specimens or parts in the soft or hardened condition may be measured by any of the several methods.

Measurements of case depth obtained by two different methods are compared in Fig. 24. One method employed chemical analysis of carbon content to the 0.40% carbon level of the carburized case; case depth was measured from the surface to the distance in from the surface at which a 0.40% carbon content was obtained. The other method employed a microhardness traverse to determine effective case depth—that is, the distance from the surface providing a minimum hardness reading of 50 HRC. The first method was used to obtain data on 25-mm (1-in.) -diam test specimens; the second method, to obtain data on a cross section of gear teeth at the junction of the involute profile and the root fillet.

Reproducibility of Results. Even when the case depth is specified unambiguously and the method of measuring it is defined, precision in results is less than for most purely physical measurements, because of the possibility of human error in determinations of case depth.

The data given in Fig. 25 correlate effective case depth to 50 HRC measured on the same gears by two different plants. Values were obtained by

Fig. 30 Carburizing conditions and carbon gradient produced using 8620H steel

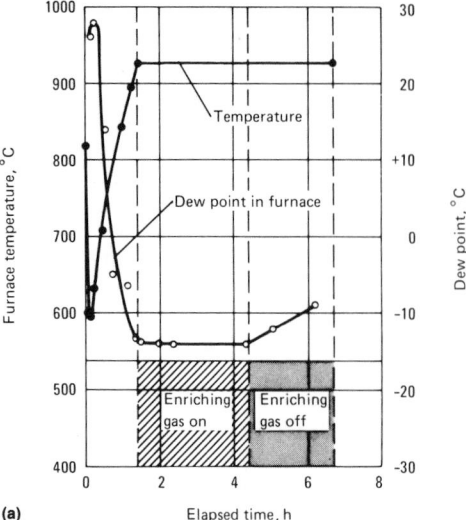

(a) Elapsed time, h

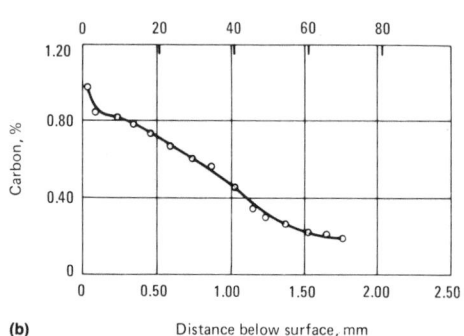

(b) Distance below surface, mm

Load: ring gears 231 kg (510 lb) net, 390 kg (860 lb) gross. **Furnace:** metallic-retort pit type, electrically heated. **Carrier gas:** 2.8 m³/h (100 ft³/h) of endothermic gas throughout cycle of 5 h and 15 min at temperature. **Enriching gas:** 0.3 m³/h (12 ft³/h) of natural gas; addition started at temperature and shut off after 3 h at temperature, or 57% of at-temperature cycle. **Generator dew point:** −4 to −3 °C (+24 to +26 °F) throughout cycle. **Heating chamber pressure:** 8.6 to 11 mm (0.34 to 0.44 in.) water column. **Carburizing temperature:** 925 °C (1700 °F). **Cooling method:** slow cool in cooling pit from 925 °C (1700 °C)

(a) Furnace temperature and atmosphere conditions for carburizing. (b) Carbon gradient produced by the cycle shown in (a)

means of the hardness traverse method, which is considered the most accurate for measuring effective case depth. Nevertheless, several discrepancies in the results obtained by the two plants are apparent.

To obtain hardness traverse readings, specimens are cut perpendicular to the carburized surface at a critical location, which for gears is normally at or near the pitch line. Specimens must be carefully ground and polished prior to hardness testing.

To compare different methods of measuring case depth on the same samples, carburized and hardened speci-

mens of 8620 steel were prepared and submitted to five different heat treating organizations for measurement by their laboratories.

These samples were prepared with reasonable care and were located within inches of one another during the carburizing and hardening operations. All of the samples were cut from the same bar. Each of the five laboratories that participated in the testing was asked to make all case depth measurements in accordance with the procedures specified in the 1960 SAE Handbook. The results are shown in Fig. 26 and 27.

These measurements (all made by trained observers in qualified laboratories) diverge somewhat more widely than might be expected. They do illustrate, perhaps in an exaggerated manner, the inherent lack of high precision in case depth measurements. With consultation among laboratories and more precise definition of the criteria, considerably closer agreement can be obtained—for instance, among different laboratories in the same multiple-plant organization or between a well-coordinated buyer and seller of carburizing services.

Commercial Carburizing Practice for Undiluted Hydrocarbon Gases or Liquids

Undiluted hydrocarbon gases or liquids normally are used only in batch-type furnaces having effective door and cover seals and fans for recirculation. Because of the infiltration of air into continuous furnaces, with which the doors must be opened and closed for charging and discharging, a large volume of gas must be fed into the furnace. Because a large flow of hydrocarbon gas or liquid would cause excessive sooting, undiluted hydrocarbon gases or liquids are not employed in continuous furnaces; combinations of carrier gas and hydrocarbon gas are used.

In the normal operation of "seasoned" batch-type units carburizing with undiluted hydrocarbon gases or liquids, first the furnace is brought to temperature with full gas or liquid flow above 760 °C (1400 °F), and then clean parts are loaded into the hot furnace. The carburizing temperature normally is between 900 and 940 °C (1650 and 1720 °F). For most steels, 925 °C (1700 °F) is used; for the more highly alloyed steels such as 3310, 9315 and 4815, 900 °C (1650 °F) is more common.

Fig. 31 Carburizing conditions and carbon gradient produced using 8620H steel

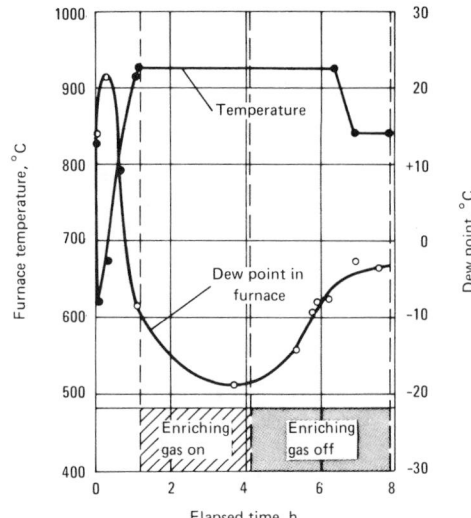

(a)   Elapsed time, h

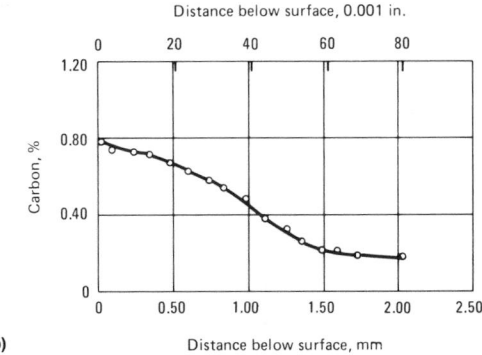

(b)   Distance below surface, mm

Load: shafts, 145 kg (320 lb) net, 260 kg (570 lb) gross. **Furnace:** metallic-retort pit type, electrically heated. **Carrier gas:** 2.8 m³/h (100 ft³/h) of endothermic gas throughout cycle of 5 h and 5 min at temperature. **Enriching gas:** 0.25 m³/h (9 ft³/h) of natural gas; addition started when load was placed in furnace, shut off after 2 h and 50 min, or 56% of at-temperature cycle. **Generator dew point:** −4 to −3 °C (+24 to +27 °F) throughout cycle. **Heating chamber pressure:** 9.9 to 12 mm (0.39 to 0.49 in.) water column. **Carburizing temperature:** 925 °C (1700 °F). **Cooling method:** quench from 845 °C (1550 °F)

(a) Furnace temperature and atmosphere conditions for carburizing. (b) Carbon gradient produced by the cycle shown in (a)

The demand for carbon by the steel surface is greatest when the work first reaches carburizing temperature. To minimize sooting, hydrocarbon flow may be decreased progressively in accordance with the decreased demand. For a cycle in which the time at carburizing temperature is 9 h, one third of the total carbon absorbed by the steel will be absorbed in the first hour, one third in the next 3 h, and one third in the last 5 h.

The efficiency of carbon utilization for typical batch-type carburizing charges, where the flow of carburizing liquid was maintained constant for 7 and 11 h, respectively, is shown in the following two examples.

Example 29. Data in Table 7, obtained from a specific batch-type installation, show how efficiency drops off to 10% or less after about the third hour of a 7-h carburizing cycle, and indicate the need for decreasing the flow of carburizing medium to obtain greater efficiency and less deposition of soot as the time cycle progresses.

Example 30. Table 8 presents data on carbon utilization efficiency for a specific installation carburizing rock bit cutters at 925 °C (1700 °F). In this operation, the load is subjected to a diffusion period between the eighth and eleventh hours, during which time efficiency is rated at 11%.

Operating conditions for two installations utilizing undiluted hydrocarbon liquids are presented in the following two examples.

Example 31. Table 9 shows details of carburizing conditions used in producing case depths of 1.5 to 3.0 mm (0.060 to 0.120 in.) on rock bit cutters made of 4815 steel. These data relate to time at 925 °C (1700 °F) and cover combined carburizing and diffusion cycles ranging from 5¼ to 21 h.

Example 32. Table 10 presents operating conditions for carburizing 9310 steel in pit-type retort furnaces that employ either propane or undiluted hydrocarbon liquids as the carburizing medium. In this particular plant, carbon potential is manually controlled for case depths under 0.9 mm (0.035 in.), while automatic control is applied for case depths of 0.9 to 2.0 mm (0.035 to 0.080 in.).

Commercial Carburizing Practice for Carrier Gas Plus Hydrocarbon Gas

Most gas carburizing furnaces use combinations of carrier gas plus hydrocarbon gas. The use of undiluted hydrocarbon gas in continuous furnaces has been virtually eliminated.

Introduction of Atmosphere. In batch furnaces, metered amounts of each gas are premixed and introduced into the carburizing chamber in quantities sufficient to ensure correct operating conditions. This includes enough flow of atmosphere to maintain positive furnace pressure, to prevent infiltration of air, and to carburize the workpieces.

The location of the atmosphere gas inlet is important. The most effective point for introduction is in the path of good gas circulation, for thorough mix-

Fig. 32 Carburizing conditions and carbon gradient produced using 8620H steel

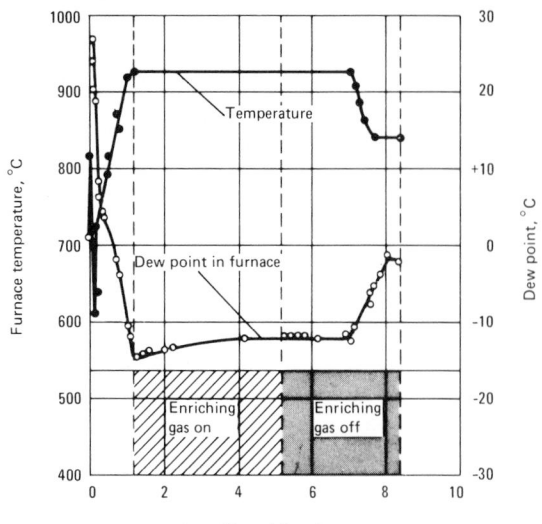

(a)

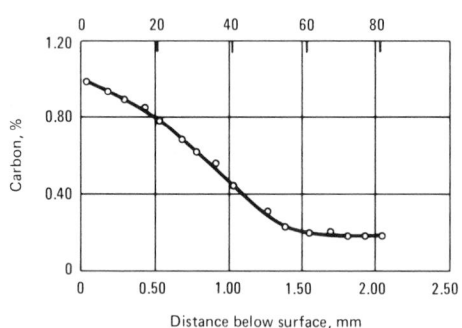

(b)

Load: shafts, 170 kg (375 lb) net, 440 kg (970 lb) gross. **Furnace:** vestibule, enclosed-quench, brick-lined heating chamber, radiant-tube heated. **Carrier gas:** 22 m³/h (770 ft³/h) of endothermic gas throughout cycle of 5 h and 53 min at temperature. **Enriching gas:** 0.28 m³/h (10 ft³/h) of propane; addition started when temperature reached 925 °C (1700 °F), shut off after 3 h and 59 min or 66% of at-temperature cycle. **Generator dew point:** + 1 to + 4 °C (+ 33 to + 39 °F) throughout cycle. **Heating chamber pressure:** 2.5 to 3.6 mm (0.10 to 0.14 in.) water column. **Carburizing temperature:** 925 °C (1700 °F). **Cooling method:** quench from 845 °C (1550 °F)

(a) Furnace temperature and atmosphere conditions for carburizing. (b) Carbon gradient produced by the cycle shown in (a)

ing, but at a spot that will avoid direct impingement of the freshly added atmosphere on the work. This usually will avoid localized sooting and nonuniform carburization.

Flow Rates. Because of the small volume and tight chamber of batch furnaces, relatively small volumes of gas are required. For a pit furnace having a work chamber 0.6 m (25 in.) in diam and 0.9 to 1.2 m (36 to 48 in.) deep, gas flows commonly used are 1.4 to 2.8 m³/h (50 to 100 ft³/h) of carrier gas and 0.2 to 0.4 m³/h (8 to 15 ft³/h) of natural gas. If

minimum sooting is desired, the flow of natural gas may be decreased as the cycle progresses, because the rate of carbon absorption at the surface of the steel decreases with elapsed time.

High rates of flow of carrier gas (with or without hydrocarbon gas) may be required for purging batch furnaces to prevent oxidation of the newly charged work. This is particularly important when the work is copper plated to prevent the carburizing of the plated areas, because the protective copper can be destroyed rapidly by an oxidizing atmosphere at high temperature.

The high flow of hydrocarbon gas during the first part of the carburizing cycle is cut down or shut off during the diffusion and cooling cycles. This "program control", either manual or automatic, is designed to maintain control of carbon potential during all stages of batch carburizing.

In continuous gas carburizing furnaces, good operating practice is not simply a matter of program control of gas ratios over a timed period. Rather it becomes a more complex arrangement of zoned control of gas ratios along the length of the furnace. To combine a carbon-saturation cycle and a diffusion cycle in a continuous furnace, varied amounts of atmosphere gas are provided along the furnace length. The mixture of carrier and active gases introduced into each zone is varied according to carbon demand along the furnace length. Figure 28 illustrates a continuous multiple-row carburizing furnace adapted for the saturation-diffusion method of carburizing. Multiple manifolding for atmosphere distribution is shown. Maintenance of different carbon potentials in the different zones, as required by the saturation-diffusion treatment, will require suitable built-in baffle arches at certain locations within the furnace. Correct directional flow of effluent gases to the ends of the furnace must be established and maintained.

In the operation of a multiple-row gas carburizing furnace, it is sometimes desirable to operate each row at a different pushing speed, in order to produce two or more different carburized case depths simultaneously. The weight in pounds per hour of the steel loaded on the different rows must be approximately balanced, to ensure all rows coming to temperature at about the same point in the furnace. Thus, slower-moving rows are not subjected to shadowing from faster-moving

Fig. 33 Production parts that were furnace cooled to quenching temperature after being carburized

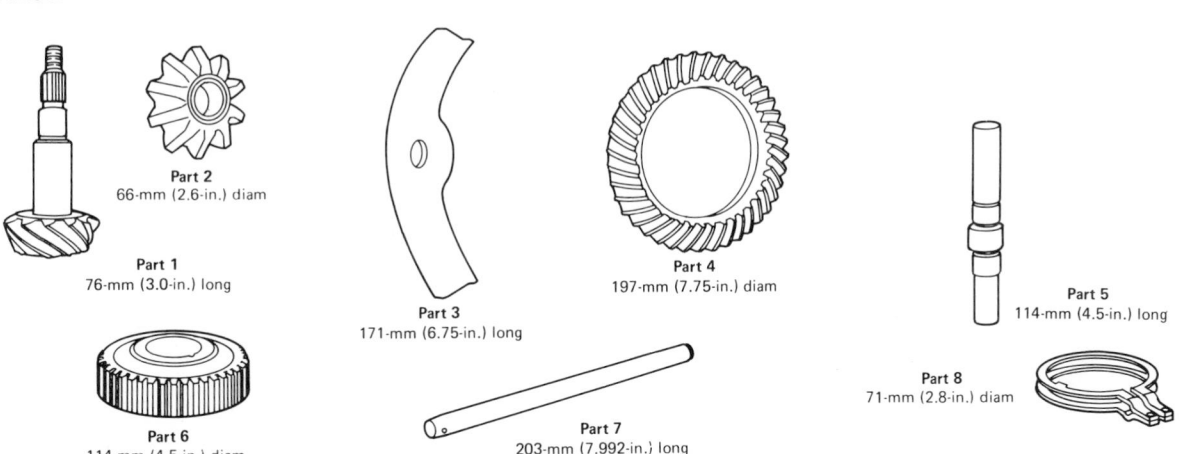

Part 2
66-mm (2.6-in.) diam

Part 1
76-mm (3.0-in.) long

Part 3
171-mm (6.75-in.) long

Part 4
197-mm (7.75-in.) diam

Part 5
114-mm (4.5-in.) long

Part 6
114-mm (4.5-in.) diam

Part 7
203-mm (7.992-in.) long

Part 8
71-mm (2.8-in.) diam

Part No.	Steel	Depth of case mm	0.001 in.	Carburizing temperature °C	°F	Oil quenched from temperature °C	°F	Tempering temperature °C	°F	Hardness, HRC
1	4118	1 to 1.25	40 to 50	885	1625	830	1525	185	365	60 min
2	1024	1 to 1.25	40 to 50	885	1625	830	1525	165	325	60 min
3	9112	1 to 1.25	40 to 50	900	1650	855	1575	440	825	30 to 45
4	4118	1 to 1.25	40 to 50	885	1625	830	1525	(a)	(a)	60 min
5	5120	1 to 1.5	40 to 60	900	1650	855	1575	165	325	60 min
6	8620	...	...	925	1700	845	1550	(a)	(a)	...
7	4322	0.9	35	915	1680	845(b)	1550(b)	(a)	(a)	...
8	1008	0.4 to 0.6	15 to 25	870	1600	800	1475	440	825	31 to 43(c)

(a) Not tempered. (b) Quenched in water. (c) HR 45-N

This procedure avoids the cost of reheating for hardening.

Fig. 34 Diametral dimensions of pinion shaft helical gear splines before and after carburizing and hardening

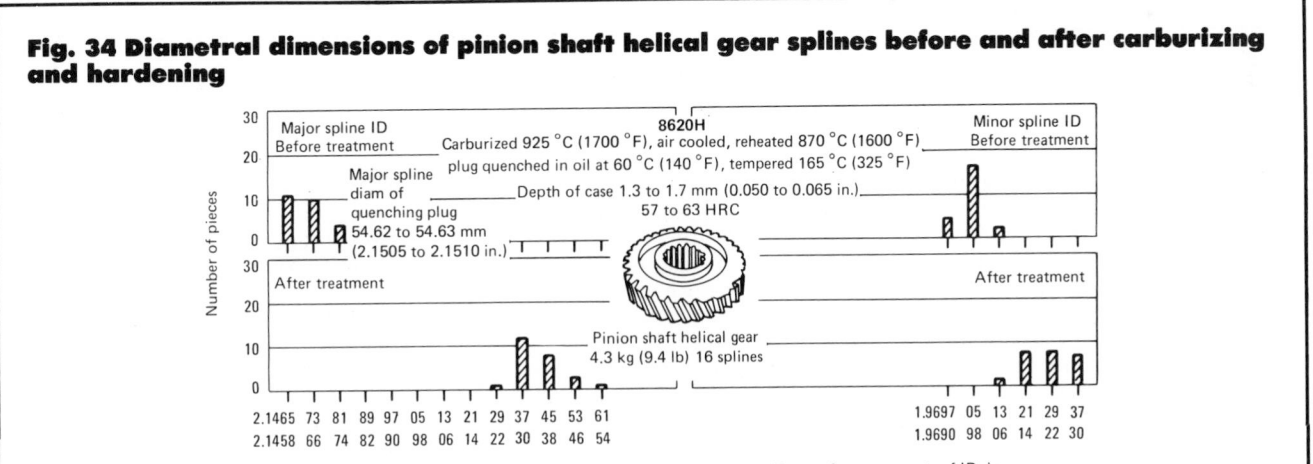

rows. Establishing a uniform zone for heating all work (regardless of differences in rate of movement through the furnace) permits control of the atmosphere in succeeding zones, so that all work, even though of different case depth from row to row, will have the same controlled carbon treatment.

It is advisable to set some practical limits on the different case depths to be obtained from row to row. The most commonly accepted limits allow for a

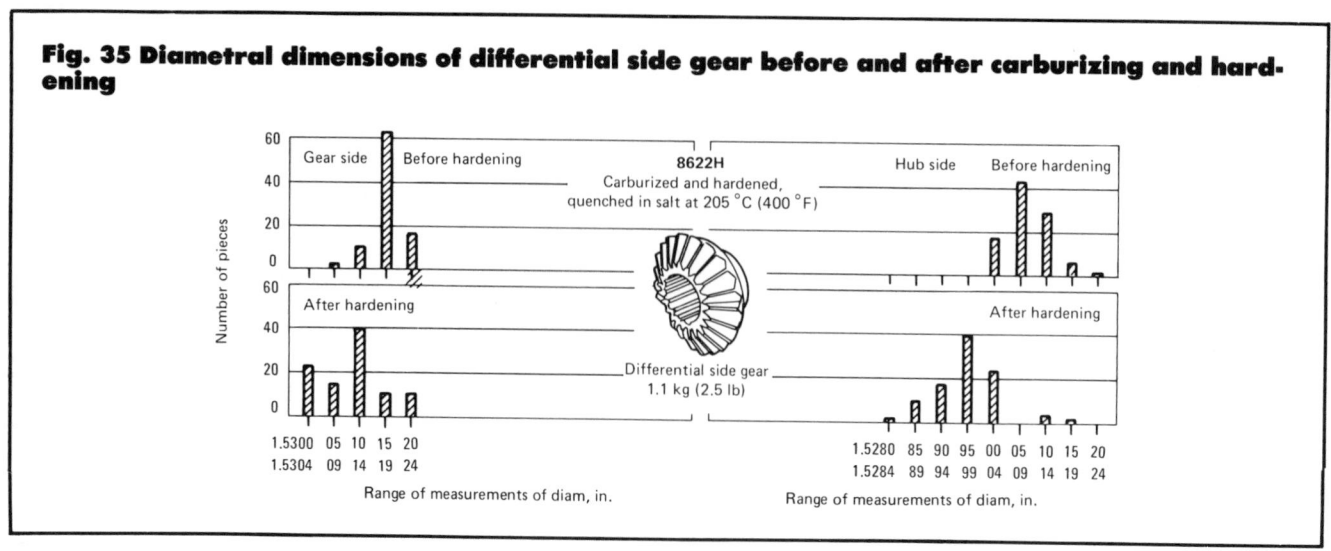

Fig. 35 Diametral dimensions of differential side gear before and after carburizing and hardening

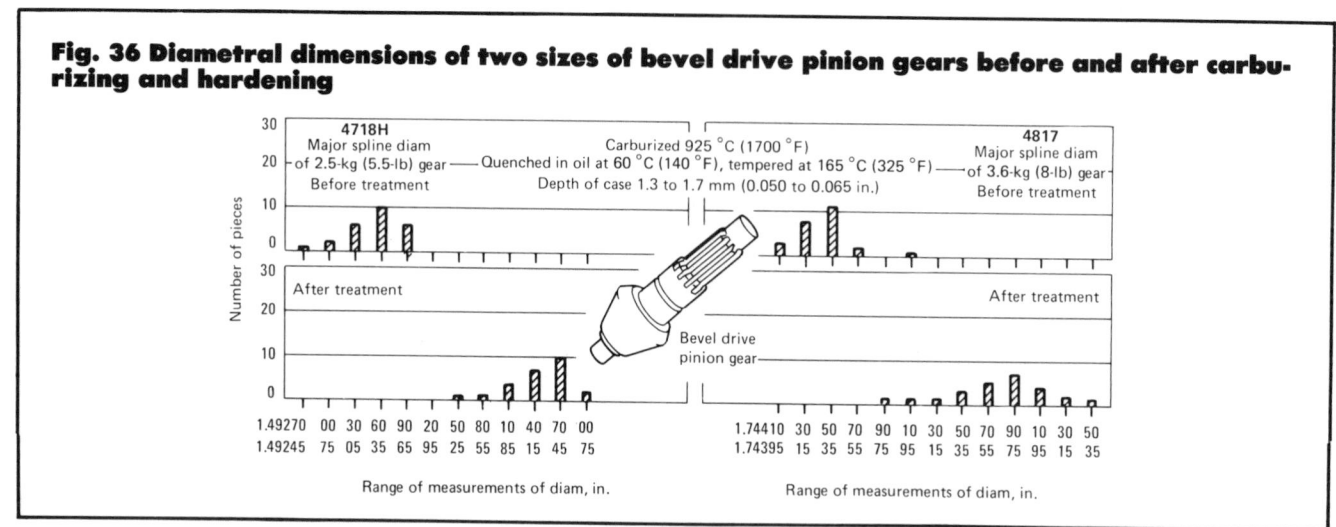

Fig. 36 Diametral dimensions of two sizes of bevel drive pinion gears before and after carburizing and hardening

depth differential of about 0.4 to 0.5 mm (0.015 to 0.020 in.) for trouble-free operation.

For large continuous furnaces, the required flow of carrier gas to the carburizing chamber in order to produce adequate furnace pressure is from one to three volume changes per hour.

Natural gas as the active addition agent will range from 3 to 15% of the carrier gas. For propane, 1 to 5% will give equivalent carbon in the atmosphere. The rate of flow of each gas is continually measured and indicated, either by visual flowmeters or by a manometer connected across an orifice of known diameter.

A dense load of large surface area will require a somewhat higher ratio of hydrocarbon gas to carrier gas to ensure uniform carbon concentrations and case depths throughout the load.

This requirement is more critical in applications where the degree of recirculation is low or nil.

Furnace pressures of 5.0 to 7.5 mm (0.2 to 0.3 in.) water column are common and can be measured most accurately with a draft gage. The presence of burning gases at the effluent ports at the top of the furnace is not an indication of positive pressure in the furnace. The hot gases are lighter than air and will rise. While they are burning out of the effluent port, air may be drawn into the furnace through leaks in the bottom casing or around fan shafts. This emphasizes the need for tightly built and maintained furnaces for continuous gas carburizing under conditions where carbon control is required.

The time interval between door openings on continuous furnaces may range from less than 5 min to 40 min or more,

depending on the length of the furnace and the case depth required. Parts such as automotive ring gears that are quenched individually in presses directly from the carburizer may require the opening of an auxiliary slot-type door every 30 s. In general, greater total gas flows and richer mixtures are required for applications where door openings are frequent. Vestibules may be swept out with natural gas immediately after the outer doors are opened and closed.

Mixing. The carrier gas almost always is mixed with hydrocarbon gas outside continuous furnaces. This may be done with individual mixers at each inlet or by one or more central mixing stations from which the mixtures are manifolded to several furnace inlets. With central mixing stations, a constant ratio of carrier gas to hydrocar-

Fig. 37 Diametral dimensions of two pinion gears before and after carburizing and hardening

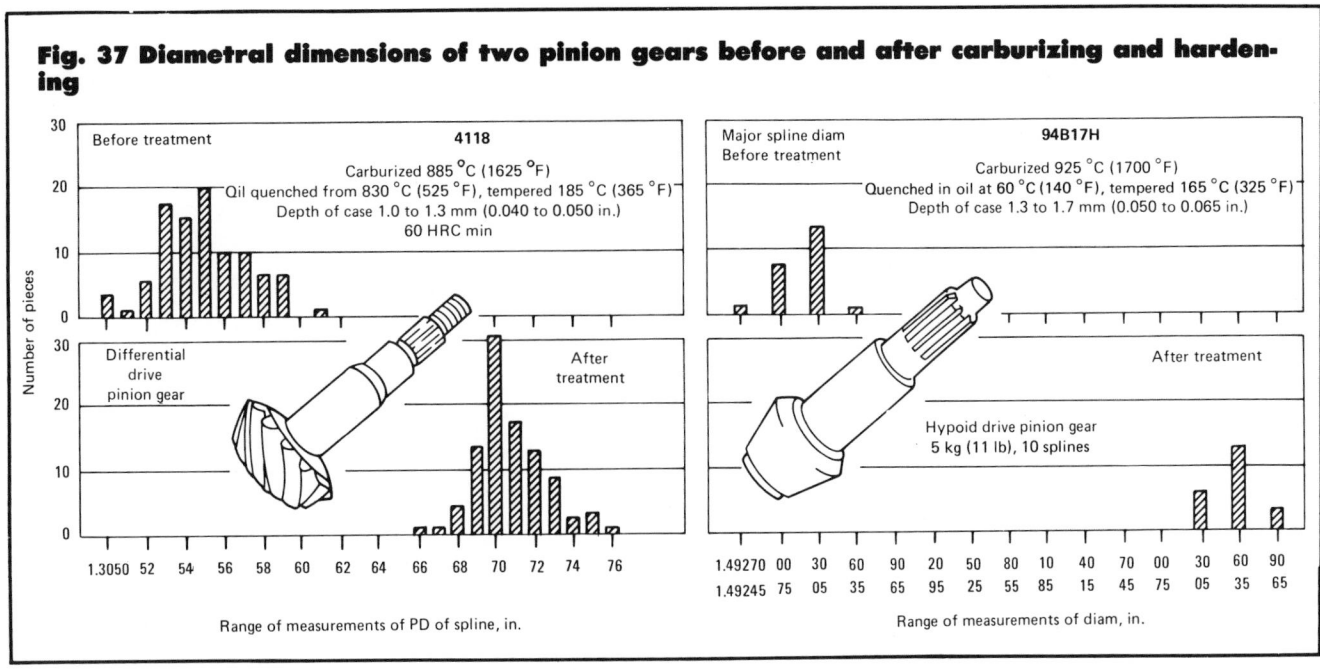

Fig. 38 Diametral dimensions of differential side gear before and after carburizing and hardening

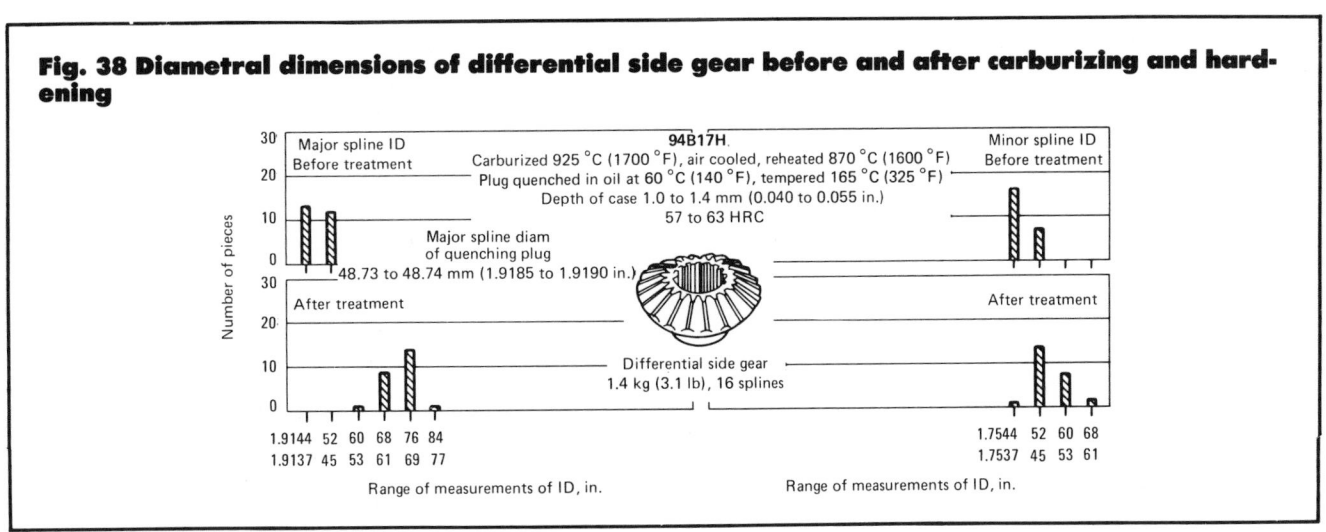

bon gas can be maintained at any or all sections of the furnace, regardless of intentional flow changes or accidental changes in line pressures of either carrier gas or hydrocarbon gas.

A constant-ratio mixing station may be an arrangement of correctly sized orifice plates in the supply lines for carrier gas and hydrocarbon gas, augmented, in the hydrocarbon gas line, by a pressure regulator that is backloaded from the carrier gas supply line. In this manner, the supply pressure of hydrocarbon gas is made dependent on the supply pressure of the carrier gas, and the respective gas flows across both orifices are maintained at constant ratio over a wide range of flow rates.

Examples of Operating Cycles. Typical operating conditions that utilize carrier gas and enriching gas are shown graphically in Fig. 29 to 32 (Examples 33 to 36). Resulting carbon gradients are included in the lower graph of each of these figures.

Example 33. Figure 29 shows details of a cycle used for gas carburizing ring gears in a pit-type furnace. The resulting carbon gradient is also shown graphically in Fig. 29. Here it is seen that surface carbon is near eutectoid composition, and that the decrease is gradual and uniform to the carbon content of the core.

Example 34. Figure 30 presents details of another cycle used for carburizing of ring gears in 230-kg (510-lb) loads (net) in a pit-type furnace. In this example and in Example 33, parts were cooled in a pit prior to reheating for hardening. With the cycle shown, the carbon content at the surface starts at 0.95%, but drops rapidly to near eutectoid composition at about 0.1 mm (0.004 in.).

Example 35. Figure 31 shows the details of a cycle and the resulting carbon gradient for carburizing shafts in a pit-type furnace. Direct quenching follows a decrease in temperature from 925 to 845 °C (1700 to 1550 °F). In this

Fig. 39 Effect of carburizing and subsequent heat treating on dimensional changes of the outside diameter of antifriction bearing races made of a 4Ni-1.5Cr steel

	Machined OD		Total case	
	mm	in.	mm	in.
1 1302		51.275	7.0	0.275
2 1050		41.350	6.4	0.250
3 908		35.750	5.7	0.225
4 907		35.725	5.7	0.225
5 869		34.225	5.1	0.200
6 718		28.250	4.6	0.180
7 592		23.290	4.6	0.180
8 566		22.275	4.1	0.160
9 562		22.130	4.1	0.160
10 525		20.675	4.1	0.160
11 370		14.550	3.3	0.130

Ten tests for each size of race

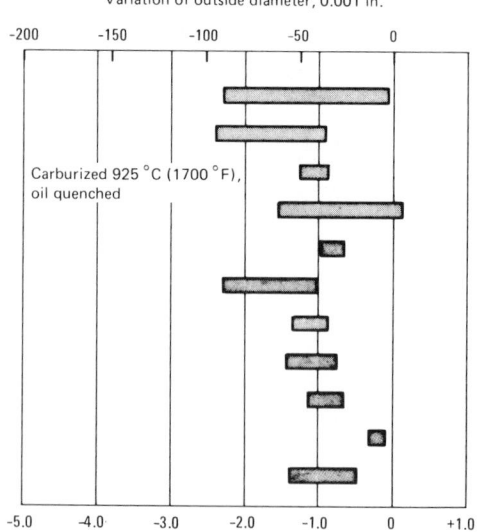

	Machined OD		Total case	
	mm	in.	mm	in.
1 1302		51.275	7.0	0.275
2 1050		41.350	6.4	0.250
3 908		35.750	5.7	0.225
4 907		35.725	5.7	0.225
5 869		34.225	5.1	0.200
6 718		28.250	4.6	0.180
7 592		23.290	4.6	0.180
8 566		22.275	4.1	0.160
9 562		22.130	4.1	0.160
10 525		20.675	4.1	0.160
11 370		14.550	3.3	0.130

Ten tests for each size of race

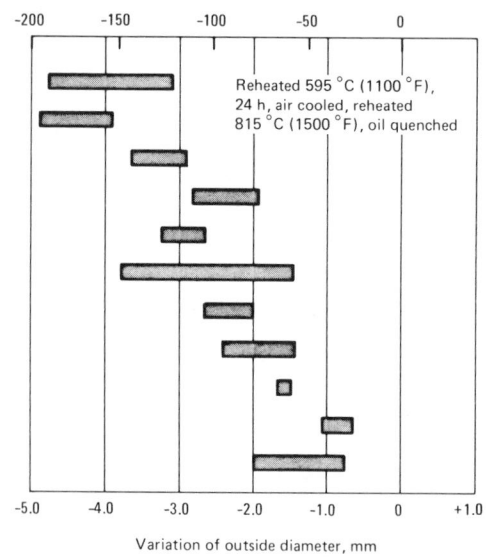

(continued)

instance, the decrease in volume of enriching gas resulted in a decrease in surface carbon to slightly below eutectoid composition.

Example 36. Figure 32 shows operating details of a cycle used for carburizing shafts in a vestibule-type furnace having an enclosed quench. Here the longer carburizing portion (66% of total time) of the carburize-diffuse cycle resulted in higher surface carbon content and greater depth to eutectoid carbon composition. It will also be noted that the proportion of enriching gas to carrier gas is much lower when propane is used.

Sooting. An excessive amount of free carbon (in the form of soot or coke) deposited on the work and in the furnace causes loss of control of the carbon potential, uneven carburizing, deterioration of furnace alloy and refractory, high cleaning costs and unpleasant working conditions. Soot may also interfere with press or plug quenching and contribute to excessive distortion.

Sooting has always been a problem in gas carburizing. The problem can be minimized by choosing a flow rate that gives a rapid rate of carburizing without excessive sooting. Another effective method includes the use of a diffusion cycle, which in many instances permits the soot to be removed from the work by reaction with the carrier gas or with regulated amounts of air introduced for that purpose. Sooting can be greatly reduced also by the use of automatic control of the enriching gas. When this is used, sediment in quenching oil, resulting from the presence of soot on quenched parts, can be reduced from a

Fig. 39 (continued)

	Machined OD		Total case	
	mm	in.	mm	in.
1	1302	51.275	7.0	0.275
2	1050	41.350	6.4	0.250
3	908	35.750	5.7	0.225
4	907	35.725	5.7	0.225
5	869	34.225	5.1	0.200
6	718	28.250	4.6	0.180
7	592	23.290	4.6	0.180
8	566	22.275	4.1	0.160
9	562	22.130	4.1	0.160
10	525	20.675	4.1	0.160
11	370	14.550	3.3	0.130

Ten tests for each size of race

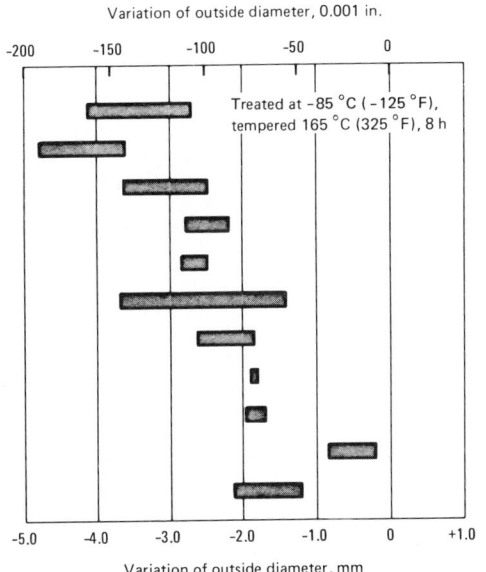

normal level of 0.1% to barely a trace.

In muffle, pit or batch furnaces, soot presents a greater problem, because the relatively smaller volume of gas is usually changed from three to five times per hour. In large continuous radiant-tube furnaces, where the ratio of gas volume to the area to be carburized is much greater, the atmosphere is changed only from one to three times per hour.

Parts are frequently processed in pit furnaces and transferred to cooling pits when finish machining is required before hardening. Under such conditions, it is sometimes advantageous to have a small amount of soot on the parts to protect them against decarburization and oxidation during transfer.

Soot deposition in endothermic generators can be minimized by operating at air-gas ratios set to produce carrier gas of dew point greater than −12 °C (+10 °F); higher temperatures, in the range −1 to +2 °C (30 to 35 °F), are recommended. Another useful step is to operate the generators at a uniform flow, allowing the surplus gas to be exhausted instead of attempting to regulate the flow to the demand, if the demand varies over a wide range.

Most continuous carburizing furnaces are "burned out" periodically (when the furnaces are empty, during a weekend) to remove soot. This is accomplished by partially opening the doors to allow air to flow through, or by introducing controlled quantities of air from the combustion blower through a series of ports along both sides of the furnace and under the furnace hearth. When starting up a furnace after such a burnout, hydrocarbon additions may have to be increased for the first 8 to 16 h of operation, to give the desired carbon potential. More hydrocarbon may also be required when new trays or other new alloy parts are first put into the furnace. When such a situation exists, the furnace is said to be underconditioned.

Direct Quenching Versus Reheating. Direct quenching is less costly than cooling the work to room temperature and then reheating to the hardening temperature prior to quenching. This is the main reason why a major portion of gas carburized parts are quenched directly from the carburizing temperature of about 925 °C (1700 °F) or from a temperature approximately 845 °C (1550 °F), without being cooled to room temperature. The decrease in temperature before quenching can be accomplished by (a) lowering the temperature in the same furnace, (b) moving the work to another zone held at the desired temperature for quenching, or (c) transferring the work to another furnace.

The type of carburizing furnace being used often has a bearing on wheth-er the work should be directly quenched or slowly cooled and reheated. Pusher-type continuous furnaces, for example, are ideally suited to direct quenching, and they can be controlled to obtain virtually any desired cycle of temperature changes. Direct quenching from a pit-type furnace, however, is both difficult and potentially hazardous.

Assume that either type of equipment is available, other factors that may influence the choice between direct quenching and reheating are: (a) the grade of steel being carburized, (b) size and shape of the part, (c) need for selective carburizing, and (d) sequence of subsequent manufacturing operations.

Some of the higher alloy steels, such as the high-nickel grades, will retain excessive amounts of austenite when quenched directly from approximately 925 °C (1700 °F). Therefore, it is common practice when processing these grades to carburize, diffuse, slow cool and reheat. However, retention of austenite has gradually become a lesser problem because of a marked decrease in the use of high-nickel carburizing steels, and the development of temperature cycles that will result in a minimum of retained austenite. For example, in some instances of carburizing high-nickel steels, temperature is first decreased below that used for quench-

Fig. 40 Effect of tempering on hardness in carburized cases

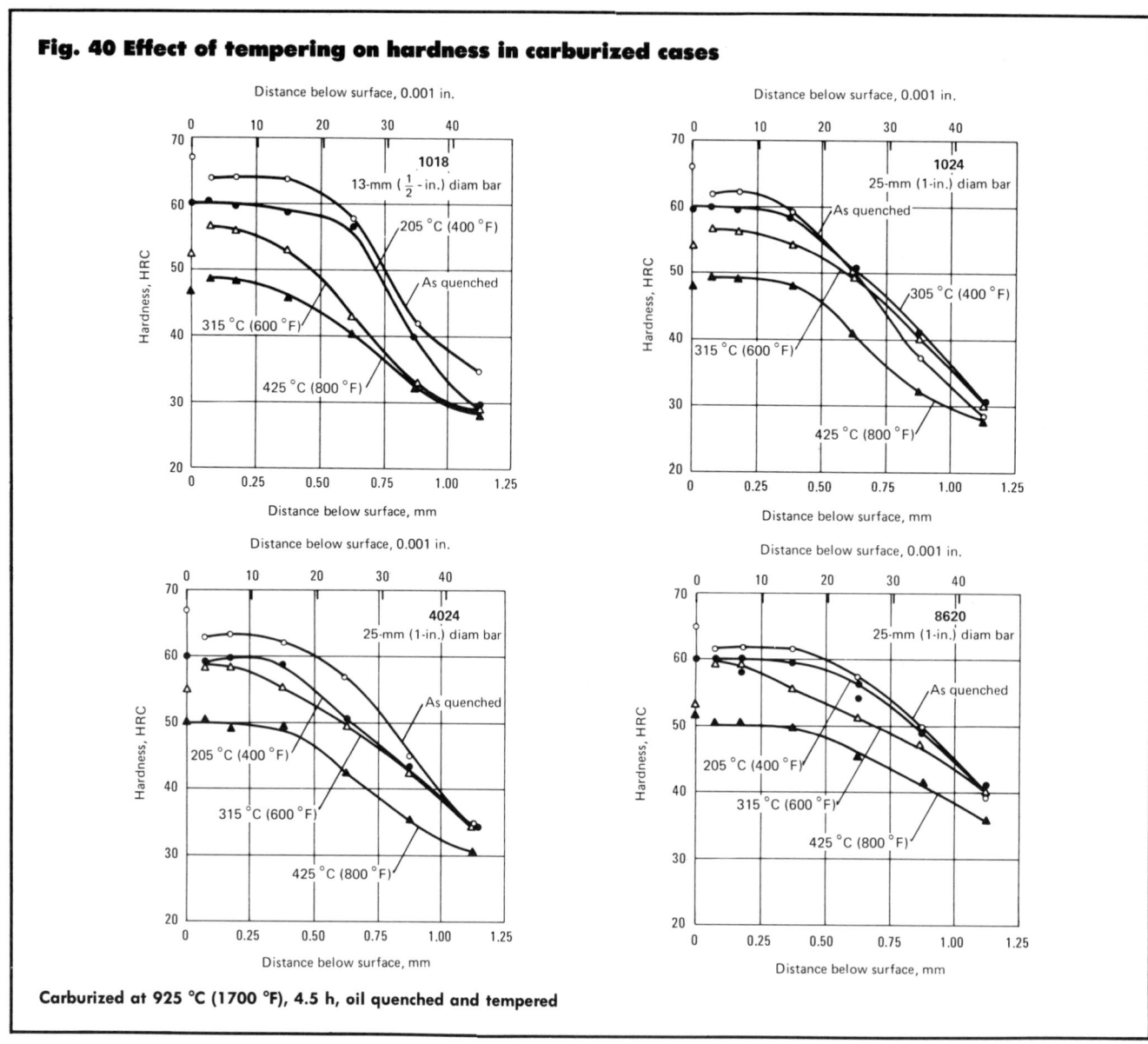

Carburized at 925 °C (1700 °F), 4.5 h, oil quenched and tempered

ing (approximately 705 °C or 1300 °F) and then increased to about 815 °C (1500 °F) prior to quenching. Some experimentation often is required to obtain optimum microstructures for high-alloy steels. This treatment may cause formation of grain-boundary carbides.

Carbon content of the case must be held to a level in which carbides will not be precipitated at grain boundaries during cooling to some reduced temperature before quenching. If boundary carbides are developed, the following difficulties are likely to result: (a) grinding checks, (b) reduced mechanical properties, (c) breaking off of case at sharp corners during mechanical clean-

ing, and (d) service failures.

Part size and shape often are factors in the selection of method, although they may not have a direct bearing; for instance, if parts are free quenched, they may exhibit less distortion when they are quenched directly than when they are reheated. However, this phenomenon must be related to specific parts and cannot be generalized. Part size and shape do become a factor in choosing between direct quenching and reheating when special quenching techniques are necessary. For instance, the gears for which data are given in Examples 37 and 43 are plug quenched, which is a relatively slow operation; therefore, reheating becomes more

practical. In many instances, parts such as ring gears (but not all ring gears) require press quenching and are therefore slow cooled, and reheated in another furnace. This practice is indicated in Examples 33 and 34. Thus, each part needs individual consideration, and if the distortion obtained in free quenching can be tolerated, there is seldom any advantage in reheating. For example, the parts shown in Fig. 33, representing a wide variety of shapes, were all direct quenched after being cooled to their hardening temperature following gas carburizing. Results in all instances were acceptable.

Machining specific areas after carburizing and prior to hardening is a

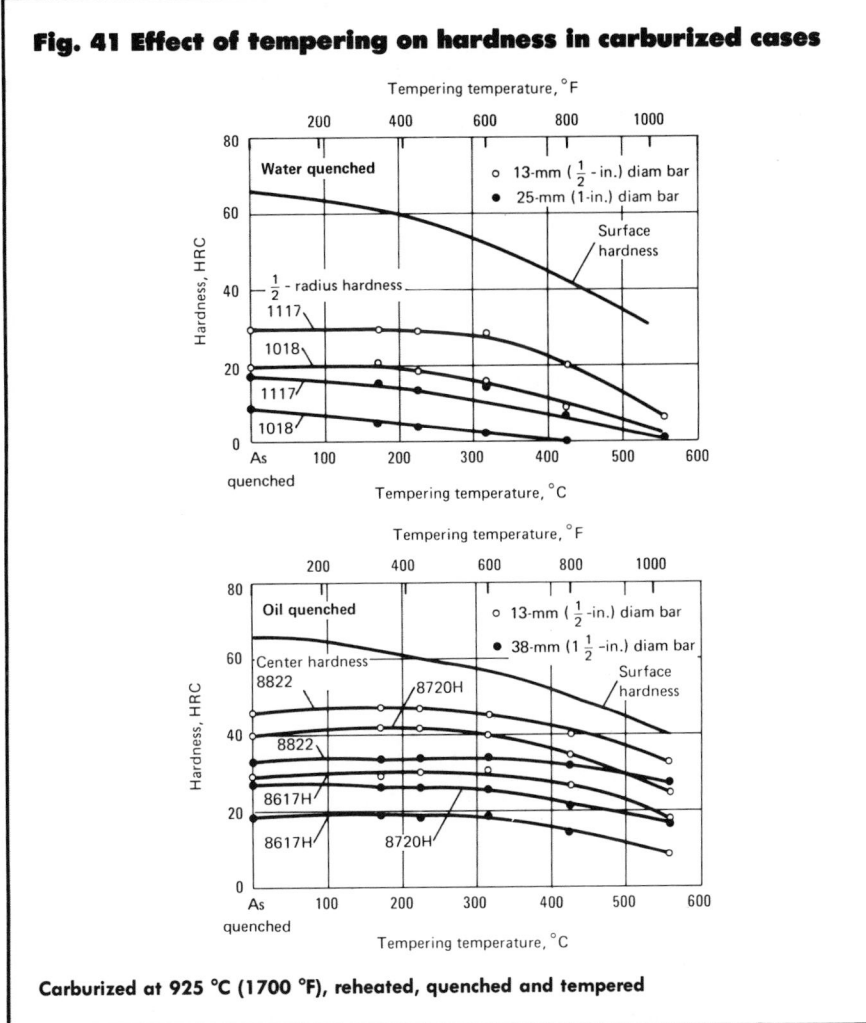

Fig. 41 Effect of tempering on hardness in carburized cases

Carburized at 925 °C (1700 °F), reheated, quenched and tempered

positive method for obtaining selective carburizing. Therefore, this is often a reason for utilizing the slow cooling and reheating procedure. Although such instances are usually special, there are sometimes other reasons for performing machining operations between carburizing and hardening, and therefore slow cooling and reheating is required.

Dimensional Control

Before being carburized, machined parts should be as near final dimensions as possible, so that heat treating times may be kept short. Nevertheless, it must be recognized that some distortion will be encountered in all carburized parts. Although the amount of distortion is influenced mainly by the size and shape of a part, other significant factors are: (a) residual stresses in the parts prior to heat treatment, (b) induced stresses caused by nonuniform heating, (c) methods of stacking and fixturing parts during heating and quenching, (d) growth of surfaces during carburizing, (e) quenching temperature, (f) severity of the quench, and (g) chemical composition of the steel.

Several means are available for decreasing the amount of distortion encountered in any specific part, but they all add to heat treating cost. Consequently, each situation should be analyzed to determine whether it is more economical to increase heat treating cost in order to decrease finishing cost, or vice versa.

Distortion determinations made on test pieces, even with identical steels and heat treating cycles, are of little value unless the test pieces simulate the shape of actual parts. Manufacturing engineers now take a realistic approach to the distortion problem by heat treating pilot lots of from 25 to 50 parts. These pilot lots are heat treated in production equipment, employing good heat treating practice, but without special loading and quenching techniques which await the development of initial data.

Significant dimensions of the parts are then measured and tabulated. Figures 34 to 39 (Examples 37 to 44) show typical results from such studies. As these examples below show, similar parts subjected to the same treatment do not distort uniformly. Therefore, the most meaningful data can be obtained by measuring identical parts only. From these data, it must be determined whether or not the maximum amount of distortion shown will require that the parts be remachined, and, if it will, whether the required machining would be impractical or prohibitively costly.

Methods of Decreasing Distortion

It may be advisable, as an alternative to subsequent machining, to attempt to decrease distortion. Various methods of decreasing distortion are available, including the following, which are discussed in order of ascending cost.

Martempering, which involves quenching in molten salt or in hot oil, is the least expensive method for decreasing distortion. Usually, the major cost it entails is related to the amortization of required capital equipment. For detailed information on the process, see the article on martempering of steel, in this volume.

Fixturing of complex parts during the heating and quenching cycles effectively decreases distortion but markedly increases heat treating costs—often by as much as 50%. A portion of the increase stems from the initial and replacement costs of specially designed and constructed fixtures; the remainder results from a decrease in "payload". Frequently, as much as half of the heating and quenching capacity of the equipment is consumed by the fixtures. Fixturing and martempering are sometimes used in combination, to further reduce distortion.

Press quenching and other means of fixture quenching, such as plug or cold die, are the most effective methods for reducing distortion. However, these methods drastically increase heat treating cost, sometimes by as much as one hundred or even several hundred percent. There are three main reasons for this increase: (*a*) equipment cost, (*b*) carburizing and hardening cycles are longer because parts usually must be slow cooled and then reheated for hardening, and (*c*) parts must be handled singly in quenching. Press quenching and other special methods are described in the article on quenching of steel. Plug quenching is frequently used to minimize dimensional change of inside diameters. The gears for which data are given in Examples 37 and 43 (Fig. 34 and 38) are typical of parts for which plug quenching has proved advantageous.

Example 37. Figure 34 shows changes in major and minor spline inside diameters of 4.3-kg (9.4-lb) pinion shaft helical gears having 16 internal splines. These gears were plug quenched after carburizing at 925 °C (1700 °F), air cooled, then reheated to 870 °C (1600 °F). Checking of 25 parts (as was done here) usually indicates the direction and magnitude of dimensional change. Parts of this shape are typical of those that are plug quenched to lessen dimensional change on inside diameters.

Example 38. Figure 35 presents dimensional change data for the 1.1-kg (2.5-lb) differential side gears representing parts that were quenched in molten salt (martempered), which is often an effective method for reducing distortion. In contrast with conventional quenching, martempering in combination with the given ratio of bore-to-wall thickness resulted in a reduction of diameter after hardening.

Example 39. Figure 36 shows changes in the major spline diameter for 25 bevel drive pinion gears. These 2.5-kg (5.5-lb) parts were carburized at 925 °C (1700 °F), directly quenched in oil, and tempered at 165 °C (325 °F).

Example 40. In Fig. 36, a somewhat wider spread of major spline diameters (compared with Example 39) is apparent when heavier bevel drive pinion gears of the same design are made from another alloy steel (4817). However, the grade of alloy steel usually is a less important factor in magnitude of distortion than part size and shape. Many tests under identical conditions are required before concluding that one grade of similar alloy steel will result in greater distortion than another.

Example 41. Measurement of 100 pieces provided the data in Fig. 37 for changes effected by carburizing, direct oil quenching and tempering of differential drive pinion gears. Although

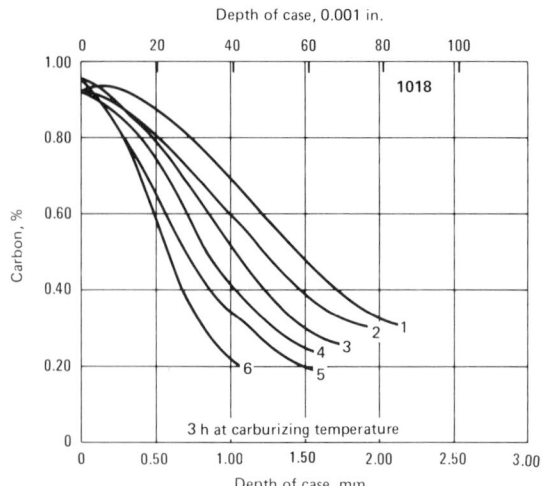

Fig. 42 Effect of increasing carburizing temperature on depth of case in 1018 steel treated for 3 h at temperature

	Carburizing temperature		Dew point	
	°C	°F	°C	°F
1	1065	1950	-22 to -21	-7 to -5
2	1040	1900	-19 to -18	-2 to 0
3	1010	1850	-17 to -16	2 to 4
4	980	1800	-14 to -13	6 to 9
5	955	1750	-12 to -11	11 to 13
6	025	1700	-10 to -9	14 to 15

Atmosphere was endothermic gas enriched with natural gas. Carbon potential was controlled automatically by the dew point method to produce surface carbon of 0.90 to 0.95%.

fewer measurements are often sufficient for practical purposes, more conclusive data are obtained from more samples, as indicated in these histograms.

Example 42. Figure 37 shows changes in major spline diameter for 5-kg (11-lb) hypoid drive pinion gears after carburizing at 925 °C (1700 °F), direct oil quenching, and tempering at 165 °C (325 °F). Dimensional change in the spline diameter of such parts probably is caused by growth during carburizing. The amount of growth is usually related closely to case depth.

Example 43. Plug quenching is often a means of minimizing dimensional change in parts like the 1.4-kg (3.1-lb) differential side gears shown in Fig. 38. They were carburized at 925 °C (1700 °F), air cooled, reheated to 870 °C (1600 °F), oil quenched over a quenching plug to minimize distortion, and tempered at 165 °C (325 °F).

Example 44. In Fig. 39, dimensional measurements on 10 pieces each of 11 sizes of bearing races indicate the effect of different heat treatments. This grade of steel (0.10C-4.0Ni-1.5Cr) retains large amounts of austenite in a direct quench, and is thus more susceptible to dimensional changes from variations in heat treatment than are steels that retain less austenite. Subzero treatments are often used for such steels, to attain greater dimensional stability. The races in this example were free quenched.

Tempering of Carburized Parts

The hardness of a carburized case decreases as the tempering temperature increases, as shown in Fig. 40 and 41. The data in Fig. 40 are from specimens carburized and quenched directly from 925 °C (1700 °F). Rockwell C hardness values shown were converted from Rockwell A for surface hardness and from Vickers for hardness traverses taken at the midlength of sectioned specimens 75 mm (3 in.) long. Each plotted point is the average of two specimens.

The data in Fig. 41 are from specimens carburized to produce case depths of 1.0 to 1.3 mm (0.040 to 0.050 in.). Specimens of the carburized 1018 and 1117 steels were reheated to 780 °C (1440 °F) and quenched vertically in agitated water. The carburized alloy steel specimens were reheated to 845 °C (1550 °F) and quenched vertically in agitated oil. After tempering, 19-mm (3/4-in.) slices were cut from each bar, and hardness readings were taken at 90° intervals. Surface hardness readings nearly coincided; therefore, only one curve is plotted for each group of steels. Rockwell C values corresponding to the J3 and J8 positions on the hardenability specimens of the alloy steels were as follows:

Steel	Hardness, HRC	
	J3	J8
8617H	30	18
8720H	41	25
8822	49	37

Many carburized and hardened parts are placed in service without tempering. However, many such parts are tempered in the range of 150 to 190 °C (300 to 375 °F); this does not greatly reduce hardness, but may increase toughness and will lessen susceptibility to grinding cracks during finishing.

For many pinions and gears for critical applications, experience has proved that tempering carburized work in the low-temperature range of 150 to 190 °C (300 to 375 °F) is beneficial. Carburized parts that will become components of aircraft are invariably tempered. In such instances, tempering is not harmful and may provide some benefit in resistance to cracking or chipping. However, in thousands of less critical applications, it has been difficult or impossible to prove the need for tempering.

Although exceptions are sometimes made to meet special requirements, carburized parts are seldom tempered at temperatures higher than approximately 190 °C (375 °F), because of the need for a hard surface.

Selective Carburizing. Some parts, because of their intended use, must be selectively carburized—that is, carburized on certain surfaces only. Many gears, for example, are carburized only on teeth and spline or bearing surfaces. In addition to satisfying the performance requirements of the part, selective carburizing may also facilitate the machining or welding of noncarburized surfaces in the hardened or hardened and tempered condition.

In selective carburizing, surfaces that are not to be carburized must be adequately protected by a coating or shield that is impervious to the carburizing atmosphere. Various means may be employed to protect or "stop-off" selected surfaces from the carburizing atmosphere. Copper plating to a minimum thickness of 13 μm (0.5 mil) is widely used for this purpose because it is relatively easy to apply, machinable and noncontaminating to furnace atmospheres. Prior to carburizing, a large, 0.9-m (36-in.) diam, 4620 steel ring gear, for example, may be copper plated on the inside-diameter flange area only to permit finish machining, after hardening, of the bore and bolt holes located in the flange. To prevent the deposition of copper on them, other surfaces are coated with a high-temperature microcrystalline wax or a chemical-resistant lacquer, which is removed prior to carburizing. After carburizing, the copper may be removed in other suitable stripping solutions. Most, if not all, of the copper plate may be removed in the course of subsequent machining operations.

Ceramic coatings, in the form of paint, also provide an economical means for protecting selected surfaces from carburizing. The surfaces must be thoroughly cleaned before ceramic paint is applied, and the first coat must be allowed to dry before a second coat is applied. The primary requirement for ceramic-paint coatings is that they adhere tightly, so as to be impervious when heated in a carburizing atmosphere. Application of ceramic paint to the bushing and button recesses of a rock bit cutter, for example, is currently employed on a production basis.

Where some leakage is permissible, blind holes can be stopped off by the insertion of copper plugs or by filling them with clay. The use of copper plugs permits a very tight fit. Therefore, pressure must be relieved or the plug will blow out from expansion of entrapped air. Through-holes may be plugged at each end to cut off circulation of the carburizing gases and thereby minimize carburization. Internal threads can be protected by the insertion of a copper capscrew; external threads, by capping with a copper nut. If a steel capscrew or nut is used, the threads should be coated with stop-off materials to facilitate unscrewing.

The success of all stop-off methods in gas carburizing depends largely on the care used in their application. Mechanical stop-offs are seldom 100% effective in production.

The most positive means of ensuring that specific areas are free from carburization is to plan the sequence of operations so that these areas are machined after carburizing and before hardening. However, in many instances, this practice is not feasible, and cost will always be greater because the work must be cooled slowly from the carburizing temperature so that it will be in the annealed condition prior to machining. Additional handling also contributes to higher cost when this practice is used.

In some instances, operations are planned so that the case is removed from certain areas by grinding the hardened workpieces, thus allowing hardening from the carburizing furnace. This practice is usually confined to special situations and generally to cases less than 1.3 mm (0.050 in.) in depth. Local softening of a carburized part by induction heating is frequently performed in the automotive industry. The heat may just temper the part, or for steels with low hardenability, may actually normalize the area to be softened.

Homogeneous Carburizing. Both the carburizing-diffusion method and the "starved carbon" (or balanced-atmosphere) method are used in production for a special carburizing application known as "homogeneous carburizing". In this process, parts of relatively small cross section, approximately 5 mm (3/16 in.) or less, are carburized throughout the section to a nearly constant carbon content. This process offers the advantage of enabling a part, such as an intricate stamping, to be made from an easily formed low-carbon steel and then carburized throughout to a carbon content that will provide desired hardness after heat treating. The balanced-atmosphere procedure may be used with parts having considerable variation in section; carburizing-diffusion is limited to parts having nearly uniform cross sections.

Example 45. One plant reports the use of homogeneous carburizing in various applications where forming, welding or assembling precludes the use of higher carbon grades of steel. For example, difficulties encountered in welding 8650 steel chain links were averted by making the links of 8620, welding them and then homogeneous carburizing to a carbon content of 0.70%. This procedure produced a continuous chain with strength equivalent to that of the 8650 steel chain, and with superior wear resistance. Maximum thickness of the links was 5 mm (³/₁₆ in.).

Example 46. In another successful application, a Belleville clutch spring was made of 1010 steel and homogeneous carburized to a carbon content of 0.65 ± 0.05%. This spring was 200 mm (8 in.) in diameter and 1.4 to 1.5 mm (0.056 to 0.059 in.) thick; producing the part from homogeneous carburized 1010 steel was less expensive than using the more costly 1065 steel, which was difficult to form and incurred a high scrap loss.

High-temperature carburizing refers to the use of carburizing temperatures above 955 °C (1750 °F). Maximum carburizing temperature in practical applications has been estimated to be approximately 1010 °C (1850 °F). The advantage of higher temperatures is that the carburizing process accelerates markedly at temperatures above 955 °C (1750 °F), thus providing a given case depth in shorter periods of time. In some applications, the advantage may be offset by a need for more frequent replacement of fixtures.

Acceleration of the process by increasing temperature is demonstrated by the curves in Fig. 42, which plot the case depths and carbon contents achieved in a 3-h period at carburizing temperatures ranging from 925 to 1065 °C (1700 to 1950 °F). These data were obtained with a 1018 steel, and carbon potential was controlled automatically throughout each cycle by the dew point method, to produce a surface carbon content of 0.90 to 0.95%.

Furnace Construction. Several problems associated with the use of higher temperatures have discouraged the widespread adoption of high-temperature carburizing. Conventional furnace materials, such as the various modifications of 35Ni-15Cr alloy, rapidly decrease in strength above 955 °C (1750 °F), while their capacity for absorbing carbon increases progressively. An increase in the oxidation rate of radiant tubes used in gas-fired furnaces is a further disadvantage. Thus, the cost of furnace maintenance increases at a rate that may more than offset the savings realized from increased production.

At higher temperatures, control of carbon potential and case depth is much more critical, and automatic control of furnace atmosphere is mandatory. Too high a carbon potential will result in the formation of complex carbides in alloy carburizing steel and will promote the retention of austenite in quenched parts. The possible adverse effects of grain growth and increased distortion above 955 °C (1750 °F) have also discouraged the use of high-temperature carburizing.

Nevertheless, the introduction of new high-temperature alloys for either cast or fabricated grids, fixtures, rails and radiant tubes has contributed to the feasibility of the process. These new alloys retain useful strength up to 1150 °C (2100 °F), and their cost has been reduced in recent years. Consequently, continuous furnaces can now be designed for carburizing at 1010 °C (1850 °F). Automatic carbon-potential and atmosphere controllers have been designed for use at temperatures up to 1095 °C (2000 °F).

At present, the application of high-temperature carburizing is limited to a few special parts, some of which are made of martensitic stainless steel. The economic feasibility of the process cannot be fully evaluated until a continuous production unit can be compared directly with a conventional unit operating at normal carburizing temperatures and processing the same part.

Grain Growth. Limited laboratory data indicate that alloy carburizing steels such as 4617, 8620 and 9315 experience little or no grain growth when they are carburized at 980 to 1040 °C (1800 to 1900 °F) for periods of 1 to 4 h. Under similar conditions, plain carbon steels exhibit appreciably more grain coarsening. The effect of carburizing temperature on grain size in eight different steels (based on data obtained from two different laboratories) is shown in Table 11.

REFERENCES
1. Predicting Carburizing Data, by F. A. Still and H. C. Child: *Heat Treatment of Metals*, Vol 5, 1978, p 67-72
2. Calculation of the Equilibrium Composition of Blended Gases for Heat Treating Furnace Atmospheres, by C. A. Stickels: *Journal of Heat Treating*, Vol 1, 1979, p 31-41
3. "JANAF Thermochemical Tables, 2nd Edition", by D. R. Stull and H. Prophet: NSRDS-NBS 37, U.S. Dept. of Commerce, Washington, DC, June 1971
4. Thermodynamics and Phase Diagram of the Fe-C System, by J. Chipman: *Metall. Trans.*, Vol 3, 1972, p 55-64

Appendix

Diffusion of Carbon and Nitrogen

Carbon and nitrogen diffusion in iron is described by Fick's laws of diffusion. Fick's first law states that the flux of the diffusing substance, carbon for example, perpendicular to a plane of unit cross sectional area is proportional to the local carbon gradient perpendicular to the plane. The constant of proportionality is the diffusion constant D, which has the units (distance)²/time. Fick's second law is a material balance about an elemental volume of the system; the flux of carbon into an elemental volume of iron less the flux of carbon out of the elemental volume equals the rate of accumulation of carbon within the volume. Combining the two laws leads to a partial differential equation that describes the diffusion process. For one-dimensional diffusion:

$$\frac{\partial C}{\partial t} = D\frac{\partial^2 C}{\partial x^2} \qquad \text{(Eq 15)}$$

where C is the concentration of the diffusing substance, t is time and x is distance. If the diffusion constant is independent of the concentration of the diffusing substance, then solutions to the diffusion equation are available for a variety of boundary conditions (Ref 1). With solutions to the diffusion equation and values of the diffusion constant, it is possible to predict the carbon gradient and depth of penetration resulting from various carburizing heat treatments.

The diffusion constant of carbon in austenitic iron, however, does depend on the carbon content. If this dependence is incorporated into the diffusion equation, exact solutions are available only for steady-state problems

$$\left(\frac{\partial C}{\partial t} = 0\right)$$

and numerical methods must be used to find the concentration profile for non-steady-state conditions. For estimating case depth resulting from various carburizing treatments, the dependence of D on carbon content can be disregarded. An approximate relation for

Fig. 43 Effective case depth after carburizing treatment

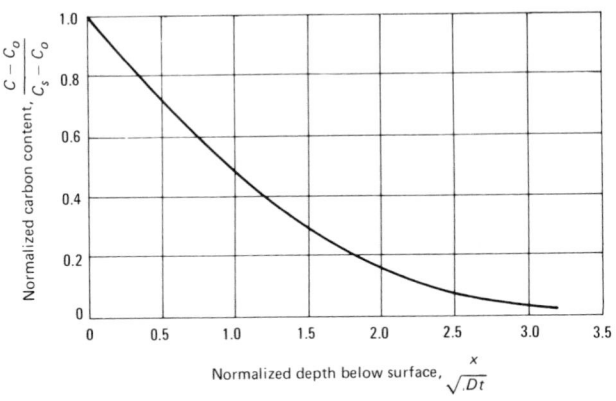

Fig. 44 Influence of surface curvature on case depth

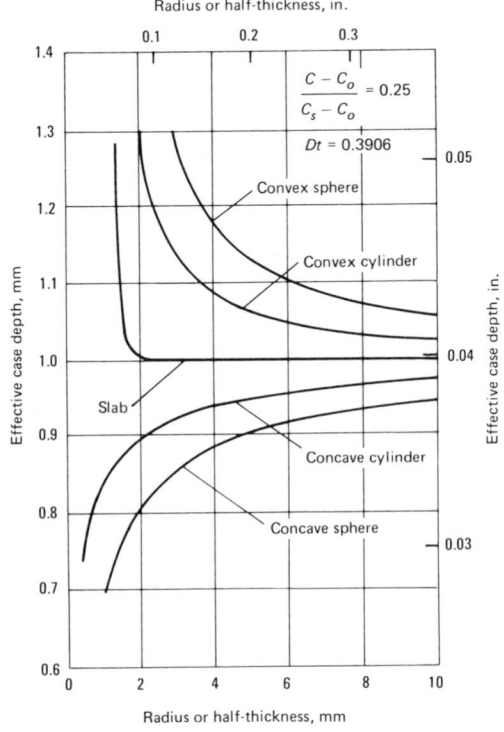

the temperature dependence of the diffusion constant for carbon in austenitic iron for temperatures from 800 to 1000 °C (1470 to 1830 °F) is

$$D(C, \gamma\text{-Fe}) = 16.2 \exp\left(\frac{-137\,800}{RT}\right) \text{mm}^2/\text{s}$$
(Eq 16)

where

$R = 8.314$ J/K·mol

and T is in degrees Kelvin (Ref 2 and 3).

The diffusion constant for carbon in ferritic iron for temperatures from 500 to 800 °C (930 to 1470 °F) given in Ref 4 is

$$D(C, \alpha\text{-Fe}) = 33.25 \exp\left(\frac{-103\,060}{RT}\right) \text{mm}^2/\text{s}$$
(Eq 17)

Rates of diffusion of nitrogen in austenite and ferrite are similar to those for carbon. For nitrogen in gamma iron, the diffusion constant given in Ref 5 is

$$D(N, \gamma\text{-Fe}) = 0.5 \exp\left(\frac{-77\,000}{RT}\right) \text{mm}^2/\text{s}$$
(Eq 18)

and for nitrogen in alpha iron (Ref 4):

$$D(N, \alpha\text{-Fe}) = 91 \exp\left(\frac{-168\,460}{RT}\right) \text{mm}^2/\text{s}$$
(Eq 19)

For carburizing, the most realistic solution to the diffusion equation is one for which it is assumed that the flux of carbon at the surface, J, is governed by a first order surface reaction:

$$J = -D\left(\frac{\partial C}{\partial x}\right)_s = k\,(C_a - C_s) \quad \text{(Eq 20)}$$

where $\left(\frac{\partial C}{\partial x}\right)_s$

is the carbon concentration gradient at the surface, C_s is the surface carbon content, C_a is the carbon potential of the furnace atmosphere and k is the surface reaction rate constant. This solution can allow for the fact that the rate of the surface reaction usually limits the rate of carbon pickup for short times, whereas the rate of carbon diffusion is rate limiting for longer times. Thus, the surface carbon content changes with time, increasing rapidly at the beginning of a carburizing treatment and increasing slowly at longer times. However, to estimate carbon gradients established after various carburizing treatments, values of k, the surface reaction rate coefficient, as well as values of D, must be known. Although some measurements of k have been made, there are no generally agreed upon values available. For estimating case depths it is sufficient to use a simpler, but less realistic, surface boundary condition for the diffusion equation. If it is assumed that the surface carbon content instantaneously takes on a value of C_s (often taken to be the maximum dissolved carbon content in austenite at the carburizing temperature) and maintains that value for the entire carburizing cycle, then the normalized carbon content:

$$C^* = \frac{C - C_0}{C_s - C_0}$$

where C_0 is the base carbon content of the alloy, is a function of the dimensionless variable

$$\left(\frac{x}{\sqrt{Dt}}\right)$$

shown in Fig. 43, where x is the distance below the surface. C^* decreases to 0.05 at a value of 2.75 for

$$\frac{x}{\sqrt{Dt}}$$

and this can be used as a definition of total case depth in mm:

Total case depth $= 2.75 \sqrt{Dt}$ (Eq 21)

where D is given by Eq 16, and t is time in seconds. Criteria similar to these were used by Harris in Ref 6 to develop his tables of total case depth as a function of time and temperature.

To determine the total case depth expected from a carburizing treatment of 8 h (28 800 s) at 927 °C (1700 °F or 1200 K) from Eq 16:

$$D = 16.2 \exp\left(\frac{-137\,800}{8.314 \times 1200}\right)$$
$$D = 1.626 \times 10^{-5} \text{ mm}^2/\text{s}$$

From Eq 21:

Total case depth
$= 2.75 \sqrt{1.626 \times 10^{-5} \times 0.288 \times 10^5}$
$= 1.88$ mm (0.074 in.)

In choosing heat treatments, one is often more concerned with effective case depth than with total case depth. In terms of carbon content, effective case depth usually is defined as the depth at which a carbon content of 0.4 wt% is reached. This can be estimated

using Fig. 43, as shown in the following example.

To determine the effective case depth resulting from a carburizing treatment of 8 h at 927 °C (1700 °F), if the base carbon content is 0.2 wt% and the surface carbon content is 1.0 wt%: when

$$C = 0.4$$

then

$$\frac{C - C_0}{C_s - C_0} = \frac{0.4 - 0.2}{1.0 - 0.2} = 0.25 = C^*$$

From Fig. 43: when

$$C^* = 0.25$$

then

$$\frac{x}{\sqrt{Dt}} = 1.6$$

because

$$D = 1.626 \times 10^{-5}$$

and

$$t = 28\,800 \text{ s}$$

therefore

$x = 1.6 \sqrt{1.626 \times 10^{-5} \times 28\,800}$
$x = 1.09$ mm (0.043 in.)

The same procedures that were used above for estimating carburized case depth can be used for estimating depth of decarburization, when steels are decarburized at temperatures where they are austenitic. For estimating depth of decarburization produced by tempering treatments, a different approach is needed because most of the carbon is combined as carbides, not dissolved in ferrite. The following expression for depth of total decarburization, that is, removal of all carbides due to tempering treatments at temperatures from 500 to 727 °C (930 to 1340 °F), is adapted from the solution to the moving boundary diffusion problem presented in Ref 1 for tarnishing reactions:

$$\tau = \sqrt{2\,D\,t\,C_1/C_2}$$ (Eq 22)

where τ is the depth of total decarburizaion in mm in tempering, D is the diffusion constant from Eq 17, t is the time in seconds, C_1 is the carbon content (dissolved) of ferrite in equilibrium with Fe_3C at the temperature of interest, and C_2 is the total carbon content of the alloy under consideration. The carbon solubility in ferrite from 400 to 727 °C (750 to 1340 °F) is

$$C_1 = 198 \exp\left(\frac{-9117.5}{T}\right) \text{ wt\%}$$ (Eq 23)

To determine the depth of total decarburization produced by tempering a 0.4% carbon steel in air for 2 h at 600 °C (1110 °F) from Eq 17:

$$D = 33.25 \exp\left(\frac{-103\,060}{8.314 \times 873}\right)$$
$$D = 2.272 \times 10^{-5} \text{ mm}^2/\text{s}$$

From Eq 23:

$$C_1 = 198 \exp\left(\frac{-9117.5}{873}\right) = 0.006 \text{ wt\%}$$

From Eq 22:

$\tau =$
$\overline{\sqrt{2 \times 2.272 \times 10^{-5} \times 7200 \times 0.006/0.4}}$
$\tau =$
0.07 mm (0.003 in.)

In atmospheres that are oxidizing as well as decarburizing, this method will overestimate the depth of decarburization present on the part after the oxide scale is removed.

The depth of penetration of nitrogen into austenitic iron during carbonitriding treatments can be estimated using Eq 18 for the diffusion constant of nitrogen and Eq 21 for total depth of penetration. The case depth formed in subcritical nitriding treatments is more difficult to predict because of the important role of nitride-forming alloying elements. For example, Eq 19, for the diffusion constant of nitrogen, and Eq 21 can be used to estimate the total depth of nitrogen penetration in plain carbon steels. Such steels are rarely of interest, however, because nitriding produces little hardening in the absence of nitride-forming alloys. For low-alloy steels such as 7140, 4140 and 8640 commonly used for nitriding, Eq 21 greatly overestimates the case depth. A better approach is to treat the diffusion as a "moving boundary" problem, such as was used for decarburization of ferrite-carbide aggregates (Ref 7). Then the estimated case depth becomes:

$$\tau, \text{ mm} = \sqrt{2\,Dt\,C_s/K}$$ (Eq 24)

where D is given by equation 19, t is time in seconds, C_s is the weight percent of dissolved nitrogen in equilibrium with Fe_4N and K is the weight percent nitrogen needed to combine with the alloying elements present. C_s and K are given by the following equations:

$$C_s = 12.3 \exp\left(\frac{-34\,700}{RT}\right) \text{ wt\%}$$ (Eq 25)

$$K = 14\left[\left(\frac{\text{wt\% Al}}{27}\right) + \left(\frac{\text{wt\% Cr}}{52}\right) + \left(\frac{\text{wt\% Mo}}{2 \times 96}\right) + \left(\frac{\text{wt\% V}}{51}\right)\right]$$

(Eq 26)

Equation 24 indicates that, as the quantity of nitride-forming elements present increases, case depth decreases; it provides a reasonable estimate of nitrided case depth for low-carbon alloy steels. However, because carbon as well as nitrogen can combine with Cr, Mo and V, Eq 24 underestimates case depth for medium- and high-carbon alloys.

Effect of Part Shape on Case Depth. The rules-of-thumb for case depth given in the previous section are derived from solutions to the diffusion equation for the concentration gradient beneath a plane surface on a solid of infinite magnitude. For solid slabs, with diffusion from both surfaces, the equations are good approximations for case depth as long as:

$$\frac{\sqrt{Dt}}{2l} < 0.2$$

where $2l$ is the thickness of the slab. Consequently, the thickness of the slab must be greater than twice the total case depth for the equations to apply. If this condition is not satisfied, the procedure outlined using Fig. 43 underestimates the effective case depth, and Eq 21 is meaningless.

The curvature of the surface influences case depth when the radius of curvature of the surfaces becomes comparable in magnitude to the case depth. For convex surfaces, case depth obtained from a given diffusion treatment is greater than that expected on plane surfaces. The case depth difference is greater for a doubly curved surface, such as a sphere, than it is on a singly curved cylindrical surface of the same radius.

Figure 44 illustrates the influence of surface curvature on case depth. For all five curves in Fig. 44, the value of Dt is the same; that is, the diffusion time and temperature are the same, and value of Dt chosen is

$$\frac{C - C_0}{C_s - C_0} = 0.25$$

at a depth of 1 mm (0.04 in.) for a slab of infinite thickness. This is approximately the effective case depth if the surface carbon content (C_s) is 1.0% and the base carbon content is 0.2%. These computed curves can be used to assess the influence of surface curvature as shown below.

Reading off the intercepts for $x = 3$ mm (0.1 in.):

	Effective case depth	
	mm	in.
Convex spherical surface, 3 mm (0.1 in.) radius	1.28	0.050
Convex cylindrical surface, 3 mm (0.1 in.) radius	1.13	0.044
Plane slab, 3 mm (0.1 in.) half-thickness	1.00	0.040
Concave cylindrical surface, 3 mm (0.1 in.) radius	0.93	0.037
Concave spherical surface, 3 mm (0.1 in.) radius	0.86	0.034

Note that, as the radius of curvature of the surfaces increases relative to the case depth, the effect of curvature on case depth diminishes.

Surface curvature is responsible for the variations in effective case depth found on carburized gears. When case depth is small compared to the curvature of tooth surfaces, case depth is uniform. However, when case depth becomes comparable in magnitude to the radius of curvature of the tooth profile, case depth is always greatest at the tip of the tooth (convex surface) and least at the root of the tooth (concave surface), with an intermediate depth at the pitch line.

Effect of Alloying Elements on Diffusion. The alloying elements present in low-alloy carburizing steels affect, to some degree, the carbon concentration gradient obtained from a given carburizing treatment. Alloying elements affect maximum carbon solubility and the diffusion constant of carbon. Some alloying elements, such as manganese and molybdenum, have little effect on the solubility of carbon in austenite. Others, such as nickel and chromium, lower the solubility; thus for a given carburizing treatment, lower carbon contents at any given case depth are found than would be present in a binary iron-carbon alloy. If carbides form at the surface during carburizing, nickel partitions preferentially to austenite and chromium partitions preferentially to carbide. As a result, the solid solubility of carbon in austenite adjacent to carbides is depressed further in nickel-bearing alloys, but is raised in chromium-bearing alloys. When steels contain more than one major alloying element, the effect on carbon solubility often is difficult to predict.

The effect of alloying elements on the diffusion constant of carbon in austenite (considered as pseudobinary diffusion) has been measured for several ternary systems in Ref 8. For most low-alloy steels, the change in diffusion constant is of the same order or less than the dependence of D on carbon content. For carburizing steels containing more than one major alloying element, there is no straightforward way to compute a correction to the diffusion constant.

To select a carburizing cycle to produce a certain desired carbon concentration gradient in a low-alloy steel, the best procedure is to choose the carburizing temperature, time and atmospheric carbon potential based on the binary iron-carbon system. Parts of the alloy steel then can be processed and analyzed, on a trial basis. Based on these results, minor alterations can be made in the processing cycle.

REFERENCES

1. *The Mathematics of Diffusion*, 2nd Ed., by J. Crank: Oxford University Press, London, 1975
2. The Diffusivity of Carbon in Iron by the Steady-State Method, by R. P. Smith: *Acta Met.,* Vol 1, 1953, p 578–587
3. Diffusion Coefficient of Carbon in Austenite, by C. Wells, W. Batz and R. F. Mehl: *Trans. AIME,* Vol 188, 1950, p 553–560
4. Diffusion of Interstitial Impurities, by D. N. Beshers: in *Diffusion,* American Society for Metals, Metals Park, OH, 1973, p 209–240
5. Kinetics of Reaction of Gaseous Nitrogen with Iron Part 1: Kinetics of Nitrogen Solution in Gamma Iron, by P. Grieveson and E. T. Turkdogan: *Trans. Met. Soc. AIME,* Vol 230, 1964, p 407–414
6. Case Depth—An Attempt at a Practical Definition, by F. E. Harris: *Metal Progress,* Vol 44, 1943, p 265–272
7. The Nitriding of Iron and Alloy Steels, by K. H. Jack: in *High Temperature Gas-Metal Reactions in Mixed Environments,* edited by S. A. Jansson and Z. A. Foroulis, Proceedings of the Symposium held in Boston, MA, May 9–10, 1972, The Metallurgical Society of AIME, NY, 1973, p 182–195
8. *Metals Reference Book,* 5th Edition, edited by C. J. Smithells: Butterworths, London, p 908–909

Carbonitriding

By the ASM Committee on
Gas Carburizing, Carbonitriding and
Nitriding*

CARBONITRIDING is a modified form of gas carburizing, rather than a form of nitriding. The modification consists of introducing ammonia into the gas carburizing atmosphere to add nitrogen to the carburized case as it is being produced. Nascent nitrogen forms at the work surface by the dissociation of ammonia in the furnace atmosphere; the nitrogen diffuses into the steel simultaneously with carbon. Typically, carbonitriding is carried out at a lower temperature and for a shorter time than gas carburizing, producing a shallower case than is usual in production carburizing.

In its effects on steel, carbonitriding is similar to liquid cyaniding. Because of problems in disposing of cyanide-bearing wastes, carbonitriding is often preferred over liquid cyaniding. In terms of case characteristics, carbonitriding differs from carburizing and nitriding in that (a) carburized cases normally do not contain nitrogen, and (b) nitrided cases contain nitrogen primarily, whereas carbonitrided cases contain both.

Carbonitriding is used primarily to impart a hard, wear-resistant case, generally from 0.075 to 0.75 mm (0.003 to 0.030 in.) deep. A carbonitrided case has better hardenability than a carburized case (nitrogen increases hardenability of steel; it also is an austenite stabilizer, and high nitrogen levels can result in retained austenite—particularly in alloy steels). Consequently, by carbonitriding and quenching, a hard-

ened case can be produced at less expense within the case-depth range indicated, using either carbon or low-alloy steel. Full hardness with less distortion can be achieved with oil quenching or, in some instances, even gas quenching, employing a protective atmosphere as the quenching medium.

Steels commonly carbonitrided include those in the 1000, 1100, 1200, 1300, 1500, 4000, 4100, 4600, 5100, 6100, 8600 and 8700 series, with carbon contents up to about 0.25%. Also, many steels in these same series with a carbon range of 0.35 to 0.50% are carbonitrided to case depths up to about 0.3 mm (0.01 in.) when a combination of a reasonably tough, through-hardened core and a hard, long-wearing surface is required (shafts and transmission gears are typical examples).

Medium-carbon parts of carbon or alloy steels are often heated for hardening in a carbonitriding atmosphere to produce a shallow case of higher hardness and greater resistance to wear than could be provided by conventional hardening alone. Steels such as 4140, 5140, 8640 and 4340 for applications like heavy-duty gearing are treated by this method at 845 °C (1550 °F).

Occasionally, parts are first carburized to depths of 1.5 to 1.8 mm (0.060 to 0.070 in.) or more, then reheated in a carbonitriding atmosphere and oil quenched. This produces an extremely hard surface with exceptionally good polishability, for long wear under heavy loading.

Applications

Although carbonitriding is a modified carburizing process, its applications are more restricted than those of carburizing. As has been stated previously, carbonitriding is largely limited to case depths of about 0.75 mm (0.03 in.) or less, while no such limitation applies to carburizing. Two reasons for this are: (a) carbonitriding is generally done at temperatures of 870 °C (1600 °F) and below, whereas, because of the time factor involved, deeper cases are produced by processing at higher temperatures; and (b) the nitrogen addition is less readily controlled than is the carbon addition, a condition that can lead to an excess of nitrogen and, consequently, to high levels of retained austenite and case porosity when processing times are too long.

The resistance of a carbonitrided surface to softening during tempering is markedly superior to that of a carburized surface. Other notable differences exist in terms of residual-stress pattern, metallurgical structure, fatigue and impact strength at specific hardness levels, and effects of alloy composition on case and core characteristics.

Various production parts that have been successfully carbonitrided are listed in Table 1. A review of this list suggests the range of applications for which carbonitriding has been found advantageous. Included in Table 1 are identifications of steels that are being carbonitrided on a production basis,

*Charles A. Stickels, *Chairman*, Principal Staff Engineer, Ford Motor Co.; Larry E. Byrnes, Manufacturing Research Manager, Federal Mogul Corp.; Jon L. Dossett, Vice President Metallurgy, U.S. Gear Corp.; Robert L. Hughes, Chief Metallurgist, Fairfield Manufacturing Co. Inc.; Kenneth D. Gladden, Senior Development Engineer, Caterpillar Tractor Co.; Dale A. Poteet, Jr., Division Engineer, John Deere Component Works; Joseph A. Riopelle, District Manager/ Chicago Office, Surface Division, Midland-Ross Corp.; Ronald D. Rogers, Chief Engineer, Materials Engineering, Rockwell International

Table 1 Typical applications and production cycles for carbonitriding

Part	Steel	Case depth mm	Case depth 0.001 in.	Furnace temperature °C	Furnace temperature °F	Total time in furnace	Quench
Carbon steels							
Adjusting yoke, 25 by 9.5 mm (1 by 0.37 in.) 1020	1020	0.05 to 0.15	2 to 6	775 and 745	1425 and 1375	64 min	Oil
Ball case, 22 g (0.8 oz.) 1010	1010	0.75	30	870	1600	5½ h	Oil
Bearing block, 64 by 32 by 3.2 mm (2.5 by 1.3 by 0.13 in.). 1010	1010	0.05 to 0.15	2 to 6	775 and 745	1425 and 1375	64 min	Oil
Brake-operating cam, 76 by 6.4 mm (3 by 0.25 in.). 1010	1010	0.05 to 0.15	2 to 6	775 and 745	1425 and 1375	64 min	Oil
Bushing, 14 g (0.5 oz.) 1113	1113	0.25 to 0.35	10 to 14	845	1550	1⅔ h	Oil
Cam, 2.3 by 57 by 64 mm (0.1 by 2.25 by 2.5 in.). 1010	1010	0.38 to 0.45	15 to 18	855	1575	2½ h	Oil
Chisel point, 16 g (0.56 oz) 1020	1020	0.13 to 0.20	5 to 8	790	1450	1 h	Oil
Cup, 13 g (0.46 oz) 1015	1015	0.08 to 0.13	3 to 5	790	1450	½ h	Oil
Distributor drive shaft, 127 mm-OD by 127 mm (5 by 5 in.) 1015	1015	0.15 to 0.25	6 to 10	815 and 745	1500 and 1375	108 min	Gas(a)
Front-axle bearing washer, 66.7-mm-OD by 3.2 mm (2.625 by 0.125 in.) 1010	1010	0.15 to 0.25	6 to 10	815 and 745	1500 and 1375	108 min	Gas(a)
Gear, 44.5-mm diam by 3.2 mm (1.75 by 0.125 in.) . . . . 1213(b)	1213(b)	0.30 to 0.38	12 to 15	855	1575	1¾ h	Oil(c)
Hex nut, 60.3 by 9.5 mm(2.4 by 0.37 in.). . . . 1030	1030	0.15 to 0.25	6 to 10	815 and 745	1500 and 1375	64 min	Oil
Hinge pin, 14-mm OD by 89 mm (0.55 by 3.5 in.). 1113	1113	0.05 to 0.15	2 to 6	775 and 745	1425 and 1375	64 min	Oil
Hood-latch bracket, 6.4-mm diam (0.25 in.). . 1015	1015	0.05 to 0.15	2 to 6	775 and 745	1425 and 1375	64 min	Oil
Link, 2 by 38 by 38 mm (0.079 by 1.5 by 1.5 in.). 1022	1022	0.30 to 0.38	12 to 15	855	1575	1½ h	Oil
Mandrel, 40 g (1.41 oz) 1117	1117	0.20 to 0.30	8 to 12	845	1550	1½ h	Oil
Ring-release lock, 23 g (0.81 oz). 1010	1010	0.25 to 0.38	10 to 15	845	1550	110 min	Oil
Segment, 1 by 64 mm (0.04 by 2.5 in.) 1010	1010	0.18 to 0.25	7 to 10	845	1550	1½ h	Gas(a)
Segment, 2.3 by 44.5 by 44.5 mm (0.09 by 1.75 by 1.75 in.) 1010	1010	0.38 to 0.45	15 to 18	855	1575	2½ h	Oil
Shaft, 4.7-mm diam by 159 mm (0.19 by 6.25 in.). . . . . 1213(b)	1213(b)	0.30 to 0.38	12 to 15	815	1500	2½ h	Gas(a)(e)
Shift collar, 59 g (2.1 oz) 1118	1118	0.30 to 0.36	12 to 14	775	1430	5½ h	Oil(f)
Slide, 1.9 by 38 by 197 mm (0.075 by 1.5 by 7.75 in.). 1010	1010	0.30 to 0.38	12 to 15	815	1500	2½ h	Gas(a)
Sliding spur gear, 66.7-mm OD (2.625 in.) . . 1018	1018	0.38 to 0.50	15 to 20	870	1600	2 h(g)	Oil(h)
Spring pin, 14.3-mm OD by 114 mm (0.56 by 4.5 in.). 1030	1030	0.25 to 0.50	10 to 20	815 and 745	1500 and 1375	144 min	Oil
Spur pinion shaft, 41.3-mm OD (1.625 in.) . . 1018	1018	0.38 to 0.50	15 to 20	870	1600	2 h(g)	Oil(j)
Transmission shift fork, 127 by 76 mm (5 by 3 in.) 1040	1040	0.25 to 0.50	10 to 20	815 and 745	1500 and 1375	162 min	Gas(a)
Transmission slide shaft, 222-mm OD by 171 mm (8.75 by 6.75 in.). . 1030	1030	0.15 to 0.25	6 to 10	815 and 745	1500 and 1375	96 min	Oil
Alloy steels							
Helical gear, 82-mm OD (3.23 in.). 8617H	8617H	0.50 to 0.75	20 to 30	845	1550	6 h(g)	Oil(h)
Hub, 50-mm OD (2 in.) 8622	8622	0.15 to 0.25	6 to 10	815 and 745	1500 and 1375	108 min	Gas(a)
Input shaft, 1.2 kg (2.6 lb) 5140	5140	0.30 to 0.35	12 to 14	775	1430	5½ h	Oil(f)
Pinion gear, 0.2 kg (0.44 lb). 4047	4047	0.30 to 0.35	12 to 14	775	1430	5½ h	Oil(f)
Ring gear, 0.9 kg (2 lb) 4047	4047	0.20 to 0.30	8 to 10	760	1400	9 h	Oil(k)
Segment, 1.4 by 83 mm (0.055 by 3.27 in.). . 8617	8617	0.18 to 0.25	7 to 10	815	1500	1½ h	Gas(a)
Spur pinion shaft, 63.5-mm OD by 203 mm (2.5 by 8 in.) 5140H	5140H	0.05 to 0.20	2 to 8	845	1550	1 h(g)	Oil(m)
Stationary gear plate, 0.32 kg (0.7 lb) 5140	5140	0.30 to 0.35	12 to 14	775	1430	5½ h	Oil(f)
Transmission main shaft sleeve, 38-mm OD by 25 mm (1.5 by 2 in.). 8622	8622	0.15 to 0.25	6 to 10	815 and 745	1500 and 1375	108 min	Gas(a)
Transmission main shaft washer, 57-mm OD by 6.4 mm (2.25 by 0.25 in.) . . . 8620	8620	0.25 to 0.50	10 to 20	815 and 745	1500 and 1375	162 min	Gas(a)

(a) Modified carbonitriding atmosphere. (b) Leaded. (c) Tempered at 190 °C (375 °F). (d) Tempered at 510 °C (950 °F). (e) Tempered at 150 °C (300 °F). (f) Tempered at 165 °C (325 °F). (g) Time at temperature. (h) Oil at 150 °C (300 °F); tempered at 150 °C (300 °F); for 1 h. (j) Oil at 150 °C (300 °F); tempered at 260 °C (500 °F) for 1 h. (k) Tempered at 175 °C (350 °F). (m) Oil at 150 °C (300 °F); tempered at 230 °C (450 °F) for 2 h.

Fig. 1 Effect of ammonia additions on nitrogen and carbon potentials determined using low carbon steel foil

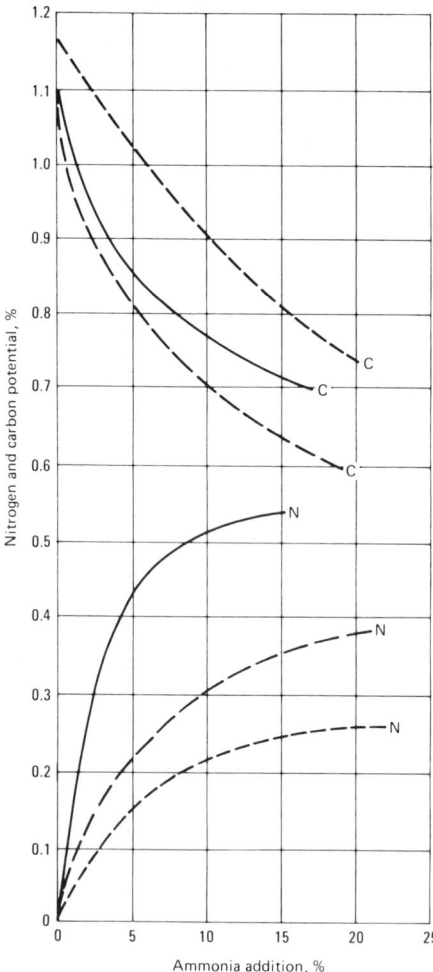

For three sets of conditions: solid lines, 3 h at 850 °C (1560 °F) and 0.29% CO₂; broken lines, 1 h at 925 °C (1695 °F) and 0.13% CO₂; dotted lines, 1 h at 950 °C (1740 °F) and 0.10% CO₂ (Ref 3)

together with carbonitriding temperatures, quenching media, tempering temperatures and case depths.

For many applications, carbonitriding the less expensive steels will provide properties equivalent to those obtained in gas carburized alloy steels. The following two examples illustrate choices of carbonitrided carbon steels for applications ordinarily requiring alloy steels.

Example 1. Carbonitrided 1117 steel was employed for a 40.8-cm- (16-in.-) long paper-cutting tool. Case depth was 0.75 mm (0.03 in.); hardness, 58 to 62 HRC.

Example 2. Camshaft eccentrics stamped from 1010 aluminum-killed steel 4 mm (0.16 in.) thick were carbonitrided to a depth of 0.25 to 0.50 mm (0.01 to 0.02 in.) and a minimum surface hardness equivalent to 58 HRC. The required metallurgical properties for this application could have been attained also with carbonitrided or gas carburized 1011 or 1016 steel. However, the 1010 steel had the advantage of facilitating the stamping operation because of its lower carbon and manganese contents.

Composition of Case

The composition of a carbonitrided case depends on temperature, time, atmosphere composition and type of steel. The higher the carbonitriding temperature, the less effective is the ammonia addition to the atmosphere as a nitrogen source, because the rate of spontaneous decomposition of ammonia to molecular nitrogen and hydrogen increases as the temperature is raised. Figure 1 shows that lower temperatures favor increased surface nitrogen concentrations. The addition of ammonia to a carburizing atmosphere has the effect of dilution by the following reaction:

$$2\,NH_3 \longrightarrow N_2 + 3\,H_2$$

Thus, as shown in Fig. 1, the carbon potential possible with a given carbon dioxide level is higher in a carburizing atmosphere than in a carbonitriding atmosphere. Dilution with nitrogen and hydrogen affects measurements of oxygen potential in a similar manner; the carbon potential possible with a given oxygen potential is higher in a carbonitriding atmosphere than in a carburizing atmosphere. Water vapor content, however, is much less affected by this dilution. Thus, the amount of dilution and its resulting effect on the atmosphere composition depends on the processing temperature, the amount of ammonia introduced, and the ratio of the total atmosphere gas flow rate to the volume of the furnace.

Carbonitriding can be carried out at such low temperatures as to produce a "compound layer", so called because iron-carbon-nitrogen compounds are formed at the surface. In certain wear applications, this type of case structure

is suitable. To produce this layer of compound, large percentages of ammonia are required. It is usually unnecessary to liquid quench parts carbonitrided in this manner. However, because the diffusion rate of nitrogen and the rate of formation of the compound are so slow at temperatures below 705 °C (1300 °F), such practice is economically applicable only to shallow cases in applications in which dimensional tolerances would be difficult to maintain if the parts were treated at higher temperatures.

Figure 2 shows carbon and nitrogen gradients and case hardness data for 1018 carbon steel and 8620 low-alloy steel that were carbonitrided for 4 h at 845 °C (1550 °F) in a batch-type radiant-tube furnace. These test data were obtained in a manufacturing plant under normal production conditions, employing a standard carbonitriding cycle. All test specimens were carbonitrided along with production loads of 22.7 kg (50 lb) of gears and shafts.

The carbonitriding atmosphere was controlled by an infrared control unit and consisted of endothermic gas at 14.2 m³/h (500 ft³/h), ammonia at 0.7 m³/h (24 ft³/h), propane at 0.007 to 0.021 m³/h (0.25 to 0.75 ft³/h), and 0.32 to 0.34% carbon dioxide. The dew point of the atmosphere was maintained at −7 to −6 °C (19 to 21 °F) throughout the carbonitriding cycle. All specimens were quenched from the carbonitriding temperature (845 °C, or 1550 °F) into warm oil at 55 °C (130 °F); they were neither tempered nor subjected to subzero treatment.

As the dew point of a carbonitriding atmosphere is increased, carbon concentration decreases and nitrogen concentration remains fairly constant. This characteristic response is demonstrated by the data in Fig. 3 for a 1020 steel carbonitrided at 845 °C (1550 °F) for 4 h, which show that, with the ammonia content of the carbonitriding atmosphere set at high (5%) and low (1%) levels, an increase in the concentration of water vapor in the inlet gas lowered the profiles of carbon concentration but did not appreciably affect nitrogen concentrations. (Although the dew point of the atmosphere in the carbonitriding furnace, rather than the dew point of the inlet gas, is the controlling factor, data based on the inlet-gas dew point demonstrate the general effects of raising and lowering dew point on case composition.)

Fig. 2 Carbon, nitrogen and hardness gradients for carburized 1018 and 8620 steels

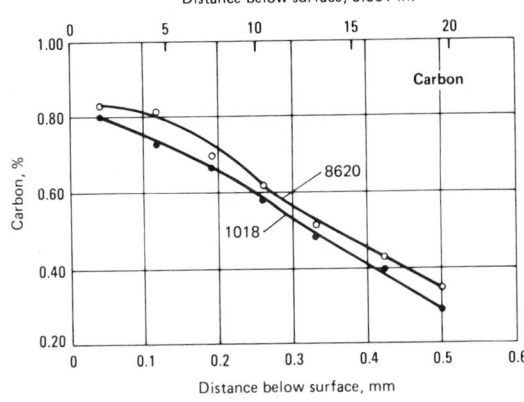

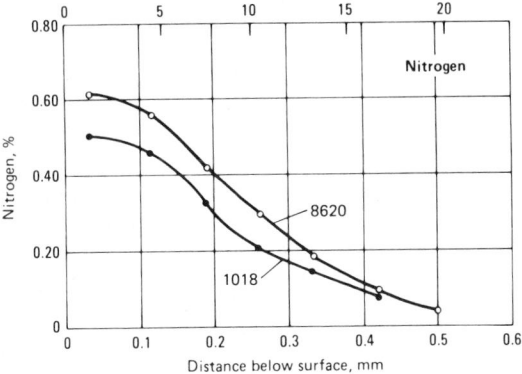

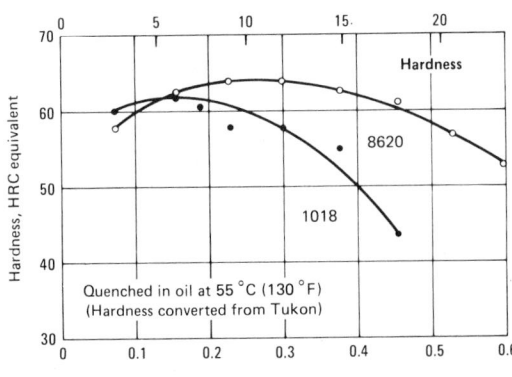

For processing details, see text.

Table 2 Effect of material/planning variables on the possibility of void formation in carbonitrided cases

All other variables held constant. Based on data presented in Fig. 4 for void formation in low-carbon foils (approximately 1010 steel) and data presented in Ref 3 and 4 on this subject

Material/processing variables	Possibility of void formation
Temperature increase	Increased
Longer cycles.	Increased
Higher case nitrogen levels	Increased
Higher case carbon levels	Increased
Aluminum killed steel.	Increased
Increased alloy content of steel .	Decreased
Severe prior cold working of material	Increased
Ammonia addition during heat up cycle	Increased

for avoiding or eliminating porosity problems.

None of the data included in this chapter were obtained from parts or test pieces containing case porosity. The successful applications listed in Table 1 show that carbonitriding can be successfully applied to a wide variety of parts and materials utilizing a wide range of processing parameters.

Hardness Gradients

Hardness at various levels in the case depends on the microstructure. Hardness gradients associated with the microstructures of 1117 steel are presented in Fig. 5. When the carbonitriding atmosphere was relatively high in ammonia (11% NH₃), the nitrogen content of the case was high, and enough austenite was retained after quenching to lower the hardness to 510 HK (48 HRC), 500-g load, at a depth of 0.025 mm (0.001 in.) below the surface. The amount of retained austenite was decreased, and hardness consequently increased, either by lowering the ammonia flow rate from 0.57 to 0.14 m³/h (20 to 5 ft³/h), which reduced the ammonia content of the furnace atmosphere from 11 to 3%, or by introducing a 15-min diffusion period at the end of the carbonitriding operation. Either treatment increased the hardness to meet or exceed a specified minimum value of

Void Formation

Subsurface voids or porosity in the case structure (Fig. 4) may occur in carbonitrided parts if the processing conditions are not adjusted properly. Although details of the mechanism of void formation are not completely understood, this problem has been related to excessive ammonia additions. Table 2 summarizes the factors that have been shown singly or in combination to contribute to void formation. No attempt has been made to quantify the interaction of the material and process variables presented in Table 2. Rather, this information should be used as a guide

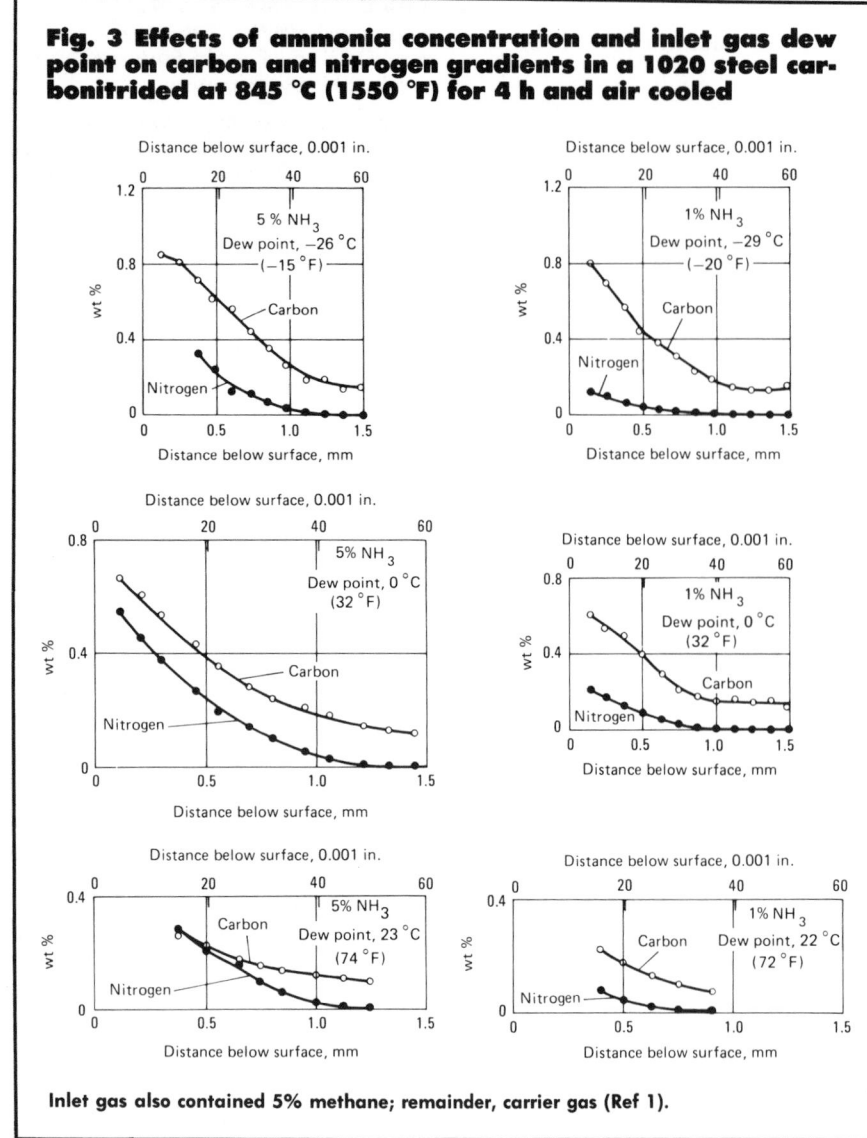

Fig. 3 Effects of ammonia concentration and inlet gas dew point on carbon and nitrogen gradients in a 1020 steel carbonitrided at 845 °C (1550 °F) for 4 h and air cooled

Inlet gas also contained 5% methane; remainder, carrier gas (Ref 1).

630 HK (55 HRC), 500-g load at 0.025 mm below the surface.

Similar data relating ammonia content to hardness for 1018 steel carbonitrided at 790 °C (1455 °F) for 2½ h and at 845 °C (1550 °F) for 2½ h are shown in Fig. 6.

Hardenability of Case

One major advantage of carbonitriding is that the nitrogen absorbed during processing lowers the critical cooling rate of the steel. That is, the hardenability of the case is significantly greater when nitrogen is added by carbonitriding than when the same steel is carburized only (Fig. 7). This permits the use of steels on which uniform case hardness ordinarily could not be obtained if they were carburized only and quenched. Where core properties are not important, carbonitriding permits the use of low-carbon steels, which cost less and may have better machinability.

Because of the hardenability effect of nitrogen, carbonitriding makes it possible to oil quench steels such as 1010, 1020 and 1113 to obtain martensitic case structures. Because of lower processing temperatures and/or the use of less severe quenches, carbonitriding may produce less part distortion and better control of dimensions than carburizing, and thus may eliminate the need for straightening or final grinding operations.

Depth of Case

Preferred case depth is governed by service application and by core hardness. Case depths of 0.025 to 0.075 mm (0.001 to 0.003 in.) are commonly applied to thin parts that require wear resistance under light loads. Case depths up to 0.75 mm (0.030 in.) may be applied to parts (such as cams) for resisting high compressive loads. Case depths of 0.63 ot 0.75 mm (0.025 to 0.030 in.) may be applied to shafts and gears that are subjected to high tensile or compressive stresses caused by torsional, bending or contact loads.

Medium-carbon steels with core hardnesses of 40 to 45 HRC normally require less case depth than steels with core hardnesses of 20 HRC or below. Low-alloy steels with medium carbon content, such as those used in automotive transmission gears, are often assigned case depths of 0.01 to 0.015 mm (0.004 to 0.006 in.).

Measurements of the case depths of carbonitrided parts may refer to effective case depth or total case depth, as with reporting case depths for carburized parts. For very thin cases, usually only the total case depth is specified. In general, it is easy to distinguish case and core microstructures in a carbonitrided piece, particularly when the case is thin and is produced at a low carbonitriding temperature; more difficulty is encountered in distinguishing case and core when high temperatures, deep cases, and medium-carbon or high-carbon steels are involved.

Effect of Time and Temperature. Based on a survey of industrial practice, Fig. 8 shows case depths for different combinations of total furnace treating time and temperature.

Figure 9(a) shows the effects of total furnace time on case depth for 1020 steel. Specimens were heated to 705, 760, 815 and 870 °C (1300, 1400, 1500 and 1600 °F) for periods of 15, 30 and 45 min. Figure 9(b) indicates the total case depths that can be obtained on an 1112 steel held for 15 min at various temperatures between about 750 and 900 °C (1380 and 1650 °F). All data in Fig. 9 were obtained in a single plant.

Case-depth uniformity in carbonitriding depends on temperature uniformity within the furnace chamber, adequate circulation and replenishment of atmosphere, and distribution of the fur-

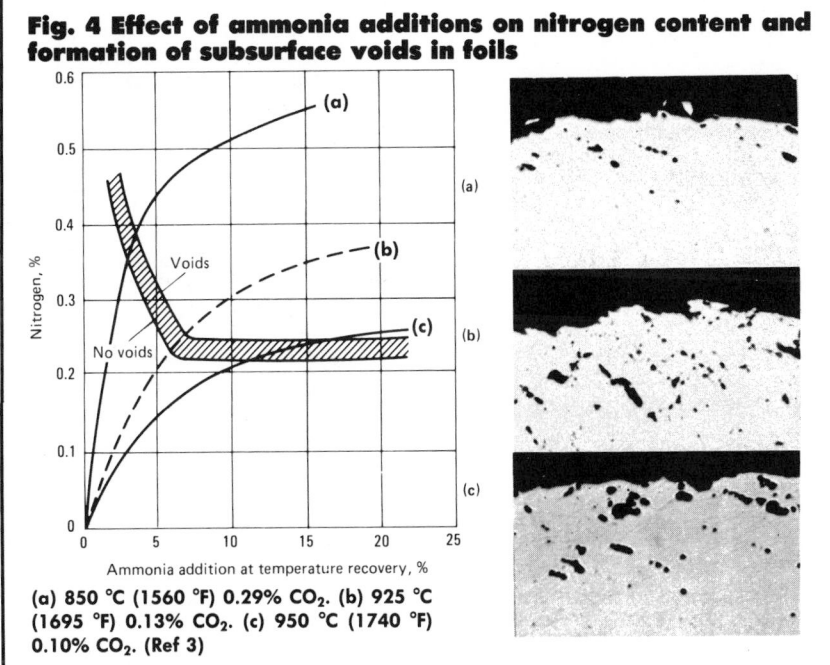

Fig. 4 Effect of ammonia additions on nitrogen content and formation of subsurface voids in foils

(a) 850 °C (1560 °F) 0.29% CO_2. (b) 925 °C (1695 °F) 0.13% CO_2. (c) 950 °C (1740 °F) 0.10% CO_2. (Ref 3)

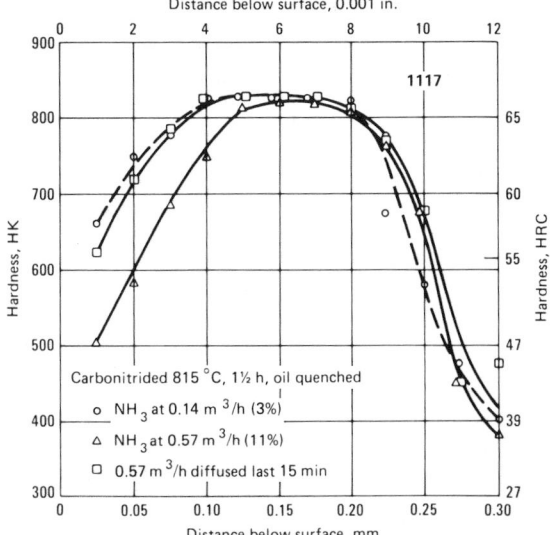

Fig. 5 Hardness gradients in 1117 steel carbonitrided at 815 °C (1500 °F) for 1½ h and quenched in oil

Carbonitrided 815 °C, 1½ h, oil quenched
o NH_3 at 0.14 m³/h (3%)
△ NH_3 at 0.57 m³/h (11%)
□ 0.57 m³/h diffused last 15 min

Required minimum hardness of 630 HK at 0.025 mm below surface was met by reducing the percentage and flow rate of ammonia or by adding a diffusion period after carbonitriding, as indicated. Atmosphere consisted of endothermic carrier gas (dew point, −1 °C) at 4.25 m³/h (150 ft³/h), natural gas at 0.17 m³/h, (6 ft³/h), and ammonia in the amounts indicated.

nace charge so that it is uniformly exposed to the atmosphere.

Example 3. Case-depth variations typical of carbonitriding at 775 to 800 °C (1425 to 1475 °F) are shown in Fig. 10. The data were obtained on two parts that were carbonitrided along with large production lots. One of these parts, a rack made of 1010 steel, was carbonitrided at 790 to 800 °C (1450 to 1475 °F) in a horizontal batch furnace equipped with an enclosed quench tank. Acceptable limits of case depth for these racks were 0.05 to 0.13 mm (0.002 to 0.005 in.).

The other part, a pinion shaft of 5140 steel, was carbonitrided at 775 °C (1425 °F) for 8 h and then quenched in oil at about 75 °C (170 °F). Attainment of acceptable case depth of 0.2 to 0.3 mm (0.008 to 0.012 in.) was 100% in 25 tests.

Although the data in Example 3 and Fig. 10 may be considered typical, they do not fully reflect the high degree of uniformity of case depth that can be achieved. For example, one plant reports total case-depth uniformity of ±0.025 mm (±0.001 in.) within a load and between loads in more than 25 000 cycles for case depths as low as 0.125 mm (0.005 in.) on small parts carbonitrided in large batch furnaces. Time, temperature and processing variables were automatically controlled.

Furnaces

Almost any furnace suitable for gas carburizing can be adapted to carbonitriding. If dense loads are to be processed, the furnace must be equipped with a fan to circulate the atmosphere. With shallow or openly spaced work loads, fan circulation of the atmosphere may not be required. For work that is to be clean and bright after quenching, the furnace must be equipped with protective-atmosphere vestibules to the quench area.

The various types of furnaces are described in the article entitled "Gas Carburizing" in this volume.

Atmosphere Constituents

The atmospheres used in carbonitriding generally comprise a mixture of carrier gas, enriching gas and ammonia. Basically, the atmospheres used in carbonitriding are produced by adding from about 2 to 12% ammonia to a standard gas carburizing atmosphere.

Ammonia used for carbonitriding is anhydrous ammonia of 99.9 + % purity. Grade designations such as "Premium", "Refrigeration" and "Metallurgical" are used to specify suitable material. Grades known as "Commercial" and "Agricultural" contain appreciable amounts of carbon dioxide, water and oils, which prohibit their use in furnace atmospheres.

Most ammonia is produced from natural gas, so costs and availability are related to natural gas supplies.

Fig. 6 Effect of ammonia content of carbonitriding gas on hardness gradient

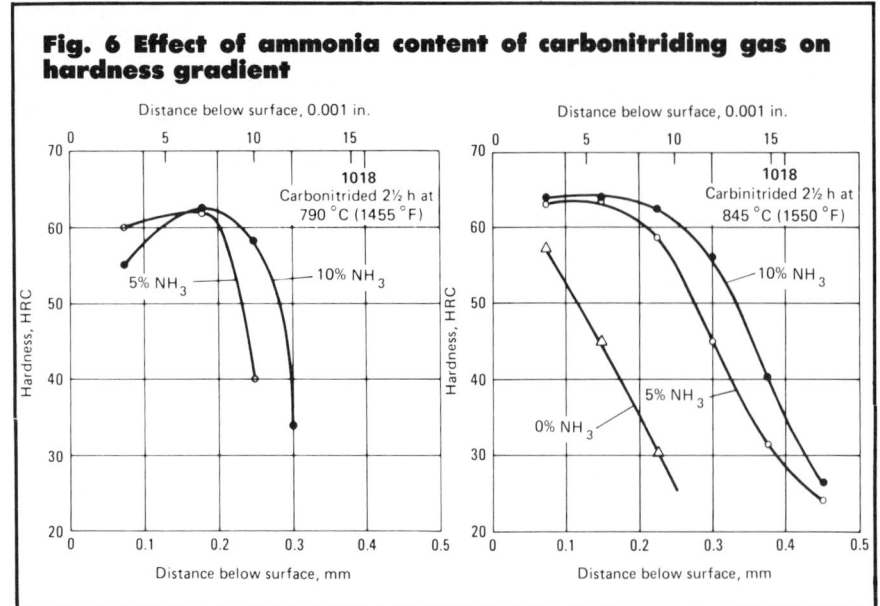

Fig. 8 Results of a survey of industrial practice regarding effects of time and temperature on depth of carbonitrided cases

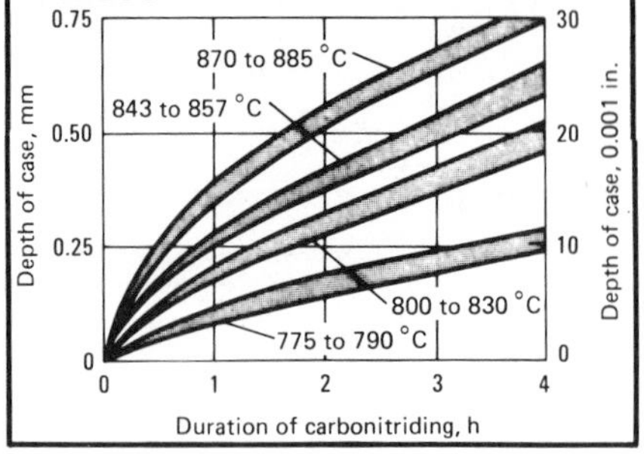

Fig. 7 End-quench hardenability curve for 1020 steel carbonitrided at 900 °C (1650 °F) compared with curve for the same steel carburized at 925 °C (1700 °F)

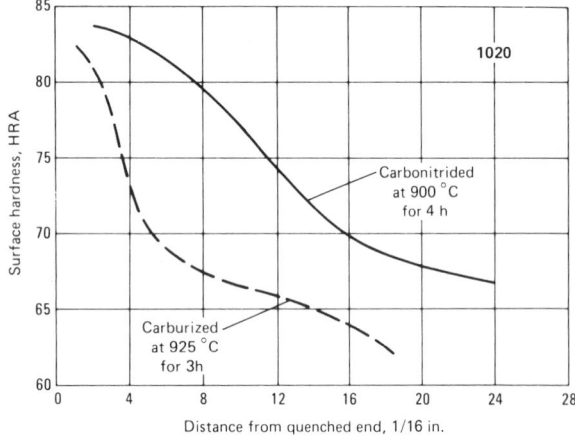

Hardness was measured along the surface of the as-quenched hardenability specimen. Ammonia and methane contents of the inlet carbonitriding atmosphere were 5%; remainder, carrier gas (Ref 2).

Control of Atmospheres

The three types of gases that comprise carbonitriding atmospheres usually are measured through flowmeters and may be premixed just before they enter the furnace. Large continuous furnaces require addition of the gas mixture at several points, to provide the desired composition within the chamber. Control of the atmosphere usually is obtained by producing a carrier gas that is as constant in chemical composition and dew point as practical, and by varying the enriching gas and ammonia either manually or automatically to give the desired carbon and nitrogen composition in the carbonitrided case. In varying the enriching gas, care must be taken not to introduce excessive amounts of high hydrocarbons into the furnace, because this will cause sooting. Heavy deposits of soot, in addition to making the work difficult to clean, have a detrimental effect on alloy furnace parts and may impede the rate of carbonitriding.

Ammonia Content. Among factors to be considered in establishing ammonia content for the influent gas are:

- *Atmosphere Turnover.* Frequently, lower ammonia percentages can be used with high rates of furnace gas change (that is, total gas flow divided by furnace chamber volume) to produce a specific microstructure for a specific furnace and load.
- *Recirculation Rate.* Generally, increases in recirculation rate, and hence in atmosphere uniformity, permit the use of lower ammonia concentrations.
- *Furnace-Cycle Time.* The rate of absorption of nitrogen by austenite decreases with time; hence, short furnace cycles and low case depths require higher ammonia percentages.
- *Furnace Temperature.* Higher ammonia percentages often must be used with higher furnace temperatures.
- *Load Size, Density and Surface Area.* As size, density and area of load increase, a higher rate of flow of ammonia is required for satisfactory hardenability.
- *Carbon Potential.* Lower ammonia percentages can normally be used with higher carbon potentials.
- *Type of Steel.* The minimum percentage of ammonia required for hardenability depends largely on steel composition. Austenite retention increases with an increase in alloy content. Therefore, lower ammonia

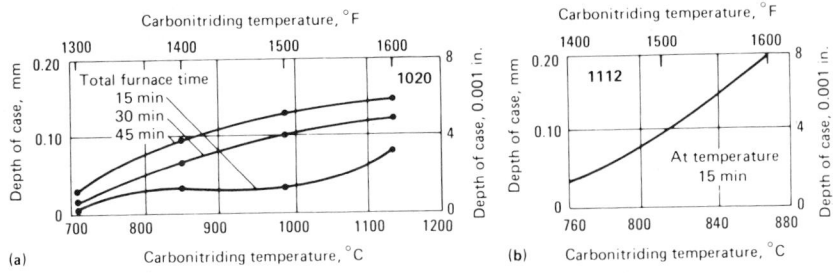

Fig. 9 Effects of temperature and of duration of carbonitriding on depth of case

Both sets of data were obtained in the same plant. (a) 1020 steel at total furnace time. (b) 1112 steel is for 15 min at temperature.

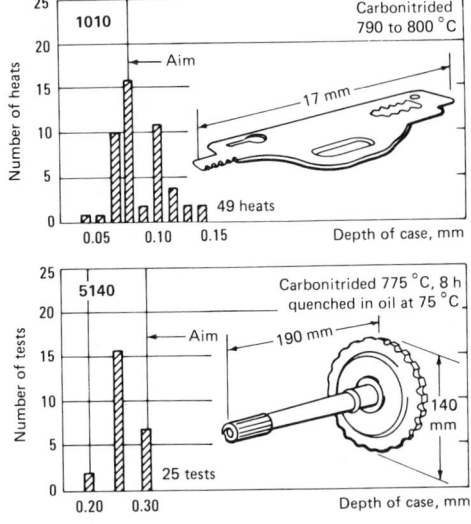

Fig. 10 Distribution of case depth of 1010 steel racks and 5140 steel pinion shafts

depending on turnover (the number of gas changes per hour). Major variables include: (a) the reactions of the gas with furnace brickwork and alloy; (b) tightness of the furnace and number of door openings (shaker-hearth and belt furnaces will have an opening between the door and hearth at all times); and (c) degree of circulation of gases, as related to load density and surface area of work. Data on the interrelationship of rate of gas flow, furnace volume, and number of gas changes per hour, as a function of type of furnace, are given in Table 3. From these data it is evident that:

- For brick-lined batch furnaces, fewer gas changes per hour are required as the furnace volume increases (ostensibly because relatively less gas leakage through furnace-door openings occurs as furnace volume increases).
- Shaker-hearth (or belt-type) furnaces require much higher gas flows per unit volume, because of large losses of atmosphere through the fixed door opening at the charge end.

Batch-Furnace Atmospheres

For maximum uniformity of case composition and case depth, the load must be heated uniformly to the carbonitriding temperature, and an adequate supply of atmosphere maintained at constant composition must be circulated throughout the load. Processing uniformity may be significantly improved by preheating parts in an atmosphere of endothermic gas for 15 to 60 min to ensure temperature uniformity within the load before adding enriching gas and ammonia for the actual carbonitriding portion of the cycle. This is a practicable procedure for batch furnaces that have sufficient heating capacity as well as an efficient work-loading pattern, good atmosphere-circulating fans, and dependable devices for mixing and controlling the addition of gases.

Control of Composition. As in gas carburizing, the composition of the atmosphere for carbonitriding should be carefully controlled for consistent results. After the rate of flow of ammonia and the temperature have been established, the ammonia flow should be held to limits of about ±10% of the predetermined flow gage reading. If simi-

contents should be used with alloy steels, particularly those containing nickel or high manganese, and with higher-manganese carbon steels such as 1117, 1118 and 1024.

Only enough ammonia should be added to sustain the carbonitriding reactions in the atmosphere; excess ammonia does not contribute to the nitrogen content of the case and can lead to void formation, as discussed above. Usually, 2.5 to 5% ammonia is sufficient to produce a satisfactory nitrogen content in the case. Ammonia contents as high as 10% or more, although relatively common, are often unnecessarily high.

Contaminants must be avoided in order to maintain an effective atmosphere in the furnace chamber. Infiltration of air and leaks in radiant tubes cause products of combustion—which are detrimental to the atmosphere—to enter the work chamber. Even the work load itself can provide a source of contamination. When significant amounts of grease, oil, cleaning agents, moisture or quenching salts are present on surfaces of the work load, these materials can seriously contaminate the atmosphere. This source of contamination can be avoided by cleaning the work with a suitable agent before charging and by the use of purging cycles.

Influence of Furnace Type. Different types of furnaces require different percentages of entering component gases and different volumes of gases,

Table 3 Influence of type of furnace on mixture and flow rate of carbonitriding gases

| Type of furnace | Constituents of gas, % by volume | | | Total gas flow, m³/h | Furnace volume, m³ | Gas changes per hour |
	Ammonia	Natural gas	Carrier gas			
Batch, brick-lined	10.4	20.8	68.8	6.8	0.29	23.5
Batch, brick-lined	4	6	90	14.5	1.42	10.2
Batch, brick-lined	8.3	8.3	83.3	10.2	2.55	4.0
Batch, brick-lined	5	20	75	11.3	8.50	1.3
Continuous, brick-lined........	3.4	2.3(a)	94.3	42.0	8.07	5.2
Continuous, brick-lined........	7	7	86	28.3	25.5	1.1
Continuous, brick-lined........	5	4	91	31.1	25.5	1.2
Shaker-hearth, metal-lined	2.5	2.5	95	6.16	0.056	110.0

(a) C_3H_8

Fig. 11 Effect on gears carbonitrided in a batch-type furnace

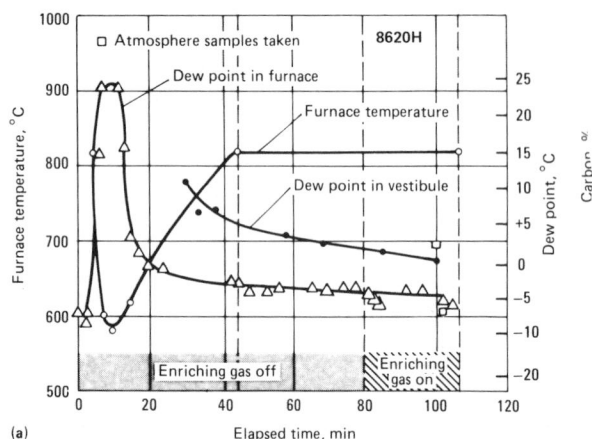

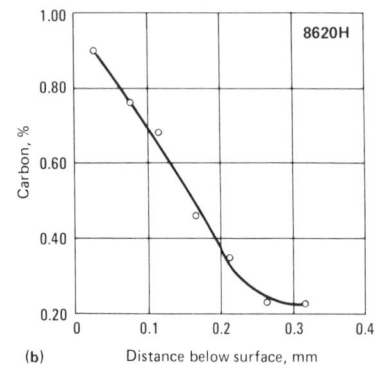

(a) Temperature and dew-point variations during carbonitriding cycle. (b) Carbon gradient produced by the cycle shown in (a)

parameters (oxygen potential, water vapor, carbon dioxide and/or methane contents) than for the same steel composition using the carburizing process.

For example, when carburizing at 925 °C (1700 °F) to obtain a case carbon content of 0.90 ± 0.05% or 0.80 ± 0.05%, it is desirable to hold the dew point within ± 0.5 °C (± 1 °F), and the carbon dioxide content within ± 0.005%, of the control values; when carbonitriding at 815 °C (1500 °F) to obtain a case carbon content of 0.80 ± 0.05%, it is desirable to hold the dew point within ± 1.5 °C (± 3 °F), and the carbon dioxide content within ± 0.03%, of the control values. The carbon dioxide levels corresponding to a certain carbon potential will not be the same for carbonitriding and carburizing, as discussed above.

Data reported for atmospheres consisting of endothermic generator gas with additions of 5% ammonia and less than 1% propane indicate dew points in the furnace chamber of −7 °C at 845 °C (+20 °F at 1550 °F) and of −4 °C at 815 °C (+25 °F at 1500 °F) for a surface carbon content of 0.80 to 0.90% with 1018 and 8620 steels. Dew points ranging from −7 to +4 °C (+20 to +40 °F) will cover most commercial carbonitriding applications at approximately 845 °C (1550 °F) with surface carbon content of about 0.80%.

Example 4. Gears (Fig. 11) made of 8620H steel were carbonitrided in batches, each of which had a net weight of 341.5 kg (755 lb) and a gross weight of 458.5 kg (1010 lb). Typically, carbonitriding was done at 815 °C (1500 °F) for 30 min, after which the gears were quenched directly from the carbonitriding temperature.

The furnace was a batch-type furnace with a brick-lined heating chamber. It was heated by radiant tubes and had a vestibule-enclosed quench. Dimensions of the furnace chamber were 1.42 m wide by 1.68 m long by 1.09 m high (56 by 66 by 43 in.) from brick floor to center of arch. The variation of furnace temperature with time during a typical carbonitriding cycle is shown in Fig. 11(a).

The atmosphere was composed of endothermic carrier gas, with enriching gas added after 33 min at temperature. The carrier gas was delivered at 21.2 m³/h (750 ft³/h) for the entire cycle, including the 63 min at temperature. The enriching gas consisted of propane, which was delivered at 0.14 m³/h (5 ft³/h), and ammonia, which was de-

lar parts are to be processed requiring the same ammonia addition, a stainless steel orifice plate can be installed in the ammonia supply line. This will ensure a constant and reproducible ammonia flow and prevent unintentional, and possibly detrimental, ammonia additions. Because of the generally lower carbon content of the carbonitrided case, the lower operating temperature (which permits higher dew points and more carbon dioxide), and the higher hardenability of the carbonitrided case, greater latitude is possible for control

livered at 1.08 m³/h (38 ft³/h). The furnace-gas analysis, determined from atmosphere samples taken (a) in the vestibule after 23 min of carbonitriding, and (b) in the furnace after 26 min of carbonitriding, is summarized in the following table:

Gas	Gas, %	
	Vestibule	Furnace
CO_2........	0.8	0.4
O_2...........	0.0	...
CO..........	22.4	20.4
CH_4........	1.2	1.2
H_2..........	34.2	34.2
N_2..........	rem	rem

The carrier-gas dew point, measured at the generator, was maintained at −15 to −14 °C (5 to 7 °F) throughout the cycle. Figure 11(a) indicates the dew points in the furnace and in the vestibule at various elapsed times during the carbonitriding cycle.

Figure 11(b) shows a typical carbon gradient in a gear resulting from the processing described.

Control of atmosphere composition begins by accurately controlling the carrier gas. With endothermic gas generators, the air-gas ratio is usually adjusted to produce a dew point of −1 to −4 °C (25 to 30 °F), which is equivalent to about 0.25 to 0.32% carbon dioxide content. When two or more generators are employed for a battery of furnaces, the generator gas can be piped to a common manifold. Depending on a wide range of operating conditions, a manual check of the dew point every 4 to 8 h and a daily or even weekly analysis of the generator gas are sufficient for most installations. However, automatic, closed-loop control based on oxygen potential, carbon dioxide or dew point is preferred for ensuring the uniformity of endothermic generator gas output. These same control systems can be used to provide accurate control of carbonitriding furnace atmospheres.

Dew-point instruments that contain a lithium chloride cell are not recommended for use with carbonitriding atmospheres, because ammonia will cause deterioration of the lithium chloride. If an infrared instrument for CO_2 is used, it is recommended that stainless steel tubing be used instead of copper to avoid the corrosive action of the ammonia. Aluminum tubing can be used successfully if care is taken to avoid any possibility of producing water condensate in the sample lines. In general, the advice of the instrument manufacturer should be obtained before using these devices with carbonitriding atmospheres.

The frequency of checking manually controlled dew point or carbon dioxide content of furnace atmospheres depends on many conditions such as type of furnace used, cycle lengths, and variations in work mix and load size. For continuous furnaces with reasonably constant work loading and work mix, a manual check of dew point and monitoring gas flows every 4 to 8 h may be sufficient for good process control. However, for batch furnaces carbonitriding widely varying materials, load sizes and part sizes, automatic control of atmosphere composition may be the only means of ensuring uniform results. The degree of process control required depends on the case depth and surface hardness tolerances required for the parts processed.

Because of the relatively light carbonitrided case, turnings from test bars are not frequently employed for determining carbon content. Metallographic examination, microhardness gradients and analysis of the furnace atmosphere are commonly used control methods.

Better control of furnace atmospheres can be obtained when the parts to be carbonitrided and the work baskets are free of dirt and oil. Sintered powder metallurgy parts containing oil must be "burned out" in a tempering furnace before they are carbonitrided.

Continuous-Furnace Atmospheres

Setting up a program for efficient and dependable control of a carbonitriding atmosphere in a continuous furnace is similar to that for a batch furnace. Good circulation of atmosphere and good control of temperature are of prime importance. It is advantageous to introduce ammonia only to those areas in which the temperature of the work is equal to the operating temperature for carbonitriding. This procedure is advisable because of the rapid action of fresh ammonia, particularly in regions of low temperature where the carburizing constituents of the atmosphere gas are not completely active with the work. If the furnace is of appreciable internal length, about 3½ m (11½ ft) or more, it is important to introduce the mixture of carburizing gas and ammonia at frequent intervals over the length of the furnace where the work is at temperature and in regions where the temperatures may be decreased for metallurgical reasons near the end of the cycle.

Safety

Gaseous carbonitriding media are highly toxic, flammable and explosive. The safety precautions that must be taken to protect equipment and personnel are essentially the same as those discussed in the article entitled "Gas Carburizing" in this volume.

The ammonia system required for carbonitriding usually consists of a number of cylinders of liquid ammonia that are connected to a common manifold. In general, ammonia from only a part of the supply is employed; the remainder is held in reserve. The flow from each cylinder should be sufficiently low to prevent freezing of the valves. Only stainless steel valves are recommended.

Outside bulk storage and vaporizing systems are much preferred to cylinder banks, even at a modest cost penalty, considering the advantage of a constant, uninterrupted source on the uniformity of work quality.

Ammonia cylinders should not be located near the furnaces, in direct sunlight, or near flammable gases or other combustibles. It is recommended that the ammonia supply be placed in a room that is well ventilated at the ceiling and separated from the work area by a fire-resistant wall.

Because ammonia is lighter than air, as well as a moderate fire hazard and a toxic material, an automatic sprinkling system is recommended. *Gas masks should be readily available, but should not be stored in the same area as the ammonia.* A sulfur stick can be used to check for ammonia leaks. Additional recommendations pertaining to safety in ammonia systems can be obtained from fire insurance companies.

Under no circumstances should combustible gas be introduced into the furnace when the furnace temperature is less than 760 °C (1400 °F). When a lower operating temperature is required, the furnace should be heated to 760 °C and purged with generator gas before the temperature is decreased. This type of operation can be very dangerous and should be done only by qualified personnel.

Fig. 12 Effect of carbonitriding temperature on dimensional stability of three 1010 steel production parts

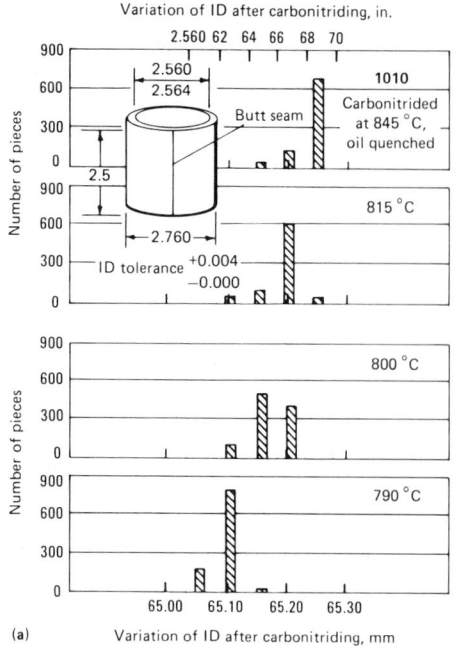

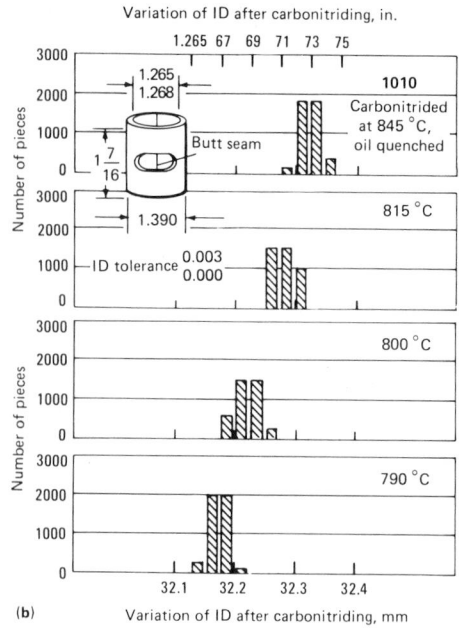

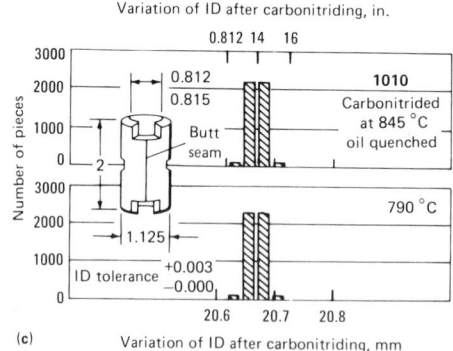

Parts were carbonitrided to produce a case depth of 0.13 to 0.20 mm (0.005 to 0.008 in.) with minimum surface hardness of 89 HR15N. Gas ratios and dew points were essentially the same for all temperatures. Time at temperature was 15 to 45 min, depending on temperature.

Temperature Selection

Choice of carbonitriding temperature is based on a number of considerations, either singly or in combination. These include steel composition, dimensional control, fatigue and wear properties, hardness, microstructural constituents, cost and equipment.

Steel Composition. Although higher temperatures permit the use of higher percentages of ammonia, ammonia content must be decreased as the alloy content of the steel increases, to minimize austenite retention. To this extent, temperature, atmosphere and alloy content are closely interrelated. *The use of temperatures near 705 °C (1300 °F) presents an explosion hazard, and results in superficial, high-nitrogen, brittle cases with low core hardness that are not suitable for most applications.* For these reasons, most carbonitriding operations are performed at 790 °C (1450 °F) or above.

Dimensional control often is the most important consideration in the selection of carbonitriding tempera-

ture. The following two examples illustrate the relation between carbonitriding temperature and dimensional stability.

Example 5. A 1010 steel rack 11.75 cm long by 1 mm thick (4.63 in. long by 0.040 in. thick) could not be kept within limits of straightness tolerance (and, because of brittleness, it could not be straightened) after it was quenched from a carbonitriding temperature of 845 °C (1550 °F) into warm oil at about 65 to 70 °C (150 to 160 °F). The part had a case-depth requirement of 0.075 mm

(0.003 in.) and a flatness requirement of 0.05 mm (0.002 in.) over its entire length. Lowering the temperature to the range of 790 to 800 °C (1450 to 1475 °F) significantly reduced distortion and increased ductility to within acceptable limits. Although temperatures below 790 °C (1450 °F) might have resulted in even less distortion and more ductility, they were not used because of equipment considerations and the increase in time required to produce the desired case depth.

Example 6. Figure 12 presents data pertaining to the effect of carbonitriding temperature on the dimensional stability of three production parts. All three parts, of 1010 commercial-quality steel sheet, were processed in batch-type equipment at temperatures ranging from 790 to 845 °C (1450 to 1550 °F). The atmospheres consisted of endothermic gas at 7.1 m³/h (250 ft³/h), enriching gas at 0.8 m³/h (30 ft³/h) and ammonia at 0.14 m³/h (5 ft³/h). Dew points were maintained at −1 to −4 °C (25 to 30 °F). Depending on temperature, furnace times were varied to produce specified case depths of 0.13 to 0.20 mm (0.005 to 0.008 in.).

Equal numbers of parts were hand loaded in baskets and processed in a batch furnace. The parts shown in Fig. 12(a) and (b) were loaded with their axes in the vertical position; parts like those in Fig. 12(c) were randomly loaded. Baskets and parts were quenched in slightly agitated oil at 60 to 70 °C (140 to 160 °F).

Data in Fig. 12 represent eight heats for the part in Fig. 12(a) and three heats for each of the two other parts. All parts were within inside-diameter tolerance before carbonitriding.

For the parts carbonitrided at four temperatures, the least distortion resulted at the lowest temperature. However, for the part with the smallest inside diameter (Fig. 12c), dimensional variation was not affected by a reduction in temperature.

Temperature also has a direct bearing on surface and core hardness. Steels treated at the higher carbonitriding temperatures and quenched from above the upper critical temperature of the core will produce higher core hardnesses. These high core hardnesses are usually desirable in applications involving high surface loads, because a strong core is needed to support the hardened case.

Special fabrication and service requirements of a particular part may restrict the choice of carbonitriding temperature. For instance, in one application, cold headed pins used in door-catch mechanisms required both riveting quality and wear resistance. A carbonitriding temperature of 790 °C (1450 °F) satisfied both requirements by producing a soft center for riveting and a thin, file-hard case for wear resistance.

Control of Retained Austenite

Nitrogen lowers the transformation temperature of austenite. Therefore, because of its nitrogen content, a carbonitrided case will, under identical post-treatment conditions, contain more retained austenite than a carburized case of the same carbon content. The low indentation hardness resulting from retained austenite is undesirable in many applications. It can be extremely detrimental in components of closefitting assemblies—for example, shaft and sleeve assemblies wherein the shaft is intended to rotate or reciprocate in the sleeve. The delayed transformation of austenite to martensite results in a volume increase that may cause moving parts to bind or "freeze" if it occurs in service.

The amount of retained austenite can be significantly decreased by cooling the quenched parts to −40 to −100 °C (−40 to −150 °F). When close-tolerance ground parts are involved, this treatment should precede finish grinding. Subzero treatment of parts that are to be tempered should precede final tempering. Subzero treatment is expensive; therefore, it is usually avoided whenever possible. It may also cause microcracks in the case structure, particularly in coarse-grain steels. Because the amount of retained austenite is normally at a maximum near the steel surface, it can be removed from symmetrical contours by grinding. However, care must be exercised in grinding high retained austenite surfaces because of the increased possibility of grinding burn or checking. If grinding is not required for any reason other than to remove retained austenite, it also may be considered an expensive operation. The most economical way to minimize retained austenite is by selection of preferred steels and control of the carbonitriding process.

Minimizing retained austenite in the carbonitrided case is assisted by modification of several processing factors:

- *Furnace Temperature.* An increase in furnace temperature will reduce the nitrogen content of the outer portions of the case, thus minimizing the amount of retained austenite. However, increasing the temperature also shifts the "nose" of the TTT diagram for a given steel, thus increasing the possibility of retained austenite. It may be better to reduce ammonia flow, rather than depend upon increased temperature to lower the nitrogen content.
- *Carbon Potential.* For most applications, hydorcarbon additions in the form of natural gas or propane should be controlled to provide a carbon potential and surface carbon concentration of 0.70 to 0.85%.
- *Ammonia content* of the carbonitriding atmosphere should be restricted to the minimum required to obtain the desired hardenability and metallurgical properties. A 5% ammonia content is usually a satisfactory starting point; a lower content decreases the rate of penetration, but may be desirable as the final choice of ammonia content.

Quenching Media and Practices

Whether carbonitrided parts are quenched in water, oil or gas depends on allowable distortion, part size and shape, steel composition, metallurgical requirements (such as hardness) and type of furnace equipment employed.

Water Quenching. Depending on allowable distortion, parts made of low-carbon steel may be quenched in water. For example, shift-lever pins made of B1212 steel are water quenched.

Water quenching usually is restricted to those furnaces in which the work is transferred from the furnace into the air prior to quenching, thus avoiding possible contamination of the furnace atmosphere by water vapor. However, water quenching from a rotary-retort furnace is feasible, provided the quench chute is equipped with gas eductors and a water-distribution system for condensing water vapor.

It should be noted that ammonia is extremely soluble in water and forms a product (NH_4OH) that is extremely corrosive to copper-base materials. In continuous operations where water is exposed to an ammonia-bearing atmosphere, brass agitators, copper tube bundles in heat exchangers, and similar copper-alloy components should be avoided.

Oil Quenching. Quenching-oil temperatures may vary from about 40 to 105 °C (100 to 220 °F). Special high-flash point oils may be used at the higher temperatures to minimize distortion; sometimes molten salt is used for the same reason. In the normal range of oil temperature (about 50 to 70 °C, or 120 to 160 °F), a mineral oil with a minimum flash point of 170 °C (335 °F) and a viscosity of 21×10^{-6} m²/s at 38 °C (21 centistokes, or 100 SUS, at 100 °F) is commonly used. Special oils containing additives for increasing the quenching rate also may be used. In general, to maintain maximum quenching effectiveness, quenching oils should have a low capacity for dissolving water. Quenching oils that dissolve even small amounts of water may lose effectiveness in three to six months; those that shed water completely may be used significantly longer.

Gas Quenching. Parts that have small mass (such as thin stampings), and that are subjected to sliding loads with low impact, may be quenched in a stream of cooled atmosphere gas. Gas or atmosphere quenching serves principally to reduce distortion and to eliminate the high costs of straightening. (Usually, however, a gas-quenched case will retain enough ductility to permit roller straightening, if required.)

In gas quenching, parts must be loaded into furnace trays carefully so that the surfaces of the parts can be cooled rapidly enough to produce desired hardness. Trays should be loaded and stacked so that the total mass of the load does not exceed that which can be satisfactorily quenched.

Tempering

Although many shallow-case carbonitrided parts are used without tempering, conditions under which the part is to be used may require it. The presence of nitrogen in the carbonitrided case increases its resistance to tempering, and the increase varies with the amount of nitrogen in the case. Thus, the maximum ammonia content of the carbonitriding atmosphere that is compatible with a fully martensitic case (minimum of retained austenite) results in the greatest resistance to tempering. Such increased resistance to tempering may be desirable where service operating temperatures are abnormally high, or where hot straightening is advantageous.

Tempering data obtained on carbonitrided cases of 1018 steel are given in Fig. 13. The data relate temper-resistance to both carbonitriding temperature and the ammonia content of the atmosphere. Figure 14 presents a summary of the effects of carbonitriding temperature and ammonia content on temper-resistance, derived from the same specimens referred to in Fig. 13.

Fig. 13 Decrease of surface hardness with increasing temperature for specimens of 1018 steel carbonitrided under the conditions indicated

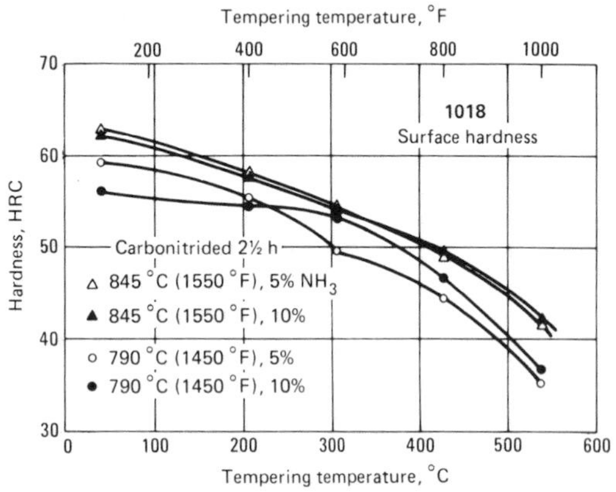

Rockwell C hardness converted from Rockwell 30-N. See also Fig. 14.

Because tempering a carbonitrided case at 425 °C (795 °F) and above results in a marked increase in notch toughness (see Table 4), parts that are to be subjected to repeated shock loading are invariably tempered to avoid impact and fatigue failures. Most carbonitrided gears are tempered at 190 to 205 °C (375 to 400 °F) to reduce surface brittleness and yet maintain a minimum case hardness of 58 HRC. Alloy steel parts that are to be surface ground are tempered to minimize grinding cracks. Low-carbon steel parts are frequently tempered at 135 to 175 °C (275 to 350 °F) to stabilize austenite and minimize dimensional variations. Tapping screws made of 1020 steel are tempered at 260 to 425 °C (500 to 795 °F) to reduce breakage in tapping holes in sheet metal. In contrast, parts that are case hardened primarily for wear re-

Table 4 Effect of tempering on Charpy V-notch impact strength of carbonitrided 1041 steel

Specimens were carbonitrided at 845 °C for 3 h in an atmosphere containing 7% ammonia, and were oil quenched from the carbonitriding temperature. Specimens were copper plated before machining of V-notch to permit exposure of the notch to the carbonitriding atmosphere.

Test	Tempering temperature °C	Tempering temperature °F	Impact strength J	Impact strength ft·lb	Hardness, HRC(a) Surface(b)	Core	Distance below surface, mm (in.) 0.075 (0.003)	0.15 (0.006)	0.25 (0.01)	0.38 (0.015)	0.64 (0.025)	1.0 (0.04)	1.4 (0.055)
1	As quenched	As quenched	1.4	1	60	53	63	64	64	63	61	61	58
2	370	700	2, 2	1.5, 1.5	47	46	57	57	55	54	49	50	50
3	425	800	29, 29	21.5, 21.5	42.5	43	57	57	56	55	49	47	47
4	480	900	69, 60	51, 44	38	38	54	54	52	50	42	38	38
5(c)	480	900	47, 52	35, 38	...	...	...	...	...	...	...	...	...
6	540	1000	78, 81	57.5, 60	35	32	49	50	50	47	36	33	32

(a) Converted from Vickers hardness. (b) Surface hardness is less than hardness at 0.075 mm below the surface because of retained austenite. (c) Tested at −18 °C (0 °F); all other tests at room temperature

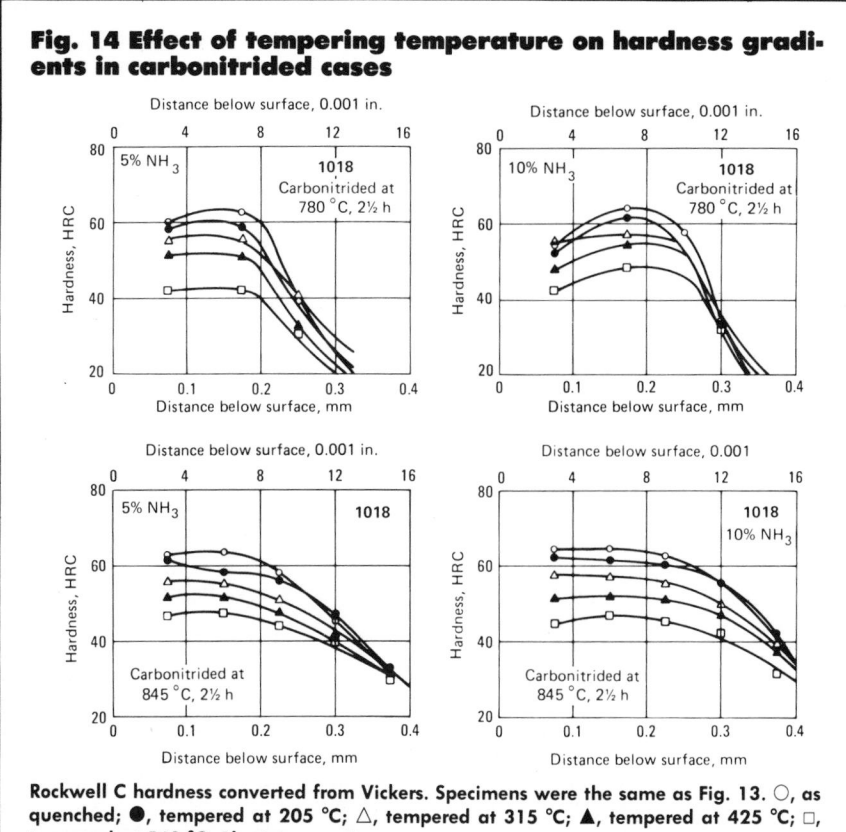

Fig. 14 Effect of tempering temperature on hardness gradients in carbonitrided cases

Rockwell C hardness converted from Vickers. Specimens were the same as Fig. 13. ○, as quenched; ●, tempered at 205 °C; △, tempered at 315 °C; ▲, tempered at 425 °C; □, tempered at 540 °C. 1h at temperature

sistance, such as dowel pins, brackets and washers, need not be tempered.

The importance of temper-resistance, whereby relatively high hardness and improved ductility are retained after tempering at the higher tempering temperatures, is illustrated by the following examples.

Example 7. A 1018 steel housing for a front-suspension ball joint required a wear-resistant bearing area and a ductile core for withstanding impact loads during service. These requirements were met by carbonitriding at 845 °C (1550 °F) to a case depth of 0.25 to 0.5 mm (0.010 to 0.020 in.), quenching from carbonitriding temperature in oil at 50 to 60 °C (120 to 140 °F), and tempering at 260 °C (500 °F). A core hardness of 85 to 95 HRB and a minimum hardness of 55 HRC in the bearing area were provided by this treatment.

Example 8. A tempering temperature of 505 °C (940 °F) was required to obtain a ductile core in carbonitrided front-suspension ball-joint studs made of 1541 steel, which are subjected to impact loads and wear during service. This temperature produced a core hard-

ness of 30 to 35 HRC and a minimum surface hardness of 50 HRC, which met the surface-wear requirements of studs carbonitrided to a case depth of 0.25 to 0.38 mm (0.010 to 0.015 in.).

Example 9. The service life of carbonitrided-and-tempered sewing-machine thread-handling parts, which operated at 12.7 m/s (2500 ft/min), and of low-carbon steel helical gears were superior to the service life of similar parts that were gas carburized and quenched. In both parts, wear was greatly reduced by the carbonitriding treatment.

Hardness Testing

The selection of a method of testing the surface hardness of carbonitrided steel depends primarily on case depth. For case depths of 0.65 mm (0.025 in.) and above, Rockwell C readings of surface hardness are generally accurate. A Rockwell C reading made on a case shallower than 0.65 mm is likely to be affected by core hardness. The surface hardness of cases deeper than 0.4 mm (0.015 in.) can be accurately measured on the Rockwell 15-N scale. When core

hardness is high enough, consistent readings can be obtained on the Rockwell 15-N scale for case depths of 0.2 to 0.4 mm (0.007 to 0.015 in.). On case depths of less than 0.2 mm, none of the Rockwell scales is reliable; testing may be done with files or a microhardness tester. Both Knoop and Vickers diamond indenters can be used at case-depth levels of 0.025 mm (0.001 in.).

In file-hardness testing, parts with a surface hardness that is less than full file hardness (64 to 68 HRC) can be tested with files that have been tempered to the desired hardness range. The file method is particularly useful in performing rapid production checks on nonsymmetrical parts. Surface hardness for such parts may be specified as Mfh 60, indicating that the part must be hardened to a mill file tempered to 60 HRC. High percentages of retained austenite will seriously affect the accuracy of file-hardness testing. A sample with high austenite content may indicate an indentation hardness of only 52 HRC and still resist a file hardened to 66 HRC.

Carbonitriding of Powder Metallurgy Parts

Carbonitriding is widely used as a process for case hardening parts made by powder metallurgy techniques from ferrous powders. Densities of the sintered compacts vary from approximately 6.5 g/cm³ up to those approaching that of wrought steel. Parts may or may not be copper infiltrated prior to carbonitriding.

Carbonitriding is extremely effective in case hardening high-density (7.2 g/cm³ minimum) sintered iron compacts. Electrolytic iron is used to obtain this high density. Three characteristics of these compacts make case hardening by other methods extremely difficult: (a) high transformation temperature, (b) very low hardenability, and (c) inherent porosity, which results in unusually high rates of diffusion.

Carbonitriding at 790 to 815 °C (1450 to 1500 °F) solves these problems; lower rates of diffusion at these temperatures permit control of case depth and allow buildup of adequate carbon in the case. The effects of nitrogen in retarding the rate of transformation provide sufficient hardenability to allow oil quenching.

File-hard cases (with microhardnesses equivalent to 60 HRC) and nor-

Fig. 15 Increase of case depth with decrease in density of iron powder metallurgy parts carbonitrided for various periods of time at 790 °C (1455 °F)

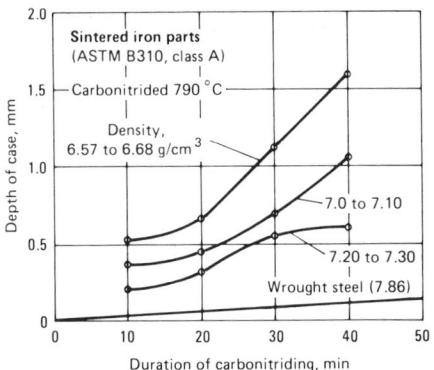

Curve for steel is based on total furnace time and represents the average of the "775 to 790 °C" (1425 to 1455 °F) band shown in Fig. 10.

mal, predominantly martensitic structures can be consistently obtained. Shallow cases are obtainable, although the allowable range of case depth must be increased over that used for wrought steels. Typical ranges of case depth are 0.08 to 0.20 mm (0.003 to 0.008 in.) and 0.15 to 0.30 mm (0.006 to 0.012 in.).

The high rate of carbon and nitrogen penetration that occurs as the result of porosity is demonstrated in Fig. 15 for parts made of iron powder conforming to ASTM B310, class A. Although the rate of penetration decreases with increasing density, case depths for the higher densities (7.20 to 7.30 g/cm^3) are much deeper than those obtained with a wrought steel (7.87 g/cm^3). Most commercial iron-powder compositions exhibit this type of response to carbonitriding; however, copper-infiltrated compacts are considerably more resistant to the penetration of carbon and nitrogen.

Tempering. Carbonitrided iron powder metallurgy parts are usually tempered, despite the fact that there is little danger of cracking untempered pieces. Tempering accomplishes the incidental result of facilitating tumbling and deburring operations. Although tempering is potentially capable of removing oil picked up and held in the pores in the part, air tempering of oil-quenched powder metallurgy parts is normally limited to temperatures not exceeding 205 °C (400 °F) because of the fire hazard at higher temperatures. Carbonitrided iron powder metallurgy parts are usually tempered at temperatures slightly higher than the temperatures used for carbonitrided wrought

steel parts. Special cleaning procedures to remove oil, thus eliminating fire hazards, are incorporated in the processing steps when the tempering temperature exceeds 205 °C.

REFERENCES

1. The Role of Water Vapor and Ammonia in Case Hardening Atmospheres, by F. A. Clarkin and M. B. Bever: *Trans ASM,* Vol 47, 1955, p 794–806
2. Carbonitriding of Plain Carbon and Boron Steels, by G. W. Powell, M. B. Bever and C. F. Floe: *Trans ASM,* Vol 46, 1954, p 1359–1371
3. A Practical Study of the Carbonitriding Process, by R. Davies and C. G. Smith: *Metal Progress,* Vol 114, No. 4, 1978, p 40–53
4. Carbonitriding: An Investigation from a Process Point of View, by J. Slycke: Linkoping Studies in Science and Technology Dissertations No. 37, Linkopings Tyckeria AB, Sweden, 1979; see also Investigation of the Carbonitriding Process, by J. Slycke: Heat Treatment '76, The Metals Society, London, 1976, p 57–63
5. Card Reader Carbonitrides Gears, by J. Dossett: *Metal Progress,* Vol 107, No. 5, 1975, p 57–58

Gas Nitriding

By the ASM Committee on Gas
Carburizing, Carbonitriding and
Nitriding*

GAS NITRIDING is a case harden-ing process whereby nitrogen is intro-duced into the surface of a solid ferrous alloy by holding the metal at a suitable temperature (below Ac_1, for ferritic steels) in contact with a nitrogenous gas, usually ammonia. Quenching is not required for the production of a hard case. The nitriding temperature for all steels is between 495 and 565 °C (925 and 1050 °F).

Application Factors

Principal reasons for nitriding are

- To obtain high surface hardness
- To increase wear resistance and anti-galling properties
- To improve fatigue life
- To improve corrosion resistance (ex-cept for stainless steels)
- To obtain a surface that is resistant to the softening effect of heat at tem-peratures up to the nitriding temper-ature.

Because of the absence of a quench-ing requirement, with attendant vol-ume changes, and the comparatively low temperatures employed in this pro-cess, nitriding produces less distortion and deformation than either carburiz-ing or conventional hardening. Some growth does occur as a result of nitrid-ing, but volumetric changes are rela-tively small.

Nitridable Steels. Of the alloying elements commonly used in commer-cial steels, aluminum, chromium, va-nadium, tungsten, and molybdenum are beneficial in nitriding, because they form nitrides that are stable at nitriding temperatures. Molybdenum, in addition to its contribution as a nitride-former, also reduces the risk of embrittlement at nitriding tempera-tures. Other alloying elements, such as nickel, copper, silicon and manganese, have little, if any, effect on nitriding characteristics.

Although at suitable temperatures all steels are capable of forming iron nitrides in the presence of nascent nitrogen, the nitriding results are more favorable in those steels that con-tain one or more of the major nitride-forming alloying elements. Because aluminum is the strongest nitride-former of the common alloying ele-ments, aluminum-containing steels (0.85 to 1.50% Al) yield the best nitrid-ing results in terms of total alloy con-tent. Chromium-containing steels can approximate these results if their chro-mium content is high enough. Unal-loyed carbon steels are not well suited to gas nitriding, because they form an extremely brittle case that spalls readi-ly, and the hardness increase in the dif-fusion zone is small.

The following steels can be gas ni-trided for specific applications:

- Aluminum-containing low-alloy steels (Table 1)
- Medium-carbon, chromium-contain-ing low-alloy steels of the 4100, 4300, 5100, 6100, 8600, 8700, 9300 and 9800 series
- Hot work die steels containing 5% chromium, such as H11, H12 and H13
- Ferritic and martensitic stainless steels of the 400 series
- Austenitic stainless steels of the 300 series
- Precipitation-hardening stainless steels, such as 17-4 PH, 17-7 PH and A-286.

Aluminum-containing steels produce a nitrided case of very high hardness and excellent wear resistance. Howev-er, the nitrided case also has low ductil-ity, and this limitation should be care-fully considered in the selection of aluminum-containing steels. In con-trast, low-alloy chromium-containing steels provide a nitrided case with con-siderably more ductility but with lower hardness. Nevertheless, these steels of-fer substantial wear resistance and good antigalling properties. Tool steels, such as H11 and D2, yield consistently high case hardness with exceptionally high core strength.

Prior Heat Treatment. All harden-able steels *must* be hardened and tem-pered before being nitrided. The tem-pering temperature must be high

*Charles A. Stickels, *Chairman*, Principal Staff Engineer, Research Staff, Ford Motor Co.; Larry E. Byrnes, Manager, Manufacturing Research, Federal Mogul Corp.; Jon L. Dossett, Vice President, Metallurgy, United States Gear Corp.; Robert L. Hughes, Chief Metallurgist, Fairfield Manufacturing Co., Inc.; Kenneth D. Gladden, Senior Development Engineer, Caterpillar Tractor Co.; Dale A. Poteet, Jr., Division Engineer, John Deere Component Works; Joseph A. Riopelle, District Manager/Chicago Office, Surface Division, Midland-Ross Corp.; Ronald D. Rogers, Chief Engineer, Materials Engineering, Automotive Operations, Rockwell International

Table 1 Nominal composition and preliminary heat treating cycles for aluminum-containing low-alloy steels commonly gas nitrided

SAE	Steel AMS	Nitralloy	C	Mn	Si	Cr	Ni	Mo	Al	Se	Austenitizing temperature(a) °C	°F	Tempering temperature(a) °C	°F
...	...	G	0.35	0.55	0.30	1.2	...	0.20	1.0	...	955	1750	565 to 705	1050 to 1300
7140	6470	135M	0.42	0.55	0.30	1.6	...	0.38	1.0	...	955	1750	565 to 705	1050 to 1300
...	6475	N	0.24	0.55	0.30	1.15	3.5	0.25	1.0	...	900	1650	650 to 675	1200 to 1250
...	...	EZ	0.35	0.80	0.30	1.25	...	0.20	1.0	0.20	955	1750	565 to 705	1050 to 1300

(a) Sections up to 50 mm (2 in.) in diameter, quenched in oil; larger sections may be water quenched.

Fig. 1 Micrographs of white nitride layers developed on vacuum melted AMS 6470 steel

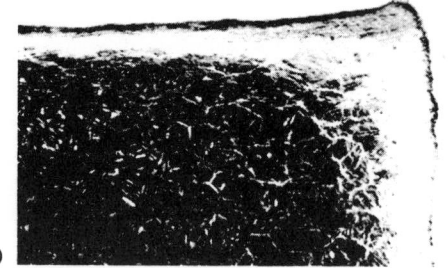 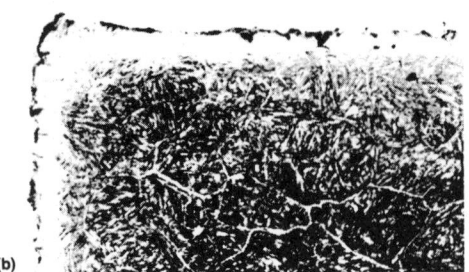

(a) 0.03-mm (0.0013-in.) white layer formed after single-stage nitriding at 525 °C (975 °F) for 60 h with 28% dissociation. (b) 0.02-mm (0.0008-in.) white layer formed after double-stage nitriding at 525 °C (975 °F) for 9½ h with 25 to 28% dissociation, then at 550 °C (1025 °F) for 50½ h and 80 to 84% dissociation. Buildup of white layer at corners during single-stage nitriding was 0.08 mm (0.0033 in.); during double-stage nitriding, 0.05 mm (0.0020 in.). Etched in 2% nital. Magnification, 150×

enough to guarantee structural stability at the nitriding temperature; the minimum tempering temperature is usually at least 30 °C (50 °F) higher than the maximum temperature to be used in nitriding.

In certain alloys, such as series 4100 and 4300 steels, hardness of the nitrided case is modified appreciably by core hardness; that is, a decrease in core hardness results in a decrease in case hardness. Consequently, in order to obtain maximum case hardness, these steels are usually provided with maximum core hardness by being tempered at the minimum allowable tempering temperature.

Applications. Examples of typical gas nitriding applications and procedures are presented in Table 2. Table 3 lists examples of parts for which nitriding eliminated production or service problems that arose when the parts were case hardened by other methods.

Two applications in which other hardening processes were more desirable than nitriding are described in the examples that follow.

Example 1. Teeth of helical speed-reducing gears were required to resist pitting and scuffing under heavy loading. The gears were formerly made of 4150 steel heat treated to a hardness of 285 HB and single-stage nitrided at 510 °C (950 °F) for 38 h. The depth of case ranged from 0.3 to 0.4 mm (0.012 to 0.015 in.). The nitrided gears failed in service because of pitting on the teeth, which was due to microscopic cracks under the case. The process was changed to progressive flame hardening; this produced a deeper case on the teeth and eliminated failures. (A deeper nitrided case was not feasible because of cost.)

Example 2. Rotating seal mating rings were required to have high stability and wear resistance to withstand service at 150 °C (300 °F) against a carbon stationary ring. The rings were originally made of SAE 7140, quenched and tempered to 25 to 35 HRC and

single-stage nitrided at 520 °C (970 °F) for 48 h. Depth of case ranged from 0.4 to 0.6 mm (0.015 to 0.025 in.); case hardness was 92 to 94 HR15-N. Rings returned from service for repairs were difficult to rework. Regrinding or relapping for flatness was not satisfactory, because of the risk of removing the hard surface layers and thus depriving parts of required surface hardness. Some rings that were outside of dimensional specifications had to be replaced. Rings subsequently were made of 52100 steel through-hardened to 60 to 64 HRC. Although inferior in wear resistance to the nitrided rings, the 52100 rings performed satisfactorily and could be reworked when necessary.

Single-Stage and Double-Stage Nitriding

Either a single- or a double-stage process may be employed when nitriding with anhydrous ammonia. In the

Table 2 Nitriding applications and procedures

Part	Dimensions or weight of part	Steel	Nitriding time, h
Single-stage nitriding			
Hydraulic barrel	50 mm (2 in.) OD, 19 mm (¾ in.) ID, 150 mm (6 in.) long	AMS 6470	48
Trigger for pneumatic hammer	...	AMS 6470	40
Governor push button	6-mm (¼-in.) diam	AMS 6470	30
Tachometer shaft	380 mm (15 in.) long	AMS 6475	25
Helical timing gear	205 mm (8 in.) OD (4.5 kg or 10 lb)	4140	24
Gear	50 mm (2 in.) OD, 6 mm (¼ in.) thick	4140	24
Generator shaft	25 mm (1 in.) OD, 355 mm (14 in.) long	4140	24
Rotor and pinion for pneumatic drill	22-mm (⅞-in.) diam	4140	9
Sleeve for pneumatic tool clutch	38-mm (1½-in.) diam	4140	9
Marine helical transmission gear	635 mm (25 in.) OD (227 kg or 500 lb)	4142	32
Oil-pump gear	50 mm (2 in.) OD, 180 mm (7 in.) long	4340	25
Loom shuttle	150 mm by 25 mm by 25 mm (6 in. by 1 in. by 1 in.)	410 stainless	8
Double-stage nitriding			
Ring gear for helicopter main transmission	380 mm (15 in.) OD, 350 mm (13.8 in.) ID, 64 mm (2.5 in.) long	AMS 6470 (a)	60(b)
Aircraft cylinder barrel	180 mm (7 in.) OD, 305 mm (12 in.) long	AMS 6470	35(c)
Bushing	10 kg (23 lb)	AMS 6470	90
Cutter spindle	3 kg (7 lb)	AMS 6470	45
Plunger	75 mm (3 in.) OD, 1525 mm (60 in.) long	AMS 6475	72
Crankshaft	205 mm (8 in.) OD (journals), 4 m (13 ft) long	4130	65
Piston ring	150 mm (6 in.) OD, 4.25 m (14 ft) long	4130	65
Clutch	1 kg (2 lb)	4140	45
Double helical gear	50 kg (108 lb)	4140	97
Feed screw	4 kg (9 lb)	4140	45
Pumper plunger	0.5 kg (1 lb)	4140	127
Seal ring	9.5 kg (21 lb)	4140	90
Stop pin	3 kg (7 lb)	4140	90
Thrust collar	3.6 kg (8 lb)	4140	90
Wear ring	40 kg (87 lb)	4140	90
Clamp	7 kg (15 lb)	4150	90
Die	21 kg (47 lb)	4340	90
Gib	10 kg (23 lb)	4340	49
Spindle	122 kg (270 lb)	4340	90
Torque gear	62.5 kg (138 lb)	4340	90
Wedge	1.8 kg (4 lb)	4340	42
Pumper plunger	1.4 kg (3 lb)	420 stainless	127

(a) Vacuum melted. (b) 9 h at 525 °C (975 °F), 51 h at 545 to 550 °C (1015 to 1025 °F). (c) 6 h at 525 °C (975 °F), 29 h at 565 °C (1050 °F)

single-stage process, a temperature in the range of about 495 to 525 °C (925 to 975 °F) is used, and the dissociation rate ranges from 15 to 30%. This process produces a brittle, nitrogen-rich layer, known as the "white nitride layer", at the surface of the nitrided case.

The double-stage process, known also as the Floe process (U.S. Patent 2 437 249), has the advantage of reducing the thickness of the white nitrided layer.

The first stage of the double-stage process is, except for time, a duplication of the single-stage process. The second stage may proceed at the nitriding temperature employed for the first stage, or the temperature may be increased to from 550 to 565 °C (1025 to 1050 °F); however, at either temperature, the rate of dissociation in the second stage is increased to from 65 to 85% (preferably, 80 to 85%). Generally, an external ammonia dissociator is necessary for obtaining the required higher second-stage dissociation.

The principal purpose of double-stage nitriding is to reduce the depth of the white layer produced on the surface of the case. Except for a reduction in the amount of ammonia consumed per hour, there is no advantage in using the double-stage process, unless the amount of white layer produced in single-stage nitriding cannot be tolerated on the finished part or unless the amount of finishing required after nitriding is substantially reduced. In Fig. 1, the amount of white layer formed during a 60-h double-stage nitriding treatment is compared to that formed

Table 3 Examples of parts for which nitriding proved superior to other case hardening processes for meeting requirements

Part	Requirement	Material and process originally used
Gear	Good wear surface and fatigue properties	Carburized 3310 steel 0.4 to 0.6-mm (0.017 to 0.025-in.) case
High-speed pinion (on gear motor)	Provide teeth with minimum (equivalent) hardness of 50 HRC	8620 steel gas carburized at 900 °C (1650 °F) to 0.5-mm (0.02-in.) case, direct quenched from 845 °C (1550 °F), and tempered at 205 °C (400 °F)
Bushings (for conveyor rollers handling abrasive alkaline material)	High surface hardness for abrasion resistance; also, resistance to alkaline corrosion	Carburized bushings
Spur gears (in train of power gears; 10-pitch, tip modified)	Sustain continuous Hertz stress of 1035 MPa (150 ksi) (overload of 1550 MPa or 225 ksi), continuous Lewis stress of 275 MPa (40 ksi) (overload of 725 MPa or 105 ksi)(c)	Carburized AMS 6260

Part	Requirement	Resultant problem	Solution
Gear	Good wear surface and fatigue properties	Difficulty in obtaining satisfactory case to meet a reliability requirement	AMS 6470 substituted for 3310 and double-stage nitrided for 25 h
High-speed pinion (on gear motor)	Provide teeth with minimum (equivalent) hardness of 50 HRC	Distortion in teeth and bore caused high rejection rate	4140 steel, substituted for 8620, was heat treated to 255 HB; parts were rough machined, finish machined, nitrided(a)
Bushings (for conveyor rollers handling abrasive alkaline material)	High surface hardness for abrasion resistance; also, resistance to alkaline corrosion	Service life of bushings was short because of scoring	Substitution of Nitralloy 135 type G (resulfurized) heat treated to 269 HB and nitrided(b)
Spur gears (in train of power gears; 10-pitch, tip modified)	Sustain continuous Hertz stress of 1035 MPa (150 ksi) (overload of 1550 MPa or 225 ksi), continuous Lewis stress of 275 MPa (40 ksi) (overload of 725 MPa or 105 ksi)(c)	Gears failed because of inadequate scuff resistance, also suffered property losses at high operating temperatures	Substitution of material of H11 type, hardened and multiple tempered (3 h + 3 h) to 48 to 52 HRC, then double-stage nitrided(d)

(a) Single-stage nitrided at 510 °C (950 °F) for 38 h. Cost increased 5%, but rejection rate dropped to zero. (b) Single-stage nitrided at 510 °C (950 °F) for 38 h. Case depth was 0.46 mm (0.018 in.), and hardness was 94 HR15-N; parts had three times the service life of carburized parts. (c) Must withstand operating temperatures to 290 °C (550 °F). (d) 15 h at 515 °C (960 °F) (15 to 25% dissociation); then 525 °C (980 °F) (80 to 83% dissociation). Effective case depth (to 60 HRC), 0.25 to 0.4 mm (0.010 to 0.015 in.); case hardness, 67 to 72 HRC (converted from Rockwell 15-N scale).

during a 60-h single-stage nitriding of the same material.

Figure 2 shows the effect of nitriding time on the depth of case developed on 4140 steel during double-stage nitriding at 525 °C (975 °F) for both stages. Use of a higher temperature during the second stage would have produced a deeper case of slightly lower hardness.

Hardness gradients obtained in double-stage nitriding of SAE 7140 (AMS 6470) are shown in Fig. 3. The hardness results shown in Fig. 3(a), (b) and (c) were obtained by grinding off progressively increasing amounts of case to form steps on which HR15-N readings were made. Data for Fig. 3(d), (e) and (f) were obtained from microhardness measurements, converted to HRC equivalents. Similar data, for double-stage nitriding of AMS 6475, are shown in Fig. 4.

Operating Procedures

Surface Preparation of Parts to be Nitrided

After hardening and tempering, and before nitriding, parts should be thoroughly cleaned. Most parts can be successfully nitrided immediately after vapor degreasing. However, some machine finishing processes such as buffing, finish grinding, lapping and

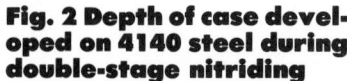

Fig. 2 Depth of case developed on 4140 steel during double-stage nitriding

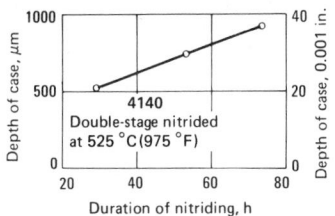

Numbers indicate hours of nitriding at 15 to 25% dissociation. Remainder of cycle at 83 to 85% dissociation

Fig. 3 Hardness gradients for double-stage nitrided SAE 7140 (AMS 6470) steel

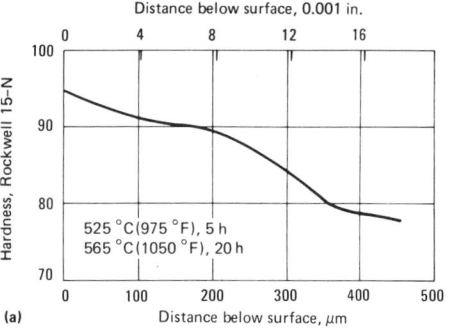

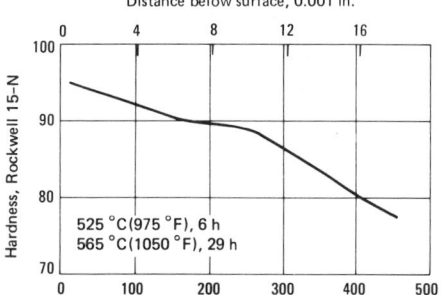

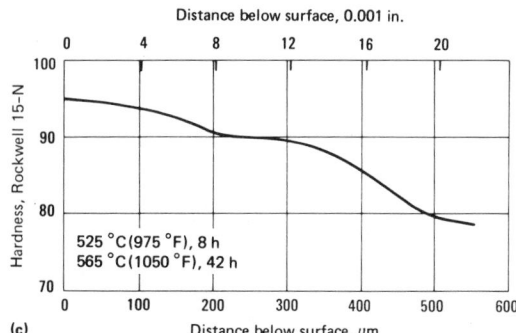

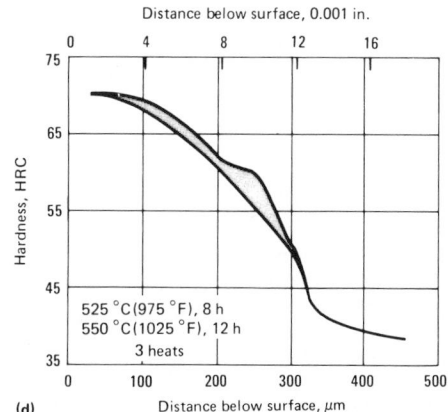

(a)-(c) Material nitrided with dissociation rates of 15 to 20% during first stage and 60 to 70% during second stage. (d)-(f) material nitrided with dissociation rates of 25 to 28% during first stage and 75 to 80% in second stage, material produced by consumable-electrode vacuum-arc-remelting method

burnishing may produce surfaces that retard nitriding and result in uneven case depth and distortion. There are two methods by which the surfaces of parts finished by such methods may be successfully conditioned before nitriding.

One method consists of vapor degreasing parts and then abrasive cleaning them with aluminum oxide grit immediately prior to nitriding. Any residual grit must be brushed off before parts are loaded into the furnace. Parts should be handled with clean gloves.

The second method is to apply a light phosphate coating. One procedure for applying such a coating is:

- Degrease.
- Rinse in cold water for ½ to 1 min.
- Dip in oxalic acid bath for 10 to 30 s. Bath is made up by adding 0.073 kg (0.16 lb) of oxalic acid crystals per gallon of water. Acid strength should be maintained at 30 points* and checked semimonthly. Bath should be discarded when yellow-green sludge adheres to parts being treated.

- Rinse in cold water for ½ to 1 min.
- Rinse in warm water for ½ to 1 min. Rinse bath should be maintained at 65 to 80 °C (150 to 180 °F), and flow regulated to keep water from becoming contaminated.
- Treat in phosphate solution at 80 °C (180 °F) for 4 min. Phosphate solution meeting requirements of MIL-C-490A, Grade I, is used. Dilute with water, approximately 1.3 kg/l (0.75 lb/gal) to strength of 30 points.* Maintain at 30 points by daily analysis.

*Strength in points is the number of millilitres of 0.1N solution of NaOH titrated against a 10-ml sample. Phenolphthalein is indicator.

- Rinse in cold water for 1 min.
- Rinse in warm water for 1 min (bath conditions as indicated above).
- Blow dry.
- Store in clean container until loaded in nitriding furnace.

Furnace Purging

After loading and sealing the furnace at the start of the nitriding cycle, it is necessary to purge the air from the retort before the furnace is heated to a temperature above 150 °C (300 °F). This prevents oxidation of parts and furnace components and, when ammonia is used as the purging atmosphere, *avoids production of a potentially explosive mixture.* Nitrogen is preferred in place of ammonia for purging, but the same precautions should be taken to avoid oxidation of parts.

A typical purging cycle, using anhydrous ammonia, follows:

- Close furnace and start flow of anhydrous ammonia gas at as fast a flow rate as is practical.
- Simultaneously set furnace temperature control at 150 °C (300 °F). Heat furnace to this temperature but do not exceed.
- When the furnace has been purged to the degree that 10% or less air and 90% or more ammonia are present in the retort, the furnace may be heated to the nitriding temperature.

Purging is employed also at the conclusion of the nitriding cycle when the furnace is cooled from the nitriding temperature. It is common practice to dilute the ammonia remaining in the retort with air, to reduce the amount of ammonia that would otherwise be released into the immediate area when the load is removed. Dilution of the ammonia lessens the discomfort to employees working near the furnace. The introduction of air into the retort must be delayed until the nitrided parts have cooled to below 150 °C (300 °F). *The formation of an explosive mixture of hydrogen and oxygen is a possibility if air is introduced before the hydrogen content of the furnace atmosphere has been reduced to a safe level (dissociation reading below 6%). Consideration must always be given to the possibility that a spark might be generated by the accidental contact of an object with the circulating fan.*

Emergency Purging. If the supply of ammonia is cut off during the nitriding cycle, or if a break occurs in the supply line, there is great danger that air will be sucked into the furnace by the contraction of gases within the furnace. This danger is greatest during the cooling cycle. To guarantee a positive pressure within the furnace, it is general practice to provide an emergency purging system that will pump dry nitrogen or an oxygen-free generated gas into

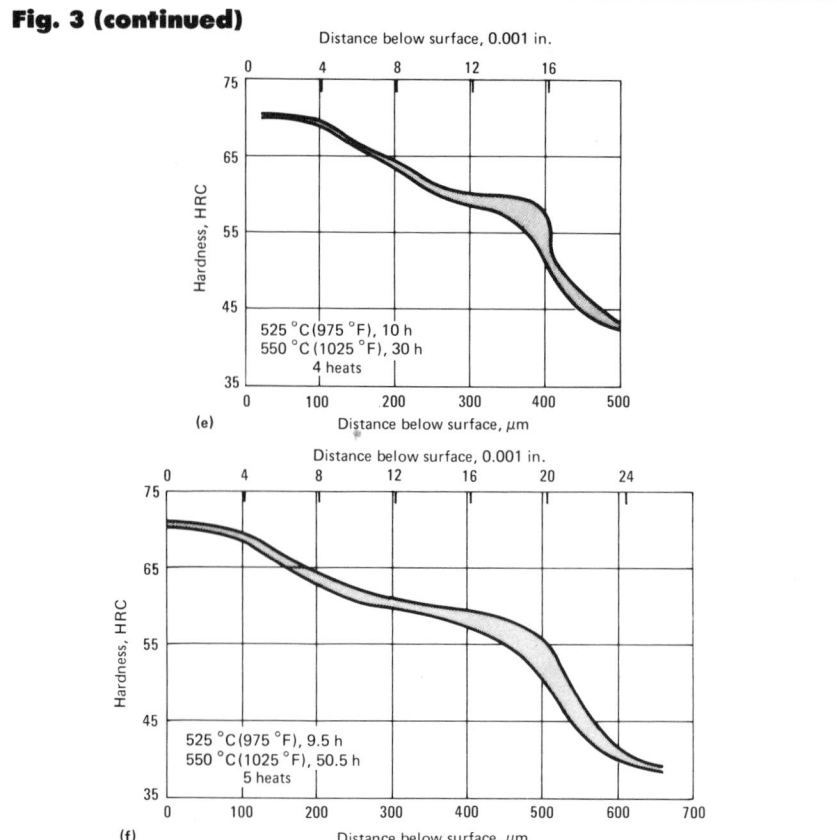

Fig. 3 (continued)

(e) 525 °C (975 °F), 10 h / 550 °C (1025 °F), 30 h / 4 heats

(f) 525 °C (975 °F), 9.5 h / 550 °C (1025 °F), 50.5 h / 5 heats

(a)-(c) Material nitrided with dissociation rates of 15 to 20% during first stage and 60 to 70% during second stage. (d)-(f) material nitrided with dissociation rates of 25 to 28% during first stage and 75 to 80% in second stage, material produced by consumable-electrode vacuum-arc-remelting method

Fig. 4 Hardness gradient obtained for double-stage nitrided AMS 6475 steel

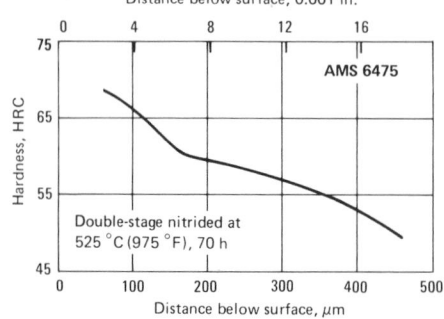

AMS 6475

Double-stage nitrided at 525 °C (975 °F), 70 h

HRC hardness numbers were obtained by conversion from diamond pyramid hardness measurements. Core hardness after nitriding was 41.5 HRC. Data represent one air-melted heat of AMS 6475.

Fig. 5 Hardness gradients and case depth relations for single-stage nitrided aluminum-containing SAE 7140 steel

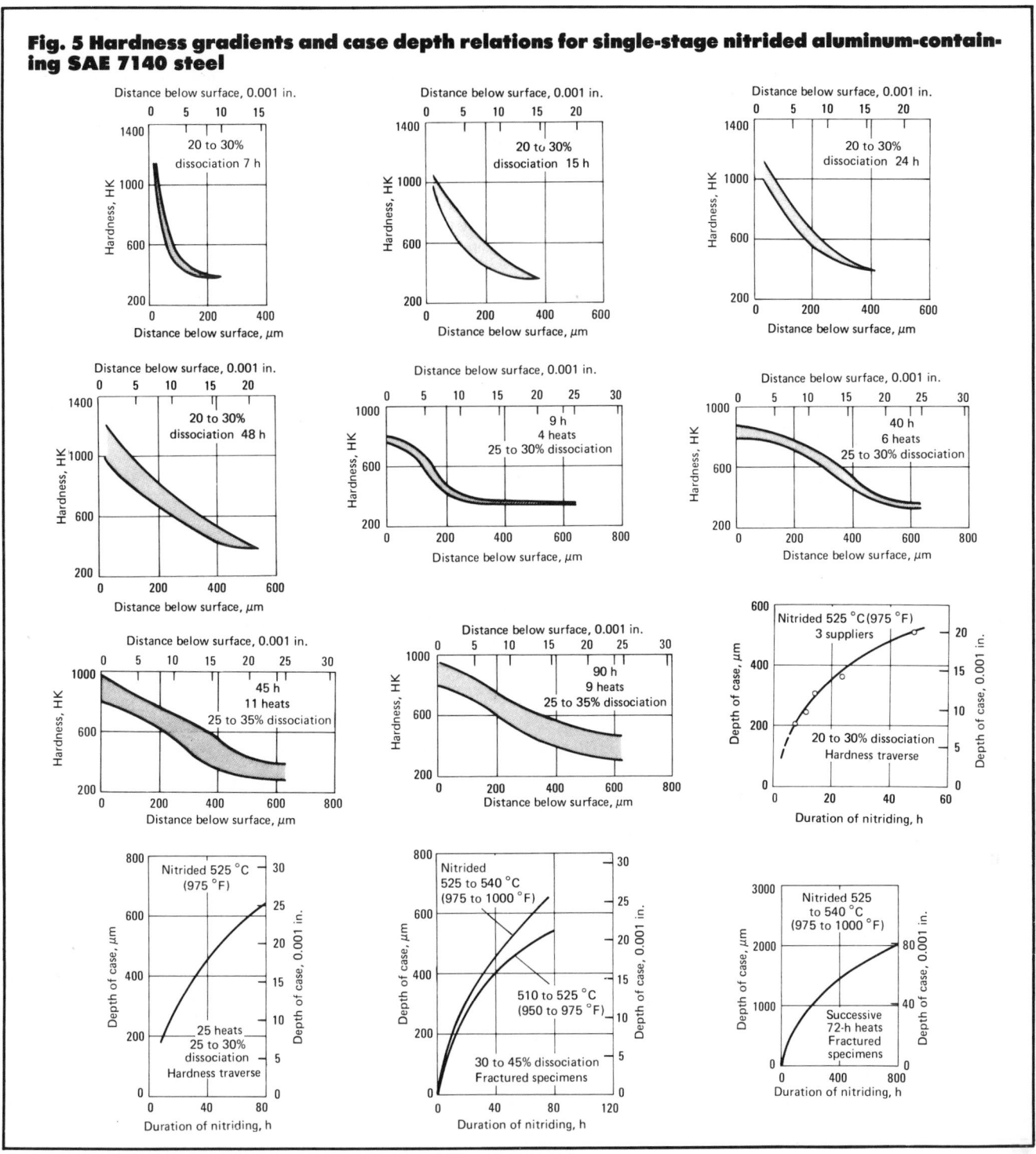

the furnace and maintain a safe pressure. If a generated gas is used, it should be as moisture-free as possible. These purge gases are released directly into the ammonia line by a special system of valves.

If emergency purging facilities are not available, further cooling of the furnace should be prevented. Exhaust and inlet lines should be closed, and if no air has entered the retort, the temperature of the furnace and load should be increased slightly to maintain a positive pressure within the retort. Tem-

perature should be maintained until the break is repaired and the flow of ammonia is restored.

Nitrogen Versus Ammonia for Purging. Advantages of nitrogen as a purging gas include its safety, ease of handling and ease of control. The use of nitrogen, however, requires additional equipment, including piping.

Ammonia requires no additional equipment and is relatively safe when properly handled; *mixtures of 15 to 26% ammonia in air, however, are explosive if ignited by a spark.*

Dissociation Rates. The nitriding process is based on the affinity of nascent nitrogen for iron and certain other metallic elements. Nascent nitrogen is produced by the dissociation of gaseous ammonia when it contacts hot steel parts. Although various rates of dissociation can be used successfully in nitriding, it is important that the nitriding cycle begin with a dissociation rate of about 15 to 30% and that this rate be maintained for 4 to 10 h, depending on duration of the total cycle; temperature should be maintained at about 525 °C (975 °F). This initial cycle develops a shallow white layer from which diffusion of nitrogen into the main case structure proceeds.

In most nitriding cycles, dissociation rates vary somewhat, even though the controlling factors—ammonia flow rate, surface area and nitriding temperature—remain constant. Characteristically, the dissociation rate gradually increases as the cycle proceeds at a constant ammonia flow rate. This increase, however, usually is not enough to affect nitrided case characteristics significantly.

When nitriding with a dissociation rate of 15 to 30%, it is normal to control this rate entirely by the flow rate of ammonia. At a dissociation rate of 80 to 85%, however, it is necessary to introduce completely dissociated ammonia from an external dissociator to ensure adequate positive flow within the furnace.

Furnace Cooling

Most nitriding furnaces are equipped with a heat exchanger that will accelerate cooling of the furnace and work load at the conclusion of the nitriding cycle. When an external water-cooled heat exchanger is used, the furnace heating elements are turned off when the nitriding cycle is completed, and the furnace temperature is allowed to drop approximately 55 °C (100 °F). At this point, the ammonia flow is approximately doubled, and the cooling water is turned on in the heat exchanger. The circulating blower of the heat exchanger also is turned on, and a gate valve is opened to permit circulation of the furnace atmosphere through the heat exchanger. Extreme care must be exercised to ensure a positive gas flow through the furnace as evidenced by the exit gas bubbles. When gas flow through the furnace has been stabilized, the flow may be reduced to the minimum required for positive pressure. After cooling to 150 °C (300 °F) or below, the furnace may be opened.

Bell-type furnaces may be cooled with a cooling bell that is placed over the sealed nitriding retort after the heating bell has been removed. The following is a typical procedure for cooling a bell-type furnace with either raw ammonia or dissociated ammonia:

- Place cooling bell in position on base.
- Insert plug of cooling bell into receptacle.
- Turn on bell cooling fan.
- Cool furnace and load to not less than 315 °C (600 °F), as indicated on base recorder. Turn off flow of dissociated ammonia, and increase flow of raw ammonia to approximately 1.4 m³/h (50 ft³/h).
- When burette reading indicates a dissociation rate of 5% or less and temperature is 120 °C (250 °F) or below, shut off flow of ammonia to furnace and open air valve to furnace. If dissociation rate is 5% or less before temperature falls to 120 °C (250 °F), flow of raw ammonia can be reduced to 1.1 m³/h (40 ft³/h) for remainder of cooling time.
- Remove cooling bell.
- Open air valve to meter; allow 4.2 m³/h (150 ft³/h) of air into furnace, and open exhaust valve wide. (Level on manometer will drop nearly to zero.)
- Turn off base fan when burette reading increases to 65% dissociation. (*A mixture of 16 to 25% ammonia in air is explosive. Therefore, fan must be shut off when ammonia level reaches 35%, or 65% burette reading.*) This eliminates hazard from sparks that may be generated by a moving fan.
- Continue to purge furnace until burette reading is 95% or higher. (This is not a safety precaution but is done to minimize discomfort to personnel nearby when furnace is opened.)
- Close air valve.
- Drain oil seal.
- Raise retort about 0.3 m (1 ft) and wipe off seal oil from retort lip before removing retort completely.
- Remove the charge from the furnace.

Control of Case Depth

Case depth and case hardness, the two criteria most commonly referred to in the control of case properties, vary not only with the duration and other conditions of nitriding, but also with steel composition, prior structure and core hardness.

Aluminum-Containing Steels. Of the aluminum-containing nitriding steels, the most widely used is SAE 7140 (AMS 6470). Figure 5 indicates the hardness gradients and case depths obtained with this steel, as a function of cycle time and nitriding conditions. Results were obtained in single-stage nitriding for various lengths of time up to 800 h and at temperatures ranging from 510 to 540 °C (950 to 1000 °F); several different dissociation rates are represented. The 800-h specimens were nitrided in 11 consecutive 72-h periods. It is apparent that the rate of nitriding decreases over extended periods of time; the case depth obtained after 800 h is only about three times that obtained in 100 h.

Chromium-Containing Low-Alloy Steels. Data relating case depth to nitriding time and conditions for chromium-containing low-alloy steels (principally, 4140, 4337, 4340 and 8640) are given in Fig. 6 and 7. Of these steels, 4140 exhibits the best nitriding characteristics because of its higher chromium content and nickel-free composition. Although 4340 develops a heavier case than 8640 in the first 24 h of nitriding, this difference begins to decrease at the end of a 48-h cycle (Fig. 7).

Data for the AMS equivalents of 4337 and 4140 (AMS 6412 and 6382, respectively) in Fig. 6 are of particular interest because they demonstrate the effect of core hardness on the hardness of the nitrided case. Core hardnesses as low as 21 to 23 HRC, and as high as 36 to 37 HRC, are considered.

Chromium-containing tool steels, such as H11, H12, H13 and D2, provide high core strength with high case hardness, an excellent combination for applications involving severe impact or very high unit loading. Use of these steels is limited primarily by high cost and fabricating difficulties. Case depth results for these steels in single-stage nitriding at 525 °C (975 °F) and at 525 to 540 °C (975 to 1000 °F) are given in Fig. 7. The relatively shallow case depths obtained reflect the retarding effect of increased chromium con-

Fig. 6 Hardness gradients for nitrided chromium-containing low-alloy steels

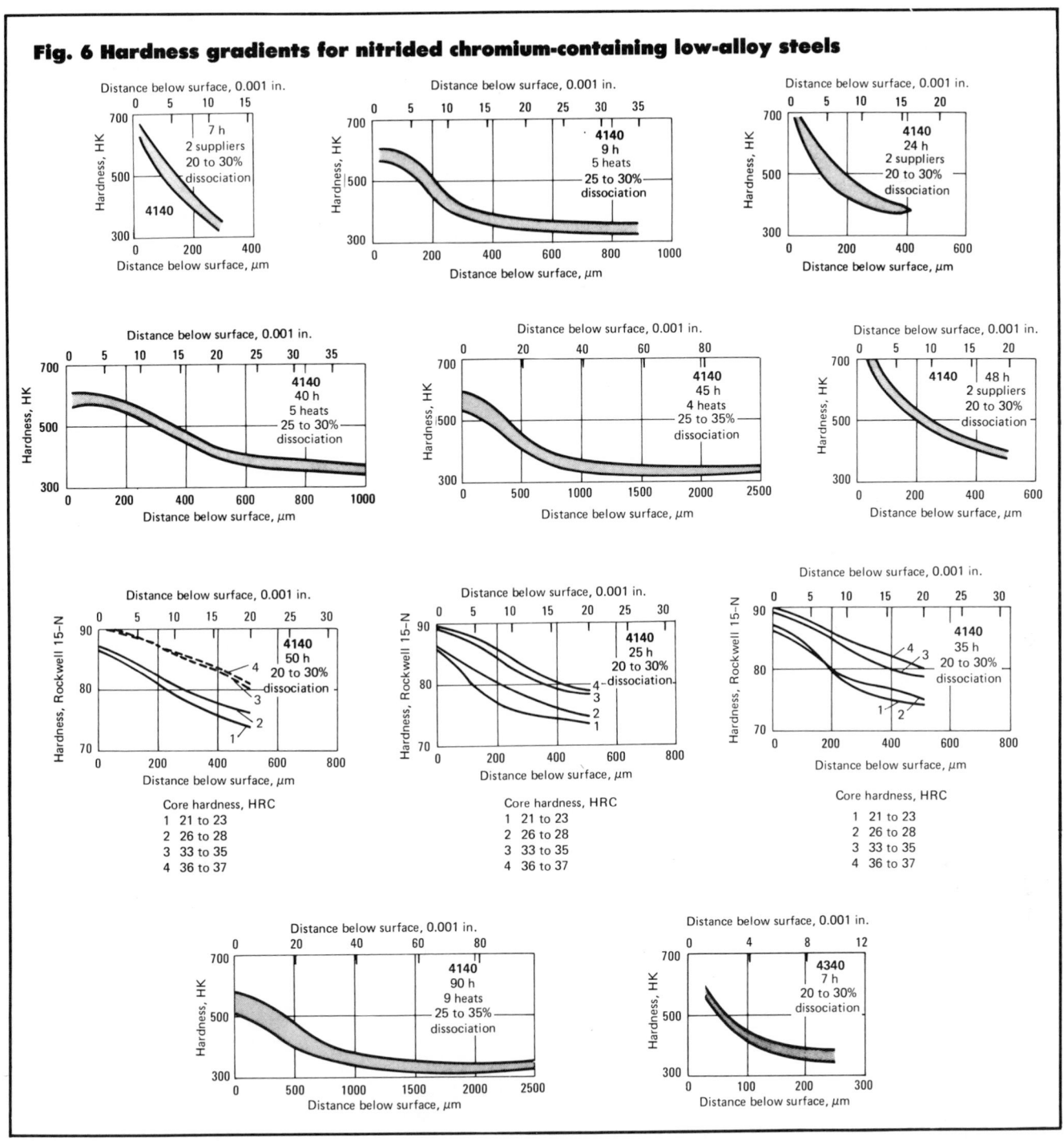

tent on the penetration of nitrogen. Hardness gradients for the same steels are shown in Fig. 8.

Dimensional Changes

During nitriding, parts increase slightly in size because of the increase in volume that occurs in the case. This change causes a stretching of the core, which results in tensile stresses that are balanced by compressive stresses in the case after the parts have cooled to room temperature. The magnitude of the permanent set in the core and case is affected by yield strength of the material, thickness of the case, and by the amount and nature of the nitrides formed. Hence, growth and distortion in nitrided parts are governed largely by composition, tempering temperature, time and temperature of nitriding, relative thickness of case and core, and shape of the part. Growth also is affected when some areas of the part are masked to prevent nitriding.

The amount of growth is usually con-

Fig. 6 (continued)

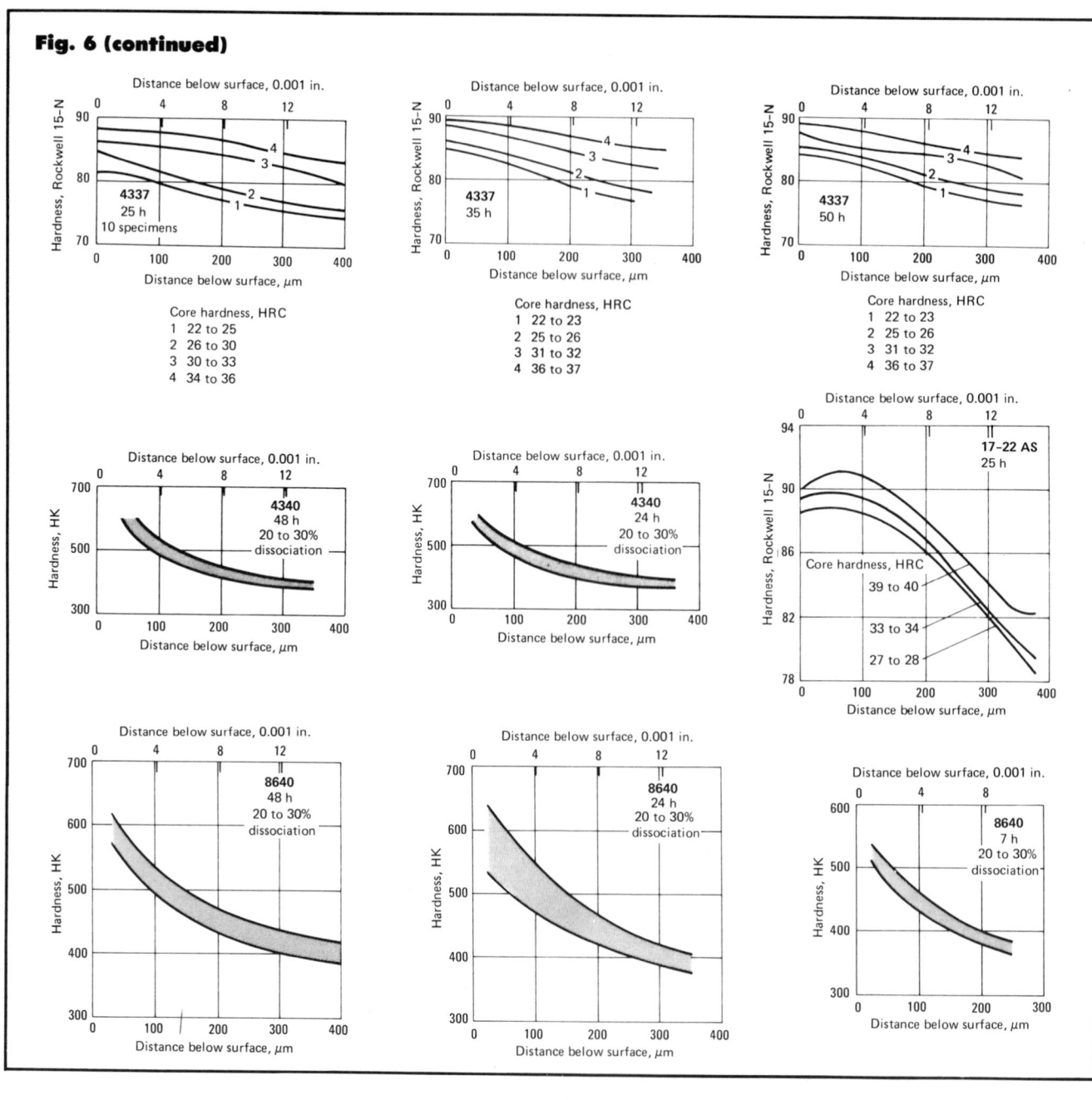

stant for identical parts nitrided in different batches by a fixed processing cycle. Thus, after the amount of growth for a particular part has been determined experimentally, allowance for it can be made during final machining prior to nitriding. Before experiments are conducted to determine growth, parts must be thoroughly stress relieved.

An example of growth as a function of the wall thickness of hollow cylinders made of Nitralloy 135 is shown in Fig. 9. These data may be used as an approximation in estimating growth when nitriding by the double-stage process. They should be used as a guide for determining size changes only with respect to parts of this design, however; the growth that occurs in solid rounds of bars is of the order of 0.04 mm (0.0015 in.) increase in diameter.

In some parts, the dimensional changes during nitriding involve both internal and external surfaces. For example, the bore diameter of a 305-mm (12-in.) -diam spur gear decreased as much as 0.025 mm (0.001 in.); whereas the over-all gear dimension increased up to 0.1 mm (0.004 in.).

Sharp corners or edges should be avoided on parts to be nitrided, because the projections formed at sharp corners, as a result of the growth that takes place, are high in nitrogen content and susceptible to chipping. Similarly, sharp edges nitride throughout the section and have no supporting core. When sharp corners are unavoidable, brittleness may be reduced by nitriding one side only, if the other side is not a wearing surface. Frequently, the problems

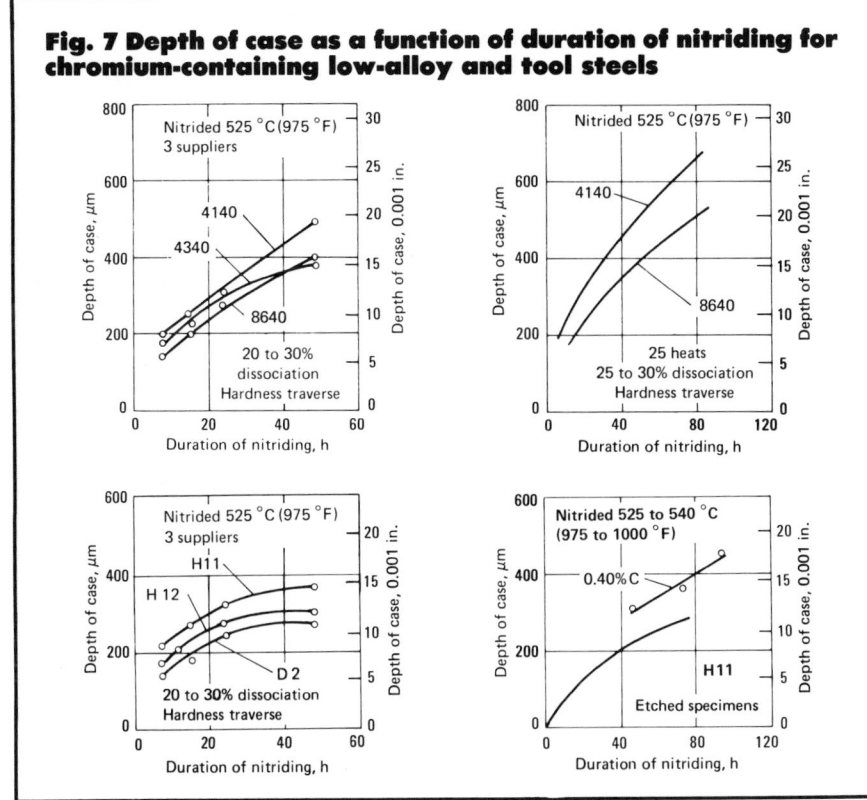

Fig. 7 Depth of case as a function of duration of nitriding for chromium-containing low-alloy and tool steels

of growth are eliminated by nitriding only those surfaces that will be subject to wear in service.

Stabilizing Treatment. In nitrided parts, there is a balance between compressive stresses in the case and tensile stresses in the core. If this balance is upset by grinding off a part of the case, slow dimensional changes may occur as the stresses approach equilibrium. (In some instances, slow dimensional changes resulting from stress redistribution during grinding have been erroneously attributed to wear.) To prevent these changes, nitrided parts are first ground almost to the final dimensions, then heated to 565 °C (1050 °F) for 1 h, and finally finish ground or lapped. Parts nitrided and not ground after nitriding have excellent dimensional stability.

Changes in Helix Angle. A sensitive indicator of dimensional changes resulting from heat treating helical gears is the change in helix angle. Accurate and reproducible measurements of helix angle can be made with electronic measuring instruments.

In general, the change in helix angle is greatest on gears with few teeth and negligible on gears with many teeth.

For example, the lead of a 13-tooth 5-pitch pinion gear might change as much as 0.0005 mm/mm (0.0005 in./in.) of face in nitriding, while the lead of the mating 67-tooth gear would not change. Consequently, when gears with few teeth are to be nitrided, the amount of helix angle change should be compensated for in machining the gear teeth.

Example 3. The amount of lead change encountered with a nitrided 13-tooth 5-pitch helical pinion gear made of 4142 steel varied from batch to batch; the depth of case also varied. It was determined that these variations resulted from the use of different nitriding time cycles from batch to batch in an attempt to accommodate different nitriding specifications in the same furnace load. These variations were sufficient to throw the lead dimension of the gears out of tolerance and necessitated reworking of the gears by a costly lapping operation.

Subsequently, when the nitriding cycle was held constant, the gears were produced regularly within the required tolerance. Figure 10 illustrates the effect of case depth on the lead change of these gears.

Lead measurements also can be used to determine whether distortion in nitrided gears has been caused by residual stress in the raw material. When there is a sharp increase in the standard deviation, or "scatter", of lead measurements after nitriding, the difficulty usually can be attributed to stress relief during the nitriding cycle.

Example 4. Shaft pinions made of mill heat treated 4150 bar stock were scrapped because of a pronounced increase in the deviation of lead measurements after nitriding. An investigation of this problem consisted of machining gears from one third of a lot of material in the as-received condition, stress relieving another third of the material at a temperature 30 °C (50 °F) higher than the nitriding temperature before machining it into gears, and oil quenching from 845 °C (1550 °F) and tempering the last portion before machining; gears of all three prior conditions were then nitrided. Figure 11(a) illustrates the improvement obtained by proper heat treatment prior to machining and nitriding.

Stress Relieving. Many standard procedures require that parts be rough machined, stress relieved and finish machined before being nitrided. For many parts, this lengthy procedure may not be required.

Example 5. Figure 11(b), which presents data relating to the lead change in 4140 steel gears after nitriding, illustrates one instance in which little benefit was obtained by the stress relief of rough-machining stresses before nitriding.

In general, it has been found that stress relieving after rough machining is required only for slender parts or parts with thin wall sections. When distortion is caused by the removal of induced machining stresses during nitriding, stress relieving at 620 °C (1150 °F) for 4 h prior to nitriding will lessen or eliminate this problem.

Design Changes. Sometimes components that have been fully stress relieved distort during nitriding. This is usually the result of the high compressive stresses induced by the volume change occurring from the nitriding action itself. Minor design changes or modification of the production planning of the part to improve the stress balance may prove helpful, as indicated for the parts illustrated in Fig. 12.

Finishing Costs. The amount of distortion resulting from nitriding is small compared to that resulting from

Fig. 8 Hardness gradients for chromium-containing tool steels

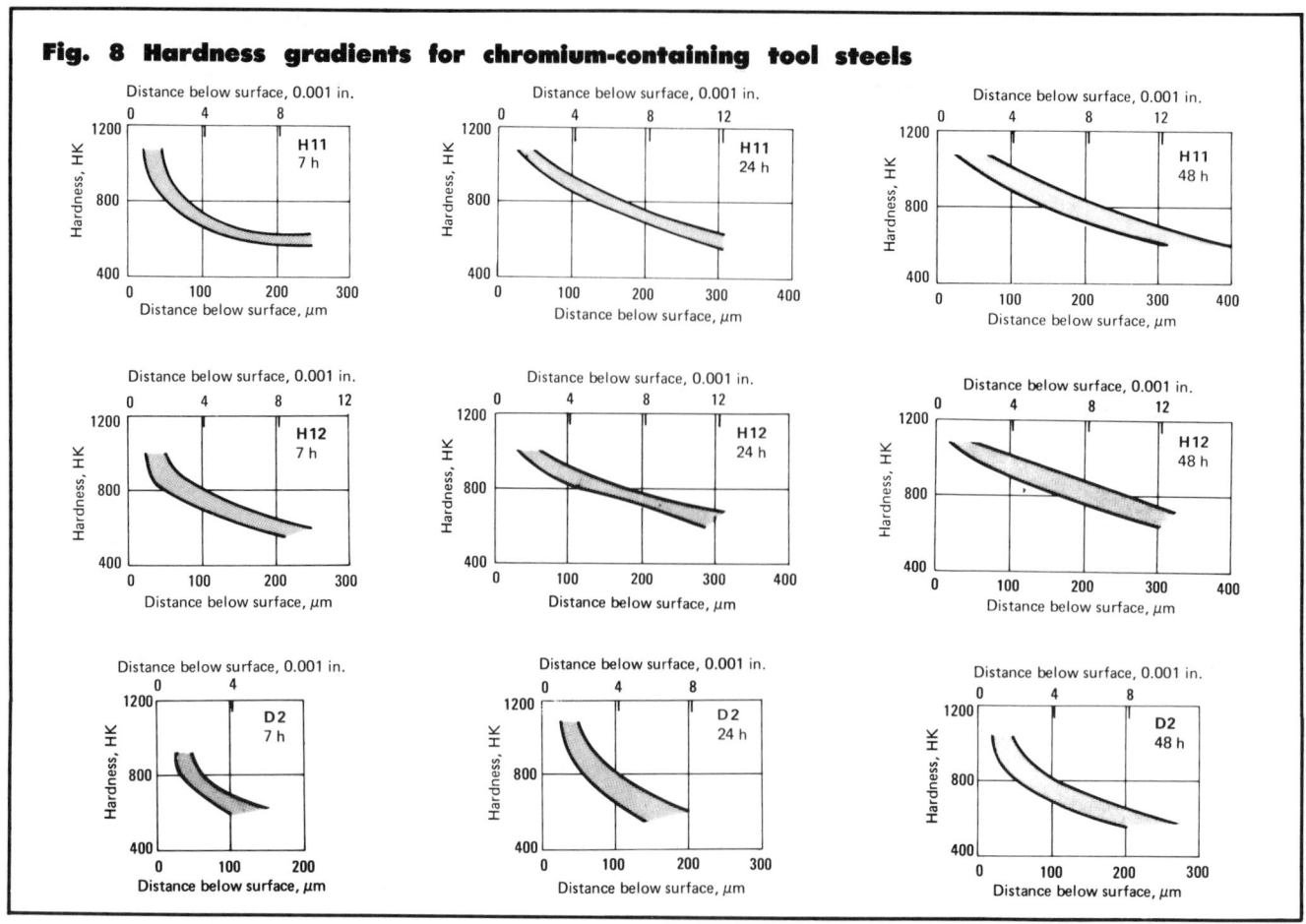

Table 4 Distortion data for carburized and for nitrided gears of five sizes

| | | Gear dimensions | | | | | | Helix error | | | | Material | |
| | L | F_c | | OD | | Diametral | Carburizing 4130 | | Nitriding AMS 6475 | | | required per gear | |
mm	in.	mm	in.	mm	in.	pitch	mm	10^{-4} in.	mm	10^{-4} in.	kg	lb
76	3.0	30	1.2	30	1.2 ...	20	0.08	32	0.025	10	0.5	1.2
84	3.3	33	1.3	36	1.4 ...	16	0.09	37	0.030	12	1.0	2.2
95	3.8	41	1.6	41	1.6 ...	14	0.12	48	0.033	13	1.3	2.8
138	5.5	58	2.3	66	2.6 ...	9	0.16	62	0.046	18	4.7	10.3
160	6.3	71	2.8	76	3.0 ...	7	0.20	78	0.066	26	7.0	15.4

other case hardening processes, which involve quenching to form martensite. Consequently, the increased cost of the nitriding operation and of steels suitable for nitriding often can be offset by the savings resulting from finishing to size prior to nitriding.

Example 6. One manufacturer realized considerable savings in the cost of producing gears by changing from 4130 steel to AMS 6475 (Nitralloy N). Gears made of 4130 steel were carburized in salt and marquenched in salt at 260 °C (500 °F), and had to be lapped after being treated to eliminate distortion; the nitralloy gears were nitrided, and

lapping was not required. Although the 4130 steel cost less than the nitralloy, the cost of the lapping required for the 4130 gears substantially offset the lower cost of this material. See Table 4 for distortion data for carburized and for nitrided gears of five sizes.

Equipment

Furnaces of several designs are in common use in gas nitriding installations. Most of these are batch furnaces, which incorporate certain essential features, including:

- A means of sealing the charge, to

exclude air and other contaminants while containing the controlled atmosphere

- An inlet line for introducing atmosphere and an outlet line for exhausting used atmosphere
- A means of heating and appropriate temperature controls
- A means, such as a fan, for circulating atmosphere and equalizing temperature throughout the work load.

The vertical retort furnace (Fig. 13) is stationary; parts to be nitrided are loaded into a work basket, which is lowered into the heating chamber. The lid rests on an asbestos gasket and dips

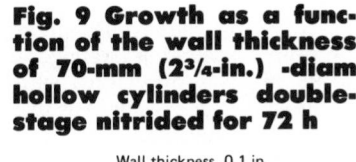

Fig. 9 Growth as a function of the wall thickness of 70-mm (2³⁄₄-in.) -diam hollow cylinders double-stage nitrided for 72 h

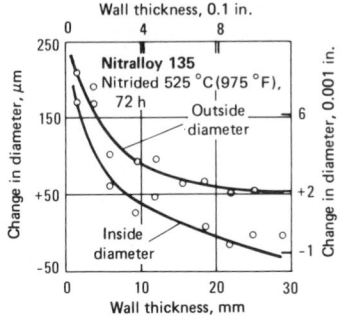

Fig. 10 Increase in lead change with depth of nitrided case for 13-tooth 5-pitch helical pinion gear

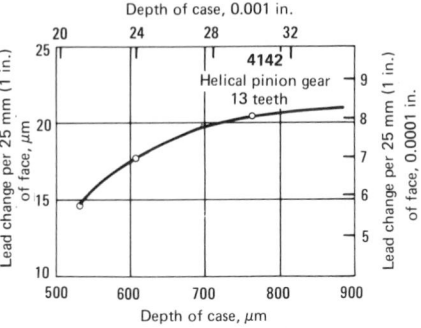

Gears were hardened, tempered at 565 °C (1050 °F), and nitrided in two stages at 525 °C (975 °F) using ammonia dissociation rates of 15 to 25% for the first stage and 83 to 85% for the second stage.

into an oil-filled trough, thus effecting the seal. Atmosphere enters at the top and leaves at the bottom of the furnace. Cooling is achieved by starting a fan and opening a valve in a water-jacketed cooling manifold. Furnaces of similar design, but without the water-jacketed manifold, are used when rapid cooling as a means of increasing furnace output is not required; the quality of nitriding achieved is equivalent to that of manifolded furnaces.

The bell-type movable furnace (Fig. 14) has a stationary base and is equipped with atmosphere inlet and outlet, control thermocouple, circulating fan, and outlets for electric power and controls. Parts to be nitrided are loaded into work baskets, which are placed on a work support at the furnace base. A retort is lowered over the base and dips into an oil-filled trough, effecting the seal. Heat is provided by a heating bell, which is lowered over the retort and rests at the bottom on a flat portion of the retort. Heat passes through the retort walls and is transferred to the work load by radiation and convection as the atmosphere is circulated. Cooling is achieved by replacing the heating bell with a cooling bell, which draws air up around the retort walls and out at the top of the bell. This air movement is accomplished by a fan in the top of the cooling bell.

It is customary to provide more bases than heating and cooling bells. This permits more efficient utilization of the bells. A cooling bell is not essential where there is no demand for rapid cooling as a means of increasing output of the base; the quality of nitriding achieved is equivalent to that achieved using a cooling bell, but the heat radiated from the retort can be a source of discomfort to persons working in the immediate area.

Box Furnaces. A box-type movable furnace with two stationary locations for work baskets also has been used for nitriding. Each location is equipped with atmosphere inlet and outlet, control thermocouple, circulating fan and separate controls. Parts to be nitrided are loaded into baskets, which are placed on a heavy metal plate at each location. A cover is placed over the charge and settles into a groove in the plate that is filled with fine chrome ore, thus effecting the seal. Lugs on the cover fit into U-shape holders through which pins are driven to secure the cover to the plate.

The furnace then is moved into position over the charge on rails, after which sliding doors at each end of the furnace are lowered. Heat passes through the walls of the cover and is transferred to the load by radiation and convection as the atmosphere is circulated. When the box furnace has been rolled into position over the work in the basket on the second plate, cooling is achieved for the work on the first plate by transfer of heat to the surroundings, either by natural or forced circulation of air. Usually, natural air circulation is rapid enough to permit cooling and recharging before the box furnace is again available.

Box furnaces of similar design are in use in which the furnace is stationary and the locations for the work loads are movable. The cover that closes over the load is similar to the retort used with the bell-type furnace.

Tube Retorts. In nitriding the inside diameters of tubes, the tube itself may act as the retort after it has been sealed (usually by welded-on covers) at both ends. A calculated volume of ammonia is sealed in the tube (after the tube has been purged of air), and the tube is heated in a suitable furnace. After the heating cycle, the tube is cooled in still or circulated air, and the covers are removed. Individual parts may be sealed in tube retorts and processed in this same manner, see the section of this article on pressure nitriding.

Temperature Control. Close control of nitriding temperatures is essential to prevent uneven heating and distortion of parts, many of which are finish machined prior to nitriding. Nitriding furnaces are equipped with two thermocouples: one to control and indicate temperature within the load, and another to control the heat source so that it does not exceed a maximum temperature—usually 5 to 15 °C (10 to 25 °F) above the nitriding temperature. These two thermocouples, in conjunction with fan circulation, normally result in control of furnace and load temperature to within ±3 to ±6 °C (±5 to ±10 °F). The override thermocouple arrangement also reduces the likelihood of overheating if the control couple or recorder fails.

Fixtures for nitriding are similar in design to those used in gas carburizing. Under nitriding conditions, ammonia and dissociation products can react chemically with materials in retorts, fans, work baskets and fixtures. This reaction contributes to further dissociation of ammonia, robs the work of atomic nitrogen, and produces an excess of hydrogen. To reduce this reaction to a minimum, furnace parts and fixtures usually are made of alloys containing high percentages of nickel and chromium (see Table 5). Under certain conditions or after extended use, even these alloys develop a surface that interferes with normal processing; their usefulness can be restored, however, by heating them in an air atmosphere and holding them for a period of time at elevated temperature, followed by sandblasting to remove scale.

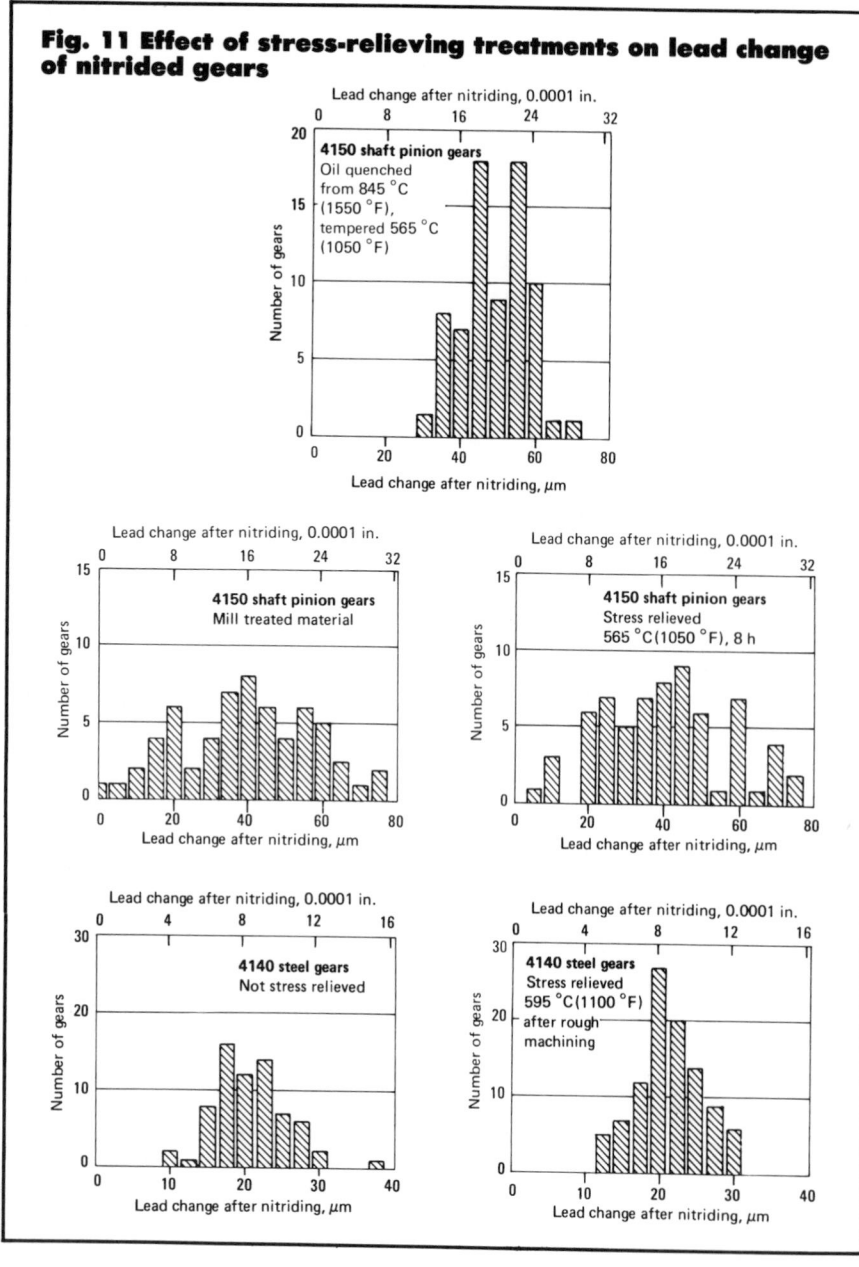

Fig. 11 Effect of stress-relieving treatments on lead change of nitrided gears

Ammonia Supply

Gas nitriding makes use of anhydrous liquid ammonia (refrigeration grade, 99.98% NH_3 by weight), which is available either in cylinders or in bulk (tank truck, trailer transport and tank car). A typical storage-tank installation with 1050-kg (2300-lb) capacity is shown in Fig. 15. Such a tank is replenished directly from a tank truck or tank car. Layouts for ammonia installations and engineering data pertaining to their operation and maintenance may be obtained from suppliers of ammonia.

Usually a storage tank is situated outside the building in which the nitriding equipment is located. At moderate outdoor temperatures, the liquid ammonia will absorb enough heat from the atmosphere to vaporize and fulfill gas requirements. On very hot days, the pressure of the gas may build up enough to actuate the pressure-relief valves. On the other hand, when temperatures are below −7 °C (+20 °F) or when very large volumes of gas are being used, an additional heat source is needed. This heat may be supplied by an electric immersion heater automatically actuated by gas pressures. Such a heater is started when gas pressure falls below 690 kPa (100 psi) and is stopped when a pressure of 1035 kPa (150 psi) is attained.

Special Precautions. To avoid leaks, exceptionally good pipe-fitting practice must be followed. Specific pipe-joint compounds must be used. One type of compound contains fine powdered lead, which is mixed in an insoluble, non-setting lubricant; another type is an oxychloride mixture with graphite, which in setting, expands to form a very hard seal. When properly applied, certain high-strength, corrosion-resistant tapes also are satisfactory, as are welded joints.

Materials used for valves, piping, gages, regulators and flow-measuring devices are similar for all installations; only iron, steel, stainless steel, and aluminum can be used, because ammonia corrodes zinc, brass and bronze. Piping should be made of extra-heavy black iron (except for vent lines, which may be made of standard-weight black iron or galvanized iron). Fittings should be made of extra-heavy malleable iron or forged steel. Valves should be made of steel and should be of the high-pressure, back-seating type.

Pressure Regulation. Ammonia

Enamel-coated carbon steel containers are satisfactory as long as the coating remains intact. Some alloys have been coated with high-temperature glass to extend their usefulness.

Low-carbon steel is unsatisfactory as a container material because it absorbs nitrogen, and the nitrided inside surface of the container becomes embrittled. Besides embrittling the container, the nitrided surface also catalyzes the decomposition of the ammonia in contact with it and thereby interferes with the proper nitriding of workpieces.

Equipment Requirements. Typi-

cal relations between production requirements and equipment requirements for gas nitriding loads of parts weighing less than 0.5 kg (1 lb) each and of parts weighing from 0.5 to 5 kg (1 to 10 lb) each are given in Table 6. These data apply to single-stage nitriding at 525 °C (975 °F) for periods of 24 and 48 h, respectively.

Example 7. Table 7 lists processing details, and correlates production and equipment requirements, for the single-stage nitriding of 5.3-kg (11.7-lb) transmission ring gears to a depth of 0.2 mm (0.008 in.).

Table 5 Recommended materials for parts and fixtures in nitriding furnaces

Materials are recommended on basis of maximum operating temperature of 565 °C (1050 °F).

	Material	
Part	Wrought	Cast
Retorts(a)....................	Type 330; Inconel 600	Not usually cast
Fans.......................	Type 330; Inconel 600	35-15 or equivalent
Trays, baskets, fixtures....................	Types 310, 330; Inconel 600	35-15 or equivalent
Thermocouple protection tube..............	Type 330; Inconel 600	Not usually cast

(a) Periodic inspection of austenitic stainless steel retorts is mandatory, because of embrittlement after long exposures to nitriding. Retorts of 18-8 stainless steel lined with high-temperature glass have been used successfully.

Table 6 Typical equipment requirements for single-stage gas nitriding

Item	Parts less than 0.5 kg (1 lb) each	Parts from 0.5 to 4.5 kg (1 to 10 lb) each
Production requirements		
Pieces per cycle..........................	800	300
Weight of pieces, kg (lb)..............	170 (375)	910 (2000)
Purging time, h.........................	2	2½
Heating time, h.........................	1	1¼
Nitriding time, h........................	24	48
Cooling time, h.........................	2	2½
Total cycle, h..........................	29	54¼
Equipment requirements		
Size of furnace..................	0.2 m³ (6.7 ft³)	0.5 m³ (17 ft³)
Retort dimensions..............	610-mm (24-in.) diam by 660 mm (26 in.) deep	710-mm (28-in.) diam by 1220 mm (48 in.) deep
Temperature....................	525 °C (975 °F)	525 °C (975 °F)
Electric supply:		
Elements........................	30 kW	48 kW
Motor, hp..........................	1	3

gas from the supply tank or cylinder bank is under pressures up to 1380 kPa (200 psi), depending on the temperature of the gas. This pressure is reduced to about 14 to 105 kPa (2 to 15 psi) by means of pressure regulators.

Another reduction must be made just ahead of each furnace or dissociator to about 255 to 1015 mm (10 to 40 in.) water column; this is adequate to supply as much as 4 m³/h (150 ft³/h) of ammonia, more than ample for most supply needs. Such supply lines are arranged to feed from a common line operating in manifold fashion at pressures not exceeding about 10 kPa (1.5 psi). Equipment to obtain this last reduction may be furnished with the dissociator or furnace.

The flow of gas into furnaces or dissociators is regulated by a suitable needle valve and is measured by a device such as a flowmeter. This device also serves to permit a visible check that gas is moving through the lines. Flow and pressure may be monitored by contact points that close and sound an alarm at predetermined settings.

Exhaust Gas. Depending on the stage of the cycle, the exhaust gas may contain air, air and ammonia, or ammonia plus hydrogen and nitrogen. Because of the variable composition of exhaust gas and the customary use of only a single exhaust line, the exhaust gas should be conducted to the outside atmosphere and released at as high an elevation as practicable. Under no circumstances should the exhaust line terminate inside a building, in a closed area, or in a container of water. Note that environmental considerations may dictate a more sophisticated approach to handling exhaust gas.

To provide a slight back pressure within the furnace, an oil-containing bubble bottle may be installed in the exhaust line. By means of appropriate valving, samples of the exhaust gas may then be withdrawn through a dissociation burette. As an alternative, a throttle valve, installed in the exhaust line, may be used to restrict the flow of exhaust gases and maintain a slight back pressure in the furnace. This pressure is indicated on a manometer (water-type) and maintained at about 25 to 50 mm (1 to 2 in.) water column. Suitable piping will permit a continuous flow of exhaust gas through a dissociation burette and past the throttle valve. Gas from the furnace may be piped into the burette from a connection just ahead of the throttle valve and returned to the exhaust line at a location just past the valve.

Safety Precautions

Anhydrous ammonia is flammable within a narrow range; *concentrations of 15 to 26% ammonia in air produce explosive mixtures.* Ammonia is classified as a nonflammable compressed or liquefied gas by the Interstate Commerce Commission and is shipped under a green label. Because of the high coefficient of expansion of liquid ammonia, all containers must be filled in accordance with Department of Transportation (DOT) regulations, to allow for this expansion in the event of temperature rise.

Dry ammonia is not corrosive to iron or steel and therefore entails no problems of internal corrosion in storage containers or piping. Moist ammonia in contact with air, however, is corrosive, and leaks in any portion of the system must be avoided. All storage containers, valves and piping should be examined periodically for signs of external corrosion. Corrosion-preventive coatings should be applied to all parts of an ammonia storage or distribution system.

Ammonia gas is not harmful at low concentrations, and because of its pungent odor, leaks are readily noticed. Leak detection, using sulfur dioxide or sensitized papers, is simple and positive.

Ammonia constitutes a potential panic hazard. Because of the discomfort resulting from traces of ammonia in air, adequate ventilation and exhaust facilities should be employed, particularly in enclosed areas. A gas mask approved for use in ammonia atmospheres always should be available for use in the event of bad leaks. Protective clothing, such as gloves, hats and gog-

Fig. 12 Minor design changes and modification of production planning to eliminate distortion during nitriding

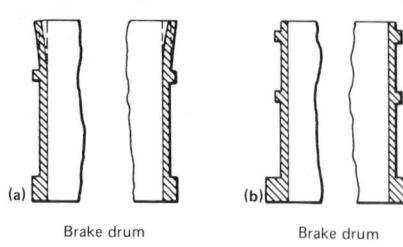

(a) Brake drum   (b) Brake drum

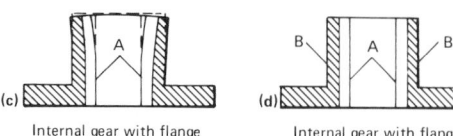

(c) Internal gear with flange   (d) Internal gear with flange

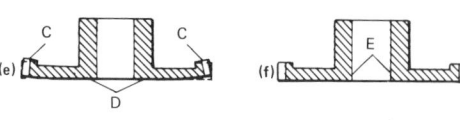

(e) Flat spur gear   (f) Flat spur gear

(a) Nitriding of ID caused excessive taper past shoulder. (b) Location of original shoulder was changed, and second shoulder was added. (c) Compressive stresses caused gear teeth A to taper during nitriding. (d) Gear teeth A and external surface B were nitrided. (e) Nitriding of gear teeth C and surface D caused warpage. (f) Entire part was nitrided except surface E, which was finished after nitriding.

Table 7 Requirements for nitriding transmission ring gears to a depth of 0.2 mm (0.008 in.)

Cycle

Purge with raw ammonia	1.5 h
Heat to 525 °C (980 °F)	3.0 h
Nitride at 525 °C (980 °F) (40% dissociation)	32.0 h
Purge with ammonia and cool	2.0 h
Purge with air and continue cooling	1.5 h
Total cycle time	40.0 h

Production requirements

Load weight	1340 kg (2950 lb)
Weight of each piece	5.3 kg (11.7 lb)
Total fixture weight	670 kg (1470 lb)
Pieces processed per hour	7½ (avg)

Furnace requirements

Furnace	Electric bell-type batch
Hearth size	1525-mm (60-in.) diam, 1800-mm (71-in.) height
Heat input rate	360 000 kJ/h (340 000 Btu/h) (360 MJ or 100 kW)
Temperature	530 °C (980 °F) (650 °C or 1200 °F max)

Atmosphere equipment

Ammonia dissociator capacity	2.8 m³/h (100 ft³/h)

Source: 3785 l (1000-gal) tank for liquid NH_3 vaporizer

Average ammonia consumption

Purging	4.2 m³/h (150 ft³/h)
Nitriding	1.75 m³/h (62 ft³/h)

gles, also should be provided for emergencies.

Ammonia is highly soluble in water. In case of severe leaks, spraying equipment is effective in carrying away the fumes. The gas is lighter than air and will rise; in emergencies, it should be remembered that the area closest to the floor will be lowest in ammonia content.

Hydrogen Hazard. *Although anhydrous ammonia is classed as a nonflammable gas, it produces considerable amounts of hydrogen (which is flammable) upon cracking. Cracking, or complete dissociation, does not occur in the nitriding furnace, but there is enough hydrogen contained in the exhaust gases to constitute a potential hazard. Because of the concentrations of hydrogen and ammonia in exhaust gases, these gases must be vented to the outside atmosphere and not into an enclosed area. The exhaust line should never be terminated in a container of water, and it is not good practice to attempt to burn the exhaust gases indoors or outdoors.*

Because of the presence of hydrogen in the nitriding furnace, the furnace should never be opened while it is heated up to nitriding temperature. If it is necessary to remove the work before the furnace has cooled to below 150 °C (300 °F), the furnace must be thoroughly purged with an inert gas, such as nitrogen. Even at 150 °C (300 °F) or below, the furnace should be thoroughly purged with air before it is opened.

Common Nitriding Problems

Some of the problems commonly encountered in nitriding are

- Low case hardness or shallow case
- Discoloration of workpieces
- Excessive dimensional changes
- Cracking and spalling of nitrided surfaces
- Variations in percentage of ammonia dissociation
- White layer deeper than permitted
- Plugging of exhaust lines and pipette lines.

A knowledge of the causes of these problems should be of assistance in avoiding, preventing or correcting them. A number of possible causes are indicated below.

Low case hardness or shallow case may be caused by:

Characteristics of the steel

- Composition unsuitable for nitriding

Fig. 13 Vertical retort nitriding furnace

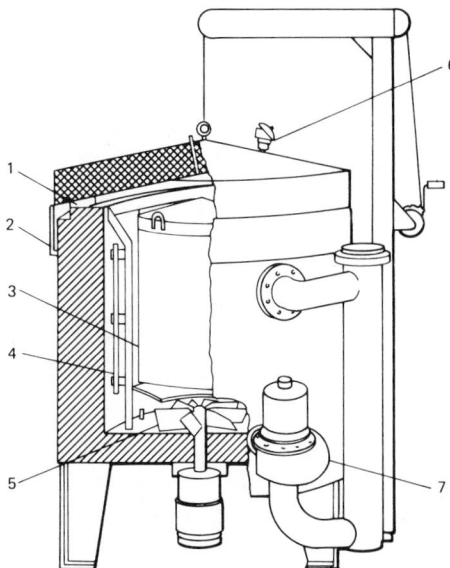

1, gasket; 2, oil seal; 3, work basket; 4, heating elements; 5, circulating fan; 6, thermocouple; and 7, cooling assembly. At end of cycle, a valve is opened and fan (not shown) incorporated in the external cooler circulates atmosphere through the water-jacketed cooling manifold.

- Improper microstructure
- Failure to quench and temper prior to nitriding
- Low core hardness
- Surface passivation, from machining, inadequate cleaning, or foreign matter

Faulty processing

- Excessively low or high nitriding temperature
- Insufficient ammonia flow
- Nonuniform circulation or temperature in furnace
- Prolonged exposure of furnace parts and work baskets to nitriding conditions such as ammonia (burnout required); see section on fixtures
- Insufficient time at temperature.

On the other hand, low case hardness or shallow case may only be apparent—occurring as the result of inaccuracies in testing, due to faulty adjustment of equipment, improper preparation or positioning of the test specimen, or the use of a test load excessive for the case depth.

Discoloration of workpieces may be caused by:

- Improper or inadequate prior surface treatment, including etching, washing, degreasing and phosphate coating

Oil in the retort, because of

- Inadequate cleaning of parts, especially those with deep holes and recesses
- Loss of pressure at seal, or overheating of seal
- Leakage at the base or other parts of the furnace

Moisture in the retort, because of

- Leakage from the cooling chamber
- Water being sucked in from water bottle during rapid cooling with inadequate gas flow

Air in the retort, because of

- Inadequate seal
- Leakage due to inadequate sealing around pipes or thermocouple
- Introduction of air to purge ammonia while charge is at or above 175 °C (350 °F).

Excessive dimensional changes may be caused by:

- Inadequate stress relieving prior to nitriding
- Inadequate support of parts during nitriding
- Inappropriate design of parts, including nonsymmetry of design, wide variations in section thickness

Unequal cases on various surfaces of parts, resulting from

- Nonuniform conditions, created by furnace design or manner in which parts are arranged in load
- Variations in absorptive power of surfaces, resulting from stop-off practices or from variations in surface metal removed, surface finishing technique, or in degree of cleanness.

Cracking and spalling of nitrided surfaces may be caused by dissociation in excess of 85% and also (especially for aluminum-containing steels) by:

- Design (particularly sharp corners)
- Excessively thick white layer
- Decarburization of surface in prior heat treatment
- Improper prior heat treatment.

Variations in percentage of ammonia dissociation may be caused by:

- Charge being too small for furnace area
- Overactive surface of furnace parts and fixtures
- Leakage or loss of sample from burette
- Change in gas flow, caused by build-up of pressure in furnace
- Variations in furnace temperature.

White layer deeper than permitted may be caused by:

- Nitriding temperature being too low
- Percentage of dissociation being below recommended minimum (15%)
- Fast purging with raw ammonia, instead of cracked ammonia, above 480 °C (900 °F) during slow cooling.

Plugging of exhaust lines and pipette lines is caused by precipitates that are formed by the reaction of ammonia with many of the various chemical compounds commonly present in ordinary domestic water. These precipitates may plug lines and prevent proper sampling, or cause pressure to build up in the furnace by plugging exhaust lines or restricting valve openings.

Enlarging lines or treating them periodically with a dilute acid solution will correct this, especially if the solution is trapped in a low spot and drained. (The use of distilled water, or water of similarly low impurity, also will eliminate this difficulty.)

Fig. 14 Schematic of bell-type furnace showing stationary base surmounted by bell

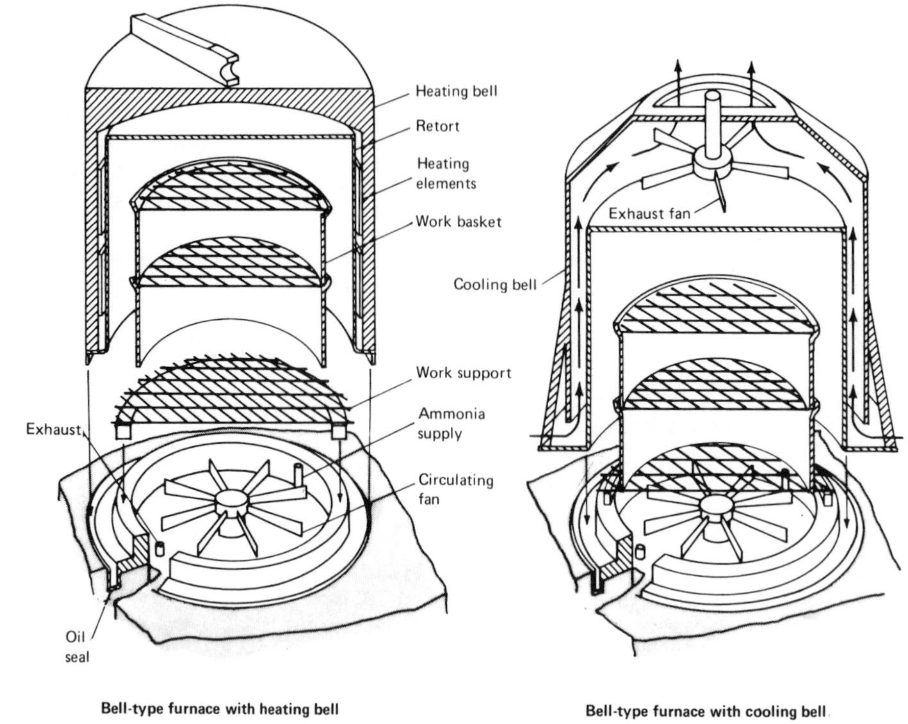

Heating bell
Retort
Heating elements
Work basket
Exhaust fan
Cooling bell
Work support
Ammonia supply
Circulating fan
Exhaust
Oil seal

Bell-type furnace with heating bell   Bell-type furnace with cooling bell

Fig. 15 Typical anhydrous ammonia storage-tank installation of 1045-kg (2300-lb) capacity

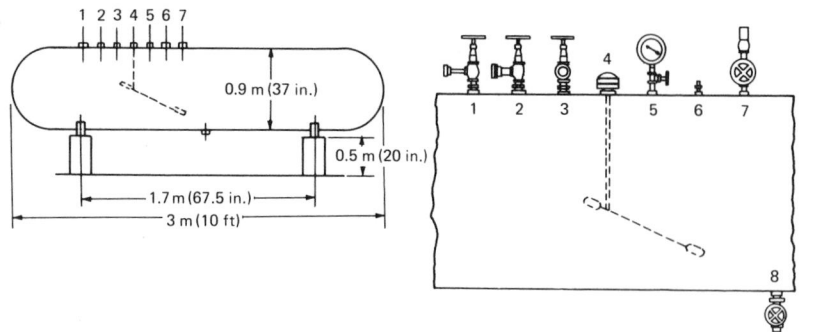

1 2 3 4 5 6 7

0.9 m (37 in.)

0.5 m (20 in.)

1.7 m (67.5 in.)

3 m (10 ft)

1, pressure-equalizing valve; 2, liquid inlet valve; 3, gas outlet valve; 4, liquid level float gage; 5, pressure gage; 6, fixed level gage; 7, pressure relief valves (2); and 8, liquid outlet valve

In some installations, water from pipettes can leak down into exhaust lines, flushing scale and other foreign material into low spots or restrictions and thus plugging the lines. A drop leg to trap such products will reduce trouble from this source, as will reduction of right-angle bends and elimination of pipes smaller than 19 mm (³/₄ in.) in diameter, where possible.

Selective Nitriding

Many coatings are available as stop-offs to prevent gas nitriding of selected areas. The success of a coating depends on such variables as density and thickness of the coating, adhesion of coating to steel, surface finish of the part and degree of leakage permitted.

Proprietary paints are effectively used in commercial heat treating operations. They also are used to touch up other coatings that have been inadvertently removed or damaged during processing. These paints usually consist of a tin base suspended in a vehicle of lacquer, aromatic hydrocarbon, or water glass. It is important that the constituents be mixed in the proper proportions (thick coatings may run, and thin coatings are not completely effective) and that the paints be applied to uniform thickness. The surface to be painted must be very clean.

Plated deposits of bronze or copper are the most common stop-off coatings. Nickel (including electroless nickel), chrome, and silver are effective also, but their higher cost restricts their use to special applications.

Thickness and density of plated coatings are important in determining their effectiveness as stop offs. Minimum thickness of bronze or copper plate should be 18 μm (0.7 mil) for ground surface finishes of 1.6 μm (64 micro-in.) or smoother, 25 μm (1.0 mil) for finishes between 1.6 and 3.2 μm (64 and 125 micro-in.), and 38 μm (1.5 mil) for finishes of 3.2 μm (125 micro-in.) and rougher. Compared to copper and bronze, nickel is a more effective stop-

Fig. 16 Hardness range as a function of depth of case for four stainless steels that were annealed prior to nitriding

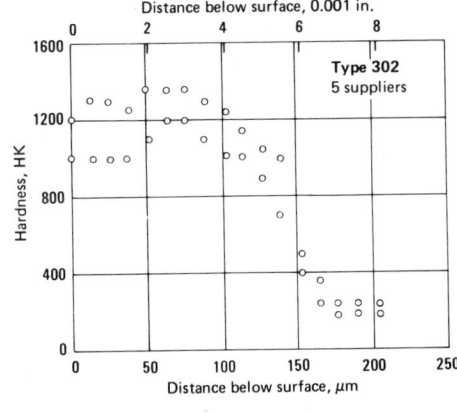

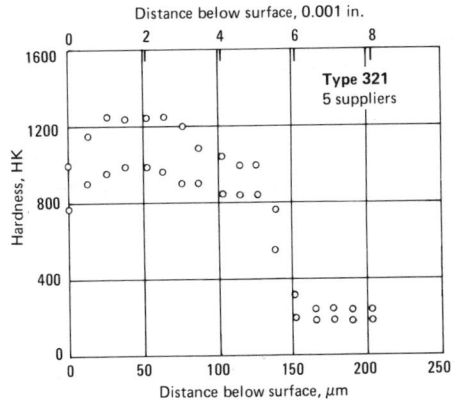

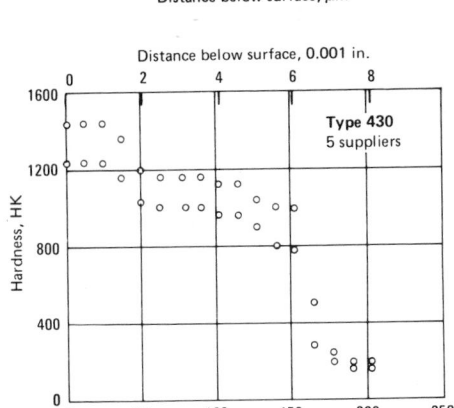

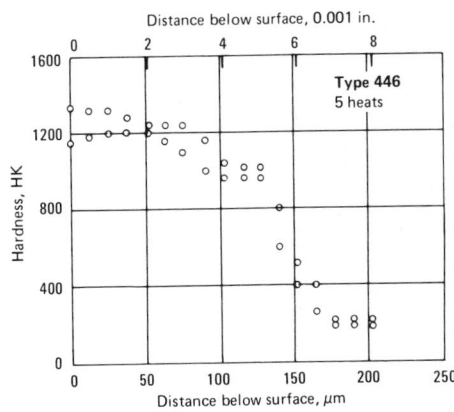

Annealing temperatures: type 302 and type 321, at 1065 °C (1950 °F); type 430, at 980 °C (1800 °F); and type 446, at 900 °C (1650 °F)

off; therefore, a thinner coating is permitted.

Electroplated silver is 100% effective when the plate thickness is a minimum of 38 μm (1.5 mil); it is 95% effective, even during long nitriding cycles, when as little as 25 μm (1.0 mil) of plate is used.

Surface finish of the basis metal also influences the thickness of the coating. A finish of 3 μm (120 micro-in.) will require a thicker coating than a finish of 1.5 μm (60 micro-in.). Usually, a finish of 1.5 μm (60 micro-in.) or smoother is recommended.

Processing Procedures. Several processing procedures are employed to accomplish selective nitriding. One of the most widely used consists of rough machining, plating, machining or grinding areas to be nitrided, nitriding, then finish machining or grinding wherever required. In another procedure, the areas to be nitrided are masked to prevent plating. When

masking is difficult, the plating material is applied to all surfaces and then selectively stripped from the areas to be nitrided.

Fine threads (external or internal) on precision parts can be protected by a tin-lead solder. The threads should be cleaned and coated with a flux containing a tinning compound, then heated slowly until both solder and flux are melted. The excess solder and flux are blown out with compressed air, leaving a coating thin enough so that it does not run during nitriding and does not require cleaning or stripping after nitriding.

When the application does not permit the retention of any protective plate on the finished part after nitriding, selection of the coating is important from the standpoint of subsequent stripping. Copper and silver are the easiest to strip; bronze is more difficult. Nickel is very difficult to remove without detrimentally affecting the part.

Nitriding of Stainless Steels

Because of their chromium content, all stainless steels can be nitrided to some degree. Although nitriding adversely affects corrosion resistance, it increases surface hardness and provides a lower coefficient of friction, thus improving abrasion resistance.

Austenitic and Ferritic Alloys. Austenitic stainless steels of the 300 series are the most difficult to nitride; nevertheless, types 301, 302, 303, 304, 308, 309, 316, 321 and 347 have been successfully nitrided. These nonmagnetic alloys cannot be hardened by heat treating; consequently, core material remains relatively soft, and the nitrided surface is limited as to the loads it can support. This is equally true of the nonhardenable ferritic stainless steels. Alloys in this group that have been satisfactorily nitrided include types 430 and 446. With proper prior treatment, these alloys are somewhat easier to nitride than the 300 series alloys.

Hardenable Alloys. The hardenable martensitic alloys are capable of providing high core strength to support the nitrided case. Hardening, followed by tempering at a temperature that is at least 15 °C (25 °F) higher than the nitriding temperature, should precede the nitriding operation. Precipitation-hardening alloys, such as 17-4 PH, 17-7 PH and A-286, also have been successfully nitrided.

Prior Condition. Before being gas nitrided, 300 series steels and nonhardenable ferritic steels should be annealed and relieved of machining stresses. The normal annealing treatments generally employed to obtain maximum corrosion resistance are usually adequate. Microstructure should be as nearly uniform as possible. Observance of these prior conditions will prevent flaking or blistering of the nitrided case. Martensitic steels, as previously noted, should be in the quenched and tempered condition.

A special pretreatment for 410 stainless is hardening from a lower-than-normal temperature; this results in a very unifrom nitrided case with reduced internal stresses. Cracking or spalling of the case is avoided; formation of brittle grain-boundary carbonitrides is suppressed. Austenitizing at 680 °C (1580 °F), followed by tempering at 595 °C (1100 °C), uniformly distributes carbides and provides low residual stress. Case growth is accommodated by a hardness of about 25 HRC.

Surface Preparation. The nitriding of stainless steels requires certain surface preparations that are not required for nitriding low-alloy steels. Primarily, the film of chromium oxide that protects stainless alloys from oxidation and corrosion must be removed. This may be accomplished by wet blasting, pickling, chemical reduction in a reducing atmosphere, or submersion in molten salts, or by one of several proprietary processes. Surface treatment must precede placement of the parts in the nitriding furnace. If there is any doubt of the complete and uniform depassivation of the surface, further reduction of the oxide may be accomplished in the furnace by means of a reducing hydrogen atmosphere or halogen-based proprietary agents. Of course, hydrogen must be dry—free of water and oxygen.

Before being nitrided, all stainless parts must be perfectly clean and free of embedded foreign particles. After depassivation, care should be exercised to avoid contaminating stainless surfaces with fingerprints. Sharp corners should be replaced with radii of not less than 1.6 mm (1/16 in.).

Nitriding Cycles. In general, stainless steels are nitrided in single-stage cycles at temperatures from about 525 to 550 °C (975 to 1025 °F) for periods ranging from 20 to 48 h, depending on the depth of case required. Dissociation rates for the single-stage cycle range from 20 to 35%. Thus, except for the

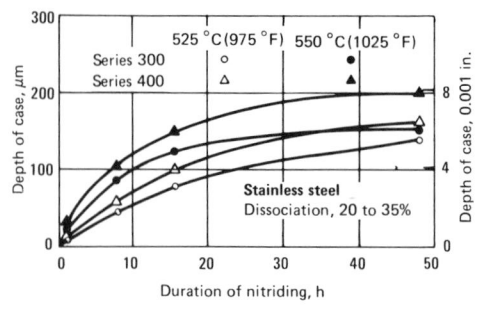

Fig. 17 Comparison of nitriding characteristics of series 300 and 400 stainless, single-stage nitrided at 525 and 550 °C (975 and 1025 °F)

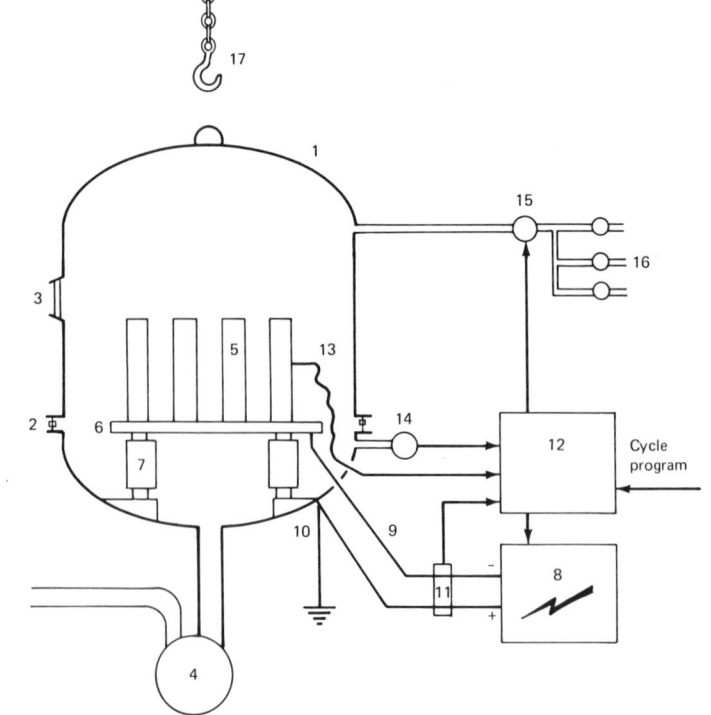

Fig. 18 Typical bell-type ion nitriding installation

1, vacuum vessel; 2, O-ring seal; 3, view port; 4, mechanical vacuum pump with 1 Pa capability; 5, workpieces; 6, load plate; 7, support insulators; 8, direct current power supply with arc suppression device; 9, current lead-in to workpieces (−); 10, ground/anode connection to vessel (+); 11, current/voltage analog network; 12, control and regulating console; 13, thermocouple; 14, vacuum gage; 15, regulating valve for process gas; 16, process gas manifold; and 17, handling equipment for load/vessel. Simplified schematic, showing main components

prior depassivation of the metal surface, the nitriding of stainless steels is similar to the single-stage nitriding of low-alloy steels.

Nitriding Results. Hardness gradients are given in Fig. 16 for types 302, 321, 430 and 446. These data are based on a 48-h nitriding cycle at 525 °C (975 °F), preceded by suitable annealing treatments. A general comparison of the nitriding characteristics of series 300 and 400 steels is presented in Fig. 17; the comparison reflects the superior results that are obtained with series 400 steels, as well as the effects of nitriding temperature on depth of case. Data are plotted for single-stage nitriding at temperatures of 525 and 550 °C (975 and 1025 °F). For steels of both series, greater case depths were obtained at the higher nitriding temperature.

Applications. Although nitriding increases the surface hardness and wear resistance of stainless steels, it decreases general corrosion resistance by combining surface chromium with nitrogen to form chromium nitride. Consequently, nitriding is not recommended for applications in which the corrosion resistance of stainless steel is of major importance. For example, a hot-air valve made of cast type 347 and used in the cabin-heating system of a jet plane was nitrided to improve its resistance to wear by the abrading action of a sliding butterfly. When the valve remained in the closed position for an extended period, the corrosive effects of salt air froze the valve into position so that it could not be opened.

In contrast, a manufacturer of steam-turbine power-generating equipment has successfully used nitriding to increase the wear resistance of types 422 and 410 stainless steel valve stems and bushings that operate in a high-temperature steam atmosphere. Large quantities of these parts have operated for 20 years or more without difficulty. In a few instances, a light-blue oxide film has formed on the valve stem diameter, causing it to "grow" and thus reduce the clearance between stem and bushing; the growth condition, however, was not accompanied by corrosive attack.

Nitrided stainless is also being used in the food-processing industry. In one application, nitrided type 321 was used to replace type 302 for a motor shaft used in the aeration of orange juice. Because the unhardened 302 shaft wore at the rubber-sealed junction of

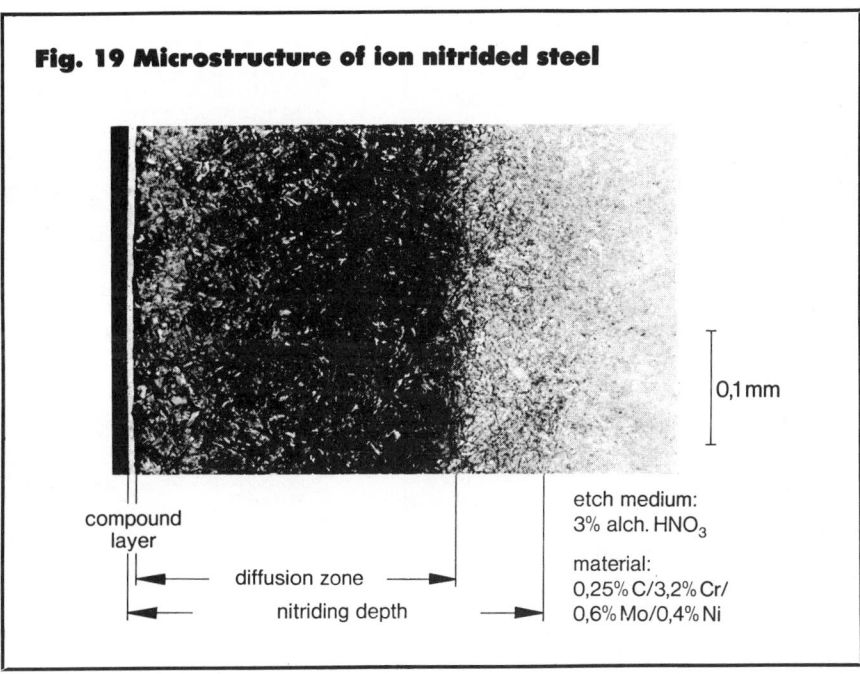

Fig. 19 Microstructure of ion nitrided steel

compound layer

diffusion zone

nitriding depth

0,1 mm

etch medium: 3% alch. HNO$_3$

material: 0,25% C/3,2% Cr/ 0,6% Mo/0,4% Ni

the motor and the juice, leaks developed within three days. The nitrided 321 shaft ran for 27 days before wear at the seal resulted in leakage. In machinery used in the preparation of dog foods, nitrided type 420 gears have replaced gears made of an unhardened stainless and have exhibited a considerable increase in life.

Modern synthetic fibers, several of which are highly abrasive, have increased the wear of textile machinery. Mechanical parts in textile machines are subjected to high humidity, absence of lubrication, high-speed movements with repeated cycling, and the abrasive action of fibers traveling at high speeds. A shear blade made of hardened, 62 to 64 HRC, 1095 steel experienced a normal life of about one million cuts (four weeks of service) in cutting synthetic fibers at the rate of 90 cuts per minute. In contrast, a nitrided type 410 blade with 0.04-mm (0.0015-in.) case depth showed less wear after completion of five million cuts.

With nitrided stainless steels, the case almost always has lower corrosion resistance than the basis material; nevertheless, the corrosion resistance of the case can be adequate for certain applications. For example, nitrided types 302 and 410 stainless steel resist attack from warp conditioner and size in the textile industry but do not resist attack from the acetic acid used in dyeing liquors.

Nitrided stainless is not resistant to mineral acids and is subject to rapid

corrosion when exposed to halogen compounds. However, a nitrided type 302 piston lasted for more than five years in a liquid-ammonia pump; it replaced a piston made of an unnitrided 300 series alloy that lasted approximately six months. Nitrided 17-4 PH impellers have performed satisfactorily and without corrosion in various types of hydraulic pumps.

Pressure Nitriding

Pressure nitriding (U.S. Patents 2 596 981, 2 779 697 and 2 986 484) differs from conventional gas nitriding in that it requires the use of a sealed retort capable of withstanding high pressures to contain the parts being nitrided. Nevertheless, it has been determined that within practical limits, depth and quality of case obtained in pressure nitriding depend less on pressure than on the ratio of the available mass of ammonia to the area of the surface presented for reaction with the gas.

Procedure. Surfaces to be nitrided are cleaned and placed in a carbon steel retort that is first evacuated of air and then filled with ammonia to a predetermined pressure. The pressure chosen depends on the total surface area of parts to be nitrided and the volume of the retort. Approximately 50 to 100 g (1.8 to 3.5 oz) of ammonia is supplied per square metre of surface to be nitrided. When only the inside surface of a part is to be hardened, as with carbon steel tubing for bottom-hole oil-well pumps, the tube can act as its own retort. The retort is then heated in any furnace whose temperature can be controlled for the required time cycle, after which the retort can be air cooled, vented and opened. Precise temperature control is not highly critical.

Advantages. Pressure nitriding provides a convenient method for nitriding part shapes that are difficult to handle by other methods. By varying the amount of ammonia added initially, the thickness of the white layer can be controlled.

Disadvantages include the following: (a) retort sealing is not always convenient, (b) after 45 h of operation, the ammonia content is about 50% expended, and further development of the case proceeds at a very slow rate, (c) to restrict the depth of the white layer to 0.00025 to 0.00050 mm (9.8 to 20 microin.), case depth must not exceed 0.50 to 0.63 mm (0.02 to 0.025 in.), and (d) in filling the welded retort with ammonia, dangerous pressures can develop if a sufficient quantity of ammonia is allowed to condense. This hazard can be avoided by keeping the retort warmer than the ammonia supply tank; however, a safety disk should be provided.

Bright Nitriding

Bright nitriding (U.S. Patents 3 399 085 and 3 684 590) is a modified form of gas nitriding, employing ammonia and hydrogen gases. Atmosphere gas is continually withdrawn from the nitriding furnace and passed through a temperature-controlled scrubber containing a water solution of sodium hydroxide (NaOH). Trace amounts of hydrogen cyanide (HCN) formed in the nitriding furnace are removed in the scrubber, thus improving the rate of nitriding. The scrubber also establishes a predetermined moisture content in the nitriding atmosphere, reducing the rate of cyanide formation and inhibiting the cracking of ammonia to molecular nitrogen and hydrogen. By this technique, control over the nitrogen activity of the furnace atmosphere is enhanced, and nitrided parts can be produced with little or no white layer at the surface. If present, the white layer will be composed of only the more ductile Fe_4N (gamma prime) phase.

Pack Nitriding

Pack nitriding (U.S. Patent 4 119 444), which is a process analo-

Fig. 20 Photomicrographs showing γ′ and ε compound layers

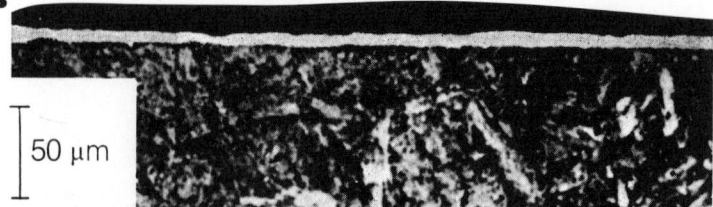

50 μm

Monophased γ′-compound zone Fe_4N

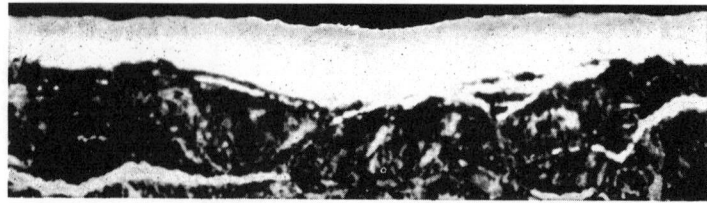

Monophased ε-compound zone $Fe_{2-3}N$

(a) Single-phase γ′ compound zone Fe_4N. (b) Single-phase ε compound zone Fe_2N-Fe_3N

Fig. 21 Influence of nitriding on fatigue strength

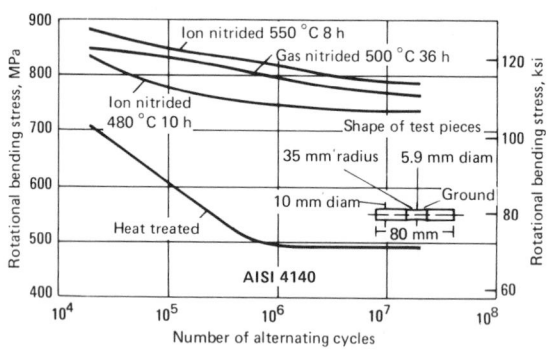

Fig. 22 Fatigue strengths of ion nitrided, quenched, and tempered steel specimen (unnotched rotating beam 6 mm or 0.24 in. diam)

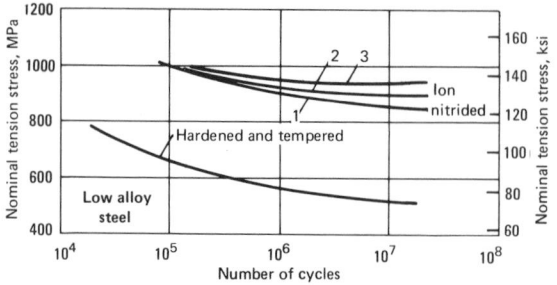

1, compound zone thickness, 5 μm (0.2 mil); thickness of nitride-free zone,* 60 μm (2.4 mil); 2, compound zone thickness, 1 μm (0.04 mil); thickness of nitride-free zone,* 80 μm (3.1 mil); 3, compound zone thickness, 0 μm (0 mil); thickness of nitride-free zone,* 95 μm (3.7 mil)

*Near surface zone free of carbonitride precipitates at grain boundaries

gous to pack carburizing, employs certain nitrogen-bearing organic compounds as a source of nitrogen. Upon heating, the compounds used in the process form reaction products that are relatively stable at temperatures up to 570 °C (1060 °F). Slow decomposition of the reaction products at the nitriding temperature provides a source of nitrogen. Nitriding times of 2 to 16 h can be employed. Parts are packed in glass, ceramic or aluminum containers with the nitriding compound, which is often dispersed in an inert packing media. Containers are covered with aluminum foil and heated by any convenient means to the nitriding temperature.

Ion Nitriding

Since the mid-1960's, nitriding equipment utilizing the glow-discharge phenomenon has been commercially available. Initially termed "glow-discharge" nitriding, the process is now generally known as ion nitriding. The term plasma nitriding is gaining acceptance.

Ion nitriding is an extension of conventional nitriding processes using plasma-discharge physics. In vacuum, high-voltage electrical energy is used to form a plasma, through which nitrogen ions are accelerated to impinge on the workpiece. This ion bombardment heats the workpiece, cleans the surface and provides active nitrogen.

Metallurgically versatile, the process provides excellent dimensional control and retention of surface finish. Ion nitriding can be conducted at temperatures lower than those conventionally employed. Control of white-layer composition and thickness enhances fatigue properties. The span of ion nitriding applications includes conventional ammonia-gas nitriding, short-cycle nitriding in salt bath or gas, and the nitriding of stainless steels.

Ion nitriding lends itself to total process automation, ensuring repetitive metallurgical results. The absence of pollution and insignificant gas consumption are important economic and public policy factors. Moreover, selective nitriding accomplished by simple masking techniques may yield significant economies.

A vacuum process utilizing electrical energy, plasma nitriding possesses the features of the "no-idle" cost and cleanliness that are characteristic of vacuum-hardening furnaces. Energy utilization is approximately equivalent to conventional ammonia-gas nitriding

operations, when cold wall ion nitriding vessels are used.

Hot-wall construction is energy efficient; holding a nitriding load at 570 °C (1060 °F) requires about 20% of the power that would be used by a similar cold-wall vessel. Adequate surface activation of high-alloy or stainless steels may not be obtained with hot-wall construction, however. The higher current densities required for cold-wall operation produce workpiece surfaces more receptive to nitriding.

Glow Discharge Process. If a sufficiently high voltage is applied between two electrodes in an evacuated chamber containing a low-pressure gas, an electric current will flow and light will be emitted. Fluorescent and neon lamps are common examples of gas-discharge tubes. In ion nitriding, parts to be processed serve as the cathode, that is, connected to the negative side of the direct-current power supply. The metallic vacuum vessel interior serves as the anode. The voltage drop between the electrodes is not linear. A sharp drop in voltage near the cathode, known as cathode fall, confines a large part of the discharge energy near the cathode, where the glow layer or glow seam is seen enveloping the cathode surface.

The glow layer near the cathodic workpiece is, in part, a plasma of positive ions and electrons. Positive ions are accelerated by the applied electrical field toward the cathode; similarly, electrons move toward the anode. The impact of positive ions heats the workpiece to nitriding temperature and cleans the surface by the ejection of surface atoms. The process of removing surface atoms by ion bombardment is termed "sputtering". The collision of electrons with neutral diatomic nitrogen molecules (N_2) produces active monatomic nitrogen atoms or ions required for nitriding.

Equipment. Figure 18 shows the main components of a typical ion nitriding installation. The vacuum vessel could be constructed with a single wall as shown in the schematic, with cooling water provided at the O-ring seal. Usually, the vacuum vessel itself is water jacketed. Where conventional vacuum heat treating furnaces are modified for ion nitriding use, reflective baffles and resistance heating elements are retained. Resistance-heating elements may be used to bring the work load to nitriding temperature; at that temperature, the elements are then shut off. Plasma discharge alone provides the

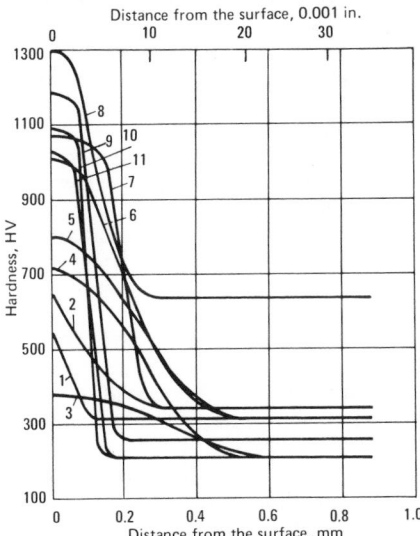

Fig. 23 Hardness profiles for various ion nitrided materials

1, Grey cast iron; 2, Ductile cast iron; 3, AISI 1040; 4, Carburizing steel; 5, Low alloy steel; 6, Nitriding steel; 7, 5% Cr hot work steel; 8, Cold work die steel; 9, Ferritic stainless steel; 10, AISI 420 stainless steel; 11, 18–8 stainless steel

necessary energy to maintain the work at temperature.

Vessel configurations include: bell-type, pit-type, and combination vessels that function as either a pit or a bell. The height capacity of combination vessels can be increased by stacking the cylindrical sections. Horizontal loading vessels are typically either small units or modifications of vacuum-hardening furnaces. Pit furnaces, which facilitate the hanging of parts, are preferred when distortion control is critical.

One design of hot-wall construction uses an external insulating wrap of the vacuum shell plus two internal reflective shields. No resistance-heating elements are incorporated. Heat is extracted by a cooling coil located between the reflective shields and vacuum vessel interior. A circulating fan, in conjunction with nitrogen backfilling to about 50 kPa (0.5 bar), provides effective cooling of the work load.

Water-jacketed vessels with few or no reflective baffles usually are not provided with a circulating fan. Without a circulating fan, backfilling alone cools very slowly. Where two or more vessels are alternately connected to one or more power supplies, rapid cooling is not important.

Load plates or other support elements of industrial ion nitriding equipment are engineered to support loads of as much as 1000 kg (22 000 lb). Support insulators are designed to avoid short circuits caused by sputtered material deposited as a conductive coating.

Mechanical vacuum pumps capable of producing a vacuum of 1 Pa (0.01 mbar, 0.008 torr) are commonly used for ion nitriding. Some equipment employs diffusion pumping to produce an initial vacuum of 10^{-3} Pa (10^{-5} torr) to vaporize oils and soaps from the surface of the work. This step, prior to glow discharge, minimizes carbon contamination.

Power supplies, ordinarily limited to 1000 V, provide maximum currents from 20 to 1000 A. Work loads of practical size require large amounts of power to reach nitriding temperature; power used to maintain the glow discharge usually is the only source of process heat. Resistance-heating elements, if available, are not used at the nitriding temperature. During nitriding, voltage is commonly in the 400- to 600-V range. A current density of 1 mA/cm² (0.93 A/ft²) of cathode is a representative value for cold-wall vessels.

Arc Suppression. A rapid-acting, current-interrupting device incorporated in the power supply to prevent arcing is highly desirable. Large power inputs result in current densities approaching the transition into arcing. Arcing occurs when the normal ohmic relation between current and voltage changes to an unstable condition in which the voltage drops with increasing current. Severe damage to the work surface, similar to arc welding, results. Adequate control of electrical power to prevent arcing is perhaps the single most important equipment requirements.

Control units may provide for complete automation of ion nitriding. Control data on temperature, gas composition and gas-pressure (vacuum) cycles are entered into the control units prior to processing. Part temperature follows the programmed cycle by varying current input to the vessel. Pressure is programmed to increase as temperature increases. Temporary extinction of the glow may be provided to avoid operation at pressures where "hole discharge" would occur. This phenomenon, known also as "hollow-cathode discharge", is described later. Additionally, the control unit monitors malfunctions and provides for automatic shutdown.

Workpiece temperature is usually measured directly by an inserted

sheathed thermocouple that is electrically isolated. Observing the work through the view port, with the glow extinguished, provides confirmation of the thermocouple measurement. At temperatures close to the limit of visibility, 500 to 510 °C (930 to 950 °F), visual estimates are quite accurate.

Optical pyrometers have been developed that respond only to infrared radiation but do not "see" the visible glow discharge. This permits the glowing surfaces to be used for aiming. These instruments are useful over the entire range of ion nitriding temperatures and can be used for control purposes.

Process Gas Composition and Pressure. Process gas is ordinarily a mixture of hydrogen and nitrogen. If a metallurgical requirement for carbon exists, methane is added. Hydrogen may be used for cleaning purposes or where the glow is used for heating without nitriding. During glow operation, gas pressures are regulated to a value within the range of 30 to 400 Pa (0.3 to 4 mbar, 0.2 to 3 torr).

Processing. The ion nitriding process is logically divided into several stages: (a) precleaning, (b) fixturing, (c) pump down, (d) cleaning by ion bombardment, (e) nitriding, (f) cooling, and (g) acceptance testing.

Work to be ion nitrided must be very clean. Even small amounts of oil, rust or other foreign material can greatly extend processing time. Vapor degreasing is generally used to remove oils. It is important that all recesses, which may be difficult to clean, are thoroughly cleansed. While small amounts of rust or scale can be cleaned by ion bombardment, the time required may make this method too costly. Chemical or abrasive cleaning usually is preferred.

Fixture design is influenced by the requirement for electrical connection to each workpiece and the characteristics of the glow layer. Thickness of the glow layer during processing determines a minimum spacing between parts. Typically, this distance is 35 to 50 mm (1.3 to 2 in.). Glow-layer thickness is determined primarily by gas pressure. When the glow layer, on opposing cathode surfaces, becomes thick enough to meet and coalesce, the current density increases markedly. This condition, known as hollow-cathode discharge or hole discharge, produces intense local heating that may damage the workpiece.

Fixture assemblies should be constructed to avoid the formation of small cavities. Hole discharge may occur in

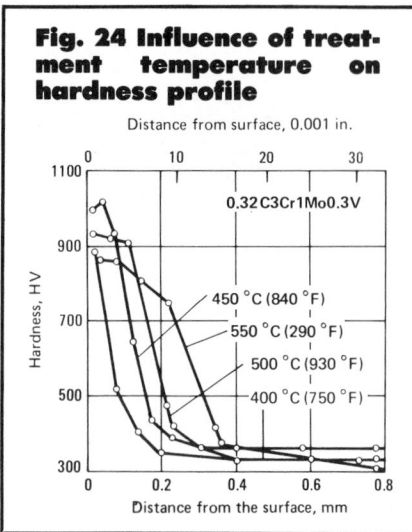

Fig. 24 Influence of treatment temperature on hardness profile

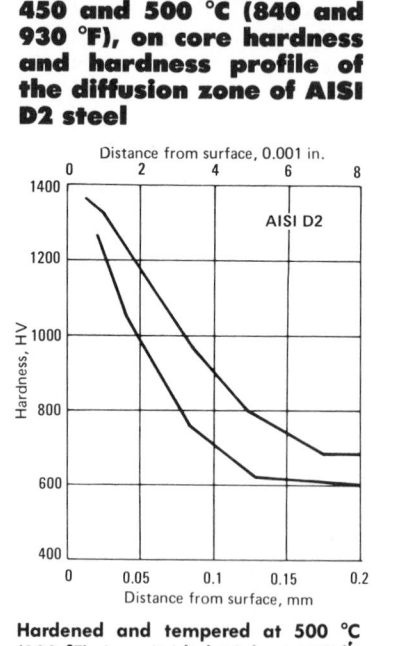

Fig. 25 Influence of the ion nitriding temperature, 450 and 500 °C (840 and 930 °F), on core hardness and hardness profile of the diffusion zone of AISI D2 steel

Hardened and tempered at 500 °C (930 °F); ion nitrided 12 h at 500 °C (930 °F); ion nitrided 30 h at 450 °C (840 °F)

slots, small holes and similar geometries. Internal acute angles are particularly troublesome because a continuous variation of gap size is presented. Potential hole-discharge problems can be foreseen and avoided by drawing upon test data and operating experience. Hollow-cathode discharge also can be avoided by utilizing knowledge of the glow phenomena to modify part geometry or to select a suitable gas pressure.

Fig. 26 Pure iron nitrocarburized by the glow-discharge technique

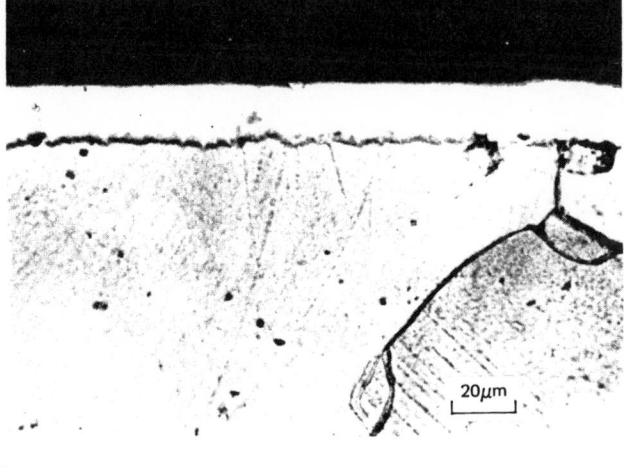

(a)

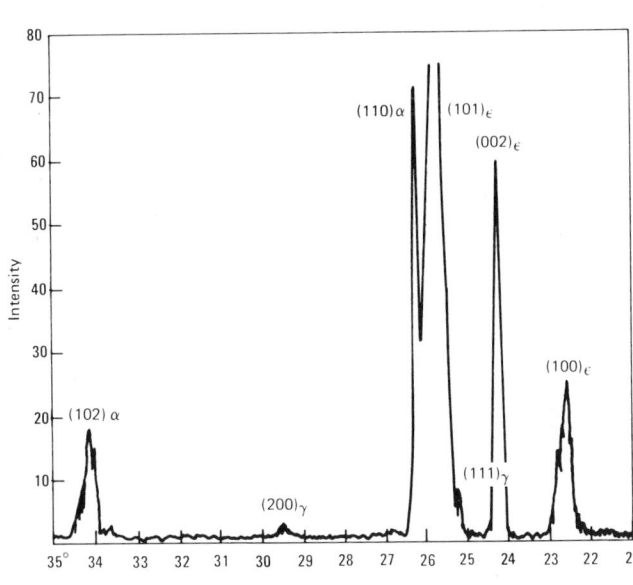

(b)

(a) Epsilon carbonitride layer formed. (b) Diffractometer tracing of compound layer shown in (a), demonstrating presence of epsilon phase.

Part geometry can be altered to eliminate susceptible areas such as holes or slots. Cavities to be protected are filled with conductive plugs, usually made of low-carbon steel. This method prevents nitriding of the covered surfaces. Alternatively, gas pressure can be held to a low value, producing a thick glow layer that spans the cavity. The result is that both hole discharge and nitriding of the cavity are avoided.

At very high pressures, the resulting thin glow layer can conform to surface contours within a cavity. Hole discharge is avoided and the interior surfaces are nitrided. Successful nitriding of bores as small as 3 mm (0.12 in.) requires a pressure of about 900 Pa (9 mbar, 7 torr). For example, a blind 3 mm (0.12 in.) bore can be nitrided uniformly to an axial depth of 40 mm (1.6 in.). Inability of the glow to enter small cavities allows close fitting shields, usually made of low-carbon steel, to be used as simple mechanical masks. Such masks can be reused many times. A maximum gap of 0.5 mm (0.02 in.) between the mask and part is effective at typical operating pressures.

After pumping down to an initial vacuum, typically about 10 Pa (0.1 mbar, 0.08 torr), process gas is admitted, raising the pressure to about 30 Pa (0.3 mbar, 0.2 torr). Electrical power is then applied to begin a period of cleaning by ion bombardment. Short-duration currents (sparks) concentrated at the sites of contamination remove the contaminant by sputtering.

Interrupted by the high-speed breaker, the glow is rather irregular. Power consumption and heating rate are low. The duration of cleaning depends on the surface area and degree of cleanliness provided by the precleaning operation. Cleaning by ion bombardment typically requires ½ to 2 h. At the end of the period, the glow has become quite steady. During this initial period of glow discharge, the thermocouple is observed to determine that it is electrically isolated, that is, not glowing.

Starting from about 100 °C (212 °F) at the end of cleaning, the work temperature follows a preselected heating ramp to nitriding temperature. Typically 1 to 5 h is required. During this time, the glow is observed for uniformity and possible hole discharge. Periodically, the glow is extinguished and the load inspected for possible visibly hot areas. As the work reaches the visible temperature range, the glow is extinguished to permit visual confirmation of the measured temperature and satisfactory temperature distribution within the load.

Temperatures employed for ion nitriding are usually in the 350 to 580 °C (660 to 1075 °F) range. Upon reaching set-point temperature, timing of the nitriding period begins. Ion nitriding times are mostly within the range of 0.2 to 36 h. Time required to obtain a given case depth is about 60 to 70% of the time needed for conventional gas nitriding. Process gas pressure during ion nitriding is typically 130 to 400 Pa (1.3 to 4 mbar, 1 to 3 torr).

At the end of the nitriding period, the glow is extinguished and cooling begins. Cooling proceeds quite slowly, even with nitrogen backfilling, unless a circulating fan and heat-exchange surface are also provided. Two or more vessels may be operated sequentially from a single power supply to minimize the need for accelerated cooling.

Acceptance Inspection. Acceptance procedures usually include (a) visual inspection for arc damage and surface finish, (b) surface hardness testing with loads of 2 to 15 kg (4.4 to 33 lb), (c) core hardness measurements, usually on masked areas, (d) micro-

Fig. 27 Effect of oxygen additions upon thickness of compound layer formed by a 2-h nitrocarburizing treatment

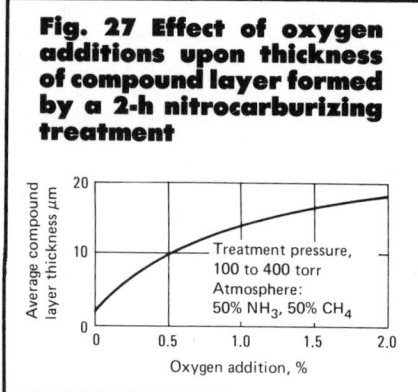

Fig. 28 The metallographic appearance of AISI 1015 material after a 2-h vacuum nitrocarburizing treatment in an ammonia/methane mixture with 1% oxygen addition

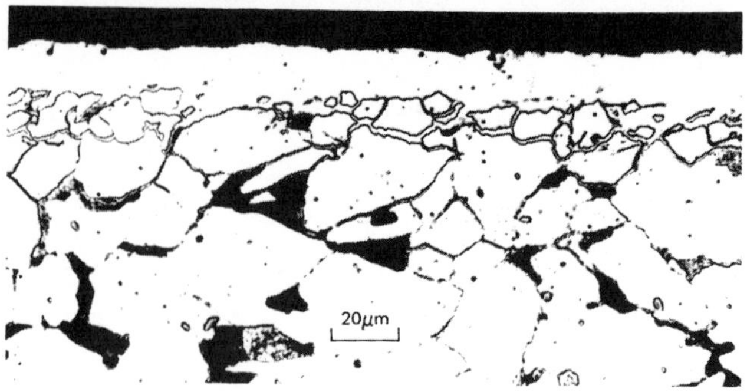

hardness profile on sectioned specimen to determine case depth, which is often defined as the depth where hardness is 2 points HRC scale equivalent above core hardness, and (e) metallographic measurement of compound zone thickness.

Structure and Properties of Ion Nitrided Steel. Ion nitriding, like other nitriding processes, produces several distinct structural zones as shown in Fig. 19. A light etching layer of iron-nitride compounds at the surface; a gradient zone of fine iron/alloy nitrides, Fe_4N, that constitutes the bulk of the case depth; and a gradient zone of interstitial nitrogen that extends to the parent material.

The light etching surface layer, commonly termed "white layer", has more recently been appropriately named "compound zone". The ion nitriding process offers the possibility of forming a single-phase compound zone with the structure Fe_4N, the gamma prime phase shown in Fig. 20(a). Depth of the gamma prime compound zone is inherently process limited to about 10 μm (0.0004 in.) maximum. Steels with alloy contents greater than 6 to 8% inherently form compound zones with only trace thickness.

Process-gas mixtures free of carbonaceous material are required to form compound zones having the gamma prime structure. In the limiting condition, a diffusion zone is formed without an overlying compound zone. Gas compositions with less than the commonly used 25% nitrogen can completely suppress compound zone formation.

A shallow gamma prime compound zone, with an underlying diffusion zone, is the desired structure for the majority of ion nitriding applications, particularly where good fatigue properties are important.

Constructional alloy steels, nitriding steels and tool steels containing ni-

tride-forming alloying elements are used to fabricate workpieces. Nitride-forming alloying elements are aluminum, chromium, molybdenum, vanadium, tungsten, titanium and niobium. Hardening and tempering is performed prior to nitriding. In common with other nitriding methods, this allows quenching distortion and stresses to be corrected or removed prior to nitriding. Hardened and tempered, the steel has useful core strength and usually is machinable.

Single-phase epsilon iron-nitride compound zones having a Fe_2N-Fe_3N structure, as shown in Fig. 20(b), are formed when the process gas includes a carbonaceous component such as methane. The epsilon structure is slightly harder and less ductile than gamma prime.

Thickness of the epsilon compound zone is not process limited; a zone 50 μm (2 mil) deep can be formed. Industrially, zones 10 to 20 μm (0.4 to 0.8 mil) deep are applied to carbon steels and cast irons where core hardness is usually low. Applications with light loads or broad area contact predominate. In addition to providing increased mechanical strength, the thicker compound zone is a good barrier against corrosion.

Treatment time is typically 2 to 4 h at 570 °C (1060 °F), similar to other short-cycle nitrocarburizing processes. The compound zone is, however, pore free with low surface roughness.

Comparison of Ion Nitriding and Ammonia-Gas Nitriding Compound Zone Structures. Ammonia-gas nitriding produces a compound zone that is a mixture of both epsilon and gamma prime structures. High internal stresses result from differ-

ences in volume growth associated with the formation of each phase. The interfaces between the two crystal structures are weak. Thicker compound zones, formed by ammonia-gas nitriding, limit accommodation of the internal stresses resulting from the mixed structure. Thickness, internal stresses and weak crystal boundaries allow the white layer to be fractured by small applied loads.

Under cyclic loading, cracks in the compound zone can serve as initiation points for the propagation of fatigue cracks. The single-phase gamma-prime compound zone, which is thin and more ductile, exhibits superior fatigue properties, as shown in Fig. 21. Reducing the thickness of the ion nitrided compound zone further improves fatigue performance. Maximization occurs at the limiting condition, where compound zone depth equals zero (Fig. 22).

Case Hardness. The bulk of the thickness of the nitride case is the diffusion zone where fine iron/alloy nitride precipitates impart increased hardness and strength. Compressive stresses are also developed, as in other nitriding processes. Hardness profiles resulting from ion nitriding are similar to ammonia-gas nitriding (Fig. 23), but near-surface hardness may be greater with ion nitriding, a result of lower processing temperatures.

The concentration and size of alloy nitride precipitates formed, together with parent material hardness, determine the hardness observed in a nitrided case. Figure 24 shows the results of ion nitriding a 0.32C-3Cr-1Mo-0.3V alloy steel at several temperatures, with time held constant. Case depth increases with temperature and near

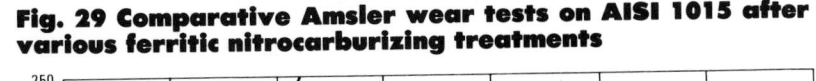

Fig. 29 Comparative Amsler wear tests on AISI 1015 after various ferritic nitrocarburizing treatments

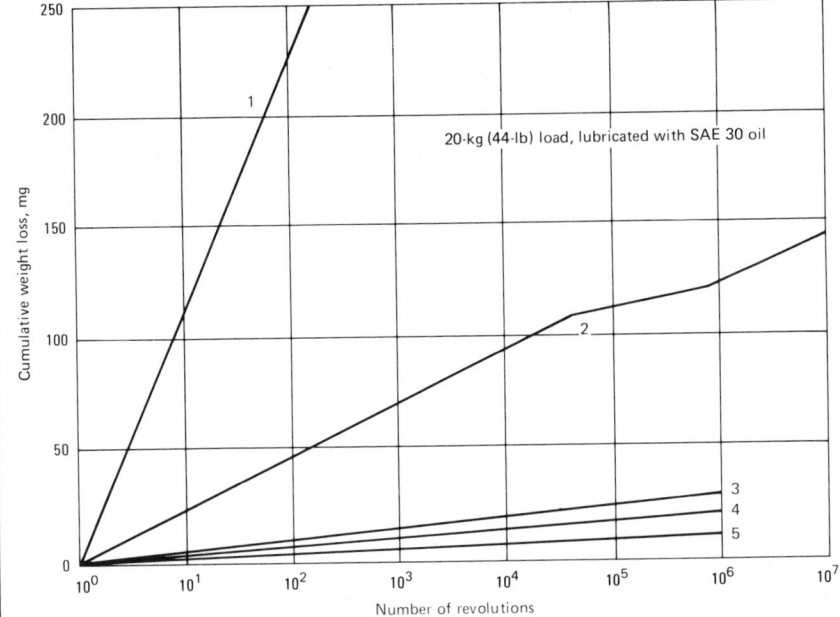

20-kg (44-lb) load, lubricated with SAE 30 oil

1, untreated; 2, cyanide-base salt-bath nitrocarburizing with sulfur; 3, subatmospheric oxynitrocarburizing; 4, gaseous nitrocarburizing; and 5, cyanide-base salt-bath nitrocarburizing (treatment 1)

surface hardness is maximized near 450 °C (840 °F). Figure 25 shows a similar effect on high-carbon high-chrome (D2) tool steel, hardened and subsequently tempered at 500 °C (930 °F). Processing at temperatures just above the hardness maximum offers several advantages: (a) higher core hardness can be retained by reducing tempering temperatures, (b) the possibility of distortion is reduced, (c) parts with low surface roughness remain virtually unchanged, and (d) at 500 to 510 °C (930 to 950 °F), where, with the plasma extinguished, the work becomes visible, and temperature can be verified by direct visual observation.

Advantages and Disadvantages of Ion Nitriding. Ion nitriding achieves repetitive metallurgical results and complete control of the nitrided layers. This control results in superior fatigue performance, wear resistance and hard layer ductility. Moreover, the process ensures high dimensional stability, eliminates secondary operations, offers low operating temperature capability, and produces parts that retain surface finish. Among operating benefits are

- Total absence of pollution
- Efficient use of gas and electrical energy

- Total process automation
- Selective nitriding by simple masking techniques
- Process span that encompasses all subcritical nitriding
- Reduced nitriding time.

The limitations of ion nitriding include high capital cost, need for precision fixturing with electrical connections, long processing times compared to other short cycle nitrocarburizing processes, and lack of feasibility of liquid quenching for carbon steels.

Applications. Among general applications requiring metallurgical properties obtainable by ion nitriding are

- Structural elements subject to cyclic loading
- Workpieces requiring precision dimensions
- Components subject to sliding wear
- Parts exposed to mild corrosion.

Metallurgical properties required by these applications are used frequently in combination for such products as: plastics processing machinery; automotive engine, transmission, chassis and accessory components; cold-forming tools; and hot-forming tools.

Screws and cylinders used to extrude plastics require close dimensional tol-

erances. In service, they are subject to high mechanical loads and severe sliding wear. The hot plastic creates abrasive, corrosive and erosive conditions at various locations along the length. Nitriding steel, pretreated for strength and toughness, receives a hard ductile layer by ion nitriding to meet this demanding service.

Side and middle housings of Wankel engines, made of gray iron, are stress relieved and finish machined prior to ion nitriding. Water-passage areas are covered with sheet-metal shields, so that only the rotor contact surfaces will be nitrided. Dimensional changes are extremely low, permitting direct use without a refinishing operation.

Similar to the Wankel engine housings are side plates for rotary automotive air-conditioning compressors. Also made of cast iron, they must be extremely flat, with good surface finish, and must be free of contamination. The epsilon layer produced by ion nitriding prevents seizure resulting from adversely hot operation.

Synchronizer components for transmissions are ion nitrided by several manufacturers to meet close dimensional tolerances. Conventional techniques such as carbonitriding fail to meet dimensional requirements. A 10-μm (0.0004-in.)-thick epsilon layer, with a superficial hardness of 550 HV5 is typically produced. Ion nitriding has proven a satisfactory substitute for more expensive chrome plating on automobile shock-absorber rods. The required wear and corrosion resistance is provided by an epsilon layer about 10 μm (0.0004 in.) thick.

Reduction gears for marine steam turbines were an early application of ion nitriding, now firmly established as the preferred nitriding method. Dimensional accuracy and fatigue properties are superior to ammonia-gas nitriding. Nitriding is confined to the tooth area by masking with sheet metal covers. Significant labor economies are achieved.

Deep-drawing punches made of high-carbon high-chrome steel are subjected to high compressive stresses in service. A core hardness of 62 HRC or higher is required after ion nitriding. Lowering the ion nitriding temperature to 470 °C (880 °F) allows retention of the core hardness while increasing the hardness of the surface layer to about 1200 HV. The considerably reduced coefficient of friction results in a great increase in service life.

Hot-forging dies are an important

application of ion nitriding. Die life is increased by improved resistance to thermal and mechanical cracking. The surface layer formed reduces the sticking of scale and inclusion of oxides.

Vacuum Nitrocarburizing

There are two main approaches to subatmospheric pressure thermochemical processing. One is known as the glow discharge method, and the other involves use of conventional cold-wall vacuum furnaces. Because the processes are closely related to ion nitriding, they are discussed in this section. Further data on other types of nitrocarburizing processes are contained in the articles on gaseous ferritic nitrocarburizing.

Ion nitriding, which is being used increasingly as an alternative to conventional gas nitriding in ammonia atmospheres, was the first thermochemical treatment to use the glow-discharge technique. In glow-discharge nitrocarburizing, which is a simple development of the ion nitriding process, the components become the cathode of an electrical circuit. They are subsequently subjected to a glow discharge generated by applying a critical voltage between the furnace chamber, which acts as anode, and the components. Consideration of glow-discharge nitriding conditions indicates that a pressure in the range of 20 to 2000 Pa (0.15 to 15 torr) and a critical applied voltage of between 400 and 1000 V should be used. The metallographic structure of pure iron after treatment in a nitrocarburizing atmosphere with an atomic concentration ratio C/N of 0.37 is shown in Fig. 26(a). A corresponding x-ray diffraction analysis is given in Fig. 26(b), where the predominance of the epsilon carbonitride phase is evident.

Cold-Wall Vacuum Furnaces. Extension of the use of cold-wall vacuum furnaces from purely thermal treatments, such as annealing and sintering, to thermochemical treatments was a natural development following introduction of such furnaces with oil quench facilities. Although vacuum carburizing has received some attention in the literature, nitrocarburizing in a vacuum furnace is a more recent development. Although there was some evidence that the presence of oxygen gave a marginal improvement in the antiscuffing behavior of the epsilon carbonitride compound layer, the significance of the role of oxygen in the kinetics of compound layer growth was not at all clear.

Table 8 Hardness of nitrocarburized specimens

Material	Treatment	Applied load g	Applied load oz	Microhardness of compound layer, HV Center region	Microhardness of compound layer, HV Inner region	Average
Low carbon steel	Toxic salt	...	...	...	...	500 to 600
AISI 1015	Toxic salt	2500	90	...	...	536
En41 Nitriding steel	Toxic salt	2500	90	...	...	803
Pure iron	Toxic salt	...	...	480 to 680	820 to 990	...
Low carbon steel	Nontoxic salt	15	0.5	340 to 450	900 to 1100	...
En40c (a)	Gaseous	200	7	...	...	820
AISI 1015	Gaseous	200	7	...	...	620
Pure iron	Gaseous	200	7	...	...	600
AISI 1015	Gaseous	25	1	...	...	600 to 900
Pure iron	Gaseous	...	...	...	...	1000 to 1200
Pure iron	Gaseous	...	...	400 to 950	780 to 850	...
AISI 1015	Subatmospheric pressure	25	1	...	...	600

(a) 3% Cr, 17% Mo nitriding steel

It had been shown that oxygen was necessary to improve the carbon mass transfer characteristics of hydrocarbon carburizing gases, and that increasing the partial pressure of oxygen in the atmosphere speeded the kinetics of carbon exchange. Similarly, the presence of oxygen had been shown to increase the rate of compound layer formation during conventional gas nitriding. Consequently, with the advent of the capability of absolute atmosphere control using vacuum furnaces, investigations of vacuum nitrocarburizing were also able to evaluate the significance of oxygen in relation to the kinetics of compound layer formation. On the basis of earlier studies on gaseous nitrocarburizing treatments, a basic atmosphere of 50% ammonia/50% methane, containing controlled oxygen additions of up to 2%, was used for experiments in a laboratory vacuum furnace. The experiments were reproduced later using a semicontinuous industrial vacuum furnace. The furnace was evacuated to a pressure of 13 Pa (0.1 torr), and then heated to the process temperature under a low flow rate of ammonia balanced against a rotary vacuum pump to give a pressure of 3900 Pa (30 torr). Once the operating temperature of 570 °C (1060 °F) was reached, the hot zone was re-evacuated, and the premixed gases were introduced into the hot zone at a pressure of 52 kPa (400 torr). To avoid either fixed flow patterns or stagnant gas pockets, a pulsed atmosphere technique was employed for the duration of the treatment. This involved pressure cycling between 13 to 52 kPa (100 to 400 torr), with 10 min dwells at 52 kPa (400 torr) throughout the treatment, after which the specimens were

oil quenched. As a result of this treatment, a compound layer is formed that has a carbon/nitrogen ratio of about 1:7; it consists predominantly of the oxygen-bearing epsilon carbonitride phase.

Layer thickness formed after a 2 h treatment varies with the oxygen content of the atmosphere. Figure 27 shows the compound zone thickness as a function of the oxygen content of the atmosphere in a nominal 50% methane/50% ammonia atmosphere. The metallographic appearance after a typical subatmospheric treatment with a 1% oxygen addition is shown in Fig. 28. The compound zone shows very little porosity, and its over-all metallographic appearance is very similar to the layer formed by gaseous nitrocarburizing in an atmosphere consisting of 50% ammonia and 50% endothermic gas at a treatment temperature of 570 °C (1060 °F). When oxygen levels of about 2% are used, however, the compound zone is quite porous. Wear properties of AISI 1015 materials treated in this manner have been evaluated by standard Amsler wear tests, the results being compared to results of similar tests with other nitrocarburizing treatments (Fig. 29). Three of the treatments confer similar wear improvement characteristics on low-carbon steels after a treatment time of 2 h at 570 °C (1060 °F).

Hardnesses of compound layers produced by subatmospheric pressure nitrocarburizing are compared (in Table 8) with hardnesses of layers resulting from other treatments. Hardness levels are high considering the ductile nature of the compound zone, and the layer

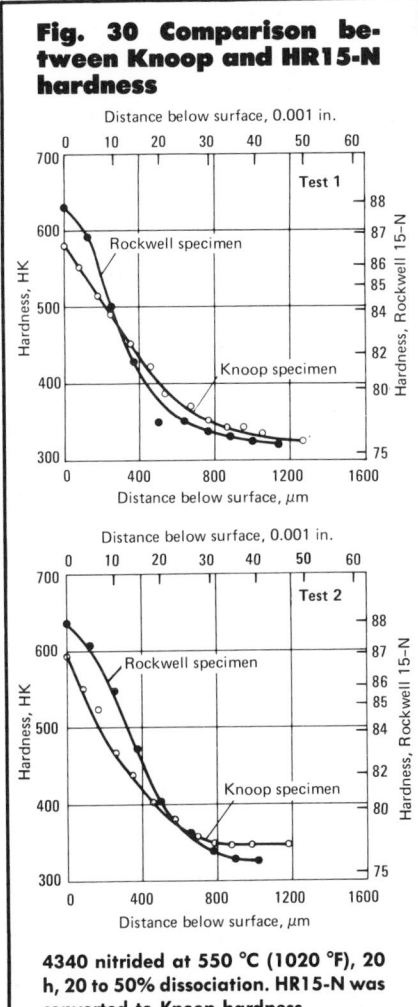

Fig. 30 Comparison between Knoop and HR15-N hardness

4340 nitrided at 550 °C (1020 °F), 20 h, 20 to 50% dissociation. HR15-N was converted to Knoop hardness.

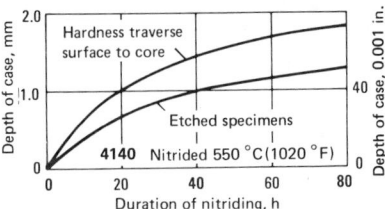

Fig. 31 Comparison of depth of nitrided case determined by hardness traverse and by etching specimens in 2% nital

Before being nitrided, the 4140 steel was oil quenched from 845 °C (1550 °F) and tempered at 595 °C (1100 °F).

hardness appears to be higher on alloy steel than on plain carbon material.

Appendix

Analysis of Exhaust Gas from Gas Nitriding Operations

Ammonia gas is completely soluble in water. When water is introduced into the dissociation burette, any ammonia present dissolves instantly, reducing the pressure within the burette. Water continues to enter the burette until it occupies a volume equivalent to that previously occupied by the ammonia. The remainder of the exhaust gas, being insoluble in water, collects at the top of the burette. The height of the water level is read directly from the scale of graduations, and this reading indicates the percentage of non-water-soluble hydrogen-nitrogen gas in the sample. If the sample was generated solely by the breakdown of ammonia, this reading is correctly called percent dissociation. When air is present at the start or end of a cycle, however, no ammonia is being dissociated and the resulting reading is not percent dissociation, but percent air. Accordingly, it is proper to subtract the reading from 100% and refer to the remainder as percent ammonia present in the sample.

Inspection and Quality Control

Indentation tests of the hardness of a nitrided case should be made using relatively light loads, regardless of case depth. These indentation methods include the superficial HR15-N (and to a limited extent, the HR30-N), the Knoop and Vickers (diamond pyramid hardness, DPH) microhardness tests. The superficial Rockwell test is made on a surface that is ground prior to and only polished lightly after nitriding; whereas the Knoop and Vickers microhardness tests are generally performed on cross sectional specimens that have been metallographically polished. Microhardness tests are generally made with loads of 100 to 500 g (0.2 to 1 lb.)

Utilizing lighter loads than the superficial Rockwell test, but greater than are commonly used for microhardness testing, the Vickers test is used extensively in Europe for quality control. Superficial measurements with 2-, 5- or 10-kg (4-, 11- or 22-lb) loads are made directly on nitrided surfaces. Loads in this range accurately reflect surface hardness while requiring only minimal surface preparation. Good accuracy results from optical measurement necessitated by the small impression.

Occasionally, both HR15-N and HR30-N tests are used for quality control of light cases; however, if the case is too shallow, the higher load used in the HR30-N test may cause penetration into the softer core, resulting in a composite case-core reading lower than obtained in the HR15-N test.

Because of metal flow or spalling, it is difficult to determine accurately by microhardness methods the case hardness at depths of less than 0.025 mm (0.001 in.) from the surface of a cross section specimen, even when light loads are applied. For this reason, surface hardness is commonly measured by the HR15-N test; the values obtained may be converted to Knoop or Vickers values in accordance with the conversion table given in ASTM E140. If a value exceeds the range of this table, conversion can be made using the tables prepared by V. E. Lysaght, which appear in his book, *Indentation Hardness Testing*, Reinhold Publishing Corp., New York, 1949.

Comparison of Hardness Measurements. When results of HR15-N and Knoop tests are compared, Knoop hardness is generally found to be higher than the equivalent HR15-N hardness for the higher-hardness portion of the depth of case measured from the nitrided surface, whereas the opposite is experienced for the lower-hardness portion of the case (see Fig. 30).

Evaluation of case depth may be accomplished by preparing a cross section of the case, etching with a suitable agent, and microscopically measuring the depth from the surface to a point of contrast between the case and core. Suitable etchants may be one of the following: (a) distilled water (250 cm³), ammonium persulfate (109 g), sodium alkyl aryl sulfonate (1 g) and saturated solution of sodium thiocyanate (10 drops); (b) 4% nital; or (c) 3% picral plus 1% benzalkonium chloride (zephiran chloride).

Case depth may be determined also by microhardness testing an unetched cross section of the case, using either the Vickers or Knoop tester. Measurement consists of making a hardness survey from near the nitrided surface to the basis metal (total case), or to a depth at which a predetermined hardness value (such as 60 HRC) is measured (effective case).

Fig. 32 Comparison of the effect of a 3% picral—1% benzalkonium chloride solution on the etching characteristics of nitrided cases produced on 2 AMS steels

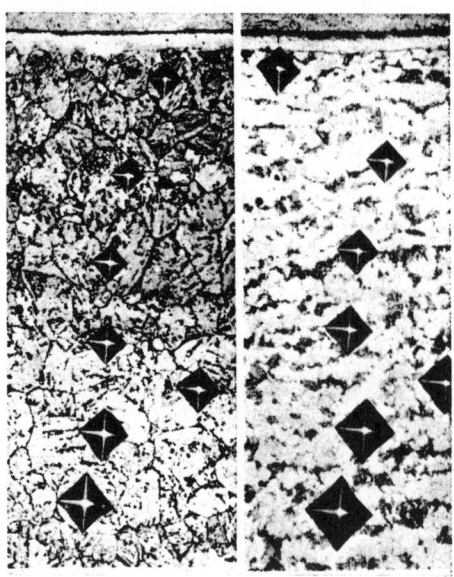

(a) 6470 steel, nitrided for 9 h. (b) 4130 steel, nitrided for 5 h. Magnification, 250×

Fig. 33 Wedge specimen for determining case depth when facilities for microhardness or etchant tests are not available and fracture specimen for determining case depth

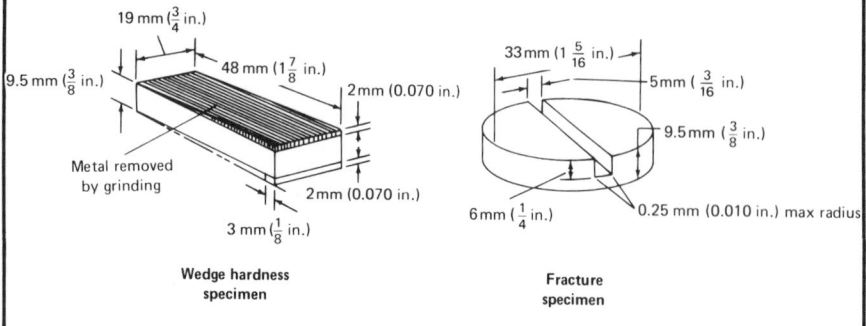

Wedge hardness specimen

Fracture specimen

In general, case depth measurements determined by microhardness tests are more accurate and reproducible than those made by visual examination of etched specimens. Frequently, the depth of case determined by examination of an etched specimen is less than that indicated by a microhardness survey, as shown by the data in Fig. 31. Also, etchants react differently on different steels. For example, the 3% picral- and 1% benzalkonium chloride solution darkens the case of aluminum-bearing steels but does not have this effect on 4100 series (see Fig. 32).

When facilities for microhardness or etchant tests are not available, a tapered-wedge control specimen may be used in conjunction with the HR15-N tester to determine case depth of the nitrided parts. Such a specimen is of the same grade of steel as the parts being nitrided and has over-all dimensions of 48 by 19 by 10 mm (1⅞ by ¾ by ⅜ in.). It is heat treated with the parts to obtain the proper hardness. After removal of 3.2 mm (⅛ in.) of material from all surfaces, a 1.8-mm (0.070-in.) taper is ground by placing a 1.8 by 3 by 19 mm (0.070 by ⅛ by ¾ in.) shim under one end of the specimen as indicated in Fig. 33.

The tapered specimen is then nitrided with the parts, after which it is reground to remove the taper so that hardness measurements can be taken at right angles to the surface. (Heat generated from grinding must be kept at a minimum, to prevent a change in hardness of the case.) This procedure results in a tapered cross section of the case. Thus, when superficial HR15-N hardness measurements are taken at 3.2-mm (⅛-in.) increments on this ground surface, each 3.2-mm (⅛-in.) increment represents a 0.13-mm (0.005-in.) increment in case depth. Results of this technique are biased, inasmuch as the relatively high load of 15 kg (33 lb) results in a series of case-core composite values.

Test Coupons. Quality control of nitrided parts is normally best accomplished by treating test coupons with each furnace load. One type of test coupon in common use is the fracture specimen illustrated in Fig. 33. Coupons must be of the same material heat treated to the same core hardness as the parts, and should be placed in locations that are representative of the nitriding conditions of the furnace. Thus, when the material and heat treating are constant, any changes in the nitriding process that may develop may be easily detected.

After nitriding, the test coupons are fractured or sectioned for determination of case depth by means of a Brinell microscope or a hardness survey. They also are used in determining the depth of the white layer, the core hardness of parts that have been nitrided all over, and the case hardness of areas that are not accessible to a hardness test. However, when possible, actual parts should be used for hardness tests of the case and core.

The data obtained from test coupons should be recorded and filed with the furnace records. Furnace temperature charts should include the dissociation readings taken during the nitriding treatment of each load.

Measurement and Removal of White Layer. Normally, the surface of nitrided parts will contain a layer of iron nitride (white layer). This white layer ranges in thickness from 0.005 to 0.05 mm (0.0002 to 0.0020 in.) depending on the length of the cycle and

Fig. 34 Nitrided surface before and after removal of white layer by immersing in an alkaline solution and blasting with 200-mesh aluminum oxide

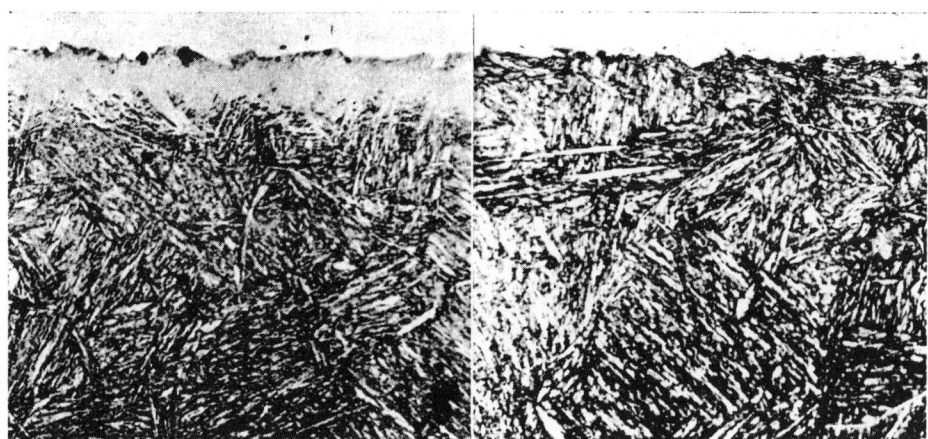

Nital etch. 500×

Table 9 General guide to amount of stock removal required for elimination of white layer from nitrided parts

| Nitriding cycle, h | Maximum amount of stock removal | | | |
| | Single-stage nitriding | | Double-stage nitriding | |
	mm	10^{-4} in.	mm	10^{-4} in.
12	0.01	5	0.01	5
24	0.03	10	0.03	10
36	0.04	15	0.03	10
48	0.05	20	0.03	10
60	0.06	25	0.04	15
72	0.08	30	0.05	20

whether single-stage or double-stage nitriding was employed. The thickness of the white layer is measured principally by metallographic methods. A prepared cross section of the nitrided surface is etched with an etchant that darkens the case but not the iron nitride layer; thus, this layer appears white and can be measured microscopically.

The white layer produced by single-stage nitriding is hard and brittle, and should be carefully removed. Double-stage nitriding produces a shallower, softer and more ductile white layer. For some applications this type of white layer is beneficial; in certain gear systems, for example, it provides a good wear-in surface. The amount of stock removal required for elimination of the white layer should be determined by testing actual parts; however, Table 9 may be used as a guide.

The amount of expensive finish grinding or lapping required for removing white layer is significantly less for parts that are double-stage nitrided than for parts that are single-stage nitrided. The white layer formed during double-stage nitriding usually can be held to a maximum thickness of 0.019 mm (0.00075 in.), although this may still be excessive for certain applications.

One method (U.S. Patent 3 069 296) for totally removing the white layer utilizes a simple alkaline solution that decomposes the iron nitride, making it friable and easily removable by light blast cleaning. A 200-mesh aluminum oxide grit is recommended for blasting. Depending on surface finish requirements, either liquid-abrasive blasting or peening with glass beads may be substituted for grit blasting. The procedure does not harm the surface finish

and has the added advantage of removing copper plate (during immersion in the alkaline solution) from parts plated for selective nitriding. Tests have shown no decrease in hardness, fatigue strength or impact strength; etching or pitting of the nitrided surface does not occur. The process does not require close control. Figure 34 shows micrographs of a nitrided surface before and after removal of the white layer by this method.

Another method (U.S. Patent 2 960 421) removes the iron nitride white layer by a diffusion process. Parts are copper plated all over and then heated to and held at about 525 °C (975 °F) for periods up to 40 h, depending on the thickness of white layer to be removed.

Parts that have been processed for the removal of white layer may be inspected by observing the reaction of nitrided surfaces after swabbing them with a 5% solution of nital or a 10% solution of ammonium persulfate. Areas from which the white layer has been removed will etch dark, whereas areas where the white layer is still present in substantial quantity will not etch. This procedure is not absolutely accurate, because areas that etch dark may still contain some white layer; however, the amount of white layer remaining is usually insufficient to affect the service performance of the part.

Pack Carburizing

By the ASM Committee on
Gas Carburizing, Carbonitriding and
Nitriding*

PACK CARBURIZING is a process in which carbon monoxide derived from a solid compound decomposes at the metal surface into nascent carbon and carbon dioxide. The nascent carbon is absorbed into the metal, and the carbon dioxide immediately reacts with carbonaceous material present in the solid carburizing compound to produce fresh carbon monoxide. The formation of carbon monoxide is enhanced by energizers or catalysts, such as barium carbonate ($BaCO_3$), calcium carbonate ($CaCO_3$), and sodium carbonate (Na_2CO_3), that are present in the carburizing compound. These energizers facilitate the reduction of carbon dioxide with carbon to form carbon monoxide. Thus, in a closed system, the amount of energizer does not change. Carburizing continues as long as enough carbon is present to react with the excess carbon dioxide.

Advantages. The continued use of pack carburizing is largely the result of improvements that have been made in the process, together with inherent advantages obtained in certain applications. Among the principal advantages of pack carburizing are the following:

- It can make use of a wide variety of furnaces because it produces its own contained environment.

- It does not require extensively trained or experienced personnel for successful operation.
- It is ideally suited for slow cooling of work from the carburizing temperature, a procedure that may be advantageous for parts that are to be finish machined after carburizing and before hardening.
- Compared to gas carburizing, it offers a wider selection of stop-off techniques for selective carburizing.

Disadvantages. By its nature, pack carburizing is less clean and less convenient than other carburizing processes. Other disadvantages generally associated with pack carburizing include the following:

- It is not well suited to production of shallow case depths where strict case-depth tolerances are required.
- It cannot provide the degree of flexibility and accuracy of control over surface carbon content and carbon gradient that can be obtained in gas carburizing.
- It is not well suited for direct quenching or quenching in dies; thus, extra handling and processing are required for the hardening operation.
- More processing time is required for pack carburizing than for gas or liquid carburizing because of the necessity of heating and cooling the extra

thermal mass associated with the compound and the container.

Carburizing Compounds

The common commercial carburizing compounds are reusable and contain 10 to 20% alkali or alkaline earth metal carbonates bound to hardwood charcoal or to coke by oil, tar or molasses. Barium carbonate is the principal energizer, usually comprising about 50 to 70% of the total carbonate content. The remainder of the energizer usually is made up of calcium carbonate, although sodium carbonate also may be used.

Hardwood charcoal is more reactive than coke as a source of carbon for pack carburizing. Nevertheless, coke offers certain advantages, such as minimum shrinkage, good hot strength and good thermal conductivity. More active carburizing compounds therefore contain both charcoal and coke, with typical compounds containing a greater percentage of coke. When nonburning properties are important, an all-coke base is used.

Addition Rate. Because of losses associated with the use of pack-carburizing compounds, new compound usually is added to the used compound before it is returned to service. The loss in energizer normally is somewhat

*Charles A. Stickels, *Chairman*, Principal Staff Engineer, Ford Motor Co.; Larry E. Byrnes, Manager-Manufacturing Research, Federal Mogul Corp.; Jon L. Dossett, Vice President-Metallurgy, U.S. Gear Corp.; Robert L. Hughes, Chief Metallurgist, Fairfield Manufacturing Company, Inc.; Kenneth D. Gladden, Senior Development Engineer, Caterpillar Tractor Co.; Dale A. Poteet, Jr., Division Engineer, John Deere Component Works; Joseph A. Riopelle, District Manager, Chicago Office, Surface Division, Midland-Ross Corp.; Ronald D. Rogers, Chief Engineer, Materials Engineering, Automotive Operations, Rockwell International

higher than loss of the rest of the compound. Therefore, somewhat larger percentages of new compound are used to ensure that the energizer level does not drop below approximately 5 to 8%. When direct quenching or severe mechanical handling methods are used, the addition rate may be as high as one part new compound to two parts used compound. When furnace cooling and careful handling methods are used, the addition rate may be one part new compound to three to five parts used compound.

Used compound often is screened to remove fines. The compound is then thoroughly mixed with the make-up material. Because many compounds, particularly those of the coated-charcoal type, are extremely friable, they require careful handling to minimize losses due to formation of dust or fines.

Special compounds have been developed to resist burning when in contact with air at carburizing temperatures. These compounds are particularly useful in operations where the work is quenched directly from the carburizing container. As previously noted, the nonburning compounds are principally coke-base materials. In the nonburning compounds, the energizer usually is distributed throughout the entire pellet by impregnating with a water-soluble salt of barium.

Carburizing compounds have been made from various combinations of organic materials, such as charred leather, peach pits, burned bone, raw bone, and bone with charcoal. In addition to a source of carbon, many of these materials contain carbonates or other catalytic materials. When applied properly, these compounds give satisfactory results. However, they are relatively expensive and, if no energizer is used, provide slow carburizing action.

Selection of Carburizing Compound. In selection of a carburizing compound, many factors should be considered. Both handling equipment and quenching methods are important. When the workpieces are cooled in a well-designed box where the compound is not exposed to air while at red heat, little consideration need be given to burning characteristics. Where there is appreciable quenching from the box, however, the rate of burning is an important economic factor and should be considered in selecting a compound.

Because the addition rate is determined by the amount of energizer retained in the used compound, the method of incorporating the energizer is important. Any evaluation of the cost of a compound should be based on the weight of steel that can be satisfactorily carburized per unit weight of compound.

Coated compound is at a disadvantage when compared with impregnated compound except where handling and cooling conditions are exceptionally favorable. However, if no attempt is made to take advantage of a more stable compound by reducing the addition rate, the more expensive, stable compound is of little advantage.

Process Control

In pack carburizing, as in other carburization processes, the carbon-concentration gradient obtained is a function of carbon potential, carburizing temperature and time, and the chemical composition of the steel. Two process-control attributes peculiar to pack carburizing are:

- There may be a variation in case depth within a given furnace load due to dissimilar thermal histories within the carburizing containers.
- Distortion of parts during carburizing may be reduced, because the compound can be used to support the parts.

Carbon Potential. The carbon potential of the atmosphere generated by the carburizing compound, as well as the carbon content obtained at the surface of the work, increase directly with an increase in the ratio of carbon monoxide to carbon dioxide. Thus, more carbon is made available at the work surface by the use of energizers and carburizing materials that promote formation of carbon monoxide.

In pack carburizing, the rate of evolution of carbonaceous gas is fixed and is almost always in excess of the rate required to supply the necessary carbon for a saturated surface layer. If the energizer is deliberately controlled to reduce the carburizing effect (ordinarily, only a bone compound can be controlled easily), it is possible to obtain a lower-carbon case than that which would be obtained under ordinary conditions of austenite saturation.

Temperature. Pack carburizing normally is performed at temperatures from 815 to 955 °C (1500 to 1750 °F). In recent years, the upper limits have been steadily raised, and carburizing temperatures as high as 1095 °C (2000 °F) have been used. Steelmaking processes have improved to the extent that fine grain size is maintained at temperatures approaching or exceeding 1040 °C (1900 °F). Above this temperature, the coarsening effect occurs only after prolonged periods of time, allowing high-temperature treatment without excessive grain coarsening.

The rate at which the carburized case is formed increases rapidly with temperature. If a factor of 1.0 is representative of 815 °C (1500 °F), the factor increases to 1.5 at 870 °C (1600 °F) and to more than 2.0 at 925 °C (1700 °F). Improved containers, fine-grain steels and other improvements now permit the use of a wide variety of temperatures.

Unless specially formulated low-carbon compounds are used, the surface carbon content is approximately that of the saturation limit for carbon in austenite. At operating temperature, the desired average carbon level throughout the case is directly dependent upon the carburizing temperature. When eutectoid cases are desired, the carburizing temperature is normally about 815 °C (1500 °F). As more carbon is required in the case, the temperature is increased. Although the composition of the compound can be varied to establish carbon potential and to limit surface carbon, control of temperature serves the same purpose and is easier to achieve.

Time. The rate of change in case depth at a particular carburizing temperature is proportional to the square root of time. The rate of carburization is thus highest at the beginning of the cycle and gradually diminishes as the cycle is extended (see Fig. 1).

Steel Composition. Any carburizing grade of carbon or alloy steel is suitable for pack carburizing. It is generally agreed that the diffusion rate of carbon in steel is not markedly influenced by the chemical composition of the steel.

Chemical composition does have an effect on the activity of carbon and thus can affect the carbon level at saturation for a particular temperature. The effects of alloy composition on austenite saturation for some Fe-C-X ternary alloys can be found in the ternary phase diagrams presented in Volume 8 of the 8th Edition of this Handbook.

Depth of Case. Even with good process control, it is difficult to obtain parts with total case-depth variation of less than 0.25 mm (0.010 in.) from maximum to minimum in a given furnace load, assuming a carburizing temperature of 925 °C (1700 °F). Commercial tolerances for case depths obtained in pack carburizing begin at ±0.25 mm (±0.010 in.) and, for deeper case depths, increase to ±0.8 mm (±0.03 in.). Lower carburizing temperatures provide some reduction in case-depth variation, because variation in the time required for all parts of the load to reach carburizing temperature becomes a smaller percentage of total furnace time. Because of the inherent variation in case depth and the cost of packing materials, pack carburizing normally is not used on work requiring a case depth of less than 0.8 mm (0.03 in.).

Typical pack-carburizing temperatures selected to produce different case depths on a variety of production parts are given in Table 1.

Distortion normally becomes more pronounced as processing temperature is increased. In some instances, carburizing temperature is selected on the basis of the maximum amount of distortion that can be tolerated. In any case, following proper container packing procedures will help minimize distortion.

Furnaces for Pack Carburizing

The suitability of a furnace for pack carburizing depends on its ability, at reasonable cost, to: (a) provide adequate thermal capacity and temperature uniformity (furnaces must be controllable to within ±5 °C, or ±9 °F, and must be capable of uniform through heating to within ±8 to ±14 °C, or ±14 to ±25 °F); and (b) provide adequate support for containers and workpieces at the required temperatures.

Modern heating systems and furnace construction provide ample heating capacity and temperature uniformity over a wide range of temperatures. A variation of ±8 °C (±14 °F) throughout the entire working section of a large furnace can be easily maintained. Many furnaces incorporate automatic compensation for heat losses at doors or other connection points. Combustion systems that maintain constant pressure or constant flow permit close temperature control on variable loads. Zoning is also a major contributor to control. To maintain good uniformity, it is necessary to load the furnace as uniformly as possible and to allow adequate space between containers—50 to 100 mm (2 to 4 in.) or more—to permit circulation of the heating gases.

The three types of furnaces most commonly used for pack carburizing are the box, car-bottom and pit types. Box furnaces are loaded by mechanical

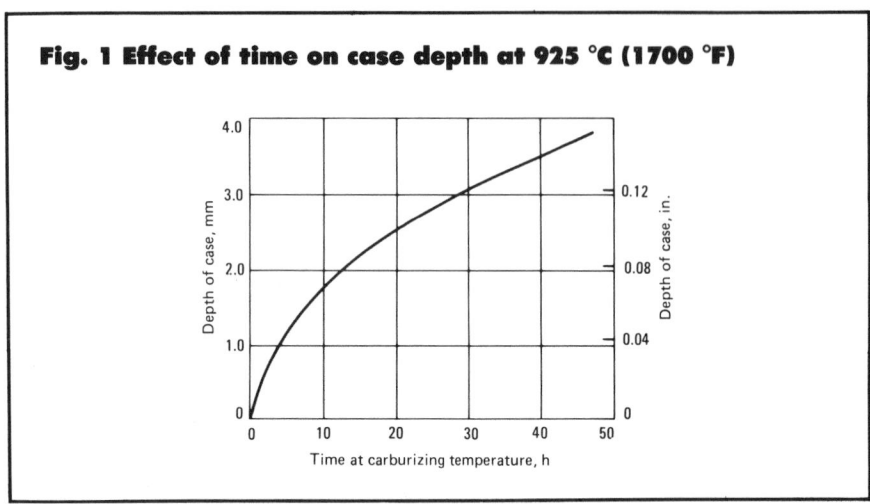

Fig. 1 Effect of time on case depth at 925 °C (1700 °F)

Table 1 Typical applications of pack carburizing

Part	Dimensions(a) OD cm	OD in.	OA cm	OA in.	Weight kg	Weight lb	Steel	Case depth to 50 HRC mm	in.	Temperature °C	°F
Mine-loader bevel gear	10.2	4.0	7.6	3.0	1.4	3.1	2317	0.6	0.024	925	1700
Flying-shear timing gear	21.6	8.5	9.2	3.6	23.6	52.0	2317	0.9	0.036	900	1650
Crane-cable drum	60.3	23.7	256.5	101.0	1 792	3 950	1020	1.2	0.048	955	1750
High-misalignment coupling gear	30.5	12.0	15.2	6.0	38.5	84.9	4617	1.2 c	0.048	925	1700
Continuous-miner drive pinion	12.7	5.0	12.7	5.0	5.4	11.9	2317	1.8	0.072	925	1700
Heavy-duty industrial gear	61.8	24.3	10.2	4.0	150	331	1022	1.8	0.072	940	1725
Motor-brake wheel	45.7	18.0	22.5	8.9	104	229	1020	3.0	0.120	925	1700
High-performance crane wheel	66.0	26.0	15.2	6.0	335	739	1035	3.8	0.150	940	1725
Calender bull gear	215.9	85.0	61.0	24.0	5 885	12 975	1025	4.0	0.160	955	1750
Kiln-trunnion roller	76.2	30.0	40.6	16.0	1 035	2 280	1030	4.0	0.160	940	1725
Leveler roll	9.5	3.7	79.4	31.3	36.7	80.9	3115	4.0	0.160	925	1700
Blooming-mill screw	38.1	15.0	332.7	131.0	2 950	6 505	3115	5.0	0.200	925	1700
Heavy-duty rolling-mill gear	91.4	36.0	403.8	159.0	11 800	26 015	2325	5.6	0.220	955	1750
Processor pinch roll	22.9	9.0	538.5	212.0	1 700	3 750	8620	6.9	0.270	1050	1925

(a) OD, outside diameter; OA, over-all (axial) dimension

devices or by in-plant transportation equipment. Car-bottom furnaces provide for convenient loading of heavy units. A car-bottom furnace with a car at each end allows a second car to be loaded while the furnace is in use, which minimizes heat loss and downtime between batches. Pit furnaces are general-purpose furnaces that may be used for carburizing and other heat treating operations and that require minimum floor space.

Adequate support of containers and workpieces does much to minimize distortion. It also helps maintain the shape and extend the life of carburizing containers. Three or more points of support should be used in car-bottom furnaces. The container should be blocked above the hearth to allow circulation around, and proper shimming of, the container. In box-type furnaces, silicon carbide and certain other hearth materials provide excellent wear resistance to maintain the shape of the hearth. Their high thermal conductivity helps promote temperature uniformity.

Furnaces for pack carburizing have a minimum number of parts that are subject to high wear or that require frequent maintenance. Very few alloy parts inside the furnace are subjected to thermal fatigue, and a minimum of auxiliary equipment is needed. The personnel who operate these furnaces do not need extensive technical training.

Carburizing Containers

Materials. Carburizing containers are made of carbon steel, of aluminum-coated carbon steel or of iron-nickel-chromium heat-resisting alloys. Although uncoated carbon steel boxes scale severely during carburizing and have short lives, they often are the most economical for processing odd lots and unusual shapes.

Aluminum coating can significantly extend the life of a carbon steel contain-er, making this material potentially the lowest in cost per hour per unit weight carburized.

In the long run, heat-resisting alloys are the most economical container materials for carburizing large numbers of moderate-size parts. However, because heat-resisting alloys are considerably higher in initial cost than plain or aluminum-coated carbon steel, they must be used continuously if they are to approach the lowest possible prorated cost.

Design and Construction. For containers of all three materials, the trend has been toward lighter construction from sheet or plate, rather than the heavier cast construction. These lighter containers require ribbing, corrugating or other bracing methods to make them rigid enough to withstand long periods at high temperature. Containers often are equipped with braced lifting eyes or hooks, special lid-receiving sections, and test-pin openings.

A carburizing container should be no larger than necessary. If possible, it should be narrow in at least one dimension, to promote uniform heating of the contents. A properly designed box will provide a cooling rate high enough to minimize formation of a carbide network in the case but low enough to avoid distortion or excessive hardening.

Lid Construction. Lids for carburizing containers vary from simple sheet-metal plates to built-up lids of metal and refractory material. The lid may add rigidity to the container. It must be tight enough to prevent air from entering and burning the compound, yet not so tight as to prevent easy expulsion of excess gas generated within the container. Lids must be capable of venting the container, and the venting means must be able to withstand the intense heat liberated by combustion of flammable gas. Lids that fit too loosely can be partly sealed with clay-base cements.

Conditioning. Before new alloy carburizing containers are placed in ser-vice, they may be conditioned by "precarburizing" without a work load. This pretreatment eliminates the possibility of the container, rather than the work load, being carburized during the first production carburizing cycle.

Packing

Intimate contact between compound and workpiece is not necessary; however, when properly packed, the compound will provide good support for the workpiece. The layer of compound surrounding the work must be heavy enough to allow for shrinkage and to maintain a high carbon potential during the entire cycle, but not so heavy as to unduly retard heating of the workpiece to carburizing temperature. If the container can be designed to conform to the shape of the workpiece, the compound will be of both uniform and minimum thickness. When compounds that contain exothermic materials are used, the thickness of the compound layer is less critical.

Work-load density—that is, net weight (piece weight) divided by gross weight (weight of the carburizing container, compound and workpieces)—is an important factor in the efficiency of pack carburizing, because it affects heating and cooling time. The smaller this percentage, the lower the relative efficiency of the process. Table 2 shows work-load densities for three different carburized parts.

Procedure. Packing of workpieces in a compound is a dusty and disagreeable operation. For this reason, grouping of boxes, workpieces and compound should be carefully planned so as to minimize handling of the compound. If possible, workpieces should come to the packer already stacked and sorted, preferably on open trays or in pans.

First, a layer of compound from 13 to 50 mm (½ to 2 in.) deep is placed in the empty box. The part or parts are then stacked in the container, and, if necessary, metal or ceramic supports or spac-

Table 2 Work-load densities in pack carburizing

Part	Dimensions(a) OD cm	in.	OA cm	in.	Weight per piece Net kg	lb	Total(b) kg	lb	Net weight, % of gross weight
Roll	7.6	3.0	122	48	37	82	72	159	51
Crane wheel	56	22	12.7	5.0	130	287	150	331	87
Gear	66	26	20.3	8.0	285	628	440	970	65

(a) OD, outside diameter; OA, over-all (axial) dimension. (b) Total weight of work plus packing material plus container, divided by number of pieces in pack

ers are applied and internal container supports are inserted.

Whenever possible, workpieces should be packed with the longest dimension vertical to the base of the container. This is extremely important in processing long parts such as shafts and rolls, because it minimizes the tendency of these parts to sag. Suspension of the work within the container or within the furnace is useful in minimizing distortion in relatively thin or delicate parts. For applications where small teeth or small holes are to be uniformly carburized, a 6- or 8-mesh material should be used to ensure good filling.

After the compound is sufficiently tamped, a final layer is placed on top of the parts. The thickness of the top layer varies according to the type of work, depth of case, type of container, and shrinkage rate of the compound, but it should be adequate to ensure that the work will be covered after shrinkage and other movements have occurred. A minimum depth of 50 mm (2 in.) is recommended. In the final step, the lid is put in place.

Process-Control Specimens. In order to control and evaluate the carburizing process, test pins or shims normally are included in the charge. To provide valid results, section sizes and locations of test specimens must closely approximate those of the workpieces. Placing a test pin close to a workpiece often will produce a thermal history identical to that of the workpiece.

For control purposes, many containers are equipped with a test-pin section that can be removed from the load during the carburizing cycle. After the pins have been quenched and fractured, case-depth readings made on them aid in evaluating whether satisfactory carburizing results are being obtained and in determining when the prescribed case depth has been attained.

Selective Carburizing

Stop-off techniques described in the article on gas carburizing in this volume apply to selective carburization by pack carburizing. In addition, it may be possible to permit any portion of a part that is not to be carburized to protrude from the carburizing container. Alternatively, an inert or slightly oxidizing material may be packed around those areas of a part that are not to be carburized.

Liquid Carburizing and Cyaniding

By the ASM Committee on Liquid Carburizing[*]

LIQUID CARBURIZING is a process used for case hardening steel or iron parts. The parts are held at a temperature above Ac_1 in a molten salt that will introduce carbon and nitrogen, or carbon alone, into the metal. Diffusion of the carbon from the surface toward the interior produces a case that can be hardened, usually by fast quenching from the bath. Carbon diffuses from the bath into the metal and produces a case comparable with one resulting from gas carburizing in an atmosphere containing some ammonia. However, because liquid carburizing involves faster heat-up (due to the superior heat-transfer characteristics of salt-bath solutions), cycle times for liquid carburizing are shorter than those for gas carburizing.

Most liquid carburizing baths contain cyanide, which introduces both carbon and nitrogen into the case. One type of liquid carburizing bath, however, uses a special grade of carbon, rather than cyanide, as the source of carbon. This bath produces a case that contains only carbon as the hardening agent.

Liquid carburizing may be distinguished from cyaniding (which is performed in a bath containing a higher percentage of cyanide) by the character and composition of the case produced. Cases produced by liquid carburizing are lower in nitrogen and higher in carbon than cases produced by cyaniding. Cyanide cases are seldom applied to depths greater than 0.25 mm (0.010 in.); liquid carburizing can produce cases as deep as 6.4 mm (0.250 in.). For very thin cases, liquid carburizing in low-temperature baths may be employed in place of cyaniding.

Cyanide-Containing Liquid Carburizing Baths

"Light case" and "deep case" are arbitrary terms that have been associated with liquid carburizing in baths containing cyanide. There is necessarily some overlapping of bath compositions for the two types of case. In general, the two types are distinguished more by operating temperature than by bath composition. Hence, the terms "low temperature" and "high temperature" are preferred.

Both low-temperature and high-temperature baths are supplied in different cyanide contents to satisfy individual requirements of carburizing activity (carbon potential) within the limitations of normal dragout and replenishment. In many instances, compatible companion compositions are available for starting the bath or for bath make-up, and for regeneration or maintenance of carburizing potential.

Low-temperature cyanide-type baths (light-case baths) are those usually operated in the temperature range from 845 to 900 °C (1550 to 1650 °F), although for certain specific effects this range is sometimes extended to 790 to 925 °C (1450 to 1700 °F). Low-temperature baths are best suited to formation of shallower cases. Low-temperature baths are generally of the accelerated cyanogen type containing various combinations and amounts of the constituents listed in Table 1 and differ from cyaniding baths in that the case produced by a low-temperature bath consists predominantly of carbon. Low temperature baths are usually operated with a protective carbon cover; however, when the carbon cover on a low-temperature bath is thin, the nitrogen content of the carburized case will be relatively high. Cyaniding baths produce cases that are about 0.13 to 0.25 mm (0.005 to 0.010 in.) deep and that contain appreciable amounts of nitrogen.

[*]Donald R. Wensing, *Chairman,* Chief Metallurgist, SKF Industries, Inc.; Donald R. Barber (retired), President, Heatbath Corp.; Robert W. Foreman, Director—Research & Development, Park Chemical Co.; George Greene, Metallurgist, Industrial Truck Division, Eaton Corp.; Karl Heinz Kopietz, Marketing Manager, Heat Treating Products, Henry E. Sanson & Sons, Inc.; W. James Laird, Jr., Vice President—Marketing, Research & Development, Upton Industries, Inc.; Quentin D. Mehrkam, Senior Vice President, Ajax Electric Co.; Burton R. Payne, Jr., President, Payne Chemical Co.; John A. Swift, Consultant, Metallurgy, Henry E. Sanson & Sons, Inc.; William G. Wood, Vice President, Technology/Research & Development, Kolene Corp.; Theodore B. Wilk, Vice President, Sales & Technology, The A. F. Holden Co.

Table 1 Operating compositions of liquid carburizing baths

Constituent	Composition of bath, %	
	Light case, low temperature 845 to 900 °C (1550 to 1650 °F)	Deep case, high temperature 900 to 955 °C (1650 to 1750 °F)
Sodium cyanide	10 to 23	6 to 16
Barium chloride	...	30 to 55(a)
Salts of other alkaline earth metals(b)	0 to 10	0 to 10
Potassium chloride	0 to 25	0 to 20
Sodium chloride	20 to 40	0 to 20
Sodium carbonate	30 max	30 max
Accelerators other than those involving compounds of alkaline earth metals(c)	0 to 5	0 to 2
Sodium cyanate	1.0 max	0.5 max
Density of molten salt	1760 kg/m^3 at 900 °C (110 lb/ft^3 at 1650 °F)	2000 kg/m^3 at 925 °C (125 lb/ft^3 at 1700 °F)

(a) Proprietary barium chloride–free deep-case baths are available. (b) Calcium and strontium chlorides have been employed. Calcium chloride is more effective, but its hygroscopic nature has limited its use. (c) Among these accelerators are manganese dioxide, boron oxide, sodium fluoride and sodium pyrophosphate.

Table 2 Relation of operating temperature to sodium cyanide content in barium-activated liquid carburizing baths

Temperature		Sodium cyanide, %		
°C	°F	min	Preferred	max(a)
815	1500	14	18	23
845	1550	12	16	20
870	1600	11	14	18
900	1650	10	12	16
925	1700	8	10	14
955	1750	6	8	12

(a) The maximum limits are based on economy. If 30% NaCN is exceeded, there is danger that NaCN will break down, with production of carbon scum and attendant frothing. To correct such a condition, the bath temperature should be lowered and the NaCN content should be adjusted to the preferred value.

In a low-temperature cyanide-type bath, several reactions occur simultaneously, depending on bath composition, to produce various end products and intermediates. These reaction products include the following: carbon (C), alkali carbonate (Na_2CO_3 or K_2CO_3), nitrogen (N_2 or 2N), carbon monoxide (CO), carbon dioxide (CO_2), cyanamide (Na_2CN_2 or $BaCN_2$), and cyanate (NaNCO).

Two of the major reactions believed to occur in the operating bath are the "cyanamide shift" and the formation of cyanate:

$$2NaCN \leftrightarrow Na_2CN_2 + C \qquad (1)$$

and either

$$2NaCN + O_2 \rightarrow 2NaNCO \qquad (2)$$

or

$$NaCN + CO_2 \leftrightarrow NaNCO + CO \qquad (3)$$

Reactions that influence cyanate content proceed as follows:

$$NaNCO + C \rightarrow NaCN + CO \qquad (4)$$

and either

$$4NaNCO + 2O_2 \rightarrow 2Na_2CO_3 + 2CO + 4N \qquad (5)$$

or

$$4NaNCO + 4CO_2 \rightarrow 2Na_2CO_3 + 6CO + 4N \qquad (6)$$

Reactions 5 and 6 deplete the activity of the bath and lead to an eventual loss of carburizing effectiveness unless suitable replenishment practice is followed. Reactions 1 and 3 are at least partly reversible. Reactions that produce either carbon monoxide or carbon are beneficial in producing the desired carburized case, as for example:

$$Fe + 2CO \rightarrow Fe[C] + CO_2 \qquad (7)$$

and

$$Fe + C \rightarrow Fe[C] \qquad (8)$$

Low-temperature (light-case) baths are usually operated at higher cyanide contents than high-temperature (deep-case) baths. The preferred operating cyanide contents shown in Table 2 provide a case that is essentially eutectoidal (>0.80% C). If a hypoeutectoid (<0.80% C) case is desired, the bath is operated at the lower end of the temperature/cyanide range. Conversely, operation at the higher end of the suggested range favors formation of a hypereutectoid surface carbon content.

High-temperature cyanide-type baths (deep-case baths) are usually operated in the temperature range from 900 to 955 °C (1650 to 1750 °F). This range may be extended somewhat, but at lower temperatures the rate of carbon penetration decreases, and at temperatures higher than about 955 °C, deterioration of the bath and equipment is markedly accelerated. However, rapid carbon penetration can be obtained by operating at temperatures between 980 and 1040 °C (1800 and 1900 °F).

High-temperature baths are used for producing cases 0.5 to 3.0 mm (0.020 to 0.120 in.) deep. In some instances, even deeper cases are produced (up to about 6 mm, or 0.250 in.), but the most important use of these baths is for the rapid development of cases 1 to 2 mm (0.040 to 0.080 in.) deep. These baths consist of cyanide and a major proportion of barium chloride (Table 1), with or without supplemental acceleration from other salts of alkaline earth metals. Although the reactions shown for low-temperature liquid carburizing salts apply in some degree, the principal reaction is the so-called cyanamide shift. This reaction is reversible:

$$Ba(CN)_2 \leftrightarrow BaCN_2 + C \qquad (9)$$

In the presence of iron, the reaction is:

$$Ba(CN)_2 + Fe \rightarrow BaCN_2 + Fe[C] \qquad (10)$$

Cases produced in high-temperature liquid carburizing baths consist essentially of carbon dissolved in iron. However, sufficient nascent nitrogen is available to produce a superficial nitride-containing skin, which aids in resisting wear and which also resists softening during tempering.

Combination Treatment. It is not uncommon for the carburizing cycle to be initiated in a high-temperature bath and then for the work load to be transferred to a low-temperature carburizing bath. Not only does this practice provide a maximum rate of carburizing, but quenching the work from the low-temperature bath reduces distortion and minimizes retained austenite.

Cyaniding (Liquid Carbonitriding)

Cyaniding, or salt-bath carbonitriding, is a heat treating process that produces a file-hard, wear-resistant surface on ferrous parts. When steel is heated above Ac_1 in a suitable bath containing alkali cyanides and cyanates, the surface of the steel absorbs both carbon and nitrogen from the molten bath. When quenched in mineral oil, paraffin-base oil, water or brine, the steel develops a hard surface layer, or case, that contains less carbon and more nitrogen than the case developed in activated liquid carburizing baths.

Because of greater efficiency and lower cost, sodium cyanide is used instead of the more expensive potassium cyanide. The active hardening agents of cyaniding baths—carbon monoxide and nitrogen—are not produced directly from sodium cyanide. Molten cyanide decomposes in the presence of air at the surface of the bath to produce sodium cyanate, which in turn decomposes in accordance with the following chemical reactions:

$$2NaCN + O_2 \rightarrow 2NaNCO \qquad (11)$$

$$4NaNCO \rightarrow Na_2CO_3 + 2NaCN + CO + 2N \qquad (12)$$

$$2CO \rightarrow CO_2 + C \qquad (13)$$

$$NaCN + CO_2 \rightarrow NaNCO + CO \qquad (14)$$

The rate at which cyanate is formed and decomposes, liberating carbon and nitrogen at the surface of the steel, determines the carbonitriding activity of the bath. At operating temperatures, the higher the concentration of cyanate, the faster the rate of its decomposition. Because the rate of cyanate decomposition also increases with temperature, bath activity is greater at higher operating temperatures. A fresh cyaniding bath must be aged for about 12 h at a temperature above its melting point to provide a sufficient concentration of cyanate for efficient carbonitriding activity. For the aging cycle to be effective, any carbon scum formed on the surface must be removed. To eliminate scum, the cyanide content of the bath must be reduced to the 25-to-30% range by addition of inert salts (sodium

chloride and sodium carbonate). At the bath aging temperature—usually about 700 °C (1290 °F)—the rate of cyanate formation is high and the rate of its decomposition is low.

Bath Composition. A sodium cyanide mixture such as grade 30 in Table 3, containing 30% NaCN, 40% Na_2CO_3 and 30% NaCl, is generally used for cyaniding on a production basis. This mixture is preferable to any of the other compositions given in Table 3. The inert salts sodium chloride and sodium carbonate are added to cyanide to provide fluidity and to control the melting points of all mixtures. The 30% NaCN mixture, as well as the mixtures containing 45, 75 and 97% NaCN, may be added to the operating bath to maintain a desired cyanide concentration for a specific application.

The carbon content of the case developed in cyanide baths increases with an increase in the cyanide concentration of the bath, thus providing considerable versatility. A bath operating at 815 to 850 °C (1500 to 1560 °F) and containing 2 to 4% cyanide may be used to restore carbon to decarburized steels with a core carbon content of 0.30 to 0.40% C, while a 30% cyanide bath at the same temperature will yield a 0.127-mm (0.005-in.) case containing 0.65% C at the surface in 45 min. Sodium cyanide concentration also has some effect on case depth, as shown for 1020 steel in Table 4.

Noncyanide Liquid Carburizing

Liquid carburizing can be accomplished in a bath containing a special grade of carbon instead of cyanide as the source of carbon. In this bath, carbon particles are dispersed in the mol-

Table 3 Compositions and properties of sodium cyanide mixtures

Constituent or property	96-98(a)	Grade 75(b)	Grade 45(b)	Grade 30(b)
Composition, %				
Sodium cyanide	97	75	45.3	30.0
Sodium carbonate	2.3	3.5	37.0	40.0
Sodium chloride	Trace	21.5	17.7	30.0
Melting point, °C (°F)	560 (1040)	590 (1095)	570 (1060)	625 (1155)
Specific gravity				
At 25 °C (75 °F)	1.50	1.60	1.80	2.09
At 860 °C (1580 °F)	1.10	1.25	1.40	1.54

(a) Appearance: white crystalline solid. This grade also contains 0.5% sodium cyanate (NaNCO) and 0.2% sodium hydroxide (NaOH); sodium sulfide (Na_2S) content, nil. (b) Appearance: white granular mixture

Table 4 Influence of sodium cyanide concentration on case depth in 1020 steel bars(a)

25.4-mm diam (1.0-in. diam), cyanided 30 min at 825 °C (1500 °F)

NaCN in bath, %	Depth of case mm	Depth of case in.
94.3	0.15	0.0060
76.0	0.18	0.0070
50.8	0.15	0.0060
43.0	0.15	0.0060
30.2	0.15	0.0060
20.8	0.14	0.0055
15.1	0.13	0.0050
10.8	0.10	0.0040
5.2	0.05	0.0020

ten salt by mechanical agitation, which is achieved by means of one or more simple propeller stirrers that occupy a small fraction of the bath.

The chemical reaction involved is not fully understood, but is thought to involve adsorption of carbon monoxide on carbon particles. The carbon monoxide is generated by reaction between the carbon and carbonates, which are major ingredients of the molten salt. The adsorbed carbon monoxide is presumed to react with steel surfaces much as in gas or pack carburizing.

Operating temperatures for this type of bath are generally higher than those for cyanide-type baths. A range of about 900 to 955 °C (1650 to 1750 °F) is most commonly employed. Temperatures below about 870 °C (1600 °F) are not recommended, and may even lead to decarburization of the steel. The case depths and carbon gradients produced are in the same range as for high-temperature cyanide-type baths (see Fig. 1 to 3), but there is no nitrogen in the case.

Temperatures above 955 °C (1750 °F)

Fig. 1 Carbon gradients produced by liquid carburizing of carbon and alloy steels

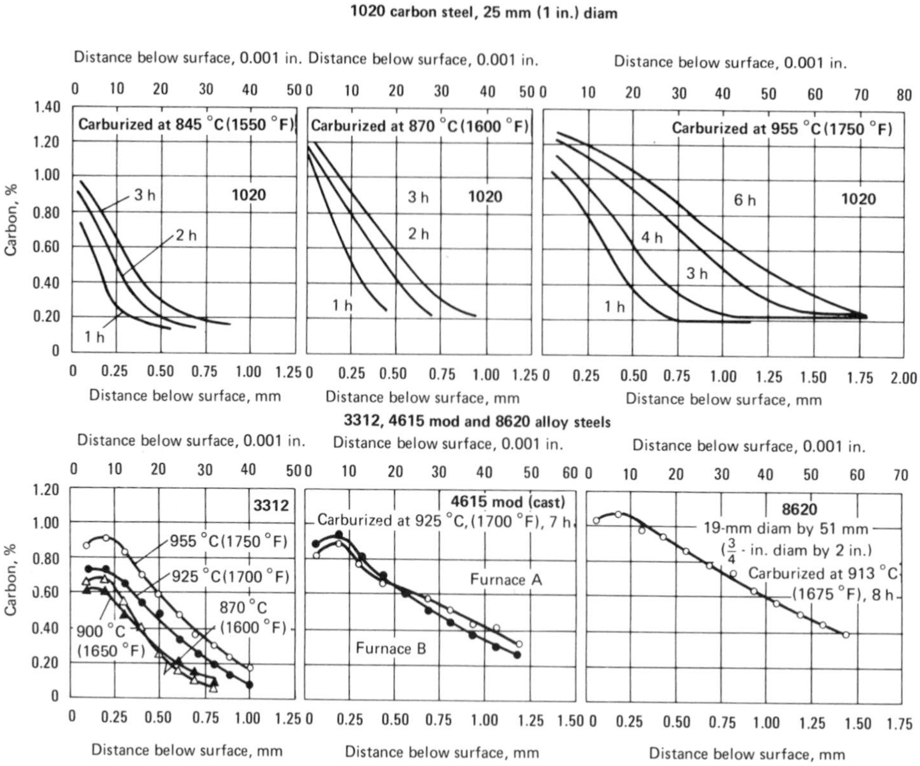

1020 carbon steel, 25 mm (1 in.) diam

3312, 4615 mod and 8620 alloy steels

Carbon gradients produced by liquid carburizing carbon and alloy steels in low-temperature and high-temperature baths. The 1020 carbon steel bars were carburized at 845, 870 and 955 °C (1550, 1600 and 1750 °F) for the periods shown. The data on 3312 alloy steel show the effect of four different carburizing temperatures on carbon gradient (time constant at 2 h). The data on modified 4615 steel castings indicate the slight differences in gradients obtained in two furnaces employing the same carburizing conditions (7 h at 925 °C, or 1700 °F). These data and the data on 8620 steel parts show a decrease in carbon content near the surface caused by diffusion of carbon during reheating to austenitizing temperature.

produce more rapid carbon penetration and do not adversely affect noncyanide baths, because no cyanide is present to break down and cause carbon scum or frothing. Equipment deterioration is the chief factor that limits operating temperature.

Parts that are slowly cooled following noncyanide carburization are more easily machined than parts slowly cooled following cyanide carburization, because of the absence of nitrogen in noncyanide-carburized cases. For the same reason, parts that are quenched following noncyanide carburization contain less retained austenite than parts quenched following cyanide carburization.

Carbon Gradients

Figure 1 shows carbon gradients produced by liquid carburizing 1020 steel bars at 845, 870 and 955 °C (1550, 1600 and 1750 °F) for various lengths of time

at carburizing temperature. Carbon-gradient data for two wrought alloy steels (3312 and 8620) and one cast alloy steel (4615 mod) are also shown. After carburizing, the 8620 steel parts were austenitized at 840 °C (1540 °F) and quenched in oil at 55 °C (130 °F). The 4615 steel parts were austenitized at 790 °C (1450 °F), quenched in salt at 190 °C (375 °F) for 3 min, and cooled in air.

Carbon penetration (case depth) in liquid carburizing is determined primarily by the carburizing temperature and the duration of the carburizing cycle. A simple formula for estimating total case depths (measured to base carbon level) obtainable in liquid carburizing is:

$$d = k \sqrt{t}$$

where d is case depth, k is a constant that represents the penetration in the first hour at temperature, and t is the time at temperature in hours. Typical

values of k at three different temperatures are 0.30 mm at 815 °C (0.012 in. at 1500 °F), 0.46 mm at 870 °C (0.018 in. at 1600 °F), and 0.64 mm at 925 °C (0.025 in. at 1700 °F).

Hardness Gradients

A hardness gradient or variation in hardness at different depths below the surface is established in parts that are quenched following liquid carburization. The data in Fig. 2 show typical hardness gradients obtainable with carbon and low-alloy steels, and illustrate the influence of carburizing temperature, duration of carburizing, quenching temperature and quenching medium. Data on 1020, 4620 and 8620 steels are plotted for cycles of 2, 4, 8, 15, 20 and 40 h. These specimens were air cooled from carburizing temperatures of 870, 900 and 925 °C (1600, 1650 and 1700 °F), reheated in neutral salt at 845 °C (1550 °F), and quenched in mol-

Fig. 2 Case-hardness gradients for two carbon steels and four low-alloy steels, showing effects of carburizing temperature and time

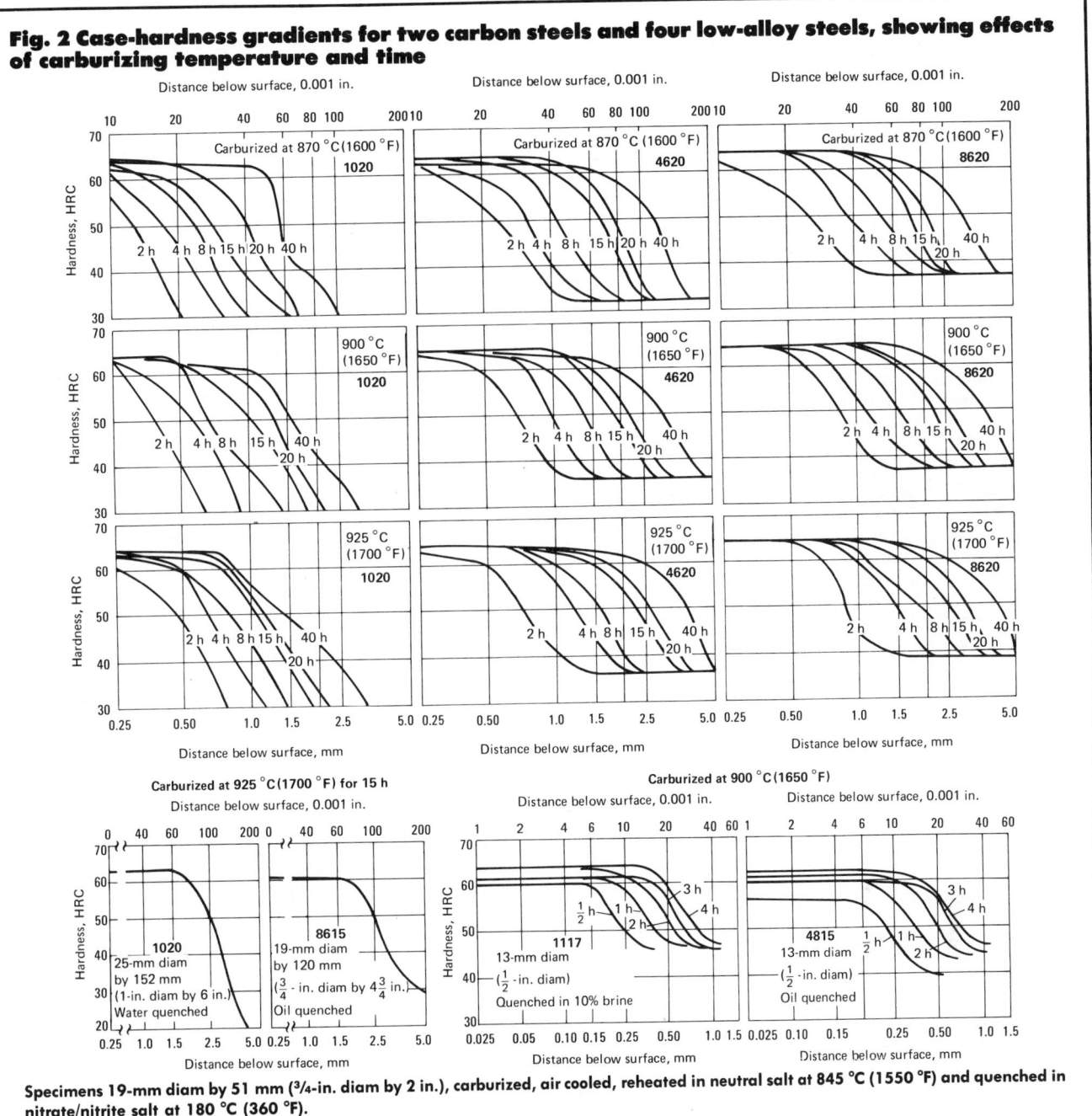

Specimens 19-mm diam by 51 mm (3/4-in. diam by 2 in.), carburized, air cooled, reheated in neutral salt at 845 °C (1550 °F) and quenched in nitrate/nitrite salt at 180 °C (360 °F).

ten salt at 180 °C (360 °F). Although the depth of case at maximum hardness is progressively extended in the alloy steels with increases in time and temperature, increases in carburizing temperature have the effect of foreshortening the curves plotted for the 1020 steel. Differences between the responses of 1020 and 8615 steels are shown to be less pronounced after carburizing at 925 °C (1700 °F) for 15 h and quenching directly from the carburiz-

ing temperature. A final series of curves indicates the results obtained with 1117 and 4815 steels after carburizing at 900 °C (1650 °F) for periods ranging from 1/2 to 4 h. The 4815 steel was quenched in oil, and the 1117 steel was quenched in a 10% brine solution.

The indentation hardness data presented in Fig. 3 for five different steels indicate the effects of normal variations in practice on the hardness gradi-

ent. The shaded bands represent the scatter in results obtained from multiple tests of each steel. Although similar surface hardnesses are obtained with all five steels, depth of hardness varies with the alloy content of the steel. A comparison among the hardnesses of these five steels at a depth of 1 mm (0.040 in.) illustrates this variation. Although a minimum case hardness of 60 HRC cannot be maintained to a depth of 1 mm (0.040 in.) with 1020

(0.30 to 0.60% Mn) steel, it can sometimes be achieved with 1113 (0.70 to 1.00% Mn) steel and can almost always be achieved with 1117 (1.00 to 1.30% Mn), 4615 and 8620 steels.

Cyaniding Time and Temperature

Bath operating temperatures for cyanide hardening vary between 760 and 870 °C (1400 and 1600 °F). Temperatures near the lower end of this range are favored for minimizing distortion during quenching from the bath temperature. Higher temperatures are selected to exceed the Ac₃ point of the steel, to achieve faster penetration, and, depending on alloy content, to produce a fully hardened core after quenching.

When low-carbon and alloy steels are to be cyanided to produce a surface capable of resisting high contact loads, the surface usually must be backed up with a fine-grain, tough core. This requires an operating bath temperature of about 870 °C (1600 °F).

The high file hardness of salt bath cyanided steel parts is a combined effect of carbon and nitrogen absorption in the carbonitrided case (see Table 5). Usually, immersion times range from 30 min to 1 h and produce case depths and surface carbon and nitrogen concentrations corresponding to those in Table 5. Lower temperatures will provide results proportionately lower than those given in Table 5.

Furnaces and Equipment

Liquid carburizing is carried out in a salt-bath furnace that may be heated either externally or internally. In an externally heated furnace, heat is introduced into an annular space between the salt pot and the surrounding insulation, which usually is made of firebrick. In an internally heated furnace, heat is introduced directly into the salt. Both internally and externally heated furnaces generally have insulated lids that slide to open the bath and allow workpieces and fixtures to be positioned, usually with an overhead crane or with similar mechanized lifting equipment.

Externally Heated Furnaces

Externally heated furnaces may be fired by gas or oil, or may be heated by means of electrical-resistance elements.

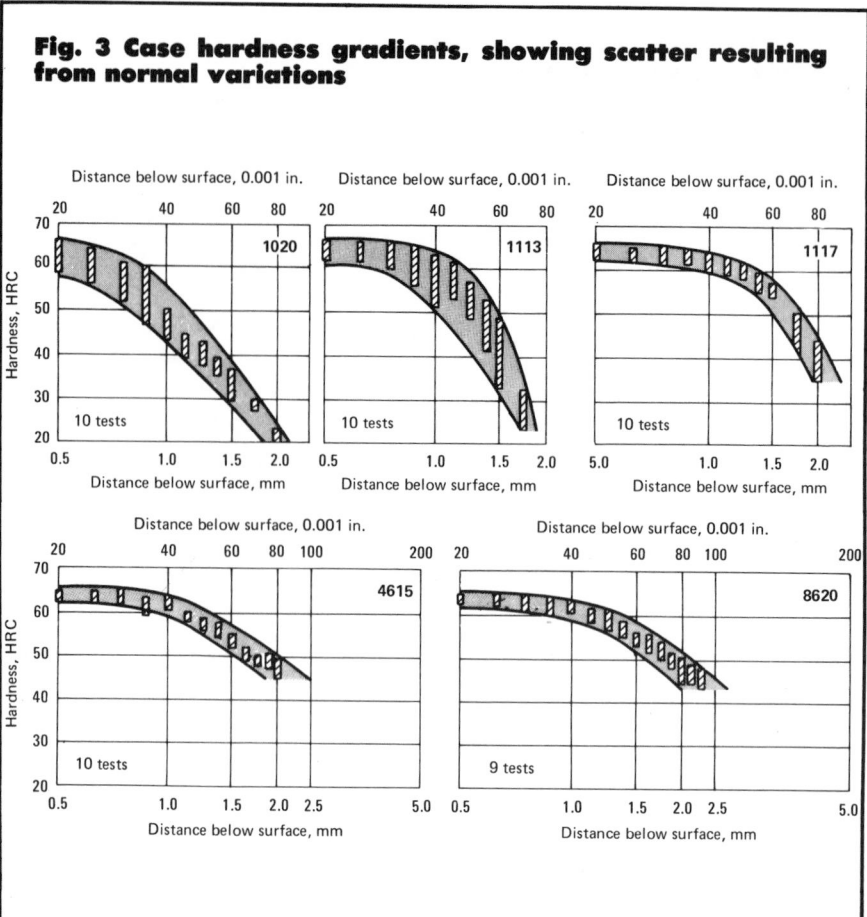

Fig. 3 Case hardness gradients, showing scatter resulting from normal variations

Table 5 Effect of cyaniding temperature and time on case depth and surface carbon and nitrogen contents

Material thickness, 2.03 mm (0.080 in.); cyanide content of bath, 20 to 30%

Steel	Case depth, mm (in.), after cyaniding for:		Analysis after 100 min at temperature(a)	
	15 min	100 min	Carbon, %	Nitrogen, %
Cyanided at 750 °C (1400 °F)				
1008	0.038 (0.0015)	0.152 (0.006)	0.68	0.51
1010	0.038 (0.0015)	0.152 (0.006)	0.70	0.50
1022	0.051 (0.0020)	0.203 (0.008)	0.72	0.51
Cyanided at 845 °C (1550 °F)				
1008	0.076 (0.0030)	0.203 (0.008)	0.75	0.26
1010	0.076 (0.0030)	0.203 (0.008)	0.77	0.28
1022	0.089 (0.0035)	0.254 (0.010)	0.79	0.27

(a) Carbon and nitrogen contents were determined from analysis of the outermost 0.076 mm (0.003 in.) of cyanided cases.

Gas-fired or oil-fired furnaces similar in design to the one shown in Fig. 4(a) are commonly used for liquid carburizing. These furnaces are generally lower in initial cost than electrode or resistance-heated furnaces and are simple to install and operate. To contain the molten carburizing salt, fuel-fired furnaces employ a steel or alloy pot, which may be either round or rectangular. Heat is applied by two or more self-cooling burners that fire tangen-

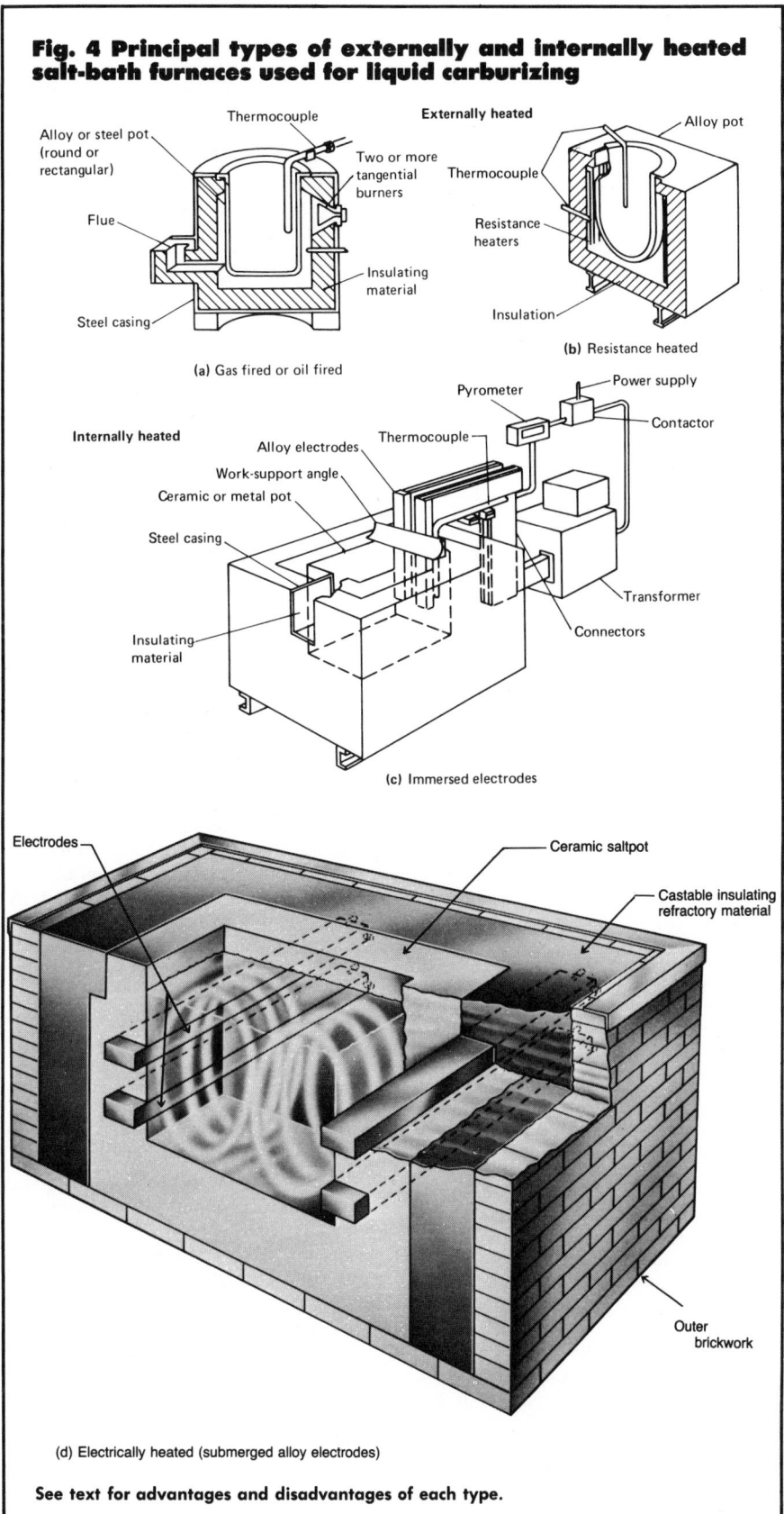

Fig. 4 Principal types of externally and internally heated salt-bath furnaces used for liquid carburizing

Externally heated

Alloy or steel pot (round or rectangular)
Thermocouple
Two or more tangential burners
Flue
Steel casing
Insulating material

(a) Gas fired or oil fired

Alloy pot
Thermocouple
Resistance heaters
Insulation

(b) Resistance heated

Internally heated

Pyrometer
Power supply
Contactor
Alloy electrodes
Thermocouple
Work-support angle
Ceramic or metal pot
Steel casing
Transformer
Connectors
Insulating material

(c) Immersed electrodes

Electrodes
Ceramic saltpot
Castable insulating refractory material
Outer brickwork

(d) Electrically heated (submerged alloy electrodes)

See text for advantages and disadvantages of each type.

tially between the outer pot wall and the inner surface of the furnace lining. The hot gases are vented through a flue, which is located near the top for atmospheric-type burners, and near the bottom for pressure-type burners and for atmospheric burners for which the flue is connected to a stack 1 to 2 m (3 to 6 ft) high to maintain negative pressure in the firing chamber. The combustion chamber is lined with firebrick, and with additional insulation if required. A steel casing completely surrounds all sides of the furnace housing and provides adequate safety in the event of pot failure.

Electrical-resistance furnaces for liquid carburizing, such as that shown in Fig. 4(b), are less widely used than furnaces fired by gas or oil. They are heated by a series of resistance heaters surrounding the salt pot. With this type of furnace, pot failure may result in the total destruction of the electrical heating elements; to guard against this possibility, carburizing temperatures below 900 °C (1650 °F) are preferred.

Salt Pots. Because the salt pot ordinarily is supported from a flange, pot size is limited by the strength of the material used. Round pots for furnaces fired by gas or oil normally range from 25 to 90 cm (10 to 36 in.) in diameter and from 20 to 75 cm (8 to 30 in.) in depth; they are about 10 mm (³⁄₈ in.) thick. Larger sizes have been built for special applications and have performed successfully. Pots larger than 35 cm (14 in.) in diameter and 45 cm (18 in.) deep are rarely used for electrical-resistance furnaces. Although it is possible to support the bottom of a large pot on a refractory pier, this may result in excessive temperature gradients.

Pots may be press formed from a single piece of low-carbon steel or iron-nickel-chromium alloy; a composition of Fe-35Ni-15Cr is usually preferred for the latter. Less-expensive welded pots may be fabricated from either of these materials.

In a well-designed furnace, life of round alloy pots will vary with maximum operating temperature as follows:

845 °C (1550 °F) . . . . . 9 to 12 months
870 °C (1600 °F) . . . . . . 6 to 9 months
900 °C (1650 °F) . . . . . . 3 to 6 months
925 °C (1700 °F) . . . . . . 1 to 3 months

In one installation, placement of an additional control thermocouple in the combustion chamber to prevent the temperature of the chamber from ex-

Fig. 5 Work-holding fixtures and wiring techniques used in liquid carburizing

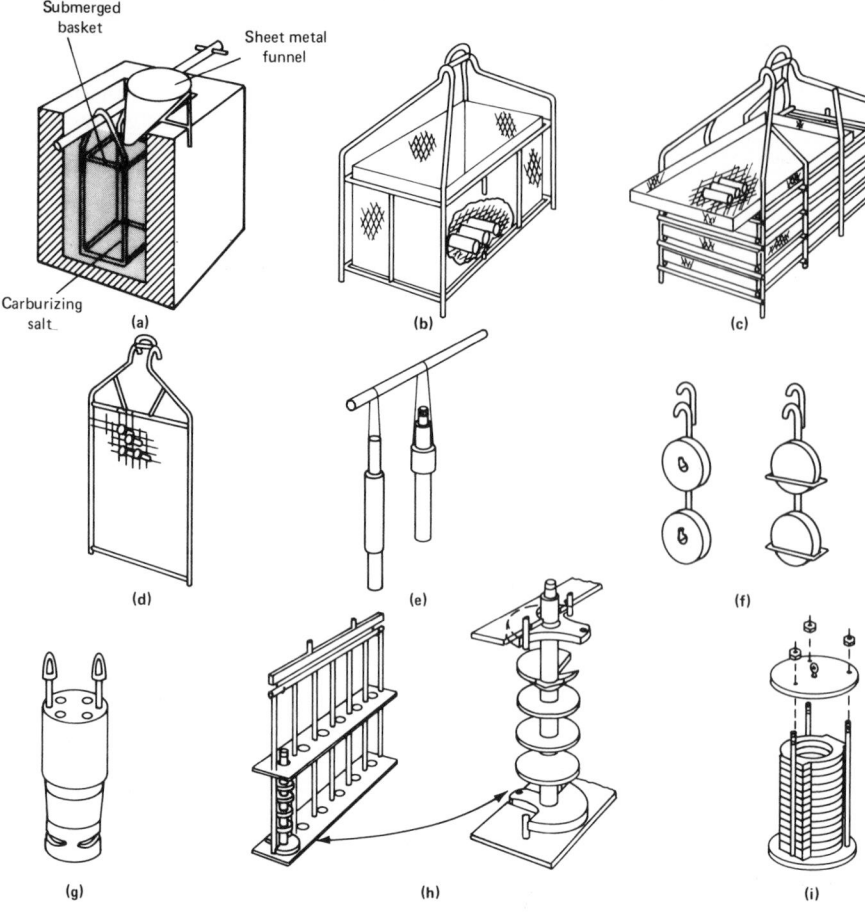

(a) Typical holding basket for small parts, equipped with a funnel for loading parts into the basket without splashing. The funnel, which is made of sheet metal, also ensures that parts are coated with salt before they are nested together. Basket may be made of carbon or alloy steel rod and steel wire mesh. Work must be free from oil, or the parts will stick together. Parts must be dry. (b) Inconel basket of simple design. Upper loop of handle is for lifting; lower loop accomodates a rod that supports the basket over the furnace. (c) Simple basket with trays, intended for small parts. Trays provide maximum loading space without adversely affecting circulation. Entire fixture is made of Inconel. (d) Method of running flat parts. (e) Method of supporting small parts. Black annealed steel wire is used for parts weighing less than 4.5 kg (10 lb); annealed stainless wire is used for heavier parts. (f) Hooks, made of nickel alloy rod, for holding circular parts. (g) Method for holding large parts in which tapped handling holes are available or can be provided. Nickel alloys are used for such fixtures because of the need for high-temperature strength. Resistance to oxidation is not a factor, because liquid carburizing salts are reducing. (h) Rack for holding six small crankshafts; exploded view shows a crankshaft in position. (i) Special rack for carburizing the outer surfaces of bearing races. Holding plates are made of mild steel; rods, of Inconel.

ceeding 1095 °C (2000 °F) served to extend the life of heat treating alloy pots to two years (previous life had been six months). Pot temperature was maintained at 900 °C (1650 °F) during a work week of 120 h (24 h per day, 5 days per week).

Temperature of the carburizing salt is measured and indicated by a thermocouple and suitable pyrometer. Externally fired furnaces, which are operated at temperatures from 790 to 925 °C (1450 to 1700 °F), may vary as much as 10 °C (20 °F) above or below the set temperature when on-off or high-low control systems are used. This is considered acceptable for many applications. Where closer control of temperature is required, a proportional control system, which can hold temperature variations to less than ±5 °C (±10 °F), should be used.

Design and Operating Factors. In the design of fuel-fired furnaces, it is important to provide ample space for combustion so that the flame does not impinge on the pot. If flame impingement is unavoidable, the pot should be rotated slightly at least once a week. Rotation of the pot and use of a sleeve reduce local deterioration in the region of flame impingement and thus prolong pot life. The combustion-chamber atmosphere also has important effects on pot life. A system controlled to range from high fire to low fire is preferable to an on-off system for two reasons: (a) the latter permits air to enter the combustion chamber during the "off" portion of the cycle, thereby accelerating scaling of the outer surfaces of the pot; and (b) closer control of temperature can be achieved.

Electrical-resistance-heated furnaces should be provided with a second pyrometer controller having its thermocouple in the heating chamber. This will prevent overheating of the resistance elements, particularly during salt meltdown, when the thermocouple that controls the temperature of the main bath is insulated by unmelted salt. Because heating elements and refractories are severely attacked by salt, it is mandatory that all salt be kept out of the combustion chamber. For this purpose, a mixture of high-temperature refractory cement, with a ceramic fiber for strength, may be used to seal joints where the pot flange rests on the retaining ring at the top of the furnace.

Regardless of the heating means, externally heated pots should be started on low fire (low heat input). After melting of the salt is observed around the top or side, of the pot, the heat can be gradually increased to complete the meltdown. Excessive heating of the sidewalls or pot bottom during start-up may create pressures sufficient to expel salt violently from the pot. For added safety, the pot should be covered during meltdown with either a cover or an unfastened steel plate.

Waste heat from flue gases may be fed to an adjacent chamber and used for preheating of work. Flue gases should always be visible to the operator; the presence of bluish white or white fumes at the vent is an indication that salts have entered the combustion chamber, and such situations require prompt corrective action.

Advantages and Disadvantages. Because of the ease with which they can be restarted, externally heated furnaces are well suited to intermittent operations. Another advantage of furnaces of this type is that a single furnace can be used for a variety of applications; a separate pot, containing the proper salt, can be used for each application.

Externally heated furnaces have several characteristics, however, that limit their usefulness in certain carburizing operations. They are less adaptable to close and uniform temperature control because they dissipate heat by convection, creating temperature gradients in the bath. Also, the temperature lag of the thermocouple and the recovery time of the furnace may result in overshooting or undershooting a desired temperature control point by as much as 14 °C (25 °F). In addition to requiring an exhaust system for flue gases, these furnaces may overheat at the bottom and sidewalls in restarting, creating in the thermally expanding molten salt a pressure buildup that may cause an explosion. Finally, externally heated furnaces are seldom practical for continuous high-volume production because of the limitations of pots with respect to size and maximum operating temperature. High maintenance cost is also a factor.

Immersed-Electrode Furnaces

The immersed-electrode furnace has greatly extended the useful range and capacity of molten carburizing baths. The electrodes can be removed and replaced without bailing out the furnace, and this design is suitable for both cyanide and noncyanide carburizing processes. In this type of furnace, the molten salt is contained in a steel or ceramic pot surrounded by suitable insulating materials that separate it from an exterior casing of heavy-gage steel. The salt is heated by passing alternating current through it by means of immersed electrodes. Heat is generated by the passage of current through the salt. This heat is quickly dissipated by a downward stirring action created by current flows. The electrodes are attached by copper connectors to a transformer that converts the line voltage to a much lower voltage (9 to 18 V) across the electrodes. Bath temperature is automatically controlled by a thermocouple-activated system that regulates the input of electric power. A typical immersed-electrode furnace for liquid carburizing is shown in Fig. 4(c).

The depth of salt pots for immersed-electrode carburizing furnaces is usually limited to about 2 m (6 ft) for metal pots. Ceramic pot depth has no limit. Furnaces with pots up to 4.5 m (15 ft) long, and with power input of 360 kW, are presently in operation. They have heating capacities of about 320 kg/h (700 lb/h). In contrast, smaller units with salt pots having work spaces measuring 23 by 18 by 35 cm deep (9 by 7 by 14 in. deep), and with 15-kW power input, can heat about 23 kg/h (50 lb/h) to 925 °C (1700 °F).

Advantages and Disadvantages. Immersed-electrode furnaces do not require iron-chromium-nickel alloy pots. Carbon steel pots of welded construction, set in insulating brick but not cemented in place, have given service life as follows:

Operating temperature		Service life, years
°C	°F	
845 (a)	1550 (a)	2 to 3
870	1600	1½ to 2
900	1650	1 to 1½
925	1700	1
(a) max		

These furnaces require minimum floor space and maintenance and can be used with all types of carburizing salts. Electrodes made of alloy steel should have an average service life equivalent to that indicated for steel pots in the above table. Worn electrodes can be replaced while the furnace is in operation.

Depending on positioning of electrodes, temperature control to within ±3 °C (±5 °F) is easily obtained with these furnaces; heat is generated within the bath, and overshooting is readily avoided. The furnaces lend themselves to mechanization and are suitable for high-volume production at operating temperatures from 815 to 955 °C (1500 to 1750 °F).

Maximum pot size is not restricted; pots may vary in length and width to suit requirements, and multiple pairs of electrodes can be installed to furnish the necessary heating capacity. Several batches of work may be carburized simultaneously and removed after different periods of time to produce a variety of case depths. Because the salt bath melts from the top downward, this type of furnace does not present a starting problem or an explosion hazard.

The immersed-electrode furnace is not recommended for intermittent operation. Depending on furnace size, reheating the salt charge may require a day or more. Pots are not intended to be interchangeable. Pot removal usually involves replacement of the surrounding insulation.

If an immersed-electrode-heated salt bath has been shut down completely and the salt has solidified, the salt between the electrodes must be melted with a torch before heating can be resumed. Insertion of an electric-resistance coil into the bath prior to salt solidification provides another means of remelting.

Submerged-Electrode Furnaces

Figure 4(d) illustrates the arrangement of components of a submerged-electrode furnace. The frame is made of heavy angle iron, and a steel plate is placed at the base beneath the brickwork. The outer brickwork consists of hollow ceramic tile or common building brick. The salt pot is made of burned alumina firebrick. The space between the sidewalls and the ceramic pot is filled with castable insulating refractory.

When salt is melted in the pot, it penetrates the refractory until it becomes cool enough to freeze. The resulting shell of solidified salt retains the liquid salt in the furnace. If the refractory develops a crack, bath temperature must be lowered to permit salt to solidify in the crack.

Water-cooled electrodes are submerged in liquid salt in the pot and are sealed in the refractory walls by frozen salt. Current travels between the electrodes, which are flush with the sidewalls. The path of current travel extends a few inches above the top of the electrodes.

Start-up and Shutdown. A submerged-electrode furnace can be started by adding molten salt from another furnace or by using a gas-fired torch or electric starting coil to melt a pool of salt that will wet both electrodes and provide molten salt for the current path. After the current path has been established in the molten salt between the electrodes, salt may be added to bring the bath up to working level. Additional salt will be required to maintain this level because a small amount will seep into the brickwork and freeze.

If the furnace must be shut down, the molten salt should be bailed from the furnace before it freezes. However, if the salt is allowed to remain in the furnace, a resistance-heated starting coil should be submerged in the bottom of the furnace while the salt is still molten. This coil remains in the frozen salt and is connected to the transformer leads to start up the furnace.

Advantages and Disadvantages. In common with immersed-electrode furnaces, submerged-electrode furnaces require minimum floor space and maintenance, and are highly adaptable to mechanization. Because the submerged-electrode furnace employs water for cooling of the electrodes and transformer, it may be operated at a 50% overload without overheating the transformer, whereas the immersed-electrode furnace, being air cooled, should not be operated above a 10% overload unless designed for overload.

Because a ceramic pot is used, pot life can be 1 to 3 years. The electrodes are usually first to fail. The furnace can be rebuilt on a planned schedule during annual shutdowns.

Because of the erosive effects that water-soluble salts with high sodium carbonate or sodium cyanide contents have on ceramic pots, submerged-electrode furnaces can be used only with low-cyanide, low-carbonate salts. Baths with high cyanide or carbonate salt require a modified basic brick. The furnace shown in Fig. 4(d) with modified brick and submerged alloy electrodes will give many years of service in both cyanide and noncyanide operation.

Furnace Parts and Fixtures

Table 6 lists wrought and cast materials used for furnace parts and fixtures. As indicated, more than one material may be safely selected for a specific furnace part or fixture. Both cost of material and length of service usually increase with alloy content. Although length of service is influenced by both operating temperature and type of carburizing salt employed, the materials listed may be used in both the low and high temperature ranges—that is, at any temperature from 845 to 955 °C (1550 to 1750 °F).

Unless parts can be suspended in the salt bath by simple wiring or by placing them in a basket or similar container,

Table 6 Materials used for furnace parts and fixtures for liquid carburizing

For temperatures between 845 and 955 °C (1550 and 1750 °F)(a)

Part	Material Wrought(b)	Cast
Pots, externally heated	35-18(c)	HT
	Inconel	HX
Pots, internally heated(d)	Carbon steel	(e)
Electrodes, immersed	446	HT
	35-18(c)	
	Inconel	
Electrodes, submerged	Carbon steel	(e)
	35-18(c)	
Thermocouple protection tubes	446	(e)
	35-18(c)	
	Inconel	
Fixtures	35-18(c)	(e)
	Inconel	
Baskets	35-18(c)	HT
	Inconel	HX

(a) When more than one material is recommended for a specific part, each has proved satisfactory in service. Multiple choices are listed in order of increasing alloy content. Cost and expected service life usually increase as alloy content increases. (b) Carbon steel has been used for most of the parts listed. (c) Refers to a series of alloys generally of the 35Ni-15Cr type or modifications that contain from 30 to 40% Ni and 15 to 23% Cr and include RA-330, 35–19, Incoloy and other proprietary alloys. (d) For immersed-electrode furnaces. Pots for submerged-electrode furnaces are made of burned alumina firebrick. (e) These types of parts are not usually cast.

some form of fixturing is required. Some typical work-holding fixtures are shown in Fig. 5, together with methods of wiring small parts. The weights of specific parts may influence both the design of the fixture and the selection of material from which it is made.

Holding fixtures and supports used in salt-bath carburizing should be kept as simple as possible. Fixture weight should be minimized to conserve heat by lessening the load to be heated. Whenever possible, fixture components should be pinned or riveted rather than welded. This permits freedom of movement of the fixture during heating and quenching, thereby extending its life. Although weldments are not significantly affected by the carburizing bath, they are subject to the stresses imposed by cyclic heating and cooling and thus are susceptible to cracking. Finally, riveting and pinning permit easy replacement of those fixture components that can be replaced.

Automatic and Semiautomatic Lines

Figure 6(a) shows a fully automatic ("jackrabbit") mechanism used for salt-bath carburizing and hardening. The mechanism has synchronized, continuous-chain conveyors that carry the work through the various operations. Work suspended from horizontal bars

is moved through baths at the proper speed by a main conveyor. Transfer conveyors carry the work from bath to bath. Completed work is picked up by a third conveyor and is dried by warm air in the enclosed upper portion of the structure while it is being returned to the loading point.

This mechanism can be used only for work having similar requirements. It does not permit the time cycle of any one operation to be varied without affecting the cycles of the other synchronized operations.

Where part requirements vary, a semiautomatic mechanism, such as that shown in Fig. 6(b), may be used. Work is transferred between operations by an overhead monorail and is automatically advanced through the carburizing and tempering furnaces by means of a push-pull mechanism. This mechanism consists of two parallel beams with reciprocating push bars. Driven either hydraulically or electrically, the bars carry the dogs, which are spaced to advance the fixtures at the center of the furnace only a part of the stroke while advancing the end fixtures through the entire stroke. By closer spacing of the work at the center of the furnace and wider spacing at the ends, high productive capacity is achieved with ease of loading and unloading.

A semiautomatic line of this type permits the time cycle for any one oper-

Fig. 6 Fully automatic and semiautomatic production lines for liquid carburizing and related operations (reheating, quenching, tempering, washing)

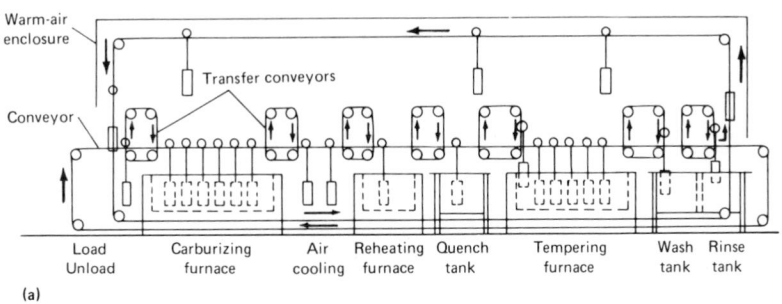

(a)

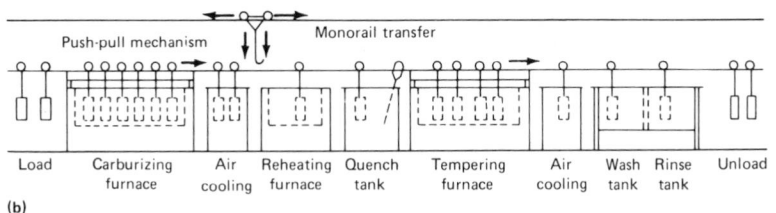

(b)

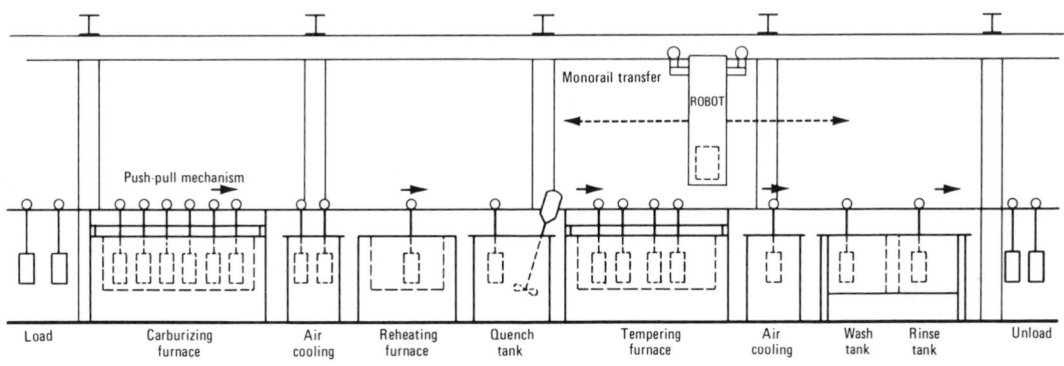

(a) Fully automatic carburizing line. (b) Semiautomatic carburizing line. (c) Fully automatic programmed-hoist carburizing line. See text for details.

ation to be varied without affecting other operations, and is less likely to require modification if work requirements change. Figure 6(c) shows a fully automatic programmable-hoist carburizing line. One or more hoists travel simultaneously back and forth, automatically advancing the fixtures that carry the work through the required stations.

The hoist movement may be controlled by a solid-state programmable control, which provides functions that normally would require extensive wiring and hundreds of relays, counters and switches. Once programmed, the controller will perform the functions specified by the user. Functions such as time cycles, sequences, drills and skips are entered easily or changed to meet metallurgical requirements. For example, some parts could be programmed to be carburized, air cooled, washed, rinsed and returned for unloading. A pushbutton command would return the controller to the standard program.

Parts suitable for fully automatic or semiautomatic installations are those that can be fixtured by wiring, racking, or placing in baskets, and that do not present problems of either buoyancy or drainage. Case-depth requirements ranging from 0.25 to 3.2 mm (0.010 to 0.125 in.) can be satisfied.

Reference to a specific liquid carburizing application in which both production and equipment requirements are provided in complete detail is often of value in estimating requirements for

Table 7 Production requirements and equipment for liquid carburizing and subsequent operations

	Total depth of case	
	0.51 to 0.64 mm (0.020 to 0.025 in.)	1.5 mm (0.060 in.)
Production requirements		
Type of steel(a)................................	8615H	8615H
No. of pieces per hour..........................	300	24
Weight of pieces, kg (lb)........................	0.09-0.4 (0.2-0.9)	7 (15)
Equipment for carburizing		
Salt-pot dimensions, cm (in.).....................	61×38×46 deep (24×15×18 deep)	452×102×86 deep (178 × 40×34 deep)
Salt-pot construction............................	Open-top iron	Open-top iron
Type of electrode..............................	Immersed	90-cm (35-in.) vertical leg (12 pairs)(c)
Operating temperature, °C (°F)....................	925 (1700)	925 (1700)
Maximum temperature deviation, °C (°F)...........	±3 (±5)	±3 (±5)
Thermocouple.................................	Chromel-Alumel	Chromel-Alumel
Thermocouple-protection tube....................	Inconel	Inconel
Location of thermocouple........................	Center of pot, rear	Center of pot, 25-cm (10-in.) immersion
Transformer, kW...............................	50(b)	60(d)
Heating capacity of furnace, kg/h (lb/h)...........	122 (270)(b)	...
Sodium cyanide limit, %.........................	16-20	8-9(e)
Carburizing period, h..........................	2½	7
Equipment for reheating after carburizing		
Salt-pot dimensions, cm (in.).....................	61×38×46 deep (24×15×18 deep)	74×114×86 deep (29×45×34 deep)
Salt-pot construction............................	Open-top iron	Open-top steel
Type of electrode..............................	Immersed	92-cm (36-in.) vertical leg (4 pairs)(c)
Operating temperature, °C (°F)....................	840 (1540)	825 (1520)
Maximum temperature deviation, °C (°F)..........	±3 (±5)	±3 (±5)
Thermocouple.................................	Chromel-Alumel	Chromel-Alumel
Thermocouple-protection tube....................	Inconel	Inconel
Location of thermocouple........................	Center of pot, rear	End, 59-cm (24-in.) immersion
Transformer, kW...............................	50(b)	70(f)
Heating capacity of furnace, kg/h (lb/h)...........	122 (270)(b)	...
Sodium cyanide limit, %.........................	10 min(g)	2-3(h)
Equipment for quenching		
Quenching medium.............................	Oil(j)	Nitrate-nitrite salt(k)
Salt-pot dimensions, cm (in.).....................	...	107×178×104 deep (42×70×41 deep)
Salt-pot construction............................	...	Open-top steel(m)
Heaters......................................	...	Electric immersion
Operating temperature, °C (°F)....................	...	205 (400)
Maximum temperature deviation, °C (°F)..........	...	±3 (±5)
Thermocouple.................................	...	Iron-constantan
Thermocouple-protection tube....................	...	Metal
Location of thermocouple........................	...	Center rear
Power requirements, kW........................	...	60(n)
Equipment for tempering		
Type of salt..................................	Neutral	Nitrate-nitrite
Salt-pot dimensions, cm (in.).....................	91×58×66 deep (36×23×26 deep)	137×102×86 deep (54×40×34 deep)
Salt-pot construction............................	Open-top steel	Open-top steel
Heaters......................................	Electric immersion	Electric immersion
Operating temperature, °C (°F)....................	190 (375)	165 (325)
Maximum temperature deviation, °C (°F).........	±3 (±5)	±3 (±5)
Thermocouple.................................	Iron-constantan	Iron-constantan
Thermocouple-protection tube....................	Metal	Metal
Location of thermocouple........................	Center rear	Center rear
Power requirements, kW........................	...	30(p)

(a) Cold drawn, magnetic tested. (b) At operating temperature of 925 °C (1700 °F). (c) Electrodes spaced 5 cm (2 in.) apart between faces below salt level. (d) Six 60-kW transformers, or total of 360 kW, at operating temperature of 925 °C (1700 °F). Transformers on 230 V, 60 Hz, three phase with carburizing salt. Normal (full-load) operating current of 159 A on each 60-kW transformer, 9- to 13-V tap. (e) Barium-base high-temperature salt. (f) Two 70-kW transformers, or total of 140 kW, at operating temperature of 790 to 870 °C (1450 to 1600 °F). Transformers on 230 V, 60 Hz, three phase with neutral and cyanide salt. Normal (full-load) operating current of 185 A on each 70-kW transformer, 9- to 13-V tap. (g) Work must not be quenched directly in nitrate-nitrite salt from this bath; prior washing in neutral salt bath is mandatory. (h) Neutral chloride salt plus bail-out from carburizing bath. (j) Oil quench tank 244 by 122 by 122 cm deep (96 by 48 by 48 in. deep) contains 3275 L (865 gal) of high-speed quenching oil heated to 50 to 55 °C (120 to 130 °F) by steam coil. Water wash tank is 169 by 107 by 102 cm deep (66.5 by 42 by 40 in. deep) and contains 1640 L (433 gal) of water heated by live steam from manifold. Steam furnishes agitation and makeup water for overflow tank. (k) Water (approx. 11 L, or 3 gal, per hour) added to salt. (m) Furnace is equipped with agitator for uniform temperature throughout salt, and with chamber around pot with automatically controlled fan to prevent overheating. Solenoid valve and timers control amount of water addition to salt. Furnace also contains filtering unit to remove high-temperature salt. (n) At operating temperature of 205 °C (400 °F); 230 V, 60 Hz, three phase with tempering salt. Each of the nine heaters for tempering salt rated at 5 kW, 230 V. Each of the three heaters in salt-separating chamber rated at 5 kW, 230 V. (p) At operating temperature of 165 °C (325 °F); 230 V, 60 Hz, three phase with nitrate salt. Furnace has six immersion heaters; each is rated at 5 kW, 230 V.

other parts that are to be carburized; the following example gives details of two deep-case processes.

Example 1. Table 7 provides detailed equipment requirements for liquid carburizing small parts ranging in weight from 0.09 to 0.4 kg (0.2 to 0.9 lb) and larger parts weighing 7 kg (15 lb) each. These parts were made of 8615H steel and were carburized to two different case depths: 0.51 to 0.64 mm (0.020 to 0.025 in.) for the small parts and 1.5 mm (0.060 in.) for the larger parts. Equipment used in reheating after carburizing, in interrupted quenching, in tempering and in washing is also described in Table 7.

Process Control

Externally heated salt baths can be held within closer temperature limits (±8 °C, or ±14 °F) when a proportional control system employing electronic instrumentation is used. Control by means of valves (on-off or high-low control) requires mechanical instrumentation and is less accurate, although for a majority of applications it is entirely adequate.

Internally heated salt baths (immersed or submerged electrodes) may be regulated to ±5 °C (±9 °F) with either mechanical or electronic on-off controllers. In either type, the temperature-control instrument operates a relay that actuates a large circuit breaker that in turn connects or disconnects the 440-V power to the step-down transformer. Welded thermocouples may be used in installations that employ electrode heating. For safety, two thermocouples are recommended—one for temperature control and one for excess temperature cutoff.

Control of Bath Composition. Control of sodium cyanide content is the most important factor in maintaining the effectiveness of a liquid carburizing bath.

Analysis of a noncyanide liquid carburizing bath is achieved by a rapid performance test in which a 1008 steel wire 1.6 mm (1/16 in.) in diameter is immersed for 3 min in the bath, then is water quenched and mechanically bent through 90°. The bath is well activated if the wire breaks before reaching the full 90° bend. A more reliable test of activity can be made by running a 1012 silicon-killed test bar for 1 h, water quenching, and measuring Rockwell C

hardness. Readings above 58 HRC indicate a well-activated bath.

Graphite Cover. A graphite cover must be maintained on the surface of a cyanide bath for efficient operation at 870 to 955 °C (1600 to 1750 °F) and during idling. Either natural flake or artificial graphite powders may be used. The former provides a more fluid cover that has less tendency to cling to the work. However, because natural graphite has a higher ash content, it introduces more impurities into the bath, which can be a problem—particularly at low operating temperatures. Furthermore, to avoid corrosion of parts, natural graphite that contains sulfur should not be used.

A noncyanide liquid carburizing bath also must have a graphite cover. A higher rate of graphite consumption, compared with a cyanide bath, is characteristic. Frequent replenishment (commonly every hour) is necessary for maintenance of proper bath activity.

Daily maintenance routines for liquid carburizing furnaces, whether fuel-fired or electrode-heated, differ in only a few details. The following items, with exceptions as noted, comprise a typical daily maintenance schedule for all types of salt-bath equipment:

- Check temperature-control system, using an auxiliary pyrometer and thermocouple. An indicating potentiometer with a long extension wire can be mounted near the furnaces and will provide accurate temperature checks faster than will a laboratory-type instrument.
- Check color of exhaust smoke from the combustion chamber of fuel-fired furnaces. A bluish white or white smoke indicates salt leakage.
- Remove sludge from bottom of pot while furnace is still at idling temperature, which normally is 705 to 730 °C (1300 to 1350 °F). The electrodes of internally heated furnaces should be scraped clean, and electric power should be shut off during the sludging and cleaning operation.
- Add fresh salt to make-up to compensate for dragout losses. If required, make room for addition of fresh salt by bailing.
- To help maintain bath composition and reduce surface heat losses, add graphite cover material to provide a thin but continuous cover.
- Check bath activity by testing for cyanide content or by quenching and bending a steel wire.

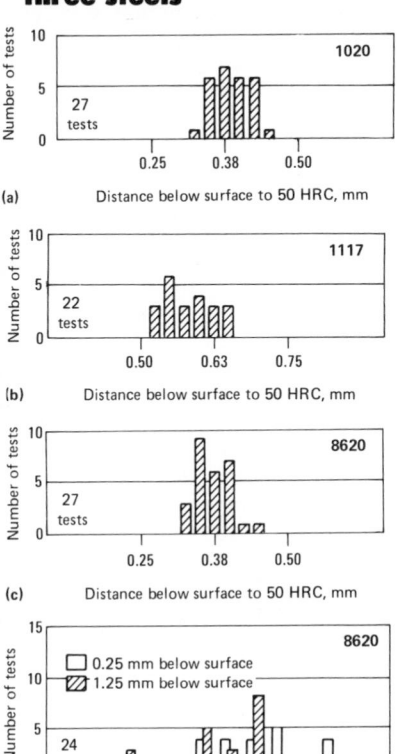

Fig. 7 Comparative case-depth and case-hardness data obtained in liquid carburizing process-control specimens made of three steels

(a) Data are for 11-mm diam by 6.4 mm (0.4375-in. diam by 0.25 in.) specimens carburized 2 h at 855 °C (1575 °F), brine quenched, and tempered at 150 °C (300 °F). (b) Data are for 15.9-mm diam (0.625-in. diam) specimens carburized 2 h at 900 °C (1650 °F) and brine quenched. (c) Data are for 12.7-mm diam by 6.4 mm (0.50-in. diam by 0.25 in.) specimens carburized 2 h at 855 °C (1575 °F), oil quenched, and refrigerated to −85 °C (−120 °F). (d) Data are for 19-mm diam by 51 mm (0.75-in. diam by 2 in.) specimens carburized 2.5 h at 915 °C (1675 °F) and water quenched.

- If possible, rotate the pot of a fuel-fired furnace at least once each week to minimize the effects of flame impingement and thus extend pot life.
- If a salt pot is leaking and the salt is still active, remove the salt and place it in sturdy steel containers. This salt may be broken up and re-used in

starting another pot (however, it is not recommended that such salt be used thereafter for replenishment).

- Prior to replacement of a pot in a resistance-heated or fuel-fired furnace, the combustion chamber should be rebuilt if contaminated with salt, to avoid rapid pot failure.
- Consult operating and maintenance instructions provided by the furnace manufacturer and salt supplier.

Shutdown and Restarting. For shutdowns of 2 days or longer, externally heated furnaces need not be idled; the heat may be shut off completely. During cooling and reheating, however, the pot should be covered to guard against violent expulsion of salt. The cover recommended by the manufacturer should be used.

It is generally advisable to idle electrode furnaces at 705 to 730 °C (1300 to 1350 °F), even over shutdown periods of one to two weeks. This simplifies restarting and eliminates possible damage to power transformers from condensation of moisture on the windings. For noncyanide carburizing furnaces with steel liners, idling above 845 °C (1550 °F) is recommended. During the idling period, the bath should be protected with a heavy carbon cover. The bath does not fume at idling temperature; therefore, ventilating air usually is not required. Excessive ventilating air should be avoided, because it will accelerate burn-off of the carbon cover. During the idling period, the transformer tap switch should be set at low voltage or idling tap. This will guard against possible overheating in the event that control-circuit difficulties arise while the equipment is unattended.

Control of Case Depth

The degree of uniformity of case depth obtained in normal production operations is indicated in Fig. 7 by data on 1020, 1117 and 8620 steels. Figures 7(a), (b) and (c) are based on information obtained with process-control specimens, show depth of case as a function of distance below the surface, in terms of a hardness of 50 HRC or higher. These data indicate that variations in case depth can be held within narrow limits when controlled carburizing procedures are employed. At a carburizing temperature of 900 °C (1650 °F), the 1117 steel produced a deeper case to 50 HRC than did the 1020 and 8620 steels,

which were carburized at 855 °C (1575 °F). Nevertheless, the total spread in case depth for any one of these steels did not exceed 0.13 mm (0.005 in.). Data presented in Fig. 7(d) indicate the range of hardnesses obtained at depths of 0.25 and 1.25 mm (0.010 and 0.050 in.) below the surface of liquid carburized 8620 steel. These data, based on 24 tests, indicate a slightly larger spread in hardness at 0.25 mm (0.010 in.) than at 1.25 mm (0.050 in.) below the surface.

Whereas the information in Fig. 7 deals with carburizing cycles of 2 and 2.5 h at temperatures ranging from 855 to 915 °C (1575 to 1675 °F), the data presented in Fig. 3 pertain to a much longer carburizing time (9½ h) at 925 °C (1700 °F). The spread in case depth at 50 HRC is considerably wider than for the light-case work on which Fig. 7 is based.

Additional data on case depth as a function of time and temperature are given for ten steels in Fig. 8. These data also reflect various criteria that have been applied to evaluate case depth—for example, data relating case depth to minimum hardness, carbon content and pearlite content.

Dimensional Changes

All parts undergo dimensional changes as a result of carburizing and hardening. From a production standpoint, it is important to know the nature and amount of dimensional change, or distortion, that can be anticipated, and the corrective action that may be taken to hold dimensional changes to a minimum. The following examples relate dimensional change to several shapes that vary in complexity.

Example 2. Figure 9(a) presents data on carburized and hardened bushings made from seamless low-carbon steel. Tolerances on the outside diameter of the bushing prior to heat treatment were set at −0.00, +0.05 mm (−0.000, +0.002 in.). The bushings were wired in clusters of seven, liquid carburized for 1 h at 900 °C (1650 °F), and oil quenched. After heat treatment, the bushings developed an out-of-round condition, ranging from 0.025 to 0.18 mm (0.001 to 0.007 in.). The data in Fig. 9(a) relate out-of-roundness with minimum and maximum outside-diameter measurements after heat treatment.

Example 3. The small gear shown

in Fig. 9(b) closed in along the bore from a minimum dimension of 17.22 mm (0.6780 in.) prior to heat treatment to a minimum of 17.14 mm (0.6750 in.) after heat treatment. In contrast, only slight contraction of the outer bearing surface occurred. The gears, made of 8615H steel, were carburized at 915 °C (1675 °F) to a depth of 0.51 to 0.64 mm (0.020 to 0.025 in.), reheated to 840 °C (1540 °F), quenched in oil at 55 °C (130 °F), and then tempered at 190 °C (375 °F) to a surface hardness of 60 to 62 HRC.

Example 4. Another gear, also made of 8615H steel, and subjected to the same heat treatment as the gear in Example 3, grew dimensionally along the bore and consequently shrank at a keyway slot. As indicated in Fig. 9(c), maximum contraction at the keyway slot amounted to 0.064 mm (0.0025 in.). Bore dimensions before and after heat treatment also are shown in Fig. 9(c).

Example 5. The bearing race shown in Fig. 9(d) was subjected to more elaborate processing to minimize dimensional variations before and after carburizing. This 8620 steel forging was normalized and stress relieved prior to being carburized. After being rough ground, it was liquid carburized for 14 h at 925 °C (1700 °F), to produce a minimum case depth of 2.3 mm (0.090 in.). It was air cooled, reheated to 845 °C (1550 °F), and salt bath quenched at 180 °C (360 °F). After being cooled to room temperature, it was tempered for 2 h at 175 °C (350 °F). Final case hardness was 61 to 63 HRC; core hardness was 40 to 43 HRC.

To minimize distortion, which was excessive when these bearing races were wired, a fixture similar to that shown in Fig. 5(i) was used throughout the heat treating cycle. As indicated by the dimensional data, the combination of fixturing and elaborate processing produced favorable results in terms of out-of-roundness and flatness. Dimensional discrepancy was held to 0.10 mm (0.004 in.) max, and in several instances, distortion was held to 0.025 mm (0.001 in.).

Example 6. The 4815 steel pinion shown in Fig. 9(e) was liquid carburized for 2½ h at 900 °C (1650 °F), quenched in still oil and then tempered for ½ h at 205 °C (400 °F) to achieve a surface hardness of 58 to 60 HRC. Runout was then checked on the pitch diameter (71.96 mm, or 2.833 in.) of the pinion. Maximum runout for one pinion amounted to 0.25 mm (0.010

Fig. 8 Effect of time and temperature on case depth of liquid carburized steels

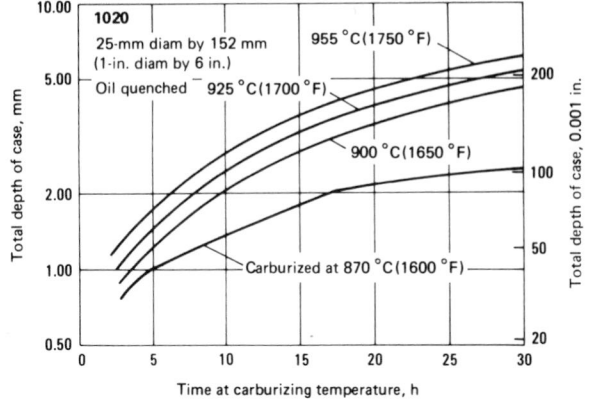

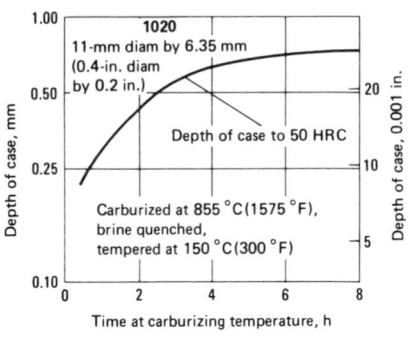

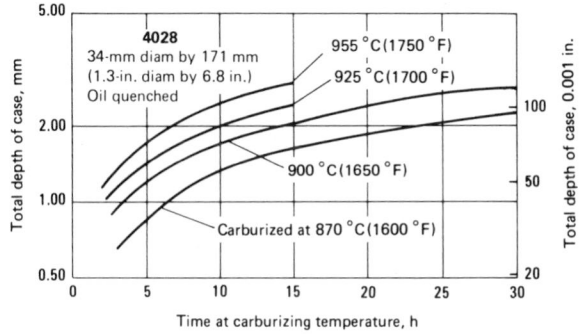

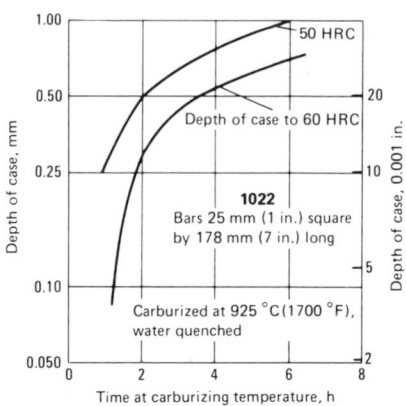

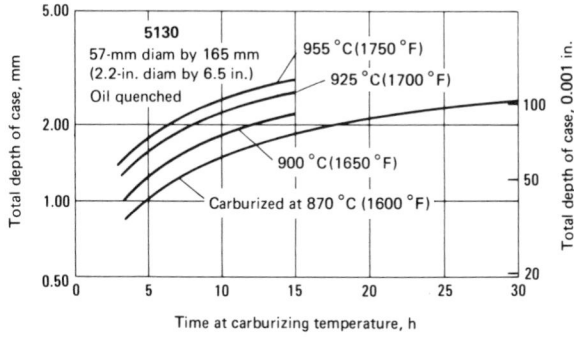

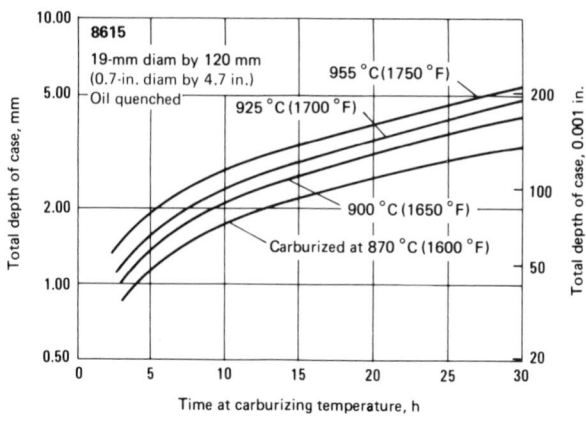

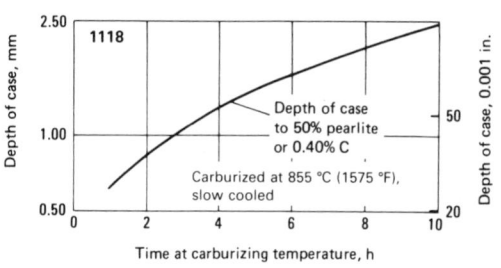

Fig. 8 (continued)

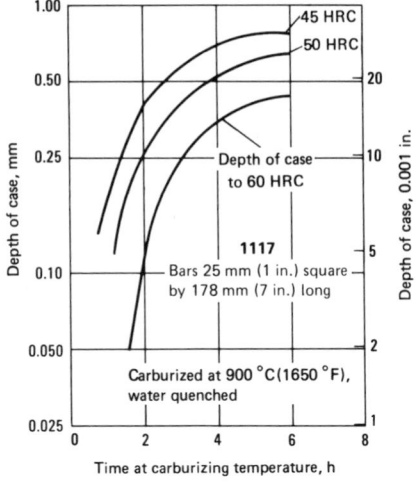

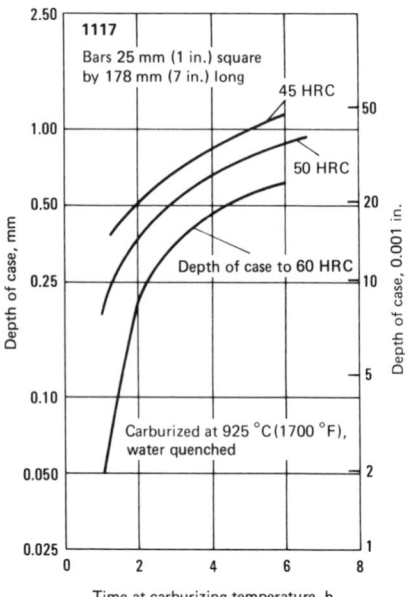

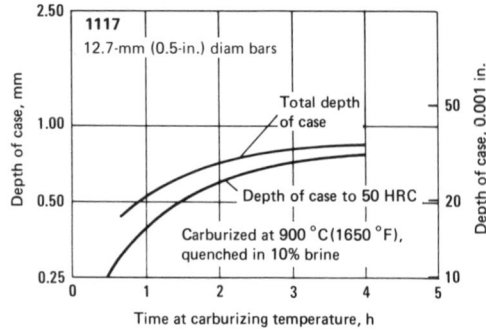

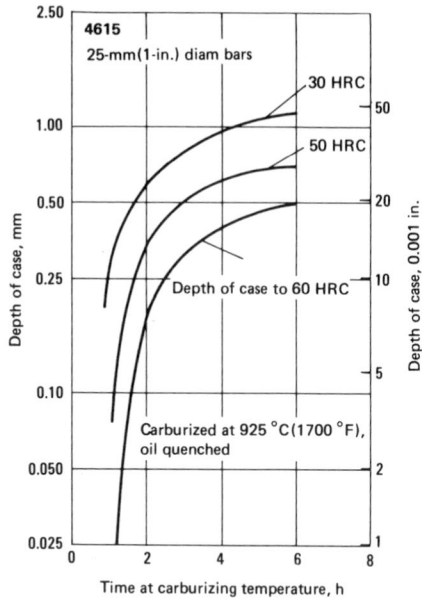

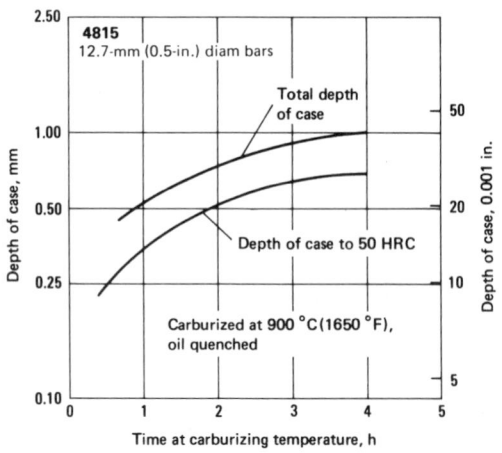

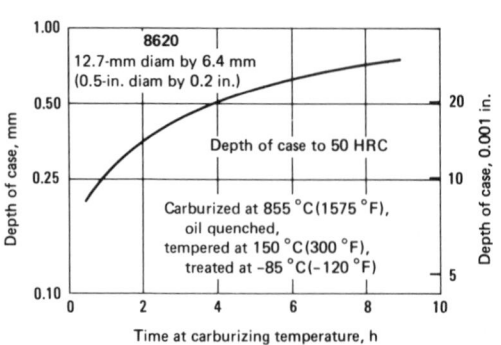

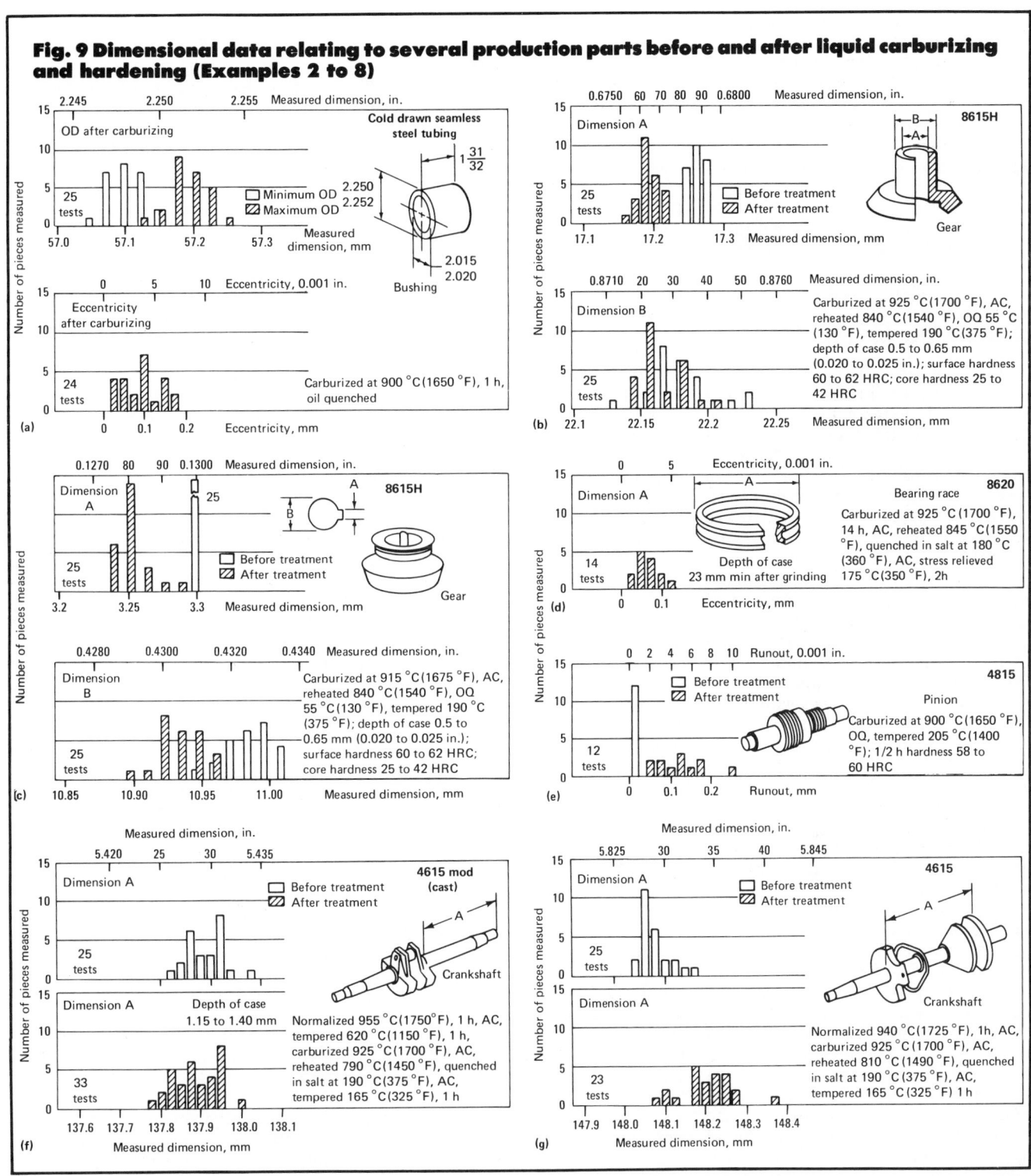

Fig. 9 Dimensional data relating to several production parts before and after liquid carburizing and hardening (Examples 2 to 8)

in.), although the majority of pinions measured indicated a runout ranging from 0.10 to 0.18 mm (0.004 to 0.007 in.). During carburizing, all of the parts were suspended vertically in the bath by means of a wire placed through a threaded hole at one end of the shaft.

Example 7. The crankshaft shown in Fig. 9(f), a shell-mold casting made of boron-modified 4615 steel, was initially normalized for 1 h at 955 °C (1750 °F) and then tempered for 1 h at 620 °C (1150 °F). After being machined, the part was liquid carburized at 925 °C or

1700 °F (case depth, 1.15 to 1.40 mm, or 0.045 to 0.055 in.), air cooled, reheated to 790 °C (1450 °F), quenched for 5 min in salt at 190 °C (375 °F), air cooled, and tempered for 1 h at 165 °C (325 °F). The dimensional data, which apply to a length measurement at one end of the shaft only, indicate a high degree of dimensional stability with a slight tendency in the direction of shrinkage.

Example 8. The data relating to the forged crankshaft in Fig. 9(g) indicate a different trend. Apparently, the checks at the throws of the crankshaft opened up, causing an increase in the A dimension. This crankshaft also was made of 4615 steel and was normalized at 940 °C (1725 °F) after forging. The carburizing and hardening procedures to which it was subjected were identical to those for the crankshaft in Example 7, except

Quenching Media

Most conventional quenching media, including water, brine, caustic solution, oil and molten salts, are suitable for quenching parts that have been liquid carburized. However, the suitability of each medium must be related to specific parts and depends primarily on the hardenability of the steel, surface and core hardness requirements, and the amount of allowable distortion.

Water and brine are the quenchants most commonly used for carbon steels. A water quench is usually maintained at 20 to 30 °C (70 to 90 °F) and agitated. Water helps to dissolve the film of carburizing salt and thus creates a localized brine that suppresses the vapor phase. With continuous use, salt concentration (chlorides, carbonates and cyanides) increases, and fresh water must be added periodically to control the concentration of contaminants and maintain a desired temperature. Sodium chloride brine (5 to 10%) and caustic (3 to 5%) solutions are used to obtain more drastic quenching. The noncyanide liquid carburizing salt provides a brine-type quench when maintained around 10% concentration by water addition. The effectiveness of brine and caustic can be severely curtailed by an excessive accumulation of contaminants. When a caustic solution is used for quenching, care must be taken to ensure that racks, baskets and fixtures are washed free from caustic and dried before being returned to the carburizing bath. Small amounts of caustic carried back to the bath will lower its cyanide content significantly.

A water-soluble polymer is sometimes used to modify the quenching rate of a water quench. However, such additives should be avoided in a quenchant used with a liquid carburizing line, unless frequent replacement or continuous salt removal by ultrafiltration can be employed. The polymers may be precipitated by salt carried into the quench, or salt buildup in the quench may render their effect variable. Either condition is undesirable.

Oil quenching is less drastic than water quenching and produces less distortion. It is often desirable to fortify the mineral oil with non-saponifiable additives that increase its quenching effectiveness and lengthen its useful life. To minimize distortion, special oils are available that can be used at temperatures as high as 175 °C (350 °F). Normally, liquid carburized parts are quenched directly into oil maintained within the range from 25 to 70 °C (80 to 160 °F).

Quenching oil should be kept free of moisture and should be agitated by propellers or impeller-type pumps. Compressed air should not be used for agitation. Because some salt will inevitably precipitate in the oil bath, periodic desludging is necessary. Screens should be placed in front of the lines leading to pumps to prevent entry of sludge.

Salt-bath quenching in a nitrate-nitrite bath further minimizes distortion. Salt quench baths are compatible with cyanide as well as noncyanide liquid carburizing baths. *However, parts should never be transferred directly from a carburizing bath containing more than 5% cyanide to a nitrate-nitrite quench bath, because this will result in a violent reaction and may cause explosion.* To avoid such reactions, immersion in a neutral salt bath (45 to 55% NaCl, 45 to 55% KCl) held at the desired temperature must precede quenching in a nitrate-nitrite bath. The neutral bath should be tested periodically for sodium cyanide content; it is general practice to limit cyanide content to a level of less than 5%. This level is never reached, as a rule, because of oxidation of the cyanide by oxygen in the air. The neutral stabilizing bath can be used alternatively for through hardening of carbon and alloy steels, provided that complete cyanide oxidation has not occurred.

All traces of nitrate should be removed from quenching fixtures before they are reimmersed in a carburizing bath. This can be accomplished by rinsing in hot water.

The buildup of high-temperature chlorides in a nitrate-nitrite bath impairs its quenching severity. It is desirable, therefore, to remove the chlorides as fast as they are being delivered. Various means of chloride removal are available, and the selection depends on furnace design. Where chloride is allowed to settle to the bottom of the quench area or an area provided for gravity separation, the chlorides can be collected in sludge pans; periodically, either the pans are removed or the bottoms of the pans are manually desludged. Some furnace designs employ continuous filtration of chlorides as the suspended crystals pass through filter baskets; the operator removes the baskets periodically to dump the collected chlorides and then returns them to the furnace.

Maintenance of Quenching Baths. Although a limited amount of dissolved salt increases the efficiency of a water quench bath, amounts in excess of 10% retard the quenching rate. Controlled addition of fresh water to the bath, together with a continuous overflow, serves to keep salt concentration at an acceptably low level. It may be required that the overflow be chemically treated in a special reservoir prior to disposal, in order to eliminate cyanide pollution (see subsequent section on disposal of cyanide wastes). For this reason, changing the water quench bath at scheduled intervals may be more convenient in small operations. For water tanks that are vigorously agitated, it is recommended that a false bottom—in the form of a perforated plate—be used to permit settling of heavier solids, which can be removed during periods of downtime.

Carryover of liquid carburizing salts into brine quench tanks actually helps maintain brine concentration. However, salt concentration should not exceed 10%. The same control of salt content applies to caustic tanks; concentration of caustic must be maintained by additions of sodium hydroxide to control the quench rate of the solution.

Some of the precautions that must be observed in the use of oil baths have already been discussed. It should be recognized that liquid carburizing salts do not dissolve in, or combine with, mineral quenching oils. Salt sludge must be removed periodically, either by mechanical means or by filtering through screens.

Proper maintenance of salt quench baths also requires sludging of contaminants. Use of separating chambers to collect these contaminants has already been discussed. Another technique involves continuous filtering out of high-

Table 8 Typical applications of liquid carburizing

Part	Weight kg	lb	Steel	Depth of case mm	in.	Temperature °C	°F	Time, h	Quench	Subsequent treatment	Hardness, HRC
Carbon steel											
Adapter............ 0.9		2	CR	1.0	0.040	940	1720	4	AC	(a)	62–63
Arbor, tapered...... 0.5		1.1	1020	1.5	0.060	940	1720	6.5	AC	(a)	62–63
Bushing............ 0.7		1.5	CR	1.5	0.060	940	1720	6.5	AC	(a)	62–63
Die block.......... 3.5		7.7	1020	1.3	0.050	940	1720	5	AC	(a)	62–63
	1.1	2.5	CR	1.3	0.050	940	1720	5	AC	(a)	59–61
Disk.............. 1.4		3	1020	1.3	0.050	940	1720	5	(b)	(b)	56–57
Flange........... 0.03		0.06	1020	0.4-0.5	0.015-0.020	845	1550	4	Oil	(c)	55 min(d)
Gage rings,											
knurled......... 0.009		0.2	1020	1.5	0.060	940	1720	6.5	AC	(a)	62–63
Hold-down											
block 0.9		2	CR	1.0	0.040	940	1720	4	AC	(a)	62–63
Insert, tapered...... 4.75		10.5	1020	1.3	0.050	940	1720	5	AC	(a)	62–63
Lever............. 0.05		0.12	1020	0.13-0.25	0.005-0.010	845	1550	1	Oil	(c)	(e)
Link............. 0.007		0.015	1018	0.13-0.25	0.005-0.010	845	1550	1	AC	...	...
Plate 0.007		0.015	1010	0.25-0.4	0.010-0.015	845	1550	2	Oil	(c)	(e)
Plug............. 0.7		1.6	CR	1.5	0.060	940	1720	6.5	AC	(a)	62–63
Plug gage 0.45		1	1020	1.5	0.060	940	1720	6.5	AC	(a)	62–63
Radius-cutout											
roll............. 7.7		17	CR	1.5	0.060	940	1720	6.5	AC	(a)	62–63
Torsion-bar cap..... 0.05		0.1	1022	0.02-0.05	0.001-0.002	900	1650	0.12	Caustic	(f)	45–47
Resulfurized steel											
Bushing............ 0.04		0.09	1118	0.25-0.4	0.010-0.015	845	1550	2	Oil	(c)	(e)
Dash sleeve 3.6		8	1117	1.1	0.045	915	1675	7	AC	(g)	58–63
Disk.............. 0.0009		0.002	1118	0.13-0.25	0.005-0.010	845	1550	1	Brine	(c)	(e)
Drive shaft........ 3.6		8	1117	1.1	0.045	915	1675	7	AC	(h)	58–63
Guide bushing...... 0.2		0.5	1117	0.75	0.030	915	1675	5	(j)	...	58–63
Nut 0.04		0.09	1113	0.13-0.25	0.005-0.010	845	1550	1	Oil	(c)	(e)
Pin.............. 0.003		0.007	1119	0.13-0.25	0.005-0.010	845	1550	1	Oil	(c)	(e)
Plug............. 0.007		0.015	1113	0.075-0.13	0.003-0.005	845	1550	0.5	Oil	(c)	(e)
Rack.............. 0.35		0.75	1113	0.13-0.25	0.005-0.010	845	1550	1	Oil	(c)	(e)
Roller............. 0.01		0.03	1118	0.25-0.4	0.010-0.015	845	1550	2	Oil	(c)	(e)
Screw............. 0.003		0.007	1113	0.075-0.13	0.003-0.005	845	1550	0.5	Oil	(c)	(e)
Shaft 0.08		0.18	1118	0.25-0.4	0.010-0.015	845	1550	2	Oil	(c)	(e)
Spring seat........ 0.01		0.02	1118	0.25-0.4	0.010-0.015	845	1550	2	Oil	(c)	(e)
Stop collar........ 0.9		2	1117	1.1	0.045	925	1700	6.5	AC	(g)	60-63
Stud............. 0.007		0.015	1118	0.13-0.25	0.005-0.010	845	1550	1	Oil	(c)	(e)
Valve bushing...... 0.02		0.05	1117	1.3	0.050	915	1675	8	AC	(g)	58–63
Valve retainer...... 0.45		1	1117	1.1	0.045	915	1675	7	(j)	...	58-63
Washer 0.007		0.015	1118	0.25-0.4	0.010-0.015	845	1550	2	Oil	(c)	(e)

(continued)

(a) Reheated at 790 °C (1450 °F), quenched in caustic, tempered at 150 °C (300 °F). (b) Transferred to neutral salt at 790 °C (1450 °F), quenched in caustic, tempered at 175 °C (350 °F). (c) Tempered at 160 °C (325 °F). (d) Or equivalent. (e) File-hard. (f) Tempered at 205 °C (400 °F). (g) Reheated at 845 °C (1550 °F), quenched in salt at 175 °C (350 °F). (h) Reheated at 775 °C (1425 °F), quenched in salt at 195 °C (380 °F). (j) Quenched directly in salt at 175 °C (350 °F). (k) Tempered at 160 °C (325 °F) and treated at −85 °C (−120 °F).

er-melting-point salts by pumping the contaminated quench salts through a filter maintained at a lower temperature. The contaminants are deposited on a wire-mesh basket, and the usable salts are returned to the quench tank.

Quenching Cyanided Parts. Cyanided steel parts are quenched either in fast-quenching oils, in water or in aqueous salt solutions. Selection of the quenchant depends on the composition of the steel, the required as-quenched hardness and the shape of the workpiece.

Water or aqueous salt solutions should be as free as possible of dissolved gases, which may cause soft spots. For this reason, pumps or impellers should be used to agitate the quenching water or brine. Compressed air should not be used as the primary means of agitation; mechanical agitation is preferred.

For maximum hardness, the quenchant should be as cold as is feasible and should be well agitated. Typical quenchant temperatures range from room temperature to about 25 °C (75 °F)

for plain water and up to about 50 °C (120 °F) for 5-to-10% aqueous salt solutions, including solutions of sodium chloride, sodium hydroxide or proprietary salt mixtures that provide corrosion protection. Use of higher temperatures with water-base quenchants causes insufficient hardness or soft spots.

Quenching oils are commonly used at temperatures from 50 to 85 °C (120 to 185 °F). Only petroleum-base quenching oils should be used for quenching cyanided parts.

Table 8 (continued)

Part	Weight kg	lb	Steel	Depth of case mm	in.	Temperature °C	°F	Time, h	Quench	Subsequent treatment	Hardness, HRC
Alloy steel											
Bearing races.......	0.9-36	2-80	8620	2.3	0.090	925	1700	14	AC	(g)	61-64
Bearing rollers......	0.20	0.5	8620	2.3	0.090	925	1700	14	AC	(g)	61-64
Coupling	0.03	0.06	8620	0.25-0.4	0.010-0.015	845	1550	2.	Oil	(c)	(e)
Crankshaft	0.9	2	8620	1.0	0.040	915	1675	6.5	AC	(h)	60-63
Gear...............	0.35	0.75	8620	1.0	0.040	915	1675	6	AC	(g)	60-63
	0.03	0.06	8620	0.075-0.13	0.003-0.005	845	1550	0.5	Oil	(c)	(e)
Idler shaft.........	0.45	1	8620	0.75	0.030	915	1675	5	(j)	...	58-63
Pintle.............	4.5-86	10-190	8620	1.5	0.060	925	1700	12	(j)	...	58-63
Piston	0.20	0.5	8620	1.3	0.050	915	1675	8	AC	(g)	60-63
Plunger	0.45-82	1-180	8620	1.3	0.050	915	1675	8	(j)	...	58-63
Ram...............	2.3-23	5-50	8620	1.1	0.045	915	1675	7	(j)	...	58-63
Retainer	0.0009	0.002	9317	0.1-0.2	0.004-0.008	845	1550	0.33	Oil	(k)	(e)
Spool	0.45-54	1-120	8620	1.3	0.050	925	1700	7	(j)	...	58-63
Thrust cup	0.20	0.5	8620	1.1	0.045	915	1675	7	(j)	...	58-63
Thrust plate........	5.4	12	8620	2.3	0.090	925	1700	14	AC	(g)	60-64
Universal socket	1.8	4	8620	1.5	0.060	915	1675	10	AC	(g)	58-63
Valve..............	0.01	0.03	8620	0.4-0.5	0.015-0.020	845	1550	4	Oil	(k)	60 min(d)
Valve seat.........	0.20	0.5	8620	1.1	0.045	915	1675	7	AC	(g)	60-63
Wear plate	0.45-3.6	1-8	8620	1.3	0.050	915	1675	7	AC	(g)	60-63

(a) Reheated at 790 °C (1450 °F), quenched in caustic, tempered at 150 °C (300 °F). (b) Transferred to neutral salt at 790 °C (1450 °F), quenched in caustic, tempered at 175 °C (350 °F). (c) Tempered at 160 °C (325 °F). (d) Or equivalent. (e) File-hard. (f) Tempered at 205 °C (400 °F). (g) Reheated at 845 °C (1550 °F), quenched in salt at 175 °C (350 °F). (h) Reheated at 775 °C (1425 °F), quenched in salt at 195 °C (380 °F).(j) Quenched directly in salt at 175 °C (350 °F). (k) Tempered at 160 °C (325 °F) and treated at −85 °C (−120 °F).

Table 9 Typical applications of liquid carburizing in noncyanide baths

Part	Weight kg	lb	Steel	Case depth mm	in.	Temperature °C	°F	Time, h	Quench	Subsequent treatment	Hardness, HRC
Production tools......	0.5-2.0	1.1-4.4	1018	0.375	0.015	925	1700	0.5-1.0	Brine	...	50-60
Bicycle forks........	51.4	3.1	1017(a)	0.05-0.08	0.002-0.003	925	1700	0.085	Brine	Temper at 425 °C (795 °F)	60
Shift lever and ball...............	~1.5	~3.3	1040, 1017(b)	0.25	0.010	925	1700	0.67	Air cool 30 s in brine	...	File hard
Screw machine spindles	0.8	1.8	4620, 8620	0.875	0.035	(c)	(c)	6.0	Molten salt, 205 °C (400 °F)	...	60-63
Clock screws and studs	0.005	0.011	1006, 1113	0.08-0.10	0.003-0.004	955	1750	0.2	Brine	...	62-64
Flat head screws	0.015	0.033	1122	0.15	0.006	925	1700	0.33	Molten salt, 290 °C (550 °F)	...	56

(a) Partial immersion. (b) Carburizer brass braze. (c) Preheat at 840 °C (1545 °F); carburize at 920 °C (1690 °F).

Salt Removal (Washing)

The ease or difficulty with which salt can be removed from liquid carburized parts depends primarily on how simply or intricately shaped the parts are and whether they were quenched in water or in oil. To some extent, removal of salt may be complicated by the presence of insoluble residues. Water-quenched parts of simple design and with no blind holes or deep recesses usually are easy to clean. They may be rinsed thoroughly in water at about 80 °C (180 °F) and then coated with a rust-preventive fluid or soluble oil. Parts that are rinsed free of cyanide by immersion in a chloride salt and then isothermally quenched in a nitrate-nitrite salt are easily cleaned by agitated hot-water washing and rinsing. It is also possible to reclaim the nitrate-nitrite salt from the wash water.

Oil-quenched parts are more difficult to clean because of the adherence of salts, some of which may be insoluble. Use of power washers with hot water or

Fig. 10 Typical parts selectively carburized by partial immersion

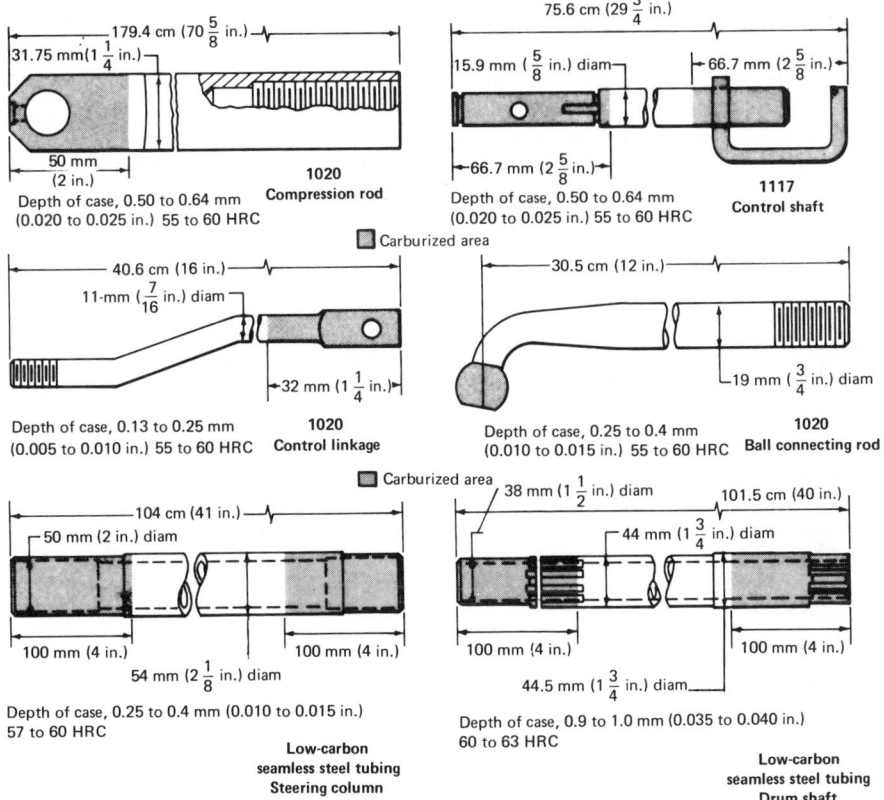

Only the portion that is to be carburized is immersed in the bath. Area to be carburized is shaded.

Typical Applications

The applicability of liquid carburizing is evidenced by the variety of parts listed in Tables 8 and 9, all of which were heat treated on a production basis. For ease of reference, the parts in Table 8 have been separated according to type of steel (carbon, resulfurized or alloy), and the parts in each group have been arranged in alphabetical order. Tables 8 and 9 also provide details, wherever they were available, regarding case depth, carburizing temperature and cycle time, method of quenching, subsequent treatment, and surface hardness.

The parts listed in Table 8 were carburized in cyanide-type baths. Noncyanide carburizing baths can be used with slight adjustments in operating conditions to do much of the carburizing described in Table 8. Noncyanide carburizing is particularly applicable to parts treated at temperatures above 900 °C (1650 °F). Some specific applications for noncyanide liquid carburizing of production parts are listed in Table 9.

In general, liquid carburizing is best suited to small and medium-size parts. Very large parts, such as rock-bit drill rods 6 m (20 ft) long and rings 2 m (7 ft) in diameter, are too large to be conveniently processed in salt, and are commonly carburized by pack methods. Because of the problems associated with salt removal, carburizing in salt baths is not recommended for parts containing small holes, threads or recessed areas that are difficult to clean.

Stop-Offs and Selective Carburizing

Selective carburizing can be accomplished in liquid carburizing baths by stopping off carbon penetration with either copper plate or copper-base paint. Because cyanide-base salts can dissolve copper, salt baths with relatively low cyanide contents are necessary. One successful formulation operates at 8 to 10% sodium cyanide with approximately 45% barium chloride energizer. Noncyanide carburizing salts will not dissolve copper.

When copper plate is employed to prevent carbon penetration, the copper layer should be fine-grained, dense, and without pinholes or other porosity. Smooth surfaces require lower plating thicknesses than do rough surfaces. Copper-plate thicknesses recommend-

with emulsion cleaners is effective. An economical cleaning procedure begins with soaking of parts in hot water to float off the oil and remove the soluble salts. The parts may then be transferred to a hot agitated dilute alkaline cleaner having high sequestering properties. (Silicated cleaners and those containing carbonates or phosphates are not recommended, because of the formation of insoluble barium compounds in the presence of barium-containing salts.) If a white, powdery overlay of barium carbonate remains on the parts, it may be removed—following removal of all cyanide—by immersion in a dilute solution of acetic or inhibited hydrochloric acid.

Complex parts with blind holes and recesses are difficult to clean, particularly if they have been oil quenched. Liquid carburizing of parts that contain blind holes for which the depth exceeds twice the diameter is not rec-ommended unless such holes can be plugged. Agitated hot water is probably the best solvent for salt held in recesses, crevices and blind holes. Normally, it will remove all soluble salts and will soften insoluble residues. When part shape and tolerances permit, tumbling for 10 to 30 min in a mild alkali and a small quantity of sand is most effective in removing insoluble surface residues.

Washing of Cyanided Parts. Cyanide-hardened parts are easy to wash, even after oil quenching, because cyanide and sodium carbonate are good detergents and because all the components of the salt bath are water soluble. The work may be soaked in a tank of agitated boiling water, rinsed in clean hot water, and then rustproofed (if required). Power spray washers, using hot water in a two-stage system, also give satisfactory results.

ed for protection against liquid carburizing for various times are as follows:

Low-temperature salts
Up to 1 h . . . . 0.013 mm (0.0005 in.)
1 to 5 h . . . . 0.020 mm (0.0008 in.)
High-temperature salts
Up to 7 h . . . . 0.025 mm (0.0010 in.)
7 to 15 h . . . . 0.040 mm (0.0015 in.)
15 to 30 h . . . . 0.050 mm (0.0020 in.)

Partial Immersion. Another method for selective carburizing in liquid carburizing baths entails partial immersion of the workpiece in the salt bath so that only the immersed areas are carburized. With this method, unless a clean-cut breakoff between carburized and noncarburized areas is required, the use of copper plate or copper-base paint is unnecessary.

Oxidation of the work at the bath surface can be reduced if the parts are initially immersed an inch or two deeper than required, to coat them with salt, and are then withdrawn to the required depth. A piece of plain carbon steel pipe with the bottom closed by welding can be inserted in a corner of the bath to displace salt if a precise adjustment of the bath level is necessary. Typical parts that are well suited to carburizing by the partial-immersion technique are shown in Fig. 10.

Combined Carburizing and Brazing

It is possible to braze and carburize steel parts simultaneously in either a cyanide or a noncyanide liquid carburizing bath, provided that the operating temperature of the bath is high enough to cause the brazing alloy to flow. First, the parts are cleaned and degreased, and then components are assembled with brazing alloy enclosed in the joints. One suitable brazing alloy, in the form of wire or thin strip, contains 55% Cu and 45% Zn, melts at 880 °C (1620 °F), and makes a sound joint at 900 to 925 °C (1650 to 1700 °F). No flux is required.

The assembly is immersed in the liquid carburizing bath for a time long enough to produce the desired case depth and at a temperature high enough to cause the brazing alloy to flow. It is then quenched to harden the steel and complete the braze. Press-fit assemblies with carefully designed lap joints are ideal for this application.

Precautions in the Use of Cyanide Salts

Cyanides cause violently poisonous reactions if allowed to come into contact with scratches or wounds (on the hands, for example); they are fatal if taken internally. Also, fatally poisonous fumes are evolved when cyanides are brought into contact with acids. The white deposits that form on hoods and cooler furnace parts consist mainly of sublimed sodium carbonate, with small amounts of sodium, potassium, and barium salts, but may contain some cyanide as the result of splashing.

When cyanide salts are removed from the packing container, the container should be opened in the room in which the cyanide is to be used. The salts should be removed from the container with a metal scoop or gloved hands, or by being dumped out as required. When not in use, the container should be covered with its original cover or with a metal substitute cover.

The precautions that should be observed in handling cyanide-type carburizing salts are the same as for any other cyanide mixture. Work material must be clean and dry, and the bath must be enclosed and well ventilated. Even the slight amount of moisture that may be deposited on parts and fixtures as a result of atmospheric humidity will cause spatter in contact with molten salt. Accordingly, operators should be equipped with long protective gloves, protective aprons, and safety glasses or face shields. When adequate precautions are observed, carburizing salts should not present serious hazards to health or safety.

Remelting a frozen cyanide bath in furnaces of other than the immersed-electrode type is potentially hazardous because of the expansion of gases as the salts are heated. This hazard is not encountered with immersed-electrode furnaces because the salts melt from the top down. If remelting is done in other types of furnaces, however, the following precaution should be observed: a steel or cast iron wedge should be inserted in the center of the bath before the bath freezes. One end of the wedge should make contact with the bottom of the pot; the other end should extend at least 10 cm (4 in.) above the surface of the bath. Before the bath is remelted, the wedge should be tapped with a hammer, loosened and

removed. The space previously occupied by the wedge will provide a vent for expanding gases during remelting. No attempt should be made to remove a wedge from a bath that is not completely solidified, because molten salt may be forcibly blown out through the opening created.

Disposal of Cyanide Wastes

Cyanide wastes, whether dissolved in quench water or in the form of solid salt from pots, pose a serious disposal problem. The cyanide contents of these wastes must be chemically altered to render the material nonpoisonous before it is discharged into sewers or streams. Because of the toxicity of cyanide wastes, local ordinances and pollution authorities should be consulted regarding the proper disposal of wastes.

Chemical Treatment. The simplest treatment consists of oxidizing the cyanide in an alkaline solution to which is added either chlorine gas or its equivalent in the form of a hypochlorite compound, such as sodium or calcium hypochlorite (bleaching powder). The choice between gas or powder depends on the quantity of cyanide to be treated, on the availability of facilities and experienced personnel for handling the oxidizing agents, and on economics. For small quantities of cyanide solutions, it may be more practical to use a hypochlorite compound than to use chlorine gas.

Depending on the oxidizing agent employed, several reactions take place when cyanide is converted into a disposable form. One reaction with chlorine gas is as follows:

$$2NaCN + 4NaOH + 2Cl_2 + 2H_2O \rightarrow (NH_4)_2CO_3 + Na_2CO_3 + 4NaCl$$

This reaction indicates that, for each kilogram (2.2 lb) of sodium cyanide, 1.42 kg (3.13 lb) of chlorine gas and 1.6 kg (3.5 lb) of sodium hydroxide are required. Because of probable side reactions, however, practical experience indicates an actual requirement of slightly more than 2 kg (4.4 lb) of chlorine for each kilogram of sodium cyanide present in the waste solution. When a hypochlorite compound is used, the amount of powder required may be estimated on the basis of available chlorine content in the compound.

Solid cyanide wastes must be dissolved in water before they can be treated. A tank of suitable capacity, equipped with a coarse screen set well above the bottom, will facilitate solution of the solid material. The tank should also have an agitator, and, for chlorine gas, a perforated pipe placed well below the solution level is required.

When cyanide wastes are to be treated with chlorine gas, the cyanide content must first be determined and the proper amount of caustic added. The gas is then introduced slowly while the temperature of the solution is kept below 50 °C (120 °F). If a sodium hypochlorite solution is used, it is only necessary that sufficient caustic be added to raise the pH of the cyanide solution above 8.5. The reaction between cyanide and the oxidizing agent should continue until a slight excess of chlorine is present in the solution. This can be determined by testing with starch iodide paper or with a solution of potassium iodide and starch. Both the iodide paper and the starch solution will turn blue in the presence of free chlorine.

Electrochemical Treatment. Although chemical treatment may be entirely adequate to meet local regulations, a recently developed electrochemical process destroys free cyanide with efficiency and economy superior to those of the chemical conversion processes. In the electrochemical process, cyanide wastes in aqueous solution are circulated through an electrochemical reactor. Within the reactor, an applied dc potential oxidizes the free cyanide and cyanate according to the reactions:

$$2CN^- + 8OH^- \longrightarrow$$
$$2CO_2 + N_2 + 4H_2O + 10e^-$$

$$2CNO^- + 4OH^- \longrightarrow$$
$$2CO_2 + N_2 + 2H_2O + 6e^-$$

Free cyanide and cyanate are converted to the nontoxic gases carbon dioxide and nitrogen, which are allowed to escape freely from the vented storage tank into which the reacted solution is circulated.

The electrochemical process is most effective at high cyanide-ion concentrations. With continual recirculation between storage tank and reactor, cyanide can be reduced to 1 ppm or less in about 100 to 150 h. By combining electrochemical and chemical treatments, effective treatment can usually be achieved at minimum cost. Electrochemical removal is used to reduce cyanide concentration to about 200 ppm, and then chemical treatment is used to complete the reduction.

Electrochemical treatment offers the following advantages:

- The process uses only electricity—no chemicals are required.
- Cost per unit weight of cyanide treated is low, depending only on the cost of electricity (about 6.6 kW·h/kg, or 3 kW·h/lb of free CN^-).
- Capital investment is relatively low.

- The process is simple to control, requiring only periodic determination of cyanide concentration.
- There are no toxic or otherwise harmful reaction products.
- Upon reaching a concentration of 1 ppm cyanide, the oxidized effluent usually may be drained directly into a sewer.
- The process also can be used to convert nitrite into nitrate.

The only significant disadvantage is that the process is time-consuming when levels of cyanide below 200 ppm must be achieved. Increasing the number of reactors decreases process times.

Safety and Disposal of Noncyanide Carburizing Salts

Noncyanide carburizing salts are safe to dispose of directly into municipal or natural water if first diluted to acceptable dissolved-solids levels. There are no significant chemical hazards in the use of these salts. They are somewhat alkaline and should be washed from the skin or eyes if contact is made. When they are used as molten salts, the usual precautions apply: avoid introduction of moisture into the bath, and prevent the hot salt from contacting the body.

Further information is available from OSHA and EPA publications.

Liquid Nitriding

By the ASM Committee on Liquid
Carburizing*

LIQUID NITRIDING (nitriding in a molten salt bath) employs the same temperature range as gas nitriding—that is, 510 to 565 °C (950 to 1050 °F). The case hardening medium is a molten nitrogen-bearing, fused-salt bath containing both cyanides and cyanates. Unlike liquid carburizing and cyaniding, which employ baths of similar compositions, liquid nitriding is a subcritical case hardening process; thus, processing of finished parts is possible because dimensional stability can be maintained. Also, liquid nitriding adds more nitrogen and less carbon to ferrous materials than that obtained through higher temperature diffusion treatments.

The liquid nitriding process has several proprietary modifications and is applied to a wide variety of carbon, low-alloy and tool steels.

Principal Uses. Liquid nitriding processes are used primarily to improve wear resistance of surfaces and to increase the endurance limit in fatigue. For many steels, resistance to corrosion is improved. These processes are not suitable for many applications requiring deep cases and hardened cores, but they have successfully replaced other types of heat treatment on a performance or economic basis. In general, the uses of liquid nitriding and gas nitriding are similar, and at times identical. Gas nitriding may be preferred in applications where heavier case depths and dependable stop-offs are required. Both processes, however, provide the same advantages: improved wear resistance and antigalling properties, increased fatigue resistance, and less distortion than other case hardening processes employing heating at higher temperatures. Four examples of parts for which liquid nitriding was selected over other case hardening methods appear in Table 1.

The degree to which various properties are affected by liquid nitriding may vary with the process used and the chemical control maintained. Thus, critical specifications should be based on prior test data or documented information.

Liquid Nitriding Systems

The term liquid nitriding has become a generic term for a number of different fused salt processes, all of which are performed at subcritical temperature. Operating at these temperatures, the treatments are essentially chemical diffusion operations, and they influence metallurgical structures primarily through absorption and reaction of nitrogen rather than through the minor amount of carbon that is assimilated. Although the different processes are represented by a number of commercial trade names, the basic subclassifications of liquid nitriding are those presented in Table 2.

A typical commercial bath for liquid nitriding is composed of a mixture of sodium and potassium salts. The sodium salts, which comprise 60 to 70% (by weight) of the total mixture, consist of 96.5% NaCN, 2.5% Na_2CO_3 and 0.5% NaCNO. The potassium salts, 30 to 40% (by weight) of the mixture, consist of 96% KCN, 0.6% K_2CO_3, 0.75% KCNO and 0.5% KCl. The operating temperature of this salt bath is 565 °C (1050 °F). With aging (a process described subsequently), the cyanide content of the bath decreases, and the cyanate and carbonate contents increase. (The cyanate content in all nitriding baths is responsible for the nitriding action, and the ratio of cyanide to cyanate is critical.) This bath is widely used for nitriding tool steels, including high speed steels, and a variety of low-alloy steels, including the aluminum-containing nitriding steels.

Another bath for nitriding tool steels has a composition as follows:

NaCN . . . . 30.00% max
Na_2CO_3 or K_2CO_3 . . . . . 25.00% max
Other active
ingredients . . . . 4.00% max
Moisture . . . . 2.00% max
KCl . . . . rem

*Donald Wensing, *Chairman,* Chief Metallurgist, SKF Industries, Inc.; Don Barber (retired), President, Heatbath Corp.; Robert W. Foreman, Director, Research & Development, Park Chemical Co.; George Green, Metallurgist, Industrial Truck Division, Eaton Yale & Towne Mfg. Co.; Karl H. Kopietz, Marketing Manager, Heat Treating Products, Henry E. Sanson & Sons, Inc.; W. James Laird, Jr., Vice President, Marketing, Research & Development, Upton Industries, Inc.; Quentin D. Mehrkam, Senior Vice President, Ajax Electric Co.; Burton R. Payne, Jr., President, Payne Chemical Co.; John A. Swift, P.E., Consultant, Metallurgy, Henry E. Sanson & Sons, Inc.; William G. Wood, Vice President, Technology/Research & Development, Kolene Corp.; Theodore B. Wilk, Vice President, A. F. Holden Co.

Table 1 Examples of parts for which liquid nitriding proved superior to other case hardening processes for meeting service requirements

Requirement	Material and process originally used	Resultant problem	Solution
Example 1—thrust washer			
Withstand thrust load without galling and deformation	Bronze; carbonitrided 1010 steel	Bronze galled, deformed; steel warped	1010 steel nitrided 90 min in cyanide-cyanate bath at 570 °C (1060 °F), water quenched(a)
Example 2—shaft			
Resist wear on splines and bearing area	Induction harden through areas	Required costly inspection	Nitride for 90 min in cyanide-cyanate salt bath at 570 °C (1060 °F)
Example 3—seat bracket			
Resist wear on surface	1020 steel, cyanide treated	Distortion; high loss in straightening(b)	1020 nitrided 90 min in cyanide-cyanate salt bath and water quenched(c)
Example 4—rocker arm shaft			
Resist wear on surface; maintain geometry	SAE 1045 steel, rough ground, induction hardened, straightened, finish ground, phosphate coated	Costly operations and material	SAE 1010 steel liquid nitrided 90 min in noncyanide fused salt at 570 to 580 °C (1060 to 1075 °F)(d)

(a) Resulted in improved product performance and extended life, with no increase in cost. (b) Also, brittleness. (c) Resulted in less distortion and brittleness, and elimination of scrap loss. (d) Eliminated finish grinding, phosphatizing and straightening

Table 2 Liquid nitriding processes

Process identification	Operating range composition	Chemical nature	Suggested post treatment	Operating temperature °C	°F	U.S. patent number
Aerated cyanide-cyanate	Sodium cyanide (NaCN), potassium cyanide (KCN) and potassium cyanate (KCNO), sodium cyanate (NaCNO)	Strongly reducing	Water or oil quench; nitrogen cool	570	1060	3 208 885
Casing salt	Potassium cyanide (KCN) or sodium cyanide (NaCN), sodium cyanate (NaCNO) or potassium cyanate (KCNO) or mixtures	Strongly reducing	Water or oil quench	510 to 650	950 to 1200	· · ·
Pressure nitriding	Sodium cyanide (NaCN), Sodium cyanate (NaCNO)	Strongly reducing	Air cool	525 to 565	975 to 1050	· · ·
Regenerated cyanate-carbonate	Type A—Potassium cyanate (KCNO), potassium carbonate (K$_2$CO$_3$)	Mildly oxidizing	Water, oil, or salt quench	580	1075	4 019 928
	Type B—Potassium cyanate (KCNO), potassium carbonate (K$_2$CO$_3$), 1–10 ppm, sulfur (S)	Mildly oxidizing	Water or oil quench, or salt quench	540 to 575	1000 to 1070	4 006 043

A proprietary nitriding salt bath has the following composition by weight: 60 to 61% NaCN, 15.0 to 15.5% K$_2$CO$_3$ and 23 to 24% KCl.

Special Processes. Several special liquid nitriding processes employ proprietary additions, either gaseous or solid, intended to serve several purposes, such as accelerating the chemical activity of the bath, increasing the number of steels that can be processed, and improving the properties obtained as a result of nitriding.

Cyanide-free liquid nitriding compositions have also been introduced, and their acceptance has contributed substantially to reduction of pollution problems.

Three processes, liquid pressure nitriding, aerated bath nitriding and aerated noncyanide nitriding, are described in the sections that follow.

Liquid Pressure Nitriding

Liquid pressure nitriding is a proprietary process in which anhydrous ammonia is introduced into a cyanide-cyanate bath. The bath is sealed and

Fig. 1 Results of liquid pressure nitriding on type 410 stainless steel

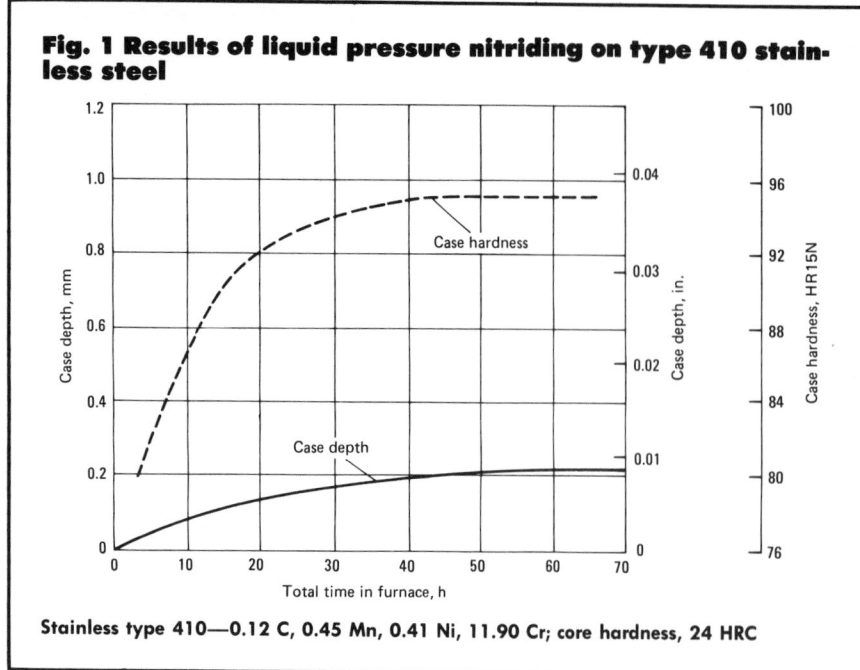

Stainless type 410—0.12 C, 0.45 Mn, 0.41 Ni, 11.90 Cr; core hardness, 24 HRC

Fig. 2 Results of liquid pressure nitriding on AISI type D2 steel

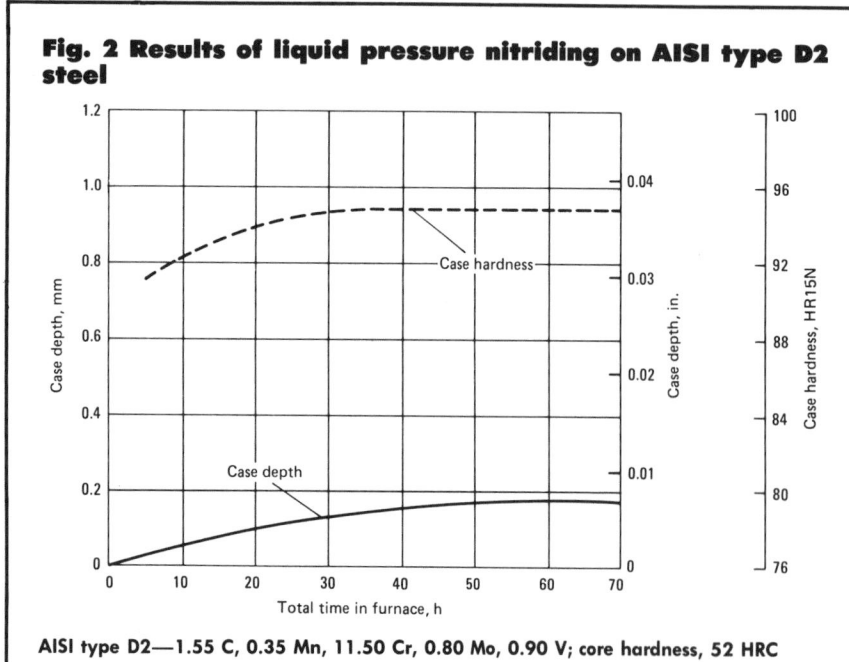

AISI type D2—1.55 C, 0.35 Mn, 11.50 Cr, 0.80 Mo, 0.90 V; core hardness, 52 HRC

ammonia addition, which serves continuously to counteract bath depletion.

The retort cover may be opened without causing complete interruption of the nitriding process. Loss of pressure within the retort results in a reduction in the nitriding rate. However, when the retort is sealed and pressure is reinstated through the resumption of ammonia gas flow, nitriding proceeds at the normal rate.

Depth of case depends on time at temperature. The average nitriding cycle is 24 h, although total cycle time may vary between 4 and 72 h. To stabilize core hardness, it is recommended that all parts be tempered at a temperature at least 28 °C (50 °F) higher than the nitriding temperature before they are immersed in the nitriding bath.

Hardness gradients and case depths resulting from pressure nitriding of 410 stainless steel, AISI Type D2 and SAE 4140 are shown in Fig. 1, 2 and 3.

Aerated Bath Nitriding

Aerated bath nitriding is a proprietary process (U.S. Patent 3 022 204; 1962) in which measured amounts of air are pumped through the molten bath. The introduction of air provides agitation and stimulates chemical activity. The cyanide content of this bath, calculated as sodium cyanide, is preferably maintained at about 50 to 60% of the total bath content, and the cyanate is maintained at 32 to 38%. The potassium content of the fused bath, calculated as elemental potassium, is between 10 and 30%, preferably about 18%. The potassium may be present as the cyanate or the cyanide, or both. The remainder of the bath is sodium carbonate.

This process produces a nitrogen-diffused case 0.3 mm (0.012 in.) deep on plain carbon or low-alloy steels in a 1½-h cycle. The surface layer (0.005 to 0.01 mm or 0.0002 to 0.0005 in. deep) of the case is composed of epsilon Fe_3N and a nitrogen-bearing Fe_3C; the nitrided case does not contain the brittle Fe_2N constituent.

Beneath the outer layer, Fe_3N is formed in a diffusion zone extending into the steel. Depth of nitrogen diffusion in 1015 steel as a function of nitriding time at 565 °C (1050 °F) is shown in Fig. 4. The outer compound layer provides wear resistance, while

maintained under a pressure of 7 to 205 kPa (1 to 30 psi). The ammonia is piped to the bottom of the retort and is caused to flow vertically. The percentage of nascent nitrogen in the bath is controlled by maintaining the ammonia flow rate at 0.5 to 1 m³/h (20 to 40 ft³/h). This results in ammonia dissociation of 15 to 30%.

The bath contains sodium cyanide and other salts, which permits an operating temperature of 525 to 565 °C (975 to 1050 °F). Because the molten salts are diffused with anhydrous ammonia, a new bath does not require aging and may be put into immediate operation employing the recommended cyanide-cyanate ratio, namely, 30 to 35% cyanide and 15 to 20% cyanate. Except for dragout losses, maintenance of the bath within the preferred ratio range is greatly simplified by the anhydrous

Fig. 3 Results of liquid pressure nitriding on SAE 4140 steel

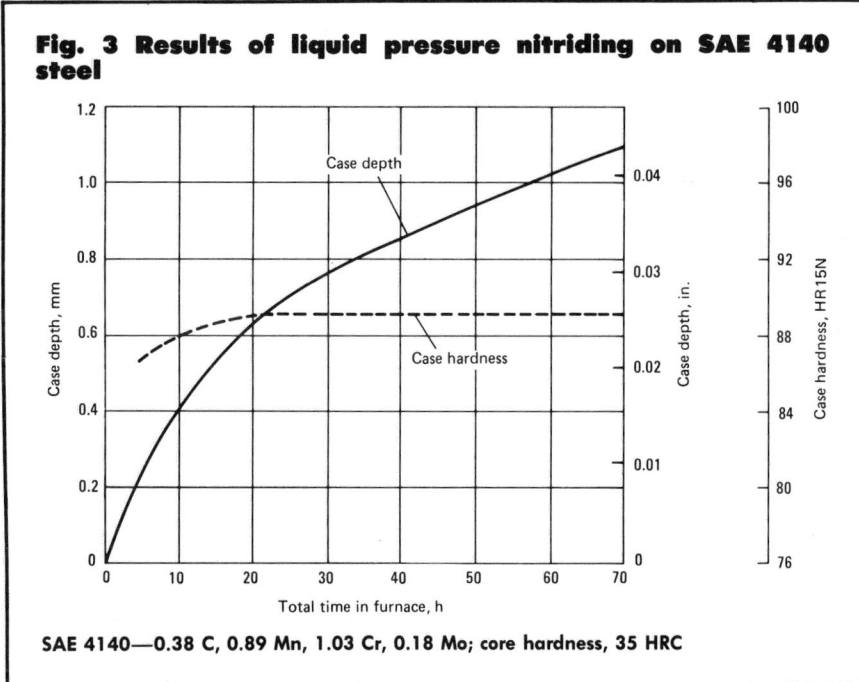

SAE 4140—0.38 C, 0.89 Mn, 1.03 Cr, 0.18 Mo; core hardness, 35 HRC

Fig. 4 Nitrogen gradients in 1015 steel as a function of time of nitriding at 565 °C (1050 °F), using the aerated bath process

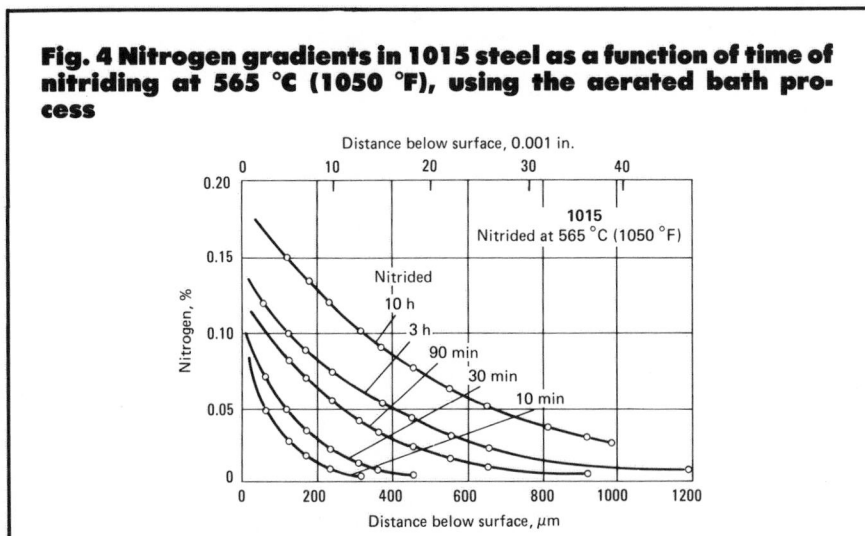

Table 3 Improvement in fatigue properties of low-temperature liquid nitrided ferrous materials

Steel type	Property improvement, %
Low carbon steels	80 to 100
Medium carbon steels	60 to 80
Stainless steels	25 to 35
Low carbon, chrome manganese steels	25 to 35
Chrome alloy, medium carbon steels	20 to 30
Cast irons	20 to 80

steels treated by this process may reach 900 HK as shown in AMS 2755B, a portion of which is reproduced in Appendix A of this article.

Aerated Noncyanide Nitriding. Environmental concerns have led to the development of cyanide-free processes for liquid nitriding. In these proprietary processes, the base salt is supplied as a cyanide-free mixture of potassium cyanate and a combination of sodium carbonate and potassium carbonate or sodium chloride and potassium chloride. It is possible to develop minor percentages of cyanide when processing heavy loads in these compositions. The problem is overcome in one process (U.S. Patent 4 019 928) by quenching in an oxidizing quench salt that destroys cyanate compounds (which have pollution capabilities) and produces less distortion than that resulting from water quenching. An alternate method utilized by U.S. Patent 4 006 643 is the incorporation of minute amounts of sulfur (1–10 ppm) in the base salt to control cyanide formation.

These noncyanide processes have been shown in tests to produce the same results as those developed in the previously mentioned liquid nitriding processes. The diffusion curves and case depths are quite similar to those shown in Fig. 1, 2 and 3. Because a high cyanate (65 to 75% KCNO) level in the absence of cyanide would be expected to produce iron nitride compound zones slightly lower in carbon and slightly higher in nitrogen, it is good practice to develop new tests and operational data when converting to one process from another. Excerpts from the AMS 2753 specification developed for noncyanide liquid salt bath nitriding are shown in Appendix B.

Case Hardness. According to AMS

the transition zone improves fatigue strength.

Another aerated bath process for liquid nitriding is a high-cyanide, high-cyanate system that is proprietary (U.S. Patent 3 208 885). The cyanide content of the fused salt is maintained in the range of 45 to 50% calculated as potassium cyanide, and the cyanate content is maintained in the range of 42 to 50% calculated as potassium cyanate. Make-up salt consists of a precise mixture of sodium and potassium cyanides that are oxidized by aeration to the mixed cyanate. The ratio of sodium

ions to potassium ions is important in duplicating the integrity of the compound zone and the diffusion zone.

The process is performed in a titanium-lined container, and it produces a compound zone of epsilon iron nitride to a depth of 0.010 to 0.015 mm (0.0004 to 0.0006 in.) and a diffusion zone of 0.356 to 0.457 mm (0.014 to 0.018 in.) in plain carbon steels with a 90-min treating time, as shown in Fig. 5. The surface hardness of the compound zone may vary between 300 HK and 450 HK if carbon or low-alloy steels are being treated. Surface hardness of stainless

Fig. 5 Nitrided case and diffusion zone produced by cyanide-cyanate liquid nitriding

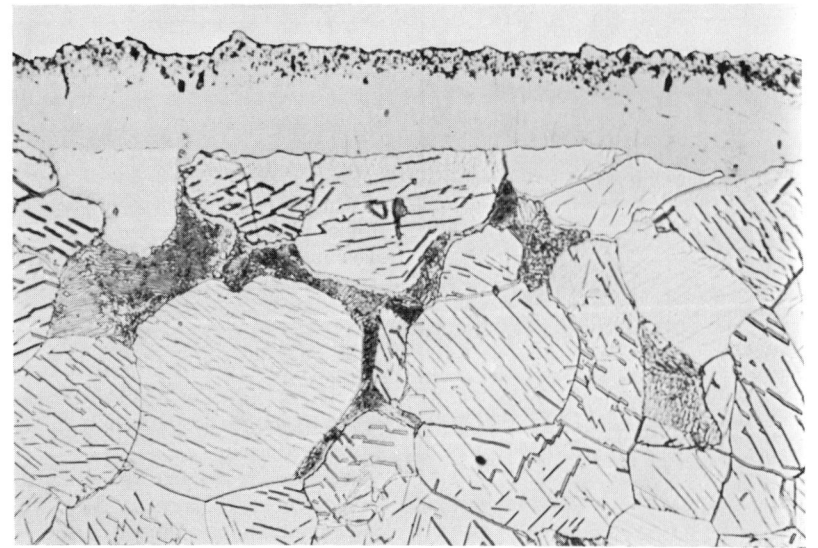

The characteristic needle structure is only seen after a 300 °C (570 °F) aging treatment.

Fig. 6 Effect of carbon content in carbon steels on the nitrogen gradient obtained in aerated bath nitriding

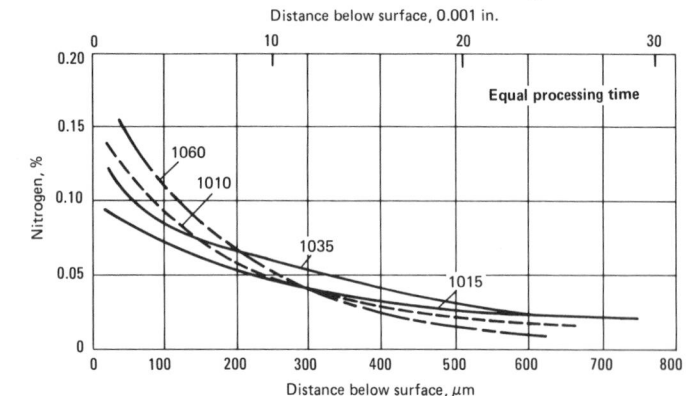

sion zone shown by the Fe₄N needles in Fig. 5 can be measured under the microscope to a depth of approximately 0.40 mm (0.016 in.), actual nitrogen penetration can be measured up to 1.02 mm (0.040 in.) as shown in Fig. 8. This nitrogen is in solution, is under stress, and is precipitated as Fe_4N. It is responsible for the fatigue improvement resulting from liquid nitriding. The improvement is more apparent in plain carbon steels, resulting in the substitution of these steels for high carbon and low-alloy steels in many applications (Table 3).

Case Depth and Case Hardness

Data indicating depth of case obtained in liquid nitriding various steels in a conventional bath at 525 °C (975 °F) for up to 70 h are shown in Fig. 9. The steels include three chromium-containing low-alloy steels (4140, 4340 and 6150), two aluminum-containing nitriding steels (SAE 7140 and AMS 6475), and four tool steels (H11, H12, M50 and D2). All were nitrided in a salt bath with an effective cyanide content of 30 to 35% and a cyanate content of 15 to 20%. Case depths were measured visually on metallographically prepared samples that were etched in 3% nital. Before being nitrided, samples were tempered to the core hardnesses indicated.

Figure 10 presents data on case hardness obtained in liquid pressure nitriding the following alloy steels and tool steels: SAE 7140, AMS 6475, 4140, 4340, medium-carbon H11, low-carbon H11, H15, and M50. The various core hardnesses, nitriding temperatures and cycle times were as noted in the graphs in Fig. 10. Case depth and hardness results are comparable to those obtained in single-stage gas nitriding.

High Speed Steels. Compared to gas nitriding of high speed steel cutting tools, liquid nitriding can produce a more ductile case with a lower nitrogen content. Nitrided case hardness data, together with details of liquid nitriding these materials, are given in the section of this volume on Heat Treating of Tool Steels.

Operating Procedures

Among the important operating procedures in liquid nitriding are the initial preparation and heating of the salt

2755, case hardness varies markedly with the alloy being nitrided. Hardness and other requirements of this specification are summarized in Appendix A.

Effects of Steel Composition. Although the properties of alloy steels are improved by the compound and diffusion layers, relatively greater improvement is achieved with plain carbon steels of low and medium carbon content. For example, although the improvement in fatigue strength of unnotched test bars of 1015 steel nitrided by this process for 90 min at 565 °C (1050 °F) and water quenched (to further enhance fatigue properties) is roughly 100%. Improvement obtained with similarly treated test bars made of 1060 steel is about 45 to 50%.

The diffusion of nitrogen in carbon steels is directly affected by carbon content, as shown in Fig. 6. The nitride-forming alloying elements also inhibit nitrogen diffusion. For example, the inhibiting effect of chromium on diffusion is shown in Fig. 7, which compares nitrogen diffusion in a low-carbon steel (1015) and a chromium-containing low-alloy steel (5115).

Although the visible nitrogen diffu-

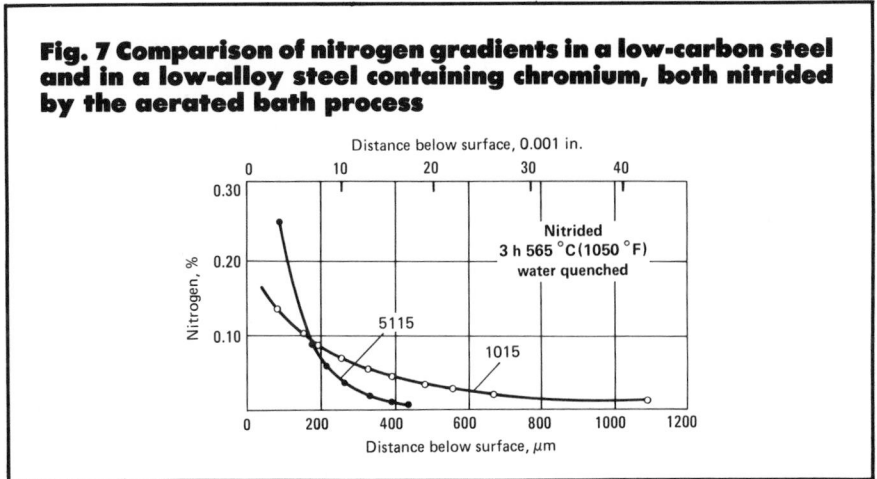

Fig. 7 Comparison of nitrogen gradients in a low-carbon steel and in a low-alloy steel containing chromium, both nitrided by the aerated bath process

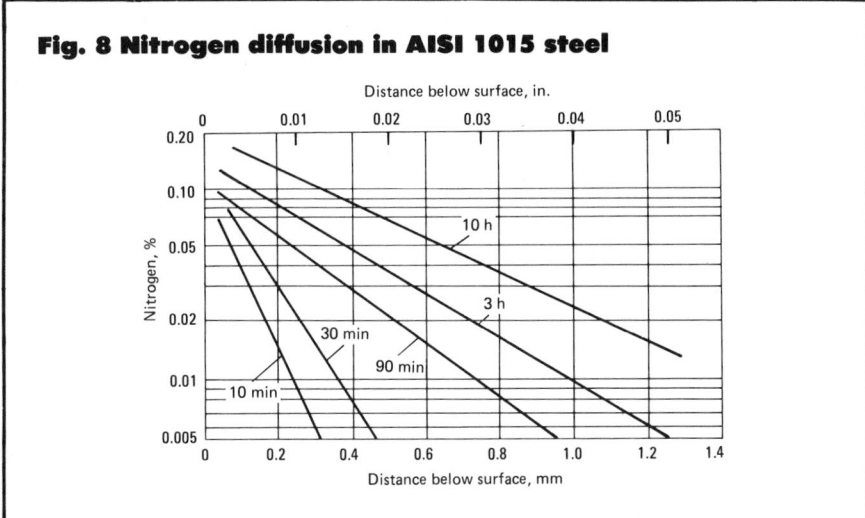

Fig. 8 Nitrogen diffusion in AISI 1015 steel

bath, aging of the molten salts (when required), and analysis and maintenance of salt bath composition. Virtually all steels must be quenched and tempered for core properties before being nitrided or stress-relieved for distortion control. So prior heat treatment may be considered an essential part of the operating procedure.

Prior Heat Treatment. Alloy steels usually are given a prior heat treatment similar to that preferred for gas nitriding. Maintenance of dimensional and geometric stability during liquid nitriding is enhanced by hardening of parts prior to nitride treatment. Tempering temperatures should be no lower than the nitriding temperature and preferably slightly above. Exceptions are those steels that have been strain hardened at elevated temperature.

Depending on steel composition, the effect of core hardness is similar to that encountered in gas nitriding.

Starting the Bath. Case-producing salt compositions may vary with respect to manufacturers, but they are basically sodium and potassium cyanides. Cyanide, the active ingredient, is oxidized to cyanate by aging as described below. The commercial salt mixture (60 to 70% sodium salts, 30 to 40% potassium salts) is melted at 540 to 595 °C (1000 to 1100 °F). *During the melting period, a cover should be placed over the retort to guard against spattering or explosion of the salt, unless the equipment is completely hooded and vented. It is mandatory that the salts be dry before they are placed in the retort; the presence of entrapped moisture may result in an eruption when the salt mixture is heated.*

Aging the Bath. Liquid nitriding compositions that do not contain a substantial amount of cyanate in the original melt must be aged before use in production. Molten salts in conventional baths should be "aged" by being held at 565 to 595 °C (1050 to 1100 °F) for at least 12 h, and no work should be placed in the bath during the aging treatment. Aging decreases the cyanide content of the bath and increases the cyanate and carbonate contents. Before nitriding is begun, a careful check of the cyanate content should be made. Nitriding should not be attempted until the cyanate content has reached at least the minimum operating level recommended for the bath.

Bath Maintenance. To protect the bath from contamination and to obtain satisfactory nitriding, all work placed in the bath should be thoroughly cleaned and free of surface oxide. An oxide-free condition is especially important when nitriding in cyanide-free salts. These compounds are not strong reducing agents and therefore are incapable of producing a good surface on any oxidized work. Either acid pickling or abrasive cleaning is recommended prior to nitriding. Finished clean parts should be preheated before being immersed in the bath, to rid them of surface moisture.

A high cyanate content (up to about 25%) will provide good results, but carbonate content should not exceed 25%. Carbonate content can be readily lowered by cooling the bath to 455 °C (850 °F) and allowing the precipitated salt to settle to the bottom of the salt pot. It can then be spooned from the bottom by means of a perforated ladle.

To minimize corrosion at the air-salt interface, salts should be completely changed every three or four months (replacement of salt is usually far more economical than replacement of the pot). When the bath is not in use, it should be covered; excessive exposure to air causes a breakdown of cyanide to carbonate and adversely affects pot life.

The ratio of cyanide content to cyanate content varies with the salt bath process and the composition of the bath. The commercial NaCN-KCN bath, after aging for one week, achieves a ratio of 21 to 26% cyanide to 14 to 18% cyanate. The bath used in liquid pressure nitriding operates with a cyanide content of 30 to 35% and a cyanate content of 15 to 20%. The aerated bath is con-

Fig. 9 Depth of case for several chromium-containing low-alloy steels, aluminum-containing steels and tool steels after liquid nitriding

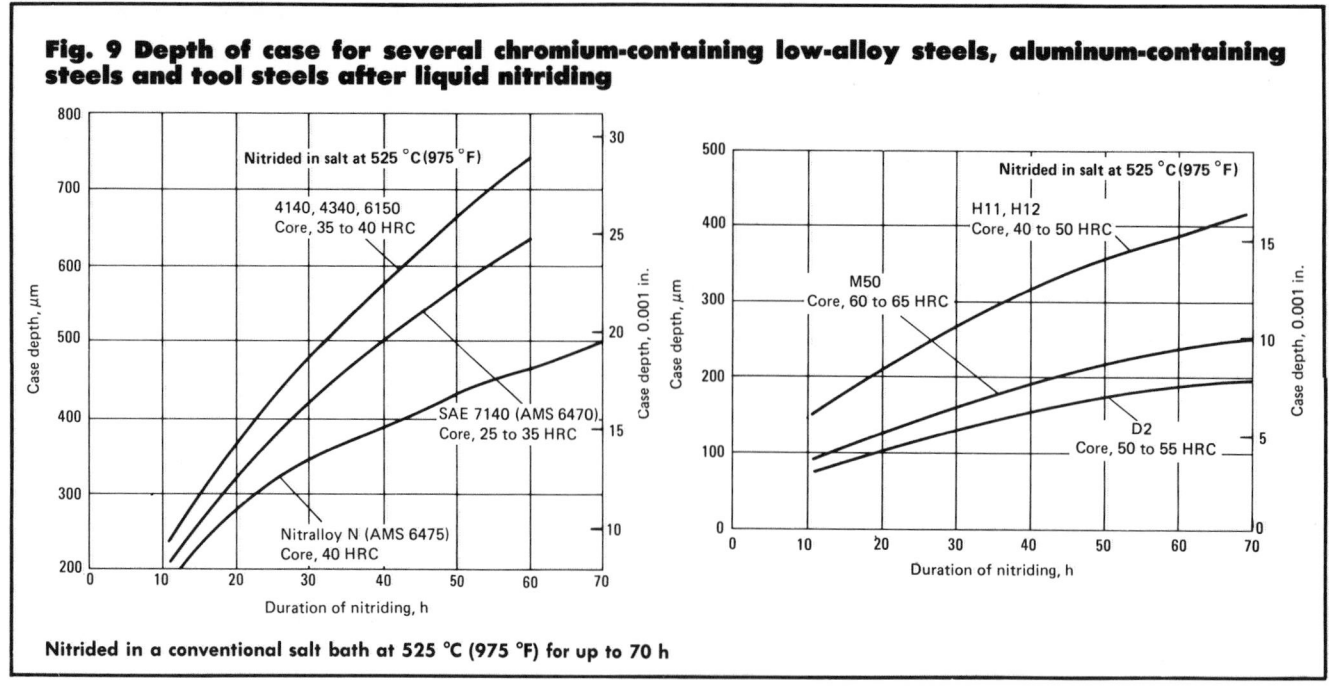

Nitrided in a conventional salt bath at 525 °C (975 °F) for up to 70 h

Fig. 10 Hardness gradients for several alloy and tool steels nitrided in salt by the liquid pressure process

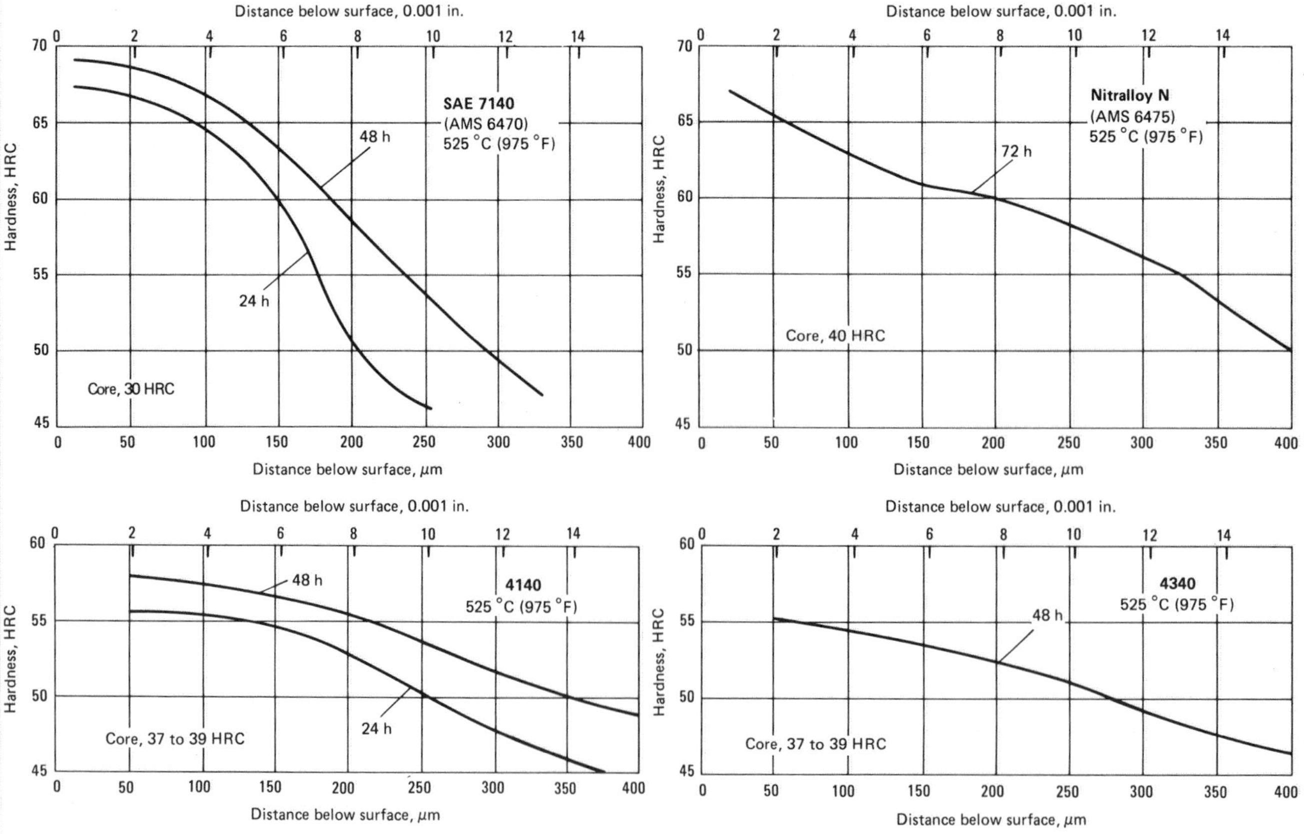

Rockwell C hardness values are converted from Knoop hardness measurements made using a 500-g load. Temperatures are nitriding temperatures.

Fig. 10 (continued)

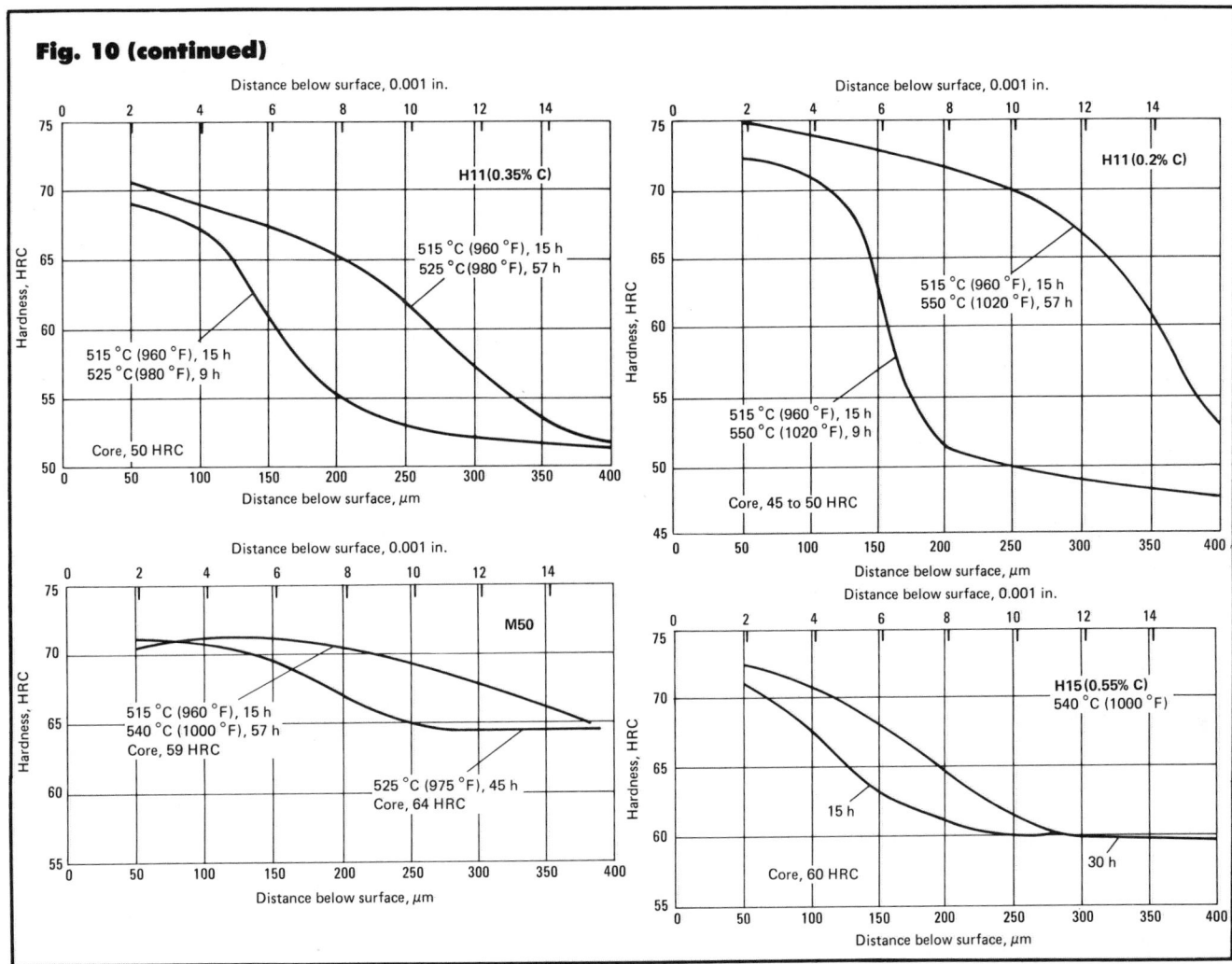

trolled to a ratio of 50 to 60% cyanide to 32 to 38% cyanate. The aerated non-cyanide nitriding process is controlled to a ratio of 36 to 38% cyanate to 17 to 19% carbonates.

All baths must be periodically relieved of oxidation products, which promote unfavorable temperature gradients. In normal operation, overheating of any bath (above 595 °C or 1100 °F) should be avoided.

Safety. Some of the compositions employed in liquid nitriding processes contain sodium cyanide or potassium cyanide, or both. These compounds can be handled safely with proper equipment and neutralized by chemical means before discharge. *The compounds are highly toxic, however, and great care should be exercised to avoid taking them internally or allowing them to be absorbed through skin abrasions. Contact between the compounds and mineral acids also generates another hazard—the formation of hydrogen cyanide (HCN) gas, an extremely toxic product. Exposure to hydrogen cyanide can be fatal.*

Equipment

Salt bath furnaces used for nitriding may be heated by gas, oil or electricity, and are essentially similar in design to salt bath furnaces used for other processes. Although batch installations are most common, semicontinuous and continuous operations are feasible. Generally, the same furnace equipment can be used for other heat treating applications by merely changing the salt. (Further details on specific types of furnaces may be found in the article "Liquid Carburizing", in this volume.)

A variety of materials are used for the pots, electrodes, thermocouple protection tubes, and fixtures employed in salt bath nitriding, depending primarily on the salt mixture and process. For example, low-carbon steel is sometimes used for furnace liners although titanium is recommended for one of the processes (U.S. Patent 3 208 885). Inconel 600 is presently being applied to the noncyanide process described in U.S. Patent 4 019 928. Type 430 stainless steel is recommended for a noncyanide process described in U.S. Patent 4 006 643. Cast HT alloy is a satisfactory fixture material, and type 446 stainless steel has been used for fixtures and thermocouple protection tubes. One plant reports the successful use of Inconel pots in liquid pressure nitriding; the same plant reports also that electrodeposited nickel performs satisfactorily as a stop-off in the liquid pressure bath. In general, however, nickel-bearing materials are not recommended for nitriding salt baths.

Maintenance Schedules

Daily

- Check temperature-measuring instruments.
- Check flowmeters, if these are required for air or anhydrous ammonia.
- Check surface condition of work for desired steel-gray color and possible pitting.
- Check case depth and case hardness, to determine operating condition of the bath.

Weekly

- Analyze salt bath composition at least once a week; a semiweekly analysis is preferred. Make necessary additions.
- Check air-salt interface on pot for undercutting. Remove salts and recharge whenever undercutting is observed.
- Check bath for nickel content. To remove traces of nickel, a steel plateout panel should be placed in the bath overnight.
- Contamination in the form of $Na_4Fe(CN)_6$ (a complex ferrocyanide that forms in cyanide-type baths) should be removed from the bath by holding the bath at 650 °C (1200 °F) for about 2 h to settle out the compound in the form of sludge.

Safety Precautions

- Operating personnel must be carefully instructed in handling the poisonous cyanide-containing salts.
- All chemical containers must be clearly marked to indicate contents.
- Personnel should be provided with facilities for washing their hands thoroughly, to prevent contamination by the cyanide salts.
- Shields, gloves, aprons and eye protection should be worn by operating personnel.
- Parts should be preheated to drive off any moisture that may be present, before they are immersed in the molten salt bath.
- Proper venting of furnace and rinse tanks to the outdoors is recommended, to provide safety against fumes and spattering and to minimize corrosion in the work area.
- *Nitrate-nitrite salts must not come in contact with nitriding salts in the molten state. Contact will result in an explosion.*

Liquid Nitrocarburizing

In liquid nitrocarburizing processes, both carbon and nitrogen are absorbed into the surface. High-cyanide nitrocarburizing baths have been in use since the late 1940s. Initially, the sulfur-containing variant was used to produce a wear-resistant surface of iron sulfide (Process 2). A sulfur-free high-cyanide bath was developed in the mid 1950s, now known as "aerated bath nitriding" (Process 1). This process and a low-cyanide variant of it (Process 4) now have extensive industrial use throughout the world.

Both Processes 1 and 2 are similar in that components are typically preheated to about 350 to 480 °C (660 to 900 °F), and then transferred to the nitrocarburizing salt bath at 570 °C (1060 °F). The major components of the baths for both processes are normally alkali metal cyanide and cyanate. Salts are predominately potassium, with sodium.

Process 1—High Cyanide Without Sulfur. At the treatment temperature, 570 °C (1060 °F) the process is controlled largely by two reactions—an oxidation reaction and a catalytic reaction. The oxidation reaction involves transformation of cyanide to cyanate:

$$4 \, NaCN + 2O_2 \longrightarrow 4 \, NaCNO \quad \text{(Eq 1)}$$

and

$$2 \, KCN + O_2 \longrightarrow 2 \, KCNO \quad \text{(Eq 1a)}$$

Though this reaction can proceed by natural oxidation of the cyanide bath, eventually leading to the desired cyanate content, the mechanism is largely uncontrollable. To provide agitation and stimulate chemical activity, therefore, dry air is introduced into the bath.

The catalytic reaction involves breaking down cyanate in the presence of the steel components being treated, thus supplying carbon and nitrogen to the surface:

$$8 \, NaCNO \longrightarrow 2 \, Na_2CO_3 + 4 \, NaCN + CO_2 + (C)_{Fe} + 4 \, (N)_{Fe} \quad \text{(Eq 2)}$$

and

$$8 \, KCNO \longrightarrow 2K_2CO_3 + 4 \, KCN + CO_2 + (C)_{Fe} + 4(N)_{Fe} \quad \text{(Eq 2a)}$$

As a result of this treatment, a wear-resistant compound zone, rich in nitrogen and carbon, is formed on component surfaces (Fig. 11).

Process 2—High Cyanide With Sulfur. The same basic oxidation and catalytic reactions of Process 1 also occur in this process. In addition, further reactions take place because of sulfites in the melt. These sulfites are reduced to sulfides, in conjunction with the oxidation of the cyanide to cyanate, as follows:

$$Na_2SO_3 + 3 \, NaCN \longrightarrow Na_2S + 3 \, NaCNO \quad \text{(Eq 3)}$$

and

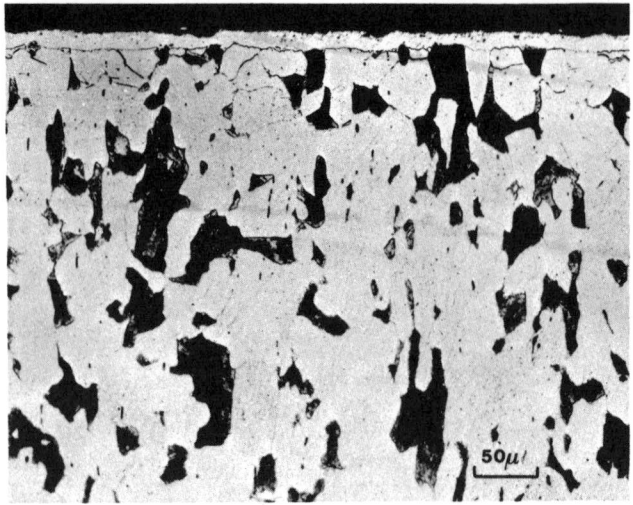

Fig. 11 Metallographic appearance of salt-bath nitrocarburized mild steel after 1.5 h at 570 °C (1060 °F) followed by water quenching

$$K_2SO_3 + 3\ KCN \rightarrow K_2S + 3\ KCNO$$
$$(Eq\ 3a)$$

Thus, the sulfur present in the bath acts as an accelerator, with the result that the cyanate is produced more readily than if the sulfur compounds were absent. Consequently, external aeration is not normally used in the process. Potassium and sodium cyanates produced by the reactions in Eq 1 and 3 catalytically decompose at the surface of ferrous materials to liberate carbon monoxide and nascent nitrogen. The carbon monoxide dissociates to liberate active carbon. The carbon, in conjunction with the nascent nitrogen, diffuses into the material being treated to form the compound zone.

The exact mechanism by which sulfur is impregnated into the material is not clear. Various sulfides react with the component being treated to form iron sulfide; this is the black deposit observed on the surface of components after treatment.

The compound layer formed on mild steel after a 90 min treatment, followed by water quenching, is shown in Fig. 12. The compound layer formed by cyanide salt-bath nitrocarburizing treatments, and, in particular, by the sulfur-containing high-cyanide process, contains an outer region of microporosity. These pores, which readily absorb oil, may assist the antiscuffing

properties of treated components under lubrication conditions.

Although little systematic investigation has been done to establish the optimum thickness of the compound layer for maximum improvement in wear and antiscuffing properties, it is believed that comparable results are obtained provided the layer is 10 to 20 μm (0.0004 to 0.0008 in.) thick.

Composition and Structural Analysis of the Compound Layer. X-ray diffraction investigations into the structure of the compound layer formed by the two high-cyanide salt-bath nitrocarburizing processes have indicated a variety of carbon and nitrogen-base phases.

One study of cyanide nitrocarburizing treatments indicated that the best antiscuffing properties were obtained when the compound layer consisted mainly of a close-packed hexagonal phase of variable carbon and nitrogen concentration. Examination of the appropriate isothermal section of the Fe-C-N ternary phase diagram (Fig. 13) indicates that this phase is the epsilon carbonitride phase. Furthermore, it is believed that provided the epsilon phase was predominant within the compound layer, small amounts of other phases, particularly Fe_4N and Fe_3C, had no serious adverse effects on antiscuffing behavior. It has been shown that with Process 1, compound layers

with less than about 2% carbon and less than about 6% nitrogen contained a mixture of the epsilon iron carbonitride and Fe_4N. With processing times in excess of 3 h, the proportion of Fe_4N was found to decrease. Furthermore, when more than 2% carbon was in the compound layer, a compound with the structure of cementite, $Fe_3(CN)$, could also be detected.

In samples treated by Process 1, a high level of oxygen within the compound layer has been reported. But whether the presence of oxygen, which is known to accelerate the formation of the compound layer by promoting the cyanide-to-cyanate oxidation reaction, is essential for improved frictional properties has not been rigorously established.

Similarly, the question arises as to whether or not sulfur, present in Process 2, contributes significantly to enhanced antiscuffing properties. The predominant presence of an epsilon carbonitride phase is required for enhanced antiscuffing properties. Electron probe microanalysis of the compound layers formed by the two processes are presented in Fig. 14.

Nontoxic Salt-Bath Nitrocarburizing Treatments

Environmental considerations and the increased cost of detoxification of cyanide-containing effluents have led to development of low-cyanide salt-bath nitrocarburizing treatments.

Cyanates are the active nitriding constituent of both high-cyanide and low-cyanide nitrocarburizing baths. Reduction of the cyanide content permits markedly higher cyanate concentrations in the low-cyanide baths; this results in greatly increased nitriding activity. Unlike the reducing high-cyanide baths, the nominal cyanate and carbonate composition of the low-cyanide baths is oxidizing. The baths are composed of primarily potassium salts with some sodium salts. During nitriding, cyanates yield nitrogen to the steel and form carbonates. Cyanate concentration is maintained by the use of organic regenerators, which supply nitrogen to reform cyanates from carbonates.

Process 3—Low Cyanide, With Sulfur. This patented process confers sulfur, nitrogen, and presumably, car-

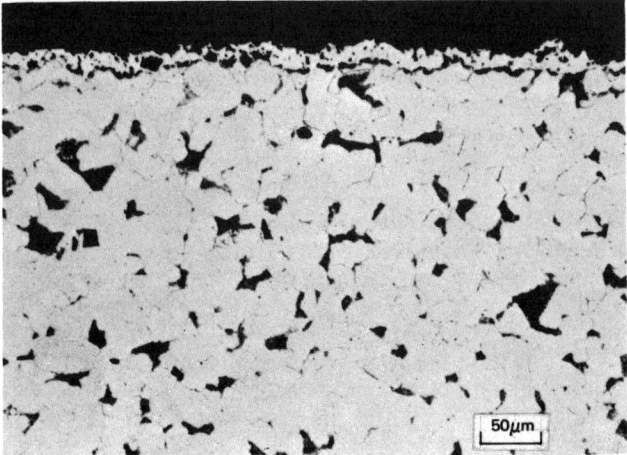

Fig. 12 Metallographic appearance of mild steel after similar treatment to Fig. 11

Iron-sulfide inclusions in the outer region of the compound zone is apparent after this treatment, in which sulfur acts as an accelerator.

50μm

Fig. 13 Phase diagram at 575 °C (1065 °F) of the ternary Fe-C-N system

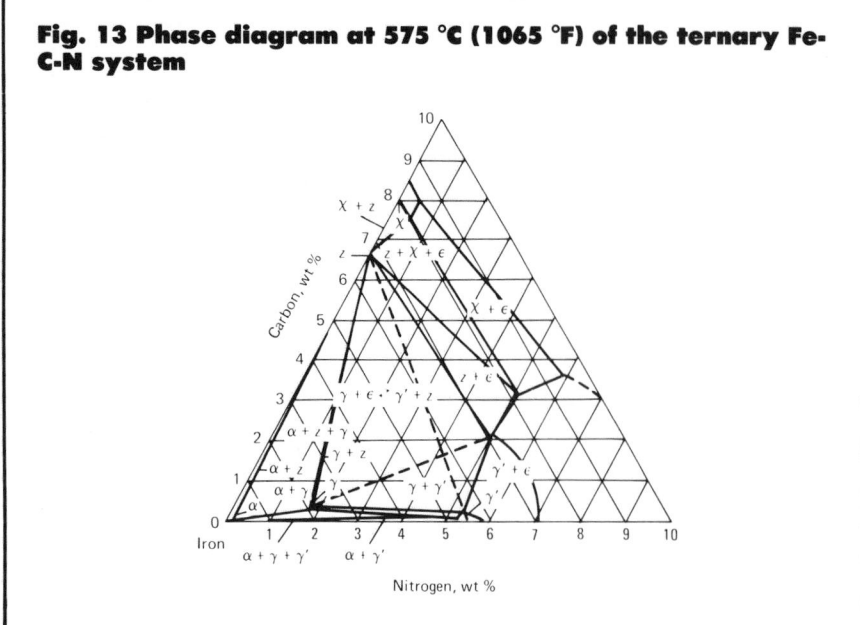

Fig. 15 Sample of plain carbon steel after low-cyanide salt-bath nitrocarburizing treatment (Process 3)

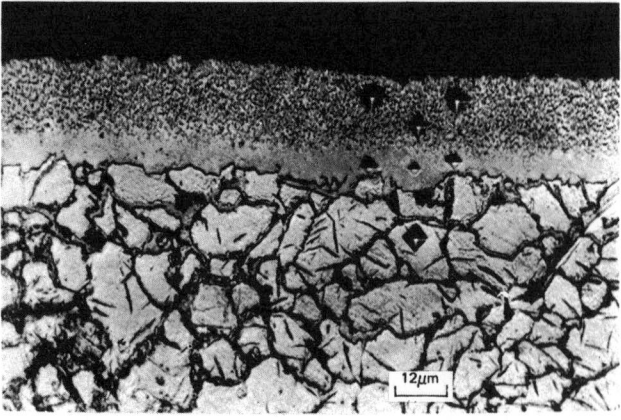

The high level of apparent porosity is a characteristic of high sulfur content in the compound zone; dark areas are actually iron-sulfide nodules, not voids.

Fig. 14 Electron microprobe traces

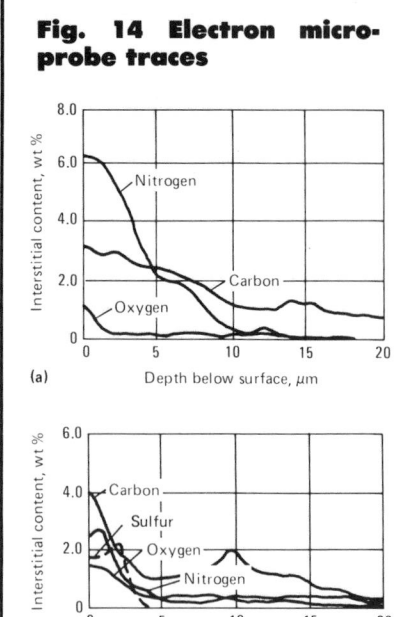

(a) Nitrogen, carbon, and oxygen in the compound layer formed by Process 1. (b) Nitrogen, carbon, oxygen, and sulfur in the compound layer formed by Process 2. Both treatments, 90 min

bon and oxygen to surfaces of ferrous materials. The process is unique in that lithium salts are incorporated in the bath composition. Cyanide is held to very low levels: 0.1 to 0.5%. Sulfur species, present in the bath at concentrations of 2 to 10 ppm, cause sulfidation to occur simultaneously with nitriding. Sulfur levels near 10 ppm result in an apparently porous compound zone (Fig. 15); the dark areas are actually iron sulfide nodules, not voids. This compound zone is similar to the high-cyanide, sulfur-containing nitrocar-

burizing process which has, however, columnar iron-sulfide inclusions.

Bath composition can be adjusted to lower sulfur levels (2 ppm) to form a less porous layer with a lower iron sulfide content.

A compound layer 20 to 25 μm (0.0008 to 0.001 in.) thick forms in 90 min at 570 °C (1060 °F) on AISI 1010 steel, compared with the 8 to 10 μm (0.0003 to 0.0008 in.) layer formed by the high-cyanide sulfur-bearing nitrocarburizing process in the same time. Figure 16 shows the thickness of the

compound layer as a function of the treatment time for the nontoxic and cyanide-base treatments.

Process 4—Low Cyanide, Without Sulfur. A low-cyanide alternative to the cyanide-based Process 2 has been developed. This process, like Process 3, is a cyanate bath with no lithium or sulfur compounds and very low cyanide levels (2 to 3%). Melon, an organic polymer, is used for bath regeneration.

When water quenching is employed, the low level of cyanide permits easier detoxification. Alternatively, quenching into a caustic-nitrate salt bath at 260 to 425 °C (500 to 795 °F) may be used for cyanide/cyanate destruction.

Processing temperature for Process 4 is 570 to 580 °C (1060 to 1080 °F); the rate of compound zone formation is comparable to that of Process 3. Metallurgical results are virtually identical with the cyanide-based Process 2.

Wear and Antiscuffing Characteristics of the Compound Zone Produced in Salt Baths

The resistance to scuffing after salt-bath nitrocarburizing treatments has

Fig. 16 Comparison of compound zone thickness produced by low-cyanide and cyanide-base treatments containing sulfur

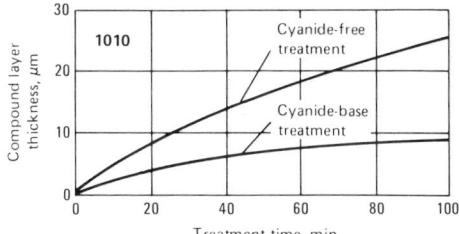

Fig. 17 Faville-Levally Falex lubricant testing machine

been frequently tested with a Faville-Levally, Falex lubricant testing machine (Fig. 17). A 32 by 6.4 mm (1.25 by 0.25 in.) test piece is attached to the main drive shaft by means of a shear pin, and two anvils or jaws having a 90° V-notch fit into holes in the lever arms. During testing, the jaws are clamped around the test piece, which rotates at 290 rpm, and the load exerted by the jaws is gradually increased. Both test pieces and jaws can be immersed totally in a small tank containing lubricant or other fluid, or tests can be carried out dry.

Table 4 lists results of a few representative Falex tests for plain low-carbon steels both before and after cyanide salt-bath nitrocarburizing treatments. The untreated low-carbon steel specimens do not show any significant scuffing resistance even when tested under oil lubricated conditions. After treatment, however, even when tested dry, there is a considerable improvement in antiscuffing properties. Specimens tested in the dry condition after salt-bath nitrocarburizing generate so much heat that they eventually become red hot, and are extruded under the applied load. Untreated test pieces seize at relatively low loads before becoming red hot, whereas treated samples, even after extrusion, show no signs of scuffing (Fig. 18a). During testing in oil, the specimens become highly polished (Fig. 18b). Similar Falex test results are reported for low-cyanide salt-bath nitrocarburizing treatments.

Appendix A*

Liquid Salt Bath Nitriding

Nitriding salts shall consist of a

*Adapted from AMS 2755B (1974).

mixture of sodium and potassium cyanide and other salts.

Salt Bath. The cyanate, cyanide and iron contents of the bath shall be controlled within the following percentages by weight:

Table 4 Results of wear testing

Condition of test pieces and jaws	Testing Medium(a)	Applied load		Condition of test pieces	Material
		kg	lb		
Untreated	SAE 30 oil	320	700	Scuffed	En32 (0 to 15% C)
Untreated	Water	270	600	Badly scuffed	En32 (0 to 15% C)
Untreated	Air	320	700	Scuffed	En32 (0 to 15% C)
Untreated	Air	205	450	Scuffed	AISI 1045
Treated (b)	SAE 30 oil	Limit of gage, 1150	Limit of gage, 2500	No scuffing	En32 (0 to 15% C)
Treated (b)	Water	450	1000	Scuffed	En32 (0 to 15% C)
Treated (b)	Air	760	1675	No scuffing, became hot and extruded	En32 (0 to 15% C)
Treated (c)	Air	660	1450	Extruded	AISI 1045

(a) Falex scuffing tests at 290 rpm in EN8 (0.4% C) jaws, 90 min running time. (b) Treatment 2, cyanide nitrocarburizing salt bath, with sulfur present as an accelerator. (c) Treatment 1, cyanide nitrocarburizing salt bath

	%	
	min	max
Cyanate determined as KNCO	42–50	
Cyanide determined as KCN	45–50	
Iron determined as $Na_4Fe(CN)_6$	0.20	

Fig. 18 Falex test pieces

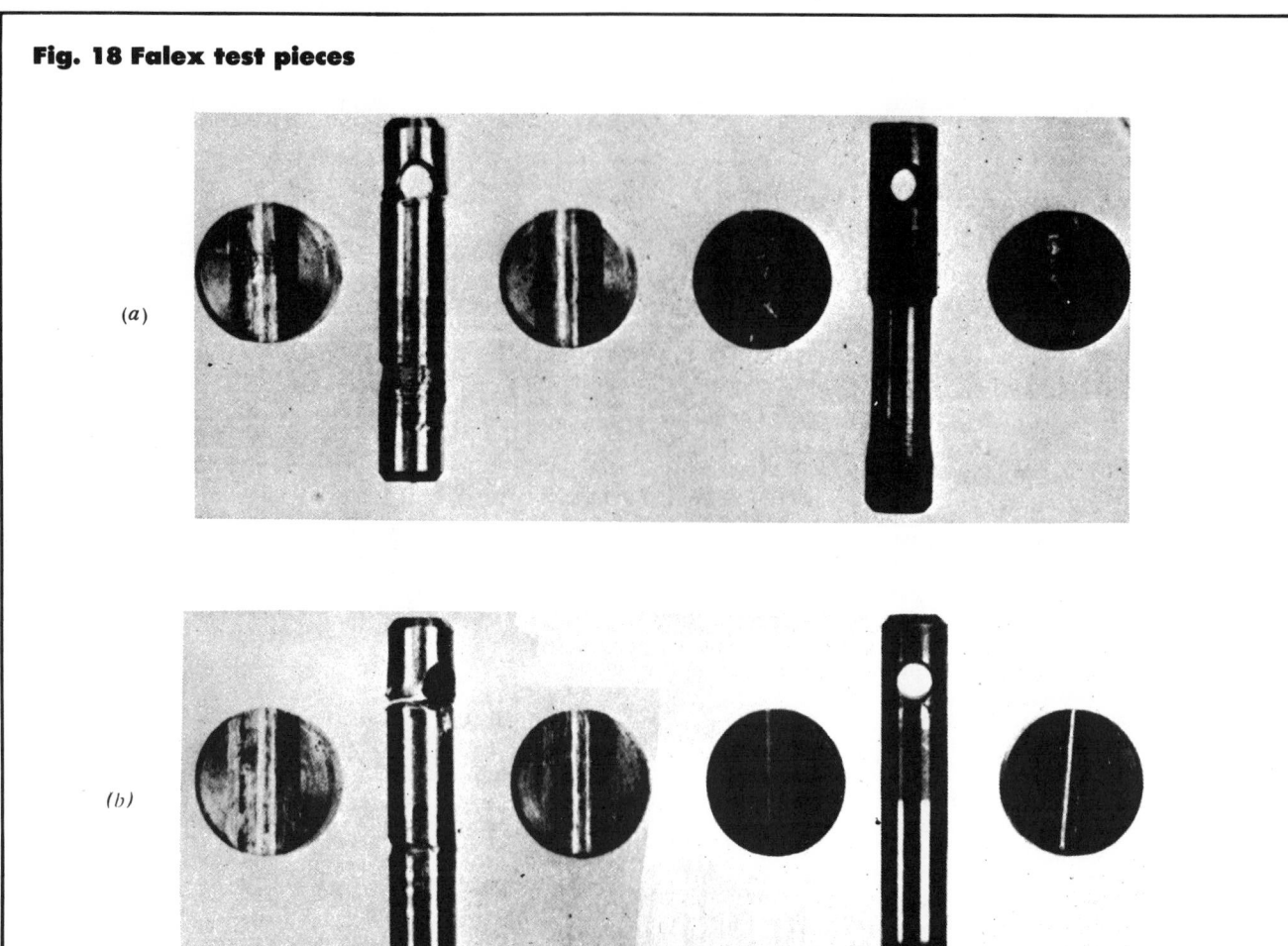

(a) Untreated and salt-bath nitrocarburized mild steel, tested dry. (b) Untreated and salt-bath nitrocarburized mild steel, tested in oil

Nitriding. Parts shall be immersed in an aerated cyanide-cyanate bath as indicated in Table 5.

Quenching. Following treatment, the parts shall be quenched in water, oil, soluble oil solution, or air. Parts, except those made of air-hardening tool steels, may be cooled to 285 to 400 °C (550 to 750 °F) prior to actual quenching, when permitted by the purchaser.

Depth of case shall be determined in accordance with SAE J423 (microscopic method) at 500X magnification as indicated in Table 6.

Case Quality. Any surface porosity present shall not extend deeper than one-half the observed depth of the compound layer, determined by examining specimens metallographically at 500X magnification.

Case hardness shall be determined by microhardness measurements in accordance with ASTM E384 on the nitrided surface or on metallographically prepared cross sections of the nitrided case using Vickers, Knoop, or other appropriate hardness tester, as agreed upon by purchaser and vendor:

Table 5 Recommended nitriding procedures

Material	Recommended time min	max	Temperature °C	°F
Carbon and low-alloy steels ...... 1 h		2 h	570 ± 6	1060 ± 10
Tool and die steels (structural).................... 30 min		3 h	535–570	1000–1060
Tool steels (cutting) .... 5 min		1 h	535–570	1000–1060
Corrosion and heat resistant steels........................ 1 h		2 h	570 ± 6	1060 ± 10
Ductile, malleable and gray cast iron...................... 1 h		4 h	570 ± 6	1060 ± 10
Powder metal products (ferrous) .... 30 min		2 h	570 ± 6	1060 ± 10

	Hardness, min (HK200)
Plain carbon steels	300
Low-alloy steels	450
Tool and die steels	700
Corrosion and heat resistant steels	900
Ductile, malleable, and gray cast iron	600
Powder metal products (ferrous)	600

Appendix B*

Liquid Salt Bath Nitriding-Noncyanide Baths

Hardening. Parts requiring core hardness shall be heat treated to the required core hardness before processing. Tempering to produce the specified core hardness shall be at a temperature not lower than 590 °C (1090 °F) except when tempering is conducted in conjunction with nitriding.

Stress Relief. Parts in which residual stresses may cause cracking or excessive distortion because of thermal shock or dimensional change because of metallurgical transformations during nitriding shall be stress relieved prior to final machining. Stress relieving shall be performed at a temperature not lower than 590 °C (1090 °F).

Cleaning. Parts, at the time of nitriding, shall be clean and free of scale or oxide, entrapped sand, core material, metal particles, oil, and grease and shall be completely dry.

Preheating. Parts shall be preheated in air at 260 to 345 °C (500 to 650 °F) to maintain bath temperature and to avoid thermal shock upon immersion in the nitriding salt.

Nitriding. Parts shall be immersed in an aerated cyanate bath as indicated in Table 7.

Quenching. Following treatment, parts shall be quenched in fused salts, water, oil, soluble oil solution, or air. Parts, except those made of air-hardening tool steels, may be cooled to 290 to 400 °C (550 to 750 °F) prior to actual quenching, when permitted by the purchaser.

Depth of compound layer shall be determined in accordance with SAE J423, microscopic method, at 500X magnification as indicated in Table 8.

Quality of Compound Layer. Any continuous surface porosity present

*Adapted from AMS 2753 (1978).

shall not extend deeper than one-half the observed depth of the compound layer, determined by examining specimens metallographically at 500X magnification.

Hardness of compound layer shall be determined by microhardness measurements in accordance with ASTM E384 on the nitrided surface or on metallographically prepared cross-sections of the nitrided case using Knoop or other appropriate hardness tester, as agreed upon by purchaser and vendor as follows:

Table 6 Depth of case measurements

	Case depth			
	mm		in.	
	min	max	min	max
Plain carbon and low-alloy steels	0.0038	0.03	0.00015	0.001
Tool and die steels (structural)	0.003	0.013	0.0001	0.0005
Tool and die steels (cutting)	0.003		0.0001	
Corrosion and heat resistant steels	0.0038	0.03	0.00015	0.001
Ductile, malleable, and gray cast iron	0.0038	0.03	0.00015	0.001
Powder metal products (ferrous)	0.0038	0.03	0.00015	0.001

Material	Hardness, min (HK/200)
Carbon steels	300
Low-alloy steels	450
Tool and die steels	700
Corrosion and heat resistant steels	900
Ductile, malleable, and gray cast iron	600
Powder metal products (ferrous)	600

Table 7 Recommended nitriding procedures

Material	Recommended time		Temperature	
	min	max	°C	°F
Carbon and low-alloy steels	1 h	2 h	580 ± 5	1075 ± 10
Tool and die steels (structural)	30 min	3 h	540–580	1000–1075
Tool steels (cutting)	5 min	1 h	540–580	1000–1075
Corrosion and heat resistant steels	1 h	2 h	580 ± 5	1075 ± 10
Ductile, malleable, and gray cast iron	1 h	4 h	580 ± 5	1075 ± 10
Powder metal products (ferrous)	30 min	2 h	580 ± 5	1075 ± 10

Table 8 Depth of compound layer

	Case depth			
	mm		in.	
Material	min	max	min	max
Carbon and low-alloy steels	0.0038	0.03	0.00015	0.001
Tool and die steels (structural)	0.003	0.013	0.0001	0.0005
Tool and die steels (cutting)	0.003		0.0001	
Corrosion and heat resistant steels	0.0038	0.03	0.00015	0.001
Ductile, malleable, and gray cast iron	0.0038	0.03	0.00015	0.001
Powder metal products (ferrous)	0.0038	0.03	0.00015	0.001

Gaseous Ferritic Nitrocarburizing

By T. Bell
Hanson Professor of
  Industrial Metallurgy
Department of Metallurgy &
  Materials
University of Birmingham
and the ASM Committee on
  Nitrocarburizing*

FERRITIC NITROCARBURIZING PROCESSES are thermochemical treatments that involve diffusional addition of both nitrogen and carbon to the surface of ferrous materials at temperatures completely within the ferrite-phase field. Cycle times are usually less than 3 h; these processes are termed "short-cycle" nitriding. The primary object of such treatments is usually to improve antiscuffing characteristics of ferrous engineering components by providing the surface with a compound layer—really, a surface zone—exhibiting good wear/friction-resistant properties. In addition, fatigue characteristics can be considerably improved, particularly when nitrogen is retained in solid solution in the "diffusion zone" beneath the compound layer. This retention normally is achieved by quenching in oil or water from the treatment temperature. Corrosion resistance provided by the compound zone is an important secondary benefit. Some distinction should be drawn between two different

processes with similar names—ferritic nitrocarburizing and carbonitriding. Carbonitriding is performed with the steel in an austenitic phase, above 760 °C (1400 °F). Ferritic nitrocarburizing is performed in the ferritic range, below 675 °C (1250 °F).

Nitrocarburizing is used on a wide range of engineering components, such as rocker-arm spacers, textile machinery gears, pump cylinder blocks and jet nozzles, which are treated for wear resistance. Components such as crankshafts and drive shafts are treated to improve fatigue properties.

Ferritic nitrocarburizing has been successfully applied to most ferrous materials, including wrought and sintered plain carbon and alloy steels, stainless steels and cast irons. The most marked improvement in both antiscuffing and fatigue properties, relative to untreated material, is found with plain low-carbon steels. Consequently, these materials form the pri-

mary basis for discussion throughout this article.

Early nitrocarburizing processes used molten cyanide-base salts to confer property improvements. Concern about over-all environmental aspects of heat treat processing with cyanide-base salts created intense interest in development of cyanide-free nitrocarburizing treatments as technically and economically viable alternatives to cyanide-base processes. Detailed metallurgical studies of materials treated by the cyanide nitrocarburizing processes led to the introduction of a variety of gaseous, vacuum and cyanide-free salt bath nitrocarburizing treatments. Many of these processes are described in other articles. This article concentrates on gaseous ferritic nitrocarburizing.

Preliminary Treatments. The surface to be nitrocarburized must be free of contaminants such as oxides, scales, oil and decarburization, if opti-

*Kenneth D. Gladden, Senior Development Engineer, Caterpillar Tractor Co.; Donald N. Guy, Project Manager, Technology Center, Lindberg Heat Treating Co.

Fig. 1 Typical furnace for gaseous nitrocarburizing

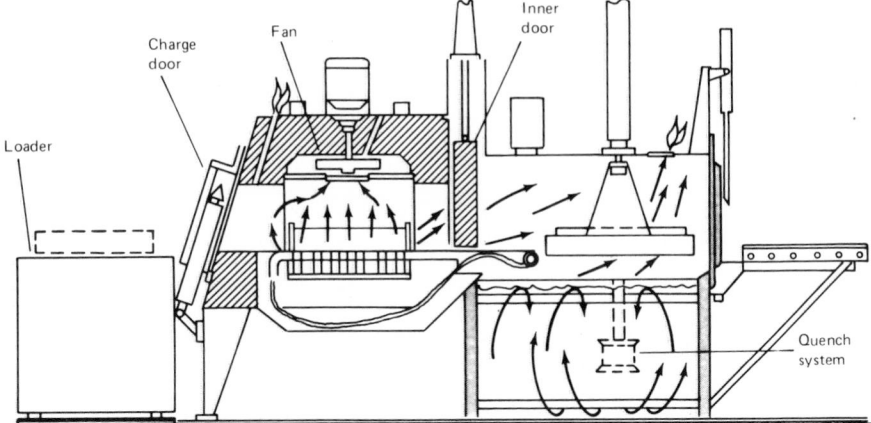

Fig. 2 AISI 1015 material after 3 h of gaseous nitrocarburizing in an ammonia/endothermic gas mixture at 570 °C (1060 °F) followed by oil quenching

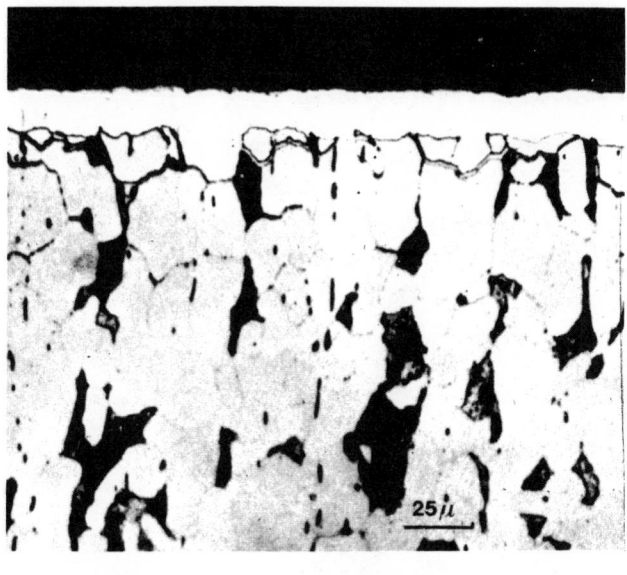

25 μ

ation of core properties during the nitrocarburizing process.

Gaseous Nitrocarburizing

Gaseous nitrocarburizing commonly employs sealed-quench batch furnaces of the same design used for carburizing and carbonitriding (Fig. 1). Furnace operating temperatures are low enough to maintain steels in the ferritic condition. The atmosphere employed consists of ammonia diluted with a carrier gas. In one process, the atmosphere is formed from equal amounts of ammonia and endothermic gas, American Gas Association (AGA) type 302. In another process, a typical atmosphere consists of 35% ammonia and 65% refined exothermic gas (AGA type 201, nominally 97% nitrogen), which may be enriched with a hydrocarbon gas. High-purity nitrogen is used as a diluent in a variant of this process. Gaseous nitrocarburizing is performed near 570 °C (1060 °F), a temperature just below the austenite range for the Fe-N system. Treatment times generally range from 1 to 5 h.

The properties produced by gaseous nitrocarburizing are similar to those produced by salt bath nitrocarburizing, and the process can be applied to most ferrous alloys including carbon steels, cast iron, stainless steels and tool steels. An advantage of the gaseous nitrocarburizing processes over salt bath processes is the elimination of environmental problems associated with the handling and disposal of toxic cyanide waste salts. Also, the annoyance of salt entrapment in holes is avoided.

The objective of the process is to produce a thin layer of iron carbonitride and nitrides, the "white layer" or compound zone, with an underlying diffusion zone containing dissolved nitrogen and iron (or alloy) nitrides. The white layer enhances surface resistance to galling, corrosion and wear. The diffusion zone increases the fatigue endurance limit significantly, especially in carbon and low-alloy steel.

The white layer is composed primarily of the epsilon carbonitride phase, with other nitrides and oxides. The exact composition is a function of the nitride-forming elements in the material and the composition of the atmosphere.

The white layer has a reduced tendency to spall compared to the white

mum results are to be obtained. Vapor degreasing is adequate for most applications. It may be necessary, especially on high chromium materials and cast irons that have been burnished or otherwise highly finished, to grit blast with fine abrasive, to apply a light phosphate coating, or both, before nitrocarburizing. Highly finished surfaces are often associated with superficial metal flow at the surface, with the result that the initiation of nitriding is difficult. Surfaces having very low sur-

face roughness can respond well to nitriding, provided that surface burnishing has been avoided. Finishing techniques with good cutting action are required.

Preliminary heat treatments range from simple stress relieving in order to control distortion to hardening and tempering in order to increase the core strength of the material. Stress relief and tempering temperatures should be at least 25 °C (45 °F) above the nitrocarburizing temperature to prevent alter-

Fig. 3 Electron microprobe traces of nitrogen, carbon, and oxygen in compound layer formed by 3 h of gaseous nitrocarburizing in ammonia/endothermic gas mixture

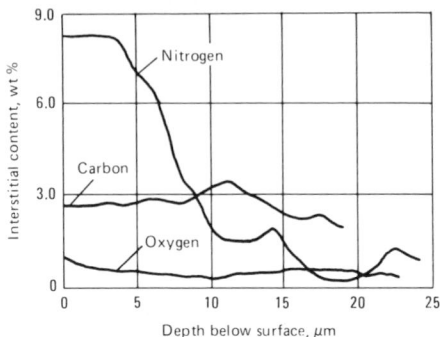

Fig. 4 Modified four-ball wear tests on AISI 1015 material

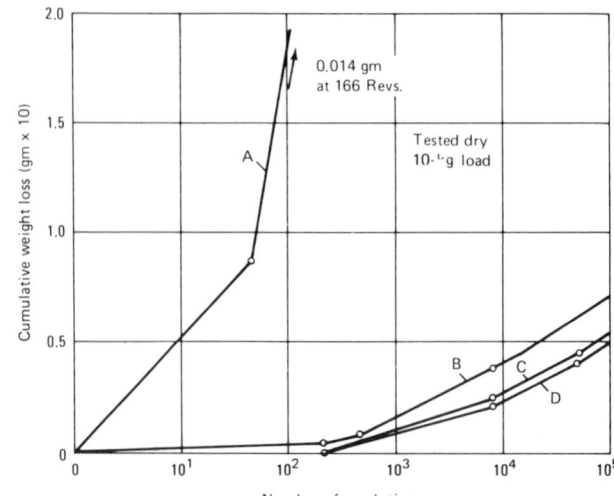

(a) Untreated. (b) Gaseous nitrocarburizing for 3 h, atmosphere cooled. (c) Gaseous nitrocarburizing for 3 h, oil quenched. (d) Gaseous nitrocarburizing for 8 h, oil quenched

layer formed during conventional gas nitriding with ammonia. This has been attributed by most investigators to an essentially single-phase white layer of the less brittle epsilon phase, but is also due in part to the fact that white layers are generally thinner in nitrocarburizing. The atmosphere conditions required to produce the less brittle phases appear to be affected by both carburizing gases such as CO and CH_4 and oxidizing gases such as air, CO_2 and H_2O.

Furnace Condition and Safety Precautions. Batch furnaces with integral oil quenches are ideally suited for performing gas nitrocarburizing; however, the over-all condition of the furnace is somewhat more critical than when the same furnace is used for other heat treating operations such as hardening, carburizing and carbonitriding. The hot chamber temperature should be controllable within ±5 °C (±10 °F) at 570 °C (1060 °F) throughout the entire volume. Thermocouple and instrument systems designed to operate at the higher temperatures of other heat treating processes are not always adequate for lower temperatures. Gas leaks in the furnace and around doors must be minimized. *Safety precautions must be carefully considered and rigorously enforced, because the process involves a combustible atmosphere that is explosive when operated below the self-ignition temperature.* Only minor gas leaks can be allowed, and double pilots should be provided at all doors. An interlock between door operation and pilot function provides added safety. Precautions must be taken to ensure

Fig. 5 Blanked gear shift gates, made from low carbon steel, after gaseous nitrocarburizing

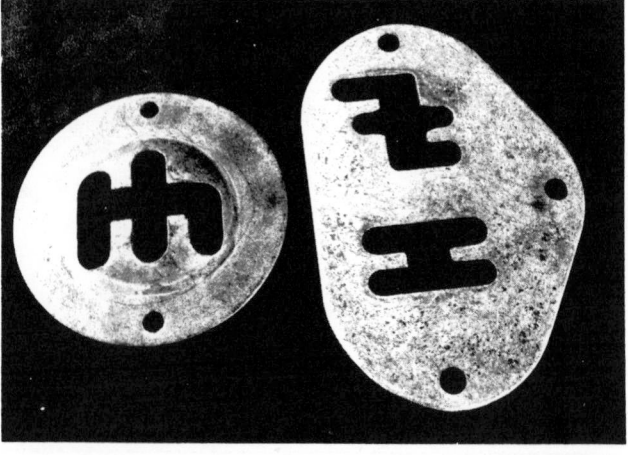

that gas burn-off ports are properly sized, free of clogging, well vented from the building and equipped with dependable pilots. All outside doors must be equipped with flame screens of sufficient capacity to cover the entire door opening.

The usual precautions also must be practiced with the quench oil: (a) adequate extinguisher equipment, (b) assurance that the oil is free from water, and (c) maintenance of adequate temperature controllers and over-tempera-

ture devices to ensure that the oil does not become overheated. Steps also must be taken to ensure that atmosphere flow is sufficient to maintain positive pressure in the furnace during quenching, to prevent the egress of air through burn-off or small leaks. At start-up, atmosphere gas may be introduced by heating the furnace above self-ignition temperature—760 °C (1400 °F)—and introducing atmosphere in the usual manner or by purging all the air from the furnace with nitrogen before introducing the reactive atmosphere. On shutdown, the furnace should be heated above self-ignition temperature of 760 °C (1400 °F) before burning the furnace out, or it should be purged with nitrogen.

Process. The basic consideration behind all gaseous nitrocarburizing processes is the type of atmosphere that can be used to cause carbon and nitrogen to be added simultaneously to the surface of ferrous materials and so produce the desired epsilon phase.

Ammonia, the most readily available source of active nitrogen, catalytically dissociates at 570 °C (1060 °F) on ferrous surfaces according to the following reaction:

$$NH_3 \rightarrow (N)_{Fe} + \frac{3}{2} H_2 \qquad \text{(Eq 1)}$$

The nitrogen in the active condition diffuses into the work being treated. After nucleation of the compound layer, saturation of the epsilon (ϵ) carbonitride can be described by the reaction:

$$NH_3 \rightarrow (N)_{\epsilon} + \frac{3}{2} H_2 \qquad \text{(Eq 2)}$$

The nitriding potential of the atmosphere is given by the expression:

$$\theta_N = \frac{pNH_3}{(pH_2)^{3/2}} \qquad \text{(Eq 3)}$$

where pNH_3 and pH_2 are the partial pressures of ammonia and hydrogen, respectively, in the nitrocarburizing atmosphere. Changes in nitrogen content of the epsilon carbonitride are related not only with changes in the value of θ_N, but also are affected by the activity of nitrogen within the carbonitride phase:

$$\theta_N = \frac{{}^a(N)}{K} \qquad \text{(Eq 4)}$$

where K is the equilibrium constant for the nitriding reaction.

The activity of nitrogen is reduced by incorporating carbon into the lattice. If reduction in activity is minimal, the value of θ_N, the nitriding potential of the atmosphere, can be assumed to dictate the nitrogen content of the epsilon carbonitride phase. However, neither the range of suitable θ_N values nor the optimum value for specific ferrous materials have been established.

A particularly suitable source of carbon for gaseous nitrocarburizing is endothermic gas, the composition of which can be readily adjusted to provide a high degree of flexibility over the control of the carburizing potential. It

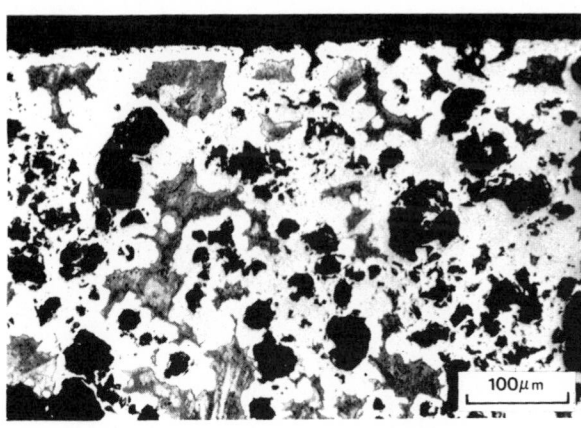

Fig. 6 Metallographic appearance of sintered iron after a 3 h gaseous controlled nitrocarburizing treatment followed by oil quenching

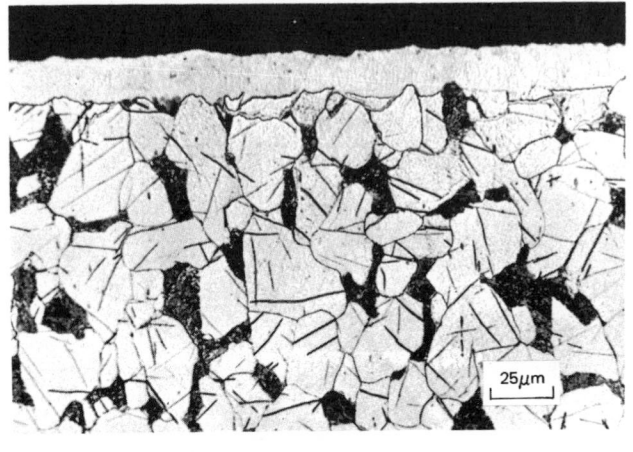

Fig. 7 Structure of mild steel after 2 h of gaseous nitrocarburizing followed by atmosphere cooling

also is widely used in heat treating for gas carburizing operations. Endothermic gas contains sufficient free oxygen to assist the rate of conversion of the epsilon nitride phase into an oxygen-bearing carbonitride structure. The reaction:

$$2 CO \rightarrow (C)_{\epsilon} + CO_2 \qquad \text{(Eq 5)}$$

controls saturation of this epsilon phase. The activity of carbon in the compound layer is given by:

$${}^a(C)_{\epsilon} = \frac{(pCO)^2}{pCO_2} \cdot K_1 \qquad \text{(Eq 6)}$$

where ${}^a(C)_{\epsilon}$ is the activity of carbon in the carbonitride phase, pCO and pCO_2

are the partial pressures of carbon monoxide and carbon dioxide, respectively, and K_1 is the equilibrium constant for the reaction. In the absence of quantitative data on the effect of nitrogen on the activity of carbon, the carburizing potential:

$$\theta_C = \frac{(pCO)^2}{pCO_2} \qquad \text{(Eq 7)}$$

has to be assumed to dictate the carbon content of the epsilon carbonitride phase.

Testing Results. The compound layer formed on AISI 1015 steel by gaseous nitrocarburizing in an ammonia/endothermic gas mixture is shown in

Fig. 8 Needle zone precipitate depth for salt-bath and gaseous nitrocarburized unalloyed low-carbon steel, compared to nitrogen diffusion depth

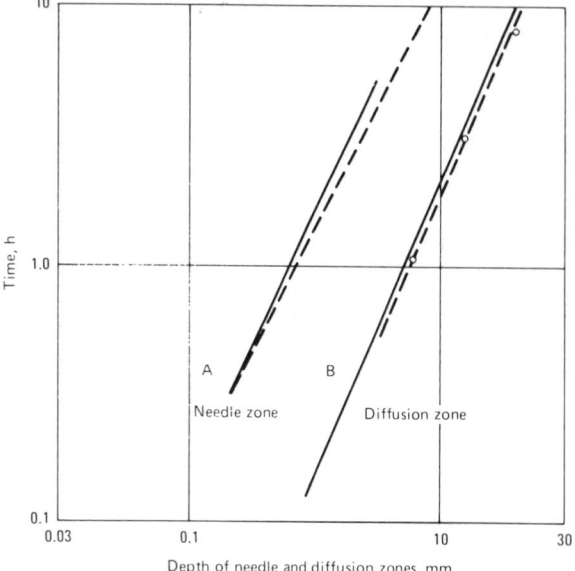

—treatment 1, cyanide-base salt-bath treatment, low-carbon steel; — — — gaseous nitrocarburized AISI 1015; A, optical metallography; B, nitrogen analysis; and o, from microhardness plot AISI 1015.

Fig. 9 Microhardness profiles of the diffusion zone for a series of steels after gaseous nitrocarburizing

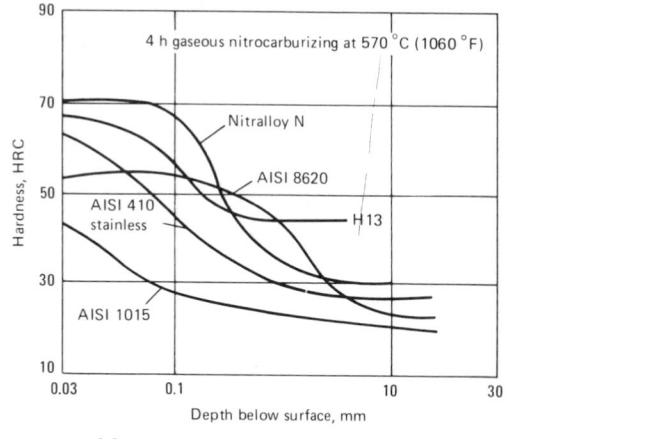

HRC values were converted from microhardness values.

Fig. 2. The layer is somewhat more dense than that formed by salt bath nitrocarburizing. X-ray diffraction analysis has confirmed the predominance of epsilon carbonitride phase in the compound layer, while microprobe analysis (Fig. 3) confirms that carbon and nitrogen contents of the compound layer are essentially within the epsilon phase limits. See the article on liquid nitriding in this volume.

A number of reports clearly demonstrate the improvement in wear resistance and antiseizure properties resulting from gaseous nitrocarburizing treatments involving the use of ammonia and endothermic gas. This improvement in wear characteristics un-der dry running conditions relative to untreated material is illustrated in Fig. 4, which shows results of modified four-ball wear tests on gaseous nitrocarburized AISI 1015 steel after various treatments.

Figure 5 illustrates stamped gear shift gates of low carbon steel, which have been nitrocarburized in an atmosphere consisting of 50% ammonia and 50% endothermic gas to prevent wear and fretting of the unfinished blanked edges.

Controlled Nitrocarburizing. A possible limitation of the process employing 50% ammonia and 50% endothermic gas is that optimum processing conditions for all classes of material, including cast irons, tool steels and stainless steels, are not ensured with a single ammonia/endothermic gas input ratio. A further, and perhaps more serious, limitation is that reproducibility may be impaired with variable loads and from furnace to furnace. These difficulties can be overcome largely by use of an infrared monitoring and control system.

With controlled nitrocarburizing, variable loads and materials can be accommodated, as well as the requirement for reproducible component growth characteristics. Controlled treatment involves processing in an atmosphere with a predetermined nitriding and carburizing potential, and controlling the nitrocarburizing potential within the furnace by infrared gas analysis of the ammonia content. Carburizing potential is controlled by a similar analyzer. An example of sintered iron treated by this process is presented in Fig. 6. The epsilon compound zone is present both on the outermost surface and also on the internal surfaces of the pores.

Vacuum Nitrocarburizing. This process is described in the article on gas nitriding.

Diffusion Zone Characteristics and Fatigue Properties

Diffusion zone characteristics are essentially independent of the type of nitrocarburizing media. During any ferritic nitrocarburizing treatment, only nitrogen diffuses inward from the carbonitride compound zone because the ferrite is normally already at its equilibrium concentration with respect to carbon. The diffusion zone beneath the compound layer on oil-quenched

Fig. 10 Increase in endurance limit of unnotched test pieces

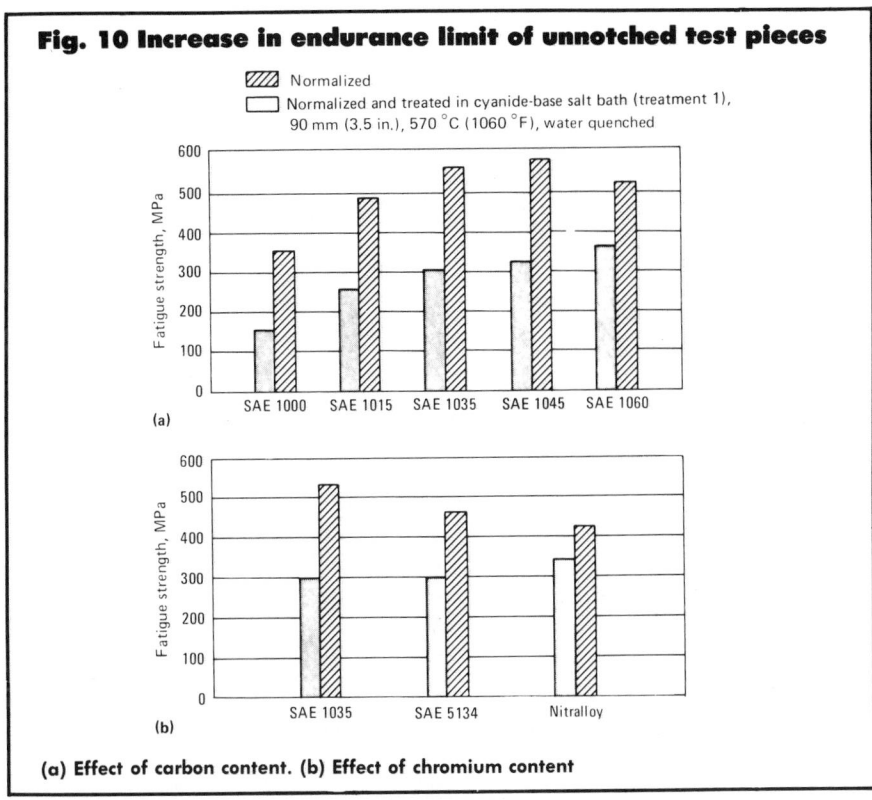

Normalized

Normalized and treated in cyanide-base salt bath (treatment 1), 90 mm (3.5 in.), 570 °C (1060 °F), water quenched

(a)

(b)

(a) Effect of carbon content. (b) Effect of chromium content

Table 1 Endurance limits of nitrocarburized steel

| Section size | | Endurance limit after: | | | | | |
| | | Salt bath treatment(a) | | Gaseous treatment(b) | | Vacuum treatment(c) | |
mm	in.	MPa	ksi	MPa	ksi	MPa	ksi
5	0.2	480	70	490	71	480	70
8	0.3	450	65	435	63	455	66
13	0.5	400	58	340	49	355	51

Note: Tests were run at 10^7 cycles on AISI 1015 steel specimens nitrocarburized at 570 °C (1060 °F) for 2 h. (a) Salt bath treatment, water quench. (b) Gaseous treatment in endothermic ammonia gas mixture, oil quench. (c) Subatmospheric pressure treatment, methane/ammonia gas mixture, oil quench

The depth of the diffusion zone lessens as the level of nitride-forming elements in the material increases, because of a drop in the rate of diffusion of nitrogen in the parent lattice. At the same time, however, the hardness of the diffusion zone rises as the alloy content increases. This rise is due to submicroscopic precipitation of alloy nitrides in a manner identical to conventional gas nitriding. These features are illustrated in Fig. 9, for a series of steels that have received a 4 h gaseous nitrocarburizing treatment at 570 °C (1060 °F) followed by oil quenching.

The improvement in fatigue strength of nitrocarburized materials, as determined with unnotched Wöhler test specimens, depends on the hardness and depth of the diffusion zone. The potential for improvement in fatigue strength lessens with increasing carbon and alloy content (Fig. 10).

As would be expected, salt, gaseous and vacuum treatments all increase fatigue resistance to the same extent, for similar treatment times and section sizes, provided the quench rate is sufficiently rapid to retain the nitrogen in solid solution. Table 1 shows this condition is fulfilled for AISI 1015 in section sizes of up to 8 mm (0.3 in.). Above this size, properties are controlled by the nature of the quenching media, in which instance vacuum nitrocarburizing produces comparable properties to gaseous nitrocarburizing followed by atmosphere cooling. Quench rate has no effect on fatigue properties conferred by alloy nitride precipitates.

To take advantage of the antigalling resistance conferred by the white layer produced by nitrocarburizing, contact stresses cannot be so high as to exceed the yield strength of the metal under the nitride layer. When contact stresses are high, the underlying metal must be strengthened, either by increasing nitrided case depth or by employing another case hardening method, such as carbonitriding. Because wear processes are complex phenomena, tests on actual parts or part assemblies are usually necessary to determine whether case hardening by nitrocarburizing will produce satisfactory results.

samples is indistinguishable from the original matrix material. After any form of nitrocarburizing, however, atmosphere cooling results in formation of needle-like precipitates of Fe_4N nitride within the diffusion zone. Figure 7 illustrates the compound zone and diffusion zone of a gaseous nitrocarburized mild steel sample that has been atmosphere cooled from the treatment temperature. The depth of the needle precipitate zone fails to reflect the true depth of nitrogen penetration. Consequently, a microhardness profile, as a function of depth on a quenched material, gives the best indication of the diffusion zone thickness in the absence of detailed nitrogen layer analysis. In Fig. 8, the needle zone precipitate depth for salt bath and gaseous nitrocarburized unalloyed low-carbon steel is compared with the true nitrogen diffusion distance as measured by detailed chemical analysis and (in gaseous nitrocarburized material) by microhardness. It can be seen that the true depth of the diffusion zone in these low-carbon steels is several times greater than the visible needle zone depth.

Vacuum Carburizing

By the ASM Committee on Gas Carburizing, Carbonitriding and Nitriding*

VACUUM CARBURIZING is a high-temperature gas carburizing process that is carried out at pressures below atmospheric pressure (below 100 kPa, or 760 torr). Vacuum carburizing temperatures typically range from 980 to 1050 °C (1800 to 1925 °F); in some cases, however, the range is extended from about 900 to 1095 °C (1650 to 2000 °F). The furnace atmosphere usually consists solely of an enriching gas such as natural gas, pure methane or propane. Nitrogen is sometimes used as a carrier gas. During the carburizing portion of the cycle, furnace pressures are maintained in the range of about 7 to 55 kPa (50 to 400 torr). The atmosphere may flow through the furnace constantly while a constant pressure is maintained; alternatively, the furnace may be repeatedly backfilled and evacuated.

Vacuum carburizing proceeds by the dissociation of hydrocarbon gas at the steel surfaces and by the direct absorption of carbon; hydrogen gas is liberated. The reaction with methane is

$$CH_4 + Fe \leftrightarrows Fe(C) + 2H_2 \qquad (Eq\ 1)$$

At normal carburizing temperatures, there is a relatively large driving force for the reaction in Eq 1 to proceed from left to right. The reaction favors rapid dissociation of the gas and rapid absorption of carbon by the hot steel sur-

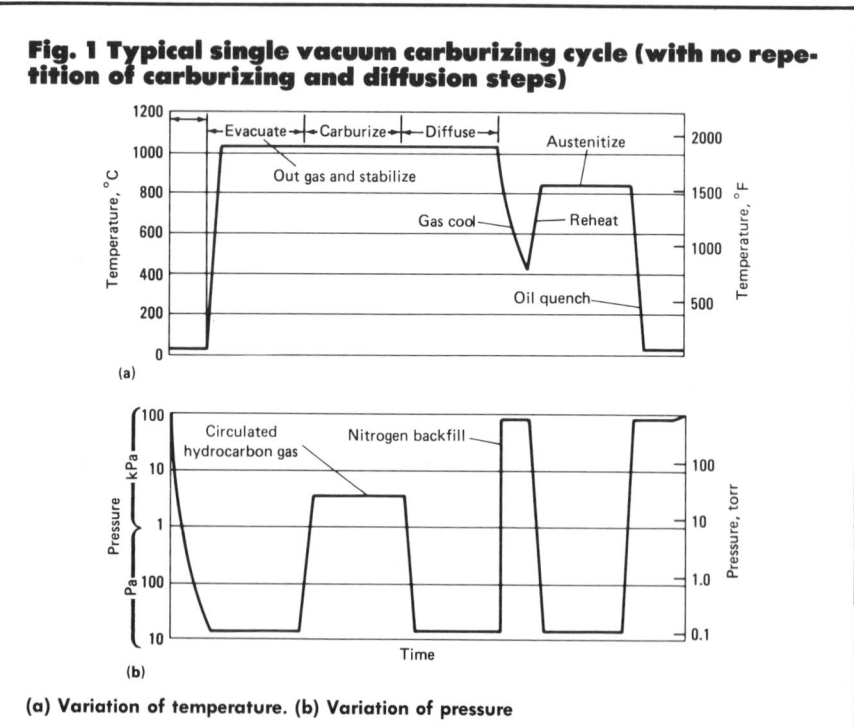

Fig. 1 Typical single vacuum carburizing cycle (with no repetition of carburizing and diffusion steps)

(a) Variation of temperature. (b) Variation of pressure

faces. Because this reaction does not normally reach equilibrium, equilibrium carbon potential control normally is not used in vacuum carburizing. Vacuum carburizing employs the so-called boost-diffuse method, a two-step process. In the first step, carburizing is carried out at saturation. This step is followed by a diffusion step, in which the carbon diffuses into the steel at low pressure, causing the surface carbon to decrease and the case depth to increase.

*Charles A. Stickels, *Chairman,* Principal Staff Engineer, Ford Motor Co.; James G. Conybear, Product Manager, Thermal Systems Technical Center, Midland-Ross Corp.; Larry E. Byrnes, Manager-Manufacturing Research, Federal Mogul Corp.; Jon L. Dossett, Vice President, Metallurgy, U.S. Gear Corp.; Robert L. Hughes, Chief Metallurgist, Fairfield Manufacturing Company, Inc.; Kenneth D. Gladden, Senior Development Engineer, Caterpillar Tractor Co.; Dale A. Poteet, Jr., Division Engineer, John Deere Component Works; Joseph A. Riopelle, District Manager, Chicago Office, Midland-Ross Corp.; Ronald D. Rogers, Chief Engineer, Materials Engineering, Rockwell International

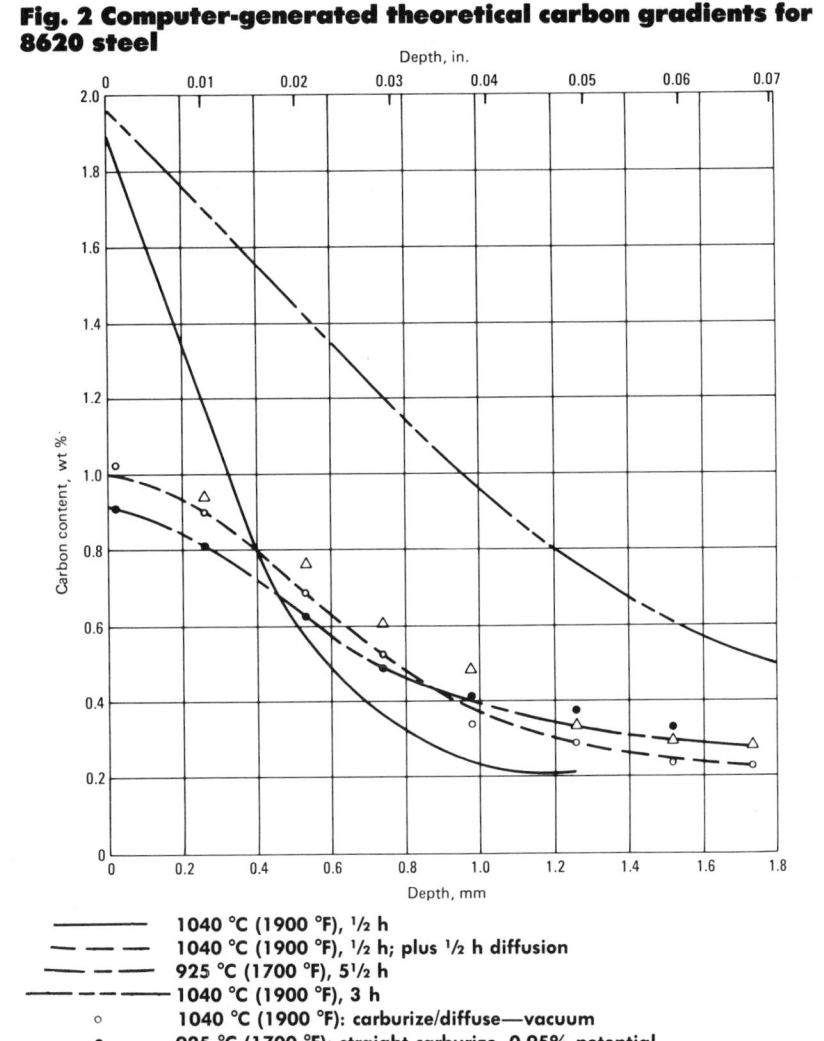

Fig. 2 Computer-generated theoretical carbon gradients for 8620 steel

————————	1040 °C (1900 °F), ½ h
– – – – –	1040 °C (1900 °F), ½ h; plus ½ h diffusion
— – — –	925 °C (1700 °F), 5½ h
– · – · –	1040 °C (1900 °F), 3 h
○	1040 °C (1900 °F): carburize/diffuse—vacuum
●	925 °C (1700 °F): straight carburize, 0.95% potential
△	925 °C (1700 °F): carburize/diffuse—atmosphere

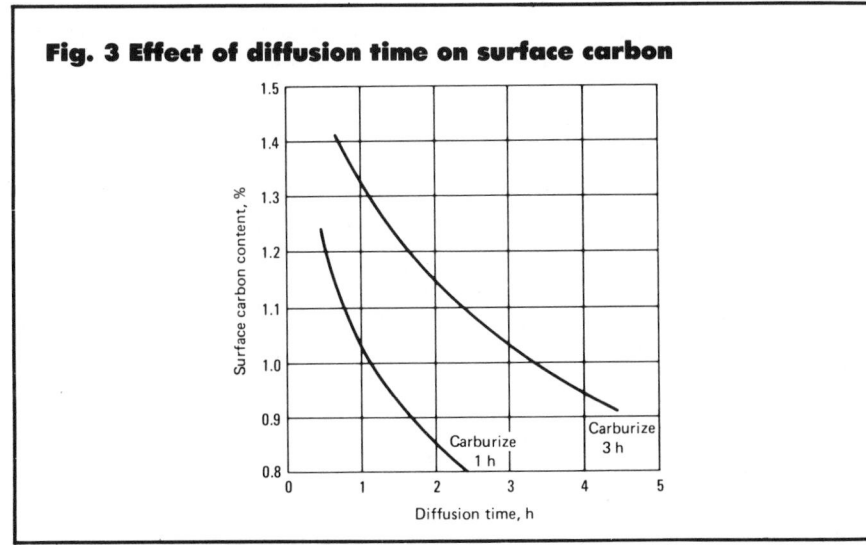

Fig. 3 Effect of diffusion time on surface carbon

The carburizing step is carried out at 7 to 55 kPa (50 to 400 torr), but most commonly at 300 torr. The carburizing gas is used to maintain the pressure. The diffusion step is carried out at normal vacuum pressures of 100 μm Hg or less.

Process Cycles

Process cycles usually start with the loading of workpieces that are relatively free of surface contaminants onto fixtures. The loads are then transferred to the cold furnace. The furnace is evacuated, usually with a mechanical pump, and the parts are heated to carburizing temperature. The combination of vacuum and heat removes most surface contaminants, leaving clean surfaces that can readily absorb carbon. Natural gas is admitted to raise the furnace pressure to about 15 to 55 kPa (100 to 400 torr). After the natural gas atmosphere is circulated for a predetermined and closely controlled length of time, the furnace is evacuated for the diffusion cycle, which is carried out under soft vacuum conditions.

Carbon absorption by the metal occurs very readily after introduction of the natural gas, and the surface carbon content quickly approaches that of saturated austenite. The rate of carbon absorption is a direct function of the partial pressure of the carburizing gas, provided that the temperature of the absorbing surface remains constant. Total carbon added to the steel is controlled by the time the heated surface is exposed to the carburizing gas. Thus, the length of the carburizing step is determined by desired case depth and surface carbon content.

After the carburizing gas has been evacuated, the diffusion cycle is timed to produce the desired surface carbon content and case depth. In certain instances, the carburizing and diffusion steps are alternated one or more times to obtain the desired carbon gradient.

Following diffusion, the carburized parts normally are treated by backfilling the furnace with nitrogen and cooling the work with fans to about 425 °C (800 °F) to refine the grains. The furnace then is immediately evacuated, and the work is reheated to about 845 °C (1550 °F) for austenitizing. The load is subsequently transferred to an internal cooling zone and lowered into an integral oil quench. Direct quenching from the carburizing temperature is seldom practical due to the high tem-

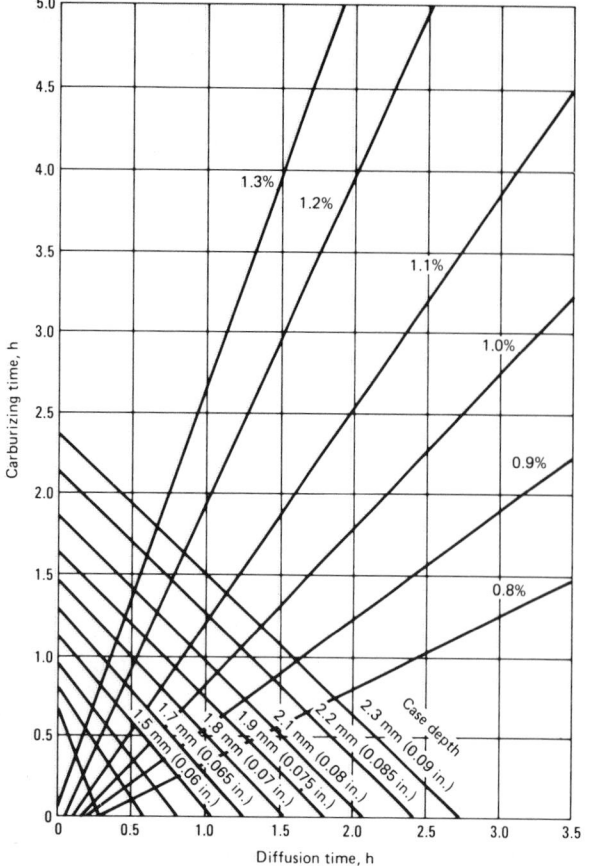

Fig. 4 Carburizing time vs diffusion time for total case and surface carbon at 1040 °C (1900 °F)

can be used to produce case depths to the desired specifications. This relationship is illustrated in Fig. 4.

Carbon Gradient Control

To achieve desired case depth and surface carbon, two factors are very important. The first is an adequate circulation of gas to achieve uniform carbon supply to the workpiece. Because carburizing temperatures above 980 °C (1800 °F) are often used in vacuum carburizing and because the pressure is reduced, the problem of achieving the necessary circulation is more difficult than in atmosphere furnaces. Furnace designs have been developed that use conventional fans, high momentum jets, pulsating gas techniques or combinations of these to overcome this difficulty.

The second factor that must be considered is that the surface carbon at the end of carburizing must be precisely known to predict the final surface carbon after diffusion. It is not difficult to determine surface carbon at high processing temperatures, above 1010 °C (1850 °F), for example, where the surface carbon reaches saturation in less than 5 min. At low temperatures, saturation requires a longer time; at 820 °C (1510 °F), the carbon profile is difficult to predict at times less than about 20 min. This effect is illustrated in Fig. 5 for carbon absorption at 845 °C (1550 °F). In the event such low temperatures are used, compensation must be made in the carburizing-diffusion relationship to account for the slow case buildup in the beginning of the cycle.

The advantage of vacuum processing, however, is that carburizing time can be predicted exactly, because no carburizing gas is admitted to the furnace until the load is uniformly at temperature. The possibility for precise carbon uniformity is illustrated in Fig. 6.

Carburizing Costs

Carburizing costs are affected greatly by carburizing time and furnace payload. As the load capacity increases, or the carburizing time decreases, the cost per pound processed decreases. Vacuum carburizing, because it is a high-temperature process, will be most cost effective in production of parts with deep cases. Because graphite, rather than alloy steel, is used for structural parts within most vacuum furnaces,

perature at which carburizing is performed. If required, parts may be tempered in the vacuum furnace, or the work may be degreased and tempered in an ordinary draw furnace.

When the entire sequence of heat processing is done under vacuum or in an inert, nonoxidizing atmosphere, the work comes out clean and bright. A typical single carburizing-diffusion cycle, with no repetition of the two steps, is illustrated in Fig. 1. Figure 1(a) shows the variation of temperature with time and Fig. 1(b) the variation of furnace pressure with time.

Carbon Gradients

In the first steps of carburizing, the surface carbon is allowed to increase without limit to the saturation value of carbon in austenite for the given temperature. This value and the resultant carbon gradient profile are predictable from classical carburizing theory. Because carbon is not lost to the vacuum

during the diffusion portion, the rate of case buildup and reduction of surface carbon are also predictable. This is illustrated by the carbon gradient curves in Fig. 2.

The final surface carbon achieved after a given time is related to the carbon profile existing in the steel after the carburizing cycle and at the start of the diffusion cycle. Figure 3 shows this effect and the relationship between surface carbon and diffusion time for two carburizing times. Even with a constant diffusion time, the surface carbon will increase as the total case depth, as determined by carburizing time, is increased. For this reason, the vacuum carburizing process differs from an atmosphere process in that the surface carbon-control parameter is time at temperature rather than an atmospheric variable such as CO_2 content or atmospheric dew point.

By combining carburizing time and diffusion time, a matrix can be developed for surface carbon and time, which

Fig. 5 Effect of vacuum carburizing time at 845 °C (1550 °F) on carbon profile for 8620 steel

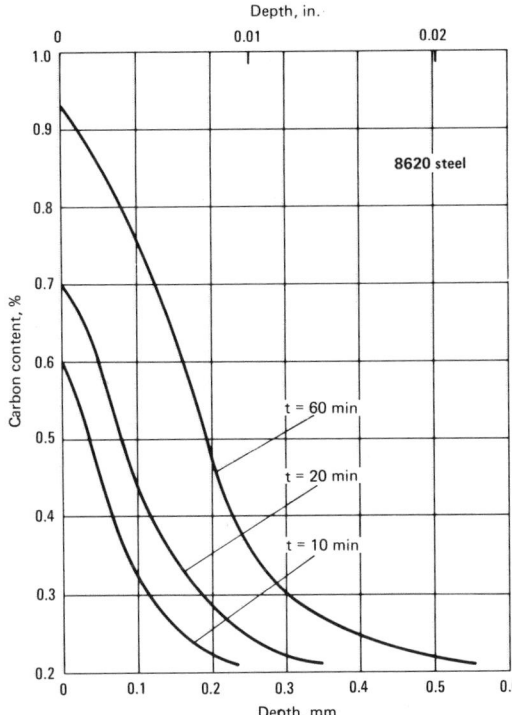

Depth, in.

8620 steel

t = 60 min

t = 20 min

t = 10 min

Carbon content, %

Depth, mm

Carburized only, 300 torr

the ratio of part weight to weight of fixturing is higher than in conventional atmosphere furnaces. Thus, less energy is required to heat the fixturing, and parts may be brought to temperature rapidly.

The operating costs for vacuum carburizing are often comparable to the costs for conventional batch carburizing, depending on local utility rates. The capital costs for vacuum furnaces may be comparable to conventional batch furnaces when the cost of an endothermic gas generator is included. Any such analysis is valid only for the particular furnaces involved, however. An important factor in a cost analysis

is the reduced strength at operating temperatures of the alloy materials used in atmosphere furnaces, with the resulting decrease in the load capacity of the furnace as temperature increases. This penalty cancels some of the effect of cycle reduction. Most vacuum furnaces, utilizing graphite as a structural material, do not have this load restriction.

Advantages of Vacuum Carburizing

In addition to reducing total processing time, which results primarily from

operating at temperatures higher than those normally used in gas carburizing, vacuum carburizing offers several other benefits:

- Because the process can be carried out using only natural gas, the need for a carrier-gas generator is eliminated. If a carrier gas is used to dilute the carburizing gas, bottled or bulk nitrogen is usually preferred.
- Preheating and postcarburizing heat treatment may be done under vacuum, which results in exceptionally clean parts and which may reduce or eliminate much of the usual precleaning and postcleaning.
- Alloying elements, such as Cr, Mn and Si, are oxidized during long carburizing cycles in conventional carburizing atmospheres, resulting in a loss of surface hardenability. This does not occur in vacuum carburizing, because of the absence of atmospheric constituents that will oxidize.
- Equipment can be started or shut down in a few minutes, and electricity and gas are used only during the carburizing cycle. These advantages can result in substantial energy savings.
- Exhaust from vacuum carburizing units is lower in substances that contribute to air pollution than exhaust from conventional gas carburizing equipment.
- Process variables are controlled with a precision that results in exceptional uniformity and repeatability among similar lots processed under the same conditions.
- Other high-temperature processes, such as brazing and sintering, can be carried out in the same furnace without delays required to condition an atmosphere furnace.
- Processes can be combined; for example, powder metallurgy parts can be sintered, carburized and hardened in a single process cycle without need for transfer of parts to a conditioned furnace for carburizing.

Fig. 6 Carbon uniformity produced by vacuum carburizing at 925 °C (1700 °F)

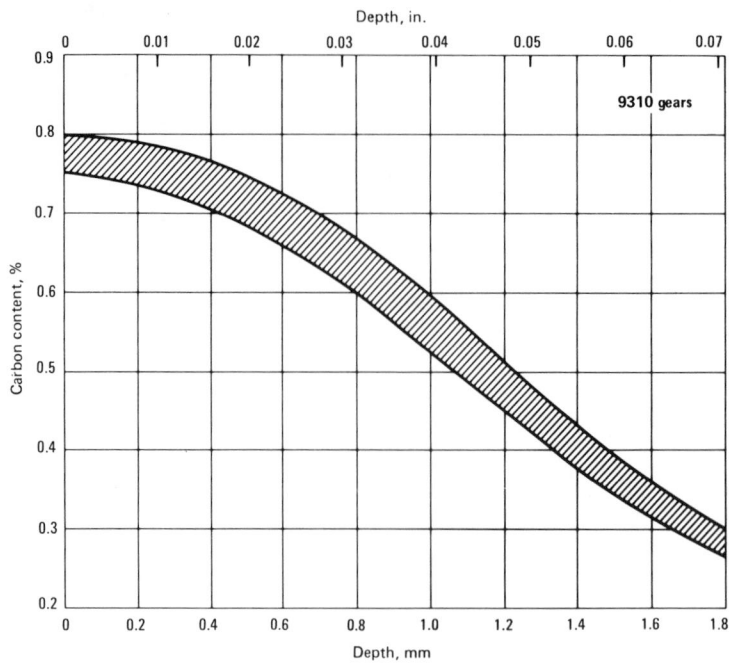

Size of load: 91 × 122 × 61 cm (36 × 48 × 24 in.)
Material: 9310 steel (gears)
Total cycle: 11 h, 300 torr
Carburizing cycle: 4.5 h
Required carbon content: 0.75 to 0.8% at surface; 0.2 to 0.35% at 1.5 mm (0.060 in.) below surface

Methods of Measuring Case Depth

CASE HARDENING may be defined as a process by which a ferrous material is hardened in such a manner that the surface layer, known as the case, becomes substantially harder than the remaining material, known as the core. Case hardening processes include carburizing, nitriding, carbonitriding, cyaniding, and induction and flame hardening. In every instance, case hardening affects chemical composition or mechanical properties, or both.

An accurate and repeatable method of measuring case depth is essential for quality control of the case hardening process and for evaluation of workpieces for conformance with specifications, such as might be done during a failure analysis. This article (sources: Ref 1 and 2) describes various methods for measuring the depth to which a change has been made in either composition or mechanical properties. Each method has its own area of application established through proven practice, and no single method is recommended for all purposes. Relationships among case depths determined by the different methods can vary extensively. Some of the factors that affect these relationships are case characteristics, parent-steel composition and quenching conditions.

Because measurements made by the various methods are not necessarily taken at the same location in a case, confusion and misunderstanding can result if the method of measurement is not specified. Specific descriptions,

such as "total case depth", "effective case depth to 50 HRC" and "case depth to 0.40% carbon" will help to avoid misunderstandings.

Typical hardness surveys taken on cross sections at the pitch line, root fillet and root land of a tooth in a carburized and hardened gear made of 8620H steel are shown in Fig. 1. These data illustrate the importance of well-defined specifications by showing that

there are variations in effective case depth even among three areas of the same gear tooth.

The methods employed for measuring case depth are chemical, mechanical or visual, and the specimens or parts may be subjected to testing in either the soft or the hardened condition. The measured case depth then may be reported as either effective or total case depth for hardened speci-

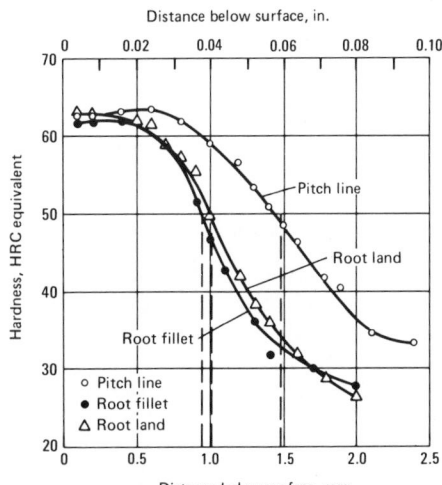

Fig. 1 Variation in hardness with distance below surface for a carburized and hardened 8620H steel gear

Effective case depths to 50 HRC: 0.94 mm (0.037 in.) at root fillet; 1.02 mm (0.040 in.) at root land; 1.46 mm (0.057 in.) at pitch line

mens and as either total case depth or equivalent effective case depth for unhardened specimens.

Effective case depth is the perpendicular distance from the surface of a hardened case to the deepest point at which a specified level of hardness is maintained. The hardness criterion, except when otherwise specified, is 50 HRC.

Total case depth may be defined as the perpendicular distance from the surface of a hardened or unhardened case to the point at which differences in chemical or physical properties of the case and core can no longer be distinguished. Total case depth sometimes is considered to be the distance from the surface to the deepest point at which the carbon content is 0.04% higher than the carbon content of the core.

Chemical Method

The chemical method of measuring case depth generally is used only for carburized cases but may be used for cyanided or carbonitrided cases as well. This method consists of determining the carbon content (and, when applicable, the nitrogen content) by chemical analysis at incremental depths below the surface. The chemical method is considered to be the most accurate method of measuring total case depth. One of two common methods is used to analyze for carbon content: combustion analysis or spectrographic analysis. Combustion carbon analysis currently is the most widely employed.

Procedure for Carburized Cases. If test specimens are used they should be of the same grade of steel as that of the parts being carburized. Specimens may be actual parts, rings or bars, and the carburized surface should be flat or otherwise suitable for accurate machining to obtain chips for subsequent carbon analysis. To ensure maximum uniformity of the carburizing process among various types of furnaces, large heat treatment facilities often use test specimens. These specimens are often standardized with respect to alloy and configuration to establish carburizing schedules for various case depths and to ensure maximum uniformity among various furnaces. Case depths of actual parts then can be correlated to the standard test specimen.

Test specimens should be carburized with actual parts or in a manner representative of the procedure to be used for actual parts. In cooling of test speci-

mens after carburizing, care should be exercised to avoid distortion and decarburization. When parts and test specimens are quenched after being carburized, they should be tempered at approximately 600 to 650 °C (1100 to 1200 °F). Time at temperature should be minimized to avoid excessive carbon diffusion. The parts and specimens should be straightened to 0.038-mm (0.0015-in.) max total indicator reading (TIR) before machining is attempted.

Test specimens must have clean surfaces and should be machined dry, taking the necessary precautions to avoid burning, in predetermined increments of depth. The increments of depth usually chosen vary from 0.05 to 0.25 mm (0.002 to 0.010 in.) depending on desired accuracy and expected case depth.

The furnace load containing the test specimens should approximate actual production conditions in terms of load density, configuration and surface area to be carburized. These three variables affect atmosphere flow, temperature uniformity and carbon demand. Differences in these conditions between production loads and the load that contains the specimens can lead to errors in correlation of case depths. A typical procedure for obtaining specimens for carbon analysis is as follows:

1 Prepare a bar of suitable material in the configuration shown in Fig. 2. Identify the bar in some manner, such as by stamping a number on the end.
2 Carburize and then quench or cool the bar as required. If bar is slowly cooled, steps 3 through 7 can be omitted.
3 Wash bar with soap and water. Rinse with methanol and dry.
4 Cut a section from the 25-mm-diam end for examination of microstructure.
5 Record as-quenched surface hardness of large-diameter end.
6 Temper bar for the time and at the temperature specified for the part with which the test bar was carburized. Record as-tempered hardness of large-diameter end.
7 Temper for 1½ h at 650 °C (1200 °F).
8 Grit blast lightly, clean centers, and straighten bar to 0.03-mm (0.001-in.) TIR taken in three places.
9 Wash bar with soap and water. Rinse with methanol and dry.
10 For case depths less than 5.0 mm

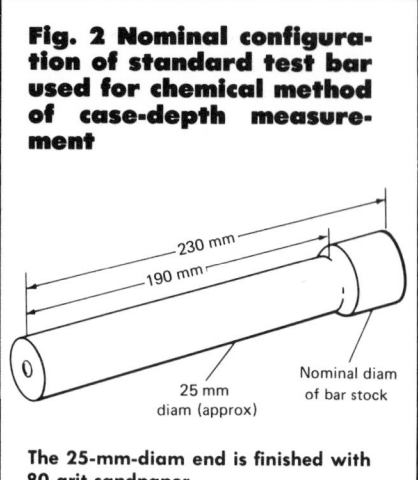

Fig. 2 Nominal configuration of standard test bar used for chemical method of case-depth measurement

230 mm
190 mm
25 mm diam (approx)
Nominal diam of bar stock

The 25-mm-diam end is finished with 80-grit sandpaper.

(0.200 in.), machine approximately 3.8 mm (0.15 in.) from the 25-mm-diam end to a depth of 5.0 mm, to ensure that the case on the end does not contaminate the specimens for carbon analysis.
11 Machine bar. Before each machining operation, record diameter of bar as measured with a micrometer. Maximum allowable taper of machined area is 0.03 mm (0.001 in.) on the radius. Machine a maximum of 0.05 mm (0.002 in.) from the radius to clean the surface. Save chips for analysis. Next, machine increments of 0.13 mm (0.005 in.) from the radius to a depth of 0.25 mm (0.010 in.) below the maximum expected case depth. Take three more increments of 0.25 mm (0.010 in.) from the radius. Save chips from each increment for separate analysis. Take precautions to ensure that chips from each cut are not burnt or contaminated by dirt, paper, oil or chips from preceding cuts.
12 Calculate and plot the carbon-gradient curve. A sample data sheet and a carbon-gradient curve are presented in Table 1 and Fig. 3, respectively.

Spectrographic Analysis. Carbon content may be determined accurately by spectrographic analysis. This method makes use of a vacuum spectrometer, which permits measurement of spectral lines in the ultraviolet region where air would ordinarily absorb much of the emitted radiation.

Many critical items must be assessed for surface carbon content to ensure uniform properties after heat treat-

Table 1 Sample data sheet for computing case-depth values for a carbon-gradient plot

Data are for 8620H steel, carburized at 925 °C (1700 °F) in a 19-tray continuous pusher furnace with infrared control of carbon dioxide content in zones 2, 3 and 4. See text for explanation of procedure, and see Fig. 3 for plot of carbon gradient.

Cut No.	D_l	D_r	Dimensional factor, mm							Carbon, %
			A_l	A_r	C_l	C_r	X	M	P	
0.....	25.35	25.36	...	...	...	...	...	...	...	...
1.....	25.20	25.23	0.15	0.13	0.15	0.13	0.07	0.03	0.03	0.987
2.....	24.98	24.99	0.22	0.24	0.37	0.37	0.18	0.06	0.13	0.953
3.....	24.76	24.76	0.22	0.23	0.59	0.60	0.30	0.06	0.24	0.918
4.....	24.49	24.47	0.27	0.29	0.86	0.89	0.44	0.07	0.37	0.871
5.....	24.22	24.22	0.27	0.25	1.13	1.14	0.57	0.06	0.50	0.818
6.....	23.94	23.91	0.28	0.31	1.41	1.45	0.71	0.07	0.64	0.787
7.....	23.69	23.65	0.25	0.26	1.66	1.71	0.84	0.06	0.77	0.717
8.....	23.41	23.38	0.28	0.27	1.94	1.98	0.98	0.07	0.91	0.675
9.....	23.10	23.10	0.31	0.28	2.25	2.26	1.13	0.08	1.05	0.583
10....	22.80	22.78	0.30	0.32	2.55	2.58	1.28	0.08	1.21	0.583
11....	22.49	22.48	0.31	0.30	2.86	2.88	1.43	0.08	1.36	0.540
12....	22.19	22.17	0.30	0.31	3.16	3.19	1.59	0.08	1.51	0.483
13....	21.87	21.87	0.32	0.30	3.48	3.49	1.74	0.08	1.67	0.444
14....	21.59	21.56	0.28	0.31	3.76	3.80	1.89	0.07	1.81	0.401
15....	21.25	21.27	0.34	0.29	4.10	4.09	2.05	0.08	1.97	0.365
16....	20.80	20.75	0.45	0.52	4.55	4.61	2.29	0.12	2.17	0.328
17....	20.27	20.24	0.53	0.51	5.08	5.12	2.55	0.13	2.42	0.283
18....	19.72	19.68	0.55	0.56	5.63	5.68	2.83	0.14	2.69	0.245

Fig. 3 Carbon gradient for carburized test bar of 8620H steel

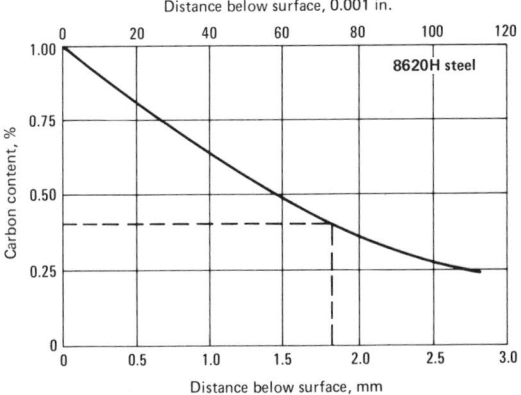

A test bar similar to the one shown in Fig. 2 was carburized at 925 °C (1700 °F) in a 19-tray continuous pusher furnace with infrared control of carbon dioxide content in zones 2, 3 and 4. Effective case depth to 0.40% carbon is 1.82 mm (0.0715 in.), and is indicated by dashed lines. See text for explanation of procedure for calculating plot points, and see Table 1 for sample data sheet containing data for this figure.

ment. The spectrographic carbon method normally uses flat test specimens that can be taper ground, step ground or reground incrementally after each carbon determination. A very small amount of material is ground from the surface (to remove oxides). Successive cuts are made and analyses are performed after each cut. Each test takes less than 2 min.

Special care must be taken for accurate measurement of the depth corresponding to each carbon determination. Case depth determined on flat or round test specimens will often be different from case depth determined directly on workpieces, because of the difference in shape.

Whereas carbon determination by the combustion method provides an average carbon content for the amount of material removed by machining, the spectrograph determines the local carbon content of the specimen to a depth of 0.03 mm (0.001 in.). A comparison of carbon values obtained from five specimens by spectrographic methods is presented in Table 2.

Mechanical Method

In the mechanical method of measuring case depth, hardness traverses are taken on the case and core of a specimen that has been prepared by one of three procedures. It is considered the most accurate method of measuring effective case depth (depth to 50 HRC). This method also is preferred for measuring total depth of thin cases (0.25 mm or less).

For measurement of effective case depth, read to point of specified hardness, which is 50 HRC (or approved equivalent) except for selectively hardened cases, for which the following values are recommended:

Carbon content, %	Case hardness, HRC
0.28 to 0.32.....................	35
0.33 to 0.42.....................	40
0.43 to 0.52.....................	45
0.53 and over....................	50

Hardness testers that produce small, shallow impressions should be used for all of the following procedures, so that the hardness values obtained will be representative of the surface or area being tested. Testers that produce Vickers or Knoop microhardness numbers with loads of at least 0.5 kg are recommended, although testers using heavier loads (such as Rockwell superficial) can be used in some instances.

Considerable care should be exercised during preparation of specimens for case-depth determination by the mechanical method to prevent grinding or cutting burn. The use of an etchant for burn detection is recommended as a general precaution, because of the serious error that can be introduced by the presence of metal whose metallurgical condition has been altered during specimen preparation.

Cross-Section Procedure. Cut specimens perpendicular to the hardened surface at a critical location, being careful to avoid any cutting or grinding practice that would affect the original hardness.

Table 2 Carbon contents of shim stock, and of surfaces of workpieces concurrently processed, as determined by spectrographic and combustion analysis

Both shim stock and workpieces were heat treated in a continuous-belt furnace with an endothermic-base atmosphere (class 301; dew point, −9 to −1 °C, or +15 to +30 °F).

	Amount of carbon present, %		
	Shim shock		Workpiece surface (spectrographic analysis)
Specimen No.	Spectrographic analysis	Combustion analysis	
1.	0.36	0.36	0.38
2	0.24	0.27	0.25
3	0.22	0.24	0.225
4	0.35	0.35	0.34
5	0.30	0.30	0.305

Fig. 4 Cross-sectioned specimen for hardness-traverse method of measuring depth of light and medium cases

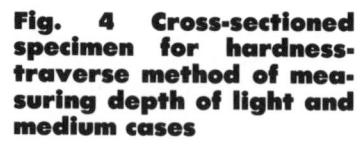

Dots show locations of hardness-indenter impressions.

Fig. 5 Cross-sectioned specimen for hardness-traverse method of measuring depth of medium and heavy cases

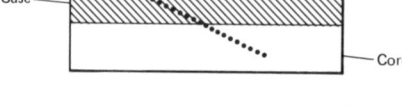

Dots show locations of hardness-indenter impressions.

Grind and polish the specimen. The surface of the area to be traversed should be polished finely enough so that hardness impressions are unaffected (the lighter the indenter load, the finer the polish necessary).

The procedure illustrated in Fig. 4 is recommended for measurement of light and medium cases. The alternative procedure illustrated in Fig. 5 is recom-

Fig. 6 Taper-ground specimen for hardness-traverse method of measuring depth of light and medium cases

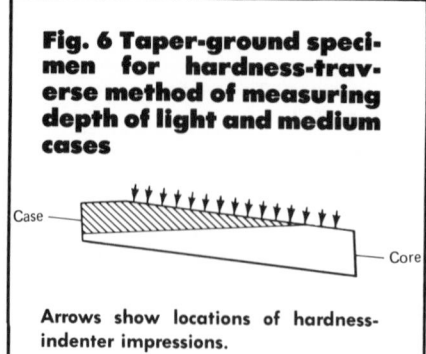

Arrows show locations of hardness-indenter impressions.

Fig. 7 Step-ground specimen for hardness-traverse method of measuring depth of medium and heavy cases

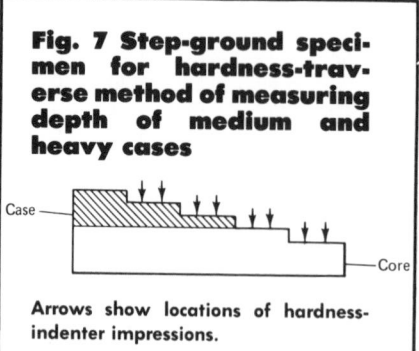

Arrows show locations of hardness-indenter impressions.

mended for measurement of medium and heavier cases.

The hardness traverse should be started far enough below the surface to ensure proper support from the metal between the center of the impression and the surface. Subsequent impressions are spaced far enough apart so as not to distort hardness values. The distance from the surface of the case to the center of the impression is measured on a calibrated optical instrument, micrometer stage, or other suitable measuring device.

Taper-Grind Procedure. This procedure, illustrated in Fig. 6, sometimes is used for measurement of light and medium cases.

A shallow taper is ground through the case, and hardness measurements are made along the surface thus prepared. The angle is chosen so that readings spaced equal distances apart will represent the hardnesses at the desired increments below the surface of the case.

Step-Grind Procedure. This procedure, illustrated in Fig. 7, is recommended for measurement of medium and heavy cases. It is essentially the same as the taper-grind procedure, with the exception that hardness readings are made on steps that are known distances below the surface.

A variation on this procedure is the step-grind method in which two predetermined depths are ground. If the hardness is greater than 50 HRC on the shallow step and less than 50 HRC on the deep step, the effective case depth to 50 HRC lies somewhere between the two steps. This variation frequently is used to ensure that the effective case depth is within specified limits.

Visual Methods

These methods employ any visual procedure, with or without the aid of magnification, for reading the depth of case produced by any of the various processes. Specimens may be prepared by combinations of fracturing, cutting (with water cooling to prevent burning), grinding and polishing. Etching with a suitable reagent normally is required to produce a contrast between the case and the core. Nital (concentrated nitric acid in alcohol) of various strengths frequently is used as the reagent for producing this contrast.

Visual methods have been classified into two general categories: macroscopic and microscopic. In macroscopic procedures, specimens normally are ground no finer than through No. 000 metallographic emery paper, and magnifications usually do not exceed 20 diameters. The Brinell glass, a hand-held optical instrument with retical markings at intervals of 0.1 mm (about 0.004 in.) and 20-diameter magnification, is a convenient tool for macroscopic measurement. In microscopic procedures, complete metallographic polishing and etching generally are required, and case depths normally are read at a magnification of 100 diameters.

Macroscopic Visual Procedures

Macroscopic methods for measuring case depth are recommended for routine process control, primarily because of the short time required for determinations and because of the minimum of specialized equipment and trained personnel that are needed. Although these methods normally are applied to hardened specimens, they have the additional advantage of being applicable to measurement of unhardened cases as well. However, the accuracy of such measurements can be improved by correlation with results of other methods. A variety of etchants may be employed with equal success, but the following procedures are typical and widely used:

- *Fracture:* Prepare part or specimen by fracturing. Examine at a magnification not exceeding 20 diameters, with no further preparation.
- *Fracture and etch:* Water quench part or specimen directly from the carburizing temperature. Fracture, then etch in 20% nitric acid in water for a time established to develop maximum contrast. Rinse in water, and read while wet.
- *Fracture or cut, and rough grind:* Prepare specimen by either fracturing or cutting (with water cooling), and then rough grinding. Etch in 10% nital for a time established to provide a sharp line of demarcation between case and core. Examine at a magnification not exceeding 20 diameters (Brinell glass), and read all of the darkened area for approximate total case depth.
- *Fracture or cut, and polish or grind:* Prepare specimen by fracturing or cutting (with water cooling). Polish, or grind through No. 000 or finer metallographic emery paper, or both. Etch in 5% nital for approximately 1 min. Rinse in two clean alcohol or water rinses. Examine at a magnification not exceeding 20 diameters (Brinell glass), and read all of the darkened zone. After correlation, effective case depth can be determined by reading from external surface of specimen to a selected line of the darkened zone. An alternative etching procedure is to etch in 25% nital for 30 s, wash in concentrated picral, rinse in alcohol, blow dry, and read as described above.
- *M_s method:* This method of case-depth measurement utilizes the fact that the martensite-start temperature (M_s point) varies with carbon content. Quenching and holding the steel for a short time at the M_s point corresponding to a given carbon content will temper the martensite formed at all lower carbon levels. Subsequent water quenching transforms austenite at all higher carbon levels to untempered martensite. Then polishing and etching of the test piece will reveal a sharp line of demarcation between tempered and untempered martensite; this line is normally read at 20-diameters magnification (Brinell glass) to a precision of ± 0.05 mm (± 0.002 in.).

The case depth is not sensitive to small temperature changes in the quenching bath. Final selection of quenching temperature usually is done statistically to produce an equal plus-and-minus distribution of error about known carbon curves.

The main factors that affect the accuracy of this method are pearlite formation during quenching to the M_s point, and time at M_s temperature. The specimen size should be sufficiently small to ensure that the severity of quench will transform all austenite of lower carbon levels to martensite without any formation of pearlite. (Specimen size may be critical for low-hardenability steels.) The time at M_s temperature should be sufficiently short so as not to allow formation of bainite, which interferes with the sharpness of the line of demarcation upon etching and can obliterate it completely.

For additional information on the M_s technique, see Ref 3.

Microscopic Visual Procedures

Microscopic methods generally are used for laboratory measurement of case depth and require complete metallographic polishing and etching suitable for the material and the process. The most common magnification used for examination is 100 diameters.

Carburized Cases. Microscopic methods may be used for laboratory determinations of total and effective case depths of material in the hardened condition. When the specimen is annealed properly, the total case depth can be determined quite precisely. For certain applications involving alloy steels of moderate to high hardenability that contain 0.4 to 0.8% carbon, the M_s method of determining case depth to specific carbon level has been found effective.

Procedure for hardened condition

1 Fracture or cut specimen (water cool when cutting) at right angles to the surface.
2 Prepare specimen for microscopic examination and etch in 2 to 5% nital.
3 For effective case depth, read from surface to metallographic structures that have been shown to be equivalent to 50 HRC. Often, the structure that is equivalent to 50 HRC consists of about 85% martensite and 15% intermediate quench products.

4 For total case depth, read to the line of demarcation between the case and the core. In alloy steels that have been quenched from a high temperature, the line of demarcation is not sharp. Read all of the darkened zone that indicates a difference in carbon content from that of the uniform core structure.

Procedure for annealed condition
(for specimens previously hardened or not cooled under controlled conditions)

1 The specimen to be annealed may be protected by copper plating or any other suitable means of preventing carbon loss.
2 Anneal specimens in a protective atmosphere, or pack in a small thin-wall container with a suitable material such as charcoal, spent chips or pitch coke.
3 Heat specimens to a temperature about 25 to 50 °C (about 50 to 100 °F) above the upper critical temperature (Ac₃) for the core. (Generally, an annealing temperature of 870 to 925 °C, or 1600 to 1700 °F, is satisfactory.) Hold specimen at temperature only long enough to transform completely to austenite; otherwise, excessive diffusion of carbon may lead to inordinately high estimates of actual total case depth.
4 Cool from the annealing temperature at the following rates:
Carbon steels: A normally satisfactory cooling rate for most plain carbon steels such as 1010, 1015 and 1018 is 150 °C/h (270 °F/h) from the annealing temperature to 425 °C (800 °F). For high-manganese steels (1500 series), boron steels, and steels with high residual alloy contents, cooling may have to be slower. Cool as desired below 425 °C.
Alloy steels: For most alloy steels, best results are obtained from isothermal transformation. For some steels, however, a low cooling rate such as 75 °C/h (135 °F/h) from the annealing temperature to 425 °C (800 °F) is satisfactory. If martensite is retained in the structure, better contrast after etching may be obtained by tempering the specimen at 540 to 590 °C (1000 to 1100 °F). Cool as desired after tempering.
5 Section, prepare and etch specimen as described under "Procedure for hardened condition".

Approximate equivalent hardness numbers for steel

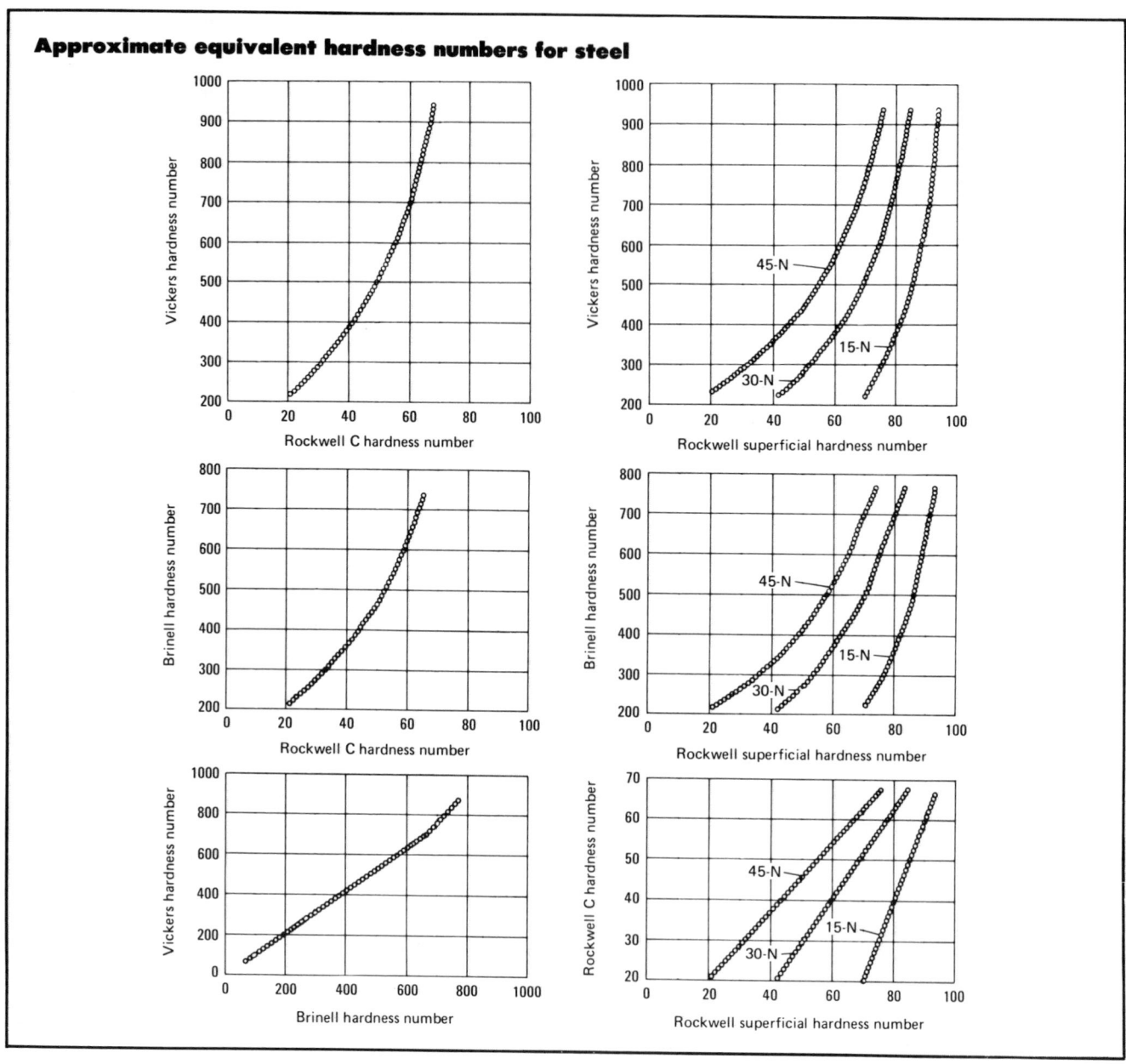

6 For total case depth, read the depth at which no further change in microstructure occurs.

Production carburizing schedules often have cooling rates similar to those described above under "Procedure for annealed condition". Specimens treated in this manner may be prepared and examined without being reheated after carburizing, and the results can be accurately correlated to a standard specimen.

Carbonitrided cases are measured for total case depth in the hardened condition. High quenching temperatures, high alloy content of the steel and high carbon content of the core

decrease the accuracy of readings obtained by this method.

Procedure

Section, prepare, etch and read as described above for carburized cases, under "Procedure for hardened condition".

Cyanided cases are thin, and only the microscopic method is recommended for accurate case-depth measurement. The usual cyanided case contains a light-etching layer followed by a totally martensitic constituent, which in turn is followed by martensite accompanied by increasingly extensive networks of other constituents, de-

pending on the type of steel. Cyanided cases are read in the hardened condition only, and results are reported as total case depth.

Procedure

1 Section, prepare and etch specimen as described above for carburized cases, under "Procedure for hardened condition".
2 Read to the line of demarcation between the case and the core. (When a sharp line of demarcation does not exist, a hardness traverse such as that described under "Mechanical Method" is recommended.)

Nitrided Cases. For measuring the depths of nitrided cases, the microscopic method is used chiefly in those situations where the available sample cannot readily be prepared for the more desirable hardness-traverse method.

Procedure

1 Section and prepare specimen as described above for carburized cases, under "Procedure for hardened condition".
2 Etch in 10% nital.
3 Read all of the darkened zone for total case depth.

Selectively Hardened Cases. Because no compositional change occurs in selective hardening (induction hardening, for example), readings must be taken on material in the hardened or the hardened-and-tempered condition

only. A procedure for reading effective case depth may be established by correlating microstructures with a hardness-traverse method. A minimum hardness of 50 HRC is commonly used, but some other value may be selected or required—for example, in lower-carbon steels that do not reach 50 HRC when fully hardened (see the in-text table correlating carbon content with effective-case-depth hardness, under "Mechanical Method"). The microstructure at the selected location will vary depending on steel composition, prior treatment, and hardness level chosen.

Procedure

1 Section, prepare and etch specimen as described above for carburized cases, under "Procedure for hardened condition".
2 For total case depth, read the entire zone containing structures hardened by the process.
3 For effective case depth, read to selected microstructure correlated with specified hardness.

REFERENCES

1. Methods of Measuring Case Depth: SAE Recommended Practice J423a, Society of Automotive Engineers, Warrendale, PA, 1963 (reaffirmed 1970)
2. Measurement of Case Depth: Chapter 4 in *Carburizing and Carbonitriding,* prepared under the direction of the ASM Committee on Gas Carburizing, American Society for Metals, Metals Park, OH, 1977
3. The Application of M_s Points to Case Depth Measurement, by E. S. Rowland and S. R. Lyle: *Transaction of ASM,* Vol 37, 1946, p 27

Heat
Processing
Equipment

Types of Heat Treating Furnaces

By John W. Smith
Manager of Engineering
Holcroft Division
Thermo Electron Corp.

FURNACES commonly used in heat treating are classified in two broad categories, batch furnaces and continuous furnaces. In batch furnaces, workpieces normally are loaded and unloaded manually. A continuous furnace has an automatic conveying system that provides a constant workload, in kilograms (pounds) per hour, to the unit.

Batch Furnaces

The basic batch furnace normally consists of an insulated chamber with an external reinforced steel shell, a heating system for the chamber, and one or more access doors to the heated chamber. Standard batch furnaces such as box, car-bottom, and pit types are most commonly used when a wide variety of heat-hold-cool temperature cycles are required.

With the addition of powered work-handling systems—integral quench tanks, slow-cool chambers and some automatic controls—the basic box-type batch furnace is upgraded to a semicontinuous batch furnace, which is a commonly used piece of heat treating equipment. The design of a semicontinuous batch furnace is shown in Fig. 1.

The use of batch equipment for heat treating usually requires considerable labor for loading, handling and unloading of the work. High labor costs dictated by the process must always be considered in the selection of batch equipment.

Uses of Batch Furnaces. Batch furnaces are normally used:

- To heat treat low volumes of parts, in terms of kilograms (pounds) per hour
- To handle special parts for which it would be difficult to adapt a conveying system for continuous handling (long drill rods processed in a pit furnace, for example)
- To process large parts in small numbers, e.g., stress relief or annealing of large weldments or castings in a car-bottom-type furnace
- To carburize parts that require heavy case depths and long cycle times; for example, work done in integral-quench batch carburizers, such as gears, or rock bits, or work done in pit-type carburizers such as drill rods or bearing races
- To process various parts requiring a wide range of heat treat cycles that can readily be changed, either manually or automatically.

Batch processing is especially appropriate when the work must be heated from room temperature to a maximum temperature at controlled rates, held at temperature, and cooled at controlled rates. For example, car-type furnaces are used for critical stress-relief work or carbon baking in saggers.

Continuous Furnaces

Continuous furnaces consist of the same basic components as batch furnaces: an insulated chamber, heating system and access doors. Figure 2 shows one configuration of a continuous furnace. Three common types of continuous furnaces are pusher, belt conveyor and roller hearth. With pusher furnaces, the total workload, in kilograms (pounds) per hour, consists of the work, trays, and any fixtures required.

In belt-conveyor furnaces, both the conveyor and work must be heated unless belt is returned inside the furnace. Roller-hearth furnaces require that both tray and work be heated. Less common types of continuous furnaces are shaker hearth, rotary hearth and walking beam.

Continuous furnaces normally are employed when high part volumes are processed in consistent cycles. In addition to providing more uniform part quality, continuous furnaces greatly reduce the manual labor content inherent in batch systems. With a continuous flow of work through the furnace, the equipment can be integrated readily into the material flow required in a manufacturing process. Continuous furnace systems also are readily adaptable to automation.

Uses of Continuous Furnaces. Continuous furnaces are normally used:

- To heat treat work to a consistent heat treat cycle and work quality. (Once a continuous furnace is properly adjusted for such variables as cycle time, zone temperatures and atmosphere integrity, the work produced will be of consistent quality. All work passing through the various furnace zones and chambers are subject to the same conditions.)
- To process large volumes of parts, in kilograms (pounds) per hour, although some continuous furnaces are designed to handle relatively low volumes. These furnaces are employed when continuous processing is required, regardless of the kilograms per hour. (A small sintering furnace rated at or below 45 kg/h (100 lb/h) is one example. Continuous furnaces, however, are considered high-production units.)
- To maintain uniform in-plant material flow. Many times, a continuous furnace can be integrated with other elements in the manufacturing process. If loading and unloading of a product can be easily automated, an automatic system can be designed to receive the parts, load them into the continuous equipment, process the parts, unload them, and return them to the manufacturing process in a specific or random manner.

Pusher Furnaces

A pusher furnace uses the "tray-on-tray" concept to move work through the furnace, as shown schematically in Fig.

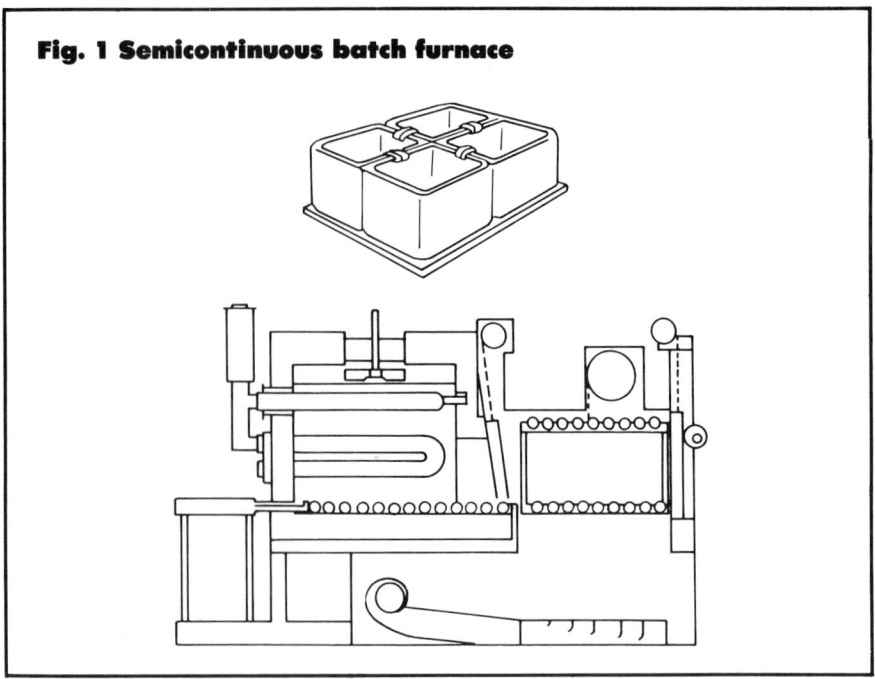

Fig. 1 Semicontinuous batch furnace

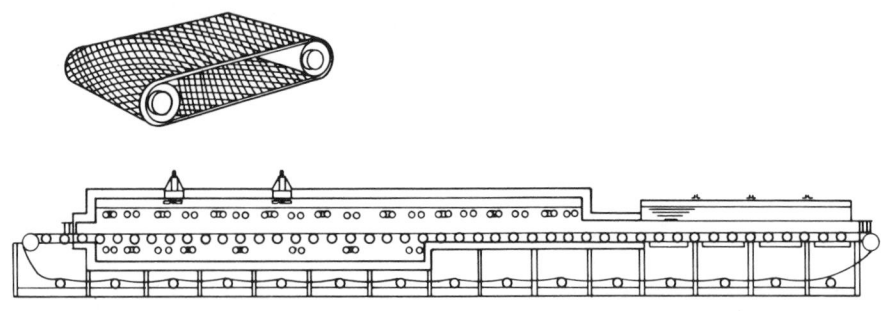

Fig. 2 Continuous furnace

3. A pusher mechanism pushes a solid row of trays from the charge end until a tray is properly located and proven at the discharge end for removal. Figure 4 shows a typical pusher mechanism. On a timed basis, the trays are successively moved through the furnace. Cycle time through the furnace is varied by changing push intervals.

Skid-Rail Furnaces. In a skid-rail pusher furnace, the work is placed on flat, normally reversible cast-alloy grid trays. The trays in turn are supported through the furnace on skid rails. The total gross load on the tray determines the number of rails necessary to minimize wear by maintaining the bearing pressure between the tray and the rail within acceptable limits. In certain applications, particularly when an endo-

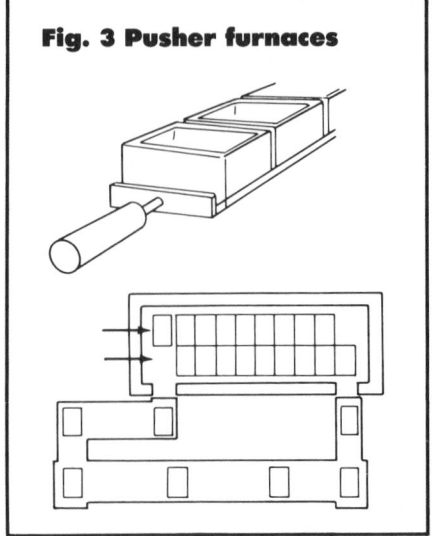

Fig. 3 Pusher furnaces

thermic or an enriched endothermic atmosphere is used, the skid rails are lubricated by the atmosphere and the coefficient of friction is reduced, decreasing wear and increasing tray and skid-rail life. Skid rails are used normally for light-to-moderate tray loadings.

Skid rails cast from nickel-chrome heat-resisting alloy have been commonly used in pusher furnaces. The skid rails normally are supported by and anchored to a series of cast-alloy pier caps and pedestals.

Because of the ever-increasing cost of nickel-chrome alloys, cast-alloy skid rails are being replaced where possible by less expensive silicon carbide refractory rails. Silicon carbide skid rails are molded and prefired into various rectangular shapes and are then bricked into the lower piers or furnace floor. With the rails resting on edge, the cast-alloy grid tray is thus supported on two or more rail faces, each normally 64 mm (2½ in.) wide. Because silicon carbide in contact with a cast alloy has a lower coefficient of friction than alloy on alloy, it makes excellent skid rails. However, in designing these rails, precautions should be taken to eliminate severe temperature gradients or thermal shock, both of which could cause the rails to fracture.

Roller Rails. Roller rails are used to support and guide the trays as they are moved through the furnace. They are supported and anchored in a manner similar to that of cast-alloy skid rails, but the mechanical advantage of the wheel and axle reduces the pushing force required to move the load.

In certain instances, rails, wheels and axles are all made of a cast alloy. The use of a dissimilar material for the axle, however, can reduce the coefficient of friction and also can eliminate the natural "galling" effect of alloy on alloy.

The cast-alloy roller rail tray usually has one or more runners and guides on the underside to keep the tray centered on the roller rails. Because the tray underside is not flat, it cannot be moved readily at 90° to its normal travel, and a transfer carriage would be used. Also, the tray normally is not reversible.

Because required pushing force is reduced, roller rails are used mostly for heavy tray loads and long pushes. To keep wear in the wheel or axle to an acceptable minimum, the total load per wheel (line pressure), in kilograms

(pounds) per inch, must be kept within acceptable limits.

Buggy trays make use of the mechanical advantage of the wheel and axle in a manner similar to that of the roller rail. Wheels and axles attached to the underside of the tray are guided through the furnace in refractory or alloy troughs.

The tray cannot readily be moved at 90° to its normal travel, and a carriage would be required for such movement. Although buggy trays can be used for

extremely heavy loads, the maximum recommended load per wheel should not be exceeded, to keep wear within acceptable limits.

Walking-Beam Furnaces

The hearth through a walking-beam furnace consists basically of two sets of support rails. One such design employs two sets of rails, one stationary and the other movable. With this system, the moving rails lift the work from the sta-

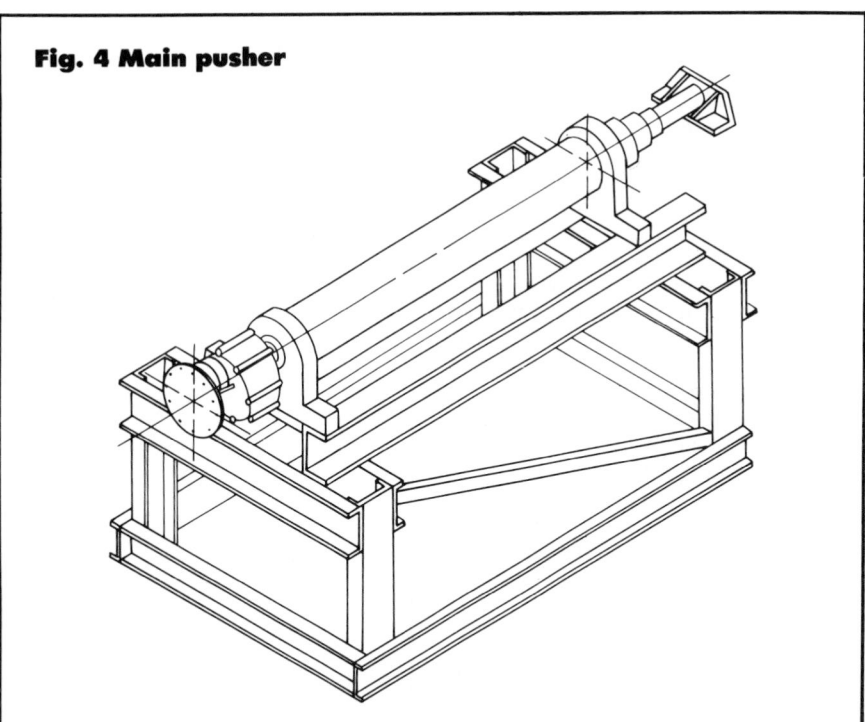

Fig. 4 Main pusher

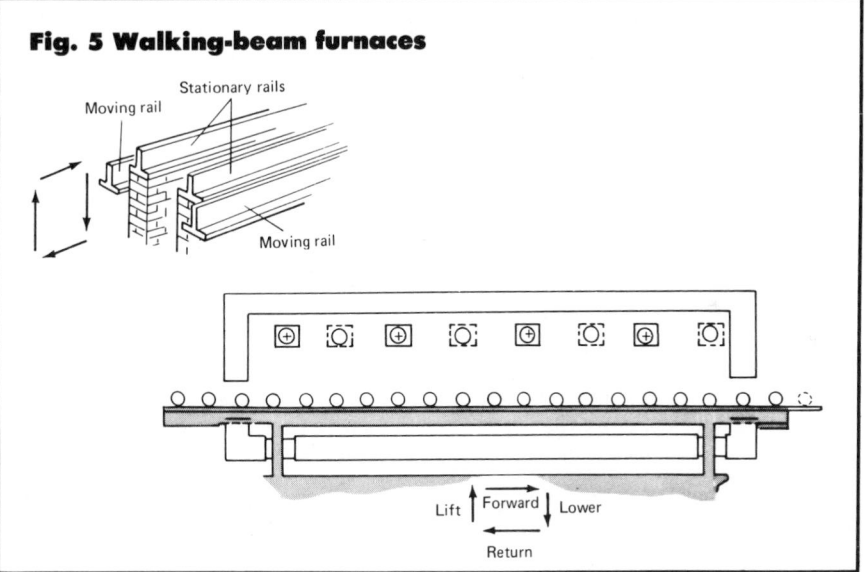

Fig. 5 Walking-beam furnaces

Moving rail
Stationary rails
Moving rail

Lift — Forward — Lower
Return

tionary rails, move it forward and then lower it back onto the stationary rails. The moving rails then return to the starting position. Figure 5 shows a typical walking-beam furnace design. In this system, the moving rail can move in an elliptical or rectangular pattern. The frequency of lift and length of stroke determine the total processing time. In another type of walking-beam furnace, both sets of rails move. One set of rails moves up and down, and the second set moves forward and backward. This system is known as a true "rectilinear" motion walking beam.

The sequence normally is as follows: the lifting beam moves up and the traveling beam moves in reverse, then the lifting beam moves down and the traveling beam moves forward. The work is thus sequenced through the furnace.

Uses of Walking-Beam Furnace. Walking beams traditionally have been used in steel mills in reheat furnace hearth systems for slabs and billets. Walking-beam systems can be built ruggedly to move extremely heavy loads. In heat treating operations, walking beams have been used successfully, with flat-top beams to carry such work as flat plates or trays or with pocketed-top beams to carry unstable parts such as rollers or shafts.

Advantages of Walking-Beam Furnaces. In this type of furnace, only the work being processed has to be heated, because normally trays or fixtures are not needed, as in a pusher furnace. Because the work is never skidded, friction is reduced for heavy loads.

The system can be loaded and unloaded automatically. A part can be picked from a specific spot and placed in a specific spot by using the walking-beam mechanism. Equipment is self-emptying on shutdown.

Disadvantages of Walking-Beam Furnaces. The mechanisms are usually much more expensive than for pusher-type systems. On large high-temperature slab or billet reheat furnaces, there is a dramatic increase in holding losses and fuel consumption due to the water-cooled insulated walking-beam rail system.

Car Furnaces

The car furnace, also called a "bogie hearth", is normally considered an extremely large batch furnace. The bottom (or floor) of the furnace is constructed as an insulated movable car

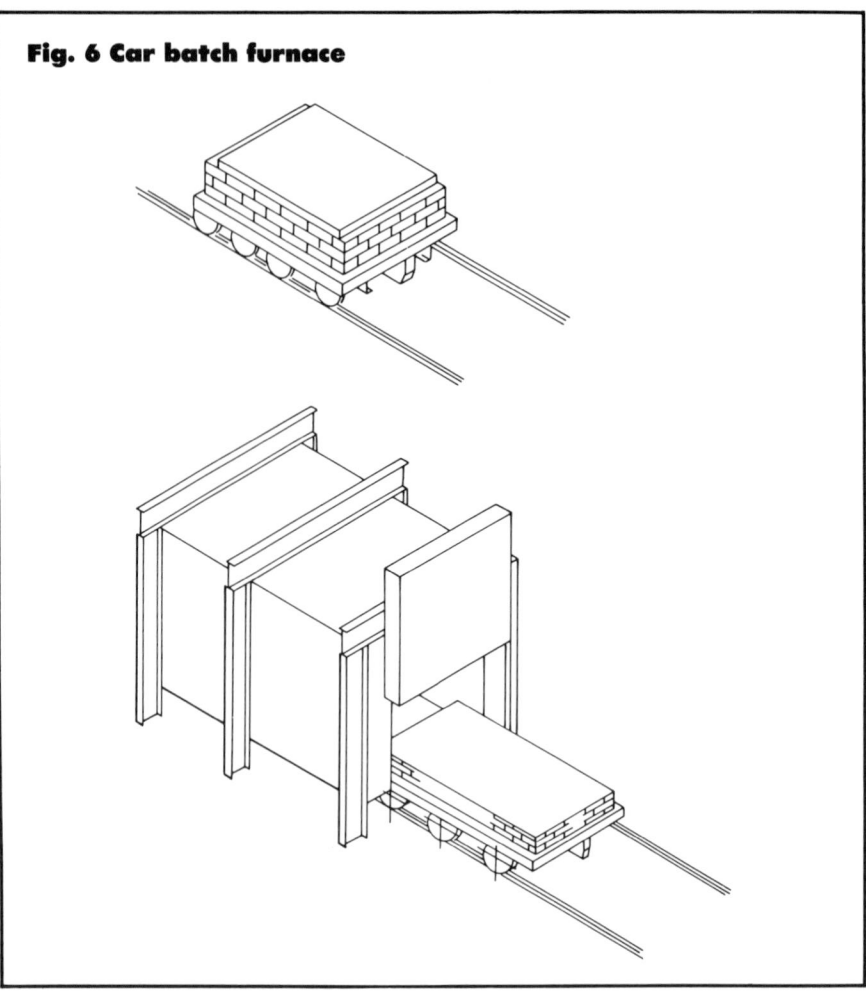

Fig. 6 Car batch furnace

that is moved out of the furnace for loading and unloading, as shown in Fig. 6. When in position inside the furnace, the car is sealed to the furnace structure with sealing troughs or solid seals. Furnace cars can be self-propelled with the motor drive mounted directly on the car, or they can be moved in and out by a floor-mounted drive with a continuous chain or a rack-and-pinion drive. Most car furnaces are nonatmosphere type due mainly to the difficulty in sealing the car.

Heating systems normally are either direct-fired or electrically heated with resistance elements. With direct-fired systems, it has proven advantageous to design a pressure-control system controlling the flues. With the large difference in fuel burned during the heat-up portion of the cycle as compared to the soak portion of the cycle, it is extremely difficult to maintain a minimum acceptable furnace pressure with constant flues. Most car furnaces are heated from room temperature with the

load already in the furnace. A typical cycle would be to heat from room temperature to a control temperature at a specific rate, hold at the control temperature for a specified time, and then slow cool to discharge temperature at a specified rate. Many packaged temperature-control systems on the market today are capable of allowing a variable programmed cycle.

The use of ceramic fiber insulation in a car furnace allows greater control of furnace temperature when following a programmed cycle. Because ceramic fiber has minimal heat storage capacity compared to hard refractories, it will heat and cool at faster rates. Also, less total heat is required to bring the furnace to the desired temperature. Further, continuous cyclic heating and cooling has no effect on ceramic fiber lining.

Car furnaces are used from the lower stress-relieving ranges around 540 °C (1000 °F) to temperatures of over 1095 °C (2000 °F) for certain applications.

Because many of the larger car furnaces are installed outdoors, increased allowances should be made for thermal holding losses caused by winds and other changes in ambient conditions.

Direct-Fired Furnace Equipment

With direct-fired furnace equipment, work being processed is directly exposed to the products of combustion, normally referred to as flue products. To minimize the scaling (oxide) effect on the work, the flue products can be controlled or varied by adjusting the fuel-air ratio of the combustion system. Although fuel-air adjustments can be made manually, more precise control can be achieved automatically by a wide variety of fuel-air ratio control systems on the market today. When direct-fired burner equipment is used in a heat treating furnace, the parts being processed often are in some primary or intermediate stage of manufacture. The oxide formed is not detrimental to the part because it will be removed later in the manufacturing process. For example, cold forged parts are annealed between successive drawing operations, and rough castings are annealed prior to machining.

Gas-Fired Equipment. Gaseous fuel used in heat treating furnaces can be natural gas, straight propane, a propane-air mix or a relatively low-Btu manufactured gas.

With the proper selection of burners, controls, orifices and pipe sizes, a combustion system can be designed to operate on 2500-Btu propane gas, 1000-Btu natural gas or 160-Btu producer gas. The number refers to the quantity of Btu's contained in a cubic foot of the gas. Manual adjustments are required for conversion from one gas to another.

Oil-Fired Equipment. Almost any grade of oil that can be satisfactorily atomized can be burned in direct-fired equipment. Lower viscosity oils such as diesel fuel and No. 2 fuel oil can be easily atomized with 28-g (16-oz) cold (room temperature) air. These are probably the fuel oils most commonly used for heat treating. Even with easily atomized oils, caution should be employed in using them on flame-supervised in furnaces operating below 760 °C (1400 °F) with interrupted pilots. At low oil flows and excess air conditions, nuisance shutdowns can occur from the flame supervision devices. In certain

instances, as dictated by the National Fire Prevention Association, "constant pilots" may be used to eliminate the shutdowns. Insurance carriers must approve the use of constant pilots for the particular application, however.

Heavier grades of oil must be atomized by a method other than low-pressure 28-g (16-oz) air. Normally, high-pressure air and steam are used. Burners that can be fired by either gas or oil are available. In most instances, oil is used as the standby fuel, to be used in peak periods when natural gas supplies are curtailed. Oil is considered desirable by some in the forging industry because it creates a "softer" scale on the billet, which is more easily removed in forging. High vanadium or sulfur content in fuel oil burned in the direct-fired process can reduce the useful life of various furnace components, especially the nickel-chrome heat-resisting alloys.

Advantages of Fuel-Fired Furnaces. These furnaces are the easiest to adjust or alter the connected input. Normally, a simple orifice change is all that is required. Recuperator heat-saving devices can be added easily and controlled cooling can be initiated easily with proper design of combustion systems. These systems also have faster heat-up times because excess control factors can be added inexpensively.

Disadvantages of Fuel-Fired Systems. The following disadvantages are common to fuel-fired furnaces:

- Requires extensive ventilation systems
- Potential explosion or fire hazard
- Requires more manpower for start-up and shutdown
- Adjustment more difficult to maintain, resulting in excessive fuel use.

Electrically Heated Furnace Equipment

Electrically heated furnaces are commonly found in all temperature ranges: from low-temperature tempering furnaces, through the heat treating range and up to forging temperatures. The basic consideration in selecting the type of heating element is to determine whether the elements are to be the "open" type, which are exposed to the furnace environment, or the "indirect" type, which are protected from the furnace environment by some means such as a radiant tube, muffle or retort. Factors affecting this decision are furnace atmosphere, need to protect the ele-

ment from mechanical damage and space required for placement of the element.

Material Selection. Almost all furnace atmospheres other than air will in some way affect the over-all performance and subsequent life of each type of heating-element material. Manufacturers of heating-element materials provide charts that allow designers to predict the material performance with any of the given atmospheres. Each heating-element material can be exposed to the different furnace atmospheres with varying degrees of success.

The notable exception is with a carburizing–type atmosphere. The conventional nickel-chrome strip heating element does not perform well in a carburizing atmosphere because the element itself carburizes, affecting element performance. Generally, in a carburizing atmosphere, heating elements are placed inside radiant tubes. However, some alternative elements are designed specifically to operate when exposed in a carburizing atmosphere.

The selection of "open" or "indirect" elements is a choice also determined by the need to protect the element against mechanical damage from parts being heated, from accumulations of metallic scale or from broken refractories. In furnaces where bottom heat is mandatory and scale can be formed readily on the parts, or where parts may fall from a tray or conveyor, electric elements should be protected in radiant tubes below the hearth. Open elements may be used throughout the upper portion of the furnace.

In some furnace designs, the physical space available determines the design of the element.

A further consideration is whether to use an element material other than nickel-chrome strip or rod. Silicon carbide (Globor) elements or molybdenum disilicide rod elements have been used with success when directly exposed to various atmospheres, although the former normally is not recommended for use in a carburizing atmosphere. Silicon carbide elements have been used on occasion inside radiant tubes for protection against carburizing atmospheres.

Metallic Resistance Heating Elements. The following are general types of furnaces, with a description of the kinds of heating elements used in each.

Low-Temperature Furnaces with

"Open" Elements. The temperature range of this type of furnace varies from approximately 150 to 675 °C (300 to 1250 °F), and the furnace is normally a recirculated-wind type. The simplest type of heating element is a commercially available duct heater, usually full-voltage, 440 V or 220 V heaters. These are quite useful when the designer can stay within the manufacturer's limitations. Maximum use temperature is normally limited to 400 °C (750 °F).

The heater should be large enough to cover the entire recirculated wind-duct cross section, but designs are limited to the heater sizes available. The design watt density for this type of duct heater is normally 34 000 W/m^2 (22 W/in.^{2}). Watt density is the expression commonly used for the connected power of each element in watts divided by its total surface area in square inches. Watts per square meter (square inch) is an important design consideration, and the allowances vary greatly with temperature, type of element and furnace.

As an alternative to the commercial unit, a custom-built duct heater can be used. A steel or alloy frame can then be designed to completely fill the air-duct cross section, and such a unit can be removed easily through a sidewall bulkhead.

The nickel-chrome ribbon material is attached to ceramic insulators mounted in tiers. A common element material is 35Ni-18Cr-44Fe, and it is normally selected in the lighter gage thicknesses and narrower widths. A typical cross section for the ribbon material would be 13 by 0.8 mm (0.50 by 0.030 in.).

Variable with temperature and wind flow, the design watt density would be in the 23 250 to 46 500 W/m^2 (15 to 30 W/in.^{2}) range. This heater is also normally designed to operate at full-line voltage of 440 or 220 V.

High Temperature Furnace with "Open" Elements. The temperature range of this type of furnace varies from approximately 675 to 955 °C (1250 to 1750 °F), and the furnace is normally a radiant-heating type.

Where large wall areas are available inside a furnace, a common method of mounting the nickel-chrome strip element is to attach it in a serpentine pattern from insulated alloy or ceramic anchors normally on the vertical walls only. With this design, especially at the higher temperatures, the structural strength of the element material and configuration must be considered. The element must support itself at operating temperatures without excessive droop or warping. Sufficient warping to cause the element to touch at various points could shorten the element's effective length, decrease the resistance and cause premature failure due to excessive currents and watt densities.

On larger furnaces with accessible wall areas, maintenance on this design element, although done from inside the "cold" furnace, is relatively easy. On smaller furnaces, accessibility for replacement of wall elements becomes a problem.

Other types of modular or drawer-type elements are available, which makes element removal and maintenance much easier.

The element strip material used with open elements is generally one of the following types: 80Ni-20Cr, 68Ni-20Cr, or 35Ni-18Cr-44Fe. The nickel-chrome element is generally selected in the heavier gage thicknesses and wider widths. A typical cross section range would be from 1.3 to 2.3 mm (0.050 to 0.090 in.) thick and 19 to 38 mm (3/4 to 1½ in.) wide. Variable with temperature and location of elements, the design watt density would be 12 400 to 23 250 W/m^2 (8 to 15 W/in.^{2}).

An alternative to the nickel-chrome strip element is the cast nickel-chrome heating element. This element has good structural strength, stability and resistance to atmospheric attack. The quality control necessary in the manufacture of this element has made it slightly less flexible and popular than nickel-chrome strip. The casting must have uniform density and cross section to ensure a guaranteed resistance without danger of hot spots. Castings made with the "lost-wax" process generally meet the quality requirements. Low voltages at the element and high currents tend to make the control hardware and wiring quite expensive. A common cast element material would be 35Ni-15Cr, and the watt-density range again would be 12 400 to 15 500 W/m^2 (8 to 10 W/in.^{2}).

Nonmetallic Resistance Heating Elements. In general, the nonmetallic heating elements are used in furnaces operating above 1010 °C (1850 °F). Silicon carbide elements are generally used in temperature ranges of 1010 °C (1850 °F) and above. They tend to be very fragile, so care should be taken in the design to allow for proper support and freedom to expand and contract.

Silicon carbide elements undergo resistance increase with age; thus, to maintain constant power over the life of these elements, it is necessary to have a voltage adjustment available, usually with a step-transformer. The useful life of a silicon carbide element is usually established at the point when its resistance has increased four times. To maintain constant power would mean that a total voltage demand of twice the initial voltage would be required, because power is equal to E/R^2.

Silicon carbide elements are provided in various diameters and lengths, with published "hot" resistances. Design watt densities vary with such factors as temperature and atmosphere, but with silicon carbide elements, conservatively designed watt densities result in better element life.

In a sintering furnace operating at 1150 °C (2100 °F) with an endothermic atmosphere, a design watt density of 31 000 to 46 500 W/m^2 (20 to 30 W/in.^{2}) would be considered appropriate. Molybdenum disilicide elements are commonly formed in U-shaped rod configurations and normally are mounted vertically. The published maximum temperature-use range in air is above 1650 °C (3000 °F), which covers all furnace temperatures up through the forging range.

Element location is an important consideration because these elements are designed to operate at very high watt densities and related high thermal heads. These elements undergo a high resistance change from cold to hot, because resistance increases with temperature. The control hardware and wiring must be properly designed to handle the high initial currents. A typical selection for a 955 °C (1750 °F) furnace with an endothermic atmosphere would be a 9-mm (0.35-in.) rod element, rated at 244 900 W/m^2 (158 W/in.^{2}), with an element temperature of 1430 °C (2610 °F).

Advantages of Electrically Heated Furnaces. The following are advantages associated with electrically heated furnaces:

- Systems are clean and free of the pollution normally found with fuel-fired systems.
- Cooler plant environment without exhaust stacks and hoods
- Quieter because of the absence of blowers and combustion noise
- More uniform heat pattern from grid elements or side-to-side uniformity on electric radiant tubes

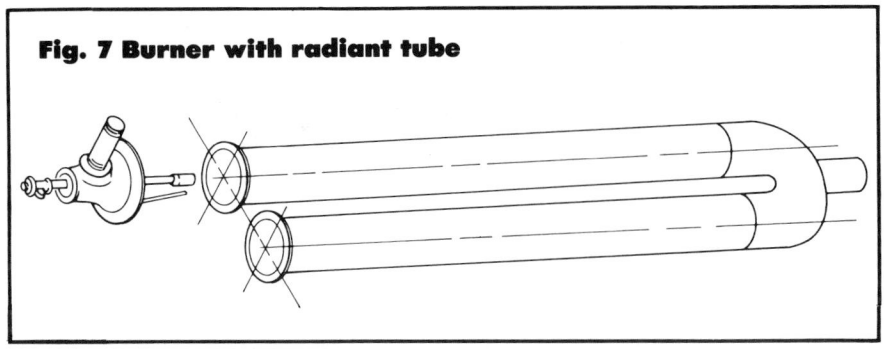

Fig. 7 Burner with radiant tube

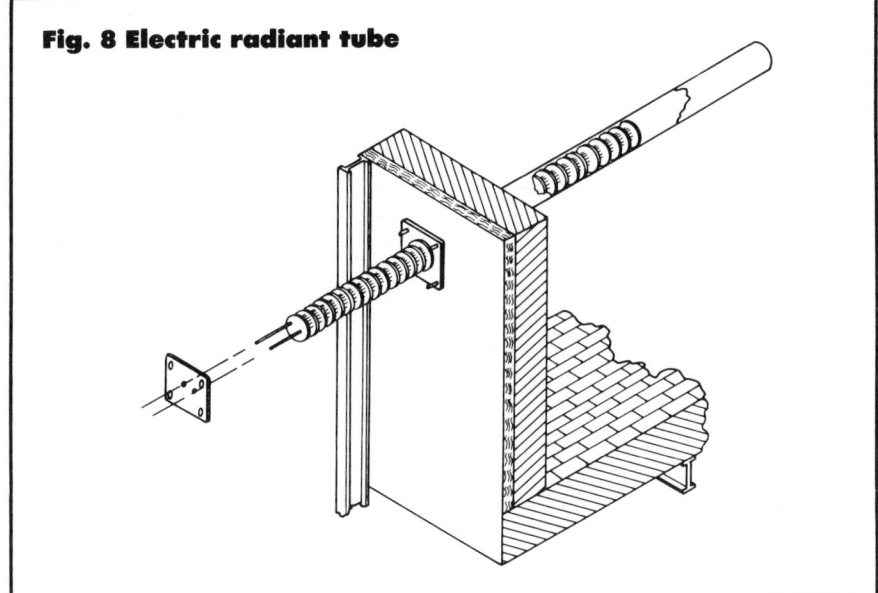

Fig. 8 Electric radiant tube

- No exhaust system required to affect building air pressures; no make-up air system required
- Does not require purge or flame safety systems
- Electrical power availability.

Disadvantages of Electrically Heated Furnaces. The following are disadvantages of electrically heated furnaces:

- Inflexible system makes changing connected heating capacities or varying individual element capacities in the same zone difficult
- High initial equipment costs
- With numerous electric furnaces in a plant and no form of peak-demand control, the user pays at a high demand rate for all power.
- Higher operating costs
- Higher maintenance costs
- Cool down times are longer because no combustion air is available.

Radiant-Tube-Heated Furnace Equipment

With fuel-fired, radiant-tube-heated furnaces, the work chamber is protected from the products of combustion. With a radiant-tube furnace, the work chamber normally contains a manufactured controlled atmosphere as dictated by the process. There are, however, cases where the chamber remains filled with air, and the only purpose of the radiant tubes is to protect the work from the high dew point flue products. Electrically heated radiant tubes normally are used to protect the heating-element material from attack by the furnace atmosphere.

Gas-fired radiant tubes, as shown in Fig. 7, are by far the most common type of fuel-fired indirect method of heating. This is mainly due to the wide availability of natural gas. The proper selection of furnace parts

such as burner components, controls, orifices and piping, will allow the same radiant tube to be fired with a wide variety of gases: natural gas, propane-air mix, straight propane and certain low-Btu manufactured gases.

Radiant-tube burners are of two basic types, sealed-head and open-type burners. A particular advantage of the sealed-head burner is that it is readily recuperable, by using the products of combustion and an air-to-flue-gas heat exchanger to preheat the combustion air prior to its entering the burner. This advantage can result in considerable fuel savings. Radiant tubes are normally constructed from centrifugally cast tubing of various diameters, with wall thicknesses of 6 and 8 mm (1/4 and 5/16 in.). In many cases, tubes fabricated from 3 and 5 mm (1/8 and 3/16 in.) wrought alloy are used.

With the high metal temperatures attained in radiant tubes in most heat treating furnaces, the alloys commonly used are the higher-grade cast alloys HT, HK, and NA22H or the wrought alloys 330, 601 and Incoloy 800.

Oil-Fired Radiant Tubes. Straight oil-fired radiant tubes are somewhat uncommon and are used mainly where an adequate supply of gaseous fuel is not available. Radiant tubes equipped to burn both oil and gas are more common, with the oil, usually No. 2 fuel oil, used as the standby fuel.

High vanadium and sulfur content in oil has a great effect on the useful life of nickel-chrome heat-resisting alloys. With the high temperatures attained inside the radiant tubes acting as a catalyst, attack on the alloy is accentuated.

When oil-fired radiant tubes are used, the construction and types of materials used are similar to those described for the gas-fired radiant tubes. Recuperation is also possible with certain sealed-head oil burners. The burner manufacturer should be consulted regarding possible damage to or problems with the atomizing system from the preheated combustion air, normally supplied at 370 to 540 °C (700 to 1000 °F).

Electrically Heated Radiant Tubes. With this design, the radiant tube protects the resistance heating element from the furnace atmosphere. A common design, as shown in Fig. 8, uses nickel-chrome alloy rod inside the radiant tube. These rods are formed into hairpin shapes supported and contained by ceramic spacer discs.

With the heating element contained in a tube, it is very important to conservatively select the heating element watt density. The design watt density is a direct function of furnace temperature and varies from 18 600 to 46 500 W/m² (12 to 30 W/in.²). Some designs use nickel-chrome strip material rather than rod in similar hairpin patterns.

If a gas-fired "U"-tube furnace is to be converted to electrically heated radiant tubes, it is desirable to replace the "U" tube with two straight tubes for improved uniformity and reduced watt density. Thus, the conversion from natural gas to electricity can be accomplished without loss in production capacity.

Other types of electric radiant tubes are available, where the radiant tube itself becomes the resistive heating element. As with cast elements, however, the quality and condition of the radiant tube will determine whether it functions properly as a resistance element. Silicon carbide elements have also been used inside radiant tubes to protect them from carburizing atmospheres.

General Furnace Maintenance

Maintenance on a furnace should be performed on a regular basis to prevent unscheduled shutdowns. In most plants, major maintenance is performed once each year, normally during the plant vacation period.

Because unscheduled maintenance is very disruptive to production, especially in plants without backup heat treating capabilities, some manufacturers prefer a number of smaller furnaces rather than a single large furnace capable of handling all production.

Many components of furnaces must be considered as consumable items, although lifespan normally can be predicted from accurate maintenance records. Many furnace owners regularly inspect and change internal furnace components such as radiant tubes, retorts and electric heating elements. Trying to get a few more months of life from a certain internal component could result in an unscheduled shutdown, and an extended loss of production can occur if a replacement part is not readily available. Most furnace equipment manufacturers provide a recommended list of spare parts that the furnace owner should maintain in stock to ensure reasonably uninterrupted production.

In addition to the consumable items that have to be regularly replaced, many furnace components must be adjusted and/or calibrated at regular intervals to maintain the efficiency and accuracy of the heat treating operation—once each shift, daily, weekly, monthly or annually. Components that require regular monitoring for adjustment and calibration are mainly those that control the quality of the heat treating process, such as thermocouples, temperature-control instruments and gas analyzers. For example, some furnace operators change all thermocouples at specified intervals to avoid the gradual deterioration that occurs prior to indiscriminate failure.

In addition, maintenance required to minimize wear and thus prolong component life must also be considered. This form of maintenance usually consists of a well-planned and documented lubrication schedule. Lubrication can be accomplished manually or with an automatic system.

Care should be taken in the proper selection of the various greases and oils, to ensure compatibility with the various furnace components. Excessive greasing should be avoided, as it will limit the life of the bearing seals. Excessively long grease supply lines should also be avoided because grease may harden before it reaches the point of use.

On manual systems, lubrication points should be carefully coded. Many special greases are incompatible with each other, and lubricating components with the wrong grease can have disastrous results.

For major maintenance such as complete refractory replacement or complete rebuilding of mechanisms and systems, it is normally best to consult the original-equipment manufacturer. Because this type of maintenance usually occurs when the equipment is quite old, the manufacturer will normally make recommendations that will improve the equipment and upgrade the system to the present state of the art and to comply with current recognized safety standards.

Salt Bath Equipment

By W. James Laird, Jr.
Vice President-Marketing
Research & Development
Upton Industries, Inc.

SALT BATHS are used in a wide variety of commercial heat treating operations, including cyaniding, liquid carburizing, liquid nitriding, austempering, martempering and tempering applications. Salt bath equipment is well adapted to heat treatment of tool steels, as well as to treatment of nonferrous alloys. Advantages of using salt bath equipment include thermal control and rapid heating rates. Applications of the various furnace designs and auxiliary equipment to specific heat treating processes are described in other articles in this volume.

Externally Heated Furnaces

Externally heated salt bath furnaces may be fired by gas or oil, or heated by means of electrical resistance elements. A typical gas- or oil-fired furnace that is commonly used in liquid carburizing applications is shown in Fig. 4a in the article, "Liquid Carburizing and Cyaniding", in this Handbook. These furnaces are generally lower in initial cost than electrode or resistance heated furnaces and are simple to install and operate. To contain the molten carburizing salt, fuel-fired furnaces employ a steel or alloy pot, which may be either round or rectangular. Heat is applied by two or more self-cooling burners that fire tangentially

between the outer wall of the pot and the inner surface of the furnace lining. The hot gases are vented through a flue located near the top for atmosphere-type burners, or near the bottom for pressure-type burners and for atmosphere-type burners for which the flue is connected to a stack about 1 to 2 m (3.3 to 6.6 ft) high, to maintain negative pressure in the firing chamber. The combustion chamber is lined with firebrick and additional insulation is required. A steel casing completely surrounds all sides of the furnace housing and provides adequate safety in the event of pot failure.

Electrical resistance furnaces for neutral heating or liquid carburizing are less widely used than furnaces fired by gas or oil. (See Fig. 4b in the article, "Liquid Carburizing and Cyaniding".) These furnaces are heated by a series of resistance heaters surrounding the salt pot. For this reason, pot failure may result in the total destruction of the electrical heating elements. Operating temperatures below 900 °C (1650 °F) are used to prevent pot failure.

Salt pots are usually supported from a flange; consequently, pot size is limited by the strength of the flange material. Round pots for gas- or oil-fired furnaces normally range from 250 to 900 mm (9.8 to 35.4 in.) in diameter and from 200 to 750 mm (7.9 to 29.5 in.) in depth; they are about 10 mm (0.4 in.) thick. Larger sizes have been built for special applications and have operated successfully. Pots larger than about 350 mm (13.8 in.) in diameter and 450 mm (17.7 in.) deep are rarely used for electrical resistance furnaces. Although it is physically possible to support the bottom of a large pot on a refractory pier, excessive temperature gradients may result.

Pots may be press formed from a single piece of low-carbon steel or iron-nickel-chromium alloy; a composition of Fe-35Ni-15Cr is usually preferred for the latter. Less expensive welded pots may be fabricated from either of these materials.

In a well-designed furnace, the life of round alloy pots will vary with maximum operating temperature about as follows:

Temperature		Service life, months
°C	°F	
840	1550	9 to 12
870	1600	6 to 9
900	1650	3 to 6
920	1690	1 to 3

In one installation, the placement of an additional control thermocouple in the combustion chamber to prevent the temperature of the chamber from exceeding 1095 °C (2000 °F) served to extend the life of high-temperature (HT) alloy pots to 2 years (previous life had been 6 months). Pot temperature was maintained at 900 °C (1650 °F) during a work week of 120 h (24 h/day, 5 days/week).

Temperature of the salt is measured and indicated by a thermocouple and suitable pyrometer. Operating within the range from 790 to 920 °C (1455 to 1690 °F), the externally fired furnace may vary as much as 10 °C (18 °F) above and below the set temperature when using on-off or high-low control systems. This is considered acceptable for many applications. Where closer control of temperature is required, a proportional control system, which will hold temperature variations to less than ±5 °C (±9 °F) should be used.

Design and Operating Factors. In the design of fuel-fired furnaces, ample space must be provided for combustion so that the flame will not impinge on the pot. If flame impingement is unavoidable, the pot should be rotated slightly at least once a week. Rotating the pot and using a sleeve reduce local deterioration in the region of flame impingement and prolong service life of the pot. The combustion-chamber atmosphere also has important effects on pot life. A system with a control range from high fire to low fire is preferable to an on-off system, because the latter allows air to enter the combustion chamber during the "off" portion of the cycle, thereby increasing the rate of sealing of the outer surfaces of the pot.

Electrical resistance heated furnaces should be equipped with a second pyrometer controller whose thermocouple is in the heating chamber. This will prevent overheating of the resistance elements, particularly during melt-down, when the thermocouple that controls the temperature of the main bath is insulated by unmelted salt. Because heating elements and refractories are severely attacked by salt, all salt must be kept out of the combustion chamber. For this purpose, a mixture of high-temperature refractory cement, with long-fiber asbestos for strength, may be used to seal joints where the pot flange rests on the retaining ring at the top of the furnace.

Externally heated pots should be started on low fire—low heat input— regardless of the method of heating. Once the salt appears to melt around the top, heat can be gradually increased to high fire to complete melt-down. *Excessive heating of the sidewalls or pot bottom during start-up may cre-* *ate pressures sufficient to expel salt violently from the pot.* For added safety, the pot should be covered during melt-down, either with a cover or with an unfastened steel plate.

The waste heat of flue gases may be fed to an adjacent chamber and used to preheat work. Flue gases should always be visible to the operator. The presence of bluish white or white fumes at the vent are an indication that salts have entered the combustion chamber; prompt corrective action is required.

Advantages and Disadvantages. Because of the ease with which they can be restarted, externally heated furnaces are well suited to intermittent operations. Another advantage of furnaces of this type is that a single furnace can be used for a variety of applications by simply changing the pot for one containing the proper salt composition.

Externally heated furnaces have several characteristics, however, that limit their usefulness in certain operations. They are less adaptable to close and uniform temperature control because they dissipate heat by convection, creating temperature gradients in the bath. Also, the temperature lag of the thermocouple and the recovery time of the furnace may result in overshooting and undershooting of a desired temperature control point by 14 °C (25 °F). In addition to requiring an exhaust system for generated flue gases, externally heated furnaces may overheat at the bottom and sidewalls in restarting creating a pressure buildup in the thermally expanding molten salt that may cause an explosion. Finally, externally heated furnaces are seldom practical for continuous high-volume production, because of the limitations of pots with respect to size and maximum operating temperature. High maintenance cost is also a factor.

Immersed-Electrode Furnaces

Introduction of the immersed-electrode furnace greatly extended the useful range and capacity of molten carburizing baths. The electrodes can be removed and replaced without bailing the furnace. This design is also suitable for neutral heating, as well as for cyanide and noncyanide carburizing processes. The molten salt is contained in a steel or ceramic pot surrounded by suitable insulating materials, which separate it from an exterior casing or framework of heavy-gage steel. The salt is heated by passing alternating current through it with immersed electrodes. As a result of the resistance built up to passage of current through salt, heat is generated within the salt itself. This heat is quickly dissipated by a downward stirring action created by the electrodes. The electrodes are attached by copper connectors to a transformer that converts the line voltage of the plant to a much lower secondary voltage (approximately 9 to 12 V) across the electrodes. Temperature is measured and automatically controlled by a system containing a thermocouple, pyrometer, relay and magnetic contactor. A typical immersed-electrode furnace is shown in Fig. 4c in the article, "Liquid Carburizing and Cyaniding", in this Handbook.

The depth of salt pots for immersed-electrode furnaces is usually limited to 0.6 m (2 ft) for metal pots; ceramic pot depth is nearly unrestricted. Furnaces with pots up to 4.5 m (14.8 ft) in length, and with a power input of 360 kW, are presently in operation. They have a capacity to heat a workload of about 320 kg/h (705 lb/h). In contrast, smaller units with salt pots having a work space measuring 230 by 180 by 350 mm (9 in. by 7 in. by 14 in.) deep, and with 15-kW power input, can heat about 22 kg/h (50 lb/h) to 920 °C (1690 °F).

Advantages and Disadvantages. Immersed-electrode furnaces do not require the use of iron-chromium-nickel alloy pots. Under normal operating conditions, the life of a carbon steel pot is usually 1 year or more. Carbon steel pots of welded construction, set into insulating brick but not cemented in place, give the following service life:

Maximum operating temperature		Service
°C	°F	life, years
840	1545	2 to 3
870	1600	1½ to 2
900	1650	1 to 1½
920	1690	1

These furnaces require minimum floor space and maintenance and can be used for all types of carburizing salts. Electrodes made of alloy steel should have an average service life equivalent to that indicated for steel pots in the above table. Worn electrodes can be replaced while the furnace is in operation.

Depending on the positioning of electrodes, temperature control to within

±3 °C (±5 °F) is easily obtained with immersed-electrode furnaces. Heat is generated within the bath, and overshooting is readily avoided. These furnaces lend themselves to mechanization and are suitable for high-volume production in the range of 815 to 1300 °C (1500 to 2370 °F).

Maximum pot size is not restricted. Pots may vary in length and width to suit requirements, and multiple pairs of electrodes can be installed to furnish the necessary heating capacity.

The immersed-electrode furnace is not recommended for intermittent operation. Depending on furnace size, reheating the salt charge may require a day or more. Pots are not intended to be interchangeable. Removal of the pot usually involves replacement of the surrounding insulation.

Steel Pot Furnaces

Some metal treating processes are performed in salt compounds that cannot be contained in a ceramic liner. For these applications, furnace manufacturers make use of a welded steel pot with immersed electrodes. This type of furnace is suitable for special applications such as case hardening in straight cyanide baths, tempering, and marquenching.

Construction. The steel pot has a sloped back wall, which produces a "bottom heating" effect resulting in better circulation and uniform temperature. This is accomplished by sloping the electrodes as shown in Fig. 1 and 2. As the current passes through the salt between the electrodes, the salt is heated, decreasing its density and causing it to rise toward the bath surface. Control of the rate of rise of the salt is effectively obtained by decreasing the distance from the electrodes to the steel pot. At the lower extremity of the electrode, the current enters the metal pot upon leaving the electrode to follow a shorter path to the other electrode. This arrangement ensures current flow through the salt along the entire electrode length. Due to the close proximity of the lower portion of the electrode to the pot, most of the heating is done in the lower part of the bath. This is the desired method of heating any liquid.

The metal pots are made of either plain steel or hot-dipped aluminized steel depending on the application. Thicknesses range from 12 to 38 mm (½ to 1½ in.). Reinforcing members for light plate, usually angular in shape, are welded around the top. Where

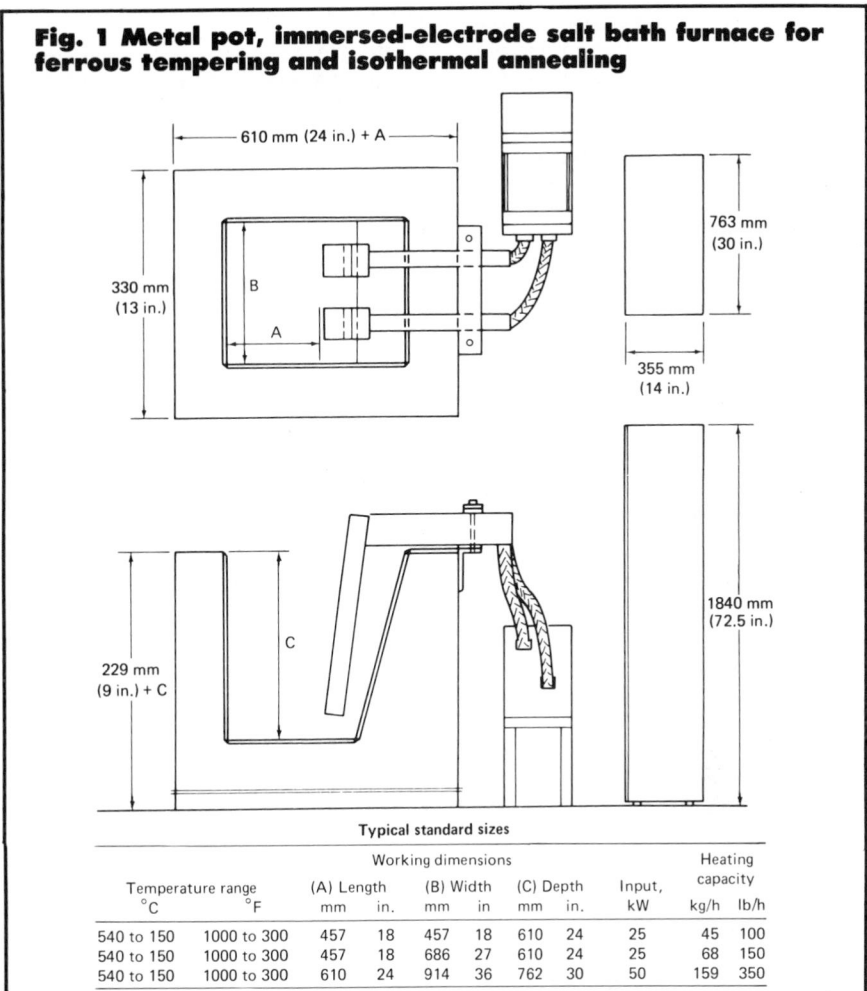

Fig. 1 Metal pot, immersed-electrode salt bath furnace for ferrous tempering and isothermal annealing

610 mm (24 in.) + A

330 mm (13 in.)

B

A

763 mm (30 in.)

355 mm (14 in.)

229 mm (9 in.) + C

C

1840 mm (72.5 in.)

Typical standard sizes

| Temperature range | | Working dimensions | | | | | | Input, | Heating capacity | |
°C	°F	(A) Length mm	in.	(B) Width mm	in	(C) Depth mm	in.	kW	kg/h	lb/h
540 to 150	1000 to 300	457	18	457	18	610	24	25	45	100
540 to 150	1000 to 300	457	18	686	27	610	24	25	68	150
540 to 150	1000 to 300	610	24	914	36	762	30	50	159	350

depth of the pot so requires, additional members are used at the midsection.

The pot is housed in an insulated 229-mm- (9-in.-) thick wall furnace with either a brick outside wall contained in a rigid welded steel frame or in a steel clad frame, depending on personal preference. In either type of construction, the frame is self supporting on a lattice formed by welding channels or beams to the underside of a steel base plate. The pot is supported on an insulated refractory pedestal.

Electrode Arrangement. Immersed electrodes are made of either mild steel or an alloy "hot" leg welded to a mild steel "cold" leg. As previously mentioned, these are shaped to follow approximately the slope of the pot wall. The portion of the electrode that crosses over the top of the salt bath and is connected to the plant power source is referred to as the "cold" leg. This is welded to the "hot" leg, or the portion of the electrode that is immersed in the

bath, with sufficient weld cross section to provide the necessary current conductor capacity (see Fig. 1). The shanks are drilled and tapped at the tinned terminal connection end for water cooling when necessary. If the latter is not required, the electrical connection is water cooled. Suitable clamping devices are used to facilitate electrode replacement.

- *Single-phase operation:* Several electrode arrangements can be used, depending on the size of the bath. If only two electrodes are required, they are normally positioned on the sloped wall side and at least 127 mm (5 in.) apart. Three electrodes are usually placed so that the center electrode, equal in size to two of the other electrodes, is used as a common conductor with equal current paths to each of the outer electrodes. More than three electrodes would be arranged in multiple groups.
- *Three-phase operation:* Three elec-

trodes are used and spaced in a manner similar to the spacing described above. They are connected to three single-phase transformers that have "Y" connected secondaries and "delta" connected primaries. The current flows from the electrodes to the pot, which is the neutral point. Several variations of three-phase connections are used, depending on the type of furnace and load requirements.

All accessories such as starting units, transformers, sludging tools and secondary connectors are the same for steel pot immersed-electrode furnaces as for ceramic furnaces.

Submerged-Electrode Furnaces

In a typical electrically heated submerged-electrode furnace, the frame is made of heavy angle iron with a steel plate at the base beneath the brickwork. The outer brickwork consists of hollow ceramic tile or common building brick. The salt pot is made of burned alumina firebrick. Castable insulating refractory fills the space between the sidewalls and the ceramic pot. An electrically heated submerged-electrode furnace is shown in Fig. 4d in the article, "Liquid Carburizing and Cyaniding", in this Handbook.

When salt is melted in the pot, it penetrates the refractory until it becomes cool enough to freeze. The resulting shell of solidified salt retains the liquid salt in the furnace. If the refractory develops a crack, bath temperature must be lowered to permit salt to solidify in the crack.

Water-cooled electrodes are in contact with liquid salt in the pot and are sealed in the refractory walls by frozen salt. Current travels between the electrodes, which are flush with the sidewalls. The path of current travel extends a few inches above the top of the electrodes.

Start-up and Shutdown. The submerged-electrode furnace can be started by adding molten salt from another furnace or by using a gas-fired torch or electric starting coil to melt a pool of salt that will wet both electrodes and provide molten salt for the current path. After the current path has been established in the molten salt between the electrodes, salt may be added to bring the bath up to working level. Additional salt will be required to maintain this level because a small

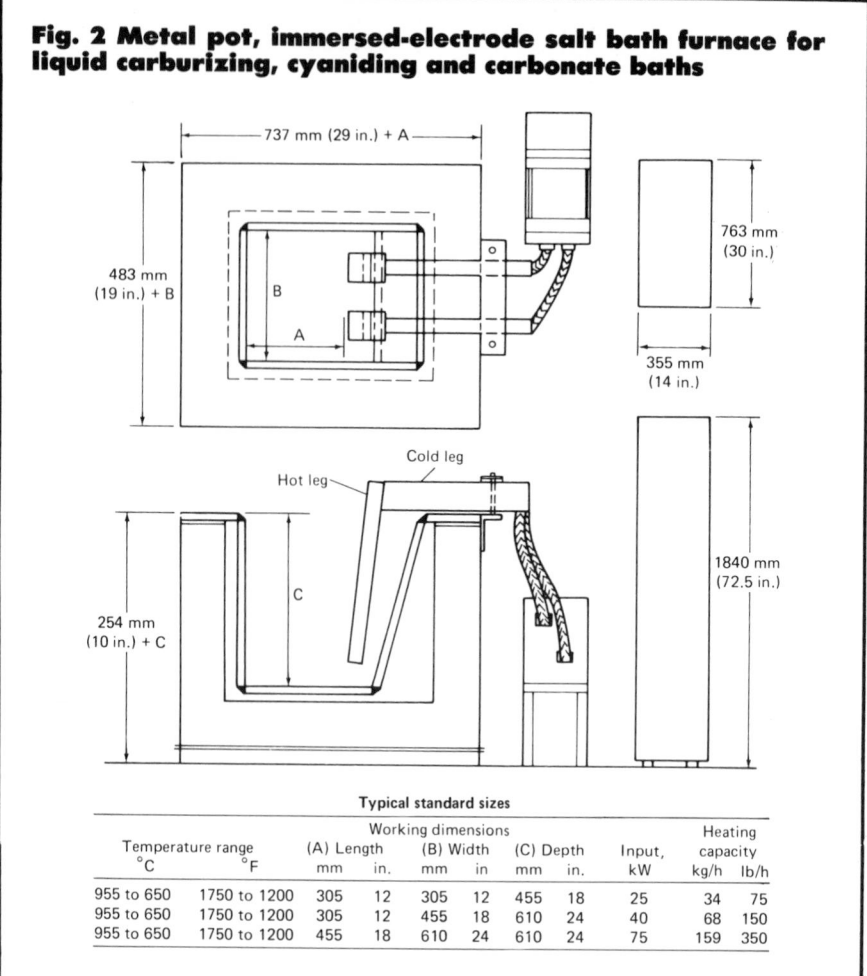

Fig. 2 Metal pot, immersed-electrode salt bath furnace for liquid carburizing, cyaniding and carbonate baths

Typical standard sizes

Temperature range °C	°F	Working dimensions (A) Length mm	in.	(B) Width mm	in	(C) Depth mm	in.	Input, kW	Heating capacity kg/h	lb/h
955 to 650	1750 to 1200	305	12	305	12	455	18	25	34	75
955 to 650	1750 to 1200	305	12	455	18	610	24	40	68	150
955 to 650	1750 to 1200	455	18	610	24	610	24	75	159	350

amount, ≃5%, will seep into the brickwork and freeze.

If the furnace must be shut down, the molten salt should be bailed from the furnace before it freezes. However, if the salt is allowed to remain in the furnace, a resistance heated starting coil should be submerged in the bottom of the furnace while the salt is still molten. This coil remains in the frozen salt, and it is connected to the transformer leads to start up the furnace.

Advantages and Disadvantages. In common with immersed-electrode furnaces, submerged-electrode furnaces require minimum floor space and maintenance and are highly adaptable to mechanization.

Because the submerged-electrode furnace employs water to cool the electrodes and transformer, it may be operated at 50% overload without overheating the transformer, whereas the immersed-electrode furnace, being air cooled, should not be operated at an overload above 10%.

Because a ceramic pot is used, unexpected pot failure is rare with submerged-electrode furnaces, and the furnaces can be rebuilt on a planned schedule during annual shutdowns. In common with other electrical equipment, submerged-electrode furnaces are at a disadvantage where electric power rates are high, but this can be overcome to some extent by working the furnace in non-peak periods when lower power rates are applicable.

Because of the erosive effects on ceramic pots of water-soluble salts with high sodium carbonate or high sodium cyanide contents, submerged-electrode furnaces can be used only with low-cyanide, low-carbonate salts. Baths with high cyanide or carbonate salt require a modified basic brick. The furnace with modified brick and submerged alloy electrodes provides many years of service in noncyanide and cyanide operations. To increase furnace life, the furnace shown in Fig. 3 is recommended. This furnace has a modified basic

Fig. 3 Electrically heated submerged graphite electrodes

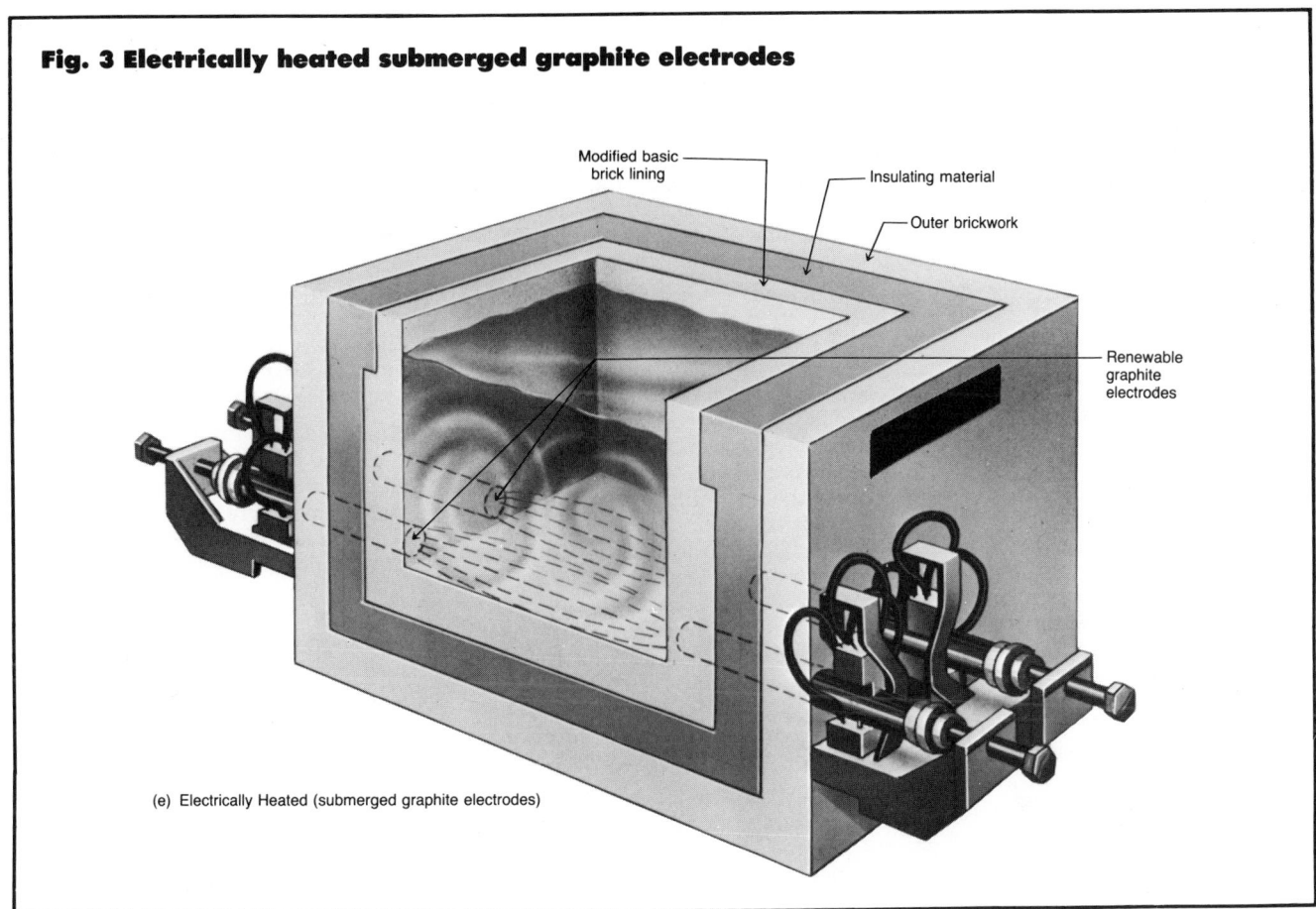

Modified basic brick lining

Insulating material

Outer brickwork

Renewable graphite electrodes

(e) Electrically Heated (submerged graphite electrodes)

brick lining for use with the basic carburizing salts. The alloy electrodes are replaced with continuing graphite electrodes. The electrodes are renewed as they become consumed without disconnecting them or even shutting off the power.

Automatic and Semiautomatic Lines

A fully automatic "jackrabbit" mechanism used for salt bath carburizing and hardening is shown in Fig. 6a in the article "Liquid Carburizing and Cyaniding", in this Handbook. The mechanism has synchronized, continuous chain conveyors that carry the work through the various operations. Work suspended from horizontal bars is moved through baths at the proper speed by a main conveyor. Transfer conveyors carry the work from bath to bath. Completed work is picked up by a third conveyor and is dried by warm air in the enclosed upper portion of the structure while it is being returned to the loading point. One operator loads and unloads the work.

This mechanism can be used only for work having similar requirements. It does not permit varying the time cycle of any one operation without affecting the cycles of the other synchronized operations.

Where part requirements are different, a semiautomatic mechanism may be used. (See Fig. 6b in the article, "Liquid Carburizing and Cyaniding".) Work is transferred between operations by an overhead monorail and is automatically advanced through the carburizing and tempering furnaces by means of a push-pull mechanism. This mechanism consists of two parallel beams with reciprocating push bars. Driven either hydraulically or electrically, the bars carry the dogs, which are spaced to advance the fixtures at the center of the furnace only a part of the stroke, while the end fixtures are advanced through the entire stroke. By closer spacing of the work at the center of the furnace and wider spacing at the ends, high productive capacity is achieved with ease of loading and unloading. A semiautomatic line of this type permits variations in the time cycling of any one operation without affecting other operations, and is less likely to require modification if the work requirements change.

A fully automatic programmable hoist carburizing line is shown in Fig. 6c in the article, "Liquid Carburizing and Cyaniding". The use of automated hoists makes possible the combination of austempering, martempering, and tempering or carburizing in one line. One or more hoists travel simultaneously back and forth automatically advancing the fixture carriers of work through the required stations.

The hoist movement is controlled by a solid state programmable control with functions that would normally require hundreds of relays, counters, switches and extensive wiring. Once programmed, the controller performs the desired commands and functions specified by the user. Time cycles, sequences, drills, and skips are easily entered or changed to meet metallurgical requirements. For instance, parts can be programmed to be carburized, air cooled, washed, rinsed, and returned for unloading. A push-button command then returns the program to standard processing.

Parts suitable for fully automatic or semiautomatic installations are those that can be fixtured by wiring, racking or placing in baskets and that do not present problems in either buoyancy or drainage.

Isothermal Quench Furnaces

Isothermal quench furnace systems were designed to eliminate the occurrence of chloride carry-over from the austenitizing bath to the quench bath, through salt separation and uniform vertical lamellar flow agitation. The three most common approaches to alleviating the salt concentration are chemical, temperature and gravity separation.

Chemical Precipitation. Chemical agents have been used to attempt to lower the solubility of the salts that precipitate into the quench. When the salts settle to the bottom of the quench tank, they are removed as sludge. This method offers little success in that the precipitate that forms is fine-textured and buoyant and thus tends to remain in suspension rather than precipitating out.

Temperature Precipitation. The elimination of carry-over salts has also been attempted by continuously pumping salt through a small auxiliary chamber whose temperature is maintained at a lower level than the main chamber. As the salt is processed through the auxiliary chamber, chlorides are continuously precipitated out.

Although this method appears practical, a fundamental error exists in its application. The salt is cooled by air blown through a space between the pot and outer shell of the precipitation chamber. Air is blown through this space to maintain the temperature lev-

els of the main chamber and precipitation chambers. The moving air cools the pot walls below the salt precipitation point so that the salt freezes and cakes to the sides. Salt buildup continues until the bath is unusable. Consequently, depending on the level of salt concentration, the bath would have to be shut down, possibly after only a few weeks of operation, to remove the remaining molten salt and chip away the caked salt.

Gravity Separation. This system of carry-over salt removal also uses a two-chamber design. The caking problem is eliminated by heavily insulating the pot walls at all points and using an internal air-water heat exchanger. Because the pot walls and salt are at the same temperature, there is no caking action. The chloride salts settle into an easily removable shallow pan at the bottom of the precipitating chamber, or if they are fine textured and buoyant, the salts float to the top of the tanks and are easily skimmed off.

The main advantages of two-chamber gravity equipment separation include:

- Easily removable variable speed propeller-type agitator with suitable baffling to provide vertical lamellar flow within the quench area, therefore, ensuring maximum quench power and minimum distortion
- Separate chloride precipitation chamber with adjustable weirs to maintain a low chloride level and subsequently high quenching power
- Easily removable internal heat exchanger to maintain quench temperature and precipitate chlorides
- Easily removable settling pan to ensure maximum efficiency in removal of chlorides
- Heavily insulated pot and precipitation chamber to eliminate salt caking on walls.

Furnace Heating. Generally, either gas or electricity may be used for heating isothermal quenching furnaces. When gas heating is desired, immersion tubes are recommended because the tubes are usually made of mild steel and provide long service life.

Further, if the pot should develop a leak, the insulation and outer shell will contain the salt. *If a furnace with an externally heated pot were to develop a leak, the nitrate-nitrite salt would drip on the hot refractory and may cause a fire hazard.* One or more immersion tubes normally are used, depending on bath size. Generally, they will have nozzle-mix sealed-in burners and will be available to FM or FIA specification.

Electric heating may be by one of the following methods depending on the maximum operating temperature:

- *Sheathed resistance strip heaters* are mounted externally to the side walls near the bottom. Maximum operating temperature is 425 °C (800 °F). They are easily removable through an insulated plug-type door. Protection against overshooting is achieved by locating a sensing device in close proximity to the heaters. The sensors operate directly on line voltage.
- *Sheathed resistance immersion heaters* have a maximum operating temperature of 425 °C (800 °F). They can operate without a transformer, but are susceptible to premature burnout due to sludge accumulation or operator tampering and abuse.
- *Immersed-electrode heaters* operate in the same manner as electrode pot furnaces for carburizing and tempering.

Furnace Construction. The pot is fabricated from firebox-quality steel plate double welded inside and out and properly supported to maintain its shape. Steel plate offers adequate resistance to chemical attack by the standard alkaline nitrate-nitrite salts at normal austempering and martempering temperatures. The pot is insulated with 102 to 152 mm (4 to 6 in.) of slab-type mineral insulation to prevent the chloride saturated nitrate salt from freezing to the sidewalls or bottom. The insulation is externally contained by a continuously welded outer steel shell. The shell is reinforced to ensure retention of the original shape and dimensions throughout its designed operating temperature range.

Fluidized-Bed Equipment

Edited by William L. James
President
Fennell Corp.

FLUIDIZED-BED TECHNIQUES are not new. A 19th century American patent describes the roasting of minerals under fluidized-bed conditions. Other established applications include potter's clay and miner's hydraulic slurries. Systems of fluidized solid particles, such as quicksand, occur in nature.

Early attempts to use fluidized beds in heat treatment of metals were limited in the temperatures that could be employed. Electrically heated furnaces capable of maintaining fluidized beds at temperatures up to 500 °C (930 °F) could be produced commercially, but difficulties were encountered when attempts were made to attain higher temperatures. A principal problem was the high rate at which refractory distributors, which distribute the hot fluidizing gases, were consumed.

In early gas-fired fluidized-bed furnace design, gas entered the base of the container after being mixed with air to make it ignitable at the point of entry. With newer designs, the mixtures are introduced separately and so cannot be ignited accidentally. This design eliminates the danger of explosion at the point of entry. The surface of the bed is heated first, and the heating of surface particles causes progressive ignition downward through the container until the entire contents of the bed achieves uniform heat treating temperature.

Newer furnace designs extend fluidized-bed technology into the higher temperature ranges required for most common heat treatments.

Principles of Fluidized-Bed Heat Treating

In fluidization, a bed of dry, finely divided particles, typically aluminum oxide in the heat treating context, is made to behave like a liquid by a moving gas fed upward through the bed. A gas-fluidized bed is considered a dense-phase fluidized bed when it exhibits a clearly defined upper limit or surface. At a sufficiently high fluid-flow rate, however, the terminal velocity of the solids is exceeded, the bed goes into motion, and the upper surface of the bed disappears. This state constitutes a disperse, dilute or lean-phase fluidized bed with pneumatic transport of solids. The general types of fluidized beds are shown in Fig. 1. The majority of beds used for heat treatment are of the aggregative or bubbling type.

Although the properties of solid and fluid alone determine the quality of fluidization (that is, whether smooth or bubbling fluidization occurs), many factors influence the rate of solid mixing, the size of bubbles and the extent of heterogeneity in the bed. These factors include bed geometry; gas-flow rate; type of gas distributor; and internal-vessel features such as screens, baffles and heat exchangers.

Determination of Fluidization Velocity. In determining the quality of fluidization, a diagram of pressure drop (Δp) versus velocity (μo) is useful as a rough indication when visual observation is not possible. A well-fluidized bed will behave as shown in

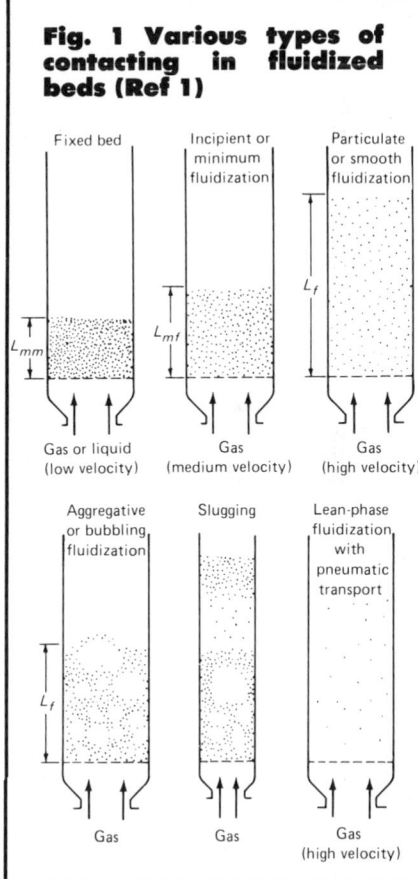

Fig. 1 Various types of contacting in fluidized beds (Ref 1)

the diagram in Fig. 2, which has two distinct zones. In the first, at relatively low flow rates in a packed bed, the pressure drop is approximately proportional to gas velocity, usually reaching a

Fig. 2 Pressure drop versus gas velocity for a bed of uniform-size particles (Ref 1)

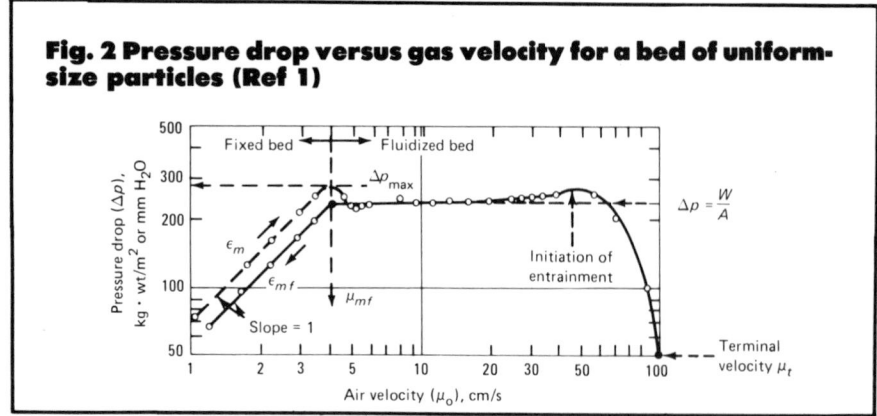

Fig. 4 Quality of fluidization as influenced by type of gas distributor (Ref 1)

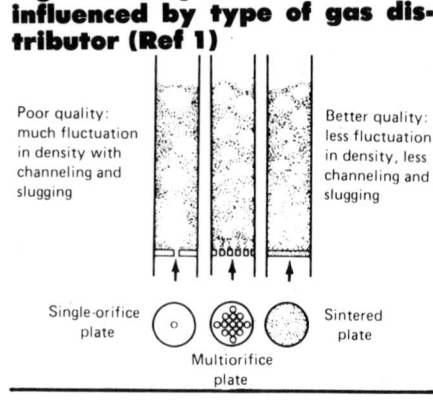

Poor quality: much fluctuation in density with channeling and slugging

Better quality: less fluctuation in density, less channeling and slugging

Single-orifice plate

Multiorifice plate

Sintered plate

Fig. 3 Pressure-drop diagrams for poorly fluidized beds (Ref 1)

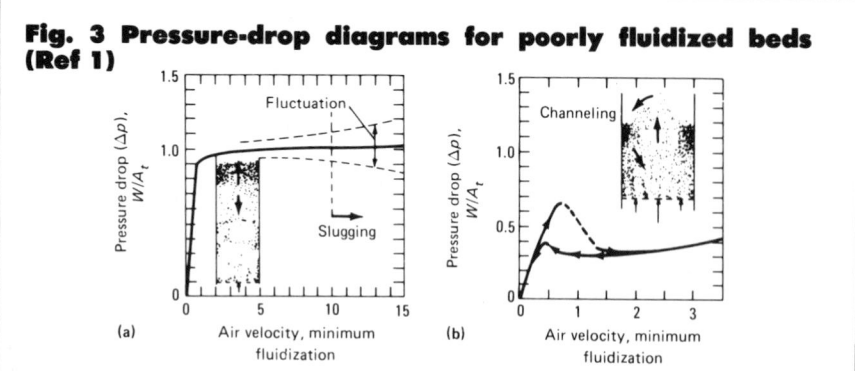

Fig. 5 Effect of temperature on the flow corresponding to minimum fluidization for particles 0.1 mm (0.004 in.) in diameter having an apparent density of 2 (Ref 1)

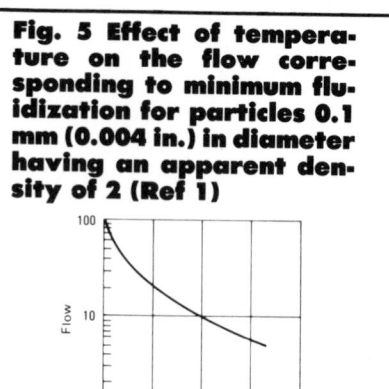

maximum value (Δp_{max}) slightly higher than the static pressure of the bed. With a further increase in gas velocity, the packed bed suddenly "unlocks" and becomes fluid-like.

When gas velocity increases beyond minimum fluidization (μ_{mf}), the bed expands and gas bubbles rise, resulting in a heterogeneous bed. This is the second zone, in which, despite a rise in gas flow, the pressure drop remains practically unchanged. The dense gas-solid phase is well aerated and can deform easily without appreciable resistance. In its hydrodynamic behavior, the dense phase can be likened to a liquid. If a gas is introduced into the bottom of a tank containing a liquid of low viscosity, the pressure required for injection is roughly the static pressure of the liquid and is independent of the flow rate of the gas. The constancy in pressure drop in both the bubbling liquid and the bubbling fluidized bed may be taken intuitively to be analogous.

The diagrams in Fig. 3 show poorly fluidized beds. The large pressure fluctuations in Fig. 3(a) suggest a slugging bed. In Fig. 3(b), an absence of the characteristic sharp change in slope at minimum fluidization and the abnor-

mally low pressure drop suggest incomplete contacting, with particles only partly fluidized.

One of the most important factors influencing the quality of fluidization is the type of distributor. Figure 4 illustrates this schematically.

Temperature Effect on Minimum Fluidization Velocity. One of the most important parameters of a fluidized bed is the minimum fluidization velocity. In simplified terms, minimum fluidization velocity (μ_{mf}) approximates to a function of the square of the particle diameter (d) and a linear function of particle mass (p) as follows:

$$\mu_{mf} \simeq d^2 p$$

In the design of heat treating furnaces, the effect of temperature must be considered. Figure 5 shows how the flow of gas required for fluidization decreases rapidly with increases in temperature.

Defluidization. One of the common misconceptions about fluidized beds is that, because of their principle of operation, they are not well suited for large solid parts with horizontal surfaces that remain stationary in the bed. With parts of this type, a cap of nonfluidized

particles collects on the horizontal surfaces, forming a thermal screen. The higher the temperature of operation, however, the greater the energy and agitation of the bed, and the smaller the likelihood that the bed will collapse. Moreover, various methods can be used to overcome this apparent disadvantage, and these are designed into most fluidized beds. These methods are:

- Movement of the part being treated
- Introduction of additional agitation in the zone of fluidization around the part, either by localized injection of fluidizing gas or by careful design of the outline of the basket that holds the parts
- Increased fluidizing velocity.

Selective Heat Treatment. Bed collapse can be turned to advantage for special heat treatments where one area must be hard and tough while the remainder must be soft and more ductile, as in the case of the engineered parts of the shape described above. In this case, after uniform heating, the

Fig. 6 Relative heat-transfer rates (Ref 1)

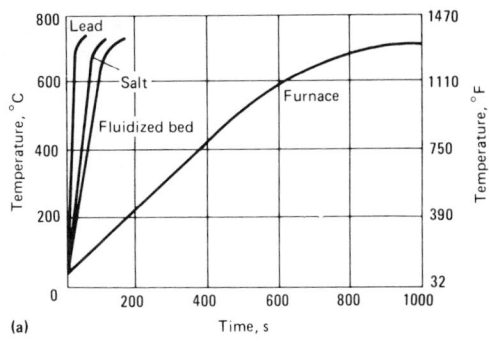

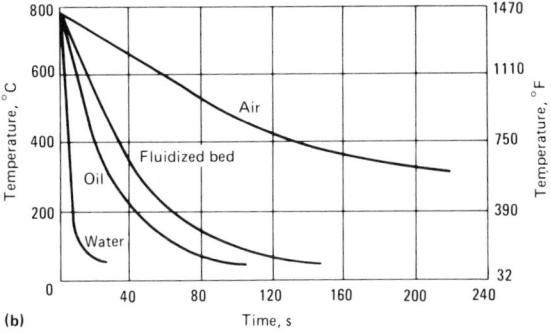

(a) Heating rates for 16-mm (0.6-in.) -diam steel bars in lead, in salt, in a fluidized bed and in a conventional furnace. (b) Quenching rates for 16-mm (0.6-in.) -diam steel bars in air, in oil, in water and in a fluidized bed

Fig. 7 Recovery rates for 25-mm (1-in.) -diam steel parts in a 0.3-m³ (10-ft³) fluidized bed (Ref 1)

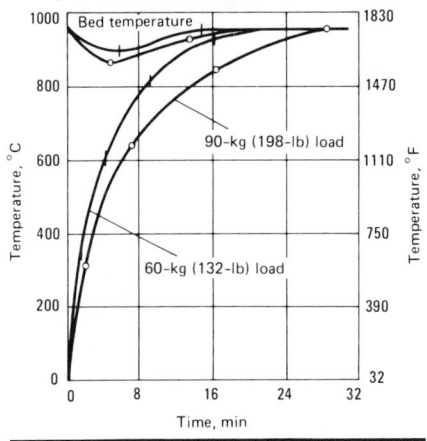

Table 1 Comparison of the effects of hardening and isothermal quenching of type D3 tool steel in salt baths and in fluidized beds

Heating or cooling medium	Diameter of test pieces mm	in.	Preheating temperature °C	°F	Total time for final heating and holding at 960 °C (1760 °F), min	Hardness, HRC At surface	At center
Salt bath.	80	3.2	500	930	44	65.5	65
Fluidized bed(a).	80	3.2	490	915	51	65	65
Salt bath.	40	1.6	540	1000	36	64.5	64
Fluidized bed.	40	1.6	500	930	41	64.5	64

(a) Small parts of the same steel but with a diameter of 8 mm (0.3 in.) were treated at the same time; hardness of these parts was 66 HRC.

product is submerged in a fluidized quenching bed, with the part to be hardened facing downwards. The top horizontal surface becomes covered with a cap of particles that form a thermal screen, which retards the vigorous cooling caused by the fluidized bed.

Heat Transfer in Fluidized Beds

An important characteristic of fluidized beds is high-efficiency heat transfer. The turbulent motion and rapid circulation of the particles in the fluid furnace provide a heat-transfer efficiency comparable with that of conventional salt- or lead-bath equipment.

The heat-transfer coefficient of a fluidized bed is typically between 120 and 1200 W/m²·°C (21 and 210 Btu/ft²·h·°F). The turbulent motion and rapid circulation rate of the particles and the extremely high solid-gas interfacial area account for this feature. The following factors are important in heat transfer:

Particle Diameter. Of all the pa-rameters that affect the heat-transfer coefficient in fluidized beds, particle diameter exerts the greatest influence. Particle diameter generally should be as small as possible; however, below a certain size, entrainment and carry-out effects may cause problems.

Bed Material. The governing physical property of any bed material is its density. There appears to be an optimum density for bed materials—about 1280 to 1600 kg/m³ (80 to 100 lb/ft³). High-density materials tend to produce lower heat-transfer coefficients and also require more power for fluidization. Carry-out problems occur with low-density materials. Other properties, such as thermal conductivity and specific heat, are less important.

Fluidization Velocity of Gas. It is essential to use the optimum flow rate that provides maximum heat-transfer rates for a particular particle density and diameter. Generally, this flow rate is considered to be between two and three times the minimum fluidization velocity. Too high a velocity leads to particle entrainment, high consumption of fluidizing gas and poor heat transfer; too low a velocity leads to poor heat transfer and lack of uniformity in processing.

Heating Rates. Relative heating rates of a 16-mm (0.6-in.) steel bar in salt, in lead, in a fluidized bed and in a conventional furnace are illustrated in Fig. 6(a); relative cooling rates for air, oil, water and a fluidized bed are shown in Fig. 6(b). Figure 7 presents heating and recovery rates for a fluidized bed. Results of both hardening and isothermal quenching of type D3 tool steel

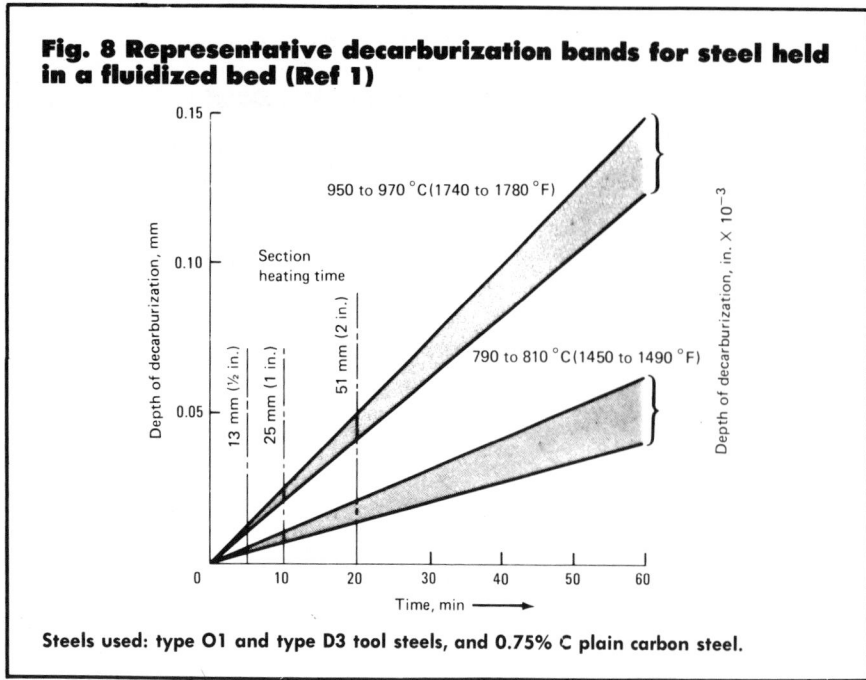

Fig. 8 Representative decarburization bands for steel held in a fluidized bed (Ref 1)

Steels used: type O1 and type D3 tool steels, and 0.75% C plain carbon steel.

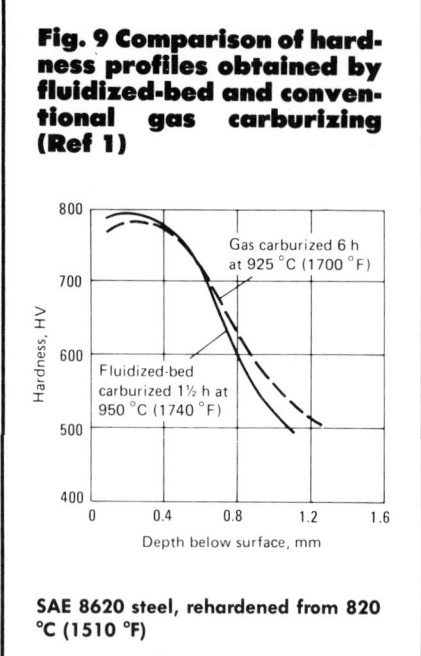

Fig. 9 Comparison of hardness profiles obtained by fluidized-bed and conventional gas carburizing (Ref 1)

SAE 8620 steel, rehardened from 820 °C (1510 °F)

with salt baths and with fluidized beds are given in Table 1. The difference between the two installations in total time for final heating and holding resulted from a difference in preheating conditions.

Control of Atmospheres

A full range of atmospheres can be used within the work zones of fluidized beds. The volume of gas used is clearly dictated by particle size, temperature of operation and optimum fluidization velocity. However, it can be shown that, with careful design and use of low-cost carrier gases such as nitrogen, even low-temperature surface treatments can be both effective and economical. In addition, one of the major advantages of a fluidized bed is that expensive gas need not be consumed while there is no work in the bed. Atmosphere conditioning is rapid: within about 30 to 60 s after an inert gas is introduced into the bed, the purity of the atmosphere is equivalent to that of the gas supply. In internal-combustion fluidized beds, two types of atmospheres can be obtained, as discussed below.

Reducing or Oxidizing Atmosphere. Adjustment of the gas/air mixture to the bed so that it is either gas-rich or oxidizing causes some decarburization or oxidation reactions in the materials being processed (the gas-

rich mixture produces somewhat less severe reactions). However, these are time-dependent reactions, and, because of the rapid heating rates of parts being processed and the subsequent short immersion times needed to obtain the correct structure and through hardness, little surface effect other than discoloration and slight scaling is exhibited in section sizes up to 25 mm (1 in.). For larger sizes, the user must be aware of surface reactions that can occur, particularly as the processing temperature increases. Figure 8 illustrates the relative decarburization bands for steels held in a fluidized bed.

Neutral Hardening and Carburizing. In electrically heated or special gas-heated beds, atmospheres for neutral hardening of tool steels or carburizing of low-carbon steels can be used for bed flotation. This will allow oxygen-free heating of tool steels. However, caution must be exercised during transport of workpieces to the quench tank, to prevent decarburization or oxidation. For carburizing, the atmosphere can be collected, rejuvenated and recycled for efficient usage of the carbon-carrying vehicle.

Surface Treatments

Fluidized beds, using atmospheres composed of ammonia, endothermic gas and oxygen, or similar combinations, are capable of performing low-tempera-

ture nitrocarburizing treatments equivalent to conventional salt-bath processes or other atmosphere processes. High-speed tools oxynitrided in a fluidized bed are comparable to similar tools treated by the more conventional gaseous process. Carburizing and carbonitriding in a fluidized bed can yield results similar to those achieved in conventional atmosphere furnaces.

Mixtures of propane and air produced the results shown in Fig. 9, which compares the case depths obtained on SAE 8620 steel bearing rings carburized in a fluidized bed and by the conventional atmosphere process. An effective case depth of 1 mm (0.04 in.) was achieved in 1.5 h using the fluidized-bed technique.

Fundamental work on this process is still being performed, but sufficient knowledge exists to compare the mechanisms of conventional gas carburizing and the fluidized-bed process.

Conventional Gas Carburizing. Carburizing occurs through catalytic decomposition of CO according to

$$CO + H_2 \rightarrow C_{Fe} + H_2O$$

Propane enrichment aids this reaction according to

$$C_3H_8 + 3CO_2 \rightarrow 6CO + 4H_2$$

and

$$C_3H_8 + 3H_2O \rightarrow 3CO + 7H_2$$

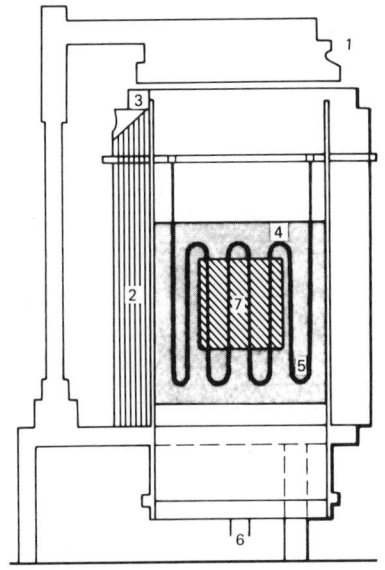

Fig. 10 Fluidized-bed furnace with internal heating by electrical-resistance elements (Ref 1)

(1) Pivoting cover in two parts. (2) Insulating lagging. (3) Refractory material. (4) Fluidized bed. (5) Heating elements. (6) Intake for fluidizing gas. (7) Parts to be treated

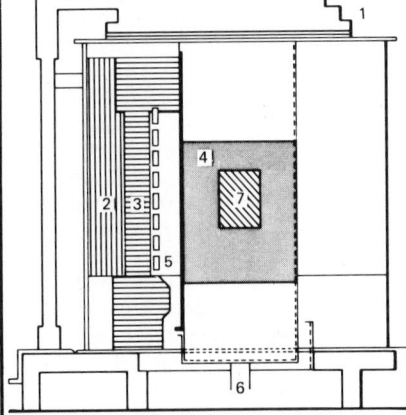

Fig. 11 Fluidized-bed furnace with external heating by electrical-resistance elements (Ref 1)

(1) Pivoting cover. (2) Insulating lagging. (3) Refractory material. (4) Fluidized bed. (5) Resistance elements. (6) Intake for fluidizing gas (air or nitrogen). (7) Parts to be treated

Fluidized-Bed Carburizing. The relatively large volumes of propane consumed during fluidized-bed carburizing, together with high gas velocities, favor carburization by the thermal decomposition of propane to precipitate carbon in accordance with

$$C_3H_8 \rightarrow C \downarrow + 2CH_4$$

The amount of carbon precipitated is proportional to the number of carbon atoms in the hydrocarbon fuel gas; that is, propane forms more carbon than methane. In addition, the purity of propane is important, especially with respect to unsaturated hydrocarbon content, which increases its carbon-forming capability.

The precipitated carbon reacts instantaneously with the oxidizing products of combustion:

$$C_3H_8 + 5O_2 \rightleftharpoons 3CO_2 + 4H_2O$$

to form carbon monoxide and hydrogen:

$$C + H_2O \rightarrow CO + H_2$$

and

$$C + CO_2 \rightarrow 2CO$$

Carburization then proceeds by the catalytic decomposition of CO by H_2 as in conventional carburizing. It is possible that carburization is further complemented by thermal dissociation of the methane formed during carbon precipitation:

$$CH_4 \rightarrow C_{Fe} + 2H_2$$

The carbon potential of the atmosphere varies with the air-to-gas ratio. For each type of hydrocarbon gas, a relationship can be established among air-to-gas ratio, temperature and carbon potential. Control of the reaction and carbon potential of the atmosphere by conventional gas analysis is possible, and fluidized-bed furnaces are equipped with sample ports and probes so that suitable measurements can be taken.

Types of Furnaces for Heat Treating With Fluidized Beds

The type of fluidized bed most widely used for heat treatment is the dense-phase type, although units have been constructed based on the dispersed-phase bed, with particle circulation for heat treatment of long, thin metal parts such as shafts and plates. In a typical dense-phase fluidized bed, the parts to be treated are submerged in a bed of fine solid particles held in suspension, without any particle entrainment, by an upward flow of gas.

Liberation of adequate quantities of heat within fluidized beds is a prime consideration in adapting them for metal processing. Because transfer of heat from the bed to the workpiece is usually much more efficient than transfer of heat from the heat source to the fluidizing medium, the greatest difficulty is encountered in transferring suitable quantities of heat to the fluidizing medium. In addition, the major part of the heat loss from any practical fluidized system is the heat content of the spent fluidizing gas. In instances in which thermal efficiency is unduly influenced by this factor, recirculation of the fluidizing gas or installation of a recuperative system may be justified, and each has been used in practical applications.

Heat input to a fluidized bed can be achieved by several different methods which are described in the following paragraphs.

Internal-Resistance-Heated Fluidized Beds. In this type of unit, the

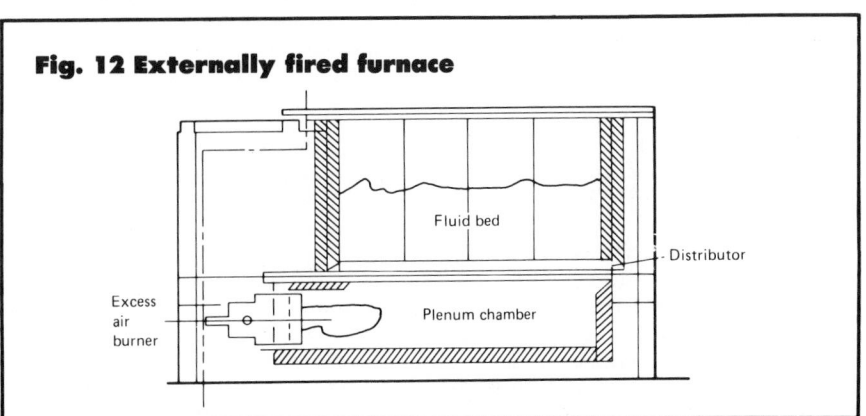

Fig. 12 Externally fired furnace

Fluid bed

Distributor

Excess air burner

Plenum chamber

gas and particles are heated by suitably sheathed internal-resistance-heated elements (see Fig. 10). For high-temperature operation between 500 and 1000 °C (930 and 1830 °F), silicon carbide elements can be used, but they must be sheathed to prevent reactions between the elements and the bed material. At lower temperatures, a mineral-insulated heater with an integral metal sheath can be used, particularly where a heater with greater structural integrity is required. With a system of this type, it is essential to ensure that there are no areas of poor fluidization close to the element. Such areas cause localized overheating and buildup of fused material on the heater, resulting in failure. This problem normally can be avoided, and such heaters provide a simple method of heating fluidized beds up to approximately 500 °C (930 °F). However, heat-up rates (from cold to operating temperature) and heat-recovery rates are relatively slow in this type of bed, particularly for higher temperatures.

External-Resistance-Heated Fluidized Beds. A fluidized bed contained in a heat-resisting pot can be heated by external resistance elements, as shown in Fig. 11. Waste-heat recovery can be used to increase thermal efficiency, and the fluidizing gas can be maintained at any desired composition. Although this appears to be a good method of applying heat to the fluidized bed, there is a severe limitation on the rate of heat input that can be achieved through the wall of the pot. Heat-up rates from cold to operating temperatures of 700 to 800 °C (1290 to 1470 °F) can be as long as 5 to 6 h.

Direct-resistance-heated fluidized beds employ an electrically conducting material such as carbon powder or silicon carbide as the bed material. Power is applied directly to the bed by means of electrodes, and heat is generated within the bed by direct passage of an electric current. However, the current distribution is influenced by a metallic workpiece situated in a region of current flow. This heating method has been shown to operate satisfactorily at temperatures up to 1300 °C (2370 °F).

External-Combustion-Heated Fluidized Beds. A fluidized bed contained in a heat-resisting pot can be heated by external gas firing (see Fig. 12). In this arrangement, a fuel/air mixture is introduced through a standard commercial burner. The burner

Fig. 13 Controlled-atmosphere fluidized-bed furnace heated by submerged combustion (Ref 1)

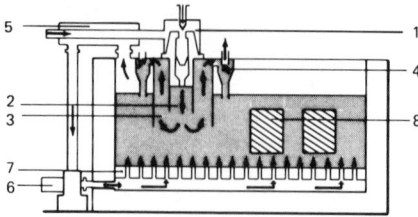

(1) Burner. (2) Combustion tube. (3) Tube through which combustion gases and particles rise. (4) Particle separators. (5) Heat exchanger. (6) Gas recycle compressor for fluidization. (7) Distributor plate. (8) Parts to be treated

can be controlled very accurately down to low temperatures for low-temperature tempering. The products of combustion are passed through a perforated metallic plate to achieve flotation of the bed. By introduction of tempering air after mixing and burning of the fuel, the burner can be operated stoichiometrically for optimum efficiency and prevention of aldehyde formation.

Submerged-Combustion Fluidized Beds. The technique of submerged combustion consists of passing the combustion products directly through the mass to be heated. This method provides an excellent rate of heat transfer and is now well established for a wide range of liquid heating applications, from heating of swimming pools to concentration of acid solutions. Application of this method to the heating of a fluidized bed requires that the burner be used in such a way as to provide strong agitation of the suspended particles, thereby achieving the desired properties of outstandingly good heat transfer and uniformity of bed temperature.

Equipment developed for this purpose consists essentially of a burner, two concentric tubes and a particle separator. A suitable gas mixture is fed through the burner into the central tube, where it is ignited. The flame develops in the tube and the combustion products escape at its lower end, where they impart heat to the suspended particles before moving up through the annular space between the two tubes. As they rise, a quantity of particles is entrained. These are separated from the gas stream by the deflector plate and fall back into the bed by vir-

Fig. 14 Gas-fired fluidized-bed furnace with internal combustion (Ref 1)

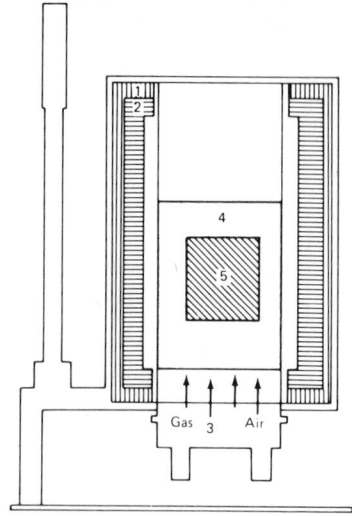

(1) Insulating lagging. (2) Refractory material. (3) Air and gas distribution box. (4) Fluidized bed. (5) Parts to be treated

tue of gravity. Figure 13 illustrates a system that incorporates submerged combustion together with a controlled atmosphere for low-temperature treatment of metals.

Internal-Combustion Gas-Fired Fluidized Beds. A major development in the heating of fluidized beds occurred when an air/gas mixture was used for fluidization and ignited in the bed, generating heat by internal combustion. Prior to this breakthrough, many technical difficulties prevented use of this mode of fluidized-bed heating. A typical furnace design incorporating this technique is shown in Fig. 14.

The advantage of this system is that the bed is fluidized by burning gases, and thus the heat is generated within the bed. In gas-fired fluidized beds, the supporting gas or fluidizing medium is a near-stoichiometric mixture of gas and air. This combustible mixture is ignited above the bed and quickly imparts its heat to the particles, which in turn heat the incoming gas further down the bed. After a period, combustion takes place spontaneously within the bed, being complete within the first 25 mm (1 in.) of the diffuser once the spontaneous combustion temperature for the gas being used is reached. This temperature commonly varies between 600 and 800 °C (1110 and 1470 °F). If the vessel is well insulated, the bed

Fig. 15 Two-stage, gas-fired, internal-combustion fluidized beds (Ref 1)

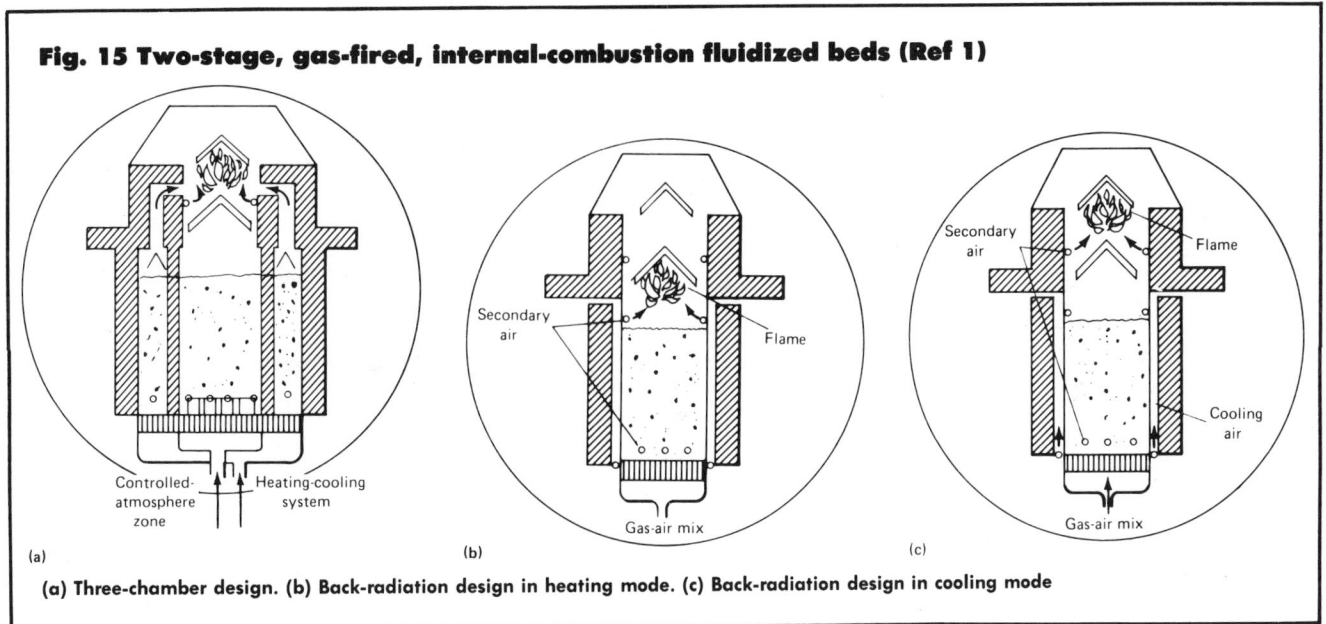

(a) Three-chamber design. (b) Back-radiation design in heating mode. (c) Back-radiation design in cooling mode

temperature can rise to a theoretical combustion temperature, and heat-up rates from cold to 800 °C (1470 °F) are typically between 1 and 1½ h. However, the following problems are inherent to the basic technique and must be considered:

- The bed is fluidized by burning gases. To obtain good temperature control and optimum fluidizing conditions, however, it is desirable that the fuel-input rate and fluidizing velocity be independently variable.
- Combustion is somewhat unstable below the spontaneous combustion temperature.
- Very high temperatures can occur in the immediate vicinity of the distributor/diffuser tile. When the bed is incorrectly fluidized, so that this heat cannot be removed from the top of the distributor, theoretical flame temperatures are achieved with consequent deterioration of the distributor. The thermal stresses of expansion and contraction on the distributor tile at these high temperatures tend, even with the best fixing techniques available, to cause failure of joints, which enhances the problem.

Two-Stage Internal-Combustion Gas-Fired Fluidized Beds. The basic problem of separating control of heat input from control of fluidizing velocity has been overcome in two alternative designs, shown in Fig. 15. In both designs, the initial heat-up from cold to

operating temperature is carried out by two-stage internal combustion. A noncombustible mixture of gas and air is introduced beneath the distributor tile. Secondary air is added to make up a stoichiometric or slightly gas-rich mixture immediately above the tile by means of jet holes drilled into heat-resisting tubes. This is done to reduce the possibility of explosion and to avoid high flame temperatures at the surface of the tile. The technique has an adverse effect on good fluidization, but this is unimportant during initial heat-up, in which the prime objective is to raise the bed to operating temperature as quickly as possible.

Once the bed has reached operating temperature, the remaining problem is that of isolating heat-up control from control of the fluidizing velocity. This is achieved in two ways:

- *Three-Chamber Design:* In this design (Fig. 15a), the heat-control outer chambers are separated from the treatment zone by a muffle. The fluidizing velocity and atmosphere are independently controlled in the inner chamber, while the outer two zones are still supplying heat by internal combustion. To achieve adequate heat input, fluidization levels in these outer chambers are above the optimum for heat transfer and surface reactions, but this fact is relatively unimportant.
- *Back-Radiation Design:* When fuel-rich gases are permitted to burn by the injection of secondary air imme-

diately above the control chamber of the fluidized bed, a back-radiation effect causes a rise in bed temperature. This design (shown operating in the heating/controlling and cooling modes in Fig. 15b and c) makes use of this effect and, at the same time, utilizes heat that is normally dissipated when gases are burned outside the furnace. It is therefore more economical in fuel usage. In principle, the gas-rich mixture is supplied to the central chamber, and extra air is added to produce stoichiometric conditions during initial heating of the bed. When cold work is loaded for treatment, the extra air is injected above the bed to produce a radiating flame and recover bed temperature. Should bed temperature exceed set temperature, the extra air is switched to the outside of the furnace wall to provide cooling and finally is mixed with the rich gas/air to produce combustion at the top of the specially constructed hood.

Applications of Fluidized-Bed Furnaces

The potential applications of fluidized-bed technology to heat treating are many. Figure 16 specifies those applications in which fluidized beds can compete with conventional furnaces.

Applications of fluidized-bed furnaces to heat treatment of metals include continuous units for all types of

Fig. 16 Fluidized-bed applications; decision model (Ref 1)

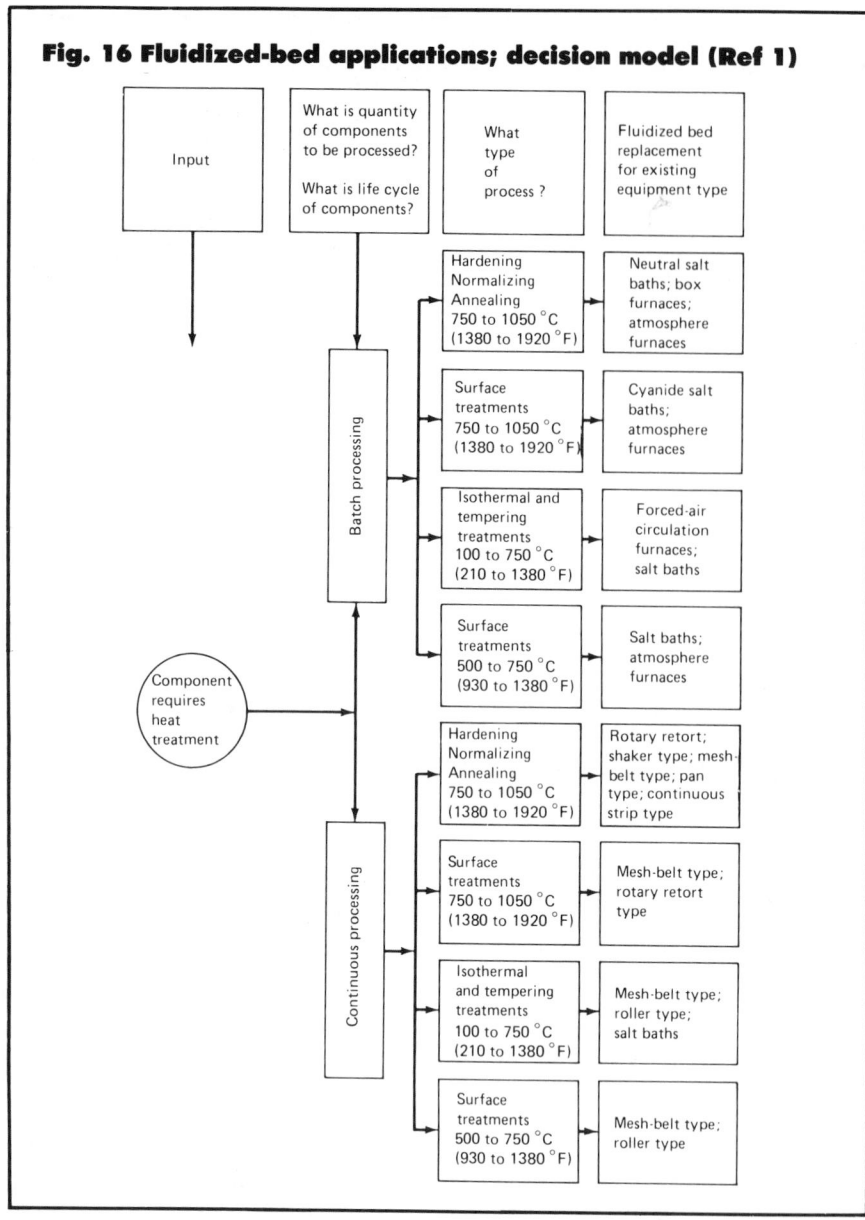

difficulties in meeting the 7- to 14-day turnaround of dies required by customers. Previously, hardening, case hardening and tempering had been done by salt-bath immersion. After studying alternatives, the firm decided to employ the latest fluidized-bed technology. Approximately one year later, the firm installed a second fluidized-bed furnace and made available its surplus capacity to other firms on a subcontract basis.

Operational Safety

As for all forms of gas heating, normally accepted safety devices are incorporated on the majority of beds presently manufactured. The "flexible tile" concept ensures that any failure of joints does not influence the performance of the bed.

Parts carrying surface oil or moisture do not create an explosion risk because the contaminants simply vaporize and are removed with the waste gas, as in conventional furnaces.

Cleaning Operations

Fluidized solids are nonabrasive and noncorrosive, and they do not wet immersed objects. There is some drag-out loss of the aluminum oxide, however, because some particles accumulate on flat surfaces occurs as workloads are removed from the fluidized bed. These particles can be removed in part by agitation, bouncing, or blowing with an air pipe. Particles can be reused by being dried, sieved and returned to the bed. When parts already scaled or preoxidized are placed in a fluidized bed, particles tend to adhere to the scale to a greater degree than if the workpieces were clean. These particles can be removed by water spraying.

REFERENCE

R. W. Reynoldson, "Controlled Atmosphere Fluidised Beds for the Heat Treatment of Metals," *Heat Treatment of Metals*, University of Aston in Birmingham, 1976

wire and strip processing (patenting, austenitizing, annealing, tempering, quenching, etc.); continuous rotary types for fasteners, bearings and other small parts; and all configurations of batch-type units for general heat treating applications. A typical batch-type unit with an output of approximately 150 kg/h (330 lb/h) is available as a standard furnace. Using mechanical handling equipment, it can be auto-mated into a continuous heat treatment line. The following example describes how one firm decided to install fluidized-bed furnaces for heat treatment.

Example 1. A company specializing in design and production of aluminum extrusion dies had relied on subcontract heat treatment facilities for hardening of dies. The decision to install in-house facilities came as a result of

Heat Treating in Vacuum Furnaces and Auxiliary Equipment

By Ed R. Byrnes, Jr.
Manager
Vacuum Equipment Marketing
Ipsen Industries
and
Roger C. Anderson
Manager
Research & Development
    Laboratories
Ipsen Industries

VACUUM HEAT TREATING is a relatively new development in metallurgical processing. Vacuum heat processing consists of thermally treating metals in heated enclosures that are evacuated to partial pressures compatible with the particular metals and processes. Vacuum is substituted for the more commonly used protective gas atmospheres during either part or all of the heat treatment. Furnace equipment used in vacuum heat treatment differs widely in size, shape, construction and method of loading.

Although originally developed for the processing of electron tube materials and refractory metals for aerospace applications, vacuum furnaces are now employed in brazing, sintering, heat treating and diffusion bonding of metals. Vacuum furnaces also are used for annealing, carburizing, heating and quenching, tempering, and stress relieving. Furnaces for vacuum heat treating are equipped for workloads ranging from several pounds up to 100 tons, and heated working chambers range in size from 0.03 m³ (1 ft³) up to hundreds of cubic feet. Although most vacuum furnaces are batch-type installations, continuous vacuum furnaces with multiple zones for purging, preheating, high-temperature processing and cooling by gas or liquid quenching also are used. Vacuum heat treating furnaces also:

- Prevent surface reactions, such as oxidation or decarburization on workpieces, thus retaining a clean surface intact
- Remove surface contaminants such as oxide films and residual traces of lubricants resulting from fabricating operations
- Add a substance to the surface layers of the work (carburization, for example)
- Remove dissolved contaminating substances from metals, using the degassing effect of a vacuum (removal of hydrogen and oxygen from titanium, for example)
- Join metals by brazing or diffusion bonding.

Vacuum Measurements

A theoretical or ideal vacuum is an empty space that does not contain either vapors, particles, gases or other matter and consequently has no atmospheric pressure. Because this condition does not exist, even in outerspace, an ideal vacuum cannot be achieved. Therefore, a manufactured vacuum is expressed in relative terms of pressure compared with the standard atmospheric pressure surrounding the earth. Standard atmosphere pressure at sea level, 45° latitude, and 0° C has the following values: 760 mm Hg, 760 000 μ or μm Hg, 29.921 in. Hg, or 14.696 psi.

The most common units of pressure used in measuring vacuum or partial pressure are as follows: millimetres of mercury (mm Hg); microns (μ Hg) or micrometres of mercury (μm Hg); and torr (1 torr equals 1 mm Hg). Microns (μ) are established and well-known units; however, micrometres (μm) are exactly equal, and consistent with metric practice.

These pressure or vacuum values refer to the height of a mercury column sustained by the differential between atmospheric pressure and an attained level of vacuum or partial pressure. Figure 1 provides a comparison of vacuum ranges in relation to atmosphere. Note the normal operating range for vacuum heat treating.

Comparison of Vacuum and Atmosphere Furnace Processing

Unlike conventional atmosphere heat treatment, vacuum heat treatment does not require control of carbon potential of prepared atmospheres and related furnace conditioning requirements. In atmosphere heat treating, the water-vapor content, or dew point, of a protective gas atmosphere is often the most critically controlled variable in addition to the temperature and time of processing. Time and temperature must be controlled in vacuum furnaces as well; however, unlike atmosphere heat treating furnaces, a vacuum furnace contains only very small amounts of residual gas or atmosphere. In a leakproof vacuum furnace, analysis of the residual atmosphere at a vacuum of about 10^{-3} torr indicates that less than 0.1% of the original air remains. The residual gases consist mainly of water vapor, with the remainder comprised of primarily organic vapors from the seals, vacuum greases and vacuum oils. The oxygen content at 10^{-3} torr is less than 1 ppm. If all of the residual gas in the vacuum furnace were converted to water vapor, the water vapor content would be approximately 1.5 ppm or equal to that of a gas with a dew point

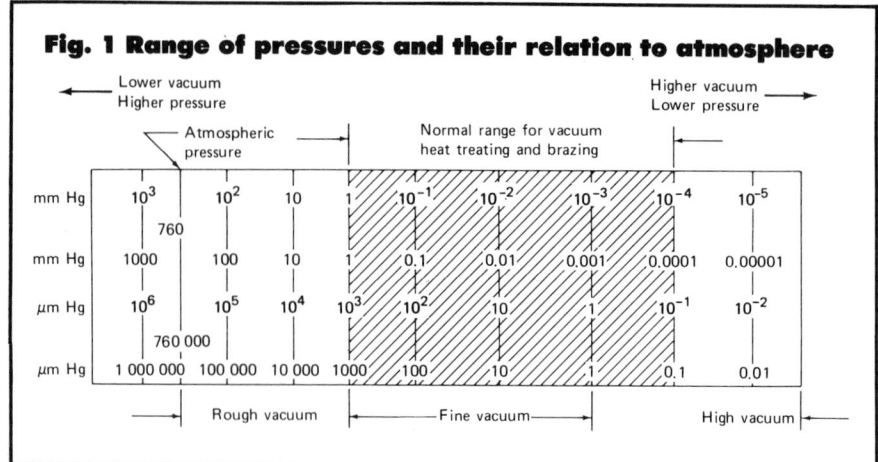

Fig. 1 Range of pressures and their relation to atmosphere

of about −80 °C (−110 °F). At a vacuum level of 10^{-4} torr, the equivalent dew point of gas is estimated to be approximately −90 °C (−130 °F) or less.

These low dew point equivalents compare favorably with the driest inert gases available from highly efficient gas-dehydration equipment. With suitable pumping systems, the concentration of oxygen and water vapor can be reduced to lower levels than that achieved in inert or reducing-gas atmospheres.

After a vacuum heat treating furnace has been evacuated, gaseous reactions such as those encountered with atmosphere heat treatment are virtually eliminated. Moreover, the vacuum extracts many gases, surface contaminants and processing lubricants that would be difficult and costly to remove by any other method. Gases drawn from the metal surface into the vacuum surrounding the charge are trapped by the vacuum pumps and exhausted from the system. This advantage of a vacuum system is of greater significance when parts with complex shapes, blind holes or deep recesses are being heat treated. A complete purging of such parts in a protective atmosphere requires an extended purging period. Even long-time purging, however, may not ensure complete removal of entrapped air and other contaminants.

When more thorough purging is required, the furnace can be evacuated with a simple vacuum system, and the enclosure or retort can be backfilled with the desired protective atmosphere. This method markedly reduces the amount of protective atmosphere and time required to produce satisfactory results.

Volatilization and Dissociation

In vacuum technology, the reactions are related principally to the physical chemistry of the materials and the laws of physics, whereas atmosphere processes are controlled largely by laws governing physical chemistry. Therefore, vapor pressures of elements and the required vacuum or backfill pressure levels become important considerations in the use of a vacuum furnace.

Vapor pressure, which is the gas pressure exerted when a substance is in equilibrium with its own vapor, increases rapidly with temperature, because the amplitude of molecular vibration increases with temperature. Some molecules in the outer surface of the solid material have higher energies than others, and they escape as free molecules or vapor. If a solid substance is contained in an enclosure devoid of any other material, molecules will continue to escape from the solid surface until their rate of escape is exactly balanced by the rate of condensation or recapture of the gaseous molecules. The equilibrium pressure developed is the vapor pressure of the substance at that temperature. Vapor pressure of a metal is dependent on temperature only.

It is normally desirable to use a vacuum-temperature combination that accelerates desorption of gases without producing vaporization of more volatile alloy constituents. Alloys with high concentrations of volatile elements, such as brass, are not heat treated in vacuum furnaces.

If brass were heated in a vacuum at a temperature of 540 °C (1000 °F) and a vacuum level on the order of 0.1 μm

Fig. 2 Vapor pressure versus temperature for carbon and various pure metals

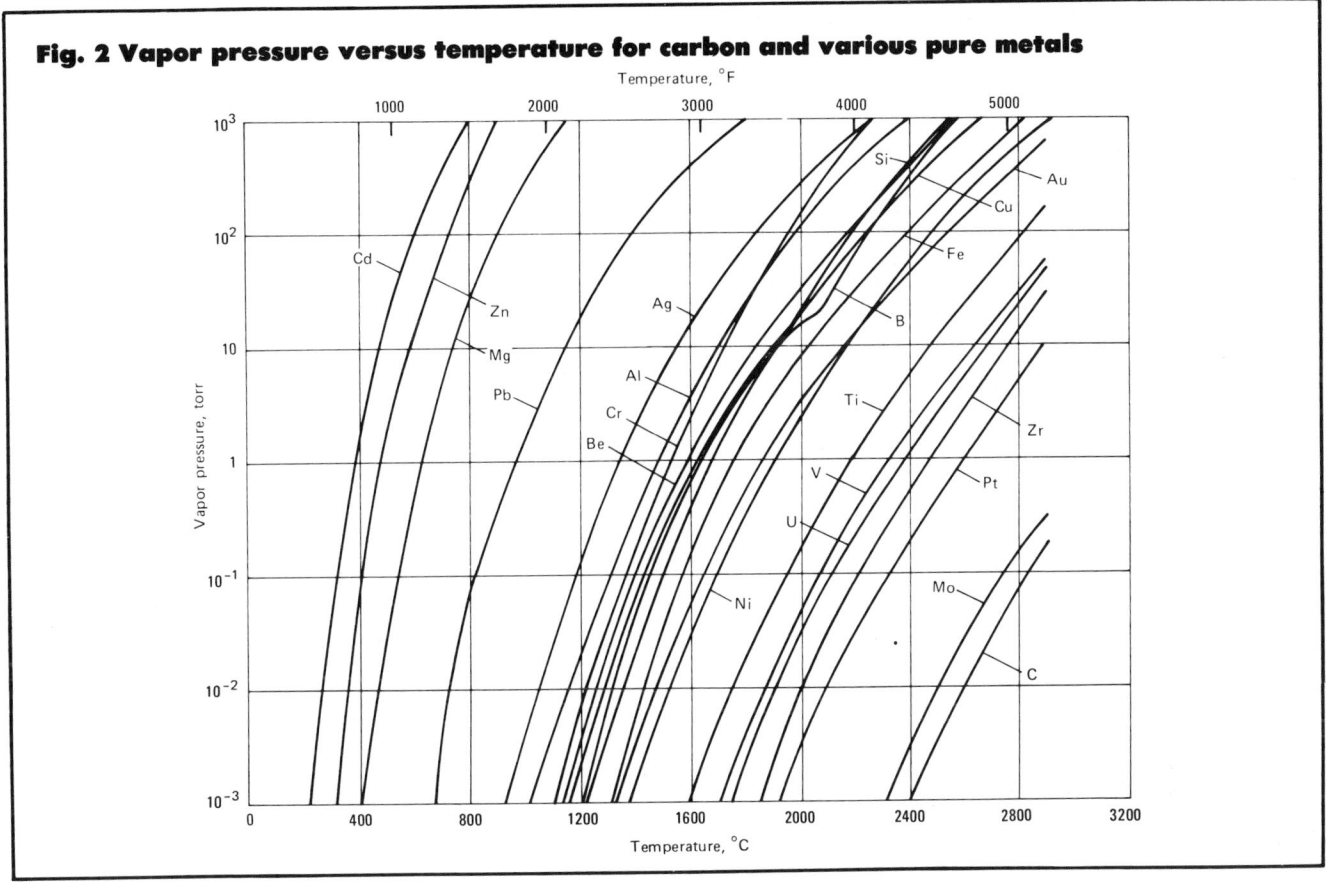

Hg, the zinc component will vaporize (volatilize), and the brass will be soon converted to copper sponge.

Metals such as aluminum, lead, zinc and magnesium have relatively high vapor pressures; if heated above a temperature at which the vapor pressure of the element exceeds the pressure in the furnace, they will evaporate or sublime rapidly. Thus, high-vacuum heat treatment is not applicable to some metals and alloys. To handle certain metals and alloys properly, the pressure must be limited to the "fine" vacuum range (Fig. 1), or a backfill partial pressure must be employed.

Alloys with lower concentrations of volatile elements can be processed in vacuum by using a backfill pressure that exceeds the sublimation pressure of the element at the temperature involved. A backfill pressure of a few hundred μm Hg at temperatures about 980 °C (1800 °F) precludes the vaporization of elements such as chromium, copper and manganese from steels processed at these higher temperatures.

For example, if pure manganese were heated to approximately 790 °C

(1455 °F) at a pressure of 10^{-4} torr, it would vaporize (Fig. 2). If the material was held at a higher temperature or lower pressure for an adequate period of time, the metal would become depleted, and eventually disappear, and the vapors would condense on the colder areas of the furnace and/or pumping system. Backfill or higher partial pressures greatly slow the rate of evaporation or volatilization. The vapor pressures of carbon and selected pure metals, as related to temperature, are shown in Fig. 2 and Table 1.

Alloy Vapor Pressures. Vapor pressures of pure metals are constant, well-established values; the vapor pressure of a given alloy is variable depending upon conditions. The vapor pressure of an alloy is governed in part by a law analogous to Dalton's law of partial pressures: the total vapor pressure of an alloy, under ideal conditions, is equal to the sum of the partial vapor pressures of its constituents. However, the partial pressure of each element in the alloy is lower than its normal vapor pressure and is proportional to its concentration.

In processing at temperatures where the vapor pressures of more volatile minor constituents are still in the micron range, alloys behave in accordance with Dalton's law. For example, if pure manganese is heated to 790 °C (1455 °F), its vapor pressure reaches 10^{-4} mm Hg; it will be impossible to evacuate to lower pressures without evaporating all of the manganese. However, when manganese is alloyed with other elements, as a solid solution in iron, for example, its effective vapor pressure is lowered. The total vapor pressure for the alloy is the sum of vapor pressures of each individual element multiplied by its concentration in the alloy. The vapor pressure of manganese in a 1% Mn alloy at 790 °C (1455 °F) is about 10^{-6} mm Hg.

Many metals form compounds by reaction with oxygen, hydrogen and nitrogen. These reactions are usually exothermic, and the possibility for dissociation of the resulting compound increases with higher temperatures. Some oxides, such as water, vaporize at temperatures so low that dissociation occurs only in the vapor phase. For an

Table 1 Vapor pressures of various elements

Element	10⁻⁴ mm Hg 0.1 μm		10⁻³ mm Hg 1.0 μm		Vapor pressure at: 10⁻² mm Hg 10 μm		10⁻¹ mm Hg 100 μm		760 mm Hg 760 000 μm	
	°C	°F	°C	°F	°C	°F	°C	°F	°C	°F
Aluminum	808	1486	890	1632	997	1825	1124	2053	2058	3733
Antimony	525	977	595	1103	678	1251	780	1434	1441	2624
Arsenic	...	...	220	428	...	...	310	590	610	1130
Barium	544	1011	626	1157	717	1321	830	1524	1404	2557
Beryllium	1030	1884	1131	2066	1247	2275	1396	2543	...	...
Bismuth	537	997	609	1128	699	1290	721	1328	1421	2588
Boron	1141	2084	1240	2262	1356	2471	1490	2712	...	...
Cadmium	180	356	220	428	264	507	321	610	766	1409
Calcium	463	865	528	982	605	1121	701	1292	1488	2709
Carbon	2290	4150	2473	4480	2683	4858	2928	5299	4831	8721
Cerium	1092	1996	1191	2174	1306	2381	1440	2622	...	...
Caesium	74	165	110	230	153	307	207	405	691	1274
Chromium	993	1818	1091	1994	1206	2201	1343	2448	2484	4500
Cobalt	1363	2484	1495	2721	1650	3000	1834	3331	...	...
Copper	1036	1895	1142	2086	1274	2323	1433	2610	2764	5003
Gallium	860	1578	966	1769	1094	1999	1249	2278	...	...
Germanium	997	1825	1113	2034	1252	2284	1422	2590	...	...
Gold	1191	2174	1317	2401	1466	2669	1647	2995	2999	5425
Indium	747	1375	841	1544	953	1746	1089	1990	...	...
Iridium	2156	3909	2342	4244	2558	4633	2813	5092	...	...
Iron	1196	2183	1311	2390	1448	2637	1604	2916	2737	4955
Lanthanum	1126	2057	1243	2268	1382	2518	1550	2820	...	...
Lead	548	1018	620	1148	718	1324	821	1508	1745	3171
Lithium	378	711	439	822	514	957	608	1125	1373	2502
Magnesium	331	628	380	716	443	829	515	959	1108	2025
Manganese	792	1456	878	1612	981	1796	1021	1868	2153	3904
Molybdenum	2097	3803	2297	4163	2535	4591	3011	5448	5573	10056
Nickel	1258	2295	1372	2500	1511	2750	1680	3054	2734	4950
Niobium	2357	4271	2541	4602	...	...	...	...	...	...
Osmium	2266	4107	2453	4444	2669	4833	2922	5288	...	...
Palladium	1272	2320	1406	2561	1567	2851	1760	3198	...	...
Platinum	1745	3171	1905	3459	2092	3794	2295	4159	4411	7965
Potassium	123	253	161	322	207	405	265	509	643	1189
Rhodium	1816	3299	1973	3580	2151	3900	2359	4274	...	...
Rubidium	88	190	123	253	165	329	217	423	679	1254
Ruthenium	3704	3736	2232	4046	2433	4408	2668	4831	...	...
Scandium	1162	2122	1283	2340	1424	2593	1596	2903	...	...
Silicon	1117	2041	1224	2233	1344	2449	1486	2705	2289	4149
Silver	848	1558	921	1688	1048	1917	1161	2120	2214	4014
Sodium	195	383	238	460	291	556	356	673	893	1638
Strontium	413	775	475	887	549	1020	639	1182	1385	2523
Tantalum	2601	4710	2822	5108	...	...	...	...	...	...
Thallium	461	862	500	932	607	1123	661	1220	1458	2655
Thorium	1833	3328	2000	3630	2198	3985	2433	4408	...	...
Tin	923	1692	1011	1850	1190	2172	1271	2318	2272	4118
Titanium	1250	2280	1385	2523	1547	2815	1725	3168	...	...
Tungsten	2769	5013	3019	5461	3312	5988	...	...	5932	10701
Uranium	1586	2885	1731	3146	1899	3448	2099	3808	...	...
Vanadium	1587	2887	1726	3137	1889	3430	2081	3774	...	...
Yttrium	1363	2484	1495	2721	2967	3000	1834	3331	...	...
Zinc	248	478	290	554	343	649	405	761	908	1665
Zirconium	1661	3020	1818	3301	2003	3634	2214	4014	...	...

Note: The vapor pressure of metals is fixed with probable values at a given temperature, and the temperature at which the solid is in equilibrium with its own vapor descends as the pressure to which it is exposed descends. For example: iron must be heated to 2737 °C (4955 °F) at atmosphere before its vapor pressure is greater than atmosphere (760 mm Hg); this point is reached at 1311 °C (2390 °F) at a pressure of 1 μm (10⁻³ mm Hg).

oxide, nitride or hydride that remains a solid over a wide range of tempera- tures, a dissociation pressure exists at any temperature that represents an equilibrium between the compound, the gas and the metal.

All metallic compounds decompose into constituent elements when heated to sufficiently high temperatures. However, many of the metal oxides are quite stable, requiring low pressures at high temperatures to effect dissociation. It is impractical to dissociate many of these compounds due to the combination of vacuum level and temperature required. When a metal oxide dissociates, the metal remains, and the oxygen is evacuated.

For example, chromium oxide will dissociate in a 10^{-5} torr vacuum at 1300 °C (2370 °F). Dissociation of a metal oxide usually depends more on temperature than on pressure. Most oxides can be dissociated under normal operating vacuum levels at approximately their reduction temperature in a highly reducing hydrogen atmosphere.

The nitrides and hydrides often have higher dissociation pressures, making many of them unstable when heated in a vacuum. For this reason, vacuum heat treating can be used both to dissociate these compounds and to remove the evolved gas without disturbing the base metal.

When oxidized surfaces brighten during vacuum heat treating, the mechanism involved is not believed to be simply thermal dissociation of the oxide. Bright surfaces do not discolor, or become brighter, when they are exposed to a vacuum atmosphere that is theoretically oxidizing. A metal surface can be maintained almost free from visible oxidation at a partial pressure several decades higher than suggested by theoretical calculations. The following theories have been proposed to explain this apparent anomaly:

- The solution and diffusion rate for oxygen exceeds its surface absorption rate.
- Oxide nucleation occurs at discrete sites rather than as a continuous film.
- The effective concentration of oxygen is reduced by carbon and hydrogen in the solid metal and by the vacuum atmosphere.

Furnace Equipment

Although conventional atmosphere furnaces can be adapted for vacuum heat treating by adding a vacuum-tight retort connected to a suitable pumping system, furnace equipment developed especially for vacuum heat treating is generally used.

Vacuum furnaces can be grouped into three basic designs: top-loading, or pit furnaces; bottom-loading, or bell furnaces; and horizontal-loading, or box furnaces. Furnace designs can be varied to fit a wide variety of processing requirements by changing the chamber length, adding internal doors, circulating fans or recirculating gas systems and internal quenching systems.

Every vacuum furnace, regardless of its end use and basic hot or cold wall design, requires (a) heating elements controlled to proper processing temperatures and cooling rates, (b) suitable vacuum enclosures with access openings, (c) vacuum pumping system, and (d) instrumentation to monitor and display critical processing data. Production furnaces may be single-chamber units, batch-type units or multichamber, semicontinuous units.

Hot Wall Vacuum Furnaces

Vacuum furnaces are classified according to the location of the heating and insulating components. Hot wall furnaces were the first type to be designed. Because of the demand of the heat treating industry for higher temperatures, lower pressures, rapid heating and cooling capabilities and higher production rates, hot wall vacuum furnaces have become essentially obsolete and have largely been replaced by cold wall vacuum furnaces.

The entire vacuum vessel is heated by external heating elements in the hot wall construction. The heat is contained by insulation materials similar to the materials used in electrically heated atmosphere heat treating furnaces. Hot wall furnaces have limited utilization because of slow heating and cooling capabilities. They are also limited in temperature because strength of materials is reduced at elevated temperature. However, hot wall equipment is readily adaptable to low-temperature operations not exceeding 980 °C (1800 °F), with moderate sized chambers.

The double-pump modification of the hot wall furnace permits the construction of larger vessels and the use of higher operating temperatures. This system incorporates a second vacuum vessel outside the vacuum retort to maintain a roughing vacuum during the heating cycle. This removes the

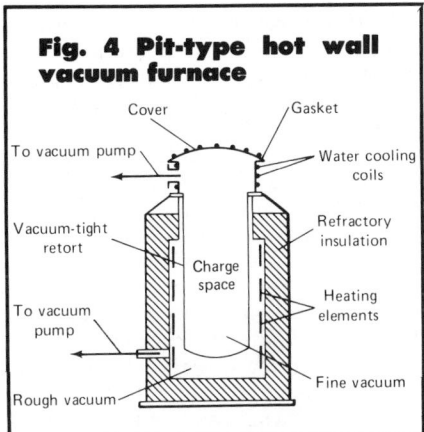

Fig. 3 Bell-type hot wall vacuum furnace

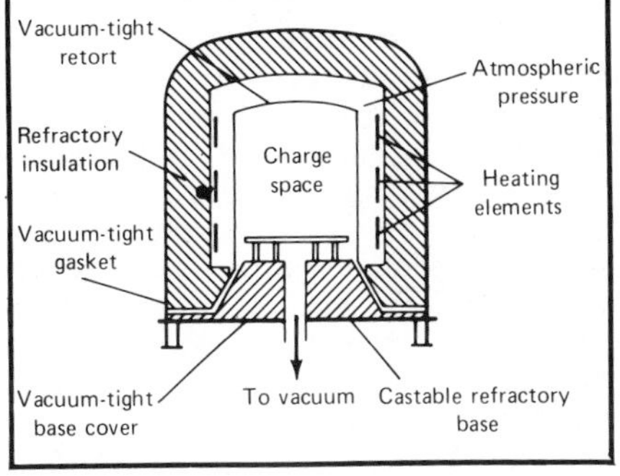

Fig. 4 Pit-type hot wall vacuum furnace

stress of the atmospheric pressure on the heated retort or vacuum vessel.

Bell-Type Furnace. A bell-type hot wall furnace is shown in Fig. 3. The workload is placed on an elevated refractory metal hearth that rests on an insulated base clad with an alloy plate material. A water-cooled circumferential flange and vacuum gasket are located on the vacuum-tight base cover adjacent to the heated zone but in an unheated area. A retort made with a heavy walled heat-resisting alloy covers the work load. A flange at the bottom of the retort fits on top of the base gasket to provide a vacuum-tight enclosure. The bell-shaped furnace equipped with internal electrical heating elements is lowered into position over the retort by a vertical hoist. The vacuum pumping system is connected through the insulated base.

Because this furnace cannot be heated or cooled rapidly, even when the bell furnace is removed, production rates and the number of thermal cycles within a given time period are limited. Moreover, because the hot retort must support the entire pressure of the external atmosphere, its wall must be quite heavy. Practical operating temperatures for a furnace of this type generally are limited to approximately 925 °C (1700 °F.).

Pit-Type Furnace. Figure 4 shows a schematic drawing of a pit-type hot wall furnace. The workload is placed in a top-loading muffle or retort made from a heat-resisting alloy. The upper end of the retort is provided with a water-cooled flange and vacuum gasket that interlock with a flange on the upper part of the furnace above the heated zone. The muffle is lowered into the furnace by an overhead hoist, providing vacuum connections for the furnace and retort.

With this construction, the space between the muffle and heating furnace can be evacuated by a roughing pump so that the pressure at the exterior surface of the muffle is essentially zero. This evacuation permits the use of a muffle with a much thinner wall and raises the maximum operating temperature of the system to approximately 1175 °C (2150 °F). When the heating cycle is completed, inert gas is bled into the retort, and air is bled simultaneously into the heating furnace so that the pressures remain balanced on both sides of the retort. The retort then can be removed from the furnace to a cooling stand, and another retort can be inserted in the hot furnace. This construction increases heating and cooling flexibility, which in turn increases cycle frequency and production.

Horizontal and vertical two-zone hot-wall vacuum furnaces are shown in Fig. 5 and 6. In both types, the heat-resisting alloy muffle is extended much further beyond the heating section of the furnace. This extended section has a water-cooled jacket to provide accelerated cooling. In the horizontal furnace, the charge is carried on an alloy hearth that can be moved in and out of the heated zone by a push rod extending through a seal in the outer fixed end of the muffle. This hearth has vertical heat baffles or multiple radiation shields at each end to confine the heat to the heated portion of the muffle. An equivalent means of transferring the workload is necessary in the vertical furnace as well, although this is not shown in Fig. 6. By using this technique, the charge can be cooled much faster, because it is only necessary to remove the heat from the hearth and workload and not from the hot end of the muffle. This increased cooling rate permits hardening of air-hardening

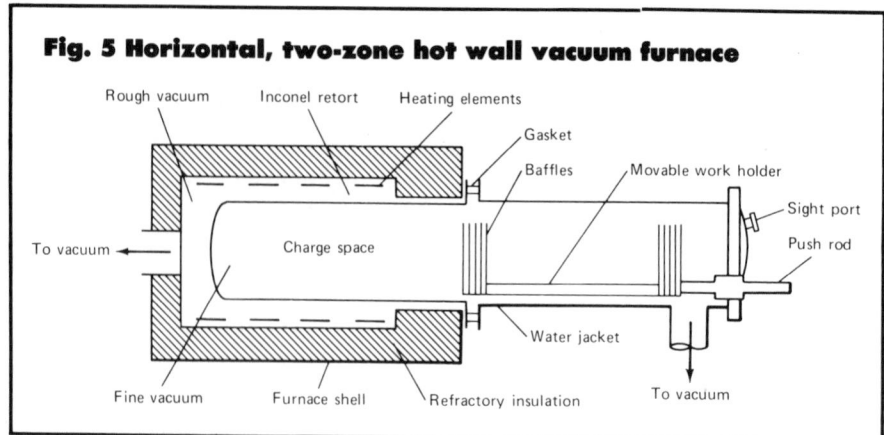

Fig. 5 Horizontal, two-zone hot wall vacuum furnace

Rough vacuum — Inconel retort — Heating elements — Gasket — Baffles — Movable work holder — Sight port — Push rod — Charge space — Water jacket — To vacuum — Fine vacuum — Furnace shell — Refractory insulation — To vacuum

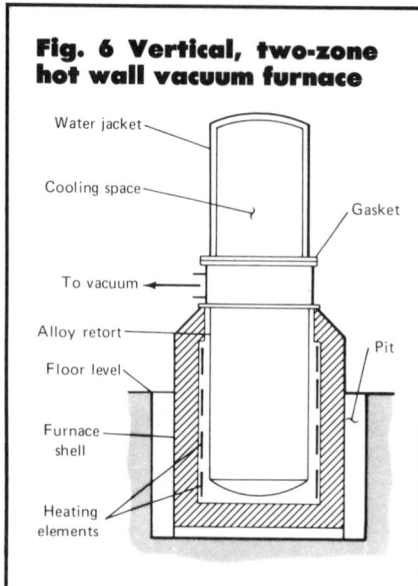

Fig. 6 Vertical, two-zone hot wall vacuum furnace

Water jacket — Cooling space — Gasket — To vacuum — Alloy retort — Pit — Floor level — Furnace shell — Heating elements

steels and is adaptable to certain solution treatments not possible with other hot wall furnaces.

Cold Wall Furnaces

Cold wall furnace units consist of a water-cooled vacuum vessel maintained near ambient temperature during high-temperature operations. Consequently, because the operating temperature does not affect the strength of the vessel material, large units can be constructed for use at high operating temperatures.

In the cold wall design, the water-cooled vacuum vessel contains and supports the internal insulation, the electrical heating elements and the hearth upon which the workload rests. The vacuum acts as (a) a substitute for the normal heat treating atmosphere to protect the workload; (b) an insulating medium in the furnace, because the thermal conductivity of a vacuum is essentially zero; (c) an effective protective coating around the heating elements, the heat shields and the supporting hearth.

The use of a vacuum as the insulating medium has permitted the use of multiple radiation shields of very low mass or special lightweight ceramic insulations that facilitate rapid heating and cooling. Rapid rates of heating and cooling are important because each treatment cycle is started at ambient temperature and must be cooled to ambient temperature at completion. As the protective medium, vacuum has permitted use of materials such as graphite, tungsten, molybdenum or tantalum, for heating elements and hot furnace structures. Such materials normally cannot be used in other furnace constructions without more elaborate, expensive and sometimes hazardous protective atmosphere environments.

Batch Vacuum Furnaces. The workload remains stationary inside the furnace during heating in batch operation; however, in semicontinuous vacuum furnaces with multiple chambers, the workload is moved within the vacuum usually after completion of a processing step or segment. By using high melting-point materials in furnace structures, extremely rapid rates of heating and high temperatures can be attained. Heat is transferred to the workload almost entirely by radiation.

Radiation cooling in the hot furnace is extremely slow, however. To reduce furnace time and shorten quenching rates, pressure in the vacuum chamber usually is increased to either just below atmospheric pressure or up to five times the atmospheric pressure by introducing a pure inert gas such as nitrogen or argon. This gas is rapidly recirculated within the furnace, through cooling coils or through an external heat exchanger and back into the furnace, with high-powered large-capacity gas pumps.

Advantages and Disadvantages

Advantages of batch cold wall vacuum furnaces compared to atmosphere furnaces include:

- Bright, oxide-free treatment of most metals and alloys
- Outgassing and purging of entrapped volumes of gas
- Retained surface finish
- Uniform heating
- Removal of surface volatiles
- No heat added to local environment
- No chemical effect on furnace or work during treatment.

Disadvantages of batch cold wall vacuum furnaces compared to atmosphere furnaces include:

- Vaporization of alloying elements with high vapor pressure
- Inert gas required to effect quench cycle.

Other advantages of batch cold wall vacuum furnaces include:

- Instant pushbutton start from cold
- Low pollution
- Wide operating temperature range
- Easy maintenance
- Safe operation
- Fully automatic processing
- Wide range of programmable heating and cooling rates
- Minimal distortion of treated work
- High heating rates and temperatures resulting from use of high-melting point materials
- Complete shutdown when not in use; no need to maintain heat to maintain low dew point
- Blanketing gas usually not required during heating.

Other disadvantages of batch cold wall vacuum furnaces include:

- Furnace must withstand atmospheric pressure and be leaktight
- High capital cost.

Types of Cold Wall Vacuum Furnaces

Bottom-Loading Furnaces. As shown in Fig. 7, the furnace is stationary and elevated well above floor level.

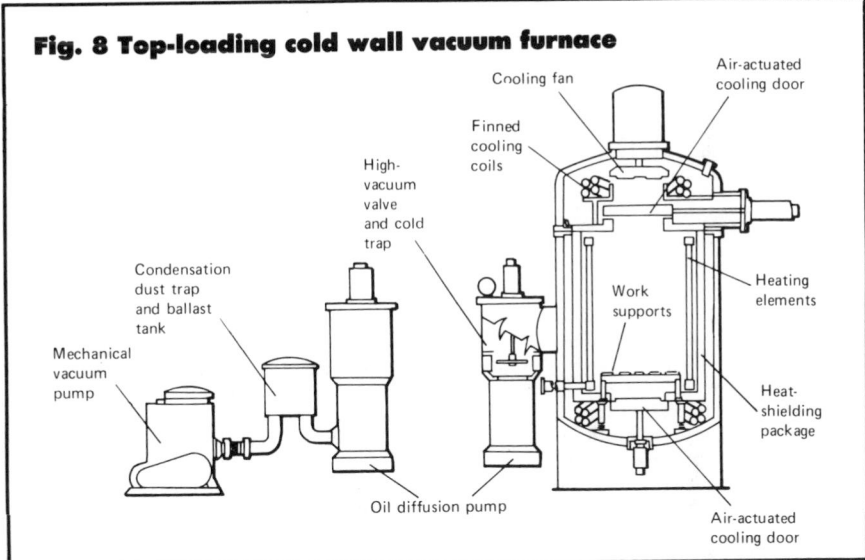

Fig. 7 Bottom-loading cold wall vacuum furnace

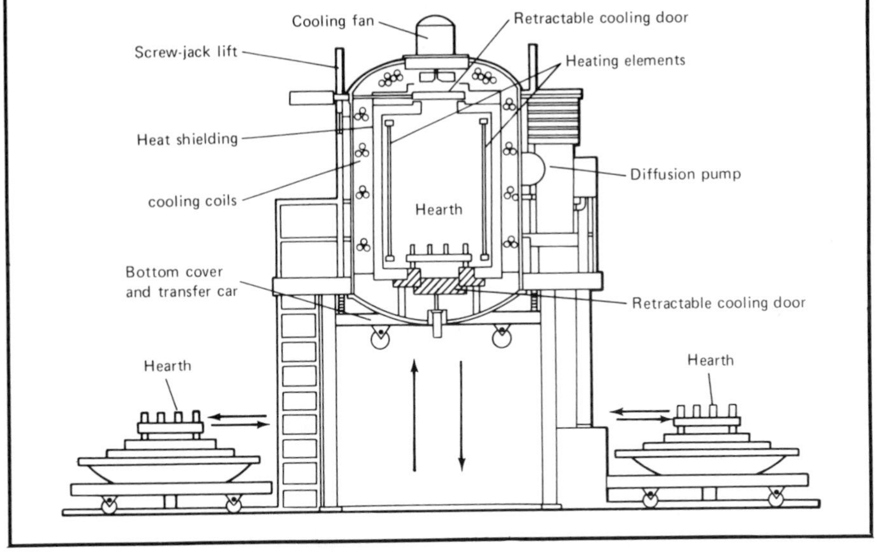

Fig. 8 Top-loading cold wall vacuum furnace

The bottom descends to floor level for ease of loading. The work is loaded on trays that are placed on the hearth by a fork lift when the bottom is in the lowered position. Such furnaces are built to handle large heavy loads and are cooled rapidly by a high-velocity internal or external circulating gas system.

Top-loading furnaces, as shown in Fig. 8, are not as widely used as the bottom-loading furnace. However, they are useful in processing long and relatively thin workpieces such as slender shafts. Workpieces are suspended vertically from hangers attached to the removable top of the furnace. Adequate head room and a vertical hoist are required. These furnaces are cooled in the same way as the bottom-loading units.

Horizontal-Loading Furnaces. A box-type horizontal furnace consists of a gastight cylindrical shell with circular convex end plates. In some designs, both of the end plates are hinged to permit easy access to the furnace interior. Furnaces also are constructed with a stationary rear plate or with a second hinged access door at the front that is smaller than the main front plate. The cylindrical shell and the end plates are water cooled by copper coils soldered to the exterior surface or through the use of double-wall construction. Depending on the intended use, the gastight shell may be made of stainless or carbon

Fig. 9 Horizontal vacuum furnace configurations

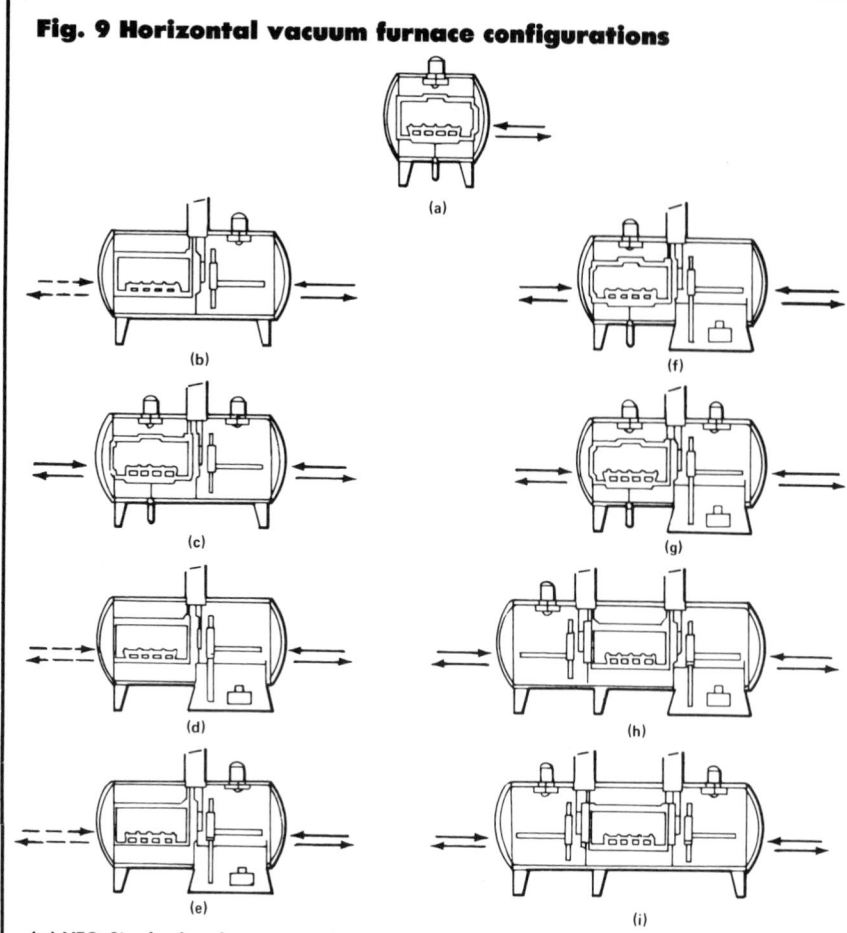

(a) VFC: Single-chamber vacuum furnace with gas/fan quenching. (b) RVF: Two-chamber vacuum furnace; one chamber for heating with integral second chamber for gas/fan quenching only and mechanism for internal in and out transverse movement of workload to and from heat chamber. Unit can be loaded and unloaded only from cooling or quenching chamber. (c) RVFCF: Same as (b) but gas/fan cooling is also included in heat chamber and unit can be loaded or unloaded from either chamber. (d) RVOQ: Two-chamber vacuum furnace; one chamber for heating with integral second chamber for oil quenching only and mechanism for internal movement of workload to and from heat zone. Unit can be loaded/unloaded from oil quenching zone only. (e) RVFOQ: Same as (d) but gas/fan cooling is also included only in quenching or cooling chamber. (f) RVFCOQ: Same as (d) but gas/fan cooling is also included only in the heating chamber and unit can be loaded or unloaded from either chamber. (g) RVFCFOQ: Same as (d) but gas/fan cooling is included in both chambers and unit can be loaded or unloaded from either chamber. (h) FRVOQ: Three-chamber vacuum furnace; middle chamber is for heating only. One end chamber contains gas/fan quenching only and internal mechanism for work movement to and from heat zone. Other end chamber contains oil quenching only and also an internal mechanism for work movement to and from heat zone. Unit can be loaded or unloaded from either end chamber. (i) FRVF: Three-chamber vacuum furnace with middle chamber for heating only. End chambers are both alike in containing gas/fan quenching only and internal mechanism for workload movement to and from heat zone. Unit can be unloaded or loaded from either end.

Fig. 10 Radiation-shield cold wall vacuum furnace

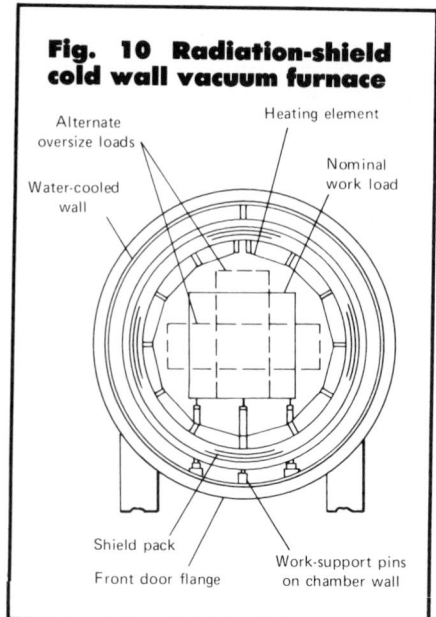

Alternate oversize loads

Heating element

Nominal work load

Water-cooled wall

Shield pack

Front door flange

Work-support pins on chamber wall

aligned with the heating chamber. A hydraulic fork lift raises the work basket, and the truck moves forward to transfer the basket into the furnace where it is lowered onto the hearth or work pedestal. Such a mechanism avoids possible damage to the interior of the furnace, which could occur if transfer is attempted without controlled movement.

Horizontal-loading furnaces may have multiple chambers, depending on the heat treating operation to be performed (Fig. 9). Special systems have been designed to transfer workloads inside multiple-chamber furnaces. The roller-hearth and pusher-type furnace designs can be adapted for vacuum furnaces.

The hearth is supported on wheels that roll on rails which are installed below and outside the heated zone and are protected by movable heat baffles. The longitudinal motion can be supplied by a sealed push rod extending through the furnace wall to an air or hydraulic cylinder or by an internal chain driven conveyor in the cool area.

Another method of work transfer within a horizontally loaded furnace uses an internal rack-and-pinion drive. An overhead chain-driven conveyor can also be used. The work trays may also be transferred longitudinally to the second chamber hearth by a lifting mechanism installed in the unheated chamber that is only exposed to the furnace heat during the short transfer time. Rack-and-pinion drives and

steel. The movable end plates are sealed by O-rings at the end faces of the cylindrical section. The pressure of the outside atmosphere on the convex ends supplies the pressure for vacuum sealing. Usually, auxiliary clamps are provided to supply sealing pressure and to prevent the door from becoming un-

sealed when positive pressure inside the furnace is used during inert gas quenching.

Many horizontal-loading furnaces are equipped with a special lifting and transfer truck that is stationed in front of the furnace. Frequently, this truck rolls on rails so that it is permanently

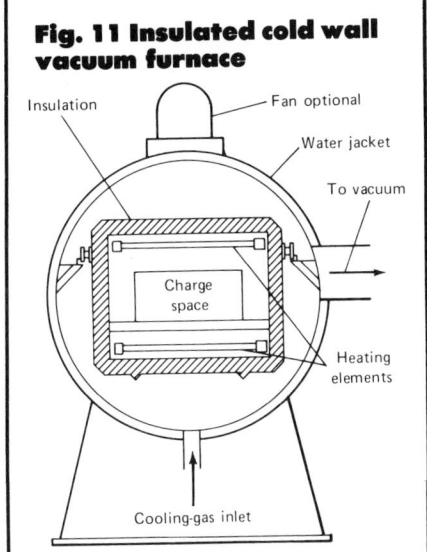

Fig. 11 Insulated cold wall vacuum furnace

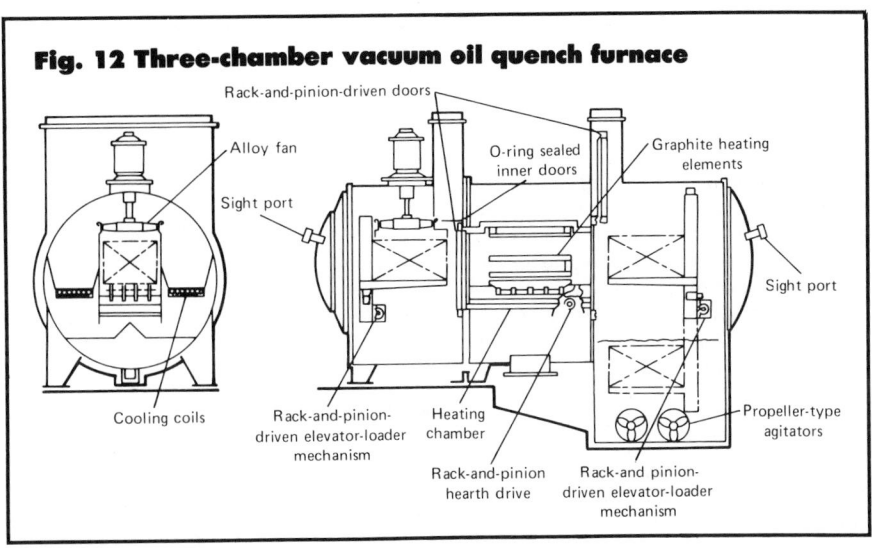

Fig. 12 Three-chamber vacuum oil quench furnace

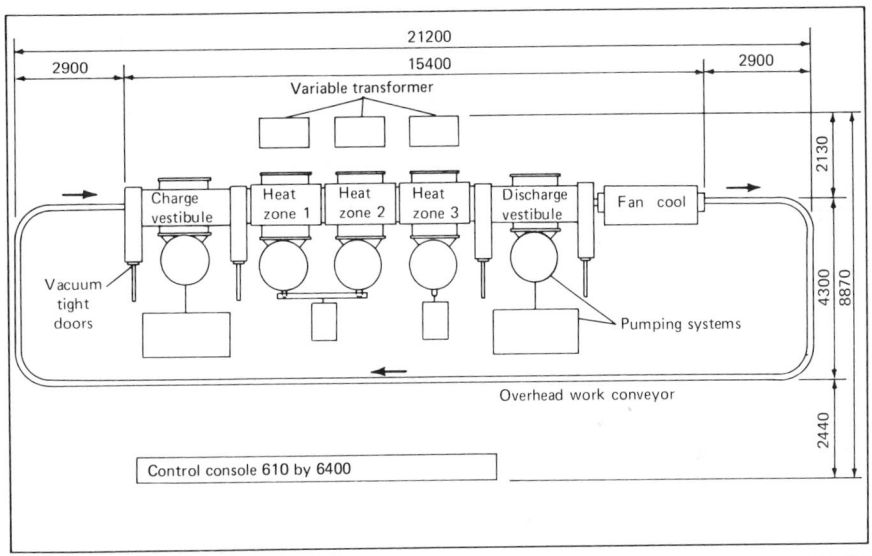

Fig. 13 Layout of an automatic vacuum furnace for fluxless aluminum brazing

pneumatic cylinders often are used to execute vertical elevator movements in and out of liquid quench tanks, as well as to open and close internal heat shields and vertical doors.

A horizontally loaded vacuum furnace equipped with radiation shields is shown, in a vertical section, in Fig. 10. The workload is exposed directly to radiation from the electrical heating elements. The multiple radiation shields are made of thin sheets of heat resisting material, such as molybdenum, in parallel layers between the heating elements and the chamber shell. An alternative construction, where thermal insulating material is used instead of radiation shields is shown in Fig. 11. This insulation may be a fiber fill, graphite felt or special low-density fiber ceramic material.

Cross-sectional views of a three-chamber oil-quench furnace are shown in Fig. 12. The front chamber is equipped with internal cooling coils and a circulating fan for accelerated gas cooling. The center chamber is the heating chamber, which can be sealed at both ends during the heating cycle by internal moving heat shields and doors equipped with O-rings. The third chamber contains the oil quench and the vertical transport system required to immerse the work in the circulated quenching oil.

Semicontinuous furnaces are constructed of modular units of three or more vacuum chambers. Each unit has a work carrier transfer system, an internal assembly of heating elements and shield package, pumping system and temperature-controlled power sys-

tem. Isolation locks or doors at each end of the vacuum heating or brazing environment separate these modular units from the entry and exit vacuum vestibules. These vestibules, in turn, have doors for access to and from atmospheric pressure and other assembly operations. Usually included with the furnace are fans for fast cooling at the exit end of the furnace and an overhead conveyor system to transfer work carriers from the exit to the furnace entrance. Work carrier loading and unloading stations are incorporated in the external overhead conveyor system. A high-volume production semicontinuous vacuum furnace is shown in Fig. 13.

These furnaces are used for fluxless brazing of aluminum heat exchangers at production rates of 100 to 250 parts per hour.

For high-volume production and easy flow of workloads, semicontinuous vacuum furnaces are equipped with electrical controls that can also be computer programmed for automatic operation. The internal and external work carrier transfer systems, door operations, pumping systems, heating systems, gas backfill, and external cooling, loading and unloading stations are all electrically interlocked and controlled for completely automatic operation. Automatically controlled mechanical loading and unloading of the work carriers can also be incorporated into the complete system.

Heating Elements

Resistance heating and induction heating are the two most common methods of heating within the cold wall furnace. When vacuum furnaces are heated inductively, a graphite cylinder is used as a susceptor; the graphite is heated by induction and radiates the heat to the work inside the cylinder. When heating is provided by the more common resistance elements, the heat transfer is also completed by radiation; therefore, the active heating surface should be large enough to effect a rapid transfer of heat.

Essentially all vacuum furnaces employ three-phase 60-Hz power supplies. Three types of power supplies and controls are used:

- Controllable variable autotransformers
- Saturable core reactors
- Silicon-controlled rectifiers (SCR).

Low voltage should be employed in the vacuum chamber because a high electrical potential can produce short circuiting of the elements by ionizing the residual gases within the chamber.

Resistance heating elements operating in a vacuum do not require oxidation-resistant properties equal to those required in oxidizing atmospheres. To improve operating efficiency, resistance heating elements are heated to higher temperatures than are the elements used in conventional furnaces because the transfer of radiant energy is proportional to the fourth power of the absolute temperature. Higher temperatures require heating elements with low vapor pressures to ensure long life. Materials meeting these requirements are

- Refractory metals, such as tungsten, molybdenum and tantalum
- Pure solid graphite in the form of bar, rod or tube
- Pure graphite cloth woven from fine filaments of pyrolyzed graphite
- Chromium-nickel elements for operating temperatures less than 980 °C (1800 °F).

Properties of these materials are compared with iron in Table 2.

Table 2 Characteristics of heating elements used in vacuum furnaces

| Material | Melting point | | Upper operating temperature limit | | Vapor pressure, torr, at: | |
	°C	°F	°C	°F	1600 °C (2910 °F)	1800 °C (3270 °F)
Molybdenum	2625	4760	1700	3100	10^{-8}	10^{-6}
Tantalum	2996	5425	2500	4530	10^{-11}	10^{-9}
Tungsten	3410	6170	2800	5070	10^{-13}	10^{-11}
Graphite	3700	6700	2500	4530	10^{-13}	10^{-10}
Iron	1535	2800	...	...	10^{-1}	1

Refractory Metals

Tungsten is capable of withstanding higher operating temperatures than the other refractory metals (see Table 2). As a heating element material, it is used as a thin sheet or as sections of woven wire screen. Wire screen is less subject to damage from thermal stresses that occur during heating or cooling.

Molybdenum, in the form of solid rod, strip or thin sheet material, is the most widely used metallic element. Material in sheet form is normally preferred because the electrical power density (watts per square inch of radiating surface) is low compared to cylindrical rod, resulting in lower operating temperatures and thus longer service life for the elements. Also, thermal expansion and contraction and the resulting stresses are handled more easily from a design standpoint. However, thin sheets are subject to mechanical damage. Hangers and supports for metallic heating elements must have good insulating properties and must be chemically stable at the temperatures and pressures encountered in service. Heating elements must not be restrained by the support system from free movement during the thermal cycle.

Molybdenum undergoes a large change in electrical resistance between room temperature and the normal operating temperatures; consequently, the design of the power supply controls the current during the early stages of heating to avoid damaging the elements, as well as furnace heating rates.

Solid Graphite Heaters

All metals lose some strength when heated, whereas crystalline carbon in the form of graphite increases in strength as the temperature increases. Pure graphite in the form of flat bar and rod is less expensive than other high-temperature metallic resistors. Graphite also has a much lower heat expansion coefficient and is more resistant to thermal shock than most metallic materials.

As shown in Table 2, graphite also has a high melting point and a low vapor pressure; thus, it is an excellent choice for a vacuum furnace heating element material. In processes where possible minute concentrations of carbonaceous material in the vacuum atmosphere will not have an adverse effect on workpieces, graphite resistors are commonly chosen. Moreover, the presence of incandescent carbon may provide additional cleansing or gettering action with respect to oxygen or water molecules in the residual vacuum atmosphere. This subtle action is not provided by any of the metallic resistors.

Replacement of graphite resistors may be less expensive than for metallic resistors, because elements often can be replaced without disturbing the surrounding insulation and because electrical connections are mechanical and require no welding.

Graphite Cloth Heaters

A third type of material used for vacuum heating elements is a cloth composed of fine graphite fibers. This material is made from rayon cloth pyrolyzed at high temperature to convert the carbon in the rayon to crystalline graphite. The cloth is strong and very flexible. It can be cut with ordinary scissors to the desired size and shape. Because the cloth is flexible, the supporting system can be simplified considerably. The ends usually are clamped in graphite electrodes. This graphite cloth also is available as a hollow cylinder with solid graphite ends of high electrical conductivity to form a self-supporting electrode with a large radiating surface area.

Fig. 14 Relation of heat shield efficiency and emissivity of sheet metal for various numbers, N, of sheets

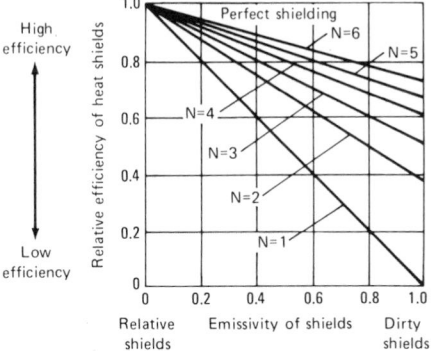

Heat Insulation

Part of the insulation in a cold wall vacuum furnace is provided by the vacuum itself. Where space is essentially void, there can be virtually no heating by convection or conduction. Radiant energy emitted by the resistors in all directions generally is confined to the desired heating zone by one of three designs of radiation shielding.

Metallic Shielding. Multiple concentric layers of thin heat-resisting sheet metal reflect the energy back into the heating chamber of the furnace. Approximately 6 mm (1/4 in.) spacing is maintained between these sheets. Thin wire coil springs are sometimes used as spacers between adjacent surfaces. Inner shields are sometimes fabricated from molybdenum, because of its superior heat-resisting qualities; the next two layers may be stainless steel, with the outer layers composed of nickel steel or plain carbon steel, depending on the maximum operating temperature of the furnace. As the number of the shield layers increases, the efficiency of the shielding increases; however, the insulating effect of each added shield decreases. Consequently, no more than five or six shields or layers are used. Approximately 85% efficiency can be achieved in this manner, even with a material of 0.5 emissivity. The effect on efficiency of the number of insulating layers and the emissivity of the sheeting material is illustrated in Fig. 14. Efficiency declines as the originally bright clean metal surfaces become affected by the deposition of oxide or sooty films. This decline must be con-

sidered during the design of the furnace system.

Although slightly faster evacuation rates and higher ultimate vacuum levels are usually obtainable with designs incorporating radiation shields, these advantages are realized only if the furnace is subsequently used for a singular process application. These advantages readily revert to disadvantages when radiation-shielded units are used for multiprocess purposes.

Sandwich construction resembles the wall of a conventional low-temperature furnace and has proven to be an efficient design. One or more layers of graphite felt or high-temperature refractory fiber fill are placed between the inner and outer sheet metal walls. For instance, the inner wall may be molybdenum and the outer wall may be stainless steel. A vacuum between the fibers reduces thermal conductivity of the wool packing to a fraction of what it would be in an oxidizing atmosphere. Thermally insulated cold wall vacuum furnaces are better suited to conventional vacuum heat treating involving multiple process and temperature requirements. Comparable evacuation rates and ultimate vacuum levels are realized with much less frequent need for vacuum outgassing or hydrogen reduction cycles compared to the radiation shield design.

Multilayer Graphite. This design utilizes multiple layers of high-purity graphite felt fastened to an outer cage of refractory metal by molybdenum clips. This material has a density of only about 80 kg/m^3 (5 lb/ft^3), and it functions as a thermal insulator in much the same way as refractory fiber fill. Because it is a felted material with inherent cohesion, it does not require support by sheet metal walls as does the fiber fill. Its emissivity is near unity—about 0.98 or better. Graphite felt is relatively inexpensive, compared to molybdenum, and is easy to replace. This method, like sandwich construction, relies upon an insulating material with low conductivity in a wall heated from one side.

Insulation Maintenance. All types of insulation can be accidentally contaminated. In many cases, foreign materials such as lubricants used in deep drawing operations, volatile metals or even cotton gloves may be inadvertently carried into vacuum furnaces with the workload. Contamination can result in extensive damage to radiation shields and considerable downtime if

mechanical cleaning of the shields is necessary. In an insulated furnace that does not depend on reflectivity, a thorough outgassing will normally return the furnace to operating condition within several hours.

The efficiency of multiple radiation shields decreases rapidly as the shields become dirty and nonreflective and also as emissivity increases. One method of cleaning molybdenum shields consists of heating the furnace to above the operating temperature after backfilling with dry hydrogen to a pressure of 1 torr. A second method consists of carefully cleaning them with clean compressed air.

An effective method of protecting the innermost radiation shield is to overlay the molybdenum with a layer of graphite foil. This material is essentially pure graphite that has been rolled into a thin sheet by a special process with dry hydrogen. It is held in place by mechanical fasteners. Any contamination from splattering of brazing filler metal or evaporation of volatile materials will be deposited on this inert material instead of the molybdenum shield. Graphite foil can be applied to almost any other metal lining surface as well.

Gas Quenching

The inability to obtain the high cooling rates necessary for many metallurgical operations severely limited early hot wall vacuum furnaces. Development of a retort that could be transferred from the hot furnace to a special cooling stand provided a somewhat inconvenient solution.

With the development of the cold wall vacuum furnace, the cooling problem was overcome to some extent by radiation shielding of multiple thin metal sheets, thermal insulation of very low mass such as refractory fiber, or special foam refractory brick or graphite fiber felt. Because of the low heat content of these insulating media, most of the heat to be dissipated is contained in the workload itself. Cooling of the work remains quite slow because of the absence of convective heat transfer in a vacuum. Backfilling the chamber with an inert gas is necessary to promote conductive transfer of heat from the work to the water-cooled shell. Fans made from stainless steel or other heat-resisting alloys can be installed above the insulated work area to promote the convective circulation of the

Fig. 15 Inert-gas recirculating system

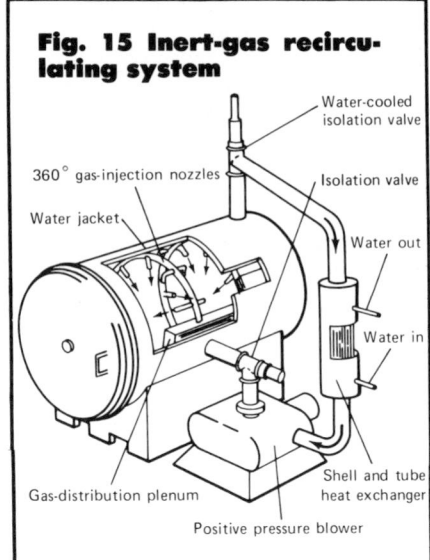

Fig. 16 Comparison of cooling rates and backfill pressures

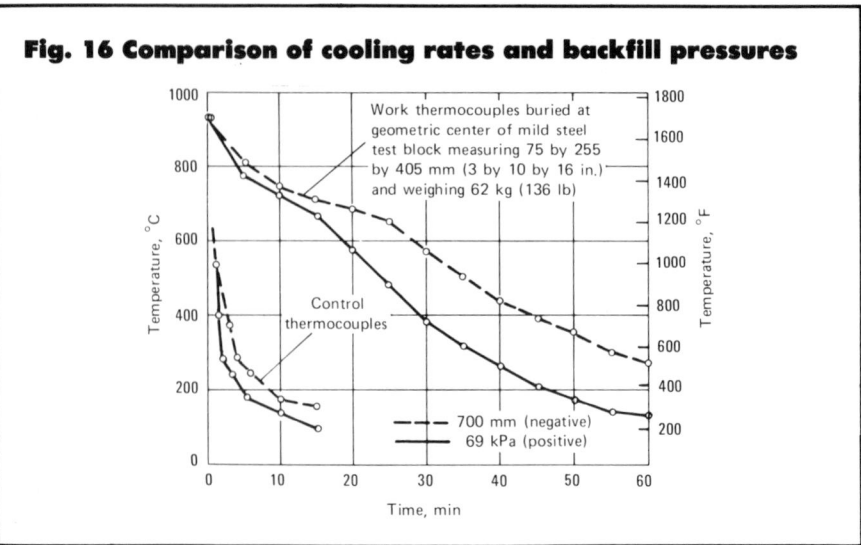

backfill atmosphere. Baffles or bungs of insulating material usually must be moved from above and below the workload to improve the access of the cooling atmosphere to the hot work.

Finned copper coils cooled by circulating water can be installed between the insulating medium and the shell of the furnace to aid in the heat transfer, rather than relying entirely upon the cold furnace shell as the transfer surface.

Circulation by an internal fan is often used because it provides the fastest gas cooling, but it is not likely to provide uniform cooling throughout the load. An inert gas circulating system with an internal manifold is shown in Fig. 15. To increase the rate of heat removal, the mass-flow coefficient must be increased. Mass flow is the product of the mass of gas moved times its velocity. To increase this coefficient, the velocity of the gas circulating through the load can be increased or the gas pressure in the system can be increased so that a greater number of gas molecules are present per unit volume.

Internal pressures used during inert gas quenching range from 100 to 650 torr on furnaces that do not have positive pressure clamps on the doors. For even more rapid gas cooling with the proper furnace construction, positive pressures up to five times atmospheric pressure have been used.

Because of technical advances in fan motors, door-sealing methods and related equipment, it has become common to cool loads with gas fans at pressures up to 1275 torr in conventional units and to several times atmosphere pressure in furnaces with turbine recirculation systems. Because the cooling rate is proportional to absolute pressure, it is increased considerably by positive pressure cooling.

By backfilling to pressures above atmospheric pressure, for example, 70 kPa (10 psi) gage, rather than slightly below atmospheric pressure, the cooling time can be shortened by as much as 30% (see Fig. 16). This development has greatly enhanced the use of vacuum equipment for air-hardening tool steels with decreased over-all cycle time.

The proportional increase of quenching speed with increasing gas pressure up to two times atmospheric pressure also applies to 5 atm, if the time delay in reaching this pressure and starting the fast gas circulation is minimal. Fast backfilling is promoted by a compact design and a special turbine system that builds up the forced circulation very rapidly. Figure 17 illustrates the effect of pressure on quenching speeds from a hardening temperature to 550 °C (1010 °F).

An increase in pressure from 1 to 2 atm will decrease cooling time by 60 s or 50%. An increase from four to five times atmospheric pressure would give a cooling time decrease of 6 s or 20%.

Liquid Quenching

When even faster cooling rates are needed, furnaces are available with liquid quench capabilities. This usually requires an arrangement where the quenching is done in a chamber isolat-

Fig. 17 Relationship between gas pressure and quenching rate

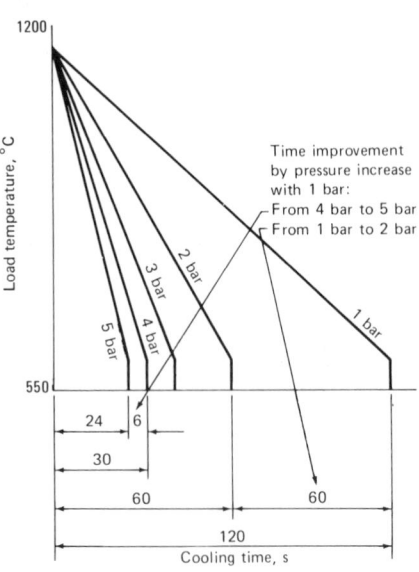

ed from the heating chamber. In one such design, a horizontal loading furnace has a rear heating chamber, with a forward cooling chamber, with a vacuum sealing door between them. When the heating cycle is completed, the work is transferred to the front chamber. After the sealing door is closed, the workload is lowered into the quench tank by an elevator. The quenching liquid is agitated vigorously by propellers, and the heat absorbed by the quenching medium is removed by a heat exchanger similar to that used in many atmosphere heat treating furnaces. The

Table 3 Typical analysis of backfill gases

Gas	Purity, %	Impurity, ppm						Carbon-aceous gas	Dew point		Thermal conductivity at 0 °C (32 °F) (a), K × 10³	Relative cooling rate (b)
		O₂	N₂	CO₂	CO	H₂	Hydro-carbons		°C	°F		
Argon	99.9995	2	2	1	...	...	1	...	−79	−110	0.04	0.74
Nitrogen	99.9993	3	...	...	...	...	...	1	−79	−110	0.06	1.0
Helium	99.998	1	10	...	...	1	1	...	−62	−80	0.34	1.03
Hydrogen	99.9	10	1500	1	2	...	25	...	−59	−75	0.42	1.4

(a) Cal·cm/°C·cm²·s. (b) Relative to nitrogen as 1.0

chamber that houses the quench tank also can be equipped with a circulating fan and cooling coils to provide forced convection gas cooling or liquid quenching. Quenching is usually completed at slightly below atmospheric pressure within an inert gas backfill.

Considerable care must be used to ensure that trays and fixtures that carry the workload into the quenchant are thoroughly cleaned before reuse. Organic residues on trays and workload will continually outgas and destroy the quality of the vacuum if not removed.

Gases Used for Backfilling

Gases used for backfilling vacuum furnaces are argon, nitrogen, helium and, in rare cases, hydrogen. Argon and nitrogen can be obtained as compressed gases or as condensed liquids stored in cryogenic containers. Helium and hydrogen are available only as compressed gases. If appreciable amounts of argon or nitrogen are to be used, it is much less expensive to purchase the gas in liquid form. Very high purity gases are available, and some typical analyses of these gases are listed in Table 3.

When a gas is purchased in a liquid state, it is stored under pressure in a large tank with a safety pressure relief valve that allows gas to escape to the atmosphere if the internal pressure exceeds a set value. Liquid gas leaves the tank through a vaporizer where ambient heat supplies the necessary heat of vaporization. Frequently, this vaporizer is a series of finned coils used to increase heat transfer. Because of the excellent insulation provided for the tank, usually very little gas is lost by venting. Sometimes a reservoir or storage tank of appreciable volume is used to store the vaporized gas at a nominal pressure to avoid sudden surges of pressure in the cryogenic system and to ensure that sufficient vol-

ume is available when backfilling is in progress.

Workload Support

The workload in most vacuum furnaces is placed on a tray or in a basket to facilitate loading. Because these fixtures are heated and cooled at rather rapid rates during processing, the material and design must allow for the cycles of thermal stresses. Molybdenum is often used for support fixtures. If moderately high temperatures are used, as with many tool and die steels, austenitic stainless steel trays may be used. However, work baskets of Inconel alloys are often used with sufficiently long service life in processing high-speed tool steels.

The work-supporting tray usually rests on a graphite or metallic hearth. Frequently, this hearth consists of three or four horizontal molybdenum or graphite bars supported by heat-resisting piers from the furnace shell below. Some hearths use ceramic bars and others use ceramic alumina rolls on which the work tray can be rolled. The end bearings for such rolls should be located outside the intense heat zone of the furnace.

Graphite hearths will react with stainless steel and Inconel alloys to form a eutectic melting at approximately 1125 °C (2060 °F). To prevent a graphite hearth from reacting with a work basket or a workload placed directly upon the hearth, a thin sheet of a ceramic material capable of resisting high temperatures can be placed between the hearth and the workpiece. Some graphite hearth blocks have a longitudinal groove at the top. An alumina ceramic rod or tube is placed in the groove and supports the work or the work basket, thus separating the graphite and the metallic workpiece.

Molybdenum will react with nickel to form a eutectic melting at approximately 1315 °C (2400 °F); therefore, it

is necessary to separate nickel-bearing alloys from a molybdenum hearth at this temperature. Slabs of honeycomb alumina ceramic are available for this purpose. Nickel and titanium form a eutectic melting at approximately 955 °C (1750 °F). Therefore, alloys of these metals must be kept separated if temperatures this high are contemplated.

Vacuum Chambers

The prime requisites in the design, fabrication and operation of a vacuum chamber are prevention of leakage from the outside atmosphere and the assurance that the planned material processing will be completed without damage to the chamber or product. Leakage is a serious consideration even though the pressure within the vessel may be held at the required vacuum, because the continual entrance of air into the evacuated chamber can be harmful to the product or internal components. If a high-capacity pump is used, the pump may well be able to maintain a low pressure inside the furnace regardless of a substantial air leak constantly bringing oxygen into the furnace, which can react with the surface of the workload. Many operators of vacuum equipment routinely conduct leak tests before energizing the heating elements. This is particularly useful when the workload consists of very expensive material. For example, a furnace is evacuated to 10^{-4} torr for at least 1 h. The vacuum valve connecting the furnace to the pump is closed, and the rise in internal furnace pressure in a specified time interval is checked. It is difficult to set specific limits on leakage tolerance, because this tolerance depends upon the type of material being processed and the duration of the heating cycle. Common industrial practice with vacuum furnaces is to establish permissible leakage rates in terms of microns or micrometres per hour. A rate of rise between 10 and 25 μm Hg per hour generally is an acceptable specification for most industrial work, but some more stringent applications can require less.

Generally, equipment manufacturers specify rate-of-rise measurements of 10 μm Hg per hour or less for an empty, clean, cold and thoroughly outgassed system. This measurement is usually obtained by checking the furnace after a vacuum cooling from a heat-up to a temperature sufficient to dry out or bake out the internal components. This procedure eliminates almost completely the outgassing effects of workload and furnace components on rate-of-rise measurements.

Once a vacuum furnace has been properly installed and put into operation, the heat treater usually needs only to be concerned with preventing and correcting the leakage of air at seals around the doors, sight ports, pumping ports, electrical and water feed lines, mechanical rotating and sliding seals for introducing mechanical force and motion into the chamber, and any other point on the vessel that is subject to opening and closing.

Unless a vacuum furnace has been built for service at vacuum levels of 10^{-6} torr or less, typical elastomer seals used in most applications will be worn out if subjected to overheating, to repeated openings and closings or to rotating or sliding motions. When seals are replaced, all flanges must be properly aligned to ensure complete seating of the seal, and sharp edges or burrs must be removed to prevent physical damage. A good grade of vacuum grease lubricant should be used sparingly to facilitate placement of the seals and to improve the sealing.

The best and most expensive vacuum chambers are constructed of an oxidation-resistant material such as a 300-series stainless steel. This material must be used for vacuum service in the pressure range of 10^{-6} torr or less or when the vacuum chambers must operate as a hot wall or retort unit. A less expensive carbon steel, adequately coated, is effective in cold wall chambers for most heat treating applications. However, unprotected carbon steel should not be used in construction of the vacuum chamber. When carbon steel is exposed to a vacuum, it can be so thoroughly degassed that oxidation can occur very rapidly when re-exposed to atmospheric conditions. Oxidation that occurs in carbon steel chambers is permanent and progressive because the

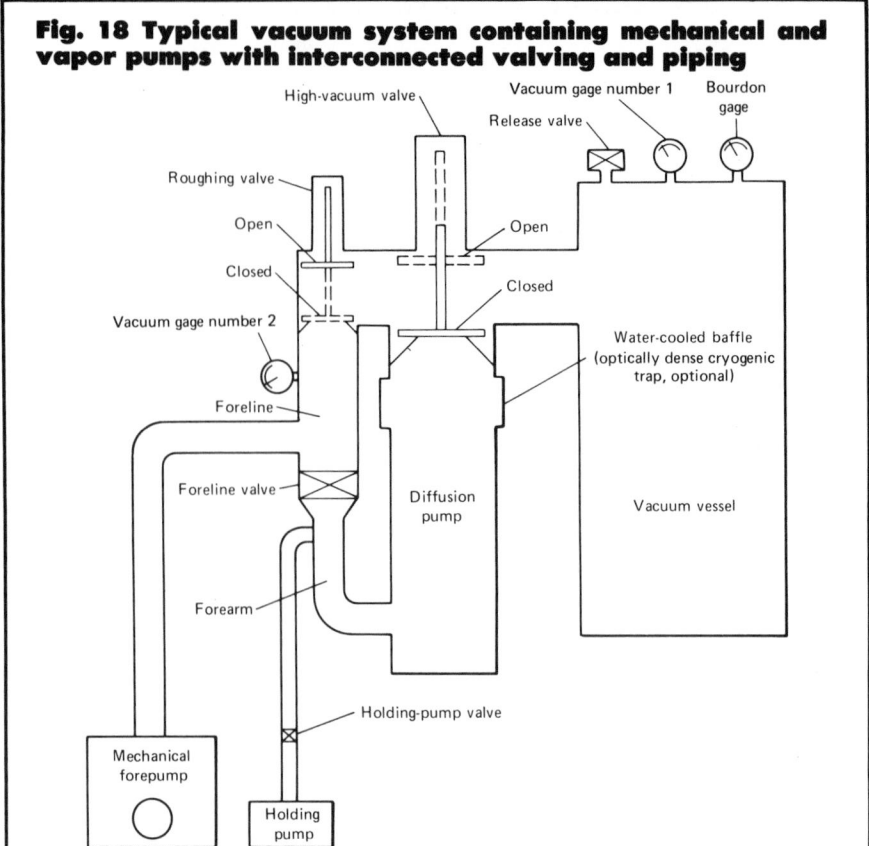

Fig. 18 Typical vacuum system containing mechanical and vapor pumps with interconnected valving and piping

vacuum chamber is completely water-jacketed and does not achieve temperatures necessary for surface cleaning.

The efficiency of the pumping system can be affected by the material used to build the vacuum chamber. For example, a stainless steel chamber with a total volume of 0.3 m³ (10 ft³) will require 10 min for evacuation to a pressure of 10^{-5} torr by a 152-mm (6-in.) diffusion pump. A chamber with the same volume constructed of unprotected carbon steel can require evacuation times of up to 30 min to achieve the same pressure. If the carbon steel surfaces are oxidized, evacuation times of up to 2 h can be required.

Pumping Systems

Vacuum vessels are evacuated by various types of pumping systems that depend, to a great extent, on the pressure range needed for processing. An adequate vacuum pumping system must attain the specified pressure and must have sufficient capacity to handle the processing gas load, not only at the ultimate pressure but at all intermediate pressures during the pumpdown cycle. Pumping systems are usually divided into two subsystems: the roughing pump and the high-vacuum pump. For certain requirements, a single pumping system is sufficient for the entire range and cycle. Pumps usually are classified as mechanical pumps or diffusion pumps. The choice of pump or combination of pumps depends largely on the pressure and gas volume or pumping rates required for a specific process and size of vacuum vessel. The vacuum system shown in Fig. 18 consists of a mechanical forepump that can be connected directly to the vacuum vessel by closing the high-vacuum valve, opening the roughing valve and closing the foreline valve. This procedure isolates the diffusion pump from the rest of the system. The diffusion pump interior can then be pumped free of air by its mechanical holding pump.

When the pressure as shown on vacuum gauge No. 2 has been reduced to a level where the diffusion pump can operate efficiently, the roughing valve is closed, and the foreline valve and the high-vacuum valve are both opened. The residual gas in the vacuum vessel can then expand continuously into the region of the diffusion pump, through the foreline to the forepump and then to the atmosphere. The water-cooled baffle or other optically dense trap atop the diffusion pump prevents diffusion pump oil from diffusing backward into the vacuum system. Vapors from the system are condensed in the liquid nitrogen cryogenic trap, if one is used, when a very high vacuum (very low pressure) of 10^{-6} torr or below is desired. Vacuum gauge No. 1 measures the pressure in the vacuum vessel. It should be a more sensitive gauge than No. 2. The release valve controls backfilling of the system to atmospheric pressure up to the high-vacuum valve.

Mechanical pumps operate on the fluid-flow principle and are primarily positive displacement pumps with suitable seals to permit operation at low pressures. Piston pumps or rotary blowers in various pumping speed ratings are available. Vacuum system levels down to 25 μmHg can be obtained with oil-sealed rotary mechanical pumps. Depending on the application, they may be called roughing or forepumps. They can discharge directly into the air against normal atmospheric pressure. A portion of air in the closed system expands into and is trapped within a chamber of the pump. This volume of air is then compressed by the movement of vanes or a piston in the interior of the pump and expelled through a port equipped with a check valve. This process is repeated and with each cycle, a portion of the remaining air volume in the closed system is expelled. When the back leakage through the pump or leaks in the wall of the container equal the rate of air removal by the pump, the closed-vessel internal pressure remains constant.

A nonmechanical limitation of the ultimate pressure of a mechanical pump is the vapor pressure of the oil itself. Most commercial pump oils are controlled grades of petroleum type SAE 20, which have a vapor pressure of about 1 μm Hg.

When air containing moisture is compressed in the interior of the pump, water may condense and contaminate the oil, consequently affecting the ultimate pressure attainable. One method used to prevent condensate is to introduce enough pure dry air at the start of the cycle to ensure that the resulting moisture level of the trapped air is below the level at which compressive

condensation can occur. Called gas ballasting, this method alleviates the moisture condensation problem.

Diffusion Pumps. When the pressure in the vacuum chamber becomes so low and the molecules so few that the path typically traversed by a gas molecule exceeds the dimensions of the chamber, the remaining gas molecules collide more often with the walls than with each other. At higher pressures, the constant and frequent collisions of adjacent gas molecules and the resulting elastic rebounding effectively scatter and expand the gas to quickly fill any new volume created. At lower pressures, this effect nearly disappears, and the remaining gas is difficult to pump with fluid-flow positive-displacement mechanical pumps.

At these pressures, it is necessary to allow molecules to diffuse randomly into the throat of the pump, then to impart a preferred direction of motion to the molecules by momentum transfer. For pumping at a vacuum system level below 10^{-3} torr, a vapor diffusion pump generally is used. Pumping action is directed by a high-velocity stream of heavy molecules in the form of a pump fluid, usually oil. The heavy molecules strike the gas molecules and push them in the desired downward direction toward the outlet of the pump. To be effective, the inlet pressure of the diffusion pump should be below 1 μm Hg so that the vapor stream is operat-

ing in nearly empty space except for the occasional diffusing gas molecule.

A schematic diagram of a three-nozzle vapor-diffusion pump is shown in Fig. 19. Vapor from a liquid held in a closed boiler heated at the bottom is forced upward inside the boiler. The vapor passes quickly through a narrow circumferential opening in the nozzles at a downward angle. Molecules of gas that stray from the vacuum chamber above the pump toward the vapor jet streaming from the nozzles encounter the downward-directed stream of heavy molecules. The over-all effect is to compress the gas molecules and force them downward to a point where they can be removed by the mechanical forepump.

Pump vapor is condensed on the cooled inner walls of the pump and returns as a liquid to the boiler. Maximum velocity is imparted to the gas molecules by using a liquid composed of heavy molecules in the boiler. Efficiency is improved by using several nozzles in line, one above the other. Special highly stable liquids with very low vapor pressure are required for such diffusion pumps.

Backstreaming is the movement of molecules of the pump fluid above the inlet flange of the diffusion pump and in the direction of the vacuum chamber. The rate of backstreaming increases rapidly as the inlet pressure exceeds 1 μm Hg and depends on the size and type of diffusion pump. Back-

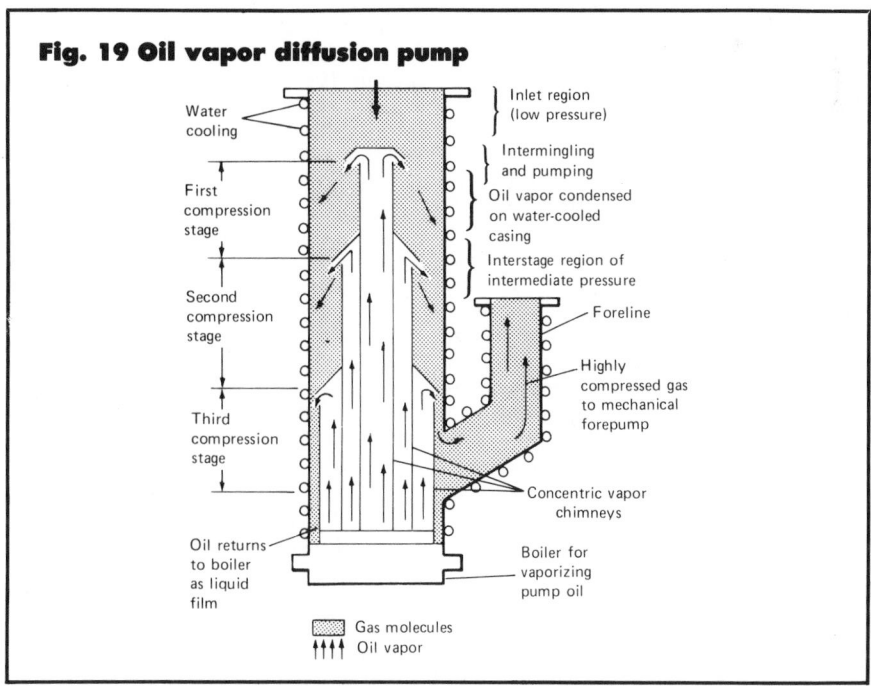

Fig. 19 Oil vapor diffusion pump

Water cooling

First compression stage

Second compression stage

Third compression stage

Oil returns to boiler as liquid film

Inlet region (low pressure)

Intermingling and pumping

Oil vapor condensed on water-cooled casing

Interstage region of intermediate pressure

Foreline

Highly compressed gas to mechanical forepump

Concentric vapor chimneys

Boiler for vaporizing pump oil

▨ Gas molecules
↟↟↟↟ Oil vapor

streaming also can be caused by exceeding the fore pressure limit, which is about 500 μm Hg, of most diffusion pumps. Backstreaming can be reduced by interposing opaque baffles between the diffusion pump throat and the vacuum chamber. Most backstreaming originates at the top jet of the pump and can be reduced by placing a water-cooled cap above the jet. The oil vapor condenses on the cap and drips back into the pump.

By using a cold trap placed above the throat of the diffusion pump, it is possible to reduce oil backstreaming further. The cold trap consists of an arrangement of baffles and water-cooled or refrigerant-cooled walls that provide an opaque path through the trap. A refrigerant such as liquid nitrogen may be used. Molecules of oil condense on the surface of the cooled baffle and remain trapped. Such traps also attract condensable vapors that may be present in the vacuum chamber. Water vapor, the most common contaminant present in high-vacuum systems, may be removed successfully in this way. The ability of the trap to serve as a pump for condensable vapors increases as the temperature of the refrigerant decreases.

Other Pumping Systems. Many types of vacuum pumps may be used depending on the pressure to be maintained for a given vacuum process. The rotary mechanical pump and the oil vapor-diffusion pump are commonly used for most vacuum metallurgical processes. Other types of pumps include steam ejector pumps, oil booster pumps and cryogenic pumps. These pumps are used in vacuum processing for such functions as degassing and drying in conjunction with a rotary pump.

Steam ejector pumps have an operating range from 1 to 10 μm Hg. Although steam ejector pumps eliminate the need for a mechanical pump, they require a large volume of steam to operate. Because their principal advantage is the capability of removing large volumes of gas vapor at low vacuum levels, they are suitable for laboratory work and for very large vacuum vessels, especially if noxious vapors are to be pumped.

Oil booster pumps have an operating range from 1000 to 1 μm Hg and require a backing pump. These pumps tend to introduce excessive backstreaming of oil vapor into the vacuum

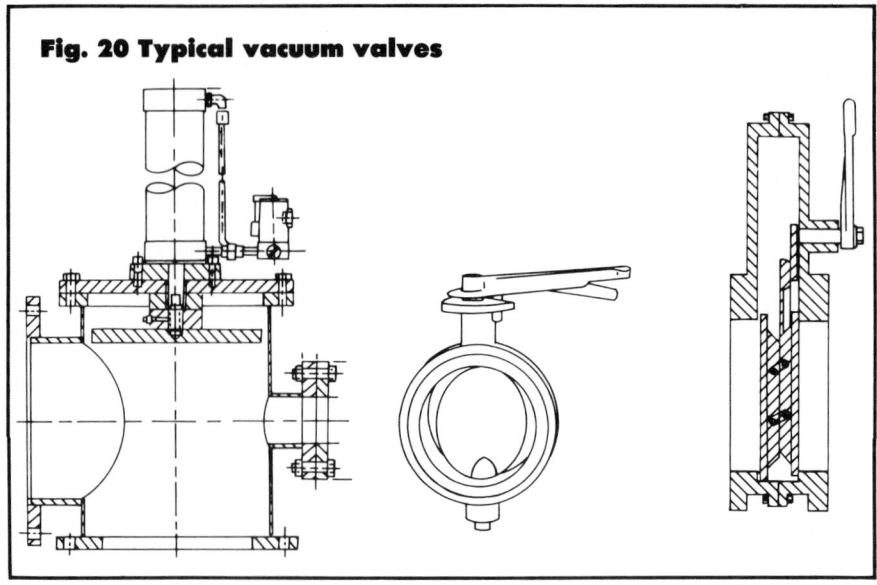

Fig. 20 Typical vacuum valves

vessel and are used mainly when high pumping capacities are desired above 10 μm Hg. They can be used in conjunction with an oil diffusion pump to achieve lower pressures while maintaining high pumping capacities.

Cryogenic pumps condense gas molecules on refrigerated surfaces at −195 °C (−320 °F) or lower. These pumps are regenerated periodically by heating the condensing surfaces to vent the accumulation of condensed gases. Cryogenic pumps are often used instead of diffusion pumps to pump high volumes of water vapor and other condensable gases in the 10^{-3} to 10^{-8} mm Hg range.

Valves. Gas flow of the vacuum pumping system is controlled by three specific types of valves, as shown in Fig. 20. High-vacuum valves isolate the oil diffusion pumps from the vessel, and butterfly and gate valves are used as roughing, foreline and holding-line valves. All of these valves have apertures that provide a minimum gas flow impedance and special seals for vacuum service.

Control Instruments

Except for vacuum gaging instruments, the instruments employed in vacuum equipment processing are similar to those used in other heat treating operations. Once vacuum control has been attained, heating cycles can be initiated either manually or automatically by the temperature recording instrument.

Temperature control systems consist of a primary sensing device, the

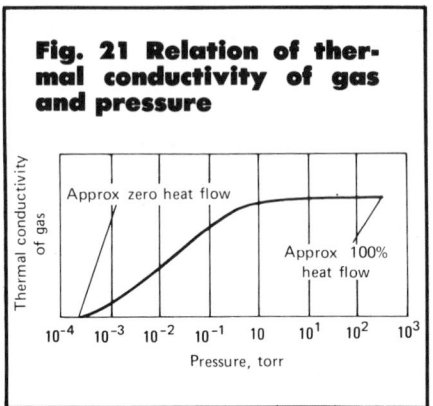

Fig. 21 Relation of thermal conductivity of gas and pressure

control instrument and the final control element. The optical pyrometer is seldom used, because it is not adaptable for automatic control.

The total radiation pyrometer is adaptable to automatic control, but its accuracy may be seriously impaired by intervening media such as gas, smoke or a discolored sight-glass window. Sublimation of materials within the furnace may cause deposition of metallic vapors on the sight-glass window, thereby reducing the radiation reaching the detector.

Thermocouples are typically used as temperature-sensing devices, although their performance varies from a vacuum to an oxidizing atmosphere. Unless the hot junction of the thermocouple is attached to the part being measured, heat transfer to the thermocouple is based almost completely on radiation. In air or other furnace atmosphere, a thermocouple receives heat by conduction and convection. For this reason, the response time of a thermocouple in a vacuum is much slower than in air. A change in air gap of 0.3 mm (0.001 in.) between the hot junction of the thermocouple and the part being measured can change the response time of the thermocouple significantly.

Control thermocouples made from nickel-nickel molybdenum are satisfactory for many heat treating applications up to 1290 °C (2350 °F). Noble metal thermocouples such as platinum-platinum rhodium are used to 1650 °C (3000 °F), and tungsten-rhenium thermocouples are used to 2205 °C (4000 °F). The thermocouples are sometimes used with unsheathed (bare wire) hot junctions, which reduces lag time tremendously. In many applications, particularly with platinum, thermocouples are sheathed in an adequate ceramic or metal protection tube. Below 1095 °C (2000 °F), sheathed base-metal thermocouples can be used to measure different locations in the load.

Because rapid heating rates can be achieved in most vacuum furnaces, it is important to determine temperatures at various locations in the workload. For instance, a large die may overheat in certain locations within the furnace. Information on uniformity of heating also is required when processing smaller parts loaded in baskets.

To bring thermocouple lead wires through the exterior shell of the furnace, special feed-through fixtures are provided. These fixtures are electrically insulated and vacuum tight, and they constitute a permanent portion of the furnace. The ends of the lead wires inside the furnace have ceramic-insulated quick-disconnecting terminals to which workpiece thermocouples can be connected. Chromel-alumel thermocouples often are used to monitor workpieces if the temperature to be measured is not above 1175 °C (2150 °F) and the time at temperatures above 980 °C (1800 °F) is not excessively long.

Pressure Control. Instruments used to measure and record the pressure inside a vacuum-processing chamber fall into two classifications: those that measure the pressure hydrostatically, and those that sense some physical characteristic of the gas which bears a definite relationship to the pressure. A Bourdon gage accurately and continuously indicates the pressure from 1 atm to approximately 0.01 atm and can be used effectively to mon-

itor the roughing cycle and the performance of the roughing pump.

Pressures of a few torr down to 10^{-4} torr can be measured periodically with a McLeod gage, which samples the gas and compresses it to a small calibrated volume. It then registers the ratio of its initial and final volume, and this ratio is an indication of the pressure at which the gas sample was taken. Special expensive diaphragm gages are available for measuring pressures continuously from atmospheric down to the range of 10^{-4} torr.

The second classification of gages includes those that sense the thermal and electrical conductivity of the gas. These physical characteristics bear a direct relationship to pressure.

The thermal conductivity gages are widely used for most vacuum metallurgical processing because they are relatively inexpensive and can monitor vacuum levels continuously between 1 and 1000 μm Hg.

The thermal conductivity, or convective heat transfer, of a gas is essentially constant as pressure is reduced until a pressure of about 1 torr (1000 μm Hg) is reached. From that point, the conductivity declines until, at a pressure somewhat less than 1 μm Hg, there is almost no heat transfer by the gas molecules to the surface of the gage walls, and the thermal conductivity becomes virtually zero. This trend for a typical common gas is shown in Fig. 21.

Thermal conductivity varies considerably from 1 to 1000 μm Hg (10^{-3} to 1 torr), but varies little above that pressure; this behavior constitutes the operational principle of the thermal conductivity gage. Two gages utilize this principle, the thermocouple and the

Pirani. The radiation or thermocouple gage is based on the principle that, as the amount of gas in a vessel decreases, the temperature of a constantly heated wire increases because less heat is radiated to the surrounding environment.

The wire temperature is measured by a fine wire thermocouple attached to the midpoint of the heated wire. The maximum temperature of the wire is about 115 °C (240 °F), and this temperature is reached at pressures of 1 μm Hg or less. A typical thermocouple gage is shown in Fig. 22. Advantages of this gage are

- It is comparatively inexpensive for both circuit and tubes.
- It samples continuously.
- Because of the relatively low wire temperature, it can be exposed to air for years without damage or danger of burnout.
- The signal can be used to activate relays or other remote controls.

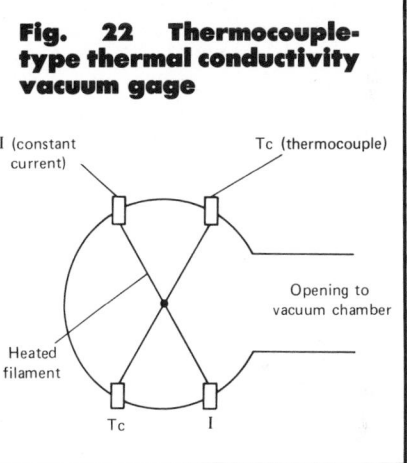

Fig. 22 Thermocouple-type thermal conductivity vacuum gage

I (constant current)

Tc (thermocouple)

Opening to vacuum chamber

Heated filament

Tc

I

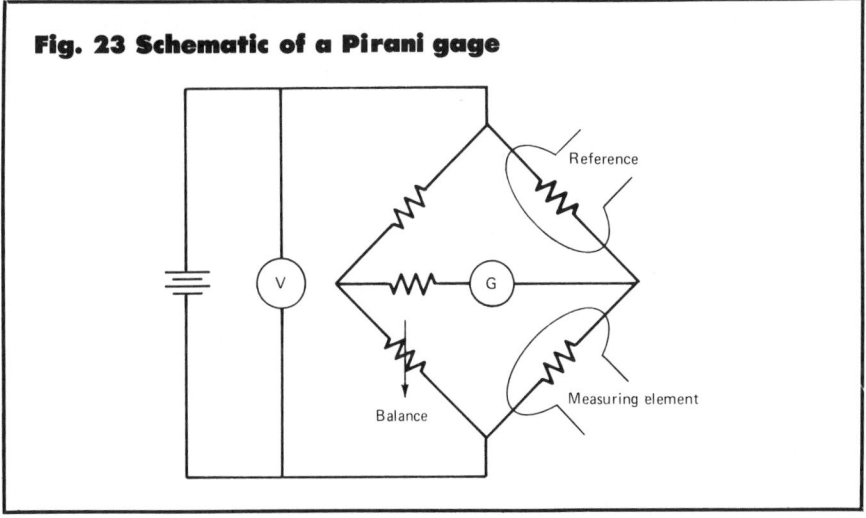

Fig. 23 Schematic of a Pirani gage

Reference

V

G

Balance

Measuring element

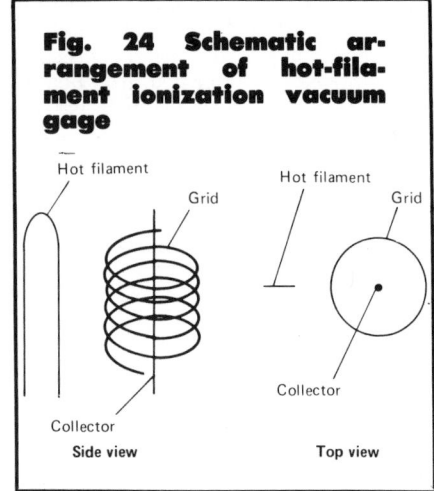

Fig. 24 Schematic arrangement of hot-filament ionization vacuum gage

Hot filament

Grid

Hot filament

Grid

Collector

Collector

Side view

Top view

Disadvantages of this gage are

- Because different gases have different thermal conductivity values, the gage requires calibration for each type of gas.
- The scale is markedly nonlinear; at low pressures, the marks on the scale are spread widely; in higher ranges, the marks are closely spaced (Fig. 21).
- Output depends somewhat upon the ambient temperature of the tube.
- The useful range is from 1 to 1000 μm Hg.

The Pirani gage (Fig. 23) uses the same thermal conductivity principle as the radiation or thermocouple gage. The wire filament is exposed to the vacuum and is heated by a constant current source. As the temperature of the wire varies, its electrical resistance varies. This change in resistance upsets the balance of a Wheatstone bridge, of which the gage is one arm. The vacuum pressure is thus measured in terms of the bridge unbalance. The system is somewhat more expensive and complicated than the thermocouple gage, with essentially the same advantages and disadvantages.

The thermocouple and Pirani gages do not measure pressures of less than 1 μm Hg; therefore, a different type of gage is necessary for systems operating below this level. For vacuums from 1 μm Hg (10^{-3} torr) to 1pm Hg (10^{-6} μm Hg or 10^{-9} torr), the hot filament ionization gage can be used. The sensing element of the gage resembles a triode vacuum tube. Figure 24 shows the three elements of a hot filament ionization gage. A heated filament emits electrons that are attracted to the loosely wound and positively charged grid. The electrons strike residual molecules of gas in the vacuum. These collisions may remove an electron from the gas molecule, resulting in a positively charged gas ion, which is attracted to the collector. The positive ion current flow from the collector to ground is a measure of the gas pressure or vacuum.

The glass ionization gage, which originated in the radio tube industry, usually is calibrated with dry nitrogen. Its response varies slightly with other gases, which must be considered when great accuracy is required.

Another type of ion gage is the cold cathode, which depends on the measurement of an ion current produced by a high-voltage discharge. The cathode in the sensing element releases electrons that spiral through a magnetic field toward the anode. This spiralling motion lengthens the distance that electrons travel between cathode and anode, thus increasing the probability of collision with gas molecules and formation of positive ions. Formation of ions varies linearly with pressure, and the ion current indirectly indicates the pressure. These rugged gages are well suited for production applications, but are more expensive than hot filament gages. They cannot be degassed as easily, however, and they are more readily contaminated and less accurate than hot filament gages. Output is linear below 10^{-3} torr, and the usable range for measuring vacuum is between 10^{-2} to 10^{-6} torr.

SELECTED REFERENCES

- *Vacuum Heat Processing*: by C. H. Junge, Metals Engineering Institute, Course 6 Lesson 7, American Society for Metals, Metals Park, OH
- *Developments in Vacuum Furnaces*: by C. H. Luiten and F. Limque, *Metallurgia*, March 1980
- *Quenching in Vacuum Furnaces*: by T. W. Ruffle & E. R. Byrnes, Jr., *Heat Treatment of Metals*, No. 4, 1979
- *Equipment and Process Developments in Atmosphere and Vacuum Heat Treatment*: by C. H. Luiten, K. H. Edler, W. Goring, and F. Limque, Tratermal 1978, Barcelona, Spain, May 1978
- *The Vacuum Heat Treatment of Tool Steels*: by R. W. Reynoldson and K. C. Harris, *Metal Treating*, Oct-Nov 1970
- *Sub-atmospheric Pressure or Vacuum Heat Treatment Processing*: by T. Bell, AIM Lecture, Linkoping University, Sweden, May 1973
- *Vacuum Heat Treating Optimizes Magnetic Properties*: by G. Koves, *Research and Development*, March 1968
- *The Heat Treatment of Engineering Components*: by J. E. McIntier, ISI Publication, No. 124, 1970
- *Vacuum Processing for Business Machines*: by R. L. VanDerHeyden, *Metals Engineering*, May 1971
- *Scientific Foundations of Vacuum Techniques*: by Saul Dushman, John Wiley & Sons Inc., 1962

Heat-Resistant Alloys for Furnace Parts, Trays and Fixtures

By the ASM Committee on Trays and Fixtures*

TRAYS AND FIXTURES made of heat-resistant alloys are among the many parts used in industrial heat treating furnaces that operate at temperatures from 540 to 980 °C (1000 to 1800 °F). A complete list of these materials may be found in Vol 3 of this Handbook, together with room-temperature and high-temperature mechanical properties.

Basic Metallurgy and Product Forms

A partial list of typical products can be divided into two categories; parts that go through the furnaces and are, therefore, subjected to thermal and/or mechanical shock include trays, fixtures, conveyor chains and belts, and quenching fixtures. Parts that remain in the furnace with less thermal or mechanical shock, include support beams, hearth plates, combustion tubes, roller and skid rails, conveyor rolls, walking beams, rotary retorts, pit-type retorts, muffles, and drive and idler drums.

These heat-resistant alloys are supplied in either wrought or cast forms and, in some situations, may be a combination of the two. The properties and costs of the two forms vary, even though their chemical compositions are similar. Because there are many foundries and fabricators who are experienced in the design and application of these products, it is important to seek their advice when purchasing high-alloy parts.

There are five types of heat-resistant alloys listed in Vol 3 of this Handbook:

- Fe-Cr alloys
- Fe-Cr-Ni alloys
- Fe-Ni-Cr alloys
- Nickel-base alloys
- Cobalt-base alloys

The great majority of heat treating furnaces use only the second and third types, because the Fe-Cr alloys do not have sufficient high-temperature strength to be useful, and the nickel- and cobalt-base types (except Inconel) are generally too expensive except for very special applications. Therefore, this discussion will be limited to the use and properties of the Fe-Cr-Ni and the Fe-Ni-Cr grades.

Room-temperature mechanical properties have limited value in selecting materials or designing for high-temperature use, but they may be useful in checking the quality of the alloys. These properties are shown in Vol 3 and also may be found in ASTM specification A297. The useful high-temperature properties of these alloys are summarized in Table 1 for castings and Table 2 for wrought products. They include nominal composition, stress to produce a creep rate of 1% in 10 000 h, and stress to rupture in 10 000 h and

*Dennis M. Wagen, *Chairman,* Executive Vice President, Stanwood Corp.; Douglas J. Cleary, Vice President and General Manager, Stanwood Corp.; Francis Fahrenwald, Fahrenwald Consulting; Norman O. Kates, Vice President-Technology, Lindberg Corp.; James Kelly, Director of Technology, Rolled Alloys; Fred W. Klag, Vice President, The Alloy Engineering Co.; Arthur L. LaMasters, Marketing Director, Cleveland Alloy Casting Co.; Ross B. Shingledecker, Metallurgical Director, Manufacturing Services, Ladish Co.; John E. Stein, Vice President, Manufacturing, Resisto-Loy Company, Inc.; Donald J. Tillack, Technical Sales Engineer, Huntington Alloys, Inc.

Table 1 Composition and elevated-temperature properties of cast heat-resistant alloys

Grade	Approximate composition, %			Temperature		Creep stress to produce 1% creep in 10 000 h		Stress to rupture in 10 000 h		Stress to rupture in 100 000 h	
	C	Cr	Ni	°C	°F	MPa	ksi	MPa	ksi	MPa	ksi
Fe-Cr-Ni alloys											
HF........ 0.20 to 0.40		19 to 23	9 to 12	650	1200	124	18.0	114	16.5	76	11.0
				760	1400	47	6.8	42	6.1	28	4.0
				870	1600	27	3.9	19	2.7	12	1.7
				980	1800	...	...	...	...	...	...
HH ...... 0.20 to 0.50		24 to 28	11 to 14	650	1200	124	18.0	97	14.0	62	9.0
				760	1400	43	6.3	33	4.8	19	2.8
				870	1600	27	3.9	15	2.2	8	1.2
				980	1800	14	2.1	6	0.9	3	0.4
HK ...... 0.20 to 0.60		24 to 28	18 to 22	650	1200	...	...	...	...	...	...
				760	1400	70	10.2	61	8.8	43	6.2
				870	1600	41	6.0	26	3.8	17	2.5
				980	1800	17	2.5	12	1.7	7	1.0
Fe-Ni-Cr alloys											
HN .... 0.20 to 0.50		19 to 23	23 to 27	650	1200	...	...	...	...	...	...
				760	1400	...	...	...	...	...	...
				870	1600	43	6.3	33	4.8	22	3.2
				980	1800	16	2.4	14	2.1	9	1.3
HT........ 0.35 to 0.75		15 to 19	33 to 37	650	1200	...	...	...	...	...	...
				760	1400	55	8.0	58	8.4	39	5.6
				870	1600	31	4.5	26	3.7	16	2.4
				980	1800	14	2.0	12	1.7	8	1.1
HV .... 0.35 to 0.75		17 to 21	37 to 41	650	1200	...	...	...	...	...	...
				760	1400	59	8.5	...	...	...	...
				870	1600	34	5.0	23	3.3	...	...
				980	1800	15	2.2	12	1.8	...	...
HX .... 0.35 to 0.75		15 to 19	64 to 68	650	1200	...	...	...	...	...	...
				760	1400	44	6.4	...	...	...	...
				870	1600	22	3.2	...	...	...	...
				980	1800	11	1.6	...	...	...	...

Note: Some stress values are extrapolated.

100 000 h at temperatures of 650, 760, 870 and 980 °C (1200, 1400, 1600 and 1800 °F). A design stress figure commonly used for uniformly heated parts not subjected to thermal or mechanical shock is 50% of the stress to produce 1% creep in 10 000 h, but this should be used carefully and should be verified with the supplier.

In general, these materials contain iron, nickel and chromium as the major alloying elements. Carbon, silicon and manganese also are present and affect the foundry pouring and rolling characteristics of these alloys, as well as their properties at elevated temperature. Nickel influences primarily high-temperature strength and toughness. Chromium increases oxidation resistance by the formation of a protective coating of chromium oxide on the surface. An increase in carbon content increases strength.

All of the alloys commonly used in castings for furnace parts have essentially an austenitic structure. The Fe-Cr-Ni alloys (HF, HH, HI, HK and HL) may contain some ferrite, depending on composition balance. If exposed to a temperature in the range of 540 to 900 °C (1000 to 1650 °F), these compositions may convert to the embrittling sigma phase. This can be avoided by using the proper proportions of nickel-chromium, carbon and associated minor elements. Chromium and silicon promote ferrite, whereas nickel, carbon and manganese favor austenite. Use of the Fe-Cr-Ni types should be limited to applications where the temperatures are steady and are not within the sigma-forming temperature range. Transformation from ferrite to sigma phase at elevated temperature is accompanied by a change from ferromagnetic to nonferromagnetic material and from a soft to a very hard and brittle material.

All heat-resistant alloys of the Fe-Ni-Cr group are wholly austenitic and are not as sensitive to composition balance as the Fe-Cr-Ni group. They also contain large primary chromium carbides in the austenitic matrix and, after exposure to service temperature, show fine, precipitated carbides. This characteristic makes these alloys useful for carburizing trays and fixtures. They are considerably stronger than the Fe-Cr-Ni alloys and may be less expensive per part if the increased strength is considered when designing for a known load.

Life expectancy of trays and fixtures is best measured in cycles rather than hours, particularly if the parts are quenched. It may be cheaper to replace all trays after a certain number of cycles to avoid expensive shutdowns caused by wrecks in the furnace. Chains or belts that cycle from room temperature to operating temperature several times a shift will not last as

Table 2 Composition and elevated-temperature properties of wrought heat-resistant alloys

Grade	C	Approximate composition, % Cr	Ni	Other	Temperature °C	°F	Creep stress to produce 1% creep in 10 000 h MPa	ksi	Stress to rupture in 10 000 h MPa	ksi
Fe-Cr-Ni alloys										
309	0.08 max	17 to 20	34 to 37	. . .	650	1200	48	7.0	. . .	. . .
					760	1400	14	2.0	. . .	. . .
					870	1600	3	0.5	10	1.45
					980	1800	. . .	. . .	3	0.5
310	0.08 max	24 to 26	19 to 22	. . .	650	1200	63	9.2	. . .	. . .
					760	1400	17	2.5	. . .	. . .
					870	1600	9	0.3	13.5	1.95
					980	1800	. . .	. . .	4	0.6
Fe-Ni-Cr alloys										
330	0.08 max	17 to 20	34 to 37	. . .	760	1400	25	3.6	30	4.4
					870	1600	13	1.9	12	1.8
					980	1800	3.5	0.52	4.5	0.65
330HC	0.4 max	17 to 22	34 to 37	. . .	760	1400	47	6.8	54	7.8
					870	1600	18	2.6	18	2.6
					980	1800	5	0.7	5	0.7
333	0.08 max	24 to 27	44 to 47	3 Mo, 3 Co, 3 W	760	1400	43	6.2	65	9.4
					870	1600	21	3.1	21	3.1
					980	1800	6	0.9	7	1.05
800	0.1 max	19 to 23	30 to 35	0.15-0.60 Al, 0.15-0.60 Ti	760	1400	19	2.8	23	3.3
					870	1600	4	0.61	12	1.7
					980	1800	1	0.23	6	0.8
802	0.2 to 0.5	19 to 23	30 to 35	. . .	760	1400	83	12.0	79	11.5
					870	1600	30	4.4	33	4.8
					980	1800	8	1.1	11.5	1.65
Ni-base alloys										
600	0.15 max	14 to 17	72 min	. . .	760	1400	28	4.1	41	6.0
					870	1600	14	2.0	16	2.3
					980	1800	4	0.56	8	1.15
601	0.10 max	21 to 25	58 to 63	1.0-1.7 Al	760	1400	28	4.0	42	6.1
					870	1600	14	2.0	19	2.7
					980	1800	5.5	0.79	8	1.2

long as stationary parts that do not fluctuate in temperature. Parts for carburizing furnaces will not last as long as those used for straight annealing.

Finally, alloy parts represent a sizable portion of the total cost of a heat treating operation. Alloys should be selected carefully, designed properly and operated with good controls throughout to keep costs at a minimum.

Material Comparison for Heat-Resistant Cast and Wrought Components

The selection of a cast or fabricated component for furnace parts and fixtures depends primarily on the operating conditions associated with heat treating equipment in the specific processes, and secondarily on the stresses that may be involved. The factors of temperature, loading conditions, work volume, the rate of heating and furnace cooling or quenching need to be examined for the operating and economy trade-offs. Other factors, such as the furnace and fixture design, type of furnace atmosphere, length of service life, and pattern availability or justification, enter into the selection.

Some of the factors affecting service life of alloy furnace parts, not necessarily in order of importance, are alloy selection, design, maintenance procedures, furnace and temperature control, atmosphere, contamination of atmosphere or work load, accidents, number of shifts operated, thermal cycle and overloading. High-alloy parts may last anywhere from a few months to many years, depending on operating conditions. In the selection of the heat-resistant alloy for a given application, all properties should be considered in relation to the operating requirements to obtain the most economical life.

If either cast or wrought alloy fabrications can be practically used, both should be considered. Similar alloy compositions in cast or wrought form may have varying mechanical properties, different initial costs and inherent advantages and disadvantages. Castings are more adaptable to complicated shapes and fabrications to similar parts, but a careful comparison should be made to determine the over-all costs of cast versus fabricated parts. Initial costs, including pattern or tooling costs, maintenance expenses and estimated life are among the factors to be included in such a comparison. Lighter weight trays and fixtures will use less fuel in heating. Cast forms are stronger than wrought forms of similar chemical composition. They will deform less rapidly than wrought shapes but may

Table 3 Recommended materials for furnace parts and fixtures for hardening, annealing, normalizing, brazing and stress relieving

When more than one material is recommended for a specific part and operating temperature, each has proved satisfactory in service. Multiple choices are listed in order of increasing alloy content.

Retorts, muffles(a)		Radiant tubes(a)		Mesh belts, wrought	Chain link		Sprockets, rolls, guides, trays	
Wrought	Cast	Wrought	Cast		Wrought	Cast	Wrought	Cast
595 to 675 °C (1100 to 1250 °F)								
430	HF	430	HF	430	430	HF	430	HF
304		304		304	304		446	
							304	
675 to 760 °C (1250 to 1400 °F)								
304	HF	347	HF	309	309	HF	304	HF
347	HH	309	HH			HH	316	HH
309(b)							309	
760 to 925 °C (1400 to 1700 °F)								
310	HH	310(c)	HH	314	314	HH	310	HH
35-18(d)	HT(e)	35-18(d)	HK	35-18(d)	35-18(d)	HL	35-18(d)	HK
Inconel	HW(e)(f)	Inconel	HL			HT		HL,HT
925 to 1010 °C (1700 to 1850 °F)								
35-18(d)	HK	Inconel	HK	314	314	HL	310	HL
Inconel	HT		HL	35-18(d)	35-18(d)	HT	35-18(d)	HT
	HW		HT	Inconel	Inconel		Inconel	
1010 to 1095 °C (1850 to 2000 °F)								
35-18(d)	HK	Inconel	HL	35-18(d)	35-18(d)	HL	35-18(d)	HL
	HL		HX	80-20	80-20	HT	Inconel	HX
Inconel	HW							
	HX							
	NA22H							
1095 to 1205 °C (2000 to 2200 °F)								
Hastelloy X	HL	Inconel	HL	35-18(d)	35-18(d)	HX	Inconel	HL
Inconel	HU		HX	80-20	80-20			HX
	HX							

(a) Temperature gradients of 40 to 95 °C (100 to 200 °F) are assumed between heat-source side and work zone side of retorts, muffles and radiant tubes. (b) The stabilized grade 309S is recommended for applications involving mechanical or thermal shock. (c) Recommended for vertical mounting only. (d) A series of alloys generally of the 35Ni-15Cr type or modifications that contain from 30 to 40% Ni and 15 to 23% Cr and include RA-330, 35-19, Incoloy and other proprietary alloys. (e) HK or HL is recommended where greater strength is needed. (f) Recommended for applications requiring shock resistance, such as shaker hearths

crack more rapidly under conditions of fluctuating temperatures. Selection should be based on the practical advantages with all facts considered.

General Considerations. Both cast and wrought alloys are well accepted by the designers and users of furnaces requiring high-temperature furnace load-carrying components. There are certain advantages for each type of manufactured component; often, the compositions are similar, if the carbon and silicon levels in the castings versus the wrought material are ignored. In general, the specifications of the wrought grades have carbon contents below 0.25%, and many are nominally near 0.05% carbon. In contrast, the cast alloys have from 0.25 to 0.50% carbon. This difference has an effect on hot strength. The difficulty in hot working the higher carbon alloys accounts for their scarcity in the wrought series. Castings and fabricated parts are not always competitive; each product has advantages, which include:

Advantages of cast alloys

- Initial cost: A casting is essentially a finished product as cast; its cost per pound is frequently less than a fabricated item.
- Strength: Similar alloy compositions are inherently stronger at elevated temperatures than wrought alloys.
- Shape: Some designs can be cast that may not be available in wrought form; they also may not be economically fabricated if wrought material was available.
- Composition: Some alloy compositions are available only in castings; they may lack sufficient ductility to

be worked into wrought material configurations.

Advantages of wrought alloys

- Section size: There is practically no limit to section sizes available in wrought form.
- Thermal fatigue resistance: Ductility of the fine grain microstructure of wrought alloys may promote better thermal fatigue resistance.
- Soundness: Wrought alloys normally are free of internal or external defects; they have smoother surfaces that may be beneficial in avoiding local hot spots.
- Availability: Wrought alloys are frequently available from stock in many forms.

Shape, complexity and number of duplicate parts (eventually affecting

Table 4 Recommended materials for parts and fixtures for carburizing and carbonitriding furnaces

When more than one material is recommended for a specific part and operating temperature, each has proved satisfactory in service. Multiple choices are listed in order of increasing alloy content.

Part	Operating temperature of part(b)	
	Wrought	Cast
Retorts(a)	...	HK
	35-18(c)	HT
	Inconel	
Muffles(a)	35-18(c)	HT
	Inconel	
Radiant tubes(a)	35-18(c)	HT
	Inconel	HU
		HX
Structural parts	35-18(c)	HT
	Inconel	
Pier caps, rails	35-18(c)	HT
	Inconel	
Tube supports	35-18(c)	HT
	Inconel	
Trays, baskets, fixtures (not quenched)	35-18(c)	HT
	Inconel	
	35-18(c)	HT(Nb)
	Inconel	HU
		HU(Nb)
Trays, baskets, fixtures (oil quenched)	35-18(c)	HT
	Inconel	HT(Nb)
		HU
		HU(Nb)
		HW

(a) Temperature gradients of 40 to 95 °C (100 to 200 °F) difference in temperature between heat-source side and work zone side of retorts, muffles and radiant tubes. (b) 815 to 1010 °C (1500 to 1850 °F). (c) A series of alloys generally of the 35Ni-15Cr type or modifications that contain from 30 to 40% Ni and 15 to 23% Cr and include RA-330, 35-19, Incoloy and other proprietary alloys

Fig. 2 Typical HT alloy carburizing furnace trays

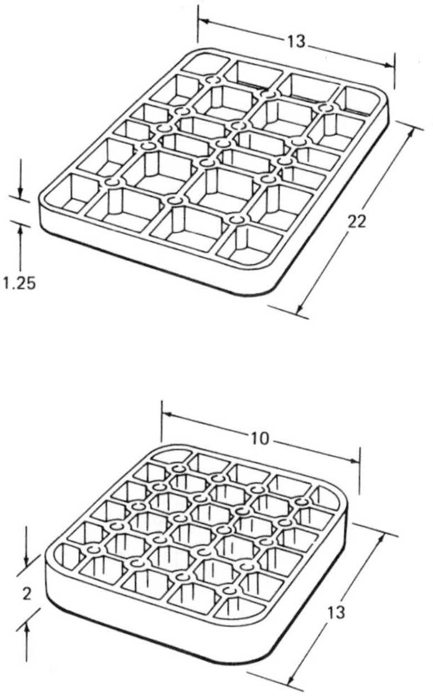

Dimensions are in inches

Fig. 1 Articulated tray for roller-rail furnace

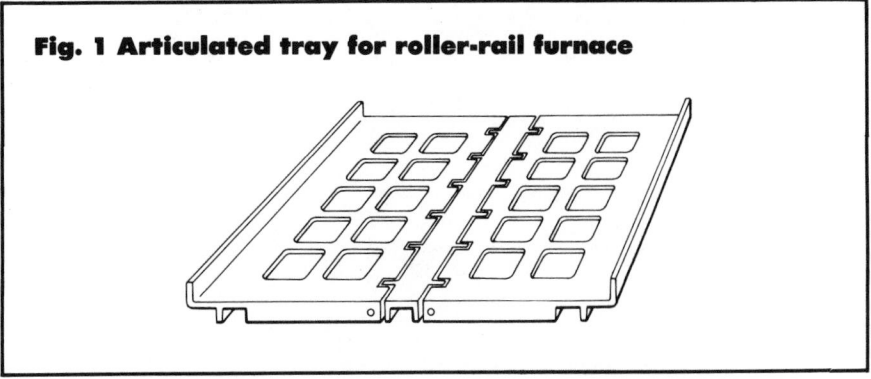

cost) usually determine the choice between casting or fabricated part. Where section thickness and configuration permit, castings are usually cheaper. The cost of the metal per pound of casting is comparable to that of a fabricated part. Total cost of the fabrication usually would be higher because the cost of forming, joining and/or assembly must be added to the cost of material. However, when only one or two types of parts are to be made, the pattern cost precludes the use of a casting.

In energy-intensive heat treating industries, use of wrought fabrications enables fuel savings through reduced heat treating time cycles. At the present level of energy costs, wrought fabrications may be economically preferable because of improvements in thermal efficiency.

Fabrications are preferred for thin sections and for parts where less weight or greater heat transfer may be required. Where thick walls are necessary for strength, or where heavy loads are transported or pushed, the cost of fabricated sections may be prohibitive. Wrought materials have a greater degree of acceptance in fabricated baskets used under carburizing or carbonitriding conditions.

A factor that must be considered in evaluating castings and fabrications is the importance of good welding techniques, particularly for parts that are used in case hardening atmospheres. Castings have replaced fabricated products because of weld failures in multiwelded fabrications.

Although cast alloys exhibit greater high-temperature strength, it is possible to place too much emphasis on this characteristic in materials selection. Strength rarely is the only requisite and frequently is not the major one. More failures occur due to brittle fracture from thermal fatigue than do from stress rupture or creep. However, high-temperature strength is important where severe thermal cycling is required.

Specific Applications. Recommended applications for alloys for parts and fixtures for various types of heat treating furnaces, based on atmosphere and temperature, are summarized in Tables 3, 4 and 5. Where more than one alloy is recommended, each has proved adequate, although service life varies in different installations because of differences in exposure conditions.

Typical Applications

Trays and Grids. Many parts to be heat treated are irregular in shape and as such must be conveyed through the continuous heat treating furnaces or loaded and unloaded from the batch furnaces on grids or trays (Fig. 1). These trays or grids must withstand exposure to the same furnace conditions as the product and as such are subject to heating and cooling, detrimental and beneficial atmospheres, as well as compression and tensile loading. Heat-resistant alloys are used extensively for these parts, although there are instances where dispensible carbon or low-alloy steel fabricated trays are employed. In this instance, the choice is based on the economics of the particular situation, taking into account the cost of materials as well as the service life expected.

Two-thirds of the approximately 15 common heat-resistant alloy compositions find application in the heat treating industry. Of these, half are recommended for use in trays and grids. The particular alloy chosen should be selected on the basis of required strength at temperature, ductility and corrosion resistance.

Trays and grids made of alloying materials using approximately 10% nickel may find an application at furnace temperatures of 650 to 870 °C (1200 to 1600 °F), but as the use temperature goes up, for example, to 1040 to 1150 °C (1900 to 2100 °F), one would probably select an alloy with twice as much nickel. If the tray or fixture is to be subjected to the thermal shock of rapid heating and cooling, one would probably specify an even higher nickel content. The particular atmosphere surrounding the trays necessitates consideration of varied amounts of chromium addition to enhance resistance to oxidation or high-temperature corrosion. If trays are to be used in an atmosphere with very high sulfur, one would select an alloy with rather high chromium and moderate nickel. Some alloys contain relatively large amounts of silicon to fortify against carburization in carburizing applications (Fig. 2).

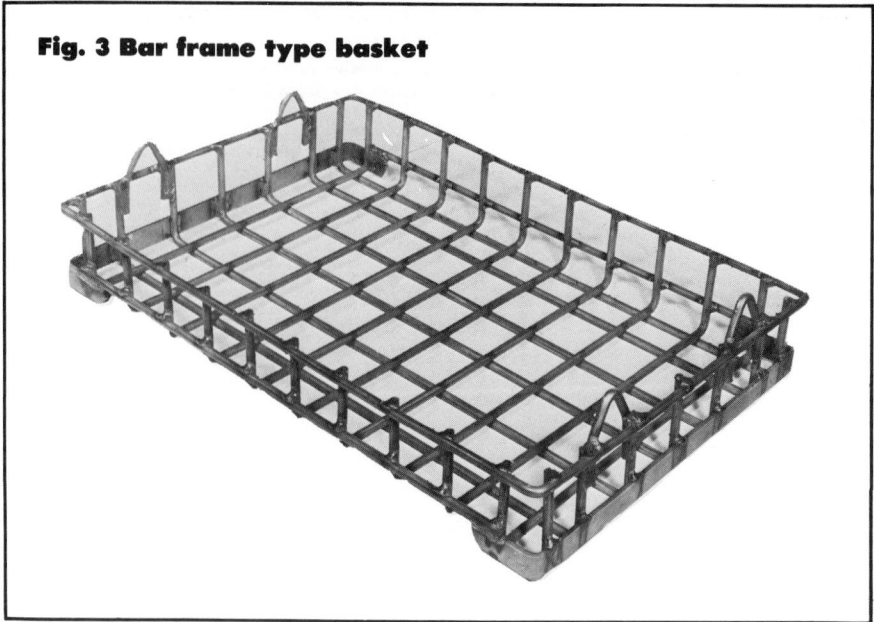

Fig. 3 Bar frame type basket

Fig. 4 Large pit furnace basket

Families of commercially available heat-resistant alloys provide sufficient selection so that an alloy that is optimized for each application and use can be specified. Alloy producers as well as vendors of trays and grids, both cast and fabricated, are an invaluable source of information regarding service life, design considerations and fabrication. Generally, a tray or grid should be of sufficient section size to provide reasonable service life under specified

Fig. 5 Tray/fixture assembly for carburizing pinions

Table 5 Recommended materials for parts and fixtures for salt baths

Where more than one material is recommended for a specific part and operating temperature, each has proved satisfactory in service. Multiple choices are listed in order of increasing alloy content (except ceramic parts).

Process and temperature range	Electrodes	Pots	Thermocouple protection tubes
Salt quenching, 205 to 400 °C (400 to 750 °F)	Low-carbon steel	Low-carbon steel	Low-carbon steel, 446
Tempering, 400 to 675 °C (750 to 1250 °F)	Low-carbon steel, 446, 35-18(a)	Aluminized low-carbon steel, 309	Aluminized low-carbon steel, 446
Neutral hardening, 675 to 870 °C (1250 to 1600 °F)	446, 35-18(a)	35-18(a), HT, HU, Ceramic Inconel	446, 35-18(a)
Carburizing, 870 to 940 °C (1600 to 1720 °F)	446, 35-18(a)	Low-carbon steel(b), 35-18(a), HT	446, 35-18(a)
Tool steel hardening, 1010 to 1315 °C (1850 to 2400 °F)	Low-carbon steel(c), 446	Ceramic	446, 35-18(a), Ceramic

(a) A series of alloys generally of the 35Ni-15Cr type or modifications that contain from 30 to 40% Ni and 15 to 23% Cr and include RA-330, 35-19, Incoloy and other proprietary alloys. (b) Immersed electrode furnaces only. (c) Low-carbon steel is recommended for completely submerged electrodes only.

loading conditions. An overly heavy tray may prolong service life, but the added energy cost to heat the tray through each cycle may offset any cost savings realized through added life. It is sometimes possible to combine materials in trays to provide sufficient strength yet maintain minimum weight. For example, in an articulated tray used in an extremely long pusher furnace, the tray grid that is subject to the compressive force of the pusher bar is of a higher nickel content than the vertical load supports that must bear the compressive load on a per tray basis. This dual alloy tray represents a compromise among weight, cost and service life. In addition, service life is greatly affected by the tray cooling process, and in general, uniform section size throughout a tray is highly desirable to minimize stress during cooling.

All service conditions should be considered when selecting alloy for trays and grids. Unlike furnace structural parts, a tray is subject to alternate heating and cooling during each use cycle. The cooling can be rapid as in quenching work, or relatively slow as in furnace-cool applications. Selection of a proper alloy will ensure adequate service life if all service conditions are known and considered.

Baskets and Fixtures. In many situations, parts being heat treated are of a size that does not permit them to be loaded directly on a furnace hearth, tray or grid. They require some type of container, such as a basket; design of baskets varies because each product is developed for a specific application and loading and must function with a specific type of furnace equipment.

Baskets and fixtures can be produced from cast alloys or fabricated using wrought alloys. Fabrications are used in light to medium loading applications, intricate designs, complex shapes and generally with lighter metal sections. Typically, these include the bar frame type basket (Fig. 3) or corrugated box or shroud. In applications involving heavy loading and/or simple shapes and designs, cast alloys are commonly selected; typically, they are large pit furnace baskets (Fig. 4).

In specific applications, a part may require special positioning. This is accomplished by utilizing a fixture that is generally adaptable to an existing tray or grid or, in some instances, placed directly into a basket or container. These components can range from simple shapes, such as round, square, rectangular or fluted bars, to extremely intricate shapes. Figures 5, 6 and 7 are examples of such fixtures. Figure 5 is a tray/fixture assembly used for carbu-

rizing pinions; Fig. 6 was designed for heat treating lawn mower blades. Figure 7 was designed for heat treating shafts.

In most applications involving operating temperatures of 790 to 1010 °C (1450 to 1850 °F), the product is generally manufactured with a material having a nominal composition of 35Ni-15Cr, which provides a fully stable austenitic structure virtually free from any embrittling phases. In addition, it provides a reasonable cost/life ratio in applications involving endothermic, exothermic and inert atmospheres even with properly controlled enrichments of natural gas, air or ammonia, typically used for gas carburizing or gas carbonitriding. For quenching, a 35Ni-15Cr alloy provides acceptable life; however, in applications of severe quenching, higher nickel alloys or carbon steel may be considered, depending on the cost/life ratio of the product. If applications involve higher temperatures, excessive oxidation or carburization, consideration should be given to increase the Ni-Cr content of the alloy. For nitriding, higher nickel content provides the best cost/life ratio.

Fig. 6 Fixture designed for heat treating lawn mower blades

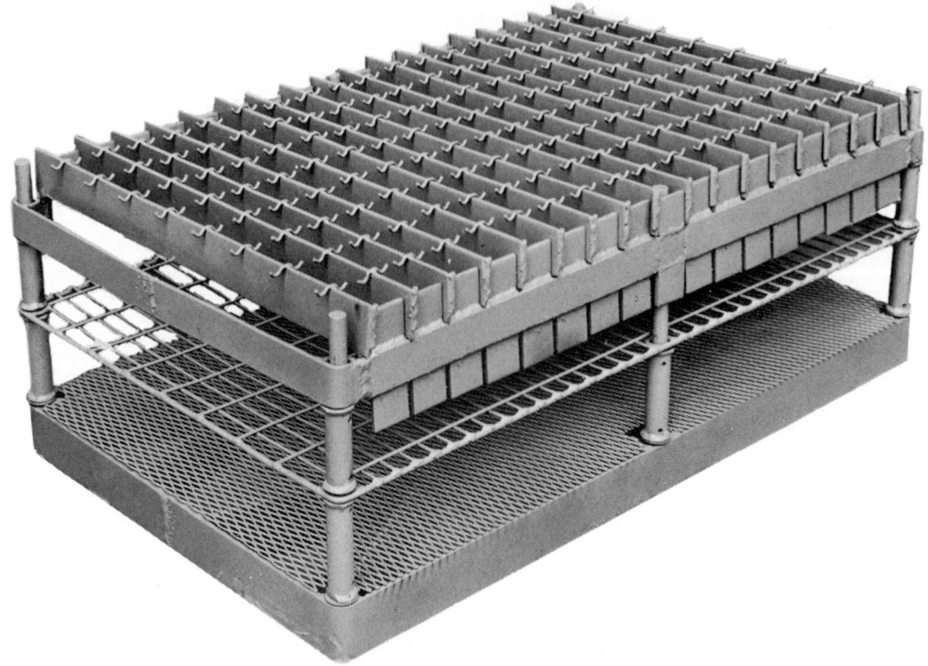

Fig. 7 Fixture designed for heat treating shafts

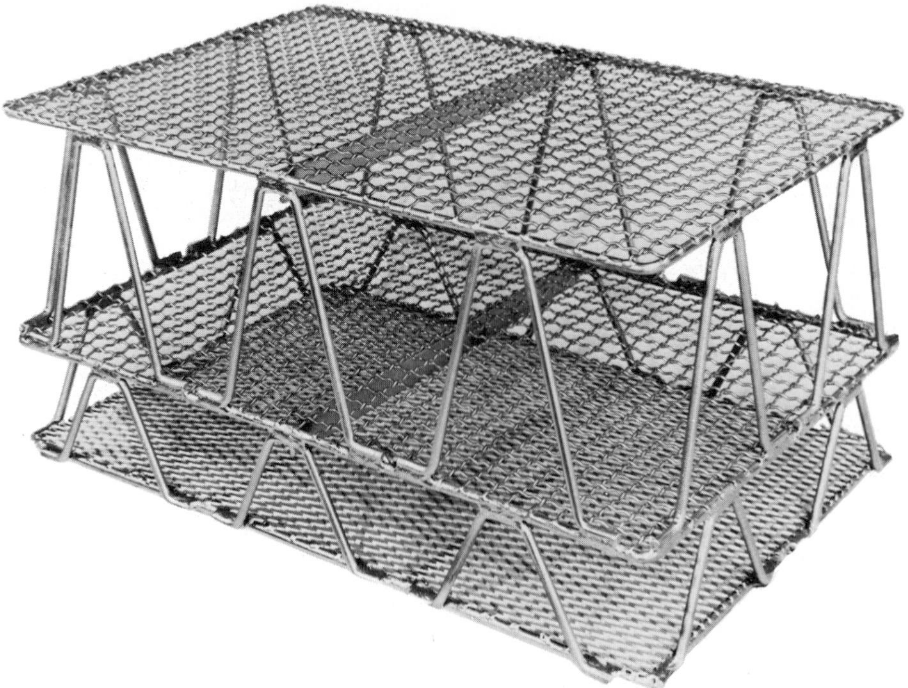

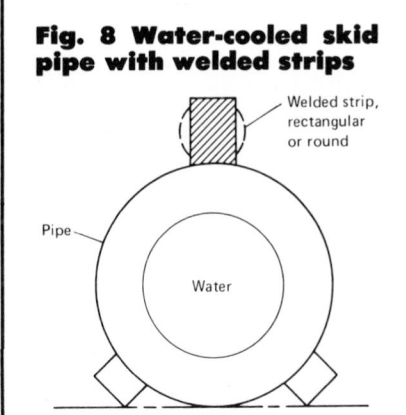

Fig. 8 Water-cooled skid pipe with welded strips

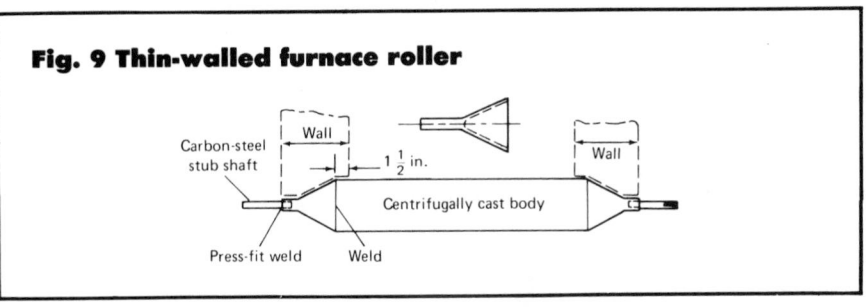

Fig. 9 Thin-walled furnace roller

Fig. 10 Roller hearth roller

In vacuum furnace applications, conventional heat-resistant alloys such as 35Ni-15Cr perform adequately. Caution should be taken to prevent vaporization of any element within these alloys. If a specific application has operating parameters that will not allow the use of a conventional alloy, molybdenum fabrications may be used.

For baskets and fixtures that may be restructured to lower temperature operations of 260 to 595 °C (500 to 1100 °F), materials such as 304, 309 and 310 stainless steel may be acceptable. If the application involves temperatures between 595 and 815 °C (1100 and 1500 °F), caution should be taken because of the potential formation of sigma, primarily in grades 309 and 310 stainless steel. In addition, when grade 304 is exposed to this temperature range, some embrittling from carbide participation results. Therefore, if the operating temperature is between 595 and 815 °C (1100 and 1500 °F), a 35Ni-15Cr alloy is generally selected because the extended life may easily justify the additional cost.

It should be noted that in the application of baskets and fixtures, periodic straightening and re-welding can greatly enhance the product's life and improve the cost/life ratio.

Skid Rails, Hearth Components and Rollers. Certain furnace parts are subject to an additional service condition that must be considered when opting for a particular design or alloy selection. This group of parts includes components of the conveyance system in a continuous furnace that is subject to wear as a result of interfacing with product or trays. Furthermore, this interfacing or wear occurs at elevated

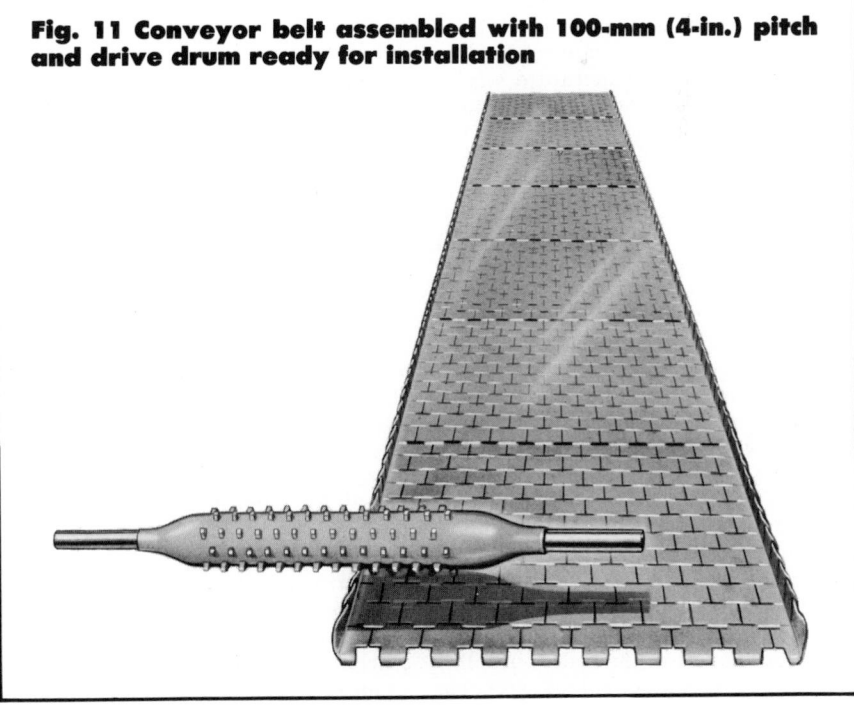

Fig. 11 Conveyor belt assembled with 100-mm (4-in.) pitch and drive drum ready for installation

Fig. 12 U-shape radiant tube

Fig. 13 Straight radiant tube

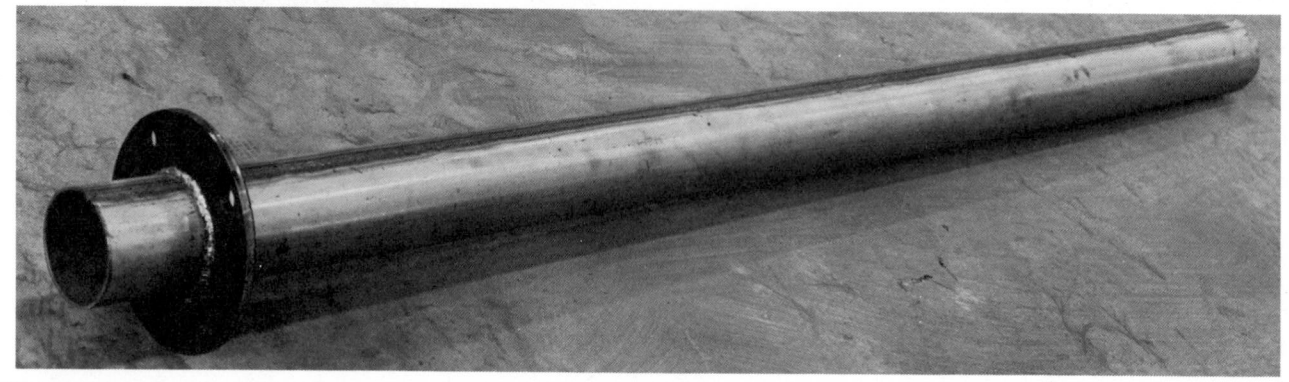

temperatures where alloy strength is diminished. Proper selection of an alloy for a specific high-temperature service involves consideration of many factors. One important factor is not to select the same composition for components that have sliding or rolling contact so as to minimize the possibility of galling or seizing. For example, when selecting an alloy to make skid rails (Fig. 8), one must consider (a) whether the rail will be cooled and the method of cooling, (b) whether adequate expansion space has been specified, (c) the amount of contact area present at the interface, and (d) how the rail will be supported and at what intervals. This implies that the design of skid rails and selection of the alloy is an integral part of furnace design; the same applies to rollers and hearth components.

Perhaps the greatest single factor affecting a roller in a heat treating furnace application is the actual bearing or roller support of the roller. In roller hearth furnaces, the rollers protrude

through the furnace walls, and the roller bearings can operate in a relatively reduced ambient temperature (Fig. 9). However, in some roller-tray furnaces, the individual rollers operate within the furnace heated area and the roller spindle or shaft must rotate on a roller support without aid of a precise, lubricated bearing (Fig. 10). Hearth components usually are nonrotating or nonmoving parts and, in most situations, are well supported by refractory piers and/or ledges. They are almost always subjected to compressive loading, although they could on occasion be subject to lateral thrust and/or bending. When selecting an alloy for these applications, one should consider the elevated-temperature mechanical properties required for the anticipated loading.

Belting. Conveyor belts are used extensively in furnace designs for brazing, sintering and hardening of carbonitriding applications. The woven belts or mesh belting are commonly used for

light duty loading, whereas the cast link belts are designed for heavy loading requirements. Figure 11 shows an assembled conveyor belt with a 100-mm (4-in.) pitch and the drive drum ready for installation.

When mesh belting is required for applications between 260 and 790 °C (500 and 1450 °F), medium carbon steel (grades 1040 to 1055) can be selected for application up to 540 °C (1000 °F). For higher temperatures, materials containing 1 to 5% chromium may be selected, or type 430 stainless is acceptable. Types 304, 309 and 316 stainless steel tend to be susceptible to carbide participation or the formation of sigma phase within this temperature range and, therefore, are not frequently selected. If stainless steel is required, type 347, which is stabilized with niobium and virtually free from carbide participation, may be selected.

Alloys commonly used for mesh belts in the temperature range of 790 to 1205 °C (1450 to 2200 °F) are 35Ni-15Cr,

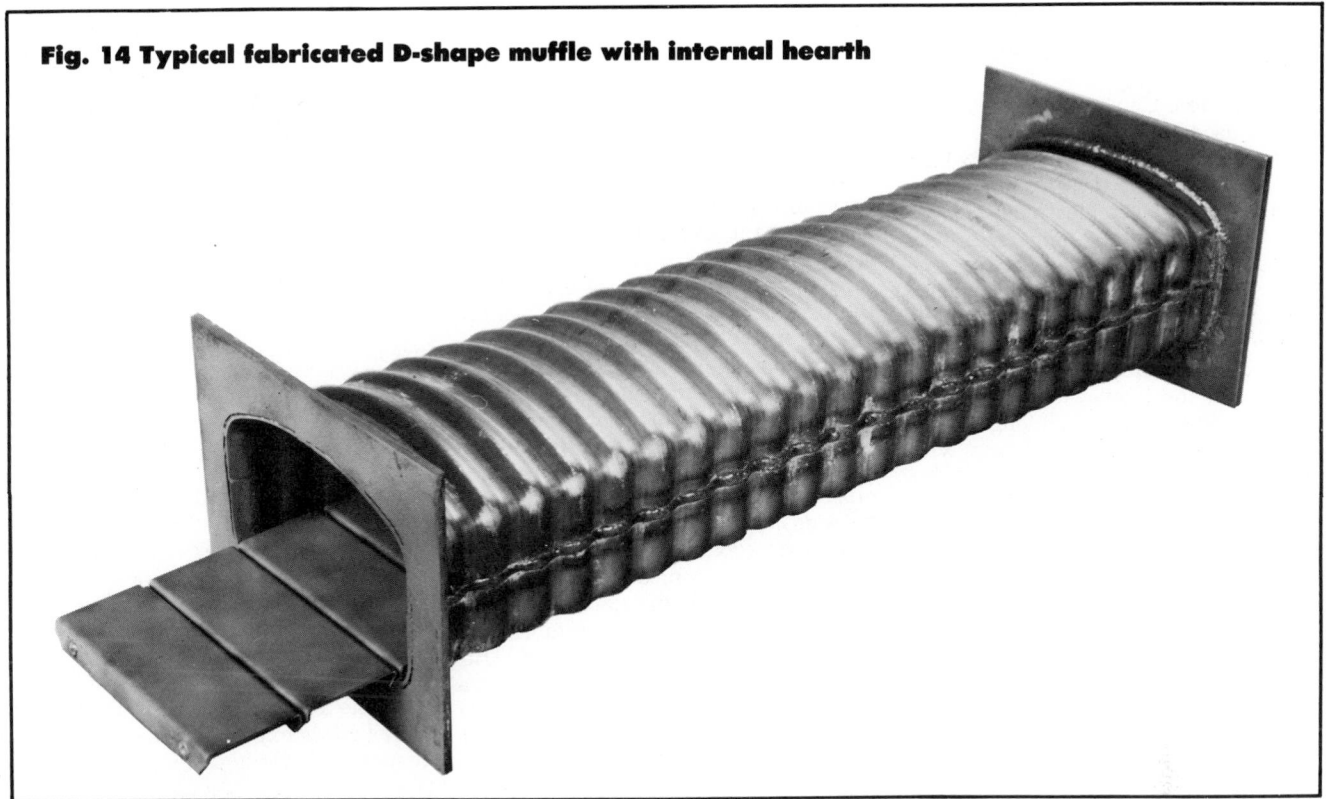

Fig. 14 Typical fabricated D-shape muffle with internal hearth

80Ni-20Cr and type 314 stainless steel. The selection of the proper alloy is based on temperature, atmosphere, possible process contaminants and cost/life ratios of the application. In addition to material selection, other key considerations for mesh belt applications are belt support, drive system, proper tension and control of side travel.

In applications involving heavier loading, the cast link belt is often used. These applications tend to be in the temperature range of 790 to 1095 °C (1450 to 2000 °F) and not in the low temperature range 260 to 790 °C (500 to 1450 °F). Materials, therefore, are similar to the high-temperature mesh belting alloys, except 35Ni-15Cr is the most popular alloy. It provides acceptable service in most conventional heat treating applications, such as hardening, gas carburizing and gas carbonitriding. The cast links generally are assembled using a wrought 35Ni-15Cr alloy with a higher carbon level. In the application of cast link belts, consideration should be given to support, drive systems, tension and side travel.

Radiant tubes can be manufactured from cast alloys or fabricated with wrought alloys and, in most applications, can be selected interchangeably depending on cost/life ratios. Fab-

rications may be selected because of the direct savings in fuel resulting from reductions in weight (fabricated tubes can weigh as much as one-quarter of the equivalent cast tubes). Also, the smooth surface of a fabricated tube is beneficial in avoiding focal points of concentrated or accelerated corrosion. The sound, smooth interior of the wrought tube permits optimum design stresses and helps to prevent the build-up of soot deposit. Figure 12 shows a typical U-shape radiant tube used in carburizing furnaces. Some furnaces use the straight radiant tube shown in Fig. 13.

A nominal 35Ni-15Cr alloy provides acceptable service life in most applications, and the material and proper application are basically the same as outlined in the discussion of baskets and fixtures. In addition to temperature and atmosphere, consideration should be given to tube design for proper expansion and contraction, support for horizontal mounting, and burner positioning to prevent flame impingement. These considerations as well as dissipation rates affect service life as severely as material selection.

Pots. Furnace design is the most important consideration in the selection of material for pots holding molten

lead or salt. Externally heated pots act as a muffle or barrier between the heating and work zones. This type of service is severe because of the great difference between outside and inside temperatures, especially while the furnace is being heated to the operating temperature, when the outside of the pot is subjected to maximum heat input and the lead or salt it contains is still solid.

When the furnace is heated by immersed or submerged electrodes, the pot is completely sealed from the outside air, and the inside of the pot is protected by the molten bath. A pot in this type of installation lasts much longer than an externally heated pot. Because of environmental reasons, salt operations, such as those using cyanide salts, have become extremely limited. The most popular operations remaining generally involve neutral salt and lead. The specific alloy selected for pots used in salt operations is directly related to salt composition.

Pots are available in both cast alloys or fabricated wrought alloys. However, the availability has become somewhat poor for cast pots, and therefore, fabricated pots are more widely used. Carbon steel pots can be used within a temperature range of 260 to 540 °C (500 to 1000 °F). For applications between 540 and 815 °C (1000 and 1500 °F), 309 stainless, 35Ni-15Cr and higher nickel alloys can be applied.

Electrodes. Choice of heat-resistant alloys used for electrodes depends chiefly on the type of furnace in which they are used. The most popular alloy for neutral salt pot electrodes is type 446 stainless steel. Immersed electrodes deteriorate rapidly along the line where the surface of the salt bath comes in contact with them. This is known as "air-line attack". Submerged electrodes, entering the bath through the side of the furnace, are never exposed to air and last much longer. This type of electrode is used only with ceramic pots.

Electrodes deteriorate badly at the salt line during the start-up period. Better service is obtained by maintaining them at a temperature just above the freezing point of the salt during short periods of inoperation. This practice not only prolongs the life of electrodes, but eliminates the tedious task of starting a cold bath. Very little power is required to hold a well-insulated, unused furnace at about 705 °C (1300 °F).

Retorts and muffles are used in heat treating furnaces to separate materials being heated from the products of combustion and, in some instances, to contain atmospheres that would otherwise escape through more porous containment vessels. In most situations, a muffle could be made either of metallic or nonmetallic materials. A typical D-shape muffle with internal hearth is shown in Fig. 14.

Muffles are treated as a separate category of HT alloy applications because an important and different set of constraints apply, in that the heat necessary to raise the inside of a muffle to the proper process temperature is applied from without. Materials and designs must be selected that will not only withstand the rigors of furnace temperature and corrosion conditions, but will, in addition, not act as a significant barrier to heat transfer. Designs must (a) provide for expansion and contraction, (b) be atmosphere tight, and (c) provide maximum area for radiating surfaces because most muffles do not include internal recirculation features. For this reason, many cast or fabricated muffles are corrugated in design. This design increases the radiating area while assisting in accommodating expansion and contraction as the muffle is cycled to and from operating temperature. Heat is transmitted to the inner wall radiating surface of a muffle by conduction, and in order to transfer heat, there must be a temperature drop across the wall of the muffle. The temperature drop is directly proportional to the thickness of the muffle wall. With heavy wall construction, the outside temperature must be raised to effect a given temperature within the muffle. Muffle material should be selected to provide a balance between alloy content, which represents strength, cost and wall thickness.

Cost of any specific furnace part or fixture increases as the alloy content increases, although not necessarily in the same proportion as the base cost of the alloy. Some cost items will be about the same regardless of the type of alloy used.

To be meaningful, computations of cost for furnace parts and fixtures must be based on the number of hours of operation. In many instances, the more expensive alloys will prove more economical when computed on this basis. For example, service comparisons show that HU may be less expensive than HT for oil quenched carburizing trays, and HW may be cheaper than HT for oil quenched carburizing fixtures. Other examples, such as brazing belts, show that the alloy of lower initial cost also may be less expensive when judged by cost per service hour.

From a practical standpoint, even cost-per-service-hour data may be incomplete. Other factors should be considered for some components, notably: (a) labor cost of replacement, (b) loss of productivity during downtime, and (c) the possibility of damage to other components when failure occurs.

Energy-Efficient Operations

By Ross Shingledecker
Metallurgical Director
Manufacturing Service
Ladish Co.

ENERGY COSTS have increased dramatically since 1973, and supplies have become less certain, both of which give heat treaters ample reason to seek out energy-efficient processes.

Considerable efforts have been made by various agencies to advise all energy consumers, including overseas users, of ways of reducing energy usage by up to 15% in comparison with 1973 levels. The readers of this article are referred to these agencies and their publications for assistance. Savings can be achieved in a variety of ways, including the introduction of more advanced technological approaches whereby expenditure of the necessary capital can yield corresponding energy savings.

Theoretical Evaluation of Energy Requirements

Perhaps the most fundamental definition of energy relates various types of energy to the amount of work being done or to the amount of potential work that energy in a particular form is capable of doing. Heat is a form of energy that, when applied to a body, increases the energy of that body. Because molecules are in constant motion, they possess kinetic energy. Because neither increases in potential energy nor increases in kinetic energy can be measured for that body as a whole, it is concluded that heat energy must have been imparted to the molecules of the body. Moreover, almost all solids and liquids expand when heated. This means that work (energy) must have been exerted on the molecules to separate them in opposition to the forces of cohesion. Heat is, therefore, a form of energy that exists as kinetic energy, especially in gases. In solids and liquids, heat also includes potential energy due to expansion.

Kinetic energy can be demonstrated by heating one end of a steel rod. The entire rod warms as heat is conducted along the rod. According to the kinetic theory of heat, the molecules of the metal in the heated end of the rod are set into rapid motion. These molecules strike neighboring molecules and impart kinetic energy to them, and this process continues throughout the length of the rod. The amount of kinetic energy is greatest at the heated end and least at the opposite end.

The potential theory of heat is demonstrated by the expansion and contraction of solids with temperature changes. In many instances, this property of metals is put to work in products and in production processes. The application of heat energy to a body can produce other dramatic effects, although these effects may not be as visible as expansion or contraction. In some instances, the temperature of an object does not increase when the object is heated, although the over-all internal energy of the body is increased. Basic examples of this phenomenon are the formation of steam when water is heated continuously at 100 °C (212 °F) or the formation of liquid when heat is applied to ice at 0 °C (32 °F). Thus, when heat is applied to an object at a specific temperature, a change of state can occur without an increase in temperature—a change from solid to liquid or from liquid to vapor, for example.

Heat treatment of metals deals with a material in its solid state. Thus, energy requirements to produce changes of state, although substantial in quantity, are of little consequence in a normal heat treating process. Further, except in the broadest concepts of heat treatment, the energy requirements related to expansion or contraction of the solid are not major concerns. In heat treating, the desired changes in properties or conditions result from changes in structure. Energy is used to change residual stress within a structure. More frequently, the changes desired in heat treating involve changes in the nature, form, size or distribution of the structural constituents. Changes in the nature of the constituents result from the effect of temperature on phase equilibrium or from changes in the composition of material, as in carburizing or nitriding. The forms of constituents are affected by the conditions under which they separate from the solid solution and by their tendencies to assume certain shapes at temperatures that permit diffusion and rearrangement, as in spheroidizing. Changes in the arrangement or distribution of constituents can be brought about by solution and reprecipitation.

The energy requirements of any particular heat treatment are finite, predictable and usually unalterable. The requirements are proportional to the temperature level at which that treatment is carried out. Further, the quantity of heat that a substance can contain or hold for each degree of temperature rise is known as its heat capacity. The ratio of this heat capacity to that of water is known as the specific heat of a substance. By use of this value, the level of energy (temperature) and the quantity of energy (heat content or thermal capacity) in a process can be measured. By use of these energy requirements in conjunction with the time to achieve the energy level (temperature) desired, the time the material is held at that level awaiting change of structure, and the prescribed time to achieve cooling of the substance, the total energy requirements for a particular heat treating process can be calculated.

The potential energy due to any given phase change or other structural change is calculable. Although such a change may require extended time to be accomplished, the kinetic energy due to the specific heat capacity of the material and the temperature level of the process is calculable and is also not time-related. The function of time relates to the over-all energy requirement because of thermal inefficiencies in heat treating processes. The inherent inefficiencies of a process continue throughout the time required to complete changes during the heat treatment. Generally, the inefficiency of a particular heat treatment, and the amount of energy consumed, is proportional to the treatment temperature.

An equipment designer or heat treater cannot alter the energy requirement of a specific heat treatment, but a great deal of latitude exists in selection of heat treatment processes. In this choice lies the opportunity for energy conservation and cost control.

Accounting Practice

Because it is an intensive user of energy, the heat treating industry remains a virtual bonanza of energy savings awaiting energy-conscious heat treatment operators. Part of the reason for inaction often relates to accounting systems. In many shops, cost systems do not account for the costs of process energy as a separate entity. These costs often are obscured within burden rates or machine center costs. If listed as a utility cost, the cost of process energy frequently is not separated from the costs of building heat and electricity. The energy cost for a particular heat treating operation in the past typically has been a rather insignificant part of the total product cost. In some situations, nearer the end-use product form, even the total cost of heat treatment became insignificant when viewed against the total cost of the product. Even with high energy prices, energy savings are measured in terms of cost effectiveness relative to other means of reducing costs rather than being seen as ends in themselves. Many corporations and large, privately held companies have reorganized their approaches to energy utilization, however. Through the energy manager approach, they have isolated and analyzed energy costs in great detail. Energy requirements are analyzed for each operation and for each phase of an operation.

Energy-Saving Strategies

Although many methods of reducing energy may seem elementary, experience has shown that simple or obvious solutions are often overlooked—particularly where quality and productivity are paramount goals, as in the heat treating industry. The following areas should be examined for possible improvements or changes:

- Product, process or specification restraints
- Modification of process
- Modification of equipment
- Alternate equipment

Product or Process. When faced with the task of improving any operation, an obvious question should be asked: "Why perform the operation at all?" Examples of the results of this type of analysis can be found in almost every sector of the heat treating industry. At one time, for example, all manufacturers of large gears specified that all forgings be double austenitized—that is, normalized, reaustenitized, quenched and tempered. When the mechanical properties of normalized, reaustenitized, quenched and tempered large rolled gear blanks or gear tires were compared with those of normalized, quenched, and tempered gear tires, however, absolutely no difference in properties was found. Now, large marine gear tires are normalized, quenched and tempered. The net savings in energy resulting from elimination of a heat treating operation have been sizable. In many such instances, product and process specifications were written when energy was comparatively inexpensive, and it was common practice to specify that certain operations be done twice to ensure that they were done once correctly.

Design criteria sometimes can be altered to reduce the energy required for heat treatment. For example, a case depth of about 0.75 mm (0.030 in.) can be attained in a 0.1 to 0.15% carbon steel in 3 h in a gas-carburizing atmosphere, whereas a case depth of 0.25 mm (0.010 in.) can be attained in the same steel in less than 1 h of exposure in the same atmosphere. Costly energy can be conserved through use of shallower case depths, even considering that the designer may have to specify a base material with a slightly higher carbon content and consequently higher cost. Further savings could accrue if designers would standardize carburized case depths: deep, medium and shallow, for example. Commercial heat treaters often process batch loads with less than optimum loading simply because other parts awaiting carburization require slightly different case depths. Any sort of overspecification usually results in increased use of energy and increased cost. Carburizing is just one example. Marked savings also can be achieved by changing combinations of both material and process. For instance, ball-bearing screws have been changed from carburized 4615 steel to 1045 steel with induction hardened threads.

Sometimes, energy can be saved by making changes in cleaning specifications. Considerable development work has been done recently on room-temperature cleaners, which usually are aqueous solutions for removing soil from the product without the aid of heat. Not all types of work can be cleaned acceptably with room-temperature cleaners, but all such cleaners achieve significant savings in energy.

Conservation Checklist. All areas of a heat-treatment-related specification should be examined for energy-saving opportunities. The following can be used as a checklist to help locate these opportunities:

- Are the service requirements of the part clearly known?
- In design of the part, has every effort been made to either simplify or eliminate one or more heat-processing operations?
- Is selection of material based on actual property requirements rather than on traditional selection practice?
- Have the possibilities of using alternative materials been investigated, for reasons of lowering cost or increasing availability?
- Has selective heat treating been considered?
- Has the possibility of using pretreated material been considered?
- Has the required case depth been reviewed?
- Is unnecessary normalizing being done?
- Are multiple tempering operations of questionable value being performed?
- Can tempering be eliminated without sacrificing quality?
- Is the heat treating equipment being upgraded in line with other equipment?
- Is a sound maintenance program being enforced in the heat-processing shop?
- Has quenching practice been thoroughly reviewed, and has consideration been given to use of alternative quenching media?
- If double quenching is being used, has the necessity for it been thoroughly reviewed?
- Is direct quenching used when possible?

Modification of Heat Treating Practices

A major equipment manufacturer modified his link-manufacturing operation to allow the link steel to transform fully in air at 540 °C (1000 °F). This modification permitted additional continuous operations to be performed at that temperature, thereby utilizing the heat in the link. The project was subsequently abandoned for mechanical reasons, but the change was metallurgically sound, and the energy savings were significant. Use of residual heat from a previous operation should be explored as a conservation strategy. Material can be fully transformed after a forging or casting operation and still retain substantial amounts of energy. It could be scheduled for immediate entry into a heat treating furnace as soon as the transformation has been fully completed. New process lines often are built with these features in mind, but practices in existing plants can also be changed to achieve the same energy savings. Large savings in energy can be accomplished by optimizing loading of batch furnaces. Too often, batch furnaces are cycled with only a portion of the full design load. In most batch-type furnaces without recuperative devices, 65% or more of the maximum energy input for the cycle is required for the furnace, regardless of the size of the load. Sometimes loads can be combined even with large differences in material thicknesses. The total furnace time is set for the largest cross section, but the reduction in energy consumption can be considerable, when compared with the energy required for processing two separate loads. A careful review of heat treating records may reveal that combining loads or improving load density could save three or more additional cycles per month.

Lowering of Cycle Temperatures. A heat treating operation that is carried out at the lowest temperature normally will require the least energy, and the temperature of a cycle should be reduced whenever possible. A traditional practice in metallurgy has been to normalize at a temperature 110 °C (200 °F) above the upper critical temperature and to austenitize at 55 °C (100 °F) above the upper critical, but this may not be necessary. Instead of normalizing at 110 °C above the upper critical, a temperature 55 °C above it may be sufficient for a particular job. For hardening, instead of 55 °C above the upper critical, a temperature 28 °C (50 °F) above it may be appropriate. The furnace operator must carry out instructions as to which temperature to use, but reduced furnace temperatures can be specified.

Energy can also be saved by minimizing heating times or holding times (soak times). Many specifications require holding times for austenitizing of one hour or more per inch of cross section. This is a conservative requirement, compared with other hardening processes (such as flame hardening and laser hardening, for example). From a metallurgical viewpoint, all that is required is that the metal be heated to a temperature above the hardening temperature. Unfortunately, design specifications are written to include excessive holding times. The metallurgist wants to achieve a certain microstructure or a certain combination of properties in a material. In repetitive situations, it would be worthwhile to discover the minimum time required to produce that microstructure or those properties.

Newer instruments for control of furnace temperature permit presetting of the desired heat-up rate. Not all material can or should be heated to temperature at the same rate or within the same time span. When the heat-up rate is specified, maximum energy savings can be realized by bringing the workpieces to temperature as quickly as possible within the constraints of the capability of the furnace and the necessity of avoiding deleterious metallurgical effects.

Trays and Fixtures. In any heat treating operation, the furnace charge often is considered to be limited to the parts being heat treated. The furnace charge, however, consists of all material that must be brought to temperature, including trays and fixtures. In job-shop operations, it often would be worthwhile to have several sets of trays and fixtures. Lighter parts could be treated in lighter-weight trays or baskets, and heavier parts could be handled in more substantial containers. This problem has been observed in batch treating of titanium parts where the net weight of the high-alloy furnace trays exceeded the net weight of the parts being treated. In another shop, shaft-type parts were being treated on a tray in a batch furnace with a grooved hearth when the workpieces could have been treated adequately without a tray. The most energy-efficient load is the one whose net-charge weight most nearly approaches the gross-charge weight.

Furnace Utilization. Most industrial furnaces are designed to be used at a specific production rate. At that rate, they will achieve the expected design thermal efficiency, but any departure from that specific design rate will result in a decrease in thermal efficiency. Once a furnace reaches a steady-state

Fig. 1 Relationship between furnace efficiency and heating rate in percentage of rated capacity

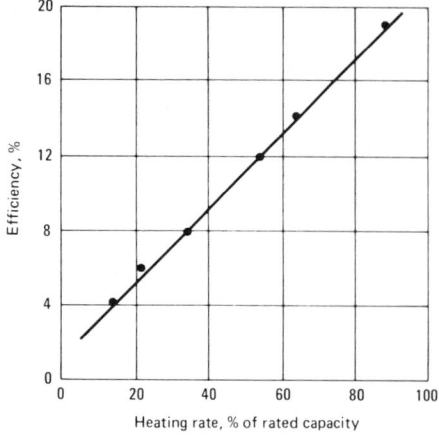

operating condition, the furnace losses will remain constant provided that the operating conditions are unchanged, regardless of whether the furnace is being operated empty or with a capacity load. This constant loss reduces the relative efficiency of the furnace when it is operated at less than its rated capacity. The curve in Fig. 1 is based on data obtained from metering of the gas used in a pusher-type heat treating furnace operating at 950 °C (1750 °F). This curve shows that furnace efficiency dropped when the furnace was operated below its design capacity.

In certain instances, heat treaters conserve energy by stacking, nesting or otherwise consolidating workpieces so as to use a furnace at its design capacity without any deleterious effect on the product. In other instances, energy can be saved by selecting the proper furnace for the job at hand. For instance, a batch furnace should never be used when a fully loaded continuous furnace can be used. Savings result not only from full utilization of a furnace but also from heating in an inherently more efficient furnace. Batch furnaces are always less efficient than continuous furnaces, in which some of the exhaust-gas heat is imparted to incoming cold workpieces.

Idling of Furnaces. Another practice that often can be modified for energy savings is idling of furnaces when not in use. For example, a batch-type furnace idling at 815 °C (1500 °F) requires 9.3 m³/h (327 ft³/h) of natural gas. Reducing the temperature to 760, 705 or 650 °C (1400, 1300 or 1200 °F)

reduces gas consumption to 6.9, 6.1 or 5.2 m³/h (245, 215 or 184 ft³/h), respectively. Thus, a 165 °C (300 °F) reduction in idling temperature reduces gas consumption and can produce a savings of 43.9% each weekend, or an annual savings of 8.6%. In this example, the furnace could produce 515 load cycles in 50 five-day weeks of operation, requiring 220 m³ (7776 ft³) of natural gas per load. If the same furnace were operated seven days a week, the same 515 loads could be processed in 30 weeks. Furthermore, the gas required per load would be reduced to 198 m³ (6995 ft³), an 11.2% savings. In addition, there would be 20 weeks of free time for repairs or for other work.

In another example, in an industrial plant with a captive heat treating facility, operators stored all flanges forged each month in front of the heat treating furnace. In the first week of the next month, the previous month's production was heat treated on a 24-h-a-day, 7-day-a-week basis and then sent to the machine shop. This energy-efficient operation was possible because this manager could stage his in-process inventory at whatever point in the process that provided the lowest over-all manufacturing cost. Although other factors, such as premium pay and urgent delivery schedules, may preclude 24-h-a-day, 7-day-a-week operation, plant managers often have introduced such a practice and have reaped significant energy savings.

If intermittent operation is necessary, the relative costs of idling a furnace and of reheating it after shutdown

can be determined with the following empirical formulas:

$$Q/Q_1 = B$$
$$\text{and}$$
$$B/K = t$$

where Q is the total amount of energy consumed in bringing the furnace from ambient temperature to operating temperature; Q_1 is an average of the combined fuel consumed per hour while idling at a given temperature and the fuel used to return from idling temperature to operating temperature; B is break-even time; K is a constant (1.7 is suggested for tube-fired furnaces, and 1.5 for direct-fired furnaces); and t is the maximum idling time in hours, that will still result in energy savings.

Thus, if it took 1223 m³/h (43 200 ft³/h) of natural gas to heat a direct-fired furnace from ambient temperature to operating temperature, and 25.2 m³/h (890 ft³/h) to idle it and to return it from idling temperature to operating temperature, the break-even point would be 1223/25.2 (43 200/890), or 48.5 h. Because the furnace is direct-fired, the optimum idling time would be 48.5/1.5, or 32.33 h. The break-even point in terms of energy savings is 48.5 h; but, considering damage to the furnace, alloy life and all other factors, the optimum idling time is 32.33 h.

Equipment Considerations

Improved Use of Equipment. In industry, if high quality and good production are being maintained, there is a natural reluctance to tamper with the equipment. But this reluctance can postpone correction of unnecessary losses of energy. Production equipment should be examined in a search for opportunities to conserve energy. With fuel-fired furnaces, the importance of proper fuel-to-air ratios cannot be overemphasized. In products of combustion, the presence of 1% CO or H_2 corresponds to a 5% loss in heat release. The problem of energy remaining in the unburned fuel can be corrected through simple adjustments of the ratio-control equipment. The effects of improper ratios (both excess fuel and excess air) are shown in Fig. 2.

If an insufficient amount of air is supplied, not all of the energy will be extracted from the fuel, as shown in Fig. 2. If too much air is supplied, additional fuel will be required to heat the

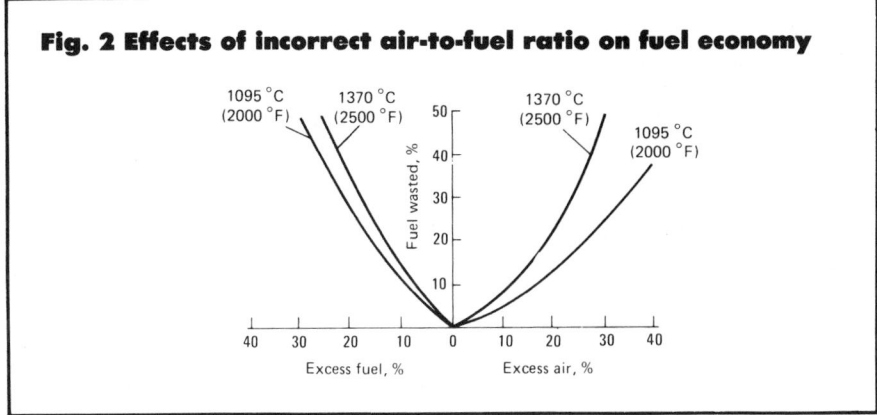

Fig. 2 Effects of incorrect air-to-fuel ratio on fuel economy

excess air. If either condition exists, a substantial loss of energy results.

Too often, furnace doors are opened too wide or left open too long, or both, when furnace charges are moved in and out. A high rate of heat loss, primarily by radiation, occurs when a furnace door is opened, and this rate of heat loss can be calculated; the following formula is often used:

$$q = 0.173 \ Ae \left[\left(\frac{T_o}{100} \right)^4 - \left(\frac{T_a}{100} \right)^4 \right]$$

where q is rate of heat loss, A is effective area of door opening, e is emissivity, T_o is absolute temperature, in furnace, and T_a is absolute temperature of air.

Examination of this formula reveals that the amount of heat lost is directly proportional to the effective area of the door opening. The advantages of limiting door openings to that area actually required for proper access to the furnace are obvious.

Convective heat losses also are proportional to the effective area of the door opening. This loss can be accentuated when both doors are opened at the same time on a continuous furnace, because a tunnel effect is created. Convective losses also result around closed doors and through other openings in fuel-fired furnaces. To conserve energy, all unintentional openings should be sealed. The number of intentional openings should be reduced to the absolute minimum, and these openings should be sealed when not in use. Even when not permitting convective losses, openings in furnaces can permit infiltration of cold air, which must be heated to exit-gas temperature. Proper control of furnace pressure will help prevent these convective losses through necessary openings.

Proper application of refractories in reworked furnaces often can reduce or eliminate constant heat losses that may occur in water-cooled members. Water-cooled skid pipes can be replaced with alloy load supports if maximum temperature of operation permits, or with skid blocks made of high-strength, high-temperature refractories. Walking beam rails sometimes can be topped with refractory shoes rather than with noninsulating alloy shoes.

Marked energy savings sometimes can be achieved by improving heat treating quality controls. Production scrap results in a twofold energy loss; not only the energy required to treat the replacement part but also the energy required to produce and process the part to the point of heat treatment. Retreatments due to poor quenching, missed draw temperatures and any other irregularity are very expensive in terms of both time and energy.

Equipment Replacement or Modification. As the cost of energy increases, the number of economically feasible opportunities to replace or alter heat treating equipment also increases. The following items are not intended to be all-inclusive, but they will point the reader in the direction of greater savings and more efficient operation.

Heat treating equipment that is sturdy and substantially built will undoubtedly provide longer life with less downtime. Such equipment may be slightly higher in cost than equipment of less substantial design, but the difference in cost often can be justified through avoidance of energy costs that continue when equipment is out of service for minor, repetitive repairs. The most efficient furnace is one requiring the least fuel per operating hour, not necessarily the one with the best hour-

ly design efficiency. Fuel cost per unit weight of product produced should be the critical measure instead of cost per furnace-cycle hour. For instance, a cold-wall vacuum furnace uses energy only during the actual heat treating cycle, whereas a conventional box furnace uses energy from the time it is lighted until it is shut down at the end of the week.

When the opportunity arises, much energy can be saved by altering the gases used in atmosphere heat treating. In some cases, an inert carrier gas can be added to the normal working gas. Perhaps a 100% substitution with an inert gas is possible during periods when working gas is not required, but the furnace must be kept "bright".

Other energy-saving opportunities may be realized by improving the insulation systems in heat treating furnaces. The energy lost through a furnace wall is a function of area, operating temperature and composition of the insulation. The first two factors are fixed, but heat flow through and heat storage in the insulating system can be reduced by addition of insulation. Heat storage, which is the amount of heat contained in the wall, can be greatly reduced by using newly developed insulating materials, principally ceramic fibers and mineral wools. At the same time, the thickness of a wall for a given heat flow (loss) also can be changed significantly. One major heat treating firm not only reduced energy requirements by using ceramic-fiber insulating materials, but also was able to heat treat larger rolls because of the decrease in required wall thickness, which in turn provided greater furnace work-zone width.

Addition of alloy fans to existing furnaces sometimes can change a stagnant atmosphere into a high-velocity stream. This increase in ambient velocity breaks up boundary layers of furnace gases that surround the workpieces and shortens the heating time. This reduced heating time is a result of the change from heating only by radiation heat transfer to a combined radiation and convection transfer. The energy savings accrue through reduced furnace time required per cycle.

Thermodynamic thermal efficiency also can be improved by use of waste combustion gases. A particular batch furnace, heated to 980 °C (1800 °F) with natural gas, with the flue gas exiting at or near that temperature, operates at only about 35% efficiency. Through use of a recuperator, the incoming combustion air can be heated by the waste products of combustion. If the combustion air is heated to 425 °C (800 °F), a 23% savings in fuel usage can be achieved. This method is perhaps the least efficient way of reusing heat from the products of combustion; sometimes, however, it is the only practical way.

Because the coefficient of heat transfer between combustion products and a solid is twice as high as between combustion products and incoming combustion air, it is better to use waste gases to preheat the workpieces, as is done in properly designed continuous direct-fired furnaces. Other commonly used recuperative methods also are employed successfully on continuous furnaces, but some of these methods employ a lower heat-transfer coefficient. Methods may vary, but increases in the cost of energy are likely to move the heat treating industry more and more toward use of waste gases.

Furnace Control Instrumentation

Temperature Control

By John E. O'Neil
Industry Manager
Fabricated Materials Market
Leeds & Northrup Co.
Unit of General Signal Corp.

TEMPERATURE INSTRUMENTATION AND CONTROL SYSTEMS used in heat treating include temperature sensors, controllers, final control elements, measurement instruments and set-point programmers. A basic control loop includes a temperature sensor, controller and final control element. The heat process temperature is detected by a temperature sensor that generates a signal proportional to the process temperature. This actual temperature is compared to the desired temperature determined by the controller set point. Based on this comparison, the controller develops an output signal that adjusts the final control element regulating heat flow to the heat treating process.

Basic Control Loop and Auxiliary Devices

The basic control loop is illustrated schematically in Fig. 1. Auxiliary devices used with this basic control loop include measurement instruments and set-point programmers (Fig. 2). The measurement instrument monitors the same temperature sensor as that used by the controller. The set-point programmer automatically varies the controller set point to provide a temperature cycle or temperature program in accordance with an established plan.

A variation of the basic control loop

frequently used in the metals industry is the recorder-controller configuration illustrated in Fig. 3. The recorder measurement of the temperature sensor signal is compared to the recorder set point. Based on this comparison signal, the integral controller regulates the process.

The basic control loop may incorporate a temperature transmitter to amplify the temperature sensor signals. This amplification is desirable, when the control instrument is remotely located from the temperature sensor. This amplification also helps to avoid interaction between two or more instruments using a common sensor. For example, the temperature signal may be transmitted to a remote central control room where it is connected to several controllers, a recorder, a data acquisition system or a digital computer.

The various measurement and control loop configurations must function properly to achieve optimum operation of the system. Although the necessity of this coordination is obvious, one or more elements of the system are commonly neglected. For example, if the temperature sensor, which may be only 1% of the instrumentation cost, is out of calibration or not properly located, the performance of the complete system is degraded. Poor input into even the most sophisticated temperature mea-

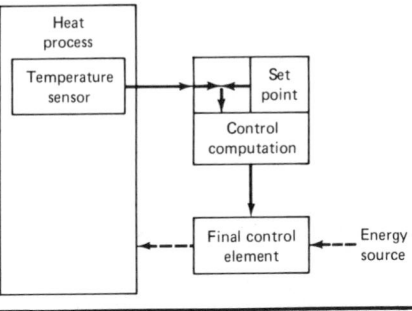

Fig. 1 Basic control loop

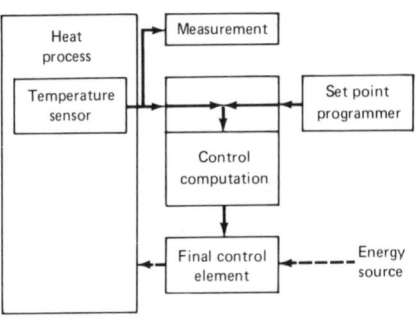

Fig. 2 Basic control loop with auxiliary devices

surement and control systems will produce poor output.

Temperature is one of the metric sys-

Fig. 3 Basic control loop with recorder-controller

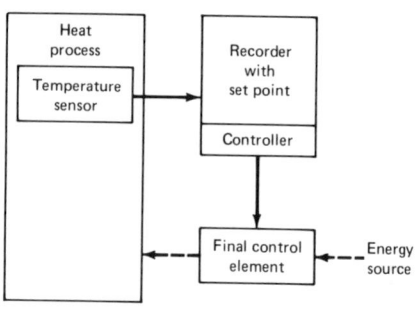

tem's seven basic units. According to Maxwell, "the temperature of a body is its thermal state considered with reference to its ability to communicate heat to other bodies". Methods or modes of communicating heat are convection, conduction and radiation. An understanding of these modes of heat transfer is essential to measurement and control of temperature.

The majority of heat processing applications in the metals industry—including virtually all heat treating—are in the temperature range of −100 to 1370 °C (−150 to 2500 °F).

Important temperature-related properties of metals include such factors as thermal conductivity, thermal expansion, electrical resistivity, ductility, hardness and oxidation resistance. In heat treating applications, temperature, time and atmosphere determine the metallurgical properties obtained. Consequently, these parameters require precise control to achieve optimum properties in the final product. If the temperature is too high, for example, excessive heat is lost, which wastes energy. If the product is not permanently damaged by excessively high temperatures, it may have to be reprocessed, thus doubling energy consumption. Therefore, precise temperature control in heat processing is mandatory for producing acceptable products as well as for conserving energy.

Temperature Sensors

Temperature sensors are classified as either contact or noncontact, and as either electrical or nonelectrical. For example, the mercury thermometer is a noncontact, nonelectrical type. Because of high-temperature requirements and the trend toward electrical instrumentation, most heat treating applications require electrical sensors. When selecting a contact sensor, certain variables should be considered, such as cost, temperature range, useful life, accuracy, size and response speed. Selection of noncontact sensors involves similar considerations, plus those related to radiation factors. With an optical pyrometer, for example, target size, surface emissivity, focal length and sighting path interference could be examined.

Thermocouples and resistance temperature detectors (also known as resistance thermometers) are the most important contact-type electrical temperature sensors used in the metals industry. Well over 90% of the sensors used in this industry are estimated to be thermocouples. A thermocouple is rugged, inexpensive, accurate, covers wide temperature ranges, and is fast in response. A resistance thermometer is more accurate and stable than a thermocouple. However, the resistance thermometer is more expensive, slower in response, and limited to lower temperatures, typically 540 °C (1000 °F).

Thermocouples

Thermocouples consist of two dissimilar wires that are metallurgically homogeneous. They are joined at one end, called the measuring or hot junction. The other end, which is connected to the copper wire of the measuring instrument circuitry, is called the reference or cold junction. The electrical signal output in millivolts is proportional to the difference in temperature between the measuring junction (hot) and the reference junction (cold). The different types of thermocouples, classified by their metallurgical compositions, have differing output signal calibrations.

The thermocouple wire combination, with or without insulators, is called a thermocouple element. A complete thermocouple assembly includes the element, and a protection tube or well, to protect the element from contamination and provide mechanical strength (see Fig. 4). In addition, the assembly provides mounting fittings and a terminal junction in a head assembly. Extension lead wires, which match the thermoelectric characteristics of the thermocouple elements, are used to connect these terminal junctions to the instrumentation. Figure 5 shows a simple thermocouple with lead wires and reference junction. The lower sketch of this figure shows details of a complete thermocouple assembly in a protection tube.

The operating life of a thermocouple depends on its operating temperature, time at the operating temperature, ambient atmosphere, and number of experienced high-to-low temperature cycles. An erroneous low output signal from a defective thermocouple indicates a measured temperature lower than the actual process temperature. Thus, if the temperature control is based on this faulty measurement, the process temperature will be too high. This mistake wastes energy and could damage both the work in process and the furnace equipment.

Heat treating operations that rely on thermocouples require numerous precautionary measures. For example, thermocouple error due to contamination normally is characterized by a low output signal. Contaminated wires do not meet the requirement for metallurgical homogeneity. Contamination comes from touching the wires with bare hands during assembly, dirty insulators, oils and dirt inside the protection tube or the atmosphere used in the process. Consequently, cleanliness is important in assembling thermocouples. Insulators that become contaminated in normal service should not be reused. Even the cutting oils used to cut threads on protection tubes contain sulfur, which can cause contamination. These tubes should be thoroughly degreased before use. In high-temperature applications, the furnace atmospheres penetrate protection tubes and sheath materials. Consequently, routine replacement of the contaminated thermocouple elements is required. Frequency of replacement depends largely on operating temperature and the environment.

Thermocouple wire and extension leadwire are annealed for metallurgical homogeneity. Therefore, the wire should not be cold worked by flexing it, as in making sharp bends. The extension leadwire should not be strained with excessive pulling when installed in electrical conduit. A separate metal conduit is often necessary, because it provides mechanical protection and shields the extension leadwire from picking up electrical signals from high-voltage electrical wiring and machinery.

Thermocouple Location. Location of thermocouple measuring junctions must allow exposure to the correct heat

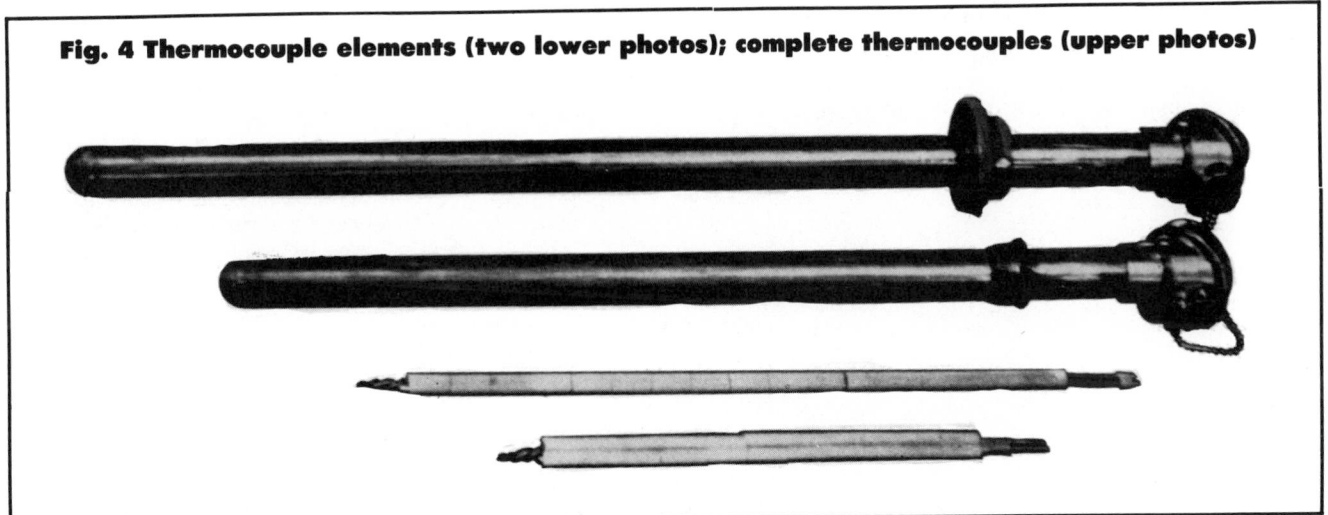

Fig. 4 Thermocouple elements (two lower photos); complete thermocouples (upper photos)

process temperature. Normally, the tip of the thermocouple (measuring junction) is placed near the work in the furnace. Thus, it is exposed to the same heating environment as the workpieces.

A common error occurs when the thermocouple assembly is not inserted far enough into the furnace. Under this condition, heat is conducted by the thermocouple assembly away from the measuring junction, toward the colder furnace wall. Consequently, the thermocouple indicates a temperature lower than the actual work temperature. One way to determine correct location is to place an adjustable flange on the thermocouple protection tube. The correct insertion depth then can be determined by adjusting the flange to achieve the maximum temperature and the minimum insertion depth, while the furnace is in operation. However, thermocouples never should be located near burners or heating elements, because radiant heat may cause erroneous readings.

The speed of response of the thermocouple to process temperature changes should be as fast or faster than that of the work in the furnace. If the workpieces are small and the thermocouple is contained in a heavy protection tube, the thermocouple temperature will lag behind the actual work temperature. Consequently, workpieces will be overheated when they are initially brought up to control temperature.

Furnace temperature uniformity is important and should be checked before installing the permanent control thermocouple. If an air recirculating system is used, an average temperature in the duct work can be measured by installing three or more thermocouples across the duct width connected in parallel. A single thermocouple in the center of the duct typically will indicate a higher temperature because this location is normally the hottest.

Types of Thermocouples. Thermocouples most commonly used in heat processing applications are listed in Table 1. Table 2 gives temperature limits for thermocouples in oxidizing atmospheres and Table 3 contains limits of error as a percentage of the temperature readings for these thermocouple types.

Selection of a specific kind of thermocouple is based primarily on the required temperature range and cost. However, the choice also is affected by its metallurgical properties. For example, type J is superior to type K in a reducing atmosphere, and type K is superior to type J in an oxidizing atmosphere. Type K is very susceptible to sulfur contamination, such as sulfur dioxide in combustion gases. Type K should be used in a large protection tube or one with a ventilated design. If the ambient atmosphere has a low oxygen concentration, the resulting green-rot corrosion results in a low output signal. Green-rot is caused by preferential oxidation of chromium at low oxygen concentrations. A new thermocouple has been designed that attempts to overcome some of the limitations of the type K device.

The two major thermocouple design classifications are (a) the compacted ceramic, and (b) the protection tube, or well-type, with integral element insulators.

Compacted ceramic thermocouples use a ceramic powder to insu-

Fig. 5 Simple thermocouple (upper view); and cut-away of a thermocouple assembly (lower view)

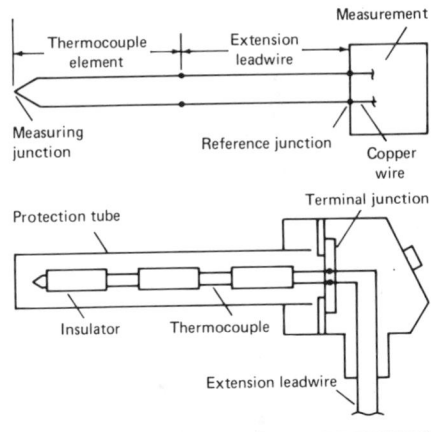

late the thermocouple wires. These wires are compacted inside a metal sheath providing protection from the environment. This compacted ceramic thermocouple is small in outside diameter, provides fast response and can be bent easily to suit installation requirements. It can be used at high pressures and high temperatures. Two types of compacted ceramic thermocouples, grounded and ungrounded, are shown in Fig. 6. The metal sheath has a plug of the same metal welded in place as the tip. In the grounded type, the end of the thermocouple is permitted to touch this plug, providing more rapid response. The ungrounded type is used when required by the measurement and/or control circuit electronics.

Protection Tubes and Wells. Thermocouple elements with insula-

Table 1 Comparison of thermocouple types

Type	Usable temperature range		Advantages	Restrictions
	°C	°F		
J (iron-constantan)	−185 to 870	−300 to 1600	Comparatively inexpensive; suitable for continuous service to 870 °C (1600 °F) in neutral or reducing atmospheres	Maximum upper limit in oxidizing atmosphere is 760 °C (1400 °F), due to the oxidation of the iron; protection tubes should be used above 480 °C (900 °F); protection tubes should always be used in a contaminating medium
K (nickel, chromium-nickel, aluminum)	−20 to 1370	0 to 2500	Suitable for oxidizing atmospheres; in higher temperature ranges, provides a more mechanically- and thermally-rugged unit than platinum, rhodium-platinum, and longer life than iron-constantan	Especially vulnerable to reducing atmospheres, requiring substantial protection when used
T (copper-constantan)	−185 to 370	−300 to 700	Resists atmosphere corrosion; applicable in reducing or oxidizing atmospheres below 315 °C (600 °F); its stability makes it useful at subzero temperatures; has high conformity to published calibration data	Copper oxidizes above 315 °C (600 °F)
E (nickel, chromium-constantan)	−185 to 870	−300 to 1600	Has high thermoelectric power; both elements are highly corrosion-resistant, permitting use in oxidizing atmospheres; does not corrode at subzero temperatures	Stability is unsatisfactory in reducing atmospheres
S (platinum, 10% rhodium-platinum) R (platinum, 13% rhodium-platinum)	−20 to 1480	0 to 2700	Usable in oxidizing atmospheres; provides a higher usable range than type K; frequently more practical than noncontact pyrometers; has high conformity to published calibration data	Easily contaminated in other than oxidizing atmospheres
B (platinum, 30% rhodium-platinum, 6% rhodium)	870 to 1650	1600 to 3000	Better stability than types S or R; increased mechanical strength; usable to higher temperatures than types S or R; reference-junction compensation is not required if junction temperature does not exceed 65 °C (150 °F)	Available in standard grade only; high temperature limit requires the use of alumina insulators and protection tubes; easily contaminated in other than oxidizing atmospheres

tors and compact ceramic elements are used inside protection tubes and wells as shown in Fig. 4 and 7. The thermocouple elements with insulators are less expensive than the compacted ceramic elements and can be manufactured by users. Protective tubes and wells are used to provide mechanical support and protect the thermocouple from the process environment. Protection tubes are made from metallic or ceramic materials and are used in air or other gaseous atmospheres (Table 4). These atmospheres are near atmospheric pressure, such as those used in heat treating furnaces.

Wells are metallic and made of either drilled metal bar stock (one-piece construction) or from welded pipes. Most often, wells are used to protect the thermocouple in liquid or gaseous environments with pressures greater than 330 kPa (50 psi) for the drilled design. Typically, they are used in quench tanks, hydraulic systems or steam lines.

If the insulation between the thermocouple wires in the element or extension leadwire breaks down from heat, contamination or mechanical damage, a low output signal results. This signal indicates an erroneously low temperature. This type of error results in process operation at an excessive temperature. Valuable energy is lost, and the work can be damaged or destroyed.

Thermocouple Applications. A typical thermocouple construction includes the thermocouple wire element with insulators mounted in a protection tube. The thermocouple head provides terminals for a junction between the thermocouple element and lead-

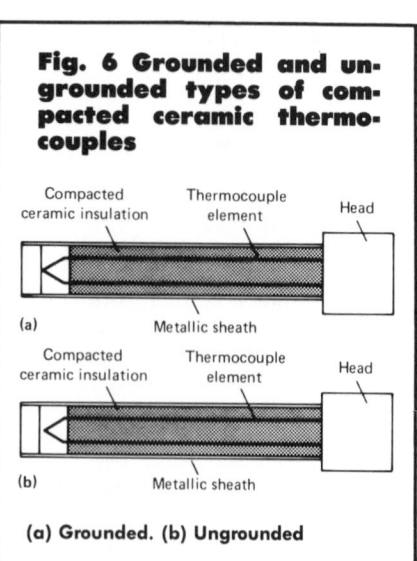

Fig. 6 Grounded and ungrounded types of compacted ceramic thermocouples

(a) Grounded. (b) Ungrounded

Table 2 Temperature limits for thermocouples in oxidizing atmospheres

Upper temperature limits are a function of wire diameter; because temperature tends to have deleterious effects on thermocouples, the larger the amount of material in thermocouple cross-section, the longer it can be expected to last

Type	Thermocouple Material	Condition	AWG (American Wire Gage) 8 °C	8 °F	14 °C	14 °F	16 °C	16 °F	20 °C	20 °F	24 °C	24 °F	30 °C	30 °F
J .... Iron-constantan		Bare	650	1200	480	900	480	900	425	800	345	650	315	600
		Protected	760	1400	595	1100	595	1100	480	900	370	700	370	700
K.......... Nickel, chromium-nickel, aluminum		Bare	1095	2000	925	1700	925	1700	870	1600	760	1400	705	1300
		Protected	1260	2300	1095	2000	1095	2000	980	1800	870	1600	815	1500
T .... Copper-constantan		Bare	315	600	315	600	260	500	205	400	205	400	205	400
		Protected	370	700	370	700	315	600	260	500	205	400	205	400
E .... Nickel, chromium-constantan		Bare	760	1400	595	1100	595	1100	480	900	370	700	370	700
		Protected	870	1600	650	1200	650	1200	540	1000	425	800	425	800
S and R ..... Platinum, rhodium-platinum		Protected	...	...	...	...	...	...	1540	2800	1480	2700	1315	2400
B .... Platinum, 30% rhodium-platinum, 6% rhodium		Protected	...	...	...	...	...	...	...	...	1705	3100	...	...

Table 3 Limits of error for thermocouples

ANSI type	Type of thermocouple	Temperature range °C	°F	Limits of error(a) Standard	Special
J .... Iron and constantan		−190 to −75	−310 to −100	...	±2%
		−75 to +315	−100 to +600	±2 °C (±4 °F)	±1 °C (±2 °F)
		315 to 425	600 to 800	±2 °C (±4 °F)	±0.33%
		425 to 760	800 to 1400	±0.75%	±0.33%
K .... Nickel, chromium and nickel, aluminum		0 to 275	32 to 530	±2 °C (±4 °F)	±1 °C (±2 °F)
		275 to 1260	30 to 2300	±0.75%	±0.38%
T .... Copper and constantan		−185 to −60	−300 to −75	...	±1%
		−100 to −60	−150 to −75	±2%	±1%
		−60 to +95	−75 to +200	±1 °C (±1.5 °F)	±0.5% °C (±0.75 °F)
		95 to 370	200 to 700	±0.75%	±0.38%
E .... Nickel, chromium and constantan		0 to 315	32 to 600	±2 °C (±3 °F)	±1 °C (±2 °F)
		315 to 870	600 to 1600	±0.5%	±0.38%
S .... Platinum, 10% rhodium and platinum		−15 to +540	0 to 1000	±1.5 °C (±2.5 °F)	±1 °C (±1.5 °F)
		540 to 1480	1000 to 2700	±0.25%	±0.15%
R .... Platinum, 13% rhodium and platinum		−15 to +540	0 to 1000	±1.5 °C (±2.5 °F)	...
		540 to 1480	1000 to 2700	±0.25%	...
B .... Platinum, 30% rhodium and platinum, 6% rhodium		870 to 1705	1600 to 3100	±0.5%	...

(a) When expressed as a percentage, the limit of error is a percentage of the temperature reading, not of the range.

wire used for connection to instrumentation.

A typical application is a car-bottom annealing furnace used up to 1095 °C (2000 °F) with an air atmosphere. An 8-gage type K thermocouple in a ceramic protection tube is selected because it is useful up to 1260 °C (2300 °F) in an oxidizing atmosphere and is less expensive than the noble metal types R, S and B. A 16-gage type K leadwire with asbestos insulation is selected because it matches the type K element up to 205 °C (400 °F), suitable for a heat treating area, and is strong enough for pulling through a conduit. Type K should be used for all the furnace measurements to maintain uniformity; that is, recording and control of furnace temperatures, plus the high limit safety control and work temperatures.

The furnace temperature thermocouple element should be 8 gage for maximum life and long enough so that its tip or measuring junction can be inserted well into the furnace cavity while the head is far enough back from the fur-

Table 4 Protection tube and thermowell materials

Material(a)	Recommended maximum temperature °C	°F	Description
Metal and metal-ceramic materials			
Carbon steel	540	1000	Satisfactory in any except corrosive atmospheres
Yoloy (Ni-Cu alloy steel)	705	1300	Resistant to corrosion in both oxidizing and reducing environments; ideally suited for use in condensate return lines, salt-water and brine solutions, condenser water lines, vent and waste piping, or corrosive water lines
Cast iron	705	1300	Generally more useful than Yoloy in the chemical industry; resistant to concentrated sulfuric acid and caustic solutions; can be used to 870 °C (1600 °F) in reducing atmospheres
304 stainless steel (18Cr-8Ni)	870	1600	Resistant to oxidation and corrosion; generally used in wet-process applications such as steam lines, oil refineries and chemical solutions; resists nitric acids well, halogen acids poorly, and the sulfuric acids moderately
316 stainless steel (18Cr-8Ni-2Mo)	870	1600	Superior to 304 stainless steel in corrosion resistance; resists pitting in phosphoric and acetic acids
446 stainless steel (28Cr, Fe)	1095	2000	Excellent corrosion resistance at high temperatures; used extensively in general-purpose alloy tubes; highly resistant to sulfur attack
Nickel	1095	2000	Resistant to attack by many chemicals at high temperatures; principally used for hot caustic and molten-salt baths; should not be used where sulfur is present
Inconel 600 (80Ni-15Cr)	1150	2100	For general high-temperature use; has greater mechanical strength than 446 stainless steel; should not be used in sulfur atmospheres
R-Monel (67Ni-30Cu)	480	900	Used where high strength and resistance to corrosion are required, such as sea water, dilute sulfuric acid and strong caustic solutions
F-11 ($1\frac{1}{4}$Cr-$\frac{1}{2}$Mo)	595	1100	Generally used in power plants, on water and steam line applications; available in thermo-wells only
F-22 ($2\frac{1}{4}$Cr-1Mo)	595	1100	Same as F-11, except has better oxidation resistance
LT-1 ($77Cr-23Al_3O_4$)	1370	2500	For high-temperature applications; intermediate between metal and ceramic tubes, has good thermal and mechanical shock resistance (continued)

(a) Materials are, in general, arranged in the order of their increasing resistance to oxidation, increasing mechanical strength at elevated temperatures, and increasing limit of useful temperatures.

nace wall to prevent its rising above 205 °C (400 °F). A ceramic protection tube will meet the process temperature requirements and is strong enough not to sag if mounted horizontally. An adjustable flange permits changing the depth of insertion in the furnace to the optimum depth. For example, when the furnace is at operating temperature, the thermocouple may be slowly withdrawn until its temperature indicator starts to drop, showing it is too close to the cold furnace wall. The adjustable flange may then be locked in place on the protection tube. Location of thermocouples depends on furnace design but, in any event, must be inserted where representative temperatures exist. These are a few typical considerations in selecting thermocouple specifications.

Resistance Temperature Detectors

Resistance temperature detectors are contact-type sensors. Their electrical resistance is proportional to temperature. Typical detector materials are platinum, copper and nickel. They are more stable and accurate than thermocouples, but even the platinum detectors have an upper temperature limit of approximately 540 °C (1000 °F), which reduces their usage in the metals industry. Because their signal is relatively large, resistance temperature detectors can be used with instruments having short temperature ranges of 5 to 10 °C (10 to 20 °F).

Resistance temperature detectors are normally larger in size and slower in response than thermocouples. However, the new thin-film deposited detectors minimize this disadvantage, which characterizes the conventional wire-wound detectors.

Resistance temperature detectors are used typically in quenching systems, low-temperature annealing furnaces and tempering furnaces. Because resistance temperature detectors are very accurate and stable, temperature errors are usually due to incorrect installation or damaged detectors.

Noncontact Sensors

Radiation sensors are noncontact-type temperature sensors used with radiation pyrometers. One type of radiation pyrometer is the optical pyrometer (see Fig. 8). Temperature measure-

Table 4 (continued)

Material(a)	Recommended maximum temperature °C	°F	Description
Ceramic materials			
L&N Fyrestan	1540	2800	A highly refractory porcelain, normally used for primary protection and often for secondary protection. L&N Fyrestan tubes can be substituted in many applications for nickel or chromium-alloy tubes, and although not as strong as metal alloys, are fully as gas-tight. Having a softening point of 1650 °C (3000 °F), these tubes can be mounted vertically at a temperature of 1540 °C (2800 °F). When mounted horizontally, the recommended maximum temperatures are 1510, 1455, 1370 and 1260 °C (2750, 2650, 2500 and 2300 °F) for unsupported lengths of 75, 150, 300 and 450 mm (3, 6, 12 and 18 in.), respectively. The tubes are impervious to air to 1650 °C (3000 °F) and to dry hydrogen to 1400 °C (2550 °F). Their low rate of thermal expansion gives good thermal shock resistance
High-purity alumina	1870	3400	A highly refractory 99 + % pure alumina, normally used for primary or secondary protection. Alumina tubes have excellent deformation resistance and superior resistance to reducing atmospheres and chemical reactions at temperatures as high as 1870 °C (3400 °F). (If the tube is mounted horizontally, it must be supported at temperatures above 1650 °C (3000 °F). Its excellent resistance to oxidizing and reducing atmospheres, as well as its exceptional stability at elevated temperatures, makes this material the ideal choice for protection of precious-metal thermocouples—especially type B. These tubes are highly resistant to corrosive alkaline vapors and aluminum chloride vapors, and are stable to acids, alkalis, molten metals, most glass fluxes, molten salts and slags. The tubes are also impermeable to most gases under the conditions found in industrial furnaces
Silicon-carbide refractory	1650	3000	Highly resistant to the cutting action of flames and gases and the corrosive effects of SO_2, tubes of silicon carbide refractory can be used for primary as well as secondary protection for temperatures up to 1650 °C (3000 °F). Ceramic-bonded silicon carbide features high thermal conductivity, high thermal-shock resistance, high strength and low permeability at differential pressures less than 50 to 75 mm (2 to 3 in.) H_2O

(a) Materials are, in general, arranged in the order of their increasing resistance to oxidation, increasing mechanical strength at elevated temperatures, and increasing limit of useful temperatures.

Table 5 Radiation wavelengths for ultraviolet, visible and infrared light

Radiation pyrometry

High temperature ← → Low temperature

← Ultraviolet (UV)	Visible vbgyor(a)	Infrared (IR)			
		Near	Middle	Far	Extreme →
← 0.4	0.75	3	6	15	→

microns

Detectors

Rayotube	Thermopile	Total radiation	UV through IR window 4 µm quartz 10 µm CaF
Spectray(b)	Silicon	90 65	UV to 0.9 µm UV to 0.8 µm
Two color	Silicon	0.85 ± 0.05 µm 1.0 ± 0.05 µm	
Optical	Eye	Red filter	0.65 µm

(a) Violet, blue, green, yellow, orange, red. (b) Spectray superior to Rayotube if interference CO_2, CO, H_2O

ments made with these devices are based on different properties of the radiant energy emitted from a hot target source. These measurements, which compare the apparent brightness of a standard electrical filament with the target, or the unknown, are commonly called the disappearing filament type. The human eye makes the comparison. A manual adjustment of the filament current is made until the apparent brightness of the filament matches that of the target; that is, it disappears. The measuring circuitry of the pyrometer converts the electrical filament current value to an indicated temperature value. This measurement, made in the visible spectrum at 0.655 µm (microns), is a narrow band measurement. Optical pyrometers generally are used to calibrate radiation sensors.

Radiation sensor detectors respond to radiant energy. They are clas-

Fig. 7 High speed steam temperature thermocouple mounted in a well

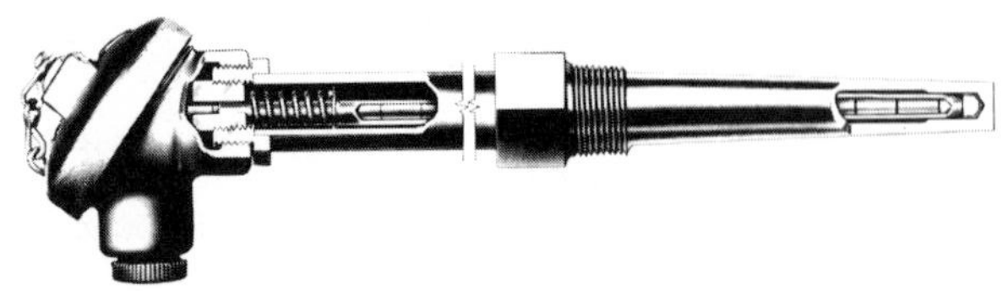

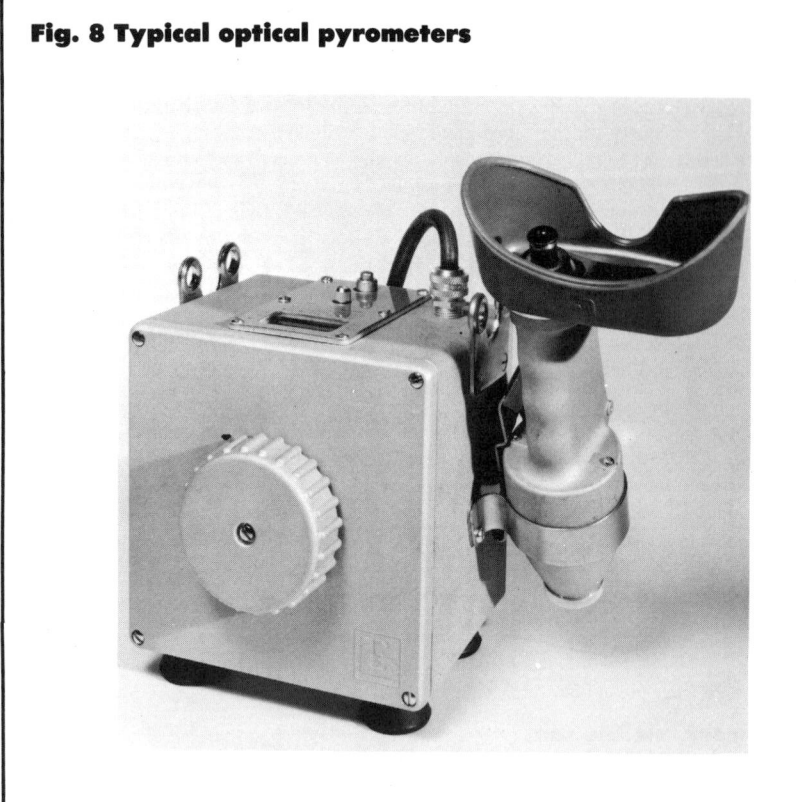

Fig. 8 Typical optical pyrometers

sified as total-radiation (wide-band) or narrow-band types, depending on the width of the radiation wavelength band to which they respond. Total radiation sensors use thermal detectors, and narrow-band sensors use photoelectric detectors. Radiant energy from metals is characterized at lower temperatures as red hot (with longer wavelengths). At higher temperatures, it is characterized as white hot (with shorter wavelengths). Radiation wavelengths in microns are shown at the top of Table 5. The bottom of the table lists detectors and the wavelengths for which they are used.

The total amount of energy transmitted from the hot body to the cold body detector is a function of the difference between the fourth powers of their absolute temperatures. The hot body may radiate or reflect the energy it receives, depending on its characteristics. If the hot body radiates 100% of the received energy, it is called a black body, with an emissivity of 1. If it radiates only 50% of the energy, the hot body has an emissivity of 0.5. Consequently, the emissivity of the hot body affects the measurements of its temperature when using radiation sensors.

One of the factors that affects the energy per unit of time emitted from a material is its surface characteristics. For example, a heated brick wall will emit 98% of its energy related to its temperature (emissivity of 0.98), whereas polished aluminum, at the same temperature, emits only 8% of its energy (emissivity of 0.08). At 150 °C (300 °F), clean gray steel emits 60% of its energy, but steel that has been buffed to a bright finish emits only 11%. Thus, the emissivity factor must be taken into consideration when using radiation sensors. In addition, energy in the infrared region can be absorbed by gases or dust particles between the hot body and the sensor, including dirt on the sensor window. Thus, smoke, water vapor, carbon dioxide and other gases will affect the measurement. When a radiation measurement is made, these factors must be either eliminated or minimized.

Radiation devices are available in various ranges such as 425 to 980 °C, 760 to 1650 °C and 1010 to 2200 °C (800 to 1800 °F, 1400 to 3000 °F and 1850 to 4000 °F, respectively). They also are made with varying response speeds to meet the different application demands. For example, detectors having a high response speed (0.01 s) are useful when the target object is moving or when the temperature of the target object varies rapidly. Slow radiation sensor response speeds are useful for filtering out rapid or erratic changes in temperature, such as those given by swirling flames.

Total radiation sensors are sensitive to broad bands of radiation wavelengths of approximately 0.3 to 10.5 μm (microns). In these wavelengths, the upper limit is determined by the window used in the radiation sensor, in this instance, calcium fluoride. If a quartz window is used, the transmission of longer wavelengths is limited, and the span is 0.3 to 4 μm. The radiant energy is focused on the center of a thermopile detector, which consists of one or more thermocouples in series. The measuring junctions of the thermocouples are at the center of the thermopile. The thermopile converts the radiant energy received from the hot body target to an electrical signal. This signal is proportional to temperature as a fourth power function, a nonlinear output signal. The sensor signal output can be increased by using thermopile detectors with more thermocouples. However, this increased mass reduces the detector response to temperature changes. Thermopile designs, using thin-film deposited thermopiles, have minimized this limitation and increased the signal output. Significant

Fig. 9 Significant components of a total radiation sensor compared in size to eye of needle

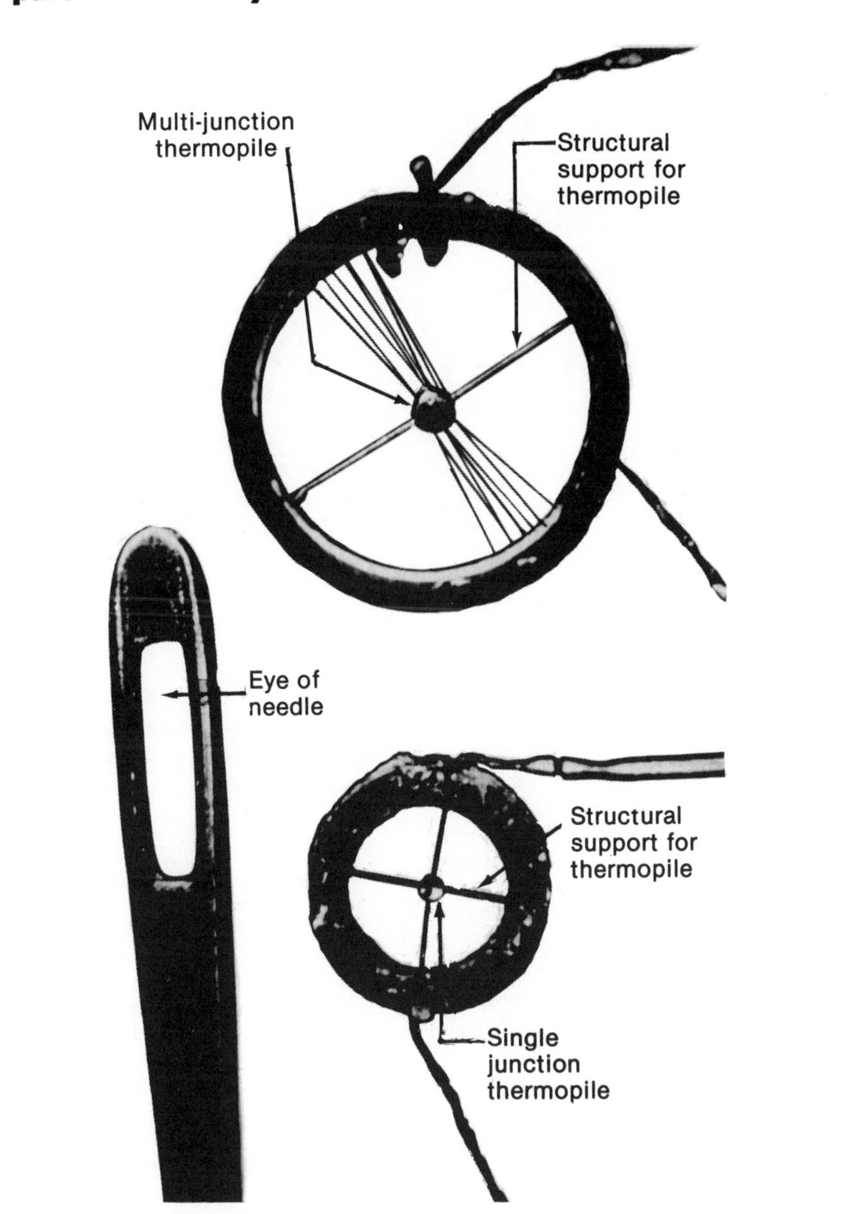

Multi-junction thermopile

Structural support for thermopile

Eye of needle

Structural support for thermopile

Single junction thermopile

Fig. 10 Sensor comparison of T^{12} versus T^4 outputs

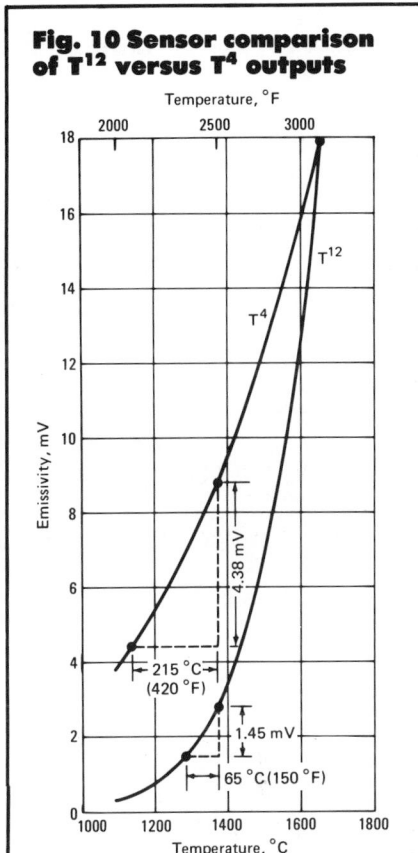

components of a total radiation sensor are shown in Fig. 9.

Narrow-band radiation sensors of the photovoltaic type, using silicon diode detectors, respond to radiant energy in the range of 0.3 to 1.0 μm. They are characterized by a large signal output, fast response speed (0.001 s) and a very nonlinear output signal (12th power) because of their narrow-band response.

A variety of sensor designs are used for specific applications to compensate for the characteristics of specific detectors. Consequently, the following comparison between total-radiation (wide-band) and narrow-band sensors is a generalization. Wide-band sensors respond to both high and low temperatures, while narrow band sensors respond only to high temperatures. Wide-band sensors are slower in response, which is sometimes a disadvantage. This slower response is advantageous when signal averaging is desired. Signal averaging facilitates

control or reading a temperature indication on a recorder or digital display. Wide-band sensors have low signal outputs, which limits the use of optical filters needed to make them spectrally selective. This low signal output characteristic also limits the minimum target size that can be used.

The signal output characteristics, 12th power, or narrow-band, versus 4th power, or wide-band sensors, are an important consideration. If the target emissivity is reduced, or if energy is lost due to interference in the optical path, the error will be less for the narrow-band sensor.

For example, referring to Fig. 10, describing the black body curves (emissivity 1.0) for T^{12} and T^4 sensors, both produce 18 mV at 1650 °C (3000 °F). At a 1370 °C (2500 °F) target temperature, with emissivity changing from 1.0 to 0.5, 50% less radiant energy and an apparent temperature loss results. Under black body conditions at 1370 °C (2500 °F), the T^{12} sensor produces 2.90 mV, and the T^4 sensor produces 8.75 mV. Loss of millivolts, due to the emittance change, produces apparent temperatures of 1290 °C (2350 °F) for the

Fig. 11 Typical digital displays (upper); typical strip chart recorders, analog displays (lower)

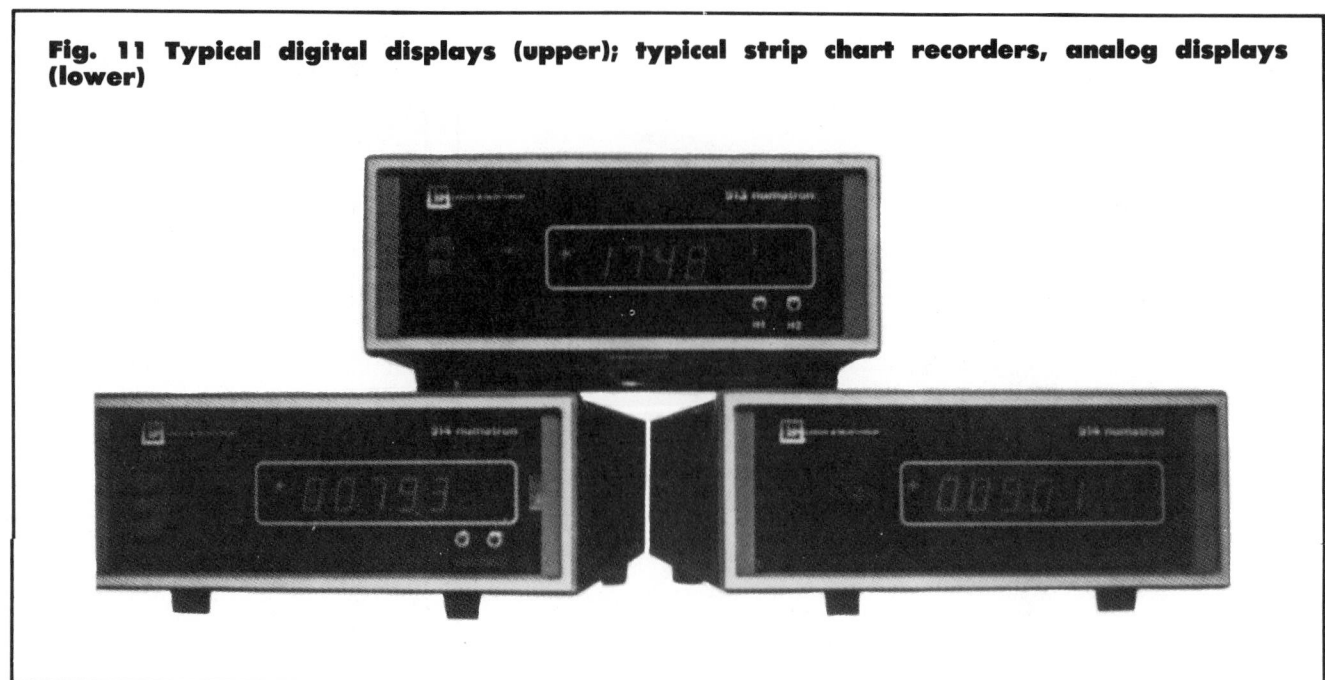

T^{12} sensor and 1140 °C (2080 °F) for the T^4 sensor. As shown in Fig. 10, the ratio of these changes is 150/421, or approximately one-third. Therefore, the error for the narrow-band sensor is one-third that of the wide-band sensor.

Erroneous temperature measurements, especially low ones, result in high-temperature operation, which wastes energy. Low-temperature errors in measuring by radiation methods can result from conditions of the target source and from optical path interference.

Target source measurement errors occur primarily due to emissivity factors or the size of the target. For example, a given metal with an unoxidized (polished) surface will have a lower emissivity than an oxidized surface, resulting in a lower indicated temperature than the actual temperature. Loose scale has a high emissivity but is lower in temperature than the metal to which it is attached, because it is a poor conductor. If the target size does not fill the radiation sensor's optical field, a lower than actual temperature is indicated. For example, the target alignment of hot rod stock in a mill may vary as it passes the sensor.

Optical path interference can be produced in several ways. Radiant energy is absorbed by water vapor and

carbon dioxide gas, steam from water quenching operations, or products of combustion from the heating process. These processes are frequently located near high-temperature measurement systems. For these conditions, the narrow-band sensor is not affected because the absorption bands for water vapor and carbon dioxide are at wavelengths longer than 1.0 μm.

Other gases may cause interference, depending on their spectral absorption bands compared with the sensor's spectral response.

Radiant energy also is lost because of dust and smoke in the optical path. One way to minimize such interference is to use an open-end target tube purged with dry air. Another method is the use of a closed-end target tube with its tip at the temperature of the work.

Radiation sensors are used typically to control annealing furnaces as well as brazing and forge furnaces. They also are used for blast furnace stoves, salt pots, checker bricks, rolling and strip mills, and induction-heating processes. One or more of the following conditions justify the use of noncontact sensors instead of contact sensors.

- Temperatures are too high for contact sensors.
- Work is moving too fast for detection by contact sensors.

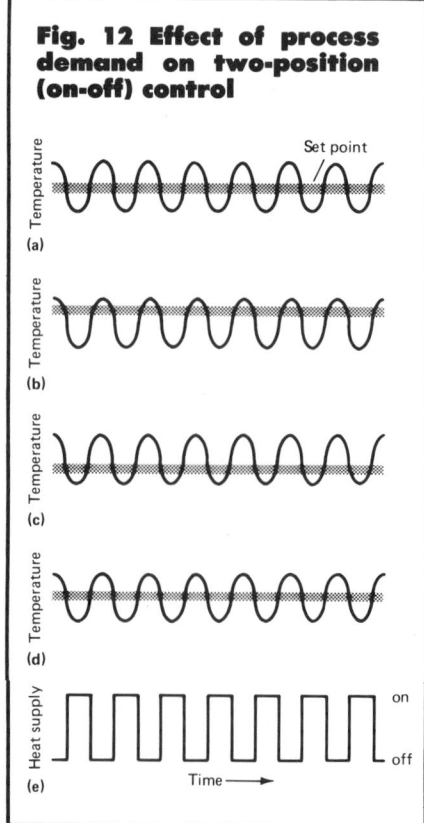

Fig. 12 Effect of process demand on two-position (on-off) control

- Required response rate is too fast for contact sensors.

Selection of the best radiation detec-

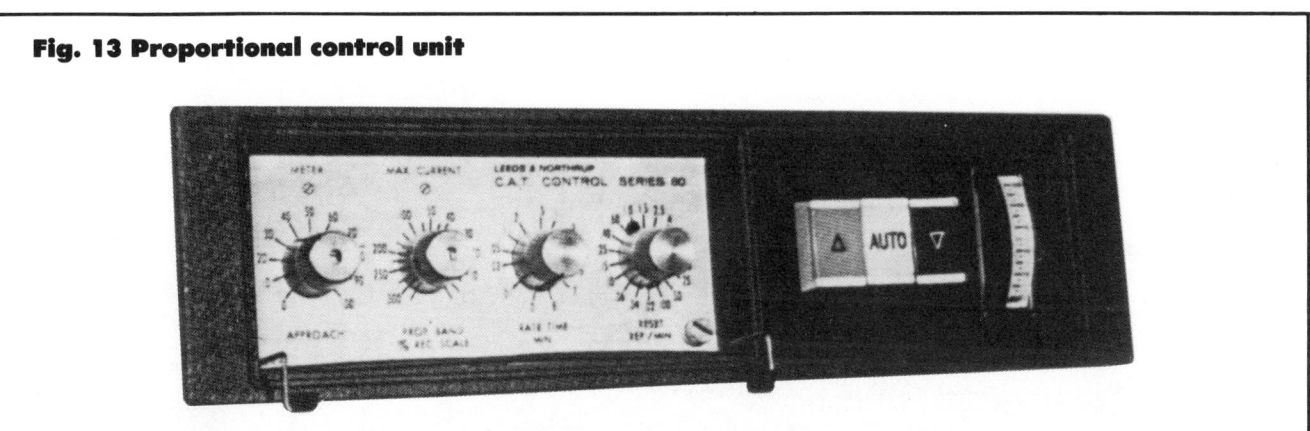

Fig. 13 Proportional control unit

tor is normally a compromise based on specific application requirements. However, when emissivity varies, the narrow band is generally the best choice because the amount of error is decreased (see Fig. 10). For instance, hot slabs of steel, which pass through the sighting field of a radiation sensor, provide a target with variable emissivity. Under these conditions, a narrow-band sensor will minimize the error. In other instances, as when temperatures are lower or emissivity is more nearly constant, a wide-band sensor is a better choice. Temperature measurement and control is a highly specialized field, and technical assistance usually can be obtained from the instrument manufacturers.

Measurement and Control Instruments

Measurement instruments measure the output signal of the temperature sensor and convert it to a temperature indication or recording in engineering units. Transmitters are used in some measurement systems to amplify and condition the temperature signal. The accuracy of the measurement depends greatly on the accuracy of the temperature sensor and the connecting leadwire. The accuracy of the measurement instrument is defined in its specifications under referenced conditions for its power supply, ambient conditions (temperature and humidity), electrical noise rejection and maximum source impedance. The accuracy of the transmitter has similar qualifications.

Measurement instruments are classified by their displays, analog or digital, and whether they are recording or nonrecording types. Analog displays

include meters and motor-driven pointers. Analog strip chart or round chart recorders include analog temperature indication. Digital displays are available with or without digital printers. In general, digital equipment is more accurate than analog equipment, but specifications must be checked in each case. The analog versus digital choice in displays and recording usually depends on user preference and specific applications. Typical digital and analog displays are shown in Fig. 11.

Temperature measurement instruments incorporate reference junction compensation in their circuitry for thermocouple measurements and emissivity compensation for noncontact radiation sensors. Reference junction compensation automatically adjusts the measurement depending on the temperature at the junction between the thermocouple wire and the copper wire of the instrument's measuring circuit. Radiation-type temperature measurements require an emissivity compensator. This compensator makes a calibration adjustment by comparing measurement of the same target with an optical pyrometer or calibrated thermocouple. The emissivity compensator is adjusted to make the radiation pyrometer measurement indication agree with the reference calibration instrument.

In small installations, the resolution and clarity of digital displays is often preferable. In larger installations, where many analog displays can be scanned easily, analog records often are preferred. Sometimes, analog records are required because trends are easier to observe. Digital displays are convenient when doing furnace surveys, such as checking temperature uniformity. Analog records provide

Fig. 14 Curves showing effect of toggle action on flow when using a nonlinear valve

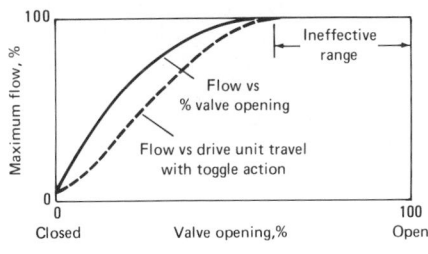

Fig. 15 Schematic view of linkage arrangement for maximum toggle action

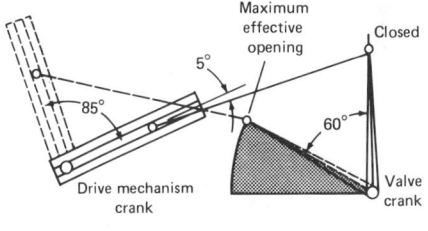

compact historical temperature data showing precisely how the process has operated over a long time cycle. These data are useful in monitoring the performance of the control instrumentation and the total process. They are also useful for making a process analysis before and after energy-saving procedures have been implemented.

Controllers. A temperature controller should provide just enough energy

Fig. 16 Digital programmer for set point and logic control

to satisfy process requirements, even though operating conditions vary. Variations include changes in process load, fuel characteristics and ambient temperature. Thus, controller requirements are more stringent when process requirements are demanding and especially when operating conditions vary significantly.

In operation, the controller set point that represents the desired temperature is compared with the process or actual temperature. The controller's stability and its sensitivity to the difference between desired and actual temperatures are critical. Based on this comparison, the controller regulates the energy flow to the process.

The two basic types of control are the two-position, or on-off type, and the proportioning, or modulating type. These two basic types exist in many variations.

On-off types of controllers are inexpensive, simple to operate and easy to maintain. On-off types, however, use energy inefficiently and increase the cost of heater and brickwork maintenance, particularly when controlling high-temperature processes where heat transfer lags are long. The process tem-

perature cycles above and below the controller set point as the controller turns the heat input on and off. During start-up, the heat remains on until the temperature reaches the controller set point and then turns off. Consequently, the thermal inertia of the process, equipment and work load, causes an overshoot beyond the set point, and energy is wasted.

After the temperature has stabilized in a characteristic on-off cycle, energy is still wasted because of the effects of heat transfer. Assuming that the temperature cycles are equally distributed above and below the controller set point, the average actual temperature will equal the set-point temperature. However, heat transfer losses or wasted energy, which result from losses above the set point, exceed the savings that result from temperatures below the set point. Thus, a net loss is produced, even though the average actual temperature is equal to the set point. These losses increase as operating temperature increases.

The magnitude of the temperature cycle, that is, the magnitude of the deviation from the set point, is directly

Fig. 17 Heat treating cycle for manganese steel castings

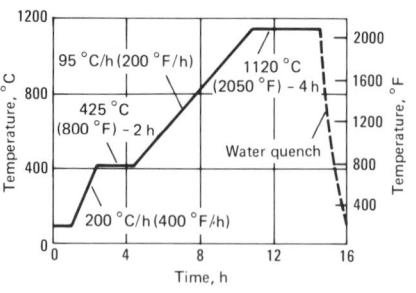

related to the heat transfer characteristics and measuring lags of a particular process. Therefore, if the temperature measurement responds rapidly to changes in heat input, departure from the set point will be small. Conversely, when response is slow, substantial temperature deviations will result.

The effect of process demand on two-position (on-off) control is illustrated in Fig. 12. The curve in Fig. 12(a) shows temperature cycles evenly above and below the set point. In other words, the heater on and off times are equal. For a

Fig. 18 Cycle for hardening tool steel parts in a vacuum furnace

Fig. 19 Single-loop proportioning controller

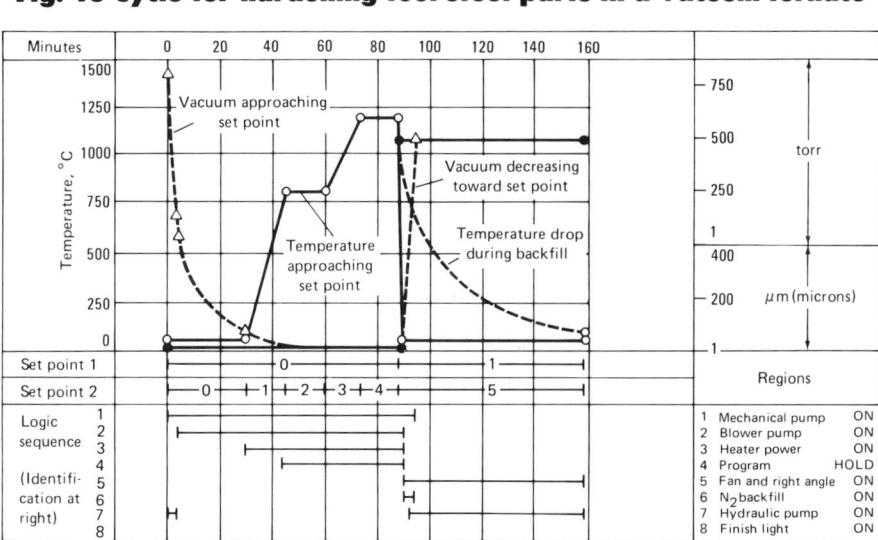

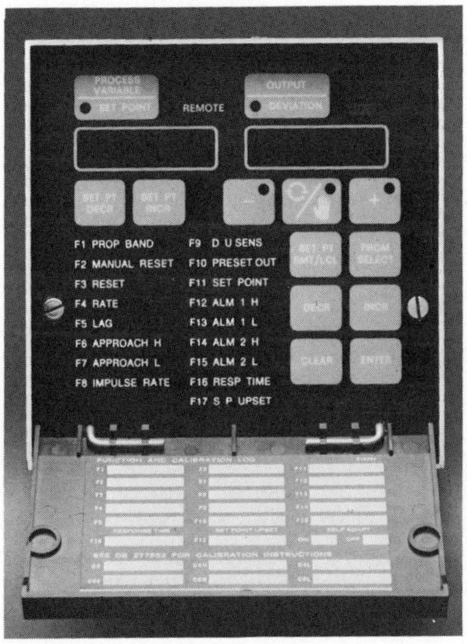

higher process demand, such as more heat required to maintain a given temperature, the response is shown by the curve in Fig. 12(b). In this case, the on time is greater. The temperature cycles are at a lower average value in the curve in Fig. 12(c). The average value is higher than the set point. The curve in Fig. 12(d) shows the effects reversed for a lower process demand, which again brings the average to very near the set point. The relationship between heat supply, the controlled variable, and the set point at about one-half load conditions also is shown in Fig. 12(e).

Cyclic process temperatures cause cyclic process pressures in enclosed gas atmosphere furnaces and ovens. This condition can cause breathing, or inspiration of ambient air, resulting in additional energy losses.

In summary, on-off controllers have the advantage of low cost and simple installation. However, they should be considered only for relatively noncritical processes.

Proportioning or Modulating Controllers. When process temperatures are critical, or when large quantities of energy are consumed, proportioning control should be considered. Compared to on-off controllers, proportioning or modulating controllers are more expensive and sophisticated. They are less expensive to operate, however, because they use energy more efficiently. The process temperature is

consistently maintained at the controller set-point temperature, because the controller adjusts the heat input to match the heat demand of the process. Any changes in the process demand temperature will be matched by corresponding changes of the heat input to offset the temperature changes and maintain the desired temperature. Proportioning controllers can be designed to minimize or eliminate set-point overshoot, which is caused by the thermal inertia of the furnace and work load when the process is initially brought from ambient to operating temperature. Normally the process heat head, if not controlled, causes the temperature to overshoot.

The process-control response characteristics vary significantly for various processes. They even can vary for the same process when different operating conditions exist. For example, a furnace might be idled at 120 °C (250 °F) and operated at 1095 °C (2000 °F). The proportioning controller, in response to the error signal comparison (set-point temperature versus actual temperature), performs a computation to solve the control equation. This computation includes proportional, integral and derivative terms; thus, the controller is referred to as a PID or three-mode controller. The proportional term is proportional band or gain, and its units are percent proportional band or gain. The integral term is reset, and its units are repeats per minute. The derivative

term is rate, and its units are minutes. These tuning adjustments (proportional band, reset and rate) match the controller response to the process dynamics. Proportional band relates to the size of the error signal. Reset relates to the error signal size, and the length of time it has existed. Rate relates to the error signal speed of change. Consequently, the controller, when properly tuned, in response to these error-signal characteristics, continuously adjusts the energy input to the process, to maintain the desired controller set-point temperature. A properly tuned proportional controller will reduce energy losses significantly by closely maintaining the desired temperature.

A typical proportional control unit is shown in Fig. 13. Proportional controllers, which are classified according to their outputs, include current-adjusting type (CAT), position-adjusting type (PAT), and duration-adjusting type (DAT). The CAT may be used with an electric power regulator, such as a silicon-controlled rectifier (SCR), to control electrically heated processes. It also can be used with an electropneumatic converter to regulate an air-operated valve. The PAT may be used with an electric motor drive unit to regulate a lever-operated valve. The DAT may be used with an electrical contactor, electric motor drive unit-valve combination or solenoid-operated valve.

Typical applications include the use of a CAT controller to regulate an SCR. The SCR, in turn, regulates electrical power to heater elements in a furnace. A PAT controller typically is used with electric-motor drive to operate a valve on a gas-fired furnace. The DAT controller is used with an electric contactor to regulate electric power to heater elements. Electric contactors are less expensive than SCR's and are adequate for slow-response, high-temperature furnaces. DAT control also is used to operate gas valves on radiant tube-fired furnaces to achieve uniform heating throughout the tube.

The process temperature, even with PID control, tends to overshoot the control set point during start-up. This action is because of the system's thermal inertia. If the temperature sensor is of heavy construction, its slow response speed will compound this condition. A feature called Approach Action may be specified in the controller to minimize this problem, if the controller is properly tuned. This function also can be achieved with a feature called *reset limiting*.

Final Elements. The final element of the control loop responds to the controller output by regulating the energy flow. It may consist of one or more components, such as valves, dampers, power regulators, heater elements, converters or actuators. Some desirable characteristics include fast response speed, linear response, high sensitivity, adequate power and reliability. The final element should have a linear response over its full range; that is, a change in the final element should produce a corresponding change in temperature. The sizing of the valves, power regulators and heaters should match the process requirements to optimize control performance. The final element may provide feedback devices, such as slidewires and limit switches, for use in the control system. The final element also may be designed to drive to a limit or to lock in its last position, in the event of a power failure.

The following list includes some final elements used with electric controllers:

- An electric motor-valve combination used to regulate flow of steam or hot water
- An electric motor-valve-burner combination used to regulate fuel combustion

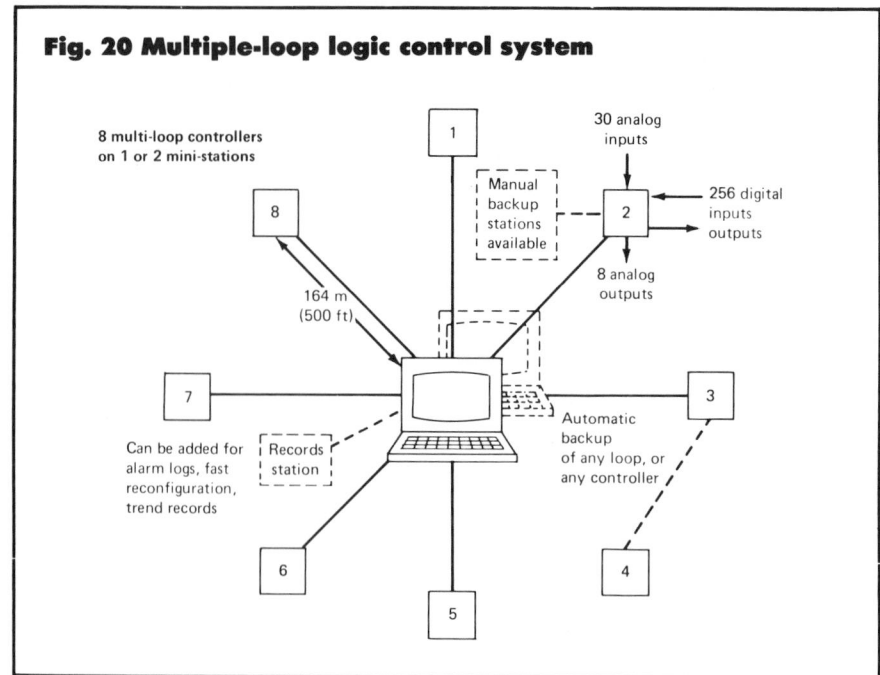

Fig. 20 Multiple-loop logic control system

8 multi-loop controllers on 1 or 2 mini-stations

30 analog inputs

256 digital inputs outputs

Manual backup stations available

8 analog outputs

164 m (500 ft)

Can be added for alarm logs, fast reconfiguration, trend records

Records station

Automatic backup of any loop, or any controller

- Pneumatic valves used to regulate fuel. An electropneumatic converter may be used to convert the output signal from the controller.
- Electrical resistance-type heating elements may be used with contactors, saturable core reactors, silicon-controlled rectifiers and triacs.

Design of the final element in any energy-efficient control system is extremely critical, because it regulates the energy flow. For example, butterfly valves are frequently selected instead of proportioning valves, because the butterfly valves cost less. However, their flow characteristics are nonlinear, which makes control over the entire range more difficult. In butterfly valves, a large flow change occurs near the closed valve position, and a small change occurs near the open position. A nonlinear toggle linkage between the electric motor drive unit and valve can reduce this nonlinearity and provide improved control.

The curves in Fig. 14 show the effect of toggle action on flow when using a nonlinear valve, compared with flow without toggle linkage. The nonlinear toggle linkage between the electric motor drive unit and the valve can reduce this nonlinearity, thus providing better control. The nonlinear linkage provides small increments of travel in the lower portion of its range. Greater increments of travel are provided in the

upper portion of its range, for equal increments of drive mechanism travel.

Maximum toggle action is obtained if the drive mechanism crank and link rod are within 5° of dead center when the controlled device crank arm and link rod approach a 90° angle. This situation is illustrated in Fig. 15, in which a valve serves as the controlling device. The control slidewire limits the drive mechanism crank movement to only 85°. Reduced toggle effect is obtained when the drive mechanism crank arm link rod angle is increased approximately 5 to 30°.

Set-Point Programmers. A temperature cycle, or temperature program is required in many heat treating applications. As the temperature changes, the metallurgical structure of the work being processed passes through phase changes or structure changes. Because these changes occur slowly, the process temperature should be changed slowly to avoid internal stresses, which cause distortion or cracking of the metal parts. Automatic control of the temperature versus time cycle is achieved by using a set-point programmer. This device provides the set-point signal to a three-mode proportioning controller. This automatic control of the complete temperature cycle ensures that the process uses a minimum amount of energy.

Fig. 21 Distributed control system

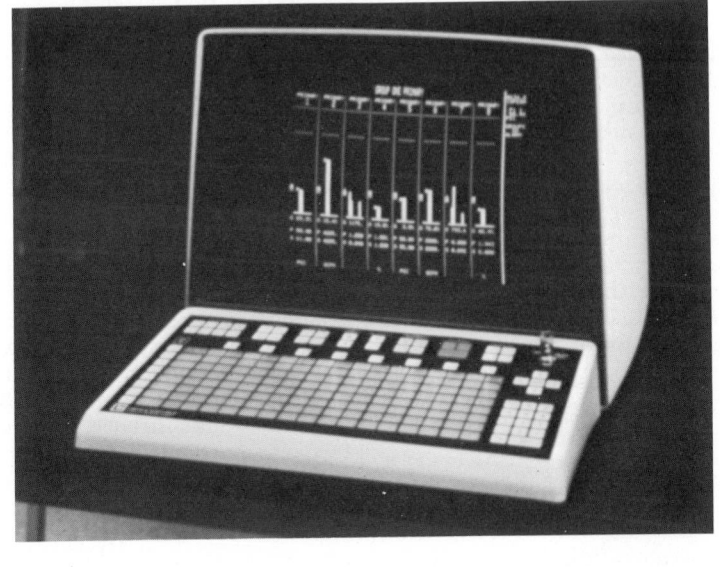

Most temperature programs specify linear heating and cooling rates with holding periods, or constant temperature plateaus. The set points may be motorized, cam actuated, curve follower or digital types. The cam actuated and curve follower types also may be used for nonlinear programs. Equipment selection is dictated by the complexity of the desired temperature program and the need for specific features, including degree of accuracy and economics.

To provide an automated temperature program, the three-mode controller set point is provided by the programmer set point. The controller computes the proper control action by comparing the programmer set point (desired temperature) to the process (actual temperature).

The accuracy of the set-point programmer is very important when evaluating its energy saving potential. However, flexibility, convenience, simplicity and cost are also significant factors in the selection of a programmer.

The digital set-point programmer provides exceptional flexibility, accuracy and resolution based on microprocessor technology. Microprocessor systems are compact, reliable, inexpensive computer systems, custom designed for specific applications. Program control in heat treating applications provides the temperature program. The logic control regulates the sequence of events program, such as operating doors, conveyors, fans and atmosphere gas additions, in the desired sequence and for the desired time periods. Digital set-point programmers provide more expanded set-point control capabilities, and their logic control eliminates the inflexible and costly hard-wired cabinet of relays used in the past.

A powerful digital programmer providing two set point outputs and logic control is shown in Fig. 16. This type of powerful programmer, working with such signals as contact closures from the heat treating equipment, can completely control the complex processing cycles in accordance with established programs.

The combination of a programmable set point and a programmable logic control provides a flexible method to completely control processes that require precise time-temperature control. Additional energy efficiency can be gained by programming the fuel-to-air ratio for fuel-fired furnaces. The temperature and fuel-to-air ratio can be programmed from a digital programmer with two set-point outputs.

Typical heat treating cycles include annealing, normalizing, hardening, carburizing, tempering and nitriding. Heating and cooling cycles can be relatively elaborate for these processes. A heat treating cycle for heating, holding and quenching manganese steel castings is shown graphically in Fig. 17. This process was conducted in a three-zone, car-bottom furnace.

Figure 18 is a graphic presentation of a programmed cycle for hardening tool steel parts in a vacuum furnace. The temperature and vacuum conditions are shown in the upper portion of the figure. Set points 1 and 2 are shown as regions just below the center of the figure. Segments of the logic sequence numbered 1 to 8 are shown as bars which indicate the time for each. The events of the sequence are identified at the lower right of Fig. 18.

Characteristics of the Process, Product and Production

A control system regulates the energy input to the process to achieve desired heating and cooling rates and maintains soak temperatures by matching heat input to heat losses. If these heat losses were always constant, a preset heat supply could keep temperature in balance indefinitely.

In practice, this never happens. Furnaces must be started up, idled, and operated at varying production rates and temperatures.

To handle such changes efficiently, control systems must constantly readjust heat input so that the work temperature is maintained despite changing operating conditions.

Two principal characteristics of the instrumentation and the process equipment influence the selection of a control system—lag and thermal inertia.

In a heat treating furnace, refractory, muffles or retorts through which the heat must flow represent heat barriers. The greater their number, the greater the heat transfer resistance or lag, with the result that there is a longer delay between the heat input change and a change in product temperature. After this mass is heated—refractory, muffles, retorts—it also represents thermal inertia.

Thermal inertia of a process is its capacity to continue a temperature change in the same direction after the heat supply is altered, before it begins to move in the opposite direction. The time it takes to turn a process around determines the extent of its thermal inertia. A furnace with a large, heavy muffle has a high heat storage capacity. Heat heads will build up rapidly during heat input periods, and the momentum from the latent heat continues to push product temperature up after the heat is turned off. The converse is also true. Product temperature will continue to drop after the heat is turned on. In short, the furnace is sluggish and the temperature swings are wide and slow.

In addition to the resistance to heat transfer, and thermal inertia of the product and process equipment, there are five characteristics of the product to be considered in selection of a temperature control system. The first characteristic of the product is its state. This is important in deciding the degree of acceptable temperature tolerances. For example, a solid product such as steel or ceramics may have a temperature tolerance of $\pm 3, 6, 8$ °C ($\pm 5, 10, 15$ °F) or even more.

The second characteristic is the material of the product, whether it is ferrous or nonferrous. Most nonferrous and high-alloy steel products have temperature tolerances of ± 3 or 6 °C (± 5 or 10 °F). On the other hand, low-carbon steel and other ferrous metals may have a temperature tolerance of ± 8 to 14 °C (± 15 to 25 °F).

The third characteristic—assuming the furnace is properly designed for the job—is the mass and shape of the product. This determines the responsibility the control system must assume for uniform heat penetration. The mass and shape of a large roll of steel influences the selection of not only the fur-

nace design but also the further selection of the control system required to maintain uniform heat from top to bottom and side to side.

The fourth characteristic of the product is its critical temperature. The nearer the product is heated to its critical temperature, the more accurate and responsive must be the control system.

The fifth and last characteristic of the product is the quality of the product after heating. In heat treating, for example, the effect of furnace atmosphere varies with temperature. Oxidation and carburizing occur faster at higher temperatures. For parts of uniform quality, temperature must be held uniformly throughout the furnace load.

Instrumentation Trends

Digital instrumentation is replacing analog instrumentation because its superior performance, flexibility and simplicity provide a more powerful tool, useful to the operator and management. Reduced costs for microprocessor technology has made digital instrumentation a practical choice. A completely automated process requires interactive loop and logic control, which is ideally suited to the capabilities of digital instrumentation. A trend toward centralized control rooms for heat

treating installations with computer supervision and management information systems favors digital control and data acquisition systems.

Individual applications and installations determine the best choice of equipment. A single-loop proportioning controller (Fig. 19) provides digital implementation of PID control. In addition, this particular controller can change its tuning constants (P, I and D) in response to changes for optimum control because it "learns" from its experiences. For large installations, a multiple loop and logic control system designed to interface with computers (Fig. 20) could control a complete heat treating department. Individual multiple loop and logic controllers controlling individual furnaces or groups of furnaces are connected to a central video display and control station. This configuration of instrumentation is called a distributed control system. An operator can view all the loops, logic and alarms on the video display and make changes from the control station (Fig. 21).

Distributed control systems are consistent with the concept of the completely automated plant, monitored and supervised with a central computer system. This type of installation integrates all plant operations, as required, to achieve optimum productivity in response to demands.

Atmosphere Control

By the ASM Committee on
Atmosphere Control*

THE PURPOSE of atmosphere control is to maintain consistent levels of constituents in a protective atmosphere and/or to determine the changes in those levels that are required in order to produce a desired result under a given set of conditions. Controls are required for various heat treating operations in many different atmospheres. All methods of atmosphere control can effectively be divided into two groups: those involving control of the atmosphere-generating system, and those involving control of the atmosphere once it is inside the furnace. Both are important in maintaining a controlled condition throughout the heating process.

Most operations performed in the heat processing industry can be done under endothermic or exothermic protective atmospheres. Effective control of the generators that produce such atmospheres may be accomplished by means of certain types of controllers—namely, combustibles controllers, infrared CO_2 controllers and oxygen probes. The required accuracy of control is directly proportional to the economics of the individual atmosphere system.

Most common among combustibles controllers is one that operates on the principle of catalytic combustion. A sample drawn from the gas generator or from the furnace is mixed with air (to supply oxygen) and then is passed over a detector. Any combustibles in the gas burn catalytically, raising the temperature of the detector and increasing the electrical resistance. The detectors in the catalytic chamber make up half of a balanced electrical circuit. A corresponding imbalance in the bridge resistance develops, and the resultant electrical output voltage proportional to the concentration of total combustibles in the sample is read out on the panel meter.

An infrared-type analyzer can be utilized for specific analysis of simple or complex mixtures of vapors and liquids and can be used to detect any components that absorb infrared energy. An infrared-type analyzing system could be of a design utilizing two separate or twin infrared radiation cells. One is referred to as a reference source, and gives a known signal in response to the reference gas. The other cell is used as a sample cell, and would vary from the known reference cell depending upon the elements in the gas contained inside that cell. When the gases in both cells are the same, they are in balance. If the gas absorbing infrared radiation in the sample cell is increased, more infrared radiation is absorbed, and consequently the beams become unequal. Movement of a membrane within the detector varies a condenser's electrical capacity, in turn resulting in an electrical signal proportional to the difference between the two beams. The signal is amplified and fed into an indicating meter.

These same principles are utilized for both furnace control and atmosphere-generator control. Infrared analyzers may be used to control systems other than carbon monoxide and carbon dioxide in furnace applications. For example, another form of atmosphere control would be dew-point control (water vapor). Because many heat treating systems are based on carbon dioxide analysis, infrared control is very common. Carbon dioxide control and/or dew-point control can be used for determining carbon potential in carburizing atmospheres. Infrared analyzers can also be used to monitor ammonia in nitriding atmospheres and carbon monoxide in other applications.

Some applications involving heat processes require the use of analyzers to determine the oxygen content of a given processed gas. The most common practice for oxygen analysis in heat treating furnaces is to determine the lack of oxygen. Depending on the level of oxygen desired or permitted in the processed gas, specific ranges of concentration may have to be determined.

The magnetic oxygen analyzer is a common analyzing system used with heat treating systems. Oxygen has an affinity for magnetic fields; most other gases do not. By adding a sample gas to a magnetic field and a detector, the change in resistance can be measured in a cell. The output from this cell is fed to a meter or a recorder and quite often is used as a permissive circuit in an atmosphere-control system.

As mentioned previously, use of a

*Paul L. Huber, *Chairman,* Manager of Sales, Sunbeam Equipment Corp.; Jeffrey W. Boswell, Project Engineer, Sunbeam Equipment Corp.; G. B. Zuber, Sales Manager, Instrument Division, Mine Safety Appliance Co.

dew-point measuring system is another method of controlling the carbon potential of a furnace atmosphere. It can also be utilized to determine the water-vapor content in any given atmosphere. One type of dew-point recorder consists of an element with two parallel electrical conductors helically wound on an insulated tube and connected to electrical power. The insulation of the tube is saturated with a hydroscopic salt and absorption of the moisture by the salt permits the electrical current to flow between the conductors, causing heating to an equilibrium point where moisture absorption ceases. That point where equilibrium exists is fed to an indicator and corresponds to the dew point of the major sample.

The above-mentioned control systems can be modified and incorporated together to form an automatic control system for any given set of conditions, utilizing normal instrumentation and final control devices. The system is ultimately a product of the needs of the user and justifiable economics. As metallurgical reproducibility of any given process becomes more important, the need for atmosphere control to produce chemical stability becomes a necessity. A second imperative not to be overlooked in the application of atmosphere-control devices is the safety aspects of operating atmosphere generators and industrial furnaces.

Analysis and Control of Endothermic and Exothermic Atmospheres

For success in metallurgical atmosphere control, gas analysis instrumentation must be applied where the atmosphere starts—at the gas generator. This is true for endothermic as well as exothermic atmosphere applications. Analyzers and associated control systems applied to a furnace should never be expected to correct deficiencies in the basic atmosphere produced by the gas generator. Instrumentation dedicated to the furnace is designed to provide information and control relative to operation of the furnace and to the product being processed. Likewise, instrumentation dedicated to the generator is designed to provide information and control relative to operation of the generator and to the atmosphere produced by it.

Because the heat treated product emerges from the furnace, there is a tendency to emphasize furnace atmosphere control and overlook the generator. In many cases, automatic atmosphere control of the gas generator will resolve heat treating atmosphere problems, simplify control of the furnace atmosphere, and in some cases reduce furnace monitoring to occasional spot checks.

Endothermic gas is used in heat treating furnaces as a protective atmosphere for hardening, stress relieving, carbon restoration and carburizing. The most recognized guides to endothermic generator operation are CO_2 and dew point. The ultimate guide is CO_2 and dew point. Any deviation from the relationship between the two indicates a generator problem, such as a leak or a carbon-coated catalyst. Figure 1 shows proper carbon dioxide/dew point relationship.

The nominal operating ranges for endothermic-atmosphere generators are -7 to $+16$ °C (20 to 60 °F) dew point and 0.2 to 0.7% CO_2. Generally, maintenance of an endothermic generator can be reduced by operating the generator near the lower end of the CO_2 range or near the higher end of the dew-point range within the limits required by the furnace process. As a rule of thumb, the dew point will rise approximately 6 °C (10 °F) between a sample taken at the generator and a sample taken at the furnace before enriching gases are admitted to produce the desired carbon potential within the furnace chamber.

The following list describes typical endothermic generator problems that can be identified, corrected, or avoided through application of proper analysis and control:

- Carbon on catalyst
- Worn or clogged carburetor or mixer
- Poor temperature control
- Combustion products leaking into retort
- Barometric pressure change
- Humidity change
- Hydrocarbons from quench oil or engine exhaust drawn into generator with air
- Change in natural gas composition
- Sticking regulators
- Dirty flowscope orifices.

When it is determined that carbon is present on the catalyst, a "burnout" procedure is followed to restore catalyst activity. Burnout is accomplished by turning off the natural gas and allowing air to pass through the catalyst. Oxygen combines with the excess carbon to form CO_2. Using the CO_2 analyzer to follow the decrease in CO_2 during burnout allows prompt determination of burnout endpoint. Operators often allow burnout to continue for excessive periods of time. Following the CO_2 decrease shows that burnout usually can be completed in less than one hour. The necessity for burning out a generator is greatly reduced when the

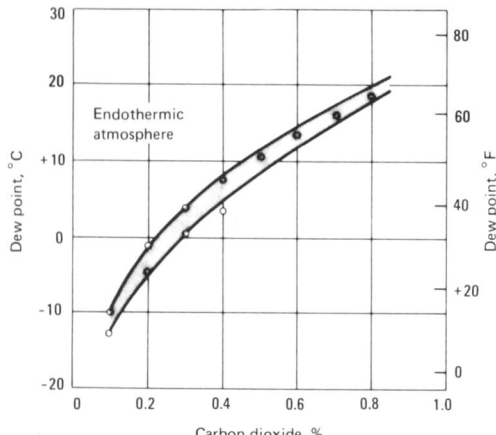

Fig. 1 Variation in the relation between dew point and carbon dioxide in generation of an endothermic atmosphere, as obtained from four plants

The generator in each plant was operated at a different temperature, all within the range from 1005 to 1095 °C (1840 to 2000 °F)

generator is controlled by CO_2 or dewpoint analysis, because large upsets commonly induced by manual adjustments are eliminated. Reports have been received indicating that endothermic generators have been controlled at 0.20% CO_2 (dew point of -4 °C or $+25$ °F) for one year without burnout. This is exceptional and somewhat contingent on generator design.

Exothermic-type atmospheres are used where inert atmospheres are required. Applications are common in the metal treating industry for bright annealing of copper, annealing of motor laminations, processing of aluminum, and annealing of coils of wire and steel sheet. Exothermic atmospheres are also used as an inert blanket in production of glass, in food processing and in rubber curing. Various chemical processes and storage facilities employ inert atmospheres produced by exothermic-gas generators.

Typical analyses of various exothermic atmospheres are listed below, along with recommended analyzers. Total combustibles analyzers are applied most frequently to monitor and/or control exothermic atmospheres. Very rich (20% or higher total combustibles) and very lean (less than 1% total combustibles) exothermic atmospheres are most accurately monitored and/or controlled by an infrared analyzer sensitized to an appropriate range of carbon monoxide.

Rich exothermic atmosphere

- 10 to 25% combustibles (approximately equal parts CO and H_2), 5% CO_2, 3% H_2O, rem N_2
- Recommended analyzers to control and/or monitor rich exothermic atmosphere:
    Infrared analyzer calibrated for 0 to 10 or 20% carbon monoxide
    Total combustibles analyzer (catalytic type) calibrated for 0 to 15%, 20 or 25% combustibles

Lean exothermic atmosphere

- 1 to 10% combustibles (approximately equal parts CO and H_2), 12% CO_2, 3% H_2O, rem N_2
- Recommended analyzers to control and/or monitor lean exothermic atmosphere:
    Total combustibles analyzer (catalytic type) calibrated for 0 to 1, 2, 5 or 10% combustibles
    Infrared analyzer calibrated for 1, 2, 5 or 10% carbon monoxide

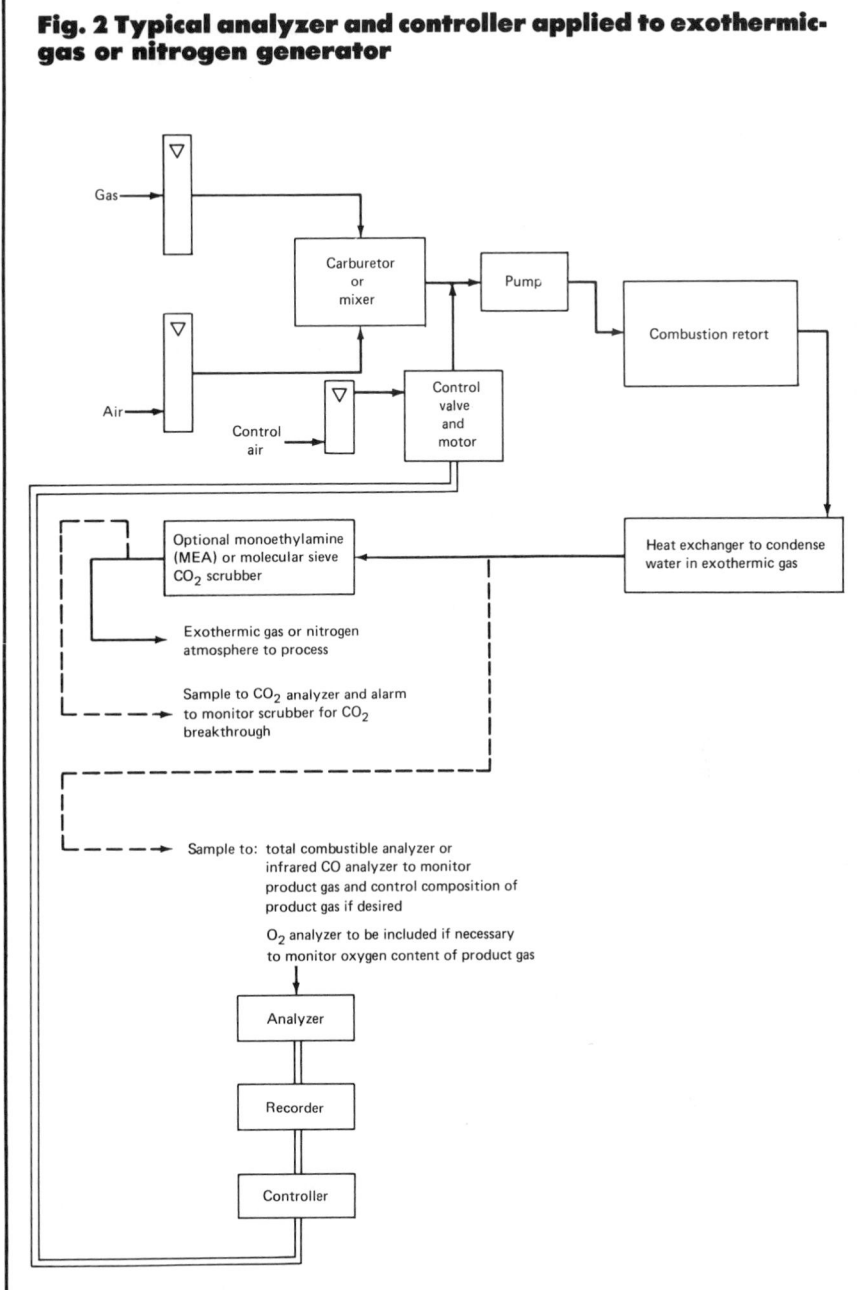

Fig. 2 Typical analyzer and controller applied to exothermic-gas or nitrogen generator

Oxygen analyzer for 0 to 1.0% or 2% oxygen to monitor lean atmospheres to ensure minimum oxygen

Nitrogen (lean exothermic) atmosphere

- 0.2% combustibles, 0.1% oxygen, 12% CO_2, 3% H_2O, rem N_2
- Recommended analyzers to control

and/or monitor nitrogen atmospheres

Recommended analyzers are similar to lean exothermic analyzers but with lower ranges specified for greater accuracy; infrared analyzers calibrated for 0 to 0.1% or 0 to 0.5% carbon monoxide are generally preferred over total combustibles analyzers when best accuracy is required

Exothermic atmospheres with CO_2 and H_2O removed

- Exothermic atmospheres may be passed through MEA (monoethyl amine) or a molecular sieve scrubber to remove CO_2 (<1000 ppm) and H_2O to produce N_2 or N_2 + combustibles atmosphere; in addition to applying the above analyzers, an infrared analyzer calibrated for 0 to 0.1% carbon dioxide provides an excellent means of monitoring CO_2 scrubber efficiency as well as prompt indication of carbon dioxide breakthrough.

Figure 2 and 3 summarize application of analyzers to exothermic- and endothermic-gas generators, indicate analyzer sample take-off locations and show proper locations for installation of control valves for automatic atmosphere control. The following adjustment procedure should be followed for proper setup of a control valve and motor when they are first installed on a gas generator for automatic control.

Control Valve and Motor Adjustment Procedure

The effectiveness of an automatic atmosphere analysis control system is directly related to the proper setup of the control valve and motor. Proper-control valve port size and driven-motor speed are important to system success. The following procedure is provided to avoid problems that are easily overlooked during initial setup of control valves and drive motors.

- Disconnect control-valve arm from motor linkage.
- Loosen control-valve locknut.
- Move control-valve arm to open position.
- Adjust control-valve port size to pass amount of air equal to 10% of air normally flowing into generator.
- Tighten control-valve locknut.
- Move control-valve arm toward close position, observe point where air flow starts to decrease, this is now considered "full open" position.
- Move control valve to locate "full close" position.
- Drive motor to "full close" position with manual control.
- Connect linkage.
- Loosen linkage arm at motor shaft.
- Rotate motor arm by hand to determine that full valve travel (close to open) corresponds to full motor travel. Adjust linkage as required to obtain above action.
- Drive motor from "full close" position

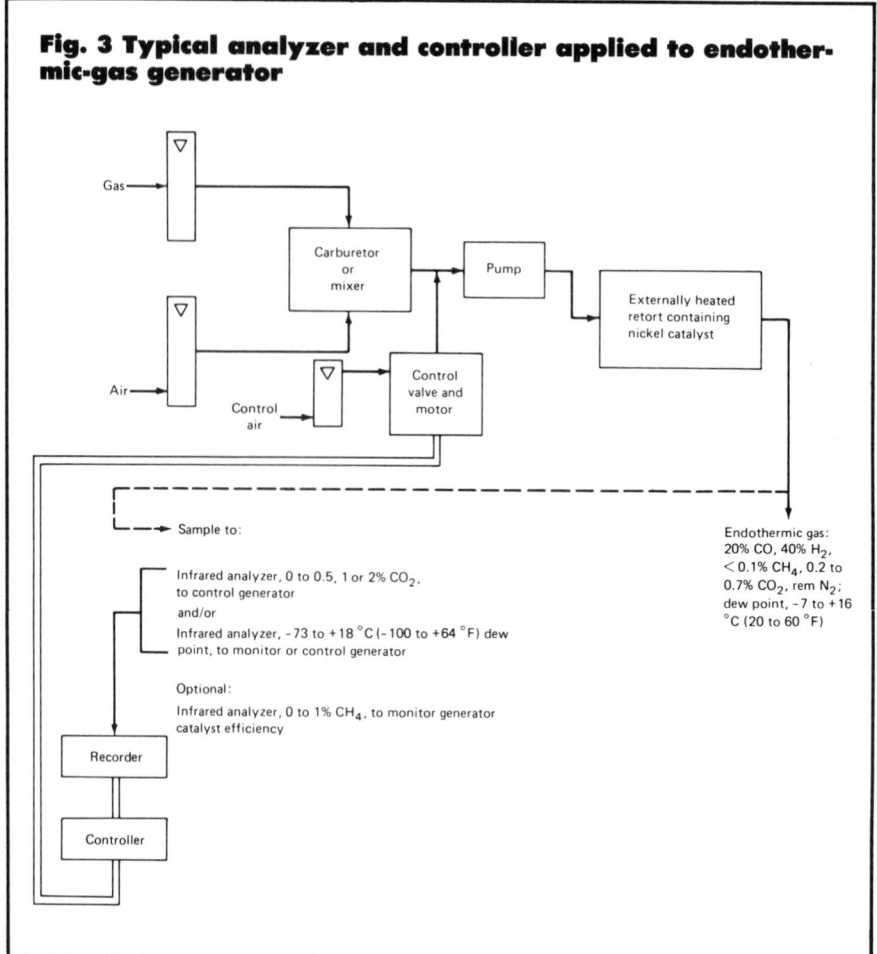

Fig. 3 Typical analyzer and controller applied to endother-mic-gas generator

Gas

Carburetor or mixer

Pump

Externally heated retort containing nickel catalyst

Air

Control air

Control valve and motor

Sample to:

Infrared analyzer, 0 to 0.5, 1 or 2% CO_2, to control generator
and/or
Infrared analyzer, -73 to +18 °C (-100 to +64 °F) dew point, to monitor or control generator

Optional:
Infrared analyzer, 0 to 1% CH_4, to monitor generator catalyst efficiency

Recorder

Controller

Endothermic gas: 20% CO, 40% H_2, <0.1% CH_4, 0.2 to 0.7% CO_2, rem N_2; dew point, -7 to +16 °C (20 to 60 °F)

toward open. Valve must start to open immediately without "dead travel".
- Continue to drive motor to "full open" position. Valve must arrive at "full open" as motor reaches end of travel.
- Drive motor from "full open" toward close. Valve must start to move immediately without "dead travel". "Dead travel" is when motor moves and valve does not move or moves in opposite direction; condition is corrected by adjusting linkage.
- Drive motor to mid-position, air flow is one-half of "full open" setting.
- Adjust generator carburetor so CO_2 or dew point is at desired set point.
- Move set point to a higher setting and observe direction in which controller drives valve when system is placed in automatic control mode.

Valve should drive open when analysis point is below set point. If set point is moved below analysis point, valve should drive closed. If valve drives in wrong direction, wires to motor must be reversed. Increasing air to endothermic generator must increase CO_2 and dew point. Increasing air to exothermic generator must decrease CO or total combustibles. During first week or two of operation, control valve on endothermic generator may keep seeking a position at full motor travel and it will be necessary for the operator to manually reset to midposition, adjust carburetor or mixer to obtain desired set point and switch back to automatic. This is a normal sequence caused by gradual cleaning and conditioning of catalyst in the endothermic generator.

Oxygen Probes

The oxygen probe is perhaps the newest technique used today in the heat processing industry for monitoring and controlling different atmospheres such as endothermic gas, exothermic gas, steam with small traces of hydrogen, direct-firing (fuel-rich) atmospheres, and hydrogen-bearing gases with low to medium water contents. Oxygen probes have been in use long enough to provide significant operating experience in heat treatment process control. Sufficient output data have been accumulated and correlated to provide practical usages that offer significant advantages over other forms of monitoring and control such as infrared and dew-point analyzers (Ref 1).

The oxygen probe is based in theory on a hot ceramic electrochemical cell. The probe will respond to oxygen, hydrogen, carbon monoxide, water and carbon dioxide and thus can determine the oxidization potential of a gas. The output of the oxygen probe is a direct measurement of the oxidation potential of the atmosphere at the process temperature of the furnace. Therefore, when the probe temperature is close to the furnace temperature, the response of the probe is a direct indication of whether the atmosphere will oxidize or reduce steel, provided that the composition of the atmosphere with regard to the proportions of carbon gases and hydrogen is known. Under such conditions, the probe will give a reliable indication of the oxidation/reduction situation for all furnace temperatures. Thus, with simple mechanical methods, additive gases or liquids can be introduced into the furnace in order to control the oxidation/reduction potential of the atmosphere.

An oxygen probe is a closed-end tube usually constructed of lime-stabilized zirconia or yttriastabilized material for temperatures up to 1600 °C (2900 °F). When such a probe is subjected to elevated temperatures, the nonporous sheath material acts as a solid electrolyte which permits the passage of oxygen ions when the inner and outer surfaces are subjected to atmospheres of different oxygen partial pressures—for example, a reference gas (such as air, because its O_2 content is constant at 20.9% by volume at sea level) and the process furnace atmosphere, respectively. The electromotive force (emf) thus generated, and measured via the electrodes attached to the sheath, is related directly to, and provides an accurate quantification of, atmosphere characteristics in terms of its oxidizing/reducing or, in some endothermic-atmosphere applications, carburizing/decarburizing tendencies at a known temperature (Ref 2). The electrodes mentioned above are in physical contact with the zirconia on both the inside and outside of the tube, and usually are constructed of platinum because of its superior chemical resistance at elevated temperatures. When the probe is subjected to elevated temperatures, there is a difference between the oxygen partial pressures on the two sides of the probe, and electricity will flow through a circuit connecting the two sides. This flow of electricity is from the higher pressure to the lower. If the oxygen pressure on one side is known, the oxygen pressure on the other side can be determined:

$$E = K \times T \log \frac{[O_2] \text{ known}}{[O_2] \text{ unknown}}$$

where T is absolute temperature, $[O_2]$ is oxygen partial pressure, K is a constant and E is the electromotive force generated. This measurement can be made using a simple panel meter without electronics (Ref 3).

Oxygen probes have been extensively used for control of carburizing furnaces. Accuracy of carbon-potential control through use of oxygen probes is estimated to be approximately ±0.05% carbon in actual practice because of the limitations of temperature control and temperature variations in typical heat treating furnaces. Because of the fast response rate, on/off control systems utilizing solenoid valves for regulating propane, natural gas, or liquid enrichment are adequate for batch furnaces. *In situ* or *ex situ* probes control carbon potential by adjusting the set point on the control instrument to the desired carbon level wanted in the furnace. The oxygen probe supplies to the control station an electrical signal related to the carbon potential. High- and low-deviation contacts are adjusted through the control instrument and either contact is made by the electrical signal from the probe. The low-deviation contact controls the solenoid valve that supplies enriching gas or liquid to the furnace. The high-deviation contact can be made to control the solenoid valve to add air or an oxidizer to the furnace. Carbon-concentration reproducibility of ±0.02% is frequently achieved using systems of this type, provided that the cycle, temperature and furnace conditions remain constant.

Continuous carburizing furnaces and straight hardening furnaces usually use proportional control in conjunction with the oxygen-probe control. Experience in control of continuous furnaces has shown that proportional control is normally required in order to cope with disturbances involving movement of work, opening of doors and the like. Compensation for these disturbances are rapid because of the rapid response rate of the oxygen probe. In the proportional-control system, the output voltage is supplied to a two-mode controller, and the controller positions a motorized or pneumatic control valve that regulates the flow of enriching gas to the furnace. Such carbon control is advantageous when short treatment cycles of 20 min or less are used, because compensations of 2 to 3 min or less are required for disturbances in such applications.

Oxygen probes often are used to monitor a furnace atmosphere while control is accomplished manually by adjusting flowmeters on enriching lines to the furnace. This is often used when the cost of control systems cannot be justified. The cost of replacing the consumable probe and the cost of simple on/off control systems must be weighed against human error, process accuracy and reproducibility.

Endothermic generators also can be controlled by use of the oxygen probe. *Ex situ* probes are generally used because of the difficulty of placing an *in situ* probe in the retort of the generator. In the case of the *ex situ* probe, a sample is taken from the output line of the generator. The electrical signal from the probe is wired to the control instrument that activates the solenoid valve located on the air-bypass line to the mixer. Thus, the air/gas ratio is automatically adjusted to give the desired properties of the endothermic gas.

Stainless steels, nickel, brass, titanium, aluminum, Incoloy and other metals and alloys are frequently annealed using exothermic gas. The reducing power of the gas necessary to protect these metals varies considerably; in general, copper alloys require a lean exothermic gas and stainless steels require a rich gas.

Usual practice for control of such a

process is to adjust the gas composition by varying the air/gas ratio prior to combustion in the gas generator. However, in practical experiments and in theoretical calculations, it is shown that it should be possible to control the air/fuel ratio closely by measurement of oxygen potential in either the gas generator or the annealing furnace (Ref 4).

By means of the oxygen probe, for any atmosphere, it should be possible to provide very early warnings of the ingress of air into the furnace from leaking seals or of ingress of water from water-cooled bearings or fans. It is also anticipated that this instrument may replace the conventional oxygen ana-lyzer as a device for providing very early warnings of possible hazardous conditions in the furnace.

REFERENCES

1. The Application of Free-Energy-Temperature Diagrams and High Temperature Electrochemical Cells in the Field of Furnace Atmospheres, by L. H. Fairbank: *Metallurgia,* Vol 79, No. 425, May 1969, p 179–185

2. Recent Developments in the Design and Use of the Oxygen Probe for Furnace Atmosphere Monitoring and Control, by L. H. Fairbank: *Heat Treatment of Metals,* Vol 4, 1977, p 1–12

3. A New Type of Gas Sensor for Combustion Work and Metal Treating Atmospheres, by D. A. Sayles and J. L. Cotter: paper presented at the 20th National ISA Iron and Steel Conference, Pittsburgh, PA, Mar 1970, and published in *Instrumentation for the Iron and Steel Industry,* Vol 20, p 57–66

4. Control of Carburizing Furnace Atmospheres Using O_2 Potential Measurements, by R. G. H. Record: *Metallurgia and Metal Forming,* Vol 39, No. 12, Dec 1972, p 413–416 and Vol 40, No. 1, Jan 1973, p 19–23

Computerized Systems for Heat Treating

By Theodore K. Thomas
Business Unit Advertising Manager
Process Control Division
Honeywell Inc.

COMPUTER MONITORING AND CONTROL of heat treating processes have recently become common practices. Consequently, this article approaches the subject from three different perspectives: an introduction to computer systems and practices that simplifies the computerized approach and compares it with traditional control methods; a glossary of computer terms; and industrial examples of computerized systems utilized effectively in industry.

Although computer technology is ever-changing, several distinct trends are in evidence, including:

• A decrease in the physical size of computer components and equipment
• A decrease in the cost of equipment
• An increase in cost of programming, or software, to run the computer.

As an illustration of the dramatic change in the physical size of computer equipment over the period of its development, a typical contemporary hand-held calculator can be compared with ENIAC, the first successful electronic digital computer. ENIAC, which was built in 1946, occupied several rooms of a large warehouse at the University of Pennsylvania, used 18 000 vacuum tubes and required a power input of 140 000 W, and yet the modern hand-held calculator far surpasses ENIAC in computing power, speed and storage capacity.

The rising cost of energy, coupled with tighter requirements for quality control in the heat treatment of metals, parallel the development of computer technology. Rising energy costs are now offsetting computer costs. It is consequently economically feasible to consider the advantages of a digital computer over conventional instrumentation in the control of furnace process variables. A computerized system provides:

• Multiple-function capabilities
• Increased speed of measurement and response to process changes
• Higher precision of measurement and control
• Greater control flexibility
• Increased accuracy in recordkeeping and reporting
• The ability to communicate electronically between computers and other elements of computerized systems.

The application of computer technology as a means of increasing furnace operation efficiency and cost effectiveness foreshadows a more productive era of heat treating.

In addition to increased production and quality standards, many scheduling problems become manageable with computer control. Greater flexibility in the use of equipment to meet production requirements is realized, as well as the ability to schedule furnace operations effectively so as to capitalize on plant services and utilities. Improvements in furnace loading and unloading are easily matched to manpower availability.

Computerized monitoring leads to more accurate and efficient reporting of problems and irregular conditions that may occur in normal heat treating practice. Timely and effective reporting of alarm conditions adds greatly to energy savings and decreases downtime. The case studies that appear in the applications section of this article describe actual industrial applications in which computers have contributed greatly to the development of more efficient operations.

Computer Basics*

Computers operate by digital methods, rather than by the analog techniques traditionally associated with industrial instrumentation. An analog instrument handles data in terms of electronic or pneumatic signals that are proportional in size to the quantity of the variable being measured, such as temperature or flow. Digital logic electronically expresses the same quanti-

*Adapted with permission from Micro, Mini, and Mainframe Basics, by P. Masucci, *Instruments & Control Systems,* June 1977

ties in terms of binary digits, or bits, each with a value of either 0 or 1. With digital logic, immense amounts of information can be accumulated and processed within limited space to very high levels of accuracy and reliability.

A computer has three main parts: a central processing unit (CPU); a data-storage unit, or memory; and input/output (I/O) equipment. The central processing unit contains an arithmetic unit, where the actual computation takes place; a control unit that dictates to the arithmetic unit which programs are to be executed; and storage registers that accumulate the numbers calculated by the arithmetic unit in accordance with the predetermined or programmed directions of the control unit.

General Computer Classifications

Computers can be generally classified as microcomputers, minicomputers and mainframes.

Microcomputers. A microcomputer contains a microprocessor, which is a central processing unit (CPU) built on a silicon chip smaller than 1 cm². By itself, the microprocessor is of little use. However, when integrated with a memory chip, a clock chip, and provisions for interfacing with a power supply and with input/output (I/O) devices such as keyboards and visual display or readout units, it becomes a microcomputer. Most microcomputers have word lengths of 4 to 8 bits and memories that contain from 256 to 65 535 words, depending on application requirements. At present (1982), some microcomputers can access 8 million words.

At the sensor level, a microprocessor or microcomputer can increase system accuracy and the ease with which signals are transmitted. Whereas analog signals may suffer from variations in amplification or other losses when transmitted over long distances, digital data are transmitted in binary units (1's and 0's), which are less subject to error. A simple microprocessor or an even simpler LSI circuit in a digital converter can gather data with minimum system loss in signal strength and quality, and also can preprocess some of the data before sending it on in binary form. Nevertheless, a microprocessor or microcomputer is "dedicated" to (designed for) a specific task or group of tasks within an application. A digital watch, for example, contains a microcomputer dedicated to computation and

display of clock time; it cannot be used to figure income tax, although it has the computing power to do so. Similarly, a digital temperature controller contains a microcomputer dedicated to measurement, display and control of temperature over a given range, using a specific type of thermocouple or other temperature-sensing device. In an industrial control system such as a fuel-fired furnace, a microcomputer can sample and evaluate temperature at several different points in the furnace, compare the results with acceptable limits, and make necessary adjustments. It also can continuously measure and regulate the flows of fuel and of combustion air to maintain a preset ratio, which yields more energy-efficient control. Conventional analog instrumentation can do all this too, but is less accurate and involves equipment that is physically much larger and that requires more power.

A microcomputer (or a minicomputer) also can integrate several instruments by acting as a data logger, preprocessor or "smart front end" to a larger computer. Whereas large mainframes may have been used a few years ago for more complex control and measurement applications, microcomputers (and minicomputers) have found successful application as such components in larger systems approaches.

Most current applications of computer technology to heat treating involve microcomputers, although minicomputers are used for supervisory control of several furnaces—for example, where each furnace is separately controlled by a microcomputer system.

Minicomputers are distinctly different from microcomputers in two ways: (a) they are physically much larger, being mounted in metal racks and cabinets; and (b) they can handle several different applications. For example, a minicomputer can measure and control many separate variables of temperature, flow or vacuum simultaneously, and at the same time can perform the supervisory function of adjusting individual control setpoints in accordance with a mathematical "optimizing" model contained in its memory. To accomplish these functions, it requires more data-storage capacity and process-interface capability than a microcomputer, hence its greater physical size.

Minicomputers have word lengths of 12 to 16 bits and a minimum memory size of 4096 (4K) words. Minicomputers

are trending toward word lengths of 16 bits or more, for product speeds (production thruputs) greater than those obtainable with microcomputers.

Mainframes, or large-scale computers, usually have a word length of 32 bits or more and a memory capacity of 16 384 words or more. A large-scale computer usually is used for after-the-fact data processing and analysis. It also can supervise or monitor the operation of a minicomputer, which is, in turn, connected to either a furnace process or its instrumentation. Within this system or network structure, the large computer can either monitor or augment the smaller device's capabilities.

Memory Systems

The memory system is the region of the computer that stores program instructions and data for instant use. Memories are of either the read-only (ROM) or the read/write type. If the data or instructions in a computer's memory do not need to be altered, read-only memories are used, in which case the computer cannot alter the contents. An example of this type of content is temperature-emf (electromotive force) tables for thermocouples, which give the millivolt output generated at a particular temperature. Read/write memories, on the other hand, can be altered and accessed by the computer. Read/write memories are also known as random-access memories (RAM) because any data location within the memory can be accessed, or interfaced with, as easily as any other. Data such as values of furnace temperature sensed by a thermocouple are temporarily stored in this type of memory.

Memories are constructed with either a ferrite core or with semiconductors. Technically, core memory is a matrix whose elements are composed of tiny toroids, or doughnut-shape devices, made of ferrite materials. These elements are magnetized to store information. Semiconductor memory is a memory matrix composed of tiny semiconductor chip circuits. All core and semiconductor read-only memories retain data whether power is off or on. Such is not the case with semiconductor random-access memories. If the power source is interrupted, memory is lost. Random access memories can be protected by connecting batteries to the memory package for "battery backup" in the event of a power failure.

Semiconductor memory is usually faster, more compact and less expensive than core memory. In applications where volatility is a key factor, battery backup can be used. Most computers use a combination of semiconductor and core memory. The type of memory used with a computer depends in part on the application. For example, a microcomputer-based controller with a limited number of functions would probably incorporate a read-only memory as the main portion of memory and a small random-access memory for temporary storage of data.

In control applications, computers must connect either to a binary-type signal (either an open or closed relay), to an analog signal or in some cases, to a digital signal. The connection between the digital computer and the binary circuit may only involve the matching of voltage levels. If an instrument or valve has a digital output, connecting it to the computer again involves only voltage or current matching. If the control has an analog output, the analog signal from the sensor to the computer must be converted to digital form. The interface, in this case, is an analog-to-digital (A/D) converter that is placed between the computer and the analog control device. If the computer must send an analog signal to a control loop, a digital-to-analog (D/A) converter acts as the interface. The digital-to-analog converter constructs an analog signal from the digital information from the computer. Analog-to-digital and digital-to-analog interfaces are integral components of even very small processes because they are usually the only possible communication interfaces between processes and analog control devices.

Software

Software provides the computer with flexibility but also accounts for many of its problems. The problems are not necessarily due to the software, but to human error. These errors result because software usually is not written in English, but in a special language that the computer can understand and translate into the binary language it uses to execute the program commands. Software language may be "machine-like" or "English-like". Machine-like languages are called assembly or *low-level* languages, and English-like languages are called *high-level* or *compiled* languages.

Assembly languages are collections of mnemonics that refer to the exact functions a computer must perform. In one particular assembly language, for example, CLA stands for "clear the accumulator", which is far easier to work with than the binary equivalent of 111110000000. Assembly languages sometimes are used with microcomputers because they conserve the microprocessor memory. Memory can be the largest hardware expense of a microcomputer system.

Whereas assembly languages are organized to match the way in which computers operate, high-level languages are organized more like spoken languages. High-level programs do not need to specify the individual steps of operations the computer performs; instead, they must only indicate what function to perform (for example, "add two fields").

While assembly languages require a special program to convert user code into binary code, high-level languages require a program to convert the high-level program into binary code. The special program, called either a compiler or an interpreter, occupies a significant portion of the memory system. This increases the size of memory required. Greater memory capacity is also required because high-level languages are not optimized for one specific machine. Consequently, they are not as efficient in their translation as are assembly languages.

Although high-level languages are not designed for specific machines, they are designed for specific functions. Of the popular languages, BASIC and FORTRAN were developed for math and scientific problem solving. FORTRAN, particularly FORTRAN IV, tends to have a greater number of features than BASIC, and thus is generally more difficult to learn. High-level languages are now in very common usage with computers of all sizes (microcomputers, minicomputers and mainframes); however, programs in these languages require greater memory capacity than programs in assembly languages. Most computers also have standard packaged operating systems written in assembly language or high-level language. With operating systems, users need to be concerned only with the application; the operating system handles file management, program location and programmers' "housekeeping" chores.

Computers have almost unlimited use. However, for a computer to provide the advantages of accuracy, flexibility and reliability, a potential user also must investigate available software, memory requirements, storage, peripherals (typers, plotters and cathode-ray tubes) and, of course, interfaces (analog-to-digital and digital-to-analog converters, etc.) to ensure purchase of a system that is compatible with a broad range of needs, as well as to determine which system will afford the greatest possibility of flexibility and expansion. With the diversity of processor types, peripherals and software, it is possible to apply computer operations effectively to almost any control or instrumentation application.

Applications of Computerized Systems to Heat Treating Processes

Computerized control systems have not been utilized with heat treating furnaces in the past because of the high cost involved. Conventional instrumentation—such as recorder-controllers with thermocouple input, and indicating controllers with special reset options for hi-limit, or excess temperature control—has been much improved by solid-state circuitry, digital setpoint indexing, digital display, and other innovations designed to increase the accuracy, response time and reliability of operation. Computer systems have been viewed as expensive and complicated, especially by furnace builders who sell control packages. Therefore, updated versions of traditional and more familiar measurement and control instrumentation have been favored by both OEM's and users of heat treating furnaces, especially for use with single-chamber furnaces performing one or two operations.

However, in response to increased quality requirements and increased production quantities, in addition to higher energy costs, computerization provides a cost-effective and flexible system for controlling these variables. For example, proper operation of a multichamber vacuum furnace requires complex temperature ramps and soaks, mass flow measurement and control of atmosphere gas admission, varying vacuum levels between chambers, sequences of door interlocks, limit alarms, load and transfer adjustment sequences, and other functions within specific time limits. Under these conditions, the advantages of a digital electronic control system—that is, computerization—not only appear desirable, but are becoming mandatory. Microprocessor-based controllers can perform all of the control functions with electronic speed, a high degree of reliability and push-button ease. The digital-display capability presents setpoint and process-variable readings in a format that is difficult to misinterpret, and provides the flexibility of one- and two-decimal place read-out capacity.

Another advantage of the new digital instrumentation is its compactness, which saves panel space and cuts installation cost. Microprocessor-based controllers and programmers can interface and communicate with a minicomputer for performance monitoring or setpoint supervisory control. All of the functions, or relay logic, for interlock sequences and digital ("on-off") inputs can be provided by digital circuitry. These functions originate in a component called the programmable logic controller (PLC) that is provided as part of the modular hardware in a microprocessor-based furnace controller package.

In more advanced control systems, operation of a multiple-furnace facility can be monitored from one supervisory computer, while each furnace is controlled independently by its own microcomputer. This is the essential concept behind the term "distributed control", which is now widely used in discussions of computerized manufacturing applications.

Computerized Control of Carburizing Furnace Atmospheres*

Gas carburizing consists of treating the steel parts to be carburized at a suitable temperature and for the required length of time in a gaseous atmosphere to bring the carbon to the surface of the steel. Carburizing proceeds by the transfer of carbon to the surface of the steel and by the diffusion of carbon from the surface to the interior of the steel. Closer control over carbon potential is a primary means of attaining more reliable results in the carburizing process. Previously, single-component control systems had provided the only approach to this goal. With recognition of the fact that variations in furnace environment can limit the

*This case study was contributed by B. K. Gupta, F. W. Fraim, and V. Jayarama and is based on the article, How Microprocessors Can Give Close Carbon Potential Control, *Heat Treating*, July 1980. Information is based on the industrial experiences at Thermo Electron Corp., Holcroft Div.

accuracy of single-component systems, efforts were directed toward developing a multicomponent system for carbon control. A microprocessor-based multicomponent analyzing system was developed based on the calculation of actual carbon potential from several measured furnace atmosphere conditions. This system has demonstrated the ability to control and monitor furnace atmospheres as a cost-effective and efficient alternative to traditional instrumentation.

Monitoring and Control Systems. The extent to which various carbon compounds transfer carbon to steel is governed by the carbon potential of the surrounding atmosphere. Carbon potential may be defined as the percent of surface carbon attained by a steel specimen in a state of equilibrium with the surrounding atmosphere. The following reactions occur in a carburizing furnace atmosphere:

$$2\,CO \rightleftharpoons C + CO_2 \qquad (Eq\ 1)$$

$$CO + H_2 \rightleftharpoons C + H_2O \qquad (Eq\ 2)$$

$$CO \rightleftharpoons C + \frac{1}{2}O_2 \qquad (Eq\ 3)$$

where C is the carbon in solution in the steel. Each chemical reaction offers a basis for measuring and controlling the carbon potential of the furnace atmosphere.

CO_2-Base Control Systems. Equation 4 may be derived from Eq 1:

$$a_c = K_1 \cdot \frac{(P_{CO})^2}{P_{CO_2}} \qquad (Eq\ 4)$$

where a_c is activity of carbon; K_1 is an equilibrium constant, P_{CO} is partial pressure of CO in the atmosphere and P_{CO_2} is partial pressure of CO_2 in the atmosphere. Activity of carbon (a_c) is related to the carbon potential; K_1 is dependent only on temperature. If temperature and P_{CO} are constant, then the carbon potential can be measured and controlled on the basis of P_{CO_2} alone.

Dew Point – Base Control Systems. From Eq 2, Eq 5 can be derived:

$$a_c = K_2 \cdot \frac{P_{CO} \cdot P_{H_2}}{P_{H_2O}} \qquad (Eq\ 5)$$

If temperature, P_{CO} and P_{H_2} are constant, carbon potential can be measured and controlled on the basis of P_{H_2O} alone.

O_2-Base Control Systems. From Eq 3, Eq 6 can be derived:

$$a_c = K_3 \cdot \frac{P_{CO}}{(P_{O_2})^{1/2}} \qquad (Eq\ 6)$$

If temperature and P_{CO} are constant, carbon potential can be measured and controlled on the basis of P_{O_2} alone.

System Limitations. Each of the monitoring/control systems described above are based on the assumption that temperature remains constant. However, this is never the case in an actual production furnace. All of these systems also assume that a constant partial pressure of CO, or CO and H_2, exists in the atmosphere, which does not occur in actual practice. Errors result because of variations in temperature and in the partial pressures of CO and H_2. Figure 1 shows the errors in carbon potential that can result in a CO_2-base system due to such variations. Any system that could accommodate for such variations and make the necessary adjustments expediently would be more accurate in monitoring and controlling carburizing furnace atmospheres.

The Computerized Approach. The microprocessor-base multicomponent system operates on the reactions given in Eq 1 and 4, and it measures both CO and CO_2. In addition, it also measures furnace temperature. Consequently, all the variables affecting carbon potential are monitored. Having determined these variables, the computer control system, through its own internal microprocessor, computes the carbon potential according to:

$$Carbon\ potential = A + \frac{B}{T}\log a_c \qquad (Eq\ 7)$$

where A and B are constants, T is reaction (part) temperature and a_c is activity of carbon.

The computerized system also has the additional capability of measuring CH_4, although CH_4 is not used in the computation of carbon potential. The capability to measure CH_4 and limit its concentration in the atmosphere (through the use of a methane limit alarm) is necessary to maintain equilibrium conditions in the furnace and to minimize soot formation.

System Components. The multicomponent system is compact in size. The major components of the system are

- Sample input
- Calibration (zero or span) gas input
- Infrared system
- Thermocouple input
- Control switches (pushbutton or thumbwheel)

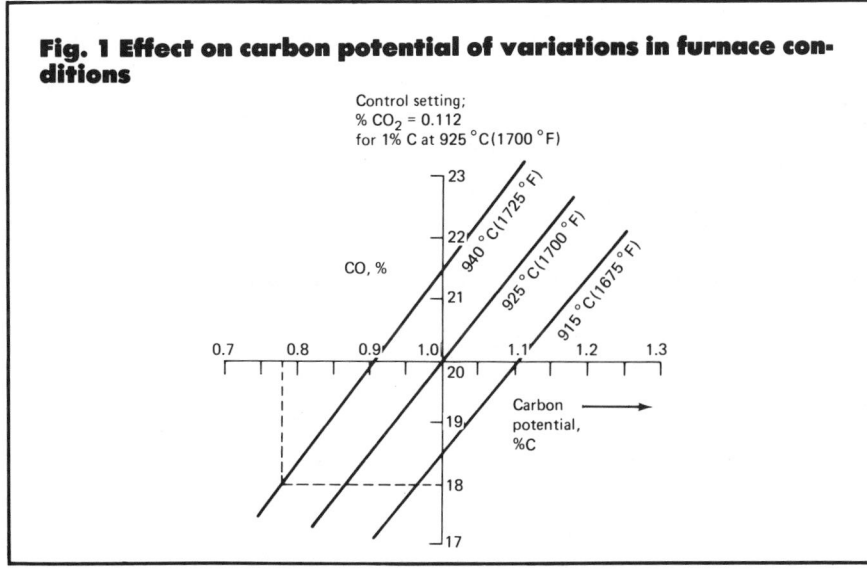

Fig. 1 Effect on carbon potential of variations in furnace conditions

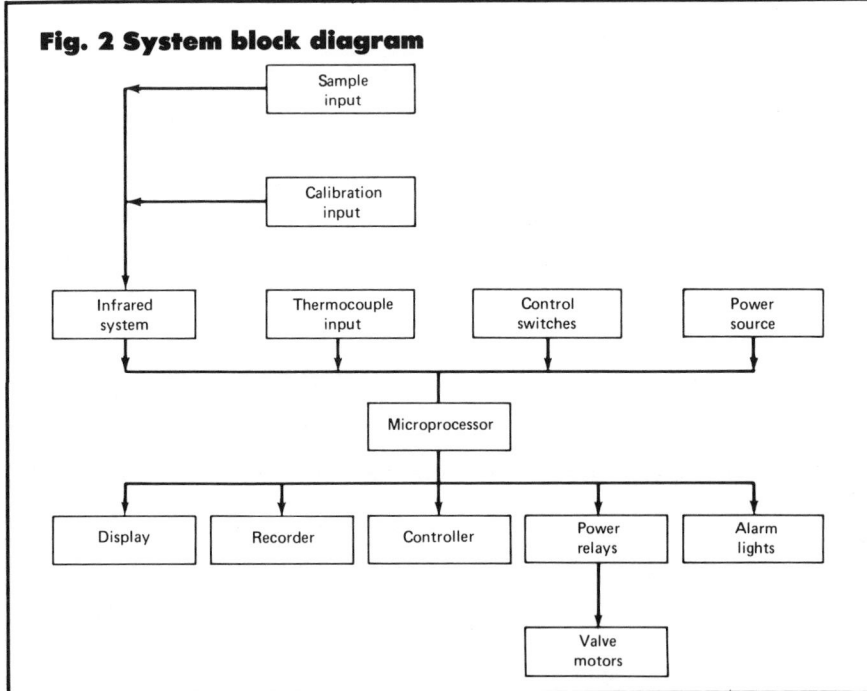

Fig. 2 System block diagram

- Power source and conditioning system
- Microprocessor
- Gas charge display
- Strip chart recorder
- Proportioning controller
- Power relays to operate valve motors
- Alarm lights.

The alarm lights are activated by automatic checks of the system to report conditions such as clogged filters or plugged sample lines. Figure 2 shows a schematic diagram of a multicomponent system. Flowmeters for the sample lines are connected to the compartments; one filter provides purge air and the other is for the sample going to the infrared analyzer system.

The computerized system is capable of monitoring and controlling six sample points, or zones of control. The unit gives the user a choice of basing the control on carbon potential, CO_2 or CH_4. For example, one or more sample points could control the atmosphere generator gas based on CO_2, while other points could control the furnace zones based on carbon potential or CH_4. All zones may be put on control, if desired. Some zones either may be on monitor or may be shut off. A choice of displays exists. The normal display shows CO, CO_2, CH_4, temperature and carbon potential. Displays may also include information on whether a zone is being monitored or controlled, and, if controlled, on what basis. A variety of other displays can be selected through use of the thumbwheel switches. A 24-point strip chart recorder shows CO, CO_2, temperature and carbon potential for each of the 6 zones of control. CH_4 is normally only on the display.

Infrared Analyzer. The infrared chamber is sealed and is continuously purged with clean, dry air. The drying is done by two heatless, regenerative-type, molecular sieve dryers, which also remove the CO_2 from the air. Controlled temperature and pressure conditions are maintained in the flow cell. The flow through the cell is also maintained at a constant level by holding a constant pressure drop across a fixed orifice; atmospheric pressure changes do not affect flow. In addition, the entire compartment is maintained at a constant temperature. These features provide the environmental stability that allows calibration adjustments to be kept to an absolute minimum.

Application of a Microprocessor-Based Control Programmer to Control of Time-vs.-Temperature Program*

Figure 3 shows a typical process profile for a heat treating furnace in which furnace temperature is controlled as a function of time in order to achieve the desired metallurgical conditioning of parts. This same type of program, however, could apply to applications for environmental chambers, weld stress relieving, or any application where process variables must be controlled as a function of time. To achieve this kind of programmed control, a separate setpoint programming source, a controller, and several interrupters and timers are needed.

The types of programming sources available are limited in their ability to provide the degree of resolution and setpoint accuracies required to achieve

*This case study was provided by Theodore K. Thomas and is based on industrial experiences at Honeywell Process Control Div.

process control that consistently produces high-quality products.

The Cam Programmer has been used extensively in the heat treating industry for many years. This instrument utilizes a rotating plexiglas disk or cam, cut to conform to a specific program, or sequence of temperature rise, level and fall. The cam moves a mechanical follower arm that is spring-loaded against it. The arm in turn adjusts the setpoint of the furnace-temperature controller to maintain the program. The Cam Programmer has several inherent limitations. The most difficult problem that arises immediately is cutting the cam to the exact program desired. This is a time consuming task. Also, rates of rise (rates at which furnace temperature is increased or decreased) are restricted due to the inability of the mechanical follower arm to follow steep cam contours.

Another method of generating setpoint-versus-time profiles involves use of a photoelectric sensor that electronically follows a curve edge drawn on a rotating disk. Drawing the curve for each program is tedious, and dirt or dust in the operating area could ultimately blind the sensor at some critical point in the process.

Capacitive sensors have also been used in generating setpoint-versus-time profiles. These electronic sensing devices follow a line scribed on a surface that rotates past the sensor. This technique also requires the drawing of curves and is susceptible to the dust, dirt and electrical noise found in many industrial areas.

All of the foregoing types of setpoint programming instruments must ultimately interface with a controller that accepts a remote setpoint input, as well as with timers and interrupters. The result in terms of process control is often marginal. Lack of characterization of the setpoint signal generated is one of the largest sources of error in programming controllers using nonlinear process sensors such as thermocouples.

Through implementation of microprocessors, however, the digital approach to solving these problems is incorporated into one instrument combining all the traditional elements of a programmed control system. Figure 4 shows a typical microprocessor-based instrument that contains the setpoint-versus-time program signal source, a digital three-mode controller with Auto-Man operating modes, and twelve

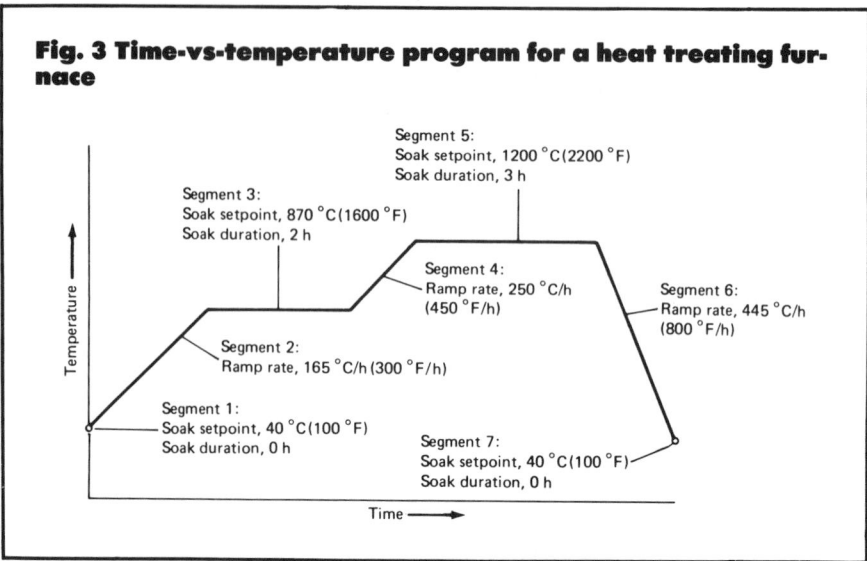

Fig. 3 Time-vs-temperature program for a heat treating furnace

Segment 5:
Soak setpoint, 1200 °C (2200 °F)
Soak duration, 3 h

Segment 3:
Soak setpoint, 870 °C (1600 °F)
Soak duration, 2 h

Segment 4:
Ramp rate, 250 °C/h (450 °F/h)

Segment 6:
Ramp rate, 445 °C/h (800 °F/h)

Segment 2:
Ramp rate, 165 °C/h (300 °F/h)

Segment 1:
Soak setpoint, 40 °C (100 °F)
Soak duration, 0 h

Segment 7:
Soak setpoint, 40 °C (100 °F)
Soak duration, 0 h

Temperature

Time

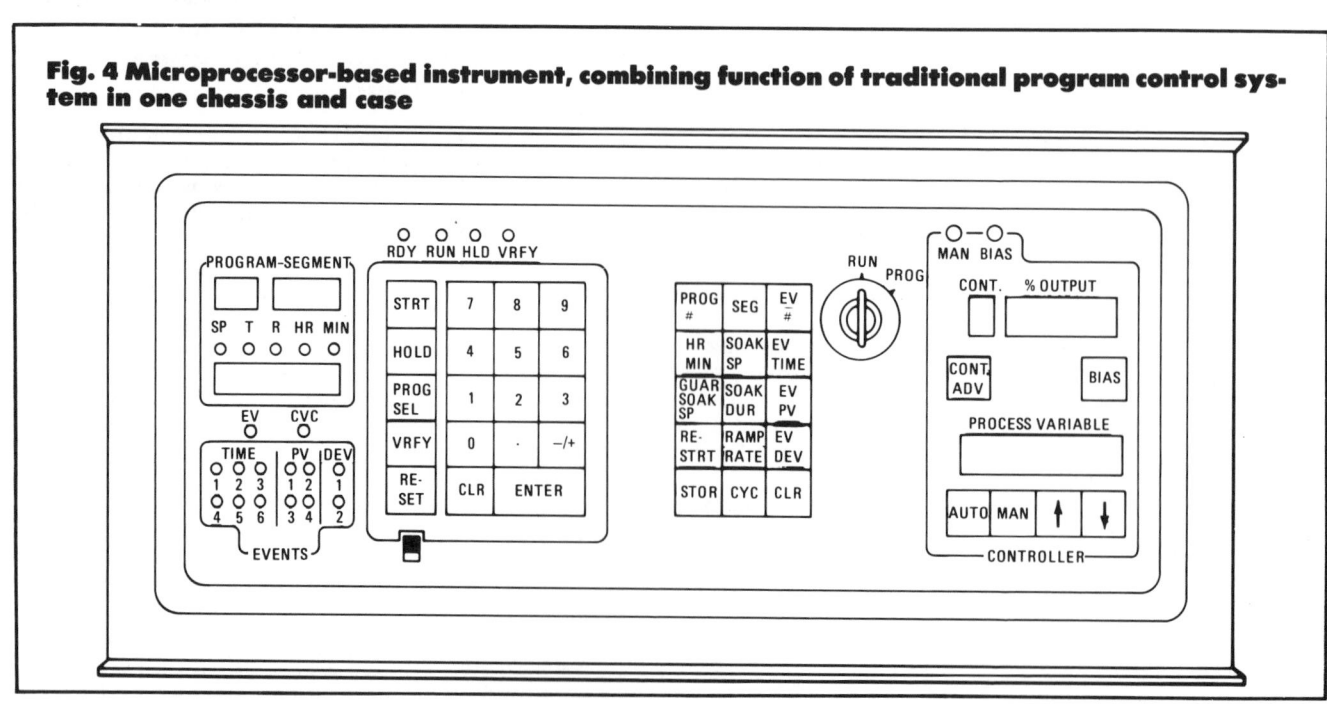

Fig. 4 Microprocessor-based instrument, combining function of traditional program control system in one chassis and case

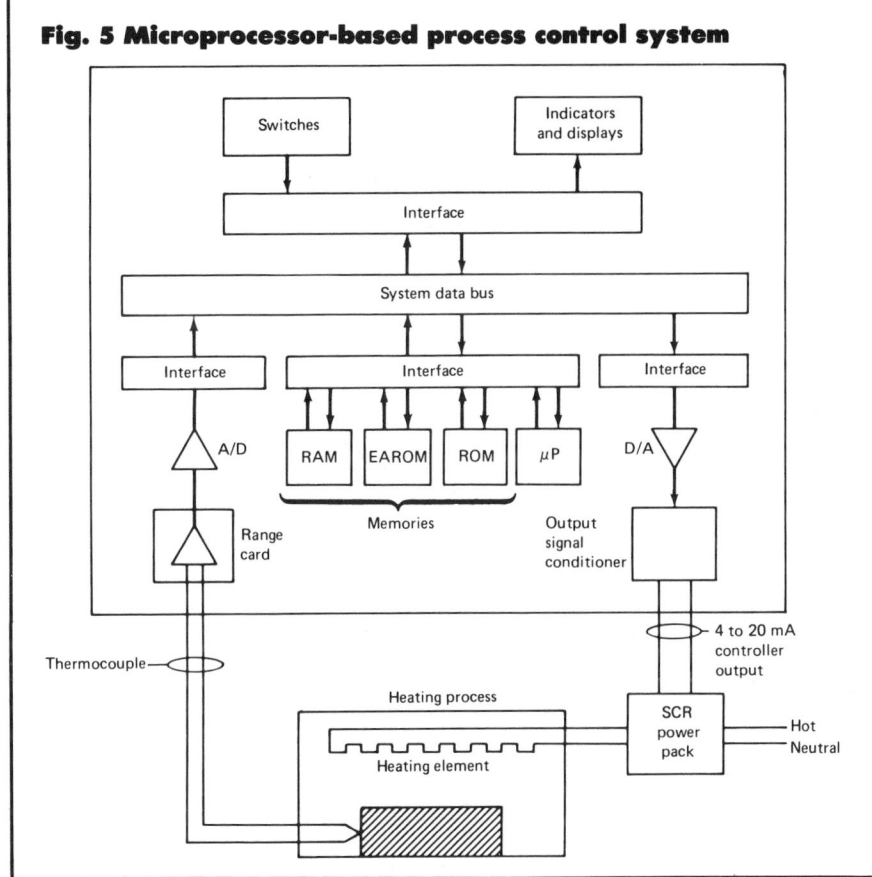

Fig. 5 Microprocessor-based process control system

A simplified description of the operation of the controller shown in Fig. 5 is as follows:

- The millivolt output from the thermocouple is conditioned and amplified in the range card to give an output of 0 to 2 Vdc.
- The analog 0-to-2-V signal is then changed to a 12-bit digital signal by the analog-to-digital converter.
- Every 300 ms, the microprocessor samples the output of the analog-to-digital converter and stores it in the RAM.
- The data in the RAM are compared with the thermocouple-versus-emf data in the ROM to determine the actual temperature being sensed by the thermocouple.
- The actual temperature is then sent to the operator's display and is also compared with the setpoint being generated by the program stored in the EAROM. The comparison and calculation of the control algorithm actually takes place in the RAM, which receives data from the ROM and EAROM and then stores the results in the EAROM.
- Every 300 ms, the microprocessor samples the results of the control algorithm and sends it to a 10-bit digital-to-analog converter whose output is 4 to 20 mA.
- The output can then be used directly to control the process or may be converted to either a time-proportioning or a position-proportioning control form if desired.

Use of Minicomputer for Job Scheduling, Temperature Control and Furnace-Pressure Control*

Production of parts for the aerospace industry includes a heat treating step after forging. Part failure could contribute to loss of life; consequently, it is extremely important that parts be heat treated properly and that accurate records be maintained concerning process control.

The critical close-tolerance heat treating is done in box furnaces, which are designed and operated to provide a high degree of temperature-uniformity. These furnaces are located in one building where two rows of furnaces face each other with oil and water quench tanks between them, as shown in Fig.

programmable event switches. All operating data (setpoint, process variable, ramp rates, percent controller output) are displayed continuously by a seven-segment digital readout. Light-emitting-diode (LED) indicators show operating mode and event-switch status. Programs are entered through the use of a front panel keyboard. These programs are then permanently stored in nonvolatile memory which requires no battery backup. If a power loss occurs, the stored programs will not be lost; reprogramming is consequently not a requirement.

Figure 5 shows a process-control loop in which a microprocessor-based instrument is used to measure, indicate and control the temperature according to a preset schedule or program. Components required to complete the system are a temperature sensor and the final control device.

Three types of memory devices are depicted in the instrument shown in Fig. 5. The read-only memory (ROM) is the permanent memory; all of its information is "burned in" by the device

manufacturer and cannot be changed. The other two types, commonly referred to as random-access memory (RAM) and electrically alterable read-only memory (EAROM), are not permanent memories. Their information is changed, as required, when the system is operating. The main difference between the two temporary types of memories is that the RAM memory is volatile (that is, it loses its memory whenever the power is turned off or fails), whereas the EAROM type is nonvolatile and does not require a battery backup to retain its memory when power is lost.

The less-expensive RAM type is used as a "scratch pad" to make temporary calculations before storing data that must be retained in EAROM. Data that must be retained include such items as:

- Temperature-versus-time programs
- Tuning constants for the algorithm
- Machine-state data, such as the current setpoint, controller output, and elapsed times.

*This case study is based on industrial experiences at the Ladish Co.

Fig. 6 Furnace area layout

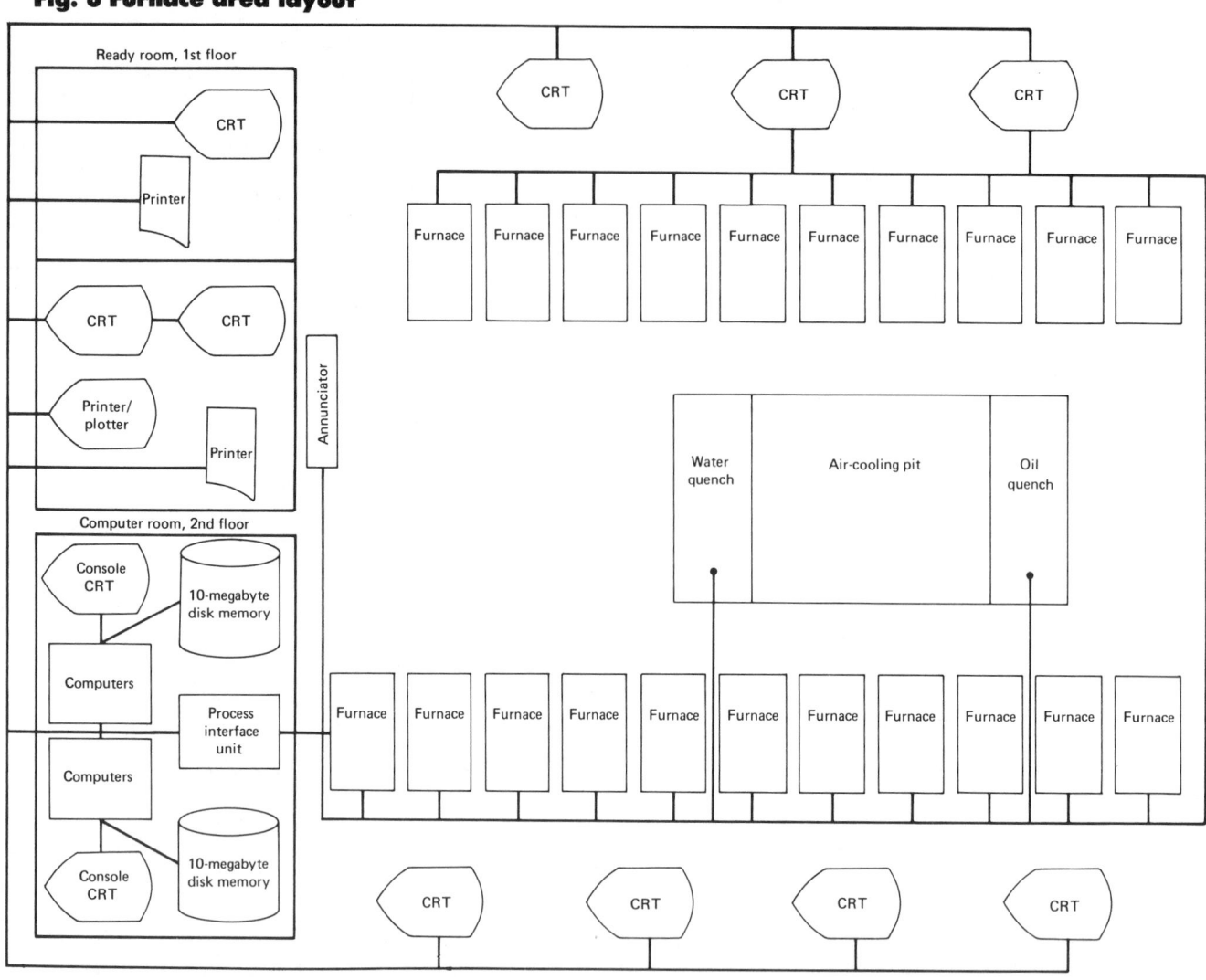

6. The furnaces are fired with excess air to meet temperature-uniformity requirements. There are five furnace thermocouples, one for control and four to indicate temperature uniformity, located in each furnace. In addition to these thermocouples, critical loads are thermocoupled in as many as nine locations and are monitored to ensure proper heat treatment. These furnaces are surveyed on a scheduled basis and are certified for very close temperature uniformity.

Computer Control. The traditional control system consisted of a controller, a circular chart recorder, and a four-point strip chart recorder for each furnace. An additional multipoint recorder was connected to the load thermocouples for critical loads. The increasing requirements for closer temperature controls, and improved recordkeeping, as well as the requirement of close tolerance quality, were the main reasons for computer control.

The main function of the computer is to provide temperature control and temperature records for the furnaces by monitoring up to 14 thermocouples in each furnace. The computer automatically adjusts the thermocouple readings for the deviation of the thermocouple wire so that all readings denote actual temperature, in either degrees Celsius or degrees Fahrenheit. The computer control system can be programmed to perform numerous functions, with several levels of response for each function. In a real-time system, the computer responds to events as they occur, with a predetermined set of priorities. If an event occurs that has top priority, such as furnace overtemperature, the computer will respond to that event first and then proceed to complete remaining lower priority commands.

System Operation. In setting up a load for heat treating, the operator enters data concerning the load, such as part and serial numbers, and information concerning the heat treating cycle, such as type of cycle, control temperature(s), heat-up rate, and length of cycle, on a cathode ray tube (CRT) terminal. Once an operator assigns a load to a furnace and instructs the computer that the furnace has been loaded, the computer executes the heat treating cycle. The system is flexible enough to allow operator changes (input) during the heat treating cycle.

The computer is programmed with several control schemes for the heat-up and hold portions of the heat treating cycle. The control scheme is automatically selected by the computer, based on information the operator has punched in at the CRT concerning the load. If an alarm condition occurs during the cycle, the computer will automatically adjust its control scheme until the problem has been corrected.

At periodic intervals, the computer will monitor and store the maximum and minimum readings of each thermocouple during the time interval. Alarm reports are also stored. At the end of each shift, all information entered by an operator concerning a load, as well as all temperature and alarm data, is transferred from disk to magnetic tape for storage as a permanent record. This data can be recalled at any time by replaying the tape.

At the end of each shift, the computer prints out a shift report showing what was heat treated during that shift. It can also generate other outputs upon operator request (such as any or all thermocouple readings, or expected completion time). Temperature-versus-time curves can be obtained using the graphics terminal and plotter. Operators are able to plot either the high and low readings for each thermocouple or the average of these readings, using an ordinate or abscissa of their choice.

Advantages of the System. Although the main purposes of the computer system are control of furnace temperature and keeping of temperature data, the system is capable of performing other functions, such as monitoring of furnace operating hours and the time until the next temperature-uniformity survey is due. The computer is also programmed to read gas meters in the plant. Through installation of additional furnace hardware, and input/output boards, the computer is able to control furnace pressure and to limit the amount of excess combustion air to the amount needed to maintain temperature uniformity. The software necessary for the computer to perform these functions is inserted in the computer before it is brought on line; however, it may be changed as heat treating needs and experience dictate.

In addition to providing improved temperature control and recordkeeping, the computer system also contributes to increased production. Many functions that are normally done at the discretion of the furnace operator are now done automatically. This in turn leads to increased energy savings.

Computer-Assisted Management System*

A commercial heat treating supplier is continuously faced with customer requests for different part configurations, use of different materials, and varying specifications, as well as close scheduling of jobs and accurate prediction of job-completion dates. The customer not only expects prompt delivery, but also high-quality parts.

One alternative to controlling costs and meeting customer service requirements is the computerized approach. The system chosen was patterned after the airline reservation system, because the goals and problems—finite capacity, generally unpredictable demand, and the need for fast delivery—were parallel. Expansion of the system to additional facilities is simple. Initial planning included the purchase of a mainframe with capabilities for handling many facilities in widely separated locations. To computerize a new plant, only a cathode ray tube, a printer and a phone line are needed to begin system operation.

A batch-type computer was in use to provide accounting, financial and managerial reports and analyses. The computer was an accepted, proven tool. The changeover from batch operations to a continuous on-line system required more modern computer hardware and a systems analyst familiar with on-line systems. Once the old accounting, financial and managerial programs were converted to the new computer, parallel systems were run for about two months. With the conversion completed, work efficiency was increased and the new computer used less space.

The on-line system was created to improve internal control of jobs progressing through the plant, to simplify paperwork, and to provide customers with fast, accurate information concerning order status. Status of jobs is maintained by heat treaters on the

*This case study was contributed by John D. Hubbard and is adapted from the article, Computer System at Hinderliter Heat Treating, *Metal Progress*, July 1980.

floor. A master job card travels with each batch of parts as it moves from work station to work station. Production information is entered on the master job card as a processing step is completed and is entered directly into the computer 24 h a day, 7 days a week.

When a customer inquires about a job status, the account number is keyed into the computer and every job the customer has in-house, including job number, purchase order number, number of pieces, weight, description, and date received, is displayed on the screen. Once the job in question is identified, the computer displays the shop routing in line-by-line detail of all operations the parts must proceed through to completion. Each individual furnace load is keyed to the actual furnace where it is being run or the exact location of that load, such as furnace number and receiving, inspection and shipping dates. From this information, the customer may be provided fast (2 min average) delivery information.

In addition to prompt and accurate customer service, shop routers are standardized; routers can be duplicated or custom built automatically. Invoices, acknowledgments and shipper forms are generated automatically, as well as special reports.

Glossary of Computer Terms

Access time: Amount of time required to access the contents of a memory location. This limitation is imposed by the speed of the memory circuitry.

Algorithm: Procedure used for performing a task

Backplane: Circuitry and mechanical elements used to connect the boards of a system

BASIC: Beginner's All-purpose Symbolic Instruction Code, oriented toward beginners rather than experienced programmers. Numerous incompatible versions exist.

Binary: Numbering system consisting of only 2 digits, either 0 or 1, as contrasted with the 10 digits, 0 to 9, of the decimal system. In electronics, the terms "binary", "two-state", and "digital" are synonymous.

BCD: Binary Coded Decimal, coding system in which each decimal digit from 0 to 9 is represented by 4 binary digits (bits)

Decimal digit	Binary code
0	0000
1	0001
2	0010
3	0011
4	0100
5	0101
6	0110
7	0111
8	1000
9	1001

Bit: Single binary digit. A bit may have two states; normally, a bit is considered "high" when its value is 1 and "low" when its value is 0.

Board: Card that contains circuitry for one or more specific functions, such as memory or interfacing

Bus: Circuitry in a backplane that allows transmission of electrical signals from one board to another

Byte: Group of 8 bits that are treated as a unit

Card frame: Enclosure that holds a system's boards in place

Chip: Integrated circuit

Clock: Device that generates electronic timing signals. Clock signals are often used to synchronize certain system operations.

COBOL: Common Business-Oriented Language, used primarily in business applications

Core memory: Type of memory that stores information on magnetically charged, doughnut-shape cores made of ferrite and lithium. Core memories have largely been superseded by semiconductor memories.

CPU: Central Processing Unit, the primary component of all computer systems. It is responsible for controlling system operations, as directed by the program it is executing.

Dedicated device: Device that is used exclusively for one function

DMA: Direct Memory Access, an arrangement where blocks of data can be transferred between main memory and a peripheral device (such as a disk drive) without processor intervention

EAROM: Electrically Alterable Read-Only Memory, a type of memory that combines the characteristics of random access memory (RAM) and read-only memory (ROM). It is nonvolatile (like read-only memory) but can be written into the processor (like RAM). EAROM, however, has a substantially longer writing time (currently about 2 ms versus 400 ns), as well as a limited number of writes (about 1 000 000) be-

fore the chip can no longer be reprogrammed.

EROM: Erasable Read-Only Memory, read-only memory that can be erased and reprogrammed. EROM is frequently erased through exposure to ultraviolet light. Also spelled EPROM.

Firmware: Part of a computer program that is incorporated, at least temporarily, as machine hardware—for example, instructions contained in a ROM

Fixed-point arithmetic: Arithmetic where the decimal point always remains at a predetermined position. Integer arithmetic is a type of fixed-point arithmetic, because the decimal point is always to the right of the mantissa.

Flag: Bit whose state signifies whether a certain condition has occurred

Flip-flop: Circuit that changes its logical state when signaled to do so by another device

Floating-point arithmetic: Arithmetic where the decimal point may occupy any position

Floppy disk: Component similar to a 45-rpm record made of flexible material and used for storing computer data

FORTRAN: Formula Translator, first high-level language. Emphasizes algebraic operations. Used primarily in scientific applications.

Gate: Circuit that performs a Boolean logic operation

General-purpose digital computer: A digital computer designed to solve a large variety of problems; that is, a computer that can be adapted to a large class of applications (as opposed to a computer that might be designed specifically to control a manufacturing process). A typical general-purpose digital computer consists of: (a) input/output (I/O) devices, which permit communications with the outside world; (b) memory, which stores data and instructions; and (c) central processing unit (CPU), which performs the arithmetic and data processing operations and provides the control that ties all of the subsystems together so that they operate in a fully automated manner.

Hard-wired logic: Group of solid-state logic modules mounted on one or more circuit boards and interconnected by electrical wiring. Logic control functions are determined by the way in which the modules are interconnected, as contrasted with a programmable controller or microprocessor, in which the logic is in program form.

Instruction: Group of bits that defines

a computer operation. An instruction may move data, do arithmetic and logic functions, control input/output devices, (typers, plotters or CRT's) or make decisions as to which instruction to execute next.

Intelligent device: Device that contains its own processor.

LSI: Large-Scale Integration, class of integrated circuits that contain the largest number of functions per chip. Microprocessors are LSI devices.

LED: Light-Emitting Diode, type of digital output display that is frequently used in calculators

Light pen: Input device used in conjunction with a video display. When the user touches the display screen with the light pen, the electronics associated with the pen will determine the coordinates of the point that the user touched. These coordinates will then be transmitted to the computer.

Line printer: Output device that prints an entire line of information at a time

Machine language: Binary code that can be directly executed by the processor, as opposed to assembly or high-level language

Mag tape: Magnetic tape, similar to that used by audio tape recorders, on which information can be stored in a computer-readable format

Mainframe: The computer itself, including the processor, main memory, input/output interfaces, and backplane

Main memory: Memory that the processor accesses directly, as opposed to peripherals such as disk and tape devices

Mass storage: Auxiliary or bulk memory, as opposed to main memory. Disk drives and tape drives are common mass storage devices.

Memory: Memory devices provide temporary and permanent storage of information. Permanent memories (ROM) instruct the microprocessor as to which logic operations to perform. Temporary memories contain information that the operator controls and has access to change.

Memory capacity: Number of bits that a memory can hold; for example, a 1K semiconductor memory can store 1000 bits (actually 1024 bits), and a 2K semiconductor memory can store 2000 bits (actually 2048 bits). Fixed memories usually contain instructions, and therefore their capacity is sometimes expressed as the number of words of a certain length that it can hold. For example, "256 × 4" means that the memory can store 256 4-bit words, which makes it a 1K memory (1024 bits). The same 1K memory could be a 128 × 8 memory. In either case, however, a 1K memory is purchases from the manufacturer, not a 256 × 4 or a 128 × 8 memory.

Microcomputer: Computer whose major sections—central processing unit (CPU), control, timing, and memory—are each contained on a single integrated (IC) chip, or at most, a few chips. In other words, a large-scale integrated (LSI) computer.

Microprocessor: Large-scale integrated (LSI) device that performs the functions of the central processing unit of a computer. It is called a microprocessor because of its extremely small size. Typically, it is contained on a single integrated chip (IC). In some cases, the microprocessor is made up of two, three or even more chips.

Minicomputer: Broad term describing any general-purpose digital computer in the low-to-moderate price range

Motherboard: Synonym for backplane

MTBF: Mean Time Between Failures, length of time for which a device can reasonably be expected to operate without malfunction

Multiplex: To combine two or more electrical signals into a single, composite signal. This may be done on a frequency basis (frequency-division multiplexing) or on a time basis (time-division multiplexing).

Off-line: Device that is not connected directly to its host computer. A keypunch is an example of an off-line device.

On-line: Device that is connected directly to its host computer

Parity check: An error-detection system in which an additional bit, the parity bit, is appended to each word or byte. Under even parity, the parity bit is 1 if there is an even number of 1's in the rest of the word. Under odd parity, the bit is 1 if there is an odd number of 1's in the rest of the word.

PASCAL: Advanced programming language, not an acronym

Peripheral: Unit, such as a communications terminal that is external to the system processor

Plotter: Hard-copy device that produces line drawings such as X/Y graphs. The coordinates of the points or lines to be plotted are normally supplied by the computer.

Port: Communication channel between a computer and another device

PCB: Printed Circuit Board, circuit board whose electrical connections are made through conductive material that is contained on the board itself, rather than with individual wires

PLA: Programmable Logic Array, device (usually an integrated circuit) containing a set of logic gates whose interconnections may be programmed

PROM: Programmable Read-Only Memory, type of read-only memory that can be programmed by the user. This programming usually requires special equipment.

RAM: Random-Access Memory, a read/write memory. A more strict definition of a RAM is a memory that stores information in such a way that each bit of information may be retrieved within the same amount of time as any other bit, as opposed to serial memory.

ROM: Read-Only Memory, memory in which information is stored permanently, such as a math function or a microprogram. An ROM is programmed according to the user's requirements during memory fabrication and cannot be reprogrammed.

Read/write memory: Memory whose contents can be continuously changed quickly and easily during system operation. It differs from a read-only memory (ROM), whose contents are fixed and not subject to change, and a reprogrammable ROM, whose contents can be changed but only periodically. A RAM is a read/write memory.

Real time: Pertains to the performance of a computation during the actual time that the related physical process occurs so that results of the computation can be used to guide the physical process

Register: Fast-access circuit used to store bits or words in a central processing unit (CPU). Registers play a key role in CPU operations. In most applications, the efficiency of the program is related to the number of registers.

Run time: Time at which the program is executed. Also, the amount of time required to execute the program.

Second source: Manufacturer who produces a product that is interchangeable with the product of another manufacturer

Semiconductor: Device (or material) with an electrical conductivity that lies between those of metal conductors and those of insulators. Integrated circuits, transistors and diodes are the most common semiconductor devices.

Semiconductor memory: Memory in which semiconductors are used as the storage elements, and characterized by low-to-moderate-cost storage and a wide range of memory operating speed, from very fast to relatively slow. Almost all semiconductor memories are volatile.

Serial memory: Memory whose contained data is accessible only in a fixed order, beginning at some prescribed reference point. Data in any particular location is not available until all data ahead of that location have been read. Such a memory is inherently slow compared with a random access memory (RAM).

Software: Coded instructions that direct the operation of a computer. A set of such instructions for accomplishing a particular task is called a program.

Solid state: Silicon or germanium semiconductor device, such as a diode, transistor or integrated circuit. May also refer to circuits, equipment or systems made from such devices.

Synchronous communication: Data transmission where the bits are transmitted at a fixed rate. The transmitter and receiver both use the same clock signals for synchronization.

Terminal: Device for communication with a computer. A typical terminal consists of a keyboard and a printer or video display.

Thermal printer: Hard-copy device that produces output on heat-sensitive paper

Volatile memory: Memory whose contents are irretrievably lost when operating power is removed. Practically all semiconductor memories are volatile.

Word length: Number of bits in a computer word. The longer the word length, the greater the precision (number of significant digits).

Furnace Safety

By the ASM Committee
on Furnace Safety*

HEAT TREATING FURNACES require safety procedures common to all industrial installations, but in addition, they have requirements specific to the use of high-temperature energy sources and potentially explosive gases and liquids used as aids to chemical processing.

Because these heat treating processes require careful control for optimum technical results as well as for safety, proper training of operating personnel is a primary consideration. Proper equipment design is also critical.

The information presented here is not intended to be interpreted as a safety standard but is offered only as a set of guidelines. Safety standards for furnaces are maintained by the National Fire Protection Association, by the U.S. Occupational Safety and Health Administration and by insurance underwriters.

All equipment should be installed and operated with awareness of the potentials for fire and explosion and the hazards to operators and equipment. Equipment designs should ensure reliable, safe operation over the expected maximum life of the equipment.

Fuel-Fired Furnaces

Fuel-fired furnaces for heat treating have several major control requirements that depend on (a) whether the process must be direct or indirect fired;

(b) whether heat treating is to be done under a particular pressure or vacuum, or in a controlled atmosphere; and (c) whether the product uses some special type of precoat or laminant. In all situations, there are fundamental control variables, and instrumentation is available to identify and control change and drift, thus achieving the desired results.

The main control elements are the three requirements for proper combustion: a source of heat, an oxidizing agent and time.

Fuel-fired furnaces can be automated to the extent that manual intervention is not required for normal operation. Many processes and operations require manual control, however; thus, furnace controls range from almost completely manual devices to highly sophisticated computer-controlled devices.

The major control variables and types of instrumentation used in each instance are described below, in sequence, from start-up through final cycle control.

Electrical Power for Fuel-Fired Furnaces

The safe use of electrical energy employed in heat treating control processes requires adherence to National Electrical Codes and to local requirements of states and communities. Good prac-

tice dictates that a circuit breaker be positioned within view of the operator. Control cabinets must be designed to ensure that operators cannot inadvertently become a path to ground. Wiring type should be based on the environment of use, and wiring for all motor and control circuits should be contained in appropriate conduits. Numbers of wires within specific conduits should be governed by the fact that elevated temperatures may be encountered. All enclosures for electrical apparatus should be designed to protect the contents from the environment. The furnace itself should be grounded for proper control.

The source of electrical power to the furnace installation should be equipped with fuses. Each motor also should be equipped individually with fuses and protected with thermal-overload elements based on operating temperatures. Motor selection should be based on such conditions of use as temperature, weather, dust, dirt, atmosphere and humidity. Manufacturers should be consulted on motor design and selection. Control-panel power should be fused and provided with externally operated switches to allow safe entry by authorized personnel.

Electronic-signal wiring from the flame-safety circuits should have its own conduit, free from the "noise" and induction present in normal power and

*Raymond Ostrowski, *Chairman*, Sales Manager, Protection Controls, Inc.; Fred J. Bartkowski, Vice President, Marshall W. Nelson & Associates, Inc.; Roger G. Blocks, President, Chem-Al, Inc.; Don G. Ensweiler, President, Heat Process Associates, Inc.; Ross Shingledecker, Metallurgical Director, Manufacturing Services, Ladish Co.

control circuits. Thermocouple wiring also should be contained in its own conduit to avoid creating induced errors from such random sources as power lines, signal wiring, motors and ballasts.

Control Circuits for Fuel-Fired Furnaces

Combustion-Air Blower Control. Combustion-air blowers must be interlocked with the combustion-limit circuits to shut down the process in the event of failure. The flow of combustion air must always be proven before and during a processing cycle with two independent sources of information. The motor should be protected from short circuits with fuses and from overheating or amperage draw with thermal breakers (heaters). The motor starter should be wired so that it will disconnect when any phase is interrupted or when the motor malfunctions. It should not be assumed that the blower is providing combustion air just because the blower motor is operating; combustion-air flow must be proven. At one time, an end switch, or rotary switch, on the motor was a common indirect method of gaining this information. A better method is to use a pressure switch in the air line for direct sensing. A sail or flag switch, although not quite as good because of the mechanical movements required, also can be used to sense air flow directly.

Gas-Pressure Control. Fuel must arrive at the burner in the correct quantity and at the correct time for safe combustion. Fuel pressure thus must be proven within an allowable range. Gas-pressure switches for both high and low gas limits are installed in the main gas lines. Visual pressure gages also are helpful to operators in setting burners and in verifying that the fuel is being supplied within the range desired and that pressure limit switches are not malfunctioning. Mercury-wetted relay pressure switches are recommended for their ease of setup and maintenance and for their reliability.

Pressure Regulators. Pressure of gaseous fuel is most commonly regulated by pressure-regulating diaphragms. Good, safe design normally requires one regulator for pilot fuel and one regulator for main-burner fuel. The pilot gas, if taken from the main fuel line, should be drawn from a point between the gas supply and the regulator for the main fuel. Thus, the pilot and main

burner can be set up optimally, safely and independently. The regulators should be vented to a safe location outside the plant to ensure safety if a regulator diaphragm is damaged in service. Good practice and manufacturers' recommendations show that diaphragm life can be substantial if regulators are shielded from thermal radiation and are used below their maximum design limits. Positive lockup regulators are recommended to prevent downstream pressure buildup during shutdown periods.

Valves. Blocking valves normally are closed valves that are energized only by the combustion-control circuits. The pilot-gas blocking valve is placed downstream of the pressure regulator and a hand-operated gas cock. A pipe union should be inserted just ahead of the electrically operated blocking valve to allow safe removal if repairs are needed. The blocking valve is opened to the pilot assembly only after the furnace is purged.

Purging of Fuel-Fired Furnaces

The furnace must be purged of any possible combustible materials. This is best accomplished by opening the furnace doors, which should be equipped with a limit switch to ensure that they are opened adequately. Once the doors are open, the combustion blower or exhaust fans can be timed to allow for a minimum of four changes of air in the combustion chamber. This purge cycle is required for safe start-up and is standard practice for all well-managed operations.

Pilot Control

Pilot assemblies can be of either the atmospheric or the blast type. The atmospheric type is similar to an atmospheric burner, in which the air is inspirated from the atmosphere by the gas stream.

In the blast type, air and gas are brought to a mixer under pressure. The gas is then reduced to atmospheric pressure and pulled into the mixer by the pressurized air stream. This is the most positive means of pilot-gas control.

Ignition

For ignition trials, a high-voltage transformer is used in conjunction with a spark plug designed for the pilot or burner assembly. The control circuit

causes the pilot valve to open and a spark to be produced. The spark continues for a short period (normally 15 s) and establishes a flame, which can be detected. If the flame is not established, because the flame or signal is inadequate, the cycle returns to the purging stage.

The voltages normally employed are approximately 5000 to 6000 V, and the high-voltage transformer is normally mounted on the furnace and grounded to it. The spark in turn is grounded to the pilot assembly, then to the furnace; hence, a well-grounded furnace is an important safety requirement.

Flame Detection

A thermocouple junction placed in intimate contact with the pilot flame is perhaps the most common means of flame detection, but thermocouples are useful only on very small pilot assemblies or burners, and they are not useful after a burner becomes hot. The flame may no longer be present, but a hot burner block or refractory may retain heat and slow the rate of thermocouple cooling. Thus, thermocouple junctions are not recommended except for quench-tank heaters of the constant-pilot, open-grid burner design or for small atmospheric burners.

Flame electrodes which are small anodes of heat-resisting alloy placed in intimate contact with the normal pilot flame, work on the principle that flame causes ionization within the burner atmosphere and thus allows a circuit to be formed to ground. The flow of a minute amount of current, at low voltage, is sufficient to sense and communicate the presence of a flame.

Flame electrodes are common on all industrial heat treating furnaces where the flame is kept on-ratio or slightly oxidizing. The flame electrode tends to become carbon coated in a reducing flame, a condition that can cause nuisance shutdowns.

Ultraviolet scanners are the third common device for sensing flame. They are normally dependable if the lens viewing the flame is kept clean. The UV scanners must not be used in any application where ultraviolet light is present from a source other than the burner in question. The UV scanner is a useful and practical device for any clean-flame, clean-furnace operation, if it is located and aimed properly. A flow of clean, filtered cooling air across the scanner face aids in keeping it clean and cool, extending scanner life appreciably.

Depending on the burner used, the application, and property-insurance requirements, it may be necessary to monitor both the main burner and the pilot flame independently.

Common and serious errors in flame detection are made by operators who circumvent flame-safety equipment rather than correcting the usually minor problems that cause nuisance shutdowns. Flame-safety equipment that uses totally enclosed relays is recommended in favor of types with accessible relays that may be kept open, for example, with a piece of paper. This point, however trivial it may seem, has been profoundly recognized by those firms who have lost operators, furnaces and product as a result of poorly designed flame-safety equipment that can be circumvented easily. Any employee found tampering with this equipment should receive disciplinary action, and all employees should be trained in use of flame-safety equipment.

Burner Operation

The main fuel supply for fuel-fired heat treating furnaces normally is natural gas, propane-air, propane, butane or one of the fuel oils. Although this discussion centers on natural gas, the same principles apply to the other gases and oils.

The main gas valve may be of the manual-reset type, requiring an operator, or may be fully automatic. The manual type usually is preferred when the furnace is run intermittently or when operators must perform some other function, such as opening doors. When the operator opens the valve, he is in effect making a conscious decision that conditions are ready for the main burner heat. The valve may be made automatic when the furnace is designed and interlocked to preclude an unsafe condition.

For furnaces with capacities greater than 422 MJ/h (400 000 Btu/h), it is recommended that a second blocking valve be inserted into the main gas line and that a normally open vent valve be installed between the main blocking valves. The vent valve should be vented out of doors away from any openings such as windows, air intakes and exhaust louvers in the building walls or roof. This "double block and bleed" arrangement prevents faulty valves or dirty valve seats from causing leakage

of fuel into the furnace between operating cycles. It is not totally foolproof: the open vent may malfunction and allow valuable fuel to leak into the atmosphere during operations, or the internal mechanisms may malfunction because their design is unsuited to operating factors such as environmental conditions. Good preventive maintenance is required. One survey showed that 1 out of every 10 blocking valves was faulty after operating for no more than 10 years.

Burner Control. The gas-air ratio ordinarily is controlled to about 10 parts air to 1 part natural gas for good combustion efficiency. There are several devices involved in control of this ratio. Typically, the amount of blower air is varied by a butterfly valve to satisfy the demands of a temperature-control device. A pulse or static pressure line is connected from the combustion air line, downstream of the butterfly valve, to a proportionator valve located in the gas line. The gas is then regulated by the ratio-control valve in proportion to the air flow, and the air-to-gas ratio remains constant throughout the firing range. The devices used to regulate the ratio fall into two broad categories: the diaphragm type, or proportionator, which uses the pulse line to keep air and gas at a specified ratio; and the mechanical-linkage type. Both are effective and common, but the diaphragm type is the more positive, because there are no linkages that can slip and require adjustment. Also, if air lines should become dirty, resulting in a lessening of air pressure, the gas pressure will follow, maintaining the correct ratio.

Ratio control alone is not sufficient to ensure safe start-up of the main burners. It is recommended that the burners be set to a low firing rate when the main burner is started. This may be done either automatically or manually. Once the main burners have been started, the furnace doors may be shut and the furnace brought up to temperature.

Temperature Control

Temperature-control devices fall into two categories: primary controls and process limiting devices. Safe operation, especially when furnace practices require long cycles and little operator attention, dictates that limits be placed on the process to ensure adequate alarm and perhaps to shut down the operation to prevent destruction of the

product, the furnace or the plant itself. Whether an analog device, a strip chart recorder, digital readout or a printout is used is a matter of operator preference and depends on the nature of the product.

The typical temperature sensor is either a thermocouple or a resistive temperature device (RTD). The thermocouple is most common. Several types of thermocouple junctions are available, with the choice depending on such factors as temperature range and furnace atmosphere. They are comparatively inexpensive and can be easily protected from atmospheres with protective "wells", which are immersion tubes that project into the furnace zone to be controlled. RTDs, although more accurate than thermocouples by factors ranging from 10 to 1 up to 50 to 1, are expensive and less rugged. For most purposes, thermocouples are satisfactory. Some firms are using heat-flow sensing to remotely ascertain interior temperatures and to provide an element of redundancy for protection of furnaces and their contents. Good temperature-sensing devices will detect failure of a thermocouple or RTD, cause the process firing rate to be reduced to its minimum rate, and perhaps provide an alarm.

Furnace temperature can be regulated by one of two very common procedures. Simple high and low firing rates are used when temperature can be allowed to vary within a fairly large range. More common, in heat treating, is the use of proportional control, wherein the temperature is held nearly constant through the use of a bridge circuit. This circuit balances the signal between the controller and the butterfly valve and holds the latter at the proper opening to maintain the desired temperature. The latter scheme, although more costly, is required for close control.

Waste-Heat Recovery

Recuperative devices used for conserving energy present special problems in safety and control instrumentation. Typically, these recuperators use the products of combustion for preheating of the combustion air. Shell and tube heat exchangers are normally used in this type of arrangement. Because preheated air becomes less dense, the air temperature must be sensed, and control of the gas-to-air ratio must be based on this temperature-density function. Experience with these devices has shown that such factors as poor design, poor gasketing, leaks in heat-exchange surfaces, overheating of burner blocks, and failure to allow for expansion and contraction have caused numerous operational problems that constitute safety hazards. Tracking of air-gas ratios can be affected by leakage of mechanical seals, and products of combustion can enter the combustion air. Thus, it is recommended that oxygen analyzers be used periodically to check the combustion air immediately ahead of the burner block. This analysis will reduce the likelihood of erroneous and perhaps hazardous conditions in the furnace, will give clues to potential design changes needed, and will give warning of part deterioration. Further, it is recommended that heat treating operators monitor the room atmosphere for carbon monoxide on a periodic basis. Although exhaust may be provided for products of combustion and sufficient air exchanges may be occurring to satisfy state regulations, there may be a temperature inversion that can cause leaking of products of combustion inside the building, resulting in a potentially dangerous situation.

Supervisory Gas-Cock System

A supervisory gas-cock system is used to ensure a safe "lightoff" procedure on a manually ignited, multiburner furnace that does not have flame-safety equipment and a programmed sequence of piloting the main burners.

The system consists of specially designed gas valves that have inlet and outlet passages for a checking pressure medium such as air or gas. Air or gas—usually air from a combustion blower—can pass only through the valve when the valve is fully closed. When the individual burner valves and the main gas line valve are closed, the air flow enters a pressure switch that closes and completes an electrical or pneumatic circuit. This allows the main gas valve, usually of the manual-reset type, to be opened. The burners are then individually manually ignited.

Supervisory gas-cock systems are used on radiant-tube furnaces and other furnaces where flame-safety systems are difficult to apply. Fewer of these systems are being used on new furnaces, because most burners now are adapted to flame safety and automatic ignition.

Electric Furnaces

Electric-furnace installations are made up of various electrical and mechanical components, many of which are water cooled and equipped with protective devices.

Furnace manufacturers generally issue instructions concerning safe practices, and these instructions should never be ignored. Potential hazards can be avoided by ensuring that operating personnel are trained thoroughly and that installations conform to safety practices and local codes.

Original equipment usually contains devices for preventing overloads and short circuits. In addition, ground detectors and surge detectors protect motor-generator units from faulty coil or transformer installations at heating stations and from breakdown of insulation in the generator windings.

Protective devices commonly used with induction-heating radio-frequency generators are as follows:

- Door interlocks
- Grounding devices to ground high-voltage circuit when furnace doors are open
- Warning lights
- Warning signs
- Circuit breaker for entire unit
- Overload relays
- Water-flow switches
- Water-temperature switches
- Time-delay relays (tube warmup)
- Grid overload relays
- Control-circuit overload relays
- Arc gaps on blocking and tank capacitors
- Surge protection
- Electronic crowbar.

Operators should become familiar with these safety devices and should inspect them periodically to ensure that they are in good working condition.

Electrical Power. Although motor generators account for the largest total power output of installed induction-heating equipment, vacuum-tube oscillators probably occur in the greatest numbers of units. Many small vacuum-tube oscillators are required to account for as many kilowatts as one 1250-kW, 3-kilocycle motor-generator set. Although many vacuum-tube oscillators for induction heating are made in small sizes, 25-kW and 50-kW outputs are also common ratings. Some have been constructed for special applications with ratings as high as 500 kW. Many

small composite (custom-built, or home-made) vacuum tube units are in use also.

Power Interlocks. Most systems produced by reputable manufacturers are designed to be completely "fail-safe". These systems also have interlocks that automatically shut down the power supply if a fault develops during operation. The system cannot be restarted until the fault is corrected. Interlocking systems also are used to increase production of induction-heating machines.

Induction-heating machines are expensive; therefore, steps are taken to keep them busy as much of the time as possible. If a hardening process, for example, requires 5 s of heating, followed by 5 s of quenching (before the part is moved from the inductor), and if another 5 s are required on the average for the operator to load and unload the part, then the generator itself is only in use one-third of the time. Production can be increased appreciably by having the same basic equipment supply two or more individual work coils (Fig. 1). This can be done by arranging the control circuit so that, if one of the work coils is demanding heat, none of the others can be started. If an operator pushes the start button at one station while another is heating, a relay withholds the actual start of heating until the first station has completed its high-frequency power demand cycle.

It is also customary in the case of interlocked multistation operation of motor-generator equipment to preset the alternator field current for the various stations. Each station has its own field-adjusting autotransformer or potentiometer, which is automatically switched into the circuit when that station has the power. This is feasible because, with only one station on at a time, it is not necessary to use the same voltage at each.

Fixtures. As introduction of automated systems increases in induction heating, the need for safety controls increases beyond the greater need for such devices as power interlocks.

For example, with highly automated induction-heating machines, a part completely foreign to the part to be treated may enter the automatic feed hopper or bin. If the fixture tries to feed this part into the coil, mechanical jamming and damage may result. If the part does get pushed into the coil itself, assorted problems may result, especial-

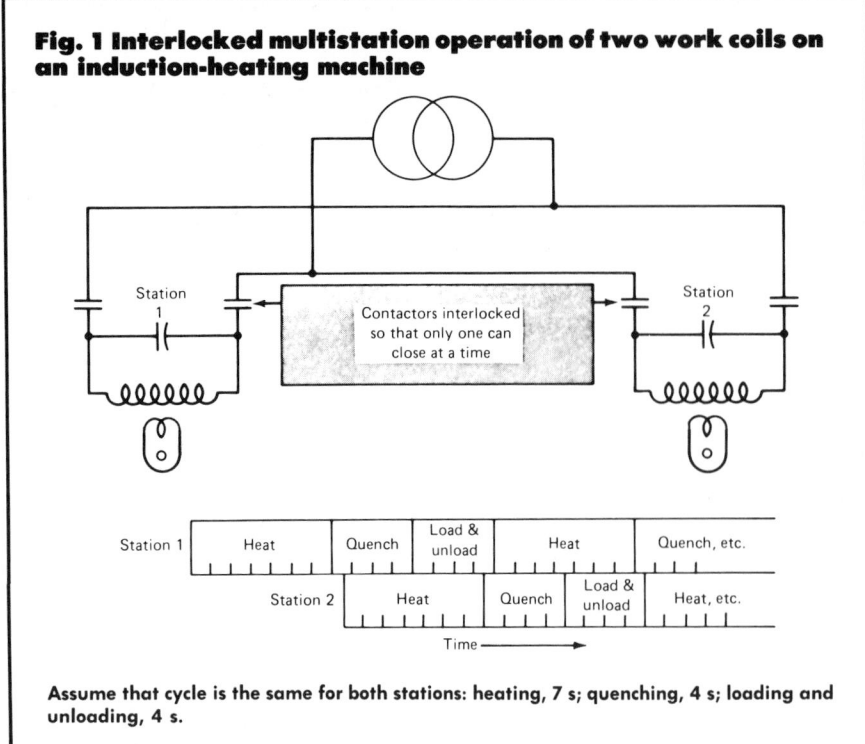

Fig. 1 Interlocked multistation operation of two work coils on an induction-heating machine

Station 1
Station 2

Contactors interlocked so that only one can close at a time

| Station 1 | Heat | Quench | Load & unload | Heat | Quench, etc. |

| Station 2 | | Heat | Quench | Load & unload | Heat, etc. |

Time ⟶

Assume that cycle is the same for both stations: heating, 7 s; quenching, 4 s; loading and unloading, 4 s.

ly if the unwanted part is larger than the inside diameter of the coil.

The greater the degree of automation, the greater the necessity for safety devices that further complicate the machine. Some automatic induction-heating machines incorporate templates through which the workpiece must pass before being fed to the more delicate parts of the mechanism. If the part touches the template, an electric circuit stops the machine until the error is manually corrected.

Spurious Radiation. Industrial heating equipment using radio-frequency energy is, in many instances, governed by rules of the Federal Communications Commission (FCC). FCC rules apply to induction-heating equipment operating at 10 kHz or above. Any operation whatsoever in frequency bands reserved for international distress equipment is strictly prohibited. One of these bands is from 490 to 510 kHz.

The best way for a user of induction-heating equipment to become familiar with the FCC rules is to acquire a copy of Part 18 of the rules and study them carefully.

Vacuum tube oscillators and some motor-generator equipment fall into

the FCC-designated categories. The rules state that operation must either be within certain narrow frequency bands (in which any amount of energy may be radiated) or be restricted in field strengths. All such equipment must be certified by a competent engineer, such certification being based upon actual measurements of field strength made around the equipment. In some circumstances, prototype models of industrial high-frequency heating equipment may be tested at the manufacturer's plant and a certificate issued to cover other equipment of the same design. Even though a specific piece of equipment may have been properly certified, and even though its spurious radiation may fall below the prescribed limits, FCC rules state that, if it interferes with communications equipment, further corrective action must be taken. The mere existence of a certificate, therefore, does not necessarily absolve the user of further responsibility.

In terms of output, induction-heating machines sometimes rival the largest communications transmitters. The frequencies, and harmonics thereof, used by many induction-heating oscillators fall within the range utilized by their

more delicate counterparts in radio and television services. If only a small portion of the power output of high-frequency heating machines were to be broadcast as unwanted (spurious) radiation, the results would be catastrophic.

Historically, dielectric heating machines have caused more interference than have induction-heating machines; they operate at higher frequencies and are more difficult to shield. However, induction-heating machines, especially vacuum tube oscillators, have also caused trouble. It is necessary to observe certain precautions in their design and operation to ensure that they do not create interference. Reputable manufacturers of induction-heating equipment take precautions to protect users of their equipment from this type of trouble. They house their equipment in heavy steel cabinets, and provide instructions which, if followed by the user, will ensure conformity with FCC rules. However, it must be emphasized that the final legal responsibility for a piece of equipment lies with the user.

Maintenance. Electrical heat treating equipment is expensive, and standby equipment generally is not maintained. Thus, preventive maintenance is critical, and ready availability of replacement parts is highly desirable.

Dust, dirt, moisture and high ambient temperatures are the primary causes of electrical equipment failures; these conditions are commonly present in industrial locations where induction-heating units are installed. In any maintenance program, warnings should be highly visible and clearly stated. The following is a typical warning:

"If the interlocks are disabled and the main circuit breaker is on with the door open, potentially lethal voltages are exposed. There is always 460 V ac present behind the control circuit breaker and on the line side of the main circuit breaker: care should be exercised at all times when the door is open. Power should be removed by opening the feed breaker or disconnect switch external to supply before working within cabinet. Solid-state circuit breaker board and isolator board are connected directly to 300 V ac. Turn off all breakers and allow one minute for capacitors to discharge before working on these boards."

Special Heat Treating Processes

Certain special heat treating processes using such systems as lasers, electron-beam heating, plasma carburizing and ion nitriding have their own unique safety requirements in addition to standard safeguards associated with high-temperature processing.

Safety of personnel is paramount, but safety of equipment often ensures personnel safety. Thus, proper care and use of equipment cannot be over stressed and frequently becomes almost synonymous with safety.

In this section, ion nitriding and plasma carburizing are used as examples of special processes and the safety precautions related to them. For all heat treating systems, however, special safety problems can be solved through sound training programs for operators and through effective and regular maintenance.

Ion Nitriding. In this system, also known as "glow-discharge" nitriding and as "ionitriding", parts are connected to the cathode for processing, and the retort is the anode. After the retort is evacuated of atmospheric gases, nitrogen and hydrogen are bled slowly into it. The glow discharge is produced when the parts are heated by electric current to approximately 500 °C (930 °F), although specified temperatures can be as high as 565 °C (1050 °F).

The retort becomes heated by radiation from the parts; additional heat is not required. The glow discharge ionizes the nitrogen, and the electrical potential accelerates the movement of the ionized nitrogen toward the parts.

Although ion nitriding is faster and produces a more ductile and fatigue-resistant case with less white layer, the extra handling and precautions it requires is an important factor in over-all cost.

Plasma Carburizing. The normal range of electrical power used for plasma-arc processing is 25 to 50 kW. Most systems are of "fail safe" design and are interlocked to shut down the power supply automatically if a fault develops during operation.

One of the most serious hazards associated with plasma-arc operation is radiation caused by electromagnetic high-electron excitation. Such radiation ranges from radio frequencies to the far ultraviolet, and it includes infrared and visible radiant-energy light rays.

The radiation produced by the plasma is capable of producing severe eye and skin burns. The plasma should never be observed with the naked eye.

Furnace Protection. The primary safety feature of a surface ion nitriding furnace is the arc-control system. Successful ion processing requires application of 300 to 1000 V dc to the workpiece. In these voltage ranges, the potential for an arc to form between the cathode and the anode is quite high during the initial part of the cycle. These arcs can be quite small; under certain circumstances, however, major arcs can occur that may be powerful enough to rupture vessel walls.

One method of controlling arcing and protecting equipment is to place a large resistance in series between the furnace vessel and the power supply. This prevents overloading of the power supply. An electrical device used to construct a type of "arc-shutdown" circuit is a Saturable Core Reactor (SCR), which operates with a low-resistance load at steady voltage.

When an arc begins to form there is an initial rapid increase in current prior to formation of the arc. Upon arc formation, the voltage drops drastically (dv/dt); the SCR senses the voltage change (dv/dt increase) and increases the resistance, thus protecting the power supply. Additionally, the SCR is normally connected to another electrical device, such as a bridge rectifier. Once the change in voltage signifying arc formation is detected, the rectifier damps the current supply. This damping is normally sufficient to shut down the arc by allowing redistribution of the energy on the portion of the part that was arcing.

This type of circuit reacts after the arc has formed, however, and the potential for ruined workpieces or holes in anodes or vessel walls remains. This is a particular problem if the unit requires operators to manually shut down the power supply after observing dead shorts, formed by misloaded parts that create short circuits.

Arc Suppression. Some equipment used in ion nitriding and plasma carburizing does not rely on an arc-shutdown circuit as the primary safety factor. Rather, true arc-suppression circuits are employed. Such a circuit senses the change in current just prior to the formation of an arc and shuts the power off completely, preventing the arc from forming. The power is then proportionally ramped back on, allowing redistribution of energy.

In one system, a counter in the controlling microprocessor tallies the number of times the power is turned off and on.

If the potential for arcing is too large, as determined by the logic preprogrammed into the microprocessor as part of the executive command package, the system shuts down, prints a fault message on both the cathode-ray tube and data logger, sounds an alarm and siren, and lights a warning light. All this occurs automatically without an arc forming and protects both the part and the equipment. The microprocessor detects a dead short by sensing and reporting the rapid frequency of shutdown/start-up cycles and prints a different "short" warning message, also with alarms, siren and lights. This sequencing is also totally automatic and requires no operator interface. The equipment has a backup system that operates on the dv/dt principle. Additionally, all leads through the vessel wall that could carry current to the power supply are triple-protected. As an example, the thermocouple has primary protection through an outer ceramic insulator. This is followed by a second powder ceramic insulator. The third protection is a high-voltage isolating amplifier between the thermocouple and the processor.

Thermocouples. The problems associated with passing a thermocouple, or any other lead, through a vessel wall in ion-processing equipment are as follows:

- The lead can become metallized, creating a pathway for catastrophic arcing. This problem becomes increasingly severe as a function of vessel use.
- The material of construction used for seals through the vessel wall is quite critical. The conditions of ion processing can affect the sealing materials, allowing increased leakage as a function of vessel use.
- Temperature uniformity throughout the entire workload in the vessel is quite critical. Proper design of thermocouple insulators that are calibrated to give true temperature readings, as well as proper design of fixturing, is necessary to achieve uniform temperatures.

Fixture Design. Design of the fixture is critical to the successful application of ion processing. Poorly designed fixtures can allow overheated or under-heated parts. If equipment without an arc-suppression circuit is used, parts can be ruined because of poor fixture design. The fixture often allows simple masking and, through proper design, can minimize or eliminate the hollow-cathode effect. This effect is signified by either (a) failure of the glow to uniformly penetrate the interior surface of a hole or cylinder, or (b) overheating and possible melting of the part, caused by overlapping glows.

Atmosphere Furnaces

Atmosphere furnaces must be considered in any discussion of furnace safety because of the potential explosion hazard produced by introduction of special flammable atmospheres.

Although many of these furnaces are supplied with "inert" gas purging and standby emergency purging, training of operators in manual "burn-out" procedures is extremely important in the event of failure of automatic controls. These emergency instructions may vary with equipment design, and thus the importance of consulting and understanding emergency procedures, as outlined by the furnace manufacturer's instructions, should not be minimized.

Protective Controls. Protective devices should be installed and interlocked and should include the following:

- A safety shutoff valve on the atmosphere supply line to the furnace
- An atmosphere gas-supply monitoring device that permits the operator to visually determine the adequacy of atmosphere gas flow at all times
- A sufficient number of temperature-monitoring devices to determine temperature in all zones of the furnace; these should be interlocked to prevent opening of the atmosphere-gas-supply safety shutoff valve until all zones are at or above 760 °C (1400 °F).
- An automatic safety shutoff valve for flame curtain burner supply gas; this should be interlocked to prevent opening of the valve when furnace temperature is below 760 °C (1400 °F).
- Audible and/or visual alarms to alert the furnace operator of abnormal conditions
- Manual door-opening facilities to permit operator control in the event of power failure.

Operator Training. The most essential safety consideration is the selection of alert and competent operators. Their knowledge and training are vital to continued safe operation. New operators should be instructed thoroughly and required to demonstrate an adequate understanding of the equipment and its operations.

Regular operators should receive scheduled retraining to maintain a high level of proficiency and effectiveness, and all operators should have ready access to operating instructions at all times. An outline of these instructions should be posted near the furnace.

Operating instructions generally are provided by the equipment manufacturer, and these instructions include schematic piping and wiring diagrams. All such instructions should include procedures for light-up, shutdown, emergencies and maintenance.

Operator training should include instructions in:

- Combustion of air-gas mixtures
- Explosion hazards
- Sources of ignition and ignition temperature
- Atmosphere gas analysis
- Handling of flammable atmosphere gases
- Handling of toxic atmosphere gases
- Functions of control and safety devices
- Purpose and basic principles of atmosphere-gas generators.

This listing is intended only to serve as a guideline; specific requirements are covered in the following standard issued for furnaces by the National Fire Protection Association: "NFPA 86C, Industrial Furnaces, Special Atmospheres, 1977".

Process Cooling

Heat treating of metals includes controlled cooling or quenching of the heated metal; metals are cooled from the specific treatment temperature in a variety of media which include air, oils, salts, water and synthetic fluids.

As a general rule, furnace equipment does not include instrumentation for control of the safety aspects of the quench media. Normal practice in layout of plant and facilities will provide for isolation of air-cooling areas, for the usual pedestrian protection at pits and for the necessary building protection should an uncontrolled conflagration

occur due to the quenching operation. However, certain equipment can be specified to ensure safe and controllable operation in specific cooling processes.

The greatest concern exists for fires associated with oil quenching. All of the ingredients for a dangerous fire—fuel, oxygen and a source of ignition—exist at the surface of an oil tank.

The most common type of fire occurs when movement of a hot workpiece is obstructed as it enters the quench oil. The result is sustained ignition and vaporization that continues as the liquid is locally heated above its flash point. Prompt immersion removes the source of local vaporization, and local flashing is extinguished by normal agitation of the oil.

A second type of fire occurs when the main body of oil is heated above the flash point because of malfunction of heating or cooling equipment, or when the quench load is greater than that for which the system was designed. When an ignition source is supplied, the resultant fire soon reaches full intensity and is very difficult to extinguish.

A third and less likely type of fire occurs because of material-handling accidents that involve spills on or near heated furnaces or cooling equipment.

Equipment is available for detection of fires and release of control media. Automatic water-spray systems are usually recommended in buildings of fire-resistant or noncombustible construction, and areas adjacent to quench-oil tanks can be protected with automatic sprinklers. Quench oil and water should not be mixed, however. High-value oils can be protected with automatic carbon dioxide or dry chemical systems. In general, these automatic systems are specified for centralized large-capacity quench-oil systems.

Special equipment is available for totally enclosed systems that operate in special atmosphere furnaces. Because water and quench oil do not mix, however, this incompatibility can be a source of trouble in these units. Water in oil rapidly turns to steam when locally heated beyond 100 °C (212 °F). This steam can cause violent boilover, increased pressure within the enclosed system, and forcible ejection of burning oil from small openings. Commercial safety equipment capable of detecting even small quantities of water in oil can be arranged to alert the furnace operator and to interrupt the quench process.

Quench-tank heating systems should be equipped with all of the safety devices normally used in conjunction with the particular heating method chosen. Overtemperature safety systems are essential. In addition, system interlocks between heating media actuators and both agitators and pumps will prevent local overheating of the bath.

Mechanical Equipment

Material handling cannot be separated from other safety considerations associated with heat treating operations. Many mechanical operations must be performed before, after, and sometimes during the actual temperature-induced transformation that usually occurs while the work is in the heat treating furnace. Doors must be opened and closed, conveyors started, rolls advanced and mechanical handling equipment activated. All of these mechanical operations constitute hazards of varying degrees of severity that must be evaluated. Some are serious enough to warrant the introduction of safety equipment to preclude serious injury of personnel and damage of expensive furnace equipment.

Furnace doors are often interlocked with other components of a facility through use of limit switches that prevent inappropriate opening or closing. For instance, in an atmosphere furnace, inner doors cannot be opened until proper vestibule ambient conditions have been restored after parts are removed. Interlocks on doors may include complex designs to prevent inadvertent opening during a power failure, on a hydrogen atmosphere furnace. There are simpler designs that prevent closing of a door on a simple normal-izing furnace before the extractor is removed. Some door interlocks are connected to pressure- or temperature-sensing devices. Such devices are used with vacuum furnaces in which inadvertent exposure of the molybdenum-graphite heating elements to air at high temperature would be disastrous. Moreover, without door interlocks, roller hearth or walking-beam furnaces could advance workpieces into an unopened discharge-zone door.

Moving parts of furnaces represent potential hazards that can be neutralized rather easily with simple time-delay relays connected to audible alarms, which in turn are connected to the start buttons of electric motors. For instance, the conveyors on some large continuous furnaces cannot be advanced until an alarm has been sounded and a timer has allowed sufficient time for workmen to stand clear.

Furnace Atmospheres and Carbon Control

Furnace Atmospheres

By the ASM Committee on Furnace Atmospheres*

CONTROL OF FURNACE ATMO-SPHERES has become increasingly critical to successful heat treating with more precise metallurgical specifications. The prevention of surface oxidation or scaling when metals are exposed to elevated temperatures remains an important task of the furnace atmosphere. In a more sophisticated view, the atmosphere within the furnace chamber is a full-fledged partner in achieving the chemical reactions that occur during heat treating.

Properly applied and controlled, furnace atmospheres provide the source of elements in some heat treating processes, surface cleansing of parts being treated in other processes, and a protective environment to guard against adverse effects of air when metals are exposed to elevated temperatures in still other processes.

Practical Flow Formula

Control of the flow of gases through a furnace chamber is important to ensure that sufficient atmosphere gas is being admitted to the furnace to seal the mechanical leaks against air infiltration or that a vestibule is being purged rapidly enough. In common practice, a chamber is considered essentially purged after five volume changes.

The simple way to adjust atmosphere flow to a furnace is through a meter that can be read directly. In some instances, however, atmosphere distribution among various inlets is required even though total atmosphere flow is metered. Flow rate is a function of the volume of a furnace chamber and time. A practical flow formula that is sufficiently accurate to deal with furnace atmosphere flows is as follows:

$$Q = 1651.25 \times A \times C \times \sqrt{\frac{h}{d}}$$

where Q is flow rate, in ft^3/h; A is area of orifice, in $in.^2$; C is coefficient of discharge, h is differential pressure; and d is specific gravity of the gas.

For a thin plate orifice in this equation, $C = 0.61$; for spuds with a tapered entrance, $C = 0.95$; and for spuds used in reverse (tapered outlet), $C = 0.9$.

Fundamentals of Gases

Gas molecules are somewhat widely separated, and they move about unceasingly in the space in which they are contained. Gases differ from liquids in two respects; gases are highly compressible and they fill any closed vessel in which they are placed. Gases resemble liquids in that both are capable of (a) flowing, (b) exerting pressure upon surfaces with which they are in contact, and (c) they exhibit flow velocity that can be measured through an orifice.

Although there is no clear distinction between gases and vapors, the term vapor usually is applied to a gas that is near its liquefying temperature. Steam and carbon dioxide usually are called vapors, because they are easy to liquefy. Air, hydrogen and nitrogen ordinarily are called gases.

Gas Pressure. The characteristic of compressibility of gases leads to a very simple relationship between the pressure of a gas and its volume. The relationship is known as Boyle's law: *The volume of a confined body of gas varies inversely as the absolute pressure, provided the temperature remains unchanged.*

The pressure of confined gases can be measured by U-shaped manometer tubes containing mercury or other liquids. The gas whose pressure is to be measured is connected to one side of the manometer, and the other leg of the tube remains open. Pressure exerted by the gas will force the liquid up the leg open to the atmosphere. The difference in height of the liquid in the two columns multiplied by the density of the liquid indicates how much the gas pressure exceeds atmospheric pressure. This pressure is known as gage pressure to distinguish it from absolute pressure, which includes the pressure of the atmosphere.

Diffusion. When two or more closed vessels, originally containing different

*Wilfred G. Shedd, *Chairman,* Co-owner and Secretary-Treasurer, Metallurgical Processing, Inc.; Douglas H. Clingner, Heat Treat Metallurgist, Fairfield Manufacturing Co., Inc.; Robert F. Gunow, President, Apt Consulting & Engineering; J. H. Kline, Heat Treat Dept. Metallurgist, Republic Steel Corp., retired; Max H. Priddy, Regional Manager, Selas Corp.; Ralph Puerta, Applied Research and Development, Air Products and Chemicals, Inc.

gases, are joined so that each gas has access to all containers, and assuming no chemical reaction takes place, molecular motion causes each gas to penetrate the entire volume of all containers. By this diffusion, the mixture eventually becomes homogeneous. Each gas expands into the total available volume as though the other gas were not present. According to Boyle's law, the absolute pressure of each gas is reduced to a lower value called its partial pressure. The pressure of the resulting mixture of gases will be equal to the sum of the partial pressures of the constituent gases. This phenomena is expressed as Dalton's law: *A mixture of several gases which do not react chemically exerts a pressure equal to the sum of the pressures which the several gases would exert separately if each were allowed to occupy the entire space alone at the given temperature.*

Avogadro's law states: *At the same temperature and pressure, equal volumes of different gases contain equal numbers of molecules.* When this law is applied to a particular quantity of a gas, the quantity is known as a mol, or gram-molecule, or gram-molecular weight. The mol of a substance is a mass in grams numerically equal to the sum of the atomic weights of the atoms in a molecule of that substance. The number of molecules in a mol of gas is known as Avogadro's Number. A mol of any gas contains 6.02×10^{23} molecules and at 0 °C (32 °F) and standard atmospheric pressure of 760 mm mercury, or sea level pressure, occupies 22.4 l (0.79 ft³) of space.

Density. The density of a gas is the amount of mass contained in a unit volume, and the density is influenced by pressure and temperature. The density and pressure increase in the same proportion, and gases expand when heated and contract when cooled. The density of air has been determined with great care. One litre (0.038 ft³) of air at standard temperature and pressure weighs 1.293 g (0.05 oz), or an air density of 1.3 kg/m³ (0.081 lb/ft³). The specific gravity of a gas is normally expressed with reference to air as standard (Table 1).

Viscosity. Molecular friction, or viscosity, is present in gases as well as in liquids, although the wide spacing of the molecules results in appreciably less viscosity in gases. This friction retards the motion of gases through such channels as tubes and ducts. The coefficients of viscosity for gases and liquids are expressed in poise or centi-

Table 1 Properties of common gases and vapors

Gas	Chemical symbol	Approximate molecular weight	Density(a) kg/m³	Density(a) lb/ft³	Specific gravity(b)
Air	...	28.97(c)	1.293	0.0807	1.000
Ammonia	NH_3	17.03	0.760	0.0474	0.588
Argon	A	39.95	1.784	0.0111	1.380
Carbon dioxide	CO_2	44.02	1.965	0.1228	1.520
Carbon monoxide	CO	28.01	1.250	0.0780	0.967
Helium	He	4.00	0.179	0.0112	0.138
Hydrogen	H_2	2.02	0.090	0.0056	0.070
Methane	CH_4	16.04	0.716	0.0447	0.552
Nitrogen	N_2	28.01	1.250	0.0780	0.968
Oxygen	O_2	32.00	1.429	0.0892	1.105
Propane	C_3H_8	44.09	1.968	0.1229	1.522
Sulfur dioxide	SO_2	64.06	2.860	0.1785	2.212

(a) Standard temperature and pressure: 0 °C (32 °F) and 760 mm mercury. (b) Relative density compared to air. (c) Because air is a mixture, it does not have a true molecular weight. This is the average molecular weight of its constituents.

poise. A rise in temperature causes a rise in the viscosity of gases and a lowering of liquid viscosity.

Temperature Effect. Charles' law states: *The volume of a fixed mass of gas and its pressure vary directly with the absolute temperature.*

Using the effect of temperature, a general gas law can be written in the form of a mathematical equation, as follows:

$$\frac{P_1 V_1}{T_1} = \frac{P_2 V_2}{T_2}$$

where P is absolute pressure, V is volume of a fixed mass, T is absolute temperature, and subscripts 1 and 2 are two different sets of conditions of pressure, volume and temperature.

When a furnace atmosphere is required to contribute an element or elements during heat treating, these fundamental gas laws become important. Many forms of control are used to develop carbon potential, or the potential of an element, within an atmosphere. Differing amounts of an element can be utilized in the heat treating chemical reaction with a constant potential reflected. A quantitative measure of an element is developed by an understanding of the chemical formula involved in the reaction and the partial pressures of the formula gases used to produce the element.

Principal Gases and Vapors

Air is an important gas in furnace atmospheres, because it comprises the atmosphere in a furnace in which no protective atmosphere is used, and be-

cause it is a major constituent in many prepared atmospheres. The chemical makeup of air is approximately 79% nitrogen and 21% oxygen with trace amounts of carbon dioxide. As an atmosphere, air behaves like an oxygen atmosphere because oxygen is the most reactive constituent in air.

Oxygen. Oxygen reacts with most metals to form oxides. In addition, oxygen reacts with carbon that is dissolved in steel to lower the carbon content of its surface.

Nitrogen. Molecular nitrogen is passive to ferrite and is entirely satisfactory for use as an atmosphere in the annealing of low-carbon steels; however, it must be completely dry to be used as a protective atmosphere for high-carbon steels, because small amounts of water vapor in the nitrogen will cause decarburization. Atomic nitrogen (created by the temperatures of normal heat treating) is not a protective atmosphere, because it combines with iron to form finely divided nitrides that impart hardness to the surface.

Carbon Dioxide and Carbon Monoxide. These two gases are important in atmospheres used in processing steel. At austenitizing temperatures, carbon dioxide reacts with surface carbon in a steel surface to produce carbon monoxide:

$$(C) + CO_2 \rightleftharpoons 2\ CO$$

in which (C) represents carbon dissolved in austenite. This reaction continues until there is no carbon dioxide available or until the steel surface is completely free of carbon—at which point, if there is a continuing supply of carbon dioxide, iron and ferrous oxide will be oxidized by the following reactions:

$$Fe + CO_2 \rightleftharpoons FeO + CO$$

$$3\,FeO + CO_2 \rightleftharpoons Fe_3O_4 + CO$$

Ferrous oxide (FeO) is the stable oxide formed above 555 °C (1030 °F), whereas the magnetic oxide (Fe$_3$O$_4$) is formed below 555 °C (1030 °F) as shown in Fig. 1.

The above reactions will proceed until equilibrium is established. These reactions progress at a rate depending on time, temperature and pressure of the system. The equilibrium conditions for carbon steels of various carbon concentrations heated in a CO-CO$_2$ atmosphere are shown in Fig. 2.

Hydrogen. Hydrogen reduces iron oxide to iron. Under certain conditions, hydrogen can act to decarburize steel. The decarburizing effect of hydrogen on steel depends on furnace temperature, moisture content (of the gas and furnace), time at temperature, and carbon content of the steel. The decarburizing effect of hydrogen at 705 °C (1300 °F) or below is negligible, but it increases markedly above this temperature. Water vapor increases the decarburizing effect because it dissociates, thus providing a supply of nascent hydrogen and oxygen. Hydrogen reacts with carbon in steel to form methane:

$$(C) + 4H \rightleftharpoons CH_4$$

Oxygen reacts with carbon in steel to form carbon monoxide as follows:

$$(C) + O \rightleftharpoons CO$$

Even a low-dew-point hydrogen atmosphere has some decarburizing effect, particularly on high carbon steels, because of the ability of the gas—even when it is not in the nascent condition—to react with carbon:

$$(C) + 2H_2 \rightleftharpoons CH_4$$

Obviously, the decarburizing potential of hydrogen in either form is markedly influenced by the carbon content of the steel and may be expected to increase directly with the increase of carbon content.

Water Vapor. Water vapor is oxidizing to iron:

$$Fe + H_2O \rightleftharpoons FeO + H_2$$

and combines with carbon in steel to form carbon monoxide and hydrogen:

$$(C) + H_2O \rightleftharpoons CO + H_2$$

It is reactive to steel surfaces at very low temperatures and low partial pressures. It is a principal cause of "blueing" during cooling cycles.

The effect of water vapor on the oxidation of iron at various temperatures is indicated by the equilibrium curves shown in Fig. 3.

Hydrocarbons. The most common hydrocarbon gases added to or found in furnace atmospheres are methane (CH$_4$), ethane (C$_2$H$_6$), propane (C$_3$H$_8$) and butane (C$_4$H$_{10}$). These gases impart a carburizing tendency to a furnace atmosphere. The chemical activity in reacting with the surface of hot steel depends on their thermal decomposition and their tendency to form nascent carbon at the steel surface, and on the temperature of the furnace chamber and work load. Thermal decomposition results in formation of soot in amounts proportional to the number of carbon atoms in the hydrocarbon; therefore, butane and propane are more likely to cause soot in the furnace chamber than are ethane and methane.

Inert Gases. The inert gases of argon and helium are frequently used during the thermal processing of reactive metals and alloys of these metals. Argon, approximately one-half the cost of helium, is generally favored. Because helium sources are limited, the price of helium is more dependent on end-use location than is argon.

Air contains approximately 0.93% argon by volume. Argon is recovered by liquefying air followed by the fractionation of liquid air. Helium is recovered by similar cryogenic methods from natural gas deposits found in western United States. A limited number of deposits in the United States and Mexico contain 1 to 8% helium. Because of the low density of helium (0.179 × 10^{-3} g/cm^3 at 20 °C, or 70 °F), much of the helium released into the atmosphere is permanently lost beyond the earth's atmosphere. As a result of this loss, the future supply of helium is threatened.

Protective atmospheres of inert gases are particularly useful during thermal processing of metals and alloys that

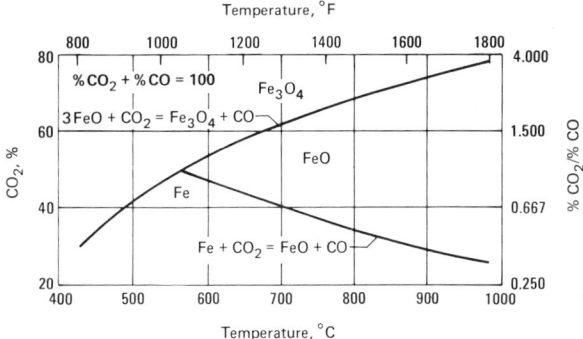

Fig. 1 Equilibrium curves for the formation of scale (FeO and Fe$_3$O$_4$) when heating iron in a CO-CO$_2$ atmosphere at different temperatures

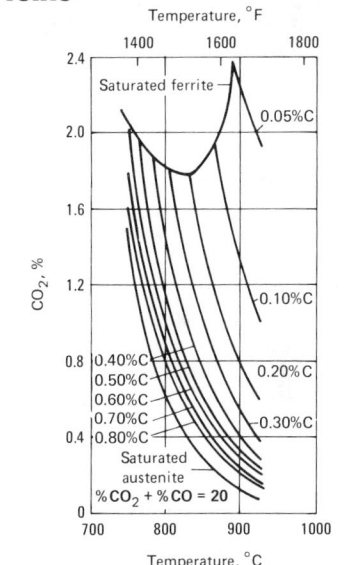

Fig. 2 Temperature and percentage of carbon dioxide for equilibrium conditions with carbon steels of various carbon contents

Fig. 3 Equilibrium curves for the formation of scale (FeO and Fe₃O₄) when heating iron in a H₂-H₂O atmosphere at different temperatures

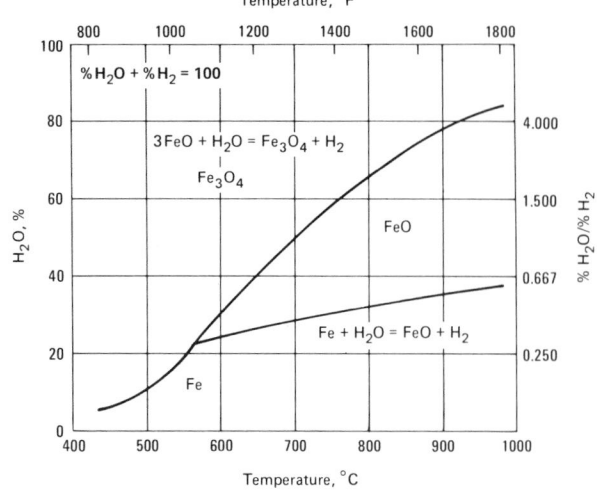

cannot tolerate the usual constituents of other protective atmospheres. Two applications of inert gases are as protective atmospheres during (a) the bright hardening of stainless steels, and (b) the heat treatment of titanium alloys. Oxygen and water vapor must be avoided during the bright hardening of stainless steels. Oxygen content of less than 0.01% and dew points below -50 °C (-60 °F) are essential. Titanium alloy heat treatments require atmospheres free of hydrogen, oxygen and carbon-bearing gases.

Careful consideration should be given to the residual entrapment of inert gases and nitrogen. When human entry into vessels and chambers is a function of operation or maintenance, special precautions are recommended. Although these gases are not toxic, asphyxiation can occur in entrapped areas. Being odorless and colorless, inert gases and nitrogen are not readily recognized as dangerous.

Helium and nitrogen can be trapped in overhead chambers of bottom-loaded furnaces, and argon and nitrogen can be trapped in vertical top-loaded furnaces. Agitation and flushing with air is recommended before entry. Other appropriate safety procedures may also be beneficial.

Furnace Atmosphere Gas Reactions

Flue gases in a direct-fired furnace are the effluent products created by combustion of hydrocarbon fuel. The composition of these gases inside a furnace contains a mixture of some, or all, of the following principal components: carbon dioxide, carbon monoxide, hydrogen, oxygen, nitrogen and water vapor.

When burners are adjusted to operate with an excess of combustion air, the products of incompletely burned fuel, carbon monoxide and hydrogen, are minimized, but measurable amounts of residual oxygen remain. Conversely, burners operating with a combustion-air deficiency consume all of the available oxygen before complete burning of the fuel. In this case, resultant oxygen is minimized, and measurable quantities of unburned carbon monoxide and hydrogen remain. Under all conditions, substantial quantities of water vapor are produced by the combustion.

Other major factors that may materially affect the over-all flue-gas furnace atmosphere are burner efficiency, tightness of the furnace and size of the door openings. The openings contribute to the rate of possible air infiltration and the amount of oxygen present. Moreover, furnaces that are equipped with multiple burners may be operating with various air-fuel ratios that can produce a mixed atmosphere of unpredictable composition.

When flue gases contain excess oxygen along with water vapor and carbon dioxide, the oxidizing potential promotes rapid formation of loose scale on

steel. Air-deficient burner operation produces much less oxygen and greater amounts of carbon monoxide and hydrogen, which are reducing constituents that can cause steel surface decarburization. Carbon dioxide and water vapor, in this instance, promote formation of a tight oxide that is not easily removed. In both instances, the amount of scale formation is a function of furnace temperature and the amount of time material is held at temperature.

Although a direct-fired furnace cannot be expected to provide a completely neutral atmosphere, specially designed direct-fired furnaces with closely controlled burner systems and specific time/temperature process cycles are capable of cost-efficient heat treating and heat processing for a variety of applications with fully acceptable results.

Carbon Dioxide Plus Hydrogen. Hydrogen will react with either carbon dioxide or oxygen to form water vapor. Water vapor has a high oxidizing potential or decarburizing potential for steel and must be controlled in a furnace atmosphere.

Water Gas Reaction. The reactions listed below by which steel or iron is oxidized at elevated temperatures are irreversible and cannot be controlled:

$$2Fe + O_2 \rightarrow 2FeO$$
$$4Fe + 3O_2 \rightarrow 2Fe_2O_3$$
$$3Fe + 2O_2 \rightarrow Fe_3O_4$$

Other oxidizing gas-metal reactions, however, *are* reversible and *can* be controlled, and hence may be employed to advantage. In the reactions:

$$Fe + H_2O \rightleftharpoons FeO + H_2$$
$$Fe + CO_2 \rightleftharpoons FeO + CO$$

water vapor and carbon dioxide are oxidizing gases, and hydrogen and carbon monoxide are reducing gases. Ultimately, the quantity of reducing gas or oxidizing gas formed may become great enough for one to cancel the effects of the other. By proper control of these reactions, a neutral, reducing, or oxidizing effect may be produced.

The opposing reactions may be controlled according to the water gas reaction, which is as follows:

$$CO + H_2O \rightleftharpoons CO_2 + H_2$$

The gases that enter into the water gas reaction react with the surface of steel to cause oxidation or reduction, depending on the equilibrium condition corresponding to the temperature and composition of the system.

Table 2 Variation of equilibrium constants, composition and dew point of an H₂-H₂O-CO-CO₂ system
Assuming that the hydrogen content remains constant at 40% and CO + CO₂ content remains constant at 20%.

Temperature		K_1	CO₂, %	K_2	H₂O, %	Dew point, °C	Dew point, °F	K_3
°C	°F							
650	1200	3.770	4.5	0.51	4.5	+32	+90	1.922
705	1300	0.942	2.6	0.66	3.4	+24	+75	0.695
760	1400	0.348	1.2	0.83	2.0	+18	+65	0.363
815	1500	0.125	0.5	1.02	1.02	+ 7	+45	0.127
870	1600	0.050	0.2	1.22	0.49	− 3	+27	0.061
925	1700	0.022	0.1	1.44	0.25	−12	+10	0.003

At 830 °C (1525 °F), the oxidizing potentials of carbon dioxide and water vapor are equal, and the reducing potentials of carbon monoxide and hydrogen are equal. At this temperature, therefore, the equilibrium constant of the water gas reaction has a value of unity. Above 830 °C, carbon dioxide is a stronger oxidizing agent than water vapor, and hydrogen is a stronger reducing agent than carbon monoxide. Below 830 °C, the reverse is true. Consider the reactions:

$$C + CO_2 \rightarrow 2CO$$
$$CO + H_2O \rightleftharpoons CO_2 + H_2$$
$$C + H_2O \rightleftharpoons CO + H_2$$

and their equilibrium constants, which are, respectively:

$$K_1 = \frac{(CO_2)}{(CO)^2}$$

$$K_2 = \frac{(CO)\,(H_2O)}{(CO_2)\,(H_2)}$$

$$K_3 = \frac{(H_2O)}{(CO)\,(H_2)}$$

Calculations for the data in Table 2 were developed, assuming that the hydrogen content of the system remains constant at 40% and the CO + CO₂ content remains constant at 20%. Figure 4 shows the equilibrium conditions for steels with carbon concentrations from 0.10 to 1.20%.

Ammonia Vapor. Ammonia dissociates in the endothermic reaction:

$$2NH_3 \rightarrow N_2 + 3H_2$$

with the dissociated ammonia containing 75% hydrogen and 25% nitrogen. Dissociation occurs when ammonia vapor is heated and passed over the proper catalyst. Dissociated ammonia then is cooled and often passed through a purifying molecular-sieve absorption system to remove undissociated ammonia and water vapor.

Ammonia vapor also may be used as an additive to suitable carbonbearing carrier atmospheres for carbonitriding or it may be used directly for nitriding processes. In these instances, partial ammonia dissociation occurs inside the furnace, allowing nitrogen to react with heated steel surfaces to form hard nitrides.

Lithium Vapor. Lithium vapor reacts with water vapor in the furnace atmosphere to form lithium oxide and hydrogen:

$$2\,Li + H_2O \rightarrow Li_2O + H_2$$

Lithium vapor also combines with any free oxygen present in the furnace atmosphere to form lithium oxide:

$$4\,Li + O_2 \rightarrow 2\,Li_2O$$

The lithium oxide formed by these reactions oxidizes some of the carbon monoxide present in the atmosphere causing the release of a certain amount of lithium:

$$2\,Li_2O + CO \rightarrow Li_2CO_3 + 2\,Li$$

The lithium liberated can then pick up more oxygen.

The practical application of this vapor as a furnace atmosphere has been in forging furnaces as a protection against scaling of the surface of steel heated to forging temperature.

Sulfurous Gases. Sulfurous gases in a furnace atmosphere are basically deleterious and are to be avoided. These gases result from the presence of sulfur compounds in industrial fuels, furnace refractories and cutting oils on work being processed.

Sulfur compounds occur as hydrogen sulfide (H_2S), sulfur dioxide (SO_2) or sulfur trioxide (SO_3), mercaptans, thiophene (C_4H_4S), and metallic sulfates. When sulfur is present in reducing atmospheres, it is generally found as

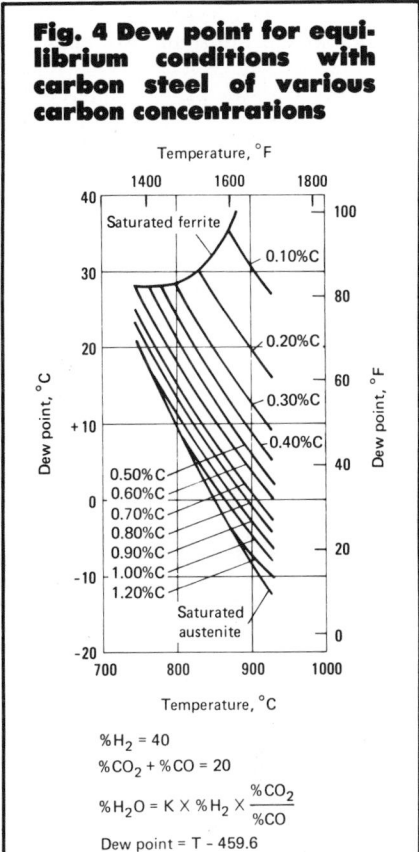

Fig. 4 Dew point for equilibrium conditions with carbon steel of various carbon concentrations

%H₂ = 40
%CO₂ + %CO = 20
%H₂O = K × %H₂ × $\frac{\%CO_2}{\%CO}$
Dew point = T − 459.6
%H₂O = $10^{8.0615 - \frac{407}{T}}$

hydrogen sulfide from the following reaction:

$$SO_2 + 3H_2 \rightleftharpoons H_2S + 2H_2O$$

When sulfur is present in an oxidizing atmosphere, the following reactions are involved:

$$C_4H_4S + 6O_2 \rightarrow 4CO_2 + 2H_2O + SO_2$$

$$2SO_2 + 3Fe + 2Ni \rightarrow$$
$$Fe_2O_3 + NiS + NiO + FeS$$

When sulfur is present in atmospheres in which high-nickel steels are heated, both nickel sulfide and nickel oxide are formed, and an "alligator" surface results. In addition to damaging parts being treated, this effect accelerates the failure of high-nickel, high-chromium heat resistant alloy furnace parts, trays and fixtures. In general, the presence of sulfur in a furnace atmosphere accelerates the rate of scaling, and this rate increases as temperature increases.

Classifications of Prepared Atmospheres

Most prepared atmospheres are commonly referred to in the field by their generic name or, in some instances, by trade names. The American Gas Association has classified the commercially important prepared atmospheres into six groups on the basis of method of preparation or the original constituents employed. These groups are designated and defined as follows:

- *Class 100—Exothermic Base.* Formed by partial or complete combustion of a gas-air mixture; water vapor may be removed to produce a desired dew point
- *Class 200—Prepared Nitrogen Base.* An exothermic base with carbon dioxide and water vapor removed
- *Class 300—Endothermic Base.* Formed by partial reaction of a mixture of fuel gas and air in an externally heated catalyst-filled chamber
- *Class 400—Charcoal Base.* Formed by passing air through a bed of incandescent charcoal
- *Class 500—Exothermic-Endothermic Base.* Formed by complete combustion of a mixture of fuel gas and air, removing water vapor, and re-forming the carbon dioxide to carbon monoxide by means of reaction with fuel gas in an externally heated catalyst-filled chamber
- *Class 600—Ammonia Base.* This can consist of raw ammonia, dissociated ammonia or partially or completely combusted dissociated ammonia with dew point regulated.

These broad areas of classification are subclassified and numerically designated to indicate variations in the method by which they are prepared. This subclassification is differentiated by replacing the two zeros of the six basic designators by the following:

- **01** indicates the use of a lean air-to-gas mixture.
- **02** indicates the use of a rich air-to-gas mixture.
- **03** and **04** indicate that preparation of the gas was completed within the furnace itself without the use of a separate machine or generator.
- **05** and **06** indicate that the original base gas was subsequently passed through incandescent charcoal before admission to the work chamber.
- **07** and **08** indicate the addition of a raw hydrocarbon fuel gas to the base gas before admission to the work chamber.
- **09** and **10** indicate the addition of a raw hydrocarbon fuel gas and raw dry anhydrous ammonia to the base gas before admission to the work chamber.
- **11** and **12** indicate the addition of a combusted mixture of chlorine, hydrocarbon fuel gas and air to the base gas before admission to the work chamber.
- **13** and **14** indicate that the base gas has had all sulfur or all sulfur and odors removed before admission to the work chamber.
- **15, 16, 17** and **18** indicate the addition of lithium vapor to the base gas before admission to the work chamber.
- **19** and **20** indicate that preparation of the gas was completed within the furnace itself with the addition of lithium vapor.
- **21** and **22** indicate some additional special treatment was given to the base gas before admission to the work chamber.
- **23** and **24** indicate the addition of steam and air in conjunction with a

catalyst within the generator to convert CO to CO_2, which is then removed.

- **25** and **26** indicate the addition of steam in conjunction with a catalyst within the generator to convert CH_4 to H_2 and CO_2, which is then removed.

This classification system provides for a large number of possibilities. In practice, only a few of the possible atmosphere classifications are industrially important. Table 3 lists significant furnace atmospheres and typical applications.

Safety Precautions. Furnace atmospheres constitute one of the major safety hazards involved in heat treating. Generally, these hazards fall into three groups:

- *Fire.* When an atmosphere contains more than 4% of combustible gases, it is classified as flammable. Included in this percentage is a practical safety margin that should never be ignored. *The combustible gases H_2, CO, CH_4 and other hydrocarbon fuel gases should never be admitted to a furnace chamber at temperatures below 760 °C (1400 °F).*
- *Explosion.* At some point, mixtures of air and combustible gas will explode when ignited. When a furnace chamber is properly gassed with the chamber temperature at or above 760 °C (1400 °F), it is likely that combustible gases will burn before creating an explosion hazard. An adjacent cold chamber or vestibule can then be flared as the atmosphere flows from the furnace to the vestibule until it is free of oxygen from the air. The vestibule can then be closed. The positive flow of atmosphere through

Table 3 Classification and application of principal furnace atmospheres

Class	Description	Common application	N₂	Nominal composition, vol % CO	CO₂	H₂	CH₄
101	Lean exothermic	Oxide coating of steel	86.8	1.5	10.5	1.2	...
102	Rich exothermic	Bright annealing; copper brazing; sintering	71.5	10.5	5.0	12.5	0.5
201	Lean prepared nitrogen	Neutral heating	97.1	1.7	...	1.2	...
202	Rich prepared nitrogen	Annealing, brazing stainless steel	75.3	11.0	...	13.2	0.5
301	Lean endothermic	Clean hardening	45.1	19.6	0.4	34.6	0.3
302	Rich endothermic	Gas carburizing	39.8	20.7	...	38.7	0.8
402	Charcoal	Carburizing	64.1	34.7	...	1.2	...
501	Lean exothermic-endothermic	Clean hardening	63.0	17.0	...	20.0	...
502	Rich exothermic-endothermic	Gas carburizing	60.0	19.0	...	21.0	...
601	Dissociated ammonia	Brazing, sintering	25.0	...	...	75.0	...
621	Lean combusted ammonia	Neutral heating	99.0	...	...	1.0	...
622	Rich combusted ammonia	Sintering stainless powders	80.0	...	...	20.0	...

the furnace and adjoining cold chamber or vestibule can then be burned. An ignited effluent from an atmosphere furnace is an immediate visual sign that a safe condition prevails.

● *Toxicity. Many of the gases making up furnace atmospheres are toxic.* Burning them at the furnace exits reduces their chemistry to the products of combustion. These products should then be vented outside the building to avoid dilution of the available oxygen supply within the building. *Ventilation of the building containing atmosphere generators and atmosphere heat treating furnaces is a major safety consideration.*

Exothermic-Base Atmospheres

Exothermic gases (class 100) have been used extensively for many years as lower-cost prepared furnace atmospheres. Exothermic atmospheres are divided into two basic classes, rich and lean. Rich exothermic atmospheres (class 102) have moderate reducing capabilities of 10 to 21% combined carbon monoxide and hydrogen, and lean exothermic atmospheres (class 101), usually with 1 to 4% combined carbon monoxide and hydrogen, have minimal reducing qualities.

Rich Exothermic Atmospheres

The principal uses of rich exothermic furnace atmospheres include clean heat treating of certain ferrous and nonferrous applications. Among these are annealing and tempering of steel, brazing of copper and silver, and sintering of powdered metals.

Reducing properties of rich exothermic atmospheres may be varied to make them suitable for specific processes. Figure 5 indicates the usual operating range of the gas generator and reflects changes (by dry volumetric measurement) in the following constituents of the product gas at any particular setting: carbon dioxide, carbon monoxide, hydrogen and unburned methane. The remainder of the mixture is nitrogen. Because these atmospheres have a carbon potential below 0.10%, steel heat treating is generally limited to processes for low carbon steels to minimize decarburizing, or processes where decarburization is unimportant. Water vapor is present in substantial quantities and may be removed partially by initial cooling and

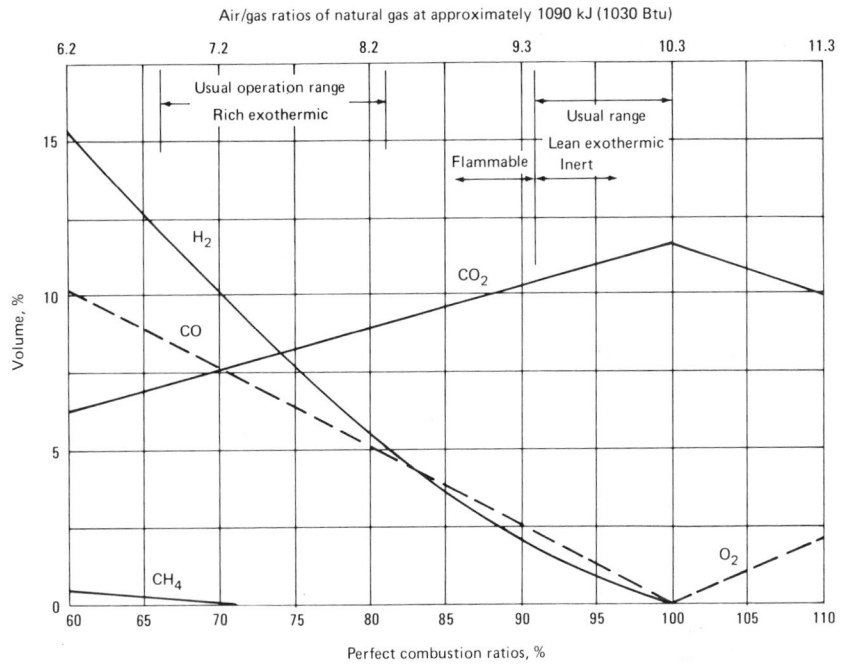

Fig. 5 Exothermic atmosphere composition versus air-fuel ratio (natural gas)

Air/gas ratios of natural gas at approximately 1090 kJ (1030 Btu)

refrigerant drying to an equivalent 5 °C (40 °F) dew point. This procedure may be followed by further dehydration with an adsorbent desiccant dryer for final dew points of −40 to −50 °C (−40 to −60 °F), as applications require.

Gas Production. Rich exothermic gas is produced by combustion of a hydrocarbon fuel such as natural gas or propane with the air-fuel ratio closely controlled. This air-gas mixture is burned in a confined combustion space to maintain a reaction temperature of at least 980 °C (1800 °F) for sufficient time to permit the combustion reaction to reach equilibrium. Heat is obtained directly from combustion. The resultant gas is then cooled to remove part of the water vapor formed by burning and to permit convenient transportation and metering. In this process, the simplified theoretical reaction of methane with air is:

$$CH_4 + 1.25O_2 + 4.75N_2 \rightarrow 0.375CO_2 + 0.625CO + 0.88H_2 + 4.75N_2 + 1.12H_2O + heat$$

where 1 volume of fuel and 6 volumes of air yield 6.63 volumes of product gas mixture, with water vapor removed. In practice, exothermic gas generators are seldom operated with an air-gas ratio lower than about 6.6 to 1, to

prevent formation of soot as a result of incomplete reaction. Trace percentages of unreacted methane also exist in the product gas.

The basic rich exothermic gas generator (Fig. 6) has a refractory-lined combustion chamber, which in some designs may be partially filled with a catalyst. The chamber incorporates a burner with a combustion-control system especially designed to provide a constant supply of air and fuel gas with a closely maintained preselected ratio. The combustion chamber is followed by a water-cooled heat exchanger for primary gas cooling and partial water vapor condensation. At this point, product atmosphere is saturated with moisture at approximately 10 °C (15 °F) above the temperature of the cooling water.

Operation is uncomplicated and continuous, incorporating input air and fuel-gas flow meters with manual or automatic control of process flow and ratio. Safety devices include sequenced burner ignition, automatic flame monitoring, pressure switches and manual-reset fuel-safety valves. Periodic operational monitoring consists of observing pressures and flow rates and taking occasional product gas samples for analysis of desired composition. Available continuous gas analyzers, either indicator-reading or recording, usually are calibrated to indicate the content of combustibles. Some analyzers may be adapted for automatic ratio control.

Operating Economics. The following are requirements needed to produce 28 m³ (1000 ft³) of rich exothermic atmosphere:

Natural gas . . . . 4.4 m³ (155 ft³)
Power . . . . 1.4 MJ (0.4 kW·h)
Cooling water . . . . 1135 l (300 gal)
Power for refrigerant
dryer . . . . 1.8 MJ (0.5 kW·h)
Power for desiccant
dryer . . . . 5.4 MJ (1.5 kW·h)

Safety Considerations. *Rich exothermic atmospheres contain sufficient carbon monoxide and hydrogen to be considered flammable gases when mixed in proper proportions with air. In handling or use as a furnace atmosphere, they should be treated with the same precautions exercised with any combustible gas.* Established furnace operating practices should be observed carefully, particularly with respect to proper purging procedures and prevention of air infiltration. Most generators are equipped with electrically programmed ignition sequencing and are monitored with additional safety devices. These devices should be periodically tested to ensure fail-safe operation. Because these atmospheres contain measurable amounts of carbon monoxide, extreme care should be taken to prevent leaking into the surrounding atmosphere. Proper venting, burn-off and ventilation facilities should be employed. Furnace safety is discussed elsewhere in this volume.

Lean Exothermic Atmospheres

Lean exothermic atmospheres generally have limited use in most heat treating applications, particularly for ferrous materials, except when these atmospheres are used as intentional surface oxidizing agents or for specialized low-temperature operations. Lean atmospheres are used to some extent for processes such as copper annealing and are employed more widely when the primary processing aim is to exclude oxygen or to provide purging and blanket gas.

Figure 5 indicates the usual operating range of the gas generator and reflects changes (by dry volumetric measurements) in the product gas at any particular setting with respect to carbon dioxide, carbon monoxide and

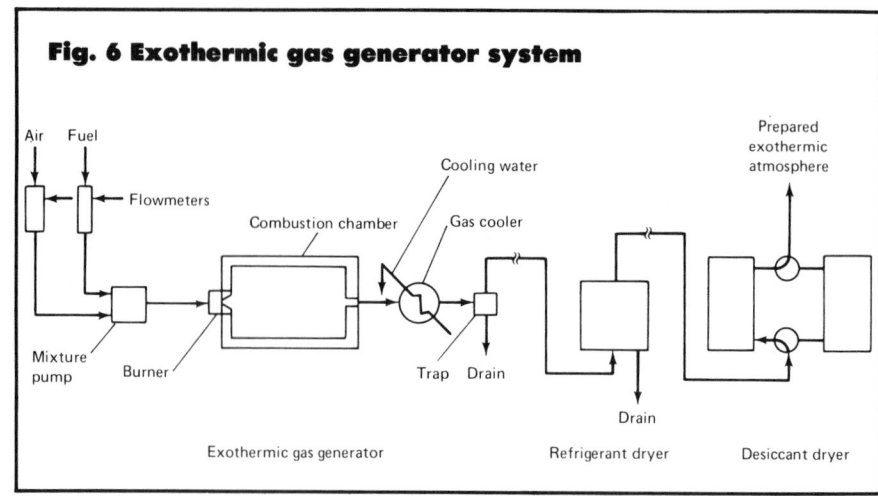

Fig. 6 Exothermic gas generator system

hydrogen. The balance of the mixture is nitrogen. Reducing capabilities are limited. Water vapor that is present in substantial quantities may be removed partially by initial cooling and refrigerant drying to an equivalent 5 °C (40 °F) dew point. This process may be followed by further dehydration with an adsorbent desiccant dryer for final dew points in the range of −40 to −50 °C (−40 to −60 °F), as applications require.

Gas Production. Lean exothermic gas is produced by combustion of a hydrocarbon fuel such as natural gas, propane, or light fuel oil with the air-fuel ratio closely controlled. This air-fuel mixture is burned in a confined combustion space for sufficient time to permit the reaction to reach equilibrium. The resultant gas is then cooled to partially condense water vapor formed by burning and to permit convenient transport and metering. In this process, the simplified theoretical reaction of methane fuel with air is:

$$CH_4 + 1.9O_2 + 7.6N_2 \rightarrow 0.9CO_2 + 0.1CO + 0.1H_2 + 7.6N_2 + 1.9H_2O + heat$$

where 1 volume of fuel and 9.5 volumes of air yield 8.7 volumes of product gas mixture, with water vapor removed. In practice, exothermic gas generators are seldom operated with an air-gas ratio higher than that required to produce a minimum of 1% total carbon monoxide and hydrogen, to avoid unwanted quantities of residual oxygen. One exception involves certain purge and blanket applications where small amounts of oxygen can be tolerated and the presence of combustible materials is not desired. In this case, the air-gas

ratio is further increased to operate with a slight amount of excess air, thus maintaining 1 to 2% oxygen in the product gas. The basic lean exothermic gas generator is shown in Fig. 6.

Operating Economics. The following are requirements needed to produce 28 m³ (1000 ft³) of lean exothermic atmosphere:

Natural gas . . . . 3.3 m³ (1000 ft³)
Power . . . . 1.4 MJ (0.4 kW·h)
Cooling water . . . . 1590 l (420 gal)
Power for refrigerant
dryer . . . . 1.8 MJ (0.5 kW·h)
Power for desiccant
dryer . . . . 5.4 MJ (1.5 kW·h)

Safety Considerations. Lean exothermic gas generators that are operated at less than 4% total combustibles are considered nonflammable and may be handled and used in the same manner as other types of inert gases. *These atmospheres, however, usually contain measurable amounts of carbon monoxide, and extreme care should be taken to avoid leakage. Proper venting and ventilation facilities should be employed. Moreover, when lean exothermic gas is used for purging and blanket applications, the chamber or vessel will lack oxygen. If it becomes necessary to enter these spaces for any reason, the chamber should be purged thoroughly with air or an air mask and a lifeline should be employed with another person outside holding the lifeline.*

Most generators are equipped with electrically programmed ignition sequencing and are monitored with additional safety devices that should be tested periodically to ensure fail-safe operation. Furnace safety is discussed in another article in this volume.

Endothermic-Base Atmospheres

Endothermic-base atmospheres are produced in generators that use air and a hydrocarbon gas as fuel. These two gases are mixed in a controlled ratio, slightly compressed, and then passed into a chamber that is filled with a nickel-bearing catalyst. This chamber has been heated externally to approximately 1040 °C (1900 °F). The gases react in this chamber to form endothermic gas. Endothermic atmospheres thus produced must be cooled rapidly to ensure the integrity of their chemical compositions. Figure 7 is a schematic diagram of an endothermic gas generator.

Endothermic gas produced from natural gas, which is primarily methane, has a typical analysis as follows: 40.4% hydrogen, 39.0% nitrogen, 19.8% carbon monoxide, 0.5% methane, 0.2% water vapor, and 0.1% carbon dioxide. Although natural gas is commonly employed, endothermic generators satisfactorily crack propane, butane or other hydrocarbon gases. The product endothermic gas chemical composition will vary when alternate hydrocarbon gases are used. When propane is used as the hydrocarbon gas, a typical analysis shows 45.3% nitrogen, 31.1% hydrogen, 23.4% carbon monoxide, 0.2% methane, <1% water vapor and virtually no carbon dioxide.

Common Applications. Endothermic atmospheres can be used in virtually all furnace processes that operate above 760 °C (1400 °F) and require strong reducing conditions. The most common use is as carrier gases in gas carburizing and carbonitriding applications. Because of the wide range of carbon equivalencies possible, however, endothermic atmospheres also are used for bright hardening of steel, for carbon restoration in forgings and bar stock, and for sintering of powder compacts that require reducing atmospheres. It is generally necessary to couple the generator with the process to ensure that the carbon equivalent being produced in the generator is appropriate for the process. However, in gas-carburizing applications, the addition of a hydrocarbon gas to the furnace will raise the carbon equivalency of the furnace gas, thus allowing a more neutral carrier gas to be produced at the generator.

Generation of Endothermic Atmospheres. In generating endothermic-base atmospheres, hydrocarbon gas and air are metered in proportions that ensure that only sufficient oxygen is admitted to form carbon monoxide and hydrogen, without any excess for the formation of carbon dioxide or water vapor. After being compressed to about 7 to 14 kPa (1 or 2 psig), the air-gas mixture is passed through a fire-check valve to the heated catalyst, which is contained in a pressure-tight retort. The retort is externally heated, usually by natural gas.

For a completely reacted gas of consistent analysis, the temperature inside the catalyst bed should be approximately 980 to 1040 °C (1800 to 1900 °F). Furthermore, the ratio of the diameter of the retort to its length must be correct for any given capacity.

After passage over the catalyst, the reaction is "frozen" by quickly chilling the gases to below 315 °C (600 °F) in a water jacket surrounding the top of the retort, thus preventing the reaction from reversing and forming carbon and carbon dioxide from carbon monoxide:

$$2CO \rightarrow C + CO_2$$

The reaction in the direction indicated predominates at temperatures of 705 to 480 °C (1300 to 900 °F). The soot-free reacted gas can be further cooled in the gas cooler for metering and distribution.

Because natural gas is composed chiefly of methane (CH_4), the over-all chemical reaction taking place in the endothermic generator using natural gas as its fuel can be expressed as follows:

$$2CH_4 + O_2 \rightarrow 2CO + 4H_2$$

neglecting 3.8 volumes of nitrogen (be-fore and after) for each 2 volumes of methane.

This reaction is not truly endothermic, as the name of the generator implies, but is exothermic. The reaction, however, takes place in two stages. In the first stage, some of the methane burns with air and generates heat. In the second stage, the surplus methane reacts with the carbon dioxide and water vapor produced in the first stage, and this later reaction is definitely endothermic. Therefore, a high temperature and a clean and sufficient catalyst bed are required to obtain a completely reacted gas and to minimize carbon dioxide, excess methane or excess water vapor.

A completely reacted gas is, in practical operations, one in which the methane content does not exceed about 0.4 to 0.8%. If the temperature is not high enough and the gas is not completely reacted, the reaction will produce soot. Once the catalyst accumulates soot, it becomes ineffective, and the gas composition will drift, resulting in more methane and higher percentages of carbon dioxide and water vapor in the products. When this occurs, the generating process cannot be controlled to maintain a definite carbon potential. Furthermore, unreacted methane will break down in the heat treating furnace to produce soot.

Clean, active catalyst is extremely important for accurate control of carbon potential. The catalyst most commonly used is of the porous refractory-base type, impregnated with nickel oxide.

The presence of hydrogen sulfide (H_2S) in natural gas will seriously affect an endothermic generator. Hy-

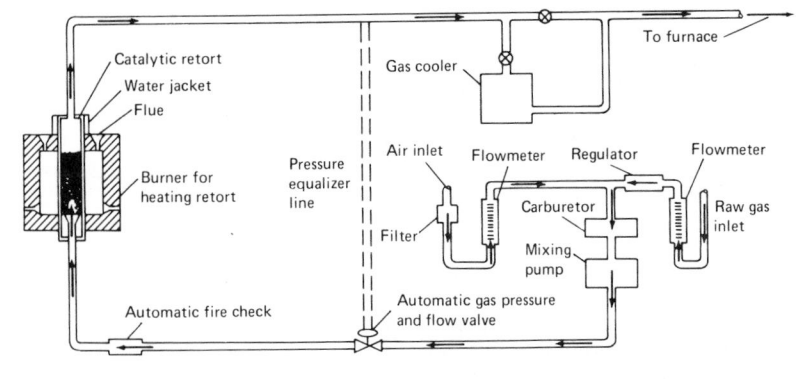

Fig. 7 Schematic flow diagram of an endothermic gas generator

drogen sulfide will cause the generator to produce high concentrations of CO_2, CH_4 and water vapor, and become unresponsive to control adjustments of the air-to-gas ratio. This causes the endothermic atmosphere being produced to have a low carbon equivalent. Specifically, any H_2S concentration over 30 ppm in natural gas of 0.6 specific gravity will have a noticeable effect on the resultant product gas.

Most endothermic generators are controlled by monitoring the dew point of the product gas. The dew point is controlled by either manual or automatic adjustment of the ratio of air and gas going into the generator. The relationship between carbon dioxide and dew point in endothermic gas is presented in Fig. 8. In broad terms, a dew point of 16 °C (60 °F) to −12 °C (10 °F) will create a product gas in equilibrium with steel containing 0.20 to 1.5% carbon at normal hardening or carburizing temperatures. Figure 9 shows the relationship between dew point and carbon content in plain carbon steels at temperatures from 815 to 925 °C (1500 to 1700 °F). Producing endothermic gas of −7 °C (20 °F) dew point or higher will ensure gas that is clean enough for continuous operation without weekend shutdowns for burnout. In some instances, percentage of carbon dioxide or the ratio of carbon monoxide to carbon dioxide is used as the controlling method of the generator.

Endothermic Generator Maintenance. The following schedule gives maintenance requirements for gas-fired endothermic-base atmosphere generators:

Weekly and/or monthly

- Burn out carbon in generator.
- Clean air filter.
- Check calibration of control instruments.
- Inspect thermocouples.

Annually, or as prescribed in operating instructions

- Test all safety controls.
- Inspect catalyst in retort and fill to proper level or replace.
- Inspect and clean burners.
- Check compressor blades and bearings in mixer pump and lubricate if necessary.
- Check motor bearings on mixer pump.
- Clean gas lines to furnaces if necessary.

Fig. 8 Relation between dew point and carbon dioxide content in the generation of an endothermic-base atmosphere

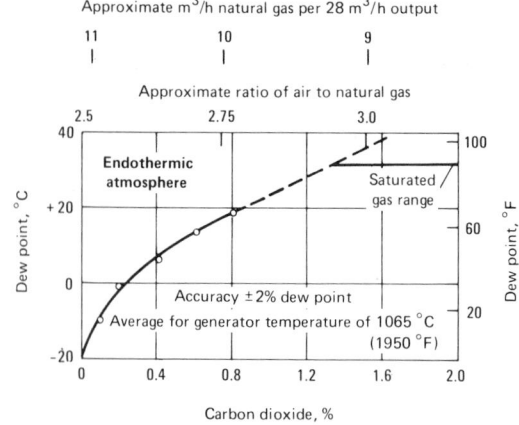

Fig. 9 Equilibrium between carbon steels and endothermic-base atmospheres

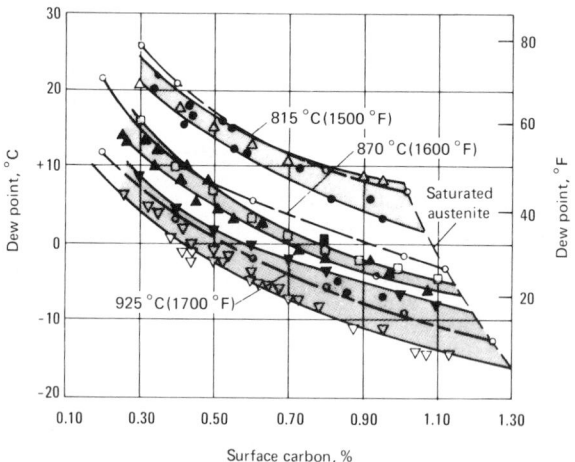

○ Groves, calculated; ● Cullen, experimental, 815 °C (1500 °F); △ Koebel, experimental, 815 °C (1500 °F); ▲ Cullen, experimental, 870 °C (1600 °F); □ Koebel, experimental, 870 °C (1600 °F); ■ production data, 870 °C (1600 °F); ▽ Cullen, experimental, 925 °C (1700 °F); ▼ Koebel, experimental, 925 °C (1700 °F); ◑ production data, 925 °C (1700 °F)

Safety Precautions. *Because endothermic gas is highly toxic and highly flammable and because it forms explosive mixtures rapidly, a safety program is imperative.* The exact program will depend on the equipment being used, local ordinances, normal operating procedures, types of emergency equipment available, plant layout and capabilities of personnel. The best safety equipment available cannot substitute for properly trained personnel.

Several areas of concern must be addressed in safety procedures and training. First, because an endothermic generator is in itself a furnace, often a gas-fired furnace, all the safety precautions normally applied to any furnace must be applied to the generator. No attempt should be made to produce endothermic gas if the generator is not at temperature. The fire-check valve, which is installed in all generators, must be in operating condition or the danger exists of an explosive gas getting into the mixer. Care must

Fig. 10 CO and H₂ versus air/gas ratio

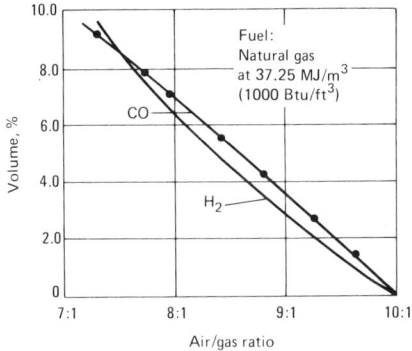

Approximate carbon-monoxide and hydrogen contents of the generated atmosphere versus the air-gas ratio of the "feed mixture".

be taken to prevent the volume of air in the mixture from increasing to a point at which the mixture becomes exothermic and hence explosive. Further, the cooling process must be maintained to prevent overheating of the supply system. *Overheating would allow a reversible reaction in the gas and create an explosive mixture.*

Prepared Nitrogen-Base Atmospheres

Prepared nitrogen-base atmospheres are exothermic atmospheres (produced by combustion of a mixture of air and fuel gas) from which almost all of the carbon dioxide and water vapor has been removed. The combination of very low dew point, approximately −40 °C (−40 °F), and the virtual absence of carbon dioxide accounts for the marked difference between the properties and applications of prepared nitrogen-base atmospheres as compared with those designated exothermic base.

The above definition conforms with the classification system used by the American Gas Association (AGA). Accordingly, the term "prepared nitrogen-base atmospheres" is not appropriate for furnace atmospheres consisting of a mixture of commercial nitrogen and other gases, nor for ammonia-base atmospheres, although both have a nitrogen base.

The high concentrations of carbon dioxide and water vapor in the products of combustion can be reduced to desirable low levels by either of two distinctly different systems. One system involves co-adsorption of both gases on a molecular sieve desiccant. The other system is a somewhat complicated miniature chemical plant that employs pumps and several heat exchangers. The carbon dioxide is absorbed by a water solution of monoethanolamine (MEA), and the water vapor is reduced by a refrigerator condenser, followed by desiccant drying.

Classification. Regardless of the method of generation, prepared nitrogen-base atmospheres are of two major types, either lean or rich. The relative richness of carbon monoxide and hydrogen in a typical rich atmosphere (class 202) is compared with a typical lean atmosphere (class 201) in Table 3.

Significantly important subtypes, which are carried in the AGA classification system as classes 223 and 224, involve the removal of almost all the carbon monoxide so that the generated atmospheres consist almost exclusively of nitrogen and hydrogen.

The methods of generator control to attain each of these two major types are described later.

Advantages and Disadvantages. The principal advantage of prepared nitrogen-base atmospheres is their applicability to a variety of heat treating operations for low-carbon, medium-carbon and high-carbon steels and for some other metals. Because of their low dew point and the virtual absence of carbon dioxide, these atmospheres (in the absence of oxygen-bearing contaminants introduced as a result of furnace operations) are neither oxidizing nor decarburizing, in contrast to the exothermic-base atmospheres. In addition, the nominal cost per unit volume of atmosphere produced are lower than for several other protective atmospheres except for the exothermic-base atmospheres.

Modified nitrogen-base atmospheres, classes 223 and 224, have the additional advantage of being nonsooting even at low furnace temperatures.

The main disadvantages of these atmospheres lie in the high initial cost of equipment, the large space requirements, and the need for more exacting maintenance and control of the generators. Further, as with any generator-based atmosphere system, extra inspection and testing of the material in process is required when generator conditions develop that force a shutdown.

Typical Applications. These atmospheres can be used in virtually all furnace applications that do not require highly reducing conditions. Because they are not decarburizing, they can be used in annealing, normalizing and hardening of medium-carbon and high-carbon steels. The low carbon monoxide contents of lean gases render them equally suitable for the thermal treatment of low-carbon steels. However, their usefulness in the heat treating of steels is predicated on their very low dew point, which can be maintained only if furnace design and operation prevent furnace atmosphere contamination with air or any other oxygen-bearing chemical compound. Consequently, in actual practice, one of the more common usages is for the annealing of steel coils in air-tight bell-type furnaces. For this application, ratios of about nine parts of air to one part of natural gas are employed to produce a dry, nondecarburizing, nonexplosive atmosphere containing about 4.0% combustibles. When these atmospheres are modified to eliminate the carbon monoxide (class 223), an additional advantage in annealing is that they do not produce soot during the period of slow cooling to relatively low temperatures.

Nitrogen-base atmospheres enriched with methane or other hydrocarbons are occasionally used as carrier gases in annealing, gas carburizing and carbon restoration, but endothermic and other protective atmospheres generally are preferred because of their higher carbon potentials and better control capability.

Lean atmospheres (with a maximum of 4.0% combustibles) are also used for purging explosive gases from furnaces and for blanketing the in-process material during furnace idle periods. While lean exothermic atmospheres are sometimes used for these purposes, the nonoxidizing, nondecarburizing nature of the nitrogen-base atmosphere is often advantageous. The rich atmospheres are not approved for use as purge gases because of the high contents of combustibles. The lean atmosphere is also used for these purposes in numerous applications within the chemical-processing industry.

Another common use of the lean atmosphere is with large semicontinuous and continuous annealing furnaces. In many instances, a lean nitrogen-base atmosphere is blended with an endothermic atmosphere to maintain adequate furnace pressure at low-

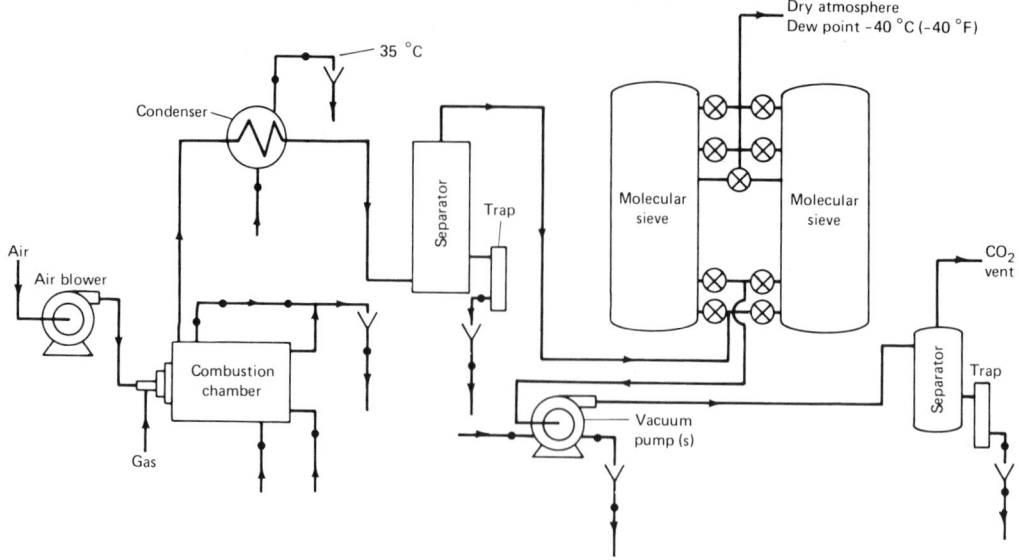

Fig. 11 Nitrogen-base atmosphere generator incorporating molecular-sieve removal of carbon-dioxide and water vapor (and using negative-pressure-regeneration of the sieves)

Gas lines are indicated by solid straight lines; water lines are shown as solid lines with closed circles. Auto-controlled valves are shown as the larger circles located between the molecular sieves.

er over-all cost. With this blend and appropriate adjustments, the higher carbon potential of the endothermic atmosphere can be maintained. With large semicontinuous furnaces, class 201 gas is often used to purge the furnace vestibules economically. The lean gas is almost always used with either type of furnace, and a part of the gas is stored under high pressure for purging the furnace when an unexpected shutdown occurs, for example, during an electrical power failure. Rich gas is avoided because it is not approved for purging for the reasons described above.

Rich nitrogen-base atmospheres may be used for the annealing or brazing of steel or copper alloys and for sintering of iron powder compacts.

Generation. In both common systems of generating prepared nitrogen-base atmospheres, molecular sieve or MEA scrubbing, the starting gas is produced by burning an appropriate mixture of air and a hydrocarbon fuel gas, which most frequently is natural gas. Combustion is accomplished within one or more refractory lined chambers. Sufficient air is provided under pressure to partially burn the fuel, and in no situation is enough air supplied to carry the burning to so-called complete combustion. Because the ratio of air to fuel for complete combustion varies widely de-

pending on the analysis of the fuel, the ratios cited are only approximate and, for ease in communication, are related to the analysis of natural gas that is commonly distributed in the midwestern areas of the United States.

The products of combustion consist of nitrogen, water vapor, carbon dioxide, carbon monoxide and hydrogen, with almost undetectable amounts of free oxygen. Figure 10 indicates the approximate pattern in which percentages of carbon monoxide and hydrogen vary in the final atmosphere versus the air-gas ratio of the feed mixture.

To maintain precise control of the air-gas ratio for the combustion process, it is common practice to use automatic analyzer/recorder/controller equipment. For simplicity, the various designs for control of air-gas ratios are not displayed in the flow diagrams of Fig. 11 and 12.

In both systems of generation, products of combustion are first passed through a water-cooled heat exchanger and then through a separator that condenses and discards most of the water of combustion. At this point, the two major systems of generation differ in the methods used for reducing the remaining water vapor and carbon dioxide to very low levels.

Molecular-Sieve Systems. A flow diagram of a typical molecular-sieve

system is shown in Fig. 11. The flow diagram for a specific generator may differ significantly depending on the design of components, although overall results are the same.

After products of combustion have passed through the heat exchanger and separator, they are piped to one tower of a dual tower system, in which both towers are filled with molecular-sieve material. While process gas is passing through the tower that is "on the line", molecular-sieve material in the other tower is being reactivated. Switching between towers is controlled by a timer that automatically triggers control valves that direct the flow. In some systems, the time for a complete cycle of a tower from "on the line" through reactivation is about 10 min.

As the gas passes through the molecular-sieve material, both carbon dioxide and water vapor are adsorbed to the extent that typically the dew point of the discharge gas is $-40\ °C\ (-40\ °F)$ or lower, and the carbon dioxide content is below 0.15%.

Molecular-sieve systems may function either at low or high operating pressures. In a low-pressure system, reactivation is accomplished by a strong vacuum and purging with some of the discharge gas of the tower that is on the line. Discharge pressure of this system is about 3 kPa (0.5 psig). In a

Fig. 12 Nitrogen-base atmosphere generator incorporating a monoethanolamine scrubbing system

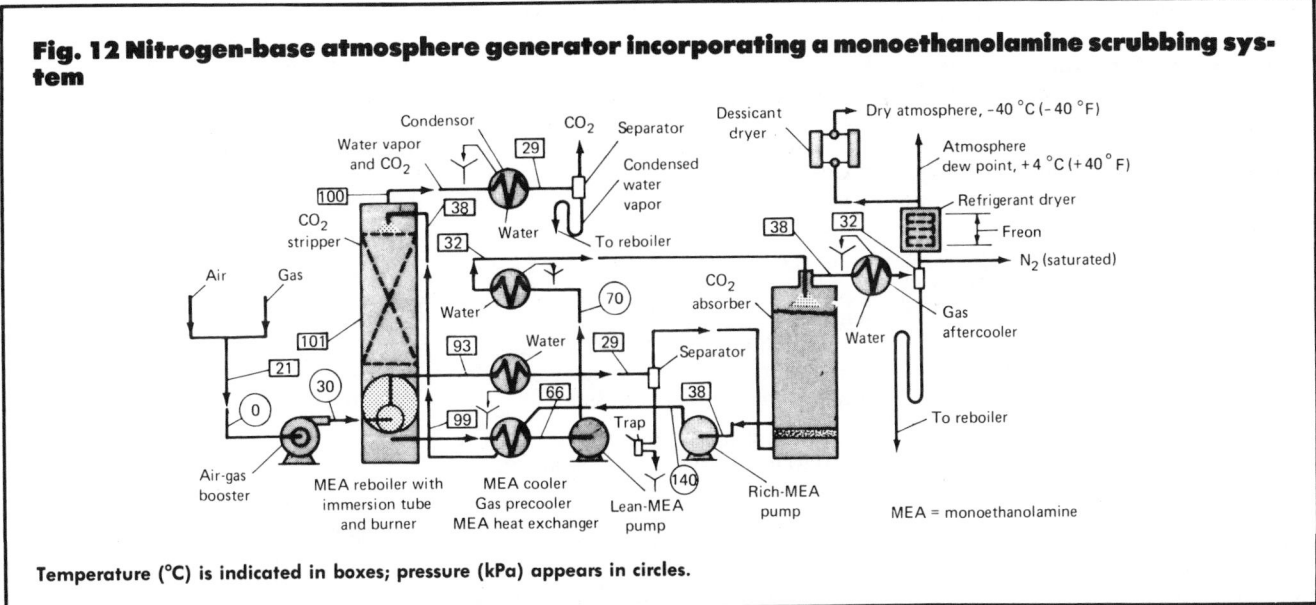

Temperature (°C) is indicated in boxes; pressure (kPa) appears in circles.

high-pressure system, operating pressure is about 590 kPa (85 psig). Reactivation consists of depressurizing to atmospheric pressure, plus purging with some of the discharge gas from the tower that is on the line.

With either type of molecular-sieve generator, heat of combustion is transferred to the water used to cool the combustion chamber. This heated water is sometimes used in other plant processes.

Because of the comparative simplicity and relative absence of operating problems, the molecular-sieve system is often preferred over the monoethanolamine system. Nevertheless, in many installations, the maintenance problems incurred with vacuum pumps are often as troublesome as the total variety of problems encountered in monoethanolamine systems.

Monoethanolamine System. Figure 12 presents a flow diagram of a typical monoethanolamine system. The flow diagram for a specific generator may differ significantly depending on design of the components, although over-all results are the same.

The combustion process is much the same as previously described except that the combustion chambers are contained in a unit identified as the reboiler-stripper. Heat of combustion is used to boil a recirculating solution of monoethanolamine. Most of the water of combustion in the gas is removed by heat exchangers and separators as described earlier. Gases are then piped to the bottom of the absorber tower, where

they pass upward through a counter flow of cooled monoethanolamine solution from the reboiler-stripper which is introduced at the top of the absorber. Intimate contact between the gas and monoethanolamine is attained with Raschig rings packed in the absorber tower. The upper portion of the reboiler-stripper shown in Fig. 12 is also packed with Raschig rings.

Because cool monoethanolamine, which is basically a water solution containing 12 to 18% monoethanolamine, absorbs carbon dioxide, the CO_2 content of the gases leaving the absorber tower is typically below 0.015%.

The monoethanolamine from the absorber is pumped through a monoethanolamine heat exchanger to the top of the reboiler-stripper. In the reboiler, the heat of combustion from the retort strips the CO_2 from the monoethanolamine. CO_2 and steam from the top of the reboiler-stripper pass through a water-cooled condenser to condense the steam before the CO_2 is vented (Fig. 12).

To reduce the moisture in the gas stream from the top of the absorber, the stream is passed through water-cooled and refrigerant-cooled heat exchangers and then through a dual-chambered desiccant drying system. While one chamber is on the line, the other is being reactivated. Reactivation is accomplished by a separately valved recirculating gas circuit. The volume and pressure of the circuit is maintained by feeding the circuit with a small portion of the discharge gas from the chamber

that is on the line. Before being passed through the chamber to be reactivated, the recirculated gas is heated to about 230 °C (450 °F) either electrically or by way of a gas heated retort. After having passed through the desiccant material in the chamber being reactivated, the water vapor that is picked up by the recirculated gas stream from the desiccant is partially condensed by way of water-cooled heat exchanger. At this point in the closed circuit, a separate gas pump is employed to ensure adequate volume and rate of flow in the recirculated gas stream.

Water vapor of the finished atmosphere is typically lower than 0.02%, or a dew point of less than −40 °C (−40 °F).

Accordingly, the prepared nitrogen-base atmosphere is virtually free of carbon dioxide and water vapor with carbon monoxide and hydrogen contents determined by the air-gas ratio fed to the combustion process.

Precautions. In the monoethanolamine process, it is essential that the air-gas ratio be maintained on the reducing side to avoid free oxygen in the prepared atmosphere and oxidation of the monoethanolamine. When oxidized, monoethanolamine is much more corrosive than a normal inhibited solution. Antifoaming agents are sometimes necessary in the monoethanolamine to avoid operating problems. Because of the very low surface tension and corrosiveness of monoethanolamine solutions, special care should be given to maintenance of monoethanolamine pumps to avoid leakage through pump-shaft seals.

Generator Maintenance. Rigid adherence to a "cookbook" recipe for maintenance of any mechanical, electrical, chemical or electronic system is frequently neither efficient nor effective. A better method is accurate observation of prescribed system performance, and adequate records of observations. The following schedules suggest minimum observation and maintenance for each of the two major systems. An adequate schedule of appropriate lubrication of all normal moving components is presumed, as well as a daily review of operator records by the immediate supervisor to detect trends and any need for corrective actions.

The following is a typical generator maintenance schedule for molecular-sieve systems:

Twice per 8-h operating shift

- Observe and record variables such as temperatures, pressures, flow rates and liquid levels.
- Observe and record details of abnormalities such as improper functioning of traps, safety switches, and excessive pump noises or leakage. Where appropriate, consult supervisor regarding the need for immediate corrective actions.
- Drain designated nonautomatic traps.

Monthly, or as prescribed by standard operating procedures

- Clean or replace all filters and strainers as required.
- With appropriate lubricants, lubricate all plug-type valves.
- Check operation of fuel safety shutoff valve and all pressure switches.
- Clean all traps, unless safety considerations override.
- Check the lens of the safety flame detector and clean as required.

Maintenance and Operating Suggestions

A complete annual overhaul of the system should include:

- Cleaning heat exchangers, for which chemical cleaning is preferred over mechanical cleaning
- Cleaning, repair, or replacement of all valves, including the safety shut-off valve and the pressure-control relief valve; pressure switches; ma-

nometers; sensing lines; traps; and flame scanner and relay

- Replacement of burner block and holder
- Rebuilding of vacuum pumps and compressor as required
- Checking accuracy of all meters, pressure controllers and pressure switches.

Do not overheat the molecular-sieve desiccant. In general, maintenance schedules for monoethanolamine generator systems coincide with those for the molecular-sieve systems except for the following major differences:

- Control amount of foaming in the absorber tower. Use antifoaming agent in monoethanolamine solution as required.
- Check monoethanolamine concentration monthly and include additions when required, along with additions of the recommended corrosion inhibitor (typically sodium metavanadate).
- The refrigerator system (for the refrigerant-cooled heat exchanger) should be checked and serviced monthly, also.
- Annual servicing should include a check of corrosion characteristics of the monoethanolamine solution. If solution needs to be replaced, the monoethanolamine circuit should be cleaned and boiled out with an appropriate caustic solution.
- Although the foregoing statements concerning vacuum pumps and molecular-sieve desiccant are not applicable here, a drying tower should not be used beyond its design cycle, contamination should be avoided, and the activated alumina desiccant should not be overheated.

Safety Precautions. In general, safety precautions outlined for exothermic-base atmospheres apply equally to prepared nitrogen-base atmosphere generators and usage. Safety precautions as outlined for endothermic-base atmosphere generators and usage are also applicable.

Atmospheres with less than 4% combustibles are underwriter-approved for purging, and are approved for furnace operating temperatures below 760 °C (1400 °F) without the requirement that the furnace-system be designed for automatic purging in case of electrical power failure or the absence of adequate furnace atmosphere pressures.

Operating Economics. The following list indicates representative utility

requirements for 28 m³ of atmosphere under so-called "standard temperature-pressure conditions" (1000 ft³ under standard temperature-pressure conditions, which is uniformly abbreviated in the English system of units as 1000 scf) for class 201 lean prepared nitrogen-base atmosphere. A comparison is drawn between the requirements for a typical monoethanolamine system, with both systems using an air-gas ratio of about 9.5 to 1.

	Molecular sieve	Monoethanolamine
Natural gas	3.77 m³ (133 scf)	3.77 m³ (133 scf)
Electric power	28 MJ (7.8 kW·h)	14.5 MJ (4.0 kW·h)
Cooling water	2780 l (735 gal)	2500 l (660 gal)

Commercial Nitrogen-Base Atmospheres

Industrial-gas nitrogen-base atmospheres have found technical acceptance in most metalworking applications. The shift of many heat treating operations to nitrogen atmospheres on some or all of the furnaces accelerated in the late 1970's as a result of shifting economics of hydrocarbon fuel supplies.

Commercial nitrogen-base atmosphere systems employed by the metalworking and heat treating industry use gases and equipment that are common among all applications. In most instances, the major atmosphere component is industrial gas nitrogen, which is supplied to the furnace from a system consisting of a storage tank, vaporizer, and a station controlling pressure and flow rate. The nitrogen serves as a pure, dry, inert gas that provides an efficient purging and blanketing function within the heat treating furnace. The nitrogen stream is often enriched with a reactive component, and the resulting composition and flow rate are determined by the specific furnace design, temperature and material being heat treated. Although there is similarity in the components of commercial nitrogen-base atmosphere systems, the flexibility of controlling atmosphere composition and flow rate independently over a wide range provides very different end use characteristics. Therefore, the classification of commercial nitrogen-base atmosphere systems is appropriately made according to three major categories of atmosphere function—protection, reactivity and carbon-control—rather than by gas or equipment components.

Protective Atmospheres. The atmosphere systems required for these applications must prevent oxidation or decarburization of the metal surface during heat treatment. Such detrimental reactions normally would occur because of residual oxygen or water vapor present in the furnace as a result of air leaks or inadequate purging. Typical applications include batch and continuous annealing of most ferrous and nonferrous metals. Atmosphere systems employed may be pure nitrogen or nitrogen blended with small quantities (usually less than 5%) of a reactive gas such as hydrogen, methane, propane or methanol vapor. Reactive gases reduce gaseous oxides present in the furnace atmosphere or reduce surface metal oxides. These systems are best employed to prevent oxidation of relatively clean metal surfaces rather than to remove gross oxidation of hot worked or forged materials.

Reactive Atmospheres. These atmosphere systems require a concentration of reactive gases greater than 5% to reduce metal oxides or to transfer a small amount of carbon to ferrous materials. The reactive components are generally hydrogen and carbon monoxide. The concentrations depend on the amount of oxide to be removed and the level of reaction products, water vapor and carbon dioxide, that are formed in the furnace atmosphere. Typical applications include brazing, powder-metal sintering and powder-metal reduction.

Carbon-Controlled Atmospheres. The main function of these nitrogen-base atmosphere systems is to react with steel in a controlled manner so that significant amounts of carbon can

be added to or removed from the steel surface. Significant change in the surface chemistry of the steel occurs in these heat treating applications. These atmosphere systems are characterized by high concentrations of reactive gases in nitrogen and by the requirement that the rate and amount of carbon transfer must be controlled by the atmosphere composition. Typical atmosphere components may include 10 to 50% hydrogen (H_2), 5 to 20% carbon monoxide (CO), and trace amounts (up to 3%) of carbon dioxide (CO_2) and water vapor (H_2O). The most common applications using a carbon-controlled atmosphere include carburizing and carbonitriding of machined parts, neutral hardening, electrical-lamination decarburization annealing, powder-metal sintering, and carbon restoration of hot worked or forged materials.

Table 4 describes heat treating applications that are presently using nitrogen-base atmosphere systems. The major components of the generated atmosphere most commonly used for each application are compared to the nitrogen-base alternatives that are presently in use.

Advantages of Commercial Nitrogen-Base Atmospheres. A nitrogen-base atmosphere system provides a technically viable substitute for generated atmospheres in most heat treating applications. Although the desired results are often the same with either a commercial nitrogen-base system or a conventional atmosphere generator, there are some significant differences in equipment, operation and function among these systems that should be understood before making the conversion from a generator to commercial nitrogen.

A generated atmosphere provides a fixed composition of output, which is

determined by the input ratio of air to hydrocarbon, the scrubbing and purification process equipment of the generator, and the relative condition of the equipment and catalysts. The commercial nitrogen-base system generally starts with elemental components stored in separate vessels, blends the desired atmosphere composition, and introduces it into the furnace. The resulting composition can be varied at different times during the cycle and at different locations within the furnace.

Most components of commercial nitrogen-base atmosphere systems are produced either by cryogenic processes or by chemical refining. As a result, the impurity levels are low, usually less than 20 ppm total impurity for liquid nitrogen, as an example. Further, because there are no by-products of air-fuel mixtures, the resulting atmosphere blend often has a dew point of $<-60\,°C$ ($<-80\,°F$) and oxygen or carbon dioxide impurity levels of <10 ppm.

The reducing or carburizing potential of an atmosphere is determined by several reactions and their associated equilibrium constants. Included among these are:

$$FeO + H_2 \rightleftarrows Fe + H_2O$$

$$Fe + 2CO \rightleftarrows FeC + CO_2$$

$$Fe + CO + H_2 \rightleftarrows FeC + H_2O$$

Because commercial nitrogen-base blends significantly reduce the amount of water vapor and carbon dioxide in the furnace atmosphere, the level of the reactive gases can also be reduced while maintaining acceptable thermodynamic potentials in the above ratios. The net result is that generated atmospheres that contain as much as 60% combustible products are often replaced with nitrogen blends containing

Table 4 Atmosphere systems and applications

Type of atmosphere	Application		Generated atmosphere Nominal composition, %			Nitrogen atmosphere	Nominal composition, %		
			N₂	H₂	CO		N₂	H₂	CO
Protective............	Annealing	Exothermic	70-100	0-16	0-11	Nitrogen-hydrogen	90-100	0-10	...
						Nitrogen-methanol	91-100	0-6	0-3
		Dissociated ammonia	25	75	...	Nitrogen-hydrogen	60-90	10-40	...
Reactive..............	Brazing	Exothermic	70-80	10-16	8-11	Nitrogen-hydrogen	95	5	...
		Dissociated ammonia	25	75	...				
	Sintering	Endothermic	40	40	20	Nitrogen-hydrogen	95	5	...
		Dissociated ammonia	25	75	...	Nitrogen-methanol	85	10	5
Carbon controlled......	Hardening	Endothermic	40	40	20	Nitrogen-methane	100	...	...
	Carburizing	Endothermic	40	40	20	Nitrogen-methanol	40	40	20
	Decarburizing	Exothermic	85	5	3	Nitrogen-hydrogen	90	10	...

less than 10% hydrogen and carbon monoxide. Many heat treating atmosphere applications can be converted from combustible to noncombustible compositions in conjunction with the switch to a commercial nitrogen-base system.

Because elemental components are available, commercial nitrogen-base heat treating practice may start with a pure nitrogen purge to eliminate oxygen from the furnace. During the heat treating cycle, a very reactive blend may be used to bright anneal heavily oxidized parts, a less reactive protective atmosphere during cooling, and a pure nitrogen purge finally to eliminate combustibles before opening the furnace.

Because the composition of a nitrogen atmosphere may be varied with time, it may also be controlled according to the functions of different zones of the heat treating furnace. For example, in a three-zone powder-metal sintering furnace, a pure nitrogen purge curtain may be used on both ends to reduce air infiltration. The presintering zone may be supplied with a reducing atmosphere with controlled ratio of H_2 and H_2O. Dew point control supplies sufficient water vapor for efficient breakdown and removal of vaporized lubricants from the powder compact. The sintering zone may use a nitrogen-hydrogen-carbon monoxide blend to provide a highly reducing atmosphere with a slight carbon potential to provide the required sintered strength and metal composition. In the cooling section, it may be necessary to add only a nitrogen protective atmosphere to bring the parts back to room temperature in a clean, nonoxidized condition.

The ability to control composition also allows flexibility of the over-all flow rate during the heat treating cycle. The most common advantage of this characteristic is the use of a high flow rate during charging of furnaces equipped with doors. When the door is opened, a microswitch initiates a purge flow of nitrogen to reduce air infiltration into the furnace. After the door is closed, the atmosphere flow rate returns to a level required to maintain adequate pressure in the closed furnace and to perform the required gas-metal surface reactions.

In commercial nitrogen-base applications where the hydrogen content of the atmosphere is significantly reduced from the generated atmosphere composition, it is often possible to achieve a

Fig. 13 Comparison of generated and nitrogen-base atmospheres

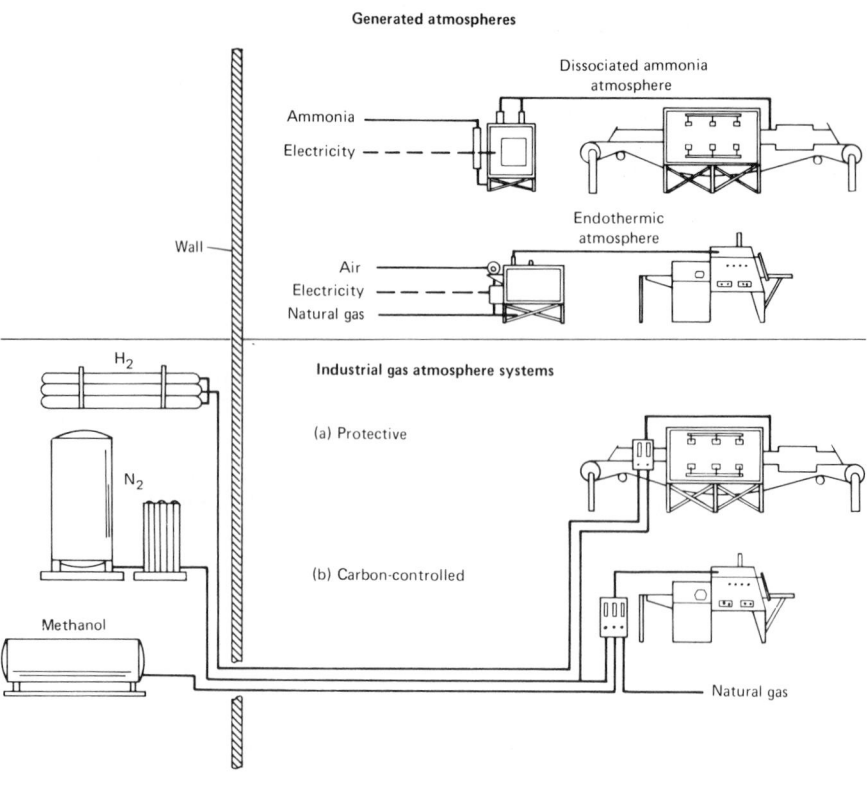

flow rate reduction because of atmosphere density. The most frequent examples of this occur in protective atmosphere applications in which endothermic gas (40% H_2) or dissociated ammonia (75% H_2) is replaced with a nitrogen-hydrogen blend of less than 10% H_2. In these instances, the atmosphere density changes from a specific gravity of 0.295 for dissociated ammonia to 0.595 for endothermic, or to 0.90 for the nitrogen-hydrogen blend. This increase in atmosphere density is responsible for increased furnace pressurization and more efficient purging of impurities from the furnace. Experience shows that an average 10 to 20% flow rate reduction may be achieved relative to the flow rates of generated atmosphere with lower densities.

Components of Commercial Nitrogen-Base Atmosphere System. The basic components of an industrial gas atmosphere system are illustrated in Fig. 13. Figure 13(a) shows a nitrogen-hydrogen protective atmosphere system used for annealing, brazing and sintering. Figure 13(b) is a nitrogen-methanol system typical of

those used for carburizing. In both instances, there are three basic parts to the commercial-nitrogen system: the storage vessels containing the elemental atmosphere components, the blend panel used to control the flow rate of each constituent gas, and the piping and wiring required for safe operation compatible with the furnace design.

Nitrogen is the main component of most commercial nitrogen-base systems. Because nitrogen is noncorrosive, special materials of construction are not required, except that they must be suitable for the temperatures of liquid nitrogen. Tanks may be spherical or cylindrical in shape. Figure 14 illustrates a typical liquid nitrogen tank. Sizes range from 2270 to 37 850 l (600 to 10 000 gal).

Hydrogen is used as a reactive reducing gas for many heat treating atmosphere applications. In commercial nitrogen-base systems, it is normally blended as a gas with nitrogen to form an atmosphere composition of 90% nitrogen, 10% hydrogen.

Liquid hydrogen is normally stored in tanks from which it is vaporized and

Fig. 14 Typical liquid nitrogen storage facility

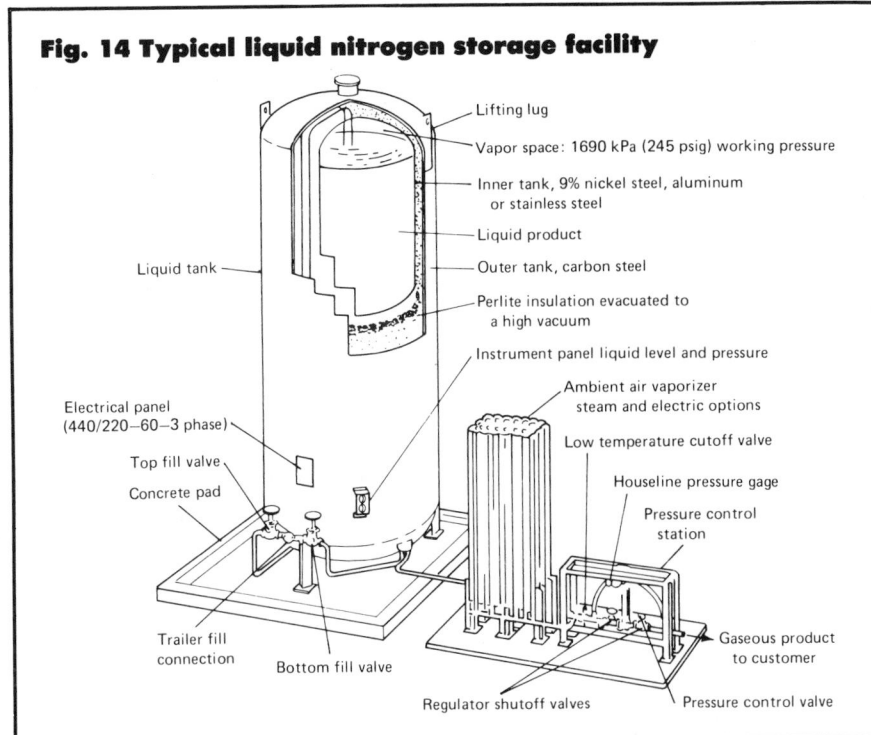

- Lifting lug
- Vapor space: 1690 kPa (245 psig) working pressure
- Inner tank, 9% nickel steel, aluminum or stainless steel
- Liquid product
- Outer tank, carbon steel
- Perlite insulation evacuated to a high vacuum
- Instrument panel liquid level and pressure
- Ambient air vaporizer steam and electric options
- Low temperature cutoff valve
- Houseline pressure gage
- Pressure control station
- Gaseous product to customer
- Pressure control valve
- Regulator shutoff valves
- Bottom fill valve
- Trailer fill connection
- Concrete pad
- Top fill valve
- Electrical panel (440/220—60—3 phase)
- Liquid tank

used as a gas. Hydrogen may also be delivered and stored as a gas. Tube trailers are available in capacities to 3570 m³ (126 000 standard ft³) of hydrogen with up to 16 960 kPa (2460 psig) pressure. Modules are available in 3 to 18 tube configurations with capacities to 4250 m³ (150 000 standard ft³) of hydrogen.

In commercial nitrogen-base atmospheres, methanol is used as a source of hydrogen and carbon monoxide for reactive and carbon controlled atmosphere systems. Methanol is delivered to the furnace as either a liquid or a vapor. When it is exposed to furnace temperatures greater than 760 °C (1400 °F), the methanol dissociates according to the following reaction:

$$CH_3OH \rightarrow 2H_2 + CO$$

Methanol is classified by the Department of Transportation (DOT) as a flammable liquid. When shipped by rail, water or highway, it must be packaged in authorized containers, and shippers must comply with all DOT regulations regarding loading, handling and labelling.

Before any connection or contact is made between the tank truck and the unloading line or other unloading equipment, the tank truck should be electrically grounded. *All containers filled from the truck should be bonded*

(electrically connected) and grounded to the truck before filling operations are started.

Air pressure should never be used for unloading tank trucks of methanol. Because the bulk storage of methanol in above- and below-ground tanks requires consideration of such factors as size of vents, diking and separation distances, which in turn depend on other variables such as nature of the tank contents, tank-wall thickness or protection, it is recommended that tank storage requirements be determined through consultation with qualified fire-protection engineers. (For guidance, refer to National Fire Protection Association pamphlet NFPA 30 or Factory Mutual Handbook of Industrial Loss Prevention.) Figure 15 illustrates an example of underground methanol storage.

Blending Equipment. Blending and flow-control systems for commercial nitrogen-base atmospheres are designed to be consistent with NFPA guidelines for generated atmospheres as well as with Factory Mutual guidelines. However, because gaseous components of the commercial nitrogen atmosphere system are independent with regard to storage, pressure and flow rate, certain differences exist compared with atmosphere generator operation. The blending system ensures

safe operation in both normal and emergency situations compatible with furnace design and processing conditions. Figure 16 illustrates generalized blending equipment schematics for two types of commercial nitrogen-base atmospheres. These schematics are consistent with the storage supply illustrations for protective and carbon-controlled atmosphere systems illustrated in Fig. 13.

In most applications, nitrogen is the predominant component, comprising 50 to 100% of the atmosphere. The enriching gas is typically hydrogen, but it may also be methane, propane or liquid methanol. Figure 16(a) illustrates a two-component blend system. Each leg consists of a pressure regulator, relief valve, pressure gage, flow meter, control valve and check valve. Nitrogen pressure is used to open a normally closed pneumatic valve on the hydrogen leg. It is, therefore, not possible for the combustible gas to flow into the furnace until the nitrogen is also available for flow. If nitrogen pressure is lost for any reason, flow of the combustible component will be interrupted automatically. An alternative method of interlocking combustibles to the nitrogen is through use of a nitrogen flow sensor. Typically this will be done by a flow switch or by a magnetic sensor on the nitrogen flow meter. The hydrogen-shutoff solenoid also should be interlocked to furnace temperature to prevent the flow of the combustible component into a furnace at less than 760 °C (1400 °F) to prevent the formation of an explosive mixture.

Figure 16(b) illustrates a nitrogen-methanol-methane blend panel typically used for carburizing applications. The design is more complex than the protective-atmosphere system because of the number of components and the more involved nature of the batch-carburizing process for which this methanol panel is designed.

Each leg typically consists of the same major components as the two-component system in Fig. 16(a): pressure regulator, relief valve, pressure gage, flow meter, flow control valve and check valve. Because the combustible components are interlocked to nitrogen pressure, a loss of nitrogen pressure will shut off the methanol and methane. The combustibles are also interlocked to furnace temperature and to the electrical power supply to the furnace and panel. A drop in furnace temperature below 760 °C (1400 °F) or a

Fig. 15 Typical underground methanol storage schematic

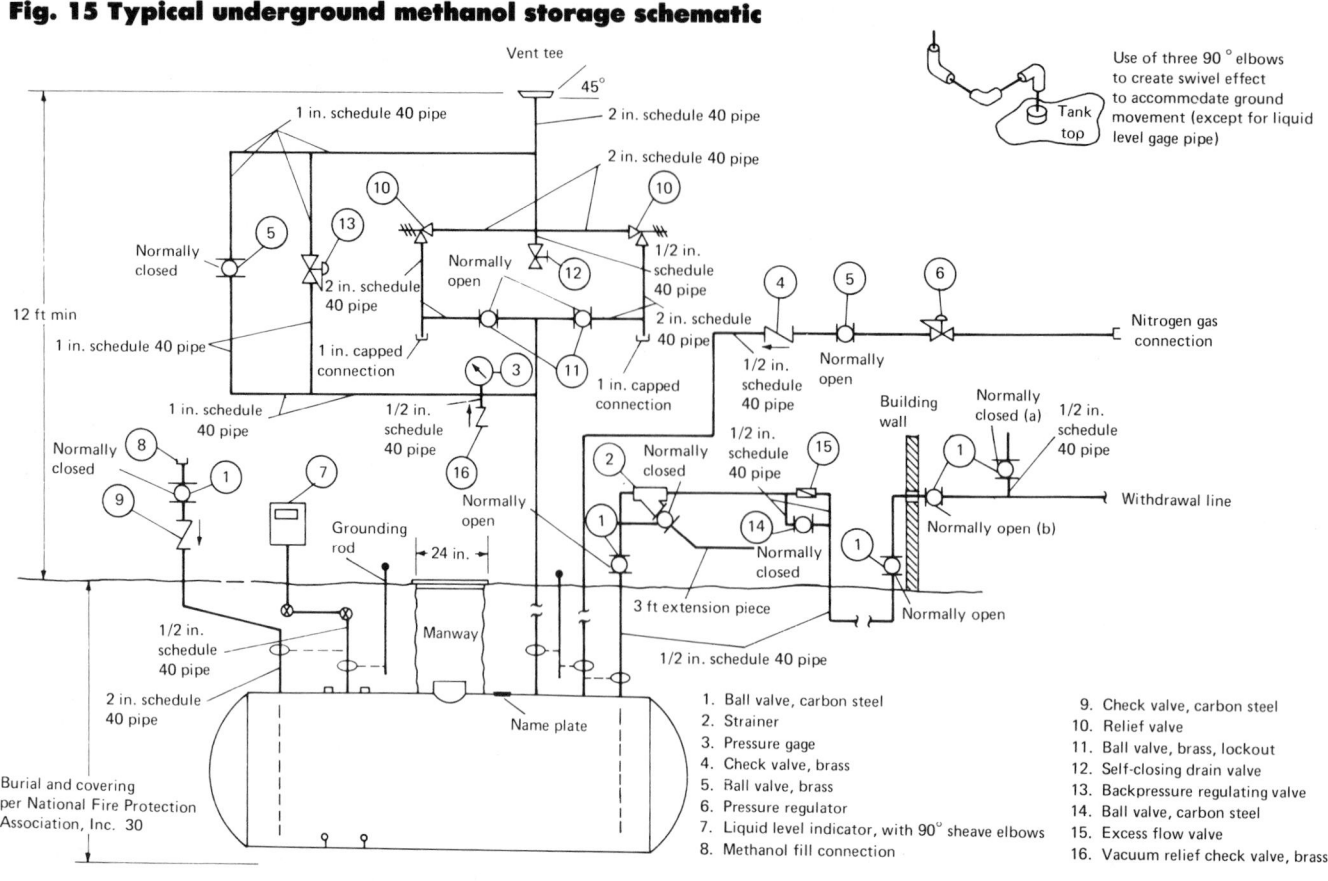

1. Ball valve, carbon steel
2. Strainer
3. Pressure gage
4. Check valve, brass
5. Ball valve, brass
6. Pressure regulator
7. Liquid level indicator, with 90° sheave elbows
8. Methanol fill connection
9. Check valve, carbon steel
10. Relief valve
11. Ball valve, brass, lockout
12. Self-closing drain valve
13. Backpressure regulating valve
14. Ball valve, carbon steel
15. Excess flow valve
16. Vacuum relief check valve, brass

(a) Vapor bleed valve to be located at the high point of the house line piping. **(b)** This valve can be eliminated if methanol piping between tank withdrawal connection and inlet to building is less than 20 ft long.

loss of electric power will shut off the combustibles.

This panel typically includes a separate nitrogen-purge system that is preset with a pressure regulator-orifice plate flow control. In the event of low furnace temperature or electric power loss, an alarm sounds, the nitrogen purge commences, and flow of combustibles is shut off. The nitrogen purge is also interlocked to the inner furnace doors through a microswitch circuit so that charging or quenching a load will initiate the high flow purge. In this mode, the purge will last from 5 to 60 min.

Commercial Nitrogen-Base Atmosphere Applications. Tables 5, 6 and 7 compare commercial nitrogen-base atmospheres with comparable generated atmospheres used in relevant heat treating processes. Shown are the generated atmospheres typically in use and the nitrogen-base alternative. Protective atmosphere applications in Table 5 indicate the potential of commercial nitrogen-base atmospheres as alternatives to exothermic, endothermic and dissociated ammonia atmospheres in a broad range of annealing operations. The nitrogen-base alternatives illustrated are pure nitrogen, nitrogen-hydrogen, nitrogen-methanol and nitrogen-hydrocarbon.

In most instances, the nitrogen atmosphere has a lower level of combustible components, carbon monoxide and hydrogen. The nitrogen atmosphere also will normally contain lower trace impurities, carbon dioxide and water vapor. In some applications, more than one system is applicable, depending on the furnace design and the grade of material processed.

Reactive atmospheres are required for such heat treating applications as brazing and powder-metal sintering. The atmosphere reacts with the metal surface to remove all metal oxides. In brazing, a chemically clean metal sur-face is required to ensure proper filler-metal flow and diffusion of the filler metal into the base metal. In powder-metal sintering, reduction of powder surface oxides is necessary to promote diffusion and thus to obtain effective bonding within the powder compact. The reactive nature of these atmospheres also applies to carbon transfer. In brazing, residual carbonaceous lubricants on the materials and organics in the brazing paste must be removed by the atmosphere for proper flow of filler metal. In carbon-steel powder-metal sintering, lubricants must be removed in one section of the furnace and carbon made available to the steel in another section to achieve proper part chemistry. Table 6 summarizes the compositions of reactive atmosphere alternatives, both with respect to major constituents and trace impurities.

Carbon-controlled atmospheres transfer carbon from the atmosphere to

Table 5 Compositions of protective atmospheres

Application	Input atmosphere	Furnace atmosphere analysis, %				Trace impurities	
		N_2	H_2	CO	CH_4	H_2O	CO_2
Carbon steel sheet, tube, wire	Exothermic-purified	80	12	8	...	0.01	0.5
	N_2–5% H_2	95	5	...	...	0.001	...
Carbon steel rod	Exothermic-purified	100	...	...	...	0.01	0.5
	Exothermic-endothermic blend	75	15	8	2	0.01	0.5
	N_2–1% C_3H_8	97	1	1	1	0.001	0.01
	N_2–5% H_2–3% CH_4	90	7	2	1	0.001	0.01
	N_2–3% CH_3OH	91	6	3	...	0.001	0.01
Copper wire, rod	Exothermic-lean	86	...	...	...	3	11
	N_2–1% H_2	99	1	...	...	0.001	...
Aluminum sheet	Exothermic-lean	86	...	...	...	3	11
	N_2	100	...	...	...	0.001	...
Stainless steel sheet, wire........	Dissociated ammonia	25	75	...	...	0.001	...
	H_2	...	100	...	...	0.0005	...
	N_2–40% H_2	60	40	...	...	0.0005	...
Stainless steel tube	Dissociated ammonia	25	75	...	...	0.001	...
	H_2	...	100	...	...	0.0005	...
	N_2–25% H_2	75	25	...	...	0.005	...
Malleable iron anneal	Exothermic-purified	98	...	2	...	0.01	0.5
	N_2–1% C_3H_8	97	1	1	1	0.001	0.2
Nickel-iron laminations..........	Dissociated ammonia	25	75	...	...	0.001	...
	N_2–15% H_2	85	15	...	...	0.001	...

Table 6 Compositions of brazing and sintering atmospheres

Application	Input atmosphere	Furnace atmosphere analysis, %				Trace impurities	
		N_2	H_2	CO	CH_4	H_2O	CO_2
Copper braze carbon steel	Exothermic-rich	70	14	11	1	0.05	4
	Endothermic	40	39	19	2	0.05	0.1
	N_2–5% H_2	95	5	...	...	0.001	...
	N_2–3% CH_3OH	91	6	3	...	0.001	0.01
Silver braze stainless steel	Dissociated ammonia	25	75	...	...	0.001	...
	N_2–25% H_2	75	25	...	...	0.001	...
Metallize ceramics	Dissociated ammonia + H_2O	25	75	...	...	3	...
	N_2–10% H_2–2% H_2O	90	10	...	...	2	...
Glass metal seal	Exothermic	75	9	7	...	3	6
	N_2–10% H_2–2% H_2O	88	10	...	...	2	...
Carbon steel sintering .... 6.4 to 6.8 g/cm³, <0.4% C	Endothermic	40	39	19	2	0.05	0.1
	N_2–5% H_2	95	5	...	...	0.001	...
Carbon steel sintering .... 6.8 to 7.2 g/cm³, >0.4% C	Endothermic	40	39	19	2	0.05	0.2
	N_2–Endothermic	87	8	4	1	0.01	0.05
	N_2–8% CH_3OH	76	16	7	1	0.005	0.05
	N_2–8% H_2–2% CH_4	90	8	1	1	0.005	0.01
Brass, bronze sintering	Dissociated ammonia	25	75	...	...	0.001	...
	Endothermic	40	39	19	2	0.05	0.3
	N_2–10% H_2	90	10	...	...	0.001	...
Stainless steel sintering........	Dissociated ammonia	25	75	...	...	0.001	...
	H_2	...	100	...	...	0.001	...
Tungsten carbide.............. Sinter	Dissociated ammonia	25	75	...	...	0.001	...
	H_2	...	100	...	...	0.001	...
Presinter	N_2–20% H_2	80	20	...	...	0.001	...
Nickel sinter	Dissociated ammonia	25	75	...	...	0.001	...
	N_2–10% H_2	90	10	...	...	0.001	...

Table 7 Compositions of carburizing atmospheres

| Application | Input atmosphere | Furnace atmosphere analysis, % | | | | | |
| | | N_2 | H_2 | CO | CH_4 | Trace impurities | |
						H_2O	CO_2
Neutral harden............	Endothermic + CH_4	39	40	19	2	0.05	0.1
	$N_2 - 2\%$ CH_4 or 1% C_3H_8	97	1	1	1	0.001	0.01
	$N_2 - 5\%$ $CH_3OH - 1\%$ CH_4	84	10	5	1	0.005	0.01
Carburize................	Endothermic + CH_4	37	40	18	5	0.05	0.1
	$N_2 - 20\%$ $CH_3OH + CH_4$	37	40	18	5	0.05	0.1
	$N_2 - 17\%$ $CH_4 - 4\%$ CO_2	70	16	7	7	0.005	0.05
	$N_2 - 20\%$ $CH_4 - 5\%$ H_2O	55	28	10	7	0.01	0.05
Carbonitride	Endothermic + CH_4 + NH_3	36	40	18	5	0.05	0.1
	$N_2 - 20\%$ $CH_3OH + CH_4 + NH_3$	36	40	18	5	0.05	0.1
	$N_2 - 17\%$ $CH_4 - 4\%$ CO_2 + NH_3	68	18	7	7	0.005	0.05
	$N_2 - 20\%$ $CH_4 - 5\%$ H_2O + NH_3	53	30	10	7	0.01	0.05
Lamination decarburize	Exothermic + H_2O	75	9	7	...	3	6
	$N_2 - 10\%$ $H_2 - 4\%$ H_2O	83	10	1	...	3	3
	$N_2 - 5\%$ $CH_3OH - 4\%$ H_2O	79	10	2	...	3	6

the surface of the metal part through controlled gas-metal reactions. The two most common reactions are:

$$2\,CO \rightleftharpoons \bar{C} + CO_2$$
$$CO + H_2 \rightleftharpoons \bar{C} + H_2O$$

Because carbon transfer is required of many of these applications, it is common to find higher concentrations of carbon monoxide and hydrogen in these atmospheres than in the protective and reactive atmospheres. The compositions of typical carbon-controlled atmosphere systems are provided in Table 7.

Carbon-controlled atmosphere systems are used in surface carburizing steel parts, in neutral hardening and in decarburize annealing of steel motor laminations. The essential requirement of all applications using these atmospheres is control. In neutral hardening, the atmosphere carbon potential must match the surface carbon of the steel to prevent decarburization or carburization. The surface carburization of steel parts requires not only the desired carbon potential in the atmosphere, but also the correct carbon availability to provide the proper case depth. Decarburization of motor and transformer laminations requires controlled carbon removal from the steel without unwanted oxidation of the steel surface. The systems presently employed in these applications are exothermic and endothermic generated atmospheres, and nitrogen-hydrocarbon and nitrogen-methanol blended atmosphere systems.

Dissociated Ammonia-Base Atmospheres

Dissociated ammonia (class 601) is a medium-cost prepared furnace atmosphere providing a dry, carbon-free source of reducing gas. Typical composition is 75% hydrogen, 25% nitrogen, less than 300 ppm residual ammonia, and less than -50 °C (-60 °F) dew point.

Principal uses of dissociated-ammonia furnace atmosphere include bright copper and silver brazing; bright heat treating of selected nickel alloys, copper alloys and carbon steels; bright annealing of electrical components; and as a carrier mixed gas for certain nitriding processes, including the "Floe" system of nitriding, which is a method of controlling white layer.

The high hydrogen content affords a strong deoxidizing potential, which is advantageous in removing surface oxides or preventing oxide formation during high temperature heat treatment. Care should be exercized, however, in selecting heat processing applications that might result in unwanted hydrogen embrittlement or surface-nitriding reactions.

Other less general uses involve dissociated ammonia fully or partially burned with air in a slightly modified exothermic gas generator. These secondary atmospheres fall into two general classifications: lean combusted ammonia (class 621), which is considered inert and nonflammable with from 0.25 to 1.0% hydrogen and the remainder nitrogen, and rich combusted ammonia (class 622), which is a moderately reducing atmosphere usually containing from 5 to 20% hydrogen and the balance nitrogen. A very high moisture content is present following combustion and primary cooling. It is often necessary to further dehydrate these gases with additional equipment in the form of refrigerant and desiccant dryers. These treated gases provide very high quality atmospheres of low-to-moderate reducing ability. Practical uses, however, are restricted because the process requires consumption of relatively costly anhydrous ammonia as the primary feedstock and because added investment is required in addition to operating costs of exothermic generators and dryers.

Required Equipment. Dissociated ammonia ($N_2 + 3H_2$) is produced from commercially supplied anhydrous ammonia (NH_3) with an ammonia dissociator. This equipment raises the temperature of ammonia vapor in a catalyst-filled retort to approximately 900 to 980 °C (1650 to 1800 °F). The gas is then cooled for metering and transport as a prepared atmosphere. At these reaction temperatures in the presence of catalyst, ammonia vapor dissociates into separate constituents of hydrogen and nitrogen. The simplified reaction is:

$$2NH_3 + heat \rightarrow N_2 + 3H_2$$

where two volumes of ammonia vapor yield four volumes of dissociated ammonia.

The basic ammonia dissociator (Fig. 17) consists of either an electrically heated or gas-fired chamber containing one or more catalyst-filled alloy retorts,

Fig. 16 Nitrogen-base atmosphere blending equipment

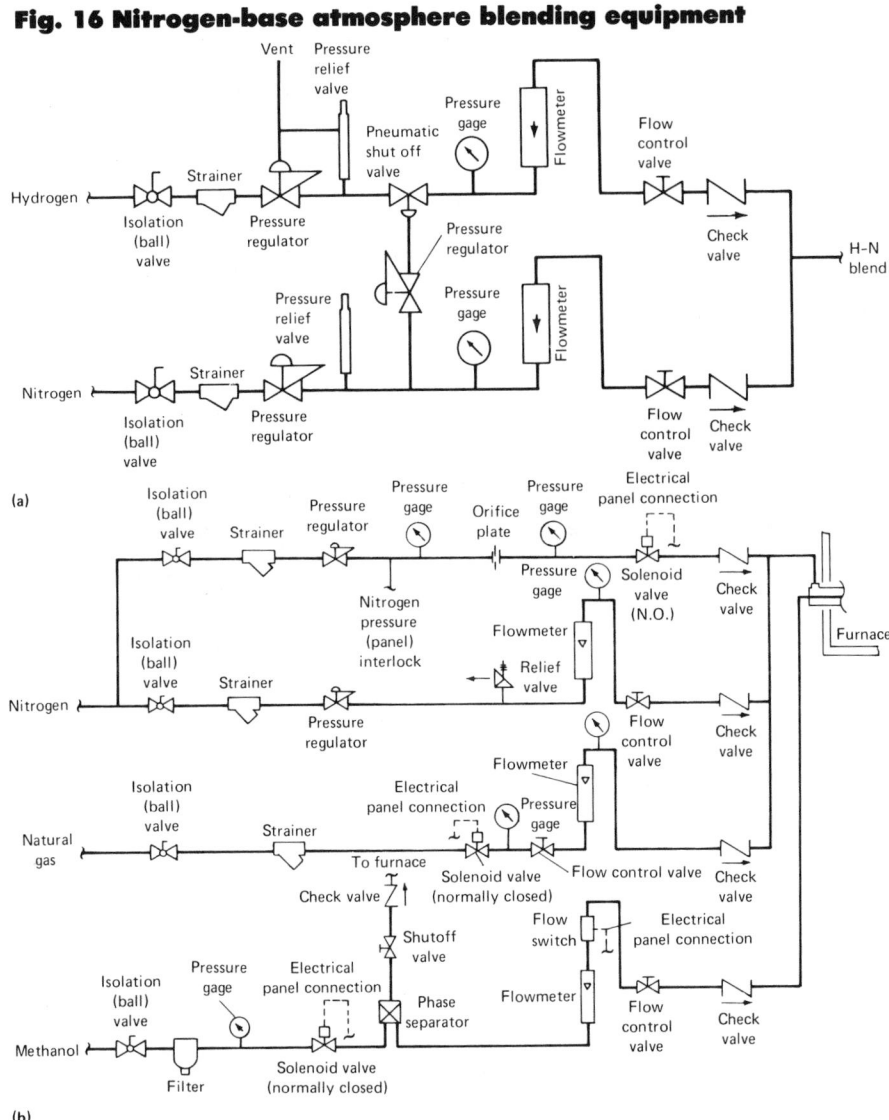

(a)

(b)

Table 8 Physiological effects of contamination of air by ammonia in various concentrations

Amount of ammonia in air, ppm	Physiological effect
53	Smallest concentration at which odor can be detected
100	Maximum concentration allowable for prolonged exposure
300 to 500	Maximum concentration allowable for short exposure (½ to 1 h)
408	Least amount causing immediate irritation to the throat
698	Least amount causing immediate irritation to the eye
1720	Least amount causing coughing
2500 to 4500	Dangerous for short exposure (½ h)
5000 to 10 000	Rapidly fatal for short exposure

followed by a water-cooled indirect heat exchanger. Operation is simple and continuous, incorporating a product gas flow meter with either manual or automatic process flow control, and suitable automatic heating-chamber temperature controls. Safety devices include items such as pressure switches, temperature-monitoring instruments and pressure-relief valves. Periodic checks of operation consist of observing pressures, temperature and flow rate, and occasional analysis of product gas for unreacted ammonia content.

Liquid anhydrous ammonia as a feedstock material is obtained by commercial delivery in pressurized cylinders by pressurized tank truck, or by rail tank car. The ammonia is transferred to a stationary on-site bulk storage vessel. Because liquid ammonia is supplied at high pressure, only pressure and flow regulation are required for conveyance to points of use.

To ensure suitable and efficient dissociator operation, liquid ammonia should first be converted to vapor, usually with a tank-mounted vaporizer. This is a chamber using an immersion type electric heater or steam coil. Some smaller dissociators may have a self-contained vaporizer using residual process heat, thus permitting operation directly from cylinder supplies.

Operating Economics. Typical operating consumptions to produce 28 m³ (1000 ft³) of dissociated ammonia are:

Anhydrous ammonia . . 10 kg (22.5 lb)
Dissociator power . . . 75 MJ (21 kW·h)
Vaporizer power . . . . . 14 MJ (4 kW·h)
Cooling water . . . . 284 l (75 gal)

To obtain the highest quality furnace atmosphere and to minimize possible dissociator maintenance, an oil-free metallurgical grade of ensured dry anhydrous ammonia should be used.

Safety Precautions. *Ammonia vapor and dissociated ammonia are flammable when mixed in certain proportions of air.* They should be treated with the same precautions exercised with any combustible gas relative to handling or to its use as a furnace atmosphere. Ammonia vapor also is extremely corrosive to certain materials such as copper and copper-bearing alloys. Care should be taken in the selection of piping materials and other devices in contact to prevent corrosion failure and resultant leaks. *Additionally, because ammonia is stored at high pressure and has a high rate of expansion, stored cylinders should not be exposed to temperatures exceeding 50 °C (120 °F), otherwise dangerously excessive pressures will result.* Ammonia piping above 25 mm (1 in.) diam should have welded steel joints with adequate pressure rating and extra heavy

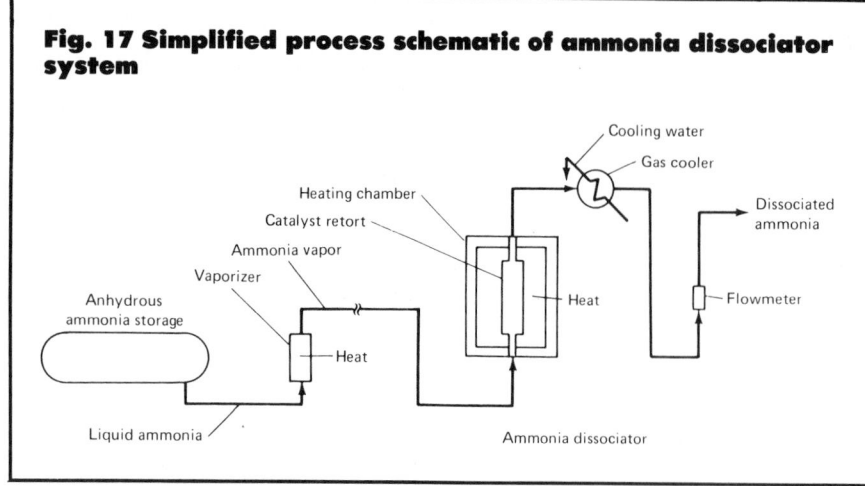

Fig. 17 Simplified process schematic of ammonia dissociator system

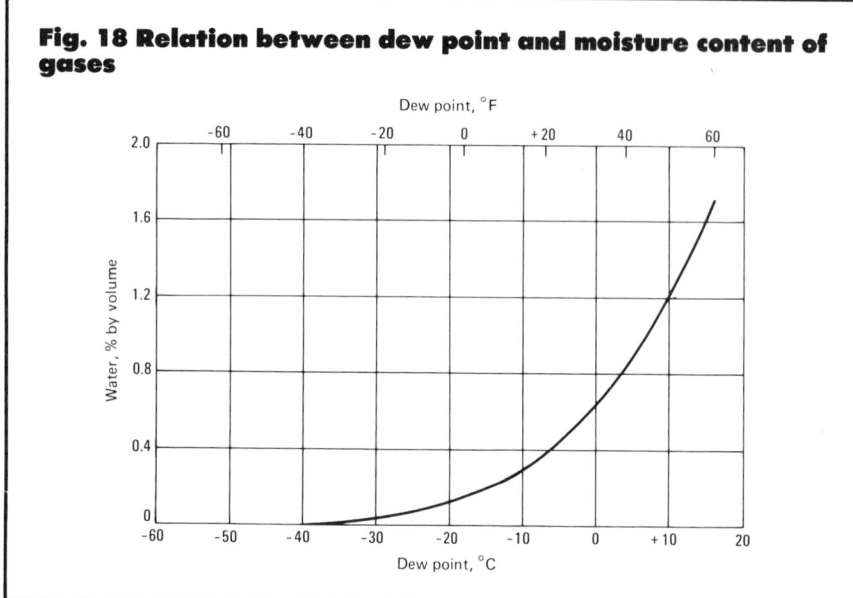

Fig. 18 Relation between dew point and moisture content of gases

(schedule 80) steel pipe should be used when joints are to be threaded.

Ammonia may also cause certain physiological effects in air concentrations over 100 ppm, resulting in eye and respiratory irritation (see Table 8). It is extremely pungent, however, and the strong odor will generally give adequate warning of any leak.

Dissociated ammonia with its high percentage of hydrogen makes it necessary to establish furnace operating practices appropriate for hydrogen atmospheres, and these should be carefully observed. Particularly important are safety procedures associated with nitrogen purging to prevent air infiltration. Adequate internal furnace circulation should be used to avoid entrapment of the low-density gas in upper portions of furnace chambers. Most ammonia dissociators are monitored with safety devices, which should be periodically tested to ensure fail-safe operation.

Hydrogen Atmospheres

Commercially available hydrogen is 98 to 99.9% pure. All cylinder hydrogen contains traces of water vapor and oxygen. Methane, nitrogen, carbon monoxide and carbon dioxide may be present as impurities in very small amounts, depending on the method of manufacture. Hydrogen is produced commercially by a variety of methods, including the electrolysis of water, the catalytic conversion of hydrocarbons, the decomposition of ammonia and the water gas-reaction. It is also obtained as a by-product in the electrolytic manufacture of sodium hydroxide and the catalytic cracking of petroleum oils.

Hydrogen is a powerful deoxidizer, and its deoxidizing potential is limited by moisture content only. Its thermal conductivity is approximately seven times that of air. Its principal disadvantage is that it is readily absorbed by most common metals, either by occlusion or by chemical combination, at elevated temperature. Absorption of hydrogen can result in serious embrittlement, especially in high-carbon steels. It may also reduce oxide inclusions in steel to form water, thus building sufficient pressure at elevated temperature to cause intergranular fracture of the steel. Dry hydrogen will decarburize high-carbon steels at elevated temperature by reacting with carbon to form methane.

Impurities. The hydrogen best suited for metallurgical purposes is made by the electrolysis of distilled water. In most heat treating procedures requiring hydrogen, water vapor and oxygen are objectionable, and the hydrogen must be purified before it can be used. Oxygen is removed by a room-temperature catalytic process that combines the oxygen with hydrogen to form water vapor. Essentially all water vapor is then removed by means of activated-alumina dryers, yielding a gas with a dew point of −50 °C (−60 °F) See Fig. 18 for a curve relating dew point and moisture content.

Applications. Dry hydrogen is used in the annealing of stainless and low-carbon steels, electrical steels and several nonferrous metals. It is used also in the sintering of refractory materials such as tungsten carbide and tantalum carbide, in the nickel brazing of stainless steel and heat-resisting alloys, and in copper brazing, direct reduction of metal ores, annealing of metal powders and the sintering of powder metallurgy compacts. Data relating to metal-oxide equilibriums in hydrogen atmospheres are given in Fig. 19.

Example 1. Stainless steel powder parts are sintered in a dry (−40 °C or −40 °F dew point) hydrogen atmosphere at 1275 °C (2325 °F), following preheating in dissociated ammonia at 425 °C (800 °F). Electric furnaces are used for both preheating and sintering. Details of equipment are given in Table 9.

Supply. Hydrogen is usually obtained in cylinders. Small quantities are supplied at a pressure of 14 MPa (2

Table 9 Equipment requirements for sintering stainless steel powder metallurgy parts in hydrogen

Production requirements

Load weight	9 kg (20 lb)
Heating cycle	40 min
Output per hour	14 kg (30 lb)

Equipment requirements

Burn-off furnace	Pusher, electrically heated; forced circulation
Size of hearth	255 by 150 by 915 mm (10 by 6 by 36 in.)
Length of cooling chamber	1.8 m (6 ft)
Power	72 MJ (20 kW)
Operating temperature	425 °C (800 °F)
Atmosphere	Dissociated ammonia, 4 m³/h (150 ft³/h)
High-heat furnace	Open chamber, electrically heated; front push, rear pull
Size of hearth	255 by 610 mm (10 by 24 in.) (preheating)
	255 by 915 mm (10 by 36 in.) (high heating)
Length of cooling chamber	2.4 m (8 ft)
Power	126 MJ (35 kW)
Operating temperature	1275 °C (2325 °F)
Atmosphere	Dry hydrogen, 10 m³/h (350 ft³/h); dew point, −40 °C (−40 °F)

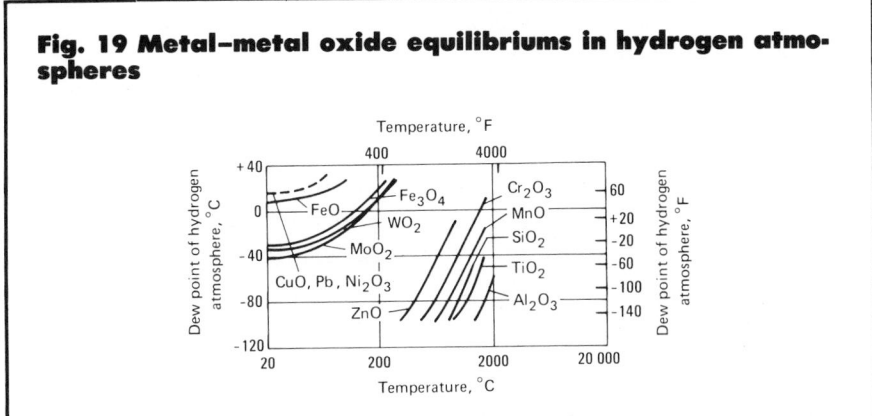

Fig. 19 Metal–metal oxide equilibriums in hydrogen atmospheres

ksi) in steel cylinders that contain 5.5 m³ (193 ft³) of hydrogen measured at standard conditions of 20 °C (68 °F) and atmospheric pressure. Larger quantities can be supplied in banks of about 12 interconnected cylinders that are mounted on portable dollies and contain a total of about 90 m³ (3200 ft³) of gas. For even larger requirements, trailers are used, containing a total of 810 m³ (28 600 ft³) of hydrogen at 17 MPa (2500 psi); the capacity of a railroad tank car is 5665 m³ (200 000 ft³).

In-Plant Generators. Very large amounts of hydrogen are supplied by generators that decompose ammonia or partially burn hydrocarbon fuels. Purification is accomplished by adsorption of unwanted gases or by diffusion of "raw" hydrogen through molecular filters that are capable of producing a gas of 99.5% purity.

Safety. *The explosive range of hydrogen in air is so great that it should be considered explosive in all ranges.* Whenever hydrogen is used in a furnace, an adequate supply of inert gas for purging should be readily available, either from storage or from a generator. Nitrogen and products of combustion are commonly used for purging.

Steam Atmospheres

Steam may be used as an atmosphere for scale-free tempering and stress re-lieving of ferrous metals in the temperature range of 345 to 650 °C (650 to 1200 °F). The steam causes a thin, hard and tenacious blue-black oxide to form on the metal surface. This oxide film, which is about 0.00127 to 0.008 mm (0.00005 to 0.0003 in.) thick, improves certain properties of various metal parts.

Effects of Steam Treating. The service life of cutting edges of high speed steel tools such as drills, reamers, taps and milling cutters is increased from 50 to 100% when the tools are steam treated after being tempered and finish ground. During machining operations, the oxide film prevents metal chips from welding to the tool and retains cutting oil to reduce heat of friction.

Steam treating decreases the porosity of sintered iron compacts and provides increased compressive strength and resistance to wear and corrosion. Steam penetrates the pores of compacts and forms the oxide internally as well as on the surface. The oxide seals the pores and partially fills the voids, thus increasing compressive strength. The hardness of the film, in addition to its ability to hold oil, increases wear resistance. The density of the film, coupled with its ability to retain oil, increases corrosion resistance.

Cast iron and steel parts steam treated at 345 °C (650 °F) or higher have increased resistance to wear and corrosion. Steam treated cast iron valve bodies also have less porosity.

Processing Considerations. Before parts are processed in steam atmospheres, their surfaces must be clean and oxide-free, to permit the formation of a uniform coating. To prevent condensation and rusting, steam should not be admitted until workpiece surfaces are above 100 °C (212 °F). Air must be purged from the furnace before the temperature exceeds 425 °C (800 °F), to prevent the formation of a brown coating instead of the desired blue-black coating.

Charcoal-Base Atmospheres

Charcoal-base atmosphere generators (AGA classes 402 to 421), which are on the way to becoming obsolete, are now used mainly by small-scale manufacturing plants that desire a generator of low initial cost for intermittent operation. The principal uses of charcoal-base atmospheres at the

Fig. 20 Flow diagram of an exothermic-endothermic gas generator

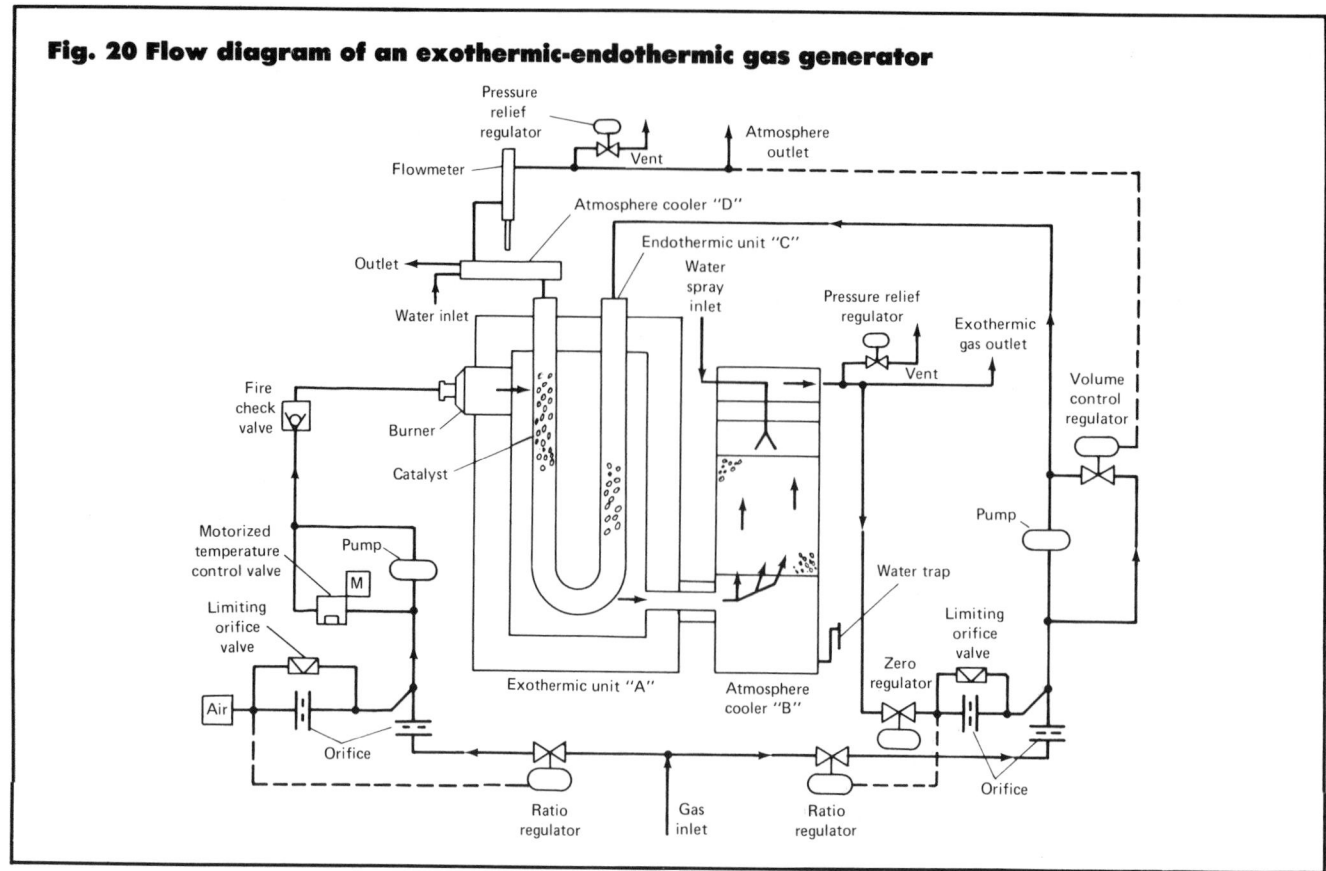

present time are for the manufacture of malleable iron castings and for providing atmosphere in small tool-room heat treating furnaces.

Composition and characteristics of a class 402 charcoal-base atmosphere are given in Table 3. These gases are produced by passing air through a bed of hot charcoal. The air is supplied by a blower and measured by a flowmeter. Incoming air burns with the charcoal at the bottom of the combustion chamber to form a mixture of nitrogen, carbon dioxide and water vapor. This reaction heats the charcoal in the upper portion of the combustion chamber to incandescence. The incandescent charcoal converts carbon dioxide to carbon monoxide and water vapor to hydrogen. These hot, dry gases pass around the charcoal-storage hopper, heating the charcoal and driving off the moisture and volatile matter through a burn-off can at the top of the hopper. The hot, dry atmosphere is taken off at the side of the generator and passed through a filter to remove fly ash.

Charcoal-base atmosphere is produced by the reaction:

$$2C + O_2 + 3.8N_2 \rightarrow 2CO + 3.8N_2$$

Theoretically, this reaction would produce 34% CO and 66% N_2. Actually, however, the following mixture is normally obtained, because of moisture and volatiles in the charcoal and the less-than-100% efficiency in converting all of the carbon dioxide.

CO_2	1.0 to 2.0%
CO	30.0 to 32.0%
H_2	1.5 to 7.0%
CH_4	0.0 to 0.5%
N_2	rem

Although the generator may operate at as low as 50% of rated capacity, lower percentages of carbon dioxide and water vapor are attained at the rated output because of the hotter charcoal bed. The type of charcoal used also influences hydrogen and water vapor content of the atmosphere.

Normally, the atmosphere is neutral to higher-carbon steels, but the carbon potential may be increased by additions of about 0.5% of natural gas in the furnace, causing further breakdown of carbon dioxide and water vapor in the atmosphere. Channeling of gases through the incandescent charcoal bed or excessive moisture content of the charcoal can cause high carbon dioxide and water vapor contents that will decarburize high-carbon steels.

Charcoal-base atmospheres may be used for hardening, annealing and normalizing high-carbon steels without scale formation or decarburization. A gray-green oxide forms on steels containing appreciable amounts of chromium. At high temperatures, the great affinity of chromium for oxygen causes a breakdown of carbon monoxide. The relatively low hydrogen content of these atmospheres makes them particularly suitable for making malleable iron castings.

The principal disadvantages of charcoal-base atmospheres are the high operating cost, the absence of suitable means for automatic control, the intermittence of operation imposed by the need to recharge charcoal and remove ashes, and the corrosion of alloy furnace parts in the area of high-temperature combustion.

Exothermic-Endothermic-Base Atmospheres

These atmospheres (classes 501 and

502), the compositions and characteristics of which are previously described under "Classification of Prepared Atmospheres", are re-formed exothermic-base atmospheres that are less reducing than conventional endothermic-base furnace atmospheres.

Applications. Potentially, exothermic-endothermic-base atmospheres could be substituted for exothermic, endothermic and nitrogen-base atmospheres in virtually every application for which any one of these three atmospheres is recommended. They may also be used as carrier gas in carburizing and carbonitriding. In practice, however, it is not practical to mix exothermic-endothermic atmospheres in existing headers with straight endothermic atmospheres because of the differences in specific gravity and chemistry. This limitation may not apply to the use of the combination exothermic-endothermic type of generator, provided one generator for one furnace can be economically employed to produce both exothermic and endothermic atmospheres for use at the same time.

Generation of exothermic-endothermic atmospheres begins with the combustion of a mixture of air and fuel gas in a refractory-lined combustion chamber. This reaction supplies the external heating required to maintain a satisfactory temperature in a secondary, or endothermic-stage, reaction. Products of combustion from the first stage are dehydrated, and a predetermined quantity of a hydrocarbon fuel is added to them. This mixture is then reacted in contact with a catalyst contained in an externally heated alloy tube.

The primary air-fuel mixture yields very close to perfect combustion. For normal operations, 0.5% CO or O_2 in the generated gas is satisfactory. This reaction, using natural gas as the fuel, yields 11 volumes of combustion products. Removal of the water formed in the reaction reduces the total to 9.225 volumes. In the secondary reaction, 7.3 volumes of combustion products and 1 volume of natural gas react to produce 10.5 volumes of atmosphere of the following nominal composition:

CO_2 . . . . 0 to 0.2%
CO 17%
H_2 . . . . 20%
N_2 . . . . rem

Equipment. Figure 20 is a schematic diagram of a typical exothermic-endothermic generator. In operation,

Table 10 Equipment requirements for hardening small parts made of 1070 steel in an exothermic-endothermic atmosphere

Production requirements

Number of parts per load	2625
Weight of each part	0.007 kg (0.015 lb)
Maximum net weight of load	18 kg (40 lb)
Production rate	7.5 loads (135 kg or 300 lb) per h

Equipment requirements

Hardening furnace	Gas-fired radiant-tube single-row pusher type with automatic quench
Size of hearth	0.9 m (3 ft) wide by 2.9 m (9½ ft) long
Heat input	790 MJ/h (750 000 Btu/h)
Operating temperature	900 °C (1650 °F)
Capacity of generator	70 m^3/h (2400 ft^3/h)
Type of atmosphere	Class 502
Capacity of oil-quench tank	1250 l (330 gal)
Type of oil	Fast, 180 °C (360 °F) flash point
Temperature of oil	70 °C (160 °F) (controlled)
Oil agitation	Medium

the flow rates of air and gas are regulated by fixed and variable orifices and balanced pressure ratio regulators to maintain the mixture at a predetermined ratio. A gas pump introduces the mixture into a refractory-lined combustion chamber (Exothermic unit A in Fig. 20). The products of combustion then pass through a waterspray cooler (Atmosphere cooler B). After the combustion products have cooled, a measured volume of a hydrocarbon fuel is blended with them. The ratio of the products to the hydrocarbon fuel is controlled by fixed and variable orifices and balanced pressure ratio regulators, and the mixture is introduced under pressure by a gas pump into a catalyst-filled U-shaped alloy retort (Endothermic unit C in Fig. 20), which is contained in the refractory-lined combustion chamber and is heated by the primary-stage combustion. From the retort, the gas mixture passes into Atmosphere cooler D. From here, it is piped through an atmosphere outlet and into the furnace.

The volume of atmosphere needed for a given work load is influenced by the size of workpieces comprising the total work load and the type of furnace used for heating. See Table 10 for production and equipment requirements for hardening small parts made of 1070 steel in an exothermic-endothermic-base atmosphere.

Operation of Generator. When the generator is used for the manufacture of endothermic-base gas, it should be operated at approximately 1010 °C (1850 °F). If the temperature is too low, reaction will be incomplete; if the temperature is too high, life of the retort will be greatly shortened.

All of the previously described precautions to be observed when operating an endothermic generator should be adhered to, when endothermic-base atmosphere is being produced.

When the same generator is used for producing an exothermic-base gas, there is no need to change the operating conditions. Therefore, by closing the gas valve in the secondary stage, the unit purges itself of endothermic gas by producing exothermic gas. When used as an exothermic generator, with about 90% perfect combustion, it will produce a gas having the following approximate composition:

CO_2 . . . . 10.0 to 11.0%
O_2 . . . . 0 to 1.0%
CO . . . . 0 to 2.0%
H_2 . . . . 0 to 2.0%
N_2 . . . . 86.0 to 89.0%

When the generator is producing endothermic gas, it can, at the same time, supply an exothermic gas volume of approximately 25% of its rated endothermic capacity.

When changing over from production of exothermic to endothermic gas, it is recommended that a period of about ½ h be allowed to establish equilibrium in the generator. Gas-analysis equipment, either manual or automatic (as described in later sections), should be employed to determine when equilibrium conditions for the desired atmosphere have been established.

Generator Maintenance. The following is a typical maintenance schedule for exothermic-endothermic atmosphere generators:

Daily

- Check flow of cooling water.
- Check dew point and secondary gas analysis, and make adjustments as necessary.
- Check instruments.

Weekly

- Check primary gas analysis and make adjustments as necessary.
- Check operation and linkage of temperature control valve.
- Lubricate as required.
- Check drive belts.

Monthly

- Check for leaks.
- Check operation of combustion and fuel safeguard equipment, such as manual reset shutoff valves, pressure switches and firecheck.
- Check and clean air filter as required.
- Check cooling distribution on atmosphere cooler B (Fig. 20).
- Check atmosphere cooler D (Fig. 20) for (a) water leaks by dew point before and after, and (b) excessive soot deposits in the after-cooler by taking readings of pressure of atmosphere before and after it passes through the cooler.

Atmospheres for Backfilling and Quenching in Vacuum

Back-filling a vacuum furnace with a cooling gas is performed to induce a more rapid cooling rate. Back-filling also is used to suppress vaporization of oil in integrated oil-quenching vacuum furnaces and to provide an atmosphere for carburizing and nitriding.

Quenching. Inert gases, nitrogen, and, in rare circumstances, hydrogen, are used for cooling. If surface integrity of workpieces is to be maintained and damage to furnace parts avoided, contaminants in the cooling gas must be kept at a low level. Commercial gases are readily available with contaminant level below 0.01%.

Back-filling and forced circulation accomplishes a more rapid cooling rate, facilitates hardening, and, in some requirements, produces annealing of metal alloys. It also enhances efficient use of the equipment. Cooling gas is usually introduced into the vacuum chamber at the end of the high-temper-

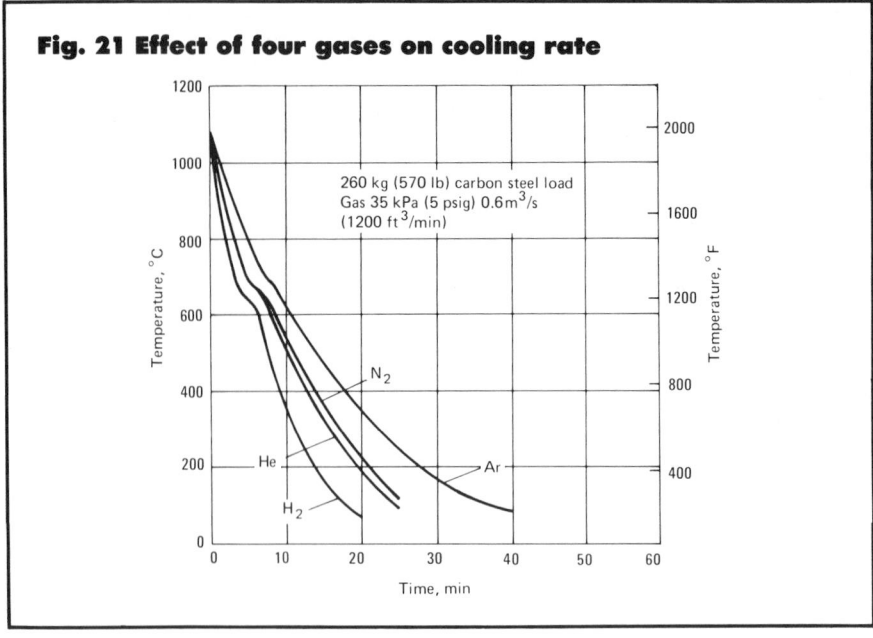

Fig. 21 Effect of four gases on cooling rate

260 kg (570 lb) carbon steel load
Gas 35 kPa (5 psig) 0.6 m³/s
(1200 ft³/min)

ature soaking period. When internal pressure reaches some predetermined level, either slightly negative or positive relative to atmospheric pressure, the gas is circulated through the load and then through appropriate heat exchangers located outside the heating zone. Design of the furnace and its structural strength dictates the maximum pressure level.

Figure 21 illustrates the effect of gas selection. Because cooling rates were established with identical loads and equipment, differences are attributed to differences in heat conductivity, heat capacity, viscosity and density of the gases tested.

Figure 22 illustrates the effect of pressure when the same cooling gas, load and equipment are used. Slightly negative pressure provides sufficient cooling to harden air-hardening materials such as air-hardening stainless steels and tool steels. Positive pressure provides a sufficient cooling rate to harden materials such as M2 tool steels. The mass of the load and part cross section dimensions also govern the cooling rate and must be recognized as cooling rate factors.

Vacuum Furnace Gas-Carburizing Atmospheres. Vacuum furnaces may be used for carburizing, by the injection of any one of several atmospheres that will induce carburizing at appropriate temperatures. Nitrogen enriched with a hydrocarbon gas is most frequently employed.

Vacuum furnace carburizing is nor-

mally carried out in the temperature range of 870 to 985 °C (1600 to 1800 °F). In the carburizing and diffusion process, surface carbon contents of 1% or more are formed initially. The high-carbon case is then diffused in vacuum until the surface carbon content and case depth are acceptable.

The diffusion process may be carried out in a series of interrupted periods during the carburizing cycle or at the end of the carburizing portion of the cycle. An interrupted carburizing cycle produces a less severe drop in carbon content and a higher hardness relative to case depth. Surface carbon content and surface hardness remain comparable in both diffusion processes.

Vacuum carburizing entails some problems with sooting and distribution of the carburizing atmosphere. Soot monitoring with subsequent enrichment control can minimize the soot problem, and forced circulation of the carburizing atmosphere assists in minimizing the distribution problem. (For more information, see the Section on vacuum carburizing in this Handbook.)

Ion-Carburizing and Ion-Nitriding Atmospheres. A hydrocarbon gas that is ionized by a high voltage system in a vacuum is a suitable atmosphere for ion carburizing. Methane is a good source of carbon and is frequently employed. With a voltage differential between an anode, which may be chamber walls, and the work load serving as the cathode, a plasma will be generated at

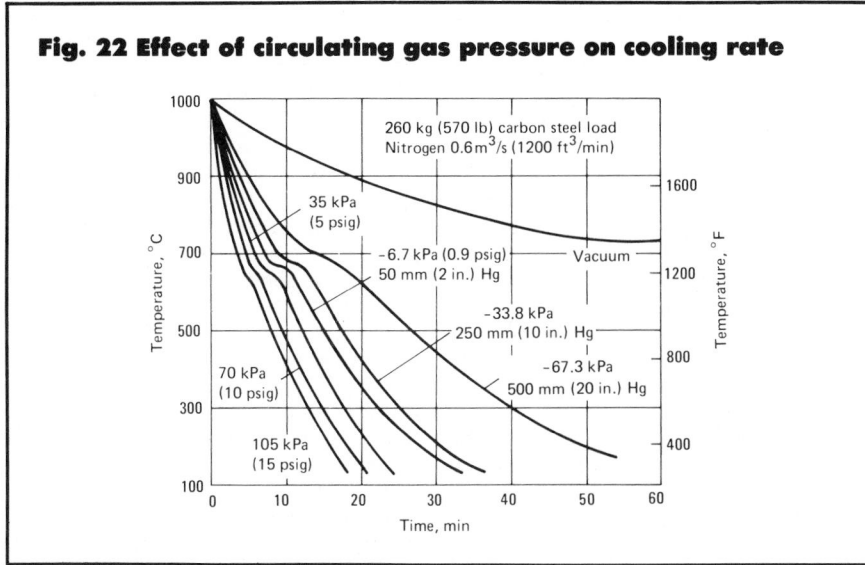

Fig. 22 Effect of circulating gas pressure on cooling rate

260 kg (570 lb) carbon steel load
Nitrogen 0.6 m^3/s (1200 ft^3/min)

35 kPa (5 psig)

−6.7 kPa (0.9 psig) 50 mm (2 in.) Hg   Vacuum

−33.8 kPa 250 mm (10 in.) Hg

70 kPa (10 psig)

−67.3 kPa 500 mm (20 in.) Hg

105 kPa (15 psig)

a critical vacuum pressure. Migration and impingement of the resulting ions on the surface of the load produce the carburizing potential. Ion carburizing is much more rapid than conventional atmosphere carburizing, and the surface-carbon content approaches saturation. Other diluting gases are added for carbon control.

An AISI 1018 steel can be ion carburized to a depth of 1.0 mm (0.040 in.) with a 10-min cycle in an atmosphere of methane at 1.3 to 2.7 kPa (10 to 20 torr) and a temperature of 1050 °C (1925 °F). Carburizing is followed by a 30-min diffusion cycle in vacuum at the same temperature. Approximately 400 V is required to produce the plasma.

Ion nitriding is similar to ion carburizing with the exception that the atmosphere gas generates nitrogen ions, and the process is carried out at lower temperatures. Ammonia or mixtures of hydrogen and nitrogen are suitable sources for the nitrogen ions.

A typical cycle of 8 h at 510 °C (950 °F) with a mixture of 75% hydrogen and 25% nitrogen at a pressure of 0.9 kPa (7 torr) and a current density of 0.8 milliamps/cm^2 will produce a nitrided case of 0.30 mm (0.012 in.) in AISI 4140 steel. Approximately 400 V is required to generate the plasma. Other alloys, such as 300 series stainless steels, M2 tool steels and nitralloy, respond to the ion nitriding process.

Partial Pressure. Nitrogen and inert gases are used in vacuum furnaces at partial-pressure vacuum levels to suppress vaporization of a constituent in the composition of the load. A gas is injected in small quantities by a needle valve. The deliberately leaked gas raises the pressure to some stable level between the ultimate vacuum of the pumping system and atmospheric pressure. For example, the vaporization of copper during the brazing of heat exchangers is suppressed by the leaking of nitrogen into the furnace and maintaining a level of 0.27 kPa (2.0 torr). For more information, see the articles on ion nitriding and ion carburizing in this Handbook.

Analyzing Atmosphere Requirements

Analysis of the atmosphere requirements of heat treating furnaces is influenced by several factors. Cost of operation is always one of the main considerations in selecting an atmosphere system. Another prime factor is the capital investment required to purchase the hardware, make the installation and satisfy the associated safety requirements. Certain circumstances dictate that all cost considerations be subordinated to achieve high levels of heat treating accuracy to satisfy stringent metallurgical specifications.

The degree of validity of any analysis is determined by the thoroughness of investigation and the definitive parameters placed on the project.

Influence of Required Processes. Many installations are built in which multiple processes are anticipated. In such instances, the atmosphere system must be capable of a reasonable response time. For example, a batch furnace may be used for carburizing in one cycle, followed by clean hardening of medium carbon steel in the next, without a long delay between heats. This processing schedule means the carbon potential will vary from 0.9 to 0.3% carbon. With an endothermic-base atmosphere, this variance means a dew point change from −12 to +18 °C (10 to 64 °F). With a nitrogen-base atmosphere, the carburizing cycle can be completed with additions of carbon-bearing material and an oxidant, and the hardening cycle can be accomplished by eliminating the additives.

In some installations, the volume of atmosphere is based on the process requirements rather than on the amount of atmosphere required to overcome air infiltration to the furnace chamber. This volume basis is used when the atmosphere provides an element in carburizing parts with large surface areas; for example, in sintering, the amount of hydrogen required becomes a function of the amount of oxide or sulfur to be reduced. Thus, distinctions need to be made between the need of an atmosphere to exclude unwanted air and the requirement that it enter into a precise chemical reaction. As the process specifications become more exacting, the need for precision of the chemical reaction increases and the need for reliable atmosphere control becomes more urgent.

Influence of Furnace Design. Furnace design and the selection of atmospheres are interdependent. In any atmosphere furnace, the degree of "tightness" is a critical factor, and among the various types of atmosphere furnaces—pusher tray, batch tray, cast-link belt, roller hearth, mesh belt, rotary retort, bell, gantry and others—the furnace designer basically has only three ways to load and four ways to unload the work.

The first way of loading and unloading an atmosphere furnace is through a slot. Sufficient atmosphere must be provided to prevent air from entering through the slot, and the amount can be significantly reduced by the use of curtains or flame screens, and by elevating the working hearth above the slot opening when possible.

The second way of loading and unloading is through a cold chamber or vestibule linked with the furnace. The most common use of a vestibule is with a pusher tray or batch tray furnace. A loaded tray is placed in the vestibule and the outer vestibule door is closed. The vestibule remains closed until sufficient time has elapsed to purge the vestibule of air. This purging occurs either by atmosphere leakage from the furnace around the inner furnace door or by admission of atmosphere gas through a direct line that is opened for a predetermined time. Sufficient atmosphere for approximately five volume changes must pass through the vestibule to be ensured that contamination will be minimized when the inner furnace door is opened to admit the work tray. Any mechanism that operates through the vestibule wall must be equipped with packing glands, and these must be maintained. The atmosphere in the vestibule also undergoes volume change when the inner door is raised and lowered, subjecting the vestibule gases to rapid temperature increases followed by decreases. The furnace door closing must not be so rapid as to cause a contraction of the gases in the vestibule greater than the flow of gas to the vestibule for normal purging.

The third way of loading and unloading is to place the work on a stationary or moveable base and to raise the work into the furnace or place the furnace over the base. The furnace chamber over the work is purged with atmosphere and the furnace is taken through a time-temperature cycle. When that cycle is completed, the work is uncovered and removed.

The fourth way of unloading a furnace is to have the work fall directly into a quench tank at the discharge end. To protect the atmosphere, an eductor must be installed to draw atmosphere from the quench chute just above the fluid line. If the quenching medium has a water base, the eductor will prevent water vapor from contaminating the atmosphere. If the quenchant is oil, the eductor will prevent oil splash and fumes from unbalancing the atmosphere. One mechanism that has been successfully used in direct quenching applications is a laminar-flow quench-fluid spray between the quench-tank eductor and the liquid quenchant in the quench chute.

An additional furnace-design consideration having a direct influence on the atmosphere is the use of fans. By increasing the flow of atmosphere gas across the faces of the work being processed, fans increase the effective concentration of the gases and accelerate chemical reactions in applications where the atmosphere is the source of a desired chemical ingredient. This is particularly significant in case hardening. A properly designed fan thoroughly mixes the atmosphere as the various chemical reactions take place, and improves temperature uniformity in dense loads. The result is more uniform part-to-part response to a process within a load.

Influence of the Availability and Dependability of Utilities. The availability and dependability of utilities influence the location of industrial plants. In heat treating operations, large amounts of energy are consumed. As the cost of energy increases, energy cost and availability receive more management attention and planning, particularly in heat treating operations that have had their energy supplies curtailed.

In the past, heat treating operations required a supply of natural gas, which was inexpensive; electricity, which was readily available; and water, which was needed for cooling purposes. These utilities not only were required for the furnaces, but also were the required ingredients for the generation of heat treating atmospheres. In many instances, natural gas was the preferred form of energy. It was used not only as fuel for furnaces but also as an acceptable feedstock for the atmosphere generators.

As energy supplies have become more expensive and less available, it has been relatively simple to go to alternate or dual fuels to heat furnaces. Finding substitute atmosphere feedstocks has been somewhat more complicated, but solutions are available. Commercial nitrogen produced by air-reduction plants provides an alternative in two forms. For large volumes, the air-reduction plant can be located at or near the using site, or nitrogen from a central air-reduction plant can be transported and stored as a liquid, at the using site. With the developing technology of utilizing methanol with nitrogen, it is now possible to establish a heat treating operation requiring carburizing atmospheres without dependence on natural gas supplies.

Influence of Estimating Atmosphere Quantities. As energy costs increase, the customary practice of employing an atmosphere generator of greater capacity than required adversely affects operating costs. This is especially true of endothermic atmosphere generation. The atmosphere feedstock is passed through a catalyst-filled heated reaction tube, which requires considerable energy to reach gas-cracking temperature. Little energy is required to maintain the reaction tube temperature over the range of minimum flow to maximum flow of reactive gases through the tube. For instance, a 68-m³/h (2400-ft³/h) endothermic generator at full flow will require a little over 1 m³/h (36 ft³/h) of natural gas per 3 m³/h (100 ft³/h) of produced gas. The same generator operating at one-third capacity or 23-m³/h (800-ft³/h) output will require a little over 1.3 m³/h (47 ft³/h) of natural gas per 3 m³/h (100 ft³/h) of produced gas. The differences on a 102-m³/h (3600-ft³/h) generator were even more pronounced. At 102-m³/h (3600-ft³/h) output, the required natural gas is 0.8 m³/h (28.7 ft³/h). At 68-m³/h (2400-ft³/h) output, the required natural gas is 0.9 m³/h (33.1 ft³/h), and at 34-m³/h (1200-ft³/h) output, the required natural gas is 1.2 m³/h (40.9 ft³/h). Because natural gas is the major cost factor in producing endothermic atmosphere, a 30% variation in operating cost is significant.

SELECTED REFERENCES

- "High Rate Carburizing in a Glo-Discharge Methane Plasma", by William Grube and Jack S. Gray: *Metallurgical Transactions, 1978,* American Society for Metals and Metallurgical Society of A.I.M.E., Vol 94, Oct 1978, p 1421
- "Ion Nitriding of Steels", by P. C. Jindal: *Journal of Vacuum Science and Technology,* Vol 15, No. 2, March/April 1978

Control of Surface Carbon Content in Heat Treating of Steel

By the ASM Committee on Control of
Surface Carbon Content*

MOST HEAT TREATING ATMO-SPHERES are gaseous mixtures containing carbon monoxide, carbon dioxide, methane, nitrogen, hydrogen and water vapor. The relative amounts of these gases depend on the type of generator gas used, on the processing temperature and on the amount of gas added during processing. For example, endothermic generator gas produced by catalytic reaction of natural gas with air results in the following composition (approximate percentages by volume): 20 CO, 40 H_2, 40 N_2, 0.1 to 0.5 CO_2, 0.2 to 1.2 H_2O, 0.2 to 0.8 CH_4. In gas carburizing, a common commercial practice is to use an endothermic gas as a carrier and to enrich it with natural or propane gas additions.

Gas Reactions. There are several possible gas reactions that may occur in a heat-treating atmosphere because of the presence of many different gas species. The most important reactions are described briefly below:

$$H_2 + CO_2 \rightleftarrows H_2O + CO \qquad \text{(Eq 1)}$$

$$H_2 + \tfrac{1}{2}O_2 \rightleftarrows H_2O \qquad \text{(Eq 2)}$$

$$CO + \tfrac{1}{2}O_2 \rightleftarrows CO_2 \qquad \text{(Eq 3)}$$

Equations 1 through 3 are homogeneous gas reactions that generally are assumed to be in equilibrium in furnace atmospheres above 790 °C (1450 °F).

Equations 4 through 6 are gas-solid reactions, where C represents carbon in solution in the steel:

$$2CO \rightleftarrows C + CO_2 \qquad \text{(Eq 4)}$$

$$CO + H_2 \rightleftarrows C + H_2O \qquad \text{(Eq 5)}$$

$$CH_4 \rightleftarrows C + 2H_2 \qquad \text{(Eq 6)}$$

When equilibrium conditions prevail during carburizing, the carbon potential in the gas phase determines the carbon concentration at the surface of the steel parts being processed. Equations 4 and 5 are much faster reactions than Eq 6, as discussed in detail later

in the section on kinetics. Thus, to maintain a given carbon potential in the gas atmosphere, the following reactions are important for replenishing CO_2 and H_2 and for maintenance of the appropriate low concentration levels of CO_2 and H_2O:

$$CH_4 + H_2O \rightarrow CO + 3H_2 \qquad \text{(Eq 7)}$$

$$CH_4 + CO_2 \rightarrow 2CO + 2H_2 \qquad \text{(Eq 8)}$$

Equations 7 and 8 usually are considered to be nonequilibrium reactions.

Control of Carbon Potential. Traditionally, control of carbon potential has been achieved, in principle, by controlling either water-vapor concentration (dew point) or carbon dioxide concentration or oxygen partial pressure.

The principle of carbon-potential control can be illustrated by the following equilibrium reaction:

$$2CO = C + CO_2 \qquad \text{(Eq 9)}$$

*James Dale, *Chairman,* Manager of Controlled Atmospheres, AIRCO Industrial Gases; Robert N. Blumenthal, Professor, Marquette University; Lewis H. Shaefer, Marketing Manager, Anarad, Inc.; Raymond L. Davis II, Carbon Control Instruments

Fig. 1 Isothermal equilibrium relationship between activity of carbon and carbon concentration in austenite for various temperatures

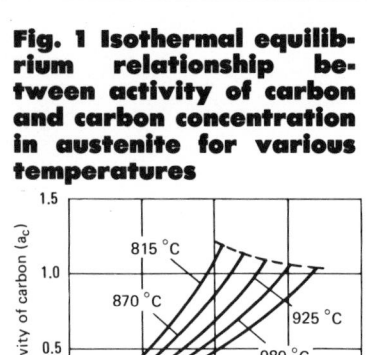

Fig. 2 Calculated equilibrium relationship between % CO_2 and wt % carbon in an endothermic-base atmosphere

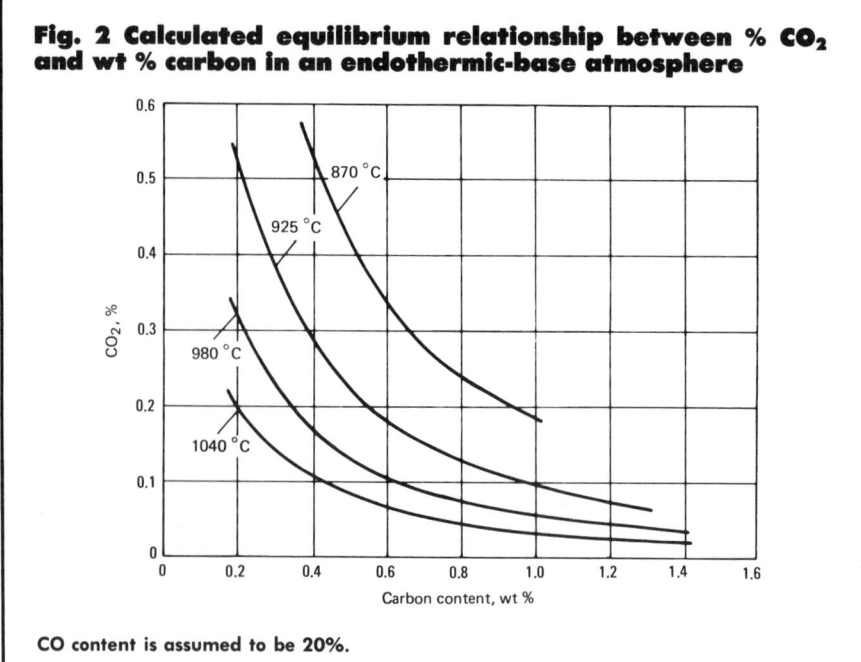

CO content is assumed to be 20%.

The equilibrium constant for Eq 9, K_1, is given by the relationship:

$$K_1 = \frac{a_c P_{CO_2}}{P_{CO^2}} \qquad \text{(Eq 10)}$$

where a_c is the activity of carbon and P_{CO_2} and P_{CO} are the partial pressures of CO_2 and CO, respectively. The expression:

$$a_c = K_1 \frac{P_{CO^2}}{P_{CO_2}} \qquad \text{(Eq 11)}$$

can be derived from Eq 10. The quantity a_c is related to the carbon potential by the equilibrium relationship shown in Fig. 1, where a_c is plotted against weight percentage of carbon in austenite for various isotherms between 815 and 1040 °C (1500 and 1900 °F). Because K_1 is temperature-dependent only and P_{CO} remains essentially constant, the carbon potential may be controlled by varying P_{CO_2}. The calculated isothermal equilibrium relationship between percentage of CO_2 and weight percentage of carbon in an endothermic-base atmosphere containing 20% CO is shown in Fig. 2. The experimentally determined relationship between percentage of CO_2 and weight percentage of carbon has been obtained by several investigators. Although the experimental values determined by different investigators are not in exact agreement with one another or with calculated values, they generally are in close enough agreement so that control of percentage CO_2 can be used as a means of controlling carbon potential.

The principle of control of carbon potential by control of H_2O vapor pressure (dew point) may be demonstrated easily by consideration of the following equations. Under equilibrium conditions, the partial pressure of H_2O is related to the partial pressure of CO_2. The following well-known water-gas reaction can be used to illustrate this relationship:

$$H_2 + CO_2 \rightleftarrows H_2O + CO \qquad \text{(Eq 1)}$$

The equilibrium constant for Eq 1, K_2, is given by:

$$K_2 = \frac{P_{H_2O} P_{CO}}{P_{H_2} P_{CO_2}} \qquad \text{(Eq 12)}$$

and the following expression for P_{CO_2} may be derived from Eq 12:

$$P_{CO_2} = \frac{P_{H_2O}}{K_2 P_{H_2}} P_{CO} \qquad \text{(Eq 13)}$$

Substitution of the right side of Eq 13 for P_{CO_2} in Eq 11 gives:

$$a_c = K_1 K_2 \frac{P_{CO}}{P_{H_2O}} P_{H_2} \qquad \text{(Eq 14)}$$

Because K_1 and K_2 are temperature-dependent only and P_{CO} and P_{H_2} remain essentially constant in the carburizing atmosphere, the carbon potential can be controlled in principle by controlling the vapor pressure of H_2O (dew point). The calculated equilibrium relationship between dew point and weight percentage of carbon for an endothermic-base atmosphere containing 20% CO and 40% H_2 is shown in Fig. 3. Although the experimental values determined by different investigators are

not in exact agreement with one another or with calculated values, they generally are in close enough agreement so that control of dew point can be used as a means of controlling carbon potential.

Based on simple thermodynamic considerations, it can be easily demonstrated that, in principle, control of oxygen partial pressure, P_{O_2}, can be used to control carbon potential. Under equilibrium conditions, P_{O_2} is related to the partial pressure of Co_2. The following well-known expression can be used to illustrate this principle:

$$CO + \frac{1}{2}O_2 \rightleftarrows CO_2 \qquad \text{(Eq 3)}$$

The equilibrium constant for Eq 3, K_3, is given by:

$$K_3 = \frac{P_{CO_2}}{P_{CO} P_{O_2^{1/2}}} \qquad \text{(Eq 15)}$$

and the following expression for P_{CO_2} may be derived from Eq 15:

$$P_{CO_2} = K_3 P_{O_2^{1/2}} P_{CO} \qquad \text{(Eq 16)}$$

Substitution of the right side of Eq 16 for P_{CO_2} in Eq 11 gives:

$$a_c = \frac{K_1 P_{CO}}{K_3 P_{O_2^{1/2}}} \qquad \text{(Eq 17)}$$

Because K_1 and K_3 are temperature-dependent only and P_{CO} remains essentially constant in the carburizing atmosphere, the carbon potential can be

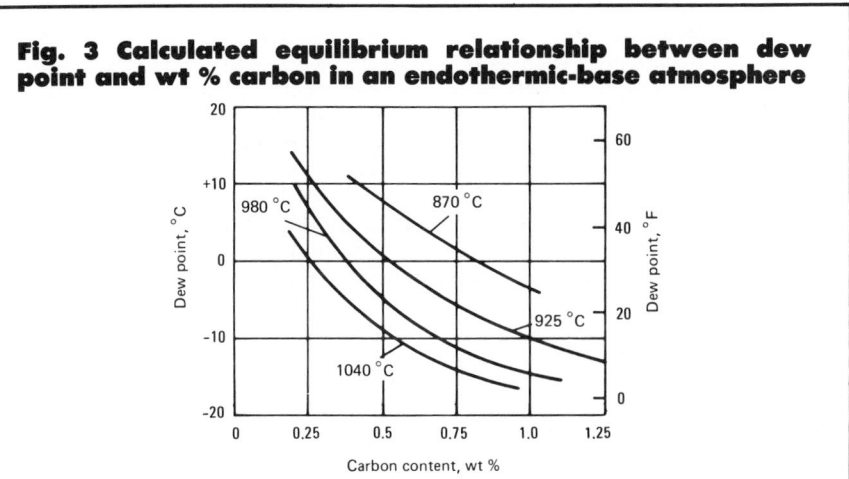

Fig. 3 Calculated equilibrium relationship between dew point and wt % carbon in an endothermic-base atmosphere

CO and H₂ contents are assumed to be 20% and 40%, respectively.

controlled, in principle, by controlling the partial pressure of oxygen.

Kinetics

As presented in the foregoing discussion, the concept of equilibrium conditions presupposes that an infinite length of time for reaction, which is necessary to ensure true equilibrium, is available. In actual heat treatment of steel, however, reaction times are important factors. Some reactions are fast and take control of the heat-treating process; these are the reactions that are of chief concern to heat treaters. Other comparatively slow reactions are not as important. Intergas reactions determine the heat-treating environment and the carbon potential it will exhibit. Gas-solid reactions are responsible for the actual carbon transfer and so determine the rate of case formation.

Homogeneous Gas-With-Gas Reactions. Some gas-with-gas reactions, such as those given in Eq 1, 2 and 3, are fast. In a well-mixed atmosphere of uniform composition, fast reactions reach equilibrium and characterize, to a large extent, the gas and its carbon potential.

Other reactions, such as those given in Eq 7 and 8, are comparatively slow. These are endothermic reactions that require considerable energy absorption but that are important as replenishment reactions. They are, however, less important in characterizing the gas, especially in terms of effective carbon potential.

Heterogeneous gas-with-solid reactions, such as those defined by Eq 4, 5 and 6, are relatively slow. These are the reactions by which carbon is transferred to steel surfaces and carbon potential is displayed and made effective. The reaction in Eq 6 is particularly slow, and this reflects the fact that CH_4 is a relatively stable molecule and does not dissociate readily. The reaction given in Eq 5 is said by some investigators to be the fastest of all carburizing reactions, and the most influential in determining the rate of case formation.

Carburization of a surface involves the following three factors not present in gas-with-gas reactions, and one of these factors becomes rate controlling.

- *Movement of molecules to the metal surface.* Reaction in contact with the metal surface is usually required. Reactions occurring away from the metal surface usually result simply in soot deposition, which hinders carburization. Turbulent mechanical circulation of the gases over the metal surfaces keeps the gases well mixed and uniform in composition, and maintains flow of the supply gases into the main stream of reacting gases. The amount of energy and time required for mixing can be significant, especially where loads are dense and gases do not have the same ready access to all parts of the load. It is difficult even to approximate flow patterns and rates, and thus calculation of kinetics for gas reactions is not feasible.

- *Reactions between gases and metal surfaces.* Carburization occurs when the proper gas molecules interact with a metal surface and transfer carbon to that surface. As the carbon is released by the gases, it is incorporated (dissolved) into the surface layer of the metal. These reactions are rate-controlling with respect to the over-all process, especially in the early hours of the process. The concentration of the active species affects the rate at which the proper molecules contact the metal surface and consequently affects the rate at which the gas mixture gives up carbon to the surface. Conversely, the concentrations of decarburizing H_2O and CO_2 molecules (relative to equilibrium concentrations) determine whether some of the carbon transferred will be reincorporated promptly into a gas molecule. Considerations such as these determine whether, at a given temperature, a given gas mixture will carburize steel at the maximum rate or at a considerably lower rate. For these reasons, increasing methane concentrations well above equilibrium values (that is, without corresponding decreases in CO_2 and H_2O) will not necessarily result in the expected increases in carburizing rates.

Dilution of the carburizing gas mixture with an inert gas such as nitrogen will slow carburization by reducing the rate at which carbon-donating molecules contact the metal surface. To this reduction is added the effect of the inert diluent on carbon potential. Generally, that effect is a drop in carbon potential. However, in the endothermic-gas technique, concentrations of inert diluent up to at least 40% do not seriously impair the carburizing power nor the carburizing rate of a well-prepared gas mixture. Despite these known factors, the mechanisms of reactions between gases and metal surfaces are too complex, and conditions at the work surfaces too difficult to evaluate, for accurate mathematical prediction of reaction rates for a given gas composition in a production carburizing furnace.

- *Diffusion of dissolved carbon from the metal surface.* When carbon has become part of the metal surface, it usually represents an increase in carbon concentration over that of the core except where the surface has been previously decarburized. With increased carbon concentration at the surface, the carbon begins to diffuse toward the leaner core at a rate that is a function of this concentration gradient and of case depth and of temperature. Initially, this diffusion rate is high; carbon diffuses inwardly at a rapid rate and surface concentration increases slowly. As the surface carbon activity approaches equilibrium with the gas carbon activity after about 2 h at 925 °C (1700 °F), the rate of inward transfer diminishes to the point at which it becomes rate controlling. Beyond that point, acceleration of carburization requires that either the carbon potential or the carburizing temperature be raised.

Kinetics in Practice. Recognizing that transfer of carbon from the gas to the steel surfaces can be rate controlling for the first several hours of a carburizing cycle, it is advisable to:

- Employ high-power circulating fans in the furnace to move the atmosphere rapidly and to keep it thoroughly mixed.
- Direct the movement of gases as much as possible to maximize circulation through the work load and to minimize short circuiting in the region of the fan.
- Allow sufficient space for movement of gas between and around workpieces.
- Use atmosphere compositions that initially will provide a carbon potential well above the concentration desired in the work surface. This will facilitate carbon transfer at a maximum rate while the case is shallow and carbon demand is high.
- Reduce carbon potential during processing in sufficient time to establish carbon concentration in the surface at the desired set point before processing is complete.
- Anticipate the total carburizing time required for a given case by reference to published data or by means of calculations based on Fick's law of diffusion. Use this as a guide in evaluating the rates of case formation obtained.

Process Control in Gas Carburizing

Control of a gas-carburizing process involves accurate measurement and adjustment of temperature, time, flow of carrier gas or atmosphere, and flow of enriching gas. Initial set points for these parameters are established from equilibrium data or from past experience. Set points may be changed as necessary to maintain required properties of the workpieces. In some installations, certain parameters such as temperature and carrier-gas or base atmosphere flow are fixed, and the time of exposure and flow of enriching gas are varied to achieve differences in case depth and surface carbon content.

Control-System Features. Regardless of the variable being controlled or the instrumentation being used, a major feature of every control system should be a set of signal devices that warn operating personnel of major malfunctions. For instance, a power failure normally renders a control system inoperative. It also shuts down the furnace. If the furnace temperature drops below 760 °C (1400 °F), there is a clear danger of an explosion. A temperature-activated battery-powered audible or visible alarm warns personnel of a decrease in temperature in sufficient time so that they can open the furnace doors and burn out or purge with an inert gas the combustible mixture before the temperature drops below the danger point.

Under normal operating conditions, control is maintained by periodic checking of the control parameters against set points or specified values, followed by adjustment of heat input, enriching-gas flow or other process variables as necessary. Case depth often is specified as effective case depth (to 0.40% carbon, or to 50 HRC), which indicates the hardened case depth. The effective case depth is influenced by steel composition and by process variables—temperature, time at temperature, carbon potential of the atmosphere and efficiency of the quench.

Accuracy. It can be shown that for a constant atmosphere composition, especially where there is no change in carbon dioxide content, a change in temperature of about 10 °C (18 °F) can produce a change of as much as 0.1% in carbon potential. Therefore, accurate control of carbon potential is achieved only when there is accurate control of temperature.

Normally, an atmosphere-control system must respond to variations of 0.01 to 0.02% in carbon dioxide content; however, this degree of control is not sufficient in high-temperature carburizing, where the set point may be below 0.1% carbon dioxide. Then, reproducibility of $\pm 0.005\%$ or better usually is recommended.

Inherently, carburization is a slow process; a part usually is at elevated temperature for several hours. Because of this, periodic fluctuations in temperature or in carbon potential usually are of little consequence. As long as average conditions are maintained, without large or prolonged excursions from setpoint conditions, it is unlikely that parts will be adversely affected. Likewise, a few minutes' difference from programmed time in the furnace usually will be of little consequence.

Only carburizing and diffusion times, not total furnace time, should be counted toward the time required to develop a given depth of case. Preheating time is disregarded because little carburizing takes place in this part of the cycle. Any equalizing time following diffusion also may be disregarded; during cooling from the diffusion temperature to the quenching temperature, any diffusion that occurs will not substantially increase case depth.

Controlling the Process. For control of carburization, there is no substitute for alert, vigilant, well-trained operators. There are many ways in which a carburizing furnace and its controls can malfunction. For instance, failure of a thermocouple or plugging of the atmosphere-sampling tube can cause the process to go out of control, and soot buildup in the furnace interferes with carburizing reactions. These and many other relatively common occurrences can be detected by operators, and corrective action can be taken before there is an adverse effect on parts in the furnace.

Moreover, considerable thought and preparation should go into the initial, preproduction planning for carburizing a new part or for putting new or renovated equipment into operation. The following is a partial checklist of items to be considered in setting up for production.

- *Temperature distribution.* Make sure that the temperature is uniform throughout the heated zones of the furnace and that thermocouples are located correctly.
- *Effects of loading.* Full trayloads of parts may have to be checked and subjected to complete metallurgical examination to determine distribution of case depth, hardness and other properties in parts throughout the load. Light loading is inefficient, but heavy loading can cause nonuniform atmosphere flow through the work. Bulk loading of parts heaped in baskets is likely to cause case depth and hardness variability.

- *Variations in carbon concentration.* Test bars for carbon determination should be placed in different furnace loads and at varying locations on trays. Results of tests should be analyzed statistically to ascertain the process capability of the furnace.
- *Variations in properties.* Effects of the positions of parts in the load on as-quenched or quenched-and-tempered properties should be analyzed statistically.
- *Atmosphere introduction.* In continuous furnaces for shallow cases and with fast pushing cycles, most of the enriching gas is piped into the furnace close to the discharge end and possibly flowed countercurrent to the work. For slow cycles and for deep cases, the main flow of enriching gas into the furnace should be nearer to the charging end. In either instance, most of the enriching gas should enter at the point where the work load has reached the carburizing temperature.
- *Dew-point characteristics.* In continuous furnaces, it should be ascertained whether the last furnace zone (diffusion zone) can develop the required dew point, with some latitude for adjustment. A small, measured quantity of air may be introduced to increase the dew point and lower the carbon potential. Some operators prefer to inject exothermic gas instead of air at the discharge end.
- *Hardenability of the steel.* Wherever possible, properties of the parts should be determined for heats of steel at both the top and bottom limits of the hardenability band.
- *Dimensional variations.* Parts requiring close tolerances should be measured before and after hardening to evaluate dimensional changes.

When a furnace is first put into operation, or when it has been overhauled or relined with new firebrick or after a long furnace shutdown, it usually is necessary to dry out and "condition" the furnace before it is used for carburizing production parts. Conditioning consists of bringing the empty furnace to the operating temperature, then admitting a carburizing atmosphere and allowing the atmosphere to react with the internal components of the furnace for at least several hours. Conditioning ensures that the internal components of the furnace will be essentially at equilibrium with the carburizing atmosphere, so that chemical reactions between the atmosphere and the work-

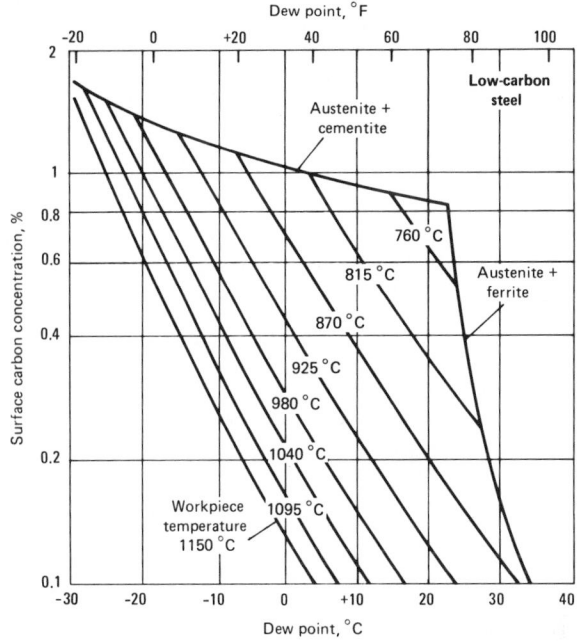

Fig. 4 Variation of carbon potential with dew point for an endothermic-base atmosphere containing 20% CO and 40% H_2 in contact with plain-carbon steel at various workpiece temperatures

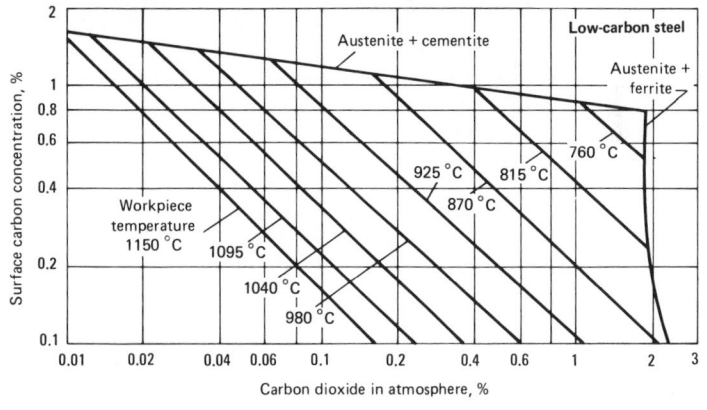

Fig. 5 Variation of carbon potential with CO_2 concentration for an endothermic-base atmosphere containing 20% CO and 40% H_2 in contact with plain-carbon steel at various workpiece temperatures

pieces will not be weakened because of reactions between the atmosphere and the furnace components.

Set points for control of the carburizing atmosphere usually are estimated by referring initially to a plot of carbon potential versus dew point (see Fig. 4) or of carbon potential versus carbon dioxide concentration (see Fig. 5). After

control has been established and the furnace is conditioned, it may become necessary to adjust set points to ensure that the desired surface carbon content is obtained consistently and reliably. Charts such as Fig. 4 and 5 also may be used to estimate the effect of a given change in set point.

Temperature control should be

maintained with an average deviation from the control set point of no more than ±3 °C (±5 °F) for good control of results. However, a control cycle of ±6 °C (±10 °F) usually is acceptable. In a continuous four-zone carburizing furnace, thermocouples should be placed near the end of the first (preheat) zone, because workpieces should be at carburizing temperature when they enter the second (carburizing) zone. Thermocouples should be placed in the centers of the second zone and the third (diffusion) zone. In the fourth (equalizing) zone, thermocouples should be close to the discharge end; this will ensure that the workpieces have reached the required temperature for subsequent quenching, which usually is about 815 °C (1500 °F).

There should be provision for two thermocouples in each furnace zone. They should be as close as possible to the workpieces. One thermocouple is for the temperature recorder/controller, and the second is for an overheat potentiometer set at the overtemperature of that zone. If a hole or crack develops in a thermocouple-protection tube, the thermocouple will come in contact with the reducing furnace atmosphere and will malfunction. When this happens, the protection tube and thermocouple must be replaced.

Circulation of the atmosphere influences temperature uniformity throughout the furnace. Without good circulation, there is no effective temperature control, and wide differences in temperature can exist between the outer sections and the inner sections of the furnace load. This is true for both continuous and batch furnaces.

Atmosphere Control. Generator gas that contains 0.1 to 0.3% carbon dioxide and that has a dew point of about −10 to +5 °C (15 to 40 °F) is ideal for use in a carburizing atmosphere. Reactions in the gas generator are controlled by regulation of the gas-to-air ratio in the mixing valve before the hydrocarbon gas is cracked in the retort tubes of the generator. Increasing the air increases the dew point, and increasing the natural gas decreases the dew point. It usually is preferable to generate a carrier gas with a composition favorable to trouble-free generator operation. Production of carrier gas with a carbon dioxide content of less than 0.1% may cause excessive sooting in the generator, requiring frequent burnouts.

At no time should the carburizing atmosphere in the furnace be controlled by varying the dew point of the generator gas. If the furnace atmosphere is too rich, raising the generator-gas dew point to decrease the carbon potential will not work effectively. The reactions in the generator will not stay in equilibrium with this method of control. If the generator-gas dew point is held constant, the reactions will stay in equilibrium; then the carbon potential of the furnace atmosphere can be controlled by varying the flow of enriching gas.

One way of controlling the carbon potential automatically is by infrared control of the furnace atmosphere composition zones. The atmosphere of each zone is assigned a predetermined set point. When the atmosphere is not on the set point, a signal is sent to a motorized valve, which either increases or decreases the flow of the enriching gas. The equalizing zone generally is controlled by admission of a predetermined amount of air to keep the carbon potential at its predetermined value.

Sampling of Atmospheres for Analysis

In carburizing of steel, three properties of the finished case are important: surface concentration of carbon, case depth and carbon gradient. Case depth depends primarily on carburizing time and temperature; it often can be presumed to depend only on time, because the temperature used in a given plant is often standardized. The other two case characteristics depend strongly on the carbon potential of the atmosphere and on accurate control of carbon potential. Carbon control is difficult unless the amounts of both carburizing and decarburizing constituents present in the atmosphere can be analyzed and controlled. The first requisite of analysis is that a representative sample of the furnace atmosphere be obtained.

The sample of gas should be taken from a point in the furnace chamber as close as possible to the work being treated. This will lessen the likelihood of obtaining a sample of stagnant gas, such as may be present near the furnace wall. The sampling point should also be as far as possible from the atmosphere inlet ports and from burner tubes. When the furnace has forced circulation through the work load, a sample taken from the atmosphere as it leaves the work will most nearly reflect its effective operating condition.

Velocity of Gas Flow. Empirical data indicate that, when carbon dioxide or water vapor is to be measured, the velocity of gas flow through the sampling probe should be at least 1.2 m/s (240 ft/min). This velocity almost completely prevents reactions between constituents by reducing the time that the gas is in the intermediate temperature range, which exists mainly in the portion of the tube that passes through the furnace wall. With lower velocities, the water-gas reaction will take place at lower temperatures and thus raise the concentration of carbon dioxide and decrease the concentration of water vapor. If the carbon potential of the atmosphere is high, carbon monoxide will decompose into carbon dioxide and soot. This will further increase the concentration of carbon dioxide and also will affect the water-vapor content through the water-gas reaction.

When sooting occurs in the probe, the sample delivered to the analyzer has a higher carbon dioxide content than actually exists in the furnace. Based on analysis of the sample, the controller would increase the flow of enriching gas, which would put the furnace out of control and compound the sooting problem in the probe. This condition can be detected by observing the flow rate of enriching gas. If it is significantly higher than the normal value for the indicated carbon dioxide content it indicates out of control furnace atmosphere. Probe sooting also may be suspected if a manual check of dew point reveals an abnormally low water-vapor content for the indicated carbon dioxide content.

An alternative method of preventing a gas reaction during cooling is water cooling of the sample probe. The water cooling prevents water-gas reaction in the sample probe. This method should not be used with dew-point analyzers when the temperature of the cooling water is below the dew point of the atmosphere. Accurate measurement of dew point will not be obtained until the sampling line is dry. For other types of analyzers, a water trap is required in the sampling line to keep condensed water out of the analyzer. Even with a trap, however, it is good practice to keep the sampling line dry and ensuring that water does not enter the atmosphere analyzer.

Probe sooting also can be eliminated by inserting a high-purity quartz liner into the probe to eliminate any catalytic effect of the probe material on chemical reactions within the probe.

If a flow rate below 1.2 m/s (240 ft/min) is used without a water-cooled sample probe, analysis will be sensitive to flow rate. Flow sensitivity can be checked by varying the flow rate from $\frac{1}{2}$ to $1\frac{1}{2}$ times the normal value and noting any change in analysis that occurs as a result.

Probe Materials and Design. The probe should be made of a heat-resisting alloy that does not react with the gas sample. Probes made of iron-chromium alloys are preferred to those made of high-nickel alloys, because nickel catalyzes the breakdown of carbon monoxide into carbon dioxide and soot.

If the sample flow rate required by the analyzer is known, the internal diameter of the sampling probe that will provide the desired velocity can be calculated. Permanent probes smaller than 6.4 mm ($\frac{1}{4}$ in.) in diameter are not recommended, because it is difficult to prevent them from becoming plugged with soot. To facilitate cleaning, a 25-mm- (1-in.-) diam pipe with tee and cap often is used as the outlet through the furnace wall. When the probe is mounted in a horizontal position, it must be made with heavy-wall tubing or must be placed inside a larger tube for increased mechanical strength.

One of the more effective methods of ensuring proper flow rate is to provide a separate sample pump for each furnace, furnace zone or carrier-gas generator being controlled. The sample pump pushes the sample gas from the furnace or generator to the analyzing instrument. This minimizes sample contamination due to leaks in the sample line, because there is always a positive pressure in the portion of the line downstream from the pump. If a sample pump should fail, only one sampling point is affected, not the entire controlled system of a multiple-point recorder-controller.

Permanent sampling probes should have a tee with a cleanout plug. Even though the instrument may have a filter of its own, it is wise to install a separate filter about 0.6 m (2 ft) from the gas sampling tube in the furnace to collect any soot or dirt. An ideal filter for this purpose is the type used for compressed-air lines, which has a replaceable porous metal filter cartridge and a transparent plastic bowl. Another common type has a glass-wool filter cartridge. The filter cartridge can be in-

spected periodically and changed when dirty, and any condensate in the filter can be drained off by a petcock on the filter. After draining, if condensate is observed in the filter cartridge, then it should be replaced with a new one.

The sample line may be of copper, aluminum or stainless steel tubing, although copper should not be used if ammonia is present in the atmosphere. A plastic tubing that will not absorb moisture also may be used, if it is protected from heat. In installations in which the sample line must pass through areas where the ambient temperature sometimes falls below the dew point, steam tracing or electrical heating of the sample line is required.

Procedures and Precautions. Before a gas sample is taken from a batch-type furnace without a charging vestibule, or from a continuous furnace from start-up, enough time should be allowed for purging of the furnace chamber and porous refractories. Otherwise, high atmosphere dew points will occur at the beginning of a cycle and will cause condensation in the sample line. For batch-type furnaces, this delay can be accomplished automatically with an electrical limit switch operated when the furnace door or cover seal is broken and a time-delay relay set for the proper purging time. For continuous furnaces, the sample pump should be turned off, or the sample line should be blocked by means of a manual valve.

In pusher-type continuous furnaces, the periodic opening of the furnace doors may cause large changes in atmosphere composition. When manual instruments are used, samples should be taken at the same time in each push cycle (preferably just before a push), in order to compare results. Use of an automatic analyzer that continuously monitors the atmosphere will permit observation of the effects brought about by venting, by resealing of doors and by changing the purge flows to the vestibules. With this as a guide, the amount of variation in atmosphere composition can be reduced.

Relationship Between Control Systems and Analysis. Usually, automatic equipment for controlling atmosphere-mixture ratios and carbon potential is more reliable than manual equipment, which varies in effectiveness with the skill of the operator. But, regardless of whether or not the process is controlled automatically, there is no substitute for alert, intelligent personnel on the operating floor. It should not

be assumed that automatic process controls are the answer to control problems. Often, equipment malfunctions have been noticed by operating personnel in time to switch over to manual control and avoid ruining of parts in the furnace.

Generator output and furnace-atmosphere composition are controlled by recording analyzers. These should have proportioning-type controls rather than on-off or two-position control because of the time lag in the sampling lines and the slow response of some analyzers. The preferred procedure for controlling carburizing atmospheres is to hold the flow of carrier gas constant and to control the flow of the enriching gas. The normal procedure for generators is to set the carburetor (or mixer) to produce a rich fuel-air ratio and then to control the flow of air in a bypass line around the carburetor.

When the carbon potential of an atmosphere is to be controlled and one constituent of the atmosphere is to be measured, the set point must be determined on the basis of chemical analysis of the carbon content of either shim stock or turnings from test bars. Equilibrium data can be used to determine the approximate set point, but equilibrium conditions normally do not exist in furnace atmospheres. Therefore, the set point must be adjusted until the work meets specifications.

Control of temperature usually is accomplished automatically, with a single-point or multiple-point recorder-controller regulating the heat input to the furnace. When several furnaces or several zones in a single furnace need to be controlled, it may be necessary to use a separate (dedicated) temperature controller for each furnace or each zone, particularly if different set points are used. Location of thermocouples that sense the control point is important, especially in continuous furnaces under zone control or where furnace doors are opened frequently.

Simple elapsed-time controllers frequently are used to trigger successive events in the carburizing cycle, such as successive pushes in a pusher furnace or the end of carburizing and the beginning of diffusion in a batch furnace. In some types of continuous furnaces, cycle time is regulated by the speed at which parts are moved through the furnace.

For batch furnaces with programmed temperatures, atmosphere set points also must be programmed. Shims removed at the end of each temperature period will serve as guides in determining correct set points.

When a furnace or generator is first placed under automatic control, it is considered good practice to monitor carbon dioxide content continuously and to correct any variations caused by equipment condition. The following list gives the interdependent variables that must be considered in order to establish and maintain automatic control of the carbon dioxide content in a carburizing atmosphere:

- Type of furnace
- Condition of equipment, including tightness and ability to maintain uniform temperature
- Size and type of typical work load
- Composition and uniformity of carrier gas
- Composition and uniformity of enriching gas
- Frequency of sampling and controlling
- Type of control—on-off or proportional
- Sensitivity of controller
- Width of proportional band on proportional controllers
- Normal and maximum response times of the motor-driven control valve
- Flow rate through fully open control valve
- Normal time for atmosphere to stabilize following a change in a controlled variable.

Dual controls offer certain advantages for improving the stability and responsiveness of automatic control systems, when the atmosphere gas is not of uniform composition, or when there are frequent large variations from normal operating conditions. Under these conditions, measurement of a single constituent of the furnace atmosphere will not necessarily lead to the desired level of process control. In these cases, control through analysis of only one atmosphere constituent results in a relatively large percentage of off-quality production. Simultaneous analysis and control of two atmosphere constituents usually is sufficient to overcome any of the more common problems.

Dual control is most often accomplished by either (a) infrared ratio control of both carbon monoxide and carbon dioxide, or (b) simultaneous independent analysis of carbon dioxide and dew point. In infrared ratio control, a two-column analyzer is used—one col-

umn sensitized to carbon monoxide and the other to carbon dioxide. The output from the two columns is sent, by means of a tapped slidewire, to a third unit that actually controls the flow of enriching gas. The use of a tapped slidewire allows the output of the analyzers to be read as the ratio $(P_{CO})^2/(P_{CO_2})$, which is directly proportional to carburizing potential.

A system for simultaneous analysis of both carbon dioxide and dew point normally consists of two independent analyzers connected to a single sampling line. In the usual arrangement, the carbon dioxide analyzer establishes primary control, and the dew-point controller activates an alarm circuit whenever the water-vapor content of the atmosphere deviates from prescribed limits.

Infrared Analyzers

Infrared analyzers are based on the principle that any compound present in the mixture will absorb infrared energy in proportion to its concentration in the mixture. The wavelengths absorbed are different for each compound. Elemental gases such as hydrogen and oxygen do not absorb infrared radiation and hence cannot be analyzed by this method.

Infrared analyzers normally are used to measure carbon monoxide, carbon dioxide and methane. For automatic control of carburizing processes where the carrier gas and enriching gas are reasonably uniform in composition, infrared analysis of carbon dioxide is considered to be the most accurate method and is the most widely used. Infrared analyzers normally are connected to a recorder-controller or a scanner-programmer-printer of either the single-point or multiple-point type, which actually performs the control function.

In multiple-point control assemblies, a single analyzer is coupled to a system of gas-sample valves and control circuits, one set of sample valves and a control circuit for each point being analyzed and controlled. After each analysis is made and the resulting control signal transmitted to the appropriate control valve, an automatic switching mechanism disconnects the sample valves and control circuit and energizes the set of sample valves and the control circuit for the next control point. A maximum of eight different furnaces, furnace zones or carrier-gas generators

Fig. 6 Elements of a positive-filtering infrared analyzer for measuring CO, CO_2 and methane contents of an atmosphere

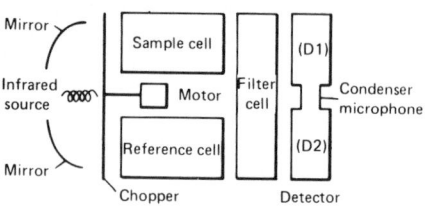

usually is recommended for a single analyzer.

An infrared analyzer responds rapidly when the analysis function is triggered by the recorder-controller. Even so, time is required for transmittal of the control signal, and for flushing of the sample cell in the analyzer so that the composition of a new sample will not be altered by remnants of the previous sample. Successive analyses can be made at intervals of about 1 min—approximately 30 s for flushing and analysis, and the remaining 30 s for controlling.

Positive-Filtering Analyzer. In this type of analyzer (illustrated schematically in Fig. 6), an electrically heated helix of nickel-chromium alloy wire is the source of infrared radiation. Radiation from this source is split into two beams by mirrors. Both beams are simultaneously interrupted by a motor-driven chopper. The resulting pulses of radiation cause alternate heating and cooling of gas in the two sides of the detector. A condenser microphone consisting of a movable metal diaphragm and a fixed metal plate is mounted between the two sides of the detector. A measurement of differential pressure can thus be obtained.

The gas stream to be analyzed is continuously passed through the sample cell. The reference cell is filled with a gas that does not absorb infrared radiation, such as nitrogen or argon. A filter cell, common to both beams, is filled with a mixture of the background gases present in the sample-gas stream; the mixture should contain all constituent gases that absorb infrared radiation other than the specific gas being measured.

To sensitize the analyzer for measurement of carbon dioxide, both sides of the detector (D1 and D2) are filled with carbon dioxide. If the radiation in

Fig. 7 Elements of a negative-filtering infrared analyzer for measuring CO, CO₂ and methane contents of an atmosphere

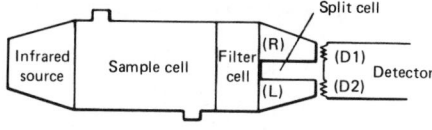

Fig. 8 Elements of a solid-state infrared analyzer for measuring CO, CO₂ and methane contents of an atmosphere

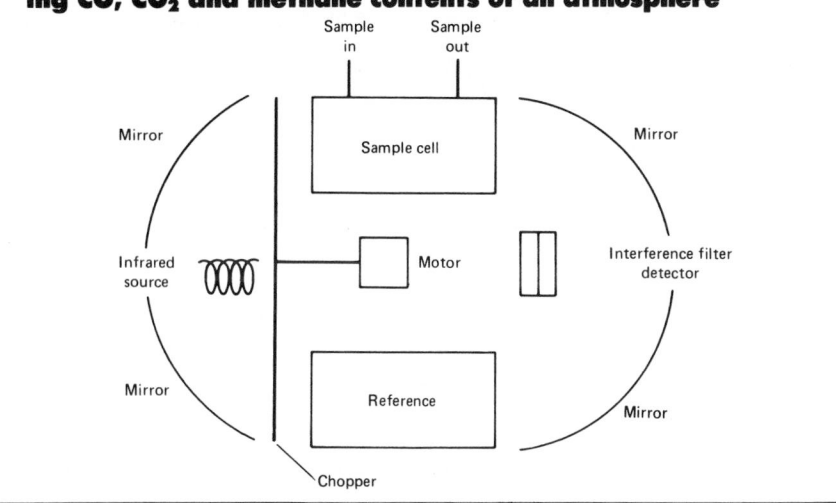

the two beams is identical, equal pulses of energy will strike both sides of the detector simultaneously, causing identical heating effects that will cancel each other and thus produce no motion of the flexible membrane. If a gas containing some carbon dioxide is admitted to the sample cell, energy pulses reaching the D1 side of the detector (Fig. 6) will be reduced. The gas in D1 will be heated less than the gas in D2, and the pressure differential will cause a movement of the flexible membrane. The higher the concentration of carbon dioxide in the sample cell, the lower the energy level reaching D1, resulting in a higher signal output.

A positive-filtering infrared analyzer has good sensitivity and accuracy—the usual stated accuracy is 2% of full scale. Its direct-current output is 0 to 100 mV. When a positive-filtering analyzer must be removed from service for repair, it can be returned to service without being conditioned prior to use.

Negative-Filtering Analyzer. In this type of analyzer (illustrated schematically in Fig. 7), an electrically heated nickel-chromium alloy filament is the source of infrared radiation. Infrared energy passes through the sample cell and a filter cell and is separated into two beams by sensitizing cones. These two beams fall on detectors D1 and D2, which are thermopiles connected in series opposition to measure the difference in energy level of the two beams.

Sensitization of the analyzer for measurement of carbon dioxide is accomplished by filling the right sensitizing cone (R in Fig. 7) with carbon dioxide, the left sensitizing cone (L) with a non-absorbing gas such as nitrogen, and the filter cell with a mixture of interfering gases. Interfering gases are any gases present in the sample-gas stream that have any absorption lines that coincide with absorption lines for carbon dioxide.

When no carbon dioxide is present in the sample flowing through the sample cell, radiation falling on detector D2 is undiminished, but radiation passing through the right sensitizing cone and falling on detector D1 is reduced to nil because of absorption by the pure carbon dioxide in the right cone. As a result, the two detectors produce a net signal that is proportional to the radiation unbalance.

If a gas sample containing some carbon dioxide is now passed through the sample cell, the radiation falling on detector D2 will be reduced by the energy absorbed in the sample cell. There will be no change on detector D1, because the pure carbon dioxide in the right sensitizing cone removes all radiation frequencies characteristic of carbon dioxide. The resulting change in net signal from both detectors is proportional to the concentration of carbon dioxide in the sample cell.

If a gas sample containing other gases that absorb infrared radiation is passed through the sample cell, an equal amount of radiation will be absorbed from the beams reaching both detectors, and the output of the analyzer will remain proportional to the amount of carbon dioxide present in the sample. If an interfering gas is present in the sample, the gas in the filter cell will ensure that its effect cannot be detected, and that the output is still proportional to the carbon dioxide content.

A negative-filtering infrared analyzer has good sensitivity, good stability and no moving parts. It has a nonli-

near output, so the scale on the recorder is expanded at the low end of the scale (where most carburizing is controlled) and compressed at the high end. The stated accuracy at the low end of the scale is ±0.002% at a set point of 0.05% carbon dioxide. When a negative-filtering analyzer is removed from service for repair, it must be conditioned for 24 h, after repairs have been made, before it can be put back into service.

Solid-State Detector Analyzer. This type of analyzer is similar in operation to the positive-filtering analyzer except that the chopper generally is used to alternate the energy through the sample cell and the reference cell. The filter cell and condensor microphone detector are eliminated and replaced by a narrow band-pass interference filter placed on top of a solid-state detector, as shown in Fig. 8. The energy from the sample cell and reference cell is collected by end mirrors and directed to the filter/detector assembly. The band pass of the interference filter is set up in such a manner that the instrument becomes specific for either CO, CO₂ or methane. Because these detectors are very small, they can be stacked side by side, and multiple components can be analyzed in a single instrument. A solid-state detector infrared analyzer has good sensitivity and accuracy. The usual stated accuracy is 2% of full scale. Direct-current outputs are available as required, and repairs tend to be more routine than with either the positive-filtering analyzer or the negative-filtering analyzer.

Advantages. The use of an infrared analyzer-control system on endothermic generators and carburizing furnaces will result in narrower deviations in gas analyses. Such a system provides more uniformity of case depth and helps keep generators and furnaces free of soot.

Examination of the daily chart record of the analyzer will reveal changes in the condition of the atmosphere. For example, work on the natural gas pipelines in the field may cause a variation in the heat content of the natural gas. This will be indicated by the analyzer, and controls automatically restrict the effect of this change.

In one plant, the chart revealed that rapid and severe increases in the carbon dioxide concentration and the dew point of the furnace atmosphere occurred with each opening of the discharge door of a continuous carburizing furnace. This problem was eliminated by improving the seal on the inner door and by adding a flame curtain to the outer door.

An infrared analyzer-control system will also compensate for the difference in the volume of natural gas required by a heavily loaded furnace on a short heating cycle and that required by a lightly loaded furnace on a long heating cycle. When a furnace or generator is being started, the system will provide a continuing check of atmosphere conditions so that equipment can be put into service as soon as proper operating levels are reached, without the necessity of frequent manual checking of dew point.

Limitations. An infrared analyzer is relatively expensive and complex, and its maintenance and repair require the skills of a trained electronics specialist. Malfunctions are not readily detectable; the unit may continue to operate despite failure of several components, but will give erroneous readings. Automatic control valves must be checked to ensure that they are operating properly and are not locked in the completely open or closed positions. The unit must be adjusted frequently to reflect changes in minor variables, such as a change in the energy content of the gas. Thus, to ensure the accuracy of the system, trained personnel must recalibrate the unit daily against a gas of known composition.

Finally, when very high dew points are encountered, enough moisture may condense in the lines and be carried to the sample cell to damage the cell or render the system inoperative until the cell has been dismantled and cleaned. Damage caused by moisture may be prevented by installation of a moisture trap or an electrical warning system ahead of the sample cell.

Dew-Point Analyzers

Dew-point analyzers measure the partial pressure of water vapor in the furnace atmosphere. Dew point is defined as the exact temperature (at a given pressure) at which a mixture of gases will begin to precipitate its moisture content. When air and gas are mixed in consistent, fixed proportions and the mixture is heated to allow chemical reactions to reach equilibrium, the dew point will reflect the chemical balance of the various components comprising the reacted products.

The use of dew point for controlling carbon potential is a fast, inexpensive and relatively simple procedure. Control of endothermic-base atmospheres by the dew-point method is widely accepted in industry.

Manually operated instruments for measuring dew point are inexpensive, rugged and simple, and serve satisfactorily under normal operating conditions. Automatic instruments are available also.

Dew-Cup Instrument. The simplest instrument for measurement of dew point is the dew cup, a schematic representation of which appears in Fig. 9. The gas sample is drawn from the furnace or generator across the outside of a polished cup made of chromium-plated copper. The cup is enclosed in a glass container so that the moisture can be seen condensing on the cup surface when the dew point is reached. The cup surface is cooled progressively by dropping small pieces of dry ice in acetone (or methanol) inside the cup until the dew point is reached, as indicated by condensation on the cup surface at a temperature indicated by a thermometer in the acetone.

The dew cup is most accurate for measurement of dew points above the freezing point of water. At dew points lower than 0 °C (32 °F), there is the possibility of supercooling, with an attendant low dew-point reading.

Use of the dew cup requires a considerable amount of skill and consistency on the part of the operator, and it is not recommended for close control. Incorrect dew-point readings may result if the atmosphere is sooty, if there are

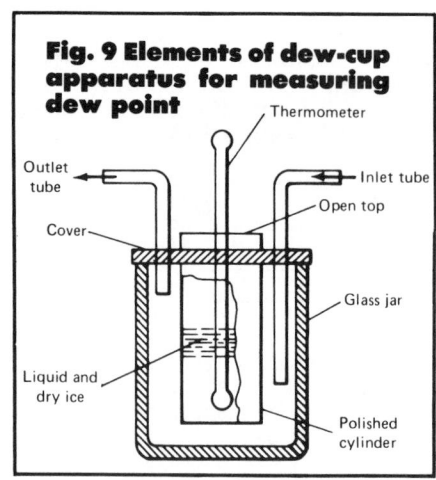

Fig. 9 Elements of dew-cup apparatus for measuring dew point

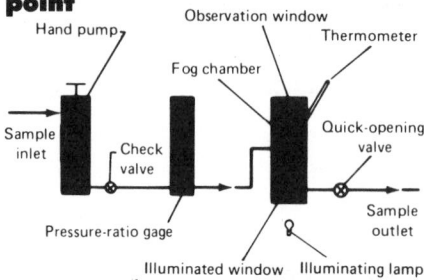

Fig. 10 Elements of fog-chamber apparatus for measuring dew point

leaks in the dew cup or sampling lines, if the flow of atmosphere is too fast, if the temperature is lowered too fast or if the lighting conditions are poor in the area where the dew point is being observed.

Fog Chamber. The fog chamber, illustrated in Fig. 10, is another type of manual instrument for measuring dew point. It is used throughout industry because it is portable and because it gives consistent and accurate readings over an extremely wide range of dew points while requiring no external cooling or mechanical refrigeration. The fog chamber operates on the principle that a rapidly expanding gas cooled adiabatically will produce a fog only when specific requirements of pressure drop, ambient temperature, and moisture content in the gas sample are satisfied.

The atmosphere sample to be tested is drawn into the apparatus and held under pressure in an observation or fog chamber by a small hand pump. A pressure-ratio gage indicates the relationship between the pressure of the furnace atmosphere sample and the ambient atmospheric pressure. The temperature is indicated by a thermometer that extends into the observation chamber. The atmosphere sample is held in the observation chamber for several seconds to stabilize the temperature, after which the quick-opening valve is depressed, releasing the pressure and creating an adiabatic cooling, which may cause a visible condensation or fog to be suspended in the chamber. The fog is easily observed by virtue of a lens system that provides a beam of light in the fog chamber when the quick-operating valve is depressed. The procedure is repeated to find the end point—the point at which the fog disappears. The dew point is then determined by referring to a chart based on the initial temperature reading of the thermometer and the pressure-ratio-gage reading at the point where the fog disappeared.

Chilled Mirror. One of the first methods devised for automatic control of dew point involves the use of the chilled-mirror instrument, a schematic representation of which is presented in Fig. 11. This method utilizes refrigeration and heating to condense and evaporate moisture from an illuminated mirror while the temperature of the mirror is being monitored. A photoelectric cell is used to detect the intensity of the light that the mirror reflects from the source of illumination. The intensity of this reflection depends on the amount of moisture present on the mirror. When the photoelectric cell registers a reflection that represents a deviation from a desired dew point, it generates an electric signal that in turn actuates appropriate auxiliary equipment at the furnace or atmosphere generator, which restores the dew point to the desired level.

This method is satisfactory for controlling relatively clean atmospheres, such as those consisting mainly of nitrogen or hydrogen, but it is not recommended for use with endothermic atmospheres unless maintenance problems created by dust and soot are solved.

Chilled-Metal Instrument. The chilled-metal method of determining dew point utilizes two refrigerated platinum electrodes that condense the moisture from a gas sample to complete an electrical circuit. The temperature of the electrodes is recorded as the circuit is completed, thus giving the dew-point temperature. A heater is employed to evaporate the moisture from the sample chamber and electrodes, to

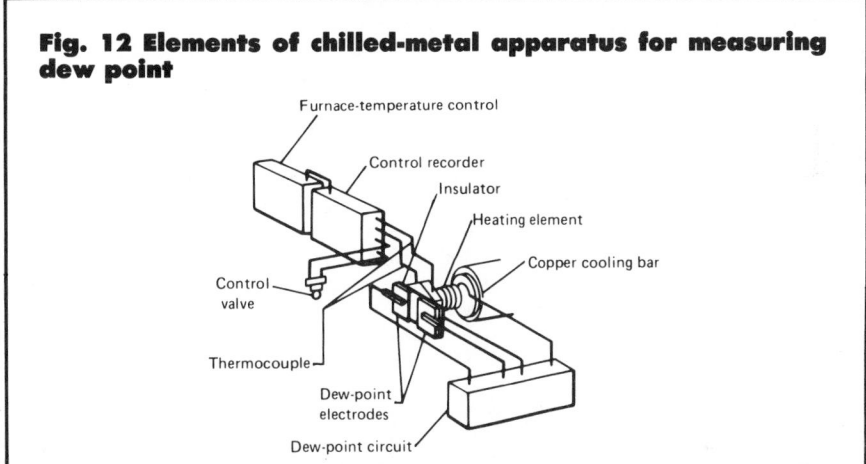

Fig. 11 Elements of chilled-mirror apparatus for measuring dew point

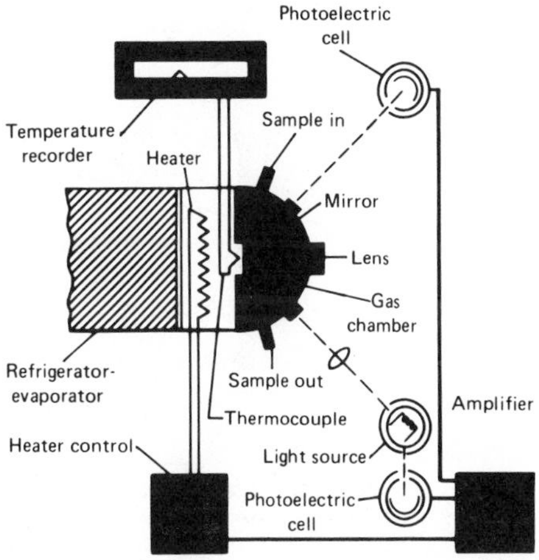

Fig. 12 Elements of chilled-metal apparatus for measuring dew point

prepare the instrument for another determination. Dirt and sooting present problems as with the chilled-mirror instrument. A schematic illustration of a chilled-metal instrument is presented in Fig. 12.

Lithium Chloride. Another type of dew-point instrument that is widely employed for continuous automatic control uses as its basis for operation the behavior of a hygroscopic salt (lithium chloride) when in contact with water vapor. Dry lithium chloride will absorb water at room temperature and dissolve, forming a saturated solution. This solution may in turn be heated to the temperature at which the evaporation tendency of the moisture just matches the absorption tendency of the salt. This temperature is directly related to the dew-point temperature.

The instrument used consists of a

thin metal tube wrapped with glass tape impregnated with dry lithium chloride. The tape is overwound with two silver wires and covered with a perforated metal guard. When the lithium chloride is exposed to moisture and becomes partly saturated, it becomes conductive, and a current is passed through the silver wire, driving off the moisture until equilibrium is achieved. The temperature is read from a sensor inside the metal tube. A limitation of this instrument is its susceptibility to contamination by raw ammonia, which makes it unacceptable for control of carbonitriding atmospheres.

Limitations of Dew-Point Analyzers. Although dew-point analyzers can reflect changes in dew point with a fair degree of accuracy, these changes do not always correlate exactly with changes in carbon potential. An error of

Fig. 13 Elements of a typical oxygen probe for controlling carburizing atmospheres

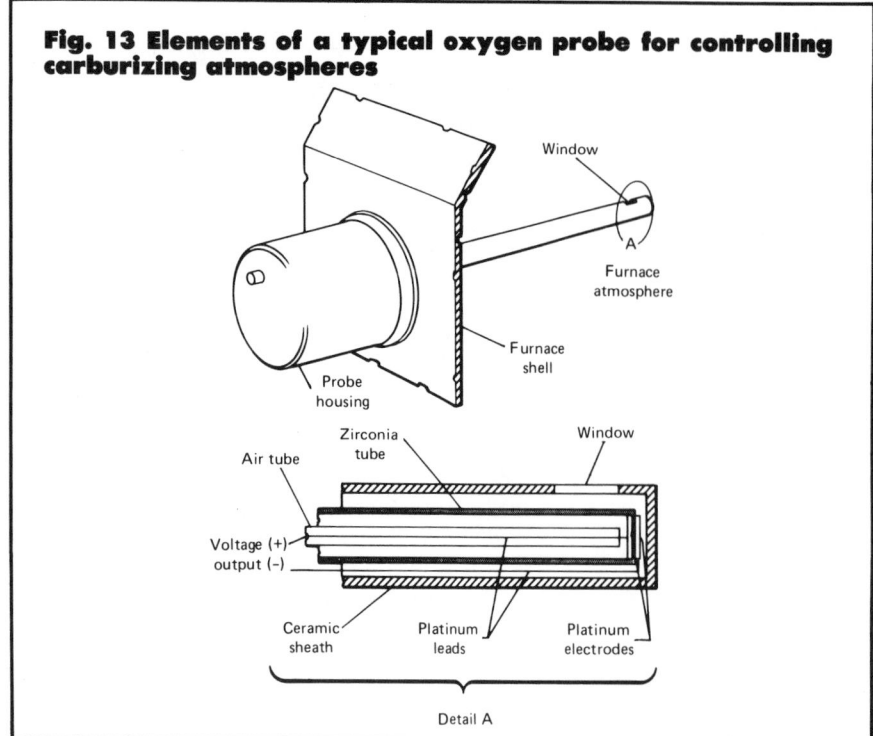

Detail A

only a few degrees Celcius can represent an error of 0.1 to 0.2% carbon in evaluation of the carbon potential of a given atmosphere, which can be as much as a 25% deviation if the desired carbon content is near the eutectoid composition.

With respect to initial investment, automatic dew-point analyzers and controllers represent one of the lower-cost approaches to automatic control of carbon potential. However, the cost of maintenance of automatic dew-point equipment is a matter of some consequence. The equipment will tolerate virtually no dirt in the sampling system and consequently requires frequent cleaning and replacement. A refrigeration system generally is used to lower the gas temperature so that dew point can be measured, and problems may attend the operation of any refrigeration system.

In any dew-point-analysis system, condensed moisture in the sampling lines or filters will result in erroneous high readings that will continue until all moisture has been evaporated by purging the sampling lines and changing the filters. It is possible for condensate to form if the temperature within the sampling lines at any point falls below the dew point of the gas being analyzed. An efficient filtering system must be used with automatic equip-ment to keep contaminants out of the sampling system of the instrument.

In most automatic systems used for checking more than one unit, rather long purge times are required between dew-point readings if the dew points of the various units are not reasonably close together. For example, if the dew points in two furnaces differ by 20 °C (36 °F), long purge times between readings may be required for accurate results. In some automatic systems, it might be impossible to record successive dew points of −18 °C and +10 °C (0 °F and 50 °F), because formation of excessive moisture on the sensing head might render the system inoperative until the moisture can be purged from the system.

Other disadvantages of a dew-point system are the lack of a method of accurate calibration, such as is available with a system for controlling carbon dioxide, and the fact that calibration is required if the system is to reflect the hydrogen content of the atmosphere.

Oxygen Probes

One of the most recently developed methods of measuring carbon potential in a furnace atmosphere is the oxygen probe, or oxygen meter. In an endothermic-base or exothermic-base carburizing atmosphere, one of the reversible reactions, given as Eq 3, is

$$CO + \frac{1}{2}O_2 \rightleftarrows CO_2$$

The carbon potential of such an atmosphere is inversely related to the square root of the partial pressure of oxygen, as indicated in Eq 17. Thus, by monitoring the concentration of oxygen, carbon potential can be defined without considering the concentrations of hydrogen, water vapor or carbon dioxide. The only atmosphere constituent that directly influences the relationship is carbon monoxide. As long as the carbon monoxide content is reasonably constant, carbon activity can be controlled by controlling oxygen content. This method of carbon-potential control has the advantage that it is less sensitive to changes in CO and/or H_2 content of the carburizing atmosphere. This can be easily ascertained by comparing Eq 17, where it is observed that a_c is proportional to P_{CO}, with Eq 11 and 14, in which a_c is proportional to P_{CO}^2 and to the product $P_{CO} P_{H_2}$, respectively.

An oxygen probe usually consists of two platinum electrodes separated by a solid electrolyte in the form of a gas-tight zirconia tube closed at one end (see Fig. 13). The probe, which usually is enclosed in a ceramic sheath, is inserted into the furnace. The furnace atmosphere enters the probe through a window in the sheath and contacts the outer electrode. The other electrode, inside the zirconia tube, is in contact with air, which serves as a reference gas of constant oxygen content. The difference between the partial pressure of oxygen in the furnace atmosphere and that in air induces an electromotive force (voltage), or emf, across the electrodes. The partial pressure of oxygen in the furnace atmosphere is determined by the voltage output (emf) of the sensor. Thus, carbon potential can be controlled by controlling the temperature in the furnace and the voltage output of the sensor, because, as shown in Fig. 14, an isothermal relationship exists between carbon content (wt %) and the emf of the sensor. The voltage output of the sensor is fed directly into an electronic control circuit. The control circuit may be designed to operate in an on-off mode, a proportional mode or a proportional-plus-reset mode, depending on the desired smoothness of response. The stated accuracy of most control systems based on the oxygen probe is ±0.05 wt % of carbon potential. Accuracies exceeding ±0.2 wt % have been reported in the literature.

The correlation between the output of an oxygen probe and actual carbon potential has been confirmed experimentally, as shown in Fig. 14.

Advantages. The response of an oxygen probe is almost instantaneous and is directly related to changes in carbon potential. There is no loss in sensitivity at high temperatures—that is, 980 to 1040 °C (1800 to 1900 °F). Calibration, cleaning, and the maintenance associated with gas-sampling systems are not required—the probe is inserted directly into the atmosphere through a hole in the furnace wall. Sudden or substantial changes in the concentrations of carbon dioxide or water vapor do not cause pronounced control problems, as such changes do with infrared analyzers. Introduction of propane into natural gas, which sometimes is done during periods of peak shaving, does not cause loss of control. Sooting has little effect on the ability to maintain control.

Disadvantages. The major disadvantage of oxygen probes is that the ceramic probe element must be replaced periodically—normally at intervals of one year or longer. Most probe failures requiring replacement occur by detachment of the outer electrode from the zirconia tube. (The outer electrode is the one exposed to the furnace atmosphere.) Another cause of failure is mechanical or thermal shock, which cracks the brittle ceramic components. In most instances, however, the increased reliability and decreased maintenance and supervision costs more than offset the expense of periodic replacement of the probe element.

Orsat Analyzer

The Orsat analyzer has for a long time been the principal tool for analysis of furnace atmospheres. Recently, however, chromatographic and infrared analyzers have been gradually displacing the Orsat analyzer because of their greater speed and sensitivity. The Orsat analyzer has a comparatively low initial cost and requires little upkeep.

Operation of this equipment consists of bubbling a sample of an atmosphere through a series of solutions, each designed to absorb one of the gaseous constituents. Hydrogen, methane, ethane and sometimes carbon monoxide are determined by controlled oxida-

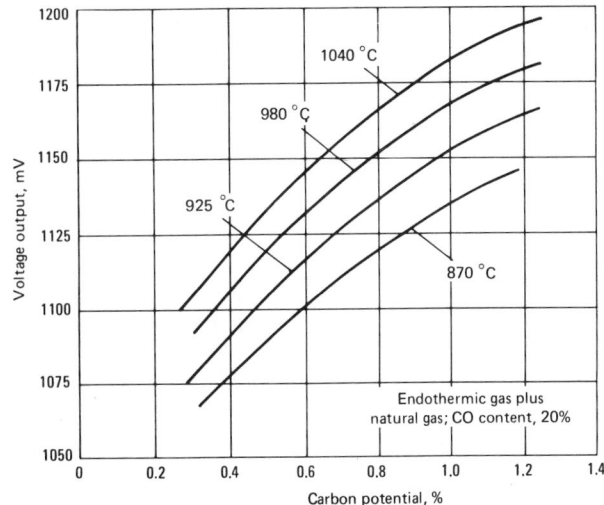

Fig. 14 Voltage across electrodes of a typical oxygen probe as a function of carbon potential at four temperatures, for endothermic gas enriched with natural gas and containing 20% CO

Courtesy of Advanced Atmosphere Control Corp.

tion to water and carbon dioxide. The Orsat analyzer employs a fixed sequence of testing for the various constituents, and this sequence must always be followed.

Advantages. The Orsat analyzer has the advantage of low initial cost, simplicity of operation, portability, and ability to analyze accurately all constituents normally encountered in heat treating atmospheres.

There are no complicated electrical circuits to maintain, and no standardization procedures are required. The operator must only be certain that solutions are reasonably fresh and that all connections are gastight. Although accuracy and reproducibility of results depend largely on the technique of the operator, satisfactory results can be obtained by nontechnical personnel after relatively short periods of training. Accuracy is actually dependent on the scale used. Higher concentrations are more accurate than lower concentrations. The ability to determine concentrations of all normal atmosphere constituents permits detection of air leaks, water leaks, or degeneration of the catalyst in the generator.

Disadvantages. The principal disadvantages of the Orsat analyzer are (a) the long analysis time required, compared with that required by the newer methods, and (b) the relatively large errors that can occur in determi-

nations of the carbon dioxide contents of most furnace atmospheres. Automatic control of the furnace atmosphere is impractical.

Determinations of carbon dioxide and oxygen contents generally can be made in about 5 min, whereas a full analysis requires 30 min or more. In contrast, a full analysis for all normal components of an atmosphere can be made in about 5 min with a gas chromatograph (see subsequent section on gas chromatography).

To achieve high carbon potentials, carbon dioxide content must be maintained at low concentrations, generally from 0.1 to 0.5%. Hence, a small error in determination of carbon dioxide content will result in a relatively large inaccuracy in the ratio of carbon monoxide to carbon dioxide, which is used for determining carbon potential. Therefore, a method that provides more accurate analysis of carbon dioxide content is preferred whenever the heat treating process employed requires close control of carbon potential.

Hot-Wire Analyzer

The hot-wire analyzer is based on the principle that carburization of steel (with the exception of high-alloy grades) is completely reversible. Moreover, the surfaces of thin and thick pieces of steel eventually come to substantially the same equilibrium with a gas of given composition and temperature. In addition, the electrical resistivity of steel is a linear function of carbon content over the range from 0.05% C to saturation. Thus, a length of fine iron wire will exhibit a resistance proportional to its average carbon content and proportional also to the carbon potential of the surrounding gas. This is the potential that would be effective in carburizing loads of production parts.

Basic Design Configuration. The U-shaped sensor of a hot-wire analyzer is a fine iron or iron alloy wire, about 0.08 mm (0.003 in.) in diameter and about 32 mm (1¼ in.) long. This wire and its supporting structure are called the sensor tip and are mounted on a long holder that provides for room-temperature connections. The tip can be readily and economically replaced.

The protection tube that houses the sensor is a ¾ IPS pipe assembly, fitted for gas and electrical connections at the cold end and equipped with a small sampling hole at the hot end. This tube is inserted through the furnace jacket so that a sample of gas from the work load can be drawn over the sensor while the sensor is held substantially at work temperature.

The measuring circuit is a Wheatstone bridge equipped with compensators for furnace temperature and for calibration of the sensor. Suitable circuitry for driving an optional recorder is provided. Also included is circuitry for operating proportionally a control solenoid valve or a motorized proportional valve.

A pump and related meters and valves are provided to ensure positive flow of sample gas over the sensor. This gear also provides protection for the sensor when the sample gas is not protective, including during standby periods.

The instrumentation is packaged in a 405-by-510-mm (16-by-20-in.) portable enclosure or in a free-standing panel 485 mm by 660 mm by 1.65 m (19 by 26 by 65 in.). The latter package is available also with a small sensor-heating furnace, in which the sensor can be located and to which the sample gas from any remote point can be brought and reheated to the preferred temperature before being passed over the sensor for measurement.

Advantages. The hot-wire analyzer can make an in-situ measurement unaffected by sampling-system disturbances, and the measurement is of

carbon potential instead of gas composition. The readout is directly in terms of percent of carbon. Accuracy is normally ±0.05% C. The equipment is simple and easy to maintain, and replacement components are readily available and inexpensive. The method is broadly applicable to any gas mixture exhibiting carbon potential, and measurements can be made either in the work furnace (in situ) or remotely in a small furnace which holds the sensor at constant temperature and reheats the sample gas before measurement.

Limitations. The hot-wire sensor cannot continuously and directly make measurements in gases rich enough to deposit soot, and it requires the protection of a good purge gas such as pure nitrogen or endothermic generator gas. The sensor is fragile and subject to breakage due to vibration of the furnace.

The speed of response below 815 °C (1500 °F) is slow for short (½-h) cycles, although speed of response is always higher than that of the work load.

In carbonitriding atmospheres, the hot-wire sensor responds also to nitrogen, and so reads out carbon plus nitrogen.

Economic Considerations. The hot-wire analyzer is among the least complicated single-point carbon-potential instruments and therefore is also among the least expensive, considering both original cost and upkeep. The most expensive component that might need periodic replacement is the alloy-protection tube.

The system requires a minimum of operator attention and needs calibration only two or three times a week.

Uses. The hot-wire method is useful for measurement and control of any atmosphere exhibiting a carbon potential. The composition of the gas mixture need not fall into a conventional pattern, and the composition can be variable, such as when nitrogen additions are made or when peak shaving is practiced. The temperature range of application is 790 to 1040 °C (1450 to 1900 °F), and the carbon range is from 0.10% C to saturation.

Operation. The sensor is installed in its work furnace port under a flow of protective gas. Interlocks usually are arranged to activate furnace sampling and furnace control during the at-heat period. Low and high limit switches enable the sensor to bypass very lean or very rich sample gas without interrup-

Fig. 15 Schematic representation of a chromatograph for atmosphere analysis

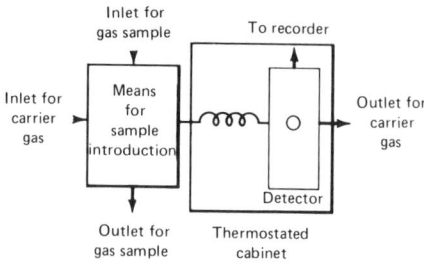

tion of furnace control. The sensor is calibrated either by decarburizing it to its minimum reading or by carburizing it with a standard gas mixture. When the protective gas is nitrogen, the sensor need not be withdrawn from its furnace port until it is to be replaced. Sensor life of three months to one year is feasible.

Hot-wire analyzers have been used to control endothermic-gas generators; pit carburizers; continuous carburizers, both pusher and belt; protected quench carburizers; bell carburizers; belt hardening furnaces; rotary carburizers; and sintering furnaces. They have been used with atmospheres made from drip feed of organic fluids, endothermic gas enriched with natural gas or propane, and natural gas alone, and with atmospheres containing various combinations of a hydrocarbon fuel with an oxygen-supplying agent.

Gas Chromatography

Gas chromatography provides a relatively fast method for measuring the concentrations of practically all the gases present in an atmosphere. A small sample of the gas mixture is inserted into the analyzer, where a steady flow of a carrier gas such as helium carries it through the analyzer column. As the sample goes through the column, individual components are separated by adsorption or partitioning. Measurement of the thermal conductivity or ionization characteristics of the gas leaving the column provides a measure of the concentration of each constituent. Various types of analyzer columns are available by which almost any gas component can be analyzed. A schematic illustration of a chromatograph is shown in Fig. 15.

With present-day chromatographs, an analysis of each constituent, except water vapor, in an atmosphere containing carbon monoxide, carbon dioxide, methane, hydrogen, water vapor and nitrogen can be made in approximately 5 min. If water vapor is analyzed, an additional 5 to 7 min is required.

The minimum full-scale ranges are 12% for hydrogen, 3% for water vapor and 1% for each of the other gases listed. Reproducibility is ±1% of the full-scale reading, except for water vapor, for which it is ±2%. Laboratory analyzers are available with more sensitive detectors that can measure concentrations in parts per million. However, they do not have the stability that is needed in an automatic process-control device.

Although several constituents may be analyzed, automatic control usually is applied to only one. Because of the intermittent nature of the analysis, the input signal to the controller can be adjusted only once for each analysis cycle. The length of the cycle limits the analyzer to single-point control.

Two limitations of gas chromatography are (a) the fact that it provides only periodic, rather than continuous, analysis of the gas stream, and (b) the difficulty of interpreting the chart record. With respect to the latter, the concentration of each constituent is recorded as a bar on the recorder chart, and each must first be identified from the recording sequence. Moreover, because the full-scale ranges differ for the various constituents, percentage of full scale must then be converted to percentage concentration.

Evaluation of Carbon Control in Processed Parts

By the ASM Committee on
Evaluation of Carbon Control
in Processed Parts*

WHEN FABRICATED STEEL PARTS are heated in a carbonaceous atmosphere that will either carburize or decarburize the surfaces of the parts, evaluation of the precise effect of the atmosphere on the parts usually is desirable. This is particularly important when the carbon content at the surface, and at significant depths below the surface, is to be controlled by adjusting the composition of the furnace atmosphere. Several methods of evaluating carbon control of processed parts are considered in the following discussion.

In addition, methods for measuring the case depth produced by carburizing or carbonitriding are described in greater detail in the article "Measurement of Case Depth", in this volume.

Hardness Tests

Hardness tests for evaluating carbon control should be used with caution.

The type of test selected should be one in which the depth of metal affected is properly related to the depth to which carbon control is desired. When hardness tests are used, one of the following methods should be adopted only after all possible sources of error in the specific application have been thoroughly investigated:

- Surface hardness measurements taken under at least two conditions of loading (Rockwell C and superficial Rockwell 15N, for example)
- Superficial Rockwell 15N tests on steps ground below the surface to significant depths
- Microhardness measurements, either on steps or, preferably, on a cross section through the carbon-control zone.

Depth-of-penetration measurements such as those made in Rockwell tests generally do not measure hardness accurately at depths of less than 0.08 mm

(0.003 in.) below the surface. Thus, if full hardness is required at the surface, as in wear applications, decarburization less than 0.08 mm (0.003 in.) deep may not be readily revealed by Rockwell testing at the surface. Such a condition would require microhardness testing or perhaps could be revealed by proper use of a file. Conversely, the surface hardness test is not an accurate indication of carbon control if the carbon control is not effective to sufficient depth to support the penetrator during the hardness test. Lower or higher carbon content at a depth of 0.3 or 0.4 mm (0.010 or 0.015 in.) below the surface can also cause false readings.

Superficial Rockwell 15N testing on steps ground below the surface to significant depths gives a better indication of the depth to which carbon control is effective. For example, if a carbon-restoration operation were being performed and the original decarburization were deeper than anticipat-

*James Dale, *Chairman*, Manager of Controlled Atmospheres, AIRCO Industrial Gases; R. N. Blumenthal, Professor, Marquette University; Raymond L. Davis II, Carbon Control Instruments; Lewis H. Shaefer, Marketing Manager, Anarad, Inc.

ed, the low carbon in the incompletely restored zone between the carbon-restored zone and the core could be detected by the step-grind and hardness-test method, whereas it might not affect a surface hardness reading.

Microhardness testing on a cross section through the carbon-control zone is the most accurate of the hardness-test methods for evaluating carbon control. With this method, each hardness impression is supported by metal of like composition. Variations in carbon content that affect hardness can be detected at any depth up to 0.06 mm (0.0025 in.). This method has the disadvantage of requiring a metallographically prepared cross section and special hardness-testing equipment that is usually available only in well-equipped laboratories.

When hardness testing is used to evaluate carbon control, it should be remembered that maximum quenched hardness is attained at about 0.80% C in plain carbon steels, and at lower carbon contents in alloy steels, depending on types and percentages of alloying elements present. If variations in carbon content above the level required to produce maximum hardness are significant, hardness measurements are not capable of evaluating carbon control. Also, low hardness in quenched parts or test specimens can be caused by insufficient quenching or by excessive retained austenite, as well as by low carbon content. Therefore, evaluation on the basis of hardness measurements alone might lead to incorrect control measures. In instances where hardness measurements do not give adequate information, one or more of the following techniques should also be used:

- Microscopic examination
- Analysis of consecutive cuts
- Analysis of shim stock
- Spectrographic analysis
- Electromagnetic testing.

Microscopic Examination

Microscopic examination is the only method of determining surface carbon variations that shows the effects of such variations on microstructure. Because of this, it will indicate which corrective action must be taken to alter surface carbon concentration. Microscopic examination is also useful in determining whether or not an improper surface carbon content is detrimental to the part.

The effect of carbon content on microstructure varies from steel to steel. It also varies for a given steel, depending on whether the steel is annealed, unannealed, hardened or tempered. The microscopic method is best used on steels with which experience has been gained in correlating results with those of other test methods of evaluating carbon control.

Microscopic examination should be used to determine the following:

Ferrite. A layer of ferrite on the surface indicates total decarburization. There is usually partial decarburization below this layer.

Carbide (Cementite). An increase in the amount of carbide indicates carburization, and a decrease indicates decarburization. A fully annealed sample is almost always best for determining depth of carburization or decarburization by observing carbide variations. Size and frequency of carbide particles can be measured in order to estimate more accurately the percentage of carbide in the microstructure. Extreme carburization can result in massive carbide at the surface.

Pearlite. When the steel is annealed, pearlite indicates eutectoid carbon content. The ratios of pearlite to free ferrite and to proeutectoid carbide can be used to estimate how far and in which direction the structure deviates from that corresponding to the eutectoid composition.

Austenite. Carburization causes some steels to retain excessive amounts of austenite at the surface after heat treating. The proper etchant must be selected that will distinguish austenite from ferrite and from massive carbide. Because both austenite and ferrite are soft, microscopic examination should be used to distinguish a sample with excessive retained austenite from one that has become decarburized.

Martensite. Variations in procedures used for etching martensitic structures (variations in metallographic preparation and in the etching reagents used) can indicate variations in carbon content. Familiarity with the microstructure of the steel is important for correct interpretation of variations. Coarsening of martensite near the surface can indicate a variation in surface carbon.

As indicated, annealing of the sample is often desirable in order that the best information may be obtained; however, annealing must be carried out in the proper atmosphere, or severe errors may result from carburization or decarburization during the annealing cycle.

Use of an inert atmosphere or of a copper-plated sample is best. A neutral atmosphere will tend to correct variations in surface carbon content, which makes it less desirable than an inert atmosphere.

The microscopic method is more sensitive than hardness testing, but it cannot detect very slight variations in carbon content and does not yield quantitative data suitable for plotting curves of atmosphere variables. Also, this type of evaluation is destructive, because cross sections must be cut perpendicular to the surface. When expensive parts are involved, appropriate test pieces should be processed along with the parts.

Analysis of Consecutive Cuts

The consecutive-cuts method of analysis can be used for accurate evaluation of carbon control at any significant depth below the surfaces of parts. Because of the extremely accurate machining operations required to obtain reliable information, this type of evaluation usually is performed on cylindrical test bars treated with the work. After a test bar has undergone the heat treatment to be evaluated, consecutive cuts of 0.03 to 0.3 mm (0.001 to 0.020 in.) or more are turned from the surface of the test bar, and the turnings from each cut are analyzed for carbon concentration.

The consecutive-cuts method is used most frequently on carburizing grades of steel to evaluate control of case carbon content. Other methods of evaluation usually are adequate for medium-carbon and high-carbon steels that have been subjected to annealing, carbon restoration or hardening.

Considerable doubt has been cast on the reliability of the data representing the first 0.03 to 0.1 mm (0.001 to 0.004 in.) cut from the surface. This reliability, however, is a function of the accuracy with which the test is performed and the care exercised in preparing and machining the test specimen.

The test bar should be made of the same grade of steel as that of the workpieces. It should be accurately machined on centers to true cylindrical shape. The diameter should be such that the section of the test bar is representative of the critical section of the workpieces to be evaluated. The length of the test bar need only be sufficient to allow enough turnings for a carbon analysis and a check analysis.

The test bar is loaded into the furnace with the workload and is subjected to exactly the same cycle. If the work is slow cooled from the furnace, the bars are:

- Slow cooled with the load
- Scrubbed with soap and water and rinsed with clean water
- Rinsed with benzene or ethyl alcohol to displace the water, and dried so that no rust forms
- Lightly blasted with shot or grit
- Straightened to 0.03 mm (0.001 in.) total indicator runout
- Scrubbed with soap and water and rinsed with clean water
- Rinsed with benzene or ethyl alcohol
- Machined.

If the work is quenched at the end of the cycle, and if the test bars are of a steel such as 1117 or 8620, the same sequence of operations can be followed except that the bars are tempered at 600 °C (1100 °F), preferably in nitrogen, between the third and fourth steps of the foregoing procedure.

Figure 1(a) shows the furnace temperature and atmosphere conditions employed in carburizing ring gears made of 8620H steel. Figure 1(b) indicates the resultant carbon gradient, as revealed by analysis of consecutive cuts on the test bars that were carburized with the gears. The test bars were prepared from 28.5-mm- (1 1/8-in.-) diam 8620H cold drawn bar stock. The bar stock was cut to 250-mm (10-in.) lengths, which were centered and accurately machined to a nominal diameter of 25 mm (1 in.).

The consecutive-cuts method of evaluating control of case carbon content provides an accurate means of determining the variations in case carbon composition that occur when different heat treating cycles are used. In most instances, the details of these variations would be difficult to determine by any other conventional method of evaluating case carbon content.

Analysis of consecutive cuts provides an accurate method of determining the cycles to be used for carburizing, for continued checks on current cycles, and for determining whether or not simpler process-control methods are adequate. Variations in factors such as cycle time, gas flows, net and gross weights, loading method and furnace condition can be evaluated. Because it provides information about the case on the finished workpiece, this method is useful for

Fig. 1 Furnace temperature and atmosphere conditions for carburizing (a), and resulting carbon gradient (b)

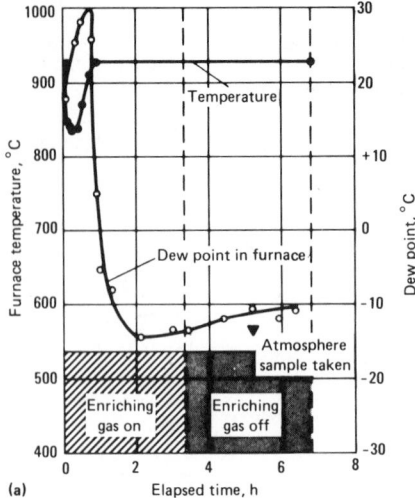

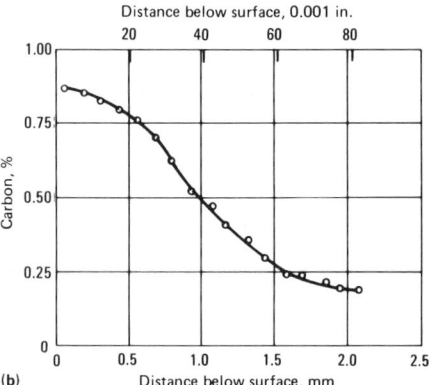

Load. Ring gears, 230 kg (510 lb) net, 390 kg (860 lb) gross
Furnace. Electric, metallic-retort pit furnace
Carrier Gas. Endothermic gas at 2.8 m³/h (100 ft³/h) throughout the cycle, including 5 h and 59 min at temperature
Enriching Gas. Addition of 0.34 m³/h (12 ft³/h) of natural gas started when load was placed in furnace; flow of natural gas stopped after 2½ h at temperature, or after 42% of at-temperature cycle.
Generator Dew Point. −6 to −4 °C (+22 to +25 °F) throughout the cycle
Heating-Chamber Pressure. 5.1 to 7.9 mm (0.20 to 0.31 in.) of water column
Carburizing Temperature. 925 °C (1700 °F)
Cooling Method. Slow cooled from 925 °C (1700 °F) in cooling pit
Atmosphere Analysis Near End of Cycle. 20.8% CO, 0.4% CO_2, 34.0% H_2, 0% O_2, 0.8% CH_4

final evaluation of all conventional types of control, such as atmosphere analysis, temperature-cycle analysis or addition of enriching gas.

There are certain disadvantages of the consecutive-cuts method of evaluating carbon control. It is a slow analytical procedure. Information cannot be obtained until the particular workpieces represented have been processed; thus, it is too late for corrections

if results are not satisfactory. Accuracy is required in machining and in analytical equipment and techniques.

Analysis of Shim Stock

Thin-gage shim stock can be used effectively for determination of the carbon potential of an atmosphere. The success of this method is due to rapid through carburizing of the thin test strip, which eliminates carbon diffusion as a variable in the atmosphere evaluation. The accuracy of this test is adequate for application to carbon control of commercial carburizing, carbon restoration and carbonitriding.

In practice, a strip of annealed 1010 steel, 32 by 75 by 0.1 to 0.15 mm (1¼ by 3 by 0.004 to 0.006 in.), is weighed before and after 1 h of exposure to the furnace atmosphere. The gain in weight, as determined on an analytical balance, is used to calculate the carbon potential of the atmosphere as follows:

$$\text{Carbon potential} =$$
$$\left(\frac{\text{Gain in weight} \times 100}{\text{Final weight}}\right) +$$
$$\text{\% original carbon}$$

Total test time varies from ½ to 1½ h.

Figure 2 illustrates the carbon-potential rod employed in the shim-stock test and procedures for exposure of shim stock to the furnace atmosphere, namely:

(a) Assembly of rod and test strip
(b) Attachment of rod to furnace wall preparatory to insertion of test strip into furnace atmosphere
(c) Insertion of test strip into furnace in unobstructed stream of furnace atmosphere above the workload. Expose for 1 h.
(d) Removal of test strip from furnace atmosphere. The same position indicated in step (b) is used for cooling the strip after it has been withdrawn from the furnace. Rod should be supported for a few minutes after withdrawal, to avoid severe bending. Test strip is cooled for 15 min.

The following precautions should be observed in shim-stock analysis:

• For furnace operating temperatures of 815 °C (1500 °F) or lower, the exposure time should be increased to 1½ h, to ensure through carburizing of the test piece.
• To prevent oxidation of the test strip during cooling, the strip should be transferred from the furnace into the pipe nipple at a time when all furnace doors are closed.
• The test strip should not be rolled or wrapped tightly around the carbon-potential rod. This permits only partial exposure of the test strip, caus-

ing low results, and also complicates removal of the strip for reweighing. Proper anchoring of the strip on the rod is accomplished by simply crimping the corners of the strip over the rod, leaving the center section open to the atmosphere.
• In continuous furnaces, the carbon potential indicated by the test strip represents only the zone or area tested, and not the entire furnace atmosphere.
• The test strip should always be exposed to an unobstructed flow of furnace atmosphere. This is easily accomplished by placing the strip above the workload, as shown in Fig. 2(c).
• A properly exposed and cooled strip containing 1.30% carbon or less should be clean and bright.
• Slight oxide discoloration is caused by improper technique. The carbon content determined on such a strip would be slightly high because of the increase in weight due to the iron oxide.
• A properly exposed and cooled strip containing about 1.40% carbon or more will have a dull matte finish. Depending on actual carbon content, the steel surface may be coated with loose, dry carbon. This is a normal appearance caused by an atmosphere of high carbon potential. The surface carbon deposit should be wiped off before the shim stock is reweighed.

The shim-stock method permits a quick evaluation of atmosphere performance; unnecessary delays, such as the time required for completion of a carburizing cycle, are avoided. Generally, this test is supplemented by analysis of test bars carburized with the work for the entire furnace cycle. Correlation of strip carbon and surface carbon on parts is then established to arrive at definite set of operating conditions.

The shim-stock method determines the maximum carbon available in the atmosphere for batch or continuous furnaces. Generally, the atmosphere carbon so determined is higher than the surface carbon of parts made of most SAE steels; exceptions are steels that contain high percentages of one or more carbide-forming elements, such as 3310, 52100 and 9310. Parts made of these or equivalent steels form surface carbides and consequently have higher surface carbon than that of shim stock.

This test is also applicable to carbonitriding atmospheres. Although the gain in weight is produced by both carbon and nitrogen, for all practical purposes it can be treated as if it were due to carbon only, permitting the same calculations as for carburizing atmospheres.

Shim stock exposed to the furnace atmosphere also may be analyzed for carbon by combustion carbon analysis. This method eliminates the necessity for weighing the sample of shim stock before and after exposure.

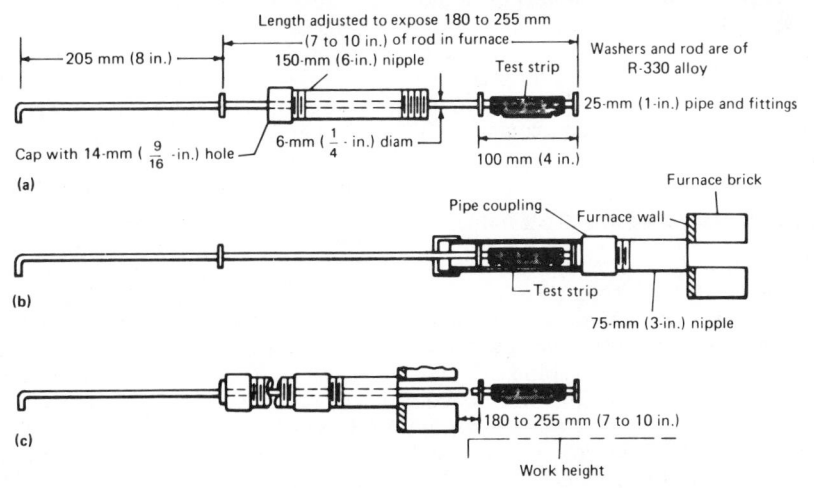

Fig. 2 Arrangements for exposing shim stock to furnace atmosphere for evaluation of carbon potential (see text)

(a) Assembly of test strip for determination of carbon potential. (b) Assembly test strip on rod attached to furnace wall preparatory to insertion into furnace (same position used for cooling of test strip.) (c) Test strip inserted into furnace above work in unobstructed stream of furnace atmosphere.

Analysis of Rolled Wire

Rolled wire also may be used for measuring carbon potential. Wire is flattened into strip and is then used for determining carbon potential in much the same manner as shim stock is used, except that there are small differences in the design of the fixture on which the rolled wire is placed.

Spectrographic Analysis

Carbon content can be determined accurately by spectrographic analysis. This method makes use of a vacuum spectrometer, which permits measurement of spectral lines in the ultraviolet region where air would ordinarily absorb much of the emitted radiation.

Spectrographic analysis normally is performed on flat test specimens that can be taper ground, step ground or reground incrementally after each carbon determination. A very small amount of material is ground from the surface (to remove oxides). Successive cuts are made, and analyses are performed after each cut.

Special care must be taken to ensure that the depth corresponding to each carbon determination is measured accurately. Case depth determined on flat or round test specimens often will be different from case depth determined directly on workpieces, because of differences in shape.

Whereas carbon determination by the combustion method provides an average carbon content for the amount of material removed by machining, spectrographic analysis determines the local carbon content of the specimen to a depth of 0.03 mm (0.001 in.) below the surface. Table 1 compares analyses performed on five samples of shim stock by the spectrographic and combustion methods.

Electromagnetic Testing

Two types of electromagnetic nondestructive tests have been used to evaluate the cases of case-hardened parts: one compares magnetic properties of a part with those of a test standard, and the other measures coercive force, which is then converted to case depth through use of a calibration chart. For a discussion of the principles of electromagnetic testing, the reader is referred to Volume 11 of the 8th Edition of this Handbook—in particular, pages 54 to 55 and 93 to 105.

Table 1 Carbon content of shim stock and workpiece surfaces as determined by spectrographic and combustion analyses

Shim stock and workpieces were heat treated, in the same load, in a 915-mm (36-in.) continuous belt furnace with an endothermic-base atmosphere (class 301; dew point, -9 to -1 °C or $+15$ to $+30$ °F).

| | Amount of carbon present, % | | Workpiece surface |
Sample No.	Spectrographic analysis	Combustion analysis	(spectrographic analysis)
1	0.36	0.36	0.38
2	0.24	0.27	0.25
3	0.22	0.24	0.225
4	0.35	0.35	0.34
5	0.30	0.30	0.305

Magnetic-Comparator Testing

One electromagnetic test is performed by placing the part to be tested in an induction coil. A reference part of known electromagnetic response is placed in a second coil. Both parts are simultaneously subjected to identical electromagnetic fields, and their responses to these fields are compared by an electronic balancing circuit. Any imbalance between responses is indicated by a meter. The imbalance is a function of the following properties of the test object: chemical composition, microstructure, case depth, surface flaws, residual stress and work hardening.

Many electromagnetic instruments are capable of breaking down an electromagnetic response into inductive components, resistive components, third-harmonic amplitude, and phase differences. The user must correlate these variables with the property or properties to be evaluated.

Standards and Test Procedures.

Electromagnetic (eddy-current) testing can be used only as a comparison test; its accuracy and usefulness depend on proper development of standards and test procedures. Acceptance and rejection standards are required for each part design to be evaluated. Once standards are developed and instrument settings are selected, production parts can be tested by comparison with the standards. Sufficient destructive testing must be carried out to supply the data required for construction of a chart or graph by which meter readings can be converted to case depths with reasonable confidence. Periodic destructive testing should be carried out to reconfirm this correlation.

In a production situation involving many parts and wide differences in specified case depth, it is difficult and costly to establish standards for all parts. To eliminate this problem, a procedure has been developed in which a standard test specimen is processed with each heat. Case depth of the test specimen is determined by magnetic comparison. The standards for carburizing a test piece are developed using the procedure described above, and are correlated with actual parts by periodic destructive testing. The standard test specimen is made from a cold finished bar of coarse-grain, silicon-killed 1018 modified steel with residual elements held to low levels. The test piece is 11 mm ($^7/_{16}$ in.) square by 75 mm (3 in.) long. Testing of the standard specimen may yield results that are different from those obtained with actual parts, but once a correlation is established, the standard test piece may be used reliably instead of actual parts.

Effectiveness and Limitations.

Test reliability for determination of case depth is as follows:

Average error	0.10 mm	(0.004 in.)
Maximum error (3% of the time)	0.44 mm	(0.017 in.)
Minimum error (18% of the time,	0.00 mm	(0.000 in.)

The magnetic-comparator test will indicate whether a production lot is acceptable or questionable. Destructive tests must then be conducted on questionable lots to determine what variables are out of specification and to assist in devising corrective measures.

Case depths of up to 5 mm (0.200 in.) may be measured. The method has been found to be more reliable for induction-hardened cases than for carburized cases. The chief variable that can adversely affect measurements is the case-to-core transition zone, which is much wider in carburized cases than in induction-hardened cases.

Coercive-Force Testing

A second type of electromagnetic instrument measures the coercive force of parts. A probe is placed on the part and the coercive-force value is taken. The case depth is read from a calibration curve previously prepared by destructive testing of samples with shallow and deep cases.

The equipment is easy to set up and to use. Results can be correlated to case depths of carburized or induction-hardened parts, usually of carbon steels. In some difficult applications, only a general trend of case depth can be given, but this is usually sufficient for production auditing. The test is extremely sensitive to core-property variations accompanying abnormal quenching.

Coercive-force testing can be applied to cases up to about 20 mm (0.800 in.) deep. The instrument readings are not affected by minor differences in steel composition, but different calibration curves must be developed for different grades of carbon and alloy steels.

Case-Depth Variation

In commercial practice, case depth is controlled within certain tolerances, which depend on the intended service of the part, the type and condition of the carburizing furnace, the cycles employed, and the limitations of control equipment. The end use of the carburized part is the principal determinant of how wide or narrow a range of case depths is acceptable. Parts to be used in less-critical applications do not require close control of case depth, and the increased cost of ensuring close control would be unjustified.

The data shown in Fig. 3 provide a summary of the reproducibility of case depths within prescribed limits for test specimens and parts made of 8620 steel. Operating details are given in the descriptions that follow.

The case variation shown in Fig. 3(a) was determined from a carbon-gradient curve secured by carburizing 25-mm (1-in.) rounds of 8620 steel with production parts at 925 °C (1700 °F) for 1¾ h in a horizontal batch-type furnace, then reducing the temperature to 845 °C (1550 °F) before quenching in oil. A diffusion type of carburizing cycle was used. Dew point was −15 °C (+5 °F) for the carburizing portions of the cycle at 925 °C (1700 °F), −12 °C (+10 °F) for the diffusion portion at 925 °C and −3 °C (+27 °F) for the equalizing portion at 845 °C (1550 °F). The carburizing atmosphere was endothermic gas enriched with natural gas, without addition of air.

After being quenched, the test pieces were tempered in lead at 650 °C (1200 °F) and then wire brushed and liquid-abrasive cleaned. The target depth to 0.40% carbon was 0.51 to 0.76 mm (0.020 to 0.030 in.). With 0.64 mm (0.025 in.) as the average, the distribution was:

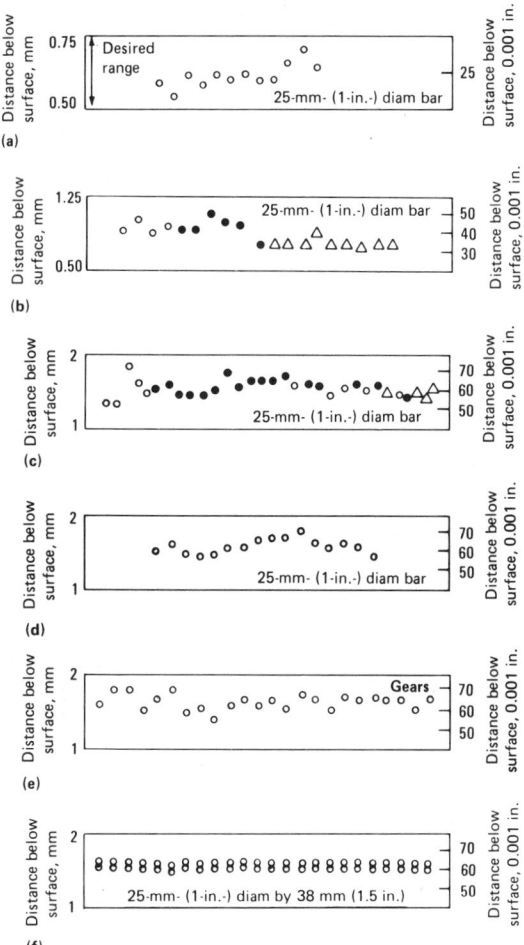

Fig. 3 Reproducibility of case depth in 8620 steel, using the criterion of depth to 0.40% carbon

Data points represent individual heats, and are plotted in chronological order (left to right). Data were collected in studies in two different plants. Legend for parts (b) and (c): ○ Furnace A; ● Furnace B; ▲ Furnace C. See text for discussion.

0.64 ± 0.03 mm (0.025 ± 0.001 in.)	75%
0.64 ± 0.05 mm (0.025 ± 0.002 in.)	92%
0.64 ± 0.08 mm (0.025 ± 0.003 in.)	100%

Test pieces similar to those described above were used to obtain case-depth data (to 0.40% carbon) from three different batch-type furnaces using endothermic gas enriched with straight natural gas, without addition of air. These data are summarized in Fig. 3(b). The loads were carburized at 925 °C (1700 °F), using a diffusion-type cycle. Dew point was maintained at −15 °C (+5 °F) for 3 h at 925 °C (1700 °F), at −12 °C (+10 °F) for the diffusion portion of the

cycle, and at −3 °C (+27 °F) for the 1 h at 845 °C (1550 °F) prior to quenching.

The three different furnaces had the following characteristics:

Furnaces A and B

Size: 76 by 122 cm by 46 cm high (30 by 48 in. by 18 in. high)

Atmosphere control: dew cell; manual reset

Rated gross load: 68 kg (1500 lb)

Furnace C

Size: 91 by 183 cm by 61 cm high (36 by 72 in. by 24 in. high)

Atmosphere control: infrared analyzer controlling carbon dioxide; automatic reset

Rated gross load: 1588 kg (3500 lb)

The maximum variation recorded in 5 years was 0.91 to 1.19 mm; target, 1.02 mm (0.036 to 0.047 in.; target, 0.040 in.). Distribution of case depth was:

1.016 ± 0.025 mm
(0.040 ± 0.001 in.) 32%
1.016 ± 0.051 mm
(0.040 ± 0.002 in.) 47%
1.016 ± 0.076 mm
(0.040 ± 0.003 in.) 90%
1.016 ± 0.102 mm
(0.040 ± 0.004 in.) 95%
1.016 +0.178, −0.102 mm
(0.040 + 0.007, −0.004 in.) 100%

For a required case depth (to 0.40% carbon) of 1.40 to 1.78 mm (0.055 to 0.070 in.), the same three furnaces used to provide the data for Fig. 3(b) produced the results plotted in Fig. 3(c). Operation and means of testing were the same as for the shallower depths, except that the total cycle time at 925 °C (1700 °F) was 10.5 h. Total cycle time was 5 h with a dew point of −15 °C (+5 °F) and 5½ h with a dew point of −12 °C (+10 °F). Equalizing time at 845 °C (1550 °F) was 1 h, and dew point was −3 °C (+27 °F). Maximum case-depth variation for 4 years was 1.422 to 1.676 mm (0.056 to 0.066 in.), with the following distribution:

1.524 ± 0.025 mm
(0.060 ± 0.001 in.) 22%
1.524 ± 0.051 mm
(0.060 ± 0.002 in.) 48%
1.524 ± 0.076 mm
(0.060 ± 0.003 in.) 75%
1.524 ± 0.102 mm
(0.060 ± 0.004 in.) 89%
1.524 ± 0.127 mm
(0.060 ± 0.005 in.) 97%
1.524 ± 0.152 mm
(0.060 ± 0.006 in.) 100%

Reproducibility of Results

The data given in Fig. 4 correlate effective case depth to 50 HRC measured on the same gears in two different plants. Values were obtained by means of the hardness-traverse method, which is considered the most accurate for measuring effective case depth. Nevertheless, several discrepancies in the results obtained at the two plants are apparent.

To compare hardness and microscopic methods of measuring case depth on the same specimens, carburized and

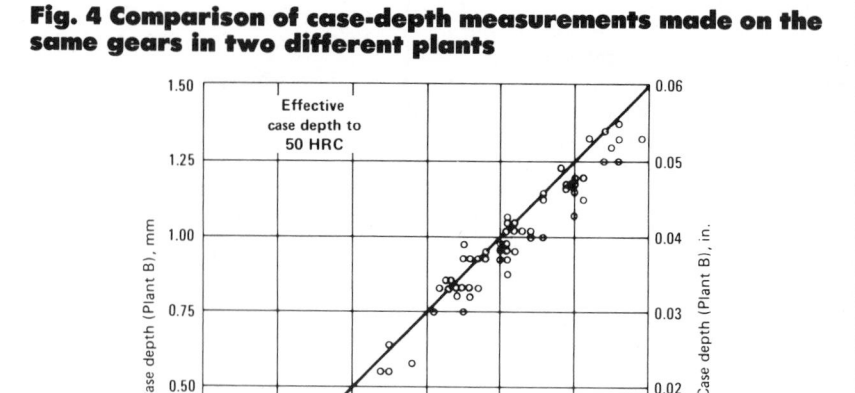

Fig. 4 Comparison of case-depth measurements made on the same gears in two different plants

hardened specimens of 8620 steel were prepared and submitted to five different heat treating organizations for measurement by their laboratories. These specimens were prepared with reasonable care and were located within a short distance of one another during carburizing and hardening. All of the specimens were cut from the same bar. Each of the five laboratories that participated in the testing was asked to make all case-depth measurements in accordance with identical industry-standard procedures. The results are shown in Fig. 5.

These measurements (all made by trained observers in qualified laboratories) diverge somewhat more widely than might be expected. They illustrate, perhaps in an exaggerated manner, the inherent lack of high precision in case-depth measurements. With consultation among laboratories and more precise definition of criteria, considerably closer agreement can be obtained—for instance, among different laboratories in the same multiple-plant organization or between closely cooperating vendors and customers.

Carbon-Concentration Gradients and Surface Carbon Content

Carbon-concentration gradients of carburized parts are influenced by the carburizing temperature and time, the type of cycle (various combinations of carburizing and diffusion times), the carbon potential of the furnace atmosphere, and the original composition of the steel.

The term "carbon gradient" not only encompasses the rate of change of carbon content with depth below the surface, but also alludes to the absolute value of carbon content in any layer except the surface layer.

Carbon Gradients. A single potential carbon gradient is one produced by carburizing at a single temperature and constant atmosphere composition for the entire cycle. In some instances, single-potential carburizing is done to produce saturated austenite at the surface of the steel. Figure 6 illustrates the influence of carburizing temperature on the carbon gradient for single potential carburizing of 1020 steel in a batch furnace. Comparable data are also given for 8620 steel carburized for 7½ h in an atmosphere containing 12% methane. Carburizing temperatures for both steels were 870, 900 and 925 °C (1600, 1650 and 1700 °F).

Figure 7 presents carbon-gradient curves obtained in a batch furnace at a carbon potential corresponding to saturated austenite for four steels after carburizing for 4 h at 870 and 925 °C (1600 and 1700 °F).

Typical influences of methane content, carbon potential and time on the single-potential carbon gradient in 1022 steel are shown in Fig. 8 and 9.

To develop optimum mechanical

Fig. 5 Comparison of case-depth measurements made on the same two samples of 8620 steel in five different laboratories

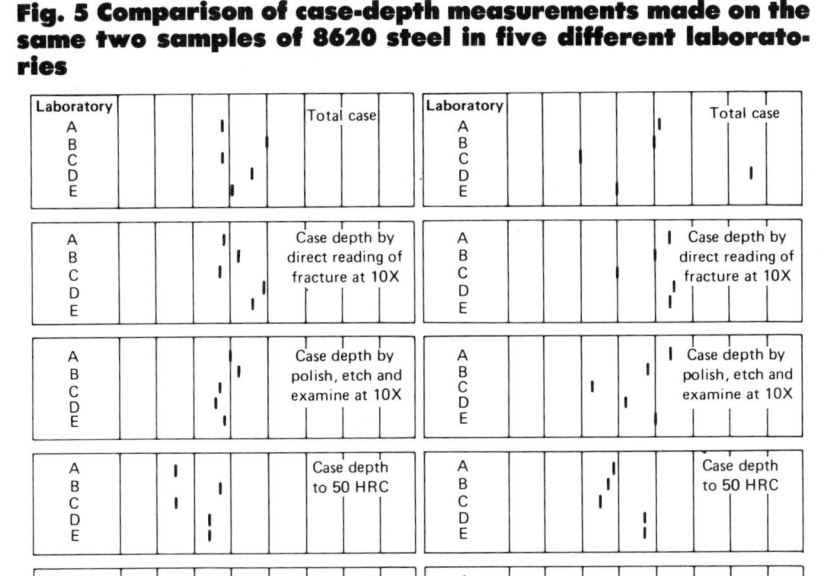

(a) Sample 1: 8620 with core hardness of 30 HRC and a surface carbon of 0.90%. (b) Sample 2: 8620 with core hardness of 40 HRC and a surface carbon of 0.90%

properties in the case, it is common practice to use carburizing cycles that consist of two or more combinations of temperature, time and atmosphere composition (carbon potential). The main objective of using a multiple carbon-potential gradient is to decrease total cycle time.

The most widely used carburizing cycles are constant-temperature cycles starting with an atmosphere carbon potential that approaches the value for carbon saturation in austenite. Part way through the cycle, the flow of the enriching gas is reduced, which lowers the carbon potential of the atmosphere. During the remainder of the time at temperature, known as the diffusion cycle, the atmosphere is maintained at a carbon potential equal to the final desired surface carbon content. During the diffusion period, the surface layer is partly decarburized because it is in an environment of lower carbon activity. It must not be assumed that all of the carbon added during the carburizing cycle will be diffused inward during the diffusion cycle; most of it does diffuse inward, but some returns to the atmosphere.

Multiple carbon-potential gradients are used for the twofold purpose of reducing the amount of retained austenite that would normally occur on quenching, and increasing the depth of the effective case and making its properties more uniform. The two examples that follow describe carbon gradients that are typical of those produced in continuous furnaces with zone control.

Example 1. Comparison of Carbon Gradients Produced With and Without Zone Control (Fig. 10). Two types of carburized cases were produced on 4027 steel by gas carburizing in the same continuous furnace using identical time-temperature cycles but different conditions of atmosphere control. For both cycles, total time in the furnace was 10.3 h, and the zone temperatures were 900, 925, 925 and 845 °C (1650, 1700, 1700 and 1550 °F). The carbon gradients produced by the two cycles are shown in Fig. 10. One case was produced without zone control—that is, without regard to the fact that there is a high demand for carbon early in the carburizing cycle and a low demand during the final part of the cycle. A uniformly distributed mixture

of carrier gas plus hydrocarbon gas was admitted continuously at various points along the length of the furnace. The ratio of carrier gas to hydrocarbon gas was high enough to ensure that only a minimum amount of soot would be deposited during the carburizing and cooling portions of the cycle. As can be seen from the curve labeled "Without zone control" in Fig. 10, the case had a high carbon content at the surface and a slightly concave curvature to the carbon gradient.

The other case (see curve labeled "With zone control" in Fig. 10) was produced under identical conditions, except for the distribution of the atmosphere within the furnace. Hydrocarbon gas was admitted only within the carburizing zones, with the greatest amount added in the first of these zones. The total amount of hydrocarbon gas was equal to that for the case produced without zone control. Carrier gas without enrichment was admitted to the furnace in the diffusion and cooling zones. The case had a lower surface carbon content and a greater effective depth than the case produced without zone control, and the carbon-gradient curve was convex in the near-surface region of the case.

Example 2. Variation of Carbon Gradient With Processing Conditions in a Three-Zone Continuous Furnace Under Manual Control (Fig. 11). The carbon gradients shown in Fig. 11 are for 4815 steel that was carburized in a three-zone, continuous pusher furnace. Carbon potential was manually controlled. For all three cycles, the temperature in zones 1 and 2 was 920 °C (1690 °F) and the atmosphere was carrier gas enriched with methane; in zone 3, the temperature was 910 °C (1670 °F) and the atmosphere was carrier gas diluted slightly with air. The three cycles differed in the carbon potentials established in the third zone, and in the time that parts were exposed in each of the three zones.

Cycle 1 was 35 h long and consisted of 8 h in zone 1 at 1.25% carbon potential, 14 h in zone 2 at 1.25% carbon potential, and 13 h in zone 3, where the carbon potential varied from 0.80% at the center of the zone to 0.55% at the discharge end. This cycle was designed to produce a total case depth of 2.8 mm (0.110 in.) at 0.25% carbon.

Cycle 2 was 25 h long and consisted of 6 h in zone 1 at 1.25% carbon potential, 10 h in zone 2 at 1.25% carbon

Fig. 6 Carbon gradients for 1020 and 8620 steels carburized at three temperatures

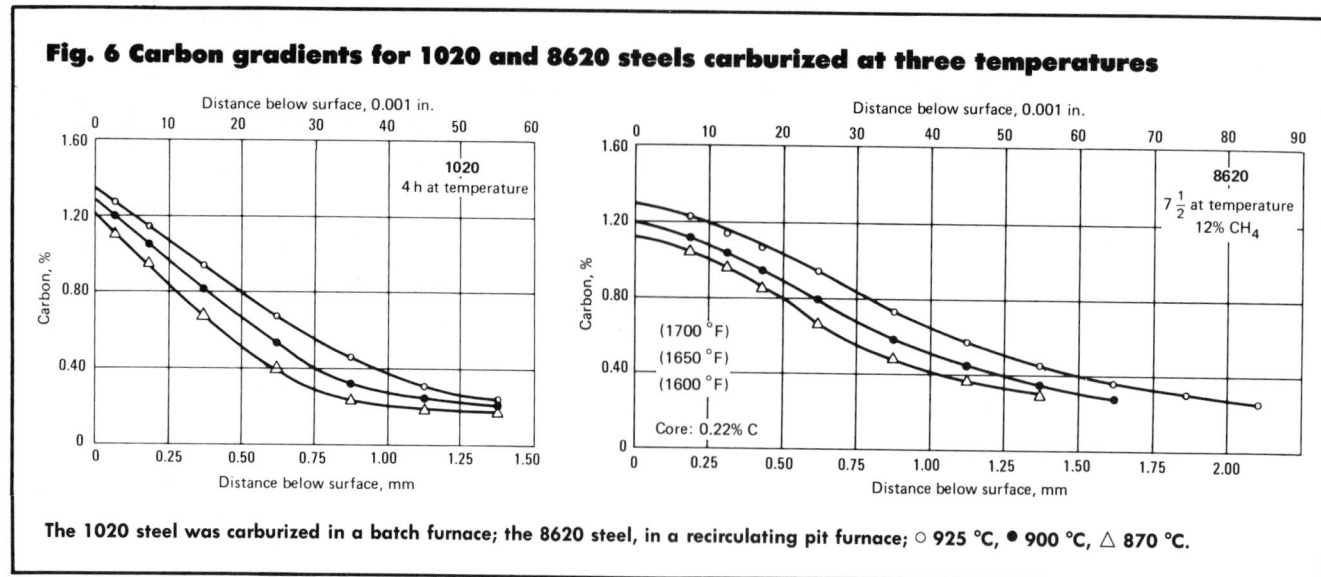

The 1020 steel was carburized in a batch furnace; the 8620 steel, in a recirculating pit furnace; ○ 925 °C, ● 900 °C, △ 870 °C.

Fig. 7 Carbon gradients for four steels

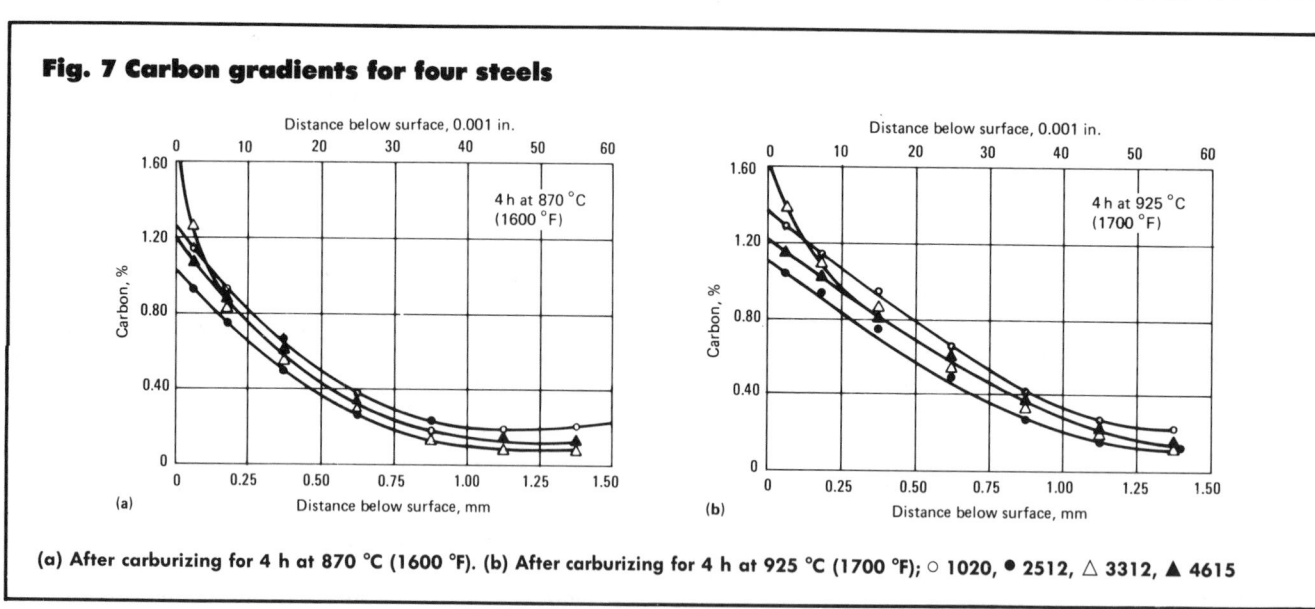

(a) After carburizing for 4 h at 870 °C (1600 °F). (b) After carburizing for 4 h at 925 °C (1700 °F); ○ 1020, ● 2512, △ 3312, ▲ 4615

potential, and 9 h in zone 3, where the carbon potential was 0.80% at the center and 0.60% at the discharge end. Cycle 2 produced a total case depth of 2.3 mm (0.090 in.) at 0.25% carbon. The carbon gradient exhibited a slightly higher surface carbon content and a narrower zone of constant carbon content in the near-surface region than did the gradient resulting from cycle 1.

Cycle 3 consisted of 4 h in zone 1 and 7 h in zone 2, both at 1.25% carbon potential, plus 6 h in zone 3, where the carbon potential was 0.8% at the center and 0.75% at the discharge end, for a total carburizing-plus-diffusion time of 17 h. This cycle produced a total case depth to 0.25% carbon of 1.8 mm (0.070 in.). The carbon gradient showed slightly higher surface carbon content than either cycle 1 or 2 (Fig. 11) and a slightly shallower plateau of constant carbon content in the near-surface region of the case.

Surface Carbon Content. As indicated in Fig. 10, carbon-potential control also affects surface carbon concentration. Surface carbon content has a pronounced effect on the properties of the steel and must be controlled for optimum results.

The amount of carbon in solution affects the amount of austenite that will be retained after quenching, and the amount of retained austenite affects the hardness of the quenched case.

Carbon potential, carburizing cycle, diffusion cycle, quenching temperature and rate, and steel composition determine the amount of austenite retained after quenching.

Effect of Carbon Gradient on Case Properties. Carbon gradient is related to case hardness, as shown by the data in Fig. 12 for specimens of 1024 steel. The specimens were carburized for 2¼ h at 900 °C (1650 °F) in an atmosphere containing 10% methane. They were oil quenched from 845 °C (1550 °F) and achieved a maximum surface hardness of about 62 HRC. However, a hardness of about 66 HRC was obtained at a depth of 0.25 mm (0.010 in.) below the surface, indicating the

Fig. 8 Carbon gradients for carburized 1022 steel test bars

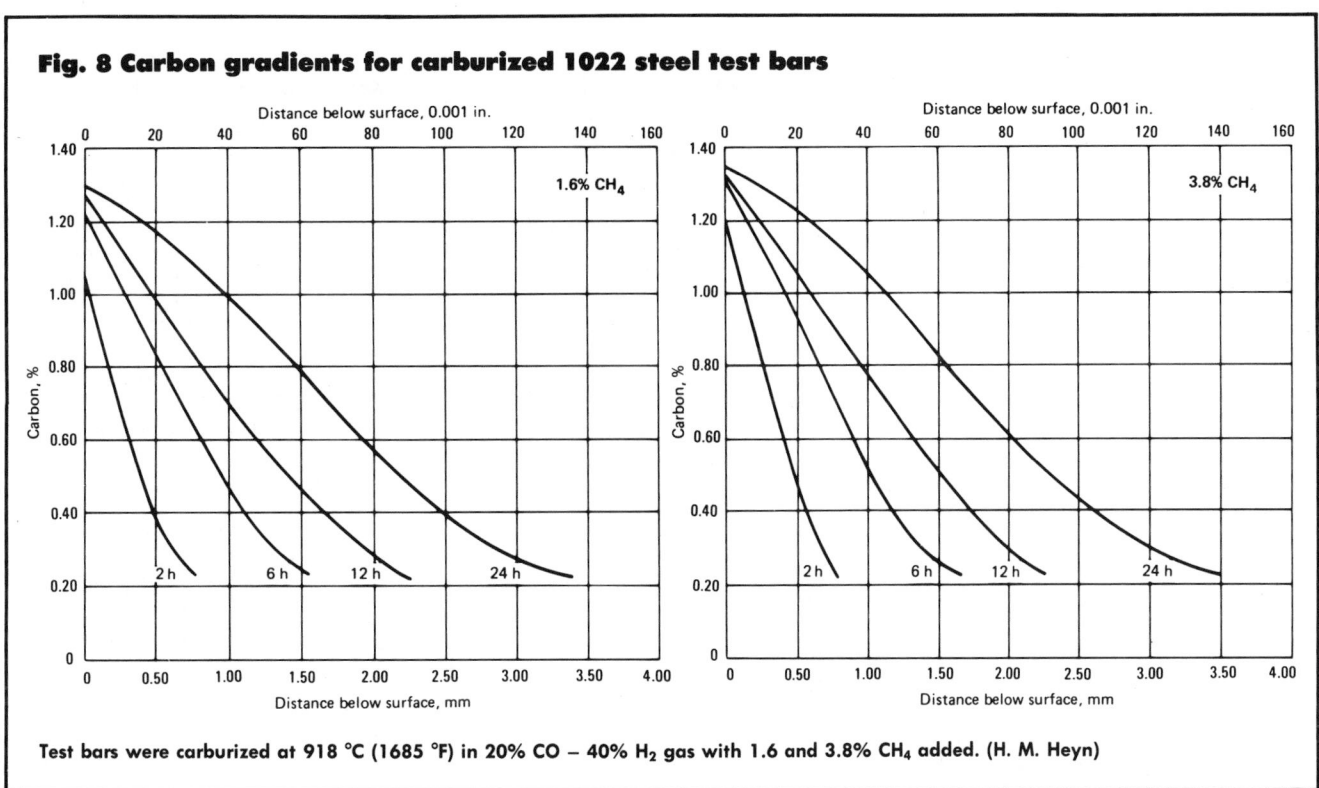

Test bars were carburized at 918 °C (1685 °F) in 20% CO — 40% H_2 gas with 1.6 and 3.8% CH_4 added. (H. M. Heyn)

Fig. 9 Carbon gradients for carburized 1022 steel

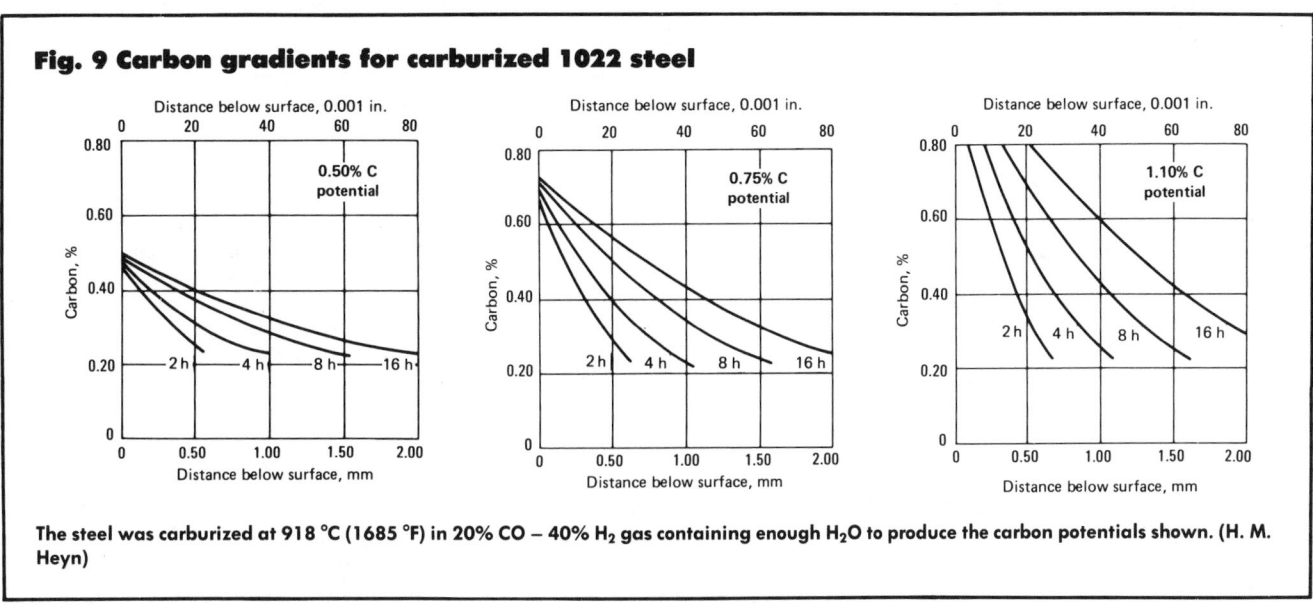

The steel was carburized at 918 °C (1685 °F) in 20% CO — 40% H_2 gas containing enough H_2O to produce the carbon potentials shown. (H. M. Heyn)

presence of some retained austenite at the surface. Retained austenite is frequently considered a normal constituent in the microstructure, corresponding to the hypereutectoid portion of the carbon gradient. Alloy steels are more likely to exhibit retained austenite than are carbon steels.

In many instances, finely dispersed retained austenite in amounts up to 30% is not detrimental to pitting-fatigue strength, whereas much lower amounts are often harmful if the austenite is not finely dispersed. But with finely dispersed austenite, small amounts of retained austenite apparently allow mating surfaces to conform slightly faster and to spread the load more evenly, thus reducing local areas of high stress.

The effect of carbon content on pitting fatigue strength has been subject to controversy for many years. Undoubtedly, other microstructural changes have clouded the issue. Rolling contact fatigue tests have indicated that pitting-fatigue strength increases with increasing hardness for medium-carbon and high-carbon steels.

The effect of carbon content on re-

Fig. 10 Effect of zone control on carbon gradient for 4027 steel carburized in a continuous furnace (Example 1)

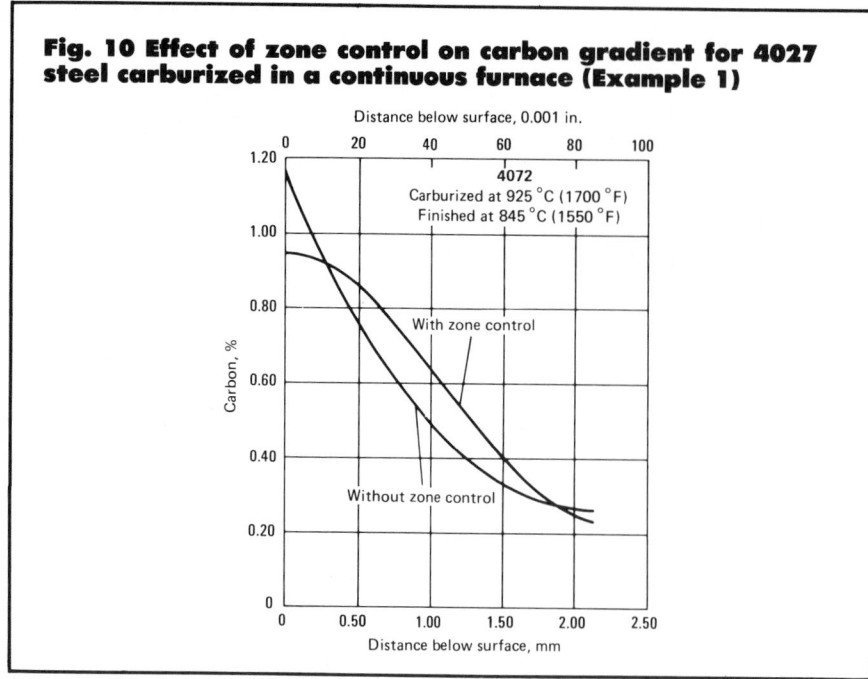

Fig. 11 Carbon gradients from three-zone, continuous pusher-type carburizing furnace with manual control of carbon potential (Example 2)

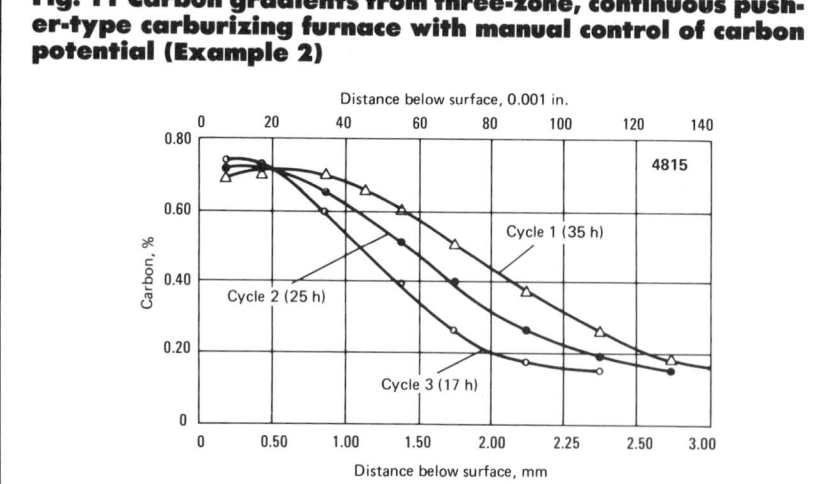

Fig. 12 Carbon and hardness gradients for 1024 steel

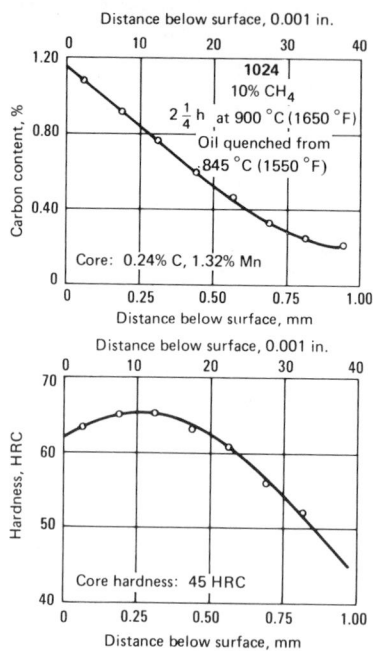

The steel was carburized in a recirculating batch furnace under conditions that produced saturated austenite at the surface. Note effect of retained austenite on surface hardness. Specimens were 25 mm (1 in.) in diameter by 150 mm (6 in.) long.

sistance to wear and scuffing is too complicated to warrant a general conclusion. Excessive retained austenite permits surfaces to deform plastically under heavy loads, resulting in ripples, or "orange peel".

Low Surface Carbon. Most steels can be carburized by the carbon saturation – carbon diffusion type of cycle to produce surface carbon content well below saturation (for example, from 0.90 to 1.00% carbon).

There is a strong preference in present carburizing practice for surface carbon concentrations of eutectoid com-

position or slightly higher. With the lean-alloy steels that are most often used, it is especially important to utilize the full hardenability of the steel. Maximum hardenability of the lean-alloy carburizing steels is obtained at carbon concentrations near eutectoid composition. The excess carbides formed at high carbon concentrations promote transformation to products other than martensite, and may also remove part of the carbide-forming elements from the austenite, thus decreasing the effective hardenability.

Low carbon concentrations at the

surface also permit direct quenching of work from the carburizing temperature, which is more economical than reheating before quenching. Direct quenching of carburized parts having high surface carbon concentrations favors retention of austenite. Such austenite is undesirable because it lowers the indentation hardness of the case and promotes secondary hardening with the formation of untempered martensite, which may change the dimensions of the finished parts as well as embrittle them.

When surface carbon concentrations near eutectoid composition are desired, a multiple-manifold arrangement may be used by which part of the carrier gas and all of the hydrocarbon gas are introduced through ports at the front of the carburizing zone, where the carbon demand is high. Only carrier gas is supplied to the other zones, and the carbon potential is adjusted to give the desired final surface carbon concentration.

Because, for a given carbon content of the core, the diffusion rate of carbon

Fig. 13 Variation in carbon content 0.25 mm (0.010 in.) below the surface for 1020 steel carburized in three similar batch furnaces

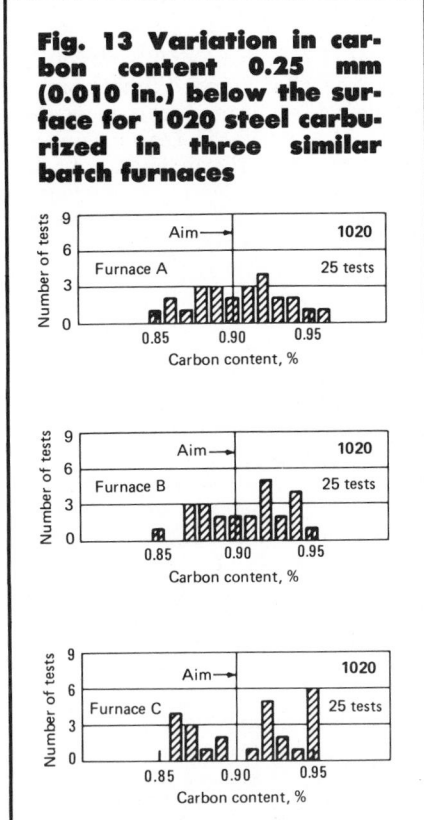

Fig. 14 Variation in surface (or near-surface) carbon content for alloy steels carburized under different conditions (see text)

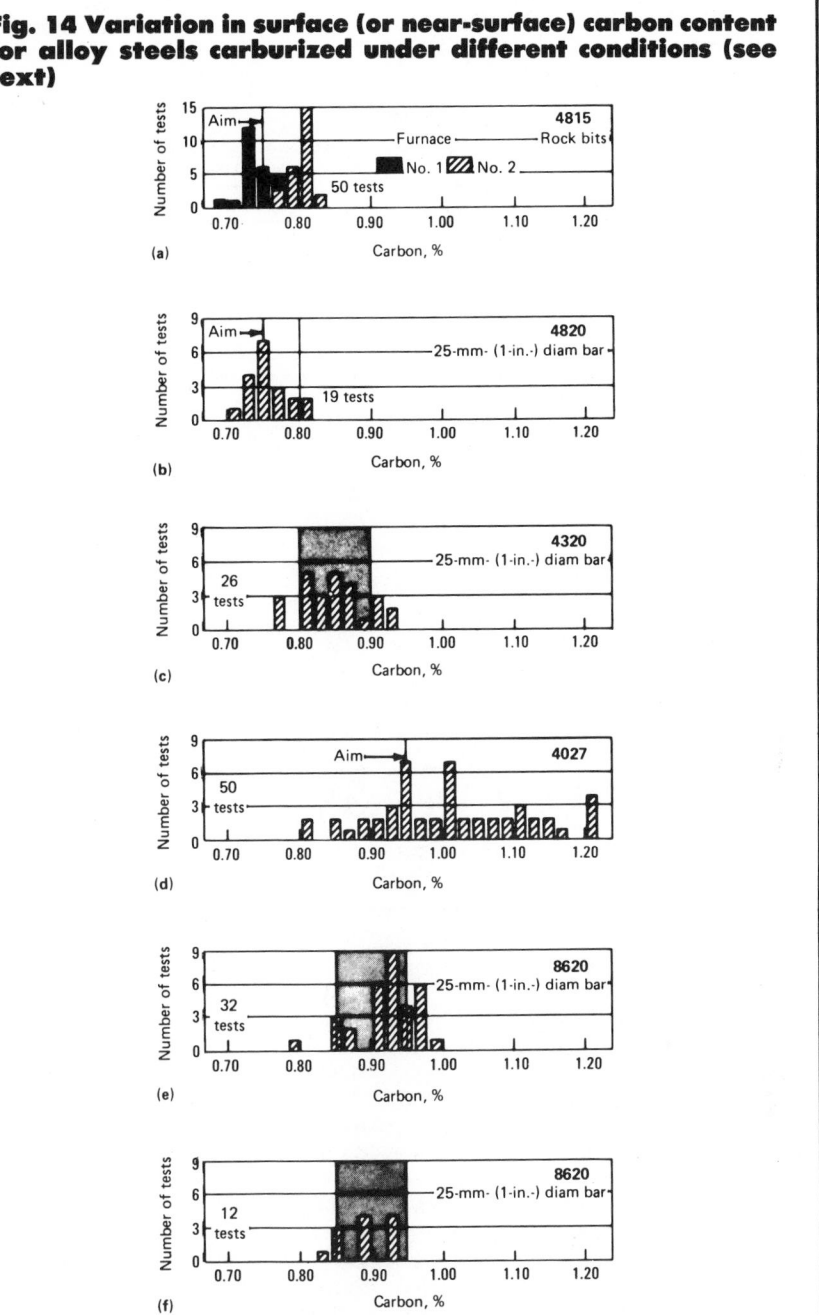

in austenite decreases with a lowering of surface carbon concentration, any "starvation" method of carburizing in which the workpiece surfaces are never carburized above the desired final value requires a longer cycle. Also, good recirculation of gases is essential. The atmosphere adjacent to the work quickly becomes depleted of carbon, and, if stagnant areas are allowed, excessively low surface carbon concentrations and shallow case depths will result on the work in these areas. This potential detriment to part quality is the chief reason why most continuous carburizing furnaces in use today are equipped with effective recirculating fans.

Variability in Surface Carbon. Statistical data relating to control of surface carbon content are given in Fig. 13 to 16 for several plain carbon and low-alloy steels. These data represent practice in eight different plants, thus involving several different types of furnaces, steel compositions and "aim" carbon contents. Control varied from good to poor, as the graphs indicate.

The data summarized in Fig. 13 compare carbon concentrations 0.25 mm (0.010 in.) below the surface, obtained

in gas carburizing 25 batches each in three similar batch furnaces. The 75 tests show a large majority of the concentrations to be within ±0.05% of the 0.90% carbon aim. Parts were carburized at 925 °C (1700 °F). The carbon potential was controlled through dew point. A carrier gas with a dew point of 1.5 °C (35 °F) was enriched with natural gas to maintain a dew point of −5.5 °C (+22 °F) in the furnace.

The data presented in Fig. 14(a) compare surface carbon content in rock bit cutters carburized in two pit furnaces to a desired carbon content of 0.75%. Furnaces were operated simultaneously, using the same carrier-gas generator. Each furnace was about 1 m (3.3 ft) in diameter by 2 m (6.6 ft) deep and contained a load of about 1360 kg (3000 lb). A carburize-diffuse cycle was used. An automatic dew-point controller was

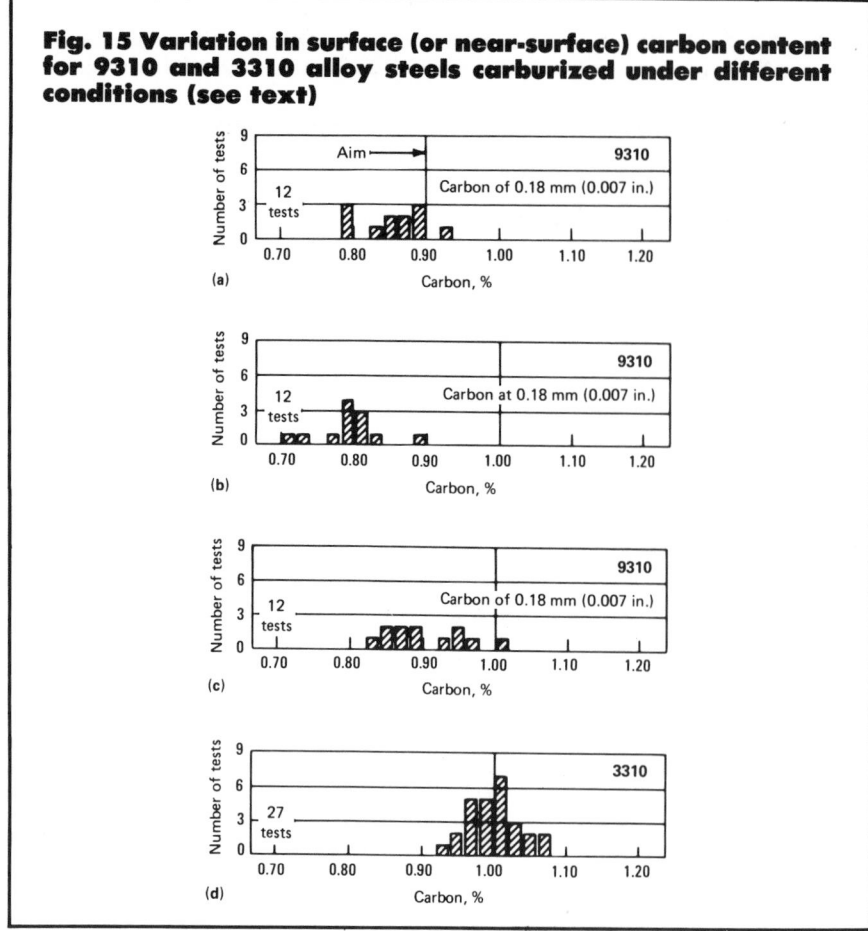

Fig. 15 Variation in surface (or near-surface) carbon content for 9310 and 3310 alloy steels carburized under different conditions (see text)

employed on the carrier-gas generator but not on the furnace.

The results plotted in Fig. 14(b) show surface carbon contents (carbon content of first 0.075-mm, or 0.003-in. cut) for 19 specimens of 4820 steel carburized along with loads of production parts in a two-row, pusher-type continuous furnace. The desired carbon content was 0.75 to 0.80%, with a specification of 0.70 to 0.90% carbon. All results were within specification, and most were within desired limits. Specimens were carburized at 925 °C (1700 °F) using a diffusion cycle, quenched from 815 °C (1500 °F) in oil at 60 °C (140 °F), tempered in lead at 620 °C (1150 °F) for 5 min, wire brushed and liquid-abrasive cleaned. The atmosphere at the charge end of the furnace was automatically controlled at a dew point of −15 °C (+5 °F), and at the discharge end at a dew point of +3 °C (+37 °F). Endothermic gas enriched with straight natural gas was used as the carburizing medium; air was added at the discharge end.

The data presented in Fig. 14(c) for 4320 steel bars were obtained under the same carburizing and test conditions as for Fig. 14(b) for 4820 steel bars, except that the dew point was −1 °C (+30 °F) at the discharge end and that specimens were tempered at 650 °C (1200 °F). The desired surface carbon content was 0.85 ± 0.05%.

The results of 50 tests that represent variation in surface carbon content for continuous carburizing of 4027 steel are given in Fig. 14(d). These results were obtained over a period of 3 to 4 years. An automatic dew-point controller was used.

The data given in Fig. 14(e) for 8620 steel bars were obtained under the same carburizing and test conditions as for carburizing of 4320 steel (Fig. 14c). The desired surface carbon content was 0.90 ± 0.05%.

The data shown in Fig. 14(f) were based on the surface carbon content (first 0.075-mm, or 0.003-in., depth of cut) of 25-mm (1-in.) rounds. The rounds were carburized with production parts in a horizontal batch furnace, quenched in oil from 845 °C (1550 °F), tempered in lead at 650 °C (1200 °F), wire brushed and liquid-abrasive cleaned. The desired carbon content was 0.90 ± 0.05%. The carburizing medium was endothermic gas enriched with natural gas; no air was added. A dew point of −15 °C (+5 °F) was used throughout the 925 °C (1700 °F) carburizing cycle, −12 °C (+10 °F) for the diffusion cycle (also at 925 °C), and −3 °C (+27 °F) for the 845 °C cycle that preceded quenching.

The data in Fig. 15(a) represent carbon content at 0.18 mm (0.007 in.) below the surface. Actual surface carbon ranged from 0.85 to 1.03%. The steel was carburized in a batch-type furnace using a hot-wire analyzer for automatic control of atmosphere. Actual case depth to 50 HRC was 1.10 to 1.32 mm (0.043 to 0.052 in.); required case depth was 1.15 to 1.25 mm (0.045 to 0.050 in.); total case depth ranged from 1.93 to 2.05 mm (0.076 to 0.081 in.).

The data summarized in Fig. 15(b) also represent carbon content at 0.18 mm below the surface. Actual surface carbon ranged from 0.84 to 1.19%. The steel was carburized in a batch-type furnace with manual control of atmosphere. Actual case depth to 50 HRC was 0.38 to 0.58 mm (0.015 to 0.023 in.); required case depth was 0.38 to 0.64 mm (0.015 to 0.025 in.); total case depth ranged from 0.64 to 0.80 mm (0.025 to 0.031 in.).

Carburizing and test conditions for the data shown in Fig. 15(c) were the same as those for the data in Fig. 15(b). In this instance, surface carbon ranged from 0.89 to 1.48%, and case depth to 50 HRC was 0.64 to 0.94 mm (0.025 to 0.037 in.). The specified case depth was 0.75 to 0.90 mm (0.030 to 0.035 in.). Total depth ranged from 0.86 to 1.35 mm (0.034 to 0.053 in.). The same batch-type furnace was used to obtain the data for Fig. 15(a), (b) and (c).

The data shown in Fig. 15(d) represent surface carbon contents of 27 samples taken at 8-h intervals; parts were carburized in trays in a continuous pusher furnace with dew-point control of the atmosphere. Desired surface carbon was 1.0%.

The carbon-concentration data for 8620 and 3310 steels that are shown in Fig. 16(a) and (b) were obtained on round-bar test specimens placed at several locations throughout each charge of production parts of the same steel

Fig. 16 Variation in near-surface (0.13 to 0.25 mm, or 0.005 to 0.010 in.) carbon content for alloy steels carburized in different types of furnaces

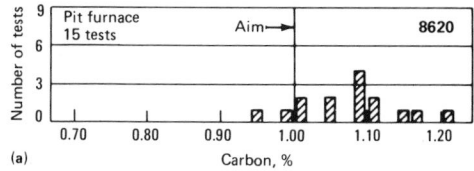

(a)

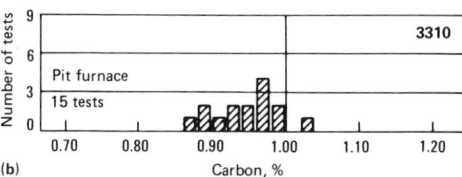

(b)

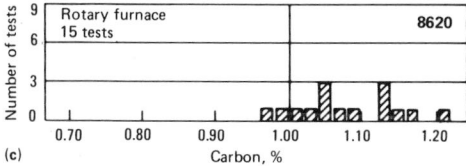

(c)

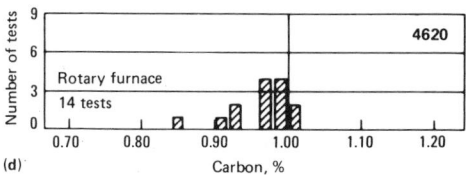

(d)

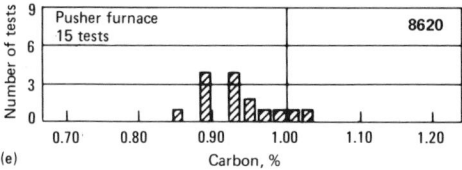

(e)

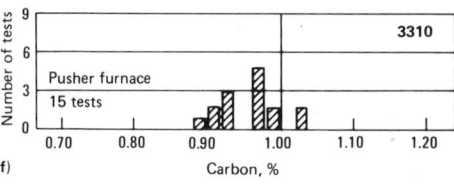

(f)

burized at 955 °C (1750 °F) in a rotary-retort furnace under an atmosphere of endothermic gas plus methane. A diffusion cycle, in which the methane was shut off for the last third of the carburizing cycle, was used for 8620 steel. The 4620 steel was carburized without a diffusion cycle, and the carbon content ranged predominantly below the desired 1.0%. Results for the 8620 steel were largely above the desired carbon content.

The data in Fig. 16(e) and (f) for 8620 and 3310 steels were obtained on round test specimens distributed throughout loads in continuous pusher furnaces. The atmospheres were endothermic gas enriched with methane. The 8620 steel specimens were carburized at 925 °C (1700 °F) in a furnace with three fans, without automatic atmosphere control or a diffusion cycle. The 3310 steel specimens were carburized at 925 °C in a furnace equipped with two fans, with automatic atmosphere control but without a diffusion cycle.

Surface carbon content may vary significantly as a function of location in the furnace. Figure 17 indicates a variation in surface carbon on 4620 steel bearing races from top to bottom of the pit furnace used. The pit furnace was about 0.75 m (30 in.) in diameter by 0.90 m (36 in.) deep. An open load of races was carburized for 7 h in a natural gas atmosphere, and this step was followed by a 3½-h diffusion cycle. Desired surface carbon was 1.00%. Each bar on the chart represents 14 heats of steel. It is apparent that, in this instance, surface carbon content was consistently higher in the bottom portion of the furnace, possibly indicating a slightly higher operating temperature there.

A horizontal batch furnace—operated the same as described for Fig. 3(b) and (c) except that, for a dew point of −12 °C (+10 °F), diffusion time was 6 h—produced the distribution of case depth shown graphically in Fig. 3(d) and tabulated as follows:

1.5 ± 0.03 mm (0.061 ± 0.001 in.) = 31%
1.5 ± 0.05 mm (0.061 ± 0.002 in.) = 50%
1.5 ± 0.08 mm (0.061 ± 0.003 in.) = 63%
1.5 ± 0.10 mm (0.061 ± 0.004 in.) = 81%
1.5 ± 0.13 mm (0.061 ± 0.005 in.) = 100%.

composition. Specimens were carburized in a vertical pit furnace having a single circulating fan. The atmosphere was endothermic gas enriched with methane. The 8620 steel was carburized at 955 °C (1750 °F), without automatic atmosphere control, using a diffusion cycle in which only endothermic gas was introduced during the last third of the carburizing cycle. The 3310 steel was carburized at 925 °C (1700 °F), with automatic atmosphere control but no diffusion cycle.

Round bars were used to obtain the data shown for 8620 and 4620 steels in Fig. 16(c) and (d). Both steels were car-

Fig. 17 Variation in surface carbon content with position in a pit furnace for 4620 steel bearing races

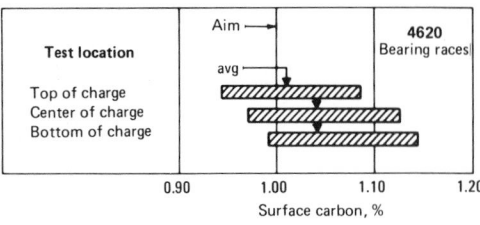

Table 2 Effectiveness and conditions of carbon restoration for four plain carbon steels

Test cut No.(a)	1020(b) Before	After	1038 Before	After	1070(c) Before	After	1095 Before	After
1	0.11	0.20	0.12	0.37	0.16	0.66	0.16	1.04
2	0.19	0.18	0.08	0.34	0.49	0.63	0.62	1.02
3	0.18	0.18	0.17	0.36	0.63	0.62	0.82	0.95
4	0.16	0.19	0.27	0.37	0.69	0.65	0.94	0.95
5	0.18	0.18	0.33	0.39	0.71	0.65	1.00	0.95
6	0.17	0.19	0.39	0.38	0.72	0.68	1.00	1.01
7	0.16	0.19	0.39	0.39	0.72	0.70	1.02	1.01
8	...	...	...	...	0.72	0.70	1.03	1.01
Shim stock, 0.10% C(d)	0.10	0.25	0.10	0.40	0.10	0.70	0.10	1.16
Shim stock, 1.30% C(d)	1.30	0.25	1.30	0.42	1.30	0.68	1.30	1.81

Conditions of treatment	1020(b)	1038	1070(c)	1095
Restoration temperature, °C (°F)	870 (1600)	870 (1600)	870 (1600)	870 (1600)
Time, h	2	1	2½	3
Dew point, °C (°F):				
Furnace	20 (68)	+7 (45)	−1 (30)	−4 (25)
Generator	23 (74)	−1 (30)	−9 (15)	−7 (20)
Atmosphere composition, %(e):				
CO_2	1.1	0.9	0.2	0.2
O_2	0.0	0.0	0.0	0.0
H_2	36.6	38.5	40.6	40.6
CO	18.0	18.5	19.4	19.6
N_2	44.1	41.9	39.4	39.2
CH_4	...	0.2	...	0.4

(a) Carbon in steel determined by analysis of consecutive cuts: depth of cuts, 0.13 mm (0.005 in.). (b) Represents four duplicate tests. (c) Represents five duplicate tests. (d) Treated for control. (e) Analysis of atmosphere in furnace

In this instance, the target depth (to 0.040% carbon) was 1.4 to 1.8 mm (0.055 to 0.070 in.). Test pieces like those described for Fig. 3(a) were used in obtaining the data.

A case depth of 1.4 to 1.8 mm (0.055 to 0.070 in.) (to 0.040% carbon) was specified for gears that were carburized at 925 °C (1700 °F) in a continuous furnace. The variation in case depth is likely to be greater in a continuous furnace than in a batch furnace, as indicated by results of 24 tests summarized in Fig. 3(e). These data were obtained from actual gears, one per week for 24 weeks.

Batch-type carburizing in the same plant that case hardened the gears produced case depths within an extremely close range, as indicated in Fig. 3(f). Case depths were determined by carburizing test pieces 25 mm (1 in.) in diameter by 38 mm (1½ in.) in length along with 175-to-205-kg (390-to-450-lb)

charges of gears, using dew-point control. The test pieces were cut in half, polished, etched in nital and then examined microscopically.

As indicated in the preceding discussion, routine carburization of parts to a case-depth variability not exceeding ±0.2 mm (±0.008 in.) is quite within reason. This degree of maximum variation can be expected over long periods of time, provided that there is adequate process control, supplemented with frequent destructive examination (several times a week) of either actual parts or production-control specimens from each furnace or integrated production line.

Carbon Restoration

Hot worked steel products usually become more or less decarburized during heating for hot rolling, forging, extruding and spinning. The depth of decarburization depends on hot working temperature, time at temperature, furnace atmosphere, reduction in area from bloom to finished size, and type of steel.

Decarburization may be corrected by carbon restoration—that is, by carburizing enough to restore the carbon content to a value equal to the original or intended content. This is particularly important when the formed product must be quenched and tempered to develop high resistance to wear or fatigue. For instance, cold heading stock intended for use in antifriction bearings or fasteners must be virtually free from decarburization. The micrographs in Fig. 18 illustrate restoration of carbon, in two stages, to decarburize 1035 steel heading stock.

Four types of atmospheres are generally used for carbon restoration:

- Nitrogen–dissociated methanol
- Nitrogen enriched with hydrocarbon (usually, methane or propane)
- Dry, purified exothermic
- Endothermic.

Of the four types, endothermic composition atmospheres are the best understood, the easiest to handle, and by far the most widely used in commercial practice. Nitrogen-methanol atmospheres can be considered as direct replacements for generated endothermic gas, because these two types of atmospheres have essentially the same composition.

When decarburization is shallow, surface carbon content can be corrected

Fig. 18 Restoration of carbon in 1035 steel bar stock

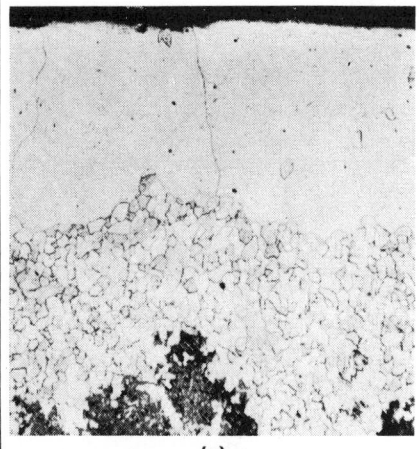 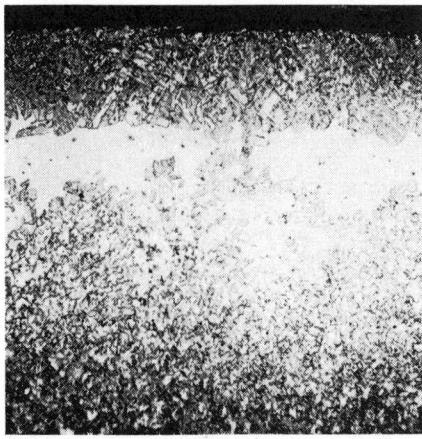 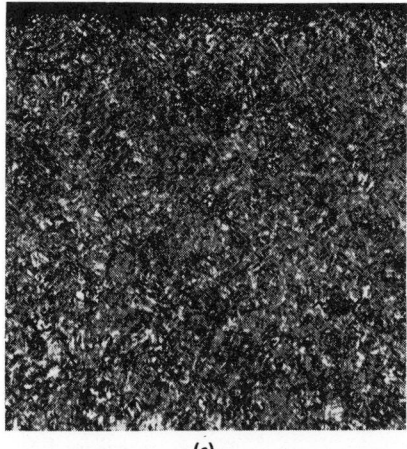

(a)   (b)   (c)

(a) As-received steel with a carbon-free depth of 0.25 to 0.30 mm (0.010 to 0.012 in.); maximum affected depth, 0.38 to 0.64 mm (0.015 to 0.025 in.). **(b):** Same steel as that shown at left, after carbon restoration at 850 °C (1560 °F) for 1 h followed by water quenching. **(c)** Same steel as that shown at left, after carbon restoration at 850 °C (1560 °F) for 2½ h followed by water quenching. All bar stock was 21.8 to 21.9 mm (0.859 to 0.862 in.) in diameter. All micrographs 100×

during the annealing or heat treating cycle. For controlled carbon restoration, the annealing temperature must be above the control temperature for the steel being processed. If decarburization is deep, it may be necessary to use a higher temperature to obtain the necessary diffusion of carbon into the steel within practical time limits.

The process of restoring carbon can be applied to steels of low, medium or high carbon content. Table 2 shows the conditions used in treatments for restoring carbon to four plain carbon steels of different carbon contents, and indicates the effectiveness of the treatment. For control, both low-carbon and high-carbon shims were treated with these steels, and results for these shims are also indicated in Table 2. Note that the low-carbon shims picked up carbon, whereas the high-carbon shims, with the exception of those treated along with the 1095 steel, lost carbon (indicating that equilibrium was established or closely approached).

Successful restoration of carbon depends on the following conditions (listed in order of importance):

- Composition of furnace atmosphere
- Composition of the steel
- Processing temperature
- Surface condition of material undergoing carbon restoration
- Type of furnace

- Furnace zone separation
- Atmosphere-tightness of the furnace
- Distribution of the load
- Flow rate and degree of circulation of the furnace atmosphere.

It is frequently expedient to control the dew point of an endothermic atmosphere by adding small amounts of air or methane to the furnace rather than by adjusting the atmosphere generator. Addition of a small quantity of air will increase the dew point, whereas addition of methane will decrease it. Addition of nitrogen to endothermic gas lowers CO_2, O_2 or dew-point concentrations by *dilution*. This is seldom done, however, because the carburizing components of the endothermic gas (CO and H_2) are also diluted, resulting in a loss of carbon attainment rate and atmosphere control. If the CO content of an endothermic gas deviates widely from the normal 20% (± 2%), this may be expected to have adverse effects on carbon-potential control where traditionally applied techniques are used (that is, C vs dew point, CO_2 content and O_2 potential).

It is difficult to control the carbon potential of atmospheres so that reproducible carbon contents are developed in the surface of the part undergoing carbon restoration. A surface carbon level slightly below that of the base

material is usually much less harmful than overcarburization, especially in fatigue applications. Consequently, if fatigue life is important to the serviceability of the part, it would be better not to risk excessive carbon restoration.

The cost of restoring carbon to parts prior to final machining operations depends on such factors as cost of steel, cost of carbon restoration, and amount of final machining required. In general, if little or no machining is required for finishing a part, there may be an economic gain in restoring carbon. Conversely, if a part requires machining all over, which would remove most or all of the carbon-depleted area, carbon restoration may be unnecessary.

Quality-Control Plans

It is considered axiomatic that, to be effective and consistent, quality-control plans for any facility must be properly conceived, formally documented and rigorously followed. However, a single plan is not universally applicable—policies and procedures that work in one plant may be ineffective in another plant. The same is true of quality-control plans for a carburizing facility.

Sampling Plans. Among commercial firms doing good-quality carburizing on a routine basis, the consensus is that only when samples are taken continually for destructive evaluation is it possible to ensure close control of case properties. "Samples" may mean production parts, pieces of production parts, or specially designed test samples. "Destructive examination" usually means, as a minimum, etching a cross section and directly measuring case depth. In many instances, measurement of hardness gradient or examination of microstructure, or both, also are specified.

In one plant, where 44xx and 86xx steel gears of various sizes are carburized to a single set of requirements in tray-type, four-zone, continuous pusher furnaces, one gear is selected at random daily from each furnace line, after the direct-quench step. Quality is determined by taking microhardness measurements on a cross section through one of the teeth. Effective case depth to 50 HRC at the pitch line is reported, and the hardness profile is compared with that of previously produced gears. As an ongoing check, each part is tested with a mill bastard file immediately after quenching in oil to ensure that a file-hard case has been achieved.

In another plant, where a wide variety of gears is processed to various specifications and where both batch and continuous pusher furnaces are used, one sample part from each production lot is cut for destructive examination. Microstructure is observed, and effec-

tive case depth to 50 HRC is determined from a plot of microhardness (Knoop) versus depth. The operation of the furnaces is such that it makes little difference what type or size of gear is being carburized. To evaluate furnace conditions, carbon gradients are measured for each furnace about every two to five working days. A standard carbon-gradient bar is run along with production parts, and then the bar is step machined and carbon content of chips determined spectrographically. For each plot of carbon gradient, the shape of the curve, case depth to 0.40% carbon, and surface carbon level are evaluated in relation to past performance.

In two other plants, part quality is evaluated by fracturing or cutting small standard test bars, etching the fractured or cut surfaces, and reading case depths with a Brinell glass. One of these plants uses test bars 11 mm square by 75 mm long (about 7/16 in. square by 3 in. long) made of coarse-grain, cold drawn, silicon-killed 1022 steel. In selecting the test-bar material, special narrow limits on composition, grain size and hardenability are imposed so that normal heat-to-heat variations in response to heat treatment are minimized. In batch furnaces, at least one test bar is processed with each batch. (Sometimes, more than one is used—for example, in the top, middle and bottom trays of a load to be carburized in a pit furnace.) In continuous furnaces, a test bar is placed in about every third tray in each row. In both

plants, satisfactory case depths on carburized or carbonitrided test bars justify release of the parts into the normal production stream. If case depths on test bars are borderline or unsatisfactory, a sample part from the production lot is destructively examined and the decision to accept or reject the parts is based on evaluation of this part.

Disposition of Rejected Parts. Experience and problem analyses have proved that nonconforming (rejected) parts must be handled by proper, strictly enforced procedures. Nonconforming parts must be identified so that they can be further investigated, and then reworked or discarded. It is extremely important that questionable parts be removed from the production area and placed in a quarantine area. Scrapped parts must be visibly mutilated to warn plant personnel against returning them to the process stream. Repairs should be made only with the knowledge of all departments concerned. Repaired components must be kept away from the normal process stream until fully tested and proved to be of adequate quality.

Most rejections of parts result from furnace, generator, gas-supply or power failures. Good preventive maintenance greatly minimizes such problems. Fortunately, nonconforming components often can be reprocessed. An efficient shop, it should be emphasized, produces only a small amount of scrap.

Localized
Heat Treating

Induction Hardening and Tempering

By Nelson Stevens
Sales Manager
Lindberg/Cycle-Dyne
A Unit of General Signal Corp.

ELECTROMAGNETIC INDUCTION is one method of generating heat within a part for hardening or tempering a steel or cast iron part. Any electrical conductor can be heated by electromagnetic induction. As alternating current from the converter flows through the inductor, or work coil, a highly concentrated, rapidly alternating magnetic field is established within the coil. The strength of this field depends primarily on the magnitude of the current flowing in the coil. The magnetic field thus established induces an electric potential in the part to be heated, and because the part represents a closed circuit, the induced voltage causes the flow of current. The resistance of the part to the flow of the induced current causes heating.

The pattern of heating obtained by induction is determined by the (a) shape of the induction coil producing the magnetic field, (b) number of turns in the coil, (c) operating frequency, (d) alternating-current power input, and (e) nature of the workpiece. Four examples of magnetic fields and induced currents produced by induction coils are shown in Fig. 1.

The rate of heating obtained with induction coils depends on the strength of the magnetic field to which the part is exposed. In the workpiece, this becomes a function of the induced currents and of the resistance to their flow.

The depth of current penetration depends upon workpiece permeability, resistivity, and the alternating current frequency. Since the first two factors vary comparatively little, the greatest variable is frequency. Depth of current penetration decreases as frequency increases. High-frequency current is generally used when shallow heating (thin case) is desired; intermediate and low frequencies are used in applications requiring deeper heating.

Most induction surface-hardening applications require comparatively high power densities and short heating cycles in order to restrict heating to the surface area. The principal metallurgical advantages that may be obtained by surface hardening with induction include increased wear resistance and improved fatigue strength.

Hardening for Wear Resistance

A shallow hardened case, ranging in depth from 0.25 to 1.5 mm (0.010 to 0.060 in.), provides a part with good wear resistance in applications involving light to moderate loading. For shallow hardening, heating may be limited

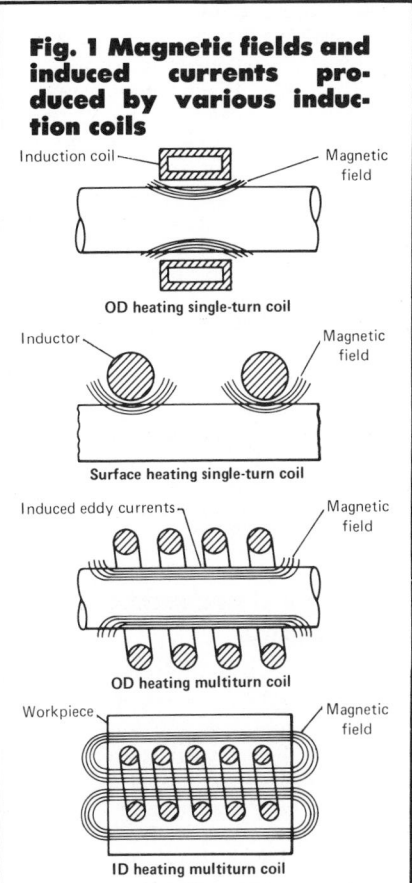

Fig. 1 Magnetic fields and induced currents produced by various induction coils

Induction coil — Magnetic field

OD heating single-turn coil

Inductor — Magnetic field

Surface heating single-turn coil

Induced eddy currents — Magnetic field

OD heating multiturn coil

Workpiece — Magnetic field

ID heating multiturn coil

to the desired depths by use of equipment in the 10-kHz to 2-MHz frequency range. Rocker-arm shafts, sucker-rod couplings, and pump shafts are typical parts requiring a shallow case depth. However, in applications where parts are subjected to heavy or impact-type loading, case depth must be increased over a range from 1.5 to 6.4 mm (0.060 to 0.250 in.) for adequate support and wear resistance. To obtain these case depths, a frequency range of from 1 to 10 kHz is recommended. Gears, track pins, heavy crankshaft bearings, camshafts and bearing races are typical parts requiring these greater case depths.

Improving Fatigue Strength

The induction case hardening of bars and shafts to depths of 3.2 to 12.7 mm (⅛ to ½ in.) has resulted in greatly improved torsional and bending fatigue strength. The process has also made it possible, in some applications, to reduce cost by substituting carbon steels for deeper hardening alloy steels.

Long bars and shafts are usually passed through the inductor coil and are hardened by being progressively heated and quenched. These parts are usually rotated to obtain more uniform results in processing. Truck, tractor and automobile axle shafts and hydraulic piston rods are typical of production parts that are processed in this manner. Equipment in the frequency range of 1 to 10 kHz is recommended for processing such parts.

Critically stressed areas of steering knuckles, flanged axle shafts and similar parts may be selectively case hardened by induction to improve torsional and bending fatigue properties. Depending on the specific application and desired case depth, frequencies ranging from 3 to 450 kHz have been effectively employed for selective hardening of these parts.

Through Hardening and Tempering

Through hardening by induction heating requires low power densities and low frequencies, particularly when applied to heavy-wall tubing or solid bars. Progressive induction heating and quenching are common in processing long sections. Low frequencies (60 to 180 Hz) provide maximum depth of heating on large cross-sections. Dual

Table 1 Characteristics of line-frequency induction heating equipment

60-Hz output, kW	Phase	60-Hz input Voltage
55	1	220/440
75	1	220/440
85	3	220/440
150	3	220/440
185	3	220/440
195	3	220/440 2300/4160
300	3	220/440 2300/4160
370	3	440/2400 4160/4800
425	3	440/2400 4160/4800
575	3	2300/4160/4800
650	3	2300/4160/4800
775	3	2300/4160/4800
875	3	2300/4160/4800

frequencies may be used to decrease the heating time when power requirements exceed 2000 kW. In dual-frequency heating, the higher-frequency power source may be operated at from 1 to 10 kHz, and the low-frequency source at 60 to 180 Hz. The inductor coil must be proportioned to apply heat at a rate that does not exceed that at which it can be conducted inwardly.

Tempering of induction hardened surfaces may be done in conventional furnace equipment or in an induction coil at lower power densities.

Characteristics of Commercial Equipment

Induction hardening is generally done at frequencies of 1000 Hz or higher. The types of commercially available low-frequency and high-frequency equipment and pertinent data are shown in Tables 1 to 5.

Line-frequency units consist essentially of a control circuit and power-factor correction capacitors, to which a transformer capable of matching low-impedance, low-voltage coils to a high-voltage power supply may be added. When a low-voltage power supply is employed to heat ferrous alloys, a transformer is not required. In larger units, the power-factor correction capacitors are mounted on separate capacitor racks; the capacitors for smaller units are usually grouped with the work-handling and coil equipment.

Because these units do not convert

Table 2 Characteristics of static frequency converters for induction heating

Output(a), kW	Input(b) V	A	Cooling
100	440 up	215	Water
200	440 up	400	Water
250	440 up	515	Water
300	440 up	585	Water
400	440 up	785	Water
500	440 up	975	Water
700	2300 up	255	Air/water
1000	2300 up	350	Air
1250	2300 up	440	Air
1500	4160 up	292	Air
2000	4160 up	395	Air

(a) 180 Hz, 750 V. (b) Three-phase

frequency, power losses are negligible, but use of the equipment is limited to through-heating applications. For these applications, line-frequency systems are capable of producing maximum temperature uniformity. Operating characteristics of line-frequency induction heating equipment are given in Table 1.

Static Frequency Converters. The most common types are the frequency tripler and the solid state converter. Basically, the tripler consists of an iron core transformer with three sets of windings, one for each input phase. Each phase voltage is applied to its own reactor and to the load in series. Operation of the tripler is based on the summation of pulses obtained from saturable reactors. Thus, by adding the pulses, a 60-Hz supply current is multiplied to 180 Hz. The operating characteristics of three-phase static frequency converters appear in Table 2.

A static frequency converter of the type described provides a highly efficient means for through heating steel bars of a given diameter, depending on output frequency. For example, at a frequency of 180 Hz, maximum heating efficiency can be achieved with diameters ranging from approximately 9 to 15 cm (3½ to 6 in.). Normally, diameters of 15 to 23 cm (6 to 9 in.) can be heated more efficiently at 60 Hz, whereas a diameter of 8 to 9 cm (3 to 3½ in.) can be heated more efficiently when output frequency is about 1 kHz.

Medium Frequency Motor-Generator Units. Motor-generator units consist of a high-frequency generator driven by a motor. Induction motors, which may be mounted integrally with the generator or separately on a

Table 3 Characteristics of motor-generator equipment for induction heating

Generator rating(a)		Motor rating(b)		Full-load current (440 V), A	Locked-rotor current (440 V), A	Type(c)	Cooling
kW	Voltage	hp	Voltage				
400 Hz (nominal)							
30	150/220	50	220/440			H	A
50	100/200	75	220/400			H	A
1000 Hz (nominal)							
100	400/800	125	220/440			H	A or W
100		150		182	840		
150	660/1320	235	440 up			V	W
175	400/800	250	440 up	315	1540	V or H	A or W
250	400/800	400	440 up	465	2700	V or H	A or W
300	400/800	500	440 up			V	W
350	400/800	600	440 up			V	W
350		560		645	4060		
470	880/1760	700	440 up			H	W
500				180 (2300 V)	1230 (2300 V)		
650	1250/2500	940	440 up			H	W
700	800	1000	440 up			H	A or W
1000	800	1500	440 up			H	A or W
1250	800	1950	440 up			H	A or W
3000 Hz (nominal)							
15	220/440	25	220/440	32	194	V	A or W
20	220/440	30	230/460			V	A or W
20		40		40	325		
30	200/400	50	220/440	60	350	V	A or W
50	220/440	75	220/440			V or H	A or W
50		85		92	581		
60	165/330	90	440			V	W
100	400/800	150	220/440			V or H	A or W
100		160		187	1076		
150		235		273	1528		
150	400/800	250	440 up			V or H	A or W
200	400/800	300	440 up	356	1850	V or H	A or W
250	400/800	400	440 up			V	W
300		460		545	2700		
300	400/800	500	440 up			V or H	A or W
500	800	800	440 up			H	A, H₂ or W
500				180 (2300 V)	1230 (2300 V)		
600	600/1200	870	440 up			H	W
700	800	1000	440 up			H	A or W
1000	800	1500	440 up			H	A or W
1250	800	1950	440 up			H	A or W
1500	800	2100	440 up			H	H₂
2500	800	3750	440 up			H	H₂

(continued)

Note: A, air; W, water; V, vertical; H, horizontal.
(a) Generator may be given higher ratings for intermittent service. Multiple ratings indicate that a choice of voltage is available for each kilowatt rating. For example, one facility indicates 100-kW generator available with voltage ratings of 400 and 800; another facility indicates 100-kW generator available with voltage ratings of 440 or 800 each with the same frequency. (b) Motor generators of the same frequency and design can generally be operated in parallel when larger power supplies are required. Generator manufacturer should be consulted before attempting parallel operation. (c) Horizontal or vertical construction

common base, are used with 1, 3 and 10 kHz generators. The integrally mounted units can be either vertical or horizontal. The motors for generators of these frequencies may be of either the induction or synchronous type. Units using a synchronous motor are mounted horizontally and separately from the generator. The data in Table 3 cover ratings for units used with a 60-Hz power supply. Some units may be used with 50-Hz sources with proportionate decreases in generator frequency and slight changes in ratings.

Medium frequency solid state units are used to produce frequencies from 180 Hz to 50 kHz. While several different types of power circuits are used, all convert the 60-Hz main-line frequency into single-phase high frequency at relatively high efficiency. Either fixed-frequency or variable-frequency output units are available.

The fixed-frequency outputs operate load-match in a manner similar to that of motor generators, working at a fixed, specific frequency—for example, 10

Table 3 continued

Generator rating(a) kW	Voltage	Motor rating(b) hp	Voltage	Full-load current (440 V), A	Locked-rotor current (440 V), A	Type(c)	Cooling
4200 Hz (nominal)							
20	220/440	30	230/460			V	W
20		40		40	325		
30	220/440	50	230/460	62	325	V	W
50	220/440	75	230/460	105	550	V	W
100	220/440 or 400/800	175	230/460	215	975	V	W
10 000 Hz (nominal)							
7.5	220/440	15	220/440	18	97	V	A or W
15	220/440	30	220/440			V	A or W
15		25		32	194		
20	220/440	40	220/440	41	214	V	A or W
30	220/440	50	220/440	60	350		
50	220/440 or 300/600	75	220/440			V or H	A or W
50		85		100	725		
75	400/800	125	220/440	145	775	V or H	A or W
100	400/440/800	175	220/440			V or H	A or W
100		170		196	1076		
150	440/440/800	250	440 up			V or H	A or W
150		265		295	2070		
175	400/800	300	440 up	353	2200	V	A or W
200	400/600/800	325	440 up			V	A or W
250	400/800	400	440 up	475	2700	V or H	A or W
350	800	500	440 up			H	H$_2$
375				140 (2300 V)	900 (2300 V)		
700	800	1000	440 up			H	A, H$_2$ or W

Note: A, air; W, water; V, vertical; H, horizontal.
(a) Generator may be given higher ratings for intermittent service. Multiple ratings indicate that a choice of voltage is available for each kilowatt rating. For example, one facility indicates 100-kW generator available with voltage ratings of 400 and 800; another facility indicates 100-kW generator available with voltage ratings of 440 or 800 each with the same frequency. (b) Motor generators of the same frequency and design can generally be operated in parallel when larger power supplies are required. Generator manufacturer should be consulted before attempting parallel operation. (c) Horizontal or vertical construction

kHz. Variable-frequency units operate over a range of frequencies for most effective load-matching. The power and tuning of a variable-frequency unit can be changed by adjusting the frequency during operation.

Solid state units can be made to provide the same heating results as motor generator units. Selection of frequency, power and duration of power should be made independent of whether a motor generator or solid state induction heater is to be used.

Solid state units have the advantage of a conversion efficiency greater than 90%, as compared with 75 to 80% of a motor generator. Solid state units also do not have the starting in-rush of a motor generator. Finally, through built-in logic in the control circuits, the solid state units can match load impedances through frequency shift, providing better load input power without shifting capacitor or transformer taps during heating (Table 4).

Vacuum Tube Units. Electronic tube units consist of a power supply section and an oscillator section. The power section provides the high voltage for the oscillator tube after rectification to a pulsating direct current, usually by solid state diodes. The oscillator tube and a tank circuit consisting of a matched inductor coil and capacitor comprise the oscillator section. The oscillator tube controls the amount of electrical energy delivered to the tank circuit from which the energy is removed by the coupled load. A small and proportionate amount of the power in the tank circuit is fed back into the grid of the oscillator tube to control the current that is delivered to the tube, and it in turn controls the amount of electrical energy entering the tank circuit. The frequency developed in the converter is determined by the inductance of the tank coil and the capacitor, which form a parallel tuned circuit. A load-matching network electrically coupled to the tank circuit is used to transmit tank circuit energy to the work. The power ratings and operating characteristics of vacuum tube equipment are listed in Table 5.

Selection of Frequency, Power and Duration of Heating

The distribution of induced current in a part is maximum on the surface and decreases rapidly within the part; the effective penetration of current increases with a decrease in the frequency. The distribution of induced current is influenced also by the magnetic and electrical characteristics of the part being heated; and since these properties change with temperature, the current distribution will change as the work is heated.

Because the heat rapidly progresses to the interior by conduction as soon as the surface is heated, the actual depth

Table 4 Characteristics of medium-frequency (1 to 50 kHz) solid state equipment for induction heating

Output, kW	Input, kV·A	Amperage at 480 V, input 30
5	6	7
10	12	14
25	29	36
50	58	73
75	87	110
100	116	146
150	174	219
200	232	292
250	290	365
300	349	438
400	465	584
500	581	730
600	698	876
800	930	1168
1000	1163	1460
1500	1744	2190
2000	2326	2920

Table 5 Characteristics of vacuum tube equipment for induction heating(a)

kW	Output Frequency(b)	Input voltage	Cooling
Single-phase input			
0.5	450 kHz	110	Air(c)
1.0	400 kHz, 2.5 MHz, 25 MHz	110/220	Air or Water(c)
1.1	500 kHz	220/440	Water(c)
1.5	450 kHz	110/220	Water(c)
2.5	450 kHz, 2.5 MHz, 5 MHz, 10 MHz	220/440	Air or Water(c)
3.0	450 kHz, 27 MHz	220/440	Air(c)
4.0	450 kHz	220/440	Air(c)
Single-phase or three-phase input			
5.0	250 kHz, 450 kHz, 600 kHz, 3.5 MHz, 4 MHz, 50 MHz	220/440	Water
7.5	250 kHz, 450 kHz, 600 kHz, 3 MHz, 5 MHz	220/440	Water
Three-phase input			
10	450 kHz, 3 MHz, 4 MHz, 5 MHz	220/440	Water
15	450 kHz, 1.4 MHz, 3 MHz, 4 MHz, 5 MHz	220/440	Water
20	180 kHz, 450 kHz, 600 kHz, 3 MHz, 5 MHz	220/440	Water
25	180 kHz, 400 kHz, 450 kHz	220/440	Water
30	180 kHz, 400 kHz, 450 kHz, 1.4 MHz	220/440	Water
35	450 kHz	220/440	Water
40	180 kHz, 450 kHz	220/440	Water
50	150 kHz, 400 kHz, 450 kHz	220/440	Water
60	180 kHz, 450 kHz	440	Water
70	450 kHz	440	Water
75	180 kHz, 400 kHz	440	Water
100	160 kHz, 180 kHz, 400 kHz, 450 kHz	440	Water
140	450 kHz	440	Water
150	180 kHz, 450 kHz	440	Water
200	450 kHz	440	Water
250	180 kHz, 450 kHz	440	Water
280	450 kHz	440	Water
300	450 kHz	2300	Water
450	450 kHz	2300	Water
560	450 kHz	440	Water
600	200 kHz	2300	Water

(a) Current requirements vary with equipment. Kilovolt-ampere input of twice the output is average. For example, an input of approximately 2 kV·A is required for 10-kW output. (b) Other output frequencies are available on special order from most manufacturers. (c) Work coils are generally water-cooled, whereas oscillator vacuum tubes may be either air- or water-cooled.

of heating is determined by the duration of heating and the power density (kilowatts per square inch of surface exposed to the inductor), as well as by the frequency. Maximum power density, minimum duration of heating, and high frequency produce a minimum depth of heating.

Selection of Frequency

In analyzing the frequency and power required for a specific application, it is desirable to consider the frequency first. Primary considerations are the depth of heating and the size of the part. Table 6 lists the frequencies and power sources most commonly used in induction hardening. As shown in this tabulation, the lower frequencies are more suitable as the size of the part and the case depth increase. However, because power density (kilowatts per square inch of surface within the work coil or inductor) also has an important influence on the depth to which the part is heated, wide deviations from this chart may be made with successful results, because higher densities of power compensate for lower frequencies. This is particularly important where equipment of a given frequency already exists in the plant or where equipment is required for several different applications. In some instances, the determining factor in selecting the frequency is the power required to provide power density sufficient for successful hardening, as lower-frequency induction equipment is available with higher power ratings.

Efficiency. The use of wrong frequency will result in a decrease in electrical efficiency; sometimes in failure to maintain a minimum case depth where shallow cases are required; or in failure to heat uniformly throughout the piece where through hardening is required. Figure 2 illustrates the decrease in transfer of energy or heating efficiency that could result if the incorrect frequency is selected. While the efficiency for all frequencies is similar with large bars, maximum efficiency is achieved with small bars by using higher frequencies. In parts of small diameter, suitable hardening temperatures cannot be obtained with the frequencies shown in Fig. 2. For such parts, MHz-frequencies are desirable, or the thin sections may sometimes be hardened successfully with normal frequencies by using a combination of induction heating and radiation from an induction heated liner that surrounds the part, except for a slit along its length to prevent electromagnetic shielding.

Table 6 Selection of power source and frequency for various applications of induction hardening

| Depth of hardening | | Section size | | Power lines, 50 or 60 Hz | Frequency converter, 180 Hz | Solid state or motor generator | | | Vacuum tube, over 200 kHz |
mm	in.	mm	in.			1000 Hz	3000 Hz	10 000 Hz	
Surface hardening									
0.38-1.27	0.015-0.050	6.35-25.4	1/4-1	. . .	. . .	. . .	. . .	. . .	Good
1.29-2.54	0.051-0.100	11.11-15.88	7/16-5/8	. . .	. . .	. . .	. . .	Fair	Good
		15.88-25.4	5/8-1	. . .	. . .	. . .	. . .	Good	Good
		25.4-50.8	1-2	. . .	. . .	. . .	Fair	Good	Fair
		Over 50.8	Over 2	. . .	. . .	Fair	Good	Good	Poor
2.56-5.08	0.101-0.200	19.05-50.8	3/4-2	. . .	. . .	. . .	Good	Good	Poor
		50.8-101.6	2-4	. . .	. . .	Good	Good	Fair	. . .
		Over 101.6	Over 4	. . .	. . .	Good	Fair	Poor	. . .
Through hardening									
Through hardening............		1.59-6.35	1/16-1/4	. . .	. . .	. . .	. . .	. . .	Good
		6.35-12.7	1/4-1/2	. . .	. . .	. . .	. . .	Fair	Good
		12.7-25.4	1/2-1	. . .	. . .	. . .	Fair	Good	Fair
		25.4-50.8	1-2	. . .	. . .	Fair	Good	Fair	. . .
		50.8-76.2	2-3	. . .	. . .	Good	Good	Poor	. . .
		76.2-152.4	3-6	Fair	Good	Good	Poor	Poor	. . .
		Over 152.4	Over 6	Good	Fair	Poor	Poor	Poor	. . .

Good indicates frequency that will most efficiently heat the material to austenitizing temperature for the specified depth. For through hardening, **Good** indicates the amount of material approximately equal to the mg/J shown in Fig. 3. **Fair** indicates a frequency that is lower than optimum but high enough to heat the material to austenitizing temperature for the specified depth. With this frequency, the current penetration relative to the section size causes current cancellation and lowered efficiency. For through hardening, **Fair** indicates an amount of material less than the pounds per kilowatt-hour shown in Fig. 3. **Fair** may also indicate a frequency higher than optimum that can overheat the surface at high-energy inputs. Converters cost more per kilowatt-hour than the converters of optimum frequency. With some equipment, the efficiency may be lower. **Poor** indicates a frequency that will overheat the surface unless low-energy input is used. Efficiency and production are low and capital cost of converters per kilowatt-hour is high.

Selection of Power

The size of the converter or the power required should be determined on the basis of power density, section size, heating method and production requirements, as indicated below.

- *Requirements for power density:* Presented in kilowatts per square inch, this information may be available from previous experience, from experiment, or from data such as those in Tables 7 and 8.
- *Size of section to be heated:* In surface hardening, the area heated at one time, multiplied by power density, indicates the total power input (kilowatts). This area is obtained by multiplying the perimeter of the part by the length of the inductor. To calculate the power required for through heating, divide the desired production load by the values for pounds per kilowatt-hour given in Fig. 3.
- *Method of heating:* When the static method of heating is used, the entire area to be hardened must be heated at one time. When the part is scanned, or progressively hardened, it is necessary to heat only a small segment or band at one time. Thus, a smaller converter can be used if the parts are progressively hardened.
- *Production requirements:* Power densities, as indicated in Tables 7 and 8, may be adjusted within limits to meet certain production requirements. Beyond this, production rate may be increased by one or more of the following modifications. For hardening by the static method: (*a*) fixturing may be improved to reduce handling time, (*b*) work may be

Fig. 2 Efficiency of energy transfer at several frequencies to type 1045 steel bars

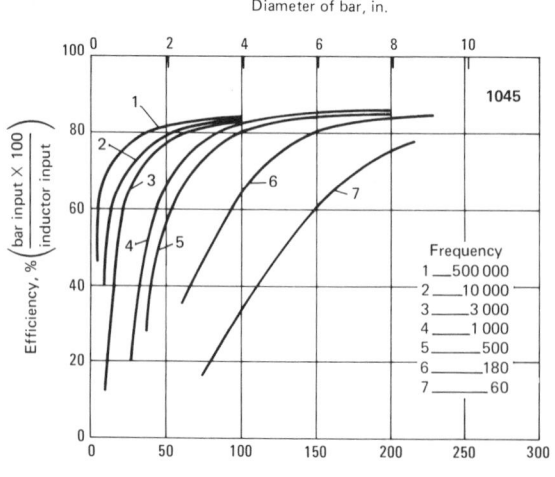

Frequency	
1	500 000
2	10 000
3	3 000
4	1 000
5	500
6	180
7	60

Bars are of various diameters heated to 1095 °C (2000 °F) with inside diameters of inductors 28.6 mm (1.125 in.) larger than outside diameters of bars.

Table 7 Power density required for surface hardening(a)

Frequency, kHz	Depth of hardening(c) mm	in.	W/mm² Low (d)	Optimum (e)	High (f)	kW/in.²(b) Low (d)	Optimum (e)	High (f)
500	0.38-1.14	0.015-0.045	10.9	15.5	18.6	7	10	12
	1.14-2.29	0.045-0.090	4.7	7.8	12.4	3	5	8
10	1.52-2.29	0.060-0.090	12.4	15.5	24.8	8	10	16
	2.29-3.05	0.090-0.120	7.8	15.5	23.3	5	10	15
	3.05-4.06	0.120-0.160	7.8	15.5	21.7	5	10	14
3	2.29-3.05	0.090-0.120	15.5	23.3	26.35	10	15	17
	3.05-4.06	0.120-0.160	7.8	21.7	24.8	5	14	16
	4.06-5.08	0.160-0.200	7.8	15.5	21.7	5	10	14
1	5.08-7.11	0.200-0.280	7.8	15.5	18.6	5	10	12
	7.11-9.14	0.280-0.360	7.8	15.5	18.6	5	10	12

(a) This table is based on use of proper frequency and normal over-all operating efficiency of equipment. Values in table may be used for static and progressive methods of heating; however, for some applications, higher inputs can be used when hardening progressively. (b) Kilowattage is read as maximum during heat cycle. (c) For greater depth of hardening, a lower kilowatt input is used. (d) Low kilowatt input may be used when generator capacity is limited. These kilowatt values may be used to calculate largest part hardened (single-shot method) with a given generator. (e) For best metallurgical results. (f) For higher production when generator capacity is available

Table 8 Approximate power density required for through heating of steel(a)

Frequency, Hz(b)	W/mm² 150-425 °C 300-800 °F	425-760 °C 800-1400 °F	760-980 °C 1400-1800 °F	980-1095 °C 1800-2000 °F	1095-1205 °C 2000-2200 °F	kW/in.² 150-425 °C 300-800 °F	425-760 °C 800-1400 °F	760-980 °C 1400-1800 °F	980-1095 °C 1800-2000 °F	1095-1205 °C 2000-2200 °F
60	0.09	0.23	(d)	(d)	(d)	0.06	0.15	(d)	(d)	(d)
180	0.08	0.22	(d)	(d)	(d)	0.05	0.14	(d)	(d)	(d)
1000	0.06	0.19	0.78	1.55	2.17	0.04	0.12	0.5	1.0	1.4
3000	0.05	0.16	0.62	0.85	1.09	0.03	0.10	0.4	0.55	0.7
10 000	0.03	0.12	0.47	0.70	0.85	0.02	0.08	0.3	0.45	0.55

(a) For hardening, tempering or forming operations. (b) Table is based on use of proper frequency and normal over-all operating efficiency of equipment. (c) In general, these power densities are for section sizes of 12 to 50 mm (1/2 to 2 in.). Higher inputs can be used for smaller section sizes, and lower inputs may be required for larger section sizes. (d) Not recommended

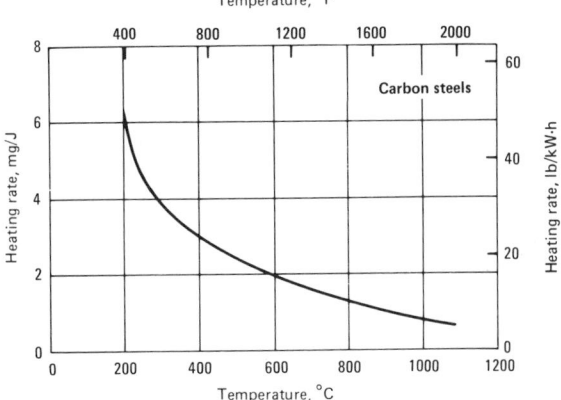

Fig. 3 Heating rate for through heating of carbon steels by induction

Temperature, °F

Carbon steels

Heating rate, mg/J

Heating rate, lb/kW·h

Temperature, °C

For converted frequencies, the total power transmitted by the inductor to the work is less than the power input to the machine, because of converter losses.

quenched outside the inductor to permit immediate reloading, and (c) two or more inductors may be used, if sufficient power is available. For hardening by the scanning method, it may be possible to increase the inductor length or to use more than one inductor, depending on available power.

If the production rate, as based on the size of converter, does not meet production requirements, two or more units of like capacity (or a single unit having two, three or four times the capacity determined) may be used to harden two, three or four pieces simultaneously in a multiple coil.

Selection of Duration of Heating

When the frequency and power density have been selected, the duration of the heating cycle becomes a fixed value for a specific set of conditions. To calculate duration of heating for surface hardening by the static method, divide the value for kilowatt seconds per square inch by power density (kilowatts per square inch). The value of kilowatt seconds per square inch is affected by case depth requirements, type of steel, and prior structure, and may be derived by experiment, as were

Fig. 4 Operating efficiency for surface hardening steels to various depths

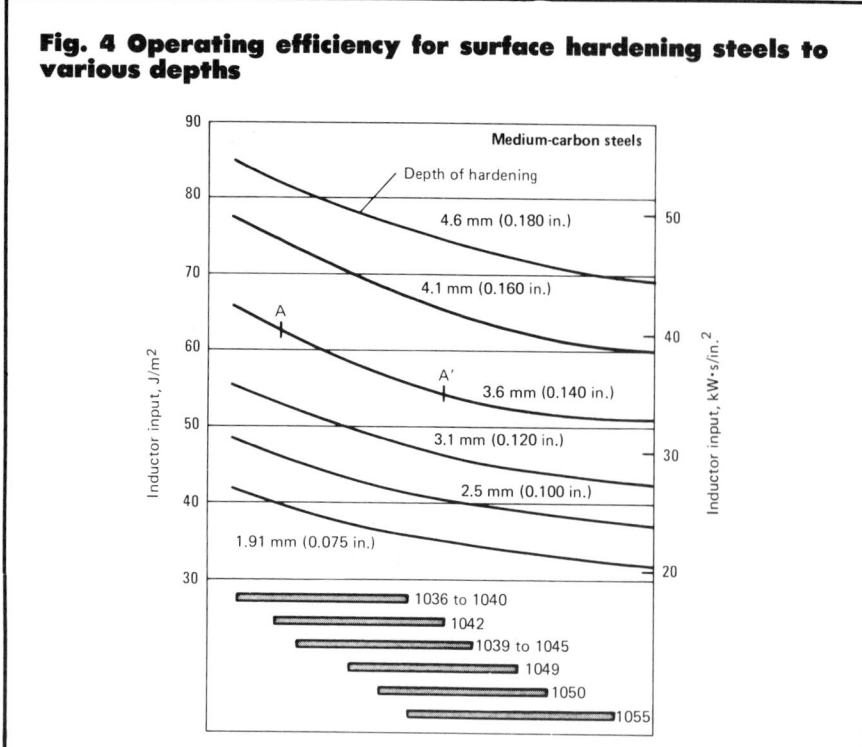

Table 10 Determination of size of smallest generator and duration of heating

Frequency	10 Hz
Power density using low input . .	5 kW/in.^{2}
Area.	7.363 in.^{2}
Power input = power density × area = 5 × 7.363	37 kW
kW·s/in.^{2}	28

Table 3 indicates that the smallest 10-Hz generator available commercially for supplying 37 kW is a 50-kW unit. Assuming that it is possible to match the impedance to utilize the full 50 kW, the maximum power density (kW/in.^{2}) using this generator can be recalculated by dividing the power (50 kW) by the area (7.363).

Power density kW/in.^{2} = 50/7.363	6.79 kW/in.^{2}
Duration of heating = kW·s/in.^{2}/kW/in.^{2} = 28/6.79	4.123 s

Table 11 Determination of frequency, inductor length and power input for scanning

Frequency	10 Hz
Power density using optimum input	10 kW/in.^{2}
kW·s/in.^{2}	27
Scanning time = $\dfrac{3600 \text{ s/h} \times 0.8 \text{ efficiency}}{200 \text{ pieces per h} \times 6 \text{ in. length}}$	2.4 s/in.
Inductor length = $\dfrac{\text{kW·s/in.}^2}{\text{kW/in.}^2 \times \text{s/in.}}$ = 27/(10 × 2.4) = 27/24	1.125 in.
Area = diameter × pi × length = 1 × 3.1416 × 1.125	3.534 in.^{2}
Power input = kW/in.^{2} × area = 10 × 3.534	35.34 kW

the data given in Fig. 4, or based on previous experience. To calculate heating time for surface hardening by the scanning method, divide kilowatt second per square inch by power density and inductor length.

Data in Fig. 4 are for a single-turn inductor of 3000- or 10 000-Hz frequency and a 1.5- to 2-mm (0.060- to 0.080-in.) air gap. Larger air gaps reduce efficiency, thus requiring more power. The values given are intended only as a guide, as different installations vary in efficiency, metering, wave form, voltage regulation and other electrical characteristics.

Example 1. A journal on a shaft is induction hardened by the static method. Diameter of the journal is 32 mm (1.25 in.) and the length to be hardened is 38 mm (1.50 in.). The steel selected is 1045, and minimum hardness of 50 HRC is required at a depth of 2.5 mm (0.100 in.). Determine the optimum frequency, kilowatt input, and duration of heating. Table 9 shows the solution.

Example 2. It is required that the shaft in Example 1 be hardened by equipment employing the smallest generator. Determine the size of this generator (in kilowatts) and the duration of heating (in seconds). Table 10 shows the solution.

Table 9 Determination of frequency, power input and duration of heating

Frequency	10 Hz
Power density	10 kW/in.^{2}
Area = perimeter × inductor length = 1.25 (3.1416) × (1.50 + 0.375)	7.363(a)
Power input = power density × area = 10 × 7.363	74 kW
kW·s/in.^{2}	28
Duration of heating = kW·s/in.^{2}/kW/in.^{2} = 28/10	2.8 s

(a) Note that inductor length equals hardened length (1.50 in.) plus 0.375 in.

Example 3. It is required to harden the full length of a 1045 steel shaft, 25 mm (1 in.) in diameter by 152 mm (6 in.) long, at the rate of 200 pieces/h. A minimum hardness of 50 HRC is required at a depth of 2.3 mm (0.090 in.). Determine the optimum frequency, inductor length, and power input (kilowatts). Table 11 shows the solution.

Example 4. It is required to harden the full length of a 1045 steel shaft, 50 mm (2 in.) in diameter by 255 mm (10 in.) long, to a minimum hardness of 50 HRC at a depth of 2.3 mm (0.090 in.). A 3000-Hz 100-kW solid state unit is

available, including a suitable work station with a scanning-type inductor 54 mm (2.125 in.) ID by 38 mm (1.5 in.) long. Determine if the available equipment can be used for hardening these shafts and what production rate can be expected. Table 12 shows the solution.

Selection of Coil Design

The success of many induction heating applications is related to the selection and design of the proper work coil, or inductor. This design is influenced by a number of factors, including the dimensions and configuration of the part to be heated, the heat pattern desired, whether the part is heated throughout its length at the same time or progressively, the number of parts to

Table 12 Determination of equipment suitability and production rate for scanning

Frequency. In Table 6, 3000 Hz at 2-to-3-in. diam, 0.090-in. case is rated	Good
Area = diam × pi × length = 2 × 3.1416 × 1.5	9.425 in.^{2}
Available kW/area = 100/9.425 = 10.6 kW/in.^{2} which is the power density if all available power is used. Comparing with Table 7, 10 kW/in.^{2} input is the low limit for 3 hertz at 0.090 in. depth. Therefore, all available power (100 kW) can be used.	
Power density (kW/in.^{2}).	10.6
kW·s/in.^{2}.	27
Scanning time (s/in.)	
$= \dfrac{27 \text{ kW·s/in.}^2}{10.6 \text{ kW/in.}^2 \times 1.5 \text{ in.}}$	1.7
Time (seconds) per piece = s/in. × length of part = 1.7 × 10 = 17 seconds per piece	
Production rate	
$= \dfrac{3600 \text{ s/h} \times 0.8 \text{ efficiency}}{17 \text{ s/piece}}$	169 pieces/h

Fig. 5 Typical work coils for high-frequency units

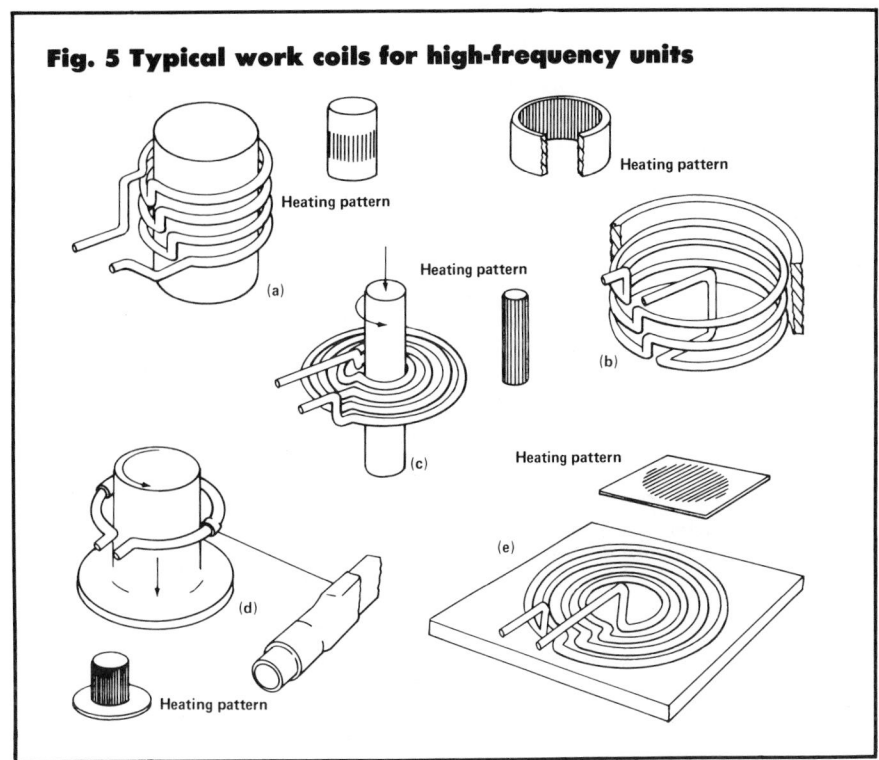

be heated at one time and the amount of power available.

The strength of the magnetic field within the inductor, or work coil, is the basic factor that determines the rate of heating. For most rapid heating rates, therefore, inductors are designed to provide the maximum flow of current in the inductor, and the closest coupling (distance between inductor inside diameter and the part) permissible, after consideration of work-handling features and arcing between the work and the coil. In practice, considerable variation exists in the design of coils for tube, spark-gap, solid state, static frequency converters and line frequencies. At all frequencies, however, the coils are generally of copper, because of its high conductivity and wide availability at moderate cost. The copper may be in the form of tubing or solid busbar, or a combination of both if required.

Basic Designs

Five basic designs of work coils for use with high-frequency (over 200 kHz) units and the heat patterns developed by each are shown in Fig. 5(a) through (e); these basic shapes are (a) a simple solenoid for external heating, (b) a coil to be used internally for heating bores, (c) a pie-plate type of coil designed to

provide high current densities in a narrow band for scanning applications, (d) a single-turn coil for scanning a rotating surface, provided with a contoured half-turn that will aid in heating the fillet, (e) a pancake coil for spot heating. Solenoid coils for external heating are most efficient and should be used whenever possible.

Commercial copper tubing may be used for such coils. The tubing must be large enough to permit an adequate flow of water for cooling. With machines of very low power, the tubing may be as small as 3 mm ($^1/_8$ in.) in diameter, but for units of 20 to 50 kW, it is usually 4.8 or 6.4 mm ($^3/_{16}$ or $^1/_4$ in.) in diameter. There is available, however, a 3-mm- ($^1/_8$-in.-) OD tubing with a wall thickness of only 0.46 mm (0.018 in.) that is reported to be able to carry about 50 kW when the volume of pressurized cooling water is adequate.

Coupling. Coil turns are normally spaced 1.6 to 2.4 mm ($^1/_{16}$ to $^3/_{32}$ in.) apart with a 1.6- to 3-mm ($^1/_{16}$- to $^1/_8$-in.) space (coupling) between the workpiece and the coil. Considerable adjustment in the heat pattern is possible by varying either the spacing between turns or the coupling for individual turns.

Inductors for medium frequencies may be considered as being of two

basic design types: (a) a single-turn inductor for high current densities (20 to 50 times the generator output current) is employed to confine heating to a comparatively narrow band or segment; and (b) a multiturn inductor for low current densities (1 to 5 times the generator output current) is employed to heat through the part or heat a wider or larger area. A step-down transformer is ordinarily utilized between single-turn inductors and the generator, while multiturn inductors are connected to the generator, directly or through another transformer.

Line-frequency (60-Hz) coils, like coils for high-frequency units, are almost invariably constructed of copper tubing through which water is passed to maintain a safe working temperature. For an equivalent voltage, many more turns are required in the low-frequency coils, but it is feasible to use multilayer construction. Optimum efficiency is usually achieved by keeping the radial dimensions of the tubing smaller than about 7.6 mm (0.300 in.). Rectangular tubing provides the most efficient use of space and strongest construction. With the insulating and lining materials now available, it is practicable to build coils that are capable of operating at up to 600 V.

Design Variables

The thickness of the copper conductor used with medium-frequency units is important. For efficient operation, the following minimum thicknesses of wall may be used as a guide in constructing multiturn coils:

Minimum wall thickness		Frequency,
mm	in.	Hz
3.0	0.120	1000
1.8	0.070	3000
1.0	0.040	10 000

Whenever possible in the construction of single-turn coils, with which current densities may be considerably higher, these minimums should be increased three or four times.

Cooling. Many surface hardening applications lend themselves to the use of single-turn inductors of busbar construction. Because of the high current densities and the extremely thin cross section in which the current confines itself, artificial cooling is required. This is usually accomplished by circulating water through channels provided for the purpose. These channels may be made by drilling connecting holes or milling out a path to make a completed loop around the bore of the inductor, and then plugging the exposed ends of the holes or brazing a copper sheet over the milled passage to make a continuous watertight cooling channel. Cross-sectional areas of the cooling passages of 32.25 to 80.6 mm² (0.050 to 0.125 in.²) provide adequate cooling at water pressures of 275 to 345 kPa (40 to 50 psi) for power inputs from 30 to 150 kW.

Quenching Orifices. The single-turn inductor often can include another water chamber with a suitable pattern or series of orifices directed for spray quenching the heated area. Such a spray quench is effective for single-shot hardening of work that is stationary while being heated and quenched. For progressive hardening of a narrow band or area, the orifices must be directed at an angle so that the quenching medium will strike the heated work as it moves out of the inductor. A satisfactory angle, which eliminates backwash of the quench, is about 30° from the work.

Bore Length. For hardening areas on cylindrical parts using a single-shot technique, the inductor bore should be longer than the area to be hardened, the actual overlap decreasing with an increase in power density. With intermediate power densities, the inductor bore should be approximately 9.5 mm (0.375 in.) longer than the area to be hardened if that area does not extend to the end of the part. If the area to be hardened is at the end of a shaft or protrudes from the shaft as on an eccentric, cam or gear, the inductor bore should be 6.4 mm (0.250 in.) longer than the area. For efficient transfer of energy the inductor bore may be 3.1 to 5.1 mm (0.120 to 0.200 in.) larger in diameter than the part. As the bore is increased, the work will approach a weaker portion of the magnetic field, heating time will be increased, and the heating pattern will be deeper unless the power density is increased correspondingly.

Scanning. With the cylindrical areas described, single-shot inductors are generally operated at power densities of 7.75 to 23.25 W/mm² (5 to 15 kW/in.²). When the length of the hardened area is such that the power density obtainable is lower than that required for a specified maximum case depth, it is logical to consider an inductor for progressive hardening, if the work is adaptable. The shorter inductor will usually permit the power density requirement to be satisfied. An inductor bore 25 mm (1 in.) long is satisfactory for shafts of 19- to 76-mm (¾- to 3-in.) diam. Bore may be shorter if power is limited.

Irregular Shapes

In constructing inductor coils for heating irregular shapes, it should be remembered that the portions of the work closest to the inductor will usually be in a stronger magnetic field and will heat more rapidly. If the irregularity is in the transverse plane of the part and does not have sharp corners, thin sections, or sharp re-entrant surfaces, a hardness pattern generally following the contour of the part may be expected with a circular inductor. If this is not satisfactory, an inductor should be constructed with a uniform air gap so that the bore will follow the irregularity. It may still be found necessary to relieve or increase the air gap around the section having least mass, in order to reduce the strength of the magnetic field and reduce the heating rate. This might be necessary for a cam or an automotive camshaft.

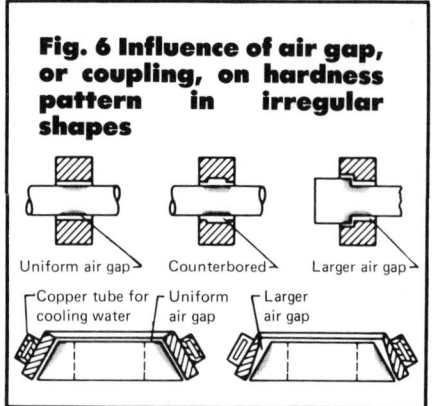

Fig. 6 Influence of air gap, or coupling, on hardness pattern in irregular shapes

Uniform air gap   Counterbored   Larger air gap

Copper tube for cooling water   Uniform air gap   Larger air gap

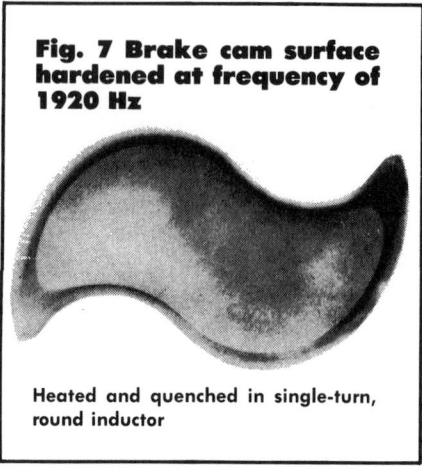

Fig. 7 Brake cam surface hardened at frequency of 1920 Hz

Heated and quenched in single-turn, round inductor

If the irregularity is in a longitudinal plane parallel to the axis of the inductor bore, as in a shaft with a bearing surface and an adjoining shoulder to be hardened, the inductor may again be machined initially to follow the shape of the part to produce a uniform air gap. The current density, however, will be strongest at the smallest bore dimension, because the current will follow the path of least impedance; therefore it will be found necessary to increase the dimension of the air gap at this point in order to decrease the heating rate.

The effect of the air gap on the hardness pattern produced for a number of irregular shapes is shown schematically in Fig. 6. In practice, it may be difficult to avoid overheating at sharp corners.

Contour Heating. Frequently, it is necessary to compromise in the design of an inductor. The following is an example of such a compromise.

Example 5. In this application, a cam was heated and quenched in a single-turn round inductor at 1920 Hz (see Fig. 7). The relatively low frequency employed shows its influence in the relatively deep hardness pattern. In the tips of the cam the current followed the shortest possible path, thus eliminating the extreme tips, which probably were not heated by induction, but by conduction of heat from other parts of the cam. The sections of the cam nearest to the inductor show deeper hardening than the rest of the part. A more uniform pattern could have been secured by using a shaped inductor and higher frequency; however, this example illustrates the results obtained from a simple round inductor shape when used with contoured parts.

Transmission Cables

Recommendations for the selection of transmission line to be used from generator to station or between heating stations are given in Table 13. The data (which apply only to true frequency generators and not to harmonic generators) cover frequencies ranging from 1 to 10 kHz and power ratings from 7.5 to 350 kW. The cable size listed is the smallest capable of carrying the rated current.

Coil Coolants

Water is commonly used for cooling inductors, although in some applications oil, modified water, or a plastic quench may be employed to serve the dual purpose of cooling the inductor and quenching the workpiece in a continuous heating and quenching operation. Generally, the water should have a hardness of less than 10 grains/gal. If the water-cooling passages are small relative to the current load carried by the inductor, it may be necessary to use distilled or deionized water to avoid a deposit buildup that could eventually stop circulation. Preferably, the water should be filtered to remove foreign particles that might clog small passageways, especially when intricately designed inductors are being used. The water should have an inlet temperature below 35 °C (95 °F), and flow should be sufficient to prevent the outlet temperature from rising above 66 °C (150 °F).

For high-voltage radio-frequency equipment and solid-state equipment, it is necessary to supply water that meets certain electrical-resistivity requirements, as specified by the equipment manufacturer. An electrical resistivity value greater than 50 $\Omega \cdot$m is a typical specification.

Table 13 Transmission lines for use from generator to station or between stations with sine-wave output

Input ratings		Single-conductor cable(a)					
		⊕ ⊖ One per leg	Maximum run		⊕ ⊖ Two per leg ⊖ ⊕	Maximum run	
A	kW	Size	m	ft	Size	m	ft
10 kHz, 220 V							
34	7.5	10	38.7	127	14	93.6	307
68	15	6	19.2	63	10	39.0	128
136	30	4/0	12.8	42	6	23.8	78
227	50	...	...	...	4/0	19.8	65
10 kHz, 400 V							
375	150	...	...	...	500 MCM	24.4	80
10 kHz, 440 V							
68	30	6	37.8	124	10	77.7	255
114	50	1/0	30.2	99	6	57.3	188
227	100	...	...	...	4/0	39.0	128
10 kHz, 800 V							
188	150	350 MCM	38.7	127	1/0	83.2	273
219	175	...	...	...	4/0	71.6	235
313	250	...	...	...	350 MCM	57.6	189
3 kHz, 400 V							
75	30	6	116.1	381	10	179.8	590
250	100	500 MCM	52.1	171	1/0	112.8	370
375	150	...	...	...	250 MCM	80.2	263
500	200	...	...	...	500 MCM	65.2	214
3 kHz, 800 V							
125	100	2	164.6	540	6	350.5	1150
188	150	4/0	109.1	358	4	247.5	812
250	200	500 MCM	104.9	344	1/0	226.8	744
375	300	...	...	...	250 MCM	160.0	525
1 kHz, 800 V							
313	250	500 MCM	243.8	800	1/0	356.6	1170
438	350	...	...	...	250 MCM	585.2	1920

(a) Cable sizes are designated by American Wire Gage (AWG) number where applicable, or by actual size in thousands of circular mils (MCM). The size listed is the smallest cable that will carry the rated current. This cable is varnished cambric insulated, single braid, and standard strand. The maximum length listed was calculated for the following conditions: The voltage across the condensers is ≤1.1 times the generator terminal voltage, and the generator output power factor is 0.9 leading.

Matching of Impedance

Efficient transfer of energy available from the high-frequency converter to the workpiece depends on good matching of impedance as well as on good coil design. This applies to all types of equipment for induction heating and is particularly important when work coils (inductors) have low impedance (one and two turns). Matching equipment consists of autotransformers, fixed-ratio step-down transformers, variable-ratio step-down transformers, and capacitors.

Motor-Generator Units and Solid State Units

Motor-generator units are rated according to maximum power, amperage and voltage. The maximum rated power cannot be attained unless the maximum ratings of amperage and voltage are reached at the same time.

Example 6. A generator rated at 50 kW and 440 V would require 114 A to secure maximum power. If a load were connected that would draw 60 A at 440 V with a unity power factor, the actual power in kilowatts would be (60 × 440) /1000, or 26.4 kW, or slightly more than half the 50 kW available. A load that drew 113 A at 200 V with unity power factor would result in (113 × 200) /1000, or 22.6 kW. The 113-A load prohibits the generator from using enough voltage to attain maximum rated power.

Solid state units are rated similarly, except that the frequency range within which rated power can be made is specified. Depending on tuning, rated power can extend over a range of frequencies. Care must be taken with variable frequency units that enough capacitance is present to make full power at the lowest rated frequency.

In addition to the conditions described in the example, single-turn inductors frequently used with medium-frequency equipment have such low inductance that considerable capacitance must be provided for in the circuit in order to maintain a rated power factor.

Fixed-Ratio Transformers. The simplest but least flexible method of matching impedance is that of the fixed-ratio step-down transformer with a capacitor bank between generator and inductor. This method is the least expensive and can be designed for a particular load, but only one type of load can be run at the rated power of the generator when properly matched. Other types of load may be run but at somewhat lower power. A more flexible method is to connect an autotransformer, capacitors and a fixed-ratio transformer between generator and inductor. This is often used when the generator voltage is high (800 V) to permit selection of a lower fixed-ratio step-down transformer.

Variable-Ratio Transformers. Two other methods are similar to those mentioned above except that they employ variable-ratio rather than fixed-ratio step-down transformers. In each of these methods considerably greater flexibility is attained.

Many variables obviously affect the ultimate selection of the transformer ratio for best matching of impedance in a given application. While no simple formulas are available for determining or selecting this ratio, certain basic observations derived from experience are helpful; for example:

• As the length of the inductor increases, an *increase* in the number of turns on the primary of the transformer is necessary to maintain constant power and constant generator voltage and current.

• As the diameter of the part increases, a *decrease* in the number of turns on the primary of the transformer is necessary to maintain constant power and constant generator voltage and current.

• In order to draw the same power when inductors and parts of the same size and shape are involved, an increase in turns on the primary of the transformer will require a higher generator voltage, and vice versa.

• With constant generator voltage and size and shape of inductor and part, a decrease in transformer ratio will increase the power and the amperage to be used, and vice versa.

• For a given generator voltage and size and shape of inductor and part, an increase in frequency will require a decrease in transformer ratio, or vice versa, to maintain the same power.

Capacitors

Capacitors are required to effect proper matching of a circuit. These capacitors are connected across the transformer primary winding and are used for power factor correction. By adding the proper amount of capacitance to the circuit, the generator can be operated at close to unity power factor, thereby supplying more useful power to the load at rated output voltage. When a job is initially balanced out, it is best to start with a minimum amount of capacitance in the circuit. If the meters indicate a lagging power factor, the generator voltage should be removed and a small amount of capacitance added to the circuit. This procedure is followed until the meters indicate a kilovar reading of zero, or as close to zero as can be obtained.

When multiturn coils are used, the coil for large parts can often be designed so that no transformer is needed and load matching is accomplished by the usual complement of capacitors alone. In tests of loaded coil impedance, it is usual to check the impedance matching with the generator in operation and the work cold. Starting with a low value of capacitance, capacitors are added to the circuit to secure unity power factor with the maximum values of voltage or amperage within the rating of the generator. If the voltage is low but the current is high, the load impedance presented to the generator is not great enough, and coil turns must be increased or the coil more loosely coupled to the load. If the current is low but the voltage is high, excessive impedance is indicated, and better matching may be accomplished by removing coil turns or coupling more closely to the work.

If changes in coil design and capacitance do not result in full loading of the generator (rated voltage and amperage at unity power factor), a transformer of the proper ratio is needed as with single-turn inductors.

Vacuum Tube Units

Vacuum tube units usually require an output transformer with a primary winding of high impedance to match the high impedance in the tube output circuit. Secondary windings are usually 1, 2, 3, 4 or 5 turns, depending on the size of the heating coil and the work load. Voltage output across the coil increases with the increase in secondary turns in the transformer. For certain applications, capacitors are sometimes used in the load circuits (primary or secondary) to afford some measure of correction of the power factor, but must be chosen with care because of the higher voltages encountered in these circuits.

When heating coils are used with an output transformer, the leads between the transformer and the coils must be kept as short and as close together as possible, so that inductance in the leads is of minimum value. Widely spaced or long leads have as much inductance as the heating coil itself and can result in an inability to draw full power from the equipment, and also may cause radio and television interference.

Influence of Temperature*

Because the magnetic permeability and electrical resistivity of ferrous loads change as heating occurs, the impedance of the work circuit will change as the work temperature increases, and therefore a perfect impedance match throughout the entire heating cycle is impossible.

If the impedance is matched while the work is cold, the demand for power will usually increase at first and then decrease as the work increases in temperature. This may be an advantage in avoiding overheating if sufficient power is available to continue heating at this stage of the heating cycle, but it may be a disadvantage if limited power is available and the change results in loss of efficiency at higher temperatures.

Matching the impedance with work that is hot will result in better transfer of energy at the higher temperatures but will cause slower initial heating of the cold work, or with tube converters, overloading. The change of impedance

*Data does not apply to variable frequency equipment.

as the work becomes heated may be partially compensated for by the timed addition of capacitance or inductance to the output circuit.

Selection of Accessory Equipment

Integration of coil design, quench arrangement and fixtures, and handling equipment frequently allows induction heat treating operations in high-speed production lines. On the other hand, manually operated or partially mechanized fixtures and work-handling arrangements are used successfully with minimum initial cost.

Work-Handling Equipment

The principles of machine design and the design of work-handling equipment can be readily adapted to induction heating accessory equipment.

The electric and electromagnetic characteristics of materials to be used in the immediate vicinity of the induction heating coil and quenching arrangements must be considered, as well as their resistance to corrosion and heat. Aluminum, brass, and austenitic stainless steels are useful because they are less readily heated by induction. Stainless steel, chromium plated steel, and brass resist corrosive quenching mediums such as water, caustic and brine. Nonmetallic materials such as ceramics, laminates, plastics, rubber and Transite are used for their corrosion resistance and electrical and thermal properties.

Typical Arrangements. Work-handling equipment is available in a number of standard units. Typical arrangements include hopper and magazine feeding of small parts, or conveyor and rotary-table feeding of larger parts to the work coil and thence into the quench. Parts may be rotated to provide greater uniformity of temperature.

In one automotive plant, gears are automatically fed into a loading chute from a storage unit. The loading chute, together with a loading cylinder, maintains a bank of parts ahead of two heat-station shuttle mechanisms, which are synchronized with all subsequent movements. Station shuttle cylinders alternately feed parts into a pickup station. Locators position each part in the induction coil, and, after being heated and quenched, parts are returned to the pickup position. The entrance of each new part into the line discharges a heat

Fig. 8 Basic arrangements for quenching induction heated parts

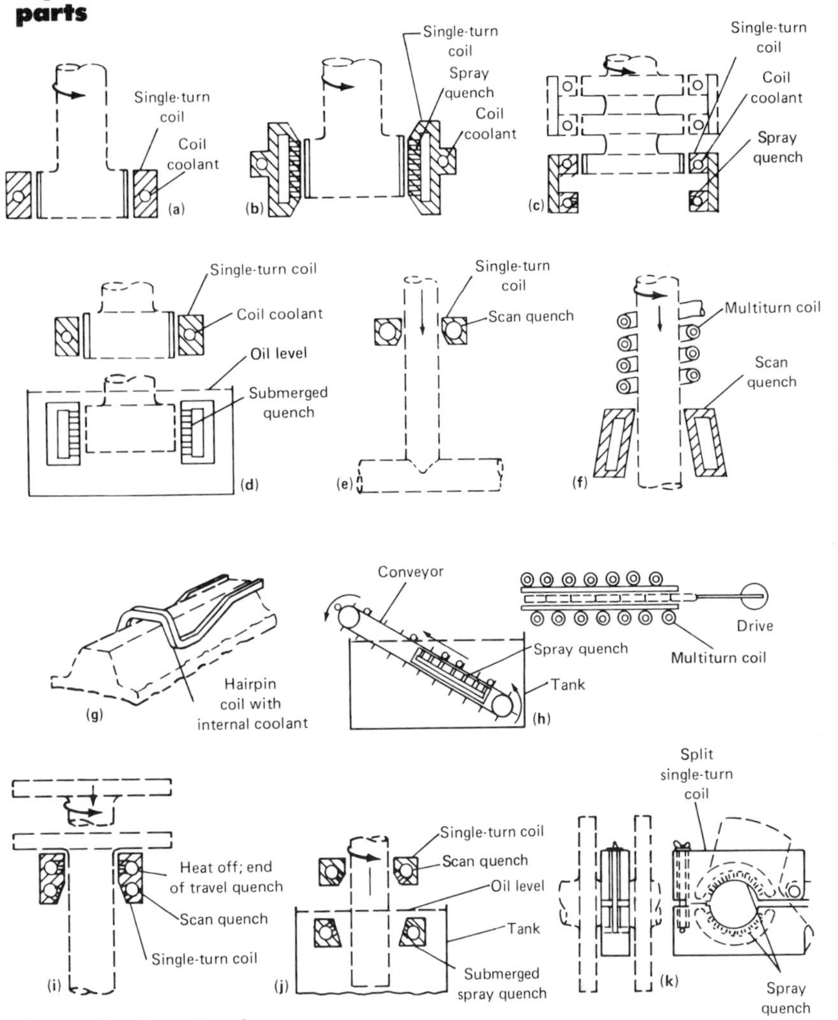

See text for discussion of parts.

treated part onto a discharge shuttle mechanism. Finally, a steel conveyor belt automatically feeds the heat treated parts into a tempering furnace.

Scanning Devices

Vertical and horizontal scanning fixtures are often designed to provide rotation of the part and controlled movement of the coil and quench arrangement along the part. Such fixtures normally have arrangements for surface hardening either the whole length or desired sections of the part.

For selective hardening, parts are frequently heated in a multiple coil, which consists of two or more connected work coils. In such an arrangement, one multiple-coil unit may be reloaded

while a second unit goes through a hardening cycle. Upon completion of the heating cycle, parts may be dropped into the quench automatically, quenched in the heating position, rotated during quenching, or held and indexed into the quenching medium in either a horizontal or a vertical position.

Integration With Machine Tools

The automation of induction heating equipment is only the first step in combining automatic machining operations with automatic induction heat treating. The multispindle automatic screw machine, the rotating-dial type of fixture for multispindle machining, the modern process line or transfer line and the common conveyor all offer possibilities for including automatic induction heating.

Parts machined in stages on automatic machines can be hardened on the same equipment. After machining the surface to be hardened, the tooling is arranged to include an inductor on one station. It is preferable to use the inductor alone on this station, but it may be combined with machine tooling on adjacent areas, if necessary.

It is often advantageous to integrate induction heating equipment with two or more other machines in order to achieve a continuous line operation. The manufacture of automotive shock absorber shafts provides a typical example of successful integration. In making these parts, bar stock is cut to length and hopper fed through a first centerless grinder for sizing to near finish tolerance. After the "green grind", the shafts are continuously fed by spin-feed rollers through guide tubes, an induction coil, and a quenching ring. The hardened shafts then proceed automatically through four tandem centerless grinders for final sizing and surface finishing.

Quenching

A great many induction hardening applications employ water as the quenching medium. Other media such as conventional quenching oil, water modified by organic polymer, and compressed air are occasionally used. Water is easiest to handle, simple to install and maintain, and generally less hazardous than other mediums. The modified-water compounds are compounds with organic polymers that are soluble in water. The temperature and concentration determine the quenching rate. The type of quench used will depend upon the metallurgical considerations.

Basic Systems for Quenching

Eleven basic arrangements for quenching induction hardened parts are shown schematically in Fig. 8, items (a) through (k). In correlation with the lettering there, these arrangements are briefly described as follows:

(a) Heat in coil; manually lift part out of coil; submerge part in tank of agitated quench medium. Used where limited production does not warrant the cost of an automated quench

(b) Heat and quench in one position; quench by means of integral quench chamber in inductor. Called single-shot method

(c) Heat in coil with part stationary; quench ring moves in place. Single-shot adaptation of scanning method

(d) Part is hydraulically lowered into quench tank after single-shot heating. Quench medium is agitated by submerged spray ring or propeller

(e) Vertical or horizontal scanning with integral spray quench. Single-turn inductor. Used for shallow hardening

(f) Vertical or horizontal scanning with multiturn coil and separate multirow quench ring. Used for deep-case or through hardening

(g) Coil scans and heats workpiece; self-quench or compressed air quench. Used in special applications with high-hardenability steels

(h) Horizontal cam-fed parts are pushed through coil, then dropped onto submerged quench conveyor

(i) Vertical scanning with single-turn inductor in combination with integral dual quench: one quench ring for scan hardening: the second for stationary quenching when the scanning travel stops. Used for parts having a diameter or a flange section too large to travel through the inductor, wherein it is desired to harden up to the shoulder or flange

(j) Vertical scanning with single-turn inductor with integral spray quench and submerged quench in tank

(k) Split inductor and integral split quench ring. Used for hardening crankshaft bearing surfaces

A water spray applied through a separate quench ring or from hollow inductors (for heating and quenching) has been used successfully in most applications involving plain carbon and low-alloy constructional steels; oil is specified for steels of higher hardenability, and for parts with nonuniform sections when difficulty with cracking and distortion is anticipated. Quenching in water or oil may also be done by submersion in an agitated bath or by a combination of a spray quench ring and submersion in a tank on completion of the heating cycle. Submersion in a brine tank or water-polymer tank may be specified for steels of very low hardenability to prevent occurrence of soft

Table 14 Typical specifications for oils used in quenching induction heated parts

General-purpose quenching oil (stock paraffin oil)

Viscosity at 40
°C (100 °F).... 70 to 85 sus
Flash point .... 165 °C (330 °F) min
Fire point ..... 175 °C (350 °F) min
Quenching temperature...... 50 to 60 °C (120 to 140 °F)

Fast quenching oil

Viscosity at 40
°C (100 °F).... 75 to 110 sus
Flash point .... 175 °C (350 °F) min
Fire point ..... 200 °C (390 °F) min
Quenching temperature...... 50 to 60 °C (120 to 140 °F)

Soluble-oil–water mixture

Water-emulsifiable oil, 10 to 12% oil in water

spots on the surface of the hardened part.

A considerable tonnage of work is handled by vertical scanning, wherein additional quenching action results from the quench medium clinging to and washing down the shaft below the initial quench ring. Satisfactory quenching occurs in horizontal scanning, even though the quench medium does not cling very long to the shaft but runs off the under side. If necessary, a secondary or auxiliary quench ring, following the initial quench ring, may be employed to assure more uniform quenching and control of residual heat.

In spray quenching in scanning operations wherein a single-row angular spray quench is employed, the quench flow must be sufficient to completely wet the surface to the point where the quench medium does not boil off after the part progresses away from the point of initial quench impingement. Otherwise, incomplete quenching will occur.

Retractable steady rests are often employed in scanning operations on long shafts to minimize warpage and the subsequent cost of straightening.

The liquid quenching mediums associated with furnace hardening—oil, water, water-base solutions, and molten salt—may be used for quenching induction heated parts. However, applications involving spray quenching through orifices in the inductor or through a separate spray quench ring are normally limited to the use of water, water-base solutions, or oil; a caustic or brine quench would not be sprayed through the inductor or quench ring because of corrosion and handling problems, but would be used for tank-immersion quenching of a heated part whose low hardenability dictated the use of a more drastic quench.

Water Quenching

There are no stringent specifications for water used in quenching systems except that a reasonably clean and adequate supply of controlled-temperature water should be available. Extremely hard water may produce an excessive coating of deposits on quench orifice surfaces exposed to radiant heat.

Quench-water systems are usually separate from the supply of water for cooling the induction heating unit, because it is desirable to have a wider range of temperature control and variable pressure control for different quenching arrangements. Public water supplies and deep wells or, if necessary, water-recirculating systems with cooling towers are usually satisfactory. Recirculating systems consist of storage tanks, pumps, and heat exchangers with adjustable controls; they are frequently protected with alarms and cut-offs for regulating pressure, temperature and flow. Pressures in water quenching vary from 205 to 585 kPa (30 to 85 psi); temperatures vary from 15 to 40 °C (60 to 105 °F).

Polyglycol Alcohol Solutions

An innovation in the field of water-base quenching media is the organic polymer solution in water which has been used with success on parts for which the steel selected has borderline hardenability—that is, where a spray oil quench does not provide a fast enough cooling rate to develop the hardness required and where a change to water would produce quench cracking.

The solution is made up by mixing from 1 to 20% solutions of organic polymers with water. As diluted for quenching, the solution is nonflammable, does not produce objectionable fumes, and does not cause skin irritation. System design is the same as for oil.

Oil Quenching

Despite the potential fire hazard that attends its use, oil spray quenching has a great many successful applications. A change from water to oil is made pri-

marily to eliminate the quench cracking that often results in hardening irregular sections such as splines, gear teeth, shoulders, keyways and oil holes. Obviously, the change is also dependent upon whether or not the hardenability of the steel is sufficient to develop the required depth of hardening.

Typical specifications for oils used in quenching induction heated parts are given in Table 14.

Oil quenching systems consist of storage tanks, pumps, and heat exchangers for heating and cooling, with adjustable automatic control. Oil quench pressures vary from 205 to 620 kPa (30 to 90 psi); oil temperatures vary from 25 to 65 °C (80 to 150 °F) and the volume is designed for ample flow and agitation without foaming at the point of quench. For example, a vertical scanning operation involving spray oil quenching a 53.98-mm- (2¹⁄₈-in.-) diam splined drive shaft at a scanning rate of 5 mm/s (12 in./min) to produce a 6.4 mm (0.250 in.) case, required a quench flow of 1 l/s (15 gal/min) at 25 to 30 °C (80 to 90 °F) inlet temperature. The oil was supplied through 2.4-mm- (³⁄₃₂-in.-) diam holes in a separate barrel-type quench ring located several inches below the single-turn heating coil. The important consideration in spray-type quenching arrangements is that sufficient quantities of controlled-temperature oil contact the heated surface rapidly enough to prevent vaporization. An excessive spray velocity with insufficient quantity of oil can cause atomization or vaporization, a condition that constitutes a severe fire hazard.

In equipment for scanning, the quench ring and work coil (inductor) move as a unit along the part. Obviously, the distance between the coil and the quench ring determines the delay between heating and quenching and must be controlled. Where spray quenching is achieved through the inductor, the angle of spray is 30 to 50° away from the heated area, striking the piece to give full coverage without excessive splashing and foaming.

Self-Quenching

A special quenching technique sometimes employed on parts with sufficient mass, such as large, coarse-pitch gears, is sometimes referred to as "self-quenching", because most of the heat at the surface of the part is rapidly absorbed by the unheated mass of metal below the surface.

Example 7. A large internal gear,

290 cm (114 in.) OD, having 127-mm (5-in.) face width, 108 teeth, 1 diametral pitch, and made of 4340 steel, is quenched in this manner. A single-turn inductor coil automatically lowers over a tooth. Power is applied for 10 s to bring the surfaces of the tooth up to 900 °C (1650 °F). The inductor automatically rises, and the gear is indexed to repeat the cycle on the next tooth. No liquid quench is employed.

This type of quenching was adopted after it was determined that a liquid quench produced a higher hardness than was necessary and occasionally produced quench cracks. Because of the high hardenability of the steel and the relatively large mass of unheated core, enough quenching action resulted from heat-loss by radiation to the surrounding atmosphere, and from the internal cooling effect of the unheated core of the tooth, to produce the desired surface hardness of 42 to 47 HRC.

Forced-Air Quenching

A modification of self-quenching that has been successfully applied to smaller-pitch gears involves a progressive or scanning-type operation and uses a blast of compressed air fed through orifices in the inductor similar to those used with a liquid spray quench. This technique has been successfully applied to an external training gear made of 4340 steel, with an outside diameter of 1.8 m (72 in.), a face width of 76 mm (3 in.), and 144 teeth. The hardening operation made use of a double hairpin-type coil scanning across the tooth face at the rate of 19 mm/s (45 in./min). Quenching was accomplished by supplying compressed air at 690 kPa (100 psi) through twelve 1.58-mm-(1/16-in.-) diam holes in a quench loop fastened to the end of the inductor to give the effect of an angular spray quench following the progressive heating. The operation involved automatic indexing at the end of each scanning pass. The entire gear was hardened in ½ h. Compressed air was used because the self-cooling action of the unheated core did not produce a quenching rate fast enough to develop the desired hardness.

Process Control

Charts or graphs, representing the relationship between depth of hardening and heating time for various power densities and for various steels, can often be used to obtain a reasonably accurate estimate of the required cycle. Other estimates can be made from results for similar parts or by careful observation of the part itself as it is heating. However, the final cycle is usually determined by experimentation.

After a part has been hardened with one experimental arrangement, it is checked for hardness and examined for cracks. It is then sectioned with a rubber-bonded cutoff wheel, checked accurately for specified hardness at a given depth and prepared for a thorough examination of the microstructure. Should hardness, hardened depth or microstructure fail to meet desired standards, or if cracks are found, the cycle is adjusted accordingly.

This procedure may be facilitated considerably with the aid of a high-speed recorder-controller together with either a thermocouple or special radiation devices that are available commercially. These units are designed to provide a minimum of inertia in response, and therefore to record and control with suitable accuracy the temperature at the surface of parts heated at normal speeds.

Once the hardening cycle has been determined, production control of the process is achieved by repetition of the experimental cycle automatically, with respect to heating time or temperature, power density, delay between heating and quenching, and the quenching cycle. In addition, parts are checked periodically by metallurgical examination for structure, case depth and hardness. The paragraphs under the four headings that follow include descriptions of equipment required to control each of these variables in a way that will ensure depth of hardening accurate to within $\pm 10\%$ and suitable structure and hardness in a medium-carbon steel.

Heating Time

Heating time of parts heated in a fixed position relative to the inductor for 10 s or less should be controlled to within ± 0.1 s. For heating times greater than 10 s but less than 60 s, this control should be within ± 0.2 s, and for heating times of 60 s or more, within ± 1.0 s.

A multiple-circuit timer of the synchronous-motor-driven type, properly operated and maintained, will control the induction hardening process within the limits mentioned. A timer that is easily adjusted should be selected if the

equipment is used for a variety of parts. Electronic timers are being used more frequently. For equipment requiring infrequent changes of time cycle, however, a drum type of timer will serve adequately.

For scanning operations, the rate of travel of the part relative to the inductor should be controlled within $\pm 1\%$. Electronic motor controls will effect tolerances within $\pm 1\%$ even though line voltages vary by $\pm 5\%$. Oil hydraulic drives are reliable after the oil has warmed up to operating temperature, and good variable-speed transmission drives are available.

Power Density

Power density, as well as heating time, must be controlled carefully if heating is to be uniform from part to part. For a given part, power density will be affected by the electrical characteristics of the high-frequency converter, by variations in line voltage and in coupling, and by part symmetry.

A uniform density of power may be achieved with automatic voltage regulation of the source of power for the induction heating unit. With short heating cycles or large voltage fluctuations ($\pm 5\%$), greater uniformity of results justifies the cost of automatic voltage regulators. When it is intended to cut one or more loads in and out of the circuit while heating others, automatic voltage regulation should always be used. In such an application it is necessary to keep the load well balanced in order to get uniform results. Rotating and electronic voltage regulators are available that are capable of controlling the generator line voltage within $\pm 1\%$.

To secure consistent results, constant coupling is necessary in induction hardening. For a given part and coil, the coupling will be determined primarily by the ability of the fixture to position the part consistently to provide a constant air gap between the coil and the part. Some variation in coupling can be tolerated when the inductor completely surrounds the workpiece by one or more turns. For face heating with a pancake-type coil or modifications thereof, the coupling is critical, and should be maintained within 0.4 mm ($1/64$ in.) if a highly restricted hardness pattern is required.

If the part to be hardened is of irregular size and shape, uneven heating will result if power density is not uniform. However, special inductors incor-

porating a laminated iron core to alter the power density, or copper plugs to fill in oil holes or keyways, will often alter the heating pattern enough so that the resultant depth-hardness characteristics are acceptable. Under these conditions, reasonable duplication of results requires careful positioning of the part in the inductor and careful positioning and fitting of copper plugs or inserts.

In scanning parts that incorporate a change in section or diameter, it may be necessary to change both the rate of travel and power density. Cam-operated electronic or hydraulic controls are available for this purpose and can be expected to give reasonable duplication of results.

Delay Time

Delay time for single-shot operations is the interval between the time when the power is turned off and the time when the quenching medium contacts the workpiece. The delay is controlled easily by using an additional circuit of the same timer that controls the heating cycle. However, in the arrangement of quench lines and manifold, variation should be avoided in the time lapse between the opening of the quench valve and the striking of the quench medium on the part. When the volume of the quench is small, solenoid-operated valves work satisfactorily, but when quench lines are larger than 76 mm (3 in.), particular care should be taken to select a quench valve that will open and close quickly and that will be trouble-free and maintain uniformity in operation.

For scanning operations the delay time is a function of the rate of travel and the distance between the inductor and the point where the quenching medium makes contact with the part. Here again, the equipment used to regulate the rate of travel for heating is accurate enough for reproducing the delay time required for scanning.

Quenching Cycle

Control of quenching cycle involves accurate control of the quenching time, the temperature of the quenching medium, and, in spray quenching, the quenching pressure, velocity and direction. The quenching pressure should be measured as close to the discharge orifice as possible, and a standard dial-type pressure gage may be used for this purpose. A standard valve may be employed to regulate the pressure of the quenching medium.

One circuit of a multiple timer used to control the heat cycle will generally control the quench cycle with sufficient accuracy. Some industrial plants utilize the residual heat in the part (not quenching the part to the temperature of the quenching medium) to make subsequent tempering unnecessary. This practice requires careful control of quenching time and other variables for uniform results.

Cooling of Equipment

The volume, pressure and temperature of water supplied for the cooling of tube, solid state or motor-generator units should be controlled within certain limits. Meters or gages are used to indicate these limits, and high- and low-pressure switches, flow switches, and temperature-controlled switches act to stop operation of the induction heating unit if the preset limits are exceeded.

The water-cooling system should have adequate volume to give uniform control of both heating and cooling, and may be supplied with filters to remove metallic particles and sludge. Where water hardness exceeds 10 to 12 grains/gal, closed recirculation systems are recommended.

Salt or acid additives that decrease the electrical resistance of the water should be avoided. Some manufacturers of tube units specify the use of distilled cooling water, to provide a minimum resistivity of 50 to $100\,\Omega\cdot$m. This standard can be met by a commercial grade of double-distilled water. With distilled water, it is advisable to install a recirculating system.

For equipment that is air-cooled, building air is normally satisfactory. In locations where air temperatures exceed 40 °C (100 °F) at certain seasons of the year, some means of cooling the equipment must be provided; usually this is done by piping air from a cooler location through a bank of filters.

Maintenance of Equipment

Induction hardening equipment is expensive, and generally it is impossible to provide standby equipment. A program of preventive maintenance is therefore imperative, and availability of certain critical parts for ready replacement is highly desirable.

Dust, dirt, moisture and high ambi-

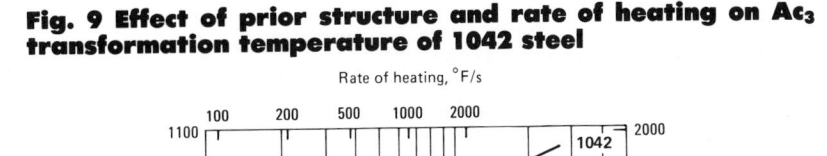

Fig. 9 Effect of prior structure and rate of heating on Ac₃ transformation temperature of 1042 steel

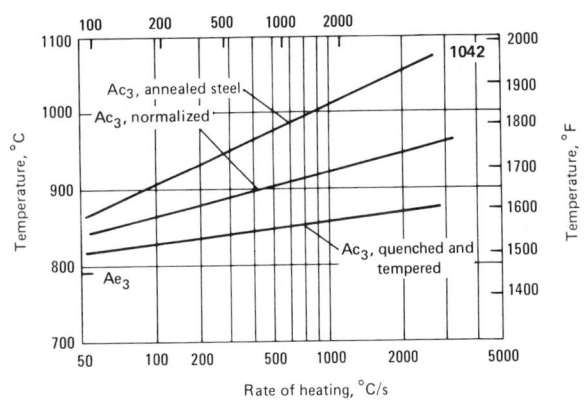

ent temperatures are the primary causes of failure in electrical equipment, and one or all of these conditions are usually present in industrial locations where induction heating units are installed. In addition, cooling with air or water is necessary for almost all induction heating units, and some problems develop in regard to suitable water-cooling systems.

Precautions and Safe Practices

Induction heating installations comprise various electrical and mechanical components, many of which are water-cooled and are equipped with protective devices. Instructions concerning safe practices and safety precautions are generally issued by the equipment manufacturer. These instructions should not be disregarded.

Many potential hazards can be avoided by entrusting installations to qualified personnel who follow in detail the instructions provided. Installations should always conform to the safety practices and codes prevailing in a particular locality.

Devices for preventing overload and short circuits are usually built in as part of original equipment. In addition to these, there are ground detectors and surge detectors designed to protect motor-generator units from faulty coil or transformer setups at the heating stations, or from breakdown of insulation in the generator windings. Also, fire-prevention devices are available to start or stop the quench in case of general power failure or malfunctioning of

other protective devices. All protective devices should be checked periodically to make certain they are in good working condition.

Protective Devices

Protective devices most commonly employed with induction heating equipment are as follows:

● Overload relays
● Bearing-protector relays
● Water-pressure switches
● Water-flow switches
● Temperature-controlled relays for generator airstream protection
● Generator surge protectors
● Ground detectors
● Door switches and interlocks
● High-low line voltage trip circuits
● Voltage and frequency limit circuits
● Hot water temperature switches
● Door-operated, gravity-type, high-voltage discharge mechanisms.

Selection of Steel for Induction Hardening

Selection of steel should be based on the total of all processing factors as well as on the engineering requirements of service. Because the processing and engineering factors are interdependent, final selection is seldom determined by one item alone.

Selection of Prior Microstructure

The microstructure of the steel before induction hardening is important in selecting the heating cycle to be used. Steels most readily hardened have carbides that are small and uniformly dispersed, as obtained by quenching and tempering. Such structures are readily austenitized. Accordingly, minimum case depths can be developed with maximum surface hardness while using very rapid rates of heating.

Pearlite-ferrite structures, typical of normalized, hot rolled and annealed steels containing 0.40 to 0.50% carbon, also respond successfully to induction hardening. Much of the induction hardened steel has such structures.

Spheroidized Structures. Steels containing more than 0.50% carbon are frequently spheroidized for improved machinability. Such spheroidized structures have the poorest response to induction hardening; the larger the carbide particles, the poorer the response. Coarsely spheroidized structures may require hardening temperatures 165 °C (300 °F) or more above the transformation temperature to redissolve the carbide and obtain uniform hardening results. Such high temperatures at the surface may lead to coarse austenitic grain size, coarse martensitic structures and significant quantities of retained austenite, which may be detrimental to fatigue resistance and may promote galling or seizing.

When a coarsely spheroidized structure is encountered in production, parts may be run through the hardening cycle twice, to achieve maximum surface hardness at a sacrifice of production time and minimum distortion.

Nonhomogeneous Structures. Any microstructure which is nonhomogeneous can cause difficulty in obtaining a completely austenitic structure. Examples of this are parts that have decarburized surfaces as found with cold-rolled material, some forgings, castings, or parts heat treated in decarburizing atmospheres. The effects of segregation and rolling can produce nonuniform chemistry and ferrite banding.

The curves in Fig. 9 indicate the effects of prior structure and heating rate on the Ac$_3$ temperature of 1042 steel. An annealed structure requires markedly higher austenitizing temperatures as the heating rate is accelerated. In contrast, a quenched and tempered structure heated at an extremely rapid rate 2775 °C (5000 °F) per s requires an austenitizing temperature that is approximately the same as that required by an annealed structure

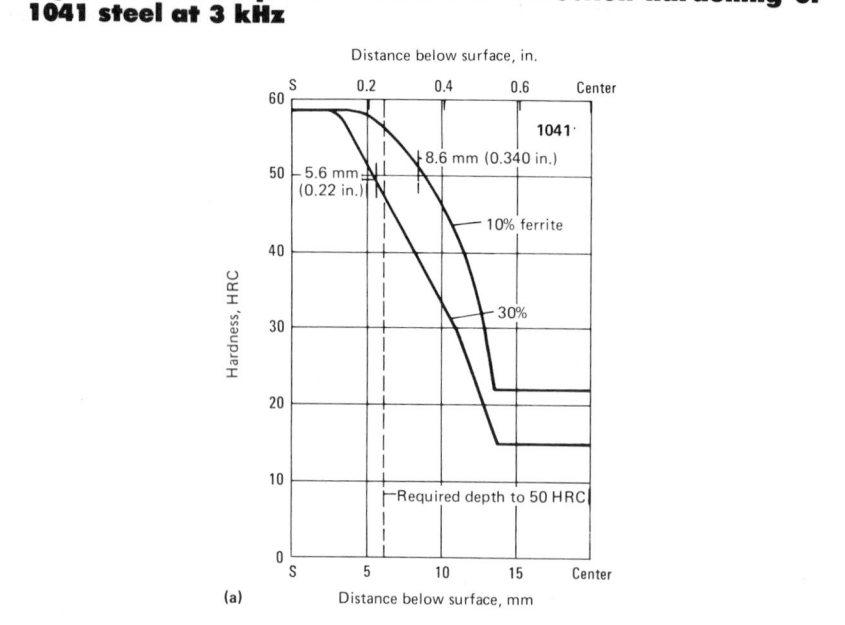

Fig. 10 Effect of prior structure on induction hardening of 1041 steel at 3 kHz

heated at a slow rate 55 °C (100 °F) per s. Variations in heating rates and austenitizing temperatures result in pronounced changes in effective case depth to 50 HRC, as described below:

Example 8. Two axles made from the same heat of 1041 steel were furnace heated for normalizing under identical conditions. However, one axle was floor cooled to a hardness of 22 HRC and contained 10% ferrite. The second axle was stack cooled to a hardness of 15 HRC and contained 30% ferrite. When both axles were later subjected to the same induction hardening treatment, the axle with the lower ferrite content exhibited a much higher effective case depth to 50 HRC, as shown in Fig. 10(a).

Details of the induction hardening

cycle were as follows: Both axles measured 41 mm (1.63 in.) in diameter and were heated with 75-kW input at an operating frequency of 3 kHz. Coil length was 38 mm (1.5 in.), and the scanning speed employed was 5 mm/s (0.20 in./s). Thus, the total heating time was 7.5 s at a heating rate of 110 °C (200 °F) per s. The axles were quenched in a 10% soluble oil maintained at 25 °C (80 °F).

Effective case depth necessarily depends on the depth of heat penetration at a given temperature, the significant temperature in all cases being the minimum austenitizing temperature required by the prior structure of the steel. Thus, as shown in Fig. 10(b), the annealed 1041 steel of Example 8 required an austenitizing temperature of

915 °C (1675 °F), while the normalized steel required an austenitizing temperature of only 865 °C (1585 °F). The difference in austenitizing temperatures 915 versus 865 °C (1675 versus 1585 °F) accounts for the difference in effective case depths 5.6 versus 8.6 mm (0.220 versus 0.340 in.).

Example 9. Prior to induction hardening, three 25-mm- (1-in.-) diam shafts made of 1070 steel were heat treated to produce three different structures—quenched and tempered at 425 °C (800 °F), normalized, and annealed. Hardness gradients obtained with these three prior structures were heated at a frequency of 450 kc. The hardening cycles are given in Fig. 11(a). Figure 11(b) shows the difference in minimum austenitizing temperature (as defined by Ac_3 points) for the three prior microstructures.

Then shafts representing each of the three prior structures were heated with an input of 25 kW at an operating frequency of 450 kHz. Coil length was 12 mm (0.5 in.), and the scanning speed was 12.7 mm/s (0.5 in./s). Heating time was 1 s at a rate of 945 °C (1700 °F) per s. Axles were quenched in water maintained at 25 °C (80 °F).

Selection of Hardening Temperature

Table 15 gives recommended hardening temperatures for a number of steels and irons commonly hardened by induction. For plain carbon steels and alloy steels not containing carbide-forming alloying elements, exhibiting a suitable prior structure, temperatures 25 °C (50 °F) higher than those used in furnace hardening are suitable, as indicated. This higher temperature compensates for the very short heating times.

Alloy Steels. When carbide-forming elements (chromium, molybdenum, vanadium and tungsten) are present in constructional steels, it may be necessary to raise the hardening temperatures 55 to 110 °C (100 to 200 °F) above those shown in Table 15 for equivalent carbon contents. The exact increase in temperature depends on the length of the heating cycle, the specific alloying elements present, the specific effects desired from the alloying elements, and the surface hardness required. These higher temperatures at the surface are used to obtain adequate solution of the carbon and alloying elements in the austenite with very short heating cy-

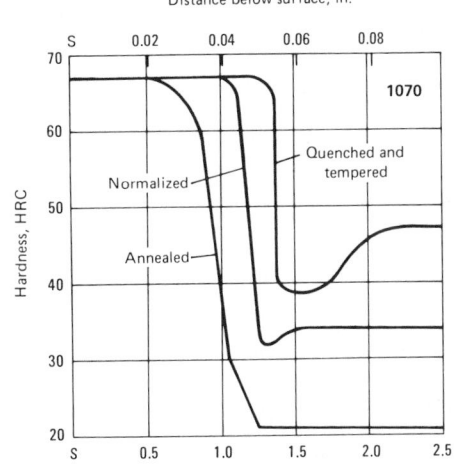

Fig. 11 Effect of prior structure on induction hardening of 1070 steel at 450 kHz

(a)

(b)

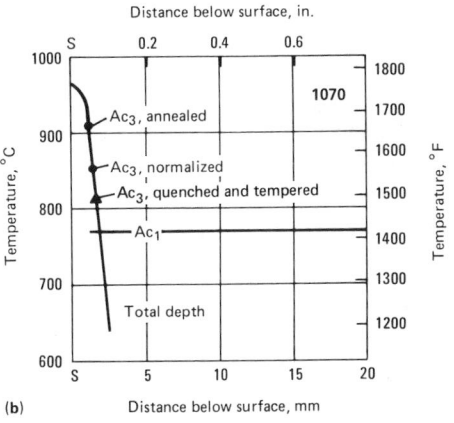

cles. As the heating cycle is lengthened to produce deep cases or through heating, hardening temperatures should approach those given in the table.

Fully annealed structures may not provide full response with the short heating times used with induction heating. The use of steels with prior heat treating and tempering may be necessary.

Grain Coarsening

With the usual induction hardening cycles, the hardening temperatures given in Table 15 will produce a fine austenitic grain size. This applies also to steels that contain carbide-forming elements and are heated 55 to 110 °C (100 to 200 °F) higher to provide alloy solution, since the alloy carbides inhib-

it austenitic grain growth as long as they remain undissolved. With longer heating cycles, there is danger of grain coarsening if the temperature is not reduced. An ASTM or fracture grain size of 6 or finer is specified for most parts hardened by induction heating.

In addition to grain coarsening, excessive hardening temperatures in induction heating may result in pronounced distortion and cracking. If specifications call for close limits of case depth and hardness pattern, overheating may also cause failure to meet specifications, actual through hardening of thin sections, or possibly hardening into a critical area where machining or drilling is yet to be done. With pronounced overheating and long heating cycles, surfaces may scale.

Control of Surface Hardness

Values given in Table 15 are minimum values for the cited surface hardnesses, which are frequently specified in commercial induction hardening. The minimum values are given to avoid needlessly close hardness specifications and correspondingly close control and high cost. Considerable variation in the hardness of the cast irons may be expected, because of a variation in the combined carbon content.

As shown in Fig. 12, the hardness obtained on a given part may be several points Rockwell C higher than the minimum values recommended in Table 15. It is also apparent from Fig. 12 that the maximum hardness obtained in hardening by induction is usually higher than that obtained by hardening the same steel from a furnace. This higher hardness is most pronounced with the short induction heating cycles characteristic for parts that are *surface* hardened; it disappears in through hardened parts, parts that are surface hardened to produce a deep case, or parts that are tempered after induction hardening.

Low hardness at the surface after induction hardening may have a number of causes, a few of which are

- Lower carbon content than specified, possibly as a result of surface decarburization before induction hardening. The usual induction hardening operations do not cause measurable decarburization, because of the short time of heating characteristic of induction treatment.

- Inadequate heating temperature or time, or both. Inadequate solution of carbon in the austenite before quenching will result in low hardness. Steels with unsuitable prior structures and alloy steels containing carbide-forming elements are most susceptible to low hardness from this source. Nonuniform performance of the induction heating unit and timer also may cause low surface hardness.

- Unsatisfactory quenching conditions. Hot water or cold oil may result in low or spotty hardness, as will low pressures of either water or oil. Orifices in the quenching equipment that are plugged because of inadequate filtering may cause the same conditions.

- Some hot-rolled steels can have different grain structures due to nonuniform deformation during hot rolling. Although these structures are rolled out linearly during hot rolling, they show up in cold drawn stock as spirals that are a result of the original hot coil being drawn and straightened into a bar. High and low hardnesses can be discovered by checking around the diameter. An examination along the diameter will show that these bands of hardness slowly spiral.

For a limited number of applications, surface hardnesses less than the minimums indicated in Table 15 are required, either to obtain greater notch toughness or to avoid quench cracking. These lower hardness values can be obtained by lessening the severity of quenching.

Table 15 Recommended induction hardening temperatures and normal expected surface hardnesses for various metals(a)

Metal	Hardening temperature °C	°F	Quench(b)	Hardness(c), HRC
Carbon and alloy steel(d)				
0.30% C	900-925	1650-1700	Water	50
0.35	900	1650	Water	52
0.40	870-900	1600-1650	Water	55
0.45	870-900	1600-1650	Water	58
0.50	870	1600	Water	60
0.60	845-870	1550-1600	Water	64
			Oil	62
	815-845	1500-1550	Water	64
			Oil	62
Cast iron(e)				
Gray iron	870-925	1600-1700	Water	45
Pearlitic malleable	870-925	1600-1700	Water	48
Nodular	900-925	1650-1700	Water	50
Stainless steel(f)				
Type 420	1095-1150	2000-2100	Oil or air	50

(a) Metals listed in this table are typical of those successfully induction hardened, and the listing is indicative rather than inclusive. (b) The quench selection depends upon the hardenability of the steel used, the diameter or cross section of the area heated, the case depth and hardness required, the need to minimize distortion, and the propensity towards quench cracking. (c) Minimum surface hardness, Rockwell C. (d) Free-machining and alloy grades with equivalent carbon contents may be induction hardened. Alloy steels containing carbide-forming elements (chromium, molybdenum, vanadium or tungsten) should be heated 55 to 110 °C (100 to 200 °F) above the temperatures indicated. (e) Combined carbon should be 0.40 to 0.50% min; hardness will vary with amount of combined carbon present. (f) Other martensitic grades of stainless steel, types 410, 416 and 440, have been induction hardened.

Fig. 12 Relation between hardness and carbon content of water-quenched steels

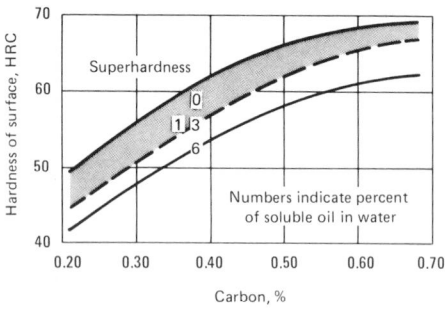

The two upper curves are for induction hardening; bottom, for furnace hardening.

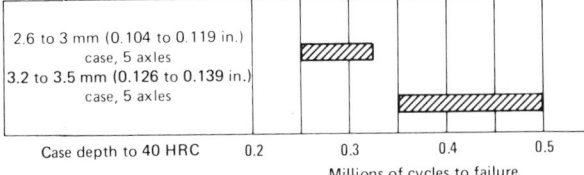

Fig. 13 Effect of case depth on fatigue life

2.6 to 3 mm (0.104 to 0.119 in.) case, 5 axles
3.2 to 3.5 mm (0.126 to 0.139 in.) case, 5 axles

Case depth to 40 HRC   0.2   0.3   0.4   0.5

Millions of cycles to failure

Automotive axle shafts made of induction hardened 1038 steel with diameters of 32 mm (1¼ in.)

Control of Case Depth and Contour

The engineering specification of depth of hardening for a given part design is based on knowledge, or estimates, of the strength required to resist service loads. Frequently, as for shafts, the analysis involves considerations of requirements for fatigue strength.

Required Case Depth

The stress pattern is an important consideration in torsional fatigue. Radial stresses of both through hardened and case hardened shafts (induction or carburized) are always tensile and relatively unimportant. Residual circumferential and longitudinal stresses are usually tensile in through hardened parts and compressive in case hardened parts. The induction hardened shaft is therefore similar to a shaft made from shallow-hardening steel or to a carburized shaft, although the residual stresses are probably lower than in a carburized case. The induction hardened case must be deep enough to provide strength across the section to exceed the applied stress, but through hardening must be avoided. A comparison of the effect of case depth in induction hardened axles is given in Example 10.

Example 10. Passenger-car axle shafts of 1038 steel were 32 mm (1¼ in.) in diameter at the reduced section, which was not machined after forging. Shafts were induction hardened to give two different case depth patterns, each having a surface hardness of 58 to 60 HRC. Shafts were not tempered. The shallower cases were 2.6 to 3 mm (0.104 to 0.119 in.) to 40 HRC and 4.5 to 5.2 mm (0.176 to 0.206 in.) to 20 HRC. The deeper cases were 3.2 to 3.5 mm (0.126 to 0.139 in.) to 40 HRC and 6.4 to 7 mm (0.253 to 0.274 in.) to 20 HRC. Five

axles of each treatment were fatigue tested under a load of 2034 J (1500 ft·lb). Comparison of the fatigue results is shown graphically in Fig. 13.

Good Operating Practice

In surface hardening carbon and alloy steels by induction, if the hardened zone is too deep for the specific section thickness, high tensile stresses are established in the surface layers and cracking may result. Excessive depth of the hardened zone can result if the quench is too severe during the austenite-to-martensite transformation. Temperature that is too high can also cause excessive scale and pitting of the surface. The frequency, power density, and time should be selected to provide the desired case depth. The quench should be selected to provide the cooling rate necessary to produce the desired hardness without cracking. The parts should be quenched to full hardness, and then tempered to specified hardness (if lower than quenched). Stress relieving either through residual heat or furnace heat is always good practice where possible.

Control Limits

Needlessly close control can result in added costs in steel, equipment and labor. The necessity for such control can start when the engineer specifies hardness depths and patterns that may be satisfactory for the design and service of the hardened part but which place unnecessary or even impracticable limitations on the size and frequency of the equipment, the tooling, fixtures, and even the steel itself.

Hardness Patterns

Hardness patterns on induction hardened parts are normally revealed by sectioning the piece with an abra-

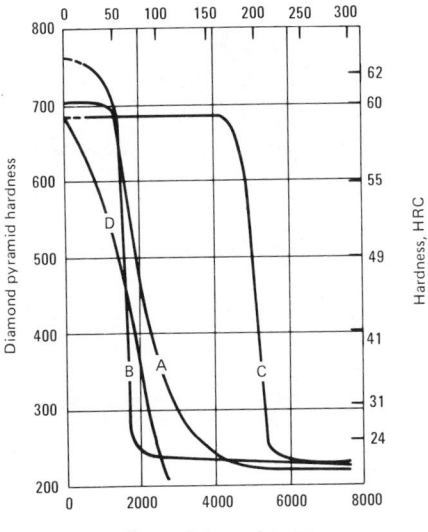

Fig. 14 Depth-hardness curves for some parts surface hardened by induction

Distance below surface, 0.001 in.

Distance below surface, μm

A is for a 47.6-mm- (1⅞-in.-) diam spindle made of 1045 steel, progressively hardened; 400 kHz, 50 kW, water spray quench. **B** is for normalized 5140 steel shown in Fig. 15; 10 kHz, 24.8 W/mm² (16 kW/in.²), 2 s. **C** is for normalized 5140 steel shown in Fig. 15; 10 kHz, 15.5 W/mm² (10 kW/in.²), 5 s. **D** is for pearlitic malleable iron; 10 kHz, 116.25 W/m² (75 kW/in.²), 1½ s, water spray quench.

sive wheel under water, rough grinding on a belt, polishing on fine emery paper, and then etching with a 2 or 5% nital solution until the pattern develops. The case depth is frequently reported as the depth from the surface at which the hardness drops below 50 HRC, although in initial studies for a given part, complete depth-hardness curves and microscopic analyses of the fully hardened and transition zones of the case are recommended. Figures 14 and 15 show depth-hardness curves and macroscopic analyses for induction hardened parts made of steel and malleable iron.

Control Factors. Basically, the depth of the hardened case obtained by induction heating followed by a suitable quench depends on the frequency, the power density, the heating time and the steel. In addition, the hardness pattern or contour depends on the shape of the part as related to coil design. The following illustrations may

Fig. 15 Macrographs illustrating the influence of heating time on depth of hardness and hardness contour for normalized 5140 steel

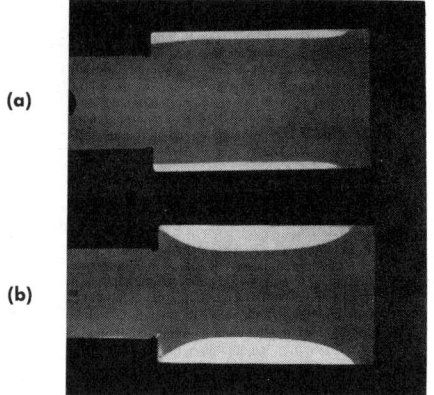

(a)

(b)

(a) 10 kHz frequency, 24.8 W/mm² (16 kW/in.²), 2 s, 1.7 mm (0.065 in.) to 50% martensite.
(b) 10 kHz frequency, 15.5 W/mm² (10 kW/in.²), 5 s, 5.2 mm (0.203 in.) to 50% martensite

Table 16 Effect of frequency on depth of case hardness

Frequency, Hz	Theoretical depth of penetration of electrical energy(a) mm	in.	Practical depth of case hardness Minimum(b) mm	in.	Working(c) mm	in.
1000	1.50	0.059	2.54	0.100	4.57-8.89	0.180-0.350
3000	0.89	0.035	1.52	0.060	3.81-5.08	0.150-0.200
10 000	0.51	0.020	1.02	0.040	2.54-3.81	0.100-0.150
120 000	0.15	0.006	0.76	0.030	1.52-2.54	0.060-0.100
500 000	0.08	0.003	0.51	0.020	1.02-2.03	0.040-0.080
1 000 000	0.05	0.002	0.25	0.010	0.25-0.76	0.010-0.030

(a) Effective initial penetration (time = 0). Values given are approximate. (b) For high power densities: 23.3 W/mm² (15 kW/in.²) min and optimum prior structure. (c) For medium power densities: 7.8 to 23.3 W/mm² (5 to 15 kW/in.²) and steel in the as-rolled condition

Fig. 16 Influence of scanning rate on depth of hardness in 28.6-mm- (1⅛-in.-) diam bars of 1045 steel

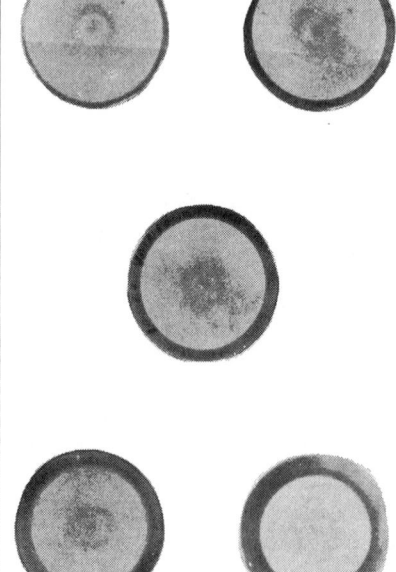

Frequency, 300 kHz; inductor input, 15 kW

Fig. 17 Nonuniform hardness pattern at end of spline, caused by overheating

Fig. 18 Distortion of hardness pattern caused by variation in size of section

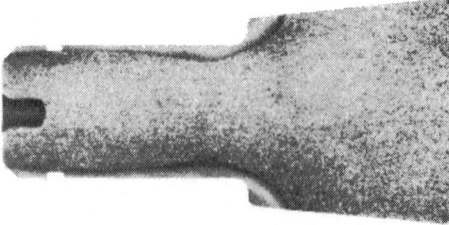

serve to relate these factors. (The influence and control of each factor have already been discussed.)

Minimum depths of case are produced by using high frequencies, short heating times as determined by high power densities, close coupling and efficient matching, and a quenched and tempered prior structure. The data in column 3 of Table 16 are presented as a guide to the minimum practicable depths of hardness possible at various frequencies, using high power density 23 W/mm² (15 kW/in.²) min and an optimum prior structure and type of steel. In practice, the depths of hardness associated with each frequency are generally greater than the minimum shown, and column 4 of Table 16 presents a guide for the practicable working depths of hardness using medium power densities 8 to 23 W/mm² (5 to 15 kW/in.²) and steel in the as-rolled or as-forged condition. Greater case depths may be obtained at each frequency by increasing the time of heating and decreasing the power density.

With suitable prior structures, short heating times and rapid quenching, the transition zone from case to core is generally sharp, as shown in Fig. 14 and 15. However, the hardness gradient may vary considerably, as illustrated in Fig. 14; the transition zone from case

to core increases with the duration of heating and with increasing coarseness of the ferrite-cementite aggregate in prior structures.

Minimum case depths are generally obtained by scanning or very high power densities such as 46.5 to 62 W/mm² (30 to 40 kW/in.²). Figure 16 illustrates the variation in case depth as well as

the uniformity of case produced at various scanning rates using a constant inductor input of 15 kW, a frequency of 300 kHz and a water spray quench. The bars of 1045 steel for this example were of 28 mm (1⅛ in.) diam.

Nonuniform Pattern. Figure 17 illustrates the nonuniformity that may occur in the hardness pattern even in symmetrically disposed pieces. If a uniform depth of hardness is necessary within a specified length of shaft, a closely coupled coil with an increased air gap at the center (see Fig. 6) may change the pattern so that it is approx-

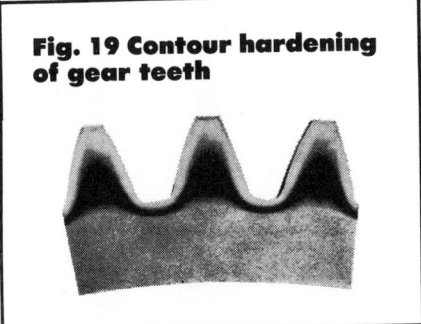

Fig. 19 Contour hardening of gear teeth

imately the same depth for the length of the inductor.

Similar effects can be obtained in the use of a multiturn coil by increasing the inside diameter of the center turns of the coil or increasing the spacing between turns at the center of the coil. The increased depth of case at the end of the spline shown in Fig. 17 is caused by overheating at the end as the coil passes. Since the heat loss to the cold section of the bar does not take place at the end, the input of energy is greater and the case deeper. If objectionable, this condition may be minimized by accelerating the scanning rate at the end of the piece.

Effect of Part Shape

The shape of the part also may have a significant influence on the heat pattern. Parts with nonuniform cross section do not develop a uniform pattern, since edges and corners heat to a higher temperature. The effect of a variation in section size, illustrated in Fig. 18, is exaggerated at higher frequencies. Though changes in the design of inductor may compensate for the distortion caused in the heat pattern by small changes in shape, if a uniform case depth is required, complex parts are difficult if not unsuitable for induction surface hardening.

Variations of the heat patterns produced in nonuniform sections are exemplified by Fig. 19, which is a macrograph of a gear. Figure 19 shows an attempt to provide single-shot contour hardening on gear teeth, that is, to provide high hardness on the surface of the teeth and at the root while maintaining a soft core in the tooth. With a uniform heating time and a uniform metallurgical structure, higher frequencies accentuate contour hardening. Because of this fact, gears—particularly, large gears—can be preheated using lower frequencies (60 to 10 000 Hz), and then

the heating can be completed with frequencies of 10 000 to 500 000 Hz.

In large gears or splines with a coarse pitch, unique hardness patterns are sometimes obtained not by controlling the heated contour but as a result of the combined influence of the hardenability of the steel and the severity of quench on the through heated gear teeth.

Residual Stress, Distortion and Cracking

When a relatively shallow outside layer is selectively heated and quenched to martensite in induction surface hardening, the volume expansion associated with martensite formation gives rise to compressive stresses at the surface and tensile stresses in the interior. The tensile stresses usually start at the junction of case and core. The stresses are compressive at the surface, irrespective of the alloy content or the case depth. Oil quenching gives weaker stress patterns than water quenching.

Stress Distribution

The final stress distribution set up by quenching depends not only on the expansion accompanying martensite formation during the quench but also, to a lesser extent, on the localized plastic deformation resulting from nonuniform thermal expansion during rapid heating to the quenching temperature and nonuniform contraction during cooling to the temperature where martensite begins to form. Thus the actual stress distribution obtained depends on the rate of heating, the depth of hardening and the steel itself.

One undesirable effect of increasing depth of hardening is an increase in the magnitude of subsurface tensile stresses. Tempering reduces the stresses near the surface.

Compressive residual stress at the surface of an induction hardened steering-knuckle pivot had a favorable influence on fatigue resistance. For this part, induction hardening to an equivalent case depth provided fatigue resistance superior to case hardening by carburizing. Apparently, increased case depth further improved the fatigue resistance in this part. High tensile stresses, however, may exist at the juncture of case and core. If the heat pattern brings the high-tensile zone near the surface in an area such as a fillet that is subject to repeated stresses

in service, early fatigue failure may result. In addition, fatigue failure may also be initiated at the juncture of the case and the core.

Distortion

Steel parts that have been surface hardened by induction generally exhibit less total distortion or distortion more readily controllable than that for the same parts quenched from a furnace. The decrease in distortion is a result of the support given by the rigid, unheated core metal, and of uniform, individual handling during heating and quenching. In scanning, distortion is controlled further by heating and quenching only a narrow band of the steel at one time. Unless a part through hardened by induction is scanned, the distortion encountered will approach the distortion that is experienced in furnace hardening.

Cracking

The forces that give rise to residual stresses and distortion will, in some instances, cause an induction heated part to crack during quenching.

Parts having severe discontinuities in configuration are particularly susceptible to cracking. The entire subject is so intimately associated with design that it can be discussed meaningfully only in relation to specific parts.

Through Hardening by Induction

The use of induction heating equipment for through hardening offers many of the same advantages over furnace hardening that it does for surface hardening. Some of these advantages are: little or no decarburization, a minimum of scale, elimination of protective atmosphere in many applications, improved working conditions, improved quality of product, practical installation of induction equipment into the machine line, and adaptability to automatic handling.

Methods

Parts can be through hardened by induction heating using the single-shot or scanning methods.

In the single-shot method, a part is positioned in the inductor coil, heated for a predetermined length of time, and quenched either in the inductor or by removal to a separate system. Advantages of this method include simplicity

Table 17 Typical operating conditions for progressive through hardening of steel parts by induction

Section size mm	in.	Material	Frequency, Hz	Power, kW(a)	Total heating time, s	Scan time s/mm	s/in.	Entering coil °C	°F	Leaving coil °C	°F
12.7 (round)	½ (round)	4130	180,	20	38	25	1	75	165	510	950
			9600(c)	21	17	25	1	510	950	925	1700
19 (round)	¾ (round)	1035 mod	180,	28.5	68.4	46	1.8	75	165	620	1150
			9600(c)	20.6	28.8	46	1.8	620	1150	955	1750
25 (round)	1 (round)	1041	180,	33	98.8	66	2.6	70	160	620	1150
			9600(c)	19.5	44.2	66	2.6	620	1150	955	1750
28.6 (round)	1⅛ (round)	1041	180,	36	114	76	3.0	75	170	620	1150
			9600(c)	19.1	51	76	3.0	620	1150	955	1750
49.2 (round)	1¹⁵⁄₁₆ (round)	14B35H	180,	35	260	178	7.0	75	165	635	1175
			9600(c)	32	119	178	7.0	635	1175	955	1750
15.9 (flat)	⅝ (flat)	1038	3000	300	11.3	38	1.5	20	70	870	1600
19 (flat)	¾ (flat)	1038	3000	332	15	50	2.0	20	70	870	1600
22.2 (flat)	⅞ (flat)	1043	3000	336	28.5	97	3.8	20	70	870	1600
25 (flat)	1 (flat)	1036	3000	304	26.3	89	3.5	20	70	870	1600
28.6 (flat)	1⅛ (flat)	1036	3000	344	36.0	122	4.8	20	70	870	1600
17.5-33.3(d)	11⁄16-1⁵⁄₁₆(d)	1037 mod	3000	580	254	61	2.4	20	70	885	1625

Section size mm	in.	Material	Production kg/h	lb/h	Operating efficiency(b) mg/J	lb/kW·h	Inductor input(b) W/m²	kW/in.²
12.7 (round)	½ (round)	4130	91.6	202	1.27	10.1	0.67	0.43
			91.6	202	1.21	9.6	1.22	0.79
19 (round)	¾ (round)	1035 mod	113.4	250	1.10	8.77	0.62	0.40
			113.4	250	1.53	12.14	0.85	0.55
25 (round)	1 (round)	1041	141.1	311	1.19	9.42	0.54	0.35
			141.1	311	2.00	15.9	0.57	0.37
28.6 (round)	1⅛ (round)	1041	153.3	388	1.18	9.39	0.52	0.34
			153.3	338	2.23	17.70	0.50	0.32
49.2 (round)	1¹⁵⁄₁₆ (round)	14B35H	194.6	429	1.55	12.3	0.29	0.19
			194.6	429	1.69	13.4	0.48	0.31
15.9 (flat)	⅝ (flat)	1038	1448.9	3194	1.34	10.6	3.61	2.33
19 (flat)	¾ (flat)	1038	1575.8	3474	1.32	10.5	3.19	2.06
22.2 (flat)	⅞ (flat)	1043	1609.4	3548	1.34	10.6	2.06	1.33
25 (flat)	1 (flat)	1036	1595.3	3517	1.46	11.6	2.25	1.45
28.6 (flat)	1⅛ (flat)	1036	1678.8	3701	1.36	10.8	2.08	1.34
17.5-33.3(d)	11⁄16-1⁵⁄₁₆(d)	1037 mod	2211.3	4875	1.06	8.4	0.40	0.26

(a) Power transmitted by the inductor at the operating frequency indicated. This power is approximately 25% less than the power input to the machine, because of losses within the machine. (b) At the operating frequency of the inductor. (c) Dual frequencies are used. (d) Irregular shape

of materials-handling equipment and ease of conversion to handle parts of different sizes. However, in the single-shot method, due to the "Curie Effect", some generators do not operate continuously at rated output and consequently there is a loss in the over-all efficiency of the system. The Curie Effect refers to the change in permeability near the temperature at which steel becomes nonmagnetic, approximately 770 °C (1420 °F). (See *Metals Handbook*, Vol. 6, 8th Ed., p. 631.)

The scanning method permits the handling of larger parts with optimum efficiency. Using this method, bars or flats can be fed continuously through one or more inductor coils and into a quench. In some installations, the bar continues to advance through a tempering inductor after quenching, emerging from one integrated line completely hardened and tempered. Another advantage of the scanning method is that distortion is minimized and can normally be held to commercial mill tolerances for bars and flats.

Power Density and Frequency

Data for surface hardening by induction are not applicable to through hardening, because through hardening requires the heating of more metal and maintenance of a minimum tempera-

ture differential between the surface and the core. Thus, lower frequency and power density are required (Tables 6 and 8). The lower frequency heats to a greater depth; the lower power density prevents the surface from overheating while the heat flows into the core. Because the initial and operating costs of large power installations decrease with a decrease in frequency, it is advantageous to select the lowest frequency that will efficiently heat the specific parts (see Table 6).

Heating Rate

In a well-designed installation a rate of 1.3 mg/J (10 lb/kW·h) can be ob-

Table 18 Determination of optimum frequency, power input and length of coil for progressive through hardening of bar stock

Frequency . . . . 3 kHz
Power density (kW/in.^{2}) is select-
ed from Table 8 for 3000 Hz and a
hardening temperature in the
range of 760 to 980 °C (1400 to
1800 °F) 0.4
Pounds per kW·h 9.0

Power input required $= \dfrac{\text{lb/h}}{\text{lb/kW·h}}$

$= \dfrac{100\ \text{ft} \times 4.17\ \text{lb/ft}}{9.0\ \text{lb/kW·h}}$ . . . . 46.3 kW

Area heated
$= \dfrac{46.3\ \text{kW}}{0.4\ \text{kW/in.}^2}$ . . . . 115.7 in.^{2}

Length of coil
$= \dfrac{\text{area}}{\text{circumference}} = \dfrac{115.7}{1.25 \times 3.14}$ · · 29.46 in.

Fig. 21 Distribution of Rockwell C hardness that was obtained in 20-mm (0.79-in.) steel shot by induction hardening

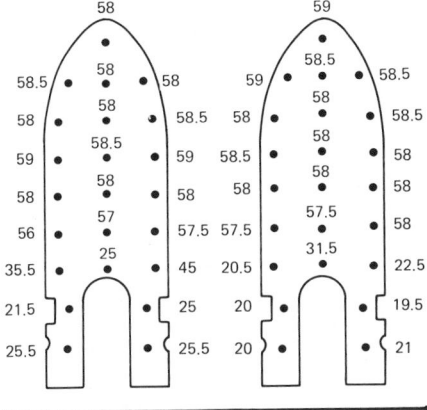

Fig. 20 Typical hardness survey on 32-mm- (1¼-in.-) diam bars of 1045, TS14B35, and 4140 steels through hardened by induction

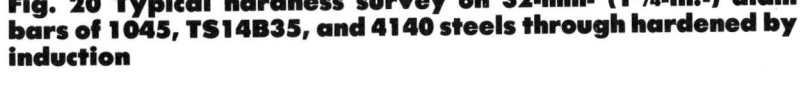

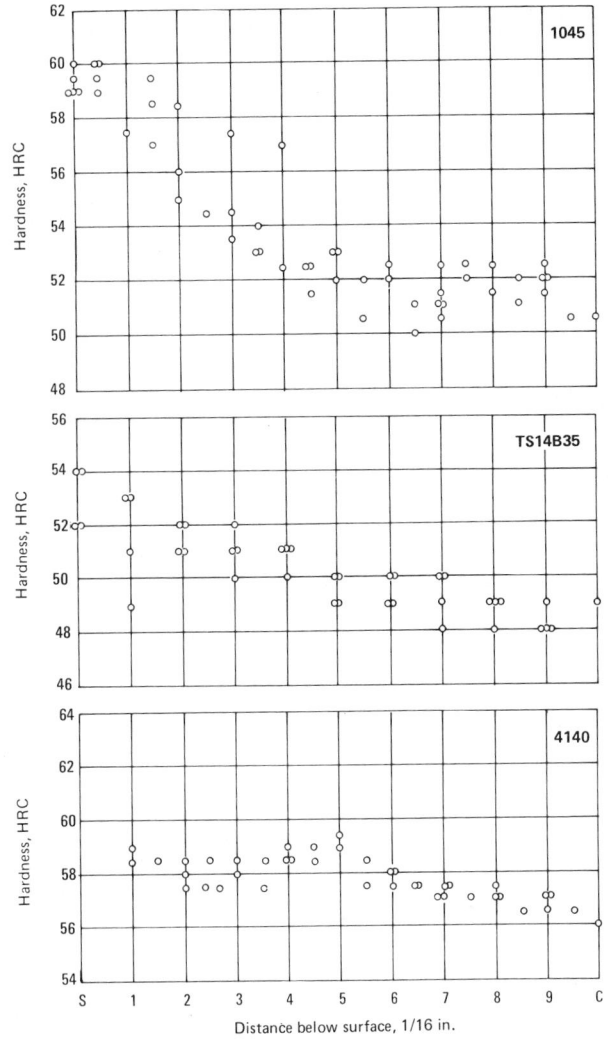

tained when continuously heating bar stock or flats to hardening temperature. This rate cannot always be maintained in production when parts of various sizes are handled on the same line. Table 17 shows typical data for installations which have been in production service for many years.

Because of the relatively short heating times involved in induction heating, it is frequently necessary, in through hardening, to heat 25 to 55 °C (50 to 100 °F) above the recommended furnace temperature to form homogeneous austenite. Fine-grained material is required for best results.

The following example illustrates the steps in determining frequency, power input and length of coil for progressive through hardening of steel bar stock.

Example 11. It is required to through harden 1041 steel bar stock, 32 mm (1¼ in.) in diameter and weighing 6.21 kg/m (4.17 lb/ft), at a rate of 8.5 mm/s (100 ft/h). Determine the frequency, power input and length of coil required. Table 18 gives the solution.

Although most induction installations for through hardening are designed for small parts, considerable work is now being done on parts 203 mm (8 in.) in diameter and larger. Recent development work in the use of lower frequencies (60 to 180 Hz), as well as general improvements in techniques, has made it practical to through heat large parts. Theoretically, parts of any size can be through heated by induction.

Figure 20 shows typical through hardening results obtained on 32-mm-

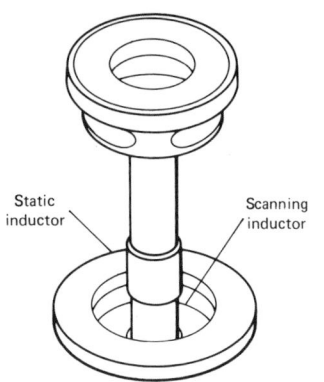

Fig. 22 Dual-frequency induction hardening of a transmission shaft

Scanning inductor employs frequency of 3000 Hz for hardening shaft to 5 mm (0.200 in.). Static inductor employs about 420 kHz for hardening flange to 0.76 mm (0.030 in.).

Table 19 Effect of carbon equivalent on surface hardness of induction hardened gray irons

| Composition, %(a) | | Carbon equivalent(b) | Hardness, HRC | | |
C	Si		As read	Rockwell 30-N	Micro-hardness
3.13	1.50	3.63	50	50	61
3.14	1.68	3.70	49	50	57
3.19	1.64	3.74	48	50	61
3.34	1.59	3.87	47	49	58
3.42	1.80	4.02	46	47	61
3.46	2.00	4.13	43	45	59
3.52	2.14	4.23	36	38	61

(a) Each iron also contained: 0.50 to 0.90 Mn, 0.35 to 0.55 Ni, 0.08 to 0.15 Cr and 0.15 to 0.30 Mo. (b) Carbon equivalent = %C + $\frac{1}{3}$% Si.

(1¼-in.-) diam bar stock of 1045, TS14B35 and 4140 steels.

Applications and Cost Relations

Hardening by induction can often contribute to cost savings by:

- Substitution of carbon steel for alloy steel
- Increased production
- Reduction in prime labor personnel
- Elimination of a machining operation
- Elimination of a straightening operation
- In-line production setup
- Less handling of parts
- Automation of the hardening operation
- Less equipment maintenance
- Less floor space
- Less energy per piece
- Less finish grinding
- Less inspection
- Less scrap.

With respect to cost relations, the applications of induction hardening may be placed in two categories: (a) where induction hardening provides a unique solution to problems posed by special requirements, and cost is secondary; (b) where induction hardening competes with other heat treating methods, and cost is paramount.

Special Requirements

Many selective hardening applications that require a restricted hardness pattern or a specified hardness gradient fall in the first category. Localized induction hardening has been specified increasingly to avoid the necessity for compromises in the mechanical properties of a given part. Two examples of this type of application follow.

Example 12. The hardness gradients produced in 20-mm steel shot illustrate another unique application of induction hardening (Fig. 21). Precise coil design is essential to the success of an operation of this type.

Example 13. This transmission shaft could be hardened properly and economically only by induction heating, employing dual frequency. As shown in Fig. 22, the 3000-Hz scanning inductor progressively hardens the shaft to a depth of about 5 mm (0.200 in.). This depth of hardening is required because of torsional loads imposed during service. When this hardening cycle has been completed, the flange indexes into the static inductor—a heating and quenching coil using radio frequency of about 420 kHz. The flange is hardened to a depth of about 0.8 mm (0.030 in.), which is sufficient for the wear resistance required. Radio frequency is necessary to control the depth of hardening; deeper hardening would create a distortion problem.

High-Frequency Resistance Hardening

In conventional induction hardening, the coil is always separate and distinct from the workpiece, and energy is transmitted from the coil to the work through an air gap, by means of induction. In high-frequency resistance hardening, a portion of the workpiece, in effect, becomes a part of the coil, and energy is transmitted to the work by means of conduction. This technique may be employed when the use of a closed loop coil is not possible or is highly inefficient.

High-frequency resistance heating usually employs frequencies in the range of 10 to 500 kHz and is capable of producing higher current densities than can be obtained with induction. The use of high frequencies minimizes problems normally encountered at the contact points in low-frequency resistance heating.

Induction Hardening of Cast Iron

Cast irons may vary significantly in combined carbon content, and results from induction hardening are nonuniform unless control is established over the combined carbon content. A minimum combined carbon content of 0.40 to 0.50% carbon (as pearlite) is recommended for cast iron to be hardened by induction with the short heating cycles that are characteristic of the induction process. Heating castings with lower combined carbon content to high hardening temperatures for relatively long periods of time may dissolve some free graphite, but such a procedure is likely to coarsen the grain structure at the surface and will result in undesirably large amounts of retained austenite in the surface layers.

Flake-Graphite Gray Iron

The surface hardness attained from induction hardening of gray iron is influenced by the carbon equivalent (%C + $\frac{1}{3}$% Si) when this hardness is measured by conventional Rockwell tests. The more graphite that is present in the microstructure, the lower the

Table 20 Operating and production data for progressive induction tempering

| Section size | | Material | Frequency, Hz | Power(a), kW | Total heating time, s | Scan time | | Work temperature Entering coil | | Leaving coil | |
mm	in.					s/mm	s/in.	°C	°F	°C	°F
12.7 (round)	1/2 (round)	4130.........	9600	11	17	0.040	1	50	120	565	1050
19 (round)	3/4 (round)	1035 mod.....	9600	12.7	30.6	0.072	1.8	50	120	510	950
25 (round)	1 (round)	1041.........	9600	18.7	44.2	0.104	2.6	50	120	565	1050
28.6 (round)	1 1/8 (round)	1041.........	9600	20.6	51	0.128	3.0	50	120	565	1050
49.2 (round)	1 15/16 (round)	14B35H.......	180	24	196	0.275	7.0	50	120	565	1050
15.9 (flat)	5/8 (flat)	1038.........	60	88	123	0.060	1.5	40	100	290	550
19 (flat)	3/4 (flat)	1038.........	60	100	164	0.080	2.0	40	100	315	600
22.2 (flat)	7/8 (flat)	1043.........	60	98	312	0.149	3.8	40	100	290	550
25 (flat)	1 (flat)	1043.........	60	85	254	0.121	3.1	40	100	290	550
28.6 (flat)	1 1/8 (flat)	1043.........	60	90	328	0.157	4.0	40	100	290	550
17.5-33.3 (irregular)	11/16 to 1 5/16 (irregular)	1037 mod.....	9600	192	64.8	0.094	2.4	65	150	550	1020
17.5-28.6 (irregular)	11/16 to 1 1/8 (irregular)	1037 mod.....	9600	154	46	0.070	1.7	65	150	425	800

| Section size | | Material | Production | | Operating efficiency(b) | | Inductor input(b) | |
mm	in.		kg/h	lb/h	mg/J	lb/kW·h	W/mm²	kW/in.²
12.7 (round)	1/2 (round)	4130.........	90	200	2.32	18.4	0.62	0.41
19 (round)	3/4 (round)	1035 mod.....	115	250	2.48	19.67	0.50	0.32
25 (round)	1 (round)	1041.........	140	310	2.09	16.6	0.54	0.35
28.6 (round)	1 1/8 (round)	1041.........	155	340	2.02	16.4	0.53	0.34
49.2 (round)	1 15/16 (round)	14B35H.......	195	430	2.26	17.9	0.31	0.20
15.9 (flat)	5/8 (flat)	1038.........	1450	3195	4.57	36.3	0.14	0.089
19 (flat)	3/4 (flat)	1038.........	1575	3475	4.37	34.7	0.13	0.081
22.2 (flat)	7/8 (flat)	1043.........	1610	3550	4.56	36.2	0.08	0.050
25 (flat)	1 (flat)	1043.........	1365	3010	4.46	35.4	0.11	0.068
28.6 (flat)	1 1/8 (flat)	1043.........	1485	3270	4.57	36.3	0.09	0.060
17.5-33.3 (irregular)	11/16 to 1 5/16 (irregular)	1037 mod.....	2210	4875	3.20	25.4	0.42	0.28
17.5-28.6 (irregular)	11/16 to 1 1/8 (irregular)	1037 mod.....	2275	5020	4.0	32	0.39	0.26

(a) Power transmitted by the inductor at operating frequency indicated. For converted frequencies, this power is approximately 25% less than the power input to the machine, because of losses within the machine. (b) At the operating frequency of the inductor

surface hardness will appear to be after hardening. Table 19 shows the surface hardness of induction hardened gray iron castings of various carbon equivalents from 3.63 to 4.23. The microstructure of these castings, which were cast in the same manner and cooled at similar rates, contained more and larger graphite flakes as the carbon equivalent increased. This resulted in lower apparent surface hardness after hardening, yet the hardened matrix was consistently 57 to 61 HRC (converted from microhardness).

Pearlite. To induction harden gray iron within these limitations, it is necessary that the carbon be present in a form that will readily dissolve in the iron at the austenitizing temperature. The combined carbon in pearlite is readily soluble, and high percentages of pearlite in the matrix are desirable. Sufficient pearlite to give a combined carbon content of 0.40 to 0.50% is satisfactory for induction hardening.

Ferrite. The prior structure and combined carbon content of as-cast gray iron are dependent on cooling rate (section size) and composition. Large sections, or slow cooling rates, may result in increased amounts of ferrite and decreased response to short heating cycles. Alloying elements, such as chromium, nickel, copper or molybdenum, will minimize the formation of ferrite on cooling.

Distortion. Induction hardening causes less distortion than would a similar quenching treatment from a furnace. Maximum warpage in the 55.9-cm (22-in.) length of a bar was found to be 0.03 mm (0.0015 in.) after induction hardening, compared with 0.17 to 0.25 mm (0.007 to 0.010 in.) for the same bars quenched from a furnace. For thin-wall cylinders, distortion is not a problem unless the thickness of the induction hardened layer exceeds 20% of the wall thickness.

For selective hardening in place or by scanning, to depths up to 3.8 mm (0.150 in.), water quenching can be safely employed. However, the hardening of keyways, cross-drilled holes, or extremely thin walls may require the use of oil, to prevent excessive distortion or cracking.

Applications. A cast frame for a baler knotter had a nominal composition of 3.30 to 3.50 C, 0.50 to 0.70 Mn,

Fig. 23 Relation between depth of induction hardening to 50 HRC and graphite nodule count in normalized and tempered nodular iron

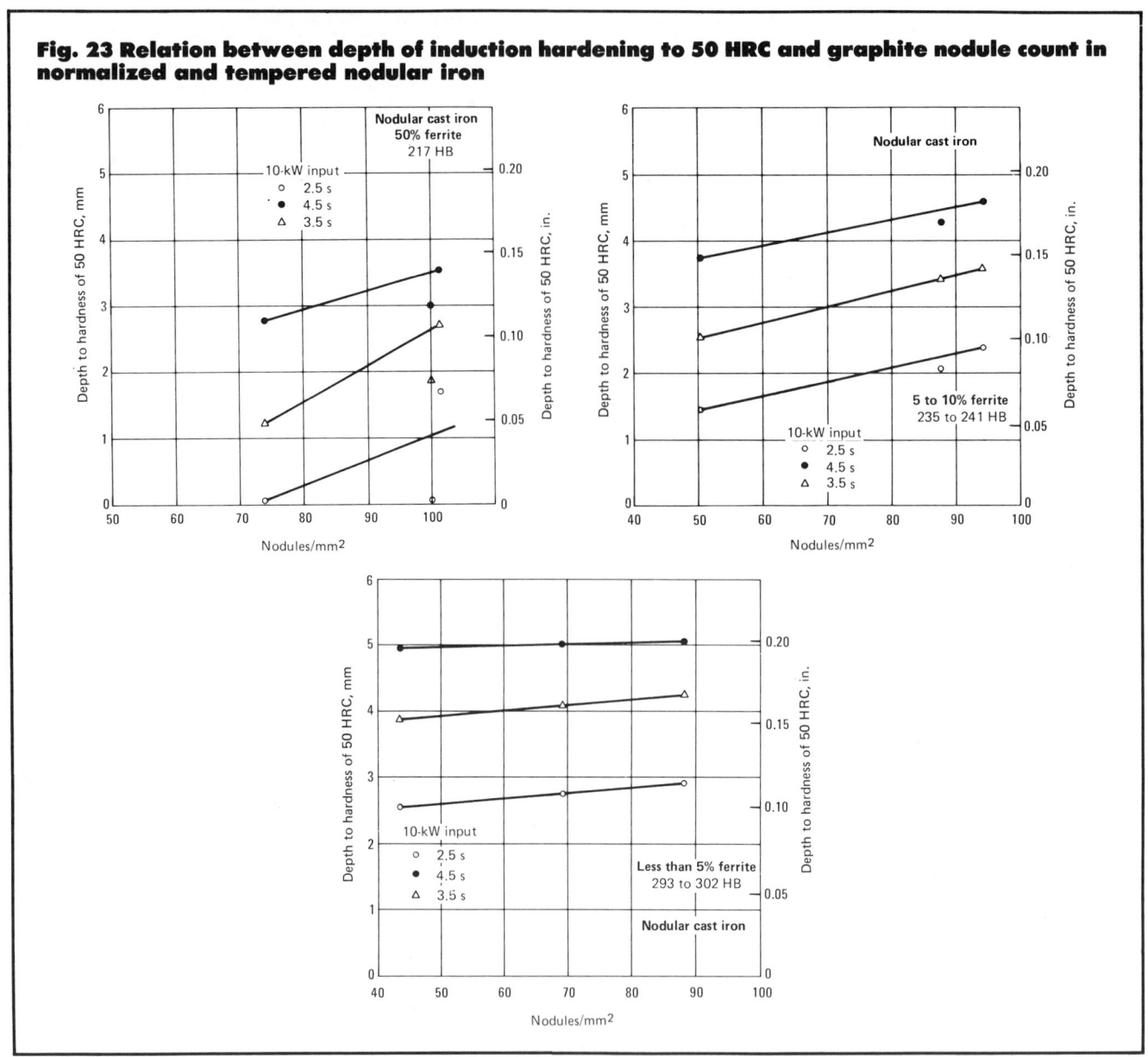

0.30 to 0.40 P, 0.15 max S, and about 2.40 to 2.60 Si. A 10.9-mm ($^7/_{16}$-in.) section of the frame was heated for 12 s with a 14-kW, 10-kHz unit, using a modified hairpin-type inductor. After water quenching for 4 s, it developed an apparent hardness of approximately 45 HRC (not indicative of the true hardness of the matrix, which can harden to 61 HRC or higher).

Ductile Iron

The response of ductile iron to induction hardening is dependent on the amount of pearlite in the matrix of as-cast, normalized, and normalized and tempered prior structures. In quenched and tempered iron, the secondary graphite nodules formed during tempering are close enough together to supply sufficient carbon to the matrix by re-solution during induction heating.

As Cast. In the as-cast condition, the amount of pearlite can vary from 10 to 95%, depending on cooling rate and silicon content. The pearlite content of the matrix increases as the cooling rate increases and as the amount of silicon decreases.

For induction heating cycles 3.5 s and longer and hardening temperatures of 955 to 980 °C (1750 to 1800 °F), 50% pearlite is considered a minimum for satisfactory hardening. Structures containing less pearlite can be hardened by using higher temperatures, but at the risk of retaining austenite, forming ledeburite, and damaging the surface. With more than 50% pearlite, hardening temperatures may be reduced to within the range of 900 to 925 °C (1650 to 1700 °F).

Prior Treatments. Because the amount of pearlite in as-cast ductile iron can fall below the minimum required for satisfactory response, a prior treatment to produce the desired microstructure is required, to ensure that all castings will respond to a given hardening cycle. Some castings, because of

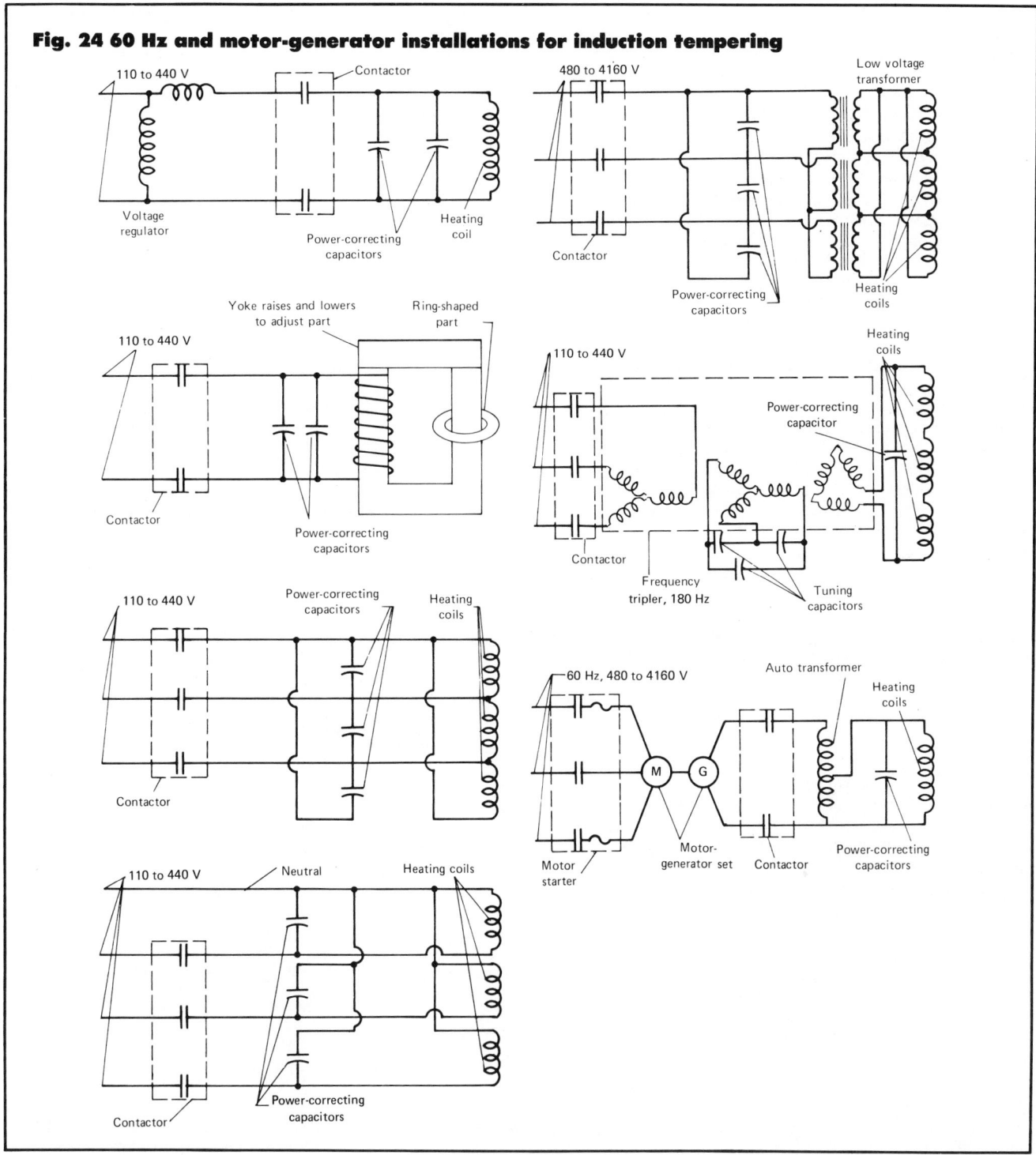

Fig. 24 60 Hz and motor-generator installations for induction tempering

predominant section size or configuration, may satisfy the minimum microstructure in the area to be hardened. It is possible, to a great extent, to relate this microstructure to a Brinell hardness measurement. At 229 HB, for example, the microstructure can vary from 20 to 50% ferrite (50% pearlite). A hardness of 229 HB would therefore be the approximate minimum for induction hardening as-cast or normalized and tempered structures.

Normalized and Tempered. Normalizing, although it produces a uniform microstructure low in ferrite and with good induction hardening characteristics, usually produces a higher hardness than is desired for machinability and ductility. Therefore, the normalizing treatment is usually followed by a tempering treatment.

To control the as-tempered hardness range within desired limits, it is necessary also to control the range of hardness developed in normalizing. This can be accomplished by providing a sufficiently fast and uniform rate of air cooling from the normalizing temperature, 900 °C (1650 °F).

For heating cycles of 3.5 s and longer, at temperatures of 955 to 980 °C (1750 to 1800 °F), a prior structure containing 50% pearlite would be considered a minimum. Examination of the normalized and tempered irons indicates that the poor response of structures of lower pearlite content is due to depletion of the matrix carbon. In the tempering operation, the carbon migrates from the pearlite matrix to the graphite nodules. In the heating cycle, carbon is reabsorbed in the matrix from the nodule; however, there is insufficient time for it to migrate throughout the ferritic areas. Another factor in the response of ductile iron is the graphite nodule count; the greater the number of nodules per unit area, the deeper the hardening for any given heat cycle. This effect is more evident as the percentage of ferrite increases (Fig. 23).

Quenched and Tempered. The response of quenched and tempered nodular iron to induction hardening is excellent over a wide range of microstructures containing up to 95% ferrite. As a prior treatment, quenching and tempering has the advantage of permitting a lower prior hardness; there is a risk of distortion and quench cracking, however.

A quenched and tempered structure that provided good response to induction hardening was obtained by oil quenching from 900 °C (1650 °F) and tempering at 620 °C (1150 °F) for 1 h. This treatment produced a hardness of 262 HB, which could have been lowered, if necessary, by increasing the tempering temperature to 675 °C (1250 °F). By induction heating to a depth of 4.7 mm (0.184 in.), a surface hardness of 54 to 56 HRC was developed, and a depth of hardness to 50 HRC of 4.2 mm (0.164 in.) was obtained.

Pearlitic Malleable Iron

Considerable research has been done on the induction hardening characteristics of pearlitic malleable iron and its capability of developing high hardness over relatively narrow surface bands. Generally, little difficulty is encountered in obtaining hardnesses in the range from 55 to 60 HRC, with the depth of penetration being controlled by the rate of heating and the tempera-

Table 21 Selection of power source and frequency for various applications of induction tempering

Size of section mm	in.	Maximum tempering temperature °C	°F	Power lines, 50 or 60 Hz	Frequency converter, 180 Hz	Solid state or medium frequency 1000 Hz	3000 Hz	10 000 Hz	Vacuum tube, over 200 kHz
3.2-6.4	1/8-1/4	705	1300	...	...	...	...	...	Good
6.4-12.7	1/4-1/2	705	1300	...	...	...	...	Good	Good
12.7-25	1/2-1	425	800	...	Fair	Good	Good	Good	Fair
		705	1300	...	Poor	Fair	Good	Good	Fair
25-50	1-2	425	800	Fair	Fair	Good	Good	Fair	Poor
		705	1300	...	Fair	Good	Good	Fair	Poor
50-152	2-6	425	800	Good	Good	Good	Fair	...	...
		705	1300	Good	Good	Good	Fair	...	...
Over 152	Over 6	705	1300	Good	Good	Good	Fair	...	...

Efficiency, capital cost and uniformity of heating are considered in the indicated selections. Indicated selections are defined as follows: **Good** indicates the optimum frequency that will heat 80% or more of the pounds per kilowatt-hour indicated in Fig. 3. **Fair** indicates a frequency that will heat 60 to 80% of the pounds per kilowatt-hour indicated in Fig. 3; or a frequency higher than optimum that increases capital cost and reduces uniformity of heating, thus requiring lower heat inputs. **Poor** indicates a frequency that will heat less than 60% of the pounds per kilowatt-hour indicated in Fig. 3; or a frequency higher than optimum that increases capital cost and reduces uniformity of heating, thus requiring lower heat inputs.

Table 22 Approximate power density required for tempering

Frequency(a), Hz	Input(b) W/mm² 150-425 °C (300-800 °F)	425-705 °C (800-1300 °F)	kW/in.² 150-425 °C (300-800 °F)	425-705 °C (800-1300 °F)
60	0.09	0.23	0.06	0.15
180	0.08	0.22	0.05	0.14
1000	0.06	0.19	0.04	0.12
3000	0.05	0.16	0.03	0.10
10 000	0.03	0.12	0.02	0.08

(a) Table is based on use of proper frequency and normal over-all operating efficiency of equipment. (b) In general, these power densities are for section sizes of 12 to 50 mm (1/2 to 2 in.). Higher inputs can be used for smaller section sizes and lower inputs may be required for larger section sizes.

ture developed at the surface of the part being hardened.

The maximum hardness obtainable in the matrix of a properly hardened part is 67 HRC. However, conventional hardness measurements made on the casting show less than 67 HRC because of the presence of the graphite particles which are averaged into the hardness. Generally, a casting with a matrix microhardness of 67 HRC will have about 62 HRC average hardness, as measured with the standard Rockwell tester.

Induction Tempering

Extensive production experience with thousands of tons of steel per month has demonstrated the commercial success of induction tempering for many applications. Metallurgically, the success of induction tempering has been related fundamentally to the possibility of compensating for short tempering times with higher tempering temperatures. Economically, induction tempering has proved particularly adaptable to automation in production lines.

Application. At present, two principal areas of application exist for induction tempering:

- Selective tempering
- Progressive tempering of bar stock previously hardened by scanning.

Many machine parts vary from one section to another with respect to load and wear requirements. Often, this variation in requirements is met by a compromise of properties obtained by uniform tempering to a single hardness level. However, it is apparent that superior performance might be expected if the mechanical properties could be adjusted to meet the particular requirements in each section by selective tempering. Within certain limitations, induction tempering is an economical means of accomplishing this. These limitations are that the parts must be

Table 23 Determination of optimum frequency, power input and length of coil for progressive tempering of 1041 steel bar stock

Frequency (selected from Table 21)	1 kHz
Power density (kW/in.^{2}) is selected from Table 8 for 1000 Hz and a tempering temperature of 535 °C (1000 °F)	0.12
Pounds per kW·h	18

Power input required $= \dfrac{\text{lb/h}}{\text{lb/kW·h}}$

$= \dfrac{100 \text{ ft} \times 4.17 = \text{lb/ft}}{18 \text{ lb/kW·h}}$ . . . . 23.2 kW

Area heated
$= \dfrac{23.2 \text{ kW}}{0.12 \text{ kW/in.}^2}$ . . . . 193.3 in.^{2}

Length of coil
$= \dfrac{\text{area}}{\text{circumference}} = \dfrac{193.3}{1.25 \times 3.14}$ . . 49.2 in.

Fig. 25 Variation of hardness with tempering temperature for furnace and induction heating

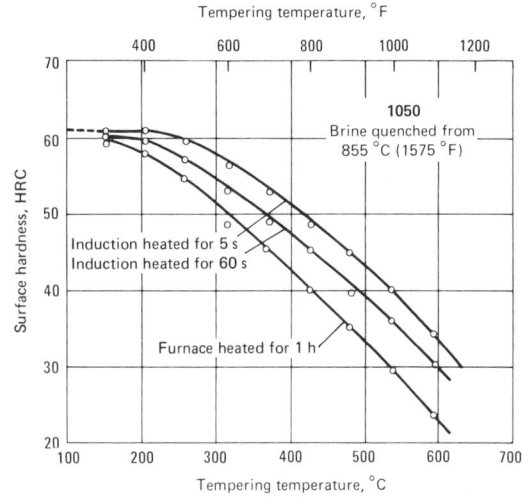

Considerable overlapping occurs in suitable frequencies, so that equipment utilized for induction hardening can often be used for the tempering treatment as well.

Power Density. Because the usual objective of induction tempering is to produce uniform hardness throughout the cross section, rather than to heat the surface, the power density is generally low from 0.08 to 0.8 W/mm^2 (0.05 to 0.5 kW/in.^{2}) of surface within the inductor. Power densities may be selected on the basis of experience, tests, or data such as those presented in Table 22 or the last column of Table 20. Furthermore, the heating time is comparatively long, to help provide uniform heating throughout the part. To meet production requirements, length of the inductor can be increased, or more than one bar can be processed at a time.

The size of the unit (in terms of kilowatts) may be determined by dividing the production rate in kilograms per hour (pounds per hour) by kilograms per kilowatt-hour (pounds per kilowatt-hour):

$$kW = \frac{kg/h}{kg/kW·h}$$

In determining the size of equipment, it is customary to apply an efficiency factor to allow for maintenance, repair, and operator fatigue. Adding 25% to the desired production, in pounds per hour, is a reasonable allowance, representing 80% efficiency.

The following example pertaining to

of such a shape and size that they can be coupled by the inductor to heat uniformly to the desired temperature in critical sections. Although this is impossible or impracticable for some parts, many can be tempered selectively by induction to obtain different degrees of hardness in the same part, with a consequent improvement in quality.

Advantage. The principal advantage of induction tempering is the possibility of integration with machine lines to avoid excessive handling of work, thereby minimizing labor cost. This is illustrated in the preparation of bar stock of specified mechanical properties before machining into cylinder-head studs and miscellaneous machine parts.

Table 20 contains pertinent data from production installations that temper thousands of tons of bar stock each month by induction.

Selection of Equipment

Because tempering is performed below the lower transformation temperature 725 °C (1333 °F), lower-frequency induction tempering installations are generally used; such installations are necessary for tempering large sections, to minimize any temperature gradient from the surface to the interior. Table 21 is presented as a guide for the selection of tempering equipment. It should be noted that line frequencies (60 Hz) may be used for tempering parts 25 to 50 mm (1 to 2 in.) in diameter or larger.

1041 bar stock illustrates the steps in determining frequency, power input and length of coil for tempering.

Example 14. It is required to induction heating to temper stock at 540 °C (1000 °F) 32-mm- (1¼-in.-) diam 1041 stock weighing 6.21 kg/m (4.17 lb·ft). Determine the frequency, power input and length of coil required. Table 23 gives the solution.

Schematic diagrams of 60-Hz installations and medium frequency equipment now being used in induction tempering operations are shown in Fig. 24. The inductors or heating coils employed are primarily of the multiturn type. For a given power density, the coils for lower-frequency installations require a greater number of turns. The application of higher voltages on the coil also requires a greater number of turns for a given power density. When using current directly from the power line at 440 V or more, the frequency is so low and the voltage so high that multiturn, multilayer coils are required for tempering operations. Such coils are expensive to build, and effective coupling is considerably more difficult. Accordingly, it may be advantageous to reduce the voltage to permit the use of a single-layer coil. The coils used with 1000-, 3000- and 10 000-Hz equipment, as well as with higher frequencies, are of the single-layer type. Among medium-frequency units, 10 000-Hz units require the fewest turns for equal applied voltages and power densities.

Fig. 26 Typical hardness survey on 19-mm- (³/₄-in.-) diam 1037, 25-mm- (1-in.-) diam 1041 and 49-mm- (1¹⁵/₁₆-in.-) diam 14B35H steel bars through hardened and tempered by induction

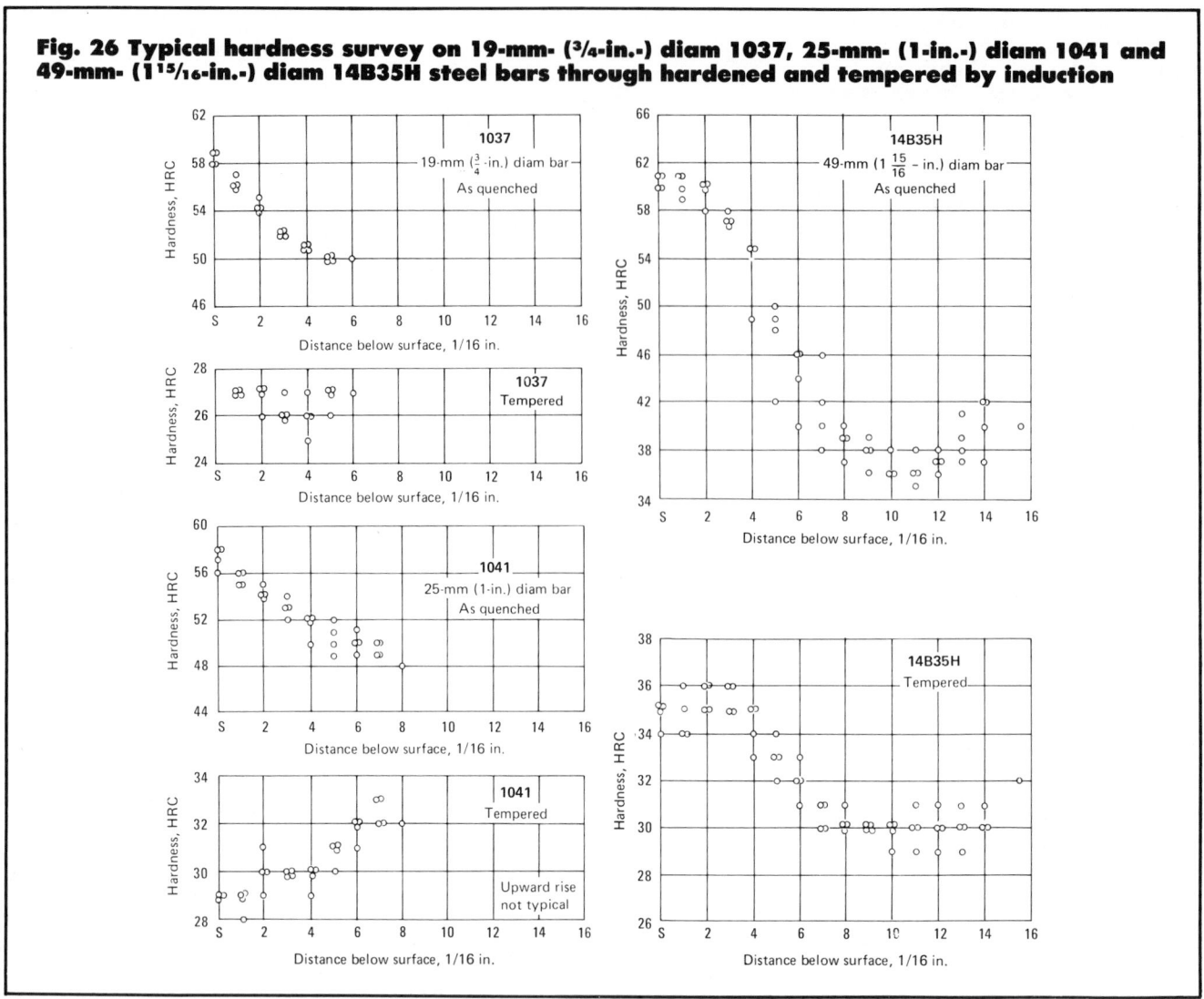

Auxiliary and Control Accessories

Because parts are cooled from the tempering temperature in air, no quenching facilities are required. However, equipment for handling is as important in induction tempering as in hardening; the fixtures already described for hardening are suitable. Frequently, the tempering operation is keyed to the hardening operation, or the same equipment may be used for both induction hardening and tempering either by merely changing the work coil or by reducing the power density and the heating time.

Radiation Pyrometer. In general, the control of induction tempering is achieved by selection of the power density and the rate of feed through the coil (scanning) on the basis of hardness tests of the tempered product. Automatic control may be obtained at tempering temperatures above 425 °C (800 °F) by use of a special radiation pyrometer and high-speed controller. This arrangement may be used to vary the speed of the scanning operation continuously or to control the power.

Voltage Regulator. With 60-Hz installations, an automatic voltage regulator in the power line ahead of the equipment is desirable if automatic control is not used. For adjustment, power input may be varied by one or more of these means:

- A continuously variable transformer
- A tapped transformer
- A saturable reactor
- Taps on the coil itself
- An SCR (silicon-controlled rectifier) controller.

With motor-generator equipment employing automatic voltage regulation, or in scanning uniform work, control of the rectified field excitation has provided good results. Solid state equipment can use either constant-voltage or constant-power regulation, which is built into the unit. SCR controllers are used with vacuum-tube equipment to give uniform results in small sections.

Selection of Induction Tempering Cycle

Fundamentally, the tempering cycle will be determined by the ultimate mechanical properties specified for the part. However, usual furnace tempering temperatures must be increased to compensate for the short heating times.

Figure 25 shows the increase in tempering temperature required to produce a given hardness as the tempering time is decreased from 1 h (furnace tempering) to 60 s and 5 s (induction tempering), in 1050 steel quenched in brine from 855 °C (1575 °F). Pieces with small cross section may be air cooled immediately upon reaching the tempering temperature, while slower heating rates or short periods of time at temperature (5 to 60 s) before cooling are desirable for larger sections, to allow penetration of heat. In scanning, of course, the power density, the rate of travel, and the length of the inductor will determine the time of tempering.

Results of Induction Tempering

Selection of the proper induction tempering cycle and of suitable equipment results in uniform hardness from the surface to center of the part. This is illustrated by Fig. 26, which shows hardness readings on the cross sections of a 19-mm- (3/4-in.-) diam bar made of 1037 steel, a 25-mm- (1-in.-) diam bar of 1041 steel, and a 49-mm- (1 15/16-in.-) diam bar of 14B35H steel. Figure 26 shows the hardness values of these steels in both the as-quenched and induction tempered conditions.

Statistical studies over several years on cross-sectional hardnesses for similar bar stock that was induction hardened and tempered by scanning reveal that 67% of all average cross-sectional hardnesses resulted in a hardness spread of ± 1.25 Rockwell C points and that approximately 95% were within ± 2.5 points. Variations in resultant hardness occur more often from differ-

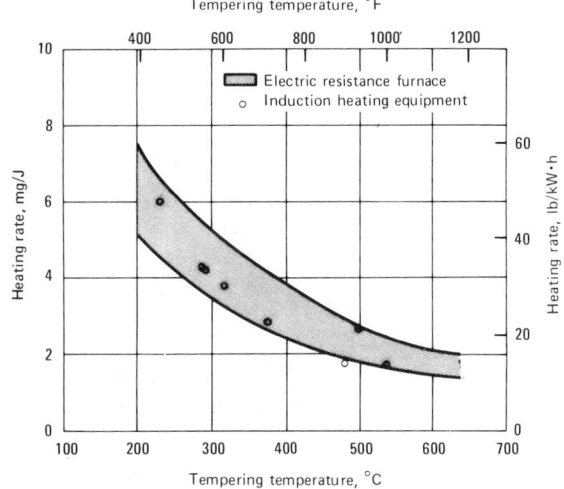

Fig. 27 Operating efficiencies for tempering by induction and electric furnace

Curves represent upper and lower efficiencies expected with electric resistance furnace. Actual operating efficiencies of continuous induction tempering installations are represented by plotted points.

ences in hardenability between respective heats of steel than from discrepancies in the induction tempering process. If each mill heat is run separately, it is possible to compensate for differences in hardenability.

Stress Distribution. In spite of the short tempering times encountered in induction tempering, results indicate that the amount of stress relief is similar for parts uniformly tempered through the cross section if an equivalent hardness is obtained by higher tempering temperatures. After differential or selective tempering, the stress distribution may be substantially different from that achieved in furnace tempering; thus, analysis or a performance test should be made before adopting induction for selective tempering.

Induction Versus Electric-Furnace Tempering. Electric furnaces were

used for tempering many years before induction heating was applied to tempering. For this reason, induction tempering is not normally used unless it proves advantageous over furnace tempering. The two examples above are not intended to imply that induction tempering is the most economical method for all parts. There are many parts which cannot be successfully induction tempered, and in many instances furnace tempering is more economical.

Figure 27 shows relative operating efficiency for electric tempering furnaces and induction tempering equipment. When making a graphical comparison between the two types of heating, allowances should be made for the somewhat higher temperatures that are required when tempering by induction to attain the same hardness range and microstructure.

Flame Hardening

FLAME HARDENING is a heat treating process in which a thin surface shell of a steel part is rapidly heated to a temperature above the critical point of the steel. After the grain structure of the shell has become austenitic, the part is quickly quenched, transforming the austenite to martensite while leaving the core of the part in its original state. In contrast, slow cooling causes transformation, as the temperature passes through the corresponding ranges, to pearlite, bainite and martensite, with the final structure being a combination of the three. The result is a relatively soft and ductile steel. To achieve hardness, therefore, the steel must be cooled rapidly so that it bypasses the first two transformation phases and transforms directly from austenite to martensite. (Ref 1 and 5)

Flame hardening employs direct impingement of a high-temperature flame or of high-velocity combustion-product gases; the part is then cooled at a rate that will produce the desired levels of hardness and other properties. The high-temperature flame is obtained by combustion of a mixture of fuel gas with oxygen or air; flame heads are used for burning the mixture. Depths of hardening from about 0.8 to 6.4 mm ($\frac{1}{32}$ to $\frac{1}{4}$ in.) or more can be obtained, depending on the fuels used, the design of the flame head, the duration of heating, the hardenability of the work material, and the quenching medium and method of quenching used. The process can be used for through hardening of work 75 mm (3 in.) or less in cross section, depending on the hardenability of the steel.

Hardening by flame differs from true case hardening, because the hardenability necessary to attain high levels of hardness is already contained in the steel, and hardening is obtained by localized heating. Although flame hardening is used mainly to develop high levels of hardness for wear resistance, the process also improves bending and torsional strength and fatigue life. One of the major advantages of flame hardening is the ability to satisfy stringent engineering requirements with carbon steels. (Ref 3)

Scope and Application

Flame hardening is applied to a wide diversity of workpieces and ferrous materials (when required mechanical properties can be provided by selective or localized hardening) for one or more of the following reasons:

- Because parts are so large as to make conventional furnace heating and quenching impracticable or uneconomical. Typical examples include large gears, machineways, and large dies and rolls.
- Because only a small segment, section or area of a part requires heat treatment, or because heat treating all over would be detrimental to the function of the part. Typical examples include the ends of valve stems and pushrods, and the wearing surfaces of cams and levers.
- Because dimensional accuracy of a part is impracticable or difficult to attain or control by furnace heating and quenching. A typical example is a large gear of complex design, for which flame hardening of the teeth would not disturb the dimensions of the gear.
- Because the use of flame hardening permits a part to be made from a less costly material, thus effecting an over-all cost saving in comparison with other technically acceptable methods. The process gives inexpensive steels the wear properties of alloyed steels, and parts can be hardened without scaling or decarburization, thereby eliminating costly cleaning operations. For example, a large carburized low-carbon alloy steel part might be made at less cost from a flame-hardened plain carbon steel. (Ref 1)

For a detailed discussion of materials suitable for flame hardening, and for a comparison of flame hardening with other methods of attaining similar results, see the sections "Selection of Process" and "Selection of Material" near the end of this article.

Methods of Flame Hardening

The versatility of flame-hardening equipment and the wide range of heat-

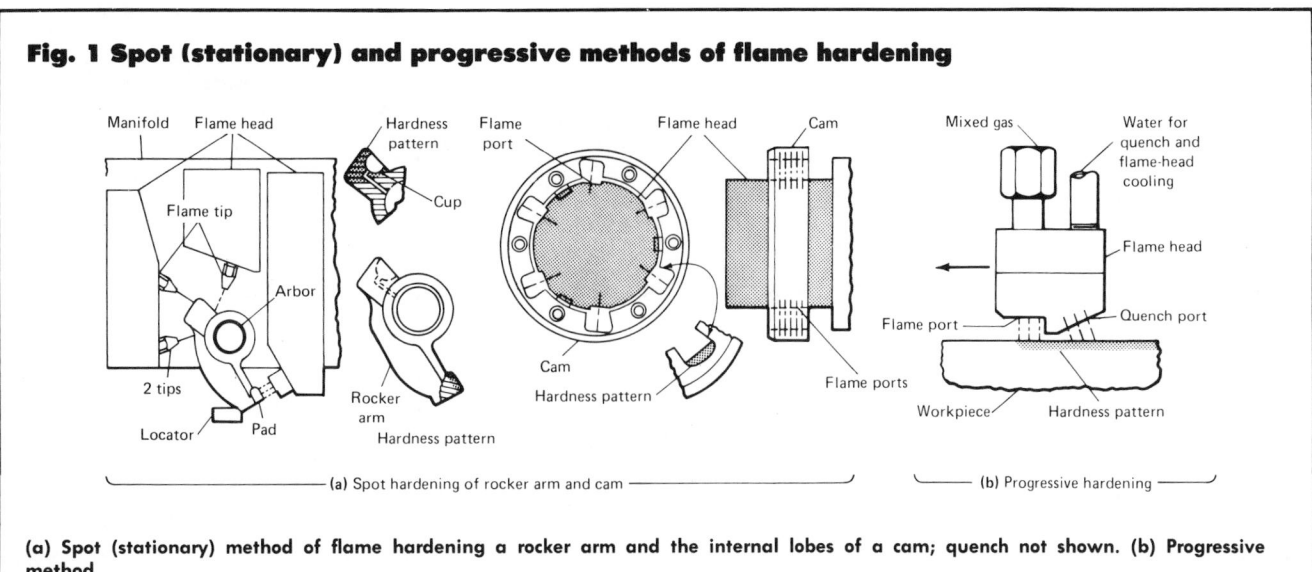

Fig. 1 Spot (stationary) and progressive methods of flame hardening

(a) Spot hardening of rocker arm and cam

(b) Progressive hardening

(a) Spot (stationary) method of flame hardening a rocker arm and the internal lobes of a cam; quench not shown. (b) Progressive method

ing conditions obtainable with gas burners often permit flame hardening to be done by a variety of methods, of which the principal ones are:

- Spot, or stationary
- Progressive
- Spinning
- Combination progressive-spinning.

Selection of the appropriate method depends on the shape, size and composition of the workpiece; the area to be hardened; the depth of case required; and the number of pieces to be hardened. In many instances, more than one method will be capable of providing the desired result; the choice will then depend on comparative costs.

The spot (stationary) method, illustrated in Fig. 1(a), consists of locally heating selected areas with a suitable flame head and subsequently quenching. The heating head may be of either single-orifice or multiple-orifice design, depending on the extent of the area to be hardened. The heat input must be balanced to obtain a uniform temperature over the entire selected area. After being heated, the parts usually are immersion quenched; however, in some mechanized operations, a spray quench may be used.

Basically, the spot method requires no elaborate equipment (except, perhaps, fixtures and timing devices to ensure uniform processing of each piece). However, the operation may be automated by indexing the heated parts into either a spray quench or a suitable quench bath.

The progressive method, illustrated in Fig. 1(b), is used to harden large areas that are beyond the scope of the spot method. The size and shape of the workpiece, as well as the volume of oxygen and fuel gas required to heat the specified area, are factors in selection of this method. In progressive hardening, the flame head is usually of the multiple-orifice type, and quenching facilities may be either integrated with the flame head or separate from it. The flame head progressively heats a narrow band that is subsequently quenched as the head and quench traverse the workpiece.

The equipment needed for flame hardening by the progressive method consists of one or more flame heads and a quenching means mounted on a movable carriage that runs on a track at a regulated speed (flame-cutting machines are adaptable to this type of flame hardening). Workpieces mounted on a turntable or in a lathe can be hardened readily by the progressive method; either the flame head or the workpiece may move. There is no practical limit on the length of parts that can be hardened by this method, because it is easy to lengthen the track over which the flame head travels. Single passes as wide as 1.5 m (60 in.) can be made; wider areas must be hardened in more than one pass.

When more than one pass is required to cover a flat surface, or when cylindrical surfaces are hardened progressively, such surfaces will exhibit soft bands because of overlapping or underlapping of the heated zones. These soft bands can be minimized, however, by closely controlling the extent of the overlapping. (Wherever overlapping occurs, the possibility of severe thermal upset and cracking should be anticipated. Tests should be conducted to determine whether overlapping will cause cracking or other harmful effects.) Simple curved surfaces may be hardened progressively by means of contoured flame heads, and some irregular surfaces may be traversed by the use of tracer-template methods.

The rate of travel of the flame head over the surface is governed mainly by the heating capacity of the head, the depth of case required, the composition and shape of the work, and the type of quench employed. Speeds ranging from 0.8 to 5.1 mm/s (2 to 12 in./min) are typical with oxyacetylene heating heads. Ordinarily, water at ambient temperature is used as a quenchant, although air sometimes is used when a less severe quench is indicated; under special conditions (particularly for quenching alloy steels), warm or hot water or a polymer-base synthetic quenchant also may be employed.

The spinning method (Fig. 2) is applied to round or semiround parts such as wheels, cams or gears. In its simplest form, the method employs a mechanism for rotating or spinning the

Fig. 2 Spinning methods of flame hardening

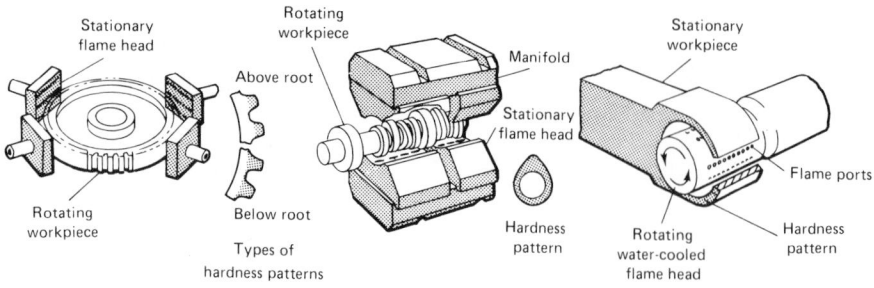

In methods illustrated at left and at center, the part rotates; in method at right, the flame head rotates. Quench not shown

Fig. 3 Combination progressive spinning method of flame hardening

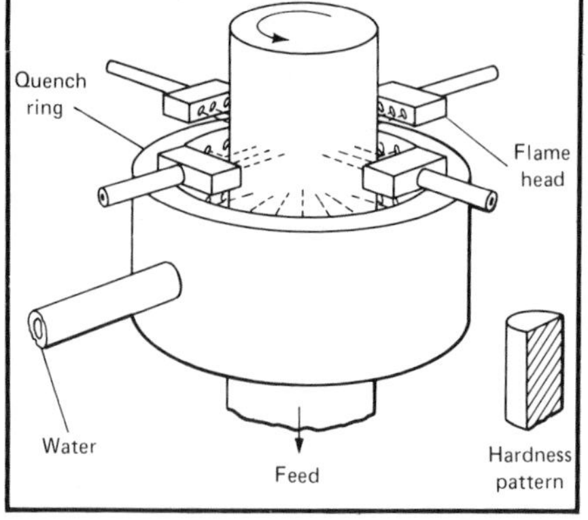

workpiece, in either a horizontal or a vertical plane, while the surface is being heated by the flame head. One or more water-cooled heating heads equal in width to the surface to be heated are employed. The speed of rotation is relatively unimportant, provided uniform heating is obtained. After the surface has been heated to the desired temperature, the flame is extinguished or withdrawn and the work is quenched by immersion or spray, or a combination of both.

The spinning method is particularly adaptable to extensive mechanization and automation; this makes it possible, for example, for all the cams on a camshaft to be hardened at the same time.

Today, fully automatic flame-heating equipment is available that can treat round components up to 1.5 m (59 in.) in diameter and up to 2 t in weight. Much of it has been designed to treat gear wheels of all types. (Ref 4)

Commercial machines have been built that provide automatic control of timing, temperature and quenching, as well as accurate control of gas flow, so that close metallurgical specifications can be met consistently. Frequently, when production is sufficient, the spinning method can be set up so that parts either are loaded manually and unloaded automatically, or are both loaded and unloaded automatically.

This method has been extended to components of irregular cross section and mass distribution. Typical are large drive wheels for tracked vehicles, cams and camshafts for marine diesel engines and crane traveling wheels. Speed, deep hardness penetration, localized hardness zone, and uniformity

of hardness pattern are the main advantages. (Ref 4)

In one technique, a rotating flame head is used for internal spin hardening of odd-shaped parts that would present handling problems if the parts themselves were rotated. Each part is positioned by a simple handling device, and the flame head rotates inside the part.

In contrast to the progressive method, in which acetylene usually is used (because of its high flame temperature and rapid heating rates), satisfactory results can be obtained in spin hardening with natural gas, propane or manufactured gas. The choice of gas depends on the shape, size and composition of the workpiece and on the depth of case required, as well as on the relative cost and availability of each gas.

A wide choice of quenchants also is possible in the spin hardening method. Because the flame is extinguished or withdrawn before the part is quenched, any appropriate quenchant may be used for immersion quenching. In spray quenching, the quenchant is usually water, a water-base liquid such as soluble oil, or a simulated oil in the form of a polymer-base quenchant; air also has been used.

The combination progressive-spinning method (Fig. 3), as the name implies, combines the progressive and spinning methods for hardening long parts such as shafts and rolls. The workpiece is rotated as in the spinning method; but in addition, the heating heads traverse the roll or shaft from one end to the other. Only a narrow circumferential band is heated progressively as the flame head moves from

one end of the work to the other. The quench follows immediately behind the heating head, either as an integral part of the head or as a separate quench ring.

This method provides a means of hardening large surface areas with relatively low gas flows. Progressive-spinning units designed to handle a broad range of diameters and lengths are available commercially.

Fuel Gases

Several different fuel gases are used in flame hardening. In selecting a fuel gas for a given application, the required rate of heating and the cost of the gas must be considered along with the initial cost of equipment and maintenance.

Flame hardening does not alter the composition of the base metal if done properly. Carburizing, neutral and oxidizing flames can be used. Oxidizing flames have high oxygen ratios and can be detrimental because they produce extremely hot temperatures that can cause decarburization and overheating. A carburizing flame can prevent some decarburization but also can introduce unwanted carbon into the surface. For best results, neutral or slightly carburizing flames should be used. (Ref 1)

A comparison of the rates of heating of fuel gases can be made when certain fundamental properties of usable mixtures with oxygen are known. A parameter that correlates well with actual heating speed is "combustion intensity", or "specific flame output". This is the product of the normal velocity of burning multiplied by the net heating value of the mixture of oxygen and fuel

Table 1 Fuel gases used for flame hardening

Gas	Heating value MJ/m³	Heating value Btu/ft³	Flame temperature With oxygen °C	°F	With air °C	°F	Usual ratio of oxygen to fuel gas	Heating value of oxy-fuel gas mixture MJ/m³	Btu/ft³	Normal velocity of burning mm/s	in./s	Combustion intensity(a) mm/s× MJ/m³	in./s× Btu/ft³	Usual ratio of air to fuel gas
Acetylene	53.4	1433	3105	5620	2325	4215	1.0	26.7	716	535	21	14 284	15 036	...
City gas........	11.2-33.5	300-900	2540	4600	1985	3605	(b)	(b)	(b)	(b)	(b)	(b)	(b)	(b)
Natural gas (methane)	37.3	1000	2705	4900	1875	3405	1.75	13.6	364	280	11	3 808	4 004	9.0
Propane........	93.9	2520	2635	4775	1925	3495	4.0	18.8	504	305	12	5 734	6 048	25.0

(a) Product of normal velocity of burning multiplied by heating value of oxy-fuel gas mixture. (b) Varies with heating value and composition.

gas. Thus, a knowledge of these two properties often permits selection of the most suitable fuel gas for a specific hardening speed and depth of case. The fuels of greatest commercial interest are ranked by combustion intensity (at metallurgically suitable ratios of mixture with oxygen) in the following order: acetylene, propane, methane. Values of normal burning velocity and the heating values of metallurgically suitable mixtures are listed in Table 1.

The time required for heat penetration is another good criterion for judging the heating qualities of a fuel provided that all other variables remain constant. Figure 4 shows comparative heating times for stabilized MAPP* (methylacetylene propadiene), acetylene and propane using an efficient coupling distance for each fuel. These curves show that a greater depth of hardness can be obtained with MAPP in a shorter length of time. (Ref 6)

The ratio of oxygen to fuel is very important in obtaining maximum heating efficiency from the fuel. However, oxygen-to-fuel ratios should not be confused with oxygen and fuel consumption rates, which vary with flame velocity, port size and heating time. Stoichiometrically, acetylene requires 2½ mols of oxygen per mol of gas for complete combustion, and MAPP and propane require 4 and 5 mols, respectively. With acetylene, however, only 1 to 1.5 volumes of oxygen is provided directly, and the remainder is drawn from the surrounding atmosphere. Neither MAPP nor propane has sufficient heat for flame hardening unless more oxygen is supplied, normally at a rate of 4 parts of oxygen per part of fuel. MAPP burns over a rather wide range of oxygen-to-fuel ratios, however, and

*MAPP is a registered trademark of The Dow Chemical Co.

thus permits a wider range of heat output while still providing high heat generation when necessary. Use of MAPP increases oxygen consumption but decreases fuel consumption, and fuel is usually more expensive than oxygen. (Ref 6)

Bulk systems of supply for oxygen and fuel gases greatly reduce their cost, but of greater importance is the elimination of cylinder handling and of the residual losses usually attending the use of gases in cylinders. Bulk systems also provide a more nearly constant supply of gas at uniform pressure.

Depth of Heating. Shallow hardness patterns (less than 3.2 mm or 0.125 in. deep) can be attained only with oxy-gas fuels. The high-temperature flames obtained with oxy-gas fuels provide the fast heat transfer necessary for effective localization of the heat pattern. Deeper hardness patterns permit the use of either oxy-gas fuels or air-gas fuels. Oxy-gas fuels will localize the heat, but care is required in their application to avoid overheating the surface during development of the deeper-seated heat. Air-gas fuels, with their slower rates of heat transfer (lower flame temperatures), minimize or eliminate surface overheating but generally extend the heat pattern beyond the desired hardness pattern. For this reason, air-gas flame hardening is generally limited to steels of shallow hardenability. In this manner, the hardness pattern is controlled by the quench rather than by the heating. The deeper-seated heat produced by air-gas flames may preclude the use of air-gas mixtures, because excessive distortion may occur. In consideration of these factors, the use of air-gas heating will depend primarily on the shape of the part insofar as the configuration favors heat localization and a lower rate of heat transfer.

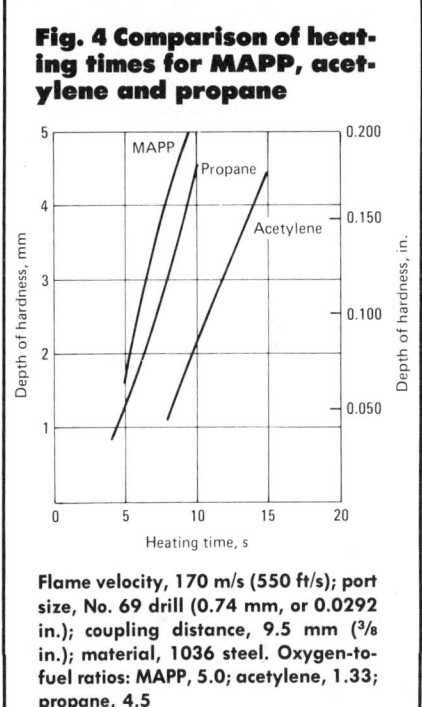

Fig. 4 Comparison of heating times for MAPP, acetylene and propane

Flame velocity, 170 m/s (550 ft/s); port size, No. 69 drill (0.74 mm, or 0.0292 in.); coupling distance, 9.5 mm (³/₈ in.); material, 1036 steel. Oxygen-to-fuel ratios: MAPP, 5.0; acetylene, 1.33; propane, 4.5

Gas Consumption, Time and Speeds. Gas consumption in flame hardening varies with the thickness of the case to be obtained; increasing or decreasing the depth of hardening increases or decreases the amount of gas used. Massive parts increase gas consumption because of their greater internal cooling effect. In order to take advantage of the maximum flame temperature from the oxy-fuel gas flame, the distance from the end of the inner cone of the flame to the work should be 1.6 mm (¹/₁₆ in.).

The speed of travel of the flame head over the work in the progressive method, as well as the time of heating for the spot and spinning methods, will

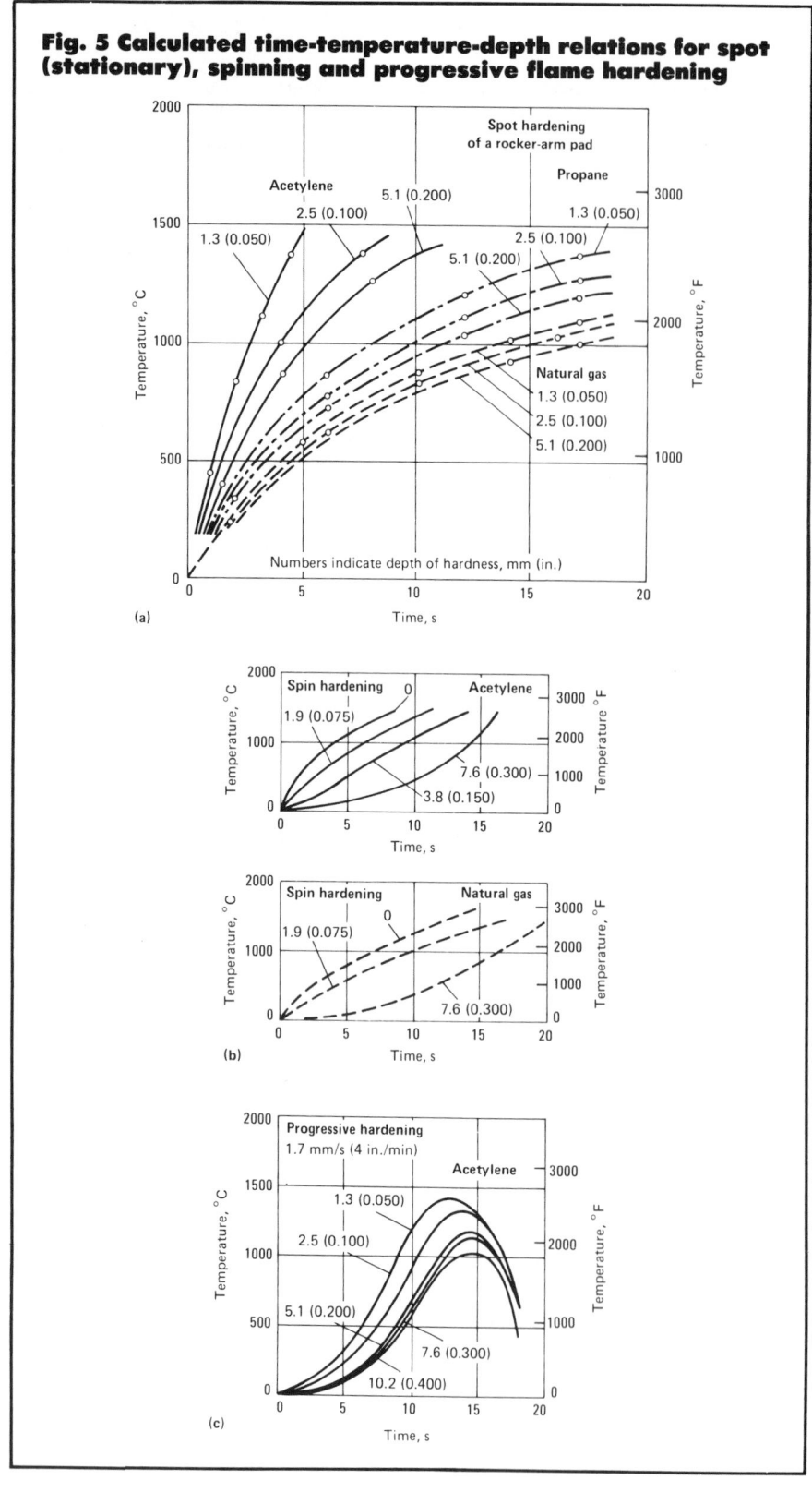

Fig. 5 Calculated time-temperature-depth relations for spot (stationary), spinning and progressive flame hardening

speeds usually vary between 0.8 and 5.1 mm/s (2 and 12 in./min) for most applications, although very thin parts may require speeds of 42 mm/s (100 in./min) or more, to avoid overheating or burning. Because of the intense heat involved, the necessity for accurate control of the rate of travel in the progressive and progressive-spinning methods cannot be overemphasized.

Time-temperature-depth relations for various fuel gases employed in the spot (stationary), spinning and progressive methods of flame hardening are shown in Fig. 5. The family of curves for the spot method (Fig. 5a) was obtained analytically by considering the flow of heat in three dimensions away from a heated spot on the surface of a rocker-arm pad. The calculations were based on heat sources of different strengths, which were varied on the basis of combustion intensity of the gases considered (acetylene, propane and natural gas). It is understood that strength of heat source will be affected also by such factors as size of tip, distance from tip to work, total gas flow, and ratio of oxygen to fuel gas; hence, these curves are intended to indicate trends in time-temperature-depth relations rather than to provide operational values for particular applications.

The curves for the time-temperature-depth relationships of fuel gases used in the spinning method, shown in Fig. 5(b), were obtained analytically by considering the flow of heat into a cylindrical body from a heat source supplying heat uniformly over the outer cylindrical surface. It was assumed that the temperature at the axis would not be raised appreciably during surface heating and that, in spin hardening, the cylinder would be rotated rapidly enough to give the effect of substantially uniform surface heating. No temperature decrease at high values of time is indicated, because in this type of process the body is quenched immediately when the surface temperature reaches a predetermined value.

The curves for the progressive method (Fig. 5c) also were obtained analytically by considering heat flow into a body from a line source moving along a flat side of the body. The heat-source strength and body configuration were chosen to be representative of progressive flame hardening. In this instance, the time variable can be correlated with travel speed if the width of the flame zone is known. For example,

vary with the thickness of case desired and the capacity of the flame head. The proximity of the quench spray to the last row of flames will affect the speed somewhat in the progressive method. Progressive and progressive-spinning

a flame zone 25 mm (1 in.) wide will pass over a point on the work surface in 15 s when the travel speed of the tip is 1.7 mm/s (4 in./min). This width of flame zone indicates the probability of a multirow tip and, if the heat source is as strong as assumed for computing the curves, will undoubtedly damage the work surface. Consequently, a higher travel speed would be used—for example, 2.5 mm/s (6 in./min), which would heat a point on the work for 10 s and result in hardening to a depth of about 2.5 mm (0.100 in.). The rapid decrease in temperature at large values of time is due to the mass quenching effect, which, in practice, would be augmented by use of water spraying or other quenching means.

Burners and Related Equipment

Burners are basic components of equipment for all methods of flame hardening. Burners vary in design, depending on whether they are fired by an oxy-fuel or an air-fuel gas mixture. Flame temperatures obtained by oxy-fuel gas combustion are 2540 °C (4600 °F) and higher. Heat transfer is by direct impingement of the flame on the surface of the workpiece. For this reason, oxy-gas burners are more commonly referred to as "flame heads". Flame temperatures obtained by air-fuel gas combustion are considerably lower (Table 1), and heat transfer is by impingement of high-velocity combustion-product gases (no direct flame) or by radiation from an incandescent refractory surface.

There is no universal flame head, and no flame head is designed specifically for one particular gas. A well-designed flame head can be used with MAPP, acetylene or propane, for example. Better flame-head design usually results in improved operation and lower gas consumption.

In general, a large number of small ports will produce a more efficient heat pattern than a few larger ports. Port spacing of 2.3 to 3.2 mm (0.090 to 0.125 in.) on centers is, in most cases, advisable. Counterboring permits higher flame velocities and frequently is advantageous or necessary when propane is used. Because acetylene has a higher flame-propagation rate, counterboring usually is unnecessary and frequently is undesirable. Counterboring reduces coupling distance and permits a stabilized flame at higher flame velocities.

For effective operation, the ratio of counterbore area to throat area should be on the order of 2 to 1. Ratios as high as 4 to 1 may be used in some cases, however. (Ref 6)

Oxy-Fuel Gas Flame Heads. Oxy-fuel gas combustion develops flame temperatures above those at which useful metals and refractory materials can survive. Accordingly, the flame head is designed in such a way as to provide a flame pattern that avoids any direct heating of its metal parts.

Generally, the flame head consists of a tube or a shell with one or more orifices drilled into it; the number and arrangement of orifices depend on the required area of heat coverage. Flame heads for use with oxy-fuel gas are shown in Fig. 6.

The drilled-face flame head has a limited range of application and usually is designed to meet the requirements of one specific part to be flame hardened in an appreciable quantity. For other applications, the flame head may be fitted with removable tips of the screw-in or insert type.

The screw-in type of removable tip shown in Fig. 6 is used widely enough so that it is available in off-the-shelf commercial flame heads or supplied as a standard item by manufacturers of flame-hardening equipment; it is also simple enough that it can be made to specification by plants applying the flame-hardening treatment.

The insert, or press-in, type of tip shown in Fig. 6 is smaller than the screw-in type and permits closer spacing of orifices, approaching that of a drilled flame head. Flame heads with removable and replaceable tips can be used over a wider range of applications by removing one or more tips and replacing them with plugs.

The heat output of the flame head is governed by the number and size of the orifices, when all other factors are equal. The individual orifices range from No. 73 to No. 51 drill size (0.6 to 1.7 mm, or 0.024 to 0.067 in., in diameter). Such small holes can readily become plugged, and when this occurs the flame head will not function properly. Flame heads with removable tips have an advantage over fixed-tip heads in that oversize or out-of-round holes, caused by mechanical or flame damage, can be corrected without replacing the head.

Integral parts of the flame system are the mixer block and mixer tube, which mix the component fuel gases

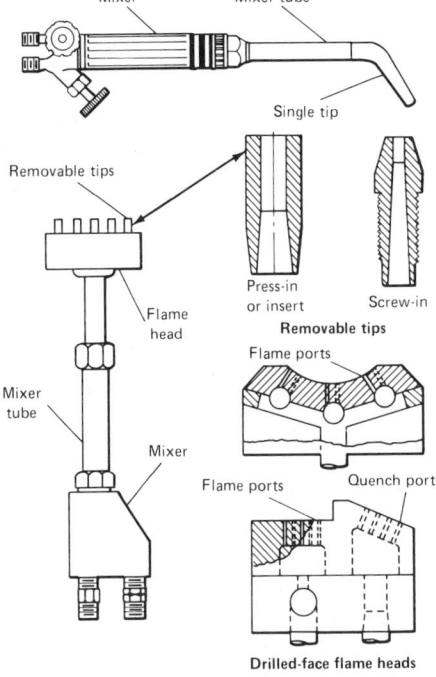

Fig. 6 Flame heads for use with oxy-fuel gas

and convey them through the orifices (Fig. 6). The capacities of mixer block and mixer tube must match the number and size of the orifices; if the mixer is too small the flame will flash back, and if the mixer is too large the flame head will not function efficiently.

To ensure identical velocity of mixed gases at all flame orifices, it is common practice to design a flame head with baffle orifices through which the gases must pass before they are burned at the flame orifices. Two rules apply to the design of baffle orifices: (a) their total area must be 1.25 to 1.50 times the area of the flame orifices, and (b) the number of baffle orifices within a single baffle should be one-fourth the total number of flame orifices.

Multiple-orifice flame heads are water cooled, because otherwise the high temperatures developed at and around the flame head would cause early deterioration. On flame heads used for progressive hardening, the quench water cools the head. On multiple-orifice flame heads used in the spinning and progressive-spinning methods, the cooling water is circulated through chambers integrated into the head. Single-orifice flame heads (welding torches, for example) generally are not water cooled.

Fig. 7 Typical flame-hardening installations using oxy-fuel gas mixtures

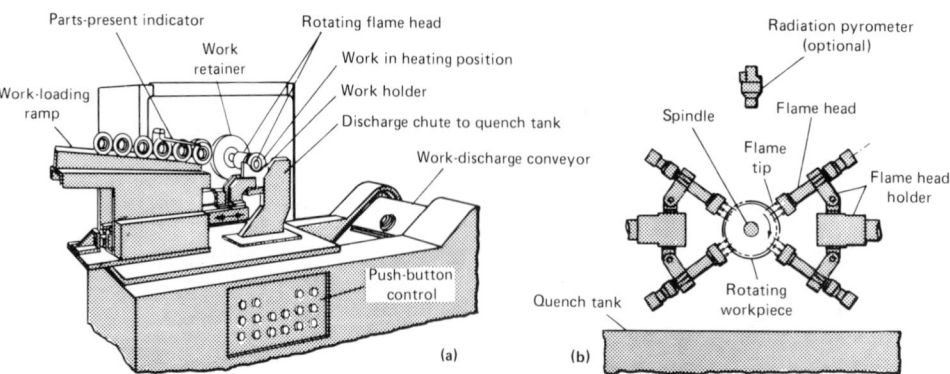

(a) Installation for high production of similar parts: hardening the 54-mm (2-¹/₈-in.) bores of hubs to a depth of about 3.2 mm (¹/₈ in.). Machine has a standard retractable spindle adapted with a rotary flame head. Spindle is driven by a variable-speed motor. Temperature of agitated quench is maintained by a water-cooled heat exchanger. (b) Installation for selective oxy-fuel gas heating of small production lots of gears, sprockets and flanges within size limits of the equipment. A radiation pyrometer is used here to control heating cycle, but many operations employ an electric timer instead. By changing work heads and spindles, equipment can be adapted to various parts.

Typical flame-hardening installations that use oxy-fuel gas mixtures are shown in Fig. 7. The equipment in Fig. 7(a) was designed to handle high production of similar parts. The equipment in Fig. 7(b) was designed to harden a variety of parts by changing the flame heads and work spindle.

New equipment configurations are being designed to handle specific problems. With conventional flame-hardening equipment, for example, it is difficult to obtain a zone of uniform temperature across the surface of the tooth gap in gears. The problem of obtaining uniform depth of hardness, not only from tooth to tooth but also across the entire face width, has been solved by development of a two-chamber burner with individual control of energy input to each chamber. The system ensures uniform heating, particularly in the tooth root, and incorporates separate control of nozzles for tooth-root and tooth-flank hardening. (Ref 4)

With this system, complex components also can be processed by spin hardening. For example, both the straight gear and bevel gear of a double pinion shaft can be simultaneously hardened, if the quenching-bath immersion depth is adequate. By distributing heating power between spur wheel and bevel wheel, austenitizing temperature can be reached under the tooth roots of both gears at the same time. This requires selection of different heating-up times for the two gears

Fig. 8 Typical burners for use with air-fuel gas

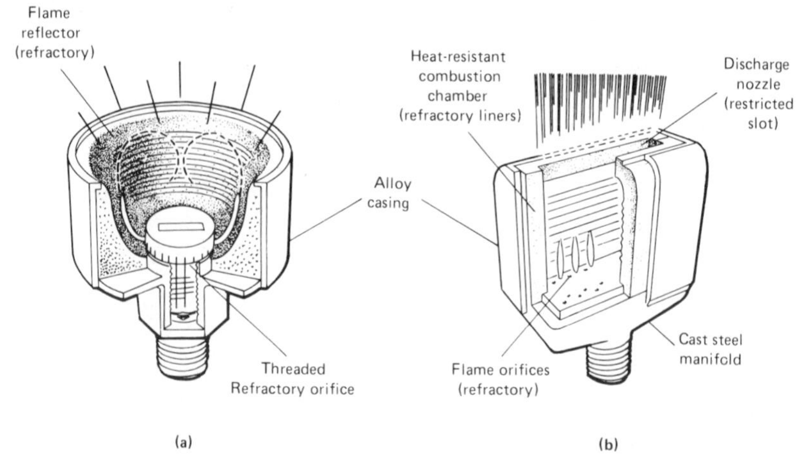

(a) Radiant type. (b) High-velocity convection type (not water cooled)

and different rates of oxygen flow to the two sets of burners. (Ref 4)

Air-Fuel Gas Burners. Air-fuel gas combustion develops lower flame temperatures that are compatible with available refractory materials. Thus, burners are designed with the aim of completely utilizing the heat generated. The burners, incorporating heat-resistant refractory liners, are of two types, generally designated as either the radiant type or the high-velocity convection type (Fig. 8).

The radiant-type burner (Fig. 8a) is essentially a refractory cup in a protective metal casing. Air-gas premixture is supplied through the pipe at the back and passes through an accurately molded ceramic tip that is screwed into the pipe and located at the bottom of the cup. With numerous narrow slots molded into its periphery, the tip functions essentially as the distributing head of a multiport burner. The many small flames wash the inner surface of the cup, making it highly incandescent for rapid heat transfer by radiation. Because combustion is completed within the cup, the burner may be positioned close to the work with no flame impingement.

The standard radiant burner employed in flame hardening is approximately 75 mm (3 in.) in cup diameter. It is particularly effective for spin hardening the teeth of large gears. A single row of burners may be arranged in a ring surrounding a gear, as shown in Fig. 9(a), or multiple rows of burners may be arranged to cover completely the surface to be hardened.

The high-velocity convection burner is basically a miniature refractory-lined furnace in which heat is released at rates as high as 414 MJ/m³·s (4 × 10⁷ Btu/ft³·h). The air-gas premixture is supplied through a pipe connection and flows through the orifices of the ceramic port plate. The burner design is such that the burning gases heat the chamber lining to a temperature approaching the theoretical flame temperature. This permits preheating of the reacting gases and accelerates combustion. In this manner, gases at approximately 1650 °C (3000 °F) are discharged through a restricted slot opening to impinge on the workpiece at velocities up to 760 m/s (2500 ft/s). High-velocity convection burners are well adapted to localized heating of parts in spin-hardening operations. Figure 9(b) shows its application for hardening the teeth of a thin gear.

Related Equipment. Burners for both the oxy-fuel gas and the air-fuel gas methods of flame hardening are implemented by pressure regulators, valves, flowmeters and protection devices. For the air-fuel gas method, a separate mixer and compressor are utilized, because the mixing function is not incorporated within the burners (Fig. 10).

Materials of Construction for Burners. Metallic components of burners may be of a variety of materials, depending on the type of burner and the service to which it is subjected. The flame heads for oxy-fuel gas flame heating are in nearly all instances made of copper or a dense grade of lead-free brass (Ref 6). These metals are relatively inexpensive, have excellent thermal conductivity and are readily machined. Because flame velocity is relatively low, there is little likelihood that the metal flame head will overheat and deteriorate as a result of the backwash of hot gases. Normally,

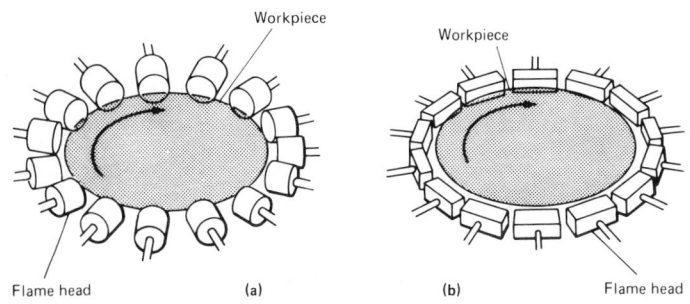

Fig. 9 Setups for flame hardening gears, idler wheels and sprockets

(a) Radiant burners. (b) High-velocity convection burners; wide-face parts can be heated with double or staggered rings of burners.

the flow of gases within the head provides sufficient cooling to maintain a safe temperature. If additional cooling is required, the head may be water cooled. Under no circumstances should the flame head be permitted to overheat to an extent that will cause damage to the burner metal, brazed burner joints, or drilled flame ports.

Life of oxy-fuel gas flame heads depends largely on the conditions of a particular application, such as containment of the heat resulting from the configuration of the piece being heated, and the degree of backwash of flame or hot gases impinging on the burner. Therefore, it is difficult to predict the exact life to be expected for a new application. Brass heads used for progressive hardening of machineways have shown an average life of 1000 to 2000 h in continuous service.

Because the melting points and ranges of heat resistance of available materials are not greatly different (compared with the large differences between flame temperatures and metal melting points), life expectancy is a relatively unimportant basis for selection of a particular metal. Availability, cost, and ease of fabrication are likely to be more important factors in material selection. A metal frequently used for flame heads is Muntz metal (60-40 lead-free brass); when cooling is a problem, however, copper is preferable.

Screw-in tips usually are made of copper, but when higher heat resistance is required, because of secondary flame, Monel K-500 can be used. Insert or press-in tips usually are made of copper, but when high wear resistance of the tip bore is desired, tips made of sintered carbide are employed.

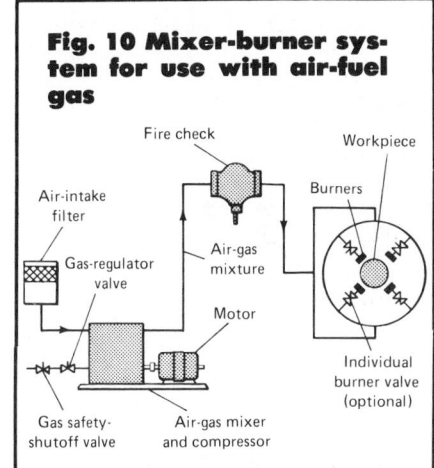

Fig. 10 Mixer-burner system for use with air-fuel gas

Cast iron, stainless steel and heat-resisting alloys are used for the casings that surround the refractory cups of radiation-type air-fuel gas burners. Heat-resisting alloys may provide reasonable life at temperatures up to 1150 °C (2100 °F). Casings of high-velocity convection burners generally are constructed of stainless steel, but may be of Inconel for more severe service. For spin hardening of gears, either type of burner may have a useful life of 10 to 15 40-h weeks of continuous operation. In many applications, longer life of casings has resulted when the operator has exercised care in setting up and operating burners. Also, water cooling of the burner casing can markedly extend its life, and may be employed provided that the added cost is justifiable and that space for the cooling arrangement is available in the heating setup.

The refractory parts of all types of

air-gas burners are formed in steel dies and prefired in kilns, and are capable of withstanding the most extreme temperatures generated in the burners. These parts generally fail from thermal or mechanical shock; hence, service life is unpredictable.

Operating Procedures and Control

The success of many flame-hardening applications depends largely on the skill of the operator. This is especially true when the volume of work is so small or so varied that the cost of automatic-control equipment is not justified.

The principal operating variables are:

- Distance from inner cone of oxy-fuel gas flames, or from air-fuel gas burner, to work surface
- Flame velocities and oxygen-to-fuel ratios (Ref 2)
- Rate of travel of flame head or work
- Type, volume and angle of quench (Ref 1).

These variables must be closely controlled to ensure duplication of desired surface hardness and depth of hardness. It is highly desirable to develop a specific procedure for each item to be flame hardened. The procedure is developed by preliminary tests on the production piece itself, if warranted, or on mock-up sections of approximately the same cross section as the production piece. After the desired contour and depth of hardened zone have been developed, the procedure is applied to production pieces and, when established, is made a part of the heat treating specification.

The critical importance of developing strict procedures is demonstrated in the following case study conducted by a flame-hardening firm. Each of five experienced flame hardeners was asked to flame harden a 1045 steel bar 25 by 50 by 450 mm (1 by 2 by 18 in.), using only experience and visual examination to guide the process. Only the traditionally and normally controlled variables were preset: coupling distance, 11 mm (⁷/₁₆ in.); water pressure, 620 kPa (90 psi); quench, water; and angle of quench, 30°. Flowmeter readings were taken, speeds of travel recorded, and flame velocities determined. After treatment, each bar was ground to determine hardness level

and depth of hardness. Results are given in Table 2. Surface hardness and depth of hardness showed little consistency, ranging from 50 to 61 HRC, and from 30 to 52 HRC, respectively. (Ref 2)

Speed of travel of the flame head, or the duration of heating, should be held constant for uniform results. In the progressive method, the flames gradually heat the workpiece in front of the flame head, and sometimes this effect must be compensated for by gradually increasing the speed of travel or by precooling. At the beginning of a pass when the progressive or progressive-spinning method is being used, the flame head or heads should be manipulated or otherwise adjusted to ensure that the beginning of the area to be hardened attains the proper temperature and depth of heating as progression begins.

Gas Pressures. Oxy-fuel gas and air-fuel gas pressures should be controlled closely for uniform input of heat. Flat oxy-fuel gas flame heads are somewhat less efficient when used on circular or curved surfaces, because each cone of flame is at a different distance from the work. Overheating causes cracking.

Oxygen-to-fuel ratio is a key factor in determining flame temperature. For example, propane produces flame temperatures of 2700 °C (4900 °F) at a 5-to-1 ratio, 2540 °C (4600 °F) at a 4-to-1 ratio, and 2340 °C (4300 °F) at a 3-to-1 ratio.

Flame velocity is one of the most important variables, because when balanced with other variables it is the main determinant of case depth. In flame hardening 1045 steel, the basic flame velocity required for establishing a case depth of 3.2 to 4.8 mm (⅛ to ³/₁₆ in.) is 152 m/s (500 ft/s). (Ref 2)

In flame hardening cast irons, where high surface temperatures are undesirable, use of lower flame velocities is distinctly advantageous. In this regard, both MAPP and propane are easy to

Table 2 Results of operator skill test

Operator	Flame velocity m/s	Flame velocity ft/s	Oxygen-to-fuel ratio	Travel speed mm/s	Travel speed in./min	Surface hardness, HRC	Hardness at 3.18 mm (0.125 in.), HRC
1..........	95	313	3.6:1	3.4	8.0	61	30
2..........	99	324	3.1:1	2.5	6.0	50	41
3..........	137	451	3.1:1	1.7	4.0	57	50
4..........	124	407	4.2:1	1.9	4.5	55	38
5..........	156	511	3.3:1	2.5	6.0	60	52

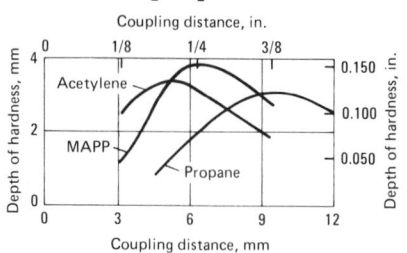

Fig. 11 Effect of coupling distance for MAPP, acetylene and propane

Flame velocity, 170 m/s (550 ft/s); port size, No. 69 drill (0.74 mm, or 0.0292 in.); material, 1052 steel. Oxygen-to-fuel ratios: MAPP, 4.5; acetylene, 1.33; propane, 4.5. Heating times: MAPP, 6 s; acetylene and propane, 8 s

control and are effective over a wide range of velocities. The ability to control the flame over a wide velocity range not only provides flexibility but also makes the operation much safer and results in better heat-pattern control. (Ref 6)

Coupling distance is another key parameter in flame hardening. The relations between coupling distance and depth of hardening for three fuel gases are illustrated in Fig. 11. In general, the coupling distance for MAPP is equal to or very slightly greater than that for acetylene. The effect of counterboring is a shortening of the coupling distance. By the same token, coupling distance increases as flame velocity increases. Efficient coupling distances for MAPP generally range from 6.4 to 9.5 mm (¼ to ⅜ in.) depending on gas velocity and port size. A counterbored port of No. 69 drill size (0.74 mm, or 0.0292 in.) will operate efficiently at a coupling distance of 4.8 mm (³/₁₆ in.). (Ref 6)

Hardening temperatures can be judged by competent operators, but to the inexperienced the heated metal will appear colder than it actually

Table 3 Procedure for spin flame hardening a small converter gear hub (Example 1)

Gear hub, of 1052 steel, shown in Fig. 15(a)

Preliminary operation

Turn on water, air, oxygen, power and propane. Line pressures: water, 220 kPa (32 psi); air, 550 kPa (80 psi); oxygen, 825 kPa (120 psi); propane, 205 kPa (30 psi). Ignite pilots.

Loading and positioning

Mount hub on spindle. Hub is held in position by magnets. Flame head previously centered in hub within 0.4 mm ($^1/_{64}$ in.). Distance from flame head to ID of gear teeth, approximately 7.9 mm ($^5/_{16}$ in.).

Cycle start

Spindle with hub advances over flame head and starts to rotate. Spindle speed, 140 rpm.

Heating cycle

Propane and oxygen solenoid valves open (oxygen flow delayed slightly). Mixture of propane and oxygen ignited at flame head by pilots. Check propane and oxygen gages for proper pressure. Adjust flame by regulating propane. Heating cycle controlled by timer. Time predetermined to obtain specified hardening depth. Propane and oxygen

Heating cycle (cont.)

solenoid valves close (propane flow delayed slightly). Spindle stops rotating and retracts. Hub stripped from spindle by ejector plate. Machine ready for recycling.

Propane regulated pressure, 125 kPa (18 psi); oxygen regulated pressure, 550 kPa (80 psi); oxygen upstream pressure, 400 kPa (58 psi); oxygen downstream pressure, 140 kPa (20 psi). Flame velocity (approx), 135 m/s (450 ft/s). Gas consumption (approx): propane, 0.02 m³ (0.6 ft³) per piece; oxygen, 0.05 m³ (1.9 ft³) per piece. Total heating time, 9.5 s.

Flame-port design: 12 ports per segment; 10 segments; port size, No. 69 (0.74 mm, or 0.0292 in.), with No. 56 (1.2-mm, or 0.0465-in.) counterbore.

Quench cycle

Hub drops into quench oil, is removed from tank by conveyor. Oil temperature, 54 ± 5.6 °C (130 ± 10 °F); time in oil (approx), 30 s.

Hardness and pattern aim

Hardness, 52 HRC min to a depth of 0.9 mm (0.035 in.) max above root of gear teeth.

Table 4 Progressive flame hardening of ring-gear teeth (Example 2)

Large bevel ring gear, sketched below, was made of 8742 steel and had 90 teeth. Diametral pitch, 1.5; face width, 200 mm (8 in.); OD, 1.53 m (60.412 in.)

Mounting

Gear mounted on holding fixture to within 0.25 mm (0.010 in.) total indicator runout.

Flame heads

Two 10-hole, double-row, air-cooled flame heads, one on each side of tooth. Flame heads set 3.2 mm (1/8 in.) from tooth.

Operating conditions

Gas pressures. Acetylene, 69 kPa (10 psi); oxygen, 97 kPa (14 psi)
Speed. 1.9 mm/s (4.5 in./min). Complete cycle (hardening pass, overtravel at each end, index time, preheat return stroke on next tooth), 2.75 min
Indexing. Index every other tooth. Index four times before immersing in coolant.
Coolant. Mixture of soluble oil and water, at 13 °C (55 °F)
Hardness aim. 53 to 55 HRC

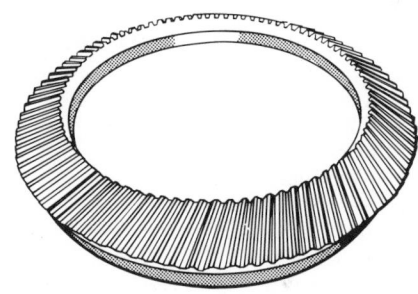

is, because of the light from the burning gases; the consequent tendency is to overheat unless the operator is equipped with didymium-tinted glasses. Radiation or optical pyrometers are often used to judge more accurately the temperatures developed. Radiation-pyrometer systems of fast response are used extensively to control work temperatures and heating times. Metallurgical examination is the best method for establishing operating conditions. Overheated spots may appear under individual flames, but with properly designed flame heads and good scanning technique, this effect is minimized and not readily detectable by microscopic examination.

Examples. The interrelation of variables can best be illustrated by specific examples of procedures used for production hardening of actual parts. The following six examples involve hardening of gears, cams, shafts and flat surfaces.

Example 1 (Table 3). A small gear hub made of 1052 steel was flame hardened by the spinning method at a spindle speed of 140 rpm. The aim was to harden to 52 HRC min to a depth of 0.9 mm (0.035 in.) max above the root of the gear teeth.

Example 2 (Table 4). A large bevel ring gear of 8742 steel was flame hardened by the progressive method. Hardness aim was 53 to 55 HRC.

Example 3 (Table 5). A free-wheel cam made of 1062 steel was flame hardened by the spot method. Aim was 60 HRC min at the surface and 59 HRC at a depth of 1.3 mm (0.050 in.), for a width of 8.8 mm (0.345 in.) on the roller surface of the cam.

Example 4 (Table 6). A 52100 steel shaft was flame hardened by the progressive method to 61 to 63 HRC.

Example 5 (Table 7). The roller path of a large ring gear was hardened according to the procedure indicated.

Hardness aim was 515 to 600 by portable Brinell, as specified in the last item in Table 7.

Example 6 (Table 8). Drop forged wear blocks of 1040 steel were flame hardened on a conveyor to 53 to 58 HRC.

Preheating

In flame hardening parts of large cross section, difficulty in obtaining the desired surface hardness and hardness penetration often can be overcome by preheating. When the available power or heat input is limited, the hardened depth can be increased by preheating. (Ref 3)

The hardness data in Fig. 12 illustrate the effectiveness of preheating in

Table 5 Spot flame hardening of a 1062 steel free-wheel cam (Example 3)

Preliminary operation

Turn on water, air, oxygen, power and propane. Line pressures: water, 205 kPa (30 psi); air, 550 kPa (80 psi); oxygen, 825 kPa (120 psi); propane, 205 kPa (30 psi). Ignite pilots.

Loading and positioning

Mount cam on flame head. Cam positioned on locating plate and two wear pads, and against three locating pins that are integral parts of flame head. Distance from flame head to cam surface, approximately 7.9 mm (5/16 in.).

Cycle start and heating cycle

Propane and oxygen solenoid valves open (oxygen flow delayed slightly). Mixture of propane and oxygen ignited at flame heads by pilots. Check propane and oxygen pressures. Adjust flame by regulating propane. Heating cycle controlled by timer. Time predetermined to obtain specified hardening depth. Propane and oxygen solenoid valves close (propane flow delayed slightly). Ejector plate (air-operated) advances and strips cam from flame head.

Propane regulated pressure, 125 kPa (18 psi); oxygen regulated pressure, 585 kPa (85 psi); oxygen upstream pressure, 425 kPa (62 psi); oxygen downstream pressure, 110 kPa (16 psi). Gas consumption (approx): propane, 0.01 m³ (0.4 ft³) per piece; oxygen, 0.04 m³ (1.3 ft³) per piece. Flame velocity (approx), 135 m/s (450 ft/s). Total heating time, 11 s.

Flame-port design: 9 ports per row; 8 rows; port size, No. 69 (0.74 mm, or 0.0292 in.), with No. 56 (1.2-mm, or 0.0465-in.) counterbore.

Quench cycle

Cam drops into quench oil, is removed from tank by conveyor. Oil temperature, 54 ± 5.6 °C (130 ± 10 °F). Time in oil (approx), 30 s.

Hardness and pattern aim

Hardness, 60 HRC min at surface, and 59 HRC min at a depth of 1.3 mm (0.050 in.) below surface, for width of 8.8 mm (0.345 in.) on cam roller surface.

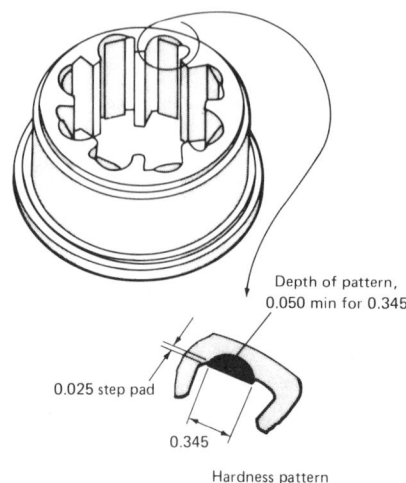

Depth of pattern, 0.050 min for 0.345

0.025 step pad

0.345

Hardness pattern

Table 6 Progressive flame hardening of a 52100 steel shaft (Example 4)

Setup

Shaft held vertically between centers to within 0.1 mm (0.005 in.) total indicator runout.

Flame heads

Three water-cooled flame heads, each with 8 holes 1.3 mm (0.052 in.) in diameter. One torch for each flame head. Flame heads set 6.4 mm (1/4 in.) from shaft.

Operating conditions

Gas pressures. Acetylene, 55 kPa (8 psi); oxygen, 76 kPa (11 psi)
Speeds. Flame-head travel, 1.3 mm/s (3 in./min); speed of shaft revolution, 20 rpm
Quench. Water nozzles directly below flame heads
Hardness aim. 61 to 63 HRC

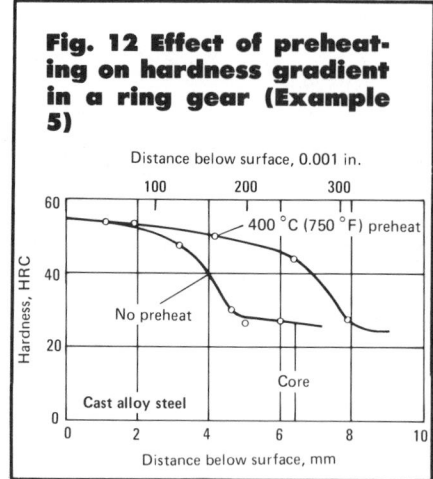

Fig. 12 Effect of preheating on hardness gradient in a ring gear (Example 5)

developing hardness penetration in the ring gear dealt with in Example 5 (sketch and operating details, Table 7).

Flame hardening of prehardened and tempered steels, especially some of the alloy steels, requires careful control of heating to avoid cracking. Preheating of the part may be advisable to minimize cracking in such steels. Hardenable cast iron also is susceptible to cracking. In one application, cast iron crane wheels had to be preheated to 480 °C (900 °F) to prevent rupture of the spokes caused by nonuniform expansion during spin heating of the tread area.

In another application, the difficulty of heat treating the internal teeth of a planetary-gear housing was overcome by spin hardening. The component had an irregular mass distribution, which led to large distortions when heat treating was done by ordinary flame-hardening or induction-hardening equipment. Case hardening of teeth also failed. The problem was solved by using a specially designed gear-wheel spin-hardening machine with close control over heating time and power and by preheating the part with the largest mass and cross section to an accurately predetermined temperature. (Ref 4)

Depth and Pattern of Hardness

In some instances, the flame-hardening procedure results in a greater depth of hardness than desired. Figure 5(a), for example, shows that spot hardening a rocker-arm pad after heating 4 s at 870 °C (1600 °F) with oxyacetylene flame produces a depth of hardness of 5.1 mm (0.200 in.). If excessive, this depth may be reduced by heating the steel to the same temperature but in a shorter time. Thus, with a reduction in heating time to 3.2 s at 870 °C, the depth of hardening will be 2.5 mm (0.100 in.) (Fig. 4a). Because of the shorter heating period, the cost of hardening each pad also will be reduced.

The problem of excessive depth resulting from a spin-hardening operation may be resolved similarly, as shown in Fig. 5(b). In this operation, heating with an oxyacetylene flame for 13.5 s at 870 °C (1600 °F) produces a hardness depth of 7.6 mm (0.300 in.).

Table 7 Flame hardening of roller path on side of large ring gear (Example 5)

Mounting on positioner

Gear should be centered on positioner in a vertical position, to approximately 1.7 mm (0.065 in.) total indicator runout on tooth OD (see sketch below).

Preheating

Flame-head gap. Set 75-mm (3-in.) max flame head at 190-mm (7½-in.) gap from flame-hardened surface and approximately 610 mm (24 in.) ahead of hardening tip.

Gas pressures. Acetylene, 89 kPa (13 psi); oxygen, 185 kPa (27 psi)

Flowmeter. 2.1 divisions for acetylene; 2.65 divisions for oxygen

Tube readings (neutral flame). Flutter, approximately ⅛ division, plus or minus

Speed setting. No. 9 notch on positioner; equivalent to 2.7 mm/s (6.42 in./min) on 5.79-m (228-in.) circumference of 1.84-m (72½-in.) flame-hardened pitch circle (approx 35½ min/rev)

Heat cutoff. Kill preheat when spread of flame toward oncoming hardened surface produces blue-purple temper color at flame-hardening starting point. (Air-block flame head may be set up for the same purpose.)

Flame hardening

Heating-head gap. (CAUTION: Must be kept parallel with flame-hardened surface.) Set and hold 125-mm (5-in.) flame head at 13-mm (½-in.) gap from flame-hardened surface and at an angle of about 35 to 40° to radius; adjust during flame hardening to follow in line with roller-path warp.

Gas pressures. Acetylene, 89 kPa (13 psi); oxygen, 185 kPa (27 psi)

Flowmeter. 2.7 divisions for acetylene; 3.2 divisions for oxygen

Tube readings (neutral flame). Normal flutter, approximately ⅛ division; watch frequently for signs of gas failure.

Water pressure (flame-hardening quench). Set pressure-reducing valve for 205 kPa (30 psi).

Speed setting. Same as for preheating. (NOTE: Double-check setting just before flame hardening.)

Heat cutoff. (No overlap permissible; aim at 13-mm (½-in.) max soft gap; use extreme caution.) As flame head approaches finish junction, back off tip 1.6 mm (¹⁄₁₆ in.) gradually during final 13 mm (½ in.) of gear travel and kill flame *abruptly* when nearest row of flames *just reaches* 6.4-mm (¼-in.) clearance from starting point of flame hardening, and immediately speed up table to fastest-speed notch.

Hardness aim (except at flame-hardening junction). 515 to 600 on portable Brinell at points 50 mm (2 in.) from ID of flame-hardened roller path (50 HRC at a depth of approximately 5.1 mm, or 0.2 in.); total depth of contour, 7.9 mm (⁵⁄₁₆ in.)

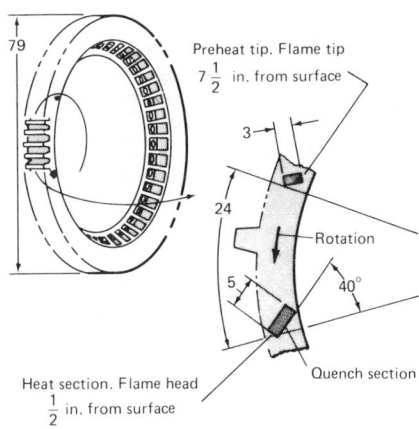

Preheat tip. Flame tip 7½ in. from surface

Rotation

40°

Quench section

Heat section. Flame head ½ in. from surface

Gear is in vertical position during hardening.

Schematic above shows ring gear and position of flame heads for hardening roller path (located on side face of ring) by procedure summarized in table. Gas cost was approximately 25% of total processing cost. Hardening extended to a depth of 4.8 mm (³⁄₁₆ in.). Hardness at surface, 530 HB. For hardness gradients, see Fig. 12.

Table 8 Flame hardening of 1040 steel wear blocks (Example 6)

Preliminary operation

Clean blocks of scale and rust, preferably by sand or shot blasting.

Loading

Load blocks on one end of conveyor belt.

Flame head

Head contains two rows of No. 54 drill size (1.4-mm, or 0.055-in.) flame holes; total of 49 holes, of which 24 are plugged; 150 mm (6 in.) between centers of end holes. Head also contains single row of water-quench holes. Head set at 16-mm (⁵⁄₈-in.) total gap; cone-point clearance of flame, 4.8 mm (³⁄₁₆ in.)

Gas pressures

Acetylene, 83 kPa (12 psi); oxygen, 150 kPa (22 psi)

Speed

Conveyor speed, 2.47 mm/s (5.83 in./min); total flame-hardening time, 1.5 min per pad

Hardness and pattern aim

Hardness, 53 to 58 HRC; total depth of hardening to core, 4 mm (⁵⁄₃₂ in.)

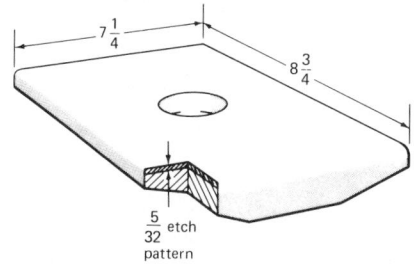

7¼

8¾

⁵⁄₃₂ etch pattern

shows the variation in depth of hardness obtained in flame hardening hubs made of 1062 steel from three different heats. The inside diameters of these hubs were flame hardened for 12 s to produce a minimum hardness of 59 HRC at a depth of 1.9 mm (0.075 in.).

Surface Hardness Pattern. The hardness pattern of any portion of a flame-hardened surface can be determined by lightly blasting the area with fine sand. The hardened portion of the surface will be less affected than the unheated area by the cutting action of the sand. This procedure also may be employed to indicate soft spots when the entire surface has been subjected to flame hardening. (Soft spots may result from nonuniform heating or the interference of scale on the surface.) Another procedure for determining surface

---

Heating at the same temperature for a period of 8 s results in a hardened depth of about 3.8 mm (0.150 in.).

In a progressive-hardening operation with a rate of travel of 1.7 mm/s (4 in./min), a hardness depth of 5.1 mm (0.200 in.) will be produced when the heating time is 12 s at 870 °C (1600 °F), as shown in Fig. 5(c). The flame head contains two rows of flame holes that

produce a heating zone 20 mm (0.8 in.) wide. If the rate of travel is increased 20% to 2 mm/s (4.8 in./min), the heating time is proportionately reduced to 10 s. This results in a hardened depth of about 2.5 mm (0.100 in.).

Variation in depth of hardness may occur in steels of the same nominal composition but from different heats. This is demonstrated in Fig. 13, which

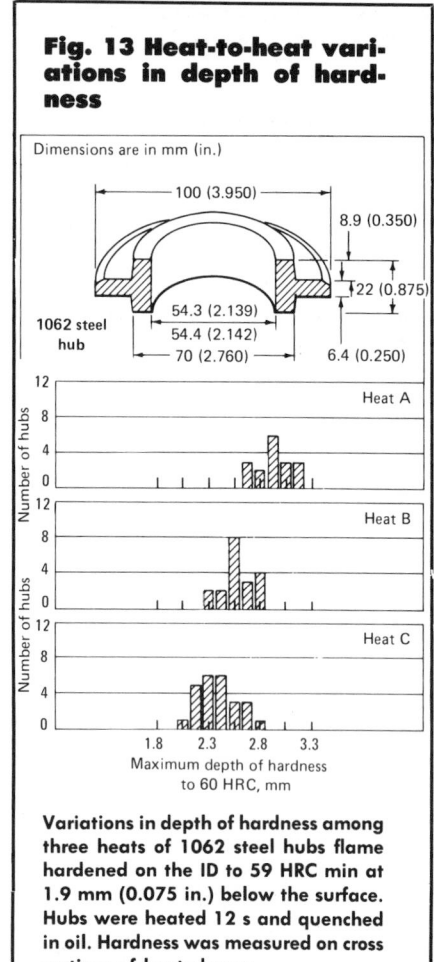

Fig. 13 Heat-to-heat variations in depth of hardness

Dimensions are in mm (in.)

Variations in depth of hardness among three heats of 1062 steel hubs flame hardened on the ID to 59 HRC min at 1.9 mm (0.075 in.) below the surface. Hubs were heated 12 s and quenched in oil. Hardness was measured on cross sections of heated area.

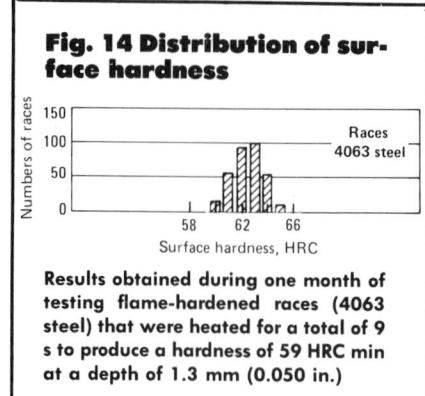

Fig. 14 Distribution of surface hardness

Results obtained during one month of testing flame-hardened races (4063 steel) that were heated for a total of 9 s to produce a hardness of 59 HRC min at a depth of 1.3 mm (0.050 in.)

hardness pattern is etching of the area with a 10% nitric acid solution. The hardened area will appear darker than the unhardened area.

These procedures are applicable also to cross sections of hardened areas for an indication of heat penetration, and they are useful as preliminary checks on the performance of the flame-hardening setup.

Variations in the hardness of the surface are exemplified by the data shown in Fig. 14. These hardness readings taken on the surface of races made of 4063 steel cover a one-month period. The races were heated for 9 s to produce a minimum hardness of 59 HRC at 1.3 mm (0.050 in.) below the surface.

Summary. Hardness of the case in flame hardening is a function of the carbon content of the steel and will range up to 65 HRC. Medium-carbon steels with 0.40 to 0.50% carbon are ideal for flame hardening, but steels with carbon contents as high as 1.50%

also can be flame hardened with special care. Normally, hardening depth ranges from 1.3 to 6.4 mm (0.05 to 0.25 in.). Heavier sections, such as large rolls and wheels, can have case depths of up to 13 mm (0.5 in.). Manganese-bearing alloys aid in the depth of hardening by decreasing the critical cooling rate which contributes to deep hardening. Therefore, manganese and free-machining grades of steel are considered excellent for flame hardening. (Ref 1)

When hardening depths are required beyond the capabilities of ordinary carbon steels (0.60 to 0.90% Mn), elevated manganese ranges such as 0.80 to 1.10%, 1.00 to 1.30% or 1.10 to 1.40% can be efficiently used. Wear resistance in many cases is not the only critical design criterion. Under high compressive loading, the hardened layer must be deep enough not only to provide the required wear life of the part, but also to contribute to the support of heavy contact loads. The case must be fully martensitic, and the material supporting the hardened layer must be of sufficient strength. However, increased hardenability may lead to cracking problems, at least with water quenching. (Ref 3)

Maintenance of Equipment

Flame Heads (Nonferrous) of the Oxy-Fuel Gas Type. Experience with nonferrous flame heads of three types of construction (drilled face, screw-in tips, and press-in tips) has shown carbon deposit, lack of proper cooling, erosion and corrosion to be the main causes of maintenance problems common to all.

Carbon Deposit. The intermittent

igniting and extinguishing of the flame causes a small deposit of carbonaceous material to build up on the sidewalls of the port, from the retrogression of the burning fuel below the orifice of the port as the flame is extinguished. Although small for each cycle, the deposit builds up gradually until it causes a restriction in the port and thus a variation in the velocity of the gas and therefore in the heating performance. Several thousand cycles may be completed before it is necessary to clean the head.

A slight amount of cleaning can be done without removing the head from the mixer tube, by pushing through the ports a wire of the proper size or a drill one size smaller than the ports. However, care must be exercised to prevent bell-mouthing or increasing the size of the port. Cleaning by this method deposits loose carbonaceous material in the fuel chamber, and this material may be moved by the velocity of the oxygen and fuel gas mixture and become lodged in other ports.

Flame heads operating on volume production should be removed from the machine and thoroughly cleaned after 40 to 120 h of operation; the interval depends on the installation, the oxygen and fuel gas mixture, and the heating cycle time.

Drilled-port and insert-port heads, when removed, can be cleaned by inserting in the ports a wire of the proper size or a drill one size smaller than the ports, and then immersing the heads in a suitable solvent or cleaner that will loosen carbonaceous deposits. Again, care must be exercised to prevent bell-mouthing or enlarging of the port. Also, the cleaner or solvent must not attack the head itself. After cleaning and rinsing in clean water, all traces of dirt and moisture should be blown from the ports, fuel chamber and water lines. Screw-in tips and heads may be cleaned in the same manner.

Erosion and Corrosion. All types of heads are subjected to the by-products of combustion, which may attack the face of the burner or the tips, causing erosion and corrosion. A flash plate of chromium on the burner face has proved helpful in reducing erosion and corrosion. In time, the face of a drilled-port burner may erode to a point at which it is necessary to reface the head by machining and reworking the counterbore. If there is enough space between ports on drilled-port heads, the port can be repaired by drilling oversize and pressing in an insert. These inserts can be removed when necessary and replaced.

Because the oxy-fuel gas type of flame head is water cooled, it may be necessary, in localities where extremely hard water is supplied, to soften the water to keep scale from forming in the cooling passages; commercial water softeners of the zeolite or polyphosphate type may be used. Recirculating systems with heat exchangers are sometimes required.

Burners of the Air-Gas Type. Failures of air-gas burner liners occur by spalling, erosion and cracking of the refractory cup or tunnel and outlet. These liners should be inspected periodically, at intervals determined by experience with the individual installation; when conditions are severe, inspection may be required as often as once every shift.

The causes of deterioration of burner liners are thermal shock from repeated and rapid heating or cooling (or both), and mechanical shock from work striking the burners. The burner casings deteriorate mainly from prolonged exposure to escaping hot gases that heat the casings locally, causing more or less progressive oxidation, cracking, growth and, infrequently, burning. If installation space permits the use of water or air cooling, these usually can be expected to increase the life of burners. Shielding, designed to conduct hot gases away from the burners and other parts of the machine, also is valuable in reducing maintenance costs.

Mechanical Components. Because both oxy-fuel and air-fuel gas systems employ pressure gages, pressure regulators, valves, flowmeters and safety devices, the maintenance problems associated with these accessories are common to both methods of flame hardening. Symptoms and probable causes of trouble with several mechanical components used in flame hardening are listed in Table 9.

Air-gas systems also have maintenance problems related to their compressors and blowers. Devices for protection against backfire and explosion should be serviced according to manufacturers' instructions.

Electrical Components. Electrical contacts should be cleaned periodically; when the heating cycle is controlled by a timer, periodic checks of the timer are advisable. Generally, if the heating cycle is inaccurate, it will be evident in the inspection of the parts. If the heating cycle is controlled by a thermopile, several items must be checked: (a) excessive flow of cooling water through the housing will cause condensation on the lens; (b) a deposit of any type on the lens will cause erratic results; (c) in high-production work, the lens should be cleaned at least twice each shift, and the temperature recorder should be balanced once each shift.

Spindle and Movable Holding Fixtures. It is important that positioning of the parts and the flame head be consistent. Therefore, holding fixtures and spindles should be periodically cleaned, checked and lubricated. Any worn part should be replaced before it becomes improperly located.

Piping. Types of piping and fittings recommended by underwriters should be used in installing the gas lines from the gas source to the site of use. In long gas lines, means should be provided for purging accumulated condensate from the line before the gas reaches the mixing chamber or torch.

Preventive Maintenance

The following is a typical schedule of preventive maintenance for flame-hardening equipment:

Daily

- Check all solenoids for leaks; clean, repair or replace, as necessary.
- Replace damaged screw-in flame tips; clean dirty or clogged tips or ports.
- Remove foreign material from surfaces of flame heads and movable machine parts.
- Check for stable pilots.
- Check for leakage of fuel gas, water, oxygen and air.
- Check level, circulation and temperature of quench oil.

- Clean lens of radiation pyrometer, and balance recorder.
- Check temperature of water-cooled parts.
- Check oil level and flow of oil to spindle.

Weekly

- Check liquid level in backflash arrestor.
- Check relief valve on backflash arrestor.
- Lubricate all movable parts as required.
- Check all fuel-gas, oxygen and air connections for leaks (use soap solution).

Bimonthly

- Remove and clean flame heads (frequency will vary with installation and use).

Semiannually or annually

- Remove, clean and repair oxygen and fuel-gas regulators.
- Remove, clean and repair all solenoids.
- Check electrical contacts and wiring.
- Clean heat exchanger and quench tank.
- Check orifice in mixing blowpipe.
- Repack motor bearings.

Safety Precautions

All fuel gases are explosive when mixed with either air or oxygen within their flammable limits. However, they are widely used throughout industry and are safe when properly handled, stored and transported in compliance with established laws of safe practice. Stabilized methylacetylene-propadiene combines the safety and handling ease of propane with the high heat energy of acetylene. (Ref 6)

The following organizations have

Table 9 Maintenance problems in flame-hardening equipment

Component	Symptom of trouble	Probable causes
Pressure gage	Erratic behavior	Mechanical damage to mechanism
Gas-supply pressure regulator	Failure to hold pressure setting at outlet	Diaphragm broken or hardened
		Diaphragm too taut
		Sticking stem or valve parts
		Damaged or dirty valve seat or plugged orifices or vents
		Broken or damaged spring
Solenoid valve	Failure to close or open	Defective valve mechanism or wiring
Fuel valve	Flame lingers after shutoff	Dirty seat or stem; damaged seat
Oxygen valve	Flame popping after shutoff	Damaged seat

published information on cylinder storage, cylinder manifolding, acetylene generators, housing and piping systems, equipment and operating procedures:

American Gas Association
1515 Wilson Blvd.
Arlington, VA 22209

American Insurance Association
(formerly National Board of Fire Underwriters)
85 John St.
New York, NY 10038

Compressed Gas Association
500 Fifth Ave.
New York, NY 10036

Industrial Risk Insurers
(formerly Factory Insurance Association)
85 Woodland St.
Hartford, CT 06102

National Fire Protection Association
470 Atlantic Ave.
Boston, MA 02210

Most state and local governments follow the recommendations of these organizations in adopting regulations pertaining to gases. A study of local regulations should be made to determine whether there is any variance from standard procedures.

Operators should be taught to keep equipment clean, free of grease and oil, and in good condition, and to avoid leaks of oxygen or fuel gas by regularly testing for them with soapy water free of oil and grease. They should open oxygen or fuel-gas valves slowly and be sure that the lines are purged before igniting. An air hose should be available for dispersing fuel gas that has been bled from the lines. Any spaces in and around the machine in which gases might accumulate should also be purged with the air hose. No flame or other source of ignition should be allowed in proximity to the area where fuel gas or oxygen is released into the air, and good ventilation is of primary importance.

Operators should recognize backfires or backflashes immediately and shut off the gases; the cause and extent of damage must be determined, and corrections made, before the torch is reignited. To prevent major explosions, a backflash arrestor must be installed in the fuel-gas line.

Hand friction lighters should not be used to ignite the gases, except for very small flame heads of low gas capacity. A flame from an acetylene torch or a burner using some other gas should be used. The pilot flame should be very close to the heating head, to prevent accumulation of gas and oxygen before ignition. For automatic flame-hardening machines, the pilot light can be permanently mounted (in which case a flame detector and alarm system should be installed to give warning when the pilot light is extinguished accidentally), or electric-spark ignition can be used. When operations are stopped for any length of time, the valves in the main supply lines should be closed. These valves must not leak, because even a small leak can cause a dangerous accumulation of gas.

Oil or grease may ignite violently in the presence of oxygen under pressure and must be kept away from cylinders, regulators, hoses and other apparatus.

The cylinders in a manifold should be secured with clamps or chain to prevent them from tipping or falling. A fall or any severe bump can detonate acetylene, with disastrous consequences. Propane is not likely to detonate, but any gas under pressure is dangerous if the valve is damaged or knocked off the cylinder. In some instances, noise can be a problem in a gas-air system. Equipment should be designed to keep the noise level within generally accepted limits, to avoid injury to nearby personnel.

Quenching Methods and Equipment

The proper application of a suitable quench in flame hardening is as important as proper heating. The quench must remove heat at a rate that will produce the desired structure and assist in controlling the depth of the hardened case.

Method of quenching and type of quenchant vary with the method of flame hardening. In spot hardening, immersion quenching is generally used, but spray quenching may be used.

Quenching After Progressive Heating. Parts heated by the progressive method usually are quenched by a spray integrated into the flame head, although for steels of high hardenability, or where it is desirable to vary or adjust the distance between the heated zone and the quench, a separate spray-quench head is sometimes used. The spray quench should issue from the head at an angle away from the flame head, to prevent interference with heating, and must provide full coverage over the heated band. The integrated type commonly sprays the quenching medium on the work 19 to 32 mm ($\frac{3}{4}$ to $1\frac{1}{4}$ in.) behind the last row of flames. When the surface to be hardened by the progressive method is vertical or overhead, additional cooling may be required after the usual spray quench.

Quenching After Spin Heating. Parts heated by the spinning method can be quenched by several different procedures. In one, the heated part is removed from the heating area and quenched by immersion in a separate quench tank. Another method integrates quenching by making the quench tank a part of the flame-hardening machine. Parts may be heated on the machine spindle or arbor and then be quenched by being dropped into a quench tank located immediately below the arbor. Parts that are either too heavy or too fragile to be dropped may be lowered into the quench by means of arms, or may be lowered into the quench while still on the arbor. Parts may be removed manually from the quench or be removed automatically by means of a conveyor belt. A spray-quenching ring sometimes is submerged in the quench bath to increase cooling rate.

Parts heated by the spinning method may be quenched also with "quench blocks" on the same plane as the flame heads. When the heating cycle is completed and the flames are extinguished, the quench is turned on. The quench blocks should cover the heated band and provide enough quenching liquid to obtain suitable and uniform hardness. There should be enough quench points around the periphery of the block to envelop the area completely with quenchant. Water or a polymer quench solution is usual for this type of quench, because of the hazard of contaminating the heating heads with oil. When oil quenching is used, the oil should not contact the heating head. This may be avoided by indexing the heated part to a separate quenching station or by retracting the heating heads into a shielded area.

When air-fuel gas mixtures are employed in spin-hardening operations, the heat pattern developed, in many instances, extends beyond the limits desired. Immersion quenching in such instances would potentially extend the hardening into areas requiring subsequent machining or contribute to an excessive amount of distortion. Localized quenching, such as that provided by quench rings, is a solution to this problem. For example, this quenching method was used in hardening the teeth of large sprockets weighing about 205 kg (450 lb). The entire tooth surface, including the root, was hardened to a depth of 4.8 mm (³/₁₆ in.). The 10-min heating cycle completely heated the teeth and backup rim. When quenching was done by rotation of the sprocket in a ring of water-spray nozzles, hardening was confined to the desired surfaces and the backup rim remained soft. A similar practice is employed for hardening of large gears.

Parts heated by the combination progressive-spinning method usually are quenched by a spray integrated in the flame heads or by separate quench blocks located below the flame heads.

Quenching Media

With spray quenching, either integrated in the flame head or by separate quench blocks, water or a dilute polymer solution is used as the quenching medium. Quenching oils should not be allowed to come in contact with oxygen or to contaminate equipment using it.

As with conventional hardening, alloy content of the steel determines the type of quench that should be used; the quenchant may be water, a brine solution, the glycol-base polymer quench, or air. (Ref 5)

By reducing the pressure of the quenchant, the rate of cooling by spray quenching can be reduced from the maximum for which the integrated or separate quench blocks are designed. Increasing the distance between the last row of flames and the point at which the quenchant impinges will allow the mass of metal below the area to be hardened to extract heat and thus will decrease the severity of quenching.

Quenching distance is another factor the operator must consider, because the quenchant must strike the heated area while it is still at the critical temperature to avoid formation of pearlite or other undesirable transformation products in the microstructure. If the spray head is angled too close to the flame, spotty hardness and blowouts can occur. If the angle of spray is too far away,

the case may not be fully hardened. (Ref 1)

Forced Air. In progressive hardening operations, forced air frequently is used as the quenchant for steels normally considered to be oil-quenching steels. Water is not used immediately after heating, because the fast cooling action would result in surface checks. Because most of these steels have relatively low Ar₃ transformation temperatures, the forced air quickly reduces the surface temperature to a point at which water can be applied without causing surface checks. The resulting hardness usually is close to that obtained with a direct oil quench. For example, 52100 steel quenched with forced air and then with water attains a surface hardness of 60 to 61 HRC.

Forced-air quenching is used also in applications where intermediate hardnesses are required. One example is the hardening of railroad rail ends to reduce "end-batter" by the impact of car wheels. The rails are indexed under the burners in four preheat stations and one high-heat station (high-temperature, high-velocity burner), which heats the rail to the hardening temperature of 870 °C (1600 °F) in 95 s. In a sixth station, air at 690 kPa (100 psi) is directed against the heated surface. This treatment results in a hardened structure of fine pearlite that provides wear resistance and sufficient ductility to withstand the impact of the moving wheels. Water is unsuitable in this application, because the high-carbon rail steel is susceptible to cracking. Rail lengths for curves, crossovers and switches are similarly hardened; however, for these, hardening is done progressively, utilizing oxy-gas burners and air quenching.

Immersion quenches vary in type in relation to the metal used, the hardness and depth desired, and the mass, design, and dimensional tolerances of the part. The quenching medium can be a caustic or brine solution, water, a soluble-oil emulsion, or any of a large variety of oils or simulated oils such as polymer quenches.

Self-Quenching. During any type of flame hardening other than through hardening, the mass of cold metal underneath the heated layer aids quenching by withdrawing heat. Thus, cooling rates are very high compared with rates for conventional quenching. (Ref 3) During progressive hardening of the teeth of gears made from medium-carbon steels such as 4140, 4150, 4340

or 4640, for example, the combination of rapid heating and the temperature gradient between the surface and the interior of the gear results in a subsequential self-quench that is equivalent to quenching in oil.

For uniform quenching, a coolant can be used at some distance from the tooth being heated. In one application involving hardening the teeth of 4150 steel gears in the size range of 4 to 12 pitch, one tooth at a time was heated and cooled by conduction. The coolant was directed at the body of the gear. With this procedure, the hardness of the teeth was 50 to 55 HRC after the gears had been tempered at 205 °C (400 °F).

Flame Hardening Problems and Their Causes

Problems that may occur in flame hardening, and their causes, are listed below.

Overheating

- Controlling pyrometer set too high
- Millivolt compensator of controlling pyrometer set incorrectly
- Heating cycle too long
- Flame heads too close to work
- Oversize flame ports
- Excess oxygen in flame
- Excessive fuel-gas pressure
- Improper pattern of flame tip

Hardness below minimum required

- Controlling pyrometer set too low
- Millivolt compensator of controlling pyrometer set incorrectly
- Heating cycle too short (underheating)
- Severity of quench too low
- Delay before quenching too long
- Part not thoroughly quenched
- Material hardenability too low for quench
- Surface decarburized

Spotty or uneven hardening

- Nonuniform heating
- Time interval between heating and quenching too short
- Quenching medium not agitated enough
- Water in quenching oil
- Scale on work
- Improper quenching medium
- Surface decarburized

Distortion

- Shape of part, or relation of portion to be hardened to remainder of section, not well adapted to flame hardening
- Metallurgically unsuitable prior structure
- Heating cycle too long
- Nonuniform heating
- Nonuniform quenching
- Excessive rate of quenching
- Material hardenability excessive

Shallow depth of hardening

- Material of low hardenability
- Excessive rate of gas flow; check for oxidizing flame
- Flame-port velocity too high
- Controlling pyrometer set too low
- Millivolt compensator of controlling pyrometer set incorrectly
- Short heating cycle or excessive scanning speed
- Severity of quench too low
- Delay before quenching too long

Excessive depth of hardening

- Low flow rate of gas; check for reducing flame
- Excess fuel gas in flame
- Controlling-pyrometer temperature set too high
- Millivolt compensator of controlling pyrometer set incorrectly
- Flame-port velocity too low

Excessive scaling

- Heating cycle too long
- Rates of gas flow too low
- Flame velocity too low
- Delay before quenching too long
- Improper arrangement of flame heads around periphery; overheating or banding.

Tempering of Flame-Hardened Parts

It is usually desirable to temper parts that have been flame hardened; the need for tempering of martensite is the same regardless of the heat treating method used to produce it. Flame-hardened steel will respond to a tempering treatment as it would if hardened to the same degree by any other method. Standard procedures, equipment and temperatures can be used. However, for work that is flame hardened because it is too large to be heated in a furnace, flame tempering may be the only feasible method of tempering available.

Flame Tempering. Large articles usually are hardened by the progressive method and can be tempered immediately by reheating the hardened surface with a flame head placed a short distance behind the quench. The reheating or tempering flame head must be designed correctly in regard to number and size of orifices or tips (flame ports) to produce the desired work temperature and temperature gradient in the flame-hardened zone at the flame-hardening speed.

Exact final adjustments can be made by varying the flow of gases through the flame head and by adjusting the distance between flame head and work surface. Tempering flame heads must have smaller heat outputs than hardening flame heads, because too-rapid heating of the hardened zone can cause cracking, and because, for tempering, the temperatures required are lower and temperature control usually is more critical.

Example 8. A 1080 steel ring, 5 m (16½ ft) in diameter, is an example of a large part that was flame tempered. Tempering was performed in a refractory-lined temporary enclosure placed around the ring, and heat was supplied by two air-and-natural-gas inspirator burners. The desired tempering temperature of 205 °C (400 °F) was measured by temperature-indicating crayons and controlled by adjusting the burners.

Simultaneous hardening and tempering is of greatest value for large workpieces, but it often can be applied economically to smaller pieces as well.

Self-Tempering. On large parts flame hardened to depths of about 6.4 mm (¼ in.) or more, the residual heat present after quenching may be sufficient to accomplish satisfactory relief of hardening stresses, and subsequent tempering in a separate operation may be unnecessary. Air-fuel gas heating, because of its lower rates of heat transfer, promotes development of the deep-seated heat required for self-tempering.

When residual heat cannot be utilized and it is desirable to eliminate the separate tempering operation, the use of a lower-carbon steel is suggested if hardness requirements permit. Preheating of heavy parts will increase the residual heat available for tempering. For this purpose, preheating may be accomplished by (a) heating the part in a furnace, if feasible; or (b) spinning the part in the burner ring with the burners firing at reduced or idling input. The choice will depend on the availability of equipment and the economics involved. The exact preheat temperature will vary, depending on the size and configuration of the part and on the degree of stress relief or temper desired. In each application, the exact schedule may be established on a trial basis as a preliminary step.

Surface Conditions

For wrought and cast steel parts, the surface conditions likely to be detrimental to successful flame hardening are, in general, those that interfere with heating or quenching, cause localized overheating, initiate cracking, or result in the presence of a soft surface skin after proper heating and quenching.

Table 10 summarizes the more common defects or conditions, their origins, and the detrimental effects to be expected when they are present on flame-hardened areas. The extent of these defects determines the amount of difficulty they may cause.

Dimensional Control

Because of its ability to heat specific areas of a part selectively, flame hardening in many applications affords greater control of dimensional stability than is obtainable by furnace heating and quenching. The magnitude of the dimensional changes that occur in flame hardening is influenced by such factors as size and shape of the part, area heated, depth of heating, hardenability, and quenching medium.

Examples of dimensional variations in production parts as a result of flame hardening are as follows:

Example 9 (Fig. 15a) pertains to a decrease in the pitch diameter of converter gear hubs made of 1052 steel. The gear teeth on the ID of the hub were hardened to a depth of 0.9 mm (0.035 in.) max above the root.

Example 10 (Fig. 15b) shows the results of flame hardening converter hubs. These hubs were made of 1062 steel from three different heats, and were hardened to 59 HRC min at a depth of 1.9 mm (0.075 in.) below the surface of the bore.

Example 11. In flywheel starter gears, dimensional control of the bore

Table 10 Surface conditions detrimental to flame hardening of steel parts

Defect or condition	Probable origin of condition	Detrimental effects to be expected on flame-hardened areas
Laps, seams, folds, fins (wrought parts)	Rolling mill or forging operations	Localized overheating (or, at worst, surface melting), with consequent grain growth, brittleness, and greater hazard of cracking
Scale (adherent)(a)	Rolling or forging; prior heat treatment; flame cutting	Insulating action against heating, with resulting underheated areas and soft spots Localized retardation of quench, causing soft spots
Rust, dirt(a)	Storage and handling of material or parts	Similar to scale as noted above Severe rusting may result in surface pitting that will remain after hardening
Decarburization	Present in as-received steel bar stock; heating for forging or prior heat treatment of parts or stock	In severely decarburized work, no hardening response will be found when parts are tested by file or other superficial means(b)
Pinholes, shrinkage (castings)	Casting defects	Localized overheating (or, at worst, surface melting), with consequent grain growth, brittleness, and greater hazard of cracking
Coarse-grain gate areas (castings)	Casting gates located in areas to be flame hardened (avoid, if possible)	Increased cracking hazard during quenching, compared with non-gated areas; shrinkage defects also likely in these areas
Improper welds	Parts welded with an alloy dissimilar to base metal	Weld-zone reaction dissimilar to base-metal reaction. Weld may separate, requiring rewelding or scrapping of the part (c)

(a) In addition to having detrimental effects on flame-hardened surfaces, scale, rust or dirt in the path of the flame may become dislodged and cause fouling of oxy-fuel gas burners or react chemically with ceramic air-fuel gas burner parts (causing rapid deterioration). When such materials enter a closed quenching system, they may clog strainers, plug quench orifices and cause excessive wear of pumps. (b) Partial decarburization lowers surface hardness as a direct function of actual carbon content of stock lost at surface, provided that steel was adequately heated and quenched. (c) To avoid these and other problems, it is mandatory that the flame hardener be given accurate and complete information on any changes in composition and past processing of the part. For example, previously hardened parts should never be flame hardened unless they have been annealed; otherwise, cracking is inevitable. (Ref 1)

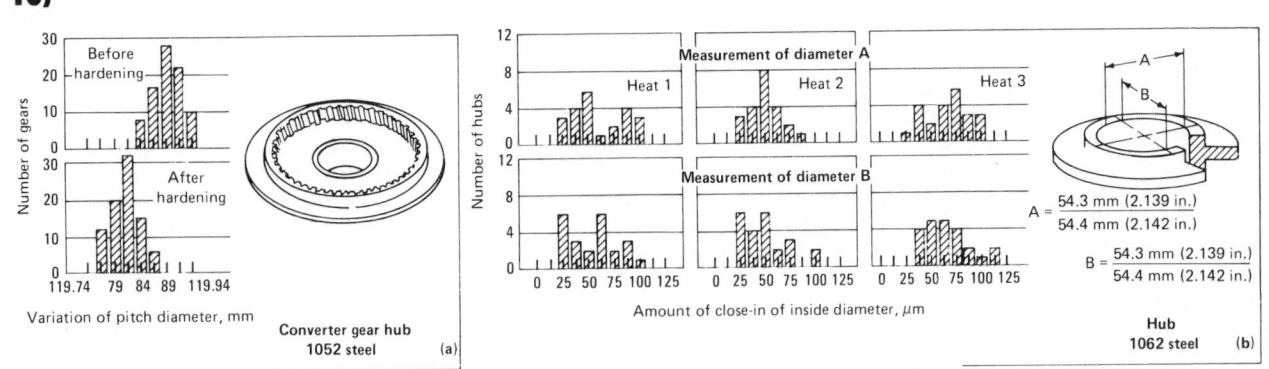

Fig. 15 Distribution of dimensional change as a result of flame hardening (Examples 9 and 10)

(a) Change in pitch diameter of converter gear hubs made of 1052 steel. Gear teeth on ID were heated for a total of 9.5 s, before being quenched in oil to provide a depth of hardness of 0.9 mm (0.035 in.) above the root. (Example 9) (b) Close-in of ID of converter hubs made of 1062 steel. ID was heated for a total of 12 s, then oil quenched to harden to 59 HRC min at a depth of 1.9 mm (0.075 in.) below the surface. ID was finish ground after hardening. (Example 10)

Table 11 Relative benefits of five hardening processes

Carburizing	Hard, highly wear-resistant surface (medium case depths); excellent capacity for contact load; good bending fatigue strength; good resistance to seizure; excellent freedom from quench cracking; low-to-medium-cost steels required; high capital investment required
Carbonitriding	Hard, highly wear-resistant surface (shallow case depths); fair capacity for contact load; good bending fatigue strength; good resistance to seizure; good dimensional control possible; excellent freedom from quench cracking; low-cost steels usually satisfactory; medium capital investment required
Nitriding	Hard, highly wear-resistant surface (shallow case depths); fair capacity for contact load; good bending fatigue strength; excellent resistance to seizure; excellent dimensional control possible; good freedom from quench cracking (during pretreatment); medium-to-high-cost steels required; medium capital investment required
Induction hardening	Hard, highly wear-resistant surface (deep case depths); good capacity for contact load; good bending fatigue strength; fair resistance to seizure; fair dimensional control possible; fair freedom from quench cracking; low-cost steels usually satisfactory; medium capital investment required
Flame hardening	Hard, highly wear-resistant surface (deep case depths); good capacity for contact load; good bending fatigue strength; fair resistance to seizure; fair dimensional control possible; fair freedom from quench cracking; low-cost steels usually satisfactory; low capital investment required

was of greater importance than control of the pitch diameter. Before the gear teeth are hardened, the ID is machined to 0.1 mm (0.003 in.) larger than final size for gears with 380-mm (15-in.) bores. Boring tolerance is ±0.05 mm (±0.002 in.); final tolerance, ±0.1 mm (±0.0045 in.). Out-of-roundness is not a problem, because the ring section is not hardened.

Selection of Process

Flame hardening is basically adaptable to surface hardening and to through hardening of selected areas. Manufacturing specifications often call for heat treatment of an entire part, but in many cases this is not necessary. Crane wheels, mill rolls and gears are examples of items for which flame hardening can be used to harden only the working surfaces. Dies often require hardening only along parting lines or in forming areas. (Ref 1)

Other hardening processes that may be applicable for accomplishing the same purposes include: (a) induction hardening for surface or through hardening; (b) carburizing, nitriding and other furnace processes in which surface composition is altered to permit hardening of the case; (c) application of hard facing materials to surfaces; and (d) through hardening of selected sections by partial immersion in molten lead or salt baths. Table 11 lists the different processes in use and describes the benefits realized from each. (Ref 3)

Induction hardening and flame hardening are the most effective methods for selectively hardening portions of a part (gear teeth, for example) without affecting other areas. A 4150 steel shaft, 760 mm (30 in.) long, with a 25-mm- (1-in.-) long gear at one end, can be core hardened to 26 to 32 HRC all over (a good machining hardness), and the gear can be hardened to 55 to 60 HRC by flame or induction hardening. If carburizing or nitriding were used instead, the main shaft would have to be masked with copper, which would have to be stripped away after hardening.

The differences among these three basic types of surface-hardening processes can be small enough for certain applications so that their feasibilities overlap. For example, a ring gear used in almost identical applications has been made for four customers using four different surface-heating procedures: (a) induction hardening a 4150 steel gear, with or without separate core hardening, depending on the specific application; (b) flame hardening the teeth of a 1045 steel gear; (c) carburizing an 8620 or 9310 steel gear, then finish grinding; and (d) core hardening

a 4150 steel gear to 32 to 34 HRC, then nitriding.

Careful analysis, however, would no doubt show that one of the four variations would be most economical. Examples of false economy are common. A company might save 20¢/kg on steel but have to pay twice that amount for hot or cold straightening to correct the distortion caused by the lower heat treatability of the lower-cost steel.

In selecting a flame-hardening process, consideration must be given to the method of applying the flame to the work and to the choice of the gas mixture. Oxy-gas equipment differs significantly from air-gas equipment in the design of the systems that control, deliver and burn the gas mixtures.

The heating characteristics of the combustion mixture largely determine the degree to which the hardening and heat-affected zones can be localized in the workpiece. Shallow hardness patterns (less than 3.2 mm or 1/8 in. deep) will be obtained only with oxy-gas mixtures, because the higher-temperature flames produced provide sufficiently fast heat transfer to effectively localize the heat pattern. Deeper hardness patterns may permit the use of either oxy-gas or air-gas mixtures. With oxy-gas mixtures the heat is localized, but care is required in controlling the heat-release rate to avoid overheating the surface during development of the deeper-seated heat.

Air-gas mixtures, because of the lower rate of heat transfer obtained with them, reduce the tendency toward surface overheating but generally extend the heat-affected zone well beyond the desired hardness pattern. For this reason, air-gas flame hardening is generally limited to those applications in which the portion to be hardened may be through heated or in which materials of shallow hardenability are employed. In the latter instance, the depth of the hardness pattern is primarily controlled by quenching rather than by heating intensity. Because of potential excessive distortion, the deeper-seated heat developed in a part by air-gas flames may preclude the use of this heating medium.

In view of these considerations, the success of air-gas flame hardening depends largely on the configuration of the part insofar as that configuration favors heat localization with its lower rates of heat transfer. In selective through hardening of gear and sprocket teeth, flanges, ribs, edges and similar projections, air-gas flame heating may be applied. On the other hand, surface hardening of gear teeth, rolls, journals, shaft areas, machineways, wear areas of forming dies, inside and outside diameters of hubs, and massive sections is usually done with oxy-fuel flame equipment. Small parts frequently require the very small, easily controllable flame characteristics obtainable with oxy-fuel burning equipment.

When more than one mode of flame application or combustion-gas mixture can be used for a flame-hardening operation, the selection of process and equipment becomes primarily an economic consideration. The equipment then may be chosen on the basis of expected cost to meet immediate and anticipated production requirements.

Whether or not a flame-hardening process is selected in preference to other hardening processes for a specific application is commonly determined on the basis of (a) suitability of the process to produce the required results, (b) control of distortion, and (c) the obvious factor of cost. Selection of flame hardening or an alternate method of hardening is discussed in the two examples that follow.

Example 12. Progressive flame hardening was used to harden the V-grooved alloy steel rim of a measuring wheel. The aluminum hub and spokes were not affected by this method of heating, and the resulting small amount of distortion of the rim was not objectionable.

Example 13 (Fig. 16). Because of service failures, and to avoid a design change, a sprocket manufacturer changed from induction hardening to flame hardening of sprocket teeth. When induction hardened, the sprockets were failing in the web area, and failure was attributed to high stress concentration resulting from induction through hardening of the teeth and rim. Continued use of induction hardening would have required major redesign of the part and remaking of a costly forging die.

For flame hardening, the sprocket was indexed on a small positioner, and standard flame-hardening tips were moved across the face of the tooth by means of a small flame-cutting dolly. Warm-water quenching was employed to eliminate the slight surface checks that had developed with cold-water quenching. Figure 16 shows the hardness pattern developed by the standard flame tips. Failures were eliminated; sprockets hardened by this method had excellent wear qualities; and cost of hardening was not increased.

Selection of Material

Application of flame hardening is limited to hardenable steels (wrought or cast) and cast irons. Typical hardnesses obtained for various grades of these materials by flame heating and quenching in air, oil or water are given in Table 12.

Maximum hardness is not the sole criterion used in selecting flame hardening as a heat treatment. Proper steel

Table 12 Response of steels and cast irons to flame hardening

Material	Air (a)	Oil (b)	Water (b)
		Typical hardness, HRC, as affected by quenchant	
Plain carbon steels			
1025 to 1035	. . .	. . .	33 to 50
1040 to 1050	. . .	52 to 58	55 to 60
1055 to 1075	50 to 60	58 to 62	60 to 63
1080 to 1095	55 to 62	58 to 62	62 to 65
1125 to 1137	. . .	. . .	45 to 55
1138 to 1144	45 to 55	52 to 57(c)	55 to 62
1146 to 1151	50 to 55	55 to 60	58 to 64
Carburized grades of plain carbon steels(d)			
1010 to 1020	50 to 60	58 to 62	62 to 65
1108 to 1120	50 to 60	60 to 63	62 to 65
Alloy steels			
1340 to 1345	45 to 55	52 to 57(c)	55 to 62
3140 to 3145	50 to 60	55 to 60	60 to 64
3350	55 to 60	58 to 62	63 to 65
4063	55 to 60	61 to 63	63 to 65
4130 to 4135	. . .	50 to 55	55 to 60
4140 to 4145	52 to 56	52 to 56	55 to 60
4147 to 4150	58 to 62	58 to 62	62 to 65
4337 to 4340	53 to 57	53 to 57	60 to 63
4347	56 to 60	56 to 60	62 to 65
4640	52 to 56	52 to 56	60 to 63
52100	55 to 60	55 to 60	62 to 64
6150	. . .	52 to 60	55 to 60
8630 to 8640	48 to 53	52 to 57	58 to 62
8642 to 8660	55 to 63	55 to 63	62 to 64
Carburized grades of alloy steels(d)			
3310	55 to 60	58 to 62	63 to 65
4615 to 4620	58 to 62	62 to 65	64 to 66
8615 to 8620	. . .	58 to 62	62 to 65
Martensitic stainless steels			
410 and 416	41 to 44	41 to 44	. . .
414 and 431	42 to 47	42 to 47	. . .
420	49 to 56	49 to 56	. . .
440 (typical)	55 to 59	55 to 59	. . .
Cast irons (ASTM classes)			
Class 30	. . .	43 to 48	43 to 48
Class 40	. . .	48 to 52	48 to 52
Class 45010	. . .	35 to 43	35 to 45
50007, 53004, 60003	. . .	52 to 56	55 to 60
Class 80002	52 to 56	56 to 59	56 to 61
Class 60-45-15	. . .	. . .	35 to 45
Class 80-60-03	. . .	52 to 56	55 to 60

(a) To obtain the hardness results indicated, those areas not directly heated must be kept relatively cool during the heating process. (b) Thin sections are susceptible to cracking when quenched with oil or water. (c) Hardness is slightly lower for material heated by spinning and combination progressive-spinning methods than for material heated by progressive or stationary methods. (d) Hardness values of carburized cases containing 0.90 to 1.10% C.

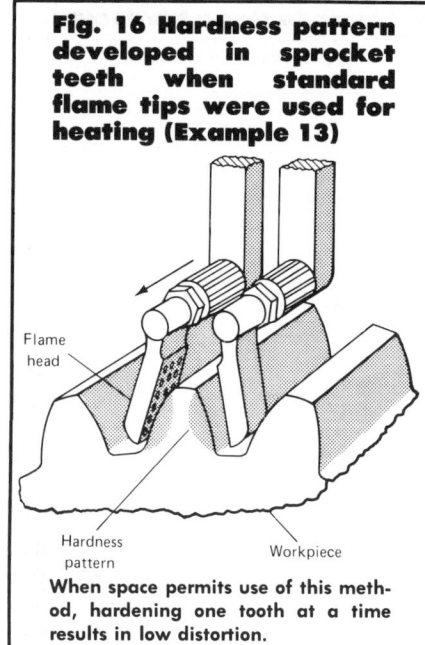

Fig. 16 Hardness pattern developed in sprocket teeth when standard flame tips were used for heating (Example 13)

Flame head

Hardness pattern

Workpiece

When space permits use of this method, hardening one tooth at a time results in low distortion.

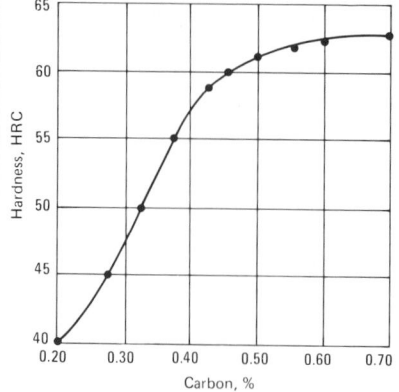

Fig. 17 Relation of carbon content to minimum surface hardness attainable by flame or induction heating and water quenching

Practical minimum carbon contents can be determined from this curve. Because actual carbon content is the criterion, the Check Analysis Tolerances shown in the AISI Steel Products Manuals should be considered in selecting grades of steel.

selection is essential to minimize distortion, for example. Plain carbon steels should be used, if possible, instead of steels whose deep hardening characteristics are more likely to incur higher internal stresses.

Some flame hardeners feel it is important to stress relieve all alloy steels and other steels with more than 0.40% carbon at 175 to 245 °C (350 to 475 °F), depending on customer specifications. This low-temperature tempering decreases hardness, but it also removes internal stress and restores toughness and ductility. (Ref 1)

Selective heating has the disadvantage of developing residual tensile stresses in the surface. As one area of a piece of metal is heated while the remainder stays cold, the hot metal expands; if restraint is sufficient, the heated metal will upset itself.

Upon cooling, this upset metal becomes short. As it cools to room temperature, it will often stabilize in a state of tension, which can be high enough to crack the part. When a part is to be induction or flame hardened, the materials engineer should work closely with the designer to keep the level of hardness, and necessary carbon, as low as possible while still meeting engineering requirements. Carbon content is the most important factor determining the level of hardness that can be attained in steels by induction or flame heating. It controls hardness level, the tendency of the part to crack, and the magnitude of residual surface stresses.

The practical level of minimum surface hardnesses attainable with water quenching for various carbon contents is shown in Fig. 17. The curve is applicable for induction hardening as well as for flame hardening. It applies also for alloy steels, except those containing stable carbide formers such as chromium and vanadium.

For best results, steels to be induction or flame hardened should be as-rolled, normalized (particularly from a high temperature) air-blast quenched, or quenched and tempered. These preferred heat treatments result in microstructures conducive to rapid and complete austenitization and full hardening. In selecting steels for either induction or flame hardening, it is important that the necessary steps be taken to ensure that the areas to be hardened are free of decarburization. Depending on stock size, steel grade, producing mill and several other factors, the depth of decarburization for as-rolled bar may run from near zero to 3.2 mm (0.125 in.). It should not be assumed that turned and polished bar is free of decarburization unless it is specifically ordered with this requirement. Carbon-restored and cold finished bar is available from mills in various carbon and alloy grades.

When maximum resistance to fatigue is desired, the hardened surface should contain residual compressive stresses; a recommended level is 172 MPa (25 ksi). Because surfaces hardened to depths of less than 1.9 mm (0.075 in.) are commonly residually stressed in tension, it is suggested that depth of hardening be at least 2.7 mm (0.105 in.) to ensure that residual stresses are compressive. This depth is particularly appropriate for manufacturers not equipped with residual stress-measuring equipment. Further, microstructure should be at least 90% martensite with no ferrite visible at a magnification of 500×.

Carbon Steels. Plain carbon steels in the range of 0.37 to 0.55% C are the most widely used for flame-hardening applications. They can be through hardened in sections up to 13 mm (½ in.). This response permits the use of carbon steel for selectively flame-hardened small gears, shafts and other parts of small cross section in which uniform properties are needed throughout the section. These same steels can be used for larger parts in which hardness is necessary only to shallow depths from 0.8 to 6.4 mm (1/32 to 1/4 in.).

Carbon steels 1035 to 1053 are suitable for flame hardening; 1042 and 1045 are the most widely available and are recommended for all flame-hardening applications except where they would be incapable of meeting requirements, as in the following examples:

- Failure of a 1045 steel part to harden with a given quench would necessitate the use of a steel of higher hardenability—for example, one with higher carbon or manganese, or both, or possibly an alloy steel.
- If increased depth of hardening is required, 1042 or 1045 may be inadequate where heavy sections are progressively hardened; therefore, substitution of 1541, 1552 or an alloy steel would be necessary.
- In applications in which wear resistance is of prime importance, it might be advisable to use a steel with 0.60% C or more to produce maximum surface hardness. Steels this high in carbon content often are quenched in oil or simulated oil to avoid the hazard of cracking due to water quenching. Thus, greater hardenability may be needed with the higher carbon content.
- When a severe quench in brine or caustic is required for hardening 1042 or 1045 steel, and such quench-

Fig. 18 Cast iron microstructures suitable for flame hardening

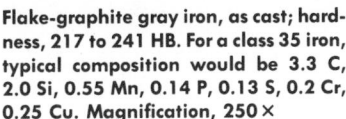

Flake-graphite gray iron, as cast; hardness, 217 to 241 HB. For a class 35 iron, typical composition would be 3.3 C, 2.0 Si, 0.55 Mn, 0.14 P, 0.13 S, 0.2 Cr, 0.25 Cu. Magnification, 250×

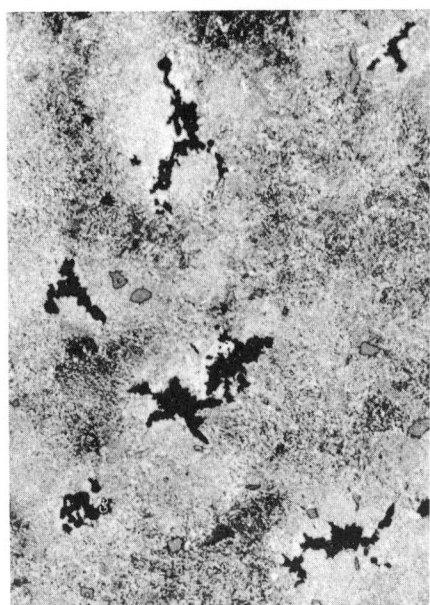

Pearlitic malleable iron, air quenched and tempered; hardness, 197 to 241 HB. Typical composition: 2.45 C, 1.45 Si, 0.50 Mn, 0.04 P, 0.10 S, 0.03 Cr. Magnification, 250×

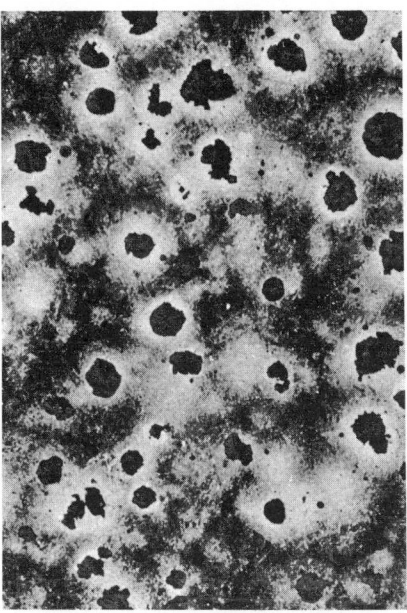

Ductile iron, air quenched and tempered; hardness, 197 to 241 HB. Typical composition: 3.8 C, 2.6 Si, 0.50 Mn, 0.02 P, 0.015 S, 0.05 Cr, 0.1 Cu, 0.05 Mg, 0.50 Ni. Magnification, 250×

ing causes cracking, a steel of higher hardenability—either carbon or alloy— should be selected, which can be hardened by a less severe quench.

Alloy Steels. The use of alloy steels for flame-hardening applications is justified only for the following reasons:

- Because high core strength is required (through heat treatment before flame hardening), and because carbon steels are inadequate to achieve this strength in the section sizes involved
- Because the mass and shape of a part, restrictions on distortion, or the hazard of cracking preclude the use of carbon steel quenched in water
- Because certain alloy grades may be more readily obtainable than carbon grades appropriate for the application. This situation often precludes use of the higher-manganese carbon grades such as:

1036	1048	1137	1141
1037	1051	1138	1144
1039	1052	1139	1145
1041	1053	1140	1151

Steels such as 4135H, 4140H, 6150H, 8640H, 8642H and 4340H are typical of the more readily obtainable alloy steels.

Carbon and alloy steel castings are widely used for flame-hardening applications, and selection of a specific composition or grade is made on much the same basis as for wrought carbon and alloy steels.

Cast Iron. Gray cast irons, ductile irons and pearlitic malleable irons having combined carbon contents of 0.35 to 0.80% can be flame hardened and will respond the same as steel.

Cast irons having less than 0.35% combined carbon will not respond readily to flame hardening because of the inability of austenite to dissolve graphite during the extremely rapid heating that occurs during flame hardening. Flame hardening of these irons produces a typical hardness of about 40 HRC. Malleable iron, in which all the carbon is in graphitic form, is not amenable to flame hardening for this reason.

Cast irons having combined carbon

contents greater than 0.80% are difficult to flame harden because of their inherent brittleness and susceptibility to cracking when heated and quenched rapidly. The low melting point of cast iron and the presence of graphite in the microstructure make cast iron susceptible to "burning" or even localized melting during flame hardening; therefore, extreme care must be taken to reduce the rate of heating when cast iron is hardened using equipment designed specifically for flame hardening of steel. For example, the distance between the inner cone and the workpiece can be increased, or the flame velocity can be decreased. Use of a flame head with smaller orifices also will decrease the heating rate.

Perhaps the most significant factor in the response of cast irons to flame hardening is prior microstructure. Micrographs showing favorable microstructures for flame hardening are presented in Fig. 18.

Other Materials. Flame hardening can be applied to other hardenable ferrous materials—for example, alloy cast irons, martensitic stainless steels and tool steels. The nature of flame hardening, however—especially the relatively high temperature gradients and higher-than-normal surface temperatures may cause retention of excessive amounts of austenite in many of the highly alloyed materials, with possible low hardness and transformation to untempered martensite in service, which is accompanied by brittleness.

Carburized parts made of plain carbon or alloy steels can be flame hardened to provide a hard case of high carbon content. The depth of carburized case may vary from a few hundredths of a millimetre to 1.6 mm (1/16 in.) or more. The flame-hardening procedure is adjusted to heat the carburized case to its full depth for hardening. Because the core of low-carbon steel so treated does not harden substantially, the method provides a means of accurately controlling the depth of hardened case.

Flame Annealing

The chemical reaction between oxy-gen and steel during flame cutting evolves intense heat, raising the temperature of the cut edge to the melting point on the surface and above the transformation temperature to a shallow depth. If the carbon or alloy content of the steel is high enough, hardening results when the heated zone is cooled rapidly.

Carbon steels containing 0.30% C or less do not harden sufficiently to prevent the use of flame-cut pieces in structures or in subsequent fabrication where bending is involved. When cut by flame, steels containing alloying elements and steels containing more than 0.30% C may harden sufficiently to prevent use of flame-cut pieces for some purposes.

When hardening is likely to occur along a cut edge, oxyacetylene flames, applied by suitable equipment, can be used either to prevent hardening or to soften an already hardened cut surface. The term "flame annealing" is applied to this process and to selective flame softening of areas of hardened steel parts.

Tool steels and certain highly alloyed steels will crack during flame cutting, or very soon after, unless they are heated to 205 to 425 °C (400 to 800 °F) before being cut.

REFERENCES

1. Flame Hardening, by N. J. Fulco: *Heat Treating,* Aug 1974, p 14-17
2. Programmable Flame Hardening Through Flow Control, by G. D. Orr and G. M. Kampitch: *Heat Treating,* Sept 1975, p 37-40
3. Selecting Steels for Heat-Treated Parts, Part II—Case Hardenable Grades, by R. F. Kern: *Metal Progress,* Dec 1968, p 71-81
4. Surface Hardening Gets Better: *Iron Age Metalworking International,* Dec 1969, p 34-35
5. How to Get the Most From Alloy Steels, by D. B. Mazer: *Machine Design,* 20 Apr 1972, p 175-179
6. Fuel Gases for Flame Hardening, by G. M. Corbett: *Welding Research Supplement,* Oct 1965, p 476-479

Laser Surface Transformation Hardening

By Ole Sandven
Chief Metallurgist
Avco Everett Metalworking Lasers

SURFACE TRANSFORMATION HARDENING of ferrous materials is an established process widely used to enhance the mechanical properties of highly stressed machine parts, such as gears and bearings. Surface hardening increases the wear resistance of the material and, under favorable circumstances, increases the fatigue strength due to residual compressive stresses that are induced in the workpiece surface by the transformation hardening process. The surface hardening process is not fundamentally different from conventional through hardening of ferrous materials. In both processes, increased hardness and strength are obtained by quenching the material from the austenite region to form hard martensite. Surface hardening differs from conventional through hardening in that only a thin surface layer is heated to austenitization temperatures prior to quenching, leaving the interior of the workpiece essentially unaffected.

Because ferrous materials are fairly good heat conductors, it is necessary to use very intense heat fluxes to heat the surface layer to austenitization temperatures without unduly affecting the bulk temperature of the workpiece.

This heat input is commonly obtained by the use of very hot flames or by high frequency induction heating. By selectively heating the workpiece surface to austenitization temperatures, desired surface hardening is obtained by application of a quench medium to the hot surface, or by self quenching.

Self quenching occurs when the cold interior of the workpiece constitutes a sufficiently large heat sink to quench the hot surface by heat conduction to the interior at a rate high enough to allow martensite to form at the surface.

In recent years, industrial lasers have become available for metalworking uses, including surface hardening. A laser can generate very intense energy fluxes at the workpiece surface and the resulting temperature profiles in the workpiece usually can be made steep enough to negate the need for external quench media. The laser beam is a beam of light, which is essentially independent of the workpiece, easily controlled, requires no vacuum, and generates no combustion products. It is ideally suited, therefore, for this purpose.

Fundamentals of Laser Surface Hardening

When a laser beam impinges on a surface, part of its energy is absorbed as heat at the surface. If the power density of the laser beam (usually given in watts per square centimetre) is sufficiently high, heat will be generated at the surface at a rate higher than heat conduction to the interior can remove it, and the temperature in the surface layer will increase rapidly. In a very short time, a thin surface layer will have reached austenitizing temperatures, whereas the interior of the workpiece is still cool. Even with a relatively moderate power density of 500 W/cm^2 (3300 W/in.^{2}), temperature gradients of 500 °C/mm (25 °F/mil) can be obtained. By moving the laser beam over the workpiece surface (see Fig. 1), a point on the surface within the path of the beam is rapidly heated as the beam passes. This area is subsequently cooled rapidly by heat conduction to the interior after the beam has passed. By selecting the correct power density and speed of the laser spot, the material will harden to the desired depth.

A relatively broad area beam, usually in the shape of a square or a rectangle, is used in the laser hardening process. The power density of a focused laser beam used for hardening is much lower than the power density of the small, intense focused spots used for welding and cutting. The power density is typically in the 1000 to 2000 W/cm^2 (6400 to 13 000 W/in.^{2}) range, occasionally as high as 5000 or as low as 500 W/cm^2 (32 000 to 3200 W/in.^{2}).

The resulting depth of case will depend on the hardening response of the material, but it will rarely be more than 2.5 mm (0.1 in.). For steel with low hardenability, such as plain carbon steel, the depth of case obtainable is much smaller, varying from perhaps 0.25 mm (0.01 in.) in mild steels to 1.3 mm (0.05 in.) in a medium carbon steel. Because of the very high heating and cooling rates obtainable, it is possible to harden steels not normally considered hardenable, such as SAE 1018. For the same reason, the hardness obtainable by the laser hardening process can, in some instances, be slightly higher than that considered possible with conventional methods.

Ferrous materials are not good absorbers of infrared and far-infrared electromatic radiation. For instance, polished pure iron has an absorptivity of about 4% at room temperature. Grit-blasted cast iron has an absorptivity of 25% at room temperature, and this value increases to about 40% at 800 °C in an inert atmosphere. The formation of oxides on the ferrous surface can increase absorptivity beyond these values, but efficient use of the laser energy demands the introduction of a controlled high-value absorbing coating on the material surface. Chemical coatings, such as manganese phosphate and paints of graphite, silicon, and carbon, have all been used successfully. Some of these coatings may burn off during the heating process, and some may leave a residue which in itself can be an indicator of the maximum surface temperature reached. In any event, the absorptivity of these coatings at the beginning of the heating cycle is high (90% or better) and continues to be higher than that of the bare material throughout the temperature excursion.

For shorter wavelength (1.06 μm near-infrared) radiation, as would be the case in (yttrium-aluminum-garnet) (YAG) laser transformation hardening, the room temperature absorptivity may be as high as 60%, but an absorbing coating is still recommended for efficient use of the laser energy.

The major advantages of laser surface hardening include: close control of the power input with modern metalworking lasers; the laser can provide high power density, which in turn minimizes the total energy input and, thereby, dimensional distortion; and, the ability of the laser to reach normally inaccessible areas on the workpiece surface. Because no vacuum or protective atmosphere enclosure is needed, and the distance from the workpiece to the last optical element of the laser system can be quite long, it is possible to process very large or irregular-shaped workpieces.

On the negative side, the depth of case obtainable is limited to about 2.5 mm (0.1 in.), usually less than half of this, and the capital cost of the equipment may be high. Therefore, careful analysis of a potential application for laser hardening is needed to ascertain the cost-effectiveness of the process.

Metallurgy of Laser Hardening

As stated earlier, the surface hardening process is not fundamentally different from the conventional hardening processes because the same metallurgical reactions occur. In surface hardening, however, the heating portion of the process cycle must, out of necessity, be much shorter than that of conventional hardening. This is particularly true for laser hardening, where heating to the austenitizing temperatures occurs within seconds or even fractions of a second. In fact, in laser surface hardening the heating period is frequently shorter than the cool-down time; hence, the expression "up quench" is often used.

The response of ferrous materials to rapid cool-down from the austenite region has been studied in great detail for many decades and is well understood. The same cannot be said for the heating period of the process. Basically, the problems associated with these high heating rates are that the formation of austenite as well as the redistribution of carbon, necessary to form a homogeneous γFe-C solid solution, are processes that require small but finite time intervals. The kinetics of these processes under very high heating rates are still somewhat uncertain, making the design of a laser surface hardening process more difficult than that of a con-

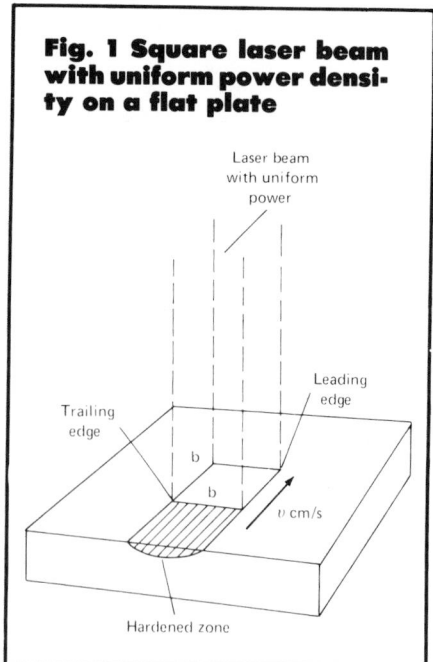

Fig. 1 Square laser beam with uniform power density on a flat plate

ventional through hardening procedure.

The basic reaction taking place during the heating period is the transformation of the bcc (α) iron into fcc (γ) iron. This occurs by nucleation and growth of the new phase in the matrix of the old phase. In slow heating, the process will start at A$_1$ (723 °C or 1335 °F) in a carbon steel and will be complete at the A$_3$ line. However, when the heating rate is high, the system is far from equilibrium conditions and the A$_3$ line will tend to be displaced upward to higher temperatures. Thus, although the temperature may be sufficiently high to form austenite under condition of slow heating, the same temperature level may be insufficient even to initiate austenitization under high heating rates. Laser hardening parameters are, therefore, usually designed to give peak temperatures well above those employed in conventional hardening to ensure austenitization.

The equilibrium room temperature structure of iron and steel will contain carbon in the form of iron carbide or graphite as a separate phase. To bring about hardening upon quenching, this carbon must be uniformly dissolved in the austenite. To do so, the carbon must be redistributed by diffusion into areas that originated from practically carbon-free ferrite.

This is a time-dependent process even at the high temperatures used in laser hardening and, under certain con-

Fig. 2 Structure of laser-hardened ductile cast iron at magnification of 250 ×

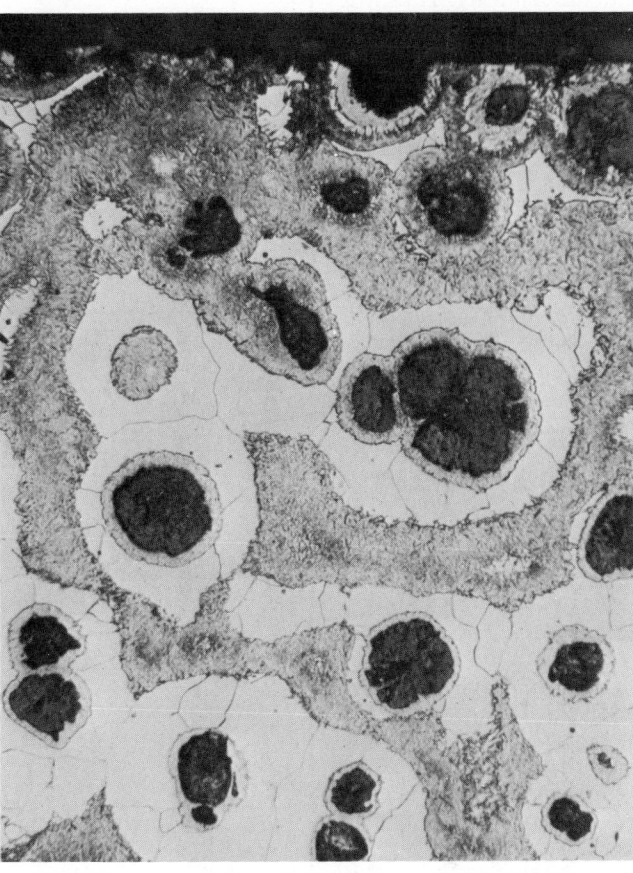

Fig. 3 Influence of processing parameters on heat penetration in laser surface transformation hardening

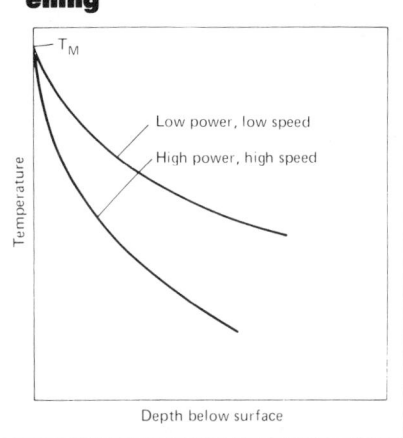

ditions, laser processing may occur too rapidly to allow for complete diffusion. This problem is obviously most prevalent where the carbon distribution in the starting material is nonuniform, such as in coarse pearlitic structures, structures containing proeutectoid cementite, spheroidized materials, and cast irons, particularly cast iron with a high content of free ferrite. Figure 2 shows the structure of a ductile cast iron after laser processing. The extent of carbon diffusion that occurred during the processing is clearly revealed by the region of martensite surrounding the graphite nodules.

The intrinsically high heating rates associated with laser hardening, combined with the need to allow sufficient time above the A_3 point to form homogeneous austenite, result in high peak temperatures. Under such conditions, there may be a tendency to form a thin surface layer, 5 to 20 μm (195 to 785 micro-in.) thick, of a phase that has yet to be categorized accurately. Retained austenite is found in laser hardened materials as revealed by x-ray analysis, but it is not clear how variation in laser processing parameters influence the amount of retained austenite.

The upper limit for the surface temperature in laser processing is set by the melting point of the material, because surface melting is undesirable in most instances. On simple plane surfaces, this is easy to avoid, but on workpieces with more complex design the surface temperature may vary across the beam-impingement area, even if the power density over the beam cross section is uniform. This can occur when the surface is curved, and in particular, when the beam strikes an area with abrupt angles. The projected power density will then be nonuniform, and the heat flow uneven, so that part of the illuminated surface may reach the melting point before the remaining area has reached sufficient temperatures to harden to the required depth. Sharp edges and corners can be troublesome, because they tend to concentrate the heat flow and lead to blunting of the edges by melting.

Another problem associated with some laser surface hardening applications is the necessity to overlap hardening passes. This may occur at the closure of a path around a cylindrical workpiece, or in the overlap zone between parallel paths being processed sequentially. Because lateral heat flow from the moving laser spot is unavoidable, backtempering will take place by the heat flow into areas already hardened by previous passes or, in the case of closed paths, at the start of the pass. The lower the processing speed, the more pronounced the effect, because relatively more heat will have time to diffuse into the previously hardened area. Even at high speeds and with the application of external heat sinks or water cooling, this effect cannot be entirely eliminated. The use of special optics may eliminate the necessity of forming overlap areas. However, this will always entail illuminating the entire width of the work area with laser radiation, and the available power of the laser limits the size of the workpiece that can be processed by the use of such optics. In many applications, the formation of a backtempered zone is not detrimental, provided that the processing is done in such a way that the backtempered area is located in a position of low service stresses or wear. Thus, for example, the interior walls of internal combustion engines may be hardened by straight or spiral laser passes, leaving unhardened material between the passes without harmful effects.

Heat Flow in Laser Hardening

In laser surface transformation hardening, thermal energy is generated by absorption of the laser radiation at the surface. The increase of temperature in the interior of the workpiece is by way of conduction only; no sources of thermal energy exist below the surface. Thus, if the rate of absorbed power and the thermal properties of the material are known, it is possible, at least in principle, to calculate the temperature distribution in the workpiece. This is of considerable value, because it is then possible to predict the results of laser processing in advance and to calculate the optimum processing parameters, such as power density, processing speed and spot size.

The long wavelength electromagnetic radiation (infrared) from a typical carbon dioxide laser is not efficiently absorbed by ferrous metals at room temperatures. It is, therefore, necessary to coat the workpiece with a substance that will aid in absorbing the laser energy. Commonly used coating materials are manganese phosphate, graphite or carbon-black paint. The paint, in the form of flat, black spray paint, is by far the most convenient coating to use. It is easy to apply and is fairly insensitive to variation in coating thickness.

When laser energy is absorbed at the surface at a rate of 500 W/cm^2 (3200 W/in.^{2}) or more, surface temperature rises very rapidly because the conduction of heat to the interior cannot keep up with the influx of energy to the surface. The higher the input flux, the more rapidly the temperature rises in the surface layer, and as a consequence, the temperature gradient in the workpiece will be steeper. The maximum surface temperature allowable is the melting point of the material, although in practical applications, temperatures should be held well below this value. This clearly acts as a constraint on the depth of austenitization that can be achieved. If lower laser power density is used and the processing speed is correspondingly decreased, the surface temperature will rise more slowly, and the temperature gradient will be less steep. This allows

austenitization to a greater depth, as shown in Fig. 3. However, the rate of cooling by self quenching will be slower, and it may be insufficient to allow the material to harden fully. Thus, the combined effects of melting temperature and hardenability act to impose a limit on the obtainable depth of case regardless of the power available for the processing. To obtain good self quenching, it is generally necessary to use high power density and high processing speed for steels of low hardenability. Steels of high hardenability can be processed to greater depth by relatively low power densities and low processing speed. For many such steels, the slow processing rate will be necessary to give the material time to form homogeneous austenite. Thus, deep case depth in plain carbon steels may be difficult to achieve. It is possible to use external quench procedures, thereby obtaining deeper case depths in steels of low hardenability. This is achieved at the cost of greater dimensional distortion, because the total power input increases with decreasing speed for constant maximum surface temperature. If the absorbed power density, laser spot dimensions, processing speed and thermal properties of the material are known, the temperature distribution in the workpiece can be calculated by means of several expressions found in the literature. The simplest expression is obtained if it is assumed that heat only flows normal to the workpiece surface, that is one-dimensional heat flow. If the workpiece is large in this direction, the temperature is given by:

$$T = T_o + 2Q/K \sqrt{\alpha t_D} \ \text{ierfc} \ \frac{\delta}{\sqrt{\alpha t_D}}$$
(Eq 1)

where T is temperature in centigrade, T_o is room temperature in centigrade, Q is absorbed power in W/cm^2, K is thermal conductivity in W/cm C°, α is thermal diffusivity in cm^2/s, δ is depth below the surface in cm, and t_D is dwell time in s.

Dwell time is the amount of time a given spot on the surface will be exposed to the laser beam. This is, therefore, equal to the length of the laser spot in the direction of travel divided by the speed of travel. The expression ierfcx is the integrated complementary error function defined by:

$$\text{ierfcx} = \int_x^\infty \text{erf} \times \text{dy} = \int_x^\infty (1\text{-erfx}) \ \text{dy}$$

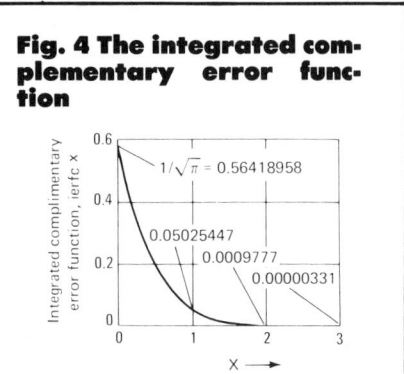

Fig. 4 The integrated complementary error function

where erfx is the error function:

$$\text{erfx} = \frac{2}{\sqrt{\pi}} \int_0^x e^{-Y^2} \ \text{dy}$$

These functions are tabulated in many sources and can easily be evaluated. Figure 4 shows a plot of ierfcx.

If we apply Eq 1 to a moving laser spot on a plane surface, it becomes clear that as the leading edge of the spot reaches a given spot on a plane surface, the temperature will rapidly start to increase, and the maximum temperature will be reached at the trailing edge of the spot, as shown in Fig. 5. A similar, but smaller, temperature increase will be experienced at points below the surface. If Eq 1 is solved for various values of the depth δ under the trailing edge of the spot, a temperature profile can be constructed representing the maximum temperature conditions in the vicinity of the surface (see Fig. 6).

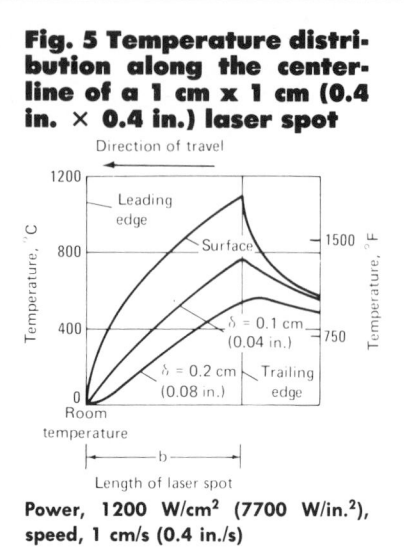

Fig. 5 Temperature distribution along the centerline of a 1 cm x 1 cm (0.4 in. × 0.4 in.) laser spot

Power, 1200 W/cm^2 (7700 W/in.^{2}), speed, 1 cm/s (0.4 in./s)

Fig. 6 Predicting depth of case in laser transformation hardening

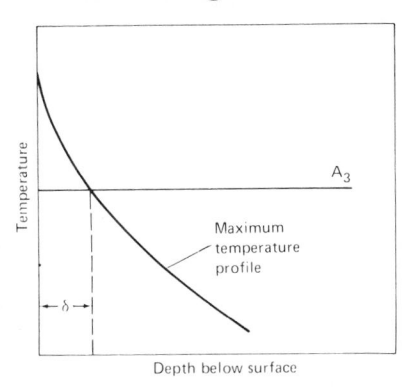

Table 1 Thermal properties of ferrous materials

Material	Transformation temperature °C	°F	Thermal diffusivity(a) cm²/s	in.²/s	Thermal conductivity(a) W/cm·°C	W/in.²·°F
1025	850	1560	0.073	0.011	0.39	0.085
1045	800	1470	0.074	0.011	0.38	0.083
1078	725	1335	0.065	0.010	0.36	0.079
4140	790	1455	0.070	0.011	0.36	0.079
5130	780	1435	0.066	0.010	0.36	0.079
Gray cast iron	···	···	0.099 to 0.148	0.015 to 0.023	0.46 to 0.57	0.10 to 0.0125

(a) To obtain values in SI units, mm²/s and W/m·K, multiply by 100.

Knowing the transformation temperature A_3, it is then a simple matter to determine the largest obtainable depth of hardened case δ.

In reality, heat diffuses in all directions into the workpiece from the moving laser spot, not only normal to the surface. Furthermore, heat is lost by radiation from the surface, and the thermal properties are not constants but are temperature dependent. Finally, in many instances, the workpiece is not large enough to be considered infinite, as Eq 1 assumes. All of these factors can contribute to make the estimates obtained from Eq 1 unreliable. Nevertheless, at high processing speeds (that is, low dwell time, t_D) and for reasonably thick specimens and large laser spots, Eq 1 gives a good estimate for depth of obtainable case. For a laser spot with dimensions 1.27 by 1.2 cm (0.5 by 0.47 in.), Eq 1 can be safely used down to a processing speed of 1 cm/s (0.4 in./s) provided that the workpiece is at least 0.6 cm (0.2 in.) thick. It is possible to develop expressions that take into account the three-dimensional nature of the heat flow, but such expressions become very tedious and time consuming in use, and it is often not worth the extra work.

By a relatively simple expansion of Eq 1, the cooling rate of any given point in the workpiece can be estimated if only one-dimensional heat flow is considered. This is done by means of the expression:

$$T = T_o + 2Q/K\left\{\sqrt{\alpha t}\ \text{ierfc}\ \frac{\delta}{2\sqrt{\alpha t}} - \sqrt{\alpha\,(t - t_D)}\ \text{ierfc}\ \frac{\delta}{2\sqrt{\alpha\,t - t_D}}\right\}$$

(Eq 2)

In this equation, the symbols have the same meaning as in Eq 1 except that t is the time elapsed since the leading edge passed over the spot for which we need to know the temperature; and $t - t_D$ is the time elapsed since the trailing edge passed over the spot, that is, the time elapsed after cooling by self quenching began. When using this equation, it is very important that the workpiece be large enough to provide an adequate heat sink.

Equation 2 can be used in conjunction with C-T diagrams to predict whether or not the cooling rate by self quenching will be adequate for hardening under a given set of circumstances. To aid in the use of Eq 1 and 2, Table 1 gives the value of the conductivity and the diffusivity for commonly used materials. The listed values are the average values from room temperature up to 1000 °C (1830 °F), and are given in units convenient for calculation of temperatures by Eq 1 and 2 when the power input (Q) is in W/cm². It should be noted that Q is not the applied power density but the estimated absorbed power in W/cm².

Processing Parameters

Many factors influence the results of laser surface transformation hardening including size of the laser spot, power density, processing speed, thermal properties of the material and laser hardenability of the material. The last factor encompasses the material's response to rapid heating and quenching and is, in part, dependent on the starting condition of the material, that is, whether the material is normalized, annealed, etc.

The parameters of principal interest are power density and processing speed. It is obvious that processing speed should be as high as possible to attain high production rates. However, the speed at which the laser spot moves over the surface is not a good measure of the production rate by itself, because the dimensions of the spot normal to the direction of travel are equally important in determining the area coverage rate. Furthermore, another important factor in determining the results of processing is the time under the beam that a given spot on the surface experiences, that is, the spot dimension in the traveling direction divided by the speed, commonly referred to as dwell time. The relative dimensions of a rectangular spot do not influence the coverage rate as long as the power density stays constant. Therefore, the area of the laser spot that will give a specific result is limited by the available power. An exception to this is when the spot is very narrow in the direction of travel and/or when the speed is very low (or equivalent, the dwell time, t_D, very long). Under such conditions, lateral heat losses become large and the spot dimensions influence the results. The conditions under which this effect becomes noticeable is:

$$B = vb/4\alpha < 3.5$$

where b is the spot dimension in the direction of travel, v is the speed of the spot and α is the diffusivity, all in cgs units.

The following are general guidelines for choice of processing conditions:

- Usable power densities in laser surface hardening are in the 500 to 5000 W/cm² (3200 to 32 000 W/in.²) range. Corresponding dwell times are in the range 0.1 to 10 s. For carbon steels, the power density is usually from 1000 to 1500 W/cm² (6400 to 9600 W/in.²), and the dwell time 1 to 2 s.
- Materials with high hardenability can be processed at low power density and high dwell time (low speed), whereas materials with low harden-

- ability should be processed at high power density and low dwell times.
- Rectangular or square laser spots with uniform power density are most suitable in obtaining uniform hardened case.
- High power density and low dwell time give shallow case, but high cooling rates. The reverse is true for low power densities.
- Maximum surface temperature is proportional to the square root of the speed. Hence, a doubling of the power density requires a quadrupling of the speed to obtain equivalent maximum surface temperatures.
- Increasing the power density results in lower total energy input for the same maximum surface temperature.
- Steel with normalized, annealed or spheroidized structures, steel with proeutectoid cementite, cast irons and steels with stable alloy carbides require longer dwell times than steels that have been hardened and tempered.
- Small workpieces will require higher power densities and lower dwell times than large pieces, unless external quenching media are used.

Metalworking Lasers

Several models of metalworking lasers of both domestic and foreign manufacture are commercially available. The majority of these are of either the neodymium, YAG solid state type or the carbon dioxide gas type. These lasers may have pulsed or continuous output power. Both types, whether pulsed or continuous wave, can be used for transformation surface hardening.

The power output of metalworking lasers at the present time is in the 50-W to 15-kW range, and more powerful systems will be available in the future. Because the cost of a laser system is approximately proportional to its maximum power output, the minimum size of a laser system for a potential industrial application should be determined on the basis of such factors as required production rate, minimum practical spot size and power density necessary to achieve the desired results. It should be noted, however, that the laser is a versatile tool and can be used for a number of other metalworking processes such as welding, cutting and hardfacing. If a laser system is to be used in a variety of applications, it may be ben-

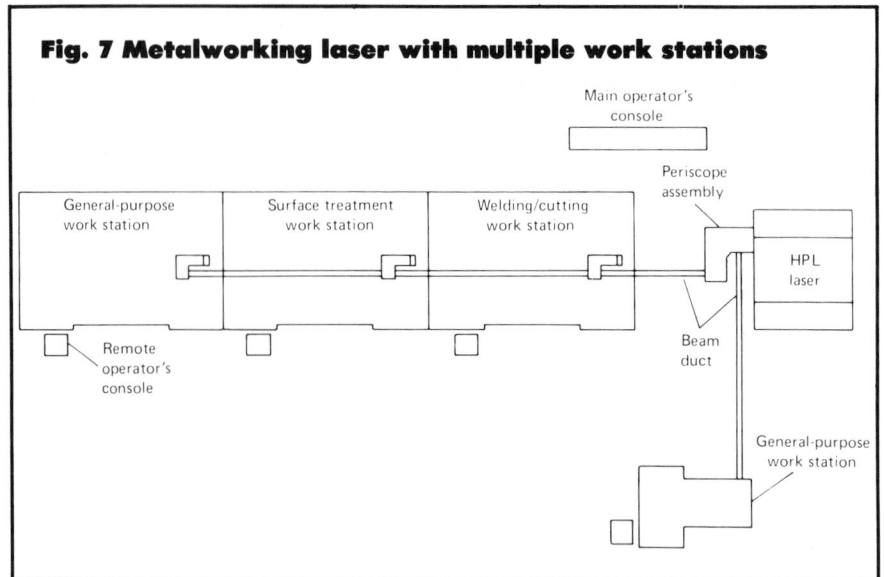

Fig. 7 Metalworking laser with multiple work stations

eficial to have as much power available as possible.

Lasers emit electromagnetic radiation in the infrared portion of the spectrum, and the laser beam from these machines is invisible. The carbon dioxide laser emits radiation with 10.6-μm wavelength, and this radiation is easily absorbed in a variety of nonmetallic substances. A thin sheet of lucite between the operator and the beam/workpiece interaction zone is sufficient to absorb potentially harmful stray radiation. The output of the YAG laser has a much shorter wavelength (1.064 μm), and therefore, the operator is required to wear special colored protective eyeglasses. YAG lasers are limited to relatively low power levels. Therefore, in metalworking processes requiring more than 500 W of power, the carbon dioxide laser is usually employed because it can deliver much higher continuous output.

The primary output beam from the laser rarely is used in metalworking applications. Instead, the output beam is directed and shaped by optical systems to generate a laser spot of the desired size and shape on the workpiece surface. Such an arrangement allows substantial flexibility in the use of a laser system. Because the coherent radiation from a laser has low loss of power with distance, the laser itself can be situated at a considerable distance from the work area. Furthermore, different types of metalworking applications can be performed with the same laser by changing the optical system. Several jobs can be performed simulta-

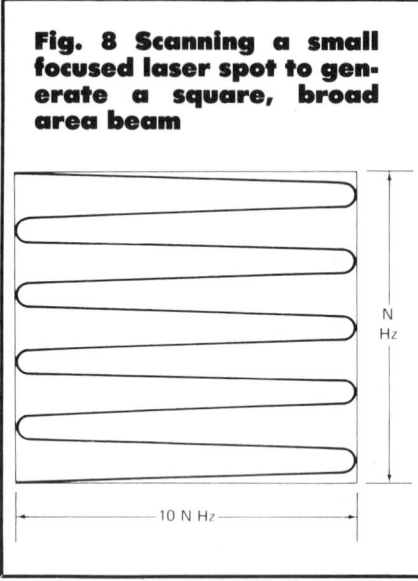

Fig. 8 Scanning a small focused laser spot to generate a square, broad area beam

neously with one laser by using several work stations, each with its own optical system designed for a specific application. Figure 7 shows a typical laser system arrangement. In this instance, each work station has its own control console, allowing manual or automatic control of beam power, duration of power delivery to the workpiece, rate of increase of power at the start of the run, rate of decrease at the end (ramp-up and ramp-down) and manipulation of the workpiece fixture protective gas flow. The laser power is automatically maintained at the desired level by a feedback control device in the laser. Positive feedback from temperature sensors, monitoring the temperature in the beam workpiece interaction zone, also can be used. The entire operation can be controlled by microprocessors. By directing the laser beam to the individual work stations in sequence and utilizing the time when the power is used at another station for workpiece manipulation, maximum usage of the laser can be achieved.

Optical Systems

In laser welding and cutting operations, a focused beam of intense power is used. For laser transformation hardening, this focused spot is replaced with a broad beam of much lower power density, partly because the hardening process requires lower power densities and partly to obtain reasonable rates of area coverage.

The simplest way to obtain such a spot is to position the workpiece surface in such a way that it intercepts the beam some distance from the focal plane of a converging beam. However, the power density distribution of a spot obtained in this manner is rarely sufficiently uniform and at the same time large enough to give satisfactory results.

The most suitable laser spot for surface hardening is a square or rectangular spot with uniform power density, sometimes referred to as top hat profile (see Fig. 1). Such a laser spot requires reshaping of the output laser beam that is attained by the use of various optical systems.

Transmission optical elements, such as lenses and windows, can be used. However, because of the long wavelength of the radiation typical for CO_2 lasers, these elements must be made from special materials such as zinc selenide to avoid excessive absorption.

Reflective optical components are less fragile and better adapted to industrial environment. They consist of flat, spherical or parabolic mirrors made from copper or molybdenum, which have excellent reflective characteristics to laser radiation. By the use of these mirrors, the beam can be reshaped and redirected to suit the particular application requirement.

The simplest way to form a square or rectangular beam is to use mirrors or lenses to form a focused spot and then to scan this spot rapidly back and forth in two perpendicular directions, as shown in Fig. 8. If scanning is done at a sufficient rate, the result will be equivalent to spreading the power of the focused spot over the area being

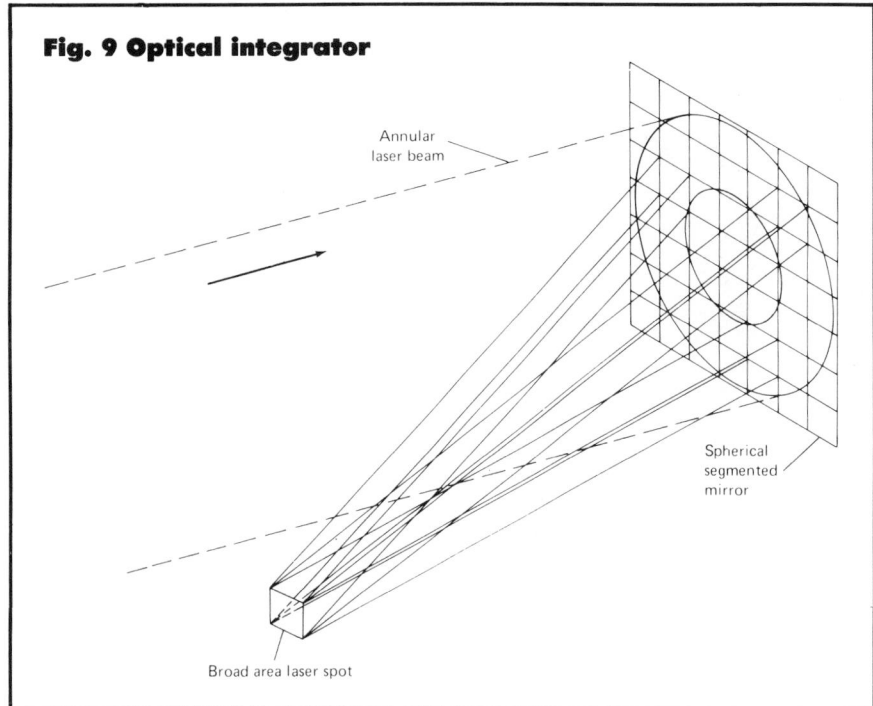

Fig. 9 Optical integrator

Annular laser beam

Spherical segmented mirror

Broad area laser spot

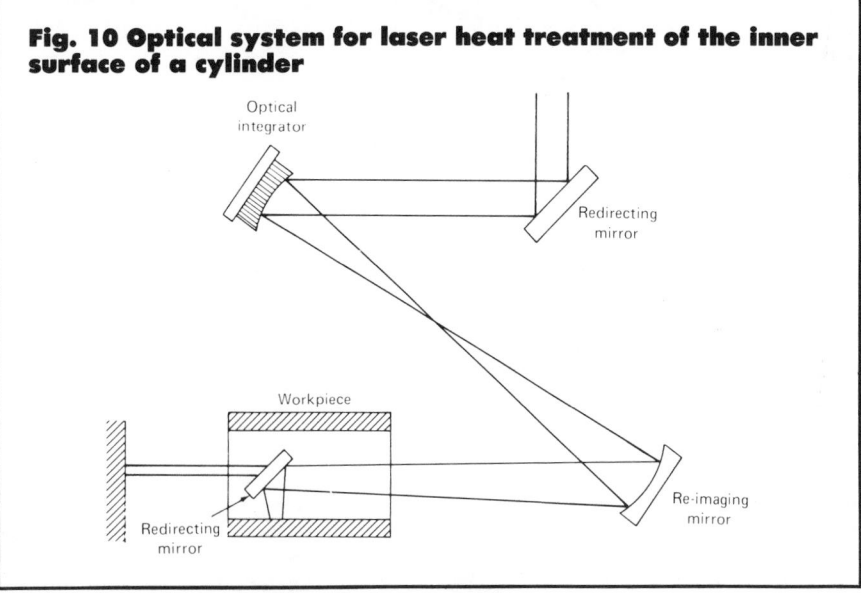

Fig. 10 Optical system for laser heat treatment of the inner surface of a cylinder

Optical integrator

Redirecting mirror

Workpiece

Re-imaging mirror

Redirecting mirror

Table 2 Laser hardening of 1045 plate

Speed mm/min	in./min	Power, kw	Measured case mm	in.	Calculated case mm	in.
510	20	2500	0.52	0.020	0.04	0.002
510	20	3000	1.02	0.040	0.82	0.030
510	20	3600	1.37	0.055	1.56	0.060
760	30	3000	0.24	0.010	0	0
760	30	3600	0.66	0.025	0.60	0.024
760	30	4150	1.24	0.050	1.08	0.043

scanned, because the thermal lag of the metal will act to even out the power input to a uniform power density. The

scanning can be performed conveniently by electromechanical vibration of one or more of the mirrors in the optical

Fig. 11 Toric mirrors for treating cylinders

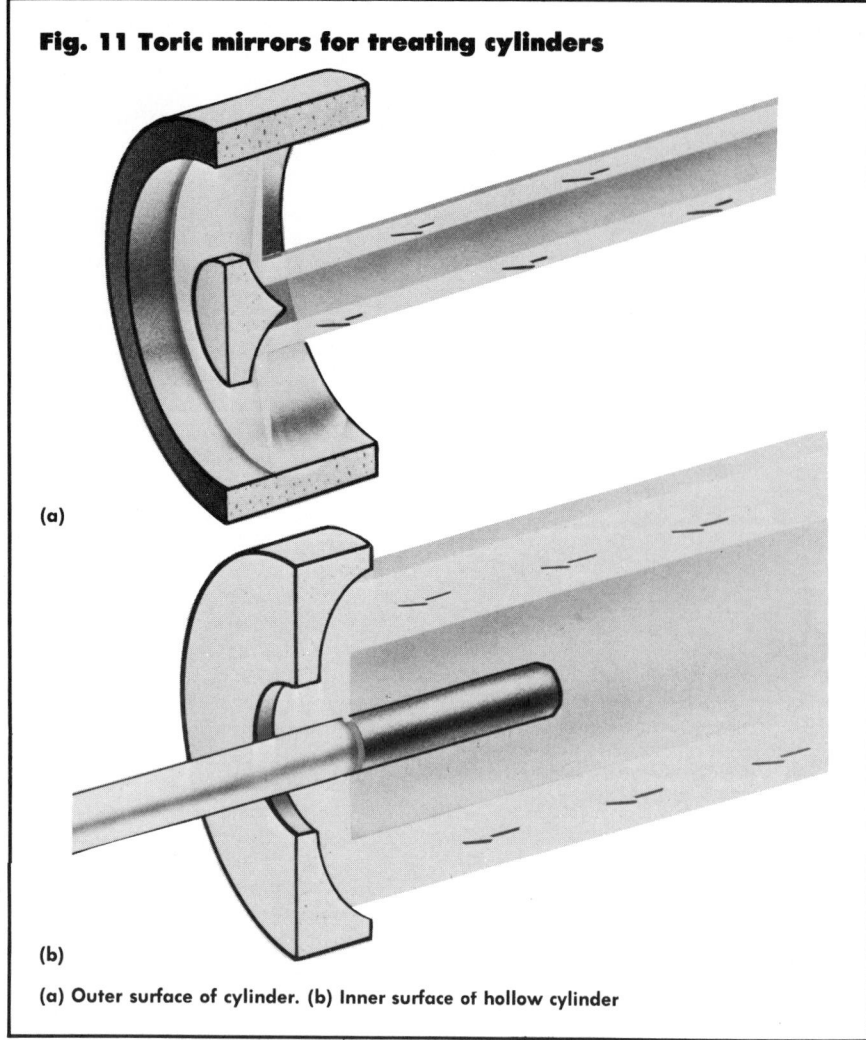

(a)

(b)

(a) Outer surface of cylinder. (b) Inner surface of hollow cylinder

Fig. 12 Using two equal laser beams to surface harden a semicircular groove

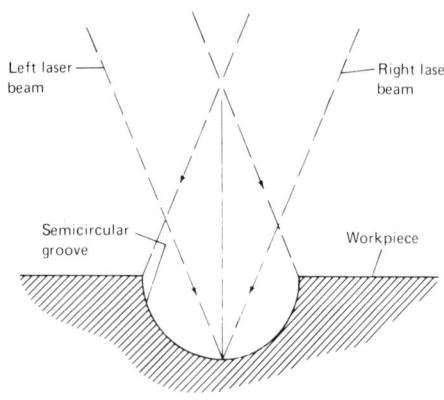

Left laser beam

Right laser beam

Semicircular groove

Workpiece

Fig. 13 Laser heat treating SAE 1045

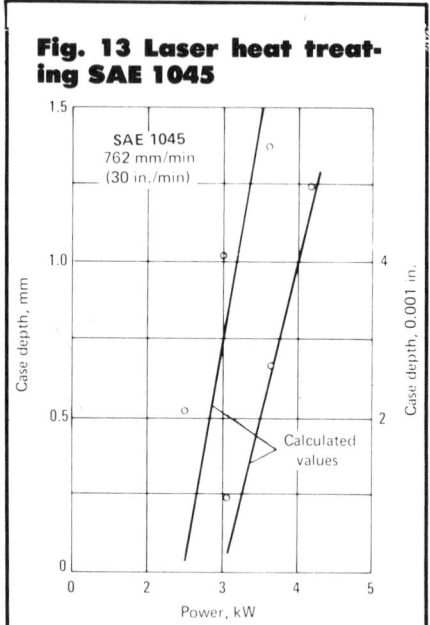

SAE 1045
762 mm/min
(30 in./min)

Case depth, mm

Case depth, 0.001 in.

Calculated values

Power, kW

system. In this way, it is possible to generate spots of desired size and shape on the workpiece surface.

Another method of forming a square or rectangular beam is to use a device known as an optical integrator. This device, shown in Fig. 9, is an array of flat mirrors mounted on a spherical surface placed to intercept the output laser beam. Each mirror forms an image of the part of the output beam that it intercepts, and all of the images from the array will be formed in the same position. In this way, the output beam is reshaped to form the desired square or rectangular spot in the focal plane of the integrator, having uniform power density over the illuminated area on the workpiece surface.

Both scanning optics and integrator optics form beams that can be redirected by flat mirrors or refocused by appropriately curved mirrors. In this way, the laser spot can be generated in the exact shape at desired location. Figure 10 shows how a spot can be projected on the inner wall of a tube.

For workpieces of cylindrical design, special optical components called toric mirrors can be used. These mirrors take advantage of the fact that the output beam from the laser can be made hollow or annular in cross section. Directing the beam by suitable mirrors onto the workpiece causes a continuous band of laser irradiated surface to form around the periphery of the workpiece, either on the inner surface of a hollow cylinder or on the outer surface of a cylinder, as shown in Fig. 11. The complete surface can be laser treated by moving the workpiece in the axial direction through the ring-shape laser spot. No backtempering is encountered because no start/stop or parallel spiral zones are generated. The relatively large area of the ring-shape laser spot limits the size of workpieces that can be processed with this technique to about 25 mm (1 in.) diameter for each 8 kW of power available.

Another technique that can be used for nonplanar surfaces is beam splitting. The laser beam is split into two equal parts by a copper prism, and the individual parts of the beam are directed at different angles to the workpiece surface by suitable reflective mirrors. Figure 12 shows the application of this technique to heat treating a wide semicircular groove. If only a single beam, normal to the bottom of the groove, had been used, the angle of incidence of the laser beam at the left and right top of the groove would have been too shallow to generate sufficient heat. By using

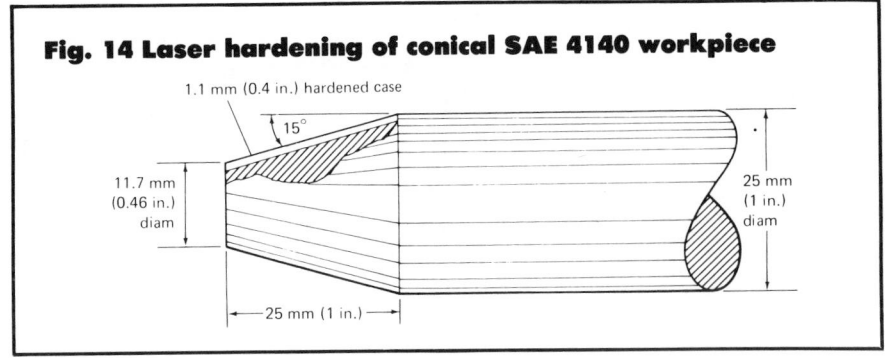

Fig. 14 Laser hardening of conical SAE 4140 workpiece

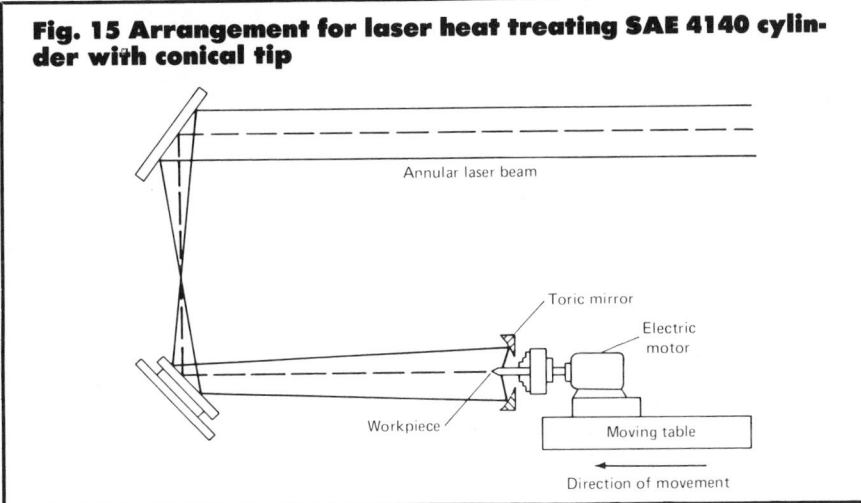

Fig. 15 Arrangement for laser heat treating SAE 4140 cylinder with conical tip

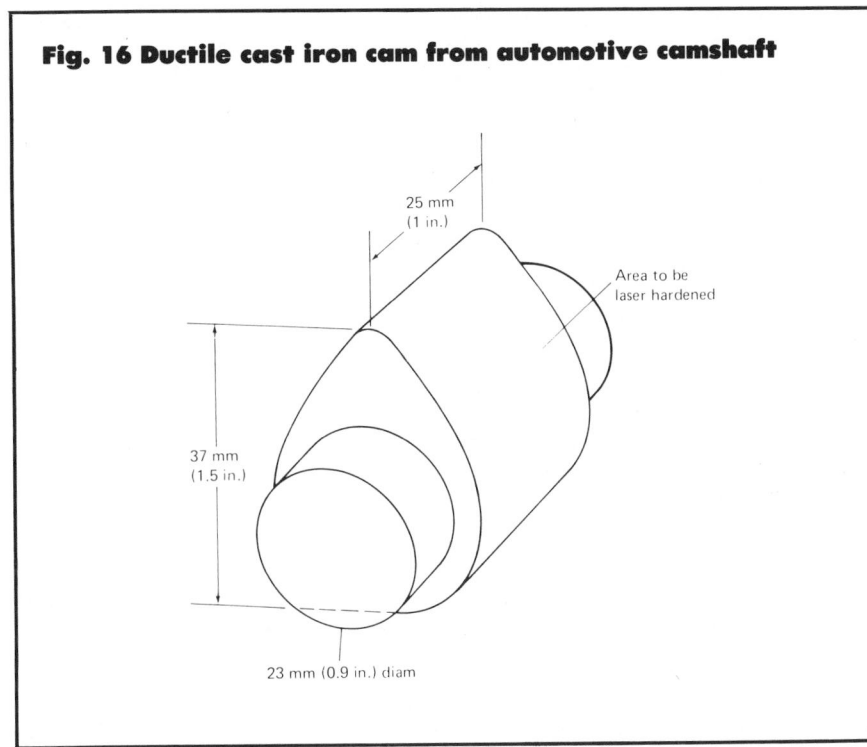

Fig. 16 Ductile cast iron cam from automotive camshaft

two beams as shown, this problem can be overcome.

Laser Surface Hardening of 1045 Steel Plate. The steel plate was 12.5 mm (0.5 in.) thick. Prior to laser processing, the material had been hardened and tempered to a hardness of 30 to 32 HRC. The desired case depth was 1 mm (0.04 in.); hardness at this depth should be 45 HRC. Single hardened band, 18 mm (0.7 in.) wide, was required. A 15-kW carbon dioxide laser was used, delivering a square beam with dimensions 18 mm by 18 mm (0.7 in. by 0.7 in.) to the workpiece surface by an optical integrator. Flat, black spray paint was used as the energy absorbing coating and applied to the workpiece surface after removal of surface oxide scale by sandblasting. No protective atmosphere was used, except for a fan blowing across the interaction zone to remove smoke. The workpiece was moved with respect to the stationary laser beam by mounting it on a controlled motion table in the work station.

The approximate range of processing parameters was calculated from Eq 1, using 0.38 W/cm·°C (0.083 W/in.·°F) for the thermal conductivity and 0.074 cm²/s (0.01 in.²/s) for the thermal diffusivity of 1045. Energy absorption was assumed to be 85% of incident power. The room temperature was 20 °C (68 °F), and the transformation temperature of 1045 was taken to be 800 °C (1470 °F). The results obtained are given in Table 2. Hence, this workpiece could be laser processed to the required case depth at a speed of 762 mm/min (30 in./min) (coverage rate 136 cm²/min or 21 in.²/min) at a power of 4.1 kW.

The relationship between calculated and measured case depth is shown in Fig. 13. As expected, using the simple one-dimensional Eq 1, the correlation between calculated and observed results are better at the higher speed.

Laser Surface Hardening of 4140 Cylinder with Conical Top. The object was to surface harden the conical part of the workpiece shown in Fig. 14 to increase the wear resistance. The minimum required case depth was 1.1 mm (0.044 in.); the hardness at this depth should be 45 HRC. The workpiece had been hardened and tempered to 40 to 42 HRC prior to laser processing to obtain strength and ductility.

Because the workpiece was cylindrical and had a maximum diameter of 25 mm (1 in.), a toric mirror could be used in conjunction with a 15-kW CO_2 laser.

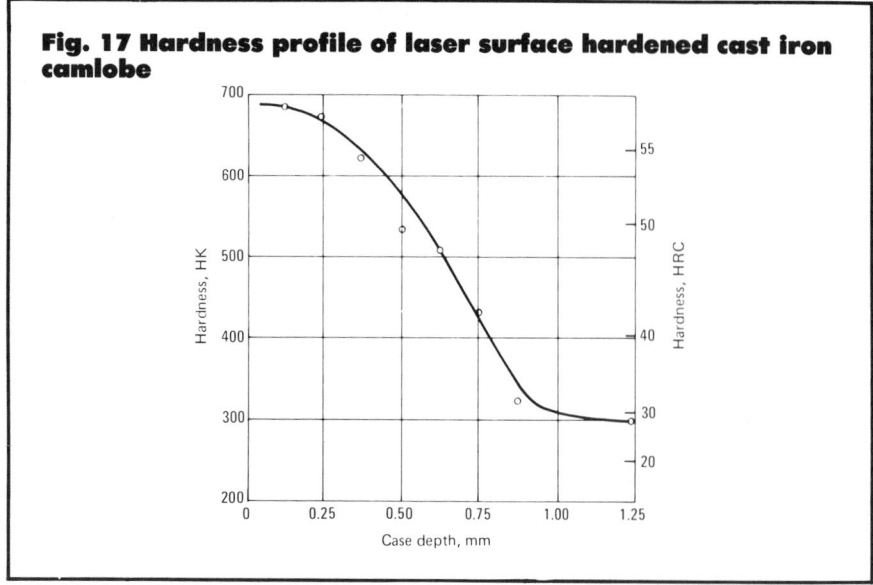

Fig. 17 Hardness profile of laser surface hardened cast iron camlobe

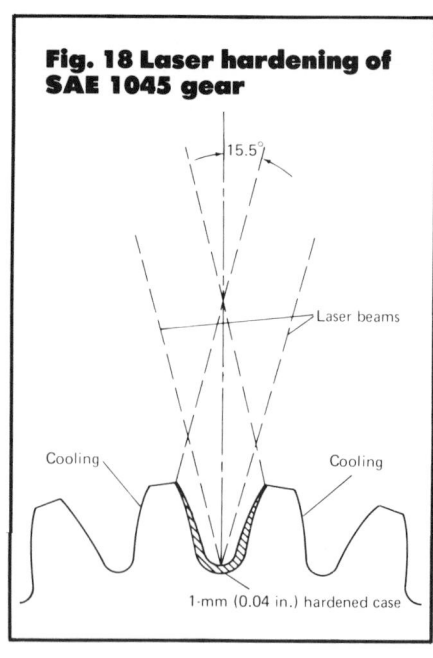

Fig. 18 Laser hardening of SAE 1045 gear

The processing arrangement is shown in Fig. 15. In this fixture, the workpiece was rotated at 1300 rpm under the ring-shape laser spot formed by the toric mirror. This was done to ensure uniform power density around the periphery of the workpiece. The ring-shape spot had a width of 5 mm (0.2 in.), and hardening of the desired area could be obtained by moving the workpiece in the axial direction. In this way, the entire area could be covered without forming any overlap zones.

Because the area under the ring-shape laser spot will increase as the beam sweeps from the tip of the cone to the cylindrical part of the workpiece, the power input had to be increased linearly with the distance in order to maintain constant power density.

Theoretical calculations, using Eq 1 with a thermal conductivity of 0.35 W/cm·°C (0.083 W/in.·°F) and a thermal diffusivity of 0.070 cm²/s (0.01 in.²/s), showed that at a processing speed of 229 mm/min (9 in./min) and a power density of 1620 W/cm² (10 400 W/in.²) would give a surface temperature of 1360 °C (2480 °F). This is only an estimate, as Eq 1 assumes a plane workpiece rather than a cylindrical one. To process at a slower rate and at lower power density would give deeper heat penetration, but the self-quenching rate would be lower. On the other hand, processing at higher speed and power would lead to shallower heat penetration.

The workpiece was coated with flat, black spray paint and processed at the following parameters:

- Axial speed: 221 mm/min (8.7 in./min); or 3.8 mm/s (0.15 in./s)
- Power: 3500 W at the tip, increasing to 7600 W at the cylindrical portion
- Rate of power increase: 620 W/s

The hardened case obtained ranged from 1.65 mm (0.06 in.) at the tip to 2 mm (0.08 in.) at the cylindrical portion of the workpiece. The surface hardness was 58 to 59 HRC.

Laser Surface Hardening of Cast Iron Camshaft Lobes. The surface of the lobes of an automotive camshaft made from ductile cast iron (see Fig. 16) was to be surface hardened to increase wear resistance. The desired case depth, defined as the depth where the hardness was 50 HRC, was 0.5 to 1.0 mm (0.02 to 0.04 in.).

A 15-kW CO_2 laser was used for the processing. The optical system delivered a focused spot, with a diameter of 10 mm (0.4 in.) to the workpiece. This spot was scanned over a distance of 22 mm (0.9 in.) normal to the direction of processing and 25 mm (1 in.) in the direction of processing. The frequency of scanning was 125 Hz in the normal direction and 700 Hz in the processing direction, forming a rectangular spot 22 mm by 25 mm (0.9 in. by 1.0 in.) on the camlobe surface.

To obtain an even hardened case around the periphery of the camlobe, it was necessary to vary the angular speed of rotation of the lobe under the laser beam. The reason is that the angle of incidence of the laser beam to the workpiece changed during rotation, from nearly normal incidence at the

Fig. 19 Dual beam optics for laser gear heat treating

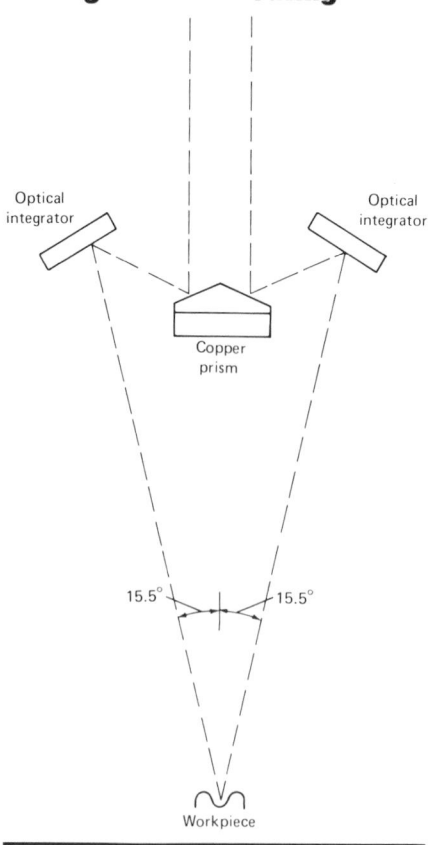

cylindrical portion of the lobe to a grazing incidence of only 20 to 30° at the flat portion. Furthermore, at constant rotational speed, the linear speed of processing would vary as the lobe rotated.

This was obtained by mounting the workpiece on a rotary table. The speed of rotation was varied by means of an electromechanical controller in a predetermined manner.

The camlobe was laser hardened, using a manganese phosphate coating to increase energy absorption. Because of the design of the workpiece, it was difficult to predict the optimum processing parameters by calculations and the parameters were, therefore, evaluated by trial and error. The results were:

- Power input.................. 9 kW
- Power density .... 1600 W/cm^2 (10 300 W/in.^{2})
- Linear speed of processing .... 760 mm/min (30 in./min) at the cylindrical portion
  180 mm/min (7 in./min) at the flat portion
- Depth of case.. 0.55 mm (0.022 in.)

The hardness profile of the surface layer of the camlobe is shown in Fig. 17.

Laser Surface Hardening of a Large Gear. The gear, made from SAE 1045 steel, had a diameter of 28 cm (11 in.) and a thickness of 10 cm (4 in.). The gear had 34 teeth and a diametral pitch of 3.35. To increase fatigue and wear properties, a hardened case of 1 mm (0.04 in.) was desired, extending in a continuous manner from the tip of one tooth to the tip of an adjacent tooth, as shown in Fig. 18.

In this application, two laser beams had to be directed toward the surface of the gear teeth at an angle of 15.5° from the normal to the root area, as shown in Fig. 18. These two beams must be abutting but not overlapping at the root between the teeth. A single beam with normal incidence to the root area could not be used because the angle of incidence at the adjacent fillets would be too shallow to generate a hardened case in this critical area, where fatigue cracks are likely to originate.

The output beam from a 15-kW CO_2 laser was split into two equal parts by a reflective copper wedge. Each beam was then directed to the workpiece surface in the form of a 12.5 mm by 12.5 mm (0.5 in. by 0.5 in.) spot by two optical integrators, as shown in Fig. 19. The projected area of each beam on the workpiece surface was approximately 12.5 mm by 25.4 mm (0.5 in. by 1 in.); thus, the total irradiated area was 3.18 cm^2 (0.5 in.^{2}).

After surface preparation with flat, black spray paint, the gear was laser processed at a total power input of 8.8 kW and with a translation speed of 500 mm/min (20 in./min). The calculated over-all power density was 1380 W/cm^2 (8900 W/in.^{2}), but the actual power density varied somewhat over the interaction area because of the variation of incidence angle to the curved surface.

Sequential runs were made on teeth 120° apart to minimize heat distortion. Water cooling was used on tooth flanks adjacent to the processing area to prevent backtempering of previously processed teeth. The obtained case depth (45 HRC) was 1.2 to 1.3 mm (0.046 to 0.052 in.) on mid flank and at the root, and 1 mm (0.04 in.) at the fillets. Surface hardness was 59 to 60 HRC.

Electron-Beam Heat Treating

By Carl Fiorletta
Supervisor, Electron Beam
  Heat Treating Systems
Sciaky Bros., Inc.

ELECTRON-BEAM HEAT TREAT-ING is a selective hardening process in which the surface of a hardenable ferrous alloy is heated rapidly above the transformation temperature of the alloy by direct bombardment or impingement of an accelerated stream of electrons. At the end of a heating cycle of 0.5 to 2.5 s, the flow of electrons is stopped abruptly to allow the part or workpiece being processed to self-quench and to form a martensitic structure with a compressive stress on the surface of the hardened area. The electron-beam hardening process normally is applied to finish-machined or ground surfaces. Because the buildup of energy is rapid and well controlled, postheat treatment operations such as grinding or straightening are not required in most instances.

Application Criteria

A part or workpiece to be heat treated is considered a suitable candidate for electron-beam heat treating if it meets the following criteria:

- The material must contain adequate carbon to produce satisfactory case hardness. Figure 1 presents relationships between carbon content and case hardness for plain carbon steels.
- The stream of electrons must have line-of-sight access to the area requiring heat treatment and a beam-impingement angle of at least 25°. Guidelines for acceptable part configurations are shown in Fig. 2.
- The component being heat treated may be processed in a vacuum envelope or chamber, or at pressures up to 1 atm in air or inert gas. Vacuum chambers in high-production heat treating systems typically have interior volumes of 0.02 to 0.11 m³ (0.8 to 4.0 ft³).
- The surface to be heat treated should be machined or ground to final dimensions. If grinding is required after heat treating, removal of 0.05 to 0.25 mm (0.002 to 0.010 in.) of stock normally is sufficient.
- To prevent magnetic interaction or unintentional deflection of the electron beam, the component being heat treated must be demagnetized prior to hardening. Demagnetization normally is required if the part has been fixtured with magnetic clamps or chucks in operations prior to heat treating.
- The mass of the part must be suffi-

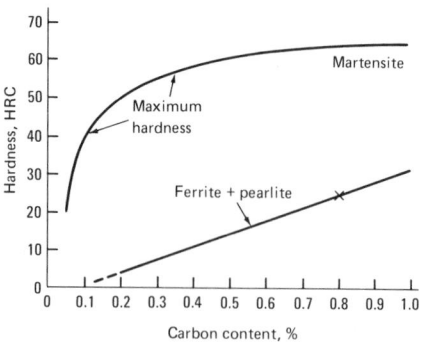

Fig. 1 Relationship between maximum hardness and carbon content for plain carbon steels

cient to self-quench the heat treated area. The ability of a part to self-quench is determined by its composition as well as its configuration. Steels of high hardenability (such as AISI 4150) can be through hardened in many instances. For plain carbon steels, however, five to eight units of mass beneath the heated area are required for each unit of mass of hardened case. The self-quench phe-

Fig. 2 Workpiece configurations and heating patterns for electron-beam heat treating

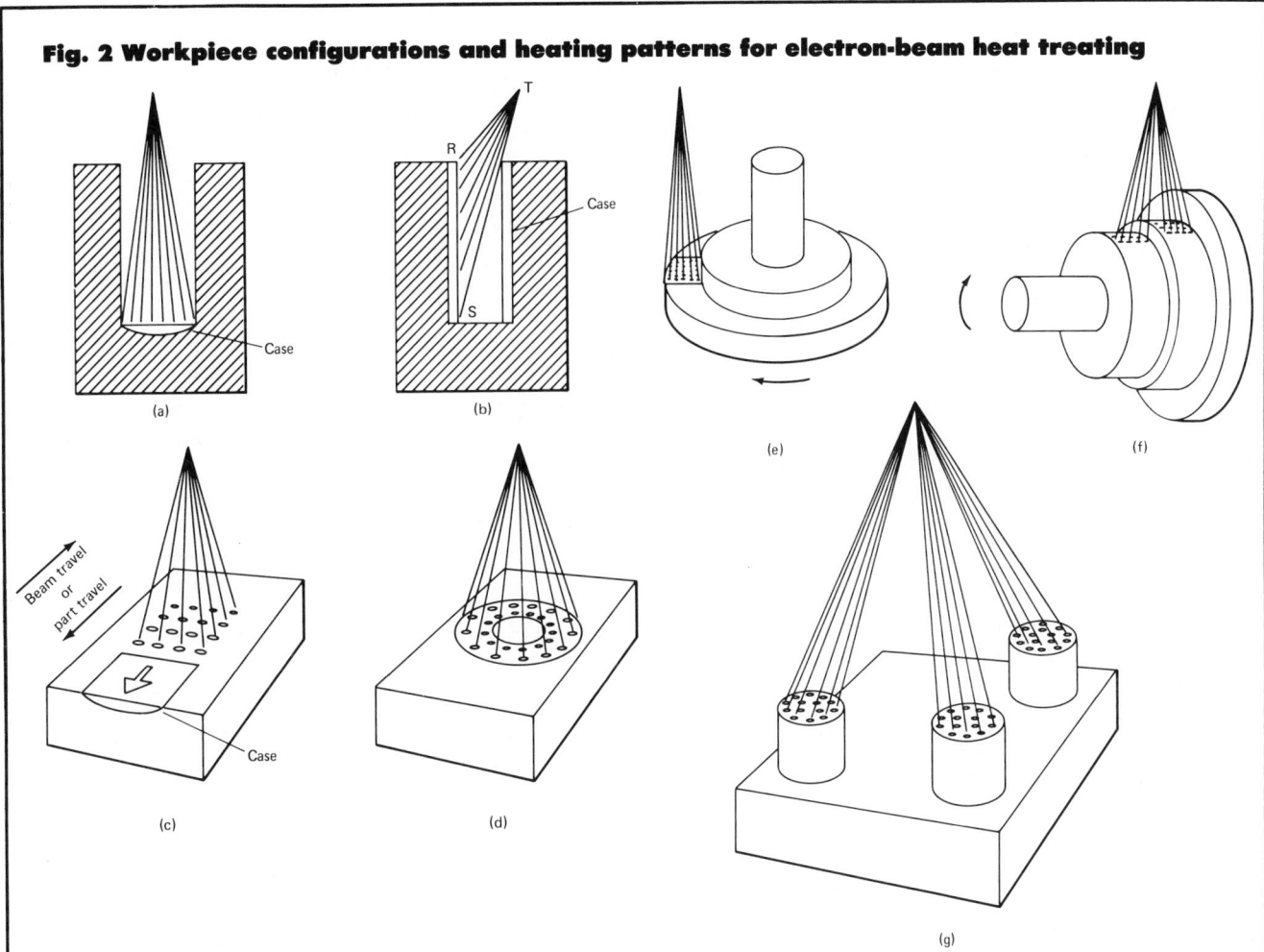

(a) Display static pattern within cavity in workpiece. (b) Maintain angle of workpiece rotation, RST, at 25° minimum. (c) Display static pattern and move the pattern or the workpiece to heat treat large areas. (d) Display static pattern; this annular pattern has well-defined inside and outside diameters. (e) Display static pattern and rotate workpiece. (f) Display more than one pattern and rotate workpiece. (g) Display multiple patterns on one workpiece or on a small group of workpieces for simultaneous hardening; patterns may be similar or dissimilar in geometric shape.

nomenon is described in detail later in this article.

Electron-Beam Equipment

In electron-beam heat treating, a highly concentrated beam of high-velocity electrons is used to heat selective surface areas. The electron beam is produced by an electron gun such as the one schematically represented in Fig. 3. Depicted is a 42-kW gun used for welding as well as for heat treating.

Beam Control. Free electrons escape from the filament when it is resistance heated to a very high tempera-ture. These electrons are accelerated and collimated into a dense, extremely energetic beam by the accelerating potential between the cathode and the anode. The high-energy beam thus formed passes through a small-diameter hole in the anode. Because of the mutual repulsion among neighboring electrons, the beam requires further collimation below the anode. This additional collimation is controlled with a focus coil that allows variation of the distance from the gun to the workpiece. A deflection coil deflects the reconverging beam to a designated location on the workpiece.

A high vacuum is needed in the region where the electrons are emitted and accelerated, both to protect the emitter from oxidation and to prevent interference with the electrons while they are still at low velocity. Therefore, the electron-gun housing is pumped and maintained at a vacuum of 10^{-5} torr. The workpieces are contained in an enclosure under a vacuum of approximately 5×10^{-2} torr. An intermediate vacuum level provides short evacuation times and higher production rates. Treating at one atmosphere does not require any evacuation time.

In electron-beam heat treating, the energy exchange is simply a matter of the electrons in the beam transferring their kinetic energy to the atomic structure of the target material in the

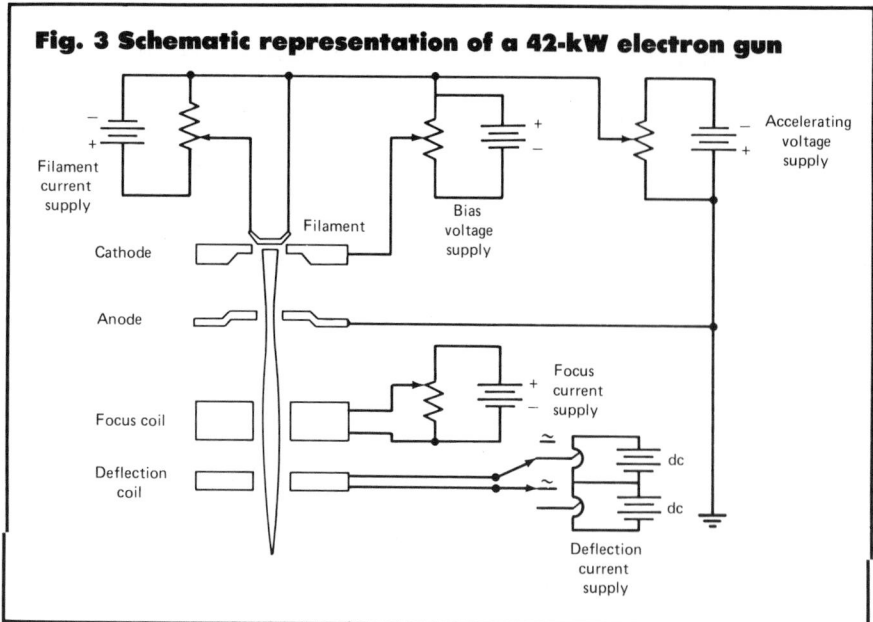

Fig. 3 Schematic representation of a 42-kW electron gun

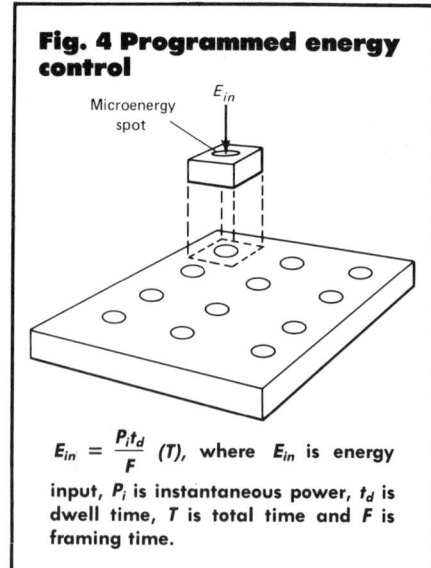

Fig. 4 Programmed energy control

$$E_{in} = \frac{P_i t_d}{F}\ (T),$$ where E_{in} is energy input, P_i is instantaneous power, t_d is dwell time, T is total time and F is framing time.

form of heat. The electron beam, when sharply focused for welding, is capable of impingement power densities on the order of 10 MW/cm² (65 MW/in.²). Because this powerful concentration of energy is easily controllable in power magnitude, power density and beam position, it is well suited for surface hardening as well. These power densities are much too high for nondestructive heat treating, however. Destructive heat treating in this context refers to controlled remelting of ferrous and nonferrous materials.

An energy concentration of 3.1 kW/cm² (20 kW/in.²) is more suitable for selective heat treating. To reduce the beam energy to this level, a single electron beam is programmed through a group of discrete beam positions referred to as a raster pattern.

Programmable Raster Pattern. The advancement that made electron-beam heat treating practical was the development of a programmable dedicated computer-control system coupled with electron-beam equipment. The control system generates a raster pattern by magnetically maneuvering the electron beam accurately from one programmed location to another while controlling energy input at each location, as shown in Fig. 4.

The power of the electron beam is distributed over the surface to be hardened at points spaced so that no "hot spots" exist when a defocused beam is applied. These locations may be spaced as required to obtain the necessary power-density distribution. The electron beam is positioned successively at each specific location, dwelled for an adjustable time interval (20 × 10⁻⁶ s minimum) and then translated at high speed to the next coordinate point. Patterns that conform to the contours of each surface to be hardened are thus produced. These patterns are repeated throughout the entire heat treating cycle, typically at a rate of 500 times per second.

Electron-beam power is also controlled as a function of time. Initial power density is sufficient to heat the surface rapidly to an austenitizing temperature just below the melting point of the ferrous alloy being hardened. A thermal gradient is generated from the surface to the interior, and the depth of the gradient increases as a function of time. As the thermal gradient gets deeper, the surface temperature tends to rise. To keep the surface temperature just below the melting point, the temperature rise is countered by decreasing the surface power density with respect to time, using the computer capability to adjust electron-beam power rapidly.

The austenitizing temperature is held just below the melting point of the alloy to take advantage of rapid carbon migration. Typical minimum times required for austenitization of the structure and dissolution of the iron carbide (Fe₃C) are on the order of ⅓ to ½ s. In this length of time, material to a depth of only about 0.25 mm (0.010 in.) can be heated sufficiently so that it will be hardened when the electron-beam en-

ergy is turned off and the workpiece allowed to self-quench. Soaking for longer times prepares deeper layers for hardening.

When the desired depth of carbon-bearing austenite is prepared, the electron-beam power is turned off. Heat flow to the interior of the metal continues until the temperature becomes equalized throughout the workpiece. The rate usually is high enough to exceed the critical cooling rate and to convert the austenite to hard martensite. A typical relationship among these various factors is shown in Fig. 5.

Raster-Pattern Control. Although part configuration must be compatible with the electron-beam process, as indicated in Fig. 2, flexibility is possible through use of static or traveling raster patterns.

Static patterns are raster patterns whose geometric shapes may be varied infinitely. The patterns are displayed on the part with no relative movement between the part and the pattern.

Traveling patterns are used when heat treating of large areas is required. A static pattern of some geometric shape is displayed on the part, and motion is applied either to the raster pattern or to the part.

Control of Case Depth

Hardening depth is simply a function of time. Because the heat applied to the surface travels through the material by conduction, the longer the surface is maintained at a temperature above the critical value, the deeper the heat will penetrate, and thus the greater will be

Fig. 5 Electron-beam energy input for heat treating

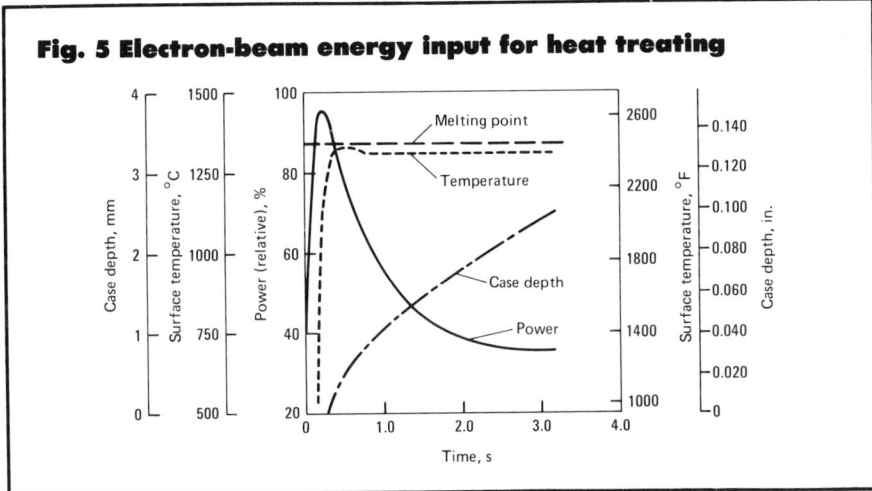

the case depth. Moreover, if the temperature of the surface is raised fast enough and if the temperature is sufficiently high to enable the carbon to go quickly into solution, it is possible to obtain a reasonably deep case with very little heat-affected zone.

Quench Media

To complete the surface-hardening process, the area heated must be quenched rapidly to develop full hardness. This limits the majority of electron-beam applications to those materials that are self-quenching. There must be sufficient mass beneath the surface being hardened to conduct the heat from the surface to the core of the workpiece rapidly. The transfer of heat must be rapid enough to prevent formation of pearlite and to reduce the surface temperature below the M_s (martensite-start) temperature before bainite transformation begins. For carbon steels, eight units of mass generally are required for each unit of case at the desired hardness.

In such a self-quenching process, quenching of the austenite occurs from the inside to the outside, which means that the inner layers harden first. Because transformation of ductile austenite to hard martensite causes expansion, the inner layer expands and plastically extends the surface layers while the surface is still austenitic. When the surface material finally transforms and expands, it generates moderate biaxial compressive forces in conjunction with the hard, virtually unyielding underlayer. This is an ideal residual-stress condition for fatigue resistance, similar to the benefits derived from controlled shot peening.

Advantages of Electron-Beam Heat Treating

Electron-beam heat treating is a precise process that allows the heat treater control over every aspect of the flow of energy into a workpiece for selective surface hardening.

Low Energy Usage. Because the flow of energy is very rapid and the area being heated is held to a selective minimum, the core and the surface area not heat treated will not be adversely affected.

Minimal Part Distortion. When energy flow is controlled and rapid and is restricted to highly selective areas, minimum distortion may be realized.

Small Heat-Affected Zone. The combination of high energy, rapid flow of energy and rapid cooling generally yields very small heat-affected zones.

Flexible Surface Hardening. By moving the pattern over the part or moving the part beneath the pattern, large areas may be heat treated in short periods of time. Depending on case depth, these traveling speeds may be in excess of 150 cm/min (60 in./min).

Compressive Surface Stress. Due to the self-quench phenomenon, ductile austenite forms hard martensite from the subsurface to the surface. When the volume of the surface layer expands during formation of martensite, this expansion is resisted by the adjacent unyielding layer.

Maximum Hardness. Case hardness attained by means of the electron-beam process follows textbook relationships of hardness and carbon content, as illustrated in Fig. 1. The ultimate hardness achieved by this process may be slightly greater than that obtained by conventional heat treating methods because of the high efficiency of self-quenching, which is more effective and much faster than quenching with external quenchants.

Short Heat Treating Time. An accelerated electron beam delivers high-level energy to the surface being modified. Due to this high energy level and to the relatively low thermal conductivities of ferrous materials, heating times for electron-beam heat treating normally are on the order of 0.5 to 2.5 s for static patterns. For most applications, heating times range from 0.5 to 1.0 s.

Minimal Workpiece Oxidation. Carbon-bearing materials heat treated in an electron-beam system may be protected in a vacuum environment and thus may be treated without surface discoloration.

Heat Treating of Cast Irons

Introduction to Heat Treating of Cast Irons

CAST IRONS may be compared with steels in their reactions to hardening. However, because cast irons (except white iron) contain graphite and substantially higher percentages of silicon, they require higher austenitizing temperatures. The graphitizing effect of silicon is so powerful that an unalloyed gray iron may become completely graphitized below the A_1 temperature during heating for austenitizing. Thus, some high-silicon irons require longer intervals at the austenitizing temperature for reabsorption of the desired carbon content in austenite. This interval is extended because silicon retards the absorption of carbon in austenite. The lower-silicon irons respond best to heat treatment.

The response of the cast iron matrix to hardening depends on the carbon and alloy content of austenite at the austenitizing temperature. The higher the austenitizing temperature above A_1, the more carbon is dissolved in austenite, and the higher the as-quenched hardness. This increase in hardness and dissolved carbon above A_1, and the decrease with graphitization during heating below A_1, are shown in Table 1.

Hardness Measurements

Conventional hardness measurements on cast irons always indicate

Table 1 Effects of quenching temperature on combined carbon content and as-quenched hardness of gray irons
Specimens, 30-mm- (1.2-in.-) diam bars 50 mm (2 in.) long, were quenched in oil from temperatures shown after holding for 1 h

Condition	Unalloyed iron(a) Combined carbon, %	Hardness, HB	Cr-Ni-Mo iron(b) Combined carbon, %	Hardness, HB
As cast	0.69	217	0.70	255
Quenched from:				
650 °C (1200 °F)	0.54	207	0.65	250
675 °C (1250 °F)	0.38	187	0.63	241
705 °C (1300 °F)	0.09	170	0.59	229
730 °C (1350 °F)	0.09	143	0.47	217
760 °C (1400 °F)	nil	137	0.45	197
790 °C (1450 °F)	0.05	143	0.42	207
815 °C (1500 °F)	0.47	269	0.60	444
845 °C (1550 °F)	0.59	444	0.69	514
870 °C (1600 °F)	0.67	477	0.76	601

(a) Total carbon content, 3.19%. (b) Total carbon content, 3.10%

lower values than the true hardness of the metal matrix. This discrepancy, which is more pronounced in gray iron than in ductile and malleable irons, occurs because conventional hardness readings are composite values that reflect the hardnesses of both the matrix material and soft graphite.

Provided that the test impression is made on a smooth, flat surface of adequate size, a Brinell hardness tester employing a 10-mm ball penetrator and a 3000-kg load will determine the average hardness of malleable irons with reasonable accuracy. The hardness value, by virtue of the size of the ball penetrator, is representative of the gross structure under test.

In determining the hardness of small castings, it is often impossible to use a Brinell tester, and a Rockwell tester must be employed. Conversions have been developed to compensate for the discrepancies inherent in hardness measurements made on malleable irons with the smaller Rockwell penetrator. Figure 1(a) and (b) show conversions from Brinell (HB) to Rockwell B and G (HRB and HRG) scales for malleable and pearlitic malleable irons,

Fig. 1 Hardness conversions

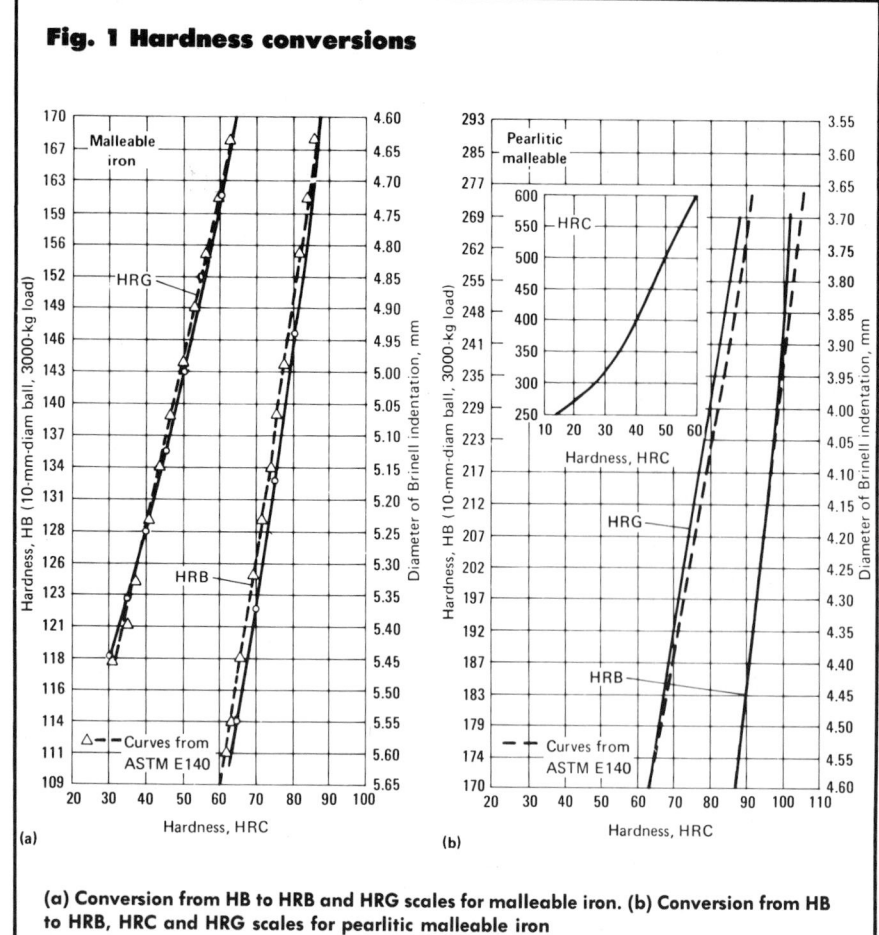

(a) Conversion from HB to HRB and HRG scales for malleable iron. (b) Conversion from HB to HRB, HRC and HRG scales for pearlitic malleable iron

Table 2 Comparative hardness values for quenched and tempered ductile irons

Iron	HB(a)	HRC converted from HB(b)	Observed HRC(c)	Microhardness, HV(d)	HRC converted from HV(b)	HRC converted from HV minus observed HRC
1	415	44.5	44.4	527	50.9	6.5
2	444	47.2	45.0	521	50.6	5.6
3	444	47.2	45.7	530	51.1	5.4
4	444	47.2	47.6	593	54.9	7.3
5	461	48.8	46.7	595	55.0	8.3
6	461	48.8	48.3	560	53.0	4.7
7	461	48.8	49.1	581	54.2	5.1
8	477	50.3	49.6	572	53.7	4.1
9	477	50.3	50.1	618	56.2	6.1
10	555	55.6	53.4	637	57.2	3.8

(a) Average of three readings for each iron. (b) Values based on SAE-ASM-ASTM hardness conversions for steel. (c) Average of five readings for each iron. (d) Average of a minimum of five readings for each iron; 100-kg load

respectively. Figure 1(b) also shows HRC equivalents for HB values, for pearlitic malleable. These conversions are generally accepted by producers of malleable iron.

Matrix hardness is determined more accurately by Brinell than by Rockwell hardness tests, although both require correction to obtain true matrix hardness. The microhardness test is the most accurate method for determining matrix hardness.

Comparative hardness readings obtained on ten quenched and tempered ductile irons are given in Table 2. Observed HRC readings ranged from 3.8 to 8.3 points lower than those converted from microhardness readings.

Figure 2(a) shows the relation between observed HRC readings and those converted from microhardness values for five gray irons of different carbon equivalents. Hardness measurements were taken at two laboratories after quenching and after tempering of each iron. The data in Fig. 2(a) show why the observed values obtained by conventional hardness testing may be misleading, and help to explain the good wear resistance of gray irons with apparently low hardness. Note that there is a correlation with carbon equivalent for all five irons tested and that the discrepancy between observed and converted hardness values diminishes at the lower hardness level.

Another comparison between observed and converted HRC values for gray and ductile irons is shown in Fig. 2(b). These irons were quenched in water from 900 °C (1650 °F) and tempered at 425 °C (800 °F) for 2 h.

Heat Treating Equipment

Heat treating equipment for hardening cast iron should be selected with an understanding of the differences between the hardening characteristics of cast iron and those of steel. For example, cast iron (especially gray iron) is more likely than steel to crack on heating if it is not heated uniformly. In the austenitizing range, metallurgical reactions within cast irons are more complex than those that occur in steels; thus, heat treating of cast iron requires more accurate temperature control. In properly balanced alloys, the hardenability of cast iron increases with alloy content, particularly manganese, vanadium and molybdenum contents. Chromium content increases hardenability, but it also complicates casting production.

For simplicity of processing, and to avoid the costs associated with the use of alloy iron, quenching facilities and procedures should, wherever possible, be arranged so as to take maximum advantage of the hardening characteristics of low-alloy or unalloyed irons. Finally, cast irons increase in volume during hardening to a much greater extent than do steels, and this added growth contributes to heat treating problems.

Table 3 summarizes the principal advantages and disadvantages of three types of heating equipment that may be

Fig. 2 Relations between observed and converted hardness values for gray and ductile irons

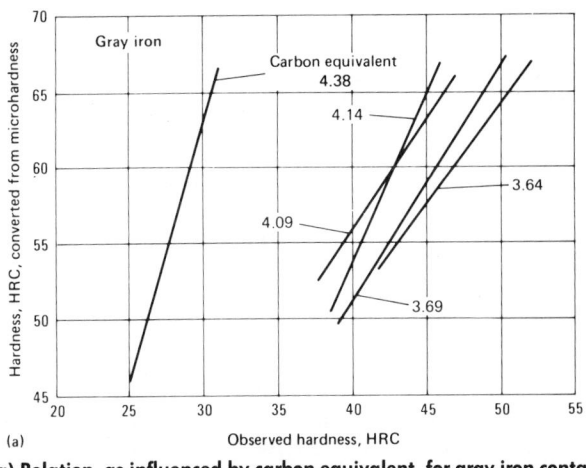
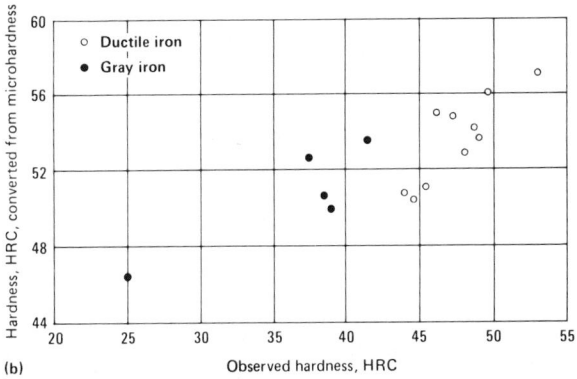

(a) Relation, as influenced by carbon equivalent, for gray iron containing type 3 graphite. (b) Relation for gray and ductile irons quenched in water from 900 °C (1650 °F) and tempered 2 h at 425 °C (800 °F)

Table 3 Advantages and disadvantages of equipments used for heat treating of cast iron

Advantages	Disadvantages
Direct-fired equipment	
Lowest initial, operating and maintenance costs	Forms oxides, making cleaning difficult
Comparatively easy to operate and mechanize	Cannot be used for machined castings
Controlled-atmosphere equipment	
Lower operating cost than salt bath	Entails higher operating costs than direct-fired equipment
Atmospheres can be adjusted more readily than salt composition, to suit work	Requires specially trained personnel to operate and maintain the atmosphere-generating equipment
No contamination of quench medium	
Machined castings can be heat treated	Equipment vulnerable to atmosphere leaks
Comparatively easy to mechanize	
Prevents decarburization and scaling	Fixturing needed to prevent distortion
Salt bath equipment	
High heating rate	High operating costs
Fixturing for preventing distortion less essential than for other methods	Salt composition less readily changed than atmosphere, to suit work
Machined castings can be heat treated	Contamination of quench medium
Salt greatly aids in removal of dirt and sand	Salt must be removed after treatment
Prevents decarburization and scaling	Not easily mechanized

considered for use in heat treating cast irons.

Advantages and disadvantages must be weighed also in selecting methods and equipment for work handling. Belt furnaces provide more uniform heating and quenching than is obtained in processes employing baskets or trays. Baskets and trays, however, permit easier fixturing for control of distortion, particularly for small parts. A major disadvantage of trays and baskets is the high cost of the alloy materials of which they must be made because of the high temperatures to which they are subjected during austenitizing for hardening. They also require more energy because they are removed from the furnace for reloading and permitted to cool down.

Heating Rate. In general, it is recommended that all sections of a casting be heated uniformly, to minimize the thermal stresses that can cause cracking. The danger of cracking is greatest at temperatures up to 425 °C (800 °F) and diminishes thereafter. The rate of heating that can be tolerated for gray iron depends on the complexity of the casting and the amount of residual stress in the as-cast condition.

Continuous furnaces with zoned temperatures, and preheating furnaces, are used to handle complex gray iron castings. Batch furnaces, in which the work is immediately exposed to high temperature, have been used for simpler castings whose shapes and section sizes ensure uniform heating. Salt baths have been used satisfactorily for castings of uniform wall thickness, such as cylinder sleeves, but the rate of heating in molten salt (about five or more times the rate achieved in air) is too fast for gray iron castings of complex shape.

Temperature Control. To avoid thermal gradients, both furnace temperature and heating rate must be uniform. Large castings, in particular, require a high degree of temperature uniformity to avoid cracking; such uniformity is best provided by furnaces with indirect firing and good circulation of atmosphere. Direct firing on a casting has been known to cause cracking and may make that area of the casting softer.

All cast irons except white iron contain graphite, some of which will dissolve in the matrix at elevated temperature. In heating for hardening, iron carbide breaks down into ferrite plus additional graphite at temperatures in the range of about 650 to 790 °C (1200 to 1450 °F). From 790 to about 845 °C (1450 to about 1550 °F), the iron achieves a state of equilibrium if held at any one temperature, and is made up of ferrite, austenite with a fixed content of dissolved carbon, and graphite. At temperatures above about 845 °C (1550 °F), all ferrite is transformed to austenite with a dissolved carbon content that increases with increasing temperature.

Although, with reasonable temperature uniformity, steel will produce the same structures over a wide temperature range, cast iron may undergo structural alterations with every temperature increment. Therefore, furnaces for hardening of cast iron should have adequate atmosphere circulation and good proportioning control to ensure uniform furnace temperatures and a narrow control range. Limiting temperature controls also are desirable.

Heating Media. Air, controlled atmosphere and molten salt are in general use for hardening of cast iron. Because cast iron is subject to decarburization and scaling at heat treating temperatures, controlled atmospheres or molten salt should be used if either of these conditions is objectionable. Salt-bath equipment is less costly to install than controlled-atmosphere equipment, but is generally more costly to operate. In one high-production martempering installation, the decrease in operating costs that was realized by switching from high-temperature salt to controlled atmosphere offset the increase in equipment costs in less than one year.

Growth during hardening is directly influenced by the heating medium used. In general, permanent growth is

Table 4 Effect of heating medium on growth of gray iron cylinders

Cylinders were 120 mm (4³/₄ in.) in diameter. Composition of iron: 3.38 total C, 2.10 Si, 0.65 Mn, 0.40 Cr, 0.55 Cu + Ni

Test(a)	Atmosphere or heating medium	Heating time, min	Temperature °C	°F	Average growth, mm/mm (in./in.)
1	Charcoal-base(b)	80	845	1550	0.00048
2	Lean endothermic(c)	60	870	1600	0.0019
3	Lean endothermic(c)	60	870	1600	0.0021
4	Air	60	870	1600	0.0026
5	Air	60	870	1600	0.0026
6	Molten salt(d)	10	845	1550	0.0027

(a) Test specimens were taken from three castings: specimens for tests 1 and 6 were taken from the first casting, 2 and 4 from the second, and 3 and 5 from the third. Specimens were fully annealed before hardening; final hardness range of oil-quenched specimens, 45 to 50 HRC. (b) Approximate composition: 64 N_2, 35 CO, 1 H_2; dew point, −29 °C (−20 °F). (c) Approximate composition: 45 N_2, 20 CO, 35 H_2, 0.4 CO_2, 0.3 CH_4; dew point, 10 °C (50 °F). (d) Composition: 48 to 52 NaCl, 48 to 52 KCl

attributed to the breakdown of iron carbide into iron and graphite, with resultant volume expansion, and to internal oxidation of the iron. Even when heating and cooling rates are fairly high, and the iron is held for only short times at the temperatures at which these effects occur, the heating medium may still exert some influence on the amount of growth that takes place. The growth in different heating media of specimens from three different 120-mm- (4³/₄-in.-) diam gray iron cylinder castings of the same composition is given in Table 4. Because these castings were ferritic prior to hardening, dimensional results were not affected by the breakdown of iron carbide during heat treatment.

From the data in Table 4 it is apparent that the oxidizing characteristics of gaseous media have the greatest influence on growth of cast iron heated in these media. It is significant that the growth that occurred in the nonoxidizing charcoal-base atmosphere was similar to that of steel during hardening.

The high growth in molten salt may be attributed to thermal upsetting caused by rapid heating. Characteristically, thermal upsetting reduces the length of a casting and increases its girth.

Quenching Facilities. Selection of a quenching medium is closely related to the complexity of the shape to be quenched, which largely determines the probability of quench cracking. Because many castings are of complex design, hot oil and molten-salt quenching are used much more frequently than water quenching. When castings with heavy sections are being quenched, agitation of the quenching bath may be necessary (for data on the effect of agitation on cooling rate, see the article on quenching of steel, in this volume).

When poor quenching facilities are used, increased alloy additions may be necessary to achieve uniform hardening. However, alloy additions may complicate production of castings in the foundry and may increase casting costs.

Heat Treating of Gray Irons

THE HEAT TREATMENT most frequently applied to gray iron, with the possible exception of stress relieving, is annealing. Annealing of gray iron consists of heating the iron to a temperature high enough to soften it, and/or to minimize or eliminate massive eutectic carbides, thus improving its machinability. This heat treatment reduces mechanical properties substantially, however. It will reduce the grade level approximately to the next lower grade; for example, the properties of a class 40 gray iron will be diminished to those of a class 30 gray iron. Figure 1 shows the effect of full annealing on tensile strength of class 30 gray iron arbitration bars. The degree of reduction of properties depends on the annealing temperature, the time at temperature, and the alloy composition of the iron.

Annealing

Gray iron commonly is subjected to one of three annealing treatments, each of which involves heating to a different range of temperature. These treatments are ferritizing annealing, medium (or "full") annealing and graphitizing annealing.

Ferritizing Annealing. For an unalloyed or low-alloy cast iron of normal composition, when the only result desired is the conversion of pearlitic carbide to ferrite and graphite for improved machinability, it generally is

unnecessary to heat the casting above the transformation range. Up to approximately 595 °C (1100 °F), the effect of short times at temperature on the structure of gray iron is insignificant. As the temperature increases above 595 °C (1100 °F), the rate at which iron carbide decomposes to ferrite plus graphite increases markedly, reaching a maximum at the lower transformation temperature (about 760 °C, or 1400 °F, for unalloyed or low-alloy iron). This is indicated in Fig. 2, which shows the structure of unalloyed gray iron in the as-cast condition and after being held for 1 h at 760 °C (1400 °F) per inch of section. Heating to a higher temperature for the same period of time may be detrimental to the annealing process if it causes partial or complete transformation to austenite.

For most gray irons, a ferritizing annealing temperature between 705 and 760 °C (1300 and 1400 °F) is recommended. The furnace temperature profile must be such that castings are sure to reach the set temperatures. Precise temperatures within this range depend on the exact composition of the iron. When machining properties are of primary importance, it is advisable to anneal several samples at various temperatures between 705 and 760 °C in order to determine the temperature that yields the lowest final hardness.

The casting must be held at temperature long enough to allow the graphi-

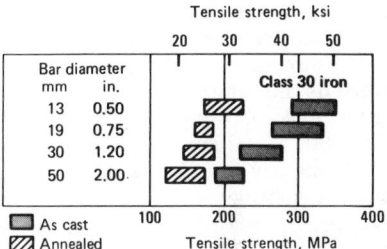

Fig. 1 Effect of annealing on tensile strength of class 30 gray iron

Tensile strength, ksi

Bar diameter				
mm	in.			Class 30 iron
13	0.50			
19	0.75			
30	1.20			
50	2.00			

As cast
Annealed

Tensile strength, MPa

Specimens were arbitration bars from 31 heats. Bars were annealed at 925 °C (1700 °F) for 2 h, plus 1 h per 25 mm (1 in.) of section over 25 mm, and cooled at a maximum rate of 160 °C/h (285 °F/h) from 925 to 565 °C (1700 to 1050 °F). Cooling continued from 565 °C at a maximum rate of 130 °C/h (230 °F/h) to 200 °C (390 °F); bars were then air cooled to room temperature.

tizing process to proceed to completion. At temperatures below 705 °C (1300 °F), an excessively long holding time usually is required. At temperatures between 705 and 760 °C (1300 and 1400 °F), holding time varies with chemical composition, and may be as short as 10 min for unalloyed irons. If an unusually low rate of cooling is employed, the

Fig. 2 Conversion of as-cast pearlitic structure of unalloyed gray iron to ferrite and graphite by annealing

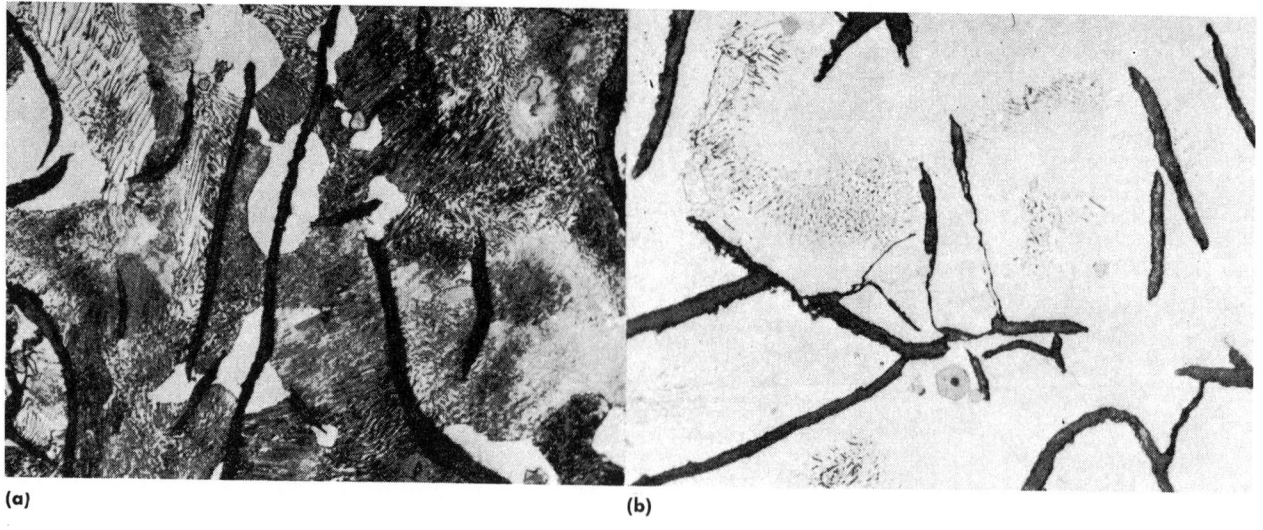

(a)   (b)

(a) As cast; 180 HB. (b) Annealed 1 h at 760 °C (1400 °F); 120 HB. Magnification, 500X

time at temperature may be still further reduced.

Although the rate of cooling *per se* is not of great importance to the annealing process, slow cooling is recommended if the stress relief that automatically occurs during annealing is to be retained as the casting cools to room temperature. A cooling rate as high as 110 °C/h (200 °F/h) down to 290 °C (550 °F) is satisfactory for all except the most complex castings.

Medium ("full") annealing is usually performed at temperatures between 790 and 900 °C (1450 and 1650 °F). This treatment is used when a ferritizing anneal would be ineffective because of the high alloy content of a particular iron. It is recommended, however, that the efficacy of temperatures at or below 760 °C (1400 °F) be tested before a higher annealing temperature is adopted as part of a standard procedure.

Holding times comparable to those used in ferritizing annealing are usually employed. When the high temperatures of medium annealing are used, however, the casting must be cooled slowly through the transformation range, from about 790 to 675 °C (1450 to 1250 °F).

Graphitizing Annealing. If the microstructure of gray iron contains massive carbide particles, higher an-nealing temperatures are necessary. Graphitizing annealing may have the purpose simply of converting massive carbide to pearlite and graphite, although in some applications it may be desired to carry out a ferritizing annealing treatment to provide maximum machinability.

Production of free carbide that must later be removed by annealing is, except with pipe and permanent mold castings, almost always an accident resulting from inadequate inoculation or the presence of excess carbide formers, which inhibit normal graphitization; thus, the annealing process is not considered to be part of the normal production cycle.

To break down massive carbide with reasonable speed, temperatures of at least 870 °C (1600 °F) are required. With each additional 55 °C (100 °F) increment in holding temperature, the rate of carbide decomposition doubles; consequently, it is general practice to employ holding temperatures of 900 to 955 °C (1650 to 1750 °F). However, at 925 °C (1700 °F) and above, the phosphide eutectic present in irons containing 0.10% P or more may melt.

The holding time at temperature may vary from a few minutes to several hours. The chill (white iron) in some high-silicon, high-carbon irons can be eliminated in as little as 15 min at 940 °C (1720 °F). In all applications, unless a controlled-atmosphere furnace is used, the time at temperature should be as short as possible, because at these high temperatures gray iron is susceptible to scaling if moisture is present in the furnace atmosphere.

The cooling rate chosen depends on the final use of the iron. If the principal object of the treatment is to break down carbides, and it is desired to retain maximum strength and wear resistance, the casting should be air cooled from the annealing temperature to about 540 °C (1000 °F), to promote formation of a pearlitic structure. If maximum machinability is the object, the casting should be furnace cooled to 540 °C (1000 °F), and special care should be exerted to ensure slow cooling through the transformation range. In both instances, cooling from 540 °C (1000 °F) to about 290 °C (550 °F) at not more than 110 °C/h (200 °F/h) is recommended, to minimize residual stresses.

Effect of Alloy Content on Time at Temperature. Certain elements, such as carbon and silicon, will accelerate decomposition of pearlite and massive carbide at annealing temperatures; therefore, when these elements are present in sufficient percentages, the time at annealing temperature may be reduced. In an investigation of the decomposition of pearlite at various temperatures in irons containing 1.93 and 2.68% Si, it was determined that the pearlite always broke down more rapidly in the higher-silicon iron, and that this iron could be effectively annealed over a greater temperature range. For example, at an annealing temperature of 750 °C (1380 °F), complete breakdown of pearlite occurred in the higher-silicon iron in 10 min, whereas 45 min was required for the lower-silicon iron. This shows the pronounced effect of silicon as an aid to diffusion of carbon to the flakes present in the iron.

On the other hand, both the carbide-stabilizing elements (vanadium, chromium and manganese) and the pearlite-stabilizing elements (phosphorus, nickel and copper) retard the rate of pearlite decomposition. The percentage increases in the time required to decompose pearlite that are effected by 0.10% additions of five of these elements are as follows:

Manganese 60%
Nickel. 30%
Copper 30%
Chromium 200%
Phosphorus 30%

Normalizing

Gray iron is normalized by being heated to a temperature above the transformation range, held at this temperature for a period of about 1 h per inch of maximum section thickness, and cooled in still air to room temperature. Normalizing may be used to enhance mechanical properties, such as hardness and tensile strength, or to restore as-cast properties that have

been modified by another heating process, such as graphitizing or the preheating and postheating associated with repair welding.

The temperature range for normalizing gray iron is approximately 885 to 925 °C (1625 to 1700 °F). Heating temperature has a marked effect on microstructure and on mechanical properties such as hardness and tensile strength. This is demonstrated in Table 1, which presents data on specimens 30 mm in diameter by 178 mm (1.2 in. in diameter by 7 in.) of unalloyed and alloy gray irons that were heated for 1 h at temperatures ranging from 540 to 980 °C (1000 to 1800 °F) in increments of 55 °C (100 °F). Specimens of both irons were cooled in still air from each of these temperatures. As indicated by the separation of the table between 760 and 815 °C (1400 and 1500 °F), air cooling from 760 °C or below softens the as-cast structure. Temperatures of 815 °C and higher are above the transformation temperature; hence, the matrix of the as-cast iron is converted to austenite on heating and is transformed during air cooling to pearlite that varies in fineness depending on the maximum temperature (the normalizing temperature) and the alloy content. For the alloy iron, the higher the normalizing temperature, the harder and stronger was the normalized structure. For the unalloyed iron, all normalizing temperatures produced the same hardness and strength, and all structures produced were softer than the as-cast material. Thus, normalizing is a function of alloy content, which affects pearlite spacing.

Some control of hardness can be exer-

cised during normalizing by allowing castings to cool in the furnace to a temperature below the normalizing temperature. Figure 3 shows the results obtained with gray iron rings that were heated to 955 °C (1750 °F), then furnace cooled to different temperatures before being removed from the furnace and cooled in air. These data also indicate that annealing can be accomplished by cooling castings in the furnace to 650 °C (1200 °F) and then air cooling. However, if stress-free castings are desired, they should be cooled in the furnace to below 455 °C (850 °F) before being removed.

Additional effects of normalizing, as a function of alloy content and carbon equivalent, are shown in Table 2. Bars 1, 3, 4, 6 and 7 are essentially free of alloying elements, except for residual amounts. Bars 1 and 3, characterized by high as-cast strength and low carbon equivalent, virtually regained their as-cast strength as a result of normalizing for 1½ h at 900 °C (1650 °F), air cooling, and stress relieving at 540 °C (1000 °F). The same treatment lowered the strength of bars 4, 6 and 7, all of which had higher carbon equivalents. Bar 2 showed an increase in strength because of the high stabilizing effect of the molybdenum and nickel contents. Bar 5, despite a high carbon equivalent, greatly exceeded as-cast strength because of its chromium, molybdenum and nickel contents.

The effect of alloy content on hardness after normalizing is shown in Fig. 4 for two alloy irons with different carbon equivalents and nickel and chromium contents. Again it is evident that alloy content has a stabilizing effect in

Table 1 Effect of air cooling from various temperatures on typical properties of gray irons

Condition(a)		Unalloyed iron(b)			Alloyed iron(c)			Combined carbon, %
		Hardness, HB	Tensile strength MPa	ksi	Hardness, HB	Tensile strength MPa	ksi	
As cast		207	265	38.1	212	265	38.7	0.84
Air cooled from:								
°C	°F							
540	1000	202	210	30.4	212	275	39.7	0.82
595	1100	190	255	36.8	210	275	40.0	0.86
650	1200	138	195	28.2	202	265	38.6	0.80
705	1300	125	180	26.4	187	265	38.4	0.81
760	1400	131	190	27.4	170	235	34.2	0.60
815	1500	152	205	29.7	212	295	42.6	0.76
870	1600	152	205	29.7	217	305	44.2	0.81
925	1700	152	205	29.6	223	290	42.4	0.82
980	1800	152	210	30.1	255	340	49.0	0.80

(a) Specimens 30 mm in diameter by 180 mm (1.2 in. in diameter by 7 in.) were held at temperature for 1 h before being cooled in still air to room temperature. (b) As-cast composition: 3.15 total C, 0.54 combined C, 2.59 Si, 0.09 P, 0.135 S, 0.88 Mn, 0.01 Cr, 0.10 Ni. (c) As-cast composition: 3.33 total C, 0.84 combined C, 2.27 Si, 0.076 P, 0.122 S, 0.72 Mn, 0.44 Cr, 0.36 Ni, 0.28 Mo

Table 2 Influence of alloy content and carbon equivalent on typical properties of gray irons before and after normalizing

Specimens 30 mm (1.2 in.) in diameter were normalized at 900 °C (1650 °F), then stress relieved at 540 °C (1000 °F).

Bar	C	Si	P	S	Composition, % Mn	Cr	Ni	Mo	Cu	Carbon equiv-alent, %	As cast Tensile strength MPa	ksi	Hard-ness, HB	Normalized Tensile strength MPa	ksi	Hardness, HB
1	2.71	2.00	0.13	0.031	0.46	0.076	0.061	0.059	...	3.37	405	58.9	241	380	55.0	241
2	3.25	2.03	0.02	0.031	0.67	0.085	0.80	0.30	0.22	3.93	380	54.8	241	425	62.0	255
3	2.66	1.90	0.03	0.018	0.63	0.063	0.092	0.042	...	3.27	400	57.7	255	385	55.7	241
4	3.15	2.20	0.38	0.018	0.44	0.074	0.071	0.071	0.39	3.88	295	43.0	229	235	34.3	179
5	3.45	2.16	0.09	0.077	0.84	0.39	1.21	0.50	0.10	4.17	250	36.2	248	405	58.5	311
6	3.31	2.10	0.39	0.070	0.41	0.069	0.08	0.055	0.44	4.01	275	39.9	212	200	29.2	163
7	3.42	2.44	0.42	0.058	0.56	0.063	0.058	0.057	0.108	4.23	215	31.1	187	180	26.0	143

Fig. 3 Room-temperature hardness of gray iron after normalizing

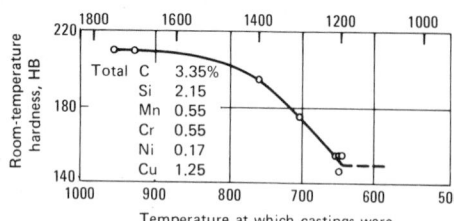

Effect of temperature at start of air cooling on hardness of normalized gray iron rings 120 mm (4³/₄ in.) in OD, 38 mm (1¹/₂ in.) high and 13 mm (¹/₂ in.) thick

the graphitizing annealing range and serves to increase hardness when the austenitizing temperature ranges from about 790 to 980 °C (1450 to 1800 °F).

Thus it can be concluded that normalizing serves to restore as-cast properties to gray iron—or, if carbon equivalent is sufficiently low, even causes these properties to be exceeded—and that the alloying elements chromium, molybdenum and nickel enhance the strengthening effect of normalizing.

Hardening and Tempering

Gray irons are hardened and tempered to improve their mechanical properties, particularly strength and wear resistance. After being hardened and tempered, these irons usually exhibit wear resistance approximately five times as great as that of pearlitic gray irons.

Furnace or salt-bath hardening can be applied to a wider variety of gray irons than either flame or induction hardening. In flame and induction hardening, a relatively large content of combined carbon is required because of the extremely short period available for solution of carbon in austenite. In furnace or salt-bath hardening, however, the casting can be held at a temperature above the transformation range for as long as necessary, and even an iron initially containing no combined carbon can be hardened.

Unalloyed gray iron of low combined carbon content must be austenitized for a relatively long time. Figure 5 shows the relation of hardness to holding time for small quenched specimens. With increased time, more carbon is dissolved in austenite, and hardness after quenching is increased.

Because of its higher silicon content, an unalloyed gray iron with a combined carbon content of 0.60% exhibits higher hardenability than a carbon steel with the same carbon content. However, because of the effect of silicon in reducing the solubility of carbon in austenite,

unalloyed irons with higher silicon contents necessarily require higher austenitizing temperatures to attain maximum hardenability.

Manganese increases hardenability. Approximately 1.50% Mn was found to be sufficient for through hardening of a 38-mm (1¹/₂-in.) section in oil, or for through hardening of a 64-mm (2¹/₂-in.) section in water.

Manganese, vanadium and molybdenum are the recognized elements for increasing the hardenability of gray iron; chromium and nickel are used to accentuate the effect of molybdenum. Although chromium, by itself, does not influence the hardenability of gray iron, its contribution to carbide stabilization is important, particularly in flame hardening. Figure 6 demonstrates the effects of various combinations of elements on hardenability.

Austenitizing. In hardening gray iron, the casting is heated to a temperature high enough to promote formation of austenite, held at that temperature until the desired amount of carbon has been dissolved, and then quenched at a suitable rate. Heating for austenitizing may be accomplished in a salt bath or in an electrically heated, gas-fired or oil-fired furnace.

The temperature to which the casting must be heated is determined by the transformation range of the particular gray iron of which it is made. The transformation range can extend more than 55 °C (100 °F) above the A_1 (transformation-start) temperature. A formula for determining the approximate A_1 transformation temperature of unalloyed gray iron is:

$$°C: 730 + 28.0 \ (\% \ Si) - 25.0 \ (\% \ Mn)$$
$$°F: 1345 + 50.4 \ (\% \ Si) - 45.0 \ (\% \ Mn)$$

Chromium raises the transformation

range of gray iron. In high-nickel, high-silicon irons, for example, each percent of chromium raises the transformation range by about 40 °C (72 °F). Nickel, on the other hand, lowers the critical range. In a gray iron containing from 4 to 5% Ni, the upper limit of the transformation range is about 710 °C (1310 °F).

Provided that recommended limits are not exceeded, the higher the casting is heated above the transformation range, the greater will be the amount of carbon dissolved in the austenite (Fig. 7) and the higher will be the hardness of the casting after quenching

(Table 3). In practice, temperatures as much as 95 °C (175 °F) higher than the calculated A₁ transformation temperature are used to ensure full austenitizing. However, excessively high temperatures should be avoided, because quenching from such high temperatures increases the danger of distortion and cracking and promotes retention of austenite, particularly in alloyed irons.

Castings should be heated through the lower temperature range slowly, in order that cracking may be avoided. Above a range of 595 to 650 °C (1100 to 1200 °F)—that is, above the stress-

relieving range—heating may be as rapid as desired. In fact, time may be saved by heating the casting slowly to about 650 °C in one furnace and then transferring it to a second furnace and bringing it rapidly up to the austenitizing temperature. As little as 20 min per inch of section may be sufficient time at temperature.

Quenching. Oil is the quenching medium most frequently used for gray iron. Water generally is not a satisfactory quenching medium for furnace-heated gray iron; it extracts heat so rapidly that distortion and cracking are likely in all except small parts of simple design. Recently developed water-soluble polymer quenches can provide the convenience of water quenching along with lower cooling rates, which can minimize thermal shock.

The least severe quenching medium is air. Unalloyed or low-alloy gray iron castings usually cannot be air quenched, because the cooling rate is not high enough to form martensite. However, for irons of high alloy content, forced-air quenching is frequently the most desirable cooling method.

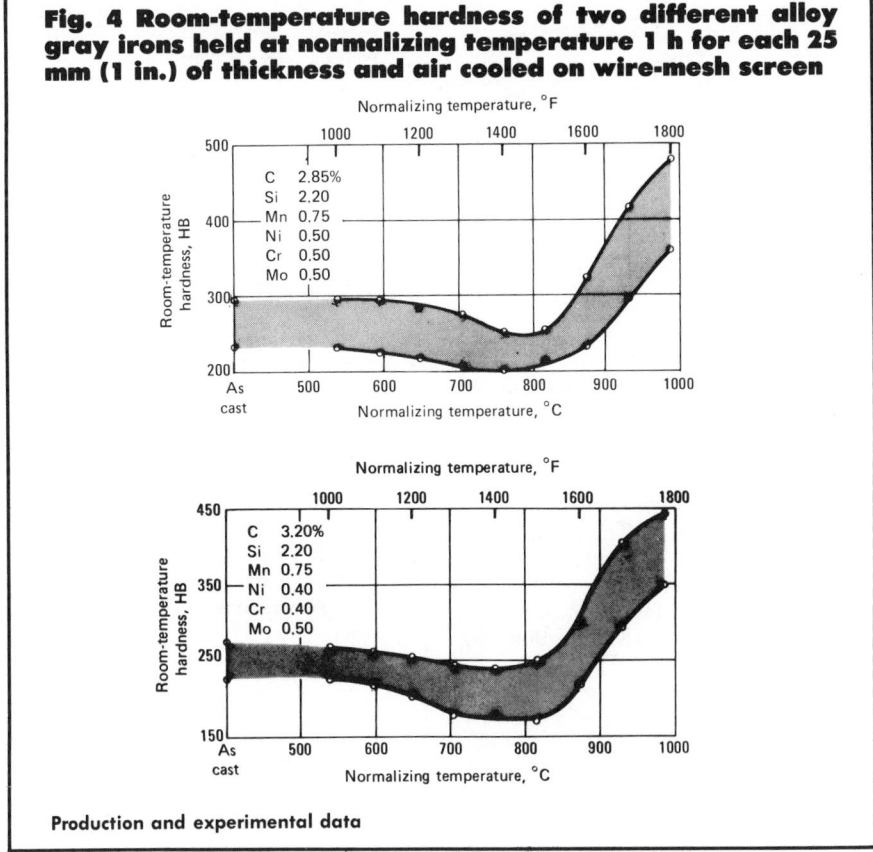

Fig. 4 Room-temperature hardness of two different alloy gray irons held at normalizing temperature 1 h for each 25 mm (1 in.) of thickness and air cooled on wire-mesh screen

Production and experimental data

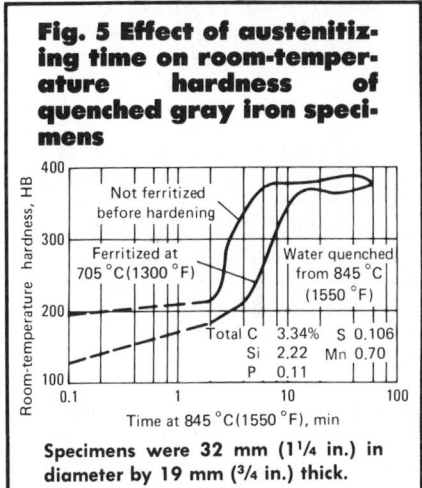

Fig. 5 Effect of austenitizing time on room-temperature hardness of quenched gray iron specimens

Specimens were 32 mm (1¼ in.) in diameter by 19 mm (¾ in.) thick.

Table 3 Effect of austenitizing temperature on hardness of various oil-quenched gray irons

			Composition, %						Hardness, HB, of as-cast iron	Hardness, HB, after oil quenching from			
TC(a)	CC(b)	Si	P	S	Mn	Cr	Ni	Mo		790 °C (1450 °F)	815 °C (1500 °F)	845 °C (1550 °F)	870 °C (1600 °F)
3.19	0.69	1.70	0.216	0.097	0.76	0.03	...	0.013	217	159	269	444	477
3.10	0.70	2.05	...	...	0.80	0.27	0.37	0.45	255	207	444	514	601
3.20	0.58	1.76	0.187	0.054	0.64	0.005	Trace	0.48	223	311	477	486	529
3.22	0.53	2.02	0.114	0.067	0.66	0.02	1.21	0.52	241	355	469	486	460
3.21	0.60	2.24	0.114	0.071	0.67	0.50	0.06	0.52	235	208	487	520	512
3.36	0.61	1.96	0.158	0.070	0.74	0.35	0.52	0.47	235	370	477	480	465

(a) Total carbon. (b) Combined carbon

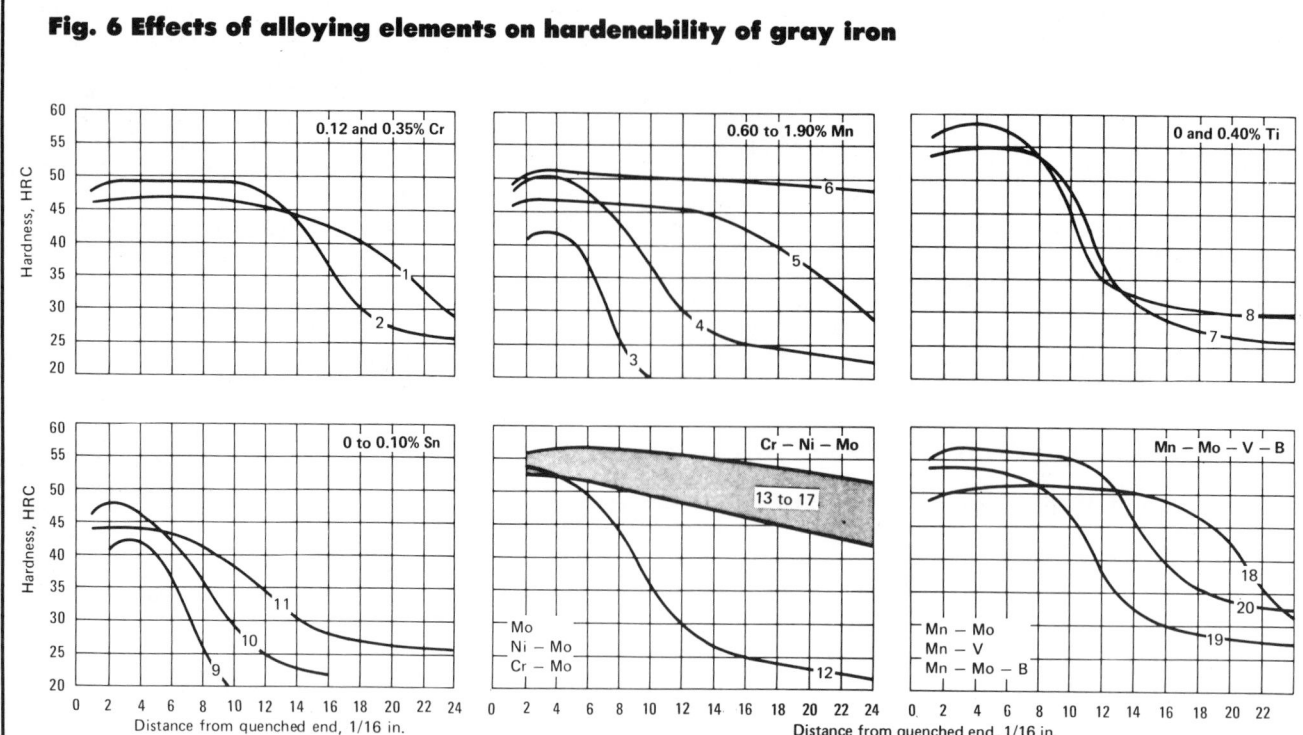

Fig. 6 Effects of alloying elements on hardenability of gray iron

Irons 1 through 20 were induction-furnace melted and poured into standard dry sand molds 30 mm (1.2 in.) in diameter. Alloy constituents for irons 21 through 32 were added to 205-kg (450-lb) ladles, from which the metal was poured into automotive-camshaft production molds. Standard end-quench hardenability specimens were machined from these castings, austenitized for 1 h at the temperatures indicated, and then water quenched.

A casting of nonuniform section should be quenched in such a way that the heavier section enters the quenching bath first. During quenching, agitation is desirable because it ensures even temperature distribution in the bath and improves quenching efficiency. Because as-quenched castings at room temperature are extremely sensitive to cracking, they should be removed from the quench bath as soon as their temperature falls to about 150 °C (300 °F) and tempered immediately.

Tempering. After quenching, castings usually are tempered at temperatures well below the transformation range for about 1 h per inch of thickest section. As the quenched iron is tempered, its hardness decreases, whereas it usually gains in strength and toughness (Fig. 8 and 9).

Applications. Examples of the quenching and tempering temperatures employed, and results obtained, for four different production parts made of gray iron are as follows:

Example 1. Unalloyed gray iron valve guides were heated in an atmosphere-controlled furnace and held for 1 h at 885 °C (1625 °F). The guides were then quenched in oil maintained at about 60 °C (140 °F). An as-quenched hardness of 45 to 50 HRC was obtained. After tempering at 480 °C (900 °F), hardness was 30 to 34 HRC.

Example 2. Valve guides made of gray iron containing 3.40 C, 2.40 Si, 0.21 Cr and 0.50 Cu were heated to 870 °C (1600 °F) and held at this temperature for 1 h. After being oil quenched from 870 °C, the guides were tempered for 1 h at 495 °C (925 °F). Hardness distributions after quenching and after tempering, for 25 guides, are shown in Fig. 10.

Example 3. Automotive valve lifter castings made of gray iron containing 3.10 to 3.30 C, 2.10 to 2.40 Si, 1.00 to 1.25 Cr, 0.40 to 0.70 Ni and 0.50 to 0.70 Mo were heated in a gas-fired radiant-tube conveyor furnace. A controlled atmosphere of endothermic generator gas was used to prevent decarburization of the machined surfaces. The castings were held for 45 min at 855 °C (1570 °F), quenched in oil at 55 °C

Fig. 7 Increase in combined carbon with increase in austenitizing temperature for gray iron

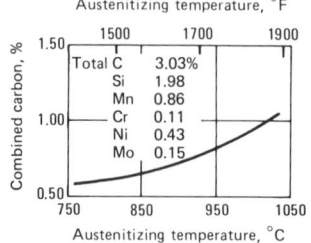

Specimens were furnace heated and water quenched. Combined carbon by difference.

(130 °F), then tempered for 3 h at 150 °C (300 °F) to a hardness of 55 to 61 HRC. The hardness readings were made 1.6 mm (1/16 in.) off center of the tappet face. A distribution of hardness

Fig. 6 (continued)

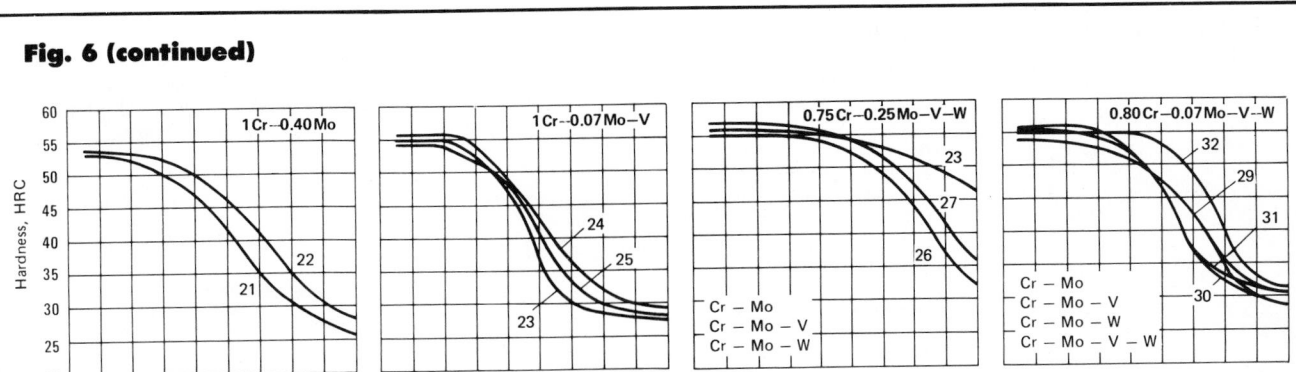

Iron	Total C	Si	P	S	Composition, % Mn	Cr	Ni	Mo	V	Other	Austenitizing temperature °C	°F
1.......	3.30	1.40	0.116	0.10	1.47	0.12	...	...	...	...	840	1540
2.......	3.30	1.90	0.116	0.10	1.43	0.35	...	...	...	...	840	1540
3.......	3.15	2.05	0.124	0.112	0.60	0.06	...	...	...	...	840	1540
4.......	2.97	2.31	0.116	0.116	0.92	0.06	...	...	...	...	840	1540
5.......	3.42	1.90	0.116	0.100	1.47	0.12	...	...	...	...	840	1540
6.......	3.13	2.29	0.116	0.018	1.90	0.08	...	...	...	...	840	1540
7.......	3.00	2.00	0.15	0.10	1.25	...	...	...	...	...	840	1540
8.......	3.00	2.00	0.15	0.10	1.25	...	...	...	...	0.40 Ti	840	1540
9.......	3.15	2.05	0.124	0.112	0.60	...	...	...	...	...	840	1540
10.......	3.10	2.25	0.120	0.160	0.65	...	...	...	...	0.05 Sn	840	1540
11.......	3.10	2.25	0.120	0.160	0.65	...	...	...	...	0.10 Sn	840	1540
12.......	3.19	1.70	0.216	0.097	0.76	0.03	...	0.013	...	...	855	1575
13.......	3.22	1.73	0.212	0.089	0.75	0.03	...	0.47	...	...	855	1575
14.......	3.20	1.76	0.187	0.054	0.64	0.005	Trace	0.48	...	...	855	1575
15.......	3.22	2.02	0.114	0.067	0.66	0.02	1.21	0.52	...	...	855	1575
16.......	3.21	2.24	0.114	0.071	0.67	0.50	0.06	0.52	...	...	855	1575
17.......	3.36	1.96	0.158	0.070	0.74	0.35	0.52	0.47	...	...	855	1575
18.......	3.21	2.01	0.15	0.10	1.53	0.40	...	0.13	...	...	840	1540
19.......	3.20	2.00	0.15	0.10	1.25	0.40	...	...	0.05	...	840	1540
20.......	3.10	2.09	0.15	0.10	1.46	0.44	...	0.14	...	0.095 B	840	1540
21.......	3.22	2.10	0.108	0.088	0.68	0.97	...	0.40	...	...	845	1550
22.......	3.20	2.15	0.108	0.093	0.70	1.00	...	0.41	...	...	845	1550
23.......	3.19	2.55	0.092	0.090	0.71	0.96	...	0.054	0.16	...	845	1550
24.......	3.17	2.20	0.094	0.092	0.66	0.95	...	0.069	0.081	...	845	1550
25.......	3.19	2.20	0.092	0.092	0.68	0.93	...	0.075	0.27	...	845	1550
26.......	3.17	1.90	0.080	0.094	0.65	0.73	...	0.19	...	...	845	1550
27.......	3.25	1.85	0.074	0.092	0.65	0.77	...	0.30	0.13	...	845	1550
28.......	3.21	1.90	0.069	0.100	0.70	0.75	...	0.28	...	0.40 W	845	1550
29.......	3.20	2.20	0.096	0.090	0.68	0.94	...	0.047	0.13	0.75 W	845	1550
30.......	3.12	1.80	0.074	0.090	0.69	0.75	...	0.064	...	...	845	1550
31.......	3.18	1.80	0.073	0.090	0.68	0.77	...	0.091	0.12	...	845	1550
32.......	3.14	1.70	0.079	0.090	0.69	0.77	...	0.071	...	0.37 W	845	1550

values representing castings from 25 heats is given in Fig. 10.

Example 4. Ring-shape castings 64 mm (2.5 in.) in outside diameter by 19 mm (0.75 in.) high by 6.4 to 13 mm (0.25 to 0.5 in.) in wall thickness, made of a hardenable gray iron, were heat treated in groups and sampled for response to heat treatment. These cast-ings were heat treated by holding for ½ h at 850 °C (1560 °F) followed by oil quenching. Hardness data for 20 consecutive lots representing 40 heats are shown in Fig. 10. Each casting was tested at and opposite the gate.

Strength. Tempering increases the tensile strength of hardened gray iron. Although the tempering temperature that will yield maximum strength depends on chemical composition and increases with higher alloy content, the high-strength tempering range for both unalloyed and alloyed gray irons is approximately 370 to 510 °C (700 to 950 °F).

Other Properties. Tempering temperatures near 370 °C (700 °F) increase

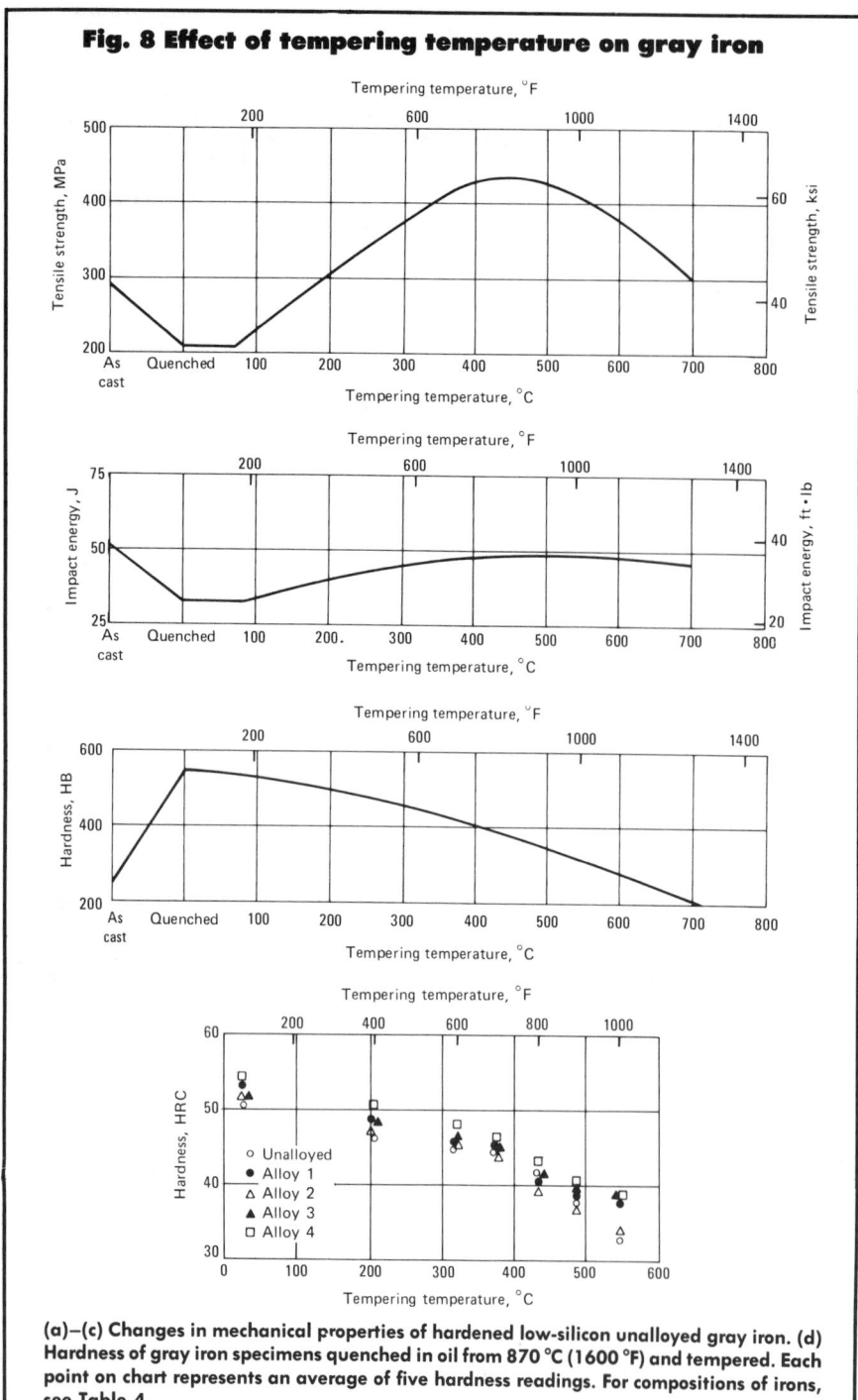

Fig. 8 Effect of tempering temperature on gray iron

(a)–(c) Changes in mechanical properties of hardened low-silicon unalloyed gray iron. (d) Hardness of gray iron specimens quenched in oil from 870 °C (1600 °F) and tempered. Each point on chart represents an average of five hardness readings. For compositions of irons, see Table 4.

Fig. 9 Influence of alloy content on hardness of quenched and tempered gray iron test castings

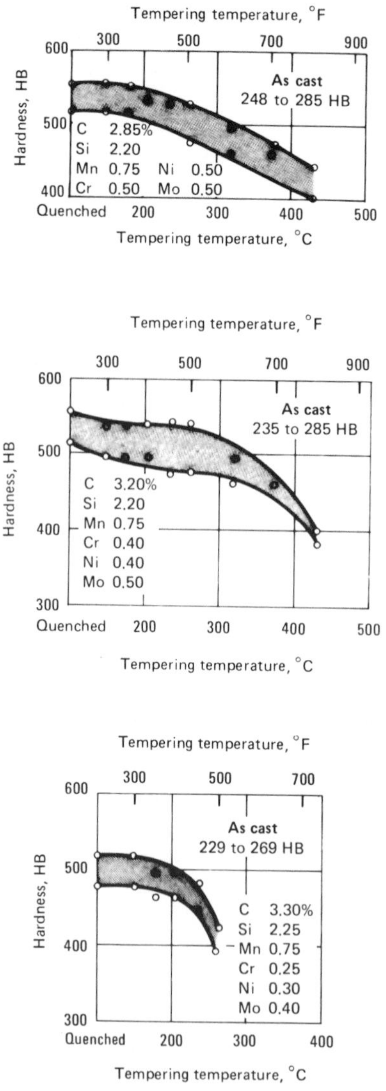

Castings were normalized to the same hardness range before being austenitized for hardening, and were oil quenched from 850 °C (1560 °F).

the impact resistance of hardened low-silicon unalloyed gray iron (Fig. 8). If alloying elements are present, higher tempering temperatures may be necessary.

Although considerable improvements in tensile strength may be brought about by hardening and tempering, proportional improvements in fatigue properties are rarely obtained.

The modulus of elasticity of a low-alloy (1.32% Ni, 0.44% Cr) gray iron was found to be moderately improved by quench hardening and tempering:

Condition	Modulus of elasticity	
	GPa	ksi
As cast.................	122	17 700
Oil quenched	112	16 200
Tempered at 300 °C (570 °F)	131	19 000

The modulus of rupture increased markedly with the use of higher tempering temperatures. A similar increase was noted for the transverse breaking load and deflection. In an alloy iron containing 4.5% Ni and 1.5% Cr, transverse breaking load was increased substantially, even with tempering temperatures as low as 205 °C (400 °F).

Austempering

In austempering, the microstructural end product of the gray iron matrix formed below the pearlite range but above the martensite range is an acicular or bainitic ferrite plus varying amounts of retained austenite depending on the transformation temperature. As shown in Fig. 11(a), the iron is quenched from a temperature above the transformation range in a hot quenching bath and is maintained in the bath at constant temperature until transformation has ceased.

In all hot quenching processes, the temperatures to which castings must be heated for austenitizing and the required holding times at temperature prior to quenching in the hot bath correspond to the temperatures and times used in conventional hardening—that is, temperatures between 760 and 900 °C (1400 and 1650 °F) and holding times of 20 min per inch of thickness. A characteristic curve for hardness as related to isothermal transformation temperature is shown in Fig. 11(b).

Gray iron is usually quenched in salt, oil or lead baths at 230 to 425 °C (450 to 800 °F) for austempering. When high hardness and wear resistance are the ultimate aim of this treatment, the temperature of the quench bath is usually held between 230 and 290 °C (450 and 550 °F).

The holding time required for maximum transformation is determined by the temperature of the quenching bath and the composition of the iron. The influence of austenite grain size in gray iron on holding time is usually small, as shown in Fig. 11(c). The effect of iron composition on the holding time may be considerable. Alloy additions, such as nickel, chromium and molybdenum, increase the time required for transformation.

The section thickness and shape of the casting often limit the use of austempering, because cooling must be fast enough to prevent any transformation of austenite until the casting

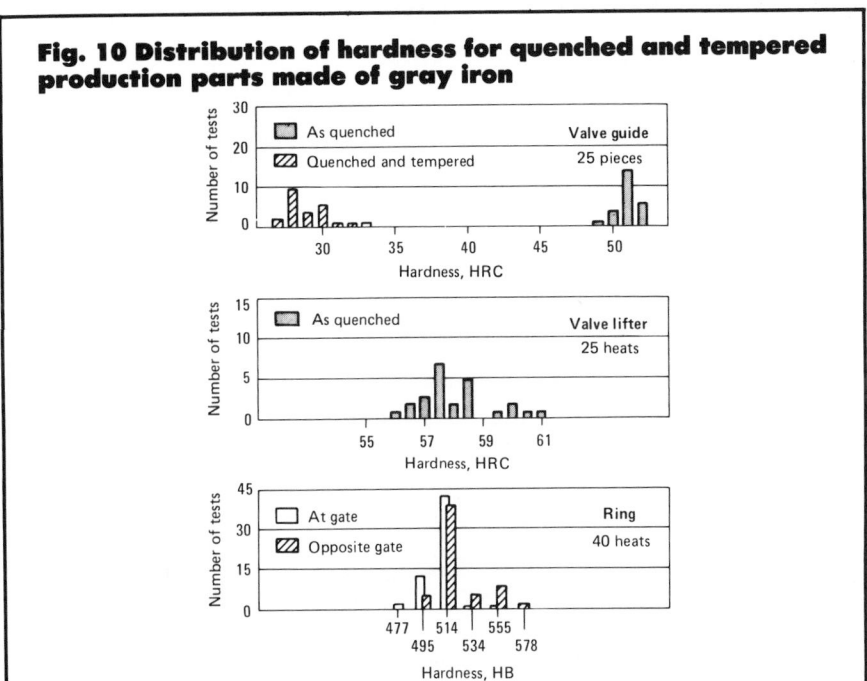

Fig. 10 Distribution of hardness for quenched and tempered production parts made of gray iron

Table 4 Composition of gray irons in Fig. 8(d)

Iron	TC(a)	CC(b)	Composition, % Si	Cr	Ni	Mo
Unalloyed	3.20	0.80	2.43	0.13	0.05	0.17
1	3.23	0.70	2.55	0.58	0.06	0.12
2	3.21	0.83	1.90	0.08	0.78	0.27
3	3.29	0.79	2.58	0.24	0.10	0.55
4	3.02	0.75	2.38	0.40	0.07	0.43

(a) Total carbon. (b) Combined carbon

reaches the temperature of the bath. Gray iron cylinder liners containing a minimum of 1.5% total Cr, Ni, Mo and Cu have been austempered at about 260 °C (500 °F), at rates of 35 to 75 liners per hour, depending on size.

Martempering

Martempering is used to produce martensite without developing the high stresses that usually accompany its formation. It is similar to conventional hardening, except that distortion is minimized. Nevertheless, the characteristic brittleness of the martensite remains in a gray iron casting after martempering, and martempered castings are nearly always tempered. As shown in Fig. 12, the casting is quenched from above the transformation range in a salt, oil or lead bath, held in the bath at a temperature slightly above the range at which martensite forms (205 to 260 °C or 400 to 500 °F for unalloyed irons) only until

the casting has reached the bath temperature, and then cooled to room temperature.

If a wholly martensitic structure is desired, the casting must be held in the hot quench bath only long enough to permit it to reach the temperature of the bath. Thus, the size and shape of the casting dictate the duration of martempering.

The following example illustrates the use of martempering.

Example 5. Cylinder liners (sleeves) for diesel and heavy-duty gasoline engines, made of gray iron containing 3.25 to 3.50 C, 2.00 to 2.25 Si, 0.25 max P, 0.12 max S, 0.55 to 0.80 Mn and 0.30 to 0.40 Cr, were fully annealed at 870 °C (1600 °F), rough machined, and then austenitized for martempering at 870 °C (1600 °F) in a neutral atmosphere for a total of 1 h (see Table 5 and Fig. 13). They were quenched for 1 min in molten nitrate-nitrite salt at 245 °C (475 °F) and air cooled for 30 min. After being washed to remove adhering salt,

Fig. 11 Austempering of gray iron

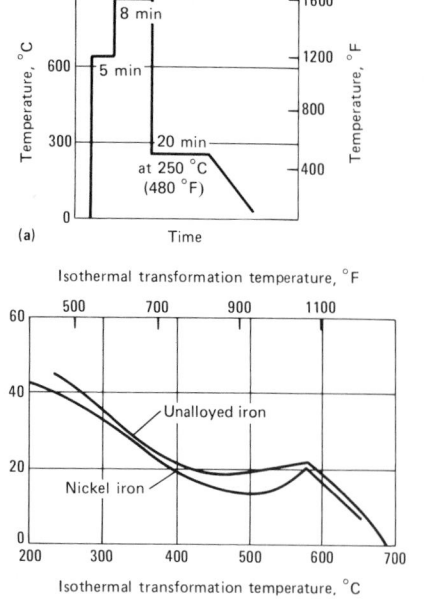

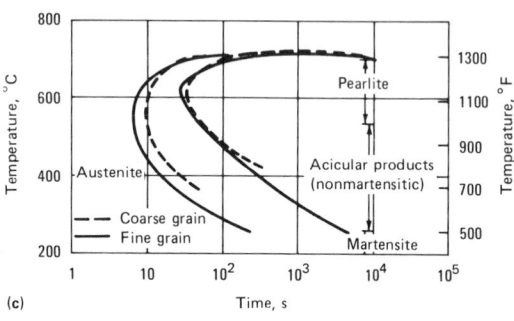

(a) Schematic representation of one austempering cycle. (b) Effect of isothermal transformation temperature on hardness of austempered gray irons. Holding times were sufficient to complete transformation. (c) Time-temperature transformation curves for gray iron containing 3.75 C, 2.9 Si, 0.4 P, 0.065 S and 0.55 Mn

Fig. 12 Schematic representation of one martempering and tempering cycle for gray iron

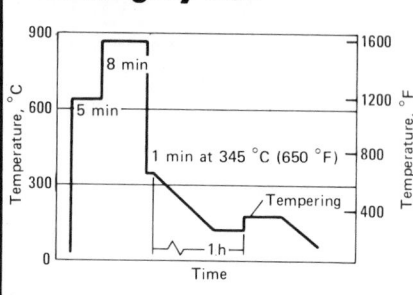

the castings were tempered for 2 h at 205 °C (400 °F).

Austenitizing for martempering was accomplished in a pusher-type continuous furnace. Sixteen sleeves were loaded on each rack, which rests on a furnace tray. On emerging from the furnace, the rack was pushed from the tray onto a quench elevator that immediately descended into the salt bath; thus, the furnace tray did not come in contact with the molten salt. The sleeves and rack were automatically raised from the salt and remained above the bath for drainage until the next rack was pushed into position, at which point the drained rack was pushed onto a slowly moving conveyor for the 30-min cooling cycle. After cooling, the rack and sleeves were moved onto the washer conveyor, and after being washed, they were deposited on the tempering-furnace conveyor. Equipment details are summarized in Table 5.

The as-cast sleeves had a graphitic structure predominantly of type A and a graphite flake size of 4 to 6. During the hardening cycle, the matrix dissolved about 0.70% C, and a martensitic structure resulted. The hardness range after hardening was 45 to 50 HRC; microhardness of the matrix was more than 60 HRC.

Martempering replaced conventional oil quenching, during which each cylinder liner had been placed on a mandrel in order to preserve shape. Despite the use of the mandrel, out-of-roundness after hardening (Fig. 13) and size variations were about twice as great for oil-quenched liners as for those that were martempered. Martempering also eliminated the small percentage of cracked sleeves previously encountered and reduced the amount of grinding stock required for finishing.

If final dimensional accuracy is important for martempered parts, allowance for growth must be made prior to heat treatment. The uniformity of growth and the allowance required depend on the condition of the iron prior to hardening. Annealing the cylinder sleeves discussed in Example 5 at 870 °C (1600 °F) and furnace cooling prior to martempering resulted in growth of 0.03 to 0.38 mm/mm (0.0010 to 0.0015 in./in.). This was only half as much growth as was encountered when as-cast or normalized pearlitic structures were martempered. Time at temperature during the annealing cycle was not critical in the range of 2 to 8 h. The effect of annealing temperature on growth of 305-mm (12-in.) -long cast specimens during subsequent martempering is shown in Fig. 14.

Some distortion, as differentiated from growth, occurs during martempering. Conditions that may promote excessive distortion are: residual stresses from casting, machining or rapid cooling during prior heat treatments; insufficient time for establishing equilibrium at the austenitizing temperature; and drafts during air cooling after the castings have been removed from the quench bath.

Austempering Versus Martempering. The maximum hardness obtained by austempering is usually less than that obtained by martempering, although this difference may be largely canceled during the tempering treatment that usually is necessary following martempering.

Example 6. A high-strength, low-silicon processed iron martempered for 1 min at 260 °C (500 °F) exhibited a hardness of 555 HB, whereas the same iron austempered for 30 min at 275 °C (530 °F) showed a hardness of only 444 HB. After being tempered for 30 min at 390 °C (730 °F), the martempered material had about the same hardness as the austempered material, but its impact resistance (Charpy test, 15-mm-square specimen, knife edges 70 mm apart) was only 14.2 J or 10.5 ft·lb, whereas that of the austempered material was 23.7 J or 17.5 ft·lb (the as-cast material had an impact resistance of 19.7 J or 14.5 ft·lb, and a hardness of 255 HB).

Both austempering and martempering can result in less distortion and growth than conventional oil quenching and tempering.

Example 7. When quenched in oil at 95 °C (200 °F) and tempered at 290 °C (550 °F), diesel-engine cylinder liners exhibited an average distortion of 0.26 mm (0.0101 in.) and cracked occasionally. When austempered, the same liners had an average distortion of 0.07 mm (0.0026 in.), and when martempered they displayed an average distortion of 0.06 mm (0.0025 in.). The average maximum growth for these liners was as follows:

Oil quenched and tempered	0.71 mm (0.0279 in.)
Austempered	0.25 mm (0.0099 in.)
Martempered	0.23 mm (0.0089 in.)

The austempering cycle consisted of preheating for 5 min at 650 °C (1200 °F), austenitizing for 8 min at 870 °C (1600 °F), quenching to 250 °C (480 °F) and holding for 20 min, and cooling in air. The martempering cycle comprised preheating for 5 min at 650 °C (1200 °F), austenitizing for 8 min at 855 °C (1575 °F), quenching for 1 min in agitated salt at 345 °C (650 °F), cooling in air for a minimum of 1 h, and tempering at 175 °C (350 °F).

Flame Hardening

Flame hardening is the method of surface hardening most commonly applied to gray iron. The mechanics of the process are dealt with in the article "Flame Hardening".

After flame hardening, a gray iron casting consists of (a) a hard, wear-resistant outer layer of martensite and (b) a core of softer gray iron, which during treatment did not reach the A_1 transformation temperature (in fact, the unhardened metal immediately below the hardened case, which has been to some extent heated by the flame, may even be partially annealed during flame hardening if it is unalloyed).

Recommended Composition. Both unalloyed and alloyed gray irons can be successfully flame hardened. However, some compositions yield much better results than others. One of the most important aspects of composition is the combined carbon content, which should be in the range of 0.50 to 0.70%, although irons with as little as 0.40% combined carbon can be flame hardened. In general, flame hardening is not recommended for irons that contain more than 0.80% combined carbon, because such irons (mottled or white irons) may crack in surface hardening.

The stability of the microconstituents from which the carbon precipitates is a factor in determining the hardness of flame-hardened iron.

Example 8. One user has observed that the stability of microconstituents containing combined carbon is indicated by the hardness of test specimens annealed at 845 °C (1550 °F). A test of four pearlitic irons from different sources, containing 3.30 max C, 1.40 to 1.60 Si, 0.80 to 1.10 Mn, and 0.15 to 0.20 Cr, gave the following results:

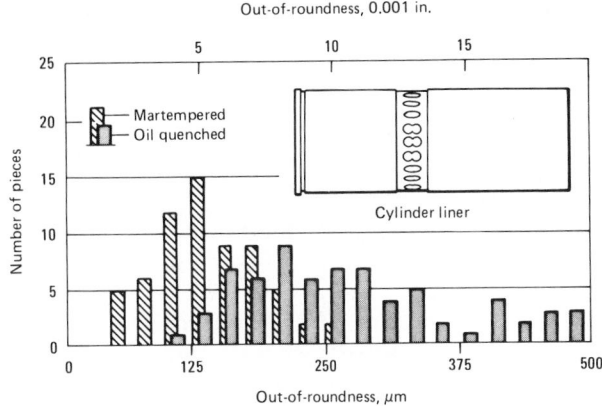

Fig. 13 Distortion in gray iron cylinder liners after martempering and after conventional oil quenching

Before being measured, liners were furnace tempered for 2 h at 205 °C (400 °F).

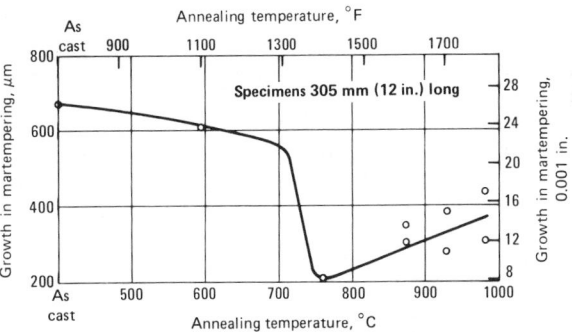

Fig. 14 Effect of annealing temperature on growth of gray iron during martempering

Heat treatment consisted of holding at the annealing temperature for 2 h and cooling in the furnace, reheating to 870 °C (1600 °F) in a neutral endothermic atmosphere, quenching for 1 min in molten salt at 245 °C (475 °F), air cooling for 30 min and tempering at 205 °C (400 °F) for 2 h. As-cast specimens were not annealed before hardening.

Fig. 15 Typical hardness gradient produced in gray iron by flame hardening

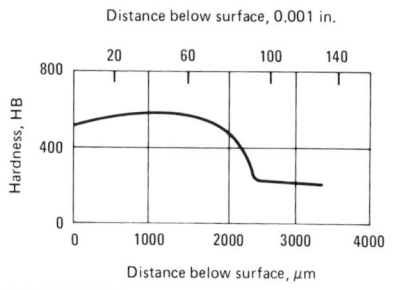

	Hardness, HB		
Iron	As cast	Annealed	Hardness(a), HRC
A	220	200	47–50
B	190	180	47–50
C	200	140	35–38
D	200	120	20–25

(a) After flame hardening as-cast iron

To obtain better stabilization of these microconstituents, 0.15 to 0.30% Mo was added, and the chromium and manganese contents were increased.

For maximum hardness, it is advisable to employ an iron containing as small an amount of total carbon as is consistent with the production of sound castings free from any danger of cracking. The coarse graphite flakes typical of high-carbon irons should be avoided because they may be burned out during flame heating, thereby producing a porous and unattractive surface.

Because silicon promotes formation of graphite and of a low combined carbon content, a relatively low silicon content also is advisable. Several users report that silicon content should not exceed 2% in any iron submitted to a flame-hardening operation, although alloy irons containing 2.4% Si can be hardened successfully. They also recommend that manganese content be held in the range of 0.80 to 1.00%, to enhance depth of hardening.

Gray iron to be flame hardened should be as free as possible from porosity and from foreign matter such as sand or slag, because porosity and even small inclusions of foreign matter can produce a rough surface or result in cracking after hardening. Rough casting surfaces should be sand or shot blasted prior to heat treatment, because skin or scale on the surface acts as a heat insulator and reduces the

Table 5 Equipment requirements for martempering cast iron cylinder liners

Cylinder liners were austenitized in a radiant-tube gas-fired single-row pusher furnace containing 27 U-type radiant tubes and accommodating 13 trays per cycle. Each tray measured 610 by 710 mm (24 by 28 in.). Austenitizing temperature was 870 °C (1600 °F). An endothermic atmosphere of 34 m³/h (1200 ft³/h) was controlled to 0.2 to 0.4% CO_2 by Orsat analysis.

Production requirements

Weight of each piece . . . . 3.8 kg (8.3 lb)
Number of pieces per push 16
Number of pushes per hour 11
Weight of empty rack . . . . 32 kg (70 lb)
Total weight of material heated and quenched per hour . . . . 1010 kg (2231 lb)

Martempering furnace requirements

Salt pot, 1.83 by 1.98 by 1.83 m (72 by 78 by 72 in.) Immersion heated, 6.6 m³ (234 ft³)(a)
Type of salt . . . . Nitrate-nitrite(b)
Operating temperature of salt . . . . 245 °C (475 °F)
Agitation, by 230-mm (9-in.) -diam propeller(c) . . . . One 7½-hp variable speed motor
Cooling system . . . . Three ½-hp fans(d)

(a) Pot, containing about 11.8 t (13 tons) of molten salt, was heated by 24 immersion rods, each rated at 5 kW. (b) Melting point of salt, 145 °C (290 °F). No chloride-separation chamber was required, because parts were transferred to salt from a controlled-atmosphere hardening furnace. (c) Propeller was in a 255-mm (10-in.) -diam pipe connected to the quench funnel. (d) Fans force outdoor air through ducts in furnace walls.

Fig. 16 Effects of stress relieving on tensile strength and hardness of gray iron

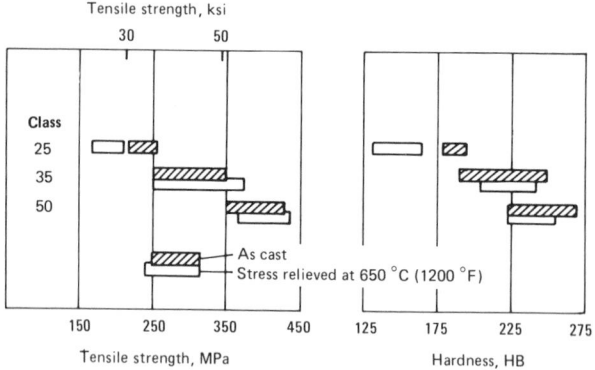

Gray iron bars, 30 mm (1.2 in.) in diameter, were stress relieved for 6 h at 650 °C (1200 °F) in a car-bottom furnace approximately 13 by 4 by 3 m (42 by 13 by 9 ft); total furnace time was 43¾ h.

effectiveness of flame hardening.

Effects of Alloying Elements. In general, alloy gray irons can be flame hardened with greater ease than can unalloyed irons—partly because alloy irons can be heat treated successfully over a wider temperature range, thus minimizing the necessity for precise pyrometric control. Also, the hardenability, or depth of martensitic case, is increased by addition of most alloying elements.

Final hardness also may be increased by alloying additions. The maximum

hardness obtainable by flame hardening an unalloyed gray iron containing approximately 3.0% total carbon, 1.7% Si and 0.60 to 0.80% Mn ranges from 400 to 500 HB. This is because the Brinell hardness value for gray iron is an average of the hardness of the matrix and that of the relatively soft graphite flakes. Actually, the matrix hardness on which wear resistance depends approximates 600 HB. With the addition of 2.5% Ni and 0.5% Cr, an average surface hardness of 550 HB can be obtained. The same result has been achieved using 1.0 to 1.5% Ni and 0.25% Mo. Small additions of chromium are particularly valuable in preventing softening and ensuring retention of a high content of combined carbon during austenitizing for hardening. Automotive camshafts containing 1% Cr, 0.50% Mo and 0.8% Mn are easily flame hardened to 52 HRC to required depth. These parts are not tempered or stress relieved.

Stress Relieving. Whenever practicable or economically feasible, flame-hardened castings should be stress relieved at 150 to 205 °C (300 to 400 °F) in a furnace, in hot oil, or by passing a flame over the hardened surface. Such a treatment will minimize distortion or cracking and will increase the toughness of the hardened layer.

Stress relieving at 150 °C (300 °F) for 7 h was found to remove 25 to 40% of the residual stresses in a flame-hardened casting, while reducing the hardness of the surface by only 2 to 5 points on the HRA scale. Although stress relieving is desirable, it can often be safely omitted.

Hardness. The surface of flame-hardened gray iron typically has a somewhat lower hardness than the metal immediately below the surface (Fig. 15). This decrease in hardness may be caused by retention of relatively soft austenite at the surface. Surface hardness often can be raised by heating in the range from 195 to 250 °C (380 to 480 °F).

The depth and microstructure of the hardened layer depend on two other factors in addition to temperature: (a) the amount of carbon and alloying elements in solution when the flame-hardened surface is quenched, and (b) the efficiency of quenching. If softness is due to the presence of austenite in the microstructure, subjecting the part to −40 °C (−40 °F) for 1 h will transform the austenite and increase the hardness.

Fatigue strength usually is increased by surface hardening, because the treatment induces compressive stresses at the surface. For example, flame or induction heating followed by water quenching induces high compressive stresses (over 205 MPa, or 30 ksi) in the fillet areas of crankshafts; however, if self-quenching instead of water quenching is used, undesirable tensile stresses may result at the surface. This degree of improvement in fatigue strength cannot be produced by through hardening.

Fig. 17 Effect of stress-relieving temperature on hardness of gray irons

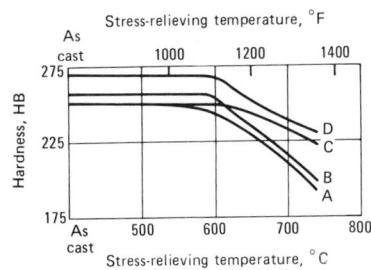

Iron	TC(a)	CC(b)	Composition, % Si	Cr	Ni	Mo
A	3.20	0.80	2.43	0.13	0.05	0.17
B	3.29	0.79	2.58	0.24	0.10	0.55
C	3.23	0.70	2.55	0.58	0.06	0.12
D	3.02	0.75	2.38	0.40	0.07	0.43

(a) Total carbon. (b) Combined carbon

Bar specimens 30 mm (1.2 in.) in diameter were held for 1 h at indicated temperatures and then air cooled.

Quenching. The various methods of flame hardening influence selection of the quenching medium. In the progressive method, only nonflammable media such as water, soluble-oil mixtures, and solutions of polyvinyl alcohol in water can be employed. Conventional oil cannot be used because of the fire hazard. In spot hardening or spinning methods, in which the flame head is withdrawn from the part before quenching, parts are quenched conventionally by immersion in hot oil.

When quenching is done with water, the water should be at about 32 °C (90 °F) for best results. Lower quenching rates, such as those obtained with 5 to 15% soluble-oil mixtures, compressed air, or compressed air and water at low pressure, are employed to prevent cracking. Air quenching is especially suited to highly alloyed cast irons because of their susceptibility to cracking.

Induction Hardening

Gray iron castings can be surface hardened by the induction method when the number of castings to be processed is large enough to warrant the relatively high equipment cost and the need for special induction coils. For details of this process and its application to gray iron, see the article "Induction Hardening and Tempering".

Fig. 18 Effect of stress-relieving temperature on residual stress in gray iron

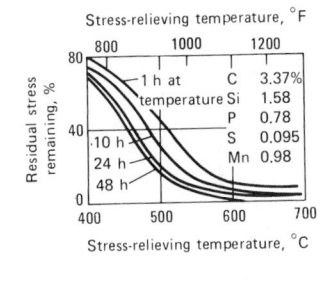

Stress Relieving

Gray iron in the as-cast condition contains residual stresses (unless the iron is cooled in the mold, in which case, much of the solidification stress is removed), because cooling (and therefore contraction) proceeds at different rates throughout various sections of a casting. The resultant residual stresses may reduce strength, cause distortion, and in some extreme cases even result in failure or cracking. The magnitudes of these stresses depend on the shape and section size of the casting, on the casting technique employed, on the composition and properties of the material cast, and on whether the casting has been stress relieved.

Temperature of stress relieving usually is well below the range for transformation of pearlite to austenite.

The effects of stress relieving at 650 °C (1200 °F) for 6 h on the tensile strength and hardness of gray irons of classes 25, 35 and 50 are plotted in Fig. 16. As indicated, the properties of the class 25 iron were affected considerably more than were those of the class 35 and class 50 irons.

Figure 17 shows the effect of stress-relieving temperature on hardness of unalloyed and alloyed gray irons.

For maximum relief of stress with minimum decomposition of carbide in unalloyed irons, a temperature range of 540 to 565 °C (1000 to 1050 °F) is desirable. Figure 18 indicates that from 75 to 85% of the residual stress can be removed by holding for 1 h in this range. Other investigations (for example, see Fig. 19) indicate that the curves in Fig. 18 are applicable to gray irons over a wide range of composition.

When almost complete stress relief (over 85%) is required in unalloyed iron, a minimum temperature of 595 °C (1100 °F) can be employed; however, some sacrifice in strength, hardness and wear resistance is likely. Fortunately, the soft irons of higher carbon equivalent normally exhibit lower level of residual stress and comparatively low creep resistance, which facilitates stress relief. In fact, irons of high carbon equivalent can be satisfactorily stress relieved at the lower end of the suggested temperature range—that is, at about 510 °C (950 °F). Low-alloy gray irons usually require higher stress-relieving temperatures—on the order of 595 to 650 °C (1100 to 1200 °F), depending on alloy content.

Quantitative data concerning the effects of alloying elements on optimum stress-relieving temperature are meager. However, it has been reported that in one instance the addition of as little as 0.14% Cr to a 3.20 C, 2.01 Si iron permitted exposure of the iron to a temperature of about 650 °C (1200 °F) for 1 h without sacrifice in room-temperature tensile strength. Figure 20 shows the effect of temperature and time on relief of stresses for seven low-alloy irons, and the tabulation below the

Fig. 19 Effect of stress-relieving temperature and time on residual stress in gray iron

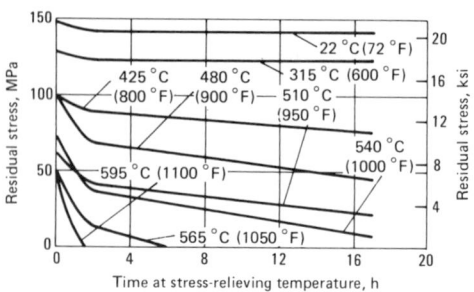

Composition of iron: 2.72 C, 1.97 Si, 0.141 P, 0.080 S, 0.51 Mn

Table 6 Effect of shakeout practice and stress relieving on residual stress in 225-kg (500-lb) diesel-engine cylinder blocks

Practice	Close-in		Stress relief, %
	mm	in.	
1 .. Shakeout after 6 h; cores in place while cooling to 27 °C (80 °F); total time, 16 h	4.1	0.160	Basis for evaluation
2 .. Same as above, except cores also shaken out at 6 h	1.9	0.076	52
3 .. Cooled in mold 16 h before shakeout	1.6	0.064	60
4 .. Practice 1, followed by stress relieving at 540 °C (1000 °F) for 2 h, furnace cooling to 370 °C (700 °F)	1.5 to 1.9	0.060 to 0.076	52 to 62
5 .. Practice 1, followed by stress relieving at 620 °C (1150 °F) for 2 h, furnace cooling to 370 °C (700 °F)	0.3 to 0.4	0.012 to 0.015	91

graphs indicates that these irons, depending on shakeout time, can be stress relieved for 8 h at 620 °C (1150 °F) with no adverse effect on hardness.

The following stress-relieving temperatures, based on normal shakeout times in the foundry, can be recommended:

Unalloyed, or alloyed
  without Cr . . . . 510 to 565 °C
                     (950 to 1050 °F)
0.15 to 0.30% Cr . . . . . 595 to 620 °C
                     (1100 to 1150 °F)
Over 0.30% Cr . . . . . . 620 to 650 °C
                     (1150 to 1200 °F)

If the service requirements of a casting demand a particularly low residual stress, temperatures about 28 °C (50 °F) above those listed may be used. When these higher temperatures are used, it is advisable, if hardness and strength are critical, to check the hardness of the stress-relieved casting to determine whether an unacceptable decrease in hardness or strength has taken place.

Rate of Heating. The rate at which gray iron castings are heated for stress relief depends on the shape and size of the part but, except for the most complex shapes, is not especially critical. When a batch-type furnace is employed, it is of the utmost importance that furnace temperature does not exceed 95 °C (200 °F) at the time of loading. After the furnace is loaded, the heating rate may be fairly high. For example, it is common practice to heat to 620 °C (1150 °F) in about 3 h, hold at temperature for 1 h, and cool to 315 °C (600 °F) in about 4 h, before removing castings from the furnace and permitting them to cool in air. These conditions apply also to continuous furnaces in which the various temperature zones can be controlled to avoid introducing additional thermal stress in the castings. It is imperative that flame im-

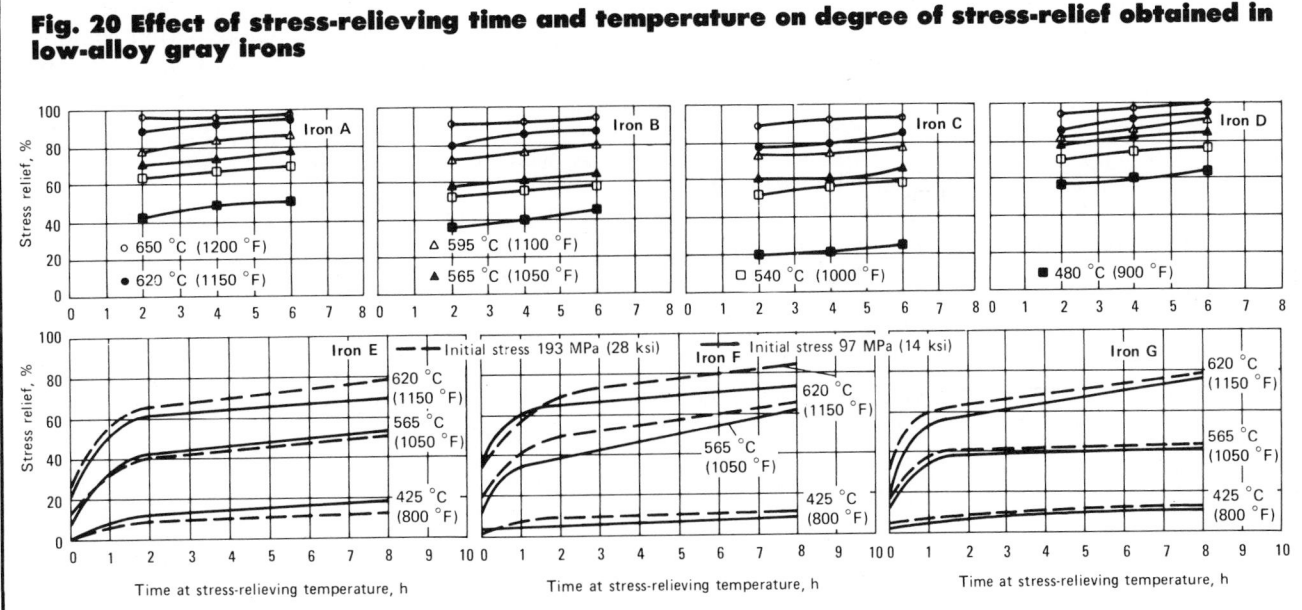

Fig. 20 Effect of stress-relieving time and temperature on degree of stress-relief obtained in low-alloy gray irons

Iron				Composition, %							Hardness, HRB	
	C	Si	P	S	Mn	Ni	Cr	Mo	Cu	V	Before stress relieving	After stress relieving for 8 h at 620 °C (1150 °F)
A	2.93	2.14	0.110	0.57	0.47	0.35	0.10	. . . .	. . . .	. . . .	98	94
B	3.43	2.12	0.104	0.70	0.81	0.34	0.18	0.23	. . . .	. . . .	98	94
C	3.24	2.55	0.107	0.62	0.87	0.51	0.20	0.22	. . . .	. . . .	95	95
D	3.91	1.43	0.54	0.25	0.32	1.56	0.06	. . . .	. . . .	. . . .	82	80
E	3.18	2.13	0.73	0.125	0.70	1.03	0.33	0.65	. . . .	. . . .	98	98
F	3.12	1.76	0.075	0.097	0.78	1.02	0.41	0.58	. . . .	. . . .	94	95
G	2.78	1.77	0.065	0.135	0.55	0.36	0.10	0.33	0.46	0.04	96	96

Compositions are given in table above, which also shows negligible effect of maximum stress-relieving conditions on hardness.

pingement on castings, which may result in variations in hardness, be avoided.

Rate of Cooling. If a casting is allowed to cool rapidly from the stress-relieving temperature to room temperature, new stresses may be developed, and the object of maximum stress elimination will not be fully achieved. For this reason, slow cooling from the stress-relieving temperature, at least in the upper temperature range, is an essential part of stress relieving.

It is generally recommended that castings be furnace cooled to 315 °C (600 °F) or lower before being allowed to cool in air; for castings of intricate design, it may even be advisable to continue furnace cooling until a temperature of about 95 °C (200 °F) has been reached. Most commercial furnaces cool slowly enough to meet all requirements.

Applications. The following examples illustrate the use of stress relieving to eliminate distortion and cracking.

Example 9. A flatness tolerance of 0.1 mm (0.004 in.) could not be maintained after machining as-cast clutch plates. These parts were made of gray iron containing 3.40 C (0.70 to 0.90 combined carbon), 1.30 to 1.80 Si, 0.25 to 0.40 Cr, 0.30 to 0.40 Mo, and nickel and copper as required for a minimum tensile strength of 275 MPa (40 ksi) and a hardness of 207 to 255 HB. After these clutch plates were stress relieved at 620 °C (1150 °F) for 2 h, the tolerance could be held.

Example 10. Necessitated by a rush order, a change in procedure from a slow cool in the mold to a fast shakeout resulted in high residual stress in cast transmission cases. These parts were made of gray iron containing 3.10 to

3.40 C, 2.15 to 2.35 Si and 0.20 to 0.40 Cr. Some of the first castings produced after the change in procedure developed cracks up to 1.3 mm (0.050 in.) wide during machining; others, although they survived machining without cracking, developed cracks during a final hot washing operation. This problem was eliminated for the remaining pieces from this order by stress relieving them at 620 °C (1150 °F) for 2 h. A return to the original cooling and shakeout practice on subsequent orders eliminated cracking and the need for stress relieving.

Example 11. Hot tears and out-of-roundness occurred in as-cast rings, and additional distortion in the flat plane of these rings occurred during machining. A change in the method of gating these castings equalized temperature differentials and controlled shrinkage distribution, thus eliminating the hot tears and the as-cast distortion. The distortion occurring during machining was eliminated by stress relieving the rings at 370 °C (700 °F) for 2 h.

Example 12. Table 6 shows the results of an evaluation of different cooling methods and stress-relieving treatments on residual stresses in 225-kg (500-lb) diesel-engine cylinder blocks. These blocks were made of gray iron containing 3.25 C, 2.20 Si, and 0.30 Cr, and having a tensile strength of 240 MPa (35 ksi).

Two vertical lines were scribed on one end of each block after shakeout.

The distance between these lines was measured before and after vertical saw cuts (parallel to the scribed lines) were made through the end of each block to the first cylinder bore. The amount of close-in between the two vertical lines after sawing was a relative measure of the residual stress.

The results (see Table 6) indicate that the complete shakeout (including cores), allowing inner sections to cool faster, materially reduced stresses, as did the procedure of leaving the casting in the mold to slow down the cooling of external sections. This latter procedure is effective if the mass of metal is sufficient to keep the external sections at a

high temperature throughout—620 °C (1150 °F) or higher—while equilibrium between outer and inner sections is taking place. As indicated, stress relieving at 540 °C (1000 °F) was substantially less effective than stress relieving at 620 °C (1150 °F).

Other examples of castings that require stress relief are pump volutes, scrolls and casings. These parts require two stress-relieving treatments: first, in the as-cast condition, to minimize distortion during machining; and second, after rough machining, to minimize distortion during final machining.

Heat Treating of Ductile Irons

By Kenneth E. Spray
Metallurgy Section Head
Material Corporate Laboratory
Clark Equipment Co.

WHEN MAXIMUM DUCTILITY and good machinability are desired and high strength is not required, ductile iron castings are generally given a full ferritize anneal. The microstructure is thus converted to ferrite and spheroidal graphite (Fig. 1). Manganese and phosphorus, and alloying elements such as chromium, nickel and molybdenum, should be as low as possible if superior machinability is desired, because these elements retard the annealing process.

Two different annealing cycles may be used satisfactorily. Selection of one or the other will depend on the type of heat treating equipment to be used:

- Hold at 900 to 955 °C (1650 to 1750 °F) for 1 h plus 1 h or more per inch of section thickness. For thin-section castings containing 2.20 to 2.70% Si, holding at 955 °C (1750 °F) for 1 to 3 h is sufficient. In heavy-section castings where chill has formed on corners, holding at 955 °C (1750 °F) for 3 to 8 h may be required. Cool to 690 °C (1275 °F) in any convenient manner (but uniformly, if residual stress is to be avoided), and hold at 690 °C (1275 °F) for 5 h plus 1 h per inch of casting section.

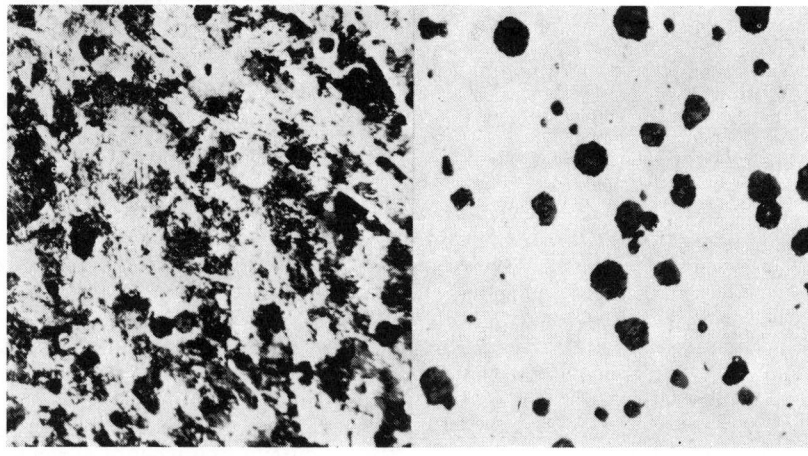

Fig. 1 Microstructure of a 4-mm (⁵/₃₂-in.) section of ductile iron as cast and after annealing

As cast   Annealed

Treatment consisted of holding at 900 °C (1650 °F) for 4 h, furnace cooling to 690 °C (1275 °F) and holding for 5 h, then furnace cooling to room temperature. Etched in picral. Magnification, 250X

- Hold at 900 to 955 °C (1650 to 1750 °F) as above, but furnace cool to 650 °C (1200 °F) so that the cooling rate between 790 and 650 °C (1450 and 1200 °F) does not exceed 20 °C/h (35 °F/h).

A shorter, subcritical annealing cycle can be used when carbides can be tolerated and maximum impact properties are not required. A reduction of as much as 50% in impact strength and an increase of 30 °C (50 °F) in transition temperature will result if prior austenitizing heat treatment is omitted. In this treatment, castings are heated to 705 °C (1300 °F) and held for 5 h plus 1 h per inch of maximum thickness. Castings should be furnace cooled to at least 595 °C (1100 °F). The influence of subcritical annealing at 705 °C (1300 °F) for various periods of time on the hardness of four ductile irons is shown in Fig. 2.

The effect of time at temperature on decomposition of primary carbide for 6-mm (¼-in.) plate castings with a nominal composition of 3.15% TC, 2.66% Si, 0.40% Mn, 0.15% P, 0.10% Cu and 0.04% Mg is given in Fig. 3. These carbides usually break down readily, in a manner similar to castings that have been post-inoculated during manufacture.

Certain carbide-forming elements, mainly chromium, form primary carbides that are very difficult, if not impossible, to decompose. For example, the presence of 0.26% chromium results in primary carbides that cannot be broken down in 2 to 20 h heat treatments at 925 °C (1700 °F). The resulting matrix after pearlite breakdown is ferritic-carbidic with only 5% elongation. Other carbide stabilizers are molybdenum contents over 0.3%, copper over 1.0%, 0.002% or more boron, and vanadium and tungsten.

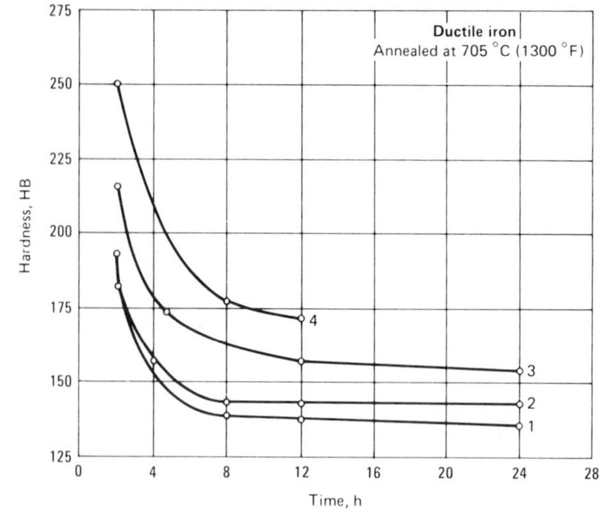

Normalizing of Ductile Iron

Normalizing can result in a considerable improvement in tensile properties and may be used in the production of ductile iron of types 100-70-03 and 120-90-02. The microstructure obtained by normalizing will depend on the composition of the castings and the cooling rate. The cooling rate depends on the mass of the casting, but it also may be influenced by the temperature and movement of the surrounding air during cooling. Normalizing generally produces a homogeneous structure of fine pearlite, if the metal is not too high in silicon content and has at least a moderate manganese content (0.3 to 0.5% or higher). Heavier castings should contain alloying elements such as nickel, molybdenum and additional

Fig. 2 Influence of time at subcritical annealing temperature on hardness

manganese for development of a fully pearlitic structure after normalizing. Lighter castings made of alloyed iron may be martensitic after normalizing.

The normalizing temperature is usually between 870 and 940 °C (1600 and 1725 °F). The standard time at temperature of 1 h per inch of section thickness or 1 h minimum is satisfactory, but shorter times often can be used successfully. The following temperatures and minimum holding times are recommended for the normalizing of unalloyed ductile iron:

Section			Time,
mm	in.	Temperature	h
Under 13	Under ½	870 °C min (1600 °F min)	1
13 to 25	½ to 1	940 °C (1725 °F)	1
Over 25	Over 1	940 °C (1725 °F)	2

These temperatures and times will vary with composition especially with silicon and chromium contents.

As shown in Fig. 4(a), the temperature of ductile iron when it is removed from the furnace for air cooling during normalizing strongly affects hardness. Because of increasing amounts of ferrite in the resultant structures, hardness decreases markedly as the normalizing temperature drops below 845 °C (1550 °F).

Alloying elements have an important function in ductile iron castings that are to be normalized. Moderate

Fig. 3 Effect of time at temperature on decomposition of primary carbide

additions increase hardenability sufficiently so that a microstructure of fine pearlite may be obtained in normalized castings with sections up to 150 mm (6 in.) thick, providing the cooling rate has been adequate. Nickel or copper are preferred as alloying elements because they strengthen the iron in all section thicknesses and do not form carbides. Molybdenum, in combination with nickel, is effective in increasing hardness and strength. The influence of various nickel contents and several combinations of alloying elements on the hardness obtained by normalizing various sections is shown in Fig. 4(b).

Normalizing is commonly followed by tempering to attain the desired

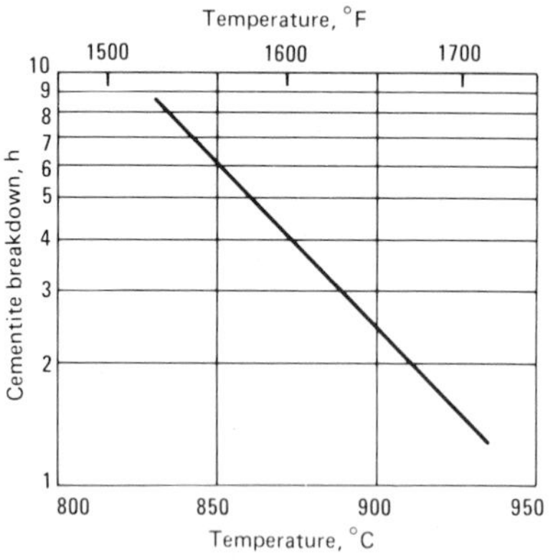

Fig. 4 Effect of variables on hardness after normalizing and after normalizing and tempering of ductile iron

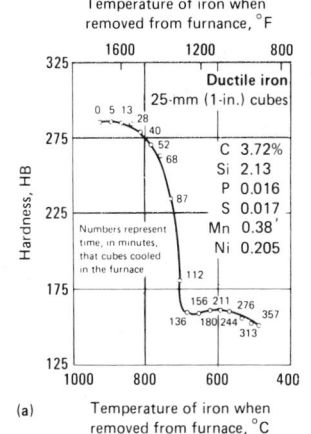

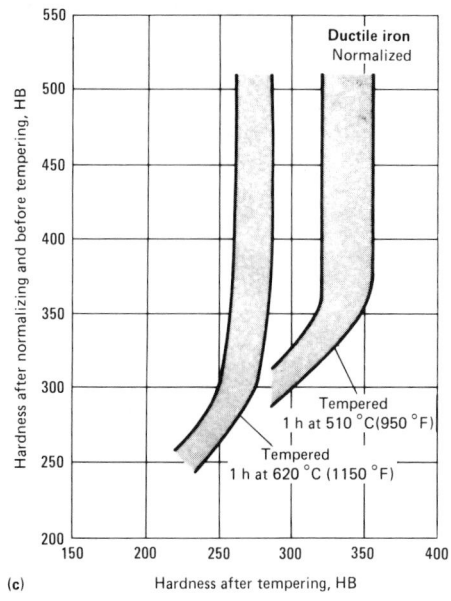

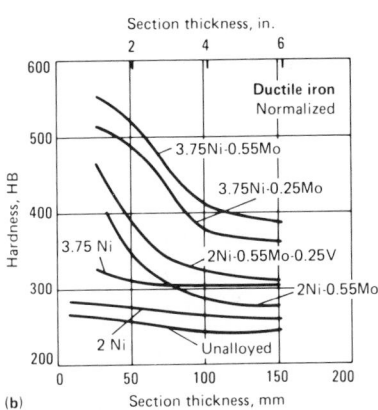

(a) Effect of temperature from which ductile iron was air cooled during normalizing. Specimens (25-mm or 1-in. cubes) were held at 925 °C (1700 °F) for 3 h and then cooled in the furnace to various temperatures in the times indicated before being cooled in air. (b) Effect of alloy content and section thickness. (c) Effect of tempering after normalizing, for unalloyed, nickel and nickel-molybdenum ductile irons

Fig. 5 Hardness of normalized ductile iron tempered at various temperatures

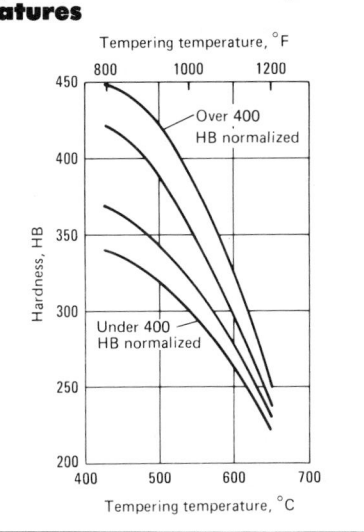

Fig. 6 Hardness of oil-quenched ductile iron tempered for 1 h

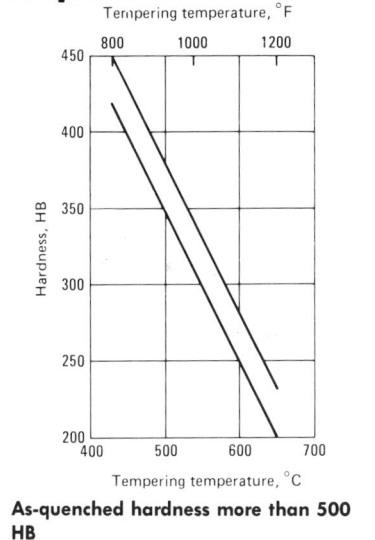

As-quenched hardness more than 500 HB

hardness and relieve residual stresses that develop upon air cooling when various parts of a casting with different section sizes cool at different rates. The effect of tempering on hardness and tensile properties depends on the composition of the iron and on the hardness level that was obtained in normalizing. In general, pearlitic structures, such as result from normalizing, soften less than the harder martensitic structures that are obtained by liquid quenching. Figure 4(c) shows the effect of tempering 25-mm (1-in.) cubes for 1 h at 510 and 620 °C (950 and 1150 °F) on the hardness of a series of unalloyed, nickel, and nickel-molybdenum ductile irons after normalizing.

Tempering after normalizing is also used to obtain high toughness and impact resistance along with high tensile properties. This consists of reheating to temperatures of 425 to 650 °C (800 to 1200 °F) and holding at the desired temperature for 1 h per inch of cross section. These temperatures are varied within the above range to meet specification limits. Figure 5 shows the effect of tempering temperature on hardness of normalized ductile iron.

Hardening and Tempering of Ductile Iron

A temperature of 845 to 925 °C (1550 to 1700 °F) is normally used for austenitizing commercial castings and produces the highest as-quenched hardness. Oil is preferred as a quenching medium, to minimize stresses, but water or brine may be used for simple shapes. Complicated castings may have

Figure 6

Fig. 7 Influence of austenitizing temperature on hardness of ductile iron

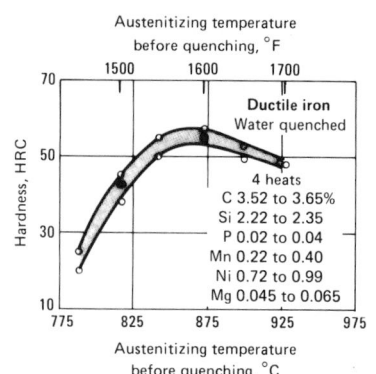

Each value represents the average of three hardness readings. Specimens (13-mm or ½-in. cubes) were heated in air for 1 h and water quenched.

to be quenched in oil at 80 to 100 °C (180 to 210 °F) to avoid cracks.

To relieve quenching stresses, castings should be tempered immediately after quenching. Tempered hardness depends on as-quenched hardness level, alloy content, and tempering time as well as temperature. Precise data on ductile irons is not available to draw precise tempering curves such as those for steels. Figure 6 can be used as a first approximation by the heat treater. More accurate control of hardness can be attained by close control of material composition and heat treating cycle.

Severity of quench is shown in Table 1 in Grossman's H values. The higher the H value, the faster the quench.

Table 1 Comparison of quenching mediums to water(a)

| Type of circulation | Air | Grossman H values | | |
		Oil	Water	Brine
No circulation of liquid or agitation of piece	0.02	0.25-0.30	0.9-1.0	2.0
Mild circulation	...	0.30-0.35	1.0-1.1	2.0-2.2
Moderate circulation	...	0.35-0.40	1.2-1.3	...
Good circulation	...	0.40-0.50	1.4-1.5	...
Strong circulation	...	0.50-0.80	1.6-2.0	...
Violent circulation	...	0.80-1.10	4.0	5.0

(a) At 15 °C (65 °F)

Fig. 8 Effect of varying amounts of carbon, silicon and manganese on hardenability of ductile irons

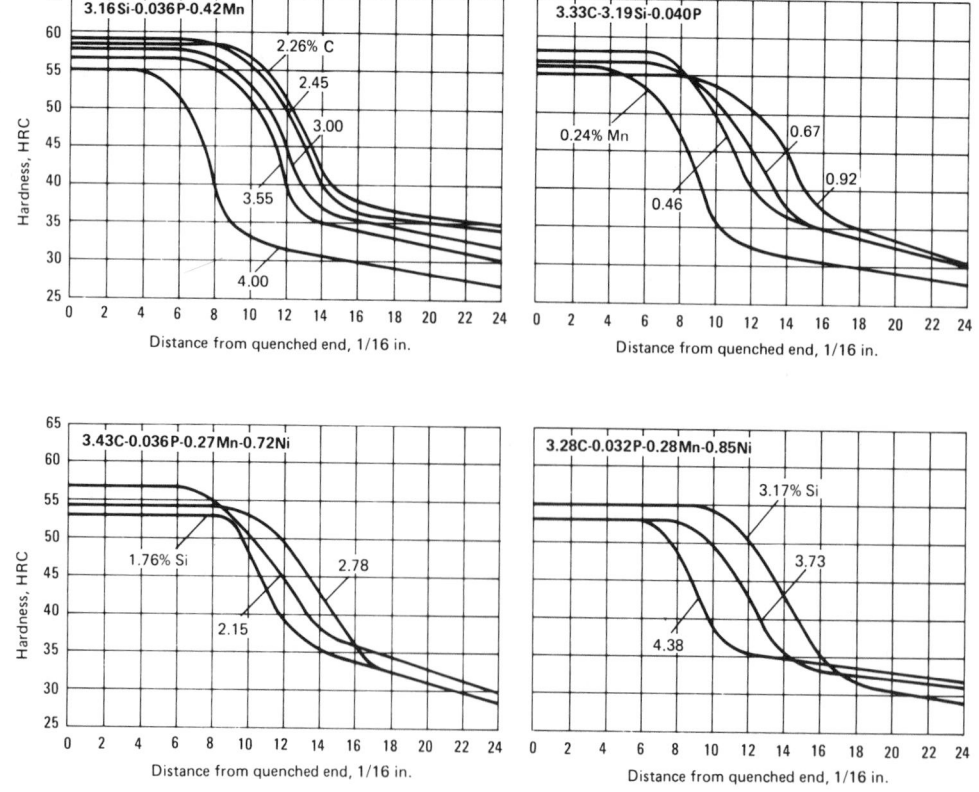

Fig. 9 Effect of alloying elements on end-quench hardenability of ductile iron

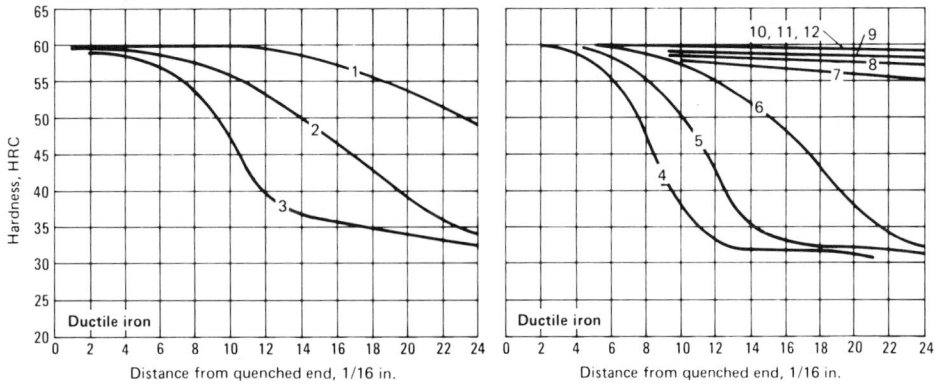

Iron	TC	Si	Mn	Cr	Ni	Mo	Other
1	3.52	2.74	0.37	0.10	...	0.41	0.061 Mg
2	3.47	2.69	0.37	0.10	...	0.23	0.050 Mg
3	3.54	2.55	0.37	0.09	...	0.022	0.049 Mg
4	3.4	2.6	0.26	...	...	...	...
5	3.4	2.6	0.26	...	1.0	...	...
6	3.4	2.6	0.26	...	2.0	...	...
7	3.4	2.6	0.26	...	4.0	...	...
8	3.4	2.6	0.26	...	2.0	0.25	...
9	3.4	2.6	0.26	...	2.0	0.55	...
10	3.4	2.6	0.26	...	3.75	0.25	...
11	3.4	2.6	0.26	...	3.75	0.55	...
12	3.4	2.6	0.26	...	2.0	0.55	0.25 V

Fig. 10 Influence of tempering temperature on mechanical properties of ductile iron quenched from 870 °C (1600 °F), tempered 2 h

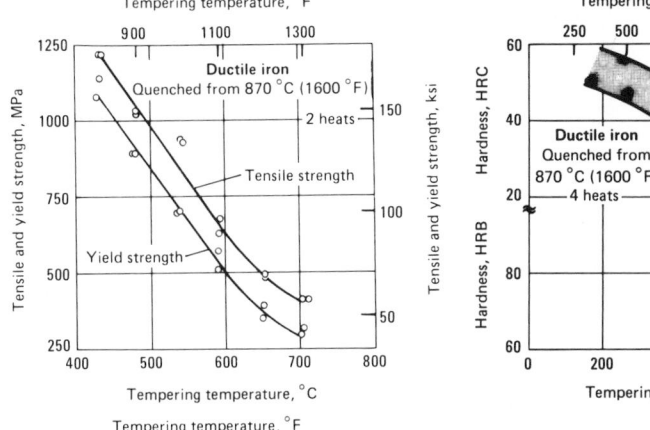

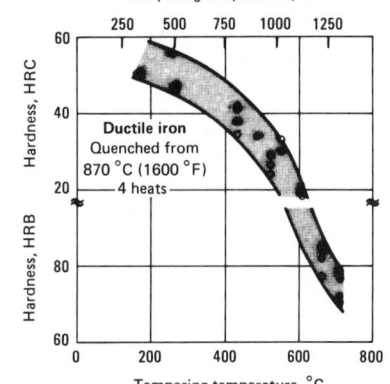

Data represent irons from four heats with composition ranges of: 3.52 to 3.68% C, 2.28 to 2.35% Si, 0.02 to 0.04% P, 0.22 to 0.41% Mn, 0.69 to 0.99% Ni, and 0.045 to 0.065% Mg. Data for tensile strength, tensile yield strength and elongation are for irons (from two of these heats) that contained 0.91 and 0.99% Ni.

Geometry of the part will generally dictate the quenching medium because of stress-induced cracking during quenching. Grossman's H values are based on the diameter of round bar steel that will be half martensitic at the center after a specific quench. The values are used as a measure of hardenability.

The influence of the austenitizing temperature on the hardness of water-quenching 13-mm (½-in.) cubes of ductile iron is shown in Fig. 7. These data show that the highest range of hardness (55 to 57 HRC) was obtained with austenitizing temperatures between 845 and 870 °C (1550 and 1600 °F). Specimens quenched from 925 °C (1700 °F) contained enough retained austenite to lower the hardness to 47 HRC.

Time at austenitizing temperature also is important for obtaining full hardness during quenching. This was indicated by further tests with 13-mm (½-in.) cubes of ductile iron; both as-cast and annealed cubes were heated in salt at 870 °C (1600 °F) and quenched in water. Cubes that were heated for a total of 2 min contained 30 to 35% ferrite in the microstructure and developed a hardness of 32 to 45 HRC. After

Fig. 11 Hardness distribution for quenched and tempered ductile iron parts

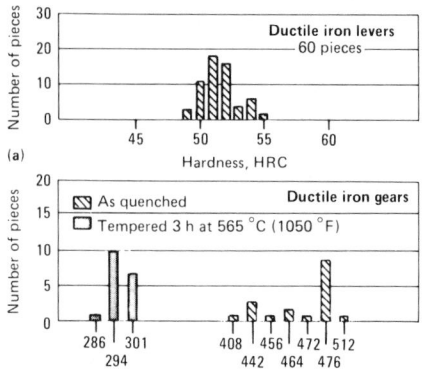

(a)

(b)

(a) Levers austenitized in salt at 870 °C (1600 °F) for ½ h then tempered in air at 175 °C (350 °F) for 2 h. Levers were made from 2 heats of electric-furnace iron containing 3.5% C, 2.3% Si, 0.03% Mn, 0.75% Ni, and 0.06% Mg. Data represent average of three hardness readings on each piece tested. (b) Gears quenched in oil from 885 °C (1625 °F) then tempered at 565 °C (1050 °F) for 3 h. Hardness of the as-cast gears ranged from 170 to 228 HB.

4 min of heating, 12 to 15% ferrite was retained, and the hardness varied from 44 to 51 HRC. When the cubes were heated for a total of 10 min or more, the microstructure did not contain ferrite, and full hardness (53 to 57 HRC) was developed.

Within their normal ranges, the total carbon and silicon contents have a minor influence on hardenability. The combined carbon in the matrix depends on the austenitizing temperature; manganese has a moderate influence. The effect of varying amounts of carbon, silicon and manganese on the hardenability of ductile irons is shown in Fig. 8.

A wide variety of hardenabilities can be obtained by the addition of alloying elements, either singly or in combination. The effects of nickel and molybdenum on Jominy end-quench curves for ductile iron are shown in Fig. 9.

After quenching, ductile iron castings are usually tempered for 1 h plus 1 h per inch of section thickness to obtain the desired hardness. Figure 10 shows the influence of tempering temperature on the hardness and tensile properties of ductile irons that have been quenched from 870 °C (1600 °F). Higher-alloy irons require the use of

Fig. 12 Isothermal transformation diagram for an unalloyed ductile iron containing 3.3% C, 2.52% Si

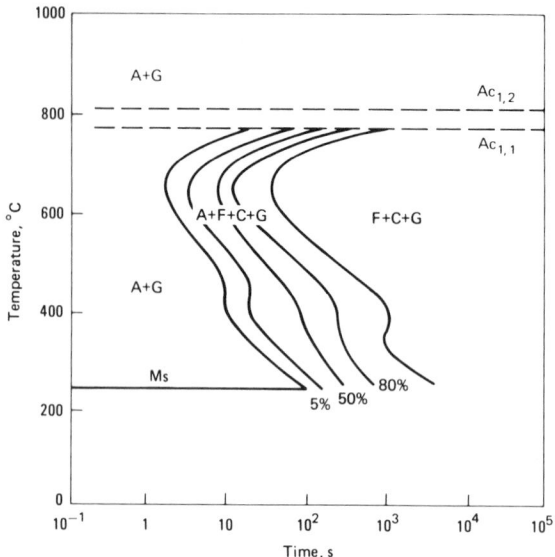

A, austenite; G, graphite nodules; F, ferrite; C, carbides

Fig. 13 Stress relief obtained in ductile iron held at three temperatures ½ to 8 h

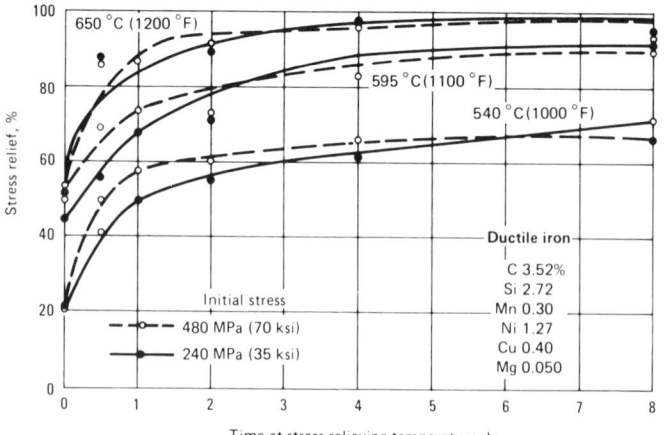

Initial hardness was 102 to 103 HRB. Hardness after holding at 540, 595 and 650 °C (1000, 1100 and 1200 °F) for 8 h was 102 to 104, 101 to 103, and 90 to 93 HRB, respectively.

somewhat higher tempering temperatures in order to obtain similar results.

The distribution of hardness readings obtained on quenched and tempered ductile iron levers is shown in Fig. 11(a). Figure 11(b) shows the hardness distribution for ductile iron gears as quenched and as tempered.

Austempering and martempering practices are based on isothermal transformation diagrams. The diagram for the unalloyed ductile iron in Fig. 12 provides information about the austenite transformation process. Transformation of austenite to bainite in unalloyed ductile iron is completed in about 2 h at austempering temperature.

The effects of alloying elements on the isothermal transformation of ductile iron can be expected to be similar to their effects in eutectoid steel. However, when carbide-forming alloys are used in large quantities, a part of any alloying element may be present as an insoluble carbide and is not effective in increasing the hardenability.

Surface Hardening of Ductile Iron

Ductile iron responds readily to surface hardening by flame or induction processes. Because of the short heating cycle in these processes, the pearlitic types of ductile iron, 80-60-03 and 100-70-03, are preferred. Irons without free ferrite in their microstructure respond almost instantly to flame or induction heating and require very little holding time at the austenitizing temperature in order to be fully hardened.

With a moderate amount of free ferrite, the response may be satisfactory, but an entirely ferritic matrix, typical of the grades with high ductility, requires several minutes at 870 °C (1600 °F) to be fully hardened by subsequent cooling. A matrix microstructure of fine pearlite, readily obtained by normalizing, has quick response to surface hardening and provides excellent core support for the hardened case.

Castings that are predominantly pearlitic in the as-cast condition (grade 80-55-06), should be given a relatively high temperature stress relief for a short time, for example, tempering at 595 to 650 °C (1100 to 1200 °F) for 1 h per inch of section thickness. This treatment will remove virtually all of the internal or residual stresses from the casting before it undergoes a localized hardening treatment.

With proper technique and control of temperature between 845 and 870 °C (1550 and 1600 °F), the following ranges of surface hardness of ductile iron can be expected in commercial production:

- Ductile iron, fully annealed (ferritic), water quenched behind the flame, 35 to 45 HRC
- Ductile iron, predominantly ferritic (partly pearlitic), stress relieved prior to heating and self quenched, 40 to 45 HRC

Table 2 Fatigue results on heat treated ductile irons

Type	Hardness, HB	Ultimate tensile strength MPa	ksi	Elongation, %	Fatigue strength MPa	ksi	Endurance ratio
As-cast............	255	664	96	3.0	278	40	0.42
Quenched and tempered(a)......	295	927	134	4.0	340	49	0.37
Quenched and tempered(b)......	350	1029	149	1.5	340	49	0.33
As-cast............	267	659	96	2.0	301	44	0.46
Normalized(c).....	325	1041	151	4.0	340	49	0.33

(a) Oil quenched from 900 °C (1650 °F); tempered 2 h at 600 °C (1110 °F). (b) Oil quenched from 900 °C (1650 °F); tempered 2 h at 550 °C (1020 °F). (c) Normalized from 900 °C (1650 °F)

- Ductile iron, predominantly ferritic (partly pearlitic), stress relieved prior to heating, water quenched, 50 to 55 HRC
- Ductile iron, mostly pearlitic, stress relieved before heating, water quenched, 58 to 62 HRC.

Heating time and temperature, amount of dissolved carbon, section size, and rate of quench help to determine final hardness values.

Flame or induction hardened ductile iron castings have been used for heavy-duty applications such as rolls for cold working titanium, ring gears for paper-mill drives, crankshafts, and large sprockets for chain drives.

For a more detailed discussion of surface hardening processes, and of the application of these processes to ductile iron, the reader may refer to the articles "Flame Hardening" and "Induction Hardening and Tempering".

Stress Relieving of Ductile Irons

When not otherwise heat treated, complex engineering castings of ductile iron may be stress relieved at 510 to 675 °C (950 to 1250 °F). Temperatures at the lower end of this range are satisfactory for many applications; temperatures at the higher end will eliminate virtually all residual stress (Fig. 13), but will also effect some reduction in hardness and tensile strength. Recommended ranges of stress-relieving temperature for various types of ductile iron are as follows:

- Unalloyed.......... 510 to 565 °C (950 to 1050 °F)
- Low-alloy.......... 565 to 595 °C (1050 to 1100 °F)
- High-alloy.......... 595 to 650 °C (1100 to 1200 °F)
- Austenitic.......... 620 to 675 °C (1150 to 1250 °F)

The required time at temperature will depend on the temperature used, the complexity of the casting, and the completeness of stress relief that is desired (Fig. 13), but 1 h plus 1 h per inch of section thickness is good practice.

Cooling should be uniform, to avoid reintroducing stresses. Castings should be furnace cooled to 290 °C (550 °F), after which they can be air cooled (in most instances; however, austenitic iron can be uniformly air cooled from the stress-relieving temperature).

Effect of Heat Treatment on Fatigue Strength

The long life fatigue properties of as-cast ductile irons are improved by heat treatment, but not in the same proportion as the monotonic properties. This is demonstrated by the endurance ratios (fatigue strength/ultimate tensile strength) shown in Table 2. In heat treating to improve properties, the proper composition and temperature must be selected to ensure the optimum improvement. For example, one cause of the low endurance ratios in quenched and tempered ductile irons is the precipitation of secondary graphite throughout the matrix on tempering. The amount of secondary graphite can be controlled by composition (primarily carbon and silicon) and tempering temperature (increases with temperature).

Heat Treating of Malleable Irons

By Edward F. Ryntz, Jr.
Senior Staff Research Engineer
General Motors Research Laboratories

FERRITIC AND PEARLITIC malleable irons are both produced by annealing white iron of controlled composition. Thus, annealing is an essential part of the manufacturing process for these irons and, as such, is discussed in detail in the article entitled "Malleable Iron", in Volume 1 of this Handbook.

The annealing treatment involves three important steps. The first causes nucleation of graphite and is initiated during heating to a high holding temperature and occurs very early during the holding period.

The second step consists of holding at 900 to 970 °C (1650 to 1780 °F); this step is called first-stage graphitization (FSG). During FSG, massive carbides are eliminated from the iron structure. At this point, the iron is rapidly cooled to 725 to 740 °C (1340 to 1360 °F) prior to entering second-stage graphitization.

The third step in the annealing treatment consists of slow cooling through the allotropic transformation range of the iron; this step is called second-stage graphitization (SSG). During SSG, a completely ferritic matrix free of pearlite and carbides is obtained when the cooling rate is 2 to 17 °C/h (3 to 30 °F/h). This cooling rate, which depends on the

silicon content of the iron and the temper carbon nodule count, may be increased to 85 °C/min (150 °F/min) to form a pearlitic matrix. Oil quenching from the first-stage graphitization temperature following completion of that step will produce a martensitic matrix.

Hardening and Tempering of Pearlitic Malleable Iron

A typical procedure for producing a hardened pearlitic malleable iron consists of (a) air quenching castings after first-stage annealing, which results in retention of about 0.75% combined carbon in the matrix; (b) reheating and holding for 1 h at 845 to 870 °C (1550 to 1600 °F) to reaustenitize the matrix; and then (c) quenching in heated (80 to 105 °C; 180 to 220 °F) and agitated oil, thus developing a matrix consisting of martensite and bainite with a hardness of 555 to 627 HB. Figure 1 shows the effects of austenitizing temperature, quenching medium and manganese content on the hardness of ferritic and pearlitic malleable iron both before and after heat treating. If direct oil quenching is used, caution must be exercised

to prevent cracking due to high combined carbon.

Increasing the austenitizing temperature increases the amount of dissolved carbon, which is measured as combined carbon in the matrix after quenching to room temperature (Fig. 2). Higher austenitizing temperatures (900 to 930 °C; 1650 to 1700 °F) result in a more homogeneous austenite, which is desirable for a more uniform martensite but which can result in a greater tendency toward distortion or cracking. Temperature and time of tempering to attain a specified hardness may be selected from curves such as those shown in Fig. 3. (For an example of hardenability data for pearlitic malleable iron, see Volume 1 of this Handbook.)

Hardened and tempered pearlitic malleable iron can be produced also from fully annealed ferritic malleable iron, the matrix of which is essentially carbon-free; graphite can be dissolved in austenite by holding at 845 to 870 °C (1550 to 1600 °F) for a time sufficiently long for production of an austenite matrix of uniform carbon content. In general, the combined carbon content of the matrix produced by this procedure is slightly lower than that of a pearlitic malleable made by air-quenching directly from the first-stage graphitization temperature (arrested annealing), and therefore the final tempering temperatures required for the development of specific hardnesses are lower (Fig. 4).

Tempering treatments consist of cycles of no less than 2 h at temperature to ensure uniformity of product. Tempering times must also be adjusted for section thickness and quenched microstructures. Fine pearlite and bainite require longer tempering times than that for martensite. In general, final hardness is controlled with process controls approximately the same as those encountered in the heat treatment of medium-carbon and higher-carbon steels. This is particularly true when

the specification requires final hardnesses in the range from 241 to 321 HB.

Example 1. Figure 5 shows the control of hardness attained by one foundry producing quenched and tempered sleeve yokes of grade 80002 pearlitic malleable iron. These yokes were heated at 870 °C (1600 °F) for 30 min, then quenched in oil at 65 °C (150 °F) and tempered at 650 °C (1200 °F) for 2 h.

The effects of tempering on the hardness of alloyed and unalloyed malleable irons are shown in Fig. 6; these data illustrate the beneficial effects of alloying on as-quenched hardness and stability at elevated temperature. During all tempering treatments, carbide has a tendency to decompose, with conse-

quent deposition of graphite on existing temper carbon nodules. This tendency is least at the lower tempering temperatures or in suitably alloyed pearlitic malleable irons.

Martempering and tempering develops mechanical properties similar to those resulting from conventional oil quenching and tempering—typical tensile strength, 862 MPa (125 ksi); yield strength, 758 MPa (110 ksi); and hardness, 300 HB.

Pearlitic malleable iron castings that are susceptible to cracking when quenched in warm oil (40 to 95 °C; 100 to 200 °F) from the austenitizing tem-

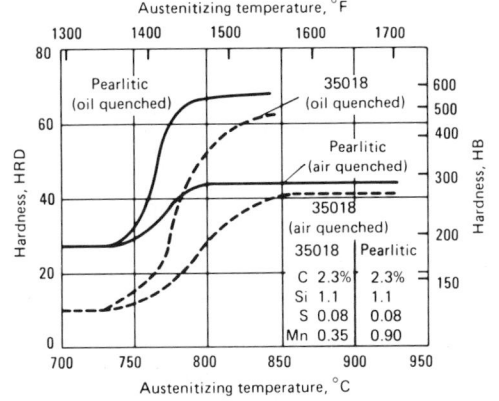

Fig. 1 Effects of austenitizing temperature, quenching medium and manganese content on hardness of as-quenched malleable iron

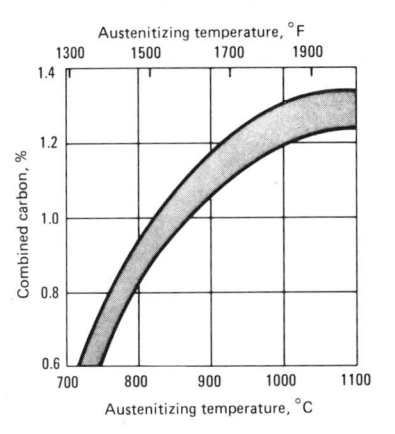

Fig. 2 Effect of austenitizing temperature on combined carbon content of pearlitic malleable iron

Fig. 3 Influence of time and tempering temperature on room-temperature hardness of pearlitic malleable iron

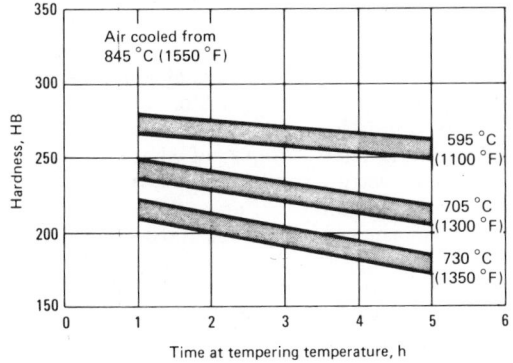

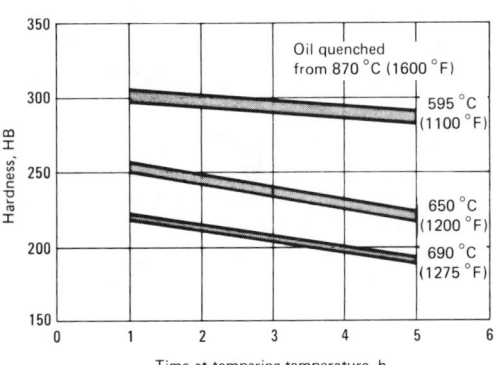

Composition: 2.35 to 2.45 C, 1.45 to 1.55 Si, 0.03 P, 0.06 to 0.15 S, 0.38 to 0.50 Mn, and less than 0.003 Cr

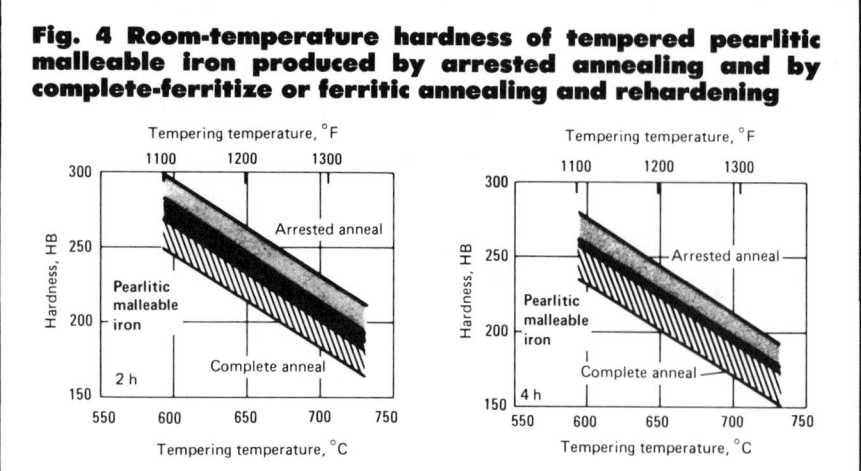

Fig. 4 Room-temperature hardness of tempered pearlitic malleable iron produced by arrested annealing and by complete-ferritize or ferritic annealing and rehardening

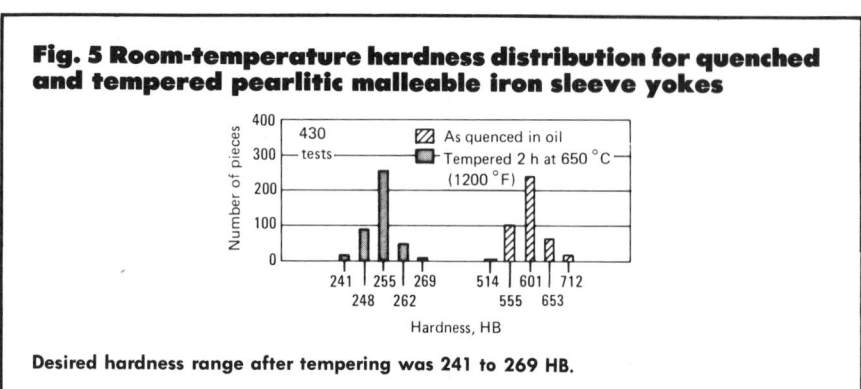

Fig. 5 Room-temperature hardness distribution for quenched and tempered pearlitic malleable iron sleeve yokes

Desired hardness range after tempering was 241 to 269 HB.

perature may be safely quenched in salt or oil at about 205 °C (400 °F).

Elevator camshafts varying in length from 0.3 to 0.45 m (12 to 18 in.) and various sizes of wear-chain components are examples of martempered pearlitic malleable iron.

Bainitic Heat Treatment of Pearlitic Malleable Iron

Both upper and lower bainite can be formed in pearlitic malleable iron with a marked increase in tensile strength and hardness but with a decrease in ductility. A pearlitic malleable iron (2.6C-1.4Si-0.5Mn-0.11S), annealed at 930 °C (1700 °F) for 16 h, air quenched, and tempered at 680 °C (1250 °F) for 4 h, developed an ultimate tensile strength of 650 MPa (94.2 ksi), a yield strength of 460 MPa (66.5 ksi), and a

3.4% elongation at 217 HB. This same iron austenitized at 900 °C (1650 °F) in molten salt for 1 h, quenched in molten salt at 295 °C (560 °F) for 3 h, and air cooled, gave an ultimate strength of 995 MPa (144.2 ksi), a yield strength of 920 MPa (133.4 ksi), and a 1% elongation at 388 HB.

Surface Hardening of Pearlitic Malleable Iron

Fully pearlitic malleable iron may be surface hardened by either induction heating and quenching or flame heating and quenching. Laser and electron-beam techniques also have been used for hardening selected areas on the surface of pearlitic and ferritic malleable iron castings that are free from decarburization. Generally, hardness in the range from 55 to 60 HRC is attainable,

with the depth of penetration being controlled by the rate of heating and by the temperature developed at the surface of the part being hardened. In induction hardening, this is accomplished by close regulation of power output, operating frequency, heating time, and alloy content of the iron.

The maximum hardness obtainable in the matrix of a properly hardened part is 67 HRC; however, as has been mentioned previously, conventional hardness measurements show less than the true matrix hardness because of the temper carbon nodules that are averaged into the hardness. Generally, a casting with a matrix microhardness of 67 HRC will have about 62 HRC average hardness, as measured with the standard Rockwell tester.

Rocker arms and clutch hubs are examples of automotive production parts that are surface hardened by induction. Flame hardening is not employed for this application because it would result in distortion that would interfere with their operation. The two examples that follow describe successful application of induction and flame hardening to other production parts.

Example 2. Grade 45010 pearlitic malleable iron was used for tools for crimping electrical connectors. The jaws of these tools as originally designed were inserts made of hardened tool steel. Shell mold casting these jaws and heat treating them to pearlitic malleable iron made it possible for the jaws to be integrally cast to the desired intricate contour, and to be hardened by induction heating and water quenching to provide required wear resistance. Hardening was thus restricted to the jaws, and the strength and toughness inherent in grade 45010 were maintained in the arms of the handles.

Example 3. Flame hardening has been used effectively on pearlitic malleable iron pinion spacers that support the cups of roller bearings. The ends of the pinion spacers were flame hardened to file hard (58 HRC or more) to a depth of about 2.4 mm (3/32 in.). This treatment eliminated service failures.

For a more complete discussion of surface hardening processes and their application to pearlitic malleable iron, see the articles "Flame Hardening" and "Induction Hardening and Tempering".

Fig. 6 Effect of tempering temperature and time on hardness of ferritic and pearlitic malleable irons in the as-received and the reheated-and-quenched conditions

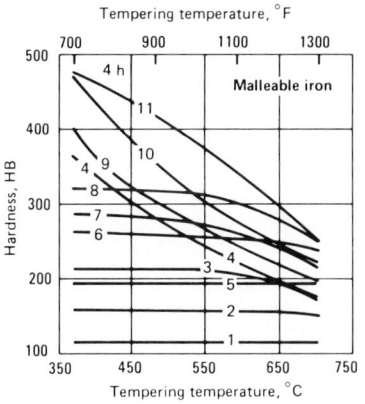

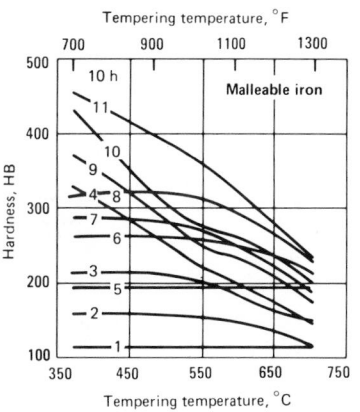

| Iron | Material | Composition, % | | | | | Alloying and prior heat treatment | Hardness, HB |
		TC	Si	S	Mn	Mo		
1	Standard (ferritic) grade 32510	2.40	1.80	0.072	0.30	⋯	Unalloyed; fully malleablized	116
2	Pearlitic malleable iron, grade 45007	2.40	1.80	0.072	0.30	⋯	Unalloyed; air quenched from 925 °C (1700 °F), tempered 8 h at 695 °C (1280 °F)	156
3	Pearlitic malleable iron, grade 60003	2.40	1.80	0.072	0.30	⋯	Unalloyed; oil quenched from 870 °C (1600 °F), tempered 3 h at 650 °C (1200 °F)	212
4	Oil-quenched malleable iron	2.40	1.80	0.072	0.30	⋯	Unalloyed; oil quenched from 870 °C (1600 °F), not tempered	444
5	Pearlitic malleable iron, grade 45010	2.40	1.80	0.076	0.90	⋯	Alloyed (Mn); air quenched from 940 °C (1720 °F), tempered 34 h at 715 °C (1320 °F)	192
6	Pearlitic malleable iron, grade 80002	2.40	1.80	0.072	0.90	0.45	Alloyed (Mn and Mo); air quenched from 940 °C (1720 °F), tempered 12 h at 620 °C (1150 °F)	262
7	Air-quenched alloyed malleable iron	2.40	1.80	0.079	0.90	⋯	Alloyed (Mn); air quenched from 925 °C (1700 °F), not tempered	285
8	Air-quenched alloyed malleable iron	2.40	1.80	0.076	1.10	⋯	Alloyed (Mn); air quenched from 925 °C (1700 °F), not tempered	321
9	Oil-quenched alloyed malleable iron	2.40	1.80	0.079	0.90	⋯	Alloyed (Mn); oil quenched from 830 °C (1525 °F), not tempered	514
10	Oil-quenched alloyed malleable iron	2.40	1.80	0.076	1.10	⋯	Alloyed (Mn); oil quenched from 830 °C (1525 °F), not tempered	578
11	Air-quenched alloyed malleable iron	2.40	1.80	0.072	0.90	0.45	Alloyed (Mn and Mo); air quenched from 940 °C (1720 °F), not tempered	514

Heat Treating of Austenitic Irons

By Richard L. Ward
Metallurgist and Manager of
  Quality Control
Bendix Industrial Group
Division of Warner & Swasey Co.

AUSTENITIC CAST IRONS are employed in a variety of applications because the alloying elements in these irons develop metallurgical properties superior to those of conventional gray and ductile irons. Nickel, chromium and copper promote corrosion resistance, and nickel and chromium improve wear resistance.

Corrosion resistance is improved because the austenitic phase is stabilized at room temperature by the alloying elements and is more corrosion resistant than pearlite. Austenite is stable at high temperatures and normally transforms into pearlite and carbide phases during cooling.

Flake-Graphite Corrosion-Resistant Austenitic Cast Irons

These cast irons exhibit properties that include (a) resistance to corrosion by alkalis, acids, salts, oils and foods; (b) high-temperature oxidation resistance; (c) high electrical resistivity; (d) nonmagnetic characteristics; (e) abrasive wear resistance; (f) uniform thermal expansion; and (g) moderate strength and toughness. Many of the irons combine several of these proper-

Table 1 Compositions of flake-graphite corrosion-resistant austenitic cast irons

Type	TC(a)	Si	Mn	Composition, % Ni	Cu	Cr
1(b)	3.00 max	1.00-2.80	0.50-1.50	13.50-17.50	5.50-7.50	1.50-2.50
1b	3.00 max	1.00-2.80	0.50-1.50	13.50-17.50	5.50-7.50	2.50-3.50
2(c)	3.00 max	1.00-2.80	0.50-1.50	18.00-22.00	0.50 max	1.50-2.50
2b	3.00 max	1.00-2.80	0.50-1.50	18.00-22.00	0.50 max	3.00-6.00(d)
3	2.60 max	1.00-2.00	0.50-1.50	28.00-32.00	0.50 max	2.50-3.50
4	2.60 max	5.00-6.00	0.50-1.50	29.00-32.00	0.50 max	4.50-5.50
5	2.40 max	1.00-2.00	0.50-1.50	34.00-36.00	0.50 max	0.10 max(e)
6(f)	3.00 max	1.50-2.50	0.50-1.50	18.00-22.00	3.50-5.50	1.00-2.00

(a) Total carbon. (b) Type 1 is recommended for applications in which the presence of copper offers corrosion-resistance advantages. (c) Type 2 is recommended for applications in which copper contamination cannot be tolerated, such as handling of foods or caustics. (d) Where some machining is required, 3.0 to 4.0 Cr is recommended. (e) Where increased hardness, strength and heat resistance are desired, and where increased expansivity can be tolerated, Cr may be increased to 2.5 to 3.0%. (f) Type 6 also contains 1.0% Mo.

ties, which result from the high-temperature austenite phase; this phase is stabilized at room temperature by the high alloy content and contains a uniform dispersion of carbides. The compositions of flake-graphite corrosion-resistant austenitic cast irons are given in Table 1; typical mechanical properties of these irons are presented in Table 2.

These alloys are susceptible to work hardening during machining and require careful cooling from the casting operation and/or subsequent heat treating operations to minimize the initial stresses and the rate of work hardening during metal-removal operations. Castings that have not been heat treated may cause "chattering" during machining.

Stress Relieving. For most applications, it is recommended that austenitic cast irons be stress relieved at 620 to 675 °C (1150 to 1250 °F), for 1 h per inch of section, to remove residual stresses

Table 2 Typical mechanical properties of flake-graphite corrosion-resistant austenitic cast irons

Type	Tensile strength(a) MPa	ksi	Hardness, HB(b)
1	170	25	131-183
1b	205	30	149-212
2	170	25	118-174
2b	205	30	171-248
3	170	25	118-159
4	170	25	149-212
5	140	20	99-124
6	170	25	124-174

(a) Minimum. (b) 3000-kg load

Table 3 Compositions of nodular-graphite corrosion-resistant austenitic cast irons

Type	TC(a)	Si	Composition, % Mn	P	Ni	Cr
D-2	3.00 max	1.50 to 3.00	0.70 to 1.25	0.08 max	18.0 to 22.0	1.75 to 2.75
D-2b	3.00 max	1.50 to 3.00	0.70 to 1.25	0.08 max	18.0 to 22.0	2.75 to 4.00
D-2c	2.90 max	1.00 to 3.00	1.80 to 2.40	0.08 max	21.0 to 24.0	0.50 max
D-3	2.60 max	1.00 to 2.80	1.00 max	0.08 max	28.0 to 32.0	2.50 to 3.50
D-3a	2.60 max	1.00 to 2.80	1.00 max	0.08 max	28.0 to 32.0	1.00 to 1.50
D-4	2.60 max	5.00 to 6.00	1.00 max	0.08 max	28.0 to 32.0	4.50 to 5.50
D-5	2.60 max	1.00 to 2.80	1.00 max	0.08 max	34.0 to 36.0	0.10 max
D-5b	2.40 max	1.00 to 2.80	1.00 max	0.08 max	34.0 to 36.0	2.00 to 3.00

(a) Total carbon

Table 4 Typical mechanical properties of nodular-graphite corrosion-resistant austenitic cast irons

Type	Minimum tensile strength MPa	ksi	Minimum yield strength MPa	ksi	Minimum elongation(a), %	Hardness, HB
D-2	400	58	205	30	8	139 to 202
D-2b	400	58	205	30	7	148 to 211
D-3	380	55	205	30	6	139 to 202
D-4	415	60	...	...	...	202 to 273
D-5	380	55	205	30	20	131 to 185

(a) In 50 mm or 2 in.

resulting from casting or machining, or both. Stress relieving should follow rough machining, particularly for castings that must conform to close dimensional tolerances, that have been extensively welded, or that are to be exposed to high stresses in service.

Holding of castings at 480 °C (900 °F) for 1 h per inch of thickness will remove about 60% of the stress; stress relieving at 675 °C (1250 °F) will remove almost 95%. It is usually acceptable to cool castings in air at a rate of 1 to 2 h per inch of section thickness, although furnace cooling produces maximum stress relief. Stress relieving does not affect tensile strength, hardness or ductility.

Spheroidize Annealing. Castings with hardnesses above 190 HB may be softened by heating to 980 to 1040 °C (1800 to 1900 °F) for ½ to 5 h, except those alloys containing 4% or more chromium. Excessive carbides cause this high hardness and may occur in rapidly cooled castings and thin sections. Annealing dissolves or spheroidizes carbides. Although it lowers hardness, spheroidize annealing does not adversely affect strength.

High-Temperature Stabilization. Except for castings of alloy type 1, which are not recommended for service above 430 °C (800 °F), castings used for either static or cyclic service at 480 °C (900 °F) or above should be given a stabilization heat treatment. This treatment consists of holding at 760 °C (1400 °F) for 4 h minimum or at 870 °C (1600 °F) for 2 h minimum, furnace cooling to 540 °C (1000 °F), and then cooling in air. This treatment stabilizes the microstructure and minimizes growth and distortion at high temperatures as carbon diffuses from the austenite and from dissolving carbides and

reprecipitates as larger-volume graphite flakes. Thus, it is usually advisable to stabilize castings prior to final machining.

Dimensional Stabilization. This treatment normally is limited to castings that require true dimensional stability, such as those used in precision machinery or scientific instruments. The treatment is not applicable to castings of type 1 alloys. Other alloys may be dimensionally stabilized by the following treatment:

- Heat to 870 °C (1600 °F), and hold for 2 h minimum plus 1 h per inch of section.
- Furnace cool, at a maximum rate of 50 °C/h (100 °F/h), to 540 °C (1000 °F).
- Hold at 540 °C (1000 °F) for 1 h per inch of section, and then cool uniformly in air.
- After rough machining, reheat to 455 to 480 °C (850 to 900 °F) and hold for 1 h per inch of section, and cool uniformly in air.
- Finish machine and reheat to 260 to 315 °C (500 to 600 °F), and cool uniformly in air.

Solution Treating. Although this treatment is seldom used, high-temper-

ature quenching is capable of producing higher-than-normal strength levels and slightly higher hardnesses by dissolving some carbon in austenite at elevated temperatures and by preventing precipitation of the carbon by rapid cooling. This treatment consists of heating to 925 to 1010 °C (1700 to 1850 °F) and quenching in oil or water. Because no metallurgical phase change occurs, the possibility of cracking is lessened.

Nodular-Graphite Corrosion-Resistant Austenitic Cast Irons

These irons, the compositions of which are given in Table 3, are similar to those listed in Table 1 but are treated with magnesium to produce spheroidal (nodular) graphite, which results in higher values of strength and ductility. To ensure satisfactory graphite nodularity, sulfur content is held to 0.02% or less, and residual magnesium content should exceed 0.03%. The procedures and temperatures of the heat treatments for these alloys are identical to those described for flake-graphite corrosion-resistant austenitic cast irons. The mechanical properties of several of these nodular austenitic irons are listed in Table 4.

Abrasion-Resistant Nickel-Chromium White Irons

These special white irons possess outstanding abrasion resistance, and are widely used in the mining, power, cement, paint, steel and foundry industries. As shown in Table 5, these irons contain less nickel, silicon and manganese than the austenitic alloys. This composition results in an entirely carbidic structure.

Because of their carbidic structures, abrasion-resistant Ni-Cr white irons must be stress relieved at 200 to 230 °C (400 to 450 °F) before being placed in service. This treatment relieves the transformation stresses present in castings after cooling to room temperature and increases strength and impact resistance with no loss in abrasion resistance.

High-Chromium, Low-Carbon White Irons

The chemical composition ranges for high-chromium, low-carbon white

Table 5 Compositions of abrasion-resistant nickel-chromium white irons

Element	Composition range, %	
	Type 1 (regular)	Type 2 (high-strength)
Total carbon	3.00-3.60	2.90 max
Silicon	0.40-0.70(a)	0.40-0.70(a)
Manganese	0.40-0.70	0.40-0.70
Sulfur	0.15 max	0.15 max
Phosphorus	0.40 max	0.40 max
Nickel	4.00-4.75	4.00-4.75
Chromium	1.40-3.50	1.40-3.50

(a) Silicon content may exceed 0.70% in thin-wall castings or in composite castings poured over gray iron cores.

irons are given in Table 6. These abrasion-resistant irons are used in the manufacture of ceramics and of equipment for dredging, oil-well drilling, coke and ore handling, and shot blasting.

Annealing. Although castings of these irons usually are finished primarily by grinding, they can be annealed to a hardness sufficiently low to permit simple machining. The annealing treatment consists of heating slowly to 760 to 790 °C (1400 to 1450 °F),

Table 6 Composition ranges for high-chromium, low-carbon white irons

Element	Composition range, %
Carbon	2.25 to 2.85
Silicon	0.25 to 1.00
Manganese	0.50 to 1.25
Chromium	24.0 to 30.0
Vanadium(a)	0.25 to 2.00
Molybdenum(a)	0 to 6.50

(a) Optional

holding at temperature for 12 to 24 h, and then furnace cooling in still air. Hardnesses as low as 350 to 450 HB can be obtained by this treatment.

Austenitizing. For maximum wear resistance under conditions of extreme rubbing or scouring, an austenitizing heat treatment is effective. The room-temperature austenite produced by this treatment readily work hardens by rubbing or mild impingement. This transformation is somewhat analogous to the behavior of austenitic high-manganese steel. Austenitizing consists of heating to 1095 to 1105 °C (2000 to 2020 °F) and cooling in air.

Heat Treating of Tool Steels

Introduction to Heat Treating of Tool Steels

By the ASM Committee on
Heat Treating of Tool Steels*

TOOL STEELS are high-quality steels made to close compositional and physical tolerances; they are used to make tools for cutting, forming or shaping a material into a part or component adapted to a definite use. The earliest tool steels were simple, plain carbon steels, but beginning in 1868, and to a greater extent early in the 20th century, many complex, highly alloyed tool steels were developed. These complex alloy tool steels, which contain, among other elements, relatively large amounts of tungsten, molybdenum, vanadium and chromium, make it possible to meet increasingly severe service demands and to provide greater dimensional control and freedom from cracking during heat treating. Many alloy tool steels also are widely used for machinery components and structural applications where particularly severe requirements must be met, such as high-temperature springs, ultrahigh-strength fasteners, special-purpose valves, and bearings of various types for elevated-temperature service. This article discusses procedures and process control requirements for heat treating the principal types of tool steel. It also provides a review of heat treating processes that are applied to tool steels and the specific applicability of these processes to the various types of tool steels.

In service, most tools are subjected to extremely high loads that are applied rapidly. They must withstand these loads a great number of times without breaking and without undergoing excessive wear or deformation. In many applications, tool steels must provide this capability under conditions that develop high temperatures in the tool. No single tool material combines maximum wear resistance, toughness, and resistance to softening at elevated temperatures. Consequently, selection of the proper tool material for a given application often requires a trade-off to achieve the optimum combination of properties. Table 1 gives the classifications and nominal compositions of various tool steels.

Most tool steels are wrought products, but precision castings can be used to advantage in some applications. The powder metallurgy process also is used in making tool steels; this process provides (a) more uniform carbide size and distribution in large sections and (b) special compositions that are difficult or impossible to produce by melting and casting and then mechanically working the cast product.

For typical wrought tool steels, raw materials (including scrap) are carefully selected, not only for alloy content but also for qualities that ensure cleanness and homogeneity in the finished product. Tool steels are generally melted in small-tonnage electric-arc furnaces to achieve composition tolerances economically, good cleanness and precise control of melting conditions. Special refining and secondary remelting processes have been introduced to satisfy particularly difficult demands on tool steel quality and performance. The medium-to-high alloy contents of many tool steels require careful control

*W. James Laird, Jr. *Chairman,* Vice President-Marketing, Research and Development, Upton Industries, Inc.; Carl J. Oxford, Jr., Vice President-Technology, National Twist, Drill & Tool, Division of Lear-Siegler, Inc.; Percy Rawcliffe, Manager, Metallurgical and Physical Laboratories, Morse Cutting Tools, Division of Gulf & Western Manufacturing; Carl Reichel, Plant Metallurgist, Drill & End Mill Division, TRW, Inc.; Ronald F. Spitzer, Chief Metallurgist, Bearings Division, TRW, Inc.; Daniel S. Zamborsky, Corporate Metallurgist, Warner & Swasey Co.

of forging and rolling, which often results in a large amount of process scrap. Semifinished and finished bars are given rigorous in-process and final inspection. This inspection can be so extensive that both ends of each bar may be inspected for macrostructure (etch quality), cleanness, hardness, grain size, annealed structure and hardenability. The inspection may require that the entire bar be subjected to magnetic and ultrasonic inspections for surface and internal discontinuities. It is important that finished tool steel bars have limited decarburization, which requires that annealing be done by special procedures under closely controlled conditions.

Such precise production practices and stringent quality controls contribute to the high cost of tool steels, as do the expensive alloying elements they contain. Insistence on quality in the manufacture of these specialty steels is justified, however, because tool steel bars generally are made into complicated cutting and forming tools worth many times the cost of the steel itself. Although some standard constructional alloy steels resemble tool steels in composition, they are seldom used for expensive tooling because, in general, they are not manufactured to the same rigorous quality standards as are tool steels.

The performance of a tool in service depends on proper design of the tool, accuracy with which the tool is made, selection of the proper tool steel and application of the proper heat treatment. A tool can perform successfully in service only when all four of these requirements have been fulfilled.

With few exceptions, all tool steels must be heat treated to develop specific combinations of wear resistance, resistance to deformation or breaking under high loads, and resistance to softening at elevated temperatures. A few simple shapes may be obtained directly from tool steel producers in correctly heat treated condition. However, most tool steels first are formed or machined to produce the required shape and then are heat treated either by the tool manufacturer or by the ultimate user. Improper finishing after heat treatment—principally grinding—can metallurgically damage most tool steels.

Processing information and service characteristics of tool steels are presented in Tables 2 to 4. This information is essential in understanding the problems involved in selection, processing and application of tool steels. Tool

Table 1 Classification and nominal compositions of various tool steels

Steel	C	Mn	Si	Composition, % W	Mo	Cr	V	Other
Water-hardening tool steels								
W1	0.60-1.40(b)	...	...	...	...	...	...	...
W2	0.60-1.40(b)	...	...	...	...	...	0.25	...
W3 (a)	1.00	...	...	...	...	...	0.50	...
W4	0.60-1.40(b)	...	...	...	...	0.25	...	...
W5	1.10	...	...	...	...	0.50	...	...
W6 (a)	1.00	...	...	...	...	0.25	0.25	...
W7 (a)	1.00	...	...	...	...	0.50	0.20	...
Shock-resisting tool steels								
S1	0.50	...	...	2.50	...	1.50	...	...
S2	0.50	...	1.00	...	0.50	...	...	...
S3 (a)	0.50	...	...	1.00	...	0.75	...	...
S4	0.55	0.80	2.00	...	...	...	...	...
S5	0.55	0.80	2.00	...	0.40	...	...	...
S6	0.45	1.40	2.25	...	0.40	1.50	...	...
S7	0.50	...	...	...	1.40	3.25	...	...
Oil-hardening cold work tool steels								
O1	0.90	1.00	...	0.50	...	0.50	...	...
O2	0.90	1.60	...	...	...	...	...	...
O6	1.45	...	1.00	...	0.25	...	...	...
O7	1.20	...	...	1.75	...	0.75	...	...
Medium-alloy air-hardening cold work tool steels								
A2	1.00	...	...	...	1.00	5.00	...	...
A3	1.25	...	...	...	1.00	5.00	1.00	...
A4	1.00	2.00	...	...	1.00	1.00	...	...
A5	1.00	3.00	...	...	1.00	1.00	...	...
A6	0.70	2.00	...	...	1.00	1.00	...	...
A7	2.25	...	...	1.00(c)	1.00	5.25	4.75	...
A8	0.55	...	...	1.25	1.25	5.00	...	...
A9	0.50	...	...	...	1.40	5.00	1.00	1.50 Ni
A10	1.35	1.80	1.25	...	1.50	...	...	1.80 Ni
High-carbon high-chromium cold work tool steels								
D1	1.00	...	...	...	1.00	12.00	...	...
D2	1.50	...	...	...	1.00	12.00	...	...
D3	2.25	...	...	...	...	12.00	...	...
D4	2.25	...	...	...	1.00	12.00	...	...
D5	1.50	...	...	...	1.00	12.00	...	3.00 Co
D6 (a)				Now included with D3 by AISI				
D7	2.35	...	...	...	1.00	12.00	4.00	...
Chromium hot work tool steels								
H10	0.40	...	...	...	2.50	3.25	0.40	...
H11	0.35	...	...	...	1.50	5.00	0.40	...
H12	0.35	...	...	1.50	1.50	5.00	0.40	...
H13	0.35	...	...	...	1.50	5.00	1.00	...
H14	0.40	...	...	5.00	...	5.00	...	...
H15 (a)	0.40	...	...	...	5.00	5.00	...	...
H16	0.55	...	...	7.00	...	7.00	...	...
H19	0.40	...	...	4.25	...	4.25	2.00	4.25 Co

(continued)

(a) These steels were not included in the March 1978 AISI Steel Products Manual, *Tool Steels*, in the main table of compositions nor in tables of heat treating practice, because of their less common use. (b) Available with various carbon contents, in increments of 0.10% within this range. (c) Optional. (d) Intermediate high speed steels M50 and M52, which are lower in alloy content than standard high speed steels, have been employed successfully in applications requiring greater abrasion resistance than plain carbon steels, but less red hardness than high speed steels. Typical uses include woodworking tools and hack saw blades. M50 and M52 steels meet the criteria promulgated by the American Society for Testing and Materials for "intermediate" high speed steels but do not meet the more stringent criteria for "standard" high speed steels.

Table 1 (continued)

Steel	C	Mn	Si	Composition, % W	Mo	Cr	V	Other
Tungsten hot work tool steels								
H20	0.35	...	...	9.00	...	2.00	...	...
H21	0.35	...	...	9.00	...	3.50	...	...
H22	0.35	...	...	11.00	...	2.00	...	...
H23	0.30	...	...	12.00	...	12.00	...	...
H24	0.45	...	...	15.00	...	3.00	...	...
H25	0.25	...	...	15.00	...	4.00	...	...
H26	0.50	...	...	18.00	...	4.00	1.00	...
Molybdenum hot work tool steels								
H41	0.65	...	...	1.50	8.00	4.00	1.00	...
H42	0.60	...	...	6.00	5.00	4.00	2.00	...
H43	0.55	...	...	...	8.00	4.00	2.00	...
Tungsten high speed tool steels, standard group								
T1	0.75	...	...	18.00	...	4.00	1.00	...
T2	0.80	...	...	18.00	...	4.00	2.00	...
T3 (a)	1.05	...	...	18.00	...	4.00	3.00	...
T4	0.75	...	...	18.00	...	4.00	1.00	5.00 Co
T5	0.80	...	...	18.00	...	4.00	2.00	8.00 Co
T6	0.80	...	...	20.00	...	4.50	1.50	12.00 Co
T7 (a)	0.75	...	...	14.00	...	4.00	2.00	...
T8	0.75	...	...	14.00	...	4.00	2.00	5.00 Co
T9 (a)	1.20	...	...	18.00	...	4.00	4.00	...
T15	1.50	...	...	12.00	...	4.00	5.00	5.00 Co
Molybdenum high speed tool steels, standard group								
M1	0.85	...	...	1.50	8.50	4.00	1.00	...
M2	0.85 or 1.00	...	...	6.00	5.00	4.00	2.00	...
M3 Cl 1	1.05	...	...	6.00	5.00	4.00	2.40	...
M3 Cl 2	1.20	...	...	6.00	5.00	4.00	3.00	...
M4	1.30	...	...	5.50	4.50	4.00	4.00	...
M6	0.80	...	...	4.00	5.00	4.00	1.50	12.00 Co
M7	1.00	...	...	1.75	8.75	4.00	2.00	...
M8 (a)	0.80	...	...	5.00	5.00	4.00	1.50	1.25 Nb
M10	0.85 or 1.00	...	...	...	8.00	4.00	2.00	...
M15 (a)	1.50	...	...	6.50	3.50	4.00	5.00	5.00 Co
M30	0.80	...	...	2.00	8.00	4.00	1.25	5.00 Co
M33	0.90	...	...	1.50	9.50	4.00	1.15	8.00 Co
M34	0.90	...	...	2.00	8.00	4.00	2.00	8.00 Co
M35 (a)	0.80	...	...	6.00	5.00	4.00	2.00	5.00 Co
M36	0.80	...	...	6.00	5.00	4.00	2.00	8.00 Co
M41	1.10	...	...	6.75	3.75	4.25	2.00	5.00 Co
M42	1.10	...	...	1.50	9.50	3.75	1.15	8.00 Co
M43	1.20	...	...	2.75	8.00	3.75	2.60	8.25 Co
M44	1.15	...	...	5.25	6.25	4.25	2.20	12.00 Co
M46	1.25	...	...	2.00	8.25	4.00	3.20	8.25 Co
M47	1.10	...	...	1.50	9.50	3.75	1.25	5.00 Co

(continued)

(a) These steels were not included in the March 1978 AISI Steel Products Manual, *Tool Steels,* in the main table of compositions nor in tables of heat treating practice, because of their less common use. (b) Available with various carbon contents, in increments of 0.10% within this range. (c) Optional. (d) Intermediate high speed steels M50 and M52, which are lower in alloy content than standard high speed steels, have been employed successfully in applications requiring greater abrasion resistance than plain carbon steels, but less red hardness than high speed steels. Typical uses include woodworking tools and hack saw blades. M50 and M52 steels meet the criteria promulgated by the American Society for Testing and Materials for "intermediate" high speed steels but do not meet the more stringent criteria for "standard" high speed steels.

steel suppliers provide more specific information on the properties developed by specific heat treatments in the steels produced by their companies. They should be consulted as to the type of steel and heat treatment best suited to meet all service requirements at the least over-all cost.

Physical properties—density, thermal expansion and thermal conductivity—of selected tool steels are given in Tables 5 and 6.

Normalizing

Normalizing requires slow and uniform heating above the transformation range to dissolve excess constituents, then cooling in still air. Normalizing breaks up nonuniform structures, relieves residual stresses, and produces greater uniformity in grain size—thus counteracting undesirable results of unequal reductions for different sections during forging, differences in temperature between varying thicknesses of sections, and the subsequent irregular cooling rates. Normalizing also conditions the steel for subsequent spheroidizing, annealing or hardening.

Applicability. Many tool steels harden even when cooled in still air; normalizing these steels for the purpose of refining a structure is not recommended. Tool steels that should not be normalized include all high speed steels, all shock-resisting steels, all hot work steels, cold work steels of types A and D (except A10), and the mold steel P4.

For other types of tool steel, normalizing is most commonly applied after forging and before annealing. Normalizing also may be used before full annealing for parts that are being hardened for a second time.

Standard practice consists of heating to the normalizing temperature and then cooling in still air. Workpieces usually are held at temperature only long enough to be heated through. No special equipment is required, but the work should be protected against decarburization during heating.

Annealing

Tool steels usually are received from the supplier in the annealed condition. However, if they are subjected to hot or cold forming, often they must be fully annealed again before subsequent operations. If a tool is to be rehardened, it should first be thoroughly annealed. This procedure is important with the steels of higher alloy content; otherwise, irregular grain growth occurs and a mixed grain size (sometimes called "fish scale") will result.

Full annealing involves heating the steel slowly and uniformly to a temperature above the transformation range, holding it at the temperature for from 1 to 4 h (which is generally long enough for complete penetration of the heat), and cooling slowly at a controlled rate. For most tool steels, controlled cooling should continue to 540 °C (1000 °F) or lower.

Atmosphere furnaces, salt baths, vacuum furnaces or lead pots may be used for annealing. Requirements of the heating equipment include reasonably accurate temperature control and a means of preventing decarburization. In box or roller-hearth furnaces, surface protection often is accomplished by packing the workpieces in pipes, in which they are then surrounded by nondecarburizing material, such as spent charcoal and mica, or cast iron chips. Furnaces with prepared atmospheres frequently are used for the annealing of tool steels.

Cracking from thermal shock can be minimized by loading the furnace at a relatively low temperature (room temperature or a few hundred degrees Fahrenheit). Following the soak at annealing temperature, the workpieces (and container, if used) should be cooled in the furnace at 8 to 22 °C/h (15 to 40 °F/h) to 540 °C (1000 °F) or lower. Below about 540 °C, the cooling rate for most tool steels is no longer critical, and the work may then be cooled in air.

Isothermal annealing is an alternative method of cooling that consists of rapidly cooling the workload in the furnace from the annealing temperature to a temperature just below the transformation range and holding the load and furnace at this temperature for one or more hours. Following this period of soaking at just below the transformation range, the load may be safely air cooled. This process, known as isothermal annealing, is best suited for applications in which full advantage can be taken of the rapid cooling to the transformation temperature, and from this temperature to room temperature. Thus, for small parts that can be handled in salt or lead baths, or for light loads in batch furnaces, isothermal annealing makes possible large savings in time, as compared with the conventional slow furnace cooling. It can also be adapted conveniently to continuous annealing cycles where adequate equipment is available.

Isothermal annealing offers no particular advantage for applications (such as the batch annealing of large furnace loads) in which the rate of cooling at the center of the load may be so slow as to preclude any rapid cooling to the transformation temperature. For such applications, conventional full annealing usually offers a better assurance of obtaining the desired properties.

Stress Relieving

Stress relieving removes or reduces residual stress induced in tools by heavy machining or other cold working, and thereby decreases the probability of distortion or cracking during hardening of the tool.

The ground surface of a hardened tool may be highly stressed after grinding but not cracked. The high stress may, however, cause cracks to develop immediately after grinding, before use or during use. Ground tools with high residual stress can often be salvaged by stress relieving, immediately after grinding, at or just below the temper-

Table 1 (continued)

Steel	C	Mn	Si	W	Mo	Cr	V	Other
High speed tool steels, intermediate group(d)								
M50	0.85	...	...	...	4.00	4.00	1.00	...
M52	0.90	...	...	1.25	4.00	4.00	2.00	...
Low-alloy special-purpose tool steels								
L1	1.00	...	...	...	...	1.25	...	...
L2	0.50-1.10(b)	...	...	...	...	1.00	0.20	...
L3	1.00	...	...	...	...	1.50	0.20	...
L4 (a)	1.00	0.60	...	...	...	1.50	0.25	...
L5 (a)	1.00	1.00	...	...	0.25	1.00	...	...
L6	0.70	...	...	...	0.25(c)	0.75	...	1.50 Ni
L7	1.00	0.35	...	...	0.40	1.40	...	...
Carbon-tungsten special-purpose tool steels								
F1	1.00	...	...	1.25	...	...	...	...
F2	1.25	...	...	3.50	...	...	...	...
F3	1.25	...	...	3.50	...	0.75	...	...
Mold steels								
P1	0.10	...	...	...	...	...	...	...
P2	0.07	...	...	...	0.20	2.00	...	0.50 Ni
P3	0.10	...	...	...	...	0.60	...	1.25 Ni
P4	0.07	...	...	...	0.75	5.00	...	...
P5	0.10	...	...	...	...	2.25	...	...
P6	0.10	...	...	...	...	1.50	...	3.50 Ni
P20	0.35	...	...	...	0.40	1.25	...	...
P21	0.20	...	...	...	...	...	...	4.00 Ni, 1.20 Al
Other alloy tool steels								
6G	0.55	0.80	0.25	...	0.45	1.00	0.10	...
6F2	0.55	0.75	0.25	...	0.30	1.00	0.10(c)	1.00 Ni
6F3	0.55	0.60	0.85	...	0.75	1.00	0.10(c)	1.80 Ni
6F4	0.20	0.70	0.25	...	3.35	...	...	3.00 Ni
6F5	0.55	1.00	1.00	...	0.50	0.50	0.10	2.70 Ni
6F6	0.50	...	1.50	...	0.20	1.50	...	...
6F7	0.40	0.35	...	...	0.75	1.50	...	4.25 Ni
6H1	0.55	...	...	...	0.45	4.00	0.85	...
6H2	0.55	0.40	1.10	...	1.50	5.00	1.00	...

(a) These steels were not included in the March 1978 AISI Steel Products Manual, *Tool Steels*, in the main table of compositions nor in tables of heat treating practice, because of their less common use. (b) Available with various carbon contents, in increments of 0.10% within this range. (c) Optional. (d) Intermediate high speed steels M50 and M52, which are lower in alloy content than standard high speed steels, have been employed successfully in applications requiring greater abrasion resistance than plain carbon steels, but less red hardness than high speed steels. Typical uses include woodworking tools and hack saw blades. M50 and M52 steels meet the criteria promulgated by the American Society for Testing and Materials for "intermediate" high speed steels but do not meet the more stringent criteria for "standard" high speed steels.

Table 2 Normalizing and annealing temperatures of tool steels

Type	Normalizing treatment/ temperature(a) °C	°F	Annealing(b) Temperature °C	°F	Rate of cooling, max °C/h	°F/h	Hardness, HB
Molybdenum high speed steels							
M1, M10 ....Do not normalize			815 to 870	1500 to 1600	22	40	207 to 235
M2 ....Do not normalize			870 to 900	1600 to 1650	22	40	212 to 241
M3, M4 ....Do not normalize			870 to 900	1600 to 1650	22	40	223 to 255
M6 ....Do not normalize			870	1600	22	40	248 to 277
M7 ....Do not normalize			815 to 870	1500 to 1600	22	40	217 to 255
M30, M33, M34, M36, M41, M42, M46, M47 ....Do not normalize			870 to 900	1600 to 1650	22	40	235 to 269
M43 ....Do not normalize			870 to 900	1600 to 1650	22	40	248 to 269
M44 ....Do not normalize			870 to 900	1600 to 1650	22	40	248 to 293
Tungsten high speed steels							
T1 ....Do not normalize			870 to 900	1600 to 1650	22	40	217 to 255
T2 ....Do not normalize			870 to 900	1600 to 1650	22	40	223 to 255
T4 ....Do not normalize			870 to 900	1600 to 1650	22	40	229 to 269
T5 ....Do not normalize			870 to 900	1600 to 1650	22	40	235 to 277
T6 ....Do not normalize			870 to 900	1600 to 1650	22	40	248 to 293
T8 ....Do not normalize			870 to 900	1600 to 1650	22	40	229 to 255
T15 ....Do not normalize			870 to 900	1600 to 1650	22	40	241 to 277
Chromium hot work steels							
H10, H11, H12, H13 ....Do not normalize			845 to 900	1550 to 1650	22	40	192 to 229
H14 ....Do not normalize			870 to 900	1600 to 1650	22	40	207 to 235
H19 ....Do not normalize			870 to 900	1600 to 1650	22	40	207 to 241
Tungsten hot work steels							
H21, H22, H25 ....Do not normalize			870 to 900	1600 to 1650	22	40	207 to 235
H23 ....Do not normalize			870 to 900	1600 to 1650	22	40	212 to 255
H24, H26 ....Do not normalize			870 to 900	1600 to 1650	22	40	217 to 241
Molybdenum hot work steels							
H41, H43 ....Do not normalize			815 to 870	1500 to 1600	22	40	207 to 235
H42 ....Do not normalize			845 to 900	1550 to 1650	22	40	207 to 235
High-carbon high-chromium cold work steels							
D2, D3, D4 ....Do not normalize			870 to 900	1600 to 1650	22	40	217 to 255
D5 ....Do not normalize			870 to 900	1600 to 1650	22	40	223 to 255
D7 ....Do not normalize			870 to 900	1600 to 1650	22	40	235 to 262
Medium-alloy air-hardening cold work steels							
A2 ....Do not normalize			845 to 870	1550 to 1600	22	40	201 to 229
A3 ....Do not normalize			845 to 870	1550 to 1600	22	40	207 to 229
A4 ....Do not normalize			740 to 760	1360 to 1400	14	25	200 to 241
A6 ....Do not normalize			730 to 745	1350 to 1375	14	25	217 to 248
A7 ....Do not normalize			870 to 900	1600 to 1650	14	25	235 to 262
A8 ....Do not normalize			845 to 870	1550 to 1600	22	40	192 to 223
A9 ....Do not normalize			845 to 870	1550 to 1600	14	25	212 to 248
A10	790	1450	765 to 795	1410 to 1460	8	15	235 to 269
Oil-hardening cold work steels							
O1	870	1600	760 to 790	1400 to 1450	22	40	183 to 212
O2	845	1550	745 to 775	1375 to 1425	22	40	183 to 212
O6	870	1600	765 to 790	1410 to 1450	11	20	183 to 217
O7	900	1650	790 to 815	1450 to 1500	22	40	192 to 217

(continued)

(a) Time held at temperature varies from 15 min for small sections to 1 h for large sizes. Cooling is done in still air. Normalizing should not be confused with low-temperature annealing. (b) The upper limit of ranges should be used for large sections and the lower limit for smaller sections. Time held at temperature varies from 1 h for light sections to 4 h for heavy sections and large furnace charges of high-alloy steel. (c) For 0.25 Si type, 183 to 207 HB; for 1.00 Si type, 207 to 229 HB. (d) Temperature varies with carbon content: 0.60 to 0.75 C, 815 °C (1500 °F); 0.75 to 0.90 C, 790 °C (1450 °F); 0.90 to 1.10 C, 870 °C (1600 °F); 1.10 to 1.40 C, 870 to 925 °C (1600 to 1700 °F). (e) Temperature varies with carbon content: 0.60 to 0.90 C, 740 to 790 °C (1360 to 1450 °F); 0.90 to 1.40 C, 760 to 790 °C (1400 to 1450 °F).

Table 2 (continued)

Type	Normalizing treatment/ temperature(a) °C	°F	Annealing(b) Temperature °C	°F	Rate of cooling, max °C/h	°F/h	Hardness, HB
Shock-resisting steels							
S1 . . . . Do not normalize			790 to 815	1450 to 1500	22	40	183 to 229(c)
S2 . . . . Do not normalize			760 to 790	1400 to 1450	22	40	192 to 217
S5 . . . . Do not normalize			775 to 800	1425 to 1475	14	25	192 to 229
S7 . . . . Do not normalize			815 to 845	1500 to 1550	14	25	187 to 223
Mold steels							
P2 . . . . Not required			730 to 815	1350 to 1500	22	40	103 to 123
P3 . . . . Not required			730 to 815	1350 to 1500	22	40	109 to 137
P4 . . . . Do not normalize			870 to 900	1600 to 1650	14	25	116 to 128
P5 . . . . Not required			845 to 870	1550 to 1600	22	40	105 to 116
P6 . . . . Not required			845	1550	8	15	183 to 217
P20	900	1650	760 to 790	1400 to 1450	22	40	149 to 179
P21	900	1650	Do not anneal				
Low-alloy special-purpose steels							
L2	871 to 900	1600 to 1650	760 to 790	1400 to 1450	22	40	163 to 197
L3	900	1650	790 to 815	1450 to 1500	22	40	174 to 201
L6	870	1600	760 to 790	1400 to 1450	22	40	183 to 212
Carbon-tungsten special-purpose steels							
F1	900	1650	760 to 800	1400 to 1475	22	40	183 to 207
F2	900	1650	790 to 815	1450 to 1500	22	40	207 to 235
Water-hardening steels							
W1, W2	790 to 925(d)	1450 to 1700(d)	740 to 790(e)	1360 to 1450(e)	22	40	156 to 201
W5	870 to 925	1600 to 1700	760 to 790	1400 to 1450	22	40	163 to 201

(a) Time held at temperature varies from 15 min for small sections to 1 h for large sizes. Cooling is done in still air. Normalizing should not be confused with low-temperature annealing. (b) The upper limit of ranges should be used for large sections and the lower limit for smaller sections. Time held at temperature varies from 1 h for light sections to 4 h for heavy sections and large furnace charges of high-alloy steel. (c) For 0.25 Si type, 183 to 207 HB; for 1.00 Si type, 207 to 229 HB. (d) Temperature varies with carbon content: 0.60 to 0.75 C, 815 °C (1500 °F); 0.75 to 0.90 C, 790 °C (1450 °F); 0.90 to 1.10 C, 870 °C (1600 °F); 1.10 to 1.40 C, 870 to 925 °C (1600 to 1700 °F). (e) Temperature varies with carbon content: 0.60 to 0.90 C, 740 to 790 °C (1360 to 1450 °F); 0.90 to 1.40 C, 760 to 790 °C (1400 to 1450 °F).

ing temperature in order to maintain the specified tool hardness.

Tools also develop high residual stress in use. It is sometimes advantageous to relieve this stress at each redressing of the tool by retempering at an appropriate temperature. This temperature should not exceed the tempering temperature; otherwise, undesirable softening will occur.

Procedure. Stress relieving is most commonly performed in air furnaces or salt baths used for tempering. Neither the heating nor the cooling rate is critical, although cooling should be slow enough to prevent the introduction of new stress. Protection against scaling or decarburization is seldom required, unless the stress-relieving temperature is above 650 °C (1200 °F). Under some conditions, vacuum or inert atmosphere furnaces may be required to prevent scaling or discoloration.

After stress relieving, it may be necessary to correct certain dimensions before hardening, because the relief of stress causes some dimensional change. Precision tools usually are stress relieved after machining and before hardening; it is often desirable to stress relieve after rough machining but before finish machining.

Austenitizing

Austenitizing is the most critical of all heating operations performed on tool steels. Excessively high austenitizing temperatures or abnormally long holding times may result in excessive distortion, abnormal grain growth, loss of ductility, and low strength; this is especially true for high speed steels, which are frequently austenitized at a temperature close to that at which melting begins. Underheating may result in low hardness and low wear resistance. At the time of quenching, if the center of a tool is cooler than the exterior, spalling or fracturing of the corners may result, particularly with water-hardening steels.

Equipment for austenitizing tool steels is chosen on the basis of steel composition, size and shape of workpieces, amount of stock removal after hardening, and production requirements. Vacuum furnaces, atmosphere furnaces, and salt baths have proved satisfactory for service over the entire austenitizing temperature range of 760 to 1300 °C (1400 to 2375 °F). Lead pots are suitable for the temperature range of about 760 to 930 °C (1400 to 1700 °F).

Workpieces must be supported during austenitizing. Lead and salt provide some of the support, but in atmosphere furnaces, special attention must be given to prevent workpieces from sagging or making contact with the furnace brickwork.

During austenitizing, continuous control of the furnace internal environment must be maintained to prevent workpieces from becoming carburized or decarburized. Salt baths must be rectified; atmospheres must be controlled for proportion of gases and dew point; lead baths must be kept free of contamination. Vacuum furnaces must be maintained at low leak rates and par-

Table 3 Hardening and tempering of tool steels

Type	Rate of heating	Preheat temperature °C	Preheat temperature °F	Hardening temperature °C	Hardening temperature °F	Time at temperature, min	Quenching medium(a)	Tempering temperature °C	Tempering temperature °F
Molybdenum high speed steels									
M1, M7, M10	Rapidly from preheat	730 to 845	1350 to 1550	1175 to 1220	2150 to 2225(b)	2 to 5	O, A or S	540 to 595(c)	1000 to 1100(c)
M2	Rapidly from preheat	730 to 845	1350 to 1550	1190 to 1230	2175 to 2250(b)	2 to 5	O, A or S	540 to 595(c)	1000 to 1100(c)
M3, M4, M30, M33, M34 ..	Rapidly from preheat	730 to 845	1350 to 1550	1205 to 1230(b)	2200 to 2250(b)	2 to 5	O, A or S	540 to 595(c)	1000 to 1100(c)
M6	Rapidly from preheat	790	1450	1175 to 1205(b)	2150 to 2200(b)	2 to 5	O, A or S	540 to 595(c)	1000 to 1100(c)
M36	Rapidly from preheat	730 to 845	1350 to 1550	1220 to 1245(b)	2225 to 2275(b)	2 to 5	O, A or S	540 to 595(c)	1000 to 1100(c)
M41	Rapidly from preheat	730 to 845	1350 to 1550	1190 to 1215(b)	2175 to 2220(b)	2 to 5	O, A or S	540 to 595(d)	1000 to 1100(d)
M42	Rapidly from preheat	730 to 845	1350 to 1550	1190 to 1210(b)	2175 to 2210(b)	2 to 5	O, A or S	510 to 595(d)	950 to 1100(d)
M43	Rapidly from preheat	730 to 845	1350 to 1550	1190 to 1215(b)	2175 to 2220(b)	2 to 5	O, A or S	510 to 595(d)	950 to 1100(d)
M44	Rapidly from preheat	730 to 845	1350 to 1550	1200 to 1225(b)	2190 to 2240(b)	2 to 5	O, A or S	540 to 625(d)	1000 to 1160(d)
M46	Rapidly from preheat	730 to 845	1350 to 1550	1190 to 1220(b)	2175 to 2225(b)	2 to 5	O, A or S	525 to 565(d)	975 to 1050(d)
M47	Rapidly from preheat	730 to 845	1350 to 1550	1180 to 1205(b)	2150 to 2200(b)	2 to 5	O, A or S	525 to 595(d)	975 to 1100(d)
Tungsten high speed steels									
T1, T2, T4, T8	Rapidly from preheat	815 to 870	1500 to 1600	1260 to 1300(b)	2300 to 2375(b)	2 to 5	O, A or S	540 to 595(c)	1000 to 1100(c)
T5, T6......	Rapidly from preheat	815 to 870	1500 to 1600	1275 to 1300(b)	2325 to 2375(b)	2 to 5	O, A or S	540 to 595(c)	1000 to 1100(c)
T15	Rapidly from preheat	815 to 870	1500 to 1600	1205 to 1260(b)	2200 to 2300(b)	2 to 5	O, A or S	540 to 650(d)	1000 to 1200(d)
Chromium hot work steels									
H10........	Moderately from preheat	815	1500	1010 to 1040	1850 to 1900	15 to 40(e)	A	540 to 650	1000 to 1200
H11, H12....	Moderately from preheat	815	1500	995 to 1025	1825 to 1875	15 to 40(e)	A	540 to 650	1000 to 1200
H13........	Moderately from preheat	815	1500	995 to 1040	1825 to 1900	15 to 40(e)	A	540 to 650	1000 to 1200
H14........	Moderately from preheat	815	1500	1010 to 1065	1850 to 1950	15 to 40(e)	A	540 to 650	1000 to 1200
H19........	Moderately from preheat	815	1500	1095 to 1205	2000 to 2200	2 to 5	A or O	540 to 705	1000 to 1300
Molybdenum hot work steels									
H41, H43....	Rapidly from preheat	730 to 845	1350 to 1550	1095 to 1190	2000 to 2175	2 to 5	O, A or S	565 to 650	1050 to 1200
H42........	Rapidly from preheat	730 to 845	1350 to 1550	1120 to 1220	2050 to 2225	2 to 5	O, A or S	565 to 650	1050 to 1200

(continued)

(a) O, oil quench; A, air cool; S, salt bath quench; W, water quench; B, brine quench. (b) When the high-temperature heating is carried out in a salt bath, the range of temperatures should be about 15 °C (25 °F) lower than given in this line. (c) Double tempering recommended for not less than 1 h at temperature each time. (d) Triple tempering recommended for not less than 1 h at temperature each time. (e) Times apply to open-furnace heat treatment. For pack hardening, a common rule is to heat 30 min/in. of cross section of the pack. (f) Preferable for large tools to minimize decarburization. (g) Carburizing temperature. (h) After carburizing. (j) Carburized case hardness. (k) P21 is a precipitation-hardening steel having a thermal treatment which involves solution treating and aging rather than hardening and tempering. (m) Recommended for large tools and tools with intricate sections

Table 3 (continued)

Type	Rate of heating	Preheat temperature °C	Preheat temperature °F	Hardening temperature °C	Hardening temperature °F	Time at temperature, min	Quenching medium(a)	Tempering temperature °C	Tempering temperature °F
Tungsten hot work steels									
H21, H22....	Rapidly from preheat	815	1500	1095 to 1205	2000 to 2200	2 to 5	A or O	595 to 675	1100 to 1250
H23........	Rapidly from preheat	845	1550	1205 to 1260	2200 to 2300	2 to 5	O	650 to 815	1200 to 1500
H24........	Rapidly from preheat	815	1500	1095 to 1230	2000 to 2250	2 to 5	O	565 to 650	1050 to 1200
H25........	Rapidly from preheat	815	1500	1150 to 1260	2100 to 2300	2 to 5	A or O	565 to 675	1050 to 1250
H26........	Rapidly from preheat	870	1600	1175 to 1260	2150 to 2300	2 to 5	O, A or S	565 to 675	1050 to 1250
Medium-alloy air-hardening cold work steels									
A2.........	Slowly	790	1450	925 to 980	1700 to 1800	20 to 45	A	175 to 540	350 to 1000
A3.........	Slowly	790	1450	955 to 980	1750 to 1800	25 to 60	A	175 to 540	350 to 1000
A4.........	Slowly	675	1250	815 to 870	1500 to 1600	20 to 45	A	175 to 425	350 to 800
A6.........	Slowly	650	1200	830 to 870	1525 to 1600	20 to 45	A	150 to 425	300 to 800
A7.........	Very slowly	815	1500	955 to 980	1750 to 1800	30 to 60	A	150 to 540	300 to 1000
A8.........	Slowly	790	1450	980 to 1010	1800 to 1850	20 to 45	A	175 to 595	350 to 1100
A9.........	Slowly	790	1450	980 to 1025	1800 ro 1875	20 to 45	A	510 to 620	950 to 1150
A10........	Slowly	650	1200	790 to 815	1450 to 1500	30 to 60	A	175 to 425	350 to 800
Oil-hardening cold work steels									
O1.........	Slowly	650	1200	790 to 815	1450 to 1500	10 to 30	O	175 to 260	350 to 500
O2.........	Slowly	650	1200	760 to 800	1400 to 1475	5 to 20	O	175 to 260	350 to 500
O6.........	Slowly	...	...	790 to 815	1450 to 1500	10 to 30	O	175 to 315	350 to 600
O7.........	Slowly	650	1200	790 to 830; 845 to 885	W:1450 to 1525; O:1550 to 1625	10 to 30	O or W	175 to 290	350 to 550
Shock-resisting steels									
S1.........	Slowly	...	...	900 to 955	1650 to 1750	15 to 45	O	205 to 650	400 to 1200
S2.........	Slowly	650(f)	1200(f)	845 to 900	1550 to 1650	5 to 20	B or W	175 to 425	350 to 800
S5.........	Slowly	760	1400	870 to 925	1600 to 1700	5 to 20	0	175 to 425	350 to 800
S7.........	Slowly	650 to 705	1200 to 1300	925 to 955	1700 to 1750	15 to 45	A or O	205 to 620	400 to 1150

(continued)

(a) O, oil quench; A, air cool; S, salt bath quench; W, water quench; B, brine quench. (b) When the high-temperature heating is carried out in a salt bath, the range of temperatures should be about 15 °C (25 °F) lower than given in this line. (c) Double tempering recommended for not less than 1 h at temperature each time. (d) Triple tempering recommended for not less than 1 h at temperature each time. (e) Times apply to open-furnace heat treatment. For pack hardening, a common rule is to heat 30 min/in. of cross section of the pack. (f) Preferable for large tools to minimize decarburization. (g) Carburizing temperature. (h) After carburizing. (j) Carburized case hardness. (k) P21 is a precipitation-hardening steel having a thermal treatment which involves solution treating and aging rather than hardening and tempering. (m) Recommended for large tools and tools with intricate sections

Table 3 (continued)

Type	Rate of heating	Preheat temperature °C	°F	Hardening temperature °C	°F	Time at temperature, min	Quenching medium(a)	Tempering temperature °C	°F
Mold steels									
P2	...	900 to 925(g)	1650 to 1700(g)	830 to 845(h)	1525 to 1550(h)	15	O	175 to 260	350 to 500
P3	...	900 to 925(g)	1650 to 1700(g)	800 to 830(h)	1475 to 1525(h)	15	O	175 to 260	350 to 500
P4	...	970 to 995(g)	1775 to 1825(g)	970 to 995(h)	1775 to 1825(h)	15	A	175 to 480	350 to 900
P5	...	900 to 925(g)	1650 to 1700(g)	845 to 870(h)	1550 to 1600(h)	15	O or W	175 to 260	350 to 500
P6	...	900 to 925(g)	1650 to 1700(g)	790 to 815(h)	1450 to 1500(h)	15	A or O	175 to 230	350 to 450
P20	...	870 to 900(h)	1600 to 1650(h)	815 to 870	1500 to 1600	15	O	480 to 595(j)	900 to 1100(j)
P21(k).......	Slowly	Do not preheat		705 to 730	1300 to 1350	60 to 180	A or O	510 to 550	950 to 1025
Low-alloy special-purpose steels									
L2	Slowly	...	...	W: 790 to 845 / O: 845 to 925	W: 1450 to 1550 / O: 1550 to 1700	10 to 30	O or W	175 to 540	350 to 1000
L3	Slowly	...	...	W: 775 to 815 / O: 815 to 870	W: 1425 to 1500 / O: 1500 to 1600	10 to 30	O or W	175 to 315	350 to 600
L6	Slowly	...	...	790 to 845	1450 to 1550	10 to 30	O	175 to 540	350 to 1000
Carbon-tungsten special-purpose steels									
F1, F2	Slowly	650	1200	790 to 870	1450 to 1600	15	W or B	175 to 260	350 to 500
Water-hardening steels									
W1, W2, W3	Slowly	565 to 650(m)	1050 to 1200(m)	760 to 815	1400 to 1550	10 to 30	B or W	175 to 345	350 to 650
High-carbon, high-chromium cold work steels									
D1, D5	Very slowly	815	1500	980 to 1025	1800 to 1875	15 to 45	A	205 to 540	400 to 1000
D3	Very slowly	815	1500	925 to 980	1700 to 1800	15 to 45	O	205 to 540	400 to 1000
D4	Very slowly	815	1500	970 to 1010	1775 to 1850	15 to 45	A	205 to 540	400 to 1000
D7	Very slowly	815	1500	1010 to 1065	1850 to 1950	30 to 60	A	150 to 540	300 to 1000

(a) O, oil quench; A, air cool; S, salt bath quench; W, water quench; B, brine quench. (b) When the high-temperature heating is carried out in a salt bath, the range of temperatures should be about 15 °C (25 °F) lower than given in this line. (c) Double tempering recommended for not less than 1 h at temperature each time. (d) Triple tempering recommended for not less than 1 h at temperature each time. (e) Times apply to open-furnace heat treatment. For pack hardening, a common rule is to heat 30 min/in. of cross section of the pack. (f) Preferable for large tools to minimize decarburization. (g) Carburizing temperature. (h) After carburizing. (j) Carburized case hardness. (k) P21 is a precipitation-hardening steel having a thermal treatment which involves solution treating and aging rather than hardening and tempering. (m) Recommended for large tools and tools with intricate sections

tial pressure control at austenitizing temperatures, above 1095 °C (2000 °F).

Preheating for Austenitizing

Preheating tool steels before austenitizing is sound practice, but it is not always required. For small pieces of simple shape, preheating may be eliminated. Preheating normally is employed as a safeguard against the cracking and extreme distortion resulting from the thermal shock undergone by a cold workpiece when it is exposed to the high temperature of the austenitizing furnace.

Preheating is especially beneficial for the highly alloyed hot work and high speed steels, because it gives them a greater length of time to reach thermal equilibrium and eliminates most of the risk of prolonged exposure to austenitizing temperatures.

Tools that are to be austenitized in salt are usually preheated in salt, but they may be preheated in an atmo-

Table 4 Processing and service characteristics of tool steels

AISI designation	Resistance to decarbur- ization	Hardening and tempering				Machin- ability	Fabrication and service		
		Harden- ing response	Amount of distortion(a)	Resistance to cracking	Approxi- mate hard- ness(b), HRC		Tough- ness	Resistance to softening	Resistance to wear
Molybdenum high speed steels									
M1.......	Low	Deep	A or S, low; O, medium	Medium	60-65	Medium	Low	Very high	Very high
M2.......	Medium	Deep	A or S, low; O, medium	Medium	60-65	Medium	Low	Very high	Very high
M3 (class 1 and class 2).......	Medium	Deep	A or S, low; O, medium	Medium	61-66	Medium	Low	Very high	Very high
M4.......	Medium	Deep	A or S, low; O, medium	Medium	61-66	Low to medium	Low	Very high	Highest
M6.......	Low	Deep	A or S, low; O, medium	Medium	61-66	Medium	Low	Highest	Very high
M7.......	Low	Deep	A or S, low; O, medium	Medium	61-66	Medium	Low	Very high	Very high
M10.....	Low	Deep	A or S, low; O, medium	Medium	60-65	Medium	Low	Very high	Very high
M30.....	Low	Deep	A or S, low; O, medium	Medium	60-65	Medium	Low	Highest	Very high
M33.....	Low	Deep	A or S, low; O, medium	Medium	60-65	Medium	Low	Highest	Very high
M34.....	Low	Deep	A or S, low; O, medium	Medium	60-65	Medium	Low	Highest	Very high
M36.....	Low	Deep	A or S, low; O, medium	Medium	60-65	Medium	Low	Highest	Very high
M41.....	Low	Deep	A or S, low; O, medium	Medium	65-70	Medium	Low	Highest	Very high
M42	Low	Deep	A or S, low; O, medium	Medium	65-70	Medium	Low	Highest	Very high
M43.....	Low	Deep	A or S, low; O, medium	Medium	65-70	Medium	Low	Highest	Very high
M44.....	Low	Deep	A or S, low; O, medium	Medium	62-70	Medium	Low	Highest	Very high
M46.....	Low	Deep	A or S, low; O, medium	Medium	67-69	Medium	Low	Highest	Very high
M47.....	Low	Deep	A or S, low; O, medium	Medium	65-70	Medium	Low	Highest	Very high
Tungsten high speed steels									
T1	High	Deep	A or S, low; O, medium	High	60-65	Medium	Low	Very high	Very high
T2	High	Deep	A or S, low; O, medium	High	61-66	Medium	Low	Very high	Very high
T4	Medium	Deep	A or S, low; O, medium	Medium	62-66	Medium	Low	Highest	Very high
T5	Low	Deep	A or S, low; O, medium	Medium	60-65	Medium	Low	Highest	Very high
T6	Low	Deep	A or S, low; O, medium	Medium	60-65	Low to medium	Low	Highest	Very high
T8	Medium	Deep	A or S, low; O, medium	Medium	60-65	Medium	Low	Highest	Very high
T15	Medium	Deep	A or S, low; O, medium	Medium	63-68	Low to medium	Low	Highest	Highest

(continued)

(a) A, air cool; B, brine quench; O, oil quench; S, salt bath quench; W, water quench. (b) After tempering in temperature range normally recommended for this steel. (c) Carburized case hardness. (d) After aging at 510 to 550 °C (950 to 1025 °F). (e) Toughness decreases with increasing carbon content and depth of hardening.

Table 4 (continued)

AISI designation	Resistance to decarburization	Hardening and tempering Harden-ing response	Amount of distortion(a)	Resistance to cracking	Approxi-mate hard-ness(b), HRC	Machin-ability	Fabrication and service Tough-ness	Resistance to softening	Resistance to wear
Chromium hot work steels									
H10...... Medium	Deep	Very low	Highest	39-56	Medium to high	High	High	Medium
H11...... Medium	Deep	Very low	Highest	38-54	Medium to high	Very high	High	Medium
H12...... Medium	Deep	Very low	Highest	38-55	Medium to high	Very high	High	Medium
H13...... Medium	Deep	Very low	Highest	38-53	Medium to high	Very high	High	Medium
H14...... Medium	Deep	Low	Highest	40-47	Medium	High	High	Medium
H19...... Medium	Deep	A, low; O, medium	High	40-57	Medium	High	High	Medium to high
Tungsten hot work steels									
H21...... Medium	Deep	A, low; O, medium	High	36-54	Medium	High	High	Medium to high
H22...... Medium	Deep	A, low; O, medium	High	39-52	Medium	High	High	Medium to high
H23...... Medium	Deep	Medium	High	34-47	Medium	Medium	Very high	Medium to high
H24...... Medium	Deep	A, low; O, medium	High	45-55	Medium	Medium	Very high	High
H25 ..... Medium	Deep	A, low; O, medium	High	35-44	Medium	High	Very high	Medium
H26...... Medium	Deep	A or S, low; O, medium	High	43-58	Medium	Medium	Very high	High
Molybdenum hot work steels									
H42...... Medium	Deep	A or S, low; O, medium	Medium	50-60	Medium	Medium	Very high	High
Air-hardening medium-alloy cold work steels									
A2...... Medium	Deep	Lowest	Highest	57-62	Medium	Medium	High	High
A3...... Medium	Deep	Lowest	Highest	57-65	Medium	Medium	High	Very high
A4...... Medium to high	Deep	Lowest	Highest	54-62	Low to medium	Medium	Medium	Medium to high
A6...... Medium to high	Deep	Lowest	Highest	54-60	Low to medium	Medium	Medium	Medium to high
A7...... Medium	Deep	Lowest	Highest	57-67	Low	Low	High	Highest
A8...... Medium	Deep	Lowest	Highest	50-60	Medium	High	High	Medium to high
A9...... Medium	Deep	Lowest	Highest	35-56	Medium	High	High	Medium to high
A10...... Medium to high	Deep	Lowest	Highest	55-62	Medium to high	Medium	Medium	High
High-carbon, high-chromium cold work steels									
D2...... Medium	Deep	Lowest	Highest	54-61	Low	Low	High	High to very high
D3...... Medium	Deep	Very low	High	54-61	Low	Low	High	Very high
D4...... Medium	Deep	Lowest	Highest	54-61	Low	Low	High	Very high
D5...... Medium	Deep	Lowest	Highest	54-61	Low	Low	High	High to very high
D7...... Medium	Deep	Lowest	Highest	58-65	Low	Low	High	Highest
Oil-hardening cold work steels									
O1...... High	Medium	Very low	Very high	57-62	High	Medium	Low	Medium
O2...... High	Medium	Very low	Very high	57-62	High	Medium	Low	Medium
O6...... High	Medium	Very low	Very high	58-63	Highest	Medium	Low	Medium
O7...... High	Medium	W, high; O, very low	W, low; O, very high	58-64	High	Medium	Low	Medium

(continued)

(a) A, air cool; B, brine quench; O, oil quench; S, salt bath quench; W, water quench. (b) After tempering in temperature range normally recommended for this steel. (c) Carburized case hardness. (d) After aging at 510 to 550 °C (950 to 1025 °F). (e) Toughness decreases with increasing carbon content and depth of hardening.

sphere furnace if more convenient. Tools that are to be austenitized in an atmosphere furnace are preheated in a gaseous atmosphere.

Procedure. Preheating is usually done in a furnace adjacent to the austenitizing furnace, although it is possible to preheat and austenitize in the same furnace. In the latter procedure, once the workpiece is heated through to the preheat temperature, the furnace temperature is raised to the austenitiz-

Table 4 (continued)

AISI designation	Resistance to decarbur- ization	Hardening and tempering					Fabrication and service		
		Harden- ing response	Amount of distortion(a)	Resistance to cracking	Approxi- mate hard- ness(b), HRC	Machin- ability	Tough- ness	Resistance to softening	Resistance to wear
Shock-resisting steels									
S1	Medium	Medium	Medium	High	40-58	Medium	Very high	Medium	Low to medium
S2	Low	Medium	High	Low	50-60	Medium to high	Highest	Low	Low to medium
S5	Low	Medium	Medium	High	50-60	Medium to high	Highest	Low	Low to medium
S6	Low	Medium	Medium	High	54-56	Medium	Very high	Low	Low to medium
S7	Medium	Deep	A, lowest; O, low	A, highest; O, high	45-57	Medium	Very high	High	Low to medium
Low-alloy special-purpose steels									
L2	High	Medium	W, low; O, medium	W, high; O, medium	45-63	High	Very high(c)	Low	Low to medium
L6	High	Medium	Low	High	45-62	Medium	Very high	Low	Medium
Low-carbon mold steels									
P2	High	Medium	Low	High	58-64(c)	Medium to high	High	Low	Medium
P3	High	Medium	Low	High	58-64(c)	Medium	High	Low	Medium
P4	High	High	Very low	High	58-64(c)	Low to medium	High	Medium	High
P5	High	. . .	W, high; O, low	High	58-64(c)	Medium	High	Low	Medium
P6	High	. . .	A, very low; O, low	High	58-61(c)	Medium	High	Low	Medium
P20	High	Medium	Low	High	28-37	Medium to high	High	Low	Low to medium
P21	High	Deep	Lowest	Highest	30-40(d)	Medium	Medium	Medium	Medium
Water-hardening steels									
W1	Highest	Shallow	High	Medium	50-64	Highest	High(e)	Low	Low to medium
W2	Highest	Shallow	High	Medium	50-64	Highest	High(e)	Low	Low to medium
W5	Highest	Shallow	High	Medium	50-64	Highest	High(e)	Low	Low to medium

(a) A, air cool; B, brine quench; O, oil quench; S, salt bath quench; W, water quench. (b) After tempering in temperature range normally recommended for this steel. (c) Carburized case hardness. (d) After aging at 510 to 550 °C (950 to 1025 °F). (e) Toughness decreases with increasing carbon content and depth of hardening.

ing temperature and the workpiece is thus brought to the austenitizing temperature without leaving the furnace. The practicality of this one-furnace procedure depends on the difference between preheating and austenitizing temperatures for the type of steel being treated and on production requirements. Some experts do not recommend this procedure for high speed steels, especially where high-inertia furnaces are used, because the difference between preheating and austenitizing temperatures for these steels may be as much as 485 °C (875 °F). In high-volume operations, where preheating is frequently performed solely to shorten production time, separate furnaces can be used for preheating and austenitizing.

Quenching

Quenching from the austenitizing temperature may be done in water, brine, oil, salt or air, depending on composition and section thickness. The quenching medium must cool the workpiece rapidly enough to obtain full hardness; it is poor practice, however, to use a quenching medium with a cooling capacity that exceeds requirements, because of the possibility that cracking may occur.

Tool steels that will harden during air cooling are frequently hot quenched to the range 540 to 650 °C (1000 to 1200 °F) after austenitizing. Quenching time is long enough to stabilize the steel at the quenching temperature but not long enough for decomposition of austenite to begin. After hot quenching, the steels are air cooled or oil quenched to ambient temperature. Hot quenching minimizes distortion without adversely affecting hardness and spalls away or prevents the hard scale from forming on most air-hardening steels during air cooling.

Martempering is often utilized to minimize distortion without sacrifice of hardness in oil-hardening tool steels or in extremely thin sections of water-hardening tool steels. Workpieces are quenched from the austenitizing temperature in an agitated bath of oil or salt. Bath temperature should be near the M_s temperature, usually about 14 °C (57 °F) above it. Time in the bath should be just sufficient for temperature to equalize throughout the workpieces, which are then air cooled to room temperature prior to tempering.

Tempering

Tempering modifies the properties of quench-hardened tool steels to produce a more desirable combination of strength, hardness and toughness than

Table 5 Thermal conductivity of selected tool steels

Temperature °C	°F	Thermal conductivity W/m·K	Btu/ft·h·°F
Type W1			
100	200	48.3	27.9
260	500	41.5	24.0
400	750	38.1	22.0
540	1000	34.6	20.0
675	1250	29.4	17.0
815	1500	24.2	14.0
Type H11			
100	200	42.2	24.4
260	500	36.3	21.0
400	750	33.4	19.3
540	1000	31.5	18.2
675	1250	30.1	17.4
815	1500	28.6	16.5
Type H13			
215	420	28.6	16.5
350	660	28.4	16.4
475	890	28.4	16.4
605	1120	28.7	16.6
Type H21			
100	200	27.0	15.6
260	500	29.8	17.2
400	750	29.8	17.2
540	1000	29.4	17.0
675	1250	29.1	16.8
Type T1			
100	200	19.9	11.5
260	500	21.6	12.5
400	750	23.2	13.4
540	1000	24.7	14.3
Type T15			
100	200	20.9	12.1
200	500	24.1	13.9
400	750	25.4	14.7
540	1000	26.3	15.2
Type M2			
100	200	21.3	12.3
200	500	23.5	13.6
400	750	25.6	14.8
540	1000	27.0	15.6
675	1250	28.9	16.7

obtained in the quenched steel. The as-quenched structure of tool steel is a heterogeneous mixture of retained austenite, untempered martensite, and carbides. More than one tempering cycle may be necessary to produce an optimum structure. It is normally desirable to transform all retained austenite. This can be more nearly accomplished by two or more shorter tempering cycles than by a single and longer cycle.

In the higher-alloy tool steels, a small amount of untempered martensite is formed from retained austenite during the cool-down from the first tempering cycle. It is good practice to double temper to ensure more nearly complete transformation of retained austenite and to temper freshly formed martensite. For some highly alloyed grades of tool steel, triple or quadruple tempering is recommended.

The changes that take place in the microstructure during tempering of hardened tool steels are time-temperature dependent. Time at tempering temperature should not be less than 1 h for any given cycle.

Most manufacturers of high speed steels recommend multiple tempers of 2 h or more each to attain the desired microstructure and properties. Maintaining recommended tempering times, temperatures, and number of tempers (a minimum of two) ensure attainment of consistent tempered martensitic structures and overcomes uncertainties caused by variations in the amount of retained austenite in the as-quenched condition. These variances are functions of differences in heat chemistry, prior thermal history, hardening temperatures and quenching conditions. Other factors that influence the tempering requirements of high speed steels are

- Increasing the free (matrix) carbon content increases the amount of retained austenite in the as-quenched condition.
- The amount of retained austenite significantly affects the rate of transformation, particularly for short tempering cycles. Multiple tempering is more important to attain an acceptable structure if short tempering times are used.
- Cobalt in alloys such as M42 reduces the amount of retained austenite in the as-quenched condition and accelerates the transformation of the retained austenite during tempering.

Enough time should be allowed during tempering for the temperature to be distributed uniformly through the tools before time at temperature is counted. This is especially true for low tempering temperatures and for tools that have large sections. Table 6 indicates the time needed for various section sizes to reach uniform temperature in different kinds of furnaces. If not enough time is allowed for the tool to reach the tempering temperature, the

result will be nonuniform tempering and possible damage to the tool. Color of the oxide film should not be used as a guide in tempering, because these temper colors indicate only the surface temperature of the tool, not the internal temperature. Grinding cracks in hardened tools may be caused by inadequate tempering.

Proper tempering depends on the accurate determination of the temperature of the load and on proper spacing of workpieces in the load to ensure that it is uniformly heated. The most common mediums used for tempering tools are oil, lead or salt baths and recirculating-atmosphere furnaces, where the "atmosphere" may be flue gas, nitrogen, argon, or even a partial vacuum. Regardless of the medium used for tempering, accurate means of temperature control are mandatory for reproducible results.

Procedure. Before a tool steel part is tempered, it should be cooled in the quenching medium or in air until it can be held in the hand without discomfort (near 50 °C, or 120 °F for most steels). For particularly large or intricate parts, it is essential to temper as soon as possible after the quench, to prevent cracking.

Heating to the tempering temperature should be slow, to obtain uniform distribution of temperature within the tool and to prevent the nonuniform relief of hardening stress that could cause cracking or warping. Satisfactory results may be obtained by charging the tools into a freely circulating medium at the desired tempering temperature and then permitting them to reach this temperature. If tempered in a liquid medium, the tools should be placed in a basket and not permitted to come in contact with the hot walls or bottom of the pot or tank. Heat transfer is most rapid for molten lead baths, less rapid for salt and oil baths, and slowest in still air.

Cooling after tempering should be relatively slow in order to prevent development of residual stress in the steel. Still air cools at a satisfactory rate.

Equipment. Recirculating-air furnaces have several advantages over most other types of equipment used for tempering. For example, such a furnace can be cooled rapidly between batches of tempering temperature so that successive work loads may enter the furnace safely at a low temperature. Another advantage of the recircu-

Table 6 Approximate heating time to attain tempering temperature

							Time required to reach furnace temperature(a):					
		In hot air oven without circulation(b)						In circulating air oven or an oil bath(c)				
Tempering temperature		Cubes or spheres		Squares or cylinders		Average flats		Cubes or spheres		Squares or cylinders		Average flats
°C	°F	min/mm	min/in.	min/mm	min/in.	min/mm	min/in.	min/mm	min/in.	min/mm	min/in.	min/mm	min/in.
120	250		30		55		80		15		20		30
150	300		30		50		75		15		20		30
175	350		30		50		70		15		20		30
205	400		25		45		65		15		20		30
260	500		25		40		60		15		20		30
315	600		25		40		55		15		20		30
370	700		20		35		50		15		20		30
425	800		20		30		45		15		20		30
480 and up	900 and up		20		30		40		15		20		30

(a) Data are given in minutes per millimetre, and in minutes per inch, of diameter or thickness, with furnace maintained at the temperature indicated in column 1. Data may be used as a guide for charges of irregular shapes and quantities by estimating total size of charge and applying the above allowance to the number of inches from outside to center of charge. (b) Times indicated are for tools with dark or scaled surfaces. If surfaces are finish ground, or otherwise brightened, twice as much time should be allowed in a still hot air oven. No extra allowance need be made for bright surfaces in a circulating oven or in an oil bath. (c) Oil baths are usually not used above 205°C (400°F).

lating-air furnace is its relatively low heat transfer rate, which permits the load to be brought to temperature more slowly. As-quenched tools heated too rapidly may develop cracks. Recirculating-air furnaces also usually afford a wider range of useful tempering temperatures than other tempering mediums, with no hazards of fire or burns from splashing of hot liquids.

Other Treatments

Carburizing of tool steels is usually restricted to special applications. Mold steels, however, are commonly carburizing and then case hardened. A marked increase in surface carbon renders most tools too brittle for their intended uses. However, tools made of shock-resisting steel, hot work steel and especially the lower-carbon types of high speed steel are sometimes carburized to advantage for use in certain die applications involving severe wear. All the common methods of carburizing (gas, pack and liquid) have been employed for these special applications. Case depths are shallower, about 0.25 mm (0.010 in.), rather than the 0.75 to 1.5 mm (0.030 to 0.060 in.) that is common on conventional carburizing steels.

Nitriding of finished high speed steel tools in cyanide-base salt baths at 510 to 565 °C (950 to 1050 °F) is a common method of increasing tool life. Nitriding time ranges from 15 min to 2 h, resulting in case depths up to about 0.05 mm (0.002 in.).

Nitriding, by either the salt bath, the ammonia gas process, or ion nitriding, has also proved advantageous for increasing the life of tools made from the A, D and H types. Before these steels are nitrided, they are usually tempered at close to or slightly above the nitriding temperature; if the temperature is higher, the base hardness may be lower than desired.

Cold Treating. The main purpose of cold treating tool steels (to −75 °C, or 100 °F or lower) is to transform retained austenite in the unfinished tool and thus to provide dimensional stability in subsequent finishing operations. The use of cold treatment on properly heat treated cutting tools does not affect tool performance. When used, cold treatment should be performed between the first and second tempering operations.

Vacuum and Ion Nitriding. Vacuum furnaces, especially vacuum tempering furnaces, can be used for nitriding of tools. The tempering furnace is ideally suited for this purpose, because they are designed to operate in the nitriding temperature range. For high speed steels and other tools tempered at 455 to 595 °C (850 to 1100 °F), the nitriding cycle can be incorporated into the final tempering cycle, by introducing a partial pressure of ammonia during the final temper. In this way, the vacuum furnace operates similar to a normal nitriding furnace, except that nitrogen or the vacuum itself acts as the dilutant for nitriding, replacing dissociated ammonia. Cycle times would be the same as for gas nitriding.

Another process used for nitriding tool steels is ion nitriding. This process is also a vacuum process, but employs the principle of glow discharge to provide energy for heating and nitriding. Because it relies on electrical energy to dissociate gases, activate surfaces, and to provide energy for reaction, ordinary nitrogen at pressures in the range 1 to 10 mm of mercury are all that is required for nitriding. In addition, by adjusting the amount of nitrogen, the surface white layer can be closely controlled or eliminated. As in the other methods the cycles are very short. In addition, the temperature range is greatly expanded, to as low as 350 °C (660 °F), because the glow discharge reaction is not dependent on ammonia breakdown as in gas nitriding. This permits greater flexibility in choice of nitriding temperature so that surface hardness, case depth and core hardness can be optimized for a given steel.

Processes and Furnace Equipment for Heat Treating Tool Steels

By the ASM Committee on Heat Treating of Tool Steels*

MOLTEN SALTS of various compositions are well adapted to all operations in the heat treatment of tool steels. For tools that cannot be ground after hardening, or for tools that require the best possible surface condition and the maintenance of sharp edges, salt bath heating provides the best results. Table 1 lists various salt bath compositions and processing temperatures for heat treating tool steels. The salt bath method of hardening tools—particularly high speed steel tools—has greatly expanded with increased use of molybdenum steels. With correct operating conditions, tools can be heat treated without carburization, decarburization and scaling. The surface will be fully hard with minimum distortion. Three salt baths are generally used—preheating, high-temperature, and quenching baths. The function of the quenching bath is to equalize the temperature as well as to ensure a clean surface after heat treatment.

Salt Baths

Most tools heat treated in salt baths are fully hard from surface to core regardless of section thickness. Because salt baths provide temperature uniformity in preheating, high-temperature heating and in quenching, distortion and residual stresses are minimized.

Tools that are heat treated in molten salt baths are heated by conduction, the molten salt provides a ready source of heat as required. Although steels come in contact with heat through the tool surfaces, the core of a tool rises in temperature at approximately the same rate as its surface. Heat is quickly drawn to the core from the surface. Salt baths provide heat at a rate equal to the heat absorption rate of the total tool. Convection or radiation heating methods are unable to maintain the rate of heating to reach equilibrium with the rate of heat absorption. The ability of a molten salt bath to supply heat at a rapid rate accounts for the uniform, high quality of tools heat treated in salt baths. Heat treating times are also shortened; for example, a 25-mm- (1-in.-) diam bar can be heated to temperature equilibrium in 4 min in a salt bath, while 20 to 30 min would be required to obtain the same properties in convection or radiation furnaces.

Salt baths are the most efficient method of heat treating tool steels;

*W. James Laird, Jr., *Chairman,* Vice President—Marketing, Research & Development, Upton Industries, Inc.; Carl J. Oxford, Jr., Vice President—Technology, National Twist Drill & Tool Division, Lear-Siegler, Inc.; Percy Rawcliffe, Manager, Metallurgical and Physical Laboratories, Morse Cutting Tools, Division, Gulf & Western Manufacturing; Carl Reichel, Drill & End Mill Division, Inc.; TRW, Inc.; Ronald Spitzer, Chief Metallurgist, Bearings Division, TRW, Inc., Daniel S. Zamborsky, Corporate Metallurgist, Warner & Swasey Co.

Table 1 Typical compositions and recommended working temperature ranges of salt mixtures used in heat treating tool steels

| Salt mixture No. | Composition, % | | | | | | Melting point | | Working range | |
	Barium chloride	Sodium chloride	Potassium chloride	Calcium chloride	Sodium nitrate	Potassium nitrate	°C	°F	°C	°F
Austenitizing salts (high heat)										
1.............	98-100	...	...	...	...	...	950	1742	1035-1300	1895-2370
2.............	80-90	10-20	...	...	...	...	870	1598	930-1300	1705-2370
Preheat salts										
3.............	70	30	...	...	...	...	335	635	700-1035	1290-1895
4.............	55	20	25	...	...	...	550	1022	590-925	1095-1700
Quench and temper salts										
5.............	30	20	...	50	...	...	450	842	500-675	930-1250
6.............	...	...	...	...	55-80	20-45	250	482	285-575	545-1065

about 93 to 97% of the electric power consumed with a salt bath operation goes directly into heating. In atmosphere furnaces, 50% of the energy consumed goes for heating, and the remaining 50% is released up the furnace stack as waste. Tool steels that are heat treated in molten salts typically are processed in ceramic-lined furnaces with submerged or immersed electrodes containing chloride-base salts.

Immersed-Electrode Furnaces

Ceramic-lined furnaces with immersed (over-the-side) electrodes have greatly extended the useful range and capacity of molten salt equipment when compared with externally heated pot furnaces. The most important of these technical advantages are

- The electrodes can be replaced without bailing out the furnace.
- Immersed electrodes allow more power capacity to be put into the furnace, thus increasing production.
- Immersed electrodes permit easy start-up when the bath is solid. A simple gas torch is used to melt a liquid path between the two electrodes, thus allowing the electrodes to pass current through the salt to obtain operating temperatures.

Immersed electrode furnaces are not as energy efficient as submerged-arc furnaces, however. The area in which the immersed electrodes enter the salt bath allow additional heat loss, through increased surface area. As can be seen in Table 2, the surface area of the salt bath (A) in the submerged-electrode furnace is smaller than the surface area plus the immersed electrodes (A + B) in the immersed electrode furnace. However, a good cast ceramic and fiber insulated cover placed over the bath and electrodes will reduce surface radiation losses up to 60%.

The immersed-electrode furnace has a pot made of high-temperature fireclay brick, surrounded on five sides by approximately 12.7 cm (5 in.) of castable and insulating brick. Figure 1 is a schematic drawing of an immersed-electrode furnace with interlocking tiles and removable electrodes. The removable electrodes enter the furnace from the top, and a seal tile is located in front of the electrodes to protect them from exposure to air at the air-bath interface. This protection helps prolong electrode life. Table 2 compares service life of electrodes and refractories for four basic furnace designs.

Over-the-top electrodes are usually built with laminated cold legs, and water cooling is always required. A typical life expectancy for electrodes operating in such a furnace at 840 °C (1550 °F) is approximately 6 months to 2 yr for over-the-top electrodes, compared with 4 to 8 years for submerged electrodes.

Submerged-Electrode Furnaces

Submerged-electrode furnaces have the electrodes placed beneath the working depth for bottom heating. Figure 2 is a schematic of a typical submerged-electrode furnace. Many submerged-electrode furnaces are designed for specific production requirements and are

Fig. 1 Salt bath furnace for neutral heating

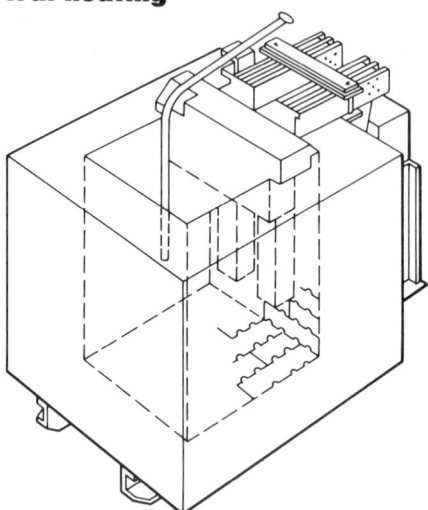

Furnace features a ceramic pot and over-the-top (immersed) electrodes.

equipped with patented features, which offer certain economical and technical advantages:

- *Maximum work space with minimum bath area:* The electrodes do not occupy any portion of the bath, so that only the alloy comes in contact with the salt. Bath size is consequently smaller, and electrode life increases many times over by incorporating unidirectional wear and eliminating excessive deterioration at the air-bath interface.
- *Circulation-convection currents:* Bottom heating provides more uniform bath temperatures and bath move-

Table 2 Service life of electrodes and refractories

Operating temperature		Service life, years	
°C	°F	Electrodes	Refractories
Submerged-electrode furnace: Furnace A			
535-735	1000-1350	10-20	10-20
735-955	1350-1750	4-8	4-8
955-1175	1750-2150	3-4	3-4
1010-1285	1850-2350	1-3	1-3
Immersed-electrode furnaces			
Furnace B			
535-735	1000-1350	2-4(a)	4-5
735-955	1350-1750	1-2(a)	2-3
955-1175	1750-2150	½-1(a)	1-2
1010-1285	1850-2350	¼-½(a)	1-1½
Furnace C			
535-735	1000-1350	2-4(a)	4-5
735-955	1350-1750	1-2(a)	2-3
955-1175	1750-2150	½-1(a)	1-2
1010-1285	1850-2350	¼-½(a)	1-1½
Furnace D			
535-735	1000-1350	2-4(a)	4-5
735-955	1350-1750	1-2(a)	2-3
955-1175	1750-2150	½-1(a)	1-2
1010-1285	1850-2350	¼-½(a)	1-1½

Furnace A

Furnace B

Furnace C

Furnace D

Note: Service life estimates are based on the assumption that proper rectification of chloride salts is being done, as well as routine unit maintenance and care. (a) Hot leg only

ment through the use of natural convection currents.

- *Triple-layer ceramic wall construction:* The temperature gradient through the wall causes any salt penetrating the wall to solidify before it can penetrate the cast refractory material that forms the center portion of the wall construction. This design requires from 5 to 8% of the initial salt charge to fill the ceramic pot. By comparison, in some designs, 140 to 150% of the initial charge is needed to seal the ceramic walls of furnaces

built with two layers of ceramic brick, backed up and supported by a steel plate. Salt penetrates the ceramic walls of any furnace and distorts the geometry of the walls. Reducing the amount of salt allowed to penetrate the ceramic walls aids in maintaining dimensions and in promoting longer furnace life.

- *Electrode placement:* Enclosing the electrode in a clear rectangular box free of any protruding obstructions eliminates potential hazards to operating personnel during cleaning. Any sludge formed in the furnace is removed easily by operating personnel.

Frame Construction. A typical submerged-electrode furnace is made of brick and ceramic material assembled, regardless of size, in a rigid, self-supporting welded steel frame. This frame consists of supporting channels or beams welded to the underside of a heavy steel plate that forms the frame base. To this base are welded lengths of heavy angle iron around the outside and on top of the plate. These pieces are notched to permit welding of the heavy angle-iron posts to the plate and vertical sides of the base-plate angle iron. Lengths of heavy angle iron are welded similarly to the top of the posts. When required, additional vertical reinforcing members are welded between the bottom and top pieces of angle iron, and prestressed horizontal members also are used to ensure that the refractory material cannot move after the furnace has been brought to operating temperature.

Brick Construction. Three types of refractory materials commonly are used in submerged-electrode furnaces. Submerged-electrode furnace liners are constructed with 23-cm (9-in.) thick high-temperature burned bricks. Consisting of approximately 42% alumina and 52% silica, the brick material is used in standard brick sizes such as 6 by 11 by 23 mm (2½ by 4½ by 9 in.) and in various brick shapes such as straights, flat backs and splits. The bricks are laid with a high-quality airsetting mortar that resists abrasion, erosion and chemical attack by chloride, fluoride and nitrate/nitrite salts. The mortar offers sufficient wear and corrosion resistance to be economically used with some salts containing cyanide. For straight cyanide or carbonate salts, a welded steel pot is used.

The outer wall of the salt bath fur-

Fig. 2 Typical submerged-electrode salt bath furnace

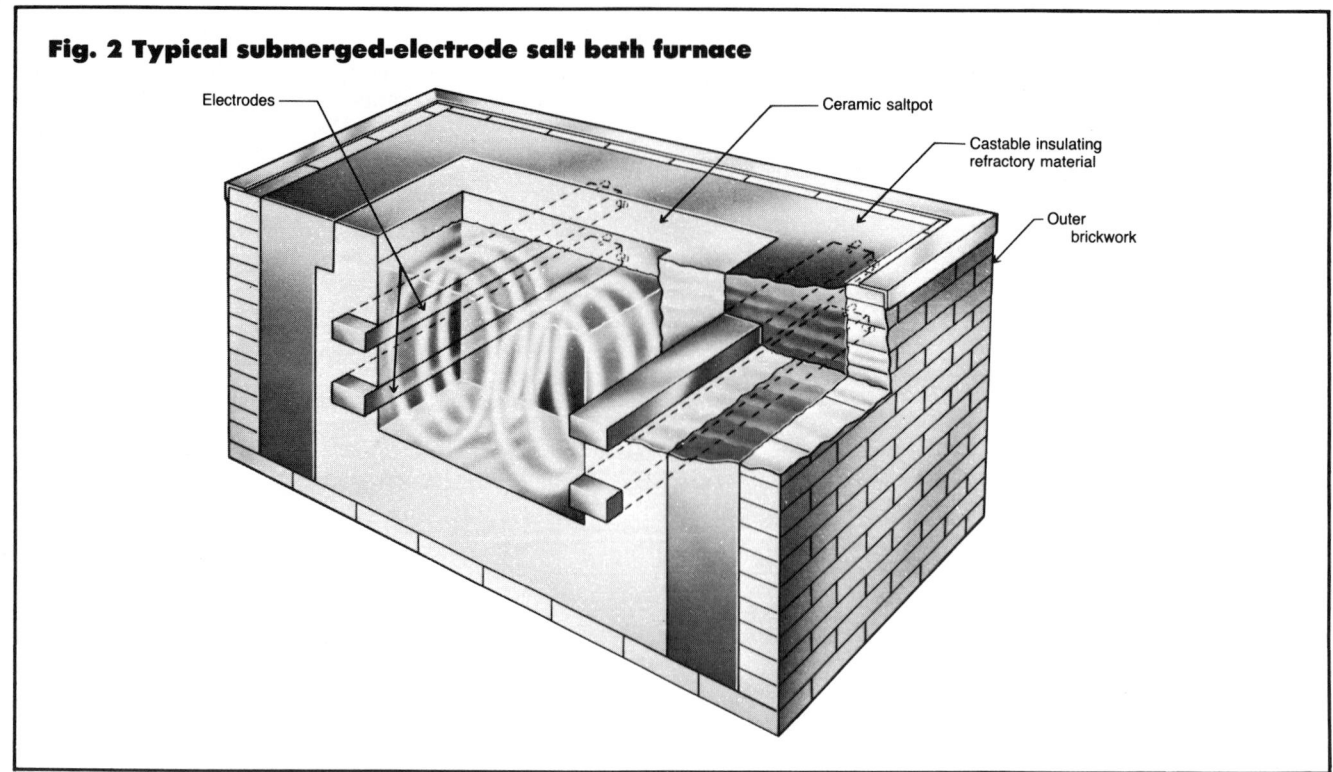

Electrodes — Ceramic saltpot — Castable insulating refractory material — Outer brickwork

nace is made of a second-quality firebrick with the same dimensions as brick used for the liner. The important qualities of this brick are the strength of material and uniformity of size and shape.

The inner castable wall is constructed with a mixture of refractory cement and aggregate that is poured between the liner and outer wall to form a 50-cm (9 ½-in.) thick monolithic wall structure. This dimension provides a temperature gradient sufficient to cause the salt to freeze in the wall, thus making the wall self-sealing. With this design, salt penetration into the wall amounts to less than 8% of the bath volume. The maximum temperature of the outside wall during furnace operation is 60 °C (140 °F).

Electrode Construction. The electrodes used in submerged-electrode salt bath furnaces vary widely in size and shape, depending on the geometry of the furnace and power requirements. All of the electrodes are located near the bottom of the bath and are built into the wall so that only one face of the electrode is in contact with the salt. This placement leaves the bath area free of obstruction for ease of cleaning and eliminates the possibility of touching the electrodes with the work.

Table 3 Relative process times and temperatures for automated heat treating of tool steels

Process stage	Operating temperature		Total time in furnace(a)
	°C	°F	
First preheat	650-870	1200-1600	X
Second preheat	760-1035	1400-1900	X
High heat	1010-1285	1850-2350	X
Isothermal quench	535-705	1000-1300	X
Air cool	Room temperature		6X, 12X, 24X
Wash, hot water	80-95	180-200	6X
Rinse, hot water	80-95	180-200	X

(a) See Table 4 for drill sizes and times in the high heat indicated by an "X" in this table.

Alloy electrodes are made by welding a 1612.9-mm² (2½-in.²) alloy material to a mild steel backing, or by welding a 127-by-127-mm (5-by-5-in.) alloy material directly to the mild steel shank.

The durability of typical electrode and ceramic components of submerged electrode furnaces is described in Table 1.

Automatic Heat Treating of Tool Steels

Figure 3 illustrates three different heat treating arrangements for production heat treatment of tool steels. Table 3 gives relative process times and tem-peratures for heat treating, and Table 4 gives process times for twist drills. The systems are equipped for cycles ranging from less than 1 min to 10 min. The parts are suspended on tong-type fixtures and are carried through the process by a chain conveyor on carrier bars. To facilitate rapid transfer of the tool steels, rotary transfer arms are placed between the preheat and the high heat units and between the high heat and the quench units. Transfer-arm placement is chiefly governed by production rate; however, transfer arms are always required between the high heat and the quench units to satisfy metallurgical conditions. The lines

Table 4 Time cycles for heat treating twist drills

Diameter		Time
mm	in.	
2.54-4.77	0.100-0.188	1 min 30 s
4.80-8.08	0.189-0.318	1 min 40 s
8.10-12.90	0.319-0.508	1 min 50 s
12.91-18.24	0.509-0.718	2 min 0 s
18.26-23.32	0.719-0.918	2 min 20 s
23.34-38.10	0.919-1.500	2 min 40 s
102-mm (4-in.) diam cups		6 min
64-mm (2½-in.) diam end mills		7 min
76-mm (3-in.) diam end mills		10 min

Pieces in high heat on smaller diameters

2.54 mm (0.100 in.) = 160 pieces/tong = 480 pieces in bath = 1.2 kg (2.6 lb)
4.77 mm (0.188 in.) = 85 pieces/tong = 255 pieces in bath = 3.4 kg (7.65 lb)
6.50 mm (0.256 in.) = 63 pieces/tong = 188 pieces in bath = 5.5 kg (12.3 lb)
8.08 mm (0.318 in.) = 25 pieces/tong = 75 pieces in bath = 3.8 kg (8.6 lb)
12.90 mm (0.508 in.) = 16 pieces/tong = 48 pieces in bath = 8.2 kg (18.2 lb)

Table 5 Ranges of endothermic-atmosphere dew point for hardening tool steels(a)

Furnace dew point; AGA class 302 atmosphere

Steel	Furnace temperature(b)		Dew point range	
	°C	°F	°C	°F
W2, W3	800	1475	7 to 12	45 to 55
S1	925	1700	4 to 7	40 to 45
S2	870	1600	4 to 16	40 to 60
O1	800	1475	7 to 12	45 to 55
O2	775	1425	7 to 12	45 to 55
O7	855	1575	−4 to 2	25 to 36
D2, D4	995	1825	−7 to −1	20 to 30
D3, D6	955	1750	−7 to −1	20 to 30
H11, H12, H13	1010	1850	1 to 7	35 to 45
T1	825	2350	−18 to −12	0 to 10
M1	1205	2200	−15 to −12	5 to 10
F2, F3	830	1525	−5 to 1	23 to 34

(a) For short times at temperature. (b) Approximate midrange of austenitizing temperatures for the specific types of tool steels

also have areas above the furnaces to accommodate air cooling of the tools. In special cases, lines will be made with a station for an isothermal nitrate quench after the neutral salt quench. This additional stage allows rapid reduction of the temperature of the tools and reduces the air cooling time from 24 times to 6 times the time at the high heat temperature. **Caution: if as little as 600 ppm of nitrate salts are allowed to enter the high heat furnace, extreme surface damage can be done to the tool being heat treated.**

Rectification of Salt Baths

Neutral salts used for austenitizing steel become contaminated with soluble oxides and dissolved metals during use, resulting from a reaction between the oxide layers present on fixtures and workpieces and the chloride salts. Because the buildup of resulting oxides and dissolved metals renders the bath oxidizing and decarburizing toward steel, the bath must be rectified periodically.

Baths of salts such as salt mixtures No. 1 and 2 in Table 1 can be rectified with silica, methyl chloride or ammonium chloride. The higher the temperature of operation, the more frequent the need for rectification. Baths in which the electrodes protrude above the surface require daily rectification with either ferrosilicon or silicon carbide. Baths operated above 1080 °C (1975 °F) require rectification once daily or more. During rectification of a bath, the silica combines with the dissolved metallic oxides to form silicates. Although these

silicates settle out as a viscous sludge that can be removed, sufficient soluble silicates can remain to cause the bath to become decarburizing. If the bath is not rectified, it becomes more viscous than water.

Methyl chloride bubbled through the bath or the submerging of ammonium chloride pellets in a perforated cage in the bath are more effective methods of rectifying salt baths. The ammonium chloride pellets react with the oxides to regenerate the original neutral salt without sludge formation or bath thickening. To remove dissolved metals from high-temperature baths, graphite rods are introduced at operating temperature. The graphite reduces any oxides to metal, which adheres to the rod. The metal can be scraped off and the rod reused.

To control the decarburizing tendency of high-temperature baths, test specimens frequently should be hardened by quenching in oil or brine. A file-soft surface indicates a need for more rectification. This test may be supplemented by analysis of the bath. High-heat baths containing in excess of 0.5% barium oxide are likely to be decarburizing to steel.

One method of rectifying austenitizing baths such as salt mixtures No. 2 and 3 of Table 1 is as follows:

● Add 57 g (2 oz) of boric acid for each 45 kg (100 lb) of salt, after every 4 h of operation.
● Insert a 76-mm (3-in.) graphite rod into the bath for 1 h for every 4 h of operation.

Atmospheres

In selecting an atmosphere that will protect the surface of tool steel against addition or depletion of carbon during heat treatment, it is desirable to choose one that requires no adjustment of composition to suit various steels. An ammonia-base atmosphere (AGA class 601) meets this requirement and has the advantage of being sufficiently reducing to prevent oxidation of high-chromium steels. In the range of dew points generally found in this gas, −40 to −50 °C (−40 to −60 °F), there is no serious depletion of carbon, because the decarburizing action is slow and any loss of carbon at the surface is partially replaced by diffusion from the interior. For applications in which high superficial hardness is important, a carburized surface can be obtained by the

Fig. 3 Process designs for automated salt bath furnaces for heat treating high speed tool steels

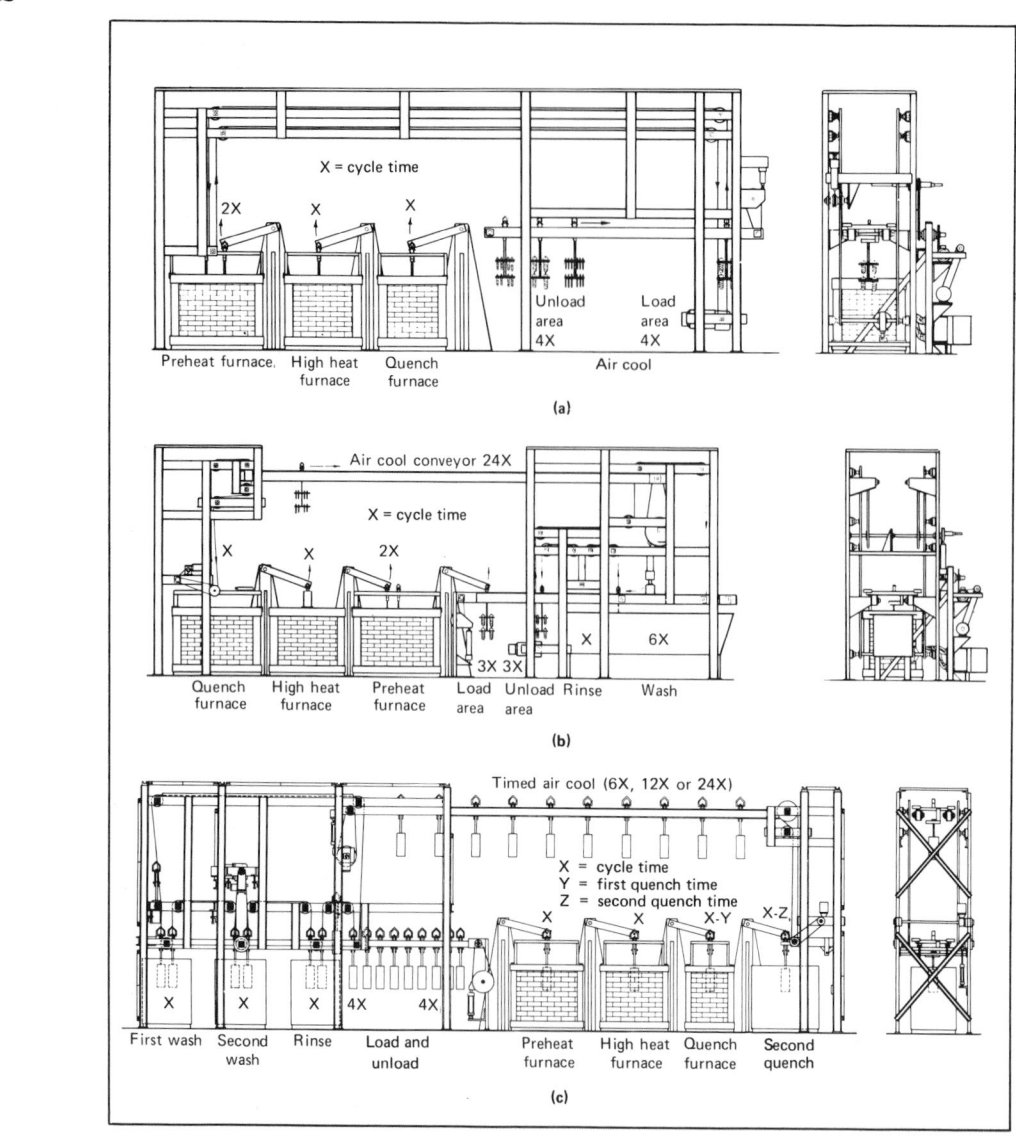

Installations can be custom designed to meet specific customer requests. (a) Does not include wash and rinse. (b) Similar to (a), but includes wash and rinse operation necessitating relocation of load and unload operations. (c) Similar to (b), but includes second quench and a variation in wash cycles as specified by customer.

addition of about 1% methane to the atmosphere. Although ammonia-base atmosphere costs more than endothermic gas, this seldom becomes important because tool treating furnaces generally are comparatively small and therefore require a correspondingly small quantity of gas.

Endothermic-base atmospheres are often used for the protection of tool steel during heat treatment. Suggested ranges of dew point for an AGA class 302 endothermic atmosphere when used for hardening some common tool steels are listed in Table 5. Relatively short heating times for hardening small tools allow treatment to be carried out with the theoretical carbon balance of the atmosphere varying over a rather wide range. However, for the hardening of large die sections, the particular composition of the die steel being treated requires careful control of the atmosphere if carburization or decarburization is to be avoided during the relatively long heat treating cycle.

Specific Classes of Tool Steels

By the ASM Committee on Heat Treating of Tool Steels*

HEAT TREATING PROCEDURES vary significantly among classes of tool steels and with intended application. The preferred heat treating and hardening procedures, as well as mechanical properties and applications are discussed in this article with respect to: water- and air-hardening tool steels, oil-hardening and high-carbon, high-chromium cold work steels, low-alloy and special purpose high-speed tool steels, and shock-resisting tool steels. Specific examples of heat treating procedures for specific applications for hot work tools are given.

Water-Hardening Tool Steels

Water-hardening tool steels containing 0.90 to 1.00% carbon are the most widely used. Carbon content affects heat treating temperatures as indicated in Table 1, which outlines recommended heat treating practices for these steels.

As a class, water-hardening tool steels are relatively low in hardenability, although they are arbitrarily classified and available as shallow-hardening, medium-hardening and deep-hardening types. Their low hardenability is frequently an advantage, because it allows tough core properties in combination with high surface hardness. Low cost and adaptability to simple heat treatment are additional advantages offered by these steels.

Water-hardening tool steels are so termed because they are most commonly quenched in an aqueous medium. There are exceptions, however; for example, thin sections may be satisfactorily quenched in oil with less distortion and danger of cracking than if quenched in water or brine.

Example 1. In one plant, it was desirable to harden small-diameter punches in oil to reduce breakage and consequent downtime of the presses. A study was made to determine the maximum diameters of water-hardening tool steels that could be fully hardened to a minimum of 60 HRC by oil quenching. Results of the study, indicating the relationship between austenitizing

temperature, type of steel and punch diameter, are shown in Fig. 1.

Experimentation proved that a greater degree of uniformity was obtained if the punches were normalized prior to hardening. Normalizing temperatures applied were: 870 °C (1600 °F) for punches up to 6.5 mm (¼ in.) in diameter; 900 °C (1650 °F) for those over 6.5 mm (¼ in.) in diameter. As indicated in Fig. 1, austenitizing temperature varied from 790 to 900 °C (1450 to 1650 °F), depending on punch diameter. The punches were austenitized by being heated vertically in a neutral salt bath. They were also quenched vertically, in a compounded oil containing additives. The quenching oil was maintained at 50 to 60 °C (120 to 140 °F) and circulated up and around workpieces at 190 l/min (50 gal/min).

Normalizing. Except in special instances where experience has proved it beneficial (as in the preceding example), normalizing is not recommended for water-hardening tool steels as received from the supplier. Normalizing

*W. James Laird, Jr., *Chairman,* Vice President-Marketing, Research & Development, Upton Industries, Inc.; Carl J. Oxford, Jr., Vice President-Technology, National Twist Drill & Tool Division, Lear-Siegler, Inc.; Percy Rawcliffe, Manager-Metallurgical and Physical Laboratories, Morse Cutting Tools, Division of Gulf & Western Manufacturing; Carl Reichel, Plant Metallurgist, Drill & End Mill Division, Augusta Plant, TRW, Inc.; Ronald F. Spitzer, Chief Metallurgist, Bearings Division, TRW, Inc.; Daniel S. Zamborsky, Corporate Metallurgist, Warner & Swasey Co.

Table 1 Recommended heat treating practice for water-hardening tool steels

| Temperature | | | Hardness | |
°C	°F	Carbon content	after treatment	Procedure
Normalizing				
815	1500	0.60 to 0.75	. . .	Heat through uniformly; hold for 15 min
790	1450	0.75 to 0.90		(light sections) to 1 h (heavy sections),
870	1600	0.90 to 1.10		then air cool
870 to 925	1600 to 1700	1.10 to 1.40		
Annealing				
740 to 760	1360 to 1400	0.60 to 0.90	156 to 201 HB	Heat through uniformly; hold for 1 to 4
760 to 790	1400 to 1450	0.90 to 1.40		h(a); furnace cool to 510 °C (950 °F) at 20 °C/h (40 °F/h), then air cool
Hardening(b)				
790 to 845	1450 to 1550	0.60 to 0.80	65 to 68 HRC	Hold at austenitizing temperature for 10 to
775 to 845	1425 to 1550	0.85 to 1.05		30 min; quench in water or brine (very
760 to 830	1400 to 1525	1.10 to 1.40		small pieces may be oil quenched)

(a) Holding times vary from about 1 h, for light sections and small furnace charges, to about 4 h, for heavy sections and large furnace charges. (b) For large tools and tools with intricate sections, preheating at 565 to 650 °C (1050 to 1200 °F) is recommended.

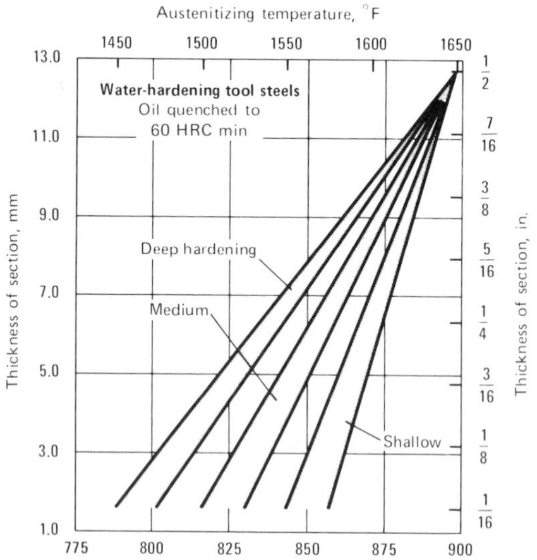

is recommended for these steels after forging or before reheat treatment, for refining the grain and producing a more uniform structure. Recommended normalizing temperatures are given in Table 1; as indicated, optimum temperature varies with carbon content.

Decarburization during air cooling will be minimized if parts are heated in a protective atmosphere or a neutral salt bath. Parts heated in salt are additionally protected during the cooling period by the film of salt that adheres to their surfaces when they are removed from the salt bath. After parts have cooled, the film of salt can be easily removed (except from recesses such as tapped holes) by a water rinse.

Annealing. Tool steels of the W types are received from the supplier in the annealed condition. Thus, annealing by the user is usually unnecessary. Annealing is applied to forged or cold worked carbon tool steel to soften it for easier machining, relieve residual stress, and produce a structure suitable for hardening. Annealing may be done in an atmosphere furnace (provided the furnace is of a type that can be cooled slowly to below 540 °C or 1000 °F), in a vacuum, or in an ordinary air furnace after the piece has been protected against surface decarburization by being packed in a suitable container with an inert material. Protection against decarburization (but not against oxidation) may be obtained also by copper plating the surface or by applying a surface-protecting paint. (Not all of these paints are equally effective, and some are difficult to remove; the prospective user should investigate such a

Fig. 1 Maximum section thicknesses of 3 classes of water-hardening tool steel that will develop minimum hardness of 60 HRC when oil quenched from various austenitizing temperatures (Example 1)

paint by trying it under his conditions of operation and then inspecting the treated part for decarburization.) The workpiece should be heated to the annealing temperature (Table 1) and held at temperature for from 1 h, for thin sections, to about 4 h, for heavy sections. When the steel has been placed in a pack to prevent surface reactions, a general rule of thumb is to allow the assembly to soak at temperature for 1 h per inch of pack cross section. Work should then be cooled in the furnace at a rate not exceeding 22 °C/h (40 °F/h),

to 510 °C (950 °F). Below this temperature, cooling rate is not critical. Hardness after annealing should be in the range of 156 to 201 HB.

Stress relieving prior to hardening is sometimes employed to minimize distortion and cracking. The procedure consists of heating the work to 650 to 720 °C (1200 to 1325 °F) and cooling in air. Usually, stress relieving of water-hardening tool steel is limited to complex or severely cold worked parts.

Example 2. A piston of W2 steel for a pneumatic clay digger varied in section thickness by as much as 6 to 1. Cracking occurred in the cupped end section when the pistons were hardened by conventional practice. Stress relieving or preheating at 675 °C (1250 °F) prior to hardening eliminated this difficulty.

In most instances, stress relieving after hardening and grinding is not employed. Periodic stress relieving of tools that have been in service will reduce the stresses imposed by such service, and is believed to be beneficial in extending service life. Temperatures used for this purpose should never exceed those used for tempering the steel after hardening.

Austenitizing temperatures for water-hardening tool steels normally vary from 760 to 845 °C (1400 to 1550 °F), as indicated in Table 1. Higher temperatures are sometimes used for special purposes (Fig. 1). Hardenability increases as austenitizing temperature increases. The optimum time at austenitizing temperature is from 10 to 30 min. Preheating is unusual except for very large tools or those with intricate cross sections (such as the W2 piston cited in Example 2).

If surfaces are to be protected against scaling or decarburization, an atmosphere furnace, lead bath or salt bath is required. It is particularly important to protect shallow-hardening steels against scaling and decarburization. Severe scaling can interfere with heat transfer during quenching and slow the required high rate of cooling. Decarburization will produce a soft surface on any tool steel, but in a deep-hardening steel it can be ground off until the underlying hard high-carbon area is reached. Grinding a shallow-hardening steel will frequently expose the soft core.

Atmospheres. Excellent results are obtained by austenitizing water-hardening tool steels in a slightly oxidizing atmosphere, as the data in Fig. 2, obtained in tests on type W2, indicate. Oxidizing atmospheres are inexpensive, and are usually produced by controlled direct-firing burners. The light scale that is produced is removed by the vigorous water or brine quench.

Endothermic atmospheres also are used, but close control is necessary to match the carbon potential of the atmosphere to the carbon content at the surface (Fig. 2). Also, endothermic installations are more expensive than the controlled-burner technique mentioned above.

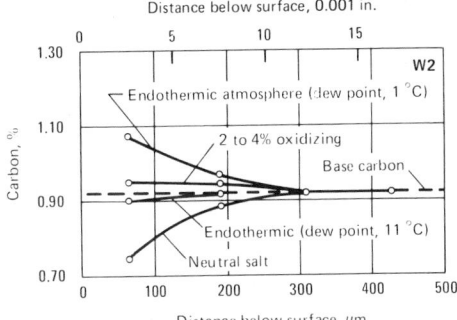

Fig. 2 Effect of furnace atmosphere on the surface carbon content of W2 tool steel

Specimens were heated at 790 °C (1450 °F) for 1 h, quenched in brine, annealed in lead at 705 °C (1300 °F), and machined in 0.13-mm (0.005-in.) cuts for analysis.

Salt baths are widely used and frequently preferred over other heating mediums for hardening type W tool steels.

Example 3. In one plant, salt baths were found to be superior to atmosphere furnaces for heat treating die sections of W1 and W2 (0.90 to 1.05% C) because: (a) die sections could be hardened in limited areas by being suspended and only partly immersed in the salt bath, and (b) long sections, such as die wiper plates measuring 25 by 100 by 760 mm (1 by 4 by 30 in.), could be hardened in a salt bath with less distortion.

Salt baths are usually lower in initial cost than endothermic-atmosphere installations. Neutral salts such as No. 3 in Table 2 of the article entitled "Furnace Processes and Equipment for Heat Treating of Tool Steels" in this volume, are commonly used. A salt of this type will operate satisfactorily in either steel-lined or ceramic-lined pot furnaces, but maintenance cost will be less with ceramic linings. Immersed-electrode heating of these furnaces is recommended.

High-temperature salt baths will cause severe decarburization (see Fig. 2) if not closely controlled. A recommended method of rectification for control of these baths is indicated in the article, "Furnace Processes and Equipment for Heat Treating of Tool Steels" in this volume.

Other disadvantages of salt baths are that: (a) salt dragout necessitates frequent replenishment of the bath, particularly when many small parts are being treated; and (b) salt is sometimes difficult to remove from parts having complex shapes or tapped holes.

Lead baths also are used for austenitizing water-hardening steels and have advantages and limitations paralleling those of the salt bath, specifically item (b) above. Both OSHA and EPA have stringent regulations to avoid lead poisoning.

Quenching. To produce maximum depth of hardness in water-hardening tool steels, it is essential that they be quenched as rapidly as possible. In most instances, water or a brine solution consisting of 10% NaCl (by weight) in water is used. Occasionally, for an even faster quench, an iced brine solution is employed. Cooling rate is a function of size of workpiece as well as of quenching medium; for this reason, small pieces can be quenched in oil (Fig. 1). This is particularly useful when heat treating thin-section tools in an atmosphere furnace containing an integral oil quench.

Tempering. Water-hardening tool steels should be tempered immediately after hardening, preferably before they reach room temperature; about 50 °C (120 °F) is optimum. Salt baths, oil baths and air furnaces are all satisfactory for tempering. However, working temperatures for both oil and salt are limited; the minimum for salt is about 165 °C (325 °F), and the maximum for oil is usually about 205 °C (400 °F).

All parts made of these steels should be tempered at temperatures not lower than 175 °C (350 °F). One hour at temperature is usually adequate; additional soaking time will further lower hardness. Figure 3 shows the effect of tempering temperature on hardness of

Fig. 3 Effect of tempering temperature on surface hardness of water-hardening tool steels austenitized at 3 different temperatures and quenched in brine

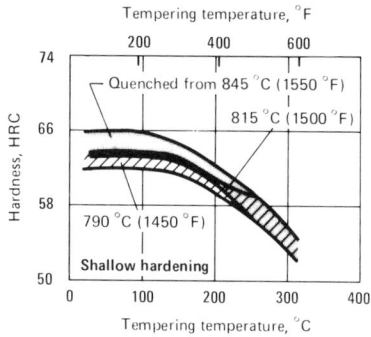 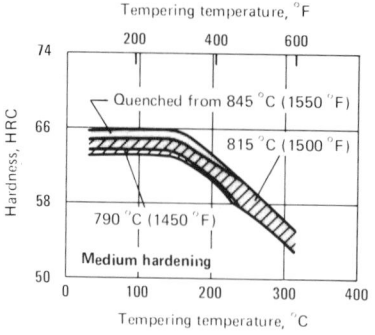 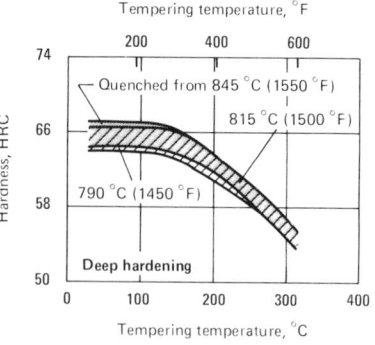

Specimens held for 1 h at the tempering temperature in a recirculating-air furnace. Cooled in air to room temperature. Data represent 20 25-mm (1-in.) diam specimens for each steel. Compositions of steels: shallow hardening, 0.90 to 1.00 C, 0.18 to 0.22 Mn, 0.20 to 0.22 Si, 0.18 to 0.22 V; medium hardening, 0.90 to 1.00 C, 0.25 Mn, 0.25 Si, no alloying elements; deep hardening, 0.90 to 1.00 C, 0.30 to 0.35 Mn, 0.20 to 0.25 Si, 0.23 to 0.27 Cr

water-hardening tool steels austenitized at 790, 815 and 845 °C (1450, 1500 and 1550 °F) and quenched in brine.

Tools should be placed in a warm 95 to 120 °C (200 to 250 °F) furnace immediately after quenching and then be brought to the tempering temperature with the furnace. This is particularly necessary when quenched tools are being accumulated for tempering in a single batch. Allowing quenched tools to stand at room temperature or placing them in a cold furnace will lead to cracking. Except for large pieces, the work will heat at about the same rate as the furnace. The low temperatures used in tempering eliminate the need for atmosphere control. A double temper is frequently used to temper any martensite that may have formed from retained austenite during quenching, and in the first tempering cycle.

The resistance to fracture by impact initially increases with tempering temperature to about 180 °C (360 °F) but falls off rapidly to a minimum at about 260 °C (500 °F). This is known as "500 °F embrittlement". For tools subjected to impact loading, tempering temperature should be selected to give an optimum combination of hardness and impact resistance.

Shock-Resisting Tool Steels

Recommended heat treating practices for shock-resisting tool steels are outlined in Table 2. These steels may

be obtained with several variations in composition, for specific applications (for example, S1 steel is available with 0.30 or 0.50% Mo or with up to 0.90% Si). The user of these nonstandard compositions should: (a) obtain from the manufacturer information as to the modifications required in heat treatment, or (b) select a heat treatment recommended for the shock-resisting tool steel of standard composition that most closely resembles the modified steel. The latter procedure should be followed only after the treatment has been tried on test samples.

Normalizing is not recommended for the shock-resisting tool steels.

Annealing. The high-silicon types (S2, S4, S5 and S6) are susceptible to graphitization and decarburization. Annealing these types at temperatures higher than those indicated in Table 2 may produce a softer structure, but it will also increase the danger of graphitization. The silicon types should not be soaked at temperature. Surfaces should be protected against decarburization by heating in a protective atmosphere or a vacuum furnace, by the use of pack annealing, or by the application of proprietary paints.

Pack annealing consists of surrounding parts with inert material inside a closed container, heating the container to the recommended temperature, and slow cooling. The selection of a packing medium for use with shock-resisting tool steels is difficult; the same general practice has produced different results in different plants. Dry

silica sand is usually satisfactory for type S1, and a combination of new and used carburizing compound is usually satisfactory for S2, S4 and S5. Burned-off cast iron chips, spent pitch coke, lime and mica are sometimes used, also. Cast iron chips decrease in carbon content and should not be used indefinitely; lime and mica should be used carefully, because they are insulators. Excessive thicknesses of inert material should not be packed around parts, because this complicates handling and lengthens heating time. Wrapping parts tightly in brown paper before surrounding them with packing material helps to keep the surfaces clean.

Proprietary paints are available that are intended to protect steel surfaces from decarburization during annealing. The use of such paints is simpler than the use of a pack anneal, but not all of these paints are effective. Moreover, considerable difficulty may be experienced in removing such paints after heat treatment. The prospective user should test any such paint on a sample of steel prior to adopting it in practice.

Stress relieving before hardening is seldom required for shock-resisting tool steel, except for extremely intricate parts of widely varying section thickness (to minimize distortion and cracking) and parts subjected to excessive stock removal (to relieve stresses induced by machining). Treatment of such parts, which involves no microstructural transformation, consists of heating them to 650 °C (1200 °F) (soak-

Table 2 Recommended heat treating practice for shock-resisting tool steels

Steel	Normalizing	Annealing Temperature(a) °C	°F	Cooling rate(b) °C/h	°F/h	Annealed hardness, HB	Hardening Preheat Temperature °C	°F	Austenitize °C	°F	Holding time, min	Quenching medium	Quenched hardness, HRC
S1	Not rec	790-815	1450-1500	22	40	183-229	...	...	900-955	1650-1750	15-45	O	57-59
S2	Not rec	760-790	1400-1450	22	40	192-217	650(c)	1200(c)	845-900	1550-1650	5-20	B, W	60-62
S4	Not rec	760-790	1400-1450	22	40	192-229	650	1200	870-925	1600-1700	5-20	B, W	61-63
									900-925	1650-1700	5-20	O	57-59
S5	Not rec	775-800	1425-1475	14	25	192-229	760	1400	870-925	1600-1700	5-20	O	58-61
S6	Not rec	800-830	1475-1525	14	25	192-229	760	1400	915-955	1675-1750	10-30	O	56-60
S7	Not rec	815-845	1500-1550	14	25	187-223	650-705	1200-1300	925-955	1700-1750	15-45(d)	A, O	60-61

(a) Lower limit of range should be used for small sections, upper limit for large sections. Holding time varies from about 1 h, for light sections and small furnace charges, to about 4 h, for heavy sections and large charges; for pack annealing, hold for 1 h per inch of pack cross section. (b) Maximum. Rate is not critical after work is cooled to about 510 °C (950 °F). (c) Preferable for large tools, to minimize decarburization. (d) For open furnace heat treatment. For pack hardening, hold for ½ h per inch of pack cross section.

ing should be avoided), furnace cooling to about 510 °C (950 °F), and then removing them from the furnace to cool in air.

Stress relieving of tools after tempering is seldom done. In some instances, however, increased tool life has been obtained by removing tools from service and stress relieving them (at a temperature no higher than the original tempering temperature) before returning them to service.

Example 4. In one plant, shock-resisting steel tools used for swaging stainless steel would sink a definite amount after a time in service. If kept in service, these tools would crack after swaging about 40 000 parts. However, by withdrawing the tools after sinking had ceased and stress relieving them at 230 °C (450 °F) for 1 h per inch of cross section, tool life was more than doubled.

Austenitizing temperatures for shock-resisting tool steels vary from 815 to 955 °C (1500 to 1750 °F). Preheating is not mandatory, but it is sometimes desirable for large tools, to minimize distortion, shorten time at the austenitizing temperature and speed up production.

These steels may be austenitized in electric or fuel-fired furnaces or in salt or lead baths. Generally, for austenitizing temperatures below 870 °C (1600 °F), a slightly oxidizing atmosphere is best, whereas above 870 °C (1600 °F) a reducing atmosphere is required. If a semimuffle fuel-fired furnace is used, the desired atmosphere can be obtained at low cost by adjustment of burners. However, if electrically heated or full-muffle fuel-fired furnaces are used, a prepared atmosphere from an external source is required.

Neutral salt baths are a practical means of heating the type S steels. A salt mixture such as No. 3 in Table 2 of the article, "Furnace Processes and Equipment for Heat Treating of Tool Steels" in this volume, is satisfactory for types S2, S4 and S5, whereas a mixture such as No. 2 in that table will be more suitable for heating S1. A recommended method of controlling these salts to prevent decarburization of the work is indicated in the discussion of rectification in the article, "Furnace Processes and Equipment for Heat Treating of Tool Steels".

If atmosphere furnaces or neutral salt baths are not available, the shock-resisting steels can be heated in a pack of neutral material such as burned pitch coke or cast iron chips. Packing mediums must be free of oil or other contaminants. Before being placed in the pack, tools should be wrapped with heavy brown paper, to prevent packing material from adhering to them as they are removed for quenching.

Types S2, S4 and S5 should be quenched soon after they reach the austenitizing temperature; types S1 and S7 are soaked at temperature for 15 to 45 min before being quenched (Table 2). Types S1 and S7 have the highest hardenability of these steels. The other types, although lower in hardenability than S1 and S7, are higher in hardenability than the W steels.

Tempering. Both the tungsten and the silicon types of shock-resisting tool steel resist softening from tempering to a greater degree than carbon tool steels. Secondary hardening does not occur in these steels, except to a minimal degree in some compositions of the tungsten type.

The effect of tempering temperature on the hardness of various types and compositions of the S steels is shown graphically in Fig. 4.

Tools made of shock-resisting steel should be tempered immediately after quenching, or cracking is likely to result, especially if they are quenched in water or brine.

Example 5. One plant made an extensive study on how much time could be safely permitted between quenching and tempering of tools made of shock-resisting steels. Results of this study are given in Table 3.

In this same plant, tool records indicate that double tempering is beneficial for tools made from the S steels. The first tempering operation is done at a temperature 30 to 55 °C (50 to 100 °F) lower than that of the second and final tempering operation.

Surface treatments such as carburizing and carbonitriding are often applied to S1 steel. Types S4 and S5 do not take an effective carburized case.

Oil-Hardening Cold Work Tool Steels

Recommended heat treating practice for oil-hardening cold work tool steels is summarized in Table 4.

Normalizing is desirable and sometimes necessary for parts that have been forged or heated previously to temperatures much higher than the proper austenitizing temperature, because it produces a more uniformly refined grain structure. Recommended normalizing temperatures are given in Table 4. Work should be held at temperature for 15 min to 1 h, depending on section size; prolonged soaking is not desirable. When tools are to be hardened after normalizing, precautions must be taken in order to avoid decarburization during normalizing. If tools are to be subsequently machined, annealing is recommended in preference to normalizing.

Annealing. Finished or semifinished tools made from oil-hardening cold work steels should be protected from decarburization or carburization during annealing. This can be accomplished by the use of dry exothermic furnace atmospheres. More often, however, it is accomplished by pack annealing, wherein work to be annealed is packed in a box and surrounded with inert protective material, such as clean cast iron chips or 6-to-8 mesh spent pitch coke. Pack annealing permits the use of an open furnace; also, slow heating and cooling occur naturally in the packed box. However, it is important that the work be soaked long enough to permit it to reach the annealing temperature. Recommended annealing temperatures, cooling rates and expected hardness values are given in Table 4.

Type O1 steel may also be cycle annealed (Table 5). Cycle annealing offers little advantage for large loads, but with individual tools that can be conveniently handled in liquid baths or other conventional furnaces, it enables substantial savings in time.

Stress Relieving. In most instances, stress relieving of finished tools prior to final hardening does not noticeably lessen distortion during hardening. If extreme dimensional accuracy after hardening is required, tools should be stress relieved after rough machining but before final light machining. A recommended stress-relieving treatment consists of heating to 650 to 675 °C (1200 to 1250 °F), holding at temperature for 1 h per inch of thickness, and then air cooling.

Preheating of the O steels will minimize distortion during subsequent hardening. It is almost always required for tools that are to be austenitized in liquid baths. Recommended preheating temperatures are listed in Table 4. Open furnaces can be used for preheating, but if scale-free and oxide-free hardening is required, preheating must be done with atmosphere control.

Austenitizing. Recommended austenitizing temperatures for oil-hardening tool steels are given in Table 4. Work that has been preheated may either be transferred to an austenitizing furnace or be heated to the austenitizing temperature in the same furnace in which it was preheated.

Decarburization and scaling can be effectively minimized in liquid salt or lead baths, and in furnaces with controlled atmospheres (such as endothermic gas, dissociated ammonia, and argon or other inert gases). However, in all of these there is some danger of decarburization if conditions are not controlled. Oxides in the molten baths or excess water vapor in the various gases will cause decarburization. The atmospheres of gas-fired or oil-fired semimuffle furnaces can be adjusted to contain from 2 to 4% oxygen, a condition that will eliminate decarburization but not oxidation. Types O1 and O2 can be satisfactorily austenitized in such an atmosphere, but it is not recommended for types O6 and O7. All type O steels may be austenitized in semimuffle furnaces if packed in inert materials such as spent pitch coke and clean cast iron chips. Adequate time must be allowed to ensure that packed work reaches prescribed temperature. If salt baths are used, a salt mixture such as No. 3 in Table 2 of the article, "Furnace Processes and Equipment for Heat Treating of Tool Steels", is recommended. For a suitable method of controlling this bath, see the discussion of rectification in that article.

Quenching. The optimum temperature range for quenching baths consisting of conventional oils is 40 to 60 °C (100 to 140 °F); agitation is recommended. Quenching oils that contain additives ("fast" oils) increase the cooling rate of the steel and permit more

Table 3 Allowable time between quenching and tempering of shock-resisting tool steels, for prevention of cracking

As determined by an extensive study conducted in one plant. Allowable time may vary significantly with size and shape of part.

Steel	Austenitizing temperature °C	°F	Quenching medium	Allowable time prior to tempering, min
S1	900	1650	Oil	30
S1	980	1800	Oil	15
S2	845	1550	Brine	10
S2	900	1650	Brine	5
S3	815	1500	Brine	10
S3	870	1600	Brine	5
S4	870	1600	Brine	10
S4	925	1700	Brine	5
S4	900	1650	Oil	30
S4	955	1750	Oil	15
S5	870	1600	Oil	30
S5	925	1700	Oil	15

Table 4 Recommended heat treating practice for oil-hardening cold work tool steels

Steel	Normalizing temperature(a) °C	°F	Annealing Temperature(b) °C	°F	Cooling rate(c) °C/h	°F/h	Annealed hardness, HB	Preheat °C	°F	Hardening Temperature Austenitize °C	°F	Holding time, min	Quenching medium	Quenched hardness, HRC
O1	870	1600	760-790	1400-1450	20	40	183-212	650	1200	790-815	1450-1500	10-30	Oil	63-65
O2	845	1550	745-775	1375-1425	20	40	183-212	650	1200	760-800	1400-1475	5-20	Oil	63-65
O6	870	1600	765-790	1410-1450	10	20	183-217	...	...	790-815	1450-1500	2-5	Oil	63-65
O7	900	1650	790-815	1450-1500	20	40	192-217	650	1200	790-830	1450-1525	10-30	Water	64-66
										845-885	1550-1625	10-30	Oil	64-66(d)

(a) Holding time, after uniform through heating, varies from about 15 min, for small sections, to about 1 h, for large sections. Work is cooled from temperature in still air. (b) Lower limit of range should be used for small sections, upper limit for large sections. Holding time varies from about 1 h, for light sections and small furnace charges, to about 4 h, for heavy sections and large charges; for pack annealing, hold for 1 h per inch of pack cross section. (c) Maximum. Rate is not critical after cooling to below 540 °C (1000 °F). (d) Sections larger than 38 mm (1½ in.) will be softer.

Fig. 4 Effect of tempering temperature on hardness of shock-resisting tool steels

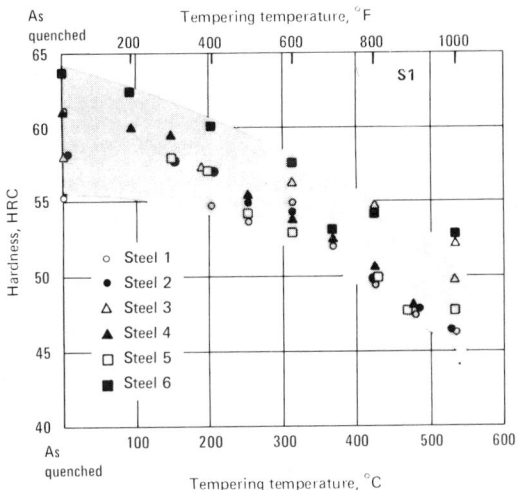

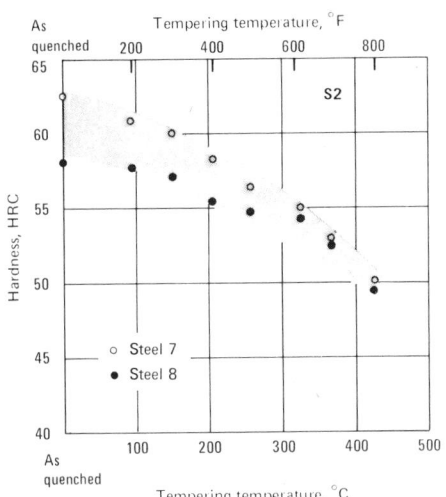

Steel		Composition, %					Quenching		
No.	Type	C	Si	W or Mo	Cr	V	°C	°F	Medium
1	S1	0.43	...	2.00 W	1.30	0.25	955	1750	...
2	S1	0.53	...	2.00 W	1.65	0.25	900	1650	...
3	S1	0.50	...	2.75 W	1.25	0.20	925	1700	...
4	S1	0.55	...	2.50 W	1.50	0.35	925	1700	...
5	S1	0.50	0.75	2.50 W	1.15	0.20	955	1750	Oil
6	S1	0.58	0.95	2.25 W	1.25	0.25	925	1700	Oil
7	S2	0.50	1.10	0.50 Mo	...	0.20	855	1575	Water
8	S2	0.50	1.10	0.50 Mo	...	0.20	900	1650	Oil

latitude in the operating temperature of the bath. Tools may be quenched in these oils at 80 °C (180 °F) without loss of hardness.

Martempering. If control of distortion is particularly important, martempering is sometimes advantageous. In martempering, the work is quenched in a bath of oil or molten salt that is usually held about 15 to 30 °C (25 to 50 °F) above the M_s temperature, and is held in the bath long enough to allow it to attain substantially equalized temperature throughout. The work is then removed from the bath and air cooled. The slow cooling through the martensitic transformation range permits the transformation of austenite to martensite to take place uniformly throughout the piece, thus minimizing distortion. Figure 5 presents a comparison of the dimensional changes in tools made of O1 steel that were oil quenched with those in tools of the same steel martempered at 230 °C (450 °F) for 10 min; the martempered tools exhibited markedly less distortion.

Tempering. The O steels should be tempered immediately after quenching (preferably before they quite reach room temperature). These steels usually are not tempered below 120 °C (250 °F) or above 540 °C (1000 °F); the most commonly used temperature range is from 175 to 205 °C (350 to 400 °F). Tempering times vary with section size. Often, a time at temperature of 1 h per inch of thickness (minimum dimension of heaviest section) or per inch of diameter, with a minimum of 1 h, is used.

Typical hardness values obtained with various tempering temperatures for oil-hardening tool steels are given in Fig. 6. The upper curve in each graph represents results from austenitizing at the higher side of the range of temperatures indicated, and the lower curve represents results from austenitizing at the lower side.

Conventional tools made from the O steels are seldom subjected to multiple tempering or subzero treatment. However, for some special tools, such as gages, where dimensional stability is critical, multiple tempering is desirable. In such instances the workpieces should be cooled to below 65 °C (150 °F) prior to each retempering. Subzero cooling to −75 °C (−100 °F) or lower is also helpful in achieving dimensional stability.

Fig. 4 (continued)

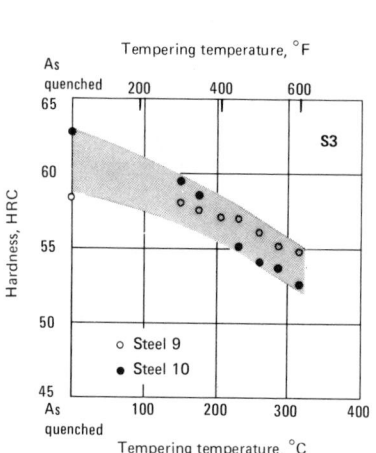

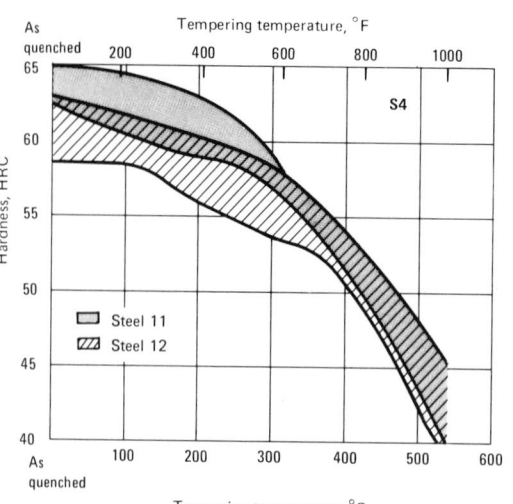

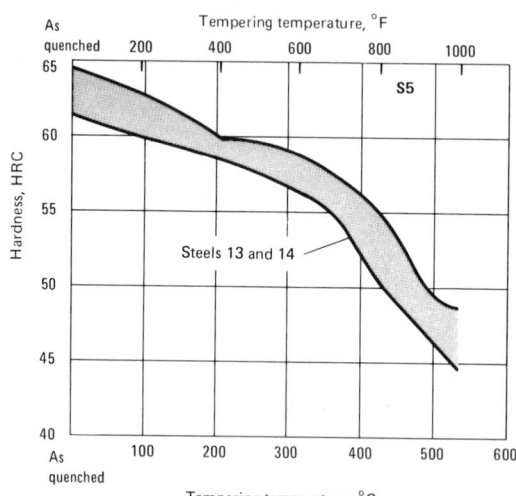

| Steel | | Composition, % | | W or | | | Quenching Temperature | | |
No.	Type	C	Si	Mo	Cr	V	°C	°F	Medium
9	S3.......	0.50	...	1.00 W	1.00	...	870	1600	Oil
10	S3.......	0.50	...	1.00 W	1.00	...	815	1500	Water
11	S4.......	0.54 to 0.60	1.90 to 2.00	...	0 to 0.34	0 to 0.25	845 to 900	1550 to 1650	Water
12	S4.......	0.54 to 0.60	1.90 to 2.00	...	...	...	870 to 955	1600 to 1750	Oil
13	S5.......	0.50	1.60	0.25 Mo	...	...	870	1600	Oil
14	S5.......	0.60	2.00	0.40 Mo	0.28	0.20	900	1650	Oil

Medium-Alloy Air-Hardening, and High-Carbon High-Chromium, Cold Work Tool Steels

Recommended heat treating practice for medium-alloy air-hardening cold work tool steels (group A) and high-carbon high-chromium cold work tool steels (group D) is summarized in Table 6.

Normalizing. Except for type A10 (see Table 6), normalizing is not recommended for any of the steels in groups A and D.

Annealing. These steels are usually supplied in the annealed condition by the manufacturer. However, they should be annealed after forging and prior to rehardening. Annealing is required also for previously hardened or welded tools that are to be reworked.

Recommended annealing temperatures for the various types are given in Table 6. Tools should be heated slowly and uniformly to the annealing temperature. Slow heating is particularly important if a hardened tool is being annealed.

Cycle, or isothermal, treatments may be employed for annealing some A and D steels (Table 5).

Stress Relieving. Tools made of A and D steels that cannot be ground after hardening are sometimes stress relieved after rough machining. This is particularly advisable for delicate tools and tools that vary markedly in cross section. Stress relieving is used also on tools that are machined to final shape, if these tools can be straightened after stress relieving and before final heat treatment. There is little advantage in stress relieving completed tools if they cannot be straightened prior to hardening, because a good preheat will relieve stresses, and the distortion which occurs in either case will remain uncorrected. Recommended temperatures for stress relieving are:

A2, A7.... 650 to 675 °C (1200 to 1250 °F)
A4, A5, A6 675 to 705 °C (1250 to 1300 °F)
D1 to D7.. 675 to 705 °C (1250 to 1300 °F)

Usually, tools can be stress relieved at these temperatures without surface protection. Tools are commonly held at temperature for 1 h per inch of cross section (minimum of 1 h) and then air cooled.

Preheating. Steels of the A and D groups are usually preheated before being austenitized for hardening. Preheating reduces subsequent distortion in the hardened parts by minimizing nonuniform dimensional changes during austenitizing. Preheating simpler tools made of grades A4, A5, A6 and A10 can often be eliminated if they are austenitized in a furnace instead of a liquid bath, because these steels are austenitized at lower temperatures.

Recommended preheating temperatures are listed in Table 6. Holding time at temperature is usually 1 h per inch of maximum cross section. Pre-

Table 5 Cycle annealing treatments for four types of tool steel

Steel	Treatment
O1	Heat to 730 °C (1350 °F), hold for 4 h; heat to 780 °C (1440 °F), hold for 2 h; cool to 690 °C (1275 °F), hold for 6 h; air cool
A2	Heat to 900 °C (1650 °F), hold for 2 h; cool to 760 °C (1400 °F), hold for 6 h; air cool
A6	Heat to 815 °C (1500 °F), hold for 2 h; cool to 650 °C (1200 °F), hold for 6 h; air cool
D2	Heat to 900 °C (1650 °F), hold for 2 h; cool to 775 °C (1425 °F), hold for 6 h; air cool

Table 6 Recommended heat treating practice for medium-alloy air-hardening, and high-carbon high-chromium, cold work tool steels

Steel	Normalizing temperature(a) °C (°F)	Annealing Temperature(b) °C	Annealing Temperature(b) °F	Cooling rate(c) °C/h	Cooling rate(c) °F/h	Annealed hardness, HB	Preheat °C	Preheat °F	Austenitize °C	Austenitize °F	Holding time, min	Quenching medium	Quenched hardness, HRC
Medium-alloy air-hardening cold work tool steels													
A2	Not rec	845-870	1550-1600	22(d)	40(d)	201-229	790	1450	925-980	1700-1800	20-45(e)	A	62-65(f)
A3	Not rec	845-870	1550-1600	22	20	207-229	790	1450	955-1010	1750-1850	25-60(e)	A	...
A4	Not rec	740-760	1360-1400	14(g)	25(g)	200-241	675	1250	815-870	1500-1600	15-90	A	61-64(f)
A5	Not rec	740-760(h)	1360-1400(h)	14	25	229-255	595	1100	790-845	1450-1550	15-45	A	62-63(f)
A6	Not rec	730-745	1350-1375	14	25	217-248	650	1200	830-870	1525-1600	20-45	A	59-63(f)
A7	Not rec	870-900	1600-1650	14(d)	25(d)	235-262	815	1500	955-980	1750-1800	30-60(e)	A	64-67(f)
A8	Not rec	845-870	1550-1600	22	40	192-223	790	1450	980-1010	1800-1850	20-45(e)	A	60-62(f)
A9	Not rec	845-870	1550-1600	14	25	212-248	790	1450	980-1025	1800-1875	20-45(e)	A	56-58(f)
A10	790(1450)	765-795	1410-1460	8	15	235-269	650	1200	790-815	1450-1500	30-60	A	62-64(f)
High-carbon high-chromium cold work tool steels													
D1	Not rec	870-900	1600-1650	22	40	207-248	815	1500	970-1010	1775-1850	15-45(e)	A	61
D2	Not rec	870-900	1600-1650	22	40	217-255	815	1500	980-1025	1800-1875	15-45(e)	A	64
D3	Not rec	870-900	1600-1650	22	40	217-255	815	1500	925-980	1700-1800	15-45(e)	O	64
D4	Not rec	870-900	1600-1650	22	40	217-255	815	1500	970-1010	1775-1850	15-45(e)	A	64
D5	Not rec	870-900	1600-1650	22	40	223-255	815	1500	980-1025	1800-1875	15-45(e)	A	64
D7	Not rec	870-900	1600-1650	22	40	235-262	815	1500	1010-1065	1850-1950	30-60(e)	A	65

(a) Holding time, after uniform through heating, varies from about 15 min, for small sections, to about 1 h, for large sections. Work is cooled from temperature in still air. (b) Lower limit of range should be used for small sections, upper limit for large sections. Holding time varies from about 1 h, for light sections and small furnace charges, to about 4 h, for heavy sections and large charges; for pack annealing, hold for 1 h per inch of pack cross section. (c) Maximum rate, to 540 °C (1000 °F) unless footnoted to indicate otherwise. (d) To 705 °C (1300 °F). (e) For open furnace heat treatment. For pack hardening, hold for ½ h per inch of pack cross section. (f) Hardness varies with austenitizing temperature. (g) To 650 °C (1200 °F). (h) One manufacturer recommends cooling from 760 to 540 °C (1400 to 1000 °F), then reheating to 730 °C (1350 °F) and cooling.

Fig. 5 Dimensional changes in O1 tools

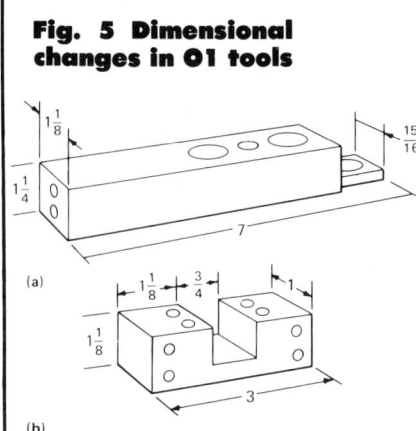

(a)

(b)

Dimensions are in inches

Tools sketched, made of O1 steel, were used for comparison of dimensional changes resulting from martempering at 230 °C (450 °F) for 10 min, and from oil quenching. Five tools of each design, processed by each method, were measured on 5 different days (a) Maximum change in flatness along the 180-mm (7-in.) dimension was 0.25-mm (0.010-in.) after oil quenching and 0.005-mm (0.0002-in.) after martempering. (b) Maximum change of the 19-mm (³/₄-in.) slot width was 0.1-mm (0.0039-in.) after oil quenching and 0.03-mm (0.0012-in.) after martempering.

heating temperatures of 790 to 815 °C (1450 to 1500 °F) are used for tools made from A2, A3, A7, A8 or A9, or from any of the D steels. For these higher temperatures, a liquid bath or a protective furnace atmosphere is required in order to prevent scaling and decarburization.

Austenitizing. Steels of groups A and D can be austenitized in molten salt baths or in various types of furnaces using gaseous atmospheres. Because of their lower austenitizing temperatures, types A4, A5, A6 and A10 may also be austenitized in molten lead, or in open furnaces with oxidizing atmospheres. However, the latter methods are not satisfactory for the other A steels or for the D steels, because of their higher austenitizing temperatures.

If salt baths are used, salt mixtures such as No. 2 or No. 3 in Table 2 of the article, "Furnace Processes and Equipment for Heat Treating of Tool Steels", are recommended; the choice between the two depends on required working temperature range. These mixtures

Table 7 Procedures for salt bath and endothermic atmosphere austenitizing of D2 inserts

Bending die inserts (Example 7)(a)

Salt bath(b) . . . . Preheat in air furnace at 650 °C (1200 °F) for 1½ h(c); austenitize at 1010 °C (1850 °F) for 35 min(c); air cool; remove salt

Endothermic atmosphere(d) . . . . . Charge directly into furnace at 705 °C (1300 °F) and preheat for 1½ h(c); austenitize at 1010 °C (1850 °F) for 2 h; air cool(c)

Trim die inserts (Example 8)(e)

Salt bath(b) . . . . Preheat in air furnace at 650 °C (1200 °F), 4 h(c), then in salt bath at 845 °C (1550 °F), 1 h(c)(f); austenitize at 1010 °C (1850 °F) for 1 h(c); air cool(c); remove salt

Endothermic atmosphere(d) . . . . . Charge directly into furnace at 705 °C (1300 °F), and preheat for 4 h(c); raise furnace temperature, and austenitize at 1010 °C (1850 °F) for 4 h; air cool(c)

(a) After austenitizing, inserts 200 by 305 by 38 mm (8 by 12 by 1½ in.) were double tempered (2 h at 510 °C or 950 °F, air cool; 2 h at 510 °C or 950 °F, air cool) and then nitrided for 48 h at 510 °C (950 °F). (b) Salt bath furnace was immersed-electrode type, 380 by 760 by 915 mm (15 by 30 by 36 in.) deep. (c) Manual loading requires 1½ min per piece. (d) Furnace was radiant-tube type, 610 by 915 by 455 mm (24 by 36 by 18 in.) high. (e) After austenitizing, die inserts were double tempered (4 h at 190 °C or 375 °F, air cool; 4 h at 190 °C or 375 °F, air cool). (f) Second preheat was necessary because of faster heating rate of salt bath.

Table 8 Hardenability in still air of several A and D tool steels

Steel	Center hardening Size of section that fully hardens at center	Hardness, HRC	Surface hardening Size of section that fully hardens at surface	Hardness, HRC
A2, A4	75-mm (3-in.) diam	59 to 61	100-mm (4-in.) diam	59 to 61
A5	. . .	. . .	100-mm (4-in.) diam	62 to 63
A6	180-mm (7-in.) cube	59 to 60	180-mm (7-in.) cube	60 to 61
D1, D2, D5	100-mm (4-in.) cube	60 to 61	125 by 125 by 255-mm (5 by 5 by 10-in.)	61 to 62

may be rectified (for the prevention of decarburization) by the method indicated in that article.

Procedures for austenitizing two different parts made of D2 steel by salt bath and by endothermic atmosphere furnace processes are shown in Table 7.

In some instances, austenitizing will cost more with atmosphere furnaces than with salt baths, and in other instances the reverse will be true. Atmospheres that have proved suitable for austenitizing the A and D steels are endothermic, dry dissociated ammonia and dry hydrogen.

Endothermic gas produced by catalytic combination of air and fuel gas is the most widely used atmosphere. This relatively inexpensive gas can be adjusted for desired carbon potential and controlled by dew point.

Dry dissociated ammonia (dew point, −50 °C or −60 °F) and dry hydrogen

(dew point, −75 °C or −100 °F) are used in applications in which complete freedom from discoloration is required. Vacuum, which excludes all atmosphere, can also be used to austenitize the A and D steels; it is particularly suitable for these steels because their air-hardening characteristics permit slow cooling rates.

Like the O steels, the A and D steels may be packed and then austenitized in semimuffle furnaces. The packing materials and heat treating procedures employed are similar to those described in the previous section on austenitizing of the O grades.

Steels of groups A and D must be held at their austenitizing temperatures long enough to obtain required carbide solution if they are to attain maximum hardness. However, hardening from excessively high austenitizing temperatures will increase the retained austenite. Although retained

Fig. 6 Hardness as a function of tempering temperature, for oil-hardening cold work tool steels

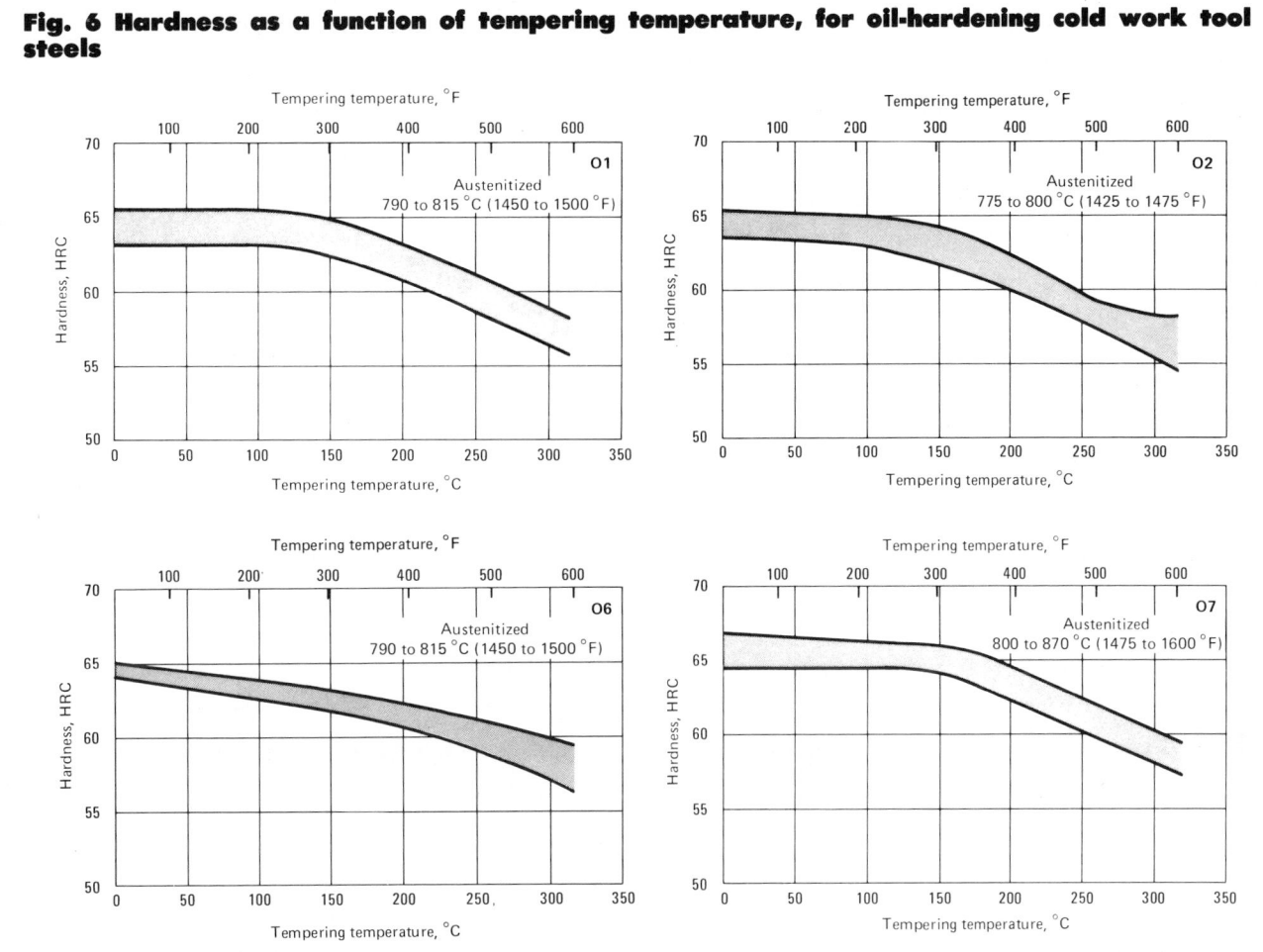

Steels O1, O2 and O6 were austenitized at the temperatures indicated, and then oil quenched. For O7 steel, large, uniform sections were austenitized at 800 to 830 °C (1475 to 1525 °F) and water quenched, and other sections were austenitized at 830 to 870 °C (1525 to 1600 °F) and oil quenched. Duration of tempering was 1 h.

austenite can be decreased by repeated tempering or subzero cooling (or both), it should be avoided.

Quenching. Steels of groups A and D, except D3, will attain maximum hardness by cooling in still air, unless sections are extremely large. However, the hardenability of these steels varies with different types, as indicated in Table 8.

Depending on section size, hardenability, and complexity of shape, the following methods are used to obtain increasingly accelerated cooling of nominally air-hardening steels:

- Cool in "still air"—that is, atmospheric air undisturbed by artificial circulation.
- Cool in "fan air"—that is, the current of air discharged from a fan.

- Cool in "air blast"—that is, the discharge from a high-pressure line.
- "Oil quench to black"—that is, quench in oil until the steel is below the temperature at which it glows dull red, then cool to room temperature in air.
- Oil quench by conventional practice.

Tempering practices for A and D steels parallel those described for O steels in the preceding section. Tempering is usually begun when the work reaches a temperature of about 50 to 65 °C (120 to 150 °F). However, these steels retain some austenite at this temperature range. To maximize transformation of austenite to martensite, cooling to room temperature, or to subzero temperature, is sometimes applied.

Opinions vary greatly as to the merits of subzero cooling, because it increases the probability of cracking during the cooling cycle. The more usual practice is to begin tempering when parts reach about 50 to 65 °C (120 to 150 °F) and then double or triple temper. Multiple tempering is effective in decreasing the amount of austenite retained in A and D steels and is a common practice in heat treating them. The general precautions and tempering practices outlined for O steels in the preceding section are followed for the A and D steels. However, because most of the steels in groups A and D (except A4, A5 and A6) soften less rapidly than the group O steels with an increase in tempering temperature (Fig. 7 and 8), higher tempering temperatures can be used for the A and D steels. A minimum tempering temperature of 205 °C

Fig. 7 Effect of tempering temperature on hardness of medium-alloy air-hardening steels

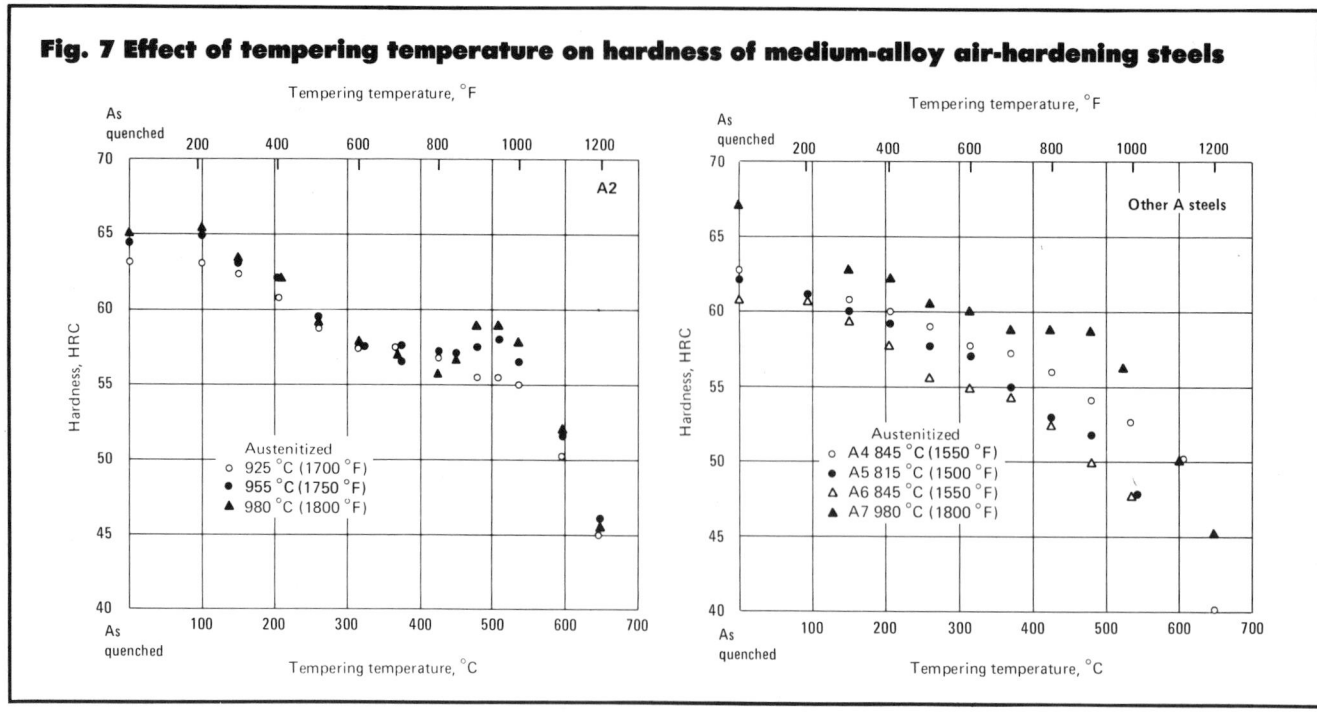

Fig. 8 Relation between tempering temperature and hardness for D2 and D3 tool steels

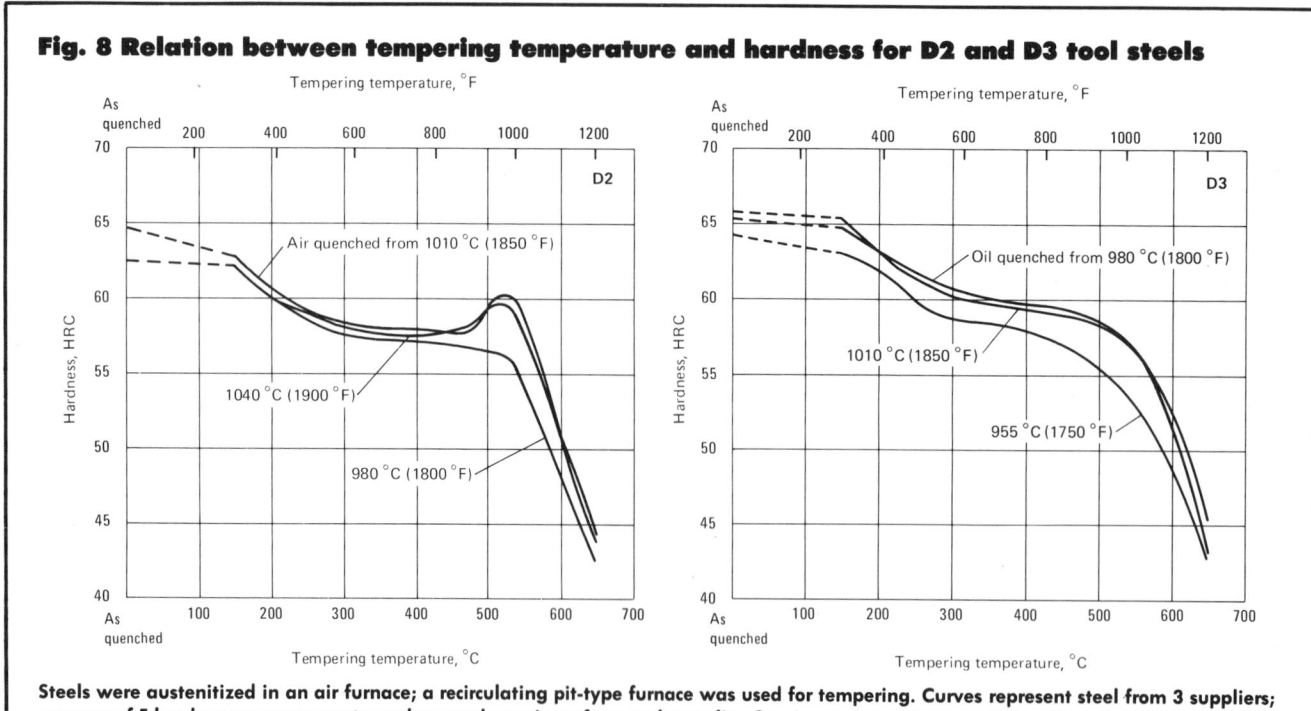

Steels were austenitized in an air furnace; a recirculating pit-type furnace was used for tempering. Curves represent steel from 3 suppliers; average of 5 hardness measurements made on each specimen from each supplier. Specimens were 25 mm (1 in.) in diam and 38 mm (1½ in.) long.

(400 °F) is a common requirement for A2, A7 and D steels. Tempering temperatures as high as 550 °C (1025 °F) are frequently used, and even higher temperatures are used for special requirements.

It will be noted in Fig. 7 and 8 that certain steels (notably A2 and D2) exhibit higher hardness after being tempered at about 540 °C (1000 °F) than after being tempered at temperatures 55 to 110 °C (100 to 200 °F) lower.

This reversal in the usual relationship is known as secondary hardening, and is caused by transformation of retained austenite during tempering at the higher temperatures, near 540 °C (1000 °F). When a steel can be tempered to the same hardness at more than one temperature (for instance, D2 to 58 to 59 HRC), it is advisable to select the highest tempering temperature that will produce the desired hardness. This will yield added toughness and may prevent tool breakage in service.

Nitriding. The A steels (particularly A2 and A7) and the D steels are often nitrided after being hardened and tempered. Nitriding may be done either in a salt bath or in an atmosphere of dissociated ammonia. High tempering temperatures of 510 to 540 °C (950 to 1000 °F) are used on steels that are to be nitrided. Excessively high nitriding temperatures, recommended range is 510 to 540 °C (950 to 1000 °F), will reduce hardness of the base metal and should not be used. Austenitizing at a higher temperature when hardening prior to nitriding will minimize loss of hardness during nitriding of some D steels (note graph for D2 in Fig. 8). For details, see the article on gas nitriding.

Hot Work Tool Steels

Nominal compositions of chromium, tungsten and molybdenum types of hot work tool steels are given in Table 1 of the article entitled "Heat Treating of Tool Steels". The steels in the group denoted in Table 1 as "Other Alloy Tool Steels" are included here in the discussion of hot work tool steels, because they are also used extensively for hot work applications. Table 9 summarizes the heat treating practices commonly employed for this composite group of tool steels.

Normalizing. Because these steels as a group are either partially or completely air-hardening, normalizing is not recommended except for the high-nickel steel 6F7. After forging or before reheat treating, 6F7 may be normalized by heating to 845 to 870 °C (1550 to 1600 °F), preferably in a controlled atmosphere, and cooling in still air.

Annealing. Recommended annealing temperatures, cooling practice and expected hardness values are given in Table 9. Heating for annealing should be slow and uniform to prevent cracking, especially when annealing hardened tools. Heat losses from the furnace usually determine the rate of cooling; large furnace loads will cool at a slower rate than light loads. For most of these steels, furnace cooling to 425 °C (800 °F), at 22 °C max (40 °F max) per h, and then air cooling, will suffice.

For types 6F2, 6F3 and 6H1, an isothermal anneal (Table 9) may be employed to advantage for small tools that can be handled in salt or lead baths or for small loads in batch-type furnaces; however, isothermal annealing has no advantage over conventional annealing for large die blocks or large furnace loads of these steels.

To minimize scaling and decarburization, small parts are usually pack annealed, while large and heavy die blocks are more commonly annealed in controlled-atmosphere furnaces.

Packing material should preferably be spent cast iron chips or spent pitch coke–petroleum coke heated to 1205 °C (2200 °F) in a semiclosed container to drive off gas and moisture. Lime, sand or mica is sometimes used, but under such material if mixed with a small amount of charcoal or other carburizing material, the steel may be decarburized. Packing material should be dry and free of all oxidizing materials, should separate all metal surfaces, and should fill the container. Containers should be sealed after packing. Holding time at the annealing temperature is 1 h per inch of container thickness. The H steels must have a neutral packing material, because they are extremely susceptible to both carburization and decarburization.

In controlled-atmosphere furnaces, the work should be supported so that it does not touch the bottom of the furnace. This will ensure uniform heating and permit free circulation of the atmosphere around the work. Workpieces should be supported so that they will not sag or distort under their own weight.

Grades 6F4 and 6F7 may be annealed without packing or controlled atmosphere if light scaling is not objectionable, because they are annealed at lower temperatures (Table 9).

Stress Relieving. It is sometimes advantageous to stress relieve tools made of hot work steel after rough machining but prior to final machining, by heating them to 650 to 730 °C (1200 to 1350 °F). This treatment minimizes distortion during hardening, particularly for dies or tools that have major changes in configuration or deep cavities. However, closer dimensional control can be obtained by hardening and tempering after rough machining and prior to final machining, provided that the final hardness obtained by this method is within the machinable range.

Preheating prior to austenitizing is nearly always recommended for all hot work steels except 6G, 6F2, 6F3 and 6F5. These four steels may or may not require preheating, depending on size and configuration of the workpieces. Recommended preheating temperatures for all the other types are given in Table 9.

Die blocks or other tools for open furnace treatment should be placed in a furnace that is not over 260 °C (500 °F). Work that is packed in containers may be safely placed in furnaces at 370 to 540 °C (700 to 1000 °F). Once the workpieces (or container) have attained furnace temperature, they are heated slowly and uniformly, at 85 to 110 °C (150 to 200 °F) per h, to the preheating temperature (Table 9) and held for 1 h per inch of thickness (or per inch of container thickness, if packed). Thermocouples should be placed adjacent to the pieces in containers. Controlled atmospheres or other protective means must be used above 650 °C (1200 °F) to minimize scaling and decarburization. A slightly reducing atmosphere is especially recommended for preheating of H41.

For certain parts—for example, intricate die-casting dies—preheating is omitted. Distortion of such parts is sometimes lessened by packing them and heating them slowly and uniformly throughout the entire range to the quenching temperature.

Austenitizing temperatures recommended for the hardening of hot work tool steels are given in Table 9. Rapid heating from the preheating temperature to the austenitizing temperature is preferred for types H16 through H43 and for type 6F4.

Except for steels H10 through H14 (see Table 9), time at the austenitizing temperature should only be sufficient to heat the work completely through; prolonged soaking is not recommended. Time cycles for several specific conditions are indicated in the next section of this article entitled "Examples of Heat Treating Procedure for Hot Work Tools".

The equipment and method employed for austenitizing are frequently determined by the size of the workpiece. For tools weighing less than about 227 kg (500 lb), any of the methods would be suitable. However, larger tools or dies would be difficult to handle in either a salt bath or a pack.

Tools or dies made of hot work steel

Table 9 Recommended heat treating practice for hot work tool steels

Steel	Normalizing temperature(a) °C	°F	Annealing Temperature(b) °C	°F	Cooling rate(c) °C/h	°F/h	Annealed hardness, HB	Hardening Preheat Temperature °C	°F	Austenitize °C	°F	Holding time, min	Quenching medium	Quenched hardness, HRC
Chromium hot work tool steels														
H10	Not recommended		845-900	1550-1650	22	40	192-229	815	1500	1010-1040	1850-1900	15-40(d)	A	56-59
H11	Not recommended		845-900	1550-1650	22	40	192-229	815	1500	995-1025	1825-1875	15-40(d)	A	53-55
H12	Not recommended		845-900	1550-1650	22	40	192-229	815	1500	995-1025	1825-1875	15-40(d)	A	52-55
H13	Not recommended		845-900	1550-1650	22	40	192-229	815	1500	995-1040	1825-1900	15-40(d)	A	49-53
H14	Not recommended		870-900	1600-1650	22	40	207-235	815	1500	1010-1065	1850-1900	15-40(d)	A	55-56
H16	Not recommended		870-900	1600-1650	22	40	212-241	815	1500	1120-1175	2050-2150	2-5	A, O	55-58
H19	Not recommended		870-900	1600-1650	22	40	207-241	815	1500	1095-1205	2000-2200	2-5	A, O	52-55
Tungsten hot work tool steels														
H20	Not recommended		870-900	1600-1650	22	40	207-235	815	1500	1095-1205	2000-2200	2-5	A, O	53-55
H21	Not recommended		870-900	1600-1650	22	40	207-235	815	1500	1095-1205	2000-2200	2-5	A, O	43-52
H22	Not recommended		870-900	1600-1650	22	40	207-235	815	1500	1095-1205	2000-2200	2-5	A, O	48-57
H23	Not recommended		870-900	1600-1650	22	40	212-255	815	1500	1205-1260	2200-2300	2-5	O	33-35(e)
H24	Not recommended		870-900	1600-1650	22	40	217-241	815	1500	1095-1230	2000-2250	2-5	A, O	44-55
H25	Not recommended		870-900	1600-1650	22	40	207-235	815	1500	1150-1260	2100-2300	2-5	A, O	46-53
H26	Not recommended		870-900	1600-1650	22	40	217-241	870	1600	1175-1260	2150-2300	2-5	A, O, S	63-64
Molybdenum hot work tool steels														
H41	Not recommended		815-870	1500-1600	22(f)	40(f)	207-235	730-845	1350-1550	1095-1190	2000-2175	2-5	A, O, S	64-66
H42	Not recommended		845-900	1550-1650	22	40	207-235	730-845	1350-1550	1120-1220	2050-2225	2-5	A, O, S	54-62
H43	Not recommended		815-870	1500-1600	22(g)	40(g)	207-235	730-845	1350-1550	1095-1190	2000-2175	2-5	A, O, S	54-58
Other alloy tool steels														
6G	Not recommended		790-815	1450-1500	22(h)	40(h)	197-229	Not required	Not required	845-855	1550-1575	…	O(j)	63 min(k)
6F2	Not recommended		780-795	1440-1460	22(m)	40(m)	223-235	Not required	Not required	845-870	1550-1600	…	O(j)	63 min(k)
6F3	Not recommended		760-775	1400-1425	22(n)	40(n)	235-248	Not required	Not required	900-925	1650-1700	…	A(p)	63 min(k)
6F4	Not recommended		705	1300	22(q)	(q)	262-285	815	1500	1010-1020	1850-1870	…	O, A	38-41(e)
6F5	Not recommended		845	1550	22(r)	(r)	262-285	Not required	Not required	870	1600	…	O, A	58-59
6F6	Not recommended		845 (pack)	1550 (pack)	22(s)	(s)	196	650-705(t)	1200-1300(t)	925-955(t)	1700-1750(t)	…	O(u)	(v)
6F7	845-870	1550-1600	670	1240	22	40	260-300	730	1350	915	1675	…	A	54-55
6H1	Not recommended		845	1550	22(w)	40(w)	202-235	760-790	1400-1450	900-940	1650-1725	…	A	48-49
6H2	Not recommended		815-845	1500-1550	22	40	202-235	705-760	1300-1400	980-1065	1800-1950	…	O, A	52-55

Note: A, air; O, oil; S, salt. (a) Holding time, after uniform through heating, varies from about 15 min, for small sections, to about 1 h, for large sections. Work is cooled from temperature in still air. (b) Lower limit of range should be used for small sections, upper limit for large sections. Holding time varies from about 1 h, for light sections and small furnace charges, to about 4 h, for heavy sections and large charges; for pack annealing, hold for 1 h per inch of pack cross section. (c) Maximum rate, to 425 °C (800 °F) unless footnoted to indicate otherwise. (d) For open furnace heat treatment. For pack hardening, hold for ½ h per inch of pack cross section. (e) Temper to precipitation harden. (f) To 540 °C (1000 °F), hold for 4 h, furnace cool to 650 °C (1200 °F), then air cool. (g) To 480 °C (900 °F), then air cool. (h) To 370 °C (700 °F). (j) To 205 to 175 °C (400 to 350 °F), then air cool. (k) Temper immediately. (m) For isothermal annealing, furnace cool to 425 °C (800 °F), then air cool. (n) For isothermal annealing, furnace cool to 670 °C (1240 °F), hold for 4 h, furnace cool to 425 °C (800 °F), then air cool. (p) Cool with forced-air blast to 205 to 175 °C (400 to 350 °F), then cool in still air. (q) Air cool from annealing temperature. (r) Furnace cool, at 20 °C (40 °F), (max) per h, to 425 °C (800 °F), reheat to 595 ± 14 °C (1100 ± 25 °F), then air cool. (s) Furnace cool to 425 °C (800 °F), then air cool. (t) Heat in pack or in controlled atmosphere. (u) To 50 °C (125 °F). (v) Pack heating, 59 to 60 HRC; atmosphere heating, 54 to 55 HRC. (w) For isothermal annealing, hold at 845 °C (1550 °F) for 2 h, furnace cool to 745 °C (1375 °F), hold for 4 to 6 h, then air cool.

must be protected against carburization and decarburization when being heated for austenitizing. Carburized surfaces are highly susceptible to heat checking. Decarburization causes decreased strength, which may result in fatigue failures; and on die-casting dies, the molten casting metal will "weld on" to decarburized surfaces and may cause "washout" because of poor wear resistance of the decarburized surface. However, the principal detrimental effect of decarburization is to mislead the heat treater as to the actual hardness of the die. To obtain specified hardness of the decarburized surface, the die is tempered at too low a temperature. The die then goes into operation at excessive internal hardness and breaks at the first application of load.

An endothermic atmosphere produced by a gas generator is probably the most widely used protective medium. The dew point is normally held from 2 to 7 °C (35 to 45 °F) in the furnace, depending on carbon content of the steel and operating temperature. A dew point of 3 to 4 °C (38 to 40 °F) is ideal for most steels of type H11 or H13 when austenitized at 1010 °C (1850 °F).

The packing of work in spent pitch coke before heating it for austenitizing has been used extensively in small shops where it has not been feasible to invest in special equipment. This procedure is generally used for small dies. New pitch coke is generally heated to 1040 to 1205 °C (1900 to 2200 °F) to burn off any combustibles that may be present as well as to remove any excessive moisture. The spent pitch coke is then sifted to remove the fines (the coke should also be sifted before re-use). Normal procedure for this method is to wrap the workpiece in plain brown wrapping paper and place it in a heat-resistant metal box in the bottom of which is about 50 mm (2 in.) of spent pitch coke. The workpiece should be covered and surrounded with approximately 50 to 100 mm (2 to 4 in.) of spent pitch coke. The cover is then placed on the box and sealed with an asbestos paste. The box is then ready to be placed in a furnace, which need not be provided with controlled atmosphere.

Quenching. Hot work steels range from high to extremely high in hardenability. Most of them will achieve full hardness by cooling in still air; however, even with those types having the highest hardenability, sections of die blocks may be so large that insufficient hardening results. In such instances, an air blast or an oil quench is required to achieve full hardness. Hot work steels are never water quenched. Recommended quenching media are listed in Table 9.

If blast cooling is used, air should be blasted uniformly on the surface to be hardened. All air must be dry. When being air quenched, dies or other tools should not be placed on concrete floors or in locations where water vapor may strike them.

Some of the hot work steels (especially the tungsten and molybdenum types) will scale considerably during cooling to room temperature in air. An interrupted quench reduces this scaling by eliminating the long period of contact with air at elevated temperature, but it also increases distortion. The procedure is best carried out by quenching from the austenitizing temperature in a salt bath held at 595 to 650 °C (1100 to 1200 °F), holding in the quench until the workpiece reaches the temperature of the bath, and then withdrawing the piece and allowing it to cool in air. An alternative, but less precise, procedure is to quench in oil at room temperature or slightly above and judge by color (faint red) when the workpiece has reached 595 to 650 °C (1100 to 1200 °F); the piece is then quickly withdrawn and permitted to cool to room temperature in air. While cooling, the pieces should be placed in a suitable rack, or be supported by wires, in such a manner that air is permitted to come in contact with all surfaces.

Steel H23 requires a different type of interrupted quench, because ferrite precipitates rapidly in this steel at 595 °C (1100 °F) and M_s is below room temperature. Type H23 should be quenched in molten salt at 165 to 190 °C (325 to 375 °F) and then air cooled to room temperature. This steel will not harden in quenching but will do so by secondary hardening during the tempering cycle.

Parts quenched in oil should be completely immersed in the oil bath, held until they have reached bath temperature, and then transferred immediately to the tempering furnace. Oil bath temperatures may range from 55 to 150 °C (130 to 300 °F), but should always be below the flash point of the oil. Oil baths should be circulated and kept free of water.

Tempering. Hot work tool steels should be tempered immediately after quenching, even though sensitivity to cracking in this stage varies considerably among the various types (for example, air-quenched 6F4 may be safely kept at room temperature for several hours before tempering, whereas 6G, 6F2 and 6F3 are susceptible to cracking if they are cooled substantially below 175 °C or 350 °F before tempering).

Hot work steels are usually tempered in air furnaces of the forced-convection type. Salt baths are used successfully for smaller parts, but for large complex parts salt bath tempering may induce too severe a thermal shock and cause cracking. The effect of tempering temperature on hardness of hot work tool steels is shown in Fig. 9.

Multiple tempering ensures that any retained austenite that transforms to martensite during the first tempering cycle is tempered before a tool is placed in service. Multiple tempering also minimizes cracks due to stress originating from the hardening operation.

Multiple tempering has proved particularly advantageous for large or sharp-cornered die blocks that are not permitted to reach room temperature before the first tempering operation.

Example 6. In one plant where many die blocks are heat treated, standard practice is as follows. When the dies have air cooled to 52 °C (125 °F), they are placed in a tempering furnace maintained at 565 °C (1050 °F). After the dies have reached furnace temperature, they are soaked for 1 h per inch of thickness. The dies are then air cooled to room temperature. Second and third tempering operations are carried out in the same manner, except that temperature may be increased as required in order to obtain desired hardness.

Most of the hot work steels have secondary hardening characteristics; H23 is the most pronounced in this respect (Fig. 9). As with A2 and D2 (discussed previously), these secondary-hardening hot work steels should be tempered at the highest temperature at which the desired hardness can be produced.

Surface Hardening. Although tools and dies made of the hot work steels usually have sufficient hardness to perform the tasks for which they were designed, they are occasionally surface hardened to acquire improved resistance to wear or heat for special applications. The two principal processes that have been used for this purpose are carburizing and nitriding.

Fig. 9a Effect of temperature on hardness of chromium hot work tool steels

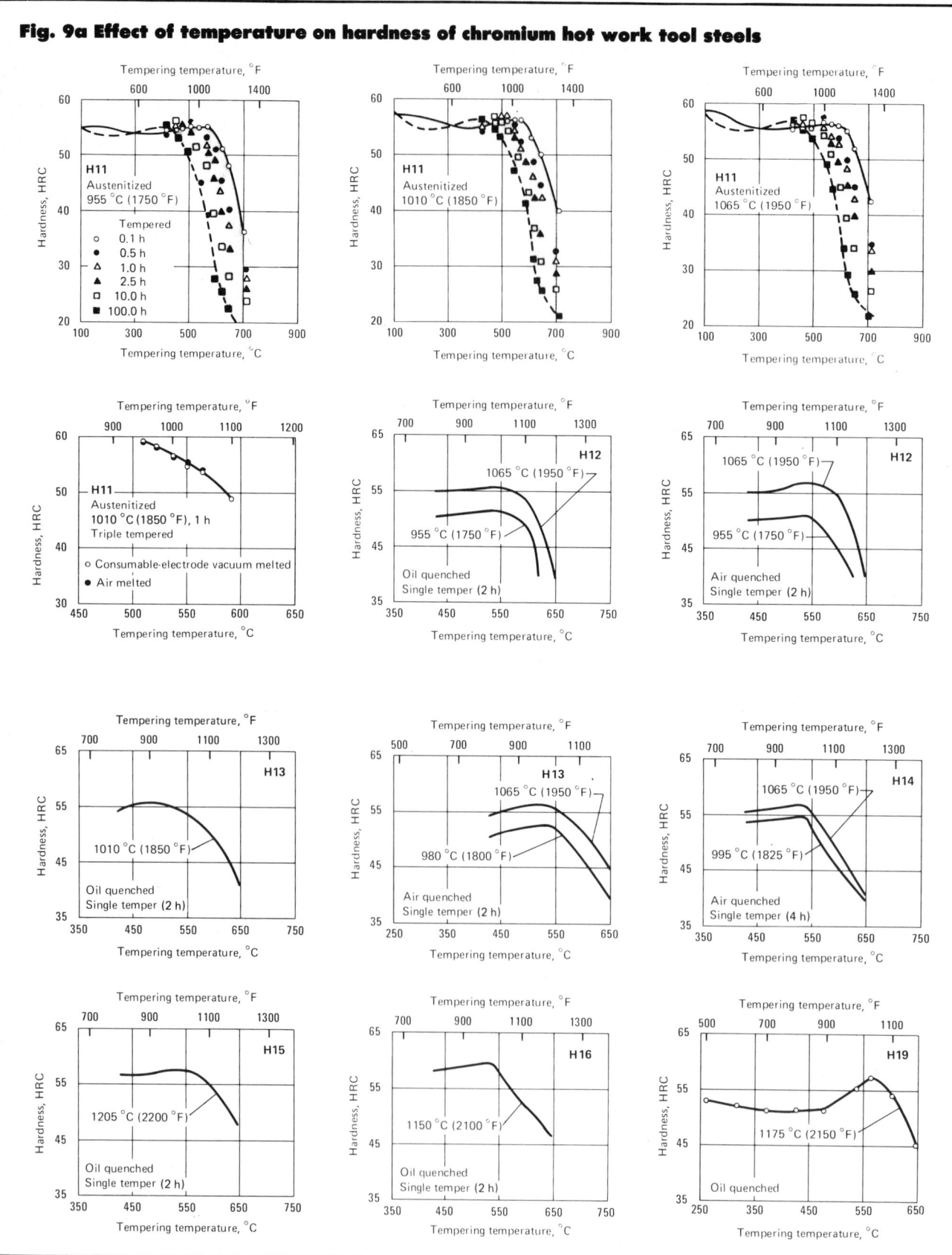

Fig. 9b Effect of temperature on hardness of tungsten hot work tool steels

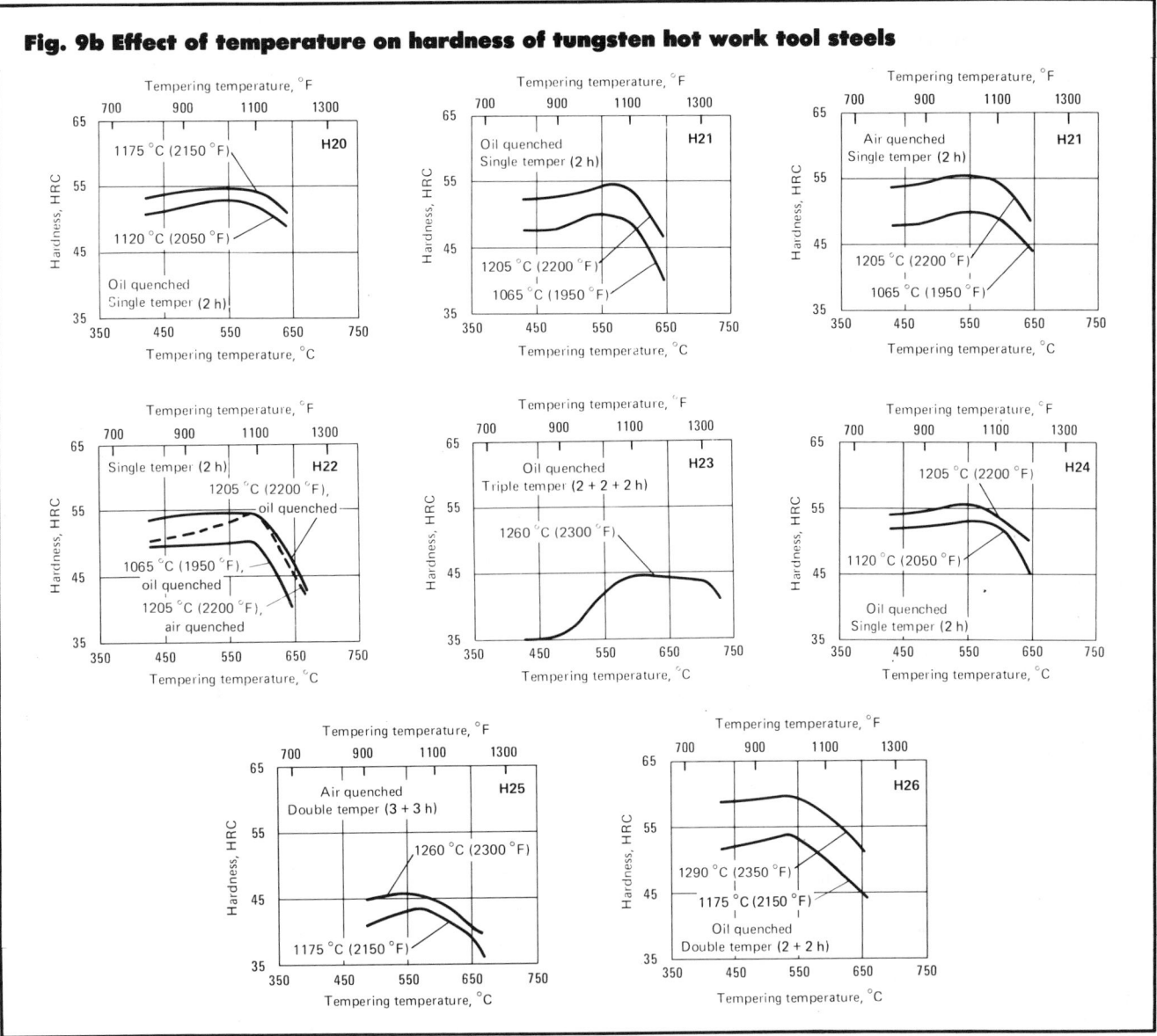

Carburizing is usually limited to hot work steels having a carbon content of 0.35% or lower. Type H12 has been reported to achieve a carburized surface hardness of 60 to 62 HRC. The carburized case should be shallow—for example, 0.4 mm (0.015 in.) max—or severe embrittlement will occur. The greater the thermal shock (or gradient) present in service—as in die casting—the shallower the case must be.

Gas or liquid nitriding is sometimes applied to the hot work steels to increase resistance to heat or wear, or both. For instance, dies for hot extrusion are sometimes nitrided to increase service life. One disadvantage of nitriding, however, is the difficulty it imposes on the reworking of tools or dies.

Another disadvantage is that it may accentuate heat checking. Hot work steels should be hardened and tempered before being nitrided, but should be neither decarburized nor carburized.

The quality and depth of the nitrided case are influenced by the chemical composition of the steel and by the time and temperature of nitriding. The presence of nitride-forming elements such as chromium and vanadium is helpful to the attainment of a satisfactory case. The fact that most of the hot work steels reach a secondary hardening peak when tempered in the vicinity of 540 °C (1000 °F) is beneficial, because nitriding is usually accomplished in a range of 510 to 540 °C (950 to 1000 °F)

over a period of 15 to 24 h. The nitrided case, in addition to being very hard, may be brittle. Brittleness increases with depth of case; hence, shallow, 0.08 to 0.2 mm (0.003 to 0.008 in.), nitrided cases are usually applied.

Examples of Heat Treating Procedure for Hot Work Tools

Tools and dies made of hot work steel extend over an extremely wide range of sizes and weights (sometimes up to several tons, as in the largest die blocks). Therefore, details of heat treating techniques may vary considerably. The following examples give details of proce-

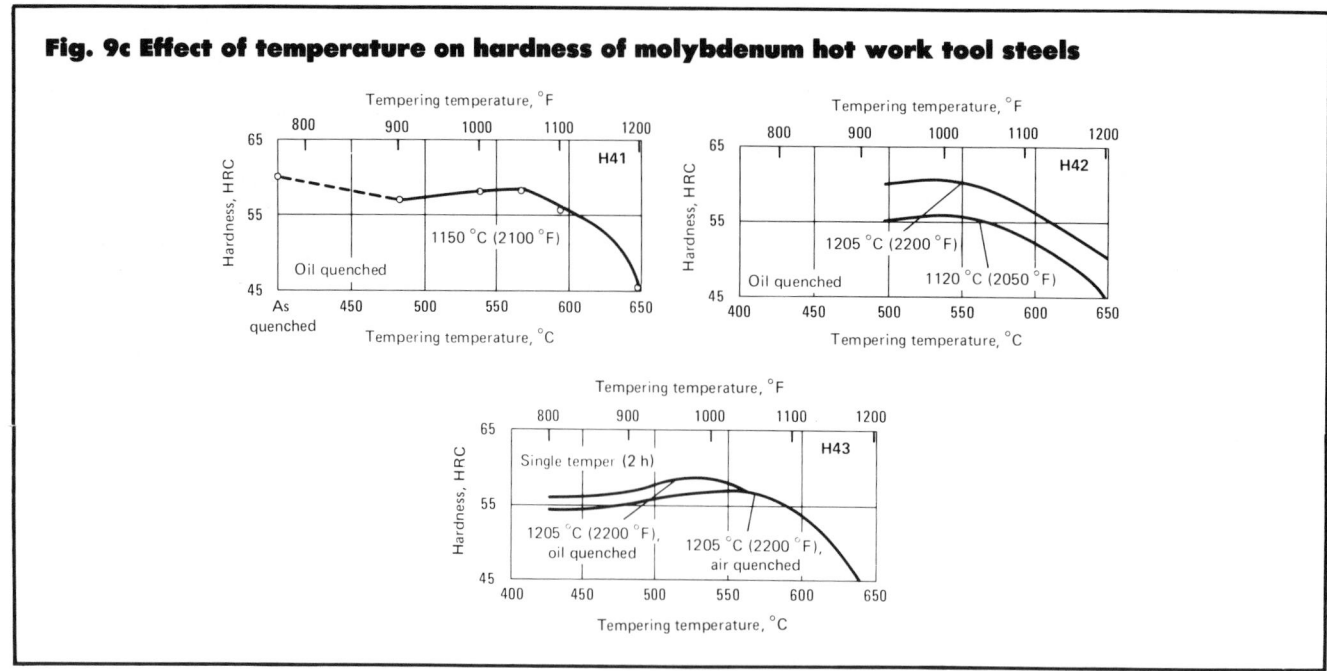

Fig. 9c Effect of temperature on hardness of molybdenum hot work tool steels

dures that have proved successful in practice:

Example 7 (H21 Hot Extrusion Die). A typical method for heat treating a 75-mm- (3-in.-) thick, 200-mm- (8-in.-) OD, 75-mm (3-in.) hole, hot extrusion die made of H21 steel comprises the following:

• Preheat at 815 to 845 °C (1500 to 1550 °F), either in a slightly oxidizing atmosphere or in neutral salt.
• Transfer to furnace (6 to 12% reducing atmosphere or neutral salt bath) operating at 1175 °C (2150 °F). Hold in furnace for approximately 20 min after the die has reached 1175 °C (2150 °F).
• Cool in still air to about 65 °C (150 °F).
• Temper at 565 °C (1050 °F) for 4 h.
• Cool to near room temperature.
• Retemper at 650 °C (1200 °F) for 4 h.
• Air cool.

Example 8 (H11 Mandrel). Mandrels made of H11, used in conjunction with the H21 die in Example 7, above, are heat treated as follows:

• Preheat at 760 °C (1400 °F) in a slightly oxidizing atmosphere.
• Transfer to atmosphere furnace (1 to 3% excess O_2) operating at 1010 °C (1850 °F) and hold for 20 min plus 5 min for each inch of thickness.
• Air cool to near room temperature (oil quenching can also be used).
• Temper (or, preferably, double temper) for desired hardness.

Example 9 (H13 Die Block). One plant employs the following procedure for heat treating die blocks made of H13 that weigh less than 23 kg (50 lb):

1 Insert eyebolt to facilitate handling.
2 Wrap die block in waxed paper and place in a heat-resistant container on a bed of spent pitch coke 75 to 100 mm (3 to 4 in.) deep.
3 Seal cover on container with asbestos paste.
4 Place container in furnace (not atmosphere-controlled) operating at 760 °C (1400 °F); bring to furnace temperature and hold for 4 h.
5 Raise furnace temperature at 30 °C (50 °F) per h to 1010 °C (1850 °F) and hold charge at this temperature for 6 h.
6 Remove die block from container by use of eyebolt.
7 Cool in still air to 345 °C (650 °F) (temperature-indicating crayons may be used), then place in furnace operating at 345 °C (650 °F) and cool in furnace at 30 °C (50 °F) per h to 95 °C (200 °F). (If the die block has no sharp corners or major changes in configuration, the interrupted cooling may be omitted.)
8 Remove from furnace and cool in air to 40 °C (100 °F).
9 Place in tempering furnace operating at 565 °C (1050 °F), bring to furnace temperature and hold for 8 h,

air cool to room temperature and check hardness.
10 Repeat step 9, except that it may be necessary to increase tempering temperature so that final hardness will be 46 to 49 HRC.

Example 10 (H13 Die Block). The following procedure has proved successful for heat treating large die blocks, 1590 kg (3500 lb), made of H13 steel:

1 Load die block into electrically heated bell-type furnace. The sequence of operations begins when furnace temperature reaches 95 °C (200 °F).
2 Raise furnace temperature at 30 °C (50 °F) per h to 370 °C (700 °F).
3 Introduce nitrogen atmosphere to furnace and increase furnace temperature at 55 °C (100 °F) per h to 790 °C (1450 °F); hold for 1 h, then shut off nitrogen, introduce endothermic atmosphere—dew point, 3 to 4 °C (38 to 40 °F)—and hold for an additional 5 h.
4 Increase furnace temperature at 55 °C (100 °F) per h to 1040 °C (1900 °F) and hold for 6 h.
5 Remove die block and air cool to 65 °C (150 °F).
6 Place die block in tempering furnace operating at 205 °C (400 °F); bring to furnace temperature and hold for 7 h.
7 Increase furnace temperature at 40

°C (100 °F) per h to 565 °C (1050 °F) and hold for 16 h.

8 Air cool to room temperature. Hardness is about 46 to 48 HRC.

9 Temper die block a second time, repeating steps 6, 7 and 8 but increasing final temperature to 580 °C (1075 °F) because a finished hardness of 42 to 43 HRC is desired.

10 Temper a third time, repeating steps 6, 7 and 8 without modification.

Example 11 (6F3 Forging Die). In one plant, dies, 495 by 215 by 150 mm (19½ by 8½ by 6 in.), used for forging pitman arms are heat treated to a final hardness of 40 to 42 HRC by the following procedure:

1 Preheat at 260 °C (500 °F).

2 When dies have attained furnace temperature, raise temperature at 55 to 85 °C (100 to 150 °F) per h to 915 °C (1675 °F); use controlled atmosphere above 760 °C (1400 °F).

3 Hold at 915 °C (1675 °F) for 6 h.

4 Air-blast cool to 175 °C (350 °F) (temperature-indicating crayons used).

5 Place in tempering furnace operating at 175 to 205 °C (350 to 400 °F). When dies have attained furnace temperature, raise temperature at 85 °C (150 °F) per h to 595 °C (1100 °F) and hold for 9 h.

6 Air cool to room temperature; check hardness.

7 Retemper, repeating steps 5 and 6 except for final temperature, which will depend on hardness obtained from first tempering.

The following three examples indicate the procedures employed in one plant for heat treating 6F2, 6F4 and H12 components used for hot upset forging of pinions.

Example 12 (6F2 Heading Tool and Gripper Die). Final hardness of 40 to 42 HRC is obtained on this tool by preheating, austenitizing, quenching, tempering and retempering as follows:

1 Preheat at 260 °C (500 °F).

2 When dies have attained furnace temperature, raise temperature at 55 to 85 °C (100 to 150 °F) per h to 855 °C (1575 °F).

3 Hold at 855 °C (1575 °F) for 1 h per inch of thickness.

4 Quench in oil at 55 °C (130 °F), to 175 °C (350 °F) (temperature-indicating crayons used); transfer as quickly as possible to tempering furnace.

5 Place in tempering furnace operating at 175 to 205 °C (350 to 400 °F). When dies have attained furnace temperature, raise temperature at 55 to 85 °C (100 to 150 °F) per h to 595 °C (1100 °F) and hold for 1 h per inch of thickness.

6 Cool in still air to room temperature; check hardness.

7 Retemper, repeating 5 and 6 except for final temperature, which depends on hardness obtained from first tempering.

Example 13 (6F4 Slab Insert). These inserts, requiring final hardness of 39 to 41 HRC, are heat treated as follows:

1 Preheat at 260 °C (500 °F).

2 When inserts have attained furnace temperature, raise temperature at 55 to 85 °C (100 to 150 °F) per h to 815 °C (1500 °F) (use controlled atmosphere above 760 °C or 1400 °F).

3 Solution treat at 1020 °C (1870 °F) for 1 h per inch of thickness.

4 Cool in still air to room temperature. **Note:** When these inserts were quenched to only 95 °C (200 °F), threads broke out of the die during tapping.

5 Precipitation harden by heating at 260 °C (500 °F) until temperature of insert equals furnace temperature, raising furnace temperature at 55 to 85 °C (100 to 150 °F) per h to 450 °C (840 °F), holding at 450 °C (840 °F) for 3 h plus 1 h per inch of thickness (minimum time, 4 h), and air cooling to room temperature.

Example 14 (H12 Punch Insert). Heat treating to a desired final hardness of 40 to 42 HRC comprises the following procedure:

1 Preheat at 260 °C (500 °F).

2 When inserts have attained furnace temperature, raise temperature at 55 to 85 °C (100 to 150 °F) per h to 870 °C (1600 °F) (use controlled atmosphere above 760 °C or 1400 °F).

3 Raise furnace temperature to 1010 °C (1850 °F); hold inserts at 1010 °C (1850 °F) for 1 h per inch of thickness.

4 Quench inserts in still air until they are cool enough to be hand-held. Transfer immediately to the tempering furnace.

5 Place in tempering furnace operating at 260 °C (500 °F). When inserts have attained furnace temperature, raise temperature at 55 to 85 °C (100 to 150 °F) per h to 595 °C (1100 °F).

Fig. 10 Effect of austenitizing temperatures on the as-quenched hardness of M2 steel

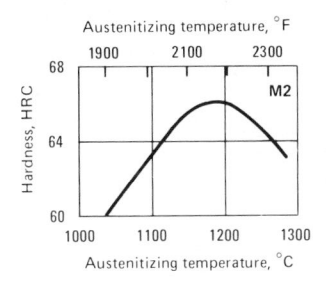

Hold at 595 °C (1100 °F) for 1 h per inch of thickness.

6 Cool in still air and check hardness.

7 Retemper, repeating steps 5 and 6 except for final temperature, which will depend on the hardness that was obtained from the first tempering cycle.

8 Temper for a third time, if time permits.

High Speed Tool Steels

High speed tool steels are used primarily for cutting tools, such as broaches, chasers, cutters, drills, hobs, reamers and taps. Nominal compositions of these steels are given in Table 1 of the article, "Heat Treating of Tool Steels", in this volume. Recommended heat treating practice is summarized for two standard groups of high speed steels and one intermediate group in Table 10 of this article; note that normalizing of high speed tool steels is not recommended.

Steels in the intermediate group, M50 and M52, are less expensive than standard high speed steels and may occasionally be used in place of standard high speed steels.

The intermediate high speed steels do not resist tempering to the same extent as M2, and therefore, they cannot be expected to perform as well as high speed steels in applications where red hardness is important. For example, in such applications as lathe tools and drills, where the tool is in continuous contact with the workpiece and high surface temperatures are the rule, M50 and M52 steels would not be expected to perform as well as standard high speed steels. When contact with the workpiece is intermittent or surface temperatures are low, in such

Table 10 Recommended heat treating practice for high speed tool steels

Steel	Normalizing	Annealing Temperature(a), °C	Annealing Cooling rate(b) °C/h	Annealed hardness, HB	Temperature Preheat, °C	Temperature Austenitize(c), °C	Hardening Holding time, min	Quenching medium	Quenched hardness, HRC
Tungsten high speed tool steels, standard group									
T1	Not recommended	870-900	20	217-255	815-870	1260-1300	2-5	O, A, S	63-65
T2	Not recommended	870-900	20	223-255	815-870	1260-1300	2-5	O, A, S	64-66
T4	Not recommended	870-900	20	229-269	815-870	1260-1300	2-5	O, A, S	64-66
T5	Not recommended	870-900	20	235-285	815-870	1275-1300	2-5	O, A, S	64-66
T6	Not recommended	870-900	20	248-302	815-870	1275-1300	2-5	O, A, S	64-66
T8	Not recommended	870-900	20	229-255	815-870	1260-1300	2-5	O, A, S	64-66
T15	Not recommended	870-900	20	241-277	815-870	1205-1260	2-5	O, A, S	65-67
Molybdenum high speed tool steels, standard group									
M1	Not recommended	815-870	20	207-235	730-845	1175-1220	2-5	O, A, S	64-66
M2	Not recommended	870-900	20	212-241	730-845	1190-1230	2-5	O, A, S	65-66
M3	Not recommended	870-900	20	223-255	730-845	1205-1230	2-5	O, A, S	64-66
M4	Not recommended	870-900	20	223-255	730-845	1205-1230	2-5	O, A, S	64-66
M6	Not recommended	870	20	248-277	790	1175-1205	2-5	O, A, S	63-66
M7	Not recommended	815-870	20	217-255	730-845	1175-1220	2-5	O, A, S	64-65
M10	Not recommended	815-870	20	207-255	730-845	1175-1220	2-5	O, A, S	64-66
M30	Not recommended	870-900	20	235-269	730-845	1205-1230	2-5	O, A, S	64-66
M33	Not recommended	870-900	20	235-269	730-845	1205-1230	2-5	O, A, S	64-66
M34	Not recommended	870-900	20	235-269	730-845	1205-1230	2-5	O, A, S	64-66
M36	Not recommended	870-900	20	235-269	730-845	1220-1245	2-5	O, A, S	64-66
M41	Not recommended	870-900	20	235-269	730-845	1190-1215	2-5	O, A, S	63-66
M42	Not recommended	870-900	20	235-269	730-845	1165-1190	2-5	O, A, S	63-66
M43	Not recommended	870-900	20	248-269	730-845	1150-1175	2-5	O, A, S	63-66
M44	Not recommended	870-900	20	248-285	730-845	1200-1225	2-5	O, A, S	63-66
M46	Not recommended	870-900	20	235-269	730-845	1190-1220	2-5	O, A, S	63-66
M47	Not recommended	870-900	20	235-269	730-845	1175-1205	2-5	O, A, S	63-66
High speed tool steels, intermediate group									
M50	Not recommended	830-845	20	197-235	730-845	1095-1120	2-5	O, A, S	63-65
M52	Not recommended	830-845	20	197-235	730-845	1120-1175	2-5	O, A, S	63-65

Note: O, oil; A, air; S, salt. (a) Pack annealing is recommended, for minimum decarburization. Steels should be held at temperature for 1 h per inch of thickness of the container. (b) Maximum. Rate is not critical after work (in pack, if employed) has been furnace cooled to 650 °C (1200 °F). (c) If steels are austenitized in a salt bath, austenitizing temperatures should be 14 °C (25 °F) lower than those in the ranges given.

applications as hack and band saw blades, blanking dies and some special woodworking tools, M50 and M52 steels may perform adequately. M50 steel is also used in ball and roller bearing races for elevated temperatures. Other applications include woodworking tools, hydraulic pump assemblies, pump pistons and pump vanes. If greater abrasion resistance is required, but not as much as afforded by standard high speed steels, then M52 may be a logical choice.

Annealing. High speed steel must be fully annealed after forging or when rehardening is required. To minimize decarburization, pack annealing in tightly closed containers is recommended. The packing material can be dry sand or lime to which a small amount of charcoal has been added; burned cast iron chips also are satisfactory. Because the packing material acts to insulate the container and thereby slow down heating, the container should be filled in such a way with the steel to be annealed that a minimum amount of packing material is required.

After the steel has reached the annealing temperature range (Table 10), it should be held at temperature for 1 h per inch of thickness of the container and should then be slowly cooled in the furnace (at a rate not exceeding 20 °C or 40 °F per h) until it reaches a temperature of 650 °C (1200 °F), when a faster rate of cooling is permissible.

Preheating. Austenite begins to form at about 760 °C (1400 °F), and preheating for hardening to slightly above this temperature will minimize stresses that might be set up because of the transformation. If the prevention of partial decarburization is important, a preheating temperature of 705 to 790 °C (1300 to 1450 °F) generally will be used. When this is not a problem, preheating at 815 to 900 °C (1500 to 1650 °F) is satisfactory.

Double preheating—in one furnace at 540 to 650 °C (1000 to 1200 °F) and in another at 845 to 870 °C (1550 to 1600 °F)—is often recommended, to minimize thermal shock.

If a single preheat is used, the T types of high speed steels are preferably preheated at 815 to 870 °C (1500 to 1600 °F), and the remaining M types at 730 to 845 °C (1350 to 1550 °F). It is common practice to preheat for twice the length of time required at the austenitizing temperature. Accordingly, to ensure a uniform flow of work, the capacity of the preheating installation is generally twice that of the austenitizing installation.

Although preheating is recommended for all high speed steels, small tools and those that do not incorporate sharp notches or abrupt changes in section, such as small tool bits and solid drill rod blanks, may be placed directly into the austenitizing furnace with reasonable safety. If consumable carbonaceous muffles are used, the preheating temperature must not exceed about 650 °C (1200 °F), because the type of atmosphere they provide is ineffective in preventing decarburization at higher temperatures.

Austenitizing. High speed steels depend on the solution of various complex alloy carbides during austenitizing to develop their heat-resisting qualities and cutting ability. These carbides do not dissolve to an appreciable extent unless the steel is heated to temperatures near the melting point (Table 10). Therefore, exceedingly accurate temperature control is required in austenitizing high speed steel. Steels containing about 3% or more vanadium may be held at the austenitizing temperature approximately 50% longer than the lower-vanadium types. The relatively pure vanadium carbide phase inherent in the microstructure of the high-vanadium steels is virtually insoluble at temperatures below the melting point and acts to restrict grain growth, thus permitting longer soaking times without detriment. However, the recommended austenitizing temperatures for these steels should not be exceeded.

Single-point tools intended for heavy-duty cutting often can be effectively austenitized at 10 to 15 °C (15 to 30 °F) above the nominal austenitizing temperature. The higher temperature increases alloy solution, temper resistance and hot hardness, but it also results in some sacrifice in toughness. To impart added toughness, fine-edged tools, such as taps and chasers, may be hardened at temperatures 15 to 30 °C (25 to 50 °F) below the nominal austenitizing temperature. Punches and dies that do not require maximum hardness may be austenitized for maximum toughness at temperatures 55 to 110 °C (100 to 200 °F) below the nominal temperature.

Other adjustments in austenitizing temperature depend on the type of heating equipment employed. Full-muffle furnaces employing a controlled atmosphere rich in carbon monoxide are usually operated at the higher temperature of the recommended range. Salt baths usually are operated 15 to 30

Fig. 11 Effect of austenitizing and tempering temperatures on impact strength of M2

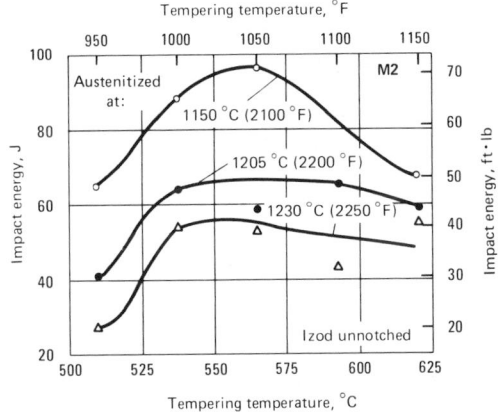

Fig. 12 Effect of austenitizing temperature on high-temperature hardness of M2 steel

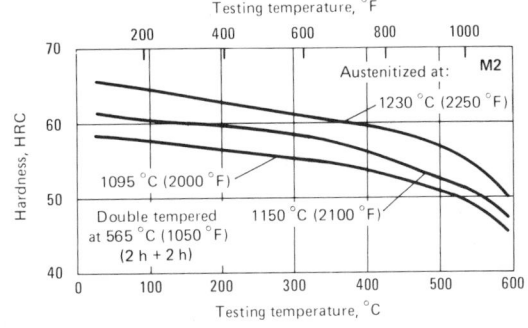

°C (30 to 50 °F) below the top of the range.

The effect of austenitizing temperature on the as-quenched hardness of M2 steel is shown in Fig. 10. Below 1175 °C (2150 °F), M2 cannot develop full hardness on quenching, because of insufficient carbide solution. At temperatures above approximately 1230 °C (2250 °F), the as-quenched hardness of M2 decreases because of too much carbon and alloy solution and an excess of retained austenite in the as-quenched steel.

Figure 11 illustrates the improved toughness of M2, as measured by the Izod unnotched impact test, that results from the use of lower-than-normal austenitizing temperatures. Numerous investigators have shown that the optimum means for attaining maximum toughness in high speed steel is through reduced austenitizing temperatures rather than by full aus-

tenitizing and overtempering to an equivalent hardness level.

Figure 12 illustrates the sacrifice in high-temperature hardness of M2 that results from the use of reduced austenitizing temperatures.

Quenching. High speed steels can be quenched in air, oil or molten salt. However, except for thin tools, which are air quenched between plates to keep them straight, it is customary to quench in oil from muffle or semimuffle furnaces and in molten salt from a high-temperature salt bath. After its temperature has been equalized in the salt quench, the tool is air cooled. For large cutters heated in a furnace, an interrupted oil quench is often used to minimize quenching strains and prevent cracking. This consists of cooling the cutters in the oil only until they lose color (about 540 °C or 1000 °F) and then cooling in air.

After quenching, high speed steel tools usually possess high residual stress, and to prevent cracking, it is good practice to transfer them from the quenchant to a tempering furnace before they have cooled to below 65 °C (150 °F). This is particularly important for large or intricate tools, for which a delay between quenching and tempering or permitting the work to cool to too low a temperature will usually induce cracking. If the work cannot be transferred to a tempering furnace at once, it should be put in a holding furnace maintained at 120 to 205 °C (250 to 400 °F) until a tempering furnace is available.

Bainitic hardening has been used in a few applications. To produce a "primary bainitic" structure, this treatment is performed by arresting the quench from the austenitizing temperature at approximately 260 °C (500 °F), holding for 4 h, then cooling to room temperature. This produces a structure with about 55% bainite and the remainder retained austenite. Subsequent tempering at normal tempering temperature transforms the retained austenite and tempers the bainite to a Rockwell C hardness 1 to 3 points lower than normal for the selected tempering temperature.

Partial Hardening to Improve Machinability. Annealed high speed steel may be partially hardened to approximately 270 to 300 HB to improve machinability. At these hardnesses, high speed steels, including the sulfurized types, are less likely to tear in shaving or back-off operations. Typical heat treating to achieve this result consists of heating to 855 to 870 °C (1575 to 1600 °F), holding for at least 1 h, quenching in oil, and tempering at 635 to 665 °C (1175 to 1225 °F) to obtain the desired hardness. If the austenitizing temperature does not exceed 870 °C (1600 °F), this treatment will not cause grain coarsening in the final hardening operation.

Certain machining operations, such as drilling and rough milling, should be performed in the annealed condition to obtain maximum tool life.

Tempering. As shown in Fig. 13 for an M2 steel austenitized at 1220 °C (2225 °F), the hardness of high speed steel is directly affected by tempering temperature and time. From the slope of the curves in Fig. 13, it can be seen that M2 undergoes secondary hardening at temperatures above approximately 370 °C (700 °F) and that sec-

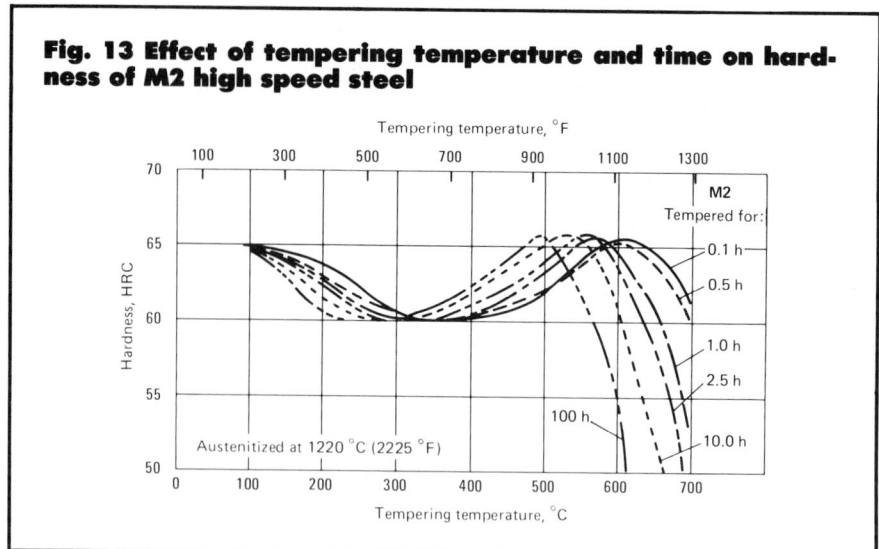

Fig. 13 Effect of tempering temperature and time on hardness of M2 high speed steel

Table 11 Effects of single and double tempering on mechanical properties of T1

Time at tempering temperature	Hardness, HRC	Bend strength		Torsion-impact strength	
		MPa	ksi	J	ft·lb
Single tempering at 565 °C (1050 °F)					
6 min	65.1	2150	312	22	16
1 h	65.7	1860	270	41	30
2½ h	65.0	2810	408	65	48
5 h	64.5	2590	376	65	48
Double tempering at 565 °C (1050 °F)					
2½ h + 2½ h	64.5	3130	454	85	63

ondary hardening proceeds at higher temperatures up to about 595 °C (1100 °F), depending on time at temperature. These temperatures approximate the practical limits for most tempering operations; lower temperatures do not evoke the secondary hardening response, and higher temperatures produce hardnesses considerably lower than are usually desired.

Emphasizing the practical time-temperature range, the response of several M and T types of high speed steel to tempering at 425 to 705 °C (800 to 1300 °F) for periods ranging from ½ to 10 h is indicated in Fig. 14.

The effect of austenitizing temperature on the tempering characteristics of several high speed steels tempered from 480 to 675 °C (900 to 1250 °F) is shown graphically in Fig. 15. For all of the steels for which data are plotted, the highest austenitizing temperature results in maximum solution of alloy carbides—which, during subsequent

tempering, produces the maximum response to secondary hardening.

High speed steels normally are subjected to a minimum of two separate tempering treatments within the range of 540 to 595 °C (1000 to 1100 °F). The duration of each treatment is usually 2 h or more at temperature. This process ensures attaining consistent martensitic structures, because the amount of retained austenite in the as-quenched condition will vary significantly because of variations in heat chemistry, prior thermal history, hardening temperature and quenching conditions.

It is essential that the time-temperature combination of the first tempering operation be adequate to condition the retained austenite. Consequently, the first tempering treatment is sometimes longer and at a slightly higher temperature than the second, because the latter is used to temper the freshly formed martensite that develops on cooling from the first

Fig. 14 Effect of tempering at temperatures from 425 to 705 °C (800 to 1300 °F) for periods of ½ to 10 h on hardness of high speed steels

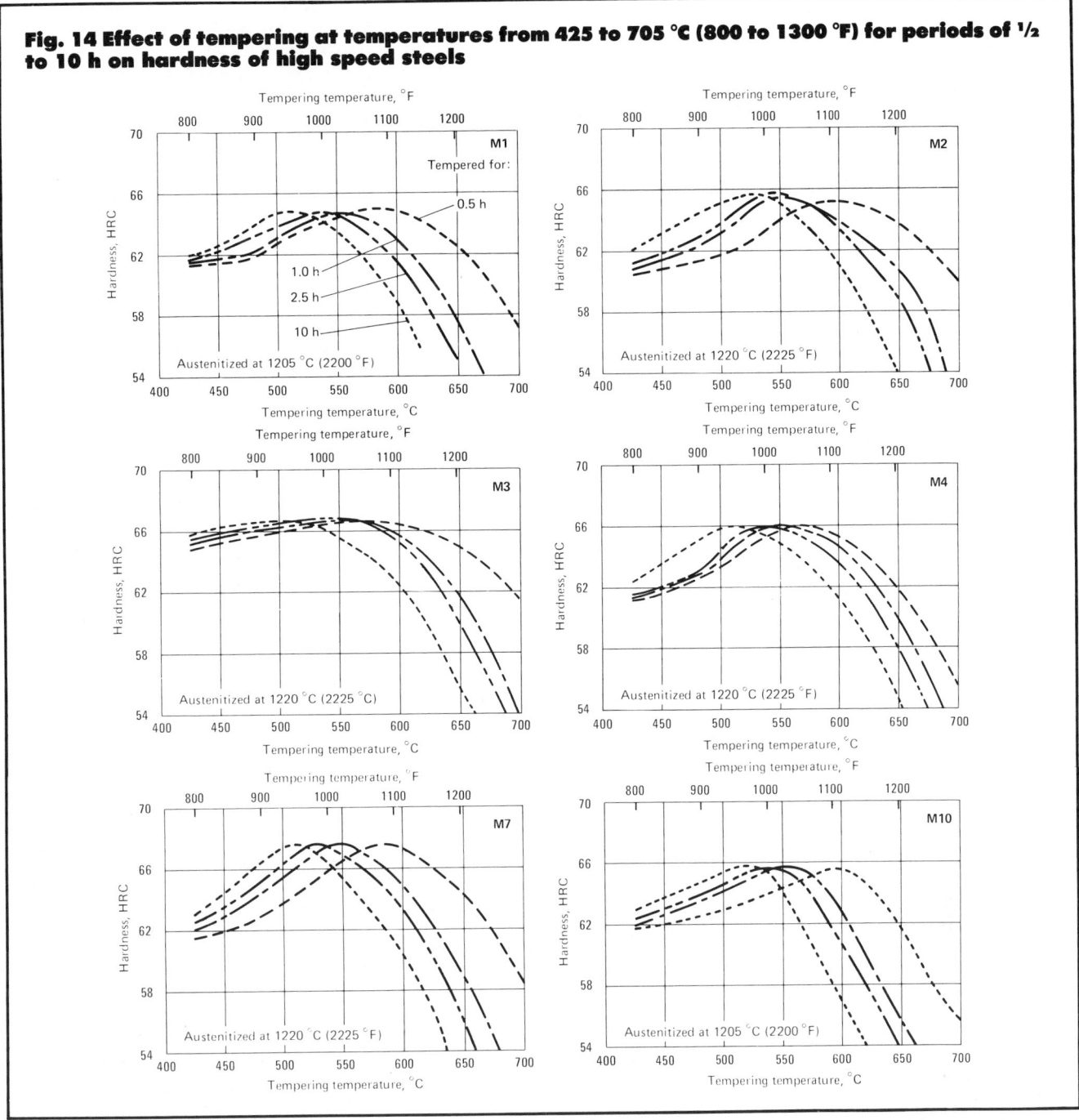

temper. Moreover, multiple tempering gains in importance in attaining an acceptable structure if short tempering times are used. The hardness of single and double tempered M2 steel austenitized at various temperatures, as affected by tempering temperature, is shown in Fig. 16.

Tempering at too low a temperature or for too short a time, or both, may not adequately condition the 20 to 30% retained austenite present after initial quenching, and the steel will still retain abnormally large quantities of austenite after cooling from the initial temper. This austenite will not transform until the steel is cooled from the second temper, and a third temper is then required to temper the martensite so formed. In order to carry these reactions as near to completion as possible, high speed steel should always be cooled to room temperature between tempers. The beneficial effect of multi-

ple tempering on mechanical properties of T1 high speed steel is shown in Table 11.

Forced-air furnaces are generally conceded to be the most desirable for tempering high speed steel, because the heat is transmitted from the heating elements to the work by convection; consequently, the transfer of heat is gradual, and there is little danger of the work cracking as the result of thermal shock. It is advisable to place the

Fig. 14 (continued)

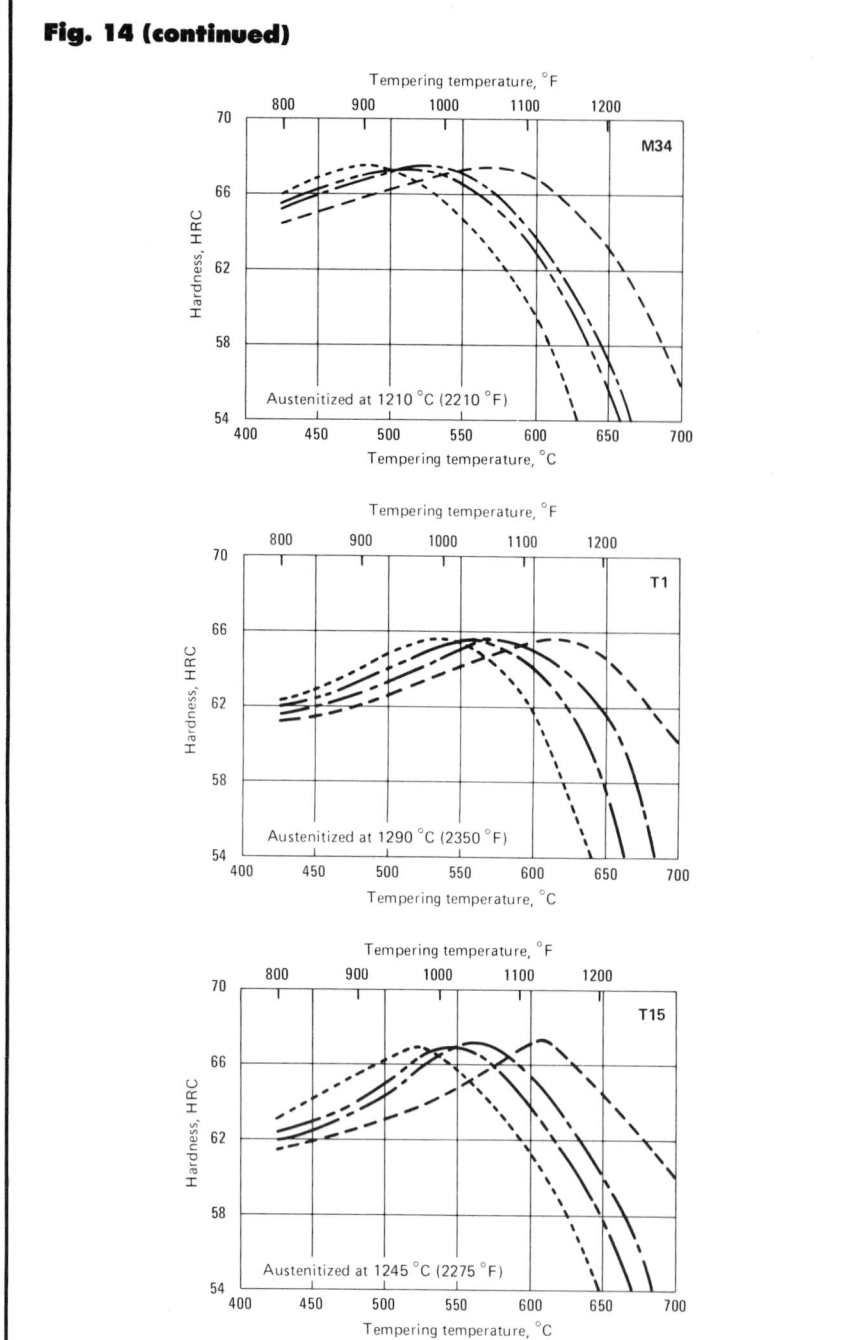

Table 12 Effect of nitriding time on surface nitrogen content of T1

Nitrogen content of first 0.025-mm (0.001-in.) layer

Time at 565 °C (1050 °F)	Nitrogen, %
3 min	0.06
10 min	0.093
30 min	0.15
90 min	0.26
3 h	0.58
6 h	1.09

Table 13 Carbon content of nitrided T1

Carbon content of the first 0.025-mm (0.001-in.) surface layer of steel originally containing 0.705% C. Some of the carbon was in pits on the surface, rather than diffused into the steel.

Nitriding Temperature °C	°F	Time, min	Surface carbon, %
455	850	30	0.85
510	950	30	0.99
565	1050	30	1.18
565	1050	360	1.63

work in a tempering chamber maintained in the temperature range of 205 to 260 °C (400 to 500 °F) and to bring the work up to the tempering temperature slowly with the furnace. This is particularly important for large or intricate tools, because too rapid a heating rate may lead to cracking.

The very rapid heating rates of molten lead or salt baths, and the attendant thermal shock, usually militate against their successful use for tempering high speed steel tools of other than simple shape and design, unless they are preheated to about 315 °C (600 °F) before being introduced into the bath.

Refrigeration treatment may be employed to transform retained austenite. The hardened or hardened and tempered tool is cooled to at least −85 °C (−120 °F) and then tempered or retempered at normal tempering temperatures. Carburized surfaces will respond satisfactorily to the −85 °C (−120 °F) treatment, even when they have been tempered prior to refrigeration.

Nitriding. Liquid nitriding is preferred to gas nitriding for high speed steel cutting tools because it is capable of producing a more ductile case with a lower nitrogen content.

Although any of the liquid nitriding baths or processes may be used to nitride high speed steel, the commercial bath consisting of 60 to 70% sodium salts and 30 to 40% potassium salts is most commonly employed. The nitriding cycle for high speed steel is of relatively short duration, seldom exceeding 1 h; in all other respects, however, the procedures and equipment are similar to those used for low-alloy steels.

The cyanide baths employed in liquid nitriding introduce both carbon and nitrogen into the surface layers of the nitrided case. Normally, the highest percentages of both elements are found in the first 0.025-mm (0.001-in.) surface layer. For carbon and nitrogen gradients, see the section on liquid nitriding.

Fig. 15 Effect of austenitizing and tempering temperatures on hardness of high speed steels

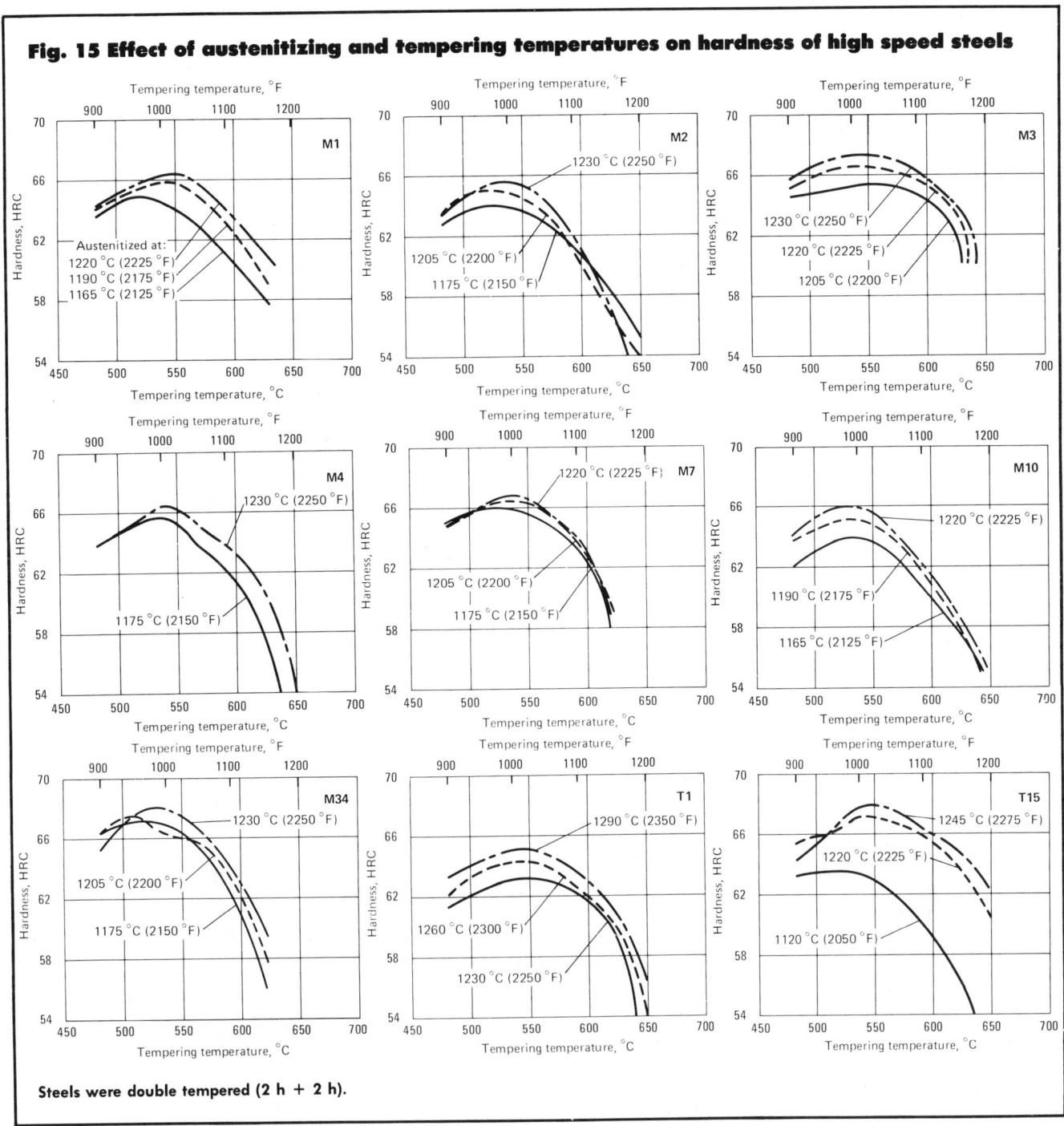

Steels were double tempered (2 h + 2 h).

The effect of time in a liquid nitriding bath at 565 °C (1050 °F) on the nitrogen content of the first 0.025-mm (0.001-in.) surface layer of a T1 high speed steel is shown in Table 12. A nitrogen content of 0.06% was obtained in the first 3 min at temperature, and it gradually increased to 1.09% at the end of a 6-h cycle at this temperature.

As shown in Table 13, carbon also was absorbed by the steel, at nitriding temperatures as low as 455 °C (850 °F). In a 30-min nitriding cycle, the carbon content of the first 0.025-mm (0.001-in.) surface layer increased with an increase in the nitriding temperature. However, it was reported that only a portion of the carbon was absorbed by the steel, most of the carbon being mechanically attached to the surface, filling microscopic pits. (This pitting is not dangerous under normal condi-

tions, because the pits are shallower than ordinary grinding or machining marks.)

High speed steel tools that are nitrided in fresh baths or for short times show steep nitrogen and hardness gradients. To avoid these steep gradients, which are believed responsible for the brittleness of the case after such treatments, the use of longer immersion time, higher temperature, or a thor-

Fig. 16 Effect of austenitizing temperature and tempering conditions on hardness of M2 speed steel

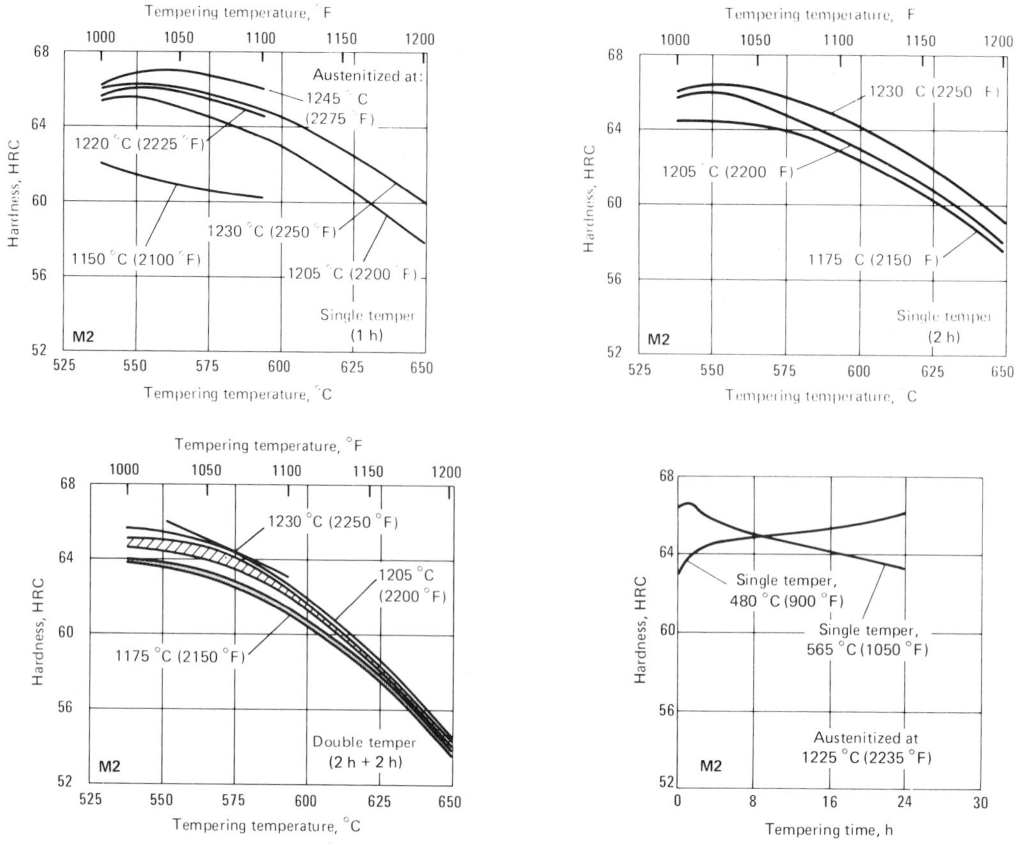

[Top left chart]
Tempering temperature, °F (1000, 1050, 1100, 1150, 1200)
Hardness, HRC (52–68)
Austenitized at:
1245 °C (2275 °F)
1220 °C (2225 °F)
1230 °C (2250 °F)
1150 °C (2100 °F)
1205 °C (2200 °F)
Single temper (1 h)
M2
Tempering temperature, °C (525, 550, 575, 600, 625, 650)

[Top right chart]
Tempering temperature, °F (1000, 1050, 1100, 1150, 1200)
Hardness, HRC (52–68)
1230 °C (2250 °F)
1205 °C (2200 °F)
1175 °C (2150 °F)
Single temper (2 h)
M2
Tempering temperature, °C (525, 550, 575, 600, 625, 650)

[Bottom left chart]
Tempering temperature, °F (1000, 1050, 1100, 1150, 1200)
Hardness, HRC (52–68)
1230 °C (2250 °F)
1205 °C (2200 °F)
1175 °C (2150 °F)
Double temper (2 h + 2 h)
M2
Tempering temperature, °C (525, 550, 575, 600, 625, 650)

[Bottom right chart]
Hardness, HRC (52–68)
Single temper, 480 °C (900 °F)
Single temper, 565 °C (1050 °F)
Austenitized at 1225 °C (2235 °F)
M2
Tempering time, h (0, 8, 16, 24, 30)

oughly aged bath is recommended. To avoid brittleness of case when relatively short immersion times are used, the cyanate content of the bath should exceed 6%. These conditions often will lower the surface hardness as well as the hardness gradient.

Figure 17 compares the hardness gradients obtained on specimens of T1 high speed steel nitrided at 565 °C (1050 °F) for 90 min in a new bath and for various lengths of time in an aged bath.

Nitriding of decarburized high speed steel tools should be avoided, because it results in a brittle surface condition. For those surfaces that have been softened from grinding, nitriding is frequently employed as an offsetting corrective measure.

Liquid nitriding provides high speed steel tools with high hardness and wear resistance and a low coefficient of friction. These properties enhance tool life in two somewhat related ways. The high hardness and wear resistance lower the abrading action of chips and work on the tool, and the low frictional characteristics serve to create less heat at and behind the tool point, in addition to assisting in the prevention of "chip pickup".

Steam treating produces a uniform, soft layer of iron oxide on the surface of finished high speed steel tools. This layer, approximately 0.005 mm (0.0002 in.) thick, has lubricant-retaining and antigalling properties, and in some applications will improve tool life by reducing tool-edge buildup. The oxide layer is removed from the tool after a short interval of operation; during this interval, the cutting surfaces of the tool develop a burnished surface that adds further to antigalling characteristics.

Steam treatment requires a special furnace with a sealed retort from which all air can be displaced by steam, which is admitted at controlled rates. The

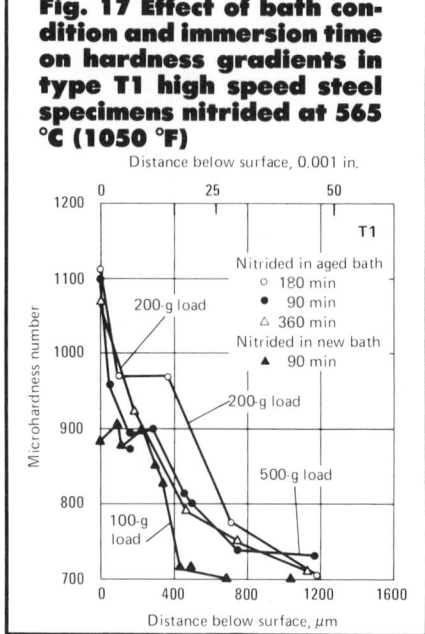

Fig. 17 Effect of bath condition and immersion time on hardness gradients in type T1 high speed steel specimens nitrided at 565 °C (1050 °F)

Distance below surface, 0.001 in. (0, 25, 50)
Microhardness number (700–1200)
T1
Nitrided in aged bath
○ 180 min
● 90 min
△ 360 min
Nitrided in new bath
▲ 90 min
200-g load
100-g load
500-g load
Distance below surface, µm (0, 400, 800, 1200, 1600)

Fig. 18 TTT diagram for M50 steel

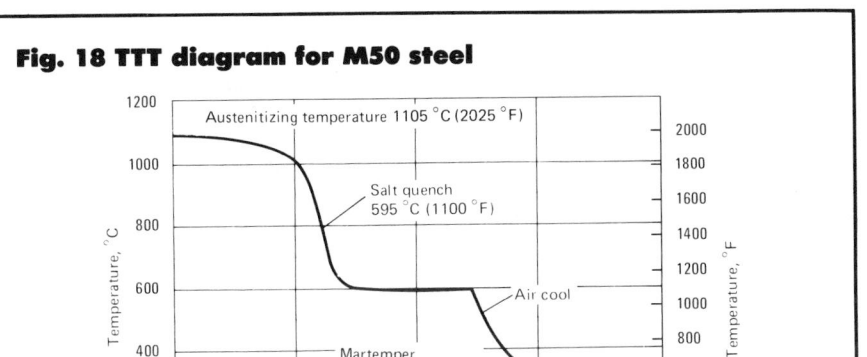

presence of any moisture in the furnace prior to the admission of the steam will cause rusting and an unsatisfactory surface finish.

A typical processing cycle involves placing the work in the special furnace, heating to approximately 370 °C (700 °F) and equalizing. After a suitable equalizing time, which depends on the load, the steam is admitted at controlled rates for approximately ½ h. The furnace is then partly sealed to develop positive steam pressure, and the temperature is raised to 525 °C (975 °F). The steam can then be shut off and the work removed from the furnace and cooled normally.

The treatment produces a blue-black film whose appearance is improved by subsequent dipping in oil. This treatment may sometimes be combined with normal tempering treatments, because the type of film produced is relatively insensitive to temperature up to approximately 580 °C (1075 °F). Steam treating offers an additional advantage for tools hardened in salt baths, because it effectively reduces the pitting that can result from adhering salt.

Carburizing is not recommended for high speed steel cutting tools because of the extreme brittleness of the case so produced. However, it is suitable for applications requiring extreme wear resistance in the absence of impact or highly concentrated loading, such as are encountered with certain types of cold work dies made from high speed steel. At the same level of hardness, the carburized layer does not have the heat resistance of normal high speed steel because carbides in the

microstructure are predominantly Fe_3C, rather than the complex alloy carbides characteristic of high speed steel.

Carburizing cycles for high speed steel consist of packing in a carburizing medium, heating to approximately 1040 to 1065 °C (1900 to 1950 °F) long enough to develop the depth of case desired, and air cooling. The usual holding time at carburizing temperature is from 10 to 60 min, to produce a case 0.05 to 0.25 mm (0.002 to 0.010 in.) deep. Deeper cases should be avoided because of the extreme brittleness developed. This treatment carburizes the surface and serves as the austenitizing treatment for hardening the entire piece. The carburized layer will harden to 65 to 70 HRC at the surface.

Hardening of Specific High Speed Steel Tools

Broaches require maximum edge hardness because of the continuous cutting action and light chip load to which they are subjected. This indicates a minimum hardness of 65 HRC for the standard grades and 66 HRC for the premium grades of high speed steel.

Broaches should be suspended vertically in the hardening furnace to avoid undue distortion, and should be quenched under controlled and uniform cooling conditions. Broaches should be straightened while still warm from the hardening operation, and should be cooled to at least 65 °C (150 °F) before tempering. These precautions are particularly important for large diameters.

Chasers, because they usually are quite small, present no particular problem in hardening with regard to straightness or residual stress. Hardness recommendations for chasers depend largely on the type of application and the pitch of the thread. For cutting steel, the following hardnesses are commonly used:

Fine-pitch threads .... 61 to 63 HRC
Coarse-pitch threads .. 64 to 65 HRC
Acme threads .... 60 to 62 HRC

For cutting cast iron or plastics, chasers should be heat treated to the maximum attainable hardness, because these materials are cut without any significant cutting force but require maximum abrasion resistance. For Acme threads, however, it is sometimes advisable to underharden.

Milling Cutters. Fine-tooth cutters and those with fragile forms should be hardened to 63 to 64 HRC. Heavy-duty milling cutters and cutters for use on soft, abrasive materials should be hardened to the maximum hardness obtainable for the particular type of steel.

Drills. Hardening techniques for drills vary, depending on the diameter of the drill. Straightness of these tools is extremely important. Various jigging methods are employed, but it is usually advisable to heat treat drills vertically suspended by their shanks in order to reduce distortion in the hardening operation. Straightening is best accomplished in the as-hardened condition before tempering. In tempering, the tempering furnace must not be overloaded, and all drills must receive the correct tempering temperature and time at temperature.

Specific recommendations for the hardness of drills for cutting steel are as follows:

- Most drills 5 mm (³/₁₆ in.) in diameter and smaller are usually hardened to 63 to 65 HRC. (Drills of this size used for plastics, aluminum or magnesium may have hardness as high as 65 HRC.)

- Drills over 5 mm (³/₁₆ in.) in diameter, to 63 to 65 HRC

- Heavy-duty drills normally use grades of high speed steel providing hardnesses equal to or higher than those noted above. (These drills generally are designed for maximum rigidity and require maximum abrasion resistance.)

Table 14 Recommended heat treating practice for low-alloy special-purpose tool steels

Steel	Normalizing temperature(a)		Annealing Temperature(b)		Annealing Cooling rate(c)		Annealed hardness, HB	Austenitizing temperature(d)		Hardening Holding time, min	Quenching medium	Quenched hardness, HRC(e)
	°C	°F	°C	°F	°C/h	°F/h		°C	°F			
L1	900	1650	775-800	1425-1475	22	40	179-207	790-845	1450-1550	10-30	O, W	64
L2	870-900	1600-1650	760-790	1400-1450	22	40	163-197	790-845	1450-1550	10-30	W	63
L3	900	1650	790-815	1450-1500	22	40	174-201	845-925	1550-1700	10-30	O	63
								775-815	1425-1500	10-30	W	64
								815-870	1500-1600	10-30	O	64
L6	870	1600	760-790	1400-1450	22	40	183-212	790-845	1450-1550	10-30	O	62
L7	900	1650	790-815	1450-1500	22	40	183-212	815-870	1500-1600	10-30	O	64

Note: O, oil; W, water. (a) Holding time, after uniform through heating, varies from about 15 min, for small sections, to about 1 h, for large sections. Work is cooled from temperature in still air. (b) Lower limit of range should be used for small sections, upper limit for large sections. Holding time varies from about 1 h, for light sections and small furnace charges, to about 4 h, for heavy sections and large charges; for pack annealing, hold for 1 h per inch of pack cross section. (c) Maximum. Rate is not critical after cooling to below 540 °C (1000 °F). (d) These steels are seldom preheated. (e) Typical average values; subject to variations depending on austenitizing temperature and quenching medium.

Taps, like drills, are slender in section and require hardening techniques that minimize distortion. This generally means hardening in the vertical position suspended in suitable jigs. Taps should be straightened in the as-hardened condition before tempering. Tempering of these tools must be carefully controlled to allow adequate heating time. Specific hardness recommendations for taps that are to be used to cut steel are as follows:

Fine-pitch taps .... 63 to 65 HRC
Coarse-pitch taps ..... 63 to 65 HRC
Acme taps .... 62 to 64 HRC
Pipe taps .... 62 to 64 HRC

Reamers encounter a minimum chip load but require maximum wear resistance. For this reason, they are always hardened to the maximum hardness attainable for each grade of steel.

Form tools of all types also should have maximum hardness. In general, a minimum of 65 HRC is necessary, and for the premium grades hardnesses ranging from 68 to 70 HRC are frequently desirable.

Hobs. Because of their shaving action, hobs require maximum edge hardness. They may become oval in shape if they are not placed in the hardening furnace in the vertical position. Such placement may require special fixtures. Techniques and temperatures in both hardening and tempering must be accurately controlled if tools of this type are to be produced successfully and economically.

The hardness of fragile tooth forms may have to be reduced to 62 to 64 HRC to avoid breakage, although the lower

Fig. 19 Hardness of low-alloy special-purpose tool steels after tempering for 2 h

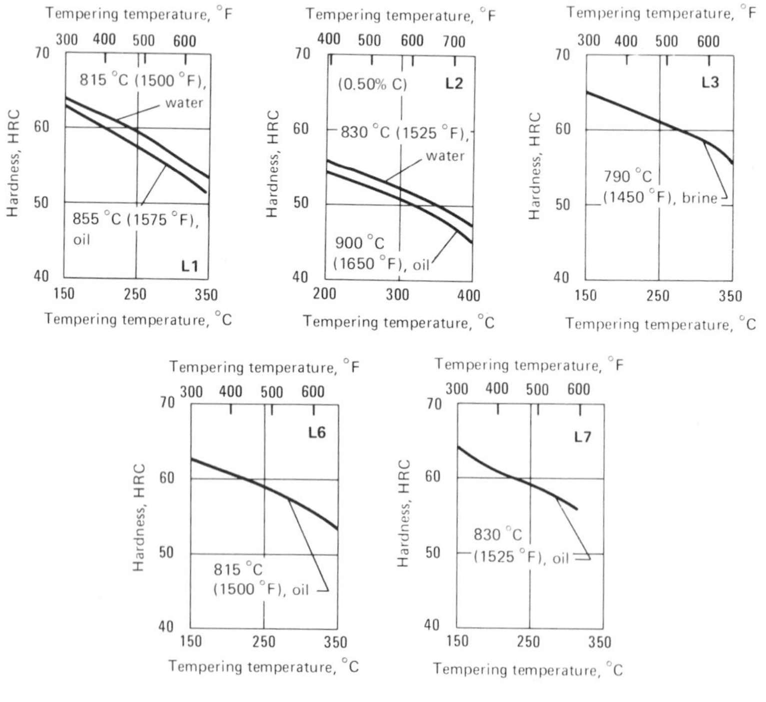

hardness results in a shorter production life.

Thread rolling dies are usually made of A2 or D2 steel, although dies made of high speed steel frequently afford superior results, particularly in rolling the harder materials. For fragile thread forms, thread rolls should be hardened to 60 to 62 HRC. For heavier thread forms and those used to roll high-strength materials, hardnesses of 63 to 65 HRC are recommended; however, at these higher hardnesses, dies are more susceptible to breakage.

Threading Dies. Most threading dies are made of carbon steel; however, button and acorn dies justify the use of high speed steel. The relation between hardness and thread form for threading dies is the same as that recommended for taps and chasers.

Tool Bits. Standard tool bits, as well

Table 15 Recommended heat treating practice for carbon-tungsten special-purpose tool steels

Steel	Normalizing temperature(a) °C	°F	Annealing Temperature(b) °C	°F	Cooling rate(c) °C/h	°F/h	Annealed hardness, HB	Preheat °C	°F	Hardening Temperature Austenitize °C	°F	Holding time, min	Quenching medium	Quenched hardness, HRC(d)
F1......	900	1650	760-800	1400-1475	20	40	183-207	650	1200	790-870	1450-1600	15	W, B	64
F2......	900	1650	790-815	1450-1500	20	40	207-235	650	1200	790-870	1450-1600	15	W, B	66
F3......	900	1650	790-815	1450-1500	20	40	212-248	650	1200	790-870	1450-1600	15	W, B, O	66

Note: W, water; B, brine; O, oil. (a) Holding time, after uniform through heating, varies from about 15 min, for small sections, to about 1 h, for large sections. Work is cooled from temperature in still air. (b) Lower limit of range should be used for small sections, upper limit for large sections. Holding time varies from about 1 h, for light sections and small furnace charges, to about 4 h, for heavy sections and large charges; for pack annealing, hold for 1 h per inch of pack cross section. (c) Maximum cooling rate. Rate is not critical after steel has been cooled to below 540 °C (1000 °F). (d) Typical average hardness values; subject to variations depending on austenitizing temperature and quenching medium employed.

as cheeking tools, offset-head bits and other special types, all require maximum hardness. Standard-duty tool bits should be hardened to 65 to 66 HRC, whereas tool bits made from the higher-alloy high speed steels should be hardened to 67 to 69 HRC when possible.

Bearing Components. The heat treatment of M50 high speed steel bearing components for aerospace applications must be capable of producing a part with high hardness, uniformly fine grain size and dimensional stability over a wide temperature range.

M50 steel has a nominal composition of 0.83C-4.0Cr-4.0Mo-1.0V with a M_s temperature of approximately 163 to 165 °C (325 to 330 °F). The time-temperature transformation (TTT) diagram for M50 is illustrated in Fig. 18.

Virtually any cooling rate capable of cooling the austenitized part to 205 °C (400 °F) or below in 15 min will produce high hardness. To minimize distortion, residual stress and crack susceptibility, a cooling similar to the "idealized" rate shown in Fig. 18 is desirable.

The following practices and procedures are recommended for heat treating M50 bearing components to provide optimum bearing properties:

- M50 can be satisfactorily heat treated in vacuum or protective atmosphere furnace. However, most bearing manufacturers prefer to heat treat these bearing components in a neutral molten salt bath or baths.
- Parts should be preheated prior to the austenitizing cycle to minimize the required soak time at the high austenitizing temperature. If a single preheat is employed, a bath temperature of 815 to 870 °C (1500 to 1600 °F) with a cycle of 5 to 15 min is recommended. If multiple preheat baths are available, recommended bath temperatures and cycles are as follows:

Cycles	Temperature °C	°F	Time(a), min
Two preheat baths			
1	675-730	1250-1350	10-15
2	815-870	1500-1600	5-15
Three preheat baths			
1	675-730	1250-1350	10-15
2	815-870	1500-1600	5-15
3	955-1010	1750-1850	5-10

(a) Time predicated on relative load size/bath capacity

- The high-temperature bath cycle is the most critical operation in heat treating M50 steel. Following preheating, parts should be austenitized at 1105 to 1120 °C (2025 to 2050 °F) for 3 to 10 min, depending on cross section and gross load weight. Optimum cycles in the austenitizing bath may be established empirically by varying the soak cycle in the high temperature bath in ½ min increments and evaluating resultant grain size and hardness. Grain size is more easily measured on as-quenched samples; however, hardness should be checked on parts subsequent to final tempering operations. Ideally, the cycle will be as short as possible to minimize grain growth while producing desired hardness.
- Following austenitizing, parts should be quenched in 540 to 595 °C (1000 to 1100 °F) molten salt for 5 to 10 min. The quench minimizes internal stresses and the core-to-surface thermal differential prior to subsequent air cooling and martempering operations.
- Parts should be subjected to a 175 to 190 °C (350 to 375 °F) martemper bath for 5 to 15 min following quench or quench/air cool operations. The martemper bath, which operates between 15 and 30 °C (25 and 50 °F) above the M_s temperature for M50,

equalizes core-to-surface thermal differentials and facilitates subsequent transformation of austenite into martensite with minimal residual stress, distortion or cracking potential. To avoid undesirable intermediate transformation products, the interval between austenitizing and martempering should not exceed 15 min.

- Following martempering, parts should be air cooled to room temperature prior to washing, tempering or subzero treatment. The air-cooling equipment and conditions should provide uniform cooling of parts from the 175 to 190 °C (350 to 375 °F) martempering bath to room temperature within 30 to 60 min. Shorter cooling rates may result in increased residual stress, distortion or susceptibility to stress cracking.
- M50 steel requires multiple tempers to provide maximum toughness and dimensional stability. Parts should be subjected to a minimum of three tempers of 540 to 550 °C (1000 to 1025 °F) for 2 to 4 h, with cooling to room temperature between each temper. Failure to cool to below 55 °C (100 °F) between tempers may result in retained austenite. Tempering may be performed either in neutral molten salts or in atmosphere or air furnaces.
- Subjection to subzero temperatures prior to and/or after initial tempering enhances transformation of retained austenite to martensite. Common deep-freeze cycles for M50 are −70 to −85 °C (−90 to −120 °F) for 2 to 4 h. Use of lower temperatures provides little if any added benefit. The deep-freeze cycle provides maximum benefit when employed before tempering; however, it is not recommended for parts not subjected to martempering or parts susceptible to cracking. When parts are subzero treated before tempering, caution

should be exercised to ensure that the total elapsed time between mar-tempering and tempering does not exceed 5 h. Use of prior stress-relief cycles reduces effectiveness of deep-freeze operation. When equipment, time constraints or part design are unfavorable for performing deep freezing prior to tempering, the parts should be subjected to deep freeze between the first and second temper-ing operations.

- Parts requiring re-treating should be annealed prior to rehardening to minimize susceptibility to develop-ing duplex/nonuniform grain.

Low-Alloy Special-Purpose Tool Steels

Nominal compositions of the low-alloy special-purpose tool steels are given in Table 1 of the article entitled "Heat Treating of Tool Steels". These steels are similar in composition to the water-hardening tool steels, except that the addition of chromium and other elements provides the L steels with greater wear resistance and hard-enability. Types L1, L3, L4 and L7 are similar to the production steel 52100 and are used for similar applications.

Because of their relatively low aus-tenitizing temperatures, the L steels are easily heat treated. Recommended heat treating practice is summarized in Table 14.

Normalizing should follow forging or any other operation in which the steel has been exposed to temperatures substantially above the transformation range. For the L steels, normalizing consists of through heating to 870 to 900 °C (1600 to 1650 °F) (Table 14) and cooling in still air. The use of a protec-tive atmosphere is recommended.

Annealing must follow normalizing and precede any rehardening opera-tion. Recommended annealing temper-atures and cooling rates, as well as expected as-annealed hardness values, are given in Table 14.

Stress relieving prior to hardening may be advantageous for complex tools, to minimize distortion during harden-ing. A common practice for complex tools is to rough machine, heat to 620 to 650 °C (1150 to 1200 °F) for 1 h per inch of cross section, cool in air, and then finish machine prior to hardening.

Austenitizing temperatures recom-mended for hardening the L steels are listed in Table 14; preheating is seldom employed for steels in this group.

Salt or lead baths and atmosphere furnaces are all satisfactory for austen-itizing these steels. A neutral salt, such as No. 3 in Table 2 of the article enti-tled "Furnace Processes and Equip-ment for Heat Treating of Tool Steels", is recommended. This salt may be deox-idized, for control of decarburization, by the method indicated earlier in this article in the section on rectification of salt baths.

Quenching. Oil is the quenching medium most commonly used for the L steels. Water or brine may be used for simple shapes, or for large sections that do not attain full hardness by oil quenching. Rolling-mill rolls made of L7 are an example of parts for which water or brine quenching is used. These steels respond well to martempering.

Tempering. Tools made of the L steels should be quenched only to a temperature at which they can be han-dled with bare hands, about 50 °C (125 °F), and should be tempered immedi-ately thereafter; otherwise, cracking is likely to occur.

The tempering characteristics of these steels are plotted in Fig. 19. For most applications, the S steels are used at near-maximum hardness. It is rec-ommended that tools made of any of these low-alloy steels be tempered at a minimum of 120 °C (250 °F), even though maximum hardness is desired. Double tempering also is recommend-ed.

Carbon-Tungsten Special-Purpose Tool Steels

Nominal compositions of carbon-tungsten special-purpose tool steels are given in Table 1 of the article entitled "Heat Treating of Tool Steels". Recom-mended heat treating practice for these steels is summarized in Table 15.

As a group, these steels are shallow hardening and usually are quenched in water or brine. Steel F3, because of the chromium addition, is the highest in hardenability.

Normalizing and Annealing. These steels should be normalized after they have been forged or otherwise sub-jected to temperatures above their hardening temperatures. Normalizing and annealing practices are essentially the same as those recommended in the preceding section on the heat treating of the low-alloy special-purpose steels. Recommendations for normalizing and

annealing the F steels are given in Table 15.

Stress relieving as outlined previ-ously for the low-alloy special-purpose steels may be advantageously applied also to the F steels. The same procedure as that described for the L steels would be used.

Austenitizing. Preheating and aus-tenitizing temperatures recommended for the carbon-tungsten special-pur-pose tool steels are given in Table 15. Equipment and practice are gen-erally the same as those previously described for the low-alloy special-purpose steels.

Quenching. Water or brine quench-ing causes high distortion in parts made of the F steels. This is often used to advantage in the rehardening of worn dies that have been used for cold drawing of bars and tubes. Such dies are "flush quenched"—that is, a spout of water is directed into the bore, thus causing shrinkage and allowing fur-ther use of dies for the same product size.

Tempering. Because tools made of the F steels (cold drawing dies, for example) are used mainly for applica-tions requiring wear resistance, they are usually placed in service at or near their maximum hardness. Therefore, tempering temperatures higher than 205 °C (400 °F) are seldom used. Tem-pering temperature versus hardness for the F steels is shown graphically in Fig. 20.

Mold Steels

The principal use of these type P steels is for plastic molds. However, some steels, such as P4, P20 and P21, are used also for die-casting dies. The several types vary widely in composi-tion (Table 1 of the article entitled "Heat Treating of Tool Steels"), from the unalloyed "hubbing iron" P1, to P4, P6 and P21, which contain over 5% total alloying elements.

The wide variations in composition, method of forming the mold cavity, molding method, and material to be molded are major influences on choice of mold material as well as method of heat treating. The two most common methods of heat treating the mold steels are (a) preharden the steel (or partially machined mold or die) to about 30 to 36 HRC, finish machine, and use at this hardness level; and (b) case harden by carburizing. Nitrided molds have proved successful in some instances, but nitriding is not used extensively.

When molds are carburized or nitrided, the same procedures are used as for production steels.

Heat treating practice for the mold steels is summarized in Table 16. P21 is a special type heat treated by the manufacturer and delivered ready for the user to machine and place in operation without further treatment. As noted in Table 16, this steel is hardened by solution treating and aging.

Annealing temperatures and expected resulting hardness values are indicated in Table 16. For some types, such as P1, the annealing temperature is not critical. A more important factor

is surface protection, especially if the mold cavities will be formed by hubbing. If surfaces are allowed to carburize, even slightly, during annealing, subsequent hubbing will be impaired.

Usually, parts are packed in an inert material such as spent pitch coke and are held at annealing temperature only long enough to become heated through; they are then cooled in the pack to below 540 °C (1000 °F), after which they may be removed from the pack. If hubbing is to follow, it is usually preferable to use the lower side of the annealing temperature range to minimize the danger of carburizing, even though annealing at the higher side of the range will result in slightly lower

hardness. Atmosphere-controlled furnaces that can be programmed for slow cooling can also be used for annealing. For hubbing deep cavities, two or more in-process anneals are sometimes required.

When cavities will be formed entirely by machining (sometimes a combination of hubbing and machining is used), annealing usually is neither necessary nor desirable, because slightly harder structures can be machined more easily. Steels as received from the manufacturer are usually suitable for machining. If hardened molds require reworking, they can be annealed as recommended in Table 16.

Variations in heat treatment, as necessitated by differences in composition, properties and intended use, are discussed in the following sections for steels P1 to P20.

P1 steel, although shown in Table 1 of the article entitled "Heat Treating of Tool Steels" as containing no alloying elements, may contain about 0.10% V, which promotes a finer grain after carburizing, with no apparent sacrifice in hubbability. This steel usually is used only for hubbed molds for injection molding of general-purpose plastics.

P1 steel can be carburized by any of the regular practices. Whether the steel is reheated to the austenitizing temperature or quenched from a programmed furnace depends on equipment used. Full hardness (Table 16) can be achieved only by water or brine quenching. Practice varies as to working hardness range.

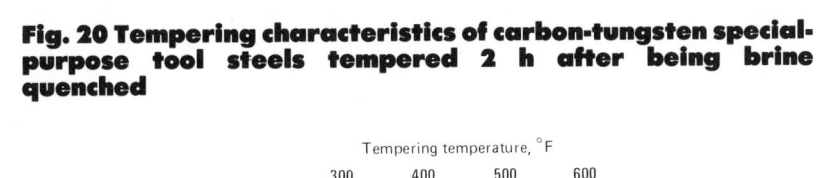

Fig. 20 Tempering characteristics of carbon-tungsten special-purpose tool steels tempered 2 h after being brine quenched

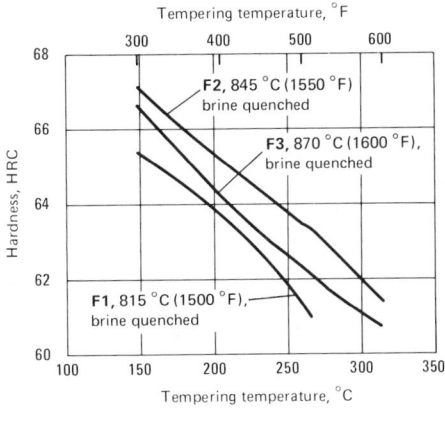

Table 16 Recommended heat treating practice for mold steels

Steel	Normalizing temperature(a), °C	Temperature(b) °C	Annealing Cooling rate(c) °C/h	Annealed hardness, HB	Carburizing temperature °C	Hardening (after carburizing) Austenitizing temperature, °C	Holding time, min	Quenching medium	Quenched hardness, HRC
P1	Not required	730-900	20	81-101	900-925	790-800	15	W, B	62-64
P2	Not required	730-815	20	103-123	900-925	830-845	15	O	62-65
P3	Not required	730-815	20	109-137	900-925	800-830	15	O	62-64
P4	Not recommended	870-900	15	116-128	970-995	970-995	15	A	62-65
P5	Not required	845-870	20	105-116	900-925	845-870	15	O, W	62-65
P6	Not required	845	8	183-217	900-925	790-815	15	A, O	60-62
P20	900	760-790	20	149-179	870-900(d)	815-870	15	O	58-64
P21	900	Not recommended			Hardened by solution treating and aging(e)				

Note: W, water; B, brine; O, oil; A, air. (a) Holding time, after uniform through heating, varies from about 15 min, for small sections, to about 1 h, for large sections. Work is cooled from temperature in still air. (b) Lower limit of range should be used for small sections, upper limit for large sections. Holding time varies from about 1 h, for light sections and small furnace charges, to about 4 h, for heavy sections and large charges; for pack annealing, hold for 1 h per inch of pack cross section. (c) Maximum. Rate is not critical after cooling to below 540 °C (1000 °F). (d) When applicable. (e) Solution treatment: Hold at 705 to 730 °C (1300 to 1350 °F) for 1 to 3 h, quench in air or oil; approximate solution treated hardness, 24 to 28 HRC. Aging treatment: Reheat to 510 to 550 °C (950 to 1025 °F); approximate aged hardness, 40 to 30 HRC.

Fig. 21 Tempering characteristics of carburized mold steels

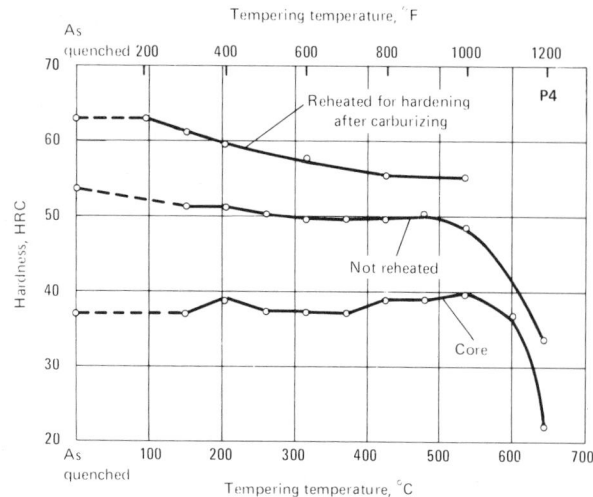
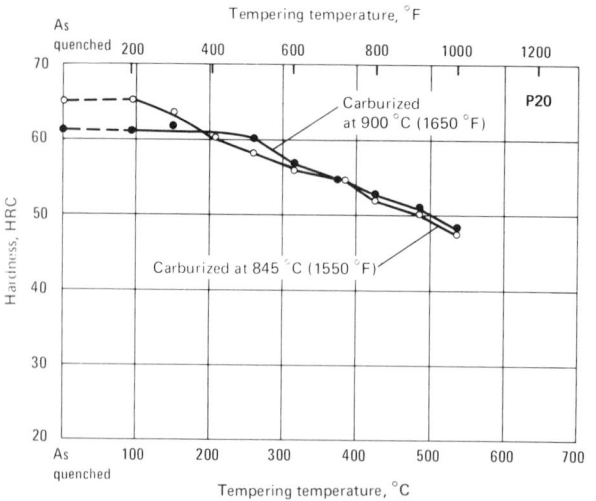

(a) Upper curve represents steel carburized in hardwood charcoal at 915 to 925 °C (1675 to 1700 °F) for 8 h, air cooled in pack, reheated at 940 to 955 °C (1725 to 1750 °F), cooled in air and tempered. Middle curve represents steel carburized in cast iron chips at 940 to 955 °C (1725 to 1750 °F), removed from pack, cooled in air and tempered. (b) Surface hardness after heating at temperature for 2 h in carburizing compound, oil quenching, and tempering

A minimum tempering temperature of 175 °C (350 °F) is recommended. This will retain a finished surface hardness of 60 HRC or slightly higher. However, a more commonly desired hardness range is 54 to 58 HRC, which is obtained by tempering at 260 to 315 °C (500 to 600 °F). If the distortion encountered from water quenching cannot be tolerated for a particular mold design, a type of mold steel that can be hardened by oil quenching must be used instead of P1.

P2 steel also is a hubbing steel, although it is less easily hubbed than P1. Carburizing and hardening practice and the working hardness range are the same as for P1, except that the alloy content of P2 increases hardenability so that full hardness can usually be obtained by oil quenching, thus minimizing distortion.

P3 steel is also hubbed, but it is less easily hubbed than P1 or P2. Except that P3 is usually oil quenched, the carburizing and hardening practice for it is essentially the same as that outlined above for P1. The operating hardness range may vary from 54 to 64 HRC, but common practice is to temper at about 315 °C (600 °F) to achieve a final hardness of 54 to 58 HRC.

P4 steel is sometimes used hubbed, but because of its resistance to cold deformation it is more often used for machined molds or dies. Of all the steels in this group, P4 is the most resistant to wear and to softening by tempering. Because of these properties, it is commonly used for injection molding of plastics that require high curing temperatures and for dies used for die casting low-melting alloys. For the latter application, a common practice is to carburize P4 in cast iron chips to obtain a slight increase in carbon content at the surface. The effect of carburizing practice, as well as case and core hardness values after tempering, is shown in Fig. 21.

Because of its high alloy content, P4 steel can be hardened by air cooling. However, it is sometimes quenched in oil to minimize scaling during cooling. For use in plastic molds, the most common working range is 56 to 60 HRC, which may be obtained by tempering the carburized and hardened molds at 205 to 315 °C (400 to 600 °F) (Fig. 21).

P5 steel, in which chromium is the major alloying element, approaches P1 in ease of hubbing and has a core strength equivalent to that of P3. After carburizing, a surface hardness of 65 HRC can be achieved by water quenching, or slightly lower values by oil quenching. Choice of quenching medium depends on mold configuration, allowable distortion, and required hardness. A common working range is 54 to 58 HRC; this can be obtained by tempering at about 260 °C (500 °F).

P6 steel, because it can seldom be annealed to a hardness of less than 183 HB (Table 16), is difficult to hub, and hence it is usually used for machine-cut cavities. It can be carburized by conventional practice. Because of its hardenability, heavy sections of P6 can be oil quenched to full hardness from 790 to 815 °C (1450 to 1500 °F). The as-quenched surface hardness is not quite so high as for some other types, because the high nickel content of P6 promotes retention of austenite. Some of this retained austenite is transformed in tempering, with the result that after tempering up to about 260 °C (500 °F) the hardness will be little or no lower than that obtained after quenching. By tempering at 315 °C (600 °F), the most common working hardness range (54 to 58 HRC) is obtained. In some plants, a working hardness range of 58 to 61 HRC, obtained by tempering at 260 °C (500 °F), is considered preferable.

P20 steel is a popular mold material for either injection or compression molding, and also for die casting low-melting alloys.

For injection molding of the general-purpose plastics or die casting of low-melting alloys, P20 is usually used in the prehardened condition. It is available at hardness levels of about 300 HB or slightly higher. In this condition, cavities are machined and the dies or molds placed in service without further heat treatment. Annealed molds or dies can be austenitized at 845 to 870 °C (1550 to 1600 °F), oil quenched, and tempered at 540 °C (1000 °F), to obtain a hardness of about 300 HB.

Type P20 is often carburized for molds used in compression molding, particularly for molding the more abrasive plastics. Carburizing temperatures no higher than 900 °C (1650 °F) are recommended for this steel, because higher temperatures may impair polishability; otherwise, conventional carburizing practice is used, and molds may be quenched in oil directly from the carburizing temperature. A common working range is 54 to 58 HRC.

Tempering characteristics for P20 carburized at two different temperatures are given in Fig. 21.

This steel is sometimes nitrided for special applications. Conventional nitriding practice is employed. Before being nitrided, P20 should first be quenched and tempered to about 300 HB as outlined above, and cavities should be machined; following this sequence will ensure freedom from carburization or decarburization.

Control of Distortion in Tool Steels*

Edited by Daniel S. Zamborsky
Corporate Metallurgist
Warner & Swasey Co.

DISTORTION in tool steel parts includes all irreversible changes in size and shape that result from processing, from heat treatment, and from temperature variations and loading in service. A basic understanding of distortion is important for two reasons. First, most tool steel parts must interact with other parts in service, and excessive distortion may prevent them from interacting in the intended manner. Second, finishing operations for correcting distortion not only are expensive but also may destroy some desirable properties and introduce others that are undesirable.

Changes in size or shape of tool steel parts may be either reversible or irreversible. Reversible changes are those caused by stressing in the elastic range or by temperature variations that neither cause changes in the metallurgical structure nor induce stresses that exceed the elastic range. Under such conditions, the initial dimensional values can be restored by a return to the original state of stress or temperature.

The upper limit of reversible dimensional change in a tool steel is determined by the stress required to initiate deformation (that is, the elastic limit corresponding to a preselected value of plastic strain), the elastic deformation per unit stress (modulus of elasticity), the effect of temperature on these properties, the coefficient of thermal expansion and the temperature-time combinations at which stress relief and phase changes occur.

For practical purposes the modulus of elasticity of all tool steels, regardless of composition or heat treatment, is 210 GPa (30×10^6 psi) at room temperature. Therefore, if a tool steel part deforms excessively under service loading but returns to its original dimensions when the load is removed, a change in grade or type of tool steel or in heat treatment will not be useful. To counteract excessive elastic distortion it is necessary to (a) reduce the applied stress by increasing the section size or (b) use a tool material with a higher modulus of elasticity (such as cemented tungsten carbide).

Irreversible changes in size or shape of tool steel parts are those caused by stresses that exceed the elastic limit or by changes in metallurgical structure (most notably, phase changes). Such irreversible changes sometimes can be corrected by thermal processing (annealing, tempering or cold treating) or by mechanical processing to remove excess material or to redistribute residual stresses.

Nature and Causes of Distortion

Distortion is a general term encompassing all irreversible dimensional changes. There are two main types: size distortion, which involves expansion or contraction in volume or linear dimensions without changes in geometrical form; and shape distortion, which entails changes in curvature or angular relations, as in twisting, bending, and/or nonsymmetrical changes in dimensions. Frequently, both size distortion and shape distortion (illustrated schematically in Fig. 1) occur during a heat treating operation.

Size distortion is the result of a change in volume produced by a change in metallurgical structure during heat

*This article has been condensed from the more complete discussion of this complicated subject provided by Bernard S. Lement in his book Distortion in Tool Steels (Ref 1).

treatment. Shape distortion results from either residual or applied stresses. Residual stresses developed during heat treatment are caused by thermal gradients within the metal (producing differing amounts of expansion or contraction) by nonuniform changes in metallurgical structure and by nonuniformity in the composition of the metal itself, such as that due to segregation.

Changes in metallurgical structure during heat treatment of tool steels are produced by the three steps described below.

The first step involves heating an annealed structure (usually consisting of ferrite and spheroidal carbides, commonly called spheroidite) to about 800 °C (1450 °F) or higher to change the ferrite to austenite and to dissolve all or most of the spheroidal carbides to the austenite. For plain carbon or low-alloy tool steels, austenitizing results in a contraction in volume. The extent of volumetric contraction decreases with increasing amounts of carbon present in the composition. This can be approximated as follows:

$$V_{SA} = -4.64 + 2.21 \,(\%C) \qquad \text{(Eq 1)}$$

where V_{SA} is the volume change in percent that occurs when spheroidite transforms to austenite. By use of this equation, it can be estimated that, if heated to a temperature high enough to dissolve all of the carbon in the austenite, a 0.50% carbon tool steel would exhibit a volume change of −3.53%, a common type containing 1% carbon would exhibit a change of −2.43%, and a very-high-carbon type containing 1.5% carbon would exhibit a change of −1.33%. However, tool steels having carbon contents higher than that of the eutectoid composition are normally austenitized at temperatures only high enough to dissolve the eutectoid amount of carbon. Under these circumstances, 1% carbon and 1.5% carbon tool steels would exhibit changes in volume of −2.77 and −2.53%, respectively, after austenitizing. These percentages are less than that calculated directly from Eq 1 because an allowance must be made for the volume occupied by undissolved carbides, which is about 3.5% for the 1.0% carbon steel and about 12% for the 1.5% carbon steel.

The second step involves cooling fast enough to cause the austenite to transform to martensite. The steel expands on transformation, the amount of expansion being in inverse proportion to

the amount of carbon in solution in the austenite:

$$V_{AM} = 4.64 - 0.53 \,(\%C) \qquad \text{(Eq 2)}$$

where V_{AM} is the percent volume change that occurs when austenite transforms to martensite. By use of Eq 2, it can be estimated that a 0.5% carbon tool steel would exhibit a volume increase for this transformation of 4.37% and that 1.0 and 1.5% carbon steels would exhibit increases of 4.07% and 3.71%, respectively, if austenitized at the normal austenitizing temperature (only 0.8% carbon, the eutectoid amount of carbon, in solution and again allowing for the volume occupied by undissolved carbides).

Equations 1 and 2 can be used to calculate the net change in dimensions in a tool steel when it is heat treated to transform it from an annealed to a fully hardened (martensitic) state. For the examples referred to above, normal heat treatment would produce net volume increases of −3.53 + 4.37 = 0.84% in the 0.5% carbon tool steel, −2.77 + 4.07 = 1.30% in the 1.0% carbon steel and −2.53 + 3.71 = 1.18% in the 1.5% carbon steel. Net changes in linear dimensions would be about one third the corresponding net changes in volume.

The third step involves reheating the freshly formed martensite to relatively low temperatures (tempering) to increase toughness and reduce lattice

stress. Tempering produces various changes in metallurgical structure, depending on temperature and time at temperature.

After very long times at room temperature or shorter times at temperatures up to 200 °C (400 °F), the high carbon martensite in plain carbon and low-alloy tool steels decomposes into low-carbon martensite (about 0.25% carbon) plus epsilon carbide, with an accompanying contraction in volume. At higher tempering temperatures, 200 to 430 °C (400 to 800 °F), the martensite decomposes into ferrite plus cementite.

Transformation of the maximum amount of austenite to martensite on quenching usually requires continuous cooling to below the martensite-finish temperature (M_f), which for a eutectoid tool steel is about −50 °C (−60 °F). To prevent cracking of very large or very intricate pieces, it is common practice to remove the tool from the quenching medium and begin tempering it while it is still slightly too warm to hold comfortably in the bare hands (about 60 °C, or 140 °F). Under these conditions, a substantial proportion of the structure (10% or more) may still be austenite. Most alloying elements lower the M_f temperature. Consequently, more austenite is retained at room temperature in the more highly alloyed tool steels. On tempering at increasing temperatures in the range 120 to 260 °C (250 to

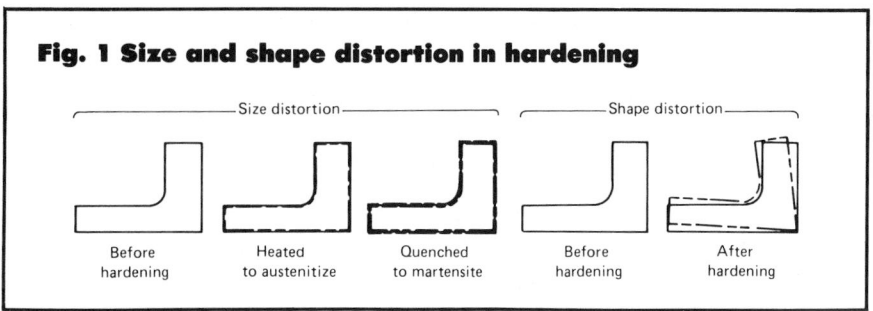

Fig. 1 Size and shape distortion in hardening

Size distortion · Before hardening · Heated to austenitize · Quenched to martensite · Shape distortion · Before hardening · After hardening

Table 1 Microconstituents in various tool steels after hardening

Steel	Hardening treatment	As-quenched hardness, HRC	Martensite, vol %	Retained austenite, vol %	Undissolved carbides, vol %
W1	790 °C (1450 °F), 30 min; WQ	67.0	88.5	9	2.5
L3	845 °C (1550 °F), 30 min; OQ	66.5	90	7	3.0
M2	1225 °C (2235 °F), 6 min; OQ	64	71.5	20	8.5
D2	1040 °C (1900 °F), 30 min; AC	62	45	40	15

Table 2 Typical dimensional changes in hardening and tempering

Tool steel	Hardening treatment Temperature °C	°F	Quenching medium	Total change in linear dimensions, % after quenching	150 300	205 400	260 500	315 600	370 700	425 800	480 900	510 950	540 1000	565 1050	595 1100
O1	815	1500	Oil	0.22	0.17	0.16	0.18	...	...	...	...	...	...	...	...
O1	790	1450	Oil	0.18	0.09	0.12	0.13	...	...	...	...	...	...	...	...
O6	790	1450	Oil	0.12	0.07	0.10	0.14	0.10	0.00	−0.05	−0.06	...	−0.07	...	...
A2	955	1750	Air	0.09	0.06	0.06	0.08	0.07	...	0.05	0.04	...	0.06	...	...
A10	790	1450	Air	0.04	0.00	0.00	0.08	0.08	0.01	0.01	0.02	...	0.01	...	0.02
D2	1010	1850	Air	0.06	0.03	0.03	0.02	0.00	...	−0.01	−0.02	...	0.06	...	...
D3	955	1750	Oil	0.07	0.04	0.02	0.01	−0.02	...	...	...	...	...	...	...
D4	1040	1900	Air	0.07	0.03	0.01	−0.01	−0.03	...	−0.4	−0.03	...	0.05	...	...
D5	1010	1850	Air	0.07	0.03	0.02	0.01	0.00	...	0.3	0.03	...	0.05	...	...
H11	1010	1850	Air	0.11	0.06	0.07	0.08	0.08	...	0.3	0.01	...	0.12	...	...
H13	1010	1850	Air	−0.01	...	...	...	...	...	...	0.00	...	0.06	...	...
M2	1210	2210	Oil	−0.02	...	...	...	...	...	...	...	−0.06	0.10	0.14	0.16
M41	1210	2210	Oil	−0.16	...	...	...	...	...	...	...	−0.17	0.08	0.21	0.23

500 °F), increasing amounts of this retained austenite transform to bainite for some tool steel compositions with an accompanying expansion in volume.

Depending on the alloy content of the tool steel, all, some, or none of the retained austenite will transform during tempering. In some highly alloyed tool steel compositions, cementite redissolves at tempering temperature of 540 to 595 °C (1000 to 1100 °F) to form alloy carbides, which induces an additional expansion in volume. The formation of alloy carbides during tempering is characteristic of tool steels containing large amounts of carbide-forming elements such as chromium, molybdenum and tungsten, which are found in high-speed tool steels.

Size Distortion in Tool Steels

Typical volume percentages of martensite, retained austenite and undissolved carbides are given in Table 1 for four different tool steels quenched from their recommended austenitizing temperatures.

Typical changes in linear dimensions for several tool steels are given in Table 2. As shown in this table, some tool steels such as A10 show very little size change when hardened and tempered over the entire range from 150 to 600 °C (300 to 1100 °F).

Other types, such as the M2 and M41 high speed steels, expand about 0.2% (2 mm/m, or 0.002 in./in.) when hardened and tempered in the temperature range of 540 to 595 °C (1000 to 1100 °F) to develop full secondary hardness. Although the information in Table 2 is useful in comparing size distortion in

several tool steels, the factor of shape distortion makes it impossible to use these data alone to predict dimensional changes of a particular tool made from any of these steels. Densities and thermal expansion characteristics for several classes of tool steels are presented in Table 3.

Shape Distortion in Tool Steels

The strength of any tool steel decreases rapidly above about 600 °C (1100 °F). At the austenitizing temperature, the yield strength is so low that plastic deformation often occurs simply from the stresses induced in the part by gravity. Therefore, long parts, large parts and parts of complex shape must be properly supported at critical locations to prevent sagging at the hardening temperature.

Rapid heating increases shape distortion, especially in large tools and in complex tools containing both light and heavy sections. If the rate of heating is high, light sections will increase in temperature much faster than heavy sections. Likewise, the outer surfaces in heavy sections will increase in temperature much faster than the interior. Differences in thermal expansion due to the differences in temperature between light and heavy sections or between surface and interior in heavy sections will be enough to set up large stresses in the material. Under these stresses, the hotter regions will deform plastically, to relieve the thermally induced stress.

Eventually, the hotter portions will reach the furnace temperature, while the cooler portions will continue to

increase in temperature. At this point, a decrease in thermal differential begins, which will cause a partial reversal in thermal stress which produced plastic deformation when the temperature differential was high. This may cause the part to undergo further plastic deformation, but to a lesser extent than the deformation caused by the initial high temperature differential. Such deformation will occur in a different direction.

Slow heating minimizes distortion by keeping temperature differentials low and thermal stresses within the elastic range of the material throughout the heating cycle. Ideally, all heat treatment of tool steel parts should start from a cold furnace to provide the greatest freedom from shape distortion during heating. Starting from a cold furnace is neither very practical nor energy efficient unless heat treating is being done in a vacuum furnace. When heat treating in fused salt or an atmosphere furnace, preheating the parts at an intermediate temperature prior to heating them to the austenitizing temperature provides the best compromise.

During quenching, large temperature differences between surface and interior, and between light and heavy sections, can cause severe shape distortion, because of thermal stress and mechanical stress produced by a martensitic transformation. This problem is most severe if the hardenability of the steel is so low that a fast cooling rate is required to obtain full hardness. In such a situation, especially when making a large or complex part, it may be best to substitute a high-hardenability, air-hardening tool steel, which re-

Table 3 Density and thermal expansion of selected tool steels

Type	Density Mg/m³	lb/in.³	μm/m·K from 20 °C to: 100 °C	205 °C	425 °C	540 °C	650 °C	μin./in. °F from 68 °F to: 200 °F	400 °F	800 °F	1000 °F	1200 °F
W1	7.84	0.283	10.4	11.0	13.1	13.8(a)	14.2(b)	5.76	6.13	7.28	7.64(a)	7.90(b)
W2	7.85	0.283	...	...	14.4	14.8	14.9	...	...	8.0	8.2	8.3
S1	7.88	0.255	12.4	12.6	13.5	13.9	14.2	6.9	7.0	7.5	7.7	7.9
S2	7.79	0.281	10.9	11.9	13.5	14.0	14.2	6.0	6.6	7.5	7.8	7.9
S5	7.76	0.280	...	...	12.6	13.3	13.7	...	...	7.0	7.4	7.6
S6	7.75	0.280	...	...	12.6	13.3	...	...	...	7.0	7.4	...
S7	7.76	0.280	...	12.6	13.3	13.7(a)	13.3	...	7.0	7.4	7.6(a)	7.4
O1	7.85	0.283	...	10.6(c)	12.8	14.0(d)	14.4(d)	...	5.9(c)	7.1	7.8(d)	8.0(d)
O2	7.66	0.277	11.2	12.6	13.9	14.6	15.1	6.2	7.0	7.7	8.1	8.4
O7	7.8	0.283	...	...	...	...	...	...	...	...	...	...
A2	7.86	0.284	10.7	10.6(c)	12.9	14.0	14.2	5.96	5.91(c)	7.2	7.8	7.9
A6	7.84	0.283	11.5	12.4	13.5	13.9	14.2	6.4	6.9	7.5	7.7	7.9
A7	7.66	0.277	...	...	12.4	12.9	13.5	...	...	6.9	7.2	7.5
A8	7.87	0.284	...	...	12.0	12.4	12.6	...	...	6.7	6.9	7.0
A9	7.78	0.281	...	...	12.0	12.4	12.6	...	...	6.7	6.9	7.0
D2	7.70	0.278	10.4	10.3	11.9	12.2	12.2	5.8	5.7	6.6	6.8	6.8
D3	7.70	0.278	12.0	11.7	12.9	13.1	13.5	6.7	6.5	7.2	7.3	7.5
D4	7.70	0.278	...	...	12.4	...	...	...	...	6.9	...	...
D5	...	...	...	...	...	12.0	...	...	...	...	6.7	...
H10	7.81	0.281	...	...	12.2	13.3	13.7	...	...	6.8	7.4	7.6
H11	7.75	0.280	11.9	12.4	12.8	12.9	13.3	6.6	6.9	7.1	7.2	7.4
H13	7.76	0.280	10.4	11.5	12.2	12.4	13.1	5.8	6.4	6.8	6.9	7.3
H14	7.89	0.285	11.0	...	...	...	6.1	...	...	...	...	...
H19	7.98	0.288	11.0	11.0	12.0	12.4	12.9	6.1	6.1	6.7	6.9	7.2
H21	8.28	0.299	12.4	12.6	12.9	13.5	13.9	6.9	7.0	7.2	7.5	7.7
H22	8.36	0.302	11.0	...	11.5	12.0	12.4	6.1	...	6.4	6.7	6.9
H26	8.67	0.313	...	...	...	12.4	...	...	...	...	6.9	...
H42	8.15	0.295	...	...	...	11.9	...	...	...	...	6.6	...
T1	8.67	0.313	...	9.7	11.2	11.7	11.9	...	5.4	6.2	6.5	6.6
T2	8.67	0.313	...	...	...	...	...	...	...	...	...	...
T4	8.68	0.313	...	...	...	11.9	...	...	...	...	6.6	...
T5	8.75	0.316	11.2	...	...	11.5	...	6.2	...	...	6.4	...
T6	8.89	0.321	...	...	...	...	...	...	...	...	...	...
T8	8.43	0.305	...	...	...	...	...	...	...	...	...	...
T15	8.19	0.296	...	9.9	11.0	11.5	...	...	5.5(c)	6.1	6.4	...
M1	7.89	0.285	...	10.6(c)	11.3	12.0	12.4	...	5.9(c)	6.3	6.7	6.9
M2	8.16	0.295	10.1	9.4(c)	11.2	11.9	12.2	5.6	5.2(c)	6.2	6.6	6.8
M3, class 1	8.15	0.295	...	...	11.5	12.0	12.2	...	...	6.4	6.7	6.8
M3, class 2	8.16	0.295	...	...	11.5	12.0	12.8	...	...	6.4	6.7	7.1
M4	7.97	0.288	...	9.5(c)	11.2	12.0	12.2	...	5.3(c)	6.2	6.7	6.8
M7	7.95	0.287	...	9.5(c)	11.5	12.2	12.4	...	5.3(c)	6.4	6.8	6.9
M10	7.88	0.255	...	...	11.0	11.9	12.4	...	...	6.1	6.6	6.9
M30	8.01	0.289	...	...	11.2	11.7	12.2	...	...	6.2	6.5	6.8
M33	8.03	0.290	...	...	11.0	11.7	12.0	...	...	6.1	6.5	6.7
M36	8.18	0.296	...	...	...	...	...	...	...	...	...	...
M41	8.17	0.295	...	9.7	10.4	11.2	...	...	5.4	5.8	6.2	...
M42	7.98	0.288	...	...	...	...	...	...	...	...	...	...
M46	7.83	0.283	...	...	...	...	...	...	...	...	...	...
M47	7.96	0.288	10.6	11.0	11.9	...	12.6	5.9	6.1	6.6	...	7.0
L2	7.86	0.284	...	...	14.4	14.6	14.8	...	...	8.0	8.1	8.2
L6	7.86	0.284	11.3	12.6	12.6	13.5	13.7	6.3	7.0	7.0	7.5	7.6
P2	7.86	0.284	...	...	13.7	...	...	...	...	7.6	...	...
P5	7.80	0.282	...	...	...	...	...	...	...	...	...	...
P6	7.85	0.283	...	...	...	...	...	...	...	...	...	...
P20	7.85	0.283	...	...	12.8	13.7	14.2	...	...	7.1	7.6	7.9

(a) From 20 to 500 °C (68 to 930 °F). (b) From 20 to 600 °C (68 to 1110 °F). (c) From 20 to 260 °C (68 to 500 °F). (d) From 38 °C (100 °F)

quires only a slow cooling rate to fully harden.

However, if lower-hardenability steels requiring liquid quenching are used, fixturing and pressure die quenching can help minimize distortion. Long symmetrical parts should be fixtured, and should be quenched in the vertical position with vertical agitation of the quench mediums.

Special Techniques for Controlling Shape Distortion

Special quenching procedures such as martempering and austempering may also be useful for controlling distortion in parts which have an appropriate configuration and been made of material having appropriate hardenability. In martempering, parts are quenched in hot molten salt fast enough to avoid transformation to high-temperature transformation products such as ferrite or pearlite. The parts are held at a bath temperature in the range from slightly above to slightly below the M_s just long enough to equalize the interior and surface temperatures. The parts are then removed from the bath and allowed to air cool to room temperature. Slow cooling through the martensitic transformation range reduces distortion as compared with rapid quenching. Martempered tools must be given the usual tempering treatment.

Austempering can be used to reduce distortion if a hardness no higher than 57 HRC is acceptable for the application. In austempering, parts are also quenched in hot molten salt but by temperature selection are forced to transform into bainite rather than martensite. Bainite forms at temperatures above those at which martensite forms. The parts must be held long enough at a temperature above the M_s (usually about 230 °C, or 450 °F) to permit the austenite to transform to lower bainite. When air cooled to room temperature, austempered tools exhibit less shape distortion and generally require no subsequent tempering.

Besides being reduced through control of rates of heating and cooling, shape distortion can be reduced by employing a localized method of heating and quenching such as flame hardening, induction hardening, electron beam or laser hardening to treat only that portion of the tool that must be hardened.

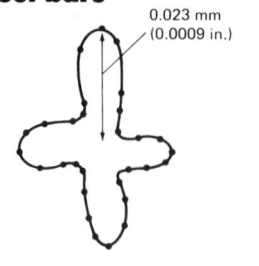

Fig. 2 Typical diameter changes during heat treatment for high speed steel bars

0.023 mm (0.0009 in.)

(a) Conventional process

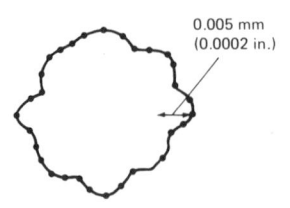

0.005 mm (0.0002 in.)

(b) Special process

Drawings produced by calculation from precision measurements of diameter. Charts are plots on polar coordinates depicting variations in diameter after heat treatment for a bar that was round within ±1.25 µm (±0.00005 in.) before heat treatment. (Courtesy of Latrobe Steel Co., Latrobe, Pennsylvania)

Controlling out-of-roundness is important for certain precision applications, such as class C and D cutting hobs made of high speed steels. Class C and D hobs must be held to close size limits because they are not ground to size after heat treatment, but rather are used in the unground condition.

Normal size distortion in hardening and tempering can be accommodated by making the tool slightly oversize or slightly undersize, as required, before heat treating. High speed steel bars, however, have been observed to go out-of-round as much as 0.05 mm (0.002 in.) during heat treatment. The pattern of size distortion shown in Fig. 2(a) can occur. It appears to be related to the initial shape of the cast ingot and to the specific primary-mill processing used to reduce the ingot into bars. By changing steelmaking, forging and rolling procedures, out-of-roundness has been reduced to the smaller differential pattern shown in Fig. 2(b), where the difference between high and low points is

only 0.005 mm (0.0002 in.). High speed steel bars made this way are marketed by a few tool steel producers as "close tolerance hob stock". An even better method of combating out-of-roundness is to use high speed tool steel bars made from hot isostatically pressed powders, which maintain the best possible symmetry during conventional heat treatment.

Stabilization involves reducing the amount of retained austenite in heat treated material. Retained austenite that can slowly transform and produce distortion if the material is later heated or subjected to stress. Stabilization also reduces internal (residual) stress, which makes distortion in service due to stress relaxation less likely to occur. Stabilization is most important for tools that must retain their exact size and shape over long periods of time (i.e., gauges, blocks).

If the tool steel chosen provides the required hardness after tempering at a relatively high temperature, it is possible to reduce the amount of retained austenite and the internal stress by multiple tempering. Initial tempering reduces internal stress and conditions the retained austenite so that it can transform to martensite on cooling from the tempering temperature. Usually, a second or third retempering is necessary to reduce the internal stress set up by the transformation of retained austenite.

Single or repeated cold treatment to a temperature below M_f will cause most of the retained austenite to transform to martensite in plain carbon or low-alloy tool steels that must be tempered at low temperatures to achieve the hardness required. Cold treatment may be applied either before or after the first temper. If, however, the tools tend to crack because of the additional stress induced by dimensional expansion during cold treatment, it is generally prudent to apply cold treatment after the tools have been tempered the first time. When cold treatment is applied after the first temper, the amount of retained austenite that transforms during the cold treatment may be considerably less than desired because some of the austenite may have been stabilized by tempering prior to cold treating. Cold treatment is usually done in a commercial refrigeration unit capable of attaining −70 to −95 °C (−100 to −140 °F). Tools must be retempered promptly after return to room temperature following cold treatment to reduce internal stress and increase the toughness of the newly formed martensite.

For some tools, a small percentage of retained austenite is desirable to improve toughness and provide a favorable internal stress pattern that will help the tool withstand service stresses. For these tools, a full stabilizing treatment may actually result in tools that are unfit to perform their required functions.

Powder Metallurgy Steels

In recent years, tool steels with improved properties have been produced by the powder metallurgy (P/M) process. In this process, molten metal alloy is solidified as a fine powder by spraying it into a chamber filled with inert gas. Steel containers are filled with these powder particles, evacuated of gases, sealed, heated, and white hot, isostatically pressed to full density. The resulting compact is rolled or forged to size on conventional steel-mill equipment or, in some instances, is used as compacted to make tools.

P/M tool steels have two major advantages: complete freedom from macrosegregation and porosity and uniform distribution of extremely fine carbides. These characteristics provide deeper hardening and faster response to hardening conditions (see Fig. 3). The latter is important, particularly for molybdenum high speed steels, which tend to decarburize rapidly at austenitizing temperatures. P/M products also show less out-of-roundness distortion in large-diameter bars (see Table 4).

When sulfur is added to P/M tool steels, they exhibit a very fine homogeneous distribution of sulfides. This uniform sulfide distribution promotes better machinability. After heat treating, the refined hardened and tempered P/M tool steels grindability and greater toughness than conventionally processed (cast and wrought) tool steels.

As of 1979, the following AISI compositions of high speed steels were available in P/M form: M2, M2 with high sulfur and high carbon, M3 class 2 with sulfur, M4, M35 with sulfur, M42 and T15. P/M steels can be substituted for their conventional counterparts in all applications, and are particularly advantageous when heavy sections are required.

The freedom from gross segregation provided by the P/M process makes it possible to readily fabricate new higher-

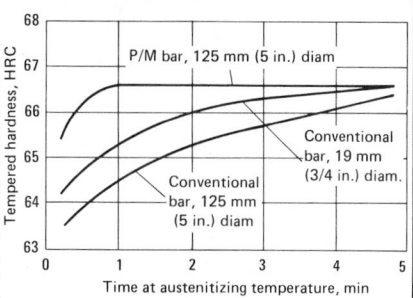

Fig. 3 Comparison of response to hardening for P/M and conventionally produced bars of M2S (HC) tool steel

Hardness at mid-radius was evaluated for bars oil quenched from 1200 °C (2200 °F) and tempered 2 + 2 + 2 h at 550 °C (1025 °F).

Table 4 Out-of-roundness distortion in large-diameter bars of M2S tool steel

Bar diameter mm	in.	Production method	Typical out-of-roundness(a) mm	in.
75	3	P/M	0.008	0.0003
		Conventional	0.020	0.0008
125	5	P/M	0.013	0.0005
		Conventional	0.033	0.0013
190	7.5	P/M	0.015	0.0006
		Conventional	0.051	0.0020

(a) Maximum diameter minus minimum diameter after normal hardening treatment

alloy tool steels compositions. One type now available, which contains 1.50 C, 3.75 Cr, 3.00 V, 10 W, 5.25 Mo and 9.00 Co, is reported to have the highest hot hardness of any high speed steel.

Surface Treatments

In many applications, service life of tool steels can be increased by surface treatments.

Oxide coatings, provided by treatment of the finish-ground tool in an alkali-nitrate bath or by steam oxidation, prevent or reduce adhesion of the tool to the workpiece. Oxide coatings have doubled tool life—particularly in machining of gummy materials such as soft copper and non-free-cutting low-carbon steels.

Plating of finished high speed steel tools with 0.0025 to 0.0125 mm (0.1 to

0.5 mil) of chromium also prolongs tool life by reducing adhesion of the tool to the workpiece. Chromium plating is relatively expensive, and precautions must be taken to prevent tool failure in service due to hydrogen embrittlement.

Electroless nickel plating has been used successfully as a replacement for chromium plating, both in routine production and for salvage plating operations on tool steel parts. Because plating by this method is accomplished by means of chemical reduction, it does not depend on any galvanic coupling between dissimilar metals, and there is no electrolysis involved. Therefore, there is no danger of hydrogen embrittlement. Plated hardness is in the high Rockwell 50's range, with good, uniform plated thickness on all surfaces, and the plated surfaces have a low coefficient of friction.

Carburizing is not recommended for high speed steel cutting tools because the cases on such tools are extremely brittle. However, carburizing is useful for applications such as cold work dies that require extreme wear resistance and that are not subjected to impact or highly concentrated loading. Carburizing is done at 1040 to 1065 °C (1900 to 1950 °F) for short periods of time (10 to 60 min) to produce a case 0.05 to 0.25 mm (0.002 to 0.010 in.) deep. The carburizing treatment also serves as an austenitizing treatment for the whole tool. A carburized case on high speed steels has a hardness of 65 to 70 HRC, but does not have the high resistance to softening at elevated temperatures exhibited by normally hardened high speed steel.

Nitriding successfully increases the life of all types of high speed steel cutting tools. However, gas nitriding in dissociated ammonia produces a case that is too brittle for most applications. Liquid nitriding for about 1 h at 565 °C (1050 °F) provides a light case, increasing both surface hardness and resistance to adhesion. For nitrided high speed steel taps, drills and reamers used in machining annealed steel, fivefold increases in life have been reported, with average increases of 100 to 200%. Obviously, if this nitrided case is removed when the tool is reground, the tool must then be retreated, which reduces the cost advantage of the process.

In addition, special surface-treatment processes, such as aerated nitriding baths, improve resistance to adhesive wear without producing excessive brittleness. Sulfur-containing nitriding baths provide a high-sulfur surface layer for additional resistance to seizing.

Sulfide Treatment. A low-temperature (190 °C, or 375 °F) electrolytic process using sodium and potassium thiocyanate provides a seizing-resistant iron sulfide layer. This process can be used as a final treatment for all types of hardened tool steels without much danger of overtempering.

Maraging Steels

Certain high-nickel maraging steels are being used for special noncutting tool applications; 18Ni(250) is the type most frequently used. For more demanding applications, the higher-strength 18Ni(300) is often preferred. Where maximum abrasion resistance is required, any of the maraging steels can be nitrided.

Maraging steels achieve full hardness—nominally 50 HRC for 18Ni(250), 54 HRC for 18Ni(300) and 58 HRC for 18Ni(350)—by means of a simple aging treatment, usually 3 h at about 480 °C (900 °F). Because the development of hardness does not depend on cooling rate, full hardness can be developed uniformly in massive sections, with almost no distortion. Decarburization is of no concern in these alloys, because they contain very little carbon and because their aging temperature is relatively low. However, if the long-time service temperature exceeds the aging temperature, maraging steels overage and undergo significant drop in hardness.

The 18Ni(300) grade is used for aluminum die-casting dies and cores, aluminum hot forging dies, dies for molding plastics, and various support tooling used in extrusion of aluminum. In die casting of aluminum, maraging steel dies can be used at higher hardness than is possible with dies made of 1113 tool steel because maraging steel is not as prone to heat checking. Because the aging process results in very little size change, it is possible to machine the intricate impressions for plastic molding dies to final size prior to final hardening.

For molding extremely abrasive types of plastics, the higher surface hardness provided by 18Ni(350) maraging steel is desirable.

REFERENCE

1. *Distortion in Tool Steels,* by B. S. Lement: American Society for Metals, 1959

SELECTED REFERENCES

- "Tool Steels", (a Steel Products Manual): American Iron and Steel Institute, March 1978
- *Source Book on Industrial Alloys and Engineering Data:* American Society for Metals, Metals Park, OH, 1978, p 251–292
- *The Metallurgy of Tool Steels,* by P. Payson: John Wiley & Sons, Inc., 1962
- *Metallurgy and Heat Treatment of Tool Steels,* by R. Wilson: McGraw-Hill, London, 1975
- *Tool Steels,* by G. A. Roberts and R. A. Cary: American Society for Metals, 1980
- *Tool Steel Simplified,* Revised Ed., by F. R. Palmer et al.: Chilton Book Co., Radnor, PA, 1978

Heat Treating of Stainless Steels and Heat-Resisting Alloys

Heat Treating of Stainless Steels

By the ASM Committee on Stainless
Steels and Heat-Resisting Alloys*

HEAT TREATING OF stainless steel serves to produce changes in physical condition, mechanical properties and residual stress level, and to restore maximum corrosion resistance when that property has been adversely affected by previous fabrication or heating. Frequently, a combination of satisfactory corrosion resistance and optimum mechanical properties is obtained in the same heat treatment.

Austenitic Stainless Steels

The austenitic stainless steels may be divided into five groups: (a) the conventional austenitics, such as types 301, 302, 303, 304, 305, 308, 309, 310, 316 and 317; (b) the stabilized compositions, primarily types 321, 347, and 348; (c) the low-carbon grades, such as types 304L, 316L and 317L; (d) the high-nitrogen grades, such as AISI types 201, 202, 304N and 316N, and the Nitronic series of alloys; and (e) the highly alloyed austenitics, such as

317LM, 317LX, JS700, JS777, 904L, AL-4X, 2RK65, Carpenter 20Cb-3, Sanicro 28, AL-6X and 254 SMO.

In furnace loading, the high thermal expansion of austenitic stainless steels (about 50% higher than that of a mild carbon steel) should be considered. The spacing between parts should be adequate to accommodate this expansion. Stacking, when necessary, should be employed judiciously, to avoid deformation of parts at elevated temperature.

Conventional austenitics cannot be hardened by heat treatment but will harden as a result of cold working. These steels are usually purchased in an annealed or cold worked state. Following welding or thermal processing, a subsequent re-anneal may be required for optimum corrosion resistance, softness and ductility. During annealing, chromium carbides, which markedly decrease resistance to intergranular corrosion, are dissolved. Annealing temperatures, which vary somewhat with the composition of the steel, are given in Table 1 for wrought

alloys and in Table 7 for the corresponding cast alloys.

Because carbide precipitation can occur at temperatures between 425 and 900 °C (800 and 1650 °F), it obviously is desirable that the annealing temperature should be safely above this limit. Moreover, because all carbides should be in solution before cooling begins, and because the chromium carbide dissolves slowly, the highest practical temperature consistent with limited grain growth is selected. This temperature is in the vicinity of 1095 °C (2000 °F).

Cooling from the annealing temperature must be rapid, but it must also be consistent with limitations of distortion. Whenever considerations of distortion permit, water quenching is used, thus ensuring that dissolved carbides remain in solution (because it precipitates carbides more rapidly, type 310 invariably requires water quenching). Where practical considerations of distortion rule out such a fast cooling rate, cooling in an air blast is

*Daniel Rapoport, *Chairman*, Technical Center, Howmet Turbine Components Corp.; Ed R. Byrnes, Jr., Manager - Vacuum Equipment Marketing, Ipsen Industries; Kenneth L. Crooks, Senior Staff Engineer, Research & Technology, Armco, Inc.; Ranes P. Dalal, Senior Development Engineer, Technical Center, Howmet Turbine Components Corp.; Matthew J. Donachie, Jr., Senior Assistant Materials Project Engineer, Pratt Whitney Aircraft Group, United Technologies Corp.; Dennis Macha, Supervisor-Metallurgical Services, Technical Center, Howmet Turbine Components Corp.; Robert N. Peterson, Manager of Metallurgical Services, Enduro Division, Republic Steel Corp.; Arthur D. Schwartz, Manager of Materials Performance Engineering, Aircraft Engine Business Group, General Electric Corp.
Reviewed by Douglas G. Frick, Associate Metallurgist, Corrosion & Welding Research, Carpenter Steel Division, Carpenter Technology Corp.; C. W. Kovach, Technical Director, Stainless Steels, Crucible Materials Research Center, Colt Industries; Jack Maurer, Market Development Manager, Allegheny-Ludlum; James D. Redmond, Manager, Stainless Steel Development, Climax Molybdenum Co.

used. With some thin-section parts, even this intermediate rate of cooling produces excessive distortion, and parts must be cooled in still air. If cooling in still air does not provide a rate sufficient to prevent carbide precipitation, maximum corrosion resistance will not be obtained. A solution to this dilemma is the use of a stabilized grade or the low-carbon alloys.

Stabilized austenitic alloys, namely types 321, 347, 348 and Carpenter 20Cb-3, contain controlled amounts of titanium or of niobium, which render the steel nearly immune to intergranular precipitation of chromium carbide and its adverse effects on corrosion resistance. Nevertheless, these alloys may require annealing to relieve stresses, increase softness and ductility, or provide additional stabilization.

To obtain maximum softness and ductility, the stabilized grades are annealed at the temperatures shown in Table 1. Unlike the unstabilized grades, these steels do not require water quenching or other acceleration of cooling from the annealing temperature to prevent subsequent intergranular corrosion; air cooling is generally adequate.

When maximum corrosion resistance of the stabilized austenitic grades is required, it may be necessary to employ a heat treatment known as a stabilizing anneal. The treatment consists of holding at 845 to 900 °C (1550 to 1650 °F) for up to 5 h, depending on section thickness. It may be applied either prior to or in the course of fabrication, and it may be followed by short-time stress relieving at 705 °C (1300 °F) without danger of harmful carbide precipitation.

Carpenter 20Cb-3 stainless steel is unlike the conventional stabilized austenitics (types 321, 347 and 348) because of its higher alloy content and improved corrosion resistance. This alloy normally is stabilized and annealed at 925 to 955 °C (1700 to 1750 °F). For special applications, the alloy can be annealed at higher temperatures (up to 1150 °C or 2100 °F), but this is premissible only if the alloy will not be subject to welding or heating temperatures over 540 °C (1000 °F).

Certain restrictions on furnace atmosphere are mandatory. Furnace combustion must be carefully controlled to eliminate carburizing or excessively oxidizing conditions. Because the properties of the stabilized steels are based on their original carbon content, carbon absorption cannot be tolerated. Excessively oxidizing conditions cause the formation of a scale that is difficult to remove in subsequent descaling operations. Direct impingement of flame on the work must be prevented. The sulfur content of the furnace atmosphere, particularly in oil-fired furnaces, must be kept low; natural gas should be used, not producer gas.

Low-carbon austenitics are intermediate in tendency to precipitate chromium carbides to the stabilized and unstabilized grades. Carbon content (0.03% max) is low enough to reduce precipitation of intergranular carbides. This characteristic of limited sensitization is of particular value in welding, flame cutting, and other hot working operations. They do not require the quenching treatment that unstabilized grades require to retain carbon in solid solution. Nevertheless, the low-carbon alloys are not satisfactory for long-time service in the sensitizing temperature range of 540 to 760 °C (1000 to 1400 °F) because they are

Table 1 Recommended annealing temperatures for austenitic stainless steels

UNS No.	Designation	Temperature(a) °C	°F
Conventional grades			
S30100, S30200, S30215 ..	301, 302, 302B	1010 to 1120	1850 to 2050
S30300, S30323	303, 303Se	1010 to 1120	1850 to 2050
S30400, S30500, S30800 ..	304, 305, 308	1010 to 1120	1850 to 2050
S30900, S30908	309, 309S	1040 to 1120	1900 to 2050
S31000, S31008	310, 310S	1040 to 1065	1900 to 1950
S31600	316	1040 to 1120	1900 to 2050
S31700	317	1065 to 1120	1950 to 2050
Stabilized grades			
S32100	321	955 to 1065	1750 to 1950
S34700, S34800	347, 348	980 to 1065	1800 to 1950
N08020	Carpenter 20Cb-3	925 to 955	1700 to 1750
Low-carbon grades			
S30403	304L, 304LN	1010 to 1120	1850 to 2050
S31603, S31703	316L, 316LN, 317L	1040 to 1110	1900 to 2025
High-nitrogen grades			
S20100, S20200	201, 202	1010 to 1120	1850 to 2050
S30451	304N	1010 to 1120	1850 to 2050
S31651	316N	1010 to 1120	1850 to 2050
S24100	Nitronic 32, Carpenter 18Cr-2Ni-12Mn	1010 to 1065	1850 to 1950
S24000	Nitronic 33	1040 to 1095	1900 to 2000
S21904	Nitronic 40, Carpenter 21Cr-6Ni-9Mn	980 to 1175	1800 to 2150
S20910	Nitronic 50, Carpenter 22Cr-13Ni-5Mn	1065 to 1120	1950 to 2050
S21800	Nitronic 60	1040 to 1095	1900 to 2000
S28200	Carpenter 18-18 PLUS	1040 to 1095	1900 to 2000
Highly alloyed grades			
... .	317LM, 317LX, 317L PLUS, 317LMO, 7L4	1120 to 1150	2050 to 2100
...	JS700, JS777	1065 to 1150	1950 to 2100
N08904	904L, AL-4X, 2RK65	1075 to 1125	1965 to 2055
N08028	Sanicro 28	...	...
N08366	AL-6X	1205 to 1230	2200 to 2250
S31254	254 SMO	1150 to 1205	2100 to 2200

(a) Temperatures given are for annealing a composite structure. Time at temperature and method of cooling depend on thickness. Light sections may be held at temperature for 3 to 5 min per 2.5 mm (0.10 in.) of thickness, followed by rapid air cooling. Thicker sections are water quenched. For many of these grades, a postweld heat treatment is not necessary. For proprietary alloys, alloy producers may be consulted for details. Although cooling from the annealing temperature must be rapid, it must also be consistent with limitations of distortion.

Table 2 Recommended annealing treatments for ferritic stainless steels

| UNS No. | Designation | Treatment temperature | |
		°C	°F
Conventional ferritic grades			
S40500 405		650 to 815	1200 to 1500
S40900 409		870 to 900	1600 to 1650
S43000 430		705 to 790	1300 to 1450
S43020 .... 430F		705 to 790	1300 to 1450
S43400 434		705 to 790	1300 to 1450
S44600 446		760 to 830	1400 to 1525
Low-interstitial ferritic grades			
S43035 439		870 to 925	1600 to 1700
S44400 444		955 to 1010	1750 to 1850
S44626 .... E-BRITE		760 to 955	1400 to 1750
S44660 .... SEA-CURE, SC-1		1010 to 1065	1850 to 1950
.... AL 29-4C		1010 to 1065	1850 to 1950
S44800 .... Al 29-4-2		1010 to 1065	1850 to 1950
S44635 .... MONIT		1010 to 1065	1850 to 1950

Postweld heat treating of the low interstitial ferritic stainless steels is generally unnecessary and frequently undesirable. Any annealing of these grades should be followed by water quenching or very rapid cooling.

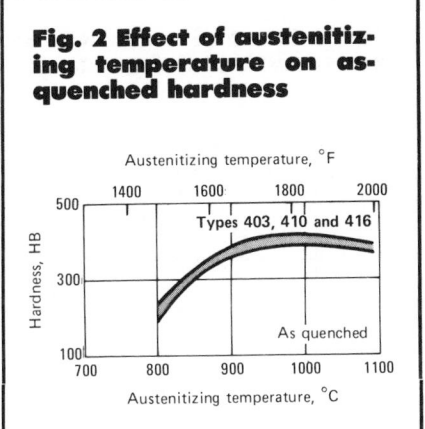

Fig. 2 Effect of austenitizing temperature on as-quenched hardness

Specimens were martensitic stainless steels containing 0.15% max carbon.

Fig. 1 Effect of heat treatments on the hardness of martensitic stainless steels

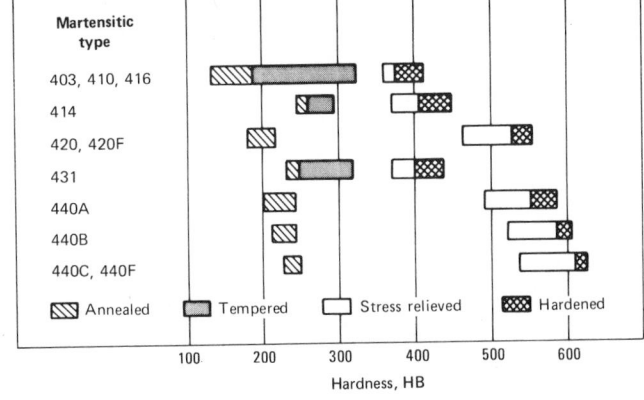

not completely immune to the formation of carbides deleterious to corrosion resistance. Recommended annealing temperatures for the low-carbon alloys are given in Table 1.

The effects of sensitization and susceptibility to general corrosion vary among the low-carbon alloys, depending on their chemical composition. Because they contain molybdenum, types 316L and 317L are susceptible to sigma phase formation as a result of long-time exposure at 650 to 870 °C (1200 to 1600 °F). However, the corrosion resistance of these grades can be improved by employing a stabilizing treatment (ASTM A262C), consisting of holding at 885 °C (1625 °F) for 2 h, prior to stress relieving at 675 °C (1250 °F). After receiving the stabilizing heat treatment, these alloys pass the sulfuric acid-copper sulfate test (ASTM A262, Practice E) for freedom from intergranular carbide precipitation.

Magnetic Permeability. The low-carbon alloys are frequently used in the production of articles requiring low magnetic permeability. These materials are nonmagnetic in the fully annealed condition, with permeabilities below 1.02 max at 200 gausses, but may develop ferromagnetic qualities as a result of cold working during fabrication. Cold working may generate some low-carbon martensite, which is strongly magnetic. Fusion welding with a low-nickel filler rod is another possible cause of magnetism. Magnetism due to any of these causes can be eliminated by a full anneal to restore the alloy to its fully austenitic condition.

Bright Annealing. All grades of austenitic stainless steel can be bright annealed in either pure hydrogen or dissociated ammonia, provided the dew point of the atmosphere is less than −50 °C (−60 °F) and the workpieces, upon entering the furnace, are dry and scrupulously clean. The furnaces used in bright annealing must be clean, moisture-free and tight if low dew points are to be maintained. If a low dew point is not maintained, a thin greenish oxide will form on the work. This oxide is very difficult to remove in subsequent descaling operations.

To maintain close control of dew point, atmosphere samples should be withdrawn from the furnace at frequent intervals and tested or continuously monitored, as in most commercial operations. Traces of oxygen in hydrogen gas can be removed before the gas enters the furnace by passing the gas through a catalytic tower that causes excess oxygen to combine with the hydrogen to form water vapor. The gas is then passed through activated alumina to remove moisture.

In using dissociated ammonia, it is important that maximum dissociation be obtained before the gas enters the furnace. The presence of any undissociated ammonia will result in objection-

Fig. 3 Effect of variations in austenitizing temperature on hardness and impact strengths of martensitic stainless steels

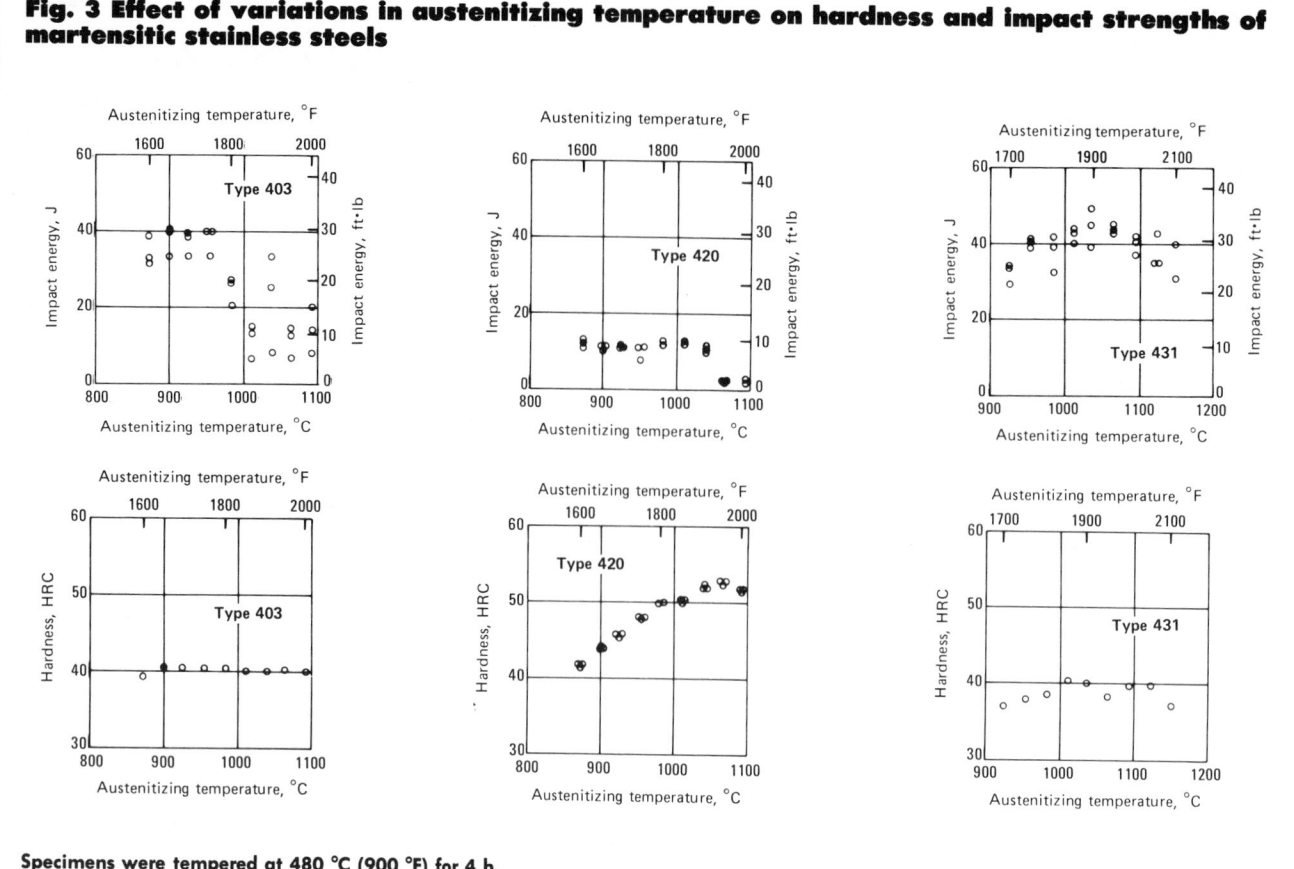

Specimens were tempered at 480 °C (900 °F) for 4 h.

able nitriding action. Because the undissociated gas is entirely soluble in water, its removal can be easily accomplished. However, the remaining fully dissociated product must be processed through drying towers to restore the required low dew point.

High-nitrogen austenitic stainless steels are heat treated in the same manner and are subject to the same problems (carbide precipitation and distortion) as conventional austenitics. They cannot be hardened by heat treatment but will harden by cold working. High-nitrogen austenitics are annealed to ensure maximum corrosion resistance, softness and ductility. Rapid cooling is preferred. Annealing temperature ranges are listed in Table 1.

Highly alloyed austenitic stainless steels contain large amounts of molybdenum to provide very good resistance to chloride corrosion. They usually are produced with low carbon to avoid sensitization and may contain copper for increased acid resistance. These alloys are austenitic in the mill-

annealed condition but may form sigma or delta ferrite phases under certain conditions of heat treatment or service. Those phases may be detrimental to corrosion resistance and mechanical properties. Annealing temperatures are confined to a narrow range to avoid sigma phase formation at lower temperature, or delta ferrite at higher temperature. Recommended annealing temperatures are given in Table 1. Rapid cooling following annealing is usually advisable, especially in heavy sections. Stress relief treatments may be used below the annealing temperature range, but holding times should be held to a minimum to avoid sigma phase and sensitization problems.

Ferritic Stainless Steels

The ferritic stainless steels may be divided into two groups, the conventional ferritics such as types 409, 430, 434 and 446, and the low-interstitial ferritics such as types 439, 444, E-BRITE, SEA-CURE, AL 29-4C and AL

29-4-2. The ferritic stainless steels are not hardened by quenching but rather develop minimum hardness and maximum ductility, toughness and corrosion resistance in the annealed and quenched condition. Therefore, the only heat treatment applied to the ferritics is annealing. This treatment relieves stresses developed during welding or cold working and provides a more homogeneous structure by dissolving transformation products formed during welding. Postweld heat treatment of the low-interstitial ferritic stainless steels is generally unnecessary and is frequently undesirable. Table 2 summarizes current annealing practices for the ferritic grades.

Austenite-Martensite Embrittlement. When grades such as 430 and 434 are cooled rapidly from above 925 °C (1700 °F), they may become brittle from austenite transforming to as much as 30% martensite. This may be corrected by a tempering treatment such as 650 to 790 °C (1200 to 1450 °F), which softens the alloy.

Table 3 Procedures for hardening and tempering wrought martensitic stainless steels to specific strength and hardness levels

Type	Austenitizing(a) Temperature(b) °C	°F	Quenching medium(c)	Tempering temperature(d) °C min	max	°F min	max	Tensile strength MPa	ksi	Hardness, HRC
403, 410.......	925-1010	1700-1850	Air or oil	565	605	1050	1125	760-965	110-140	25-31
				205	370	400	700	1105-1515	160-220	38-47
414	925-1050	1700-1925	Air or oil	595	650	1100	1200	760-965	110-140	25-31
				230	370	450	700	1105-1515	160-220	38-49
416, 416(Se)	925-1010	1700-1850	Oil	565	605	1050	1125	760-965	110-140	25-31
				230	370	450	700	1105-1515	160-220	35-45
420	985-1065	1800-1950	Air or oil(e)	205	370	400	700	1550-1930	225-280	48-56
431	985-1065	1800-1950	Air or oil(e)	565	605	1050	1125	860-1035	125-150	26-34
				230	370	450	700	1210-1515	175-220	40-47
440A..........	1010-1065	1850-1950	Air or oil(e)	150	370	300	700	...	...	49-57
440B..........	1010-1065	1850-1950	Air or oil(e)	150	370	300	700	...	...	53-59
440C, 440F	1010-1065	1850-1950	Air or oil(e)	...	160	...	325	...	...	60 min
				...	190	...	375	...	...	58 min
				...	230	...	450	...	...	57 min
				...	355	...	675	...	...	52-56

(a) Preheating to a temperature within the process annealing range (see Table 5) is recommended for thin-gage parts, heavy sections, previously hardened parts, parts with extreme variations in section or with sharp re-entrant angles, and parts that have been straightened or heavily ground or machined, to avoid cracking and minimize distortion, particularly for types 420, 431, and 440A, B, C and F. (b) Usual time at temperature ranges from 30 to 90 min. The low side of the austenitizing range is recommended for all types subsequently tempered to 25 to 31 HRC; generally, however, corrosion resistance is enhanced by quenching from the upper limit of the austenitizing range. (c) Where air or oil is indicated, oil quenching should be used for parts more than 6.4 mm (¼ in.) thick; martempering baths at 150 to 400 °C (300 to 750 °F) may be substituted for an oil quench. (d) Generally, the low end of the tempering range of 150 to 370 °C (300 to 700 °F) is recommended for maximum hardness, the middle for maximum toughness, and the high end for maximum yield strength. Tempering in the range of 370 to 565 °C (700 to 1050 °F) is not recommended, because it results in low and erratic impact properties and poor resistance to corrosion and stress corrosion. (e) For minimum retained austenite and maximum dimensional stability, a subzero treatment −75 °C ±10 °C (−100 °F ±20 °F) is recommended; this should incorporate continuous cooling from the austenitizing temperature to the cold transformation temperature.

Fig. 4 Effect of variations in austenitizing time on hardness and impact strengths of martensitic stainless steels

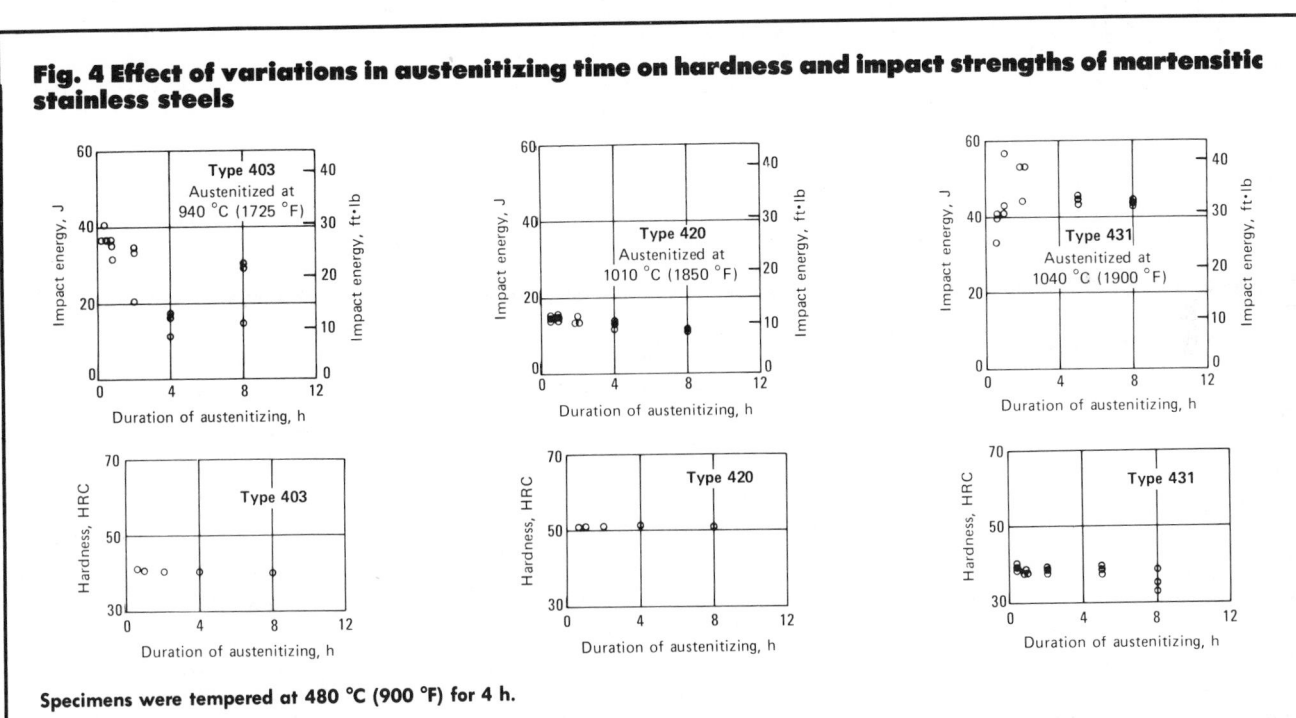

Specimens were tempered at 480 °C (900 °F) for 4 h.

475 °C (885 °F) Embrittlement. A potentially harmful form of embrittlement common to the ferritic grades can develop from prolonged exposure to, or slow cooling within, the temperature range from about 370 to 540 °C (700 to 1000 °F) with the maximum rate of embrittlement occurring at about 475 °C (885 °F). The embrittlement is caused by precipitation of the α′ phase, and the effects of embrittlement increase rapidly with chromium content. Certain heat treatments must be controlled to avoid embrittlement. The brittle condition can be eliminated by

Fig. 5 Effect of variations in tempering temperature on hardness and impact strengths of martensitic stainless steels

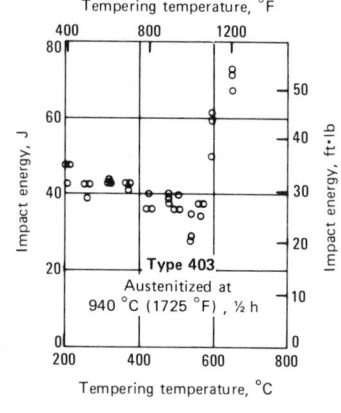
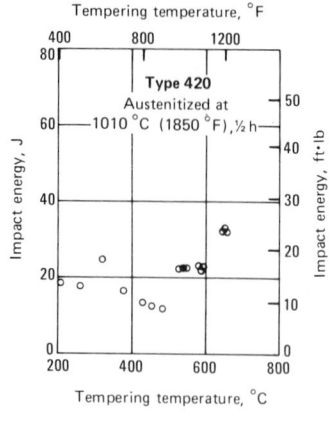
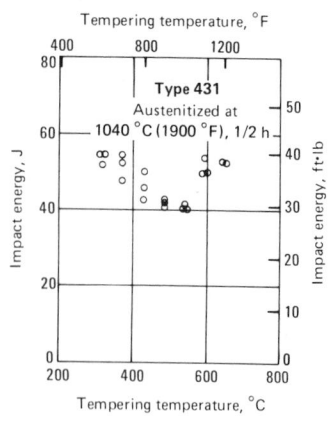

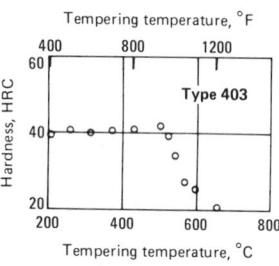
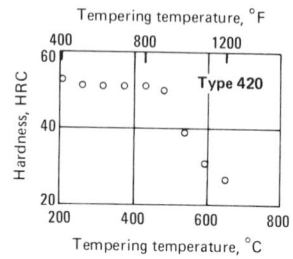
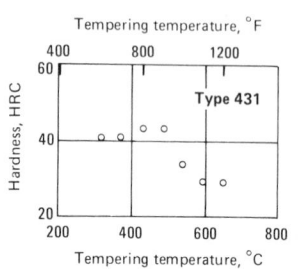

Table 4 Annealing temperatures and procedures for wrought martensitic stainless steels

Type	Process (subcritical) annealing Temperature(a), °C	Hardness	Full annealing Temperature(b)(c), °C	Hardness	Isothermal annealing(c) Procedure (d)	Hardness
403, 410 650-760		86-92 HRB	830-885	75-85 HRB	Heat to 830 to 885 °C; hold 6 h at 705 °C	85 HRB
414 650-730		99 HRB-24 HRC	Not recommended		Not recommended	
416, 416(Se) 650-760		86-92 HRB	830-885	75-85 HRB	Heat to 830 to 885 °C; hold 2 h at 720 °C	85 HRB
420 675-760		94-97 HRB	830-885	86-95 HRB	Heat to 830 to 885 °C; hold 2 h at 705 °C	95 HRB
431 620-705		99 HRB-30 HRC	Not recommended		Not recommended	
440A 675-760		90 HRB-22 HRC	845-900	94-98 HRB	Heat to 845 to 900 °C; hold 4 h at 690 °C	98 HRB
440B 675-760		98 HRB-23 HRC	845-900	95 HRB-20 HRC	Same as 440A	20 HRC
440C, 440F 675-760		98 HRB-23 HRC	845-900	98 HRB-25 HRC	Same as 440A	25 HRC

(a) Air cool from temperature; maximum softness is obtained by heating to temperature at high end of range. (b) Soak thoroughly at temperature within range indicated; furnace cool to 790 °C; continue cooling at 15 to 25 °C/h to 595 °C; air cool to room temperature. (c) Recommended for applications in which full advantage may be taken of the rapid cooling to the transformation temperature and from it to room temperature. (d) Preheating to a temperature within the process annealing range is recommended for thin-gage parts, heavy sections, previously hardened parts, parts with extreme variations in section or with sharp re-entrant angles, and parts that have been straightened or heavily ground or machined to avoid cracking and minimize distortion, particularly for types 420 and 431, and 440A, B, C and F.

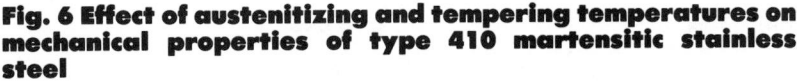

Fig. 6 Effect of austenitizing and tempering temperatures on mechanical properties of type 410 martensitic stainless steel

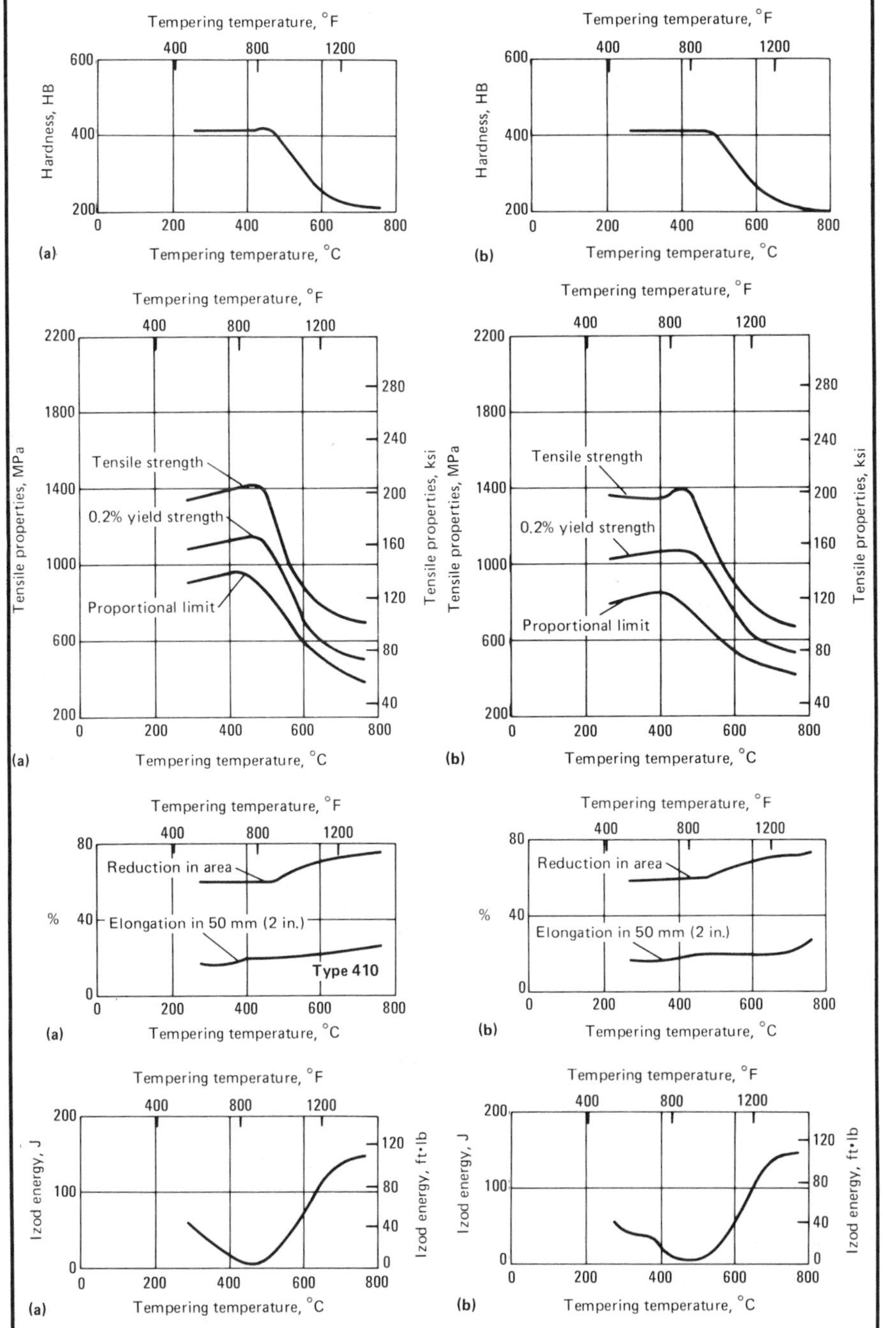

Austenitized 30 min; oil quenched to 65 to 95 °C (150 to 200 °F); double stress relieved at 175 °C (350 °F) for 15 min and water quenched; tempered 2 h. (a) Quenched from 925 °C (1700 °F). (b) Quenched from 1010 °C (1850 °F)

the treatments listed in Table 2, using temperatures clearly above the upper boundary of embrittlement, followed by rapid cooling to prevent a recurrence.

Intermetallic Phase Embrittlement. Intermetallic phases, such as sigma, chi and Laves, may form at elevated temperatures in ferritic stainless steels containing more than about 14% chromium. These intermetallic phases increase hardness (sometimes usefully) and decrease ductility, notch toughness, and corrosion resistance. The temperature range over which these phases form is approximately 600 to 1000 °C (1100 to 1830 °F).

Intermetallic phase embrittlement is primarily a service problem where long exposures at elevated temperatures are involved. These phases can be dissolved by heating to above 1000 °C (1830 °F).

Duplex stainless steels consist of a mixed microstructure of austenite and ferrite. Some duplex stainless steels are rich in ferrite, others in austenite and others are equally balanced. Compared with type 316, the annealed duplex alloys provide improved resistance to chloride stress corrosion cracking. Although the duplex grades are generally not as resistant as the low-interstitial ferritics, they are normally available in heavier section thicknesses. Another useful characteristic of the duplex grades is that they typically have yield strengths more than twice that of the conventional austenitic stainless steels.

Duplex stainless steels, such as SAF 2205, AF 22, DP 3 and Ferralium alloy 225, are alloyed with 0.15 to 0.20% nitrogen. This minimizes alloy element segregation between the ferrite and austenite, thereby improving the as-welded corrosion resistance compared with the type 329 alloy. Recommended annealing temperatures for duplex stainless steels are listed below:

UNS No.	Designation	Temperature °C (°F)
S32900	329, 7-Mo	925-955 (1700-1750)
S31500	3RE60	975-1025 (1785-1875)
... .	SAF 2205, AF 22	1020-1100 (1870-2010)
... .	DP 3	1065-1175 (1950-2150)
S32550	Ferralium alloy 225	1065-1175 (1950-2150)

Note: Cooling from the annealing temperature must be rapid, but it also must be consistent with limitations of distortion.

Fig. 7 Effect of austenitizing and tempering temperatures on typical mechanical properties of type 414 martensitic stainless steel

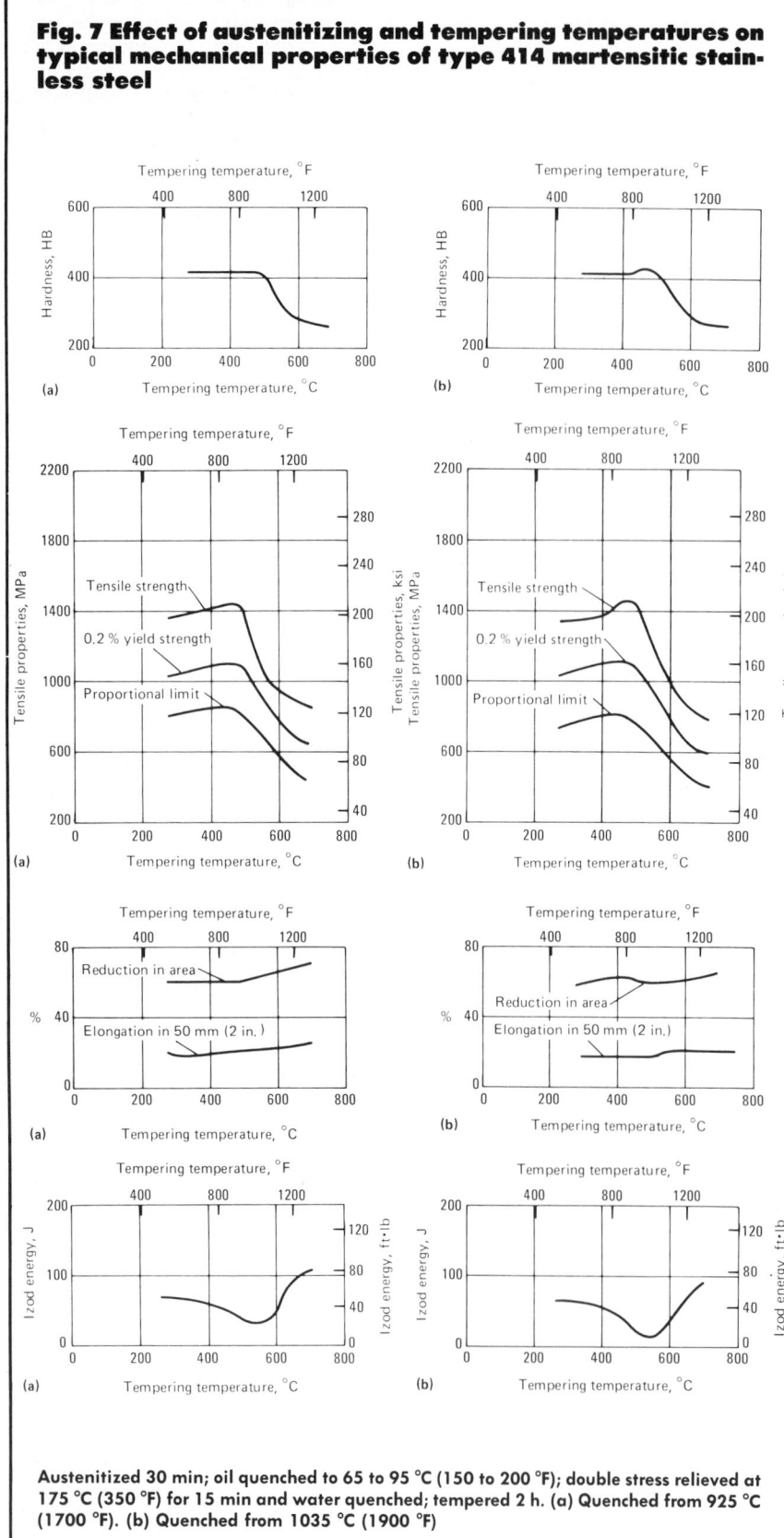

Austenitized 30 min; oil quenched to 65 to 95 °C (150 to 200 °F); double stress relieved at 175 °C (350 °F) for 15 min and water quenched; tempered 2 h. (a) Quenched from 925 °C (1700 °F). (b) Quenched from 1035 °C (1900 °F)

Martensitic Stainless Steels

The heat treating of martensitic stainless steel is essentially the same as for plain carbon or low-alloy steels, in that maximum strength and hardness depend chiefly on carbon content. The principal metallurgical difference is that the high alloy content of the stainless grades causes the transformation to be so sluggish, and the hardenability to be so high, that maximum hardness is produced by air cooling in the center of sections up to approximately 30.5 cm (12 in.) thick.

Surface hardness ranges for the various heat treated conditions from fully annealed to fully hardened are given in Fig. 1.

The martensitic stainless steels are more sensitive to heat treating variables than are carbon and low-alloy steels; rejection rates due to faults in heat treating are correspondingly high.

Prior Cleaning. To avoid contamination, all parts and heat treating fixtures must be cleaned thoroughly before they are placed in the furnace. Proper cleaning is particularly important when the heat treatment is to be performed in a protective atmosphere. Grease, oil, and even location lines made by an ordinary lead pencil, can cause carburization. Perspiration stains from fingerprints are a source of chloride contamination and may cause severe scaling in oxidizing atmospheres. Furthermore, a protective atmosphere cannot be effective unless it is permitted to make unobstructed contact with metal surfaces.

Preheating. Martensitic stainless steels normally are hardened by being heated to the austenitizing range of 925 to 1065 °C (1700 to 1950 °F) and then cooled in air or oil.

The thermal conductivity of stainless steels is characteristically lower than that of carbon and alloy steels. Accordingly, high thermal gradients and high stresses during rapid heating may cause warpage and cracking in some parts. To avoid these problems, preheating is usually recommended in the treatment of martensitic stainless steels. In annealing or hardening, the following parts should be preheated: (a) heavy section parts, (b) parts with both thin and thick sections, (c) parts with sharp corners and re-entrant angles, (d) heavily ground parts, (e) parts machined with heavy deep cuts, (f) parts

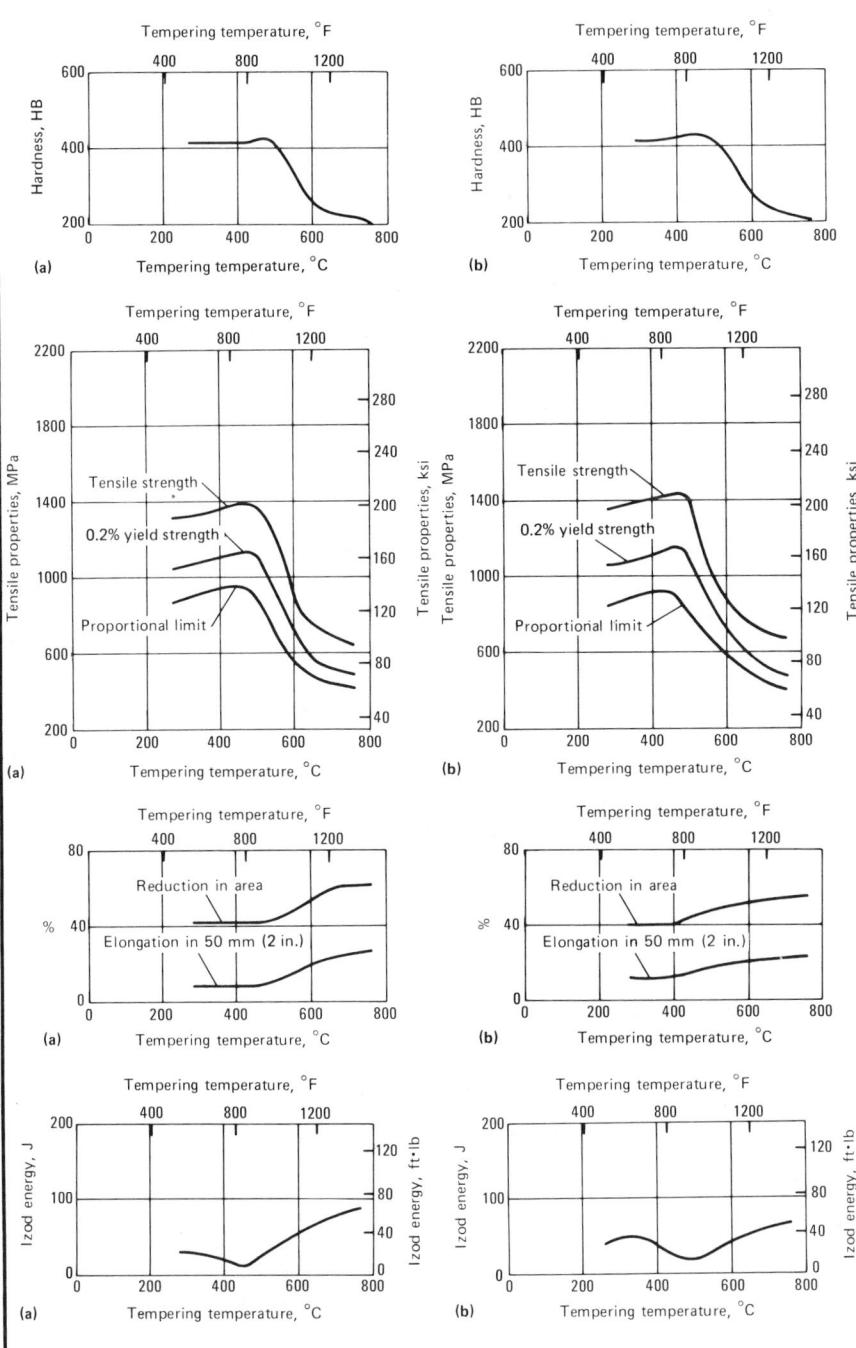

Fig. 8 Effect of austenitizing and tempering temperatures on typical mechanical properties of type 416 martensitic stainless steel

Austenitized 30 min; oil quenched to 65 to 95 °C (150 to 200 °F); double stress relieved at 175 °C (350 °F) for 15 min and water quenched; tempered 2 h. (a) Quenched from 925 °C (1700 °F). (b) Quenched from 980 °C (1800 °F)

that have been cold formed or straightened, and (g) previously hardened parts that are being reheat treated.

Preheating is normally accomplished at 760 to 790 °C (1400 to 1450 °F), and heating need be continued only long enough to ensure that all portions of each part have reached the preheating temperature. Large heavy parts are sometimes preheated at approximately 540 °C (1000 °F) prior to the 790 °C (1450 °F) preheat. Types 403, 410 and 416 require less preheating than the higher-carbon types 414, 431, 420, and the 440 grades.

Austenitizing temperatures, soaking times, and quenching mediums are summarized in Table 3. When maximum corrosion resistance and strength are desired, the steel should be austenitized at the high end of the temperature range. For alloys that are to be tempered above 565 °C (1050 °F), the low side of the austenitizing range is recommended, because it enhances ductility and impact properties.

The effect of austenitizing temperature on the as-quenched hardness of three martensitic grades is shown in Fig. 2. The hardness increases with increasing austenitizing temperature to about 980 °C (1800 °F), then decreases because of austenite retention and (occasionally) the formation of delta ferrite.

Certain anomalies in these steels that should be considered before specifying a heat treating procedure are exemplified in the opposing injurious effects of the high and low extremes of austenitizing temperature, depending on the subsequent tempering temperature. For example, type 431 showed the following Izod impact properties that are caused by retained austenite:

Temperature, °C Austenitizing	Tempering	Izod impact, J
980	315	20.3-33.9
1065	315	40.7-81.3
980	595	74.6-108.5
1065	595	61.0-74.6

Soaking times employed in the hardening of martensitic stainless steels represent a compromise between (a) achieving maximum solution of chromium-iron carbides for maximum strength and corrosion resistance and (b) avoiding decarburization, excessive grain growth, retained austenite, brittleness, and quench cracking. For sections 13 mm (½ in.) thick and under, a soaking time of 30 to 60 min is sometimes recommended. For most parts, adding 30 min for each additional inch of thickness or fraction thereof has proven adequate. However, soaking times should be doubled if parts to be hardened have been fully annealed or isothermally annealed.

The effect of soaking time at austenitizing temperature, and of other variables, on the impact strength and room-temperature hardness of types 403, 420 and 431 is plotted in Fig. 3, 4 and 5.

Quenching. Because of their high hardenability, martensitic stainless steels can be quenched in either oil or air. Some decrease in corrosion resistance and ductility, resulting from air quenching, may occur in these grades. These steels may precipitate carbides in grain-boundary areas if heavy sections are cooled slowly through the temperature range of about 870 to 540 °C (1600 to 1000 °F). Too slow a cooling rate in bright annealing these alloys may impair their corrosion resistance. Although oil quenching is preferred, air cooling may be required for large or complex sections, to prevent distortion or quench cracking.

Martempering is particularly easy with these steels because of their high hardenability.

Retained Austenite. The higher-carbon martensitic grades, such as 440C, and the higher-nickel type 431, are likely to retain large amounts of untransformed austenite in the as-quenched structure, frequently as much as 30% by volume. Stress relieving at about 150 °C (300 °F) has little effect. Delayed transformation, particularly in type 440C, may occur as a result of temperature fluctuations in service, thus resulting in embrittlement and unacceptable dimensional changes.

Subzero Cooling. A portion of the austenite retained in quenching may be transformed by subzero cooling to about −75 °C (−100 °F) immediately after quenching. To obtain maximum transformation of retained austenite, double tempering may be necessary. Parts should be air cooled to room temperature between the tempering cycles.

Subzero cooling is frequently included in the hardening treatment of parts such as the slides and sleeves of slide valves, and bearings requiring maximum dimensional stability.

Fig. 9 Effect of austenitizing and tempering temperatures on typical mechanical properties of type 420 martensitic stainless steel

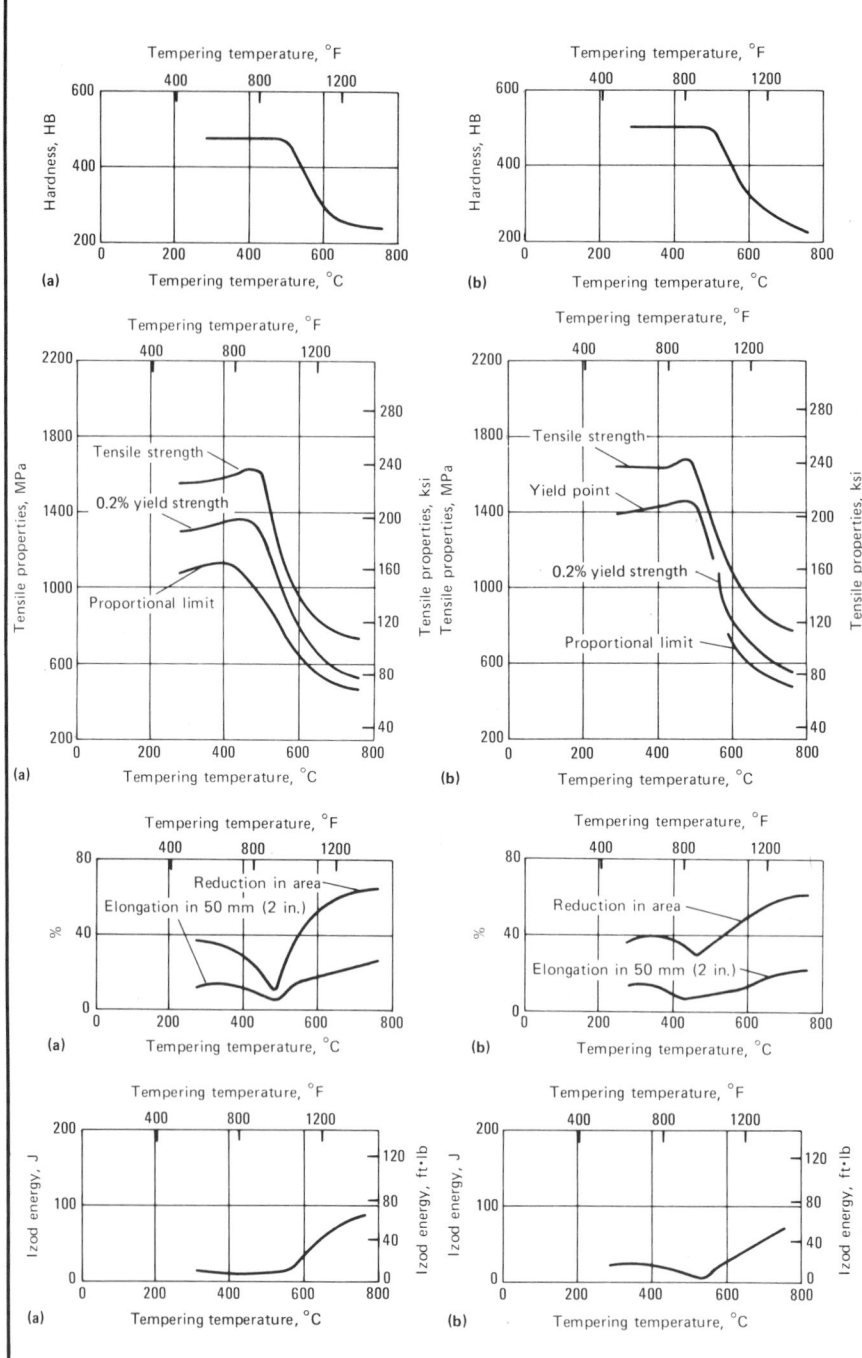

Austenitized 30 min; oil quenched to 65 to 95 °C (150 to 200 °F); double stress relieved at 175 °C (350 °F) for 15 min and water quenched; tempered 2 h. (a) Quenched from 925 °C (1700 °F). (b) Quenched from 1025 °C (1875 °F)

Reheating. For fully hardened steels, increasing degrees of recovery are achieved by:

- *Stress relieving* at 150 to 370 °C (300 to 700 °F) after hardening, to reduce transformation stresses without significantly affecting microconstituents or mechanical properties
- *Tempering* at intermediate temperatures, to modify properties
- *Subcritical annealing* (variously called process, mill, or low annealing) in the upper portion of the ferritic range, just below the lower critical Ac_1 temperature, to achieve maximum softening without the complications of re-entering the gamma or austenitic field
- *Full annealing* for maximum softening by a return to the austenitic range, followed by slow cooling.

Although the strength, elongation and hardness curves shown in Fig. 6 through 11 appear to have the same general form as those of low-alloy steel, the increase in tensile strength and hardness between 400 and 510 °C (750 and 950 °F) may be associated with a serious decrease in notch toughness, while tempering on the high side of the temperature range generally coincides with a decrease in corrosion resistance. The tempering temperatures most frequently employed to achieve desired hardness and other mechanical properties are included in Table 3.

The Izod impact curves in Fig. 6 through 11 reveal a loss of impact strength when parts are tempered within the range of 370 to 650 °C (700 to 1200 °F). Tempering within this range also results in decreased corrosion resistance, particularly resistance to stress-corrosion cracking (Fig. 12). Double tempering (cooling parts to room temperature after the first tempering treatment) also is beneficial for resistance to stress corrosion.

Annealing. Temperatures and resulting hardnesses for process (subcritical) annealing, full annealing and isothermal annealing are given in Table 4. Full annealing is an expensive and time-consuming treatment; it should be used only when required for subsequent severe forming. Types 414 and 431 do not respond to full or isothermal annealing procedures within a reasonable soaking period.

Isothermal annealing is recommended where maximum softening is required and adequate facilities for controlled slow cooling are not available.

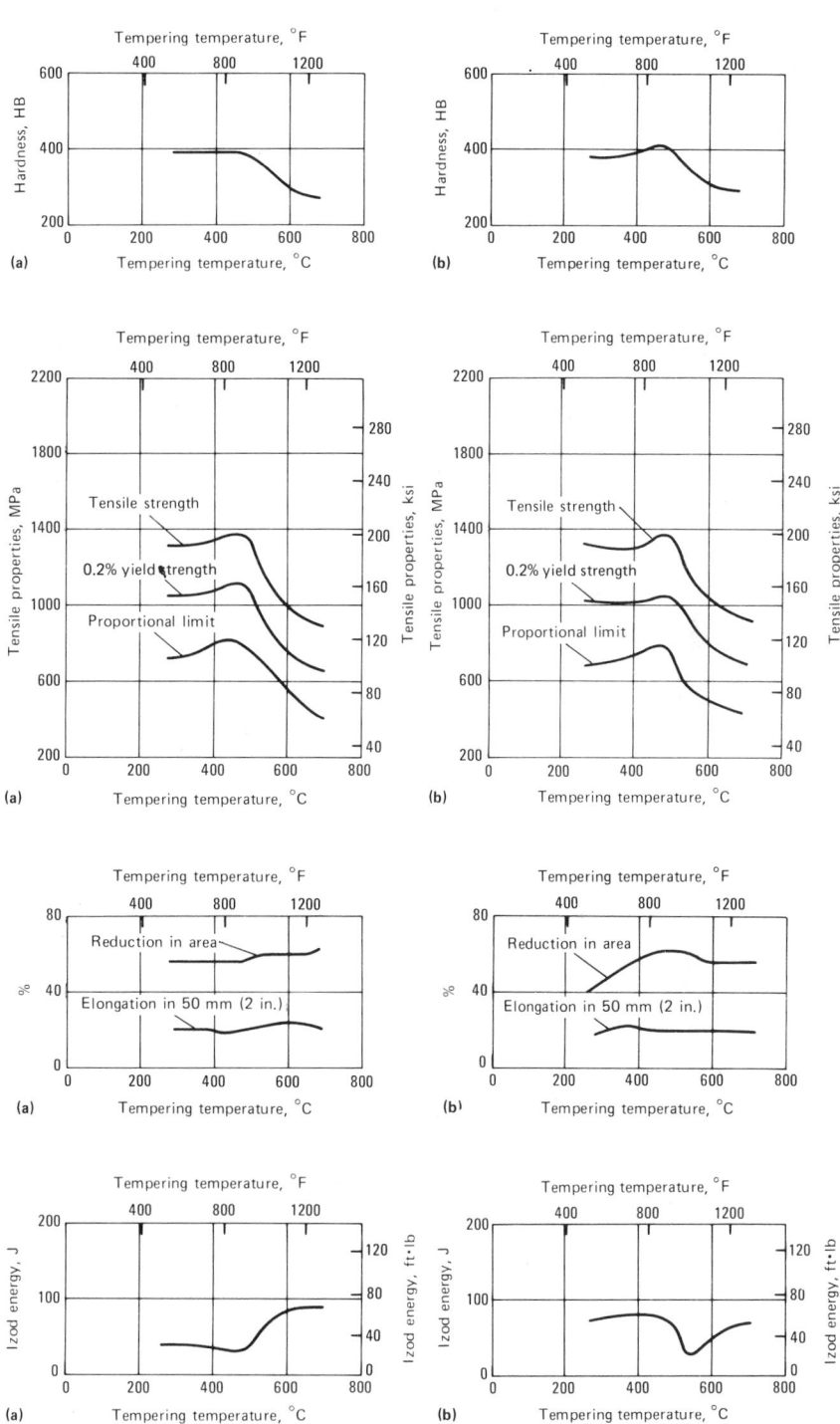

Fig. 10 Effect of austenitizing and tempering temperatures on typical mechanical properties of type 431 martensitic stainless steel

Austenitized 30 min; oil quenched to 65 to 95 °C (150 to 200 °F); double stress relieved at 175 °C (350 °F) for 15 min and water quenched; tempered 2 h. (a) Quenched from 925 °C (1700 °F). (b) Quenched from 1040 °C (1900 °F)

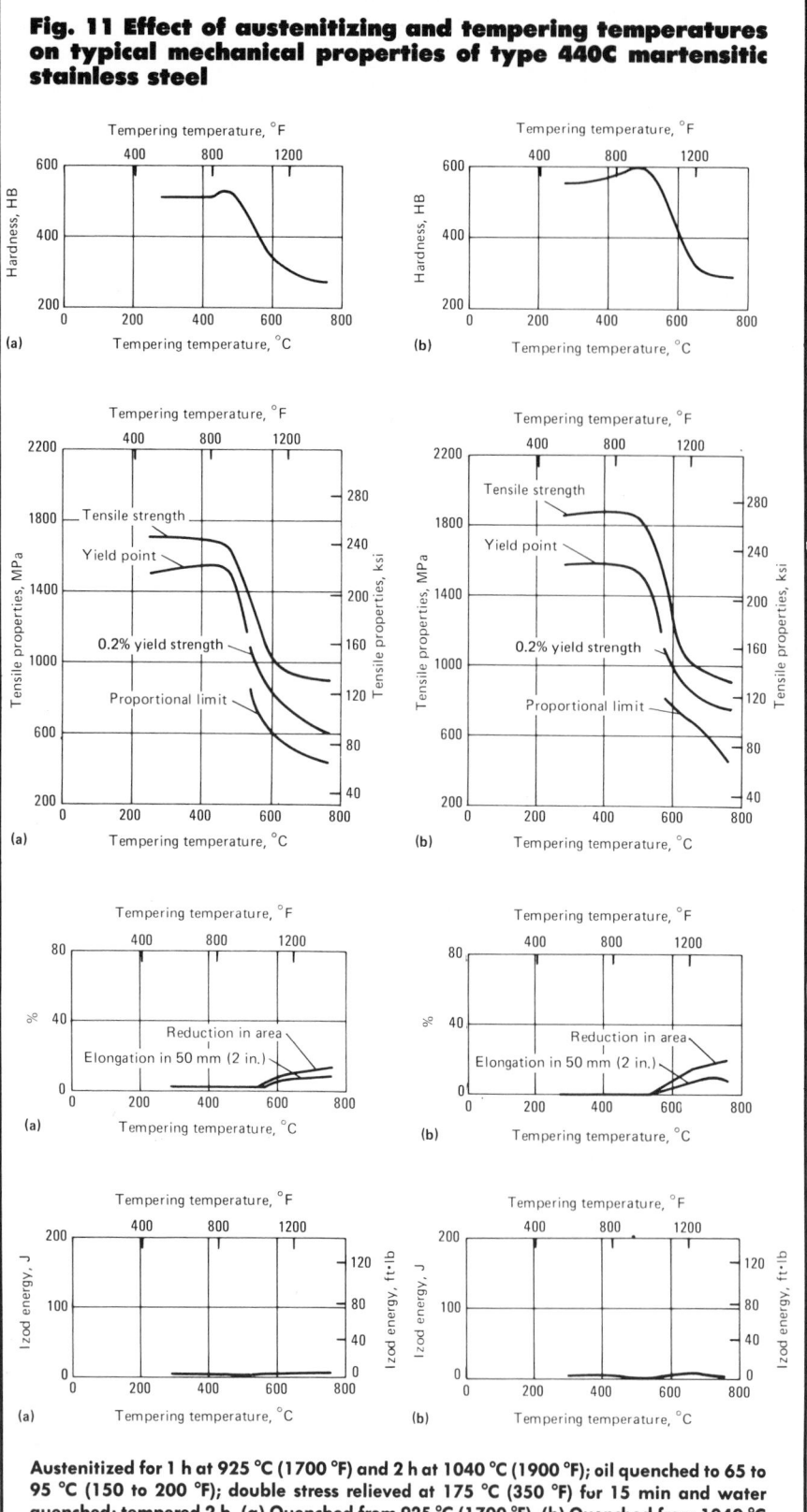

Fig. 11 Effect of austenitizing and tempering temperatures on typical mechanical properties of type 440C martensitic stainless steel

Austenitized for 1 h at 925 °C (1700 °F) and 2 h at 1040 °C (1900 °F); oil quenched to 65 to 95 °C (150 to 200 °F); double stress relieved at 175 °C (350 °F) for 15 min and water quenched; tempered 2 h. (a) Quenched from 925 °C (1700 °F). (b) Quenched from 1040 °C (1900 °F)

Full annealing, isothermal annealing, and especially repeated process annealing promote the formation of coarse carbides that take longer to dissolve at austenitizing temperatures. Subcritical annealing is recommended for all applications that do not require maximum softness.

Salt Baths. Many stainless steel parts are heat treated in molten salt, with excellent results. The baths usually employed consist of barium chloride with 5 to 35% sodium or potassium chloride. Alkaline-earth and other metallic oxides build up in these baths through use, but these oxides are not harmful to low-carbon stainless steels. However, if these salt baths are to be used also for hardening other alloy steels, to avoid surface decarburization it is necessary to rectify the baths with graphite to remove the metallic oxides and with methyl chloride gas to convert the alkaline-earth oxides back to chlorides. A bath treated with methyl chloride will carburize low-carbon stainless steels unless it is aged for at least 24 h before the stainless is treated in it. To avoid this problem, stainless steel parts should be heat treated in a salt bath reserved exclusively for stainless steels.

Protective Atmospheres. Argon or helium, if used as protective atmospheres, should be exceptionally dry (with a dew point below −50 °C or −60 °F). Because they are expensive and cannot be generated, they are rarely used. Exothermic and endothermically generated gas can be used with excellent results. These require dew point or infrared control so as not to carburize or decarburize the stainless grade being heat treated. Endothermic gas containing approximately 40% hydrogen can embrittle martensitic stainless steels that are oil quenched.

An exothermic gas ratio of 6.5 or 7 to 1 is satisfactory for grades of stainless containing not more than 0.15% carbon. When an endothermic atmosphere is used, dew points for specific steels and austenitizing temperatures should be:

Alloy	Austenitizing temperature, °C	Dew point, °C
420	1010	10-12
403, 410, 414, 416, 431	980	16-18
440C	1040	2-4

Table 5 Recommended conditions for abrasive blast cleaning of precipitation-hardening stainless steels prior to heat treatment(a)

Material	Abrasive Grit No.	Nozzle Size mm	in.	Angle, °	Air pressure KPa	psi	Cleaning speed mm²/s	in.²/min
Alumina (dry)	30	6.4	¼	45 to 60	170 to 655(b)	25 to 95(b)	130 to 215	12 to 20
Garnet or alumina (dry)	36	9.5	⅜	60	240	35	645	60
Wet blasting	220	6.4	¼	45 to 60	170 to 655(b)	25 to 95(b)	65 to 110	6 to 10

(a) All abrasive must be removed by thorough scrubbing. (b) Depending on metal thickness

Fig. 12 Effect of tempering temperature on the stress-corrosion characteristics of two martensitic stainless steels at high stress

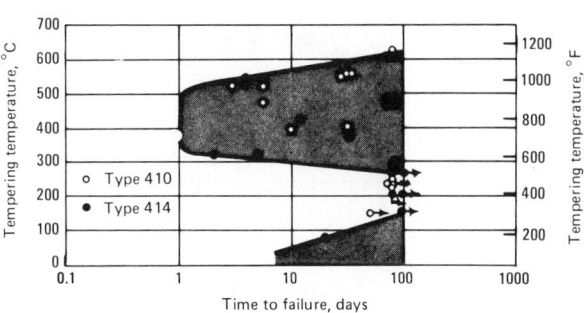

Data apply to a stress level of 550 MPa (80 ksi) for tests in a salt fog cabinet.

Table 6 Recommended heat treating procedures for semi-austenitic precipitation-hardenable stainless steels

UNS S17400

Homogenization. 1175 ± 15 °C (2150 ± 25 °F), 2 h + 30 min per 25 mm (1 in.)(a)

Austenite conditioning (solution treatment). 1040 ± 15 °C (1900 ± 25 °F), 30 min + 30 min per 25 mm (1 in.)(a)

Transformation cooling. To below +30 °C (+90 °F)

Precipitation hardening. To obtain minimum tensile strengths shown, use the following treatments for wrought alloys(b):

MPa	ksi		
1310	190	480 ± 5 °C	(900 ± 10 °F) .... 1 h
1170	170	495 ± 5 °C	(925 ± 10 °F) .... 4 h
1070	155	550 ± 5 °C	(1030 ± 10 °F) .... 4 h
1030	150	565 ± 5 °C	(1050 ± 10 °F) .... 4 h
1000	145	580 ± 5 °C	(1075 ± 10 °F) .... 4 h
930	135	620 ± 5 °C	(1150 ± 10 °F) .... 4 h

Comparable treatments for cast materials(b):

MPa	ksi		
1240	180	480 ± 5 °C	(900 ± 10 °F) .... 4 h
1170	170	495 ± 5 °C	(925 ± 10 °F) .... 4 h
1030	150	540 ± 5 °C	(1000 ± 10 °F) .... 4 h
900	130	595 ± 5 °C	(1100 ± 10 °F) .... 4 h

(continued)

(a) To prevent cracking and ensure uniform properties, cool as follows: 75 mm (3 in.) and less, oil quench or air cool; 75 to 150 mm (3 to 6 in.), air cool; 150 mm (6 in.) and over, air cool under cover. *All parts must be cooled to below +30 °C (+90 °F)* prior to the precipitation-hardening cycle. (b) If hardness exceeds maximum specified, reheat treat at a slightly higher temperature for a minimum of 30 min. (c) Air cool to room temperature; *do not reheat before transformation cooling.* (d) Time at heat is dependent upon section size. Normally, a 1 h hold at temperature is suggested.

Hydrogen embrittlement can become an important concern in the martensitic grades, generally increasing with hardness and carbon content. (It is variable and less acute in ferritic steels, and is virtually unknown in the austenitic grades.)

The embrittling hydrogen may be acquired as a result of the melting process, a heat treating atmosphere, or chemical and electrochemical processes such as pickling and electroplating.

Most heat treating atmospheres contain hydrogen in the form of (a) moisture, (b) hydrocarbons, or (c) elemental hydrogen as an atmosphere or a dissociation product. The use of pure hydrogen or dissociated ammonia for bright annealing in one plant was associated with cracking of wire coils of types 431 and 440C, although other plants have reported no similar difficulty. Nevertheless, it is possible that some loss in ductility may result from the bright annealing of any of the martensitic stainless steels.

Embrittlement has been found in oil-quenched types 403, 410, 414 and 431. Air quenching or the subsequent tempering of oil-quenched material releases the entrapped hydrogen, and ductility is restored. If a green oxide appears on the oil-quenched alloy, this must be removed prior to tempering or it will interfere with the release of entrapped hydrogen.

Precipitation-Hardening Stainless Steels

In the heat treating of precipitation-hardening stainless steels, areas of primary interest include: (a) cleaning prior to heat treatment; (b) furnace atmospheres; (c) time-temperature cycles; (d) effect of variations in cycles; and (e) scale removal after heat treating.

Prior Cleaning. All parts must be cleaned thoroughly prior to heat treating. Because the chemical composition of these steels is delicately balanced, failure to remove drawing lubricants, cutting oils and grease can lead to surface carburization and improper response to heat treatment. As a secondary benefit, thorough cleaning promotes the formation of a uniform surface scale that is readily removable.

The recommended cleaning procedure comprises vapor degreasing or solvent cleaning, followed by mechanical scrubbing with a mild abrasive alkaline cleaner to remove insoluble soils. All traces of cleaners should be removed by thoroughly rinsing with warm water.

Wet or dry abrasive blasting may be substituted for the above procedures. Recommended grits and operating details for blasting are given in Table 5. After blasting, all traces of abrasive must be removed from the work by scrubbing thoroughly.

In some applications, cleaning prior to heat treating may be accomplished by closely controlled pickling in a 10% HNO$_3$-2% HF aqueous solution at 45 to 60 °C (110 to 140 °F). Time should be limited to 2 or 3 min. This method is not recommended for cleaning severely formed or previously heat treated parts.

Furnaces fired with oil or natural gas are not entirely satisfactory for the heat treatment of these steels where finished surfaces are not to be subsequently machined. In such units, it is difficult to control combustion contaminants and to eliminate flame impingement on the parts being treated. Electric furnaces or gas-fired radiant-tube furnaces are generally used for heat treating precipitation-hardening stainless steels.

Furnace Atmospheres. Air is a satisfactory furnace atmosphere for austenite-conditioning and annealing operations. Controlled reducing atmospheres, such as dissociated ammonia or bright-annealing gas, introduce the potential hazard of nitriding or carburizing, either of which has a deleterious effect on mechanical properties.

Bright annealing may be done in hydrogen, argon or helium atmospheres, provided a dew point of −55 °C (−65 °F) or lower is maintained. The cooling rate from the annealing temperature must be approximately equal to that of cooling in still air. Austenite-conditioning treatments at temperatures as high as 925 to 955 °C (1700 to 1750 °F) may also be performed in dry hydrogen, argon or helium, maintaining the same low dew point. A scale-free surface will be obtained.

Table 6 (continued)

UNS S17700

Full annealing. 1065 ± 15 °C (1950 ± 25 °F), 3 min + 1 min per 0.25 mm (0.01 in.); air cool

RH treatments

Austenite conditioning. 955 ± 15 °C (1750 ± 25 °F), 10 min + 1 min per 0.25 mm (0.01 in.); air cool(c)

Transformation cooling. To below −70 °C (−90 °F), 8 h

Precipitation hardening. To obtain minimum tensile strengths shown, the following treatments are recommended(b):

MPa	ksi
1450	210 .... 510 ± 5 °C (950 ± 10 °F) .... 1 h
1240	180 .... 565 ± 5 °C (1050 ± 10 °F) .... 1 h
1170	170 .... 580 ± 5 °C (1075 ± 10 °F) .... 1 h
1030	150 .... 595 ± 5 °C (1100 ± 10 °F) .... 1 h

TH treatments

Austenite conditioning. 760 ± 15 °C (1400 ± 25 °F), 1½ h; cool within 1 h to below 15 °C (60 °F) but above 0 °C (32 °F), and hold at least ½ h before precipitation hardening

Precipitation hardening. To obtain minimum tensile strengths shown, the following treatments are recommended(b):

MPa	ksi
1240	180 .... 565 ± 5 °C (1050 ± 10 °F) .... 1½ h
1170	170 .... 580 ± 5 °C (1075 ± 10 °F) .... 1½ h
1030	150 .... 595 ± 5 °C (1100 ± 10 °F) .... 1½ h

UNS S15700

Full annealing. Same as for 17-7 PH

RH treatments

Austenite conditioning. Same as for 17-7 PH

Transformation cooling. Same as for 17-7 PH

Precipitation hardening. To obtain minimum tensile strengths shown, the following treatments are recommended(b):

MPa	ksi
1550	225 .... 510 ± 5 °C (950 ± 10 °F) .... 1 h
1310	190 .... 565 ± 5 °C (1050 ± 10 °F) ... 1 h

TH treatments

Austenite conditioning. 760 ± 15 °C (1400 ± 25 °F), 1½ h; cool within 1 h to below 15 °C (60 °F) but above 0 °C (32 °F), and hold at least ½ h before precipitation hardening

Precipitation hardening. For minimum tensile strength of 1310 MPa (190 ksi) (b): 565 ± 5 °C (1050 ± 10 °F), 1½ h

(continued)

(a) To prevent cracking and ensure uniform properties, cool as follows: 75 mm (3 in.) and less, oil quench or air cool; 75 to 150 mm (3 to 6 in.), air cool; 150 mm (6 in.) and over, air cool under cover. *All parts must be cooled to below +30 °C (+90 °F) prior to the precipitation-hardening cycle.* (b) If hardness exceeds maximum specified, reheat treat at a slightly higher temperature for a minimum of 30 min. (c) Air cool to room temperature; *do not reheat before transformation cooling.* (d) Time at heat is dependent upon section size. Normally, a 1 h hold at temperature is suggested.

The lower austenite-conditioning temperatures, such as 760 °C (1400 °F), present difficulties in achieving scale-free surfaces in dry hydrogen, argon or helium. An air atmosphere is generally used at these temperatures. For complete freedom from scale or discoloration at the lower temperatures, a vacuum furnace is required.

Final hardening of these steels is performed at relatively low temperatures, and an air atmosphere is acceptable for these treatments.

Heat Treating Procedures. Rec-

Table 6 (continued)

UNS S35000

Full annealing. Wrought materials only: 1065 ± 15 °C (1950 ± 25 °F), 3 min + 1 min per 0.25 mm (0.01 in.); air cool

Austenite conditioning. 930 ± 5 °C (1710 ± 10 °F), 10 min + 1 min per 0.25 mm (0.01 in.); air cool(c)

Transformation cooling. To −75 ± 5 °C (−100 ± 10 °F), 3 h (minimum)

Precipitation hardening. To obtain minimum tensile strengths shown, the following treatments are recommended(b):

MPa	ksi		
1270	185	.... 455 ± 5 °C (850 ± 10 °F)	.... 3 h
1170	170	.... 510 ± 5 °C (950 ± 10 °F)	.... 3 h
1140	165	.... 540 ± 5°C (1000 ± 10 °F)	.... 3 h

UNS S35500

Homogenization. Castings only: 1095 ± 15 °C (2000 ± 25 °F), 2 h; air cool (water quench sections over 50 mm, or 2 in.). Bar and forgings: 1050 ± 15 °C (1925 ± 25 °F), 1 to 3 h; water quench. Cool all forms to below −70 °C (−90 °F), hold 3 h minimum.

Full annealing. 1025 ± 15 °C (1875 ± 25 °F), 1 h per 25 mm (1 in.); water quench

Machinability treatment. 760 ± 15 °C (1400 ± 25 °F), 3 h; air cool. Refrigerate to −70 °C (−90 °F), hold for 3 h. Reheat to 565 ± 15 °C (1050 ± 25 °F), 3 h

For −70 °C (−90 °F) transformation:

Austenite conditioning. Castings: 980 ± 15 °C (1800 ± 25 °F), 2 h; air cool (oil quench sections over 3 mm, or 0.125 in.). Wrought materials: 930 ± 15 °C (1710 ± 25 °F), 15 min per in.; air cool (oil quench sections over 3 mm or 0.125 in.)

Transformation cooling. To −75 ± 5 °C (−100 ± 10 °F), 3 h

Precipitation hardening. To obtain minimum tensile strengths shown, use the following treatments for wrought alloys(b):

MPa	ksi		
1310	190	.... 455 ± 5 °C (850 ± 10 °F)	.... 3 h
1170	170	.... 540 ± 5 °C (1000 ± 10 °F)	.... 3 h

For castings, to obtain minimum tensile strength of 1240 MPa (180 ksi)(b): 455 ± 5 °C (850 ± 10 °F), 2 h

UNS S45000

Solution annealing. 1040 ± 15 °C (1900 ± 25 °F), 1 h at heat (d), water quench

Precipitation hardening. Typical tensile strengths shown may be obtained by the following treatments:

MPa	ksi		
1345	195	.... 480 ± 5 °C (900 ± 10 °F)	.... 4 h, air cool
1170	170	.... 540 ± 5 °C (1000 ± 10 °F)	.... 4 h, air cool
1105	160	.... 565 ± 5 °C (1050 ± 10 °F)	.... 4 h, air cool
965	140	.... 620 ± 5 °C (1150 ± 10 °F)	.... 4 h, air cool

(continued)

(a) To prevent cracking and ensure uniform properties, cool as follows: 75 mm (3 in.) and less, oil quench or air cool; 75 to 150 mm (3 to 6 in.), air cool; 150 mm (6 in.) and over, air cool under cover. *All parts must be cooled to below +30 °C (+90 °F) prior to the precipitation-hardening cycle.* (b) If hardness exceeds maximum specified, reheat treat at a slightly higher temperature for a minimum of 30 min. (c) Air cool to room temperature; *do not reheat before transformation cooling.* (d) Time at heat is dependent upon section size. Normally, a 1 h hold at temperature is suggested.

ommended procedures for full annealing, austenite conditioning, transformation cooling and age-tempering (precipitation hardening) are given in Table 6.

17-4 PH (UNS S17400) is a precipitation hardening steel that has an essentially martensitic structure and limited formability when supplied in the solution treated condition. Fabrication is followed by hardening in the range of 480 to 620 °C (900 to 1150 °F) (Table 6).

This alloy should not be put into service in any application in the solution treated condition, because in this condition its ductility can be relatively low and its resistance to stress-corrosion cracking is poor. Hardening to any of the strength levels shown in Table 6 improves both toughness and resistance to stress corrosion.

17-7 PH (UNS S17700) is normally supplied in the annealed condition (condition A), in which it is soft and formable. Heat treatment is accomplished through the use of the TH or RH procedures indicated in Table 6. The choice of method of heat treatment is in most instances dictated by the ease with which the particular sequence fits into the user's production techniques.

This alloy is also supplied in the cold rolled condition (condition C). Here, transformation has been achieved by cold rolling, and heat treatment is reduced to a single step: 480 °C (900 °F) for 1 h. Although strength and stress-corrosion resistance are greatly increased by this treatment, ductility is reduced and formability is limited.

PH 15-7 Mo (UNS S15700) is a high-strength modification of 17-7 PH. It is supplied in the same condition as 17-7 PH and requires identical heat treating procedures. Table 6 illustrates the strength levels obtainable with PH 15-7 Mo.

15-5 Ni alloys (UNS S15500) are normally supplied in the annealed condition. They should not be put into service in the annealed condition because of susceptibility to stress corrosion cracking. The alloy can be hardened by heating to a temperature in the range of 480 to 620 °C (900 to 1150 °F) for 1 to 4 h, depending on the temperature, then air cooling. Table 6 illustrates typical strength levels obtainable.

13-8 Mo alloys (UNS S13800) are normally supplied in the annealed condition. They can be hardened to high strength levels by a single low temperature treatment. Table 6 illustrates typical strength levels versus hardening procedures.

AM-350 and PYRO MET 350 (UNS S35000) are normally purchased in the fully annealed condition; however, after severe forming or cold working, it may require a second annealing treatment. The annealing temperature limits, indicated in Table 6, are critical.

High temperatures reduce strength; lower temperatures adversely affect formability.

After annealing and conditioning at 930 °C (1710 °F), AM-350 is usually subzero cooled, and then aged at 455 °C (850 °F) for 3 h; this treatment produces maximum strength. Maximum toughness is achieved by aging in the range of 480 to 540 °C (900 to 1000 °F). The recommended temperature for subzero cooling must be carefully observed. Cooling to much lower temperatures, such as −195 °C (−320 °F), results in incomplete transformation, as does failure to hold at the recommended temperature for at least 3 h.

AM-355 and PYRO MET 355 (UNS S35500) flat products are supplied in either the solution treated or solution treated and cold rolled condition, whereas bar products are usually supplied in the equalized and overtempered condition for best machinability. Most castings are supplied in the as-cast condition.

Although, as indicated in Table 6, the homogenizing and austenite-conditioning treatments applied to castings differ from those applied to wrought materials (higher temperatures being used for castings in both treatments), subzero cooling is required for all forms, to obtain maximum toughness and corrosion resistance. The full annealing treatment shown in Table 6 would normally be applicable only to flat products. The machinability treatment is required for obtaining good machining characteristics in this alloy.

Wrought materials should be aged at 455 °C (850 °F) for maximum strength, and at 540 °C (1000 °F) for maximum ductility and toughness. The usual aging treatment for castings consists of holding at 455 °C (850 °F) for 2 h.

Custom 450 (UNS S45000) stainless is normally supplied in the annealed condition, requiring no further heat treatment for many applications. It is easily fabricated in the annealed condition. A single-step hardening treatment develops higher strength with good ductility and toughness.

The recommended minimum hardening temperature of 480 °C (900 °F) produces the optimum combination of strength, ductility, and toughness. Hardening at temperatures up to 620 °C (1150 °F) increases the ductility and decreases strength (see Table 6).

Custom 455 (UNS S45500) stainless is normally in the annealed condition.

Table 6 (continued)

UNS S45500

Solution annealing. 830 ± 15 °C (1525 ± 25 °F), 1 h at heat (d), water quench
Precipitation hardening. Typical tensile strengths shown may be obtained by the following treatments:

MPa	ksi
1725	250 .... 480 ± 5 °C (900 ± 10 °F) .... 4 h, air cool
1620	235 .... 510 ± 5 °C (950 ± 10 °F) .... 4 h, air cool
1450	210 .... 540 ± 5 °C (1000 ± 10 °F) .... 4 h, air cool
1310	190 .... 565 ± 5 °C (1050 ± 10 °F) .... 4 h, air cool

UNS S15500

Solution annealing. 1040 ± 15 °C (1900 ± 25 °F), 1 h (d), water quench
Precipitation hardening. Typical tensile strengths shown may be obtained by the following treatments:

MPa	ksi
1380	200 .... 480 ± 5 °C (900 ± 10 °F) .... 1 h, air cool
1170	170 .... 550 ± 5 °C (1025 ± 10 °F) .... 4 h, air cool
1000	145 .... 620 ± 5 °C (1150 ± 10 °F) .... 4 h, air cool

H1150M condition (after annealing). 760 ± 8 °C (1400 ± 15 °F), 2 h, air cool + 620 ± 5 °C (1150 ± 10 °F), 4 h, air cool
Typical tensile strength. 860 MPa (125 ksi)

(continued)

(a) To prevent cracking and ensure uniform properties, cool as follows: 75 mm (3 in.) and less, oil quench or air cool; 75 to 150 mm (3 to 6 in.), air cool; 150 mm (6 in.) and over, air cool under cover. *All parts must be cooled to below +30 °C (+90 °F)* prior to the precipitation-hardening cycle. (b) If hardness exceeds maximum specified, reheat treat at a slightly higher temperature for a minimum of 30 min. (c) Air cool to room temperature; *do not reheat before transformation cooling.* (d) Time at heat is dependent upon section size. Normally, a 1 h hold at temperature is suggested.

It is relatively soft and easily formable in the annealed condition. A single-step hardening treatment develops exceptionally high yield strengths with good ductility and toughness. Harden by heating in the range of 480 to 565 °C (900 to 1050 °F). See Table 6.

Variations in Heat Treating Cycles. One of the principal advantages of these materials is their versatility. Although certain heat treatments have been classified as standard, there are applications where deviations from these standards are desirable. The series of curves in Fig. 13 through 17 illustrates how these deviations affect mechanical properties.

Scale Removal After Heat Treating. The amount and nature of scale vary with the degree of cleanness of the work being treated, the furnace atmosphere, and the temperature and duration of heat treatment. In the following discussion, it will be assumed that all heat treating operations are performed in an air atmosphere. A variety of descaling methods may be employed; the choice depends on the type of steel and the facilities available.

In removing scale formed during homogenization or full annealing, the use of a 10% HNO₃ - 2% HF aqueous solution at 45 to 60 °C (110 to 140 °F) has been effective. Exposure to the acid solution should be limited to a period of 3 min. Removal of loosened scale may be facilitated by the use of high-pressure water or steam. A uniform surface is evidence of a well-cleaned part.

The austenite-conditioning treatments produce a scale that is best removed by mechanical means. Acids should be avoided, because they are a possible source of intergranular attack. Wet grit blasting processes have been widely used to remove these scales and have been found to be highly satisfactory.

The final step in heat treating (precipitation hardening) produces a discoloration of heat tint. It is desirable to use mechanical means to remove this oxide from 17-7 PH, PH 15-7 Mo, AM-350 and AM-355. The HNO₃ - HF solution has been used on these steels, but extreme care is required to prevent intergranular attack. The acid solution

Table 6 (continued)

UNS S13800

Annealing. 925 ± 8 °C (1700 ± 15 °F), 1 h (d), air cool or oil quench

Precipitation hardening. Typical tensile strengths shown may be obtained by the following treatments:

MPa	ksi				
1550	225	.... 510 ± 5 °C	(950 ± 10 °F)	....	4 h, air cool
1310	190	.... 565 ± 5 °C	(1050 ± 10 °F)	....	4 h, air cool
1000	145	.... 620 ± 5 °C	(1150 ± 10 °F)	....	4 h, air cool

H1150M condition (after annealing). 760 ± 8 °C (1400 ± 15 °F), 2 h, air cool + 620 ± 5 °C (1150 ± 10 °F), 4 h, air cool

Typical tensile strength. 895 MPa (130 ksi)

(a) To prevent cracking and ensure uniform properties, cool as follows: 75 mm (3 in.) and less, oil quench or air cool; 75 to 150 mm (3 to 6 in.), air cool; 150 mm (6 in.) and over, air cool under cover. *All parts must be cooled to below +30 °C (+90 °F) prior to the precipitation-hardening cycle.* (b) If hardness exceeds maximum specified, reheat treat at a slightly higher temperature for a minimum of 30 min. (c) Air cool to room temperature; *do not reheat before transformation cooling.* (d) Time at heat is dependent upon section size. Normally, a 1 h hold at temperature is suggested.

Table 7 Annealing of ferritic and austenitic stainless steel castings

Type	Minimum temperature °C	Minimum temperature °F	Quench(a)	Typical ultimate tensile strength(b) MPa	ksi
For full softness					
CB-30	790	1450	FC + A(c)	660	95
CC-50	790	1450	A	670	97
For maximum corrosion resistance					
CE-30	1095	2000	W, O, A	670	97
CF-3, CF-3M	1040	1900	W, O, A	530	77
CF-8, CF-8C (d)	1040	1900	W, O, A	530	77
CF-8M, CF-12M (e)	1040	1900	W, O, A	550	80
CF-16F, CF-20	1040	1900	W, O, A	530	77
CH-20	1095	2000	W, O, A	610	88
CK-20	1095	2000	W, O, A	520	76
CN-7M	1120	2050	W, O, A	480	69

(a) FC, furnace cool; W, water; O, oil; A, air. (b) Approximate. (c) Furnace cool to 540 °C (1000 °F), then air cool. (d) CF-8C may be reheated to 870 to 925 °C (1600 to 1700 °F), then air cooled, for precipitation of niobium carbides. (e) CF-12M should be quenched from a temperature above 1095 °C (2000 °F).

may be used satisfactorily with 17-4 PH. To a lesser extent, electropolishing has also been used to remove the final heat tint resulting from precipitation hardening.

Stainless Steel Castings

The heat treatment of stainless steel castings follows closely in purpose and procedure the thermal processing of comparable wrought materials. However, the differences in detail warrant separate consideration here.

Because they are not cold worked or cold formed, castings of the older, con-ventional martensitic grades CA-15 and CA-40 (UNS J91150 and J91153) do not require subcritical annealing to remove the effects of cold working. However, in work-hardenable ferritic alloys, machining and grinding stress-es are relieved at temperatures from about 260 to 540 °C (500 to 1000 °F). Casting stresses in the martensitic castings noted above should be relieved by subcritical annealing prior to fur-ther heat treatment. When these hard-ened martensitic castings are stress relieved, the stress-relieving tempera-ture must be kept below the final tem-pering or aging temperature.

An improved, cast martensitic alloy, CA-6NM (UNS-J91540), possesses bet-ter casting behavior, improved weld-ability and equals or exceeds all of the mechanical, corrosion and cavitation resistance properies of CA-15 and, as a result, has largely replaced the older alloy. Both CA-6NM and CA-15 cast-ings are normally supplied in the nor-malized condition at 955 °C (1750 °F) minimum and tempered at 595 °C (1100 °F) minimum. However, when it is necessary or desirable to anneal CA-6NM castings, a temperature of 790 to 815 °C (1450 to 1500 °F) should be used. The alloy should be furnace cooled or otherwise slow cooled to 595 °C (1100 °F), after which it may be cooled in air. When stress relieving is required, CA-6NM may be heated to 620 °C (1150 °F) maximum, followed by slow cooling to prevent the formation of martensite.

Homogenization. Alloy segrega-tion and dendritic structures may occur in castings, and may be particularly pronounced in heavy sections. Because castings are not subjected to the high-temperature mechanical reduction and soaking treatments entailed in the mill processing of wrought alloys, it is fre-quently necessary to homogenize some alloys at temperatures above 1095 °C (2000 °F) to promote uniformity of chemical composition and microstruc-ture. Full annealing of martensitic castings results in recrystallization and maximum softness, but it is less effective than homogenization in elimi-nating segregation. Homogenization is a common procedure in the heat treat-ment of precipitation-hardening cast-ings.

Ferritic and Austenitic Alloys. The ferritic, austenitic, and mixed fer-ritic-austenitic alloys are not harden-able by heat treatment. They can be heat treated to improve their corrosion resistance and machining characteris-tics. The ferritic alloys CB-30 and CC-50 (UNS J91803 and J92615) are an-nealed to relieve stresses and reduce hardness by being heated above 790 °C (1450 °F) (Table 7).

The austenitic alloys achieve maxi-mum resistance to intergranular corro-sion by the high-temperature heating and quenching procedure known as solution annealing (Table 7). As-cast structures, or castings exposed to tem-peratures in the range from 425 to 870 °C (800 to 1600 °F), may contain com-plex chromium carbides precipitated preferentially along grain boundaries in wholly austenitic alloys. This micro-

Fig. 13 Effect of variations in annealing temperature on typical mechanical properties of 17-7 PH sheet, strip and plate

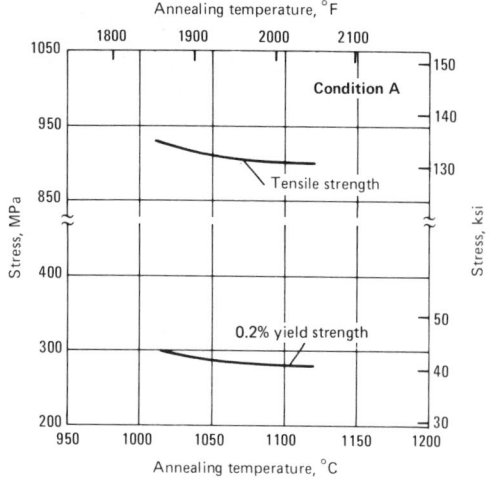

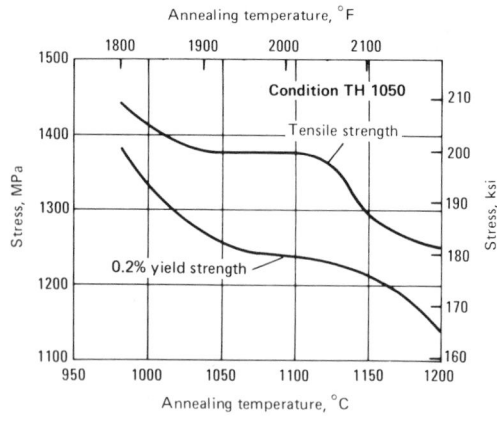

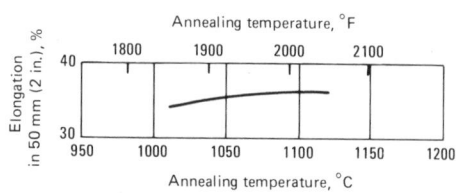

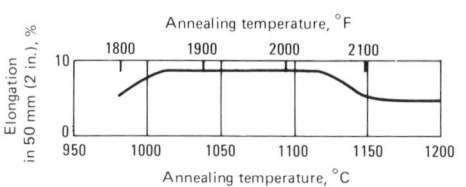

Fig. 14 Effect of variations in transformation treatment temperature and time on typical mechanical properties of 17-7 PH sheet, strip and plate

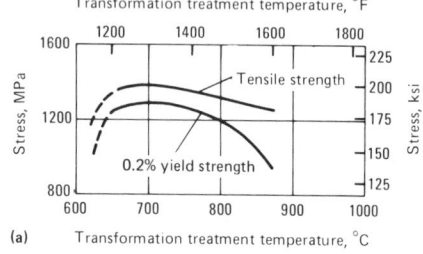

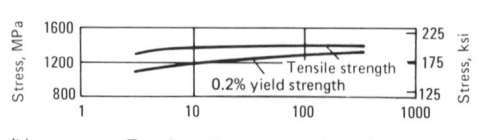

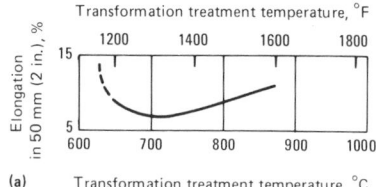

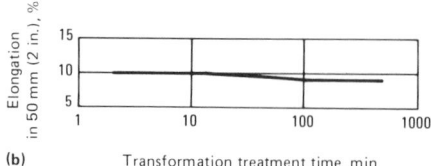

(a) Heated for 90 min; cooled to 15 °C (60 °F); hardened at 565 °C (1050 °F). (b) Heated at 760 °C (1400 °F); hardened at 565 °C (1050 °F), 90 min

Table 8 Effects of four methods of heat treatment on typical mechanical properties of cast CA-15 (a)

Heat treatment(b)	Ultimate tensile strength		Yield strength		Elonga-tion in 50 mm (2 in.), %	Reduc-tion in area, %
	MPa	ksi	MPa	ksi		
Treatment 1	1230	178	1010	146	9.0	13.0
Homogenize: 1 h at 1040 °C (1900 °F), AC	1250	181	970	141	12.5	28.0
Solution anneal: ½ h at 955 °C (1750 °F), OQ	1280	185	990	143	7.0	14.0
Temper: 3 h at 300 °C (575 °F), AC	1320	191	1020	148	8.0	12.5
Treatment 2	1260	183	1120	162	6.5	9.5
Anneal: 1 h at 900 °C (1650 °F), FC	1300	188	1130	164	5.5	16.0
Solution anneal: 1¼ h at 1010 °C (1850 °F), OQ	1340	194	1070	155	9.0	23.0
Temper: 3 h at 370 °C (700 °F), OQ	1380	200	1050	152	12.0	42.0
Treatment 3(c)	790	115	480	70	15.5	60.0
Anneal: 1 h at 900 °C (1650 °F), FC	810	117	630	91	16.5	37.0
Solution anneal: 1¼ h at 1010 °C (1850 °F), OQ	830	120	680	98	9.5	23.0
Temper: 2 h at 620 °C (1150 °F), AC	860	125	590	85	12.5	32.0
Treatment 4(d)	680	99	520	76	21.0	65.0
Anneal: 1 h at 900 °C (1650 °F), FC	710	103	540	79	20.5	56.0
Solution anneal: 1½ h at 995 °C (1825 °F), FAC	710	103	540	79	18.5	61.5
Temper: 2 h at 705 °C (1300 °F), AC	720	104	550	80	20.5	60.0

(a) Specimens taken from shell mold cast keel blocks; data indicate results obtained on four specimens treated by each method. (b) Each treatment comprised three processes as listed. AC, air cool; OQ, oil quench; FC, furnace cool; FAC, forced-air cool. (c) AMS 5351-B. (d) MIL-S-16993

Fig. 15 Effect of variations in austenite-conditioning temperature and time on typical mechanical properties of 17-7 PH sheet, strip and plate

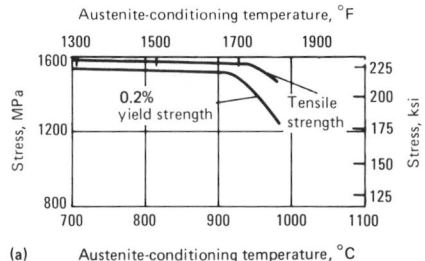

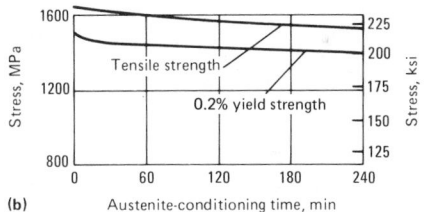

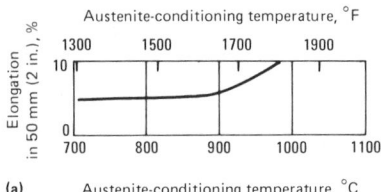

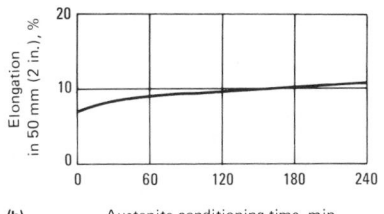

(a) Heated for 10 min; air cooled, liquid cooled to −75 °C (−100 °F), 8 h; hardened at 510 °C (950 °F), 1 h. (b) Heated at 955 °C (1750 °F); air cooled; liquid cooled to −75 °C (−100 °F), 8 h; hardened at 510 °C (950 °F), 1 h

Fig. 16 Effect of variations in hardening temperature and time on typical mechanical properties of 17-7 PH sheet, strip and plate

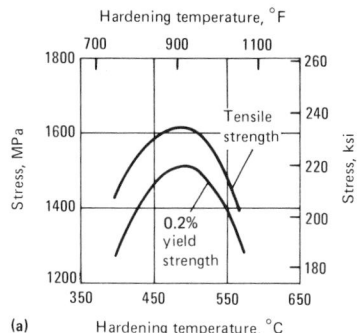

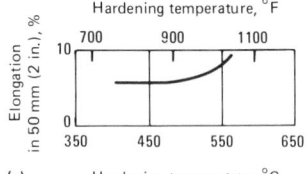

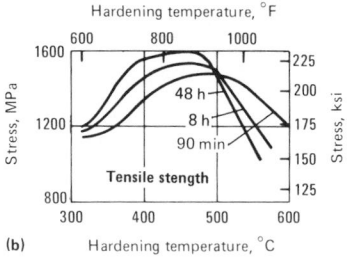

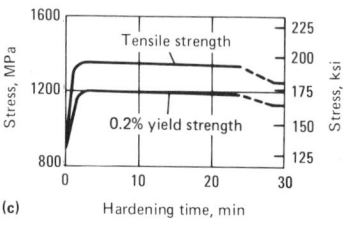

(a) Heated at 955 °C (1750 °F), 10 min; air cooled, liquid cooled to −75 °C (−100 °F), 8 h; hardened for 1 h. (b) Heated at 760 °C (1400 °F), 90 min; air cooled to room temperature, water quenched to 15 °C (60 °F); hardened as indicated. Elongation data not available. (c) Heated at 760 °C (1400 °F), 90 min; air cooled to room temperature, water quenched to 15 °C (60 °F); hardened at 565 °C (1050 °F). Elongation data not available

Table 9 Heat treatment of martensitic stainless steel castings

Annealing temperature(a)		Hardening treatment				Typical ultimate tensile strength(c)	
		Austenitizing temperature(b)		Tempering temperature			
°C	°F	°C	°F	°C	°F	MPa	ksi
CA-15 alloy							
845-900	1550-1650	···	···	···	···	550	80
···	···	925-1010(d)	1700-1850(d)	370 max(e)	700 max(e)	1380	200
···	···	925-1010(d)	1700-1850(d)	595-760	1100-1400	690-930	100-135
CA-40 alloy							
845-900	1550-1650	···	···	···	···	620	90
···	···	980-1010	1800-1850	315 max(e)	600 max(e)	1520	220
···	···	980-1010	1800-1850	595	1100	1030	150
···	···	980-1010	1800-1850	650	1200	970	140
···	···	980-1010	1800-1850	760	1400	760	110
CA-6NM alloy							
760-815	1450-1500	···	···	···	···	550	80
···	···	950-980	1750-1800	595-620	1100-1150	830	120

(a) Annealing for maximum softness; slow furnace cool from temperature. (b) Quench in oil or air. (c) Approximate. (d) Hold at temperature for a minimum of 30 min. (e) Tempering at 370 to 595 °C (700 to 1100 °F) is not recommended because low impact ductility results.

Table 10 Effect of temperature of 1-h aging treatment on typical properties of investment cast 17-4 PH stainless steels(a)

Aging temperature		Ultimate tensile strength		Yield strength, 0.2%		Elongation in 50 mm (2 in.),%	Hardness, HRC
°C	°F	MPa	ksi	MPa	ksi		
Alloy with 0.15 to 0.40% niobium							
As cast		1070	153	770	112	3.5	···
480	900	1380	200	1070	153	15	44
510	950	1360	197	1080	157	13	42
540	1000	1130	164	970	141	14	39
565	1050	1120	163	1040	151	16	35
595	1100	1120	162	990	143	16	34
650	1200	1010	147	860	125	15	30
Alloy without niobium							
As cast		1120	162	990	143	2.7	38
480	900	1360	198	1140	166	12	43
510	950	1250	182	1110	161	13	42
540	1000	1280	186	1100	159	14	38
565	1050	980	142	910	132	16	35
595	1100	1080	157	840	122	16	34
650	1200	1050	153	900	130	12	32

(a) As determined on cast test bars. Before aging, specimens were homogenized (1½ h at 1150 °C or 2100 °F, air cool) and solution annealed (½ h at 1040 °C or 1900 °F, oil quench); subzero transformation not employed. After 1 h at aging temperature, specimens were air cooled.

structure is susceptible to intergranular corrosion, especially in oxidizing solutions. (In partially ferritic alloys, carbides tend to precipitate in the discontinuous ferrite pools; thus, these alloys are less susceptible to intergranular attack.) The purpose of solution annealing is to ensure complete solution of carbides in the matrix and to retain these carbides in solid solution.

Solution annealing procedures for all austenitic alloys are similar, and consist of heating to a temperature of about 1095 °C (2000 °F), holding for a time sufficient to accomplish complete solution of carbides, and quenching at a rate fast enough to prevent reprecipitation of the carbides—particularly while cooling through the range from 870 to 540 °C (1600 to 1000 °F). Tem-

Table 11 Effect of aging time at 480 °C (900 °F) on typical properties of investment cast 17-4 PH stainless steels(a)

Aging time, h	Ultimate tensile strength		Yield strength		Elongation in 25 mm (1 in.), %	Hardness, HRC
	MPa	ksi	MPa	ksi		
Alloy with 0.15 to 0.40% niobium						
½	1380	201	1280	185	7	45
1	1380	200	1070	155	15	44
2	1340	194	1060	153	13	45
4	1300	188	1080	156	9	43
Alloy without niobium						
½	1380	201	1080	156	10	43
1	1360	198	1130	164	12	43
2	1390	202	1080	157	12	44
4	1180	171	980	142	16	38
8	1130	164	970	141	14	37

(a) Treatment prior to aging: 1½ h at 1150 °C (2100 °F), air cool; 1 h at 1040 °C (1900 °F), oil quench

Table 12 Effects of tempering temperature on typical properties of shell mold cast AM-355 (a)

Condition	Ultimate tensile strength		Yield strength(b)		Elongation in 50 mm (2 in.),%	Reduction in area, %
	MPa	ksi	MPa	ksi		
Annealed	1290	187	480	70	6	3.5
Subzero transformed	1400	203	960	140	6	2.5
Tempered 3 h at:						
480 °C (900 °F)	1440	209	1170	170	20	9
540 °C (1000 °F) ...	1320	192	1100	159	34	13
595 °C (1100 °F) ...	1190	173	940	136	35	14
650 °C (1200 °F) ...	1010	147	590	86	33	15

(a) Treatment prior to tempering: 1½ h at 1095 °C (2000 °F), furnace cool to 980 °C (1800 °F); soak at 980 °C (1800 °F) for 1½ h, water quench; subzero cool at −85 °C (−120 °F) for 6 h. (b) 0.2% offset

Table 13 Effect of heat treating and welding sequence on typical properties of AM-355 castings

Sequence	Ultimate tensile strength		Yield strength, 0.2%		Elongation in 50 mm (2 in.), %
	MPa	ksi	MPa	ksi	
Heat treated after welding					
A...........	1450	210	1100	160	15
B...........	1380	200	1120	162	12
C...........	1410	205	1070	155	12
D...........	1410	205	1100	160	8
Not heat treated after welding					
E...........	1070	155	830	120	11

Note: Heat treating and welding procedures and sequences were as follows: A—2 h at 1095 °C (2000 °F), air cool; 2 h at 1010 °C (1850 °F), water quench; 3 h at −75 °C (−100 °F); 3 h at 455 °C (850 °F), air cool; 3 h at 510 °C (950 °F), air cool; helium-shielded arc welding; repeat heat treatment. B—2 h at 1095 °C (2000 °F), air cool; 2 h at 1010 °C (1850 °F), water quench; 3 h at −75 °C (−100 °F); 3 h at 455 °C (850 °F), air cool; 3 h at 510 °C (950 °F), air cool; 2 h at 1010 °C (1850 °F), water quench; helium-shielded arc welding; repeat heat treatment through 510 °C (950 °F) tempering. C—Helium-shielded arc welding; 2 h at 1095 °C (2000 °F), air cool; 2 h at 1010 °C (1850 °F), water quench; 3 h at −75 °C (−100 °F); 3 h at 455 °C (850 °F), air cool; 3 h at 510 °C (950 °F), air cool. D—2 h at 1095 °C (2000 °F), air cool; 2 h at 1010 °C (1850 °F), water quench; helium-shielded arc welding; 2 h at 1095 °C (2000 °F), air cool; 2 h at 1010 °C (1850 °F), water quench; 3 h at −75 °C (−100 °F); 3 h at 455 °C (850 °F), air cool; 3 h at 510 °C (950 °F), air cool. E—2 h at 1095 °C (2000 °F), air cool; 2 h at 1010 °C (1850 °F), water quench; 3 h at −75 °C (−100 °F); 3 h at 455 °C (850 °F), air cool; 3 h at 510 °C (950 °F), air cool; helium-shielded arc welding.

Fig. 17 Effect of hardening temperature on typical room-temperature properties of 17-4 PH

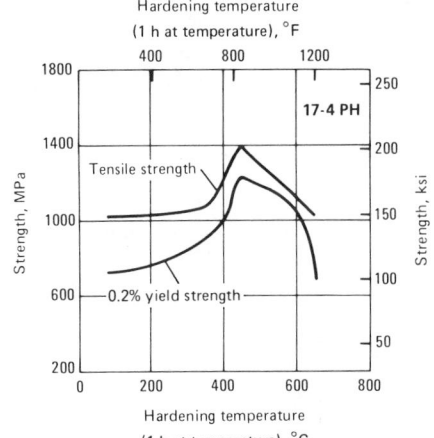

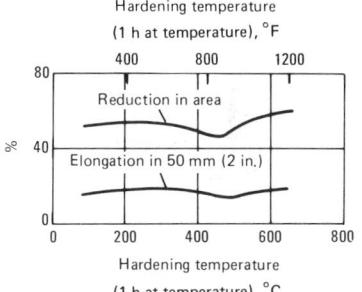

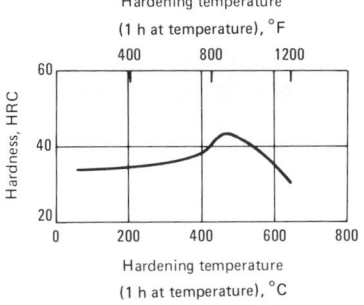

Data are average values for bars 25 to 89 mm (1 to 3½ in.) in diameter from four heats.

peratures to which castings should be heated prior to quenching vary somewhat, depending on the alloy (Table 7).

As shown in Table 7, a two-step heat treating procedure may be applied to the niobium-containing CF-8C (UNS J92710) alloy. The first treatment con-

Fig. 18 Effect of tempering temperature on the mechanical properties of a CA-6NM standard keel block

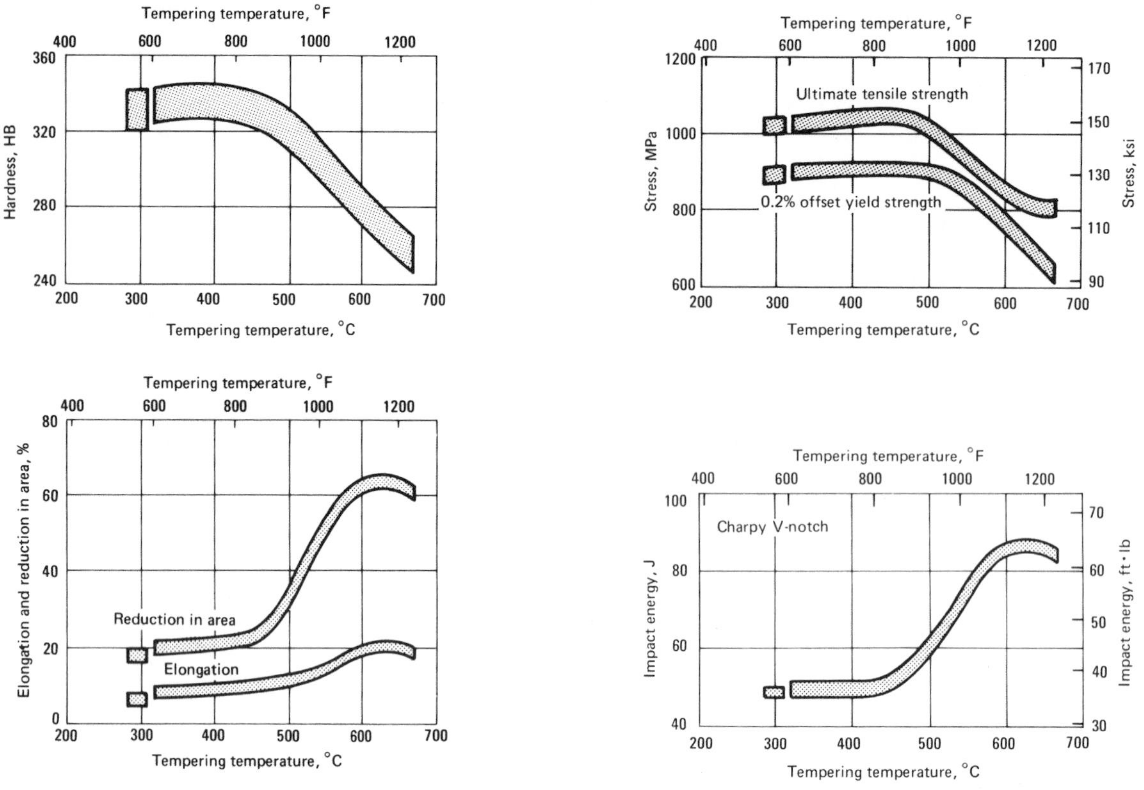

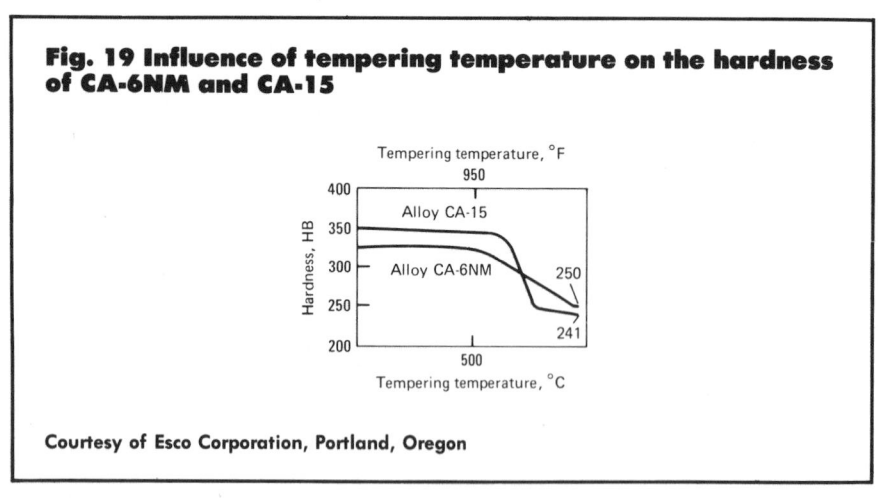

Courtesy of Esco Corporation, Portland, Oregon

Fig. 19 Influence of tempering temperature on the hardness of CA-6NM and CA-15

Courtesy of Esco Corporation, Portland, Oregon

sists of solution annealing. This is followed by a stabilizing treatment at 870 to 925 °C (1600 to 1700 °F), which precipitates niobium carbides, prevents formation of the damaging chromium carbides, and provides maximum resistance to intergranular attack.

Because of their low carbon contents, CF-3 and CF-3M (UNS J92700 and J92800) as cast do not contain enough chromium carbides to cause selective intergranular attack, and hence may be used in some corrodents in this condition; for maximum corrosion resistance, however, these grades require solution annealing.

Martensitic Alloys. Castings of the CA-6NM composition should be hardened by air cooling or oil quenching from a temperature of 1010 to 1065 °C (1850 to 1950 °F). Even though the carbon content of this alloy is lower than that of CA-15, this fact in itself and the addition of molybdenum and nickel enable the alloy to harden completely without significant austenite retention when cooled as suggested.

The choice of cooling medium is determined primarily by the maximum section size. Section sizes in excess of 125 mm (5 in.) will harden completely when cooled in air. CA-6NM is not prone to cracking during cooling from elevated temperatures. For this reason, no problem should arise in the air cool-

Fig. 20 Effect of tempering temperature on typical room-temperature mechanical properties of CA-15 castings

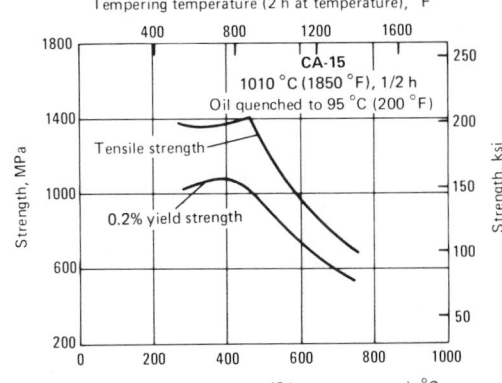

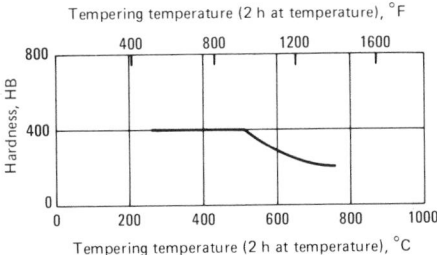

The minor loss of toughness and ductility that does occur is associated with the lesser degree of tempering which takes place at the lower temperature and not with embrittlement, as might be the situation with other 12% chromium steels that contain no molybdenum. The addition of molybdenum to 12% chromium steels makes them unusually stable thermally and normally not susceptible to embrittlement in the annealed or annealed and cold worked conditions, even when exposed for long periods of time at 370 to 480 °C (700 to 900 °F). There are no data currently available on such steels in the quenched and tempered or normalized and tempered conditions.

Another significant practical advantage of CA-6NM is its relative freedom from the rapid drop in hardness when tempered above about 510 °C (950 °F). Figure 19 shows clearly that a given increase in tempering temperature produces a much more gradual decrease in hardness as compared with CA-15. This makes heat treating much easier and cheaper and decreases the frequency of rejects and/or the necessity for reheat treatment.

The hardening procedures for CA-15 castings are similar to those used for the comparable wrought alloy (type 410). Austenitizing consists of heating to 955 to 1010 °C (1750 to 1850 °F), and soaking for a minimum of 30 min; the high side of this temperature range is normally employed. Parts are then cooled in air or quenched in oil. To reduce the probability of cracking in the brittle, untempered martensitic condition, tempering should take place immediately after quenching.

Tempering is performed in two temperature ranges: (a) up to 370 °C (700 °F) for maximum strength and corrosion resistance, and (b) from 595 to 760 °C (1100 to 1400 °F) for improved ductility at lower strength levels. Tempering in the range of 370 to 595 °C (700 to 1100 °F) is normally avoided, because of the resultant low impact strength. Figure 20 shows the nominal mechanical properties obtained in CA-15 castings as a function of tempering temperature. Additional data on mechanical properties are given in Table 8. These data are based on several heats of shell cast CA-15 alloy. The standard heat treating procedures for CA-15, CA-40, and CA-6NM are given in Table 9. In the hardened and tempered condition, CA-40 provides higher tensile strength and lower ductility

ing or oil quenching of configurations that include thick as well as thin sections.

A wide choice of mechanical properties is available through the choice of tempering temperature. Castings of CA-6NM are normally supplied normalized and tempered at 595 to 620 °C (1100 to 1150 °F). Reaustenitizing occurs upon tempering above 620 °C (1150 °F), the amount of reaustenitization increasing with increasing temperature. Depending on the amount of this transformation, cooling from such tempering temperatures may adversely affect both ductility and toughness through the transformation to untempered martensite.

Even though the alloy is characterized by a decrease in impact strength

when tempered in the range of 370 to 595 °C (700 to 1100 °F), the minimum reached is significantly higher than that of CA-15. This improvement in impact toughness results from the presence of molybdenum and nickel in the composition and from the lower carbon content. The best combination of strength with toughness is obtained when the alloy is tempered above 510 °C (950 °F).

Figure 18 describes the effect of tempering temperature on the hardness, strength, ductility and toughness properties of CA-6NM and illustrates that strengths even higher than those considered typical can be obtained by tempering at lower temperatures without a disturbing loss of ductility or toughness.

than CA-15 tempered at the same temperature. Both alloys can be annealed by cooling slowly from the range 845 to 900 °C (1550 to 1650 °F).

Precipitation-Hardening Alloys. It is desirable to subject precipitation-hardenable castings to a high-temperature homogenization treatment to reduce alloy segregation and to obtain more uniform response to subsequent heat treatment. Even investment castings that are cooled slowly from the pouring temperature exhibit more nearly uniform properties when they have been homogenized. Recommended homogenizing treatments for precipitation-hardening alloys 17-4 PH and AM-350 are included in Table 6.

17-4 PH Castings. When 17-4 PH (ASTM CB-7Cu-1 and CB-7Cu-2) is cast in plastic-bonded shell molds, the surface is carburized by decomposition of the binder. The added carbon prevents proper heat treating response of the casting surface. Satisfactory response is obtained when surface carbon is removed prior to the homogenization treatment.

In addition to homogenization, other heat treating procedures for 17-4 PH castings include solution annealing and precipitation hardening. Details of these procedures are given in Table 7. The preferred temperature range for precipitation hardening is 480 to 595 °C (900 to 1100 °F). The mechanical properties obtained at different aging temperatures are given in Table 10.

The tendency of 17-4 PH castings to overage is reduced by the addition of about 0.25% combined niobium plus tantalum to the alloy. The effect of time at aging temperature on the mechanical properties of niobium-free and niobium-containing 17-4 PH investment castings is shown in Table 11.

AM-350 and AM-355. Although investment castings made of these alloys do not necessarily require a homogenizing treatment, this treatment provides a more uniform response to subsequent heat treatment. Shell mold and sand castings made of AM-355 that were extremely brittle without homogenization regained ductility after homogenizing at 1095 °C (2000 °F) for 2 h minimum. Heat treating procedures and effects of tempering temperatures up to 650 °C (1200 °F) on mechanical properties of AM-355 shell mold castings are given in Table 12.

When AM-355 castings are welded, maximum mechanical properties are obtained when the castings are fully heat treated after welding (Table 13). Heat treatments prior to welding have little effect on properties when a complete heat treatment follows welding.

Stress Relieving of Austenitic Stainless Steels

AUSTENITIC STAINLESS STEEL has good creep resistance; consequently, it must be heated to about 900 °C (1650 °F) to attain adequate stress relief. In some instances, heating to the annealing temperature may be desirable. Holding at a temperature lower than about 870 °C (1600 °F) results in only partial stress relief. The most effective stress-relieving results are achieved by slow cooling. Quenching or other rapid cooling, as is normal in the annealing of austenitic stainless steel, will usually reintroduce residual stresses.

Selection of Treatment

Selection of an optimum stress-relieving treatment is difficult, because heat treatments that provide adequate stress relief can impair the corrosion resistance of stainless steel, and heat treatments that are not harmful to corrosion resistance may not provide adequate stress relief. To avoid specifying a heat treatment that might prove harmful, ASME Code neither requires nor prohibits stress relief of austenitic stainless steel.

Metallurgical characteristics of austenitic stainless steels that may affect the selection of a stress-relieving treatment are discussed below.

- *Heating in the range from 480 to 815 °C (900 to 1500 °F):* Chromium carbides will precipitate in the grain boundaries of wholly austenitic unstabilized grades. In partially ferritic cast grades, the carbides will precipitate initially in the discontinuous ferrite pools rather than in a continuous grain-boundary network. After prolonged heating such as is necessary for heavy sections, however, grain boundary carbide precipitation will occur. For cold worked stainless, carbide precipitation may occur as low as 425 °C (800 °F); for types 309 and 310, the upper limit for carbide precipitation may be as high as 900 °C (1650 °F). In this condition, the steel is susceptible to intergranular corrosion. By using stabilized or extra-low-carbon grades, these intergranular precipitates of chromium carbide can be avoided.

- *Heating in the range from 540 to 925 °C (1000 to 1700 °F):* The formation of hard, brittle sigma phase may result, which can decrease both corrosion resistance and ductility. During the times necessary for stress relief, sigma will not form in fully austenitic wrought, cast or welded stainless. However, if the stainless is partly ferritic, the ferrite may transform to sigma during stress relief. This is generally not a problem in wrought stainless steels, because they are fully austenitic; however, some wrought grades—particularly types 309, 309Cb, 312 and 329—may contain some ferrite. Furthermore, the composition of most austenitic stainless welds and castings is intentionally adjusted so that ferrite is present as a deterrent to cracking. The niobium (columbium)-containing cast grade CF-8C normally contains 5 to 20% ferrite, which is more likely to transform to sigma than the niobium (columbium)-free ferrite in the unstabilized CF-8 grade.

- *Slow cooling an unstabilized grade (other than an extra-low-carbon grade):* Through either of the above temperature ranges, slow cooling may allow sufficient time for these detrimental effects to take place.

- *Heating at 815 to 925 °C (1500 to 1700 °F):* The coalescence of chromium carbide precipitates or sigma phase will occur, resulting in a form less harmful to corrosion resistance or mechanical properties.

- *Heating at 955 to 1120 °C (1750 to 2050 °F):* This annealing treatment

Fig. 1 Effect of stress relieving on corrosion rate of type 347 stainless steel

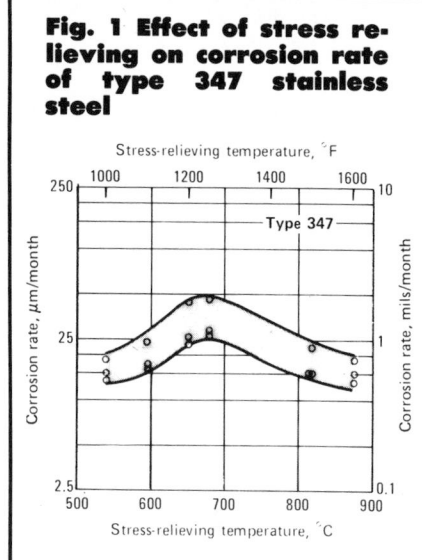

Fig. 2 Stress relief obtained in type 347 stainless steel, as a function of temperature, initial stress, and time at temperature

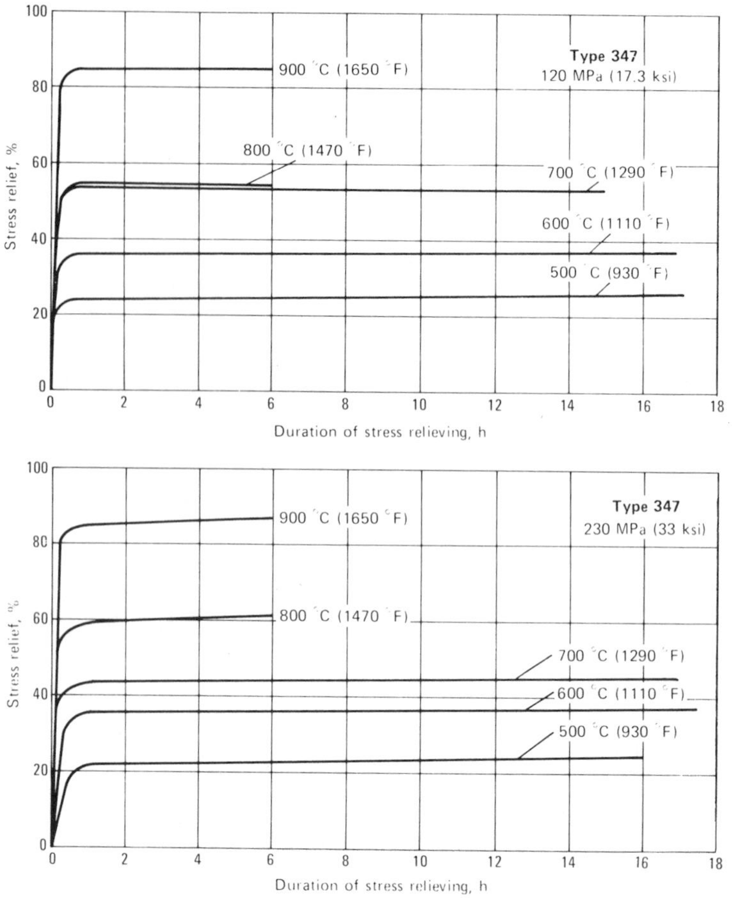

causes all grain-boundary chromium carbide precipitates to redissolve and transforms sigma back to ferrite, as well as fully softening the steel.

- *Stress relieving to improve the notch toughness:* Unlike carbon and alloy steels, austenitic steels are not notch sensitive. Consequently, stress relieving to improve notch toughness would be of no benefit. Notch-impact strength may actually be decreased if the steel is stress relieved at a temperature at which chromium carbide is precipitated or sigma phase forms.

Although stabilized alloys do not require high-temperature annealing to avoid intergranular corrosion, the stress-relieving temperature exerts an influence on the general corrosion resistance of these alloys. Figure 1 shows the effect of stress relieving for 2 h at various temperatures on the corrosion rate of type 347 stainless steel in boiling 65% nitric acid. The corrosion resistance of type 347 in boiling nitric acid is better when the material is treated at 815 to 870 °C (1500 to 1600 °F) than when treated at 650 to 705 °C (1200 to 1300 °F).

Figure 2 shows how the percentage of stress relief increases with an increase in stress-relieving temperature for type 347 stainless steel. These data also demonstrate the relative unimportance of holding time.

General Recommendations. In the selection of the proper stress-relieving treatment, consideration must be given also to the specific material used, fabrication procedures in-

volved, and to the design and operating conditions of the equipment. Stress relieving generally is not advisable unless the service environment is known or suspected to cause stress corrosion. If stress relieving seems warranted, due regard should be given the metallurgical factors and their effect on the steel in the intended service. The use of stabilized or extra-low-carbon grades is advantageous in view of the greater latitude allowed in stress relieving.

Table 1 gives suggested stress-relieving treatments for service applications and environments. Because of the varying degrees of stress relief that may be required, number of different grades of stainless in use, many fabricating procedures that may be employed, and the multitude of service requirements, many alternative treatments are indicated in Table 1 to allow selection of the stress-relieving treat-

ment best suited to particular circumstances.

Results Obtained by Various Treatments

Inadequate Stress Relief. Austenitic stainless steels have in many instances been stress relieved at temperatures normally used for carbon steels (540 to 650 °C, or 1000 to 1200 °F). Although at these temperatures virtually all residual stress is relieved in carbon steel, only 30 to 40% of the residual stress is relieved in austenitic stainless (Fig. 2). Because the treatment does not provide adequate stress relief, stainless stress relieved in this temperature range is often susceptible to stress corrosion. Table 2 shows the residual stresses remaining in solid austenitic stainless steels after being stress relieved for various times at tem-

Table 1 Stress-relieving treatments for austenitic stainless steels

Application or desired characteristics	Suggested thermal treatment(a)		
	Extra-low-carbon grades, such as 304L and 316L	Stabilized grades, such as 318, 321 and 347	Unstabilized grades, such as 304 and 316
Severe stress corrosion	A, B	B, A	(b)
Moderate stress corrosion	A, B, C	B, A, C	C(b)
Mild stress corrosion	A, B, C, E, F	B, A, C, E, F	C, F
Remove peak stresses only	F	F	F
No stress corrosion	None required	None required	None required
Intergranular corrosion	A, C(c)	A, C, B(c)	C
Stress relief after severe forming	A, C	A, C	C
Relief between forming operations	A, B, C	B, A, C	C(d)
Structural soundness(e)	A, C, B	A, C, B	C
Dimensional stability	G	G	G

(a) Thermal treatments are listed in order of decreasing preference. A: anneal at 1065 to 1120 °C (1950 to 2050 °F), slow cool. B: stress relieve at 900 °C (1650 °F), slow cool. C: anneal at 1065 to 1120 °C (1950 to 2050 °F), quench(f) or cool rapidly. D: stress relieve at 900 °C (1650 °F), quench or cool rapidly. E: stress relieve at 480 to 650 °C (900 to 1200 °F), slow cool. F: stress relieve at below 480 °C (900 °F), slow cool. G: stress relieve at 205 to 480 °C (400 to 900 °F), slow cool (usual time, 4 h per inch of section). (b) To allow the optimum stress-relieving treatment, the use of stabilized or extra-low-carbon grades is recommended. (c) In most instances, no heat treatment is required, but where fabrication procedures may have sensitized the stainless steel the heat treatments noted may be employed. (d) Treatment A, B or D also may be used, if followed by treatment C when forming is completed. (e) Where severe fabricating stresses coupled with high service loading may cause cracking. Also, after welding heavy sections

peratures ranging from 595 to 1010 °C (1100 to 1850 °F).

Annealing and Water Quenching. Numerous instances have been reported in which satisfactory service was obtained for vessels and parts that were stress relieved by being annealed (at 1065 to 1120 °C, or 1950 to 2050 °F) and water quenched. However, it is unlikely that these products were subjected to service environments conducive to severe stress corrosion, because a water quench will almost always reintroduce high residual stresses. One instance in which stress corrosion was caused by annealing and quenching is described below.

Example 1. Thirty type 316 stainless vessels were quench annealed and placed in the same type of service at different locations. All but two of the vessels gave many years of excellent service. These two failed within 2 months because of the presence of chlorides in the environment at their particular locations; chlorides were absent at the locations of the other vessels.

Intergranular Corrosion. In a number of instances, partially stress-relieved stainless steel parts have failed through intergranular corrosion.

Example 2. Type 316 stainless steel hardware used in coastal steam stations was partially stress relieved at 620 to 650 °C (1150 to 1200 °F). Failure by intergranular attack in seawater occurred in less than 6 months.

Example 3. A type 304 stainless steel heat exchanger was partially stress relieved at 650 °C (1200 °F) for 2

Table 2 Stresses in austenitic stainless steel after various treatments

Treatment Temperature		Time,	Residual stress	
°C	°F	h	MPa	ksi
After welding 23.5-cm (9.25-in.-) OD, 16.5-cm (6.5-in.) ID pipe				
As welded			205 to 175	30.0 to 25.7
595	1100	16	140	20.0
595	1100	48	140	20.0
595	1100	72	160	23.0
650	1200	4	150 to 165	21.5 to 24.0
After welding 12.7-cm-(5-in.-) OD, 10.2-cm-(4-in.-) ID pipe				
As welded			125 to 100	18.5 to 14.7
650	1200	4	95 to 105	13.7 to 15.3
650	1200	12	110	16.0
650	1200	36	108	15.6
900	1650	2	nil	nil
1010	1850	1	nil	nil

h and furnace cooled. Failure by intergranular attack occurred in 7 days.

Prevention of Stress Corrosion by Stress Relieving. A number of instances have been recorded in which beneficial effects were derived from an adequate stress-relief treatment.

Example 4. Heaters made of type 316L failed after a few weeks of service in contact with acid organic chloride and ammonium chloride, whereas heaters that had been stress relieved at 955 °C (1750 °F) were completely free of stress-corrosion cracking after 4 years of service under the same conditions.

Example 5. Two type 316L stainless steel vessels were used in 85% phosphoric acid service. One was not stress relieved and experienced extensive stress corrosion. Stress relieving the

other vessel at 540 °C (1000 °F) completely prevented the stress corrosion. This illustrates that, even though a stainless steel component may not be completely stress relieved, reducing the stress level may prevent stress corrosion.

Stress relief of unstabilized grades of stainless at 900 °C (1650 °F) will result in some intergranular carbide precipitation, and in some instances a small amount of intergranular attack may be encountered. However, failure after a few years by intergranular attack is preferable to failure within a few weeks by stress-corrosion cracking. Moreover, the intergranular attack probably could be avoided by using an extra-low-carbon or stabilized grade of austenitic stainless steel.

Heat Treating of Heat-Resisting Alloys

By the ASM Committee on
Stainless Steels
and Heat-Resisting Alloys*

PROCEDURES AND PROCESS CONTROL for heat treating the principal types of heat-resisting alloys and refractory metals are discussed in this article. Tables 1, 2, 3 and 4 list representative wrought and cast materials, and include primarily those alloys most frequently used in the aerospace industry.

The first part of this article reviews the heat treating processes that are applied to heat-resisting alloys. Later sections describe the applicability of these processes to specific alloys.

Stress Relieving

Stress relieving of heat-resisting alloys and refractory metals frequently entails a compromise; the desirability of maximum relief of residual stress must be weighed against possible effects deleterious to high-temperature properties and corrosion resistance.

True stress relieving of wrought material usually is confined to alloys that are not age-hardenable. Thus, the time and temperature cycles may vary considerably, depending on the metallurgical characteristics of the alloy and on the type and magnitude of residual stresses developed by previous fabricating processes.

Stress-relieving temperatures are usually below the annealing or recrystallization temperatures. Typical cycles for wrought alloys are listed in Table 5; temperatures at least 25 °C (50 °F) higher or lower than those listed are usually satisfactory.

Some heat-resisting alloy castings are placed in service in the as-cast condition. However, castings may be stress relieved: (a) when they are of an extremely complex shape that might crack during the initial heating-up period in service, (b) when their dimensional tolerances are stringent, and (c) after they have been welded. It is not possible to tabulate the stress relief cycles for cast alloys because they are particularly dependent on geometry and prior processing. Stress relief cycles can be developed by empirical studies of stress decay with time and temperature, as determined by nondestructive means such as x-ray diffraction for many alloys. This is not an effective technique for superalloys, where extensive material testing of critical properties and subsequent data analysis must be performed to determine the efficacy of a given cycle. Particular care must be given to such time-dependent effects as LCF hold time, crack growth, and creep rate.

Annealing

When applied to heat-resisting alloys, annealing implies full annealing—that is, complete recrystallization and the attainment of maximum softness. The practice is usually applied to wrought alloys of the nonhardening type. For a majority of the hardenable alloys, annealing cycles are the same as those used for solution treating. However, the two treatments serve different purposes. Annealing is used mainly to increase ductility (and reduce hardness) to facilitate forming or machin-

*Daniel Rapoport, *Chairman*, Technical Center, Howmet Turbine Components Corp.; Ed R. Byrnes, Jr., Manager—Vacuum Equipment Marketing, Ipsen Industries; Kenneth L. Crooks, Senior Staff Engineer, Research & Technology, Armco, Inc.; Ranes P. Dalal, Senior Development Engineer, Technical Center, Howmet Turbine Components Corp.; Matthew J. Donachie, Jr., Senior Assistant Materials Project Engineer, Pratt Whitney Aircraft Group, United Technologies Corp., Dennis Macha, Supervisor—Metallurgical Services, Technical Center, Howmet Turbine Components Corp.; Robert N. Peterson, Manager of Metallurgical Services, Enduro Division, Republic Steel Corp.; Arthur D. Schwartz, Manager of Materials Performance Engineering, Aircraft Engine Business Group, General Electric Corp.

Table 1 Nominal compositions of wrought heat-resisting alloys

Alloy	UNS number	Composition, %										
		Cr	Ni	Co	Mo	W	Nb	Ti	Al	Fe	C	Other
Iron-base and iron-nickel-chromium solid-solution alloys												
16-25-6	...	16.0	25.0	...	6.00	...	...	...	...	50.7	0.06	1.35 Mn; 0.70 Si; 0.15 N
17-14CuMo	...	16.0	14.0	...	2.50	...	0.4	0.3	...	62.4	0.12	0.75 Mn; 0.50 Si; 3.0 Cu
19-9DL	K63198	19.0	9.0	...	1.25	1.25	0.4	0.3	...	66.8	0.30	1.10 Mn; 0.60 Si
Carpenter 20Cb-3	N08020	20.0	34.0	...	2.50	...	1.0 max	...	...	42.4	0.07 max	3.5 Cu
Incoloy 800, 800H	N08800	21.0	32.5	...	...	...	...	0.38	0.38	45.7	0.05	...
Incoloy 801	N08801	20.5	32.0	...	...	...	...	1.13	...	46.3	0.05	...
Incoloy 802	...	21.0	32.5	...	...	...	...	0.75	0.58	44.8	0.35	...
Incoloy 825	...	21.5	42.0	...	3.0	...	...	0.9	0.1	30.0	0.03	0.5 Mn; 2.25 Cu
N-155	R30155	21.0	20.0	20.0	3.00	2.5	1.0	...	...	32.2	0.15	0.15 N; 0.02 La; 0.02 Zr
RA330	N08330	19.0	36.0	...	...	...	...	...	...	45.1	0.05	...
Cobalt-base solid-solution alloys												
Haynes 25 (L-605)	R30605	20.0	10.0	50.0	...	15.0	...	...	...	3.0	0.10	1.5 Mn
Haynes 188	R30188	22.0	22.0	37.0	...	14.5	...	...	...	3.0 max	0.10	0.90 La
S-816	R30816	20.0	20.0	42.0	4.0	4.0	4.0	...	...	4.0	0.38	...
Stellite 6B	...	30.0	1.0	61.5	...	4.5	...	...	...	1.0	1.0	...
UMCo-50	...	28.0	...	49.0	...	...	...	...	...	21.0	0.12 max	...
Nickel-base solid-solution alloys												
Hastelloy B	N10001	1.0 max	63.0	2.5 max	28.0	...	...	...	...	5.0	0.05 max	0.03 V
Hastelloy B-2	N10665	1.0 max	69.0	1.0 max	28.0	...	...	...	...	2.0 max	0.02 max	...
Hastelloy C-4	N06455	16.0	63.0	2.0 max	15.5	...	...	0.7 max	...	3.0 max	0.015 max	...
Hastelloy C-276	N10276	15.5	59.0	...	16.0	3.7	...	...	...	5.0	0.02 max	...
Hastelloy N	N10003	7.0	72.0	...	16.0	...	...	0.5 max	0.2	5.0 max	0.06	...
Hastelloy S	...	15.5	67.0	...	15.5	...	...	...	0.2	1.0	0.02 max	0.02 La
Hastelloy W	N10004	5.0	61.0	2.5 max	24.5	...	...	...	...	5.5	0.12 max	0.6 V
Hastelloy X	N06002	22.0	49.0	1.5 max	9.0	0.6	...	...	2.0	15.8	0.15	...
Inconel 600	N06600	15.5	76.0	...	...	...	...	...	...	8.0	0.08	0.25 max Cu
Inconel 601	N06601	23.0	60.5	...	...	...	...	...	1.35	14.1	0.05	0.5 max Cu
Inconel 617	...	22.0	55.0	12.5	9.0	...	...	...	1.0	...	0.07	...
Inconel 625	N06625	21.5	61.0	...	9.0	...	3.6	0.2	0.2	2.5	0.05	...
Inconel 690	...	30.0	60.0	...	...	...	...	...	...	9.5	0.03	...
NA-224	...	27.0	48.0	...	...	6.0	...	...	...	18.5	0.50	...
Nimonic 75	...	19.5	75.0	...	...	...	...	0.4	0.15	2.5	0.12	0.25 max Cu
RA-333	N06333	25.0	45.0	3.0	3.0	3.0	...	...	...	18.0	0.05	...
Iron-base precipitation-hardening alloys												
A-286	K66286	15.0	26.0	...	1.25	...	...	2.0	0.2	55.2	0.04	0.005 B; 0.3 V
Discaloy	K66220	14.0	26.0	...	3.0	...	...	1.7	0.25	55.0	0.06	...
Haynes 556	...	22.0	21.0	20.0	3.0	2.5	0.1	...	0.3	29.0	0.10	0.50 Ta; 0.02 La; 0.002 Zr
Incoloy 903	...	0.1 max	38.0	15.0	0.1	...	3.0	1.4	0.7	41.0	0.04	...
Pyromet CTX-1	...	0.1 max	37.7	16.0	0.1	...	3.0	1.7	1.0	39.0	0.03	...

(continued)

(a) No longer active, but shown here for reference. (b) Also known as Rolls Royce C-263

Table 1 (continued)

Alloy	UNS number	Composition, %										
		Cr	Ni	Co	Mo	W	Nb	Ti	Al	Fe	C	Other
V-57	...	14.8	27.0	...	1.25	...	...	3.0	0.25	48.6	0.08 max	0.01 B; 0.5 max V
W-545	K66545	13.5	26.0	...	1.5	...	...	2.85	0.2	55.8	0.08	0.05 B
Cobalt-base precipitation-hardening alloys												
AR-213	...	19.0	0.5 max	65.0	...	4.5	...	...	3.5	0.5 max	0.17	6.5 Ta; 0.15 Zr; 0.1 Y
MP-35N	R30035	20.0	35.0	35.0	10.0	...	...	...	...	...	...	...
MP-159	...	19.0	25.0	36.0	7.0	...	0.6	3.0	0.2	9.0	...	...
Nickel-base precipitation-hardening alloys												
Astroloy	N09979;	15.0	56.5	15.0	5.25	...	...	3.5	4.4	<0.3	0.06	0.03 B; 0.06 Zr
D-979	K66979 (a)	15.0	45.0	...	4.0	4.0	...	3.0	1.0	27.0	0.05	0.01 B
IN 100	N13100	10.0	60.0	15.0	3.0	3.0	...	4.7	5.5	<0.6	0.15	1.0 V; 0.06 Zr; 0.015 B
IN 102	N06102	15.0	67.0	...	2.9	3.0	2.9	0.5	0.5	7.0	0.06	0.005 B; 0.02 Mg; 0.03 Zr
Incoloy 901	N09901	12.5	42.5	...	6.0	...	...	2.7	...	36.2	0.10 max	...
Inconel 702	...	15.5	79.5	...	...	...	...	0.6	3.2	1.0	0.05	0.5 Mn; 0.2 Cu; 0.4 Si
Inconel 706	N09706	16.0	41.5	...	...	...	...	1.75	0.2	37.5	0.03	2.9 (Nb + Ta); 0.15 max Cu
Inconel 718	N07718	19.0	52.5	...	3.0	...	5.1	0.9	0.5	18.5	0.08 max	0.15 max Cu
Inconel 721	...	16.0	71.0	...	...	...	...	3.0	...	6.5	0.04	2.2 Mn; 0.1 Cu
Inconel 722	...	15.5	75.0	...	...	...	1.0	2.4	0.7	7.0	0.04	0.5 Mn; 0.2 Cu; 0.4 Si
Inconel 751	N07750	15.5	72.5	...	...	...	1.0	2.3	1.2	7.0	0.05	0.25 max Cu
Inconel X-750	...	15.5	73.0	...	...	...	...	2.5	0.7	7.0	0.04	0.25 max Cu
M252	N07252	19.0	56.5	10.0	10.0	...	...	2.6	1.0	<0.75	0.15	0.005 B
Nimonic 80A	N07080	19.5	73.0	1.0	...	...	...	2.25	1.4	1.5	0.05	0.10 max Cu
Nimonic 90	N07090	19.5	55.5	18.0	...	...	...	2.4	1.4	1.5	0.06	...
Nimonic 95	...	19.5	53.5	18.0	...	...	...	2.9	2.0	5.0 max	0.15 max	+ B; + Zr
Nimonic 100	...	11.0	56.0	20.0	5.0	...	...	1.5	5.0	2.0 max	0.30 max	+ B; + Zr
Nimonic 105	...	15.0	54.0	20.0	5.0	...	...	1.2	4.7	...	0.08	0.005 B
Nimonic 115	...	15.0	55.0	15.0	4.0	...	...	4.0	5.0	1.0	0.20	0.04 Zr
Nimonic 263 (b)	...	20.0	51.0	20.0	5.9	...	...	2.1	0.45	0.7 max	0.06	...
Pyromet 860	...	13.0	44.0	4.0	6.0	...	...	3.0	1.0	28.9	0.05	0.01 B
Refractory 26	...	18.0	38.0	20.0	3.2	...	6.5	2.6	0.2	16.0	0.03	0.015 B
René 41	N07041	19.0	55.0	11.0	10.0	...	...	3.1	1.5	<0.3	0.09	0.01 B
René 95	...	14.0	61.0	8.0	3.5	3.5	3.5	2.5	3.5	<0.3	0.16	0.01 B; 0.05 Zr
René 100	...	9.5	61.0	15.0	3.0	...	...	4.2	5.5	1.0 max	0.16	0.015 B; 0.06 Zr; 1.0 V
Udimet 500	N07500	19.0	48.0	19.0	4.0	...	...	3.0	3.0	4.0 max	0.08	0.005 B
Udimet 520	...	19.0	57.0	12.0	6.0	1.0	...	3.0	2.0	...	0.08	0.005 B
Udimet 630	...	17.0	50.0	...	3.0	3.0	6.5	1.0	0.7	18.0	0.04	0.004 B
Udimet 700	...	15.0	53.0	18.5	5.0	...	...	3.4	4.3	<1.0	0.07	0.03 B
Udimet 710	...	18.0	55.0	14.8	3.0	1.5	...	5.0	2.5	...	0.07	0.01 B
Unitemp AF2-1DA	...	12.0	59.0	10.0	3.0	6.0	...	3.0	4.6	<0.5	0.35	1.5 Ta; 0.015 B; 0.1 Zr
Waspaloy	N07001	19.5	57.0	13.5	4.3	...	...	3.0	1.4	2.0 max	0.07	0.006 B; 0.09 Zr

(a) No longer active, but shown here for reference. (b) Also known as Rolls Royce C-263

Table 2 Compositions of nickel-base heat-resistant casting alloys

Alloy designation	C	Ni	Cr	Co	Mo	Fe	Al	B	Ti	W	Zr	Others
							Nominal composition, %					
B-1900	0.1	64	8	10	6	. . .	6	0.015	1	. . .	0.10	4 Ta(a)
Hastelloy X	0.1	50	21	1	9	18	. . .	. . .	. . .	1	. . .	. . .
IN-100	0.18	60.5	10	15	3	. . .	5.5	0.01	5	. . .	0.06	1 V
IN-713C	0.12	74	12.5	. . .	4.2	. . .	6	0.012	0.8	. . .	0.1	1.75 Ta, 0.9 Nb
IN-713LC	0.05	75	12	. . .	4.5	. . .	6	0.01	0.6	. . .	0.1	4 TA
IN-738	0.17	61.5	16	8.5	1.75	. . .	3.4	0.01	3.4	2.6	0.1	2 Nb
IN-792	0.2	60	13	9	2.0	. . .	3.2	0.02	4.2	4	0.1	2 Nb
Inconel 718	0.04	53	19	. . .	3	18	0.5	. . .	0.9	. . .	. . .	0.1 Cu, 5 Nb
Inconel X-750	0.04	73	15	. . .	. . .	7	0.7	. . .	2.5	. . .	. . .	0.25 Cu, 0.9 Nb
M-252	0.15	56	20	10	10	. . .	1	0.005	2.6	. . .	. . .	. . .
MAR-M 200	0.15	59	9	10	. . .	1	5	0.015	2	12.5	0.05	1 Nb(b)
MAR-M 246	0.15	60	9	10	2.5	. . .	5.5	0.015	1.5	10	0.05	1.5 Ta
MAR-M 247	0.15	59	8.25	10	0.7	0.5	5.5	0.015	1	10	0.05	1.5 Hf, 3 Ta
NX 188 (DS)	0.04	74	. . .	. . .	18	. . .	8	. . .	. . .	. . .	. . .	. . .
René 77	0.07	58	15	15	4.2	. . .	4.3	0.015	3.3	. . .	0.04	. . .
René 80	0.17	60	14	9.5	4	. . .	3	0.015	5	4	0.03	. . .
René 100	0.18	61	9.5	15	3	. . .	5.5	0.015	4.2	. . .	0.06	1 V
TRW-NASA VIA	0.13	61	6	7.5	2	. . .	5.5	0.02	1	6	0.13	0.4 Hf, 0.5 Nb, 0.5 Re, 9 Ta
Udimet 500	0.1	53	18	17	4	2	3	. . .	3	. . .	. . .	. . .
Udimet 700	0.1	53.5	15	18.5	5.25	. . .	4.25	0.03	3.5	. . .	. . .	. . .
Udimet 710	0.13	55	18	15	3	. . .	2.5	. . .	5	1.5	0.08	. . .
Waspaloy	0.07	57.5	19.5	13.5	4.2	1	1.2	0.005	3	. . .	0.09	. . .
WAZ-20 (DS)	0.20	72	. . .	. . .	. . .	. . .	6.5	. . .	. . .	20	1.5	. . .

(a) B-1900 + Hf also contains 1.5% Hf. (b) MAR-M 200 + Hf also contains 1.5% Hf.

Table 3 Compositions of cobalt-base heat-resistant casting alloys

Alloy designation	C	Co	Cr	Ni	Al	B	Fe	Ta	W	Zr	Others
						Nominal composition, %					
AiResist 13	0.45	62	21	. . .	3.4	. . .	. . .	2	11	. . .	0.1 Y
AiResist 213	0.20	64	20	0.5	3.5	. . .	0.5	6.5	4.5	0.1	0.1 Y
AiResist 215	0.35	63	19	0.5	4.3	. . .	0.5	7.5	4.5	0.1	0.1 Y
FSX-414	0.25	52.5	29	10	. . .	0.010	1	. . .	7.5	. . .	. . .
Haynes 21	0.25	64	27	3	. . .	. . .	1	. . .	. . .	. . .	5 Mo
Haynes 25; L-605	0.1	54	20	10	. . .	. . .	1	. . .	15	. . .	. . .
J-1650	0.20	36	19	27	. . .	0.02	. . .	2	12	. . .	3.8 Ti
MAR-M 302	0.85	58	21.5	. . .	. . .	0.005	0.5	9	10	0.2	. . .
MAR-M 322	1.0	60.5	21.5	. . .	. . .	. . .	0.5	4.5	9	2	0.75 Ti
MAR-M 509	0.6	54.5	23.5	10	. . .	. . .	. . .	3.5	7	0.5	0.2 Ti
MAR-M 918	0.05	52	20	20	. . .	. . .	. . .	7.5	. . .	0.1	. . .
NASA Co-W-Re	0.40	67.5	3	. . .	. . .	. . .	. . .	. . .	25	1	2 Re, 1 Ti
S-816	0.4	42	20	20	. . .	. . .	4	. . .	4	. . .	4 Mo, 4 Nb, 1.2 Mn, 0.4 Si
V-36	0.27	42	25	20	. . .	. . .	3	. . .	2	. . .	4 Mo, 2 Nb, 1 Mn, 0.4 Si
WI-52	0.45	63.5	21	. . .	. . .	. . .	2	. . .	11	. . .	2 Nb + Ta
X-40	0.50	57.5	22	10	. . .	. . .	1.5	. . .	7.5	. . .	0.5 Mn, 0.5 Si

ing, prepare for welding, relieve stresses after welding, produce specific microstructures, or soften age-hardened structures by re-solution of second phases. Solution treating is intended to dissolve second phases to produce maximum corrosion resistance or to prepare for aging. Additionally, it will homogenize microstructure prior to aging.

Annealing practices vary considerably among different plants. Representative annealing temperatures, holding times, and cooling procedures are given in Table 5. Experience with specific parts for known requirements often indicates advantageous modifications of temperature, time or cooling method.

Most wrought heat-resisting alloys can be cold formed but are more difficult to form than austenitic stainless steels. Severe cold forming may require several intermediate annealing operations. Full annealing must be followed by fast cooling.

Annealing of weldments should immediately follow welding of the age-hardenable alloys where highly restrained joints are involved. If the configuration of the weldment does not permit annealing, aging can be used for stress relieving the joints. Contact with

Table 4 Compositions of commercially important refractory alloys

Designation	Nominal composition, %
Molybdenum alloys	
Mo-0.5Ti	Mo-0.5Ti-0.02W
TZM	Mo-0.5Ti-0.1Zr-0.02W
Mo-30W	Mo-30W
Niobium alloys	
Nb-1Zr	Nb-1Zr
FS-80	Nb-0.75Zr
FS-82	Nb-33Ta-0.75Zr
FS-85	Nb-27.5Ta-11W-1Zr
SCb-291	Nb-10Ta-10W
Cb-752	Nb-10W-2.5Zr
B-66	Nb-5Mo-5V-1Zr
C-103	Nb-10Hf-1Ti
C-129Y	Nb-10W-10Hf-0.15Y
Tantalum alloys	
"63" Metal	Ta-2.5W-0.15Nb
Ta-10W	Ta-10W
T-111	Ta-8W-2Hf
T-222	Ta-10W-2.5Hf-0.01C
Tungsten alloys	
W-ThO$_2$	W-1ThO$_2$; W-2ThO$_2$
W-Mo alloys	Various Mo contents; W-2Mo and W-15Mo are most common
W-Re alloys	Various Re contents up to 26%; W-1.5Re, W-3Re and W-25Re are most common
Doped W	50 ppm Si, 90 ppm K, 15 ppm Al, 35 ppm O

copper chill blocks should be avoided during welding. Nickel plating is desirable.

Reheating for hot working is an annealing practice whose aim is to promote adequate formability of the metal being deformed. Temperatures vary widely depending on alloy and working practice. Control of temperature can be critical to resultant properties, as varying degrees of recrystallization may be desired. In most standard operations, heating or reheating for hot working is a full annealing step with recrystallization and dissolution of all or most secondary phases. Occasionally, reheating for hot working is restricted to temperatures that do not dissolve all secondary phases so that the remaining phases can be used to limit grain growth.

Heat Treatment for High Strength

The strengthening of heat-resisting alloys usually requires solution treating and aging; typical cycles for wrought alloys are given in Table 6, and for casting alloys, in Table 7. Note that the cooling rate from the solution temperature is critical for some alloys. Alternate heat treatments may be used to improve specific properties.

Solution Treating. In some instances, the solution treating temperature employed will depend on the properties desired. This is indicated in Table 6 for alloys A-286, Inconel 718, René 41, Udimet 700, and Waspaloy. A higher temperature is specified for optimum creep and creep-rupture properties; a lower temperature, for optimum short-time tensile properties at elevated temperature. The higher solution treating temperature will result in some grain growth and more extensive solution of carbides in wrought alloys. The principal objective is to put gamma prime type phases into solution and dissolve some carbides.

After aging, the resulting microstructure of these wrought alloys consists of large grains that contain the principal aging phases and of a heavy concentration of carbides in the grain boundaries. The lower solution treating temperature dissolves the principal aging phases without grain growth or significant high-temperature carbide solution.

For some wrought alloys, for example, Nimonic 80A and Nimonic 90 in Table 6, an intermediate solution treating temperature is selected to produce a compromise of the properties. For other alloys, such as Udimet 500 and Udimet 700, the intermediate temperature aging treatment is used to tailor the grain boundaries for improved creep-rupture properties.

Quenching. The purpose of quenching heat-resisting alloys is to maintain, at room temperature, the supersaturated solid solution obtained during solution treating. Quenching permits a finer gamma prime particle size to be achieved on aging. Cooling methods commonly used are indicated in Table 6 for wrought alloys and in Table 7 for casting alloys.

Aging treatments strengthen age-hardenable alloys by causing the precipitation of one or more phases from the supersaturated matrix that is developed by solution treating and retained by rapid cooling from the solution treating temperature.

Factors that influence the selection of number of aging steps and aging temperature include (a) type and number of precipitating phases available, (b) anticipated service temperature, (c) precipitate size, (d) the combination of strength and ductility desired and heat treatment of similar alloys.

Principal aging phases in the heat-resisting alloys usually include one or more of the following: gamma prime Ni$_3$Al or Ni$_3$(Al, Ti), eta (Ni$_3$Ti), or gamma double prime, Ni$_3$Nb. Secondary phases that may be present include: carbides (M$_{23}$C$_6$, M$_7$C$_3$, M$_6$C and MC), nitrides (MN), carbonitrides (MCN), and borides (M$_3$B$_2$), as well as Laves phase (M$_2$Ti) and delta phase (Ni$_3$Nb). The above phases occur principally in nickel-base alloys. The primary phases in cobalt-base alloys are M$_{23}$C$_6$, M$_7$C$_3$, M$_6$C, and MC. The primary phases in iron-base alloys will be similar to those in nickel alloys although eta is more apt to be found because of the generally higher Ti to Al ratio in iron alloys than in nickel alloys.

When more than one phase is capable of precipitating from the alloy matrix, judicious selection of a single aging temperature may result in obtaining optimum amounts of multiple precipitating phases. Alternatively, a double aging treatment that produces different sizes and types of precipitate at different temperatures may be employed. The aging temperature also determines not only the type but also the size distribution of precipitate. Examples of coprecipitation and of double aging are described in the section of this article concerned with nickel-base alloys.

Exposure to temperatures higher than the optimum aging temperature results in a decrease in strength through the process of overaging; at still higher temperatures, re-solution may occur. High aging temperatures will produce coarser gamma prime particles than lower temperatures and result in higher creep-rupture properties. For optimum short-time elevated-temperature properties, small, finely dispersed particles of gamma prime precipitate are desired. Therefore, lower final aging temperatures are employed than those used to obtain high creep-rupture properties. For all gam-

Table 5 Typical stress relieving and annealing cycles for wrought heat-resisting alloys

Alloy	Stress relieving Temperature °C	°F	Holding time per inch of section, h	Annealing(a) Temperature °C	°F	Holding time per inch of section, h
Iron-base and iron-nickel-chromium alloys						
RA-330.............	900	1650	1(b)	1110(c)	2025(c)	¼(d)
19-9 DL	675(e)	1250(e)	4	980	1800	1
A-286	(f)	(f)	...	980	1800	1
Discaloy	(f)	(f)	...	1035	1900	1
Nickel-base alloys						
Astroloy	(f)	(f)	...	1135	2075	4
Hastelloy B	(f)	(f)	...	1175	2150	1
Hastelloy C	(f)	(f)	...	1215	2225	1
Hastelloy W........	(f)	(f)	...	1175	2150	1
Hastelloy X	(f)	(f)	...	1175	2150	1
Incoloy 800	870	1600	1½	980	1800	¼
Incoloy 800H........	...	...	...	1175	2150	...
Incoloy 825	...	...	...	980	1800	...
Incoloy 901	(f)	(f)	...	1095	2000	2
Inconel 600	900	1650	1	1010	1850	¼(d)
Inconel 601	...	...	...	980	1800	...
Inconel 625	870	1600	1	980	1800	1
Inconel 690	...	...	...	1040	1900	½
Inconel 718	(f)	(f)	...	955	1750	1
Inconel X-750	880(g)	1625(g)	...	1035	1900	½
Nimonic 80A........	(f)	(f)	...	1080	1975	2
Nimonic 90	(f)	(f)	...	1080	1975	2
René 41	(f)	(f)	...	1080	1975	2
Udimet 500	(f)	(f)	...	1080	1975	4
Udimet 700	(f)	(f)	...	1135	2075	4
Waspaloy...........	(f)	(f)	...	1010	1850	4
Cobalt-chromium-nickel-base alloys						
L-605 (HS-25)	(h)	(h)	...	1230	2250	1
N-155 (HS-95).......	(h)	(h)	...	1175	2150	...
S-816..............	(h)	(h)	...	1205	2200	1
Refractory metals(j)						
Ta-10W............	1205(k)	2200(k)	1	1425(k)	2600(k)	1
FS-80	1095(k)	2000(k)	1	1315(k)	2400(k)	1
FS-82	1095(k)	2000(k)	1	1315(k)	2400(k)	1
Mo-0.5 Ti	1095(m)	2000(m)	½	1315(m)(n)	2400(m)(n)	1
TZM	1205(m)	2200(m)	1	1425(m)(n)(p)	2600(m)(n)(p)	1

(a) Minimum hardness is achieved by cooling rapidly from the annealing temperature, to prevent precipitation of hardening phases. Water quenching is preferred, and is usually necessary for heavy sections; air cooling is preferred for heavy sections of Waspaloy, Udimet 500, Udimet 700 and Inconel X-750, because water quenching causes cracking. However, for complex shapes subject to excessive distortion, oil quenching is often adequate and more practical. Rapid air cooling usually is adequate for parts formed from strip or sheet. Rapid cooling from the annealing or solution treating temperature does not suppress the aging reaction of some alloys, such as Astroloy; these alloys become harder and stronger. (b) Time given is minimum; some plants use as long as 3 h per inch. (c) Nominal temperature; 1035 to 1175 °C (1900 to 2150 °F) is commonly used. (d) Short time is required for prevention of grain coarsening. (e) Nominal temperature; 650 to 705 °C (1200 to 1300 °F) is permissible. (f) Full annealing is recommended, because intermediate temperatures cause aging. (g) Used only for stress equalizing of warm worked grades. (h) Full annealing is recommended if further fabrication is performed; otherwise, material can be stress relieved at approximately 55 °C (100 °F) below annealing temperature. (j) Annealing temperatures depend on prior plastic deformation, degree of cold work, alloy content and interstitial purity. Annealing temperatures given are those most frequently used for cold worked sheet or plate; in many instances, more precise determination of the recrystallization temperature is necessary for a specific application. (k) Heat and cool in vacuum or inert-gas atmosphere. (m) Heat and cool in hydrogen or vacuum. (n) Seldom used as finished product in annealed condition, because recrystallization raises the ductile-brittle transition temperature, resulting in brittleness at low temperatures. (p) For vacuum-arc-cast material with a minimum of 50% cold work.

ma prime dispersions, care must be taken to ensure the correct carbide distribution. A principal reason for two-step aging sequences, in addition to gamma prime or gamma double prime control, is the need to precipitate or control grain boundary carbide morphology.

Surface Contamination

Age-hardenable heat-resisting alloys may have good oxidation resistance within their normal range of service temperature, at or above their aging temperatures—in a range of 760 to 980 °C (1400 to 1800 °F)—depending on the alloy. Others may require coatings, particularly blade alloys for turbojet engines and compressors. However, at temperatures used for solution treating, these alloys are susceptible to intergranular oxidation, a defect that adversely affects thermal fatigue.

Table 6 Typical solution treating and aging cycles for wrought heat-resisting alloys

Alloy	Solution treating Temperature °C	°F	Time, h	Cooling procedure	Aging Temperature °C	°F	Time, h	Cooling procedure
Iron-base alloys								
A-286	980	1800	1	Oil quench	720	1325	16	Air cool
Discaloy	1010	1850	2	Oil quench	730	1350	20	Air cool
					650	1200	20	Air cool
N-155	1175	2150	1	Water quench	815	1500	4	Air cool
Nickel-base alloys								
Astroloy	1175	2150	4	Air cool	845	1550	24	Air cool
	1080	1975	4	Air cool	760	1400	16	Air cool
Hastelloy B	1175	2150	1/2	(a)	(b)	(b)	...	...
Hastelloy B-2	1065	1950	1/2	Rapid quench	...	...	...	...
Hastelloy C-4	1065	1950	1/2	Rapid quench	...	...	...	...
Hastelloy C-276	1120	2050	1/2	Rapid quench	...	...	...	...
Hastelloy N	1175	2150	1/2	Rapid quench	...	...	...	...
Hastelloy S	1065	1950	1/2	Rapid quench	...	...	...	...
Hastelloy C	1220	2225	1	(a)	(b)	(b)	...	...
Hastelloy W	1175	2150	1	(a)	(b)	(b)	...	...
Hastelloy X	1175	2150	1	(a)	...	...	...	...
Inconel 901	1095	2000	2	Water quench	790	1450	2	Air cool
					720	1325	24	Air cool
Inconel 600	1120	2050	2	Air cool	...	...	...	...
Inconel 601	1150	2100	1	Air cool	...	...	...	...
Inconel 617	1175	2150	2	(a)	...	...	...	...
Inconel 625	1150	2100	2	(a)	...	...	...	...
Inconel 706	925-1010	1700-1850	...	...	845	1550	3	Air cool
					720	1325	8	Furnace cool
					620	1150	8	Air cool
	925-1010	1700-1850	...	...	730	1350	8	Furnace cool
					620	1150	8	Air cool
Inconel 718	980	1800	1	Air cool	720	1325	8	Furnace cool
					620	1150	8	Air cool
Inconel X-750 (AMS 5667)	855	1625	24	Air cool	705	1300	20	Air cool
Inconel X-750 (AMS 5668)	1150	2100	2	Air cool	845	1550	24	Air cool
					705	1300	20	Air cool
Nimonic 80A	1080	1975	8	Air cool	705	1300	16	Air cool
Nimonic 90	1080	1975	8	Air cool	705	1300	16	Air cool
René 41	1065	1950	1/2	Air cool	760	1400	16	Air cool
Udimet 500	1080	1975	4	Air cool	845	1550	24	Air cool
					760	1400	16	Air cool
Udimet 700	1175	2150	4	Air cool	845	1550	24	Air cool
	1080	1975	4	Air cool	760	1400	16	Air cool
Waspaloy	1080	1975	4	Air cool	845	1550	24	Air cool
					760	1400	16	Air cool
Cobalt-base alloys								
Haynes 25; L-605	1230	2250	1	Rapid air cool	(b)	(b)	...	...
Haynes 188	1175	2150	1/2	Rapid air cool	...	...	...	...
Haynes 556	1175	2150	1/2	Rapid air cool	...	...	...	...
S-816	1175	2150	1	(a)	760	1400	12	Air cool
Stellite 6B	1230	2250	1	Air cool	...	...	...	...

Note: Alternate treatments may be used to improve specific properties. (a) To provide an adequate quench after solution treating, it is necessary to cool below about 540 °C (1000 °F) rapidly enough to prevent precipitation in the intermediate temperature range. For sheet metal parts of most alloys, rapid air cooling will suffice. Oil or water quenching is frequently required for heavier sections that are not subject to cracking. (b) Aging occurs in service at elevated temperatures.

Intergranular oxidation is measured optically as depth of intergranular penetration. Figure 1(a) shows the depth of intergranular oxidation that occurred in René 41 heated in air. Oxidation data as determined by weight change for several age-hardenable alloys and type 310 stainless steel are plotted in Fig. 1(b).

Oxidation resistance is enhanced by chromium and aluminum additions as well as by certain other elements. Chromium allows the formation of a protective chromium oxide and is most beneficial at intermediate temperatures and below $\leq$870 °C ($\leq$1600 °F), whereas aluminum forms aluminum oxide which is more protective at the higher temperature of $\leq$870 °C ($\leq$1600 °F). The principal mode of intergranular attack involves the preferential oxidation of chromium, aluminum, titanium, zirconium and boron. Molybdenum increases susceptibility to intergranular attack in age-hardenable alloys.

In relation to intergranular oxidation, aluminum is preferable to titanium as a hardening element for two reasons: (a) increased aluminum reduces formation of the eta phase from gamma prime; and (b) aluminum oxide provides a denser and less permeable barrier to the diffusion of oxygen.

Carbon pickup can occur if the solution treating atmosphere has a carburizing potential. For instance, the carbon content of the surface of A-286 alloy has been observed to increase from 0.05 to 0.30%. The added carbon forms a stable carbide (TiC), thus removing titanium from solid solution and preventing normal precipitation hardening in the surface layers. TiN can be formed in the same manner as a result of nitrogen contamination.

Miscellaneous Contaminants. All exposed surfaces of heat-resisting alloy parts should be kept free of dirt, fingerprints, oil, grease, forming compounds, lubricants and scale. Lubricants or fuel oils that contain sulfur-bearing compounds are particularly active in corroding the metal surface by first forming Cr_2S_3, and then, as the attack progresses, also forming a Ni-Ni_3S_2 eutectic that melts at 645 °C (1190 °F), particularly at low pressures of less than 10^{-4} torr in vacuum.

Scale and slag from furnace hearths are another source of contamination. Contact with steel scale, slag and furnace spallings should be avoided; low-

Table 7 Typical solution treating and aging cycles for heat-resisting casting alloys

Alloy	Solution treating				Aging			
	Temperature(a)		Time, h	Cooling procedure	Temperature(b)		Time, h	Cooling procedure
	°C	°F			°C	°F		
A-286	1095	2000	2	Rapid cool	720	1325	16	Air cool
B-1900	As cast		...	...	...	...	...	...
FSX-414	1150	2100	4	Rapid cool	980	1800	4	Air cool
Hastelloy B	1175	2150	2	Rapid cool	(c)	(c)	...	...
Hastelloy C	1220	2225	1	Rapid cool	(c)	(c)	...	...
HS-31 (X-40)	As cast		...	...	...	...	...	...
IN-100	As cast		...	...	...	...	...	...
IN-713C	As cast		...	...	...	...	...	...
IN-738	1120	2050	2	Air cool	845	1550	24	Air cool
IN-792	1120	2050	2	Air cool	845	1550	24	Air cool
IN-939	1160	2120	4	Air cool	850	1560	16	Air cool
Inconel 718	1095	2000	1	Air cool	620	1150	10	Air cool
MAR-M 200	...	...	...	...	870	1600	50	Air cool
MAR-M 200 DS	1230	2250	4	Air cool	870	1600	32	Air cool
MAR-M 246	...	...	...	...	845	1550	50	Air cool
MAR-M 247	...	...	...	...	870	1600	16	Air cool
MAR-M 302	As cast		...	...	...	...	...	...
MAR-M 509	As cast		...	...	...	...	...	...
René 41	1095	2000	1/2	Rapid cool	900	1650	4	Air cool
René 80	1220	2225	2	Air cool	1095	2000	4	Air cool
					1055	1925	4	Air cool
					845	1550	16	Air cool
Udimet 700	1150	2100	2	Air cool	760	1400	16	Air cool

(a) Furnace temperature tolerance of $\pm$15 °C ($\pm$25 °F) is satisfactory. (b) Furnace temperature tolerance of $\pm$10 °C ($\pm$15 °F) is recommended. (c) Aging occurs in service at elevated temperature. Use a vacuum or protective atmosphere for heat treating at temperatures above 1040 °C (1900 °F) and subsequent cooling.

melting constituents can form on the metal surface and promote corrosion.

Protective Atmospheres

Protective atmospheres are used in annealing or solution treating if heavy oxidation cannot be tolerated. If oxidation can be tolerated, because of subsequent stock removal, heat-resisting alloys can be solution treated in air or in the normal mixture of air and combustion products found in gas-fired furnaces. However, refractory metals must always be heat treated in a vacuum or in an inert-gas atmosphere (argon, helium, or an ArHe mixture) or hydrogen. In some cases, ceramic coatings are used to prevent surface attack.

Exothermic Atmosphere. A lean and dilute exothermic atmosphere is relatively safe and economical. The surface scale formed in such an atmosphere can be removed by pickling or by salt bath descaling and pickling. Such an atmosphere, formed by burning fuel gas with air, contains about 85% nitrogen, 10% carbon dioxide, 1.5% carbon monoxide, 1.5% hydrogen,

and 2% water vapor. This atmosphere will produce a scale rich in chromium oxides.

Endothermic atmospheres prepared by reacting fuel gas with air in the presence of a catalyst are not recommended, because of their carburizing potential. Similarly, the endothermic mixture of nitrogen and hydrogen formed by dissociating ammonia is not used, because of the probability of nitriding.

Dry hydrogen (dew point, −50 °C (−60 °F) or lower) is used in preference to dissociated ammonia for bright annealing of heat-resisting alloys. If the hydrogen is prepared by catalytic gas reactions instead of by electrolysis, residual hydrocarbons, such as methane, should be limited to about 50 ppm, to prevent carburizing. Hydrogen is not recommended for bright annealing of alloys containing significant amounts of elements (such as aluminum or titanium) that form stable oxides not reducible at normal heat treating temperatures and dew points. Hydrogen is not recommended for annealing or solution treating alloys that contain boron, because of the danger of deboroni-

Fig. 1 Effect of time and temperature on oxidation

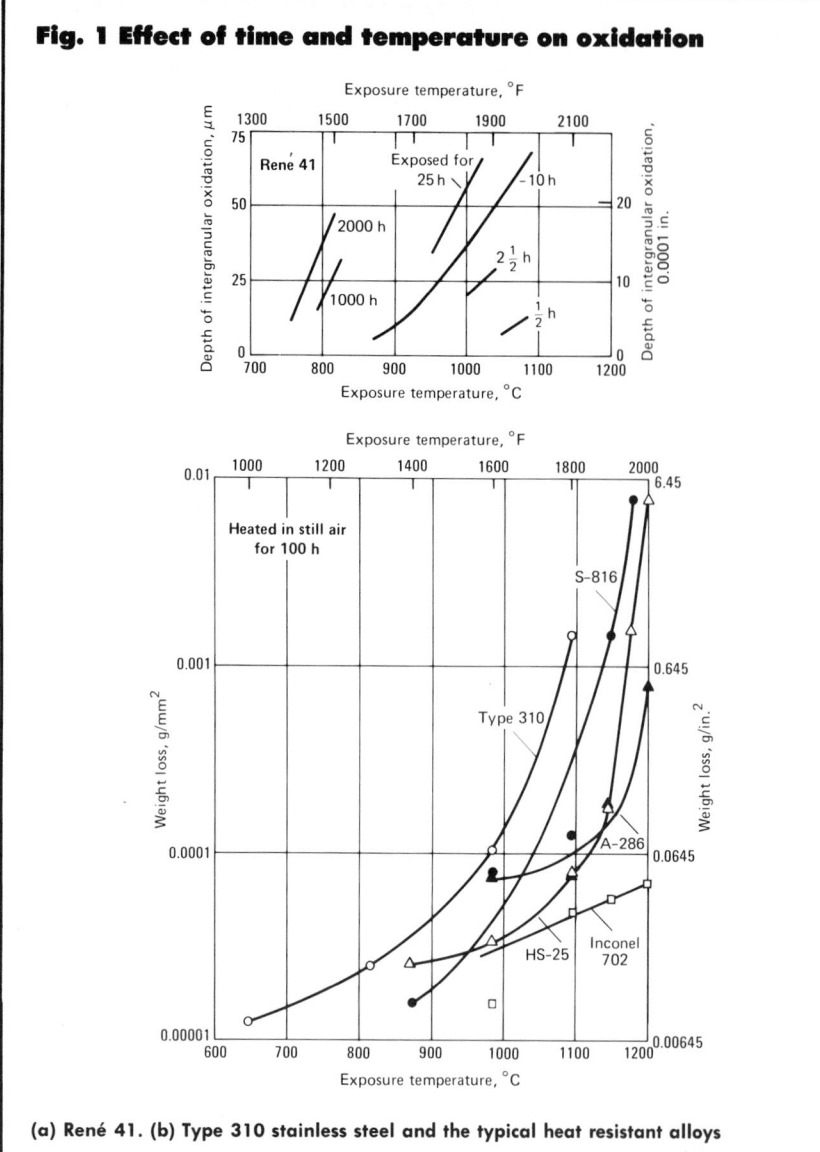

(a) René 41. **(b)** Type 310 stainless steel and the typical heat resistant alloys

zation through formation of boron hydrides. Nor can hydrogen be used for heat treating niobium and tantalum, because of its embrittling effect.

Dry argon (dew point: −50 °C, or −60 °F or lower) should be used if no oxidation can be tolerated. It is mandatory that this type of atmosphere be used in a sealed retort or sealed furnace chamber. A purge of at least ten times the volume of the retort is recommended before the retort is placed in the furnace. The argon must be kept flowing continually during and after the treatment until the workpieces have cooled nearly to room temperature, to prevent the formation of an oxide film.

Heat-resisting alloys containing stable-oxide formers such as aluminum and titanium, with or without boron, must be bright annealed in a vacuum or in a chemically inert gas such as argon. If used, argon must be pure and dry—dew point, −50 °C (−60 °F) or lower. If the argon has a slightly higher dew point—but not over −40 °C (−40 °F)—oxidation will be limited to a thin surface film that can usually be tolerated.

Vacuum Atmosphere. Vacuum atmosphere, generally below 20 μm (2 × 10^{-3} torr) is commonly used for heat-resisting alloys above 815 °C (1500 °F). It is particularly desirable when parts are at or close to final dimensions.

Atmospheres for Aging. Air is the most common aging atmosphere. The smooth, tight oxide layer that is formed is usually unobjectionable on the finished product. However, if this oxide layer must be minimized, a lean exothermic gas (air-gas ratio, about 10 to 1) can be employed. It will not entirely prevent oxidation, but the oxide layer will be very light. *The use of gases containing hydrogen and carbon monoxide for aging cycles is dangerous because of the explosion hazard at temperatures below 760 °C (1400 °F).*

Furnace Equipment

Basic equipment considerations seldom differ from those influencing the selection of furnaces for heat treating steel. In general, the temperature-control limits are ±14 °C (±25 °F) and temperature may range up to about 1290 °C (2350 °F). Belt conveyor furnaces, although widely used for production annealing, are less gastight than roller-hearth furnaces. Consequently, atmosphere costs for a belt conveyor furnace are likely to be higher than for a roller-hearth furnace of the same volume. Batch heating for annealing or solution treating is usually done in box furnaces. These may have provisions for purging, preheating and quenching, if the high-temperature compartment is supplemented by other chambers.

Aging of heat-resisting alloys, commonly in the range of 650 to 900 °C (1200 to 1650 °F), is usually done in box furnaces, with or without protective atmospheres. The usual operating-temperature tolerance is ±14 °C (±25 °F) for wrought alloys and ±8 °C (± 15 °F) for casting alloys. Continuous furnaces are seldom used, because of the long aging cycles. Salt baths are not recommended, because reaction could occur between chloride in the bath and the alloy surface during the long-time immersion that would be required for aging.

Vacuum furnaces are used for heat treating of niobium and tantalum, and other heat-resisting alloys. Heating may be accomplished by resistance elements or by induction. Furnace design dictates a batch operation. Cooling can be accomplished in a vacuum retort pressurized with an inert gas that provides conductive cooling after heating is discontinued. Further details are provided in the article on heat treating in vacuum furnaces.

Fixturing

Fixtures for holding finished parts or assemblies during heat treatment may be of either the support type or the restraint type. For alloys that must be cooled rapidly from the solution treating temperature, the best practice is to employ minimum fixturing during solution treating and quenching and to control dimensional relations by the use of restraining fixtures during aging.

Support fixtures are used when restraint is not required or when the part itself provides sufficient self-restraint. A support fixture also functions as an aid in handling parts and helps the part to support its own weight. Long narrow pieces, such as tubes or bolts, are most easily fixtured by hanging vertically. Parts that have a large flat surface can be placed on a flat furnace tray or plate. Examples of such parts are rings, cylinders and beams. For parts of slightly asymmetrical shape, special supports can be built up from a flat tray. If these supports are fabricated by welding, they must be stress relieved prior to use.

Asymmetrical parts can be supported in several ways. One method is to lay the part on a tray of sand, making certain that most of the bottom area is well supported. Alluvial garnet sand is most commonly used as the supporting medium. Another method of support is the use of a ceramic casting formed to the shape of the part. However, this method is costly and subject to size limitations. Turbine blades and asymmetrical ducting are examples of parts that can be supported either in a sand tray or by ceramic castings.

Restraint fixtures are generally more complicated than support fixtures and may require machined grooves, lugs or clamps to hold parts to a given shape.

To maintain symmetry and roundness in an A-286 frame assembly during aging, the assembly was processed on a flat plate into which grooves had been machined. These grooves accepted the rims on the outer and inner shrouds and held them in restraint during heat treatment. To prevent the center hub from rising or dropping in relation to the outer shroud, both the hub and the shroud were clamped to the grooved-plate fixture.

It is possible to perform some straightening of parts in aging fixtures of the type described above. A slightly distorted part can be forced into the fixture and clamped. Some stress relieving will occur along with aging. However, fixtures for hot sizing are not always successful for heat-resisting alloys, because of the high creep strength of these alloys at the aging temperature.

The use of threaded fasteners for clamping is not recommended, because they are difficult to remove after heat treatment. A slotted bar held in place by wedges is preferred.

Usually, the coefficient of expansion of both the fixture and part should be nearly the same. However, in some applications, the fixture is purposely made from a material having different expansion characteristics, in order to apply pressure to the part as the temperature increases.

For example, for securing a René 41 turbine blade in place for brazing and solution treatment, sand or cast-ceramic techniques could not be used because of the rapid heating and cooling rates required. Instead, a guillotine-type fixture was constructed from Hastelloy X, with 11 contour stations conforming to the correct shape for the blade. Inconel tubes standing along the edges of the blade were secured to the fixture with molybdenum wire. The wire passed through a hole in the blade and fixture and through the tube. The wire in each tube was looped and tied around a short tube placed horizontally atop the standing tube. Thus, during heating, the tubes, which have a greater expansion coefficient than the wire, produced a clamping effect on the edges of the blade.

Relation of Fabrication to Heat Treating

A uniformly fine-grained microstructure can be produced in the age-hardenable heat-resisting alloys if final deformation is carried out in the lower part of the hot working range. This type of structure can be solution treated to a uniform grain size. If the final working temperature is too low, or if the metal is not reduced enough during final working, a duplex structure is produced during solution treating. Alloys deformed in the upper hot working range have a coarser structure that cannot be refined by solution treating.

Cold working is usually performed on alloys in the solution treated condition because of the markedly lower strength and increased ductility of the material before aging (see Table 8). Cold working itself affects mechanical properties, through its influence on (a) grain growth during subsequent solution treatment, and (b) the reaction kinetics of aging.

Effect of Grain Size. The age-hardenable heat-resisting alloys are susceptible to critical grain growth if they are solution treated after small amounts of cold or hot work. Larger amounts of cold work refine the grain size. Excessive grain growth may have deleterious effects on creep and fatigue properties in all heat-resisting alloys. Therefore, on parts subjected to cold or hot work prior to solution treating, the critical amount of work (about 1 to 6% cold work, depending on the alloy, or 10% hot work) must be exceeded in all areas, to avoid the growth of abnormally large grains. This rule applies to items such as cold headed bolts, spun or stretch-formed sheets, and parts formed by simple bending.

Effect on Aging. Cold working accelerates the aging reaction, causing the early appearance of precipitates at normal aging temperature and the appearance of a precipitate at temperatures below the normal aging temperature if the alloy is cold worked sufficiently and held at temperature long enough. In other words, cold working renders the material more readily overageable at the normal aging temperature. Properly charted information cor-

Table 8 Typical effects of aging on room-temperature mechanical properties of solution treated heat-resisting alloys

| Alloy | Yield strength(a) | | | | Elongation in 50 mm (2 in.), % | |
| | Not aged | | Aged | | | |
	MPa	ksi	MPa	ksi	Not aged	Aged
A-286	240	35	760	110	52	33
René 41	620	90	1100	160	45	15
Inconel X-750	410	60	650	92	45	24
HS-25	480	69	480	70	55	45

(a) At 0.2% offset

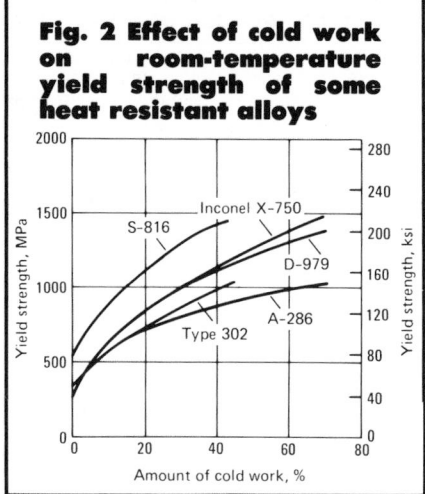

Fig. 2 Effect of cold work on room-temperature yield strength of some heat resistant alloys

relating cold work with aging response, however, can be used to shorten the required aging period or to lower the normal aging temperature.

Rates of Work Hardening. In cold forming operations such as sheet rolling, tube drawing, pressing, brake and stretch forming, and deep drawing, the governing material property is the rate of hardening in relation to the amount of cold deformation. Materials such as A-286, Nimonic 75, and Hastelloy X can be reduced as much as 90% before a softening treatment is required. For the more complex alloys such as René 41 and Udimet 500, incremental reductions of 35 to 40% can be achieved without intermediate annealing.

Figure 2 shows the relative work hardening rates of four heat-resisting alloys compared to type 302 stainless steel. The iron-base A-286 alloy work hardens at about the same rate as type 302. The highest work hardening rates are those of the cobalt-base alloys, such as S-816.

In general, the greater the work hardening rate of a given alloy, the less the amount of cold deformation possible before annealing is required. When a large amount of total reduction is required, as in the draw swaging of tubes, the metal is formed in small incremental reductions with interstage anneals between successive reductions.

Welding of age-hardenable heat-resisting alloys requires precise control of prewelding and postwelding heat treatments. A typical welding sequence for wrought alloys is as follows:

• Material received in mill-annealed

condition (the specific condition is critical for certain alloys)
• Forming by various methods (in-process anneals used if required)
• Welding
• Solution treating of weldment
• Aging of weldment.

Mill annealing cycles must dissolve precipitated phases without causing excessive grain growth and must not dissolve M_6C in alloys such as René 41. Large grain size and dissolved carbides may produce a continuous grain-boundary carbide precipitate that promotes cracking during welding. A maximum grain size of ASTM 3 is usually recommended for welding.

Fusion welds in age-hardenable heat-resisting alloys should be solution treated after welding to eliminate the uncontrolled aging that occurs in the heat-affected zone of the base metal during welding, and to relieve residual stresses. Special procedures, such as double aging, have been used in large and complex components containing restrained welds, for which solution treatment is not practical.

Cast non-age-hardenable cobalt-base superalloy turbine vanes are repair welded, and heat treatment procedures, such as preheat and post-heat, have been developed to minimize cracking and surface contamination in these alloys. Age-hardenable cast nickel-base alloys are not normally repair welded with the exception of the weldable superalloy Inconel 718. Cracking and dimensional control are problems encountered with the welding of normal cast nickel-base age-hardenable alloys.

Service requirements of weldments must be considered in the selection of heat treatment. If service temperatures do not exceed the aging temperature, no appreciable change in properties is likely in service. However, if the service temperature is above the normal aging temperature, strength decreases and ductility increases in service.

The heat treatments that may be applied to weldments in nonhardenable heat-resisting alloys are limited to stress relieving and, in some instances, stabilizing. Stress relieving must be done at relatively high temperatures (815 to 980 °C, or 1500 to 1800 °F) to be effective. Because the stress-relieving temperature range coincides with the carbide-precipitation range, the composition of the alloys should be controlled

to avoid precipitating carbides in harmful amounts.

Heat Treating of Iron-Base Alloys

The iron-base heat-resisting alloys (Table 1) include some that are hardenable by solution treating and aging and some that are not hardenable by heat treatment. Heat treating temperatures for selected alloys are given in Tables 5, 6 and 7.

Stress Relieving and Annealing. Alloys (such as 19-9 DL) that depend entirely on hot and cold work to develop high strength frequently require stress relieving (Table 5). However, these alloys cannot be stress relieved when maximum corrosion resistance is required in service, because the recommended stress-relieving temperatures are within the sensitizing range of the alloys. If these alloys are to be exposed to acid or steam in service, intergranular cracking can be minimized or eliminated by annealing instead of stress relieving.

Iron-base heat-resisting alloys, such as A-286 and Incoloy 901, which rely on gamma prime type hardening, although not susceptible to sensitization at stress-relieving temperatures, still cannot be stress relieved, because the intermediate temperatures of stress relief result in aging. Consequently, the restoration of ductility and reduction of stresses in cold formed parts and weldments made of these alloys is achieved by heating rapidly to the annealing temperature. This procedure will result in partial solutioning of the gamma prime phase, reduction in residual stress and subsequent reprecipitation of gamma prime on cooling. In forging these alloys, the finishing temperature is usually above 925 °C (1700 °F), and stress relieving is not required for as-forged parts. These alloys usually are solution treated and aged after forging.

Castings. Iron-chromium-nickel alloys, identified by Alloy Casting Institute designations HA through HX, are usually exposed to service temperatures of 650 to 1250 °C (1200 to 2200 °F) in process equipment such as heat treating furnaces, oil-refinery equipment, and cement kilns. Normally, they are placed in service in the as-cast condition, but when experience indicates the advisability of stress relieving, the operation is performed by placing the castings in a cold furnace and

and heating slowly to 980 to 1040 °C (1800 to 1900 °F), holding at temperature for 1 h per inch of section thickness, and then furnace cooling.

Castings usually are not annealed, because the need for restoring ductility seldom exists, and slow cooling through the temperature range of 705 to 900 °C (1300 to 1650 °F) may cause formation of sigma phase in some grades, such as HD. Therefore, heat-resisting alloy castings are annealed only for special reasons. Some of these reasons are discussed in the following paragraphs. However, in some cases, stress relieving operations are performed to relieve residual stresses.

Iron-chromium casting alloys containing 9 to 30% chromium and little or no nickel vary widely in response to heat treatment, depending on composition. Most castings of alloys in this group are used in the as-cast condition. Type HA (9% Cr, 1% Mo) is an exception; it transforms from austenite to ferrite on air cooling from temperatures above 815 °C (1500 °F). For maximum softness, castings of this alloy are annealed by furnace cooling from 885 °C (1625 °F) minimum (at about 30 °C/h [50 °F/h]) to below 705 °C (1300 °F).

The iron-chromium-nickel and iron-nickel-chromium casting alloys containing 20 to 30% chromium and 10 to 20% nickel (such as HE, HF and HL), as well as those containing about 12 to 20% chromium and 25 to 67% nickel (such as HT, HU and HX), are usually used in the as-cast condition. However, for applications that require maximum thermal fatigue properties, annealing is sometimes beneficial. Castings of alloy HT (35% Ni, 15% Cr), for example, may give improved performance by being heated to 1040 °C (1900 °F), held at temperature for about 12 h, and furnace cooled, prior to being placed in service.

Solution Treating and Aging. The principal hardenable iron-base alloys are A-286 and Incoloy 901. Discussion and examples here will be confined to these two alloys, which contain titanium and aluminum, and which are hardened mainly by precipitation of gamma prime. The gamma prime is metastable and transforms to eta phase if the aluminum to titanium ratio is too low.

Iron-base alloys such as A-286 are relatively insensitive to the rate of cooling from the solution treating temperature (Table 9). Thus, section thickness

Table 9 Effect of cooling rate from 980 to 535 °C (1800 to 1000 °F) on short-time properties of A-286(a)

Cooling rate		Yield strength(b)		Tensile strength		Elongation, %	Reduction in area, %
°C/min	°F/min	MPa	ksi	MPa	ksi		
Tested at 20 °C (70 °F)							
Oil quench		680	99	1030	149	24	39
15	27	690	100	1030	150	22	42
7	15	680	99	1020	148	24	39
5	9	700	102	990	143	22	36
1	2	660	96	1030	149	23	33
Tested at 650 °C (1200 °F)							
Oil quench		650	94	780	113	8	10
5	27	650	94	770	112	6	7
7	15	630	92	770	112	9	16
5	9	590	86	740	108	7	17
1	2	220	90	740	108	11	18

(a) After cooling from the solution temperature, all samples were aged at 720 °C (1325 °F) for 16 h and air cooled. (b) At 0.2% offset

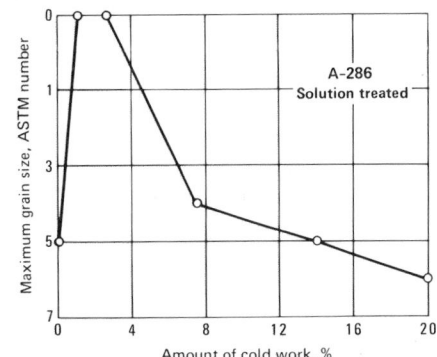

Fig. 3 Effect of cold work

Influence of cold work on grain size of A-286 alloy solution treated at 900 °C (1650 °F) for 1 h and oil quenched

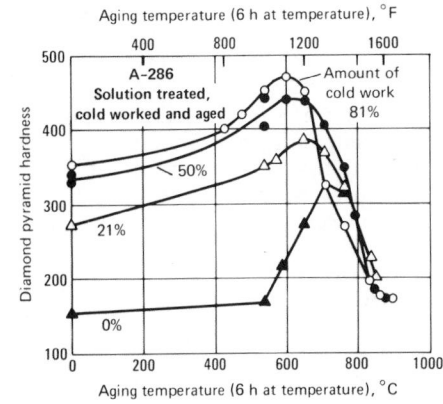

Fig. 4 Effect of cold work and aging on diamond pyramid hardness on A-286

is not of major concern in relation to cooling practice. The effect of varying amounts of cold work on grain growth in A-286 during solution treating is

Table 10 Typical uniformity of properties obtained in double aging of Incoloy 901 turbine disks

Location	Ultimate tensile strength MPa	ksi	Yield strength(a) MPa	ksi	Elongation in 50 mm (2 in.), %	Reduction in area, %
ID, tangential	1170	170	870	127	16	19
Midradius, radial	1180	171	890	129	16	19
OD, tangential	1180	171	880	128	17	20
Specified (min)	1140	165	830	120	12	15

(a) At 0.2% offset

Table 11 Effect of intermediate aging on typical properties of Incoloy 901

	Ultimate tensile strength MPa	ksi	Yield strength(a) MPa	ksi	Elongation in 50 mm (2 in.), %	Reduction in area, %	Creep-rupture life, h
Tested at 20 °C (70 °F)							
No intermediate aging(b):							
Heat A	1050	152	790	115	12	13	...
Heat B	1080	157	790	114	17	16	...
With intermediate aging(c):							
Heat A	1040	151	730	106	12	15	...
Heat B	1040	151	710	103	12	13	...
Tested at 650 °C (1200 °F)							
No intermediate aging(b):							
Heat A	...	...	...	...	1.0	...	76
Heat B	...	...	...	...	1.5	...	118
With intermediate aging(c):							
Heat A	...	...	...	...	11	...	45
Heat B	...	...	...	...	7	...	54

(a) At 0.2% offset. (b) Heat treatment: 1120 °C (2050 °F) for 2 h, water quench; 745 °C (1375 °F) for 24 h, air cool. (c) Heat treatment: 1120 °C (2050 °F) for 24 h, water quench; 815 °C (1500 °F), 4 h; air cool; 745 °C (1375 °F), 24 h; air cool

illustrated in Fig. 3. The initial material was in the solution treated condition with a maximum grain size of ASTM 5. Cold working 14% and re-solution treating reproduced this same grain size. However, cold working in the range of 1 to 5% caused excessive grain growth during subsequent solution treating.

Cold working affects the properties after aging, as illustrated in Fig. 4 for A-286. The hardness after cold working and before aging, as well as the peak hardness after aging, increases greatly with increasing amounts of cold rolling. The temperature at which peak hardness is attained decreases with increasing amounts of reduction. The temperature at which softening occurs also decreases with increasing amounts of deformation. (Note, in Fig. 4, that the material cold worked 81% and then aged at 760 °C or 1400 °F for 16 h is softer than material that was not cold worked before being aged.)

To obtain uniformity of properties in parts that have been cold worked non-uniformly, higher-than-normal aging temperatures are sometimes used, followed by a second aging cycle at lower-than-normal aging temperature. For example, in A-286 parts with varying amounts of cold work, double aging at 760 and 705 °C (1400 and 1300 °F), respectively, provides more uniform hardness, short-time tensile properties, and creep-rupture properties than the normal 720 °C (1325 °F) aging treatment. The higher aging temperature also improves the structural stability of the part in service.

The uniformity of properties after double aging is illustrated by the following example.

Turbojet turbine disks made of Incoloy 901, each weighing 170 kg (375 lb), were heat treated in a gas-fired, box-type furnace by the following procedure:

- *Solution treating:* Load 10 disks on tray in single layer (total weight of load, 1700 kg or 3750 lb). Heat to 1080 °C (1975 °F), hold 2 h at temperature ±10 °C (±20 °F), quench rapidly in water.
- *First aging treatment (stabilizing):* Load 20 disks on tray, stacking in two layers (total weight of load, 3400 kg or 7500 lb). Heat to 775°C (1425 °F), hold 2 h at temperature ±5 °C (±10 °F), cool in air.
- *Second aging treatment:* Load as in first aging. Heat to 720 °C (1325 °F), hold 2 h at temperature ±5 °C (±10 °F), cool in air.

Uniformity of properties obtained with this procedure is shown in Table 10.

Practical Heat Treatment. Acceptable mechanical properties are not always accomplished as a result of the solution treating and aging procedure first tried for a given alloy or application. To develop specified properties, changes of the following kinds are often required:

- Adjust the solution temperature or time.
- Adjust the aging temperature.
- Add an intermediate (stabilizing) aging treatment at a temperature higher than that of the final aging.
- Add a second (final) aging treatment at a lower temperature.
- Adjust one or both aging temperatures in a double aging cycle.
- Add a third aging treatment.

These modifications of practice are illustrated for A-286 and Incoloy 901 in the following eight examples. Note that for forging alloys, protective atmosphere is usually not required as the surface is subsequently machined off. However, the surface of precision investment casting must be protected during heat treatment with protective atmosphere as argon or vacuum.

Example 1 (A-286). The treatment first specified—900 °C (1650 °F) for 2 h, oil quench; 705 °C (1300 °F) for 16 h, air cool—resulted in borderline values of yield strength (610 to 630 MPa, or 89 to 91 ksi for two heats). Adding a second aging treatment (650 °C, or 1200 °F, for 16 h) increased the yield strength to 630 and 700 MPa (92 and 101 ksi) for the same two heats of material.

Example 2 (Incoloy 901). In one application, the alloy failed to meet a specification requiring 830 MPa (120

Table 12 Effect of single and double aging on room-temperature properties of Incoloy 901

	Ultimate tensile strength		Yield strength		Elongation in 50 mm (2 in.), %	Reduction in area, %
	MPa	ksi	MPa	ksi		
Specification...............	1140	165	827	120	12	15
Single aging(a)...........	1150 to 1160	167 to 169	800 to 810	116 to 118	20 to 23	24 to 29
Double aging(b)...........	1190 to 1210	173 to 175	830 to 890	121 to 129	18 to 22	24 to 29

Single and double aging data reflect the results of four tests. (a) Solution treated at 1085 °C (1985 °F) for 2 h, water quenched; aged at 770 °C (1450 °F) for 2 h, air cooled; aged at 720 °C (1325 °F) for 24 h, air cooled. (b) Re-aged at 650 °C (1200 °F) for 12 h and air cooled

Table 13 Effects of adding a third aging treatment on properties of Incoloy 901

Ultimate tensile strength		Yield strength		Elongation in 50 mm (2 in.), %	Reduction in area, %	Creep-rupture elongation(a)
MPa	ksi	MPa	ksi			
Specification						
1140	165	830	120	12	15	4% in 23 h(b)
Double aging(c)						
1160 to 1210(d)	169 to 175(d)	810 to 900(d)	118 to 131(d)	22 to 23(d)	25 to 30(d)	4.9% in 31 h to 2.8% in 85 h(e)
Triple annealing(f)						
1200 to 1240(d)	174 to 180(d)	850 to 930(d)	123 to 135(d)	18 to 20(d)	23 to 29(d)	7% in 64 h to 6.3% in 74 h(g)

(a) At 650 °C (1200 °F) and 620 MPa (90 ksi). (b) Minimum values. (c) Solution treated at 1085 °C (1985 °F) for 2 h, water quenched; aged at 775 °C (1425 °F) for 2 h, air cooled; aged at 720 °C (1325 °F) for 24 h, air cooled. (d) Seven tests. (e) Three tests. (f) Solution treated at 1085 °C (1985 °F) for 2 h; water quenched; aged at 790 °C (1450 °F) for 2 h, air cooled; aged at 720 °C (1325 °F) for 24 h, air cooled; aged at 650 °C (1200 °F) for 12 h, air cooled. (g) Two tests

ksi) minimum yield strength after double aging—1095 °C (2000 °F) for 2 h, water quench; 790 °C (1450 °F) for 2 h air cool; 720 °C (1325 °F) for 2 h, air cool. The specification was met by adding a third aging cycle at 595 °C (1100 °F) for 24 h.

Example 3 (Incoloy 901). Creep-rupture ductility was improved, although at some sacrifice of yield strength and creep-rupture life, by the addition of intermediate aging to the heat treating procedure. Table 11 compares the effects of the two treatments on mechanical properties.

Example 4 (Incoloy 901). Single aging failed to provide sufficient yield strength to meet specifications. An additional aging treatment at 650 °C (1200 °F) for 12 h provided required properties. Table 12 lists specification requirements and compares mechanical properties obtained by the two treatments.

Example 5 (Incoloy 901). A problem of low yield strength and low creep-rupture ductility was solved by increasing the temperature of the first aging (stabilizing) treatment and adding a third (final) aging treatment. Table 13 lists the mechanical properties obtained from the original and revised heat treatments.

Example 6 (Incoloy 901). Increas-

Table 14 Effect of revision in aging treatment on creep-rupture properties of Incoloy 901

	Creep rupture(a)			
	Original treatment(b)		Revised treatment(c)	
				Elongation in 50 mm (2 in.), %
Test No.	Life, h	Elongation, %	Life, h	
1	72	4	74	13
2	126	4	115	12
3	161	4	160	13
4	111	4	110	9
5	127	4	84	9
6	76	4	84	8
7	127	4	98	9

(a) At 650 °C (1200 °F) and 552 MPa (80 ksi), specified minimums; life, 23 h; elongation, 5%. (b) Solution treated at 1085 °C (1985 °F) for 2 h, cooled; aged at 715 °C (1325 °F) for 24 h, air cooled. (c) Same conditions as in (b), except that temperature of first aging was 810 °C (1490 °F)

ing the temperature of the first aging treatment in a double aging cycle provided specified minimum creep-rupture properties. Results of seven tests made on specimens heat treated by the two procedures are shown in Table 14, with details of heat treating procedures.

Eaxmple 7 (A-286). Table 15 shows the effects of alternative approaches to solving the problem of insufficient notch ductility in heat treated A-286 alloy—namely, the use of an additional aging treatment and the substitution of a higher aging temperature for single aging. Although both approaches pro-

vided required creep-rupture properties, the use of the higher aging temperature (with no increase in the solution treating temperature) is recommended over the additional aging treatment.

Example 8 (A-286 weldment). For control of dimensions during heat treatment, a turbine combustion case weldment (Fig. 5) was loaded on radial spokes of a spider made from an HT alloy casting. Sizing was done mechanically between solution heat treating and aging, and allowed for the linear shrinkage of about 0.1% during aging.

A roller-hearth furnace, utilizing a rich unpurified exothermic atmosphere, was used for solution treating at 980 °C (1800 °F) for 1 h. Cooling was controlled by a water-jacketed cooler that maintained a rate comparable to that of cooling in still air. Parts were aged at 720, 730, 745 or 760 °C (1325, 1350, 1375 or 1400 °F) in a pit furnace without protective atmosphere. Exact aging temperature depended on test results from specimens of each heat.

The following examples illustrate the use of interstage annealing or softening with welding and forming operations.

Example 9. A part for a turbine shaft (Fig. 6) was machined from an A-286 alloy forging, then annealed at 900 °C (1650 °F) prior to forming the cone section, which was shear spun in one pass from the initial thickness of 6.4 mm (0.250 in.) to final thickness of 2 mm (0.080 in.). After forming, the part was solution treated at 900 °C (1650 °F) and then aged at 720 °C (1325 °F).

Example 10. An exhaust header-half stamping was made from 2.1-mm-(0.082-in.-) thick 19-9 DL sheet (Fig. 6). The part required interstage annealing at 980 °C (1800 °F) between forming operations and was annealed again after being welded into an assembly.

Heat Treating of Nickel-Base Alloys

As indicated in Tables 1 and 2, most of the nickel-base heat-resisting alloys contain titanium and aluminum, which make the material hardenable by aging. Heat treating temperatures for selected alloys are given in Tables 5, 6 and 7.

Stress Relieving and Annealing. These processes are generally applied to wrought nickel alloys of the solid solution strengthened type. Nickel-base alloy castings are seldom stress relieved or annealed. Such parts are used in the as-cast, solution treated, or solution treated and aged condition. The age hardenable nickel-base alloys (such as René 41) are more crack-sensitive than the iron-base alloys and must be annealed during fabrication to relieve forming and welding stresses. As with other wrought alloys, grain size of nickel alloys after annealing depends on the amount of prior cold or hot work and the annealing temperature. This effect is illustrated for Nimonic 90 in Fig. 7.

Table 15 Effects of alternative aging treatments on rupture properties of A-286

Treatment	Creep-rupture(a) Life, h	Elongation in 50 mm (2 in.), %	Location of failure
Original heat treatment(b)			
900 °C (1650 °F), 2 h, oil quench; 720 °C (1325 °F), 16 h, air cool	7 to 69	...	Notch
Revised heat treatments(c)			
Original plus additional age at 650 °C (1200 °F), 12 h, air cool	74 to 142	5.6 to 7.7	Smooth bar
900 °C (1650 °F), 2 h, oil quench; 730 °C (1350 °F), 16 h, air cool	24 to 82	4.9 to 7.7	Smooth bar

(a) Specification requirements for creep-rupture at 650 °C (1200 °F) and 450 MPa (65 ksi); life, 23 h min; no failure in notch permitted. (b) Ten specimens tested. (c) Five specimens for each treatment

Holding time at annealing temperature ranges from 1 to 2 h for most alloys. The more highly alloyed wrought materials such as Astroloy, Udimet 500, and Udimet 700 require longer holding times for dissolving microconstituents. For other alloys, such as Inconel 600, time at temperature is critical because of grain growth, and a holding time of about 15 min per inch of section thickness is commonly employed.

Room-temperature yield strength is sensitive to grain size, as illustrated below for René 41 forged bar:

ASTM grain size	Average 0.2% yield strength MPa	ksi
3	840	122
5	900	130
7	1010	147
8	1060	154

For some wrought alloys, annealing practice can have a marked effect on subsequent response to solution treating and aging. This is shown for René 41, one of the more sensitive alloys, in the following example. Parts formed from René 41 sheet showed strain age cracking after being solution treated at 1080 °C (1975 °F) for ½ h, air cooled, and then aged at 760 °C (1400 °F) for 16 h. Cracking was attributed to a carbide network in the grain boundaries. Cause of the carbide network was traced to an in-process anneal at 1175 °C (2150 °F). It was found that at 1175 °C (2150 °F) the M_6C carbide was dissolved. Subsequent exposure to temperatures between 760 and 870 °C (1400 and 1600 °F) produced $M_{23}C_6$ carbide network in the grain boundaries. This

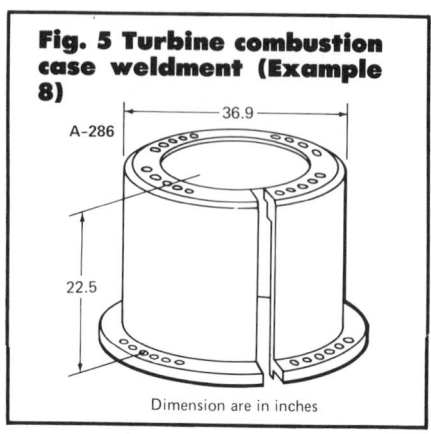

Fig. 5 Turbine combustion case weldment (Example 8)

A-286
36.9
22.5

Dimension are in inches

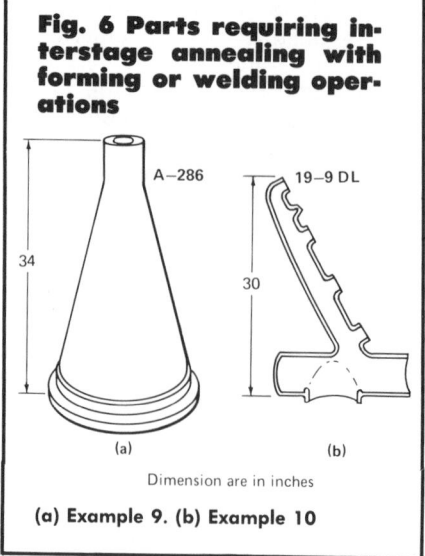

Fig. 6 Parts requiring interstage annealing with forming or welding operations

A-286
19-9 DL
34
30
(a)
(b)

Dimension are in inches

(a) Example 9. (b) Example 10

network reduced ductility to an unacceptable level.

If the annealing temperature was kept below 1095 °C (2000 °F), M_6C was not dissolved and ductility was improved. A similar effect can occur in weldments of nickel-base alloys if they are annealed at temperatures above 1095 °C (2000 °F).

A problem similar to that described in the preceding example occurred in René 41 bar stock. Grain-boundary carbide network reduced ductility and caused difficulty (sometimes cracking) during forming and welding. Investigation of the cause of the grain-boundary network indicated that the bar stock was produced with a final rolling temperature of 1175 °C (2150 °F). Light reductions were taken during final rolling to ensure proper size for the finished bar stock and to eliminate the possibility of surface tearing. This high rolling temperature, coupled with relatively light reductions (in the range of 2 to 3%), produced boundary network because (a) the M_6C carbides were dissolved at the rolling temperature and (b) slow cooling through the range of 870 to 760 °C (1600 to 1400 °F) produced $M_{23}C_6$ in an unfavorable morphology (grain boundary carbide network). Rolling temperatures of 1150 °C (2100 °F) max, coupled with a final reduction in rolling of at least 10 to 15%, eliminated the network and produced bar that could be welded and formed.

Multiple shaping and forming operations on nickel-base alloys may require numerous in-process anneals. Fabrication of the 90-cm- (35-in.-) long René 41 tubular truss member shown in Fig. 8 exemplifies fabrication by draw swaging to obtain the end reduction, and hydraulic forming to obtain the expanded center section. Both operations were necessary because the structural design of the tube exceeded the limits of the tube-reduction process alone. Shaping involved (a) roll forming and welding to produce the tube blank, (b) three draw swaging steps with four anneals, (c) rotary swaging, and (d) four hydraulic expanding steps with three anneals. Each of the seven anneals consisted of holding at 1080 °C (1975 °F) in an argon atmosphere for 30 min and then water quenching.

This method has also been applied successfully to the fabrication of HS-25 and Hastelloy X tubular parts.

Atmospheres. Nickel-base alloys are susceptible to attack by sulfur-bearing atmospheres. Therefore, the gas used in a semimuffle furnace or for manufacturing exothermic atmosphere must have a very low sulfur content. Thin sheet is especially likely to be damaged by sulfur-bearing atmospheres or by oxidation in air. Argon inert environment atmospheres frequently are used to preclude excessive oxidation. Vacuum furnaces may be used provided the necessary cooling rates can be obtained.

Example 11. Low creep-rupture life was experienced with 0.38-mm- (0.015-in.-) thick Waspaloy sheet solution treated in air at 995 °C (1825 °F). Rupture life of 260 MPa (37.5 ksi) at 815 °C (1500 °F) was increased from 15 h (heat treated in air) to 28 h merely by using argon as a protective atmosphere. However, for thicker sheet 1.25 mm (0.050 in.) of the same alloy, there was no significant difference between specimens solution treated in air and in argon.

Solution Treating. Solution treat-

Fig. 7 Effect of cold work and annealing on grain size

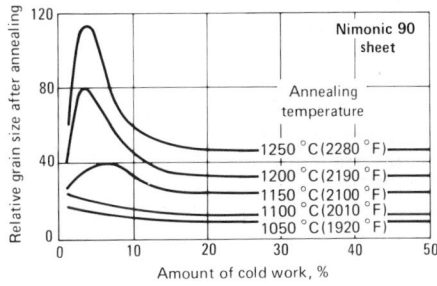

Nimonic 90 sheet was cold rolled in steps from 1.8 to 0.9 mm (0.072 to 0.036 in.) thick and annealed at five temperatures.

Fig. 8 Interstage annealing

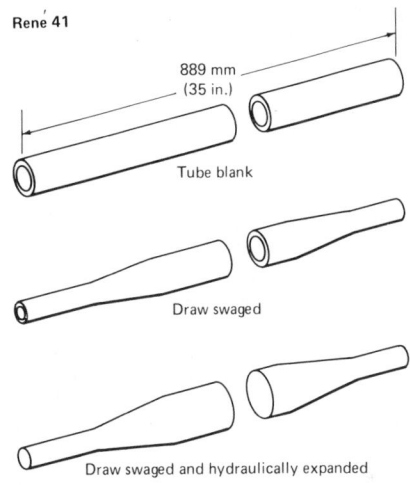

	Original		Dimension(a) Draw swaged			Expanded	
	mm	in.	mm	in.		mm	in.
Tube diam	31.8	1.25	19.1	0.75	(−40%)	57	2.25 (+80%)
Wall	1.0	0.040	1.5	0.059	(+48%)	0.50	0.020 (−50%)
Length	889	35.0	895.4	35.25	(+0.71%)	895.4	35.25 (+0.71%)
Swaged transition area		0°		10°			10°

(a) Numbers in parentheses indicate percentage change from original dimension.

ing temperatures of nickel-base alloys are determined primarily by the properties desired after subsequent aging. In wrought alloys, for optimum creep-rupture properties, higher solution treating temperatures are used than when optimum short-time tensile yield strength and elongation at elevated temperature are desired. For instance, René 41 and Udimet 500 are solution treated at 1175 °C (2150 °F) for optimum creep-rupture properties and at 1080 °C (1975 °F) for optimum short-time tensile properties. For cast alloys, solution treatment temperatures primarily are a function of incipient melting temperatures as well as furnace capability. Protective atmosphere or vacuum is necessary.

Because wrought nickel-base alloys are often solution-annealed below the gamma prime solubility limit and cast nickel-base alloys frequently are solutioned very close to the incipient melting temperature, control of furnace temperature is important when these materials are being heat treated. Variations in furnace temperature can result in failure to achieve specified properties. Should this occur because of an inadequate solution heat treatment, a reheat treatment sometimes is effective for producing desired results. When incipient melting has occurred, reheat treatments usually are not able to re-establish adequate properties in the affected part.

Table 16 indicates how reheat treating provided turbine wheels of wrought René 41 alloy with sufficient yield strength to exceed a minimum specification of 896 MPa (130 ksi), which was not consistently met by a single two-stage treatment under identical conditions. Re-treatment also increased both tensile strength and reduction of area. Elongation was not affected.

Higher solution heat-treat temperatures and a faster quench yield more favorable properties for forming René 41 (Fig. 9). However, exposure to temperatures at which the carbides dissolve promotes subsequent formation of brittle films at the grain boundaries.

In welded tubular components of René 41 solution annealed at 1150 °C (2100 °F), cracking is often a serious problem, even though the high-temperature annealing is followed by a solution heat treatment at 1080 °C (1975 °F) prior to welding. Solution treating at 1080 °C (1975 °F) after a 1150 °C (2100 °F) anneal precipitates M$_6$C along grain boundaries, resulting in

Table 16 Effect of reheat treatment on mechanical properties(a) of René 41 turbine wheels

Treatment	Ultimate tensile strength MPa	ksi	Yield strength(b) MPa	ksi	Elongation in 50 mm (2 in.), %	Reduction in area, %
Heat treated(c)	1310-1380	190-200	850-920	123-133(d)	16-22	17-21
Re-heat treated(e)	1340-1420	195-206	920-1030	134-150(f)	16-22	18-24

(a) Data represent eight tests for each treatment; properties at 20 °C (68 °F). (b) Specified minimum. (c) Solution treated at 1065 °C (1950 °F), air cooled; aged at 760 °C (1400 °F) for 16 h, air cooled. (d) Average, 880 MPa (128 ksi). (e) Re-heat treated repeating procedure in (c). (f) Average, 980 MPa (142 ksi)

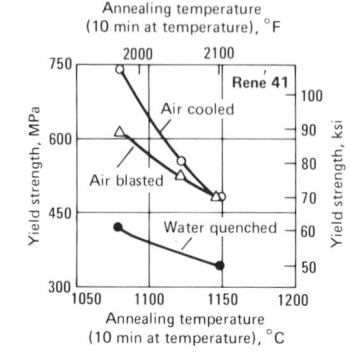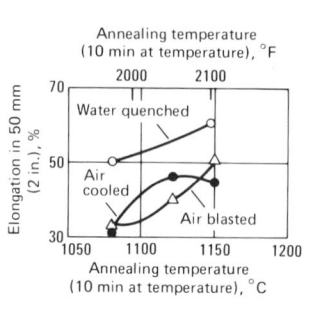

Fig. 9 Effects of annealing temperature and quenching rate on typical room-temperature ductility and yield strength of René 41

cracking when the alloy is welded and subsequently aged. The M$_6$C carbides begin to dissolve between 1105 and 1150 °C (2025 and 2100 °F). Consequently, a solution annealing temperature has been established at 1105 ±15 °C (2025 ± 25 °F) instead of 1150 °C (2100 °F). In addition, a modified aging cycle has been developed to prevent cracking in welded components without adversely affecting mechanical properties. Welding and aging tests were performed on material solution treated at 1080 °C (1975 °F). Several microcracks were observed in welded specimens that were aged between 760 and 845 °C (1400 and 1550 °F). A modified heat treatment was developed in which welded components are first aged for 1 h at 900 °C (1650 °F), then for 10 h at 760 °C (1400 °F). Parts subjected to mild forming, and which are not welded, are aged for 16 h at 760 °C (1400 °F).

Shrinkage during Aging. For design purposes, the contraction of components made of René 41 must be taken into consideration. An average contraction of 0.033 mm/mm (0.0013 in./in.)

can be expected after aging for 16 h at 760 °C (1400 °F) or for 1 h at 900 °C (1650 °F) followed by 10 h at 760 °C (1400 °F). This average contraction value reflects both longitudinal and transverse measurements on three heats of René 41, and the following treatments prior to aging:

● Solution treated at 1080 °C (1975 °F)
● Solution treated at 1080 °C (1975 °F), annealed at 1105 °C (2025 °F), re-solutioned at 1080 °C (1975 °F).

Dilatometric measurements indicate that contraction is essentially complete within 6 h after the aging temperature is reached.

Sheet structures experiencing moderate cold working are generally spot welded and double aged in air furnaces without a prior re-solution treatment. Although desirable, re-solutioning introduces distortion effects that are detrimental to the structure and that outweigh any benefits obtained by the treatment. Air atmosphere is used during aging to provide a tight, adherent, dark scale useful for emissivity purposes in actual flight conditions.

Tubular Components. In fabricating tubular components that are reduced at the ends by swaging, multiple annealing is frequently required to remove cold work. Tubes are annealed in an argon atmosphere having an oxygen content of not more than 400 ppm and a dew point not higher than −20 °C (−5 °F). The annealing temperature is held just below the carbide solubility temperature (1105 ± 15 °C or 2025 ± 25 °F) for 5 min. The tubes are then quenched in water in less than 3 s. The total intergranular attack on the material after three interstage anneals is generally less than 0.3 mil. Tubular components are aged in an air atmosphere.

Effect of Rate of Cooling from the Solution Treating Temperature. Nickel-base alloys (such as René 41, Waspaloy and Astroloy) that contain large amounts of hardening elements cannot be uniformly softened throughout heavy sections even by water quenching from the solution temperature. Variations in hardness from 180 to 240 HB, caused by precipitation during cooling, have been noted from surface to center in water-quenched 968-mm²- (1½-in.²-) bars of René 41. In air-cooled Astroloy forgings, higher strength and ductility values are produced in 100- to 125-mm (1- to 2-in.) sections than in 102- to 127-mm (4- to 5-in.) sections because of the uncontrolled overaging that occurs in the heavier sections.

In order to obtain more uniform hardness after solution treating and more uniform response to aging, heavy sections of these alloys are often air cooled from the solution temperature, despite the partial aging that occurs as a result of slower cooling.

Occasionally, a time delay may be used before quenching, particularly when parts may be susceptible to quench cracking. Component hardness will vary with time delays to some extent. The effect of time delay before quenching on the hardness of René 41 sheet is shown in Fig. 10. Hardening is caused by precipitation of gamma prime and carbides.

Example 12. High hardness (35 HRC) of semifabricated parts made of Waspaloy and René 41 was experienced after the parts were cooled from 1095 °C (2000 °F) by an air blast. Water quenching at first produced no improvement, because, owing to the loca-

tion of the quench tank, a time delay of 10 s occurred in the quenching operation. The problem was solved by moving the quench tank closer to the furnace, which allowed quenching to black heat in less than 4 s. The parts were then about 20 HRC as quenched and could be easily formed.

Occasionally, experimentation must be conducted to determine quenching mediums and procedures that yield an acceptable combination of cooling rate, distortion control, and cost. In some applications, combined quenches, such as water immersion and water spray, are used simultaneously on a part having a heavy, thick base and a thin web.

Figure 11 shows a semifinished forged turbine blade of René 41 and schematically illustrates the method by which it was simultaneously spray and immersion quenched with water after being solution treated for in-process annealing prior to cold forming. The blade was required to be as soft as possible after annealing (maximum acceptable hardness of 18 HRC), and to have a minimum of distortion and oxidation.

Parts to be heated were hand-loaded into an electrically heated full-muffle box furnace, in which an argon atmosphere was used. Processing consisted of the following:

- Degrease, alkaline clean, and dry. Handle parts with gloves to avoid contamination by fingerprints.
- Place in handling fixture.
- Place in furnace operating at 1080 °C (1975 °F) with argon flowing at 0.03 m³/min (1 ft³/min); muffle volume, 0.03 m³ (1 ft³).
- Reduce argon flow to 0.01 m³/min (⅓ ft³/min) when furnace load reaches 1080 °C (1975 °F).
- Hold parts at 1080 °C (1975 °F) for 15 min.
- Place parts in quench (Fig. 10) within 2 s after withdrawal from furnace.
- When parts are cool, remove from quench.

The annealed blades had an average hardness of 95 HRB (18 HRC). As the result of the special quenching technique, distortion was held to an acceptable minimum. Oxidation consisted only of a dark-blue surface discoloration.

Air cooling rates for nickel-base blade and vane alloys generally are sufficient to produce a supersaturated

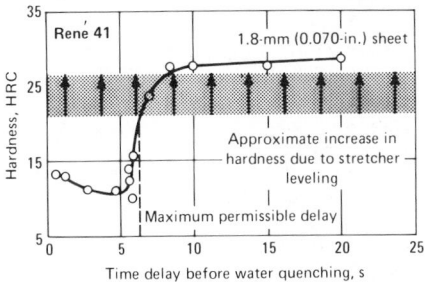

Fig. 10 Effect of time delay in quenching on hardness of René 41

Effect of delay before water quenching on the hardness of René 41 sheet solution heat treated at 1080 °C (1975 °F). Shaded band shows the increase in hardness due to stretcher leveling (about 2% strain) after quenching.

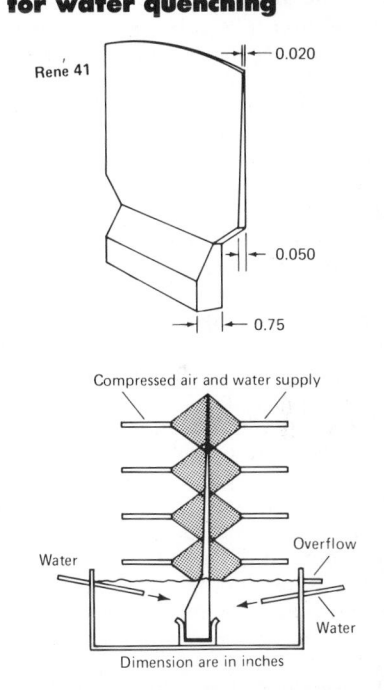

Fig. 11 René 41 forged turbine blade and immersion-spray arrangement for water quenching

gamma matrix for subsequent development of optimum hardness by aging. Occasionally, as noted for thick section parts such as turbine wheels, air cooling may be insufficient to produce the desired structure. Even for blade and vane components which are relatively small, cooling rates may be inhibited by other considerations, as for example in the pack aluminizing of airfoil sur-

Fig. 12 Precipitation in aging

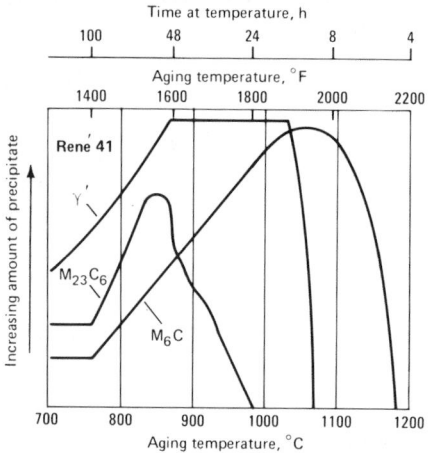

Relative amounts of precipitates resulting from aging René 41 at various times and temperatures after solution treating at 1205 °C (2200 °F) and water quenching

Table 17 Effect of eliminating intermediate aging on typical room-temperature mechanical properties of Udimet 500

	Ultimate tensile strength		Yield strength (0.2%)		Elongation in 50 mm (2 in.), %	Reduction in area, %
	MPa	ksi	MPa	ksi		
Specified min	1030	150	690	100	10	15
Obtained with intermediate aging(a)						
Test 1............	1030	149	830	120	7	11
Test 2............	970	141	810	118	4	5
Obtained without intermediate aging(b)						
Test 1............	1170	170	800	116	14.5	17
Test 2............	1230	179	850	123	14	16

(a) Heat treatment: 4 h at 1080 °C (1975 °F), air cool; 24 h at 845 °C (1550 °F), air cool (intermediate aging); 16 h at 760 °C (1400 °F), air cool. (b) Same as (a), but without intermediate aging

faces or the hot isostatic pressure closing of internal casting porosity. In such cases, either forced argon cooling may be used or the components may be slow cooled from the pack or HIP temperature and reheated, followed by rapid cooling from a casting diffusion treatment or solution treatment temperature.

Aging. When more than one phase is capable of precipitating from the alloy matrix, judicious selection of a single aging temperature may result in obtaining optimum properties in the component. Although single age temperatures can be determined for many alloys to produce near optimum properties, the need to produce an appropriate gamma prime/gamma double prime intergranular dispersion with an appropriate carbide dispersion in the grain boundary militates against a single age step. Figure 12 indicates the range of carbide and gamma prime precipitation in an older wrought alloy, René 41. Laboratory experiments coupled with the results of Fig. 12 suggested that a single-step aging temperature which would produce the best carbide ($M_{23}C_6$) and gamma prime morphologies was 870 °C (1600 °F). In all likelihood, a two-step aging sequence would be better and, in fact, almost all wrought nickel-base alloys are given a two- to four-step aging sequence.

One alternative to single aging is a double aging treatment. In Nimonic 80A, initial aging at 900 °C (1650 °F)

Fig. 13 Light-gage (0.81-mm, or 0.032-in., wall thickness) welded pressure vessel of Inconel X-750 alloy in a nonrestrictive rack for heat treating

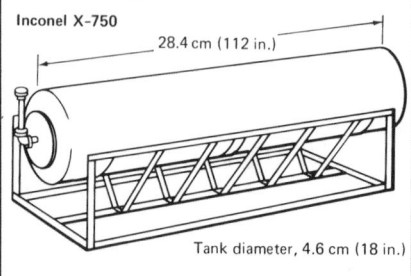

Inconel X-750

28.4 cm (112 in.)

Tank diameter, 4.6 cm (18 in.)

precipitates $M_{23}C_6$ in the grain boundaries and gamma prime within the grains. In a second aging treatment at 700 to 760 °C (1290 to 1400 °F), additional precipitation of a finer gamma prime occurs within the grains, to provide an optimum combination of strength and ductility.

Udimet 500 is typical of nickel-base age-hardenable alloys that are strengthened by gamma prime precipitation and also contain TiC and $M_{23}C_6$. The alloy is normally double aged. Solution treating at 1065 to 1175 °C (1950 to 2150 °F) is followed by air cooling. All phases except TiC are dissolved as a result of the solution treatment. Appreciable hardening takes place during cooling because of rapid precipitation of gamma prime. Intermediate aging at 845 °C (1550 °F) precipitates $M_{23}C_6$ in a discontinuous form, mainly at the grain boundaries. Final aging at 760 °C (1400 °F) determines the ulti-

mate hardness by controlling the size of the gamma prime precipitate. The presence of $M_{23}C_6$ in grain boundaries increases the creep-rupture life; however, a continuous carbide network should be avoided because it markedly decreases ductility. Higher ductility at room temperature can be obtained by omitting the 845 °C (1550 °F) aging treatment, but undesirable cellular precipitation of $M_{23}C_6$ in the grain boundaries will cause low ductility in creep-rupture tests at elevated temperatures.

Elimination of the intermediate aging treatment proved beneficial for increasing the room-temperature strength and ductility of Udimet 500 forgings that would not be subjected to service temperatures above 730 °C (1350 °F). Unacceptable room-temperature mechanical properties were obtained by solution treating at 1080 °C (1975 °F) for 4 h and air cooling, aging at 845 °C (1550 °F) for 24 h and air cooling, and aging at 760 °C (1400 °F) for 16 h and air cooling. Acceptable properties were produced by eliminating the 845 °C (1550 °F) treatment only. Table 17 compares specified minimum properties with those obtained by the two treatments.

Weldments are heat treated by a variety of special methods.

Example 13. To meet final requirements for an Inconel X-750 welded pressure vessel, 285 cm long by 46 cm (9 ft 4 in. long by 18 in.) in diameter, for a missile application, a procedure was established that incorporated the following special techniques:

1 Cradle vessel in a lightweight, nonrestrictive rack (Fig. 13).

2 Purge interior of vessel with argon.

3 Place cradled vessel in furnace (car-bottom, elevator type) and heat at 165 to 220 °C/h (300 to 400 °F/h) to 955 °C (1750 °F), to relieve forming and welding stresses. Hold at 955 °C (1750 °F) for 15 min; maintain argon at 21 kPa (3 psi) and −50 °C (−60 °F) dew point within the interior of vessel, to prevent sagging and oxidation; maintain furnace atmosphere of lean exothermic gas at +20 °C (+65 °F) dew point and approximate composition of 10.5% CO_2, 1.5% CO, 1 to 2% H_2, rem N_2.

4 Remove cradled vessel from furnace and air cool.

5 Reheat cradled vessel to 705 °C (1300 °F), and hold at temperature for 6 h under the same atmospheric conditions as described in step 3, to age that portion of the weld that was effectively solution treated in cooling from the welding temperature.

6 Air or furnace cool in cradle.

René 41 is solution treated at 1080 °C (1975 °F) before fusion welding. A higher solution treating temperature, such as 1175 °C (2150 °F), permits grain-boundary precipitation of $M_{23}C_6$ to occur in the heat-affected zone during welding, and this precipitate causes catastrophic age cracking after welding or during subsequent heat treatment of highly restrained weldments.

When it is possible to re-solution treat, at 1065 to 1080 °C (1950 to 1975 °F) after welding, the standard aging treatments are used—that is, for maximum tensile strength, 760 °C (1400 °F) for 16 h, or, for improved creep-rupture properties, 900 °C (1650 °F) for 4 h.

The re-solution treating of large and complex structural components containing restrained fusion welds is seldom feasible, on account of distortion. Single aging of René 41 fusion welds, particularly those under high restraint, without re-solution treating, has resulted in a high incidence of strain age cracking in the heat-affected zone. One fabricator overcame this difficulty by using a duplex aging treatment of 900 °C (1650 °F) for 1 h plus 760 °C (1400 °F) for 10 h.

Weld repairs of weldments may be employed to remedy defects or to repair weld casting defects in metals or to repair weld base metal cracks originating from service application. In the case of weld repair of weldments of wrought age-hardening alloys, the repair may originate either before or after aging. For the case of René 41, for example, if weld repair precedes aging it can be followed by the double aging treatment cited above. If weld repair follows aging, the problem is more severe. Assemblies that require weld repair after aging may be classified as follows:

- Simple assemblies that can be water quenched
- Complex assemblies that cannot be water quenched because of low-restraint welds
- Complex assemblies that cannot be water quenched because of high-restraint welds.

The first type of assembly is simply re-solution treated at 1080 °C (1975 °F) and water quenched prior to repair welding; the repair is then followed by double aging. For the second type of assembly, postweld heat treatments at 980 °C (1800 °F) for 1 h and at 760 °C (1400 °F) for 6 h are used. Re-solution treating and air cooling before repair welding is the correct procedure for the third type of assembly; the repair is then followed by double aging. In all repair procedures, parts subjected to aging after weld repair must be cooled slowly from the aging temperature, either in the furnace or in an insulating material.

Resistance welding of René 41 requires stringent process controls. For applications in which the service temperature exceeds the aging temperature, or for weldments that do not require maximum ductility, the resistance welded assembly can be aged after welding. For all other conditions, however, René 41 should be welded in the solution treated and aged condition. Aging after spot welding decreases the ductility of the spot weld zone. Because spot welds can develop high stress concentrations in service, good ductility is essential.

Heat Treating of Cobalt-Base Alloys

Cobalt-chromium-nickel-base alloys can be full annealed or stress relief annealed but are rarely solution treated because the gamma prime type precipitates generally do not participate in strengthening of cobalt alloys. Aging heat treatments may be given but the purpose of these would be to modify the carbide distribution. Heat treating temperatures for selected alloys are given in Tables 5, 6 and 7.

Full annealing, rather than stress relieving, is recommended for most of the wrought cobalt-chromium-nickel-base alloys whenever high residual stresses are developed during fabrication. When fabrication is completed, however, these alloys can be safely stress relieved at about 55 °C (100 °F) below the annealing temperature (Table 5) provided that excessive distortion does not occur.

Annealing of the cast cobalt-base alloys is often avoided, because of the undesirable precipitation of coarse carbides that occurs during slow cooling after annealing. Investment-cast gas-turbine buckets and guide vanes that are made to close dimensional tolerances and must exhibit excellent resistance to stress cracking are occasionally stress relieved at 870 to 900 °C (1600 to 1650 °F) for 2 to 4 h. There is no apparent effect of cooling rate on the properties of annealed or stress relieved cobalt-base alloys since gamma prime type precipitates are not involved in the hardening process.

Aging reactions of hardenable cobalt-base alloys such as HS-31 (X-40) are based on the random precipitation of various complex carbides and intermetallics. HS-31, a cast alloy containing 0.4% C, may contain $Cr_{23}C_6$, Cr_3C_2, MC or M_6C (or both), as well as mu (M_7W_6) phase. The ratio of chromium to carbon in the alloy determines which of the chromium carbides will appear during aging; a high ratio favors the precipitation of $Cr_{23}C_6$ (cobalt, iron, tungsten and nickel can replace part of the chromium in $Cr_{23}C_6$). The $Cr_{23}C_6$ can be precipitated in different forms, represented by discrete, acicular or lamellar shapes. Discrete particles are obtained by aging at temperatures near 760 °C (1400 °F). Aging at higher temperatures causes the appearance of the acicular precipitate. The third form, lamellar, resembles pearlite and is nucleated by cooling from a solution treating temperature between 1205 °C (2200 °F) and the temperature range where $Cr_{23}C_6$ precipitates.

During elevated-temperature testing of turbine blades made from vacuum melted X-40 alloy, it was observed that the blades were elongating as a result of tensile overstress (Fig. 14). To increase the tensile properties, the material was subjected to aging treatments applicable to air-melted X-40 alloy, but these were not sufficiently beneficial. Special aging treatments were then

Fig. 14 Effect of aging treatments on typical tensile properties of vacuum melted X-40 alloy

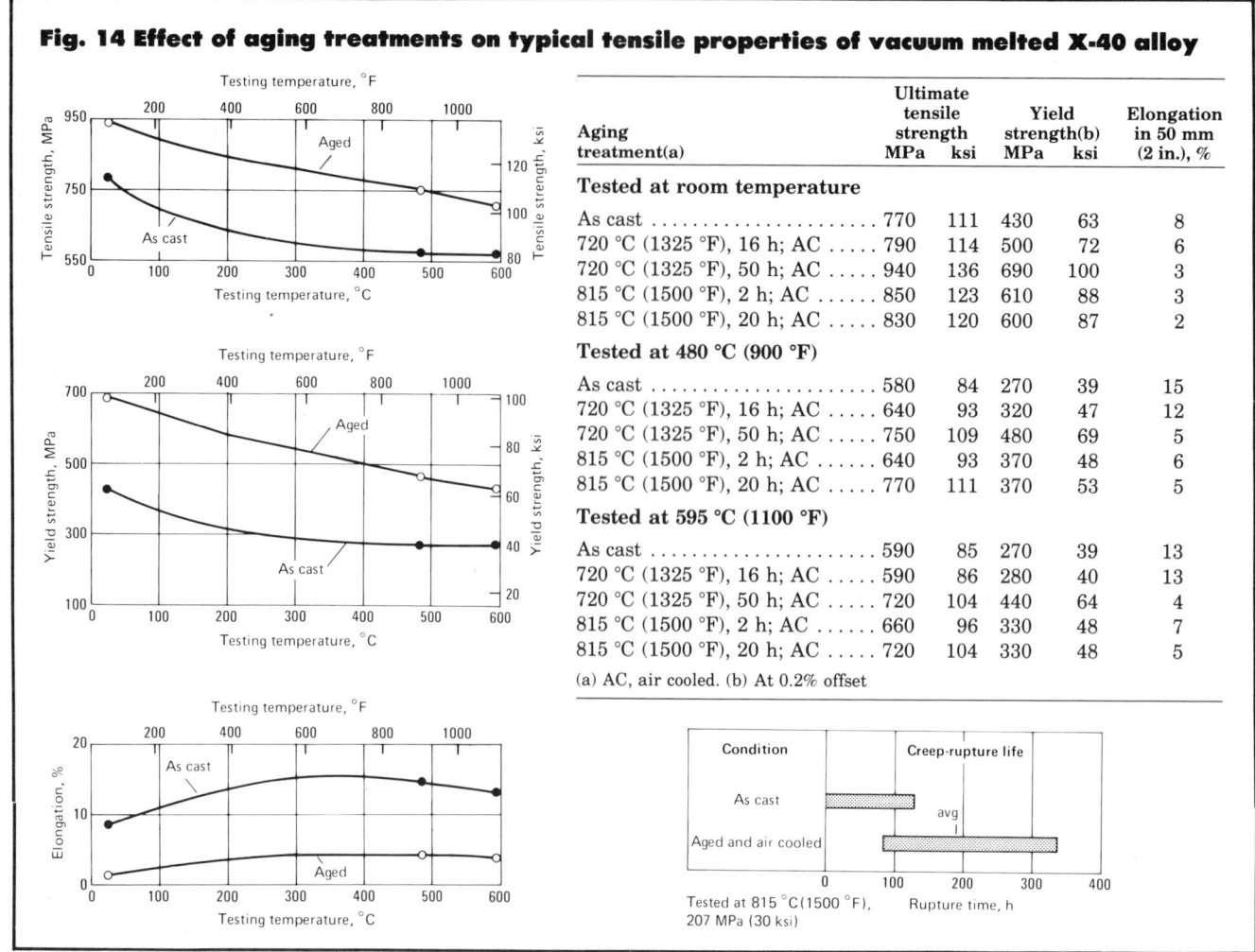

Aging treatment(a)	Ultimate tensile strength		Yield strength(b)		Elongation in 50 mm (2 in.), %
	MPa	ksi	MPa	ksi	
Tested at room temperature					
As cast	770	111	430	63	8
720 °C (1325 °F), 16 h; AC	790	114	500	72	6
720 °C (1325 °F), 50 h; AC	940	136	690	100	3
815 °C (1500 °F), 2 h; AC	850	123	610	88	3
815 °C (1500 °F), 20 h; AC	830	120	600	87	2
Tested at 480 °C (900 °F)					
As cast	580	84	270	39	15
720 °C (1325 °F), 16 h; AC	640	93	320	47	12
720 °C (1325 °F), 50 h; AC	750	109	480	69	5
815 °C (1500 °F), 2 h; AC	640	93	370	48	6
815 °C (1500 °F), 20 h; AC	770	111	370	53	5
Tested at 595 °C (1100 °F)					
As cast	590	85	270	39	13
720 °C (1325 °F), 16 h; AC	590	86	280	40	13
720 °C (1325 °F), 50 h; AC	720	104	440	64	4
815 °C (1500 °F), 2 h; AC	660	96	330	48	7
815 °C (1500 °F), 20 h; AC	720	104	330	48	5

(a) AC, air cooled. (b) At 0.2% offset

developed, and these provided satisfactory tensile properties. Average values resulting from these treatments are given in the table accompanying Fig. 14.

The best of these treatments (50 h at 720 °C, 1325 °F, air cool) resulted in increases of well over 50% in the yield strength up to 595 °C (1100 °F), compared to the tensile properties of as-cast material, and in a corresponding decrease in ductility (Fig. 13). Creep-rupture tests of material aged by the best treatment showed an average increase of 290% in the creep-rupture life of the alloy, compared to as-cast material (Fig. 14). Data from tests conducted at 480 °C (900 °F) showed no significant change in the fatigue curve.

Cobalt-base alloys generally are not susceptible to attack by furnace atmospheres and air environments can be used for heat treatment. At higher tem-peratures or in the case of vacuum melted alloys which contain reactive elements such as zirconium (for example, MAR-M 509), a vacuum, inert or almost neutral environment may be required.

Cobalt-base alloys are welded, either as fabrications or to repair weld cracks in base alloys after service operation. Stress relief heat treatments are applied after welding but there is no uniform preweld heat treatment practice.

Heat Treating of Refractory Metals

Presently available refractory metals and alloys are not heat treated to achieve high strength, but are used in the stress-relieved or annealed (recrystallized) condition.

Stress relieving is recommended for molybdenum, molybdenum alloys, and unalloyed tungsten when cold work in processing exceeds about 50%. Stress relieving is performed below the recrystallization temperature to retain the toughness acquired during cold working. Lowering the level of residual stress permits further fabrication with less danger of cracking and delamination. Processing history markedly affects the optimum stress-relieving temperature.

Unalloyed tantalum, Ta-10W, FS-80 and FS-82 are stress relieved to increase ductility and to reduce residual stress resulting from welding and cold work. These materials may be placed in service in either the stress-relieved or the recrystallized (fully annealed) condition, because they do not lose ductility during recrystallization as do molybdenum alloys and unalloyed molybdenum and tungsten.

Table 18 Causes, effects, prevention and correction of contamination of refractory metals and alloys during heat treatment

Contaminants		Preventive action	Corrective action	Remarks
Type and quantity	General effect			
Unalloyed molybdenum or tungsten: TZM alloy				
Oxygen and nitrogen in all quantities, including minute concentrations	Formation of surface oxides and nitrides. Diffusion into surface causes embrittlement.	Use of hydrogen atmosphere, inert gas, or vacuum	Pickle in molten caustic or caustic solution to remove surface contamination. Acid pickle to remove subsurface contamination zone.	Dry hydrogen is virtually always the atmosphere selected for heat treating molybdenum and tungsten.
FS-80; FS-82; Ta-10 W; unalloyed tantalum				
Oxygen, nitrogen and hydrogen in all quantities, including minute concentrations in inert gases or moderate vacuum	Formation of surface and subsurface oxides, nitrides and hydrides. Rapid diffusion at elevated temperature causes embrittlement.	Use of high vacuum (10-mm Hg, min) or high-purity inert gas (preferably argon)	Remove hydrogen contamination by high-vacuum thermal treatment. Acid pickle to remove surface layers contaminated by oxygen and nitrogen. No corrective action is possible when excessive diffusion of oxygen or nitrogen occurs.	Diffusion of oxygen, nitrogen and hydrogen is rapid. Observable surface effects are negligible, but extreme loss of ductility occurs when impure inert atmosphere or poor vacuum is used.

Recrystallization annealing of unalloyed molybdenum, Mo-0.5Ti, TZM, and unalloyed tungsten raises the ductile-brittle transition temperature of the materials and results in a nearly complete loss of ductility at room temperature. For this reason, mill products of these materials are supplied in the stress-relieved condition. Unalloyed tantalum, Ta-10W, and FS-80 and FS-82 niobium alloys retain low ductile-brittle transition temperatures after recrystallization. Mill products of these materials are supplied in either the stress-relieved or annealed condition, depending on end use or fabrication requirements.

Annealing temperatures for refractory metals depend on prior processing history, chemical composition, amount of cold work, and sequence of in-process anneals. Therefore, annealing temperatures are not standardized, although those given in Table 5 may be considered typical.

Unalloyed niobium and tantalum and many of their alloys may be cold formed in either the stress-relieved or the annealed (recrystallized) condition. They are seldom formed in the as-rolled condition because of their susceptibility to cracking. A stress relief treatment at about 55 °C (100 °F) below the recrystallization temperature reduces residual stress and improves ductility enough to permit some forming. However, annealing should precede more severe operations, such as spinning, flow turning and deep drawing.

Recrystallization greatly improves the ductility of niobium and tantalum alloys but lowers their strength by removing the effects of strain hardening. However, when the material is used at a temperature above the recrystallization temperature, the effects of strain hardening are not utilized in service. Therefore, it is usually preferable to do all forming in the annealed condition.

Heat treatment after forming may consist of either stress relieving or annealing. When the formed parts are given an oxidation-resistant coating prior to use, the thermal treatment involved in the coating process usually provides adequate stress relief.

Unalloyed molybdenum, molybdenum alloys, and unalloyed tungsten are nearly always formed in the stress-relieved condition. Recrystallization results in a marked decrease in the ductility at or near room temperature.

Surface Contamination. Tungsten, molybdenum, and the alloys of molybdenum are less susceptible to surface contamination during heat treatment than are niobium and tantalum alloys. However, atmospheric contamination of the Mo-0.5Ti alloy is known to occur in sheet, with the probable formation of TiO_2 and TiN.

Absorption of oxygen, nitrogen and hydrogen by niobium and tantalum increases as temperatures increase above about 650 °C (1200 °F). The hard, brittle surface layers formed are detrimental to further forming and machining operations, and to performance characteristics. Stress relieving or annealing of material with a contaminated surface, even in an inert atmosphere, will result in diffusion and usually in a stronger, less ductile alloy having a higher transition temperature from ductile to brittle failure.

Atmospheres. Refractory metals require a protective atmosphere during heat treatment, to avoid degradation of properties from surface contamination. Molybdenum and tungsten have an extremely low solubility for hydrogen; hence, hydrogen atmospheres are almost always used for treating these metals.

Because niobium and tantalum are severely embrittled by hydrogen, oxygen and nitrogen, these metals must be heat treated in a vacuum or an inert atmosphere, such as argon with an extremely low dew point (−55 °C, or −70 °F or lower). Argon is preferred over helium because argon is cheaper and its greater density provides more positive gas coverage of parts. Table 18 lists important factors in the prevention of contamination from environment during heat treatment of refractory metals.

Heat Treating of Nonferrous Metals

Heat Treating of Aluminum Alloys

By the ASM Committee on Heat
Treating of Aluminum Alloys*

HEAT TREATING, in its broadest sense, refers to any of the heating and cooling operations that are performed for the purpose of changing the mechanical properties, the metallurgical structure or the residual stress state of a metal product. When the term is applied to aluminum alloys, however, its use frequently is restricted to the specific operations employed to increase strength and hardness of the precipitation-hardenable wrought and cast alloys. These usually are referred to as the "heat treatable" alloys, to distinguish them from those alloys in which no significant strengthening can be achieved by heating and cooling. The latter, generally referred to as "non-heat-treatable" alloys, when in wrought form depend primarily on cold work to increase strength. Heating to decrease strength and increase ductility (annealing) is used with alloys of both types; metallurgical reactions may vary with type of alloy and with degree of softening desired. Except for the low-temperature stabilization treatment sometimes given 5xxx series alloys (which is a mill treatment and not discussed in this article), complete or partial annealing treatments are the only ones used for non-heat-treatable alloys.

One essential attribute of a precipitation-hardening alloy system is a temperature-dependent equilibrium solid solubility characterized by increasing solubility with increasing temperature. Although this condition is met by most of the binary aluminum alloy systems, many exhibit very little precipitation hardening, and these alloys ordinarily are not considered heat treatable. Alloys of the binary Al-Si and Al-Mn systems, for example, exhibit relatively insignificant changes in mechanical properties as a result of heat treatments that produce considerable precipitation.

The solubility-temperature relationship required for precipitation hardening is illustrated by the Al-Cu system (see Fig. 1). The equilibrium solid solubility of copper in aluminum increases as temperature increases—from about 0.20% at 250 °C (480 °F) to a maximum of 5.65% at the eutectic melting temperature of 548 °C (1018 °F). (It is considerably lower than 0.20% at temperatures below 250 °C.) For Al-Cu alloys containing from 0.2 to 5.6% Cu, two distinct equilibrium solid states are possi-

ble. At temperatures above the lower curve in Fig. 1 (solvus), the copper is completely soluble, and when the alloy is held at such temperatures for sufficient time to permit needed diffusion, the copper will be taken completely into solid solution. At temperatures below the solvus, the equilibrium state consists of two solid phases: solid solution, α, plus an intermetallic-compound phase, θ (Al_2Cu). When such an alloy is converted to all solid solution by holding above the solvus temperature and then the temperature is decreased to below the solvus, the solid solution becomes supersaturated and the alloy seeks the equilibrium two-phase condition; the second phase tends to form by solid-state precipitation.

The preceding description is a gross oversimplification of the actual changes that occur under different conditions even in simple binary Al-Cu alloys. A variety of different nonequilibrium precipitate structures is formed at temperatures below the solvus. In alloys of the Al-Cu system, a succession of precipitates is developed from a rapidly cooled supersaturated solid solution (SSS). These precipitates develop sequentially either with in-

*James T. Staley, *Chairman,* Technical Manager, Alloy Technology Division, Alcoa Technical Center, Alcoa Laboratories; Sven E. Axter, Specialist Engineer, The Boeing Commercial Airplane Co.; Larry J. Barker, Senior Research Associate, Kaiser Aluminum and Chemical Corp.; John H. Hull, Chief Metallurgist, Grafton Plant, Wyman-Gordon Co.; Robert B. Leholm, Materials Engineer, McDonnell-Douglas Corp.; Paul C. Wilson, Chief Engineer, Arwood Corp.

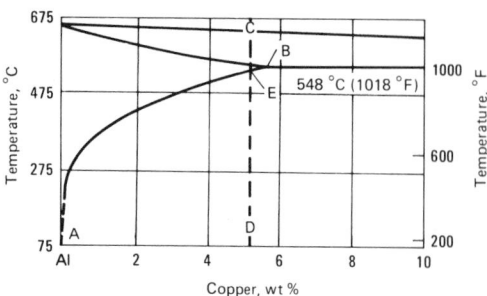

Fig. 1 Aluminum-rich end of the aluminum-copper equilibrium diagram

Line AB represents the increase in solubility of copper in solid aluminum with increasing temperature. See text for discussion.

creasing temperature or with increasing time at temperatures between room temperature and the solvus. The several stages are identified by the following notation:

$$SSS \longrightarrow GP\ zones \longrightarrow \theta'' \longrightarrow$$
$$\theta' \longrightarrow \theta\ (Al_2Cu)$$

At temperatures in the natural aging range (about -20 to $+60$ °C, or 0 to 140 °F), the distribution of copper atoms changes with time from random to the disklike planar aggregates termed GP zones, which form on particular crystallographic planes of the aluminum matrix. These aggregates create coherency strain fields that increase resistance to deformation, and their formation is responsible for the changes in mechanical properties that occur during natural aging. At higher temperatures, transition forms of approximate composition Al_2Cu develop and further increase strength. In the highest strength condition, both the θ'' and θ' transition precipitates may be present. When time and temperature are increased sufficiently to form high proportions of the equilibrium θ, the alloy softens and is said to be "overaged".

Commercial alloys whose strength and hardness can be significantly increased by heat treatment include 2xxx, 6xxx and 7xxx series wrought alloys (except 7072) and 2xx.0, 3xx.0 and 7xx.0 series casting alloys. Some of these contain only copper, or copper and silicon, as the primary strengthening alloy addition(s). Most of the heat treatable alloys, however, contain combinations of magnesium with one or more of the elements copper, silicon and zinc. Characteristically, even small amounts of magnesium in concert with these elements accelerate and

accentuate the strength changes attributable to precipitation hardening.

In the heat treatable wrought alloys, with some notable exceptions (2024, 2219 and 7128), such solute elements are present in amounts that are within the limits of mutual solid solubility at temperatures below the eutectic temperature (lowest melting temperature). In contrast, some of the casting alloys of the 2xx.0 series and all of the 3xx.0 series alloys contain amounts of soluble elements that far exceed solid-solubility limits. In these alloys, the phase formed by combination of the excess soluble elements with the aluminum will never be dissolved, although the shapes of the undissolved particles may be changed by partial solution.

Most of the heat treatable aluminum alloy systems exhibit multistage precipitation and undergo accompanying strength changes analogous to those of the Al-Cu system. The structural changes from the solid solution to the high-strength tempers involve precipitate particles that are too small to be resolved by light microscopy. In some instances, the presence of precipitation can be detected by etching, but definitive determination of the details of precipitate structure responsible for the changes in mechanical and other properties generally requires electron microscopy and special x-ray techniques.

In some alloys, sufficient precipitation occurs in a few days at room temperature to yield stable products with properties that are adequate for many applications. These alloys sometimes are precipitation heat treated to provide increased strength and hardness in wrought or cast products. Other alloys with slow precipitation reactions

at room temperature are always precipitation heat treated before being used.

Precipitation heat treatments generally are low-temperature, long-term processes. Temperatures range from 115 to 190 °C (240 to 375 °F); times vary from 5 to 48 h.

Choice of time-temperature cycles for precipitation heat treatment should receive careful consideration. Larger particles of precipitate result from longer times and higher temperatures; however, the larger particles must, of necessity, be fewer in number with greater distances between them. The objective is to select the cycle that produces optimum precipitate size and distribution pattern. Unfortunately, the cycle required to maximize one property, such as tensile strength, is usually different from that required to maximize others, such as yield strength and corrosion resistance. Consequently, the cycles used represent compromises that provide the best combinations of properties.

To recap, heat treatment to increase strength of aluminum alloys is a three-step process:

1 Solution heat treatment: dissolution of soluble phases
2 Quenching: development of supersaturation
3 Aging: precipitation of solute atoms either at room temperature (natural aging) or elevated temperature (artificial aging or precipitation heat treatment).

Temper designations for heat treatable aluminum alloys are presented at the end of this article. Typical solution and precipitation heat treatments for mill products are given in Tables 1 and 2, and treatments for castings are given in Table 3.

Solution Heat Treating

To take advantage of the precipitation-hardening reaction, it is necessary first to produce a solid solution. The process by which this is accomplished is called solution heat treating, and its objective is to take into solid solution the maximum practical amounts of the soluble hardening elements in the alloy. The process consists of soaking the alloy at a temperature sufficiently high and for a time long enough to achieve a nearly homogeneous solid solution.

Nominal commercial solution heat treating temperature is determined by the composition limits of the alloy and

Table 1 Typical solution and precipitation heat treatments for aluminum alloy mill products

These times and temperatures are typical for various forms, sizes and methods of manufacture and may not exactly describe optimum treatments for specific items.

Alloy	Product form	Solution heat treatment(a) Metal temperature(b) °C	°F	Temper designation	Precipitation heat treatment Metal temperature(b) °C	°F	Time(c), h	Temper designation
2011	Rolled or cold finished rod and bar 525		975	T3(d)	160	320	14	T8(d)
				T4	...	...	...	...
				T451(e)	...	...	...	...
2014(f)	Flat sheet 500		935	T3(d)	160	320	18	T6
				T42	160	320	18	T62
	Coiled sheet 500		935	T4	160	320	18	T6
				T42	160	320	18	T62
	Plate................. 500		935	T42	160	320	18	T62
				T451(e)	160	320	18	T651(e)
	Rolled or cold finished wire, rod and bar 500		935	T4	160(g)	320(g)	18	T6
				T42	160(g)	320(g)	18	T62
				T451(e)	160(g)	320(g)	18	T651(e)
	Extruded rod, bar, shapes and tube.............. 500		935	T4	160(g)	320(g)	18	T6
				T42	160(g)	320(g)	18	T62
				T4510(e)	160(g)	320(g)	18	T6510(e)
				T4511(e)	160(g)	320(g)	18	T6511(e)
	Drawn tube............ 500		935	T4	160(g)	320(g)	18	T6
				T42	160(g)	320(g)	18	T62
	Die forgings .... 500(h)		935(h)	T4	170	340	10	T6
	Hand forgings and rolled rings................ 500(h)		935(h)	T4	170	340	10	T6
				T452(j)	170	340	10	T652(j)
2017	Rolled or cold finished wire, rod and bar 500		935	T4	...	...	...	...
				T42	...	...	...	...
				T451(e)	...	...	...	...
2018	Die forgings .... 510(k)		950(k)	T4	170	340	10	T61
2024(f)	Flat sheet 495		920	T3(d)	190	375	12	T81(d)
				T361(d)	190	375	8	T861(d)
				T42	190	375	9	T62
					190	375	16	T72

(continued)

(a) Material should be quenched from the solution-treating temperature as rapidly as possible and with minimum delay after removal from the furnace. When material is quenched by total immersion in water, unless otherwise indicated, the water should be at room temperature, and should be suitably cooled so that it remains below 38 °C (100 °F) during the quenching cycle. Use of high-velocity, high-volume jets of cold water also is effective for some materials. (b) The nominal temperatures listed should be attained as rapidly as possible and maintained within ± 6 °C (± 10 °F) of nominal during the time at temperature. (c) Approximate time at temperature. The specific time will depend on the time required for the load to reach temperature. The times shown are based on rapid heating, with soak time measured from the time the load reaches a temperature within 6 °C (10 °F) of the applicable temperature. (d) Cold working subsequent to solution heat treatment and prior to any precipitation heat treatment is necessary to attain the specified properties for this temper. (e) Stress relieved by stretching to produce a specified amount of permanent set subsequent to solution heat treatment and prior to any precipitation heat treatment. (f) These heat treatments also apply to alclad sheet and plate in these alloys. (g) An alternative treatment of 8 h at 177 °C (350 °F) also may be used. (h) Solution heat treatment is followed by quenching in water 60 to 82 °C (140 to 180 °F). (j) Stress relieved by 1 to 5% cold reduction subsequent to solution heat treatment and prior to precipitation heat treatment. (k) Solution heat treatment is followed by quenching in water at 100 °C (212 °F). (m) Solution heat treatment is followed by quenching in room-temperature air blast. (n) By suitable control of extrusion temperature, product may be quenched directly from extrusion press to provide specified properties for this temper. Some products may be adequately quenched in room-temperature air blast. (p) See U.S. Patent 4 082 578. (q) Applicable to tread plate only. (r) An alternative treatment of 8 h at 171 °C (340 °F) also may be used. (s) Cold working subsequent to precipitation heat treatment is necessary to attain the specified properties for this temper. (t) An alternative treatment of 3 h at 182 °C (360 °F) also may be used. (u) An alternative treatment of 6 h at 182 °C (360 °F) also may be used. (v) No solution heat treatment; 72 h at room temperature following press quench, followed by two-stage precipitation heat treatment comprised of 8 h at 107 °C (225 °F) plus 16 h at 149 °C (300 °F). (w) Aging practice varies with product, size, nature of equipment, loading procedures and furnace-control capabilities. The optimum practice for a specific item can be ascertained only by actual trial treatment of the item under specific conditions. Typical procedures involve a two-stage treatment comprised of 3 to 30 h at 121 °C (250 °F) followed by 15 to 18 h at 163 °C (325 °F) for extrusions. An alternative two-stage treatment of 8 h at 99 °C (210 °F) followed by 24 to 28 h at 163 °C (325 °F) also may be used. (x) Aging of aluminum alloys 7050, 7075, 7175 and 7475 from any temper to the T73 or T76 temper series requires closer-than-normal controls on aging variables such as time, temperature, heatup rate, etc., for any given item. In addition, when material in a T6-type temper is reaged to a T73- or T76-type temper, the specific condition of the T6 material (such as property levels and other effects of processing variables) is extremely important and will affect the capability of the reaged material to conform to the requirements specified for the applicable T73- or T76-type temper. (y) Two-stage treatment comprised of 6 to 8 h at 107 °C (225 °F) followed by: 24 to 30 h at 163 °C (325 °F) for sheet and plate; 8 to 10 h at 177 °C (350 °F) for rolled or cold finished rod and bar; 6 to 8 h at 177 °C (350 °F) for extrusions and tube; 8 to 10 h at 177 °C (350 °F) for forgings in the T73 temper and 6 to 8 h at 177 °C (350 °F) for forgings in the T7352 temper. (z) An alternative two-stage treatment comprised of 4 h at 96 °C (205 °F) followed by 8 h at 157 °C (315 °F) also may be used. (aa) For sheet, plate, tube and extrusions, an alternative two-stage treatment comprised of 6 to 8 h at 107 °C (225 °F) followed by 14 to 18 h at 168 °C (335 °F) may be used, provided that a heatup rate of approximately 14 °C/h (25 °F/h) is employed. For rolled or cold finished rod and bar, the alternative treatment is 10 h at 177 °C (350 °F). (bb) An alternative three-stage treatment comprised of 5 h at 99 °C (210 °F), 4 h at 121 °C (250 °F) and then 4 h at 149 °C (300 °F) also may be used. (cc) 7175-T736 and -T73652 heat treatments are directed to specific results, may vary from supplier to supplier and are either proprietary or patented. (dd) Must be preceded by soak at 466 to 477 °C (870 to 890 °F). See U.S. Patent 3 791 880.

Table 1 (continued)

Alloy	Product form	Solution heat treatment(a) Metal temperature(b) °C	°F	Temper designation	Precipitation heat treatment Metal temperature(b) °C	°F	Time(c), h	Temper designation
2024(f)	Coiled sheet 495		920	T4	...	...	...	...
				T42	190	375	9	T62
					190	375	16	T72
	Plate................. 495		920	T351(e)	190	375	12	T851(e)
				T361(d)	190	375	8	T861(d)
				T42	190	375	9	T62
	Rolled or cold finished wire, rod and bar 495		920	T4	190	375	12	T6
				T351(e)	190	375	12	T851(e)
				T36(d)	190	375	8	T86(d)
				T42	190	375	16	T62
	Extruded rod, bar, shapes and tube............. 495		920	T3	190	375	12	T81
				T3510(e)	190	375	12	T8510(e)
				T3511(e)	190	375	12	T8511(e)
				T42	190	375	16	T62
	Drawn tube........... 495		920	T3(d)	...	...	...	...
				T42	...	...	...	...
2025	Die forgings 515		960	T4	170	340	10	T6
2036	Sheet 500		930	T4	...	...	...	...
2117	Rolled or cold finished wire and rod 500		935	T4	...	...	...	...
				T42	...	...	...	...
2218	Die forgings .... 510(k)		950(k)	T4	170	340	10	T61
		510(m)	950(m)	T41	240	460	6	T72
2219(f)	Flat sheet 535		995	T31(d)	175	350	18	T81(d)
				T37(d)	165	325	24	T87(d)
				T42	190	375	36	T62
	Plate................. 535		995	T31(d)	175	350	18	T81(d)
				T37(d)	175	350	18	T87(d)
				T351(e)	175	350	18	T851(e)
				T42	190	375	36	T62
	Rolled or cold finished wire, rod and bar 535		995	T351(e)	190	375	18	T851(e)

(continued)

(a) Material should be quenched from the solution-treating temperature as rapidly as possible and with minimum delay after removal from the furnace. When material is quenched by total immersion in water, unless otherwise indicated, the water should be at room temperature, and should be suitably cooled so that it remains below 38 °C (100 °F) during the quenching cycle. Use of high-velocity, high-volume jets of cold water also is effective for some materials. (b) The nominal temperatures listed should be attained as rapidly as possible and maintained within ± 6 °C (± 10 °F) of nominal during the time at temperature. (c) Approximate time at temperature. The specific time will depend on the time required for the load to reach temperature. The times shown are based on rapid heating, with soak time measured from the time the load reaches a temperature within 6 °C (10 °F) of the applicable temperature. (d) Cold working subsequent to solution heat treatment and prior to any precipitation heat treatment is necessary to attain the specified properties for this temper. (e) Stress relieved by stretching to produce a specified amount of permanent set subsequent to solution heat treatment and prior to any precipitation heat treatment. (f) These heat treatments also apply to alclad sheet and plate in these alloys. (g) An alternative treatment of 8 h at 177 °C (350 °F) also may be used. (h) Solution heat treatment is followed by quenching in water 60 to 82 °C (140 to 180 °F). (j) Stress relieved by 1 to 5% cold reduction subsequent to solution heat treatment and prior to precipitation heat treatment. (k) Solution heat treatment is followed by quenching in water at 100 °C (212 °F). (m) Solution heat treatment is followed by quenching in room-temperature air blast. (n) By suitable control of extrusion temperature, product may be quenched directly from extrusion press to provide specified properties for this temper. Some products may be adequately quenched in room-temperature air blast. (p) See U.S. Patent 4 082 578. (q) Applicable to tread plate only. (r) An alternative treatment of 8 h at 171 °C (340 °F) also may be used. (s) Cold working subsequent to precipitation heat treatment is necessary to attain the specified properties for this temper. (t) An alternative treatment of 3 h at 182 °C (360 °F) also may be used. (u) An alternative treatment of 6 h at 182 °C (360 °F) also may be used. (v) No solution heat treatment; 72 h at room temperature following press quench, followed by two-stage precipitation heat treatment comprised of 8 h at 107 °C (225 °F) plus 16 h at 149 °C (300 °F). (w) Aging practice varies with product, size, nature of equipment, loading procedures and furnace-control capabilities. The optimum practice for a specific item can be ascertained only by actual trial treatment of the item under specific conditions. Typical procedures involve a two-stage treatment comprised of 3 to 30 h at 121 °C (250 °F) followed by 15 to 18 h at 163 °C (325 °F) for extrusions. An alternative two-stage treatment of 8 h at 99 °C (210 °F) followed by 24 to 28 h at 163 °C (325 °F) also may be used. (x) Aging of aluminum alloys 7050, 7075, 7175 and 7475 from any temper to the T73 or T76 temper series requires closer-than-normal controls on aging variables such as time, temperature, heatup rate, etc., for any given item. In addition, when material in a T6-type temper is reaged to a T73- or T76-type temper, the specific condition of the T6 material (such as property levels and other effects of processing variables) is extremely important and will affect the capability of the reaged material to conform to the requirements specified for the applicable T73- or T76-type temper. (y) Two-stage treatment comprised of 6 to 8 h at 107 °C (225 °F) followed by: 24 to 30 h at 163 °C (325 °F) for sheet and plate; 8 to 10 h at 177 °C (350 °F) for rolled or cold finished rod and bar; 6 to 8 h at 177 °C (350 °F) for extrusions and tube; 8 to 10 h at 177 °C (350 °F) for forgings in the T73 temper and 6 to 8 h at 177 °C (350 °F) for forgings in the T7352 temper. (z) An alternative two-stage treatment comprised of 4 h at 96 °C (205 °F) followed by 8 h at 157 °C (315 °F) also may be used. (aa) For sheet, plate, tube and extrusions, an alternative two-stage treatment comprised of 6 to 8 h at 107 °C (225 °F) followed by 14 to 18 h at 168 °C (335 °F) may be used, provided that a heatup rate of approximately 14 °C/h (25 °F/h) is employed. For rolled or cold finished rod and bar, the alternative treatment is 10 h at 177 °C (350 °F). (bb) An alternative three-stage treatment comprised of 5 h at 99 °C (210 °F), 4 h at 121 °C (250 °F) and then 4 h at 149 °C (300 °F) also may be used. (cc) 7175-T736 and -T73652 heat treatments are directed to specific results, may vary from supplier to supplier and are either proprietary or patented. (dd) Must be preceded by soak at 466 to 477 °C (870 to 890 °F). See U.S. Patent 3 791 880.

Table 1 (continued)

Alloy	Product form	Solution heat treatment(a) Metal temperature(b) °C	°F	Temper designation	Precipitation heat treatment Metal temperature(b) °C	°F	Time(c), h	Temper designation
2219(f)	Extruded rod, bar, shapes and tube.............. 535		995	T31(d)	190	375	18	T81(d)
				T3510(e)	190	375	18	T8510(e)
				T3511(e)	190	375	18	T8511(e)
				T42	190	375	36	T62
	Die forgings and rolled rings................ 535		995	T4	190	375	26	T6
	Hand forgings......... 535		995	T4	190	375	26	T6
				T352(j)	175	350	18	T852(j)
2618	Forgings and rolled rings................ 530		985	T4	200	390	20	T61
4032	Die forgings .... 510(h)		950(h)	T4	170	340	10	T6
6005	Extruded rod, bar, shapes and tube.............. 530(n)		985(n)	T1	175	350	8	T5
6009(p)	Coiled sheet 555		1030	T4	175	350	8	T6
6010(p)	Coiled sheet 565		1050	T4	175	350	8	T6
6053	Die forgings 520		970	T4	170	340	10	T6
6061(f)	Sheet 530		985	T4	160	320	18	T6
				T42	160	320	18	T62
	Plate................. 530		985	T4(q)	160	320	18	T6(q)
				T42	160	320	18	T62
				T451(e)	160	320	18	T651(e)
	Rolled or cold finished wire, rod and bar 530		985	T4	160(r)	320(r)	18	T6
					160(r)	320(r)	18	T89(d)
					160(r)	320(r)	18	T93(s)
					160(r)	320(r)	18	T913(s)
					160(r)	320(r)	18	T94(s)
				T42	160(r)	320(r)	18	T62
				T451(e)	160(r)	320(r)	18	T651(e)
	Extruded rod, bar, shapes and tube.............. 530(n)		985(n)	T4	175	350	8	T6
				T4510(e)	175	350	8	T6510(e)
				T4511(e)	175	350	8	T6511(e)
		530	985	T42	175	350	8	T62

(continued)

(a) Material should be quenched from the solution-treating temperature as rapidly as possible and with minimum delay after removal from the furnace. When material is quenched by total immersion in water, unless otherwise indicated, the water should be at room temperature, and should be suitably cooled so that it remains below 38 °C (100 °F) during the quenching cycle. Use of high-velocity, high-volume jets of cold water also is effective for some materials. (b) The nominal temperatures listed should be attained as rapidly as possible and maintained within ± 6 °C (± 10 °F) of nominal during the time at temperature. (c) Approximate time at temperature. The specific time will depend on the time required for the load to reach temperature. The times shown are based on rapid heating, with soak time measured from the time the load reaches a temperature within 6 °C (10 °F) of the applicable temperature. (d) Cold working subsequent to solution heat treatment and prior to any precipitation heat treatment is necessary to attain the specified properties for this temper. (e) Stress relieved by stretching to produce a specified amount of permanent set subsequent to solution heat treatment and prior to any precipitation heat treatment. (f) These heat treatments also apply to alclad sheet and plate in these alloys. (g) An alternative treatment of 8 h at 177 °C (350 °F) also may be used. (h) Solution heat treatment is followed by quenching in water 60 to 82 °C (140 to 180 °F). (j) Stress relieved by 1 to 5% cold reduction subsequent to solution heat treatment and prior to precipitation heat treatment. (k) Solution heat treatment is followed by quenching in water at 100 °C (212 °F). (m) Solution heat treatment is followed by quenching in room-temperature air blast. (n) By suitable control of extrusion temperature, product may be quenched directly from extrusion press to provide specified properties for this temper. Some products may be adequately quenched in room-temperature air blast. (p) See U.S. Patent 4 082 578. (q) Applicable to tread plate only. (r) An alternative treatment of 8 h at 171 °C (340 °F) also may be used. (s) Cold working subsequent to precipitation heat treatment is necessary to attain the specified properties for this temper. (t) An alternative treatment of 3 h at 182 °C (360 °F) also may be used. (u) An alternative treatment of 6 h at 182 °C (360 °F) also may be used. (v) No solution heat treatment; 72 h at room temperature following press quench, followed by two-stage precipitation heat treatment comprised of 8 h at 107 °C (225 °F) plus 16 h at 149 °C (300 °F). (w) Aging practice varies with product, size, nature of equipment, loading procedures and furnace-control capabilities. The optimum practice for a specific item can be ascertained only by actual trial treatment of the item under specific conditions. Typical procedures involve a two-stage treatment comprised of 3 to 30 h at 121 °C (250 °F) followed by 15 to 18 h at 163 °C (325 °F) for extrusions. An alternative two-stage treatment of 8 h at 99 °C (210 °F) followed by 24 to 28 h at 163 °C (325 °F) also may be used. (x) Aging of aluminum alloys 7050, 7075, 7175 and 7475 from any temper to the T73 or T76 temper series requires closer-than-normal controls on aging variables such as time, temperature, heatup rate, etc., for any given item. In addition, when material in a T6-type temper is reaged to a T73- or T76-type temper, the specific condition of the T6 material (such as property levels and other effects of processing variables) is extremely important and will affect the capability of the reaged material to conform to the requirements specified for the applicable T73- or T76-type temper. (y) Two-stage treatment comprised of 6 to 8 h at 107 °C (225 °F) followed by: 24 to 30 h at 163 °C (325 °F) for sheet and plate; 8 to 10 h at 177 °C (350 °F) for rolled or cold finished rod and bar; 6 to 8 h at 177 °C (350 °F) for extrusions and tube; 8 to 10 h at 177 °C (350 °F) for forgings in the T73 temper and 6 to 8 h at 177 °C (350 °F) for forgings in the T7352 temper. (z) An alternative two-stage treatment comprised of 4 h at 96 °C (205 °F) followed by 8 h at 157 °C (315 °F) also may be used. (aa) For sheet, plate, tube and extrusions, an alternative two-stage treatment comprised of 6 to 8 h at 107 °C (225 °F) followed by 14 to 18 h at 168 °C (335 °F) may be used, provided that a heatup rate of approximately 14 °C/h (25 °F/h) is employed. For rolled or cold finished rod and bar, the alternative treatment is 10 h at 177 °C (350 °F). (bb) An alternative three-stage treatment comprised of 5 h at 99 °C (210 °F), 4 h at 121 °C (250 °F) and then 4 h at 149 °C (300 °F) also may be used. (cc) 7175-T736 and -T73652 heat treatments are directed to specific results, may vary from supplier to supplier and are either proprietary or patented. (dd) Must be preceded by soak at 466 to 477 °C (870 to 890 °F). See U.S. Patent 3 791 880.

Table 1 (continued)

Alloy	Product form	Solution heat treatment(a) Metal temperature(b) °C	°F	Temper designation	Precipitation heat treatment Metal temperature(b) °C	°F	Time(c), h	Temper designation
6061(f)	Drawn tube............ 530		985	T4	160(r)	320(r)	18	T6
				T42	160(r)	320(r)	18	T62
	Die and hand forgings... 530		985	T4	175	350	8	T6
	Rolled rings........... 530		985	T4	175	350	8	T6
				T452(j)	175	350	8	T652(j)
6063	Extruded rod, bar, shapes							
	and tube.............. (n)		(n)	T1	205(t)	400(t)	1	T5
		520(n)	970(n)	T4	175(u)	350(u)	8	T6
		520	970	T42	175(u)	350(u)	8	T62
	Drawn tube........... 520		970	T4	175	350	8	T6
					175	350	8	T83(d)(n)
					175	350	8	T831(d)(n)
					175	350	8	T832(d)(n)
				T42	175	350	8	T62
6066	Extruded rod, bar, shapes							
	and tube.............. 530		990	T4	175	350	8	T6
				T42	175	350	8	T62
				T4510(e)	175	350	8	T6510(e)
				T4511(e)	175	350	8	T6511(e)
	Drawn tube.......... 530		990	T4	175	350	8	T6
				T42	175	350	8	T62
	Die forgings 530		990	T4	175	350	8	T6
6070	Extruded rod, bar, shapes							
	and tube.............. 545(n)		1015(n)	T4	160	320	18	T6
				T42	160	320	18	T62
6151	Die forgings 515		960	T4	170	340	10	T6
	Rolled rings........... 515		960	T4	170	340	10	T6
				T452(j)	170	340	10	T652(j)
6262	Rolled or cold finished							
	wire, rod and bar 540		1000	T4	170	340	8	T6
					170	340	12	T9(s)
				T451	170	340	8	T651(e)
				T42	170	340	8	T62

(continued)

(a) Material should be quenched from the solution-treating temperature as rapidly as possible and with minimum delay after removal from the furnace. When material is quenched by total immersion in water, unless otherwise indicated, the water should be at room temperature, and should be suitably cooled so that it remains below 38 °C (100 °F) during the quenching cycle. Use of high-velocity, high-volume jets of cold water also is effective for some materials. (b) The nominal temperatures listed should be attained as rapidly as possible and maintained within ± 6 °C (± 10 °F) of nominal during the time at temperature. (c) Approximate time at temperature. The specific time will depend on the time required for the load to reach temperature. The times shown are based on rapid heating, with soak time measured from the time the load reaches a temperature within 6 °C (10 °F) of the applicable temperature. (d) Cold working subsequent to solution heat treatment and prior to any precipitation heat treatment is necessary to attain the specified properties for this temper. (e) Stress relieved by stretching to produce a specified amount of permanent set subsequent to solution heat treatment and prior to any precipitation heat treatment. (f) These heat treatments also apply to alclad sheet and plate in these alloys. (g) An alternative treatment of 8 h at 177 °C (350 °F) also may be used. (h) Solution heat treatment is followed by quenching in water 60 to 82 °C (140 to 180 °F). (j) Stress relieved by 1 to 5% cold reduction subsequent to solution heat treatment and prior to precipitation heat treatment. (k) Solution heat treatment is followed by quenching in water at 100 °C (212 °F). (m) Solution heat treatment is followed by quenching in room-temperature air blast. (n) By suitable control of extrusion temperature, product may be quenched directly from extrusion press to provide specified properties for this temper. Some products may be adequately quenched in room-temperature air blast. (p) See U.S. Patent 4 082 578. (q) Applicable to tread plate only. (r) An alternative treatment of 8 h at 171 °C (340 °F) also may be used. (s) Cold working subsequent to precipitation heat treatment is necessary to attain the specified properties for this temper. (t) An alternative treatment of 3 h at 182 °C (360 °F) also may be used. (u) An alternative treatment of 6 h at 182 °C (360 °F) also may be used. (v) No solution heat treatment; 72 h at room temperature following press quench, followed by two-stage precipitation heat treatment comprised of 8 h at 107 °C (225 °F) plus 16 h at 149 °C (300 °F). (w) Aging practice varies with product, size, nature of equipment, loading procedures and furnace-control capabilities. The optimum practice for a specific item can be ascertained only by actual trial treatment of the item under specific conditions. Typical procedures involve a two-stage treatment comprised of 3 to 30 h at 121 °C (250 °F) followed by 15 to 18 h at 163 °C (325 °F) for extrusions. An alternative two-stage treatment of 8 h at 99 °C (210 °F) followed by 24 to 28 h at 163 °C (325 °F) also may be used. (x) Aging of aluminum alloys 7050, 7075, 7175 and 7475 from any temper to the T73 or T76 temper series requires closer-than-normal controls on aging variables such as time, temperature, heatup rate, etc., for any given item. In addition, when material in a T6-type temper is reaged to a T73- or T76-type temper, the specific condition of the T6 material (such as property levels and other effects of processing variables) is extremely important and will affect the capability of the reaged material to conform to the requirements specified for the applicable T73- or T76-type temper. (y) Two-stage treatment comprised of 6 to 8 h at 107 °C (225 °F) followed by: 24 to 30 h at 163 °C (325 °F) for sheet and plate; 8 to 10 h at 177 °C (350 °F) for rolled or cold finished rod and bar; 6 to 8 h at 177 °C (350 °F) for extrusions and tube; 8 to 10 h at 177 °C (350 °F) for forgings in the T73 temper and 6 to 8 h at 177 °C (350 °F) for forgings in the T7352 temper. (z) An alternative two-stage treatment comprised of 4 h at 96 °C (205 °F) followed by 8 h at 157 °C (315 °F) also may be used. (aa) For sheet, plate, tube and extrusions, an alternative two-stage treatment comprised of 6 to 8 h at 107 °C (225 °F) followed by 14 to 18 h at 168 °C (335 °F) may be used, provided that a heatup rate of approximately 14 °C/h (25 °F/h) is employed. For rolled or cold finished rod and bar, the alternative treatment is 10 h at 177 °C (350 °F). (bb) An alternative three-stage treatment comprised of 5 h at 99 °C (210 °F), 4 h at 121 °C (250 °F) and then 4 h at 149 °C (300 °F) also may be used. (cc) 7175-T736 and -T73652 heat treatments are directed to specific results, may vary from supplier to supplier and are either proprietary or patented. (dd) Must be preceded by soak at 466 to 477 °C (870 to 890 °F). See U.S. Patent 3 791 880.

Table 1 (continued)

Alloy	Product form	Solution heat treatment(a) Metal temperature(b) °C	°F	Temper designation	Precipitation heat treatment Metal temperature(b) °C	°F	Time(c), h	Temper designation
6262	Extruded rod, bar, shapes and tube	540(n)	1000(n)	T4	175	350	12	T6
				T4510(e)	175	350	12	T6510(e)
				T4511(e)	175	350	12	T6511(e)
		540	1000	T42	175	350	12	T62
	Drawn tube	540	1000	T4	170	340	8	T6
					170	340	8	T9(s)
				T42	170	340	8	T62
6463	Extruded rod, bar, shapes and tube	(n)	(n)	T1	205(t)	400(t)	1	T5
		520(n)	970(n)	T4	175(u)	350(u)	8	T6
		520	970	T42	175(u)	350(u)	8	T62
6951	Sheet	530	985	T4	160	320	18	T6
				T42	160	320	18	T62
7001	Extruded rod, bar, shapes and tube	465	870	W	120	250	24	T6
					120	250	24	T62
				W510(e)	120	250	24	T6510(e)
				W511(e)	120	250	24	T6511(e)
7005	Extruded rod, bar and shapes	...	...	...	...	...	...	T53(v)
7050	Plate	475	890	W51(e)	(w)	(w)	(w)	T7651(x)
					(y)	(y)	(y)	T73651(x)
	Extrusions	475	890	W510(e)	(w)	(w)	(w)	T76510(x)
				W511(e)	(w)	(w)	(w)	T76511(x)
	Die and hand forgings	475	890	W	(y)	(y)	(y)	T736(x)
				W52(e)	(y)	(y)	(y)	T73652(x)
7075(f)	Sheet	480	900	W	120(z)	250(z)	24	T6
					120(z)	250(z)	24	T62
					(w)	(w)	(w)	T76(x)
					(y)(aa)	(y)(aa)	(y)(aa)	T73(x)
	Plate	480	900	W	120(z)	250(z)	24	T62
				W51(e)	(y)(aa)	(y)(aa)	(y)(aa)	T7351(e)(x)
					120(z)	250(z)	24	T651(e)
					(w)	(w)	(w)	T7651(x)

(continued)

(a) Material should be quenched from the solution-treating temperature as rapidly as possible and with minimum delay after removal from the furnace. When material is quenched by total immersion in water, unless otherwise indicated, the water should be at room temperature, and should be suitably cooled so that it remains below 38 °C (100 °F) during the quenching cycle. Use of high-velocity, high-volume jets of cold water is effective for some materials. (b) The nominal temperatures listed should be attained as rapidly as possible and maintained within ± 6 °C (± 10 °F) of nominal during the time at temperature. (c) Approximate time at temperature. The specific time will depend on the time required for the load to reach temperature. The times shown are based on rapid heating, with soak time measured from the time the load reaches a temperature within 6 °C (10 °F) of the applicable temperature. (d) Cold working subsequent to solution heat treatment and prior to any precipitation heat treatment is necessary to attain the specified properties for this temper. (e) Stress relieved by stretching to produce a specified amount of permanent set subsequent to solution heat treatment and prior to any precipitation heat treatment. (f) These heat treatments also apply to alclad sheet and plate in these alloys. (g) An alternative treatment of 8 h at 177 °C (350 °F) also may be used. (h) Solution heat treatment is followed by quenching in water 60 to 82 °C (140 to 180 °F). (j) Stress relieved by 1 to 5% cold reduction subsequent to solution heat treatment and prior to precipitation heat treatment. (k) Solution heat treatment is followed by quenching in water at 100 °C (212 °F). (m) Solution heat treatment is followed by quenching in room-temperature air blast. (n) By suitable control of extrusion temperature, product may be quenched directly from extrusion press to provide specified properties for this temper. Some products may be adequately quenched in room-temperature air blast. (p) See U.S. Patent 4 082 578. (q) Applicable to tread plate only. (r) An alternative treatment of 8 h at 171 °C (340 °F) also may be used. (s) Cold working subsequent to precipitation heat treatment is necessary to attain the specified properties for this temper. (t) An alternative treatment of 3 h at 182 °C (360 °F) also may be used. (u) An alternative treatment of 6 h at 182 °C (360 °F) also may be used. (v) No solution heat treatment; 72 h at room temperature following press quench, followed by two-stage precipitation heat treatment comprised of 8 h at 107 °C (225 °F) plus 16 h at 149 °C (300 °F). (w) Aging practice varies with product, size, nature of equipment, loading procedures and furnace-control capabilities. The optimum practice for a specific item can be ascertained only by actual trial treatment of the item under specified conditions. Typical procedures involve a two-stage treatment comprised of 3 to 30 h at 121 °C (250 °F) followed by 15 to 18 h at 163 °C (325 °F) for extrusions. An alternative two-stage treatment of 8 h at 99 °C (210 °F) followed by 24 to 28 h at 163 °C (325 °F) also may be used. (x) Aging of aluminum alloys 7050, 7075, 7175 and 7475 from any temper to the T73 or T76 temper series requires closer-than-normal controls on aging variables such as time, temperature, heatup rate, etc., for any given item. In addition, when material in a T6-type temper is reaged to a T73- or T76-type temper, the specific condition of the T6 material (such as property levels and other effects of processing variables) is extremely important and will affect the capability of the reaged material to conform to the requirements specified for the applicable T73- or T76-type temper. (y) Two-stage treatment comprised of 6 to 8 h at 107 °C (225 °F) followed by: 24 to 30 h at 163 °C (325 °F) for sheet and plate; 8 to 10 h at 177 °C (350 °F) for rolled or cold finished rod and bar; 6 to 8 h at 177 °C (350 °F) for extrusions and tube; 8 to 10 h at 177 °C (350 °F) for forgings in the T73 temper and 6 to 8 h at 177 °C (350 °F) for forgings in the T7352 temper. (z) An alternative two-stage treatment comprised of 4 h at 96 °C (205 °F) followed by 8 h at 157 °C (315 °F) also may be used. (aa) For sheet, plate, tube and extrusions, an alternative two-stage treatment comprised of 6 to 8 h at 107 °C (225 °F) followed by 14 to 18 h at 168 °C (335 °F) may be used, provided that a heatup rate of approximately 14 °C/h (25 °F/h) is employed. For rolled or cold finished rod and bar, the alternative treatment is 10 h at 177 °C (350 °F). (bb) An alternative three-stage treatment comprised of 5 h at 99 °C (210 °F), 4 h at 121 °C (250 °F) and then 4 h at 149 °C (300 °F) also may be used. (cc) 7175-T736 and -T73652 heat treatments are directed to specific results, may vary from supplier to supplier and are either proprietary or patented. (dd) Must be preceded by soak at 466 to 477 °C (870 to 890 °F). See U.S. Patent 3 791 880.

Table 1 (continued)

Alloy	Product form	Solution heat treatment(a) Metal temperature(b) °C	°F	Temper designation	Precipitation heat treatment Metal temperature(b) °C	°F	Time(c), h	Temper designation
7075(f)	Rolled or cold finished wire, rod and bar 490		915	W	120	250	24	T6
					120	250	24	T62
					(y)(aa)	(y)(aa)	(y)(aa)	T73(x)
				W51(e)	120	250	24	T651(e)
	Extruded rod, bar, shapes				(y)(aa)	(y)(aa)	(y)(aa)	T7351(e)(x)
	and tube.............. 465		870	W	120(bb)	250(bb)	24	T6
					120(bb)	250(bb)	24	T62
					(y)(aa)	(y)(aa)	(y)(aa)	T73(x)
					(w)	(w)	(w)	T76(x)
				W510(e)	120(bb)	250(bb)	24	T6510(e)
					(y)(aa)	(y)(aa)	(y)(aa)	T73510(e)(x)
					(w)	(w)	(w)	T76510(x)
				W511(e)	120(bb)	250(bb)	24	T6511(e)
					(y)(aa)	(y)(aa)	(y)(aa)	T73511(e)(x)
					(w)	(w)	(w)	T76511(x)
	Drawn tube............ 465		870	W	120	250	24	T6
					120	250	24	T62
					(y)(aa)	(y)(aa)	(y)(aa)	T73(x)
	Die forgings .... 470(h)		880(h)	W	120	250	24	T6
					(y)	(y)	(y)	T73(x)
				W52(j)	(y)	(y)	(y)	T7352(j)(x)
	Hand forgings......... 470(h)		880(h)	W	120	250	24	T6
					(y)	(y)	(y)	T73(x)
				W52(j)	120	250	24	T652(j)
					(y)	(y)	(y)	T7352(j)(x)
	Rolled rings........... 470		880	W	120	250	24	T6
7175	Die forgings .... (cc)		(cc)	W	(cc)	(cc)	(cc)	T66(cc)
		(cc)	(cc)	W	(cc)	(cc)	(cc)	T736(x)(cc)
		(cc)	(cc)	W52(j)	(cc)	(cc)	(cc)	T73652(j)(x)(cc)
	Hand forgings......... (cc)		(cc)	W	(cc)	(cc)	(cc)	T736(x)(cc)
		(cc)	(cc)	W52(j)	(cc)	(cc)	(cc)	T7365(j)(x)(cc)

(continued)

(a) Material should be quenched from the solution-treating temperature as rapidly as possible and with minimum delay after removal from the furnace. When material is quenched by total immersion in water, unless otherwise indicated, the water should be at room temperature, and should be suitably cooled so that it remains below 38 °C (100 °F) during the quenching cycle. Use of high-velocity, high-volume jets of cold water also is effective for some materials. (b) The nominal temperatures listed should be attained as rapidly as possible and maintained within ± 6 °C (± 10 °F) of nominal during the time at temperature. (c) Approximate time at temperature. The specific time will depend on the time required for the load to reach temperature. The times shown are based on rapid heating, with soak time measured from the time the load reaches a temperature within 6 °C (10 °F) of the applicable temperature. (d) Cold working subsequent to solution heat treatment and prior to any precipitation heat treatment is necessary to attain the specified properties for this temper. (e) Stress relieved by stretching to produce a specified amount of permanent set subsequent to solution heat treatment and prior to any precipitation heat treatment. (f) These heat treatments also apply to alclad sheet and plate in these alloys. (g) An alternative treatment of 8 h at 177 °C (350 °F) also may be used. (h) Solution heat treatment is followed by quenching in water 60 to 82 °C (140 to 180 °F). (j) Stress relieved by 1 to 5% cold reduction subsequent to solution heat treatment and prior to precipitation heat treatment. (k) Solution heat treatment is followed by quenching in water at 100 °C (212 °F). (m) Solution heat treatment is followed by quenching in room-temperature air blast. (n) By suitable control of extrusion temperature, product may be quenched directly from extrusion press to provide specified properties for this temper. Some products may be adequately quenched in room-temperature air blast. (p) See U.S. Patent 4 082 578. (q) Applicable to tread plate only. (r) An alternative treatment of 8 h at 171 °C (340 °F) also may be used. (s) Cold working subsequent to precipitation heat treatment is necessary to attain the specified properties for this temper. (t) An alternative treatment of 3 h at 182 °C (360 °F) also may be used. (u) An alternative treatment of 6 h at 182 °C (360 °F) also may be used. (v) No solution heat treatment; 72 h at room temperature following press quench, followed by two-stage precipitation heat treatment comprised of 8 h at 107 °C (225 °F) plus 16 h at 149 °C (300 °F). (w) Aging practice varies with product, size, nature of equipment, loading procedures and furnace-control capabilities. The optimum practice for a specific item can be ascertained only by actual trial treatment of the item under specific conditions. Typical procedures involve a two-stage treatment comprised of 3 to 30 h at 121 °C (250 °F) followed by 15 to 18 h at 163 °C (325 °F) for extrusions. An alternative two-stage treatment of 8 h at 99 °C (210 °F) followed by 24 to 28 h at 163 °C (325 °F) also may be used. (x) Aging of aluminum alloys 7050, 7075, 7175 and 7475 from any temper to the T73 or T76 temper series requires closer-than-normal controls on aging variables such as time, temperature, heatup rate, etc., for any given item. In addition, when material in a T6-type temper is reaged to a T73- or T76-type temper, the specific condition of the T6 material (such as property levels and other effects of processing variables) is extremely important and will affect the capability of the reaged material to conform to the requirements specified for the applicable T73- or T76-type temper. (y) Two-stage treatment comprised of 6 to 8 h at 107 °C (225 °F) followed by: 24 to 30 h at 163 °C (325 °F) for sheet and plate; 8 to 10 h at 177 °C (350 °F) for rolled or cold finished rod and bar; 6 to 8 h at 177 °C (350 °F) for extrusions and tube; 8 to 10 h at 177 °C (350 °F) for forgings in the T73 temper and 6 to 8 h at 177 °C (350 °F) for forgings in the T7352 temper. (z) An alternative two-stage treatment comprised of 4 h at 96 °C (205 °F) followed by 8 h at 157 °C (315 °F) also may be used. (aa) For sheet, plate, tube and extrusions, an alternative two-stage treatment comprised of 6 to 8 h at 107 °C (225 °F) followed by 14 to 18 h at 168 °C (335 °F) may be used, provided that a heatup rate of approximately 14 °C/h (25 °F/h) is employed. For rolled or cold finished rod and bar, the alternative treatment is 10 h at 177 °C (350 °F). (bb) An alternative three-stage treatment comprised of 5 h at 99 °C (210 °F), 4 h at 121 °C (250 °F) and then 4 h at 149 °C (300 °F) also may be used. (cc) 7175-T736 and -T73652 heat treatments are directed to specific results, may vary from supplier to supplier and are either proprietary or patented. (dd) Must be preceded by soak at 466 to 477 °C (870 to 890 °F). See U.S. Patent 3 791 880.

Table 1 (continued)

Alloy	Product form	Solution heat treatment(a) Metal temperature(b) °C	°F	Temper designation	Precipitation heat treatment Metal temperature(b) °C	°F	Time(c), h	Temper designation
7475	Sheet .... 515(dd)		960(dd)	W	120	250	3	
				plus	155	315	3	T61(dd)
					(w)	(w)	(w)	T761(x)(dd)
	Plate .... 510(dd)		950(dd)	W51(e)	120	250	24	T651(dd)
					(w)	(w)	(w)	T7651(x)(dd)
					(y)	(y)	(y)	T7351(x)(dd)
Alclad 7475	Sheet 495		920	W	120	250	3	
				plus	155	315	3	T61(dd)
					(w)	(w)	(w)	T761(x)(dd)

(a) Material should be quenched from the solution-treating temperature as rapidly as possible and with minimum delay after removal from the furnace. When material is quenched by total immersion in water, unless otherwise indicated, the water should be at room temperature, and should be suitably cooled so that it remains below 38 °C (100 °F) during the quenching cycle. Use of high-velocity, high-volume jets of cold water also is effective for some materials. (b) The nominal temperatures listed should be attained as rapidly as possible and maintained within ± 6 °C (± 10 °F) of nominal during the time at temperature. (c) Approximate time at temperature. The specific time will depend on the time required for the load to reach temperature. The times shown are based on rapid heating, with soak time measured from the time the load reaches a temperature within 6 °C (10 °F) of the applicable temperature. (d) Cold working subsequent to solution heat treatment and prior to any precipitation heat treatment is necessary to attain the specified properties for this temper. (e) Stress relieved by stretching to produce a specified amount of permanent set subsequent to solution heat treatment and prior to any precipitation heat treatment. (f) These heat treatments also apply to alclad sheet and plate in these alloys. (g) An alternative treatment of 8 h at 177 °C (350 °F) also may be used. (h) Solution heat treatment is followed by quenching in water 60 to 82 °C (140 to 180 °F). (j) Stress relieved by 1 to 5% cold reduction subsequent to solution heat treatment and prior to precipitation heat treatment. (k) Solution heat treatment is followed by quenching in water at 100 °C (212 °F). (m) Solution heat treatment is followed by quenching in room-temperature air blast. (n) By suitable control of extrusion temperature, product may be quenched directly from extrusion press to provide specified properties for this temper. Some products may be adequately quenched in room-temperature air blast. (p) See U.S. Patent 4 082 578. (q) Applicable to tread plate only. (r) An alternative treatment of 8 h at 171 °C (340 °F) also may be used. (s) Cold working subsequent to precipitation heat treatment is necessary to attain the specified properties for this temper. (t) An alternative treatment of 3 h at 182 °C (360 °F) also may be used. (u) An alternative treatment of 6 h at 182 °C (360 °F) also may be used. (v) No solution heat treatment; 72 h at room temperature following press quench, followed by two-stage precipitation heat treatment comprised of 8 h at 107 °C (225 °F) plus 16 h at 149 °C (300 °F). (w) Aging practice varies with product, size, nature of equipment, loading procedures and furnace-control capabilities. The optimum practice for a specific item can be ascertained only by actual trial treatment of the item under specific conditions. Typical procedures involve a two-stage treatment comprised of 3 to 30 h at 121 °C (250 °F) followed by 15 to 18 h at 163 °C (325 °F) for sheet and plate; 8 to 10 h at 177 °C (350 °F) for extrusions. An alternative two-stage treatment of 8 h at 99 °C (210 °F) followed by 24 to 28 h at 163 °C (325 °F) also may be used. (x) Aging of aluminum alloys 7050, 7075, 7175 and 7475 from any temper to the T73 or T76 temper series requires closer-than-normal controls on aging variables such as time, temperature, heatup rate, etc., for any given item. In addition, when material in a T6-type temper is reaged to a T73- or T76-type temper, the specific condition of the T6 material (such as property levels and other effects of processing variables) is extremely important and will affect the capability of the reaged material to conform to the requirements specified for the applicable T73- or T76-type temper. (y) Two-stage treatment comprised of 6 to 8 h at 107 °C (225 °F) followed by: 24 to 30 h at 163 °C (325 °F) for sheet and plate; 8 to 10 h at 177 °C (350 °F) for rolled or cold finished rod and bar; 6 to 8 h at 177 °C (350 °F) for extrusions and tube; 8 to 10 h at 177 °C (350 °F) for forgings in the T73 temper and 6 to 8 h at 177 °C (350 °F) for forgings in the T7352 temper. (z) An alternative two-stage treatment comprised of 4 h at 96 °C (205 °F) followed by 8 h at 157 °C (315 °F) also may be used. (aa) For sheet, plate, tube and extrusions, an alternative two-stage treatment comprised of 6 to 8 h at 107 °C (225 °F) followed by 14 to 18 h at 168 °C (335 °F) may be used, provided that a heatup rate of approximately 14 °C/h (25 °F/h) is employed. For rolled or cold finished rod and bar, the alternative treatment is 10 h at 177 °C (350 °F). (bb) An alternative three-stage treatment comprised of 5 h at 99 °C (210 °F), 4 h at 121 °C (250 °F) and then 4 h at 149 °C (300 °F) also may be used. (cc) 7175-T736 and -T73652 heat treatments are directed to specific results, may vary from supplier to supplier and are either proprietary or patented. (dd) Must be preceded by soak at 466 to 477 °C (870 to 890 °F). See U.S. Patent 3 791 880.

an allowance for unintentional temperature variations. Although ranges normally listed allow variations of ±6 °C (±10 °F) from the nominal, some highly alloyed, controlled-toughness, high-strength alloys require that temperature be controlled within more restrictive limits. Broader ranges may be allowable for alloys with greater intervals of temperature between their solvus and eutectic melting temperatures.

Overheating. Care must be exercised to avoid exceeding the initial eutectic melting temperature. If appreciable eutectic melting occurs as a result of overheating, properties such as tensile strength, ductility and fracture toughness may be degraded. Specifications generally categorize as unacceptable those materials that exhibit mi-

crostructural evidence of overheating. Figure 2 shows the grain-boundary melting that occurs when the eutectic melting temperature of the alloy is exceeded. The material becomes brittle and nonsalvageable. This dangerous condition usually is not detectable by either visual examination or nondestructive testing.

Although maximum temperature must be restricted to avoid melting, the lower limit should, when possible, be above the temperature at which complete solution occurs (solvus). In the 5.25% Cu alloy represented by line CD in Fig. 1, these temperatures would be 548 and 536 °C (1018 and 996 °F), respectively. However, under production conditions, a range of 537 to 546 °C (999 to 1014 °F) would probably be

used, to provide a 2.2 °C (4 °F) margin to safeguard against eutectic melting and a 1.7 °C (3 °F) cushion on the low side for increased solution and diffusion rates.

For certain alloys such as 2024 and 2219, which can be represented on Fig. 1 as alloys containing more than 5.65% Cu, complete solution can never occur. For these alloys, the minimum solution heat treating temperature is established so that it is as close as practical to the eutectic temperature while providing a margin of safety commensurate with the capability of the equipment. The proximity of typical solution-treating temperature ranges to eutectic melting temperatures for three common alloys is shown in the following table:

Alloy	Solution-treating temperature °C	°F	Eutectic melting temperature °C	°F
2014	496–507	925–945	510	950
2017	496–507	925–945	513	955
2024	488–499	910–930	502	935

Nonequilibrium Melting. When high heating rates are employed, the phenomenon of nonequilibrium melting must be considered. This phenomenon can also be explained with the help of the Al-Cu phase diagram (Fig. 1). The room-temperature microstructure of an F-temper product containing 4% Cu consists of a solid solution of copper in aluminum and particles of Al_2Cu. When this product is heated slowly, the Al_2Cu begins to dissolve, and if heating is slow enough, all of the Al_2Cu is dissolved when temperatures above the solvus (500 °C, or 932 °F) are reached. When the heating rate is high, however, much of the Al_2Cu remains undissolved. If a material with this microstructure is heated at or above the eutectic temperature of 548 °C (1018 °F), melting will begin at the interface between the Al_2Cu and the matrix. With sufficient time above the eutectic temperature, this metastable liquid will dissolve to form a solid solution and will leave no trace provided that hydrogen gas has not condensed at the interface to form a void. If the product is quenched before the liquid has time to equilibrate, however, it will solidify and form fine eutectic rosettes. This nonequilibrium melting should not be confused with true equilibrium melting, which would occur in an alloy containing more than 5.65% Cu. In such an alloy, eutectic melting is equilibrium melting. No matter how long such an alloy is held above the eutectic temperature, the liquid will never solidify. In commercial alloys, which usually are ternaries or quaternaries of the major alloying elements, the situation is more complex. Different phases have different solvus temperatures, and nonequilibrium melting may occur at different temperatures depending on composition, size of precipitates and rate of heating. When new solution heat treating equipment (which provides higher heating rates) is employed, careful examination of alloy microstructures should be included as part of the certification process.

Underheating. When the temperatures attained by the parts or pieces being heat treated are appreciably below the normal range, solution is incomplete, and strength somewhat lower than normal is expected. The shallow slope of the solvus at its intersection with the composition line (typified by point E on line AB in Fig. 1) indicates that a slight decrease in temperature below point E will result in a large reduction in the concentration of the solid solution and a correspondingly significant decrease in final strength. The effect of solution-treating temperature on the strength of two aluminum alloys is illustrated by the following data:

Solution treating temperature °C	°F	Tensile strength MPa	ksi	Yield strength MPa	ksi
6061-T6 sheet 1.6 mm (0.064 in.) thick					
493	920	301	43.7	272	39.4
504	940	316	45.8	288	41.7
516	960	333	48.3	305	44.3
527	980	348	50.5	315	45.7
2024-T4 sheet 0.8 mm (0.032 in.) thick					
488	910	419	60.8	255	37.0
491	915	422	61.2	259	37.5
493	920	433	62.8	269	39.0
496	925	441	63.9	271	39.3

Table 2 Soak times and maximum quench delays for solution treatment of wrought aluminum alloys

See Table 1 for solution-treating temperatures.

Thickness(a), mm (in.)	Soak time, minutes Air furnace(b) min	max(d)	Salt bath(c) min	max(d)	Maximum quench delay, s
Thru 0.41 (0.016) 20		25	10	15	5
0.51 (0.020) 20		30	10	20	7
0.64 (0.025) 25		35	15	25	7
0.81 (0.032) 25		35	15	25	7
1.02 (0.040) 30		40	20	30	10
1.27 (0.050) 30		40	20	30	10
1.35 (0.053) 30		40	20	30	10
1.80 (0.071) 35		45	25	35	10
2.03 (0.080) 35		45	25	35	10
2.29 (0.090) 35		45	25	35	10
2.54 (0.100) 40		55	30	45	15
3.18 (0.125) 40		55	30	45	15
4.06 (0.160) 50		60	35	45	15
4.57 (0.180) 50		60	35	45	15
6.35 (0.250) 55		65	35	45	15
Over 6.35 (0.250) thru 12.7 (0.500) 65		75	45	55	15
For each additional 12.7 (1/2) or fraction .... +30		+30	+20	+20	(e)
Rivets (all)........................ 60		...	30	...	5

(a) Minimum dimension of thickest section. (b) Soak time begins when all pyrometer instruments recover to original operating temperature. (c) Soak time begins at time of immersion except when a heavy charge causes bath temperature to drop below specified minimum, in which case soak time begins when bath regains minimum temperature. (d) Applicable to alclad materials only. (e) Increases in thickness above 12.7 mm (1/2 in.) do not affect maximum quench delay, which remains constant at 15 s.

In the tabulation above, note especially the effects of small increments of temperature, within the normal range, on the properties of 0.8-mm (0.032-in.) 2024-T4 sheet.

Solution Treating Time. The time at the nominal solution heat treating temperature ("soak time") required to effect a satisfactory degree of solution of the undissolved or precipitated soluble phase constituents and to achieve good homogeneity of the solid solution is a function of microstructure before heat treatment. This time requirement can vary from less than a minute for thin sheet to as much as 20 h for large sand or plaster-mold castings. Guideline information for soak times required for wrought products of various section thicknesses is given in Table 2. Similar guidelines for castings are presented in Table 3. The time required to heat a load to the treatment temperature in furnace heat treatment also increases with section thickness and furnace loading, and thus total cycle time increases with these factors.

Soak time for alclad sheet and for parts made from alclad sheet must be held to a minimum, because excessive diffusion of alloying elements from the core into the cladding reduces corrosion

Table 3 Typical heat treatments for aluminum alloy sand and permanent mold castings

Alloy	Temper	Type of casting(a)	Solution heat treatment(b) Temperature(c) °C	°F	Time, h	Aging treatment Temperature(c) °C	°F	Time, h
201.0	T6	S.......	510-515; 525-530	950-960; 980-990	2 14-20	155	310	20
	T7	S.......	510-515; 525-530	950-960; 980-990	2 14-20	190	370	5
204.0	T4	S or P.....	520	970	10	...	...	...
208.0	T55	S........	...	...	...	155	310	16
222.0	O(d)	S........	...	...	...	315	600	3
	T61	S........	510	950	12	155	310	11
	T551	P........	...	...	...	170	340	16-22
	T65	... .	510	950	4-12	170	340	7-9
242.0	O(e)	S........	...	...	...	345	650	3
	T571	S........	...	...	...	205	400	8
		P........	...	...	...	165-170	330-340	22-26
	T77	S........	515	960	5(f)	330-355	625-675	2 (min)
	T61	S or P.....	515	960	4-12(f)	205-230	400-450	3-5
295.0	T4	S........	515	960	12	...	...	...
	T6	S........	515	960	12	155	310	3-6
	T62	S........	515	960	12	155	310	12-24
	T7	S........	515	960	12	260	500	4-6
296.0	T4	P........	510	950	8	...	...	...
	T6	P........	510	950	8	155	310	1-8
	T7	P........	510	950	8	260	500	4-6
319.0	T5	S........	...	...	...	205	400	8
	T6	S........	505	940	12	155	310	2-5
		P........	505	940	4-12	155	310	2-5
328.0	T6	S........	515	960	12	155	310	2-5
332.0	T5	P........	...	...	...	205	400	7-9
333.0	T5	P........	...	...	...	205	400	7-9
	T6	P........	505	940	6-12	155	310	2-5
	T7	P........	505	940	6-12	260	500	4-6
336.0	T551	P........	...	...	...	205	400	7-9
	T65	P........	515	960	8	205	400	7-9
354.0	...	(g).......	525-535	980-995	10-12	(h)	(h)	(h)
355.0	T51	S or P.....	...	...	...	225	440	7-9
	T6	S........	525	980	12	155	310	3-5
		P........	525	980	4-12	155	310	2-5
	T62	P........	525	980	4-12	170	340	14-18
	T7	S........	525	980	12	225	440	3-5
		P........	525	980	4-12	225	440	3-9
	T71	S........	525	980	12	245	475	4-6
		P........	525	980	4-12	245	475	3-6
C355.0	T6	S........	525	980	12	155	310	3-5
	T61	P........	525	980	6-12	Room temperature 155	310	8 (min) 10-12
356.0	T51	S or P.....	...	...	...	225	440	7-9
	T6	S........	540	1000	12	155	310	3-5
		P........	540	1000	4-12	155	310	2-5
	T7	S........	540	1000	12	205	400	3-5
		P........	540	1000	4-12	225	440	7-9
	T71	S........	540	1000	10-12	245	475	3
		P........	540	1000	4-12	245	475	3-6

(continued)

(a) S, sand; P, permanent mold. (b) Unless otherwise indicated, solution treating is followed by quenching in water at 65 to 100 °C (150 to 212 °F). (c) Except where ranges are given, listed temperatures are ±6 °C or ±10 °F. (d) Stress relieve for dimensional stability as follows: hold 5 h at 413 ± 14 °C (775 ± 25 °F); furnace cool to 345 °C (650 °F) over a period of 2 h or more; furnace cool to 230 °C (450 °F) over a period of not more than 1/2 h; furnace cool to 120 °C (250 °F) over a period of approximately 2 h; cool to room temperature in still air outside the furnace. (e) No quench required; cool in still air outside furnace. (f) Air-blast quench from solution-treating temperature. (g) Casting process varies (sand, permanent mold or composite) depending on desired mechanical properties. (h) Solution heat treat as indicated, then artificially age by heating uniformly at the temperature and for the time necessary to develop the desired mechanical properties. (j) Quench in water at 65 to 100 °C (150 to 212 °F) for 10 to 20 s only. (k) Cool to room temperature in still air outside furnace.

Table 3 (continued)

Alloy	Temper	Type of casting(a)	Solution heat treatment(b) Temperature(c) °C	°F	Time, h	Aging treatment Temperature(c) °C	°F	Time, h
A356.0	T6	S........	540	1000	12	155	310	3-5
	T61	P........	540	1000	6-12	Room temperature		8 (min)
						155	310	6-12
357.0	T6	P........	540	1000	8	175	350	6
	T61	S........	540	1000	10-12	155	310	10-12
A357.0	...	(g).......	540	1000	8-12	(h)	(h)	(h)
359.0	...	(g).......	540	1000	10-14	(h)	(h)	(h)
A444.0	T4	P........	540	1000	8-12	...	...	...
520.0	T4	S........	430	810	18(j)	...	...	...
535.0	T5(d)	S........	400	750	5	...	...	...
705.0	T5	S........	...	...	...	Room temperature or		21 days
						100	210	8
		P........	...	...	...	Room temperature or		21 days
						100	210	10
707.0	T5	S........	...	...	...	155	310	3-5
		P........	...	...	...	Room temperature or		21 days
						100	210	8
	T7	S........	530	990	8-16	175	350	4-10
		P........	530	990	4-8	175	350	4-10
710.0	T5	S........	...	...	...	Room temperature		21 days
711.0	T1	P........	...	...	...	Room temperature		21 days
712.0	T5	S........	...	...	...	Room temperature or		21 days
						155	315	6-8
713.0	T5	S or P.....	...	...	...	Room temperature or		21 days
						120	250	16
771.0	T53(d)	S........	415(k)	775(k)	5(k)	180(k)	360(k)	4(k)
	T5	S........	...	...	...	180(k)	355(k)	3-5(k)
	T51	S........	...	...	...	205	405	6
	T52	S........	...	...	...	(d)	(d)	(d)
	T6	S........	590(k)	1090(k)	6(k)	130	265	3
	T71	S........	590(e)	1090(e)	6(e)	140	285	15
850.0	T5	S or P.....	...	...	...	220	430	7-9
851.0	T5	S or P.....	...	...	...	220	430	7-9
	T6	P........	480	900	6	220	430	4
852.0	T5	S or P.....	...	...	...	220	430	7-9

(a) S, sand; P, permanent mold. (b) Unless otherwise indicated, solution treating is followed by quenching in water at 65 to 100 °C (150 to 212 °F). (c) Except where ranges are given, listed temperatures are ±6 °C or ±10 °F. (d) Stress relieve for dimensional stability as follows: hold 5 h at 413 ± 14 °C (775 ± 25 °F); furnace cool to 345 °C (650 °F) over a period of 2 h or more; furnace cool to 230 °C (450 °F) over a period of not more than ½ h; furnace cool to 120 °C (250 °F) over a period of approximately 2 h; cool to room temperature in still air outside the furnace. (e) No quench required; cool in still air outside furnace. (f) Air-blast quench from solution-treating temperature. (g) Casting process varies (sand, permanent mold or composite) depending on desired mechanical properties. (h) Solution heat treat as indicated, then artificially age by heating uniformly at the temperature and for the time necessary to develop the desired mechanical properties. (j) Quench in water at 65 to 100 °C (150 to 212 °F) for 10 to 20 s only. (k) Cool to room temperature in still air outside furnace.

protection. For the same reason, reheat treatment of alclad sheet less than 0.76 mm (0.030 in.) thick generally is prohibited, and the number of reheat treatments permitted for thicker alclad sheet is limited.

The soak times for wrought alloys take into account the normal thermal lag between furnace and part and the difference between surface and center temperatures for commercial equipment qualified to the standards of MIL-H-6088. The rapid heating rates of salt baths permit all immersion time to be counted as soak time unless the bath temperature drops below the minimum of the range. Even then, soak time begins as soon as the bath temperature returns to the minimum. In air furnaces, soak time does not begin until all furnace instruments return to their original set temperature—that is, the temperature reading before insertion of the load.

In air furnaces, thermocouples may also be attached to, or buried in, parts located in the load in such a manner as to represent the hottest and coldest temperatures in each zone. In this way, it is possible to ensure that adequate soaking is obtained.

Special consideration is given also to establishing soak times for hand and die forgings; soak time in some specifications is extended to complete solution and homogenization in areas that received marginal reduction during forging. Considerable variation exists in the amount of soak time added; some specifications call for an arbitrary addition, such as one hour, and others require one hour per inch of thickness of the original forging.

In air furnaces, careful attention should be given to arrangement of the load. Air flow and natural temperature distribution within the furnace should be arranged (a) to offer minimum resis-

Fig. 2 Micrograph showing eutectic melting caused by slight overheating during solution heat treating of 2024-T4 sheet

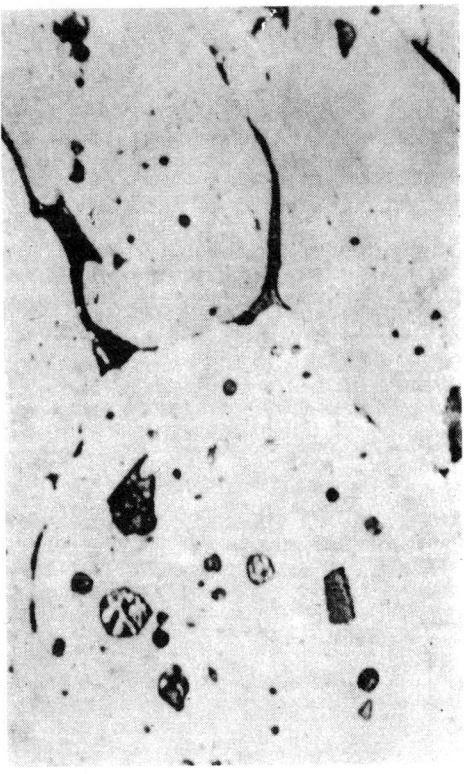

The rosettes (arrows) and heavy grain boundaries are manifestations of material that melted during solution treating and resolidified during subsequent quenching. The as-cast structure in these areas is extremely brittle and, because of the nearly continuous grain-boundary films, brittleness is imparted to the gross structure. Overheated material of this kind cannot be salvaged by reheat treating. Keller's reagent; magnification, 1000X

tance to air flow, (b) to produce the least disturbance in the natural temperature distribution, and (c) to afford constant replenishment of the envelope of air around each part. It is common practice to specify a minimum spacing of 50 mm (2 in.) between parts, but large complex shapes may require considerably greater spacing. Many operators have found conservative loading practices to be more economical in the long run than heavier loading, because with lighter loads heating rates are higher and fewer rejections and service failures are encountered.

High-Temperature Oxidation. There is a condition, commonly but erroneously known as HTO or high-temperature oxidation, which can lead to deterioration of properties in aluminum alloys.

Moisture in contact with aluminum at high temperature serves as a source of nascent hydrogen, which diffuses into the metal. Foreign materials, such as sulfur compounds, function as decomposers of the natural oxide surface film, eliminating it as a barrier either between the moisture and the aluminum or between the nascent hydrogen and the aluminum. The most common manifestation of high-temperature oxidation is surface blistering, but occasionally the only manifestations are internal discontinuities or voids, which can be detected only by careful ultrasonic inspection or by metallographic techniques.

It is important to recognize that the symptoms of high-temperature oxidation are identical to those of unsoundness or high gas content in the original ingot or of other improper mill practice. Blisters resulting from ingot defects, improper extrusion or improper rolling may be lined up in the direction of working. However, it usually is impossible to distinguish among defect sources, and therefore the possibility that a contaminated atmosphere is the cause of the defects must be checked.

Not all alloys and product forms are equally vulnerable to this type of attack. The 7xxx series alloys are most susceptible, followed by the 2xxx alloys. Extrusions undoubtedly are the most susceptible form; forgings are probably second. Low-strength alloys and alclad sheet and plate are relatively immune to high-temperature oxidation. (Blistering of alclad material as a result of inadequate bonding is not the same as the blistering caused by high-temperature oxidation.)

If the protective oxide film formed during mill operations is removed from the mill product by a subsequent mechanical conditioning operation, such as sanding, the conditioned surface will be more susceptible to high-temperature oxidation than those from which the film was not removed.

Moisture can be minimized by thoroughly drying parts and racks before they are charged. Drain holes often are needed in racks of tubular construction, to avoid entrapment of water. Another common requirement is adjustment of the position of the quench tank with respect to furnace doors and air intake. Because it is unlikely that all moisture can be eliminated from the atmosphere in a production heat treating furnace, it is extremely important to eliminate all traces of other contaminants from both the parts and the furnace atmosphere.

The most virulent contaminants in attacking aluminum are sulfur compounds. Residues from forming or machining lubricants, or from a sulfur dioxide protective atmosphere used in prior heat treatment of magnesium, are potential sources of sulfur contamination. In one plant, surface contamination resulted from sulfur-containing materials in tote boxes used to transport parts. In another, an epidemic of blistering was cured by rectifying a "sour" degreaser. In a third instance, it was found that a vapor-degreasing operation was not completely removing a thin, hard waxy residue, and an alkaline cleaning operation was added.

Very often, the source of contamination is obscure and difficult to detect, and the problem must be combated in another way. The most common of the alternative methods is use of a protective fluoborate compound in the furnace. Such a compound usually is effective in minimizing the harmful effects of moisture and other undesirable contaminants because it forms a barrier layer or film on the aluminum surface. The additive is not a universal solution; in some applications, high-temperature oxidation has occurred even though a fluoborate compound was employed. Also, the use of such compounds, particularly ammonium fluoborate, may present a hazard to personnel if used in poorly sealed furnaces or in furnaces that discharge their atmospheres into enclosed areas.

Protective fluoborate compounds accentuate staining or darkening of the parts being heat treated. (At times, this attack, particularly on parts located near the protective-compound container during heat treatment, has been severe enough to be termed "corrosion".) Although this minor nuisance might be considered a small price to pay for solution of a problem of high-temperature oxidation, the residual compound in the furnace dissipates slowly. Therefore, subsequent loads of alloys and product forms whose end uses require bright surfaces, and that are not susceptible to high-temperature oxidation, may be detrimentally affected.

Successful use of fluoborate protective compounds appears to depend on specifying the right amount for each furnace; this must be established on a trial-and-error basis. One aircraft manufacturer adds 4 g/m^3 (0.004 oz/ft^3) of furnace chamber to each load. Another adds 0.45 kg (1 lb) per shift to a metal container hung on the furnace chamber wall, thus avoiding loss of the compound during quenching.

A second method of combating high-temperature oxidation is to anodize the work before it is heat treated. The resultant aluminum oxide film prevents attack by contaminants in the furnace atmosphere. The only deterrents to the use of anodizing are its cost (in money and time) and the slight surface frostiness which results from the subsequent stripping operation.

The usual objection to the blistered surface produced by high-temperature oxidation is its unsightly appearance. This often can be improved (for salvage purposes) by applying local pressure to flatten each blister and then finishing by a mechanical process such as polishing, buffing, sanding or abrasive blasting. In general, the effect of HTO on static properties and fatigue strength is slight. However, if a void resulting from HTO is located close to another stress concentration, such as a hole, much greater degradation of fatigue strength is likely. In critical aluminum alloy forgings, any blistering must be evaluated carefully for its effect on the integrity of the part. Any "cosmetic" salvage should be performed only after it has been established that the blisters are superficial and will not remain in the finished product.

Quenching

In most instances, to avoid those types of precipitation that are detrimental to mechanical properties or to corrosion resistance, the solid solution formed during solution heat treatment must be quenched rapidly enough (and without interruption) to produce a supersaturated solution at room temperature—the optimum condition for precipitation hardening. The resistance to stress-corrosion cracking of certain Cu-free Al-Zn-Mg alloys, however, is improved by slow quenching. Most frequently, parts are quenched by immersion in cold water or, in continuous heat treating of sheet, plate or extrusions in primary fabricating mills, by progressive flooding or high-velocity spraying with cold water. However, parts of complex shape, often with both thin and thick sections (such as die forgings, most castings, impact extrusions, and components formed from sheet) are commonly quenched in a medium that provides somewhat slower cooling. This medium may be water at 65 to 80 °C (150 to 180 °F), boiling water, an aqueous solution of polyalkaline glycol, or some other fluid medium such as forced air or mist.

If appreciable precipitation during cooling is to be avoided, two requirements must be satisfied. First, the time required for transfer of the load from the furnace to the quenching medium must be short enough to preclude slow precooling into the temperature range where very rapid precipitation takes place. For alloy 7075, this range was determined to be 400 to 290 °C (750 to 550 °F), and some sources quote this range (or a slightly different range) as the most critical range for quenching of any aluminum alloy. Later work has shown that the most critical range is alloy-dependent, and as will be discussed in detail under "Quench-Factor Analysis", significant errors can result from the assumption that precipitation is negligible outside of a so-called "critical range".

The second requirement for avoidance of appreciable precipitation during quenching is that the volume, heat-absorption capacity and rate of flow of the quenching medium be such that little or no precipitation occurs during cooling. Any interruption of the quench that might allow reheating into a temperature range where rapid precipitation can occur must be prohibited.

For maximum dimensional stability, some forgings and castings are fan cooled or still-air cooled. In such instances, precipitation-hardening response is limited, but satisfactory values of strength and hardness are obtained. Extrusions produced without separate solution heat treatment can be air or mist quenched, but thicker sections may require water quenching by immersion or spraying. Alloys that are relatively dilute, such as 6063 and 7005, are particularly well suited to air quenching, and their mechanical properties are not greatly affected by its low cooling rate.

Lower quenching rates are employed for forgings, castings and complex shapes to minimize warpage or other distortion and the magnitude of residual stresses developed as a consequence of temperature nonuniformity from surface to interior. The effects of these low rates on mechanical properties vary with alloy composition and temper. The effects on yield strength of average quenching rates through the range from 400 to 290 °C (750 to 550 °F) are shown for four alloys in Fig. 3. For alloys relatively high in sensitivity to quenching rate, such as 7075, rates of about 300 °C/s (540 °F/s) or higher are required in order to obtain near-maximum strength after precipitation heat treatment. The other alloys in Fig. 3 maintain their strengths at cooling rates as low as about 100 °C/s (180 °F/s). As cautioned previously, significant precipitation can occur outside this temperature range, and thus curves such as those in Fig. 3 are useful mainly for comparison of relative sensitivities to quench rate. Moreover, for high-strength alloys, toughness and corrosion resistance may be impaired without significant loss of tensile strength. Where such properties are important, further restrictions on minimum quench rate may be necessary.

Delay in Quenching. Whether the transfer of parts from the furnace to the quench is performed manually or me-

Fig. 3 Effect of quenching rate through critical temperature range on final yield strength of four wrought aluminum alloys

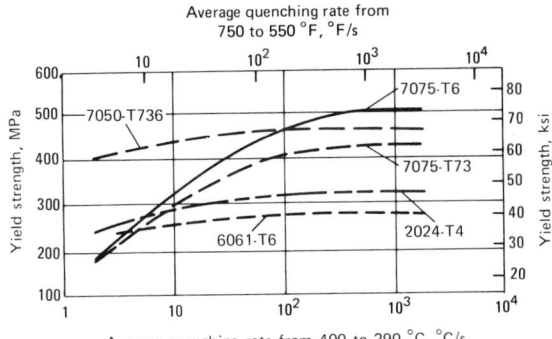

Tempers shown are those after aging.

Fig. 4 Cooling curves for alclad and nonclad aluminum products cooled from 495 °C (920 °F) in forced air

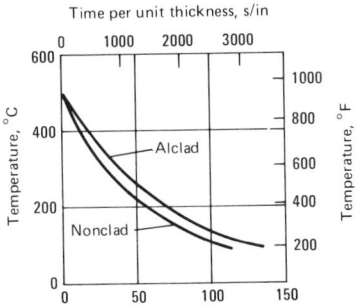

| Thickness | | Maximum quench delay, s | |
mm	in.	Alclad	Nonclad
0.41	0.016	6.4	4.4
0.51	0.020	8.0	5.5
0.64	0.025	10.0	6.8
0.81	0.032	12.8	8.8
1.02	0.040	20.0	11.0

Air temperature, 25 °C (80 °F); air velocity, 2.3 m/s (450 ft/min). Tabulated values of quench delay (maximum delay before the material being quenched has cooled below 400 °C, or 750 °F) were determined from cooling curves shown.

chanically, it must be completed in less than the specified maximum time. The maximum allowable transfer time or "quench delay" varies with the temperature and velocity of the ambient air and the mass and emissivity of the parts. From cooling curves such as those illustrated in Fig. 4, maximum quench delays (see table accompanying Fig. 4) can be determined that will ensure complete immersion before the parts cool below 400 °C (750 °F). MIL-H-6088 specifies maximum quench delays for high-strength alloys of 5, 7, 10, and 15 s for thickness ranges of (a) up to 0.016 in. (0.41 mm), (b) 0.017 to 0.031 in. (0.43 to 0.79 mm), (c) 0.032 to 0.090 in. (0.81 to 2.29 mm), and (d) over 0.090 in., respectively. Quench delay is conservatively defined as commencing "when the furnace door begins to open or the first corner of a load emerges from a salt bath" and ending "when the last corner of the load is immersed in the water quench tank". Recommended maximum quench-delay times are listed in Table 2. However, exceeding the maximum delay time is permitted if temperature measurements of the load prove that all parts are above 415 °C (775 °F) when quenched.

It is relatively easy to control quench delay in day-to-day operations by using a stopwatch or, if necessary, by attaching thermocouples to parts. However, although the cooling rate between 400 and 260 °C (750 and 500 °F) is most critical and must be extremely high for many high-strength alloys, it cannot be directly measured in production operations. It is usual to rely on standardized practices, augmented by results of tension tests and tests of susceptibility to intergranular corrosion.

Water-immersion quenching normally is controlled in practice by stipulating maximum quench-delay time and maximum water temperature. The first requirement controls the cooling rate during transfer and, for high-strength alloys, often is based on the criterion of complete immersion before the metal cools below 415 °C (775 °F). The second requirement controls the cooling rate during immersion. MIL-H-6088 specifies that for water-immersion quenching, except quenching of forgings and castings, the temperature of the water shall not exceed 38 °C (100 °F) upon completion of quenching. This requirement controls both the temperature of the quench water prior to immersion and the ratio of the combined mass of load and rack to the volume of water. However, to ensure adequate quenching effectiveness, it is necessary also that the cooling fluid flow past all surfaces of each part during the first few seconds after immersion. Before parts enter the furnace, their placement in racks or baskets should be compatible with this requirement. During the first few seconds of quenching, agitation of the parts or the water should be sufficient to prevent local increases in temperature due to the formation of steam pockets.

In one application, it was found that 2024-T4 plates 13 by 760 by 760 mm (½ by 30 by 30 in.), quenched singly into a large volume of still water, were quite susceptible to intergranular corrosion. This susceptibility disappeared completely when the quenching practice was modified by adding sufficient agitation to break up the insulating blanket of steam that formed on the surface of the hot metal. Quenching practices for small parts such as fasteners and hydraulic fittings have been modified for the same reason. Dumping in bulk from baskets has been replaced by methods, such as the use of shaker hearth furnaces or special racking, which permit parts to be quenched singly.

Spray Quenching. For spray quenching, the quench rate is controlled by the velocity of the water and by volume of water per unit area per unit time of impingement of the water on the workpiece. Rate of travel of the workpiece through the sprays is an important variable.

Local increases in temperature that occur within the first few seconds of quenching, caused by a phenomenon such as plugged spray nozzles, are particularly deleterious. The remaining "internal heat" may be sufficient to reheat the surface region. When this happens, a large loss in strength occurs at the previously quenched surface. The loss of strength in the affected area of a heavy part is much more severe than that caused by an inadequate quenching rate alone. This is illustrated for 75-mm- (3-in.-) thick 7075-T62 plate in Fig. 5, which compares, at various depths, the properties of a plate for which quenching was interrupted on one side after 3 s with those of a plate that was quenched from one side only.

Quench-Factor Analysis. Although useful as first approximations, average quenching rates and critical temperature ranges are too qualitative to permit accurate prediction of the effects of quenching rate when the rate of cooling does not change smoothly. For such instances, a procedure known as "quench-factor analysis" has been developed. To understand this procedure, one must recognize that the rate of precipitation during quenching depends on two competing factors: supersaturation and diffusion. At high temperatures, supersaturation is low, and so precipitation rate is low despite the high diffusion rate. At low temperatures, diffusion rate is low, and thus precipitation rate is low despite the

high degree of supersaturation. At intermediate temperatures, precipitation rate is highest. Consequently, times to produce equal amounts of precipitation follow a C-shape pattern. Using this C-curve, a quench curve can be analyzed to predict quantitatively how certain properties will be affected. For example, it has been shown that the strengths of several alloys, and the type

of corrosion that will attack alloy 2024-T3, can be predicted by means of this method.

Precipitation kinetics for continuous cooling can be described by the equation:

$$\zeta = 1 - \exp(k\tau) \qquad \text{(Eq 1)}$$

where ζ is the fraction transformed, k is a constant, and:

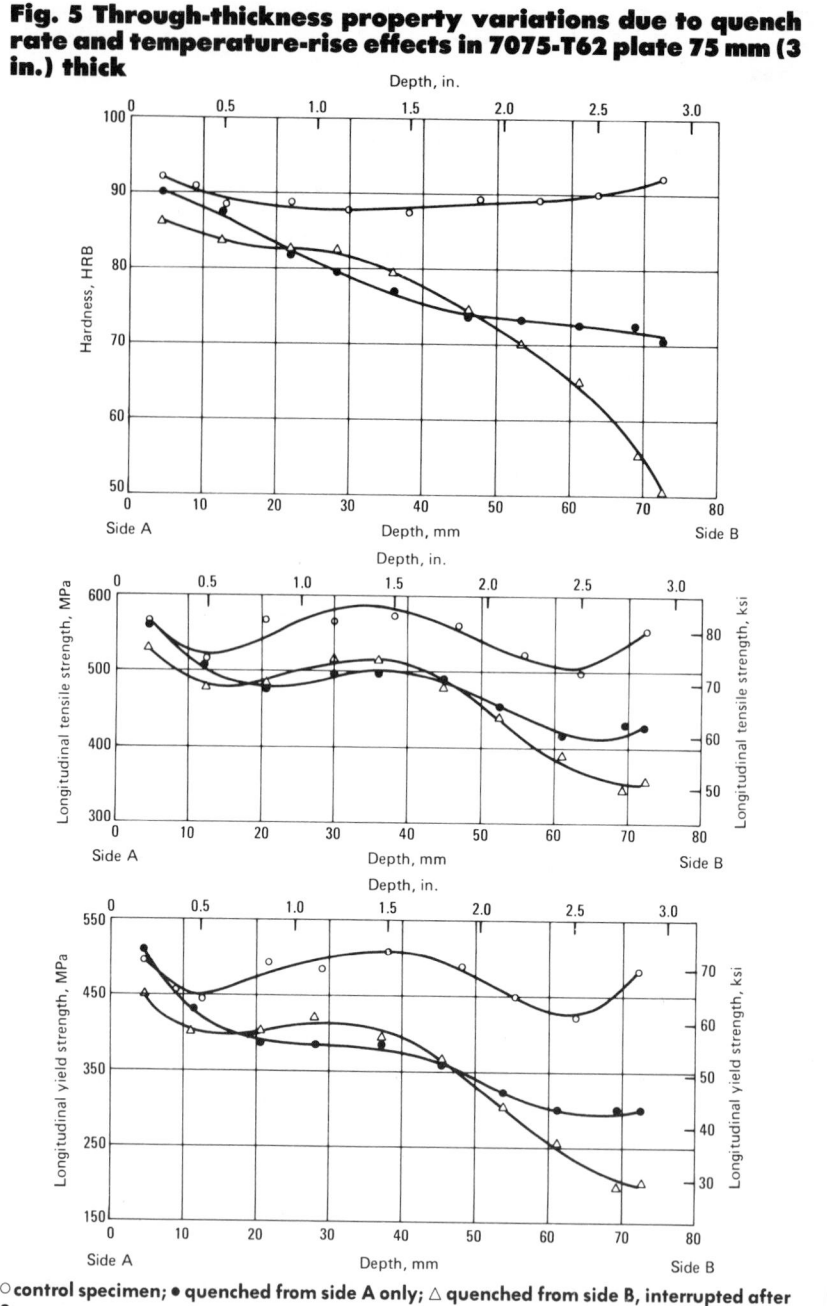

Fig. 5 Through-thickness property variations due to quench rate and temperature-rise effects in 7075-T62 plate 75 mm (3 in.) thick

○ control specimen; ● quenched from side A only; △ quenched from side B, interrupted after 3 s

Fig. 6 Method of determining quench factor, τ, using a cooling curve and a C-curve

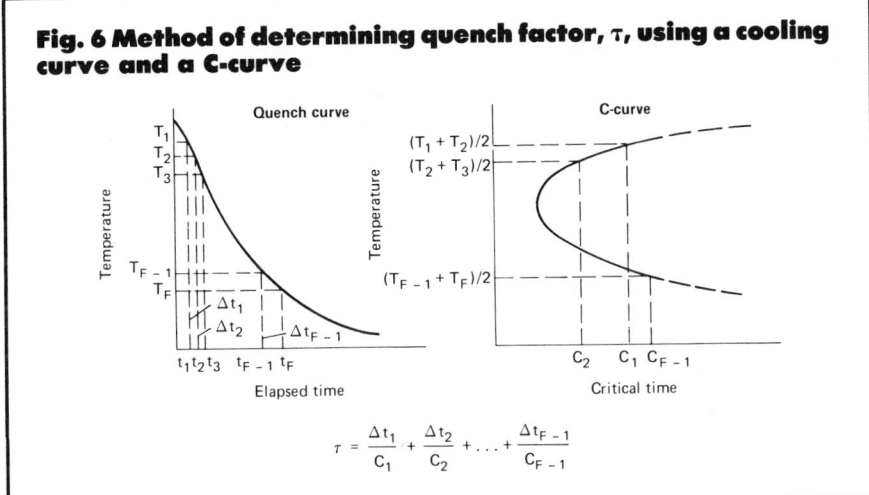

$$\tau = \frac{\Delta t_1}{C_1} + \frac{\Delta t_2}{C_2} + \ldots + \frac{\Delta t_{F-1}}{C_{F-1}}$$

Fig. 7 C-curve indicating type of corrosion attack on 2024-T4 sheet

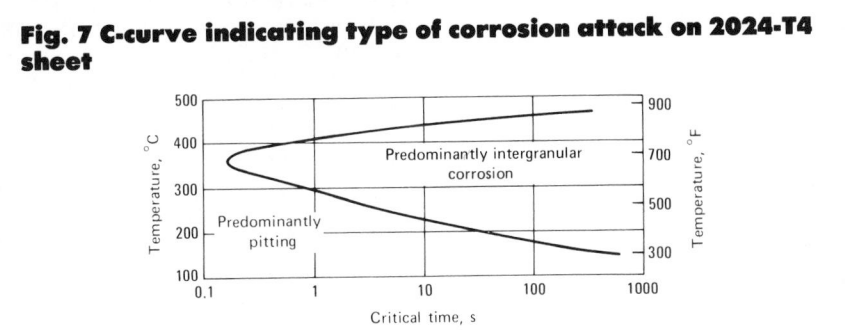

$$\tau = \int \frac{dt}{C_t}$$

where t is time and C_t is critical time as a function of temperature; the locus of critical times is the C-curve. When $\tau = 1$, the fraction transformed, ζ, equals the fraction transformed value designated by the C-curve. This integral can be graphically integrated using the method illustrated in Fig. 6 and discussed below.

Predicting Corrosion Behavior. Alloy 2024-T4, for example, is susceptible to intergranular corrosion when a critical amount of solute is precipitated during quenching, but will corrode in the less severe pitting mode when lesser amounts are precipitated. For predicting the effects of proposed quenching conditions on the corrosion characteristics of 2024-T4, the postulated quench curve is drawn and the quench factor is calculated using the C-curve in Fig. 7. Corrosion characteristics are predicted from the plot in Fig. 8. When the quench factor, τ, is less

than 1.0, continuously quenched 2024-T4 will corrode by pitting.

These relationships are applied to studies of effects of proposed changes in quench practice on design of new quenching systems. For example, consider that the goal of a proposed quenching system for 2024-T4 sheet products is to minimize warpage while preventing susceptibility to intergranular corrosion. Warpage occurs when the stresses imposed by temperature differences across the parts exceed the flow stress. As quenching rate decreases, the tendency for large differences in temperature to occur decreases but the tendency for intergranular corrosion to occur increases.

The C-curve in Fig. 7 indicates that quenching rate can be decreased near the solution heat treating temperature and near room temperature without greatly sacrificing corrosion characteristics, but this information does not provide a quantitative answer. Simple calculations, however, can reveal a

multitude of hypothetical cooling curves that provide slow quenching during a large portion of the quench cycle but sufficiently rapid quenching where critical times are short so that desirable corrosion characteristics are obtained.

As an example, one-, two- and three-step quench curves that would ensure acceptable corrosion behavior in 2024-T4 sheet (quench factor, 0.99) were calculated. Some of these curves are plotted in Fig. 9. This illustration shows that 2024 can be quenched at a rate of 470 °C/s (850 °F/s) or higher and still develop acceptable corrosion characteristics if the quenching rate is linear from the solution temperature to 150 °C (300 °F). If sheet 3.2 mm (0.125 in.) thick is air-blast quenched (rate of heat removal, 5.68 W/m²·°C) to 395 °C (740 °F), however, the quenching rate from 395 to 150 °C must be at least 945 °C/s (1700 °F/s) to maintain the acceptable corrosion behavior. It may also be air-blast quenched to 395 °C, spray quenched at 3300 °C/s (6000 °F/s) to 250 °C (480 °F), then air-blast quenched to 150 °C.

Other curves could be drawn, of course, but the important points are that air-blast quenching cannot be continued to more than a few degrees below 395 °C (740 °F) and cannot be initiated at more than a few degrees above 270 °C (520 °F) even if infinite quenching rates are attained from 395 to 270 °C.

Predicting yield strength is more complex than predicting corrosion behavior, and requires some knowledge of the relationship between extent of precipitation and loss in ability to develop property. Because attainable strength of precipitation-hardening aluminum alloys is a function of the amount of solute remaining in solid solution after quenching, relationships between strength, σ, attainable after continuous cooling, and quench factor, τ, can be expressed as follows:

$$\sigma = \sigma_{max} \exp(k_1\tau) \qquad \text{(Eq 2)}$$

where σ_{max} is the strength attainable with an infinite quenching rate, k_1 is 0.005013, and:

$$\tau = \frac{dt}{C_{t99.5}}$$

where t is time and $C_{t99.5}$ is the C-curve for $\sigma_{99.5}$—that is, critical time as a function of temperature to reduce attainable strength to 99.5% of σ_{max}.

The advantage of predicting yield

Fig. 8 Type and depth of attack on 2024-T4 sheet vs quench factor

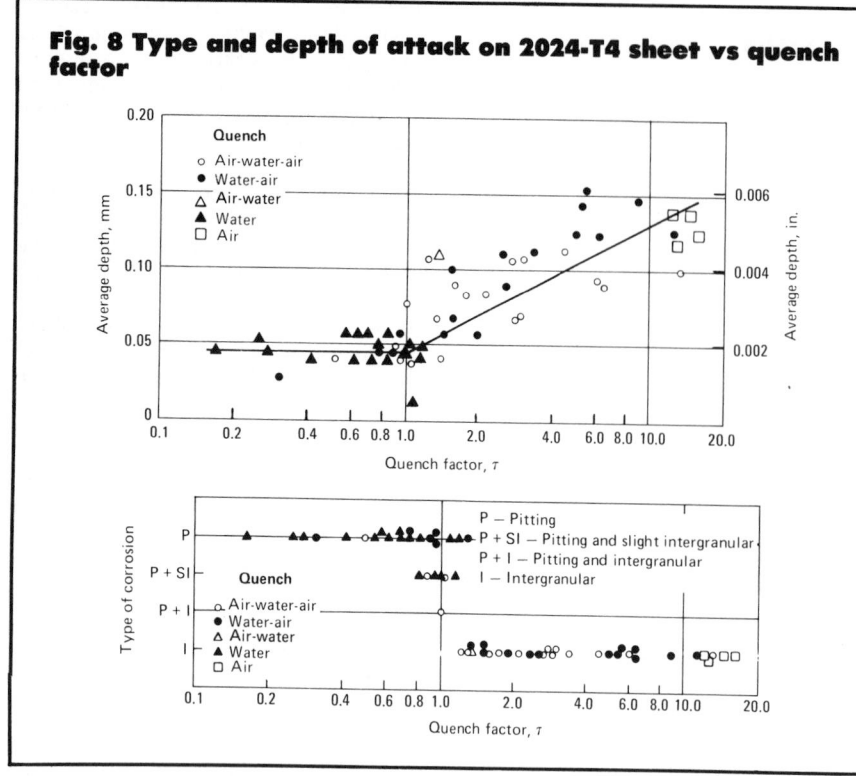

Fig. 9 Quench curves for 2024-T4 sheet, to eliminate susceptibility to intergranular corrosion

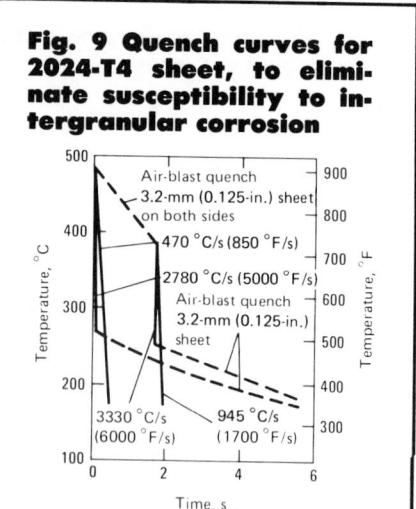

Fig. 10 Cooling curves for 7075-T6 sheet

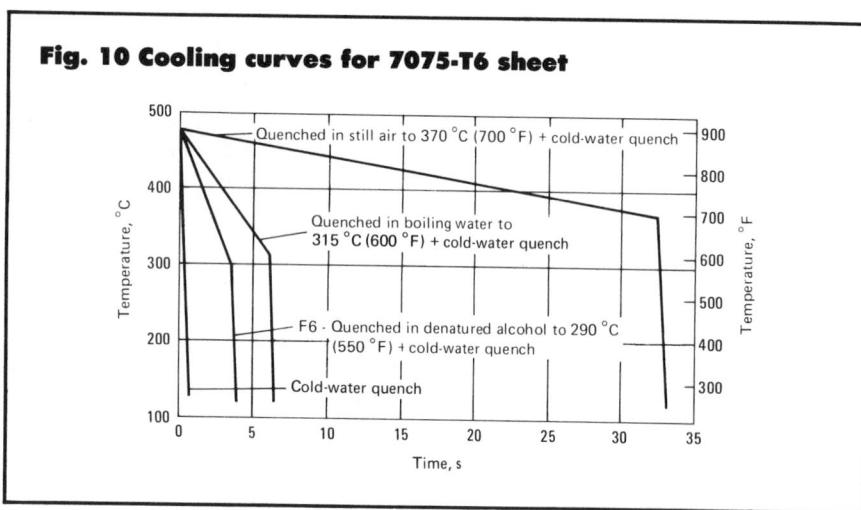

strength from quench factor instead of from average quenching rate is illustrated by the following comparison. Four specimens of alloy 7075-T6 quenched by various means (see Fig. 10) were selected. Yield strengths were predicted both from average quenching rate between 400 and 290 °C (750 and 550 °F) and from quench factor. Quench factor was calculated using the C-curve for 99.5% maximum yield strength for 7075-T6 (Fig. 11), and yield strength was estimated from the above equation defining the quench factor, τ (see Fig. 12).

A comparison of predicted yield strength with actual yield strength is given in Table 4. Yield strengths predicted from quench factor agree very well with measured yield strengths for all specimens, the maximum error being 19.3 MPa (2.8 ksi). Yield strengths predicted from average quenching rates, however, differ from measured values by as much as 226 MPa (32.8 ksi).

The advantage of using the quench factor for predicting yield strength from cooling curves is apparent. Cooling curves that have long holding times

either above or below the critical temperature range from 400 to 290 °C (750 to 550 °F) cannot be used to predict yield strength from average quenching rate. In such instances, prediction of yield strength on the basis of quench factor is particularly advantageous.

An underlying assumption of both quench-factor analysis and average-cooling-rate estimation is that the only effect of temperature is on the kinetics of precipitation. Recent evidence indicates that this assumption is not valid, however, when portions of the metal are quenched locally but reheated significantly before quenching is complete.

Quenching To Minimize Residual Stress and Warpage. Although cold-water immersion or flushing is most common, because it produces the most effective quench (and has been required by MIL-H-6088 for 2014, 2017, 2024, 2117, 7075 and 7178 alloys except forgings), it presents problems involving residual stress and warpage.

Residual stresses in heavy sections of aluminum alloys originate from differential thermal expansion during quenching—that is, the still-warm central material contracts, pulling in the already cooled outer shell. The magnitude of stresses increases with section size, as shown in Fig. 13.

The distribution pattern of residual stresses in as-quenched parts (compression in the outer layers and tension in the central portion) is usually desirable in service. Compressive stresses inhibit failure by fatigue and stress corrosion—two mechanisms that initiate in the outer fibers. Unfortunately, metal-removal operations required after heat

Table 4 Yield-strength values for 7075-T6 sheet predicted from cooling curves using average quench rate and quench factor

Quench	Average quench rate from 400 to 290 °C (750 to 550 °F)		Quench factor, τ	Measured yield strength		Yield strength predicted from average quench rate		Yield strength predicted from quench factor	
	°C/s	°F/s		MPa	ksi	MPa	ksi	MPa	ksi
Cold water 935		1680	0.464	506	73.4	499	72.4	498	72.3
Denatured alcohol to 290 °C (550 °F), then cold water 50		90	8.539	476	69.1	463	67.2	478	69.4
Boiling water to 315 °C (600 °F), then cold water 30		55	15.327	458	66.4	443	64.2	463	67.1
Still air to 370 °C (700 °F), then cold water 5		9	21.334	468	67.9	242	35.1	449	65.1

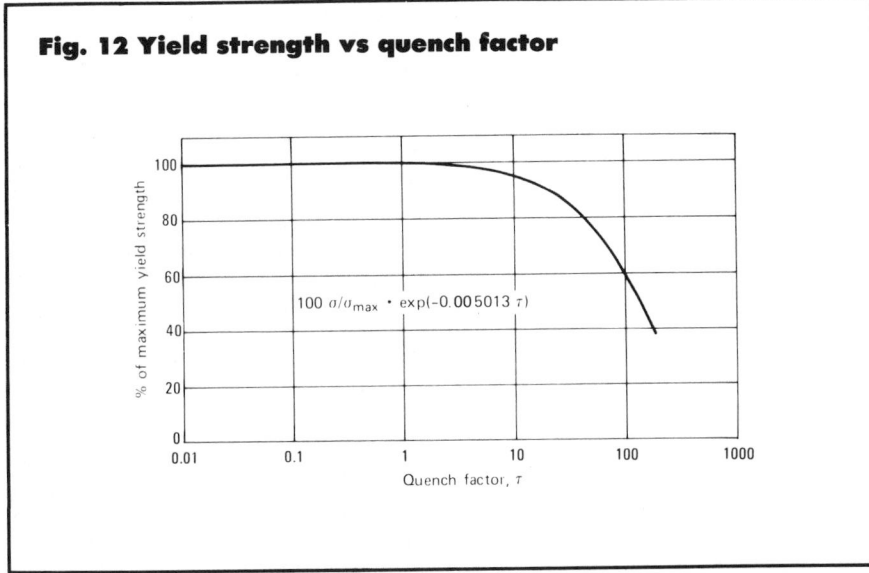

Fig. 11 C-curve for 99.5% maximum yield strength of 7075-T6 sheet

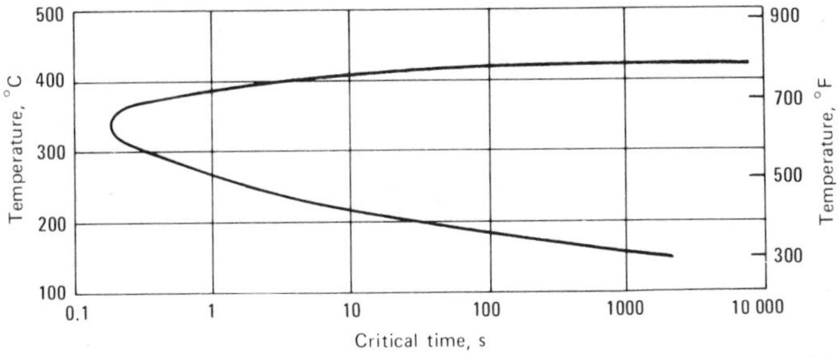

Fig. 12 Yield strength vs quench factor

$$100 \, \sigma/\sigma_{max} \cdot \exp(-0.005013\,\tau)$$

treating often expose material that is stressed in tension. Also, metal-removal operations that are asymmetrical (with respect to residual stresses) cause distortion by redistributing residual stresses. When close-tolerance parts are being fabricated, the resulting warpage can be costly and difficult to correct.

Although service performance is sometimes a factor, the major incentive for reducing residual stress differen-

tials has been a reduction in warpage during machining or an improvement in shape before machining.

One approach to reducing the cooling-rate differential between surface and center is the use of a milder quenching medium—water that is hotter than that normally used or water-glycol solutions. Boiling water, which is the slowest quenching medium used for thick sections, is sometimes employed for quenching wrought products even though it lowers mechanical properties and corrosion resistance. Quenching of castings in boiling water, however, is standard practice, and is reflected in design allowables.

Another approach to the minimization of residual stresses that is generally successful consists of rough machining to within 3.2 mm (0.125 in.) or less of finish dimensions, heat treating and then finish machining. This procedure is intended to reduce the cooling-rate differential between surface and center by reducing thickness; other benefits that accrue if this technique is used to reduce or reverse surface tension stresses in finished parts are improvements in strength, fatigue life, corrosion resistance, and reduced probability of stress-corrosion cracking.

Several factors (especially quenching warpage) sometimes preclude general use of this procedure. The thinner and less symmetrical a section, the more it will warp during quenching, and the residual stresses resulting from straightening of warped parts (plus straightening costs) often are less desirable than the quenching stresses. Holding fixtures and die quenching may be helpful, but precautions must be taken to ensure that they do not retard quenching rates excessively.

Fig. 13 Effect of quenching from 540 °C (1000 °F) on residual stresses in solid cylinders of alloy 6151

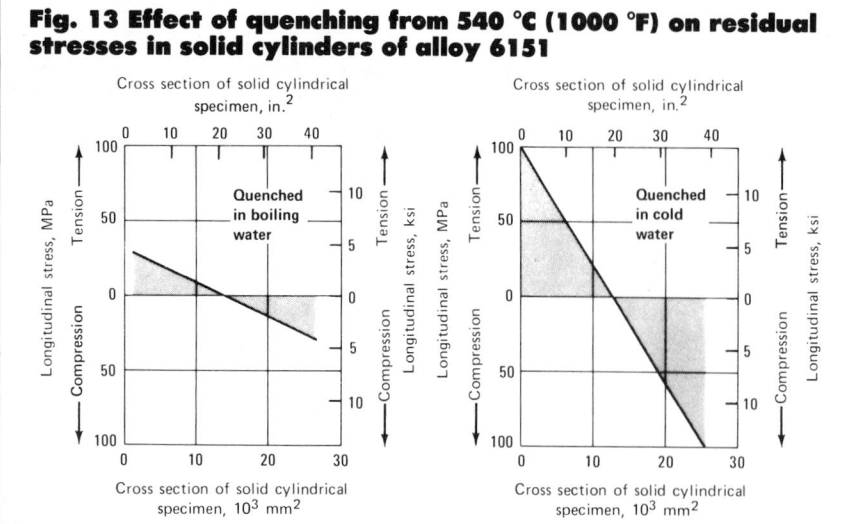

Fig. 14 Effect of quenching medium on strength of 6061-T6 sheet

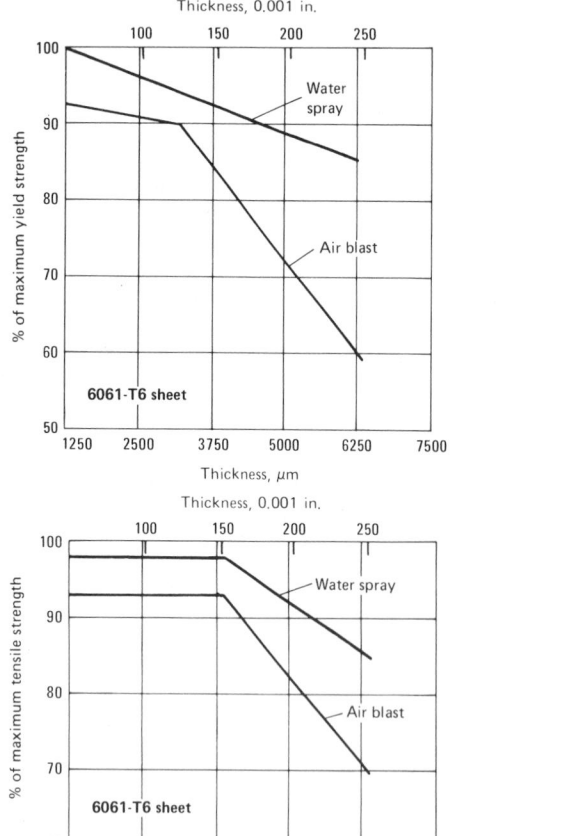

Water-immersion quench equals 100%. Control of coolant flow will minimize decrease in mechanical properties.

Other factors that must be considered are the availability of heat treating facilities and whether or not the advantages of such a manufacturing sequence offset the delay and cost entailed in a double-machining setup.

Warpage of thin sections during quenching is also a problem. Even in the same load, symmetry of cooling usually varies significantly among identical parts and the resultant inconsistent warpage usually requires costly hand straightening. Consequently, a significant amount of effort has been devoted to reducing or eliminating warpage by changing racking positions to achieve symmetry of cooling.

For sheet-metal parts, one manufacturer uses a double screen floor in the quenching rack to reduce the force of initial contact between water and parts. Others allow parts to "free fall" from rack to quench tank. Spacing and positioning on the rack are carefully controlled so that parts will enter the water with minimum impact. With this technique, water turbulence must be avoided, because it will often cause parts to float for a few seconds, greatly reducing their cooling rate.

Because of the difficulties encountered with quenching in cold water, milder quenchants have been employed. Indiscriminate use of milder quenchants can have catastrophic effects; however, when their use is based on sound engineering judgment and a metallurgical knowledge of the effects on the specific alloy, significant cost savings or performance improvements can be realized. The most frequent advantage is the reduction in costly straightening operations and in resultant uncontrolled residual stresses. For example, one aircraft manufacturer utilizes water-spray and air-blast quenching for weldments and complex formed parts made from 6061, an alloy whose corrosion resistance is insensitive to quenching rate. Straightening requirements are negligible and, through careful control of racking and coolant flow, the decrease in mechanical properties is minimized, as shown by the data in Fig. 14. Another development for reducing straightening costs is quenching in water-polymer solutions. Quenching of formed sheet-metal parts in aqueous solutions of polyalkylene glycol or in similar inversely soluble media has significantly reduced the cost of straightening these parts after quenching. The Aerospace Metals Engineering Committee has recommend-

Fig. 15 Effects of flattening on fatigue characteristics of alclad 2024-T4 sheet

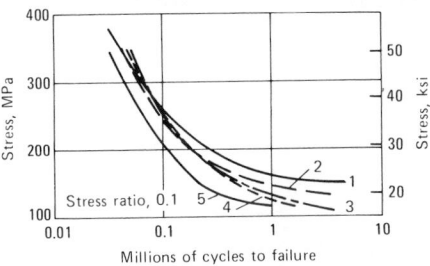

Sheet 1.02 mm (0.040 in.) thick was annealed, solution heat treated and quenched, and then fatigue tested. The sheet represented by curve 1 was not bent. All other sheet was bent 90° in the annealed condition. Flattening (unbending) was done in either the annealed condition (curve 2) or the solution heat treated and quenched condition (curves 3, 4 and 5). Details of bending and flattening were as follows: (1) Not bent. (2) Bend radius, 3.18 mm (1/8 in.); flattened in annealed condition. (3) Bend radius, 3.18 mm (1/8 in.); flattened in quenched condition after 3 days of storage at −18 to −12 °C (0 to 10 °F). (4) Bend radius, 3.18 mm (1/8 in.); flattened in quenched condition after 14 days of storage at −18 to −12 °C (0 to 10 °F). (5) Bend radius, 1.59 mm (1/16 in.); flattened in quenched condition after 3 days of storage at −18 to −12 °C (0 to 10 °F).

Table 5 Limits for quenching in glycol-water solutions

Data are for wrought aluminum alloy products other than forgings.

Glycol concentration, vol %	Alloys	Maximum thickness	
		mm	in.
12 to 16	2014, 2017, 2117, 2024, 2219	2.03	0.080
	7075, 7175	25.4	1.000
17 to 22	2014, 2017, 2117, 2024, 2219	1.80	0.071
	7075, 7079, 7175, 7178, 6061	12.7	0.500
23 to 28	2014, 2017, 2117, 2024, 2219	1.60	0.063
	7075, 7079, 7175, 7178, 6061	9.53	0.375
29 to 34	2014, 2017, 2117, 2024, 2219	1.02	0.040
	7075, 7079, 7175, 7178, 6061	6.35	0.250
35 to 40	7075, 7079, 7175, 7178, 6061	2.03	0.080

ed, for several alloys, maximum thicknesses that can be quenched in solutions of specific concentrations while maintaining acceptable property levels. Typical parameters for quenching wrought products (other than forgings) in glycol-water solutions are presented in Table 5.

Treatments that Precede Precipitation Heat Treating

Immediately after being quenched, most aluminum alloys are nearly as ductile as they are in the annealed condition. Consequently, it is often advantageous to form or straighten parts in this temper. Moreover, at the mill level, controlled mechanical deformation is the most common method of reducing residual quenching stresses. Because precipitation hardening will occur at room temperature, forming or straightening usually follows as soon after quenching as possible. In addition, maximum effectiveness in stress relief is obtained by working the metal immediately after quenching.

In some alloys, notably those of the 2xxx series, cold working of freshly quenched material greatly increases its response to later precipitation heat treatment. Mills take advantage of this phenomenon by applying a controlled amount of rolling (sheet and plate) or stretching (extrusion, bar and plate) to produce higher mechanical properties. However, if the higher properties are used in design, reheat treatment must be avoided.

Forming and Straightening. These operations vary in degree from minor corrections of warpage to complete forming of complex parts from solution-treated flat blanks. Particular value is gained when enough forming can be done at this stage of processing to eliminate the distortion caused by quenching. However, production operations must be adjusted so that most of the plastic deformation is accomplished before an appreciable amount of precipitation hardening takes place.

Although the most severe forming operations may have to be arranged to avoid natural aging, it often is desirable to allow some natural aging to occur and thus avoid formation of Lüders lines. This condition of nonuniform deformation is most likely to occur shortly after quenching and diminishes significantly after a few hours of natural aging. Complete freedom from Lüders lines, however, may require one or two days of natural aging prior to forming. Thus, the forming operation may have to be timed so as to obtain the most appropriate trade-off of these characteristics for the specific parts involved. Lüders lines also can be reduced by employing low strain rates or by forming at temperatures of 150 to 175 °C (300 to 350 °F).

Residual stresses in sheet-metal parts formed in the quenched condition are higher than those in parts formed in the annealed condition. Consequently, forming in the quenched condition should be selected judiciously for parts that are critical in fatigue (Fig. 15) or stress corrosion.

Re-solution heat treatment of parts formed after quenching often causes excessive grain growth in critically strained regions and thus is not recommended.

Mechanical Stress Relief. Deformation consists of stretching (bar, extrusions and plate) or compressing (forgings) the product sufficiently to achieve a small but controlled amount (1 to 3%) of plastic deformation. If the benefits of mechanical stress relieving are needed, the user should refrain from reheat treating.

Figure 16 illustrates the beneficial effect of 3% permanent deformation in compression on a large forging.

These methods are most readily adaptable to mill and forge shop products and require equipment of greater capacity than that found in most manufacturing plants. Application of these methods to die forgings and extrusions usually requires construction of special dies and jaws. Stretching generally is limited to material of uniform cross section; however, it has been applied successfully to stepped extrusions and to a 3-by-14-m (10-by-47-ft) aircraft wing skin roll-tapered to a thickness range of 7.1 to 3.2 mm (0.280 to 0.125 in.).

Specific combinations of the supplemental digits are used to denote the tempers produced when mechanical deformation is used primarily to relieve residual stresses induced during the quenching operation. For products stress relieved by stretching, the digits 51 follow the basic Tx designation (T451, for example). For products stress relieved by compressive deformation, the supplementary digits are 52.

Fig. 16 Effect of 3% permanent deformation in compression (T652 treatment) on distribution of stress in a large forging

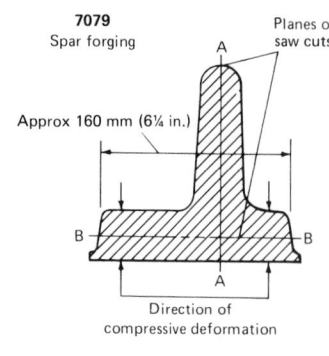

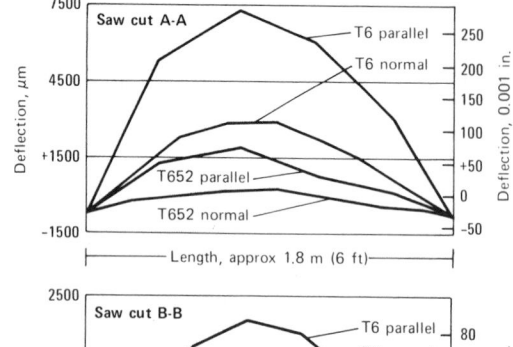

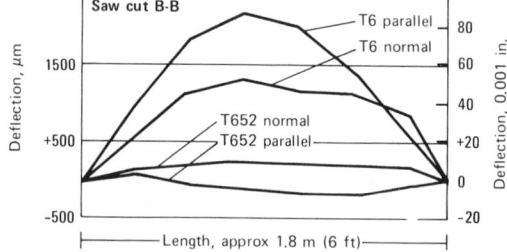

"Parallel" and "normal" refer to warpage directions with respect to the plane of the saw cut.

Fig. 17 Effectiveness of various uphill quenching treatments in reducing residual quenching stresses in 2014 plate

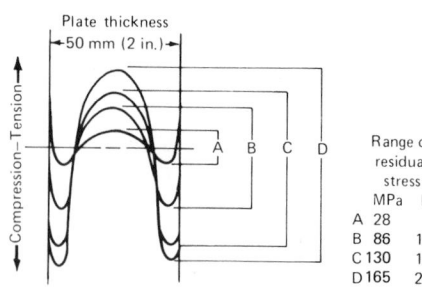

A: cooled to −195 °C (−320 °F), then uphill quenched in a steam blast. B: cooled to −75 °C (−100 °F), then uphill quenched in a steam blast. C: cooled to −75 or −195 °C (−100 or −320 °F), then uphill quenched in boiling water. D: standard specimen, quenched and aged to T6 temper in conventional manner with no further treatment. Note: uphill quenching treatments (single cycle only) were applied from ½ to 1½ h after quenching from the recommended solution-treating temperature. All specimens were aged to the T6 temper after uphill quenching.

An additional digit is added to designations for extrusions: an added zero specifies that the product has not been straightened after final stretching; an added one indicates that straightening may have been performed after final stretching.

Thermal Treatments for Relieving Stresses. Numerous attempts have been made to develop a thermal treatment that will remove, or appreciably reduce, quenching stresses. Normal precipitation heat treating temperatures are generally too low to provide appreciable stress relief. Exposure to higher temperatures (at which stresses are relieved more effectively) results in lower properties. However, such treatments sometimes are utilized when even moderate reduction of residual stress levels is important enough so that some sacrifice in mechanical properties can be accepted. The T7 temper for castings is a typical example of this kind of treatment and is discussed later.

Other thermal stress-relief treatments, known as "subzero treatment" and "cold stabilization", involve cycling of parts above and below room temperature. The temperatures chosen are those that can be readily obtained with boiling water and mixtures of dry ice and alcohol—namely, +100 and −73 °C (+212 and −100 °F)—and the number of cycles ranges from one to five. The maximum reduction in residual stress that can be effected by these techniques is about 25%. The maximum effect can be obtained only if the subzero step is performed first, and immediately after quenching from the solution-treating temperature while yield strength is low. No benefit is gained from more than one cycle.

A 25% reduction in residual stress is sometimes sufficient to permit fabrication of a part that could not be made without this reduction. However, if a general reduction is needed, as much as 83% relief of residual stress is possible by increasing the severity of the "uphill quench"—that is, more closely approximating the reverse of the cooling-rate differential during the original quench. This may be accomplished by a patented process that involves extending the subzero step to −195 °C (−320 °F) and then very rapidly "uphill quenching" in a blast of live steam (Fig. 17). The rate of reheating is extremely critical, and therefore, to ensure proper application of the steam blast, a special fixture usually is required for each part.

This process will not solve all problems of warpage in machining. It may reduce warpage internally but increase warpage of the extreme outer layers, although in the opposite direction (Fig. 18). Also, the effect of the altered residual-stress pattern on performance

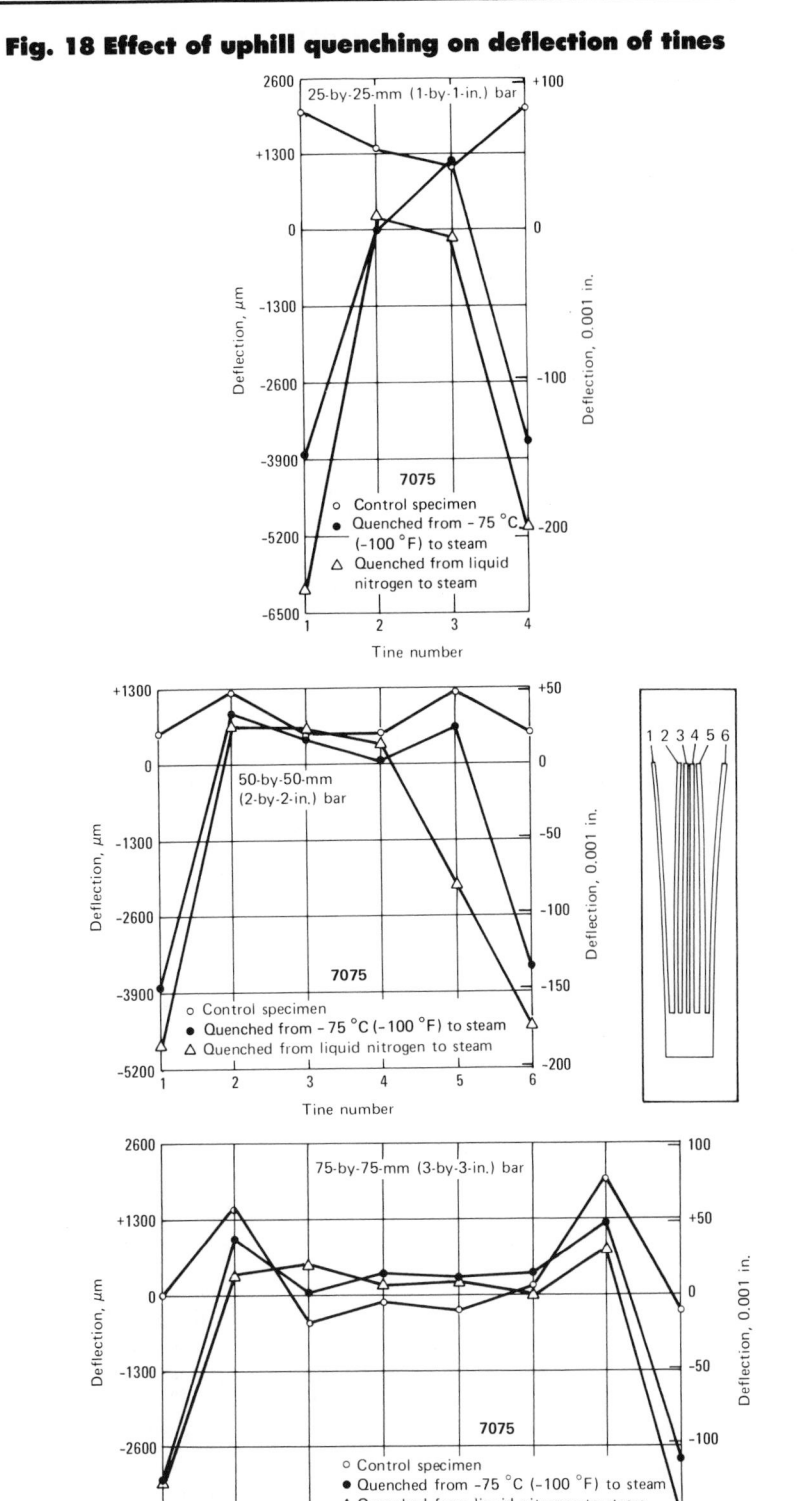

Fig. 18 Effect of uphill quenching on deflection of tines

Six-tine specimen shown was machined from 50-by-50-mm (2-by-2-in.) bar. Similar specimens machined from 25-by-25-mm (1-by-1-in.) and 75-by-75-mm (3-by-3-in.) bars had four and eight tines, respectively.

must be evaluated carefully for each part. This is particularly important for parts subjected to cyclic loading or exposed to corrosive environments such as marine atmospheres, especially if the process is introduced after the start of production and original performance tests are not repeated. Further disadvantages are the cost and hazard involved in handling liquid nitrogen and live steam.

Natural Aging. The more highly alloyed members of the 6xxx wrought series, the copper-containing alloys of the 7xxx group, and all of the 2xxx alloys are almost always solution heat treated and quenched. For some of these alloys—particularly the 2xxx alloys—the precipitation hardening that results from natural aging alone produces useful tempers (T3 and T4 types) that are characterized by high ratios of tensile to yield strength and high fracture toughness and resistance to fatigue. For the alloys that are used in these tempers, the relatively high supersaturation of atoms and vacancies retained by rapid quenching causes rapid formation of GP zones, and strength increases rapidly, attaining nearly maximum stable values in four or five days. Tensile-property specifications for products in T3- and T4-type tempers are based on a nominal natural aging time of four days. In alloys for which T3- or T4-type tempers are standard, the changes that occur on further natural aging are of relatively minor magnitude, and products of these combinations of alloy and temper are regarded as essentially stable after about one week.

In contrast to the relatively stable condition reached in a few days by 2xxx alloys that are used in T3- or T4-type tempers, the 6xxx alloys and to an even greater degree the 7xxx alloys are considerably less stable at room temperature and continue to exhibit significant changes in mechanical properties for many years. The differences in rate and duration of changes in tensile yield strength of representative alloys of the three types are illustrated in Fig. 19. Because of the relative instability of the 7xxx alloys, the naturally aged temper (after solution heat treatment and quenching) is designated by the suffix letter W. For specific description of this condition, the time of natural aging should be included (example: 7075-W, 1 month).

Aging characteristics vary from alloy to alloy with respect to both time to ini-

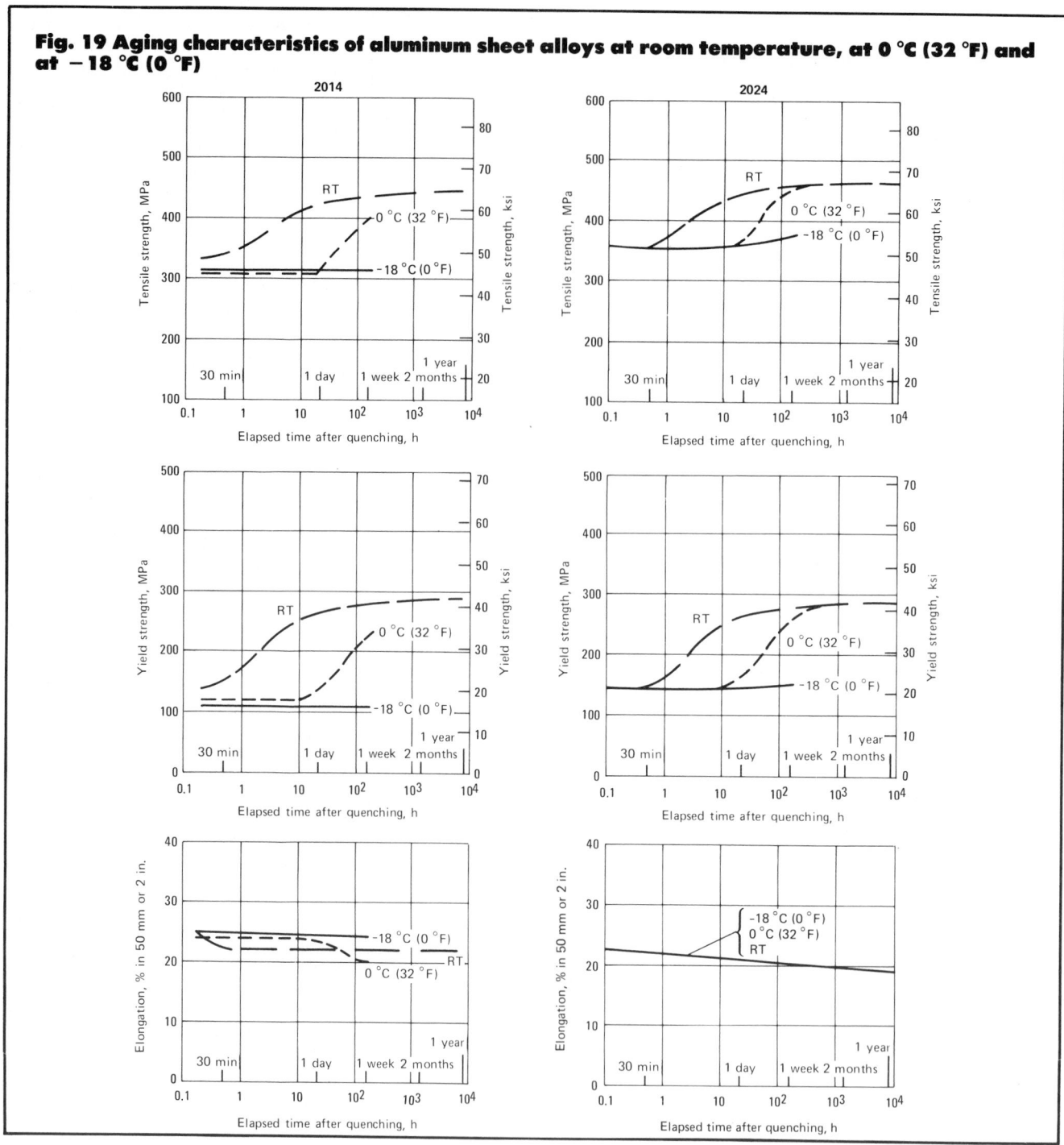

Fig. 19 Aging characteristics of aluminum sheet alloys at room temperature, at 0 °C (32 °F) and at −18 °C (0 °F)

tial change in mechanical properties and rate of change, but aging effects always are lessened by reductions in aging temperature (see Fig. 19). With some alloys, aging can be suppressed or delayed for several days by holding at a temperature of −18 °C (0 °F) or lower. It is usual practice to complete forming and straightening before aging changes mechanical properties appre-

ciably. When scheduling makes this impractical, aging may be avoided in some alloys by refrigerating prior to forming. It is conventional practice to refrigerate alloy 2024-T4 rivets to maintain good driving characteristics. Full-size wing plates for current-generation jet aircraft have been solution heat treated and quenched at the primary fabricating mill, packed in dry

ice in specially designed insulated shipping containers and transported by rail about 2000 miles to the aircraft manufacturer's plant for forming.

Unanticipated difficulties may arise as a result of failure to control refrigerator or part temperature closely enough. If opening of the cold box to insert or remove parts is done too frequently, the cooling capacity of the

Fig. 19 (continued)

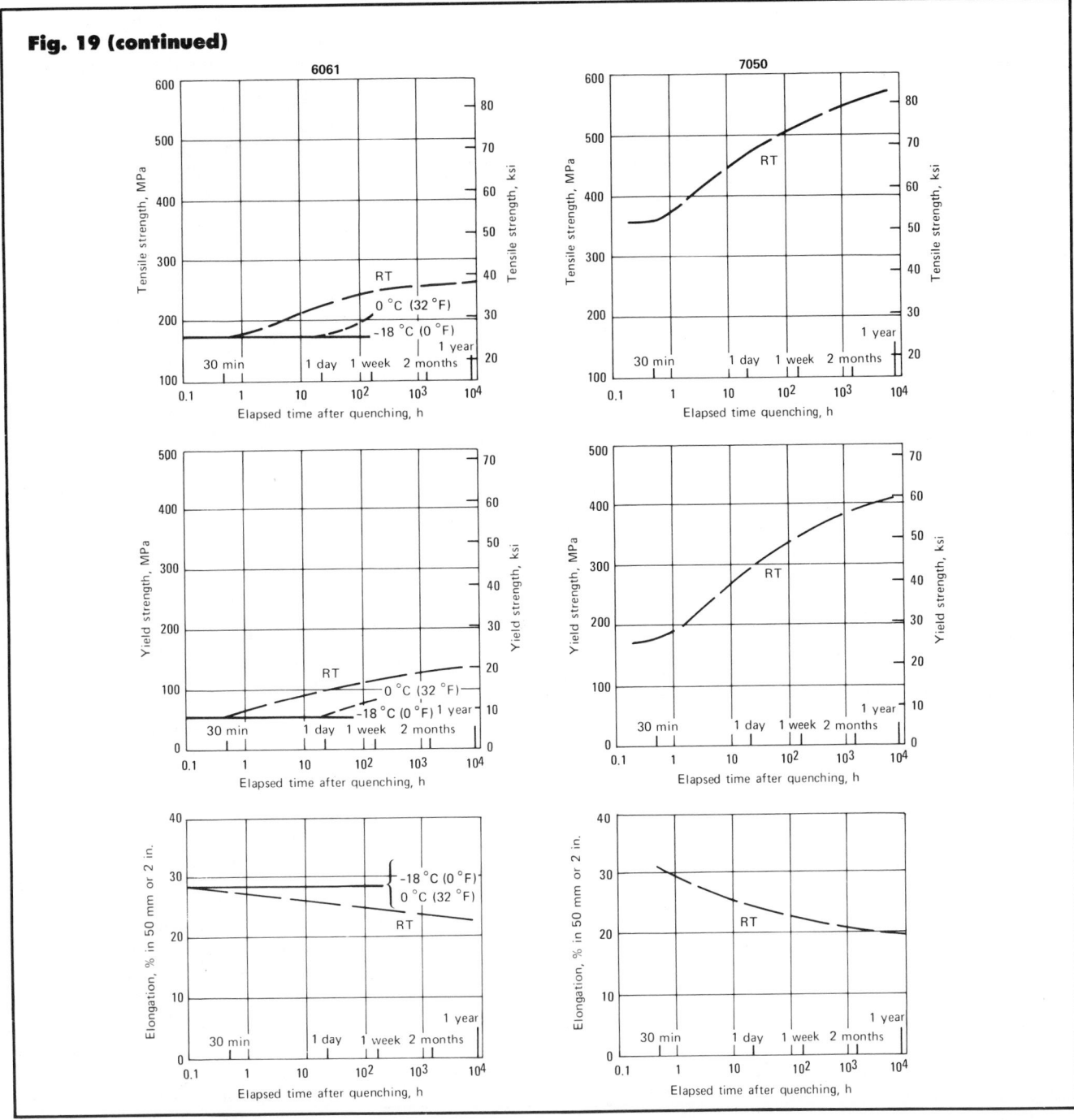

refrigerator may be exceeded. At times, the rate at which heavy-gage parts can be cooled in a still-air cold box has been found to be insufficient. This problem has been solved in one plant by immersing parts in a solvent at −40 °C (−40 °F) before placing them in the refrigerator.

Cold Work Strain Hardening. The T3-type tempers are distinguished from T4-type tempers by significant mechanical-property differences re-

sulting from cold work strain hardening associated with certain mechanical operations performed after quenching. Roller or stretcher leveling to achieve flatness or straightness introduces modest strains (on the order of 1 to 4%) that cause changes in mechanical properties (primarily, increases in strength). Further increases in strength can be obtained by cold rolling, additional stretching, combinations of these operations or, for products such as hand

forgings, compressive deformation. The tempers produced by these operations followed by natural aging alone (no precipitation heat treatment) are classified as T3-type tempers, and an additional digit is used to indicate a variation in strain hardening that results in significant changes in properties. In the most recently introduced 2xxx aircraft alloy, 2324, high strength is achieved by cold rolling plate to a T39 temper.

Fig. 19 (continued)

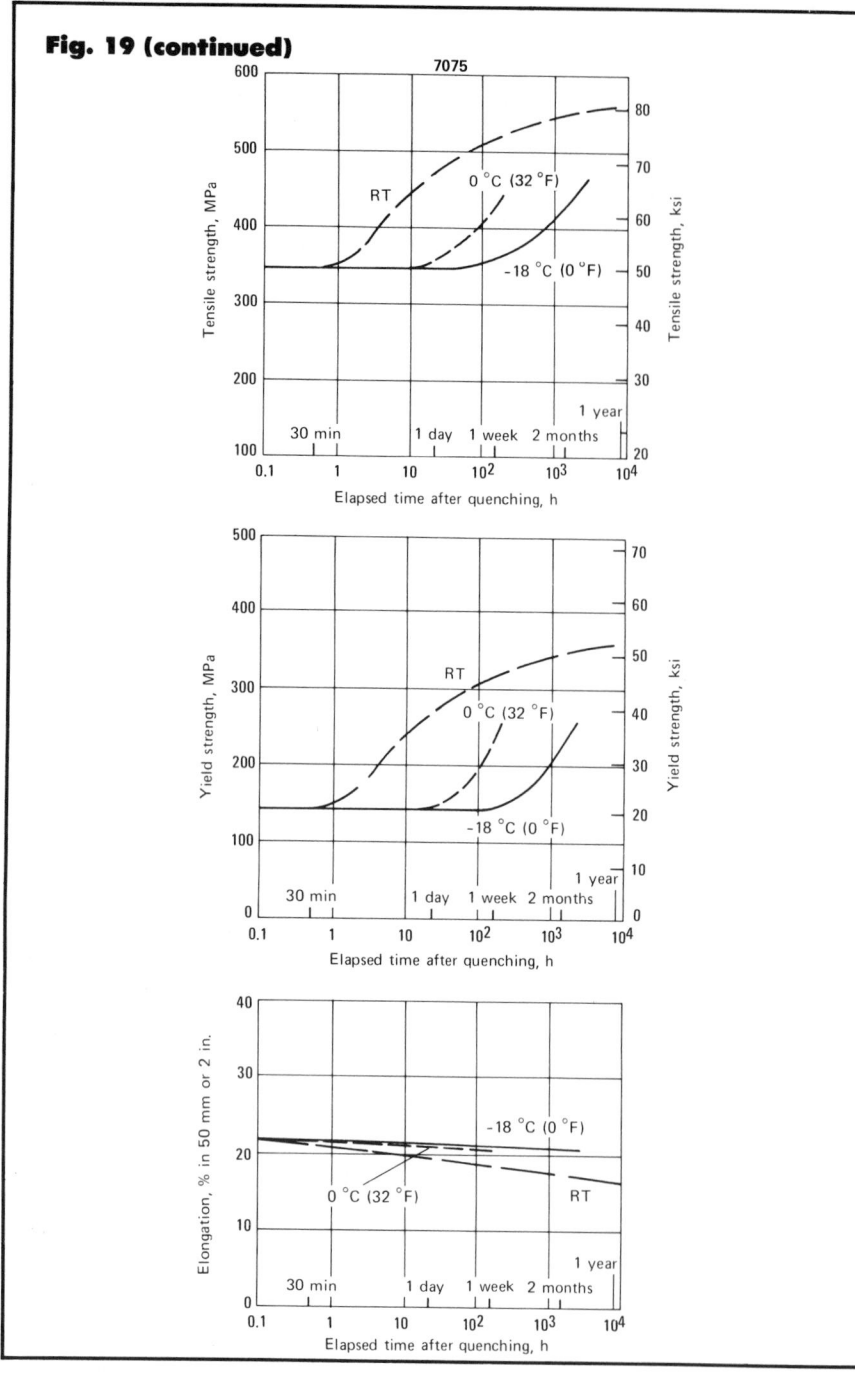

Precipitation Heat Treatment.
Production of material in T5- through T10-type tempers (see the section on temper designations near the end of this article) necessitates precipitation heat treating at elevated temperatures (artificial aging). Although the hardening precipitate developed by this operation is submicroscopic, structures before and after precipitation heat treatment often can be distinguished by etching metallographic specimens. In aluminum alloys in the solution heat treated and quenched condition, coloration contrast between grains of differing orientation is relatively high, particularly in 2xxx series wrought alloys and 2xx.0 series casting alloys. This contrast is noticeably decreased by precipitation heat treatment.

Differences in type, volume fraction, size and distribution of the precipitated particles govern properties as well as the changes observed with time and temperature, and these are all affected by the initial state of the structure. The initial structure may vary in wrought products from unrecrystallized to recrystallized and may exhibit only modest strain from quenching or additional strain from cold working after solution heat treatment. These conditions, as well as the time and temperature of precipitation heat treatment, affect the final structure and the resulting mechanical properties.

Because mechanical properties and other characteristics change continuously with time and with temperature, as shown in Fig. 20 (a), (b) and (c) by typical curves for three wrought alloys, treatment to produce a combination of properties corresponding to a specific alloy-temper combination requires one or more rather specific and coordinated combinations of time and temperature, with both parameters being subject to practical limitations. Recommended commercial treatments often are compromises between (a) time and cost factors and (b) the probability of obtaining the intended properties, with consideration of allowances for variables such as composition within specified range and temperature variations within the furnace and load. Use of higher temperatures may reduce treatment time; but if the temperature is too high, characteristic features of the precipitation-hardening process reduce the probability of obtaining the required properties.

T6 and T7 Tempers. Precipitation heat treatment following solution heat treatment and quenching produces T6- and T7-type tempers. Alloys in T6-type tempers generally have the highest strengths practical without sacrifice of the other properties and characteristics found by experience to be satisfactory and useful for engineering applications. Alloys in T7 tempers are "overaged", which means that some degree of strength has been sacrificed or "traded off" to improve one or more other characteristics. Strength may be sacrificed to improve dimensional stability, particularly in products intended for service at elevated temperatures, or to lower residual stresses in order to reduce warpage or distortion in machining. T7-type tempers frequently are specified for cast or forged engine parts. Precipitation heat treating temperatures used to produce these tempers generally are higher than those used to

Fig. 20a Aging characteristics of alloy 2014 sheet

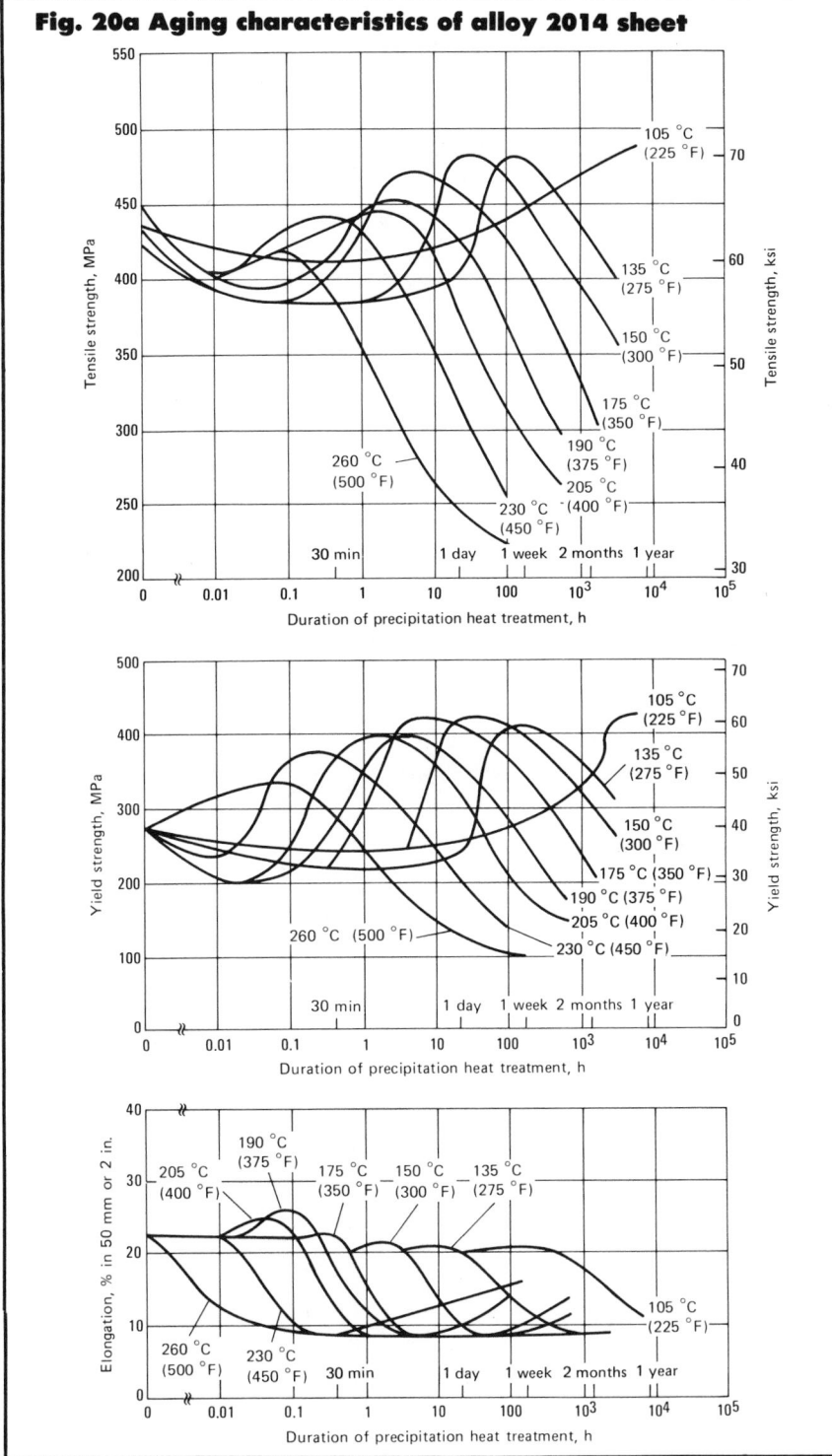

per has greatly minimized stress-corrosion cracking of large and complex machined parts made of these alloys, which occasionally occurred with T6-type tempers. The precipitation heat treatments used to produce the T73- and T76-type tempers consist either of a two-stage isothermal precipitation heat treatment or of heating at a controlled rate to a single treatment temperature. The microstructural/electrochemical relationships that are required in order to achieve the desired corrosion-resisting characteristics can be developed by using only a single-stage precipitation heat treatment above about 150 °C (300 °F), but higher strength is obtained by preceding this with a lower-temperature stage or with a slow, controlled heatup. Extended natural aging can provide the same results, but the times required at room temperature are impractical. Either during the preliminary stage or during slow heatup, a fine, high-density dispersion of GP zones is nucleated. Either the time and temperature of the first step or the rate of heating must be controlled to produce GP zones that will not dissolve but will transform to the η′ precipitate when heated to the aging temperature above 150 °C (300 °F). The aging practice that produces the results in the shortest time depends on the GP-zone solvus temperature. This temperature, in turn, depends on vacancy concentration, a factor influenced by solution heat treating temperature and quench rate, and on composition. If first-step aging time is too short, if first-step aging temperature is too far below the GP-zone solvus, or if heating rates are too high, the GP zones will dissolve above 150 °C (300 °F), and the resultant coarse and widely distributed precipitate will provide lower strength. The T76-type treatments have the same operational sequence but employ second-stage heating only long enough to develop a resistance to exfoliation corrosion higher than that provided by the T6-type tempers. Materials in the T73-type temper also have high resistance to exfoliation corrosion.

Recommended treatments to produce T5- and T6-type tempers, and those of the T7-type employed for dimensional and property stabilization, provide adequate tolerance for normal variations encountered with good operating practices. On the other hand, the T73, T736 and T76 tempers for alloys 7049, 7050, 7075, 7175 and 7475 involve changes in strength that occur significantly more

produce T6-type tempers in the same alloys.

Two important groups of T7-type tempers—the T73 and T76 types—have been developed for the wrought alloys of the 7xxx series, which contain more than about 1.25% copper. These tempers are intended to improve resistance to exfoliation corrosion and stress-corrosion cracking, but as a result of overaging, they also increase fracture toughness and, under some conditions, reduce rates of fatigue-crack propagation. The T73-type tem-

rapidly at the temperatures employed in the second stage of the T7x precipitation heat treatment cycle compared to the changes occurring at the temperatures employed to produce the T6 temper.

As illustrated in Fig. 21, variations in soak time of several hours and variations in soak temperature of up to 11 °C (20 °F) from the nominal aging practice of 24 h at 120 °C (250 °F) affect the strength of 7075-T6 by as much as 28 MPa (4 ksi). In contrast, similar variations in second-step soak time and temperature for 7075-T73—that is, variations from 24 h at 165 °C (325 °F)—affect strength by up to 150 MPa (22 ksi).

Consequently, control of both temperature and time to achieve the mechanical properties and corrosion resistance specified for these tempers is more critical than the control required in producing the T6 temper. Moreover, rate of heating from the first to the second aging step must be considered, because precipitation occurs during this period.

Heat treaters attempt to adjust to these new problems by empirically modifying soak times to compensate for precipitation during heating and for effects of soaking at temperatures above or below the nominal. A method has been developed recently, however, that permits quantitative compensation for the effects of precipitation during heating and of soaking either above or below the recommended temperature. For overaging, these effects can be described by the following equation:

$$YS = Y \exp -\left(\frac{t_c}{F_{YS}} + \theta\right) \quad \text{(Eq 3)}$$

where YS is yield strength; Y is a term having units of strength that is dependent on alloy, fabrication and test direction; t_c is time at soak temperature; F_{YS} is a temperature-dependent term; and

$$\theta = \frac{dt}{F_{YS}}$$

where t is time during heating.

Equation 3 provides the basis for selection of a nominal aging time that will result in the desired yield strength and gives the furnace operator a method of compensating for heating rate and for differences between desired and attained soak temperatures.

Specifics will be illustrated using data for alloy 7050. The value of F_{YS} for

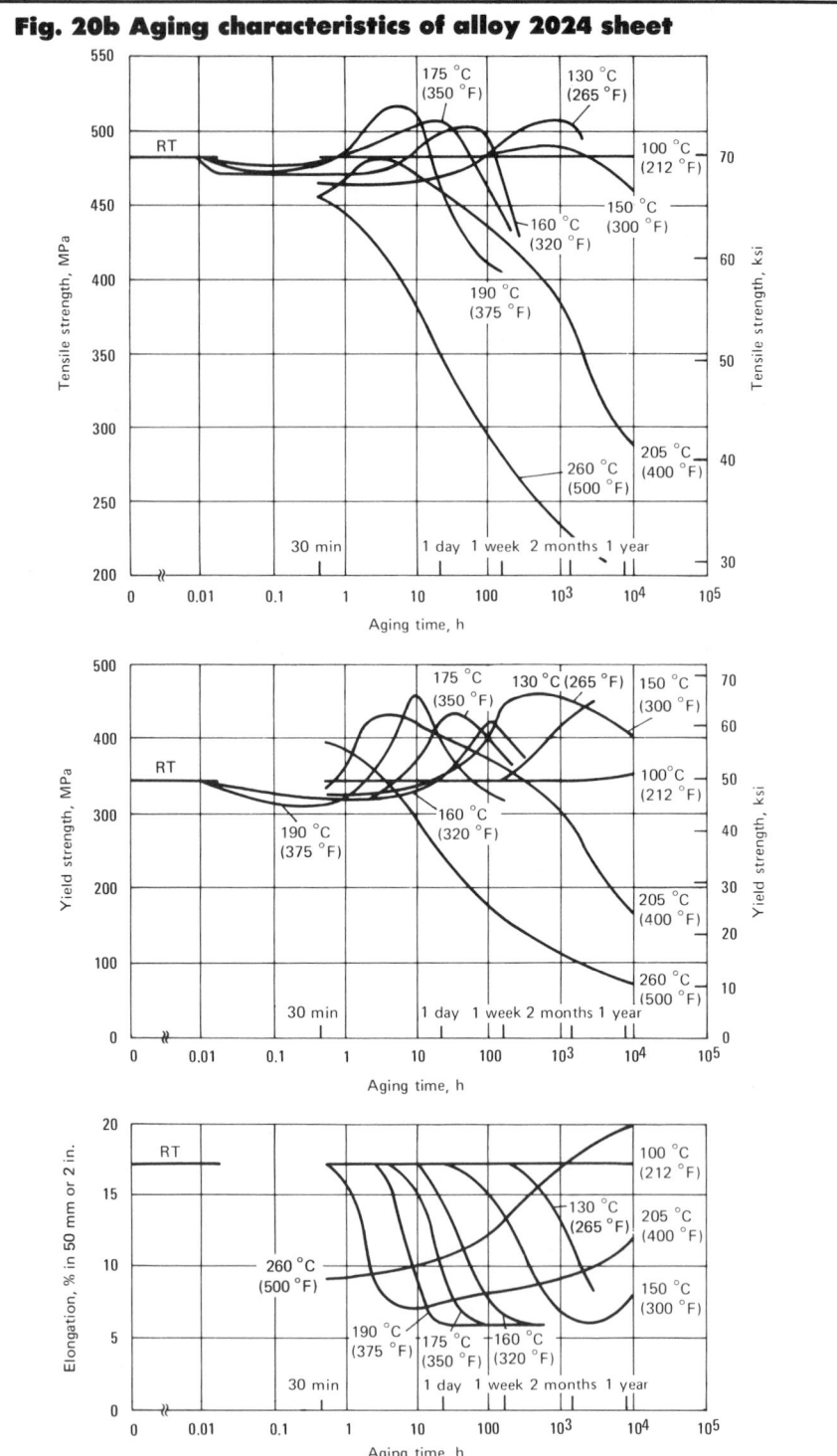

Fig. 20b Aging characteristics of alloy 2024 sheet

7050 can be calculated by use of the following equation:

$$F_{YS} = 1.45 \times 10^{-16} \exp\left(\frac{32\,562}{T_F + 460}\right) h$$

(Eq 4)

where T_F is temperature in °F, or

$$F_{YS} = 1.45 \times 10^{-16} \exp\left(\frac{18\,090}{T_K}\right) h$$

where T_K is temperature in K.

In one experiment, lengths of 7050-

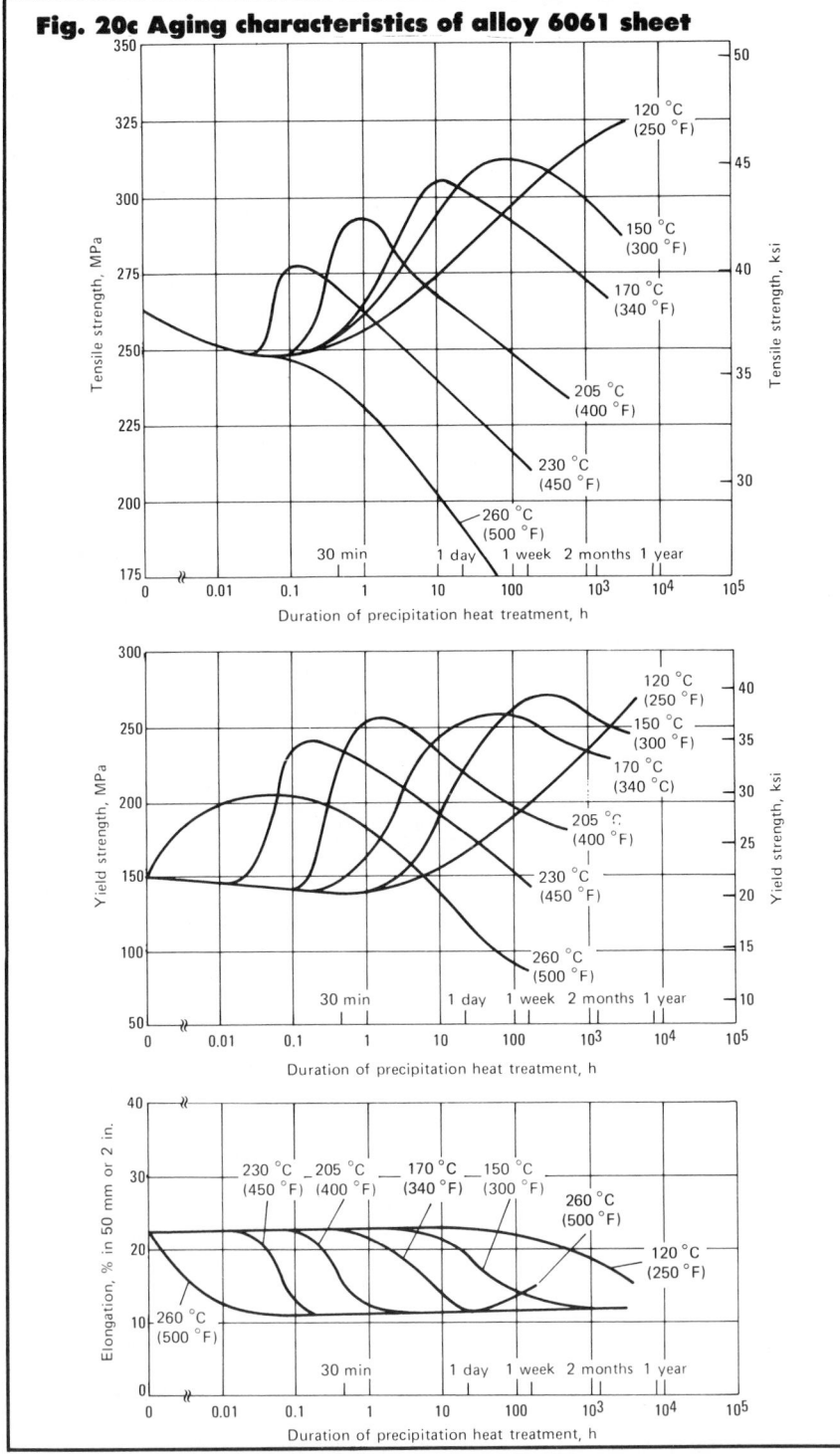

Fig. 20c Aging characteristics of alloy 6061 sheet

sate for soaking at temperatures other than the nominal can be large (Fig. 23). For example, the calculated difference in strength between alloy 7050 extrusions soaked 29 h at 160 °C (320 °F) and at 165 °C (325 °F) is about 50 MPa (about 7 ksi), and the calculated difference in strength between 7050 extrusions soaked 29 h at 155 °C (315 °F) and at 170 °C (335 °F) is about 100 MPa (about 14 ksi).

Neglecting to compensate for time spent heating the work to the soak temperature will increase the variability. Strength loss attributed to heatup was 14 MPa (2 ksi).

These kinetic relationships also can assist in selection of equivalent aging times for alternate second-step aging temperatures. Equations 3 and 4 can be rearranged to yield the following equation:

$$t_{T_2} = t_{T_1} \exp \frac{32\,562}{T_1 + 460} - \frac{32\,562}{T_2 + 460} \qquad \text{(Eq 5)}$$

where t_{T_1} is aging time at temperature T_1, t_{T_2} is aging time at temperature T_2 that will provide equivalent yield strength, and T_1 and T_2 are in °F. For example, the time at 175 °C (350 °F) equivalent to aging alloy 7050 for 29 h at 165 °C (325 °F) is calculated as follows:

$$t_{350} = 29/\exp\,(1.28) = 29/3.6 = 8\text{ h}$$

T8-Type Tempers. In some alloys—particularly certain alloys of the 2*xxx* series—strain introduced by cold working after solution heat treatment and quenching nucleates a finer, denser precipitate dispersion that considerably increases strength, as illustrated for alloy 2024 in Fig. 24. This effect is the basis for the higher-strength T8-type tempers of alloys 2011, 2024, 2124, 2219 and 2419, which are produced by applying controlled amounts of cold rolling, stretching, or combinations of these operations.

Alloys 2024, 2124 and 2219 in T8-type tempers are particularly well suited for supersonic and military aircraft; alloy 2219 in such tempers, and alloy 2014-T65, were the principal materials for the fuel and oxidizer tanks (which also served as the primary structure) of the Saturn V space vehicles. Re-solution heat treatment of mill products supplied in these tempers can result in grain growth and in substantially lower strength than is nor-

W (4 days) extrusions were aged 24 h at 120 °C (250 °F) plus the equivalent of 3 to 42 h at 165 °C (325 °F). For the second step, a logarithmic heatup was used in which 10 h were required for the load to reach 155 °C (315 °F), and nominal soak temperature was 165 °C (325 °F). Figure 22 indicates that yield

strength generally agreed with values predicted using Equation 3. The deviation of the curve for short-transverse strength at the short aging times indicates that the method is inadequate for predicting strength on the underaging side of the aging curve.

The effects of neglecting to compen-

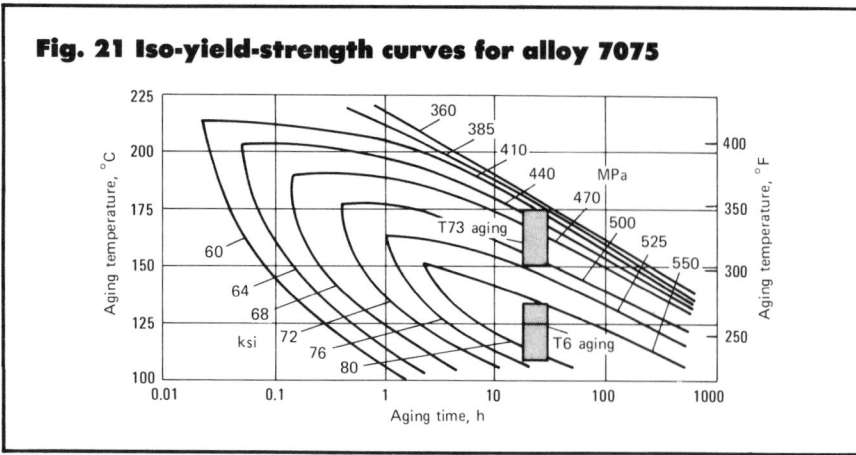

Fig. 21 Iso-yield-strength curves for alloy 7075

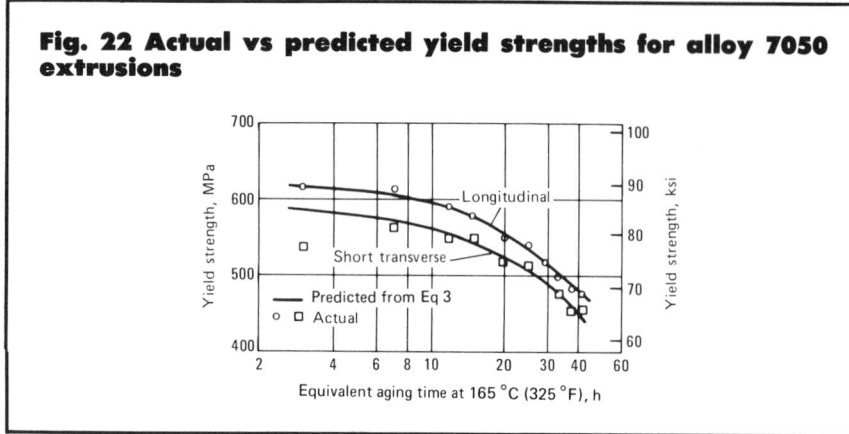

Fig. 22 Actual vs predicted yield strengths for alloy 7050 extrusions

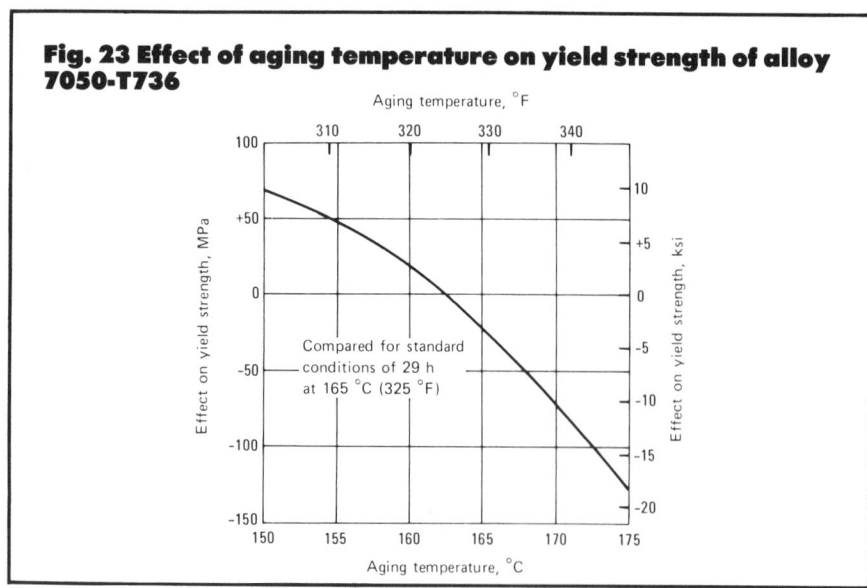

Fig. 23 Effect of aging temperature on yield strength of alloy 7050-T736

of material precipitation heat treated to T6-type tempers. On the other hand, these operations have measurable detrimental effects on final strength when T73-, T736- or T76-type tempers are produced, particularly in the direction opposite the direction of cold work. Accordingly, specification properties are somewhat lower for the stress-relieved versions of these tempers. Decreasing the overaging time to compensate for the loss in strength is not advisable, because this would impair development of the desired corrosion characteristics.

Temperature control and uniformity present essentially the same problems in precipitation heat treating as they do in solution heat treating.

Good temperature control and uniformity throughout the furnace and load is required for all precipitation heat treating. Recommended temperatures are generally those that are least critical and that can be used with practical time cycles. Except for 7xxx alloys in T7x tempers, these temperatures generally allow some latitude and should have a high probability of meeting property specification requirements. Furnace radiation effects seldom are troublesome except in those few furnaces that are used for both solution and precipitation heat treating. Generally, such situations should be avoided, because the high heat capacity needed for the higher temperatures may be difficult to control at normal aging temperatures.

Soak time in precipitation heat treating is not difficult to control; the specified times carry rather broad tolerances. Heavier loads with parts racked closer together, and even nested, are not abnormal. The principal hazard is undersoaking due to gross excesses in loading practices. Some regions of the load may reach soak temperature long after soak time has been called. Placement of load thermocouples is critical, and limiting the size and spacing of a load may be necessary for aging to the T73 and T76 tempers. As discussed above, soak time is not as critical for peak aged (T6 and T8) tempers.

Hardening of Cast Alloys

In hardening of cast alloys, selection of a specific cycle that will produce the optimum combination of properties in all configurations is not possible. Foundry practice (chills, gating, type of

mal for the original temper. Such reheat treatment is not recommended.

Alloys of the 7xxx series do not respond favorably to the sequence of operations used to produce T8-type tempers, and no such tempers are standard for these alloys. The strains associated with stretching or compressive stress 999rel of 7xxx alloys have relatively little effect on the mechanical properties

Fig. 24 Effects of cold work after quenching and before aging on tensile properties of alloy 2024 sheet

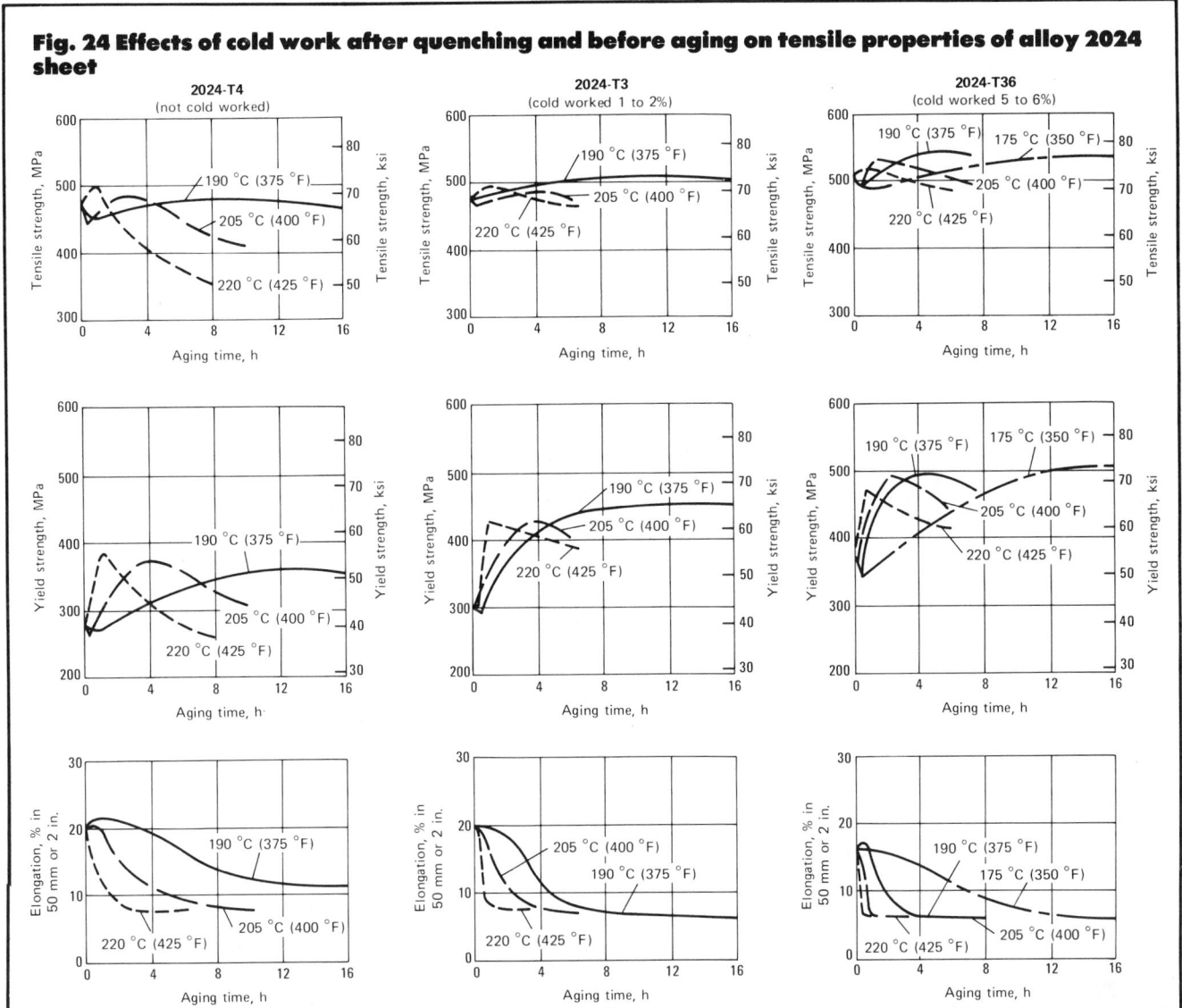

mold) plays an important role in the response of a casting, or a portion of a casting, to heat treatment. Hence, the heat treating practice used for each casting usually is based on preliminary testing.

Premium-quality casting specifications such as MIL-A-21180 can require different strengths and ductility levels in the same casting. Yield strength is largely controlled by the limiting hardening-element level, and thus is not controllable inside a given casting. This leaves the heat treater only the option of hardening to the highest specified yield strength. Ductility, however, is controlled for a given yield strength by soundness and microstructural fineness, and is thus determined in the foundry and not by the heat treater.

Tensile strength is not a fundamental property in this sense, but rather the result of the ductility at a given yield strength.

In general, the principles and procedures for heat treating wrought and cast alloys are similar. The major differences between solution-treating conditions for castings and those for wrought products are found in soak times and quenching media. Solution of the relatively large microconstituents present in castings requires longer soaking periods than those used for wrought products (Table 3). When heat treatment of castings must be repeated, solution times become similar to those for wrought products, because the gross solution and homogenization has been accomplished and is irreversible under

normal conditions. Reduction of quenching stresses and distortion dictate that quenching be done in boiling water and sometimes in milder media. A commercially important variety is a mixture of polyalkylene glycol and water, which has no detrimental effect on properties for thicknesses under approximately 3.2 mm (0.125 in.).

Cast products of heat treatable aluminum alloys have the highest combinations of strength, ductility and toughness when produced in T6-type tempers. Developing T6-type tempers in cast products requires the same sequence of operations employed in developing tempers of the same type in wrought products—solution heat treating, quenching and precipitation heat treating.

Fig. 25 Comparison of the precipitation-hardening characteristics of 356.0-T4 sand and permanent mold castings

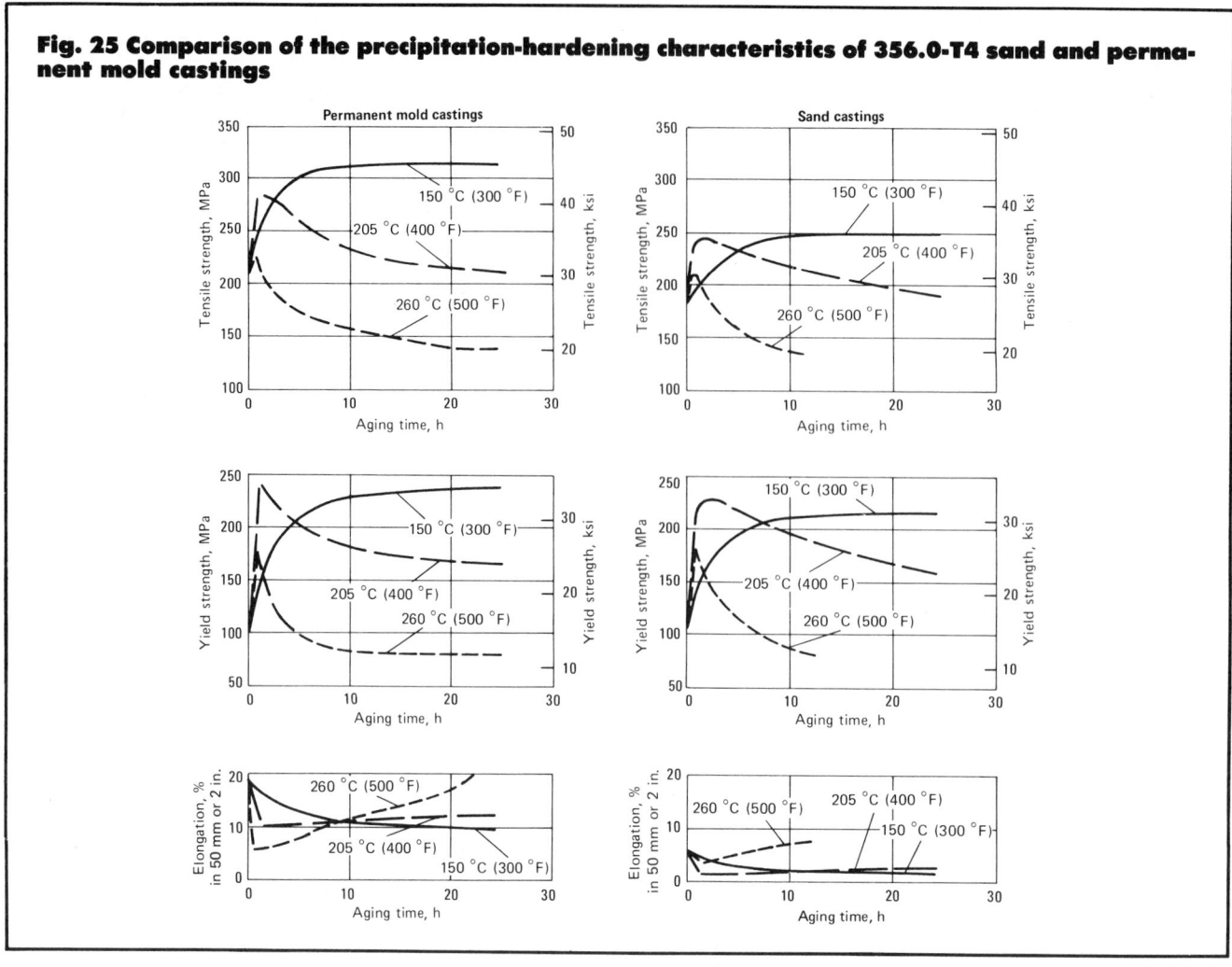

A comparison of the precipitation-hardening characteristics of sand and permanent mold castings of alloy 356.0-T4 is provided in Fig. 25. Because of the finer cast structure and higher supersaturation of the more rapidly solidified permanent mold castings, their tensile properties are superior to those of sand castings of the same composition similarly heat treated.

Among precipitation treatments unique to castings are those resulting in the T5 and T7 tempers. The T5 temper is produced merely by applying a precipitation treatment to the as-cast casting, without previous solution treatment. A moderate increase in strength is achieved without warpage and subsequent straightening. High hardness and dimensional and strength stability at elevated temperatures account for the almost universal use of materials in T5 tempers for pistons and other engine parts. Some applications demand combinations of strength, toughness and dimensional stability that cannot be met by heat treating to T5-, T6- or T8-type tempers. For these applications, T7-type tempers are developed by solution heat treating, quenching in a medium that provides a moderate cooling rate and then precipitation heat treating at a temperature higher than that used to develop T5-, T6- and T8-type tempers. Heat treating to T7-type tempers results in lower strength than that of material in T6- or T8-type tempers, develops high ductility and toughness, and carries precipitation far enough to minimize further precipitation during service.

Effects of Reheating

The precipitation characteristics of aluminum alloys must be considered frequently during evaluation of the effects of reheating on mechanical properties and corrosion resistance.

Such evaluations are necessary for determining standard practices for manufacturing operations, such as hot forming and straightening, adhesive bonding, and paint and dry-film lubricant curing, and for evaluating the effects of both short-term and long-term exposure to elevated temperatures in service.

The stage of precipitation that exists in an alloy at the time of reheating plays a significant role in the effects of reheating. Consequently, it is extremely dangerous to reheat material in a solution heat treated temper without first carefully testing the effects of such reheating. In one such test, 2024-T4 sheet was found to be very susceptible to intergranular corrosion when subjected to a 15-min drying operation at 150 °C (300 °F) during the first 8 h after quenching; no susceptibility was evident when the same drying operation was performed more than 16 h after quenching. In another test, 7075-W

(0.2 to 600 h) bar and plate were reheated for hot forming at 175 °C (350 °F) for 20 min. Strengths after aging to the T6 temper were 10 to 15% lower than those for standard 7075-T6. In contrast, similar reheating of T6 material for up to 1 h at 175 °C (350 °F) produced no detrimental effect.

If reheating is performed on material in the W or T4 condition, its effect can be estimated from families of precipitation heat treating curves such as those presented in Fig. 20. Such curves can also be used for reheating of precipitation heat treated material at the precipitation heat treating temperature. For reheating at other temperatures, other data may be needed (Fig. 26). The heat treating and reheating curves may be used as the bases for limitations on reheating (Table 6).

Annealing

Annealing treatments employed for aluminum alloys are of several types that differ in objective. Annealing times and temperatures depend on alloy type as well as on initial structure and temper.

Full Annealing. The softest, most ductile and most workable condition of both non-heat-treatable and heat treatable wrought alloys is produced by "full annealing" to the temper designated "O". Strain-hardened products in this temper normally become recrystallized, but hot worked products may remain unrecrystallized. In the case of heat treatable alloys, the solutes are sufficiently thoroughly precipitated to prevent natural age hardening. A higher maximum temperature than that used for stress-relief annealing, controlled cooling to a lower temperature and additional holding time at the lower temperature generally are employed.

For both heat treatable and non-heat-treatable aluminum alloys, reduction or elimination of the strengthening effects of cold working is accomplished by heating at a temperature from about 260 to about 440 °C (500 to 825 °F). The rate of softening is strongly temperature-dependent; the time required to soften a given material by a given amount can vary from hours at low temperatures to seconds at high temperatures.

If the purpose of annealing is merely to remove the effects of strain hardening, heating to about 345 °C (650 °F) will usually suffice. If it is necessary to remove the hardening effects of a heat treatment or of cooling from hot working temperatures, a treatment designed to produce a coarse, widely spaced precipitate is employed. This usually consists of soaking at 415 to 440 °C (775 to 825 °F) followed by slow cooling (28 °C/h, or 50 °F/h, max) to about 260 °C (500 °F). The high diffusion rates that exist during soaking and slow cooling permit maximum coalescence of precipitate particles and result in minimum hardness.

As a result of this treatment, only partial precipitation occurs in 7xxx alloys, and a second treatment (soaking at 230 ±6 °C, or 450 ±10 °F, for 2 h) is required. When the need arises for small additional improvements in formability, cooling at 28 °C/h (50 °F/h) should be extended to 230 °C (450 °F), and the material should be soaked at 230 °C for 6 h. The effects of eliminating or prolonging the 230 °C second step on the ductility of 7075-O sheet are compared with the standard treatment in Table 7.

In annealing, it is important to ensure that the proper temperature is reached in all portions of the load; therefore, it is common to specify a soaking period of at least 1 h. The maximum annealing temperature is moderately critical; it is advisable not to exceed 415 °C (775 °F), because of oxidation and grain growth. The heating rate can be critical, especially for alloy 3003, which usually requires rapid heating for prevention of grain growth. Relatively slow cooling, in still air or in the furnace, is recommended for all alloys to minimize distortion. Typical annealing conditions used for some alloys in common use are listed in Table 8.

Products that can be heated and

Fig. 26 Effects of reheating on tensile properties of alclad 2024-T81 sheet

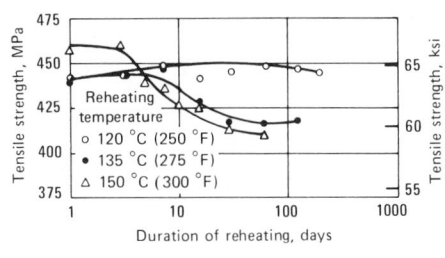

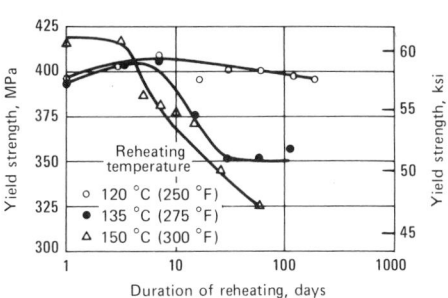

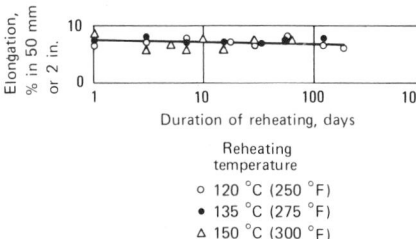

Table 6 Reheating schedules for wrought aluminum alloys
The schedules given in this table normally will not decrease strength more than 5%.

Alloy and temper	150 °C (300 °F)	165 °C (325 °F)	175 °C (350 °F)	190 °C (375 °F)	205 °C (400 °F)	220 °C (425 °F)	230 °C (450 °F)
			Reheating time at a temperature of:				
2014-T4	(a)	(a)	(a)	(a)	(a)	(a)	(a)
2014-T6	20-50 h	8-10 h	2-4 h	½-1 h	5-15 min	(b)	(b)
2024-T3, 2024-T4	(a)	(a)	(a)	(a)	(a)	(a)	(a)
2024-T81, 2024-T86	20-40 h	. . .	2-4 h	1 h	½ h	15 min	5 min
6061-T6, 6062-T6,							
6063-T6	100-200 h	50-100 h	8-10 h	1-2 h	½ h	15 min	5 min
7075-T6, 7178-T6	10-12 h	1-2 h	1-2 h	½-1 h	5-10 min	(b)	(a)

(a) Reheating not recommended. (b) Bring to temperature

Table 7 Effects of annealing treatments on ductility of 7075-O sheet

Annealing treatment	Elongation in tension(a), % in 50 mm or 2 in., for thickness of:			Bend angle(b), degrees, for thickness of:		Elongation in bending(c), % in 50 mm or 2 in., for thickness of:	
	0.5 mm (0.020 in.)	1.6 mm (0.064 in.)	2.6 mm (0.102 in.)	1.6 mm (0.064 in.)	2.6 mm (0.102 in.)	1.6 mm (0.064 in.)	2.6 mm (0.102 in.)
Treatment 1(d) 12	12	12	82	73	48	50
Treatment 2(e)..........................14	14	14	91	76	58	57
Treatment 3(f)..........................16	16	. . .	92.5	84	56	60

(a) Uniform elongation of gridded tension specimens. (b) Bend angle at first fracture. (c) Elongation in bend test for 1.3-mm (0.05-in.) gage spanning fracture. (d) Soak 2 h at 415 ± 14 °C (775 ± 25 °F); furnace cool to 260 °C (500 °F) at 30 °C/h (50 °F/h); air cool. (e) Soak 2 h at 425 °C (800 °F), air cool; soak 2 h at 230 °C (450 °F), air cool. (f) Soak 1 h at 425 °C (800 °F); furnace cool to 230 °C (450 °F) at 30 °C/h (50 °F/h); soak 6 h at 230 °C (450 °F), air cool

Table 8 Typical full annealing treatments for some common wrought aluminum alloys

These treatments, which anneal the material to the "O" temper, are typical for various sizes and methods of manufacture and may not exactly describe optimum treatments for specific items.

Alloy	Metal temperature °C	°F	Approximate time at temperature, h	Alloy	Metal temperature °C	°F	Approximate time at temperature, h
1060 345		650	(a)	5457 345		650	(a)
1100 345		650	(a)	5652 345		650	(a)
1350 345		650	(a)	6005 .... 415(b)		775(b)	2-3
2014 .... 415(b)		775(b)	2-3	6009 .... 415(b)		775(b)	2-3
2017 .... 415(b)		775(b)	2-3	6010 .... 415(b)		775(b)	2-3
2024 .... 415(b)		775(b)	2-3	6053 .... 415(b)		775(b)	2-3
2036 .... 385(b)		725(b)	2-3	6061 .... 415(b)		775(b)	2-3
2117 .... 415(b)		775(b)	2-3	6063 .... 415(b)		775(b)	2-3
2124 .... 415(b)		775(b)	2-3	6066 .... 415(b)		775(b)	2-3
2219 .... 415(b)		775(b)	2-3	7001 .... 415(c)		775(c)	2-3
3003 415		775	(a)	7005 .... 345(d)		650(d)	2-3
3004 345		650	(a)	7049 .... 415(c)		775(c)	2-3
3105 345		650	(a)	7050 .... 415(c)		775(c)	2-3
5005 345		650	(a)	7075 .... 415(c)		775(c)	2-3
5050 345		650	(a)	7079 .... 415(c)		775(c)	2-3
5052 345		650	(a)	7178 .... 415(c)		775(c)	2-3
5056 345		650	(a)	7475 .... 415(c)		775(c)	2-3
5083 345		650	(a)	**Brazing sheet**			
5086 345		650	(a)				
5154 345		650	(a)	No. 11 and 12 345		650	(a)
5182 345		650	(a)	No. 21			
5254 345		650	(a)	and 22 345		650	(a)
5454 345		650	(a)	No. 23			
5456 345		650	(a)	and 24 345		650	(a)

(a) Time in the furnace need not be longer than necessary to bring all parts of the load to annealing temperature. Cooling rate is unimportant. (b) These treatments are intended to remove the effects of solution treatment and include cooling at a rate of about 30 °C/h (50 °F/h) from the annealing temperature to 260 °C (500 °F). Rate of subsequent cooling is unimportant. Treatment at 345 °C (650 °F), followed by uncontrolled cooling, may be used to remove the effects of cold work, or to partly remove the effects of heat treatment. (c) These treatments are intended to remove the effects of solution treatment and include cooling at an uncontrolled rate to 205 °C (400 °F) or less, followed by reheating to 230 °C (450 °F) for 4 h. Treatment at 345 °C (650 °F), followed by uncontrolled cooling, may be used to remove the effects of cold work, or to partly remove the effects of heat treatment. (d) Cooling rate to 205 °C (400 °F) or below is less than or equal to 30 °C/h (50 °F/h).

cooled very rapidly, such as wire, are annealed by continuous processes that require a total heating and cooling time of only a few seconds. Continuous annealing of coiled sheet is accomplished in a total time of a few minutes. For these extremely rapid operations, maximum temperature may exceed 440 °C (825 °F).

Although material annealed from the precipitation-hardened condition usually has sufficient ductility for most forming operations, this ductility often is slightly lower than that of material that has not been subjected to prior heat treatment—that is, material annealed at the producing source. Therefore, when maximum ductility is required, annealing of a previously heat treated product is sometimes unsuccessful.

Partial Annealing. Annealing of cold worked non-heat-treatable wrought alloys to obtain intermediate mechanical properties (H2-type tempers) is referred to as "partial annealing" or "recovery annealing". Temperatures used are below those that produce extensive recrystallization, and incomplete softening is accomplished by substructural changes in dislocation density and rearrangement into cellular patterns (polygonization). Benda-

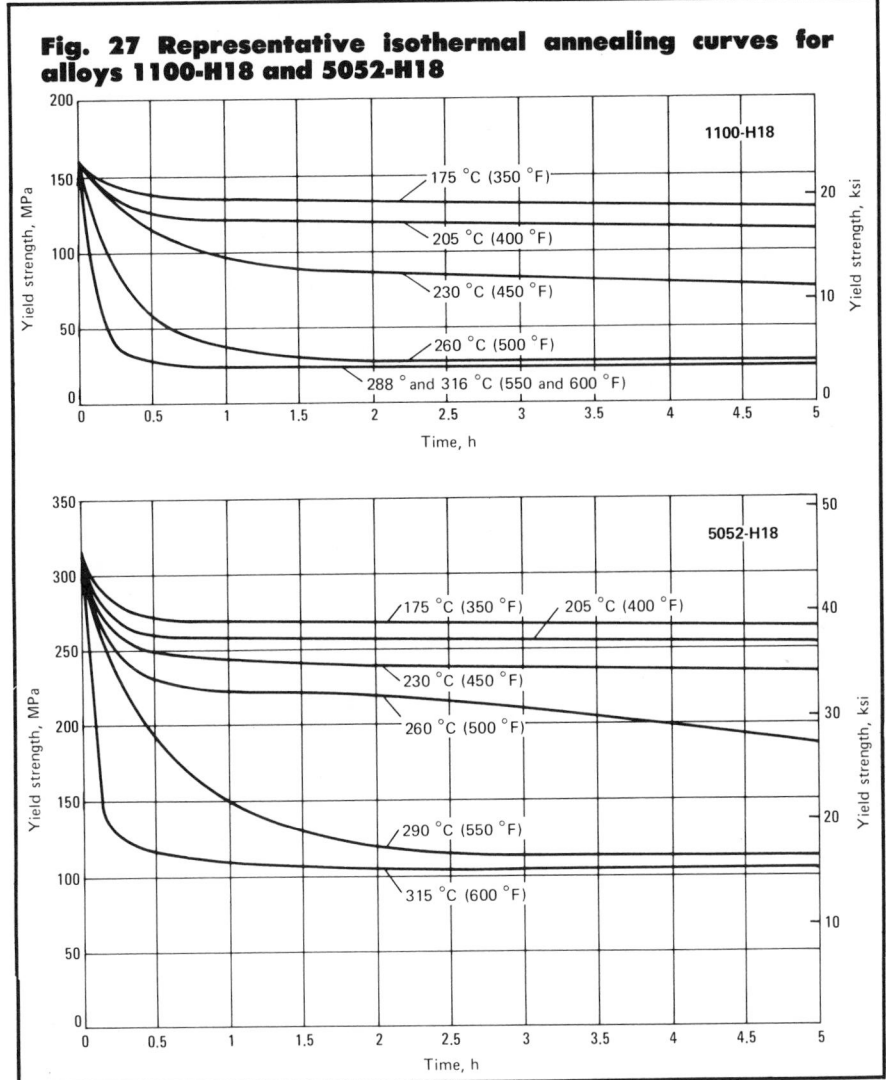

Fig. 27 Representative isothermal annealing curves for alloys 1100-H18 and 5052-H18

is used for heat treatable alloy products that subsequently will be inspected ultrasonically. The product is heated to its normal solution heat treating temperature, then cooled in still air to room temperature. This temper is referred to as the O1 temper.

Controlled-Atmosphere Annealing and Stabilizing. Aluminum alloys that contain even very small amounts of magnesium will form a surface magnesium oxide unless the atmosphere in the annealing furnace is free of moisture and oxygen. Examples include alloy 3004, which is used for cooking utensils, and alloys of the 5xxx series.

Another problem that control of the annealing atmosphere helps to overcome or avoid is "oil staining" by oil-base roll lubricants that do not burn off at lower annealing temperatures. If the oxygen content of the furnace atmosphere is kept very low during such annealing, the oil will not oxidize and stain the work.

Temperature control for full and partial annealing is somewhat more critical than for stress-relief annealing; the temperatures and times specified are selected to produce recrystallization and, in the case of heat treatable alloys, a precipitate of maximum size; for this the cooling rate must be closely controlled. Even allowing the load to cool in the furnace may result in an excessively high rate. Similarly, lowering the furnace-control instrument by 28 °C (50 °F) each hour may produce stepped cooling, which is not satisfactory for severe forming operations. For maximum softening, a continuous cooling rate of not more than 28 °C/h is recommended.

Annealing of castings for 2 to 4 h at temperatures from 315 to 345 °C (600 to 650 °F) provides the most complete relief of residual stresses and precipitation of the phases formed by the excess solute retained in solid solution in the as-cast condition. Such annealing treatments provide maximum dimensional stability for service at elevated temperatures. The annealed temper is designated "O". (This temper was designated "T2" prior to 1975.)

Grain Growth

Many of the aluminum alloys in common use are subject to grain growth during solution treatment or annealing. This phenomenon can occur during or after recrystallization of material

bility and formability of an alloy annealed to an H2-type temper generally are significantly higher than those of the same alloy in which an equal strength level is developed by a final cold working operation (H1-type temper). Treatments to produce H2-type tempers require close control of temperature to achieve uniform and consistent mechanical properties.

Figure 27 shows changes in yield strength as functions of temperature and time for sheet of two non-heat-treatable alloys (1100 and 5052) initially in the highly cold worked condition (H18 temper). From these curves, it is apparent that, by selection of appropriate combinations of time and temperature, mechanical properties intermediate to those of cold worked and fully annealed material can be obtained. It is also evident that yield

strength depends much more strongly on temperature than on time of heating.

Stress-Relief Annealing. For cold worked wrought alloys, annealing merely to remove the effects of strain hardening is referred to as "stress-relief annealing". Such treatments employ temperatures up to about 345 °C (650 °F), or up to 400 ± 8 °C (750 ± 15 °F) for 3003 alloy, and cooling to room temperature. No appreciable holding time is required. Such treatment may result in simple recovery, partial recrystallization or full recrystallization. Age hardening may follow stress-relief annealing of heat treatable alloys, however, because a concentration of soluble alloying elements sufficient to cause natural aging remains in solid solution after such treatments.

A special form of stress-relief temper

that has been subjected to a small critical amount of prior cold work. It is usually manifested by surface roughening during subsequent fabrication operations and frequently results in rejections for appearance or functional reasons. Less frequently, some deterioration of mechanical properties is encountered, and this is undesirable regardless of surface-roughening effects.

Degree of susceptibility to grain growth varies with alloy, structure, and chemical-composition variation, and from one product form to another. The critical range of cold work is ordinarily about 5 to 15%. Usually, temperatures of 400 °C (750 °F) and above must be reached before grain growth occurs, but some growth has been encountered at temperatures as low as 345 °C (650 °F). Grain growth that occurs during initial recrystallization is more a function of composition, structure and degree of cold work than of temperature *per se*; temperatures in excess of 455 °C (850 °F) in common alloys can lead to secondary-recrystallization grain-growth problems. The common symptom indicating moderately large-grain material is roughening or "orange peel" on the external surfaces of bends. Severe growth of grains to fingernail size and larger sometimes is evident in parts made from annealed (O temper) material by stretch forming and then thermal treating or similar operations. This type of grain growth often is detected during subsequent anodizing, etching, and chemical milling operations.

Cracking during welding or brazing is another characteristic which may indicate that severe grain growth has occurred. In such instances, cracks propagate along grain boundaries that provide little obstruction to their progress.

If the surface roughening is objectionable from either an appearance or a functional aspect, the desirability of surface-smoothing operations, such as sanding or buffing, must be evaluated. If reductions in mechanical properties are suspected, these must be established by test and evaluated in relation to the anticipated service.

In one application, a part that had been made by stretch forming O-temper 2-mm (0.080-in.) sheet and heat treating exhibited significantly lower tensile and yield strengths in portions where severe grain growth had occurred than in portions having normal grain size:

Test	Grain structure	Tensile strength MPa	ksi	Yield strength MPa	ksi
Transverse					
1	Coarse	265	38.5	247	35.8
2	Coarse	263	38.2	241	35.0
3	Fine	311	45.1	261	37.8
Longitudinal					
1	Coarse	259	37.6	243	35.3
2	Coarse	269	39.0	245	35.6
3	Fine	305	44.2	270	39.1

In other similar investigations, no detrimental effects have been discovered, and in many cases such parts have served satisfactorily in critical applications.

When a grain-growth problem is discovered, it is too late to change the condition of the parts in question, but several possible methods are available for preventing recurrence of the difficulty. The simplest of these is relieving the causative stress by interjecting a stress-relief anneal into the manufacturing sequence immediately prior to the solution-treating or full-annealing cycle in which the grain growth occurred. This approach is usually successful and practical. Another possibility is to adjust the amount of stress present in the part immediately prior to the critical heat treatment so that the stress level is outside the critical range. This may be done by adding a cold working operation before forming, such as prestretching of blanks, or by forming in multiple stages with a stress-relief anneal before each stage.

A third method that is sometimes successful consists of increasing the heating rate during the critical heat treatment by reducing the size of furnace loads or by changing from an air furnace to a salt bath. In one application, severe grain growth was found during bending of alloy 1100 rectangular tubing. The roughening of the inside surfaces of the parts, which occurred during forming of the large-grain material, impaired their functioning as radar waveguides. Investigation disclosed that, to minimize handling marks, the material was procured in the strain-hardened (H14) temper and was stress-relief annealed at 345 °C (650 °F) immediately prior to forming. Grain growth occurred during annealing as a result of the moderate amount of cold work introduced at the mill. The problem was eliminated by changing the stress-relieving operation to a 5-min heating period in an air furnace operating at 540 °C (1000 °F). The explanation advanced for the success of this treatment was that, due to the rapid heating rate, the temperature of the material was raised through the recrystallization range for the less severely cold worked grains before the critically cold worked grains had time to grow appreciably.

Furnace Equipment and Accessories

Both molten salt baths and air-chamber furnaces are suitable for solution heat treating of aluminum alloys. The choice of furnace equipment depends largely on the alloy and the configuration of the parts to be processed. Both heating media have advantages and disadvantages. Oil- and gas-fired furnaces, in designs that allow the products of combustion to come in contact with the work, are usually unsatisfactory because they promote high-temperature oxidation.

Salt baths heat the work faster (see Table 2) than air furnaces, provided that the amount of work introduced at any one time is controlled to prevent the temperature of the bath from falling below the desired range. If the temperature is permitted to fall below the minimum limit, much of the advantage of the salt bath is lost, because of the necessity for reheating the large mass of salt.

Salt baths are also more readily adapted to the introduction, at any time, of small amounts of work requiring different soaking periods. (Economical utilization of air furnaces usually dictates accumulation of a large load of parts of similar thickness before charging.) Also, the buoyant effect of the salt reduces distortion during heating, and the large reservoir of heat facilitates temperature control and uniformity.

Salt bath operation entails special housekeeping requirements. Dragout is costly and unsightly. Because residual salt on parts may result in corrosion, all salt must be completely removed, including that from crevices and blind holes. In addition, salt residue from the quench water must be kept to a minimum by a constant water overflow or by providing a fresh-water rinse for all parts after quenching. When these provisions are impractical, corrosion can be inhibited by adding 14 g (½ oz) of sodium or potassium dichromate to each 45 kg (100 lb) of the molten salt.

Precautions. Molten salt baths are

potentially hazardous and require special precautions. Operators must be protected from splashing and dripping of the hot salt. Because heated nitrates are powerful oxidizing agents, they must never be allowed to come in contact with combustibles and reducing agents, such as magnesium and cyanides. Most authorities advise against inserting aluminum alloys containing more than a few percent of magnesium into molten nitrate. To avoid exposure of personnel to nitrous fumes produced during decomposition of nitrates, good ventilation is essential.

When molten nitrates are being used, the possibilities of explosions resulting from both physical and chemical reactions must be avoided. The former result from rapid expansion of gases entrapped beneath the surface of the bath. Hence, parts entering the bath must be clean and dry; they must also be free of pockets or cavities that contain air or other gases. Chemical-reaction explosions result from rapid breakdown of the nitrates due to overheating or reaction with the pot material. Stainless steel pots (preferably of type 321 or 347) are more resistant to scaling than those made of carbon steel or cast iron and therefore present a lower probability of local overheating. Sludge or sediment accumulations in bottom-heated pots can also lead to local overheating. Overheat controls are essential to ensure against temperatures exceeding 595 °C (1100 °F).

It is vitally important that water be kept away from a nitrate tank. In controlling a nitrate fire, do not use water or any fire extinguisher containing water. The best extinguisher is dry sand, a supply of which should be kept near the tank.

Extra sacks of salt should be stored in a dry place, distant from the tank. If the fresh salt being added to the bath is even slightly damp, it should be added very slowly or when the bath is frozen.

Air furnaces are used more widely than salt baths because they permit greater flexibility in operating temperature. When production schedules and the variety of alloys requiring heat treatment necessitate frequent changes in temperature, the time and cost of adjusting the temperature of a large mass of salt makes the use of an air furnace almost mandatory. However, waiting periods are often required to allow the walls of air furnaces to stabilize at the new temperature before parts are introduced. Otherwise, parts may radiate heat to colder walls or absorb radiant heat from hotter walls, and the temperature indicated by the control instrument will not reflect actual metal temperature in the usual manner. Air furnaces are also more economical when the product mix includes a few rather large parts; holding the temperature of a large volume of salt in readiness for an occasional large part is far more expensive than heating an equal volume of air.

Temperature Control

The importance of close temperature control in solution treating has been noted in a previous section of this article.

Each control zone of each furnace should contain at least two thermocouples. One thermocouple, with its instrument, should act as a controller, regulating the heat input; the other should act independently as a safety cutoff, requiring manual reset if its set temperature (usually the maximum of the specified range) is exceeded during the solution-treating cycle.

Safety cutoffs are mandatory for salt baths, to guard against explosions, and often have paid for themselves in air furnaces by saving a load of parts or even the furnace itself. It is important, however, that they be tested periodically (by deliberately overshooting the empty furnace) to guard against "frozen" or corroded contacts resulting from prolonged periods of idleness.

At least one of the instruments for each zone should be of the recording type, and both instruments should have restricted scales—for instance, 400 to 600 °C, rather than 0 to 600 °C. This is required for maximum accuracy because manufacturers' guarantees are specified in terms of percent of scale.

In the placement of instruments, exposure to extremes in ambient temperature, humidity, vibration, dust and corrosive fumes should be avoided. Ambient temperatures between 5 and 50 °C (40 and 120 °F) are satisfactory, but temperature changes of 6 °C/h (10 °F/h) or more should be avoided. It is also essential that instruments and thermocouple circuits be shielded from electromagnetic fields commonly associated with the leads of high-amperage furnace heating elements.

Temperature-sensing elements must be capable of responding more rapidly to temperature changes than the materials being processed. Therefore, thermocouple wire diameter should not exceed 1½ times the thickness of the minimum-gage material to be heat treated, and should in no case exceed 14 gage. Thermocouples for salt baths should be enclosed in suitable protection tubes. Air-furnace thermocouples should be installed in open-end protection tubes, with the thermocouple junction extending sufficiently beyond the tube to prevent any loss in sensitivity.

Temperature-sensing elements should be located in the furnace work chamber, not in ducts and plenums, and should be as close as possible to the working zone. Specification MIL-H-6088C restricts distance between the sensing element and the working zone to a maximum of 102 mm (4 in.). The safety-cutoff thermocouple should be located to reflect the highest temperature in the working zone. The control thermocouple should be located in a position where it will read a temperature approximately halfway between the hottest and coldest temperatures.

Probe Checks. After the temperature-measurement equipment is properly installed, it must be checked frequently for accuracy. This is accomplished by inserting a calibrated probe thermocouple into the furnace adjacent to each furnace thermocouple and comparing its reading on a calibrated test potentiometer with that indicated by the furnace instrument. Correction factors should be applied to the furnace instruments after each probe check, but if the correction required exceeds 3 °C (5 °F), the source of the deviation should be corrected. MIL-H-6088C recommends that this check be made weekly, but many operators make the check as frequently as once each shift.

Temperature-Uniformity Surveys. In controlling the temperature of parts that are being heat treated it must first be determined that the temperature indicated by the furnace instruments truly represents the temperature of the nearby air or salt. Second, the uniformity of temperature within the working zone must be shown to be within a range of 11 °C, or 20 °F (6 °C, or 10 °F, for precipitation heat treatment of alloy 2024). This is accomplished by measuring the temperature at several test locations, using calibrated test thermocouples and a calibrated test potentiometer, and reading furnace instruments nearly simultaneously. MIL-H-6088 recommends

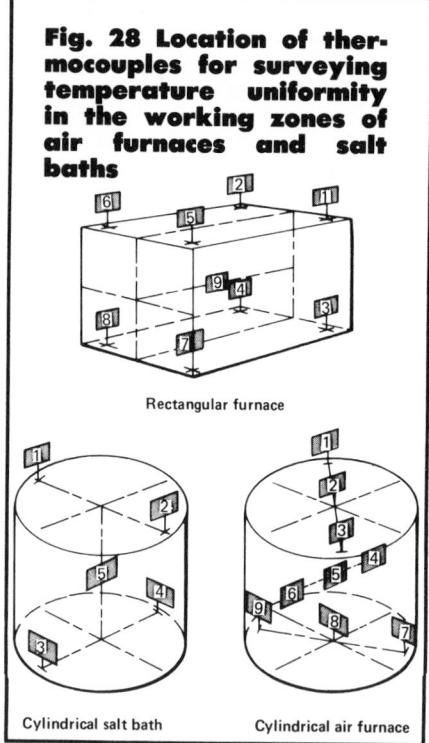

Fig. 28 Location of thermocouples for surveying temperature uniformity in the working zones of air furnaces and salt baths

Rectangular furnace

Cylindrical salt bath   Cylindrical air furnace

monthly surveys with one test location per 1.1 m³, or 40 ft³ (0.7 m³, or 25 ft³, for air furnaces on initial survey), but with a minimum of nine test locations distributed as shown in Fig. 28. Despite the large size of some furnaces, rather surprising temperature uniformities have been reported. In one instance, the initial survey of an air furnace measuring 12.5 by 1.2 by 3.0 m (41 by 4 by 10 ft) showed maximum temperature variations of $+1.7$, -1.1 °C ($+3$, -2 °F). When a partition 0.3 m (1 ft) thick was lowered, converting the furnace to two chambers 6.1 by 1.2 by 3.0 m (20 by 4 by 10 ft) each, the spread was $+1.1$, -0.6 °C ($+2$, -1 °F) in one section and $+0.6$, -1.1 °C ($+1$, -2 °F) in the other.

For each furnace load, one thermocouple (the "cold" couple) should be placed in the coldest area of the furnace and another (the "hot" couple) in the hottest area. In addition to these two thermocouples, a load thermocouple should be installed. The load couple should be of approximately the same gage as the sheet or other product being heat treated. If heavy plate, forgings or castings are being heat treated, a similar discarded item should be used at the controlling load couple. The thermocouple should be placed in a drilled hole and packed to hold it firmly in place

during the heat treating cycle. In some instances, the items being heat treated can be used as the load couples. The thermocouples can be placed in holes drilled in areas that will be removed in making the finished article.

It is important that items of different thicknesses—1-mm (0.040-in.) sheet and 25-mm (1-in.) plate, for example—not be heat treated in the same furnace load.

In salt baths, uniformity surveys usually are made by holding a probe thermocouple in each location until thermal equilibrium is reached; in air furnaces, a mock heat treating cycle is required. First, the air furnace is stabilized at the test temperature. Then a rack containing the test thermocouples is inserted into the furnace. By using multiple switches or a multipoint recording instrument, all test thermocouples and furnace instruments are read every 5 min. As the temperature approaches the test range, it is advisable to increase the frequency of readings to detect possible overshooting. After thermal equilibrium is reached, readings should be continued until the recurrent temperature pattern is established.

Surveys of salt baths generally are considered acceptable whether they are made while the bath is empty or filled with work. It is controversial whether surveys of air furnaces should be made with or without a load. Undoubtedly, recovery overshoots are most likely to occur with a very light load and would not be detected if a heavier load were used. Certainly, if all loads are essentially alike, surveys should be made with typical loads. With widely varying loads, the optimum approach is to make several surveys initially, including one with an empty furnace, and then to make succeeding surveys with an empty furnace to ensure against changes in furnace characteristics. If any changes are made in the furnace that might affect temperature distribution, such as repair of vanes or louvers, several surveys should be repeated.

Another aspect of the problem of temperature control in air furnaces is the necessity of ensuring that the temperature of the parts is the same as that of the surrounding air. Furnace components whose temperature differs from the air temperature must be suitably shielded to prevent radiation to or from the parts being heat treated. In a furnace used for solution heat treating of rivets, unshielded heating elements

have been known to produce part temperatures as much as 19 °C (35 °F) higher than the control temperature, resulting in eutectic melting and cracking. In two other instances, reradiation through inadequate shielding produced a radiation effect of as much as 11 °C (20 °F). One of these problems was solved by painting the shield with reflective aluminum paint and the other by adding a 12.7-mm- (½-in.-) thick layer of asbestos to the 1.6-mm (1/16-in.) stainless steel shield.

Furnace-wall temperatures that differ appreciably from the temperature of the parts also must be avoided. Consequently, when the operating temperature of an air furnace is changed, waiting periods are required after the furnace instrument indicates stability, to allow the furnace walls to stabilize at the new temperature. The magnitude of this limitation is directly proportional to the efficiency of the furnace as an insulated chamber, but possibilities of such radiation should be recognized even in thin-wall furnaces.

Radiation effects are potentially dangerous because they often cannot be detected by ordinary thermocouples. Specially prepared radiation panels with thermocouples attached are used, and their readings are compared with adjacent free thermocouples. These panels normally are made of material of the same gage as the thinnest parts to be heat treated and should have a single surface area of about 650 cm² (100 in.²). A thermocouple is attached to the center of the panel by welding or peening. In order to detect the maximum effect, panel surfaces should be darkened so that their emissivity is at least as high as that of any material to be processed. During the test, the panel surfaces should be parallel to the suspected source or recipient of radiation. As an example of the number of panels required, several aerospace companies specify one panel for every 1.5 linear metres (5 linear feet) of furnace wall.

Instrument Calibration. All instruments and thermocouples must be accurately calibrated, and it is essential that the calibrations be traceable directly to the National Bureau of Standards. The chain of traceability should consist of not more than four links for sensing elements and three links for measuring elements. To illustrate, if the article calibrated by the National Bureau of Standards is called a primary standard, then the chain of traceability of measuring elements should consist of primary standard, test potentiometer, and furnace instrument. Similarly, the chain for sensing elements should consist of primary standard, secondary standard, test thermocouple, and furnace thermocouple. Every effort should be made to ensure that the temperature indicated by the furnace instruments is as close as possible to the actual temperature. To achieve this, it is necessary to apply correction factors obtained during calibration to the next lower echelon of accuracy. Even then, if all errors inherent in the chain are in the same direction, a considerable difference will exist between the measured and actual temperatures. Therefore, it is advisable to operate as close to the mean of the desired range as possible.

Dimensional Changes During Heat Treatment

In addition to the completely reversible changes in dimensions that are simple functions of temperature change and are caused by thermal expansion and contraction, dimensional changes of more permanent character are encountered during heat treatment. These changes are of several types, some of mechanical origin and others caused by changes in metallurgical structure. Changes of mechanical origin include those arising from stresses developed by gravitational or other applied forces, from thermally induced stresses or from relaxation of residual stresses. Dimensional changes also accompany recrystallization, solution and precipitation of alloying elements.

Solution Heat Treatment. Distortion as a result of creep during solution heat treatment should be avoided by proper loading of parts in baskets, racks or fixtures or by provision of adequate support for long pieces of plate, rod, bar and extrusions heat treated in horizontal roller hearth furnaces. Sheet is provided with air-pressure support in continuous heat treating furnaces to avoid scratching, gouging and distortion. If parts are to be solution heat treated in fixtures or racks made of materials (such as steel) with coefficients of thermal expansion lower than that of the aluminum being treated, allowance should be made for this differential expansion to ensure that expansion of the aluminum is not restricted. Straightening immediately

after solution heat treating may be preferable to fixturing.

Solution of phases formed by major alloying elements causes volumetric expansion or contraction, depending on the alloy system, and this may have to be taken into account in heat treatment of long pieces. For example, solution heat treatment and quenching of alloy 2219 causes lengthwise contraction of about 2 mm/m (0.002 in./in.). Solution heat treatment and quenching of alloys of the 7xxx series is accompanied by lengthwise expansion—about 0.6 mm/m (0.0006 in./in.) for alloy 7075 rod or plate.

Quenching. The most troublesome changes in dimensions and shape are those that occur during quenching or that result from stresses induced by quenching. Due to its nonuniform cooling, quenching may produce warpage or distortion, particularly in thin material and in thin sections of parts that contain variations in thickness. For thick-section products or parts, changes in external shape may be small because of rigidity, but the interior-to-surface temperature gradients that form with rapid cooling create residual stresses; these stresses normally are compressive at the surfaces and tensile in the interior.

As previously discussed, warpage or distortion of thin-section material can be reduced by using a quenching medium that provides slower cooling; however, cooling must be sufficient to produce the required properties. Slower quenching can also reduce the magnitude of residual stresses in thicker parts or pieces, as shown in Fig. 13 for cylindrical specimens of alloy 6151 quenched in cold or boiling water. Stress range (maximum tensile stress plus maximum compressive stress) for a cylinder with a radius of 89 mm (3.5 in.) is about 205 MPa (30 ksi) when the cylinder is quenched in cold water but less than 70 MPa (10 ksi) when it is quenched in boiling water. The effects of average cooling rate through the temperature range from 400 to 290 °C (750 to 550 °F) on longitudinal stress ranges developed in alloy 2014 cylinders 75 mm (3 in.) in diameter are shown in Fig. 29.

High stresses induced by rapid quenching generally are reduced only modestly by the precipitation heat treatments used to produce T6- or T8-type tempers. Consequently, for the alloys that require rapid cooling to develop the properties of these tempers,

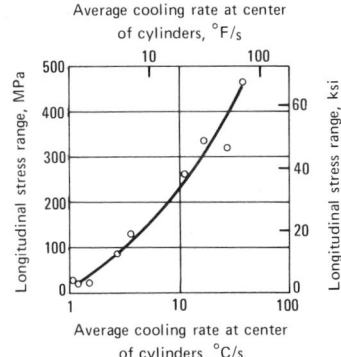

Fig. 29 Effect of quenching rate on longitudinal stress ranges in alloy 2014-T4 cylinders quenched in various media

Cylinders were 75 mm (3 in.) in diameter by 230 mm (9 in.) long. Cooling rate was measured from 400 to 290 °C (750 to 555 °F). Stress range is maximum tensile stress plus maximum compressive stress.

those incorporating mechanical stress relief (Tx51, Tx52) usually are specified when substantial metal must be removed to produce final shapes. Other T8-type tempers, such as T86 and T87, also have low residual stress as a result of the stretching required to produce them.

Heat Treatments for Precipitation and Stabilization. The most significant dimensional changes associated with precipitation heat treatments and stabilizing heat treatments arise from concurrent dilution of the solid solution (which changes lattice parameter) and formation of precipitate. Changes in density and specific volume resulting from these changes in metallurgical structure are the reverse of those caused by solution of the alloy phases. However, because the strongest tempers are those in which the precipitate is present in nonequilibrium transition forms, the amount of change during precipitation heat treatment does not totally compensate for the previous (and opposite) change that occurred during solution heat treatment. Most of the heat treatable alloys expand (grow) during precipitation heat treatment. Exceptions are alloys of the 7xxx wrought series and the 7xx.0 casting series, which exhibit contraction.

In alloys of the 2xxx series, the amount of growth decreases with increasing magnesium content. Thus, growth of about 1.5 mm/m (0.0015 in./in.) can be expected during precipitation heat treatment of alloy 2219-T87, about 0.5 mm/m (0.005 in./in.) for treatment of alloy 2014-T6 and less than 0.1 mm/m (0.0001 in./in.) for treatment of alloy 2024-T851. Alloys 7050 and 7075, on the other hand, contract about 0.3 mm/m (0.0003 in./in.) on precipitation heat treating from the W temper to the T6 temper and about 0.7 mm/m (0.0007 in./in.) on treating from the W temper to the T73 temper.

Stabilizing T7-type treatments cause greater amounts of growth than the T5-, T6- or T8-type treatments for the same alloys. This increased growth is associated either with formation of increased amounts of transition precipitates or with transformation of transition precipitates to equilibrium phases.

Dimensional Stability in Service

Dimensional stability of heat treated parts in service depends on alloy, temper and service conditions. Of the latter, excluding mechanical conditions such as applied loads, the most important is service temperature range relative to the range in which precipitation occurs. Residual stresses constitute another source of dimensional changes. Stress relief minimizes changes due to residual stresses, and most mill products usually are supplied in tempers that include stress relief. Potential dimensional change as a result of further precipitation in parts that operate at elevated temperatures is minimized for wrought products by use of T7-type stabilizing treatments and for castings by use of T5-type treatments. However, components of high-precision equipment, such as instruments for aerospace guidance systems and optical and telescopic devices, may require special supplementary treatments during manufacture to further reduce stresses or subsequent precipitation. (These treatments are discussed below, under "Stability of Precision Equipment".)

The T3- and T4-type tempers are the least stable dimensionally because of possible precipitation in service. Alloys 2024 and its variants have the smallest dimensional change in aging; the total change from the quenched to the average state is of the order of 0.06 mm/m (0.00006 in./in.), less than the change due to a temperature variation of 3 °C

(5 °F). These alloys therefore can be used in the T3- and T4-type tempers, except for precision equipment. For all other alloys, T6- or T8-type tempers should be used, because in these tempers all the alloys have good dimensional stability.

Stability of Precision Equipment. Proper maintenance of high-precision devices, such as gyros, accelerometers and optical systems, requires use of materials in which dimensional changes from metallurgical instability are limited to 10 μm/m (10 μin./in.). Several laboratory investigations and considerable practical experience have shown that wrought alloys 2024 and 6061 and casting alloy 356.0 are well suited and generally preferred for such applications. Dimensional changes were no greater than 10 μm/m when alloys 2024-T851 and -T62, 6061-T651 and -T62, and 356.0-T51, -T6 and -T7 were tested for more than a year at room temperature and for several months at 70 °C (160 °F) and when the same alloys were tested with repeated thermal cycling between +20 and −70 °C (+68 and −94 °F).

Because stresses applied or induced by acceleration in such devices generally are not high, strength levels lower than those of the highest-strength tempers frequently are satisfactory. To increase precision of machining to intended dimensions, as well as to promote maximum stability, it is common practice to apply additional thermal treatments for stress relief and precipitation of 1 to 2 h at temperatures of 175 to 205 °C (350 to 400 °F) after rough machining. These additional treatments sometimes are repeated at successive stages of processing, and even after final machining. In addition, it has been claimed that one or two cyclic treatments consisting of cooling to −100 °C (−150 °F), holding for 2 h, heating to 232 to 240 °C (450 to 465 °F) and again holding for 2 h can improve dimensional stability of 356-T6 castings.

Quality Assurance

Quality-assurance criteria that heat treated materials must meet always include minimum tensile properties and, for certain alloys and tempers, adequate fracture toughness and resistance to detrimental forms of corrosion (such as intergranular or exfoliation attack) or to stress-corrosion cracking. All processing steps through

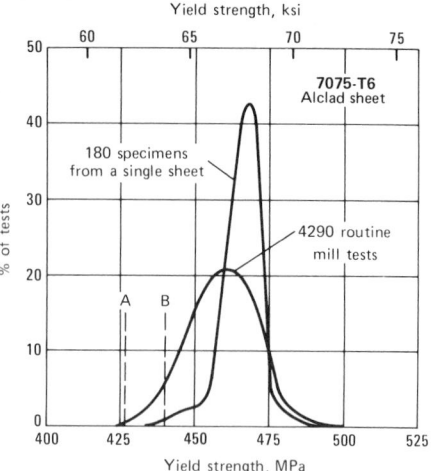

Fig. 30 Comparison of distribution of yield strength in heat treated 7075-T6 clad sheet product with distribution in a single sheet

A is 95% probability that not more than 1% of all material will fall below this value; B is 95% probability that not more than 10% of all material will fall below this value. (A and B refer only to curve representing 4290 routine mill tests.)

heat treatment must be carefully controlled to ensure high and reliable performance.

Tensile Tests. In general, the relatively constant relationships among various properties allow the use of tensile properties alone as acceptance criteria. The minimum guaranteed strength is ordinarily that value above which it has been statistically predicted with 95% probability that 99% or more of the material will pass. The inherent variability within lots and among specimens from a given piece is shown in Fig. 30. Testing provides a check for evidence of conformance; process capability and process control are the foundations for guaranteed values.

Published minimum guaranteed values are applicable only to specimens cut from a specific location in the product, with their axes oriented at a specific angle to the direction of working as defined in the applicable procurement specification. In thick plate, for example, the guaranteed values apply to specimens taken from a plane midway between the center and the surface, and with their axes parallel to the width dimension (long transverse). Different properties should be expected in specimens taken from other locations, or in specimens whose axes were parallel to the thickness dimension (short transverse). However, the specified "referee" locations and orientations do provide a useful basis for lot-to-lot comparisons, and constitute a valuable adjunct to other process-control measures.

Tensile tests can be used to evaluate the effects of changes in the process, provided specimens are carefully selected. A variation in process that produces above-minimum properties on test specimens, however, is not necessarily satisfactory. Its acceptability can be judged only by comparing the resulting properties with those developed by the standard process on similarly located specimens. Finally, variations in heat treating procedure are likely to affect the relationships among tensile properties and other mechanical properties. In applications where other properties are more important than tensile properties, the other properties should be checked also.

Hardness tests are less valuable for acceptance or rejection of heat treated aluminum alloys than they are for steel. Nevertheless, hardness tests have some utility for process control. Typical hardness values for various alloys and tempers are given in Table 9. Figure 31 shows the general relationship between longitudinal tensile strength and hardness for aluminum alloys.

Intergranular-Corrosion Test. The most common test for susceptibility to intergranular corrosion is carried out as follows:

- Use a specimen that has at least 19 cm² (3 in.²) of surface area.
- Remove any cladding by filing or etching.
- Clean the specimen by immersing it for 1 min in a solution containing 5% concentrated nitric acid and 0.5% hydrofluoric acid at a temperature of 95 °C (200 °F); rinse in distilled water. Immerse for 1 min in concentrated nitric acid at room temperature; rinse in distilled water.
- Immerse the specimen for 6 h in a freshly prepared solution containing 57 g of sodium chloride and 10 ml of 30% hydrogen peroxide per litre of water at a temperature of 30 ± 5 °C (86 ± 9 °F). More than one specimen may be corroded in the same container provided that at least 4.6 ml of solution is used for each square cen-

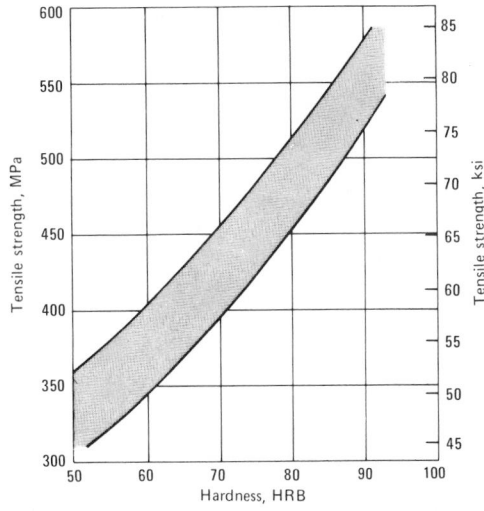

Fig. 31 Tensile strength vs hardness for various aluminum alloys and tempers

timetre (30 ml/in.²) of specimen surface and that the specimens are electrically insulated from each other.

- After the immersion period, wash the specimen with a soft-bristle brush to remove any loose corrosion product. Cut a cross-sectional specimen at least 19 mm (³⁄₄ in.) long through the most severely corroded area; mount and metallographically polish this specimen.
- Examine the cross-sectional specimen microscopically at magnifications of 100× and 500× both before and after etching with Keller's reagent.
- Describe the results of the microscopic examination in terms of the five degrees of severity of intergranular attack illustrated in Fig. 32, as follows: x areas of severity (a), y areas of severity (b), z areas of severity (c), and so on.

Electrical Conductivity. For control of the corrosion and stress-corrosion characteristics of certain tempers, notably the T73 and T76 types, the materials must meet combination criteria of yield strength plus electrical conductivity. Although these criteria are based on indirect measurements of properties, their validity for ensuring the intended corrosion and stress-corrosion resistance has been firmly established by extensive correlation and testing.

Low tensile strengths may be accompanied by high levels of electrical con-

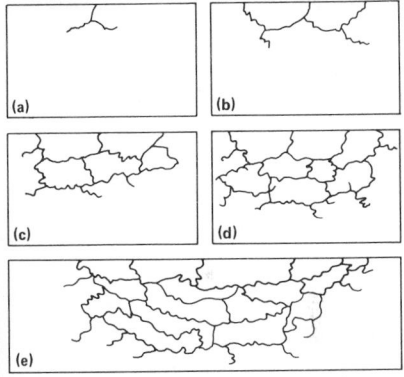

Fig. 32 Five degrees of severity of intergranular attack

Severity of intergranular attack (schematic), as observed microscopically in transverse sections after test for susceptibility to intergranular corrosion. Top of each area shown in surface exposed to corrosive solution. See text.

ductivity, so electrical conductivity is sometimes used as a quality-assurance diagnostic tool. However, because the correlation between strength and electrical conductivity is strongly a function of chemical composition and fabricating practice, use of electrical conductivity is not recommended except for rough screening. This screening must be followed by hardness testing, and then by tensile testing if the hardness tests indicate that the heat treatment was suspect.

Table 9 Typical acceptable hardness values for wrought aluminum alloys

Acceptable hardness does not guarantee acceptable properties; acceptance should be based on acceptable hardness plus written evidence of compliance with specified heat treating procedures. Hardness values higher than the listed maximums are acceptable provided that the material is positively identified as the correct alloy.

Alloy and temper	Product form(a)	Hardness HRB	HRE	HRH	HR15T
2014-T3, -T4, -T42	All..........................	65-70	87-95	...	...
2014-T6, -T62, -T65	Sheet(b).....................	80-90	103-110	...	...
	All others...................	81-90	104-110	...	...
2014-T61	All..........................	...	100-109	...	...
2024-T3	Not clad(c)...................	69-83	97-106	111-118	82.5-87.5
	Clad, thru 1.60 mm (0.063 in.).................	52-71	91-100	109-116	80-84.5
	Clad, over 1.60 mm (0.063 in.).................	52-71	93-102	109-116	...
2024-T36	All..........................	76-90	100-110	...	85-90
2024-T4, -T42(d)	Not clad....................	69-83	97-106	111-118	82.5-87.5
	Clad, thru 1.60 mm (0.063 in.).................	52-71	91-100	109-116	80-84.5
	Clad, over 1.60 mm (0.063 in.).................	52-71	93-102	109-116	...
2024-T6, -T62	All..........................	74.5-83.5	99-106	...	84-88
2024-T81	Not clad....................	74.5-83.5	99-106	...	84-88
	Clad........................	...	99-106	...	...
2024-T86	All..........................	83-90	105-110	...	87.5-90
6053-T6	All..........................	...	79-87	...	74.5-78.5
6061-T4(d)	Sheet......................	...	60-75	88-100	64-75
	Extrusions; bar..............	...	70-81	82-103	67-78
6061-T6	Not clad, 0.41 mm (0.016 in.)..................	...	...	...	75-84
	Not clad, 0.51 mm (0.020 in.) and over.......................	47-72	85-97	...	78-84
	Clad........................	...	84-96	...	...
6063-T5	All..........................	...	55-70	89-97	62.5-70
6063-T6	All..........................	...	70-85	...	...
6151-T6	All..........................	...	91-102	...	...
7075-T6, -T65	Not clad(e)..................	85-94	106-114	...	87.5-92
	Clad: Thru 0.91 mm (0.036 in.).................	...	102-110	...	86-90
	Over 0.91 thru 1.27 mm (over 0.036 thru 0.050 in.)..............	78-90	104-110	...	...
	Over 1.27 thru 1.57 mm (over 0.050 thru 0.062 in.)..............	76-90	104-110	...	...
	Over 1.57 thru 1.78 mm (over 0.062 thru 0.070 in.)..............	76-90	102-110	...	...
	Over 1.78 mm (0.070 in.).................	73-90	102-110	...	...
7079-T6, -T65	All(e)......................	81-93	104-114	...	87.5-92
7178-T6	Not clad(f)..................	85 min	105 min	...	88 min
	Clad: Thru 0.91 mm (0.036 in.).................	...	102 min	...	86 min
	Over 0.91 thru 1.57 mm (over 0.036 thru 0.062 in.)..............	85 min	...	...	...
	Over 1.57 mm (0.062 in.).................	88 min	...	...	...

(a) Minimum hardness values shown for clad products are valid for thicknesses up to and including 2.31 mm (0.091 in.); for heavier-gage material, cladding should be locally removed for hardness testing or test should be performed on edge of sheet. (b) 126 to 158 HB (10-mm ball, 500-kg load). (c) 100 to 130 HB (10-mm ball, 500-kg load). (d) Alloys 2024-T4, 2024-T42 and 6061-T4 should not be rejected for low hardness until they have remained at room temperature for at least three days following solution treatment. (e) 136 to 164 HB (10-mm ball, 500-kg load). (f) 136 HB min (10-mm ball, 500-kg load)

Indices of Fracture Toughness. In products of the newer high-strength alloys 2124, 2224, 2324, 7050, 7149, 7150, 7175 and 7475, which were developed to provide high fracture toughness, minimum values of the applicable indices, K_{Ic} or K_c, are being established by accumulation of statistical data from production lots as a basis for guaranteed minimum values. These indices and their attendant values indicate the resistance that a material possesses in regard to crack propagation. Existing specifications for certain products, alloys and tempers contain minimum values of these fracture-toughness indices.

Some specifications allow use of the less-expensive notch tensile test as a basis for release of high-toughness alloy products. In these instances, correlations between K_{Ic} and the ratio of the notch tensile strength and the yield strength of a companion smooth specimen were used to establish the appropriate notch-yield ratio as a lot-release criterion.

Temper Designations for Heat Treatable Aluminum Alloys

The temper designations used in the United States for heat treatable aluminum alloys are part of the system that has been adopted as an American National Standard (ANSI H35.1). Used for all wrought and cast product forms except ingot, the system is based on the sequences of mechanical or thermal treatments, or both, used to produce the various tempers. The temper designation follows the alloy designation and is separated from it by a hyphen. Basic temper designations consist of individual capital letters. Major subdivisions of basic tempers, where required, are indicated by one or more digits following the letter. These digits designate specific sequences of treatments that produce specific combinations of characteristics in the product. Variations in treatment conditions within major subdivisions are identified by additional digits. The conditions during heat treatment (such as time, temperature and quenching rate) used to produce a given temper in one alloy may differ from those employed to produce the same temper in another alloy.

Designations for the common heat treated tempers, and descriptions of the sequences of operations used to produce those tempers, are given in the following paragraphs. (For the entire aluminum alloy temper designation system, including designations for non-heat-treatable alloys, see pages 24 to 27 in Volume 2 of this Handbook.)

Basic Temper Designations

O Annealed. Applies to wrought products that are annealed to obtain lowest strength temper, and to cast products that are annealed to improve ductility and dimensional stability. The O may be followed by a digit other than zero.

W Solution heat treated. An unstable temper applicable to any alloy that naturally ages (spontaneously ages at room temperature) after solution heat treatment. This designation is specific only when the period of natural aging is indicated—for example, W ½ h. (See also the discussion of the Tx51, Tx52 and Tx54 tempers, in the section below on subdivision of the T temper.)

T Heat treated to produce stable tempers other than O. Applies to products that are thermally treated, with or without supplementary strain hardening, to produce stable tempers. The T is always followed by one or more digits, as discussed below.

Subdivision of T Temper

In T-type designations, the T is followed by a number from 1 to 10; each number denotes a specific sequence of basic treatments, as follows:

T1 Cooled from an elevated-temperature shaping process and naturally aged to a substantially stable condition. Applies to products that are not cold worked after an elevated-temperature shaping process such as casting or extrusion, and for which mechanical properties have been stabilized by room-temperature aging. If the products are flattened or straightened after cooling from the shaping process, the effects of the cold work imparted by flattening or straightening are not recognized in specified property limits.

T2 Cooled from an elevated-temperature shaping process, cold worked, and naturally aged to a substantially stable condition. Applies to products that are cold worked specifically to improve strength after cooling from a hot working process such as rolling or extrusion, and for which mechanical properties have been stabilized by room-temperature aging. The effects of cold work, including any cold work imparted by flattening or straightening, are recognized in specified property limits.

T3 Solution heat treated, cold worked, and naturally aged to a substantially stable condition. Applies to products that are cold worked specifically to improve strength after solution heat treatment, and for which mechanical properties have been stabilized by room-temperature aging. The effects of cold work, including any cold work imparted by flattening or straightening, are recognized in specified property limits.

T4 Solution heat treated and naturally aged to a substantially stable condition. Applies to products that are not cold worked after solution heat treatment, and for which mechanical properties have been stabilized by room-temperature aging. If the products are flattened or straightened, the effects of the cold work imparted by flattening or straightening are not recognized in specified property limits.

T5 Cooled from an elevated-temperature shaping process and artificially aged. Applies to products that are not cold worked after an elevated-temperature shaping process such as casting or extrusion, and for which mechanical properties or dimensional stability, or both, have been substantially improved by precipitation heat treatment. If the products are flattened or straightened after cooling from the shaping process, the effects of the cold work imparted by flattening or straightening are not recognized in specified property limits.

T6 Solution heat treated and artificially aged. Applies to products that are not cold worked after solution heat treatment, and for which mechanical properties or dimensional stability, or both, have been substantially improved by precipitation heat treatment. If the products are flattened or straightened, the effects of the cold work imparted by flattening or straightening are not recognized in specified property limits.

T7 Solution heat treated and stabilized. Applies to products that have been precipitation heat treated to the extent that they are overaged. Stabilization heat treatment carries the mechanical properties beyond the point of maximum strength to provide some special characteristic, such as enhanced resistance to stress-corrosion cracking or to exfoliation corrosion.

T8 Solution heat treated, cold worked, and artificially aged. Applies to products that are cold worked specifically to improve strength after solution heat treatment, and for which mechanical properties or dimensional stability, or both, have been substantially improved by precipitation heat treatment. The effects of cold work, including any cold work imparted by flattening or straightening, are recognized in specified property limits.

T9 Solution heat treated, artificially aged, and cold worked. Applies to products that are cold worked specifically to improve strength after they have been precipitation heat treated.

T10 Cooled from an elevated-temperature shaping process, cold worked, and artificially aged. Applies to products that are cold worked specifically to improve strength after cooling from a hot working process such as rolling or extrusion, and for which mechanical properties or dimensional stability, or both, have been substantially improved by precipitation heat treatment. The effects of cold work, including any cold work imparted by flattening or straightening, are recognized in specified property limits.

When it is desirable to identify a variation of one of the ten major T tempers described above, additional digits, the first of which cannot be zero, may be added to the designation.

The following specific sets of additional digits have been assigned to stress-relieved wrought products:

Tx51 Stress relieved by stretching. Applies to the following products when stretched to the indicated amounts after solution heat treatment or after cooling from an elevated-temperature shaping process:

Product form	Permanent set, %
Plate	1½ to 3
Rod, bar, shapes, extruded tube	1 to 3
Drawn tube	½ to 3

Applies directly to plate and to rolled or cold finished rod and bar. These products receive no further straightening after stretching. Applies to extruded rod, bar, shapes and tubing, and to drawn tubing, when designated as follows:

Tx510 Products that receive no further straightening after stretching

Tx511 Products that may receive minor straightening after stretching to comply with standard tolerances

Tx52 Stress relieved by compressing. Applies to products that are stress relieved by compressing after solution heat treatment, or after cooling from a hot working process to produce a permanent set of 1 to 5%

Tx54 Stress relieved by combining stretching and compressing. Applies to die forgings that are stress relieved by restriking cold in the finish die. (These same digits—and 51, 52 and 54—may be added to the designation W to indicate unstable solution heat treated and stress-relieved tempers.)

The following temper designations have been assigned to wrought products heat treated from the O or the F temper to demonstrate response to heat treatment:

T42 Solution heat treated from the O or the F temper to demonstrate response to heat treatment, and naturally aged to a substantially stable condition

T62 Solution heat treated from the O or the F temper to demonstrate response to heat treatment, and artificially aged

Temper designations T42 and T62 also may be applied to wrought products heat treated from any temper by the user when such heat treatment results in the mechanical properties applicable to these tempers.

Subdivision of O Temper

In temper designations for annealed products, a digit following the O indicates special characteristics. For example, O1 denotes that a product has been heat treated according to a time/temperature schedule approximately the same as that used for solution heat treatment, and then air cooled to room temperature, to accentuate ultrasonic response and provide dimensional stability; this designation applies to products that are to be machined prior to solution heat treatment by the user.

Heat Treating of Copper Alloys

By the ASM
Committee on Heat
Treating of Copper and
Copper Alloys*

HEAT TREATING PROCESSES that are applied to copper and copper alloys include homogenizing, annealing, stress relieving, solution treating, precipitation (age) hardening, and quench hardening and tempering.

Homogenizing

Homogenizing is a process wherein prolonged high-temperature soaking is used to reduce the extent of chemical segregation, or coring, which occurs as a natural result of solidification in some alloys. It is applied to copper alloys to improve the hot and cold ductility of cast billets for mill processing, and occasionally to castings to meet specified hardness, ductility or toughness requirements.

Homogenization is required most frequently for alloys with wide freezing ranges, such as tin (phosphor) bronzes, copper nickels and silicon bronzes. Although coring occurs to some extent in alpha brasses, alpha aluminum bronzes and copper-beryllium alloys, these alloys survive primary mill processing and become homogenized during normal process working and an-

nealing. There is rarely a necessity to apply homogenization to finished or semifinished mill products.

The time and temperature required for the process varies with the alloy, the cast grain size and the desired degree of homogenization. Typical soak times vary from 3 to over 10 h. Temperatures normally are above the upper annealing range, to within 50 °C (90 °F) of the solidus temperature.

The normal precautions that apply to annealing should be used for homogenization of any particular alloy. The furnace atmosphere should be selected so as to control both surface and internal oxidation. Where there is appreciable danger of liquefying segregated phases, the materials, particularly castings, should be well supported and heated slowly through the final 100 °C (180 °F).

Typical applications of homogenization are:

- Alloy C71900 (Cu-Ni-Cr) billets: 1040 to 1065 °C (1900 to 1950 °F) for 4 to 9 h, to prevent cracks, seams and excessive wood-fiber structure in extrusions
- Alloy C52100 and C52400 (phosphor

bronzes, 8 and 10% tin): 775 °C (1425 °F) for 5 h, to reduce embrittlement in billets and slabs that are to be cold rolled.

Annealing

Annealing is a heat treatment intended to soften, and increase the ductility and/or toughness of, metals and alloys. It is applied to wrought products during and after mill processing, and to castings. The process includes heating, holding and cooling, and a proper process description should include heating rate, temperature, time at temperature, atmosphere, and cooling rate where each may affect results.

Wrought Products

Annealing of cold worked metal is accomplished by heating to a temperature that produces recrystallization and, if desirable, by heating beyond the recrystallization temperature to produce grain growth. Temperatures commonly used for annealing cold worked coppers and copper alloys are given in Table 1.

Annealing is primarily a function of

*Patrick A. Tully, *Chairman*, Manager, Metallurgy, Ampco Pittsburgh Corp.; Carl J. Gaffoglio, Manager, Applications Engineering, Copper Development Association, Inc.; David S. Hibbard, Development Engineer, Research & Technical Center, Anaconda Industries; Theodore J. Louzon, Metallurgical Engineer, Bell Laboratories; J. Howard Mendenhall,Technical Associate, Olin Brass, Olin Corp.; Richard G. O'Rourke, Senior Quality Engineer, Brush Wellman, Inc.; Donald G. Schmidt, Consultant, R. Lavin & Sons, Inc.; Robert F. Schmidt, Technical Director, Colonial Metals Co.; Stanley Shapiro, President, Revere Research, Inc.

Table 1 Annealing temperatures for cold worked coppers and copper alloys

Alloy	Common name	Annealing temperature °C	°F
Wrought coppers			
C10200....................................	Oxygen-free copper	425-650	800-1200
C11000....................................	Electrolytic tough pitch copper	250-650	500-1200
C11300, C11400, C11500, C11600............................	Silver-bearing tough pitch copper	400-475	750-900
C12000....................................	Phosphorus-deoxidized copper, low residual phosphorus	325-650	600-1200
C12200....................................	Phosphorus-deoxidized copper, high residual phosphorus	375-650	700-1200
C14500....................................	Phosphorus-deoxidized, tellurium-bearing copper	425-650	800-1200
Wrought copper alloys			
C17000, C17200, C17500....................................	Beryllium copper	775-925(a)	1425-1700(a)
C21000....................................	Gilding metal	425-800	800-1450
C22000....................................	Commercial bronze	425-800	800-1450
C22600....................................	Jewelry bronze	425-750	800-1400
C23000....................................	Red brass	425-725	800-1350
C24000....................................	Low brass	425-700	800-1300
C26000....................................	Cartridge brass	425-750	800-1400
C26800, C27000, C27400....................................	Yellow brass	425-700	800-1300
C28000....................................	Muntz metal	425-600	800-1100
C31400....................................	Leaded commercial bronze	425-650	800-1200
C33000, C33500..........................	Low-leaded brass	425-650	800-1200
C33200, C34200, C35300....................................	High-leaded brass	425-650	800-1200
C34000, C35000..........................	Medium-leaded brass	425-650	800-1200
C35600....................................	Extra-high-leaded brass	425-650	800-1200
C36000....................................	Free-cutting brass	425-600	800-1100
C36500, C36600, C36700, C36800............................	Leaded Muntz metal	425-600	800-1100
C37000....................................	Free-cutting Muntz metal	425-650	800-1200
C37700....................................	Forging brass	425-600	800-1100
C38500....................................	Architectural bronze	425-600	800-1100
C44300, C44400, C44500....................................	Inhibited admiralty	425-600	800-1100
C46200....................................	Naval brass	425-600	800-1100
C48200, C48500..........................	Leaded naval brass	425-600	800-1100
C50500....................................	Phosphor bronze	475-650	900-1200
C51000, C52100, C54200....................................	Phosphor bronze	475-675	900-1250
C53200, C53400, C54400....................................	Free-cutting phosphor bronze	475-675	900-1250
C60600, C60800..........................	Aluminum bronze	550-650	1000-1200
C61000....................................	Aluminum bronze	600-675	1100-1250
C61300, C61400..........................	Aluminum bronze	750-875	1400-1600
C61800, C61900 C62400....................................	Aluminum bronze	600-650(b)	1100-1200(b)
C63000....................................	Aluminum bronze	650-700(c)	1200-1300(c)
C63200....................................	Aluminum bronze	675-725(c)	1250-1350(c)
C64200....................................	Aluminum bronze	Above 650	Above 1200
C65100....................................	Low-silicon bronze	475-675	900-1250
C65500....................................	High-silicon bronze	475-700	900-1300
C67000, C67500..........................	Manganese bronze	425-600	800-1100
C68700....................................	Aluminum brass	425-600	800-1100
C70600....................................	Copper nickel, 10%	600-825	1100-1500
C71500....................................	Copper nickel, 30%	650-825	1200-1500
C75200, C75700, C77000....................................	Nickel silver	600-825	1100-1500

(a) Solution-treating temperature; see Table 5 for temperatures for specific alloys. (b) Cool rapidly (cooling method important in determining result of annealing). (c) Air cool (cooling method important in determining result of annealing).

metal temperature and time at temperature. Except for multiphase alloys and certain precipitation-hardening alloys, or alloys susceptible to fire cracking, rates of heating and cooling are relatively unimportant. Source and application of heat, furnace design, furnace atmosphere and shape of workpiece are important, because they affect finish, cost of annealing and uniformity of results obtained.

Because of the multiplicity of influential variables, it is difficult to tabulate a definite annealing schedule that will result in completely recrystallized metal of a specific grain size. The effects of annealing temperature on the tensile strength, elongation and grain size of hard drawn (63%) C27000 (yellow brass) wire annealed for 1 h and the effect of annealing time on the grain size of C27000 strip are illustrated in Fig. 1.

The annealing response of Alloy C26000 (cartridge brass) strip after a reduction of 40.6% by cold rolling is shown in Fig. 2. Time at temperature was 1 h. The actual increases in hardness and tensile properties shown at temperatures below the recrystallization range are typical of alloys such as brasses, nickel silvers, phosphor bronzes and alpha aluminum bronzes. Depending on the individual alloy, these increases are attributable to phenomenon of the strain-aging and/or lattice-ordering type.

An increased amount of cold work prior to annealing lowers the recrystallization temperature. The lower the degree of prior deformation, the larger the grain size after annealing. For fixed temperature and duration of annealing, the larger the original grain size before working, the larger the grain size after recrystallization.

In commercial mill practice, copper alloys usually are annealed at successively lower temperatures as the material approaches the final anneal, with intermediate cold reductions of at least 35% and as high as 50 to 60% in single or multiple passes wherever practicable. The higher temperatures initially used accelerate homogenization, and the resulting large grains permit more economical reduction during the early working operation.

During subsequent anneals, the grain size should be decreased gradually to approximate the final grain size required. This point usually is reached one or two anneals before the final anneal. With such a sequence and with sufficiently severe intermediate reductions, it is then possible to produce a

Fig. 1 Effects of annealing temperature and time on characteristics of C27000 wire and strip

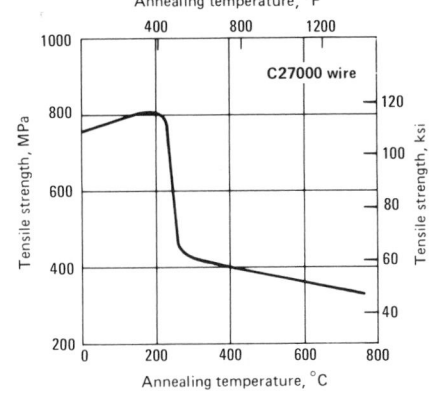
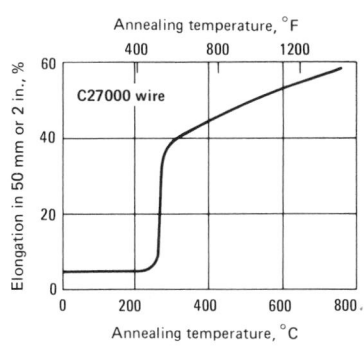
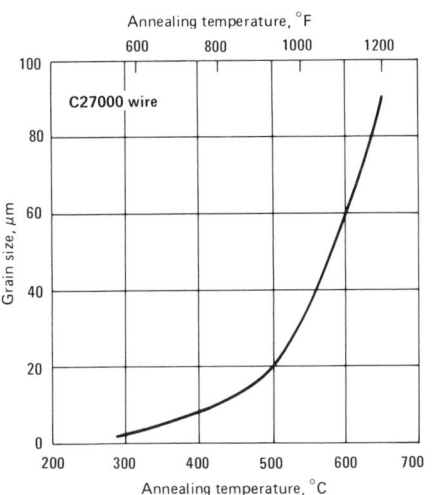
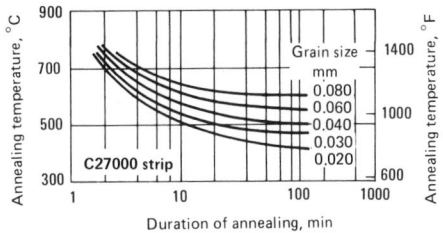

Effects of annealing temperature (annealing time: 1 h) on tensile strength, elongation and grain size of C27000 wire hard drawn 63%; effect of annealing time on grain size of C27000 strip 1.3 mm (0.050 in.) thick

uniform final grain size within a lot and from lot to lot.

The grain size and mechanical properties required for further cold working vary considerably with the alloy and with the amount and kind of further cold work to be done. The object of annealing for cold working is to obtain the maximum combination of ductility and strength. However, when press-

drawn parts are to be finished by polishing and buffing, the grain size should be as fine as practicable to keep the surface texture smooth and thus to avoid the need for excessive buffing and the attendant excessive finishing costs. The anneal must be governed by definite specifications and coordinated with cold working operations, to yield the desired finished properties.

Fig. 2 Annealing data on alloy C26000

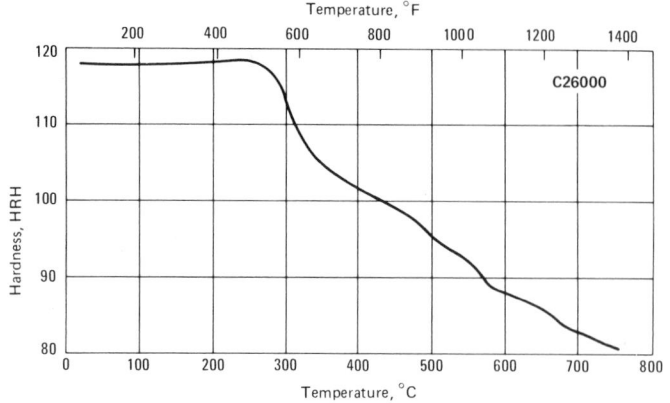

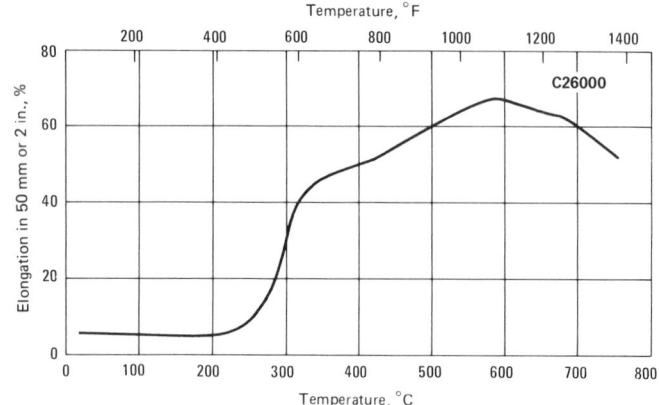

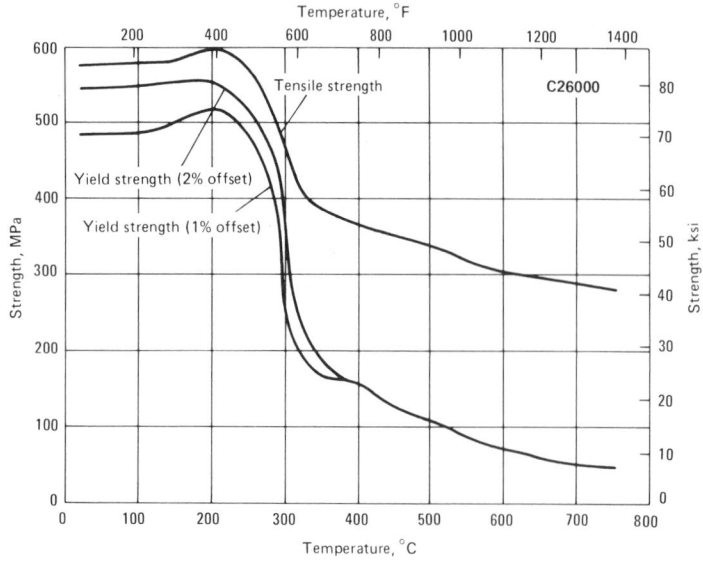

Finish rolling reduction 40.6%

Annealing to Specific Properties

Although specific properties are most frequently produced by controlled cold working of annealed material, there are occasions in which annealing to temper is necessary or advantageous. In hot rolling of copper alloy plate—particularly large patterns—the finishing temperature may not be consistent or controllable, and varying degrees of work hardening may occur. Also, small quantities and/or odd sizes of required drawn or roll tempered materials may not be readily available, while appropriate stocks of harder material are. Thin-gage strip (0.25 mm or 0.010 in. thick) for radiator fabrication produced by annealing to temper is more closely controlled and superior for fabrication than strip in cold worked tempers. In each case, an anneal is used to alter hardness and tensile properties to levels between the hard and fully annealed tempers with reasonably predictable results. For most copper alloys, the rapid drops in tensile properties and hardness that occur with an increase in temperature in the annealing range necessitate very close control of the annealing process to produce the desired results. Temperatures in the lower annealing range are used, with special precautions to avoid any overheating. The resultant microstructures may indicate incomplete recrystallization for the harder tempers, and grain sizes generally up to 0.025 mm (0.001 in.) for softer tempers. Tensile strengths and hardness levels similar to those of $\frac{1}{8}$, $\frac{1}{4}$ and $\frac{1}{2}$ hard cold worked tempers can be produced by annealing hard worked brasses, nickel silvers and phosphor bronzes. While the yield strength for a given final hardness tends to be lower for alloys annealed to temper than for those cold worked to temper, the fatigue resistance of some phosphor bronze spring materials in annealed $\frac{1}{2}$ hard tempers appears to be superior to that of cold worked material. Table 2 gives typical properties of annealed-to-temper mill materials. The successful use of annealing to provide specific tempers in mill products requires well-regulated working and annealing schedules designed to produce homogeneous material with controlled grain size, such that the final anneal can produce a uniform result throughout a material lot.

General Precautions

For best results in annealing copper

Fig. 2 (continued)

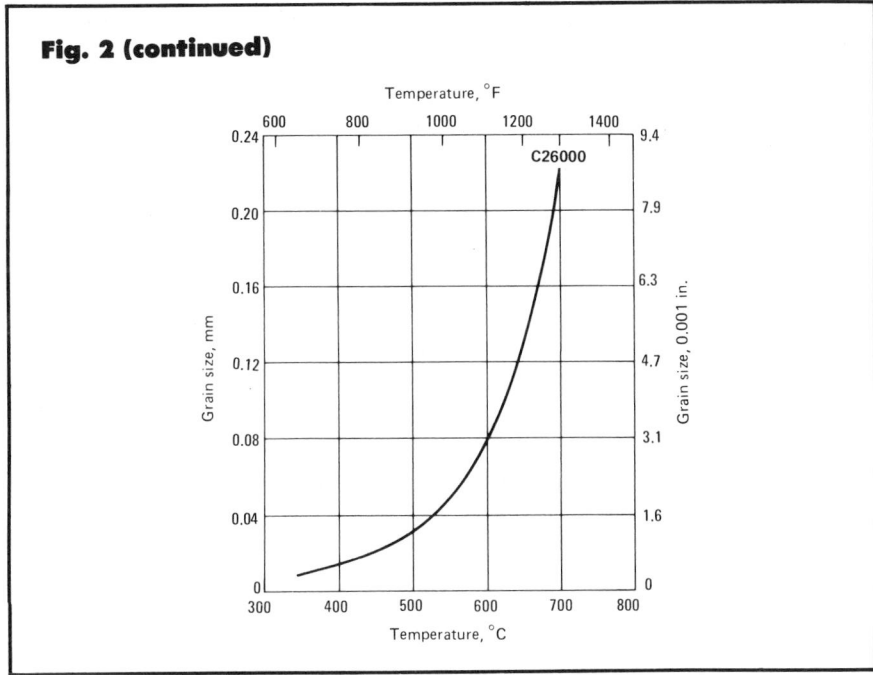

Table 2 Typical properties of copper alloys annealed to temper

Alloy	Common name	Annealed temper	Tensile strength MPa	ksi	Approximate hardness, HR30T
C26000	Cartridge brass	¼ hard	340-405	49-59	43-51
		½ hard	395-460	57-67	56-66
C51100, C53200,					
C53400, C54400	Phosphor bronze	½ hard	380-485	55-70	57-73
C75200	Nickel silver	¼ hard	400-495	58-72	49-67
		½ hard	455-550	66-80	62-72

and copper alloys, the following precautions should be observed.

Sampling and Testing. Test specimens must represent the extreme conditions of the furnace load. For copper alloys that do not contain grain-growth inhibitors, the best and most accurate test for the extent of annealing is the size of the average grain. Grain size is usually the basis for acceptance or rejection of the material. This determination requires special equipment not always available in the plants of consumers or fabricators. For convenience in testing, Rockwell-type hardness testers are used to approximate the grain size; ASTM specifications correlate Rockwell hardness with grain-size values for many copper alloys.

Effect of Pretreatment. Because the amount of cold working and the anneal prior to cold working greatly affect the results of annealing after cold working, any schedule that is set up must take this pretreatment into account. Once a schedule has been estab-

lished, both the anneal and the pretreatment must be adhered to for consistent results.

Effect of Time. In most furnaces, there is an appreciable difference between the temperature of the metal and that of the furnace; consequently, time in the furnace greatly affects the final temperature of the metal. For a fixed anneal and furnace temperature, time must vary with type of work load.

Oxidation should be held to a minimum, to reduce loss of metal and costs of pickling, and to improve finish. In some instances, specially prepared atmospheres are used to produce a bright or clean annealed material. Control of furnace atmosphere usually also results in better furnace economy.

Effect of Lubricants. Lubricants on metal to be annealed may cause staining that is difficult to remove. Regardless of the type of furnace or the article to be annealed, it is advisable to eliminate as much of the lubricant as possible before the metal is heated.

Hydrogen Embrittlement. When copper that contains oxygen is to be annealed, the hydrogen in the furnace atmosphere must be kept to a minimum. This reduces the embrittlement caused by the combination of the hydrogen in the atmosphere with the oxygen in the copper, forming water vapor under pressure and resulting in minute porosity in the metal. For temperatures lower than about 480 °C (900 °F), the hydrogen content of the atmosphere preferably should not exceed 1%, and as the temperature is increased, hydrogen content should approach zero.

Impurities. Occasionally, it is difficult to obtain proper grain growth by annealing under standard conditions that previously have resulted in the desired grain size. This difficulty sometimes may be traced to impurities in the alloy.

Loading. It usually is inadvisable to anneal a variety of different sizes or kinds of material in the same charge, because of the different rates of heating and the resulting final metal temperatures.

Fire cracking occurs when some alloys that contain residual stresses are heated too rapidly. Leaded alloys are particularly susceptible to fire cracking. The remedy is to heat slowly until the stresses are relieved. Special types of cold deformation, such as "springing" (flexing or reeling through a straightener), aid considerably in preventing fire cracking by inducing countervailing mechanical stresses.

Thermal shock or fatigue occurs when rapid and extreme changes in temperatures occur. Stresses that result in thermal shock are influenced by thermal expansion, themal conductivity, strength, toughness and the rate of change of the temperature and the condition of the material. Brasses containing lead, lead and tin or lead and certain impurities including bismuth or tellurium may be hot short. If they are repeatedly subject to extreme temperature changes, they may be subject to thermal shock, especially if highly stressed in tension on the surface.

Cooling. Alpha brasses containing less than 70% copper may contain beta that is formed during casting or during heat treatment above 600 °C (1110°F), especially if the metal section is massive. Quenching rapidly will entrap beta in the brass. Slow cooling will permit the time and temperature to convert the beta to alpha brass.

Sulfur Stains. Excessive sulfur in the fuel or lubricant will cause discoloration of the metal; red stains appear on yellow brass, and black or reddish-brown stains on copper-rich alloys.

Castings

Annealing is applied to castings of some duplex alloys, such as manganese bronzes and aluminum bronzes, where it is intended to correct the effects of mold cooling. The extremely slow cooling of sand and plaster castings, or the rapid cooling of permanent mold or die castings, can produce microstructures resulting in high hardness and/or low ductility, and occasionally inferior corrosion resistance. Typical annealing treatments for castings are in the range of 580 to 700 °C (1075 to 1300 °F) for 1 h at temperature. For aluminum bronzes, rapid cooling by water quenching or high-velocity air is advisable.

Stress Relieving

Stress relieving is a process intended to relieve internal stress in materials or parts without appreciably affecting their properties. Stress-relieving heat treatments are applied to copper alloys as one means of accomplishing this objective.

During processing or fabrication of copper alloys by cold working, strength and hardness increase due to plastic strain. Because plastic strain is accompanied by elastic strain, residual stresses remain in the resultant product. If allowed to remain in sufficient magnitude, residual surface tensile stresses can result in stress-corrosion cracking of material in storage or service, unpredictable distortion of material during cutting or machining, and hot cracking of materials during processing, brazing or welding. In brasses that contain more than 15% zinc, stress-corrosion cracking, or "season cracking", can occur if both sufficient amounts of residual tensile stress and small amounts of atmospheric ammonia are present. Other copper alloys, such as cold worked aluminum bronzes and silicon bronzes, may also suffer stress corrosion cracking under more severe environments.

Although mill practice for stress relief frequently involves mechanical means such as flexing, cross-roll straightening or shot peening, stress-relief heat treatments are employed for some tubular products and odd shapes. Thermal stress relief is also used for formed parts and fabrications made by material users. It is important to recog-

nize that thermal stress relief reduces residual stress by eliminating part of the residual elastic strain, whereas mechanical stress relief merely redistributes residual stress into a less detrimental pattern.

Stress-relief heat treatments are carried out at temperatures below those normally used for annealing. Typical process stress-relieving temperatures for 19 wrought copper alloys are given in Table 3. Temperatures for treatment of cold formed or welded structures are generally 50 to 110 °C (90 to 200 °F) higher.

From a practical standpoint, higher-temperature/shorter-time treatments are preferable. However, to guarantee the preservation of mechanical properties, lower temperatures and longer times are sometimes necessary. The optimum cycle produces adequate stress relief without adversely affecting properties. As illustrated in Fig. 2, some alloys may undergo slight increases in properties during stress-relief heat treatment.

To detect the presence of significant residual stress, and to evaluate the effectiveness of stress-relieving treatments, samples of material may be tested with mercurous nitrate solutions as described in ASTM B154. Because of the hazards of mercurous salts, tests in high concentrations of moist ammonia have also been used. Warping of rod or tube during longitudinal saw slitting has also been used as a crude field test for residual stress.

Table 3 Typical stress-relieving temperatures for 19 wrought copper alloys

Alloy	Common name	Stress-relieving temperature(a) °C	°F
C21000	Gilding metal	190	375
C22000	Commercial bronze	205	400
C23000	Red brass	230	450
C24000	Low brass	260	500
C26000	Cartridge brass	260	500
C27000	Yellow brass	260	500
C28000	Muntz metal	205	400
C36000	Free-cutting brass	245	475
C44300, C44400, C44500	Inhibited admiralty	290	550
C51000, C52100	Phosphor bronze	205	400
C61300, C61400	Aluminum bronze	345	650
C65500	High-silicon bronze	345	650
C70600, C71500	Copper nickel	260	500
C75200	Nickel silver	260	500

(a) Time at temperature, 1 h

Hardening

Copper alloys that are hardened through heat treatment are of two general types: those that are softened by high-temperature quenching and hardened by lower-temperature treatments, and those that are hardened by quenching from high temperatures through martensitic-type reactions. Alloys that harden during low-to-intermediate-temperature treatments following solution quenching include precipitation-hardening, spinodal-hardening and order-hardening types. Quench-hardening alloys comprise aluminum bronzes, nickel-aluminum bronzes, and a few copper-zinc alloys. Quench-hardened alloys normally are tempered to improve toughness and ductility and reduce hardness in a manner similar to that for alloy steels.

Low-Temperature-Hardening Alloys

For purposes of comparison, Table 4 lists examples of the various types of low-temperature-hardening alloys, as well as with typical heat treatments and attainable property levels for these alloys. Additional details are given in the three subsections that follow.

Precipitation-Hardening Alloys. Most copper alloys of the precipitation-hardening type find use in electrical- and heat-conduction applications. The heat treatment must therefore be designed to develop the necessary mechanical strength and electrical con-

Table 4 Typical heat treatments and resulting properties for several low-temperature-hardening alloys

Alloy	Solution-treating temperature(a)		Aging treatment Temperature		Time, h	Hardness	Electrical conductivity, % IACS
	°C	°F	°C	°F			
Precipitation hardening							
C15000 980		1795	500-550	930-1025	3	30 HRB	87-95
C17000, C17200, C17300 760-800		1400-1475	300-350	575-660	1-3	35-44 HRC	22
C17500, C17600 900-950		1650-1740	455-490	850-915	1-4	95-98 HRB	48
C18000(b), C81540 900-930		1650-1705	425-540	800-1000	2-3	92-96 HRB	42-48
C18200, C18400, C18500, C81500 980-1000		1795-1830	425-500	800-930	2-4	68 HRB	80
C94700 775-800		1425-1475	305-325	580-620	5	180 HB	15
C99400 885		1625	482	900	1	170 HB	17
Spinodal hardening							
C71900 900-950		1650-1740	425-760	800-1400	1-2	86 HRC	4-4
C72800 815-845		1500-1550	350-360	660-680	4	32 HRC	···

(a) Solution treating is followed by water quenching. (b) Alloy C18000 (81540) must be double aged—typically, 3 h at 540 °C (1000 °F) followed by 3 h at 425 °C (800 °F) (U.S. Patent No. 4 191 601)—to develop the higher levels of electrical conductivity and hardness.

ductivity. The resulting hardness and strength depend on both the effectiveness of the solution quench and the control of the precipitation (aging) treatment. Note: "age hardening" or "aging" is used in heat treating practice as substitutes for the terms "precipitation" or "spinodal hardening." Copper alloys harden by elevated temperature treatment rather than ambient temperature (natural) aging as in the case of some aluminum alloys. As solute atoms proceed through the coagulation, coherency, and precipitation cycle in the quenched alloy lattice, the hardness increases, reaches a peak, and then decreases with time. Electrical conductivity increases continuously with time until some maximum is reached, normally in the fully precipitated condition. The optimum condition generally preferred results from a precipitation treatment of temperature and duration just beyond those that correspond to the hardness aging peak. Cold working prior to precipitation aging tends to improve the heat treated hardness. In the case of lower-strength wrought alloys such as C18200 (Cu-Cr) and C15000 (Cu-Zr), some heat treated hardness may be sacrificed to attain increased conductivity, with final hardness and strength being enhanced by cold working. Two precipitation treatments are necessary in order to develop maximum electrical conductivity and hardness in alloy C18000 (Cu-Ni-Si-Cr) because of two distinct precipitation mechanisms.

Problems in producing anticipated properties in precipitation-hardening alloys may be diagnosed in accordance with the following guidelines:

Problem	Diagnosis
Low hardness	Solution temperature too low; solution quench delayed or cooling rate too low; aging temperature too low and/or time too short (underaging), or temperature too high and/or time too long (overaging)
Low hardness; low conductivity	Inadequate solution treatment and/or underaging
Low hardness; high conductivity	Inadequate solution treatment and/or overaging
High hardness; low conductivity	Underaging; contaminated material

When precipitation hardening is performed at the mill, further treatment following fabrication of parts is not required. However, it may be desirable to stress relieve parts to remove stresses induced during fabrication, particularly for highly formed cantilever-type springs and intricate machined shapes that require maximum resistance to relaxation at moderately elevated temperatures.

Spinodal-Hardening Alloys. Alloys that harden by spinodal decomposition are hardened by a treatment similar to that used for precipitation-hardening alloys. The soft and ductile spinodal structure is generated by a high-temperature solution treatment followed by quenching. The material can be cold worked or formed in this condition. A lower-temperature spinodal-decomposition treatment, commonly referred to as aging, is then used to increase the hardness and strength of the alloy. Spinodal-hardening alloys are basically copper-nickel alloys with chromium or tin additions. The hardening mechanism is related to a miscibility gap in the solid solution and does not result in precipitation. The spinodal-hardening mechanism results in chemical segregation of the alpha crystal matrix on a very fine (Angstrom) scale, and requires the use of the electron microscope to discern the metallographic effects. Since no crystallographic changes take place, spinodal-hardening alloys retain excellent dimensional stability during hardening.

Order-Hardening Alloys. Certain alloys, generally those that are nearly saturated with an alloying element dissolved in the alpha phase, will undergo an ordering reaction when highly cold worked material is annealed at a relatively low temperature. Alloys C61500, C63800, C68800 and C69000 are examples of copper alloys that exhibit this behavior. Strengthening is attributed to short-range ordering of the solute atoms within the copper matrix, which greatly impedes the motion of dislocations through the crystals.

The low-temperature order-annealing treatment also acts as a stress-relieving treatment, which raises yield strength by reducing stress concentrations in the lattice at the focuses of dislocation pileups. As a result, order-annealed alloys exhibit improved stress-relaxation characteristics.

Order annealing is done for relatively short times at relatively low temperatures, generally in the range from 150 to 400 °C (300 to 750 °F). Because of the low temperature, no special protective atmosphere is required. Order hardening is frequently done after the final fabrication step to take full advantage of the stress-relieving aspect of the treatment, especially where resistance to stress relaxation is desired.

Quench Hardening and Tempering

Quench hardening and tempering (also referred to as "quench-and-temper hardening") is used primarily for aluminum bronze and nickel aluminum bronze alloys, and occasionally for some cast manganese bronze alloys with zinc equivalents of 37 to 41%. Aluminum bronzes with 9 to 11.5% Al, and nickel aluminum bronzes with 8.5 to 11.5% Al, respond in a practical way to quench hardening by a martensitic-type reaction. Alloys higher in aluminum content generally are too susceptible to quench cracking, whereas those with lower aluminum contents do not contain enough high-temperature beta phase to respond to quench treatments.

Heat Treating Equipment

Although basic furnace design is similar for all copper alloys, consideration must be given to the annealing temperature range and method of cooling. Solid-solution alloys that do not precipitation harden usually are annealed at temperatures below 760 °C (1400 °F) and may be cooled at any convenient rate. Precipitation or spinodal hardenable alloys are solution treated at temperatures up to 1040 °C (1900 °F) and require rapid quenching to ambient temperatures.

Batch-type atmosphere furnaces may be heated electrically or by oil or gas. When nonexplosive atmospheres are used, electrically heated furnaces permit the atmosphere to be introduced directly into the work chamber.

Furnaces that are heated by gas or oil and that employ protective atmospheres sometimes use a muffle to contain the atmosphere and protect the work from the direct fire of the burners. A properly constructed and safely operated muffle that prevents infiltration of air by maintaining positive pressure is always required when explosive atmospheres, such as hydrogen, are used.

When protective atmospheres are

Fig. 3 Continuous conveyor furnace for heat treating copper alloys in a controlled atmosphere

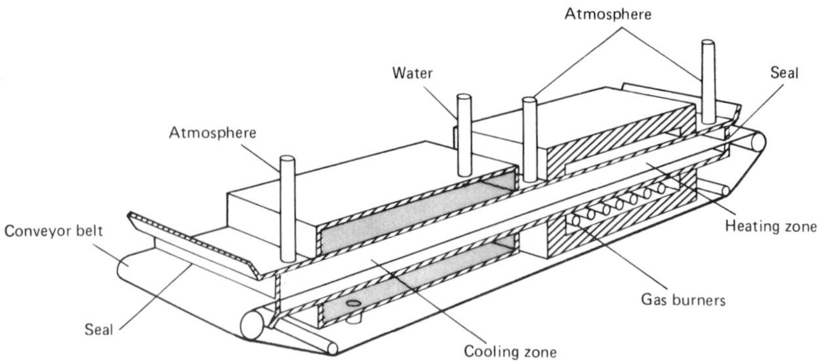

used during annealing, the work must be cooled in the atmosphere nearly to room temperature, to prevent surface scale or discoloration. Metal temperatures above 65 °C (150 °F) in air may result in light tarnishing. If some degree of surface oxidation and discoloration can be tolerated, direct-natural-gas-fired furnaces may be used. The products of combustion from the gas-air burners are controlled to yield reducing combustion products similar in composition to manufactured protective atmospheres. Parts annealed in reducing atmospheres developed by control of the furnace air/gas ratio require cleaning to restore luster.

Continuous atmosphere furnaces (Fig. 3) offer versatility for solution heat treating of a wide variety of products. Usually, the furnace consists of a vestibule that provides a seal for the atmosphere and in some instances preheats the work, a heating chamber of sufficient length to ensure complete solution treating, and a cooling or quenching chamber that also serves as an atmosphere seal.

Because the work usually is conveyed at a fixed rate through the furnace, moderate temperature gradients are less harmful than in batch furnaces. When long heating chambers are required, the furnace may be divided into individual temperature-controlled heating zones. Thus, it is practical to develop a high temperature in the entrance zone to facilitate heating of the work to the desired temperature. The cooling chamber may be either a long tunnel through which cool protective atmosphere is circulated or a water-quench zone supplied with a protective atmosphere.

Products such as stampings, ma-

chined shapes, castings and small assemblies are conveyed through the furnace on an endless belt or conveyor chain. Long sections such as tubing, bar and flat products, or heavy sections that permit stacking on trays, may be conveyed on a roller hearth. In rolling-mill operations, the product is uncoiled at the entrance of the furnace and pulled through the furnace by terminal equipment at the exit end; thus, there are no moving parts within the furnace. For wire products, annealing is carried out in bell furnaces, with the wire reel-wound, or in-line resistance annealing is performed upon exit of the product from the drawing machine prior to reel-winding.

Salt Baths. Molten neutral salts may be used for annealing, stress relieving, solution heat treating or aging of copper alloys. The composition of the salt mixture depends on the temperature range required. For heating between 705 and 870 °C (1300 and 1600 °F), mixtures of sodium chloride and potassium chloride are commonly used. Various mixtures of barium chloride with sodium and potassium chlorides are used for a wider temperature range (595 to over 1095 °C, or 1100 to over 2000 °F). The latter mixtures are compatible with each other and are commonly used in multiple-furnace operations when it is advantageous to preheat the work in one mixture at a low temperature and then transfer the work to a high-temperature bath. The least common neutral salts are mixtures of calcium chloride, sodium chloride and barium chloride. They have an operating temperature range of 540 to 870 °C (1000 to 1600 °F) but usually are operated between 540 and 650 °C (1100 and 1200 °F).

Fig. 4 Temperature variations in two types of furnaces (Example 1)

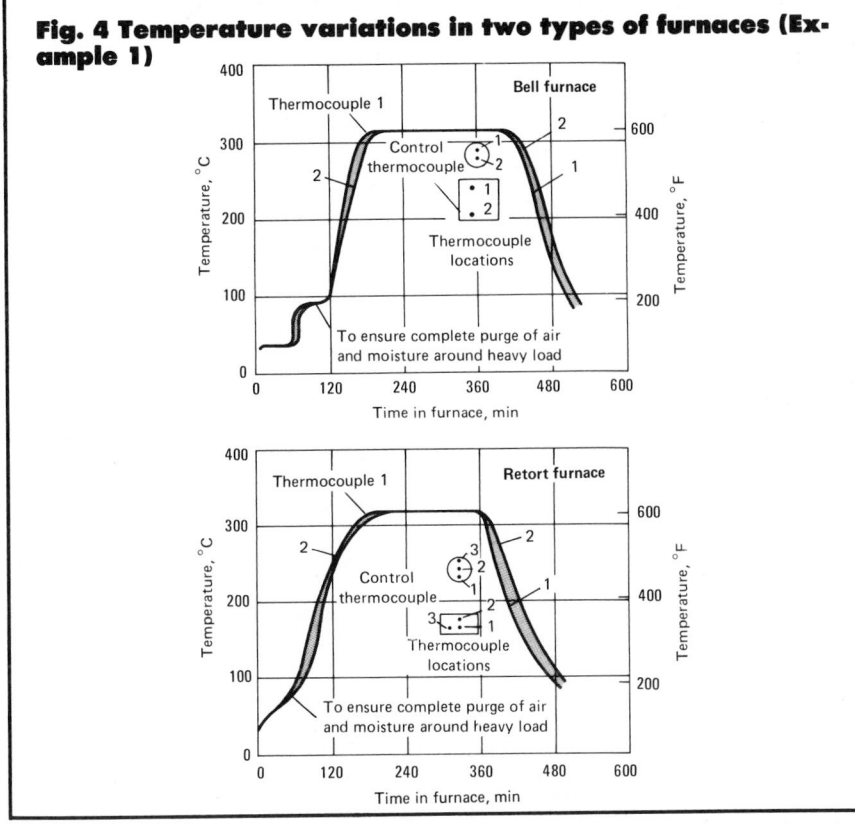

Fig. 5 Effects of metal thickness and heating medium on aging time required to develop maximum strength in C17200 strip

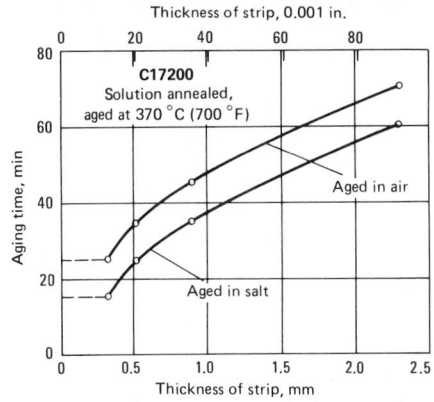

The sodium chloride – carbonate mixtures (not true neutral salts) are used between 595 and 925 °C (1100 and 1700 °F), primarily for annealing.

For operating temperatures below 540 °C (1000 °F), the only practical mixtures are the nitrate-nitrite salts.

Cyanide-base salts have limited application for heating of copper alloys. Although copper is soluble in cyanide, these salts can be used, with precaution, when a very bright finish is required.

None of the above salt mixtures is applicable for solution treating of standard beryllium copper alloys, because of intergranular attack, pitting or discoloration.

Aging and stress-relieving operations require furnace equipment that can be controlled to within 3 °C (5 °F) throughout the work zone. Unless cleaning after heating is permissible, it may be necessary to use controlled-atmosphere or vacuum equipment.

Because of the necessity for close temperature control, forced-convection (recirculating-air) and salt-bath furnaces are commonly used for aging and stress relieving. Forced-convection furnaces may be of box, bell or pit type. Each is equipped with a fan that recir-culates the constant-temperature atmosphere over the work. When forced-convection furnaces are fired by gas or oil and protective atmosphere or vacuum is used, the work must be contained in a properly operating muffle chamber or retort to seal off all products of combustion and to prevent infiltration of air. Temperature variations and heating and cooling times are compared in the following example.

Example 1. A comparison was made of temperature variations in a bell furnace and in a pit retort furnace during heat treating of small flat springs made of beryllium copper (see Fig. 4). Both furnaces were rated at 30 kW. The load in each furnace weighed 90 kg (200 lb) and contained 55 000 to 60 000 springs. An exothermic gas, produced by a generator using an air/gas ratio of 6.75 to 1 (capacity, 10 m³/h, or 350 ft³/h), was used as the protective atmosphere. The composition of the atmosphere was 6.5% CO, 6% CO_2, 10% H_2, rem N_2; dew point was 2 °C (35 °F) after refrigeration (18 to 21 °C, or 65 to 70 °F, as generated).

Salt baths can reduce total furnace time by up to 30%, as compared to that required with atmosphere furnaces (Fig. 5). Salt baths are particularly valuable when the age-hardening time is of short duration and precise control of time at the aging temperature is required.

Commercially available nitrate-nitrite salt mixtures (40 to 50% sodium nitrate, remainder sodium or potassium nitrite) that melt at 143 °C (290 °F) are used for aging and stress relieving. All material to be heated in salt should be properly cleaned and dried before being immersed in the molten salt; any organic substance (such as oil or grease) will react violently with the nitrate-nitrite salt.

Protective Atmospheres

Selection of protective atmospheres for heat treating of copper and copper alloys is influenced by the temperature used in the heat treating process.

Heating Above 705 °C (1300 °F). An exothermic atmosphere is the least expensive protective atmosphere for heat treating of copper alloys. The air/gas ratio is adjusted to produce a combusted gas that contains 2 to 7% hydrogen for use in muffle furnaces operating at 705 to 995 °C (1300 to 1825 °F). This atmosphere is used successfully for solution treating of alloys such as beryllium coppers, chromium coppers, zirconium coppers and copper-nickel-silicon alloys.

Usually, combusted gases are dried with a surface cooler, using tap water to keep the water/hydrogen ratio reducing throughout the heating and cooling cycle. It may be necessary to lower the dew point further by refrigerating the gas. If the furnace atmosphere is not sufficiently reducing, or if the muffle leaks air, a "subscale", or internal oxidation of the hardening elements below the surface of the metal, results. Subscale formation can occur rapidly above 845 °C (1550 °F) if the atmosphere becomes oxidizing.

Dissociated ammonia is used primarily for annealing and brazing operations. The gas is very flammable, and can explode if air enters the furnace while at an elevated temperature or if the furnace is improperly purged before reaching the elevated temperature.

Dissociated ammonia can be partly or completely burned with air to reduce cost and flammability. The hydrogen content can be controlled within a range of 1 to 24%, the remainder being nitrogen saturated with water vapor. Water must be removed to maintain a reducing atmosphere.

Hydrogen is highly reducing to copper oxide at elevated temperatures and is recommended for elevated-temperature bright annealing and brazing.

Commercial hydrogen contains about 0.2% oxygen, which if not removed may cause internal oxidation of the reactive alloying elements in the copper.

When mixed with air, hydrogen is explosive at elevated temperature; therefore, the furnace must be purged before being heated to high temperature, and air must not enter the furnace.

Heating Below 705 °C (1300 °F).
Combusted gas (lean exothermic atmosphere) is the most widely used protective atmosphere for annealing of copper and copper alloys. Because of its low sulfur content, natural gas is the preferred fuel for production of combusted gas. The air/gas ratio is adjusted to produce a hydrogen content of 0.5 to 1%. Combusted gas is dried before entering the furnace, to prevent discoloration and staining of the metal by water vapor during the cooling cycle.

Steam is the cheapest atmosphere for protecting copper alloys during annealing. Although the annealed metal is not as bright as when heated in a combusted-fuel-gas atmosphere, it is satisfactory for some applications. For products such as tightly wound coils of strip, steam can be used during the heating cycle and combusted fuel gas can be used during cooling.

Inert gases, dissociated ammonia burned with air, and vacuum are more expensive and are not in common use

Table 5 Solution treating and precipitation hardening of copper-beryllium alloys

Alloy	Solution treatment(a) Temperature °C	°F	Time(b), h	Aging treatment Temperature °C	°F	Time, h
C17000	775-800	1425-1475	½-3	300-330	575-625	1-3
C17200	775-800	1425-1475	½-3	300-330	575-625	1-3
C17300	775-800	1425-1475	½-3	300-330	575-625	1-3
C17500	900-925	1650-1700	½-3	455-480	850-900	1-3
C17510	900-925	1650-1700	½-3	455-480	850-900	1-3

(a) All alloys are cooled immediately and rapidly from the solution-treating temperature. Thin sections such as strip can be cooled in circulating atmosphere; heavier sections require water quenching. (b) Shorter times may be desirable to minimize grain growth, particularly for thin sections.

for annealing of copper alloys. A major disadvantage of vacuum is that heating and cooling are slow because heat is transferred by radiation only.

Copper-Beryllium Alloys

Solution Treating

Wrought copper-beryllium alloy mill products generally are supplied solution treated or solution treated and cold worked. Material in these conditions can be fabricated without further heat treatment. Thus, solution treating typically is not a part of the fabricating process unless it is necessitated by a special requirement such as softening of the material for additional forming or is employed as a salvage operation for parts that have been incorrectly heated for precipitation hardening.

Solution treating must be carefully controlled to produce desired grain size, dimensional tolerances and mechanical properties, and to prevent surface oxidation. Table 5 gives recommended schedules for solution treating and precipitation hardening of the five major copper-beryllium alloys that are produced in wrought form. Optimum mechanical and physical properties for specific applications can be attained by varying these schedules, but the temperatures and times given in this table constitute the most conventional practice and typically provide maximum tensile strength. In addition to the wrought copper-beryllium alloys, there is a wide variety of copper-base casting alloys (C81300 through C82800) that contain beryllium. Appropriate solution-treating and aging schedules for these alloys are dictated by levels of beryllium and other additives.

Solution-treating temperature limits must be adhered to if optimum properties are to be obtained from the precipitation-hardening treatment. Exceeding the upper limit causes grain

coarsening in wrought material and overheating in wrought and cast materials. Coarse grain size impairs formability; overheating results in a brittle material that does not respond fully to precipitation hardening.

Solution treating below the specified minimum temperature results in insufficient solution of the beryllium-rich phase. This results in lower hardness after precipitation hardening (Fig. 6).

Effect of Time. The time at the solution-treating temperature depends on the amount of beryllium-rich phase that must be dissolved. Solution of this phase must be complete to produce maximum strength after precipitation hardening.

In cast products, the as-cast structure usually contains a large amount of microsegregation within the dendritic pattern. Therefore, castings must be heated for a time sufficient to homogenize the structure. A minimum of 3 h at temperature is recommended for this purpose.

Solution treating of wrought material also removes the effects of cold working and permits additional forming. Some grain growth will occur during softening for additional forming, because the solution-treating temperature is above the recrystallization temperature. Therefore, to minimize grain growth, excessive time at temperature must be avoided. It is recommended that wrought alloys be held at temperature 1 h for each inch or fraction of an inch of section thickness. The optimum time for a specific application must be determined by mechanical testing and microscopic examination of the alloy.

Effect of Oxidation. When copper-beryllium alloys are solution treated in air or in an oxidizing atmosphere, two types of oxidation are encountered. A continuous and tenacious oxide surface layer forms on alloys with high beryllium contents. Low-beryllium alloys

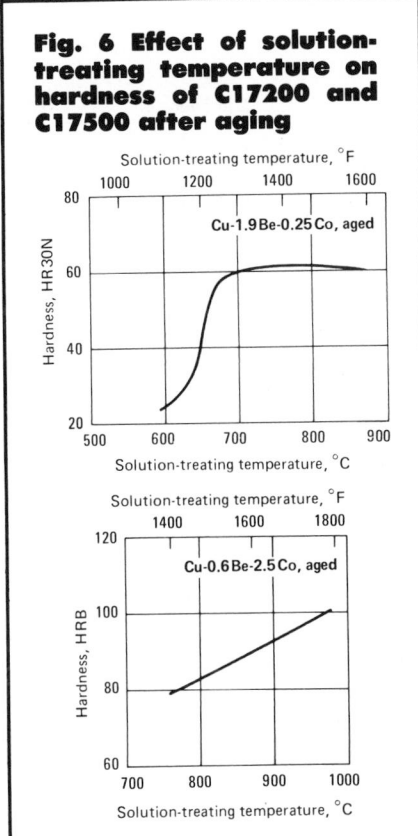

Fig. 6 Effect of solution-treating temperature on hardness of C17200 and C17500 after aging

Fig. 7 Time/temperature relationships in aging of C17200 strip

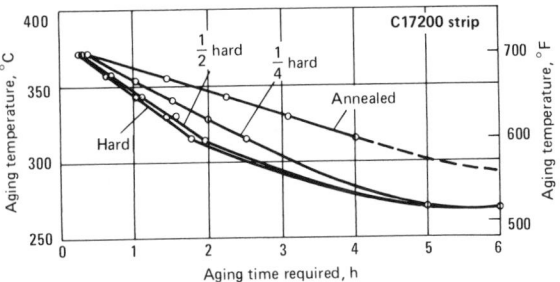

Aging time required for development of maximum strength in annealed, ¼-hard, ½-hard and hard C17200 strip aged at various temperatures in a recirculating-air furnace

form a loosely adhering scale and are subject to internal oxidation.

The oxide layer on high-beryllium alloys does not significantly affect the mechanical properties of the precipitation-hardened material, but it is abrasive and causes severe wear of tools and dies if not removed. In addition to the abrasive effect of the oxide, oxidation of low-beryllium alloys decreases mechanical properties. This is caused by the surface layer of internal oxidation, which reduces the effective section thickness of the material.

The oxides on both types of alloys may be removed by chemical or abrasive cleaning methods.

Quenching is a critical phase of the solution-treating process. Successful treatment requires that the material be quenched immediately, and at the highest possible rate, after being removed from the furnace. Any time lapse during transfer from the furnace to the quenching medium will permit some cooling and will cause precipitation. Precipitation is rapid at elevated temperatures, and its occurrence significantly affects the properties obtained during subsequent precipitation hardening. The maximum allowable delay before quenching depends on the mass of the load, the size of the parts and the transfer equipment. Mechanical testing and microscopic examination of the structure should be used to evaluate the effectiveness of the quenching operation.

Water quenching is the most common method of retaining the solid-solution condition in both wrought and cast products; however, because of their shape, some castings may crack as the result of the rapid cooling. Such castings may be quenched in oil or forced air; however, the slower cooling rates may cause some precipitation. Thin-gage strip is typically cooled in forced air.

Precipitation Hardening

Cold working of solution-treated copper-beryllium alloys influences the strength attainable through subsequent aging; the greatest response to aging occurs in material in the cold rolled hard temper. In general, work hardening offers no advantages beyond the hard temper, because formability is poor and control of the precipitation-hardening treatment for maximum strength is critical. For some applications, however, wire is drawn to higher levels of cold work prior to precipitation hardening.

Table 6 lists the properties usually specified for mill products of the common copper-beryllium alloys, and Fig. 7 shows the time required to develop maximum tensile strength in one of these alloys aged at various temperatures. The aging times in Fig. 7 vary slightly from those given in Table 6 for the same alloy; the latter are primarily for acceptance-test purposes.

Special combinations of properties can be obtained by varying either the aging time or the aging temperature. Table 7 shows the age-hardening response from underaging to overaging for cold rolled material in various tempers. Note that as tensile strength increases, elongation decreases and does not recover substantially with overaging but electrical conductivity continues to increase. The response of alloy C17200 (1.9Be-0.2 Co/Ni) at other aging temperatures within the hardening range (290 to 400 °C, or 550 to 750 °F) will be similar to the response at 370 °C (700 °F), but the corresponding time cycles will vary.

Recommended precipitation-hardening cycles for solution-treated copper-beryllium castings are presented in Table 8.

Effect of Temperature. Close control of temperature is critical in the conventional aging of copper-beryllium alloys. As indicated by the data in Tables 6 and 7, a change in temperature affects the time required for development of maximum properties. Also, the higher temperatures can result in lower property values. Normal commercial control of ± 6 °C (±10 °F) is adequate for temperatures in the range from 315 to 370 °C (600 to 700 °F).

Problems involving temperature usually arise when test data are translated into production-control data. A strip specimen of the material tested at the mill may just meet the minimum specification requirements; however, when the fabricator heat treats a large mass of parts made of the same material, tests may indicate properties below the minimum requirements. The low properties may result from heat treatment of too large a mass of parts rela-

Table 6 Properties and precipitation treatments usually specified for copper-beryllium alloys

Initial condition	Standard aging treatment Time, h	Temperature °C	°F	Tensile strength MPa	ksi	Yield strength(a) MPa	ksi	Elongation(b), %	Hardness(c)	Electrical conductivity, % IACS
C17200										
Flat products:										
Annealed	None	...	...	415-540	60-78	195-380	28-55	35-60	45-78 HRB	17-19
¼ hard	None	...	...	515-605	75-88	415-550	60-80	10-40	68-90 HRB	16-18
½ hard	None	...	...	585-690	85-100	515-655	75-95	10-25	88-96 HRB	15-17
Hard	None	...	...	690-825	100-120	620-770	90-112	2-8	96-102 HRB	15-17
Annealed(d)....	3	315	600	1140-1345	165-195	965-1205	140-175	4-10	35-40 HRC	22-25
Annealed	½	370	700	1105-1310	160-190	895-1205	130-175	3-10	34-40 HRC	22-25
¼ hard(d)......	2	315	600	1205-1415	175-205	1035-1275	150-185	3-6	37-42 HRC	22-25
¼ hard	⅓	370	700	1170-1380	170-200	965-1275	140-185	2-6	36-42 HRC	22-25
½ hard(d)......	2	315	600	1275-1485	185-215	1105-1345	160-195	2-5	39-44 HRC	22-25
½ hard	¼	370	700	1240-1450	180-210	1070-1345	155-195	2-5	38-44 HRC	22-25
Hard(d)........	2	315	600	1310-1575	190-220	1140-1415	165-205	1-4	40-45 HRC	22-25
Hard	¼	370	700	1275-1480	185-215	1105-1415	160-205	1-4	39-45 HRC	22-25
Rod, bar and plate:										
Annealed	None	...	...	415-585	60-85	185-205	20-30	35-60	45-85 HRB	17-19
Hard	None	...	...	585-895	85-130	515-725	75-105	10-20	88-103 HRB	15-17
Annealed(d)....	3	315	600	1140-1345	165-200	1000-1205	145-175	3-10	36-41 HRC	22-25
Hard(d)........	2	315	600	1205-1550	175-225	1035-1380	150-200	2-5	39-45 HRC	22-25
Wire(e):										
Annealed	None	...	...	450-590	65-85	185-240	20-35	35-55	...	17-19
¼ hard	None	...	...	620-795	90-115	485-655	70-95	10-35	...	15-17
½ hard	None	...	...	760-930	110-135	620-760	90-110	4-10	...	15-17
¾ hard	None	...	...	895-1070	130-155	760-930	110-135	2-8	...	15-17
Annealed(d)....	3	315	600	1140-1310	165-190	1000-1205	145-175	3-8	...	22-25
Annealed	½	370	700	1105-1310	160-190	930-1205	135-175	3-8	...	22-25
¼ hard(d)......	2	315	600	1205-1415	175-205	1105-1310	160-190	2-5	...	22-25
¼ hard	¼	370	700	1170-1415	170-205	1035-1310	150-190	2-5	...	22-25
½ hard(d)......	1½	315	600	1310-1480	190-215	1205-1380	175-200	1-3	...	22-25
½ hard	¼	370	700	1275-1480	185-215	1170-1380	170-200	1-3	...	22-25
¾ hard(d)......	1	315	600	1345-1585	195-230	1245-1415	180-205	1-3	...	22-25
¾ hard	¼	370	700	1310-1585	190-230	1205-1415	175-205	1-3	...	22-25

(continued)

(a) At 0.2% offset. (b) In 50 mm or 2 in. (c) Rockwell B and C hardness values are accurate only if metal is at least 1 mm (0.040 in.) thick. (d) Heat treatment that provides optimum strength. (e) For wire diameters greater than 1.3 mm (0.050 in.)

tive to the capacity of the furnace, from inadequate time at the proper temperature, or from use of a higher temperature to gain production speed or to obtain fixture conformity when the parts are in fixtures.

Effect of Grain Size. The effect of grain size on the properties of heat treated material is less significant for copper-beryllium alloys than for solid-solution alloys such as brass and phosphor bronze. The relatively high temperatures required for solution treating of copper-beryllium alloys usually override the effects of cold work and time at temperature. Low solution-treating temperatures will result in fine grain size, but, if the temperature is too low to dissolve the beryllium-rich phase, response to aging will be affected adversely and the benefits obtained

from the fine grain size will be nullified. For this reason, grain sizes below about 0.015 mm (0.0006 in.) are not practical for most beryllium copper products, regardless of dimensions. With normal commercial practice, and depending on the product, the grain size of solution treated material will range from about 0.015 to 0.060 mm.

Fixturing for Close Tolerances. Excellent dimensional accuracy can be achieved by properly supporting beryllium copper parts during aging. Usually, overaging is necessary to hold close tolerances. Fixture design should be based on the following principles:

- Fixtures should be of minimum weight.
- Excessive clamping pressure should be avoided, to prevent stripping of

clamping-screw threads and warping of fixtures.
- Parts should be held only at critical locations.
- A maximum number of parts should be held by a minimum number of clamping screws.
- Design should minimize warping of the fixture, thus maximizing its service life.
- When configuration of the part permits, the fixture should be designed so that parts can be stacked. Parts having no more than two planes often can be stacked (see Fig. 8), provided that no burrs are present.

Example 2. A comparison was made of the dimensional variations that occurred during aging of solenoid guides heated as loose pieces and as fixtured

Table 6 (continued)

Initial condition	Standard aging treatment Time, h	Temperature °C	°F	Tensile strength MPa	ksi	Yield strength(a) MPa	ksi	Elonga- tion(b), %	Hardness(c)	Electrical conduc- tivity, % IACS
C17000										
Flat products:										
Annealed	None	...	...	415-540	60-78	170-365	25-55	35-60	45-78 HRB	17-19
¼ hard	None	...	...	515-605	75-88	310-515	45-75	10-40	68-90 HRB	16-18
½ hard	None	...	...	585-690	85-100	450-620	65-90	10-25	88-96 HRB	15-17
Hard	None	...	...	690-825	100-120	550-760	80-110	2-8	96-102 HRB	15-17
Annealed	3	315	600	1035-1240	150-180	895-1105	130-165	4-10	33-39 HRC	22-25
Annealed(d)	3	345	650	1105-1275	160-185	860-1140	125-165	4-10	34-40 HRC	22-25
¼ hard	2	315	600	1105-1310	160-190	860-1140	135-170	3-6	34-40 HRC	22-25
¼ hard(d)	3	330	625	1170-1345	170-195	895-1170	130-170	3-6	36-41 HRC	22-25
½ hard	2	315	600	1170-1380	170-200	895-1170	145-175	2-5	36-41 HRC	22-25
½ hard(d)	2	330	625	1240-1380	180-200	965-1240	140-180	2-5	38-42 HRC	22-25
Hard	2	315	600	1240-1450	180-210	965-1240	155-180	2-5	38-42 HRC	22-25
Hard(d)	2	330	625	1275-1415	185-205	1070-1345	155-195	2-5	39-43 HRC	22-25
Rod and bar:										
Annealed	None	...	...	415-585	60-85	185-205	20-30	35-60	45-85 HRB	17-19
Hard	None	...	...	585-895	85-130	515-725	75-105	10-20	88-103 HRB	15-17
Annealed	3	315	600	1035-1240	150-180	860-1070	125-155	4-10	32-39 HRC	22-25
Annealed(d)	3	345	650	1105-1275	160-185	930-1140	135-165	4-10	34-40 HRC	22-25
Hard	2	315	600	1140-1380	165-200	930-1140	135-165	2-5	36-41 HRC	22-25
Hard(d)	2	345	650	1205-1415	175-205	965-1170	140-170	2-5	38-42 HRC	22-25
C17500, C17510										
Rod, bar, plate and flat products:										
Annealed	None	...	...	240-380	35-55	185-205	20-30	20-35	20-43 HRB	25-30
Hard	None	...	...	515-585	75-85	380-550	55-80	3-10	78-88 HRB	20-30
Annealed	3	480	900	690-760	100-120	550-690	80-100	10-20	92-100 HRB	45-60
Annealed(d)	3	455	850	725-825	105-120	550-725	80-105	8-12	93-100 HRB	45-52
Hard	2	480	900	760-860	110-130	690-825	100-120	8-15	95-103 HRB	45-60
Hard(d)	2	455	850	795-930	115-135	725-860	105-125	5-8	97-104 HRB	45-52

(a) At 0.2% offset. (b) In 50 mm or 2 in. (c) Rockwell B and C hardness values are accurate only if metal is at least 1 mm (0.040 in.) thick. (d) Heat treatment that provides optimum strength. (e) For wire diameters greater than 1.3 mm (0.050 in.)

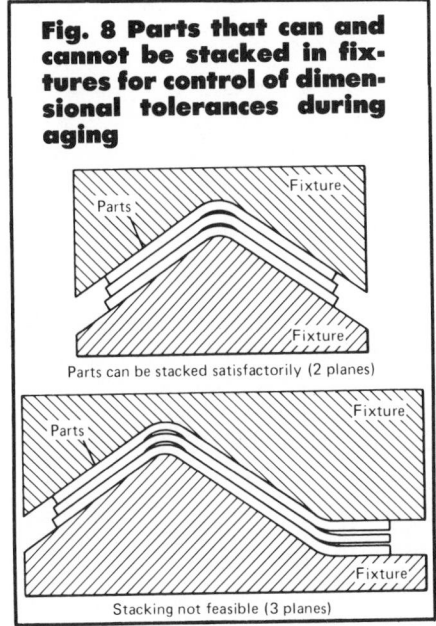

Fig. 8 Parts that can and cannot be stacked in fixtures for control of dimensional tolerances during aging

Fixture
Parts
Fixture

Parts can be stacked satisfactorily (2 planes)

Fixture
Parts
Fixture

Stacking not feasible (3 planes)

pieces. The fixture that was used supported, with adequate pressure, all inside and outside surfaces of the part. Although the total cost of aging the guides as fixtured pieces was nearly 2½ times the cost of aging them as loose pieces (total cost per piece, based on aging of 100 000 pieces per year), a significant improvement (reduction) in dimensional variations was achieved by the use of fixtures (see Table 9).

An understanding of the behavior of beryllium copper during the hardening treatment is helpful in design and use of fixtures for parts with a controlled gap or opening. These may be U-shaped parts, circular parts, or clips of various designs. During heating for hardening, the material will move toward the direction in which it was plastically formed or elastically deflected. If both conditions occur before hardening, movement due to the elastic effect pre-

vails. These phenomena may be illustrated by the following example.

Example 3. Spring clips like the one shown in Fig. 9 were press formed to a slightly undersize ID (24.9 to 25.2 mm, or 0.980 to 0.990 in.) so that they would fit snugly on a mandrel 25.4 ± 0.03 mm (1 ± 0.001 in.) in diameter. However, during aging for ½ h at 350 °C (660 °F), the metal moved in the direction of elastic deflection, which caused the clips to be loose on the mandrel. Because precipitation hardening of beryllium copper is a cumulative time/temperature reaction, the hardening treatment can be interrupted. Therefore, the clips (without being placed on the mandrel) were partially aged for 5 min at 350 °C (this decreased the original dimension to 24.8 to 25.0 mm, or 0.975 to 0.985 in.) and then were placed on the mandrel and aged at 350 °C for the remaining 25 min.

Following this aging treatment, none

Table 7 Effects of special precipitation-hardening treatments on mechanical properties and electrical conductivity of Cu-1.9Be strip

Initial condition	Aging treatment Time, min	Temperature °C	Temperature °F	Tensile strength MPa	Tensile strength ksi	Yield strength(a) MPa	Yield strength(a) ksi	Elongation(b), %	Electrical conductivity, % IACS	Fatigue strength(c) MPa	Fatigue strength(c) ksi	Modulus of elasticity GPa	Modulus of elasticity 10⁶ psi
Alloy C17200													
Annealed	None	...	...	465	67.5	250	36	49	18.0	205	30	115	16.5
	5	370	700	855	124	695	101	18	19.5	...	...	120	17.5
	15	370	700	1195	173	1055	153	10	22.0	...	...	125	18.0
	30	370	700	1260	182.5	1060	153.5	6	23.0	...	...	125	18.0
	60	370	700	1240	180	1055	153	5	25.5	255	37	130	18.5
	120	370	700	1195	173.5	1040	151	6	26.0	...	...	130	18.5
	240	370	700	1150	167	980	142	6	26.5	...	...	130	19.0
¼ hard	None	...	...	570	82.5	485	70.5	21	17.0	220	32	115	17.0
	5	370	700	1115	162	945	137	9	18.5	...	...	125	18.0
	15	370	700	1250	181	1115	162	6	20.5	...	...	130	18.5
	30	370	700	1290	187	1125	163.5	4	23.5	290	42	130	18.5
	60	370	700	1230	178.5	1060	154	3	25.5	...	...	130	18.5
	120	370	700	1185	172	1000	145	4	26.5	...	...	130	19.0
	240	370	700	1155	167.5	970	141	6	27.0	...	...	130	19.0
½ hard	None	...	...	605	87.5	555	80.5	17	16.0	230	33	115	17.0
	3	370	700	1010	146.5	885	128	11	18.0	230	33	125	18.0
	5	370	700	1280	186	1110	161	3	21.0	295	43	125	18.0
	15	370	700	1310	190	1175	170.5	2	23.0	305	44	130	18.5
	30	370	700	1325	192.5	1180	171	2	24.5	305	44	130	18.5
	60	370	700	1280	185.5	1105	160	2	25.0	295	43	130	18.5
	120	370	700	1200	174	1040	150.5	3	26.0	275	40	130	18.5
	240	370	700	1185	172	1035	150	3	27.0	275	40	130	19.0
	420	370	700	1010	146.5	860	125	10	27.0	200	29	130	19.0
Hard	None	...	...	730	106	690	100	5	15.0	270	39	120	17.5
	5	370	700	1300	188.5	1125	163	3	18.0	...	...	125	18.0
	15	370	700	1360	197	1195	173	2	21.0	...	...	130	18.5
	30	370	700	1310	190	1170	170	1	24.5	315	46	130	19.0
	60	370	700	1295	188	1105	160	1	26.5	...	...	130	19.0
	120	370	700	1240	180	1090	158	2	27.5	...	...	130	19.0
	240	370	700	1215	176	1055	153	2	27.5	250	36	130	19.0
Alloy C17500													
Annealed	None	...	...	350	51	170	25	30	25	...	...	110	16.3
	120	425	800	805	117	625	91	14	44	...	...	135	19.3
	120	455	850	835	121	675	98	14	48	...	...	140	20.0
	120	480	900	805	116.5	625	91	14	48	...	...	140	20.0
	120	510	950	795	115	600	87	16	48.5	215	31	140	20.0
Hard	None	...	...	440	63.5	425	61.5	2	27.8	...	...	125	18.3
	120	425	800	985	142.5	860	125	11	44.0	...	...	140	20.0
	120	455	850	915	133	800	116	13	45.0	...	...	140	20.0
	120	480	900	850	123	760	110.5	13	47.5	250	36	140	20.0
	120	510	950	800	116	705	102	12	49.0	...	...	140	20.0

(a) At 0.2% offset. (b) In 50 mm or 2 in. (c) 10⁷ cycles

Table 8 Recommended precipitation-hardening schedules and resulting properties for solution-treated copper-beryllium castings

Alloy	Solution treatment Temperature °C	°F	Time, min	Aging treatment Temperature °C	°F	Time, min	Tensile strength MPa	ksi	Yield strength(a) MPa	ksi	Elonga-tion(b), %	Hardness	Electrical conduct-ivity, % IACS
C81300	980–1010	1800–1850	60	480	900	120	365	53	250	36	11	89 HB(c)	60
C81700	900–925	1650–1700	60	455	850	180	635	92	470	68	8	217 HB(d)	48
C81800	900–925	1650–1700	60	480	900	180	705	102	515	75	8	92 HRB	45
C82000	900–925	1650–1700	180	480	900	180	690	100	515	75	8	195 HB(d)	45
C82100	900–925	1650–1700	60	455	850	180	635	92	470	68	8	217 HB(d)	48
C82200	900–925	1650–1700	60	445–455	835–850	120	655	95	515	75	8	96 HRB	45
C82400	785–850	1450–1560	60	345	650	180	1035	150	965	140	1	34 HRC	25
C82500	785–800	1450–1475	60	345	650	180	1105	160	795	115	1	40 HRC	20
C82600	785–800	1450–1475	60	345	650	180	1105	160	1035	150	1	40 HRC	19
C82700	785–800	1450–1475	180	345	650	180	1070	155	895	130	0	39 HRC	20
C82800	785–800	1450–1475	60	345	650	180	1140	165	1000	145	1	42 HRC	18

(a) At 0.2% extension under load. (b) In 50 mm or 2 in. (c) 500-kg load. (d) 3000-kg load

Table 9 Dimensional variations in beryllium copper solenoid guides aged with and without fixtures (Example 2)

Dimension	Dimensional variation(a) Loose pieces	Fixtured pieces
A(b)	Up to 3/16 in.	1/32 in. max
B	−1/8 to +1/16 in.	±0.010 in.
C	−4° to +2°	±1/2°
D	−4° to +2°	±1/2°

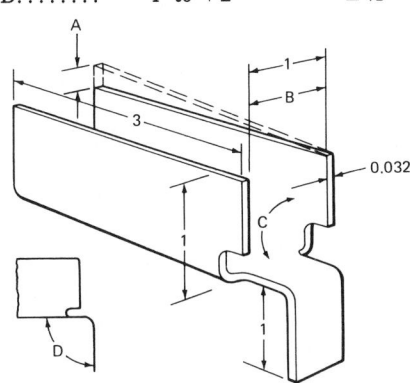

Linear dimensions are in inches.

(a) Assuming no deviations from the die operation.
(b) Twist in 3 in.

Table 10 Suggested methods for measuring the hardness of heat treated C17200

Thickness mm	in.	Hardness test method	Load, kg
0.03-0.08	0.001-0.003	Diamond pyramid	0.2
0.08-0.38	0.003-0.015	Diamond pyramid	0.5
0.38-0.51	0.015-0.020	Rockwell superficial 15N	15
0.51-1.02	0.020-0.040	Rockwell superficial 15N	30
1.02 & up	0.040 & up	Rockwell B or C	Std

Fig. 9 Problem in dimensional control (Example 3)

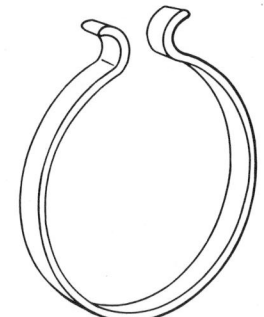

Spring clip representing a type of beryllium copper part that should be partially aged without a fixture and then placed on a mandrel for final aging, for control of diameter

Fig. 10 Hardness variation in beryllium copper springs (Example 4)

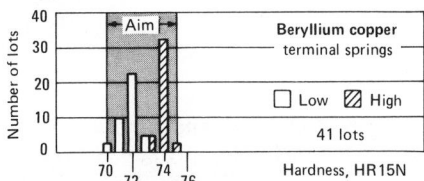

Variation in hardness for 41 lots of heat treated beryllium copper terminal springs tested over a three-month period; high and low values in each lot are plotted.

Inspection and Quality Control. In most instances, the completeness of aging can be verified by hardness testing. Exceptions are tension testing of specimens taken from large parts and simulated service testing to determine elastic performance.

Hardness measurements should always be made with the method and load most suitable for the thickness of the material and the normal level of hardness expected. Table 10 indicates suggested methods for testing various thicknesses of a hardened beryllium copper alloy. Variations obtained with

of the clips was larger than the diameter of the mandrel; the actual final dimension was governed by the time and temperature of aging—that is, by the amount of stress relief that occurred.

Selection of the correct time and temperature for the first step of the interrupted aging sequence is important but not critical. If the time is inadequate, the direction of movement will not be established; if excessive, there will not be sufficient time for heating the parts on the fixture to allow enough stress

relief for close conformity. Duration of the preliminary treatment usually should be about 15% of the total aging time. Allowance must be made also for the mass effect of the load on the time required for it to reach furnace temperature. Interrupted aging procedures also can be used to restrict increases in the outside diameters of various parts.

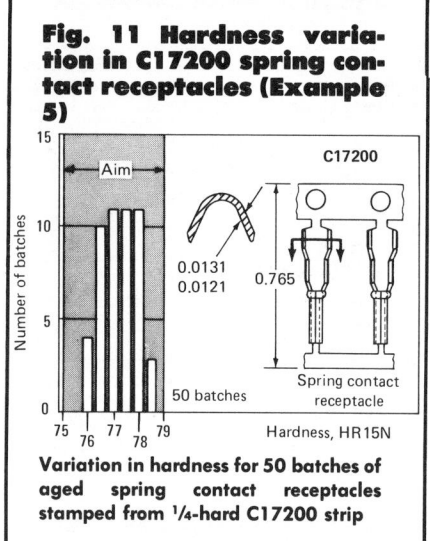

Fig. 11 Hardness variation in C17200 spring contact receptacles (Example 5)

Variation in hardness for 50 batches of aged spring contact receptacles stamped from ¼-hard C17200 strip

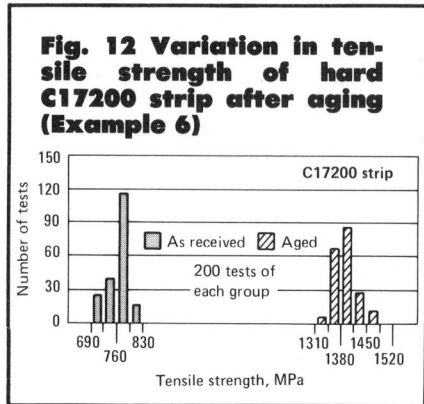

Fig. 12 Variation in tensile strength of hard C17200 strip after aging (Example 6)

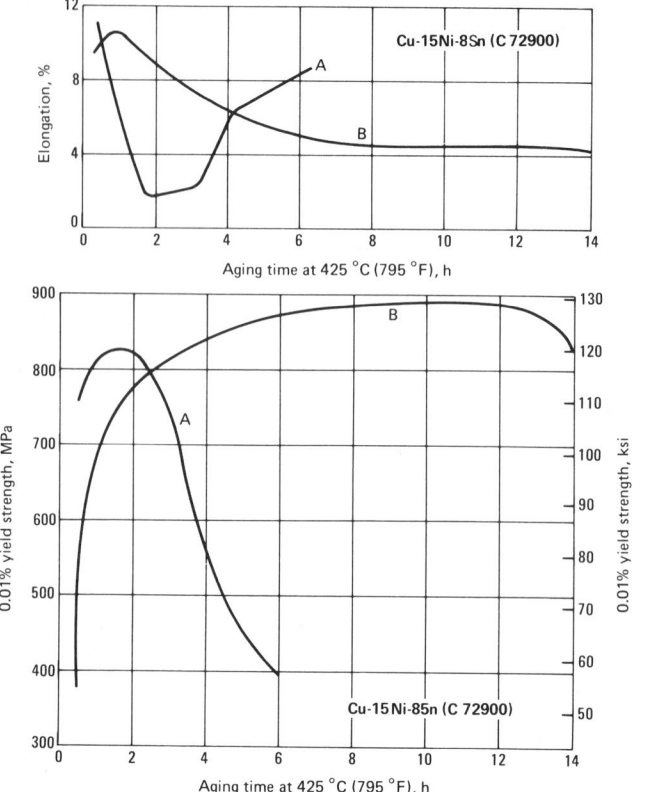

Fig. 13 Aging response of Cu-15Ni-8Sn (C72900) solution treated above and below (microduplexed) the single phase boundary

A, water quench for 30 min. at 825 °C (1520 °F); B, water quench for 1 h at 725 °C (1340 °F), microduplexed

heat treated parts and with strip are presented in the following examples.

Example 4. Terminal springs made of a beryllium-copper alloy (0.8 to 1% Be, 0.85% max Ni + Co + Fe, 3% max Zn + Sn) were solution annealed, then aged for 5 h at 343 ± 6 °C (650 ± 10 °F) in a batch-type recirculating-air furnace (see Fig. 10). The furnace load consisted of 280 000 pieces. Hardness tests were conducted on specimens of strip material representative of the parts. Specifications required a hardness of 70 to 75 HR15N. Results of hardness tests on 41 production lots within a three-month period are presented in Fig. 10.

Example 5. Spring contact receptacles stamped from ¼-hard beryllium copper strip (alloy C17200) were aged in a salt bath at 300 °C (575 °F) for 30 min. The salt pot, 915 by 840 by 760 mm (36 by 33 by 30 in.), contained 1090 kg (2400 lb) of nitrate-nitrite salt (see Fig. 11).

A strip of about 1000 of these parts was wound on a "birdcage" reel 265 mm (10½ in.) in diameter and 305 mm (12 in.) wide. Each receptacle weighed about 0.6 g (0.02 oz); therefore, the total weight of the receptacles on each reel was about 565 g (20 oz). Each reel weighed about 0.9 kg (2 lb), and one furnace load consisted of eight reels.

Specified hardness for these parts was 75 to 79 HR15N. Figure 11 shows the variation in hardness for 50 batches.

Example 6. The variation in tensile strength obtained after aging hard-temper strip (alloy C17200) is illustrated in Fig. 12. The strip, 0.17 to 0.19 mm (0.0065 to 0.0075 in.) thick and 15.9 mm (0.625 in.) wide, was aged 3 h at 315 °C (600 °F).

Copper-Chromium Alloys

Copper-chromium alloys of 0.5 to 1.0 % Cr are solution treated, in molten salt or in controlled-atmosphere furnaces to avoid scaling, at 980 to 1010 °C (1800 to 1850 °F) and rapidly quenched. Solution-treated chromium copper is soft and ductile; therefore, it can be cold worked in a manner similar to that used for unalloyed copper.

After being solution treated, the material may be aged for several hours at 400 to 500 °C (750 to 930 °F) to produce special mechanical and physical properties. A typical aging cycle is 4 h or more at 455 °C (850 °F).

Typical effects of heat treatment and cold work on the properties of chromium copper are shown in Table 11. The hard drawn specimens were obtained by reducing the cross-sectional area of solution-annealed specimens by approximately 40%.

Copper-Zirconium Alloys

Solution treatment of zirconium copper (99.7 min Cu, 0.13 to 0.30 Zr) con-

Table 11 Typical effects of heat treatment and cold work on properties of copper-1% chromium alloys

Condition	Ultimate tensile strength MPa	ksi	Yield strength(a) MPa	ksi	Elongation(b), %	Hardness	Electrical conductivity, % IACS
Alloy C18200							
Solution treated................	240	35	105	15	42	50 HRF	35-42
Solution treated and aged......................	350	51	275	40	15	90 HB(c)	75-82
Solution treated and drawn 40%...............	415	60	310	45	15	65 HRB	40
Solution treated, hard drawn and aged..........	435	63	385	56	18	68-75 HRB	80
Solution treated, aged and drawn 30%...............	480	70	425	62	18	75-80 HRB	80
Alloy C81500							
Cast, solution treated and aged.....................	350	51	275	40	17	105 HB(c)	75-80

(a) At 0.5% extension under load. (b) In 50 mm or 2 in. (c) 500-kg load

Table 12 Effect of heat treatment and cold work on properties of copper-zirconium alloy C15000

Solution treating temperature(a) °C	°F	Amount of cold work, %	Aging Temperature °C	°F	Time, h	Tensile strength MPa	ksi	Yield strength MPa	ksi	Elongation(b), %	Hardness, HRB	Electrical conductivity, % IACS
900	1650	20	475	885	1	310	45	260	38	25	48	85 min
900	1650	80	425	795	1	425	62	380	55	12	64	85 min
980	1795	None	...	...	...	200	29	41(c)	6(c)	54	...	64
980	1795	20	...	...	...	270	39	250(c)	36(c)	26	37	64
980	1795	80	...	...	...	440	64	420(c)	61(c)	19	73	64
980	1795	None	500	930	3	205	30	90	13	51	...	87
980	1795	None	550	1025	3	205	30	90	13	49	...	95
980	1795	20	400	750	3	330	48	260	38	31	50	80
980	1795	20	450	840	3	330	48	275	40	28	57	92
980	1795	85	400	750	3	495	72	440	64	24	79	85
980	1795	85	450	840	3	470	68	425	62	23	74	91

(a) Hold 30 min, water quench. (b) In 50 mm or 2 in. (c) 0.5% extension under load

Table 13 Typical heat treating schedules and resulting properties for precipitation hardened miscellaneous alloys

Alloy	Solution treatment Temperature °C	°F	Time, min	Tempering treatment Temperature °C	°F	Time, min	Tensile strength MPa	ksi	Yield strength(a) MPa	ksi	Elongation(b), %	Hardness, HB(c)
C94700	775-800	1425-1475	120	305-325	580-620	300	585	85	415	60	10	180
C94800	...	...	...	305-325	580-620	360-1000	415	60	205	30	8	120
C96600	995	1825	60	510	950	180	760	110	485	70	7	230
C99400	885	1625	60	480	900	60	545	79	370	54	...	170
C99500	885	1625	60	480	900	60	595	86	425	62	8	196

(a) At 0.2% extension under load for C96600; at 0.5% extension under load for all other alloys. (b) In 50 mm or 2 in. (c) 3000-kg load

sists of heating to 900 to 980 °C (1650 to 1795 °F) and quenching in water. The material may then be precipitation hardened for 1 to 4 h at 500 to 550 °C (930 to 1020 °F). If cold working is done prior to aging, the aging temperature is reduced to 370 to 480 °C (700 to 900 °F) for 1 to 4 h.

Time at the solution-treating temperature should be minimized, to limit grain growth and possible internal oxidation by reaction of the zirconium with the furnace atmosphere. Because solution and diffusion of the zirconium occur rapidly at the solution-treating temperature, holding at temperature is not required.

Maximum mechanical properties

and resistance to softening are developed with maximum solution of zirconium. If material containing 0.15% Zr or more is heated above 980 °C (1795 °F), the Cu_3Zr phase will begin to melt. A slight amount of melting will not affect mechanical properties; but if excessive melting occurs, ductility of the alloy will decrease.

Normally, as the solution temperature is increased from 900 to 980 °C (1650 to 1795 °F), the aging temperature should also be increased to maintain high electrical conductivity. The following aging treatments will produce the best combination of mechanical properties and electrical conductivity:

Condition	Aging treatment Temperature	Time, h
Solution treated at 900 °C (1650 °F)	500 °C (930 °F)	3
Solution treated at 900 °C and cold worked	400 °C (750 °F)	3
Solution treated at 980 °C (1795 °F)	(550 °C (1020 °F)	3
Solution treated at 980 °C and cold worked	450 °C (840 °F)	3

The increase in strength of zirconium copper depends primarily on cold work. Although aging results in some increase in strength, its chief effect is to increase electrical conductivity. The properties developed by various combinations of heat treatment and cold work are given in Table 12.

Miscellaneous Precipitation-Hardening Alloys

Other alloys which can be age hardened are the nickel-tin bronze alloys C94700 and C94800, copper-nickel-beryllium alloy C96600, and the complex special alloys C99400 and 99500. The solution treating and precipitation hardening treatments for these alloys are shown in Table 13.

A protective atmosphere of exothermic gas or dissociated ammonia is recommended during aging of these alloys to produce bright surfaces. Parts heated in gas-fired or oil-fired furnaces in which the products of combustion are used as a protective atmosphere may tarnish slightly and require cleaning. Aging in an oxidizing atmosphere results in scaling.

Alloys C19000 and C19100 (Cu-4%Ni - 0.25%P) also respond to precipitation hardening. The alloys are solution treated at 705 to 790 °C (1300 to 1450 °F). A reducing or neutral atmosphere should be used to prevent internal oxidation, especially on thin sections. Water quenching is preferred, while rapid air cooling may be adequate for separate small parts.

Precipitation hardening is accomplished by aging the alloy at 425 to 480 °C (800 to 900 °F) for 1 to 3 h.

While annealing is required to facilitate cold working prior to aging, temperatures as low as 620 °C (1150 °F) followed by normal air cooling are adequate.

Spinodal-Hardening Alloys

Table 14 lists the recommended solution treatment for the Cu-Ni-Cr and Cu-Ni-Sn spinodal hardening alloys. It is important to maintain control within the limits shown in order to obtain proper heat treating response in the subsequent spinodal aging treatment. Exceeding the upper limit may result in excessive grain growth in wrought materials, which could impair formability in the solution annealed condition. Overheating cast material may cause incipient melting, resulting in brittle material that does not respond to spinodal hardening, particularly in the Cu-Ni-Sn alloys. Solution treating below the minimum temperature results in incomplete solution and failure of the material to fully harden during the spinodal aging treatment.

Wrought materials of some of these alloys can be extensively cold worked, with up to 90% reduction, after an effective solution treatment and quench. Solution anneals used between working schedules must be controlled toward the lower end of the temperature range and minimum time to minimize grain growth, because the solution temperature is above the recrystallization temperature. However, wrought alloys should be solution treated for approximately ½ h for each inch of section thickness (or fraction thereof).

Homogenization

Cast microstructures of spinodal hardening alloys generally require homogenization to encourage uniform distribution of hardening elements and adequate response to the hardening treatments. The C71900 alloy may be homogenized by a prolonged solution treating temperature. The Cu-Ni-Sn alloys have a tendency to develop porosity at higher temperatures, and therefore homogenization is best accomplished by first heating the alloys to 725 °C (1335 °F) and holding long enough to spheroidize the gamma phase (3 to 12 h). The temperature is then increased to the regular solution temperature in preparation for the quench.

Oxidation

When spinodal Cu-Ni-Sn alloys are solution treated in air or oxidizing atmospheres, oxidation is encountered which can be extremely abrasive to tools although it will not substantially affect the mechanical properties of the spinodal hardened material. The oxide generally does not penetrate far below the surface of these alloys and it can be removed by mechanical, chemical or abrasive cleaning methods.

Quenching

Quenching is a critical step of the solution process for Cu-Ni-Sn alloys. Successful heat treating requires that the material be quenched at the highest possible rate from the solution treating temperature. It is therefore necessary to provide both rapid transfer from furnace to quench tank, and an

Table 14 Recommended solution heat treating temperatures and times for spinodal alloys

Alloy	CDA No.	Solution heat treating temperature °C	°F	Time at temperature, h
Cu-30Ni-3Cr	C71900	900 to 950	1650 to 1740	½ to 2
Cu-4Ni-4Sn	C72600	700 to 760	1300 to 1400	½ to 2
Cu-9Ni-6Sn	C72700	730 to 790	1350 to 1450	½ to 2
Cu-10Ni+8Sn+0.2Nb	C72800	805 to 845	1480 to 1550	½ to 2
Cu-15Ni-8Sn	C72900	815 to 860	1500 to 1575	½ to 2

Table 15 Typical strengths and recommended aging times for various spinodal alloys(a)

Alloy	CDA No.	Solution treated and cold worked temper	Aging cycle	Tensile strength		Yield strength		Elongation, %
				MPa	ksi	MPa	ksi	
Cu-4Ni-4Sn	C72600	TD 02(½H)	90 min at 350 °C (660 °F)	635-690	92-100	495-570	72-83 (0.05)	12
Cu-4Ni-4Sn	C72600	TD 06(XH)	90 min at 350 °C (660 °F)	690-725	100-105	565-620	82-90 (0.05)	9
Cu-4Ni-4Sn	C72600	TD 08(S)	90 min at 350 °C (660 °F)	705-795	102-115	565-655	82-95 (0.05)	7
Cu-9Ni-6Sn	C72700	TD 04(H)	90 min at 350 °C (660 °F)	860-1035	125-150	760-895	110-130 (0.05)	8
Cu-9Ni-6Sn	C72700	TD 14(SS)	90 min at 350 °C (660 °F)	1055-1145	153-166	930-985	135-143 (0.05)	...
Cu-10Ni-8Sn-0.2Nb ..	C72800	TB 00 cast and solution treated	4-6 h at 350 °C (660 °F)	830-965	120-140	550-690	80-100 (0.01)	3
Cu-10Ni-8Sn-0.2Nb ..	C72800	TB 00 hot work and solution treated	3-5 h at 350 °C (660 °F)	965-1070	140-155	690-825	100-120 (0.01)	6-14
Cu-10Ni-8Sn-0.2Nb ..	C72800	TD 01(¼H)	3 h at 350 °C (660 °F)	1140-1240	165-180	895-930	130-135 (0.01)	7
Cu-10Ni-8Sn-0.2Nb ..	C72800	TD 04(H)	3 h at 350 °C (660 °F)	1205-1380	175-200	930-1000	135-145 (0.01)	7
Cu-10Ni-8Sn-0.2Nb ..	C72800	TD 06(XH)	3 h at 350 °C (660 °F)	1205-1380	175-200	965-1035	140-150 (0.01)	5
Cu-10Ni-8Sn-0.2Nb ..	C72800	TD 08(S)	3 h at 350 °C (660 °F)	1240-1380	180-200	1000-1070	145-155 (0.01)	4
Cu-10Ni-8Sn-0.2Nb ..	C72800	TD 14(SS)	90 min at 350 °C (660 °F)	1240-1380	180-200	1070-1140	155-165 (0.01)	2.5
Cu-15Ni-8Sn	C72900	TD 14(SS)	90 min at 350 °C (660 °F)	1140-1380	165-200	1035-1170	150-170 (0.05)	3
Cu-30Ni-3Cr	C71900	Hot extruded	90 min at 760 °C (405 °F)	550	80	345	50 (0.20)	25

(a) At 350 °C (660 °F)

efficient quenching medium. Delays sufficient to cause loss of temperature before the quench could allow the Cu-Ni-Sn intermetallic gamma phase to form, reducing the effectiveness of the entire heat treatment. The quenching media, which is generally water, must be sufficiently cold and agitated to maintain a rapid cooling rate to below the 200 °C (400 °F) level to ensure that no premature spinodal hardening occurs.

In some circumstances, particularly with the low tin alloys (C72600 and C72700), oil, air, or cold gaseous medium quenching may be used for small parts or very thin sections, but a careful evaluation of these methods should be conducted to insure the adequacy of the quench. Both mechanical testing of fully treated materials and metallographic examination of the microstructure should be employed in evaluating the quenching practice.

The quenching of spinodal hardening C71900 (Cu-Ni-Cr) alloy is somewhat less critical as to the over-all cooling rate, but it is equally important to begin the quench at a temperature above the 950 °C (1650 °F) temperature.

Spinodal Hardening (Aging)

Table 15 illustrates typical aging treatments and some resultant properties of various spinodal hardening alloys. The effect of cold working between the solution quench and aging steps is also illustrated for the Cu-Ni-Sn alloys.

The Cu-Ni-Cr (C71900) alloy can be hardened by slow cooling from the solution treating (full annealing) range thru 760 °C (1400 °F), or by a spinodal aging treatment in the 425 to 760 °C (800 to 1400 °F) range after a solution treatment. The slow cool from solution temperature tends to produce slightly greater ductility.

The Cu-Ni-Sn alloys are hardened by treating in a rather narrow temperature range of 350 to 360 °C (660 to 680 °F). Development of the maximum properties requires careful control of temperature and time at temperature. The use of hardness alone to evaluate results may not be adequate, since high hardness may be maintained where excessive aging causes a decrease in elastic properties. Variations in tensile properties of 68.9 to 103.35 MPa (10 to 15 ksi) are possible without significant hardness change.

Microduplexing

The Cu-Ni-Sn spinodal alloys can be treated using a combination of cold working and heat treatment called microduplexing. The alloys are cold worked to significant reductions (40 to 60% is typical), and given a partial solution treatment below the single phase boundary, typically at 725 °C (1335 °F). The alloy is then aged at a higher spinodal hardening temperature level of 425 °C (800 °F) for an extended time. The higher aging temperature is used to keep the aging time within reason, as microduplex age

hardening takes place much more slowly than full spinodal hardening. This treatment results in moderately high tensile properties and significantly greater ductility. The curves in Figure 13 illustrate the effect of the microduplex treatment compared to conventional spinodal treatment using a 425 °C (800 °F) age.

Effect of Grain Size

The temperatures required for solution treatment are well into the recrystallization/growth range for spinodal hardening alloys. The normal grain size obtained in properly treated wrought products will be in the 0.010 to 0.090 mm range, but coarser grain sizes have been used without adverse results. Maintaining low solution temperatures in an attempt to minimize grain growth is good practice, but is secondary to ensuring a complete solution treatment and adequate quench.

An extremely fine grain size (2 to 4 μm) may be produced in the microduplexing treatment or Cu-Ni-Sn alloys because of the much lower solution treating temperature and nucleating effects of undissolved gamma phase particles.

Fixturing in Close Tolerances

Excellent dimensional reproducibility can be achieved without fixturing during heat treatment. This is because the spinodal alloys do not undergo a crystallographic change during hardening. The chemical segregation (spi-

Table 16 Typical heat treatments and resulting properties for complex (alpha-beta) aluminum bronzes

Alloy	Typical condition(a)	Tensile strength MPa	ksi	Yield strength(b) MPa	ksi	Elonga- tion(c), %	Hard- ness, HB
C62400	As forged or extruded	620-690	90-100	240-260	35-38	14-16	163-183
	Solution treated at 870 °C (1600 °F) and quenched, tempered 2 h at 620 °C (1150 °F)	675-725	98-105	345-385	50-56	8-14	187-202
C63000	As forged or extruded	730	106	365	53	13	187
	Solution treated at 855 °C (1575 °F) and quenched, tempered 2 h at 650 °C (1200 °F)	760	110	425	62	13	212
C95300	As cast	495-530	72-77	185-205	27-30	27-30	137-140
	Solution treated at 855 °C (1575 °F) and quenched, tempered 2 h at 620 °C (1150 °F)	585	85	290	42	14-16	159-179
C95400	As cast	585-690	85-100	240-260	35-38	14-18	156-179
	Solution treated at 870 °C (1600 °F) and quenched, tempered 2 h at 620 °C (1150 °F)	655-725	95-105	330-370	48-54	8-14	187-202
C95500	As cast	640-710	93-103	290-310	42-45	10-14	183-192
	Solution treated at 855 °C (1575 °F) and quenched, tempered 2 h at 650 °C (1200 °F)	775-800	112-116	440-470	64-68	10-14	217-234

(a) As-cast condition is typical for moderate sections shaken out at temperatures above 540 °C (1000 °F) and fan cooled; or mold cooled, annealed at 620 °C (1150 °F) and fan (rapid) cooled. (b) At 0.5% extension under load. (c) In 50 mm or 2 in.

nodal decomposition) is not accompanied by a second phase which can cause distortion during the aging process. Therefore, stability is greatly increased over precipitation hardening alloys. Fixturing may still be necessary for extremely tight dimensional control but is not generally necessary.

Copper-Aluminum (Aluminum Bronze) Alloys

The microstructures and consequent heat treatabilities of aluminum bronzes vary with aluminum content much the same as these characteristics vary with carbon content in steels. Unlike steels, aluminum bronzes are tempered above the normal transformation temperature, typically in the range from 565 to 675 °C (1050 to 1250 °F). In selection of tempering temperatures, consideration must be given both to required properties and to hardness obtained on quenching. Normal tempering time is 2 h at temperature. Moreover, heavy or complex sections should be heated slowly to avoid cracking. After the tempering cycle has been

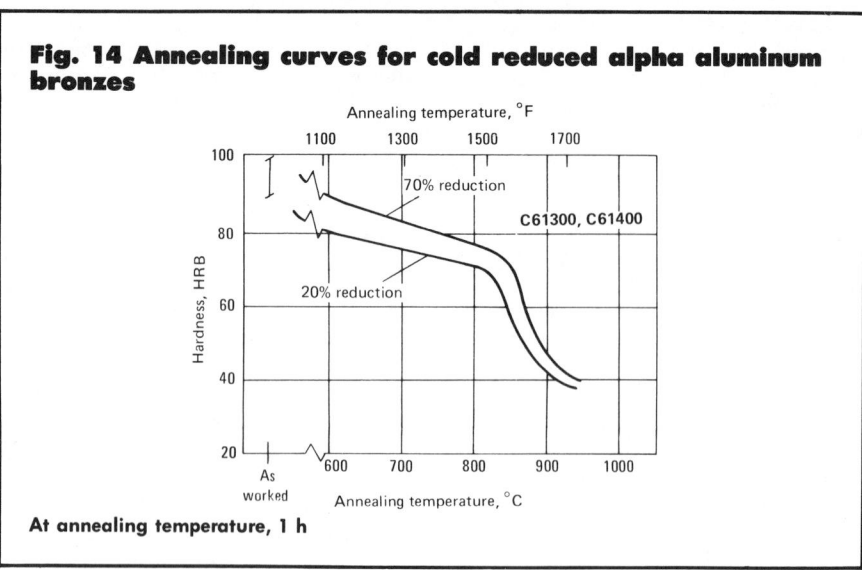

Fig. 14 Annealing curves for cold reduced alpha aluminum bronzes

completed, it is important that aluminum bronzes be cooled rapidly, using water quenching, spray cooling or fan cooling. Slow cooling through the range from 565 to 275 °C (1050 to 530 °F) can cause the residual tempered martensit-

ic beta phase to decompose, forming the embrittling alpha-gamma eutectoid. The presence of appreciable amounts of this eutectoid structure can result in low tensile elongation, low energy of rupture, severely reduced impact values, and reduced corrosion resistance in some media. For adequate protection against detrimental eutectoid transformation, cooling after tempering should bring the alloy to a temperature below 370 °C (700 °F) within about 5 min, and to a temperature below 275 °C (530 °F) within 15 min. Normally, the danger of eutectoid transformation is much smaller in nickel aluminum bronzes, and these alloys can be air cooled after tempering.

The normal precautions used in heat treating of steel have been found to be applicable to aluminum bronze, with critical cooling rates being somewhat lower than those for steel. Soaking time should be at least 45 min at temperature, and longer times may be used without fear of excess surface oxidation. Oil quenching is used on heavy, complex sections to avoid quench cracking, particularly in nickel aluminum bronzes such as C63000.

Alpha aluminum bronzes are those aluminum bronzes that contain less than 9% aluminum, or less than 8.5% aluminum with up to 3% iron. They are essentially single-phase alloys, except for fine iron-rich particles in those alloys that contain iron. For alpha aluminum bronzes, effective strengthening can be attained only by cold work, and annealing and/or stress relieving are the only heat treatments of practical use. The most prevalent alloys of this group are C60600, C61000, C61300 and C61400. In addition, alloys containing up to 9.6% Al, with microstructures containing small amounts of beta phase at high temperatures, have such limited heat treatability that they also can be hardened only by cold work.

Annealing of alpha aluminum bronzes is carried out at temperatures from about 540 to about 870 °C (1000 to 1600 °F), with the iron-containing alloys requiring temperatures nearer the high end of this range. Alloys of intermediate composition (containing small amounts of beta phase), such as C61900, normally are annealed at 595 to 650 °C (1100 to 1200 °F). Figure 14 presents annealing curves for typical alpha aluminum bronze alloys C61300 and C61400.

Complex (alpha-beta) aluminum bronzes are those aluminum bronzes whose normal microstructures contain more than one phase to the extent that beneficial quench-and-temper treatments are possible. These copper-aluminum alloys, with and without iron, are heat treated by procedures somewhat similar to those used for heat treatment of steel, and have isothermal transformation diagrams that resemble those of carbon steels. For these alloys, the quench-hardening treatment is essentially a high-temperature soak intended to dissolve all of the alpha phase into the beta phase. Quenching results in a hard room-temperature beta martensite structure; and subsequent tempering reprecipitates fine alpha needles in the structure, forming a tempered beta martensite.

Table 16 gives typical tensile properties and hardnesses of alpha-beta aluminum bronzes after various stages of heat treatment.

Heat Treating of Lead and Lead Alloys

By William B. Hampshire
Technical Service/Metallurgy
Lead Industries Association, Inc.

LEAD normally is considered to be unresponsive to heat treatment. Yet, some means of strengthening lead and lead alloys may be required for some applications. Lead alloys for battery components, for example, can benefit from improved creep resistance in order to retain dimensional tolerances for the full service life. Battery grids also require improved hardness to withstand industrial handling.

The absolute melting point of lead is 327.4 °C (621.3 °F). Therefore, in applications in which lead is used, recovery and recrystallization processes and creep properties have great significance. Attempts to strengthen the metal by reducing the grain size or by cold working (strain hardening) have proved unsuccessful. Lead-tin alloys, for example, may recrystallize immediately and completely at room temperature. Lead-silver alloys respond in the same manner within two weeks.

Transformations that are induced in steel by heat treatment do not occur in lead alloys, and strengthening by ordering phenomena, such as in the formation of lattice superstructures, has no practical significance.

Despite these obstacles, however, attempts to strengthen lead have met with some success.

Solid-Solution Hardening

In solid-solution hardening of lead alloys, the rate of increase in hardness generally improves as the difference between the atomic radius of the solute and the atomic radius of lead increases.

Specifically, in one study of possible binary lead alloys, it was found that the following elements, in the order listed, provided successively greater amounts of solid-solution hardening: thallium, bismuth, tin, cadmium, antimony, lithium, arsenic, calcium, zinc, copper and barium. Unfortunately, these elements have successively decreasing solid-solution solubilities, and thus the most potent solutes have the most limited solid-solution hardening effects. Within the midrange of this series, however, are elements that, when alloyed with lead, produce useful strengthening.

A useful level of strengthening normally requires solute additions in excess of the room-temperature solubility limit. In most lead alloys, homogenization and rapid cooling result in a breakdown of the supersaturated solution during storage. Although this breakdown produces coarse structures in certain alloys (lead-tin alloys, for example), it produces fine structures in others (lead-antimony alloys, for example). In alloys of the lead-tin system, the initial hardening produced by alloying is quickly followed by softening as the coarse structure is formed.

In lead-antimony alloys, at suitable solute concentrations, the structure may remain single phase with hardening by Guinier-Preston (GP) zones formed during aging. At higher concentrations, and in certain other systems, aging may produce precipitation hardening as discrete second-phase particles are formed.

Alloys that exhibit precipitation hardening typically are less susceptible to overaging and thus are more stable with time than alloys hardened by GP zones. Lead-calcium and lead-strontium alloys have been observed to age harden through discontinuous precipitation of a second phase—Pb_3Ca in lead-calcium alloys and Pb_3Sr in lead-strontium alloys—as grain boundaries move through the structure.

Solution Treating and Aging

Useful strengthening of lead can be attained by adding sufficient quantities of antimony to produce hypoeutectic lead-antimony alloys. Small amounts of arsenic have particularly strong effects on the age-hardening response of such alloys, and these effects are enhanced by solution treating and rapid quenching prior to aging.

An investigation (Ref 1) was conducted on the effects of additions of 0.15% arsenic on the age-hardening behavior of five hypoeutectic lead-antimony alloys. Accurately weighed quantities of commercially pure lead, antimony and arsenic were melted under a nitrogen atmosphere to produce the alloys listed below:

Alloy	Antimony, %	Arsenic, %
200	2.0	. . .
215	2.0	0.15
400	4.0	. . .
415	4.0	0.15
600	6.0	. . .
615	6.0	0.15
800	8.0	. . .
815	8.0	0.15
1000	10.0	. . .
1015	10.0	0.15

To minimize segregation during solidification, the melts, each weighing about 100 g (3.5 oz), were chilled rapidly by casting into a shallow horizontal steel boat. Pieces weighing about 15 g (0.5 oz) were cut from the castings and were used as hardness-test specimens after being subjected to one of two heat treatments. One group of specimens, designated group A, was air cooled from the liquid state and quenched in ice water as soon as solidification was complete. Another group, designated group B, was solution treated for 4 h at 250 °C (480 °F) and then quenched in ice water.

In view of the rapid age hardening of the arsenical alloys, it was desired that the hardness tests be commenced as soon after quenching as possible, and one minute was chosen as a reasonable period of time. For group B specimens, this was easily achieved by solution treating the specimens, which had previously been ground with 600 grade paper, in a stream of nitrogen in a vertical tube furnace. The specimens then were dropped into ice water and transferred after 20 s to water at room temperature. Hardness of these specimens was measured with a Vickers hardness tester using a 2.5-kg load.

Because the specimens in group A could not be prepared for hardness testing before they were quenched, a different procedure was necessary to permit testing to commence after one minute. These specimens were remelted in small flat-base cylindrical cups punched from thin aluminum foil. After quenching, the foil was peeled from the specimen and the hardness measurement was made on the flat surface (after the top surface had been quickly smoothed on a file to ensure proper seating in the test machine).

Cooling curves were obtained for the specimens in group A by means of Chromel-Alumel thermocouples, which were thinly sheathed with mild steel. After remelting, the thermocouple was removed before the beginning of solidification. Quenching then was timed to follow complete solidification on the basis of the previous cooling curves. The group A specimens were cooled at a rate of 1 °C/s (1.8 °F/s) to solidification at 232 °C (450 °F). These specimens showed evidence of surface segregation.

Hardness tests were continued for up to two years, with the specimens being stored at ambient temperature, which for the majority of the time was between 20 and 22 °C (68 and 72 °F). To investigate variations in hardness, tests were made on some of the specimens using a Reichert microhardness tester (5.3-g load, applied for 10 s).

Test Results. Results of hardness testing showed that lead-antimony alloys of commercial purity demonstrate significant age hardening, particularly after solution treating, as shown in Fig. 1 and 2. In the quench-cast alloys of group A, the hardness at any time increased with antimony content. For the solution-treated specimens of group

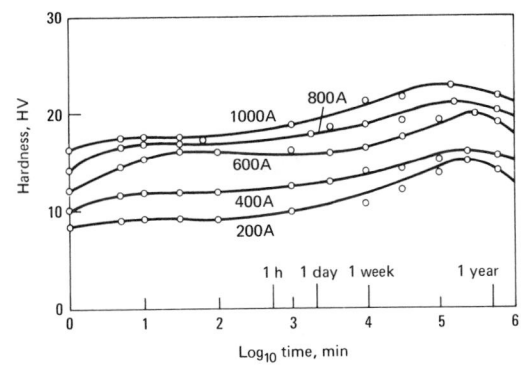

Fig. 1   Age hardening of Pb-Sb alloys, solidified and water quenched

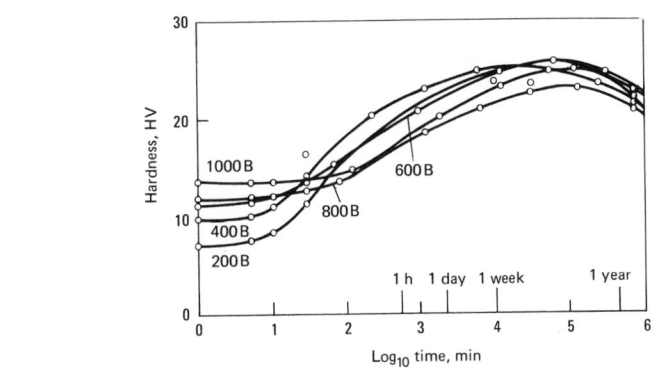

Fig. 2   Age hardening of Pb-Sb alloys, solution treated 4 h at 250 °C (480 °F) and water quenched

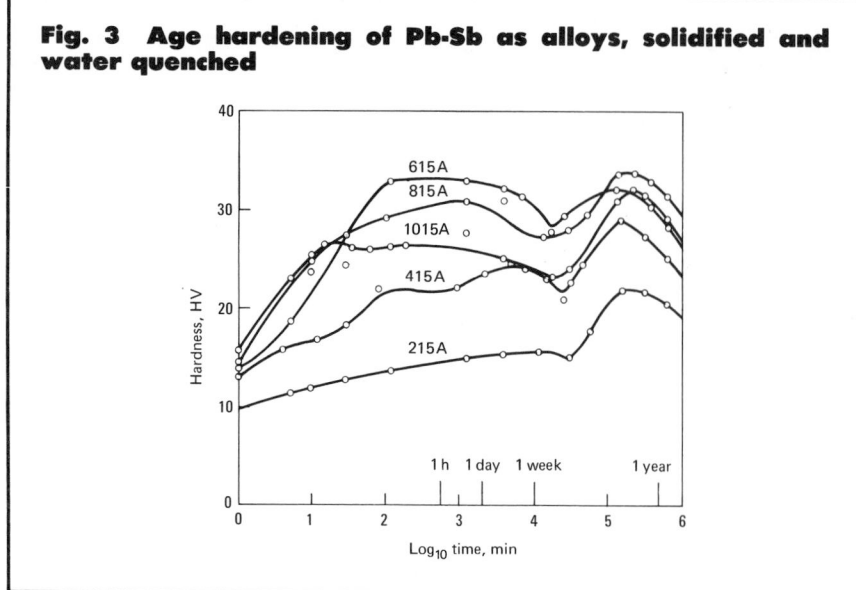

Fig. 3   Age hardening of Pb-Sb as alloys, solidified and water quenched

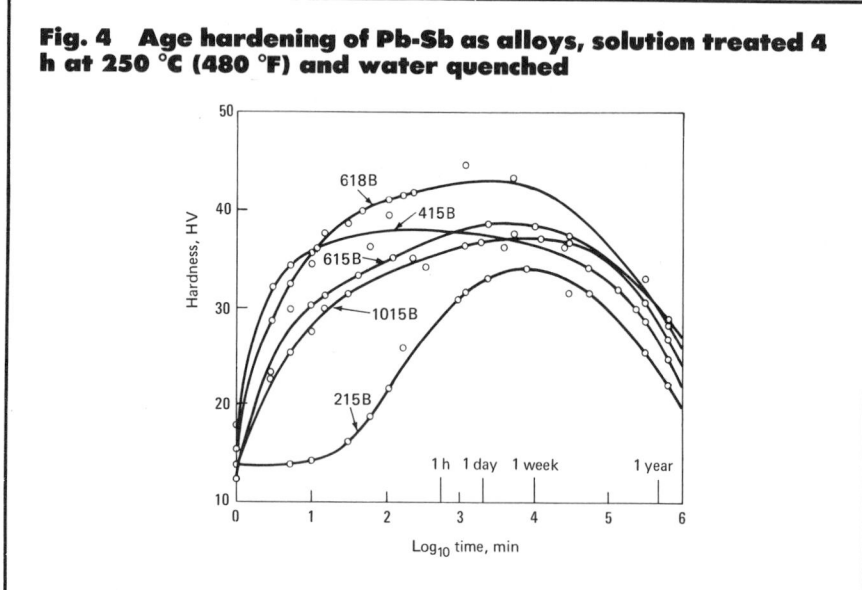

Fig. 4   Age hardening of Pb-Sb as alloys, solution treated 4 h at 250 °C (480 °F) and water quenched

B, aging was more effective for alloys with lower antimony contents.

Figures 3 and 4 show that both the rate and the extent of hardening are increased by addition of 0.15% arsenic. As shown in Fig. 4, increases in hardness of the solution-treated hypoeutectic alloys are pronounced in the first 10 min.

Hardness Stability. For any given alloy, both heat treatments result in hardnesses after 1 min and after 2 years that are somewhat comparable, as indicated in Fig. 1 through 4. For most of the two-year period, the solution-treated specimens were hard-er than the quench-cast specimens. Other investigations have also shown that alloys cooled slowly after casting are always softer than quenched alloys. As shown in Fig. 3, the alloys with 2 and 4% antimony harden comparative-ly slowly, and the alloy containing 6% antimony appears to undergo optimum hardening.

Application. To reduce the antimo-ny contents of the positive plates in lead-acid storage batteries, because of antimony's detrimental effect on charge retention, there has been a trend toward replacing eutectic alloys with a Pb-6Sb-0.15As alloy. Battery grids made of this arsenical alloy will age harden slowly after casting and air cooling. However, storing grids for several days constitutes unproductive use of floor space and results in unde-sirable interruptions in manufacturing sequences.

Although large-scale solution treat-ment of battery grids might be difficult to justify economically or to achieve without some distortion, quenching of grids cast from arsenical lead-antimony alloys offers an attractive alternative method of effecting improvements in strength. The suitability of quench-cast grids can be assessed by comparing the values given in Fig. 3 with the hard-ness level that battery grids require in order to withstand industrial handling (about 18 HV, the hardness of the eutectic alloy). The alloy containing 2% antimony clearly does not respond suf-ficiently well to be considered as a pos-sible alternative. The 4% antimony alloy, however, attains a hardness of 18 HV after 30 min, and the alloys that contain 6, 8, and 10% antimony could be handled almost immediately. Fur-thermore, the values given in Fig. 3 for hardness after two years are superior in all instances to those for air-cooled alloys of similar composition. Hardness curves decline steadily after two years, and full evaluation of these alloys for use in the battery industry would re-quire battery performance tests.

Dispersion Hardening

Another mechanism for strengthen-ing of lead alloys involves elements that have low solubilities in solid lead, such as copper and nickel. Alloys that contain these elements can be pro-cessed so that no homogenization re-sults; most of the strengthening that occurs is developed through dispersion hardening, with some solid-solution hardening taking place as a secondary effect. The resulting structure is more stable than those developed by other hardening processes. Dispersion strengthening also has been achieved through powder metallurgy methods in which lead oxide, alumina or similar materials are dispersed in pure lead.

Fabrication

Although alloy selection is impor-tant, care must be taken in fabrication as well. Castings should be cooled rap-idly to a temperature below that at which the structure breaks down, or a coarse structure will be obtained. Age-hardening alloys should be extruded at a temperature above the breakdown temperature, and extrusions should not be allowed to cool slowly. Rolled alloys often are processed at insufficient temperatures; when this occurs, homogenization after rolling is required if age hardening is to produce a beneficial response.

Cold Storage

Cold storage has been shown to improve the response of lead-antimony alloys to age hardening. Cooling a homogenized Pb-2Sb alloy to -10 °C ($+15$ °F) and holding for one or two days prior to room-temperature aging results in increases in both the rate of age hardening and the maximum hardness attained. This behavior has been explained as the result of a reduction in the mobility of quenched-in free vacancies and a consequent reduction in their annihilation. The process allows the vacancies to form complexes with sol-ute atoms, and these complexes improve the efficiency of nucleation during aging.

Service Temperatures

Service temperatures for lead alloys must be kept low to prevent overaging. Some cable-sheathing alloys, for example, have retained most of their creep resistance for up to 20 years, but exposure to elevated temperatures could have reduced this performance substantially. Even the normally stable age-hardened lead-antimony and lead-calcium alloys can be altered detrimentally by high service temperatures or excessive working.

REFERENCE

1. The Effect of Arsenic on the Age-Hardening of Lead-Antimony Alloys, by J. D. Williams: *Metallurgia,* Vol 74, No. 443, Sept 1966, p 105–108

SELECTED REFERENCES

For additional information on heat treatment of lead and lead alloys, refer to the following sources and their bibliographies:

- *Lead and Lead Alloys,* by W. Hofmann: Springer-Verlag, Berlin, 1970, p 262–267

- Effect of low-temperature treatment on the aging of lead antimony alloys, by J. J. Regidor *et al:* paper presented at LEAD '71, The Fourth International Conference on Lead, Hamburg, Sept 1971

- Structural control of non-antimonial lead alloys via alloy additions, heat treatment and cold work, by R. D. Prengaman: paper presented at Pb80, The Seventh International Lead Conference, Madrid, May 1980

Heat Treating of Magnesium Alloys

By the ASM Committee on
Heat Treating of Magnesium Alloys*

MAGNESIUM ALLOYS usually are heat treated either to improve mechanical properties or as a means of conditioning for specific fabricating operations. The type of heat treatment selected depends on alloy composition and form (cast or wrought), and on anticipated service conditions.

Solution heat treatment improves strength and results in maximum toughness and shock resistance. Artificial aging (precipitation heat treatment) subsequent to solution treatment gives maximum hardness and yield strength, but with some sacrifice of toughness. As applied to castings, artificial aging without prior solution treatment or annealing is a stress-relieving treatment that also somewhat increases tensile properties. Annealing of wrought products lowers tensile properties considerably and increases ductility, thereby facilitating some types of fabrication. Modifications of these basic treatments have been developed for specific alloys, to obtain the most desirable combinations of properties. For example, increasing the aging time for some magnesium

alloy castings considerably increases yield strength (although with some sacrifice of ductility). Also, combinations of solution treating, strain hardening and artificial aging are applied to alloy HM21A sheet to improve mechanical properties over those attainable by solution treating and artificial aging alone.

For certain magnesium alloys, development of properties depends almost entirely on heat treatment. In magnesium-zirconium alloys, however, the extremely pronounced grain-refining effect of the zirconium also plays a very important role in improving mechanical properties.

The basic temper designations outlined in Table 1 for magnesium alloys are used throughout this article to indicate the various types of heat treatment. For a detailed explanation of these designations, which are the same as those applied to aluminum alloys, see "Temper Designation System for Aluminum and Aluminum Alloys", on pages 24 to 27 in Volume 2 of this Handbook.

The mechanical properties of most

magnesium casting alloys can be improved by heat treatment. Casting alloys can be grouped into six general classes of commercial importance on the basis of composition, as follows:

- Magnesium-aluminum-manganese (example: AM100A)
- Magnesium-aluminum-zinc (examples: AZ63A, AZ81A, AZ91C, AZ92A)
- Magnesium-zinc-zirconium (examples: ZK51A, ZK61A)
- Magnesium-rare earth metal-zinc-zirconium-(examples: EZ33A, ZE41A)
- Magnesium-rare earth metal-silver-zirconium, with or without thorium (examples: QE22A, QH21A)
- Magnesium-thorium-zirconium, with or without zinc (examples: HK31A, ZH62A, HZ32A).

In most wrought alloys, maximum mechanical properties are developed through strain hardening, and these alloys generally are either used without subsequent heat treatment or merely aged to a T5 temper. Occasionally, however, solution treatment, or a

*Alan H. Braun, *Chairman*, Manager Research and Development, Wellman Dynamics Corp.; Emmett Bossing, Chief Metallurgist, Teledyne Cast Products; Sidney L. Couling, Senior Research Metallurgist, Battelle Columbus Laboratories; Henry J. Profitt, Technical Director, The Light Metal Testing Division, Haley Industries Ltd.

Table 1 Basic temper designations

For a more complete explanation of the designations outlined here, see pages 24 to 27 in Volume 2 of this Handbook.

F As fabricated
O Annealed, recrystallized (wrought products only)
H Strain hardened (wrought products only):
   H1 Strain hardened only
   H2 Strain hardened and partially annealed
   H3 Strain hardened and stabilized
W Solution heat treated; unstable temper
T Heat treated to produce stable tempers other than F, O, or H:
   T2 Annealed (cast products only)
   T3 Solution heat treated and cold worked
   T4 Solution heat treated
   T5 Artificially aged only
   T6 Solution heat treated and artificially aged
   T7 Solution heat treated and stabilized
   T8 Solution heat treated, cold worked and artificially aged
   T9 Solution heat treated, artificially aged and cold worked
   T10 Artificially aged and cold worked

Table 2 Heat treatments commonly applied to magnesium alloys

Alloy	Heat treatment(a)	Alloy	Heat treatment(a)
Casting alloys			
AM100A	T4, T5, T6, T61(b)	ZE41A	T5
		ZE63A	T6(c)
AZ63A	T4, T5, T6	ZH62A	T5
AZ81A	T4	ZK51A	T5
AZ91C	T4, T6	ZK61A	T4, T6
AZ92A	T4, T6		
EZ33A	T5	**Wrought alloys**	
HK31A	T6	AZ80A	T5
HZ32A	T5	HM21A	T5, T8, T81(d)
QE22A	T6		
QH21A	T6	HM31A	T5
		ZK60A	T5

(a) Indicated by temper designations (see Table 1). (b) Same as T6 except aged for longer time to increase yield strength. (c) Thermal treatment must include hydriding. (d) Mill modification of T8 to improve mechanical properties

Table 3 Annealing temperatures for wrought magnesium alloys

Alloy	Original temper	Annealing temperature(a) °C	°F
AZ31B	F, H10, H11, H23, H24, H26	345	650
AZ31C	F	345	650
AZ61A	F	345	650
AZ80A	F, T5, T6	385	725
HK31A	H24	400	750
HM21A	T5, T8, T81	455	850
HM31A	T5	455	850
ZK60A	F, T5, T6	290	550

(a) Time at temperature, 1 h or more

combination of solution treatment with strain hardening and artificial aging, will substantially improve mechanical properties. Wrought alloys that can be strengthened by heat treatment are grouped into four general classes according to composition:

- Magnesium-aluminum-zinc (example: AZ80A)
- Magnesium-thorium-zirconium (example: HK31A)
- Magnesium-thorium-manganese (examples: HM21A, HM31A)
- Magnesium-zinc-zirconium (example: ZK60A).

Types of Heat Treatment

The heat treatments commonly used for various magnesium alloys, both cast and wrought, are indicated by temper designations in Table 2.

Annealing. Wrought magnesium alloys in various conditions of strain hardening or temper can be annealed by being heated at 290 to 455 °C (550 to 850 °F), depending on alloy, for one or more hours (Table 3). This procedure usually will provide a product with the maximum anneal that is practical.

Because most forming operations on magnesium are done at elevated temperature, the need for fully annealed wrought material is less than with many other metals.

Stress Relieving of Wrought Alloys. Stress relieving is used to remove or reduce residual stresses induced in wrought magnesium products by cold and hot working, shaping and forming, straightening, and welding.

Table 4 gives the recommended stress-relieving times and temperatures for wrought magnesium alloys to obtain assemblies with maximum freedom from stress. When extrusions are welded to hard rolled sheet, the lower stress-relieving temperature and the longer time should be used to minimize distortion—for example, 150 °C (300 °F) for 60 min, rather than 260 °C (500 °F) for 15 min.

Stress Relieving of Castings. The precision machining of castings to close dimensional limits, the necessity of avoiding warpage and distortion, and the desirability of preventing stress-corrosion cracking in welded magnesium-aluminum casting alloys make it mandatory that cast components be substantially free from residual stresses. Although magnesium castings do not normally contain high residual stresses, the low modulus of elasticity of magnesium alloys means that comparatively low stresses can produce appreciable elastic strains.

Residual stresses may arise from contraction due to mold restraint during solidification, from nonuniform cooling after heat treatment, or from quenching. Machining operations also can result in residual stress and require intermediate stress relieving prior to final machining.

Weld repairs may introduce severe stresses and should be followed by some type of heat treatment to prevent subsequent movement and cracking, as discussed in the section of this article concerning heat treatment of repair-welded castings.

The following heat treatments for castings will provide stress relief without significantly affecting mechanical properties:

Alloy	Temper	Heat treatment
Mg-Al-Mn	All	1 h at 260 °C (500 °F)
Mg-Al-Zn	All	1 h at 260 °C (500 °F)
ZK61A	T5	2 h at 330 °C (625 °F), then 48 h at 130 °C (265 °F)
ZE41A	All	2 h at 330 °C (625 °F)

Table 4 Recommended stress-relieving treatments for wrought magnesium alloys

Alloy	Sheet Annealed Temperature °C	Sheet Annealed Temperature °F	Sheet Annealed Time, min	Hard rolled Temperature °C	Hard rolled Temperature °F	Hard rolled Time, min	Extrusions and forgings Temperature °C	Extrusions and forgings Temperature °F	Extrusions and forgings Time, min
AZ31B	345	650	120	150	300	60	...	...	...
AZ31B-F	...	...	...	...	...	...	260	500	15
AZ61A	345	650	120	205	400	60	...	...	...
AZ61A-F	...	...	...	...	...	...	260	500	15
AZ80A-F	...	...	...	...	...	...	260	500	15
AZ80A-T5	...	...	...	...	...	...	205	400	60
HK31A	345	650	60	290	550	30	...	...	...
HM21A-T5	...	...	...	...	...	...	370	700	30
HM21A-T8	...	...	...	370	700	30	...	...	...
HM21A-T81	...	...	...	400	750	30	...	...	...
HM31A-T5	...	...	...	...	...	...	425	800	60
ZK60A-F	230	450	180	...	...	...	260	500	15
ZK60A-T5	...	...	...	...	...	...	150	300	60

Note: Stress relieving after welding, to prevent stress-corrosion cracking, is necessary only for alloys that contain more than 1.5% aluminum.

Solution Treating and Aging. Schedules for solution treating and aging of magnesium alloy castings are summarized in Table 5. In solution treating of magnesium-aluminum-zinc alloys, parts should be loaded into the furnace at approximately 260 °C (500 °F) and then raised to the appropriate solution-treating temperature slowly, to avoid fusion of eutectic compounds and resultant formation of voids. The time required to bring the load from 260 °C to the solution-treating temperature is determined by the size of the load and by the composition, size, weight and section thickness of the parts, but 2 h is a typical time. All other heat treatable magnesium alloys can be loaded into the furnace at the solution-treating temperature. For alloy HK31A, it is important to bring the load to temperature as rapidly as possible, to avoid grain coarsening.

During aging, magnesium alloy parts should be loaded into the furnace at the treatment temperature, held for the appropriate period of time, and then cooled in still air. As indicated in Table 5, there is a choice of artificial aging treatments for some alloys; results are closely similar for the alternative treatments given.

Reheat Treating. Under normal circumstances, when mechanical properties are within expected ranges and the prescribed heat treatment has been carried out, reheat treating is seldom necessary. However, if the microstructures of heat treated castings indicate too high a compound rating or if the castings have been aged excessively by slow cooling after solution treating, reheat treating is called for. Most magnesium alloys can be reheat treated with little danger of germination (excessive grain growth). When reheat treating of alloy HK31A is necessary, however, the castings should be checked carefully for evidence of germination. To prevent germination in Mg-Al-Zn alloys, solution reheat treating time should be limited to 30 min (assuming proper solution treatment of thick sections during prior heat treatment).

Effects of Major Variables

Casting size and section thickness, relation of casting size to volume capacity of the furnace, and arrangement of castings in the furnace are mechanical considerations that can affect heat treating schedules for all metals.

Section Size and Heating Time. There is no general rule for estimating time of heating per unit of thickness for magnesium alloys. However, because of the high thermal conductivity of these alloys, combined with their low specific heat per unit volume, parts reach soaking temperature quite rapidly. The usual procedure is to load the furnace and then begin the soaking period when the loaded furnace reaches the desired temperature.

The heat treating times given in Table 5 have been found to be satisfactory for normal furnace loads and for castings of moderate section thickness. In the heat treating of magnesium alloy castings with thick sections (occasionally as low as 25 mm or 1 in. but usually over 50 mm or 2 in.), a good rule is to double the time at the solution treating temperature. For example, the usual solution treatment for AZ63A castings is 12 h at about 385 °C (725 °F), whereas 25 h at about 385 °C is suggested for castings with section thicknesses greater than 50 mm (2 in.). Similarly, the suggested solution-treating schedule for preventing excessive grain growth in AZ92A castings is 6 h at about 405 °C (765 °F), 2 h at about 350 °C (665 °F) and 10 h at about 405 °C; but for castings with sections more than 50 mm (2 in.) thick, it is recommended that the last soak at 405 °C be extended from 10 h to 19 h. The best way to determine whether or not additional solution-treating time is required is to cut a section through the thickest portion of a scrap casting and examine the center of the section microscopically: if heat treatment is complete, this examination will reveal a low compound rating.

Heat Treating Time and Temperature. As demonstrated by the data in Fig. 1, 2 and 3, the mechanical properties of magnesium alloys can be varied within wide limits by varying the heat treating times and temperatures recommended in Table 5. (Although, as illustrated in Fig. 1, the highest mechanical properties in test bars of QE22A-T6 were obtained by solution treating for 4 h at 540 °C, or 1005 °F; less distortion due to sagging is experienced in production castings solution treated for 8 h at 527 °C, or 980 °F.) The risk of incipient melting can also occur at the higher temperature.

Protective Atmospheres. Although magnesium alloys can be heat treated in air, protective atmospheres are almost always used for solution treating. Government specification MIL-M-6857, for heat treating of magnesium castings, requires a protective atmosphere for solution treating above 400 °C (750 °F). Protective atmospheres serve the dual purpose of preventing surface oxidation (which, if severe, can

Table 5 Recommended solution-treating and aging schedules for magnesium alloy castings

For castings up to 51 mm (2 in.) in section thickness; heavier sections may require longer times at temperature.

Alloy	Final temper	Aging(a) Temperature °C, ±6(b)	°F, ±10(b)	Time, h	Solution treating(c) Temperature °C, ±6(b)	°F, ±10(b)	Time, h	Maximum temperature °C	°F	Aging after solution treating Temperature °C, ±6(b)	°F, ±10(b)	Time, h
Magnesium-aluminum-zinc alloys(d)												
AM100A	T5	232	450	5	...	...	...	...	...	...	...	...
	T4	...	...	...	424(e)	795(e)	16–24(e)	432	810	...	...	...
	T6	...	...	...	424(e)	795(e)	16–24(e)	432	810	232	450	5
	T61	...	...	...	424(e)	795(e)	16–24(e)	432	810	218	425	25
AZ63A	T5	260(f)	500(f)	4(f)	...	...	...	...	...	...	...	...
	T4	...	...	...	385	725	10–14	391	735	...	...	...
	T6	...	...	...	385	725	10–14	391	735	218(f)	425(f)	5(f)
AZ81A	T4	...	...	...	413(e)	775(e)	16–24(e)	418	785	...	...	...
AZ91C	T5	168(g)	335(g)	16(g)	...	...	...	...	...	...	...	...
	T4	...	...	...	413(e)	775(e)	16–24(e)	418	785	...	...	...
	T6	...	...	...	413(e)	775(e)	16–24(e)	418	785	168(h)	335(h)	16(h)
AZ92A	T5	260	500	4	...	...	...	...	...	...	...	...
	T4	...	...	...	407(j)	765(j)	16–24(j)	413	775	...	...	...
	T6	...	...	...	407(j)	765(j)	16–24(j)	413	775	218	425	5
Magnesium-zirconium alloys												
EZ33A	T5	216(k)	420(k)	5(k)	...	...	...	...	...	...	...	...
HK31A(m)	T6	...	...	...	566	1050	2	571	1060	204	400	16
HZ32A	T5	316	600	16	...	...	...	...	...	...	...	...
QE22A(n)	T6	...	...	...	527	980	4–8	538	1000	204	400	8
QH21A(n)	T6	...	...	...	527	980	4–8	538	1000	204	400	8
ZE41A	T5	329(p)	625(p)	2(p)	...	...	...	...	...	...	...	...
ZE63A(q)	T6	...	...	...	480	895	10–72	491	915	141	285	48
ZH62A	T5	329	625	2	...	...	...	...	...	...	...	...
	plus:	177	350	16	...	...	...	...	...	...	...	...
ZK51A	T5	177(r)	350(r)	12(r)	...	...	...	...	...	...	...	...
ZK61A	T5	149	300	48	...	...	...	...	...	...	...	...
	T6	...	...	...	499(s)	930(s)	2(s)	502	935	129	265	48

(a) Aging of castings to the T5 temper is done from the as-cast condition. (b) Except where quoted differently. (c) After solution treatment and before subsequent aging, castings are cooled to room temperature by fast fan cooling, except where otherwise indicated. Use carbon dioxide or sulfur dioxide atmosphere above 400 °C (750 °F). (d) For solution treating, Mg-Al-Zn alloys are loaded into the furnace at 260 °C (500 °F) and brought to temperature over a 2-h period at a uniform rate of temperature increase. (e) Alternative treatment, to prevent germination (excessive grain growth): 6 h at 413 ± 6 °C (775 ± 10 °F), 2 h at 352 ± 6 °C (665 ± 10 °F), 10 h at 413 ± 6 °C (775 ± 10 °F). (f) Alternative treatment: 5 h at 232 ± 6 °C (450 ± 10 °F). (g) Alternative treatment: 4 h at 216 ± 6 °C (420 ± 10 °F). (h) Alternative treatment: 5–6 h at 216 ± 6 °C (420 ± 10 °F). (j) Alternative treatment, to prevent germination (excessive grain growth): 6 h at 407 ± 6 °C (765 ± 10 °F), 2 h at 352 ± 6 °C (665 ± 10 °F), 10 h at 407 ± 6 °C (765 ± 10 °F). (k) Alternative treatment, which can be used where maximum resistance to creep at elevated temperature is not of prime importance: 2 h at 343 ± 6 °C (650 ± 10 °F). (m) Alloy HK31A castings must be loaded into the furnace already at temperature and brought back to temperature as quickly as possible. (n) Quench from solution-treating temperature either in water at 65 °C (150 °F) or in other suitable quenching medium. (p) This treatment is adequate for development of satisfactory properties; it may be followed by 16 h at 177 ± 6 °C (350 ± 10 °F), to provide very slight improvements in mechanical properties. (q) Alloy ZE63A must be solution treated in a special hydrogen atmosphere, because its mechanical properties are developed through hydriding of some of its alloying elements. Hydriding time depends on section thickness; as a guide, 6.4-mm (1/4-in.) sections require approximately 10 h, and 19-mm (3/4-in.) sections require about 72 h. Following solution treatment, ZE63A should be quenched in oil, water spray or air blast. (r) Alternative treatment: 8 h at 218 ± 6 °C (425 ± 10 °F). (s) Alternative treatment: 10 h at 482 ± 6 °C (900 ± 10 °F).

decrease strength) and of preventing active burning should the furnace exceed proper temperature.

The two gases normally used are sulfur dioxide and carbon dioxide. Inert gases also may be used; however, in most instances, these gases are not practical because of higher cost. Sulfur dioxide is available bottled, while carbon dioxide may be obtained either bottled or as the product of recirculated combustion gases from a gas-fired furnace. A concentration of 0.7% (0.5% min) sulfur dioxide will prevent active burning to a temperature of 565 °C (1050 °F), provided that melting of the alloy has not occurred. Carbon dioxide in a concentration of 3% will prevent active burning to 510 °C (950 °F), and a carbon dioxide concentration of 5% will provide protection to about 540 °C (1000 °F).

Although sulfur dioxide is more expensive per unit volume, the volume of sulfur dioxide required for a protective atmosphere is only 1/6 the volume of carbon dioxide required. Thus, the cost of producing a protective atmosphere with bottled gases is less using sulfur dioxide. Where gas-fired furnaces are used and the atmosphere is obtained by recirculating combustion gases, a carbon dioxide atmosphere has the lower cost.

The use of sulfur dioxide requires frequent cleaning of furnace controls and fixtures and replacement of furnace parts, because of the formation of corrosive sulfuric acid in the furnace system. When it is desired to heat treat both magnesium and aluminum alloy castings in the same furnace, a carbon dioxide atmosphere is required, because sulfur dioxide is harmful to aluminum. Government regulations regarding the use of sulfur dioxide also suggest the use of carbon dioxide.

Fig. 1 Tensile properties of alloy QE22A-T6 as functions of solution-treating temperature

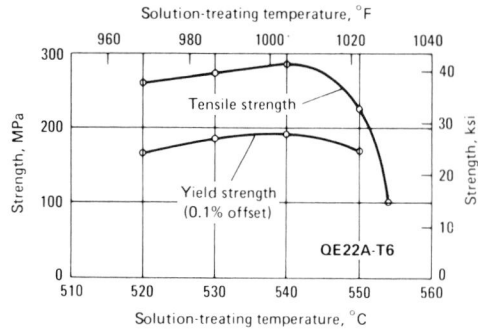

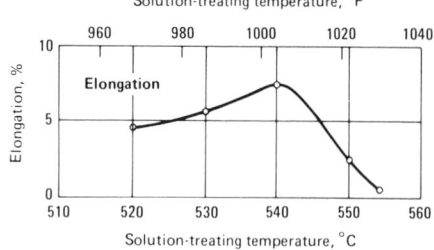

Data were obtained from test bars of casting alloy QE22A-T6 machined from 25-mm-(1-in.-) diam cast specimens. The bars were held at temperature for 4 h.

Table 6 Effect of quenching medium on average tensile properties of QE22A-T6

Quenching medium	Tensile strength		Yield strength(a)		Elonga-tion(b),
	MPa	ksi	MPa	ksi	%
Still air(c)	232	33.6	158	22.9	3.8
Air blast(c)	250	36.2	182	26.4	3.5
Water at 65 °C (150 °F)(c)	270	39.2	190	27.5	3.0
30% glycol at room temperature(d)	269	39.0	190	27.5	3.0

(a) At 0.2% offset. (b) In 50 mm or 2 in. (c) Properties determined on bars machined from 25-mm- (1-in.-) diam separately cast specimens. (d) Properties determined on bars machined from castings

Equipment and Processing

In solution treating and artificial aging of magnesium alloys, it is standard practice to use an electrically heated or gas-fired furnace equipped with a high-velocity fan or comparable means for circulating the atmosphere and promoting uniformity of temperature. However, because the atmosphere for solution treating sometimes contains sulfur dioxide, only furnaces that are gastight and that provide an inlet for introducing protective atmosphere are suitable.

The atmosphere is circulated past the heating elements and through the load in the basket. A rapid rate of circulation is necessary for maintaining an even distribution of temperature throughout the load. The minimum rate of circulation varies with furnace design and loading practice; one manufacturer recommends a rate equivalent to about 45 changes of atmosphere per minute.

Furnace Loading. Loading of furnaces used for heat treatment of magnesium parts is an important consideration. Parts must be clean and free from grinding dust, shavings, chips, sawings and other fines; this is particularly important at the higher temperatures used for solution treatment. A furnace load should be comprised of only one alloy composition because of the variation in nonequilibrium fusion points among various alloys. Furnaces should be loaded in an orderly manner that will not interfere with air circulation and cause uneven heating.

Temperature Control. A high degree of temperature control is required for heat treating of magnesium alloys. As indicated in Table 5, the maximum allowable temperature variation is ±6 °C (±10 °F) for the solution treating operation.

The safest and most suitable temperature-control system for solution and precipitation treating of magnesium alloys consists of three types of furnace controls. The first control detects the temperature within the heating chamber and controls the source of heat so as to attain and hold a desired predetermined temperature. This control is composed of a temperature-sensing device (usually a thermocouple) and a recording controller.

The second control is a furnace-charge-temperature sensor and is used to determine when the load has reached temperature. It consists of a thermocouple strategically located in the charge and a temperature-indicating device; an indicator-controller may be used to actuate an alarm, signal, or timing device to indicate that the charge has reached temperature. This control is useful for preventing overshooting of the desired maximum temperature because of high heat input or because the load is small in comparison with the size of the furnace.

The third control is a safety device and is used to prevent serious overheating of the load or a possible magnesium fire. It consists of a thermocouple, usually located in the roof of the furnace, and an indicating controller that will turn off the source of heat to the furnace. This controller should require resetting before the furnace will heat again.

Quenching Media. Magnesium alloy products normally are quenched in air following solution treatment. Still air usually is sufficient; forced-air cooling is recommended for dense loads or for parts that have very thick sections.

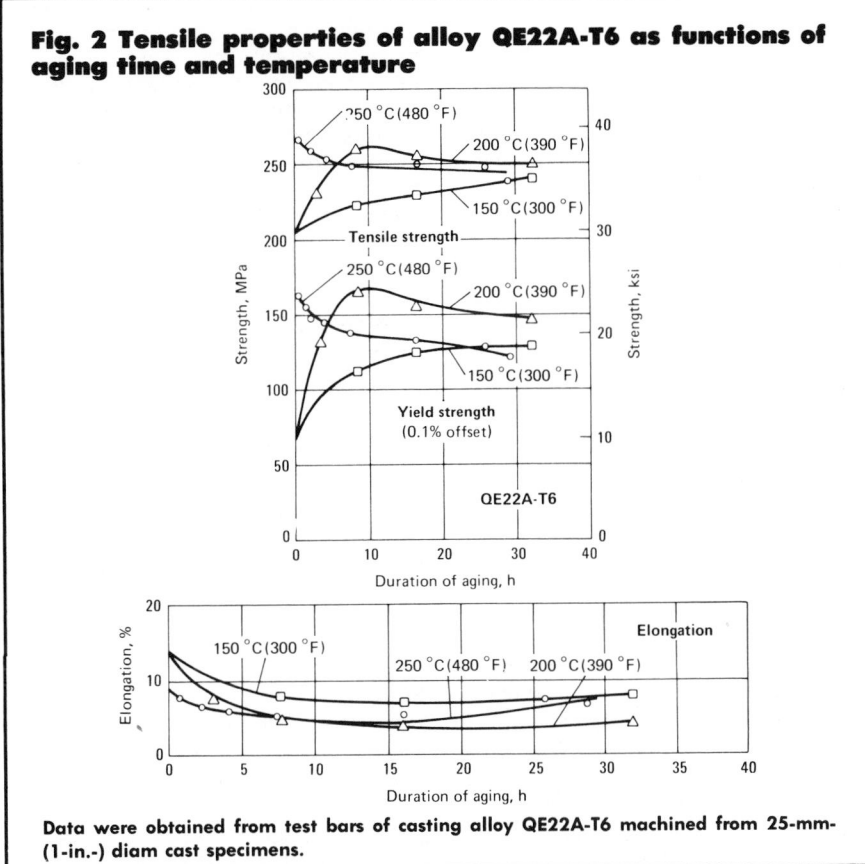

Fig. 2 Tensile properties of alloy QE22A-T6 as functions of aging time and temperature

Data were obtained from test bars of casting alloy QE22A-T6 machined from 25-mm-(1-in.-) diam cast specimens.

ing alloys exhibit good dimensional stability and can be considered free from additional dimensional changes.

Some cast magnesium-aluminum-manganese and magnesium-aluminum-zinc alloys in certain tempers exhibit slight permanent growth after relatively long exposure to temperatures exceeding 95 °C (Fig. 4 and 5). This growth, although slight, can give rise to problems:

Example. An aircraft-engine cover plate, sand cast of alloy AZ63A, was used in the as-cast condition (F temper), and the operating temperature of the engine was high enough to cause growth. As a result of this growth, it was necessary to pry off the cover plate when the engine was taken down for overhauling, and the plate could not be replaced because the holes did not line up properly with mating studs. This problem could have been eliminated by proper choice of temper (T5 or T6) prior to placing the cover in service.

In contrast to the growth characteristics of the magnesium-aluminum-zinc alloys are those of the magnesium alloys containing thorium, rare earth metals and zirconium as major alloying elements. These alloys normally are used in the T5 or T6 temper, and they shrink, rather than grow, on exposure to elevated temperature (Table 8).

The only common exceptions to air quenching are alloys QE22A and QH21A, for which water quenching at 60 to 95 °C (140 to 200 °F) is used to develop the best mechanical properties. Glycol or oil quenchants will produce similar properties with reduced distortion. For QE22A products subject to distortion as a result of the severity of water quenching, air cooling may be used provided that the cooling rate exceeds 3 °C/s (5 °F/s). The effects of quenching in still air, in forced air and in water at 65 °C (150 °F) on the tensile properties of separately cast test bars of QE22A-T6 are indicated by the data in Table 6.

Control of Distortion. The strength of cast magnesium alloys decreases at elevated temperature to such an extent that it is often necessary during solution treating to prevent intricate castings from sagging from their own weight and to keep flat castings from warping as a result of the relief of casting stresses. To accomplish these ends, tie bars are made an integral part of the casting, simple fixturing is used, or complicated cast or machined fix-

tures are produced. The method used depends on the complexity and quantity of castings to be made, and on the degree of dimensional control required. Whatever the method, it should not interfere unnecessarily with the free circulation of heat about the castings. Although fixturing decreases warpage of castings, some castings still require straightening after solution treating. Straightening is most readily done after solution treating, prior to aging.

Prevention of Heat Treating Problems

Six common problems that may be encountered in heat treating magnesium alloys are oxidation, fusion voids, warpage, grain coarsening, germination, and inconsistent properties. Causes and prevention are discussed in Table 7.

Dimensional Stability

In normal service up to approximately 95 °C (200 °F), all magnesium cast-

Heat Treatment of Weld-Repaired Castings

Magnesium sand castings that have been reclaimed or repaired by welding may be subsequently heat treated to relieve residual stresses. Heat treatment also may be necessary for restoring the mechanical properties of the casting when these have been impaired or modified by preheating prior to welding, or for heat treating the weld zone as such.

Table 9 shows the heat treating schedules recommended for magnesium castings after welding. The heat treating procedures shown are based on both the temper of the casting before the welding operation and the temper desired after welding. These postwelding heat treatments are normally all that is required for adequate stress relief of castings and for optimum mechanical properties in weld areas. The solution treatments here require the use of a protective atmosphere to prevent oxidation or burning. Only the minimum time for complete solution

Table 7 Causes and prevention of problems commonly encountered in heat treatment of magnesium alloys

Oxidation

Cause: Heat treating without use of protective or inert atmosphere; can lead to local weakening of the metal part, and even to burning of the metal in the furnace.

Prevention: Heat treat in a controlled atmosphere containing about 0.5 to 1.5% SO_2 or 3 to 5% CO_2, or (less practical, because more costly) in an inert-gas atmosphere. Ensure that furnace is clean and completely dry.

Fusion voids

Cause: Use of improper rate of heating from 260 to 370 °C (500 to 700 °F) for Mg-Al-Zn alloys, or exceeding recommended temperature in solution heat treating of these alloys or of the alloys that contain zinc, thorium and rare earth metals as major alloying elements.

Prevention: Charge furnace with Mg-Al-Zn alloys at 260 °C (500 °F), then heat gradually to solution-treating temperature over a period of 2 h. Control solution temperature so as not to exceed designated temperature by more than 6 °C (10 °F).

Warpage

Cause: Lack of support of castings during heat treatment; uneven distribution of heat.

Prevention: Support long spans of thin cross section; use jigs for intricate shapes. Distribute load in furnace so as to obtain good circulation of atmosphere.

Grain coarsening

Cause: Occurs in HK31A as a result of delay in attaining solution temperature or of holding at solution temperature for an excessive period of time.

Prevention: Prior to solution treating of HK31A, furnace should be at temperature; castings should be loaded quickly, and the loaded furnace should be closed and brought to temperature as rapidly as possible. Time at temperature should be controlled.

Germination

Cause: Grain growth, which occurs in AM100A, AZ81A, AZ91C and AZ92A toward the end of the solution-treating cycle.

Prevention: Use antigermination heat treating schedules (presented as alternative treatments in footnotes e and j in Table 5).

Inconsistent properties

Cause: Insufficient or excessive furnace temperature, inadequate circulation of heat in the furnace, faulty temperature control, very slow cooling from the solution-treating temperature, or inadequate solution-treating time for heavy sections.

Prevention: Check temperature at various positions in furnace with standardized thermocouple. Distribute castings in furnace so as to provide adequate circulation of heat. Check temperature controls often, and ensure that controls are located so as to provide uniformity of temperature. Increase solution-treating time to allow complete homogenization.

(½ h) is used for the welded casting alloys AZ81A, AZ91C and AZ92A if the castings were in the T4 or the T6 condition prior to welding.

Welded Mg-Al-Zn alloy castings that do not require solution treatment after welding should be stress relieved 1 h at 260 °C (500 °F), to eliminate the possibility of stress-corrosion cracking.

Evaluation of Heat Treated Parts

The effectiveness of heat treating procedures can be determined by hardness and tensile tests and by microscopic examination.

Indentation hardness tests are rapid and usually can be performed on the parts heat treated, without the necessity for a special test specimen. Brinell and Rockwell E hardness tests are normally used, but the Rockwell 15T superficial test may be required for thin sections. Soft material with large grains should be tested by the Brinell method for best results. The strength of magnesium alloys usually increases as hardness increases, but a graph of strength values determined from hardness plotted against actual strength values from the same specimens shows so much scatter that hardness cannot be used as an index for strength. Furthermore, hardness cannot indicate how an alloy will machine, or how easily it can be fabricated by other methods; it is primarily a measure of the temper of the material, and normally it suffices for this purpose.

Tensile tests more accurately indicate the temper of magnesium alloys but require special test specimens. For casting alloys, these can be separately cast unmachined specimens, although specimens machined from castings are more representative of actual casting properties. Test bars must be machined from extrusions, forgings and sheet. Test bars from extrusions and forgings are machined round for convenience. Standard ASTM specimens and procedures are usually employed to help ensure uniform results and avoid errors from variable testing speed, presence of scratches, and other causes.

Microscopic Examination. A method has been devised for microscopic examination of heat treated magnesium alloys whereby even inexperienced personnel can polish specimens and rate microstructure as a series of numbers (P. F. George, "Numerical Rating Method for the Routine Metallographic Examination of Commercial Magnesium Alloys", ASTM Bulletin No. 125, August 1944). In this method, standard microstructures are prepared for purposes of comparison. Specimens cut from magnesium alloy parts are examined, and by comparing them with the standard micrographs, it is possible to rate the alloy for the following conditions:

- Massive compound in cast alloys
- Percentage of pearlitic type of precipitate in cast alloys
- Porosity and "burning" voids in improperly solution heat treated cast alloys
- Grain size in cast and wrought alloys
- Massive compound in extruded, forged or rolled alloys.

Determination of Temper. For alloys AZ91C and AZ92A, the following solution can be applied to heat treated parts to identify the T6 conditions: 60 parts ethylene glycol; 20 parts glacial acetic acid; 19 parts distilled water; 1 part nitric acid, 42° Be. For ZE41A, the following solution can be applied to heat treated parts to identify the T5 condition: 5 parts glacial acetic acid; 95 parts distilled water. For either of these tests, the test surface should be prepared by sanding first with 180-grit paper and then with 220- to 400-grit paper to produce a smooth area about 25 mm (1 in.) square. After the test section has been wiped clean, one drop of the solution is applied with an eyedropper. After 30 s, the test section is rinsed with water and blotted dry with a piece of soft cloth.

Fig. 3 Variation of yield strength with aging time and temperature for sand cast AZ63A and AZ92A

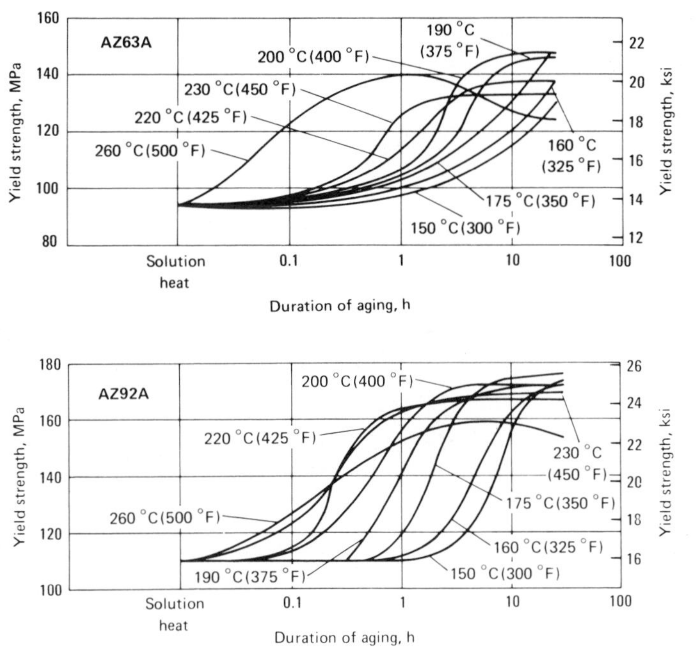

Source: T. E. Leontis and C. E. Nelson: *Transactions of AIME*, Vol 191, 1951, p 120

Table 8 Contraction of magnesium casting alloys at elevated temperatures

Temperature		Unit contraction, 100 µm/m (0.0001 in./in.), after exposure time, h, of:			
°C	°F	10	100	1000	5000
EZ33A-T5					
205	400	1.1	1.3	1.3	1.3
260	500	1.3	1.6	1.8	1.9
315	600	1.2	1.5	1.7	1.8
370	700	1.0	1.2	1.3	1.4
HK31A-T6					
205	400	0.3	0.3	0.3	0.3
260	500	0.3	0.5	0.7	0.7
315	600	0.2	0.5	1.3	1.3
370	700	0.6	1.1	1.1	1.1
HZ32A-T5					
205	400	1.1	1.3	1.4	1.4
260	500	0.8	1.0	1.1	1.2
315	600	0.8	1.0	1.1	1.2
370	700	0.6	0.8	0.9	1.0

For AZ91C and AZ92A, the surface of a part in a T6 condition will appear darker where etched. For ZE41A, the surface will appear lighter where etched if the part is in the T5 condition. This alloy will show a bluish tint in the as-cast condition, whereas alloy ZE41A in the T5 temper—either 2 h at 330 °C (625 °F) or 2 h at 330 °C plus 16 h at 175 °C (350 °F)—tends to show a brown-to-beige tint.

The inspected area should be wiped clean, and a fresh type I chrome pickle should be applied to prevent corrosion.

Prevention and Control of Magnesium Fires

Improper heat treating of magnesium alloy castings not only will result in damaged castings but also may cause fire. Only clean, dry castings should be introduced into the furnace. Castings contaminated with filings, chips, oil or other foreign matter should first be cleaned. The furnace itself should be protected from these same contaminants and should be completely dry. The charge should consist of castings of only one alloy, and the recommended heat treating cycle for that alloy should be closely followed.

Occasionally, because of equipment malfunction, carelessness or operator error, fires do occur. They are usually detectable either by an increase in furnace temperature without an increase in input of heat or by seepage of light-colored smoke from the furnace.

UNDER NO CIRCUMSTANCE SHOULD WATER BE USED TO EXTINGUISH A MAGNESIUM FIRE.

All power, fuel and protective-atmosphere lines should be shut off immediately. This in itself may extinguish small fires, which will die from lack of oxygen in a tight furnace. Should the fire continue to burn, several methods of extinguishing it are possible, depending on the nature of the fire. If the fire is small and the burning castings are readily accessible and can be *safely* removed from the furnace, they should be removed to a steel container and covered with commercially available fire-extinguishing powder intended for use with magnesium. If the burning castings are not both safely and readily accessible, a pump can be used to throw a stream of powder onto the burning castings in the furnace.

When the fire is such that neither of the above methods can safely be used for extinguishing it, another method is use of boron trifluoride (BF₃) or boron trichloride (BCl₃), both of which are available as bottled gases. As with the other methods, all power, fuel and protective-atmosphere lines should be shut off.

Boron trifluoride gas is introduced into the furnace through a pipe in the furnace door or wall so that a minimum

Fig. 4 Variation of growth with aging time and temperature for solution-treated AZ63A, AZ92A and AZ91C

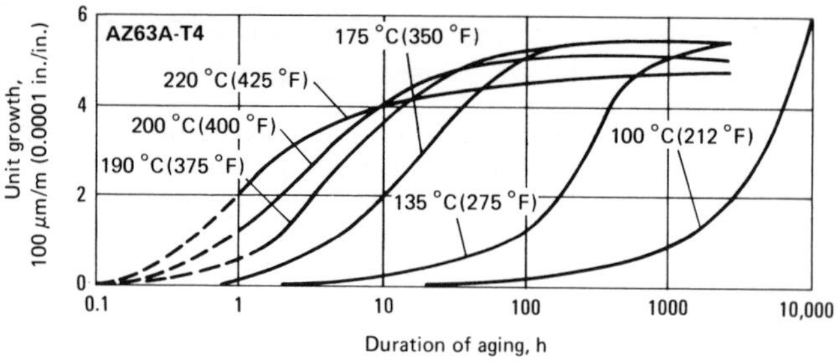

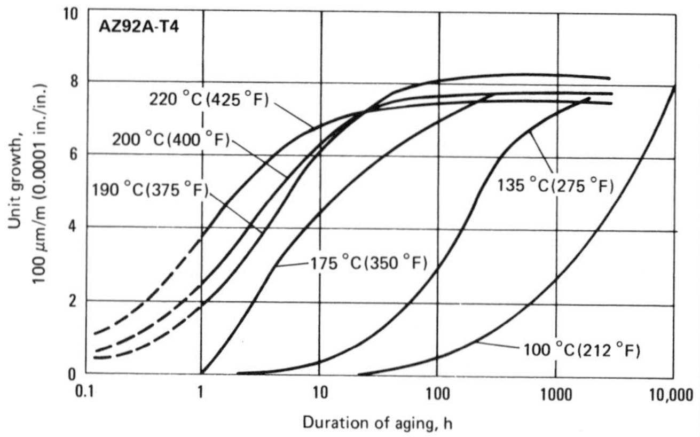

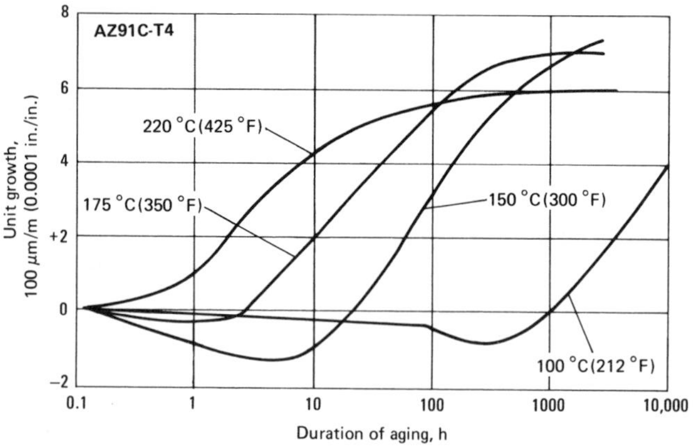

Source: T. E. Leontis and C. E. Nelson: *Transactions of AIME,* Vol 191, 1951, p 120

Fig. 5 Variation of growth with aging time and temperature for solution-treated AZ63A and AZ92A

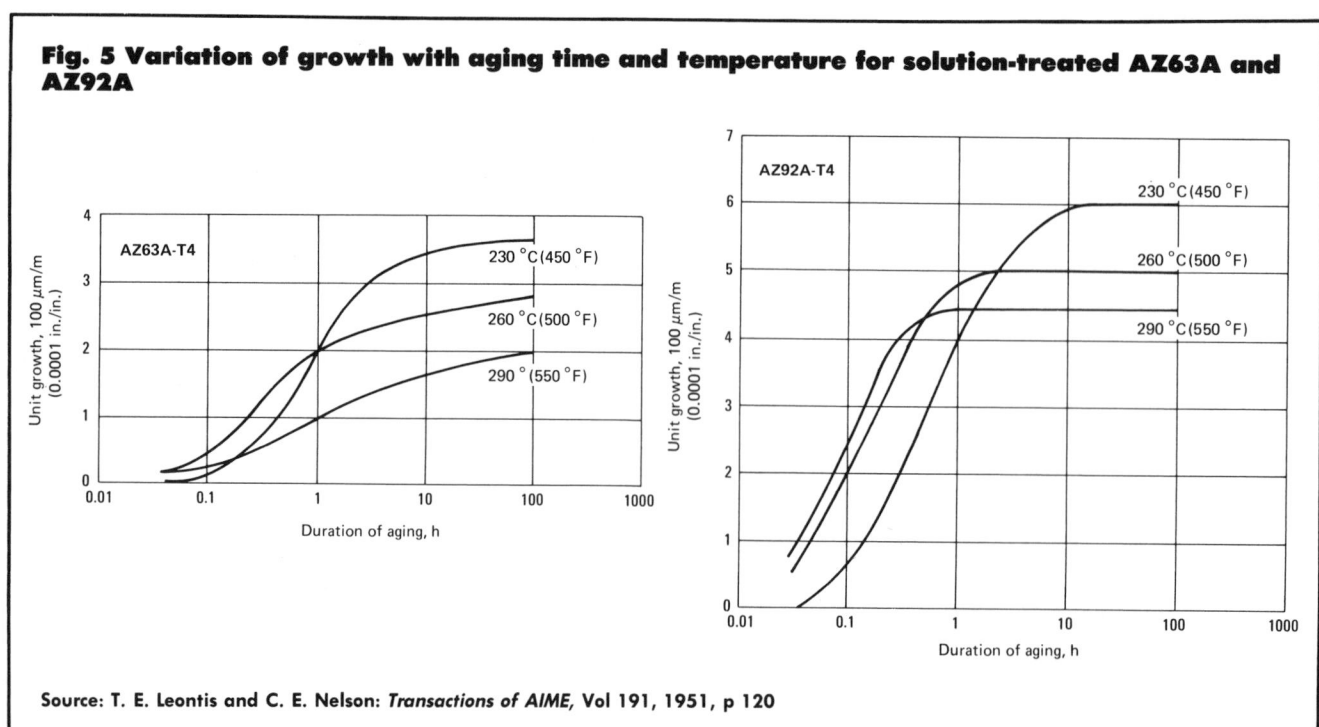

Source: T. E. Leontis and C. E. Nelson: *Transactions of AIME,* Vol 191, 1951, p 120

Table 9 Postweld heat treatments for magnesium alloy castings

Alloy	Welding rod	Temper before welding	Desired temper after welding	Postweld heat treatment
AZ63A	AZ63A or	F	T4.........	12 h at 385 ± 6 °C (725 ± 10 °F)(b)
	AZ92A(a)	F	T6.........	12 h at 385 ± 6 °C (725 ± 10 °F)(b), plus 5 h at 220 °C (425 °F)
		T4	T4.........	½ h at 385 ± 6 °C (725 ± 10 °F)
		T4 or T6	T6.........	½ h at 385 ± 6 °C (725 ± 10 °F), plus 5 h at 220 °C (425 °F)
AZ81A	AZ92A or AZ101	T4	T4.........	½ h at 413 ± 6 °C (775 ± 10 °F)(c)
AZ91C	AZ92A or	T4	T4.........	½ h at 413 ± 6 °C (775 ± 10 °F)(c)
	AZ101	T4 or T6	T6.........	½ h at 413 ± 6 °C (775 ± 10 °F)(c), plus 4 h at 215 °C (420 °F) or 16 h at 170 °C (335 °F)
AZ92A	AZ92A	T4	T4.........	½ h at 407 ± 6 °C (765 ± 10 °F)(c)
		T4 or T6	T6.........	½ h at 407 ± 6 °C (765 ± 10 °F)(c), 4 h at 260 °C (500 °F) or h at 220 °C (425 °F)
EZ33A	EZ33A	F or T5	T5.........	2 h at 345 °C (650 °F)(d), or 5 h at or 220 °C (420 or 425 °F)
HK31A	HK31A(g)	T4 or T6	T6.........	16 h at 205 °C (400 °F)(e)
HZ32A	HZ32A(g)	F or T5	T5.........	16 h at 315 °C (600 °F)
QE22A	QE22A	F or T6	T6.........	4 h at 524 ± 6 °C (975 ± 10 °F)(c); quench; 8 h at 205 °C (400 °F)
QH21A	QH21A	F or T6	T6.........	4 h at 524 ± 6 °C (975 ± 10 °F)(c); quench; 8 h at 205 °C (400 °F)
ZE41A	ZE41A(g)	F or T5	T5.........	2 h at 330 °C (625 °F)(f)
ZH62A	ZH62A(g)	F or T5	T5.........	12 h at 250 °C (480 °F)(f)
ZK51A	ZK51A(g)	F or T5	T5.........	2 h at 330 °C (625 °F), plus 16 h at 175 °C (350 °F)

(a) AZ63A rod must be used for welding AZ63A in the F temper, because 12 h at 385 °C (725 °F) causes germination in welds made with AZ92A rod; AZ92A rod normally is used for welding AZ63A in the T4 or T6 condition unless AZ63A rod is required by specifications. (b) Preheat to 260 °C (500 °F); heat to specified temperature at no more than 83 °C/h (150 °F/h). (c) Use carbon dioxide or sulfur dioxide atmosphere. (d) Heating for 2 h at 345 °C (650 °F) results in slight loss of creep strength. (e) Alternative treatment: 1 h at 315 °C (600 °F), plus 16 h at 205 °C (400 °F). (f) Alternative treatment: 2 h at 330 °C (625 °F), plus 16 h at 175 °C (350 °F). (g) Or EZ33A

concentration of 0.04% is produced. The gas flow is continued until the fire is extinguished and the temperature of the furnace drops to 370 °C (700 °F). Teflon hose is suitable for transferring the pressurized boron trifluoride gas from bottle to furnace.

Boron trichloride gas also is introduced into the furnace through a pipe in the furnace door or wall, in a concentration of about 0.4%. To ensure a sufficient volume of gas, a bank of infrared lamps or some other suitable device is required for heating the bottled gas. Boron trichloride gas reacts with hot magnesium to form a protective film over the castings. A supply of gas is maintained until the fire is extinguished and the temperature of the furnace drops to 370 °C (700 °F). In a totally enclosed furnace, it is feasible to employ the furnace fan to circulate the boron trifluoride or boron trichloride extinguishing gas around the castings. (Teflon hose can be used for transferring the boron trichloride gas from the bottle to the furnace.)

Boron trichloride predates boron trifluoride as a fire extinguisher for magnesium. Boron trifluoride is preferred, because it is effective in lower concentration, it does not require a heat source to ensure an adequate supply of gas, and its reaction products are less hazardous than those of boron trichloride. The fumes of boron trichloride are irritating and are ranked with hydrochloric acid fumes as a health hazard.

If the heat treating furnace fire includes more than several hundred pounds of magnesium parts, is well advanced prior to discovery, involves a large pool of metal on the floor of the furnace, or is in a furnace with excessive air leaks, these gases cannot be expected to extinguish the fire completely. However, both boron trifluoride and boron trichloride are effective in slowing or suppressing the fire until it can be extinguished with other solid materials.

Dry cast iron chips, graphite powder combined with heavy hydrocarbons, and (occasionally) foundry melting flux have been used for extinguishing large magnesium fires; the effectiveness of these materials is based on smothering the fire from the oxygen of the atmosphere in the furnace.

Besides the normal safety equipment used by personnel in fire fighting, colored glasses should be used when fighting a magnesium fire, for protection of the eyes against the characteristic intense white light.

Heat Treating of Nickel and Nickel Alloys

By Donald J. Tillack
Technical Sales Engineer
Huntington Alloys, Inc.
and
E. B. Fernsler (retired)
Technical Service Manager
Huntington Alloys, Inc.

NICKEL and nickel alloys may be subjected to one or more of five principal types of heat treatment, depending on chemical composition, fabrication requirements and intended service. These methods include:

Annealing—A heat treatment designed to produce a recrystallized grain structure and softening in work-hardened alloys. Annealing usually requires temperatures between 705 and 1205 °C (1300 and 2200 °F), depending on alloy composition and degree of work hardening.

Stress relieving—A heat treatment used to remove or reduce stresses in work-hardened non-age-hardenable alloys without producing a recrystallized grain structure. Stress-relieving temperatures for nickel and nick-

el alloys range from 425 to 870 °C (800 to 1600 °F), depending on alloy composition and degree of work hardening.

Stress equalizing—A low-temperature heat treatment used to balance stresses in cold worked material without an appreciable decrease in the mechanical strength produced by cold working.

Solution treating—A high-temperature heat treatment designed to put age-hardening constituents and carbides into solid solution. Normally applied to age-hardenable materials before the aging treatment.

Age hardening (precipitation hardening)—A treatment performed at intermediate temperatures (425 to 870 °C; 800 to 1600 °F) on certain alloys in order to develop maximum strength by precipitation of a dis-

persed phase throughout the matrix.

Figure 1 illustrates the effect of holding for 3 h at various temperatures on room-temperature mechanical properties of cold drawn Monel 400 rod. Heating at 700 °C (1300 °F) or above produces the soft annealed condition; heating in the range 400 to 595 °C (750 to 1100 °F) results in stress relieving; and heating in the range 260 to 480 °C (500 to 900 °F) results in stress equalization—producing an increase in the proportional limit, a slight increase in tensile strength, and no significant change in elongation.

Annealing

As applied to nickel and nickel alloys, annealing consists of heating the metal at a predetermined temperature for a definite time and then slowly or rapidly cooling it, to produce a change

Fig. 1 Effects of holding 3 h at various temperatures on room-temperature properties of cold drawn Monel 400 rod

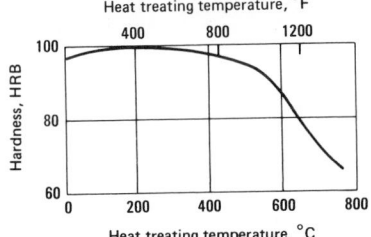

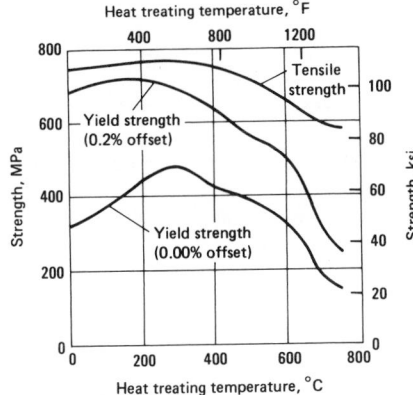

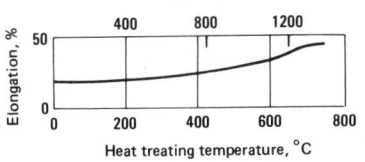

in mechanical properties—usually, a complete softening as a result of recrystallization. Nickel and nickel alloys that have been hardened by cold working operations, such as rolling, deep drawing, spinning or severe bending, require softening before cold working can be continued. The thermal treatment that will produce this condition is known as annealing, or soft annealing.

Annealing Practice and Methods. The differences in chemical composition among nickel and nickel alloys (Table 1) necessitate modifications in annealing temperatures (Table 2) as well as in furnace atmospheres. The precipitation-hardening alloys must be cooled rapidly after annealing if maximum softness is desired.

Three soft-annealing methods in general commercial use—open, closed and salt bath annealing—are described in the following paragraphs.

Open annealing is used most often. The material to be annealed is heated at the selected temperature and protected from oxidation by the products of combustion in a fuel-heated furnace, or by a reducing gas introduced into an electric furnace. Temperature control is critical, because the annealing period is short (Table 2). During preliminary heating of the furnace, the burners may be adjusted for optimum combustion; but, before charging, the air supply should be reduced to provide the excess of reducing gases required. Also, the top vents or dampers should be closed or partly closed to provide a positive gas pressure over the hearth and thereby prevent leakage of air.

Closed (box) annealing requires more

time than open annealing because of the lower temperatures used (Table 2). Temperature control is less critical than in open annealing. In most instances, the weight of the container exceeds that of the work; consequently, the amount of fuel required, heating time and cost are greater than in open annealing. Small pressed parts, rivets, coiled wire and similar work are annealed in boxes because of greater ease in handling. After the work has been charged, the boxes should be sealed with fireclay, sand or other suitable refractory material. A small quantity of charcoal, prebaked for 3 h at 650 to 815 °C (1200 to 1500 °F) to remove occluded gases, should be packed in the box at points where air could enter. Reducing gas is admitted to the box immediately upon withdrawal of the box from the furnace, or when the burners are shut off. Following annealing, the flow of gas should be regulated to maintain a positive pressure within the box; if the gas is combustible, it will burn gently at all outlets.

Salt bath annealing is used for special work with small parts. Inorganic salts, such as chlorides and carbonates of sodium, potassium and barium, which are relatively stable at temperatures considerably above their respective melting points, are fused in large metallic or refractory containers at temperatures up to about 700 °C (1300 °F); at higher temperatures, heat-resisting Fe-Ni-Cr alloy pots or refractory containers should be used. Excessive fuming of the bath is an indication of its maximum usable temperature.

Particular care must be exercised to remove all traces of sulfur from the

Table 1 Nominal compositions of nickels and nickel alloys

Material	Ni	C	Mn	Fe	Si	Composition, wt% Cu	Cr	Ti	Al	Mo	Co	Other
Nickel 200	99.5	0.06	0.25	0.15	0.05	0.05	...	...	...	...	...	...
Nickel 201	99.5	0.01	0.20	0.15	0.05	0.05	...	...	...	...	...	...
Monel 400	66.0	0.12	0.90	1.35	0.15	31.5	...	...	...	...	...	...
Monel R-405	66.0	0.18	0.90	1.35	0.15	31.5	...	...	...	...	...	...
Monel K-500	65.0	0.15	0.60	1.00	0.15	29.5	...	0.50	2.80	...	...	...
Inconel 600	76.0	0.04	0.20	7.20	0.20	0.10	15.8	...	...	...	...	...
Inconel 601	60.5	0.05	0.50	14.1	0.25	...	23.0	...	1.35	...	...	...
Inconel 617	54.0	0.07	...	...	...	...	22.0	...	1.00	9.0	12.5	...
Inconel 625	61.0	0.05	0.25	2.5	...	...	21.5	...	...	9.0	...	3.65 Nb
Inconel 718	52.5	0.04	0.20	18.0	0.20	0.10	19.0	0.80	0.60	3.00	...	5.20 Nb
Inconel X-750	73.0	0.04	0.70	6.75	0.30	0.05	15.0	2.50	0.80	...	...	0.85 Nb
Hastelloy B	64.0	0.03	...	5.0	...	...	...	...	...	28.0	2.5 max	...
Hastelloy C	56.0	0.05	...	5.5	...	...	15.5	...	...	16.0	2.5 max	4.0 W
Hastelloy X	48.0	0.10	...	18.5	...	...	22.0	...	...	9.0	1.5	0.6 W

Table 2 Annealing, stress-relieving and stress-equalizing schedules for nickels and nickel alloys

	Soft annealing							
	Open annealing				Closed annealing			
	Temperature		Time,	Cooling	Temperature		Time,	Cooling
Material	°C	°F	min	method(a)	°C	°F	h	method
Nickel 200	815 to 925	1500 to 1700	½ to 5	AC or WQ	705 to 760	1300 to 1400	2 to 6	AC
Nickel 201	760 to 870	1400 to 1600	½ to 5	AC or WQ	705 to 760	1300 to 1400	2 to 6	AC
Monel 400..................	870 to 980	1600 to 1800	2 to 10	AC or WQ	760 to 815	1400 to 1500	1 to 3	AC
Monel R-405..............	870 to 980	1600 to 1800	2 to 10	AC or WQ	760 to 815	1400 to 1500	1 to 3	AC
Monel K-500	870 to 1040	1600 to 1900	5 to 15	WQ	——————— Not applicable ———————			
Inconel 600...............	925 to 1040	1700 to 1900	15 to 30	AC or WQ	925 to 980	1700 to 1800	1 to 3	AC
Inconel 601...............	1095 to 1175	2000 to 2150	15 to 30	AC or WQ	1095 to 1175	2000 to 2150	1 to 3	AC
Inconel 617...............	1120 to 1175	2050 to 2150	15 to 30	AC or WQ	1120 to 1175	2050 to 2150	1 to 3	AC
Inconel 625...............	980 to 1150	1800 to 2100	15 to 30	AC or WQ	980 to 1150	1800 to 2100	1 to 3	AC
Inconel 718...............	955 to 980	1750 to 1800	15 to 30	AC	——————— Not applicable ———————			
Inconel X-750...........	1095 to 1150	2000 to 2100	15 to 30	AC	——————— Not applicable ———————			
Hastelloy B	1095 to 1185	2000 to 2165	5	AC or WQ	...	...	...	...
Hastelloy C	1215	2220	5	WQ	...	...	...	...
Hastelloy X	1175	2150	5 to 15	AC or WQ	1175	2150	1	AC or WQ

(a) AC, air cool; WQ, water quench

| | Stress relieving | | | | Stress equalizing | | | |
| | Temperature | | Time, | Cooling | Temperature | | Time, | Cooling |
Material	°C	°F	min	method	°C	°F	h	method
Nickel 200	480 to 705	900 to 1300	½ to 3	AC	260 to 480	500 to 900	1 to 2	AC
Nickel 201	480 to 705	900 to 1300	½ to 3	AC	260 to 480	500 to 900	1 to 2	AC
Monel 400..................	540 to 565	1000 to 1050	1 to 2	AC	230 to 315	450 to 600	1 to 3	AC
Monel R-405..............	...	...	...	...	...	...	...	...
Monel K-500	...	...	...	...	...	...	...	...
Inconel 600...............	760 to 870	1400 to 1600	1 to 2	AC	...	...	...	...
Inconel 601...............	...	...	...	...	...	...	...	...
Inconel 617...............	...	...	...	...	...	...	...	...
Inconel 625...............	...	...	...	...	...	...	...	...
Inconel 718...............	...	...	...	...	...	...	...	...
Inconel X-750...........	...	...	...	...	...	...	...	...
Hastelloy B	1095 to 1185	2000 to 2165	1/12	AC or WQ	——————Not applicable ——————			
Hastelloy C	1215	2220	1/12	WQ	——————Not applicable ——————			
Hastelloy X	...	...	...	...	——————Not applicable ——————			

(a) AC, air cool; WQ, water quench

fused salts in order to prevent embrittlement of the work. This may be accomplished in 2 to 3 h by adding to the fused chlorides and carbonates a small amount (1 lb or more) of a mixture consisting (by volume) of 3 parts powdered borax and 1 part powdered charcoal. If test pieces of nickel strip or wire do not embrittle after 3 or 4 h in the purified salt bath, the desulfurizing treatment was sufficient.

The material to be annealed is placed in the molten salts and absorbs heat rapidly. After being annealed, the work metal is quenched in water to free it from particles of the salt mixture. The annealed material will not be bright and may be flash pickled to achieve a bright surface.

Bright Annealing. The temperatures required for soft annealing of nickel and nickel alloys are sufficiently high to cause slight surface oxidation unless the materials are heated in vacuum or in a furnace provided with a reducing atmosphere. Nickel 200, Monel 400 and similar alloys will remain bright and free from discoloration when heated and cooled in a reducing atmosphere. However, nickel alloys containing chromium, titanium and aluminum will form a thin oxide film. Even if oxidation is not important, the furnace atmosphere must be suitably sulfur-free and not strongly oxidizing.

The protective atmosphere most commonly used in heating nickel and nickel alloys is that provided by controlling the ratio between the fuel and air supplied to burners firing directly into the furnace. A desirable reducing condition may be obtained by using a slight excess of fuel so that the products of combustion contain at least 2% carbon monoxide plus hydrogen (preferably, 4%) with no more than 0.05% uncombined oxygen.

Another method of maintaining desired conditions of furnace atmosphere is to introduce a prepared atmosphere into the heating and cooling chambers. This can be added to the products of combustion in a direct-fired furnace; however, introduction of prepared atmospheres is more commonly practiced with indirectly heated equipment.

Prepared atmospheres suitable for use with nickel and nickel alloys include: dried hydrogen, dried nitrogen, dissociated ammonia, and cracked or

Table 3 Prepared atmospheres suitable for annealing of nickels and nickel alloys

Atmospheres 2 through 7 can be used for bright annealing of nickel, modified nickels, and nickel-copper alloys; atmosphere 4 or atmosphere 7 must be used for bright annealing of nickel alloys that contain chromium or molybdenum, or both.

| Atmosphere | Air-to-gas ratio(a) | Composition, vol % | | | | | | Dew point (approx) | |
		H_2	CO	CO_2	CH_4	O_2	N_2	°C	°F
1 Completely burned fuel, lean atmosphere.............	10 to 1	0.5	0.5	10.0	0.0	0.0	89.0	Saturated(b)	
2 Partially burned fuel, medium-rich atmosphere	6 to 1	15.0	10.0	5.0	1.0	0.0	69.0	Saturated(b)	
3 Reacted fuel, rich atmosphere	3 to 1	38.0	19.0	1.0	2.0	0.0	40.0	+20	+70
4 Dissociated ammonia (complete dissociation)	No air	75.0	0.0	0.0	0.0	0.0	25.0	−55 to −75	−70 to −100
5 Dissociated ammonia, partially burned...............	1.25 to 1(c)	15.0	0.0	0.0	0.0	0.0	85.0	Saturated(b)	
6 Dissociated ammonia, completely burned	1.8 to 1(c)	1.0	0.0	0.0	0.0	0.0	99.0	Saturated(b)	
7 Electrolytic hydrogen, dried(d)	No air	100.0	0.0	0.0	0.0	0.0	0.0	−55 to −75	−70 to −100

(a) Based on use of natural gas containing nearly 100% methane and rated at 37 MJ/m³ (1000 Btu/ft³). For high-hydrogen manufactured gas (20 MJ/m³, or 550 Btu/ft³), ratios are about 50% of values listed. For manufactured gas with lower hydrogen and high carbon monoxide contents (17 MJ/m³, or 450 Btu/ft³), ratios are about 40% of values listed. For propane, ratios are about twice those listed. For butane, multiply listed values by three. (b) When atmosphere is cooled by tap-water heat exchangers, dew point will be about 6 to 8 °C (10 to 15 °F) above the temperature of the tap water. Dew point may be reduced to about 5 °C (40 °F) by refrigeration equipment, and to −55 °C (−70 °F) or lower by activated-absorption equipment. (c) Ratio of air to dissociated ammonia. (d) Dried to a dew point of −55 to −75 °C (−70 to −100 °F) by alumina plus molecular sieve.

partially reacted natural gas. Properties of various protective atmospheres are shown in Table 3. Nickels, modified nickels and nickel-copper alloys can be bright annealed in all of these atmospheres except the first one listed (completely burned fuel, lean atmosphere). Inconel 600 and other alloys containing chromium or molybdenum, or both, require completely dissociated ammonia or dried hydrogen for bright annealing.

The conditions for bright annealing should be maintained during heating, regardless of the method used for cooling. If facilities for cooling under protective atmosphere are not available, nickel and nickel alloys may be quenched in a 2% (by volume) solution of denatured alcohol. This will reduce the "oxide flash" formed by the oxygen of the air during transfer from the furnace to the quenching tank.

Dead-Soft Annealing. When the nickel alloys are annealed at higher temperatures and for longer periods, a condition commonly described as "dead-soft" is obtained, and hardness numbers will result that are 10 to 20% lower than those of the "soft" condition. This cannot be accomplished without increasing the grain size of the metal. Therefore, this treatment should be used only for those few applications in which grain size is of little importance.

Torch Annealing. Some large equipment is hardened locally by fabricating operations. If the available annealing furnace is too small to hold the work piece, the hardened sections can be annealed with the flames of oil or acetylene torches adjusted so as to be highly reducing. The work should be warmed gently at first, with sweeping motions of the torch, and should not be brought to the annealing temperature until sufficient preheating has been done to prevent cracking as a result of sudden release of stress. (**note:** Torch annealing is a poor method for general use, because it provides irregular and insufficient annealing and produces heavily oxidized surfaces.)

Process-Control Factors in Annealing

Among the more important process-control factors in annealing nickel and nickel alloys are selection of suitably sulfur-free fuels for heating, control of furnace temperature, effects of prior cold work and of cooling rates, control of grain size, control of protective atmospheres, and protection from contamination by foreign material.

Fuels for Heating. Nickel and nickel alloys are subject to intergranular attack when heated in the presence of sulfur or sulfur compounds. Fuels for heating must be low in sulfur content.

Gas is the best fuel for heating nickel alloys, and should be used if available. Good heating is achieved readily with gas because of the ease with which gas can be mixed with air and its supply controlled. Gaseous fuels require little combustion space, and automatic control of temperature and furnace atmosphere is easily accomplished.

Natural gas, consisting chiefly of methane (CH_4) and smaller amounts of ethane (C_2H_6), propane (C_3H_8) and butane (C_4H_{10}), and essentially free of sulfur compounds, is available in many areas.

Manufactured gases are produced from coal or oil, which may contain substantial amounts of sulfur. These gases should not be used unless the sulfur compounds are effectively removed during gas manufacture. Sulfur occurs in these manufactured gases as hydrogen sulfide (H_2S) and as organic sulfides.

Suppliers of manufactured gas endeavor to keep the total sulfur content of their product to a value below the maximum set by state regulatory agencies. Thus, manufactured gas supplied to consumers usually contains less than 6.9 g of total sulfur per 10 m³ of gas (less than 30 gr per 100 ft³). This sulfur content may vary considerably from day to day, but, where adequate maintenance of sulfur removal is observed by the gas supplier, total sulfur content will average 2.3 to 3.4 g per 10 m³ (10 to 15 gr per 100 ft³) or lower. This sulfur content is acceptable for heating nickel alloys; however, the generally accepted statutory limit (6.9 g per 10 m³) is marginal.

Two other satisfactory gaseous fuels are butane and propane, which are components of natural gas that liquefy and separate out when the gas is compressed. Both fuels are stored and shipped in tank cars that range in capacity from 30 to 57 m³ (8000 to 15 000 gal), and may be distributed throughout a plant in pipelines as liquids under their own vapor pressures. Butane can be considered in a sense as an oil fuel of high volatility, and proper means must be provided for gasifying it by heating, before it is mixed with air for combustion. Propane is more volatile and does not require the application of heat to convert it from liquid to gas. It is also obtainable in cylinders equipped with pressure regulators to control the flow of gas. These cylinders are useful for occasional work in heating small objects.

Furnace-Temperature Control. The importance of accurate control of annealing temperature cannot be overemphasized. Satisfactory indicating, controlling, recording, and controlling-recording pyrometers are available. Iron-Constantan and Chromel-Alumel thermocouples should be changed every three months (or more often, if required); noble-metal thermocouples may be used as long as their accuracy is not impaired. All thermocouples should be checked daily for accuracy.

Effect of Prior Cold Work. The greater the amount of cold work to which the material has been subjected before annealing, the lower is the temperature required to produce the same degree of softness without increasing grain size, and the shorter is the time required at any one temperature.

The amount of any type of previous cold work also has a critical influence on the ductility of nickel and nickel alloys after annealing. If only a small amount of cold work is done (for example, approximately 10% reduction), full ductility for deep drawing and spinning cannot be restored by annealing, even though the hardness is reduced to that of soft material. A minimum of approximately 20% cold working is required between anneals to ensure maximum ductility and softness following annealing.

Effect of Cooling Rate. Neither slow cooling, whether in or out of the furnace, nor rapid cooling by quenching, has any specific effect on the softness of the annealed, solid-solution nickel materials. Therefore, rapid cooling is preferable (except for heavy sec-

tions, in which it may set up excessive thermal stresses), both as a time-saver and to minimize the amount of oxidation. Some of the age hardenable alloys, such as Monel K-500, must be cooled rapidly by quenching from the annealing temperature.

Control of Grain Size. Coarse-grain material is unsuitable for most cold working operations. A coarse grain in the high-nickel materials cannot be refined by thermal treatment. It can be removed only by cold working sufficiently to effect recrystallization to a smaller grain size during a subsequent annealing treatment. Maximum workability is obtained with material that has been annealed without allowing appreciable grain growth to occur. The average grain diameter should not exceed 0.064 mm (0.0025 in.). This gives the best combination of ductility to permit extensive deformation, strength to withstand the action of tools, and surface quality to facilitate polishing.

Effect of Fluctuating Atmospheres. If nickel and nickel alloys are annealed in atmospheres that fluctuate between oxidizing (excess of air) and reducing (excess of carbon monoxide or hydrogen), severe intercrystalline attack will occur with resulting embrittlement, even though the atmosphere is sulfur-free. This type of embrittlement can be prevented by maintaining a constant and sufficient excess of a reducing atmosphere during heating and cooling. Alloys containing chromium or molybdenum are affected less than nickel and nickel-copper alloys.

Protection From Contamination by Foreign Material. Many of the lubricants used for deep drawing and spinning contain sulfur or lead. Unless these elements are removed before annealing, they will cause embrittlement. They can be removed by washing with hot trisodium phosphate solution or with a volatile organic grease-removing solvent such as trichloroethylene. Lubricants of any type should be entirely removed from the material before annealing. Paints, and other adherent substances that may contain sulfur, lead or similarly harmful ingredients, should be removed by appropriate methods before annealing.

Stress Relieving

In stress relieving, careful regulation of time and temperature is required. These variables usually are deter-

mined experimentally for each application; some typical ranges are given in Table 2. Figure 1 illustrates the effect of stress relief at temperatures from about 400 to about 600 °C (750 to 1100 °F) on the room-temperature properties of Monel 400.

Stress Equalizing

Stress equalizing is a low temperature heat treatment (Table 2) that effects what is known as "partial recovery". This recovery, which precedes any detectable microscopic structural changes, consists of a considerable increase in the proportional limit, slight increases in hardness and tensile strength, no significant change in elongation or reduction of area, balancing of stresses and return of electrical conductivity toward its characteristic value for the alloy in the annealed condition.

The temperature required for stress equalizing depends on the composition of the alloy. Figure 1 shows an optimum temperature range of about 230 to 315 °C (450 to 600 °F) for cold drawn Monel 400 rod. A temperature of about 275 °C (525 °F) is recommended for commercial use. Long treatment time at this temperature has no detrimental effect.

Stress equalizing is usually applied to coil springs, wire forms, and flat spring stampings. If coil springs are to be given a cold "set" or cold pressing after coiling, the stress equalization should be carried out *before* the setting operation, which involves stressing the material beyond the elastic limit. Any cold working stresses set up by this operation are in such a direction that they will benefit, rather than harm, the material. If stress is equalized *after* cold pressing, part of the beneficial cold working stress will be removed.

Age Hardening

Addition of magnesium, aluminum, silicon, titanium and certain other alloying elements to nickel and nickel alloys, separately or in combinations, produces an appreciable response to age hardening. The effect is dependent on both chemical composition and aging temperature; it is caused by precipitation of submicroscopic particles throughout the grains, which results in a marked increase in hardness and strength.

Prior Solution Treating. Unlike

Table 4 Solution-treating and age-hardening schedules for nickel alloys

Alloy	Solution treating Temperature °C	°F	Time, h	Cooling method(a)	Age hardening
Monel K-500	980	1800	½ to 1	WQ	Heat to 595 °C (1100 °F), hold 16 h; furnace cool to 540 °C (1000 °F), hold 6 h; furnace cool to 480 °C (900 °F), hold 8 h; air cool
Inconel 718	980	1800	1	AC	Heat to 720 °C (1325 °F), hold 8 h; furnace cool to 620 °C (1150 °F), hold until furnace time for entire age-hardening cycle equals 18 h; air cool
Inconel X-750	1150	2100	2 to 4	AC	Heat to 845 °C (1550 °F), hold 24 h; air cool; reheat to 705 °C (1300 °F), hold 20 h; air cool
	980	1800	1	AC	Heat to 730 °C (1350 °F), hold 8 h; furnace cool to 620 °C (1150 °F), hold until furnace time for entire age-hardening cycle equals 18 h; air cool
Hastelloy X	1175	2150	1	AC	Heat to 760 °C (1400 °F), hold 3 h; air cool; reheat to 595 °C (1100 °F), hold 3 h; air cool

(a) WQ, water quench; AC, air cool

precipitation hardening stainless steels and aluminum-base alloys, the nickel alloys normally do not require solution treating in the upper annealing temperature range prior to age hardening. However, solution treating may be employed to enhance special properties (Table 4). For example, Inconel X-750 may be solution treated for 2 to 4 h at 1150 °C (2100 °F) and air cooled prior to a double (high and low temperature) aging cycle to develop maximum creep, relaxation and rupture strength at temperatures above about 600 °C (1100 °F). This combination of heat treatments is considered essential for high-temperature springs and turbine blades made of Inconel X-750.

Age-hardening practices for several nickel alloys are summarized in Table 4. In general, nickel alloys are soft when quenched from temperatures ranging from 790 to 1220 °C (1450 to 2225 °F); however, they may be hardened by holding at 480 to 870 °C (900 to 1600 °F) or above and then furnace or air cooling. Quenching is not a prerequisite to aging; the alloys can be hardened from the hot worked and cold

worked conditions, as well as from the soft condition.

Hardening Techniques. Nickel alloys usually are hardened in sealed boxes placed inside a furnace, although small horizontal or vertical furnaces without boxes may be used also. The box or furnace should hold the parts loosely packed, yet afford a minimum of excess space. Electric furnaces provide the optimum temperature uniformity of ± 6 °C (± 10 °F) and the freedom from contamination required for this work. Gas-heated furnaces, particularly those of the radiant-tube type, can be made to give satisfactory results. It is difficult to obtain good results from oil heating, even with muffle furnaces. All lubricants should be removed from the work before hardening.

Because of the long time of aging and the difficulty of excluding air from the box or furnace, truly bright hardening cannot be accomplished commercially. For semibright hardening, dry hydrogen (see No. 7 in Table 3) or cracked and dried ammonia (see No. 4 in Table 3) should be used. When bright or semibright hardening is not required, other atmospheres may be used, such as ni-

trogen, cracked natural gas free of sulfur, cracked city gas, cracked hydrocarbons, or a generated gas. The use of sulfur-free gases is necessary to avoid embrittlement.

Salt baths are used occasionally for small parts. The hardened material is never bright, and must be flash pickled to restore the natural color. Inorganic salts are used, such as chlorides and carbonates of sodium and potassium, which are relatively stable at temperatures considerably above their respective melting points. It is extremely important that the salts be free of all traces of sulfur, so that the work does not become embrittled.

Heat-Resisting Alloys

Procedures for heat treating the nickel-base alloys used in turbojet engines and for other high-temperature applications are described in the article on heat treating of heat-resisting alloys, in this volume. Alloys for which heat treating data and processing examples are given include René 41, Nimonic 90, Waspaloy, Astroloy, Udimet 500 and Inconel X-750.

Annealing of Precious Metals

By Edward D. Zysk
Technical Director
Engelhard Industries

ANNEALING BEHAVIOR* of commercial fine silver and silver alloys, of commercial rhodium, of commercial platinum and palladium and of chemically pure platinum and palladium are given in Fig. 1, 2, 3 and 4.

Consolidated polycrystalline ruthenium usually is hot worked at high temperatures. For thicknesses of less than 0.5 mm (0.020 in.), previously hot worked material can be cold worked with very small reductions to thicknesses of about 0.25 mm (0.010 in.) using intermediate anneals at 1050 to 1250 °C (1920 to 2280 °F). At 20 °C (68 °F), the hardness of an annealed bar of ruthenium would be 200 to 350 HV.

Iridium, like tungsten, is initially hot worked. Subsequent fabrication, which is done warm (that is, below the recrystallization temperature), results in a fibrous structure. Recrystallization of warm worked iridium occurs at a temperature of 1000 °C (1830 °F) or higher. The hardness of a warm drawn 0.5-mm- (0.020-in.-) diam iridium wire, annealed at 1000 °C, would be 200 to 240 HV; as warm-drawn wire would have a hardness of 600 to 700 HV. Recrystallized iridium, like tungsten, is relatively brittle at room temperature.

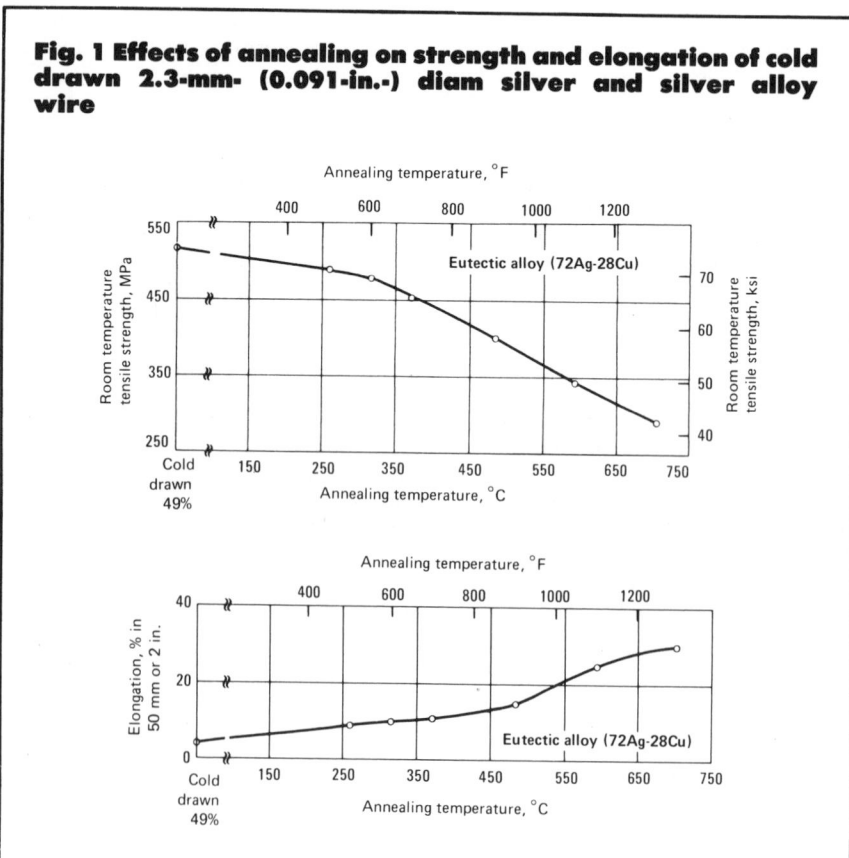

Fig. 1 Effects of annealing on strength and elongation of cold drawn 2.3-mm- (0.091-in.-) diam silver and silver alloy wire

*Source of data for silver and silver alloys: *Metals Handbook,* 8th Ed., Vol 1, 1961, p 1182 and 1183. Source of data for other metals: R. W. Douglass, C. A. Krier and R. I. Jaffee, Report from Battelle Memorial Institute to Office of Naval Research, Aug 31, 1961.

Fig. 1 (continued)

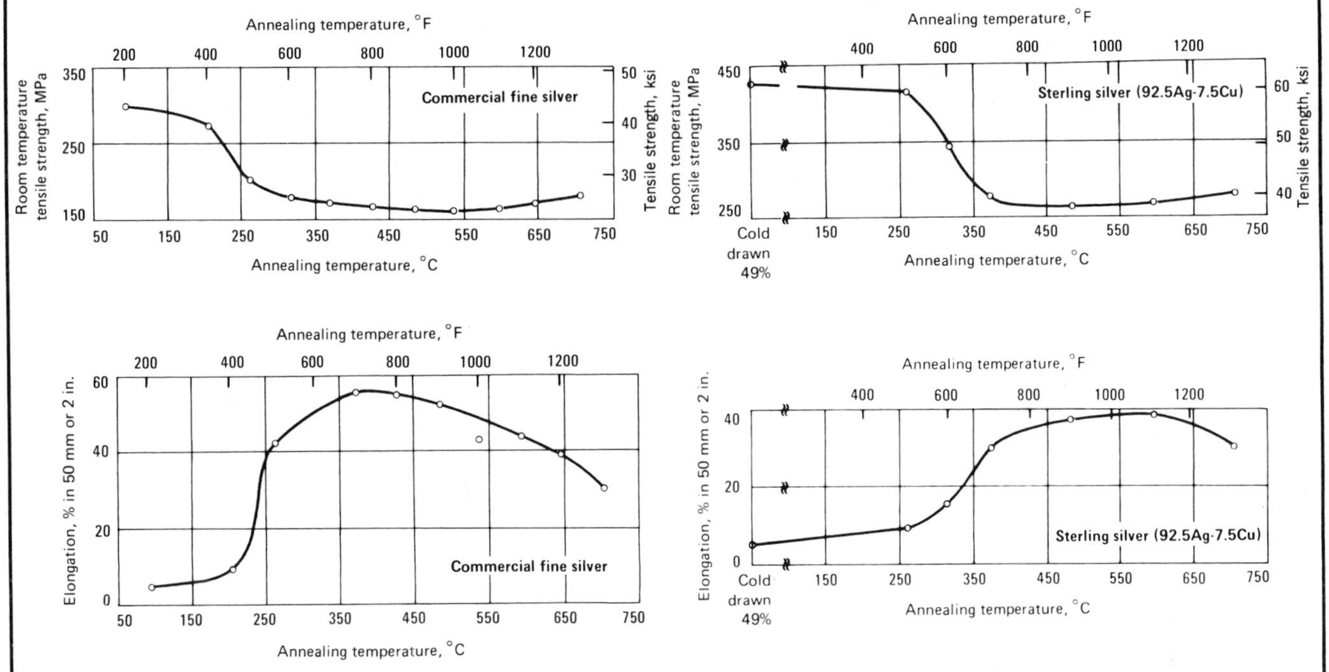

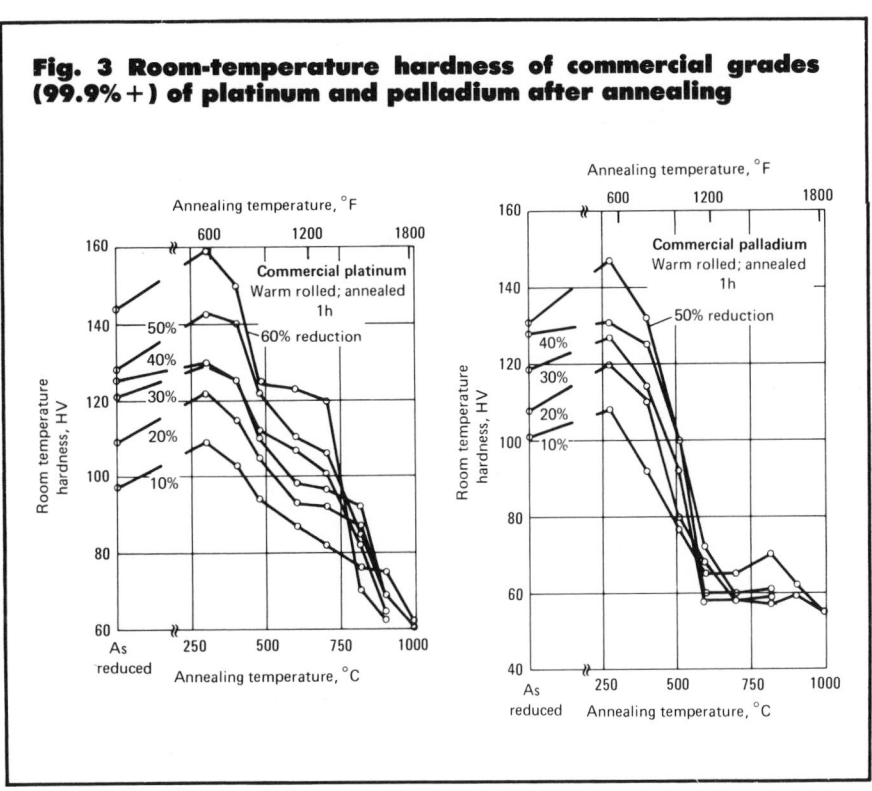

Fig. 2 Room-temperature hardness of commercial rhodium after annealing at various temperatures

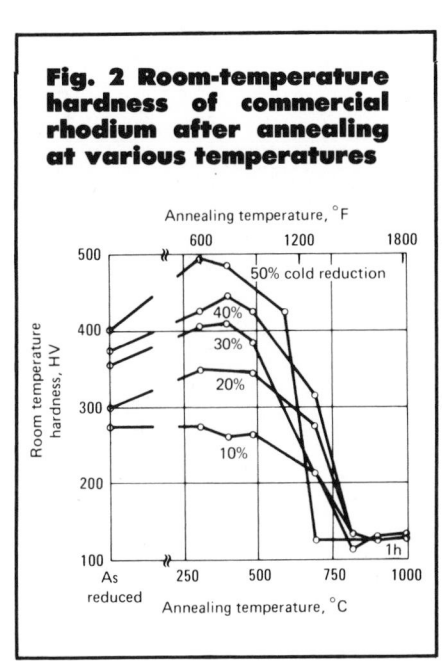

Fig. 3 Room-temperature hardness of commercial grades (99.9%+) of platinum and palladium after annealing

Fig. 4 Room-temperature hardness of chemically pure (99.99% +) platinum and palladium after annealing

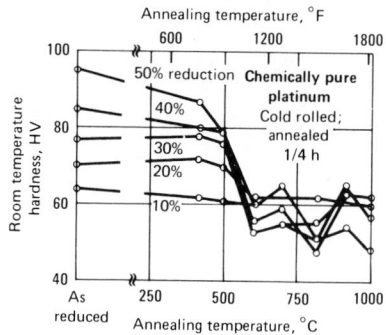

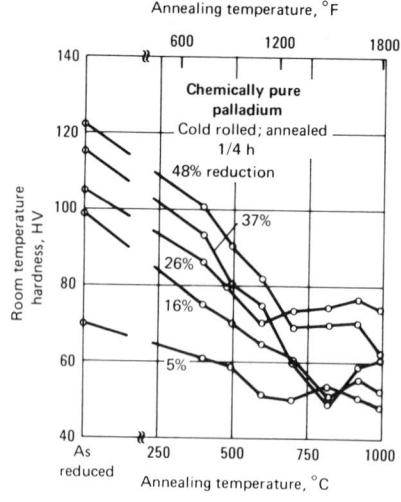

Heat Treating of Titanium and Titanium Alloys

By the ASM Committee on Titanium and Titanium Alloys*

TITANIUM AND TITANIUM AL-LOYS are heat treated for the following purposes:

- To reduce residual stresses developed during fabrication (stress relieving)
- To produce an optimum combination of ductility, machinability, and dimensional and structural stability (annealing)
- To increase strength (solution treating and aging)
- To optimize special properties such as fracture toughness, fatigue strength and high-temperature creep strength.

Various types of annealing treatments (single, duplex, beta and recrystallization annealing, for example), and solution treating and aging treatments, are imposed to achieve selected mechanical properties. Stress relieving and annealing may be employed to prevent preferential chemical attack in some corrosive environments, to prevent distortion (a stabilization treatment) and to condition the metal for subsequent forming and fabricating operations.

Response of titanium and titanium alloys to heat treatment depends on the composition of the metal. Unalloyed titanium is allotropic. Its close-packed hexagonal structure (alpha phase) changes to a body-centered cubic structure (beta phase) at 885 °C (1625 °F), and this structure persists at temperatures up to the melting point.

With respect to their effects on the allotropic transformation, alloying elements in titanium are classified as alpha stabilizers or as beta stabilizers. Alpha stabilizers, such as oxygen and aluminum, raise the alpha-to-beta transformation temperature. Nitrogen and carbon also are alpha stabilizers, but these elements usually are not added intentionally in alloy formulation. Beta stabilizers, such as manganese, chromium, iron, molybdenum, vanadium and niobium, lower the alpha-to-beta transformation temperature and, depending on the amount added, may result in retention of some beta phase at room temperature. Alloying elements such as zirconium and tin have essentially no effect on the alpha-to-beta transformation temperature. Based on the types and amounts of al-

loying elements they contain, titanium alloys are classified as alpha, near-alpha, alpha-beta or beta alloys. Near-alpha alloys are alloys with predominantly alpha stabilizer, plus limited beta stabilizers (normally, 2% or less).

Alpha and near-alpha titanium alloys can be stress relieved and annealed, but high strength cannot be developed in these alloys by any type of heat treatment. The commercial beta alloys are, in reality, metastable beta alloys. When these alloys are exposed to selected elevated temperatures, the retained beta phase decomposes and strengthening occurs. For beta alloys, stress-relieving and aging treatments can be combined, and annealing and solution treating may be identical operations.

Alpha-beta alloys are two-phase alloys and, as the name suggests, comprise both alpha and beta phases at room temperature. These are the most common and the most versatile of the three types of titanium alloys. Phase compositions, sizes and distributions can be manipulated by heat treatment within certain limits to enhance a spe-

*Walter Herman, *Chairman,* Technical Director, Viking Metallurgical Corp.; Roger V. Carter, Manager, Metals Technology, Boeing Commercial Airplane Co.; William H. Heil, Supervisor, Quality Assurance, Timet; Ralph J. Kotfila, Lead Engineer—Technology, McDonnell Aircraft Co.; Charles J. Scholl, Chief Product Metallurgist, Wyman-Gordon Co.

cific property or to attain a range of strength levels.

Not all heat treating cycles are applicable to all titanium alloys, because the various alloys are designed for different purposes. Alloys Ti-5Al-2Sn-2Zr-4Mo-4Cr (commonly called "Ti-17") and Ti-6Al-2Sn-4Zr-6Mo are designed for strength in heavy sections; Ti-6Al-2Sn-4Zr-2Mo for creep resistance; Ti-6Al-2Cl-1Ta-1Mo and Ti-6Al-4V-ELI for resistance to stress corrosion in aqueous salt solutions, and for high fracture toughness; Ti-5Al-2.5Sn for weldability; and Ti-6Al-6V-2Sn, Ti-6Al-4V and Ti-10V-2Fe-3Al for high strength at low-to-moderate temperatures.

Stress Relieving

Titanium and titanium alloys can be stress relieved without adversely affecting strength or ductility. Stress-relieving treatments decrease the undesirable residual stresses that result from (a) nonuniform hot forging deformation from cold forming and straightening, (b) asymmetric machining of plate (hogouts) or forgings, and (c) welding and cooling of castings. Removal of such stresses helps maintain shape stability and eliminates unfavorable conditions, such as the loss of compressive yield strength commonly known as the Bauschinger effect.

When symmetrical shapes are machined in the annealed condition, employing moderate cuts and uniform stock removal, stress relieving may not be required. Compressor disks made of Ti-6Al-4V have been satisfactorily machined in this manner, conforming with dimensional requirements. In contrast, thin rings made of the same alloy could be machined at a higher production rate to more stringent dimensions by stress relieving 2 h at 540 °C (1000 °F) after rough machining.

Separate stress relieving may be omitted when the manufacturing sequence can be adjusted to employ annealing or hardening as the stress-relieving process. For example, forging stresses may be relieved by annealing prior to machining. Large, thin rings have been effectively processed with minimum distortion by rough machining in the annealed state, followed by solution treating, quenching, partial aging, finish machining and final aging. Partial aging relieves quenching stresses, and final aging relieves

Table 1 Recommended stress-relief treatments for titanium and titanium alloys

Parts can be cooled from stress relief by either air cooling or slow cooling

Alloy	Temperature °C	°F	Time, h
Commercially pure Ti (all grades)	480 to 595	900 to 1100	1/4 to 4
Alpha or near-alpha titanium alloys			
Ti-5Al-2.5Sn	540 to 650	1000 to 1200	1/4 to 4
Ti-8Al-1Mo-1V	595 to 705	1100 to 1300	1/4 to 4
Ti-6Al-2Sn-4Zr-2Mo	595 to 705	1100 to 1300	1/4 to 4
Ti-6Al-2Cb-1Ta-0.8Mo	595 to 650	1100 to 1200	1/4 to 2
Ti-0.3Mo-0.8Ni (Ti Code 12)	480 to 595	900 to 1100	1/4 to 4
Alpha-beta titanium alloys			
Ti-6Al-4V	480 to 650	900 to 1200	1 to 4
Ti-6Al-6V-2Sn (Cu + Fe)	480 to 650	900 to 1200	1 to 4
Ti-3Al-2.5V	540 to 650	1000 to 1200	1/2 to 2
Ti-6Al-2Sn-4Zr-6Mo	595 to 705	1100 to 1300	1/4 to 4
Ti-5Al-2Sn-4Mo-2Zr-4Cr (Ti-17)	480 to 650	900 to 1200	1 to 4
Ti-7Al-4Mo	480 to 705	900 to 1300	1 to 8
Ti-6Al-2Sn-2Zr-2Mo-2Cr-0.25Si	480 to 650	900 to 1200	1 to 4
Ti-8Mn	480 to 595	900 to 1100	1/4 to 2
Beta or near-beta titanium alloys			
Ti-13V-11Cr-3Al	705 to 730	1300 to 1350	1/12 to 1/4
Ti-11.5Mo-6Zr-4.5Sn (Beta III)	720 to 730	1325 to 1350	1/12 to 1/4
Ti-3Al-8V-6Cr-4Zr-4Mo (Beta C)	705 to 760	1300 to 1400	1/6 to 1/2
Ti-10V-2Fe-3Al	675 to 705	1250 to 1300	1/2 to 2
Ti-15V-3Al-3Cr-3Sn	790 to 815	1450 to 1500	1/12 to 1/4

stresses developed during finish machining.

Table 1 presents combinations of time and temperature that are used for stress relieving titanium and titanium alloys. The ranges in both time and temperature indicate that more than one combination may yield satisfactory results. The higher temperatures usually are used with shorter times, and the lower temperatures with longer times, for effective stress relief. During stress relief of solution-treated and aged titanium alloys, care should be taken to prevent overaging to lower strength. This usually involves selection of a time-temperature combination that provides partial stress relief. The parts, in bulk or in fixtures, may be charged directly into a furnace operating at the stress-relief temperature. If a part is mounted in a massive fixture, a thermocouple should be attached to the largest part of the fixture.

Figure 1 illustrates the effects of stress relieving Ti-6Al-4V at five temperatures ranging from 260 to 620 °C (500 to 1150 °F) for periods of time ranging from 5 min to 50 h.

The rate of cooling from the stress-relieving temperature is not critical. Uniformity of cooling *is* critical, partic-ularly in the temperature range from 480 to 315 °C (900 to 600 °F). Oil or water quenching should not be used to accelerate cooling, however, because this can induce residual stresses by unequal cooling. Furnace or air cooling is acceptable.

Stress-relieving treatments must be based on the metallurgical response of the alloy involved. Generally, this requires holding at a temperature sufficiently high to relieve stresses without causing an undesirable amount of precipitation or strain aging in alpha-beta and beta alloys, or without producing undesirable recrystallization in single-phase alloys that rely on cold work for strength.

Stress relieving of the more highly alloyed alpha-beta compositions, and of beta alloys, should be done using a thermal exposure that is compatible with annealing, solution-treating, stabilization or aging processes.

There are no nondestructive testing methods that can measure the efficiency of a stress-relief cycle other than direct measurement of residual stresses by x-ray diffraction. No significant changes in microstructure due to stress-relieving heat treatments can be detected by optical microscopy.

Weldments. The temperatures used for stress relieving complex weldments of alpha or alpha-beta alloys should be near the high ends of the ranges given in Table 1. Complex weldments may be defined as those having multiple welds in complex configurations, possibly involving combinations of machine and manual welding. In complex weldments made with commercially pure titanium, Ti-5Al-2.5Sn alloy or Ti-6Al-4V alloy, more than 70% of the residual stress is relieved during the first hour at temperature. Simple weldments of commercially pure titanium often are used without stress relief.

Annealing

Annealing of titanium and titanium alloys serves primarily to increase fracture toughness, ductility at room temperature, dimensional and thermal stability, and creep resistance. Many titanium alloys are placed in service in the annealed state. Because improvement in one or more properties generally is obtained at the expense of some other property, the annealing cycle should be selected according to the objective of the treatment. Common annealing treatments are:

- Mill annealing
- Duplex annealing
- Triplex annealing
- Recrystallization annealing
- Beta annealing.

Recommended annealing treatments for several alloys are given in Table 2. Mill annealing is a general-purpose treatment given to all mill products. It is not a full anneal, and may leave traces of cold or warm working in the microstructures of heavily worked products (particularly sheet). Duplex and triplex annealing alter the shapes, sizes and distributions of phases to those required for improved creep resistance or fracture toughness. Both recrystallization and beta annealing treatments are used to improve fracture toughness. Beta annealing is done at temperatures above the beta transus of the alloy being annealed.

Straightening, sizing and flattening may be combined with annealing by use of appropriate fixtures. The parts, in bulk or in fixtures, may be charged directly into a furnace operating at the annealing temperature.

Either air or furnace cooling may be used, but the two methods may result

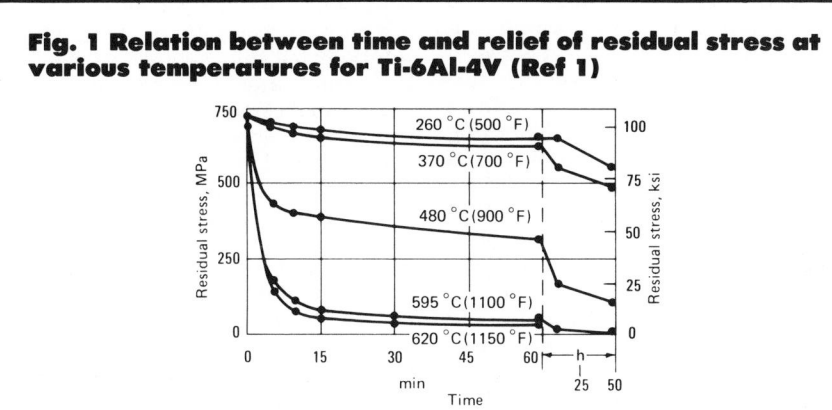

Fig. 1 Relation between time and relief of residual stress at various temperatures for Ti-6Al-4V (Ref 1)

Table 2 Recommended annealing treatments for titanium and titanium alloys

Alloy	Temperature °C	°F	Time, h	Cooling method
Commercially pure Ti (all grades)	650 to 760	1200 to 1400	1/10 to 2	Air
Alpha or near-alpha titanium alloys				
Ti-5Al-2.5Sn	720 to 845	1325 to 1550	1/6 to 4	Air
Ti-8Al-1Mo-1V	790(a)	1450(a)	1 to 8	Air or furnace
Ti-6Al-2Sn-4Zr-2Mo	900(b)	1650(b)	1/2 to 1	Air
Ti-6Al-2Cb-1Ta-0.8Mo	790 to 900	1450 to 1650	1 to 4	Air
Alpha-beta titanium alloys				
Ti-6Al-4V	705 to 790	1300 to 1450	1 to 4	Air or furnace
Ti-6Al-6V-2Sn (Cu + Fe)	705 to 815	1300 to 1500	3/4 to 4	Air or furnace
Ti-3Al-2.5V	650 to 760	1200 to 1400	1/2 to 2	Air
Ti-6Al-2Sn-4Zr-6Mo	(c)	(c)	. . .	. . .
Ti-5Al-2Sn-4Mo-2Zr-4Cr (Ti-17)	(c)	(c)	. . .	. . .
Ti-7Al-4Mo	705 to 790	1300 to 1450	1 to 8	Air
Ti-6Al-2Sn-2Zr-2Mo-2Cr-0.25Si	705 to 815	1300 to 1500	1 to 2	Air
Ti-8Mn	650 to 760	1200 to 1400	1/2 to 1	(d)
Beta or near-beta titanium alloys				
Ti-13V-11Cr-3Al	705 to 790	1300 to 1450	1/6 to 1	Air or water
Ti-11.5Mo-6Zr-4.5Sn (Beta III)	690 to 760	1275 to 1400	1/6 to 1	Air or water
Ti-3Al-8V-6Cr-4Zr-4Mo (Beta C)	790 to 815	1450 to 1500	1/4 to 1	Air or water
Ti-10V-2Fe-3Al	(c)	(c)	. . .	. . .
Ti-15V-3Al-3Cr-3Sn	790 to 815	1450 to 1500	1/12 to 1/4	Air

(a) For sheet and plate, follow by 1/4 h at 790 °C (1450 °F), then air cool. (b) For sheet, follow by 1/4 h at 790 °C (1450 °F), then air cool (plus 2 h at 595 °C or 1100 °F, then air cool, in certain applications). For plate, follow by 8 h at 595 °C (1100 °F), then air cool. (c) Not normally supplied or used in annealed condition (see Table 3). (d) Furnace or slow cool to 540 °C (1000 °F), then air cool.

in different levels of tensile properties. For example, air cooling of Ti-6Al-6V-2Sn from the mill-annealing temperature results in lower tensile strength than that obtained by furnace cooling. If distortion is a problem, the cooling rate should be uniform down to 315 °C (600 °F).

Stability. In alpha-beta titanium alloys, thermal stability is a function of beta-phase transformations. During cooling from the annealing temperature, beta may transform and, under certain conditions and in certain alloys, may form the brittle intermediate phase omega. A stabilization annealing treatment is designed to produce a stable beta phase capable of resisting further transformation when exposed to elevated temperatures in service. Alpha-beta alloys that are lean in beta, such as Ti-6Al-4V, can be air cooled from the annealing temperature without impairing their stability. Furnace (slow) cooling may promote formation of Ti₃Al, an ordering reaction that can degrade resistance to stress corrosion. Slight increases in strength (up to 34 MPa, or 5 ksi) can be gained in Ti-6Al-4V and in Ti-6Al-6V-2Sn by cooling from the annealing temperature to 540 °C (1000 °F) at a rate of 56 °C/h (100 °F/h).

To obtain maximum creep resistance and stability in the near-alpha alloy Ti-8Al-1Mo-1V and Ti-6Al-2Sn-4Zr-2Mo, a duplex annealing treatment is employed. This treatment begins with solution annealing at a temperature high in the alpha-beta range, usually 28 to 56 °C (50 to 100 °F) below the beta transus for Ti-8Al-1Mo-1V and 19 to 56 °C (35 to 50 °F) below the beta transus for Ti-6Al-2Sn-4Zr-2Mo. Forgings are held for 1 h (nominal) and then air or fan cooled depending on section size. This treatment is followed by stabilization annealing for 8 h at 595 °C (1100 °F). Final annealing temperature should be at least 56 °C (100 °F) above the maximum anticipated service temperature. Maximum creep resistance can be developed in Ti-6Al-2Sn-4Zr-2Mo by beta annealing or beta processing.

Straightening During Annealing. It may be difficult to prevent distortion of close-tolerance thin sections during annealing. Straightening of bar to close tolerances, and flattening of sheet, present major problems for titanium producers and fabricators. Because of springback and resistance to straightening at room temperature,

it is necessary to employ elevated-temperature forming. At annealing temperatures, many titanium alloys have creep resistance low enough to permit straightening during annealing. With proper fixturing, and in some instances judicious weighting, sheet-metal fabrications and thin, complex forgings have been straightened with satisfactory results. Again, uniform cooling to below 315 °C (600 °F) can improve results.

Various jigs and processing techniques have been proposed for annealing titanium in a manner that will yield a flat product. "Creep flattening" and "vacuum creep flattening" are two such techniques. Creep flattening consists of heating titanium sheet between two clean, flat sheets of steel in a furnace containing an oxidizing or inert atmosphere. Vacuum creep flattening is used to produce stress-free flat plate for subsequent machining. The plate is placed on a large, flat ceramic bed that has integral electric-heating elements. Insulation is placed on top of the plate, and a plastic sheet is sealed to the frame. The bed is slowly heated to the annealing temperature while a vacuum is pulled under the plastic. Atmospheric pressure is used to creep flatten the plate.

Solution Treating and Aging

A wide range of strength levels can be obtained in alpha-beta or beta alloys by solution treating and aging. The origin of heat treating responses of titanium alloys lies in the instability of the high-temperature beta phase at lower temperatures. Heating an alpha-beta alloy to the solution-treating temperature produces a higher ratio of beta phase. This partitioning of phases is maintained by quenching; on subsequent aging, decomposition of the unstable beta phase occurs, providing high strength. Commercial beta alloys, generally supplied in the solution-treated condition, need only be aged.

After being cleaned, titanium components should be loaded into fixtures or racks that will permit free access to the heating and quenching media. Thick and thin components of the same alloy may be solution treated together, but the time at temperature (soaking time) is determined by the thickest section. For most alloys, the rule is 20 to 30 min per inch of thickness, to get the required temperature, followed by the required soak time.

Time/temperature combinations for solution treating are given in Table 3. A load may be charged directly into a furnace operating at the solution-treating temperature. Although preheating is not essential, it may be used to minimize distortion of complex parts.

Solution Treating. To obtain high strength with adequate ductility, it is necessary to solution treat at a temperature high in the alpha-beta field, normally 28 to 83 °C (50 to 150 °F) below the beta transus of the alloy. If high fracture toughness or improved resistance to stress corrosion is required, beta annealing or beta solution treating may be desirable. A change in the solution-treating temperature of alpha-beta alloys alters the amount of beta phase and consequently changes the response to aging (see Table 4). Selection of solution-treating temperature usually is based upon practical considerations such as the desired level of tensile properties and the amount of ductility to be obtained after aging.

Because solution treating involves heating to temperatures only slightly below the beta transus, proper control of temperature is essential. If the beta transus is exceeded, tensile properties (especially ductility) are reduced and cannot be fully restored by subsequent thermal treatment. The beta transus temperatures for commercial alloys are listed in Table 5.

Beta alloys normally are obtained from producers in the solution-treated condition. If reheating is required, soak times should be only as long as necessary to obtain complete solutioning. Solution-treating temperatures for beta alloys are above the beta transus; because no second phase is present, grain growth can proceed rapidly.

Quenching. The rate of cooling from the solution-treating temperature has an important effect on strength. If the rate is too low, appreciable diffusion may occur during cooling, and decomposition of the altered beta phase during aging may not provide effective strengthening.

For alloys relatively high in beta-stabilizer content, and for products of small section size, air or fan cooling may be adequate; such slow cooling, where allowed by specified mechanical properties, is preferred because it minimizes distortion. Beta alloys generally are air quenched from the solution-treating temperature.

Water or a 5% brine or caustic soda solution is preferred for quenching

Table 3 Recommended solution treating and aging (stabilizing) treatments for titanium alloys

Alloy	Solution temperature °C	°F	Solution time, h	Cooling rate	Aging temperature °C	°F	Aging time, h
Alpha or near-alpha alloys							
Ti-8Al-1Mo-1V...................	980 to 1010(a)	1800 to 1850(a)	1	Oil or water	565 to 595	1050 to 1100	...
Ti-6Al-2Sn-4Zr-2Mo	955 to 980	1750 to 1800	1	Air	595	1100	8
Alpha-beta alloys							
Ti-6Al-4V	955 to 970(b)(c)	1750 to 1775(b)(c)	1	Water	480 to 595	900 to 1100	4 to 8
	955 to 970	1750 to 1775	1	Water	705 to 760	1300 to 1400	2 to 4
Ti-6Al-6V-2Sn (Cu + Fe)...........	885 to 910	1625 to 1675	1	Water	480 to 595	900 to 1100	4 to 8
Ti-6Al-2Sn-4Zr-6Mo	845 to 890	1550 to 1650	1	Air	580 to 605	1075 to 1125	4 to 8
Ti-5Al-2Sn-2Zr-4Mo-4Cr	845 to 870	1550 to 1600	1	Air	580 to 605	1075 to 1125	4 to 8
Ti-6Al-2Sn-2Zr-2Mo-2Cr-0.25Si	870 to 925	1600 to 1700	1	Water	480 to 595	900 to 1100	4 to 8
Beta or near-beta alloys							
Ti-13V-11Cr-3Al	775 to 800	1425 to 1475	¼ to 1	Air or water	425 to 480	800 to 900	4 to 100
Ti-11.5Mo-6Zr-4.5Sn (Beta III)......	690 to 790	1275 to 1450	⅛ to 1	Air or water	480 to 595	900 to 1100	8 to 32
Ti-3Al-8V-6Cr-4Mo-4Zr (Beta C)	815 to 925	1500 to 1700	1	Water	455 to 540	850 to 1000	8 to 24
Ti-10V-2Fe-3Al	760 to 780	1400 to 1435	1	Water	495 to 525	925 to 975	8
Ti-15V-3Al-3Cr-3Sn	790 to 815	1450 to 1500	¼	Air	510 to 595	950 to 1100	8 to 24

(a) For certain products, use solution temperature of 890 °C (1650 °F) for 1 h, then air cool or faster. (b) For thin plate or sheet, solution temperature can be used down to 890 °C (1650 °F) for 6 to 30 min, then water quench. (c) This treatment is used to develop maximum tensile properties in this alloy.

Table 4 Variation of tensile properties of Ti-6Al-4V bar stock with solution-treating temperature

Solution-treating temperature °C	°F	Room-temperature tensile properties(a) Tensile strength MPa	ksi	Yield strength(b) MPa	ksi	Elongation in 4D, %
845	1550	1025	149	980	142	18
870	1600	1060	154	985	143	17
900	1650	1095	159	995	144	16
925	1700	1110	161	1000	145	16
940	1725	1140	165	1055	153	16

(a) Properties determined on 13-mm (1/2-in.) bar after solution treating, quenching and aging. Aging treatment: 8 h at 480 °C (900 °F), air cool. (b) At 0.2% offset

alpha-beta alloys, because these quenchants provide cooling rates necessary to prevent decomposition of the beta phase obtained by solution treating, to provide maximum response to aging. The need for rapid quenching is further emphasized by short quench-delay-time requirements. Depending on the mass of the sections being heat treated, some alpha-beta alloys can only tolerate a maximum delay of 7 s, whereas more highly beta-stabilized alloys can tolerate quench delay times of up to 20 s. For example, the effect of quench delays on Ti-6Al-4V bar is shown in Fig. 2.

Less sensitive to delayed quenching are alloys such as Ti-6Al-2Sn-4Zr-6Mo and Ti-5Al-2Sn-2Zr-4Mo-4Cr, in which fan air cooling develops good strength through 100-mm (4-in.) sections.

Section size influences effectiveness of quenching and, in turn, response to aging. The amount and type of beta stabilizer in the alloy determine depth of hardening or strengthening. Thick sections exhibit lower tensile properties unless the alloy is highly alloyed with beta stabilizers. The practical significance of section size for some alloys is shown in Table 6. The effects of quenched section size on the tensile properties of Ti-6Al-4V alloy are illustrated in Fig. 3. (For additional data, see Tables 2 and 3 on page 526 of Volume 1 of the 8th Edition of this Handbook.)

Aging. The final step in heat treating titanium alloys to high strength consists of reheating to an aging temperature between 425 and 650 °C (800 and 1200 °F). Aging causes decomposition of the supersaturated beta phase retained on quenching. A summary of

aging times and temperatures is presented in Table 3. The time/temperature combination selected depends on required strength.

Aging at or near the annealing temperature will result in overaging. This condition, called solution treated and overaged, or STOA, is sometimes used to obtain modest increases in strength while maintaining satisfactory toughness and dimensional stability.

Although the aged condition is not necessarily one of equilibrium, proper aging produces high strength with adequate ductility and metallurgical stability. Heat treatment of alpha-beta alloys for high strength frequently involves a series of compromises and modifications, depending on the type of service and on special properties that are required, such as ductility and suitability for fabrication. This has become especially true where fracture toughness is important in design and strength is lowered to improve design life.

During aging of some highly beta-stabilized alpha-beta alloys, beta transforms first to a metastable transition phase referred to as omega phase. Retained omega phase, which produces brittleness unacceptable in alloys heat treated for service, can be avoided by severe quenching and rapid reheating to aging temperatures above 425 °C (800 °F). Because a coarse alpha phase forms, however, this treatment might not produce optimum strength proper-

Table 5 Beta transformation temperatures of titanium alloys

Alloy	Beta transus	
	°C, ±15	°F, ±25
Commercially pure Ti, 0.25 max O$_2$	910	1675
Commercially pure Ti, 0.40 max O$_2$	945	1735
Alpha and near-alpha alloys		
Ti-5Al-2.5Sn	1050	1925
Ti-8Al-1Mo-1V	1040	1900
Ti-6Al-2Sn-4Zr-2Mo	995	1820
Ti-6Al-2Cb-1Ta-0.8Mo	1015	1860
Ti-0.3Mo-0.8Ni (Ti code 12)	880	1615
Alpha-beta alloys		
Ti-6Al-4V	1000(a)	1830(b)
Ti-6Al-6V-2Sn (Cu + Fe)	945	1735
Ti-3Al-2.5V	935	1715
Ti-6Al-2Sn-4Zr-6Mo	940	1720
Ti-5Al-2Sn-2Zr-4Mo-4Cr (Ti-17)	900	1650
Ti-7Al-4Mo	1000	1840
Ti-6Al-2Sn-2Zr-2Mo-2Cr-0.25Si	970	1780
Ti-8Mn	800(c)	1475(d)
Beta or near-beta alloys		
Ti-13V-11Cr-3Al	720	1330
Ti-11.5Mo-6Zr-4.5Sn (Beta III)	760	1400
Ti-3Al-8V-6Cr-4Zr-4Mo (Beta C)	795	1460
Ti-10V-2Fe-3Al	805	1480
Ti-15V-3Al-3Cr-3Sn	760	1400

(a) ±20. (b) ±30. (c) ±35. (d) ±50

Fig. 2 Effects of quench delay on tensile properties of Ti-6Al-4V bar (Ref 2)

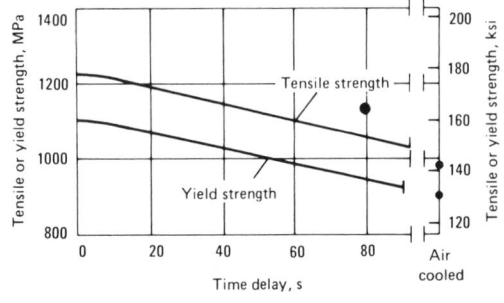

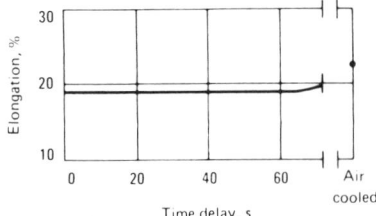

Bar, 13 mm (½ in.) in diameter, was solution treated 1 h at 955 °C (1750 °F), water quenched, aged 6 h at 480 °C (900 °F) and air cooled.

ties. An aging practice that ensures that aging time and temperature are adequate to carry out any omega reaction to completion usually is employed. Aging above 425 °C (800 °F) generally is adequate to complete the reaction.

The metastable beta alloys do not require solution treatment. Final hot working, followed by air cooling, leaves these alloys in a condition comparable to a solution-treated state. In some instances, however, solution treating at 790 °C (1450 °F) has produced better uniformity of properties after aging. Aging at 480 °C (900 °F) for 8 to 60 h produces tensile strengths of 1.10 to 1.38 GPa (160 to 200 ksi). Aging for times longer than 60 h may provide higher strengths, but will decrease ductility and fracture toughness if the alloy contains chromium and titanium-chromium compounds are formed. Short aging times can be used on cold worked material to produce a significant increase in strength over that obtained by cold working. Use of beta alloys at service temperatures above 315 °C (600 °F) for prolonged periods is not recommended, because the loss of ductility caused by metallurgical instability is progressive.

Other Special Thermal Treatments. Certain physical properties, such as notch strength, fracture toughness and fatigue resistance, can be enhanced in some alloys by special thermal treatments. Three such treatments are given below:

- *Solution treating and overaging of Ti-6Al-4V:* Heat 1 h at 955 °C (1750 °F), water quench, then 2 h at 705 °C (1300 °F), air cool. Advantages: improved notch strength, fracture toughness and creep strength at strength levels similar to those obtained by regular annealing.
- *Recrystallization annealing of Ti-6Al-4V or Ti-6Al-4V-ELI:* Heat 4 h or more at 925 to 955 °C (1700 to 1750 °F), furnace cool to 760 °C (1400 °F) at a rate no higher than 56 °C/h (100 °F/h), cool to 480 °C (900 °F) at a rate no lower than 370 °C/h (670 °F/h), air cool to room temperature. Advantages: improved fracture toughness and fatigue-crack-growth characteristics at somewhat reduced levels of strength.
- *Beta annealing of Ti-6Al-4V, Ti-6Al-4V-ELI and Ti-6Al-2Sn-4Zr-2Mo.* Ti-6Al-4V or Ti-6Al-4V-ELI: Heat 5 min to 1 h at 1010 to 1040 °C (1850 to 1900 °F), air cool to 650 °C (1200 °F)

Table 6 Relation of tensile strength of solution treated and aged titanium alloys to size

	Tensile strength of square bar in section size of:										
	13 mm (½ in.)		25 mm (1 in.)		50 mm (2 in.)		75 mm (3 in.)		100 mm (4 in.)		150 mm (6 in.)	
Alloy	MPa	ksi	MPa	ksi	MPa	ksi	MPa	ksi	MPa	ksi	MPa	ksi
Ti-6Al-4V	1105	160	1070	155	1000	145	930	135	. . .	. . .	. . .	. . .
Ti-6Al-6V-2Sn (Cu + Fe).	1205	175	1205	175	1070	155	1035	150	. . .	. . .	. . .	. . .
Ti-6Al-2Sn-4Zr-6Mo	1170	170	1170	170	1170	170	1140	165	1105	160	. . .	. . .
Ti-5Al-2Sn-2Zr-4Mo-4Cr (Ti-17) . . .	1170	170	1170	170	1170	170	1105	160	1105	160	1105	160
Ti-10V-2Fe-3Al	1240	180	1240	180	1240	180	1240	180	1170	170	1170	170
Ti-13V-11Cr-3Al	1310	190	1310	190	1310	190	1310	190	1310	190	1310	190
Ti-11.5Mo-6Zr-4.5Sn (Beta III)	1310	190	1310	190	1310	190	1310	190	1310	190	. . .	. . .
Ti-3Al-8V-6Cr-4Zr-4Mo (Beta C) . .	1310	190	1310	190	1240	180	1240	180	1170	170	1170	170

Fig. 3 Effects of quenched section size on tensile properties of Ti-6Al-4V (Ref 2)

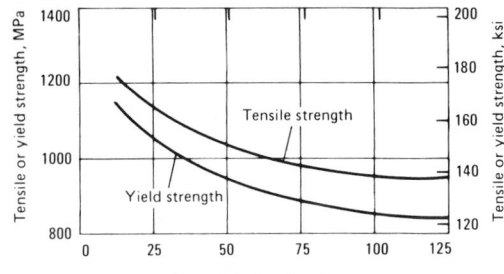
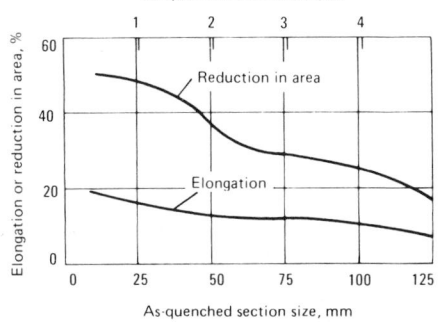

at a rate of 85 °C/min (150 °F/min) or higher, then 2 h at 730 to 790 °C (1350 to 1450 °F), air cool. Advantages: improved fracture toughness, high-cycle fatigue strength and resistance to aqueous stress corrosion. Ti-6Al-2Sn-4Zr-2Mo: Heat ½ h at 1020 °C (1870 °F), air cool, then 8 h at 595 °C (1100 °F), air cool. Advantages: improved creep strength at elevated temperatures as well as improved fracture toughness.

Post Heat Treating Requirements. Titanium reacts with the oxygen, water and carbon dioxide normally found in oxidizing heat treating atmospheres and with hydrogen formed by decomposition of water vapor. Unless the heat treatment is performed in a vacuum furnace or in an inert atmosphere, oxygen will react with the titanium at the metal surface and produce an oxygen-enriched layer commonly called "alpha case". This brittle layer must be removed before the component is put into service. It can be removed by machining, but certain machining operations may result in excessive tool wear. Standard practice is to remove alpha case by other mechanical methods or by chemical methods, or by both.

Oxidation rates of commercial titanium alloys vary, and Table 7 can be used as a guide to determine how much metal should be removed. Temperature and total time at temperature must be known. One method to check for complete removal of alpha case is to etch the component with a solution composed of 18 g of ammonium bifluoride per litre of water (2.4 oz/gal). The presence or absence of alpha case is detected by the difference in etching characteristics: light gray shows the presence of alpha case; dark gray indicates its absence. If the component has been machined, such as a forging, the ammonium bifluoride treatment must be preceded by etching in a solution consisting nominally of 5% HF, 30% min HNO_3, balance water. For other mill products, such as plate, microexamination of representative samples removed from the plate is commonly used.

Small amounts of hydrogen (100 to 200 ppm) can be tolerated in titanium alloys with the specific limiting amount determined by the type of alloy. High hydrogen content can lead to premature failure of a component. Hydrogen pickup occurs not only during heat treatment but also during pickling or chemical cleaning operations used to remove alpha case. The amount of hydrogen pickup can only be determined by chemical analysis. If high hydrogen content is found, vacuum annealing is required. A typical vacuum annealing cycle consists of heating at or close to the annealing temperature for 2 to 4 h in a vacuum of not less than 10 μm.

Hardness testing is not recommended as a nondestructive method of checking the efficiency of heat treatment. The correlation between strength and hardness is poor. Whenever verification of a property is required, the appropriate mechanical test should be used.

Contamination During Heat Treatment

Before being subjected to any thermal treatment, titanium components should be cleaned and dried. Caution:

Table 7 Minimum metal removal after thermal exposure of titanium alloys

Heat treating temperature °C	°F	Time at temperature, h	Minimum stock removal per surface(a) mm	in.
480 to 593	900 to 1100	Up to 12	0.005	0.0002
594 to 648	1101 to 1200	Up to 4	0.008	0.0003
		4 to 12	0.015	0.0006
649 to 704	1201 to 1300	Up to 1	0.013	0.0005
		1 to 8	0.020	0.0008
		8 to 12	0.025	0.0010
705 to 760	1301 to 1400	Up to 1	0.025	0.0010
		1 to 4	0.036	0.0014
		4 to 8	0.038	0.0015
		8 to 12	0.043	0.0017
761 to 787	1401 to 1450	Up to 1	0.030	0.0012
		1 to 2	0.038	0.0015
		2 to 4	0.046	0.0018
		4 to 8	0.051	0.0020
		8 to 12	0.056	0.0022
788 to 815	1451 to 1500	Up to 1/2	0.036	0.0014
		1/2 to 1	0.041	0.0016
		1 to 2	0.051	0.0020
816 to 871	1501 to 1600	Up to 1/2	0.058	0.0023
		1/2 to 1	0.066	0.0026
		1 to 2	0.076	0.0030
872 to 898	1601 to 1650	Up to 1/2	0.058	0.0023
		1/2 to 1	0.081	0.0032
		1 to 2	0.089	0.0035
899 to 926	1651 to 1700	Up to 1/2	0.086	0.0034
		1/2 to 1	0.091	0.0036
		1 to 2	0.107	0.0042
927 to 954	1701 to 1750	Up to 1/2	0.097	0.0038
		1/2 to 1	0.107	0.0042
		1 to 2	0.122	0.0048

(a) Values shown are typical; actual values may vary with alloy type.

Do not use ordinary tap water in cleaning titanium components. Oil, fingerprints, grease, paint and other foreign matter should be removed from all surfaces. Cleaning is required because the chemical reactivity of titanium at elevated temperatures can lead to its contamination or embrittlement and can increase its susceptibility to stress corrosion. After cleaning, parts should be handled with clean gloves to prevent recontamination. If a component is to be sized, straightened or heat treated in a fixture, the fixture also should be free of any foreign matter and loosely adhering scale.

Titanium is chemically active at elevated temperatures and will oxidize in air. However, oxidation is not of primary concern in heat treating of titanium, although it may be a problem in sheet-forming operations. Oxygen pickup during heat treatment results in a surface structure composed predominantly of alpha phase and causes formation of

scale. This condition is detrimental because of the brittle nature of the oxygen-enriched alpha structure, which also is very abrasive to either carbide or high speed steel machine tools. At 955 °C (1750 °F), the alpha structure can extend 0.2 to 0.3 mm (0.008 to 0.012 in.) below the surface and must be removed.

An antioxidant spray coating may be applied to clean sheet-metal parts in order to minimize oxygen pickup. Such coatings work effectively at temperatures up to about 760 °C (1400 °F), but their use does not fully eliminate the need for removing the surface structure after heat treating.

The danger of hydrogen pickup is of greater importance than that of oxidation. Current specifications limit hydrogen content to a maximum of 125 to 200 ppm, depending on alloy and mill form. Above these limits, hydrogen embrittles some titanium alloys, thereby reducing impact strength and notch

tensile strength and causing delayed cracking.

Hydrogen Pickup. With the exceptions of high vacuum, salt baths and chemically inert gases such as argon, all heat treating atmospheres contain some hydrogen at temperatures used for annealing titanium. Hydrocarbon fuels produce hydrogen as a by-product of incomplete combustion, and electric furnaces with air atmospheres contain hydrogen from breakdown of water vapor. However, because small amounts of hydrogen can be tolerated in titanium and because inert media are expensive, most titanium heat treating operations are performed in conventional furnaces employing oxidizing atmospheres with at least 5% excess oxygen in the flue gas.

An oxidizing atmosphere serves in two ways to reduce hydrogen pickup: it reduces the partial pressure of hydrogen in the surrounding atmosphere, and it provides the titanium with a protective surface oxide that retards hydrogen pickup.

Oxidation rates of titanium alloys vary considerably. A comparison of the scaling rates of commercially pure titanium and titanium alloys in air at temperatures from 650 to 980 °C (1200 to 1800 °F) is given in Fig. 4. Table 8 indicates the measurable thickness of oxide formed on commercially pure titanium after 1/2 h at various temperatures in air.

Nitrogen is absorbed by titanium during heat treatment at a much slower rate than oxygen and thus does not present a serious contamination problem. Dry nitrogen has been used successfully as a lower-cost protective atmosphere for heat treating of titanium forgings that are to be fully machined after treatment. If absorbed in sufficient quantities, however, nitrogen forms a hard, brittle compound.

Carbon monoxide and carbon dioxide decompose in the presence of hot titanium and produce surface oxidation.

Chlorides. Titanium alloys are subject to stress corrosion when parts with high residual stress are exposed to chlorides at temperatures above 290 °C (550 °F). Salt from fingerprints, and the chlorides contained in some degreasing solutions, may cause stress-corrosion cracking at temperatures above 315 °C (600 °F). Although this phenomenon is readily produced in laboratory testing, and is known to occur during heat treatment, hot-salt cracking has not

Fig. 4 Scaling rates of titanium, titanium alloys and stainless steel in air at various temperatures

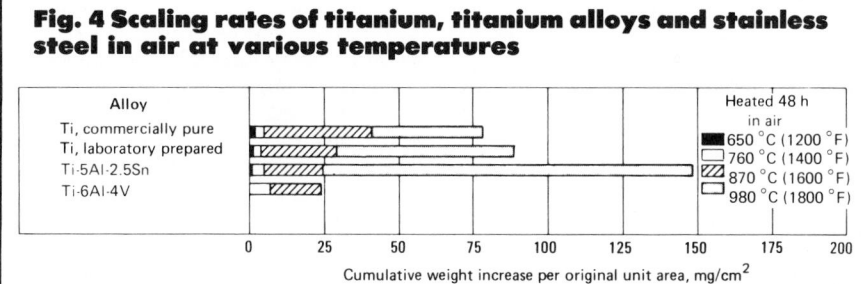

Cumulative weight increase per original unit area, mg/cm²

Table 8 Thickness of oxide on commercially pure titanium heated for ½ h in air

Temperature		Measurable thickness	
°C	°F	mm	in.
315	600	None	
425	800	None	
540	1000	None	
650	1200	<0.005	<0.0002
705	1300	0.005	0.0002
760	1400	0.008	0.0003
815	1500	<0.025	<0.001
870	1600	<0.025	<0.001
925	1700	<0.05	<0.002
980	1800	0.05	0.002
1040	1900	0.10	0.004
1095	2000	0.36	0.014

been a significant problem in service. Care is required during thermal processing to ensure freedom from chloride contamination.

Growth During Heat Treatment

Solution treating of large parts requires allowances for growth during heat treatment. The growth due to heating may be retained after cooling, and this growth may be increased either by longer holding times at solution temperature or by lower heating rates. Table 9 gives examples of net growth of Ti-6Al-4V specimens heated to 955 °C (1750 °F).

Furnace Equipment and Accessories

Atmospheres. An oxidizing atmosphere should be maintained during any thermal treatment of titanium. Furnaces normally operated with exothermic atmospheres, endothermic cracked-ammonia atmospheres or hydrogen atmospheres, because of the danger of hydrogen pickup, should be thoroughly "burned out" before being used for processing of titanium. If dimensions, shape or size do not permit removal of scale by subsequent pickling or machining, antioxidant coatings suitable for use to 760 °C (1400 °F) can be employed to minimize contamination. A vacuum or an inert gas such as argon also can be used.

Furnaces. Titanium usually is annealed or stress relieved in conventional furnaces constructed for annealing of steel. These furnaces are electric, gas fired or oil fired, in order of decreasing popularity. The temperature-control equipment for these operations should have an accuracy of ±5.5 °C (±10 °F) and should be capable of controlling and recording the desired temperature within ±14 °C (±25 °F), except where

control within ±8 °C (±15 °F) is required by MIL-H-81200.

Vacuum annealing furnaces are of either the cold-wall or the hot-wall type and may be heated by gas or electricity. Cold-wall electric vacuum furnaces are used most commonly with titanium. Maximum furnace operating temperature depends upon the heating elements and radiation shields, but usually these furnaces are designed for a maximum temperature of 980 °C (1800 °F) and are adequate for all titanium alloys. Hot-wall electric furnaces and gas-fired vacuum furnaces have been used in production. When the furnace employs a metallic retort, operating temperatures are held below 980 °C (1800 °F); higher temperatures can be achieved with ceramic retort tubes.

Laboratory vacuum annealing furnaces usually are operated at pressures of 0.1 μm or less, whereas production furnaces are designed to operate at pressures of 0.5 to 3.0 μm.

Vacuum annealing is expensive, and generally it is used only when: (a) a reduction in hydrogen content is required, (b) further hydrogen contamination is prohibited or (c) allowances that can be made for stock removal are insufficient to permit surface contamination resulting from annealing in air. Hydrogen outgassing at 705 °C (1300 °F) and below is so slow that its cost may be prohibitive. A temperature of 730 °C (1350 °F) is recommended as a minimum, and temperatures from 760 to 790 °C (1400 to 1450 °F) are preferred. At a temperature of 760 °C, removal of 100 ppm of hydrogen from 13- to 25-mm (½- to 1-in.) sections of Ti-6Al-4V alloy required approximately 2 h at a pressure of <10 μm. Actual time at temperature may vary widely depending on the capacity of the furnace to maintain a vacuum.

Solution-treating equipment can vary from a simple furnace with accurate temperature control and a water-

Table 9 Effect of heating rate and time at 955 °C (1750 °F) on growth of Ti-6Al-4V

Test conditions: 50-mm (2-in.) specimens were taken in the longitudinal direction (except where otherwise indicated) from material annealed 2 h at 705 °C (1300 °F) and air cooled. No growth was observed in specimens tested during annealing.

Mill heat(a)	Heating rate °C/min	°F/min	Holding time(b), h	Net growth(c), %
A	3.3	6	0	0.27
B	3.3	6	0	0.22
A	3.3	6	1	0.60
B	3.3	6	1	0.49
A	3.3	6	2	1.00
B	3.3	6	2	0.90(d)
B	10	18	1	0.32
B(e)	10	18	1	0.35

(a) Beta transus temperatures (determined metallographically) were 990 °C (1810 °F) for heat A and 1015 °C (1860 °F) for heat B. (b) All specimens water quenched after holding for time indicated. (c) As determined by Leitz-Wetzler dilatometer. (d) Calculated from curve. (e) Specimen taken in transverse direction.

quench tank to specialized installations for treating complex parts. Electrically heated furnaces are preferred because they minimize hydrogen pickup, although fuel-fired furnaces with slightly oxidizing conditions or with muffles that protect the metal from combustion products have been used successfully. Resistance and induction heating also have been used to reduce heating times and to minimize contamination during solution treatment. Accuracy of temperature-control equipment should be within ±2.8 °C (±5 °F), and the desired temperature should be controlled within ±14 °C (±25 °F).

To reduce distortion in long, thin

products such as sheet or extrusions, in hollow cylinders and in long forgings during immersion quenching, parts often are suspended vertically in an electrically heated drop-bottom furnace. In addition, weights usually are attached to the bottom ends of sheet to improve flatness during heating and to facilitate lowering of the sheet into the quench tank.

Quenching Media. Because rapid cooling is required after solution treating of most titanium alloys, either water or a 5% brine or caustic soda solution is most widely used as the quenching medium. Low-viscosity oil with a high flash point has been used effectively in vertical immersion quenching of sheet to reduce distortion. Quenching oils used with steel provide rapid cooling to 370 to 425 °C (700 to 800 °F), and these oils are satisfactory. Their use, however, should be limited to thin sections to avoid degradation of strength compared to that obtained by water quenching from the same solution temperature. Various concentrations of glycol in water will produce quench rates between those of water and those of oil.

Aging Furnaces. Because they do not involve combustion by-products, furnaces of the electrical-resistance type are preferable for aging titanium and its alloys. Retorts, however, may be used with oil-fired or gas-fired furnaces to avoid contamination. Aging furnaces normally are equipped with internal fans to promote circulation of air or other atmosphere throughout the work zone. Temperature-control equipment should be accurate to ±1.1 °C (±2 °F) and should be capable of controlling temperature within ±8 °C (±15 °F).

At normal aging temperatures of 480 to 595 °C (900 to 1100 °F), a protective atmosphere is not required. Aging in air produces a superficial scale that can be removed easily by mechanical or chemical means (this scale also may be left in place, because it does not affect properties).

Fixtures. In fixturing titanium components or assemblies to prevent distortion, the thermal-expansion characteristics of both the titanium alloy and the fixture itself must be considered. Ideally, both the alloy and the fixture will have equivalent thermal expansion characteristics within the intended aging-temperature range. Mild steel is commonly used because it is low in cost and can be made reasonably resistant to oxidation at aging temper-

atures through use of coatings such as electroless nickel. When mild steel fixtures are used, allowances must be made for the slight difference between the thermal expansion of the mild steel and that of titanium to avoid undesirable growth or distortion of the treated part.

In some applications, it is necessary to reduce or eliminate existing distortion in a part or assembly. This distortion may have resulted from water quenching, from relief of residual stresses during machining, from stresses induced by welding, or from uncontrollable springback after forming. Proper fixturing during aging can be used to minimize such distortion. Fixtures also must guard against sagging; for example, Ti-6Al-4V has a tendency to sag at 955 °C (1750 °F) during solution heat treating. Because titanium alloys exhibit creep behavior within the normal range of aging temperatures, it is possible to fixture and "creep form" components or assemblies to desired shape. Parts also may be sized by fixturing during aging.

Summary of Practice. Key considerations in heat treating of titanium and its alloys—practices that are to be followed and those that should be avoided—are summarized below.

- Provide sufficient stock for post-treatment metal-removal requirements (contaminated metal removal).
- Clean components, fixtures and furnaces prior to heat treatment. (Caution: Do not use ordinary tap water in cleaning of titanium components.)
- Use temperature controls with an upper cutoff to prevent temperature from exceeding beta transus.
- Charge cold components into furnaces operating at the required temperature.
- Stack and support components to allow free access of heating and quenching media.
- Observe quench-delay requirements to ensure hardening response during aging.
- Review property requirements and select optimum heat treating procedure.
- Review strength requirements and select proper aging cycle.
- Remove alpha case after all heat treating is complete.
- Check for the presence of hydrogen after all processing is complete.
- Do not nest components.

- Do not allow temperature to exceed beta transus (unless it is specified as a beta anneal process).
- Do not rely on inert atmosphere or vacuum for prevention of oxygen contamination.
- Do not rely on hardness tests for measurement of the effects of heat treatment.
- Do not pickle assemblies with faying surfaces.

Production Examples of Heat Treating Processes

The examples that follow describe applications of heat treating processes to specific titanium parts and assemblies, and indicate typical relationships among heat treating and other production operations.

Example 1 (Alpha Alloy Weldment). Because alpha alloys are not hardenable, heat treatment of welded compressor cases made of alpha alloy Ti-5Al-2.5Sn was limited to annealing and stress relieving. After the subassemblies had been formed, machined and stress relieved, the cases were assembled, using manual and mechanized inert-gas-shielded tungsten-arc welding and resistance welding. The completed assemblies were stress relieved for 1 h at 620 °C (1150 °F) in an electric muffle furnace with air atmosphere. No fixturing or protective coating was used. After being stress relieved, the cases were descaled by grit blasting and light pickling in a nitric-hydrofluoric acid solution.

Alpha-Beta Alloy Weldments. Development of maximum properties in alpha-beta alloy weldments requires solution treating followed by rapid quenching and then aging. However, if reduced strength is acceptable or higher ductility is required, workpieces may be annealed only, after welding.

Example 2 (Wing Rib). After being inert-gas-shielded tungsten-arc welded with Ti-6Al-4V filler rod, a wing rib made of annealed Ti-6Al-4V sheet was solution treated for 1 h at 900 °C (1650 °F) in an air muffle furnace and water quenched. The part was protected by a glass coating during heating. (The coating, which consisted of hydrous borax, boric acid and aluminum hydrate in a volatile carrier, was sprayed on the sheet at a dry weight of 269 to 323 g/m², or 0.88 to 1.06 oz/ft².) After being quenched, the part was inserted in an aging fixture and forced into the desired shape. It was aged in the fixture

Fig. 5 Turbojet compressor case, showing fabrication processes employed in its manufacture (Example 4)

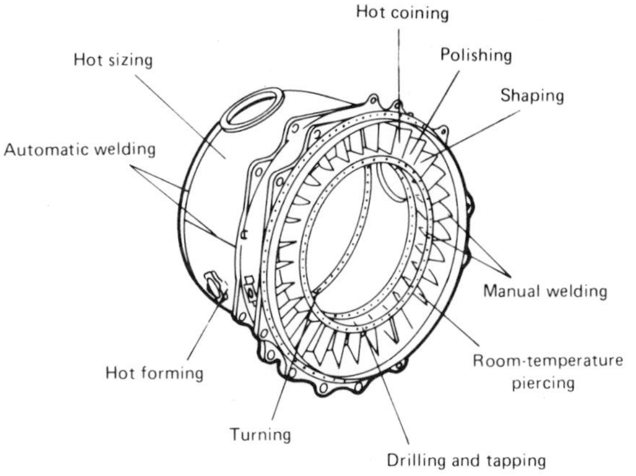

for 4 h at 540 °C (1000 °F) in an air muffle furnace. Sufficient relaxation and relief of stresses occurred during aging to bring the part into conformity with dimensional requirements. After aging, the welds had an ultimate tensile strength of 1090 MPa (158 ksi), a yield strength of 985 MPa (143 ksi), and 6% elongation in 12.7 mm (½ in.).

Example 3 (Pressure Vessel). A spherical pressure vessel 610 mm (24 in.) in diameter and 6.4 mm (¼ in.) in wall thickness was fabricated from 19-mm- (¾-in.-) thick hemispherical forgings of Ti-6Al-4V. The hemispheres were rough machined and tack welded to a special frame mounted on the lid of a cylindrical electric furnace. The rough machined hemispheres were then solution treated for 1 h in argon at 900 °C (1650 °F), water quenched and finish machined. Furnace loading and unloading were accomplished by use of an overhead crane, and the parts were quenched within 6 s after being removed from the furnace. The two hemispheres were joined by mechanical inert-gas-shielded tungsten-arc welding, using Ti-6Al-4V titanium filler metal and a reinforced weld area. Following welding, the vessel was aged for 6 h in a circulating-air furnace at 540 °C (1000 °F). No fixturing or protective coating was used. Cleaning after aging consisted of pickling in nitric-hydrofluoric acid.

Example 4 (Turbojet Compressor Case). Figure 5 shows a 760-mm- (30-in.-) diam jet-engine compressor case that was fabricated from Ti-5Al-2.5Sn. The shell of the case was roll

formed into two half-round segments after the ports had been punched. After rolling, the ports were deep drawn in resistance-heated dies at 595 °C (1100 °F). After being loaded in stainless steel fixtures, the half-round segments were stress relieved in an air muffle furnace for 1 h at 620 °C (1150 °F). The ports were then sized to final dimensions while the curvature of the segments was maintained with a fixture. After the ports had been drawn and the segments stress relieved, all components were cleaned by grit blasting and pickling. The half-round segments then were welded by the inert-gas-shielded tungsten-arc method, and the resulting weldments were stress relieved for 1 h at 620 °C without fixturing.

The shrouds were brake formed to a radius of 2.5 t at room temperature. Between brake-forming operations, the shrouds were resistance heated in air at 595 °C (1100 °F) for 5 min. Following brake forming, the shrouds were wrap formed to their final diameters at 120 °C (250 °F) in heated dies to reduce springback. They were then stress relieved 1 h at 620 °C (1150 °F) in air, and air cooled. No fixturing was necessary for maintaining tolerances.

The vanes were fabricated from machined bar stock and sheet. The annealed material was welded to form hollow vanes and then hot coined at 650 °C (1200 °F) in dies to flatten the parts and relieve welding stresses. Following coining, the vanes were cleaned and machined to an airfoil contour.

The final assembly was made by joining the various stress-relieved components by fusion welding. After assem-

bly, the entire compressor case was stress-relieved (without fixturing) in an air furnace for 1 h at 620 °C (1150 °F) and air cooled. Then the assembly was cleaned by grit blasting and pickling.

Wing section heat treating practices and applications are discussed in Examples 5 and 6.

Example 5. Skins of Ti-6Al-4V were stretch formed at room temperature in the fully annealed condition and solution treated in air for 30 min at 900 °C (1650 °F). After being water quenched, the distorted parts were placed in a steel aging frame and clamped to make them conform to the desired configuration. The skins and fixtures were aged 4 h at 540 °C (1000 °F) in a muffle furnace. Sufficient relaxation occurred during aging so that the skins were within drawing tolerances.

Example 6 (T-section extrusions). Ti-6Al-4V wing sections (1.22 m or 48 in. long) for spars were hung vertically on a rack for solution treatment in an air muffle furnace at 900 °C (1650 °F). The rack and the parts were then rapidly quenched in water. Straightening of the distorted (primarily, bowed) extrusions was accomplished by use of fixtures during aging. The parts were subjected to a 50% overbend by insert spacers and aged 4 h at 540 °C (1000 °F) before being machined. This aging cycle was repeated after each of the three machining operations required. Heat treating was performed without protective atmosphere. After the solution treatment and the first aging treatment, the parts were pickled in nitric-hydrofluoric acid. Pickling removed the contaminated surface layer and increased tool life in machining.

Example 7 (Rocket-Motor Cases). In fabrication of rocket-motor cases made of Ti-6Al-4V, the individual components were machined to a maximum wall thickness of 12.7 mm (0.500 in.), or to 7.62 mm (0.300 in.) wherever possible, allowing at least 1.3 mm (0.050 in.) of stock for cleanup after solution treating. Only cylindrical components were placed in fixtures for the solution treating operation. Solution treating was performed by holding the parts for 2 h at a temperature 19 to 33 °C (35 to 60 °F) below the beta transus (aim: 28 °C, or 50 °F, below beta transus) in a bottom-loading gantry furnace and then quenching them rapidly in a violently agitated 3% solution of sodium hydroxide.

Tensile-test coupons representative of all parts in the load were heated and quenched with the work. Before the parts themselves were aged, the test coupons were aged for 8 h at each of four temperatures—480, 510, 540 and 565 °C (900, 950, 1000 and 1050 °F)—to determine the optimum aging temperature in terms of desired mechanical properties. Then the parts were fixtured (a predetermined load was applied to promote creep forming), aged 8 h at optimum temperature, and air cooled. Use of a protective atmosphere was not required.

After being aged, the components were machined to final dimensions prior to welding, removing surface material that had been oxidized and contaminated during solution treating. Components were then welded into the final assemblies without preheating or postheating. The completed assemblies were stress relieved for 2 h at 480 °C (900 °F) and air cooled.

REFERENCES

1. "Titanium Alloy Handbook", by R. A. Wood and R. J. Favor: Report MCIC-HB-02, Battelle Memorial Institute, Columbus, OH, Dec 1972
2. "Properties and Processing Ti-6Al-4V": Timet, Apr 1980
3. How to Descale Titanium, by A. E. Durkin: *Materials and Methods,* Vol 38, Oct 1953, p 107-109
4. Properties and Structure of Titanium After 30-Min. Heating at 1200 to 2000 °F, by E. Walden and L. A. Dixon: *Metal Progress,* Vol 64, Aug 1953, p 88-89

SELECTED REFERENCES

- The Oxidation and Contamination of Ti and Ti Alloys: DMIC Memo 238, July 1968
- Production Techniques for Extruding, Drawing and Heat Treatment of Titanium Alloys: AFML-TR-68-349, Dec 1968
- "Influence of Metallurgical Factors on the Fatigue Crack Growth Rate in Alpha-Beta Titanium Alloys", by J. C. Chesnutt, A. Thompson and J. C. Williams: AFML-TR-78-68, Rockwell Science Center
- "Improvement of Reliability and the Mechanical Properties of Titanium Alloy Forgings", by T. Gurganus and G. Hall: AFML-TR-75-211, Alcoa Technical Center
- "Improved Manufacturing Methods for Producing High Integrity More Reliable Titanium Forgings", by R. Sparks and J. Long: AFML-TR-73-301, Wyman Gordon, Worcester, MA, Feb 1974
- *Titanium Science and Technology,* by R. I. Jaffe and H. M. Burte: Plenum Press, 1973
- "Residual Stresses, Stress Relief and Annealing of Titanium and Titanium Alloys", by D. J. Maykuth: DMIC Report S-23, July 1968
- Heat Treating Titanium and Its Alloys, by J. A. Burger and D. K. Hanink: *Metal Progress,* Vol 91, No. 6, June 1967, p 70-75
- "Hydrogen Contamination in Titanium and Titanium Alloys; Part IV: The Effect of Hydrogen on the Mechanical Properties and Control of Hydrogen in Titanium Alloys", by D. N. Williams *et al:* report from Battelle Memorial Institute to Wright Air Development Center, Contract AF 33(616)-2813, Mar 1957
- "Scaling of Titanium and Titanium Alloys", by H. W. Maynor, Jr., B. R. Barrett and R. E. Swift: report from University of Kentucky to Wright Air Development Center, issued as WADC Report 54-190, Part I, Mar 1955, and Part II, June 1955
- "A Study of the Air Contamination of Three Titanium Alloys", by J. E. Reynolds, H. R. Ogden and R. I. Jaffee: Titanium Metallurgical Laboratory Report 10, Battelle Memorial Institute, Columbus, OH, 1955
- Kinetics of the Reaction of Titanium with Oxygen, Nitrogen and Hydrogen, by E. A. Gulbransen and K. F. Andrew: *Journal of Metals,* Vol 1, 1949, p 741-748
- "Hydrogen in Titanium and Titanium Alloys", by D. N. Williams: Titanium Metallurgical Laboratory Report 100, Battelle Memorial Institute, Columbus, OH, May 1958
- An Experimental and Thermodynamic Investigation of the Hydrogen-Titanium System, by A. D. McQuillan: *Proceedings of the Royal Society* (London), Vol A204, Dec 1950, p 309-323
- Vacuum Degassing, by C. B. Griffith and M. W. Mallett: *Vacuum Metallurgy,* Vol 147, 1954

Heat Treating of Tin-Rich Alloys

By Joseph B. Long
Consultant

IN HEAT TREATING of tin-rich alloys, it is difficult to secure an effective and permanent degree of hardening. In most tin-base alloys, diffusion usually occurs at ambient temperatures, and therefore quenched and tempered alloys tend to overage and resoften.

Binary Alloys

Tin-antimony, tin-bismuth, tin-lead and tin-silver alloys can be temper hardened by solution treatment and aging. However, only the tin-antimony alloys can be permanently strengthened by heat treatment; all other tin-rich binary alloys will gradually soften at room temperature. The greatest improvement obtainable in binary tin-antimony alloys occurs in the alloy that contains 9% antimony; a hardness of 21 HB and a tensile strength of 51 MPa (7.4 ksi) can be increased to 26 HB and 65 MPa (9.4 ksi). This alloy is tempered for 48 h at 100 °C (212 °F) after being quenched from 225 °C (435 °F). During this tempering treatment, elongation decreases from 20 to 10% (in 50 mm or 2 in.).

Ternary Alloys

Permanent effects of heat treatment also carry over into ternary alloys of tin, antimony and cadmium. This was discovered in an early investigation of the strength and hardness of ternary alloys containing up to 43% cadmium and 14% antimony using chill cast specimens. It was found that the strengthening effect of cadmium in the terminal solution tin phase alpha is much greater than that of antimony. The presence of the sigma phase (principally SbSn) as primary cuboids has no effect on strength or hardness, but the presence of primary epsilon (CdSb) destroys the useful mechanical properties. The maximum combination of strength, ductility and hardness is obtained in alloys that have finely dispersed precipitates of the sigma and epsilon phases in an alpha matrix, or finely dispersed epsilon in a matrix of alpha with a eutectoid of alpha plus gamma (cadmium-rich solid solution). These structures typically are achieved by quenching or rapid cooling from elevated temperatures to avoid precipita-

tion of primary sigma and epsilon. In this study, the maximum stable values obtained in alloys containing 7 to 9% antimony and 5 to 7% cadmium were as follows: tensile strength, 108 MPa (15.7 ksi); elongation, 15% (in 50 mm or 2 in.); and hardness, 35 HB.

Additional heat treatment studies have been directed to a group of cold workable tin-rich alloys containing 3 to 8% cadmium and 1 to 9% antimony. Two forms of hardening were observed on quenching of these alloys from 185 to 200 °C (365 to 390 °F). One form results from the change in solubility of antimony in tin or in the beta phase. The other, which produces more intensive hardening, is analogous to hardening of binary cadmium-tin alloys by quenching and depends on suppression of eutectoid decomposition of the beta phase. Permanent improvement results in the first instance. Thus, a maximum tensile strength of 101 MPa (14.6 ksi) was achieved in a Sn-3Cd-7Sb alloy that was quenched from 190 °C (375 °F) and aged for either 24 h at 100 °C (212 °F) or 18 months at room temperature.

Further studies have been carried out on tin-base alloys containing 7 to 10% antimony and 0 to 3% cadmium in an effort to locate a bearing alloy that would be suitable at mildly elevated temperatures. In this composition range, it was found that alloys containing 0.5 to 2% cadmium (but not 3%) can be strengthened considerably by quenching and tempering.

Optimum properties (tensile strength: 92 MPa, or 13.4 ksi) were obtained in a Sn-9Sb-1.5Cd alloy quenched from 220 °C (430 °F) and then aged for 1000 h at 140 °C (285 °F). This alloy consists of finely divided sigma and epsilon phases in a matrix of alpha.

Pewter

Many pewter articles are manufactured from sheet prepared by cold reduction of cast bars or slabs. Tin-rich pewter alloys containing antimony and copper will work harden during sheet-rolling operations that involve small percentage reductions (20%). Upon standing at room temperature, the alloy will recrystallize and soften until it has reverted to the hardness of the original cast bar or slab. On the other hand, if large reductions (such as 90%) are made and the crystals are heavily worked, the alloy will work soften. Finally, as the crystals increase in size, hardness will increase slightly, but never to the level of the original cast material.

The hardness values of spun pewterware, or of other articles that have been manufactured by mechanically working the metal, can be restored by heat treatment at temperatures from 110 to 150 °C (230 to 300 °F). The time required varies from 3 h at the lower temperature to a few minutes at the higher temperature. A tin alloy containing 6% antimony and 2% copper will harden to 90% of the hardness of the as-cast material after annealing for 1 h at 200 °C (390 °F). Longer annealing times at lower temperatures have smaller but similar effects on the recovery from work softening.

Heat Treating of Special-Purpose Alloys

HEAT TREATING procedures for several special-purpose alloys used in military and aerospace applications, and in other products where special properties are required, are discussed in this article. Included are procedures for treating depleted uranium, (DU) zirconium, tantalum, niobium and alloys of these metals. Also presented are lists of references for expanded study of these metals and their alloys.

Heat Treating of Depleted Uranium and Its Alloys

By the ASM Committee on
Heat Treating of Depleted Uranium*

The major applications of depleted uranium (DU) and its alloys are those for which density is an important, if not an overriding, consideration (at 18.7 g/cm³, unalloyed depleted uranium is one of the densest of all elements). Among these applications are kinetic energy penetrators for military use, aircraft and missile counterweights, radiation shielding, gyrorotors and ballast. Dilute alloys containing 0.75 wt%

Ti or 2 wt% Mo are used in production of kinetic energy penetrators (the largest single application of DU) because superior mechanical properties can be developed in these alloys, and their corrosion resistance can be improved, by heat treatment.

Most unalloyed DU is produced in cast form and is used without heat treatment and with or without machining. However, requirements for rolled and heat treated DU may increase significantly in the future.

Phase changes in DU and its alloys are accompanied by significant changes in volume. Volumetric shrinkage from the high temperature gamma phase to the low-temperature phase is 1.8%. Changes in linear dimensions are influenced by preferred orientation. For a random orientation, linear shrinkage is 0.6%. Because shrinkage cannot be predicted accurately, these dimensional changes make it impossible to machine (DU) to final dimensions before heat treatment. In rough machining for heat treatment, an envelope at least 0.4 to 0.5 mm (0.015 to 0.020 in.) thick should be allowed for final machining.

Metallurgical Characteristics of Depleted Uranium

In processing of uranium, several considerations must be addressed: hydrogen embrittlement and hydride formation; ready oxidation in air, attack by hot water; and dissolution by acids. In addition, fine particles of uranium metal are pyrophoric and can ignite spontaneously at room temperature.

The metal uranium is obtained from uranium hexafluoride (UF_6) tailings from the uranium enrichment process that provides U-235 uranium for the nuclear industry. In typical production of depleted uranium, UF_6 tailings are reduced to uranium tetrafluoride (UF_4), called "green salt", which is further reduced to derby uranium metal by a thermite-type bomb reduction with magnesium metal. Calcium metal has also been used to reduce UF_4 to derby metal. Typical derby chemical analysis ranges (in ppm) are 5 to 50 Cu, 8 to 40 Al, 30 to 150 Fe, 10 to 50 Ni, 10 to 100 Si, 1 to 10 Mg, 10 to 50 C, 15 to 40 O, 8 to 40 N and 4 to 18 H. Minor high-vapor-pressure contaminants can be re-

*Arthur L. Geary, Senior Metallurgist, Nuclear Metals, Inc.; Nicholas C. Jessen, Jr., Group Leader, Nuclear Division, Union Carbide Corp.; Allen B. Townsend, Development Consultant, Union Carbide Corp.

Table 1 Typical chemical and gas analyses for vacuum-induction-melted DU

Element	Average analysis, ppm
Carbon	32
N$_2$	3
O$_2$	14
Bulk H$_2$	0.16
Aluminum	10
Copper	10
Iron	30
Manganese	8
Nickel	4
Lead	5
Silicon	40

moved by volatilization during subsequent vacuum melting operations, while the other listed elemental contaminants remain. These impurities should be monitored and controlled to ensure expected metallurgical response to processing. Typical chemical and gas analyses for vacuum-induction-melted depleted uranium are shown in Table 1. In general, hardness varies directly, and ductility varies inversely, with impurity content.

Phase Changes. Chemically and metallurgically, DU is identical with natural uranium. Transformation from the alpha phase (orthorhombic) to the beta phase (tetragonal) occurs at 662 °C (1224 °F). The beta phase is stable up to 773 °C (1423 °F), where it transforms to the high-temperature gamma phase (body-centered cubic). The latter is stable to the melting point, 1132 °C (2070 °F). The gamma phase in unalloyed DU cannot be retained to room temperature by commercial quench rates.

Grain Size and Orientation Control

The grains of cast DU or of DU worked in the gamma region are quite large, typically 2 to 3 mm (0.08 to 0.12 in.) in diameter. Large grain sizes are undesirable for material that is to be worked because they result in rough machined surfaces and variations in mechanical properties. Grain size can be refined considerably by multiple beta quenching. In workpieces more than 25 mm (1 in.) thick, however, refinement is limited to an outer layer of grains because the cooling rates at greater depths are too low. Generally,

the rim of fine-grain material is sufficiently thick to produce a smooth surface after subsequent working.

Beta treatment is also used as a heat treatment for DU worked in the high alpha temperature range. Beta treatment consists of heating the DU into the beta range, holding for a suitable time and cooling at a rapid rate. A common temperature range is 720 to 730 °C (1330 to 1350 °F). The purpose of this heat treatment is to eliminate the preferred orientation that develops during working.

The final alpha grain size is insensitive to temperature within the beta range and to variations in holding time. Times from 1 min to 1 h at temperatures greater than 700 °C (1290 °F) have no effect on final alpha grain size. Cooling rate, however, has a significant effect. Water-quenched material has a significantly finer grain size, and the grains have rough scalloped edges. Air-cooled grains have more uniform boundaries. Extremely low cooling rates, such as those obtained in a furnace-cool cycle, produce large alpha grains.

High cooling rates such as those achieved by water quenching produce high residual stresses in beta-treated material. In thin sections, these residual stresses can produce appreciable plastic deformation in the alpha phase. This alpha phase can be recrystallized, and sometimes grain refined, by an anneal in the alpha range. Alpha annealing after beta treatment will not produce recrystallization at the center of a thick section, because the center does not cool rapidly enough for sufficient straining of the lattice to occur.

When DU is water quenched from the beta phase, high stresses develop due to the combination of (a) volume contraction during beta-to-alpha transformation and (b) the radial temperature gradient. These stresses are compressive at the surface and tensile at the center. The tensile stresses are high enough to produce failure near the centerline. Large numbers of repetitive quenches from the beta phase can produce sponginess, cracks or holes in the center section of the workpiece.

Grain growth is extremely sensitive to orientation as well as to differences in metal purity, to prior deformation and to heat treatments that affect the dispersions of contaminant second phases. Those contaminants that are in solution tend to delay recrystallization and often result in mixed, or incom-

plete, recrystallized structures. Those elements having uranium compounds that show limited solubility, thereby existing as inclusions, do not delay recrystallization appreciably.

The amount of work that exists in uranium metal prior to heat treatment has an important effect on final grain size. A 1 to 2% strain in uranium constitutes critical strain. Recrystallization of material with this amount of strain results in very large grains. Consequently, a plate prepared for forming operations is produced with 10 to 15% warm work to ensure that no areas of critical strain exist in the final wrought product. As the amount of work in the metal to be annealed is increased, the temperature needed for uniform recrystallization is lowered. For most formed parts made from relatively pure material, recrystallization will not occur below 400 °C (750 °F) with annealing times of up to 10 h.

Cold Working

Alpha uranium is slightly softer than steel and is considered to be malleable and ductile. Alpha uranium is readily worked at room temperature; however, directionality and texture persist because of a pronounced anisotropy. Aside from the complication of directionality, the tensile strength of uranium can be greatly enhanced by cold working, as shown in Table 2; hardness also can be increased significantly, as shown in Table 3.

Annealing

Annealing of cold worked DU is similar to that in other metals. The first stage is recovery, in which there is a slight decrease in hardness, a small decrease in electrical resistivity and a pronounced sharpening of x-ray line shape. Recovery is followed by recrystallization. The variation of recrystallization temperature as a function of cold work is shown in Fig. 1 for an annealing time of 1½ h. Recrystallization begins at 400 °C (750 °F) and is complete at 450 °C (840 °F) in material cold worked 90 to 94%. Light cold working (about 4%) causes recrystallization to begin at 525 °C (975 °F), but recrystallization is not complete after 1½ h at 600 °C (1110 °F).

The grain size of cold worked and annealed DU depends upon a variety of factors: annealing time and temperature; amount and homogeneity of the cold work strain; cold working temperature; and volume, size and dispersion of inclusions. Cold or warm working of DU often results in a banded or duplex structure; this persists as a duplex grain size after recrystallization. The average grain size for material rolled to 50% reduction at 300 °C (570 °F) is illustrated in Fig. 2. This shows the effect of annealing temperature (annealing time, 1 h) on materials of various purities. Grain size is about 0.01 to 0.015 mm (0.0004 to 0.0006 in.) for impure uranium and as large as 0.04 mm (0.0015 in.) for high-purity metal.

Table 4 presents typical mechanical properties of uranium in various conditions. These data are not precise values and are intended for use only as guidelines for selection of heat treatments.

The following heat treating procedure can be followed to produce fine-grain material. It is best to start with a relatively pure material that has very small amounts of inclusion-producing impurities. Following a hot breakdown of the rolling ingot at 630 °C (1165 °F), rolling operations are performed at 300 to 400 °C (570 to 750 °F). After 40 to 60% warm work, the material is given a short recrystallization anneal (about 30 min per inch of thickness) at 630 °C. The rolling stock is cooled from the annealing temperature to below 400 °C and given further warm work. This process is repeated as often as final stock thickness will allow.

If the rolled stock is to be used in subsequent forming operations, the final rolling procedure should leave 15 to 20% warm work in the plate. Forming operations should be carried out warm at temperatures not exceeding 375 °C (705 °F). A final anneal at 630 °C (1165 °F) for 30 min per inch of thickness (minimum, 6 to 8 min) will produce parts with a grain size of ASTM 6 to 10, depending on the amount of warm rolling possible. Intermediate anneals at

temperatures below the suggested 630 °C (1165 °F) will help develop the finer grain size. Test specimens will be required to establish the times necessary at these lower temperatures to effect complete recrystallization.

Alpha recrystallization annealing will not remove the anisotropy of the crystal structure produced by the rolling and forming processes. The stock for most forming operations is produced by "square rolling", or by giving the rolled plate essentially equal reductions in the longitudinal and transverse directions. This process produces

a plate with relatively uniform properties in the plane of the plate. Equal reductions taken at 45° to the standard longitudinal and transverse directions, will produce even more uniform forming stock.

Cast Uranium. The grain size of cast uranium is difficult to define because the large cast grains all have well-organized substructures (see Fig. 3). The cast microstructure and mechanical properties can be improved by beta heat treating (see Fig. 4). In this process, the casting is heated to about 740 °C (1365 °F), water quenched, and

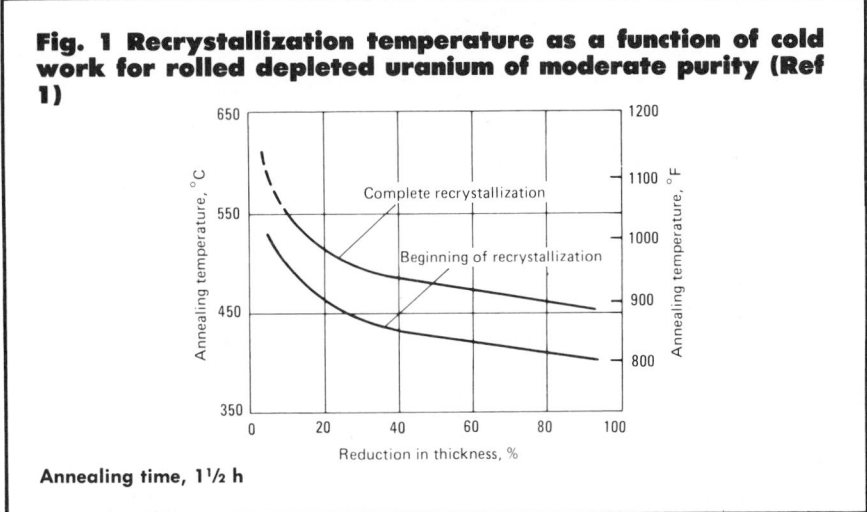

Fig. 1 Recrystallization temperature as a function of cold work for rolled depleted uranium of moderate purity (Ref 1)

Annealing time, 1½ h

Table 2 Typical mechanical properties of DU as functions of amount of cold work

Cold work, %	Ultimate tensile strength		Tensile yield strength(a)		Compressive yield strength		Elonga-tion(b), %	Hard-ness(c), HV
	MPa	ksi	MPa	ksi	MPa	ksi		
0(d)	1060	154	375	54	405	58	15	294
15	1140	165	525	76	500	72	17	352
25	1190	173	600	87	575	83	14	354
40	1280	186	660	96	605	87	13	359
55	1360	197	905	131	690	100	11	397

(a) At 0.2% offset. (b) In 50 mm or 2 in. (c) 1-kg load. (d) This material is highly directional and was not beta heat treated.

Table 3 Hardness data for cold-worked uranium rod

Cold work, %	Hardness									
	HR15N		HR30N		HR30T		HR45T		HV(a)	
	Average	Range(b)	Average	Range	Average	Range	Average	Range	Average	Range(c)
0	72	71 to 74	43	42 to 44	79	77 to 82	68	67 to 69	294	281 to 308
15	73	71 to 74	46	42 to 49	81	79 to 82	72	69 to 73	352	335 to 366
25	75	74 to 76	47	43 to 49	81	79 to 83	73	71 to 74	354	348 to 361
40	76	75 to 77	50	47 to 52	83	81 to 85	73	69 to 75	359	339 to 376
55	78	76 to 79	56	55 to 58	85	83 to 86	79	78 to 80	397	376 to 423

(a) 1-kg load. (b) Results of five indentations. (c) Results of eight indentations across the diametral cross-section of the bar

Fig. 2 Average grain size of depleted uranium as a function of annealing temperature (Ref 2)

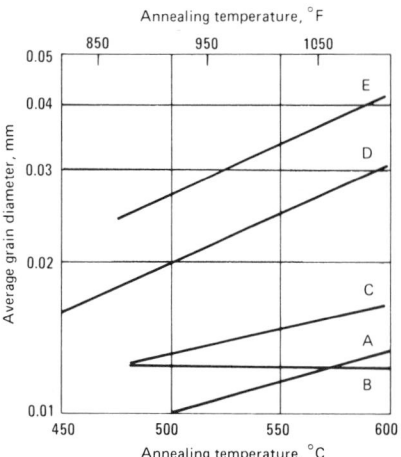

One-hour anneals after 50% reduction by rolling at 300 °C (570 °F). A and B are impure metal; C, D and E are high-purity uranium in decreasing order of submicroscopic inclusion content

then given by an alpha anneal. Grain refining occurs because of the presence of small levels of uranium-iron and uranium-silicon compounds. These compounds are put into solution by the 740 °C (or higher) beta heat treatment, kept in solution by the water quench, and then precipitated by the alpha anneal. If the alpha-anneal temperature is low, the precipitation of the compounds is fine and well dispersed. After this fine precipitate dispersion is achieved, a second beta quench can be more effective as a grain refining step. Two beta quenches and two alpha annealing cycles will produce the desired grain structure in stock as thick as 32 mm, or 1.25 in. (see Fig. 3).

The specific details of the recommended beta heat treatment—such as temperature, time, quenching procedure and furnace conditions—are dictated by the final metallurgical condition desired. When beta heat treating is followed by water quenching, the uranium lattice undergoes heavy strain as

Fig. 3 Typical grain structure of cast depleted uranium

a result of the beta-to-alpha transformation. This transformation can occur by either diffusion or martensitic mechanisms, depending on the severity of quenching. Transformation from beta to alpha is a diffusion reaction at low cooling rates, a martensitic reaction at high cooling rates, and a mixed diffusion and martensitic at intermediate cooling rates. In general, to relieve the residual stresses induced by beta quenching and to precipitate a fine dispersion of secondary compounds, an annealing temperature of 575 °C (1065 °F) is used. Typical tensile properties that can be developed by various heat treatments in cast and wrought uranium are shown in Tables 5 and 6, respectively.

Dilute Alloys of Depleted Uranium

Dilute alloys that are heat treated in larger quantities are DU-0.7 wt% Ti and DU-2 wt% Mo. Both are used as cores in kinetic energy penetrators. The ability of these alloys to age harden is related to the fact that titanium and molybdenum have extended solid solubility in the high-temperature gamma phase and essentially complete

Table 4 Typical room-temperature mechanical properties of uranium in various conditions

Fabrication history	Yield strength(a) MPa	ksi	Tensile strength MPa	ksi	Elongation(b), %	Reduction in area, %
Cast	207	30	448	65	5	10
Gamma extruded	172	25	552	80	10	12
Beta rolled	207	30	586	85	12	...
					20	
Alpha extruded, 600 °C (1080 °F)	207	30	621	90	15	
Alpha rolled at:						
300 °C (570 °F)	759	110	1172	170	7	14
500 °C (930 °F)	414	60	897	130	20	...
600 °C (1080 °F)	276	40	759	110	20	...
Annealed, after rolling at:						
300 °C (570 °F)	345	50	759	110	5	...
500 °C (930 °F)	276	40	690	100	15	...
Beta treated after alpha rolling:						
Water quenched	241	35	586	85	10	12
Slow cooled	207	30	414	60	7	...

(a) At 0.2% offset. (b) In 50 mm or 2 in.

Fig. 4 Grain of cast depleted uranium refined by beta quenching

Table 5 Typical tensile properties vs heat treating methods for cast uranium

Heat treating methods	Tensile strength MPa	ksi	Yield strength(a) MPa	ksi	Elongation(b), %	Reduction in area, %	J-integral J/mm²	in.·lb/in.²	Tearing modulus	Charpy impact energy J	ft·lb
As cast	420	61	205	30	6	...	...	...	...	...	...
Vacuum heat treated, 640 °C (1184 °F), 1 h	450	65	215	31	5	...	...	...	...	...	...
Vacuum heat treated, 650 °C (1202 °F), 2 h, then 630 °C (1196 °F), 24 h	565	82	185	27	13	...	...	...	...	...	...
Salt annealed	450	65	215	31	8	...	...	...	...	...	...
Beta quenched, vacuum annealed	785	114	295	43	22	17	0.034	192	35	~14(d)	~10(c)
							0.016	90	11	~7(c)	~5(d)

(a) At 0.2% offset. (b) In 50 mm or 2 in. (c) 21 °C. (d) 54 °C

Table 6 Typical tensile properties vs heat treating methods for wrought uranium

Method of heat treating	Tensile strength MPa	ksi	Yield strength(a) MPa	ksi	Elongation(b), %	Reduction in area, %
Vacuum heat treated	800	116	271	39	31	28
Salt annealed or short vacuum heat treated	655	95	272	39	12	12
Vacuum arc melt, vacuum heat treated	780	113	217	31	49	...
Vacuum heat treated plate	835	121	273	40	40	...
Salt annealed	885	128	213	31	20	...

(a) At 0.2% offset. (b) In 50 mm or 2 in.

Table 7 Cooling rates for DU-0.75Ti in various quench media

Media	Quench rate °C/s	°F/s
Flowing argon	3.8	6.8
Conventional or soluble oil	38–40	68–72
0.05% PVA(a)	80	145
Water	98	175
10% brine	190	340

(a) Polyvinyl alcohol

insolubility in the low-temperature alpha phase. On rapid quenching, the gamma transforms martensitically to supersaturated alpha prime. A fine dispersion of intermetallic compound develops during subsequent aging at temperatures above 300 °C (570 °F).

Figure 5 shows the microstructure of water-quenched 36-mm- (1.4-in.-) diam U-0.75Ti bar. The outside of the bar (Fig. 5a) has transformed completely to lenticular alpha prime. The fineness of this structure increases with increasing quench rate. The grain boundaries visible in the structure are those of the prior gamma grains. The U-0.75Ti alloy is shallow hardening, however. At the center of the bar, transformation to alpha phase plus U_2Ti starts at prior grain boundaries (Fig. 5b). Similar structures are found in 18-mm- (0.7-in.-) diam bar oil quenched from the gamma phase. Aging to peak hardness produces no detectable change in structure.

Solution Treating and Aging. Heat treating for improved hardness and mechanical properties in dilute DU alloys consists of solution treating in the gamma-phase temperature range of 800 to 850 °C (1470 to 1560 °F),

quenching to room temperature, and aging in the alpha temperature range. Alternatively, interrupted quenching in a molten metal or salt bath held at the appropriate temperature can be used.

The time at the solution-treating temperature is not critical. Times as short as 2 to 5 min produce a completely gamma structure. Longer solution-treating times are generally used—typically, ½ to 1 h. Excessively long times should be avoided because they lead to large gamma grain sizes.

An important consideration in the selection of conditions for gamma solution treatment is the hydrogen level required in the final product. Hydrogen is detrimental to ductility in U-0.75Ti and must be maintained at 1 ppm or less to ensure high ductility in heat treated parts. These levels have been achieved consistently in extruded 36-mm (1.4-in.) -diam rod by vacuum outgassing for 2½ h at 850 °C (1560 °F) at 10^{-5} torr. Unalloyed uranium and uranium alloys are sensitive to hydrogen and for maximum material properties require extensive outgassing. The literature should be consulted before selecting conditions for these alloys.

Quenching. Table 7 gives the rates at which DU-0.75Ti cools when quenched in various media. The test slugs used for measuring these rates were 22 mm (0.875 in.) in diameter by 21 mm (0.845 in.) long. Cooling rates for other DU alloys should be similar. Except for DU-0.75Ti, which is cooled by very slow argon gas quenching, response to subsequent aging at 350 °C (660 °F), as determined by hardness measurements, was independent of quench rates above 40 °C/s (72 °F/s). Because DU-Ti alloys are shallow hardening, higher quench rates are needed to achieve uniform hardening response in larger-diameter bars or thicker plates.

Small-diameter bars and plates can be plunge quenched, but larger diameter bars (greater than 19 mm, or 0.75 in.) develop centerline voids if plunge quenched. These voids pose a particularly serious problem. Once they are formed, there is no easy way to heal them. Void formation is related to the stresses caused by the large volume change associated with the gamma-to-alpha prime transformation and high radial thermal gradients. Centerline voids can be minimized by end quenching—that is, by lowering

Fig. 5 Microstructure of solution treated and quenched U-0.75Ti

(a)

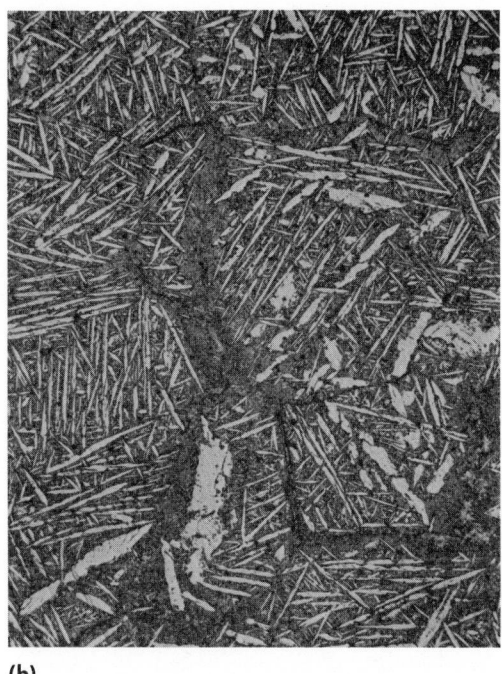

(b)

Bar 36 mm (1.4 in.) in diameter end quenched into water at 455 mm/min (18 in./min). Chrome-acetic electroetch. Magnification, 100×. (a) Edge. (b) Center

the bars, end on, at a controlled rate, into the quench media. Bars 36 mm (1.4 in.) in diameter have acceptable levels of centerline voids when quenched in this way, 18 at a time, into circulating water at 455 mm/min (18 in./min). The number and size of centerline voids, as detected by ultrasonic techniques, are substantially lower in bars end quenched at 255 mm/min (10 in./min). The aging response of the more slowly quenched bars is identical with those quenched at 455 mm/min (18 in./min).

Aging Results. Hardness and strength levels achieved on aging of dilute DU alloys are illustrated in Fig. 6, 7 and 8. The hardness curves for U-0.75Ti apply equally well to oil-quenched and water-quenched material (Fig. 6). The scatter band is caused by nominal differences in titanium and trace element contents of the alloys; iron and copper, even at low levels, contribute to the hardening response. Silicon is reported to retard hardening.

The heat treater has a wide selection of time-temperature options to achieve specified combinations of hardness and strength. For example, U-0.75Ti can be hardened to 45 HRC by any of the following treatments: 16 h at 380 °C (720 °F); 5 h at 400 °C (750 °F); or 1¾ h at 420 °C (790 °F). For production runs, conditions are selected to optimize equipment utilization.

Metastable High Alloys

Heat treatment of DU alloys with 4.0 wt% or more molybdenum or equivalent is similar to that for dilute alloys. The treatment starts with a gamma-phase solution treatment of 1 h at about 800 °C (1470 °F). This is followed by either (*a*) quenching to room temperature followed by aging in the alpha region or (*b*) quenching directly to the aging temperature and holding to achieve the desired properties. Special care must be taken in heat treating large sections. Cracking has occurred in billets of U-6 wt% Nb 205 mm (8 in.) in diameter (or larger) that were water quenched from 800 °C (1470 °F). Refer-

ences should be consulted for details regarding specific high alloys.

Processing and Equipment

Generally, three basic furnace designs are used for heating or heat treating of unalloyed uranium: molten baths, inert-atmosphere furnaces and vacuum furnaces. The type of furnace chosen depends primarily on desired final properties and material quality.

Uranium has been heated in molten lead for 50 h at 350 °C (660 °F) with no appreciable reaction. Longer periods, and/or temperatures of 800 to 1000 °C (1470 to 1830 °F), however, have caused uranium to be completely penetrated. Molten salts are the most common heating media used in industry for preheating uranium prior to fabrication operations and for final heat treatment. Table 8 shows corrosion results for six heat treating mixtures of salts. The main disadvantage of molten salt baths is the potential for hydrogen

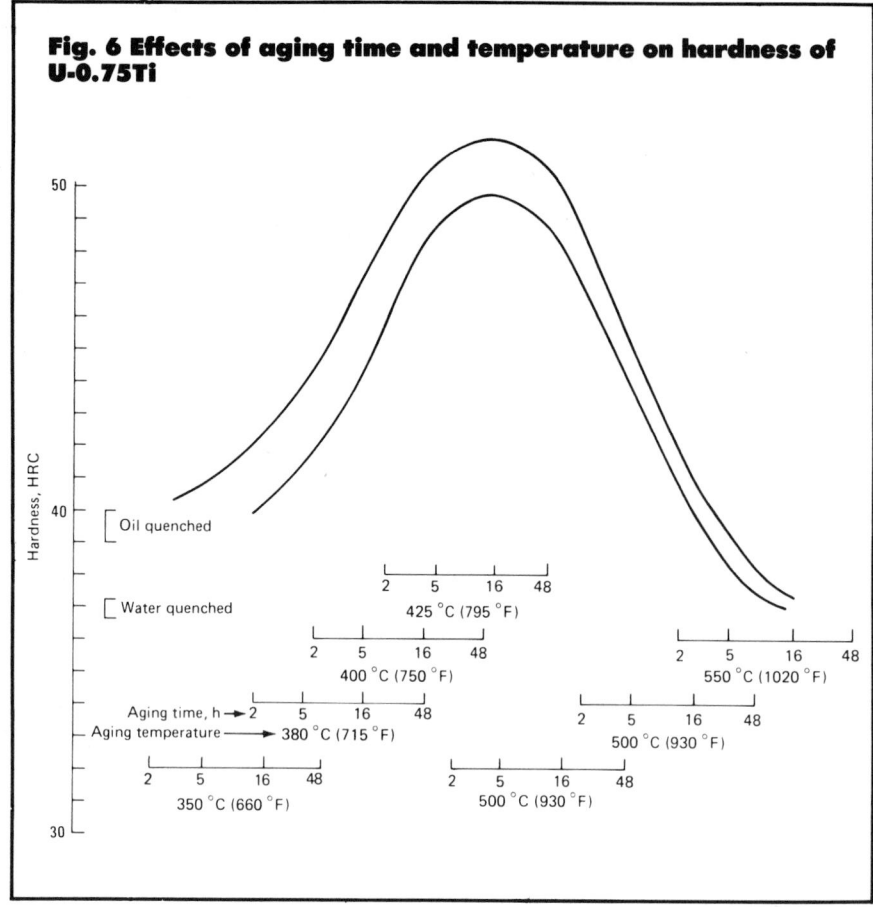

Fig. 6 Effects of aging time and temperature on hardness of U-0.75Ti

Table 8 Corrosion of uranium in molten salts at 595 °C (1100 °F)

Salt mixture	Time, h	Observed attack
$44Na_2CO_3$-$30K_2CO_3$-$26Li_2CO_3$	½ to 2	No corrosion
	4	Surface pitting
$74K_2CO_3$-$26Li_2CO_3$	½ to 2	No corrosion
	4	Pitted
$47Na_2CO_3$-$32K_2CO_3$-$21Li_2CO_3$	½ to 2	No corrosion
	2	Pits beginning
	4	Badly pitted
$53K_2CO_3$-$46.6Li_2CO_3$	½ to 1	No corrosion
	2	Surface pits
$20NaOH$-$30K_2CO_3$-$50Na_2CO_3$	½ to 4	Surface etching
$47Na_2CO_3$-$47K_2CO_3$-$6Li_2CO_3$	½	Scaling, 0.24% weight loss
	2	Scaling, 1.0% weight loss
	4	Scaling, 1.3% weight loss

contamination. Consequently, any planned use of molten salt baths for heating of uranium should include a design for removing the hydrogen from the bath, such as sparging the bath with CO_2 gas. Hydrogen, which is par-

ticularly deleterious to uranium, drastically reduces its tensile elongation, as shown in Fig. 9.

Furnace Atmospheres. No appreciable attack on uranium occurs in dry furnace atmospheres of helium, argon,

carbon monoxide, carbon dioxide or hydrocarbon gases at temperatures up to 500 °C (930 °F). However, relatively slight amounts of water vapor in any of these gases can cause extensive corrosion. Uranium reacts with water, in either liquid or vapor form, to produce UO_2 and hydrogen. The UO_2 spreads over the entire surface area as a black powder and, depending on the amount of exposure and purity of the metal, can result in excessive pitting and surface cratering. Thus, in heat treating of uranium in atmospheres of commercial inert gases, furnaces should be equipped with gas line dryers to dry the gas thoroughly before it is used.

Vacuum Treatment. Vacuum heat treating of uranium provides the best over-all environment for obtaining maximum tensile properties and high-quality metal surfaces. The principal advantage of vacuum-furnace heat treating is the potential for removing hydrogen from the metal.

Figure 10 shows the time required for vacuum heat treating uranium stock of different thicknesses to achieve maximum ductility. This illustration assumes an initial hydrogen concentration within the part of 2 ppm, which is typical of material processed through carbonate preheating baths, and it can be used for cast material.

A vacuum of 300 torr limits oxidation to an acceptable level while maintaining the hydrogen at its initial level. A vacuum of 10^{-5} torr is needed for significant lowering of the hydrogen level.

Cleaning. Surfaces of parts to be heat treated should be free of moisture, grease and cutting lubricants. Heavier oxides that form on DU and its alloys can be removed by pickling for about ½ h in a 1:1 mixture of nitric acid and water at 25 °C (75 °F). Copper, which is often used as cladding during fabrication, can also be removed with 1:1 nitric acid.

CAUTION: U-Nb and U-Zr alloys can produce explosions during pickling in nitric acid. The problem can be eliminated by adding 1 to 2 vol % of hydrofluoric acid to the pickling solution.

Quenching. The oil-quench media can be contained within the vacuum chamber. Following solution treating, the furnace is backfilled with argon or helium. The work is then transferred to a position directly over the quench tank and lowered at a controlled rate into the quench media.

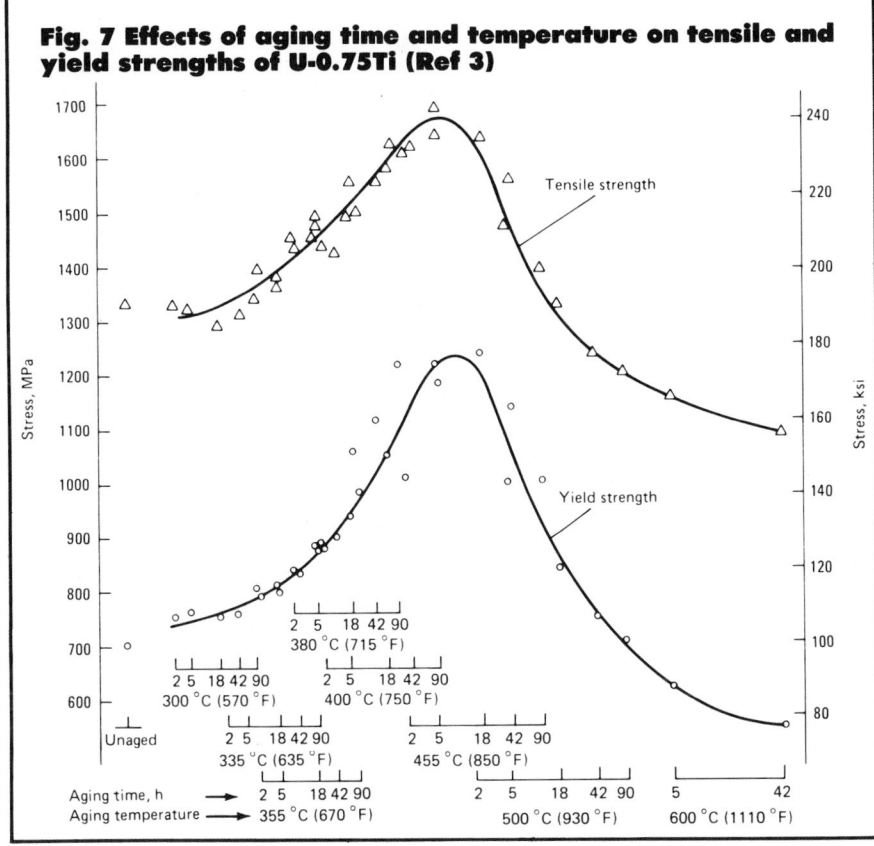

Fig. 7 Effects of aging time and temperature on tensile and yield strengths of U-0.75Ti (Ref 3)

°C (+4 °F). Following aging, the pieces should be either furnace cooled to 100 °C (212 °F) or water quenched. Oxidation of DU is exothermic, and the heat generated can make the reaction self-sustaining. *Rods 18 mm (0.7 in.) in diameter have reached red heat when they were not water quenched after being aged at 425 °C (795 °F) in a lead pot.*

Examples of Heat Treatment

The following procedures are examples of heat treatment used to meet certain specifications.

Example 1. A DU-0.75Ti alloy is to be heat treated to a hardness of 44 to 52 HRC. The heat treated material is to be machined to form a kinetic energy penetrator with a cylindrical body and a conical nose. Mechanical properties and hydrogen level are not as follows: The procedure is specified.

- Machine alpha-extruded bar stock to the approximate dimensions, allowing a 0.4-mm (0.015-in.) envelope for finish machining.
- Assemble premachined blanks in a rectangular basket with the grid-support structure; the nose ends should be pointed upward.
- Degrease to remove residual cutting fluid.
- Place the baskets in a vacuum furnace with an integral oil quenching tank. Solution treat ½ h at 850 °C (1560 °F) and at a pressure of less than 300 torr.
- Oil quench.
- Degrease.
- Age 2 h at 440 °C (820 °F) in an inert-gas recirculating furnace.
- Furnace cool.

Example 2. Bars of DU-0.75Ti alloy, 36 mm (1.4 in.) in diameter, are to be heat treated to the following specifications: hardness, 38 to 44 HRC; minimum 0.2% yield strength, 725 MPa (105 ksi); minimum elongation 12%; and maximum hydrogen content, 1 ppm. The procedure is to:

- Cut extruded bar stock to length.
- Pickle in 1:1 nitric acid to remove copper sheath.
- Rinse and air dry.
- Place rods vertically in a basket (see Fig. 7).
- Solution treat 2½ h at 850 °C (1560 °F) in a vacuum of 5×10^{-5} torr, or better.

The high vapor pressure of water precludes incorporation of water-base quench media into the furnace. Specially designed bottom-loading furnaces have been used successfully with separate quench tanks. The sequence of events is as follows. At the end of solution treating, the quench tank is moved under the furnace. The furnace is then backfilled with argon or helium, the power turned off, and the bottom door removed. The load is dropped rapidly to a position 50 to 75 mm (2 to 3 in.) above the water level and then is dropped at a controlled rate until all of the workpiece is submerged.

Fixtures. Neither DU nor its alloys are especially strong at gamma-solution treating, temperatures. Low strength coupled with a high density places special requirements on fixturing to provide proper support and thus prevent sagging. If possible, rods and irregular-shape pieces should be placed with the long axis in a vertical position and supported every 150 mm (6 in.); spans for horizontal pieces should be limited to 75 to 100 mm (3 to 4 in.). Figure 11 shows a fixture used for solution treating of bars 36 mm (1.4 in.)

in diameter by 380 mm (15 in.) long. This basket is made of Inconel 600. Copper shims are used at the points of contact between the DU and the Inconel because these two materials react at solution treating temperatures to produce a DU-Fe eutectic composition with a melting point of 725 °C (1335 °F). Molten-metal attack would result in extensive local wastage of the DU and welding to the fixture.

Salt Baths. If a salt bath is used for solution treating, the times should be kept as short as possible, not only to minimize hydrogen pickup but also to limit corrosive attack. Pieces exposed to molten chloride salts ($BaCl_2$, KCl and NaCl eutectic) and molten carbonate salts (35% Li_2CO_3 and 65% K_2CO_3) at 730 °C (1350 °F) have lost 0.08 mm (0.003 in.) of thickness in 1 h. Such attack would be significantly more severe at 850 °C (1560 °F).

Aging treatments can be carried out in recirculating inert gas furnaces, lead baths, lead-tin baths or molten salt baths. Because the aging reactions are temperature sensitive, the temperature should be controlled to within +5 °C (+9 °F) and preferably to within +2

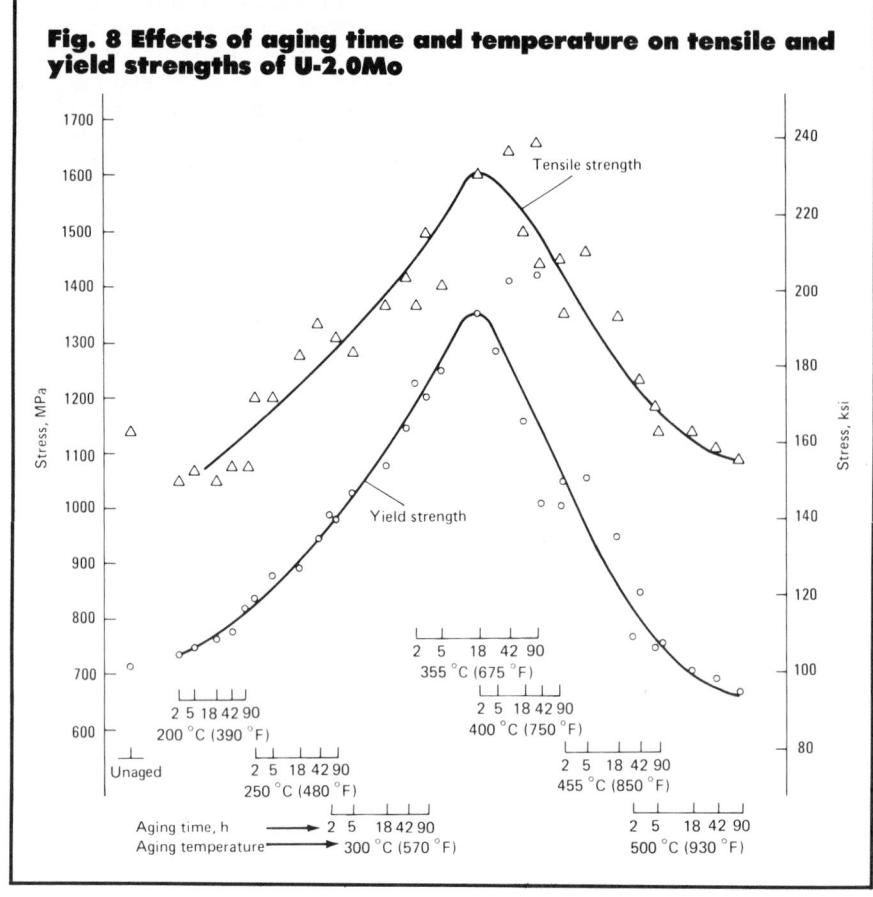

Fig. 8 Effects of aging time and temperature on tensile and yield strengths of U-2.0Mo

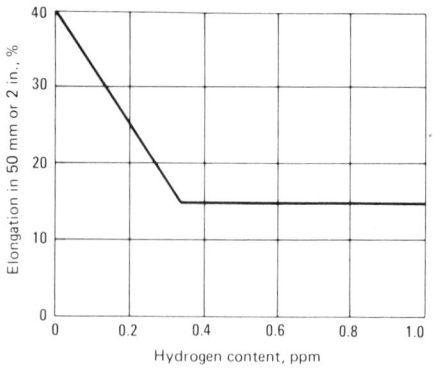

Fig. 9 Elongation vs hydrogen content for wrought depleted uranium

Fig. 10 Time required to achieve maximum ductility in depleted uranium plate of various thicknesses under a vacuum of 10^{-4} torr

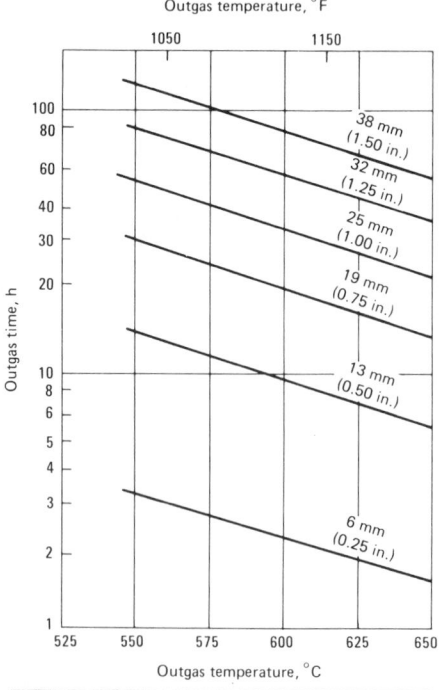

- Quench into circulating water at 455 mm/min (18 in./min).
- Air dry.
- Age 16 h at 350 °C (660 °F) in an inert gas recirculating furnace.

Licensing and Health and Safety Requirements

Possession of more than 15 lb (6.8 kg) of depleted uranium in any form requires a license from the U.S. Nuclear Regulatory Commission. Title 10, Part 40, of Federal Regulations describes the steps necessary and the requirements to obtain such a license. In addition, all other local, state and federal regulations also are effective as applicable.

The greatest potential source of contamination in the heat treating area is uranium oxide. The area should be isolated from the remainder of the plant, and everyone entering should be required to wear disposable protective footwear. Smoking and eating should be restricted.

The toxicity of depleted uranium if it *enters the blood stream, may result in poisoning similar to that caused by lead, arsenic, mercury or any other heavy metal.*

REFERENCES

1. "Recrystallization of Cold-Rolled Uranium", by E. E. Hayes: U. S. Atomic Energy Commission Report TID-2501, 1949
2. Recrystallization and Grain Growth in Uranium, by E. S. Fisher: in Reactor Technology and Chemical Processing, Vol 9 of *Proceedings of the International Conference on the Peaceful Uses of Atomic Energy*, United Nations, 1956
3. The Effect of Aging on the Mechanical Behaviors of U-0.75 wt. % Ti and U-2.0 wt. % Mo, by K. H. Eckelmeyer and F. J. Zanner: *Journal of Nuclear Materials*, Vol 62, No. 1, Oct 1976, p 37-49

SELECTED REFERENCES

- *Nuclear Reactor Fuel Elements: Metallurgy and Fabrication*, edited by A. R. Kaufmann: Interscience Publishers, a division of John Wiley and Sons, New York, 1962.
- *Physical Metallurgy of Uranium Alloys*, edited by J. J. Burke *et al*: Brook Hill Publishing Co., Chestnut Hill, MA, 1976

Fig. 11 Fixture for solution treating and end quenching of 18 depleted uranium bars

Heat Treating of Zirconium and Its Alloys

By R. Terrence Webster
Principal Metallurgical Engineer
Teledyne Wah Chang

The major application of zirconium is in the fuel cladding for water-cooled nuclear reactors and in ancillary reactor core parts such as water channels and fuel rod spacers. The major alloys for nuclear applications are the Zircaloys, which are a series of zirconium, tin, iron, chromium and nickel alloys. Another alloy used is zirconium-niobium.

Zirconium, zirconium-tin alloys and zirconium-niobium alloys are also used in corrosion resistant equipment in the chemical processing industries and in energy related applications.

Annealing Temperatures

For most of the zirconium alloys in industrial applications, strength cannot be increased by heat treatment. Consequently, the only heat treatments required are full annealing and stress-relief annealing.

Zirconium-niobium alloys can be heat treated to higher strength, but they normally are used in the annealed condition. The heat treatments for zirconium and zirconium alloys are as follows:

- Full recrystallization anneal—700 to 785 °C (1290 to 1445 °F) for 1 h per inch of thickness
- Stress relief anneal—480 to 595 °C (895 to 1100 °F) for 1 h per inch of thickness.

To avoid excessive grain growth in highly cold worked thin sheet and foil, full annealing at 595 °C (1100 °F) for 6 h may be desirable.

Processing

The nuclear reactor-grade alloys are annealed in vacuum (at pressures from 10^{-2} to 10^{-5} torr) or in an inert-gas atmosphere of argon or helium. This is done to prevent formation of an oxide layer on the surface, which can be removed only by mechanical means such as abrasion or machining.

The commercial alloys can be annealed in air because the oxide layer is fully corrosion resistant and need not be removed. Prior to heat treating, the metal surfaces should be free of oxide scale, soil and any foreign substance that can be absorbed into the metal and cause brittleness. Heavy oxides and scale must be removed by machining, abrasion or grinding. Light oxides can be removed by acid pickling in a hydrofluoric acid-nitric acid solution. Soil, grease and oils can be removed with solvents such as trichloroethylene or acetone. ANSI/ASTM Standard B614 can be used as a guide for descaling and cleaning of zirconium and zirconium alloys.

Heat Treating of Tantalum and Niobium and Their Alloys

By the ASM Committee on Heat Treating of Tantalum and Niobium and Their Alloys*

At the present time, tantalum is primarily used as a capacitor material for electronic hardware and niobium as an alloying addition to other materials. However, both are finding increased use by themselves and as base elements for alloys used in aerospace and in applications where resistance to chemical attack is important. Niobium-base alloys are widely used as superconductors.

Tantalum and niobium are sister metals, not only in the sense that they are often found together in nature, but also because their like chemical and metallurgical properties require similar procedures for melting, fabrication and heat treating. Both are chemically active, ductile metals that are readily embrittled by pickup of a few hundred parts per million of O_2, N_2, C or H_2. Surface oxidation of both metals occurs in air above 300 °C (570 °F), and the oxidation rate increases with increasing temperature. Hydrogen embrittlement occurs if either metal is cathodic in a galvanic couple or is exposed to a hydrogen atmosphere during cooling from elevated temperature.

To prevent pickup of these four interstitial elements during heat treating, it

*Robert E. Droegkamp, Manager, Research and Development, Fansteel Metals, Inc.; Louis (Ned) E. Huber, Jr., Kawecki-Berylco Industries, Inc., Division of Cabot Corp.; R. Terrence Webster, Principal Metallurgical Engineer, Teledyne Wah Chang.

is essential to use appropriate surface-cleaning procedures, furnace maintenance and operational practices.

Both metals first can be purified by electron-beam melting in a vacuum of 10^{-5} torr or better. Alloying can be done by either electron-beam or vacuum-arc melting, the latter method being chosen for better control when alloying elements tend to be volatile.

Mill products also are made from powders by resistance or induction sintering of cold isostatically compacted bars in a vacuum of 10^{-5} torr.

Conventional fabrication procedures are used; if heavy sections are heated in air, all surface contamination must be removed by machining or grinding and pickling before annealing.

It is strongly suggested that contact be made with the primary fabricated metal supplier for discussion and review of plans and procedures prior to any thermal treatment of tantalum or niobium or their alloys. Numerous expensive failures have resulted from attempts to perform thermal treatment attempts in air, in atmospheres with inadequate inert gas protection or in inadequate vacuum systems.

Cleaning Procedures

To avoid contamination of tantalum and columbium by interstitial elements and metallic contaminants, it is mandatory that the material be chemically clean before it is subjected to any heating operation such as annealing or welding. Cleaning and degreasing present no special problems (conventional methods and materials may be used, although hot caustics must be avoided). First, thorough degreasing is carried out using a detergent or solvent. Degreasing is followed by chemical etching, typically with a mixture 60 HNO_3, 20 HF and 20 H_2SO_4 (volume percentages); hot and cold water rinses in distilled water; and spot-free drying. The etching solution may be strengthened by HF additions, or weakened by water additions, to achieve the amount of stock removal necessary to ensure cleanness of the metal surface. One company eliminates H_2SO_4 because some evidence indicates that it can contribute to weld embrittlement. Nitric acid should always be present, however, because it prevents hydrogen pickup during pickling.

Elevated-temperature forgings will have an oxygen-contaminated outer layer, and this must be removed from all surfaces by machining or grinding before acid pickling.

Annealing Practice

Tantalum and niobium and their alloys are most often used in the fully recrystallized condition to achieve the best fabrication response, although some applications require stress relieved or cold worked properties. The recrystallization temperature is so highly dependent on purity, amount of cold work and prior history, that current practice is to anneal pilot samples to ensure that the correct temperatures are used. Time at temperature is typically 1 h.

Table 9 can be used as a guide for choosing pilot temperatures. Materials given heavy fabrication reductions will recrystallize to finer grain sizes at lower temperatures than will those given lighter fabrication reductions. The recrystallization annealing temperature is also somewhat dependent on interstitial purity. For example, pure tantalum containing 200 ppm oxygen requires a higher recrystallization annealing temperature than pure tantalum containing less than 50 ppm oxygen.

A typical annealing sequence is:

- Visually verify material cleanness
- Load, using tantalum, tantalum alloy or molybdenum fixtures for support, or tantalum foil for protection, as required. Tantalum, niobium and certain of their alloys exhibit low yield strengths at the necessary annealing temperatures. Sticking to fixtures and other pieces can cause damage, especially with tantalum because of its high density.
- Pump down
- Leak rate check
- Power on to temperature
- Time as required at temperature
- Power off
- When temperature drops below 1000 °C (1830 °F), backfill to 15 mm Hg with industrial high-purity (99.995% min) argon or helium.
- Before removing load from furnace, allow to cool to below 200 °C (390 °F), which can require from 3 to 15 h depending on furnace size and mass of load.

This sequence is not intended to be a detailed procedure for annealing these materials. This sequence information is intended to create awareness of the difficulties and risks of heat treating these materials to avoid repetition of past costly errors. The major risk is loss of vacuum at temperature resulting in the extremely costly destruction by oxidation not only of the parts being heat treated but also of the shielding and resistance elements.

Furnaces

Tantalum and niobium and their alloys are easily contaminated during

Table 9 Annealing temperatures for tantalum and niobium and their commercial alloys

Alloy designation	Nominal alloy additions, %	Annealing temperature			
		Stress relief		Recrystallization	
		°C	°F	°C	°F
Tantalum alloys					
Ta	None	850	1560	1000–1250	1830–2280
Ta	None(a)	1000	1830	1200–1350	2190–2460
FS63	2.5W, 0.15Nb	1000	1830	1200–1300	2190–2370
FS61	7.5W(a)	...	...	1400–1550	2550–2820
FS60	10W	1100	2010	1300–1600	2370–2910
T111	8W, 2Hf	1100	2010	1400–1650	2550–3000
T222	9W, 2.4Hf, 0.01C	1100	2010	1400–1650	2550–3000
Niobium alloys					
Nb	None	800	1470	900–1200	1650–2190
FS80	1Zr	875–1150	1610–2100	1150–1250	2100–2280
SNb 291	10Ta, 10W	1000	1830	1150–1200	2100–2190
Nb 752	10W, 2.5Zr	...	...	1300–1400	2370–2550
C 129Y	10W, 10Hf, 0.1Y	900	1650	1150–1250	2100–2280
FS85	28Ta, 11W, 0.8Zr	1150	2100	1300–1400	2370–2550
C103	10Hf, 1Ti, 0.7Zr	...	...	1250–1375	2280–2510

(a) Powder metallurgy; all other compositions are vacuum melted.

annealing, and special care must be exercised in furnace selection, cleanness of work and annealing practice used. Normally used are cold-wall radiant-heated furnaces with refractory metal heater elements, primary heat shields, permanent hearth materials and support fixtures. These furnaces operate at vacuums of 10^{-4} torr or better and have an acceptably low leak rates.

Vacuum Leak Control. Leak-rate control tends to be the key to successful heat treating of tantalum and niobium alloys, especially with products having high surface-to-volume ratios, such as low gage wire, tube and strip. A suggested maximum leak rate is 0.1 μm/ min. This expression includes chamber volume and pressure rise per unit of time, which is the only meaningful measurement.

A leak rate is only meaningful in a stabilized system—that is, one that has pumped for a period of time and is no longer outgassing. It is defined as the difference in pressure (torr) for one second in a volume of one litre.

For example, a 1200-litre vacuum chamber is isolated by closing the high-vacuum valve. The pressure rise is observed and an increment of rise is timed by stopwatch. The chamber in this example rose 2×10^{-3} torr in 5 min. The rate is calculated as:

$$\frac{Pressure\ change \times chamber\ volume}{Time} =$$

$$\frac{2 \times 10^{-3}}{300} \times 1200 =$$

$$0.8 \times 10^{-2}\ torr\ l/s$$

where pressure change is measured in torr; chamber volume in litres; and time in seconds. This rate is acceptable within the suggested limit of 10^{-2} torr l/s.

Hot-wall argon-atmosphere furnaces have been used, but adsorbed gases and metals on hot furnace walls are more likely to cause contamination. Argon must be free of H_2 and have a dew point below -50 °C (-60 °F).

Furnaces must be clean and usually must not be used for other operations or other metals unless a given practice has been found to be satisfactory. Furnaces previously used to perform brazing operations should be avoided.

Good practice dictates that furnaces be heated to a temperature 100 °C (180 °F) above the annealing temperature in the empty condition, to remove adsorbed gases.

Furnace qualification is often a customer requirement and usually includes a limit on the amount of allowable contamination as measured by increased hardness or interstitial content.

The trend toward bonding tantalum and niobium to other metals presents special problems that must be carefully reviewed from a metallurgical standpoint. The recommendations is Table 9 probably do not apply to most clad materials.

SELECTED REFERENCES

- *Tantalum and Niobium, Metallurgy of the Rarer Metals -6,* by G. L. Miller: Butterworths Scientific Publications, London, 1959
- *Columbium and Tantalum,* by Frank T. Sisco and Edward Epremian: John Wiley & Sons, Inc., London, 1963
- "The Engineering Properties of Tantalum and Tantalum Alloys", by F. F. Schmidt and H. R. Ogden: DMIC Report 189, Sept 13, 1963
- "The Engineering Properties of Columbium and Columbium Alloys", by F. F. Schmidt and H. R. Ogden: DMIC Report 188, Sept 6, 1963
- *Metals Handbook,* Vol 2, 8th ed., "Heat Treating, Cleaning and Finishing", 1979, p 269–270
- *Metals Handbook,* Vol 3, 9th ed., "Properties and Selection: Stainless Steels, Tool Materials and Special-Purpose Metals", 1980, p 321–325
- Postheating of Cb-1Zr and T111 (Ta + 8W + 2Hf) Weldments: NASA-Lewis Specification No. RM−5, June 1971
- Interactions of Refractory Metals with Active Gases in Vacua and Inert Gas Environments, by H. Inoye: in *Refractory Metals Alloys, Metallurgy and Technology,* edited by I. Machlin, R. T. Begley and E. D. Weisert, p 165

Heat Processing of Powder Metallurgy Parts

Sintering

By the ASM Committee on Sintering*

SINTERING is the process by which loose or compressed powders are bonded by heating at temperatures below the melting points of the major constituents. Densification may or may not occur. If powders of two or more different metals are heated together to a sufficiently high temperature, alloying may take place simultaneously with sintering. Sometimes a liquid phase forms and assists in consolidation, or a compact may be sintered for a short time and then infiltrated with a molten metal of lower melting point.

The processes operative in sintering include vapor and/or liquid transport, diffusion, and plastic flow. The predominant process is diffusion. A preform, usually a "green" compact, becomes a sintered part in a series of continuous stages. As the temperature increases, the interparticle contacts increase in size and strength. As diffusion progresses, neck formation occurs, which causes the pores to become rounded. With continued sintering, densification increases as a result of volume diffusion. Ultimately, pores coalesce. The total sintering process causes strength, ductility, and thermal and electrical conductivity to increase. With increased temperature, approaching the melting point, porosity becomes isolated and theoretical density is approached asymptotically.

A common production method used to achieve higher densities is re-pressing and resintering. Re-pressing closes up the larger pores mechanically, and fresh bonds are formed during resintering. The improvement obtainable by these operations is illustrated by the following example.

Example. Re-pressing and Resintering for Increased Density (Fig. 1). Densities of compacts 25 mm (1 in.) in diameter and 25 mm high were measured after pressing at pressures varying from 4.2 to 8.4 tonnes/cm^2 (1 tonne = 1 Mg, or 1 metric ton), or 30 to 60 tons/in.^{2}, and sintering for 1 h in dissociated ammonia at 1120 °C (2050 °F). The relation of compacting pressure and density, after sintering, is shown graphically in Fig. 1. Also shown in Fig. 1 is the increase in density that is effected by re-pressing at 7.0 tonnes/cm^2 (50 tons/in.^{2}) and resintering for 1 h in dissociated ammonia at 1120 °C (2050 °F).

Recent technology has been developed that utilizes high-temperature sintering to achieve densification as well. If near-theoretical density is required, other processes, such as hot pressing or re-pressing, hot isostatic pressing, hot forging and super solidus sintering, are employed.

Sintering Furnaces

The burn-off chamber of a sintering furnace used for sintering ferrous preforms is usually controlled to heat the preforms to temperatures from 500 to 800 °C (930 to 1470 °F). It is important that all lubricants, including zinc stearate, stearic acid and waxes, be volatilized and expelled from the furnace before the preforms enter the high-temperature section, and both the flow of gas and the time of heating should be sufficient to ensure that this is done. If lubricants pass into the sintering section, they will be decomposed, and the liberated products may adversely affect the sintering process, the parts and the furnace.

The sintering section of the furnace may be refractory lined or fully muffled. In small furnaces it is cheaper to use a muffle to obtain gas tightness. In larger furnaces it is more economical to make the shell gastight and eliminate the muffle. Production sintering furnaces are supplied throughout with a protective atmosphere and are divided into: (a) a burn-off section, which serves

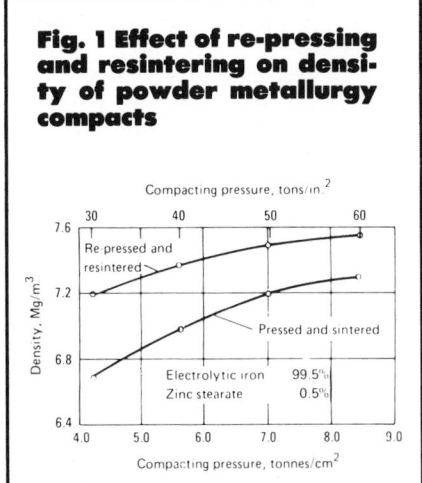

Fig. 1 Effect of re-pressing and resintering on density of powder metallurgy compacts

*Kenneth H. Moyer, *Chairman,* Product Development Engineer, Hoeganaes Corp.; J. Howard Beck, President, BTU Engineering Corp.; A. J. Craig, Jr., Chief Metallurgist, Homelite; Donald Dyke, Engineering Manager, Sintered Specialties; Donald Grendon, Sales Engineer, Drever Co.; Erhard Klar, Manager Research and Development, Glidden Metals, SCM Corp.; George Otto, Maytag Co.; Thomas Sibley, Manager, P/M Applications, Air Products & Chemicals, Inc.; Sang-Kee Suh, Senior Research Engineer, Ford Motor Co.

also for preheating; (b) a high-temperature heating section; and (c) a cooling section. These sections have typical length ratios of 1:1:2 or 1:1:3.

The high-temperature heating section must be long enough to allow sufficient time for the preforms to heat up to temperature and enough soak time at temperature for adequate sintering. Multiple-control zones are used to obtain suitable temperature gradients.

Large sintering furnaces usually are constructed with a gastight shell and electrical-resistance heating elements or gas-fired radiant tubes exposed to the heating chambers. Muffles generally are not used in larger furnaces because they are expensive to purchase and to maintain, and also introduce thermal losses. Muffle construction, however, is widely used when the dew point of the atmosphere must be kept below about -40 °C (-40 °F). By using high-purity alumina refractories, muffle-free furnaces with exposed molybdenum heating elements have been operated successfully at low dew points. Full muffle furnaces have one further advantage: they purge faster, because there is no porous brickwork in the chamber and because less purge gas is required.

The cooling section often begins with a short, insulated zone in which the preforms cool slowly enough to avoid thermal shock and to allow for carbon restoration. This is followed by a cooling section that may be a long, water-jacketed extension or a shorter, forced-convection cooling system. Automatic control of the temperature of the cooling water is most desirable. If the temperature of the walls should fall below the dew point of the protective gas, condensed water may collect on the workpieces and cause staining. If cooling time is insufficient, the sintered parts will oxidize when they emerge into the air.

Preforms commonly are conveyed through the furnace by mesh belts, roller hearths, pusher mechanisms or walking beams.

Mesh-belt conveyors typically handle a nominal loading of 480 Pa (10 lb/ft²) at 1120 °C (2050 °F). Stretching of the alloy belt limits the length of the furnace and the size of the furnace load. It is desirable to keep the temperature below 1150 °C (2100 °F) when using a mesh belt. Because each end of this type of furnace usually remains open during operation, consumption of the protective gas is high and ample gas

capacity must be provided. Flame curtains or nitrogen baffles may be used to prevent oxygen intrusion. A variable-speed drive gives flexibility in adjusting time and temperature cycles.

Roller-hearth furnaces are similar in arrangement to mesh-belt furnaces except that, in place of the belt, a series of driven rollers is fitted along the entire length of the hearth. These rollers are spaced to support and carry trays loaded with preforms. Maximum operating temperature is limited by the properties of the alloy rolls and usually is about 1120 °C (2100 °F). Depending on roll spacing, loads four to seven times greater than can be handled by a mesh-belt furnace of equal length can be conveyed on a roller hearth. Furnace doors are provided and are opened only to charge or discharge the trays. Thus, consumption of protective gas is less, and heat losses are lower, than for mesh-belt furnaces, because the atmosphere flow may be diminished.

Pusher-type furnaces are suitable for sintering of preforms that are too heavy to be carried by a mesh belt or that require sintering temperatures greater than 1150 °C (2100 °F). With this type of equipment, preforms are fed into the furnace on trays, which are advanced by mechanical pushers. Alternatively, for small batches, the trays may be pushed through manually.

Walking-beam furnaces are used for high-temperature sintering. These furnaces can move heavy loads at high temperatures. The moving mechanism is capable of four basic motions: up, forward, down and reverse. The up motion lifts the tray from the shelves; the forward motion advances the trays; the lowering motion unloads the trays from the beam, and the reverse motion returns the empty beam to the original position. Normal operation involves temperatures up to 1400 °C (2550 °F). Doors at both ends are closed except during loading and unloading, thus reducing gas consumption.

Vacuum furnaces are also used for high-temperature sintering. They may be mechanized to provide continuous production of parts. Parts are protected by the absence of reactive gases, and furnaces may be backfilled with an inert or protective gas. Burn-off generally is accomplished in a separate conventional furnace. Heating is done by means of resistance heating elements. Cooling can be accelerated by backfilling with a nonreactive gas.

Sintering Atmospheres

Protective atmospheres are used in powder metallurgy (a) to prevent oxidation and reduce oxides, (b) to control carbon contents of iron and iron alloy preforms, and (c) to flush volatilized lubricants from the furnace.

Oxidation and decarburization of iron preforms are caused by oxygen, water vapor and carbon dioxide when present in excess proportions with respect to hydrogen and carbon monoxide contents. Iron oxides are reduced by hydrogen, carbon monoxide and carbon. Carburization is caused by carbon monoxide and by hydrocarbons such as methane.

Copper and bronze preforms are susceptible to general oxidation and to scaling or discoloration by oxygen. These compacts are not adversely affected by hydrogen, carbon monoxide or carbon. Selective attack on zinc in brass compacts is caused by carbon dioxide, oxygen, sulfur and water vapor.

Vacuum is used mainly for sintering preforms of stainless steels and tool steels; soft magnetic materials; and refractory metals such as tantalum, titanium, zirconium and uranium—all of which react with most of the usual protective gases, including hydrogen. Vacuum is also being used to an increasing extent for sintering of conventional ferrous materials at high temperatures.

When the moisture content of any atmosphere must be kept very low (as in sintering of alloys containing chromium), the furnace must be operated and maintained with special care, to eliminate all leakage or back-diffusion of air that would contaminate the furnace atmosphere. One factor is often overlooked; the dew point of the gas fed into the furnace may be different from the moisture content of the gas in contact with the workpieces, owing to higher oxygen content within the porous compacts. Under vacuum, care must be exercised to prevent the pressure from dropping below the vapor pressure of the alloy constituents, which can result in depletion.

The atmospheres most commonly used for sintering are: hydrogen, dissociated ammonia, nitrogen-base exothermic gas, purified rich exothermic gas, endothermic gas, and vacuum. Each is discussed individually in the sections that follow.

Table 1 Characteristics of sintering atmospheres for powder metallurgy products

Atmosphere	Typical dew point, °C	Al	Cu	Brass	Bronze	Ni	Ag	Mo	W	Fe	Fe-Cu	Fe-C	Fe-Cu-C	Carbon steel	Stainless steel	Relative cost per unit volume(b)
Hydrogen:																
Liquid	−75	R	R	R	R	R	R	Y	Y	R	R	...	...	...	Y	9–20
Bulk gas	−70	R	R	R	R	R	R	Y	Y	R	R	...	...	...	Y	20–35
Steam-methane	−40 to −50	R	R	R	R	R	R	Y	Y	R	R	...	...	...	Y	9–14
Nitrogen base, with:																
Endothermic enrichment	−20 to −10	X	R	R	R	R	R	...	...	C2	C2	C2	C2	C2	...	1.5–6(e)
Hydrogen enrichment	−70	Y(d)	R	R	R	R	R	R	R	N	N	N	N	N	R	1.7–7(e)
Methanol enrichment	−20 to −10	X	R	R	R	R	R	...	...	C2	C2	C2	C2	C2	...	1.6–6.5(e)
Ammonia-base:																
Dissociated NH₃	−40 to −50	R	R	R	R	R	R	R	R	R	N	N	...	N	R	3.3–7.2
Burned NH₃, rich	+20 to +30(c)	R	R	...	R	R	R	...	...	D3	D3	...	...	...	...	2.3–5.1
Exothermic gas:																
Rich, saturated	+20 to +30	X	R	...	R	R	R	...	...	D3	D3	...	...	...	...	1
Medium rich, saturated	+20 to +30	X	...	...	R	R	R	...	...	...	...	...	...	...	...	0.9
Purified exothermic gas:																
Rich	−40	R	R	R	R	R	R	...	...	C1	C1	C1	C1	C1	...	1.5–2.2
Medium rich	−40	R	R	R	R	R	R	...	...	C1	C1	C1	C1	C1	...	1.5–2.2
Endothermic gas:																
Rich, dry	−20 to −10	X	R	R	R	R	R	...	...	C3	C3	C3	C3	C3	...	1.6–3.2
Fairly rich, dry	−5 to 0	X	R	R	R	R	R	...	...	C2	C2	C2	C2	C2	...	1.5–3.1
Medium rich, saturated	+20 to +30	X	R	...	R	R	R	...	...	D1	...	...	...	...	...	1.5–3
Lean, saturated	−20 to −30	X	R	...	R	R	R	...	...	D3	D3	...	...	...	...	1.5–2.5

(a) R, reducing; Y, recommended; C1, mildly carburizing; C2, carburizing; C3, strongly carburizing; N, neither carburizing nor decarburizing; X, not recommended; D1, mildly decarburized; D2, decarburizing; D3, strongly decarburizing. (b) Costs are approximate and relative to rich, saturated exothermic gas. (c) Dew point may be reduced by refrigeration or by absorbent-tower dehydration. (d) Nitrogen with no enriching gas is recommended. (e) Price range includes on-site plant production or liquid tanker deliveries.

The characteristics and applications of gas atmospheres for sintering are summarized in Table 1.

Hydrogen provided as a liquid or in bulk gaseous trailers is the most economical source for most sintering facilities. For volumes larger than 170 000 to 340 000 m³ (6 to 12 million ft³) per month, hydrogen may be produced at lower cost by steam methane reforming.

The explosiveness of hydrogen-oxygen mixtures demands that hydrogen be handled with extreme care. Its high thermal conductivity (seven times that of air) helps in increasing the rate of heat transfer in the heating and cooling chambers. Its low density causes it to diffuse rapidly outward and can allow back-diffusion of air through small openings and cracks, which may result in contamination of the atmosphere.

Hydrogen delivered as a liquid typically contains less than 0.001% impurities, with an oxygen content of less than 0.0002%. Bulk gaseous hydrogen usually has less than 0.05% impurities and less than 0.0005% oxygen. This high purity makes hydrogen suitable for sintering of stainless steels, carbides, tungsten and other refractory metals, magnetic materials, superalloys and other metals in which high reducing potential and no nitriding is desired.

Dissociated Ammonia. Cracked or dissociated ammonia is made by passing ammonia gas (from large cylinders or tanks) over a heated catalyst, and consists of a mixture of 75% hydrogen and 25% nitrogen by volume. It is dry and can be used as a substitute for pure hydrogen in nearly all sintering applications, including stainless steel, iron, brass, copper and tungsten. Its use, particularly for sintering molybdenum and ferrous materials, is sometimes avoided because of the danger of nitriding by traces of undissociated ammonia, which are nearly always present and which dissociate on contact with hot metal. This residual ammonia may be removed almost completely by passing the gas through water (and subsequently drying it), through activated alumina or through a molecular sieve.

Exothermic gas is made by burning natural gas, propane or a similar hydrocarbon gas in a refractory-lined combustion chamber with controlled amounts of air.

Rich or medium-rich exothermic gases are the atmospheres most commonly used for sintering. The richest gas is formed from a 6-to-1 ratio of air to natural gas and contains about 14% hydrogen, 10% carbon monoxide, 1% methane, 5% carbon dioxide and 70% nitrogen. It has a dew point approximately 5 °C (9 °F) above that of the cooling water and is useful for sintering preforms of copper, bronze, silver, iron, and iron copper. It usually is strongly decarburizing at the sintering temperature to ferrous metals. A medium-rich gas, made by reacting air with natural gas in a ratio of 6.75 to 1, has been used for sintering nonferrous compacts. The principal advantages of exothermic gas are low flammability and low cost (it is the most economical gas available). Exothermic gas also may be usefully employed where removal of lubricant is desired.

Like all gases derived from hydrocarbons, exothermic gas may have a carburizing or a decarburizing effect on ferrous parts, depending on the carbon content of the work and on temperature.

Purified Rich Exothermic Gas. By removal of water and carbon dioxide from exothermic gas, a stable and mildly reactive gas with a dew point below −45 °C (−50 °F) may be produced. This gas is called purified rich exothermic gas. It has virtually no carburizing or decarburizing effect on iron-graphite compacts. Purified rich exothermic gas is used for sintering preforms of iron, iron-copper, iron-carbon and iron-copper-carbon; and also for copper infiltration of iron or iron-carbon compacts.

Endothermic gas is made in an externally heated catalytic chamber by reacting a hydrocarbon with an amount of air insufficient to support combustion. The ratio of air to natural gas can be varied from 4.5:1 to 2.4:1, depending on the desired carbon potential of the atmosphere. The most common air:gas ratio is approximately 2.4:1, which produces gas of the following composition: 20% CO, 38% H_2, traces of CO_2 and CH_4, rem N_2. Dew point for practical operation is between −5 and +5 °C (23 and 41 °F). Traditionally, endothermic generators have provided low-cost carbon-controlled atmospheres. Depending on the air:gas ratio, endothermic gas can be either carburizing or decarburizing; therefore, it is used to sinter ferrous-base preforms and most nonferrous preforms.

Nitrogen-base atmospheres are made by mixing cryogenically produced nitrogen with one of several enriching gases. Liquid storage vessels are the most economical source for most sintering facilities. For larger requirements, on-site plants provide gaseous nitrogen at lower cost.

The atmosphere generally consists of 75 to 95% nitrogen, so that advantage may be taken of nitrogen's low dew point. Gaseous hydrogen, dissociated ammonia, endothermic gas or methanol is used for enrichment. Hydrogen and dissociated ammonia provide reducing atmospheres with neutral carbon potentials. Endothermic gas and methanol produce atmospheres with medium carbon potentials. Methanol is provided as a liquid that dissociates in the furnace to form essentially carbon monoxide and hydrogen. Hydrocarbons can be added to any of the atmospheres to increase carbon potential. With the proper enrichment gas, nitrogen-base atmospheres can be used for sintering and infiltrating iron, iron-carbon, iron-copper and iron-copper-carbon compacts, and for sintering nonferrous compacts. Stainless steels and refractory metals can be sintered when nitriding is not critical. For some applications, such as soft magnetic materials or highly corrosion-resistant stainless steels and some refractory metals, ni-

Table 2 Typical sintering temperatures and holding times for various metals and prealloyed powders

Material	Temperature °C	°F	Time, min
Aluminum	570–650	1060–1200	5–25
Bronze	760–870	1400–1600	10–20
Brass	760–940	1400–1725	10–45
Copper	900–1010	1650–1850	12–45
Iron, iron alloys	1010–1425	1850–2600	8–45
Nickel	1010–1150	1850–2100	30–45
Stainless steel	1095–1315	2000–2400	30–60
Alnico magnets	1205–1300	2200–2375	120–150
Ferrites	1205–1480	2200–2700	10–30
90% tungsten, 6% nickel, 4% copper	1345–1595	2450–2900	12–20
Tungsten carbide	1425–1480	2600–2700	20–30
Molybdenum	2050	3730	120 approx
Tungsten	2345	4250	480 approx
Tantalum	2400	4350	480 approx

trogen absorption occurs and should be avoided. The principal advantages of nitrogen-base atmospheres are safety and elimination of generating equipment.

Vacuum is finding increasing application as an environment for protection of materials that are reactive at elevated temperatures. It is beneficial for sintering soft magnetic preforms, stainless steels, tantalum, titanium, uranium, zirconium, refractory metals, and superalloys that react with hydrogen, carbon or nitrogen. Care must be taken to prevent alloy depletion, which can occur if the vapor pressure of an element is higher than the pressure in the vacuum vessel. Further, backfilling with reactive gases for cooling purposes can cause detrimental reactions.

Sintering Practice

Time and temperature cycles must be carefully chosen to develop the properties required in sintered preforms. The major causes of low density or low strength are insufficient temperature or insufficient time at temperature. Of the two, low temperature usually is the more critical condition. When the preform is composed of a mixture of two or more constituents, longer time at the sintering temperature is required for diffusion than is required for sintering a similar prealloyed preform.

Invariably, the optimum values of time and temperature for sintering any given compact are determined empirically. Within practical limits, higher sintering temperatures result in better

properties and are more effective than longer times. The limits of temperature are defined by the available equipment. Temperature is also limited by the increased degree of difficulty of controlling dimensions at higher temperatures. As sintering is prolonged, the measurable increase in density continues at a lower rate, until a point is reached beyond which the diminishing returns do not justify the expense of further sintering. The ranges of time and temperatures most widely used are given in Table 2.

Alloying in Iron-Base Compacts. Carbon is the most common alloying addition to iron. It usually is added, as graphite, to the original blend of powders so that alloying will occur during sintering, but it may be added from a carburizing furnace atmosphere. The reactions that occur between iron and graphite and among iron, graphite and the furnace atmosphere are fairly rapid at the sintering temperature. For this reason, it is necessary that the atmosphere be maintained at the desired carbon potential or that restoration of the carbon potential be made prior to cooling. Carbon gradients may result from improper control of carbon from the atmosphere. Final carbon content is determined by the amount of graphite added, by the carbon potential maintained and by control of restoration.

Copper often is alloyed with iron in the sintering furnace. It has a solid solubility in iron of 8 to 9% at normal sintering temperatures. As cooling occurs, some precipitation hardening is likely, depending on cooling rate. Additions of copper cause compacts to expand, which often makes it more difficult to control dimensions. For consistent compact size, it is necessary to maintain close control over raw materials and over sintering temperature, atmosphere and time.

Carbon is also alloyed with iron copper preforms. The solubility of copper in iron is lowered by the addition of carbon, and the presence of copper decreases the sensitivity of compacts to variations in the sintering atmosphere. Iron-copper-carbon alloys have good mechanical properties, and dimensions of compacts made from these materials are generally easier to control than are those of similar compacts made from iron and copper.

To obtain increased strength and density, postsintering operations such as re-pressing and resintering may be employed. When the part is to be re-pressed and resintered, it generally is not given a thorough sintering until the last pass through the furnace. The initial pass (or passes, when resintering is done more than once) is done mainly to anneal the compact and burn off the lubricant. It is desirable that alloying be kept to a minimum during initial sintering so that the compact will not be strengthened sufficiently to resist compaction in subsequent re-pressing operations. As the density of the compact approaches that of a solid material and interconnecting porosity disappears, the reaction of the compact to the atmosphere also approaches that of solid material.

Prealloyed powders provide a means of obtaining preforms with structures more uniform than those of preforms produced by diffusing particles of the constituent metals. Additional alloying may be achieved through mixing of elemental or other

Table 3 Nominal compositions and sintering temperatures for some prealloyed metal powders

| Composition | Sintering temperature | | | |
| | Hydrogen-base atmosphere | | Nitrogen-base atmosphere | |
	°C	°F	°C	°F
Copper-base alloys				
90 Cu, 0.5 P, 9.5 Zn	940	1725	. . .	. . .
90 Cu, 1.5 Pb, 8.5 Zn	900	1650	900	1650
90 Cu, 10 Zn	900	1650	900	1650
85 Cu, 15 Zn	900	1650	900	1650
78.5 Cu, 1.5 Pb, 0.3 P, 19.7 Zn	900	1650	. . .	. . .
78.5 Cu, 1.5 Pb, 20 Zn	900	1650	900	1650
70 Cu, 0.3 P, 29.7 Zn	870	1600	. . .	. . .
70 Cu, 30 Zn	900	1650	870	1600
60 Cu, 40 Zn	. . .	. . .	815	1500
64 Cu, 18 Ni, 18 Zn	980	1800	980	1800
64 Cu, 1.5 Pb, 18 Ni, 16.5 Zn	870	1600	940	1725
70 Cu, 10 Ni, 20 Zn	940	1725	. . .	. . .
Iron-base alloys				
17 Cr, 12 Ni, 2.5 Mo, rem Fe	1210	2210	. . .	. . .
12.5 Cr, rem Fe	1210	2210	. . .	. . .
50 Fe, 50 Ni	1175	2150	. . .	. . .
97 Fe, 3 Si	1260	2300	. . .	. . .
1.8 Ni, 0.5 Mo, add 0.6 graphite, rem Fe	1135	2075	. . .	. . .
1.75 Ni, 1.5 Cu, 0.5 Mo, rem Fe	1110	2030	. . .	. . .
0.4 Ni, 0.7 Mo, 2 Cu, rem Fe	1120	2050	. . .	. . .
1 Mo, 0.4 Mn, rem Fe	1120	2050	. . .	. . .

intermetallic powders to achieve more complex alloys. Table 3 lists nominal compositions of some prealloyed metal powders, and gives sintering temperatures for these materials.

Liquid-Phase Sintering. The oldest and most important example of liquid-phase sintering is production of 90Cu-10Sn bronze for self-lubricated bearings. At the sintering temperature, the elemental tin is liquid, which allows lower sintering temperatures (815 to 895 °C; 1500 to 1640 °F), and shorter sintering times (5 to 10 min) than those required for prealloyed

bronze. Powder producers supply lubricated bronze premixes with graded particle size distributions, which allow better control of dimensional change (growth) during sintering. The self-lubricated bearings are sized after sintering for further control of dimensions.

Liquid-phase sintering may be successfully employed for other alloys as well. These include iron-copper, iron-phosphorus, aluminum, iron-sulfur, tungsten carbide–nickel, tungsten carbide–cobalt and tungsten carbide–iron alloys.

Heat Treating of Ferrous Powder Metallurgy Parts

By Howard Boyer
Consultant

HEAT TREATING procedures used for powder metallurgy parts are theoretically the same as those used for wrought or cast parts of the same composition; however, some practical limitations must be considered in dealing with iron-base alloy (steel) powder metallurgy parts. If a powder compact has the same composition as its wrought or cast counterpart, it should likewise respond in the same manner to heat treatment. The principles are the same—that is, the maximum hardness that can be achieved is controlled by the content of carbon that is in solution at the austenitizing temperature, not necessarily by the total carbon content. In powder metallurgy, because alloys are produced by mixing various powders and then sintering the mixture, some of the carbon can be incorporated in a graphitic state and can remain so at the austenitizing temperature. Because graphitic carbon does not contribute to hardening, combined carbon must be known before an optimum heat treating procedure can be selected.

Furthermore, as with wrought or cast parts, the hardenabilities of powder metallurgy parts are controlled by the amounts of alloying elements they contain. It is therefore essential to know the complete composition before attempting to perform any heat treatment.

For example, if a specific powder metallurgy part is composed essentially of iron with a combined carbon content of 0.5% and if section thickness does not exceed about 4.8 mm ($^3/_{16}$ in.), full hardness should be obtained by austenitizing at about 870 °C (1600 °F) and quenching in oil.

For sections greater than about 4.8 mm, an aqueous (water-brine or synthetic polymer) quench will probably be necessary to attain full hardness. This quenching treatment may induce cracking, because powder metallurgy parts always contain some voids and thus are more likely to crack than their wrought or cast counterparts. Therefore, oil quenching is always preferred for powder metallurgy parts. Thicker sections should contain sufficient alloying elements such as manganese and/or chromium to provide the required hardenability.

Heating Media. Powder metallurgy parts for through hardening normally should be heated in a gaseous atmosphere, such as the type provided by an endothermic generator, using a carbon potential equal to that of the combined carbon in the compact. Heating of powder metallurgy parts in a molten salt bath is not recommended even for those having maximum density, because they are extremely difficult to wash free of salt. If the salt remains in the voids, the parts will corrode.

Tempering. Austenitized and quenched powder metallurgy parts should be tempered just as are their wrought counterparts. Tempering temperature should be 150 °C (300 °F) or higher if some decrease in hardness can be tolerated.

Hardness Evaluation. Accurate evaluation of hardness of powder metallurgy parts by conventional indentation methods is difficult, because the inherent porosity produces erroneous low readings. The error depends on the degree of porosity. For example, parts that have been processed by simple pressing and sintering may have a density as low as about 6.6 g/cm^3, as shown in Fig. 1. With this low density, the error in an indentation hardness test is

Fig. 1 Case depth vs time for carbonitriding powder metallurgy parts of three different densities

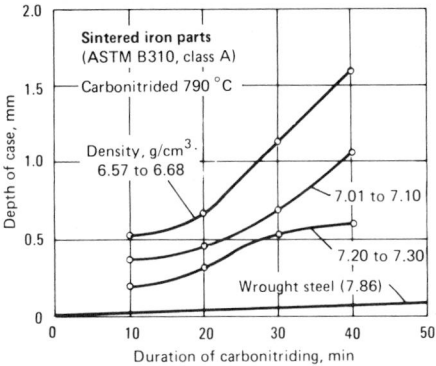

so great that test results are essentially worthless. As density is increased, usually by re-pressing and re-sintering, the error in indentation hardness testing will be reduced. However, as shown in Fig. 1, the density will still be lower than that of wrought steel and, at best, some error will remain.

The indentor of a conventional hardness tester provides an average value, termed "apparent hardness", across metal particles and voids. A more meaningful value of hardness is obtained by microhardness testing—that is, testing of individual particles. Not even this method is infallible, however, because an apparently solid area selected by the microscope may have a void directly beneath it. Consequently, a simple file test is sometimes preferred. If density and composition have been determined accurately, the approximate amount of error can be established for a specific part, thus allowing an indentation method to be used for routine control. This approach requires some "cut and try" before the degree of accuracy can be established.

Surface Hardening. Powder metallurgy parts with low combined carbon contents of 0.10 to 0.20%, for example, can be carburized by conventional gas carburizing. Liquid carburizing is not recommended, because of the difficulty of washing the parts free of salt.

Density should be known for powder metallurgy parts because the carburizing gases always penetrate the voids. The result of this penetration is that a distinct case is not achieved as with a wrought part, but rather the carbon penetration is generally deeper and relatively nonuniform. The degree of this condition varies with density. For parts that have been re-pressed and re-sintered, this condition may be tolerable. If carburizing takes place at too great a depth, however, the quenched part is likely to be excessively brittle.

A second consideration is hardenability. If the compact is of a straight carbon composition and is of any significant thickness, water quenching will be required, with the inherent danger of cracking. Therefore, use of conventional carburizing for specific powder metallurgy parts should be evaluated carefully before it is specified. A modification of carburizing that is generally more practical for surface hardening is described in the paragraphs that follow.

Carbonitriding is widely used as a process for case hardening of parts made by powder metallurgy techniques from ferrous powders. Densities of the sintered compacts vary from approximately 6.5 g/cm³ to values that approach those of wrought steel. Parts may or may not be infiltrated with copper prior to carbonitriding.

Carbonitriding is extremely effective in case hardening parts of high density (7.2 g/cm³) made from sintered iron compacts and is reasonably effective for those of lower density. The carbonitriding process overcomes, to a large degree, the disadvantages of other case-hardening processes, such as carburizing. With other processes, high transformation temperature, low hardenability and inherent porosity produce unusually high rates of diffusion. Carbonitriding at 790 to 815 °C (1450 to 1500 °F) solves these problems. Lower rates of diffusion at these temperatures permit control of case depth and allow adequate buildup of the carbon in the case. Furthermore, the effects of nitrogen in retarding the rate of transformation provide sufficient hardenability to allow oil quenching.

File-hard cases with microhardnesses equal to 60 HRC and with predominantly martensitic structures are consistently obtained, although the allowable case-depth range must be increased over that used for wrought steels. Typical ranges of case depth are 0.08 to 0.20 mm (0.003 to 0.008 in.) and 0.15 to 0.30 mm (0.006 to 0.012 in.).

The high rate of carbon and nitrogen penetration that occurs as a result of porosity is demonstrated in Fig. 1 for parts made of iron powder conforming to ASTM B310, class A. Although the rate of penetration decreases with increasing density, case depths for the higher densities (7.2 to 7.3 g/cm³) are much deeper than those obtained with wrought steel at a density of 7.87 g/cm³. Although most commercial iron-powder compositions exhibit this type of response to carbonitriding, some copper-infiltrated compacts are considerably more resistant to penetration of carbon and nitrogen.

Equipment and techniques for carbonitriding of powder metallurgy parts are basically the same as those used for wrought steel parts. One industrial plant uses carbonitriding at 790 °C (1450 °F) in a batch-type furnace for all copper-free iron parts. Lower temperatures are avoided to minimize the danger of explosion, and higher temperatures are avoided because they alter case characteristics and make control of case depth more difficult.

The processing cycle, including the composition of the atmosphere, is critical. Ammonia content profoundly influences both hardenability and dimensional changes. Because dimensional changes in heat treating are often crucial to the economic justification of producing parts by powder metallurgy, gas compositions, temperatures and quenching media must be controlled closely.

In addition to batch-type equipment, shaker-hearth and belt-type furnaces are also well suited to carbonitriding of powder metallurgy parts. In processing to a new specification, it has become common practice to make furnace adjustments while just a few of the parts are being processed, thereby sacrificing a few parts to arrive at optimum atmosphere adjustments. Such adjustments are then recorded so that they can be used when the next lot of similar parts is processed.

Tempering of Carbonitrided Parts. Powder metallurgy parts usually are tempered despite the fact that there is little danger that untempered pieces will crack. Handling during tempering may eliminate the need for tumbling and deburring operations. Although tempering is potentially capable of removing oil picked up and held by the pores in the workpieces, air tempering of powder metallurgy parts generally is limited to 205 °C (400 °F) because of the fire hazard that accompanies tempering at higher temperatures. As a rule, powder metallurgy parts are tempered at temperatures slightly higher than those used for wrought parts. Special cleaning procedures such as ultrasonic degreasing should be used for cleaning oil-quenched parts when tempering is done at temperatures that exceed 205 °C (400 °F).

Hardness Testing of Carbonitrided Parts. In addition to the reasons given in the earlier discussion of hardness evaluation, carbonitrided parts cannot be accurately tested by indentation methods because they have very thin cases. File testing is the principal method used for routine testing. For development work, the microhardness test is an invaluable tool for determining true hardness.

Système International d'Unités (SI)

SI Base Units

Quantity	Unit	Symbol	Quantity	Unit	Symbol	Quantity	Unit	Symbol
Length	metre	m	Amount of substance	mole	mol	Plane angle(a)	radian	rad
Mass	kilogram	kg				Solid angle(a)	steradian	sr
Time	second	s	Luminous intensity	candela	cd			
Electric current	ampere	A						
Thermodynamic temperature	kelvin	K						

(a) Supplementary unit

SI Derived Units(a)

Quantity	Unit	Symbol	Formula	Quantity	Unit	Symbol	Formula
Frequency (of a periodic phenomenon)	hertz	Hz	s^{-1}	Capacitance	farad	F	C/V
				Electric resistance	ohm	Ω	V/A
Force	newton	N	$kg \cdot m/s^2$	Conductance	siemens	S	A/V
Pressure, stress	pascal	Pa	N/m^2	Magnetic flux	weber	Wb	$V \cdot s$
Energy, work, quantity of heat	joule	J	$N \cdot m$	Magnetic flux density	tesla	T	Wb/m^2
Power, radiant flux	watt	W	J/s	Inductance	henry	H	Wb/A
Quantity of electricity, electric charge	coulomb	C	$A \cdot s$	Luminous flux	lumen	lm	$cd \cdot sr$
				Illuminance	lux	lx	lm/m^2
Electric potential, potential difference, electromotive force	volt	V	W/A	Activity (of radionuclides)	becquerel	Bq	s^{-1}
				Absorbed dose	gray	Gy	J/kg

(a) Derived units in this list include only those units for which special names and symbols have been approved by the General Conference on Weights and Measures (CGPM)

SI Prefixes(a)

Prefix	Multiplication factor	Symbol	Prefix	Multiplication factor	Symbol
exa	$1\ 000\ 000\ 000\ 000\ 000\ 000 = 10^{18}$	E	deci(b)	$0.1 = 10^{-1}$	d
peta	$1\ 000\ 000\ 000\ 000\ 000 = 10^{15}$	P	centi(c)	$0.01 = 10^{-2}$	c
			milli	$0.001 = 10^{-3}$	m
tera	$1\ 000\ 000\ 000\ 000 = 10^{12}$	T			
giga	$1\ 000\ 000\ 000 = 10^{9}$	G	micro	$0.000\ 001 = 10^{-6}$	μ
mega	$1\ 000\ 000 = 10^{6}$	M	nano	$0.000\ 000\ 001 = 10^{-9}$	n
			pico	$0.000\ 000\ 000\ 001 = 10^{-12}$	p
kilo	$1\ 000 = 10^{3}$	k			
hecto(b)	$100 = 10^{2}$	h	femto	$0.000\ 000\ 000\ 000\ 001 = 10^{-15}$	f
deka(b)	$10 = 10^{1}$	da	atto	$0.000\ 000\ 000\ 000\ 000\ 001 = 10^{-18}$	a

(a) Used to form multiples and decimal fractions of the base and derived SI units. (b) Normally avoided. (c) Use not recommended.

Abbreviations and Symbols

AAR Association of American Railroads

AC air cooled

A$_{cm}$ solubility limit between austenite and austenite plus cementite

Ac$_1$ temperature at which austenite begins to form on heating

Ac$_3$ temperature at which transformation of ferrite to austenite is completed on heating

Ac$_{cm}$ temperature at which cementite completes solution in austenite in hypereutectoid steel

ACI Alloy Casting Institute

Ae$_{cm}$, Ae$_1$, Ae$_3$ equilibrium transformation temperature in steel

AGA American Gas Association

AISI American Iron and Steel Institute

AMS Aerospace Materials Specification (of SAE)

ANSI American National Standards Institute

approx approximately

Ar$_1$ temperature at which transformation to ferrite or to ferrite plus cementite is completed on cooling

Ar$_3$ temperature at which transformation of austenite to ferrite begins on cooling

Ar$_{cm}$ temperature at which cementite begins to precipitate from austenite on cooling

ASME American Society of Mechanical Engineers

ASTM American Society for Testing and Materials

atm atmosphere (pressure)

AWG American Wire Gage

AWS American Welding Society

Btu British thermal unit

CH cold work hardened

cm centimetre

cph close-packed hexagonal

CQ commercial quality

dc direct current

diam diameter

DQ drawing quality

DQSK drawing quality special killed

DU depleted uranium

EC electrical conductivity

emf electromotive force

Eq equation

FC furnace cooled

fcc face-centered cubic

FIA Forging Industry Association

Fig. figure

FSG first stage graphitization

ft foot

g gram

gal gallon

GP Guinier-Preston

h hour

HB Brinell hardness

HIP hot isostatic pressure

HK Knoop hardness

hp horsepower

HR hot rolled

HRA Rockwell A hardness

HRB Rockwell B hardness

HRC Rockwell C hardness

HSc Scleroscope numbers (hardness)

HSLA high strength low alloy

HT heat treating

HTO high temperature oxidation

HV Vickers hardness

I current, intensity of magnetization

IACS International Annealed Copper Standard (electrical conductivity)

ID inside diameter

in. inch

IPS International Pipe Standard

J joules

kg kilogram

kW kilowatt

ksi kips (1000 pounds) per square inch

lb pound

LSI large scale integrated

M$_s$ temperature at which martensite starts to form from austenite on cooling

m metre

max maximum

MEA moneothanolamine

Mil military

min minimum, or minute

MJ megajoule

mm millimetre

mol moles

MPa megapascal

MW megawatt

NFPA National Fire Protection Association

No. number

OA over-all (axial) dimension

OD outside diameter

OQ oil quenched

Oxy oxygen

oz ounce

PAG polyalkylene glycols

pH negative logarithm of hydrogen activity

P/M powder metallurgy

POP printing out paper (computers)

ppm parts per million

psi pounds per square inch

psig pounds per square inch, gage

PVA polyvinyl alcohol

PVP polyvinylpurrolidone

RAM random access memory

Ref reference

rem remainder

rev revolution

RH refrigeration hardened

ROM read only memory

rpm revolutions per minute

RT room temperature

RTD resistance temperature detectors

s second

SAE Society of Automotive Engineers

scf standard cubic foot

std standard

SQ structural quality

SSG second stage graphitization

STOA solution treated and overaged

t tonne (metric)

T temperature

TH transformation hardened

TTT transformation characteristics

UNS Unified Numbering System

US United States

V volt

VA volt ampere

vol volume

vol% volume percent

vs versus

W watt

WQ water quenched

wt weight

wt% weight percent

× diameters, magnification

YAG yttrium aluminum garnet (laser)

°**C** degree Celsius (centigrade)

°**F** degree Fahrenheit

ΔT temperature difference

μm micron (micrometre)

± maximum deviation plus or minus

Index